9789811258169-2
T0417602

Unilateral Variational Analysis in Banach Spaces

Part I: General Theory

Unilateral Variational Analysis in Banach Spaces

Part I: General Theory

Lionel Thibault
University of Montpellier, France

NEW JERSEY · LONDON · SINGAPORE · BEIJING · SHANGHAI · HONG KONG · TAIPEI · CHENNAI · TOKYO

Published by

World Scientific Publishing Co. Pte. Ltd.
5 Toh Tuck Link, Singapore 596224
USA office: 27 Warren Street, Suite 401-402, Hackensack, NJ 07601
UK office: 57 Shelton Street, Covent Garden, London WC2H 9HE

British Library Cataloguing-in-Publication Data
A catalogue record for this book is available from the British Library.

UNILATERAL VARIATIONAL ANALYSIS IN BANACH SPACES
(In 2 Parts)
Part I: General Theory
Part II: Special Classes of Functions and Sets

Copyright © 2023 by World Scientific Publishing Co. Pte. Ltd.

All rights reserved. This book, or parts thereof, may not be reproduced in any form or by any means, electronic or mechanical, including photocopying, recording or any information storage and retrieval system now known or to be invented, without written permission from the publisher.

For photocopying of material in this volume, please pay a copying fee through the Copyright Clearance Center, Inc., 222 Rosewood Drive, Danvers, MA 01923, USA. In this case permission to photocopy is not required from the publisher.

ISBN 978-981-125-816-9 (Set_hardcover)
ISBN 978-981-125-817-6 (Set_ebook for institutions)
ISBN 978-981-125-818-3 (Set_ebook for individuals)

ISBN 978-981-125-494-9 (Part I_hardcover)

ISBN 978-981-125-495-6 (Part II_hardcover)

For any available supplementary material, please visit
https://www.worldscientific.com/worldscibooks/10.1142/12797#t=suppl

Printed in Singapore

To Janine and Sylvain

To the memory of my parents Arntz and Germaine

To Dennis and Sylvette

In the memory of my parents Artur and Germaine

Contents

Part I

List of Figures	xix
Preface	xxi

Chapter 1. Semilimits and semicontinuity of multimappings — 1
    1.1. Generalities on multimappings — 2
    1.2. Semilimits of multimappings — 4
    1.3. Epigraphical semilimits — 16
    1.4. Semicontinuity of multimappings — 19
    1.5. Scalarization of semicontinuity — 24
    1.6. Semicontinuity of sum and convexification — 33
    1.7. Michael continuous selection theorem — 34
    1.8. Hausdorff-Pompeiu semidistances — 36
        1.8.1. Hausdorff-Pompeiu excess and distance — 37
        1.8.2. Truncated Hausdorff-Pompeiu excess and distance — 39
        1.8.3. Hausdorff and Attouch-Wets semicontinuities — 45
        1.8.4. Hölder continuity of metric projection with convex set as variable — 47
    1.9. Further results — 48
        1.9.1. Further properties of semicontinuous multimappings — 48
        1.9.2. Hausdorff-Pompeiu distance between boundaries of sets — 50
        1.9.3. Distances between cones — 53
    1.10. Comments — 56

Chapter 2. Tangent cones and Clarke subdifferential — 65
    2.1. Clarke tangent and normal cones — 65
        2.1.1. Definitions and various characterizations — 65
        2.1.2. Interior tangent cone and calculus for Clarke tangent and normal cones of intersection — 72
        2.1.3. Epi-Lipschitz sets; their geometrical, tangential and topological properties — 76
    2.2. Clarke subdifferential — 83
        2.2.1. C-subdifferential: definition and examples — 83
        2.2.2. Differentiability and strict differentiability — 84
        2.2.3. C-subdifferential, minimizers and derivatives — 93
        2.2.4. C-subdifferential and convex functions — 95
        2.2.5. C-subdifferential and locally Lipschitz functions — 103

2.2.6. Gradient representation of C-subdifferential of locally Lipschitz functions	113
2.2.7. Clarke tangent and normal cones of epigraphs and graphs	115
2.2.8. C-subdifferential and directionally Lipschitz functions	118
2.2.9. Clarke tangent and normal cones through distance function	121
2.2.10. Rockafellar theorem for C-subdifferential of finite sum of functions	123
2.2.11. Bouligand-Peano tangent cone and Bouligand directional derivative	128
2.2.12. Tangential regularity of sets and functions	134
2.2.13. Tangent cones of inverse and direct images	140
2.2.14. Tangential regularity of sets versus tangential regularity of distance functions	148
2.2.15. Signed distance function and Clarke tangent cone	151
2.2.16. Sublevel representation of epi-Lipschitz sets	154
2.3. Local Lipschitz property of continuous convex functions	156
2.3.1. Lipschitz property of convex functions	156
2.3.2. Applications to directional Lipschitz property	161
2.3.3. Lipschitz property of continuous vector-valued convex mappings	162
2.4. Chain rule, compactly Lipschitzian mappings, supremum functions	170
2.4.1. Compactly Lipschitzian mappings and chain rule	170
2.4.2. C-subdifferential of supremum of finitely many functions, extension of Lipschitz mappings, tangent cones of sublevel sets	179
2.4.3. Clarke theorem for C-subdifferential of supremum of infinitely many Lipschitz functions	185
2.4.4. Valadier theorem for suprema of infinitely many convex functions in normed spaces	188
2.4.5. C-subgradients of distance function through metric projection; nonzero C-normals at boundary points	191
2.5. Optimization problems with constraints	194
2.5.1. Penalization principles with Lipschitz/non-Lipschitz objective functions	194
2.5.2. Minimization under a set-constraint	197
2.5.3. Ekeland variational principle and Bishop-Phelps principles	198
2.5.4. General optimization problems	202
2.6. Clarke tangent cone in terms of Bouligand-Peano tangent cones	205
2.6.1. Daneš' drop theorem	205
2.6.2. C-tangent cone and limit inferior of B-tangent cones	207
2.7. Basic tangential properties through measure theory	209
2.7.1. Points of nullity of symmetrized B-tangent cone	209

	2.7.2. Lipschitz surfaces	212
	2.7.3. Metrics on the set of vector subspaces	215
	2.7.4. Points of nullity of trace of B-tangent cone on subspace	219
	2.7.5. Tangential properties through Hausdorff measure	220
2.8.	Further results	228
	2.8.1. Intersection of C-normal cones	228
	2.8.2. Subset of a Cartesian product	230
	2.8.3. Compactly epi-Lipschitz sets	232
2.9.	Comments	236

Chapter 3. Convexity and duality in locally convex spaces — 249

3.1.	Convex functions on topological vector spaces	249
	3.1.1. Subdifferential and directional derivatives of convex functions on topological vector spaces	249
	3.1.2. Topological and Lipschitz properties of convex functions on topological vector spaces	256
	3.1.3. Lipschitz property of convex functions under growth conditions	260
	3.1.4. Coercive convex functions	263
	3.1.5. Subdifferentiability and topological properties of subdifferential of convex functions	264
	3.1.6. Subdifferential of suprema of infinitely many convex functions in locally convex spaces	268
	3.1.7. Subdifferential properties of one variable convex functions	272
3.2.	Subdifferentiability of convex functions in finite dimensions	277
	3.2.1. Subdifferentiability via the relative interior	277
	3.2.2. Subdifferentiability of polyhedral convex functions	278
3.3.	Conjugates in the locally convex setting	281
	3.3.1. General properties and examples of Legendre-Fenchel conjugate	281
	3.3.2. Pointwise supremum of continuous affine functions	293
	3.3.3. Biconjugate and Fenchel-Moreau theorem	295
	3.3.4. Dual conditions for coercivity of convex functions	301
	3.3.5. Global Lipschitz property of conjugate functions	302
3.4.	$\mathcal{B}$-differentiability and continuity of subdifferential	303
	3.4.1. $\mathcal{B}$-differentiability: Definition and intrinsic characterizations for convex functions	304
	3.4.2. $\mathcal{B}$-differentiability of convex functions and continuous selections of subdifferentials	308
	3.4.3. $\mathcal{B}$-differentiability of convex functions and continuity of their subdifferentials	310
	3.4.4. Lipschitz continuity of ε-subdifferential	316
3.5.	Asymptotic functions and cones	318
	3.5.1. Definitions and general properties	318
	3.5.2. Asymptotic functions under convexity	324
3.6.	Brønsted-Rockafellar theorem	327
3.7.	Duality for the sum with a linear function	330

3.8. Duality in convex optimization ... 332
3.9. Duality, infsup property, Lagrange multipliers ... 347
3.10. Linear optimization problem ... 349
3.11. Sum and chain rules of subdifferential under convexity ... 351
3.12. Application to chain rule for C-subgradients and C-normals ... 359
3.13. Calculus rules for normals and tangents to convex sets ... 360
3.14. Chain rule with partially nondecreasing functions ... 364
3.15. Extended rules for subdifferential of maximum of finitely many convex functions ... 366
3.16. Limiting formulas for subdifferential of convex functions ... 370
    3.16.1. Limiting sum/chain rule for subdifferential of convex functions ... 370
    3.16.2. Limiting rules for subdifferential of composition with inner vector-valued convex mapping ... 377
3.17. Subdifferential determination and maximal monotonicity for convex functions on Banach spaces ... 379
3.18. Normals to convex sublevel sets ... 381
    3.18.1. Normals to convex sublevels under Slater condition in locally convex spaces ... 381
    3.18.2. Horizon subgradients of convex functions ... 383
    3.18.3. Limiting formulas for normals to convex sublevels: Reflexive Banach space case ... 385
    3.18.4. Limiting formulas for normals to convex sublevels: General Banach space case ... 389
    3.18.5. Limiting formulas for normals to intersection of finitely many sublevels ... 391
    3.18.6. Limiting formulas for normals to vector convex sublevels ... 399
3.19. Continuity of conjugate functions, weak compactness of sublevels, minimum attainment ... 401
    3.19.1. Continuity of conjugate functions and weak compactness of sublevels ... 401
    3.19.2. Attainment of the minimum of $f - \langle x^*, \cdot \rangle$ ... 406
3.20. Subdifferential of distance functions from convex sets ... 416
3.21. Moreau envelope, strongly convex functions ... 419
    3.21.1. Moreau envelope ... 419
    3.21.2. Strongly convex functions ... 430
3.22. Gâteaux differentiability at subdifferentiability points ... 435
3.23. Further results ... 436
    3.23.1. Duality with partial conjugate ... 436
    3.23.2. Calculus for ε-subdifferential of convex functions ... 438
    3.23.3. Extended calculus for ε-subdifferential of convex functions ... 441
    3.23.4. ε-Subdifferential determination of convex functions and cyclic monotonicity ... 448
    3.23.5. Limiting subdifferential chain rule for convex functions on locally convex spaces ... 451

3.24. Comments 453

Chapter 4. Mordukhovich limiting normal cone and subdifferential 473
    4.1. Fréchet normal and subgradient 473
        4.1.1. Definitions and first properties 473
        4.1.2. Fréchet subgradients of distance functions 488
    4.2. Separable reduction principle for F-subdifferentiability 493
        4.2.1. Preparatory lemmas 494
        4.2.2. Separable reduction of Fréchet subdifferentiability 497
    4.3. Fuzzy calculus rules for Fréchet subdifferentials 501
        4.3.1. Borwein-Preiss variational principle 502
        4.3.2. Fuzzy sum rule for Fréchet subdifferential under Fréchet differentiable renorm 508
        4.3.3. Applications to convex functions and Asplund spaces 512
        4.3.4. Fuzzy sum rule for Fréchet subdifferential in Asplund space 515
        4.3.5. Fuzzy chain rule for Fréchet subdifferential 518
        4.3.6. Stegall variational principle, Fréchet derivative of conjugate function 520
    4.4. Mordukhovich limiting subdifferential in Asplund space 523
        4.4.1. Definitions, properties, calculus 523
        4.4.2. Calculus rules 527
        4.4.3. L-Subdifferential of distance function 535
        4.4.4. Mordukhovich limiting subdifferential in normed space 540
    4.5. Representation of C-subdifferential via limiting subgradients 557
        4.5.1. Horizon L-subgradient and representation of C-subdifferential 557
        4.5.2. Analytic description of horizon limiting subgradient 561
    4.6. Proximal normal cone and subdifferential 563
        4.6.1. Definition and properties of proximal subgradient 563
        4.6.2. Proximal subgradients of distance functions 573
        4.6.3. Proximal fuzzy calculus and proximal representation of the limiting subdifferential 576
    4.7. Further results 580
        4.7.1. F-normal cone to graphs of multimappings 580
        4.7.2. L-subdifferential versus C-subdifferential in the real line 582
    4.8. Comments 583

Chapter 5. Ioffe approximate subdifferential 591
    5.1. Hadamard subgradient 591
        5.1.1. General properties 591
        5.1.2. Hadamard subdifferential of sums of functions 597
    5.2. Ioffe A-subdifferential on separable Banach spaces 599
        5.2.1. Definition for Lipschitz functions and comparisons 599
        5.2.2. A-normal cone in separable Banach spaces 602

- 5.3. Ioffe A-subdifferential of Lipschitz functions on Banach spaces — 607
  - 5.3.1. Definition, properties and sum — 607
- 5.4. A-normal cone and A-subdifferential of general functions — 616
  - 5.4.1. A-normal cone in general Banach spaces — 616
  - 5.4.2. A-subdifferential of general functions — 622
  - 5.4.3. Chain rule for A-subdifferential and mean value inequality — 627
  - 5.4.4. Representation of C-subdifferential with A-subgradients — 631
  - 5.4.5. Extended A-subdifferential sum rule — 632
- 5.5. Further results — 636
  - 5.5.1. A-normals to compactly epi-Lipschitz sets — 636
  - 5.5.2. Subdifferentially pathological Lipschitz functions — 638
- 5.6. Comments — 643

Chapter 6. Sequential mean value inequalities — 647
- 6.1. Mean value inequalities with Dini derivatives — 647
  - 6.1.1. Mean value inequalities with lower/upper Dini directional derivatives — 647
  - 6.1.2. Sub-sup regularity and saddle functions — 650
  - 6.1.3. Extended gradient representations of subdifferentials — 655
  - 6.1.4. Conditions for monotonicity and other properties via Dini semiderivates — 664
  - 6.1.5. Mean value inequality for images of sets and Denjoy-Young-Saks theorem — 667
  - 6.1.6. Mean value inequality with Dini subgradients — 674
- 6.2. Zagrodny mean value inequality — 678
  - 6.2.1. Density properties for subdifferentials — 680
  - 6.2.2. Zagrodny mean value theorem — 681
  - 6.2.3. Subdifferential and tangential characterizations of Lipschitz properties — 684
  - 6.2.4. Subdifferential and tangential characterizations of monotonicity and convexity — 690
- 6.3. Approximate and sequential Rolle-type theorems — 694
- 6.4. Multidirectional mean value inequalities — 699
- 6.5. Comments — 711

Chapter 7. Metric regularity — 715
- 7.1. Aubin-Lipschitz property and metric regularity — 715
- 7.2. Openness and metric regularity of convex multimappings: Robinson-Ursescu theorem — 726
- 7.3. Criteria and estimates of rates of openness and metric regularity of multimappings — 728
- 7.4. Metrically regular transversality of system of sets — 737
- 7.5. Metric regularity of convex feasible sets, Hoffman inequality — 750
- 7.6. Metric regularity and Lipschitz additive perturbation — 756
- 7.7. Optimality conditions and calculus of tangent and normal cones under metric subregularity — 762

7.7.1. Optimality conditions under metric subregularity or other conditions	762
7.7.2. Estimates of coderivatives under metric subregularity or other conditions; regularity of nonsmooth constraints	764
7.7.3. General optimality conditions	767
7.7.4. Normal/tangent cone calculus and chain rule	769
7.8. More on subdifferential calculus for convex functions	778
7.9. Further results	780
7.9.1. Metric subregularity of polyhedral multimappings	780
7.9.2. Metric regularity/subregularity of subdifferential and growth conditions	782
7.10. Comments	810
Appendix A. Topology	817
Appendix B. Topological properties of convex sets	821
Appendix C. Functional analysis	829
Appendix D. Measure theory	835
Appendix E. Differential calculus and differentiable manifolds	837
Bibliography	843
Index	879

Part II

List of Figures	xix
Preface	xxi
Chapter 8. Subsmooth functions and sets	893
8.1. Definition and first properties of subsmooth functions	893
8.2. Directional derivatives and subdifferentials of subsmooth functions	902
8.2.1. General properties of derivatives and subdifferentials	902
8.2.2. Submonotonicity of subdifferentials	909
8.2.3. Subdifferential characterizations of one-sided subsmooth functions	917
8.3. Subsmooth sets	922
8.3.1. Definition of subsmooth sets and general properties	922
8.3.2. Subsmoothness of sets versus Shapiro property	930
8.4. Epi-Lipschitz subsmooth sets	933
8.5. Metrically subsmooth sets	939
8.6. Subsmoothness of a set and α-far property of the C-subdifferential of its distance function	950
8.7. Preservation of subsmoothness under operations	955
8.8. Metric subregularity under metric subsmoothness	967

- 8.9. Equi-subsmoothness of sets and subdifferential of their distance functions — 973
- 8.10. Further results — 978
  - 8.10.1. ε-Localization of subsmooth functions by convex functions — 979
  - 8.10.2. Metric regularity of subsmooth-like multimappings — 981
- 8.11. Comments — 985

Chapter 9. Subdifferential determination — 989
- 9.1. Denjoy function — 989
- 9.2. Subdifferentially and directionally stable functions — 993
  - 9.2.1. Subdifferentially and directionally stable functions, properties and examples — 994
  - 9.2.2. Subdifferential determination of subdifferentially and directionally stable functions — 999
- 9.3. Essentially directionally smooth functions and their subdifferential determination — 1003
  - 9.3.1. Essentially directionally smooth functions, properties and examples — 1003
  - 9.3.2. Subdifferential determination of essentially directionally smooth functions — 1009
- 9.4. Comments — 1013

Chapter 10. Semiconvex functions — 1017
- 10.1. Semiconvex functions — 1017
  - 10.1.1. Semiconvexity, moduli of semiconvexity — 1017
  - 10.1.2. Semiconvexity of diverse types of functions — 1020
  - 10.1.3. Sup-representation of linearly semiconvex functions — 1023
  - 10.1.4. Composite stability for semiconvexity and distance function — 1028
  - 10.1.5. Lipschitz continuity of semiconvex functions — 1032
- 10.2. Subdifferentials and derivatives of semiconvex functions — 1034
  - 10.2.1. Directional derivatives and subdifferentials — 1034
  - 10.2.2. Properties under linear semiconvexity and linear semiconcavity — 1042
  - 10.2.3. Subdifferential and tangential characterizations of semiconvex functions — 1045
- 10.3. Max-representation and extension of Lipschitz semiconvex functions — 1047
  - 10.3.1. Max-representation with quadratic/differentiable functions — 1048
  - 10.3.2. Max-representation in uniformly convex space — 1050
- 10.4. Semiconvex multimappings — 1063
- 10.5. Comments — 1065

Chapter 11. Primal lower regular functions and prox-regular functions — 1069
- 11.1. s-Lower regular functions — 1069
  - 11.1.1. Primal lower and s-lower regular functions — 1069

11.1.2. Convexly composite functions	1072
11.1.3. Coincidence of subdifferentials of s-lower regular functions	1078
11.1.4. Subdifferential characterization of s-lower regular functions	1081
11.2. Moreau s-envelope	1086
11.3. Moreau envelope of primal lower regular functions in Hilbert spaces	1099
11.3.1. First properties related to continuity of proximal mapping	1100
11.3.2. Differentiability properties of Moreau envelope of primal lower regular functions	1101
11.4. Subdifferential determination of primal lower regular functions	1113
11.5. Prox-regular functions	1116
11.5.1. Definition and examples	1116
11.5.2. Subdifferential characterization of prox-regular functions	1117
11.5.3. Differentiability of Moreau envelopes under prox-regularity	1122
11.6. Comments	1125
Chapter 12. Singular points of nonsmooth functions	1131
12.1. Singular points of nonsmooth mappings	1131
12.2. Singular points of convex and semiconvex functions	1140
12.3. Comments	1150
Chapter 13. Non-differentiability points of functions on separable Banach spaces	1153
13.1. Non-differentiability points of subregular functions	1153
13.2. Null sets in infinite dimensions	1153
13.2.1. Aronszajn null sets	1154
13.2.2. Porous sets	1156
13.2.3. Haar null sets	1163
13.3. Hadamard non-differentiability points of Lipschitz functions	1169
13.3.1. Hadamard non-differentiability points of Lipschitz mappings	1169
13.3.2. More on interior tangent property via signed distance function	1175
13.3.3. Non-differentiability points of one-sided Lipschitz functions	1176
13.4. Zajíček extension of Denjoy-Young-Saks theorem	1178
13.5. Comments	1184
Chapter 14. Distance function, metric projection, Moreau envelope	1187
14.1. Distance function and metric projection	1187
14.1.1. Density of points with nearest/farthest points	1187

14.1.2. Differentiability of distance functions and farthest distance functions under differentiable norms	1193
14.1.3. Genericity of points with nearest points, Lau theorem	1203
14.2. Genericity attainment and other properties of Moreau envelopes	1208
14.3. L-subdifferential by means of Moreau envelopes	1220
14.4. Comments	1223

Chapter 15. Prox-regularity of sets in Hilbert spaces ... 1227
  15.1. $\rho(\cdot)$-prox-regularity of sets ... 1227
  15.2. Uniform and local prox-regularity ... 1254
    15.2.1. Uniform prox-regularity ... 1254
    15.2.2. Uniform prox-regularity of r-enlargement and r-exterior set ... 1267
    15.2.3. Linear semiconvexity of distance function to a prox-regular set ... 1273
    15.2.4. Uniform prox-regularity of connected components ... 1277
    15.2.5. Local (r,α)-prox-regularity ... 1278
    15.2.6. Directional derivability of the metric projection ... 1290
  15.3. Change of metric ... 1295
  15.4. Prox-regularity in operations ... 1296
    15.4.1. Uniform prox-regularity in operations ... 1297
    15.4.2. Local prox-regularity in operations ... 1306
  15.5. Continuity properties of $C \mapsto P_C(u)$ ... 1312
  15.6. Further results ... 1316
    15.6.1. Representation of multimappings with prox-regular values ... 1316
    15.6.2. Continuous selections of lower semicontinuous multimappings with prox-regular values ... 1318
  15.7. Comments ... 1320

Chapter 16. Compatible parametrization and Vial property of prox-regular sets, exterior sphere condition ... 1329
  16.1. Compatible parametrization of prox-regular sets ... 1329
  16.2. Strongly convex sets and Vial property of prox-regular sets ... 1334
    16.2.1. Strongly convex sets ... 1334
    16.2.2. Vial property ... 1341
    16.2.3. Closedness of Minkowski sums and ball separations properties ... 1346
  16.3. Exterior/interior sphere condition ... 1349
  16.4. Comments ... 1359

Chapter 17. Differentiability of metric projection onto prox-regular sets ... 1363
  17.1. Further properties of (r,α)-prox-regularity ... 1363
  17.2. Differentiability of metric projection ... 1376
    17.2.1. Variational and prox-regularity properties of submanifolds ... 1376

17.2.2. Smoothness of metric projection onto prox-
regular sets with smooth boundary 1384
17.3. Characterization of epi-Lipschitz sets with smooth
boundary 1395
17.3.1. Properties of derivatives of metric projection 1395
17.3.2. Smoothness of the boundary of a set via the
metric projection 1399
17.4. Metric projection onto submanifold 1406
17.4.1. Differentiability of metric projection onto
submanifold 1407
17.4.2. Characterization of submanifolds via metric
projection 1409
17.4.3. Smoothness property of signed distance function 1414
17.5. Comments 1418

Chapter 18. Prox-regularity of sets in uniformly convex Banach spaces 1421
18.1. Uniformly convex Banach spaces 1421
18.1.1. Strictly convex normed spaces 1421
18.1.2. Uniformly convex Banach spaces 1426
18.1.3. Uniformly smooth Banach spaces 1433
18.1.4. Characterizations of uniformly convex/smooth
norms via duality mappings 1443
18.1.5. Xu-Roach theorems on moduli of convexity and
smoothness 1450
18.2. Proximal normals in normed spaces 1463
18.3. Prox-regular sets and J-plr functions 1469
18.4. Local Moreau envelopes of J-plr functions 1477
18.4.1. Fréchet differentiability of local Moreau envelope 1478
18.4.2. Uniform continuity of local proximal mappings of
J-plr functions 1480
18.5. Characterizations of local prox-regular sets in uniformly
convex Banach spaces 1484
18.5.1. Metric projection of local prox-regular sets 1484
18.5.2. Basic characterizations of local prox-regularity in
uniformly convex Banach spaces 1489
18.5.3. Tangential regularity 1493
18.6. Characterizations and properties of uniformly prox-
regular sets in uniformly convex Banach spaces 1495
18.6.1. Characterizations of uniform prox-regularity of
sets in uniformly convex Banach spaces 1495
18.6.2. Connected components of r-prox-regular sets 1501
18.6.3. Enlargements and exterior points of r-prox-
regular sets 1502
18.7. Lipschitz continuity of metric projection and radius of
prox-regularity 1504
18.8. Prox-regularity and geometric variational properties of
cones 1507
18.9. Comments 1508

Appendix A. Topology	1513
Appendix B. Topological properties of convex sets	1517
Appendix C. Functional analysis	1525
Appendix D. Measure theory	1531
Appendix E. Differential calculus and differentiable manifolds	1533
Bibliography	1539
Index	1575

List of Figures

1.1	Illustration of Example 1.78.	49
2.1	Clarke tangent cones.	68
2.2	Bouligand-Peano tangent cones.	131
2.3	Daneš drop property.	206
4.1	Fréchet normal cones.	479
4.2	Mordukhovich limiting normal cones.	526

Preface

The study in depth of *unilateral properties* in mathematical analysis apparently began at the end of the 19th century with the 1899 Baire's thesis (Thèse de doctorat ès sciences mathématiques). Given a real-valued function f of n real variables, under the term of *maximum of f at a point P_0* and with the notation $M[f, P_0]$, R. Baire introduced in his thesis [**61**, p. 4] what is nowadays called the upper limit (or limit superior) of f at P_0 usually denoted as $\limsup_{P \to P_0} f(P)$ (or $\overline{\lim}_{P \to P_0} f(P)$). Considering in [**61**] a decreasing sequence $(\rho_n)_n$ of positive reals tending to 0 and the nondecreasing sequence $(M_n)_n$ of suprema of f on $B[P_0, \rho_n]$, Baire defined $M[f, P_0]$ as the limit of M_n as $n \to \infty$. The property $M[f, P_0] = f(P_0)$, or equivalently the inequality $M[f, P_0] \leq f(P_0)$, was observed by Baire [**61**] to be equivalent to the fact that for each $\varepsilon > 0$ there exists some $\rho > 0$ such that $f(P) < f(P_0) + \varepsilon$ for all $P \in B[P_0, \rho]$. Baire then wrote: "La fonction possède donc au point P_0 l'une des deux propriétés dont l'ensemble constitue la continuité"(English translation: "The function thus has one of the two properties which together constitute the continuity"). That *unilateral* property was then called the *upper semicontinuity* of f at P_0 by Baire in [**61**, p. 6]. The *minimum* $m[f, P_0]$ *of f at P_0* and its *lower semicontinuity* at P_0 were defined similarly in [**61**], and various properties of $M[f, \cdot]$ and $m[f, \cdot]$ and of the semicontinuity notions were established. Through those concepts Baire carried out in his dissertation thesis the thorough analysis of his celebrated classification of diverse classes of discontinuous functions. By analogy, for a multimapping $M : X \rightrightarrows Y$ from a metric space X with closed values into a compact metric space Y, C. Kuratowski said in his 1932 paper [**640**, p. 148] that M is *upper* (resp. *lower*) *semicontinuous* at x_0 whenever for each sequence $(x_n)_n$ converging to x_0 one has $\operatorname{Ls} M(x_n) \subset M(x_0)$ (resp. $M(x_0) \subset \operatorname{Li} M(x_n)$). For a sequence of sets $(S_n)_n$ in a metric space, the set $\operatorname{Ls} S_n$ (resp. $\operatorname{Li} S_n$) is taken in [**640**] as the set of $\lim y_n$ with $y_n \in S_n$ for infinitely many n (resp. for large n). In a Euclidean space, the former set was considered much earlier in 1887 with the same above form under the name of *"limit"* by P. Painlevé [**779**, p. 123]. In the same year 1887, G. Peano [**782**, p. 302] instead called for a family of figures $M(x)$, in finite dimensions, *"limit"* of $M(x)$ as $x \to x_0$ the set of y such that $\lim_{x \to x_0} d(y, M(x)) = 0$, and this is nowadays known to coincide with the usual limit inferior $\operatorname{Lim\,inf}_{x \to x_0} M(x)$; the notation $\lim(M, x_0)$ was used by Peano in his other monograph [**784**, p. 302] published in 1908. In this same 1908 monograph [**784**, p. 237] Peano also employed, with the notation $\operatorname{Lm}(M, x_0)$, the set of y such that 0 is a limit point of the mapping $x \mapsto d(y, M(x))$ as $x \to x_0$, which corresponds to the limit superior $\operatorname{Lim\,sup}_{x \to x_0} M(x)$. We refer to Section 1.10 in Chapter 1 for comments. In what follows and throughout the text, instead of the notations $\operatorname{Ls} M(x_n)$ and $\operatorname{Li} M(x_n)$, we will utilize the today more usual ones $\operatorname*{Lim\,sup}_{n\to\infty} M(x_n)$ and $\operatorname*{Lim\,inf}_{n\to\infty} M(x_n)$. Those concepts are clearly *unilateral semilimit* notions for a multimapping M and they generate *unilateral semicontinuities* for the multimapping by means of the inclusions $\operatorname*{Lim\,sup}_{x\to x_0} M(x) \subset M(x_0)$ and $M(x_0) \subset \operatorname*{Lim\,inf}_{x\to x_0} M(x)$.

Previously to [**640**], instead of the classic tangent line to a regular curve or tangent plane to a regular surface, G. Bouligand employed the Lim sup to put into light another viewpoint for the concept of tangent. For a set S in a finite-dimensional Euclidean space X and a point $x_0 \in S$, Bouligand considered in his 1928 paper [**157**, p. 29] the set of elements belonging (with the preceding notation) to $x_0 + \operatorname*{Lim\,sup}_{n\to\infty} \frac{1}{t_n}(S-x_0)$ for some $t_n \downarrow 0$; the notation Δ was used there for that set. Bouligand provided an analysis of sets S without point x_0 where the corresponding set Δ coincides with the whole space X. Several other studies involving this set Δ were carried out by Bouligand in a series of works [**158, 159, 160, 161, 162**] and by others influenced in the 1930s by his works as, for example, the authors of [**217, 367, 368, 714, 826, 867**]; see comments in Section 2.9 in Chapter 2. This concept of tangent was previously considered in 1908 by G. Peano [**784**, p. 331] in the form $\operatorname{Lm}\left(x_0 + \frac{1}{t}(S-x_0), 0\right)$ for $t \downarrow 0$ (see Penao's notation Lm recalled above), and it was denoted there Tang(S, x_0). In his earlier 1887 manuscript [**782**, p. 305] Peano used another notion as tangent, namely the set of elements y such that $\lim_{t\downarrow 0} d\left(y - x_0, \frac{1}{t}(S - x_0)\right) = 0$, which corresponds, with the nowadays notation, to elements belonging to $x_0 + \operatorname*{Lim\,inf}_{n\to\infty} \frac{1}{t_n}(S - x_0)$ for every $t_n \downarrow 0$. In the 1908 manuscript [**784**, p. 335] Peano also stated in terms of Tang(S, x_0) necessary optimality conditions when x_0 is a maximizer of a function under the constraint set S. Both above concepts appeared as *unilateral* viewpoints of tangents in geometry and mathematical analysis.

Unilateral derivates of real-valued functions were already present in U. Dini's 1878 book [**339**]. Given a real-valued function f on an interval in $\mathbb{R}$ the *right-hand* (resp. *left-hand*) *derivate* of f at x is defined at the page 66 of Dini's book [**339**] as the limit in $\mathbb{R} \cup \{-\infty, +\infty\}$ of $\frac{f(x+\delta)-f(x)}{\delta}$ as δ tends to 0 with $\delta > 0$ (resp. with $\delta < 0$); Dini used there the terminology "*derivata della funzione $f(x)$ a destra* (resp. *a sinistra*) *di x*". The *right-hand upper* (resp. *lower*) *derivate* of f at x is defined in A. Denoy's paper [**327**, p. 144-145] and in S. Saks' book [**876**, p. 108] as the upper (resp. lower) limit of the above difference quotient as δ tends to 0 with $\delta > 0$; the left-hand upper (resp. lower) derivate is defined there similarly. Denjoy in [**327**, p. 146] and Saks in [**876**, p. 108] called these four quantities the *unilateral extreme derivates*, and Saks also said *unilateral Dini derivates*. For a locally Lipschitz function f on $X = \mathbb{R}^N$, if x_0 is a minimizer of f on $S \subset X$ and $\mathcal{T}$ denotes the above tangent set in the sense of Bouligand (or equivalently the Peano tangent set Tang(S, x_0)), the following "*necessary optimality condition*" holds:

$$\underline{d}_D^+ f(x_0; x - x_0) \geq 0 \quad \text{for all } x \in \mathcal{T},$$

where by $\underline{d}_D^+ f(x_0; h)$ we denote the above right-hand lower Dini derivate at 0 of the one-variable function $t \mapsto f(x + th)$. It is worth noticing that the *distributional derivative* does not allow to state an optimality condition in such a situation. In many cases in mathematics and diverse sciences, the set S arises as a closed subset

of a Banach space X in the form
$$S = \{x \in X : g_i(x) \leq 0 \ i \in I, h_j(x) = 0 \ j \in J\},$$
where I and J are finite sets. When the functions g_i are all null and the functions f, h_j are continuously differentiable, under suitable conditions the classical Lagrangian optimality condition is well-known, that is, there exist reals $(\lambda_j)_{j \in J}$ such that
$$Df(x_0) + \sum_{j \in J} \lambda_j Dh_j(x_0) = 0.$$
The great progress in this direction with nondifferentiable functions was achieved with the concept of subdifferential of a convex function $\varphi : X \to \mathbb{R}$ independently introduced in 1963 by J. J. Moreau [738] and R. T. Rockafellar [844] as given by the *unilateral* description
$$\partial \varphi(x) = \{v \in \mathbb{R}^N : \langle v, y - x \rangle \leq \varphi(y) - \varphi(x), \ \forall y \in \mathbb{R}^N\}.$$
In addition to its rich calculus similar to that of differential calculus, this subdifferential is known, and this will be seen later, to allow, for example with $S = \{x : g_i(x) \leq 0, \ i \in I\}$ in the above minimization problem, to obtain under certain conditions and the convexity of the functions f, g_j that there exist non-negative reals $(\lambda_j)_{j \in J}$ such that
$$\partial f(x_0) + \sum_{j \in J} \lambda_j \partial f_j(x_0) \ni 0.$$
Such an optimality condition, for the minimization problem with nondifferentiable nonconvex locally Lipschitz functions f, g_i, h_j, was proved with the subdifferential $\partial \varphi(x)$ (also called generalized gradient) introduced by F. H. Clarke in his 1973 thesis [232] for a nondifferentiable nonconvex function φ. As it will be presented in the treatise, this subdifferential also enjoys a full calculus and coincides with that of Moreau-Rockafellar when φ is convex. Two other fundamental subdifferentials, yielding optimality conditions as above and enjoying rich calculus, were introduced by B. S. Mordukhovich [720] and A. D. Ioffe [515, 517] respectively. All those subdifferentials offer efficient *unilateral first order variation* tools for the analysis of nondifferentiable functions, and they possess rich calculus. According to the necessity and advantages pointed out in the 1960s by Moreau and Rockafellar to work, for the analysis of unilateral properties, with nondifferentiable functions taking values in the extended line $\mathbb{R} \cup \{-\infty, +\infty\}$, the subdifferential theories of convex and nonconvex functions were carried out with such extended real-valued functions.

One of the main aims of the book is to develop the theory of the unilateral concepts: semilimits of variable sets, semicontinuities of multimappings, tangent cones, normal cones, subdifferentials. A second main aim is to present various basic classes of sets and functions which can be seen as *subregular* in a unilateral viewpoint via the aforementioned subdifferentials, and to study the smallness of sets of points which are singular relative to subdifferentials. A third main aim is to analyze in depth the very large class of prox-regular sets, that is, sets whose metric projection is continuous near a reference point inside. There exist diverse books/lectures devoted (or partially devoted) to the analysis of nondifferentiable convex functions as [35, 43, 45, 55, 72, 78, 134, 146, 155, 214, 384, 494, 529, 643, 649, 744, 801, 852, 1000]. Chapter 3 furnishes a comprehensive presentation of the general theory of extended real-valued convex functions on locally convex spaces, and many results related to these functions either appear here for the first time in a book or are established with new simple proofs. After the theory of semilimits and semicontinuities for multimappings in Chapter 1, tangent cones and Clarke subdifferential are completely studied in Chapter 2 in the setting of normed spaces. The Mordukhovich limiting subdifferential in Asplund spaces and the Ioffe approximate subdifferential in Banach spaces are developed in details in Chapter 4 and Chapter 5, respectively. A systematic exposition of mean value inequalities for nondifferentiable functions is given in Chapter 6. The theory of metric regularity in Banach spaces for mappings and multimappings is the topic of Chapter 7. Subsmooth sets and subsmooth functions are investigated in Chapter 8, and the subdifferential determination property of such functions and others is studied in Chapter 9. Three important other classes of functions and their very useful properties are developed in details: semiconvex functions in Chapter 10, and primal lower regular and prox-regular functions in Chapter 11. The smallness of sets of subdifferentially singular points of certain such functions is the subject of Chapters 12 and 13. The study of differential properties of distance functions and metric projection in Chapter 14 is made in a very large context. Many results of that chapter are employed in the systematic exposition of prox-regular sets presented in Chapters 15, 16, 17 and 18. Each chapter finishes with a section called *"Comments"* which contains discussions and references regarding papers related to the results presented in the chapter in question. When a theorem is due to authors of a same paper (that is, co-authors) this is indicated as Theorem [**Author and Author'**], while we use the form Theorem [**Author; Author'**] for the case of authors of distinct papers; the same convention is adopted for certain examples, lemmas and propositions. For convenience of the reader, we recalled in Appendix diverse basic results used in the text and related to the theories: Topology, Functional Analysis, Convex Sets in Finite Dimensions, Measure Theory, Differential Geometry. The bibliography contains most of the contributions which are pertinent to the topics in the book.

Certain chapters in the book were the subject of lectures given in the 2010s at the "Université de Montpellier", and they benefited from comments of some students. I am also indebted to my colleague Florent Nacry for many valuable suggestions.

Montpellier
September, 2021

Lionel Thibault

CHAPTER 1

Semilimits and semicontinuity of multimappings

We denote by $\mathbb{R}_+$ the set of non-negative real numbers $[0,+\infty[$ and by $\mathbb{N}$ the set of positive integers $1, 2, \cdots, n, \cdots$. By a *countable set* we will mean a set which is empty or in bijection with a nonempty subset of $\mathbb{N}$, that is, a set which is either finite or in bijection with $\mathbb{N}$ itself. Let (X, τ) be a topological space. A subset S of X is said to be *closed near* $x \in S$ whenever, for some neighborhood V of x, the set $S \cap V$ is closed in V with respect to the induced topology on V. The closure of the set $S \subset X$ will be denoted by $\operatorname{cl} S$ (also by $\operatorname{cl}_\tau S$ or $\operatorname{cl}_X S$ when the topology or the space needs to be emphasized). For the boundary of S we will employ the notation $\operatorname{bdry} S$ (also $\operatorname{bdry}_\tau S$ or $\operatorname{bdry}_X S$). It will be convenient to write $S \ni u \to x$ or $u \underset{S}{\to} x$ to mean that $u \to x$ with $u \in S$. Similarly, for a mapping ϕ from a neighborhood of $x \in X$ into a topological space Y, we will write $u \to_\phi x$ when $u \to x$ along with $\phi(u) \to \phi(x)$. Throughout the book, when the integral of a mapping with values in a normed space is considered, *it will always be the Bochner integral taken in the (unique up to isometric isomorphism) Banach space completion of the normed space.*

Given a real-valued function $g : X \to \mathbb{R}$ on a space X and a nonempty subset S of X, J. J. Moreau and R. T. Rockafellar fully exploited independently in convex duality and convex optimization the observation that minimizing (resp. the infimum of) g on the constraint set S amounts to minimizing (resp. coincides with the infimum of) the extended real-valued function g_S on the global space X, where $g_S : X \to \mathbb{R} \cup \{-\infty, +\infty\}$ is defined by $g_S(x) = g(x)$ if $x \in S$ and $g_S(x) = +\infty$ if $x \in X \setminus S$. So, they were led to introduce the so-called indicator function of S, which is the extended real-valued function Ψ_S defined on X by $\Psi_S(x) = 0$ if $x \in S$ and $\Psi_S(x) = +\infty$ if $x \in X \setminus S$. It is then clear that $g_S = g + \Psi_S$. A large part of the book will involve extended real-valued functions. Given such a function $f : X \to \mathbb{R} \cup \{-\infty, +\infty\}$, one naturally associate with it the sets

$$\operatorname{dom} f := \{x \in X : f(x) < +\infty\} \text{ and } \operatorname{epi} f := \{(x, r) \in X \times \mathbb{R} : f(x) \leq r\}$$

called the *(effective) domain* of f and the *epigraph* of f respectively. For any $r \in \mathbb{R} \cup \{-\infty, +\infty\}$ one also associates the *sublevel set*

$$\{f \leq r\} := \{u \in X : f(u) \leq r\}$$

as well as the similar sets $\{f < r\}$ etc; when the feature that f is a function needs to be emphasized it will be convenient to write $\{f(\cdot) \leq r\}$, $\{f(\cdot) < r\}$, etc. As we will see, the sets $\operatorname{dom} f$, $\operatorname{epi} f$ and $\{f \leq r\}$ will play crucial roles in the study of diverse properties of f. Many mathematical optimization problems require the study of properties of the dependence on r of the set $\{f \leq r\}$, which amounts to the study of the *multimapping* defined in this way.

Given a subset S of a vector space X, we recall that S is convex provided that

(1.1) $$(1-t)x + ty \in S \quad \text{for all } t \in]0,1[, \, x,y \in S;$$

the empty set $\emptyset$ and the global set X are the least and greatest convex sets in X. We will use the notation $\operatorname{co} S$ for the *convex hull* of the set S, that is, the intersection of all convex subsets of X containing S. Assuming that (X,τ) is a topological vector space, $\overline{\operatorname{co}}\,S$ or $\overline{\operatorname{co}}^\tau S$ will stand for the *τ-closed convex hull* of S (that is, the intersection of all τ-closed convex subsets of X containing S). According to this and what precedes, if X is a normed (or more generally, locally convex) space, the weak closure of S is denoted by $\operatorname{cl}_w S$, that is, the closure with respect to the weak topology $w(X, X^*)$, where X^* is the topological dual of X. Similarly $\operatorname{cl}_{w^*} G$ will be the weak* closure of a subset $G \subset X^*$. In addition to the closed convex hull $\overline{\operatorname{co}}\,S$ of the subset S of X, the weak* closed convex hull of the the subset G of X^* will be similarly denoted by $\overline{\operatorname{co}}^* G$.

Suppose that X is a metric space. For any $x \in X$ and any real number $\delta \geq 0$ (resp. $\delta > 0$) the closed (resp. open) ball centered at x with radius δ will be denoted by $B_X[x,\delta]$ or $B[x,\delta]$ (resp. $B_X(x,\delta)$ or $B(x,\delta)$). If X is a normed space, for $x = 0$ and $\delta = 1$, we will put $\mathbb{B}_X$ or $\mathbb{B}$ in place of $B_X[0,1]$, and $\mathbb{B}_X$ is generally called the *closed unit ball* (centered at 0) of the normed space X. For $\delta = +\infty$, we adopt the convention $\delta \mathbb{B}_X = B(x,\delta) = B[x,\delta] = X$ for all $x \in X$. The usual distance function associated with S is denoted by $d(\cdot, S)$, $d(S, \cdot)$, $d_S(\cdot)$, or $\operatorname{dist}(\cdot, S)$, that is,

$$d(x, S) := \inf_{y \in S} d(x, y)$$

along with $d(x, S) = +\infty$ whenever S is empty. It is usual for each $x \in X$ to denote $\operatorname{Proj}_S x := \{u \in X : d(x, u) = \inf_{y \in S} d(x, y)\}$, and the dependence on x of the set $\operatorname{Proj}_S x$ defines a *multimapping*.

In addition to the foregoing examples of multimappings, in the case when a real-valued function f is defined on $X \times T$, where T appears as a set of parameters, the minimization of $f(x, t)$ in x leads naturally to consider, for each $t \in T$, the set

$$\operatorname{Argmin}_X f(\cdot, t) := \operatorname{Argmin}_{x \in X} f(x,t) := \left\{ u \in X : f(u,t) = \inf_{x \in X} f(x,t) \right\}$$

of solutions of the minimization problem; so is defined the *multimapping* assigning to each $t \in T$ the set $\operatorname{Argmin}_X f(\cdot, t)$. As will be seen later in the book, several other concepts involved in the theory of variational analysis when X is a normed space, like tangent cone, normal cone, and subdifferential, induce multimappings from X into either X or its topological dual X^*. So, we start with some generalities and then with the concepts of semilimits and semicontinuities for multimappings.

1.1. Generalities on multimappings

Let T and X be two nonempty sets. One defines a *multimapping* (or *multifunction, set-valued mapping*) $M : T \rightrightarrows X$ from T into X as the correspondence which assigns to each $t \in T$ a set $M(t)$ included in X. The set $M(t)$ is called the *image of the point t under M*. If A is a subset of T, its *image under M* is defined as the set

$$M(A) := \{x \in X : x \in M(t) \text{ for some } t \in A\},$$

so $M(A) = \bigcup_{t \in A} M(t)$; the set $M(T)$ is called the *range* of M and it is denoted by Range M or $\operatorname{Im} M$. The *inverse image of a set* $B \subset X$ *under* M is defined by

$$\overset{-1}{M}(B) := \{t \in T : M(t) \cap B \neq \emptyset\}$$

and the *effective domain* of the multimapping M is

$$\operatorname{Dom} M := \{t \in T : M(t) \neq \emptyset\},$$

hence $\operatorname{Dom} M = \overset{-1}{M}(X)$. When $M(t)$ is closed (resp. compact etc) for all $t \in A$, one says that M is *closed-valued* (resp. *compact-valued* etc) on A. The set

$$\operatorname{gph} M := \{(t, x) \in T \times X : x \in M(t)\}$$

is called the *graph* of M. It is worth pointing out that the graph $\operatorname{gph} M$ is taken as a subset of $T \times X$ but not of $T \times 2^X$, where 2^X denotes the power set of X, that is, the set of all subsets of X. Doing so, we make an *important difference* between the multimapping M and the mapping $\widetilde{M} : T \to 2^X$ given by $T \ni t \mapsto M(t) \in 2^X$. The difference also appears through the definition of $M(A)$ above which is a subset of X but not the subset of 2^X given by $\{M(t) : t \in A\}$.

REMARK 1.1. Given a nonempty subset S of T and a mapping $\phi : S \to X$, it will be often convenient to associate to ϕ the multimapping $M_\phi : T \rightrightarrows X$ defined by $M_\phi(t) = \{\phi(t)\}$ if $t \in S$ and $M_\phi(t) = \emptyset$ if $t \in T \setminus S$. The obvious equality $\operatorname{Dom} M_\phi = S$ leads to also denote $\operatorname{Dom} \phi := S$. □

The *inverse* of M is defined as the multimapping $M^{-1} : X \rightrightarrows T$ which assigns to each $x \in X$ the set

$$M^{-1}(x) := \{t \in T : x \in M(t)\},$$

thus for a set $B \subset X$ it is easily seen that $M^{-1}(B)$ coincides with the inverse image $\overset{-1}{M}(B)$ as defined above, that is,

$$M^{-1}(B) = \{t \in T : M(t) \cap B \neq \emptyset\}.$$

Therefore, we will use only the symbol $M^{-1}(B)$. Obviously,

$$\operatorname{Dom} M^{-1} = \operatorname{Range} M \quad \text{and} \quad \operatorname{Range} M^{-1} = \operatorname{Dom} M,$$

so unlike the inverse image under *a (classic) mapping*, $M^{-1}(X)$ does not coincide with T; the latter fact is due to the possibility that it may happen that $M(t) = \emptyset$ for some elements $t \in T$.

Evidently $M(A) \subset M(A')$ whenever $A \subset A' \subset T$, and for a family $(A_i)_{i \in I}$ of subsets of T it is readily seen that

$$M\left(\bigcup_{i \in I} A_i\right) = \bigcup_{i \in I} M(A_i) \quad \text{and} \quad M\left(\bigcap_{i \in I} A_i\right) \subset \bigcap_{i \in I} M(A_i);$$

easy examples can be found where the latter inclusion is not an equality. The inverse image M^{-1} being a multimapping, for a family $(B_i)_{i \in I}$ of subsets of X we also have

$$M^{-1}\left(\bigcup_{i \in I} B_i\right) = \bigcup_{i \in I} M^{-1}(B_i) \quad \text{and} \quad M^{-1}\left(\bigcap_{i \in I} B_i\right) \subset \bigcap_{i \in I} M^{-1}(B_i);$$

thus, unlike the inverse image under *a (classic) mapping*, the inverse image of an intersection of sets is not equal to (but only included in) the intersection of the inverse images of these sets. Note also that for a set $B \subset X$ there is no relationship between $M^{-1}(X \setminus B)$ and $T \setminus M^{-1}(B)$.

Consider now a family $(M_i)_{i \in I}$ of multimappings from T into X and define $M, N : T \rightrightarrows X$ as the pointwise union and pointwise intersection of the family of multimappings, that is,

$$M(t) := \bigcup_{i \in I} M_i(t) \quad \text{and} \quad N(t) := \bigcap_{i \in I} N_i(t) \quad \text{for all } t \in T.$$

It is not difficult to check that

$$\operatorname{gph} M = \bigcup_{i \in I} \operatorname{gph} M_i \quad \text{and} \quad \operatorname{gph} N = \bigcap_{i \in I} \operatorname{gph} N_i.$$

Let $N : X \rightrightarrows Y$ be a multimapping. One defines the composition multimapping $N \circ M : T \rightrightarrows Y$ by

$$N \circ M(t) := N(M(t)) = \bigcup_{x \in M(t)} N(x).$$

It is an exercise to see, for $A \subset T$ and $C \subset Y$, that

$$(N \circ M)(A) = N(M(A)) \quad \text{and} \quad (N \circ M)^{-1}(C) = M^{-1}(N^{-1}(C)).$$

If X is a vector space and $Y = X$, the addition of M and N, and the scalar multiplication of M by a real λ is defined by

$$(M + N)(t) := M(t) + N(t) \quad \text{and} \quad (\lambda M)(t) := \lambda M(t) \quad \text{for all } t \in T,$$

where $C + C' := \{x + x' : x \in C, x' \in C'\}$ and $\lambda C := \{\lambda x : x \in C\}$ are, for subsets $C, C' \subset X$, the *Minkowski addition* of C and C', and the *Minkowski scalar multiplication* of C by λ.

1.2. Semilimits of multimappings

We assume, throughout this section (unless otherwise stated), that T and X are endowed with a topology τ_T and a topology τ_X respectively, and we consider a nonempty subset $T_0 \subset T$ and a point $\bar{t} \in \operatorname{cl} T_0$. We recall that we write $T_0 \ni t \to \bar{t}$ or $t \underset{T_0}{\to} \bar{t}$ to mean that $t \to \bar{t}$ with $t \in T_0$.

As usual, if T_0 is a neighborhood of $\bar{t}$, we write only $t \to \bar{t}$. Let us point out other usual particular cases where T_0 will be omitted in the convergence notation. If $T = \mathbb{R} \cup \{-\infty, +\infty\}$ is endowed with its natural topology (which is known to be metrizable) and T_0 is a subset of $\mathbb{R}$ containing an interval of the form $]a, +\infty[$, unless otherwise stated $t \to +\infty$ will be written in place of $T_0 \ni t \to +\infty$. For $\mathbb{N} \cup \{+\infty\}$ endowed with the metrizable topology induced by that of $\mathbb{R} \cup \{-\infty, +\infty\}$, unless otherwise specified we will write $n \to \infty$ instead of $\mathbb{N} \ni n \to +\infty$. Also $t \downarrow \bar{r}$ will mean $]\bar{r}, \bar{s}[\ni t \to \bar{r}$ for some $\bar{r}, \bar{s} \in [-\infty, +\infty]$ with $\bar{r} < \bar{s}$.

Let $\mathcal{N}(\bar{t}; \tau_T)$, or $\mathcal{N}(\bar{t})$ for short, denote the class of all neighborhoods of $\bar{t}$ and

$$\mathcal{N}_{T_0}(\bar{t}) := \{V \cap T_0 : V \in \mathcal{N}(\bar{t})\}.$$

In all this section, unless other specified, M is a multimapping from a subset of T containing T_0 into the topological space X.

DEFINITION 1.2. One defines the *inner limit* or *limit inferior* (resp. *outer limit* or *limit superior*) of the multimapping M at $\bar{t}$ relative to T_0 as
$$\liminf_{T_0 \ni t \to \bar{t}} M(t) := \{x \in X : \forall W \in \mathcal{N}(x), \exists V \in \mathcal{N}(\bar{t}), \forall t \in V \cap T_0, M(t) \cap W \neq \emptyset\}$$
(resp.
$$\limsup_{T_0 \ni t \to \bar{t}} M(t) := \{x \in X : \forall W \in \mathcal{N}(x), \forall V \in \mathcal{N}(\bar{t}), \exists t \in V \cap T_0, M(t) \cap W \neq \emptyset\}).$$
When both sets coincide, one says that $M(t)$ *inner-outer*, or *Painlevé-Peano* (or *Painlevé-Peano-Kuratowski*) *converges* as $T_0 \ni t \to \bar{t}$ to the common value, which is then called the *inner-outer limit*, or *Painlevé-Peano limit* (or *Painlevé-Peano-Kuratowski limit*) of M at $\bar{t}$ relative to T_0 and it is denoted by $\lim_{T_0 \ni t \to \bar{t}} M(t)$. If in addition, both inner and outer limits coincide with $M(\bar{t})$ for $\bar{t} \in T_0$, one says that M is *inner-outer*, or *Painlevé-Peano* (or *Painlevé-Peano-Kuratowski*) *continuous* at $\bar{t}$ relative to T_0.

In addition to the situation of families of sets which will be the main focus of this chapter, we will also see that the counterpart to families of epigraphs is involved in many results. Recall that the *epigraph* epi f of a function $f : X \to \mathbb{R} \cup \{-\infty, +\infty\}$ is defined by
$$\operatorname{epi} f := \{(x, r) \in X \times \mathbb{R} : f(x) \leq r\}.$$
Using the multimapping $M : T \rightrightarrows X \times \mathbb{R}$ with $M(t) := \operatorname{epi} f_t$, where $f_t : X \to \mathbb{R} \cup \{-\infty, +\infty\}$ the convergence counterpart of Definition 1.2 for families of functions is the following:

DEFINITION 1.3. Let $(f_t)_{t \in T_0}$ be a family of functions from X into $\mathbb{R} \cup \{-\infty, +\infty\}$, where T_0 is as above. One says that this family *inner-outer converges*, or *Painlevé-Peano converges* (or *Painlevé-Peano-Kuratowski converges*) to a function $f : X \to \mathbb{R} \cup \{-\infty, +\infty\}$ as $T_0 \ni t \to \bar{t}$ if (epi f_t) converges to epi f in $X \times \mathbb{R}$ as $T_0 \ni t \to \bar{t}$.

The name *Painlevé-Peano limit* (or *Painlevé-Peano-Kuratowski limit*) is attributed in Definition 1.2 because of contributions of these authors to the development of this concept of convergence when one regards M as a family of sets $(M(t))_{t \in T}$; see comments in Section 1.10.

Of course, in Definition 1.2 above the classes of neighborhoods $\mathcal{N}(x)$ and $\mathcal{N}(\bar{t})$ can be replaced by fundamental systems (or bases) of neighborhoods of x and $\bar{t}$. According to the convention above, when T_0 is a neighborhood of $\bar{t}$, we will write $t \to \bar{t}$ in place of $T_0 \ni t \to \bar{t}$ under the symbols Lim inf and Lim sup. Also, taking $\mathbb{N}$ for T_0 as subset of $\mathbb{N} \cup \{\infty\}$ we will write $n \to \infty$ under the above symbols. In this case, for a sequence $(S_n)_{n \in \mathbb{N}}$ of subsets of X, putting $M(n) := S_n$ for each $n \in \mathbb{N}$ and writing
$$\liminf_{n \to \infty} S_n \quad \text{and} \quad \limsup_{n \to \infty} S_n$$
in place of the corresponding semilimits of M we see through the natural basis of neighborhoods of ∞ in $\mathbb{N} \cup \{\infty\}$ that

(1.2) $\qquad \liminf_{n \to \infty} S_n = \{x \in X : \forall W \in \mathcal{N}(x), \exists k \in \mathbb{N}, \forall n \geq k, S_n \cap W \neq \emptyset\}$

and

(1.3) $\qquad \limsup_{n \to \infty} S_n = \{x \in X : \forall W \in \mathcal{N}(x), \forall k \in \mathbb{N}, \exists n \geq k, S_n \cap W \neq \emptyset\}.$

If both semilimits coincide, one says that the sequence $(S_n)_n$ *inner-outer converges* (also *Painlevé-Peano converges*, or *Painlevé-Peano-Kuratowwski converges*) to the common value of semilimits.

Consider the situation when T and T_0 are in the forms $T = I \times U$ and $T_0 = I_0 \times U_0$, where U is a topological space, U_0 a nonempty subset of U, I an interval of $\mathbb{R}$ and I_0 a subinterval of I. Denote by $\bar{r} \in [-\infty, +\infty[$ the left-hand extremity of I_0 and suppose that $\bar{r} \notin I$. For $\bar{u} \in \operatorname{cl}_U U_0$, it will be often convenient to write the semilimits with subscript "$r \downarrow \bar{r}, u \xrightarrow[U_0]{} \bar{u}$", that is,

(1.4)
$$\operatorname{Lim\,inf}_{r \downarrow \bar{r}, u \xrightarrow[U_0]{} \bar{u}} M(r,u) := \operatorname{Lim\,inf}_{T_0 \ni (r,u) \to (\bar{r},\bar{u})} M(r,u), \quad \operatorname{Lim\,sup}_{r \downarrow \bar{r}, u \xrightarrow[U_0]{} \bar{u}} M(r,u) := \operatorname{Lim\,sup}_{T_0 \ni (r,u) \to (\bar{r},\bar{u})} M(r,u).$$

EXAMPLE 1.4. (a) In $\mathbb{R}$ considering $S_n = \{(-1)^n\}$, $S'_n = \{1/n, (-1)^n\}$, and $S''_n = [n, n+1]$ for each $n \in \mathbb{N}$, one has

$$\operatorname{Lim\,inf}_{n \to \infty} S_n = \operatorname{Lim\,inf}_{n \to \infty} S''_n = \emptyset \text{ and } \operatorname{Lim\,inf}_{n \to \infty} S'_n = \{0\},$$

$$\operatorname{Lim\,sup}_{n \to \infty} S_n = \{-1, 1\}, \operatorname{Lim\,sup}_{n \to \infty} S'_n = \{-1, 0, 1\}, \text{ and } \operatorname{Lim\,sup}_{n \to \infty} S''_n = \emptyset.$$

(b) Let S' and S'' be subsets of X and $S_n = S'$ if n is odd and $S_n = S''$ if n is even. Then

$$\operatorname{Lim\,inf}_{n \to \infty} S_n = (\operatorname{cl} S') \cap (\operatorname{cl} S'') \text{ and } \operatorname{Lim\,sup}_{n \to \infty} S_n = (\operatorname{cl} S') \cup (\operatorname{cl} S'').$$

(c) Assume that (X, τ_X) is a topological vector space and that $M(t)$ is convex for all $t \in T_0$. Then $\operatorname{Lim\,inf}_{T_0 \ni t \to \bar{t}} M(t)$ is convex, but the example in (b) makes clear that $\operatorname{Lim\,sup}_{T_0 \ni t \to \bar{t}} M(t)$ can fail to be convex.

(d) If $(S_n)_{n \in \mathbb{N}}$ is a nondecreasing sequence of subsets of X in the sense that $S_n \subset S_{n+1}$ for every $n \in \mathbb{N}$, then one has

$$\operatorname{Lim\,inf}_{n \to \infty} S_n = \operatorname{Lim\,sup}_{n \to \infty} S_n = \operatorname{cl}\Big(\bigcup_{n \in \mathbb{N}} S_n\Big).$$

(e) Similarly, if $(S_n)_{n \in \mathbb{N}}$ is a nonincreasing sequence of subsets of X in the sense that $S_{n+1} \subset S_n$ for every $n \in \mathbb{N}$, one has

$$\operatorname{Lim\,inf}_{n \to \infty} S_n = \operatorname{Lim\,sup}_{n \to \infty} S_n = \bigcap_{n \in \mathbb{N}} \operatorname{cl} S_n.$$

$\square$

The first properties below follow directly from the definition.

PROPOSITION 1.5. *Let T' and Y be other topological spaces, $T'_0 \subset T'$ and $\bar{t}' \in \operatorname{cl} T'_0$. Let N be a multimapping from a subset of T containing T_0 into Y and M' be a multimapping from a subset of T' containing T'_0 into Y.*
(a) *One has*

$$\operatorname{Lim\,inf}_{T_0 \ni t \to \bar{t}} M(t) \times N(t) = \Big(\operatorname{Lim\,inf}_{T_0 \ni t \to \bar{t}} M(t)\Big) \times \Big(\operatorname{Lim\,inf}_{T_0 \ni t \to \bar{t}} N(t)\Big),$$

$$\Big(\operatorname{Lim\,sup}_{T_0 \ni t \to \bar{t}} M(t)\Big) \times \Big(\operatorname{Lim\,inf}_{T_0 \ni t \to \bar{t}} N(t)\Big) \subset \operatorname{Lim\,sup}_{T_0 \ni t \to \bar{t}} M(t) \times N(t),$$

$$\operatorname*{Lim\,sup}_{T_0 \ni t \to \bar t} M(t) \times N(t) \subset \left(\operatorname*{Lim\,sup}_{T_0 \ni t \to \bar t} M(t)\right) \times \left(\operatorname*{Lim\,sup}_{T_0 \ni t \to \bar t} N(t)\right),$$

and the latter inclusion with outer limit is an equality whenever $\left(\operatorname*{Lim\,sup}_{T_0 \ni t \to \bar t} N(t)\right) = \left(\operatorname*{Lim\,inf}_{T_0 \ni t \to \bar t} N(t)\right)$.

(b) In the case of separate variables the following equality is valid

$$\operatorname*{Lim\,sup}_{T_0 \times T_0' \ni (t,t') \to (\bar t, \bar t')} M(t) \times M'(t') = \left(\operatorname*{Lim\,sup}_{T_0 \ni t \to \bar t} M(t)\right) \times \left(\operatorname*{Lim\,sup}_{T_0' \ni t' \to \bar t'} M'(t')\right).$$

REMARK 1.6. In general, the inclusion concerning the outer limit (or limit superior) is not an equality. For $T_0 = \mathbb{N}$ and $M(n) = N(n) = S_n$ with $S_n = \{(-1)^n\}$ for each $n \in \mathbb{N}$, we see that

$$\operatorname*{Lim\,sup}_{n \to \infty} M(n) \times N(n) = \{(-1,-1),(1,1)\}$$

and $\left(\operatorname*{Lim\,sup}_{n \to \infty} M(n)\right) \times \left(\operatorname*{Lim\,sup}_{n \to \infty} N(n)\right) = \{-1,1\} \times \{-1,1\}$,

which confirms the claim. □

It is readily seen from the definition that the outer limit (or limit superior) of M is described as in Proposition 1.7 below as an intersection of appropriate closed sets. Considering the *grill* of $\mathcal{N}_{T_0}(\bar t)$ defined by

$$\mathcal{N}_{T_0}^{\#}(\bar t) := \{P \subset T_0 : \forall U \in \mathcal{N}_{T_0}(\bar t),\ P \cap U \neq \emptyset\},$$

a similar description can also be obtained for the inner limit (or limit inferior) as follows.

PROPOSITION 1.7. *With notation above we have*

$$\operatorname*{Lim\,inf}_{T_0 \ni t \to \bar t} M(t) = \bigcap_{P \in \mathcal{N}_{T_0}^{\#}(\bar t)} \operatorname{cl}\left(\bigcup_{t \in P} M(t)\right) \text{ and } \operatorname*{Lim\,sup}_{T_0 \ni t \to \bar t} M(t) = \bigcap_{V \in \mathcal{N}_{T_0}(\bar t)} \operatorname{cl}\left(\bigcup_{t \in V} M(t)\right),$$

so both semilimits are closed sets of X.

PROOF. Only the left equality needs to be proved. The first member of the first equality being obviously included in the second one, fix any $x \notin \operatorname*{Lim\,inf}_{T_0 \ni t \to \bar t} M(t)$. According to the definition of the inner limit of M, there exists some $W \in \mathcal{N}(x)$ such that for any $V \in \mathcal{N}_{T_0}(\bar t)$ there is $t_V \in V \cap T_0$ such that $M(t_V) \cap W = \emptyset$. Then setting $P_W := \{t \in T_0 : M(t) \cap W = \emptyset\}$ we see that $P_W \cap V \neq \emptyset$ for all $V \in \mathcal{N}_{T_0}(\bar t)$, that is, $P_W \in \mathcal{N}_{T_0}^{\#}(\bar t)$. Further, from the definition of P_W we have $W \cap \left(\bigcup_{t \in P_W} M(t)\right) = \emptyset$, which ensures $x \notin \operatorname{cl}\left(\bigcup_{t \in P_W} M(t)\right)$. So, x does not belong to the second member of the left equality of the proposition and the proof is complete. □

REMARK 1.8. In addition to the above descriptions, setting

$$B(S,\varepsilon) := \bigcup_{x \in S} B(x,\varepsilon)$$

for any subset S of a metric space, it is worth mentioning the following other descriptions which are easy consequences of Definition 1.2.
(a) If X is a metric space, then

$$\operatorname*{Lim\,sup}_{T_0 \ni t \to \bar{t}} M(t) = \bigcap_{\varepsilon > 0} \bigcap_{V \in \mathcal{N}_{T_0}(\bar{t})} \bigcup_{t \in V} B(M(t), \varepsilon),$$

$$\operatorname*{Lim\,sup}_{n \to \infty} S_n = \bigcap_{\varepsilon > 0} \bigcap_{k \in \mathbb{N}} \bigcup_{n \geq k} B(S_n, \varepsilon),$$

$$\operatorname*{Lim\,inf}_{T_0 \ni t \to \bar{t}} M(t) = \bigcap_{\varepsilon > 0} \bigcup_{V \in \mathcal{N}_{T_0}(\bar{t})} \bigcap_{t \in V} B(M(t), \varepsilon),$$

$$\operatorname*{Lim\,inf}_{n \to \infty} S_n = \bigcap_{\varepsilon > 0} \bigcup_{k \in \mathbb{N}} \bigcap_{n \geq k} B(S_n, \varepsilon).$$

(b) If X is a topological vector space, then

$$\operatorname*{Lim\,sup}_{T_0 \ni t \to \bar{t}} M(t) = \bigcap_{W \in \mathcal{N}_X(0)} \bigcap_{V \in \mathcal{N}_{T_0}(\bar{t})} \bigcup_{t \in V} (M(t) + W),$$

$$\operatorname*{Lim\,sup}_{n \to \infty} S_n = \bigcap_{W \in \mathcal{N}_X(0)} \bigcap_{k \in \mathbb{N}} \bigcup_{n \geq k} (S_n + W),$$

$$\operatorname*{Lim\,inf}_{T_0 \ni t \to \bar{t}} M(t) = \bigcap_{W \in \mathcal{N}_X(0)} \bigcup_{V \in \mathcal{N}_{T_0}(\bar{t})} \bigcap_{t \in V} (M(t) + W),$$

$$\operatorname*{Lim\,inf}_{n \to \infty} S_n = \bigcap_{W \in \mathcal{N}_X(0)} \bigcup_{k \in \mathbb{N}} \bigcap_{n \geq k} (S_n + W).$$

$\square$

More generally the semilimits can be defined along a filter basis. Recall that a family $\mathcal{B}$ of subsets of a set S is called a *filter basis* whenever all $B \in \mathcal{B}$ are nonempty and, for any $B_1, B_2 \in \mathcal{B}$ there is $B_3 \in \mathcal{B}$ with $B_3 \subset B_1 \cap B_2$. Given a multimapping $M : S \rightrightarrows X$, one defines the *inner limit* or *limit inferior* (resp. *outer limit* or *limit superior*) of the multimapping M along the filter basis $\mathcal{B}$ as

$$\operatorname*{Lim\,inf}_{\mathcal{B}} M := \{x \in X : \forall W \in \mathcal{N}(x), \exists B \in \mathcal{B}, \forall t \in B,\ M(t) \cap W \neq \emptyset\}$$

and

$$\operatorname*{Lim\,sup}_{\mathcal{B}} M := \{x \in X : \forall W \in \mathcal{N}(x), \forall B \in \mathcal{B}, \exists t \in B,\ M(t) \cap W \neq \emptyset\}.$$

When both sets coincide, their common value $\operatorname*{Lim}_{\mathcal{B}} M$ is the inner-outer (or Painlevé-Peano, or Painlevé-Peano-Kuratowski) limit of M along the filter basis $\mathcal{B}$.

REMARK 1.9. (a) Defining the grill $\mathcal{B}^\#$ of $\mathcal{B}$ as

$$\mathcal{B}^\# := \{P \subset S : \forall B \in \mathcal{B},\ P \cap B \neq \emptyset\},$$

and proceeding as in Proposition 1.7, one obtains

$$\operatorname*{Lim\,inf}_{\mathcal{B}} M = \bigcap_{P \in \mathcal{B}^\#} \operatorname{cl}\left(\bigcup_{t \in P} M(t)\right) \quad \text{and} \quad \operatorname*{Lim\,sup}_{\mathcal{B}} M = \bigcap_{B \in \mathcal{B}} \operatorname{cl}\left(\bigcup_{t \in B} M(t)\right),$$

so both semilimits are closed sets of X.
(b) If X is a topological vector space, one easily sees as in Remark 1.8 that

$$\operatorname*{Lim\,sup}_{\mathcal{B}} M = \bigcap_{W \in \mathcal{N}_X(0)} \bigcap_{B \in \mathcal{B}} \bigcup_{t \in B} (M(t) + W),$$

$$\operatorname*{Lim\,inf}_{\mathcal{B}} M = \bigcap_{W \in \mathcal{N}_X(0)} \bigcup_{B \in \mathcal{B}} \bigcap_{t \in B} (M(t) + W).$$

$\square$

The following corollary is easily derived from Proposition 1.7.

COROLLARY 1.10. *Let $M, N : T \rightrightarrows X$ be multimappings. Then the following hold.*
(a) *If $M(t) \subset N(t) \subset \operatorname{cl} M(t)$ for all t near $\bar{t}$, then*

$$\operatorname*{Lim\,inf}_{T_0 \ni t \to \bar{t}} M(t) = \operatorname*{Lim\,inf}_{T_0 \ni t \to \bar{t}} N(t) \quad \text{and} \quad \operatorname*{Lim\,sup}_{T_0 \ni t \to \bar{t}} M(t) = \operatorname*{Lim\,sup}_{T_0 \ni t \to \bar{t}} N(t).$$

(b) *If $\bar{t} \in T_0$, then*

$$\operatorname*{Lim\,inf}_{T_0 \ni t \to \bar{t}} M(t) \subset \operatorname{cl}(M(\bar{t})) \subset \operatorname*{Lim\,sup}_{T_0 \ni t \to \bar{t}} M(t).$$

PROOF. The equalities in (a) are direct consequences of Proposition 1.7 above. Concerning the inclusions in (b) under the assumption $\bar{t} \in T_0$, according to Proposition 1.7 again, the second inclusion follows from the relation $\bar{t} \in V \cap T_0$ for all $V \in \mathcal{N}(\bar{t})$, and the first inclusion follows from the observation $\{\bar{t}\} \in \mathcal{N}_{T_0}^{\#}(\bar{t})$. $\square$

The case of iterated outer semilimits will occur in various chapters.

PROPOSITION 1.11. *With the above notation the following inclusion holds*

$$\operatorname*{Lim\,sup}_{T_0 \ni t \to \bar{t}} \operatorname*{Lim\,sup}_{T_0 \ni \tau \to t} M(\tau) \subset \operatorname*{Lim\,sup}_{T_0 \ni t \to \bar{t}} M(t),$$

and the inclusion is an equality whenever $\bar{t} \in T_0$.

PROOF. Fix any x in the left-hand side, and take any open neighborhood W of x in X and any open neighborhood V of $\bar{t}$ in T. By Definition 1.2 there is some $t \in V \cap T_0$ such that $W \cap \operatorname*{Lim\,sup}_{T_0 \ni \tau \to t} M(\tau) \neq \emptyset$. Choosing $u \in W \cap \operatorname*{Lim\,sup}_{T_0 \ni \tau \to t} M(\tau)$ and noting that W is a neighborhood of u and V is a neighborhood of t, Definition 1.2 again furnishes some $\tau \in V \cap T_0$ such that $W \cap M(\tau) \neq \emptyset$. We deduce through Definition 1.2 that $x \in \operatorname*{Lim\,sup}_{T_0 \ni t \to \bar{t}} M(t)$. Therefore, (a) is justified.

Now assume that $\bar{t} \in T_0$ and put $\Gamma(t) := \operatorname*{Lim\,sup}_{T_0 \ni \tau \to t} M(\tau)$ for each $t \in T_0$. By Corollary 1.10(b) we have $\Gamma(\bar{t}) \subset \operatorname*{Lim\,sup}_{T_0 \ni t \to \bar{t}} \Gamma(t)$, which is exactly the converse inclusion of the previous one. This confirms the desired equality. $\square$

When the topology τ_X is metrizable, that is, generated by a distance d, the semilimits can be characterized through the usual lower and upper limits of the distance function as stated in the next proposition. We limit ourselves to the situation either involving $\bar{t} \in \operatorname{cl}(T_0)$ or with $\mathbb{N}$; the case related to a filter basis is similar. Given a sequence $(r_n)_{n \in \mathbb{N}}$ of extended reals in $\mathbb{R} \cup \{-\infty, +\infty\}$ we know

that its *limit inferior* or *lower limit* and its *limit superior* or *upper limit* are defined as
$$\liminf_{n\to\infty} r_n := \sup_{n\in\mathbb{N}} \inf_{k\geq n} r_k \quad \text{and} \quad \limsup_{n\to\infty} r_n := \inf_{n\in\mathbb{N}} \sup_{k\geq n} r_k,$$
and we also know that $\liminf_{n\to\infty} r_n$ (resp. $\limsup_{n\to\infty} r_n$) is the minimum (resp. maximum) of limits in $\mathbb{R} \cup \{-\infty, +\infty\}$ of convergent (in $\mathbb{R} \cup \{-\infty, +\infty\}$) subsequences of $(r_n)_{n\in\mathbb{N}}$. Regarding an extended real-valued function $\varphi : T_0 \to \mathbb{R} \cup \{-\infty, +\infty\}$ defined on the subset T_0 of the topological space (T, τ_T), recall that its usual *limit inferior* or *lower limit* and *limit superior* or *upper limit* are defined as
$$\liminf_{T_0 \ni t \to \bar{t}} \varphi(t) := \sup_{V \in \mathcal{N}_{T_0}(\bar{t})} \inf_{t \in V} \varphi(t) \quad \text{and} \quad \limsup_{T_0 \ni t \to \bar{t}} \varphi(t) := \inf_{V \in \mathcal{N}_{T_0}(\bar{t})} \sup_{t \in V} \varphi(t);$$
the above situation of sequences is then a particular case with $T_0 = \mathbb{N}$ and $T = \mathbb{N} \cup \{+\infty\}$ equipped with its canonic topology. When the topological space (T, τ_T) is metrizable, it is known (and not difficult to see) that the following sequential descriptions hold true:
$$\liminf_{T_0 \ni t \to \bar{t}} \varphi(t) = \min \left\{ \lim_{n\to\infty} \varphi(t_n) : T_0 \ni t_n \to \bar{t} \right\}$$
and
$$\limsup_{T_0 \ni t \to \bar{t}} \varphi(t) = \max \left\{ \lim_{n\to\infty} \varphi(t_n) : T_0 \ni t_n \to \bar{t} \right\},$$
that is, the lower (resp. upper) limit is the least (resp. greatest) of limits $\lim_{n\to\infty} \varphi(t_n)$ such that the sequence $(t_n)_{n\in\mathbb{N}}$ in T_0 converges to $\bar{t}$.

PROPOSITION 1.12. *Assume that (X, d) is a metric space and let $x \in X$. Then, the following hold:*
(a) *One has $x \in \operatorname*{Liminf}_{T_0 \ni t \to \bar{t}} M(t)$ if and only if $\limsup_{T_0 \ni t \to \bar{t}} d(x, M(t)) = 0$, or equivalently $\lim_{T_0 \ni t \to \bar{t}} d(x, M(t)) = 0$.*
(b) *One has $x \in \operatorname*{Limsup}_{T_0 \ni t \to \bar{t}} M(t)$ if and only if $\liminf_{T_0 \ni t \to \bar{t}} d(x, M(t)) = 0$.*

PROOF. It is enough to invoke the above definition of upper (resp. lower) limit of real-valued functions and to use in the definition of semilimits of M the basis of neighborhoods formed by the open balls centered at x. □

Proposition 1.12 allows us to prove the following sequential characterization for a sequence of sets.

PROPOSITION 1.13. *Assume that X is a metric space and let $(S_n)_{n\in\mathbb{N}}$ be a sequence of subsets of X. Then, given $x \in X$ the following hold:*
(a) *$x \in \operatorname*{Liminf}_{n\to\infty} S_n$ if and only if there exists a sequence $(x_n)_{n\in\mathbb{N}}$ converging to x with $x_n \in S_n$ for n large enough.*
(b) *$x \in \operatorname*{Limsup}_{n\to\infty} S_n$ if and only if there exist a sequence $(x_n)_{n\in\mathbb{N}}$ converging to x and an increasing mapping $s : \mathbb{N} \to \mathbb{N}$ such that $x_n \in S_{s(n)}$ for all $n \in \mathbb{N}$.*

PROOF. (a) Only the direct implication needs to be proved for (a) since the reverse one is immediate. So, fix $x \in \operatorname*{Liminf}_{n\to\infty} S_n$. First, we observe that this inclusion entails that $S_n \neq \emptyset$ for n large enough, say $n \geq N$. If for each integer

$n \geq N$ we set $r_n := d(x, S_n) + \frac{1}{n}$, there exists some $x_n \in S_n$ such that $d(x_n, x) < r_n$. Taking any $x_n \in X$ for $n < N$, it appears that $x_n \to x$ since $r_n \to 0$ according to the assertion (a) of Proposition 1.12, and this translates the desired implication.

(b) Again, it is enough to establish the direct implication. Let $x \in \underset{n \to \infty}{\operatorname{Lim\,sup}} S_n$. From (b) of Proposition 1.12, there is an increasing mapping $s : \mathbb{N} \to \mathbb{N}$ such that $\lim_{n \to \infty} d(x, S_{s(n)}) = 0$ along with $d(x, S_{s(n)}) < +\infty$ for all $n \in \mathbb{N}$. Then, it suffices to put $r_n := d(x, S_{s(n)}) + \frac{1}{n}$ for all $n \in \mathbb{N}$ and to argue like in (a) above. □

When both T and X are metrizable spaces, sequential descriptions can also be given for multimappings.

PROPOSITION 1.14. *Assume that T and X are metric spaces. Then, given $x \in X$ the following sequential characterizations hold.*

(a) $x \in \underset{T_0 \ni t \to \bar{t}}{\operatorname{Lim\,inf}} M(t)$ *if and only if for any sequence $(t_n)_{n \in \mathbb{N}}$ in T_0 converging to $\bar{t}$ there exists a sequence $(x_n)_{n \in \mathbb{N}}$ in X converging to x and such that $x_n \in M(t_n)$ for n large enough. This means that $x \in \underset{T_0 \ni t \to \bar{t}}{\operatorname{Lim\,inf}} M(t)$ if and only if for any sequence $(t_n)_{n \in \mathbb{N}}$ in T_0 converging to $\bar{t}$ one has $x \in \underset{n \to \infty}{\operatorname{Lim\,inf}} M(t_n)$.*

Another sequential characterization of the inclusion $x \in \underset{T_0 \ni t \to \bar{t}}{\operatorname{Lim\,inf}} M(t)$ is that for any sequence $(t_n)_{n \in \mathbb{N}}$ in T_0 converging to $\bar{t}$ there exist a subsequence $(t_{s(n)})_{n \in \mathbb{N}}$ and a sequence $(x_n)_{n \in \mathbb{N}}$ in X converging to x and such that $x_n \in M(t_{s(n)})$ for all $n \in \mathbb{N}$.

(b) $x \in \underset{T_0 \ni t \to \bar{t}}{\operatorname{Lim\,sup}} M(t)$ *if and only if there exist a sequence $(t_n)_{n \in \mathbb{N}}$ in T_0 converging to $\bar{t}$ and a sequence $(x_n)_{n \in \mathbb{N}}$ converging to x in X and such that $x_n \in M(t_n)$ for all integers n.*

PROOF. (a) Fix $x \in \underset{T_0 \ni t \to \bar{t}}{\operatorname{Lim\,inf}} M(t)$ and consider any sequence $(t_n)_{n \in \mathbb{N}}$ in T_0 converging to $\bar{t}$. From Proposition 1.12 we have $\lim_{n \to \infty} d(x, M(t_n)) = 0$ and for large n, say $n \geq N$, we can choose $x_n \in M(t_n)$ with $d(x, x_n) < d(x, M(t_n)) + \frac{1}{n}$. Taking any $x_n \in X$ for $n < N$, the sequence $(x_n)_{n \in \mathbb{N}}$ fulfills the desired property.

Conversely, suppose that $x \notin \underset{T_0 \ni t \to \bar{t}}{\operatorname{Lim\,inf}} M(t)$. There exists a neighborhood $W \in \mathcal{N}(x)$ such that for each integer $n \in \mathbb{N}$ there exists $t_n \in T_0 \cap B(\bar{t}, 1/n)$ with $W \cap M(t_n) = \emptyset$. The sequence $(t_n)_{n \in \mathbb{N}}$ fulfills the negation of the property of the right-hand member of the first equivalence of (a). This finishes the proof of the first sequential characterization in (a).

Concerning the subsequential property, we observe that it can be translated by saying that for any sequence $(t_n)_n$ in T_0 converging to $\bar{t}$ there exists some subsequence $(t_{s(n)})_n$ such that $\lim_{n \to \infty} d(x, M(t_{s(n)})) = 0$, which is easily seen to be equivalent to $\underset{T_0 \ni t \to \bar{t}}{\lim} d(x, M(t)) = 0$. So, the subsequential property is a characterization of the inclusion $x \in \underset{T_0 \ni t \to \bar{t}}{\operatorname{Lim\,inf}} M(t)$, and this finishes the proof of (a).

(b) By Proposition 1.12 again the inclusion $x \in \underset{T_0 \ni t \to \bar{t}}{\operatorname{Lim\,sup}} M(t)$ holds if and only if $\underset{T_0 \ni t \to \bar{t}}{\operatorname{lim\,inf}} d(x, M(t)) = 0$, and it is not difficult to see that the latter equality is equivalent to the property of the right-hand member of the assertion in (b). □

Characterizations similar to those in Proposition 1.14 when T and X are general topological spaces involve nets in place of sequences. Recall that a *net* $(t_j)_{j \in J}$ of the set T is a family of elements of T where $(J, \preceq_J)$ is a *directed set*, that is, $\preceq_J$ is a preorder on the nonempty set J such that for any pair $j_1, j_2 \in J$ there exists some $j_3 \in J$ with $j_1 \preceq_J j_3$ and $j_2 \preceq_J j_3$. It is often convenient to write $j_3 \succeq_J j_2$ in place of $j_2 \preceq j_3$. A *subnet* of $(t_j)_{j \in J}$ is any net of the form $(t_{s(i)})_{i \in I}$ where $(I, \preceq_I)$ is a directed set and $s : I \to J$ is a directed mapping, in the sense that for each $j_0 \in J$ there exists some $i_0 \in I$ such that $s(i) \succeq_J j_0$ for all $i \succeq_I i_0$. Endowing T with the topology τ_T, the net $(t_j)_{j \in j}$ converges to some $\bar{t} \in T$ provided for each neighborhood V of $\bar{t}$ there is some $j_V \in J$ such that $t_j \in V$ for all $j \succeq_J j_V$. Clearly, any subnet of a net converging to $\bar{t}$ converges also to $\bar{t}$.

EXAMPLE 1.15. (a) Let $\mathcal{B}$ be a basis of neighborhoods of $\bar{t} \in T$. Taking some $t_B \in B$, for each $B \in \mathcal{B}$, and endowing $\mathcal{B}$ with the directed preorder $\supset$, it is easily seen that $(t_B)_{B \in \mathcal{B}}$ is a net converging to $\bar{t}$.
(b) Similarly, if $\mathcal{B}$ is a filter basis of T converging to $\bar{t}$ (that is, for each neighborhood V of $\bar{t}$ there is $B \in \mathcal{B}$ with $B \subset V$), then the net $(t_B)_{B \in \mathcal{B}}$ (constructed as above) converges to $\bar{t}$.
(c) More generally, let $(J, \preceq_J)$ be a directed set and $\mathcal{B}$ be a filter basis of T converging to $\bar{t}$, for example, a basis of neighborhoods of $\bar{t} \in T$. Endow the set $J \times \mathcal{B}$ with the product preorder of $\preceq_J$ and $\supset$. If $t_{j,B} \in B$ for all $j \in J$ and $B \in \mathcal{B}$, then it is not difficult to see that $(t_{j,B})_{(j,B) \in J \times \mathcal{B}}$ is a net converging to $\bar{t}$.
(d) Let $(S_j)_{j \in J}$ be a net of subsets of the topological space X. Its inner and outer semilimits are defined as those of the multimapping $J \ni j \mapsto S_j$ along the (canonical) filter basis $\{\Sigma_j : j \in J\}$, where $\Sigma_j := \{i \in J : i \succeq_J j\}$. So, we have the following descriptions similar to those of sequences:
$$\operatorname*{Lim\,inf}_{j \in J} S_j = \{x \in X : \forall W \in \mathcal{N}(x), \exists i \in J, \forall j \succeq_J i, W \cap S_j \neq \emptyset\},$$
$$\operatorname*{Lim\,sup}_{j \in J} S_j = \{x \in X : \forall W \in \mathcal{N}(x), \forall i \in J, \exists j \succeq_J i, W \cap S_j \neq \emptyset\}.$$
(e) More generally, let $(M_j)_{j \in J}$ be a net of multimappings from T into the topological space X. The semilimits $\operatorname*{Lim\,inf}_{T_0 \ni t \to \bar{t}, j \in J} M_j(t)$ and $\operatorname*{Lim\,sup}_{T_0 \ni t \to \bar{t}, j \in J} M_j(t)$ are defined as those of the multimapping $T \times J \ni (t, j) \mapsto M_j(t)$ along the filter basis $\{V \times \Sigma_j : V \in \mathcal{N}_{T_0}(\bar{t}), j \in J\}$, where as above $\Sigma_j := \{i \in J : i \succeq_J j\}$. $\square$

In addition to the features exposed in Proposition A.1 in Appendix, we recall the following fundamental properties:
- A mapping $f : T \to X$ is continuous at $\bar{t} \in T$ if and only if, for any net $(t_j)_{j \in J}$ of T converging to $\bar{t}$, the net $(f(t_j))_{j \in J}$ converges to $f(\bar{t})$.
- For a subset Q of T, an element $t \in \operatorname{cl} Q$ if and only if there exists some net $(t_j)_{j \in J}$ in Q converging to t.
- A subset Q of the topological space T is compact if and only if every net in Q has a subnet converging to some element in Q.

In the context of the general topological space T the lower and upper limits of an extended real-valued function $\varphi : T_0 \to \mathbb{R} \cup \{-\infty, +\infty\}$ are known to be characterized as follows:
$$\liminf_{T_0 \ni t \to \bar{t}} \varphi(t) = \min \left\{ \lim_{j \in J} \varphi(t_j) : T_0 \ni t_j \to \bar{t} \right\}$$

and
$$\limsup_{T_0 \ni t \to \bar{t}} \varphi(t) = \max \left\{ \lim_{j \in J} \varphi(t_j) : T_0 \ni t_j \to \bar{t} \right\},$$
that is, the lower (resp. upper) limit is the least (resp. greatest) of limits $\lim_{j \in J} \varphi(t_j)$ such that the net $(t_j)_{j \in J}$ in T_0 converges to $\bar{t}$.

PROPOSITION 1.16. *Assume T and X are general topological spaces. Given $x \in X$ the following hold:*
(a) $x \in \underset{T_0 \ni t \to \bar{t}}{\text{Lim inf}} M(t)$ *if and only if any net $(t_j)_{j \in J}$ in T_0 converging to $\bar{t}$ admits a subnet $(t_{s(i)})_{i \in I}$ for which there exists a net $(x_i)_{i \in I}$ in X converging to x and such that $x_i \in M(t_{s(i)})$ for all $i \in I$.*
(b) $x \in \underset{T_0 \ni t \to \bar{t}}{\text{Lim sup}} M(t)$ *if and only if there exist a net $(t_j)_{j \in J}$ in T_0 converging to $\bar{t}$ and a net $(x_j)_{j \in J}$ converging to x in X and such that $x_j \in M(t_j)$ for all $j \in J$.*

PROOF. (a) Suppose that $x \in \underset{T_0 \ni t \to \bar{t}}{\text{Lim inf}} M(t)$ and take a net $(t_j)_{j \in J}$ in T_0 converging to $\bar{t}$, where $\preceq$ denotes the preorder of J. By Definition 1.2, for each $W \in \mathcal{N}(x)$, there exists some $V_W \in \mathcal{N}_{T_0}(\bar{t})$ such that $W \cap M(t) \neq \emptyset$ for all $t \in T_0 \cap V_W$. Fix for each $j \in J$ some $s(j, W) \in J$ satisfying $j \preceq s(j, W)$ and $t_{s(j,W)} \in V_W$, and hence we may choose $x_{j,W} \in W \cap M(t_{s(j,W)})$. Let $I := J \times \mathcal{N}(x)$ be endowed with the product preorder $\preceq_I$ of $\preceq$ and $\supset$. This preorder is directed and the above mapping $s : I \to J$ is also directed since for each $j_0 \in J$, fixing $W_0 \in \mathcal{N}(x)$ and considering any $(j, W) \in I$ with $(j_0, W_0) \preceq_I (j, W)$ one has $j_0 \preceq j \preceq s(j, W)$. Consequently, $(t_{s(i)})_{i \in I}$ is a subnet of $(t_j)_{j \in J}$. Since $x_{j,W} \in W$ for each $W \in \mathcal{N}(x)$, it follows that $(x_i)_{i \in I}$ converges to x (see Example 1.15(c) above).

Conversely suppose that $x \notin \underset{T_0 \ni t \to \bar{t}}{\text{Lim inf}} M(t)$. There is by Definition 1.2 some $W \in \mathcal{N}(x)$ such that for each $V \in \mathcal{N}_{T_0}(\bar{t})$ there exists $t_V \in V$ with $W \cap M(t_V) = \emptyset$. Then, taking the directed preorder $\supset$ on $\mathcal{N}_{T_0}(\bar{t})$ the net $(t_V)_{V \in \mathcal{N}_{T_0}(\bar{t})}$ satisfies the negation of the right-hand property of (a) of the proposition. The equivalence in (a) is then established.

(b) Fix $x \in \underset{T_0 \ni t \to \bar{t}}{\text{Lim sup}} M(t)$. By Definition 1.2, for each $W \in \mathcal{N}(x)$ and each $V \in \mathcal{N}_{T_0}(\bar{t})$, we can choose $t_{V,W} \in V$ and $x_{V,W} \in W \cap M(t_{V,W})$. Considering $J := \mathcal{N}_{T_0}(\bar{t}) \times \mathcal{N}(x)$ endowed with the product preorder of $\supset$ and $\supset$, the nets $(t_j)_{j \in J}$ and $(x_j)_{j \in J}$ converge to $\bar{t}$ and x respectively (see Example 1.15(b) above), and further $x_j \in M(t_j)$ for all $j \in J$, which is the property of the right-hand side of (b). This finishes the proof since the converse implication follows directly from Definition 1.2. $\square$

When the topological spaces T and X are not metrizable, taking into account Propositions 1.14 and 1.16, it is often useful (as we will see in Chapter 4) to define the *sequential limit inferior* or *sequential inner limit* of M at $\bar{t}$ relative to T_0 by

(1.5)
$$^{\text{seq}}\underset{T_0 \ni t \to \bar{t}}{\text{Lim inf}} M(t) := \left\{ \begin{array}{c} x \in X : \forall \text{ sequence } T_0 \ni t_n \to \bar{t},\ \exists \text{ a sequence } x_n \to x, \\ \text{with } x_n \in M(t_n) \text{ for } n \text{ large enough} \end{array} \right\},$$

that is, $x \in {}^{\text{seq}}\underset{T_0 \ni t \to \bar{t}}{\text{Lim inf}} M(t)$ if and only if for any sequence $(t_n)_{n \in \mathbb{N}}$ in T_0 converging to $\bar{t}$ there exists a sequence $(x_n)_{n \in \mathbb{N}}$ in X converging to x and such that $x_n \in M(t_n)$

for n large enough. Similarly, the *sequential limit superior* or *sequential outer limit* is defined by

$$(1.6) \qquad {}^{\text{seq}}\operatorname*{Lim\,sup}_{T_0 \ni t \to \bar{t}} M(t) := \left\{ \begin{array}{c} x \in X : \exists \text{ sequences } T_0 \ni t_n \to \bar{t},\ x_n \to x, \\ \text{with } x_n \in M(t_n) \text{ for all } n \end{array} \right\},$$

that is, $x \in {}^{\text{seq}}\operatorname*{Lim\,sup}_{T_0 \ni t \to \bar{t}} M(t)$ if and only if there exist a sequence $(t_n)_{n \in \mathbb{N}}$ in T_0 converging to $\bar{t}$ and a sequence $(x_n)_{n \in \mathbb{N}}$ converging to x in X and such that $x_n \in M(t_n)$ for all integers n.

The next two propositions will be used in Chapter 5. The first one is related to an important case where the topological outer limit is the closure of the sequential one.

PROPOSITION 1.17. *Assume that the topology τ_T is metrizable and that any point in X admits a basis of τ_X-closed neighborhoods. Assume also that there is a sequentially τ_X-compact set K in X such that $M(t) \subset K$ for all $t \in T_0$ near $\bar{t}$. Then one has the equality*

$$\operatorname*{Lim\,sup}_{T_0 \ni t \to \bar{t}} M(t) = \operatorname{cl}_{\tau_X} \left({}^{\text{seq}}\operatorname*{Lim\,sup}_{T_0 \ni t \to \bar{t}} M(t) \right).$$

PROOF. By Proposition 1.16(b) the right-hand side is obviously contained in the left-hand side. To prove the converse inclusion, let V be a neighborhood of $\bar{t}$ such that $M(V \cap T_0) \subset K$. Let also $(\varepsilon_n)_n$ be a decreasing sequence in $]0, +\infty[$ tending to 0 with $B(\bar{t}, \varepsilon_n) \subset V$ for all $n \in \mathbb{N}$, where $B(\bar{t}, \varepsilon_n)$ denotes the open ball relative to a distance generating the topology τ_T. Fix any x in the left-hand side and fix also any τ_X-closed neighborhood U of x. By the definition of outer limit, for each $n \in \mathbb{N}$ there is some $x_n \in U \cap M(T_0 \cap B(\bar{t}, \varepsilon_n))$. Since $x_n \in K$, there is a subsequence $(x_{s(n)})_n$ τ_X-converging to some u. It results that u belongs to both sets U and ${}^{\text{seq}}\operatorname*{Lim\,sup}_{T_0 \ni t \to \bar{t}} M(t)$, so U meets the latter set. This being true for any τ_X-closed neighborhood U of x, it ensues that x belongs to the right-hand side of the statement. □

The second proposition considers a situation when both topological and sequential outer limits coincide.

PROPOSITION 1.18. *Assume that the topology τ_T is metrizable and that X is a reflexive Banach space. Assume also that there is a real $\gamma \geq 0$ such that $M(t) \subset \gamma \mathbb{B}_X$ for all $t \in T_0$ near $\bar{t}$. Then, for X endowed with the weak topology one has the equality*

$$\operatorname*{Lim\,sup}_{T_0 \ni t \to \bar{t}} M(t) = {}^{\text{seq}}\operatorname*{Lim\,sup}_{T_0 \ni t \to \bar{t}} M(t).$$

PROOF. Fix any $x \in \operatorname{cl}_w \left({}^{\text{seq}}\operatorname*{Lim\,sup}_{T_0 \ni t \to \bar{t}} M(t) \right)$. The latter set being bounded (since it is included in $\gamma \mathbb{B}_X$), the Eberlein-Šmulian theorem gives a sequence $(x_n)_n$ in ${}^{\text{seq}}\operatorname*{Lim\,sup}_{T_0 \ni t \to \bar{t}} M(t)$ converging weakly to x. Then for each $n \in \mathbb{N}$ (by the definition of sequential outer limit) there is a sequence $(x_{n,k})_k$ converging weakly to x_n and a sequence $(t_{n,k})_k$ in T_0 converging to $\bar{t}$ such that, for all $k \in \mathbb{N}$ one has $x_{n,k} \in M(t_{n,k})$ and $M(t_{n,k}) \subset \gamma \mathbb{B}_X$. The set $C := \operatorname{cl}_w \{x_{n,k} : (n,k) \in \mathbb{N}^2\}$ is then compact and separable relative to the topology induced by $w(X, X^*)$, so this induced topology is generated by a distance d_C. This allows us to construct an increasing mapping $s : \mathbb{N} \to \mathbb{N}$ such that, for all $n \in \mathbb{N}$ both inequalities $d_C(x_{n,s(n)}, x_n) < 1/n$ and $d_T(t_{n,s(n)}, \bar{t}) < 1/n$ are satisfied, where d_T denotes a distance on T generating the topology τ_T. It results as $n \to \infty$ that $x_{n,s(n)} \to x$ weakly and $t_{n,s(n)} \to \bar{t}$ relative to d_T, which guarantees that x belongs to $^{\text{seq}}\operatorname*{Lim\,sup}_{T_0 \ni t \to \bar{t}} M(t)$. The latter set is then weakly closed and this combined with Proposition 1.17 justifies the equality of the proposition. $\square$

In addition to properties in Proposition 1.12 the next proposition establishes with the help of that proposition and Proposition 1.16 other properties related to the distance function.

PROPOSITION 1.19. *Assume that (X, d) is a metric space and S is a subset of X. One has the following properties:*
(a) *The inclusion $S \subset \operatorname*{Lim\,inf}_{T_0 \ni t \to \bar{t}} M(t)$ holds if and only if, for all $y \in X$,*
$$d(y, S) \geq \limsup_{T_0 \ni t \to \bar{t}} d(y, M(t)).$$
(b) *The inclusion $S \subset \operatorname*{Lim\,sup}_{T_0 \ni t \to \bar{t}} M(t)$ holds if and only if, for all $y \in X$,*
$$d(y, S) \geq \liminf_{T_0 \ni t \to \bar{t}} d(y, M(t)).$$
(c) *If X is a finite-dimensional normed space, then for all $y \in X$,*
$$d\left(y, \operatorname*{Lim\,sup}_{T_0 \ni t \to \bar{t}} M(t)\right) = \liminf_{T_0 \ni t \to \bar{t}} d(y, M(t)).$$

PROOF. (a) If the inequality property in (a) is satisfied, then for any $x \in S$ we have $\limsup_{T_0 \ni t \to \bar{t}} d(x, M(t)) = 0$, hence $x \in \operatorname*{Lim\,inf}_{T_0 \ni t \to \bar{t}} M(t)$ by Proposition 1.12(a). This proves the implication $\Leftarrow$ in (a). Conversely, suppose that $S \subset \operatorname*{Lim\,inf}_{T_0 \ni t \to \bar{t}} M(t)$ with $S \neq \emptyset$. Fix $y \in X$ and take any $x \in S$. For any real $\varepsilon > 0$, by definition of limit inferior of multimapping there is some $V \in \mathcal{N}_{T_0}(\bar{t})$ such that, for all $t \in V$, one has $M(t) \cap B(x, \varepsilon) \neq \emptyset$. For each $t \in V$, it ensues that
$$d(y, M(t)) \leq d(y, x) + d(x, M(t)) \leq d(y, x) + \varepsilon.$$
From this we see that $\limsup_{T_0 \ni t \to \bar{t}} d(y, M(t)) \leq d(y, x)$. The latter inequality being true for all $x \in S$, it results that $\limsup_{T_0 \ni t \to \bar{t}} d(y, M(t)) \leq d(y, S)$.

(b) As above, the implication $\Rightarrow$ follows from Proposition 1.12(b). Conversely, fix any $y \in X$, and suppose that S is a nonempty subset of $\operatorname*{Lim\,sup}_{T_0 \ni t \to \bar{t}} M(t)$ and take any x in this set. For any $\varepsilon > 0$ and $V \in \mathcal{N}_{T_0}(\bar{t})$, the definition of outer limit of multimapping provides some $t_V \in V$ such that $M(t_V) \cap B(x, \varepsilon) \neq \emptyset$, which implies
$$d(y, M(t_V)) \leq d(y, x) + d(x, M(t_V)) < d(y, x) + \varepsilon.$$
From this and the definition of lower limit of extended real-valued function it follows that $\liminf_{T_0 \ni t \to \bar{t}} d(y, M(t)) \leq d(y, x) + \varepsilon$, which guarantees as desired that the

right-hand side of (b) is not greater than the left-hand one.

(c) Now assume that X is a finite-dimensional normed space. Fix any real $r > \liminf\limits_{T_0 \ni t \to \bar{t}} d(y, M(t))$ (since the desired inequality is obvious if the latter lower limit is $+\infty$). Choose a net $(t_j)_{j \in J}$ in T_0 tending to $\bar{t}$ such that $\liminf\limits_{T_0 \ni t \to \bar{t}} d(y, M(t)) = \lim\limits_{j \in J} d(y, M(t_j))$. There exists some $j_0 \in J$ such that, for each $j \succeq j_0$, there is $x_j \in M(t_j)$ such that $r > \|y - x_j\|$, and hence in particular $\{x_j : j \succeq j_0\}$ is bounded. Taking a subnet of $(x_j)_{j \in J}$ converging to some x, we have $x \in \operatorname*{Lim\,sup}\limits_{T_0 \ni t \to \bar{t}} M(t)$ according to Proposition 1.16(b). So,

$$r \geq \|y - x\| \geq d\left(y, \operatorname*{Lim\,sup}\limits_{T_0 \ni t \to \bar{t}} M(t)\right),$$

which completes the proof. □

REMARK 1.20. Even in $X = \mathbb{R}^2$, the inequality

$$d\left(y, \operatorname*{Lim\,inf}\limits_{T_0 \ni t \to \bar{t}} M(t)\right) \leq \limsup\limits_{T_0 \ni t \to \bar{t}} d(y, M(t))$$

fails in general. Indeed, endow $\mathbb{R}^2$ with the Euclidean norm $\|\cdot\|$ and put $S_n := \mathbb{R} \times \{0\}$ if the integer n is even and $S_n = \{0\} \times \mathbb{R}$ if n is odd. Then $\operatorname*{Lim\,inf}\limits_{n \to \infty} S_n = \{(0,0)\}$ hence $d\left(y, \operatorname*{Lim\,inf}\limits_{n \to \infty} S_n\right) = \|y\|$, while $\limsup_{n \to \infty} d(y, S_n) = \max\{|y_1|, |y_2|\}$ for all $y = (y_1, y_2) \in \mathbb{R}^2$. □

1.3. Epigraphical semilimits

Let us keep the topological spaces (T, τ_T) and (X, τ_X) along with a set $T_0 \subset T$ and $\bar{t} \in \operatorname{cl}(T_0)$. Consider for each $t \in T_0$ a function $f_t : X \to \mathbb{R} \cup \{-\infty, +\infty\}$ and recall that its epigraph is given by

$$\operatorname{epi} f_t := \{(x, r) \in X \times \mathbb{R} : f_t(x) \leq r\}.$$

Take any $(\bar{x}, \bar{r}) \in X \times \mathbb{R}$ and for convenience put $\mathcal{N}_0(\bar{t}) := \mathcal{N}_{T_0}(\bar{t})$, that is,

$$\mathcal{N}_0(\bar{t}) = \{U \cap T_0 : U \in \mathcal{N}(\bar{t})\}.$$

Suppose first that $(\bar{x}, \bar{r}) \in \operatorname*{Lim\,inf}\limits_{T_0 \ni \to \bar{t}}(\operatorname{epi} f_t)$. This yields

$$\forall \varepsilon > 0, \forall W \in \mathcal{N}(\bar{x}), \exists V \in \mathcal{N}_0(\bar{t}), \forall t \in V, (\operatorname{epi} f_t) \cap (W \times]\bar{r} - \varepsilon, \bar{r} + \varepsilon[) \neq \emptyset,$$

or equivalently

$$\forall \varepsilon > 0, \forall W \in \mathcal{N}(\bar{x}), \exists V \in \mathcal{N}_0(\bar{t}), \forall t \in V, \exists x \in W, \exists r \in]\bar{r} - \varepsilon, \bar{r} + \varepsilon[,\ f_t(x) \leq r,$$

thus

$$\forall \varepsilon > 0, \forall W \in \mathcal{N}(\bar{x}), \exists V \in \mathcal{N}_0(\bar{t}), \forall t \in V, \inf_{x \in W} f_t(x) < \bar{r} + \varepsilon,$$

which in turn gives

$$\forall \varepsilon > 0, \forall W \in \mathcal{N}(\bar{x}), \exists V \in \mathcal{N}_0(\bar{t}), \sup_{t \in V} \inf_{x \in W} f_t(x) \leq \bar{r} + \varepsilon.$$

It ensues that for any $\varepsilon > 0$

(1.7) $$\overline{L}_{\bar{t}}(\bar{x}) := \sup_{W \in \mathcal{N}(\bar{x})} \inf_{V \in \mathcal{N}(\bar{t})} \sup_{t \in V} \inf_{x \in W} f_t(x) \leq \bar{r} + \varepsilon,$$

hence
$$\overline{L}_{\bar{t}}(\bar{x}) \leq \bar{r}. \tag{1.8}$$

Conversely, suppose that $\overline{L}_{\bar{t}}(\bar{x}) \leq \bar{r}$, that is,
$$\sup_{W \in \mathcal{N}(\bar{x})} \inf_{V \in \mathcal{N}(\bar{t})} \sup_{t \in V} \inf_{x \in W} f_t(x) \leq \bar{r}.$$

Take any $\varepsilon > 0$ and any $W \in \mathcal{N}(\bar{x})$. By the previous inequality we have
$$\inf_{V \in \mathcal{N}(\bar{t})} \sup_{t \in V} \inf_{x \in W} f_t(x) \leq \bar{r},$$
which gives some $V \in \mathcal{N}_0(\bar{t})$ such that
$$\sup_{t \in V} \inf_{x \in W} f_t(x) < \bar{r} + \varepsilon/2,$$
hence for any $t \in V$ we obtain $\inf_{x \in W} f_t(x) < \bar{r} + \varepsilon/2$, and this furnishes some $x \in W$ such that $f_t(x) < \bar{r} + \varepsilon/2$. Putting $r := \bar{r} + \varepsilon/2$ we get
$$(x, r) \in (\operatorname{epi} f_t) \cap (W \times]\bar{r} - \varepsilon, \bar{r} + \varepsilon[), \text{ so } (\operatorname{epi} f_t) \cap (W \times]\bar{r} - \varepsilon, \bar{r} + \varepsilon[) \neq \emptyset.$$

It follows that $(\bar{x}, \bar{r}) \in \operatorname*{Lim\,inf}_{T_0 \ni \to \bar{t}}(\operatorname{epi} f_t)$. This and (1.8) ensure that
$$(\bar{x}, \bar{r}) \in \operatorname*{Lim\,inf}_{T_0 \ni \to \bar{t}}(\operatorname{epi} f_t) \Leftrightarrow \overline{L}_{\bar{t}}(\bar{x}) \leq \bar{r}.$$

Considering the function $\overline{L}_{\bar{t}} : X \to \mathbb{R} \cup \{-\infty, +\infty\}$ defined for any $u \in X$ by
$$\overline{L}_{\bar{t}}(u) := \sup_{W \in \mathcal{N}(u)} \inf_{V \in \mathcal{N}(\bar{t})} \sup_{t \in V} \inf_{x \in W} f_t(x),$$
it results that $\operatorname*{Lim\,inf}_{T_0 \ni \to \bar{t}}(\operatorname{epi} f_t)$ is an epigraphical set, more precisely
$$\operatorname*{Lim\,inf}_{T_0 \ni \to \bar{t}}(\operatorname{epi} f_t) = \operatorname{epi} \overline{L}_{\bar{t}}.$$

Now suppose that $(\bar{x}, \bar{r}) \in \operatorname*{Lim\,sup}_{T_0 \ni t \to \bar{t}}(\operatorname{epi} f_t)$. This means
$$\forall \varepsilon > 0, \forall W \in \mathcal{N}(\bar{x}), \forall V \in \mathcal{N}_0(\bar{t}), \exists t \in V, (\operatorname{epi} f_t) \cap (W \times]\bar{r} - \varepsilon/2, \bar{r} + \varepsilon/2[) \neq \emptyset,$$
or equivalently
$$\forall \varepsilon > 0, \forall W \in \mathcal{N}(\bar{x}), \forall V \in \mathcal{N}_0(\bar{t}), \exists t \in V, \exists x \in W, \exists r \in]\bar{r} - \varepsilon/2, \bar{r} + \varepsilon/2[, f_t(x) \leq r,$$
which gives
$$\forall \varepsilon > 0, \forall W \in \mathcal{N}(\bar{x}), \forall V \in \mathcal{N}_0(\bar{t}), \inf_{t \in V} \inf_{x \in W} f_t(x) < \bar{r} + \varepsilon/2.$$

Consequently, for any $\varepsilon > 0$
$$\sup_{W \in \mathcal{N}(\bar{x})} \sup_{V \in \mathcal{N}_0(\bar{t})} \inf_{t \in V} \inf_{x \in W} f_t(x) \leq \bar{r} + \varepsilon/2,$$
hence
$$\underline{L}_{\bar{t}}(\bar{x}) := \sup_{W \in \mathcal{N}(\bar{x})} \sup_{V \in \mathcal{N}_0(\bar{t})} \inf_{t \in V} \inf_{x \in W} f_t(x) < \bar{r} + \varepsilon.$$

It results that $\underline{L}_{\bar{t}}(\bar{x}) \leq \bar{r}$.

Conversely, proceeding as above one can see that the inequality $\underline{L}_{\bar{t}}(\bar{x}) \leq \bar{r}$ implies that for any $\varepsilon > 0$ and any $W \in \mathcal{N}(\bar{x})$ there exists a $V \in \mathcal{N}_0(\bar{t})$ such that for every $t \in V$
$$(\operatorname{epi} f_t) \cap (W \times]\bar{r} - \varepsilon, \bar{r} + \varepsilon[) \neq \emptyset,$$

so $(\bar{x},\bar{r}) \in \underset{T_0 \ni t \to \bar{t}}{\operatorname{Lim\,sup}}(\operatorname{epi} f_t)$.

We then obtain the equivalence
$$(\bar{x},\bar{r}) \in \underset{T_0 \ni t \to \bar{t}}{\operatorname{Lim\,sup}}(\operatorname{epi} f_t) \Leftrightarrow \underline{L_{\bar{t}}}(\bar{x}) \leq \bar{r}.$$

Considering the function $\underline{L_{\bar{t}}} : X \to \mathbb{R} \cup \{-\infty, +\infty\}$ defined for any $u \in X$ by
$$\underline{L_{\bar{t}}}(u) := \sup_{W \in \mathcal{N}(u)} \sup_{V \in \mathcal{N}(\bar{t})} \inf_{t \in V} \inf_{x \in W} f_t(x),$$

it results that $\underset{T_0 \ni t \to \bar{t}}{\operatorname{Lim\,inf}}(\operatorname{epi} f_t)$ is an epigraphical set, more precisely
$$\underset{T_0 \ni t \to \bar{t}}{\operatorname{Lim\,inf}}(\operatorname{epi} f_t) = \operatorname{epi} \underline{L_{\bar{t}}}.$$

The above development serves as a guide for introducing the following concepts.

DEFINITION 1.21. Let $\bar{t} \in \operatorname{cl}(T_0)$ and let $(f_t)_{t \in T_0}$ be a family of functions from X into $\mathbb{R} \cup \{-\infty, +\infty\}$, where X and T_0 are as above. For any $\bar{x} \in X$ the extended real
$$\underset{T_0 \ni t \to \bar{t}}{\lim\sup} \inf_{x \to \bar{x}} f_t(x) := \sup_{W \in \mathcal{N}(\bar{x})} \inf_{V \in \mathcal{N}(\bar{t})} \sup_{t \in V} \inf_{x \in W} f_t(x)$$

is called the *upper epi-limit* (or *upper Γ-limit*) at $\bar{x}$ of $f_t(x)$ as $T_0 \ni t \to \bar{t}$. The other extended real
$$\underset{T_0 \ni t \to \bar{t}}{\lim\inf} \inf_{x \to \bar{x}} f_t(x) := \sup_{W \in \mathcal{N}(\bar{x})} \sup_{V \in \mathcal{N}(\bar{t})} \inf_{x \in V} \inf_{x \in W} f_t(x)$$

is called the *lower epi-limit* (or *lower Γ-limit*) at $\bar{x}$ of $f_t(x)$ as $T_0 \ni t \to \bar{t}$.

It is worth noticing that the lower epi-limit clearly coincides to the usual lower limit as $(t,x) \to (\bar{t},\bar{x})$, that is,
$$\underset{T_0 \ni t \to \bar{t}}{\lim\inf} \inf_{x \to \bar{x}} f_t(x) = \underset{T \ni t \to \bar{t}, x \to \bar{x}}{\lim\inf} f_t(x).$$

Notice also that the similar equality fails for the upper epi-limit and the usual upper limit as $(t,x) \to (\bar{t},\bar{x})$.

DEFINITION 1.22. The extended real-valued functions from X to $\mathbb{R} \cup \{-\infty, +\infty\}$ defined by
$$u \mapsto \underset{T_0 \ni t \to \bar{t}}{\lim\sup} \inf_{x \to u} f_t(x) \text{ and } u \mapsto \underset{T_0 \ni t \to \bar{t}}{\lim\inf} \inf_{x \to u} f_t(x)$$

are called the *upper epi-limit* (or *upper Γ-limit*) and *lower epi-limit* (or *lower Γ-limit*) of the family $(f_t)_{t \in T_0}$ as $T_0 \ni t \to \bar{t}$. When the latter upper and lower epi-limits coincide, one says that the family $(f_t)_{t \in T_0}$ *epi-converges* (or *Γ-converges*) to the common function.

Both above upper and lower epi-limits are lower τ_X-semicontinuous functions on X since their epigraphs are closed sets in $X \times \mathbb{R}$ as outer and inner limits of a family of sets in $X \times \mathbb{R}$.

In the literature, certain authors adopting the terminology "epi-limit"(resp. "Γ-limit") use notation $\operatorname{epi} - \underset{t \to \bar{t}}{\lim} f_t$ (resp. $\Gamma - \underset{t \to \bar{t}}{\lim} f_t$).

The above development can then be summarized as follows:

PROPOSITION 1.23. Keep f_t and T_0 as above.
(a) The inner limit of the family $(\mathrm{epi}\, f_t)_{t\in T_0}$ is the epigraph of the upper epi-limit (or upper Γ-limit) of $(f_t)_{t\in T_0}$, both as $T_0 \ni t \to \bar{t}$.
(b) The outer limit of the family $(\mathrm{epi}\, f_t)_{t\in T_0}$ is the epigraph of the lower epi-limit (or lower Γ-limit) of $(f_t)_{t\in T_0}$, both as $T_0 \ni t \to \bar{t}$.

It is clear that

$$\underline{\rho}(V,W) := \inf_{t\in V} \inf_{x\in W} f_t(x) \leq \overline{\rho}(V,W) := \sup_{t\in V} \inf_{x\in W} f_t(x)$$

and that $V \mapsto \underline{\rho}(V,W)$ (resp. $V \mapsto \overline{\rho}(V,W)$) is nondecreasing (resp. nonincreasing) on the ordered set $(\mathcal{N}_0(\bar{t}), \supset)$. Therefore,

$$\lim_{V\in\mathcal{N}_0(\bar{t})} \underline{\rho}(V,W) \leq \lim_{V\in\mathcal{N}_0(\bar{t})} \overline{\rho}(V,W),$$

or equivalently

$$\sup_{V\in\mathcal{N}_0(\bar{t})} \inf_{t\in V} \inf_{x\in W} f_t(x) \leq \inf_{V\in\mathcal{N}_0(\bar{t})} \sup_{t\in V} \inf_{x\in W} f_t(x).$$

It results that

$$\sup_{W\in\mathcal{N}(\bar{x})} \sup_{V\in\mathcal{N}_0(\bar{t})} \inf_{t\in V} \inf_{x\in W} f_t(x) \leq \sup_{W\in\mathcal{N}(\bar{x})} \inf_{V\in\mathcal{N}_0(\bar{t})} \sup_{t\in V} \inf_{x\in W} f_t(x).$$

This corresponds to (b) in the next proposition; the assertion (a) therein is obvious.

PROPOSITION 1.24. With the above data the following hold.
(a) For every real $\lambda > 0$

$$\lim_{T_0\ni t\to\bar{t}} \inf_{x\to\bar{x}} (\lambda f_t)(x) = \lambda \lim_{T_0\ni t\to\bar{t}} \inf_{x\to\bar{x}} f_t(x)$$

and similarly

$$\lim_{T_0\ni t\to\bar{t}} \sup_{x\to\bar{x}} \inf (\lambda f_t)(x) = \lambda \lim_{T_0\ni t\to\bar{t}} \sup_{x\to\bar{x}} \inf f_t(x).$$

(b) One also has the inequality

$$\lim_{T_0\ni t\to\bar{t}} \inf_{x\to\bar{x}} \inf f_t(x) \leq \lim_{T_0\ni t\to\bar{t}} \sup_{x\to\bar{x}} \inf f_t(x).$$

1.4. Semicontinuity of multimappings

As above, throughout this section and the next ones of this chapter, unless otherwise stated, (T, τ_T) and (X, τ_X) are topological spaces and $M : T \rightrightarrows X$ is a multimapping from T into X; we will also suppose that $\bar{t}$ is a point in T. We observe by Corollary 1.10 that we have

$$M(\bar{t}) \subset \mathrm{Lim\,sup}_{t\to\bar{t}} M(t),$$

and that, whenever $M(\bar{t})$ is closed, we also have

$$\mathrm{Lim\,inf}_{t\to\bar{t}} M(t) \subset M(\bar{t}).$$

This observation thus leads to define the concepts of inner and outer semicontinuities as follows.

DEFINITION 1.25. The multimapping M is said to be *inner semicontinuous* (isc, for short) at $\bar{t} \in T$ provided

$$(1.9) \qquad M(\bar{t}) \subset \liminf_{t \to \bar{t}} M(t).$$

Similarly, one says that M is *outer semicontinuous* (osc, for short) at $\bar{t}$ when

$$(1.10) \qquad \limsup_{t \to \bar{t}} M(t) \subset M(\bar{t}).$$

When M is inner (resp. outer) semicontinuous at each point of a set $T_0 \subset T$, one says that M is inner (resp. outer) semicontinuous on T_0. If $T_0 = T$ one just says that M is inner (resp. outer) semicontinuous.

REMARK 1.26. (a) The multimapping M is outer semicontinuous at $\bar{t}$ if and only if

$$M(\bar{t}) = \limsup_{t \to \bar{t}} M(t)$$

since the reverse inclusion of (1.10) always holds (as recalled above), hence $M(\bar{t})$ is closed whenever M is outer semicontinuous at $\bar{t}$.
(b) If M is inner semicontinuous at $\bar{t}$, using Corollary 1.10 and (1.9) we see that $\operatorname{cl}(M(\bar{t})) = \liminf_{t \to \bar{t}} M(t)$.
(c) Since $\liminf_{t \to \bar{t}} M(t) = \liminf_{t \to \bar{t}} \operatorname{cl}(M(t))$ (see Corollary 1.10), M is inner semicontinuous at $\bar{t}$ if and only if the multimapping $t \mapsto \operatorname{cl}(M(t))$ is inner semicontinuous at $\bar{t}$. $\square$

The name of outer semicontinuity is mainly justified by the assertion (a) of the following proposition characterizing the outer semicontinuity property in terms of neighborhoods.

PROPOSITION 1.27. *The following assertions hold.*
(a) *The multimapping M is outer semicontinuous at $\bar{t} \in T$ if and only if, for each $\bar{x} \in X \setminus M(\bar{t})$, there exists a neighborhood $V \times W$ of $(\bar{t}, \bar{x})$ such that*

$$W \cap M(t) = \emptyset \quad \text{for all } t \in V.$$

(b) *The multimapping M is outer semicontinuous if and only if its graph $\operatorname{gph} M$ is closed in $T \times X$.*

PROOF. (a) It suffices to use the equality $\limsup_{t \to \bar{t}} M(t) = \bigcap_{V \in \mathcal{N}(\bar{t})} \operatorname{cl}(M(V))$ of Proposition 1.7.
(b) It is enough to see that the property of the second member of assertion (a) is fulfilled at any $\bar{t} \in T$ if and only if the set $(T \times X) \setminus \operatorname{gph} M$ is open in $T \times X$. $\square$

DEFINITION 1.28. Taking the characterization in (b) above into account, one also says that the multimapping M is *closed at $\bar{t}$* to translate that it is outer semicontinuous at $\bar{t}$. This also means, according to Remark 1.26(a) and Proposition 1.16, that for any net $(t_j, x_j)_{j \in J}$ of $\operatorname{gph} M$ converging to $(\bar{t}, x)$ in $T \times X$ one has $x \in M(\bar{t})$. When the latter holds with sequences in place of nets, we say that the multimapping M is *sequentially outer semicontinuous* or *sequentially closed at $\bar{t}$*. Of course both concepts are equivalent whenever both topological spaces T and X are metrizable.

Let us state now a characterization of the inner semicontinuity of M in terms of neighborhoods.

PROPOSITION 1.29. *The following characterizations of inner semicontinuity property hold.*
(a) *The multimapping M is inner semicontinuous at $\bar{t} \in T$ if and only if, for every open set U of X with $U \cap M(\bar{t}) \neq \emptyset$, there exists a neighborhood V of $\bar{t}$ in T such that*
$$U \cap M(t) \neq \emptyset \quad \text{for all } t \in V.$$
(b) *The multimapping M is inner semicontinuous if and only if the inverse image under M of any open set U of X (that is, $M^{-1}(U)$) is an open set of T.*

PROOF. The assertion in (b) follows directly from (a) and it is easily seen that (a) is a direct consequence of Definitions 1.2 and 1.25. □

Proposition 1.29 thus says that the inner semicontinuity translates, in addition to the inner property (1.9), another face of the behavior of M near $\bar{t}$. Following the literature, we then consider the terminology in the definition:

DEFINITION 1.30. One says that the multimapping M is *lower semicontinuous* (lsc, for short) at $\bar{t}$ when the property of the right-hand side of the assertion (a) of Proposition 1.29 is fulfilled.

Of course, by the above proposition there is equivalence between inner and lower semicontinuity. So, we will say equally well inner or lower semicontinuity for a multimapping. Further, taking (a) into account the inner semicontinuity can also be involved with a pair $(\bar{t}, \bar{x}) \in \operatorname{gph} M$ as follows.

DEFINITION 1.31. One says with $\bar{x} \in M(\bar{t})$ that the multimapping M is *inner (or lower) semicontinuous at $\bar{t}$ for $\bar{x}$* whenever, for any neighborhood U of $\bar{x}$, there exists a neighborhood V of $\bar{t}$ such that

(1.11) $$U \cap M(t) \neq \emptyset \quad \text{for all } t \in V.$$

Obviously, M is inner (or lower) semicontinuous at $\bar{t}$ whenever, for every $\bar{x} \in M(\bar{t})$, the multimapping M is inner (or lower) semicontinuous at $\bar{t}$ for $\bar{x}$. While through the latter characterization, the inner semicontinuity of M at $\bar{t}$ can be seen as a behavior of M related to points inside $M(\bar{t})$, the assertion (a) of Proposition 1.27 translates in contrast only an outer behavior of M under the outer semicontinuity property. So, it is often useful to consider the following other behavior of upper expansion.

DEFINITION 1.32. One says that the multimapping M is *upper semicontinuous* (usc, for short) at $\bar{t} \in T$ provided that, for every open set U of X with $M(\bar{t}) \subset U$, there exists some neighborhood V of $\bar{t}$ in T such that
$$M(t) \subset U \quad \text{for all } t \in V.$$
When the property holds at each point $\bar{t} \in T$ (resp. $\bar{t} \in T_0$, where $T_0 \subset T$), one says that M is upper semicontinuous (resp. upper semicontinuous on T_0).

REMARK 1.33. If $M(\bar{t}) = \emptyset$ and M is upper semicontinuous at $\bar{t}$, then according to the above definition $M(t) = \emptyset$ for all t in a neighborhood of $\bar{t}$.

This ensures that the (effective) domain $\operatorname{Dom} M$ of M is closed in T whenever M is upper semicontinuous on the whole space T. □

Let us state some characterizations of global upper semicontinuity property.

PROPOSITION 1.34. *The following assertions are pairwise equivalent.*
(a) *The multimapping M is upper semicontinuous;*
(b) *for every open set U of X, the set $\{t \in T : M(t) \subset U\}$ is open in T;*
(c) *the inverse image under M of any closed set C of X, that is $M^{-1}(C)$, is a closed set in T.*

PROOF. The equivalence between (a) and (b) follows from Definition 1.32, and the equivalence between (b) and (c) follows from the equality
$$\complement_T \{t \in T : M(t) \cap C \neq \emptyset\} = \{t \in T : M(t) \subset \complement_X C\}.$$
$\square$

An important result concerning images of compact sets under upper semicontinuous multimappings is the following:

THEOREM 1.35 (compactness theorem of images of compact sets by upper semicontinuous multimappings). *Let K be a compact set of T such that M is upper semicontinuous on K and $M(t)$ is compact in X for every $t \in K$. Then the set $M(K)$ is compact in X.*

PROOF. Let $(U_i)_{i \in I}$ be a family of open sets in X with $M(K) \subset \bigcup_{i \in I} U_i$. For each $t \in K$, by compactness of $M(t)$ there is a finite subset $I_t \subset I$ with $M(t) \subset \bigcup_{i \in I_t} U_i$, so the upper semicontinuity of M at t furnishes an open neighborhood V_t of t such that $M(V_t) \subset \bigcup_{i \in I_t} U_i$. By compactness of K there are $t_1, \cdots, t_m \in K$ such that $K \subset V_{t_1} \cup \cdots \cup V_{t_m}$. It results that
$$M(K) \subset \bigcup_{k=1}^{m} \bigcup_{i \in I_{t_k}} U_i,$$
which justifies the compactness of $M(K)$. $\square$

The next result on the composition is often useful. Thanks to the formula
$$(N \circ M)^{-1}(C) = M^{-1}(N^{-1}(C)),$$
it is a direct consequence of the assertion (b) of Proposition 1.29 and of the assertion (c) of Proposition 1.34.

PROPOSITION 1.36. *Let (Y, τ_Y) be another topological space and $N : X \rightrightarrows Y$ be another multimapping. If M and N are lower (resp. upper) semicontinuous on T and X respectively, then $N \circ M$ is lower (resp. upper) semicontinuous on T.*

The next proposition states two cases when the upper semicontinuity property entails the outer semicontinuity property at a point. Recall that a Hausdorff topological space is *normal* whenever, for any pair of disjoint closed sets C_1 and C_2 (that is, $C_1 \cap C_2 = \emptyset$), there exist two disjoint open sets W_1 and W_2 such that $C_i \subset W_i$, $i = 1, 2$. Any metrizable space is seen to be normal by taking the open sets $W_i := \{x \in X : d(x, C_i) < d(x, C_j)\} \supset C_i$, $i \neq j$. Further, such disjoint open sets $W_1 \supset C_1$ and $W_2 \supset C_2$ are also easily seen to exist for any pair of disjoint compact sets C_1 and C_2 of any Hausdorff topological space.

1.4. SEMICONTINUITY OF MULTIMAPPINGS

PROPOSITION 1.37. Assume that the topological space (X, τ_X) is Hausdorff and that M is upper semicontinuous at $\bar{t} \in T$. Then the multimapping M is outer semicontinuous at $\bar{t}$ under anyone of the following conditions:
(a) the set $M(\bar{t})$ is compact;
(b) the set $M(\bar{t})$ is closed and (X, τ_X) is normal (which holds in particular whenever X is metrizable).

PROOF. Fix any $\bar{x} \in X \setminus M(\bar{t})$. Under either the assumption in (a) or the assumption in (b) there exist two open sets W and U in X with $\bar{x} \in W$, $M(\bar{t}) \subset U$, and $W \cap U = \emptyset$. The upper semicontinuity property of M at $\bar{t}$ gives some neighborhood V of $\bar{t}$ such that $M(t) \subset U$ for all $t \in V$. Consequently, $V \times W$ is a neighborhood of $(\bar{t}, \bar{x})$ in $T \times X$ and $W \cap M(t) = \emptyset$ for all $t \in V$. The multimapping M is thus outer semicontinuous at $\bar{t}$ according to Proposition 1.27(a). □

Taking Proposition 1.27 into account one deduces immediately from the latter proposition the following corollary.

COROLLARY 1.38. Assume that (X, τ) is Hausdorff and M is upper semicontinuous on T. Then, under anyone of conditions (a) and (b) below, M is outer semicontinuous on T or equivalently the graph of M is closed in $T \times X$:
(a) the multimapping M is closed-valued and $\operatorname{Range} T$ is metrizable;
(b) the multimapping M is compact-valued.

We consider now a case where there is an equivalence between the outer and upper semicontinuity properties at a point.

PROPOSITION 1.39. Let the topological space (X, τ_X) be Hausdorff. Assume that there exists a neighborhood V of $\bar{t}$ in T such that M is closed-valued on V and that the set $M(V)$ is contained in a compact set of X. Then M is upper semicontinuous at $\bar{t}$ if and only if M is outer semicontinuous at $\bar{t}$.

PROOF. The direct implication $\Rightarrow$ follows from Proposition 1.37. To show the reverse implication, consider the compact set $K = \operatorname{cl}(M(V))$ and fix an open set $U \supset M(\bar{t})$. If $K \subset U$, Definition 1.32 is fulfilled with the neighborhood V of $\bar{t}$. So, suppose that the compact set $K \setminus U$ is nonempty. For each $x \in K \setminus U$ we have $x \in X \setminus M(\bar{t})$ and hence, from the outer semicontinuity of M at $\bar{t}$ and from Proposition 1.27, there exists an open neighborhood $V_x \times W_x$ of $(\bar{t}, x)$ with $V_x \subset V$ such that $W_x \cap M(V_x) = \emptyset$. The compactness of $K \setminus U$ gives $x_1, \cdots, x_k \in K \setminus U$ (with $k \in \mathbb{N}$) such that $K \setminus U \subset \bigcup_{i=1}^{k} W_{x_i}$. Consequently, $V' := \bigcap_{i=1}^{k} V_{x_i}$ is a neighborhood of $\bar{t}$ and $M(V') \subset U$. This ensures the desired implication. □

REMARK 1.40. It is worth noticing according to the above proof that the implication $\Rightarrow$ does not require the Hausdorff property of the space X. □

We can state the following important direct corollary of Proposition 1.39.

COROLLARY 1.41. Assume that the topological space (X, τ_X) is Hausdorff, that M is closed-valued, and that, for each $\bar{t} \in T$, there exists some neighborhood V of $\bar{t}$ such that $M(V)$ is included in a compact set of X. Then the following assertions are pairwise equivalent:
(a) The multimapping M is upper semicontinuous on T;
(b) the graph of M is closed in $T \times X$;
(c) the multimapping M is outer semicontinuous on T.

Under the compactness of the alone set $M(\bar{t})$ the following sequential characterization of upper semicontinuity is useful in many situations.

PROPOSITION 1.42. *Assume that T and X are two metric spaces and that $M(\bar{t})$ is a nonempty compact set. Then M is upper semicontinuous at $\bar{t}$ if and only if for any sequence $(t_n)_n$ in T tending to $\bar{t}$ and any sequence $(x_n)_n$ in X with $x_n \in M(t_n)$ for all $n \in \mathbb{N}$, there exists a subsequence of $(x_n)_n$ converging to a point belonging to $M(\bar{t})$.*

PROOF. Suppose first that M is upper semicontinuous at $\bar{t}$. Let $(t_n)_n$ in T with $t_n \to \bar{t}$ as $n \to \infty$ and let $(x_n)_n$ in X with $x_n \in M(t_n)$ for all n. For each $n \in \mathbb{N}$ we easily obtain by the upper semicontinuity of M at $\bar{t}$ some integer K_n such that
$$x_k \in M(t_k) \subset \{x \in X : d(x, M(\bar{t})) < 1/n\} \quad \text{for all } k \geq K_n.$$
This easily furnishes a subsequence $(x_{s(n)})_n$ (where $s : \mathbb{N} \to \mathbb{N}$ is an increasing function) such that, for each $n \in \mathbb{N}$ one has $d(x_{s(n)}, M(\bar{t})) < 1/n$, and hence there is some $u_n \in M(\bar{t})$ with $d(x_{s(n)}, u_n) < 1/n$. By compactness of $M(\bar{t})$ there is a subsequence $(u_{\sigma(n)})_n$ converging to some $u \in M(\bar{t})$. The subsequence $(x_{s\circ\sigma(n)})_n$ of $(x_n)_n$ then converges to $u \in M(\bar{t})$, which justifies the implication $\Rightarrow$.

Now suppose that M is not upper semicontinuous at $\bar{t}$, that is, there exists an open set $W \supset M(\bar{t})$ such that for each integer n there is a $t_n \in B(\bar{t}, 1/n)$ with $M(t_n) \not\subset W$. Note that $W \neq X$ and put
$$\varepsilon := \inf_{x \in M(\bar{t})} d(x, X \setminus W) = \min_{x \in M(\bar{t})} d(x, X \setminus W) > 0,$$
where the second equality is due to the compactness of $M(\bar{t})$. It is easily seen that $W_\varepsilon := \{x \in X : d(x, M(\bar{t})) < \varepsilon\} \subset W$, hence $M(t_n) \not\subset W_\varepsilon$ for every n. Consequently, for each $n \in \mathbb{N}$ there is some $x_n \in M(t_n)$ with $x_n \notin W_\varepsilon$, that is, $d(x_n, M(\bar{t})) \geq \varepsilon$. It ensues that $t_n \to \bar{t}$ as $n \to \infty$ along with $x_n \in M(t_n)$ for all n while no subsequence of $(x_n)_n$ can converge to a point in $M(\bar{t})$. This completes the proof of the proposition. $\square$

1.5. Scalarization of semicontinuity

We start this section with Berge's theorem on semicontinuity of value function. It is a fundamental theorem in variational analysis providing conditions ensuring a semicontinuity property of functions defined as pointwise suprema (resp. infima) over varying sets, the so-called *marginal pointwise suprema (resp. infima) value functions*. We recall that if, for an extended real-valued function h from T into a closed interval $I = [\inf(I), \sup(I)]$ of interest of $\mathbb{R} \cup \{-\infty, +\infty\}$, we take infimum and supremum of h in I over subsets of T, then
$$\inf_{t \in A} h(t) = \sup(I) \quad \text{and} \quad \sup_{t \in A} h(t) = \inf(I) \quad \text{whenever } A = \emptyset.$$
We also recall that the function h is lower (resp. upper) semicontinuous at $\bar{t} \in T$ if and only if for any real $r < h(\bar{t})$ (resp. $r > h(\bar{t})$) there exists a neighborhood V of $\bar{t}$ such that $r < h(t)$ (resp. $r > h(t)$) for all $t \in V$. It is known and readily seen that h is lower (resp. upper) semicontinuous at $\bar{t}$ if and only if $h(\bar{t}) \leq \liminf\limits_{t \to \bar{t}} h(t)$
$\left(\text{resp. } h(\bar{t}) \geq \limsup\limits_{t \to \bar{t}} h(t)\right)$.

1.5. SCALARIZATION OF SEMICONTINUITY

THEOREM 1.43 (C. Berge theorem of semicontinuity of marginal functions I).
] Let $f : T \times X \to \mathbb{R} \cup \{-\infty, +\infty\}$ be an extended real-valued function and $M : T \rightrightarrows X$ be a multimapping. Let $\varphi : T \to \mathbb{R} \cup \{-\infty, +\infty\}$ be the supremum value function defined by

$$\varphi(t) := \sup_{x \in M(t)} f(t,x) \quad \text{for all } t \in T,$$

the supremum being taken in a closed interval I of interest of $\mathbb{R} \cup \{-\infty, +\infty\}$ containing the range of f.
(a) If the multimapping M is lower semicontinuous at $\bar{t} \in T$ and if the function f is lower semicontinuous at every point of $\{\bar{t}\} \times M(\bar{t})$, then the function φ is lower semicontinuous at $\bar{t}$.
(b) If the multimapping M is upper semicontinuous at $\bar{t}$, if the function f is upper semicontinuous at every point of $\{\bar{t}\} \times M(\bar{t})$, and if the set $M(\bar{t})$ is compact, then the function φ is upper semicontinuous at $\bar{t}$.

PROOF. (a) We may suppose that $\varphi(\bar{t}) > \inf(I)$, otherwise the result is direct. Fix any real r with $\inf(I) < r < \varphi(\bar{t})$. Choose by definition of φ some $\bar{x} \in M(\bar{t})$ such that $r < f(\bar{t}, \bar{x})$. From the lower semicontinuity of f at $(\bar{t}, \bar{x})$ there exists some open set $V' \times W \ni (\bar{t}, \bar{x})$ such that, for each $t \in V'$, we have

(1.12) $$r < f(t,x) \quad \text{for all } x \in W.$$

Since $W \cap M(\bar{t}) \neq \emptyset$ (because $\bar{x} \in M(\bar{t}) \cap W$), the lower semicontinuity of the multimapping M at $\bar{t}$ gives a neighborhood $V \subset V'$ of $\bar{t}$ in T such that

$$W \cap M(t) \neq \emptyset \quad \text{for all } t \in V.$$

We deduce from this and from (1.12) that $r < \varphi(t)$ for all $t \in V$, thus the function φ is lower semicontinuous at $\bar{t}$.
(b) Suppose that $\varphi(\bar{t}) < +\infty$ and fix any real $r > \varphi(\bar{t})$. Choose a real s satisfying $r > s > \varphi(\bar{t})$. If $M(\bar{t}) = \emptyset$, then the upper semicontinuity of the multimapping M at $\bar{t}$ ensures that $M(t) = \emptyset$ for all t in a neighborhood V of $\bar{t}$, which entails, for all $t \in V$, that $\varphi(t) \leq \inf(I) \leq \varphi(\bar{t})$ hence $\varphi(t) < r$, so the function φ is upper semicontinuous at $\bar{t}$. So, suppose that $M(\bar{t}) \neq \emptyset$. Then for each $x \in M(\bar{t})$ we have $f(\bar{t}, x) < s$ and hence from the upper semicontinuity of the funtion of f at $(\bar{t}, x)$ there exist some neighborhood V_x of $\bar{t}$ and some open set $W_x \ni x$ such that $f(t, y) < s$ for all $t \in V_x$ and $y \in W_x$. By the compactness of $M(\bar{t})$ choose $x_1, \cdots, x_n \in M(\bar{t})$ such that $M(\bar{t}) \subset \bigcup_{i=1}^{n} W_{x_i} =: W$ and consider the neighborhood $V' := \bigcap_{i=1}^{n} V_{x_i}$ of $\bar{t}$. We have $f(t,y) < s$ for all $t \in V'$ and $y \in W$. By the upper semicontinuity property of the multimapping M choose a neighborhood $V \subset V'$ of $\bar{t}$ such that $M(t) \subset W$ for all $t \in V$. Then for each $t \in V$ we have

$$f(t, y) < s \quad \text{for all } y \in M(t), \quad \text{hence } \varphi(t) \leq s.$$

Consequently, $\varphi(t) < r$ for all $t \in V$, thus the function φ is upper semicontinuous at the point $\bar{t}$. $\square$

When the function f does not depend on t, another theorem avoiding the compactness of $M(\bar{t})$ is also valid.

THEOREM 1.44 (C. Berge theorem of semicontinuity of marginal functions II).
Let $f : X \to \mathbb{R} \cup \{-\infty, +\infty\}$ be a function and $M : T \rightrightarrows X$ be a multimapping. Let $\varphi : T \to \mathbb{R} \cup \{-\infty, +\infty\}$ be the supremum value function defined by
$$\varphi(t) := \sup_{x \in M(t)} f(x) \quad \text{for all } t \in T,$$
the supremum being taken as above in a closed interval I of interest of $\mathbb{R} \cup \{-\infty, +\infty\}$ containing the range of f.

If the multimapping M is upper semicontinuous at $\bar{t}$ and if the function f is upper semicontinuous on an open set $U \supset M(\bar{t})$, then the function φ is upper semicontinuous at $\bar{t}$.

PROOF. Suppose that $\varphi(\bar{t}) < +\infty$ and fix any real $r > \varphi(\bar{t})$. Choose a real s satisfying $r > s > \varphi(\bar{t})$. The set $W := \{x \in U : f(x) < s\}$ is open (by the upper semicontinuity of f on the open set U) and $M(\bar{t}) \subset W$. By the upper semicontinuity of M at $\bar{t}$ there is a neighborhood V of $\bar{t}$ such that $M(V) \subset W$. Consequently, for any $t \in V$ we obtain
$$\varphi(t) = \sup_{x \in M(t)} f(x) \leq s < r,$$
which confirms the upper semicontinuity of φ at $\bar{t}$. □

Applying Berge's theorem to the opposite of the distance function and using the equalities
$$d(x, M(t)) = \inf_{x' \in M(t)} d(x', x) = - \sup_{x' \in M(t)} \left(- d(x', x) \right),$$
where $\sup_{x' \in M(t)} \left(- d(x', x) \right) = -\infty$ if $M(t) = \emptyset$ since the supremum is taken in the interval $[-\infty, 0]$, we obtain the following result of semicontinuity of the distance function to a multimapping.

PROPOSITION 1.45. Assume that (X, d) is a metric space. Then the following hold:
(a) If the multimapping M is lower semicontinuous at $\bar{t}$, then for each $x \in X$, the function $t \mapsto d(x, M(t))$ is upper semicontinuous at $\bar{t}$.
(b) If the multimapping M is upper semicontinuous at $\bar{t}$ and if $M(\bar{t})$ is compact, then for each $x \in X$, the function $t \mapsto d(x, M(t))$ is lower semicontinuous at $\bar{t}$.

Our next step is to provide characterizations of semicontinuity of a multimapping through its pointwise support function. The concept of support function is fundamental in variational analysis and it will be used in other chapters. In the book, for a normed vector space X, characterizations of semicontinuity in terms of support functions will be applied on the vector space X endowed with either its strong topology or its weak topology. It will often be used also with the topological dual space X^* equipped with the weak* topology. To keep a unified viewpoint, we will then present the concept for locally convex spaces.

Suppose that (X, τ_X) is a Hausdorff locally convex (topological real vector) space and let X^* be its topological dual space. As usual, we denote by $w(X, X^*)$ (resp. $w(X^*, X)$) the *weak topology* of X (resp. the *weak* topology* of X^*), that is, the topology associated with the family of seminorms $(p_{u^*})_{u^* \in X^*}$ (resp. $(p_u)_{u \in X}$) defined by
$$p_{u^*}(x) := |\langle u^*, x \rangle| \; \forall x \in X, \quad \text{and} \quad p_u(x^*) := |\langle x^*, u \rangle| \; \forall x^* \in X^*.$$

So, the collection of sets

$$\{x \in X : \langle u_1^*, x\rangle < \varepsilon, \cdots, \langle u_m^*, x\rangle < \varepsilon\} \text{ with } \varepsilon > 0, m \in \mathbb{N}, u_1^*, \cdots, u_m^* \in X^*$$

(resp.

$$\{x^* \in X^* : \langle x^*, u_1\rangle < \varepsilon, \cdots, \langle x^*, u_m\rangle < \varepsilon\} \text{ with } \varepsilon > 0, m \in \mathbb{N}, u_1, \cdots, u_m \in X)$$

forms a basis of neighborhoods of zero in X (resp. in X^*). Sometimes, we will just say the w-topology (resp. w^*-topology).

DEFINITION 1.46. For any set C of the Hausdorff locally convex space X, one defines its *support function* $\sigma(\cdot, C)$ (also denoted by σ_C) as the extended real-valued function on X^* given by

(1.13) $$\sigma(x^*, C) := \sup_{x \in C} \langle x^*, x\rangle \quad \text{for all } x^* \in X^*.$$

Endowing X^* with the weak* topology, we know that the topological dual of $(X^*, w(X^*, X))$ is identified with X, so given a subset D of X^* its support function in X (or relative to X) is defined in the same way by

$$\sigma(x, D) := \sup_{x^* \in D} \langle x^*, x\rangle \quad \text{for all } x \in X.$$

In both equalities the supremum is taken in $[-\infty, +\infty]$.

Clearly, support functions are *positively homogeneous* in the sense that

$$\sigma(tx^*, C) = t\sigma(x^*, C) \text{ (resp. } \sigma(tx, D) = t\sigma(x, D)) \text{ for every real } t > 0.$$

Support functions are also extended real-valued convex functions according to the natural adaptation below of the classical concept of convexity to functions which may take infinite values.

DEFINITION 1.47. Given a nonempty convex set U of a vector space X, an extended real-valued function $f : U \to \mathbb{R} \cup \{-\infty, +\infty\}$ is called *convex* whenever its epigraph epi f is a convex set in $X \times \mathbb{R}$, which is equivalent to say that, for all $t \in [0, 1]$ and $u, u' \in U$, and for all reals $r \geq f(u)$ and $r' \geq f(u')$, we have

$$f(tu + (1-t)u') \leq tr + (1-t)r',$$

and this in turn is equivalent to require that for all $t \in [0, 1]$ and $u, u' \in U$, and for all reals $r > f(u)$ and $r' > f'(u')$

$$f(tu + (1-t)u') < tr + (1-t)r',$$

otherwise stated, the strict epigraph of f is convex. When $f(U) \subset \mathbb{R} \cup \{+\infty\}$, the convexity property may obviously be translated by

$$f(tu + (1-t)u') \leq tf(u) + (1-t)f(u') \quad \text{for all } t \in]0,1[,\ u, u' \in U.$$

In addition to its positive homogeneity and to its convexity, the support function fulfills diverse basic properties and we list below some of them which will be used next and throughout the book. First, it is clear that the convex function $\sigma(\cdot, C)$ on X^* (resp. $\sigma(\cdot, D)$ on X) is lower $w(X^*, X)$-semicontinuous (resp. lower $w(X, X^*)$-semicontinuous), and $\sigma(\cdot, C) = -\infty$ whenever $C = \emptyset$, since the supremum is taken in $[-\infty, +\infty]$. The Hahn-Banach separation theorem also ensures that a point x of X belongs to $\overline{\text{co}}\,C$ if and only if $\langle x^*, x\rangle \leq \sigma(x^*, C)$ for all $x^* \in X^*$. So, for two subsets C_1, C_2 of X (resp. D_1, D_2 of X^*) we have

(1.14) $$\overline{\text{co}}\,C_1 \subset \overline{\text{co}}\,C_2 \Leftrightarrow \sigma(\cdot, C_1) \leq \sigma(\cdot, C_2) \text{ on } X^*$$

and
(1.15) $$\overline{co}\, C_1 = \overline{co}\, C_2 \Leftrightarrow \sigma(\cdot, C_1) = \sigma(\cdot, C_2) \text{ on } X^*.$$
Similarly, for subsets of X^* the following hold:
(1.16) $$\overline{co}^*\, D_1 \subset \overline{co}^*\, D_2 \Leftrightarrow \sigma(\cdot, D_1) \leq \sigma(\cdot, D_2) \text{ on } X$$
and
(1.17) $$\overline{co}^*\, D_1 = \overline{co}^*\, D_2 \Leftrightarrow \sigma(\cdot, D_1) = \sigma(\cdot, D_2) \text{ on } X.$$

Writing $\sigma(x^*, C_1 + C_2) = \sup_{x_1 \in C_1} \langle x^*, x_1 \rangle + \sup_{x_2 \in C_2} \langle x^*, x_2 \rangle$ when C_1 and C_2 are nonempty, we also see that
(1.18) $$\sigma(x^*, C_1 + C_2) = \sigma(x^*, C_1) + \sigma(x^*, C_2)$$
and this equality still holds true even if C_1 or C_2 is empty. Analogously,
(1.19) $$\sigma(x, D_1 + D_2) = \sigma(x, D_1) + \sigma(x, D_2).$$
Given a nonempty bounded set B in the locally convex space X, the function $\sigma(\cdot, B)$ is finite on X^*, so using (1.18) and (1.14) we see that
(1.20) $$C_1 + B \subset C_2 + B \implies C_1 \subset \overline{co}\, C_2,$$
so if C_1 and C_2 are closed and convex
(1.21) $$C_1 + B = C_2 + B \implies C_1 = C_2.$$
Also, for any nonempty bounded set B' in $(X^*, w(X^*, X))$, the function $\sigma(\cdot, B')$ if finite on X and
(1.22) $$D_1 + B' \subset D_2 + B' \implies D_1 \subset \overline{co}^*\, D_2,$$
so if D_1 and D_2 are $w(X^*, X)$-closed and convex
(1.23) $$D_1 + B' = D_2 + B' \implies D_1 = D_2.$$
The results provided by (1.20), (1.21), (1.22) and (1.23) are known in the literature as Rådström cancellation laws.

Similar equalities to (1.18) are also valid for images under continuous linear mappings. We recall that the adjoint of a continuous linear mapping $A : X \to Y$ from X into another Hausdorff locally convex space Y is the linear mapping $A^* : Y^* \to X^*$ defined by
(1.24) $$A^*(y^*) = y^* \circ A \quad \text{for all } y^* \in Y^*.$$
Since $\sigma(y^*, A(C)) = \sup_{x \in C} \langle y^*, A(x) \rangle = \sup_{x \in C} \langle y^* \circ A, x \rangle$, it follows that
(1.25) $$\sigma(y^*, A(C)) = \sigma(A^*(y^*), C) \quad \text{for all } y^* \in Y^*.$$
Similarly, for a set $D \subset Y^*$ and its support function defined in X we note that $\sigma(x, A^*(D)) = \sup_{y^* \in D} \langle A^*(y^*), x \rangle = \sup_{y^* \in D} \langle y^*, A(x) \rangle$, thus
(1.26) $$\sigma(A(x), D) = \sigma(x, A^*(D)) \quad \text{for all } x \in X.$$

As said above, for $C \subset X$ (resp. $D \subset X^*$), the support function $\sigma(\cdot, C)$ (resp. $\sigma(\cdot, D)$) is a $w(X^*, X)$ (resp. $w(X, X^*)$) lower semicontinuous convex function. Corollary 3.83 will provide useful characterizations of convex functions on X^* (resp. on X) which are support functions of nonempty closed convex subsets of X^* (resp. of X).

1.5. SCALARIZATION OF SEMICONTINUITY

Let us also recall that the *polar set* of the subset C of X is defined as

(1.27) $$C^\circ := \{x^* \in X^* : \langle x^*, x\rangle \leq 1, \forall x \in C\},$$

so C° is a weak* closed convex set containing the zero of X^*, $C^\circ = X^*$ if either $C = \emptyset$ or $C = \{0\}$, and $\left(\overline{co}(\{0\} \cup C)\right)^\circ = C^\circ$. Similarly, for a subset D of X^*, one defines the *prepolar set* of D as

$$^\circ D := \{x \in X : \langle x^*, x\rangle \leq 1, \forall x^* \in D\}.$$

Since the topological dual of $(X^*, w(X^*, X))$ is (classically) identified with X (as said above), when there is no risk of ambiguity we will write in general D° instead of $^\circ D$ and call in this case D° as *the polar of D in X*.

For $C_1, C_2 \subset X$, clearly $C_2^\circ \subset C_1^\circ$ whenever $C_1 \subset C_2$, and

(1.28) $$(C_1 + C_2)^\circ \subset C_1^\circ \cap C_2^\circ \quad \text{whenever } 0 \in (\operatorname{cl} C_1) \cap (\operatorname{cl} C_2).$$

It is also clear that $\left(\bigcup_{i \in I} C_i\right)^\circ = \bigcap_{i \in I} C_i^\circ$, for $C_i \subset X$. Further, it is known and not difficult to prove (via the Hahn-Banach separation theorem for the inclusion of the left member into the second) that

(1.29) $$^\circ(C^\circ) = \overline{co}(\{0\} \cup C) \quad \text{and} \quad (^\circ D)^\circ = \overline{co}^*(\{0\} \cup D).$$

We then derive that

(1.30) $$^\circ\left(\bigcap_{i \in I} C_i^\circ\right) = \overline{co}\left(\{0\} \cup \left(\bigcup_{i \in I} C_i\right)\right).$$

For a convex set C in X and a continuous linear mapping $A : X \to Y$, it easily seen that

(1.31) $$\left(A(C)\right)^\circ = (A^*)^{-1}(C^\circ).$$

Clearly, when nonempty cones K_1, K_2 are involved in (1.28) (instead of sets C_1, C_2) one has the equality

(1.32) $$(K_1 + K_2)^\circ = K_1^\circ \cap K_2^\circ.$$

Similarly, given a closed convex cone K in a Hausdorff locally convex space Y and a continuous linear mapping $A : X \to Y$, one has

(1.33) $$^\circ\left(A^*(K^\circ)\right) = A^{-1}(K),$$

where $A^* : Y^* \to X^*$ is the adjoint linear mapping of A.

We also notice that the collection of polar sets (resp. prepolar sets) of finite subsets of X (resp. X^*) clearly forms a *basis of weak* (resp. weak) neighborhoods of zero* in X^* (resp. X).

THEOREM 1.48 (scalarization of upper semicontinuity through support functions). Assume that (X, τ_X) is a Hausdorff locally convex space.
(a) If the multimapping M is τ_X-upper semicontinuous at $\bar{t} \in T$, then for each $x^* \in X^*$, the function $t \mapsto \sigma(x^*, M(t))$ is upper semicontinuous at $\bar{t}$.
(b) If the set $M(\bar{t})$ is convex and closed and if for every $x^* \in X^*$ the function $t \mapsto \sigma(x^*, M(t))$ is upper semicontinuous at $\bar{t}$, then the multimapping M is τ_X-outer semicontinuous at $\bar{t}$ (or equivalently τ_X-closed at $\bar{t}$).
(c) Suppose that $M(\bar{t})$ is closed and the range under M of some neighborhood of $\bar{t}$ in contained in some τ_X-compact set of X. Then the multimapping M is τ_X-upper semicontinuous at $\bar{t}$ if and only if, for every $x^* \in X^*$, the function $t \mapsto \sigma(x^*, M(t))$

is upper semicontinuous at $\bar{t}$.

(d) If the set $M(\bar{t})$ is convex and $w(X, X^*)$-compact and if, for every $x^* \in X^*$, the function $t \mapsto \sigma(x^*, M(t))$ is upper semicontinuous at $\bar{t}$, then the multimapping M is $w(X, X^*)$-upper semicontinuous at $\bar{t}$.

(e) Suppose that the set $M(\bar{t})$ is convex and $w(X, X^*)$-compact. Then, the multimapping M is $w(X, X^*)$-upper semicontinuous at $\bar{t}$ if and only if, for each $x^* \in X^*$, the function $t \mapsto \sigma(x^*, M(t))$ is upper semicontinuous at $\bar{t}$.

PROOF. (a) Fixing any $x^* \in X^*$ and recalling that
$$\sigma(x^*, M(t)) = \sup_{x \in M(t)} \langle x^*, x \rangle,$$
we see that the assertion (a) directly follows from Theorem 1.44.

(b) Assume that $M(\bar{t})$ is convex and closed and that the function $t \mapsto \sigma(x^*, M(t))$ is upper semicontinuous for all $x^* \in X^*$. Take any $x \in \limsup_{t \to \bar{t}} M(t)$ (the result being obvious if the latter is empty) and take any $x^* \in X^*$. By Proposition 1.16 there exist a net $(t_j)_{j \in J}$ converging to $\bar{t}$ and a net $(x_j)_{j \in J}$ τ_X-converging to x with $x_j \in M(t_j)$ for all $j \in J$, hence $\langle x^*, x_j \rangle \leq \sigma(x^*, M(t_j))$. The upper semicontinuity of the function $t \mapsto \sigma(x^*, M(t))$ then entails $\langle x^*, x \rangle \leq \sigma(x^*, M(\bar{t}))$. Using this and the closedness of the convex set $M(\bar{t})$, we obtain $x \in M(\bar{t})$ hence $\limsup_{t \to \bar{t}} M(t) \subset M(\bar{t})$. This means that the multimapping M is τ_X-outer semicontinuous at $\bar{t}$.

(c) The statement in (c) follows from (a) and (b) and from Proposition 1.39.

(d) Now suppose that the function $t \mapsto \sigma(x^*, M(t))$ is upper semicontinuous at $\bar{t}$ for every $x^* \in X^*$. If $M(\bar{t}) = \emptyset$, then $\sigma(0, M(\bar{t})) = -\infty$, hence by the upper semicontinuity at $\bar{t}$ of the function $t \mapsto \sigma(0, M(t))$ there exists a neighborhood V of $\bar{t}$ such that $\sigma(0, M(t)) < 0$ for all $t \in V$. This entails $M(t) = \emptyset$ for all $t \in V$, thus the multimapping M is upper semicontinuous at $\bar{t}$ in the case $M(\bar{t}) = \emptyset$. So, suppose that $M(\bar{t}) \neq \emptyset$. Taking $\bar{x} \in M(\bar{t})$ and considering in place of M the multimapping $t \mapsto M(t) - \bar{x}$ (for which the function
$$t \mapsto \sigma(x^*, M(t) - \bar{x}) = -\langle x^*, \bar{x} \rangle + \sigma(x^*, M(t))$$
remains upper semicontinuous at $\bar{t}$) we may suppose without loss of generality that $0 \in M(\bar{t})$. Let W be any $w(X, X^*)$-open set containing $M(\bar{t})$. By Lemma 1.49 below, according to the $w(X, X^*)$-compactness of $M(\bar{t})$, there exists a $w(X, X^*)$-neighborhood W_0' of zero in X such that $M(\bar{t}) + 2W_0' \subset W$. Take $u_1^*, \cdots, u_m^* \in X^*$ such that
$$W_0 := \{x \in X : \langle u_i^*, x \rangle \leq 1, \forall i = 1, \cdots, m\} \subset W_0', \quad \text{hence} \quad M(\bar{t}) + 2W_0 \subset W.$$
By the $w(X, X^*)$-compactness of $M(\bar{t})$ again choose $x_1, \cdots, x_n \in M(\bar{t})$ such that $M(\bar{t}) \subset \bigcup_{i=1}^{n} (x_i + W_0)$. Observe that the closed convex set
$$C := \mathrm{co}\,\{0, x_1, \cdots, x_n\} + 2W_0$$
has its polar C° which is polyhedral because of the definition of W_0, and notice that C° is contained in $\mathrm{co}\{0, u_1^*, \cdot, u_m^*\}$ according to the inclusions $W_0 \subset C$ and (1.29). Consequently, C° is a bounded polyhedral convex set of the finite-dimensional vector space spanned by $u_1^*, \cdots, u_m^*$ (see Subsection 3.2.2, for details), so there exist $v_1^*, \cdots, v_p^* \in X^*$ such that

(1.34) $$\{x^* \in X^* : \langle x^*, x \rangle \leq 1 \,\forall x \in C\} = \mathrm{co}\,\{v_1^*, \cdots, v_p^*\}.$$

Put $S := \{x_1, \cdots, x_n\}$ and note that for each $j = 1, \cdots, p$ with $v_j^* \neq 0$ we have
$$\sigma(v_j^*, M(\bar{t})) \leq \sigma(v_j^*, S + W_0) < \sigma(v_j^*, S + 2W_0) \leq \sigma(v_j^*, C) \leq 1.$$
Then for each $j = 1, \cdots, p$ with $v_j^* \neq 0$, the inequality $\sigma(v_j^*, M(\bar{t})) < 1$ holds, hence from the upper semicontinuity at $\bar{t}$ of the function $t \mapsto \sigma(v_j^*, M(t))$ there is some neighborhood V of $\bar{t}$ such that, for each $j = 1, \cdots, p$ with $v_j^* \neq 0$, we have $\sigma(v_j^*, M(t)) < 1$ for all $t \in V$. The latter property then holds for all $v_j^*, j = 1, \cdots, p$. By (1.34) and (1.29), for all $t \in V$, we obtain $M(t) \subset C$, hence $M(t) \subset W$. This says that the multimapping M is $w(X, X^*)$-upper semicontinuous at $\bar{t}$.
(e) The statement in (e) is a direct consequence of assertions (a) and (d). □

LEMMA 1.49. *Let K be a compact set of a topological vector space X. Then, for any open set $W \supset K$, there exists a neighborhood U of zero in X such that $K + U \subset W$.*

PROOF. We may suppose $K \neq \emptyset$. For each $x \in K$, since $x \in W$, we can choose some open neighborhood U_x of zero such that $x + U_x + U_x \subset W$. According to the compactness of K there are $x_1, \cdots, x_m \in K$ such that $K \subset \bigcup_{i=1}^{m}(x_i + U_{x_i})$. For the neighborhood $U := \bigcap_{i=1}^{m} U_{x_i}$ of zero, we then have
$$K + U \subset \bigcup_{i=1}^{m}(x_i + U_{x_i}) + U \subset W.$$
□

From the above theorem we directly derive the corollary:

COROLLARY 1.50. *Assume that (X, τ_X) is a Hausdorff locally convex space and that M is a multimapping with closed convex values. Assume also that each point of T has a neighborhood whose range is contained in some τ_X-compact set of X. Then the multimapping M is τ_X-upper semicontinuous if and only if, for every $x^* \in X^*$, the function $t \mapsto \sigma(x^*, M(t))$ is upper semicontinuous.*

We study now the support function scalarization of lower semicontinuity of multimappings.

THEOREM 1.51 (scalarization of lower semicontinuity through support function). *Assume that (X, τ_X) is a Hausdorff locally convex space.*
(a) If the multimapping M is τ_X-lower semicontinuous at $\bar{t}$, then, for every $x^* \in X^*$, the function $t \mapsto \sigma(x^*, M(t))$ is lower semicontinuous at $\bar{t}$.
(b) Suppose that the range under M of some neighborhood of $\bar{t}$ in contained in some τ_X-compact set of X and that M takes on closed convex values on some neighborhood of $\bar{t}$. Then the multimapping M is τ_X-lower semicontinuous at $\bar{t}$ if and only if, for every $x^* \in X^*$, the function $t \mapsto \sigma(x^*, M(t))$ is lower semicontinuous at $\bar{t}$.

PROOF. (a) The assertion follows directly from (a) of Berge's theorem I.
(b) Let us prove the implication $\Leftarrow$. Fix a τ_X-compact set K of X and a neighborhood V_0 of $\bar{t}$ such that $M(V_0) \subset K$ and such that M takes on closed convex values on V_0. Since K is τ_X-compact in the Hausdorff topological space (X, τ_X) and since $w(X, X^*) \subset \tau_X$, the topologies induced on K by $w(X, X^*)$ and τ_X coincide. So, it suffices to show that the multimapping M is $w(X, X^*)$-lower semicontinuous at $\bar{t}$.

Let W_0 be a $w(X, X^*)$-open set with $M(\bar{t}) \cap W_0 \neq \emptyset$. Taking $\bar{x} \in M(\bar{t}) \cap W_0$ and considering in place of M the multimapping $t \mapsto M(t) - \bar{x}$ (for which the function

$$t \mapsto \sigma(x^*, M(t) - \bar{x}) = -\langle x^*, \bar{x} \rangle + \sigma(x^*, M(t))$$

remains lower semicontinuous at $\bar{t}$) we may suppose without loss of generality that $0 \in M(\bar{t}) \cap W_0$. Choose a $w(X, X^*)$-open symmetric convex neighborhood W of zero in X such that $W \subset W_0$. Then $0 \in M(\bar{t}) \cap W$. We claim that there exists a neighborhood $V \subset V_0$ of $\bar{t}$ such that

(1.35) $\qquad M(t) \cap W \neq \emptyset \quad$ for all $t \in V$.

Suppose the contrary. Then there exists a net $(t_j)_{j \in J}$ of V_0 converging to $\bar{t}$ and such that $M(t_j) \cap W = \emptyset$ for all $j \in J$. By the Hahn-Banach separation theorem, for each $j \in J$, there exists $x_j^* \in X^* \setminus \{0\}$ such that

(1.36) $\qquad \langle x_j^*, x \rangle \geq -1$ for all $x \in W$ and $\sigma(x_j^*, M(t_j)) \leq -1$.

The first inequality means that $|\langle x_j^*, x \rangle| \leq 1$ for all $x \in W$ because of the symmetric property of W. Further, the set $\{x^* \in X^* : |\langle x^*, x \rangle| \leq 1\ \forall x \in W\}$ is bounded and closed with respect to the $w(X^*, X)$-topology hence $w(X^*, X)$-compact. So, we may extract from $(x_j^*)_{j \in J}$ a subnet (that we do not relabel) converging weakly* to some $u^* \in X^*$. Put $\varepsilon = 1/6$ and take a weak neighborhood $W' \subset \varepsilon W$ of zero such that $|\langle u^*, v \rangle| < \varepsilon$ for all $v \in W'$. By the $w(X, X^*)$-compactness of K choose $u_1, \cdots, u_p \in K$ such that $K \subset \bigcup_{k=1}^{p} (u_k + W')$. Choose also $j_0 \in J$ such that $|\langle u^* - x_j^*, u_k \rangle| < \varepsilon$ for all $k = 1, \cdots, p$ and all $j \in J$ with $j_0 \preceq j$. Then for each $x \in K$ there is some $k \in \{1, \cdots, p\}$ such that $x \in u_k + W'$, thus for every $j \in J$ with $j_0 \preceq j$ we have

$$|\langle u^* - x_j^*, x \rangle| \leq |\langle u^* - x_j^*, u_k \rangle| + |\langle u^*, x - u_k \rangle| + |\langle x_j^*, x - u_k \rangle| < 3\varepsilon,$$

which ensures that, for every $j \in J$ with $j_0 \preceq j$,

$$\langle u^*, x \rangle < \langle x_j^*, x \rangle + (1/2) \quad \text{for all } x \in K,$$

and hence

(1.37) $\qquad \sigma(u^*, M(t_j)) \leq \sigma(x_j^*, M(t_j)) + 1/2 \leq -1/2,$

where the second inequality follows from (1.36). But the inclusion $0 \in M(\bar{t})$ ensures that $0 \leq \sigma(u^*, M(\bar{t}))$, thus $-1/2 < \sigma(u^*, M(\bar{t}))$. By the lower semicontinuity of the function $t \mapsto \sigma(u^*, M(t))$ there exists some $j_1 \in J$ with $j_0 \preceq j_1$ such that $-1/2 < \sigma(u^*, M(t_j))$, which contradicts (1.37). The claim (1.35) is then established. We deduce that $M(t) \cap W_0 \neq \emptyset$ for all $t \in V$ hence M is $w(X, X^*)$-lower semicontinuous at $\bar{t}$.

The reverse implication of (b) following from (a), the proof is finished. $\square$

The following result is a direct corollary of the latter proposition.

COROLLARY 1.52. *Assume that (X, τ_X) is a Hausdorff locally convex space and that M is a multimapping with closed convex values. Assume also that each point of T has a neighborhood whose range is contained in some τ_X-compact set of T. Then the multimapping M is τ_X-lower semicontinuous if and only if, for every $x^* \in X^*$, the function $t \mapsto \sigma(x^*, M(t))$ is lower semicontinuous.*

1.6. Semicontinuity of sum and convexification

This section is turned to the question of preservation of semicontinuities under sum and convexification.

PROPOSITION 1.53. Assume that (X, τ_X) is a topological vector space. Let $\alpha : T \to \mathbb{R}$ be a continuous function and $M_1, M_2 : T \rightrightarrows X$ be two multimappings. Let $M(t) = \alpha(t)M_1(t) + M_2(t)$ and $N(t) = \overline{\operatorname{co}}\, M_1(t)$ for all $t \in T$. The following hold:

(a) If M_1 and M_2 are lower semicontinuous at $\bar{t} \in T$, then the multimappings M and N are lower semicontinuous at $\bar{t}$.

(b) If $M_1(\bar{t})$ and $M_2(\bar{t})$ are compact and if M_1 and M_2 are upper semicontinuous at $\bar{t}$, then M is upper semicontinuous at $\bar{t}$.

(c) Assume that the Hausdorff topological vector space (X, τ_X) is locally convex. If $N(\bar{t})$ is compact and M_1 is upper semicontinuous at $\bar{t}$, then N is upper semicontinuous at $\bar{t}$.

PROOF. (a) Let W be any open set of X with $W \cap M(\bar{t}) \neq \emptyset$. There are $x_i \in M_i(\bar{t})$, $i = 1, 2$, such that $\alpha(\bar{t})x_1 + x_2 \in W$, and hence there is an open set $V' \ni \bar{t}$ and open sets $W_i \ni x_i$, $i = 1, 2$, such that $\alpha(V')W_1 + W_2 \subset W$. Since $W_i \cap M_i(\bar{t}) \neq \emptyset$, there exists, by the lower semicontinuity of M_i, a neighborhood $V \subset V'$ of $\bar{t}$ such that $W_i \cap M_i(t) \neq \emptyset$ for all $t \in V$ and $i = 1, 2$. Fixing $t \in V$ and choosing $a_i \in W_i \cap M_i(t)$, we obtain $\alpha(t)a_1 + a_2 \in W \cap M(t)$, hence $W \cap M(t) \neq \emptyset$. This says that M is lower semicontinuous at $\bar{t}$.

To prove the lower semicontiniuity property of N, it is enough to show the lower semicontinuity of the multimapping Q, where $Q(t) = \operatorname{co} M_1(t)$. Let W be any open set of X with $W \cap \operatorname{co} M_1(\bar{t}) \neq \emptyset$. There are $x_i \in M_1(\bar{t})$ and nonnegative reals λ_i, $i = 1, \cdots, n$, with $\lambda_1 + \cdots + \lambda_n = 1$ such that $\lambda_1 x_1 + \cdots + \lambda_n x_n \in W$. Choose open sets $W_i \ni x_i$ such that $\lambda_1 W_1 + \cdots + \lambda_n W_n \subset W$. Since $W_i \cap M_1(\bar{t}) \neq \emptyset$, there exists a neighborhood V of $\bar{t}$ such that $W_i \cap M_1(t) \neq \emptyset$ for all $t \in V$ and all $i = 1, \cdots, n$. Like above fixing $t \in V$ and choosing $a_i \in W_i \cap M_1(t)$, we obtain $\lambda_1 a_1 + \cdots + \lambda_n a_n \in W \cap \operatorname{co} M_1(t)$, hence $W \cap Q(t) \neq \emptyset$. Consequently, the multimapping Q is lower semicontinuous at $\bar{t}$.

(b) Suppose first that $\alpha(t) = 1$ for all t. If $M_1(\bar{t})$ (resp. $M_2(\bar{t})$) is empty, then $M_1(t)$ (resp. $M_2(t)$) is empty, according to the upper semicontinuity of M_1 (resp. M_2), for all t in some neighborhood of $\bar{t}$, hence so is $M(t)$. Therefore, suppose that $M(\bar{t}) \neq \emptyset$ and fix any open set $W \supset M(\bar{t})$. For each $u \in M_1(\bar{t})$ and each $v \in M_2(\bar{t})$ choose two open sets $W'_{u,v} \ni u$ and $W''_{u,v} \ni v$ such that $W'_{u,v} + W''_{u,v} \subset W$. By the compactness of $M_2(\bar{t})$ there are $v_1, \cdots, v_q \in M_2(\bar{t})$ such that $M_2(\bar{t}) \subset \bigcup_{j=1}^{q} W''_{u,v_j}$. The open sets $W''_u := \bigcup_{j=1}^{q} W''_{u,v_j}$ and $W'_u := \bigcap_{j=1}^{q} W'_{u,v_j}$ contain $M_2(\bar{t})$ and u respectively and they satisfy $W'_u + W''_u \subset W$. Choose now (by compactness of $M_1(\bar{t})$) elements $u_1, \cdots, u_p \in M_1(\bar{t})$ such that $M_1(\bar{t}) \subset \bigcup_{i=1}^{p} W'_{u_i}$ and set $W' := \bigcup_{i=1}^{p} W'_{u_i}$ and $W'' := \bigcap_{i=1}^{p} W''_{u_i}$. The open sets W' and W'' contain $M_1(\bar{t})$ and $M_2(\bar{t})$ respectively and we have

$$(1.38) \quad W' + W'' = \bigcup_{i=1}^{p}(W'_{u_i} + W'') \subset \bigcup_{i=1}^{p}(W'_{u_i} + W''_{u_i}) \subset W.$$

Take by the upper semicontinuity of M_1 and M_2 at $\bar{t}$ a neighborhood V of $\bar{t}$ such that for all $t \in V$ we have $M_1(t) \subset W'$ and $M_2(t) \subset W''$. Then $M(t) \subset W$ according to (1.38), and this translates the upper semicontinuity of M at $\bar{t}$.

The upper semicontinuity of the multimapping $t \mapsto \alpha(t)M_1(t)$ can be shown in a similar way, and then we deduce the upper semicontinuity of M in the general case of the statement of (b).

(c) We may suppose that $N(\bar{t}) \neq \emptyset$ (since otherwise $M_1(\bar{t}) = \emptyset$ and the upper semicontinuity of M_1 at $\bar{t}$ yields a neighborhood V of $\bar{t}$ with $M(t) = \emptyset$, hence $N(t) = \emptyset$ for all $t \in V$). Let an open set $W \supset N(\bar{t})$. According to the compactness of $N(\bar{t})$ and to the local convexity of τ_X there is by Lemma 1.49 an open convex neighborhood W_0 of zero in X such that

$$(1.39) \quad N(\bar{t}) + 2W_0 \subset W.$$

The set $N(\bar{t}) + W_0$ being an open set containing $M_1(\bar{t})$, the upper semicontinuity of M_1 at $\bar{t}$ gives a neighborhood V of $\bar{t}$ such that for all $t \in V$ we have $M_1(t) \subset N(\bar{t}) + W_0$, thus

$$\overline{\text{co}}\, M_1(t) \subset \text{cl}\left(N(\bar{t}) + W_0\right) \subset N(\bar{t}) + 2W_0.$$

We deduce from this and (1.39) that $N(t) \subset W$ for all $t \in V$, hence N is upper semicontinuous at $\bar{t}$. The proof is then complete. □

REMARK 1.54. It is worth mentioning that the compactness assumption of $N(\bar{t}) = \overline{\text{co}}\, M_1(\bar{t})$ is satisfied whenever $M_1(\bar{t})$ is τ-compact and τ is either the norm or the weak topology of a Banach space. Indeed, (see any classical book on normed space, e.g., [**366**, p. 416 and 484]) in any Banach it is known by Mazur Theorem (resp. Krein-Šmulian Theorem) that the closed (resp. weakly closed) convex hull of a compact (resp. weakly compact) set is compact (resp. weakly compact). □

1.7. Michael continuous selection theorem

We show in this section that lower semicontinuous multimappings admit continuous selections under suitable assumptions.

DEFINITION 1.55. Given a multimapping $M : T \rightrightarrows X$, one calls *selection* of M over a subset $U \subset T$ any mapping $f : U \to X$ such that $f(t) \in M(t)$ for all $t \in U$. When U is the whole set T, one just says that f is a selection of M.

The result on continuous selection requires two lemmas, and each one has its own interest. The first lemma is concerned with the preservation of lower semicontinuity under intersection with an open ball moving continuously.

LEMMA 1.56. Let T be a topological space, (X, d) be a metric space and $M : T \rightrightarrows X$ be a multimapping which is lower semicontinuous at $\bar{t} \in T$. Let $\rho : T \to]0, +\infty[$ and $f : T \to X$ be two mappings which are continuous at $\bar{t}$. Then the multimapping $t \mapsto M(t) \cap B(f(t), \rho(t))$ is lower semicontinuous at $\bar{t}$.

PROOF. Let V be an open set in X with $M(\bar{t}) \cap B(f(\bar{t}), \rho(\bar{t})) \cap V \neq \emptyset$ and $\bar{x}$ in this intersection. We can choose two reals r, s with $d(\bar{x}, f(\bar{t})) < s < r < \rho(\bar{t})$. Since $M(\bar{t}) \cap \big(B(\bar{x}, r-s) \cap V\big) \neq \emptyset$ (because it contains $\bar{x}$), the lower semicontinuity of M at $\bar{t}$ gives a neighborhood U of $\bar{t}$ such that $M(t) \cap B(\bar{x}, r-s) \cap V \neq \emptyset$ for all $t \in U$. By continuity of f and ρ at $\bar{t}$ and by the inequalities $d(\bar{x}, f(\bar{t})) < s$ and $r < \rho(\bar{t})$, there is a neighborhood $U_0 \subset U$ of $\bar{t}$ such that for all $t \in U_0$ one has $d(f(t), \bar{x}) < s$ and $r < \rho(t)$. Then, considering any $t \in U_0$ we obtain for every $x \in B(\bar{x}, r-s)$

$$d(x, f(t)) \leq d(x, \bar{x}) + d(\bar{x}, f(t)) < r - s + s = r < \rho(t),$$

hence $B(\bar{x}, r-s) \subset B(f(t), \rho(t))$. It follows that $M(t) \cap B(f(t), \rho(t)) \cap V \neq \emptyset$ for all $t \in U_0$, which confirms the lower semicontinuity of $t \mapsto M(t) \cap B(f(t), \rho(t))$ at the point $\bar{t}$. □

The second lemma establishes the existence of approximate continuous selection provided the first space T is paracompact.

Given a paracompact Hausdorff topological space T, we recall (see Theorem A.3 in Appendix) that for each open cover $(U_j)_{j \in J}$ of T there exists a locally finite continuous partition of unity $(\varphi_i)_{i \in I}$ subordinated to $(U_j)_{j \in J}$. This means that each φ_i is a continuous function from T into $[0, 1]$ which is null outside some U_j, each $t \in T$ has a neighborhood on which all but a finite number of φ_i are null, and $\sum_{i \in I} \varphi_i(t) = 1$ for every $t \in T$.

LEMMA 1.57. Let T be a paracompact Hausdorff topological space and X be a normed space. Let $M : T \rightrightarrows X$ be a lower semicontinuous multimapping with nonempty convex values. Then for any real $\varepsilon > 0$ there exists a continuous mapping $f : T \to X$ such

$$d(f(t), M(t)) < \varepsilon \quad \text{for all } t \in T.$$

PROOF. Fix any real $\varepsilon > 0$, and for each $x \in X$ consider the set $U_x := M^{-1}\big(B(x, \varepsilon)\big)$, which is open by the lower semicontinuity of M. Noting that $(U_x)_{x \in X}$ is an open cover of T, choose a locally finite continuous partition of unity $(\varphi_i)_{i \in I}$ subordinated to this cover. For each $i \in I$ take $x_i \in X$ such that $\varphi_i(t) = 0$ for all $t \in T \setminus U_{x_i}$. The mapping $f : T \to X$ defined by $f(t) = \sum_{i \in I} \varphi_i(t) x_i$ for all $t \in T$, is continuous. Given any $t \in T$, for every $i \in I$ with $\varphi_i(t) \neq 0$ we have $t \in U_{x_i}$, hence $d(x_i, M(t)) < \varepsilon$, so by convexity of $d(\cdot, M(t))$ it follows that $d(f(t), M(t)) < \varepsilon$. □

We are now ready to prove the result of continuous selection.

THEOREM 1.58 (E. Michael continuous selection theorem). Let T be a paracompact Hausdorff topological space and X be a Banach space. Let $M : T \rightrightarrows X$ be a lower semicontinuous multimapping with nonempty closed convex values. Then there exists a continuous selection of M.

PROOF. Let $M_0 := M$ and by Lemma 1.57 with $\varepsilon = 1/2^0$ let $f_0 : T \to X$ be a continuous mapping such that $d(f_0(t), M(t)) < 1/2^0$ for all $t \in T$. Considering the multimapping with nonempty convex values $M_1 : T \rightrightarrows X$ defined by

$$M_1(t) = M_0(t) \cap B(f_0(t), 1/2^0) \quad \text{for all } t \in T.$$

We note by Lemma 1.56 that M_1 is lower semicontinuous, so by Lemma 1.57 again we can take a continuous mapping $f_1 : T \to X$ with $d(f_1(t), M_1(t)) < 1/2^1$, thus

$\|f_1(t) - f_0(t)\| < 1/2^0 + 1/2^1$ and $d(f_1(t), M(t)) < 1/2^1$. By induction we obtain a sequence $(f_n)_{n \geq 0}$ of mappings from X into Y such that for each integer $n \geq 0$

$$\|f_{n+1}(t) - f_n(t)\| < \frac{1}{2^n} + \frac{1}{2^{n+1}} \quad \text{and} \quad d(f_n(t), M(t)) < \frac{1}{2^n} \quad \text{for all } t \in T.$$

The sequence $(f_n)_{n \geq 0}$ then converges uniformly on T to a continuous mapping $f : T \to X$ and clearly $d(f(t), M(t)) = 0$, that is, $f(t) \in M(t)$ for all $t \in T$. □

The assertion (b) in the following corollary says that, given a multimapping M as in Theorem 1.58, any point $(\bar{t}, \bar{x})$ in its graph belongs to the graph of a continuous selection.

COROLLARY 1.59. Let T be a paracompact Hausdorff topological space and X be a Banach space. Let $M : T \rightrightarrows X$ be a lower semicontinuous multimapping with nonempty closed convex values.
(a) Let S be a nonempty closed subset T and $f_0 : S \to X$ be a mapping which is continuous on S relative to the induced topology and which satisfies $f_0(t) \in M(t)$ for all $t \in S$. Then f_0 can be extended to a continuous selection of M.
(b) For any $\bar{t} \in T$ and any $\bar{x} \in M(\bar{t})$ there exists a continuous selection f of M such that $f(\bar{t}) = \bar{x}$.

PROOF. The assertion (b) clearly follows from (a). To justify (a), define the multimapping $M_0 : T \rightrightarrows X$ by $M_0(t) = \{f_0(t)\}$ if $t \in S$ and $M_0(t) = M(t)$ otherwise. This multimapping M_0 is convex-valued and by Lemma 1.60 below it is lower semicontinuous. Then the assertion (a) follows from Theorem 1.58 above. □

LEMMA 1.60. Let T, X be topological spaces and $M : T \rightrightarrows X$ be a multimapping which is lower semicontinuous. Let S be a nonempty closed subset of T and $N : S \rightrightarrows X$ be a multimapping with $N(t) \subset M(t)$ for all $t \in S$. Assume that N is lower semicontinuous relative to the topology induced on S by the one of T. Then the multimapping $M_0 : T \rightrightarrows X$, defined by

$$M_0(t) = \begin{cases} N(t) & \text{if } t \in S \\ M(t) & \text{if } t \in T \setminus S, \end{cases}$$

is lower semicontinuous.

PROOF. For any $\bar{t} \in T \setminus S$, the multimappings M_0 and M coincide on the open neighborhood $T \setminus S$ of $\bar{t}$, hence M_0 is lower semicontinuous at $\bar{t}$. Now, fix any $\bar{t} \in S$ and take any open set V in X with $M_0(\bar{t}) \cap V \neq \emptyset$. Then $N(\bar{t}) \cap V \neq \emptyset$, and this also gives $M(\bar{t}) \cap V \neq \emptyset$ since $N(\bar{t}) \subset M(\bar{t})$. We obtain two neighborhoods U_1 and U_2 of $\bar{t}$ in T such that $N(t) \cap V \neq \emptyset$ for all $t \in U_1 \cap S$ and $M(t) \cap V \neq \emptyset$ for all $t \in U_2$. For the neighborhood $U := U_1 \cap U_2$ in T, we deduce that $M_0(t) \cap V \neq \emptyset$ for all $t \in U$, so the multimapping M_0 is lower semicontinuous at $\bar{t}$. □

1.8. Hausdorff-Pompeiu semidistances

In addition to the above concepts of semilimits inferior and superior, in some cases one needs to invoke the Hausdorff-Pompeiu excess and distance.

1.8.1. Hausdorff-Pompeiu excess and distance.
We begin with the following definition.

DEFINITION 1.61. Assume that (X, d) is a metric space. Given two subsets S, S' of X, the *Hausdorff-Pompeiu excess* of the set S over the set S' is defined by

$$\mathrm{exc}(S, S') := \sup_{x \in S} d(x, S')$$

(where as already seen $d(x, S') = +\infty$ if $S' = \emptyset$ and $\sup_{x \in S} d(x, S') = 0$ if $S = \emptyset$, since infimum and supremum are taken in $[0, +\infty]$), and the *Hausdorff-Pompeiu distance* between S and S' is

$$\mathrm{haus}(S, S') := \max\{\mathrm{exc}(S, S'), \mathrm{exc}(S', S)\}.$$

When for a family of subsets $(S_t)_{t \in T_0}$ of X one has $\lim_{T_0 \ni t \to \bar{t}} \mathrm{haus}(S_t, S) = 0$, one says that the family $(S_t)_{t \in T_0}$ *Hausdorff-Pompeiu converges* to S as $T_0 \ni t \to \bar{t}$.

So, a multimapping $M : T \rightrightarrows X$ is *Hausdorff-Pompeiu continuous* at $\bar{t} \in T$, when the family of sets $(M(t))_{t \in T}$ Hausdorff-Pompeiu converges to the set $M(\bar{t})$ as $t \to \bar{t}$.

From the definition above we see that $\mathrm{exc}(S, S') = 0$ if and only if $S \subset \mathrm{cl}\, S'$, so $\mathrm{haus}(S, S') = 0$ if and only if $\mathrm{cl}\, S = \mathrm{cl}\, S'$. Further, $\mathrm{exc}(S, S') < +\infty$ for every nonempty set $S' \subset X$ if and only if S is bounded.

Let S, S' be nonempty sets in the metric space (X, d). For any set $P \supset S$ in X we observe that for any $x \in X$

$$d(x, S') - d(x, S) = \sup_{u \in S} \left(d(x, S') - d(x, u)\right) = \sup_{u \in S} \inf_{v \in S'} \left(d(x, v) - d(x, u)\right)$$
$$\leq \sup_{u \in S} \inf_{v \in S'} d(u, v) = \sup_{u \in S} d(u, S')$$
$$= \sup_{u \in S} \left(d(u, S') - d(u, S)\right) \leq \sup_{u \in P} \left(d(u, S') - d(u, S)\right),$$

thus

$$\sup_{x \in X} \left(d(x, S') - d(x, S)\right) \leq \mathrm{exc}(S, S') \leq \sup_{x \in P} \left(d(x, S') - d(x, S)\right)$$
$$\leq \sup_{x \in X} \left(d(x, S') - d(x, S)\right),$$

which implies

(1.40) $\quad \mathrm{exc}(S, S') = \sup_{x \in P} \left(d(x, S') - d(x, S)\right) = \sup_{x \in X} \left(d(x, S') - d(x, S)\right).$

Further, recalling that $\sup_{i \in I} \sup_{z \in Z} f_i(z) = \sup_{z \in Z} \sup_{i \in I} f_i(z)$, we also see from the first equality in (1.40) that

(1.41) $\quad \mathrm{exc}(S, S') = \max\{0, \sup_{x \in P} \left(d(x, S') - d(x, S)\right)\} = \sup_{x \in P} \left(d(x, S') - d(x, S)\right)^+.$

Considering now in X any set $Q \supset S \cup S'$, where S and S' are still nonempty, we deduce from (1.40) that

$$\max\{\mathrm{exc}(S,S'), \mathrm{exc}(S',S)\} = \max\left\{\sup_{x\in Q} (d(x,S) - d(x,S')), \sup_{x\in Q} (d(x,S') - d(x,S))\right\}$$

$$= \sup_{x\in Q} \left(\max\{d(x,S) - d(x,S'), d(x,S') - d(x,S)\}\right)$$

$$= \sup_{x\in Q} |d(x,S) - d(x,S')|,$$

and hence

(1.42) $\qquad \mathrm{haus}(S,S') = \sup_{x\in Q} |d(x,S') - d(x,S)| = \sup_{x\in X} |d(x,S') - d(x,S)|.$

Taking now three nonempty subsets S, S', S'' of X and considering any $x \in X$ we can write for all $x'' \in S''$

$$d(x, S') - d(x, S) \leq d(x'', S') + d(x, x'') - d(x, S)$$
$$\leq \mathrm{exc}(S'', S') + d(x, x'') - d(x, S),$$

which gives $d(x,S') - d(x,S) \leq d(x,S'') - d(x,S) + \mathrm{exc}(S'',S')$ by taking the infimum over $x'' \in S''$. The latter inequality being true for all $x \in X$, taking the supremum over $x \in X$ and using (1.40) we obtain

$$\mathrm{exc}(S, S') \leq \mathrm{exc}(S, S'') + \mathrm{exc}(S'', S'),$$

and the inequality is readily seen to be still true if anyone of the three sets is empty. From this we also obtain that

$$\mathrm{exc}(S, S') \leq \mathrm{haus}(S, S'') + \mathrm{haus}(S'', S'),$$

and hence

$$\mathrm{haus}(S, S') \leq \mathrm{haus}(S, S'') + \mathrm{haus}(S'', S').$$

The latter inequality combined with the evident symmetry of $\mathrm{haus}(\cdot,\cdot)$ tells us that $\mathrm{haus}(\cdot,\cdot)$ is *a semidistance on the set of all subsets* of the metric space X and *a distance on the set of nonempty closed bounded subsets* of X. By a *semidistance on a set* Y we mean a function δ from $Y \times Y$ into $\mathbb{R}_+ \cup \{+\infty\}$ which satisfies the properties of a distance except that it is not required that $\delta(y,y') = 0$ implies $y = y'$.

We have then proved the following results.

PROPOSITION 1.62. *Given a metric space* (X,d), *the following hold:*
(a) *The Hausdorff-Pompeiu excess satisfies the triangle inequality, and the Hausdorff-Pompeiu function* $\mathrm{haus}(\cdot,\cdot)$ *is a semidistance on the collection of all subsets of* X.
(b) *For nonempty subsets* S, S' *of* X, *one has for any set* $P \supset S$ *in* X

$$\mathrm{exc}(S, S') = \sup_{x\in P} (d(x,S') - d(x,S)) = \sup_{x\in X} (d(x,S') - d(x,S))$$

$$= \sup_{x\in P} (d(x,S') - d(x,S))^+ = \sup_{x\in X} (d(x,S') - d(x,S))^+,$$

and for any set $Q \supset S \cup S'$ *in* X

$$\mathrm{haus}(S, S') = \sup_{x\in Q} |d(x,S') - d(x,S)| = \sup_{x\in X} |d(x,S') - d(x,S)|.$$

REMARK 1.63. The equality in (b) of the above Proposition 1.62 ensures in particular that for any $x \in X$ fixed the function $S \mapsto d(x, S)$ satisfies the inequality
$$|d(x, S) - d(x, S')| \leq \text{haus}(S, S'),$$
so it is Lipschitzian with Lipschitz constant 1 with respect to the Hausdorff-Pompeiu semidistance on the collection of nonempty subsets of X. □

REMARK 1.64. When S, S' are convex, expressions for $\text{exc}(S, S')$ and $\text{haus}(S, S')$ in terms of support functions will be established later in Proposition 2.80. □

Although $\text{haus}(\cdot, \cdot)$ is a true distance merely on the set of nonempty closed bounded subsets of X, it is usual to keep the term of Hausdorff-Pompeiu distance on the set of all subsets of X.

1.8.2. Truncated Hausdorff-Pompeiu excess and distance. Easy examples give $\text{haus}(S, S') = +\infty$ very often for unbounded sets, in particular for epigraphs of functions. Further, even for the simple nonempty closed bounded sets $S_n := \{0\} \cup [n, n+1]$ in $\mathbb{R}$, for which one should expect the convergence to the set $S := \{0\}$, one observes that $\text{haus}(S_n, S) \to +\infty$ as the integer n tends to $+\infty$, while
$$\liminf_{n \to \infty} S_n = \limsup_{n \to \infty} S_n = \{0\}.$$
On the other hand, we know by Proposition 1.62 that
$$\text{exc}(S, S') = \sup_{x \in X} \left(d(x, S') - d(x, S)\right)^+ \quad \text{and} \quad \text{haus}(S, S') = \sup_{x \in X} |d(x, S') - d(x, S)|.$$
Those features lead for a normed space $(X, \|\cdot\|)$ and the distance d associated with the norm $\|\cdot\|$, to consider for any extended real $\rho \in \,]0, +\infty]$ and subsets S, S' of X the ρ-*excess* of S over S' and the *Hausdorff-Pompeiu ρ-metric* between S and S' defined by $\text{exc}_\rho(S, S') = \text{haus}_\rho(S, S') = 0$ when both sets S, S' are empty and otherwise

(1.43) $$\text{exc}_\rho(S, S') := \sup_{\|x\| \leq \rho} \left(d(x, S') - d(x, S)\right)^+$$

(where we recall that $t^+ := \max\{t, 0\}$ for $t \in \mathbb{R} \cup \{-\infty, +\infty\}$), and

(1.44) $$\text{haus}_\rho(S, S') := \max\{\text{exc}_\rho(S, S'), \text{exc}_\rho(S', S)\} = \sup_{\|x\| \leq \rho} |d(x, S') - d(x, S)|,$$

where the latter equality follows as for the similar equality in (1.42). Of course, for $\rho := +\infty$ one has $\text{exc}_\infty(S, S') = \text{exc}(S, S')$ and $\text{haus}_\infty(S, S') = \text{haus}(S, S')$. According to the above comment, when working with nonbounded sets, it will be often necessary to utilize the true truncated Hausdorff-Pompeiu excesses or distances, that is, with reals $\rho \in \,]0, +\infty[$.

Recalling for $s, t \in \mathbb{R}$ that $(s+t)^+ \leq s^+ + t^+$, it is clear that the non-negative functions $\text{exc}_\rho(\cdot, \cdot)$ and $\text{haus}_\rho(\cdot, \cdot)$ fulfill for nonempty sets the triangle inequality, that is,
$$\text{exc}_\rho(S, S') \leq \text{exc}_\rho(S, S'') + \text{exc}_\rho(S'', S')$$
and
$$\text{haus}_\rho(S, S') \leq \text{haus}_\rho(S, S'') + \text{haus}_\rho(S'', S').$$
So, $\text{haus}_\rho(\cdot, \cdot)$ is a semidistance on the set of nonempty subsets of X in the sense recalled above. Further, for reals $\rho, \rho' > 0$ the inequality $\rho \leq \rho' + |\rho' - \rho|$ ensures

that every $x \in \rho \mathbb{B}_X$ can be written as $x = x' + |\rho' - \rho|u$ with $x' \in \rho' \mathbb{B}_X$ and $u \in \mathbb{B}_X$, so for nonempty sets S, S' in X it results that
$$d(x, S') - d(x, S) \le d(x', S') - d(x', S) + 2|\rho' - \rho|,$$
which yields
$$\mathrm{exc}_\rho(S, S') \le \mathrm{exc}_{\rho'}(S, S') + 2|\rho' - \rho| \quad \text{and} \quad \mathrm{haus}_\rho(S, S') \le \mathrm{haus}_{\rho'}(S, S') + 2|\rho' - \rho|,$$
hence
$$|\mathrm{exc}_\rho(S, S') - \mathrm{exc}_{\rho'}(S, S')| \le 2|\rho - \rho'| \quad \text{and} \quad |\mathrm{haus}_\rho(S, S') - \mathrm{haus}_{\rho'}(S, S')| \le 2|\rho - \rho'|.$$
We can summarize as follows:

PROPOSITION 1.65. *Given a normed space $(X, \|\cdot\|)$ the following hold:*
(a) *For any $\rho \in {]}0, +\infty]$ the ρ-excess satisfies the triangle inequality, and the function $\mathrm{haus}_\rho(\cdot, \cdot)$ is a semidistance on the collection of nonempty subsets of X.*
(b) *For any reals $\rho, \rho' > 0$ and any nonempty sets S, S' in X, the following inequalities hold true*

(1.45)
$$|\mathrm{exc}_\rho(S, S') - \mathrm{exc}_{\rho'}(S, S')| \le 2|\rho - \rho'| \text{ and } |\mathrm{haus}_\rho(S, S') - \mathrm{haus}_{\rho'}(S, S')| \le 2|\rho - \rho'|.$$

On the collection of nonempty bounded subsets of X, the convergence with respect to (resp. the topology associated with) the family $\bigl(\mathrm{haus}_\rho(\cdot, \cdot)\bigr)_{\rho \in {]}0, +\infty[}$ of semidistances is known in the literature as the Attouch-Wets convergence (resp. topology) because of the various contributions of those authors to that convergence; see the related comments in Section 1.10. The related convergence for functions is defined via their epigraphs (like the Painlevé-Peano-Kuratowski convergence in Definition 1.3).

DEFINITION 1.66. *The family $(S_t)_{t \in T_0}$ of subsets of the normed space X converges in the sense of Attouch-Wets to a set S of X as $T_0 \ni t \to \bar{t}$ provided that for every real $\rho \in {]}0, +\infty[$*
$$\lim_{T_0 \ni t \to \bar{t}} \mathrm{haus}_\rho(S_t, S) = 0.$$
When in addition the set S is required to be closed, then it is unique whenever the family Attouch-Wets converges.

One says that a family of functions $(f_t)_{t \in T_0}$ from $X \to \mathbb{R} \cup \{-\infty, +\infty\}$ Attouch-Wets converges as $T_0 \ni t \to \bar{t}$ to a function f from X to $\mathbb{R} \cup \{-\infty, +\infty\}$ if the family of epigraphs $(\mathrm{epi}\, f_t)_{t \in T_0}$ Attouch-Wets converges in $X \times \mathbb{R}$ to $\mathrm{epi}\, f$.

Since $\mathrm{exc}_\rho(S', S'')$ and $\mathrm{haus}_\rho(S', S'')$ are nondecreasing with respect to ρ, it is clear that $(S_t)_{t \in T_0}$ Attouch-Wets converges to S if and only if there is some real $\bar{\rho} > 0$ such that
$$\lim_{T_0 \ni t \to \bar{t}} \mathrm{haus}_\rho(S_t, S) = 0 \quad \text{for every real } \rho \ge \bar{\rho}.$$
Obviously, the Hausdorff-Pompeiu convergence entails the Attouch-Wets convergence.

EXAMPLE 1.67. *With subsets of $\mathbb{R}$ given by $S_n := \{0\} \cup [n, n+1]$, we saw above that the sequence $(S_n)_{n \in \mathbb{N}}$ does not converge to $\{0\}$. Nevertheless, this sequence Attouch-Wets converges to $\{0\}$. Indeed, fixing any real $\rho > 0$ and taking any integer*

$n \geq 2\rho$, it is clear that, for all $x \in [-\rho, \rho]$, we have $d(x, S_n) = |x|$. This ensures that $\text{haus}_\rho(S_n, \{0\}) = \sup_{|x| \leq \rho} |d(x, \{0\}) - d(x, S_n)| = 0$ for all $n \geq 2\rho$.

In fact, we will see in Proposition 1.75 that the Painlevé-Peano and Attouch-Wets convergences coincide when the normed space X is *finite-dimensional*. $\square$

Now let us notice that in the context of normed spaces the excess and Hausdorff-Pompeiu distances can also be described as follows:

PROPOSITION 1.68. *In the normed space $(X, \|\cdot\|)$ one has for any subsets S, S' of X*

(1.46) $$\text{exc}(S, S') = \inf\{r > 0 : S \subset S' + r\mathbb{B}_X\}$$

and

(1.47) $$\text{haus}(S, S') = \inf\{r > 0 : S \subset S' + r\mathbb{B}_X, S' \subset S + r\mathbb{B}_X\}.$$

PROOF. It suffices to prove the equality concerning the excess. Take any sets S, S' in X and put $\eta(S, S') := \inf\{r > 0 : S \subset S' + r\mathbb{B}_X\}$. Observe that for $S = \emptyset$ (resp. $S \neq \emptyset$ and $S' = \emptyset$) the equality is trivial. So, suppose that both sets are nonempty. Note first that for any real $r > 0$ with $S \subset S' + r\mathbb{B}_X$ (if any) and any $x \in S$, we have $x \in S' + r\mathbb{B}_X$, hence $d(x, S') \leq r$. This yields

$$\text{exc}(S, S') = \sup_{x \in S'} d(x, s') \leq \inf\{r > 0 : S \subset S' + r\mathbb{B}_X\} = \eta(S, S').$$

Conversely, if $\eta(S, S') = 0$ the inequality $\eta(S, S') \leq \text{exc}(S, S')$ is trivial. Then suppose $\eta(S, S') > 0$ and take any $0 < t < \inf\{r > 0 : S \subset S' + r\mathbb{B}_X\}$. It ensues that $S \not\subset S' + t\mathbb{B}_X$, so there is a $z \in S$ with $z \notin S' + t\mathbb{B}_X$, which gives $d(z, S') \geq t$, and hence $\sup_{x \in S} d(x, S') \geq t$. This yields

$$\text{exc}(S, S') = \sup_{x \in S} d(x, S') \geq \inf\{r > 0 : S \subset S' + r\mathbb{B}_X\} = \eta(S, S'),$$

which translates the desired converse inequality. $\square$

On the basis of the above proposition it is natural, in addition to the ρ-excess $\text{exc}(S, S')$ and the Hausdorff-Pompeiu ρ-semidistance $\text{haus}_\rho(S, S')$, to consider also for any real $\rho > 0$ the expressions that we call the *ρ-pseudo-excess* of S over S' and the *Hausdorff-Pompeiu ρ-pseudo-distance* between S and S' defined by

(1.48) $$\widehat{\text{exc}}_\rho(S, S') := \inf\{r > 0 : S \cap \rho\mathbb{B}_X \subset S' + r\mathbb{B}_X\}$$
$$= \text{exc}(S \cap \rho\mathbb{B}_X, S') = \sup_{x \in S \cap \rho\mathbb{B}_X} d(x, S')$$

and

(1.49) $$\widehat{\text{haus}}_\rho(S, S') := \max\{\widehat{\text{exc}}_\rho(S, S'), \widehat{\text{exc}}_\rho(S', S)\}.$$

While the equality $\text{haus}(S, S') = 0$ entails that $\text{cl}\, S = \text{cl}(S')$, such a property fails for the equality $\widehat{\text{haus}}_\rho(S, S') = 0$. Further, even in $\mathbb{R}$ simple examples as $S := \{1\}$, $S' := \{-1\}$ and $S'' := \{-3/2, 3/2\}$ show, with $\rho := 1$, that $\widehat{\text{exc}}_\rho$ can fail the triangle inequality since $\widehat{\text{exc}}_\rho(S, S'') + \widehat{\text{exc}}_\rho(S'', S') < \widehat{\text{exc}}_\rho(S, S')$. Nevertheless, for a sequence of sets $(S_n)_n$ of X, we will see in Proposition 1.70 below that $\text{haus}_\rho(S_n, S) \to 0$ for all reals $\rho > 0$ if and only if $\widehat{\text{haus}}_\rho(S_n, S) \to 0$ for all reals $\rho > 0$. Those observations explain why we call $\widehat{\text{haus}}_\rho(\cdot, \cdot)$ the Hausdorff-Pompeiu ρ-pseudo-distance.

In order to show the equivalence between the Attouch-Wets convergence and the convergence with respect to the pseudo-distances above, let us prove the following lemma.

LEMMA 1.69. *Let S, S' be subsets of the normed space $(X, \|\cdot\|)$ and let two reals $r > 0$, $\rho > 0$.*
(a) *If $S' \neq \emptyset$ and $d(x, S') - d(x, S) < r$ for all $x \in \rho \mathbb{B}_X$, then $S \cap \rho \mathbb{B}_X \subset S' + r \mathbb{B}_X$.*
(b) *If $S' \neq \emptyset$ and C is a subset of X such that $C \subset S' + r \mathbb{B}_X$, then*
$$d(x, S') - d(x, C) \leq r \quad \text{for all } x \in X.$$
(c) *If $S \cap \rho' \mathbb{B}_X \subset S' + r \mathbb{B}_X$ for some real $\rho' > 2\rho + d(0, S)$, then*
$$d(x, S') - d(x, S) \leq r \quad \text{for all } x \in \rho \mathbb{B}_X.$$
(d) *One always has*
$$\widehat{\mathrm{exc}}_\rho(S, S') \leq \mathrm{exc}_\rho(S, S') \quad \text{and} \quad \widehat{\mathrm{haus}}_\rho(S, S') \leq \mathrm{haus}_\rho(S, S').$$
(e) *If S and S' are nonempty, then for any real $\rho' \geq 2\rho + d(0, S)$ one has*
$$\mathrm{exc}_\rho(S, S') \leq \widehat{\mathrm{exc}}_{\rho'}(S, S'),$$
and for any real $\rho' \geq 2\rho + \max\{d(0, S), d(0, S')\}$ one has
$$\mathrm{haus}_\rho(S, S') \leq \widehat{\mathrm{haus}}_{\rho'}(S, S');$$
further, if at least one of the sets S, S' is empty, then $\mathrm{haus}_\rho(S, S') = \widehat{\mathrm{haus}}_{\rho'}(S, S')$ for all pairs ρ, ρ' of positive reals.

PROOF. (a) We may suppose $S \neq \emptyset$. Then, under the assumption of (a), for any $x \in S \cap \rho \mathbb{B}_X$ we have $d(x, S') < r$, hence obviously $x \in S' + r \mathbb{B}_X$, so $S \cap \rho \mathbb{B}_X \subset S' + r \mathbb{B}_X$.
(b) Under the inclusion $C \subset S' + r \mathbb{B}_X$, we have for all $x \in X$
$$d(x, C) \geq d(x, S' + r \mathbb{B}_X) = \inf\{\|x - y - rb\| : y \in S', b \in \mathbb{B}_X\}$$
$$\geq \inf\{\|x - y\| - r : y \in S'\} = d(x, S') - r,$$
which justifies the desired inequality in (b).
(c) Assume that $\rho' > 2\rho + d(0, S)$ along with $S \cap \rho' \mathbb{B}_X \subset S' + r \mathbb{B}_X$. The inequality entails that $S \cap \rho' \mathbb{B}_X \neq \emptyset$, so $S' \neq \emptyset$ according to the former inclusion. Fix any $x \in \rho \mathbb{B}_X$. Take any $0 < \varepsilon < \rho' - 2\rho - d(0, S)$ and choose some $z \in S$ such that $\|x - z\| < d(x, S) + \varepsilon$. Then, we have
$$\|z\| \leq \|x - z\| + \|x\| \leq d(x, S) + \|x\| + \varepsilon \leq 2\|x\| + d(0, S) + \varepsilon < \rho',$$
hence $z \in S \cap \rho' \mathbb{B}_X$. This combined with the inequality $\|x - z\| < d(x, S) + \varepsilon$ implies $d(x, S \cap \rho' \mathbb{B}_X) \leq d(x, S) + \varepsilon$, thus $d(x, S \cap \rho' \mathbb{B}_X) = d(x, S)$. This and (b) guarantee that $d(x, S') - d(x, S) \leq r$.
(d) Since $\widehat{\mathrm{exc}}_\rho(S, S') = \sup_{x \in S \cap \rho \mathbb{B}_X} d(x, S')$ by (1.48), the inequality
$$\widehat{\mathrm{exc}}_\rho(S, S') \leq \mathrm{exc}_\rho(S, S')$$
is trivial if $S = \emptyset$, and otherwise it is due to the equality
$$\mathrm{exc}_\rho(S, S') = \sup_{x \in \rho \mathbb{B}_X} \left(d(x, S') - d(x, S)\right)^+$$
in (1.43).
(e) Suppose first that S, S' are nonempty. Fix any $\rho' \geq 2\rho + d(0, S)$. Take any

$\sigma \in \,]0, \rho[$, so $\rho' > 2\sigma + d(0, S)$. Considering any $t > \widehat{\text{exc}}_{\rho'}(S, S')$, we have $t > 0$ and $S \cap \rho' \mathbb{B}_X \subset S' + t\mathbb{B}_X$ according to the definition of $\widehat{\text{exc}}_{\rho'}$, thus, for every $x \in \sigma \mathbb{B}_X$, the assertion (c) gives

$$d(x, S') - d(x, S) \leq t, \quad \text{so } \bigl(d(x, S') - d(x, S)\bigr)^+ \leq t.$$

This yields $\text{exc}_\sigma(S, S') \leq t$, hence $\text{exc}_\sigma(S, S') \leq \widehat{\text{exc}}_{\rho'}(S, S')$. The latter inequality being true for every real $\sigma \in \,]0, \rho[$, the relation (1.45) guarantees that $\text{exc}_\rho(S, S') \leq \widehat{\text{exc}}_{\rho'}(S, S')$. The inequality of (e) relative to the excesses is established and it obviously implies the inequality relative to the semidistances.

Finally, when at least one of the sets is empty, the stated equality in (e) is easily verified. □

PROPOSITION 1.70. *For each $t \in T_0 \subset T$ let S_t be a subset of the normed space X and let S be a nonempty subset of X. Then the following hold:*
(a) *One has* $\lim_{T_0 \ni t \to \bar{t}} \text{exc}_\rho(S, S_t) = 0$ *for any real $\rho > 0$ if and only if*

$$\lim_{T_0 \ni t \to \bar{t}} \widehat{\text{exc}}_\rho(S, S_t) = 0 \quad \text{for any real } \rho > 0.$$

(b) *One has* $\lim_{T_0 \ni t \to \bar{t}} \text{exc}_\rho(S_t, S) = 0$ *for any real $\rho > 0$ if and only if*

$$\lim_{T_0 \ni t \to \bar{t}} \widehat{\text{exc}}_\rho(S_t, S) = 0 \quad \text{for any real } \rho > 0.$$

(c) *The family $(S_t)_{t \in T_0}$ Attouch-Wets converges to S if and only if*

$$\lim_{T_0 \ni t \to \bar{t}} \widehat{\text{haus}}_\rho(S_t, S) = 0 \quad \text{for every real } \rho > 0.$$

PROOF. (a) From the inequality in (d) of the above lemma it is evident, for each real $\rho > 0$, that $\lim_{T_0 \ni t \to \bar{t}} \widehat{\text{exc}}_\rho(S, S_t) = 0$ whenever $\lim_{T_0 \ni t \to \bar{t}} \text{exc}_\rho(S, S_t) = 0$. Conversely, suppose that $\lim_{T_0 \ni t \to \bar{t}} \widehat{\text{exc}}_r(S, S_t) = 0$ for all reals $r > 0$. Taking any real $\rho > 0$ and putting $\rho' := 2\rho + d(0, S)$, for every $t \in T_0$ close enough to $\bar{t}$ the set $S_t \cap \rho' \mathbb{B}$ is nonempty and the inequality in (e) of the above lemma gives $\text{exc}_\rho(S, S_t) \leq \widehat{\text{exc}}_{\rho'}(S, S_t)$, thus $\lim_{T_0 \ni t \to \bar{t}} \text{exc}_\rho(S, S_t) = 0$. The equivalence in (a) is proved.

(b) As above the inequality in (d) of the above lemma ensures, for each real $\rho > 0$, that $\lim_{T_0 \ni t \to \bar{t}} \widehat{\text{exc}}_\rho(S_t, S) = 0$ whenever $\lim_{T_0 \ni t \to \bar{t}} \text{exc}_\rho(S_t, S) = 0$. Conversely suppose that $\lim_{T_0 \ni t \to \bar{t}} \widehat{\text{exc}}_r(S_t, S) = 0$ for every real $r > 0$. Take any real $\rho > 0$ and put $\rho' := 4\rho + d(0, S)$. Fix any $x \in \rho \mathbb{B}_X$ and any $t \in T_0$. If $d(0, S_t) < 2\rho + d(0, S)$, then $\rho' > 2\rho + d(0, S_t)$, and hence the inequality in (e) of the above lemma guarantees that $\text{exc}_\rho(S_t, S) \leq \widehat{\text{exc}}_{\rho'}(S_t, S)$. If $d(0, S_t) \geq 2\rho + d(0, S)$, then for any $x \in \rho \mathbb{B}_X$

$$d(x, S_t) \geq d(0, S_t) - \|x\| \geq 2\rho + d(0, S) - \|x\| \geq \rho + d(0, S)$$
$$\geq \rho + d(x, S) - \|x\| \geq d(x, S),$$

which tells us that $\text{exc}_\rho(S_t, S) = 0$ according to the definition of $\text{exc}_\rho(\cdot, \cdot)$. In any case we then have $\text{exc}_\rho(S_t, S) \leq \widehat{\text{exc}}_{\rho'}(S_t, S)$, thus $\lim_{T_0 \ni t \to \bar{t}} \text{exc}_\rho(S_t, S) = 0$. This finishes the proof of the equivalence in (b).

(c) Finally, (c) follows directly from (a) and (b). □

For convex sets in a Hilbert space containing the origin, we will see next that inequalities in Lemma 1.69(d) are equalities. We need first the following lemma.

LEMMA 1.71. *Let S be a closed convex set of a Hilbert space $(H, \|\cdot\|)$ and $\bar{x} \in S$. Then, given any real $r > 0$ one has*
$$d(x, S) = d(x, S \cap B[\bar{x}, r]) \quad \text{for all } x \in B[\bar{x}, r].$$

PROOF. Clearly, we have $d(\cdot, S) \leq d(\cdot, S \cap B[\bar{x}, r])$. Now, fix any $x \in B[\bar{x}, r]$. Since S is a closed convex set, the metric projection mapping P_S is Lipschitz with constant 1, hence
$$\|P_S(x) - \bar{x}\| = \|P_S(x) - P_S(\bar{x})\| \leq \|x - \bar{x}\| \leq r.$$
It ensues that $P_S(x) \in S \cap B[\bar{x}, r]$, thus
$$d(x, S \cap B[\bar{x}, r]) \leq \|x - P_S(x)\| = d(x, S).$$
We then conclude that $d(x, S) = d(x, S \cap B[\bar{x}, r])$. □

LEMMA 1.72. *Let $(H, \|\cdot\|)$ be a Hilbert space and S, S' be two closed convex subsets containing the origin. Then for every $\rho \in \,]0, +\infty]$ one has*
$$\widehat{\mathrm{exc}}_\rho(S, S') = \mathrm{exc}(S \cap \rho \mathbb{B}, S' \cap \rho \mathbb{B}) = \mathrm{exc}_\rho(S, S')$$
along with
$$\widehat{\mathrm{haus}}_\rho(S, S') = \mathrm{haus}(S \cap \rho \mathbb{B}, S' \cap \rho \mathbb{B}) = \mathrm{haus}_\rho(S, S').$$

PROOF. We may suppose that $0 < \rho < +\infty$. By Lemma 1.71 we have on the one hand
$$\widehat{\mathrm{exc}}_\rho(S, S') = \sup_{x \in S \cap \rho \mathbb{B}} d(x, S')$$
$$= \sup_{x \in S \cap \rho \mathbb{B}} d(x, S' \cap \rho \mathbb{B}) = \mathrm{exc}(S \cap \rho \mathbb{B}, S' \cap \rho \mathbb{B}),$$
and on the other hand by Lemma 1.71 again
$$(1.50) \quad \mathrm{exc}_\rho(S, S') = \sup_{x \in \rho \mathbb{B}} \left(d(x, S') - d(x, S)\right)^+ = \sup_{x \in \rho \mathbb{B}} \left(d(x, S' \cap \rho \mathbb{B}) - d(x, S \cap \rho \mathbb{B})\right)^+$$
$$(1.51) \quad = \sup_{x \in H} \left(d(x, S' \cap \rho \mathbb{B}) - d(x, S \cap \rho \mathbb{B})\right)^+ = \mathrm{exc}(S \cap \rho \mathbb{B}, S' \cap \rho \mathbb{B}),$$
where the first equality in (1.51) follows from Proposition 1.62(b). The equalities concerning the excesses are then justified and they clearly entail the ones for the corresponding semidistances. □

For convex sets S, S' in a normed space, removing the requirement $0 \in S \cap S'$ in Lemma 1.72 above we have the following inequalities. In particular, they show that a sequence of convex sets $(S_n)_n$ Attouch-Wets converges to a convex set S if and only if $\mathrm{haus}(S_n \cap \rho \mathbb{B}, S \cap \rho \mathbb{B}) \to 0$ for large $\rho > 0$.

PROPOSITION 1.73. *Let S, S' be two nonempty convex sets in the normed space $(X, \|\cdot\|)$. Then one has for any $\rho > 2d(0, S')$*
$$\widehat{\mathrm{exc}}_\rho(S, S') \leq \mathrm{exc}(S \cap \rho \mathbb{B}, S' \cap \rho \mathbb{B}) \leq 4\,\widehat{\mathrm{exc}}_\rho(S, S'),$$
and also for any $\rho > 2\max\{d(0, S), d(0, S')\}$
$$\widehat{\mathrm{haus}}_\rho(S, S') \leq \mathrm{haus}(S \cap \rho \mathbb{B}, S' \cap \rho \mathbb{B}) \leq 4\,\widehat{\mathrm{haus}}_\rho(S, S').$$

PROOF. First, the obvious inequality $\sup_{x \in S \cap \rho \mathbb{B}} d(x, S') \leq \sup_{x \in S \cap \rho \mathbb{B}} d(x, S' \cap \rho \mathbb{B})$ for every $\rho > 0$ clearly ensures the left inequality about excesses. Now, fix any real ρ_0 with $d(0, S') < \rho_0 < \rho/2$. Take any $x \in S \cap \rho \mathbb{B}$ and consider any real $r > d(x, S')$. There exists $x' \in S'$ with $\|x - x'\| < r$. By the inequality $d(0, S') < \rho_0$, choose some $z' \in S'$ with $\|z'\| < \rho_0$. Let $y' := (1-t)z' + tx'$ with $t = (\rho - \rho_0 + r)^{-1}(\rho - \rho_0)$ and note that $y' \in S'$ by convexity of S'. Then

$$\|y'\| \leq (1-t)\|z'\| + t\|x'\| \leq (1-t)\rho_0 + t(\|x\| + r) \leq \rho_0 + t(\rho - \rho_0 + r) = \rho,$$

which gives

$$d(x, S' \cap \rho \mathbb{B}) \leq \|x - y'\| \leq \|x - x'\| + \|x' - y'\| \leq r + (1-t)\|x' - z'\|$$
$$\leq r + (1-t)(\|x'\| + \|z'\|) \leq r + (1-t)(r + \rho + \rho_0)$$
$$= r\left(1 + \frac{r + \rho + \rho_0}{\rho - \rho_0 + r}\right) = 2r\left(1 + \frac{\rho_0}{\rho - \rho_0 + r}\right),$$

and hence $d(x, S' \cap \rho \mathbb{B}) \leq 2r\left(1 + \frac{\rho_0}{\rho - \rho_0}\right) \leq 4r$. This being true for any $r > d(x, S')$, we obtain $d(x, S' \cap \rho \mathbb{B}) \leq 4 d(x, S')$ for any $x \in S \cap \rho \mathbb{B}$. It results that

$$\operatorname{exc}(S \cap \rho \mathbb{B}, S' \cap \rho \mathbb{B}) = \sup_{x \in S \cap \rho \mathbb{B}} d(x, S' \cap \rho \mathbb{B}) \leq 4 \sup_{x \in S \cap \rho \mathbb{B}} d(x, S') = 4\widehat{\operatorname{exc}}_\rho(S, S'),$$

which translates the second inequality for the excesses. The similar inequalities for the semidistances directly follow from those for the excesses. □

1.8.3. Hausdorff and Attouch-Wets semicontinuities. Taking into account Proposition 1.70 as well as (1.46), (1.47) and the definition of $\widehat{\operatorname{exc}}_\rho(\cdot, \cdot)$, we define the Hausdorff and Attouch-Wets semicontinuities:

DEFINITION 1.74. A multimapping $M : T \rightrightarrows X$ from T into a metric space X is *Hausdorff upper* (resp. *lower*) *semicontinuous* at $\bar{t} \in T$ provided

$$\lim_{t \to \bar{t}} \operatorname{exc}(M(t), M(\bar{t})) = 0 \quad (\text{resp. } \lim_{t \to \bar{t}} \operatorname{exc}(M(\bar{t}), M(t)) = 0),$$

or equivalently, for any real $\varepsilon > 0$ there is a neighborhood V of $\bar{t}$ such that, for all $t \in V$,

$$M(t) \subset \{x \in X : d(x, M(\bar{t})) < \varepsilon\} \quad (\text{resp. } M(\bar{t}) \subset \{x \in X : d(x, M(t)) < \varepsilon\}).$$

If X is a normed space, this is equivalent to say that, for any real $\varepsilon > 0$ there is a neighborhood V of $\bar{t}$ such that, for all $t \in V$,

$$M(t) \subset M(\bar{t}) + \varepsilon \mathbb{B}_X \quad (\text{resp. } M(\bar{t}) \subset M(t) + \varepsilon \mathbb{B}_X).$$

Assuming that X is a normed space, the multimapping M is said to be *Attouch-Wets upper* (resp. *lower*) *semicontinuous* at $\bar{t}$ whenever

$$\lim_{t \to \bar{t}} \operatorname{exc}_\rho(M(t), M(\bar{t})) = 0 \; \forall \rho > 0, \text{ or equivalently } \lim_{t \to \bar{t}} \widehat{\operatorname{exc}}_\rho(M(t), M(\bar{t})) = 0 \; \forall \rho > 0,$$

(resp. whenever

$$\lim_{t \to \bar{t}} \operatorname{exc}_\rho(M(\bar{t}), M(t)) = 0 \; \forall \rho > 0, \text{ or equivalently } \lim_{t \to \bar{t}} \widehat{\operatorname{exc}}_\rho(M(\bar{t}), M(t)) = 0 \; \forall \rho > 0).$$

The Attouch-Wets upper (resp. lower) semicontinuity at $\bar{t}$ then amounts to requiring that for any reals $\rho > 0$ and $\varepsilon > 0$ there is a neighborhood V of $\bar{t}$ such that, for all $t \in V$,

$$M(t) \cap \rho \mathbb{B}_X \subset M(\bar{t}) + \varepsilon \mathbb{B}_X \quad (\text{resp. } M(\bar{t}) \cap \rho \mathbb{B}_X \subset M(t) + \varepsilon \mathbb{B}_X).$$

More generally, if X is a topological vector space one says that M is *Hausdorff upper* (resp. *lower*) *semicontinuous at* $\bar{t}$ when for every neighborhood W of zero in X there exists a neighborhood V of $\bar{t}$ such that, for all $t \in V$,
$$M(t) \subset M(\bar{t}) + W \quad (\text{resp. } M(\bar{t}) \subset M(t) + W).$$
Obviously, the classic upper semicontinuity in Section 1.4 implies the Hausdorff upper semicontinuity; and the Hausdorff lower semicontinuity entails the classic lower semicontinuity in Section 1.4.

Links between classic semicontinuities and Hausdorff semicontinuities for a multimapping will be stated in Proposition 1.81 of the next section. Concerning the Attouch-Wets semicontinuities, they are linked to the outer and lower semicontinuities as it is seen from the following proposition studying the connection between the Attouch-Wets convergence with the Painlevé-Peano convergence.

PROPOSITION 1.75. *For each $t \in T_0 \subset T$ let S_t be a subset of the normed space X and let S be a nonempty subset of X. Then the following hold:*
(a) *If* $\lim_{T_0 \ni t \to \bar{t}} \mathrm{exc}_\rho(S, S_t) = 0$ *for all real numbers $\rho > 0$, then* $\mathrm{cl}\, S \subset \mathrm{Lim\,inf}_{T_0 \ni t \to \bar{t}} S_t$.
(b) *If* $\lim_{T_0 \ni t \to \bar{t}} \mathrm{exc}_\rho(S_t, S) = 0$ *for all real numbers $\rho > 0$, then* $\mathrm{Lim\,sup}_{T_0 \ni t \to \bar{t}} S_t \subset \mathrm{cl}\, S$.
(c) *So, the Attouch-Wets convergence (hence also the Hausdorff-Pompeiu convergence) of $(S_t)_{t \in T_0}$ to a nonempty closed set S implies the Painlevé-Peano convergence of the family to S.*
(d) *The implications in assertions (a) and (b) are equivalences (hence Attouch-Wets and Painlevé-Peano convergences for nonempty closed sets coincide) whenever X is a finite-dimensional normed vector space.*

PROOF. (a) Suppose that $\lim_{T_0 \ni t \to \bar{t}} \mathrm{exc}_\rho(S, S_t) = 0$ for any real $\rho > 0$ and take any $x \in S$. Then by definition of $\mathrm{exc}_\rho(\cdot, \cdot)$, for any positive real $\rho \geq \|x\|$ we have $d(x, S_t) \leq \mathrm{exc}_\rho(S, S_t)$ for all $t \in T_0$, which yields $\lim_{T_0 \ni t \to \bar{t}} d(x, S_t) = 0$. This says by Proposition 1.12 that $x \in \mathrm{Lim\,inf}_{T_0 \ni t \to \bar{t}} S_t$. Combining this with the closedness of the limit inferior of sets, we see that $\mathrm{cl}\, S \subset \mathrm{Lim\,inf}_{T_0 \ni t \to \bar{t}} S_t$.

(b) Suppose now that $\lim_{T_0 \ni t \to \bar{t}} \mathrm{exc}_\rho(S_t, S) = 0$ for any real $\rho > 0$ and take any $x \in \mathrm{Lim\,sup}_{T_0 \ni t \to \bar{t}} S_t$. By Proposition 1.16 there exist nets $(t_j)_{j \in J}$ in T_0 converging to $\bar{t}$ and $(x_j)_{j \in J}$ in X converging to x with $x_j \in S_{t_j}$ for all $j \in J$. Choose some $j_0 \in J$ and some real $\rho > 0$ such that $\rho \geq \|x_j\|$ for all $j \succeq j_0$. By definition of $\mathrm{exc}_\rho(\cdot, \cdot)$ we can then write, for all $j \succeq j_0$,
$$d(x, S) \leq d(x, x_j) + d(x_j, S) \leq d(x, x_j) + \mathrm{exc}_\rho(S_{t_j}, S),$$
so we obtain that $d(x, S) = 0$, that is, $x \in \mathrm{cl}\, S$. The assertion in (b) is thus established.

(c) The assertion in (c) follows directly from (a) and (b).

(d) Assume that X is a finite-dimensional normed space.

Suppose that $\mathrm{cl}\, S \subset \mathrm{Lim\,inf}_{T_0 \ni t \to \bar{t}} S_t$. Then for $T_1 := T_0 \cup \{\bar{t}\}$ and for $M(t) := S_t$ for $t \in T_1 \setminus \{\bar{t}\}$ and $M(\bar{t}) := \mathrm{cl}\, S$, the multimapping M is lower semicontinuous at $\bar{t}$ relative to T_1. If for some $\rho > 0$ we have that $\widehat{\mathrm{exc}}_\rho(S, S_t)$ does not converge to 0, then there exist some $\varepsilon > 0$ and some net $(t_j)_{j \in J}$ in T_0 tending to $\bar{t}$ such that for each $j \in J$ we can find some point $x_j \in (S \cap \rho \mathbb{B}_X) \setminus (S_{t_j} + 2\varepsilon \mathbb{B}_X)$. Choose by the

compactness of $\rho\mathbb{B}_X$ some subnet of $(x_j)_{j\in J}$ (that we do not relabel) converging to some point $x \in (\operatorname{cl} S) \cap \rho\mathbb{B}_X$. This yields some $j_0 \in J$ such that for all $j \succeq j_0$
$$2\varepsilon \leq d(x_j, S_{t_j}) \leq d(x, S_{t_j}) + \|x_j - x\| \leq d(x, M(t_j)) + \varepsilon.$$
On the other hand, the lower semicontinuity of the multimapping M at $\bar{t}$ entails, by Proposition 1.45, the upper semicontinuity of the function $t \mapsto d(x, M(t))$ at $\bar{t}$ relative to T_1. Consequently, taking the limit superior in the latter inequality gives
$$2\varepsilon \leq \limsup_{j\in J} d(x, M(t_j)) + \varepsilon \leq d(x, M(\bar{t})) + \varepsilon = \varepsilon,$$
which is a contradiction. This and Proposition 1.70(a) say that the reverse implication of (a) holds.

Now suppose that for some $\rho > 0$ we have that $\widehat{\operatorname{exc}}_\rho(S_t, S)$ does not converge to 0 as $T_0 \ni t \to \bar{t}$. Then there are some $\varepsilon > 0$ and some net $(t_j)_{j\in J}$ in S tending to $\bar{t}$ such that for each $j \in J$ there exists some point $x_j \in (S_{t_j} \cap \rho\mathbb{B}_X) \setminus (S + 2\varepsilon\mathbb{B}_X)$. Taking a cluster point x of $(x_j)_{j\in J}$ (according to the compactness of $\rho\mathbb{B}_X$), we obtain $x \in \underset{T_0 \ni t \to \bar{t}}{\operatorname{Lim\,sup}} S_t$ and $x \notin S + \varepsilon\mathbb{B}_X$. So $\underset{T_0 \ni t \to \bar{t}}{\operatorname{Lim\,sup}} S_t$ is not included in $\operatorname{cl} S$. This and Proposition 1.70(b) ensure the converse implication of (b), thus the proof of the proposition is finished. $\square$

In the case when $M(\bar{t})$ is a closed set of a finite-dimensional normed space, we have:

PROPOSITION 1.76. *Let T be a metric space, X be a finite-dimensional normed space and $M : T \rightrightarrows X$ be a multimapping. Assume that the multimapping M is closed at $\bar{t}$ (in the sense (see Definition 1.28) that, for any sequences $(t_n)_n$ converging to $\bar{t}$ and $(x_n)_n$ converging to x with $x_n \in M(t_n)$, one has $x \in M(\bar{t})$). Then, for any real $\rho > 0$ one has*
$$\operatorname{exc}\big(M(t) \cap \rho\mathbb{B}_X, M(\bar{t}) \cap \rho\mathbb{B}_X\big) \to 0 \quad \text{as } T \ni t \to \bar{t}.$$

PROOF. Suppose that, for some real $\rho > 0$, the property does not hold. There exist some real $\varepsilon > 0$ and some sequences $(t_n)_n$ converging to $\bar{t}$ and $(x_n)_n$ with $x_n \in M(t_n) \cap \rho\mathbb{B}_X$ such that
$$d(x_n, M(\bar{t}) \cap \rho\mathbb{B}_X) \geq \varepsilon \quad \text{for all } n \in \mathbb{N}.$$
Extracting a subsequence if necessary, we may suppose that the sequence $(x_n)_n$ converges to a certain x, hence $d(x, M(\bar{t}) \cap \rho\mathbb{B}_X) \geq \varepsilon$. On the other hand, the closedness assumption of the multimapping M at $\bar{t}$ tells us that $x \in M(\bar{t})$, and clearly we also have $x \in \rho\mathbb{B}_X$. Therefore, $x \in M(\bar{t}) \cap \rho\mathbb{B}_X$, which contradicts the above inequality $d(x, M(\bar{t}) \cap \rho\mathbb{B}_X) \geq \varepsilon$, and finishes the proof. $\square$

1.8.4. Hölder continuity of metric projection with convex set as variable. This subsection establishes the Hölder continuity of the metric projection with respect to the truncated Hausdorff distance on convex sets. Given a nonempty closed convex set S of a Hilbert space H and a point $x \in H$, we denote by $\operatorname{proj}_S(x)$ or $P_S(x)$ the unique nearest point of x in S. The mapping proj_S defined in this way is called as usual the *metric projection onto S*. In the context of Hilbert space H we know and recall that $p = \operatorname{proj}_S(x)$ if and only if
$$\langle x - p, y - p \rangle \leq 0 \quad \text{for all } y \in S.$$

For any extended real $r \in \,]0, +\infty]$ define the *open r-enlargement* of S by
$$U_r(S) := \{x \in H : d_S(x) < r\}.$$

PROPOSITION 1.77. *Let S, S' be two nonempty closed convex subsets of a Hilbert space H.*

(a) *Given $r, \rho \in \,]0, +\infty]$ for any $x \in U_r(S) \cap \rho\mathbb{B}$ and $x' \in U_r(S') \cap \rho\mathbb{B}$ one has*
$$\|\mathrm{proj}_S(x) - \mathrm{proj}_{S'}(x')\|^2$$
$$\leq \|x - x'\|^2 + 2d_S(x)\mathrm{exc}_{r+\rho}(S', S) + 2d_{S'}(x')\mathrm{exc}_{r+\rho}(S, S')$$
$$\leq \|x - x'\|^2 + 2\left(d_S(x) + d_{S'}(x')\right) \mathrm{haus}_{r+\rho}(S, S').$$

(b) *In particular, for all $x, x' \in H$ one has*
$$\|\mathrm{proj}_S(x) - \mathrm{proj}_{S'}(x')\|^2 \leq \|x - x'\|^2 + 2\left(d_S(x) + d_{S'}(x')\right) \mathrm{haus}(S, S'),$$
so for $x \in X$ fixed the mapping $S \mapsto \mathrm{proj}_S(x)$ is locally Hölderian with Hölder exponent $1/2$ with respect to the Hausdorff-Pompeiu semidistance on the collection of nonempty closed convex subsets of H.

PROOF. (a) The inequality $\|a\|^2 - \|b\|^2 \leq 2\langle a, a - b\rangle$ entails, with $p := \mathrm{proj}_S(x)$ and $p' := \mathrm{proj}_{S'}(x')$, that
$$\|p - p'\|^2 - \|x - x'\|^2 \leq 2\langle p - p', p - p' - x + x'\rangle$$
$$= 2\langle p' - p, x - p\rangle + 2\langle p - p', x' - p'\rangle.$$

Further, $\|p'\| \leq \|x' - p'\| + \|x'\| = d_{S'}(x') + \|x'\| \leq r + \rho$ and similarly $\|p\| \leq r + \rho$. On the other hand, noting for $q := \mathrm{proj}_S(p')$ that $\langle q - p, x - p\rangle \leq 0$ (since $p = \mathrm{proj}_S(x)$ and $q \in S$), we also have that
$$\langle p' - p, x - p\rangle \leq \langle p' - q, x - p\rangle \leq \|x - p\| d_S(p') \leq \|x - p\| \mathrm{exc}_{r+\rho}(S', S).$$

Interchanging, we also have $\langle p - p', x' - p'\rangle \leq \|x' - p'\| \mathrm{exc}_{r+\rho}(S, S')$, which combined with what precedes confirms the assertion (a).

(b) The inequality in (b) directly follows from (a) and the local Hölder property follows from this inequality and from Remark 1.63. □

EXAMPLE 1.78. The following example shows that the exponent $1/2$ in the above Hölder property is sharp. Consider in the usual Euclidean space $\mathbb{R}^2$ the closed bounded sets $S := \{(x, y) \in \mathbb{R}^2 : 0 \leq x \leq 1, 1 \leq y \leq 2\}$ and $S_\varepsilon := \{(x, y) \in S : x\sqrt{\varepsilon} + y \geq 1 + \varepsilon\}$ for $\varepsilon \in \,]0, 1[$ (see Figure 1.1). It is easily seen that $\mathrm{proj}_S(0, 0) = (0, 1)$, $\mathrm{proj}_{S_\varepsilon}(0, 0) = (\sqrt{\varepsilon}, 1)$, so
$$\|\mathrm{proj}_S(0,0) - \mathrm{proj}_{S_\varepsilon}(0,0)\| = \sqrt{\varepsilon}, \quad \text{whereas } \mathrm{haus}(S_\varepsilon, S) = \varepsilon/\sqrt{1+\varepsilon},$$
which justifies the sharpness of the exponent $1/2$. □

1.9. Further results

1.9.1. Further properties of semicontinuous multimappings.
Consider first the semicontinuity property of the multimapping $t \mapsto M(t) \times N(t)$.

PROPOSITION 1.79. *Let T, X, Y be topological spaces and let $M : T \rightrightarrows X$ and $N : T \rightrightarrows Y$ be multimappings. Let $\Gamma : T \to X \times Y$ be the multimapping defined by $\Gamma(t) := M(t) \times N(t)$ for all $t \in T$.*

(a) *For $X \times Y$ endowed with the product topology, Γ is lower semicontinuous at*

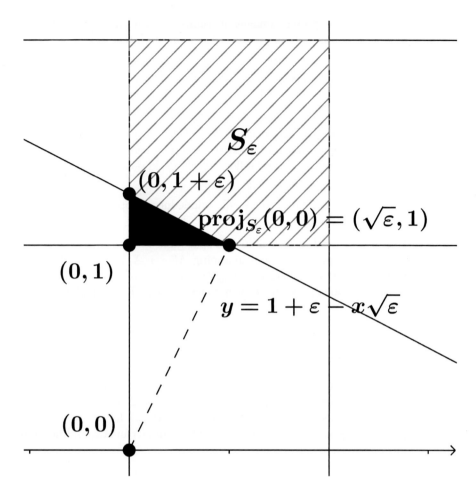

FIGURE 1.1. Illustration of Example 1.78.

$\bar{t} \in T$ if and only if M and N are lower semicontinuous at $\bar{t}$.
(b) The similar property fails for the upper semicontinuity.

EXERCISE 1.80 (Proof of the proposition). Prove the assertion (a) and provide a counterexample for (b). □

In the context of metric spaces, semicontinuities and Hausdorff semicontinuities of multimappings are linked as follows.

PROPOSITION 1.81. Let T and X be two metric spaces and $M : T \rightrightarrows X$ be a multimapping, and let $\bar{t} \in T$. The following assertions hold.
(a) If M is upper semicontinuous at $\bar{t}$ in the classic sense, then it is Hausdorff upper semicontinuous at $\bar{t}$.

This implication is an equivalence whenever $M(\bar{t})$ is compact.
(b) If M is Hausdorff lower semicontinuous at $\bar{t}$, then it is lower semicontinuous at $\bar{t}$ in the classic sense.

The implication is an equivalence whenever $M(\bar{t})$ is compact.

EXERCISE 1.82. Prove the proposition. □

As with continuous mappings, images of connected sets under semicontinuous multimappings (with connected values) preserve connectedness.

PROPOSITION 1.83. *Let T and X be two topological spaces and $M : T \rightrightarrows X$ be a mulimapping with nonempty connected values. Then the image $M(S)$ of any connected set $S \subset T$ is connected whenever M is either lower semicontinuous or upper semicontinuous on S.*

EXERCISE 1.84. Prove the proposition. (Use the characterization of connectedness of $M(S)$ with pair of disjoint closed (resp. open) sets containing $M(S)$ for upper (resp. lower) semicontinuity and follow the classic known proof for continuous mappings). □

The next proposition is related to a local boundedness property of weak∗ upper semicontinuous multimappings.

DEFINITION 1.85. *Let T be a topological space and X be a topological vector space (resp. a metric space). A multimapping $M :\rightrightarrows X$ is said to be locally bounded near a point $\bar{t} \in T$ if there is a neighbourhood U of $\bar{t}$ such that $M(U)$ is bounded in X. If M is locally bounded near every point in T, one just says that it is locally bounded.*

PROPOSITION 1.86. *Let T be a metric space, X be a Banach space and $M : T \rightrightarrows X^*$ be a multimapping with weak* compact values, which is weak* upper semicontinuous. Then M is locally bounded.*

EXERCISE 1.87. Prove the proposition. (Hint: Suppose the contrary, that is, there are $\bar{t} \in T$, a sequence $(t_n)_n$ converging to $\bar{t}$ and a sequence $(x_n^*)_n$ in X^* with $\|x_n^*\| \to +\infty$ as $n \to \infty$ and with $x_n^* \in M(t_n)$ for all $n \in \mathbb{N}$. For $K := \{\bar{t}\} \cup \{t_n : n \in \mathbb{N}\}$, argue that $M(K)$ is weak* compact in X^*, and hence one obtains a contradiction). □

The next proposition generates upper semicontinuous multimappings.

PROPOSITION 1.88. *Let T be a metric space, X be a normed space and $M : T \rightrightarrows X^*$ be a locally bounded multimapping with nonempty values. Then, there exists a unique smallest weak* upper semicontinuous multimapping $\Gamma : T \rightrightarrows X^*$ with w^*-compact values (resp. with w^*-compact convex values) such that $\Gamma(t) \supset M(t)$ for all $t \in T$.*

EXERCISE 1.89. Prove the proposition. Hint: For the desired multimapping with w^*-compact values, put $\Gamma(t) := \bigcap_{\varepsilon > 0} \mathrm{cl}_{w(X^*,X)} M(B_T(t, \varepsilon))$ and use Propositions 1.7, 1.11, 1.39. Proceed in a similar way for the multimapping with w^*-compact convex values. □

1.9.2. Hausdorff-Pompeiu distance between boundaries of sets.

Given two nonempty subsets S, S' of a normed space $(X, \|\cdot\|)$, can one compare $\mathrm{haus}(S, S')$ and $\mathrm{haus}(\mathrm{bdry}(S), \mathrm{bdry}(S'))$? We first give a counter example for the equality. Endowing $\mathbb{R}^2$ with the Euclidean norm and considering the sets $S := B[0,1]$ and $S' := \mathbb{R}^2 \setminus B(0,2)$, we see that $\mathrm{haus}(S, S') = \mathrm{exc}(S', S) = +\infty$ while

$$\mathrm{haus}\big(\mathrm{bdry}(S), \mathrm{bdry}(S')\big) = \mathrm{exc}\big(\mathrm{bdry}(S), \mathrm{bdry}(S')\big) = \mathrm{exc}\big(\mathrm{bdry}(S'), \mathrm{bdry}(S)\big) = 1.$$

Despite the latter example, the following theorem provides a general condition under which the equality holds true.

THEOREM 1.90 (M.D. Wills). Let S, S' be two nonempty bounded closed convex subsets of a normed space $(X, \|\cdot\|)$. Then one has

$$\operatorname{haus}(S, S') = \operatorname{haus}(\operatorname{bdry}(S), \operatorname{bdry}(S')).$$

The proof of the theorem requires diverse preparatory lemmas. Some of them have their own interest.

LEMMA 1.91. Let S, S' be two nonempty sets in a normed space $(X, \|\cdot\|)$. The following hold:
(a) For any $x \in \operatorname{cl}(S) \setminus \operatorname{int}(S')$ one has $d(x, \operatorname{bdry}(S')) \leq \operatorname{exc}(S, S')$.
(b) If $\operatorname{cl}(S) \cap \operatorname{int}(S') = \emptyset$, then $\operatorname{exc}(S, S') \geq \operatorname{exc}(S, \operatorname{bdry}(S'))$.

PROOF. To prove (a) fix any $x \in \operatorname{cl}(S) \setminus \operatorname{int}(S')$. If $x \in \operatorname{cl}(S')$, then $x \in \operatorname{bdry}(S')$ and the inequality is evident. So, suppose $x \notin \operatorname{cl}(S')$ and take any real $r > \operatorname{exc}(S, S')$. Since $x \in \operatorname{cl}(S)$, there is some $y \in S'$ such that $\|x - y\| < r$. Then the relations $y \in S'$ and $x \notin \operatorname{int}(S')$ furnish some $y' \in [x, y] \cap \operatorname{bdry}(S')$, so $\|x - y'\| \leq \|x - y\| < r$. This entails that $d(x, \operatorname{bdry}(S')) < r$, which justifies the inequality in the assertion (a) of the lemma.

The other assertion easily follows from the first. □

LEMMA 1.92. Let S, S' be nonempty bounded convex sets in a normed space $(X, \|\cdot\|)$ with $\operatorname{int} S \neq \emptyset$. Then one has

$$\operatorname{exc}(\operatorname{bdry}(S), \operatorname{bdry}(S')) \leq \operatorname{haus}(S, S').$$

The proof of the lemma is the purpose of the following exercise.

EXERCISE 1.93 (Proof of the lemma). Let S, S' satisfying the assumptions in the lemma and suppose X is not reduced to zero. Fix any $x \in \operatorname{bdry}(S)$. If $x \notin \operatorname{int}(S')$, one has $d(x, \operatorname{bdry}(S')) \leq \operatorname{exc}(S, S')$ by Lemma 1.91(a). Suppose that $x \in \operatorname{int}(S')$.
(a) Since $x \notin \operatorname{int}(S) \neq \emptyset$, argue that there exists some $x^* \in X^*$ with $\|x^*\| = 1$ such that $\langle x^*, x \rangle = \sigma_S(x^*)$, where we recall that σ_S denotes the support function of S.
(b) Argue that

$$d(x, \{u \in X : \langle x^*, u \rangle = \sigma_{S'}(x^*)\}) = |\sigma_{S'}(x^*) - \langle x^*, x \rangle|$$
$$= \sigma_{S'}(x^*) - \langle x^*, x \rangle = \sigma_{S'}(x^*) - \sigma_S(x^*).$$

(c) Fix any real $\varepsilon > 0$ and take by what precedes some $z \in X$ with $\langle x^*, z \rangle = \sigma_{S'}(x^*)$ such that

$$\|x - z\| < \sigma_{S'}(x^*) - \sigma_S(x^*) + \varepsilon.$$

(d) Argue that $z \notin \operatorname{int}(S')$ and that $[x, z] \cap \operatorname{bdry}(S') \neq \emptyset$.
(e) Choosing $y' \in [x, z] \cap \operatorname{bdry}(S')$, argue that

$$\|x - y'\| < \sigma_{S'}(x^*) - \sigma_S(x^*) + \varepsilon \leq \operatorname{exc}(S', S) + \varepsilon$$

(Hint: For the second inequality one can use the independent result in Proposition 2.80 in the next chapter).
Then derive the inequality in the lemma. □

LEMMA 1.94. Let S, S' be nonempty bounded sets in a normed space $(X, \|\cdot\|)$. If S' is convex, then

$$\operatorname{exc}(S, S') \leq \operatorname{exc}(\operatorname{bdry}(S), \operatorname{bdry}(S')).$$

EXERCISE 1.95 (Proof of the lemma). Let S, S' be as in the statement of the lemma and let $x \in S$. Put $\delta := \text{exc}(\text{bdry}(S), \text{bdry}(S'))$.
(I) If $x \in \text{bdry}\, S$ or $x \in \text{cl}(S')$, argue that $d(x, S') \leq \delta$.
(II) Suppose that $x \in \text{int}(S) \setminus \text{cl}(S')$ and fix any real $\varepsilon > 0$. Choose some $y \in S'$ with $\|x - y\| < d(x, S') + \varepsilon$.
(a) Argue that there are $x' \in \text{bdry}\, S$ and $t \in [0, 1]$ such that $x = (1 - t)x' + ty$.
(b) Argue that there is $y' \in \text{bdry}(S')$ with $\|x' - y'\| < \delta + \varepsilon$, and that $[y, y'] \subset \text{cl}(S')$.
(c) Argue that $y'' := (1 - t)y' + ty$ lies in $\text{cl}(S')$, and that $\|x - y''\| < \delta + \varepsilon$.
(d) Deduce that $d(x, S') < \delta + \varepsilon$ and derive the inequality in the lemma. □

EXAMPLE 1.96. Lemma 1.94 fails without the convexity of S'. In $\mathbb{R}^2$ endowed with the Euclidean norm, for $S := B[0, 1]$ and $S' := \text{bdry}\, S$, we see that
$$\text{exc}(\text{bdry}(S), \text{bdry}(S')) = 0 < 1 = \text{exc}(S, S').$$
□

For $S \subset X$ nonempty and for any real $r > 0$, put $S_r := \{u \in X : d(u, S) \leq r\}$.

LEMMA 1.97. Let S, S' be two nonempty bounded convex sets in a normed space $(X, \|\cdot\|)$. Assume that $\text{int}(S) = \emptyset$. If $x \in \text{bdry}(S) = \text{cl}\, S$ satisfies $d(x, \text{bdry}(S_r)) \to 0$ as $r \to 0$ with $r > 0$, then $d(x, \text{bdry}(S')) \leq \text{haus}(S, S')$.

EXERCISE 1.98 (Proof of the lemma). Let S, S' be as in the lemma.
(I) Apply Lemma 1.91 to obtain the inequality in the lemma if $x \notin \text{int}(S')$.
(II) Suppose that $x \in \text{int}(S')$. For each integer $n \in \mathbb{N}$, choose $x_n \in \text{bdry}(S_{1/n})$ such that $\|x_n - x\| < (1/n) + d(x, \text{bdry}(S_{1/n}))$.
(a) Argue that for n large enough, say $n \geq N$, one has $x_n \in \text{int}(S')$.
(b) Argue through Lemma 1.92 that
$$d(x_n, \text{bdry}(S')) \leq \text{haus}(S_{1/n}, S') \leq \text{haus}(S, S') + (2/n).$$
(c) Conclude that the inequality in the lemma holds true. □

LEMMA 1.99. Let S be a nonempty convex set in a normed space $(X, \|\cdot\|)$. Then for any $x \in S$ the function $r \mapsto d(x, \text{bdry}(S_r))$ is nondecreasing on $]0, +\infty[$.

EXERCISE 1.100 (Proof of the lemma). Let S be as in the lemma and let $0 < r < \rho$ in $\mathbb{R}$.
(a) Show that $\text{int}(S_\rho) = \{u \in X : d(u, S) < \rho\}$, and hence $\text{bdry}(S_\rho) = \{u \in X : d(u, S) = \rho\}$.
(b) For any $y \in \text{bdry}(S_\rho)$ argue (with the intermediate value theorem applied with $d(\cdot, S)$) that there is some $u \in [x, y]$ such that $d(u, S) = r$.
(c) Deduce that $\|x - y\| \geq \|x - u\| \geq d(x, \text{bdry}(S_r))$ and conclude that
$$d(x, \text{bdry}(S_r)) \leq d(x, \text{bdry}(S_\rho)).$$
□

LEMMA 1.101. For any nonempty bounded closed convex set S in a normed space $(X, \|\cdot\|)$ and for any $x \in \text{bdry}(S)$ one has
$$d(x, \text{bdry}(S_r)) \to 0 \quad \text{as } r \to 0 \text{ with } r > 0.$$

EXERCISE 1.102 (Proof of the lemma). Let S be as in the lemma. For any fixed $n \in \mathbb{N}$, by the inclusion $x \in \text{bdry}(S)$ choose some $y_n \in B[x, 1/n] \setminus S$.
(a) Argue that there exists some integer $k(n) \geq n$ such that $y_n \notin S_{1/k(n)}$.

(b) Argue that there are $z_n \in \text{bdry}\,(S_{1/k(n)})$ and $t_n \in [0,1]$ such that $z_n = t_n y_n + (1-t_n)x$.
(c) Derive that $d(x, \text{bdry}\,(S_{1/k(n)})) \le \|x - z_n\| \le 1/n$.
(d) From (c) and the nondecreasing property of $r \mapsto d(x, \text{bdry}\,(S_r))$ in Lemma 1.99 conclude that $d(x, \text{bdry}\,(S_r)) \to 0$ as $r \to 0$ with $r > 0$. □

PROOF OF THEOREM 1.90. Note first that the left-hand side in the theorem is not greater than the right according to Lemma 1.94. Concerning the converse inequality, it is enough to justify that

$$\text{exc}\,(\text{bdry}\,(S), \text{bdry}\,(S')) \le \text{haus}\,(S, S').$$

The case $\text{int}\,S \ne \emptyset$ follows from Lemma 1.92 and the case $\text{int}\,S = \emptyset$ is a consequence of Lemma 1.97 and Lemma 1.101. □

EXAMPLE 1.103. Theorem 1.90 (resp. Lemma 1.101) fails without the closedness property of the convex sets S, S' (resp. the convex set S). Let $\ell_\mathbb{R}^2(\mathbb{N})$ be the (real) vector space of sequences $x : \mathbb{N} \to \mathbb{R}$ such that $\sum_{n=1}^\infty |x_n|^2 < +\infty$. Endow it with its usual Hilbert norm and consider its usual basis $(e_k)_{k\in\mathbb{N}}$ defined by $e_k(n) = 1$ if $n = k$ and $e_k(n) = 0$ if $n \ne k$. Denote by E the vector subspace spanned by the set $\{e_k : k \in \mathbb{N}\}$, and put $S := E \cap B[0,1]$ and $S' := B[0,1]$. Since S is dense in $B[0,1]$, we see that $S_r = B[0, 1+r]$, so $d(0, \text{bdry}\,(S_r)) = 1 + r$ does not tend to 0 as $r \to 0$ with $r > 0$. This says that the conclusion of Lemma 1.101 fails with the convex nonclosed set S and $x = 0_{\ell^2} \in \text{bdry}\,(S) = B[0,1]$.

On the other hand, since $\text{bdry}\,S' = \{u \in \ell_\mathbb{R}^2(\mathbb{N}) : \|u\| = 1\}$ we have

$$\text{haus}\,(S, S') = 0 < 1 = \text{haus}\,(\text{bdry}\,(S), \text{bdry}\,(S')),$$

hence the conclusion of Theorem 1.90 does not hold with the convex closed set S' and the convex nonclosed set S. □

1.9.3. Distances between cones. Given two nonempty closed cones K, K' of a normed space $(X, \|\cdot\|)$, it results from the positive homogeneity of the function $d(\cdot, K)$ that, for any real $\rho > 0$, the ρ-excess of K over K' satisfies

$$\text{exc}_\rho(K, K') = \sup_{\|x\|\le\rho} \left(d(x, K') - d(x, K)\right)^+ = \sup_{\|u\|\le 1}\left(d(\rho u, K') - d(\rho u, K)\right)^+$$
$$= \rho \sup_{\|u\|\le 1} \left(d(u, K') - d(u, K)\right)^+ = \rho\,\text{exc}_1(K, K'),$$

and hence

$$\text{haus}_\rho(K, K') = \rho\,\text{haus}_1(K, K').$$

Then the Attouch-Wets convergence for nonempty closed cones amounts to convergence with respect to $\text{haus}_1(\cdot, \cdot)$, which is easily seen to be a distance on the set of nonempty closed cones of X. From the positive homogeneity of the function $d(\cdot, K)$ again we also see that

$$\text{exc}_1(K, K') := \sup_{\|x\|\le 1}\left(d(x, K) - d(x, K')\right)^+ = \sup_{\|x\|= 1}\left(d(x, K) - d(x, K')\right)^+,$$

and similarly

$$\text{haus}_1(K, K') = \sup_{\|x\|\le 1} |d(x, K) - d(x, K')| = \sup_{\|x\|=1} |d(x, K) - d(x, K')|,$$

with the convention that each above supremum on $\{x \in X : \|x\| = 1\} = \emptyset$ is zero when X is reduced to zero. It is worth noticing that $\mathrm{haus}_1(K, K')$ offers also a way to measure the aperture/opening between the cones K, K'. We also note that $\mathrm{haus}_1(K, K') \leq 1$ since, for all $x \in \mathbb{B}_X$ one has $0 \leq d(x, K) \leq 1$ because $0 \in K$.

Among other ways to measure the *aperture/opening* between nonempty closed cones K, K', consider the expressions

(1.52) $\qquad \mathrm{haus}(K \cap \mathbb{B}_X, K' \cap \mathbb{B}_X)$ and $\mathrm{haus}(K \cap \mathbb{S}_X, K' \cap \mathbb{S}_X)$,

where by convention $\mathrm{haus}(K \cap \mathbb{S}_X, K' \cap \mathbb{S}_X) = 0$ if both cones K, K' are reduced to zero and $\mathrm{haus}(K \cap \mathbb{S}_X, K' \cap \mathbb{S}_X) = 1$ if one of the two cones K, K' is null and the other not. Clearly, each one of the two above expressions defines a distance on the set of closed cones in X.

On the other hand, concerning the ρ-pseudo-excess of K over K' one has by the positive homogeneity of $d(\cdot, K')$ that

(1.53) $$\widehat{\mathrm{exc}}_\rho(K, K') = \sup_{x \in K \cap \rho \mathbb{B}_X} d(x, K') = \sup_{u \in K \cap \mathbb{B}_X} d(\rho u, K')$$
$$= \rho \sup_{u \in K \cap \mathbb{B}_X} d(u, K') = \rho \widehat{\mathrm{exc}}_1(K, K'),$$

and hence

$$\widehat{\mathrm{haus}}_\rho(K, K') = \rho \widehat{\mathrm{haus}}_1(K, K').$$

Another way to measure the aperture between the nonempty closed cones K and K' is then given by

$$\widehat{\mathrm{haus}}_1(K, K') = \max\{\widehat{\mathrm{exc}}_1(K, K'), \widehat{\mathrm{exc}}_1(K', K)\}$$
$$= \max\left\{\sup_{x \in K \cap \mathbb{B}_X} d(x, K'), \sup_{x \in K' \cap \mathbb{B}_X} d(x, K)\right\}.$$

This function $\widehat{\mathrm{haus}}_1(\cdot, \cdot)$ obviously defines a pseudo-distance on the set of nonempty closed cones in the normed space X.

The following exercise shows that $\widehat{\mathrm{haus}}_1(K, K') = \mathrm{haus}(K \cap \mathbb{B}, K' \cap \mathbb{B})$ for nonempty closed convex cones K, K' in a Hilbert space.

EXERCISE 1.104. For any two nonempty closed convex cones K, K' in a Hilbert space H, justify that

$$\widehat{\mathrm{haus}}_1(K, K') = \mathrm{haus}(K \cap \mathbb{B}_X, K' \cap \mathbb{B}_X) = \mathrm{haus}_1(K, K').$$

$\square$

In order to compare in the context of a general normed space, let us notice the following lemma.

LEMMA 1.105. Let K, K' be nonempty closed cones in a normed space X.
(a) Concerning $\mathrm{exc}_1(K, K')$ and $\widehat{\mathrm{exc}}_1(K, K')$ one has

$$\widehat{\mathrm{exc}}_1(K, K') \leq \mathrm{exc}_1(K, K') \leq 2\,\widehat{\mathrm{exc}}_1(K, K').$$

(b) For any $x \in \mathbb{B}_X$ one has

$$d(x, K') \leq d(x, K' \cap \mathbb{B}_X) \leq 2\,d(x, K').$$

(c) If the closed cone K' is not reduced to zero, then for any $x \in \mathbb{S}_X$

$$d(x, K') \leq d(x, K' \cap \mathbb{B}_X) \leq d(x, K' \cap \mathbb{S}_X) \leq 2\,d(x, K').$$

PROOF. (a) The left inequality in (a) is a direct consequence of Lemma 1.69(d). Concerning the right inequality, it suffices to notice, with $\rho = 1$ and $\rho' = 2$ in (e) in Lemma 1.69, that
$$\mathrm{exc}_1(K, K') \leq \widehat{\mathrm{exc}}_2(K, K') = 2\,\widehat{\mathrm{exc}}_1(K, K'),$$
where the latter equality is due to (1.52).

(c) For (c) we only have to justify its last inequality. Fix any $x \in \mathbb{S}_X$ and take any nonzero $y \in K'$. For $z := y/\|y\| \in K' \cap \mathbb{S}_X$, we have
$$d(x, K' \cap \mathbb{S}_X) \leq \|x - z\| \leq \|x - y\| + \|y - z\| = \|x - y\| + |1 - \|y\||$$
$$= \|x - y\| + |\|x\| - \|y\|| \leq 2\|x - y\|,$$
so $d(x, K' \cap \mathbb{S}_X) \leq 2\,d(x, K')$.

(b) For (b) we may suppose that K' is not reduced to zero, otherwise the inequalities are evident. Fix any $x \in \mathbb{B}_X$. Since X is non-null (because $K' \neq \{0\}$), there exists some $u \in \mathbb{S}_X$ and $t \in [0, 1]$ such that $x = tu$. Noticing that $d(0, K' \cap \mathbb{B}_X) = 0$, by the convexity of $d(\cdot, K' \cap \mathbb{B}_X)$ and by (c) above we have
$$d(x, K' \cap \mathbb{B}_X) \leq t\,d(u, K' \cap \mathbb{B}_X) \leq 2t\,d(u, K') = 2d(x, K'),$$
where the latter equality is due to the positive homogeneity of $d(\cdot, K')$. This finishes the proof of the lemma. □

With Lemma 1.105 at hands one can establish in the next exercise the inequalities in the following proposition translating the metric equivalence between the above distances.

PROPOSITION 1.106. For any nonempty closed cones K, K' in a normed space X show the following inequalities:
$$\widehat{\mathrm{haus}}_1(K, K') \leq \mathrm{haus}_1(K, K') \leq 2\,\widehat{\mathrm{haus}}_1(K, K')$$
$$\widehat{\mathrm{haus}}_1(K, K') \leq \mathrm{haus}(K \cap \mathbb{B}_X, K' \cap \mathbb{B}_X) \leq 2\,\widehat{\mathrm{haus}}_1(K, K')$$
$$\widehat{\mathrm{haus}}_1(K, K') \leq \mathrm{haus}(K \cap \mathbb{S}_X, K' \cap \mathbb{S}_X) \leq 2\,\widehat{\mathrm{haus}}_1(K, K').$$

EXERCISE 1.107. Prove the proposition. □

Although $\widehat{\mathrm{haus}}_1(\cdot, \cdot)$ does not enjoy in general the triangle inequality as shown in Exercise 1.109 below, we will see in Section 2.7 in the next chapter that it behaves very well in the metric point of view. For example, this will be illustrated in Theorem 2.251 establishing that, for two closed vector subspaces L, L' of a Hilbert space H

(1.54) $$\widehat{\mathrm{haus}}_1(L, L') = \|P_L - P_{L'}\|,$$

where P_L denotes the metric projection operator onto the closed vector subspace L and $\|P_L - P_{L'}\|$ is the usual norm of the continuous linear operator $P_L - P_{L'}$ from H into itself. On the other hand, as proved in the next exercise a certain function of $\widehat{\mathrm{haus}}_1$ is a distance on the set of closed cones.

EXERCISE 1.108. Let X be a normed space.
Let K, K', K'' be three nonempty closed cones in X.
(a) For each $x \in K \cap \mathbb{B}_X$, argue that for any real $\varepsilon > 0$ one can find some $x' \in K'$ with $\|x - x'\| < \widehat{\mathrm{exc}}_1(K, K') + \varepsilon$, then some $x'' \in K''$ with $\|x' - x''\| \leq$

$$\big(\widehat{\mathrm{exc}}_1(K', K'') + \varepsilon\big)\|x'\|.$$

(b) Deduce that, for x, x'' as in (a)
$$\|x - x''\| \leq \widehat{\mathrm{exc}}_1(K, K') + \varepsilon + \big(\widehat{\mathrm{exc}}_1(K', K'') + \varepsilon\big)\big(1 + \widehat{\mathrm{exc}}_1(K, K') + \varepsilon\big),$$
and hence
$$\widehat{\mathrm{exc}}_1(K, K'') \leq \widehat{\mathrm{exc}}_1(K, K') + \widehat{\mathrm{exc}}_1(K', K'') + \widehat{\mathrm{exc}}_1(K, K')\,\widehat{\mathrm{exc}}_1(K', K'').$$

(c) Derive that
$$1 + \widehat{\mathrm{haus}}_1(K, K'') \leq \big(1 + \widehat{\mathrm{haus}}_1(K, K')\big)\big(1 + \widehat{\mathrm{haus}}_1(K', K'')\big).$$

(d) Conclude that $(K, K') \mapsto \log\big(1 + \widehat{\mathrm{haus}}_1(K, K')\big)$ is a distance on the set of nonempty closed cones in X. □

The next exercise provides an example showing that, even restricted to the set $\mathrm{Svect}(X)$ of non-null closed vector subspaces of X, the function $\widehat{\mathrm{haus}}(\cdot, \cdot)$ may fail to satisfy the triangle inequality.

EXERCISE 1.109. Let $X := \mathbb{R}^2$ be endowed with the sum norm given by $\|(r, s)\| := |r| + |s|$ and, for $0 < a' < a'' < 1$, let the vector subspaces
$$L := \mathbb{R} \times \{0\}, \quad L' := \{(r, s) \in \mathbb{R}^2 : s = a'x\}, \quad L'' := \{(r, s) \in \mathbb{R}^2 : s = a''r\}.$$

(a) Check that
$$\widehat{\mathrm{haus}}(L, L') = a', \quad \widehat{\mathrm{haus}}(L, L'') = a'', \quad \widehat{\mathrm{haus}}(L', L'') = (a'' - a')/(1 + a').$$

(A picture can be made.)

(b) Check that $\widehat{\mathrm{haus}}(L, L') + \widehat{\mathrm{haus}}(L', L'') - \widehat{\mathrm{haus}}(L, L'') < 0$. □

1.10. Comments

We begin this section of comments with some faces of the origins of semilimits of sequences of sets and of semicontinuities of multimappings.

The concepts of semilimits of families of sets has a long story. Outer semilimits (also called upper semilimits) of sequences of particular sets were considered by P. Painlevé in his 1887 thesis [**779**, p. 123-124] (thèse de doctorat ès sciences mathématiques). Indeed, in [**779**, p. 123] (see also [**780**, p. B123-B124], the publication of the thesis) for a complex-valued function F of one complex variable, Painlevé used outer semilimits (with the name "*limit*") of equibounded sequences of disjoint singularity lines $(L_n)_{n \in \mathbb{N}}$ of F to obtain a suitable expansion series of the function, establishing in this way a result in the line of the well-known Mittag-Leffler theorem regarding the case of a *sequence of singularity points*. As it appears in page 8 of Zoretti's paper [**1016**], semilimits of sets were also utilized by P. Painlevé in the development of his 1902 lectures at the "Ecole Normale Supérieure de Paris". Zoretti himself benefited from these lectures for his thesis [**1016**]. Let $(E_\alpha)_{\alpha \in A}$ be a family of sets in the plane $\mathbb{R}^2$ with $A \subset \mathbb{R}$ and let α_0 be a limit point of A. In page 8 of the 1905 publication [**1016**] of his thesis (thèse de doctorat ès sciences mathématiques), L. Zoretti utilized the outer limit (or limit superior) as $\alpha \to \alpha_0$ under the name of "*limit set*". Then, Zoretti proved the following theorem: If the sets E_α are continua with $E_{\alpha_2} \subset E_{\alpha_1}$ for $\alpha_1 \leq \alpha_2$, then the "limit set" (that is, the outer limit) of the family either is a continuum or is reduced to one point. Recall that, according to G. Cantor, a continuum E is a well-chained set satisfying $E = E'$, where the *derived set* E' is the set of all limit points (also called accumulation points) of the set E. In the page 9 of [**1016**], Zoretti observed that, in the above theorem, it is enough that there exists some $a \in \mathbb{R}^2$ such that for every $\varepsilon > 0$ and every $\delta > 0$ one has $E_\alpha \cap B(a,\varepsilon) \neq \emptyset$ for every $\alpha \in A$ with $|\alpha - \alpha_0| < \delta$; otherwise stated Lim inf $E_\alpha \neq \emptyset$ in nowadays notation. Zoretti precised in the page 8 of [**1016**] that the theorem stated above was demonstrated by P. Painlevé in the aforementioned lectures that he gave in 1902 at the "Ecole Normale Supérieure de Paris". This theorem was used by Zoretti in his thesis to study the behavior of analytic functions of one-complex variable around sets of singular points. Among various related results, let us cite the following theorem [**1016**, p. 15, Théorème A]: Given a one-complex variable function admitting a singular point z_0 as an element of a *discontinuous set* of singularities, then the function is not continuous around the point z_0.

The inner limit (also called lower limit) of a family of sets was considered in 1887 by G. Peano [**782**], as pointed out by S. Dolecki and G. Greco [**342, 343**]. In the 1887 paper [**782**, p. 302], G. Peano defined in two/three dimensions the inner limit (or lower limit) of sets under the name of "limit" as follows: "Una figura si può considerare o come fissa o come variabile. Diremo *limite d'una figura variabile* F il luogo dei punti le cui distanze dalla figura F hanno per limite zero"(English translation: A figure can either be considered as fixed or as variable. We will say that the limit of a variable figure F is the locus of points whose distances from figure F have zero as limit). This concept of limit of figures corresponds to what is called nowadays the inner (or lower) semilimit of families of sets according to Proposition 1.12(a) in the manuscript. After examining (in pages 303-304) the example of figure $F(t)$ given by the equation $f(x,y,z,t) = 0$ (resp. $f(x,y,z,t) = 0$ and $g(x,y,z,t) = 0$) with continuous functions f,g, Peano wrote in page 305: "Como applicazione delle cose precedenti, tratteremo da un nuovo punto di vista le tangenti a curve i piani tangenti a superficie"(English translation: As application, we will treat from a new point of view tangents to curve and tangent planes to surface). Peano then used his above notion of "limit" to define, in page 305 of the same book [**782**], an extension of tangents in the following way: "Sia F una linea od una superficie, e P_0 un punto di essa. Si immagini la figura omotetica della F, con centro di omotetia in P_0 e con rapporto di omotetia r. Col crescere indefinitamente di r, questa figura omotetica tende nei casi più comuni verso un limite. Esso è in generale la tangente alla linea, od il piano tangente alla superficie F; ma può anche essere, in casi speciali, una figura più complicata, cui potremo dare in ogni caso il nome di *figura tangente* alla F nel suo punto P_0"(English translation: Let F be either a line or a surface, and P_0 one of its points. Consider the homothetic figure of F, through the homothety with center P_0 and constant r. As $r \to \infty$, this homothetic figure tends in the most common cases towards a limit. It is generally the tangent to the line, or the tangent plane to the surface F; but it can be, in special cases, a more complicated figure, to which we can give in any case the name of the *tangent figure* to F at the point P_0). All the pages 302-305 in [**782**] make clear that this tangent figure was the main motivation for which Peano considered his above "limit" for figures. We will see in Subsection 7.7.4 of Chapter 7 that the translation of this tangent figure with the opposite of P_0 corresponds to the *adjacent tangent cone* defined in (7.35). The concept of outer limit (or limit superior) of sets was also involved by G. Peano: In his other 1908 manuscript [**784**, p. 237], for a variable figure F, Peano denoted by $\mathrm{Lm}\,(F,x_0)$ the set of y such that 0 is a limit point of the mapping $x \mapsto d(y, F(x))$ as $x \to x_0$, and by Proposition 1.12(b) this set corresponds to the outer limit $\underset{x \to x_0}{\text{Lim sup}}\, F(x)$.

The semilimits of an infinite family of sets were also considered by S. Janiszewski in his 1911 thesis (thèse de doctorat ès sciences mathématiques) published in [558]. In finite dimensional Euclidean spaces, the inner and outer semilimits were defined in [558, p. 93, 94] in the forms in (a) and (b) in Proposition 1.12 of the book under the respective names of *"limit set"* and *"accumulation set"* of the family; note that what is named "accumulation set" by Janiszewski in [558] was called "limit set" by Zoretti [1016] as seen earlier. In the case of a sequence, Janiszewski's definitions amount to (1.2) and (1.3) with $B(x, \varepsilon)$ in place of W therein. Through these notions, Janiszewski showed with Theorem I in [558, Chap. I, p. 98] that, if the inner limit (or limit inferior) of a sequence of (equi-bounded) continua is nonempty, then the outer limit (or limit superior) is either a continuum or a singleton set. This is similar to the above theorem stated in Zoretti's thesis, but it was derived from a more general result in the lemma in page 97 of [558]; regarding this lemma we refer also to G. Rabaté's comments in [826, Remarque II, p. 59, 60]. It was also proved in Theorem VIII in [558, Chap. I, p. 107] that, given an infinite family $(E_\alpha)_\alpha$ of pairwise disjoint (equi-bounded) continua such that the outer limit (or limit superior) is nonempty, then any continuum C containing the union $\bigcup_\alpha E_\alpha$ admits a *continuum of condensation* K, that is, a continuum K which is included in the derived set of $C \setminus K$. Janiszewski also established in his thesis several other results on (or using) semilimits of families of continua of diverse types, in particular for continua of condensation and *irreducible continua*; an irreducible continuum with respect to a property is a continuum such that no proper continuum subset enjoys the property.

In his 1925 thesis [953, p. 7] (thèse de doctorat ès sciences mathématiques), the way that F. Valisesco defined the convergence of a sequence $(E_n)_n$ of nonempty closed bounded sets in the real line $\mathbb{R}$ to a subset E of $\mathbb{R}$ amounts to requiring that for each $\varepsilon > 0$ one has:

(i) for each $x \in E$ there exists an integer k (depending on x and ε) such that $E_n \cap B(x, \varepsilon) \neq \emptyset$ for all $n \geq k$;
(ii) there is an integer p (depending on ε) such that $E_n \subset E + B(0, \varepsilon)$ for all $n \geq p$.

In pages 8 and 9, it was proved in [953] that the above convergence holds if and only if for each $\varepsilon > 0$ there is some $p \in \mathbb{N}$ such that for every $n \geq p$ one has $E_n \subset E + B(0, \varepsilon)$ and $E \subset E_n + B(0, \varepsilon)$. The Cauchy property for $(E_n)_n$ was defined in the same line and the convergence in the above sense of such a sequence was shown in page 11. Various other sequential results were also provided. Although each E_n was taken as a set in $\mathbb{R}$, Valisesco knew that his study could be applied to sequences of sets in $\mathbb{R}^N$. Indeed, he wrote in page 6 of his thesis: "Nous considèrerons dans ce chapitre seulement les ensembles linéaires, les notions que nous poserons s'étendant d'elles-mêmes à un espace quelconque" (English translation: "In this chapter, we will consider only linear sets, the notions that we will use, can be applied to any space").

In addition to the study of sequences $(E_n)_n$, a large development was devoted in [953] to multimappings with values in $\mathbb{R}$. Let be given a multimapping $M : T \rightrightarrows \mathbb{R}$ from a subset T in $\mathbb{R}^N$ into the real line $\mathbb{R}$ (called a *"multiform function"* in [953,

1.10. COMMENTS

p. 12]) with closed bounded values. Vasilesco's definitions of horizontal closedness and exterior closedness at $\bar{t} \in T$ can be translated as follows:

(a) M is *horizontally closed* at $\bar{t}$ whenever $\limsup_{t \to \bar{t}} M(t) \subset M(\bar{t})$, which means in the terminology in our treatise that M is outer semicontinuous at $\bar{t}$.

(b) M is *exteriorly closed* at $\bar{t}$ provided that for each $\varepsilon > 0$ there exists $\delta > 0$ such that $M(\bar{t}) \subset M(t) + B(0, \varepsilon)$ for all $t \in T \cap B(\bar{t}, \delta)$, which in terms in the book says that M is Hausdorff lower semicontinuous at $\bar{t}$.

In the page 17, Vasilesco's definition of continuity of M at $\bar{t}$ can be rephrased in requiring that for any $\varepsilon > 0$ there is $\delta > 0$ such that $M(t) \subset M(\bar{t}) + B(0, \varepsilon)$ and $M(\bar{t}) \subset M(t) + B(0, \varepsilon)$ for all $t \in T \cap B(\bar{t}, \delta)$; this is clearly equivalent to the continuity with respect to what we call the Hausdorff-Pompeiu distance. Uniform continuity of M was defined similarly. Various related results were established, in particular a theorem of Arzelà-Ascoli type was proved in the page 43. A classification in the line of R. Baire was also developed in Chapter V of Vasilesco's thesis.

Given a sequence of nonempty closed bounded sets $(E_n)_n$ in the complex plane, D. Pompeiu defined, in his thesis (thèse de doctorat ès sciences mathématiques) published in 1905 in [**818**], the *excess* ("*écart*" in French) of E_h with respect to E_k by
$$\Delta_{hk} := \sup_{z \in E_h} \operatorname{dist}(z, E_k)$$
(see page 281 therein), where we reproduce the terminology and notation in [**818**]. Then (in page 282) he called $\Delta_{hk} + \Delta_{kh}$ the "*mutual excess*" of E_h and E_k and said that a closed set E_ω is the "*limit*" of $(E_n)_n$ if $\Delta_{n\omega} + \Delta_{\omega n} \to 0$ as $n \to \infty$. Taking a collection $\mathcal{E}$ of closed sets in $\mathbb{C}$, a "*limit-point*" of $\mathcal{E}$ was defined as the limit of a sequence of (distinct) elements in $\mathcal{E}$, and the collection of all limits of such sequences of $\mathcal{E}$ was taken as the (Cantor) "*derived set*" $\mathcal{E}'$ of $\mathcal{E}$. If $\mathcal{E}' \subset \mathcal{E}$, it was said that $\mathcal{E}$ is closed, and $\mathcal{E}$ was declared "*reducible*" provided it is countable and closed. The thesis [**818**] was mainly devoted to the following problem: Let D be an open simply connected set in $\mathbb{C}$ and $f : D \to \mathbb{C}$ be a function which is complex differentiable ("*monogène*" in Painlevé's terminology in [**779**]) at every point in $D \setminus E$ and continuous at every point in E, where E is a subset of D. Under which condition on E is the function f holomorphic on D? It was well-known that this is the case when E is either a singleton or a reducible set (in the classical sense), and Pompeiu observed that the latter case could be deduced from the former singleton case by classical arguments related to reducible sets. By the theorem in page 27 of P. Painlevé's thesis [**779**] the result was also known to hold true for any *rectifiable line* E in $\mathbb{C}$. In 4 lines of the same page 282 Pompeiu wrote and noticed that, on the basis of the above concepts, the same arguments which allow to move from the case where E contains just one point to that of a countable number of points of a reducible set, can be used to move from a rectifiable line of points to a sequence of rectifiable lines fulfilling the above *reducible property*. This was the motivation for which Pompeiu introduced the aforementioned "mutual excess" $\Delta_{hk} + \Delta_{kh}$ and devoted two pages of his thesis to its use. Indeed, at the end of page 282 Pompeiu wrote: "J'ai tenu à exposer ce procédé de démonstration parce que c'est celui qui se présente naturellement à l'esprit lorsque l'on considère des ensembles de lignes et que l'on veut établir entre ces ensembles et les ensembles de points les plus grandes analogies"(English translation: "I wanted to expose this demonstration process because it is the one that naturally comes to mind when considering sets of lines and that we wish to establish between these sets and sets of points the most great

analogies"). No property of the convergence was studied in [**818**] and no continuity of multimappings was considered there. Instead, Pompeiu noticed in page 283 that in the absence of the *reducible property*, a limit of rectifiable lines may fail to be a rectifiable line, so the above process cannot be applied in such a situation. Then, in Chapters IV and V, Pompeiu employed other methods to obtain the desired result with more general sets E.

In the comments with regard to convergence of sets in his book [**473, 474**], F. Hausdorff cited both P. Painlevé and D. Pompeiu. In fact, in page 343 of [**474**] Hausdorff referred to Pompeiu's paper [**818**] for the metric between two sets and to Painlevé's works for limits of sets as presented above. The excesses were defined in page 167 there and the distance between two sets S, S' were defined as the maximum of the two asymmetrical excesses whereas the sum was employed instead by Pompeiu. Nowadays, Hausdorff's definition with maximum is generally utilized for the distance between two sets. The related convergence (resp. limit) was called *metric convergence* (resp. *metric limit*) in [**474**]. The inner limit (or limit inferior) and the outer limit (or limit superior) as defined in (1.2) and (1.3) respectively in the book were called the *lower closed limit* and *the upper closed limit* by Hausdorff (with notation $\underline{\text{Fl}}\, E_n$ and $\overline{\text{Fl}}\, E_n$), see [**474**, p. 168], and in the case of equality the common value was called the *closed limit* there. Diverse properties of those semilimits were studied in Subsection 2 of Section 28 of Chapter VI in [**474**] and the following theorem was established: Any sequence of sets in a separable metric space contains a subsequence for which the closed limit exists. In Subsection 3 of Section 28 of Chapter VI [**474**], various comparison results were proved for the above lower/upper closed limit and a related asymmetric excess; in particular, it was shown there: In a compact metric space, a sequence of nonempty closed sets metrically converge in the above metric sense if and only if the lower and upper closed limits coincide. An extension to compact metric spaces of the aforementioned Zoretti theorem was demonstrated in Subsection 4 of Section 29 of Chapter VI in [**474**].

Many of the convergence results for sets in [**474**] were reproduced by C. Kuratowski in the two volumes [**641**] and [**642**]. Inner and outer limits were called there *limit inferior* and *limit superior*. They were defined in [**641**, p. 241 and 243] via the equalities (1.2) and (1.3) in our manuscript and they were denoted in [**641**, p. 241 and 243] (as in [**640**]) by $\text{Li}\, S_n$ and $\text{Ls}\, S_n$. Kuratowski also established in pages 241-250 in [**641**] diverse additional features for $\text{Li}\, S_n$ and $\text{Ls}\, S_n$: comparisons; calculus for union, intersection, Cartesian product; characterization in metric spaces of the inclusion $x \in \text{Li}\, S_n$ (resp. $x \in \text{Ls}\, S_n$) by $\lim d(x, S_n) = 0$ (resp. $\liminf d(x, S_n) = 0$). G. Choquet studied in his 1947 paper [**225**] inner and outer limits of a family of sets $(E_j)_{j \in J}$, where the indexed set J is endowed with a filter $\mathcal{F}$ under the names of *limit inferior* and *limit superior* and with the notations $\inf . (E_j)_{\mathcal{F}}$ and $\sup . (E_j)_{\mathcal{F}}$. Many properties were established in this framework in [**225**], in particular an extension for connected sets E_j of the results recalled earlier of Zoretti and of Janiszewski.

The term *"multimapping"* was used by J. J. Moreau [**748**, p. 205] as the translation of the French term *"multi-application"* that one can find in C. Castaing [**209, 210**]. Multimapping is also the expression used by J.-P. Penot [**795**]. Note that one also finds "multifunction", "relation", "multivalued mapping"... and "set-valued mapping" (the latter invokes a mapping assigning to each element of the first space a subset of the second space); E. Michael in [**706**, p. 361] used the name "carrier". We refer to Yu. G. Borisovich, B. D. Gelman, A. D. Myshkis and V. V. Obukhovskii [**117**] for various generalities related to multimappings.

The terminology of *inner limit* and *outer limit* in Definition 1.2 was utilized by R. T. Rockafellar and R. J-B. Wets [**865**] as well as the names of *inner semicontinuity* and *outer semicontinuity* in Definition 1.25. Instead of outer semicontinuity for the related property in Definition 1.25, the name "upper semicontinuity" was used by C. Kuratowski [**640**] and G. Choquet [**225**]. In the treatise, we employed the name upper semicontinuity for multimappings to translate the distinct property in Definition 1.32.

Now, let us provide references for the results in this chapter. The first equality in Proposition 1.7 is a particular case of a more general result of G. Choquet [**225**, p. 61, Formula (2)]; the concept of *grill* of a filter was introduced in his paper [**224**]. Equalities in Remark 1.9(a) appeared in [**225**, p. 61, Formulas (1) and (2)]. The characterizations in Proposition 1.12 probably go back to Kuratowski [**641**, p. 242-243] for sequences of sets.

Proposition 1.17 and Proposition 1.18 are particular cases of more general results by J. M. Borwein and S. Fitzpatrick [**129**] where X is taken as a weakly compactly generated Banach space, that is, a Banach space X admitting a weakly compact set K such $X = \mathrm{cl}_w(\mathrm{span}\, K)$. We followed the terminology of Rockafellar and Wets [**865**, Definition 5.4] in attributing the name of outer semicontinuity for a multimapping satisfying the property (1.10), while it is called upper semicontinuity by C. Kuratowski in [**640**, p. 148] and [**642**, p. 32] and by G. Choquet [**225**, p. 67, Définition 2]. For a multimapping, the upper semicontinuity in Definition 1.32 in the manuscript corresponds to the concept of upper semicontinuity in the "strong sense" in Choquet's paper [**225**, p. 70, Définition 2 bis]. Proposition 1.27 was observed by Choquet in [**225**, Théorème 1], and for metric spaces it was contained in the proof of the statement in Section 2 of the earlier paper [**640**] of Kuratowski since the compactness assumption of the space of values is not used there. The properties in Propositions 1.37 and 1.39 were pointed out by Choquet [**225**, p. 70].

Theorems 1.43 and 1.44 as stated are taken, for their essential parts, from C. Berge [**91**, Theorems 1 and 2, p. 121-122]. The main statements of Theorem 1.48 and Theorem 1.51 as well as the proofs are taken from the book of C. Castaing and M. Valadier [**214**, Theorems II-20 and II-21].

Theorem 1.58 is due to E. Michael [**706**, Theorem 3.2"] and was announced in his previous 1953 note [**705**]; see also Michael's footnote in [**706**, p. 364]. Lemma 1.57 corresponds to Lemma 4.1 in [**706**]. The proof of Lemma 1.57 follows Michael's arguments in [**706**, p. 368] and the proof of Theorem 1.58 follows the arguments in [**706**, p. 368, 369] and in [**89**, p. 22] and [**44**, p. 357]. Corollary 1.59 translates Proposition 1.4 in [**706**, p. 363].

The quantitative convergence furnished by the Hausdorff-Pompeiu distance is clearly not efficient for nonempty bounded subsets of a normed space X, since $\mathrm{haus}(K, K') = +\infty$ for any pair of distinct closed cones K, K' in X. The idea is to involve the Hausdorff-Pompeiu distance of truncations of such sets. With regard to the origins of such a process of truncation, we mention first, for non null vector subspaces L, L', the expression

$$(1.55) \qquad \max\left\{\sup_{x \in L \cap \mathbb{S}_X} d(x, L'),\ \sup_{x \in L' \cap \mathbb{S}_X} d(x, L)\right\}$$

utilized in the 1948 paper of M. G. Krein, M. A. Krasnoselskiĭ and D. P. Milman [**618**] as well as in the 1957 paper of I. C. Gohberg and M. G. Krein [**441**] and in the 1958 paper of T. Kato [**598**]; see more in the comments about (1.56) below and in the related historical facts in Comments Section 2.9 in Chapter 2. In a normed space X, the expression in (1.49) defining $\widehat{\operatorname{haus}}_\rho(\cdot,\cdot)$ (denoted $\widehat{dl}_\rho(\cdot,\cdot)$ in [**865**, 4(11)]) appeared for convex cones with the particular $\rho = 1$ in the 1967 paper of D. W. Walkup and R. J-B. Wets [**968**, p. 229] and for convex sets (under notation $\sigma_\rho(\cdot,\cdot)$) in the 1969 paper of U. Mosco [**752**, p. 522]. Assuming that X is reflexive and endowing the classes of closed convex cones of X and X^* with the associated distances $\widehat{\operatorname{haus}}_{\rho_0}(\cdot,\cdot)$ with $\rho_0 = 1$, it was established in Theorem 1 in [**968**] that the mapping assigning to each closed convex cone in X its polar in X^* is an isometry. In 1986, with any pair of proper functions $f, g : X \to \mathbb{R} \cup \{+\infty\}$ it was associated in [**36**] the semidistance $d_{[\lambda],\rho}(f,g) := \sup_{\|x\|\leq\rho} |f_{[\lambda]}(x) - g_{[\lambda]}(x)|$, where $f_{[\lambda]}$ denotes the *Baire envelop* defined by $f_{[\lambda]}(x) := \inf_{y\in X} \left(f(y) + \frac{1}{\lambda}\|x-y\|\right)$. So, for two subsets S, S' we see that $d_{[1],\rho}(\Psi_S, \Psi_{S'}) = \sup_{\|x\|\leq\rho} |d_S(x) - d_{S'}(x)|$, hence the semidistances $\operatorname{haus}_\rho(\cdot,\cdot)$ (in the book) constitute for convex sets particular cases of the semidistances whose diverse properties are studied for convex functions by H. Attouch and R. J-B. Wets in [**36**, p. 52-57]. Lemma 1.1 in Mosco's paper [**752**, p. 523] showed, for a sequence $(S_n)_n$ of closed convex sets of a reflexive Banach X, that the convergence $\widehat{\operatorname{haus}}_\rho(S_n, S) \to 0$ for every $\rho > 0$ implies that both inner limit and sequential weak outer limit of $(S_n)_n$ coincide with S, that is, $\operatorname{Liminf} S_n = {}^{\text{w-seq}}\operatorname{Limsup} S_n = S$; today, the property corresponding to both latter equalities is called the *Mosco convergence* of $(S_n)_n$ to S. All the aforementioned works were concerned with particular classes of closed sets. The step of a suitable quantitative convergence for the class of all closed subsets of a normed space was carried out in a series of papers [**37, 38, 39**] by H. Attouch and R. J-B. Wets by means of families of semidistances $\left(\operatorname{haus}_\rho(\cdot,\cdot)\right)_{\rho>0}$ and pseudo semidistances $\left(\widehat{\operatorname{haus}}_\rho(\cdot,\cdot)\right)_{\rho>0}$; notice that notation $\operatorname{haus}_\rho(\cdot,\cdot)$ is employed in [**36**, p. 696] in place of the one $\widehat{\operatorname{haus}}_\rho(\cdot,\cdot)$ in our manuscript. In [**36**], Attouch and Wets established: the pseudo-distance property of $\widehat{\operatorname{haus}}_\rho(\cdot,\cdot)$ in the proof of their Proposition 1.2, comparisons with $\operatorname{haus}(S \cap \rho\mathbb{B}, S' \cap \rho\mathbb{B})$ in their Proposition 1.4, comparison between $\operatorname{haus}_\rho(\operatorname{epi} f, \operatorname{epi} g)$ and $d_{[\lambda],\rho'}(f,g)$ in Section 3, comparison with Painlevé-Peano-Kuratowski convergence and with Mosco convergence in Section 4, comparison between $\operatorname{haus}_\rho(\operatorname{epi} f, \operatorname{epi} g)$ and $\operatorname{haus}_{\rho'}(\operatorname{gph} \partial f, \operatorname{gph} \partial g)$ in Section 5, calculus for certain operations in Section 6. In the second companion paper [**38**], Attouch and Wets provided Hölderian type properties with respect to the above pseudo semidistance for optima of suitably well-conditioned optimization problems, while they proved in [**39**] Lipschitzian properties for ε-approximate solutions of convex optimization problems. All these contributions speak in favor to name the related convergence as Attouch-Wets convergence, following in this way the terminology in the literature and in [**865**].

The results in Lemma 1.69 and their proofs are all mainly inspired by those in Proposition 4.37 in the book [**865**] of R. T. Rockafellar and R. J-B. Wets. The equality $\widehat{\operatorname{haus}}_\rho(S, S') = \operatorname{haus}_\rho(S, S')$ in Lemma 1.72 was proved in [**865**, Proposition 4.37] by another method. Proposition 1.73 and its proof are taken from Proposition

4.39 in [**865**]; in fact the arguments of the proofs for [**865**, Proposition 4.39] and for Proposition 1.73 in the book follow those by F. H. Clarke in [**243**, Lemma 3, p. 172] and by P. D. Loewen and R. T. Rockafellar in [**673**, Lemma 2.1].

The inequality (b) in Proposition 1.77 was shown by J. J. Moreau [**751**, Lemma p. 362, inequality (2.17)], and a previous result in this line was proved by J. W. Daniel [**308**, Theorem 2.2] but with a less accurate Hölder constant. The proof in the manuscript follows the one by M. D. P. Monteiro Marques [**716**, Proposition 4.7], see also F. Nacry and L. Thibault [**755**, Lemma 2.5] and [**757**]. Example 1.78 is due to Daniel [**308**, p. 235].

The connectedness result under upper semicontinuity in Proposition 1.83 can be found in G. Choquet's article [**225**, p. 68, Théorème 2], since the proof there still holds without the compactness of X if the upper semicontinuity in the sense of Definition 1.32 in the book is used instead of the outer semicontinuity. In the 1985 paper [**492**] by J.-B. Hiriart-Urruty another proof was given as well as various applications in optimization, calculus of variations, control theory; references can also be found in [**492**] for other earlier proofs between 1970 and 1980. Proposition 1.88 is taken from the paper [**124**] of J. M. Borwein.

Theorem 1.90 is due to M. D. Wills [**973**] and the main parts in Subsection 1.9.2 are taken from [**973**].

Regarding Subsection 1.9.3, on the set $\mathrm{Svect}(X)$ of non-null closed vector subspaces of a normed space X the expression

$$(1.56) \qquad \max\left\{\sup_{x \in L \cap \mathbb{S}_X} d(x, L'), \sup_{x \in L' \cap \mathbb{S}_X} d(x, L)\right\} = \widehat{\mathrm{haus}}_1(L, L'),$$

as already said earlier about (1.55), was probably introduced for the first time by M. G. Krein, M. A. Krasnoselskiĭ and D. P. Milman in their 1948 paper [**618**]; it provided a way to measure the aperture/opening between two vector subspaces L, L'. It is known as the *geometric opening* between L, L'. Note that the equality

$$\max\left\{\sup_{x \in L \cap \mathbb{S}_X} d(x, L'), \sup_{x \in L' \cap \mathbb{S}_X} d(x, L)\right\} = \widehat{\mathrm{haus}}_1(L, L')$$

holds true according to the equality $\sup_{x \in L \cap \mathbb{S}_X} d(x, L') = \sup_{x \in L \cap \mathbb{B}_X} d(x, L')$ due to the convexity of $d(\cdot, L')$. Previously to [**618**] another form of the geometric opening has been considered in 1947 in the context of Hilbert spaces by M. G. Krein and M. A. Krasnoselskiĭ [**617**] as

$$\theta(L, L') := \|P_L - P_{L'}\|,$$

where P_L denotes the orthogonal projection onto the closed vector subspace L; the arguments that $\theta(L, L')$ coincides with the expression in (1.56) were already present in [**618**]. Many results related to the geometric opening metric in (1.56) were established by M. G. Krein, M. A. Krasnoselskiĭ and D. P. Milman [**618**], by I. C. Gohberg and M. G. Krein [**441**], by I. C. Gohberg and A. S. Markus [**442**]. Comparisons with diverse other metrics on the space of subspaces of a Banach space were sutdied by E. R. Berkson [**92**]. As written by M. I. Ostrovskiĭ in [**778**, Notes and Remarks 3.7], the geometric opening seems to have been introduced by the authors of [**617**, **618**] as an important concept to extend the theory of Carleman-von Neumann of defect numbers of Hermitian operators to the setting of arbitrary operators in Hilbert and Banach spaces. Given a complex Banach space Y and a

linear operator A from a vector subspace $\mathrm{Dom}\, A \subset Y$ into Y, a number $\lambda \in \mathbb{C}$ is said to be of *regular type* for A whenever there exists some real $c(\lambda) > 0$ such that
$$\|(A - \lambda I)y\| \geq c(\lambda)\|y\| \quad \text{for all } y \in \mathrm{Dom}\, A,$$
where I denotes the identity mapping on Y. Using the above notion of geometric opening, it was proved in [**618**] the following: For any connected open set U in $\mathbb{C}$ made up of numbers of regular type for A, the number $\dim\left(\mathrm{Im}(A - \lambda I)\right)^{\perp}$ is a constant with respect to $\lambda \in U$.

The expression
$$\max\left\{\sup_{x \in L \cap \mathbb{S}_X} d(x, L' \cap \mathbb{S}_X), \sup_{x \in L' \cap \mathbb{S}_X} d(x, L \cap \mathbb{S}_X)\right\} = \mathrm{haus}(L \cap \mathbb{S}_X, L' \cap \mathbb{S}_X),$$
called the *spherical opening* between L, L', has been proposed by I. C. Gohberg and A. S. Markus in their 1959 paper [**442**]. The other expression
$$\max\left\{\sup_{x \in L \cap \mathbb{B}_X} d(x, L' \cap \mathbb{B}_X), \sup_{x \in L' \cap \mathbb{B}_X} d(x, L \cap \mathbb{B}_X)\right\} = \mathrm{haus}(L \cap \mathbb{B}_X, L' \cap \mathbb{B}_X),$$
known as the *ball opening* between L, L' has been used in 1965 by R. Douady [**355**] and V. I. Gurariĭ [**451**]. The properties in Exercise 1.108 and the counter example in Exercise 1.109 are taken from [**442, 778**]. For more on measures of opening between two vector subspaces as well as on topologies on $\mathrm{Svect}(X)$, we refer to the well-developed survey of M. I. Ostrovskiĭ [**778**]. The adaptation of the above measures of opening to closed convex cones of the space X has been achieved by A. Iusem and A. Seeger [**530**].

CHAPTER 2

Tangent cones and Clarke subdifferential

In classical analysis normal vectors to a smooth submanifold appear as orthogonal vectors to the tangent vector space. In unilateral variational analysis, instead of normal vector space or tangent vector space one uses local approximations with *cones*. We start here with the concept of Clarke tangent and normal cones in the analysis of nonsmooth sets.

2.1. Clarke tangent and normal cones

The Clarke tangent cone of a set S at $x \in S$ provides a tangential approximation of S near x which has the remarkable convexity property (as we will see in the first theorem below). The Clarke normal cone will be the negative polar of the Clarke tangent cone, hence there will be a duality between these two convex cones.

2.1.1. Definitions and various characterizations.

DEFINITION 2.1. Let S be a set of the normed space $(X, \|\cdot\|)$ and $x \in S$. The *Clarke tangent cone* or *C-tangent cone* of S at x is defined as the following inner limit or limit inferior of the set-difference quotient

$$T^C(S;x) := \liminf_{t\downarrow 0,\, u\xrightarrow{S} x} \frac{1}{t}(S - u),$$

where notation in (1.4) is used. Otherwise stated, according to the definition of the inner limit or limit inferior of a multimapping (see Definition 1.2), a vector $h \in X$ belongs to $T^C(S;x)$ provided that for any neighborhood V of h in X there are a neighborhood U of x in X and a real $\varepsilon > 0$ such that for all $u \in U \cap S$ and $t \in]0, \varepsilon[$ one has $V \cap t^{-1}(S - u) \neq \emptyset$, that is,

$$(u + tV) \cap S \neq \emptyset.$$

The set $T^C(S;x)$ is obviously a cone containing zero, and any element in $T^C(S;x)$ is called a *C-tangent* of S at the point x.

The *Clarke normal cone* or *C-normal cone* $N^C(S;x)$ of C at x is the negative polar $\left(T^C(S;x)\right)^\circ$ of the Clarke tangent cone, that is,

$$N^C(S;x) := \{x^* \in X^* : \langle x^*, h\rangle \leq 0,\ \forall h \in T^C(S;x)\}.$$

When $x \notin S$ we define both tangent and normal cones to be empty. Every element in $N^C(S;x)$ is called a *C-normal* of S at x.

We recall that the polar of a subset $A \subset X$ is the set (see (1.27))

$$A^\circ := \{u^* \in X^* : \langle u^*, u\rangle \leq 1,\ \forall u \in A\}.$$

When A is a *cone* K, that is, $rK \subset K$ for all reals $r > 0$, it is easily seen that

$$K^\circ = \{u^* \in X^* : \langle u^*, u\rangle \leq 0,\ \forall u \in K\},$$

thus in such a case one says that K° is the *negative polar of the cone K*.

When X is a Hilbert space H with the inner product $\langle \cdot, \cdot \rangle_H$ it is often convenient (as usual) to identify through the Riesz linear isometry $J : H \to H^*$ any element $J(u)$ of the dual space H^* with the element $u \in H$ itself. This identification makes $N^C(S; x)$ as a subset of H through the equality $\langle J(u), h \rangle = \langle u, h \rangle_H$.

The first theorem of this chapter states two useful sequential characterizations of the Clarke tangent cone. It also establishes its convexity as well as other fundamental properties of Clarke tangent and normal cones.

THEOREM 2.2 (sequential characterizations of Clarke tangent cone, convexity and properties under convexity). *Let S be a set of the normed space $(X, \|\cdot\|)$. For any $x \in S$ the following hold.*
(a) *The Clarke tangent cone has the two following sequential characterizations, that is, each one of properties (a_1) and (a_2) characterizes the inclusion $h \in T^C(S; x)$:*

   (a_1) *For any sequence $(x_n)_n$ of S converging to x and any sequence of positive reals $(t_n)_n$ converging to 0, there exists a sequence $(h_n)_n$ in X converging to h such that*
   $$x_n + t_n h_n \in S \quad \text{for all } n \in \mathbb{N} \text{ (or for all } n \text{ large enough)}.$$

   (a_2) *For any sequence $(x_n)_n$ of S converging to x and any sequence of positive reals $(t_n)_n$ converging to 0, there is an increasing function $s : \mathbb{N} \to \mathbb{N}$ and a sequence $(h_n)_n$ in X converging to h such that*
   $$x_{s(n)} + t_{s(n)} h_n \in S \quad \text{for all } n \in \mathbb{N} \text{ (or for all } n \text{ large enough)}.$$

(b) $T^C(S; x)$ *is a closed convex cone.*
(c) *For $X = X_1 \times \cdots \times X_m$ and $x_k \in S_k \subset X_k$ for all $k = 1, \cdots, m$, one has the equalities:*
$$T^C(S_1 \times \cdots \times S_m; (x_1, \cdots, x_m)) = T^C(S_1; x_1) \times \cdots \times T^C(S_m; x_m)$$
and $\quad N^C(S_1 \times \cdots \times S_m; (x_1, \cdots, x_m)) = N^C(S_1; x_1) \times \cdots \times N^C(S_m; x_m)$.
(d) *Whenever S is convex, the Clarke tangent and normal cones satisfy the equalities*
$$T^C(S; x) = \mathrm{cl}\,(\mathbb{R}_+(S-x)) \quad \text{and} \quad N^C(S; x) = \{x^* \in X^* : \langle x^*, y - x \rangle \leq 0 \ \forall y \in S\}.$$

PROOF. (a) The equivalences of (a_1) and (a_2) with the inclusion $h \in T^C(S; x)$ follow directly from the sequential characterizations in Propositions 1.14 and 1.16 of the inner limit or limit inferior of a multimapping between metric spaces.
(b) The cone $T^C(S; x)$ is closed since it is the limit inferior of a multimapping. Let us prove its convexity.

First proof. Let $h, h' \in T^C(S; x)$. Fix any neighborhood W of $h + h'$ and take two neighborhoods V and V' of h and h' respectively such that $V + V' \subset W$. From Definition 2.1 there exist on the one hand $\varepsilon' > 0$ and a neighborhood U' of x with $(u' + t'V') \cap S \neq \emptyset$ for all $t' \in \,]0, \varepsilon'[$ and $u' \in S \cap U'$, and on the other hand there are a positive real $\varepsilon < \varepsilon'$, a neighborhood $U \subset U'$ of x, and a neighborhood $V_0 \subset V$ of h such that $U +]0, \varepsilon[V_0 \subset U'$ and $(u + tV_0) \cap S \neq \emptyset$ for all $t \in \,]0, \varepsilon[$ and $u \in S \cap U$. Therefore, for all $t \in \,]0, \varepsilon[$ and $u \in S \cap U$, we can take some $v \in V_0$ such that $u + tv \in S$ and hence $u + tv \in S \cap U'$, and this yields $((u + tv) + tV') \cap S \neq \emptyset$ according to what precedes, thus $(u + tW) \cap S \neq \emptyset$. This means that the inclusion $h + h' \in T^C(S; x)$ holds, and the desired convexity of the cone $T^C(S; x)$ follows.

Second proof. Let $h, h' \in T^C(S; x)$ and let any sequences $(x_n)_n$ of S converging to x and $(t_n)_n$ converging to 0 with $t_n > 0$. By (a_1) above choose a sequence $(h_n)_n$ in X converging to h such that $x_n + t_n h_n \in S$ for all n. For $x'_n := x_n + t_n h_n$ we then have $x'_n \to x$ with $x'_n \in S$, hence by (a_1) there is a sequence $(h'_n)_n$ in X converging to h' such that, for all $n \in \mathbb{N}$, we have $x'_n + t_n h'_n \in S$, that is, $x_n + t_n(h_n + h'_n) \in S$. Since $h_n + h'_n \to h + h'$, property (a_1) again says that $h + h' \in T^C(S; x)$. This combined with the cone property of $T^C(S; x)$ easily yields its convexity.

(c) The first equality of (c) follows directly from the definition of the Clarke tangent cone (or from (a_1)) and this equality obviously ensures the equality related to the normal cones. Notice that the first equality of (c) can also be seen as a consequence of Proposition 1.5(a).

(d) We only need to prove the first equality of (d) since the second one is a direct consequence of it.

First proof. Let any $y \in S$. Take any neighborhood V of $y - x$ and write it as $V = y - x - W$ where W is a neighborhood of zero. For the neighborhood $U := x + W$ of x we have, for any $u = x + w \in U \cap S$ (with $w \in W$) and any positive real $t < 1$, that $u + t(y - x - w) = ty + (1 - t)u \in S$ according to the convexity of S, so $(u + tV) \cap S \neq \emptyset$. This translates that $T^C(S; x)$ contains $(S - x)$, and hence also $\mathrm{cl}(\mathbb{R}_+(S - x))$.

Fix now any $h \in T^C(S; x)$. Take any neighborhood V of h and consider U and $\varepsilon > 0$ as given by Definition 2.1. Fixing $t \in]0, \varepsilon[$ we have $(x + tV) \cap S \neq \emptyset$, that is, $V \cap t^{-1}(S - x) \neq \emptyset$, and hence $V \cap (\mathbb{R}_+(S - x)) \neq \emptyset$. So $h \in \mathrm{cl}(\mathbb{R}_+(S - x))$, which combined with what precedes establishes the first equality of (d).

Second proof. Take any $y \in S$. Consider any sequence $(t_n)_n$ tending to zero with $t_n > 0$ and any sequence $(x_n)_n$ in S converging to x. There exists some n_0 such that, for all $n \geq n_0$, one has $0 < t_n < 1$, hence $x_n + t_n(y - x_n) \in S$ by convexity of S. Since $y - x_n \to y - x$ as $n \to \infty$, the assertion (a_2) tells us that $y - x \in T^C(S; x)$, so $\mathrm{cl}(\mathbb{R}_+(S - x)) \subset T^C(S; x)$ since $T^C(S; x)$ is a closed cone.

Conversely let $h \in T^C(S; x)$. Taking $t_n = 1/n$, the assertion (a_1) gives a sequence $(h_n)_n$ converging to h such that, for all n, one has $x + t_n h_n \in S$ thus $h_n \in \mathbb{R}_+(S - x)$, which implies that $h \in \mathrm{cl}(\mathbb{R}_+(S - x))$. □

DEFINITION 2.3. For a convex set S of a normed space X and $x \in S$, the second members of the equalities in Theorem 2.2(d) are generally called the *tangent and normal cones* of the convex set S at x respectively. They are simply denoted by $T(S; x)$ and $N(S; x)$ respectively, that is,

$$T(S; x) = \mathrm{cl}\left(\mathbb{R}_+(S - x)\right) \quad \text{and} \quad N(S; x) = \{x^* \in X^* : \langle x^*, y - x \rangle \leq 0 \; \forall y \in S\},$$

so clearly one also has $N(S; x) = (T(S; x))^\circ$.

Theorem 2.2 then says that the Clarke tangent and normal cones of a convex set S coincide with the above standard tangent and normal cones of convex sets in Definition 2.3.

The first proofs of (b) and (d) in Theorem 2.2 as well as those of several other later results of this section allow us to see that the variational analysis theory with Clarke concepts work well even in locally convex spaces (instead of normed spaces). This is one of the reasons for which in many places of this section we give the proofs with neighborhood arguments (instead of, or in addition to, sequential ones).

REMARK 2.4. We point out that the Clarke tangent cone is neither isotone nor antitone, that is, the inclusion $S \subset S'$ with $x \in S$ does not imply either $T^C(S;x) \subset T^C(S';x)$ or $T^C(S';x) \subset T^C(S;x)$. Indeed, for $S = [0,+\infty[\times[0,+\infty[$ and $S' = \{(r,s) \in \mathbb{R} \times \mathbb{R} : s \geq -|r|\}$, we have $S \subset S'$,

$$T^C(S;(0,0)) = [0,+\infty[\times[0,+\infty[, \quad T^C(S';(0,0)) = \{(r,s) \in \mathbb{R} \times \mathbb{R} : s \geq |r|\},$$

so we see that

$$T^C(S;(0,0)) \not\subset T^C(S';(0,0)) \quad \text{and} \quad T^C(S';(0,0)) \not\subset T^C(S;(0,0)).$$

□

Figure 2.1 shows the Clarke tangent cone at the point x_i of $S \subset \mathbb{R}^2$ (or more exactly $x_i + T^C(S;x_i)$), for $i \in \{1,2,5,6,7,9\}$. At the point x_i, for $i \in \{3,4,8,10,11,12,13\}$, we have

$$T^C(S;x_3) =]-\infty,0] \times \{0\}, \quad T^C(S;x_4) = \{(0,0)\},$$

$$T^C(S;x_8) = \{(0,0)\}, \quad T^C(S;x_{10}) = \mathbb{R} \times \{0\},$$

$$T^C(S;x_{11}) = [0,+\infty[\times\{0\}, \quad T^C(S;x_{12}) = \{0\} \times \mathbb{R}, \quad T^C(S;x_{13}) = \{(0,0)\}.$$

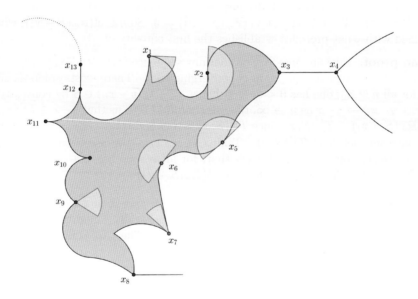

FIGURE 2.1. Clarke tangent cones.

The following proposition states a local property along with the coincidence for fitted sets.

PROPOSITION 2.5. Let S be a set of the normed space $(X, \|\cdot\|)$ and $x \in S$.
(a) The concepts of Clarke tangent and normal cones are local in the sense that

$$T^C(S;x) = T^C(S \cap U;x) \quad \text{and} \quad N^C(S;x) = N^C(S \cap U;x)$$

for any neighborhood U of x.

(b) For any set S' with $S \cap U \subset S' \cap U \subset (\mathrm{cl}\, S) \cap U$ for some neighborhood U of x, one has
$$T^C(S'; x) = T^C(S; x) \quad \text{and} \quad N^C(S'; x) = N^C(S; x).$$

(c) If $x \in S$ is not an isolated point of S, then $h \in T^C(S; x)$ if and only if, for any sequence $(t_n)_n$ in $]0, +\infty[$ tending to 0 and any sequence $(x_n)_n$ in $S \setminus \{x\}$ converging to x, there is a sequence $(h_n)_n$ in X converging to h such that $x_n + t_n h_n \in S$ for infinitely many (resp. for all) $n \in \mathbb{N}$.

PROOF. The assertion (a) follows directly from the definition of $T^C(S; x)$ as a limit inferior of sets or from the sequential characterization (a) in the above theorem. Concerning (b) take any $h \in T^C(S'; x)$. Fix any sequences $(t_n)_n$ tending to 0 with $t_n > 0$, and $(x_n)_n$ in S converging to x. Observing that for some integer N_0 we have $x_n \in S'$ for all $n \geq N_0$, by the sequential characterization (a) in the above theorem we see that there is a sequence $(h_n)_n$ in X converging to h such that $x_n + t_n h_n \in S'$ for all $n \geq N_0$, hence there is some integer $N \geq N_0$ such that $x_n + t_n h_n \in S' \cap U$ for all $n \geq N$. For each integer $n \geq N$, the inclusion $x_n + t_n h_n \in \mathrm{cl}\, S$ allows us to choose some $u_n \in S \cap B(x_n + t_n h_n, t_n/n)$, so $u_n = x_n + t_n(h_n + e_n)$ with $\|e_n\| < 1/n$. It ensues that $x_n + t_n(h_n + e_n) \in S$ for all $n \geq N$ along with $h_n + e_n \to h$, which guarantees by (a) in the above theorem that $h \in T^C(S; x)$.

Conversely, fix any $h \in T^C(S; x)$. Consider any sequences $(t_n)_n$ tending to 0 with $t_n > 0$, and $(x_n)_n$ in S' converging to x. Fix some integer N such that, for all $n \geq N$, we have $x_n \in \mathrm{cl}\, S$. Then, for each $n \geq N$, the set $S \cap B(x_n, t_n/n)$ is nonempty, so there is some $e_n \in X$ with $\|e_n\| < 1/n$ such that $x_n + t_n e_n \in S$. From the inclusion $h \in T^C(S; x)$ and from the sequential characterization (a) in the above theorem we obtain a sequence $(h_n)_n$ converging to h such that, for all $n \geq N$, we have $(x_n + t_n e_n) + t_n h_n \in S$, and hence $x_n + t_n(h_n + e_n) \in S'$ for n large enough along with $h_n + e_n \to h$. The same sequential characterization in the above theorem tells us that $h \in T^C(S'; x)$, which finishes the proof of (b).

By Theorem 2.2(a) the inclusion $h \in T^C(S; x)$ implies the property in (c) with all $n \in \mathbb{N}$, and the latter in turn obviously entails the property with infinitely many $n \in \mathbb{N}$. So, assume now that the property in (c) with infinitely many $n \in \mathbb{N}$ is fulfilled with a nonzero vector h. Take any sequence $(t_n)_n$ in $]0, +\infty[$ tending to 0. Since x is not an isolated point of S, for each n, we can choose some nonzero vector $e_n \in B(0,1)$ such that $x + (t_n/n)e_n \in S \cap B(x, t_n/n)$. Clearly, $\bigl(x + (t_n/n)e_n\bigr)_n$ is a sequence in $S \setminus \{x\}$ converging to x, hence by assumption there is a sequence $(h'_n)_n$ in X converging to h such that $(x + (t_n/n)e_n) + t_n h'_n \in S$ for infinitely many n. Putting $h_n := h'_n + (1/n)e_n$, we see that $h_n \to h$ as $n \to \infty$ and $x + t_n h_n \in S$ for infinitely many n. From this and the assumption again it easily follows that, for any sequence $(t_n)_n$ in $]0, +\infty[$ tending to 0 and any sequence $(x_n)_n$ in S converging to x, there exists some sequence $(h_n)_n$ in X converging to h with $x_n + t_n h_n \in S$ for infinitely many n. Consequently, Theorem 2.2(a_2) guarantees that $h \in T^C(S; x)$. □

In the line of the above assertion (c) we have:

PROPOSITION 2.6. Let S be a set of the normed space $(X, \|\cdot\|)$ and $x \in \mathrm{cl}\, S$. Let S' be a subset of X such that $S \cap U \subset S' \cap U \subset (\mathrm{cl}\, S) \cap U$ for some neighborhood U of x. Then, a vector $h \in T^C(S \cup \{x\}; x)$ if and only if, for any sequence $(t_n)_n$ in $]0, +\infty[$ tending to 0 and any sequence $(x_n)_n$ in S' converging to x, there is a sequence $(h_n)_n$ in X converging to h such that $x_n + t_n h_n \in S$ for all $n \in \mathbb{N}$.

PROOF. First, fix any $h \in T^C(S \cup \{x\}; x)$. Consider any sequences $(t_n)_n$ tending to 0 with $t_n > 0$, and $(x_n)_n$ in S' converging to x. Thanks to the inclusion $S' \cap U \subset (\operatorname{cl} S) \cap U$, we can fix $N \in \mathbb{N}$ such that, for all integers $n \geq N$, we have $x_n \in \operatorname{cl} S$. Then, for each integer $n \geq N$, the set $S \cap B(x_n, t_n/n)$ is nonempty, so there is some $e_n \in X$ with $\|e_n\| < 1/n$ such that $x_n + t_n e_n \in S$. Since $h \in T^C(S \cup \{x\}; x)$, Theorem 2.2($a_1$) furnishes a sequence $(h'_n)_n$ converging to h such that, for all integers $n \geq N$, we have $(x_n + t_n e_n) + t_n h'_n \in S \cup \{x\}$, so $x_n + t_n e_n + t_n h'_n \in \operatorname{cl} S$. For each integer $n \geq N$, we can choose $e'_n \in B(0, 1/n)$ such that $x_n + t_n h_n \in S$, where $h_n := h'_n + e_n + e'_n$. This justifies the implication $\Rightarrow$ of the proposition.

Conversely, suppose that the property of the proposition holds true with a vector $h \in X$. Take any sequence $(t_n)_n$ in $]0, +\infty[$ tending to 0 and any sequence $(x_n)_n$ in $S \cup \{x\}$ converging to x. For each $n \in \mathbb{N}$ the inclusion $x_n \in \operatorname{cl} S$ holds, so we can find $e_n \in B(0, 1/n)$ satisfying $x_n + t_n e_n \in S$. By the inclusion $S \cap U \subset S' \cap U$, choose $N \in \mathbb{N}$ such that $x_n + t_n e_n \in S'$ for all $n \geq N$. Then, the property of the proposition gives us a sequence $(h_n)_n$ in X converging to h such that $x_n + t_n e_n + t_n h_n \in S$ for all integers $n \geq N$. Since $e_n + h_n \to h$ as $n \to \infty$, it results that $h \in T^C(S \cup \{x\})$. □

Regarding symmetric sets we have:

PROPOSITION 2.7. *Assume that the subset S of the normed space X contains 0 and, for some neighborhood U of zero, $S \cap U$ is symmetric (with respect to the origin), that is, $-(S \cap U) = S \cap U$. Then $T^C(S; 0)$ and $N^C(S; 0)$ are closed vector subspaces of X and X^* respectively.*

PROOF. By Proposition 2.5 we may suppose that S is symmetric. Fix any $h \in T^C(S; 0)$ and consider any sequences $(t_n)_n$ in $]0, +\infty[$ tending to 0 and $(x_n)_n$ in S converging to 0. By symmetry of S we also have $-x_n \in S$ for all $n \in \mathbb{N}$, thus there is a sequence $(h_n)_n$ in X converging to h with $-x_n + t_n h_n \in S$ for all $n \in \mathbb{N}$. By symmetry of S again we obtain $x_n + t_n(-h_n) \in S$ for all $n \in \mathbb{N}$, so $-h \in T^C(S; 0)$. The closed convex cone $T^C(S; 0)$ is then symmetric with respect to 0, and hence it is a closed vector space (as easily seen) and its negative polar $N^C(S; 0)$ as well. □

The property in the latter proposition can be seen as a particular case of general results for inverse images which will be established later.

Now let us examine the boundary bdry S of a subset S of X. We first show that in the description (a_2) in Theorem 2.2 we can invoke merely sequences $(x_n)_n$ in the boundary of S.

PROPOSITION 2.8. *Let S be a subset of a normed space X and let $x \in \operatorname{bdry} S$. Let $S^c := X \setminus S$. The following are equivalent, for a vector $h \in X$:*
(a) *The vector $h \in T^C(S \cup \{x\}; x)$;*
(b) *for any sequence $(t_n)_n$ in $]0, +\infty[$ tending to 0 and any sequence $(x_n)_n$ in bdry S converging to x, there is a sequence $(h_n)_n$ in X converging to h such that $x_n + t_n h_n \in S$ for infinitely many (resp. for all) $n \in \mathbb{N}$;*
(c) *for any sequence $(t_n)_n$ in $]0, +\infty[$ tending to 0 and any sequence $(x_n)_n$ in bdry S converging to x, there is a sequence $(h_n)_n$ in X converging to h such that $x_n + t_n h_n \in \operatorname{cl} S$ for infinitely many (resp. for all) $n \in \mathbb{N}$.*

PROOF. Suppose that $h \in T^C(S \cup \{x\}; x)$. Proposition 2.6 (applied with $S' = \operatorname{cl} S$) says that, for any sequence $(t_n)_n$ in $]0, +\infty[$ tending to 0 and any sequence $(x_n)_n$ in $\operatorname{cl} S$ converging to x, there is a sequence $(h_n)_n$ in X converging to h such that, for each $n \in \mathbb{N}$, we have $x_n + t_n h_n \in S$. Since $\operatorname{bdry} S \subset \operatorname{cl} S$, we deduce that, for any sequence $(t_n)_n$ in $]0, +\infty[$ tending to 0 and any sequence $(x_n)_n$ in $\operatorname{bdry} S$ converging to x, there exists a sequence $(h_n)_n$ in X converging to h such that $x_n + t_n h_n \in S$ for all $n \in \mathbb{N}$. This justifies that (a) entails (b_1), where (b_1) denotes the property (b) with all $n \in \mathbb{N}$. Further, denoting by (b_2) the property (b) with infinitely many $n \in \mathbb{N}$ and denoting by (c_1) (resp. (c_2)) the property (c) with all $n \in \mathbb{N}$ (resp. with infinitely many $n \in \mathbb{N}$), it is obvious that the implications $(b_1) \Rightarrow (b_2) \Rightarrow (c_2)$ and $(b_1) \Rightarrow (c_1) \Rightarrow (c_2)$ hold true.

It remains to show the implication $(c_2) \Rightarrow (a)$. Take any sequence $(t_n)_n$ in $]0, +\infty[$ tending to 0 and any sequence $(x_n)_n$ in $\operatorname{cl}(S \cup \{x\}) = \operatorname{cl} S$ converging to x. Consider first the case when $x_n + t_n h \notin \operatorname{cl} S$ for large n, say $n \geq N$. For each integer $n \geq N$, there is some $r_n \in [0, t_n[$ such that $u_n := x_n + r_n h \in \operatorname{bdry} S$. Putting $u_n := x$ for $n < N$ and noting that $t_n - r_n \to 0$ with $t_n - r_n > 0$, by assumption there exists a sequence $(h'_n)_n$ in X converging to h satisfying $u_n + (t_n - r_n) h'_n \in \operatorname{cl} S$ for all $n \in K$, where K is some infinite subset of $\mathbb{N} \setminus \{1, \cdots, N\}$. For each $n \in K$, it results that

$$x_n + t_n \left(h + \frac{t_n - r_n}{t_n}(h'_n - h) \right) = u_n + (t_n - r_n) h'_n \in \operatorname{cl}(S \cup \{x\}),$$

so with $h_n := h + \frac{t_n - r_n}{t_n}(h'_n - h)$, clearly $h_n \to h$ and $x_n + t_n h_n \in \operatorname{cl}(S \cup \{x\})$ for all $n \in K$. In the remaining case, there is an infinite set $K' \subset \mathbb{N}$ such that, for each $n \in K'$, we have $x_n + t_n h \in \operatorname{cl} S$. Consequently, in any case there is an increasing function $s : \mathbb{N} \to \mathbb{N}$ and a sequence $(h_n)_n$ converging to h with $x_{s(n)} + t_{s(n)} h_n \in \operatorname{cl}(S \cup \{x\})$ for all n. This guarantees that $h \in T^C(\operatorname{cl}(S \cup \{x\}); x)$ by Theorem $2.2(a_2)$. From Proposition 2.5(b) we obtain that $h \in T^C(S \cup \{x\}; x)$, that is, the desired implication $(c_2) \Rightarrow (a)$ is justified. $\square$

COROLLARY 2.9. *Let S be a subset of a normed space X and let $\bar{x} \in \operatorname{bdry} S$. Let S' be another subset of X for which there is an open neighborhood U of $\bar{x}$ such that*

$$U \cap \operatorname{bdry} S \subset U \cap S' \subset U \cap \operatorname{cl} S.$$

Then one has the inclusion $T^C(S'; \bar{x}) \subset T^C(S \cup \{\bar{x}\}; \bar{x})$.

PROOF. Fix any $h \in T^C(S'; \bar{x})$ and take any sequence $(t_n)_n$ tending to 0 with $t_n > 0$ and any sequence $(x_n)_n$ in $\operatorname{bdry} S$ converging to $\bar{x}$. There is some integer n_0 such that for every $n \geq n_0$ one has $x_n \in U$ hence $x_n \in S'$. This furnishes a sequence $(h_n)_n$ converging to h such that $x_n + t_n h_n \in S'$ for all $n \geq n_0$. Since $x_n + t_n h_n \to \bar{x}$, there is some integer $n_1 \geq n_0$ such that for every $n \geq n_1$ one has $x_n + t_n h_n \in U$, thus $x_n + t_n h_n \in \operatorname{cl} S$. This gives that $h \in T^C(S \cup \{\bar{x}\}; \bar{x})$ according to the equivalence (c)$\Leftrightarrow$(a) in Proposition 2.8. $\square$

Another result concerning the boundary of S establishes the relationship between the C-tangent cones of $\operatorname{bdry} S$, $\operatorname{cl} S$ and $\operatorname{cl}(X \setminus S)$.

PROPOSITION 2.10. *Let S be a subset of a normed space X and let $\bar{x} \in \operatorname{bdry} S$. Then, setting $S^c := X \setminus S$ one has the equality*

$$T^C(\operatorname{bdry} S; \bar{x}) = T^C(S \cup \{\bar{x}\}; \bar{x}) \cap T^C(S^c \cup \{\bar{x}\}; \bar{x}).$$

PROOF. Since $\bar{x} \in \mathrm{bdry}\, S = \mathrm{bdry}\,(S^c)$, the inclusion of the left member into the right one follows directly from the inclusion in Corollary 2.9. Now put $\overline{S} := \mathrm{cl}\, S$ and take any h in the right member, hence in $T^C(\overline{S}; \bar{x}) \cap T^C(\overline{S^c}; \bar{x})$ by Proposition 2.5(b). Consider any convex neighborhood V of h. There exist a real $\varepsilon > 0$ and a neighborhood U of $\bar{x}$ such that, for all $t \in \,]0, \varepsilon]$ and $x \in U \cap \overline{S}$ (resp. $x \in U \cap \overline{S^c}$) we have $(x + tV) \cap \overline{S} \neq \emptyset$ (resp. $(x + tV) \cap \overline{S^c} \neq \emptyset$). Fix any $t \in \,]0, \varepsilon]$ and any $x \in U \cap \mathrm{bdry}\, S$. Noting that

$$x + tV = \big((x + tV) \cap \overline{S}\big) \cup \big((x + tV) \cap \overline{S^c}\big),$$

the convex (hence connected) set $x + tV$ which meets the closed sets $\overline{S}$ and $\overline{S^c}$, must meet their intersection $\mathrm{bdry}\, S$, that is, $(x + tV) \cap \mathrm{bdry}\, S \neq \emptyset$. This says that $h \in T^C(\mathrm{bdry}\, S; \bar{x})$. □

2.1.2. Interior tangent cone and calculus for Clarke tangent and normal cones of intersection. We have already considered the negative polar of the Clarke tangent cone to define the Clarke normal cone. For two nonempty *convex cones* K and L of X it is known from Functional Analysis (related to vector spaces) that the equality $(K \cap L)^\circ = K^\circ + L^\circ$ (or equivalently the crucial inclusion $(K \cap L)^\circ \subset K^\circ + L^\circ$) requires some condition on the cones K and L. A usual condition in Functional Analysis ensuring this crucial inclusion is that $K \cap \mathrm{int}\, L \neq \emptyset$. Indeed, taking $x^* \in (K \cap L)^\circ \setminus L^\circ$, we see for $Q := \mathrm{int}\, L$ that the open convex cone $G := \{(x, r) \in Q \times \mathbb{R} : \langle x^*, x \rangle > r\}$ is non void and disjoint from the convex cone $K \times [0, +\infty[$. Applying the Hahn-Banach separation theorem gives some $(y^*, -\lambda) \in X^* \times \mathbb{R}$ such that on the one hand $\langle y^*, y \rangle - \lambda s \leq 0$ for all $y \in K$ and $s \geq 0$ (thus $\lambda \geq 0$), and on the other hand $\langle y^*, u \rangle - \lambda r > 0$ for all $(u, r) \in G$. The assumption $K \cap Q \neq \emptyset$ easily gives $\lambda \neq 0$, and hence dividing both inequalities by $\lambda > 0$ we way suppose that $\lambda = 1$. Then, the first inequality ensures that $y^* \in K^\circ$, while the second one entails, for any $q \in Q$ and any $\varepsilon > 0$, that $\langle y^*, q \rangle - \langle x^*, q \rangle + \varepsilon > 0$ (since $(q, \langle x^*, q \rangle - \varepsilon) \in G$), hence $x^* - y^* \in Q^\circ = L^\circ$. So, for the convex cones K and L, we have established the equality

(2.1) $\qquad (K \cap L)^\circ = K^\circ + L^\circ \quad \text{whenever} \quad K \cap \mathrm{int}\, L \neq \emptyset.$

Having the above condition $K \cap Q \neq \emptyset$ in mind, the need of the formula

$$N^C(S_1 \cap S_2; x) \subset N^C(S_1; x) + N^C(S_2; x)$$

leads to the introduction of an open convex cone Q_x as a substitute for Q and such that $T^C(S; x)$ coincides with the closure of Q_x whenever $Q_x \neq \emptyset$.

DEFINITION 2.11. *Let S be a nonempty set in the normed space $(X, \|\cdot\|)$. The interior tangent cone $I(S; x)$ of S at $x \in S$ is defined by $h \in I(S; x)$ provided there are some $\varepsilon > 0$ and two neighborhoods U and V of x and h respectively such that*

(2.2) $\qquad U \cap S + \,]0, \varepsilon[\, V \subset S.$

As for the C-tangent cone, one puts $I(S; x) = \emptyset$ whenever $x \notin S$. When $I(S; x) \neq \emptyset$ (resp. $h \in I(S; x)$), the set S is said to satisfy at x (resp. at x in the direction h) the *interior tangent cone property/condition*, or the *interior tangent property/condition* for short.

The interior tangent property of S at $x \in S$ amounts to saying that there exist a neighborhood U of x, an open convex neighborhood V of h and a real $\varepsilon > 0$

such that all translates $x + [0, \varepsilon]V$ of the truncated open convex cone $[0, \varepsilon]V$ with $x \in U \cap S$ are contained in the set S, that is,

(2.3) $$x +]0, \varepsilon[V \subset S \quad \text{for all } x \in S \cap U.$$

Obviously $I(S; x)$ is an open cone included in $T^C(S; x)$ and its non-vacuity (that is, the interior tangent cone property) entails that the topological interior of S is nonempty; further $I(S; x) \neq \emptyset$ whenever $x \in \text{int } S$. Arguments similar to those for (a) of Theorem 2.2 gives the sequential characterization: $h \in I(S; x)$ if and only if, for any sequence of reals $(t_n)_n$ converging to 0 with $t_n > 0$, any sequence $(x_n)_n$ of S converging to x, and any sequence $(h_n)_n$ of X converging to h, we have

(2.4) $$x_n + t_n h_n \in S \quad \text{for } n \text{ large enough.}$$

The case of Cartesian product is a direct consequence of the above definition and can be stated as follows:

PROPOSITION 2.12. *Let $S_1, \cdots, S_m$ be subsets of normed spaces $X_1, \cdots, X_m$. For any $x = (x_1, \cdots, x_m)$ in $S_1 \times \cdots \times S_m$ one has*

$$I(S_1 \times \cdots \times S_m; x) = I(S_1; x_1) \times \cdots \times I(S_m; x_m).$$

The next theorem presents diverse fundamental relations between the interior tangent cone and the C-tangent cone.

THEOREM 2.13 (Clarke tangent and normal cones to intersection). *Let $S, S_1, \cdots, S_m$ be subsets of X. For any x in all these sets the following hold:*
(a) $I(S; x) \subset T^C(S; x)$ and $I(S; x) + T^C(S; x) = I(S; x)$.
(b) $I(S; x)$ is an open convex cone, and whenever $I(S; x) \neq \emptyset$

$$T^C(S; x) = \text{cl}\left(I(S; x)\right) \quad \text{and} \quad I(S; x) = \text{int}\left(T^C(S; x)\right).$$

(c) *The inclusions*

$$I(S_1; x) \cap \cdots \cap I(S_m; x) \subset I(S_1 \cap \cdots \cap S_m; x)$$

and $T^C(S_1; x) \cap I(S_2; x) \cap \cdots \cap I(S_m; x) \subset T^C(S_1 \cap \cdots \cap S_m; x)$
are always true, hence one has

$$T^C(S_1; x) \cap \cdots \cap T^C(S_m; x) \subset T^C(S_1 \cap \cdots \cap S_m; x)$$

whenever $T^C(S_1; x) \cap I(S_2; x) \cap \cdots \cap I(S_m; x) \neq \emptyset$.
(d) *If* $T^C(S_1; x) \cap I(S_2; x) \cap \cdots \cap I(S_m; x) \neq \emptyset$, *then*

$$N^C(S_1 \cap \cdots \cap S_m; x) \subset N^C(S_1; x) + \cdots + N^C(S_m; x).$$

PROOF. (a) The inclusion of (a) follows directly from the definitions. Since $0 \in T^C(S; x)$ the right-hand side of the equality is obviously contained in the left-hand side. The converse inclusion is obvious if $I(S; x) = \emptyset$. So, fix any $h \in T^C(S; x)$ and $h' \in I(S; x)$.

First proof. By the definition of $I(S; x)$ choose an $\varepsilon' > 0$, a neighborhood U' of x and a symmetric neighborhood V of zero in X such that $U' \cap S +]0, \varepsilon'[(h' + V + V) \subset S$. Choose now from the definition of $T^C(S; x)$ a positive $\varepsilon < \varepsilon'$, a neighborhood $U \subset U'$ of x, and a neighborhood $V' \subset V$ of zero such that $U +]0, \varepsilon[(h + V') \subset U'$ and such that for all $t \in]0, \varepsilon[$ and $u \in U$ we have $(u + t(h + V')) \cap S \neq \emptyset$, which allows us to select some $v_{t,u} \in V'$ satisfying $u + t(h + v_{t,u}) \in S$. Fix any $t \in]0, \varepsilon[$ and $u \in U \cap S$, and note that $u + t(h + v_{t,u}) \in U' \cap S$, which ensures that

$u + t(h + v_{t,u}) + t(h' + V + V) \subset S$ according to the choice of ε', U', V. We can then write

$$u + t(h+h'+V) = u+t(h+v_{t,u})+t(h'-v_{t,u}+V) \subset u+t(h+v_{t,u})+t(h'+V+V) \subset S,$$

which translates that $h + h' \in I(S;x)$.

Second proof. Take any sequences $(x_n)_n$ of S converging to x, $(t_n)_n$ converging to 0 with $t_n > 0$, and $(v_n)_n$ of X converging to $h + h'$. By the sequential characterization of $T^C(S;x)$ there exists a sequence $(h_n)_n$ in X converging to h such that $x_n + t_n h_n \in S$ for all $n \in \mathbb{N}$. Since $v_n - h_n \to h'$ and $x_n + t_n h_n \to x$ with $x_n + t_n h_n \in S$, the sequential characterization of $I(S;x)$ ensures that, for n large enough, we have $(x_n + t_n h_n) + t_n(v_n - h_n) \in S$, that is, $x_n + t_n v_n \in S$. The same sequential characterization says that $h + h' \in I(S;x)$ as desired.

(b) The assertion in (a) gives on the one hand $I(S;x) + I(S;x) \subset I(S;x)$, which corresponds to the convexity of the cone $I(S;x)$. On the other hand assuming $I(S;x) \neq \emptyset$ and choosing $h' \in I(S;x)$ the assertion in (a) yields for any $h \in T^C(S;x)$ that $h + \frac{1}{n}h' \in I(S;x)$, which implies that $h \in \mathrm{cl}\,(I(S;x))$. So, $T^C(S;x) = \mathrm{cl}\,(I(S;x))$ and this equality guarantees that $I(S;x) = \mathrm{int}\,(T^C(S;x))$ since $I(S;x)$ is an open convex set.

(c) It suffices to establish (c) for $m = 2$. Fix any $h \in I(S_1;x) \cap I(S_2;x)$ and take any sequence $(t_n)_n$ tending to 0 with $t_n > 0$, any sequence $(x_n)_n$ in $S_1 \cap S_2$ converging to x, and any sequence $(h_n)_n$ in X converging to h. For n large enough we have both $x_n + t_n h_n \in S_1$ and $x_n + t_n h_n \in S_2$, so $x_n + t_n h_n \in S_1 \cap S_2$. This entails $h \in I(S_1 \cap S_2;x)$ and justifies the first inclusion of (c).

The second inclusion of (c) being obvious if its left member is empty, fix any $h \in T^C(S_1;x) \cap I(S_2;x)$. Fix any sequence $(x_n)_n$ in $S_1 \cap S_2$ converging to x and any sequence $(t_n)_n$ tending to 0 with $t_n > 0$ for all n. From the sequential characterization of $T^C(S_1;x)$ there exists a sequence $(h_n)_n$ converging to h with $x_n + t_n h_n \in S_1$ for all n. Since $x_n \in S_2$ for all n, the sequential characterization of $I(S_2;x)$ ensures, for all n large enough, that $x_n + t_n h_n \in S_2$, thus $x_n + t_n h_n \in S_1 \cap S_2$. This justifies the inclusion $h \in T^C(S_1 \cap S_2;x)$.

Suppose now $T^C(S_1;x) \cap I(S_2;x) \neq \emptyset$ and choose $\overline{h} \in T^C(S_1;x) \cap I(S_2;x)$. Fix any $h \in T^C(S_1;x) \cap T^C(S_2;x)$. For each integer n we have $h + (1/n)\overline{h} \in T^C(S_1;x)$ since $T^C(S_1;x)$ is a convex cone, and $h + (1/n)\overline{h} \in I(S_2;x)$ by (a) above. So, by what precedes we obtain $h + (1/n)\overline{h} \in T^C(S_1 \cap S_2;x)$, thus $h \in T^C(S_1 \cap S_2;x)$ according to the closedness of the C-tangent cone, and this finishes the proof of assertion (c).

(d) Here also, it is enough to prove the assertion for $m = 2$. Under the assumption of (d), we have by (b), by the third inclusion of (c), and by (2.1)

$$\left(T^C(S_1 \cap S_2)\right)^\circ \subset \left(T^C(S_1;x) \cap T^C(S_2;x)\right)^\circ = N^C(S_1;x) + N^C(S_2;x),$$

and this finishes the proof of the theorem. $\square$

EXAMPLE 2.14. The converse inclusions in (c) and (d) of the above theorem fail. Consider in $\mathbb{R}^2$ the sets

$$S_1 := \{(r,s) \in \mathbb{R}^2 : r \leq 0 \text{ or } s \geq 0\} \text{ and } S_2 := \{(r,s) \in \mathbb{R}^2 : s \geq -|r|\}.$$

For $\overline{x} := (0,0)$ we have $I(S_1;\overline{x}) = \{(r,s) \in \mathbb{R}^2 : r < 0 \text{ and } s > 0\}$ and $I(S_2;\overline{x}) = \{(r,s) \in \mathbb{R}^2 : s > |r|\}$, hence $I(S_1;\overline{x}) \cap I(S_2;\overline{x}) \neq \emptyset$ (this intersection contains, for

example, $(-1, 2)$). However, the situation in Figure 2.1 with x_9 gives
$$T^C(S_1 \cap S_2; \overline{x}) = \{(r, s) \in \mathbb{R}^2 : s \geq \max\{r, 0\}\},$$
while $T^C(S_1; \overline{x}) = \{(r, s) \in \mathbb{R}^2 : r \leq 0 \text{ and } s \geq 0\}$ and $T^C(S_2; \overline{x}) = \{(r, s) \in \mathbb{R}^2 : s \geq |r|\}$; so the converse inclusion for C-tangent cones in (c) of the theorem is not satisfied.

In addition, $N^C(S_1 \cap S_2; \overline{x}) = \{(r, s) \in \mathbb{R}^2 : r \geq 0 \text{ and } s \leq -r\}$, while
$$N^C(S_1; \overline{x}) = \{(r, s) \in \mathbb{R}^2 : r \geq 0, s \leq 0\} \text{ and } N^C(S_2; \overline{x}) = \{(r, s) \in \mathbb{R}^2 : s \leq -|r|\};$$
so the converse inclusion for C-normal cones in (d) of the above theorem fails. $\square$

Regarding the inclusion of sets and the interior tangent cone, the following remark is useful.

REMARK 2.15. The interior tangent cone is neither isotone nor antitone. Consider in $\mathbb{R}^2$ as in Remark 2.4 the sets $S = \{(r, s) \in \mathbb{R}^2 : r \geq 0, s \geq 0\}$ and $S' = \{(r, s) \in \mathbb{R}^2 : s \geq -|r|\}$. We have $S \subset S'$, $I(S; (0, 0)) = \{(r, s) \in \mathbb{R}^2 : r > 0, s > 0\}$, and $I(S'; (0, 0)) = \{(r, s) \in \mathbb{R}^2 : s > |r|\}$, so
$$I(S; (0, 0)) \not\subset I(S'; (0, 0)) \quad \text{and} \quad I(S'; (0, 0)) \not\subset I(S; (0, 0)).$$
$\square$

With the interior tangent cone at hand we can prove the inclusion below relative to the Clarke tangent cone of intersection. The arguments are quite similar to those in Proposition 2.10.

PROPOSITION 2.16. *Let S_1 and S_2 be two closed sets of the normed space X and let $\overline{x} \in S_1 \cap S_2$. Then one has*
$$T^C(S_1; \overline{x}) \cap T^C(S_2; \overline{x}) \cap I(S_1 \cup S_2; \overline{x}) \subset T^C(S_1 \cap S_2; \overline{x}).$$

PROOF. Take any h in the left member of the deired inclusion, and by definition of the interior tangent cone choose a real $\varepsilon_0 > 0$, a neighborhood U_0 of $\overline{x}$ and a convex neighborhood V_0 of h such that $U_0 \cap (S_1 \cup S_2) +]0, \varepsilon]V_0 \subset S_1 \cup S_2$. Consider any convex neighborhood V of h with $V \subset V_0$. There exist a real $\varepsilon \in]0, \varepsilon_0]$ and a neighborhood $U \subset U_0$ of $\overline{x}$ such that, for each $i = 1, 2$ and for all $t \in]0, \varepsilon]$ and $x \in U \cap S_i$ we have $(x + tV) \cap S_i \neq \emptyset$. Fix any $t \in]0, \varepsilon]$ and any $x \in U \cap (S_1 \cap S_2)$. Then, the inclusion $x + tV \subset S_1 \cup S_2$ yields
$$((x + tV) \cap S_1) \cup ((x + tV) \cap S_2) = x + tV.$$
Consequently, the convex (hence connected set) $x + tV$, meeting the closed sets S_1 and S_2, must meet $S_1 \cap S_2$, that is, $(x + tV) \cap (S_1 \cap S_2) \neq \emptyset$. This guarantees that $h \in T^C(S_1 \cap S_2; \overline{x})$. $\square$

REMARK 2.17. (a) With $\overline{x} \in \text{bdry } S$, it is worth noticing that the inclusion
$$T^C(S \cup \{\overline{x}\}; \overline{x}) \cap T^C(S^c \cup \{\overline{x}\}; \overline{x}) \subset T^C(\text{bdry } S; \overline{x})$$
in Proposition 2.10 can also be seen as a consequence of the above proposition by taking $S_1 = \text{cl } S$ and $S_2 = \text{cl}(X \setminus S)$ and observing that $I(S_1 \cup S_2; \overline{x}) = X$ since $S_1 \cup S_2 = X$.

(b) Another situation where the above proposition can be applied is the following. Let $g : X \to \mathbb{R}$ be a real-valued continuous function and let $\overline{x} \in X$ with $g(\overline{x}) = 0$.

Since $S_1 \cup S_2 = X$ with $S_1 := \{g(\cdot) \leq 0\}$ and $S_2 := \{g(\cdot) \geq 0\}$, we obtain from the proposition

$$T^C(\{g(\cdot) \leq 0\}; \bar{x}) \cap T^C(\{g(\cdot) \geq 0\}; \bar{x}) \subset T^C(\{g(\cdot) = 0\}; \bar{x}).$$

The converse inclusion being clearly seen from Corollary 2.9, the equality

$$T^C(\{g(\cdot) = 0\}; \bar{x}) = T^C(\{g(\cdot) \leq 0\}; \bar{x}) \cap T^C(\{g(\cdot) \geq 0\}; \bar{x})$$

holds true. □

2.1.3. Epi-Lipschitz sets; their geometrical, tangential and topological properties. Theorem 2.13 above makes clear that sets S for which $I(S; x)$ is nonempty is of great interest. We establish next a useful characterisation for such a property at boundary points.

Recall that the *epigraph* epi f of an extended real-valued function $f : U \to \mathbb{R} \cup \{-\infty, +\infty\}$ on a subset U of the normed space X is defined as the subset of $X \times \mathbb{R}$ given by

$$\text{epi } f := \{(x, r) \in X \times \mathbb{R} : x \in U \text{ and } f(x) \leq r\}.$$

Recall also that, for a nonempty subset W of a metric space (Y, d_Y), a mapping g from W into a metric space (Z, d_Z) is *Lipschitz* or *Lipschitzian on* W whenever there exists some real $\gamma \geq 0$ such that

$$d_Z(g(y_1), g(y_2)) \leq \gamma d_Y(y_1, y_2) \quad \text{for all } y_1, y_2 \in W;$$

one says that γ is *a Lipschitz constant* of g over W or g is γ-*Lipschitz over* W. If the set W is open, one also says that g is *Lipschitz continuous on* W.

If W is a neighborhood of a point $y \in W$, one says that g is *Lipschitz, Lipschitzian,* or *Lipschitz continuous near y* provided there is a neighborhood $W' \subset W$ of y over which g is Lipschitz. When the set W is open and g is Lipschitz near any point of W, one says that g is *locally Lipschitz, locally Lipschitzian* or *locally Lipschitz continuous on* W. Similarly, the extended real-valued function $f : U \to \mathbb{R} \cup \{-\infty, +\infty\}$ on an open set $U \subset X$ is declared to be *Lipschitz, Lipschitzian,* or *Lipschitz continuous near a point* $x \in X$ provided that f is finite on a neighborhood of x and Lipschitz therein.

The local closedness of sets will be utilized. A set S of a normed space X is said to be *closed near a point* $x \in S$ provided there exists a neighborhood U of x such that $S \cap U$ is closed in U with respect to the induced topology. This is equivalent to saying that there is a real $\rho > 0$ such that $S \cap B[x, \rho]$ is closed in X. Clearly, closed sets and open sets are closed near any of their points.

DEFINITION 2.18. Let S be a subset of a normed space X and let $\bar{x} \in S$. The set S is called *epi-Lipschitz* (or *epi-Lipschitzian*) at $\bar{x}$ if it can be viewed around $\bar{x}$, after a linear homeomorphism, as the epigraph of a locally Lipschitz continuous function. Precisely, this means that there exist a nonzero vector $h \in X$, some closed vector hyperplane E of X with $X = E \oplus \mathbb{R}h$, some neighborhood W in X of $\bar{x} = \bar{u} + \bar{r}h$ with $\bar{u} \in E$ and $\bar{r} \in \mathbb{R}$, and some function $f : E \to \mathbb{R}$ Lipschitz continuous near $\bar{u}$ (with E endowed with the induced norm) such that $f(\bar{u}) = \bar{r}$ and

$$W \cap S = W \cap \{u + rh : u \in E, r \in \mathbb{R}, f(u) \leq r\}.$$

In such a case one says that S is *epi-Lipschitz* at $\bar{x}$ *in the direction* h.

Clearly, if the set S is epi-Lipschitz at $\bar{x}$, then it is closed near $\bar{x}$, int $S \neq \emptyset$ and $\bar{x} \in \operatorname{bdry} S$.

DEFINITION 2.19. A closed set S of a normed space is called *epi-Lipschitz* if it is epi-Lipschitz at any of its boundary points.

Below and in many other places in the book, a point in the normed space X for which a specific assumption is required will be very often denoted by $\bar{x}$.

THEOREM 2.20 (R.T. Rockafellar). Let S be a subset of a normed space $(X, \|\cdot\|)$ and $\bar{x} \in S$. Then S has the interior tangent cone property at $\bar{x}$ and is closed near $\bar{x}$ with $\bar{x} \in \operatorname{bdry} S$ if and only if it epi-Lipschitz at $\bar{x}$ in a nonzero direction h, otherwise stated, there exist some closed vector hyperplane E of X with $X = E \oplus \mathbb{R}h$, some neighborhood W of $\bar{x}$ in X, and some function $f : E \to \mathbb{R}$ Lipschitz continuous near $\pi_E(\bar{x})$ (with E endowed with the induced norm and $x = \pi_E(x) + \pi_h(x)h$ with $\pi_E(x) \in E$ and $\pi_h(x) \in \mathbb{R}$) such that $f(\pi_E(\bar{x}) = \pi_h(\bar{x})$ and
$$W \cap S = W \cap \{u + rh : u \in E, r \in \mathbb{R}, f(u) \leq r\}.$$
Given S epi-Lipschitz at $\bar{x}$, the above function $f : E \to \mathbb{R}$ can even be chosen Lipschitz on the whole space E.

PROOF. It can be checked in a straightforward way that the epi-Lipschitz condition entails that S is closed near $\bar{x}$ and satisfies the interior tangent property at $\bar{x}$. Let us prove the converse. Without loss of generality, we may suppose that the set S is closed. Let $h \in I(S; \bar{x})$. We observe that $h \neq 0$ since otherwise we would get that $\bar{x} \in \operatorname{int} S$, contradicting the assumption $\bar{x} \in \operatorname{bdry} S$. Take a real number $\delta' > 0$ and neighborhoods U' and V' of $\bar{x}$ and h in X such that
$$S \cap U' + tV' \subset S \quad \text{for all } t \in [0, \delta'].$$
Choose some $u^* \in X^*$ with $\langle u^*, h \rangle = 1$ and put $E := \operatorname{Ker} u^*$, so X appears as the topological direct sum $X = E \oplus \mathbb{R}h$ and $\bar{x} = \bar{u} + \bar{r}h$, with $\bar{u} := \bar{x} - u^*(\bar{x})h \in E$ and $\bar{r} := u^*(\bar{x}) \in \mathbb{R}$. Choose two reals $\eta > 0$ and $0 < \delta < \min\{\delta'/2, 1\}$ such that, for $U_0 := E \cap B(0, \eta)$, $U := (\bar{u} + U_0) \oplus]\bar{r} - \delta, \bar{r} + \delta[h$ and $V := U_0 \oplus]1 - \delta, 1 + \delta[h$, we have $U \subset U'$ and $V \subset V'$, hence
$$(2.5) \qquad S \cap U + tV \subset S \quad \text{for all } t \in [0, 2\delta].$$
For any $u \in U_0$ and any real t with $\bar{r} + \delta(1 - \delta) < t < \bar{r} + \delta$, writing $\bar{u} + \delta u + th = (\bar{u} + \bar{r}h) + \delta(u + \frac{t-\bar{r}}{\delta}h)$ yields through the above inclusion that $\bar{u} + \delta u + th \in S$. So, for $I :=]\bar{r} - \delta, \bar{r} + \delta[$, the function f defined (with E endowed with the induced topology) on the neighborhood $\bar{u} + \delta U_0$ of $\bar{u}$ in E by
$$f(\bar{u} + \delta u) := \inf\{r \in I : \bar{u} + \delta u + rh \in S\} \quad \text{for all } u \in U_0,$$
is real-valued and $f(\bar{u} + \delta u) = \inf\{r \in \mathbb{R} : r > \bar{r} - \delta \text{ and } \bar{u} + \delta u + rh \in S\}$.

Claim 1. $S \cap ((\bar{u} + \delta U_0) \oplus Ih) = \{\bar{u} + \delta u + rh : (\bar{u} + \delta u, r) \in ((\bar{u} + \delta U_0) \times I) \cap \operatorname{epi} f \}$.
Fix any $(u, r) \in U_0 \times I$ with $(\bar{u} + \delta u, r) \in \operatorname{epi} f$. Then $f(\bar{u} + \delta u) \leq r$, and hence for any $r' \in I$ with $r' > r$ we can find by the definition of f some $r'' \in I$ with $\bar{u} + \delta u + r''h \in S$ and such that $f(\bar{u} + \delta u) \leq r'' < r'$. Thus by (2.5) we have
$$\bar{u} + \delta u + r'h = (\bar{u} + \delta u + r''h) + (r' - r'')h \in S \cap U +]0, 2\delta[V \subset S,$$
which ensures that $\bar{u} + \delta u + rh \in S$ according to the closedness of S. So, we have proved the inclusion $\supset$ of the claim and the reverse inclusion is obvious.

Claim 2. The function f is Lipschitz continuous on $\bar{u} + \delta U_0$.
To show this Lipschitz property, it is enough to prove that

$$f(\bar{u} + \delta u + tv) \le f(\bar{u} + \delta u) + t,$$

for all $u, v \in U_0$ and $t \in [0, \delta]$ with $\bar{u} + \delta u + tv \in \bar{u} + \delta U_0$. Fix $u \in U_0$ and fix also $r \in I$ with $\bar{u} + \delta u + rh \in S$. For each $v \in U_0$ and each $t \in [0, \delta]$ with $\bar{u} + \delta u + tv \in \bar{u} + \delta U_0$, the inclusion in (2.5) gives

$$(\bar{u} + \delta u + tv) + (r + t)h = (\bar{u} + \delta u + rh) + t(v + h) \in S \cap U + tV \subset S.$$

Since $r + t \ge r > \bar{r} - \delta$, we obtain $f(\bar{u} + \delta u + tv) \le r + t$ for all $r \in I$ with $\bar{u} + \delta u + rh \in S$. It follows that $f(\bar{u} + \delta u + tv) \le f(\bar{u} + \delta u) + t$, justifying the claim of Lipschitz property of f on $\bar{u} + \delta U_0$. Take any real-valued extension on E of f outside of $\bar{u} + \delta U_0$, or even any Lipschitz extension of f on E with the same Lipschitz constant (which exists as known and shown in Proposition 2.79 below).

The choice $W = (\bar{u} + \delta U_0) \oplus Ih$ satisfies the desired equality $W \cap S = W \cap \{u + rh : u \in E, r \in \mathbb{R}, f(u) \le r\}$. Further, this equality combined with the inclusion $\bar{u} + \bar{r}h = \bar{x} \in W \cap \text{bdry } S$ ensures (by continuity of f at $\bar{x}$) that $\bar{r} = f(\bar{u}) = \bar{r}$. This finishes the proof of the theorem. $\square$

REMARK 2.21. Let H be a Hilbert space endowed with the inner product $\langle \cdot, \cdot \rangle$ and the associated norm $\|\cdot\|$. Let S be a subset of H which is closed near a point $\bar{x} \in \text{bdry } S$ and has the interior tangent cone property at $\bar{x}$ in a direction $h \in H$ with $\|h\| = 1$. The above proof shows that for the orthogonal vector space $E := (\mathbb{R}h)^\perp$ (of $\mathbb{R}h$ in H), there exist a neighborhood W of $\bar{x}$ in H and a Lipschitz continuous function $f : E \to \mathbb{R}$ such that

$$W \cap S = W \cap \{u + rh : u \in E, r \in \mathbb{R}, f(u) \le r\}.$$

We notice that $H = E \oplus \mathbb{R}h$ and that the above mappings $\pi_E : X \to E$ and $X \ni x \mapsto \pi_h(x)h \in \mathbb{R}h$ (with $x = \pi_E(x) + \pi_h(x)h$) correspond now to the orthogonal projections from X onto E and $\mathbb{R}h$ respectively. Further, for any $x \in H$ we have

$$\|x\|^2 = \|\pi_E(x)\|^2 + (\pi_h(x))^2 = \|(\pi_E(x), \pi_h(x))\|_{E \times \mathbb{R}},$$

where $\|\cdot\|_{E \times \mathbb{R}}$ denotes the canonic Hilbert norm on $E \times \mathbb{R}$ associated to the canonic Hilbert inner product on $E \times \mathbb{R}$ given for $(u_i, r_i) \in E \times \mathbb{R}$, $i = 1, 2$, by

$$\langle (u_1, r_1), (u_2, r_2) \rangle = \langle u_1, u_2 \rangle + r_1 r_2.$$

The linear mapping $A : E \times \mathbb{R} \to H$ defined by $A(u, r) = u + rh$ is then a bijective isometry and

$$W \cap S = W \cap A(\text{epi } f).$$

In the case of the Euclidean space $H = \mathbb{R}^n$, the set S can then be seen near the point $\bar{x}$ as the image by an orthogonal n-square matrix of the epigraph of a Lipschitz continuous function. $\square$

EXAMPLE 2.22. The set S in the previous Figure 2.1 is epi-Lipschitz at x_1, x_2, x_5, x_6, x_7, x_9 but it is not epi-Lipschitz at the other points x_i. $\square$

According to Definition 2.19 the following geometric characterization of epi-Lipschitz property of sets is a direct corollary of Theorem 2.20.

COROLLARY 2.23. A nonempty closed set S in a normed space X is epi-Lipschitz if and only if $I(S; x) \ne \emptyset$ for all $x \in S$.

Cartesian products of epi-Lipschitz sets furnish by Proposition 2.12 and Definition 2.18 another example of epi-Lipschitz sets. We state this in the following proposition.

PROPOSITION 2.24. *Let $S_1, \cdots, S_m$ be subsets of normed spaces $X_1, \cdots, X_m$. If each set S_i, $i = 1, \cdots, m$, has the interior tangent cone property (resp. is epi-Lipschitz) at a point $\overline{x}_i \in S_i$ in the space X_i, then the set $S_1 \times \cdots \times S_m$ has the interior tangent cone property) (resp. is epi-Lipschitz) at the point $(\overline{x}_1, \cdots, \overline{x}_m)$ in the product space $X_1 \times \cdots \times X_m$.*

The next proposition rephrases with the language of sets with the interior tangent cone condition a property in the assertion (b) of Theorem 2.13.

PROPOSITION 2.25. *Let S be a subset of the normed space X which has the interior tangent cone property at $\overline{x} \in S$. Then one has*
$$T^C(S; \overline{x}) = \mathrm{cl}\,(I(S; \overline{x})) \quad \text{and} \quad I(S; \overline{x}) = \mathrm{int}\,(T^C(S; \overline{x})).$$

Under the interior tangent cone property at a boundary point, the C-normal cone is not reduced to zero as the following corollary shows.

COROLLARY 2.26. *Let S be a subset of a normed space X which is has the interior tangent cone property at a point $\overline{x} \in S \cap \mathrm{bdry}\, S$. Then $N^C(S; \overline{x}) \neq \{0\}$.*

PROOF. By the interior tangent cone property we have $I(S; \overline{x}) \neq \emptyset$. Further, $0 \notin I(S; \overline{x})$ since otherwise we would have $\overline{x} \in \mathrm{int}\, S$ by definition of $I(S; \overline{x})$. Then, $I(S; \overline{x})$ being an open convex set, the Hahn-Banach separation theorem gives a nonzero vector $u^* \in X^*$ such that $\langle u^*, h \rangle < 0$ for all $h \in I(S; \overline{x})$, hence $\langle u^*, h \rangle \leq 0$ for all $h \in T^C(S; \overline{x})$ according to the equality $T^C(S; \overline{x}) = \mathrm{cl}\,(I(S; \overline{x}))$ in Proposition 2.25 above. This means that $u^* \in (T^C(S; \overline{x}))^\circ$, or equivalently $u^* \in N^C(S; \overline{x})$. □

We already observed in Proposition 2.5(b) the coincidence of Clarke tangent cones of sets S, S' satisfying the inclusions $S \subset S' \subset \mathrm{cl}\, S$. A similar property holds for the interior tangent cone of sets.

PROPOSITION 2.27. *Let S be a set of the normed space X such that $S \cup \{\overline{x}\}$ has the interior tangent cone property at $\overline{x} \in \overline{S}$, where $\overline{S}$ denotes the closure of S. The following hold:*
(a) There exists an open neighborhood U of $\overline{x}$ such that
$$\overline{\mathrm{int}\, S \cap U} = \overline{\mathrm{int}\, \overline{S} \cap U} = \overline{S} \cap U.$$
(b) If C is a subset of X containing $\overline{x}$ for which there exits some neighborhood U_0 of $\overline{x}$ such that
$$(\mathrm{int}\, \overline{S}) \cap U_0 \subset C \cap U_0 \subset \overline{S} \cap U_0,$$
then C has the interior tangent cone property at $\overline{x}$ and $I(C; \overline{x}) = I(S \cup \{\overline{x}\}; \overline{x})$; in particular $\overline{S}$ has the interior tangent cone property at $\overline{x}$ and $I(\overline{S}; \overline{x}) = I(S \cup \{\overline{x}\}; \overline{x})$.

PROOF. Let h be a vector in the direction of which $S \cup \{\overline{x}\}$ has the interior tangent cone property at $\overline{x}$. Take by definition a real $\varepsilon > 0$, an open neighborhood U of $\overline{x}$ and an open neighborhood V of h such that $(S \cup \{\overline{x}\}) \cap U +]0, \varepsilon[V \subset S \cup \{\overline{x}\}$, thus in particular $S \cap U +]0, \varepsilon[V \subset S \cup \{\overline{x}\}$. Since the left member of the latter inclusion is open (as the union of the open sets $u + tV$ with $u \in S \cap U$ and $t \in]0, \varepsilon[$), it ensues that $S \cap U +]0, \varepsilon[V \subset \mathrm{int}\,(S \cup \{\overline{x}\})$. Fix any $t \in]0, \varepsilon[$, any $v' \in V$ and any $x' \in \overline{S} \cap U$. Choose a sequence $(x_n)_n$ in S converging to x'. For some integer N

we have $x_n \in S \cap U$ for all $n \geq N$ since U is a neighborhood of x'. It ensues that $x_n + tv' \in \text{int}\,(S \cup \{\overline{x}\})$ for all $n \geq N$, thus $x' + tv' \in \overline{\text{int}\,(S \cup \{\overline{x}\})}$. From Lemma 2.28 below we obtain $x' + tv' \in \overline{\text{int}\,S}$, so $x' \in \overline{\text{int}\,S}$. The two latter inclusions ensure $\overline{S} \cap U \subset \overline{\text{int}\,S}$ and

(2.6) $\qquad \overline{S} \cap U +]0, \varepsilon[V \subset \overline{S}, \quad \text{hence} \quad \overline{S} \cap U +]0, \varepsilon[V \subset \overline{\text{int}\,S}.$

On the one hand, the inclusion $\overline{S} \cap U \subset \overline{\text{int}\,S}$ guarantees the equalities

$$\overline{S} \cap U = \overline{\text{int}\,S} \cap U = \overline{\text{int}\,\overline{S}} \cap U$$

in (a) since we always have $\overline{\text{int}\,S} \subset \overline{\text{int}\,\overline{S}} \subset \overline{S}$.

On the other hand, choosing an open neighborhood $U' \subset U \cap U_0$ of $\overline{x}$, a real $0 < \varepsilon' < \varepsilon$, and a neighborhood $V' \subset V$ of h such that $U' +]0, \varepsilon'[V' \subset U_0$, where U_0 is any neighborhood of $\overline{x}$, the right-hand inclusion in (2.6) entails that, for C and U_0 as in the assumption in (b),

$$C \cap U' +]0, \varepsilon'[V' = (C \cap U' +]0, \varepsilon'[V') \cap U_0 \subset (\text{int}\,\overline{S}) \cap U_0.$$

Then, by the assumption in (b) again we obtain $C \cap U' +]0, \varepsilon'[V' \subset C$, which justifies that C has the interior tangent cone property at $\overline{x}$.

From the equalities in (a) and the assumption of (b) it is easily seen that $\overline{S} \cap U' = \overline{C} \cap U'$, or equivalently $\overline{S \cup \{\overline{x}\}} \cap U' = \overline{C} \cap U'$. It ensues by Proposition 2.5(b) that $T^C(S \cup \{\overline{x}\}; \overline{x}) = T^C(C; \overline{x})$. The latter equality implies the desired equality $I(S \cup \{x\}; \overline{x}) = I(C; \overline{x})$ of (b) since $I(P; x) = \text{int}(T^C(P; x))$ whenever the set P has the interior tangent cone property at x (see Theorem 2.13). The proof is then finished. $\qquad \square$

LEMMA 2.28. *Let S be a nonempty set of the normed space X and let $\overline{x} \in X$. Then, denoting by $\overline{S}$ the closure of S one has*

$$\overline{\text{int}\,(S \cup \{\overline{x}\})} = \overline{\text{int}\,S}.$$

PROOF. We only need to show that $\text{int}\,(S \cup \{\overline{x}\}) \subset \overline{\text{int}\,S}$, and we suppose $X \neq \{0\}$. Fix any x in the left member that we suppose nonempty. There exists a real $r > 0$ such that $B(x, r) \subset S \cup \{\overline{x}\}$. If $x \neq \overline{x}$, choosing $0 < \rho < \min\{r, \|x - \overline{x}\|\}$ we see that $B(x, \rho) \subset S$, so $x \in \text{int}\,S$.

Now suppose that $x = \overline{x}$, so $B(\overline{x}, r) \subset S \cup \{\overline{x}\}$. Then, considering the nonempty set $U := B(\overline{x}, r) \setminus \{\overline{x}\}$, we derive $U \subset S$, thus $U \subset \text{int}\,S$. Since $\overline{x} \in \overline{U}$, it results that $\overline{x} \in \overline{\text{int}\,S}$, completing the proof. $\qquad \square$

In addition to the relevance of the interior tangent property in the calculus of the Clarke tangent cone of intersection in Theorem 2.13 above, Clarke tangent cones to a set with such a property and to its complement are linked as follows.

PROPOSITION 2.29. *Let S be a set of the normed space X, let $\overline{x} \in S \cap \text{bdry}\,S$, and let $S' := (X \setminus S) \cup \{\overline{x}\}$. The equality*

$$I(S'; \overline{x}) = -I(S; \overline{x})$$

holds, thus S has the interior tangent cone property at $\overline{x}$ if and only if S' has the interior tangent cone property at $\overline{x}$. So,

$$T^C(S'; \overline{x}) = -T^C(S; \overline{x})$$

whenever S has the interior tangent cone property at $\overline{x}$.

PROOF. Suppose that $I(S;\bar{x})$ is nonvoid and, for any fixed $h \in I(S;\bar{x})$, let us prove that $-h \in I(S';\bar{x})$. Take, from the definition of $I(S;\bar{x})$, a real $\varepsilon > 0$, a neighborhood U of $\bar{x}$, and an open neighborhood V of h such that $S \cap U +]0,\varepsilon[V \subset S$. Since the set $S \cap U +]0,\varepsilon[V$ is open, we have $S \cap U +]0,\varepsilon[V \subset \operatorname{int} S$. Choose a real $0 < \varepsilon' < \varepsilon$ and neighborhoods U' and V' of $\bar{x}$ and h such that $U' +]-\varepsilon',\varepsilon'[V' \subset U$. To see that $-h \in I(S';\bar{x})$ it suffices to show that $S' \cap U' +]0,\varepsilon'[(-V') \subset S'$. If the latter inclusion does not hold, there exist a real $0 < t < \varepsilon'$, and elements $x' \in S' \cap U'$ and $h' \in V'$ such that $u := x' - th' \in X \setminus S'$, hence $u \in S \cap U$. Writing $x' = u + th'$ we obtain $x' \in S \cap U + tV \subset \operatorname{int} S$, which is a contradiction since $x' \in S' = (X \setminus S) \cup \{\bar{x}\}$ and $\bar{x} \in \operatorname{bdry} S$. Therefore, $-h \in I(S';\bar{x})$ and the inclusion $I(S;\bar{x}) \subset -I(S';\bar{x})$ is justified.

Since $(X \setminus S') \cup \{\bar{x}\} = S$ we may permute S and S' to get $I(S';\bar{x}) \subset -I(S;\bar{x})$. The desired equality concerning the interior tangent cone is then established.

The above equality ensures that S is has the in interior tangent cone property at $\bar{x}$ if and only if S' has the interior tangent cone property at $\bar{x}$. Further, under the interior tangent cone property assumption the same equality guarantees the equality $T^C(S;\bar{x}) = -T^C(S';\bar{x})$ since the Clarke tangent cone is the closure of the interior tangent cone whenever the latter is nonempty (see Theorem 2.13). This finishes the proof. □

For a convex set with nonempty interior, it is well known that the interior of the set coincides with the interior of its closure. A first corollary of the previous proposition and Proposition 2.27 shows a similar relationship with a set with the interior tangent property.

COROLLARY 2.30. Let S be a set of the normed space X which has the interior tangent cone property at $\bar{x} \in S \cap \operatorname{bdry} S$. Then, there exists a neighborhood U of $\bar{x}$ such that, with $\overline{S} := \operatorname{cl} S$, one has

$$U \cap \operatorname{int} \overline{S} = U \cap \operatorname{int}(\overline{\operatorname{int} S}) = U \cap \operatorname{int} S.$$

PROOF. Clearly, we have $\bar{x} \in \operatorname{bdry}(\complement_X S)$. Further, from the above proposition we know that $(\complement_X S) \cup \{\bar{x}\}$ has the interior tangent property at $\bar{x}$. Then, Proposition 2.27 furnishes a neighborhood U of $\bar{x}$ such that

$$U \cap \operatorname{cl}(\operatorname{int}(\complement_X S)) = U \cap \operatorname{cl}(\complement_X S).$$

Writing $\operatorname{cl}(\complement_X S) = \complement_X \operatorname{int} S$ and

$$\operatorname{cl}(\operatorname{int}(\complement_X S)) = \operatorname{cl}(\complement_X \overline{S}) = \complement_X \operatorname{int} \overline{S},$$

we deduce the equality $U \cap \complement_X \operatorname{int} \overline{S} = U \cap \complement_X \operatorname{int} S$, from which we see (without difficulty) that $U \cap \operatorname{int} \overline{S} = U \cap \operatorname{int} S$. This combined with the easy double inclusion $\operatorname{int} S \subset \operatorname{int}(\overline{\operatorname{int} S}) \subset \operatorname{int} \overline{S}$ yields the desired double equality of the proposition. □

It is also worth stating another direct corollary which directly follows from Proposition 2.27 and Proposition 2.29.

COROLLARY 2.31. Let X be a normed space and S be a closed set in X with nonempty boundary. Assume that S is epi-Lipschitz. Then the nonempty closed set $S' := X \setminus \operatorname{int} S$ is also epi-Lipschitz and the following equalities hold:

$$S = \operatorname{cl}(\operatorname{int} S), \quad S' = \operatorname{cl}(\operatorname{int}(S')) = \operatorname{cl}(H \setminus S),$$
$$\operatorname{bdry} S = \operatorname{bdry}(\operatorname{int} S) = \operatorname{bdry}(S') = \operatorname{bdry}(\operatorname{int}(S')).$$

One also has the equalities

$$I(S'; x) = -I(S; x) \quad \text{and} \quad T^C(S'; x) - T^C(S; x)$$

for all $x \in \text{bdry } S$.

Given a set S with the interior tangent cone property at $\bar{x} \in S \cap \text{bdry } S$, we also note, by Propositions 2.29 and 2.10, that $T^C(\text{bdry } S; \bar{x}) = T^C(S; \bar{x}) \cap -T^C(S; \bar{x})$, and this entails that $T^C(S; \bar{x})$ is a vector space since $T^C(S; \bar{x})$ is a convex cone. We state those properties in the corollary:

COROLLARY 2.32. *Let S be a subset of the normed space X which satisfies the interior tangent cone property at $\bar{x} \in S \cap \text{bdry } S$. Then, the following equality holds*

$$T^C(\text{bdry } S; \bar{x}) = T^C(S; \bar{x}) \cap -T^C(S; \bar{x}),$$

which yields in particular that $T^C(\text{bdry } S; \bar{x})$ is a closed vector subspace of X.

For $h \in I(S; x)$, the definition of the interior tangent cone gives some real $t > 0$ and some neighborhood V of h such that $x + tV \subset S$ hence $x + th \in \text{int } S$, that is, $h \in t^{-1}(\text{int } S - x)$. So, we have the inclusion

$$I(S; x) \subset]0, +\infty[(\text{int } S - x).$$

With convex sets, the converse inclusion holds:

PROPOSITION 2.33. *Let S be a nonempty convex set of the normed space X. Then for any $x \in S$*

$$I(S; x) =]0, +\infty[(\text{int } S - x),$$

so the convex set S has the interior tangent cone property at x if and only if $\text{int } S \neq \emptyset$; in such a case S has the interior tangent cone property at any of its points. So, any open convex set of X has a Lipschitz boundary, that is, its closure is locally around every point of its boundary the epigraph of a Lipschitz function in the sense of Theorem 2.20.

PROOF. By the inclusion preceding the statement of the proposition it suffices to show that the right-hand member (if it is nonempty) is included in the left-hand one. Fix any $h \in]0, +\infty[(\text{int } S - x)$. This gives some real $s > 0$ such that $x + sh \in \text{int } S$, hence there is some neighborhood V' of h such that $x + sV' \subset S$. Choose a neighborhood U of zero and a neighborhood V of h such that $s^{-1}U + V \subset V'$. For any $x' \in x + U$ we have

$$x' + sV \subset x + s(s^{-1}U + V) \subset x + sV' \subset S,$$

thus the convexity of S yields $x' + tV \subset S$ for all $t \in]0, s[$ and all $x' \in (x + U) \cap S$. This means that $h \in I(S; x)$, justifying the desired inclusion. □

C-normals to sets with the interior tangent cone property enjoy a remarkable estimate.

PROPOSITION 2.34. *Let S a subset of a normed space $(X, \|\cdot\|)$ which has the interior tangent cone property at $\bar{x} \in S$. Then there exist a neighborhood U of $\bar{x}$, a real $\gamma > 0$ and a nonzero vector $\bar{h} \in X$ such that*

$$\|x^*\| \leq \gamma |\langle x^*, \bar{h}\rangle| \quad \text{for all } x \in S \cap U, x^* \in N^C(S; x).$$

PROOF. Fixing a nonzero $\overline{h} \in I(S;\overline{x})$, there exist an open neighborhood U of $\overline{x}$ and a real $r > 0$ such that $S \cap U +]0,r](\overline{h} + 2r\mathbb{B}) \subset S$. Fix any $x \in S \cap U$ and any $x^* \in N^C(S;x)$. Noting that $\overline{h} + r\mathbb{B} \subset T^C(S;x)$, it ensues that for every $b \in \mathbb{B}$ we have $\langle x^*, \overline{h} + rb \rangle \leq 0$, or equivalently $\langle x^*, b \rangle \leq -r^{-1}\langle x^*, \overline{h}\rangle$. It results that $\|x^*\| \leq r^{-1}|\langle x^*, \overline{h}\rangle|$. □

2.2. Clarke subdifferential

With the geometrical concepts of Clarke tangent and normal cones of sets at hand, the present section will develop the notion of Clarke subdifferential of extended real-valued functions defined on normed spaces.

2.2.1. C-subdifferential: definition and examples.
In $\mathbb{R}^m$ and $\mathbb{R}^m \times \mathbb{R}$ endowed with their canonical inner products, denoted as usual by $\langle \cdot, \cdot \rangle$, consider a smooth function $f : \mathbb{R}^m \to \mathbb{R}$ and its associated (closed) upper hypersurface, more usually called nowadays its associated epigraph,

$$\mathrm{Hyps}^{\leq}(f) := \{(x,r) \in \mathbb{R}^m \times \mathbb{R} : f(x) - r \leq 0\} = \{(x,r) \in \mathbb{R}^m \times \mathbb{R} : f(x) \leq r\} =: \mathrm{epi}\, f.$$

Given $x \in \mathbb{R}^m$ and denoting by $\nabla f(x)$ the gradient of f at x, the (closed) upper tangent half-space of f at $(x, f(x))$ is known to be given by the half-space

(2.7) $\quad \mathrm{Hypl}^{\leq}(\nabla f(x), -1) := \{(h,r) \in \mathbb{R}^m \times \mathbb{R} : \langle \nabla f(x), h \rangle - r \leq 0\}.$

This makes clear that the gradient $\nabla f(x)$ of f at $x \in \mathbb{R}^m$ is characterized, in a unilateral point of view, as the unique element $\zeta \in \mathbb{R}^m$ such that the vector $u := (\zeta, -1)$ is *perpendicular* to the upper tangent half-space of f at x, or equivalently, *normal* to the upper hypersurface $\mathrm{Hyps}^{\leq}(f)$ at $(x, f(x))$.

When the function f is nonsmooth at x it can happen that there are several or no elements $\zeta \in \mathbb{R}^m$ such that $(\zeta, -1)$ is normal to the upper hypersurface $\mathrm{Hyps}^{\leq}(f)$ at $(x, f(x))$. Taking into account both these latter situations and the property at the beginning in the presence of smoothness, one introduces the Clarke subdifferential through the Clarke normal cone as follows.

DEFINITION 2.35. Let U be a neighborhood of a point x in a normed space X and $f : U \to \mathbb{R} \cup \{-\infty, +\infty\}$ be an extended real-valued function. For any $x \in U$ where f is finite, the *Clarke subdifferential* or *C-subdifferential* $\partial_C f(x)$ of f at x is defined by

$$\partial_C f(x) := \{x^* \in X^* : (x^*, -1) \in N^C(\mathrm{epi}\, f; (x, f(x)))\},$$

where we recall that

$$\mathrm{epi}\, f := \{(x,r) \in X \times \mathbb{R} : x \in U \text{ and } f(x) \leq r\}$$

is the epigraph of f. We also put $\partial_C f(x) = \emptyset$ when $f(x)$ is not finite (observing in such a case that $(x, f(x)) \notin \mathrm{epi}\, f$). Any element in $\partial_C f(x)$ is called a *C-subgradient* of f at x.

Extending f to a function from the whole space X into $\mathbb{R} \cup \{-\infty, +\infty\}$ by putting $\overline{f}(u) = f(u)$ if $u \in U$ and $\overline{f}(u) = +\infty$ if $u \in X \setminus U$, we see that $\mathrm{epi}\, f = \mathrm{epi}\, \overline{f}$ hence $\partial_C f(x) = \partial_C \overline{f}(x)$. So, concerning the Clarke subdifferential of f, it will be very often assumed that the extended real-valued function f is defined on *the whole normed space X*.

EXAMPLE 2.36. (a) According to the cases of the points x_7 and x_9 in Figure 2.1, for $f := |\cdot|$, we see that
$$T^C(\mathrm{epi}\, f; (0,0)) = \{(r,s) : s \geq |r|\} = T^C(\mathrm{epi}\, (-f); (0,0)),$$
hence
$$\partial_C f(0) = \{r^* \in \mathbb{R} : r^* r - s \leq 0, \forall (r,s), |r| \leq s\} = \{r^* \in \mathbb{R} : r^* r \leq |r|, \forall r \in \mathbb{R}\},$$
which gives
$$\partial_C(|\cdot|)(0) = [-1,1] = \partial_C(-|\cdot|)(0).$$
(b) Similarly, the cases of the points x_{11} and x_{10} in Figure 2.1 furnishes with $f := \sqrt{|\cdot|}$
$$T^C(\mathrm{epi}\, f; (0,0)) = \{0\} \times [0, +\infty[\quad \text{and} \quad T^C(\mathrm{epi}\, (-f); (0,0)) = \{0\} \times \mathbb{R},$$
or equivalently $N^C(\mathrm{epi}\, f; (0,0)) = \mathbb{R} \times]-\infty, 0]$ and $N^C(\mathrm{epi}\, (-f); (0,0)) = \mathbb{R} \times \{0\}$. Consequently
$$\partial(\sqrt{|\cdot|})(0) = \mathbb{R} \quad \text{and} \quad \partial(-\sqrt{|\cdot|})(0) = \emptyset.$$
(c) Consider the functions $f, g : \mathbb{R} \to \mathbb{R} \cup \{+\infty\}$ defined by
$$f(r) := \begin{cases} \sqrt{r} & \text{if } r \geq 0 \\ +\infty & \text{if } r < 0 \end{cases} \qquad g(r) := \begin{cases} -\sqrt{r} & \text{if } r \geq 0 \\ +\infty & \text{if } r < 0. \end{cases}$$
It is easily seen that
$$T^C(\mathrm{epi}\, f; (0,0)) = \{0\} \times [0, +\infty[\quad \text{and} \quad T^C(\mathrm{epi}\, g; (0,0)) = [0, +\infty[\times \mathbb{R},$$
so $N^C(\mathrm{epi}\, f; (0,0)) = \mathbb{R} \times]-\infty, 0]$ and $N^C(\mathrm{epi}\, g; (0,0)) =]-\infty, 0] \times \{0\}$, hence
$$\partial_C f(0) = \mathbb{R} \quad \text{and} \quad \partial_C g(0) = \emptyset.$$
(d) Consider the modification of the first function in (c) given by $f : \mathbb{R} \to \mathbb{R}$ with
$$f(r) := \begin{cases} \sqrt{r} & \text{if } r \geq 0 \\ 0 & \text{if } r < 0. \end{cases}$$
Using the cases of the points x_7 and x_9 in Figure 2.1, it is clear that
$$T^C(\mathrm{epi}\, f; (0,0)) =]-\infty, 0] \times [0, +\infty[\quad \text{and} \quad T^C(\mathrm{epi}\, (-f); (0,0)) = [0,, +\infty[\times [0, +\infty[,$$
which is equivalent to
$$N^C(\mathrm{epi}\, f; (0,0)) = [0, +\infty[\times]-\infty, 0] \quad \text{and} \quad N^C(\mathrm{epi}\, (-f); (0,0)) =]-\infty, 0] \times]-\infty, 0].$$
It results that $\partial_C f(0) = [0, +\infty[$ and $\partial_C(-f)(0) =]-\infty, 0]$. □

Other examples with typical nonsmooth functions will be provided later; see in particular Examples 2.84, 2.66, 2.191, 2.200.

2.2.2. Differentiability and strict differentiability. Among various properties with which we will be concerned in the next subsection, we will study the link between the Clarke subdifferential and some usual concepts of derivatives. In the setting of infinite-dimensional spaces, three notions of derivatives are generally needed. For an open set U of the normed space X, a mapping G from U into a normed space Y is Fréchet (resp. Hadamard, resp. Gâteaux) differentiable at a point $x \in U$ provided there exists a continuous linear mapping $DG(x) : X \to Y$ such that

(2.8) $$\lim_{x' \to x} \frac{1}{\|x' - x\|} (G(x') - G(x) - DG(x)(x' - x)) = 0$$

(resp. for all $h \in X$

(2.9) $$\lim_{t \to 0, h' \to h} t^{-1}\big(G(x + th') - G(x)\big) = DG(x)(h),$$

$$\text{resp.} \quad \lim_{t \to 0} t^{-1}\big(G(x + th) - G(x)\big) = DG(x)(h) \;).$$

Because of the linearity of $DG(x)$, it is equivalent to require the two latter equalities with $t \downarrow 0$ in place of $t \to 0$. The continuous linear mapping $DG(x)$ which is unique, is called the Fréchet (resp. Hadamard, resp. Gâteaux) *derivative* (or *differential*) of G at x. When it is essential in a context to emphasize the type of derivative $DG(x)$, we will write $D_F G(x)$ (resp. $D_H G(x)$, resp. $D_G G(x)$). In some situations, it can be more convenient to use superscripts $D^F G(x)$ (resp. $D^H G(x)$, resp. $D^G G(x)$). Let $F : V \to Z$ be another mapping from an open set $V \supset G(U)$ of Y into a normed space Z. If G and F are Fréchet (resp. Hadamard) differentiable at x and $y := G(x)$ respectively, then it is well known and will be shown just after Proposition 2.38 that $F \circ G$ is Fréchet (resp. Hadamard) differentiable at x and $D(F \circ G)(x) = DF(y) \circ DG(x)$. Unfortunately, such a property fails with the Gâteaux differentiability, see (b) in Example 2.39 below.

If the stronger property

$$\lim_{x' \to x, x'' \to x} \frac{1}{\|x' - x''\|} \big(G(x') - G(x'') - DG(x)(x' - x'')\big) = 0$$

(resp. for all $h \in X$

$$\lim_{t \to 0, (x', h') \to (x, h)} t^{-1}\big(G(x' + th') - G(x')\big) = DG(x)(h),$$

$$\text{resp.} \quad \lim_{t \to 0, x' \to x} t^{-1}\big(G(x' + th) - G(x')\big) = DG(x)(h) \;)$$

holds, one says that G is *strictly Fréchet (resp. Hadamard, resp. Gâteaux) differentiable* at x; in the two latter equalities we can equivalently as above replace $t \to 0$ by $t \downarrow 0$. It is not difficult to show (see Proposition 2.37 below) that G is Fréchet (resp. strictly Fréchet) differentiable at x if and only if

$$t^{-1}\big(G(x + th) - G(x) - tDG(x)(h)\big) \to 0$$

$$\text{(resp. } t^{-1}\big(G(x' + th) - G(x') - tDG(x)(h)\big) \to 0\text{)}$$

uniformly with respect to $h \in \mathbb{B}_X$ as $t \downarrow 0$ (resp. as $t \downarrow 0$ and $x' \to x$). Clearly, when X is not the null space it is equivalent to have the uniform convergence with respect to the unit sphere of X, that is, with respect to $h \in X$ with $\|h\| = 1$.

Of course, the Fréchet (resp. strict Fréchet) differentiability implies the Hadamard (resp. strict Hadamard) differentiability, which itself entails the Gâteaux (resp. strict Gâteaux) differentiability. We will see in Example 2.77 that even in the infinite-dimensional space $\ell^1(\mathbb{N})$ the Hadamard (resp. strict Hadamard) differentiability at a point $\bar{x}$ does not entail the Fréchet (resp. strict Fréchet) differentiability at $\bar{x}$ (even for continuous convex functions).

PROPOSITION 2.37. *Denoting by $\mathcal{B}$ the class of all symmetric bounded (resp. strongly compact) sets of X, then G is Fréchet (resp. Hadamard) differentiable at a point $\bar{x} \in U$ if and only if there is some continuous linear mapping $A : X \to Y$ such that, for any $B \in \mathcal{B}$,*

$$t^{-1}\big(G(\bar{x} + th) - G(\bar{x}) - tA(h)\big) \to 0$$

uniformly with respect to $h \in B$ as $t \downarrow 0$.

The similar characterization also holds true for the strict Fréchet (resp. strict Hadamard) differentiability where one requires that for each $B \in \mathcal{B}$,
$$t^{-1}\big(G(u+th) - G(u) - tA(h)\big) \to 0$$
uniformly with respect to $h \in B$ as $t \downarrow 0$ and $u \to \bar{x}$.

PROOF. We prove the equivalence for the strict Hadamard differentiability and without loss of generality we may take $U = X$. Let G be strictly Hadamard differentiable at $\bar{x}$, and with $A := D_H G(\bar{x})$ let us define
$$\rho(t, x', h') := \|t^{-1}\big(G(x'+th') - G(x') - tAh\big)\| \text{ for all } t \in \mathbb{R}\setminus\{0\}, x', h' \in X.$$
Suppose, by contradiction, that the property of the proposition does not hold with that continuous linear mapping A. There exist a nonempty symmeric compact set B and some real $\varepsilon > 0$ such that
$$\inf_{n\in\mathbb{N}} \sup_{t\in\,]0,1/n[,\,x'\in B(\bar{x},1/n)} \sup_{h'\in B} \rho(t, x', h') > \varepsilon,$$
which, for each $n \in \mathbb{N}$, furnishes elements $t_n \in\,]0, 1/n[$, $x_n \in B(\bar{x}, 1/n)$, $h_n \in B$, such that $\rho(t_n, x_n, h_n) > \varepsilon$. By compactness of B, a (non-relabelled) subsequence of $(h_n)_n$ converges to some $h \in B$. Consequently, $\rho(t, x', h') \not\to 0$ as $t \to 0$ and $(x', h') \to (\bar{x}, h)$, contradicting the strict Hadamard differentiability of G at $\bar{x}$.

Conversely, suppose the property holds with a continuous linear mapping $A : X \to Y$ and define $\rho(t, x', h')$ as above. Fix any $h \in X$ and choose $t_n \to 0$ with $t_n > 0$, $x_n \to \bar{x}$ and $h_n \to h$ such that $\limsup_{t\downarrow 0,(x',h')\to(\bar{x},h)} \rho(t, x', h') = \lim_{n\to\infty} \rho(t_n, x_n, h_n)$. Putting $K := \{h\} \cup \{h_n : n \in \mathbb{N}\}$ and using the compactness of the symmetric set $B := K \cup (-K)$, the property yields that $\lim_{t\downarrow 0,x'\to\bar{x}} \sup_{h'\in B} \rho(t, x', h') = 0$. We then derive that $\lim_{n\to\infty} \rho(t_n, x_n, h_n) = 0$, hence $\lim_{t\downarrow 0,(x',h')\to(\bar{x},h)} \rho(t, x', h') = 0$. Similarly, taking the symmetry of B into account we obtain $\lim_{t\uparrow 0,(x',h')\to(\bar{x},h)} \rho(t, x', h') = 0$, completing the proof. □

Another useful characterization of Hadamard differentiability is the following.

PROPOSITION 2.38. *Let $\bar{x} \in U$ and let $G : U \to Y$ be defined as above. Then G is Hadamard differentiable at $\bar{x}$ if and only if there exists a continuous linear mapping $A : X \to Y$ such that, for any mapping $\varphi :\,]-r, r[\,\to U$ derivable at 0 with $\varphi(0) = \bar{x}$, one has*
$$t^{-1}\big(G(\varphi(t)) - G(\varphi(0))\big) \to A(\varphi'(0)) \quad \text{as } t \to 0,$$
(that is, $G \circ \varphi$ is derivable at 0 and $(G\circ\varphi)'(0) = A(\varphi'(0))$).

PROOF. There is no loss of generality in taking $U = X$.

Suppose that G is Hadamard differentiable at $\bar{x}$. Denote $A := DG(\bar{x})$ and take any mapping $\varphi :\,]-r, r[\,\to X$ derivable at 0 with $\varphi(0) = \bar{x}$. Setting $h := \varphi'(0)$ we have, for some $\varepsilon(t) \to 0$ as $t \to 0$, that
$$t^{-1}\big(G(\varphi(t)) - G(\varphi(0))\big) = t^{-1}\big(G(\bar{x} + t(h + \varepsilon(t))) - G(\bar{x})\big) \to A(h) \text{ as } t \to 0.$$
This means that the property in the proposition holds true.

Conversely, suppose that G is not Hadamard differentiable at $\bar{x}$ and note that we may take $\bar{x} = 0$ (without loss of generality). Consider any continuous linear mapping $A : X \to Y$. There are some $h \in X$, some $\varepsilon > 0$ and some sequences $t_n \downarrow 0$ and $h_n \to h$ such that

(2.10) $\qquad \left\| t_n^{-1}\big(G(\bar{x} + t_n h_n) - G(\bar{x}) - t_n A(h)\big) \right\| \geq \varepsilon \quad \text{for all } n \in \mathbb{N}.$

Taking a subsequence if necessary, we may and do suppose that the sequence $(t_n)_n$ is decreasing. Consider the odd function $\varphi :]-t_2, t_2[\to X$ defined by $\varphi(0) = 0$ and $\varphi(t) = th_n$ for all $t \in]t_{n+1}, t_n]$ with $n \geq 2$. We notice for any $t \in]t_{n+1}, t_n]$ that $t^{-1}(\varphi(t) - \varphi(0)) = h_n$, which easily ensures that $t^{-1}(\varphi(t) - \varphi(0)) \to h$ as $t \to 0$ (according to the oddness of φ). This means that φ is derivable at 0 with $\varphi'(0) = h$. Further, by (2.10)
$$\left\| t_n^{-1}\big(G(\varphi(t_n)) - G(\varphi(0)) - t_n A(h)\big) \right\| \geq \varepsilon \quad \text{for all } n \geq 2.$$
Consequently, the property in the proposition is not satisfied. This finishes the proof. $\square$

Let U and V be open sets of normed spaces X and Y respectively, let $G_1 : U \to Y$ be a mapping with $G_1(U) \subset V$, and let $G_2 : V \to Z$ be a mapping from V into a normed space Z. Let also $\bar{x} \in U$. If G_1 and G_2 are Fréchet differentiable at $\bar{x}$ and $\bar{y} := G_1(\bar{x})$ respectively, from the definition of Fréchet differentiability in (2.8) it is easily seen that the mapping $G_2 \circ G_1$ is Fréchet differentiable at $\bar{x}$ and
$$D_F(G_2 \circ G_1)(\bar{x}) = D_F G_2(\bar{y}) \circ D_F G_1(\bar{x}).$$
Clearly, it also follows from Proposition 2.38 that the Hadamard differentiability of G_1 and G_2 at $\bar{x}$ and $\bar{y}$ respectively ensures the Hadamard differentiability of $G_2 \circ G_1$ at $\bar{x}$ along with the chain rule
$$D_H(G_2 \circ G_1)(\bar{x}) = D_H G_2(\bar{y}) \circ D_H G_1(\bar{x}).$$

Proposition 2.37, in particular, tells us that, when X is finite-dimensional, the Hadamard (resp. strict Hadamard) differentiability coincides with the Fréchet (resp. strict Fréchet) differentiability. In contrast to that, even in $\mathbb{R}^2$ the Gâteaux differentiability does not imply the Hadamard differentiability, as shown in the following example. The function in the example also illustrates that Gâteaux differentiability does not imply continuity, and that the Gâteaux differentiability is not preserved under composition.

EXAMPLE 2.39. (a) Fixing two reals $\alpha_0 \neq \alpha_1$ and defining $f : \mathbb{R}^2 \to \mathbb{R}$ by
$$f(x, y) = \begin{cases} \alpha_0 & \text{if } x > 0 \text{ and } y = x^2, \\ \alpha_1 & \text{otherwise,} \end{cases}$$
the function f is Gâteaux differentiable at $(0,0)$ (with $D_G f(0,0) = 0$) whereas it is not Hadamard differentiable at $(0,0)$ since with $t_n := 1/n$, we have $(1, t_n) \to (1, 0)$ as $n \to \infty$ and
$$t_n^{-1}\Big(f\big((0,0) + t_n(1, t_n)\big) - f(0,0)\Big) = t_n^{-1}(\alpha_0 - \alpha_1) \not\to 0.$$
(b) We note that the function f is Gâteaux differentiable at $(0,0)$ while it is not continuous at this point.
(c) Through this function f we can also see that the Gâteaux differentiability is not preserved under composition. Indeed, define the mapping $G : \mathbb{R} \to \mathbb{R}^2$ by $G(x) = (x^2, x^4 \cos(1/x))$ if $x \neq 0$ and $G(0) = (0,0)$. Clearly, the mapping G is $\mathcal{C}^1$ on $\mathbb{R}$ whereas the function $f \circ G$ is not (Gâteaux) differentiable at 0 since $(f \circ G)(x) = \alpha_0$ if $x \in \{1/(2n\pi) : n \in \mathbb{Z} \setminus \{0\}\}$ and $(f \circ G)(x) = \alpha_1$ otherwise. $\square$

Despite the latter example of function f, the strict Gâteaux and strict Hadamard differentiabilities coincide whenever the first space is finite-dimensional.

PROPOSITION 2.40. *Let U be an open set of the normed space X and $G : U \to Y$ be a mapping from U into a normed space Y. Assume that X is finite-dimensional. Then, the mapping G is strictly Gâteaux differentiable at a point $\bar{x} \in U$ if and only if it is strictly Fréchet (or strictly Hadamard) differentiable at $\bar{x}$.*

PROOF. Without loss of generality, suppose that $U = X$ and X is not reduced to zero. Assume that G is strictly Gâteaux differentiable at $\bar{x}$. Let $e_1, \cdots, e_m$ be a basis of the vector space X and let $\|\cdot\|$ be the relative sum norm, that is, $\|v\| := |v_1| + \cdots + |v_m|$, for $v = v_1 e_1 + \cdots + v_m e_m$ with $v_1, \cdots, v_m \in \mathbb{R}$. Take any real $\varepsilon > 0$ and choose a real $\delta > 0$ such that, for all $x' \in B(\bar{x}, 2\delta)$ and all non-null $t \in\ ]-\delta, \delta[$,
$$\|t^{-1}(f(x' + te_j) - f(x') - tAe_j)\| \leq \varepsilon/m, \text{ for } j = 1, \cdots, m,$$
where A denotes the Gâteaux derivative of G at $\bar{x}$. Let $\mathbb{B}$ be the closed unit ball of X relative to $\|\cdot\|$ and let h be any non-zero element of $\mathbb{B}$. Write $h = h_{k(1)} e_{k(1)} + \cdots + h_{k(n)} e_{k(n)}$, where $h_{k(1)}, \cdots, h_{k(n)}$ are the non-null coordinates of h with $n \leq m$ and $k(1) < \cdots < k(n)$. Consider any $x' \in B(\bar{x}, \delta)$ and any non-null $t \in\ ]0, \delta[$. Put $u_1 := x'$ and $s_1 := th_{k(1)}$, and if $n > 1$, define u_i and s_i (depending on (x', t, h)), for $i = 2, \cdots, n$, by
$$u_i := x' + t(h_{k(1)} e_{k(1)} + \cdots + h_{k(i-1)} e_{k(i-1)}) \quad \text{and} \quad s_i := th_{k(i)},$$
so $u_i \in B(\bar{x}, 2\delta)$ and $s_i \in\ ]-\delta, \delta[$ with $s_i \neq 0$, for all $i = 1, \cdots, n$. Then,
$$t^{-1}(f(x' + th) - f(x') - tAh) = \sum_{i=1}^{n} h_{k(i)} s_i^{-1}(f(u_i + s_i e_{k(i)}) - f(u_i) - s_i A e_{k(i)}),$$
which ensures that $\|t^{-1}(f(x' + th) - f(x') - tAh)\| \leq \varepsilon$. Consequently, the mapping G is strictly Fréchet differentiable at $\bar{x}$. □

In the case of extended real-valued functions, we will declare $f : X \to \mathbb{R} \cup \{-\infty, +\infty\}$ to be Fréchet, Hadamard, or Gâteaux differentiable (resp. strictly Fréchet, Hadamard, or Gâteaux differentiable) at $x \in X$ whenever f is finite on an open set U containing x and the restriction of f to U (with values in $\mathbb{R}$) satisfies the corresponding differentiability at x.

When X is a Hilbert space H with the inner product $\langle \cdot, \cdot \rangle_H$, $Y = \mathbb{R}$ and G is differentiable at x in one of the above senses, the vector $\nabla G(x) \in H$ such that $DG(x)(h) = \langle \nabla G(x), h \rangle_H$ for all $h \in H$ is called the *gradient* of G at x in the considered sense.

If G is Hadamard differentiable at x, then we have $\lim\limits_{t \to 0, h' \to 0_X} t^{-1}(G(x + th') - G(x)) = 0$ and from the latter equality we easily see that G is continuous at x. In contrast to that, the aforementioned function $f : \mathbb{R}^2 \to \mathbb{R}$ in Example 2.39 (defined with α_0 and α_1) is not continuous at the point $(0,0)$ whereas it is Gâteaux differentiable at that point.

If instead G is strictly Hadamard differentiable at x, then (as well-known and proved below) G is Lipschitz continuous near x.

PROPOSITION 2.41. *Let U be an open set of the normed space X and $g : U \to Y$ be a mapping from U into a normed space Y. If G is strictly Hadamard differentiable at $\bar{x} \in U$, then it is Lipschitz continuous near $\bar{x}$.*

PROOF. From the definition of the strict Hadamard differentiability of G at $\bar{x}$ we see that $\lim_{t\to 0,(x',h')\to(\bar{x},0_X)} t^{-1}\bigl(G(x'+th') - G(x') - tA(h')\bigr) = 0$, where $A := DG(\bar{x})$. Then there exists some positive real $r \le 1$ such that, for all $t \in]0,r[$, $x' \in B(\bar{x},r)$ and $h' \in r\mathbb{B}_X$, we have $x', x'+th' \in U$ and
$$\|t^{-1}\bigl(G(x'+th') - G(x')\bigr) - A(h')\| \le 1, \text{ so } \|G(x'+th') - G(x')\| \le t(1+r\|A\|).$$
Fix some positive real $\alpha < r^2/2$. Take two different elements $x', x'' \in B(\bar{x},\alpha)$ and consider
$$t := r^{-1}\|x' - x''\| \quad \text{and} \quad h' := r(x''-x')/\|x'-x''\|.$$
Since $t \le 2r^{-1}\alpha < r$, we have
$$\|G(x'') - G(x')\| = \|G(x'+th') - G(x')\|$$
$$\le t(1+r\|A\|) = r^{-1}(1+r\|A\|)\,\|x''-x'\|,$$
which translates the Lipschitz continuity of G near $\bar{x}$. □

When G is γ-Lipschitz continuous near $\bar{x}$, since
$$\|t^{-1}\bigl(G(x'+th') - G(x')\bigr) - t^{-1}\bigl(G(x'+th) - G(x')\bigr)\| \le \gamma\|h'-h\|$$
for t small enough, x' near $\bar{x}$, and h' near h, we see that the Gâteaux (resp. strict Gâteaux) differentiability of G at $\bar{x}$ is equivalent to the Hadamard (resp. strict Hadamard) differentiability of G at $\bar{x}$. We formulate this equivalence in the following proposition.

PROPOSITION 2.42. *Let U be an open set of the normed space X and $G : U \to Y$ be a mapping from U into a normed space Y which is Lipschitz continuous near a point $\bar{x} \in U$. Then G is Gâteaux (resp. strictly Gâteaux) differentiable at $\bar{x}$ if and only if it is Hadamard (resp. strictly Hadamard) differentiable at $\bar{x}$.*

It is also not difficult to see that G is strictly Fréchet differentiable at $\bar{x}$ whenever G is Gâteaux differentiable near $\bar{x}$ and DG is continuous at $\bar{x}$ (see Proposition 2.44 below), where the space of continuous linear mappings $L(X,Y)$ is endowed with the usual norm. This ensures in particular that G is Gâteaux differentiable on an open set U with its Gâteaux derivative continuous on U (with $L(X,Y)$ endowed with the usual norm) if and only if G is Fréchet (resp. Hadamard) differentiable on U with $D_F G$ (resp. $D_H G$) continuous on U; in such a case one says that G is *of class $\mathcal{C}^1$* on U.

The next proposition is concerned with the continuity of strict Fréchet derivative on open sets. A much more general result is even established, and as stated it will be used later in the text. In the proposition and the one after, given a class $\mathcal{B}$ of nonempty bounded subsets of X, on the space $L(X,Y)$ we denote by $\tau_\mathcal{B}$ the topology of uniform convergence on sets in $\mathcal{B}$, that is, the topology associated with the family of seminorms $(p_B)_{B\in\mathcal{B}}$ defined, for each $B \in \mathcal{B}$, by
$$p_B(A) := \sup_{u \in B} \|A(u)\| \quad \text{for all } A \in L(X,Y).$$
Of course, $\tau_\mathcal{B}$ coincides with the usual norm topology of $L(X,Y)$ when $\mathcal{B}$ is the class of all nonempty symmetric bounded subsets of X.

PROPOSITION 2.43. *Let U be an open set of the normed space X and $G : U \to Y$ be a mapping from U into a normed space Y which is strictly Fréchet (resp. strictly Hadamard, resp. strictly Gâteaux) differentiable at $\bar{x} \in U$ and Fréchet* (resp. Hadamard, resp. Gâteaux) differentiable at each point of a subset S of U containing $\bar{x}$. Then the Fréchet (resp. Hadamard, resp. Gâteaux) derivative mapping DG (defined on S) is $\tau_\mathcal{B}$-continuous at $\bar{x}$ relative to S, where $\mathcal{B}$ is the class of nonempty symmetric bounded (resp. compact, resp. finite) subsets of X; otherwise stated, for each $B \in \mathcal{B}$

$$\lim_{S \ni x \to \bar{x}} \left(\sup_{h \in B} \|(DG(x) - DG(\bar{x}))(h)\| \right) = 0.$$

The Fréchet case amounts to saying that the strict Fréchet differentiability of G at $\bar{x}$ and its Fréchet differentiability at each point in S entails that DG is continuous at $\bar{x}$ relative to S with $L(X,Y)$ endowed with the usual norm.

PROOF. Fix any $B \in \mathcal{B}$ and consider any real $\varepsilon > 0$. By the strict differentiability choose a real $\delta > 0$ such that $B(\bar{x}, \delta) \subset U$ and for all $h \in B$, $u \in B(\bar{x}, \delta)$ and $t \in]0, \delta[$

$$\|t^{-1}(G(u + th) - G(u) - tDG(\bar{x})(h))\| \leq \varepsilon/2.$$

Take any $x \in S \cap B(\bar{x}, \delta)$ and choose, by the suitable differentiability of G at x, a real $\delta' \in]0, \delta[$ such that for all $h \in B$ and $t \in]0, \delta']$

$$\|t^{-1}(G(x + th) - G(x) - tDG(x)(h))\| \leq \varepsilon/2.$$

Then, making $u = x$ in the first inequality and $t = \delta'$ in both inequalities, we obtain $\|(DG(x) - DG(\bar{x}))(h)\| \leq \varepsilon$ for all $h \in B$. As desired, it ensues that

$$\sup_{h \in B} \|(DG(x) - DG(\bar{x}))(h)\| \leq \varepsilon \quad \text{for all } x \in S \cap B(\bar{x}, \delta).$$

The conclusion in the Fréchet case is due to the coincidence between the usual norm topology of $L(X,Y)$ and $\tau_\mathcal{B}$ when $\mathcal{B}$ is the class of nonempty symmetric bounded sets. □

The converse of the previous proposition with $S = U$ is a particular case of the following proposition.

PROPOSITION 2.44. Let U be an open set of the normed space X and $G : U \to Y$ be a mapping from U into a normed space Y which is Gâteaux differentiable at each point of U. Then G is strictly Fréchet (resp. strictly Hadamard, resp. strictly Gâteaux) differentiable at $\bar{x} \in U$ whenever $D_G G$ is $\tau_\mathcal{B}$-continuous at $\bar{x}$, where $\mathcal{B}$ is the class of nonempty symmetric bounded (resp. compact, resp. finite) subsets of X and $\tau_\mathcal{B}$ is as above.

PROOF. Let any $B \in \mathcal{B}$ and fix a real $r > \sup\{\|h\| : h \in B\}$. Let $\varepsilon > 0$ and choose a real $\delta > 0$ such that $B(\bar{x}, 2\delta) \subset U$ and

$$\sup_{h \in B} \|(DG(x) - DG(\bar{x}))(h)\| \leq \varepsilon \quad \text{for all } x \in B(\bar{x}, 2\delta),$$

where $DG(x) := D_G G(x)$. Consider a positive real $\delta' < \min\{\delta, \delta/r\}$. Fix any $h \in B$, any $x \in B(\bar{x}, \delta')$ and any $t \in]0, \delta'[$. We have

$$\|t^{-1}(G(x + th) - G(x) - tDG(\bar{x})(h))\| = \left\| \int_0^1 (DG(x + \theta th) - DG(\bar{x}))(h) \, d\theta \right\| \leq \varepsilon,$$

hence G satisfies the desired strict differentiability property at $\bar{x}$. □

As a direct consequence of both previous propositions we have:

COROLLARY 2.45. Let U be an open set of the normed space X and $G : U \to Y$ be a mapping from U into a normed space Y which is Fréchet (resp. Hadamard, resp. Gâteaux) differentiable on U. Then G is strictly Fréchet (resp. strictly Hadamard, resp. strictly Gâteaux) differentiable on U if and only if DG is $\tau_{\mathcal{B}}$-continuous on U, where $DG(x)$ is the corresponding derivative at x, $\mathcal{B}$ is the class of nonempty symmetric bounded (resp. compact, resp. finite) subsets of X and $\tau_{\mathcal{B}}$ is as above.

Proposition 2.43 ensures in particular that the Fréchet differentiability does not imply the strict Fréchet differentiability. A classical and simple function f illustrating that point is given by $f : \mathbb{R} \to \mathbb{R}$ with $f(x) := x^2 \sin(1/x)$ if $x \neq 0$ and $f(0) = 0$. This function f is differentiable at 0 but not strictly differentiable at 0 as easily seen through the choice of $x_n := (\pi(n + 1/2))^{-1}$ and $y_n := x_{n+1}$. Further, as shown in the examples below, there are mappings which are strictly Fréchet differentiable at $\bar{x}$ and not of class $\mathcal{C}^1$ near $\bar{x}$, even differentiable over no neighborhood of $\bar{x}$.

EXAMPLE 2.46. (a) Consider the continuous convex function $f : \mathbb{R} \to \mathbb{R}$ defined by $f(x) = 0$ if $x \leq 0$, $f(1/2^n) = 1/2^{2n}$ for all $n \in \mathbb{N}$, and f defined linearly otherwise. The function f is easily seen to be differentiable at 0 with $Df(0) = 0$. To see that it is strictly differentiable at 0, take any x, y with $0 < x < y < 1/2$. Choose some $k \in \mathbb{N}$ such that $1/2^{k+1} < y \leq 1/2^k$. By the convexity of f and its affinity on $[1/2^{k+1}, 1/2^k]$ we have

$$0 \leq \frac{f(y) - f(x)}{y - x} \leq \frac{f(y) - f(1/2^{k+1})}{y - 1/2^{k+1}} = 3\frac{1}{2^{k+1}} \leq 3y,$$

from which it is not difficult to deduce the strict differentiability of f at 0 (in fact, even in the general context of normed spaces we will see later that a real-valued convex function is Fréchet differentiable at a point if and only if it is strictly Fréchet differentiable at that point). Nevertheless, there is no neighborhood of 0 over which f is differentiable since it is clearly not differentiable at any point $1/2^n$.
(b) Let S be a measurable subset of $[0, 1]$ such that $\lambda(I \cap S) > 0$ as well as $\lambda(I \cap \complement_{[0,1]}S) > 0$ for every nonempty open interval I in $[0, 1]$, where λ denotes the Lebesgue measure. Consider the function $\varphi := \mathbf{1}_S$, that is, $\varphi(x) = 1$ if $x \in S$ and $\varphi(x) = 0$ otherwise. Putting $f(x) := \int_0^x \varphi(t)\,dt$ for all $x \in \mathbb{R}$, the function f is Lipschitz continuous (with $\gamma = 1$ as Lipschitz constant) and hence there is some measurable subset $A \subset]0, 1[$ with $\lambda(A) = 1$ such that f is differentiable on A with $f'(x) = \varphi(x)$ for all $x \in A$. Then, $f'(x) = 1$ if $x \in A_1 := A \cap S$ and $f'(x) = 0$ if $x \in A_0 := A \setminus S$. Noting that $\lambda(I \cap A_0) > 0$ and $\lambda(I \cap A_1) > 0$ for all nonempty open intervals I of $]0, 1[$, we see that both sets A_0 and A_1 are dense in $]0, 1[$. It follows that f is strictly differentiable at no point in $]0, 1[$ according to Proposition 2.43 above. $\square$

Unlike the above second example the next theorem provides a smallness property (in Baire category sense) of the size of the set of differentiability but non strict differentiability points of mappings between normed spaces. Recall that a subset P of a topological space T is of *first category* provided it can be written as the countable union of nowhere dense sets, that is, $P = \bigcup_{n \in \mathbb{N}} P_n$ with $\operatorname{int}(\operatorname{cl} P_n) = \emptyset$.

THEOREM 2.47 (L. Zajíček: generic strict F-differentiability of F-differentiable mappings). Let U be an open set of the normed space X and $G : U \to Y$ be a mapping from U into a normed space Y. Then the set of points in U at which G is Fréchet differentiable but not strictly Fréchet differentiable is of first category in U.

Before proving the theorem, let us establish the following lemma.

LEMMA 2.48. *Let $G : X \to Y$ be a mapping from the normed space X into a normed space Y. Let $A : X \to Y$ be a linear mapping, $z \in X$, $\varepsilon > 0$ and $\delta > 0$ such that*
$$\|G(z+h) - G(z) - A(h)\| < \varepsilon \|h\|$$
for all non-zero $h \in B(0,\delta)$. Then, for $x, y \in X$ with $x \neq y$, the inequalities $\|x - z\| < \delta$, $\|y - z\| < \delta$ and $\|x - y\| \geq \|x - z\|$ imply that
$$\|G(y) - G(x) - A(y - x)\| < 3\varepsilon\|y - x\|.$$

PROOF. Let $x, y \in X$ as in the statement. Since $\|G(u) - G(z) - A(u - z)\| \leq \varepsilon\|u - z\|$ with the points $u = x$ and $u = y$ and since the inequality is strict with at least one of these points, we have
$$\|G(y) - G(x) - A(y - x)\| < \varepsilon(\|y - z\| + \|x - z\|)$$
$$\leq \varepsilon(\|y - x\| + 2\|x - z\|) \leq 3\varepsilon\|y - x\|,$$
which is the conclusion of the lemma. □

PROOF OF THEOREM 2.47. We may suppose that $U = X$. Denote by S the set of points in X where G is Fréchet differentiable and not strictly Fréchet differentiable. For each pair of integers $n, m \in \mathbb{N}$ denote by $S_{n,m}$ the set of points $w \in S$ such that the following properties (2.11) and (2.12) are satisfied:

(2.11) $\|G(w + h) - G(w) - DG(w)(h)\| < \|h\|/m$ for all non-zero $h \in (1/n)\mathbb{B}_X$,

and for each $\delta > 0$ there exist $x, y \in B(w, \delta)$ such that

(2.12) $$\|G(y) - G(x) - DG(w)(y - x)\| > (8/m)\|y - x\|.$$

Since $S = \bigcup_{n,m\in\mathbb{N}} S_{n,m}$, it is enough to show that each set $S_{n,m}$ is nowhere dense in the space X. Suppose on the contrary that for some fixed pair n, m there are some $u \in S_{n,m}$ and some real $\eta > 0$ satisfying $B(u, \eta) \subset \operatorname{cl} S_{n,m}$. Putting $\delta := \min\{\eta/4, 1/(8n)\}$ we can choose by (2.12) two points $x, y \in B(u, \delta)$ such that

(2.13) $$\|G(y) - G(x) - DG(u)(y - x)\| > (8/m)\|y - x\|.$$

From the inclusions $x, y \in B(u, \eta/4)$ we have $B(x, \|y - x\|) \subset B(u, \eta)$, and this allows us to take some $\bar{u} \in S_{n,m} \cap B(x, \|y - x\|)$. Since $\|y - x\| < 1/(4n)$, we also have $\|x - \bar{u}\| < 1/(4n)$, and hence $\|y - \bar{u}\| < 1/(2n)$. The two latter inequalities combined with the inequality $\|x - y\| \geq \|x - \bar{u}\|$ tell us that we can apply the above lemma with $z = \bar{u}$, $\varepsilon = 1/m$, $\delta = 1/n$ and $A := DG(\bar{u})$ to obtain

(2.14) $$\|G(y) - G(x) - DG(\bar{u})(y - x)\| < (3/m)\|y - x\|.$$

Setting $h := (y - x)/(2n\|y - x\|)$ and $v := u + h$ we get from (2.11)

(2.15) $$\|G(v) - G(u) - DG(u)(v - u)\| < (1/m)\|v - u\|.$$

On the other hand
$$\|\bar{u} - u\| \leq \|\bar{u} - x\| + \|x - u\| < (1/4n) + 1/(8n) = 3/(8n)$$

and
$$\|\bar{u} - v\| \le \|\bar{u} - u\| + \|u - v\| < 3/(8n) + 1/(2n) < 1/n.$$
Further, $\|u-v\| = 1/(2n) > 3/(8n) > \|\bar{u}-u\|$, thus keeping in mind that $\bar{u} \in S_{n,m}$ we can apply again the lemma above with $z = \bar{u}$, $\varepsilon = 1/m$, $\delta := 1/n$, $A := DG(\bar{u})$, and this gives
$$\|G(v) - G(u) - DG(\bar{u})(v - u)\| < (3/m)\|v - u\|.$$
Combining this with (2.15) yields
$$\|DG(u)(v - u) - DG(\bar{u})(v - u)\| < (4/m)\|v - u\|,$$
or equivalently with $w := (y - x)/\|y - x\|$
(2.16) $$\|DG(u)(w) - DG(\bar{u})(w)\| < (4/m).$$
From (2.13) and (2.14) we also see that
$$\|DG(u)(y - x) - DG(\bar{u})(y - x)\| > (5/m)\|y - x\|,$$
otherwise stated
$$\|DG(u)(w) - DG(\bar{u})(w)\| > (5/m),$$
and this contradicts (2.16) and finishes the proof. □

2.2.3. C-subdifferential, minimizers and derivatives. In addition to the relationship analyzed below between derivatives and the Clarke subdifferential, another important point is how the Clarke subdifferential allows us to extend the classical Fermat necessary optimality condition for local minimizers of functions which are not differentiable. The following definition recalls the concepts of minimizers and local minimizers.

DEFINITION 2.49. Given a subset S of a topological space X, a point $x \in S$ is a *minimizer* (resp. *maximizer*) of a function $f : X \to \mathbb{R} \cup \{-\infty, +\infty\}$ over S provided $f(x) \le f(x')$ (resp. $f(x) \ge f(x')$) for all $x' \in S$. When S is the whole space X, one says that x is a *global minimizer* (resp. *maximizer*) or just a minimizer (resp. maximizer) of f; if instead S is a neighborhood of x, one says that x is a *local minimizer* (resp. *maximizer*) of f. When the above inequality is fulfilled on $S \cap V$ for some neighborhood V of x, one says that x is a *local minimizer* (resp. *local maximizer*) of f over S.

If $f(x) < f(x')$ (resp. $f(x) > f(x')$) for all $x' \in S$ with $x' \ne x$, the point x is said to be a *strict minimizer* (resp. *strict maximizer*) of f over S. Similarly as above, one defines the concepts of global and local strict minimizer (resp. maximizer).

With the definitions of various derivatives and of local minimizers at hand, we can now establish in the next proposition (as said above) some properties of the Clarke subdifferential at differentiability points as well as at local minimizer points. Other fundamental first properties are studied as well.

PROPOSITION 2.50. Let U be an open set of the normed space X and $f, g : U \to \mathbb{R} \cup \{-\infty, +\infty\}$ be two functions, and let any point $x \in U$ where both functions are finite.
(a) If f and g coincide near x, then $\partial_C f(x) = \partial_C g(x)$.
(b) The set $\partial_C f(x)$ is weak* closed and convex (maybe empty) in X^*.
(c) If x is a local minimizer of f with $|f(x)| < +\infty$, then $0 \in \partial_C f(x)$.
(d) If f is finite near x and Hadamard differentiable at x, then $Df(x) \in \partial_C f(x)$.

(e) If f is finite near x and strictly Hadamard differentiable at x, then one has the equality $\partial_C f(x) = \{Df(x)\}$.

(f) For any real $r > 0$ the equality $\partial_C(rf)(x) = r\partial_C f(x)$ holds.

PROOF. If f and g coincide on a neighborhood $V \subset U$ of x, then $\operatorname{epi} f \cap (V \times \mathbb{R}) = \operatorname{epi} g \cap (V \times \mathbb{R})$ and $f(x) = g(x)$. So, $N^C(\operatorname{epi} f; (x, f(x))) = N^C(\operatorname{epi} g; (x, g(x)))$ according to Proposition 2.5(b), hence the equality $\partial_C f(x) = \partial_C g(x)$ follows.

The assertion in (b) is a direct consequence of the weak* closedness and convexity of the Clarke normal cone.

Suppose now that x is a local minimizer of f with $|f(x)| < +\infty$, that is, there exists some neighborhood $V \subset U$ of x such that $f(u) \geq f(x)$ for all $u \in V$. Take any $(h, r) \in T^C(\operatorname{epi} f; (x, f(x)))$. For $t_n := 1/n$ there exist $(h_n, r_n)_n$ in $X \times \mathbb{R}$ converging to (h, r) such that $(x, f(x)) + t_n(h_n, r_n) \in \operatorname{epi} f$ and $x + t_n h_n \in V$ for n large enough, which yields

$$f(x) \leq f(x + t_n h_n) \leq f(x) + t_n r_n \quad \text{and hence } r_n \geq 0.$$

Consequently $\langle (0, -1), (h, r) \rangle = -r \leq 0$, which entails that

$$(0, -1) \in N^C(\operatorname{epi} f; (x, f(x))), \text{ that is, } 0 \in \partial_C f(x).$$

This establishes the property (c).

Let us prove (d). Suppose that f is Hadamard differentiable at x and take as above any $(h, r) \in T^C(\operatorname{epi} f; (x, f(x)))$ and, for $t_n = 1/n$, take $(h_n, r_n)_n$ in $X \times \mathbb{R}$ converging to (h, r) such that $(x + t_n h_n, f(x) + t_n r_n) \in \operatorname{epi} f$. Then, for n large enough there is some $\theta_n \to 0$ in $\mathbb{R}$ such that

$$f(x) + t_n Df(x)(h_n) + t_n \theta_n = f(x + t_n h_n) \leq f(x) + t_n r_n,$$

that is, $\langle Df(x), h_n \rangle - r_n \leq -\theta_n$. Thus, $\langle Df(x), h \rangle - r \leq 0$, which yields

$$(Df(x), -1) \in N^C(\operatorname{epi} f; (x, f(x))), \text{ that is, } Df(x) \in \partial_C f(x).$$

Concerning (e) fix any $x^* \in \partial_C f(x)$. Consider any $h \in X$ and take any sequences $(t_n)_n$ tending to 0 with $t_n > 0$ and $(x_n, r_n)_n$ in $X \times \mathbb{R}$ converging to $(x, f(x))$ with $x_n \in U$ and $r_n \geq f(x_n)$. The strict Hadamard differentiability of f at x ensures on the one hand that $s_n := t_n^{-1}(f(x_n + t_n h) - f(x_n))$ is finite for n large enough, and on the other hand that $s_n \to Df(x)(h)$. Since $(x_n, r_n) + t_n(h, s_n) \in \operatorname{epi} f$ for large n, we deduce that $(h, Df(x)(h)) \in T^C(\operatorname{epi} f; (x, f(x)))$. Consequently, it follows from the definition of $\partial_C f(x)$ that $x^*(h) - Df(x)(h) \leq 0$. So, $x^* = Df(x)$ and this combined with (d) translates the desired equality of (e).

Finally, denoting by $T(f; x)$ and $N(f; x)$ the Clarke tangent and normal cone of $\operatorname{epi} f$ at $(x, f(x))$ and fixing $r > 0$, one can check that $(h, \rho) \in T(f; x) \Leftrightarrow (h, r\rho) \in T(rf; x)$, hence

$$(x^*, -s) \in N(f; x) \Leftrightarrow (x^*, -s/r) \in N(rf; x).$$

The latter equivalence easily yields the desired equality in (f). □

In addition to the Fermat-type property (c) of the above proposition, it is worth observing that the set of strict local minimizers is of countable cardinality in many situations, especially in the case of a topological space with a countable basis of open sets $(U_n)_{n \in \mathbb{N}}$ (that is, each open set U is the union of a subfamily of the open sets $(U_n)_{n \in \mathbb{N}}$).

PROPOSITION 2.51. Let X be a topological space with a countable basis of open sets and $f : X \to \mathbb{R} \cup \{-\infty, +\infty\}$ be an extended real-valued function. Then the set of strict local minimizers (resp. maximizers) of f is countable.

PROOF. It suffices to consider the case of strict local minimizers. Fix a countable basis $(U_n)_{n \in \mathbb{N}}$ of open sets. For each $n \in \mathbb{N}$, denote by S_n the set of points $x \in U_n$ such that $f(x) < f(u)$ for all $u \in U_n$ with $u \neq x$. Obviously, each set S_n contains at most one point, hence the set of strict local minimizers of f is countable since it coincides with $\bigcup_{n \in \mathbb{N}} S_n$. □

The Clarke subdifferential of a function has been defined above through the Clarke normal cone of its epigraph. Given a set S of the normed space X and $x \in S$, there is a particular function whose Clarke subdifferential at x coincides with the Clarke normal cone of S at x. Recalling the definition of the *indicator function of S* as the extended real-valued function Ψ_S (see the presentation preceding Section 1.1 of Chapter 1) on X given by $\Psi_S(u) = 0$ if $u \in S$ and $\Psi_S(u) = +\infty$ otherwise, we have epi $\Psi_S = S \times [0, +\infty[$ hence the property of the Clarke normal to a cartesian product in (c) of Proposition 2.2 gives

$$N^C(\text{epi } \Psi_S; (x, 0)) = N^C(S; x) \times]-\infty, 0].$$

So, $(x^*, -1) \in N^C(\text{epi } \Psi_S; (x, 0))$ if and only if $x^* \in N^C(S; x)$, which means that $N^C(S; x) = \partial_C \Psi_S(x)$. We state this property in the proposition:

PROPOSITION 2.52. For any set S of the normed space X and any point $x \in S$ the equality

$$N^C(S; x) = \partial_C \Psi_S(x)$$

holds.

2.2.4. C-subdifferential and convex functions. Let us consider the case of convex functions. Real-valued convex functions are very classical. The concept has been adapted beyond, with functions which may take infinite values, in Definition 1.47.

Let us state some basic convex functions which will be utilized in the book.

EXAMPLE 2.53. (a) Positively linear combinations of convex functions (that is, $\alpha_1 f_1 + \cdots + \alpha_m f_m$ with f_i convex and $\alpha_i > 0$) are convex functions.
(b) If $f : Y \to \mathbb{R} \cup \{-\infty, +\infty\}$ is convex on a vector space Y, then for any linear mapping $A : X \to Y$ from a vector space X into Y and any element $a \in X$, the function $x \mapsto f(a + Ax)$ is convex on X.
(c) Any norm/seminorm on a vector space X is convex on X, and also positively homogeneous.
(d) Let X be a vector space and $b : X \times X \to \mathbb{R}$ be a bilinear function which is symmetric (that is $b(u, v) = b(v, u)$) and positive semidefinite in the sense $b(v, v) \geq 0$ for all $v \in X$. Let $q : X \to \mathbb{R}$ be defined by $q(x) := b(x, x)$ for all $x \in X$. For all $x, y \in X$ and $t \in [0, 1]$ one has the equality

(2.17) $\qquad q(tx + (1-t)y) = tq(x) + (1-t)q(y) - t(1-t)q(x-y),$

hence in particular q is convex.

Indeed, let $x, y \in X$ and $t \in [0, 1]$. We have
$$tq(x) + (1-t)q(y) - q(tx + (1-t)y)$$
$$= tq(x) + (1-t)q(y) - t^2 q(x) - (1-t)^2 q(y) - 2t(1-t)b(x,y)$$
$$= t(1-t)\big(q(x) + q(y) - 2b(x,y)\big) = t(1-t)q(x-y),$$
which confirms the desired equality.

(e) Let H be a Hilbert space with inner product $\langle \cdot, \cdot \rangle$ and $A : H \to H$ be a continuous linear mapping which is symmetric (that is, $\langle u, Av \rangle = \langle Au, v \rangle$) and positive semidefinite (that is, $\langle v, Av \rangle \geq 0$). It results from (d) that f, defined by
$$f(x) = \langle x, Ax \rangle + \langle b, x \rangle + c$$
with $b \in H$ and $c \in \mathbb{R}$, is a convex function.

(f) Given a convex set C of a vector space X, a convex function $f : C \to \mathbb{R}$ and a nondecreasing convex function $\varphi : I \to \mathbb{R}$ on an interval I in $\mathbb{R}$ containing $f(C)$, the function $\varphi \circ f$ is convex on C.

Indeed, take $x, y \in C$ and $t \in {]0, 1[}$. One has
$$f(tx + (1-t)y) \leq tf(x) + (1-t)f(y)$$
along with $f(x)$, $f(y)$, $tf(x) + (1-t)f(y)$, $f(tx + (1-t)y)$ all in I, hence by the nondecreasing property of φ and then by its convexity one obtains
$$\varphi\big(f(tx + (1-t)y)\big) \leq \varphi\big(tf(x) + (1-t)f(y)\big) \leq t\varphi\big(f(x)\big) + (1-t)\varphi\big(f(y)\big).$$
This translates the convexity of $\varphi \circ f$.

(g) The convexity of φ is not necessary for the convexity of $\varphi \circ f$. Indeed, the function $f = \exp$ is convex on $\mathbb{R}$ and the function $\log \circ \exp$ is convex on $\mathbb{R}$ although the function $\log$ is not convex on $]0, +\infty[$. □

The distance function from a convex set in a normed space is another fundamental example of convex functions which will be involved in several places in the book. The following result presents the convexity of such functions as a particular case of a much larger useful class of functions.

PROPOSITION 2.54. *Let X, Y be two vector spaces, C and C' be nonempty convex subsets of X and Y respectively, and $f : C \times C' \to \mathbb{R} \cup \{-\infty, +\infty\}$ be a convex function. Then, the function $x \mapsto \inf_{y \in C'} f(x, y)$ is convex on C.*

In particular, the distance function d_S from any convex set S of a normed space is convex on the space.

PROOF. Denote by φ the infimum function and fix any pair $(x_i, r_i) \in C \times \mathbb{R}$ with $r_i > \varphi(x_i)$, $i = 1, 2$. Fix also $0 < t_i < 1$ with $t_1 + t_2 = 1$. By definition of infimum, there is $y_i \in C'$ with $f(x_i, y_i) < r_i$. Thus, with $x' := t_1 x_1 + t_2 x_2 \in C$ and $y' := t_1 y_1 + t_2 y_2 \in C'$ we see that
$$\varphi(x') \leq f(x', y') < t_1 r_1 + t_2 r_2,$$
which translates the convexity of φ on C.

In the case of d_S, with S a convex set in the normed space X, it suffices to take $Y := X$, $C := X$, $C' := S$ and $f(x, y) := \|x - y\|$. □

Convex functions taking the value $-\infty$ possess some pathological properties. Recall first the concept of core of a subset of a vector space.

DEFINITION 2.55. Given a subset S of a vector space X, the *core* of S is the set
$$\operatorname{core} S := \{x \in X : \forall y \in X, \exists r > 0,\, x + ty \in S, \forall t \in [-r, r]\}.$$
Considering both vectors y and $-y$, it is clear that
$$\operatorname{core} S := \{x \in X : \forall y \in X, \exists r > 0,\, x + ty \in S, \forall t \in [0, r]\}.$$
Obviously, $\operatorname{core} S \subset S$, and $\operatorname{int} S \subset \operatorname{core} S$ in normed (or topological) vector spaces. The following example shows that the latter inclusion fails to be an equality.

EXAMPLE 2.56. Consider the set $S = C_1 \cup C_2 \cup D$ in $\mathbb{R}^2$ where D is the line $D = \{(x,y) \in \mathbb{R}^2 : x = 0\}$ and C_1 and C_2 are the two disks with radius 1 centered at $(-1, 0)$ and $(1, 0)$, that is,
$$C_1 = \{(x,y) \in \mathbb{R}^2 : (x+1)^2 + y^2 \le 1\} \quad \text{and} \quad C_2 = \{(x,y) \in \mathbb{R}^2 : (x-1)^2 + y^2 \le 1\}.$$
We have $(0,0) \in \operatorname{core} S$ whereas $(0,0) \notin \operatorname{int} S$. □

PROPOSITION 2.57. Let $f : X \to \mathbb{R} \cup \{-\infty, +\infty\}$ be an extended real-valued convex function which takes the extended value $-\infty$ at least at one point in X. If $\operatorname{core}(\operatorname{dom} f) \ne \emptyset$, then
$$f(x) = -\infty \quad \text{for every } x \in \operatorname{core}(\operatorname{dom} f).$$

PROOF. Choose $\bar{x} \in X$ with $f(\bar{x}) = -\infty$ and take any $x \in \operatorname{core}(\operatorname{dom} f)$. By Definition 2.55 above there exits some real $\tau > 1$ such that $u := \tau x + (1-\tau)\bar{x} \in \operatorname{dom} f$. Since $x = tu + (1-t)\bar{x}$ with $t = 1/\tau$ in $]0, 1[$, fixing a real $r \ge f(u)$ we have
$$f(x) \le tr + (1-t)s \quad \text{for every } s \in \mathbb{R},$$
hence it results that $f(x) = -\infty$. □

REMARK 2.58. More generally, the same arguments show that if the convex function f takes the value $-\infty$ at some point and the relative core $\operatorname{rcore}(\operatorname{dom} f) \ne \emptyset$, then
$$f(x) = -\infty \quad \text{for every } x \in \operatorname{rcore}(\operatorname{dom} f).$$
Given a convex set S of a vector space X, its *relative core* is defined by
$$\operatorname{rcore} S := \{x \in X : \forall y \in E, \exists r > 0,\, x + ty \in S, \forall t \in [-r, r]\},$$
where E denotes the vector subspace spanned by $S - S := \{u - v : u, v \in S\}$. □

Before giving a first description of the Clarke subdifferential of a convex function, we need the following lemma which has its own interest.

LEMMA 2.59 (slope inequality for convex functions). Let I be an interval of $\mathbb{R}$ and $\varphi : I \to \mathbb{R} \cup \{-\infty, +\infty\}$ be a convex function and let three reals $r_1 < r_2 < r_3$ in I. Then
$$\frac{\varphi(r_2) - \varphi(r_1)}{r_2 - r_1} \le \frac{\varphi(r_3) - \varphi(r_1)}{r_3 - r_1} \le \frac{\varphi(r_3) - \varphi(r_2)}{r_3 - r_2},$$
where in each one of the three inequalities we assume that $\varphi(r_i)$ is finite when it appears in both members.

Further, the inequality between the first member and the second (resp. the first and the third) characterizes the convexity of the function φ.

PROOF. Let us verify only the first inequality assuming $|\varphi(r_1)| < +\infty$. The inequality being trivial if $\varphi(r_3) = +\infty$, suppose that $\varphi(r_3) < +\infty$ and take any real $\rho_3 \geq \varphi(r_3)$. For $t := (r_3 - r_2)/(r_3 - r_1)$ we have $0 < t < 1$ and $r_2 = tr_1 + (1-t)r_3$, so by convexity of φ

$$\varphi(r_2) \leq \frac{r_3 - r_2}{r_3 - r_1}\varphi(r_1) + \frac{r_2 - r_1}{r_3 - r_1}\rho_3.$$

Addind $-\varphi(r_1)$ to both members of the latter inequality, we obtain the first inequality with ρ_3 in place of $\varphi(r_3)$, so the desired inequality is obtained by letting $\rho_3 \downarrow \varphi(r_3)$.

The arguments above also yield the convexity under the assumed inequalities. □

Now let us adapt the usual definition of strict convexity to extended real-valued functions by declaring that a function $f : C \to \mathbb{R} \cup \{+\infty\}$ on a convex set C of a vector space is *strictly convex* on C provided that for any points $x \neq y$ in C where f is finite and any real $t \in]0,1[$

(2.18) $$f(tx + (1-t)y) < tf(x) + (1-t)f(y).$$

Through the slope inequality in Lemma 2.59, it is an exercise to see that if I is an open interval in $\mathbb{R}$ on which the function φ on I is real-valued and differentiable, then φ is convex (resp. strictly convex) on I if and only if φ' is nondecreasing (resp. increasing) on I. If φ is twice differentiable, then the condition $\varphi''(\cdot) \geq 0$ on I is necessary and sufficient for the convexity of φ whereas the condition $\varphi''(\cdot) > 0$ is merely sufficient for its strict convexity. Given an open convex set U of a normed space $(X, \|\cdot\|)$ and a function $f : U \to \mathbb{R}$, the conditions for convexity in the next proposition directly follow by considering, for each $x \in U$ and each $v \in X$, the function $t \mapsto f(x + tv)$ on the open interval $\{t \in \mathbb{R} : x + tv \in U\}$.

PROPOSITION 2.60. *Let U be an open convex set of a normed space $(X, \|\cdot\|)$ and $f : U \to \mathbb{R}$ be a function differentiable on U.*
(A) *The following are equivalent:*
(a) *the function f is convex on U;*
(b) *the inequality $\langle Df(x) - Df(y), x - y \rangle \geq 0$ holds for all $x, y \in U$;*
(c) *one has $f(y) \geq f(x) + \langle Df(x), y - x \rangle$ for all $x, y \in U$.*
If f is twice differentiable on U, then the latter assertions are also equivalent to the inequality $D^2 f(x)(v,v) \geq 0$ for any $x \in U$ and any $v \in X$.
(B) *Regarding the strict convexity we also have the equivalence between the assertions:*
(a') *The function f is strictly convex on U;*
(b') *the strict inequality $\langle Df(x) - Df(y), x - y \rangle > 0$ holds for all points $x \neq y$ in the set U;*
(c') *one has $f(y) > f(x) + \langle Df(x), y - x \rangle$ for all points $x \neq y$ in U.*
If f is twice differentiable on U, then each one of the latter assertions is implied by the strict inequality $D^2 f(x)(v,v) > 0$ for any $x \in U$ and any nonzero $v \in X$.

EXAMPLE 2.61. We recall that the well-known convexity of the following functions is justified by the classical conditions (A-a) and (B-b') above:
- On $]0, +\infty[$ the function $x \mapsto x^s$ is convex (resp. strictly convex) if $s \geq 1$ (resp. $s > 1$).
- On $]0, +\infty[$ the function $x \mapsto 1/x^s$ is strictly convex if $s > 0$.

- On $]0,+\infty[$ the function $x \mapsto -x^s$ is convex (resp. strictly convex) if $s \in [0,1]$ (resp. $s \in]0,1[$).
- On $]0,+\infty[$ the function $x \mapsto -\log x$ is strictly convex.
- On $]0,+\infty[$ the function $x \mapsto x \log x$ is strictly convex. □

The first description of Clarke subdifferentials of convex functions will use the standard directional derivative.

DEFINITION 2.62. Given any function $\varphi : V \to \mathbb{R} \cup \{-\infty,+\infty\}$ on a nonempty convex set V of a vector space X and a point $x \in V$ with $|\varphi(x)| < +\infty$, it is usual when, for $h \in]0,+\infty[(V-x)$, the limit

$$\lim_{t \downarrow 0} t^{-1}\big(\varphi(x+th) - \varphi(x)\big)$$

exists in $\mathbb{R} \cup \{-\infty,+\infty\}$, to call it the *(standard) directional derivative, or the (right-hand side) Dini directional derivative* of φ at x in the direction h; it is denoted by $\varphi'(x;h)$ or $d_D^+\varphi(x;h)$. When $]0,+\infty[(V-x) = X$ and the latter limit exists for all $h \in X$, the function $\varphi'(x;\cdot)$ (or $d_D^+\varphi(x;\cdot)$) from X to $\mathbb{R} \cup \{-\infty,+\infty\}$ is called the *(standard) directional (or (right-hand side) Dini directional) derivative* of φ at x.

For the convex function $f : U \to \mathbb{R} \cup \{-\infty,+\infty\}$, given a point $x \in U$ where f is finite and a vector $h \in]0,+\infty[(U-x)$ such that the interval $I = \{t \in \mathbb{R} : x+th \in U\}$ is not reduced to 0, Lemma 2.59 tells us that the function

(2.19) $\quad t \mapsto t^{-1}\big(f(x+th) - f(x)\big)$ is nondecreasing on $I \setminus \{0\}$,

so the (standard) directional derivative $f'(x;h)$ exists in $\mathbb{R} \cup \{-\infty,+\infty\}$ with

(2.20) $\qquad f'(x;h) = \inf_{t>0,\, x+th \in U} t^{-1}\big(f(x+th) - f(x)\big).$

PROPOSITION 2.63. Let U be a nonempty open convex set of the normed space X, $f : U \to \mathbb{R} \cup \{-\infty,+\infty\}$ be an extended real-valued convex function, and $x \in U$ be a point where f is finite. Then one has

$$\partial_C f(x) = \{x^* \in X^* : \langle x^*, x' - x \rangle \leq f(x') - f(x),\ \forall x' \in U\}$$
$$= \{x^* \in X^* : \langle x^*, h \rangle \leq f'(x;h) := \lim_{t \downarrow 0} t^{-1}\big(f(x+th) - f(x)\big),\ \forall h \in X\}.$$

In particular, whenever the convex function f is Gâteaux differentiable at x, one has $\partial_C f(x) = \{Df(x)\}$.

PROOF. For the first equality, it is enough to observe that the convexity of $E_f := \mathrm{epi}\, f$ ensures by Proposition 2.2 that

$$N^C\big(E_f;(x,f(x))\big) = \{(x^*,s) \in X^* \times \mathbb{R} : \langle x^*, x'-x \rangle + s(r-f(x)) \leq 0\ \forall (x',r) \in E_f\},$$

and this clearly justifies the desired first equality. This equality allows us to see that $x^* \in \partial_C f(x)$ if and only if for every $h \in X$ we have

$$\langle x^*, h \rangle \leq \inf_{t>0,\, x+th \in U} t^{-1}\big(f(x+th) - f(x)\big) = \lim_{t \downarrow 0} t^{-1}\big(f(x+th) - f(x)\big),$$

where the equality follows from the analysis preceding the statement of the proposition. The second equality is then proved.

If in addition f is Gâteaux differentiable at x, the second equality above ensures that $x^* \in \partial_C f(x)$ if and only $\langle x^*, h \rangle \leq \langle Df(x), h \rangle$ for all h, which (changing h in $-h$) is equivalent to $\langle x^*, h \rangle = \langle Df(x), h \rangle$ for all $h \in X$. This means that $\partial_C f(x) = \{Df(x)\}$ whenever the convex function f is Gâteaux differentiable at the point x. □

DEFINITION 2.64. Given a *convex function* $f : U \to \mathbb{R} \cup \{-\infty, +\infty\}$ and a point x where it is finite, the set in the second member in Proposition 2.63 presents a universal character. It is called the subdifferential of the convex function f at x and as it is usual, it will be denoted $\partial f(x)$ without any subscript, that is,

$$\partial f(x) = \{x^* \in X^* : \langle x^*, x' - x \rangle \le f(x') - f(x), \ \forall x' \in U\}; \tag{2.21}$$

the equality

$$\partial f(x) = \{x^* \in X^* : (x^*, -1) \in N(\text{epi } f; (x, f(x)))\}$$

also holds according to the above proposition, to Proposition 2.52 and to the equality between $N^C(\text{epi } f; \cdot)$ and $N(\text{epi } f; \cdot)$, where for any convex set S of a normed space, $N(S; \cdot) := \partial \Psi_S$ is called the normal cone of the convex set S.

The case of positively homogeneous convex functions is sometimes of interest. Recall that a function $f : X \to \mathbb{R} \cup \{-\infty, +\infty\}$ on a vector space X is *positively homogeneous* provided

$$f(tx) = tf(x) \quad \text{for all } x \in X \text{ and } t > 0.$$

PROPOSITION 2.65. *Let $f : X \to \mathbb{R} \cup \{-\infty, +\infty\}$ be a convex function which is positively homogeneous on the normed space X with $f(0) = 0$. Then for every $x \in X$ where f is finite, one has*

$$\partial f(x) = \{x^* \in X^* : x^* \in \partial f(0) \text{ and } \langle x^*, x \rangle = f(x)\}.$$

PROOF. Fix $x \in X$ with $|f(x)| < +\infty$. Take any x^* in the right-hand member. Applying the above proposition, for any $y \in X$, one has

$$\langle x^*, y - x \rangle = \langle x^*, y \rangle - f(x) \le f(y) - f(x),$$

hence $x^* \in \partial f(x)$. Conversely, fix any $x^* \in \partial f(x)$. Then, for any real $t > 0$

$$(t - 1)\langle x^*, x \rangle = \langle x^*, tx - x \rangle \le f(tx) - f(x) = (t - 1)f(x),$$

so taking successively $t > 1$ and $0 < t < 1$ gives $\langle x^*, x \rangle = f(x)$. Further, fixing any $y \in X$ one has for every real $t > 0$

$$\langle x^*, ty - x \rangle \le tf(y) - f(x) \text{ or equivalently } \langle x^*, y \rangle - t^{-1}\langle x^*, x \rangle \le f(y) - t^{-1}f(x),$$

so we obtain $\langle x^*, y \rangle \le f(y)$ as $t \to +\infty$, which says that $x^* \in \partial f(0)$. □

In the next examples we describe the subdifferential of some convex functions which arise very often in various variational problems.

EXAMPLE 2.66. (a) Given a normed space $(X, \|\cdot\|)$ it is clear from Proposition 2.63 that

$$\partial \|\cdot\|(0) = \mathbb{B}_{X^*}. \tag{2.22}$$

So, it follows from Proposition 2.65 that

$$\partial(\|\cdot\|)(x) = \{x^* \in \mathbb{B}_{X^*} : \langle x^*, x \rangle = \|x\|\},$$

hence

$$\partial(\|\cdot\|)(x) = \{x^* \in X^* : \|x^*\| = 1, \ \langle x^*, x \rangle = \|x\|\} \quad \text{for all } x \ne 0. \tag{2.23}$$

Similarly, for any real $p > 1$,

$$\partial(\frac{1}{p}\|\cdot\|^p)(x) = \{x^* \in X^* : \langle x^*, x \rangle = \|x^*\|_* \|x\|, \ \|x^*\|_* = \|x\|^{p-1}\}. \tag{2.24}$$

For $p = 2$, the multimapping $\partial(\frac{1}{2}\|\cdot\|^2)$ is generally denoted by J and called the (normalized) *duality multimapping* or *duality mapping*, so

(2.25) $\quad J(x) := \partial(\frac{1}{2}\|\cdot\|^2)(x) = \{x^* \in X^* : \langle x^*, x\rangle = \|x^*\|_*\|x\|,\ \|x^*\|_* = \|x\|\};$

more will be said in Section 11.2. Let us verify the equality (2.24). Fix any real $p > 1$ and set $f(x) = (1/p)\|x\|^p$. It is easily seen that f is Fréchet differentiable at 0 with $Df(0) = 0$, so $\partial f(0) = \{0\}$ according to Proposition 2.63. Fix now any $x \neq 0$ and write by Proposition 2.63 again that $x^* \in \partial f(x)$ if and only if, for all $h \in X$,

$$\langle x^*, h\rangle \leq \lim_{t\downarrow 0} t^{-1}\bigl(f(x + th) - f(x)\bigr) = \lim_{t\downarrow 0} t^{-1}(\|x + th\| - \|x\|)\|x\|^{p-1}(1 + \varepsilon(t)),$$

where $\lim_{t\downarrow 0}\varepsilon(t) = 0$. It results that $x^* \in \partial f(x)$ if and only if

$$\left\langle \frac{x^*}{\|x\|^{p-1}}, h\right\rangle \leq \lim_{t\downarrow 0} t^{-1}(\|x + th\| - \|x\|) \text{ for all } h \in X,$$

which is equivalent by Proposition 2.63 to $x^*/(\|x\|^{p-1}) \in \partial(\|\cdot\|)(x)$, or equivalently $\|x^*/(\|x\|^{p-1})\| = 1$ and $\langle x^*/(\|x\|^{p-1}), x\rangle = \|x\|$ according to (2.23). The equality (2.24) is then justified.

(b) Let $\ell_{\mathbb{R}}^p(\mathbb{N})$, for a real $p \geq 1$, (resp. $\ell_{\mathbb{R}}^\infty(\mathbb{N})$) denote the set of real sequences $x : \mathbb{N} \to \mathbb{R}$ such that $\sum_{n=1}^\infty |x_n|^p < +\infty$ (resp. $\sup_{n\in\mathbb{N}} |x_n| < +\infty$) endowed with the standard norm $\|\cdot\|_p$ (resp. $\|\cdot\|_\infty$). The convex function $\frac{1}{p}\|\cdot\|_p^p$ is known to be differentiable for $1 < p < \infty$. Concerning the case $p = 1$, we have

$$\partial(\|\cdot\|_1)(x) = \left\{\zeta \in \ell_{\mathbb{R}}^\infty(\mathbb{N}) : \zeta_n = \frac{x_n}{|x_n|} \text{ if } x_n \neq 0,\ \zeta_n \in [-1,1] \text{ if } x_n = 0\right\}.$$

Indeed, from Proposition 2.65 we have that $\zeta \in \ell_{\mathbb{R}}^\infty(\mathbb{N})$ is in $\partial\|\cdot\|_1(x)$ if and only if $\|\zeta\|_\infty \leq 1$ and $\sum_{n=1}^\infty \zeta_n x_n = \sum_{n=1}^\infty |x_n|$, which (taking into account the inequality $\zeta_n x_n \leq |x_n|$ because $\|\zeta\|_\infty \leq 1$) is equivalent to $\|\zeta\|_\infty \leq 1$ and $\zeta_n x_n = |x_n|$ for all n, or equivalently $\zeta_n = x_n/|x_n|$ if $x_n \neq 0$ and $|\zeta_n| \leq 1$ if $x_n = 0$.

(c) More generally, given an uncountable set Γ let $\ell_{\mathbb{R}}^1(\Gamma)$ be the vector space of functions $x : \Gamma \to \mathbb{R}$ such that the family $(x_\gamma)_{\gamma\in\Gamma}$ is summable, endowed with the norm $\|x\|_1 := \sum_{\gamma\in\Gamma}|x_\gamma|$. Its topological dual is known (and easily seen) to coincide with $\ell_{\mathbb{R}}^\infty(\Gamma)$ the vector space of bounded functions from Γ into $\mathbb{R}$. The above arguments also show that, for any $x \in \ell_{\mathbb{R}}^1(\Gamma)$,

$$\partial(\|\cdot\|_1)(x) = \left\{\zeta \in \ell_{\mathbb{R}}^\infty(\Gamma) : \zeta_\gamma = \frac{x_\gamma}{|x_\gamma|} \text{ if } x_\gamma \neq 0,\ \zeta_\gamma \in [-1,1] \text{ if } x_\gamma = 0\right\},$$

and the above set is never a singleton since $x_\gamma \neq 0$ only for countably many indices γ. Consequently, according to Proposition 2.63 the norm in $\ell_{\mathbb{R}}^1(\Gamma)$ is not Gâteaux differentiable at any point whenever Γ is uncountable.

The case of the usual norm of the space $c_{\mathbb{R}}^0(\mathbb{N})$ will be discussed later in Example 2.200. $\quad\square$

The example of the normal cone of a convex cone $K \ni 0$ is also of great interest. Since for $x \in K$ we have $N(K; x) = \partial \Psi_K(x)$ for the positively homogeneous convex indicator function Ψ_K of K and since $\Psi_K(0) = 0$, the equality 2.26 below follows directly from Proposition 2.65.

EXAMPLE 2.67. (a) Given a convex cone K in a normed space X with $0 \in K$, one has by definition of $N(K;0)$

$$N(K;0) = K^\circ = \{x^* \in X^* : \langle x^*, h\rangle \leq 0, \forall h \in K\},$$

and more generally for any $x \in K$

(2.26) $\quad N(K;x) = \{x^* \in X^* : \langle x^*, x\rangle = 0 \text{ and } \langle x^*, h\rangle \leq 0 \ \forall h \in K\}.$

(b) For $p, q \in \mathbb{N}$ and $x \in -\mathbb{R}_+^p \times \{0_{\mathbb{R}^q}\}$ with $\mathbb{R}^+ := [0, +\infty[$ one has

(2.27) $\quad N(-\mathbb{R}_+^p \times \{0_{\mathbb{R}^q}\}; x) = \{\lambda \in \mathbb{R}^{p+q} : \forall i \in \{1, \cdots, p\}, \lambda_i \geq 0, \lambda_i x_i = 0\}.$

This equality is clearly a particular case of (a). □

Assume now that $f : X \to \mathbb{R} \cup \{-\infty, +\infty\}$ is a convex function defined on the whole normed space X. If f is finite at x, by Theorem 2.63 we can write

$$x^* \in \partial f(x) \Leftrightarrow \langle x^*, x'\rangle - f(x') \leq \langle x^*, x\rangle - f(x), \ \forall x' \in X,$$

or equivalently

$$x^* \in \partial f(x) \Leftrightarrow \sup_{x' \in X} \left(\langle x^*, x'\rangle - f(x')\right) \leq \langle x^*, x\rangle - f(x).$$

This makes clear one of the main interests of the function in x^* provided by the above supremum. Further, when X is $\mathbb{R}^n$ and f is smooth the supremum coincides with $\langle \nabla f(x), x\rangle - f(x)$, which is related to the well-known Legendre transform in Analysis; see more in the comments at the end of Chapter 3. The role of this supremum function in various aspects in Analysis, in particular in Convex Duality, will be largely discussed in Chapter 3 in the general framework of locally convex spaces.

DEFINITION 2.68. Given a function $\varphi : X \to \mathbb{R} \cup \{-\infty, +\infty\}$ on a locally convex space X, its *Legendre-Fenchel conjugate* (or *transform*) is the function $\varphi^* : X^* \to \mathbb{R} \cup \{-\infty, +\infty\}$ defined by

$$\varphi^*(x^*) := \sup_{x' \in X} \left(\langle x^*, x'\rangle - \varphi(x')\right) \quad \text{for all } x^* \in X^*.$$

Obviously, this function φ^* is always convex and weak* lower semicontinuous, and the inequality (generally called *Fenchel inequality*)

$$\langle x^*, x\rangle - \varphi(x) \leq \varphi^*(x^*)$$

holds true. If φ is finite at some point, then φ^* does not take the value $-\infty$.

Coming back to the proper convex function $f : X \to \mathbb{R} \cup \{+\infty\}$ on a normed space X, for $|f(x)| < +\infty$ we can rewrite the above equivalences as

$$x^* \in \partial f(x) \Leftrightarrow f^*(x^*) \leq \langle x^*, x\rangle - f(x) \Leftrightarrow f^*(x^*) = \langle x^*, x\rangle - f(x).$$

We state those features as follows:

PROPOSITION 2.69. Let $f : X \to \mathbb{R} \cup \{+\infty\}$ be a proper convex function on a normed space X. Then for any $x \in X$ with $|f(x)| < +\infty$ one has

$$\partial f(x) = \{x^* \in X^* : \langle x^*, x\rangle = f(x) + f^*(x^*)\} = \{x^* \in X^* : \langle x^*, x\rangle \geq f(x) + f^*(x^*)\}.$$

2.2.5. C-subdifferential and locally Lipschitz functions.

The inclusion $(x^*, -1) \in N^C(\text{epi } f; (x, f(x)))$ of the definition of x^* in the Clarke subdifferential can be translated through the Clarke tangent cone to the epigraph as: $x^* \in \partial_C f(x)$ if and only if

$$(2.28) \qquad \langle x^*, h \rangle \leq \inf\{r \in \mathbb{R} : (h, r) \in T^C(\text{epi } f; (x, f(x)))\} \quad \text{for all } h \in X.$$

We recall that the infimum over the empty set is $+\infty$, and we point out that the inclusion $(h, r) \in T^C(\text{epi } f; (x, f(x)))$ implies $(h, r') \in T^C(\text{epi } f; (x, f(x)))$ for every real $r' \geq r$, as is easily seen. Taking then Theorem 2.13 and Theorem 2.20 into account, we study below the right-hand member of the inequality (2.28) when f is finite and Lipschitz continuous near x.

THEOREM 2.70 (w* compactness of $\partial_C f(\bar{x})$ under Lipschitz continuity). Let U be an open set of the normed space X and $f : U \to \mathbb{R} \cup \{-\infty, +\infty\}$ be a function which is finite and Lipschitz continuous near a point $\bar{x} \in U$ with $\gamma \geq 0$ as a Lipschitz constant. The following hold:

(a) For all $h \in X$

$$\min\{r \in \mathbb{R} : (h, r) \in T^C(\text{epi } f; (\bar{x}, f(\bar{x})))\} = f^\circ(\bar{x}; h),$$

where

$$f^\circ(\bar{x}; h) := \limsup_{t \downarrow 0, x' \to \bar{x}} t^{-1}(f(x' + th) - f(x')).$$

(b) The function $f^\circ(\bar{x}; \cdot)$ is finite, sublinear (that is, convex and positively homogeneous), γ-Lipschitz continuous on X, hence in particular $|f^\circ(\bar{x}; h)| \leq \gamma \|h\|$ for all $h \in X$; further

$$(2.29) \qquad f^\circ(\bar{x}; h) = \limsup_{t \downarrow 0, (x', h') \to (\bar{x}, h)} t^{-1}(f(x' + th') - f(x')) \quad \text{for all } h \in X.$$

(c) For any $h \in X$ and any real $\varepsilon > 0$

$$(h, f^\circ(\bar{x}; h) + \varepsilon) \in I(\text{epi } f; (\bar{x}, f(\bar{x}))).$$

(d) The set $\partial_C f(\bar{x})$ is nonempty and weak* compact in X^*, and

$$\partial_C f(\bar{x}) = \{x^* \in X^* : \langle x^*, h \rangle \leq f^\circ(\bar{x}; h), \forall h \in X\},$$

along with

$$f^\circ(\bar{x}; h) = \max\{\langle x^*, h \rangle : x^* \in \partial_C f(\bar{x})\} \quad \text{for all } h \in X,$$

that is, $f^\circ(\bar{x}; \cdot)$ is the support function of $\partial f(\bar{x})$; further $\partial_C f(\bar{x}) \subset \gamma \mathbb{B}_{X^*}$.

PROOF. Without loss of generality we may suppose that $U = X$.
(a)-(c): Let $h \in X$ and $f^\circ(\bar{x}; h)$ defined by

$$f^\circ(\bar{x}; h) = \limsup_{t \downarrow 0, x' \to \bar{x}} t^{-1}(f(x' + th) - f(x')).$$

The Lipschitz continuity property of f near $\bar{x}$ clearly ensures the finiteness of $f^\circ(\bar{x}; h)$ as well as the inequality $f^\circ(\bar{x}; h) \leq \gamma \|h\|$. Further, choosing $0 < r < 1$ such that f is γ-Lipschitz on $B(\bar{x}, 2r)$, we have, for all $0 < t < \min\{1, r/(1 + |h|)\}$, $x' \in B(\bar{x}, r)$ and $h' \in B(h, r)$,

$$|t^{-1}(f(x' + th') - f(x' + th))| \leq \gamma \|h' - h\|,$$

and this ensures that

$$f^\circ(\bar{x}; h) = \limsup_{t \downarrow 0, (x', h') \to (\bar{x}, h)} t^{-1}(f(x' + th') - f(x')).$$

Fix now any $h \in X$ and any $\varepsilon > 0$, and take any sequences $(t_n)_n$ tending to 0 with $t_n > 0$ and $(x_n, r_n)_n$ in $X \times \mathbb{R}$ converging to $(\overline{x}, f(\overline{x}))$ with $r_n \geq f(x_n)$, and any sequence $(h_n, s_n)_n$ in $X \times \mathbb{R}$ converging to $(h, f^\circ(\overline{x}; h) + \varepsilon)$. We have

$$\limsup_n t_n^{-1}\big(f(x_n + t_n h_n) - f(x_n)\big) < f^\circ(\overline{x}; h) + \varepsilon/2,$$

which ensures for n large enough that $t_n^{-1}\big(f(x_n + t_n h_n) - f(x_n)\big) < s_n$ and this implies $(x_n, r_n) + t_n(h_n, s_n) \in \operatorname{epi} f$. This confirms the inclusion $(h, f^\circ(\overline{x}; h) + \varepsilon) \in I\big(\operatorname{epi} f; (\overline{x}, f(\overline{x}))\big)$ in (c) and entails the nonvacuity of $I\big(\operatorname{epi} f; (\overline{x}, f(\overline{x}))\big)$.

Then using the first equality $T^C(S; u) = \operatorname{cl}\big(I(S; u)\big)$ in (b) of Theorem 2.13 whenever $I(S; u) \neq \emptyset$, we deduce from the last inclusion above the inequality $\leq$ for the first equality in (a). Let us show the reverse inequality. Fix any $r \in \mathbb{R}$ with $(h, r) \in T^C\big(\operatorname{epi} f; (\overline{x}, f(\overline{x}))\big)$. Choose sequences $(x_n)_n$ in X converging to $\overline{x}$ and $(t_n)_n$ of positive reals converging to 0 such that

$$f^\circ(\overline{x}; h) = \lim_n t_n^{-1}\big(f(x_n + t_n h) - f(x_n)\big).$$

Denoting by V an open set containing $\overline{x}$ where f is finite and γ-Lipschitz continuous, we may suppose that x_n and $x_n + t_n h \in V$ for all n. By the sequential characterization in Proposition 2.2 of $T^C(\cdot; \cdot)$ there exists some sequence $(h_n, r_n) \to (h, r)$ such that $(x_n, f(x_n)) + t_n(h_n, r_n) \in \operatorname{epi} f$. We easily derive from this inclusion and from the γ-Lipschitz continuity of f on V that $f^\circ(\overline{x}; h) \leq r$, which justifies the desired reverse inequality. We have then proved the equality

$$\inf\{r \in \mathbb{R} : (h, r) \in T^C\big(\operatorname{epi} f; (\overline{x}, f(\overline{x}))\big)\} = f^\circ(\overline{x}; h),$$

and the infimum is attained since $f^\circ(\overline{x}; h)$ is finite and the set whose infimum is taken is closed.

The convexity and positive homogeneity of $f^\circ(\overline{x}; \cdot)$ follow from the latter equality and the convexity of the cone $T^C\big(\operatorname{epi} f; (\overline{x}, f(\overline{x}))\big)$; and it can also be directly checked easily from the (limit superior) definition of $f^\circ(\overline{x}; \cdot)$. This combined with the inequality $f^\circ(\overline{x}; h) \leq \gamma \|h\|$ guarantees that

$$f^\circ(\overline{x}; h) \leq f^\circ(\overline{x}; h') + \gamma \|h - h'\|,$$

hence $f^\circ(\overline{x}; \cdot)$ is γ-Lipschitz continuous on X and $|f^\circ(\overline{x}; h)| \leq \gamma \|h\|$. Assertions (a), (b), and (c) are then established.

(d) Fix $h \in X$. The function $f^\circ(\overline{x}; \cdot)$ being by (b) sublinear and continuous, there exists by the (analytic) Hahn-Banach theorem (see Theorem B.1 in Appendix) a continuous linear functional $u^* \in X^*$ such that $\langle u^*, h \rangle = f^\circ(\overline{x}; h)$ and $\langle u^*, \cdot \rangle \leq f^\circ(\overline{x}; \cdot)$, which yields the nonemptiness of $\partial_C f(\overline{x})$ and the expression of $\partial_C f(\overline{x})$ and $f^\circ(\overline{x}; h)$ in (d) according to (a), and to the characterization (2.28) above of ∂_C through the Clarke tangent cone. The inequality $f^\circ(\overline{x}; h) \leq \gamma \|h\|$ for all $h \in X$ then ensures the inclusion $\partial_C f(\overline{x}) \subset \gamma \mathbb{B}_{X^*}$, which entails the weak star compactness of the weak star closed set $\partial_C f(\overline{x})$. The proof is then complete. $\square$

DEFINITION 2.71. When f is finite and *Lipschitz continuous* near $\overline{x}$, the above function $f^\circ(\overline{x}; \cdot)$ given by

$$f^\circ(\overline{x}; h) = \limsup_{t \downarrow 0, x' \to \overline{x}} t^{-1}\big(f(x' + th) - f(x')\big)$$

is called the *Clarke directional derivative* of f at $\overline{x}$.

2.2. CLARKE SUBDIFFERENTIAL

When f is not Lipschitz near $\bar{x}$, the analytical expression of
$$\inf\{r \in \mathbb{R} : (h,r) \in T^C(\text{epi}\, f; (\bar{x}, f(\bar{x})))\}$$
(in Theorem 2.70) is much less simple and involves the so-called *upper epi-limit* or *upper Γ-limit* of a function of two (or several) variables which was developed in Section 1.3 of the previous chapter. Given two metric spaces E and F, a set $E_0 \subset E$, an element $\bar{u} \in \text{cl}\,(E_0)$, an element $h \in F$, and for each $u \in E_0$ a function $\varphi_u : F \to \mathbb{R} \cup \{-\infty, +\infty\}$, recall (see Definition 1.21) that the upper epi-limit (or upper Γ-limit) at h of φ_u as $E_0 \ni u \to \bar{u}$ is defined by

$$\lim_{E_0 \ni u' \to \bar{u}} \sup_{h' \to h} \inf \varphi(u', h') := \sup_{\eta > 0} \inf_{\varepsilon > 0} \sup_{u' \in E_0 \cap B(\bar{u}, \varepsilon)} \inf_{h' \in B(h, \eta)} \varphi(u', h').$$

Consider a function $f : U \to \{-\infty, +\infty\}$ on an open set U of the normed space X which is finite at $\bar{x}$ (but not necessarily Lipschitz near $\bar{x}$). Extend f to X with $f(x) = +\infty$ if $x \notin X$. Put $E := [0, +\infty[\times \text{epi}\, f$ and $E_0 :=\,]0, +\infty[\times \text{epi}\, f$, and for each $(t, x, s) \in E_0$ define $\varphi_{t,x,s} : X \to \mathbb{R} \cup \{-\infty, +\infty\}$ by

$$\varphi_{t,x,s}(h) = t^{-1}\bigl(f(x+th) - s\bigr) \quad \text{for all } h \in X.$$

Notice that $(0, \bar{x}, f(\bar{x})) \in \text{cl}\,(E_0)$ and that
$$\text{epi}\, \varphi_{t,x,s} = t^{-1}\bigl(\text{epi}\, f - (x, s)\bigr).$$
Therefore, Definition 2.1 and Proposition 1.23 yield that
$$T^C\bigl(\text{epi}\, f; (\bar{x}, f(\bar{x}))\bigr) = \operatorname*{Lim\,inf}_{\substack{t \downarrow 0,\, (x,s) \xrightarrow[\text{epi}\, f]{} (\bar{x}, f(\bar{x}))}} \text{epi}\, \varphi_{t,x,s} = \text{epi}\, L,$$
where for every $h \in X$
$$L(h) := \lim \sup_{\substack{t \downarrow 0,\, (x,s) \xrightarrow[\text{epi}\, f]{} (\bar{x}, f(\bar{x}))}} \inf_{h' \to h} t^{-1}\bigl(f(x + th') - s\bigr).$$
If f is lower semicontinuous at $\bar{x}$ it can be seen that
$$L(h) = \lim \sup_{t \downarrow 0,\, x \to_f \bar{x}} \inf_{h' \to h} t^{-1}\bigl(f(x + th') - f(x)\bigr).$$

Since $T^C(\text{epi}\, f; (\bar{x}, f(\bar{x})))$ is a closed convex cone, the function L is lower semicontinuous, convex and positively homogeneous, and $L(0) \in \{-\infty, 0\}$. When f is Lipschitz continuous near $\bar{x}$ we know by Theorem 2.70(a) that
$$\min\{r \in \mathbb{R} : (h, r) \in T^C(\text{epi}\, f; (\bar{x}, f(\bar{x})))\} = f^\circ(\bar{x}; h),$$
so this assures us under this Lipschitz property of f near $\bar{x}$ that
$$L(h) = f^\circ(\bar{x}; h) \quad \text{for all } h \in X.$$

Further, we have
$$x^* \in \partial_C f(\bar{x}) \Leftrightarrow (x^*, -1) \in \bigl(T^C(\text{epi}\, f; (\bar{x}, f(\bar{x})))\bigr)^\circ$$
$$\Leftrightarrow \langle x^*, h \rangle \leq r \; \forall (h, r) \in T^C(\text{epi}\, f; (\bar{x}, f(\bar{x}))),$$
which means
$$x^* \in \partial_C f(\bar{x}) \Leftrightarrow \langle x^*, h \rangle \leq L(h) \; \forall h \in X.$$
We summarize the above features in the following theorem.

THEOREM 2.72 (R.T. Rockafellar). Let $f : U \to \mathbb{R} \cup \{-\infty, +\infty\}$ be an extended real-valued function on an open set U of a normed space X and let $\overline{x} \in U$ be a point where f is finite. The following hold.

(a) The Clarke tangent cone $T^C(\text{epi } f; (\overline{x}, f(\overline{x})))$ is the epigraph of a function $f^\uparrow(\overline{x}; \cdot) : X \to \mathbb{R} \cup \{-\infty, +\infty\}$, that is,

$$T^C(\text{epi } f; (\overline{x}, f(\overline{x}))) = \text{epi } f^\uparrow(\overline{x}; \cdot)$$

or $f^\uparrow(\overline{x}; h) = \inf\{r \in \mathbb{R} : (h, r) \in T^C(\text{epi } f; (\overline{x}, f(\overline{x})))\}$ for all $h \in X$.

(b) For any $h \in X$ one has

$$f^\uparrow(\overline{x}; h) = \lim_{t \downarrow 0,\, (x,s) \xrightarrow[\text{epi } f]{} (\overline{x}, f(\overline{x}))} \sup_{h' \to h} \inf\, t^{-1}(f(x + th') - s),$$

and if f is lower semicontinuous at $\overline{x}$

$$f^\uparrow(\overline{x}; h) = \lim_{t \downarrow 0,\, x \to_f \overline{x}} \sup_{h' \to h} \inf\, t^{-1}(f(x + th') - f(x)).$$

(c) The function $f^\uparrow(\overline{x}; \cdot)$ is lower semicontinuous, convex, positively homogeneous and $f^\uparrow(\overline{x}; 0) \in \{0, -\infty\}$.

(d) One has the description

$$\partial_C f(\overline{x}) = \{x^* \in X^* : \langle x^*, h \rangle \leq f^\uparrow(\overline{x}; h),\ \forall h \in X\}.$$

(e) If f is Lipschitz continuous near $\overline{x}$

$$f^\uparrow(\overline{x}; h) = f^\circ(\overline{x}; h) \quad \text{for all } h \in X.$$

DEFINITION 2.73. Given an extended real function $f : U \to \mathbb{R} \cup \{-\infty, +\infty\}$ on an open set U of a normed space, the expression

$$f^\uparrow(\overline{x}; h) = \lim_{t \downarrow 0,\, (x,s) \xrightarrow[\text{epi } f]{} (\overline{x}, f(\overline{x}))} \sup_{h' \to h} \inf\, t^{-1}(f(x + th') - s)$$

in the assertion (b) above is called the *Rockafellar directional derivative* of f at $\overline{x}$ in the direction h.

The Rockafellar directional derivative $f^\uparrow(\overline{x}; \cdot)$ not only allows us to describe the Clarke subdifferential $\partial_C f(\overline{x})$ as in Theorem 2.72(d), but as we will see in Corollary 3.81 (in the next chapter) the nonemptiness of $\partial_C f(\overline{x})$ is characterized by the simple property $f^\uparrow(\overline{x}; 0) = 0$, or equivalently $f^\uparrow(\overline{x}; 0) > -\infty$.

Let us come back to functions which are locally Lipschitz continuous around the reference point. The assertion in (d) of Theorem 2.70 above ensures, whenever f is Lipschitz continuous near $\overline{x}$, the equalities

(2.30) $$\partial_C f(\overline{x}) = \{x^* \in X^* : \langle x^*, \cdot \rangle \leq f^\circ(\overline{x}; \cdot)\} = \partial f^\circ(\overline{x}; \cdot)(0),$$

where, in the third right-hand member, $\partial f^\circ(\overline{x}; \cdot)$ denotes the subdifferential at 0 (in the sense of Definition 2.64) of the real-valued continuous convex function $f^\circ(\overline{x}; \cdot)$. Let U be an open neighborhood of $\overline{x}$ over which the function f is locally Lipschitz continuous. Observing, for any $h \in X$, that $f^\circ(\cdot; h)$ is upper semicontinuous on U according to its definition as a limit superior, the assertion (d) of Theorem 2.70 easily yields that the Clarke subdifferential of the locally Lipschitz continuous function f is upper semicontinuous from U endowed with the topology induced by the norm of X into $(X^*, w(X^*, X))$. Recall that a multimapping $M : U \rightrightarrows Y$ from U into a topological space Y is *upper semicontinuous* at $\overline{x} \in U$ (see the previous chapter) when for any open set W of Y with $M(\bar{x}) \subset W$ there exists some neighborhood V of $\bar{x}$ such that $M(u) \subset W$ for all $u \in V$. When Y is a locally convex topological vector space and M is assigned to have nonempty weakly compact convex images, it is known (see Theorem 1.48 in the previous chapter) that M is upper semicontinuous at $\bar{x}$ with respect to the topology $w(Y, Y^*)$ if and only if for each $y^* \in Y^*$ the support function $\sigma(y^*, M(\cdot))$ is upper semicontinuous at $\bar{x}$, where $\sigma(y^*, M(u)) := \sup_{y \in M(u)} \langle y^*, y \rangle$. In our case above the locally convex space Y is $(X^*, w(X^*, X))$, so that $Y^* = X$. So, for $M(\cdot) = \partial_C f(\cdot)$ with f locally Lipschitz continuous on U, we have $\sigma(v, \partial_C f(u)) = f^\circ(u; v)$ for all $v \in X$ and $f^\circ(\cdot; v)$ is upper semicontinuous on U for all $v \in X$. The upper semicontinuity of $\partial^C f$ on U is then justified.

With $f : U \to \mathbb{R}$ locally Lipschitz on the open set U of the normed space X, the equality (2.29) in Theorem 2.70

$$f^\circ(x; h) = \limsup_{t \downarrow 0, (x', h') \to (x, h)} t^{-1} \big(f(x' + th') - f(x') \big) \quad \text{for all } x \in U, h \in X,$$

guarantees that the Clarke directional derivative is not only upper semicontinuous with respect to x but with respect to the pair (x, h), that is, $f^\circ(\cdot; \cdot)$ is upper semicontinuous on $U \times X$.

Concerning the opposite of the function f which is locally Lipschitz continuous on the open set U, we observe that, for all $x \in U$ and $h \in X$,

$$(-f)^\circ(x; h) = \limsup_{t \downarrow 0, x' \to x} t^{-1} \big(f(x') - f(x' + th) \big)$$

$$= \limsup_{t \downarrow 0, x' \to x} t^{-1} \big(f(x' + th + t(-h)) - f(x' + th) \big)$$

$$\leq \limsup_{t \downarrow 0, u \to x} t^{-1} \big(f(u + t(-h)) - f(u) \big) = f^\circ(x; -h).$$

Changing in the latter inequality f into $-f$ and h into $-h$ gives also $f^\circ(x; -h) \leq (-f)^\circ(x; h)$, hence $(-f)^\circ(x; h) = f^\circ(x; -h)$.

So we have the following properties.

PROPOSITION 2.74. Let U be an open set of the normed space $(X, \|\cdot\|)$ and $f : U \to \mathbb{R}$ be a locally Lipschitz function on U. Then the following hold:
(a) The function $f^\circ(\cdot; \cdot)$ is upper semicontinuous on $U \times X$. The multimapping $\partial_C f(\cdot)$ takes on nonempty weak* compact values on U and is upper semicontinuous from U into $(X^*, w(X^*, X))$; in particular, the multimapping $\partial_C f$ is norm-to-weak* closed (or equivalently norm-to-weak* outer semicontinuous) at any $x \in U$, that is, for any net $(x_j)_{j \in J}$ in U converging to x in $(X, \|\cdot\|)$ and any net $(x_j^*)_{j \in J}$ converging weakly star to x^* in X^* with $x_j^* \in \partial_C f(x_j)$, one has $x^* \in \partial_C f(x)$.
(b) For all $x \in U$ and $h \in X$, the following equalities are fulfilled:

$$(-f)^\circ(x; h) = f^\circ(x; -h) \quad \text{and} \quad \partial_C(rf)(x) = r \partial_C f(x) \quad \text{for all } r \in \mathbb{R}.$$

(c) For all $x \in U$ and $h \in X$, one has

$$\{\langle x^*, h \rangle : x^* \in \partial_C f(x)\} = [-f^\circ(x; -h), f^\circ(x; h)].$$

PROOF. The first part of the assertion (a) is completely proved above. The norm-to-weak* closedness property follows from Proposition 1.37(a) or can be seen directly (with $h \in X$) through the inequality $\langle x_j^*, h \rangle \leq f^\circ(x_j; h)$ and the upper semicontinuity of $f^\circ(\cdot; h)$.

Fix any $x \in U$. The left equality in (b) has been also argued in the analysis preceding the statement of the proposition. This left equality guarantees that an element $x^* \in \partial_C(-f)(x)$ if and only if for all $h \in X$ we have
$$\langle x^*, h \rangle \leq f^\circ(x; -h), \text{ that is, } \langle -x^*, -h \rangle \leq f^\circ(x; -h),$$
which means $-x^* \in \partial_C f(x)$. It results that $\partial_C(-f)(x) = -\partial_C f(x)$, so $\partial_C(\lambda f)(x) = \lambda \partial_C f(x)$ for all $\lambda \in \mathbb{R}$, since the equality is trivial for $\lambda \geq 0$.

Regarding (c), we note by (d) in Theorem 2.70 that, for every $h \in X$,
$$f^\circ(x; h) = \max\{\langle x^*, h \rangle : x^* \in \partial_C f(x)\}, \ -f^\circ(x; -h) = \min\{\langle x^*, h \rangle : x^* \in \partial_C f(x)\}.$$
This and the closedness and convexity of the set $\{\langle x^*, h \rangle : x^* \in \partial_C f(x)(h)\}$ justify the equality of (c). $\square$

REMARK 2.75. The assertion (a) above says that the norm-to-weak* topological outer limit (or limit superior) $\underset{u \to \overline{x}}{\operatorname{Lim\,sup}}\, \partial_C f(u)$ coincides with $\partial_C f(\overline{x})$ whenever the function f is Lipschitz continuous near $\overline{x}$. The result fails without the Lipschitz property. It suffices to construct, in finite dimensions, a set for which the multimapping $N^C(S; \cdot)$ is not closed at $0 \in S$ and to choose as f the indicator function Ψ_S. Consider the set
$$S := \{(x, y, z) \in \mathbb{R}^3 : z = \pm xy\}.$$

Let us first determine $T^C(S; (0,0,0))$. Let (u,v,w) in this tangent cone and let $(t_n)_n$ in $]0, +\infty[$ tending to 0. By the first sequential characterization in Theorem 2.2(a) there is a sequence $(u_n, v_n, w_n)_n$ in $\mathbb{R}^3$ converging to (u,v,w) such that, for all $n \in \mathbb{N}$, we have $t_n(u_n, v_n, w_n) \in S$ or equivalently $w_n = \pm t_n u_n v_n$. This ensures that $w = 0$. On the other hand, we claim that $(1,0,0)$ and $(0,1,0)$ belong to $T^C(S; (0,0,0))$. Indeed let any sequence $(t_n)_n$ tending to 0 with $t_n > 0$ and any sequence $(x_n, y_n, z_n)_n$ in S converging to $(0,0,0)$. Then $z_n = \pm x_n y_n$. Suppose that there is some n_0 such that $z_n = x_n y_n$ for all $n \geq n_0$. Then, for all $n \geq n_0$, we can write
$$(x_n, y_n, z_n) + t_n(1, 0, y_n) = \big(x_n + t_n, y_n, (x_n + t_n)y_n\big) \in S.$$
If $z_n = -x_n y_n$ for all n in an infinite subset J of $\mathbb{N}$, then for all $n \in J$
$$(x_n, y_n, z_n) + t_n(1, 0, -y_n) = \big(x_n + t_n, y_n, -(x_n + t_n)y_n\big) \in S.$$
The second sequential characterization in Theorem 2.2(a) of the Clarke tangent cone says that $(1,0,0) \in T^C(S; (0,0,0))$; and similarly $(0,1,0) \in T^C(S, (0,0,0))$. Putting all together and noting that $T^C(S, (0,0,0))$ is a vector space according to Proposition 2.7 (since S is clearly symmetric with respect to the origin), we obtain that
$$T^C(S; (0,0,0)) = \mathbb{R} \times \mathbb{R} \times \{0\} \quad \text{and} \quad N^C(S; (0,0,0)) = \{0\} \times \{0\} \times \mathbb{R}.$$

Now fix any $x \neq 0$ and let us show that
$$T^C(S; (x, 0, 0)) = \mathbb{R} \times \{0\} \times \{0\}. \tag{2.31}$$

Let (u, v, w) in this tangent cone and take a sequence $(t_n)_n$ in $]0, +\infty[$ tending to 0. Since $(x, t_n, t_n x) \in S$ and $t_n^2 \downarrow 0$, there is a sequence $(u_n, v_n, w_n)_n$ converging to (u, v, w) such that, for all n we have $(x, t_n, t_n x) + t_n^2 (u_n, v_n, w_n) \in S$, thus
$$x + t_n w_n = \pm(1 + t_n v_n)(x + t_n^2 u_n).$$

For n large enough, this entails that $x + t_n w_n = (1 + t_n v_n)(x + t_n^2 u_n)$ or equivalently $w_n = x v_n + t_n u_n (1 + t_n v_n)$, which gives $w = xv$. Similarly, since $(x, t_n, -t_n x) \in S$, there is a sequence $(u'_n, v'_n, w'_n)_n$ converging to (u, v, w) such that, for all n we have $(x, t_n, -t_n x) + t_n^2 (u'_n, v'_n, w'_n) \in S$, thus
$$-x + t_n w'_n = \pm(1 + t_n v'_n)(x + t_n^2 u'_n).$$
For n large enough, this implies that $-x + t_n w'_n = -(1 + t_n v'_n)(x + t_n^2 u'_n)$ or equivalently $w'_n = -x v'_n - t_n u'_n (1 + t_n v'_n)$, which yields $w = -xv$. Both equalities $w = xv$ and $w = -xv$ ensure that $w = v = 0$ (keep in mind that $x \neq 0$), which justifies the inclusion of the left member of (2.31) into the right one.

To obtain the equality in (2.31) it remains to prove that $(1, 0, 0)$ and $(-1, 0, 0)$ are in $T^C(S; x, 0, 0)$. The proof of the inclusion related to $(1, 0, 0)$ is obtained as above. Let us establish the inclusion related to $(-1, 0, 0)$ (noting that S is now not symmetric with respect to $(x, 0, 0)$). Let any sequence $(t_n)_n$ in $]0, +\infty[$ with $t_n \to 0$ and any sequence $(x_n, y_n, z_n)_n$ in S converging to $(x, 0, 0)$. Then $z_n = \pm x_n y_n$. Suppose that there is some n_0 such that $z_n = x_n y_n$ for all $n \geq n_0$. Then, for all $n \geq n_0$, we can write
$$(x_n, y_n, z_n) + t_n(-1, 0, -y_n) = (x_n - t_n, y_n, (x_n - t_n) y_n) \in S.$$
If $z_n = -x_n y_n$ for all n in an infinite subset J of $\mathbb{N}$, then for all $n \in J$
$$(x_n, y_n, z_n) + t_n(-1, 0, y_n) = (x_n - t_n, y_n, -(x_n - t_n) y_n) \in S.$$
Again the second sequential characterization in Theorem 2.2 of the Clarke tangent cone tells us that $(-1, 0, 0) \in T^C(S; (x, 0, 0))$. Putting all together we see that
$$T^C(S; (x, 0, 0)) = \mathbb{R} \times \{0\} \times \{0\} \quad \text{and} \quad N^C(S; (x, 0, 0)) = \{0\} \times \mathbb{R} \times \mathbb{R}.$$
It follows that the multimapping $N^C(S; \cdot)$ is not closed at $(0, 0, 0) \in S$. □

Through the second equality in (b) of Proposition 2.74 we can also characterize the strict Gâteaux differentiability of a function in terms of its C-subdifferential. Indeed, we note first, for f Lipschitz continuous near a point $\bar{x}$, that requiring the equality between the extreme terms $f^\circ(\bar{x}; h)$ and $-f^\circ(\bar{x}; -h)$ in (c) of Proposition 2.74 amounts to requiring that
$$\limsup_{t \downarrow 0, x' \to \bar{x}} t^{-1}(f(x' + th) - f(x')) = \liminf_{t \downarrow 0, x' \to \bar{x}} t^{-1}(f((x' - th) + th) - f(x' - th))$$
$$= \liminf_{t \downarrow 0, x' \to \bar{x}} t^{-1}(f(x' + th) - f(x')).$$
Further, the equality $f^\circ(\bar{x}; -h) = -f^\circ(\bar{x}; h)$ for all $h \in X$ entails that $f^\circ(\bar{x}; \cdot)$ is a continuous linear functional on X. Consequently, $f^\circ(\bar{x}; h) = -f^\circ(\bar{x}; -h)$ for all $h \in X$ if and only if f is strictly Gâteaux differentiable at $\bar{x}$.

Taking into account the assertion (c) of Proposition 2.74 above, we see that the strict Gâteaux differentiability characterizes the single valuedness of $\partial_C f(\bar{x})$ whenever f is Lipschitz continuous near $\bar{x}$. This property is stated in the proposition below which complements the assertion (e) in Proposition 2.50 for locally Lipschitz continuous functions.

PROPOSITION 2.76. *Assume that the function f is finite and Lipschitz continuous on an open set U of the normed space X and let $\bar{x} \in U$. Then $\partial_C f(\bar{x})$ is a singleton if and only if f is strictly Gâteaux (or equivalently, strictly Hadamard) differentiable at $\bar{x}$.*

EXAMPLE 2.77. (a) The above proposition combined with Example 2.66(b) yields that the Lipschitz convex function $\|\cdot\|_1$ in $\ell^1_\mathbb{R}(\mathbb{N})$ is Gâteaux (or equivalently strictly Gâteaux) differentiable at x if and only if $x_n \neq 0$ for all $n \in \mathbb{N}$; in that case the Gâteaux derivative is given by

$$\langle D\|\cdot\|_1(x), h\rangle = \sum_{n\in\mathbb{N}} \frac{x_n}{|x_n|} h_n \quad \text{for all } h \in \ell^1_\mathbb{N}(\mathbb{R}).$$

(b) Although (a) furnishes a large set of points in $\ell^1_\mathbb{R}(\mathbb{N})$ where the function $\|\cdot\|_1$ is strictly Hadamard differentiable, this function is Fréchet differentiable nowhere. Indeed, fix any $x \in \ell^1_\mathbb{R}(\mathbb{N})$ with $x_n \neq 0$ for all $n \in \mathbb{N}$ and define $h^k \in \ell^1_\mathbb{R}(\mathbb{N})$ by

$$h^k_n = 0 \quad \text{if } n < k \quad \text{and} \quad h^k_n = -2x_n \quad \text{if } n \geq k,$$

so $\|h^k\|_1 \to 0$ as $k \to \infty$. If $\|\cdot\|_1$ is Fréchet differentiable at x, then it Fréchet derivative coincides with the Gâteaux derivative obtained in (a), which is in contradiction with the fact that

$$(\|h^k\|_1)^{-1} \left| \|x+h^k\|_1 - \|x\|_1 - \sum_{n\in\mathbb{N}} \frac{x_n}{|x_n|} h^k_n \right| = (\|h^k\|_1)^{-1} \left| -2\sum_{n\geq k} |x_n| \right| = 1.$$

□

We turn now to the epi-Lipschitz property of sublevel sets of locally Lipschitz functions. Let S be a subset of the normed space X which is closed near $\bar{x} \in \text{bdry } S$ and epi-Lipschitz at this point in a direction $h \in X$ with $h \neq 0$. By Definition 2.18 there are a topologically complemented closed vector hyperplane E of $\mathbb{R}h$ (so, $X = E \oplus \mathbb{R}h$), a neighborhood W of $\bar{x}$ in X, and a function $f : E \to \mathbb{R}$ Lipschitz continuous near $\pi_E \bar{x}$ (with E endowed with the induced norm and $x = \pi_E x + (\pi_h x)h$ with $\pi_E x \in E$ and $\pi_h x \in \mathbb{R}$) such that

$$W \cap S = W \cap \{u + rh : u \in E, r \in \mathbb{R}, f(u) \leq r\}.$$

Consider the linear isomorphism $\pi : X \to E \times \mathbb{R}$ defined by $\pi(x) := (\pi_E x, \pi_h x)$ and note that the function $g : X \to \mathbb{R}$ with $g(x) := f(\pi_E x) - \pi_h x$ is Lipschitz near $\bar{x}$ along with $W \cap S = W \cap \{x \in X : g(x) \leq 0\}$ and

$$g^\circ(\bar{x}; v) = f^\circ(\pi_E \bar{x}; \pi_E v) - \pi_h v.$$

With $\bar{v} := \pi^{-1}(0_E, 1)$ we have $g^\circ(\bar{x}; \bar{v}) = -1 < 0$, so $0 \notin \partial_C g(\bar{x})$. This says that any set epi-Lipschitz at $\bar{x} \in \text{bdry } S$ is locally around $\bar{x}$ the sublevel set of a locally Lipschitz function g with $0 \notin \partial_C g(\bar{x})$. The next proposition shows that the converse also holds true.

PROPOSITION 2.78. Let $g : X \to \mathbb{R}$ be a function which is Lipschitz near a point $\bar{x}$ of the normed space X and let $S := \{x \in X : g(x) \leq 0\}$. Assume that $\bar{x} \in \text{bdry } S$ and $0 \notin \partial_C g(\bar{x})$. Then, the set S is epi-Lipschitz at $\bar{x}$.

In fact, the set S is epi-Lipschitz at $\bar{x}$ in a nonzero direction v in the space X whenever $g^\circ(\bar{x}; v) < 0$.

PROOF. We know by (2.29) that, for every $v \in X$,

$$g^\circ(\bar{x}; v) = \limsup_{t\downarrow 0, (x', v') \to (\bar{x}, v)} t^{-1}\big(g(x' + tv') - g(x')\big).$$

Since $0 \notin \partial g(\bar{x})$, there is some nonzero $v \in X$ such that $g^\circ(\bar{x};v) < 0$, and hence the latter equality furnishes some real $\varepsilon > 0$ such that, for all $x' \in B(\bar{x},\varepsilon)$, $v' \in B(v,\varepsilon)$, and $t \in {]}0,\varepsilon[$, we have

$$t^{-1}\big(g(x'+tv') - g(x')\big) < 0, \quad \text{that is, } g(x'+tv') < g(x').$$

So, for all $x' \in S \cap B(\bar{x},\varepsilon)$, $v' \in B(v,\varepsilon)$, and $t \in {]}0,\varepsilon[$, we obtain

$$g(x'+tv') < g(x') \leq 0$$

(the second inequality being due to the inclusion $x' \in S$). This yields that

$$S \cap B(\bar{x},\varepsilon) + {]}0,\varepsilon[\, B(v,\varepsilon) \subset S,$$

which means that S is epi-Lipschitz at $\bar{x}$ according to Theorem 2.20. This justifies the first assertion of the proposition. The arguments for the second assertion are also contained above. □

Other results representing epi-Lipschitz sets as sublevel sets of Lipschitz functions plus a condition like the one in the above proposition will be established in Theorem 2.154 and Theorem 2.155.

In many of the results above the function f is supposed to be Lipschitz continuous on a neighborhood U of x and to take its values in $\mathbb{R} \cup \{-\infty,+\infty\}$. The set U being a neighborhood of x, we have already observed that $\partial_C f(x)$ coincides with $\partial_C(f_{|U})(x)$, where $f_{|U} : U \to \mathbb{R}$ is the restriction of f to U. Of course $f_{|U}$ is Lipschitz continuous on U. The next proposition tells us in particular that we may also, when $\partial_C f(x)$ is considered, suppose without loss of generality that f is real-valued and locally Lipschitz continuous on X instead of supposing that it is Lipschitz continuous merely near x.

PROPOSITION 2.79. *Let (E,d) be a metric space, S be a nonempty set of E and $f : S \to \mathbb{R}$ be a real-valued function defined on S and γ-Lipschitz on S for some real $\gamma \geq 0$, that is, $|f(u) - f(u')| \leq \gamma d(u,u')$ for all $u, u' \in S$. Then the function f can be extended to a γ-Lipschitz function to the whole metric space E; in particular the functions $\varphi, \Phi : E \to \mathbb{R}$ are two such extensions defined, for all $x \in X$, by*

$$\varphi(x) := \sup_{u \in S}\big(f(u) - \gamma d(u,x)\big) \text{ and } \Phi(x) := \inf_{u \in S}\big(f(u) + \gamma d(u,x)\big),$$

and $\varphi(x) \leq \Phi(x)$ for all $x \in E$. Further, any γ-Lipschitz extension $F : E \to \mathbb{R}$ satisfies $\varphi \leq F \leq \Phi$ on E.

PROOF. It suffices to prove the assertion corresponding to Φ. It is readily seen that $\Phi(x) = f(x)$ whenever $x \in S$ and thus Φ is an extension of f. Fix now $x, x' \in E$. For any $u \in S$ we have

$$f(u) + \gamma d(u,x) \leq f(u) + \gamma d(u,x') + \gamma d(x,x'),$$

so that

$$\inf_{u \in S}\big(f(u) + \gamma d(u,x)\big) \leq \inf_{u \in S}\big(f(u) + \gamma d(u,x') + \gamma d(x,x')\big)$$
$$= \inf_{u \in S}\big(f(u) + \gamma d(u,x')\big) + \gamma d(x,x'),$$

otherwise stated $\Phi(x) \leq \Phi(x') + \gamma d(x,x')$. This ensures on the one hand that Φ is finite on E (since $f(u_0) = \Phi(u_0) \leq \Phi(x') + \gamma d(u_0, x')$ for a fixed point $u_0 \in S$), and on the other hand that Φ is γ-Lipschitz on the whole space E. □

Through Theorem 2.70, we can express in terms of support functions the Hausdorff-Pompeiu excess and distance defined in Definition 1.61 whenever the involved sets are convex.

PROPOSITION 2.80. Let S, S' be two nonempty convex subsets of a normed space X. The following hold:
(a) For all $x \in X$, one has $d(x, S) = \sup\limits_{x^* \in \mathbb{B}_{X^*}} (\langle x^*, x \rangle - \sigma(x^*, S))$;
(b) the excess can be expressed as
$$\mathrm{exc}(S', S) = \sup_{x^* \in \mathbb{B}_{X^*} \cap \mathrm{dom}\, \sigma(\cdot, S)} (\sigma(x^*, S') - \sigma(x^*, S)),$$
so if S is addition bounded
$$\mathrm{exc}(S', S) = \sup_{x^* \in \mathbb{B}_{X^*}} (\sigma(x^*, S') - \sigma(x^*, S)).$$
(c) If both nonempty convex sets S and S' are bounded, one has the equality
$$\mathrm{haus}(S, S') = \sup_{x^* \in \mathbb{B}_{X^*}} |\sigma(x^*, S') - \sigma(x^*, S)|.$$

PROOF. (a) Taking any $x^* \in \mathbb{B}_{X^*}$, we have, for all $y \in S$,
$$\langle x^*, x \rangle - \sigma(x^*, S) = \langle x^*, x \rangle - \sup_{y \in S} \langle x^*, y \rangle$$
$$= \inf_{y \in S} \langle x^*, x - y \rangle \leq \inf_{y \in S} \|x - y\|,$$
thus $\sup\limits_{x^* \in \mathbb{B}_{X^*}} (\langle x^*, x \rangle - \sigma(x^*, S)) \leq d_S(x)$.

To prove the reverse inequality, take by the nonemptiness of $\partial d_S(x)$ (see Theorem 2.70(d)) some $u^* \in \partial d_S(x)$, so $\|u^*\| \leq 1$ by Theorem 2.70(d). Then for all $y \in S$, the convexity of d_S yields by Proposition 2.63
$$d_S(x) = d_S(x) - d_S(y) \leq \langle u^*, x \rangle - \langle u^*, y \rangle,$$
which entails
$$d_S(x) \leq \langle u^*, x \rangle + \inf_{y \in S} (-\langle u^*, y \rangle) = \langle u^*, x \rangle - \sigma(u^*, S).$$

This combines with the previous inequality guarantees the equality in (a).
(b) Clearly, by (a) one has $d_S(x) = \sup\limits_{x^* \in \mathbb{B}_{X^*} \cap D} (\langle x^*, x \rangle - \sigma(x^*, S))$, where $D := \mathrm{dom}\, \sigma(\cdot, S)$. So, concerning the first equality in (b), it suffices to write
$$\sup_{y \in S'} d_S(y) = \sup_{x^* \in \mathbb{B}_{X^*} \cap D} \sup_{y \in S'} (\langle x^*, y \rangle - \sigma(x^*, S))$$
$$= \sup_{x^* \in \mathbb{B}_{X^*} \cap D} (\sigma(x^*, S') - \sigma(x^*, S)).$$

The second equality in (b) follows directly from the first since $\mathrm{dom}\, \sigma(\cdot, S'') = X$ for any nonempty bounded set $S'' \subset X$.
(c) Finally (c) is a direct consequence of (b). □

2.2.6. Gradient representation of C-subdifferential of locally Lipschitz functions.
Taking the second equality of (b) of Proposition 2.74 into account, for a function f which is Lipschitz continuous near x, the set $\partial_C f(x)$ is also called the *Clarke generalized gradient* of f at x. The terminology is justified in finite dimensions mainly by Theorem 2.83 below.

Before stating the theorem, let us establish the Rademacher theorem concerning differentiability points of Lipschitz continuous functions. We start by recalling the concept of absolutely continuous mappings. Let u be a maping from an interval $I \subset \mathbb{R}$ into a normed vector space $(X, \|\cdot\|)$. One says that u is *absolutely continuous* on $[\alpha, \beta] \subset I$ provided for any $\varepsilon > 0$ there exists some $\delta > 0$ such that for all $s_i < t_i$ in $[\alpha, \beta]$, $i = 1, \cdots, p$, with $t_i < s_{i+1}$ for $i = 1, \cdots, p-1$ and $\sum_{i=1}^{p}(t_i - s_i) < \delta$ one has $\sum_{i=1}^{p}\|u(t_i) - u(s_i)\| < \varepsilon$. The mapping u is locally absolutely continuous on I whenever it is absolutely continuous on any $[\alpha, \beta] \subset I$. Obviously, u is locally absolutely continuous on I whenever it is locally Lipschitz continuous on I.

We recall the classical Lebesgue theorem.

THEOREM 2.81. *Let I be an interval of $\mathbb{R}$ with nonempty interior. Then any locally absolutely continuous function $u : I \to \mathbb{R}$ is derivable Lebesgue almost everywhere on I and for all $\tau, t \in I$ one has*

$$u(t) - u(\tau) = \int_{\tau}^{t} u'(s)\,ds,$$

where $u'(s)$ denotes the derivative of u at each s between τ and t where the derivative exists.

THEOREM 2.82 (H. Rademacher). *Any real-valued locally Lipschitz continuous function on an open set U of $\mathbb{R}^m$ is differentiable Lebesgue almost everywhere on the open set U.*

PROOF. According to Proposition 2.79 for extension of Lipschitz functions we may suppose that f is a real-valued Lipschitz function on $\mathbb{R}^m$. For each nonzero h, writing $\mathbb{R}^m = E_h \oplus \mathbb{R}h$ with some $(m-1)$-dimensional vector subspace E_h, the above Lebesgue theorem combined with the Fubini theorem ensures that, for almost all $x \in \mathbb{R}^m$ the bilateral limit $df(x;h) := \lim_{t \to 0} t^{-1}(f(x+th) - f(x))$ exists for all $h \in \mathbb{Q}^m$. Let us show that for almost all $x \in \mathbb{R}^m$ the function $df(x;\cdot)$ is additive on $\mathbb{Q}^m$. Fix any $h, k \in \mathbb{Q}^m$. Consider as usual any $\mathcal{C}^1$ function φ on $\mathbb{R}^m$ with compact support and with $\int_{\mathbb{R}^m} \varphi(x)\,dx = 1$. Denoting by g the integral convolution $g := f * \varphi$, that is, $g(x) = \int_{\mathbb{R}^m} f(y)\varphi(x-y)\,dy$, the function g is $\mathcal{C}^1$ and $Dg(x)(v) = (f * D\varphi(\cdot)(v))(x)$ for all $v \in \mathbb{R}^m$.

By the Lebesgue dominated convergence theorem we have, for every $x \in \mathbb{R}^m$,

$$Dg(x)(h) = \lim_{t \to 0} t^{-1}(g(x+th) - g(x))$$
$$= \int_{\mathbb{R}^m} \varphi(y) \lim_{t \to 0} t^{-1}(f(x-y+th) - f(x-y))\,dy,$$

which means that, for the bounded measurable function $df(\cdot;h)$ which is defined almost everywhere, we have $Dg(x)(h) = \varphi * df(\cdot;h)(x)$. The latter equality being true for every vector $h \in \mathbb{Q}^m$, it ensues that $\varphi * (df(\cdot;h+k) - df(\cdot;h) - df(\cdot;k)) = 0$.

For any integer $j \in \mathbb{N}$, applying the latter equality (in place of φ) with the mollifier φ_j defined by $\varphi_j(x) := j^m \varphi(jx)$, we obtain that

$$\varphi_j * \big(df(\cdot; h+k) - df(\cdot; h) - df(\cdot; k)\big) = 0.$$

Passing to the limit as $j \to \infty$ yields that $df(\cdot; h+k) - df(\cdot; h) - df(\cdot; k) = 0$ almost everywhere according to the classical result saying that, for any bounded measurable function ψ, the sequence $(\varphi_j * \psi)_j$ converges almost everywhere to ψ.

This and the countability of $\mathbb{Q}^m$ furnish a negligible set $N \subset \mathbb{R}^m$ such that for each $x \in \mathbb{R}^m \setminus N$ the function $df(x; \cdot)$ is well defined on $\mathbb{Q}^m$ and $df(x; h+k) = df(x; h) + df(x; k)$ for all $h, k \in \mathbb{Q}^m$. Fix any $x \in \mathbb{R}^m \setminus N$. The equicontinuity (in fact equi-Lipschitz property) of the family $(q_t)_{t \in \mathbb{R} \setminus \{0\}}$ with $q_t(v) := t^{-1}\big(f(x+tv) - f(x)\big)$ guarantees that $\lim_{t \to 0} q_t(v)$ exists for all $v \in \mathbb{R}^m$. It follows that the function $df(x; \cdot)$ is well defined and additive on $\mathbb{R}^m$, thus linear according to the obvious equality $df(x; \alpha v) = \alpha df(x; v)$. The Lipschitz function f on $\mathbb{R}^m$ is then Gâteaux differentiable, hence Fréchet differentiable, at x (see Proposition 2.42). $\square$

We can now prove the gradient representation result.

THEOREM 2.83. [F. Clarke: gradient representation of C-subdifferential] Let $f : \mathbb{R}^m \to \mathbb{R} \cup \{-\infty, +\infty\}$ be a function which is finite and Lipschitz continuous near $\overline{x} \in \mathbb{R}^m$. Then, denoting by Δ_f the set of points where ∇f exists and considering any set $D_f \subset \Delta_f$ such that $\Delta_f \setminus D_f$ is Lebesgue negligible, one has

$$f^\circ(\overline{x}; h) = \max\left\{\lim_{n \to \infty} \langle \nabla f(x_n), h \rangle : D_f \ni x_n \to \overline{x}\right\}$$

and

$$\partial_C f(\overline{x}) = \operatorname{co}\left\{\lim_{n \to \infty} \nabla f(x_n) : D_f \ni x_n \to \overline{x}\right\},$$

where in each right-hand member above are considered all sequences $(x_n)_n$ in D_f converging to $\overline{x}$ and such that the appropriate limit involving ∇f exists.

PROOF. Without loss of generality we may suppose that f is finite and globally Lipschitz on $\mathbb{R}^m$ with $\gamma \geq 0$ as a Lipschitz constant. Put $\alpha(h) := \limsup_{v \to 0, \overline{x}+v \in D_f} \langle \nabla f(\overline{x}+v), h\rangle$. Obviously $\alpha(h) \leq f^\circ(\overline{x}; h)$. Let us prove the reverse inequality for any fixed h with $\|h\| = 1$. Consider any real $\varepsilon > 0$ and take some $\delta > 0$ such that $\langle \nabla f(\overline{x}+v), h\rangle \leq \alpha(h) + \varepsilon$ for all $\overline{x}+v \in D_f \cap B(\overline{x}, \delta)$. Write $\mathbb{R}^m = E \oplus \mathbb{R}h$ for some vector space E with dimension equal to $m-1$ and write any $v \in \mathbb{R}^m$ as $v = v_1 + v_2 h$ with $v_1 \in E$ and $v_2 \in \mathbb{R}$. By the Rademacher theorem above, the set N of all $v \in \mathbb{R}^m$ where the function $u \mapsto f(\overline{x}+u)$ is not differentiable is Lebesgue null. According to the Fubini theorem there exists some $(m-1)$-Lebesgue null set N_1 in E such that for each $y \in Q := E \setminus N_1$ the set of $t \in \mathbb{R}$ with $\overline{x}+y+th \in D_f$ has its complement Lebesgue null in $\mathbb{R}$, and at each such t the Lipschitz continuous function $\tau \mapsto f(\overline{x}+y+\tau h)$ is derivable with $\langle \nabla f(\overline{x}+y+th), h\rangle$ as derivative. So, for any $y \in Q$, any $r \in \mathbb{R}$ and any $s > 0$,

$$s^{-1}\big(f(\overline{x}+y+(r+s)h) - f(\overline{x}+y+rh)\big) = s^{-1}\int_0^s \langle \nabla f(\overline{x}+y+(r+t)h), h\rangle\, dt.$$

Then, for any $v \in B(0, \delta/3)$ with $v_1 \in Q$ and any $0 < s < \delta/3$, we have

$$s^{-1}\big(f(\overline{x}+v+sh) - f(\overline{x}+v)\big) = s^{-1}\big(f(\overline{x}+v_1+(v_2+s)h) - f(\overline{x}+v_1+v_2 h)\big) \leq \alpha(h) + \varepsilon,$$

and according to the continuity of f on $B(\bar{x}, \delta)$ the latter inequality still holds for every $v \in B(0, \delta/3)$ and every $0 < s < \delta/3$ since the set of $v \in B(0, \delta/3)$ with $v_1 \in Q$ is dense in $B(0, \delta/3)$. The latter inequality ensures that $f°(\bar{x}; h) \leq \alpha(h) + \varepsilon$, thus $f°(\bar{x}; h) \leq \alpha(h)$ for all h with $\|h\| = 1$, and hence for all $h \in X$. This clearly implies

$$f°(\bar{x}; h) = \sup \left\{ \lim_n \langle \nabla f(x_n), h \rangle : D_f \ni x_n \to \bar{x} \right\}$$

and the supremum is a maximum because the set whose supremum is taken is compact according to the Lipschitz property of f near $\bar{x}$. Finally, this equality combined with the compactness of the set whose supremum is taken in its right member guarantees the equality of the proposition concerning $\partial_C f(\bar{x})$ in $\mathbb{R}^m$. □

EXAMPLE 2.84. (a) Let $f : \mathbb{R} \to \mathbb{R}$ be defined by $f(0) = 0$ and $f(x) = x^2 \sin(1/x)$ if $x \neq 0$. We have

$$\nabla f(0) = 0 \quad \text{and} \quad \nabla f(x) = 2x \sin(1/x) - \cos(1/x) \text{ if } x \neq 0.$$

The function f is Lipschitz continuous and obviously $\operatorname*{Lim\,sup}_{x \to 0}\{\nabla f(x)\} \subset [-1, 1]$. Further, fixing any $a \in \mathbb{R}$ and putting $x_n := (a + 2n\pi)^{-1}$ for every $n \in \mathbb{N}$, we see that $\nabla f(x_n) \to -\cos(a)$. It follows that $\operatorname*{Lim\,sup}_{x \to 0}\{\nabla f(x)\} = [-1, 1]$, so

$$\partial_C f(0) = [-1, 1] = \operatorname*{Lim\,sup}_{x \to 0}\{\nabla f(x)\}$$

according to Theorem 2.83 above.

(b) Let U be an open set of $\mathbb{R}^m$, let $f_1, f_2, f_3 : U \to \mathbb{R}$ be three $\mathcal{C}^1$ functions and let $f(x) = \max\{\min\{f_1(x), f_2(x)\}, f_3(x)\}$ be such that $\{x \in U : f_i(x) = f_j(x)\}$ is Lebesgue negligeable in $\mathbb{R}^m$ for each pair (i, j) with $i \neq j$. Let

$$U_1 := \{x \in U : f_2(x) > f_1(x) > f_3(x)\}, \quad U_2(x) := \{x \in U : f_1(x) > f_2(x) > f_3(x)\},$$

$$U_3; = \{x \in U : f_3(x) > \min\{f_1(x), f_2(x)\}\}.$$

Then for any $\bar{x} \in \bigcap_{i=1}^{3} \mathrm{cl}\, U_i$ one has

$$\partial_C f(\bar{x}) = \mathrm{co}\,\{\nabla f_1(\bar{x}), \nabla f_2(\bar{x}), \nabla f_3(\bar{x})\}.$$

It suffices to observe that $f(x) = f_i(x)$ if $x \in U_i$ for some $i \in \{1, 2, 3\}$ and to apply the above theorem (noticing that each set U_i is open). □

2.2.7. Clarke tangent and normal cones of epigraphs and graphs. We turn now to particular properties related to epigraphs of functions and graphs of mappings. We begin with some basic properties of the second component of a C-normal vector to the epigraph of a function.

PROPOSITION 2.85. Let U be an open set of the normed space X and $f : U \to \mathbb{R} \cup \{-\infty, +\infty\}$ be a function.

(a) For any point $x \in U$ where f is finite and for any real $s \geq f(x)$, one always has $(0, 1) \in T^C(\mathrm{epi}\, f; (x, s))$, hence

$$(x^*, -r) \in N^C(\mathrm{epi}\, f; (x, s)) \Rightarrow r \geq 0.$$

(b) For any point $x \in U$ around which f is finite and Lipschitz continuous with $\gamma \geq 0$ as a Lipschitz constant, one has

$$(x^*, -r) \in N^C(\mathrm{epi}\, f; (x, f(x))) \Rightarrow \|x^*\| \leq r\gamma,$$

hence in particular
$$(x^*, 0) \in N^C\bigl(\operatorname{epi} f; (x, f(x))\bigr) \Rightarrow x^* = 0.$$

PROOF. (a) Let $(t_n)_n$ be a sequence in $]0, +\infty[$ tending to 0 and $(x_n, s_n)_n$ be a sequence in epi f converging to (x, s). We observe that $(x_n, s_n + t_n) \in$ epi f, that is, $(x_n, s_n) + t_n(0, 1) \in$ epi f. It ensues that $(0, 1) \in T^C\bigl(\operatorname{epi} f; (x, s)\bigr)$.

Take now $(x^*, -r) \in N^C\bigl(\operatorname{epi} f; (x, s)\bigr)$. The previous inclusion tells us that $\langle (x^*, -r), (0, 1) \rangle \leq 0$, that is, $r \geq 0$ as desired.

(b) Now assume that f is Lipschitz continuous near x with $\gamma \geq 0$ as a Lipschitz constant and take $(x^*, -r) \in N^C\bigl(\operatorname{epi} f; (x, f(x))\bigr)$. Fix any $h \in X$. We know by the assertions (a) and (b) of Theorem 2.70 that $(h, f^\circ(x; h)) \in T^C\bigl(\operatorname{epi}; (x, f(x))\bigr)$ and $f^\circ(x; h) \leq \gamma \|h\|$. We then have $\langle x^*, h \rangle - r f^\circ(x; h) \leq 0$ and, since $r \geq 0$ by (a), we deduce
$$\langle x^*, h \rangle \leq r f^\circ(x; h) \leq r \gamma \|h\|.$$
This being true for all $h \in X$, we obtain $\|x^*\| \leq r\gamma$. □

By definition, any C-subgradient of f at x is connected to the C-normal cone $N^C\bigl(\operatorname{epi} f; (x, f(x))\bigr)$ in a unique way by the inclusion $(x^*, -1) \in N^C\bigl(\operatorname{epi} f; (x, f(x))\bigr)$. On the other hand, by the property (a) in Proposition 2.85 above any $(x^*, -r) \in N^C\bigl(\operatorname{epi} f; (x, f(x))\bigr)$ with $r \neq 0$ is connected to $\partial_C f(x)$ by the relations $r > 0$ and $r^{-1} x^* \in \partial_C f(x)$. The situation when $r = 0$ is quite particular. Assume that $X = \mathbb{R}^m$ and that $\mathbb{R}^m \times \mathbb{R}$ is endowed with its usual inner product. If a nonzero $\zeta \in \mathbb{R}^m$ is such that $(\zeta, 0) \in N^C\bigl(\operatorname{epi} f; (x, f(x))\bigr)$, then the boundary $\operatorname{Hypl}(x^*, 0) := \{(x, r) \in X \times \mathbb{R} : \langle x^*, x \rangle = 0\}$ of the associated half-space $\operatorname{Hypl}^{\leq}(\zeta, 0)$ (see (2.7)) is a vertical hyperplane in $\mathbb{R}^m \times \mathbb{R}$. Such a vector $\zeta \in \mathbb{R}^m$ then furnishes a "generalized horizon gradient direction" of f at x.

DEFINITION 2.86. Let $f : U \to \mathbb{R}\{-\infty, +\infty\}$ be a function on an open set U of the normed space X and let $x \in U$ with $|f(x)| < +\infty$. Any $x^* \in X^*$ such that $(x^*, 0) \in N^C(\operatorname{epi} f; (x, f(x)))$ is called a *horizon C-subgradient* of f at x. The set of all horizon C-subgradients of f at x is called the *horizon C-subdifferential* of f at x and denoted by $\partial_C^\infty f(x)$. If f is not finite at x, one puts $\partial_C^\infty f(x) = \emptyset$.

It clear from this definition and Proposition 2.85(a) that
$$(2.32) \quad N^C(\operatorname{epi} f; (x, f(x))) = \mathbb{R}_+\bigl(\partial_C f(x) \times \{-1\} \cup \partial_C^\infty f(x) \times \{0\}\bigr),$$
which provides a way to recover $N^C(\operatorname{epi} f; (x, f(x)))$ from both sets $\partial_C f(x)$ and $\partial_C^\infty f(x)$. We also note by Proposition 2.85(b) above that $\partial_C^\infty f(x) = \{0\}$ whenever f is Lipschitz continuous near x. According to this and (2.32) and to the nonvacuity of $\partial_C f(x)$ when f is finite and Lipschitz near x, we deduce the following proposition.

PROPOSITION 2.87. Let $f : U \to \mathbb{R}\{-\infty, +\infty\}$ be a function on an open set U of the normed space X. If f is finite and Lipschitz continuous near $x \in U$, then
$$N^C(\operatorname{epi} f; (x, f(x))) = \mathbb{R}_+\bigl(\partial_C f(x) \times \{-1\}\bigr).$$

It is worth pointing out that, for $s > f(x)$, the inclusion $T^C\bigl(\operatorname{epi} f; (x, f(x))\bigr) \subset T^C\bigl(\operatorname{epi} f; (x, s)\bigr)$ may fail even for lower semicontinuous function. Indeed, considering the simple lower semicontinuous function $f : \mathbb{R} \to \mathbb{R}$ with $f(u) = u$ if $u \leq 0$ and $f(u) = 1$ for $u > 0$, we see that $T^C\bigl(\operatorname{epi} f; (0, f(0))\bigr) = \{(u, r) \in \mathbb{R}^2 : u \leq 0, r \geq u\}$

whereas $T^C(\text{epi } f;(0,1)) =\,]-\infty,0]\times[0,+\infty[$. However the inclusion holds true for continuous functions as it is proved below.

PROPOSITION 2.88. *Let U be an open set of the normed space X and $f: U \to \mathbb{R} \cup \{-\infty, +\infty\}$ be a function. Then, at any point $x \in U$ where f is finite and continuous one has for any real $s \geq f(x)$*

$$T^C(\text{epi } f;(x,f(x))) \subset T^C(\text{epi } f;(x,s))$$

and $\quad N^C(\text{epi } f;(x,s)) \subset N^C(\text{epi } f;(x,f(x))).$

PROOF. Let $x \in U$ be a point where f is finite and continuous. Fix any (h,r) in $T^C(\text{epi } f;(x,f(x)))$. Take any sequence $(t_n)_n$ tending to 0 with $t_n > 0$ and any sequence $(x_n, s_n)_n$ in epi f converging to (x,s). The continuity of f at x says that $f(x_n) \to f(x)$ and $f(x_n)$ is finite for n large enough, say $n \geq N$. Then, there exists some sequence $(h_n, r_n)_n$ converging to (h,r) with $(x_n, f(x_n)) + t_n(h_n, r_n)$ in epi f for all $n \geq N$. From this it is readily seen that $(x_n, s_n) + t_n(h_n, r_n)$ is in epi f for all $n \geq N$, which proves that (h,r) belongs to $T^C(\text{epi } f;(x,s))$. This translates the inclusion concerning the Clarke tangent cones and the inclusion of the Clarke normal cones is a direct consequence of the latter. □

The Clarke tangent and normal cones to the graphs of locally Lipschitz continuous mappings between normed spaces have also particular properties. They are closed vector subspaces; this extends Corollary 2.32 which can be seen as the case of graphs of locally Lipschitz real-valued functions.

PROPOSITION 2.89. *Let $G: X \to Y$ be a mapping between two normed spaces which is Lipschitz continuous near $\bar{x} \in X$. Then $(h,k) \in T^C(\text{gph } G;(\bar{x}, G(\bar{x})))$ if and only if*

$$\lim_{t\downarrow 0, u\to \bar{x}} t^{-1}(G(u+th) - G(u)) = k,$$

that is, the limit exists and is equal to k.

Consequently, the Clarke tangent and normal cones to gph G at $(\bar{x}, G(\bar{x}))$ are vector subspaces of $X \times Y$ and $X^* \times Y^*$ respectively.

PROOF. Let $h \in X$ such that $\lim_{t\downarrow 0, u\to \bar{x}} t^{-1}(G(u+th) - G(u))$ exists and is equal to k. Then, for any sequence $(x_n)_n$ converging to $\bar{x}$ and for any sequence of positive numbers $(t_n)_n$ tending to 0, setting $k_n := t_n^{-1}(G(x_n + t_n h) - G(x_n))$ we have

$$(x_n, G(x_n)) + t_n(h, k_n) \in \text{gph } G \quad \text{and} \quad (h, k_n) \to (h, k),$$

which means $(h,k) \in T^C(\text{gph } G;(\bar{x}, G(\bar{x})))$.

Suppose now that $(h,k) \in T^C(\text{gph } G;(\bar{x}, G(\bar{x})))$ and take an arbitrary sequence $(x_n)_n$ converging to $\bar{x}$ and an arbitrary sequence of positive reals $(t_n)_n$ tending to 0. The sequential characterization of the Clarke tangent cone in Theorem 2.2 gives a sequence $(h_n, k_n)_n$ converging to (h,k) such that

$$(x_n, G(x_n)) + t_n(h_n, k_n) \in \text{gph } G \quad \text{for all } n \text{ large enough}.$$

This entails the equality $k_n = t_n^{-1}(G(x_n + t_n h_n) - G(x_n))$ for n sufficiently large. Denoting by γ a Lipschitz constant of G near $\bar{x}$, we have for large n

$$r_n := \|t_n^{-1}(G(x_n + t_n h_n) - G(x_n + t_n h))\| \leq \gamma \|h_n - h\|, \quad \text{hence} \quad r_n \to 0.$$

Consequently, the inequality $\|k_n - t_n^{-1}(G(x_n + t_n h) - G(x_n))\| \leq r_n$ for large n yields that $t_n^{-1}(G(x_n + t_n h) - G(x_n)) \to k$, which finishes the proof of the first assertion.

Concerning the second assertion, the Clarke tangent cone to $\mathrm{gph}\, G$ being a convex cone (see Theorem 2.2), it is enough to show that the opposite of any vector of this tangent cone remains in this cone. Consider any $(h, k) \in T^C(\mathrm{gph}\, G; (\overline{x}, G(\overline{x})))$. Take any sequence $(x_n)_n$ converging to $\overline{x}$ and any sequence of positive reals $(t_n)_n$ tending to 0. Since, for $x_n' := x_n - t_n h$, the sequence $(x_n')_n$ converges to $\overline{x}$, the assertion above tells us that $t_n^{-1}(G(x_n' + t_n h) - G(x_n')) \to k$, which means $t_n^{-1}(G(x_n + t_n(-h)) - G(x_n)) \to -k$. Thus, the assertion above once again entails that $(-h, -k) \in T^C(\mathrm{gph}\, G; (\overline{x}, G(\overline{x})))$, completing the proof. $\square$

2.2.8. C-subdifferential and directionally Lipschitz functions.

The assertion (c) of Theorem 2.70 says in particular that the epigraph $\mathrm{epi}\, f$ is epi-Lipschitz at $\overline{x}$ whenever f is Lipschitz continuous near $\overline{x}$. We may then think that an analytic Lipschitz-like description of the epi-Lipschitz property of the epigraph of a function $f : X \to \mathbb{R} \cup \{-\infty, +\infty\}$ exists. This is confirmed by Proposition 2.90 below. Before stating the proposition, given a function f on X (resp. a set S in X) recall our notation $x \to_f \overline{x}$ (resp. $S \ni x \to \overline{x}$) means $x \to \overline{x}$ and $f(x) \to f(\overline{x})$ (resp. $x \to \overline{x}$ with $x \in S$).

PROPOSITION 2.90. *Let X be a normed space and $f : U \to \mathbb{R} \cup \{-\infty, +\infty\}$ be a function on an open set U of X which is finite at $\overline{x} \in U$ and let $h \in X$. The following assertions are equivalent:*
(a) *there exists some $r \in \mathbb{R}$ such that $(h, r) \in I(\mathrm{epi}\, f; (\overline{x}, f(\overline{x})))$;*
(b) *there exists some $\beta \in \mathbb{R}$ and $\delta, \varepsilon > 0$ such that for all $h' \in B(h, \delta)$, $t \in\,]0, \varepsilon[$, $x' \in \mathrm{dom}\, f$ with $\|x' - \overline{x}\| < \delta$ and $s' \geq f(x')$ with $|s' - f(\overline{x})| < \varepsilon$, one has $x' + th' \in U$ and*
$$t^{-1}(f(x' + th') - s') \leq \beta;$$
(c) $\displaystyle\limsup_{\substack{E_f \ni (x', s') \to (\overline{x}, \overline{s}),\, h' \to h \\ t \downarrow 0}} t^{-1}(f(x' + th') - f(x')) < +\infty$, *where $E_f := \mathrm{epi}\, f$ and $\overline{s} := f(\overline{x})$.*

If f is lower semicontinuous at $\overline{x}$, the assertions (d) and (e) below can be added to the list:
(d) *there exists some $\beta \in \mathbb{R}$ and $\delta, \varepsilon > 0$ such that for all $h' \in B(h, \delta)$, $t \in\,]0, \varepsilon[$, and $x' \in U$ with $\|x' - \overline{x}\| < \delta$ and $|f(x') - f(\overline{x})| < \varepsilon$, one has $x' + th' \in U$ and*
$$t^{-1}(f(x' + th') - f(x')) \leq \beta;$$
(e) $\displaystyle\limsup_{\substack{x' \to_f \overline{x},\, h' \to h \\ t \downarrow 0}} t^{-1}(f(x' + th') - f(x')) < +\infty.$

PROOF. We will only consider the case when f is lower semicontinuous at $\overline{x}$, the general case being similar and even easier. The equivalence between (d) and (e) is obvious. Suppose (a), that is, $(h, r) \in I(\mathrm{epi}\, f; (\overline{x}, f(\overline{x})))$ for some $r \in \mathbb{R}$. There exist some reals $\delta, \varepsilon > 0$ such that

$$(\mathrm{epi}\, f) \cap \big(B(\overline{x}, \delta) \times]f(\overline{x}) - \varepsilon, f(\overline{x}) + \varepsilon[\big) +]0, \varepsilon[(B(h, \delta) \times]r - \delta, r + \delta[) \subset \mathrm{epi}\, f.$$

For $t \in]0, \varepsilon[$, $h' \in B(h, \delta)$, and $x' \in U \cap B(\overline{x}, \delta)$ with $|f(x') - f(\overline{x})| < \varepsilon$, we then obtain

$$(x', f(x')) + t(h', r) \in \text{epi } f, \quad \text{that is, } t^{-1}(f(x' + th') - f(x')) \le r,$$

which obviously implies (d).

Suppose now that the property of (d) is fulfilled. There are $\beta \in \mathbb{R}$ and $\delta', \varepsilon > 0$ such that $B(\overline{x}, \delta') \subset U$ and for any $t \in]0, \varepsilon[$, $h' \in B(h, \delta')$, and $x' \in X$ with $\|x' - \overline{x}\| < \delta'$ and $|f(x') - f(\overline{x})| < \varepsilon$, we have $x' + th' \in U$ and

(2.33) $\quad t^{-1}(f(x' + th') - f(x')) \le \beta, \quad \text{that is, } f(x' + th') \le f(x') + t\beta.$

By the lower semicontinuity of f at $\overline{x}$, choose some positive real $\delta < \delta'$ such that $f(\overline{x}) - \varepsilon < f(x')$ for all $x' \in B(\overline{x}, \delta)$. Put $r := \beta + \delta$ and take any $t \in]0, \varepsilon[$, $(h', s') \in B(h, \delta) \times]r - \delta, r + \delta[$, and $(x', r') \in (\text{epi } f) \cap \big(B(\overline{x}, \delta) \times]f(\overline{x}) - \varepsilon, f(\overline{x}) + \varepsilon[\big)$. Then

$$f(\overline{x}) - \varepsilon < f(x') \le r' < f(\overline{x}) + \varepsilon,$$

hence by (2.33) we obtain

$$f(x' + th') \le f(x') + t\beta \le r' + ts', \quad \text{thus } (x', r') + t(h', s') \in \text{epi } f.$$

This means

$$(\text{epi } f) \cap \big(B(\overline{x}, \delta) \times]f(\overline{x}) - \varepsilon, f(\overline{x}) + \varepsilon[\big) +]0, \varepsilon[\big(B(h, \varepsilon) \times]r - \delta, r + \delta[\big) \subset \text{epi } f,$$

hence $(h, r) \in I\big(\text{epi } f; (\overline{x}, f(\overline{x}))\big)$. The equivalence between (a) and (d) is proved. $\square$

The Lipschitz-like property (b) above is translated in the following definition.

DEFINITION 2.91. Let $f : X \to \mathbb{R} \cup \{-\infty, +\infty\}$ be a function which is finite at $\overline{x} \in X$. One says that f is *directionally Lipschitz at $\overline{x}$ in the direction $h \in X$* (or *with respect to $h \in X$*) whenever there exists some $r \in \mathbb{R}$ such that $(h, r) \in I\big(\text{epi } f; (\overline{x}, f(\overline{x}))\big)$; when such a vector h exists, f is said to be *directionally Lipschitz*, or *directionally Lipschitzian*, at $\overline{x}$. The above properties hold for a function $f : U \to \mathbb{R} \cup \{-\infty, +\infty\}$ on an open set $U \ni \overline{x}$ in X if they hold for the extension of f by $+\infty$ on $X \setminus U$.

The function f is directionally Lipschitz at $\overline{x}$ in the direction h if and only if the property (b) in Proposition 2.90 holds true, or equivalently the property (d) when f is lower semicontinuous at $\overline{x}$. More precisely in the case of lower semicontinuity of f at $\overline{x}$, for the constants in (d) of the proposition, one says that f is *directionally Lipschitz at $\overline{x}$ in the direction h with the constants $\beta \in \mathbb{R}$, $\varepsilon > 0$ and $\delta > 0$*.

The equality in Proposition 2.74(b) concerning the Clarke subdifferential of the opposite of a locally Lipschitz continuous function can be extended to directionally Lipschitz functions:

PROPOSITION 2.92. *Let X be a normed space and $f : U \to \mathbb{R} \cup \{-\infty, +\infty\}$ be a function on an open set U of X which is finite at $\overline{x} \in U$ and directionally Lipschitz at $\overline{x}$. Then one has*

$$\partial_C(-f)(\overline{x}) = -\partial_C f(\overline{x}), \quad \text{hence } \partial_C(rf)(\overline{x}) = r\partial_C f(\overline{x}) \quad \text{for all } r \in \mathbb{R}.$$

PROOF. We may extend f to the whole space X by setting $f(x) = +\infty$ for all $x \in X$. Put $H_f^< := \{(u, s) \in X \times \mathbb{R} : s < f(u)\} \cup \{(\overline{x}, f(\overline{x}))\}$ and $H_f := \{(u, s) \in X \times \mathbb{R} : s \le f(u)\} \cup \{(\overline{x}, f(\overline{x}))\}$. Since $H_f^<$ is the union of $\{(\overline{x}, f(\overline{x}))\}$ and the complement in $X \times \mathbb{R}$ of epi f, by Proposition 2.29 we have $T^C(\text{epi } f; (\overline{x}, f(\overline{x}))) = -T^C(H_f^<; (\overline{x}, f(\overline{x})))$. Further, the Clarke tangent cone of $H_f^<$ at $(\overline{x}, f(\overline{x}))$ coincides with the Clarke tangent cone of H_f at $(\overline{x}, f(\overline{x}))$ by Proposition 2.5(b) since $H_f^< \subset H_f \subset \text{cl}(H_f^<)$, where the second inclusion can be seen with the fact that any pair $(u, s) \in H_f$ is the limit of the sequence of pairs $(u, s - n^{-1})_n$ of $H_f^<$. Observe also that $(h, r) \in T^C(H_f; (\overline{x}, f(\overline{x})))$ if and only if $(h, -r) \in T^C(\text{epi } (-f); (\overline{x}, -f(\overline{x})))$, as easily seen from the equivalence $(u, s) \in H_f \Leftrightarrow (u, -s) \in \text{epi}(-f)$. It results that

$$\begin{aligned} x^* \in \partial_C(-f)(\overline{x}) &\Leftrightarrow \langle x^*, h \rangle - r \leq 0, \; \forall (h, r) \in T^C(\text{epi }(-f); (\overline{x}, -f(\overline{x}))) \\ &\Leftrightarrow \langle -x^*, -h \rangle - r \leq 0, \; \forall (h, -r) \in T^C(H_f; (\overline{x}, f(\overline{x}))) \\ &\Leftrightarrow \langle -x^*, -h \rangle - r \leq 0, \; \forall (-h, r) \in T^C(\text{epi } f; (\overline{x}, f(\overline{x}))) \\ &\Leftrightarrow -x^* \in \partial_C f(\overline{x}), \end{aligned}$$

and this finishes the proof. □

REMARK 2.93. The result in the above proposition fails for non directionally Lipschitz functions even when f is continuous. For the continuous function $f : \mathbb{R} \to \mathbb{R}$ with $f(x) := \sqrt{|x|}$, we have

$$T^C(\text{epi } f; (0, 0)) = \{0\} \times [0, +\infty[\quad \text{and} \quad T^C(\text{epi }(-f); (0, 0)) = \{0\} \times \mathbb{R},$$

which gives

$$N^C(\text{epi } f; (0, 0)) = \mathbb{R} \times]-\infty, 0] \quad \text{and} \quad N^C(\text{epi }(-f); (0, 0)) = \mathbb{R} \times \{0\},$$

hence $\partial_C f(0) = \mathbb{R}$ whereas $\partial_C(-f)(0) = \emptyset$. □

We saw in Remark 2.75 examples of non-locally Lipschitz continuous functions f such that the Clarke subdifferential multimapping $\partial_C f$ is not sequentially closed at $\overline{x}$. The next proposition shows that the desired closedness property at $\overline{x}$ holds true whenever f is directionally Lipschitz at $\overline{x}$, extending in this way the closedness property in Proposition 2.74(a) relative to locally Lipschitz continuous functions.

PROPOSITION 2.94. Let $(X, \|\cdot\|)$ be a normed space, S be a nonempty subset of X, and $f : U \to \mathbb{R} \cup \{-\infty, +\infty\}$ be a function on an open set U of X. The following hold:
(a) If S has the interior tangent cone property at a point $\overline{x} \in S$, then for any net $(x_j)_{j \in J}$ in S converging to $\overline{x}$ in $(X, \|\cdot\|)$ and any net $(x_j^*)_{j \in J}$ converging weakly star to x^* with $x_j^* \in N^C(S; x_j)$, one has $x^* \in N^C(S; \overline{x})$; this means that the C-normal cone multimapping $N^C(S; \cdot)$ is norm-to-weak* closed (or outer semicontinuous) at $\overline{x}$.
(b) If f is directionally Lipschitz at a point $\overline{x} \in U$ with $|f(\overline{x})| < +\infty$, then for any net $(x_j)_{j \in J}$ converging to $\overline{x}$ in $(X, \|\cdot\|)$ with $f(x_j) \to f(\overline{x})$ and for any net $(x_j^*)_{j \in J}$ converging weakly star to x^* with $x_j^* \in \partial_C f(x_j)$, one has $x^* \in \partial_C f(\overline{x})$.

PROOF. It suffices to prove (a), since (b) is obtained by applying (a) to epi f. Consider any $h \in I(S; \overline{x})$ and observe from the definition of $I(S; \overline{x})$ that there is an open neighborhood U of $\overline{x}$ such that $h \in I(S; x)$ for all $x \in S \cap U$. Taking some $j_0 \in J$ such that $x_j \in U$ for all $j \succeq j_0$, we see for all such j that $\langle x_j^*, h \rangle \leq 0$, hence $\langle x^*, h \rangle \leq 0$. Since $T^C(S; x) = \text{cl}(I(S; \overline{x}))$ according to the interior tangent cone property of S at $\overline{x}$ (see Theorem 2.13(b)), we deduce that $\langle x^*, h \rangle \leq 0$, for all $h \in T^C(S;\bar{x})$, which means that $x^* \in N^C(S;\bar{x})$; this translates the f-graphical norm-to-weak* outer semicontinuity of the multimapping $\partial_C f$ at $\bar{x}$. □

The directional Lipschitz property under convexity will be studied in Subsection 2.3.2 as well as its relationship with the Rockafellar directional derivative $f^\uparrow(\cdot;\cdot)$.

2.2.9. Clarke tangent and normal cones through distance function. Above, several properties of the Clarke directional derivative and subdifferential of a locally Lipschitz continuous function are given. For the particular important Lipschitz continuous distance function d_S from a set S, the additional properties below are relevant in many situations. Among such properties, we have other useful descriptions of the Clarke tangent and normal cones of S.

PROPOSITION 2.95. Let S be a subset of the normed space $(X, \|\cdot\|)$ and $x \in S$. Then the following hold:
(a) $(d_S)^\circ(x;h) = \limsup_{t\downarrow 0, S \ni x' \to x} t^{-1} d_S(x' + th)$ for all $h \in X$.
(b) $h \in T^C(S;x) \Leftrightarrow (d_S)^\circ(x;h) \leq 0 \Leftrightarrow (d_S)^\circ(x;h) = 0$.
(c) The normal cone $N^C(S;x))$ is related to $\partial_C d_S(x)$ by the equality
$$N^C(S;x) = \mathrm{cl}_{w^*}\left(\mathbb{R}_+ \partial_C d_S(x)\right),$$
where cl_{w^*} denotes the weak* closure; in particular
$$\partial_C d_S(x) \subset N^C(S;x) \cap \mathbb{B}_{X^*}.$$

PROOF. (a) The right member of (a) being obviously not greater than the left one, let us prove the reverse inequality. Fix $h \in X$. For any $t > 0$ and any $u \in X$ choose some $y(t,u) \in S$ satisfying $\|u - y(t,u)\| < d_S(u) + t^2$ and observe that $y(t,u) \to x$ as $t \downarrow 0$ and $u \to x$. We then have the inequality
$$t^{-1}\bigl(d_S(u+th) - d_S(u)\bigr) \leq t^{-1}\bigl(d_S(u+th) - \|u - y(t,u)\|\bigr) + t \leq t^{-1} d_S(y(t,u) + th) + t,$$
which guarantees that $(d_S)^\circ(x;h) \leq \limsup_{t\downarrow 0, S\ni x' \to x} t^{-1} d_S(x'+th)$. So, the assertion (a) is established.

(b) Take now any $h \in T^C(S;x) = \liminf_{t\downarrow 0, u \xrightarrow{S} x} \frac{1}{t}(S - u)$. The scalarization characterization (through the distance function) of the inner limit of multimappings in Proposition 1.12(a) says that the inclusion $h \in \liminf_{t\downarrow 0, u \xrightarrow{S} x} \frac{1}{t}(S - u)$ is equivalent to $\limsup_{t\downarrow 0, S \ni u \to x} \frac{1}{t} d(u + th, S) = 0$ (since $d(h, \frac{1}{t}(S - u)) = \frac{1}{t} d(u + th, S)$), which is in turn equivalent to $(d_S)^\circ(x;h) = 0$ according to the assertion (a). The non-negativity of $(d_S)^\circ(x;h)$ (see (a)) ensures also that the equality $(d_S)^\circ(x;h) = 0$ is equivalent to $(d_S)^\circ(x;h) \leq 0$, which completes the proof of the assertion in (b).

(c) Observing for $x^* \in \partial_C d_S(x)$ that, for any $h \in T^C(S;x)$, we have $\langle x^*, h \rangle \leq (d_S)^\circ(x;h) = 0$ according to (b), we see that the right member of the equality in (c) is included in $\bigl(T^C(S:x)\bigr)^\circ = N^C(S;x)$.

Fix now any $h \in X$ in the negative polar cone of the right member of (c). Then we have $\langle x^*, h \rangle \leq 0$ for all $x^* \in \partial_C d_S(x)$, hence by (d) in Theorem 2.70 we obtain $(d_S)^\circ(x;h) \leq 0$, which entails $h \in T^C(S;x)$ according to (b). This establishes the equality in (c) and finishes the proof of the proposition because the final inclusion $\partial_C d_S(x) \subset N^C(S;x) \cap \mathbb{B}_{X^*}$ follows from the above equality and the Lipschitz property of d_S with 1 as a Lipschitz constant. □

EXAMPLE 2.96. In the Euclidean space $\mathbb{R}^2$ the following examples show that the inclusion $\partial_C d_S(x) \subset N^C(S;x) \cap \mathbb{B}_{X^*}$ in (c) of the above proposition may be strict.

(a) Let $S \subset \mathbb{R}^2$ be the graph of the function $-|\ |$, that is, $S := \{(r,s) \in \mathbb{R}^2 : s = -|r|\}$. Clearly $T^C(S;0_{\mathbb{R}^2}) = \{0_{\mathbb{R}^2}\}$, hence $N^C(S;0_{\mathbb{R}^2}) = \mathbb{R}^2$ or equivalently $N^C(S;0_{\mathbb{R}^2}) \cap \mathbb{B}_{\mathbb{R}^2} = \mathbb{B}_{\mathbb{R}^2}$.

On the other hand,

$$d_S(r,s) = \begin{cases} \sqrt{r^2+s^2} & \text{if } s \geq |r| \\ |r+s|/\sqrt{2} & \text{if } s \leq r \text{ and } r \geq 0 \\ |r-s|/\sqrt{2} & \text{if } s \leq -r \text{ and } r \leq 0, \end{cases}$$

so the gradient of d_S takes the values

$$\nabla d_S(r,s) = \begin{cases} \frac{1}{\sqrt{r^2+s^2}}(r,s) & \text{if } s > |r| \\ \frac{1}{\sqrt{2}}(1,1) & \text{if } -r < s < r \\ \frac{1}{\sqrt{2}}(-1,-1) & \text{if } s < -r \text{ and } r > 0 \\ \frac{1}{\sqrt{2}}(-1,1) & \text{if } r < s < -r \\ \frac{1}{\sqrt{2}}(1,-1)) & \text{if } s < r \text{ and } r < 0). \end{cases}$$

Consequently, by the gradient representation of Clarke subdifferential in Theorem 2.83 we have

$$\partial_C d_S(0_{\mathbb{R}^2}) = \overline{\mathrm{co}}\Big(\{-a,a,-b,b\} \cup \{\zeta_{r,s} : s > |r|\}\Big),$$

where $a = (1/\sqrt{2}, 1/\sqrt{2})$, $b = (-1/\sqrt{2}, 1/\sqrt{2})$, and $\zeta_{r,s} := (r^2+s^2)^{-1/2}(r,s)$. From the latter equality we see that

$$\partial_C d_S(0_{\mathbb{R}^2}) \neq N^C(S;0_{\mathbb{R}^2}) \cap \mathbb{B}_{\mathbb{R}^2}.$$

(b) The above set S is not epi-Lipschitz at $(0,0)$. The next example shows that the equality can fail even with epi-Lipschitz sets. Let $S' \subset \mathbb{R}^2$ be the epigraph of $-|\ |$, that is, $S' := \{(r,s) \in \mathbb{R}^2 : s \geq -|r|\}$. The set S' is obviously epi-Lipschitz and it is not difficult to see that $T^C(S';0_{\mathbb{R}^2}) = \{(h,k) \in \mathbb{R}^2 : k \geq |h|\}$, so

$$N^C(S';0_{\mathbb{R}^2}) \cap \mathbb{B}_{\mathbb{R}^2} = \{(\rho,\sigma) \in \mathbb{R}^2 : \sigma \leq -|\rho|, \rho^2 + \sigma^2 \leq 1\}.$$

On the other hand,

$$d_{S'}(r,s) = \begin{cases} 0 & \text{if } s \geq -|r| \\ |r+s|/\sqrt{2} & \text{if } s \leq -r \text{ and } r \geq 0 \\ |r-s|/\sqrt{2} & \text{if } s \leq r \text{ and } r \leq 0, \end{cases}$$

thus the gradient of $d_{S'}$ takes the values

$$\nabla d_{S'}(r,s) = \begin{cases} 0 & \text{if } s > -|r| \\ \frac{1}{\sqrt{2}}(-1,-1) & \text{if } s < -r \text{ and } r > 0 \\ \frac{1}{\sqrt{2}}(1,-1)) & \text{if } s < r \text{ and } r < 0). \end{cases}$$

It follows again from the gradient representation of Clarke subdifferential in Theorem 2.83 that

$$\partial_C d_{S'}(0_{\mathbb{R}^2}) = \mathrm{co}\{-a, -b, (0,0)\},$$

where a, b are as in (a). It is then clear that

$$\partial_C d_{S'}(0_{\mathbb{R}^2}) \neq N^C(S';0_{\mathbb{R}^2}) \cap \mathbb{B}_{\mathbb{R}^2}.$$

It is also worth observing that the above expression of $d_{S'}(r, s)$ can be written in the form $d_{S'}(r, s) = 0$ if $(r, s) \in S'$ and $d_{S'}(r, s) = ||r| + s|/\sqrt{2}$ if $(r, s) \notin S'$. □

A particular important case where the inclusion in (c) of Proposition 2.95 is an equality is that of convex sets. More general cases will be seen in Chapter 15.

PROPOSITION 2.97. *Let S be a convex subset of the normed space $(X, \|\cdot\|)$. Then, for any $x \in S$ the following equalities hold*
$$\partial d_S(x) = N(S; x) \cap \mathbb{B}_{X^*} \quad \text{and} \quad N(S; x) = \mathbb{R}_+ \partial d_S(x).$$

PROOF. Fix any $x \in S$. By Proposition 2.95(c) it suffices, for the left equality, to prove the inclusion of the right-hand side into the left. Fix any $x^* \in N(S; x) \cap \mathbb{B}_{X^*}$. Then, for any $x' \in X$ we have by Theorem 2.2(d), for all $u \in S$,
$$\langle x^*, x' - x \rangle = \langle x^*, x' - u \rangle + \langle x^*, u - x \rangle \leq \langle x^*, x' - u \rangle \leq \|x' - u\|,$$
which entails $\langle x^*, x' - x \rangle \leq d_S(x') = d_S(x') - d_S(x)$, for all $x' \in X$. This means that $x^* \in \partial d_S(x)$ as desired.

The second equality is a direct consequence of the first. □

2.2.10. Rockafellar theorem for C-subdifferential of finite sum of functions. We establish now a crucial subdifferential calculus rule for the Clarke subdifferential of the sum of two functions, one of them being locally Lipschitz continuous.

THEOREM 2.98 (R.T. Rockafellar's theorem for C-subdifferential sum rule). *Let U be an open set of the normed space X and $f_1, \cdots, f_m : U \to \mathbb{R} \cup \{-\infty, +\infty\}$ be functions on U. Assume that f_1 is finite at $\bar{x} \in U$ and $f_2, \cdots, f_m$ are finite on U and locally Lipschitz continuous near $\bar{x}$. Then*
$$\partial_C(f_1 + \cdots + f_m)(\bar{x}) \subset \partial_C f_1(\bar{x}) + \cdots + \partial_C f_m(\bar{x}).$$

EXAMPLE 2.99. The Lipschitz condition of $m - 1$ functions is crucial in the above theorem. We have seen in Example 2.36(b) that for the convex function $f_1 : \mathbb{R} \to \mathbb{R} \cup \{+\infty\}$ defined by $f_1(x) := -\sqrt{x}$ if $x \geq 0$ and $f_1(x) := +\infty$ if $x < 0$, we have $\partial f_1(0) = \emptyset$. This ensures with $f_2(x) := f_1(-x)$ for all $x \in \mathbb{R}$, we have $\partial_C f_2(0) = \emptyset = \partial_C f_1(0)$ whereas $\partial_C(f_1 + f_2)(0) = \mathbb{R}$ since $f_1 + f_2 = \Psi_{\{0\}}$. The inclusion in the above theorem then fails for these (convex) functions f_1, f_2 at the point $\bar{x} = 0$. □

It is enough to prove the theorem for $m = 2$ with $f_1 =: f$ finite at $\bar{x}$ and $f_2 =: g$ finite on U and locally Lipschitz continuous near $\bar{x}$. We may also suppose that $U = X$. The proof will use the next lemmas. To state them put
$$S := \{(u, r, s) \in X \times \mathbb{R} \times \mathbb{R} : f(u) + g(u) \leq r + s\},$$
$S_1 := \{(u, r, s) \in X \times \mathbb{R} \times \mathbb{R} : f(u) \leq r\}$ and $S_2 := \{(u, r, s) \in X \times \mathbb{R} \times \mathbb{R} : g(u) \leq s\}$. Denote by $T(S)$, $T(S_1 \cap S_2)$, and $T(S_k)$ the Clarke tangent cones at $(\bar{x}, f(\bar{x}), g(\bar{x}))$ of S, $S_1 \cap S_2$, and S_k respectively and denote also by $I(S_2)$ the interior tangent cone at $(\bar{x}, f(\bar{x}), g(\bar{x}))$ of S_2. We denote similarly by $N(S)$, $N(S_1 \cap S_2)$, and $N(S_k)$ the Clarke normal cones of the above sets at $(\bar{x}, f(\bar{x}), g(\bar{x}))$. We readily observe that we have the equality
$$I(S_2) = \{(h, \lambda, \mu) \in X \times \mathbb{R} \times \mathbb{R} : (h, \mu) \in I\big(\text{epi } g; (\bar{x}, g(\bar{x}))\big)\}$$
along with similar ones for $T(S_k)$.

LEMMA 2.100. Under the above notations, one has the implications:
$$(h, \lambda, \mu) \in T(S) \Rightarrow (h, \lambda + \mu) \in T^C\big(\operatorname{epi}(f+g); (\overline{x}, (f+g)(\overline{x}))\big)$$
and
$$x^* \in \partial_C(f+g)(\overline{x}) \Rightarrow (x^*, -1, -1) \in N(S).$$

PROOF. Fix $(h, \lambda, \mu) \in T(S)$. Consider any sequences $(t_n)_n$ tending to 0 with $t_n > 0$ and $(x_n, \rho_n)_n$ converging to $(\overline{x}, f(\overline{x}) + g(\overline{x}))$ with $(x_n, \rho_n) \in \operatorname{epi}(f+g)$, that is, $\rho_n \geq f(x_n) + g(x_n)$. We have
$$S \ni (x_n, \rho_n - g(x_n), g(x_n)) \to (\overline{x}, f(\overline{x}), g(\overline{x})),$$
hence from the inclusion $(h, \lambda, \mu) \in T(S)$ and from the sequential characterization in Theorem 2.2 there exists a sequence $(h_n, \lambda_n, \mu_n)_n$ converging to (h, λ, μ) such that for all n
$$(x_n, \rho_n - g(x_n), g(x_n)) + t_n(h_n, \lambda_n, \mu_n) \in S,$$
or equivalently
$$\rho_n + t_n(\lambda_n + \mu_n) \geq (f+g)(x_n + t_n h_n),$$
and this implies $(x_n, \rho_n) + t_n(h_n, \lambda_n + \mu_n) \in \operatorname{epi}(f+g)$. The same sequential characterization of the Clarke tangent cone ensures that $(h, \lambda + \mu) \in T^C\big(\operatorname{epi}(f+g); (\overline{x}, (f+g)(\overline{x}))\big)$, that is, the first implication is established.

Now fix $x^* \in \partial_C(f+g)(\overline{x})$, that is, $(x^*, -1) \in N^C\big(\operatorname{epi}(f+g), (\overline{x}, (f+g)(\overline{x}))\big)$. By the previous implication, for any $(h, \lambda, \mu) \in T(S)$ we have $(h, \lambda+\mu) \in T^C\big(\operatorname{epi}(f+g); (\overline{x}, (f+g)(\overline{x}))\big)$, and hence by the definition of $N^C(\operatorname{epi}(f+g); \cdot)$ we obtain $\langle x^*, h \rangle - \lambda - \mu \leq 0$. This entails, as desired, $(x^*, -1, -1) \in N(S)$ according to the equality $N(S) = \big(T(S)\big)^\circ$. $\square$

The second lemma is related to $T(S)$ and $T(S_1 \cap S_2)$.

LEMMA 2.101. The inclusion $T(S_1 \cap S_2) \subset T(S)$ holds.

PROOF. Let $(h, \lambda, \mu) \in T(S_1 \cap S_2)$. Take any sequences $t_n \downarrow 0$ and $(x_n, r_n, s_n) \to (\overline{x}, f(\overline{x}), g(\overline{x}))$ with $(x_n, r_n, s_n) \in S$, that is, $r_n + s_n \geq f(x_n) + g(x_n)$. Observing that
$$S_1 \cap S_2 \ni (x_n, r_n + s_n - g(x_n), g(x_n)) \to (\overline{x}, f(\overline{x}), g(\overline{x})),$$
and using the inclusion $(h, \lambda, \mu) \in T(S_1 \cap S_2)$ and the sequential characterization of the Clarke tangent cone we obtain a sequence $(h_n, \lambda_n, \mu_n) \to (h, \lambda, \mu)$ such that for all n

(2.34) $\qquad (x_n, r_n + s_n - g(x_n), g(x_n)) + t_n(h_n, \lambda_n, \mu_n) \in S_1 \cap S_2.$

Translating the inclusion in S_1 and then in S_2, we obtain
$$f(x_n + t_n h_n) \leq r_n + s_n - g(x_n) + t_n \lambda_n \text{ and } g(x_n + t_n h_n) \leq g(x_n) + t_n \mu_n,$$
which ensures that $r_n + s_n + t_n(\lambda_n + \mu_n) \geq (f+g)(x_n + t_n h_n)$, that is,
$$(x_n, r_n, s_n) + t_n(h_n, \lambda_n, \mu_n) \in S.$$
So, by the sequential characterization of the Clarke tangent cone again, we have $(h, \lambda, \mu) \in T(S)$, which finishes the proof of the lemma. $\square$

PROOF OF THEOREM 2.98. Fix $x^* \in \partial_C(f+g)(\bar{x})$. The first lemma says that $(x^*, -1, -1) \in N(S) = (T(S))^\circ$, hence $(x^*, -1, -1) \in N(S_1 \cap S_2)$ according to the second lemma. On the other hand, observing that $I(S_2) \cap T(S_1) \neq \emptyset$ because $(0, 0, 1) \in I(S_2) \cap T(S_1)$ by (c) in Theorem 2.70 and the equality $g^\circ(\bar{x}; 0) = 0$, we further have $N(S_1 \cap S_2) \subset N(S_1) + N(S_2)$ by (d) in Theorem 2.13. Consequently, $(x^*, -1, -1) \in N(S_1) + N(S_2)$, which is easily seen to mean that there are $u^*, v^* \in X^*$ such that $(u^*, -1, 0) \in N(S_1)$, $(v^*, 0, -1) \in N(S_2)$, and $x^* = u^* + v^*$. Obviously the two latter inclusions translate that $(u^*, -1) \in N\bigl(\text{epi } f; (\bar{x}, f(\bar{x}))\bigr)$ and $(v^*, -1) \in N\bigl(\text{epi } g; (\bar{x}, g(\bar{x}))\bigr)$, that is, $u^* \in \partial_C f(\bar{x})$ and $v^* \in \partial_C g(\bar{x})$. The proof is then complete. □

EXAMPLE 2.102. The converse inclusion in Theorem 2.98 fails. This is confirmed with the simple functions $f_1 := |\cdot|$ and $f_2 := -|\cdot|$ since $\partial_C(f_1 + f_2)(0) = \{0\}$ while $\partial_C f_1(0) + \partial_C f_2(0) = [-1, 1] + [-1, 1]$. □

A particular important case where we have the equality corresponds to the strict Hadamard differentiability of g at $\bar{x}$, since in addition to the inclusion $\partial_C(f+g)(\bar{x}) \subset \partial_C f(\bar{x}) + D_H g(\bar{x})$ from Theorem 2.98 and Proposition 2.50(e), writing f in the form $f = (f + g) + (-g)$ gives by Theorem 2.98 and Proposition 2.50(e) again that

$$\partial_C f(\bar{x}) \subset \partial_C(f+g)(\bar{x}) - D_H g(\bar{x}), \text{ or equivalently } \partial_C f(\bar{x}) + D_H g(\bar{x}) \subset \partial_C(f+g)(\bar{x}).$$

We then have the following corollary.

COROLLARY 2.103. Let U be an open set of the normed space X and $f : U \to \mathbb{R} \cup \{-\infty, +\infty\}$ be finite at $\bar{x} \in U$. If $g : U \to \mathbb{R}$ is strictly Hadamard differentiable at the point $\bar{x}$, then

$$\partial_C(f+g)(\bar{x}) = \partial_C f(\bar{x}) + D_H g(\bar{x}).$$

Assuming that $f = g + \rho$ with ρ strictly Hadamard differentiable at $\bar{x}$ and $D_H(\rho)(\bar{x}) = 0$ and applying the above corollary to $f + \rho$ give:

COROLLARY 2.104. Let $f, g : U \to \mathbb{R} \cup \{-\infty, +\infty\}$ be two extended real-valued functions on an open set U of the normed space X which are finite at $\bar{x} \in U$. If there exists a function $\rho : U \to \mathbb{R}$ which is strictly Hadamard differentiable at $\bar{x}$ with $D\rho(\bar{x}) = 0$ and such that $g = f + \rho$, then $\partial_C g(\bar{x}) = \partial_C f(\bar{x})$.

The next theorem says more than Theorem 2.98 for the subdifferential of a sum under the convexity of the functions.

THEOREM 2.105 (J.J. Moreau; R.T. Rockafellar: theorem for subdifferential of sum of convex functions on normed space). Let U be an open convex set of the normed space X and $f_1, \cdots, f_m : U \to \mathbb{R} \cup \{+\infty\}$ be extended real-valued convex functions. Assume that there is some $a \in U$ such that f_1 is finite at a and $f_2, \cdots, f_m$ are finite at a and continuous at a. Then, for all $x \in U$

$$\partial(f_1 + \cdots + f_m)(x) = \partial f_1(x) + \cdots + \partial f_m(x).$$

PROOF. Obviously the right member is included in the left by the description for convex functions in Proposition 2.63, and as above it suffices to prove the reverse inclusion with $m = 2$. We put $f := f_1$ and $g := f_2$ and we suppose that f and g are finite at x. Fix some real $\bar{r} > \max\{f(a), g(a)\}$ and by continuity of g at a choose some neighborhood $V \subset U$ of a and some real $\varepsilon > 0$ such that $g(u) < \bar{r} - \varepsilon$ for all

$u \in V$. Putting $\alpha := \min\{f(x), g(x)\}$, we have on the one hand, for all $u \in V$ and $r > \bar{r} - \varepsilon$,
$$g(x + (u - x)) = g(u) < r \leq g(x) + (r - \alpha),$$
so considering S_i as defined above, Proposition 2.33 ensures that $(a - x, \bar{r} - \alpha, \bar{r} - \alpha) \in I(S_2)$. On the other hand, writing
$$f(x + (a - x)) = f(a) \leq f(x) + (\bar{r} - \alpha),$$
we see by Theorem 2.2(d) that $(a - x, \bar{r} - \alpha, \bar{r} - \alpha) \in T(S_1)$, hence $T(S_1) \cap I(S_2) \neq \emptyset$. So, the rest of the proof follows the last part of the proof of Theorem 2.98 above. □

REMARK 2.106. Given a nonempty open convex set U of X and convex functions $f_1, \cdots, f_m : U \to \mathbb{R} \cup \{+\infty\}$ the "Moreau-Rockafellar condition" that there is some $a \in \operatorname{dom} f_1$ such that $f_2, \cdots, f_m$ are finite and continuous at a is equivalent to requiring that each function f_i with $i \in \{2, \cdots, m\}$ is continuous at some point $a_i \in \bigcap_{j=1}^{m} \operatorname{dom} f_j$. Indeed, let C be any convex subset of $\bigcap_{j=1}^{m} \operatorname{dom} f_j$ such that each f_i with $i \in \{2, \cdots, m\}$ is finite and continuous at some $c_i \in C$. The point $c := (c_2 + \cdots + c_m)/(m - 1)$ belongs to C and for each $i \in \{2, \cdots, m\}$ we have $c \in \operatorname{int}(\operatorname{dom} f_i)$, or equivalently (by Theorem 2.158 that we will see later in this chapter, or by Theorem 3.18 in Chapter 3) the convex function f_i is continuous at c since f_i is continuous at some point in $\operatorname{dom} f_i$. □

The functions f_1, f_2 involved in Example 2.99 are convex, and hence they say that the inclusion $\partial(f_1 + f_2)(\bar{x}) \subset \partial f_1(\bar{x}) + \partial f_2(\bar{x})$ (or the corresponding equality) in Theorem 2.105 does not hold in some situations. In that example the subdifferentials $\partial f_i(\bar{x})$, $i = 1, 2$, are empty sets. We provide below another example where all the involved subdifferentials are nonempty.

EXAMPLE 2.107. In $\mathbb{R} \times \mathbb{R}$ consider the closed convex sets
$$S_1 := \{(x, y) \in \mathbb{R} \times \mathbb{R} : (x + 1)^2 + y^2 \leq 1\}$$
and $S_2 := \{(x, y) \in \mathbb{R} \times \mathbb{R} : (x - 1)^2 + y^2 \leq 1\}.$

For $i = 1, 2$, denote by $f_i := \Psi_{S_i}$ the indicator function of S_i, so f_i is convex and lower semicontinuous. Clearly, for $\bar{x} := (0, 0)$
$$\partial f_1(\bar{x}) + \partial f_2(\bar{x}) = ([0, +\infty[\times \{0\}) + (] -\infty, 0] \times \{0\}) = \mathbb{R} \times \{0\},$$
while
$$\partial(f_1 + f_2)(\bar{x}) = \partial \Psi_{\{\bar{x}\}}(\bar{x}) = \mathbb{R} \times \mathbb{R}.$$
So, $\partial(f_1 + f_2)(\bar{x}) \not\subset \partial f_1(\bar{x}) + \partial f_2(\bar{x})$ whereas $\partial f_1(\bar{x})$ and $\partial f_2(\bar{x})$ are nonempty. □

As shown in the above example and in Example 2.99, the inclusion of the left-hand side into the right-hand one in Theorems 2.98 and 2.105 may fail without a certain condition. In variational analysis, given a function formed from data functions via a certain operation, any condition ensuring the inclusion of the subdifferential of that function at a reference point in terms of subdifferentials of data functions at this reference point is called a *qualification condition*.

In addition to Theorem 2.98 the case of the sum of functions with separate variables is also of interest.

2.2. CLARKE SUBDIFFERENTIAL

PROPOSITION 2.108. Let X_k, $k = 1, \cdots, m$, be normed spaces and $f_k : X_k \to \mathbb{R} \cup \{+\infty\}$ be extended real-valued functions. For $\varphi(x_1, \cdots, x_m) = f_1(x_1) + \cdots + f_m(x_m)$ one has

$$\partial_C \varphi(x_1, \cdots, x_m) \subset \partial_C f_1(x_1) \times \cdots \times \partial_C f_m(x_m).$$

PROOF. It is enough to consider $m = 2$. Put $X := X_1 \times X_2$, $x := (x_1, x_2)$, $f(x) := f_1(x_1)$, and $g(x) := f_2(x_2)$, and observe that $\varphi = f + g$. It is enough to prove the inclusion in the case $f_k(x_k) < +\infty$. With the development subsequent to the statement of Theorem 2.98, we have

$$S_1 = \{(x_1, x_2, r, s) : f_1(x_1) \leq r\}, \ S_2 = \{(x_1, x_2, r, s) : f_2(x_2) \leq s\},$$

(2.35)
$$S = \{(x_1, x_2, r, s) : f_1(x_1) + f_2(x_2) \leq r + s\}.$$

Through the sequential characterization of the Clarke tangent cone, it is easily seen from the previous descriptions of S_k that, for $T_k := T(\mathrm{epi}\, f_k; (x_k, f_k(x_k)))$, the equality

$$T(S_1 \cap S_2) = \{(u_1, u_2, \lambda, \mu) : (u_1, \lambda) \in T_1, (u_2, \mu) \in T_2\}$$

is fulfilled. Fix any $(x_1^*, x_2^*) \in \partial_C \varphi(x_1, x_2)$. From both Lemma 2.100 and Lemma 2.101 and from (2.35) we have $(x_1^*, x_2^*, -1, -1) \in N(S_1 \cap S_2)$. Using this and taking the negative polar in the latter equality yields $(x_k^*, -1) \in N^C(\mathrm{epi}\, f_k; (x_k, f_k(x_k)))$, that is, $x_k^* \in \partial_C f_k(x_k)$ as required. □

Before passing to the next subsection let us apply Corollary 2.104 to provide a chain rule for the composition of a strictly F-differentiable function with an inner locally Lipschitz vector-valued mapping.

PROPOSITION 2.109. Let $G : U \to Y$ be a mapping from an open set U of the normed space X into a normed space Y and let $g : V \to \mathbb{R}$ be a function from an open set V of Y containing $G(U)$. Assume that G is Lipschitz continuous near $\bar{x} \in U$ and g is strictly Fréchet differentiable at $\bar{y} := G(\bar{x})$, and let $\bar{y}^* := Dg(\bar{y})$. Then $g \circ G$ is Lipschitz continuous near $\bar{x}$ and

$$\partial_C(g \circ G)(\bar{x}) = \partial_C(\bar{y}^* \circ G)(\bar{x}).$$

PROOF. By strict Fréchet differentiability of g at $\bar{y}$ there exists a function $\varepsilon : V \times V \to [0, +\infty[$ with $\varepsilon(y, y') \to 0$ as $(y, y') \to (\bar{y}, \bar{y})$ and such that

$$|g(y) - g(y') - \langle \bar{y}^*, y - y' \rangle| \leq \|y - y'\| \varepsilon(y, y') \quad \text{for all } y, y' \in V.$$

Denoting by γ a Lipschitz constant of G on some neighborhood $U_0 \subset U$ of $\bar{x}$, it ensues that

$$|g(G(x)) - g(G(x')) - (\bar{y}^* \circ G(x) - \bar{y}^* \circ G(x'))| \leq \gamma \|x - x'\| \varepsilon(G(x), G(x'))$$

along with $\varepsilon(G(x), G(x')) \to 0$ as $(x, x') \to (\bar{x}, \bar{x})$. This clearly says that $g \circ G - \bar{y}^* \circ G$ is strictly Fréchet differentiable at $\bar{x}$ with a null derivative at this point. Corollary 2.104 then guarantees the equality $\partial_C(g \circ G)(\bar{x}) = \partial_C(\bar{y}^* \circ G)(\bar{x})$. □

C-subdifferentials of product and quotient furnish examples of application of the proposition.

EXAMPLE 2.110. Let $f, g : U \to \mathbb{R}$ be two locally Lipschitz functions on an open set U of a normed space X.
(a) Defining $p : \mathbb{R}^2 \to \mathbb{R}$ by $p(r, s) := rs$ and $G : U \to \mathbb{R}^2$ by $G(x) := (f(x), g(x))$ we see that the product function fg can be written as $fg = p \circ G$. So, we can clearly apply the above proposition with $p \circ G$ to obtain for any $x \in U$ that
$$\partial_C(p \circ G)(x) = \partial_C(g(x)f + f(x)g)(x).$$
From this and Proposition 2.74(b) it results for every $x \in U$ that
$$\partial_C(fg)(x) = \partial_C(g(x)f + f(x)g)(x) \subset g(x)\partial_C f(x) + f(x)\partial_C g(x).$$
(b) Assume in addition that $g(x) \neq 0$ for every $x \in U$. Considering the function $q : D_q \to \mathbb{R}$ defined $q(r, s) := r/s$ for all $(r, s) \in D_q := \{(\rho, \sigma) \in \mathbb{R}^2 : \sigma \neq 0\}$, we see that $f/g = q \circ G$ (with G as in (a) above). Repeating the arguments in (a) with q in place of p, we obtain
$$\partial_C\left(\frac{f}{g}\right)(x) = \frac{1}{(g(x))^2}\partial_C(g(x)f - f(x)g)(x) \subset \frac{1}{(g(x))^2}(g(x)\partial_C f(x) - f(x)\partial_C g(x))$$
for all $x \in U$. □

As a direct corollary of Proposition 2.109 we can extend the second equality in (b) of Proposition 2.74 as follows.

COROLLARY 2.111. Let X be a normed space and $f : X \to \mathbb{R}$ be a function which is Lipschitz continuous near $\bar{x} \in X$. If a function $\varphi : \mathbb{R} \to \mathbb{R}$ is strictly differentiable at $\bar{r} := f(\bar{x})$, then $\varphi \circ f$ is Lipschitz continuous near $\bar{x}$ and
$$\partial_C(\varphi \circ f)(\bar{x}) = \varphi'(\bar{r})\partial_C f(\bar{x}).$$

2.2.11. Bouligand-Peano tangent cone and Bouligand directional derivative.
In Theorem 2.13 under the interior tangent cone property the inclusion
$$T^C(S_1 \cap S_2; x) \subset T^C(S_1; x) \cap T^C(S_2; x)$$
is established for two subsets S_1, S_2 of the normed space X. The reverse inclusion generally requires (as a sufficient condition) a tangential regularity of the sets. To state this regularity we need the concept of Bouligand-Peano tangent vector to a set $S \subset X$ at $x \in S$. We will follow the tradition in the literature in denoting the set of such tangent vectors by $T^B(S; x)$. In addition to its use in defining the tangential regularity, this set of such tangent vectors is very important in its own right and it will be involved in many other parts of the book.

DEFINITION 2.112. The *Bouligand-Peano tangent (or contingent) cone* (also called the Bouligand tangent cone) of the set $S \subset X$ at a point $x \in S$ is defined as the following outer limit (or limit superior) of the set-difference quotient
$$T^B(S; x) := \limsup_{t \downarrow 0} \frac{1}{t}(S - x),$$
otherwise stated, a vector $h \in T^B(S; x)$ if and only if, for any neighborhood V of h in X and for any real $\varepsilon > 0$, there is some $t \in]0, \varepsilon[$ such that $V \cap (t^{-1}(S - x)) \neq \emptyset$, or equivalently for any neighborhood V of h in X and for any real $\varepsilon > 0$
$$(x +]0, \varepsilon[V) \cap S \neq \emptyset.$$
It is easily seen that $T^B(S; x)$ is a cone, and $h \in T^B(S; x)$ if and only if
(2.36)
$$U \cap (x +]0, +\infty[V) \cap S \neq \emptyset,$$

for any neighborhoods V of h and U of x.

When $x \notin S$, one puts by convention $T^B(S;x) = \emptyset$.

As for the Clarke tangent cone, the Bouligand-Peano tangent cone is evidently a local concept in the sense that

$$T^B(S;x) = T^B(S \cap U; x) \quad \text{for any neighborhood } U \text{ of } x.$$

Concerning properties (a_1) and (d) below, they follow directly from properties of outer limit (or limit superior) of a multimapping between metric spaces in the previous chapter. Further, the inclusion

$$T^B(S;x) \subset \text{cl}\left(\mathbb{R}_+(S-x)\right)$$

always holds according to the definition $T^B(S;x) = \underset{t\downarrow 0}{\text{Lim sup }} \frac{1}{t}(S-x)$, then, assuming the convexity of the set S, the assertion (e) below results from the inclusions

$$T^C(S;x) \subset T^B(S;x) \subset \text{cl}(\mathbb{R}_+(S-x)) = T^C(S;x),$$

where the latter equality above is due to Proposition 2.2(d).

All those features together provide the assertions (a_1), (d) and (e) in the following proposition.

PROPOSITION 2.113. Let S be a subset of the normed space X and let any $x \in S$. The following assertions hold.
(a) (a_1) The Bouligand-Peano tangent cone $T^B(S;x)$ is closed (maybe nonconvex) and $T^C(S;x) \subset T^B(S;x)$;
(a_2) the equality $T^B(S;x) = T^B(S \cap U; x)$ holds for any neighborhood U of x;
(a_3) for any set S' with $S \cap U \subset S' \cap U \subset (\text{cl } S) \cap U$ for some neighborhood U of x, one has $T^B(S';x) = T^B(S;x)$;
(a_4) $T^B(\cdot;x)$ is isotone in the sense that $T^B(S;x) \subset T^B(S';x)$ for any $S' \supset S$; in particular for $x \in S_1 \cap \cdots \cap S_m$ with $S_1 \subset X, \cdots, S_m \subset X$,

$$T^B(S_1 \cap \cdots \cap S_m; x) \subset T^B(S_1;x) \cap \cdots \cap T^B(S_m;x).$$

(a_5) For $x \in S_1 \cap \cdots \cap S_m$ with $S_1 \subset X, \cdots, S_m \subset X$,

$$T^B(S_1 \cup \cdots \cup S_m; x) = T^B(S_1;x) \cup \cdots \cup T^B(S_m;x).$$

(b) For $X = X_1 \times \cdots \times X_m$, $x_k \in S_k \subset X_k$ for $k = 1, \cdots, m$, one has the inclusion

$$T^B(S_1 \times \cdots \times S_m; (x_1, \cdots, x_m)) \subset T^B(S_1;x_1) \times \cdots \times T^B(S_m;x_m).$$

(c) A vector $h \in T^B(S;x)$ if and only if there are sequences $(t_n)_n$ tending to 0 with $t_n > 0$ and $(h_n)_n$ converging to h such that

$$x + t_n h_n \in S \quad \text{for all } n \in \mathbb{N}.$$

(d) The inclusion $T^C(S;x) \subset T^B(S;x)$ is fulfilled as well as $\left(T^B(S;x)\right)^\circ \subset N^C(S;x)$.
(e) If S is convex, then

$$T^B(S;x) = \text{cl}\left(\mathbb{R}_+(S-x)\right) = T^C(S;x).$$

PROOF. The assertions (a_1), (d) and (e) have been argued before the statement of the proposition. The proofs of (a_2), (a_4), (a_5) and (b) are direct consequences of the definition. Concerning (a_3), for any open neighborhood V of h, the set $x+]0,\varepsilon[V$ is open, so

$$(x+]0,\varepsilon[V) \cap \text{cl } S \neq \emptyset \Leftrightarrow (x+]0,\varepsilon[V) \cap S \neq \emptyset,$$

so $T^B(\mathrm{cl}\, S; x) = T^B(S; x)$; this yields $T^B(S; x) = T^B(S'; x)$ by (a_2) and the isotony property in (a_4).

Finally, the assertion (c) follows from the sequential characterization of the outer limit (or limit superior) of a multimapping between metric spaces in Proposition 1.14(b) in the first chapter. □

REMARK 2.114. From (c) above we easily see, for $S := \{(r, s) \in \mathbb{R}^2 : s = |r|\}$, that $T^B(S; (0,0)) = S$. So, unlike the Clarke tangent cone, the Bouligand-Peano tangent cone is not convex in general.

With x_3 in Figure 2.1 (for example) we see that the equality (a_5) does not hold for the Clarke tangent cone.

We also note that simple examples show that the reverse inclusion of the intersection in (a_4) above (which is the one usually required), that is, $T^B(S_1; x) \cap T^B(S_2; x) \subset T^B(S_1 \cap S_2; x)$, may fail to be true. Take in $\mathbb{R}$ the closed sets $S_1 := \{0\} \cup \{\frac{1}{3^n} : n \in \mathbb{N}\}$, $S_2 := \{0\} \cup \{\frac{1}{2^n} : n \in \mathbb{N}\}$, and the point $x = 0$.

With the same sets S_1 and S_2, it is not difficult to verify that $(1,1) \notin T^B(S_1 \times S_2; (0,0))$, while clearly $T^B(S_1; 0) \times T^B(S_2; 0) = [0, +\infty[\times [0, +\infty[$, so the reverse inclusion in (b) above fails in general. □

Figure 2.2 (with the same set of Figure 2.1) shows the Bouligand-Peano tangent cone at the point x_i of $S \subset \mathbb{R}^2$ (more precisely $x_i + T^B(S; x_i)$), for $i \in \{1, 2, 5, 6, 7, 9\}$. At the point x_i for $i \in \{3, 4, 8, 10, 11, 12, 13\}$ we have

$$T^B(S; x_3) = \{(h_1, h_2) \in \mathbb{R}^2 : h_1 \leq -|h_2|\} \cup ([0, +\infty[\times \{0\}),$$

$$T^B(S; x_4) = (]-\infty; 0] \times \{0\}) \cup \{(h_1, h_2) \in \mathbb{R}^2 : h_1 \geq 0 \text{ and } h_2 = \pm h_1\},$$

$$T^B(S; x_8) = \{(h_1, h_2) \in \mathbb{R}^2 : h_1 \leq 0 \text{ and } h_2 \geq -h_1\} \cup ([0, +\infty[\times \{0\}),$$

$$T^B(S; x_9) = \{(h_1, h_2) \in \mathbb{R}^2 : h_2 \leq h_1 \text{ or } h_2 \geq -h_1\}, \quad T^B(S; x_{10}) = \mathbb{R}^2,$$

$$T^B(S; x_{11}) = [0, +\infty[\times \{0\}, \quad T^B(S; x_{12}) = \{0\} \times \mathbb{R}, \quad T^B(S; x_{13}) = \{0\} \times \mathbb{R}.$$

So, for $i \in \{1, 2, 5, 6, 7, 11, 12\}$ the Clarke and Bouligand-Peano tangent cones coincide, that is, the set S is tangentially regular at those points according to Definition 2.122 below. In contrast, $T^C(S; x_i)$ and $T^B(S; x_i)$ are completely different for all other indices i.

Concerning the first inclusion in Proposition 2.113(d), a more accurate one can be obtained. Given two subsets C_1, C_2 of a vector space X, we recall that the *Minkowski difference* or *star-difference* $C_1 \overset{*}{-} C_2$ in X is defined as

$$C_1 \overset{*}{-} C_2 := \{u \in X : u + C_2 \subset C_1\} = \bigcap_{x \in C_2} (C_1 - x);$$

this set is closed whenever C_1 is closed. If K is a closed cone, it is clear from the left equality above that the set $K \overset{*}{-} K$ is stable by addition, so it is a closed convex cone included in K. Consequently, for a set S of the normed space $(X, \|\cdot\|)$ and $x \in S$, the set

$$T^B(S; x) \overset{*}{-} T^B(S; x)$$

is a closed convex subcone of $T^B(S; x)$. Further, taking any $h \in T^C(S; x)$ and any $h' \in T^B(S; x)$, there exist sequences $(t_n)_n$ in $]0, +\infty[$ tending to 0 and $(h'_n)_n$ in X converging to h' such that $x + t_n h'_n \in S$ for all $n \in \mathbb{N}$. Since $h \in T^C(S; x)$ and $x + t_n h'_n \to x$ as $n \to \infty$, there is a sequence $(h_n)_n$ in X converging to h such that, for all $n \in \mathbb{N}$, we have $(x + t_n h'_n) + t_n h_n \in S$, that is, $x + t_n(h_n + h'_n) \in S$. This

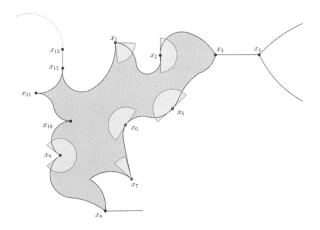

FIGURE 2.2. Bouligand-Peano tangent cones.

says that $h + h' \in T^B(S;x)$. Consequently, $T^C(S;x) + T^B(S;x) \subset T^B(S;x)$, which gives the inclusion $T^C(S;x) \subset T^B(S;x) \stackrel{*}{-} T^B(S;x)$. We state all those features in the proposition:

PROPOSITION 2.115. Let S be a subset of the normed space $(X, \|\cdot\|)$ and let any $x \in S$. Then $T^B(S;x) \stackrel{*}{-} T^B(S;x)$ is a closed convex cone and the following inclusions hold true
$$T^C(S;x) \subset T^B(S;x) \stackrel{*}{-} T^B(S;x) \subset T^B(S;x).$$

REMARK 2.116. For the set S in Figure 2.2 we note that at the point x_{10} we have $T^B(S;x_{10}) \stackrel{*}{-} T^B(S;x_{10}) = \mathbb{R}^2$ is different from $T^C(S;x_{10})$. □

Concerning the Bouligand-Peano tangent cone of the boundary of a set, an equality similar with the one (related to Clarke tangent cone) in Proposition 2.10 can be established.

PROPOSITION 2.117. Let S be a subset of the normed space X and let $\overline{x} \in$ bdry S. Then, setting $S^c := X \setminus S$ one has the equality
$$T^B(\text{bdry } S; \overline{x}) = T^B(S \cup \{\overline{x}\}; \overline{x}) \cap T^B(S^c \cup \{\overline{x}\}; \overline{x}).$$

PROOF. Put $\overline{S} := \text{cl } S$. Since $T^B(S \cup \{\overline{x}\}; \overline{x}) = T^B(\overline{S}; \overline{x})$ by Proposition 2.113(a_3), the left member is obviously included in the right. Conversely, take any h in the right member, thus $h \in T^B(\overline{S}; \overline{x}) \cap T^B(\overline{S^c}; \overline{x})$. Consider any $\varepsilon > 0$ and any convex neighborhood V of h. For $U_\varepsilon := \overline{x} +]0, \varepsilon]V$, by definition of Bouligand-Peano tangent cone, $U_\varepsilon \cap \overline{S} \neq \emptyset$ and $U_\varepsilon \cap \overline{S^c} \neq \emptyset$, and clearly $U_\varepsilon = (U_\varepsilon \cap \overline{S}) \cup (U_\varepsilon \cap \overline{S^c})$. Then, the convex (hence connected) set U_ε must meet $\overline{S} \cap \overline{S^c}$, that is, it meets bdry S. This translates that $h \in T^B(\text{bdry } S; \overline{x})$. □

REMARK 2.118. (a) Let S_1, S_2 be two closed sets of the normed space X and $\overline{x} \in S_1 \cap S_2$. Define, for a set $S \subset X$ containing $\overline{x}$, the (nonconvex) open cone $I^B(S; \overline{x})$ as the set of $h \in X$ for which there are a real $\varepsilon > 0$ and a neighborhood

V of h such that $\bar{x}+]0,\varepsilon]V\subset S$. Proceeding as in the proof of Proposition 2.16 one can claim that
$$T^B(S_1;\bar{x})\cap T^B(S_2;\bar{x})\cap I^B(S_1\cup S_2;\bar{x})\subset T^B(S_1\cap S_2;\bar{x}).$$
Indeed, take any h in the left-hand side. Choose a real $\varepsilon_0 > 0$ such that $\bar{x}+[0,\varepsilon_0]B(h,\varepsilon_0)\subset S_1\cup S_2$. For any real $\varepsilon\in]0,\varepsilon_0]$, putting $U_\varepsilon := \bar{x}+]0,\varepsilon]B(h,\varepsilon)$ we have $U_\varepsilon\cap S_1\neq\emptyset$ and $U_\varepsilon\cap S_2\neq\emptyset$. Further, since $U_\varepsilon\subset S_1\cup S_2$, we also have with the convex (hence connected) set U_ε
$$(U_\varepsilon\cap S_1)\cup(U_\varepsilon\cap S_2) = U_\varepsilon.$$
The sets S_1 and S_2 being closed, it ensues that $U_\varepsilon\cap(S_1\cap S_2)\neq\emptyset$ for any $\varepsilon\in]0,\varepsilon_0]$, so $h\in T^B(S_1\cap S_2;\bar{x})$, justifying the claim.

We note that the inclusion from the right-hand side into the left in the above proposition can also be obtained as a consequence of the claim.

(b) Given a continuous function $g: X\to\mathbb{R}$ with $g(\bar{x}) = 0$, as in the inclusion in Remark 2.17 we see from (a) that
$$T^B(\{g(\cdot)\leq 0\})\cap T^B(\{g(\cdot)\geq 0\})\subset T^B(\{g(\cdot) = 0\}).$$
Since the converse inclusion is obvious by the monotonicity of $T^B(\cdot;\bar{x})$, the equality
$$T^B(\{g(\cdot) = 0\}) = T^B(\{g(\cdot)\leq 0\})\cap T^B(\{g(\cdot)\geq 0\})$$
is established. $\square$

We analyse now the Bouligand-Peano tangent cone of the epigraph of a function $f: U\to\mathbb{R}\cup\{-\infty,+\infty\}$ on an open set U of the normed space X which is assumed to be finite at $\bar{x}\in U$. As in the analysis preceding Theorem 2.72, we may suppose $U = X$ and for each $t\in E_0 :=]0,+\infty[$ we define $\varphi_t: X\to\mathbb{R}\cup\{-\infty,+\infty\}$ by
$$\varphi_t(h) = t^{-1}\big(f(\bar{x}+th) - f(\bar{x})\big)\quad\text{for all } h\in X.$$
Notice that $0\in E := \operatorname{cl} E_0$ and that
$$\operatorname{epi}\varphi_t = t^{-1}\big(\operatorname{epi} f - (\bar{x}, f(\bar{x}))\big).$$
By Definition 2.112 and Proposition 1.23 we have that

(2.37) $\qquad T^B\big(\operatorname{epi} f; (\bar{x}, f(\bar{x}))\big) = \underset{t\downarrow 0}{\operatorname{Lim\,inf}}\operatorname{epi}\varphi_t = \operatorname{epi} L,$

where for every $h\in X$
$$L(h) := \liminf_{t\downarrow 0}\inf_{h'\to h} t^{-1}\big(f(\bar{x}+th') - f(\bar{x})\big) = \liminf_{t\downarrow 0, h'\to h} t^{-1}\big(f(\bar{x}+th') - f(\bar{x})\big).$$

DEFINITION 2.119. Given an extended real-valued function $f: U\to\mathbb{R}\cup\{-\infty,+\infty\}$ on an open set U of a normed space X and given $\bar{x}\in U$ with $|f(\bar{x})| < +\infty$, the above function L will be denoted $f^B(\bar{x};\cdot)$, that is,

(2.38) $\qquad f^B(\bar{x}; h) = \liminf_{t\downarrow 0, h'\to h} t^{-1}\big(f(\bar{x}+th') - f(\bar{x})\big)$

and called the *(one-sided) Bouligand directional derivative* or *directional B-derivative* because of (2.37). So, one has

(2.39) $\qquad T^B\big(\operatorname{epi} f; (\bar{x}, f(\bar{x}))\big) = \operatorname{epi} f^B(\bar{x};\cdot),$

or equivalently

(2.40) $\qquad \inf\{r\in\mathbb{R}: (h,r)\in T^B\big(\operatorname{epi} f; (\bar{x}, f(\bar{x}))\big)\} = f^B(\bar{x}; h),\quad\text{for all } h\in X.$

We note that the equality (2.38) tells us that the Bouligand directional derivative $f^B(\bar{x}; h)$ coincides with the *(right-hand side) Hadamard lower directional derivative* of f at $\bar{x}$ in the direction h defined by

(2.41) $$\underline{d}_H^+ f(\bar{x}; h) := \liminf_{t \downarrow 0, h' \to h} t^{-1}\big(f(\bar{x} + th') - f(\bar{x})\big).$$

The terminology "Hadamard lower directional derivative" is due to the left member in (2.9) defining the Hadamard differentiability.

Obviously, for functions $f_k : U \to \mathbb{R} \cup \{-\infty, +\infty\}$ which are finite at $\bar{x}$ and such that $f_1 + \cdots + f_m$ is well defined near $\bar{x}$ one has

$$f_1^B(\bar{x}; h) + \cdots + f_m^B(\bar{x}; h) \leq (f_1 + \cdots + f_m)^B(\bar{x}; h)$$

whenever the sum involved in the left member of the inequality is well defined.

It is also of interest to note that, if f is locally Lipschitz continuous near $\bar{x}$,

$$f^B(\bar{x}; h) = \liminf_{t \downarrow 0} t^{-1}\big(f(\bar{x} + th) - f(\bar{x})\big),$$

that is, the Bouligand directional derivative of f at $\bar{x}$ coincides with the so-called *(right-hand side) Dini lower directional derivative* of f at $\bar{x}$ defined by

(2.42) $$\underline{d}_D^+ f(\bar{x}; h) := \liminf_{t \downarrow 0} t^{-1}\big(f(\bar{x} + th) - f(\bar{x})\big)$$

for any extended real-valued function which is finite at $\bar{x}$. Of course,

(2.43) $$\underline{d}_H^+ f(\bar{x}; h) \leq \underline{d}_D^+ f(\bar{x}; h)$$

for any extended real-valued function f.

Observing for a set S in X that for $t > 0$ and $\bar{x} \in S$ one has

$$d(h, t^{-1}(S - \bar{x})) = t^{-1} d(\bar{x} + th, S) = t^{-1}\big(d_S(\bar{x} + th) - d_S(\bar{x})\big),$$

the expression above for $f^B(\bar{x}; \cdot)$ when f is Lipschitz continuous near $\bar{x}$ ensures the equality $\liminf_{t \downarrow 0} d(h, t^{-1}(S - \bar{x})) = (d_S)^B(\bar{x}; h)$. Consequently, Proposition 1.12 gives

$$h \in T^B(S; \bar{x}) \Leftrightarrow (d_S)^B(\bar{x}; h) = 0 \Leftrightarrow (d_S)^B(\bar{x}; h) \leq 0.$$

We put all those properties together in the following proposition.

PROPOSITION 2.120. *Let S be a subset of the normed space X, let any $\bar{x} \in S$, let U be an open set containing $\bar{x}$, and let $f_1, \cdots, f_m : U \to \mathbb{R} \cup \{-\infty, +\infty\}$ be functions which are finite at $\bar{x}$ and such that $f_1 + \cdots + f_m$ is well defined near $\bar{x}$. The following properties hold:*
(a) *One has*

$$f_1^B(\bar{x}; h) + \cdots + f_m^B(\bar{x}; h) \leq (f_1 + \cdots + f_m)^B(\bar{x}; h)$$

whenever the sum involved in the left member of the inequality is well defined.
(b) *One has the equivalences*

$$h \in T^B(S; \bar{x}) \Leftrightarrow (d_S)^B(\bar{x}; h) = 0 \Leftrightarrow (d_S)^B(\bar{x}; h) \leq 0.$$

We have seen in Proposition 2.88 above that for a real $s > f(x) > -\infty$, the inclusion of $T^C(\text{epi } f; (x, f(x)))$ in $T^C(\text{epi } f; (x, s))$ generally requires the continuity of f at x. The similar inclusion

$$T^B(\text{epi } f; (x, f(x))) \subset T^B(\text{epi } f; (x, s))$$

does not require any assumption on the function f. Indeed, consider any (h, r) in $T^B(\text{epi } f; (x, f(x)))$ and take a sequence of reals $(t_n)_n$ tending to 0 with $t_n > 0$ and a sequence $(h_n, r_n)_n$ converging to (h, r) such that $(x, f(x)) + t_n(h_n, r_n) \in \text{epi } f$ for all n. Obviously, $(x, s) + t_n(h_n, r_n) \in \text{epi } f$ for all n, so (h, r) is in $T^B(\text{epi } f; (x, s))$, justifying the desired inclusion. We then have the following proposition:

PROPOSITION 2.121. *Let U be an open set of the normed space X and $f : U \to \mathbb{R} \cup \{-\infty, +\infty\}$ be an extended real-valued function. Then, for any $x \in U$ with $|f(x)| < +\infty$ and for any real $s \geq f(x)$, one has*

$$T^B(\text{epi } f; (x, f(x))) \subset T^B(\text{epi } f; (x, s)).$$

2.2.12. Tangential regularity of sets and functions. As seen in Proposition 2.113 above, one always has $T^C(S; x) \subset T^B(S; x)$. The Clarke tangential regularity of the set S is then defined as follows.

DEFINITION 2.122. A set S of the normed space X is called *Clarke tangentially regular* or *C-tangentially regular* at a point $\bar{x} \in S$ when $T^C(S; \bar{x}) = T^B(S; \bar{x})$. For an open set U in X, an extended real-valued function $f : U \to \mathbb{R} \cup \{-\infty, +\infty\}$ is said to be *Clarke tangentially regular*, or *C-tangentially regular*, at a point $\bar{x} \in U$ with $|f(\bar{x})| < +\infty$ whenever its epigraph is (Clarke) tangentially regular at $(\bar{x}, f(\bar{x}))$. When there is no risk of confusion we will just say *tangentially regular* for sets as well as for functions.

Recalling that

$$\text{epi } f^{\uparrow}(x; \cdot) = T^C(\text{epi } f; (x, f(x))) \text{ and } \text{epi } f^B(x; \cdot) = T^B(\text{epi } f; (x, f(x))),$$

the tangential regularity of f yields:

PROPOSITION 2.123. *Let U be an open set of a normed space X and $f : U \to \mathbb{R} \cup \{-\infty, +\infty\}$ be a function which is tangentially regular at a point $\bar{x} \in U$ with $|f(\bar{x})| < +\infty$. Then one has*

$$f^{\uparrow}(\bar{x}; h) = f^B(\bar{x}; h) = \underline{d}_H^+ f(\bar{x}; h)$$

for every $h \in X$.

The assertion (e) in Proposition 2.113 directly ensures the following:

PROPOSITION 2.124. *Any convex set is tangentially regular at any of its point and any convex function is tangentially regular at any point where it is finite.*

The next proposition shows that the inclusion related to the Cartesian product of finitely many sets in (b) of Proposition 2.113 is in fact an equality whenever all the sets but one of them are tangentially regular. It also shows that the Bouligand directional derivative of the indicator function of a set is equal to the indicator function of the Bouligand-Peano tangent cone of this set.

PROPOSITION 2.125. *Let S be a subset of the normed space X and let any point $x \in S$.*
(a) *The set S is tangentially regular at x whenever $S \cap U$ is convex for some neighborhood U of the point x.*
(b) *For $X = X_1 \times \cdots \times X_m$, $x_k \in S_k \subset X_k$ for $k = 1, \cdots, m$, one has the equality*

$$T^B(S_1 \times \cdots \times S_m; (x_1, \cdots, x_m)) = T^B(S_1; x_1) \times \cdots \times T^B(S_m; x_m)$$

whenever all the sets S_k except one of them are tangentially regular at x_k.
(c) *The Bouligand directional derivative at x of the indicator function of the set S*

coincides with the indicator function of the Bouligand-Peano tangent cone of S at x, that is,
$$(\Psi_S)^B(x;h) = \Psi_{T^B(S;x)}(h) \quad \text{for all } h \in X.$$
(d) The set S is tangentially regular at x if and only if its indicator function Ψ_S is tangentially regular at x.

PROOF. The assertion (a) follows from Proposition 2.124 above. Let us prove (b). Assume that, for each $k = 2, \cdots, m$, the set S_k is tangentially regular at x_k and fix any $h_k \in T^B(S_k; x_k) = T^C(S_k; x_k)$. Fix also any $h_1 \in T^B(S_1; x_1)$. From the sequential characterization of $T^B(S_1; x_1)$ there are sequences $(t_n)_n$ tending to 0 with $t_n > 0$ and $(h_{1,n})_n$ converging to h_1 in X_1 such that $x_1 + t_n h_{1,n} \in S_1$ for all $n \in \mathbb{N}$. For each $k = 2, \cdots, m$, the sequential characterization of the Clarke tangent cone produces a sequence $(h_{k,n})_n$ converging to h_k in X_k such that $x_k + t_n h_{k,n} \in S_k$ for all n. Consequently, for all $n \in \mathbb{N}$
$$(x_{1,n}, \cdots, x_{m,n}) + t_n(h_{1,n}, \cdots, h_{m,n}) \in S_1 \times \cdots \times S_m,$$
hence $(h_1, \cdots, h_m) \in T^B(S_1 \times \cdots \times S_m; (x_1, \cdots, x_m))$. This guarantees the desired equality of (b) since, as said above, the reverse inclusion follows from (a) of Proposition 2.113.

For the set $S \subset X$ with $x \in S$, since epi $\Psi_S = S \times [0, +\infty[$, the assertion (b) above ensures that $T^B(\text{epi}\,\Psi_S; (x,0)) = T^B(S;x) \times [0,+\infty[$. By (2.39) the latter equality means that epi $(\Psi_S)^B(x;\cdot) = $ epi $\Psi_{T^B(S;x)}$, otherwise stated $\Psi_S^B(x;h) = \Psi_{T^B(S;x)}(h)$ for all $h \in X$. The equality in (c) is then proved.

Concerning (d), as seen in the proof of (c), we have $T^B(\text{epi}\,\Psi_S; (x,0)) = T^B(S;x) \times [0,+\infty[$. Further, by Theorem 2.2 we also have $T^C(\text{epi}\,\Psi_S; (x,0)) = T^C(S;x) \times [0,+\infty[$. Consequently, $T^C(\text{epi}\,\Psi_S; (x,0)) = T^B(\text{epi}\,\Psi_S; (x,0))$ if and only if $T^C(S;x) = T^B(S;x)$, that is, the function Ψ_S is tangentially regular at x if and only if the set S is tangentially regular at x. □

Since, for $x \in S_k \subset X$ with $k = 1, \cdots, m$, we always have $T^B(S_1 \cap \cdots \cap S_m; x) \subset T^B(S_1;x) \cap \cdots \cap T^B(S_m;x)$, under the additional assumption of tangential regularity of $S_1, \cdots, S_m$ at x, the inclusions in (c) and (d) of Theorem 2.13 become equalities and $T^C(S_1 \cap \cdots \cap S_m; x) = T^B(S_1 \cap \cdots \cap S_m; x)$, that is, the set $S_1 \cap \cdots \cap S_m$ is tangentially regular at x. This calculus rule for the Clarke tangent cone is stated in the next proposition. This proposition also provides other calculus rules in the form of equality for functions. All those results make clear the relevance of the tangential regularity in calculus rules of the Clarke normal cone of an intersection and the Clarke subdifferential of the sum of finitely many functions.

The assertion (b_2) of the proposition is related to the standard directional derivative in Definition 2.62.

PROPOSITION 2.126. Let $S_1, \cdots, S_m$ be subsets of X containing $\bar{x} \in X$, U be an open set containing $\bar{x}$, and $f_1, \cdots, f_m : U \to \mathbb{R} \cup \{-\infty, +\infty\}$ be extended real-valued functions which are finite at $\bar{x}$.
(a) If $S_1, \cdots, S_m$ are tangentially regular at $\bar{x}$ and
$$T^C(S_1; \bar{x}) \cap I(S_2; \bar{x}) \cap \cdots \cap I(S_m; \bar{x}) \neq \emptyset,$$
then $S_1 \cap \cdots \cap S_m$ is tangentially regular at $\bar{x}$ along with
$$T^C(S_1 \cap \cdots \cap S_m; \bar{x}) = T^C(S_1; \bar{x}) \cap \cdots \cap T^C(S_m; \bar{x})$$

and
$$N^C(S_1 \cap \cdots \cap S_m; \overline{x}) = N^C(S_1; \overline{x}) + \cdots + N^C(S_m; \overline{x}).$$

(b) If the function f is finite and Lipschitz continuous near $\overline{x}$, then it is tangentially regular at $\overline{x}$ if and only if anyone of the two following properties is fulfilled:

(b$_1$) the Clarke and Bouligand directional derivatives of f at $\overline{x}$ coincide, that is,
$$f^\circ(\overline{x}; h) = f^B(\overline{x}; h) \quad \text{for all } h \in X;$$

(b$_2$) the (standard) directional derivative $f'(\overline{x}; h) := \lim_{t \downarrow 0} t^{-1}(f(\overline{x} + th) - f(\overline{x}))$
in any direction h exists and
$$f^\circ(\overline{x}; h) = f'(\overline{x}; h) \quad \text{for all } h \in X.$$

(c) If $f_1, \cdots, f_m$ are tangentially regular at $\overline{x}$ and all these functions except one of them are Lipschitz continuous near $\overline{x}$, then the function $f_1 + \cdots + f_m$ is tangentially regular at $\overline{x}$ and
$$\partial_C(f_1 + \cdots + f_m)(\overline{x}) = \partial_C f_1(\overline{x}) + \cdots + \partial_C f_m(\overline{x}).$$

PROOF. Arguments for (a) are already given above. Concerning (b) we know that $T^B(\text{epi } f; (\overline{x}, f(\overline{x}))) = \text{epi } f^B(\overline{x}; \cdot)$, and $T^C(\text{epi } f; (\overline{x}, f(\overline{x}))) = \text{epi } f^\circ(\overline{x}; \cdot)$ when f is Lipschitz continuous near $\overline{x}$, see (2.40) and (a) of Theorem 2.70. This clearly ensures the equivalence between (b$_1$) and the tangential regularity of f. Since obviously
$$\liminf_{t \downarrow 0} t^{-1}(f(\overline{x} + th) - f(\overline{x})) \leq \limsup_{t \downarrow 0} t^{-1}(f(\overline{x} + th) - f(\overline{x})) \leq f^\circ(\overline{x}; h),$$
we see that (b$_2$) is equivalent to (b$_1$).

It is enough to prove (c) for $m = 2$ with $f_1 =: f$ and $f_2 =: g$, assuming that f and g are finite at $\overline{x}$ and tangentially regular at $\overline{x}$ and that g is locally Lipschitz continuous near $\overline{x}$. For the tangential regularity of $f + g$ in (c), put $E_\varphi := \text{epi } \varphi$ for any function φ, and fix any $(h, r) \in T^B(\text{epi } (f + g); (\overline{x}, f(\overline{x}) + g(\overline{x})))$, that is, $r \geq (f + g)^B(\overline{x}; h)$. Since $(f + g)^B(\overline{x}; h) \geq f^B(\overline{x}; h) + g^B(\overline{x}; h)$ (as easily seen from the definition of the Bouligand directional derivative as a limit inferior), the tangential regularity of f and g along with the Lipschitz property of g near $\overline{x}$ yields
$$r - g^\circ(\overline{x}; h) \geq f^B(\overline{x}; h) = \inf\{r' \in \mathbb{R}: (h, r') \in T^C(E_f; (\overline{x}, f(\overline{x})))\}.$$

Fix any real $\varepsilon > 0$ and take some $(h, r') \in T^C(E_f; (\overline{x}, f(\overline{x})))$ with $r - g^\circ(\overline{x}; h) + \varepsilon > r'$. We know from Theorem 2.70 that $(h, g^\circ(\overline{x}; h) + \varepsilon) \in I(E_g; (\overline{x}, g(\overline{x})))$. Consider any sequences $(x_n, \rho_n)_n$ in $X \times \mathbb{R}$ converging to $(\overline{x}, f(\overline{x}) + g(\overline{x}))$ with $(x_n, \rho_n) \in E_{f+g}$ and $(t_n)_n$ tending to zero with $t_n > 0$. We have $(x_n, \rho_n - g(x_n))_n$ converges to $(\overline{x}, f(\overline{x}))$ with $\rho_n - g(x_n) \geq f(x_n)$, hence by the sequential characterization of $T^C(E_f; (\overline{x}, f(\overline{x})))$ there exists a sequence $(h_n, r'_n)_n$ converging to (h, r') with
$$(x_n, \rho_n - g(x_n)) + t_n(h_n, r'_n) \in E_f \quad \text{for all } n.$$
The sequential characterization of $I(E_g; (\overline{x}, g(\overline{x})))$ implies that
$$(x_n, g(x_n)) + t_n(h_n, g^\circ(\overline{x}; h) + \varepsilon) \in E_g \quad \text{for large } n.$$
Both latter inclusions ensure that for n large enough
$$(f + g)(x_n + t_n h_n) \leq \rho_n + t_n(r'_n + g^\circ(\overline{x}; h) + \varepsilon) \leq \rho_n + t_n(r'_n + r - r' + 2\varepsilon)$$

because $g^\circ(\overline{x}; h) < r - r' + \varepsilon$, thus $(h, r + 2\varepsilon) \in T^C(E_{f+g}; (\overline{x}, f(\overline{x}) + g(\overline{x})))$. Consequently, $(h, r) \in T^C(E_{f+g}; (\overline{x}, f(\overline{x}) + g(\overline{x})))$, and this means that $f + g$ is tangentially regular at $\overline{x}$.

For the equality of subdifferentials in (c), it is enough to show the converse inclusion of that of Theorem 2.98. Let $u^* \in \partial_C f(\overline{x})$ and $v^* \in \partial_C g(\overline{x})$. The tangential regularity property of f (resp. g) says that the Clarke and Bouligand-Peano tangent cones of E_f at $(\overline{x}, f(\overline{x}))$ (resp. of E_g at $(\overline{x}, g(\overline{x}))$) coincide, thus (2.28) yields for any $h \in X$ that $\langle u^*, h \rangle \leq f^B(\overline{x}; h)$ and $\langle v^*, h \rangle \leq g^B(\overline{x}; h)$, which entails

$$\langle u^* + v^*, h \rangle \leq f^B(\overline{x}; h) + g^B(\overline{x}; h) \leq (f+g)^B(\overline{x}; h).$$

We deduce from this and the inclusion of the Clarke tangent cone into the Bouligand Peano one that

$$\langle u^* + v^*, h \rangle \leq \inf\{r \in \mathbb{R} : (h, r) \in T^C(E_{f+g}; (\overline{x}, f(\overline{x}) + g(\overline{x})))\},$$

hence $u^* + v^* \in \partial_C(f+g)(\overline{x})$ as required. □

Given any function $f : U \to \mathbb{R} \cup \{-\infty, +\infty\}$, we recall (see Definition 2.62) that the standard directional derivative in (b$_2$) of the above proposition is also called the *right-hand side Dini directional derivative* and denoted by $d_D^+ f(\overline{x}; \cdot)$, that is,

(2.44) $$d_D^+ f(\overline{x}; h) := f'(\overline{x}; h) := \lim_{t \downarrow 0} t^{-1}(f(\overline{x} + th) - f(\overline{x})).$$

In some situations the equality in (b$_1$) is needed only for certain directions $h \in X$.

DEFINITION 2.127. A locally Lipschitz function $f : U \to \mathbb{R}$ on an open set U of the normed space X will be called *Clarke subregular* or *C-subregular* at a point $x \in U$ in a direction $h \in X$ provided that

$$f^\circ(x; h) = f^B(x; h),$$

or equivalently (as seen in the proof of (b) in the above proposition) $f'(x; h)$ exists and

$$f^\circ(x; h) = f'(x; h).$$

When anyone of the above equivalent equalities hold for all $h \in X$, the function f will be said to be *Clarke directionally subregular* or *directionally C-subregular* at x. We will also say, for short, that f is *Clarke subregular* or *C-subregular* at x. By (b$_1$) and (b$_2$) in Proposition 2.126 the locally Lipschitz function f is C-subregular at x if and only if it is *tangentially regular* at x (see Definition 2.62). If f is Clarke subregular at any point in U, we will say that it is Clarke subregular or C-subregular (on U).

Characterizations of Clarke subregularity of a locally Lipschitz function f in terms of upper semicontinuity of the lower Dini directional derivative $\underline{d}_D^+ f(\cdot; h)$ will be established later in Proposition 6.6. The separate regularity of a bivariate function will be developed in Proposition 6.11 and Theorem 6.16.

Before continuing with the tangential regularity of some classes of sets, it worth observing that, with the (standard) directional derivative recalled in Definition 2.62, the second equality in Proposition 2.63 can be rewritten as:

PROPOSITION 2.128. Let U be an open convex set of the normed space X and $f : U \to \mathbb{R} \cup \{+\infty\}$ be an extended real-valued convex function. Then, at any $x \in U$ where f is finite the (standard) directional derivative $f'(x; \cdot)$ exists and
$$\partial f(x) = \{x^* \in X^* : \langle x^*, h \rangle \leq f'(x; h),\ \forall h \in X\}.$$

Concerning tangentially regular sets, the graph of a strictly differentiable mapping is a particular example of such a set. This is in fact a consequence of a more general result. In the statement of the result, as already recalled in (1.24), for a continuous linear operator $A : X \to Y$ between two normed spaces, $A^* : Y^* \to X^*$ denotes its *continuous linear adjoint* defined between the topological duals by
$$A^*(y^*) := y^* \circ A \quad \text{for all } y^* \in Y^*.$$
So, for a mapping $G : U \to Y$ from an open set U of X into Y which is differentiable at a point $x \in U$ in some sense, $DG(x)^*$ denotes the continuous linear adjoint of $DG(x)$, hence the image of a set $Q \subset Y^*$ under $DG(x)^*$ is
$$DG(x)^*(Q) = \{DG(x)^*(y^*) : y^* \in Q\} = \{y^* \circ DG(x) : y^* \in Q\}.$$

PROPOSITION 2.129. Let S be a subset of a normed space Y and $G : X \to Y$ be a mapping from the normed space X into Y. Let
$$M(x) = G(x) + S \quad \text{for all } x \in X.$$
(a) If G is Hadamard differentiable at a point $\bar{x} \in X$, then for every $y \in M(\bar{x})$,
$$T^B\big(\mathrm{gph}\, M; (\bar{x}, y)\big) = \{(h, k) \in X \times Y : k \in DG(\bar{x})h + T^B(S; y - G(\bar{x}))\},$$
hence in particular
$$\mathrm{gph}\, DG(\bar{x}) = T^B\big(\mathrm{gph}\, G; (\bar{x}, G(\bar{x}))\big).$$
(b) Assume that G is strictly Hadamard differentiable at $\bar{x}$. Then, for every $y \in M(\bar{x})$,
$$T^C\big(\mathrm{gph}\, M; (\bar{x}, y)\big) = \{(h, k) \in X \times Y : k \in DG(\bar{x})h + T^C(S; y - G(\bar{x}))\}$$
and
$$N^C\big(\mathrm{gph}\, M; (\bar{x}, y)\big) = \{(-DG(\bar{x})^*(y^*), y^*) \in X^* \times Y^* : y^* \in N^C(S; y - G(\bar{x}))\},$$
hence $\mathrm{gph}\, M$ is tangentially regular at $(\bar{x}, y)$ whenever S is tangentially regular at $y - G(\bar{x})$; in such a case one has
$$T^B\big(\mathrm{gph}\, M; (\bar{x}, y)\big) = T^C\big(\mathrm{gph}\, M; (\bar{x}, y)\big)$$
$$= \{(h, k) \in X \times Y : k \in DG(\bar{x})h + T^C(S; y - G(\bar{x}))\}.$$
In particular $\mathrm{gph}\, G$ is tangentially regular at $(\bar{x}, G(\bar{x}))$ and
$$\mathrm{gph}\, DG(\bar{x}) = T^C\big(\mathrm{gph}\, G; (\bar{x}, G(\bar{x}))\big) = T^B\big(\mathrm{gph}\, G; (\bar{x}, G(\bar{x}))\big)$$
$$\text{and} \quad N^C\big(\mathrm{gph}\, G; (\bar{x}, G(\bar{x}))\big) = \{(-DG(\bar{x})^*(y^*), y^*) : y^* \in Y^*\}.$$

PROOF. (a) Suppose that G is Hadamard differentiable at $\bar{x}$ and put $A := DG(\bar{x})$. Fix $y \in M(\bar{x})$ and let $(h, k) \in T^B\big(\mathrm{gph}\, M; (\bar{x}, y)\big)$. By the sequential characterization of the Bouligand-Peano tangent cone, there exist a sequence of reals $(t_n)_n$ tending to 0 with $t_n > 0$ and a sequence $(h_n, k_n)_n$ in $X \times Y$ converging to (h, k) such that for all n
$$(\bar{x}, y) + t_n(h_n, k_n) \in \mathrm{gph}\, M, \quad \text{that is,} \quad y + t_n k_n \in G(\bar{x} + t_n h_n) + S.$$

From the Hadamard differentiability of G at $\bar{x}$, there is a sequence $(e_n)_n$ in Y converging to 0 such that $G(\bar{x} + t_n h_n) = G(\bar{x}) + t_n(Ah_n + e_n)$, so

$$y + t_n k_n \in G(\bar{x}) + t_n(Ah_n + e_n) + S, \quad \text{that is,} \quad y - G(\bar{x}) + t_n(k_n - Ah_n - e_n) \in S.$$

This ensures that $k - Ah \in T^B(S; y - G(\bar{x}))$, proving, for the first equality in (a), the inclusion of the left member into the right one.

For the converse inclusion fix any (h, k) in $X \times Y$ such that $k \in Ah + T^B(S; y - G(\bar{x}))$. There are a sequence of reals $(t_n)_n$ tending to 0 with $t_n > 0$ and a sequence $(e_n)_n$ in Y converging to 0 such that $y - G(\bar{x}) + t_n(k - Ah + e_n) \in S$. The Hadamard differentiability of G at $\bar{x}$ furnishes a sequence $(e'_n)_n$ in Y converging to 0 such that $G(\bar{x} + t_n h) = G(\bar{x}) + t_n(Ah + e'_n)$, so

$$y + t_n(k + e_n + e'_n) \in G(\bar{x} + t_n h) + S, \quad \text{that is,} \quad (\bar{x}, y) + t_n(h, k + e_n + e'_n) \in \mathrm{gph}\, M.$$

This entails that $(h, k) \in T^B(\mathrm{gph}\, M; (\bar{x}, y))$. The first equality in (a) is then established and the second equality follows from the first by taking $S := \{0\}$.

(b) Now suppose that G is strictly Hadamard differentiable at $\bar{x}$ and put as above $A := DG(\bar{x})$. Fix $y \in M(\bar{x})$ and fix also any $(h, k) \in T^C(\mathrm{gph}\, M; (\bar{x}, y))$. Let $(z_n)_n$ be any sequence in Y converging to $y - G(\bar{x})$ with $z_n \in S$ for all n and let $(t_n)_n$ be any sequence of reals tending to 0 with $t_n > 0$. We observe that, for $y_n := G(\bar{x}) + z_n$, we have $y_n \in M(\bar{x})$, that is, $(\bar{x}, y_n) \in \mathrm{gph}\, M$ and $(\bar{x}, y_n) \to (\bar{x}, y)$. Then, the inclusion (h, k) in $T^C(\mathrm{gph}\, M; (\bar{x}, y))$ yields a sequence $(h_n, k_n)_n$ converging to (h, k) such that

$$(\bar{x}, y_n) + t_n(h_n, k_n) \in \mathrm{gph}\, M, \quad \text{that is,} \quad y_n + t_n k_n \in G(\bar{x} + t_n h_n) + S.$$

By the Hadamard differentiability of G, we have $G(\bar{x} + t_n h_n) = G(\bar{x}) + t_n(Ah_n + e_n)$ for some sequence $(e_n)_n$ in Y converging to 0. Therefore,

$$y_n + t_n k_n \in G(\bar{x}) + t_n(Ah_n + e_n) + S, \quad \text{that is,} \quad z_n + t_n(k_n - Ah_n - e_n) \in S,$$

so $k - Ah \in T^C(S; y - G(\bar{x}))$. This proves, for the first equality in (b), the inclusion of the first member into the second.

To prove the converse inclusion, fix (h, k) in $X \times Y$ such that $k - Ah \in T^C(S; y - G(\bar{x}))$. Consider any sequence $(x_n, y_n)_n$ in $\mathrm{gph}\, M$ with $(x_n, y_n) \to (\bar{x}, y)$. Let $(t_n)_n$ be any sequence of reals tending to 0 with $t_n > 0$. Since $y_n - G(x_n) \in S$ and $y_n - G(x_n) \to y - G(\bar{x})$ according to the continuity of G at $\bar{x}$, the inclusion of $k - Ah$ in $T^C(S; y - G(\bar{x}))$ furnishes a sequence $(v_n)_n$ converging to $k - Ah$ such that $y_n - G(x_n) + t_n v_n \in S$. For $k_n := v_n + Ah$, we have $k_n \to k$ and $y_n - G(x_n) + t_n(k_n - Ah) \in S$. From the strict Hadamard differentiability of G at $\bar{x}$, we have a sequence $(e_n)_n$ in Y converging to 0 such that

$$G(x_n + t_n h) = G(x_n) + t_n(Ah + e_n).$$

From this equality and from the latter inclusion, we obtain for all integers n

$$y_n - G(x_n + t_n h) + t_n e_n + t_n k_n \in S,$$

that is,

$$y_n + t_n(k_n + e_n) \in G(x_n + t_n h) + S = M(x_n + t_n h)$$

or equivalently $(x_n, y_n) + t_n(h, k_n + e_n) \in \mathrm{gph}\, M$. This proves that $(h, k) \in T^C(\mathrm{gph}\, M; (\bar{x}, y))$. Consequently, the first equality in (b) is proved and it is easy to see that it implies the second equality. The tangential regularity of $\mathrm{gph}\, M$ and $\mathrm{gph}\, G$ as well as the related equalities follow from the equalities in (a) and the first in (b). Finally, the last equality in (b) is obtained from the second in (b) by taking $S = \{0\}$. □

2.2.13. Tangent cones of inverse and direct images. The inverse image of a set arises in many situations; for example, contraint sets of optimization problems often appear in this form. Such sets are the objective of the next two propositions.

The first proposition is concerned with inverse image under isomorphism.

PROPOSITION 2.130. *Let $A : X \to Y$ be an isomorphism between two normed spaces X, Y and S be a subset Y. Then for any $b \in Y$ and any $x \in A^{-1}(b+S)$ one has*
$$T^C(A^{-1}(b+S); x) = A^{-1}(T^C(S; A(x) - b)),$$
$$T^B(A^{-1}(b+S); x) = A^{-1}(T^B(S; A(x) - b))$$
along with the equality
$$N^C(A^{-1}(b+S); x) = A^*(N^C(S; A(x) + b)).$$

PROOF. The tangential equalities directly follow from the sequential characterizations of $T^C(\cdot; \cdot)$ and $T^B(\cdot; \cdot)$. Concerning the equality with normal cones, putting $C := A^{-1}(b+S)$ and writing for $x^* \in X^*$
$$x^* \in N^C(C; x) \Leftrightarrow \langle x^*, h \rangle \leq 0 \ \forall h \in T^C(C; x)$$
$$\Leftrightarrow \langle x^* \circ (A^{-1}), Ah; \rangle \leq 0 \ \forall h \in T^C(C; x)$$
$$\Leftrightarrow \langle x^* \circ (A^{-1}), v \rangle \leq 0 \ \forall v \in T^C(S; A(x) + b),$$
we see that $x^* \in N^C(C; x)$ if and only if $x^* \circ A^{-1} \in N^C(S; A(x)+b)$, or equivalently $x^* \in A^*(N^C(S; A(x) + b))$ since $A^*(x^* \circ (A^{-1})) = x^*$ and A^* is bijective. This justifies the equality with normal cones. □

Recall now an inverse mapping theorem under strict Fréchet differentiability (see the strict differentiability statement in Theorem E.1 in Appendix).

THEOREM 2.131. *Let X and Y be two Banach spaces. Assume that $G : X \to Y$ is strictly Fréchet differentiable at a point $\bar{x} \in X$ and that $DG(\bar{x})$ is a bijection from X onto Y. Then there exist open neighborhoods U and V of $\bar{x}$ and $G(\bar{x})$ respectively such that the restriction $G_{|U} : U \to V$ of G to U with values in V is a bijection from U onto V whose inverse $F := (G_{|U})^{-1}$ is strictly Fréchet differentiable at $\bar{y} := G(\bar{x})$ with $DF(\bar{y}) = (DG(\bar{x}))^{-1}$.*

PROPOSITION 2.132. *Let S be a subset of a normed space Y and $G : X \to Y$ be a mapping from a normed space X into Y which is Hadamard differentiable at a point $\bar{x} \in G^{-1}(S)$. Denote by $DG(\bar{x})$ the Hadamard derivative of G at $\bar{x}$. The following hold:*
(a) $T^B(G^{-1}(S); \bar{x}) \subset DG(\bar{x})^{-1}(T^B(S; G(\bar{x})))$.
(b) *If G is strictly Hadamard differentiable at $\bar{x}$, then*
$$DG(\bar{x})^{-1}(I(S; G(\bar{x}))) \subset I(G^{-1}(S); \bar{x}).$$
(c) *If G is strictly Hadamard differentiable at $\bar{x}$ and $DG(\bar{x})^{-1}(I(S; G(\bar{x}))) \neq \emptyset$, one has*
$$DG(\bar{x})^{-1}(T^C(S; G(\bar{x}))) \subset T^C(G^{-1}(S); \bar{x}),$$

and if in addition S is tangentially regular at $\bar{x}$ then $G^{-1}(S)$ is tangentially regular at $\bar{x}$ and the latter inclusion is an equality.
(d) If X and Y are Banach spaces and if G is strictly Fréchet differentiable at $\bar{x}$ and $DG(\bar{x})$ is a bijection from X onto Y, then the inclusions in (a) and (b) are equalities, and one also has

$$T^C(G^{-1}(S);\bar{x}) = DG(\bar{x})^{-1}(T^C(S;G(\bar{x}))).$$

PROOF. (a) The proof of (a) is similar to that of the first part of (a) in the previous proposition. Take $h \in T^B(G^{-1}(S);\bar{x})$ and choose sequences $(t_n)_n$ tending to 0 with $t_n > 0$ and $(h_n)_n$ converging to h such that $\bar{x} + t_n h_n \in G^{-1}(S)$, that is, $G(\bar{x} + t_n h_n) \in S$. Since G is Hadamard differentiable at $\bar{x}$ we have $\lim_{n\to\infty} t_n^{-1}(G(\bar{x} + t_n h_n) - G(\bar{x})) = DG(\bar{x})(h)$, or equivalently $G(\bar{x}+t_n h_n) = G(\bar{x})+t_n(DG(\bar{x})(h)+e_n)$, where $e_n \to 0$ in Y. So, $G(\bar{x}) + t_n(DG(\bar{x})(h) + e_n) \in S$ for all n, and this says that $DG(\bar{x})(h) \in T^B(S;G(\bar{x}))$.
(b) For (b), put $A := DG(\bar{x})$ and fix $h \in A^{-1}(I(S;G(\bar{x})))$. Consider any sequences $(x_n)_n$ of $G^{-1}(S)$ converging to $\bar{x}$, $(t_n)_n$ tending to 0 with $t_n > 0$, and $(h_n)_n$ converging to h in X. The strict Hadamard differentiability of G at $\bar{x}$ gives some sequence $(e_n)_n$ in Y converging to 0 and such that $G(x_n + t_n h_n) = G(x_n) + t_n(A(h_n) + e_n)$. Since $G(x_n) \in S$ and $G(x_n) \to G(\bar{x})$ (according to the continuity of G at $\bar{x}$ due to the strict Hadamard differentiability at this point), from the inclusion $A(h) \in I(S;G(\bar{x}))$ and the sequential characterization of $I(S;G(\bar{x}))$ we have, for n large enough, $G(x_n) + t_n(A(h_n) + e_n) \in S$, thus $G(x_n + t_n h_n) \in S$, that is, $x_n + t_n h_n \in G^{-1}(S)$. This establishes the desired inclusion of (b).
(c) Concerning the inclusion of (c) fix $h_0 \in A^{-1}(I(S;G(\bar{x})))$ and take any $h \in A^{-1}(T^C(S;G(\bar{x})))$. For any integer n we have $A(h + \frac{1}{n}h_0) = A(h) + \frac{1}{n}A(h_0) \in I(S;G(\bar{x}))$ by (a) in Theorem 2.13, then $h + \frac{1}{n}h_0 \in I(G^{-1}(S);\bar{x})$ by (b) above. So, Theorem 2.13(b) gives $h \in T^C(G^{-1}(S);\bar{x})$. The desired inclusion then holds and, whenever S is in addition tangentially regular at $G(\bar{x})$, it is easily seen from (a) that this inclusion is an equality and that $G^{-1}(S)$ is tangentially regular at $\bar{x}$.
(d) By the inverse mapping theorem recalled above fix neighborhoods U and V of $\bar{x}$ and $\bar{y} := G(\bar{x})$ such that $G_{|U} : U \to V$ is a bijection from U onto V whose inverse $F := (G_{|U})^{-1} : V \to U$ is strictly Fréchet differentiable at $\bar{y}$ with $DF(\bar{y}) = (DG(\bar{x}))^{-1}$. Put $\Lambda := DF(\bar{y})$, $Q := G^{-1}(S)$, $Q_U := Q \cap U$, and $S_V := S \cap V$. Note that $S_V = F^{-1}(Q_U)$.
By (a) we have

$$T^B(S_V;\bar{y}) = T^B(F^{-1}(Q_U);\bar{y}) \subset \Lambda^{-1}(T^B(Q_U;\bar{x})),$$

which is equivalent to $T^B(S;\bar{y}) \subset \Lambda^{-1}(T^B(Q;\bar{x}))$, so (a) is an equality. Applying the same arguments with (b) transforms the inclusion there into an equality too.

Now let us show that $A^{-1}(T^C(S;\bar{y}))$ is included in $T^C(G^{-1}(S);\bar{x}) = T^C(Q;\bar{x})$, where $A := DG(\bar{x})$. Fix any $h \in A^{-1}(T^C(S;\bar{y}))$ and consider any sequences $(t_n)_n$ tending to 0 with $t_n > 0$ and $(x_n)_n$ converging to $\bar{x}$ with $x_n \in Q$, that is, $G(x_n) \in S$. Since $Ah \in T^C(S_V;\bar{y})$, there exists a sequence $(v_n)_n$ converging to Ah such that, for all n large enough, one has $G(x_n) + t_n v_n \in S_V = G(Q_U)$, or equivalently $F(G(x_n) + t_n v_n) \in Q_U$. The strict Fréchet differentiability of F at $\bar{y}$ gives some sequence $(e_n)_n$ in X converging to 0 such that for large n

$$F(G(x_n) + t_n v_n) = F(G(x_n)) + t_n(\Lambda v_n + \|v_n\|e_n).$$

Consequently, for n large enough, we obtain $x_n + t_n(\Lambda v_n + \|v_n\|e_n) \in Q$, so $h = \Lambda(Ah) \in T^C(Q;\bar{x})$, which justifies the desired inclusion $A^{-1}(T^C(S;\bar{y})) \subset T^C(G^{-1}(S);\bar{x})$. Using this inclusion with F in place of G gives $\Lambda^{-1}(T^C(Q_U;\bar{x})) \subset T^C(F^{-1}(Q_U);\bar{y})$, which means that $T^C(G^{-1}(S);\bar{x}) \subset A^{-1}(T^C(S;\bar{y}))$. This justifies the equality with the C-tangent cone in (d) and finishes the proof. □

A first corollary is concerned with the interior tangent cone property of inverse images.

COROLLARY 2.133. *Let S be a subset of a normed space Y and $G : X \to Y$ be a mapping from a normed space X into Y which is strictly Hadamard differentiable at a point $\bar{x} \in G^{-1}(S)$. If S has the interior tangent cone property at $G(\bar{x})$ and the Hadamard derivative $DG(\bar{x})$ of G at $\bar{x}$ is surjective, then the set $G^{-1}(S)$ has the interior tangent cone property at $\bar{x}$.*

PROOF. Assume that $A := DG(\bar{x})$ is surjective and S has the interior tangent property at $G(\bar{x})$, or equivalently $I(S;G(\bar{x})) \neq \emptyset$. By surjectivity of A, we have $A^{-1}(I(S;G(\bar{x}))) \neq \emptyset$. Then Proposition 2.132(b) ensures that $I(G^{-1}(S);\bar{x}) \neq \emptyset$, hence the set $G^{-1}(S)$ has the interior tangent property at $\bar{x}$ as desired. □

The next corollary is related to tangents and normals to submanifolds. Recall that a subset M of a Banach space X is by definition (see Definition E.8 in Appendix) a $\mathcal{C}^1$-submanifold at a point $m_0 \in M$ provided there exist a closed vector subspace E of X, an open neighborhood U of m_0 in X, an open neighborhood V of zero in X, and a $\mathcal{C}^1$-diffeomorphism $\varphi : U \to V$ from U onto V with $\varphi(m_0) = 0$ such that
$$\varphi(M \cap U) = E \cap V;$$
the pair (U,φ) is a local chart and E is a model space. Then, for such a submanifold M Proposition 2.132 gives $T^B(M;m_0) = D\varphi(m_0)^{-1}(E)$ and $T^C(M;m_0) = D\varphi(m_0)^{-1}(E)$. Further, from the latter equality it is clear that $N^C(M;m_0) = \{x^* \in X^* : \langle x^*, D\varphi(m_0)^{-1}(z)\rangle = 0, \forall z \in E\}$. So, we have the following corollary of Proposition 2.132(d).

COROLLARY 2.134. *Let X be a Banach space and M be a $\mathcal{C}^1$-submanifold at $m_0 \in M$ in X with a local chart (U,φ) and a closed vector subspace E as model space. Then one has the equalities*
$$T^B(M;m_0) = T^C(M;m_0) = D\varphi(m_0)^{-1}(E)$$
$$N^C(M;m_0) = \{x^* \in X^* : x^* \circ (D\varphi(m_0)^{-1}) \in E^{\perp_{X^*}}\},$$
where $E^{\perp_{X^*}}$ is the annihilator of E in X^*, that is, $E^{\perp_{X^*}} := \{x^* \in X^* : \langle x^*, z\rangle = 0, \forall z \in E\}$.

The following theorem translates the results in (a) and (c) in Proposition 2.132 above to functions. (Note that a direct proof can easily be given).

THEOREM 2.135 (*C*-subdifferential chain rule under strict *H*-differentiability). *Let $G : X \to Y$ be a mapping between two normed spaces X and Y which is strictly Hadamard differentiable at a point $\bar{x} \in X$, and let $g : Y \to \mathbb{R} \cup \{-\infty, +\infty\}$ be a function which is finite and Lipschitz continuous near $G(\bar{x})$. Then $g \circ G$ is finite and Lipschitz continuous near $\bar{x}$ and*
$$(g \circ G)^\circ(\bar{x};h) \leq g^\circ(G(\bar{x}); DG(\bar{x})(h)) \text{ and } \partial_C(g \circ G)(\bar{x}) \subset DG(\bar{x})^*(\partial_C g(G(\bar{x}))),$$

and the inequality and inclusion are equalities whenever in addition g is tangentially regular at $G(\bar{x})$.

PROOF. The function $g \circ G$ is evidently Lipschitz continuous near $\bar{x}$ and, for $\widehat{G} : X \times \mathbb{R} \to Y \times \mathbb{R}$ with $\widehat{G}(x,r) = (G(x), r)$, we have $\widehat{G}^{-1}(\operatorname{epi} g) = \operatorname{epi}(g \circ G)$. So, for $S := \operatorname{epi} g$, $u := (\bar{x}, g \circ G(\bar{x}))$, $A := DG(\bar{x})$ and $\widehat{A} := D\widehat{G}(u)$, we have $\widehat{A}(h, r) = (A(h), r)$ and

$$T^C(\widehat{G}^{-1}(S); u) = \operatorname{epi}(g \circ G)^\circ(\bar{x}; \cdot), \quad \widehat{A}^{-1}(T^C(S; \widehat{G}(u))) = \operatorname{epi}\big(g^\circ(G(\bar{x}); \cdot) \circ A\big);$$

similar equalities also hold with the corresponding Bouligand-Peano tangent cones and Bouligand directional derivatives. Further, the non vacuity of $\widehat{A}^{-1}(I(S; \widehat{G}(u))$ is ensured by the inclusion $\widehat{A}(h, g^\circ(G(\bar{x}); A(h)) + \varepsilon) \in I(S; \widehat{G}(u))$ for all $h \in X$ and $\varepsilon > 0$ (see Theorem 2.70). We may then apply (c) in Proposition 2.132 above to obtain the desired inequality for $(g \circ G)^\circ(\bar{x}; \cdot)$.

Similarly, we see that $g^B(G(\bar{x}); Ah) \leq (g \circ G)^B(\bar{x}; h)$ by the analogous above equalities for B-tangent cones and by (a) in Proposition 2.132. This combined with the previous inequality entails that

$$(2.45) \qquad (g \circ G)^\circ(\bar{x}; h) = g^\circ(G(\bar{x}); Ah) \quad \text{for all } h \in X$$

whenever g is tangentially regular at $G(\bar{x})$.

On the other hand, $A^*(\partial_C g(G(\bar{x})))$ is weak* closed according to the weak* compactness of $\partial_C g(G(\bar{x}))$ and its support function clearly coincides with $h \mapsto g^\circ(G(\bar{x}); Ah)$. The inclusion of the proposition then follows from the above inequality. Further, by (2.45) the inclusion is an equality when g is tangentially regular at $G(\bar{x})$. □

EXAMPLE 2.136. Let $f : U \to \mathbb{R}$ be a function on an open set U of a normed space X which is strictly Hadamard differentiable at a point $\bar{x} \in U$. Writing $|f| = |\cdot| \circ f$, we obtain

$$\partial_C|f|(\bar{x}) = [-Df(\bar{x}), Df(\bar{x})] = \{rDf(\bar{x}) : |r| \leq 1\}$$

by applying the above theorem. □

LEMMA 2.137. Let X be a normed space, S be a subset of X and U be an open set with $U \cap S \neq \emptyset$. Let $g : U \to [0, +\infty[$ be a locally Lipschitz continuous function such that $d_S(x) \leq g(x)$ for all $x \in U$ along with $g(x) = 0$ for all $x \in U \cap S$. Then one has for all $x \in U \cap S$

$$d_S^\circ(x; \cdot) \leq g^\circ(x; \cdot) \quad \text{and} \quad \partial_C d_S(x) \subset \partial_C g(x).$$

PROOF. Fix any $x \in U \cap S$. By Proposition 2.95(a) we have for every $h \in X$

$$d_S^\circ(x; h) = \limsup_{\substack{S \ni u \to x \\ t \downarrow 0}} t^{-1} d_S(u + th) \leq \limsup_{\substack{S \ni u \to x \\ t \downarrow 0}} t^{-1} g(u + th) = \limsup_{\substack{S \ni u \to x \\ t \downarrow 0}} t^{-1}[g(u + th) - g(u)].$$

This entails that

$$d_S^\circ(x; h) \leq \limsup_{\substack{u \to x \\ t \downarrow 0}} t^{-1}[g(u + th) - g(u)] = g^\circ(x; h),$$

which justifies the inequality of the lemma and finishes the proof since the inclusion of the lemma is a direct consequence of the inequality. □

PROPOSITION 2.138. Let X and Y be normed spaces, $G : X \to Y$ be a mapping which is strictly Hadamard differentiable at a point $\bar{x} \in X$ and S be a subset of Y containing $\bar{y} := G(\bar{x})$. Assume that there are a real $\gamma \geq 0$ and an open neighborhood U of $\bar{x}$ such that the metric inequality

$$d(x, G^{-1}(S)) \leq \gamma d(G(x), S)$$

holds for all $x \in U$. Then one has with $A := DG(\bar{x})$

$$\left(d(\cdot, G^{-1}(S))\right)^\circ (\bar{x}; h) \leq \gamma (d(\cdot, S))^\circ(\bar{y}; Ah) \text{ for all } h \in X,$$

$$\partial_C d\left(\cdot, G^{-1}(S)\right)(\bar{x}) \subset \gamma A^* \left(\partial_C d(\cdot, S)(G(\bar{x}))\right),$$

$$A^{-1}\left(T^C(S; \bar{y})\right) \subset T^C\left(G^{-1}(S); \bar{x}\right),$$

where $A := DG(\bar{x})$.

PROOF. Suppose that the metric inequality in the proposition is satisfied. Put $Q := G^{-1}(S)$ and fix any $h \in X$. Using the above lemma we see that $d_Q^\circ(\bar{x}) \leq \gamma (d_S \circ G)^\circ(\bar{x}; h)$. Since $(d_S \circ G)^\circ(\bar{x}, h) \leq d^\circ(\bar{y}; DG(\bar{x})h)$ according to the Lipschitz property of d_S and the strict Hadamard differentiability of G at $\bar{x}$, it follows that

$$d_Q^\circ(\bar{x}; \cdot) \leq \gamma d_S^\circ(\bar{y}; \cdot) \circ A(\cdot).$$

Considering the convex function $f := d_S^\circ(\bar{y}; \cdot)$ and the linear mapping $A := DG(\bar{x})$, and using the latter inequality along with the equality $d_Q^\circ(\bar{x}; \cdot)(0) = \gamma (f \circ A)(0)$, whose both members are null, we easily obtain that

$$\partial d_Q^\circ(\bar{x}; \cdot)(0) \subset \gamma \partial (f \circ A)(0).$$

Since the convex function f is continuous as well as the linear mapping A, the chain rule in Theorem 2.135 entails that

$$\partial_C d_Q(\bar{x}) \subset \gamma A^*(\partial f(A0)) = \gamma A^*(\partial f(0)) = \gamma A^*(\partial_C d_S(\bar{y})).$$

If $Ah \in T^C(S; \bar{y})$, we know that $d_S^\circ(\bar{y}; Ah) = 0$ (see Proposition 2.95(b)), so the above inequality $d_Q^\circ(\bar{x}; h) \leq \gamma d_S^\circ(\bar{y}; Ah)$ yields that $d_Q^\circ(\bar{x}; h) = 0$ (keep in mind that one always has $d_Q^\circ(\bar{x}; h) \geq 0$), which gives $h \in T^C(Q; \bar{x})$. This jsutifies the inclusion $A^{-1}\left(T^C(S; \bar{y})\right) \subset T^C\left(G^{-1}(S); \bar{x}\right)$ and finishes the proof. □

Another case which occurs frequently with $G^{-1}(S)$ is when G is strictly Fréchet differentiable with a surjective derivative at the point of interest. In order to examine such a case, let us establish first the so-called Graves theorem. Given a surjective continuous linear mapping $A : X \to Y$ between Banach spaces, recall that the *classical Banach-Schauder open mapping theorem* (see Theorem C.3 in Appendix) says that A is open, or equivalently there exists a real constant $c > 0$ such that $c\mathbb{B}_Y \subset A(\mathbb{B}_X)$.

THEOREM 2.139 (L.M. Graves's open mapping theorem). Let X, Y be Banach spaces and let $G : B_X(\bar{x}, \varepsilon) \to Y$ be a mapping with $\bar{x} \in X$ and $\varepsilon > 0$. Assume that there are a surjective continuous linear mapping $A : X \to Y$ and reals $c > 0$, $\alpha > 0$ and $\beta > 0$ with $c > \alpha + \beta$ such that $c\mathbb{B}_Y \subset A(\mathbb{B}_X)$ and

$$\|G(x') - G(x) - A(x' - x)\| \leq \beta \|x' - x\| \quad \text{for all } x, x' \in B_X(\bar{x}, \varepsilon).$$

Then for any $t > 0$ and $x \in X$ with $\|x - \bar{x}\| + t < \varepsilon$ one has

$$B_Y[G(x), \alpha t] \subset G(B_X[x, t]).$$

2.2. CLARKE SUBDIFFERENTIAL

PROOF. Choose a real $K > 0$ such that $c > 1/K > \alpha + \beta$. Fix any $x \in X$ and $t > 0$ with $\|x - \bar{x}\| + t < \varepsilon$, and note (replacing $G(\cdot)$ by $G(\cdot) - G(x)$) we may and do suppose that $G(x) = 0$. Observe that $B[x,t] \subset B(\bar{x}, \varepsilon)$. Take any $y \in B[0, \alpha t]$ and put $u_0 := x$ and $v_0 := y$. Let us construct inductively a sequence $(u_n, v_n)_{n \geq 0}$ in $B[x,t] \times Y$ such that for all $n \geq 1$

(2.46) $\qquad \|u_n - u_{n-1}\| \leq (\beta K)^{n-1} K \|y\|, \quad v_{n-1} = A(u_n - u_{n-1}),$

(2.47) $\qquad v_n = A(u_n - u_{n-1}) - \big(G(u_n) - G(u_{n-1})\big), \quad \|v_n\| \leq (\beta K)^n \|y\|.$

Clearly $(u_0, v_0) \in B[x,t] \times Y$. Suppose that we have constructed $(u_1, v_1), \cdots$, (u_n, v_n) fulfilling the above properties. Since $1/K < c$, we have $\mathbb{B}_Y \subset A(K\mathbb{B}_X)$, hence we can choose $u_{n+1} \in X$ such that

$$\|u_{n+1} - u_n\| \leq K\|v_n\| \quad \text{and} \quad A(u_{n+1} - u_n) = v_n,$$

and we see that $\|u_{n+1} - u_n\| \leq (\beta K)^n K \|y\|$ according to the induction assumption $\|v_n\| \leq (\beta K)^n \|y\|$ at the step n. This and the inductive inequalities assumption $\|u_j - u_{j-1}\| \leq (\beta K)^{j-1} K \|y\|$, $j = 1, \cdots, n$, imply that

(2.48) $\qquad \|u_{n+1} - x\| \leq \dfrac{K}{1 - \beta K} \|y\| \leq \alpha^{-1} \|y\| \leq t.$

Then we can define $v_{n+1} = A(u_{n+1} - u_n) - \big(G(u_{n+1}) - G(u_n)\big)$, so

$$\|v_{n+1}\| \leq \beta \|u_{n+1} - u_n\| \leq \beta K \|v_n\|, \quad \text{hence } \|v_{n+1}\| \leq (\beta K)^{n+1} \|y\|.$$

The sequence $(u_n, v_n)_{n \geq 0}$ is then inductively well-defined as above.

The inequality in (2.46) tells us that $(u_n)_{n \geq 0}$ is a Cauchy sequence in $B[x,t]$, thus it converges to some $u \in B[x,t]$. On the other hand, the equalities in (2.46) and (2.47) give $v_{n-1} - v_n = G(u_n) - G(u_{n-1})$, hence

$$y - v_n = v_0 - v_n = G(u_n) - G(u_0) = G(u_n).$$

Since $v_n \to 0$ (because $\|v_n\| \leq (\beta K)^n \|y\|$) and since G is (Lipschitz) continuous on $B[x,t] \subset B(\bar{x}, \varepsilon)$ by the inequality assumption in the theorem, we obtain $y = G(u)$. This means that $B[G(x), \alpha t] \subset G(B[x,t])$. $\qquad\square$

From the above theorem we derive the metric inequality in the next corollary. A large study will be devoted to such metric inequalities in Chapter 7.

COROLLARY 2.140. *Under the assumptions of Theorem 2.139, for every $(x,y) \in X \times Y$ satisfying $\|x - \bar{x}\| + \alpha^{-1} \|G(x) - y\| < \varepsilon$ one has*

$$d(x, G^{-1}(y)) \leq \alpha^{-1} \|y - G(x)\|.$$

PROOF. Fix $(x,y) \in X \times Y$ with $\|x - \bar{x}\| + \alpha^{-1} \|G(x) - y\| < \varepsilon$ and $y \neq G(x)$, and put $t := \alpha^{-1} \|y - G(x)\|$. By Theorem 2.139 there is some $u \in B[x,t]$ such that $y = G(u)$, hence

$$d(x, G^{-1}(y)) \leq \|x - u\| \leq t, \quad \text{so } d(x, G^{-1}(y)) \leq \alpha^{-1} \|y - G(x)\|,$$

and the latter inequality still holds true if $y = G(x)$. $\qquad\square$

Proposition 2.132 and Proposition 2.138 provided first C-tangential formulas for inverse images. From the above Corollary 2.140 we derive below C-tangential calculus rules for inverse images under $\mathcal{C}^1$-mappings with surjective derivatives. With the same assumption, calculus rules for C-normal cones as well as chain rules for C-subdifferentials will be established later in Theorem 3.174.

PROPOSITION 2.141. Let X, Y be Banach spaces and $G : X \to Y$ be a mapping which is strictly Fréchet differentiable at $\bar{x} \in X$ with $A := DG(\bar{x})$ surjective. Let $\bar{y} := G(\bar{x})$. For any subset S of Y with $\bar{y} \in S$ one has

$$A^{-1}\left(T^C(S; \bar{y})\right) \subset T^C\left(G^{-1}(S); \bar{x}\right);$$

if in addition S is tangentially regular at $\bar{y}$, then the above inclusion is an equality and $G^{-1}(S)$ is tangentially regular at $\bar{x}$.

PROOF. Fix any $h \in A^{-1}\left(T^C(S; \bar{y})\right)$ and let any sequences $(t_n)_n$ tending to 0 with $t_n > 0$ and $(x_n)_n$ in $G^{-1}(S)$ with $x_n \to \bar{x}$. Since $S \ni G(x_n) \to \bar{y}$ there exists $(v_n)_n$ converging to Ah with $G(x_n) + t_n v_n \in S$ for all $n \in \mathbb{N}$. By Corollary 2.140 there are a real $\gamma \geq 0$ and neighborhoods U and V of $\bar{x}$ and $G(\bar{x})$ respectively such that

$$d(x, G^{-1}(y)) \leq \gamma \|G(x) - y\| \quad \text{for all } x \in U,\ y \in V.$$

Take n large enough so that $x_n + t_n h \in U$ and $G(x_n) + t_n v_n \in V$. For any such n there is some $h_n \in X$ with $x_n + t_n h_n \in G^{-1}(G(x_n) + t_n v_n) \subset G^{-1}(S)$ such that

$$\|x_n + t_n h - (x_n + t_n h_n)\| \leq 2\gamma \|G(x_n + t_n h) - G(x_n) - t_n v_n\|$$

(in the case when $G(x_n + t_n h) = G(x_n) + t_n v_n$, just take $h_n = h$). By the strict differentiability of G at $\bar{x}$ we derive for some sequence $(e_n)_n$ in Y with $e_n \to 0$ that $\|h_n - h\| \leq 2\gamma \|A(h) - v_n + e_n\|$, hence $h_n \to h$ as $n \to \infty$. This ensures that $h \in T^C\left(G^{-1}(S); \bar{x}\right)$.

If S is tangentially regular at $\bar{y}$, then $G^{-1}(S)$ is tangentially regular at $\bar{x}$ and the tangential inclusion is an equality since one always has

$$T^C\left(G^{-1}(S); \bar{x}\right) \subset T^B\left(G^{-1}(S); \bar{x}\right) \subset A^{-1}\left(T^B(S; \bar{y})\right)$$

see Proposition 2.132(a). $\square$

Taking $S = \{0\}$ yields directly the following corollary.

COROLLARY 2.142 (L.A. Lyusternik's tangential equality). Let X, Y be Banach spaces and $G : X \to Y$ be a mapping which is strictly Fréchet differentiable at $\bar{x} \in X$ with $DG(\bar{x})$ surjective and $G(\bar{x}) = 0$. Then one has

$$T^B\left(G^{-1}(0); \bar{x}\right) = T^C\left(G^{-1}(0); \bar{x}\right) = DG(\bar{x})^{-1}(0).$$

Estimates for tangent cones to direct images of sets are also needed in some cases.

PROPOSITION 2.143. Let S be a subset of the normed space X and $G : X \to Y$ be a mapping from X into a normed space Y, and let $\bar{y} \in G(S)$.
(a) If G is Hadamard differentiable at a point $u \in G^{-1}(\bar{y}) \cap S$, then

$$DG(u)\left(T^B(S; u)\right) \subset T^B(G(S); \bar{y});$$

so when G is Hadamard differentiable at any point of $G^{-1}(\bar{y}) \cap S$, one has

$$\bigcup_{x \in G^{-1}(\bar{y}) \cap S} DG(x)\left(T^B(S; x)\right) \subset T^B(G(S); \bar{y}).$$

(b) Assume that S is closed and G is continuous on S and of class $\mathcal{C}^1$ on an open set containing $G^{-1}(\bar{y}) \cap S$. Assume also that there exists a neighborhood V of $\bar{y}$ and a compact set K of X such that $G^{-1}(V) \cap S \subset K$. Then

$$\bigcap_{x \in G^{-1}(\bar{y}) \cap S} DG(x)\left(T^C(S; x)\right) \subset T^C(G(S); \bar{y}).$$

2.2. CLARKE SUBDIFFERENTIAL

PROOF. (a) Let $u \in G^{-1}(\bar{y}) \cap S$ be a point where G is Hadamard differentiable and let $h \in T^B(S;u)$. There exists a sequence $(h_n)_n$ in X converging to h and a sequence $(t_n)_n$ of positive reals tending to 0 such that $u + t_n h_n \in S$ for all integers n. Then $G(u + t_n h_n) \in G(S)$ and by the Hadamard differentiability of G at u we have $G(u + t_n h_n) = \bar{y} + t_n(DG(u)(h_n) + e_n)$ for some sequence $(e_n)_n$ converging to 0 in Y. This ensures $DG(u)(h) \in T^B(G(S); \bar{y})$, proving the inclusion of (a).
(b) Fix any vector k in the left member of (b). Take any sequence $(y_n)_n$ in $G(S)$ converging to $\bar{y}$ and any sequence of positive reals $(t_n)_n$ tending to 0. Choose for each n some $x_n \in S$ with $G(x_n) = y_n$. For n large enough we have $x_n \in G^{-1}(V) \cap S \subset K$, thus some subsequence $(x_{s(n)})_n$ converges to some $x \in S$. The continuity of G at x gives $G(x) = \bar{y}$, hence $x \in G^{-1}(\bar{y}) \cap S$. Take by assumption a vector $h \in T^C(S;x)$ such that $DG(x)(h) = k$ and choose by the sequential characterization of the Clarke tangent cone a sequence $(h_n)_n$ in X converging to h such that $x_{s(n)} + t_{s(n)} h_n \in S$ for all n. By the $\mathcal{C}^1$ property of G near x, there exists a sequence $(e_n)_n$ in Y converging to 0 such that

$$y_{s(n)} + t_{s(n)}\big(DG(x)(h_n) + e_n\big) = G(x_{s(n)} + t_{s(n)} h_n) \in G(S).$$

Since $DG(x) h_n \to k$ (because $k = DG(x) h$), the characterization with subsequence in assertion (a$_2$) of Theorem 2.2 ensures that $k \in T^C(G(S); \bar{y})$, which guarantees the desired inclusion of (b). □

An estimate for interior tangent cones to direct images also hold.

PROPOSITION 2.144. *Let S be a subset of a Banach space X and $G : X \to Y$ be a mapping from X into a Banach space Y, and let S be a subset of X. Assume that G is strictly Fréchet differentiable at a point $\bar{x} \in S$ and that the derivative $A := DG(\bar{x})$ is surjective. The following hold.*
(a) *One has the inclusion*

$$A\big(I(S; \bar{x})\big) \subset I\big(G(S); G(\bar{x})\big).$$

(b) *If in addition S has the interior tangent property at the point $\bar{x}$, then $G(S)$ has the interior tangent property at the point $G(\bar{x})$.*

PROOF. The assertion (b) being a direct consequence of (a), it suffices to prove (a). By Corollary 2.140 there are neighborhoods U of $\bar{x}$ and V of $G(\bar{x})$ such that

(2.49) $\quad\quad d(u, G^{-1}(v)) \leq \gamma \|v - G(u)\| \quad$ for all $u \in U$, $v \in V$.

Fix any $h \in I(S; \bar{x})$. Take any sequence $(y_n)_n$ in $G(S)$ converging to $G(\bar{x})$, any sequence $(t_n)_n$ in $]0, +\infty[$ tending to 0 and any sequence $(v_n)_n$ in Y converging to $A(h)$. By (2.49) there is $n_0 \in \mathbb{N}$ such that for each integer $n \geq n_0$ one can find some $x_n \in G^{-1}(y_n)$ with $\|x_n - \bar{x}\| \leq 2\gamma\|y_n - G(\bar{x})\|$, so $x_n \to \bar{x}$. By (2.49) again there is an integer $N_0 \geq n_0$ such that for every integer $n \geq N_0$ one has

$$d\big(x_n + t_n h, G^{-1}(y_n + t_n v_n)\big) \leq \gamma \|y_n + t_n v_n - G(x_n + t_n h)\|$$
$$= \gamma \|y_n + t_n(v_n - A(h)) - y_n - t_n e_n\|$$
$$= \gamma t_n \|v_n - A(h) - e_n\|,$$

for some $e_n \to 0$ in Y according to the strict Fréchet differentiability of G at $\bar{x}$. Therefore, for each integer $n \geq N_0$ we obtain some h_n with $G(x_n + t_n h_n) = y_n + t_n v_n$ such that

$$\|(x_n + t_n h_n) - (x_n + t_n h)\| \leq 2\gamma t_n \|v_n - A(h) - e_n\|, \text{ i.e. } \|h_n - h\| \leq 2\gamma \|v_n - A(h) - e_n\|,$$

thus $h_n \to h$. This and the inclusion $h \in I(S;\bar{x})$ yields that for n large enough one has $x_n + t_n h_n \in S$, which in turn gives $y_n + t_n v_n \in G(S)$. The tangential characterisation (2.4) of the interior tangent cone guarantees that $A(h) \in I(G(S); G(\bar{x}))$, which confirms (a). □

The above proposition yields the following corollary related to the interior tangent cone property for the Minkowski sum of sets.

COROLLARY 2.145. *Let $S_1, \cdots, S_m$ be subsets of a Banach space X. If each set S_i, $i = 1, \cdots, m$ has the interior tangent cone property at a point $\bar{x}_i \in S_i$, then the Minkowski sum $S_1 + \cdots + S_m$ has the interior tangent cone property at the point $\bar{x}_1 + \cdots + \bar{x}_m$.*

PROOF. Assume that each set S_i, $i = 1, \cdots, m$, has the interior tangent cone property at $\bar{x}_i$, so the product set $S_1 \times \cdots \times S_m$ has the interior tangent cone property at $(\bar{x}_1, \cdots, \bar{x}_m)$ according to Proposition 2.24. Denote $A : X \times \cdots \times X \to X$ the surjective continuous linear mapping defined by $A(x_1, \cdots, x_m) = x_1 + \cdots + x_m$ for all $(x_1, \cdots, x_m)$ in $X \times \cdots \times X$, and note that $S_1 + \cdots + S_m = A(S_1 \times \cdots \times S_m)$. Then, Proposition 2.144(b) entails that $S_1 + \cdots + S_m$ has the interior tangent property at $\bar{x}_1 + \cdots + \bar{x}_m$ as stated in the corollary. □

2.2.14. Tangential regularity of sets versus tangential regularity of distance functions. In many results above the tangential regularity of sets or functions allowed us to transform inclusion calculus rules into equalities. Now our task is to compare the tangential regularity of a set with the tangential regularity of the associated distance function.

The comparaison will use the following inequalities concerning distance functions from tangent cones.

PROPOSITION 2.146. *Let S be a subset of the normed space $(X, \|\cdot\|)$ and $x \in S$.*
(a) *For all $h \in X$ one has*
$$(d_S)^B(x;h) \leq d\left(h, T^B(S;x)\right) \quad \text{and} \quad (d_S)^\circ(x;h) \leq d\left(h, T^C(S;x)\right).$$
(b) *If X is finite-dimensional, the left inequality is an equality.*

PROOF. It is easily seen that $\frac{1}{t} d_S(u + th) = d(h, \frac{1}{t}(S - u))$ for all $t > 0$ and $u \in X$. This guarantees first through (b) and (a) in Proposition 1.19 (concerning the distance of a point from a semilimit of sets) that
$$\liminf_{t \downarrow 0} \frac{1}{t} d_S(x + th) = \liminf_{t \downarrow 0} d\left(h, \frac{1}{t}(S - x)\right) \leq d\left(h, T^B(S; x)\right),$$
and
$$\limsup_{t \downarrow 0, u \xrightarrow{S} x} \frac{1}{t} d_S(u + th) = \limsup_{t \downarrow 0, u \xrightarrow{S} x} d\left(h, \frac{1}{t}(S - u)\right) \leq d\left(h, T^C(S; x)\right),$$
which justifies by Proposition 2.95(a) the inequalities of the assertion (a).

If X is finite-dimensional, (c) in Proposition 1.19 and the reasoning above ensure that the left inequality in (a) is an equality. □

We are now ready to compare in the following corollary the tangential regularities of the set S and its associated distance function. The characterization produced in the second assertion below in the finite-dimensional setting is similar to the one (already observed) saying that the set S is tangentially regular at $x \in S$ if and only if its indicator function is tangentially regular at x (see (c) of Proposition 2.125).

2.2. CLARKE SUBDIFFERENTIAL

COROLLARY 2.147. *Let S be a subset of the normed space $(X, \|\cdot\|)$ with $x \in S$.*
(a) *The tangential regularity of the distance function d_S at x entails the tangential regularity of the set S at x.*
(b) *If the vector space X is finite-dimensional, then the implication in (a) is an equivalence.*
(c) *If vector space X is finite-dimensional and the set S is tangentially regular at x, then $d_S^\circ(x; h) = d(h, T^C(S; x))$ for all $h \in X$.*

PROOF. (a) Suppose that d_S is tangentially regular at x, that is, $(d_S)^\circ(x; \cdot) = (d_S)^B(x; \cdot)$. For any $h \in T^B(S; x)$, by Proposition 2.146(a) we have $(d_S)^B(x; h) = 0$, hence $(d_S)^\circ(x; h) = 0$. The assertion (b) in Proposition 2.95 then gives $h \in T^C(S; x)$, so $T^B(S; x) \subset T^C(S; x)$, or equivalently $T^B(S; x) = T^C(S; x)$.
(b)-(c) Assume that X is finite-dimensional and that S is tangentially regular at x, that is, $T := T^B(S; x) = T^C(S; x)$. Fix any $h \in X$. From Proposition 2.146 we have $(d_S)^B(x; h) = d(h, T) \geq (d_S)^\circ(x; h)$, so $(d_S)^B(x; h) = d(h, T) = (d_S)^\circ(x; h)$ since the inequality $(d_S)^B(x; h) \leq (d_S)^\circ(x; h)$ always holds. This finishes the proof. □

In preparation for an example of a set for which the reverse implication of (a) in Corollary 2.147 above fails, let us recall a form of the Riesz classical result characterizing finite-dimensional normed spaces.

LEMMA 2.148. *Let $(X, \|\cdot\|)$ be a normed space.*
(a) *For any closed subspace $Y \neq X$ and for any $0 < r < 1$, there exists some $e \in X$ with $\|e\| = 1$ such that $d(e, Y) > r$.*
(b) *If X is infinite-dimensional, then for any $0 < r < 1$ there exists a sequence of unit vectors $(e_n)_{n \in \mathbb{N}}$ in X such that $\|e_n - e_m\| > r$ for all $n, m \in \mathbb{N}$ with $n \neq m$.*

PROOF. (a) Since Y is a vector space, there exists $x \in X \setminus Y$ such that $1 > d(x, Y) > r$. The inequality $1 > d(x, Y)$ gives some $y \in Y$ with $1 > \|x - y\|$, so putting $e := (x - y)/\|x - y\|$ (which is well defined) we obtain
$$d(e, Y) = d(x - y, Y)/\|x - y\| \geq d(x - y, Y) = d(x, Y) > r.$$
(b) Assume that X is infinite-dimensional. From (a) we obtain by induction a sequence of unit vectors $(e_n)_{n \in \mathbb{N}}$ in X such that $d(e_{n+1}, \mathrm{Vect}\{e_1, \cdots, e_n\}) > r$, hence in particular $\|e_n - e_m\| > r$ for all $n, m \in \mathbb{N}$ with $n \neq m$. □

EXAMPLE 2.149 (J.M. Borwein and M. Fabian's example). In any infinite dimensional normed space $(X, \|\cdot\|)$ the reverse implication of the assertion (a) in Corollary 2.147 fails. Indeed, given any infinite dimensional normed space $(X, \|\cdot\|)$ we even construct a closed subset S which is tangentially regular at some $\bar{x} \in S$ and such that neither the distance function d_S nor its opposite $-d_S$ is tangentially regular at $\bar{x}$.

Take $0 < r < 1$ and by the above lemma consider a sequence $e_0, e_1, e_2, \cdots$ of elements of the unit sphere of X such that $\|e_n - e_m\| > r$ for all $n \neq m$ in $\mathbb{N} \cup \{0\}$, and put

$$S := \{0\} \cup \{z_n : n \in \mathbb{N}\} \quad \text{where } z_n := 4^{-n}\left(e_0 + \frac{1}{4}e_n\right) \text{ for all } n \in \mathbb{N}.$$

Since $4^{-n}(e_0 + (1/4)e_n) \to 0$ as $n \to \infty$, the set S is compact, hence closed.

First we claim that $T^B(S; 0) = \{0\}$. Suppose (by contradiction) that there exits some nonzero vector $h \in T^B(S; 0)$, hence there are a sequence $(t_n)_n$ in $]0, +\infty[$ tending to 0 and a sequence $(h_n)_n$ of nonzero vectors of X converging to h with

$0 + t_n h_n \in S$ for all $n \in \mathbb{N}$. Since all $t_n h_n$ are different from zero and $t_n h_n \to 0$ as $n \to \infty$, the sequence $(t_n h_n)_n$ has a subsequence whose all elements are pairwise distinct. Extracting a subsequence and relabeling it, we may suppose that $t_i h_i \neq t_j h_j$ for all $i \neq j$ in $\mathbb{N}$. Now, for each $n \in \mathbb{N}$ select some $m_n \in \mathbb{N}$ such that

$$t_n h_n = z_{m_n} = 4^{-m_n}\left(e_0 + \frac{1}{4}e_{m_n}\right).$$

Using this and the fact above that $t_i h_i \neq t_j h_j$, it is immediate that $m_i \neq m_j$ for $i \neq j$ in $\mathbb{N}$. Since $t_n 4^{m_n} h_n = e_0 + (1/4)e_{m_n}$ we have $3/4 \leq t_n 4^{m_n}\|h_n\| \leq 5/4$, hence the sequence $(t_n 4^{m_n})_n$ in bounded in $\mathbb{R}$ because $\|h_n\| \to \|h\| \neq 0$. Extracting a subsequence, we may suppose that $(t_n 4^{m_n})_n$ tends to some real s and by what precedes $3/4 \leq s\|h\| \leq 5/4$, so $s > 0$. Consequently,

$$0 = \lim_{n,k \to \infty} \|h_n - h_k\|$$
$$= \lim_{n,k \to \infty} \left\|(t_n^{-1} 4^{-m_n} - t_k^{-1} 4^{-m_k})e_0 + \frac{1}{4}(t_n^{-1} 4^{-m_n} e_{m_n} - t_k^{-1} 4^{-m_k} e_{m_k})\right\|$$
$$= \frac{1}{4s} \lim_{n,k \to \infty} \|e_{m_n} - e_{m_k}\| \geq \frac{r}{4s} > 0,$$

and this contradiction confirms that $T^B(S;0) = \{0\}$. We then have $T^C(S;0) = T^B(S;0) = \{0\}$ (recall that one always has $T^C(S;x) \subset T^B(S;x)$), which justifies that the set S is then tangentially regular at 0.

To see that the distance function d_S is not tangentially regular at 0, we proceed to show that $(d_S)^\circ(0;e_0) = 1$ and $(d_S)^B(0;e_0) < 1$. Let us first prove that $(d_S)^\circ(0;\cdot) = \|\cdot\|$. Since $\|e_i\| = 1$ for all $i = 0,1,2,\cdots$, we observe, for any fixed $k, n \in \mathbb{N}$ with $n < k$, that

$$\left\|4^{-n}(e_0 + \frac{1}{4}e_n) - 4^{-k}(e_0 + \frac{1}{4}e_k)\right\| \geq 4^{-n} - 4^{-k} - \frac{1}{4}(4^{-n} + 4^{-k})$$

(2.50)
$$= \frac{3}{4}4^{-n} - \frac{5}{4}4^{-k} \geq \frac{3}{4}4^{-n} - \frac{5}{4}4^{-n-1} = 7 \cdot 4^{-n-2}.$$

Fix $n \in \mathbb{N}$ and any $v \in X$ with $\|v\| < \frac{7}{2}4^{-n-2}$. By (2.50) we have on the one hand, for $k > n$,

$$\|(z_n + v) - z_k\| \geq \|z_n - z_k\| - \|v\| \geq 7 \cdot 4^{-n-2} - \|v\|$$
$$> 7 \cdot 4^{-n-2} - \frac{7}{2}4^{-n-2} = \frac{7}{2}4^{-n-2} > \|v\|,$$

and on the other hand, for $k < n$, since $\|v\| < \frac{7}{2}4^{-n-2} < \frac{7}{2}4^{-k-2}$, we have

$$\|(z_n + v) - z_k\| \geq \|z_k - z_n\| - \|v\| \geq 7 \cdot 4^{-k-2} - \|v\|$$
$$> 7 \cdot 4^{-k-2} - \frac{7}{2}4^{-k-2} = \frac{7}{2}4^{-k-2} > \|v\|.$$

Observe also that

$$\|(z_n + v) - 0\| = \left\|4^{-n}(e_0 + \frac{1}{4}e_n) + v\right\| \geq 4^{-n} - 4^{-n-1} - \|v\|$$
$$\geq 4^{-n} - 4^{-n-1} - \frac{7}{2}4^{-n-2} = \frac{17}{2}4^{-n-2} > \|v\|,$$

and that $\|(z_n + v) - z_k\| = \|v\|$ for $k = n$. We deduce from all those relations that
$$\|v\| = \min_{x \in S} \|(z_n + v) - x\| = d_S(z_n + v) \quad \text{whenever } \|v\| < \frac{7}{2} 4^{-n-2}.$$
Consequently, for any nonzero $h \in X$, taking $t_n := \frac{7}{2} 4^{-n-2}(1 + \|h\|)^{-1}$ and noting that $\|t_n h\| < \frac{7}{2} 4^{-n-2}$ it follows that
$$(d_S)^\circ(0; h) \geq \limsup_{n \to \infty} t_n^{-1}\big(d_S(z_n + t_n h) - d_S(z_n)\big) = \limsup_{n \to \infty} t_n^{-1} \|t_n h\| = \|h\|.$$
It ensues $(d_S)^\circ(0; h) = \|h\|$ for all $h \in X$ since the inequality $(d_S)^\circ(0; h) \leq \|h\|$ always holds. This entails in particular $(d_S)^\circ(0; e_0) = 1$ as desired.

It remains to show $(d_S)^B(0; e_0) < 1$. Take now $t_n := 4^{-n}$. Since
$$\|t_n e_0 - z_n\| = \left\| 4^{-n} e_0 - 4^{-n} \left(e_0 + \frac{1}{4} e_n\right) \right\| = 4^{-n-1} = 4^{-1} t_n,$$
we have $d_S(t_n e_0) \leq 4^{-1} t_n$, hence
$$(d_S)^B(0; e_0) \leq \liminf_{n \to \infty} t_n^{-1}\big(d_S(0 + t_n e_0) - d_S(0)\big) \leq t_n^{-1} 4^{-1} t_n = 1/4 < 1,$$
as required.

Finally, the opposite of the distance function to S, that is, $-d_S$ is not either tangentially regular at 0. Indeed, on the one hand $(-d_S)^\circ(0; e_0) = 1$ since $\partial_C(-d_S)(0) = \mathbb{B}_{X^*}$ because $\partial_C d_S(0) = \mathbb{B}_{X^*}$ according to the equality $(d_S)^\circ(0; \cdot) = \|\cdot\|$ established above, and on the other hand obviously $(-d_S)^B(0; e_0) \leq 0$. □

2.2.15. Signed distance function and Clarke tangent cone. In addition to the roles of the usual distance function for the Clarke and Bouligand-Peano tangent cones (see Proposition 2.95 and Proposition 2.120), the signed distance function needs to be employed in many situations. Recall that the *signed distance function* from a subset S of a normed space X is defined by

(2.51) $\quad \operatorname{sgd}(x, S) := d(x, S) - d(x, S^c) = d(x, \overline{S}) - d(x, \overline{S^c}) \quad$ for all $x \in X$,

where $S^c := X \setminus S$ and $\overline{S} := \operatorname{cl} S$. With $S = \emptyset$ we have $\operatorname{sgd}(\cdot, S) = +\infty$, and with $S = X$ we have $\operatorname{sgd}(\cdot, S) = -\infty$. Sometimes, it is convenient to write sgd_S in place of $\operatorname{sgd}(\cdot, S)$.

Clearly, we have $\operatorname{sgd}_{S^c} = -\operatorname{sgd}_S$, $|\operatorname{sgd}_S| = \max\{d_S, d_{S^c}\}$ and
(2.52)
$$\operatorname{sgd}_S(x) = \begin{cases} d_S(x) & \text{if } x \in S^c \\ -d_{S^c}(x) & \text{if } x \in S \end{cases} \quad \text{and} \quad \operatorname{sgd}_S(x) = \begin{cases} d_S(x) > 0 & \text{if } x \in \operatorname{int}(S^c) \\ 0 & \text{if } x \in \operatorname{bdry} S \\ -d_{S^c}(x) < 0 & \text{if } x \in \operatorname{int} S. \end{cases}$$

It results from the two latter descriptions of sgd_S that
$$\operatorname{sgd}_S(x) = \begin{cases} d_S(x) = d(x, \operatorname{bdry} S) & \text{if } x \in \operatorname{cl}(S^c) \\ -d_{S^c}(x) = -d(x, \operatorname{bdry} S) & \text{if } x \in \operatorname{cl}(S) \end{cases} \quad \text{and } |\operatorname{sgd}_S| = d(\cdot, \operatorname{bdry} S),$$

and
$$x \in \operatorname{bdry} S \Leftrightarrow \operatorname{sgd}(x, S) = 0 \Leftrightarrow d(x, S) = d(x, S^c)$$
along with
$$\operatorname{int} S = \{\operatorname{sgd}_S(\cdot) < 0\}, \quad \operatorname{int}(S^c) = \{\operatorname{sgd}_S(\cdot) > 0\}$$
and
$$\operatorname{cl} S = \{\operatorname{sgd}_S(\cdot) \leq 0\}, \quad \operatorname{cl}(S^c) = \{\operatorname{sgd}_S(\cdot) \geq 0\}.$$
We recall that the set $\operatorname{bdry} S$ above denotes the boundary of S.

If $S \neq \emptyset$ and $S \neq X$, taking $x \in S$ and $x' \in S^c$, there is some $x'' \in [x, x'] \cap$ bdry S, so $\operatorname{sgd}(x'', S) = d(x'', S) = d(x'', S^c) = 0$, which entails

$$\operatorname{sgd}(x', S) - \operatorname{sgd}(x, S) = d(x', S) - d(x'', S) - d(x'', S^c) + d(x, S^c)$$
$$\leq \|x' - x''\| + \|x'' - x\| = \|x' - x\|.$$

The inequality $\operatorname{sgd}(x', S) - \operatorname{sgd}(x, S) \leq \|x' - x\|$ being evident in the case when both x, x' are in either S or S^c, we see that $\operatorname{sgd}(\cdot, S)$ is Lipschitz with constant 1 whenever $S \neq \emptyset$ and $S \neq X$. We state those properties in the assertions (a)-(c) in the following proposition.

PROPOSITION 2.150. *Let S, Q be two subsets of the normed space $(X, \|\cdot\|)$ and let S^c denote the complement $X \setminus S$ of S in X. The following hold.*
(a) *One has* $\operatorname{sgd}_{S^c} = -\operatorname{sgd}_S$ *and*

$$\operatorname{sgd}_S(x) = \begin{cases} d_S(x) = d(x, \operatorname{bdry}, S) > 0 & \text{if } x \in \operatorname{int}(S^c) \\ 0 & \text{if } x \in \operatorname{bdry} S \\ -d_{S^c}(x) = -d(x, \operatorname{bdry} S) < 0 & \text{if } x \in \operatorname{int} S. \end{cases}$$

(b) *One has*

$$\operatorname{sgd}_S(x) = \begin{cases} d_S(x) = d(x, \operatorname{bdry} S) & \text{if } x \in \operatorname{cl}(S^c) \\ -d_{S^c}(x) = -d(x, \operatorname{bdry} S) & \text{if } x \in \operatorname{cl}(S), \end{cases}$$

$|\operatorname{sgd}_S| = \max\{d_S, d_{S^c}\} = d(\cdot, \operatorname{bdry} S)$ *and*

$$x \in \operatorname{bdry} S \Leftrightarrow \operatorname{sgd}(x, S) = 0 \Leftrightarrow d(x, S) = d(x, S^c)$$

along with

$$\operatorname{int} S = \{\operatorname{sgd}_S(\cdot) < 0\}, \quad \operatorname{int}(S^c) = \{\operatorname{sgd}_S(\cdot) > 0\}$$

and

$$\operatorname{cl} S = \{\operatorname{sgd}_S(\cdot) \leq 0\}, \quad \operatorname{cl}(S^c) = \{\operatorname{sgd}_S(\cdot) \geq 0\}.$$

(c) *If $S \neq \emptyset$ and $S \neq X$, then the signed distance function is Lipschitz on X with 1 as a Lipschitz constant, that is,*

$$|\operatorname{sgd}(x, S) - \operatorname{sgd}(x', S)| \leq \|x - x'\| \quad \text{for all } x, x' \in X.$$

(d) *One has the equivalence*

(2.53) $\qquad (\operatorname{sgd}_S \leq \operatorname{sgd}_Q) \Leftrightarrow (\operatorname{cl} Q \subset \operatorname{cl} S \text{ and } \operatorname{int} Q \subset \operatorname{int} S),$

so if Q and S are closed

$$(\operatorname{sgd}_S \leq \operatorname{sgd}_Q) \Leftrightarrow (Q \subset S).$$

PROOF. According to the analysis preceding the statement of the proposition we only have to prove (d). It suffices to note that the right-hand side in (2.53) is equivalent to $\operatorname{cl} Q \subset \operatorname{cl} S$ and $\operatorname{cl}(S^c) \subset \operatorname{cl}(Q^c)$, and the property that both latter inclusions hold is itself equivalent to the property that both inequalities $d_S \leq d_Q$ and $d_{Q^c} \leq d_{S^c}$ hold. Then it follows from the left description of the signed distance in (2.52) that the equivalence in (2.53) holds true.

Finally, the other equivalence in (d) under the closedness of both Q and S is a direct consequence of (2.53). $\qquad \square$

REMARK 2.151. With $\overline{S} := \operatorname{cl} S$, it is worth pointing out that the equality $\operatorname{sgd}(\cdot, S) = \operatorname{sgd}(\cdot, \overline{S})$ fails. Indeed, in $\mathbb{R}$ taking $S := \mathbb{Q} \cap\,]-\infty, 0[$, we see that, for all $x \in \mathbb{R}$, we have
$$\operatorname{sgd}(x, S) = d(x,]-\infty, 0]) = \max\{x, 0\},$$
while $\operatorname{sgd}(x, \overline{S}) = d(x, -]\infty, 0]) - d(x, [0, +\infty[) = x.$

□

The signed distance function allows us to describe the vectors of the Clarke tangent cone $T^C(S \cup \{x\}, x)$ which belongs to the opposite of $T^C((X \setminus S) \cup \{x\}; x)$ as follows.

PROPOSITION 2.152. Let S be a subset of the normed space $(X, \|\cdot\|)$ and $x \in \operatorname{bdry} S$. Then, with $S^c := X \setminus S$ one has
$$T^C(S \cup \{x\}; x) \cap -T^C(S^c \cup \{x\}; x) = \{h \in X : (\operatorname{sgd}_S)^\circ(x; h) \leq 0\}$$
$$= \{h \in X : \langle x^*, h \rangle \leq 0, \forall x^* \in \partial_C \operatorname{sgd}_S(x)\}.$$

PROOF. Put $\overline{S} := \operatorname{cl} S$. The function $\operatorname{sgd}(\cdot, S)$ being Lipschitz, the equality between the second and third members is clear since we know that $\operatorname{sgd}(\cdot, S)^\circ$ is the support function of $\partial_C \operatorname{sgd}(\cdot, S)(x)$ (see Theorem 2.70(d)). By the sum rule in Theorem 2.98 and by Proposition 2.74(b) we also know that
$$\partial_C \operatorname{sgd}(\cdot, S)(x) \subset \partial_C d_S(x) + \partial_C(-d_{S^c})(x) = \partial_C d_S(x) - \partial_C d_{S^c}(x).$$
Then, for any h in the first member of the proposition and any $u^* \in \partial_C d_S(x)$ and $v^* \in \partial_C d_{S^c}(x)$, we have $\langle u^*, h \rangle \leq 0$ and $\langle -v^*, h \rangle \leq 0$. All together justify the inclusion of the first member of the proposition into the third.

Now fix any h in the second member of the proposition. By Proposition 2.95(a) consider a sequence $(t_n)_n$ in $]0, +\infty[$ tending to 0 and a sequence $(x_n)_n$ in $\overline{S}$ such that $d_S^\circ(x; h) = \lim_{n \to \infty} t_n^{-1} d_S(x_n + t_n h)$. If $x_n + t_n h \in \overline{S}$ for infinitely many n, then $d_S^\circ(x; h) = 0$ or equivalently $h \in T^C(\overline{S}; x)$ by Proposition 2.95. Suppose that $x_n + t_n h \notin \overline{S}$ for large n, say $n \geq N$. Then, for any integer $n \geq N$, we have
$$t_n^{-1} d_S(x_n + t_n h) \leq t_n^{-1}\big(d_S(x_n + t_n h) + d_{S^c}(x_n)\big)$$
$$= t_n^{-1}\big(\operatorname{sgd}(x_n + t_n h, S) - \operatorname{sgd}(x_n, S)\big),$$
hence again $d_S^\circ(x; h) = 0$. So in any case, we get $h \in T^C(\overline{S}; x)$ (see Proposition 2.95(b)).

On the other hand, since the inequality $\operatorname{sgd}(\cdot, S)^\circ(x; h) \leq 0$ is equivalent to $\big(-\operatorname{sgd}(\cdot, S)\big)^\circ(x; -h) \leq 0$ (see Proposition 2.74(b)), we have $\operatorname{sgd}(\cdot, S^c)^\circ(x; -h) \leq 0$ because $\operatorname{sgd}(\cdot, S^c) = -\operatorname{sgd}(\cdot, S)$. By what precedes it ensues that $-h \in T^c(\overline{S^c}; x)$. It finally follows that
$$h \in T^C(\overline{S}; x) \cap -T^C(\overline{S^c}; x),$$
and this finishes the proof since $T^C(\overline{S}; x) = T^C(S \cup \{x\}; x)$ (see Proposition 2.5(b)).

□

Recalling that $T^C(S; x) = -T^C(S^c \cup \{x\}; x)$ whenever the set S has the interior tangent cone property at $x \in S \cap \operatorname{bdry} S$ (see Proposition 2.29), we directly derive the following corollary.

COROLLARY 2.153. Let S be a subset of the normed space $(X, \|\cdot\|)$ which has the interior tangent cone property at $x \in S \cap \mathrm{bdry}\, S$. Then, the following equalities hold
$$T^C(S \cup \{x\}; x) = \{h \in X : \mathrm{sgd}(\cdot, S)^\circ(x; h) \leq 0\}$$
$$= \{h \in X : \langle x^*, h \rangle \leq 0, \forall x^* \in \partial_C \mathrm{sgd}_S(x)\}.$$

2.2.16. Sublevel representation of epi-Lipschitz sets.

In the first step of this subsection mainly devoted to sublevel representations of epi-Lipschitz sets, we characterize sets with the interior tangent cone property by means of the signed distance function. Doing so, we obtain a significant result in addition to Corollary 2.153.

THEOREM 2.154 (interior tangent property via signed distance). Let S be a subset of a normed space $(X, \|\cdot\|)$ and $\bar{x} \in S \cap \mathrm{bdry}\, S$. Then, the following assertions hold:
(a) The set S has the interior tangent property at $\bar{x}$ in a nonzero direction h if and only if
$$(\mathrm{sgd}_S)^\circ(\bar{x}; h) < 0.$$
(b) The set S has the interior tangent property at $\bar{x}$ if and only if $0 \notin \partial_C \mathrm{sgd}_S(\bar{x})$.

PROOF. Recalling that the function $(\mathrm{sgd}_S)^\circ(\bar{x}; \cdot)$ is the support function of $\partial_C \mathrm{sgd}(\bar{x})$, we see that the assertion (b) directly follows from (a). If $(\mathrm{sgd})^\circ(\bar{x}; h) < 0$, the equality $\overline{S} = \{u \in X : \mathrm{sgd}_S(u) \leq 0\}$ (where $\overline{S} := \mathrm{cl}\, S$) yields by Proposition 2.78 that $\overline{S}$ has the interior tangent property at $\bar{x}$ in the direction h, hence S has also the interior tangent property at $\bar{x}$ in the direction h according to Proposition 2.27(b).

Suppose now that the set S has the interior tangent property at the point $\bar{x}$ in the direction h and put $S^c := X \setminus S$. Proposition 2.29 tells us that $S^c \cup \{x\}$ has the interior tangent property at $\bar{x}$ in the direction $-h$, so by definition of sets satisfying the interior tangent property, there is a real $\varepsilon_0 \in\,]0, 1[$ such that, with $S' := S^c \cup \{\bar{x}\}$, we have for any $\varepsilon \in\,]0, \varepsilon_0]$

(2.54) $\quad S \cap B(\bar{x}, 3\varepsilon) +\,]0, \varepsilon[B(h, 2\varepsilon) \subset S \quad$ and $\quad S' \cap B(\bar{x}, 3\varepsilon) +\,]0, \varepsilon[B(-h, \varepsilon) \subset S'.$

According to Proposition 2.27(a) we can also choose the above real $\varepsilon_0 > 0$ such that for any real $\varepsilon \in\,]0, \varepsilon_0]$

(2.55) $\quad B(\bar{x}, \varepsilon) \cap \overline{S} = B(\bar{x}, \varepsilon) \cap \overline{X \setminus S^c} \quad$ and $\quad B(\bar{x}, \varepsilon) \cap \overline{S^c} = B(\bar{x}, \varepsilon) \cap \overline{X \setminus S}.$

Fix any real $\varepsilon \in\,]0, \varepsilon_0]$.

Claim 1. $d_S(x) \geq d_S(x + tv)$ for all $t \in\,]0, \varepsilon[$, $v \in B[h, \varepsilon]$ and $x \in B(\bar{x}, \varepsilon)$.
Fix any $x \in B(\bar{x}, \varepsilon)$, any $v \in B[h, \varepsilon]$, and any $t \in\,]0, \varepsilon[$. Choose a sequence $(y_n)_n$ in S such that $d_S(x) = \lim_{n \to \infty} \|x - y_n\|$. Since $d_S(x) \leq \|x - \bar{x}\| < \|x - \bar{x}\| + \varepsilon$, we may suppose that $\|x - y_n\| < \|x - \bar{x}\| + \varepsilon$ for every $n \in \mathbb{N}$. Then,
$$\|y_n - \bar{x}\| \leq \|y_n - x\| + \|x - \bar{x}\| < 2\|x - \bar{x}\| + \varepsilon < 3\varepsilon.$$
For every $n \in \mathbb{N}$, it results that $y_n + tv \in S$ by (2.54), hence
$$d_S(x + tv) \leq \|x + tv - (y_n + tv)\| = \|x - y_n\|,$$
so $d_S(x + tv) \leq d_S(x)$ as says the claim.

Claim 2. For all $0 < t < \frac{\varepsilon}{2(1+\|h\|)}$ and $x \in B(\bar{x}, \varepsilon/2) \cap \overline{S^c}$, one has $x - th \notin \overline{S}$ and $d_S(x - th) \geq d_S(x) + \varepsilon t/2$.

Fix any such real t with $0 < t < \frac{\varepsilon}{2(1+\|h\|)}$ and fix also any $x \in B(\bar{x}, \varepsilon/2) \cap (X \setminus \overline{S})$. Take any $v \in B[h, \varepsilon]$, and note that

$$\|x - tv - \bar{x}\| \leq \|x - \bar{x}\| + t\|v\| < \frac{\varepsilon}{2} + \frac{\varepsilon}{2(1+\|h\|)}(\varepsilon + \|h\|) < \varepsilon.$$

Thus, $x - tv \in B(\bar{x}, \varepsilon)$, which gives by Claim 1

$$d_S(x - tv) \geq d_S(x - tv + tv) = d_S(x).$$

This means that

(2.56) $\qquad d_S(y) \geq d_S(x) \quad \text{for all } y \in B[x - th, \varepsilon t].$

Choose $z_t \in S$ such that

$$d_S(x - th) \geq \|x - th - z_t\| - \varepsilon t/2.$$

Noticing that $d_S(x) > 0$ (since $x \notin \overline{S}$), the inclusion $z_t \in S$ and (2.56) imply that $z_t \notin B[x - th, \varepsilon t]$. Then, we can choose $u_t \in [x - th, z_t]$ (the line segment) such that $\|x - th - u_t\| = \varepsilon t$. It ensues that

$$d_S(x - th) \geq \|x - th - z_t\| - \frac{1}{2}\varepsilon t = -\frac{1}{2}\varepsilon t + \|x - th - u_t\| + \|u_t - z_t\|$$

$$= \frac{1}{2}\varepsilon t + \|u_t - z_t\| \geq \frac{1}{2}\varepsilon t + d_S(u_t),$$

hence by (2.56) we get $d_S(x - th) \geq d_S(x) + \varepsilon t/2$. By continuity of the function d_S and by (2.55) we deduce that the latter inequality still holds for all $x \in B(\bar{x}, \varepsilon/2) \cap (X \cap \overline{S^c})$ as stated in the claim. Clearly, the inequality also tells us that $x - th \notin \overline{S}$.

Claim 3. For all $0 < t < \frac{\varepsilon}{2(1+\|h\|)}$ and $x \in B(\bar{x}, \varepsilon/2) \cap \overline{S}$, one has $x + th \notin \overline{S^c}$ and $d_{S^c}(x + th) \geq d_{S^c}(x) + \varepsilon t/2$.

It suffices to apply Claim 2 with the set $S^c \cup \{\bar{x}\}$ and the vector $-h$ in place of S and h respectively.

Claim 4. For all $0 < t < \frac{\varepsilon}{4(1+\|h\|)}$ and $x \in B(\bar{x}, \varepsilon/4)$,

$$\operatorname{sgd}_S(x + th) - \operatorname{sgd}_S(x) \leq -\frac{\varepsilon t}{2}.$$

Fix any real t with $0 < t < \frac{\varepsilon}{4(1+\|h\|)}$ and fix any $x \in B(\bar{x}, \varepsilon/4)$.

- If $x \in \overline{S}$, by Claim 3 we have $x + th \notin \overline{S^c}$, that is, $x + th \in \operatorname{int} S$, and

$$\operatorname{sgd}_S(x + th) - \operatorname{sgd}_S(x) = (-d_{S^c}(x + th)) - (-d_{S^c}(x)) \leq -\frac{\varepsilon t}{2}.$$

- If $x \notin \overline{S}$, that is, $x \in \operatorname{int}(S^c)$, let

$$\tau = \sup\{\theta \mid \theta \leq t, \ [x, x + \theta h] \subset X \setminus S\}.$$

Then $x + \tau h \in B(\bar{x}, \varepsilon/2) \cap \overline{S^c}$ and by Claim 2,

$$\operatorname{sgd}_S(x + \tau h) - \operatorname{sgd}_S(x) = d_S(x + \tau h) - d_S(x)$$
$$= d_S(x + \tau h) - d_S(x + \tau h - \tau h)$$
(2.57) $\qquad\qquad\qquad \leq -\frac{\varepsilon \tau}{2}.$

In the case $\tau = t$, the latter inequality translates the desired claim. Now, suppose $\tau < t$. Then $x + \tau h \in \text{bdry}\, S$, so writing $x + th = x + \tau h + (t - \tau)h$, we see by Claim 3 that $x + th \notin \overline{S^c}$ and

$$\text{sgd}_S(x+th) - \text{sgd}_S(x+\tau h) = (-d_{S^c}(x+\tau h+(t-\tau)h)) - (-d_{S^c}(x+\tau h)) \leq -\frac{\varepsilon(t-\tau)}{2}.$$

By adding this and (2.57), we deduce the Claim 4.

Finally, it results from Claim 4 that

$$(\text{sgd}_S)^\circ(\overline{x}; h) \leq -\varepsilon/2.$$

This being true for any $\varepsilon \in\,]0, \varepsilon_0]$, we obtain $(\text{sgd}_S)^\circ(\overline{x}; h) \leq 0$, hence the proof is complete. $\square$

Putting Proposition 2.78 and Theorem 2.154 together yields the global sublevel representation of epi-Lipschitz sets.

THEOREM 2.155 (sublevel representation of epi-Lipschitz sets). Let S be a nonempty closed set of a normed space $(X, \|\cdot\|)$ with $S \neq X$. Then, the following are equivalent:
(a) the set S is epi-Lipschitz;
(b) there exists a Lipschitz function $g : X \to \mathbb{R}$ such that $0 \notin \partial_C g(x)$ for every $x \in \text{bdry}\, S$ and

$$S = \{x \in X : g(x) \leq 0\};$$

(c) the set S enjoys the canonical qualified sublevel representation:

$$S = \{x \in X : \text{sgd}_S(x) \leq 0\} \quad \text{with} \quad 0 \notin \partial_C \text{sgd}_S(u)\ \forall u \in \text{bdry}\, S.$$

PROOF. The implication (c) $\Rightarrow$ (b) is obvious and (b) $\Rightarrow$ (a) follows from Proposition 2.78. Finally, the last implication (a) $\Rightarrow$ (c) results from Theorem 2.154. $\square$

2.3. Local Lipschitz property of continuous convex functions

We have already noted that convex functions f are tangentially regular. So, when the convex function f is locally Lipschitz near x, then $f^\circ(x; \cdot) = f'(x; \cdot)$. In this section we will investigate Lipschitz properties of convex functions and mappings.

2.3.1. Lipschitz property of convex functions.
In preparation for some future needs we will start with the equi-Lipschitz property of families of convex functions.

LEMMA 2.156. Let U be an open convex set of a normed space X with $\overline{x} \in U$ and $(f_i)_{i \in I}$ be a family of convex functions from U into $\mathbb{R}$. Assume, for some reals $\beta \geq 0$ and $r > 0$, that $B(\overline{x}, 2r) \subset U$ and

$$f_i(x) - f_i(\overline{x}) \leq \beta \quad \text{for all } i \in I,\ x \in B(\overline{x}, 2r).$$

Then the family $(f_i)_{i \in I}$ is equi-Lipschitzian on $B(\overline{x}, r)$, more precisely, for each $i \in I$, one has for all $x, y \in B(\overline{x}, r)$,

$$|f_i(x) - f_i(y)| \leq (2\beta/r)\|x - y\| \quad \text{and} \quad |f_i(x) - f_i(\overline{x})| \leq (\beta/r)\|x - \overline{x}\|.$$

PROOF. For each $i \in I$ put $g_i(x) := f_i(x + \overline{x}) - f_i(\overline{x})$ for all $x \in -\overline{x} + U$, so $g_i(x) \leq \beta$ for all $x \in B(0, 2r)$. Further, for any $x \in B(0, 2r)$ the convexity of g_i yields $g_i(x) - g_i(0) \geq g_i(0) - g_i(-x)$ thus $g_i(x) \geq -g_i(-x) \geq -\beta$ since $g_i(0) = 0$. So $|g_i(x)| \leq \beta$ for all $i \in I$ and $x \in B(0, 2r)$. Now fix arbitrary $x, y \in B(0, r)$ and consider any $s > r^{-1}\|x - y\|$. Setting $z_s := x + s^{-1}(x - y)$ we see that $\|z_s\| < 2r$ and $x = (1 - t)y + tz_s$, where $t = (1 + s)^{-1}s < 1$. For each $i \in I$, by the convexity of g_i we have

$$(2.58) \qquad g_i(x) \leq (1 - t)g_i(y) + tg_i(z_s) = g_i(y) + t(g_i(z_s) - g_i(y)),$$

hence on the one hand $g_i(x) - g_i(y) \leq 2\beta t \leq 2\beta s$, and on the other hand with $y = 0$ in (2.58) we obtain $g_i(x) \leq \beta s$. Consequently, on the one hand $g_i(x) - g_i(y) \leq (2\beta/r)\|x - y\|$, and by symmetry it follows that $|g_i(x) - g_i(y)| \leq (2\beta/r)\|x - y\|$; on the other hand $g_i(x) \leq (\beta/r)\|x\|$ and $-g_i(x) \leq g_i(-x) \leq (\beta/r)\|x\|$, hence $|g_i(x)| \leq (\beta/r)\|x\|$. □

THEOREM 2.157 (equi-Lipschitz property of families of convex functions: local upper equi-boundedness case). *Let U be an open convex set of a normed space X and $(f_i)_{i \in I}$ be a family of convex functions from U into $\mathbb{R}$. Then the family of functions $(f_i)_{i \in I}$ is equi-Lipschitzian on a neighborhood of a point $\overline{x} \in U$ if and only if the family of functions $(f_i(\cdot) - f_i(\overline{x}))_{i \in I}$ is equi-bounded from above on some neighborhood of this point $\overline{x}$.*

PROOF. The "if part" follows directly from the above lemma. Conversely, taking a ball $B(\overline{x}, r)$ over which $(f_i)_{i \in I}$ is equi-Lipschitzian with a real $\gamma \geq 0$ as a common Lipschitz constant, it is readily seen that $f_i(x) - f_i(\overline{x}) \leq r\gamma$ for all $i \in I$ and $x \in B(\overline{x}, r)$. □

A more simple criterion for the local Lipschitz behavior of a single convex function is the assertion (a) of the following theorem.

THEOREM 2.158 (Lipschitz property for convex functions). *Let U be an open convex set of a normed space X and $f : U \to \mathbb{R} \cup \{+\infty\}$ be an extended real-valued convex function. The following properties are equivalent:*
(a) *there exists a point in U on a neighborhood of which f is bounded from above;*
(b) *the function f is finite at some point in U and continuous at this point;*
(c) $\operatorname{int}(\operatorname{dom} f) \neq \emptyset$ *and f is continuous on $\operatorname{int}(\operatorname{dom} f)$;*
(d) $\operatorname{int}(\operatorname{dom} f) \neq \emptyset$ *and f is locally Lipschitz continuous on $\operatorname{int}(\operatorname{dom} f)$.*

PROOF. We have only to prove the implication (a) $\Rightarrow$ (d) since the implications (d) $\Rightarrow$ (c) $\Rightarrow$ (b) $\Rightarrow$ (a) are obvious. Suppose (a) and denote $\overline{x}$ a point in U on a neighborhood of which f is bounded from above. Observe immediately that $\overline{x} \in \operatorname{int}(\operatorname{dom} f)$, so $\operatorname{int}(\operatorname{dom} f) \neq \emptyset$. Choose by (a) some reals $\beta \geq 0$ and $r > 0$ such that $f(x) \leq \beta$ for all $x \in B(\overline{x}, r) \subset \operatorname{dom} f$. Fix an arbirary $u \in \operatorname{int}(\operatorname{dom} f)$ and choose a real $t > 1$ such that $\overline{y} := \overline{x} + t(u - \overline{x}) \in \operatorname{int}(\operatorname{dom} f)$. For $s := 1 - (1/t)$, we have $u = s\overline{x} + (1 - s)\overline{y}$ and $0 < s < 1$. Consider any $x \in B(u, sr)$ and note that, for x' such that $x = sx' + (1 - s)\overline{y}$, we have $x' \in B(\overline{x}, r)$, hence

$$f(x) \leq sf(x') + (1 - s)f(\overline{y}) \leq s\beta + (1 - s)f(\overline{y}).$$

The latter inequality entails that f is bounded from above on $B(u, sr)$, so Lemma 2.156 tells us that f is Lipschitz on $B(u, sr/2)$ and the proof is finished. □

Lipschitz properties of locally bounded separately convex-convex and convex-concave functions will be studied in Proposition 6.9. The next result is related to convex functions on finite dimensional spaces.

COROLLARY 2.159. *Any convex function $f : X \to \mathbb{R} \cup \{+\infty\}$ on a finite-dimensional normed space X is locally Lipschitz continuous on the interior of its (effective) domain whenever the latter is nonempty.*

PROOF. Suppose $\operatorname{int}(\operatorname{dom} f) \neq \emptyset$ and fix $\bar{x} \in \operatorname{int}(\operatorname{dom} f)$. Choose $x_1, \cdots, x_m$ in $\operatorname{int}(\operatorname{dom} f)$ such that, for $U := \operatorname{int}(\operatorname{co}\{x_1, \cdots, x_m\})$, we have $\bar{x} \in U \subset \operatorname{dom} f$. Then $f(x) \leq \max\{f(x_1), \cdots, f(x_m)\}$ for all $x \in U$, so the corollary follows from Theorem 2.158 above. □

The classical Banach-Steinhaus theorem or uniform boundedness principle says that any family of continuous linear operators $(A_i)_{i \in I}$ from a Banach space into a normed space which is pointwise bounded is uniformly bounded in the sense that the family of norms $(\|A_i\|)_{i \in I}$ is bounded. The latter property is obviously equivalent to the equi-Lipschitz property of the linear operators $(A_i)_{i \in I}$. So, the next theorem can be seen as a Banach-Steinhaus type theorem for real-valued convex functions. The case of vector-valued convex mappings will be examined in Theorem 2.177. The proof of the Banach-Steinhaus type theorem for real-valued convex functions will utilize Lemma 2.160 below. The lemma involves the concept of core of a set presented in Definition 2.55.

We saw that the topological interior of a subset of a topological vector space is always included in the core of the set. While Example 2.56 showed that the two concepts are different, Lemma 2.160 provides a large class of convex sets S where $\operatorname{int} S = \operatorname{core} S$ in Banach spaces.

LEMMA 2.160. *Let U be a nonempty convex set of a Banach space X and let $f : U \to \mathbb{R} \cup \{+\infty\}$ be a lower semicontinuous convex function on U. Then the following hold.*
(a) *For each $\bar{x} \in \operatorname{core}(\operatorname{dom} f)$ if any, the function f is bounded from above on a neighborhood of $\bar{x}$.*
(b) *One has the equality*

$$\operatorname{core}(\operatorname{dom} f) = \operatorname{int}(\operatorname{dom} f).$$

PROOF. Fix any $\bar{x} \in \operatorname{core}(\operatorname{dom} f)$ (if any), put $U_0 := U - \bar{x}$ and define $g := f(\cdot + \bar{x}) - f(\bar{x})$ on U_0. Observe that $g(0) = 0$ and $\bar{x} + \operatorname{dom} g = \operatorname{dom} f$, hence $0 \in \operatorname{core}(\operatorname{dom} g)$. For each $n \in \mathbb{N}$ consider the symmetric convex set

$$C_n := \left\{ x \in U_0 : \max\left\{ g\left(-\frac{1}{n}x\right), g\left(\frac{1}{n}x\right) \right\} \leq n \right\},$$

where the convexity property is due to the convexity of the function g. We claim that $U_0 = \bigcup_{n \in \mathbb{N}} C_n$. Indeed, take any $x \in U_0$ and choose some real $r > 0$ (depending on x) such that $[-rx, rx] \subset \operatorname{dom} g$ according to the inclusion $0 \in \operatorname{core}(\operatorname{dom} g)$. Choose an integer $n \geq r$ (depending on x) such that

$$\max\{g(-rx), g(rx)\} \leq n,$$

and note by convexity of g with $g(0) = 0$ that for $y = \pm x$ there exists some $t \in {]}0, 1]$ (depending on y) such that

$$g\left(\frac{1}{n}y\right) \leq tg(ry) \leq tn \leq n, \quad \text{hence } x \in C_n.$$

This confirms that $U_0 = \bigcup_{n \in \mathbb{N}} C_n$. Further, the open set U_0 endowed with the induced metric is a Baire space and each set C_n is closed relative to U_0. Therefore, there is some $k \in \mathbb{N}$ such that $\mathrm{int}_{U_0}(C_k) \neq \emptyset$, or equivalently $\mathrm{int}_X(C_k) \neq \emptyset$. This and the symmetric convexity of C_k entails that $0 \in \mathrm{int}_X(C_k)$, which ensures in particular that the function g is bounded from above on a neighborhood of 0, or equivalently f is bounded from above on a neighborhood of $\overline{x}$. This translates the assertion (a).

Concerning (b) we note by (a) that any $\overline{x} \in \mathrm{core}(\mathrm{dom}\, f)$ (if any) possesses a neighborhood included in $\mathrm{dom}\, f$. It results that $\mathrm{core}(\mathrm{dom}\, f) \subset \mathrm{int}_X(\mathrm{dom}\, f)$, hence the desired equality in (b) holds true. □

THEOREM 2.161 (equi-Lipschitz property of families of convex functions: pointwise upper equi-boundedness case). *Let U be an open convex set of a Banach space X and $(f_i)_{i \in I}$ be a family of lower semicontinuous convex functions from U into $\mathbb{R}$. Assume that there exists a neighborhood $V \subset U$ of $\overline{x} \in U$ such that, for each $x \in V$, the family $\bigl(f_i(x) - f_i(\overline{x})\bigr)_{i \in I}$ is bounded from above. Then the family of functions $(f_i)_{i \in I}$ is equi-Lipschitzian on a neighborhood of $\overline{x}$.*

PROOF. Consider the lower semicontinuous convex function $f : U \to \mathbb{R} \cup \{+\infty\}$ defined by

$$f(x) = \sup_{i \in I}\bigl(f_i(x) - f_i(\overline{x})\bigr) \quad \text{for all } x \in U.$$

By the assumption on V we have $\overline{x} \in \mathrm{core}(\mathrm{dom}\, f)$, hence by Lemma 2.160(a), the function f is bounded from above on a neighborhood W of $\overline{x}$. This amounts to saying that the family of functions $\bigl(f_i - f_i(\overline{x})\bigr)_{i \in I}$ is equi-bounded from above over W. This and Lemma 2.156 guarantee the equi-Lipschitz property of the family $(f_i)_{i \in I}$ over a neighborhood of $\overline{x}$. □

The first corollary is concerned with useful criteria for point of continuity of lower semicontinuous convex functions on Banach spaces.

COROLLARY 2.162. *Let U be an open convex set of a Banach space and $f : U \to \mathbb{R} \cup \{+\infty\}$ be a lower semicontinuous convex function. Given $\overline{x} \in U$, the following are equivalent:*
(a) *the function f is finite at $\overline{x}$ and locally Lipschitz continuous near $\overline{x}$;*
(b) *the function f is finite at $\overline{x}$ and continuous at $\overline{x}$;*
(c) *the interiority property $\overline{x} \in \mathrm{int}(\mathrm{dom}\, f)$ is satisfied;*
(d) *the inclusion $\overline{x} \in \mathrm{core}(\mathrm{dom}\, f)$ holds.*

PROOF. The implications (a)$\Rightarrow$(b)$\Rightarrow$(c) are clear and the equivalence (c)$\Leftrightarrow$(d) is due to Lemma 2.160(b). The final inclusion (c)$\Rightarrow$(a) follows from Theorem 2.161 by taking for I a suitable singleton set there. □

The previous corollary directly yields the following.

COROLLARY 2.163. *Let U be an open convex set of a Banach space X. Then any lower semicontinuous convex function $f : U \to \mathbb{R} \cup \{+\infty\}$ is locally Lipschitz continuous on the interior of its (effective) domain whenever the latter is nonempty.*

The third corollary is concerned with a continuity result for a bivariate function which is convex with respect to the first variable.

COROLLARY 2.164. *Let U and V be open convex sets of a Banach space X and a normed space Y respectively. Let $f : U \times V \to \mathbb{R}$ be a real-valued separately continuous function which is convex with respect to the first variable. Then f is jointly continuous on $U \times V$.*

PROOF. Fix any $(\overline{x}, \overline{y}) \in U \times V$ and, for $U_0 := U - \overline{x}$ and $V_0 := V - \overline{y}$, define the separately continuous function $g : U_0 \times V_0 \to \mathbb{R}$ by $g(x, y) := f(\overline{x} + x, \overline{y} + y) - f(\overline{x}, \overline{y} + y)$ for all $(x, y) \in U_0 \times V_0$. For each $y \in V_0$, the function $g(\cdot, y)$ is convex on U_0 and $g(0, y) = 0$. Let $(x_n, y_n)_{n \in \mathbb{N}}$ be any sequence in $U_0 \times V_0$ converging to $(0, 0)$ and let any real $\varepsilon > 0$. For each $x \in U_0$ the set $\{g(x, y_n) : n \in \mathbb{N}\}$ is bounded since $\{0\} \cup \{y_n : n \in \mathbb{N}\}$ is compact in Y and included in V_0, thus by Theorem 2.161 the family $(g(\cdot, y_n))_{n \in \mathbb{N}}$ is equicontinuous at 0. Then, there exists some neighborhood of zero $W \subset U_0$ such that $|g(x, y_n)| < \varepsilon$ for all $n \in \mathbb{N}$ and $x \in W$. Choosing an integer k such that $x_n \in W$ for all $n \geq k$, it results that $|g(x_n, y_n)| < \varepsilon$ for all $n \geq k$, which guarantees the continuity of g at $(0, 0)$, or equivalently of f at $(\overline{x}, \overline{y})$. □

In addition to the above criteria for local Lipschitz continuity, the next proposition provides a quantitative global Lipschitz property for convex functions. We need to establish first a lemma showing the convexity of the signed distance function to any convex set. The lemma is clearly interesting for its own right and it will also be used in other places of the book.

LEMMA 2.165. *Let S be a subset of a normed space $(X, \|\cdot\|)$. The following hold.*
(a) For all $x \in X$,
$$\operatorname{sgd}_S(x) = \inf_{y \in S} \left(\|x - y\| - d_{S^c}(y) \right),$$
where $S^c := X \setminus S$.
(b) If S is convex, then the function d_{S^c} is concave on S, so the signed distance function sgd_S is convex on the space X.

PROOF. We may suppose $S \neq \emptyset$, since otherwise both assertions (a) and (b) are trivial.
(a) Denote by $\rho(x)$ the right-hand member in (a).

Fix any $x \in S$. On the one hand, $\rho(x) \leq \|x - x\| - d_{S^c}(x) = -d_{S^c}(x) = \operatorname{sgd}_S(x)$. On the other hand, for any $y \in S$, we have $\|x - y\| - d_{S^c}(y) \geq -d_{S^c}(x) = \operatorname{sgd}_S(x)$, hence taking the infimum over $y \in S$ gives $\rho(x) \geq \operatorname{sgd}_S(x)$. We then deduce that $\operatorname{sgd}_S(x) = \rho(x)$.

Fix now any $x \in S^c$. For any $y \in S$, obviously $\|x - y\| - d_{S^c}(y) \leq \|x - y\|$, thus taking the infimum over $y \in S$ yields $\rho(x) \leq d_S(x) = \operatorname{sgd}_S(x)$. On the other hand, for any $y \in S$ there is some $z \in [x, y] \cap \operatorname{bdry} S$, hence
$$d_S(x) + d_{S^c}(y) \leq \|x - z\| + \|y - z\| = \|x - y\|, \text{ so } \operatorname{sgd}_S(x) \leq \|x - y\| - d_{S^c}(y),$$
which entails $\operatorname{sgd}_S(x) \leq \rho(x)$. In this second case, we also obtain $\operatorname{sgd}_S(x) = \rho(x)$.

In conclusion, we have $\operatorname{sgd}_S(x) = \rho(x)$, for all $x \in X$.
(b) Assume that S is convex and put $C := S^c$. First, we show that d_C is concave on S. If $\operatorname{int} S = \emptyset$, then the function d_C is null on S (since $C = S^c$ is dense in X), hence concave on S. Suppose that $\operatorname{int} S \neq \emptyset$. Fix any $x_1, x_2 \in \operatorname{int} S$ and $t_1, t_2 \in {]0, 1[}$ with $t_1 + t_2 = 1$, and put $u := t_1 x_1 + t_2 x_2$. Suppose (by contradiction)

that $d_C(u) < r_1 + r_2$, where $r_i := t_i d_C(x_i) > 0$ with $i = 1, 2$. There exists some $z \in C$ such that $\|u - z\| < r_1 + r_2$. Since $B(u, r_1 + r_2) = B(t_1 x_1, r_1) + B(t_2 x_2, r_2)$, we can choose $z_1, z_2 \in X$ such that $z = t_1 z_1 + t_2 z_2$ and $\|z_i - x_i\| < d_C(x_i)$, $i = 1, 2$. The latter inequality tells us that $z_i \notin C$, or equivalently, $z_i \in S$. The convexity of S gives $z = t_1 z_1 + t_2 z_2 \in S$, which contradicts the inclusion $z \in C$, so d_C is concave on int S. By continuity of d_C, it follows that d_C is concave on cl(int S), hence also on the convex set S (since cl(int S) $\supset S$ by convexity of S (see Proposition B.5(b) in Appendix).

The function $(x, y) \mapsto \|x - y\| - d_{S^c}(y)$ is then convex on $X \times S$, hence (a) and Proposition 2.54 imply that the function sgd_S is convex on X. $\square$

PROPOSITION 2.166. *Let X be a normed space, $f : X \to \mathbb{R} \cup \{+\infty\}$ be a convex function, and U be a nonempty open convex set of X over which f is bounded. Then, for each real $r > 0$ such that $U_r := \{x \in U : d(x, X \setminus U) > r\}$ is nonempty, the function f is Lipschitz on U_r; as long as β is an upper bound of $|f|$ on U the real $2\beta/r$ is a Lipschitz constant of f over U_r.*

PROOF. Let $\beta \geq 0$ be a real upper bound of $|f|$ over U and let $r > 0$ be as in the statement. Consider arbitrary $x \neq y$ in U_r. Choose an integer n such that $\|x - y\|/n < r/2$ and put $x_i := x + \frac{i}{n}(y - x)$ for $i = 0, 1, \cdots, n$. For each $i = 0, \cdots, n-1$, we have on the one hand $x_{i+1} \in B(x_i, r/2)$ and $B(x_i, r) \subset U$ since from the concavity on U of $d(\cdot, X \setminus U)$ obtained in the above lemma we have

$$d(x_i, X \setminus U) \geq \frac{i}{n} d(y, X \setminus U) + \left(1 - \frac{i}{n}\right) d(x, X \setminus U) > r.$$

On the other hand, $f(u) - f(x_i) \leq 2\beta$ for all $u \in B(x_i, r)$. From the second inequality in the conclusion of Lemma 2.156 we obtain $|f(x_{i+1}) - f(x_i)| \leq (2\beta/r)\|x_{i+1} - x_i\|$. For $\gamma := (2\beta/r)$, we then conclude that

$$|f(y) - f(x)| \leq \sum_{i=0}^{n-1} |f(x_{i+1}) - f(x_i)| \leq \gamma \sum_{i=0}^{n-1} \|x_{i+1} - x_i\| = \gamma \|y - x\|.$$

$\square$

2.3.2. Applications to directional Lipschitz property. Consider a convex function $f : X \to \mathbb{R} \cup \{+\infty\}$ on a normed space X. Note that $(u, r) \in \text{int epi } f$ for some $r \in \mathbb{R}$ if and only if $u \in \text{dom } f$ and f is bounded from above on some neighborhood of u, which is equivalent to the continuity of f at $u \in \text{dom } f$ according to Theorem 2.157. Further, by Proposition 2.33 we have for any $x \in \text{dom } f$

(2.59) $\qquad I\big(\text{epi } f; (x, f(x))\big) =]0, +\infty[(\text{int}(\text{epi } f) - (x, f(x))).$

This equality tells us that f is directionally Lipschitz at some point $x \in \text{dom } f$ if and only if $\text{int}(\text{epi } f) \neq \emptyset$, which is equivalent by what precedes to the continuity of f at some point in $\text{dom } f$.

Now assume that the convex function f is directionally Lipschitz at some point in $\text{dom } f$, which (as seen above) amounts to assuming that f is continuous at some point in $\text{dom } f$, and this is equivalent to int $(\text{dom } f) \neq \emptyset$ and f is continuous on this set int $(\text{dom } f)$ by Theorem 2.158. For $x \in \text{dom } f$ the function f is directionally Lipschitz at x in a direction $h \in X$ if and only if by (2.59) there are $t \in]0, +\infty[$ and $r \in \mathbb{R}$ such that

$(h, r) \in t^{-1}\big(\text{int}(\text{epi } f) - (x, f(x))\big),$ or equivalently $(x + th, f(x) + tr) \in \text{int}(\text{epi } f).$

For $t \in]0, +\infty[$ the latter inclusion holds for some $r \in \mathbb{R}$ if and only if (by what precedes) f is continuous at $x + th$, or equivalently $x + th \in \text{int}(\text{dom } f)$ since (as seen above) $\text{int}(\text{dom } f)$ is the set of points where the directionally Lipschitz convex function f is both finite and continuous. Therefore, the directionally Lipschitz convex function f is directionally Lipschitz at $x \in \text{dom } f$ in the direction h if and only if $h \in]0, +\infty[(\text{int}(\text{dom } f) - x)$.

All those features guarantee the following proposition.

PROPOSITION 2.167. *Let $f : X \to \mathbb{R} \cup \{+\infty\}$ be a convex function on a normed space X. Then f is directionally Lipschitz at some point in $\text{dom } f$ if and only if it continuous at some point in $\text{dom } f$.*

In such a case, for any $x \in \text{dom } f$ the function f is directionally Lipschitz at x, and the set of vectors $h \in X$ with respect to which f is directionally Lipschitz at x coincides with $]0, +\infty[(\text{int}(\text{dom } f) - x)$.

Given a function f which is directionally Lipschitz at $\overline{x}$, the next proposition expresses, by means of $\text{dom } f^\uparrow(\overline{x}; \cdot)$, the set of directions with respect to which f is directionally Lipschitz at $\overline{x}$.

PROPOSITION 2.168. *Let $f : U \to \mathbb{R} \cup \{-\infty, +\infty\}$ be a function on an open set U of a normed space X which is directionally Lipschitz at a point $\overline{x} \in U$ with $|f(\overline{x})| < +\infty$. Then the set of vectors in X with respect to which f is directionally Lipschitz at $\overline{x}$ coincides with $\text{int dom } f^\uparrow(\overline{x}; \cdot)$.*

PROOF. For convenience we may extend f to the entire space X by putting $f(x) = +\infty$ for all $x \in X \setminus U$. Fix any vector $\overline{h} \in X$ with respect to which f is directionally Lipschitz at $\overline{x}$. There exist by Proposition 2.90 reals $\alpha > 0$ and $\beta > 0$ such that for any $t \in]0, \alpha]$, for any $(x', s') \in \text{epi } f$ with $x' \in B(\overline{x}, \alpha)$ and $s' \in]f(\overline{x}) - \alpha, f(\overline{x}) + \alpha[$, for any $h' \in B(\overline{h}, \alpha)$, one has
$$t^{-1}(f(x' + th') - s') < \beta.$$
This gives in particular the inequality $f^\uparrow(\overline{x}; v) \leq \beta$ for every $v \in B(\overline{h}, \alpha)$ along with $\text{int dom } f^\uparrow(\overline{x}; \cdot) \neq \emptyset$, and we also see that $\overline{h} \in \text{int dom } f^\uparrow(\overline{x}; \cdot)$.

Now take any $h \in \text{int dom } f^\uparrow(\overline{x}; \cdot)$. If $f^\uparrow(\overline{x}; h_0) = -\infty$ for some $h_0 \in X$, by Proposition 2.57 the convex function $f^\uparrow(\overline{x}; \cdot)$ takes the value $-\infty$ on the whole set $\text{int dom } f^\uparrow(\overline{x}; \cdot)$, hence choosing $\varepsilon > 0$ with $B(h, \varepsilon) \subset \text{int dom } f^\uparrow(\overline{x}; \cdot)$ we have $B(h, \varepsilon) \times \mathbb{R} \subset \text{epi } f^\uparrow(\overline{x}; \cdot)$. Suppose that $f^\uparrow(\overline{x}; \cdot)$ takes the value $-\infty$ nowhere. Since the convex function $f^\uparrow(\overline{x}; \cdot)$ is bounded from above (by β) on $B(\overline{h}, \alpha)$, it is continuous on $\text{int dom } f^\uparrow(\overline{x}; \cdot)$ according to Theorem 2.158. Keeping in mind that $h \in \text{int dom } f^\uparrow(\overline{x}; \cdot)$, it ensues that there are reals $\varepsilon > 0$ and $\gamma' > 0$ such that $f^\uparrow(\overline{x}; h') < \gamma'$ for all $h' \in B(h, \varepsilon)$, hence $B(h, \varepsilon) \times]\gamma', +\infty[\subset \text{epi } f^\uparrow(\overline{x}; \cdot)$. Then, in any case there exist reals $\varepsilon > 0$ and $\gamma > \gamma' > 0$ such that
$$B(h, \varepsilon) \times]\gamma', +\infty[\subset \text{epi } f^\uparrow(\overline{x}; \cdot) = I(\text{epi } f; (\overline{x}, f(\overline{x}))),$$
where the latter equality involving the interior tangent cone $I(\cdot; \cdot)$ is due to the interior tangent cone property of $\text{epi } f$ at $(\overline{x}, f(\overline{x}))$ (see Proposition 2.25). We deduce that f is directionally Lipschitz at $\overline{x}$ in the direction h. This finishes the proof. □

2.3.3. Lipschitz property of continuous vector-valued convex mappings. We develop now some continuity and Lipschitz properties of vector-valued

2.3. LOCAL LIPSCHITZ PROPERTY OF CONTINUOUS CONVEX FUNCTIONS

convex mappings. Such mappings will be involved in Theorem 2.189 of the next subsection and in Chapter 7.

Given a convex cone K of a vector space Y with $0 \in K$ we associate with it a *preorder* $\leq_K$ on Y defined by $y \leq_K z$ provided $z - y \in K$. Sometimes, it will be convenient to write $z \geq_K y$ instead of $y \leq_K z$. Clearly, for $y \leq_K z$ and $y' \leq_K z'$, and for any real $t > 0$ the following inequalities hold true:
$$ty \leq_K tz, \quad -z \leq_K -y, \quad y + y' \leq_K z + z'.$$

As usual we adjoin an *abstract greatest element* $+\infty_Y$ to the ordered space $(Y, \leq_K)$, and in $Y \cup \{+\infty_K\}$ we put for any $y \in Y \cup \{+\infty_K\}$
$$y + (+\infty_K) = (+\infty_K) + y = +\infty_K \quad \text{and} \quad t(+\infty_K) = +\infty_K \quad \text{for every real } t > 0.$$

A mapping f from a convex set C of a vector space X into $Y \cup \{+\infty_Y\}$ is said to be *K-convex* whenever

(2.60) $\quad f(tx + (1-t)x') \leq_K tf(x) + (1-t)f(x') \quad \text{for all } t \in \,]0,1[,\, x, x' \in C;$

it is readily seen that this amounts to saying that the *K-epigraph*
$$\text{epi}_K f := \{(x, y) \in C \times Y : f(x) \leq_K y\}$$
is convex in $X \times Y$. We define the *effective domain* of f as $\text{dom } f := \{x \in C : f(x) \in Y\}$ and we note that

$\text{dom } f \quad \text{and} \quad f(\text{dom } f) + K$ are convex in X and Y respectively.

Recalling that a linear functional $\zeta : Y \to \mathbb{R}$ is positive (with respect to K) if $\zeta(y) \geq 0$ for all $y \geq_K 0$, the Hahn-Banach separation theorem clearly yields when Y is a locally convex space, under the closedness of the convex cone K, that f is K-convex if and only if the real-valued function $\zeta \circ f : \text{dom } f \to \mathbb{R}$ is convex for all positive continuous linear functionals ζ on Y.

When Y is a normed space endowed with the norm $\|\cdot\|$, the convex cone K of Y is said to be $\|\cdot\|$-*normal* whenever there exists some real $\beta > 0$ such that

(2.61) $\quad (\mathbb{B}_Y - K) \cap (\mathbb{B}_Y + K) \subset \beta \mathbb{B}_Y.$

First, in such a case if $0 \leq_K y \leq_K 0$, then for any $n \in \mathbb{N}$
$$ny \in (-K) \cap K \subset (\mathbb{B}_Y - K) \cap (\mathbb{B}_Y + K) \subset \beta \mathbb{B}_Y,$$
which implies $\|y\| \leq (\beta/n)$, hence $y = 0$; this means that the preorder $\leq_K$ is an *order* on Y.

Further, under the normality condition on K the property

(2.62) $\quad x \leq_K y \leq_K z \Longrightarrow \|y\| \leq \beta \max\{\|x\|, \|z\|\}$

holds true. Indeed, consider first the case $\|x\| \leq \|z\|$. Supposing $z \neq 0$ we can write on the one hand
$$\frac{y}{\|z\|} \in \frac{z}{\|z\|} - K \subset \mathbb{B}_Y - K$$
and on the other hand
$$\frac{y}{\|z\|} \in \frac{x}{\|z\|} + K \subset \mathbb{B}_Y + K,$$
which yields
$$\frac{y}{\|z\|} \in (\mathbb{B}_Y - K) \cap (\mathbb{B}_Y + K) \subset \beta \mathbb{B}_Y.$$

It ensues that $\|y\| \leq \beta\|z\|$, and the latter inequality still holds if $z = 0$, since this equality entails $x = 0$ because $\|x\| \leq \|z\|$, hence $y = 0$ according to $x \leq_K y \leq_K z$ and the order property established above for $\leq_K$.

Writing $-z \leq_K -y \leq_K -x$ the above case tells us that $\|y\| \leq \beta\|x\|$ if $\|z\| \leq \|x\|$. The desired inequality is then justified.

Conversely, assume the property (2.62). Taking any $y \in (\mathbb{B} - K) \cap (\mathbb{B} + K)$, there are $b_1, b_2 \in \mathbb{B}$ and $h_1, h_2 \in K$ such that $y = b_1 + h_1 = b_2 - h_2$, hence $b_1 \leq_K y \leq_K b_2$. From this and (2.62) it ensues that $\|y\| \leq \beta$, that is, $y \in \beta\mathbb{B}$.

We have then proved:

PROPOSITION 2.169. *A convex cone K (with $0 \in K$) of the normed space $(Y, \|\cdot\|)$ is $\|\cdot\|$-normal if and only if the property (2.62) holds.*

EXAMPLE 2.170. Let $\mathcal{B}(T, \mathbb{R})$ (resp. $\mathcal{C}(T, \mathbb{R})$) be the space of real-valued bounded (resp. continuous) functions on a set (resp. on a compact topological space) T.
(a) The usual convex cones of non-negative functions in $(\mathcal{B}(T, \mathbb{R}), \|\cdot\|_\infty)$ and $(\mathcal{C}(T, \mathbb{R}), \|\cdot\|_\infty)$ are clearly $\|\cdot\|_\infty$-normal; clear is also the $\|\cdot\|_p$-normality of the cone of classes of non-negative functions in $(L^p_\mathbb{R}(S, \mathcal{S}, \mu), \|\cdot\|_p)$, where $(S, \mathcal{S}, \mu)$ is a measure space.
(b) Let $\mathcal{C}^1([0, \pi], \mathbb{R})$ be the space of continuously differentiable functions from $[0, \pi]$ into $\mathbb{R}$ endowed with the norm given by $\|\varphi\|_{\mathcal{C}^1} := \|\varphi\|_\infty + \|\varphi'\|_\infty$ for all $\varphi \in \mathcal{C}^1([0, \pi], \mathbb{R})$. Let K be the convex cone of functions $\varphi \in \mathcal{C}^1([0, \pi], \mathbb{R})$ with $\varphi(t) \geq 0$ for all $t \in [0, \pi]$. Take $\varphi_n(t) := \frac{1}{n}\sin(nt)$ and $\psi_n(t) := 1/n$, and observe that $-\psi_n \leq_K \varphi_n \leq_K \psi_n$ and $1 \leq \|\varphi_n\|_{\mathcal{C}^1}$ along with $\|\psi_n\|_{\mathcal{C}^1} = 1/n$. Consequently, (2.62) cannot hold true, so the convex cone K is not $\|\cdot\|_{\mathcal{C}^1}$-normal. $\square$

We need to extend the concept of upper semicontinuity to mappings with values in the ordered normed space Y.

DEFINITION 2.171. *Let U be an open set of a topological space (X, τ) and $f : U \to Y \cup \{+\infty_K\}$ from U into $Y \cup \{+\infty_K\}$, where Y is a normed space equipped with a convex cone K containing 0. The mapping f is said to be K-upper semicontinuous at $\bar{x} \in U$ if for any neighborhood V of zero in Y there exists a neighborhood $U_0 \subset U$ of $\bar{x}$ such that*
$$f(x) \in f(\bar{x}) + V - K \quad \text{for all } x \in U_0 \cap \operatorname{dom} f.$$

So, a family $(f_i)_{i \in I}$ of mappings from U into $Y \cup \{+\infty_K\}$ is K-upper equisemicontinuous at $\bar{x}$ if for any neighborhood V of zero in Y there exists a neighborhood $U_0 \subset U$ of $\bar{x}$ such that for every $i \in I$
$$f_i(x) \in f_i(\bar{x}) + V - K \quad \text{for all } x \in U_0 \cap \operatorname{dom} f_i.$$

PROPOSITION 2.172. *Let U be an open convex set of a normed space X and let Y be a normed space endowed with a normal convex cone K with $0 \in K$. Let $(f_i)_{i \in I}$ be a family of K-convex mappings from U into Y. The family $(f_i)_{i \in I}$ is equicontinuous at a point $\bar{x} \in U$ if and only if it is K-upper equisemicontinuous at this point $\bar{x}$.*

PROOF. It is enough to prove that the condition is sufficient. By normality of K fix a real $\beta > 0$ such that $(\mathbb{B}_X - K) \cap (\mathbb{B}_X + K) \subset \beta\mathbb{B}_Y$. Take any real $\varepsilon > 0$ and set $\varepsilon' := \varepsilon/\beta$. Choose by the upper equisemicontinuity a real $r > 0$ with $\bar{x} + r\mathbb{B}_X \subset U$ such that
$$f_i(\bar{x} + x) - f_i(\bar{x}) \in \varepsilon'\mathbb{B}_Y - K, \quad \text{for all } x \in r\mathbb{B}_X \text{ and } i \in I.$$

This and the convexity of f_i yield, for all $x \in r\mathbb{B}_X$ and $i \in I$,
$$f_i(\bar{x} + x) - f_i(\bar{x}) \in f_i(\bar{x}) - f_i(\bar{x} - x) + K$$
$$\subset \varepsilon'\mathbb{B}_Y + K + K = \varepsilon'\mathbb{B}_Y + K,$$
so it results that
$$f_i(\bar{x} + x) - f_i(\bar{x}) \in (\varepsilon'\mathbb{B}_Y - K) \cap (\varepsilon'\mathbb{B}_Y + K) \subset \varepsilon\mathbb{B}_Y.$$
This confirms the desired equicontinuity. $\square$

A first characterization of equi-Lipschitz property of families of convex vector-valued mappings is the following:

THEOREM 2.173 (equicontinuity for families of convex vector mappings). Let U be an open convex set of a normed space X and let Y be a normed space endowed with a normal convex cone K with $0 \in K$. Let $(f_i)_{i \in I}$ be a family of K-convex mappings from U into Y and let $\bar{x} \in U$. The following assertions are equivalent:
(a) the family $(f_i)_{i \in I}$ is equi-Lipschitzian around $\bar{x}$;
(b) the family $(f_i)_{i \in I}$ is equicontinuous at $\bar{x}$;
(c) there exist an element $\bar{y} \in Y$ and a family $(M_i)_{i \in I}$ of mappings from U into Y which is equicontinuous at $\bar{x}$ and such that
$$M_i(\bar{x}) \leq_K \bar{y} \; \forall i \in I, \quad \text{and} \quad f_i(x) - f_i(\bar{x}) \leq_K M_i(x) \; \forall i \in I, \forall x \in U;$$
(d) there exist an element $\bar{y} \in Y$ and a family $(M_i)_{i \in I}$ of mappings from U into Y which is K-upper equisemicontinuous at $\bar{x}$ and such that
$$M_i(\bar{x}) \leq_K \bar{y} \; \forall i \in I, \quad \text{and} \quad f_i(x) - f_i(\bar{x}) \leq_K M_i(x) \; \forall i \in I, \forall x \in U;$$
(e) the family of functions $(\|f(\cdot)_i\|)_{i \in I}$ is equibounded from above, that is, the exists a real $r > 0$ and a neighborhood $V \subset U$ of $\bar{x}$ such that $\|f_i(x)\| \leq r$ for all $i \in I$ and $x \in V$.

PROOF. By normality of the cone K fix a real $\beta > 0$ such that (see (2.61))
$$(\mathbb{B}_Y - K) \cap (\mathbb{B}_Y + K) \subset \beta\mathbb{B}_Y.$$
Put $U_0 := -\bar{x} + U$ and consider the family $(g_i)_{i \in I}$ of mappings from U_0 into Y defined for each $i \in I$ by
$$g_i(x) = f_i(x + \bar{x}) - f_i(\bar{x}) \quad \text{for all } x \in U_0,$$
and note that $g_i(0) = 0$.
(a)$\Leftrightarrow$(e). For this equivalence it suffices to show the implication $\Leftarrow$. Suppose that (e) is satisfied and take any real $\varepsilon > 0$. Then there exist a real $r > \varepsilon$ and a symmetric convex neighborhood $V \subset U_0$ of zero such that
$$\|g_i(x)\| \leq r \quad \text{for all } i \in I \text{ and } x \in V.$$
It follows from the convexity of g_i along with the equality $g_i(0) = 0$ that for any $i \in I$ and any $x \in V$
$$-\frac{\varepsilon}{r} g_i(-x) \leq_K g_i\left(\frac{\varepsilon}{r}x\right) \leq_K \frac{\varepsilon}{r} g_i(x),$$
which ensures with the above real $\beta > 0$ (see (2.62)) that
$$\left\|g_i\left(\frac{\varepsilon}{r}x\right)\right\| \leq \beta \frac{\varepsilon}{r} \max\{\|g_i(x)\|, \|g_i(-x)\|\}.$$

This and the above inequality $\|g_i(u)\| \leq r$ for every $u \in V$ yield
$$\|g_i(x) - g_i(0)\| = \|g_i(x)\| \leq \varepsilon\beta \quad \text{for all } i \in I \text{ and } x \in r^{-1}\varepsilon V,$$
which translates the equicontinuity of $(g_i)_{i \in I}$ at zero, and hence the desired equicontinuity of the family $(f_i)_{i \in I}$ at $\overline{x}$.

(a)⇔(b). Suppose (b) holds, so the family $(g_i)_{i \in I}$ is equicontinuous at zero with $g_i(0) = 0$. Choose a real $r > 0$ with $2r\mathbb{B}_X \subset U_0$ and such that $g_i(x) - g_i(x') \in \mathbb{B}_Y$ for all $x, x' \in r\mathbb{B}_X$ and $i \in I$. Fix any $x, x' \in r\mathbb{B}_X$. For each real $s > r^{-1}\|x - x'\|$ set $z_s := x' + s^{-1}(x' - x)$. Then $z_s \in 2r\mathbb{B}_X$ and with $t := (1+s)^{-1}s < 1$ we have $x' = (1-t)x + tz_s$, so the convexity of g_i gives
$$g_i(x') \in (1-t)g_i(x) + tg_i(z_s) - K = g_i(x) + t\big(g_i(z_s) - g_i(x)\big) - K,$$
which entails
$$g_i(x') - g_i(x) \in t\big(g_i(z_s) - g_i(x)\big) - K \subset t\mathbb{B}_Y - K.$$
It follows with the above real $\beta > 0$ that
$$g_i(x') - g_i(x) \in (t\mathbb{B}_Y - K) \cap (t\mathbb{B}_Y + K) \subset t\beta\mathbb{B}_Y \subset s\beta\mathbb{B}_Y.$$
This being true for every real $s > r^{-1}\|x' - x\|$, we deduce that $\|g_i(x') - g_i(x)\| \leq r^{-1}\beta\|x' - x\|$, proving the equivalence (b) ⇔ (a) since the implication (a) ⇒ (b) is obvious.

(d)⇒(b). Suppose (d) and keep g_i as defined above. Take any real $\varepsilon > 0$ and choose a real $t > 1$ such that $\|\overline{y}\| < (t-1)\varepsilon$. By the upper equisemicontinuity of $(M_i)_{i \in I}$ there exists a real $r > 0$ such that
$$M_i(\overline{x} + r\mathbb{B}_X) - M_i(\overline{x}) \subset \varepsilon\mathbb{B}_Y - K \quad \text{for all } i \in I.$$
It ensues that, for each $i \in I$ and each $x \in r\mathbb{B}_X$,
$$g_i(x) - g_i(0) = g_i(x) \in M(\overline{x} + x) - K$$
$$= \big(M_i(\overline{x} + x) - M_i(\overline{x})\big) + M_i(\overline{x}) - K$$
$$\subset \varepsilon\mathbb{B}_Y + \overline{y} - K \subset \varepsilon\mathbb{B}_Y + \varepsilon(t-1)\mathbb{B}_Y - K$$
$$= \varepsilon t\mathbb{B}_Y - K = t(\varepsilon\mathbb{B}_Y - K),$$
where the latter left equality is due to the fact that $\alpha C + \beta C = (\alpha + \beta)C$ for any convex set C and any reals $\alpha \geq 0$ and $\beta \geq 0$. Taking the convexity of g_i into account we obtain
$$g_i(t^{-1}r\mathbb{B}_X) - g_i(0) \subset \varepsilon\mathbb{B}_Y - K,$$
which by Proposition 2.172 says that $(g_i)_{i \in I}$ is equicontinuous at 0. Then, the implication (d) ⇒ (b) holds true.

Finally, the implication (c) ⇒ (d) is obvious and the implication (b) ⇒ (c) is also obvious with $M_i(x) := f_i(x) - f_i(\overline{x})$ and $\overline{y} := 0$. The proof of the theorem is then finished. □

THEOREM 2.174 (Lipschitz property for vector convex mappings). *Let U be an open convex set of a normed space X and let Y be a normed space endowed with a normal convex cone K with $0 \in K$. Let $f : U \to Y$ be a K-convex mapping from U into Y and let $\overline{x} \in U$. The following assertions are equivalent:*
(a) *the mapping f is Lipschitzian around $\overline{x}$;*
(b) *the mapping f is continuous at $\overline{x}$;*
(c) *there exists a mapping $M : U \to Y$ which is continuous at $\overline{x}$ and such that*

$f(x) \leq_K M(x)$ for all $x \in U$;
(d) there exists a mapping $M : U \to Y$ which is K-upper semicontinuous at $\bar{x}$ and such that $f(x) \leq_K M(x)$ for all $x \in U$;
(e) the mapping f is norm-bounded around $\bar{x}$, that is, there exists a real $r > 0$ and a neighborhood $V \subset U$ of $\bar{x}$ such that $\|f(x)\| \leq r$ for all $x \in V$.

PROOF. Putting $M_0(x) := M(x) - f(\bar{x})$, it is enough to observe that the statement (c) (resp. (d)) in the theorem is equivalent to saying that M_0 is (continuous) (resp. K-upper semicontinuous) at $\bar{x}$ and $f(x) - f(\bar{x}) \leq_K M_0(x)$ for all $x \in U$. So, applying Theorem 2.173 justifies the equivalences of the theorem. $\square$

The next proposition shows that such a result fails without the normality of the cone K. We need first to establish a lemma.

LEMMA 2.175. *If a closed convex cone K of a normed space Y is not $\|\cdot\|$-normal, then there exists some $a \in K$ such that the order interval $\{y \in Y : 0 \leq_K y \leq_K a\}$ is not $\|\cdot\|$-bounded.*

PROOF. For each $n \in \mathbb{N}$, the non-normality of K gives some $u_n \in K$, $y_n \in K$ such that $0 \leq_K y_n \leq_K u_n$, $\|u_n\| = 1$, and $\|y_n\| \geq 3^n$. The vector $a := \sum_{k=1}^{\infty} 2^{-k} u_k$ is well defined and $a \in K$. For each $n \geq 2$, putting $a_n := a - \sum_{k=1}^{n-1} 2^{-k} u_k - 2^{-n} y_n$, we see that
$$\|a_n\| \geq \left(\frac{3}{2}\right)^n - \left\|a - \sum_{k=1}^{n-1} 2^{-k} u_k\right\|,$$
hence the sequence $(a_n)_{n \geq 2}$ is not $\|\cdot\|$-bounded. Since
$$0 \leq_K a - \sum_{k=1}^{n} 2^{-k} u_k = a_n + 2^{-n}(y_n - u_n) \leq_K a_n \leq_K a,$$
the order interval $\{y \in Y : 0 \leq_K y \leq_K a\}$ is not $\|\cdot\|$-bounded. $\square$

PROPOSITION 2.176. *Let $(Y, \|\cdot\|)$ be a normed space and K be a closed convex cone in Y which is not $\|\cdot\|$-normal. Then there exists a K-convex mapping $\varphi : \mathbb{R} \to Y$ which is K-upper bounded near 0 but not continuous at 0.*

PROOF. By the above lemma fix $a \in K$ such that $\{y \in Y : 0 \leq_K y \leq_K a\}$ is not $\|\cdot\|$-bounded, so $\{y \in Y : \rho_1 a \leq_K y \leq_K \rho_2 a\}$ is not $\|\cdot\|$-bounded for all reals $0 \leq \rho_1 \leq \rho_2$. Fix a real $\sigma \in {]0,1[}$ and take any real r satisfying $1 < r < \sigma^{-1}(1 - \sigma + \sigma^2)$, which is possible since $1 - \sigma + \sigma^2 > \sigma$. For each $n \in \mathbb{N}$ we can choose $a_n \in K$ such that $\sigma^{2n} a \leq_K a_n \leq_K r\sigma^{2n} a$ and $\|a_n\| \geq n$. Define $\varphi : \mathbb{R} \to Y$ with $\varphi(t) = 0$ for all $t \leq 0$, $\varphi(\sigma^n) = a_n$ for each $n \in \mathbb{N}$, and φ defined linearly otherwise, that is,
$$\varphi(t) = \frac{\sigma^n}{\sigma^n - \sigma^{n+1}} a_{n+1} - \frac{\sigma^{n+1}}{\sigma^n - \sigma^{n+1}} a_n + \frac{t}{\sigma^n - \sigma^{n+1}}(a_n - a_{n+1}),$$
for all $t \in [\sigma^2, +\infty[$ if $n = 1$ and all $t \in [\sigma^{n+1}, \sigma^n]$ if $n \geq 2$. We note that $a_{n+1} \leq_K a_n$, since
$$a_n - a_{n+1} \geq_K \sigma^{2n}(1 - r\sigma^2)a \geq_K \sigma^{2n}\big(1 - \sigma(1 - \sigma + \sigma^2)\big)a$$
$$= \sigma^{2n}(1 - \sigma)(1 + \sigma^2)a \geq_K 0.$$

Further, with $c_n := (\sigma^n - \sigma^{n+1})^{-1}(a_n - a_{n+1})$ we also have $c_{n+1} \leq_K c_n$; indeed

$$\frac{a_n - a_{n+1}}{\sigma^n - \sigma^{n+1}} - \frac{a_{n+1} - a_{n+2}}{\sigma^{n+1} - \sigma^{n+2}}$$

$$\geq_K \frac{\sigma^{2n} - r\sigma^{2n+2}}{\sigma^n - \sigma^{n+1}} a - \frac{r\sigma^{2n+2} - \sigma^{2n+4}}{\sigma^{n+1} - \sigma^{n+2}} a$$

$$= \frac{\sigma^n}{1-\sigma}(1 - r\sigma^2 - r\sigma + \sigma^3)a$$

$$\geq_K \frac{\sigma^n}{1-\sigma}\bigl(1 - \sigma(1 - \sigma + \sigma^2) - (1 - \sigma + \sigma^2) + \sigma^3\bigr)a = 0.$$

On the other hand, with any positive linear functional $\zeta : Y \to \mathbb{R}$, the inequalities $\sigma^{2n} a \leq_K a_n \leq_K r\sigma^{2n} a$ give $\sigma^{2n}\zeta(a) \leq \zeta(a_n) \leq r\sigma^{2n}\zeta(a)$, hence $\zeta(a_n) \to 0$ as $n \to \infty$. The non-decreasing function $\zeta \circ \varphi$ is then continuous at 0, so continuous on $\mathbb{R}$ according to the definition of φ. This combined with the above inequalities and the definition of φ entails that $\zeta \circ \varphi$ is a convex function for every positive linear functional ζ, so the mapping φ is K-convex.

To finish the proof, it remains to see that the K-convex mapping φ is K-majorized on $[-\sigma, \sigma]$ (by a_1, for example), but φ is not $\|\cdot\|$-continuous at 0 since $\|\varphi(\sigma^n)\| \to +\infty$ (keep in mind $\|a_n\| \geq n$) and $\sigma^n \to 0$ as $n \to \infty$. □

With Proposition 2.172 and Theorem 2.173, we can state and prove a Banach-Steinhaus type theorem for vector-valued convex mappings.

THEOREM 2.177 (Banach-Steinhaus type theorem for convex mappings). Let X be a Banach space and U be an open convex set of X, and let Y be a normed space and K be a normal convex cone in Y with $0 \in K$. Let $(f_i)_{i \in I}$ be a family of continuous K-convex mappings from U into Y and let $\bar{x} \in U$. The following assertions are equivalent:
(a) the family $(f_i)_{i \in I}$ is equi-Lipschitzian around $\bar{x}$;
(b) the family $(f_i)_{i \in I}$ is equicontinuous at $\bar{x}$;
(c) there exists a neighborhood $U_0 \subset U$ of $\bar{x}$ such that, for each $x \in U_0$, the family $\bigl(\|f_i(x) - f_i(\bar{x})\|\bigr)_{i \in I}$ is bounded in $\mathbb{R}$.

PROOF. Suppose (c) and fix any real $\varepsilon > 0$. Considering $x \mapsto f_i(x + \bar{x}) - f_i(\bar{x})$ we may suppose that $\bar{x} = 0$ and $f_i(\bar{x}) = 0$. Choose a real $r > 0$ such that $r\mathbb{U}_X \subset U$ and $(\|f_i(x)\|)_{i \in I}$ is bounded from above for each $x \in r\mathbb{U}_X$, where we recall that $\mathbb{U}_X$ denotes the open unit ball of X centered at 0. Denote by g_i the restriction of f_i to $r\mathbb{U}_X$ and put

$$W := \bigcap_{i \in I} \operatorname{cl}_X g_i^{-1}(\varepsilon \mathbb{U}_Y - K) \quad \text{and} \quad V := W \cap (-W).$$

Since W is easily seen to be convex according to the K-convexity of each g_i, the set V is a closed symmetric convex set in X. Take any $x \in X$ and choose a real $t > 0$ such that $tx \in r\mathbb{U}_X$. There exists a real $s > 1$ such that $\|g_i(tx)\| < \varepsilon s$ for all $i \in I$, hence by the K-convexity of g_i we have

$$g_i(s^{-1}tx) \in s^{-1}g_i(tx) - K \subset \varepsilon \mathbb{U}_Y - K,$$

which entails $s^{-1}tx \in W$. Taking into account the definition of V, it ensues that $X = \bigcup_{n \in \mathbb{N}} nV$, which gives $\operatorname{int}_X(nV) \neq \emptyset$ for some $n \in \mathbb{N}$, thus $\operatorname{int}_X V \neq \emptyset$. It results that $0 \in \operatorname{int}_X V$ since V is a symmetric convex set. Choose some real $\rho > 0$ such that $\rho \mathbb{U}_X \subset V$, so for each $i \in I$ we have $\rho \mathbb{U}_X \subset \operatorname{cl}_X g_i^{-1}(\varepsilon \mathbb{U}_Y - K)$ and since $g_i^{-1}(\varepsilon \mathbb{U}_Y - K)$ is convex and open (by the continuity of g_i), we deduce that $\rho \mathbb{U}_X \subset g_i^{-1}(\varepsilon \mathbb{U}_Y - K)$. The latter inclusion translates the K-upper equisemicontinuity of the family $(g_i)_{i \in I}$ at 0, or equivalently the K-upper equisemicontinuity of the family $(f_i)_{i \in I}$ at $\overline{x}$. Proposition 2.172 and Theorem 2.173 complete the proof. □

Our next step with convex mappings is to show that such a mapping is $\mathcal{C}^1$ on an open set whenever it is Fréchet differentiable on that open set. Let us first establish a result concerning the strict differentiability of convex mappings.

PROPOSITION 2.178. *Let U be an open convex set of a normed space X and let Y be a normed space endowed with a normal convex cone K with $0 \in K$. Let $f : U \to Y$ be a K-convex mapping from U into Y and let $\overline{x} \in U$.*
(a) The convex mapping f is strictly Fréchet (resp. strictly Hadamard) differentiable at $\overline{x}$ if and only if it is Fréchet (resp. Hadamard) differentiable at $\overline{x}$.
(b) The convex mapping f is strictly Hadamard differentiable at $\overline{x}$ whenever it is Gâteaux differentiable at $\overline{x}$ and continuous at $\overline{x}$.

PROOF. If f is Hadamard differentiable at $\overline{x}$, we know that it is continuous at $\overline{x}$. Then, under either the Hadamard differentiability of f at $\overline{x}$ or the assumption in (b), by Theorem 2.174 there are reals $\delta > 0$ and $\gamma > 0$ such that f is γ-Lipschitz on $B[\overline{x}, \delta] \subset U$. Further, this gives under (b) that the mapping f is Hadamard differentiable at $\overline{x}$.

By the normality of the convex cone K, let a real $\beta > 0$ such that $(\mathbb{B}_Y - K) \cap (\mathbb{B}_Y + K) \subset \beta \mathbb{B}_Y$. Considering $f - Df(\overline{x})(\cdot)$ in place of f, we can and do suppose that $Df(\overline{x})$ is null. Take $B := \mathbb{B}_Y$ if f is Fréchet differentiable at $\overline{x}$, and take B as any symmetric compact subset of $\mathbb{B}_Y$ if f is Hadamard differentiable at $\overline{x}$. Fix any real $\varepsilon > 0$ and choose by the differentiability of f at $\overline{x}$ a real $r \in]0, \delta/3[$ such that

$$\sup_{h \in B} \|r^{-1}(f(\overline{x} + rh) - f(\overline{x}))\| < \varepsilon/(3\beta).$$

Take a positive real $\eta < \min\{\delta/3, \varepsilon r/(6\beta\gamma)\}$ and consider any $x \in B(\overline{x}, \eta)$ and any $t \in]0, \eta[$. Take any $h \in \mathbb{B}_Y$. From the convexity of f it results that

$$t^{-1}(f(x + th) - f(x)) = (-t)^{-1}(f(x + th - th) - f(x + th))$$
$$\in r^{-1}(f(x + th + rh) - f(x + th)) - K.$$

Writing

$$r^{-1}(f(x + th + rh) - f(x + th))$$
$$= r^{-1}\Big(\big(f(x + th + rh) - f(\overline{x} + rh)\big) + \big(f(\overline{x} + rh) - f(\overline{x})\big) + \big(f(\overline{x}) - f(x + th)\big) \Big),$$

we see with $\varepsilon' := \varepsilon/(3\beta)$ that

$$t^{-1}(f(x + th) - f(x))$$
$$\in r^{-1}\Big(\gamma \|x - \overline{x} + th\| \mathbb{B}_Y + r\varepsilon' \mathbb{B}_Y + \gamma \|x - \overline{x} + th\| \mathbb{B}_Y \Big) - K.$$

It ensues that

$$t^{-1}(f(x + th) - f(x)) \in (\varepsilon/\beta)\mathbb{B}_Y - K.$$

Similarly, we can write
$$t^{-1}\big(f(x) - f(x+th)\big) = (-t)^{-1}\big(f(x+(-t)(-h)) - f(x)\big)$$
$$\in r^{-1}\big(f(x+r(-h)) - f(x)\big) - K,$$
hence with $\varepsilon' := \varepsilon/(3\beta)$ as above
$$t^{-1}\big(f(x) - f(x+th)\big)$$
$$\in r^{-1}\Big(\big(f(x-rh) - f(\overline{x}-rh)\big) + \big(f(\overline{x}-rh) - f(\overline{x})\big) + \big(f(\overline{x}) - f(x)\big)\Big) - K$$
$$\subset r^{-1}\Big(\gamma\|x-\overline{x}\|\mathbb{B}_Y + r\varepsilon'\mathbb{B}_Y + \gamma\|x-\overline{x}\|\mathbb{B}_Y\Big) - K.$$
It ensues that $t^{-1}\big(f(x) - f(x+th)\big) \in (\varepsilon/\beta)\mathbb{B}_Y - K$. It follows that
$$t^{-1}\big(f(x+th) - f(x)\big) \in \big((\varepsilon/\beta)\mathbb{B}_Y - K\big) \cap \big((\varepsilon/\beta)\mathbb{B}_Y + K\big) \subset \varepsilon\mathbb{B}_Y,$$
which ensures that, for all $x \in B(\overline{x}, \eta)$ and $t \in\,]0, \eta[$,
$$\sup_{h \in B} \|t^{-1}\big(f(x+th) - f(x)\big)\| \le \varepsilon,$$
justifying the corresponding strict differentiability of f at $\overline{x}$. □

Combining the preceding proposition with Corollary 2.45 we obtain the desired theorem about the $\mathcal{C}^1$ property of Fréchet differentiable convex mappings.

THEOREM 2.179 ($\mathcal{C}^1$-property of Fréchet differentiable convex mappings). Let U be an open convex set of a normed space X and let Y be a normed space endowed with a normal convex cone K with $0 \in K$. A K-convex mapping $f : U \to Y$ from U into Y is of class $\mathcal{C}^1$ on U if and only if it is Fréchet differentiable on U.

2.4. Chain rule, compactly Lipschitzian mappings, supremum functions

Formulae for the Clarke subdifferential of $g \circ G$ generally requires the compactly Lipschitzian property of the vector-valued mapping G.

2.4.1. Compactly Lipschitzian mappings and chain rule.
The chain rule for composition with compactly Lipschitzian mapping will use a mean value theorem available for locally Lipschitz continuous functions. This mean value theorem has also many other applications.

THEOREM 2.180 (mean value theorem with C-subdifferential for Lipschitz functions). Let U be an open convex set of the normed space X and $f : U \to \mathbb{R}$ be a locally Lipschitz continuous real-valued function. Then for any $a, b \in U$ with $a \ne b$ there exist some $c \in\,]a, b[\,:= \{a + t(b-a) : t \in\,]0, 1[\}$ and $c^* \in \partial_C f(c)$ such that
$$f(b) - f(a) = \langle c^*, b - a \rangle.$$

PROOF. For $a, b \in U$ with $a \ne b$, set as usual
$$\varphi(t) := f(a + t(b-a)) + t\big(f(a) - f(b)\big) \quad \text{for all } t \in [0, 1].$$
Since $\varphi(0) = \varphi(1)$, the continuous function φ attains at some $s \in\,]0, 1[$ its minimum or maximum, hence $0 \in \partial_C\varphi(s)$ (because $\partial_C(-\varphi)(s) = -\partial_C\varphi(s)$). Using Theorem 2.135 and the calculus rule for sum, and putting $c := a + s(b-a)$ we obtain some $c^* \in \partial_C f(c)$ such that $0 = \langle c^*, b-a \rangle + f(a) - f(b)$, which ensures the desired equality. □

We saw in Proposition 2.76 that a locally Lipschitz continuous function $f : U \to \mathbb{R}$ on an open set of a normed space is strictly Gâteaux differentiable at a point $\bar{x} \in U$ if and only if $\partial_C f(\bar{x})$ is a singleton. Since the multimapping $\partial_C f : U \rightrightarrows X^*$ is $\|\cdot\|$-to-w^* upper semicontinuous at $\bar{x}$ (see Proposition 2.74), we can say that f is strictly Gâteaux differentiable at $\bar{x}$ if and only if $\partial_C f(\bar{x})$ is a singleton and $\partial_C f$ is $\|\cdot\|$-to-w^* upper semicontinuous at $\bar{x}$. By means of the mean value theorem above we establish a similar characterization of strict Fréchet differentiability.

PROPOSITION 2.181. *Let $f : U \to \mathbb{R}$ be a locally Lipschitz continuous function on an open set U of a normed space X and let $\bar{x} \in U$. The following are equivalent:*
(a) *the function f is strictly Fréchet differentiable at $\bar{x}$;*
(b) *$\partial_C f(\bar{x})$ is a singleton and the multimapping $\partial_C f : U \rightrightarrows X^*$ is $\|\cdot\|$-to-$\|\cdot\|_*$ upper semicontinuous, that is, for every real $\varepsilon > 0$ there is a neighborhood $U_0 \subset U$ of $\bar{x}$ such that*
$$\partial_C f(x) \subset \partial_C f(\bar{x}) + \varepsilon \mathbb{B}_{X^*} \quad \text{for all } x \in U_0;$$
(c) *there exists some $z^* \in X^*$ such that for every real $\varepsilon > 0$ there is a neighborhood $U_0 \subset U$ of $\bar{x}$ such that*
$$\partial_C f(x) \subset B_{X^*}(z^*, \varepsilon) \quad \text{for all } x \in U_0.$$

PROOF. (a)$\Rightarrow$(b) Suppose that f is strictly F-differentiable at $\bar{x}$. By Proposition 2.76 this entails in particular that $\partial_C f(\bar{x}) = \{Df(\bar{x})\}$. Take any real $\varepsilon > 0$. There is an open neighborhood $U_0 \subset U$ of $\bar{x}$ and a real $\delta > 0$ such that for any $h \in \mathbb{S}_X$, any $y \in U_0$ and any $t \in \,]0, \delta[$
$$-\varepsilon \leq t^{-1}\big(f(y+th) - f(y)\big) - \langle Df(\bar{x}), h \rangle \leq \varepsilon = \varepsilon \|h\|.$$
Fixing $x \in U_0$, and passing to the upper limit in the right-hand inequality above as $y \to x$ and $t \downarrow 0$, yields for every $h \in \mathbb{S}_X$
$$f^\circ(x; h) \leq \langle Df(\bar{x}), h \rangle + \varepsilon \|h\|,$$
and, in terms of support functions, this is equivalent to
$$\sigma\big(\cdot, \partial_C(x)\big) \leq \sigma\big(\cdot, Df(\bar{x}) + \varepsilon \mathbb{B}_{X^*}\big).$$
Since both sets $\partial_C f(x)$ and $Df(\bar{x}) + \varepsilon \mathbb{B}_{X^*}$ are convex and w^*-closed, it ensues that $\partial_C f(x) \subset Df(\bar{x}) + \varepsilon \mathbb{B}_{X^*}$ for all $x \in U_0$, which justifies the implication (a)$\Rightarrow$(b).
(b)$\Rightarrow$(c) This is trivial.
(c)$\Rightarrow$(a) Suppose (c) and let z^* be as given by (c). Take any real $\varepsilon > 0$. By (c) there is a neighborhood $U_0 \subset U$ of $\bar{x}$ such that $\partial_C f(U) \subset z^* + \varepsilon \mathbb{B}_{X^*}$. Choose a real $\delta > 0$ such that $B(\bar{x}, \delta) + \,]0, \delta[\mathbb{B}_X \subset U_0$. Take any $h \in \mathbb{S}_X$, any $x \in B(\bar{x}, \delta)$ and any $t \in \,]0, \delta[$. Since $B(\bar{x}, \delta) + \,]0, \delta[\mathbb{B}_X$ is an open convex set, by Theorem 2.180 there is some $z_{t,x} \in [x, x+th]$ along with $\zeta_{t,x} \in \partial_C f(z_{t,x})$ (both depending also on $h \in \mathbb{S}_X$) such that
$$t^{-1}\big(f(x+th) - f(x)\big) = \langle \zeta_{t,x}, h \rangle.$$
Therefore, choosing some $b^* \in \mathbb{B}_{X^*}$ (depending on t, x, h) such that $\zeta_{t,x} = z^* + \varepsilon b^*$, it ensues that
$$\left| t^{-1}\big(f(x+th) - f(x)\big) - \langle z^*, h \rangle \right| = \varepsilon |\langle b^*, h \rangle| \leq \varepsilon.$$
This ensures that
$$\limsup_{x \to \bar{x}, t \downarrow 0} \left(\sup_{h \in \mathbb{S}_X} \left| t^{-1}\big(f(x+th) - f(x)\big) - \langle Df(\bar{x}), h \rangle \right| \right) = 0,$$

which means that f is strictly F-differentiable at $\bar{x}$. The proof is finished. □

Before proceeding with composition with compactly Lipschitzian mappings, let us give a first application of the above mean value theorem to convexity of locally Lipschitz functions on normed spaces. A similar more general result for lower semicontinuous functions in the framework of Banach spaces will be established in Theorem 6.68. Given a nonempty subset S of the normed space X, recall that a multimapping $M : U \rightrightarrows X^*$ is *monotone* provided that for all $(x_i, x_i^*) \in \operatorname{gph} M$, $i = 1, 2$, one has
$$\langle x_1^* - x_2^*, x_1 - x_2 \rangle \geq 0.$$

PROPOSITION 2.182. *Let U be an open convex set of the normed space X and $f : U \to \mathbb{R}$ be a locally Lipschitz continuous real-valued function. Then f is convex if and only if the multimapping $\partial_C f$ is monotone.*

PROOF. The implication $\Rightarrow$ follows from the equality
$$\partial_C f(x) = \{x^* \in X^* : \langle x^*, y - x \rangle \leq f(y) - f(x),\ \forall y \in U\}.$$
Concerning the converse implication, suppose that $\partial_C f$ is monotone and fix any $x, y \in U$ with $x \neq y$ and any $s, t \in\ ]0, 1[$ with $s + t = 1$. Setting $z := sx + ty$, by Theorem 2.180 there are $c \in\ ]x, z[$, $d \in\ ]z, y[$, $c^* \in \partial_C f(c)$ and $d^* \in \partial_C f(d)$ such that
$$f(z) - f(x) = \langle c^*, z - x \rangle \quad \text{and} \quad f(z) - f(y) = \langle d^*, z - y \rangle.$$
Noting that $z - x = t(y - x)$ and $z - y = s(x - y)$, multiplying the first inequality by s and the second by t and adding together yield
$$f(z) - sf(x) - tf(y) = st\langle c^* - d^*, y - x \rangle = -st\lambda \langle c^* - d^*, c - d \rangle \leq 0,$$
where $\lambda \geq 0$ is so that $x - y = \lambda(c - d)$. This confirms the convexity of f. □

The chain rule for the C-subdifferential of $g \circ G$ in Proposition 2.109 or Theorem 2.135 involved the strict differentiability of the function g or the mapping G respectively. Through the mean value theorem above we can deal with a particular class of locally Lipschitz continuous mappings G without requiring any other property for the locally Lipschitz function g.

DEFINITION 2.183. *Let $\mathcal{K}$ be a collection of nonempty bounded subsets of a normed space Y with $\{0\} \in \mathcal{K}$ and let $G : U \to Y$ be a mapping from an open set U of the normed space X into Y. One says that G is $\mathcal{K}$-Lipschitzian at $x \in U$ if there exist a neighborhood $W \subset U$ of x, a multimapping $K : X \rightrightarrows X$ with images in $\mathcal{K}$, and a real-valued function $\rho :\ ]0, 1] \times W \times X \to \mathbb{R}_+$ such that*

(i) $\lim_{t \downarrow 0, x' \to x} \rho(t, x'; h) = 0$ for each $h \in X$ and ρ is bounded from above on $]0, 1] \times W \times \mathbb{B}_X$;

(ii) $K(\cdot)$ is upper semicontinuous at 0 with $K(0) = \{0\}$;

(iii) for all $x' \in W$, all $t \in\ ]0, 1]$ and all $h \in X$ with $x' + th \in W$
$$t^{-1}\big(G(x' + th) - G(x')\big) \in K(h) + \rho(t, x'; h)\mathbb{B}_Y.$$

When $\mathcal{K}$ is the collection of all nonempty compact subsets of Y, the mapping G will be said to be *compactly Lipschitzian* (or *compactly Lipschitz*) at x.

Let us list some first properties of such mappings.

2.4. CHAIN RULE, COMPACTLY LIPSCHITZIAN MAPPINGS

PROPOSITION 2.184. *Let $\mathcal{B}$ be the collection of all nonempty bounded subsets of the normed space Y and U be an open set of the normed space X. Then a mapping $G : U \to Y$ is $\mathcal{B}$-Lipschitzian at $x \in U$ if and only if it is Lipschitz continuous near the point x.*

In particular, any compactly Lipschitzian mapping at x is Lipschitz continuous near x; further when Y is finite-dimensional, a mapping $G : U \to Y$ is compactly Lipschitzian at x if and only if it is Lipschitz continuous near x.

PROOF. If G is Lipschitz continuous on a neighborhood $W \subset U$ of x with a Lipschitz constant $\gamma \geq 0$ therein, then taking $\rho(\cdot) = 0$ and $K(h) = \gamma \|h\| \mathbb{B}_Y$, we see that G is $\mathcal{B}$-Lipschitzian at x.

Now suppose that G is $\mathcal{B}$-Lipschitzian at x. Choose W, $\rho(\cdot)$, and $K(\cdot)$ as given by Definition 2.183 and choose some real upper bound $\beta \geq 0$ of ρ on $]0,1] \times W \times \mathbb{B}_X$ according to (ii). Choose some positive real $\delta < 1$ such that $x + 2\delta \mathbb{B}_X \subset W$ and $K(h) \subset \mathbb{B}_Y$ for all $h \in \delta \mathbb{B}_X$. Then by (iii) we have

$$G(x' + th') - G(x') \in t(1+\beta)\mathbb{B}_Y$$

for all $t \in]0,1]$, $x' \in x + \delta \mathbb{B}_X$, and $h' \in \delta \mathbb{B}_X$. Fix some positive real $\alpha < \delta/2$. Consider two different points $x', x'' \in x + \alpha \mathbb{B}_X$ and choose

$$t = \delta^{-1}\|x'' - x'\| \quad \text{and} \quad h' = \delta(x'' - x')/\|x'' - x'\|.$$

Then $t \leq 2\delta^{-1}\alpha < 1$, and hence

$$G(x'') - G(x') = G(x' + th') - G(x') \in t(1+\beta)\mathbb{B}_Y = \delta^{-1}(1+\beta)\|x'' - x'\|\mathbb{B}_Y,$$

which means that G is Lipschitz continuous near x.

Finally, the particular cases of the proposition follow in a direct way. □

The second property is related to the collection of all singletons of Y.

PROPOSITION 2.185. *Let $\mathcal{S}$ be the collection of all singletons of the normed space Y and U be an open set of the normed space X. Then a mapping $G : U \to Y$ is $\mathcal{S}$-Lipschitzian at x if and only if it is strictly Hadamard differentiable at x.*

In particular, any mapping which is strictly Hadamard differentiable at x is compactly Lipschitzian at x.

PROOF. Suppose that G is $\mathcal{S}$-Lipschitzian at x. Consider the mapping $A : X \to Y$ given by $\{A(h)\} = K(h)$ for all $h \in X$. Let us first prove that A is linear. Fix any $h, h' \in X$. Then by Definition 2.183 there exist $\varepsilon_i(t) \to 0$ in Y as $t \downarrow 0$ with $i = 1, 2$ such that, for all positive t small enough,

$$t^{-1}\bigl(G(x + t(h+h')) - G(x)\bigr) = A(h+h') + \varepsilon_1(t)$$

and

$$t^{-1}\bigl(G(x + t(h+h')) - G(x)\bigr)$$
$$= t^{-1}\bigl(G((x+th) + th') - G(x+th)\bigr) + t^{-1}\bigl(G(x+th) - G(x)\bigr)$$
$$= A(h') + A(h) + \varepsilon_2(t).$$

So, $A(h+h') + \varepsilon_1(t) = A(h) + A(h') + \varepsilon_2(t)$, and hence passing to the limit as $t \downarrow 0$ we get $A(h+h') = A(h) + A(h')$. Similarly, one proves that $A(\lambda h) = \lambda A(h)$ for all $\lambda \in \mathbb{R}$. Then, A is linear and condition (ii) easily ensures the continuity of the linear operator A. Further, on the one hand the proposition above says that G is

Lipschitz continuous near x, and on the other hand conditions (i) and (iii) of the definition of $\mathcal{S}$-Lipschitzian mapping imply, for each $h \in X$,

$$\lim_{t\downarrow 0, x' \to x} t^{-1}\big(G(x' + th) - G(x')\big) = A(h).$$

Consequently, the mapping G is Lipschitz near x and strictly Gâteaux differentiable at x, hence it is strictly Hadamard differentiable at x (see Proposition 2.42).

For the reverse implication, assume that G is strictly Hadamard differentiable at x, so it is γ-Lipschitz continuous on a neighborhood $W \subset U$ of x (see Proposition 2.41) and strictly Gâteaux differentiable at x. Put $A := DG(x)$ and define $\rho :]0,1] \times W \times X \to \mathbb{R}_+$ by $\rho(t, x'; h) = 0$ if $x' + th \notin W$ and

$$\rho(t, x'; h) = \|t^{-1}\big(G(x' + th) - G(x')\big) - A(h)\| \quad \text{if } x' + th \in W.$$

It is clear that $\rho(t, x'; h) \le \gamma + \|A\|$ for all $(t, x'; h) \in]0,1] \times W \times \mathbb{B}_X$. Setting $K(h) = \{A(h)\}$ it is easily seen that G satisfies the conditions of the definition of $\mathcal{S}$-Lipschitzian mapping. The proof is then complete. $\square$

The next theorem provides a sequential property which characterizes compactly Lipschitzian mappings.

THEOREM 2.186 (sequential characterization of compactly Lipschitzian mappings). *Let X, Y be normed spaces and U be an open subset of X with $x \in U$, and let a mapping $G : U \to Y$ from U into Y. The following assertions are equivalent:*
(a) *the mapping G is compactly Lipschitzian at x;*
(b) *the mapping G is Lipschitz continuous near x and, for each $h \in X$ and any sequences $(x_n)_n$ in U converging to x and $(t_n)_n$ tending to 0 with $t_n \in]0,1]$ and $x_n + t_n h \in U$, the sequence $t_n^{-1}\big(G(x_n + t_n h) - G(x_n)\big)$ has a cluster point in Y;*
(c) *the conditions (i) and (iii) in Definition 2.183 are fulfilled with a multimapping $K : X \rightrightarrows X$ taking nonempty compact values and for which there exists some real $\gamma > 0$ such that, instead of (ii) the multimapping K satisfies the condition*
(ii') $K(0) = \{0\}$ *and for all $h, h' \in X$*

$$K(h) \subset K(h') + \gamma\|h - h'\|\mathbb{B}_Y.$$

PROOF. Suppose that G is compactly Lipschitzian at x and let W, $K(\cdot)$, and $\rho(\cdot)$ as given by Definition 2.183. Proposition 2.184 says that G is Lipschitz continuous near x. Fix $h \in X$ and consider any sequences $(x_n)_n$ in U converging to x and $(t_n)_n$ tending to 0 with $t_n \in]0,1]$ and $x_n + t_n h \in U$. We may suppose without loss of generality $x_n, x_n + t_n h \in W$. According to condition (iii) of Definition 2.183, there exists $b_n \in \mathbb{B}_Y$ such that

$$t_n^{-1}[G(x_n + t_n h) - G(x_n)] - \rho(t_n, x_n; h)b_n \in K(h).$$

By compactness of $K(h)$ we can choose an increasing function $i : \mathbb{N} \to \mathbb{N}$ such that the associated subsequence of the first member of the latter inclusion converges in Y. Since $\rho(t_n, x_n; h) \to 0$ (by condition (i)), we obtain that the sequence

$$\big(t_{i(n)}^{-1}[G(x_{i(n)} + t_{i(n)}h) - G(x_{i(n)})]\big)_n$$

converges in Y. This establishes the first implication (a) $\Rightarrow$ (b) of the theorem.

Now suppose the Lipschitz continuity of G near x along with the sequential property in (b) of the theorem. For convenience we set

$$q(t, x'; h) := t^{-1}[G(x'+th) - G(x')] \quad \text{for all } (t, x', h) \in]0,1] \times U \times X \text{ with } x'+th \in U.$$

Fix $h \in X$ and put
$$K(h) := \left\{ \lim_{n \to \infty} q(t_n, x_n; h) : t_n \downarrow 0,\ U \ni x_n \to x \text{ with } x_n + t_n h \in U \right\}.$$
First we claim that $K(h)$ is compact. Let $(y_k)_k$ be any sequence in $K(h)$. Then
$$y_k = \lim_{n \to \infty} q(t_{k,n}, x_{k,n}; h)$$
for some sequence $(t_{k,n})_n$ tending to 0 with $t_{k,n} > 0$ and some sequence $(x_{k,n})_n$ of U converging to x with $x_{k,n} + t_{k,n} h \in U$, and put $y_{k,n} := q(t_{k,n}, x_{k,n}; h)$. Choose an increasing function $i : \mathbb{N} \to \mathbb{N}$ such that
$$t_{k,i(k)} + \|x_{k,i(k)} - x\| < 1/k \quad \text{and} \quad \|y_{k,i(k)} - y_k\| < 1/k.$$
Set $(t'_k, x'_k) = (t_{k,i(k)}, x_{k,i(k)})$ and $y'_k = y_{k,i(k)}$. Then $y'_k = q(t'_k, x'_k; h)$, hence, according to the sequential assumption property, there exists an increasing function $j : \mathbb{N} \to \mathbb{N}$ such that the subsequence $(y'_{j(k)})_k$ converges to some $y \in Y$, that is,
$$y = \lim_{k \to \infty} q(t'_{j(k)}, x'_{j(k)}; h),$$
which ensures that $y \in K(h)$. Since $\|y'_{j(k)} - y_{j(k)}\| < 1/j(k)$, we obtain that $(y_{j(k)})_k$ converges to y, so $K(h)$ is compact in Y. Now let us prove that conditions (i), (ii'), and (iii) are fulfilled. Fix an open neighborhood $W \subset U$ of x over which G is Lipschitz continuous with some Lipschitz constant γ. On the one hand, consider $h, h' \in X$ and take any $y \in K(h)$. Then $y = \lim_{n \to \infty} q(t_n, x_n; h)$ for some sequence $(x_n)_n$ of W converging to x and some sequence $t_n \downarrow 0$ with $x_n + t_n h \in W$ and $x_n + t_n h' \in W$. Observing, by the γ-Lipschitz property of G, that
$$q(t, x'; h) \in q(t, x'; h') + \gamma \|h - h'\| \mathbb{B}_Y$$
for all $(t, x') \in]0,1] \times W$ with $x' + th \in W$ and $x' + th' \in W$, and observing also that $(q(t_n, x_n; h'))_n$ admits a convergent subsequence (according to the sequential assumption property), we see that
$$K(h) \subset K(h') + \gamma \|h - h'\| \mathbb{B}_Y \quad \text{for all } h, h' \in X.$$
This clearly implies $K(h) \subset \gamma \|h\| \mathbb{B}_Y$ since $K(0) = \{0\}$, thus condition (ii') is obtained. On the other hand, because of the compactness of $K(h)$ we have
$$q(t, x'; h) \in K(h) + d(q(t, x'; h), K(h)) \mathbb{B}_Y \text{ for all } t \in]0,1],\ x' \in U \text{ with } x' + th \in U,$$
where $d(\cdot, K(h))$ denotes the distance function from the subset $K(h)$ of X. Let us show that
$$d(q(t, x'; h), K(h)) \xrightarrow[t \downarrow 0,\ x' \to x]{} 0.$$
If not, there exists some $\varepsilon > 0$, some sequence $(t_n)_n$ tending to 0 with $t_n \in]0,1]$, and some sequence $(x_n)_n$ of W converging to x such that $x_n + t_n h \in W$ and
$$d(q(t_n, x_n; h), K(h)) \geq \varepsilon \quad \text{for all } n \in \mathbb{N}.$$
According to the sequential assumption property, there exists some subsequence $(q(t_{i(n)}, x_{i(n)}; h))_n$ converging to some y in Y, hence $y \in K(h)$, so $d(y, K(h)) = 0$. Taking the limit in the latter inequality along the subsequence indexed by $i(n)$ gives $\varepsilon \leq d(y, K(h))$, which contradicts the above equality $d(y, K(h)) = 0$. Define $\rho :]0,1] \times W \times X \to \mathbb{R}_+$ by
$$\rho(t, x'; h) = 0 \text{ if } x' + th \notin W \text{ and } \rho(t, x'; h) := d(q(t, x'; h), K(h)) \text{ if } x' + th \in W.$$

By what precedes we have $q(t,x';h) \in K(h) + \rho(t,x';h)\mathbb{B}_Y$ for all $t \in {]}0,1]$, $x' \in W$, and $h \in X$ such that $x' + th \in W$ and $\rho(t,x';h) \to 0$ as $t \downarrow 0$ and $x' \to x$. Moreover, for $t \in {]}0,1]$, $x' \in W$, and $h \in \mathbb{B}_X$ with $x' + th \in W$, noting by γ-Lipschitz property of G on W that

$$\|q(t,x';h)\| = \|t^{-1}[G(x'+th) - G(x')]\| \le \gamma \|h\| \le \gamma,$$

we see that

$$\rho(t,x';h) = d\big(q(t,x';h), K(h)\big) \le \|q(t,x';h)\| + \sup_{y \in K(h)} \|y\| \le 2\gamma$$

since $K(h) \subset \gamma \mathbb{B}_Y$. This tells us that the function ρ is upper bounded on $]0,1] \times W \times \mathbb{B}_X$. Condition (ii') as well as conditions (i) and (iii) of Definition 2.183 are then fulfilled, so the implication (b $\Rightarrow$ (c) is proved.

The proof is then finished since the implication (c) $\Rightarrow$ (a) is obvious. $\square$

Scalarizations of compactly Lipschitzian mappings fulfill the following directional and C-subdifferential closedness properties.

PROPOSITION 2.187. *Let X, Y be normed spaces and let U be an open set of X and $G : U \to Y$ be a mapping which is compactly Lipschitzian at $x \in U$. Let $(x_j)_{j \in J}$ be a net of U converging to x and $(y_j^*)_{j \in J}$ be a bounded net in Y^* converging weakly star to y^*. Then the following hold:*
(a) *For each $h \in X$ one has $\limsup_{j \in J}(y_j^* \circ G)^\circ(x_j;h) \le (y^* \circ G)^\circ(x;h)$.*
(b) *If $(x_j^*)_{j \in J}$ is a net of X^* converging weakly star to x^* with $x_j^* \in \partial_C(y_j^* \circ G)(x_j)$, then $x^* \in \partial_C(y^* \circ G)(x)$.*
(c) *If $y^* = 0$, that is, the bounded net $(y_j^*)_{j \in J}$ converges weakly star to 0, then for any net $(x_j^*)_{j \in J}$ with $x_j^* \in \partial_C(y_j^* \circ G)(x_j)$ one has $x_j^* \to 0$ weakly star in X^*.*

PROOF. Choose some real $\gamma > 0$ such that $\|y^*\| \le \gamma$, $\|y_j^*\| \le \gamma$ for all $j \in J$, and some positive real $\delta_0 < 1$ such that G is Lipschitz continuous on $B(x,\delta_0)$. Let W, $K(\cdot)$ and $\rho(\cdot)$ as given by Definition 2.183. Fix $h \in X$ and take any $\varepsilon > 0$. Choose some $v_1, \cdots, v_m \in Y$ such that $K(h) \subset \{v_1, \cdots, v_m\} + \frac{\varepsilon}{2\gamma}\mathbb{B}_Y$. Choose also some positive real $\delta < \delta_0$ such that $x + \delta\mathbb{B}_X + [0,\delta]h \subset W$ and such that for all $t \in {]}0,\delta]$ and $x' \in B(x,\delta)$

$$t^{-1}\big(y^* \circ G(x'+th) - y^* \circ G(x')\big) \le (y^* \circ G)^\circ(x;h) + \varepsilon \text{ and } \rho(t,x';h) \le \varepsilon/(2\gamma).$$

Put $e_j^* = y_j^* - y^*$ and choose some $j_0 \in J$ such that for all $j \succeq j_0$ we have $x_j \in B(x,\delta)$ and $|e_j^*(v_k)| \le \varepsilon$ for all $k = 1, \cdots, m$. Then for all $j \succeq j_0$, $t \in {]}0,\delta]$ and $x' \in B(x,\delta)$ we deduce through the inclusion

$$t^{-1}\big(G(x'+th) - G(x')\big) \in K(h) + \rho(t,x';h)\mathbb{B}_Y$$

that $t^{-1}\big(e_j^* \circ G(x'+th) - e_j^* \circ G(x')\big) \le 3\varepsilon$, hence

$$t^{-1}\big(y_j^* \circ G(x'+th) - y_j^* \circ G(x')\big) \le (y^* \circ G)^\circ(x;h) + 4\varepsilon,$$

and the latter entails $(y_j^* \circ G)^\circ(x_j;h) \le (y^* \circ G)^\circ(x;h) + 4\varepsilon$. This establishes the inequality of (a).

(b) Concerning the assertion (b), writing $\langle x_j^*, h \rangle \le (y_j^* \circ G)^\circ(x_j;h)$ and using (a) we see that $\langle x^*, h \rangle \le (y^* \circ G)^\circ(x;h)$, thus $x^* \in \partial_C(y^* \circ G)(x)$.

(c) If $y_j^* \to 0$ weakly*, then the inequality $\langle x_j^*, h \rangle \leq (y_j^* \circ G)^\circ(x_j; h)$ and (a) give that $\limsup_{j \in J} \langle x_j^*, h \rangle \leq 0$. Combining this with the inequality obtained with $-h$ in place of h, we arrive at

$$0 \leq \liminf_{j \in J} \langle x_j^*, h \rangle \leq \limsup_{j \in J} \langle x_j^*, h \rangle \leq 0,$$

so $x_j^* \to 0$ weakly*. □

The latter proposition is used in the theorem of chain rule for C-subdifferential when the inner mapping is compactly Lipschitzian.

THEOREM 2.188 (chain rule for C-subdifferential). *Let X, Y be normed spaces and U be an open set of X. Let $G : U \to Y$ be a mapping which is compactly Lipschitzian at $x \in U$ and $g : Y \to \mathbb{R} \cup \{-\infty, +\infty\}$ be a function which is finite and Lipschitz continuous near $G(x)$. Then $g \circ G$ is Lipschitz continuous near x and*

$$(g \circ G)^\circ(x; h) \leq \max\{(y^* \circ G)^\circ(x; h) : y^* \in \partial_C g(G(x))\} \quad \forall h \in X,$$

$$\partial_C(g \circ G)(x) \subset \overline{\text{co}}^* \left(\bigcup_{y^* \in \partial_C g(G(x))} \partial_C(y^* \circ G)(x) \right).$$

PROOF. Only the inequality needs to be proved since the inclusion is a direct consequence of this inequality. Let $U_0 \subset U$ be an open convex neighborhood of x over which G is γ_G-Lipschitz continuous and such that g is γ_g-Lipschitz continuous on an open convex set V containing $G(U_0)$. Fix any $h \in X$ and choose two sequences $(x_n)_n$ converging to x and $(t_n)_n$ tending to 0 with $t_n > 0$ such that

$$(2.63) \qquad (g \circ G)^\circ(x; h) = \lim_{n \to \infty} t_n^{-1}\big(g(G(x_n + t_n h)) - g(G(x_n))\big).$$

Without loss of generality me may suppose that $x_n, x_n + t_n h \in U_0$ for all n. For each n, the mean value theorem above yields some $c_n \in [G(x_n), G(x_n + t_n h)]$ and $c_n^* \in \partial_C g(c_n)$ such that (see Theorem 2.180)

$$t_n^{-1}\big(g(G(x_n + t_n h)) - g(G(x_n))\big) = t_n^{-1} \langle c_n^*, G(x_n + t_n h) - G(x_n) \rangle,$$

and since $\|c_n^*\| \leq \gamma_g$ there is some w^*-cluster point y_h^* of $(c_n^*)_n$, and the closedness property of $\partial_C g$ at x (see Proposition 2.74) ensures that $y_h^* \in \partial_C g(G(x))$. A second application of the mean value theorem with the function $c_n^* \circ G$ gives some $z_n \in [x_n, x_n + t_n h]$ such that $t_n^{-1}\big(c_n^* \circ G(x_n + t_n h) - c_n^* \circ G(x_n)\big) \leq (c_n^* \circ G)^\circ(z_n; h)$, thus

$$t_n^{-1}\big(g(G(x_n + t_n h)) - g(G(x_n))\big) \leq (c_n^* \circ G)^\circ(z_n; h).$$

So, by (2.63) and (a) of Proposition 2.187 we have $(g \circ G)^\circ(x; h) \leq (y_h^* \circ G)^\circ(x; h)$. This ensures that $(g \circ G)^\circ(x; h)$ is not greater than the supremum of the set in the right-hand side of the first inequality of the proposition, and it is easily seen by (b) of Proposition 2.187 that this supremum is in fact a maximum. The proof is complete. □

The composition of convex mappings requires weaker assumptions than in Theorem 2.188. We give below such a statement. Others will be established later in Section 3.11 and in Section 7.8.

Let K be a convex cone of a normed space Y with $0 \in K$ and $G : X \to Y \cup \{+\infty_Y\}$ be a mapping from a normed space X into Y which is K-convex (see

(2.60)). Let $g : Y \to \mathbb{R} \cup \{+\infty\}$ be a convex function which is nondecreasing on $G(\operatorname{dom} G) + K$. We set by convention
$$g(G(x)) = +\infty \quad \text{when } G(x) = +\infty_K.$$
For $x_1, x_2 \in G^{-1}(\operatorname{dom} g)$ and $t_1, t_2 > 0$ with $t_1 + t_2 = 1$, writing by the K-convexity of G
$$t_1 G(x_1) + t_2 G(x_2) \in G(t_1 x_1 + t_2 x_2) + K,$$
it results from the nondecreasing property of the function g on $G(\operatorname{dom} G) + K$ and then from its convexity
$$g(G(t_1 x_1 + t_2 x_2)) \leq g(t_1 G(x_1) + t_2 G(x_2)) \leq t_1 g(G(x_1)) + t_2 g(G(x_2)),$$
which means that the function $g \circ G$ is convex on X.

If g is nondecreasing on both $G(\operatorname{dom} G) + K$ and $G(\operatorname{dom} G) - K$, then for any $x \in G^{-1}(\operatorname{dom} g)$ and $y^* \in \partial g(G(x))$ we have
$$-y^* \in K^\circ := \{v^* \in Y^* : \langle v^*, y \rangle \leq 0 \ \forall y \in K\}.$$
Indeed, taking $y^* \in \partial g(G(x))$ we can write for every $y \in K$
$$\langle y^*, -y \rangle \leq g(G(x) - y) - g(G(x)) \leq 0,$$
and this clearly gives $-y^* \in K^\circ$.

THEOREM 2.189. *Let X and Y be normed spaces, K be a convex cone of Y containing zero, $G : X \to Y \cup \{+\infty_Y\}$ be a K-convex mapping, and $g : Y \to \mathbb{R} \cup \{+\infty\}$ be an extended real-valued convex function which is nondecreasing on $G(\operatorname{dom} G) + K$. Then the function $g \circ G$ is convex and under either assumption (a) or (b) below one has for all $x \in X$*
$$\partial(g \circ G)(x) = \bigcup_{y^* \in (-K^\circ) \cap \partial g(G(x))} \partial(y^* \circ G)(x),$$
and if g is nondecreasing on both $G(\operatorname{dom} G) + K$ and $G(\operatorname{dom} G) - K$, the equality remains true with the inclusion $y^* \in (-K^\circ) \cap \partial g(G(x))$ replaced by only $y^* \in \partial g(G(x))$.

(a) For some $\bar{x} \in \operatorname{dom} G$ the function g is finite at $G(\bar{x})$ and continuous at this point;
(b) there exist some $\bar{x} \in X$ and $\bar{y} \in \operatorname{dom} g$ and some neighborhoods U of $\bar{x}$ and V of $\bar{y}$ such that $V \subset G(x) + K$ for all $x \in U$.

PROOF. Fix any $x \in G^{-1}(\operatorname{dom} g)$ and denote by ψ the indicator function of $\operatorname{epi}_K G := \{(x, y) \in X \times Y : G(x) \leq_K y\}$ and by $\widehat{g} : X \times Y \to \mathbb{R} \cup \{+\infty\}$ the function defined by $\widehat{g}(x', y') = g(y')$ for all $(x', y') \in X \times Y$. We observe that $x^* \in \partial(g \circ G)(x)$ means that
$$\langle x^*, x' - x \rangle \leq g(G(x')) - g(G(x)), \quad \forall x' \in \operatorname{dom} G,$$
which is equivalent, according to the nondecreasing property of g on $G(\operatorname{dom} G) + K$, to the inequality
$$\langle x^*, x' - x \rangle \leq \widehat{g}(x', y') + \psi(x', y') - \widehat{g}(x, G(x)) - \psi(x, G(x)), \quad \forall (x', y') \in X \times Y,$$
otherwise stated $(x^*, 0) \in \partial(\widehat{g} + \psi)(x, G(x)) = \partial \widehat{g}(x, G(x)) + \partial \psi(x, G(x))$, where the latter equality is due to the Moreau-Rockafellar theorem (see Theorem 2.105) under either assumption (a) or (b). This is then equivalent to the existence of some $y^* \in \partial g(G(x))$ such that $(x^*, -y^*) \in \partial \psi(x, G(x))$. Since the inequality $\langle x^*, x' - x \rangle \leq$

$\langle y^*, y' - G(x) \rangle$ is fulfilled for all $x' \in \text{dom} \, G$ and $y' \in G(x') + K$ if and only if $y^* \in -K^\circ$ and $x^* \in \partial(y^* \circ G)(x)$, the first equality in the theorem holds true.

If in addition g is also nondecreasing on $G(\text{dom} \, G) - K$, we know by the analysis preceding the theorem that $\partial g(G(x)) \subset -K^\circ$. The proof is then complete. $\square$

2.4.2. C-subdifferential of supremum of finitely many functions, extension of Lipschitz mappings, tangent cones of sublevel sets. We turn now to providing estimates for the C-subdifferential of pointwise maximum of locally Lipschitz continuous functions, the case of minimum being deduced through the opposite since $\partial_C(-g)(x) = -\partial_C g(x)$ for any locally Lipschitz continuous function g. We start first with pointwise maximum of finitely many functions.

PROPOSITION 2.190. *Let U be an open convex set of the normed space X and $(f_i)_{i \in I}$ be a finite family of locally Lipschitz continuous real-valued functions defined on U, and let $f(u) = \max_{i \in I} f_i(u)$ for all $u \in U$. For any $x \in U$ and for $I(x) := \{i \in I : f_i(x) = f(x)\}$, one has*

$$f^\circ(x; h) \leq \max_{i \in I(x)} f_i^\circ(x; h) \text{ and } f^B(x; h) \geq \max_{i \in I(x)} f_i^B(x; h),$$

for all $h \in X$, and

$$\partial_C f(x) \subset \text{co} \left(\bigcup_{i \in I(x)} \partial_C f_i(x) \right),$$

otherwise stated, for each $x^* \in \partial_C f(x)$ there exist $(x_i^*)_{i \in I}$ with $x_i^* \in \partial_C f_i(x)$ and nonnegative reals $(\lambda_i)_{i \in I}$ with $\sum_{i \in I} \lambda_i = 1$ such that

$$x^* = \sum_{i \in I} \lambda_i x_i^* \text{ and } \lambda_i = 0 \text{ if } f_i(x) < f(x).$$

Further, the inequalities and inclusion are equalities and f is tangentially regular at x whenever in addition the functions $(f_i)_{i \in I(x)}$ are tangentially regular at the point x.

PROOF. Fix $h \in X$. Choose a sequence $(t_n)_n$ tending to 0 with $t_n > 0$ and a sequence $(x_n)_n$ in U converging to x with $x_n + t_n h \in U$ and such that

$$f^\circ(x; h) = \lim_{n \to \infty} t_n^{-1} \big(f(x_n + t_n h) - f(x_n) \big).$$

Since I is finite, there exists some $k \in I$ and an infinite subset $N \subset \mathbb{N}$ (both depending on h) such that $f_k(x_n + t_n h) = f(x_n + t_n h)$ for all $n \in N$. The continuity of f and f_k at x ensures that $f_k(x) = f(x)$, that is, $k \in I(x)$. Further, the above equality concerning $f^\circ(x; h)$ implies

$$f^\circ(x; h) \leq \limsup_{N \ni n \to \infty} t_n^{-1} \big(f_k(x_n + t_n h) - f_k(x_n) \big) \leq f_k^\circ(x; h),$$

and this yields the inequality of the proposition for $f^\circ(x; h)$. The required inclusion follows from the latter inequality.

Now take any $i \in I(x)$ and note that

$$f^B(x; h) = \liminf_{t \downarrow 0} t^{-1} \big(f(x + th) - f(x) \big) \geq \liminf_{t \downarrow 0} t^{-1} \big(f_i(x + th) - f_i(x) \big) = f_i^B(x; h),$$

so $f^B(x; h) \geq \max_{i \in I(x)} f_i^B(x; h)$.

Finally, the directional regularity of f and the equalities concerning the directional derivatives and the Clarke subdifferential follow. □

EXAMPLE 2.191. (a) The result in Example 2.136 can also be obtained through the above formula. Indeed, given the function $f : U \to \mathbb{R}$ strictly Hadamard differentiable at $\overline{x}$ in the open set U with $f(\overline{x}) = 0$, applying the above proposition with $f_1 := -f$ and $f_2 := f$, we get

$$\partial_C(|f|)(\overline{x}) = [-Df(\overline{x}), Df(\overline{x})] = \{r\,Df(\overline{x}) : |r| \leq 1\}.$$

(b) Let us come back to Example 2.84(b), but here in infinite dimensions, that is, we consider three C^1 functions $f_1, f_2, f_3 : U \to \mathbb{R}$ defined on a open set U of a normed space X (instead of $\mathbb{R}^m$) and $f : U \to \mathbb{R}$ given by

$$f(x) := \max\{\min\{f_1(x), f_2(x)\}, f_3(x)\} \quad \text{for all } x \in U.$$

Consider also U_1, U_2, U_3 as defined in that example, that is,

$$U_1 := \{x \in U : f_2(x) > f_1(x) > f_3(x)\}, \; U_2(x) := \{x \in U : f_1(x) > f_2(x) > f_3(x)\},$$

$$U_3; = \{x \in U : f_3(x) > \min\{f_1(x), f_2(x)\}\},$$

and assume merely that $\overline{x} \in \bigcap_{i=1}^{3} \operatorname{cl} U_i$. We claim that

$$\partial_C f(\overline{x}) = \operatorname{co}\{Df_1(\overline{x}), Df_2(\overline{x}), Df_3(\overline{x})\}.$$

Putting $g(x) := \min\{f_1(x), f_2(x)\}$ and noting that $f(\overline{x}) = g(\overline{x})$ the above proposition tells us that $\partial_C f(\overline{x}) \subset \operatorname{co}(\{Df_3(\overline{x})\} \cup \partial_C g(\overline{x}))$. In the same way $\partial_C(-g(\overline{x})) \subset \operatorname{co}\{-Df_2(\overline{x}), -Df_1(\overline{x})\}$, or equivalently $\partial_C g(\overline{x}) \subset \operatorname{co}\{Df_2(\overline{x}), Df_1(\overline{x})\}$. It ensues that $\partial_C f(\overline{x}) \subset \operatorname{co}\{Df_1(\overline{x}), Df_2(\overline{x}), Df_3(\overline{x})\}$. On the other hand, keeping in mind that $\overline{x} \in \operatorname{cl} U_i$) we see for each $i \in \{1, 2, 3\}$ that

$$Df_i(\overline{x}) = \lim_{U_i \ni x \to \overline{x}} Df_1(x) = \lim_{U_i \ni x \to \overline{x}} Df(x) \in \partial_C f(\overline{x}),$$

It results that $\partial_C f(\overline{x}) = \operatorname{co}\{Df_1(\overline{x}), Df_2(\overline{x}), Df_3(\overline{x})\}$, and the claim is justified.
(c) Now given $\alpha, \beta \in \mathbb{R}$ consider the Lipschitz function $f : \mathbb{R}^2 \to \mathbb{R}$ defined by $f(x, y) := |\alpha|x| + \beta y|$ for all $(x, y) \in \mathbb{R}^2$. Proposition 2.190 tells us that $\partial_C f(0, 0) = \operatorname{co}(\partial_C g(0, 0), -\partial_C g(0, 0))$ where $g(x, y) := \alpha|x| + \beta y$. Since

$$\partial_C g(0, 0) = (\alpha[-1, 1]) \times \{\beta\} = [-|\alpha|, |\alpha|] \times \{\beta\},$$

it ensues that

$$\partial_C f(0, 0) \subset \operatorname{co}([-|\alpha|, |\alpha|] \times \{-\beta, \beta\}) = [-|\alpha|, |\alpha|] \times [-|\beta|, |\beta|].$$

On the other hand, each of points $\pm(\alpha, \beta), \pm(-\alpha, \beta)$, or equivalently each of points $\pm(|\alpha|, |\beta|), \pm(-|\alpha|, |\beta|)$, belongs to $\partial_C f(0, 0)$ as a limit of $\nabla f(x_n, y_n)$ as $n \to \infty$, for suitable (x_n, y_n) converging to $(0, 0)$. All together ensure that $\partial_C f(0, 0) = [-|\alpha|, |\alpha|] \times [-|\beta|, |\beta|]$. □

Extensions of real-valued Lipschitz functions defined on a subset of a metric space to the whole space have been provided in Proposition 2.79 with preservation of the Lipschitz constant. Using the max rule formula in Proposition 2.190 above we establish next an extension result for Lipschitz mappings with values in Hilbert spaces.

THEOREM 2.192 (extension of Lipschitz mappings between subsets of Hilbert spaces). *Assume that X and Y are two Hilbert spaces. Then any Lipschitz mapping from a subset S of X into Y can be extended to a Lipschitz mapping on the whole space X with the same Lipschitz constant.*

LEMMA 2.193. *Let $(r_i)_{i \in I}$ be a family of positive real numbers, and $(x_i)_{i \in I}$ and $(y_i)_{i \in I}$ be families of elements in Hilbert spaces X and Y respectively with the same index set I such that*

$$\|y_i - y_j\| \leq \|x_i - x_j\| \quad \text{for all } i, j \in I.$$

If $\bigcap_{i \in I} B[x_i, r_i] \neq \emptyset$, then $\bigcap_{i \in I} B[y_i, r_i] \neq \emptyset$.

PROOF. Taking the weak compactness of closed balls in the Hilbert space Y into account, we can suppose that the index set I is finite. If $x_{i_0} \in \bigcap_{i \in I} B[x_i, r_i]$ for some $i_0 \in I$, we see that $y_{i_0} \in \bigcap_{i \in I} B[y_i, r_i]$. So, suppose that there is some $a \in \bigcap_{i \in I} B[x_i, r_i]$ different from all x_i, $i \in I$. Putting

$$f(y) := \max_{i \in I} f_i(y), \text{ where } f_i(y) := \|y - y_i\|^2 / \|a - x_i\|^2 \quad \text{for all } y \in Y,$$

we see that the convex function f is weakly lower semicontinuous with $f(y) \to +\infty$ as $\|y\| \to +\infty$. This ensures that f has a global minimizer $b \in Y$, and hence $0 \in \partial f(b)$. We shall show that $f(b) \leq 1$, and thus $b \in \bigcap_{i \in I} B[y_i, r_i]$.

Setting $I_0 := \{i \in I : f_i(b) = f(b)\}$, Proposition 2.190 tells us that there are reals $t_i \geq 0$ with $\sum_{i \in I_0} t_i = 1$ such that $0 = 2 \sum_{i \in I_0} t_i (b - y_i)/(\|a - x_i\|^2)$. Putting $\Lambda := \sum_{i \in I_0} t_i/(\|a - x_i\|^2)$ and $\lambda_i := t_i/(\|a - x_i\|^2 \Lambda)$, it is clear that $b = \sum_{i \in I_0} \lambda_i y_i$ and $\sum_{i \in I_0} \lambda_i = 1$.

On the other hand, putting $\mu := f(b)$ we have, for any $i, j \in I_0$,

$$\|x_i - x_j\|^2 \geq \|y_i - y_j\|^2 = \|y_i - b\|^2 + \|y_j - b\|^2 - 2\langle y_i - b, y_j - b\rangle$$
$$= \mu\|x_i - a\|^2 + \mu\|x_j - a\|^2 - 2\langle y_i - b, y_j - b\rangle.$$

Summing after multiplying by $\lambda_i \lambda_j$ and noting that

$$\sum_{i,j \in I_0} \lambda_i \lambda_j \langle y_i - b, y_j - b\rangle = \sum_{i \in I_0} \lambda_i \langle y_i - b, b - b\rangle = 0,$$

it results that

$$0 \geq \sum_{i,j \in I_0} \lambda_i \lambda_j (\mu\|x_i - a\|^2 + \mu\|x_j - a\|^2 - \|x_i - x_j\|^2)$$
$$= \sum_{i,j \in I_0} \lambda_i \lambda_j (\|x_i - a\|^2 + \|x_j - a\|^2 - \|x_i - x_j\|^2)$$
$$+ (\mu - 1) \sum_{i,j \in I_0} \lambda_i \lambda_j (\|x_i - a\|^2 + \|x_j - a\|^2).$$

For $c := \sum_{i \in I_0} x_i$, we have

$$\sum_{i,j \in I_0} \lambda_i \lambda_j (\|x_i - a\|^2 + \|x_j - a\|^2 - \|x_i - x_j\|^2) = -2 \sum_{i,j \in I_0} \lambda_i \lambda_j \langle x_i - a, a - x_j \rangle$$
$$= -2\langle c - a, a - c \rangle = 2\|a - c\|^2 \geq 0,$$

which combined with what precedes ensures that $\mu - 1 \leq 0$, that is, $\mu \leq 1$ as desired. $\square$

PROOF OF THEOREM 2.192. Let $G : S \to Y$ be a Lipschitz mapping with constant γ. We may suppose $\gamma > 0$ since with $\gamma = 0$ the extension is obvious because G is null on S. Considering $\gamma^{-1} G$ in place of G, we may also suppose that $\gamma = 1$ is a Lipschitz constant for G. Fixing any $u \in X \setminus S$, we see that $\|G(x) - G(x')\| \leq \|x - x'\|$ for all $x, x' \in S$ and that $\bigcap_{x \in S} B[x, \|u - x\|] \neq \emptyset$ since u obviously belongs to this intersection. Then the above lemma guarantees that $\bigcap_{x \in S} B[G(x), \|u - x\|] \neq \emptyset$, so choosing b_u in the latter intersection and putting $G_u(u) := b_u$ and $G_u(x) := G(x)$ for all $x \in S$ it is clear that G_u is a Lipschitz extension of G to $S \cup \{u\}$ with $\gamma = 1$ as Lipschitz constant. By Zorn's lemma we deduce the desired Lipschitz extension to the whole Hilbert space X. $\square$

The subdifferential of maximum of finitely many functions in Proposition 2.190 will also be applied to estimate the Clarke normal cone of intersection of finitely many sublevel sets in Proposition 2.196. For this purpose, we need first, through Proposition 2.79, to estimate tangent cones to sublevel sets. More general estimates for tangent cones of such sets will be provided later when the theory of metric regularity will be developed in Chapter 7. One of the estimations below involves the directional continuity which amounts to saying that, given an open set U of X, a mapping $G : U \to Y$ is *directionally continuous* at a point $\overline{x} \in U$ provided for each fixed $h \in X$ the one-variable mapping $t \mapsto G(\overline{x} + th)$ is continuous at zero.

PROPOSITION 2.194. Let $(f_i)_{i \in I}$ be a finite family of functions from the normed space X into $\mathbb{R} \cup \{+\infty\}$. Let $S := \{x \in X : f_i(x) \leq 0 \ \forall i \in I\}$ and $\overline{x} \in S$ be such that $I(\overline{x}) = \{i \in I : f_i(\overline{x}) = 0\}$ is nonempty.
(a) One has
$$T^B(S; \overline{x}) \subset \{h \in X : f_i^B(\overline{x}; h) \leq 0 \ \forall i \in I(\overline{x})\}.$$

If the functions f_i are directionally continuous at $\overline{x}$ for all $i \in I \setminus I(\overline{x})$ and Hadamard differentiable at $\overline{x}$ for all $i \in I(\overline{x})$, then
$$T^B(S; \overline{x}) = \{h \in X : \langle Df_i(\overline{x}), h \rangle \leq 0 \ \forall i \in I(\overline{x})\}$$
whenever the so-called "Slater qualification condition under differentiability" is fulfilled:

there is some $\overline{h} \in X$ such that $\langle Df_i(\overline{x}), \overline{h} \rangle < 0$ for all $i \in I(\overline{x})$.

(b) If the functions f_i are continuous at $\overline{x}$ for all $i \in I \setminus I(\overline{x})$ and Lipschitz continuous near $\overline{x}$ for all $i \in I(\overline{x})$, then
$$\{h \in X : f_i^\circ(\overline{x}; h) < 0 \ \forall i \in I(\overline{x})\} \subset I(S; \overline{x}),$$
and further
$$\{h \in X : f_i^\circ(\overline{x}; h) \leq 0 \ \forall i \in I(\overline{x})\} \subset T^C(S; \overline{x})$$

whenever in addition the following "generalized Slater qualification condition" is fulfilled:

there is some $\overline{h} \in X$ such that $f_i^\circ(\overline{x}; \overline{h}) < 0$ for all $i \in I(\overline{x})$.

So, if the functions f_i, with $i \notin I(\overline{x})$ are continuous at $\overline{x}$ and if the other functions f_i, with $i \in I(\overline{x})$ are tangentially regular at $\overline{x}$ and Lipschitz continuous near $\overline{x}$, then under the above "generalized Slater condition" one has
$$T^C(S; \overline{x}) = T^B(S; \overline{x}) = \{h \in X : f_i^\circ(\overline{x}; h) \leq 0 \ \forall i \in I(\overline{x})\}.$$
(c) If for each $i \in I \setminus I(\overline{x})$ the function f_i is continuous at $\overline{x}$ and for each $i \in I(\overline{x})$ the function f_i is Lipschitz continuous near $\overline{x}$ and strictly Gâteaux differentiable at $\overline{x}$, and if the Slater qualification condition in (a) above holds, then
$$T^C(S; \overline{x}) = T^B(S; \overline{x}) = \{h \in X : \langle Df_i(\overline{x}), h \rangle \leq 0 \ \forall i \in I(\overline{x})\}.$$

PROOF. (a) Fix any $h \in T^B(S; \overline{x})$. Choose a sequence $(t_n)_n$ of positive numbers tending to 0 and a sequence $(h_n)_n$ converging to h and such that $\overline{x} + t_n h_n \in S$ for all n. Then for each $i \in I(\overline{x})$ we have
$$t_n^{-1}(f_i(\overline{x} + t_n h_n) - f_i(\overline{x})) = t_n^{-1} f_i(\overline{x} + t_n h_n) \leq 0 \quad \text{hence } f_i^B(\overline{x}; h) \leq 0,$$
which ensures the inclusion in (a).

Suppose that the functions f_i are directionally continuous at $\overline{x}$ for all $i \in I \setminus I(\overline{x})$ and Hadamard differentiable at $\overline{x}$ for all $i \in I(\overline{x})$, and suppose also the Slater condition in (a) holds. Let $h \in X$ such that $\langle Df_i(\overline{x}), h \rangle \leq 0$ for all $i \in I(\overline{x})$. For each $i \in I(\overline{x})$ and each $r > 0$ we have $\langle Df_i(\overline{x}), h + r\overline{h} \rangle < 0$, then there exists some $\varepsilon > 0$ such that for all positive reals $t < \varepsilon$ the inequality $t^{-1}(f_i(\overline{x} + t(h + r\overline{h})) - f_i(\overline{x})) < 0$ holds, that is, $f_i(\overline{x} + t(h + r\overline{h})) < 0$. For every positive real t small enough this inequality still holds for all $i \in I \setminus I(\overline{x})$ because of the continuity of $t \mapsto f_i(\overline{x} + t(h + r\overline{h}))$ at 0. In consequence, $\overline{x} + t(h + r\overline{h}) \in S$ for every positive real t small enough, thus $h + r\overline{h} \in T^B(S; \overline{x})$. The closedness of $T^B(S; \overline{x})$ entails $h \in T^B(S; \overline{x})$. So, the equality in (a) is proved.

(b) Consider any $h \in X$ with $f_i^\circ(\overline{x}; h) < 0$ for all $i \in I(\overline{x})$. Take any sequence $(x_n)_n$ in S converging to $\overline{x}$, any sequence of positive reals $(t_n)_n$ tending to 0, and any sequence $(h_n)_n$ converging to h. For each $i \in I(\overline{x})$ we then have $\limsup_n t_n^{-1}(f_i(x_n + t_n h_n) - f_i(x_n)) < 0$, hence for all n large enough $t_n^{-1}(f_i(x_n + t_n h_n) - f_i(x_n)) < 0$, which entails $f_i(x_n + t_n h_n) < f_i(x_n) \leq 0$. For each $i \notin I(\overline{x})$ by continuity of f_i at $\overline{x}$ the inequality $f_i(x_n + t_n h_n) < 0$ still holds for n large enough. Therefore, $x_n + t_n h_n \in S$ for large n, and this says that $h \in I(S; \overline{x})$, proving the first inclusion of (b).

Suppose the generalized Slater condition holds. Taking any $h \in X$ with $f_i^\circ(\overline{x}; h) \leq 0$ for all $i \in I(\overline{x})$, then for every $r > 0$ we have $f_i^\circ(\overline{x}; h + r\overline{h}) < 0$ for all $i \in I(\overline{x})$ (since $f_i^\circ(\overline{x}; h + r\overline{h}) \leq f_i^\circ(\overline{x}; h) + r f_i^\circ(\overline{x}; \overline{h})$), thus by what precedes $h + r\overline{h} \in I(S; \overline{x}) \subset T^C(S; \overline{x})$. This implies $h \in T^C(S; \overline{x})$ according to the closedness property of $T^C(S; \overline{x})$, hence the second inclusion in (b) is established.

If in addition the functions f_i, with $i \in I(\overline{x})$, are all tangentially regular at $\overline{x}$, or equivalently $f_i^\circ(\overline{x}; \cdot) = f_i^B(\overline{x}; \cdot)$ for $i \in I(\overline{x})$, then the latter inclusion combined with the first inclusion in (a) gives the desired equality in (b).

(c) Since $f_i^\circ(\overline{x}; \cdot) = \langle Df_i(\overline{x}), \cdot \rangle$, for each $i \in I(\overline{x})$, by the strict Gâteaux differentiability property of the function f_i at $\overline{x}$ along with its Lipschitz property near $\overline{x}$, the assertion (c) follows from the equality in (b). □

REMARK 2.195. (a) When the sublevel set of only one function f Lipschitz near $\overline{x}$ is considered, the "generalized Slater qualification condition" in (b) of the above proposition is clearly equivalent to $0 \notin \partial_C f(\overline{x})$.

(b) If all the functions f_i with $i \in I$ are convex and continuous at $\overline{x}$, then the "generalized Slater qualification condition" at $\overline{x}$ is equivalent to the following "Slater qualification condition for convex functions continuous at $\overline{x}$": There exists some $x_0 \in \bigcap_{i \in I} \text{int}(\text{dom } f_i)$ such that $f_i(x_0) < 0$ for all $i \in I$. Indeed, suppose first the "generalized Slater qualification condition" holds at $\overline{x}$ with some vector $\overline{h} \in X$. There exists some real $\overline{t} > 0$ such that for each $i \in I(\overline{x})$ we have $f_i(\overline{x} + t\overline{h}) < 0$ and $\overline{x} + t\overline{h} \in \text{int}(\text{dom } f_i)$ for all $t \in]0, \overline{t}[$. Further, the continuity at $\overline{x}$ of each function f_i with $i \in I \setminus I(\overline{x})$ combined with the inequality $f_i(\overline{x}) < 0$ gives some $t_0 \in]0, \overline{t}[$ such that for any $i \in I \setminus I(\overline{x})$ we have $f_i(\overline{x} + t_0\overline{h}) < 0$ and $\overline{x} + t_0\overline{h} \in \text{int}(\text{dom } f_i)$. The point $x_0 := \overline{x} + t_0\overline{h}$ then fulfills the desired condition.

Conversely, suppose the existence of x_0. For each $i \in I(\overline{x})$ and $t \in]0, 1]$ we have by convexity of f_i

$$t^{-1}[f_i(\overline{x} + t(x_0 - \overline{x})) - f_i(\overline{x})] \leq f(x_0)$$

which yields by the convexity and continuity of f_i at $\overline{x}$ and by Proposition 2.124 and Proposition 2.126(b)

$$f_i^\circ(\overline{x}; x_0 - \overline{x}) = f_i'(\overline{x}; x_0 - \overline{x}) \leq f(x_0) < 0.$$

This translates the "generalized Slater qualification condition". □

We can now state and prove the proposition concerning the Clarke normal cone of intersection of finitely many sublevel sets.

PROPOSITION 2.196. Let $(f_i)_{i \in I}$ be a finite family of functions from the normed space X into $\mathbb{R} \cup \{+\infty\}$. Let $S := \{x \in X : f_i(x) \leq 0 \; \forall i \in I\}$ and $\overline{x} \in S$ be such that $I(\overline{x}) := \{i \in I : f_i(\overline{x}) = 0\}$ is nonempty. Assume that the functions f_i are continuous at $\overline{x}$ for all $i \in I \setminus I(\overline{x})$ and Lipschitz continuous near $\overline{x}$ for all $i \in I(\overline{x})$, and assume also that the "generalized Slater qualification condition" is fulfilled:

there is some $\overline{h} \in X$ such that $f_i^\circ(\overline{x}; \overline{h}) < 0$ for all $i \in I(\overline{x})$.

Then,

$$N^C(S; \overline{x}) \subset \sum_{i \in I(\overline{x})} \mathbb{R}_+ \partial_C f_i(\overline{x}).$$

If in addition the functions f_i, with $i \in I(\overline{x})$, are tangentially regular at $\overline{x}$, then the latter inclusion is an equality.

PROOF. First consider the case of only one function f, so $S \times \{0\} = E_f \cap (X \times \{0\})$ and $f(\overline{x}) = 0$, where $E_f := \text{epi } f$. The local Lipschitz property of f and the inequality $f^\circ(\overline{x}; \overline{h}) < 0$ entail that $(\overline{h}, 0) \in I(E_f; (\overline{x}, 0))$ by Theorem 2.70(c), hence $(\overline{h}, 0) \in I(E_f; (\overline{x}, 0)) \cap T^C(X \times [0, +\infty[; (\overline{x}, 0))$. Theorem 2.13 tells us that

$$N^C(S; \overline{x}) \times \mathbb{R} \subset N^C(E_f; (\overline{x}, 0)) + \{0\} \times \mathbb{R}.$$

Fixing any nonzero $x^* \in N^C(S; \overline{x})$, we can then find $(u^*, -s) \in N^C(E_f; (\overline{x}, 0))$ with $s \geq 0$ such that $(x^*, -1) = (u^*, -s) + (0, -1 + s)$. So $u^* = x^* \neq 0$, and the local Lipschitz property of f then ensures that $s > 0$ according to Proposition 2.85(b), thus $s^{-1}u^* \in \partial_C f(\overline{x})$. Consequently, $N^C(S; \overline{x}) \subset \mathbb{R}_+ \partial f(\overline{x})$.

In the general case of $f_1, \cdots, f_m$, there exists an open set $U \ni \overline{x}$ such that $f_i(x) < 0$ for all $i \notin I(\overline{x})$ and $x \in U$. Setting $f := \max_{i \in I(\overline{x})} f_i$ and $S_0 := \{x \in X : f(x) \leq 0\}$, we see that $S \cap U = S_0 \cap U$, thus $N^C(S; \overline{x}) = N^C(S_0; \overline{x})$. The first inclusion of the proposition then results from the previous case and from Proposition 2.190.

If in addition the functions f_i, with $i \in I(\overline{x})$, are tangentially regular at $\overline{x}$, then the inclusion is an equality since the reverse inclusion is easily seen from the equality in Proposition 2.194(b). □

Through the above proposition and Theorem 2.154, the Clarke normal cone of any set with the interior tangent property can be expressed in terms of the Clarke subdifferential of the associated signed distance function, completing in this way Corollary 2.153.

COROLLARY 2.197. Let S be a subset of a normed space $(X, \|\cdot\|)$ which has the interior tangent property at $\overline{x} \in S \cap \operatorname{bdry} S$. Then, the Clarke normal cone of S at $\overline{x}$ can be described as
$$N^C(S; \overline{x}) = \mathbb{R}_+ \partial_C \operatorname{sgd}_S(\overline{x}).$$

PROOF. We recall first that $\overline{S} := \operatorname{cl} S = \{u \in X : \operatorname{sgd}_S(u) \leq 0\}$. Since $(\operatorname{sgd}_S)^\circ(\overline{x}; h) < 0$ for some $h \in X$ according to Theorem 2.154, it results from the above Proposition 2.196 that $N^C(\overline{S}; \overline{x}) \subset \mathbb{R}_+ \partial_C \operatorname{sgd}_S(\overline{x})$.

On the other hand, we know from Corollary 2.153 that
$$T^C(S; \overline{x}) = \{h \in X : \langle x^*, h \rangle \leq 0, \forall x^* \in \partial_C \operatorname{sgd}_S(\overline{x})\},$$
and from this we easily see that $\mathbb{R}_+ \partial_C \operatorname{sgd}_S(\overline{x}) \subset \left(T^C(S; \overline{x})\right)^\circ = N^C(S; \overline{x})$. Since $N^C(\overline{S}; \overline{x}) = N^C(S; \overline{x})$, the desired equality of the proposition is justified. □

2.4.3. Clarke theorem for C-subdifferential of supremum of infinitely many Lipschitz functions. This subsection studies (in complement of Proposition 2.190) the pointwise supremum of infinitely many functions.

THEOREM 2.198 (F. Clarke's theorem for C-subdifferential of suprema of infinitely many Lipschitz functions). Let T be a Hausdorff topological space, U be an open set of the normed space X and $\overline{x} \in U$. Let a function $g : U \times T \to \mathbb{R}$ and
$$f(x) := \sup_{t \in T} g(x, t) \quad \text{and} \quad T(x) := \{t \in T : g(x, t) = f(x)\}, \quad \text{for every } x \in U.$$
Assume that the following conditions (i)-(iv) are fulfilled:
(i) The functions $g_t := g(\cdot, t)$, $t \in T$, are equi-Lipschitz on some neighborhood of $\overline{x}$ and f is finite on this neighborhood;
(ii) there exists a neighborhood $U_0 \subset U$ of $\overline{x}$ and a compact set K in T such that $\bigcup_{x \in U_0} T(x) \subset K$ and $T(x) \neq \emptyset$ for all $x \in U_0$;
(iii) the function $g(\overline{x}, \cdot)$ is upper semicontinuous on T;
(iv) the restriction of the multimapping $(x, t) \mapsto \partial_C g_t(x)$ to $U_0 \times T$ has its graph which is closed in $U_0 \times T \times X^*$ relative to the weak* topology of X^*.

Then, f is Lipschitz near $\overline{x}$ and one has the inequality $f^\circ(\overline{x}; h) \leq \max_{t \in T(\overline{x})} g_t^\circ(\overline{x}; h)$ for all $h \in X$ and the inclusion
$$\partial_C f(\overline{x}) \subset \overline{\operatorname{co}}^* \left(\bigcup_{t \in T(\overline{x})} \partial_C g_t(\overline{x}) \right).$$

If in addition each function g_t, $t \in T(\overline{x})$, is tangentially regular at $\overline{x}$, then so is f and the above inequality with $f^\circ(\overline{x}; h)$ and inclusion with $\partial_C f(\overline{x})$ are equalities.

PROOF. Fix any $h \in X$. By (i) and (ii) we may suppose that there is a convex neighborhood U_0 of $\overline{x}$ satisfying (ii) and such that f is finite on U_0 and the functions g_t are γ-Lipschitz therein with some real $\gamma \geq 0$. Then f is γ-Lipschitz on U_0 too, since for all $x, x' \in U_0$, writing $g(x', t) \leq g(x, t) + \gamma \|x' - x\|$ and taking the supremum over $t \in T$ gives $f(x') \leq f(x) + \gamma \|x' - x\|$. Let $(r_n)_n$ be a sequence of positive reals tending to 0 and $(x_n)_n$ be a sequence in U_0 converging to $\overline{x}$ with $x_n + r_n h \in U_0$ and such that

$$f^\circ(\overline{x}; h) = \lim_{n \to \infty} \frac{1}{r_n} \big(f(x_n + r_n h) - f(x_n) \big).$$

For each $n \in \mathbb{N}$, choose some $t_n \in T(x_n + r_n h)$, and by the mean-value theorem (see Theorem 2.180) take some $\theta_n \in [0, 1]$ and $x_n^* \in \partial_C g_{t_n}(x_n + r_n \theta_n h)$ such that

$$\frac{1}{r_n} \big(g_{t_n}(x_n + r_n h) - g_{t_n}(x_n) \big) = \langle x_n^*, h \rangle.$$

For each $n \in \mathbb{N}$, it results that

$$\frac{1}{r_n} \big(f(x_n + r_n h) - f(x_n) \big) \leq \frac{1}{r_n} \big(g_{t_n}(x_n + r_n h) - g_{t_n}(x_n) \big) = \langle x_n^*, h \rangle.$$

From the inclusion $(t_n, x_n^*) \in K \times (\gamma \mathbb{B}_{X^*})$, take a $\tau \times w^*$ cluster point $(\overline{t}, x^*)$ of the sequence $(t_n, x_n^*)_n$, so (iv) ensures that $x^* \in \partial_C g_{\overline{t}}(\overline{x})$. Further, since $f(x_n + r_n h) = g(t_n, x_n + r_n h)$, we see that $f(x_n + r_n h) \leq g(t_n, \overline{x}) + \gamma \|x_n + r_n h - \overline{x}\|$ for all $n \in \mathbb{N}$, hence $f(\overline{x}) \leq g(\overline{t}, \overline{x})$ by (iii), or equivalently $\overline{t} \in T(\overline{x})$. Consequently,

$$f^\circ(\overline{x}; h) \leq \langle x^*, h \rangle \leq g_{\overline{t}}^\circ(\overline{x}; h) \leq \sup_{t \in T(\overline{x})} g_t^\circ(\overline{x}; h),$$

and the latter supremum is a maximum since $t \mapsto g_t^\circ(\overline{x}; h)$ is upper semicontinuous on T by (iv) and the scalarization theorem for upper semicontinuity (see Theorem 1.48), and $T(\overline{x}) \subset K$ is closed (hence compact) according to the upper semicontinuity in (iii) of $t \mapsto g_t(\overline{x})$. This justifies the inclusion and inequality of the theorem.

Now assume in addition that the functions g_t, $t \in T(\overline{x})$, are tangentially regular at $\overline{x}$. Fixing any $h \in X$ and taking any $r > 0$ with $\overline{x} + rh \in U_0$, we have for all $t \in T(\overline{x})$,

$$\frac{1}{r} \big(f(\overline{x} + rh) - f(\overline{x}) \big) \geq \frac{1}{r} \big(g_t(\overline{x} + th) - g_t(\overline{x}) \big),$$

so $f^B(\overline{x}; h) \geq \sup_{t \in T(\overline{x})} g_t^B(\overline{x}; h)$. The tangential regularity of f at $\overline{x}$ and the desired equalities concerning $f^\circ(\overline{x}; h)$ and $\partial_C f(\overline{x})$ then follow. □

REMARK 2.199. The above proof reveals that we still have

$$f^\circ(\overline{x}; h) \leq \sup_{t \in T(\overline{x})} g_t^\circ(\overline{x}; h) \quad \text{and} \quad \partial_C f(\overline{x}) \subset \overline{\mathrm{co}}^* \left(\bigcup_{t \in T(\overline{x})} \partial_C g_t(\overline{x}) \right),$$

if instead of (iii) we assume:
(iii') the multimapping $x \mapsto T(x)$ on U_0 has a multimapping selection M which is closed at $\overline{x}$.
Just take $t_n \in M(\overline{x} + r_n h)$ in the proof. The difference is that the supremum over $T(\overline{x})$ is not necessarily attained.

Further, if in addition each function $g(\cdot, t)$, with $t \in T(\overline{x})$, is tangentially regular at $\overline{x}$, then the inclusion and inequality are equalities and the supremum over $T(\overline{x})$ is attained since in this case $\overline{t} \in T(\overline{x})$ by (iii') and

$$\sup_{t \in T(\overline{x})} g_t^B(\overline{x}; h) \leq f^B(\overline{x}; h) \leq f^\circ(\overline{x}; h) \leq g_{\overline{t}}^\circ(\overline{x}; h)$$

as given by the above proof. $\square$

The subdifferential of the usual norm of the space $\ell_{\mathbb{R}}^p(\mathbb{N})$ has been described in Example 2.66(b). In the following example we express the subdifferential of the usual norm $\|\cdot\|_{c^0}$ of $c_{\mathbb{R}}^0(\mathbb{N})$ as an application of the above proposition.

EXAMPLE 2.200. Let $c_{\mathbb{R}}^0(\mathbb{N})$ be the space of sequences $x : \mathbb{N} \to \mathbb{R}$ with $\lim_n x_n = 0$. The set $T := \mathbb{N} \cup \{+\infty\}$ endowed with the topology induced by the usual topology of $\mathbb{R} \cup \{-\infty, +\infty\}$ is a compact space and we can extend each $x \in c_{\mathbb{R}}^0(\mathbb{N})$ as a continuous function on T by putting $x_\infty := 0$. For each $n \in T$, setting $f_n(x) = |x_n|$ for all $n \in T$, we see that $\|x\|_{c^0} = \max_{n \in T} f_n(x)$ and that the assumptions in Theorem 2.198 above are fulfilled. For each fixed nonzero $x \in c_{\mathbb{R}}^0(\mathbb{N})$, clearly $T(x) := \{n \in T : |x_n| = \|x\|_{c_0}\}$ is a finite subset of $\mathbb{N}$, hence the above theorem gives the equality

$$\partial \|\cdot\|_{c^0}(x) = \mathrm{co}\left\{\frac{x_n}{|x_n|}e_n^* : n \in \mathbb{N}, |x_n| = \|x\|_{c^0}\right\},$$

where e_n^* is the continuous linear functional defined on $c_{\mathbb{R}}^0(\mathbb{N})$ by $e_n^*(x) = x_n$. Identifying the topological dual of $c_{\mathbb{R}}^0(\mathbb{N})$ with $\ell_{\mathbb{R}}^1(\mathbb{N})$, for any nonzero $x \in c_{\mathbb{R}}^0(\mathbb{N})$ we can write in $\ell_{\mathbb{R}}^1(\mathbb{N})$

$$\partial \|\cdot\|_{c^0}(x) = \mathrm{co}\left\{\frac{x_n}{|x_n|}e_n : n \in \mathbb{N}, |x_n| = \|x\|_{c^0}\right\},$$

where $e_n := (0, \cdots, 0, 1, 0 \cdots)$ is the sequence whose only nonzero term is 1 in the n-th place.

Further, concerning the zero vector 0_{c^0} we have

$$\partial \|\cdot\|_{c^0}(0_{c^0}) = \mathbb{B}_{\ell^1} = \left\{\zeta \in \ell_{\mathbb{R}}^1(\mathbb{N}) : \sum_{n=1}^\infty |\zeta_n| \leq 1\right\},$$

as already seen for the subdifferential at zero of any norm of a vector space. $\square$

The following corollary of Theorem 2.198 deals with a family of continuously differentiable functions indexed by a compact set.

COROLLARY 2.201 (Danskin type formula for max-function). Let T be a compact Hausdorff topological space, U be an open set of the normed space X. Let $g : U \times T \to \mathbb{R}$ be a function which is upper semicontinuous in its second variable and let $f(x) := \sup_{t \in T} g(x, t)$ for all $x \in U$. Assume that g is differentiable on U in its first variable and that $D_1 g(\cdot, \cdot)$ is continuous on $U \times T$. Then the function f is locally Lipschitz continuous on U and for each $x \in U$

$$f'(x; h) = f^\circ(x; h) = \max_{t \in T(x)} \langle Dg_t(x), h \rangle \quad \text{for all } h \in X,$$

$$\partial_C f(x) = \overline{\mathrm{co}}^* \{Dg_t(x) : t \in T(x)\},$$

where $T(x) := \{t \in T : g(x, t) = f(x)\}$.

PROOF. Since f is clearly finite valued, by Theorem 2.198 it suffices to show that the family $(g_t)_{t \in T}$ is locally equi-Lipschitz. Fix any $\bar{x} \in U$. For each $\tau \in T$ choose a real $\beta_\tau \geq 0$, an open convex neighborhood U_τ of $\bar{x}$ in U and an open neighborhood V_τ of τ in T such that $\|D_1 g(x,t)\| \leq \beta_\tau$ for all $t \in V_\tau$ and $x \in U_\tau$. Let $\tau_1, \cdots, \tau_n$ be such that $T = V_{\tau_1} \cup \cdots \cup V_{\tau_n}$ and put $U_0 := U_{\tau_1} \cap \cdots \cap U_{\tau_n}$ and $\beta := \max\{\beta_{\tau_1}, \cdots, \beta_{\tau_n}\}$. Then $\|D_1 g(x,t)\| \leq \beta$ for all $t \in T$ and all $x \in U_0$, which clearly entails that the family $(g_t)_{t \in T}$ is equi-Lipschitz on U_0. □

As another corollary of Theorem 2.198, we can describe the C-subdifferential of a pointwise supremum over a compact space T of families of C^1 functions, in terms of Radon probability measures on an appropriate compact subset of T.

COROLLARY 2.202. Let X be a separable reflexive Banach space, U be a nonempty open subset of X and T be a compact Hausdorff topological space. Let $g : U \times T \to \mathbb{R}$ be a function which is upper semicontinuous in its second variable and let $f(x) := \sup_{t \in T} g(x,t)$ for all $x \in U$. Assume that g is differentiable on U in its first variable and that $D_1 g(\cdot, \cdot)$ is continuous on $U \times T$. Then $x^* \in \partial_C f(x)$ if and only if there exists a Radon probability measure μ on $T(x) := \{t \in T : g(t, x) = f(x)\}$ such that

$$x^* = \int_{T(x)} Dg_t(x) \, d\mu(t).$$

PROOF. Fix any $x \in U$ and note that $T_0 := T(x)$ is compact. Denote by $\mathcal{M}_+^1(T_0)$ the set of Radon probability measures on T_0 endowed with the topology τ induced by the weak* topology on the topological dual of $C(T, \mathbb{R})$ (the space of continuous real-valued functions on T_0 equipped with the supremum norm). Since $\partial_C f(x)$ is $w(X^*, X)$ compact and the mapping $\mu \mapsto \int_{T_0} Dg_t(\bar{x}) \, d\mu(t)$ is τ-to-$w(X^*, X)$ continuous from $\mathcal{M}_+^1(T_0)$ into X^*, the set

$$\left\{ \int_{T_0} Dg_t(x) \, d\mu(t) : \mu \in \mathcal{M}_+^1(T_0) \right\}$$

is weak* compact and convex in X^*. From this and Corollary 2.201 it results that, for every $h \in X$,

$$f^\circ(\bar{x}; h) = \max\{\langle Dg_t(\bar{x}), h \rangle : t \in T_0\}$$
$$= \max \left\{ \int_{T_0} \langle Dg_t(\bar{x}), h \rangle \, d\mu(t) : \mu \in \mathcal{M}_+^1(T_0) \right\},$$

which guarantees the equivalence in the corollary. □

2.4.4. Valadier theorem for suprema of infinitely many convex functions in normed spaces. In the case of convex functions, some results of other types are very useful. One of the results will need the concept of ε-subdifferential for convex functions. Given an extended real-valued convex function $f : X \to \mathbb{R} \cup \{-\infty, +\infty\}$ on a normed space X and a real $\varepsilon \geq 0$, the ε-subdifferential (in the sense of convex analysis) of f at a point $x \in X$ with $|f(x)| < +\infty$ is defined by

(2.64) $\qquad \partial_\varepsilon f(x) := \{x^* \in X^* : \langle x^*, y - x \rangle \leq f(y) - f(x) + \varepsilon, \ \forall y \in X\}.$

2.4. CHAIN RULE, COMPACTLY LIPSCHITZIAN MAPPINGS

If $f(x)$ is not finite, one puts $\partial_\varepsilon f(x) = \emptyset$. Evidently, $\partial_\varepsilon f(x)$ is a $w(X^*, X)$-closed convex set in X^* and clearly

$$\partial f(x) = \bigcap_{\varepsilon > 0} \partial_\varepsilon f(x).$$

For $\varepsilon = 0$ the ε-subdifferential $\partial_\varepsilon f(x)$ coincides with the subdifferential $\partial f(x)$. It is also readily seen that for any $a^* \in X^*$ and $\beta \in \mathbb{R}$

$$\partial_\varepsilon (f + \langle a^*, \cdot \rangle + \beta)(x) = a^* + \partial_\varepsilon f(x),$$

so in particular $\partial_\varepsilon (\langle a^*, \cdot \rangle + \beta)(x) = \{a^*\}$.

Suppose that $f(x)$ is finite. It is easily seen from the very definition that $x^* \in \partial_\varepsilon f(x)$ if and only if for all $h \in X$

$$\langle x^*, h \rangle \leq \inf_{t > 0} t^{-1}[f(x + th) - f(x) + \varepsilon].$$

This leads, for a convex function $f: X \to \mathbb{R} \cup \{-\infty, +\infty\}$ on a vector space X and for a point $x \in X$ with $|f(x)| < +\infty$, to define the function $f'_\varepsilon(x; \cdot) : X \to \mathbb{R} \cup \{-\infty, +\infty\}$, called the *directional ε-derivative* of f at x, by

(2.65) $$f'_\varepsilon(x; h) := \inf_{t > 0} t^{-1}[f(x + th) - f(x) + \varepsilon] \quad \text{for all } h \in X.$$

Clearly, $f'_\varepsilon(x; 0) = 0$ and $f'_\varepsilon(x; rh) = r f'_\varepsilon(x; h)$ for every real $r > 0$. Further, for any reals $r, s > 0$ we have with $\theta := (r + s)^{-1} rs$ (depending on r, s)

$$\frac{1}{r}[f(x + rh) - f(x) + \varepsilon] + \frac{1}{s}[f(x + sh') - f(x) + \varepsilon]$$

$$= \frac{1}{\theta}\left[\frac{s}{r+s} f(x + rh) + \frac{r}{r+s} f(x + sh') - f(x) + \varepsilon\right]$$

$$\geq \frac{1}{\theta}[f(x + \theta(h + h')) - f(x) + \varepsilon] \geq f'_\varepsilon(x; h + h'),$$

which implies that $f'_\varepsilon(x; h) + f'_\varepsilon(x; h') \geq f'_\varepsilon(x; h + h')$. The function $f'_\varepsilon(x; \cdot)$ is then a positively homogeneous convex function null at zero.

We also notice that on $]0, +\infty[$ the function $\varepsilon \mapsto f'_\varepsilon(x; h)$ is nondecreasing, so

$$\lim_{\varepsilon \downarrow 0} f'_\varepsilon(x; h) = \inf_{\varepsilon > 0} f'_\varepsilon(x; h) = \inf_{\varepsilon > 0} \inf_{t > 0} t^{-1}[f(x + th) - f(x) + \varepsilon]$$

$$= \inf_{t > 0} \inf_{\varepsilon > 0} t^{-1}[f(x + th) - f(x) + \varepsilon] = \inf_{t > 0} t^{-1}[f(x + th) - f(x)],$$

hence

$$\lim_{\varepsilon \downarrow 0} f'_\varepsilon(x : h) = f'(x; h).$$

Coming back to the context where X is a normed space, we see that

$$\partial_\varepsilon f(x) = \partial f'_\varepsilon(x; \cdot)(0).$$

From the very definition of $\partial_\varepsilon f(x)$ we also see that

$$x^* \in \partial_\varepsilon f(x) \iff f^*(x^*) \leq \langle x^*, x \rangle - f(x) + \varepsilon.$$

The ε-subdifferential will be used in Corollary 2.205 below. Let us establish first a simple feature related to supremum of convex functions.

PROPOSITION 2.203. Let T be any nonempty set and X be a normed space. Let $g : X \times T \to \mathbb{R} \cup \{+\infty\}$ be a function such that $g_t := g(\cdot, t)$ is convex for every $t \in T$ and let $f(x) := \sup_{t \in T} g(x, t)$ for all $x \in X$. Let $\bar{x}$ be a point at which the convex function f is finite and let $T_\eta(\bar{x}) := \{t \in T : g(\bar{x}, t) \geq f(\bar{x}) - \eta\}$. Then for any real $\eta \geq 0$ and any $t \in T_\eta(\bar{x})$ one has

$$\partial g_t(\bar{x}) \subset \partial_\eta f(\bar{x}).$$

PROOF. Let $t \in T_\eta(\bar{x})$ and $x^* \in \partial g_t(\bar{x})$. It suffices to write for every $x \in X$

$$\langle x^*, x - \bar{x} \rangle \leq g_t(x) - g_t(\bar{x}) \leq f(x) - f(\bar{x}) + \eta,$$

which gives as desired $x^* \in \partial_\eta f(\bar{x})$. $\square$

We pass now to one of basic descriptions of subdifferential of supremum of convex functions.

THEOREM 2.204 (M. Valadier's theorem for subdifferential of suprema of infinitely many convex functions I). Let T be any nonempty set and X be a normed space. Let $g : X \times T \to \mathbb{R} \cup \{+\infty\}$ be a function such that $g_t := g(\cdot, t)$ is convex for every $t \in T$ and let $f(x) := \sup_{t \in T} g(x, t)$ for all $x \in X$. Let $\bar{x}$ be a point such that the convex function f is finite at $\bar{x}$ and bounded from above near $\bar{x}$. Let $T_\eta(\bar{x}) := \{t \in T : g(\bar{x}, t) \geq f(\bar{x}) - \eta\}$. Then the equality

$$\partial f(\bar{x}) = \bigcap_{\eta > 0} \overline{\mathrm{co}}^* \left(\bigcup_{t \in T_\eta(\bar{x}), x \in B(\bar{x}, \eta)} \partial g_t(x) \right)$$

holds true.

PROOF. To show the inclusion $\subset$, fixing any $h \in X$, any real $\varepsilon > 0$ and any real $\eta > 0$ such that f is bounded from above on $B(\bar{x}, \eta)$, it suffices to find $\tau \in T_\eta(\bar{x})$ and $x \in B(\bar{x}, \eta)$ such that $g'_\tau(x; h) \geq f'(\bar{x}; h) - \varepsilon$. Note that $f'(\bar{x}; h)$ is finite by the continuity of the convex function f at $\bar{x}$ (see Theorem 2.158). Choose a real $s \in]0, 1]$ with $s\|h\| < \eta$ and such that

(2.66) $\qquad (2s)^{-1}[f(\bar{x} + 2sh) - f(\bar{x})] \leq f'(\bar{x}; h) + (\eta/4).$

Set $x := \bar{x} + sh$ and take $\tau \in T$ such that

(2.67) $\qquad f(x) - g_\tau(x) \leq \min\{\eta/4, s\varepsilon\}.$

Noting that $f(x) - f(\bar{x}) = [f(\bar{x} + sh) - f(\bar{x})] \geq sf'(\bar{x}; h)$, it ensues that

(2.68) $\qquad g_\tau(x) \geq f(x) - (\eta/4) \geq sf'(\bar{x}; h) + f(\bar{x}) - (\eta/4).$

Further, for $u := \bar{x} + 2sh$ we see from (2.66) (since $g_\tau(u) \leq f(u)$) that

$$g_\tau(u) \leq f(\bar{x}) + 2sf'(\bar{x}; h) + (s\eta/2) \leq f(\bar{x}) + 2sf'(\bar{x}; h) + \eta/2.$$

Combining the latter inequality and (2.68) with the (Jensen) inequality $2g_\tau(x) \leq g_\tau(\bar{x}) + g_\tau(u)$ we obtain $g_\tau(\bar{x}) \geq f(\bar{x}) - \eta$, that is, $\tau \in T_\eta(\bar{x})$. On the other hand, we have

$$g'_\tau(x; h) \geq (-s)^{-1}[g_\tau(x - sh) - g_\tau(x)] = s^{-1}[g_\tau(x) - g_\tau(\bar{x})]$$
$$\geq s^{-1}[g_\tau(x) - f(\bar{x})] \geq s^{-1}[f(x) - f(\bar{x})] - \varepsilon,$$

where the latter inequality is due to (2.67). It follows that $g'_\tau(x; h) \geq f'(\bar{x}; h) - \varepsilon$ as desired.

Let us prove the converse inclusion. Let a real $\eta_0 > 0$ such that f is bounded from above on $B(\bar{x}, 2\eta_0)$, so the family $\{g_t - g_t(\bar{x}) : t \in T_\eta, 0 < \eta \leq \eta_0\}$ is bounded from above on $B(\bar{x}, 2\eta_0)$, where $T_\eta := T_\eta(\bar{x})$. By Lemma 2.156 there is some real $\gamma \geq 1$ such that all the functions $\{g_t : t \in T_\eta, 0 < \eta \leq \eta_0\}$ are Lipschitz on $B(\bar{x}, \eta_0)$ with the same Lipschitz constant γ. Fix any $\eta \in {]0, \eta_0]}$. Take any x^* in $Q_\eta := \bigcup\{\partial g_t(x) : t \in T_\eta, x \in B(\bar{x}, \eta/(2\gamma))\}$. Choosing $\tau \in T_\eta$ and $u \in B(\bar{x}, \eta/(2\gamma))$ such that $x^* \in \partial g_\tau(u)$, it follows that for every $y \in X$

$$\langle x^*, y - \bar{x} \rangle = \langle x^*, y - u \rangle + \langle x^*, u - \bar{x} \rangle \leq g_\tau(y) - g_\tau(u) + \gamma \|u - \bar{x}\|$$
$$\leq f(y) - g_\tau(\bar{x}) + 2\gamma\|u - \bar{x}\| \leq f(y) - f(\bar{x}) + 2\eta,$$

so $x^* \in \partial_{2\eta} f(\bar{x})$. For $\eta \in {]0, \eta_0]}$ this means that $Q_\eta \subset \partial_{2\eta} f(\bar{x})$, so $\overline{\text{co}}^* Q_\eta \subset \partial_{2\eta} f(\bar{x})$. Consequently, $\bigcap_{\eta > 0} \overline{\text{co}}^* Q_\eta \subset \partial f(\bar{x})$, which finishes the proof. $\square$

The above result involves subgradients at points x near $\bar{x}$. The next corollary uses instead merely ε-subgradients at the reference point.

COROLLARY 2.205. *Under the assumptions and notation of Theorem 2.204 one also has*

$$\partial f(\bar{x}) = \bigcap_{\eta > 0} \overline{\text{co}}^* \left(\bigcup_{t \in T_\eta(\bar{x})} \partial_\eta g_t(\bar{x}) \right).$$

PROOF. Put $Q_\eta := \bigcup_{t \in T_\eta(\bar{x})} \partial_\eta g_t(\bar{x})$ for every real $\eta > 0$.

Let us show first the inclusion $\supset$. Take any real $\eta > 0$ and consider any $x^* \in Q_\eta$. Choose $\tau \in T_\eta(\bar{x})$ such that $x^* \in \partial_\eta g_\tau(\bar{x})$. For any $y \in X$ it ensues that

$$\langle x^*, y - \bar{x} \rangle \leq g_\tau(y) - g_\tau(\bar{x}) + \eta \leq f(y) - f(\bar{x}) + 2\eta,$$

which gives $x^* \in \partial_{2\eta} f(\bar{x})$. It follows that $Q_\eta \subset \partial_{2\eta} f(\bar{x})$, hence $\overline{\text{co}}^*(Q_\eta) \subset \partial_{2\eta} f(\bar{x})$. This implies the desired inclusion $\supset$ of the corollary.

Concerning the converse inclusion, fix $\eta_0 > 0$ and $\gamma \geq 1$ as in the second part of the proof of Theorem 2.204. Take any $0 < \eta < \eta_0/2$ and consider any $t \in T_\eta(\bar{x})$, any $x \in B(\bar{x}, \eta)$ and any $x^* \in \partial g_t(x)$. For any $h \in X$ with $\|h\| \leq 1$ we have $\langle x^*, \eta h \rangle \leq g_t(x + \eta h) - g_t(x) \leq \gamma \eta$, so $\|x^*\| \leq \gamma$. Then, for any $y \in X$ we can write

$$\langle x^*, y - \bar{x} \rangle \leq \langle x^*, y - x \rangle + \|x^*\| \, \|x - \bar{x}\| \leq g_t(y) - g_t(\bar{x}) + 2\gamma\eta.$$

This says that $x^* \in \partial_{2\gamma\eta} g_t(\bar{x})$, hence $x^* \in Q_{2\gamma\eta}$. From this and Theorem 2.204 we see that the left member in the corollary is included in $\bigcap_{\eta > 0} \overline{\text{co}}^*(Q_\eta)$, which finishes the proof. $\square$

2.4.5. C-subgradients of distance function through metric projection; nonzero C-normals at boundary points. We apply Valadier's theorem to compute the C-subdifferential of distance function at points outside the set. Given a nonempty set S of a metric space (E, d) and a point $x \in E$, we denote by $\text{Proj}_S x$ the set of *nearest points* of x in S, that is,

(2.69) $$\text{Proj}_S x := \{u \in S : d(u, x) = d_S(x)\}.$$

Even when S is closed, this set $\text{Proj}_S x$ may be empty for $x \notin S$ in either general metric spaces or infinite-dimensional normed spaces. In such a situation, sets of approximate nearest points can be useful. For a real $\varepsilon > 0$, one defines the set of ε-nearest points of x in S as

(2.70) $$\operatorname{Proj}_{S,\varepsilon} x := \{u \in S : d(u,x) \leq d_S(x) + \varepsilon\}.$$

Clearly, $\operatorname{Proj}_{S,\varepsilon} x \neq \emptyset$ for every $\varepsilon > 0$.

Consider now a Hilbert space H. Noting that
$$\|x - y\|^2 = \|x\|^2 - (2\langle x, y\rangle - \|y\|^2),$$
we can write
$$\frac{1}{2} d_S^2(x) = \frac{1}{2}\|x\|^2 - \sup_{y \in S}\left(\langle x, y\rangle - \frac{1}{2}\|y\|^2\right).$$
Utilizing the so-called Asplund function associated with S, defined by
$$\varphi_S(u) := \sup_{y \in S}\left(\langle u, y\rangle - \frac{1}{2}\|y\|^2\right),$$
it results that

(2.71) $$\frac{1}{2} d_S^2(x) = \frac{1}{2}\|x\|^2 - \varphi_S(x).$$

The function φ_S is obviously convex and the latter equality ensures that it is also finite-valued and locally Lipschitz. Since $\|\cdot\|^2/2$ is $\mathcal{C}^1$ with the identity on H as gradient, the same equality (2.71) gives

(2.72) $$d_S(x)\partial_C d_S(x) = x + \partial_C(-\varphi_S)(x) = x - \partial_C \varphi_S(x) = x - \partial \varphi_S(x).$$

Given $\varepsilon > 0$, putting $\eta(\varepsilon, x) := \varepsilon^2 + 2\varepsilon d_S(x)$ we observe that $u \in X$ satisfies $\varphi_S(x) \leq \langle x, u\rangle - (1/2)\|u\|^2 + \eta(\varepsilon, x)/2$ if and only if $\|x - u\|^2 \leq d_S^2(x) + \eta(\varepsilon, x)$, that is, $\|x - u\| \leq d_S(x) + \varepsilon$, which means $u \in \operatorname{Proj}_{S,\varepsilon} x$. Since $\eta(\varepsilon, x) \downarrow 0$ as $\varepsilon \downarrow 0$, applying the above Valadier theorem with $g : X \times S \to \mathbb{R}$ defined by $g(u, s) := \langle u, s\rangle - \frac{1}{2}\|s\|^2$ for all $s \in S$, we obtain

$$\partial \varphi_S(x) = \bigcap_{\varepsilon > 0} \overline{\operatorname{co}}\{u \in H : u \in \operatorname{Proj}_{S,\varepsilon} x\} = \bigcap_{\varepsilon > 0} \overline{\operatorname{co}}(\operatorname{Proj}_{S,\varepsilon} x).$$

So, for $x \notin \operatorname{cl}_H S$ we obtain from (2.72)

$$\partial_C d_S(x) = \frac{1}{d_S(x)}\left(x - \bigcap_{\varepsilon > 0} \overline{\operatorname{co}}(\operatorname{Proj}_{S,\varepsilon} x)\right).$$

We summarize in the following proposition.

PROPOSITION 2.206. *Let S be a nonempty subset of a Hilbert space H. Then for any $x \in H \setminus \operatorname{cl}_H S$ one has*

$$\partial_C d_S(x) = \frac{1}{d_S(x)}\left(x - \bigcap_{\varepsilon > 0} \overline{\operatorname{co}}(\operatorname{Proj}_{S,\varepsilon} x)\right) = \bigcap_{\varepsilon > 0} \overline{\operatorname{co}}\left(\frac{x - \operatorname{Proj}_{S,\varepsilon} x}{d_S(x)}\right).$$

Assume now that S is *ball-compact* in the sense that $S \cap r\mathbb{B}$ is compact in H for every real $r > 0$, so $\operatorname{Proj}_S x \neq \emptyset$ for all $x \in H$. Further, with the notations preceding the statement of Proposition 2.206, for any fixed point $x \in H \setminus S$, any real $r > 0$ and any $x' \in B(x, r)$, we have

$$\{s \in S : g(x', s) = \varphi_S(x')\} = \operatorname{Proj}_S x' \subset S \cap B[x, 2r + d_S(x)] =: K$$

and K is compact. All the assumptions of Theorem 2.198 are fulfilled for the fixed point x in place of $\bar{x}$, and hence
$$\partial \varphi_S(x) = \overline{\operatorname{co}}\{u \in H : u \in \operatorname{Proj}_S x\} = \overline{\operatorname{co}}(\operatorname{Proj}_S x).$$
This and the equality $d_S(x)\partial_C d_S(x) = x - \partial \varphi_S(x)$ in (2.72) justify the following result (since any ball-compact set is clearly closed).

PROPOSITION 2.207. *Let H be a Hilbert space and S be a set in H which is ball-compact. Then for any $x \in H \setminus S$ one has*
$$\partial_C d_S(x) = \frac{1}{d_S(x)}(x - \overline{\operatorname{co}}(\operatorname{Proj}_S x)) = \overline{\operatorname{co}}\left(\frac{x - \operatorname{Proj}_S x}{d_S(x)}\right).$$

In fact, the result in the above proposition will be extended to weakly closed subsets of Hilbert spaces in Proposition 4.125(a).

The next corollary shows that any closed set of a finite dimensional normed space contains nonzero C-normals at any boundary point.

COROLLARY 2.208. *Let S be a nonempty set of a finite dimensional normed space X which is closed near $\bar{x} \in S \cap \operatorname{bdry} S$. Then $\partial_C d_S(\bar{x})$ is not reduced to zero, hence in particular $N^C(S; \bar{x}) \neq \{0\}$.*

PROOF. We may suppose that S is closed. Denote by $\|\cdot\|$ a Euclidean norm of X and choose also a sequence $(x_n)_n$ in $X \setminus S$ converging to $\bar{x}$. For each $n \in \mathbb{N}$ there exists by Proposition 2.207 some $x_n^* \in \partial_C d_S(x_n)$ with $\|x_n^*\| = 1$. There exists a subsequence of $(x_n^*)_n$ (that we do not relabel) which converges to some x^* with $\|x^*\| = 1$. By outer semicontinuity of $\partial_C d_S$ (see Proposition 2.74(a)) and since $\bar{x} \in S$, we get that $x^* \in \partial_C d_S(\bar{x}) \subset N^C(S; \bar{x})$ (see Proposition 2.95(c) for the second inclusion). This finishes the proof. □

REMARK 2.209. The corollary could also been justified by applying Theorem 2.83 to the distance function d_S and using as above Proposition 2.95(c). □

The property in Corollary 2.208 fails in infinite dimensions even though the closed set is convex, as illustrated in the following example.

EXAMPLE 2.210. Consider a separable infinite-dimensional Banach space X and let $(u_n)_n$ be a dense sequence in the unit sphere $\mathbb{S}_X$. The set $K := \{0\} \cup \{\frac{1}{n} u_n : n \in \mathbb{N}\}$ is strongly compact (since $\frac{1}{n} u_n \to 0$), hence its closed convex hull $C := \overline{\operatorname{co}} K$ is also strongly compact. Therefore, $\operatorname{int} C = \emptyset$ since X is infinite-dimensional. It ensues that $0 \in \operatorname{bdry} C$. On the other hand, for any $x^* \in N(C; 0)$ we have for any $n \in \mathbb{N}$
$$\left\langle x^*, \frac{1}{n} u_n - 0 \right\rangle \leq 0, \quad \text{hence } \langle x^*, u_n \rangle \leq 0.$$
This entails that $x^* = 0$ since $(u_n)_n$ is dense in $\mathbb{S}_X$. Then, at the boundary point 0 of the closed convex set C in X we have $N(C; 0) = \{0\}$, showing that the property in Corollary 2.208 fails in infinite dimensions. (In fact, this set C is classical in Functional Analysis, as an example of closed convex set with a non-supporting point in its boundary). □

PROPOSITION 2.211. *Let $f : X \to \mathbb{R} \cup \{+\infty\}$ be a proper convex function on a finite-dimensional normed space X and let $\bar{x} \in \operatorname{dom} f$. Then $\bar{x} \in \operatorname{bdry}(\operatorname{dom} f)$ if and only if $\partial f(\bar{x})$ is either empty or unbounded.*

PROOF. If $\bar{x} \in \operatorname{int}(\operatorname{dom} f)$ we know (see Corollary 2.159) that f is Lipschitz continuous near $\bar{x}$, hence $\partial f(\bar{x})$ is nonempty and bounded according to Theorem 2.70(d). Now, suppose that $\bar{x} \in \operatorname{bdry}(\operatorname{dom} f)$. Put $S := \operatorname{cl}(\operatorname{dom} f)$ and note that $\bar{x} \in \operatorname{bdry} S$ since $\operatorname{int}(\operatorname{cl} S) = \operatorname{int} S$ (see Corollary B.10 in Appendix). Suppose that $\partial f(\bar{x}) \neq \emptyset$ and fix $x^* \in \partial f(\bar{x})$. Choose a Euclidean norm $\|\cdot\|$ on X and let d_S be the distance function from S relative to the Euclidean norm $\|\cdot\|$. Let the affine function $\varphi(\cdot) := f(\bar{x}) + \langle x^*, \cdot - \bar{x}\rangle$. For each $n \in \mathbb{N}$ consider the function $\varphi_n : X \to \mathbb{R}$ defined by $\varphi_n(x) := \varphi(x) + nd_S(x)$, and note that $\varphi_n \leq f$ with $\varphi_n(\bar{x}) = f(\bar{x})$. By the proof of Corollary 2.208 there is some $u^* \in \partial d_S(\bar{x})$ with $\|u^*\|_* = 1$, where $\|\cdot\|_*$ is the Euclidean dual norm of $\|\cdot\|$. For any $x \in X$ we see by (2.21) that

$$\langle x^* + nu^*, x - \bar{x}\rangle \leq \varphi(x) - \varphi(\bar{x}) + nd_S(x) - nd_S(\bar{x}) = \varphi_n(x) - \varphi_n(\bar{x}) \leq f(x) - f(\bar{x}),$$

hence $x^* + nu^* \in \partial f(\bar{x})$. This yields that $\partial f(\bar{x})$ is unbounded, which finishes the proof. $\square$

The next example illustrates that the equivalence in Proposition 2.211 fails in infinite dimensions.

EXAMPLE 2.212. Let C be the closed convex set of the infinite-dimensional Banach space X as in Example 2.210, with $0 \in \operatorname{bdry} C$ and $N(C; 0) = \{0\}$. Consider the convex function f given by the indicator function of C. We have $\partial f(0) = \{0\}$ since $\partial f(0) = N(C; 0)$, so $\partial f(0)$ is neither empty nor unbounded while $0 \in \operatorname{bdry}(\operatorname{dom} f)$. $\square$

We will see in Corollaries 6.65 that the implication $\Rightarrow$ in Proposition 2.211 can be extended to C-subdifferentials of lower semicontinuous functions on finite-dimensional normed spaces.

2.5. Optimization problems with constraints

Now we proceed to establish necessary optimality conditions for mathematical programming problems. We start with some penalizations for minimization over a constraint set.

2.5.1. Penalization principles with Lipschitz/non-Lipschitz objective functions. The following penalization lemma is often crucial for many optimization problems.

LEMMA 2.213 (F. Clarke penalization lemma). Let S be a nonempty subset of a metric space (E, d) and $f : E \to \mathbb{R}$ be a real-valued function which is γ-Lipschitz on E. Then

$$\inf_S f = \inf_E (f + \gamma d_S),$$

thus any minimizer of f over S (if it exists) is a minimizer of $f + \gamma d_S$ over E.

PROOF. Fix $x \in E$ and take any $y \in S$. Then $f(x) + \gamma d(x, y) \geq f(y) \geq \inf_S f$, hence taking the infimum over $y \in S$ yields $f(x) + \gamma \inf_{y \in S} d(x, y) \geq \inf_S f$, that is, $f(x) + \gamma d_S(x) \geq \inf_S f$. This means $\inf_E (f + \gamma d_S) \geq \inf_S f$, thus the desired equality is justified since we also have $\inf_E (f + \gamma d_S) \leq \inf_S (f + \gamma d_S) = \inf_S f$ (where the latter equality is due to $d_S(x) = 0$ for all $x \in S$). $\square$

REMARK 2.214. If, in addition to the assumptions of the above lemma, the set S is closed, then for any real $\gamma' > \gamma$ one has
$$\operatorname*{Arg\,min}_{E}(f + \gamma' d_S) = \operatorname*{Arg\,min}_{S} f,$$
which amounts to saying that any minimizer over E of $f + \gamma' d_S$ belongs to S. Here $\operatorname*{Arg\,min}_{S} f$ denotes the set of minimizers of f over S, that is,
$$\operatorname*{Arg\,min}_{S} f := \left\{ x \in S : f(x) = \inf_{S} f \right\}.$$
Indeed, if $\bar{x}$ is a minimizer over E of $f + \gamma' d_S$, by the above lemma we have
$$f(\bar{x}) + \gamma' d_S(\bar{x}) = \inf_{S} f = \inf_{E}(f + \gamma d_S) \leq f(\bar{x}) + \gamma d_S(\bar{x}) \leq f(\bar{x}) + \gamma' d_S(\bar{x}).$$
Then $f(\bar{x}) + \gamma' d_S(\bar{x}) = f(\bar{x}) + \gamma d_S(\bar{x})$, which ensures $(\gamma' - \gamma) d_S(\bar{x}) = 0$, hence $\bar{x} \in S$ according to the closedness assumption of the set S. □

A variant of the above lemma related to functions of two variables is also valid. It will be employed in Section 5.2 of Chapter 5.

LEMMA 2.215. Let S and T be two nonempty subsets of two metric spaces (E, d) and (F, d') respectively. Let $f : E \times F \to \mathbb{R}$ be a real-valued function and let $(\bar{x}, \bar{y}) \in S \times T$ be a minimizer of f over $S \times T$. The following hold:
(a) If there exists a real constant $\gamma \geq 0$ such that $f(\cdot, y)$ is γ-Lipschitz continuous on E for all $y \in T$, then
$$f(\bar{x}, \bar{y}) \leq f(x, y) + \gamma\, d(x, S) \quad \text{for all } x \in E,\ y \in T.$$
(b) If in addition to the assumption in (a), there is also a real constant $\gamma' \geq 0$ such that $f(x, \cdot)$ is γ'-Lipschitz continuous on F, then
$$f(\bar{x}, \bar{y}) \leq f(x, y) + \gamma\, d(x, S) + \gamma'\, d'(y, T) \quad \text{for all } x \in E,\ y \in F.$$

PROOF. (a) Fix any $x \in E$ and $y \in T$. Write, for every $x' \in S$,
$$f(\bar{x}, \bar{y}) \leq f(x', y) \leq f(x, y) + \gamma\, d(x', x),$$
so taking the infimum over $x' \in S$ gives $f(\bar{x}, \bar{y}) \leq f(x, y) + \gamma\, d_S(x)$.
(b) Putting $g(x, y) := f(x, y) + \gamma\, d_S(x)$, the assertion (a) says that $(\bar{x}, \bar{y})$ is a minimizer of g over $E \times T$. Using the additional assumption in (b), the assertion (a) again guarantees the property in (b). □

The case where f is merely lower semicontinuous is sometimes of interest. Assume that $f : E \to \mathbb{R} \cup \{+\infty\}$ is proper and such that there exist $a \in E$ and constants $\alpha \geq 0$ and $\beta \in \mathbb{R}$ such that
$$-\alpha d(x, a) + \beta \leq f(x) \quad \text{for all } x \in E.$$
Considering any $\lambda \geq \alpha$ and $x \in E$ we observe that for all $x \in E$
$$f(y) + \lambda d(y, x) \geq (\lambda - \alpha) d(y, a) - \lambda d(x, a) + \beta \geq -\lambda d(x, a) + \beta,$$
so setting $f_\lambda(x) := \inf_{y \in E}\left(f(y) + \lambda d(y, x)\right)$ the function f_λ is finite-valued on E; further, writing
$$f(y) + \lambda d(y, x) \leq \left(f(y) + \lambda d(y, x')\right) + \lambda d(x, x')$$
and taking the infimum of both members over $y \in E$ yields $f_\lambda(x) \leq f_\lambda(x') + \lambda d(x, x')$, hence
$$|f_\lambda(x) - f_\lambda(x')| \leq \lambda d(x, x').$$

It is also clear that $\lambda \mapsto f_\lambda(x)$ is nondecreasing and $f_\lambda(x) \le f(x)$ for all $x \in E$.

PROPOSITION 2.216. *Let S be a nonempty bounded subset of a metric space (E, d) and $f : E \to \mathbb{R} \cup \{+\infty\}$ be an extended real-valued function finite at some point in S for which there exist $a \in E$ and numbers $\alpha \ge 0$ and $\beta \in \mathbb{R}$ such that*

$$-\alpha d(x, a) + \beta \le f(x) \quad \text{for all } x \in E.$$

Assume that there is a topology τ on E coarser than the initial metric topology such that S is τ-compact and f is τ-lower semicontinuous. Then one has

$$\inf_S f = \lim_{\lambda \uparrow \infty} \inf_E (f + \lambda d_S) = \sup_{\lambda \ge 0} \inf_E (f + \lambda d_S).$$

PROOF. Set $f_\lambda(x) := \inf_{y \in E} (f(y) + \lambda d(y, x))$ and fix any $\lambda > \alpha$. By what precedes the proposition the function f_λ is λ-Lipschitz continuous on E, so it is bounded from below on the bounded set S. Choose some $u_\lambda \in S$ satisfying $f_\lambda(u_\lambda) \le \lambda^{-1} + \inf_S f_\lambda$ and then choose some $y_\lambda \in E$ such that $f(y_\lambda) + \lambda d(y_\lambda, u_\lambda) \le f_\lambda(u_\lambda) + \lambda^{-1}$. Then we have

$$(2.73) \qquad f(y_\lambda) \le f(y_\lambda) + \lambda d(y_\lambda, u_\lambda) \le \frac{2}{\lambda} + \inf_S f_\lambda \le \frac{2}{\lambda} + \inf_S f.$$

This ensures that

$$\beta - \alpha d(y_\lambda, a) + \lambda d(y_\lambda, u_\lambda) \le \frac{2}{\lambda} + \inf_S f,$$

which yields, with $\mu := \beta - \alpha \sup_{x \in S} d(x, a)$, that

$$\mu + (\lambda - \alpha) d(y_\lambda, u_\lambda) \le (2/\lambda) + \inf_S f.$$

It ensues that $d(y_\lambda, u_\lambda) \to 0$ as $\lambda \uparrow \infty$. By the τ-compactness of S we can and do suppose (via a subnet) that $(u_\lambda)_\lambda$ τ-converges to some $u \in S$, so $(y_\lambda)_\lambda$ τ-converges to u. Invoking the τ-lower semicontinuity of f and (2.73) we obtain

$$f(u) \le \liminf_{\lambda \uparrow \infty} f(y_\lambda) \le \liminf_{\lambda \uparrow \infty} \left(\inf_S f_\lambda\right) \le \inf_S f \le f(u).$$

According to the nondecreasing property in λ of $\inf_S f_\lambda$ we deduce that

$$f(u) = \inf_S f = \lim_{\lambda \uparrow \infty} \inf_S f_\lambda = \sup_{\lambda \ge 0} \inf_S f_\lambda.$$

Finally, it suffices to note that

$$\inf_S f_\lambda = \inf_{x \in S} \inf_{y \in E} (f(y) + \lambda d(y, x)) = \inf_{y \in E} \inf_{x \in S} (f(y) + \lambda d(y, x)) = \inf_E (f + \lambda d_S).$$

$\square$

REMARK 2.217. The equality between the two members of the first equality in the proposition is in fact a penalization result, while the equality between the first and third members can be seen as a duality result. $\square$

COROLLARY 2.218. *Let S be a nonempty weakly compact subset of a normed space X and $f : X \to \mathbb{R} \cup \{+\infty\}$ be an extended real-valued lower semicontinuous convex function which is finite at some point in S. Then one has*

$$\inf_S f = \lim_{\lambda \uparrow \infty} \inf_X (f + \lambda d_S).$$

PROOF. Clearly the convex function f is weakly lower semicontinuous since by Mazur theorem its sublevel sets are weakly closed. Taking a point $(\bar{x}, r) \notin \text{epi } f$ and applying the Hahn-Banach separation theorem, we obtain that f is bounded from below by an affine continuous function, so the minorization assumption in Proposition 2.216 above is satisfied. The corollary then follows from that proposition. □

2.5.2. Minimization under a set-constraint. We can now begin with necessary optimality conditions for optimization problems with constraints. The first problem which we deal with is that of minimization under a set-constraint. Given a set S of the normed space X and an extended real-valued function $f: X \to \mathbb{R} \cup \{-\infty, +\infty\}$, a point $\bar{x} \in S$, where f is finite, is a *local solution* of the problem

$$\begin{cases} \text{Minimize} & f(x) \\ \text{subject to} & x \in S \end{cases}$$

whenever there exists a neighborhood U of $\bar{x}$ in X such that $f(\bar{x}) \leq f(x)$ for all $x \in S \cap U$. If U is the whole space X, one says that $\bar{x}$ is a *(global) solution* of the problem.

Before establishing a first necessary optimality condition, we need a property for the distance function from a truncated set. In addition to that property, we also prove two related properties for nearest points to which we will refer in other places; see (2.69) for the definition and notation of the set of nearest points.

LEMMA 2.219. *Let S be a nonempty subset of a metric space (E, d) and $\bar{x} \in E$. Let $\bar{\delta} := d(\bar{x}, S)$. Then for any real $r > 0$, one has*

$$d(x, S \cap B(\bar{x}, \bar{\delta} + 2r)) = d(x, S \cap B[\bar{x}, \bar{\delta} + 2r]) = d(x, S) \quad \text{for all } x \in B[\bar{x}, r]$$

as well as the equalities

$$\text{Proj}(x, S \cap B(\bar{x}, \bar{\delta} + 2r)) = \text{Proj}(x, S) \quad \text{for all } x \in B(\bar{x}, r)$$

and

$$\text{Proj}(x, S \cap B[\bar{x}, \bar{\delta} + 2r]) = \text{Proj}(x, S) \quad \text{for all } x \in B[\bar{x}, r].$$

PROOF. Fix $r > 0$ and put $S' := S \cap B(\bar{x}, \bar{\delta} + 2r)$. Notice that obviously $d(\bar{x}, S) = d(\bar{x}, S')$.

Take any $x \in B[\bar{x}, r]$. For the equalities related to the distance function, it is enough to show the inequality $d(x, S) \geq d(x, S')$. For any $u \in S \setminus S'$, observing that $u \notin B(\bar{x}, \bar{\delta} + 2r)$, we see that

$$d(x, u) \geq d(u, \bar{x}) - d(x, \bar{x}) \geq \bar{\delta} + 2r - r = \bar{\delta} + r$$
$$\geq d(\bar{x}, S) + d(\bar{x}, x) = d(\bar{x}, S') + d(\bar{x}, x) \geq d(x, S').$$

This ensures that $d(x, u) \geq d(x, S')$ for all $u \in S$, and hence $d(x, S) \geq d(x, S')$. This then justifies the series of equalities for the distance function.

Now fix $x \in B(\bar{x}, r)$. From the previous equalities we obtain that

(2.74) $\quad \text{Proj}(x, S) \supset \text{Proj}(x, S \cap B[\bar{x}, \bar{\delta} + 2r]) \supset \text{Proj}(x, S \cap B(\bar{x}, \bar{\delta} + 2r)).$

Further, for any $y \in \text{Proj}(x, S)$ we have $d(y, x) = d(x, S) \leq d(\bar{x}, S) + d(x, \bar{x})$, hence

$$d(y, \bar{x}) \leq d(y, x) + d(x, \bar{x}) \leq \bar{\delta} + 2d(x, \bar{x}) < \bar{\delta} + 2r,$$

which gives $y \in S \cap B(\bar{x}, \bar{\delta} + 2r)$. This combined with the above equalities related to the distance function yields $y \in \text{Proj}(x, S \cap B(\bar{x}, \bar{\delta} + 2r))$. This and (2.74) confirm the first equality for $\operatorname{Proj}(\cdot, S)$ on the open ball $B(\bar{x}, r)$. The other similar equality on the closed ball $B[\bar{x}, r]$ is obtained with the same arguments. □

PROPOSITION 2.220. *Let S be a subset of the normed space X and $\bar{x} \in S$. Let $f : X \to \mathbb{R} \cup \{-\infty, +\infty\}$ be a function which is finite and γ-Lipschitz continuous near $\bar{x}$. If $\bar{x}$ is a local solution of the above minimization problem of f over S, then*
$$0 \in \partial_C f(\bar{x}) + \gamma \partial_C d_S(\bar{x}), \quad \text{hence} \quad 0 \in \partial_C f(\bar{x}) + N^C(S; \bar{x}).$$

PROOF. Choose a real $r > 0$ such that f is finite and γ-Lipschitz over $B(\bar{x}, 2r)$ and $f(\bar{x}) \leq f(x)$ for all $x \in S \cap B(\bar{x}, 2r)$. For $C := S \cap B(\bar{x}, 2r)$, Lemma 2.213 says that $f(\bar{x}) \leq f(x) + \gamma d_C(x)$ for all $x \in B(\bar{x}, 2r)$. For all $x \in B(\bar{x}, r)$, we have $d_C(x) = d_S(x)$ by Lemma 2.219 above, thus $f(\bar{x}) + \gamma d_S(\bar{x}) \leq f(x) + \gamma d_S(x)$. Consequently, $0 \in \partial_C f(\bar{x}) + \gamma \partial_C d_S(\bar{x})$ according to the subdifferential sum formula in Theorem 2.98.

The second condition follows from the inclusion $\partial_C d_S(\bar{x}) \subset N^C(S; \bar{x})$ because $\bar{x} \in S$ (see Proposition 2.95). □

2.5.3. Ekeland variational principle and Bishop-Phelps principles.
The approach below where inequality/equality constraints are present will use the following Ekeland variational principle.

THEOREM 2.221 (I. Ekeland variational principle). *Let (E, d) be a complete metric space and $f : E \to \mathbb{R} \cup \{+\infty\}$ be an extended real-valued proper lower semicontinuous function which is bounded from below. Let $\varepsilon > 0$ and let $a \in E$ be a point satisfying*
$$f(a) \leq \inf_E f + \varepsilon.$$
Then for any real $\lambda > 0$ there exists some $b \in E$ such that
(a) $d(b, a) \leq \lambda$;
(b) $f(b) \leq f(b) + \frac{\varepsilon}{\lambda} d(a, b) \leq f(a)$;
(c) $f(b) < f(x) + \frac{\varepsilon}{\lambda} d(x, b)$ for all $x \neq b$ in E, that is, the point b is a *strict minimizer* of the function $f + \frac{\varepsilon}{\lambda} d(\cdot, b)$ over E.

PROOF. Fix ε, a as above and fix also any real $\lambda > 0$. For each $x \in E$, put

(2.75) $$M(x) := \left\{ y \in E : f(y) + \frac{\varepsilon}{\lambda} d(x, y) \leq f(x) \right\}$$

and observe that $M(x)$ is closed according to the lower semicontinuity of f, and that

(2.76) $$x \in M(x) \quad (\text{ hence } M(x) \neq \emptyset).$$

Note also that

(2.77) $$y \in M(x) \Rightarrow M(y) \subset M(x),$$

since if we suppose $y \in M(x)$ we have for any $z \in M(y)$ that $f(z) + \frac{\varepsilon}{\lambda} d(z, y) \leq f(y)$, which ensures that $f(z) + \frac{\varepsilon}{\lambda} d(z, x) - \frac{\varepsilon}{\lambda} d(x, y) \leq f(y)$, that is, $f(z) + \frac{\varepsilon}{\lambda} d(z, x) \leq f(y) + \frac{\varepsilon}{\lambda} d(x, y)$, hence $f(z) + \frac{\varepsilon}{\lambda} d(z, x) \leq f(x)$ (because $y \in M(x)$), and the latter means that $z \in M(x)$.

Set now

(2.78) $$m(x) = \inf_{y \in M(x)} f(y)$$

and note that $m(x)$ is finite (because $-\infty < m(x)$ since f is bounded from below on E, and if $f(x) < +\infty$ then $m(x) < +\infty$, and if $f(x) = +\infty$ then $M(x) = E$ hence $m(x) < +\infty$ because $f \not\equiv +\infty$; keep in mind f is proper). Obviously by (2.75) we have for all $y \in M(x)$

$$m(x) + \frac{\varepsilon}{\lambda}d(y,x) \leq f(x), \text{ that is, } \frac{\varepsilon}{\lambda}d(y,x) \leq f(x) - m(x),$$

and hence

(2.79) $$\frac{\varepsilon}{\lambda}\text{diam}\,(M(x)) \leq 2(f(x) - m(x)).$$

Consider the sequence $(x_n)_{n\in\mathbb{N}}$ defined by

(2.80) $\quad x_1 = a \quad \text{and} \quad x_{n+1} \in M(x_n) \quad \text{with} \quad f(x_{n+1}) \leq m(x_n) + 1/2^n$

(the sequence is well defined by induction since $m(x_n) + 1/2^n > \inf_{M(x_n)} f$). By (2.77) we have $M(x_{n+1}) \subset M(x_n)$, and hence

$$m(x_n) \leq m(x_{n+1}).$$

Since $m(x) \leq f(x)$ according to (2.76) and (2.78), we may write

$$m(x_{n+1}) \leq f(x_{n+1}) \leq m(x_n) + 1/2^n \leq m(x_{n+1}) + 1/2^n$$

(the second inequality being due to (2.80)), and hence by the finiteness of $m(x_{n+1})$

$$0 \leq f(x_{n+1}) - m(x_{n+1}) \leq 1/2^n.$$

This and (2.79) imply that $\text{diam}\,(M(x_n)) \to 0$ as $n \to \infty$. Since $(M(x_n))_n$ is a nonincreasing sequence of nonempty closed subsets of the complete metric space E the convergence property $\text{diam}(M(x_n)) \to 0$ ensures that $\bigcap_{n\in\mathbb{N}} M(x_n) = \{b\}$ for some $b \in X$, which entails on the one hand

(2.81) $\quad b \in M(x_1) = M(a), \quad \text{thus} \quad f(b) + \frac{\varepsilon}{\lambda}d(b,a) \leq f(a),$

and on the other hand

$$b \in M(x_n) \Rightarrow M(b) \subset M(x_n) \Rightarrow M(b) \subset \bigcap_{n\in\mathbb{N}} M(x_n) \Rightarrow M(b) = \{b\}.$$

Consequently, for any $x \neq b$, we have $x \notin M(b)$, thus

$$f(x) + \frac{\varepsilon}{\lambda}d(x,b) > f(b).$$

Properties (b) and (c) of the theorem are then established. It remains to write by (2.81) and the assumption $f(a) \leq \inf_X f + \varepsilon$ that

$$\frac{\varepsilon}{\lambda}d(b,a) \leq f(a) - f(b) \leq \inf_X f + \varepsilon - f(b) \leq f(b) + \varepsilon - f(b) = \varepsilon$$

(the last equality being due to the fact that $f(b)$ is finite), and to deduce from this the inequality $d(b,a) \leq \lambda$. The proof of the theorem is now complete. $\square$

Translating, in the setting of a Banach space $(X, \|\cdot\|)$, property (c) above as $f(b) - r < (\varepsilon/\lambda)\|b - x\|$ for all $(x,r) \in \text{epi}\,f$ with $x \neq b$ leads through the closed convex cone

$$K_{\varepsilon,\lambda} := \left\{(x,r) \in X \times \mathbb{R} : r \geq \frac{\varepsilon}{\lambda}\|x\|\right\} = \text{epi}\,(\frac{\varepsilon}{\lambda}\|\cdot\|)$$

to see that (c) says equivalently that for all $(x,r) \in \text{epi } f$ with $x \neq b$
$$(b-x, f(b)-r) \notin K_{\varepsilon,\lambda} \Leftrightarrow (x,r) \notin (b, f(b)) - K_{\varepsilon,\lambda}.$$
Further, using the assumption $f(a) \leq \inf_X f + \varepsilon$ it is not difficult to check that properties (a) and (b) hold true simultaneously if and only if
$$(b, f(b)) \in \text{epi } f \cap \big((a, f(a)) - K_{\varepsilon,\lambda}\big).$$
Putting those features together, the Ekeland variational principle can be interpreted geometrically in the form:
$$(b, f(b)) \in \text{epi } f \cap \big((a, f(a)) - K_{\varepsilon,\lambda}\big) \text{ and } \{(b, f(b))\} = \text{epi } f \cap \big((b, f(b)) - K_{\varepsilon,\lambda}\big).$$

Given a pointed convex cone Q (that is, $Q \cap -Q = \{0\}$) of a vector space Y, it defines an order $\preceq_Q$ on Y by $y_1 \preceq_Q y_2 \Leftrightarrow y_2 - y_1 \in Q$. So, for $C \subset Y$ the equality $\{c\} = C \cap (c - Q)$ means that there is no element different from c which is below c with respect to $\preceq_Q$, that is, c is a *minimal element* of C with respect to the order $\preceq_Q$. The above geometrical interpretation then translates that $(b, f(b))$ is the unique element of epi f with $(b, f(b)) \preceq_{K_{\varepsilon,\lambda}} (a, f(a))$ which $K_{\varepsilon,\lambda}$-minimal in the epigraph epi f of f.

The next corollary of the Ekeland variational principle does not assume that a is an ε-minimizer of the function f (that is, $f(a) \leq \inf_E f + \varepsilon$).

COROLLARY 2.222. *Let (E,d) be a complete metric space and $f : E \to \mathbb{R} \cup \{+\infty\}$ be a proper lower semicontinuous function which is bounded from below. Then for any $a \in \text{dom } f$ and any real $\varepsilon > 0$ there exists some $b \in E$ such that*
(a) $f(b) + \varepsilon d(b, a) \leq f(a)$;
(b) $f(b) < f(x) + \varepsilon d(b, x)$ *for all $x \neq b$ in E.*

PROOF. Choosing a real $\varepsilon' > 0$ satisfying $f(a) \leq \inf_E f + \varepsilon'$ and putting $\lambda := \varepsilon'/\varepsilon$, it suffices to apply the Ekeland variational principle above with λ and with ε' in place of ε. □

At this stage, from the Ekeland variational principle we can derive the density of *C-subdifferentiability points* (that is, points x with $\partial_C f(x) \neq \emptyset$) in the effective domain of any proper lower semicontinuous function f.

PROPOSITION 2.223. *Let U be an open set of a Banach space X and $f : U \to \mathbb{R} \cup \{+\infty\}$ be a proper lower semicontinuous function. Then $\text{Dom } \partial_C f$ is graphically dense in dom f, that is, for any $x \in \text{dom } f$ there exists a sequence $(x_n)_n$ in $\text{Dom } \partial_C f$ such that $(x_n, f(x_n))_n$ converges to $(x, f(x))$.*

PROOF. Fix any $\bar{x} \in \text{dom } f$, any real $\varepsilon > 0$. By the lower semicontinuity of f there is a real $r > 0$ such that $B[\bar{x}, r] \subset U$ and $f(\bar{x}) - \varepsilon < f(x)$ for all $x \in B[\bar{x}, r]$, hence $f(\bar{x}) \leq \inf_{B[\bar{x},r]} f + \varepsilon$. Taking $0 < \lambda < \min\{\varepsilon, r/2\}$ there exists by the Ekeland variational principle (see Theorem 2.221) some $u \in B[\bar{x}, \lambda]$ with $f(u) \leq f(\bar{x}) \leq f(u) + \varepsilon$ and such that
$$f(u) \leq f(x) + (\varepsilon/\lambda)\|x - u\| \quad \text{for all } x \in B[\bar{x}, r].$$
Since $\|u - \bar{x}\| < r$, the point u is a local minimizer of $f + (\varepsilon/\lambda)\|\cdot - \bar{x}\|$, thus by the sum rule for C-subdifferential (see Theorem 2.98) $0 \in \partial_C f(u) + (\varepsilon/\lambda)\mathbb{B}$. It results that $\partial_C f(u) \neq \emptyset$, which justifies the proposition since $\|u - \bar{x}\| \leq \varepsilon$ and $|f(u) - f(\bar{x})| \leq \varepsilon$. □

2.5. OPTIMIZATION PROBLEMS WITH CONSTRAINTS

Before going to optimization problems with inequality/equality constraints, let us establish two results which have various consequences both in functional and variational analysis. They will be used in other chapters and they need the following definition.

DEFINITION 2.224. Given a closed convex set C of a Banach space X, a *nonzero* continuous linear functional $u^* \in X^*$ is called a *support functional of C* whenever it attains its supremum on C, that is, there exists at least some $\bar{x} \in C$ such that
$$\langle u^*, \bar{x} \rangle = \sup_{x \in C} \langle u^*, x \rangle, \quad \text{or equivalently} \quad u^* \in N(C; \bar{x}).$$
In such a case, one says that u^* supports C at $\bar{x}$ (or u^* is a support functional of C at $\bar{x}$) and that $\bar{x}$ is a *support point of C* (which is supported by u^*). Obviously, any support point of C belongs to the boundary of C since $N(C; x) = \{0\}$ for any point $x \in \text{int } C$.

If the interior of a closed convex set C is nonempty, any boundary point $\bar{x}$ of C is seen to be a support point by applying the Hahn-Banach separation theorem with $\bar{x} \notin \text{int } C$. However, there are closed bounded convex sets C with empty interior admitting boundary points which are not support points (see Example 2.210). Nevertheless, Theorem 2.227 below tells us that the set of support points is dense in the boundary of any closed convex set of a Banach space. Let us first start with the density of support functionals.

THEOREM 2.225 (E. Bishop and R.R. Phelps: variational principle for support functionals). *Let C be a nonempty bounded closed convex subset of a Banach space X. Then the set of support functionals of C is strongly dense in X^* (whenever X is not reduced to zero).*

In particular, the set of continuous linear functionals on X which attain their norm on $\mathbb{B}_X$ is strongly dense in X^*.

PROOF. Take any nonzero $u^* \in X^*$ and any real ε with $0 < \varepsilon < \|u^*\|$. The lower semicontinuous function $\langle -u^*, \cdot \rangle + \Psi_C$ being bounded from below, the second form of the Ekeland variational principle in Corollary 2.222 above gives us some $x_\varepsilon \in C$ at which the function $f := \langle -u^*, \cdot \rangle + \Psi_C(\cdot) + \varepsilon \| \cdot - x_\varepsilon \|$ attains its minimum. It follows that $0 \in \partial f(x_\varepsilon)$, which furnishes by Theorem 2.98 and Example 2.66(a) some $x^* \in N(C; x_\varepsilon)$ such that $0 \in x^* - u^* + \varepsilon \mathbb{B}_{X^*}$, that is, $\|x^* - u^*\| \leq \varepsilon$. Further, $x^* \neq 0$ since $\|x^*\| \geq \|u^*\| - \varepsilon > 0$. The first statement is then proved and the second is obtained with $C = \mathbb{B}_X$. □

REMARK 2.226. Let be given a proper lower semicontinuous function $\varphi : X \to \mathbb{R} \cup \{+\infty\}$ on the Banach space X such that $\lim_{\|x\| \to +\infty} \frac{\varphi(x)}{\|x\|} = +\infty$. For each $u^* \in X^*$, applying Corollary 2.222 as above with the function $\varphi + \langle u^*, \cdot \rangle$ which is lower semicontinuous and bounded from below on X, the proof reveals that the set of $x^* \in X^*$ such that $0 \in \partial_C(\varphi + \langle x^*, \cdot \rangle)(X)$ is strongly dense in X^*, or equivalently
$$\text{cl}_{\|\cdot\|_*}(\text{Range } \partial_C \varphi) = X^*.$$
So, if φ is in addition convex, then the set of $x^* \in X^*$ such that $\varphi + \langle x^*, \cdot \rangle$ attains its minimum is strongly dense in X^*. □

THEOREM 2.227 (E. Bishop and R.R. Phelps: variational principle for support points). *Let C be a nonempty closed convex subset of a Banach space X with $C \neq X$. Then the set of support points of C is dense in the boundary of C.*

PROOF. Consider any $\bar{x} \in \text{bdry}\, C$ and any real ε with $0 < \varepsilon < 1$. Choose a point $u \in B[\bar{x}, \varepsilon^2] \setminus C$. Then $\|\bar{x} - u\| \leq \|x - u\| + \varepsilon^2$ for all $x \in C$. According to the Ekeland variational principle in Theorem 2.221 there is some $x_\varepsilon \in C \cap B[\bar{x}, \varepsilon]$ which is a minimizer on X of the function f given by $f(x) := \|x-u\| + \Psi_C(x) + \varepsilon\|x - x_\varepsilon\|$. It results that $0 \in \partial f(x_\varepsilon)$. Since $u - x_\varepsilon \neq 0$ (keep in mind that $u \notin C$), from Theorem 2.98 and Example 2.66(a) there are $u^* \in X^*$ with $\|u^*\| = 1$ and $v^* \in \varepsilon \mathbb{B}_{X^*}$ such that $u^* + v^* \in N(C; x_\varepsilon)$, and clearly $\|u^* + v^*\| \geq 1 - \varepsilon > 0$. We conclude that x_ε is a support point of C with $\|x_\varepsilon - \bar{x}\| \leq \varepsilon$. □

2.5.4. General optimization problems. A general *mathematical minimization problem*, with inequality/equality and set constraints, generally appears in the form

$$(\mathcal{P}) \quad \begin{cases} \text{Minimize} & f_0(x) \\ \text{subject to} & f_1(x) \leq 0, \cdots, f_p(x) \leq 0, f_{p+1}(x) = 0, \cdots, f_{p+q}(x) = 0, x \in S, \end{cases}$$

where S is a subset of X and f_k is a real-valued function on X for every $k \in \{0, 1, \cdots, p+q\}$. Taking S' as the set of $x \in S$ satisfying the inequality and equality constraints, the problem $(\mathcal{P})$ amounts to the minimization problem under the constraint set S'. So, the global and local solutions of $(\mathcal{P})$ are global and local solutions of the previous problem as defined before Proposition 2.220. One says that S' is the set of *feasible* (or *admissible*) points of the problem $(\mathcal{P})$.

THEOREM 2.228 (F. Clarke: Lagrange multiplier with **slC**-subdifferential). Assume that X is a Banach space and $\bar{x}$ is a local solution of $(\mathcal{P})$. Assume also that the set S is closed and the functions f_k are Lipschitz continuous near $\bar{x}$. Then the following necessary optimality conditions hold:
There are real numbers $\lambda_0, \lambda_1, \cdots, \lambda_{p+q}$ (called *Lagrange multipliers*) not all zero such that

(a) $\lambda_0 \geq 0, \lambda_1 \geq 0, \cdots, \lambda_p \geq 0$;
(b) $\lambda_k f_k(\bar{x}) = 0$ for all $k = 1, \cdots, p$;
(c) $0 \in \sum\limits_{k=0}^{p+q} \lambda_k \partial_C f_k(\bar{x}) + N^C(S; \bar{x})$.

One also says that such a $(1+p+q)$-uple $(\lambda_0, \cdots, \lambda_{p+q})$ is a $(1+p+q)$-uple Lagrange multiplier (relative to the C-subdifferential) for the local solution $\bar{x}$.

PROOF. Put $\varepsilon_n = 1/n$ and put for all $x \in X$

$$g_n(x) := \max\{f_0(x) - f_0(\bar{x}) + \varepsilon_n^2,\, f_i(x),\, |f_j(x)| : i \in I,\, j \in J\},$$

where $I := \{1, \cdots, p\}$ and $J := \{p+1, \cdots, p+q\}$. Choose two reals $r > 0$ and $\gamma \geq 0$ such that, for all n, the function g_n is γ-Lipschitz continuous on $B[\bar{x}, 2r]$ and $\bar{x}$ is a global solution of $(\mathcal{P})$ with the constraint set $C := S \cap B[\bar{x}, 2r]$ in place of S. We observe that $\inf_C g_n \geq 0$ and $g_n(\bar{x}) = \varepsilon_n^2$. Then, for each fixed $n \in \mathbb{N}$, the Ekeland variational principle in Theorem 2.221 gives some $x_n \in C$ satisfying

$$\|x_n - \bar{x}\| \leq \varepsilon_n \quad \text{and} \quad g_n(x_n) \leq g_n(x) + \varepsilon_n\|x - x_n\| \; \forall x \in C,$$

and we have $g_n(x_n) > 0$ since otherwise x_n would be a feasible point better than $\bar{x}$ (for the problem with C in place of S). By Lemma 2.213, for large n, say $n \geq n_0$ we have $x_n \in B(\bar{x}, r)$ and

$$g_n(x) + \varepsilon_n\|x - x_n\| + \gamma d_C(x) \geq g_n(x_n) \; \forall x \in B(\bar{x}, r),$$

hence $0 \in \partial_C g_n(x_n) + \varepsilon_n \mathbb{B}_{X^*} + \gamma \partial_C d_S(x_n)$ according to Theorem 2.98 and Lemma 2.219. Fix any $n \geq n_0$. Using Proposition 2.190 and the inequality $g_n(x_n) > 0$, we obtain on the one hand real numbers $\lambda_{0,n} \geq 0$, $\lambda_{i,n} \geq 0$ for $i \in I$ with $\lambda_{i,n} = 0$ if $f_i(x_n) \leq 0$, and $\mu_{j,n} \geq 0$ for $j \in J$ with $\mu_{j,n} = 0$ if $f_j(x_n) = 0$ such that

$$\lambda_{0,n} + \sum_{i \in I} \lambda_{i,n} + \sum_{j \in J} \mu_{j,n} = 1,$$

and on the other hand $x^*_{0,n} \in \partial_C f_0(x_n)$, $x^*_{i,n} \in \partial_C f_i(x_n)$ for $i \in I$, and $u^*_{j,n} \in \partial_C |f_j|(x_n)$ for $j \in J$ such that for some $e^*_n \in \varepsilon_n \mathbb{B}_{X^*}$

$$\lambda_{0,n} x^*_{0,n} + \sum_{i \in I} \lambda_{i,n} x^*_{i,n} + \sum_{j \in J} \mu_{j,n} u^*_{j,n} + e^*_n \in -\gamma \partial_C d_S(x_n).$$

For each $j \in J$ with $f_j(x_n) \neq 0$, writing $|f_j| = | | \circ f_j$, we know by Theorem 2.188 that $\partial_C |f_j|(x_n) \subset \frac{f_j(x_n)}{|f_j(x_n)|} \partial_C f_j(x_n)$, thus for some $x^*_{j,n} \in \partial_C f_j(x_n)$ we have

$$\mu_{j,n} u^*_{j,n} = \lambda_{j,n} x^*_{j,n} \quad \text{where } \lambda_{j,n} := \mu_{j,n} f_j(x_n)/|f_j(x_n)|.$$

For each $j \in J$ with $f_j(x_n) = 0$, put $\lambda_{j,n} = 0$ and take any $x^*_{j,n} \in \partial_C f_j(x_n)$. We note that $\lambda_{0,n} + \sum_{i \in I} \lambda_{i,n} + \sum_{j \in J} |\lambda_{j,n}| = 1$ and that the sequence

$$(x^*_{0,n}, x^*_{1,n}, \cdots, x^*_{p,n}, \cdots, x^*_{p+q,n})_n$$

is bounded in $(X^*)^{1+p+q}$. We may then take a w^*-cluster point

$$(\lambda_0, \cdots, \lambda_{p+q}, x^*_0, \cdots, x^*_{p+q}) \text{ in } \mathbb{R}^{1+p+q} \times (X^*)^{1+p+q}$$

and we see that $\lambda^*_0 \geq 0$, $\lambda_i \geq 0$ for $i \in I$, and

$$\lambda_0 + \sum_{i \in I} \lambda_i + \sum_{j \in J} |\lambda_j| = 1.$$

Further, by the w^*-closedness property of the Clarke subdifferential of locally Lipschitz continuous functions (see Proposition 2.74(a)) we obtain $x^*_k \in \partial_C f_k(\overline{x})$ for all $k = 0, \cdots, p+q$ and

$$\sum_{k=0}^{p+q} \lambda_k x^*_k \in -\gamma \partial_C d_S(\overline{x}).$$

Finally, if for some $i \in I$ the inequality $f_i(\overline{x}) < 0$ holds, then for n large enough we have $f_i(x_n) < 0$ by continuity of f_i, hence $\lambda_{i,n} = 0$ ensuring in turn $\lambda_i = 0$. The proof is then complete. $\square$

Necessary optimality conditions for optimization problems with $\mathcal{C}^1$ (or more generally strictly differentiable) data follow immediately from the above theorem.

COROLLARY 2.229. *Assume that X is a Banach space and $\overline{x}$ is a local solution of $(\mathcal{P})$. Assume also that the set S is closed and the functions f_k are strictly Hadamard differentiable at $\overline{x}$. Then the following necessary optimality conditions hold:*
There are real numbers $\lambda_0, \lambda_1, \cdots, \lambda_{p+q}$ not all zero such that

(a) $\lambda_0 \geq 0, \lambda_1 \geq 0, \cdots, \lambda_p \geq 0$;
(b) $\lambda_k f_k(\overline{x}) = 0$ for all $k = 1, \cdots, p$;
(c) $0 \in \sum_{k=0}^{p+q} \lambda_k D f_k(\overline{x}) + N^C(S; \overline{x})$.

If $\lambda_0 = 0$, the objective function f_0 is not taken into account in (c) of the above corollary. So, an assumption on the data defining the constraints is generally needed to ensure $\lambda_0 \neq 0$. Such an assumption is usually called a *constraint qualification*. In the case of a $(1+p+q)$-uple multiplier $(1, \lambda_1, \cdots, \lambda_{p+q})$ (that is, $\lambda_0 = 1$) satisfying conditions (a), (b) and (c) in Theorem 2.228 the $(p+q)$-uple $(\lambda_1, \cdots, \lambda_{p+q})$ is called a *normalized Lagrange $(p+q)$-uple multiplier*. Clearly, when $\lambda_0 > 0$, dividing by $\lambda_0 > 0$ the conditions (a), (b) and (c) in Theorem 2.228 yields the $(p+q)$-uple $(1, \lambda_1/\lambda_0, \cdots, \lambda_{p+q}/\lambda_0)$ normalized Lagrange multiplier.

COROLLARY 2.230. *Assume that X is a Banach space and $\bar{x}$ is a local solution of $(\mathcal{P})$. Assume also that the functions f_k are strictly Hadamard differentiable at $\bar{x}$ and the following conditions (i) and (ii) are satisfied:*
(i) *there is a vector $\bar{v} \in T^C(S; \bar{x})$ with $Df_k(\bar{x})(\bar{v}) = 0$ for all $k = p+1, \cdots, p+q$ and such that*

$$D_k f_k(\bar{x})(\bar{v}) < 0 \quad \text{for each } k \in \{1, \cdots, p\} \text{ with } f_k(\bar{x}) = 0;$$

(ii) $\left(\sum_{k=p+1}^{p+q} \zeta_k Df_k(\bar{x}) \in -N^C(S; \bar{x}) \right) \Rightarrow \zeta_{p+1} = \cdots = \zeta_{p+q} = 0.$

Then there exists a $(p+q)$-uple normalized Lagrange multiplier for the local solution $\bar{x}$, that is, the following necessary optimality conditions hold:
There are real numbers $\lambda_1, \cdots, \lambda_{p+q}$ such that

(a) $\lambda_1 \geq 0, \cdots, \lambda_p \geq 0$;
(b) $\lambda_k f_k(\bar{x}) = 0$ for all $k = 1, \cdots, p$;
(c) $0 \in Df_0(\bar{x}) + \sum_{k=1}^{p+q} \lambda_k Df_k(\bar{x}) + N^C(S; \bar{x})$.

PROOF. Denote by K the set of $k \in \{1, \cdots, p\}$ such that $f_k(\bar{x}) = 0$. By the previous corollary there are reals $\lambda_0, \lambda_1, \cdots, \lambda_{p+q}$, not all null, satisfying

(2.82) $$0 \in \lambda_0 Df_0(\bar{x}) + \sum_{k=1}^{p+q} \lambda_k Df_k(\bar{x}) + N^C(S; \bar{x}),$$

and such that $\lambda_0 \geq 0$, $\lambda_k \geq 0$ for all $k \in \{1, \cdots, p\}$ and $\lambda_k = 0$ for all k in $\{1, \cdots, p\} \setminus K$. It suffices to show that $\lambda_0 > 0$ (since in this case the result is obtained after multiplying (2.82) by λ_0^{-1}). Suppose (by contradiction) $\lambda_0 = 0$. From (2.82) and (i) we derive that $\sum_{k \in K} \lambda_k Df_k(\bar{x})(\bar{v}) \geq 0$. Combining this with the inequalities $\lambda_k \geq 0$ and $D_k f_k(\bar{x})(\bar{v}) < 0$ for all $k \in K$, it ensues that $\lambda_k = 0$ for all $k \in K$, or equivalently $\lambda_k = 0$ for all $k \in \{1, \cdots, p\}$. Consequently, we obtain from (2.82) again the inclusion $0 \in \sum_{k=p+1}^{p+q} \lambda_k Df_k(\bar{x}) + N^C(S; \bar{x})$ with $(\lambda_{p+1}, \cdots, \lambda_{p+q}) \neq 0_{\mathbb{R}^q}$, which contradicts the assumption (ii) and finishes the proof. $\square$

Noticing that $N^C(S; \bar{x}) = 0$ for $S = X$ we obtain directly from Corollary 2.230 the following other corollary.

COROLLARY 2.231 (Optimality condition under Mangasarian-Fromovitz qualification:1). *Assume that X is a Banach space and $\bar{x}$ is a local solution of $(\mathcal{P})$ with $S = X$. Assume also that the functions f_k are strictly Hadamard differentiable at $\bar{x}$ and the so-called Mangasarian-Fromovitz constraint qualification is satisfied:*

The vectors $Df_{p+1}(\overline{x}), \cdots, Df_{p+q}(\overline{x})$ are linearly independent and there is a vector $\overline{v} \in X$ with $Df_k(\overline{x})(\overline{v}) = 0$ for all $k = p+1, \cdots, p+q$ and such that

$$D_k f_k(\overline{x})(\overline{v}) < 0 \quad \text{for each } k \in \{1, \cdots, p\} \text{ with } f_k(\overline{x}) = 0.$$

Then the following necessary optimality conditions hold:
There are real numbers $\lambda_1, \cdots, \lambda_{p+q}$ such that
  (a) $\lambda_1 \geq 0, \cdots, \lambda_p \geq 0$;
  (b) $\lambda_k f_k(\overline{x}) = 0$ for all $k = 1, \cdots, p$;
  (c) $0 = Df_0(\overline{x}) + \sum_{k=1}^{p+q} \lambda_k Df_k(\overline{x})$.

When there is no equality constraint, the Mangasarian-Fromovitz qualification condition is reduced to the Slater qualification condition. Under the latter condition, tangent and normal cones of constraint sets with merely inequalities have been described in terms of derivatives of the data in Proposition 2.194 and Proposition 2.196. We will see in Chapter 7 that the above Mangasarian-Fromovitz constraint qualification allows us also to describe, in terms of the derivatives of $f_1, \cdots, f_{p+q}$, the tangent and normal cones of the (constraint) set $S := \{x \in X : f_1(x) \leq 0, \cdots, f_p(x) \leq 0, f_{p+1}(x) = 0, \cdots, f_{p+q}(x) = 0\}$.

2.6. Clarke tangent cone in terms of Bouligand-Peano tangent cones

Let us now turn to study how and when the Clarke tangent cone can be expressed in terms of Bouligand-Peano tangent cones. In regard to that, recalling that

$$T^C(S;x) = \operatorname*{Lim\,inf}_{t\downarrow 0, u \xrightarrow{S} x} \frac{1}{t}(S-u) \quad \text{and} \quad T^B(S;x) = \operatorname*{Lim\,sup}_{t\downarrow 0} \frac{1}{t}(S-x)$$

leads us to investigate the comparison between $T^C(S;x)$ and $\operatorname*{Lim\,inf}_{u \xrightarrow{S} x} T^B(S;u)$. Daneš' theorem is an efficient way to approach the problem.

2.6.1. Daneš' drop theorem. Given a subset S of the vector space X and a point $a \in X$, one defines the *(closed) drop* $D(a, S)$ *with vertex* a *and basis* S as $D(a, S) := \overline{\operatorname{co}}(\{a\} \cup S)$, that is, the closed convex hull of $S \cup \{a\}$. It is easily seen that the equality $D(a, S) = \operatorname{co}(\{a\} \cup S)$ holds whenever the set S is convex, closed, and bounded since in such a case it is easily seen that $\operatorname{co}(\{a\} \cup S)$ is closed.

THEOREM 2.232 (J. Daneš' drop theorem). *Let S be a nonempty closed subset of a Banach space $(X \|\cdot\|)$, an element $a \in S$, and $B[b, r]$ be a closed ball centered at $b \in X$ with radius $r < d(b, S)$. Then there exists a point $c \in S \cap D(a, B[b, r])$ satisfying*

$$S \cap D(c, B[b, r]) = \{c\}.$$

PROOF. Applying, for any $\varepsilon > 0$, Corollary 2.222 of the Ekeland variational principle to the complete metric subspace $E := S \cap D(a, B[b, r])$ with the function $f := \|\cdot - b\|$ we obtain some $c \in E$ satisfying

$$\|c - b\| + \varepsilon\|c - a\| \leq \|a - b\|$$

and

(2.83) $$\|c - b\| < \|x - b\| + \varepsilon\|x - c\| \quad \text{for all } x \neq c \text{ in } E.$$

On the one hand, we can easily verify, since $c \in D(a, B[b,r])$, that

(2.84) $$D(c, B[b,r]) \subset D(a, B[b,r]).$$

On the other hand, for any $y = tc + (1-t)(b+u) \in D(c, B[b,r])$ with $\|u\| \leq r$ and $t \in [0,1[$ we have

$$\|y-b\| + \varepsilon \|y-c\| = \|t(c-b) + (1-t)u\| + \varepsilon \|(1-t)(b-c) + (1-t)u\|$$
$$\leq t\|c-b\| + \varepsilon(1-t)\|c-b\| + (1-t)(1+\varepsilon)r,$$

then in order that y does not fulfill (2.83) it suffices to choose $\varepsilon > 0$ small enough that

$$t\|c-b\| + \varepsilon(1-t)\|c-b\| + (1-t)(1+\varepsilon)r \leq \|c-b\|,$$

which is easily seen to be equivalent to $\varepsilon \leq (\|c-b\| + r)^{-1}(\|c-b\| - r)$. Since $\delta := d(b,S) \leq \|c-b\|$ and $\|c-b\| \leq \|a-b\|$, we have

$$0 < (r + \|a-b\|)^{-1}(\delta - r) \leq (r + \|c-b\|)^{-1}(\|c-b\| - r),$$

hence choosing $\varepsilon > 0$ satisfying $\varepsilon < (r + \|a-b\|)^{-1}(\delta - r)$ we deduce from what precedes that no element $y \neq c$ in $D(c, B[b,r])$ fulfills (2.83). Combining this with (2.84) yields

$$S \cap D(c, B[b,r]) = \{c\}$$

and finishes the proof. $\square$

Figure 2.3 illustrates the statement of Daneš drop property.

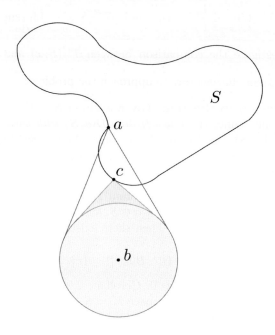

FIGURE 2.3. Daneš drop property.

2.6.2. *C*-tangent cone and limit inferior of *B*-tangent cones.
Given a set S of a normed space X and $x \in S$, the obvious inclusion $\operatorname*{Lim\,inf}_{u \to_S x} T^B(S; u) \subset T^B(S; x)$ tells us that

$$\operatorname*{Lim\,inf}_{u \to_S x} T^B(S; u) \subset T^C(S; x)$$

whenever S is tangentially regular at x.

In preparation for the extension of the latter inclusion to sets which are not tangentially regular, let us first observe the following lemma concerning the limit inferior of Bouligand-Peano tangent cones.

LEMMA 2.233. *Let S be a subset of a normed space X and let $x \in S$. If $h \in \operatorname*{Lim\,inf}_{u \to_S x} T^B(S; u)$, then for any neighborhood V of h there exists a neighborhood U of x such that for all $u \in U \cap S$ and all $\varepsilon > 0$ one has*

$$(u+]0, \varepsilon[V) \cap S \neq \emptyset.$$

PROOF. Let V be any neighborhood of zero in X and let any real $\varepsilon > 0$. Choose a neighborhood W of zero with $W + W \subset V$, and take a neighborhood U of x such that $(h + W) \cap T^B(S; u) \neq \emptyset$ for every $u \in U \cap S$. Choose for each $u \in U \cap S$ some $h_u \in (h + W) \cap T^B(S; u)$, hence by definition of $T^B(S; u)$ note that $(u+]0, \varepsilon[(h_u + W)) \cap S \neq \emptyset$. It results that $(u+]0, \varepsilon[(h + V)) \cap S \neq \emptyset$ for all $u \in U \cap S$, which translates the desired property. $\square$

Through the Daneš' theorem we show that the previous property is a characterization of Clarke's tangent cone of closed sets in Banach spaces.

PROPOSITION 2.234. *Assume that X is a Banach space and S is a subset which is closed near $x \in S$. Then a vector $h \in T^C(S; x)$ if and only if for any neighborhood V of h there exists a neighborhood U of x such that for all $u \in U \cap S$ and all $\varepsilon > 0$ one has*

$$(u+]0, \varepsilon[V) \cap S \neq \emptyset.$$

PROOF. Of course, any vector in $T^C(S; x)$ satisfies the property of the proposition. Suppose now $h \notin T^C(S; x)$. Without loss of generality we may suppose that S is closed. By Definition 2.1 of the Clarke tangent cone there exists some positive real $r < \|h\|$ such that for every $\delta > 0$, putting $\delta' := \delta/(1 + \|h\| + r)$, there are $x_\delta \in S \cap B(x, \delta')$ and $t_\delta \in {]0, \delta']}$ for which

$$S \cap (x_\delta + t_\delta B[h, 2r]) = \emptyset, \text{ or equivalently } B[x_\delta + t_\delta h, 2t_\delta r] \cap S = \emptyset.$$

Daneš' drop theorem yields some $u_\delta \in S \cap D(x_\delta, B[x_\delta + t_\delta h, t_\delta r])$ with

$$S \cap D(u_\delta, B[x_\delta + t_\delta h, t_\delta r]) = \{u_\delta\},$$

and we observe that $\|u_\delta - x\| < \delta$ since $u_\delta \in x_\delta + [0, t_\delta] B[h, r]$. We claim that

(2.85) $$S \cap B(u_\delta, t_\delta r) \cap (u_\delta + [0, 1] B[t_\delta h, t_\delta r]) = \{u_\delta\}.$$

The boundedness of $B := B[x_\delta + t_\delta h, t_\delta r]$ ensures that $D(x_\delta, B) = \operatorname{co}(\{x_\delta\} \cup B)$, then $u_\delta = (1 - \tau)b + \tau x_\delta$ for some $b \in B$ and some $\tau \in {]0, 1]}$, where the inequality $\tau > 0$ is due to the fact that $u_\delta \notin B$ since $S \cap B = \emptyset$. Then we obtain

$$B[t_\delta h, t_\delta r] = B - x_\delta = B - \frac{1}{\tau} u_\delta + \frac{1}{\tau}(1 - \tau) b = \frac{1}{\tau}(\tau B + (1 - \tau) b - u_\delta) \subset \frac{1}{\tau}(B - u_\delta).$$

So, for any u in the left member of (2.85) there are some $s \geq 0$ and $b \in B$ such that $u - u_\delta = s(b - u_\delta)$. Since $\|u - u_\delta\| < t_\delta r$ and $\|b - u_\delta\| \geq t_\delta r$ (because $S \cap (B + t_\delta r \mathbb{B}) = \emptyset$), we have $s < 1$, thus $u = (1-s)u_\delta + sb \in S \cap D(u_\delta, B) = \{u_\delta\}$, which proves the claim.

Choosing some positive real $\delta_0 < 1$ such that $[0, \delta_0]B[h, r] \subset B(0, r)$ we see that $u_\delta +]0, t_\delta \delta_0[B[h, r] \subset B(u_\delta, t_\delta r)$, which combined with (2.85) entails that

$$S \cap \left(u_\delta +]0, t_\delta \delta_0[B(h, r) \right) = \emptyset$$

since $0 \notin B(h, r)$. So, there exists $V := B(h, r)$ such that for every $\delta > 0$ there are $u_\delta \in S \cap B(x, \delta)$ and $\varepsilon_\delta := t_\delta \delta_0 > 0$ with $S \cap \left(u_\delta +]0, \varepsilon_\delta[B(h, r) \right) = \emptyset$. This means that the property of the proposition is not satisfied and the proof is finished. □

We can now state and prove the theorem concerning connexion between the limit inferior of Bouligand-Peano tangent cones and the Clarke tangent cone.

THEOREM 2.235 (limit inferior of B-tangent cones). *Assume that S is a subset of a Banach space $(X, \|\cdot\|)$ which is closed near $x \in S$. Then*

$$\operatorname*{Lim\,inf}_{u \xrightarrow{S} x} T^B(S; u) \subset T^C(S; x).$$

If in addition X is finite-dimensional, the latter inclusion is an equality.

PROOF. The inclusion follows directly from the lemma and proposition above.

Now assume that X is finite-dimensional and fix $h \in T^C(S; x)$. Take any closed bounded neighborhood V of h and by definition of Clarke tangent cone choose some neighborhood U of x and some $\varepsilon > 0$ such that for any $u \in S \cap U$ and $t \in]0, \varepsilon[$ we have $V \cap t^{-1}(S - u) \neq \emptyset$. Consider any fixed element $u \in S \cap U$ and take a sequence $(t_n)_n$ in $]0, \varepsilon[$ tending to 0. Then, for each n there exists $h_n \in V \cap t_n^{-1}(S - u)$. Denoting by h' a cluster point of the bounded sequence $(h_n)_n$, we see that $h' \in T^B(S; u) \cap V$, therefore $T^B(S; u) \cap V \neq \emptyset$. This translates by Definition 1.2 that $h \in \operatorname*{Lim\,inf}_{S \ni u \to x} T^B(S; u)$. □

EXAMPLE 2.236 (J.S. Treiman). In any infinite-dimensional Banach space X the inclusion in Theorem 2.235 above fails to be an equality. Consider a linearly independent sequence $a, b, e_1, e_2, \cdots$ in the unit sphere of X such that
(2.86)
$$d(a, \operatorname{Vect}\{b, e_1, e_2, \cdots\}) > 1/2 \text{ and } d(e_{n+1}, \operatorname{Vect}\{a, b, e_1, \cdots, e_n\}) > 1/2 \,\forall n,$$

where $\operatorname{Vect} Q$ denotes the vector space spanned (that is, generated) by a subset Q of X. For each $m \in \mathbb{N}$ put

$$S_m := \left\{ \frac{1}{m} a \right\} \cup \left(\frac{1}{m} a + \bigcup_{k \in \mathbb{N}} \left(\left[\frac{1}{k}, +\infty \right[\left\{ \frac{1}{m^2} x_k + b \right\} \right) \right).$$

It can be checked that the set

$$S := \left(\bigcup_{m \in \mathbb{N}} S_m \right) \cup \{ rb : r \geq 0 \}$$

is closed along with

$$T^C(S; 0) = \{ rb : r \geq 0 \} \quad \text{and} \quad T^B \left(S; \frac{1}{m} a \right) = \{ 0 \}.$$

The latter equality implies $\operatorname*{Liminf}_{S \ni u \to x} T^B(S;u) = \{0\}$, so the inclusion in the theorem above is not an equality for that set S. □

REMARK 2.237. (a) Assume that X is a finite-dimensional vector space and $f : X \to \mathbb{R} \cup \{+\infty\}$ be a proper lower semicontinuous function. Let $(x_n)_n$ be a sequence of points where f is Fréchet differentiable with $x_n \to x$ and $f(x_n) \to f(x)$ as $n \to \infty$ and such that $\lim_{n \to \infty} Df(x_n)$ exists. Take any $(h,r) \in T^C(\operatorname{epi} f; (x, f(x)))$. By Theorem 2.235 there exists a sequence $(h_n, r_n)_n$ converging to (h,r) with $(h_n, r_n) \in T^B(\operatorname{epi} f; (x_n, f(x_n)))$. For each $n \in \mathbb{N}$, it is easily seen from the definition of B-tangent cones that $\langle Df(x_n), h_n \rangle \leq r_n$, so

$$\left\langle \lim_{n \to \infty} Df(x_n), h \right\rangle - r \leq 0.$$

This means that $\lim_{n \to \infty} Df(x_n) \in \partial_C f(x)$, hence

$$\operatorname*{Limsup}_{u \to_f x} \{Df(u)\} \subset \partial_C f(x).$$

(b) A more general result will be established in Theorem 4.120(b). □

Although Example 2.236 shows that the inclusion in Theorem 2.235 fails in general to be an equality in infinite dimensions, the next proposition proves that the equality holds true for the class of sets with the interior tangent property.

PROPOSITION 2.238. Let S be a subset of a Banach space $(X, \|\cdot\|)$ which is closed near $x \in S$. If S satisfies the interior tangent property at x, then one has the equality

$$\operatorname*{Liminf}_{u \to_S x} T^B(S;u) = T^C(S;x).$$

PROOF. Fix any $h \in I(S;x)$. There is an $\varepsilon > 0$ such that $S \cap U +]0, \varepsilon[V \subset S$, where $U := B(x, \varepsilon)$ and $V := B(h, \varepsilon)$. For every $u \in U \cap S$, this obviously entails that $h \in T^B(S;u)$, and hence $h \in \operatorname*{Liminf}_{u \to_S x} T^B(S;u) =: K$. This means that $I(S;x) \subset K$. Using the closedness of the set K (see Proposition 1.7) and the fact that $T^C(S;x)$ is the closure of $I(S;x) \neq \emptyset$ (see Theorem 2.13), we obtain that $T^C(S;x) \subset K$, which yields the equality of the proposition according to the reverse inclusion from Theorem 2.235. □

2.7. Basic tangential properties through measure theory

It is readily seen that $T^B(S;x) = \{0\}$ whenever x is an isolated point in S. As known (and proved in Lemma 2.244 below), the set of isolated points in S is countable, hence it can be seen as a negligible set. Given a closed subspace F of the normed space X, the present section will study how negligible is the set of points $x \in S$ where $F \cap T^B(S;x)$ is reduced to zero.

2.7.1. Points of nullity of symmetrized B-tangent cone.
For the set of points x where $F \cap T^B(S;x)$ is zero, we begin with the case where F is the whole space X, which corresponds to the set of $x \in S$ where $T^B(S;x)$ is zero. In fact, we will study more generally in this case $F = X$, the negligibility of the set of points $x \in S$ where the symmetrized B-tangent cone $T^B(S;x) \cap -T^B(S;x)$ is reduced to zero.

Before proving the first proposition preparing the result of this section related to such properties, recall that a point x of a set S of a topological space X is *isolated* in S provided there exists some neighborhood V of x such that $V \cap S = \{x\}$.

PROPOSITION 2.239. *Assume that the normed space X is finite-dimensional and let S be a subset of X with $x \in S$.*
(a) *The point x is isolated in S if and only if $T^B(S;x) = \{0\}$.*
(b) *If $K(x)$ is a nonempty closed cone such that $(K(x) \setminus \{0\}) \subset X \setminus T^B(S;x)$, then the point x is an isolated point in $S \cap (x + K(x))$.*

PROOF. (a) If x is not isolated in S, we can find a sequence of points $(x_n)_n$ in S converging to x with $x_n \neq x$. Extracting a subsequence, we may also suppose that $h_n := (x_n - x)/\|x_n - x\|$ converges to some $h \in X$ with $\|h\| = 1$ and $h \in T^B(S;x)$ since, for $t_n := \|x_n - x\|$, we see that $x + t_n h_n = x_n \in S$, justifying (a). (Note that the assertion (a) can also be seen as a consequence of the assertion (b) by taking $K(x) = X$ in (b). It seemed to be better to us to give a direct proof).
(b) Suppose that $(K(x) \setminus \{0\}) \subset X \setminus T^B(S;x)$ and that $K(x) \neq \{0\}$. Consider the nonvoid compact set $Q := K(x) \cap \{h \in X : \|h\| = 1\}$. For each $h \in Q$ choose some positive integer $m(h)$ such that $(x+]0, 1/m(h)[B(h, 1/m(h))) \cap S = \emptyset$. Choose also $h_1, \cdots, h_k \in Q$ such that $Q \subset B(h_1, 1/m(h_1)) \cup \cdots \cup B(h_k, 1/m(h_k))$. Putting $p := \max\{m(h_1), \cdots, m(h_k)\}$ we obtain

$$(2.87) \qquad S \cap \left(x+]0, 1/p[\left(\bigcup_{i=1}^{k} B\left(h_i, \frac{1}{m(h_i)}\right) \right) \right) = \emptyset,$$

which entails

$$(x+]0, 1/p[Q) \cap S = \emptyset, \quad \text{hence} \quad B(x, 1/p) \cap (x + K(x)) \cap S = \{x\}.$$

The latter equality guarantees that x is an isolated point in $S \cap (x + K(x))$. □

EXAMPLE 2.240. The properties (a) and (b) in Proposition 2.239 above fail in the infinite-dimensional setting, even for Hilbert space. Consider any infinite-dimensional separable Hilbert space H and a countable orthonormal basis $(e_n)_{n \in \mathbb{N}}$ of H, and set $S := \{0\} \cup \{n^{-1} e_n : n \in \mathbb{N}\}$. Then 0 is not isolated in S while $T^B(S; 0) = \{0\}$, which says that (a) fails. We also see that (b) is not fulfilled since, for $K := X$ and $x = 0$, we have $(K \setminus \{0\}) \subset X \setminus T^B(S;x)$ whereas $S \cap (x + K) = S$ holds. □

In the finite-dimensional setting, the assertion (a) above says that $x \in S$ is isolated in S whenever $T^B(S;x)$ is reduced to zero, so the set of $x \in S$ such that $T^B(S;x) = \{0\}$ is countable (see Lemma 2.244 below). Proposition 2.243 below states that even the bigger set of $x \in S$ such that $T^B(S;x) \cap -T^B(S;x) = \{0\}$ is countable. This fact will be a consequence of the following more general result.

PROPOSITION 2.241. *Let X be a finite-dimensional normed space and S be a nonempty closed subset. Let $(K(x))_{x \in S}$ be a family of nonempty closed symmetric cones of X and let $S_0 := \{x \in S : K(x) \cap T^B(S;x) = \{0\}\}$. Then there exists a sequence of closed subsets $(S_i)_{i \geq 1}$ of S such that $S_0, S_1, \cdots$ cover the set*

$$\{x \in S : K(x) \cap T^B(S;x) \cap -T^B(S;x) = \{0\}\}$$

and such that, for each integer $i \geq 0$,

$$K(x) \cap T^B(S_i; x) = \{0\} \quad \text{for all } x \in S_i.$$

2.7. BASIC TANGENTIAL PROPERTIES THROUGH MEASURE THEORY

PROOF. Let $(U_m)_{m\in\mathbb{N}}$ be a countable basis of nonempty open sets of X and $(h_n)_{n\in\mathbb{N}}$ be a dense sequence of the unit sphere of X.

Consider first a family of nonempty closed cones $(P(x))_{x\in S}$ with $P(x) \subset K(x)$ and put
$$E := \{x \in S : P_0(x) \cap T^B(S;x) = \emptyset\},$$
where $P_0(x) := P(x) \setminus \{0\}$. Let $E_0 := \{x \in E : P(x) \neq \{0\}\}$. Fix, for a moment, any $x \in E_0$. We have $P(x) \neq \{0\}$ with $P(x) \subset K(x)$ and $\{h \in P(x) : \|h\| = 1\} \cap T^B(S;x) = \emptyset$. Thus, as in the proof of (2.87) in Proposition 2.239(b), through the compactness of $\{h \in P(x) : \|h\| = 1\}$ we obtain unit vectors $h_1, \cdots, h_k \in P(x)$ and positive integers $p, m(h_1), \cdots, m(h_k)$ such that

$$\{h \in P(x) : \|h\| = 1\} \subset \bigcup_{i=1}^{k} B\left(h_i, \frac{1}{m(h_i)}\right) \text{ and } S \cap \left(x +]0, 1/p[\left(\bigcup_{i=1}^{k} B\left(h_i, \frac{1}{m(h_i)}\right)\right)\right) = \emptyset.$$

This easily furnishes a real $r > 0$ and a finite subset $J \subset \mathbb{N}\times\mathbb{N}$ such that $P_0(x) \subset W_J$ and $B(x, r) \cap (x + W_J) \cap S = \emptyset$, where $W_J :=]0, +\infty[(\bigcup_{(i,n)\in J} B(h_i, 1/n))$. Then for each U_m in the countable basis of open sets such that $x \in U_m \subset B(x,r)$, we have $U_m \cap (x + W_J) \cap S = \emptyset$. So, a point $x \in S$ with $P_0(x) \neq \emptyset$ belongs to E_0 if and only if there are $m \in \mathbb{N}$ and a nonempty finite subset J of $\mathbb{N}\times\mathbb{N}$ such that $x \in U_m$, $P_0(x) \subset W_J$ and $U_m \cap (x + W_J) \cap S = \emptyset$, where W_J is defined as above. This leads to consider, for each $m \in \mathbb{N}$ and each nonempty finite subset J of $\mathbb{N} \times \mathbb{N}$, the set
$$A_{m,J} := \{x \in S \cap U_m : \emptyset \neq P_0(x) \subset W_J, U_m \cap (x + W_J) \cap S = \emptyset\}.$$
Denoting by Δ the set of pairs (m, J) such that $A_{m,J} \neq \emptyset$ with $m \in \mathbb{N}$ and $J \subset \mathbb{N}\times\mathbb{N}$, we have $E_0 = \bigcup_{(m,J)\in\Delta} A_{m,J}$. For each $(m, J) \in \Delta$ put
$$S_{m,J} := \{x \in \mathrm{cl}(S \cap U_m) : U_m \cap (x + W_J) \cap S = \emptyset\},$$
so $A_{m,J} \subset S_{m,J}$, hence $E \subset \Sigma_0 \bigcup (\bigcup_{(m,J)\in\Delta} S_{m,J})$, where $\Sigma_0 := \{x \in S : P(x) = \{0\}\}$. Fix any $(m, J) \in \Delta$. By Lemma 2.242 below the set $S_{m,J}$ is closed and for each $x \in S_{m,J}$ we have $(x - W_J) \cap S_{m,J} = \emptyset$, hence in particular $P(x) \cap -T^B(S_{m,J}; x) = \{0\}$.

Take now $K(x) \cap -T^B(S;x)$ as substitute for $P(x)$. For the sets $S_{m,J}$ so obtained, we have for each $x \in S_{m,J}$
$$K(x) \cap -T^B(S;x) \cap -T^B(S_{m,J};x) = \{0\},$$
or equivalently $K(x) \cap T^B(S_{m,J};x) = \{0\}$ since $K(x)$ is a symmetric set and $T^B(S_{m,J};x) \subset T^B(S;x)$. The countable family formed by S_0 and $S_{m,J}$ with $(m, J) \in \Delta$ fulfills the desired properties of the proposition. □

LEMMA 2.242. *Let U, W be nonempty open sets of a normed space X and let S, S_U be nonempty subsets of X with $S \cap U \subset S_U \subset \mathrm{cl}(S \cap U)$. Then the set*
$$C = \{x \in S_U : U \cap (x + W) \cap S = \emptyset\}$$
is closed in S_U with respect to the induced topology, and for each $x \in C$ one has $(x + W) \cap C = \emptyset$ and $(x - W) \cap C = \emptyset$.

PROOF. For $G := \{x \in X : U \cap (x + W) \cap S \neq \emptyset\}$, fix any $x \in G$ and take $u \in U \cap (x + W) \cap S$. Choosing $\delta > 0$ such that $B(u, \delta) \subset x + W$ we see, for any $x' \in B(x, \delta)$, that
$$u = x' + (u + (x - x')) - x \in x' + B(u, \delta) - x \subset x' + (x + W) - x = x' + W,$$

thus $u \in U \cap (x' + W) \cap S$, which ensures that $x' \in G$ and then G is open. Consequently, the set C is closed in S_U with respect to the induced topology.

Fix any $x \in C$. We claim that $(x + W) \cap C = \emptyset$. By definition of C, the inclusion $x \in C$ ensures that we have $(x+W) \cap (S \cap U) = \emptyset$ and this combined with the openness of $x+W$ easily yields $(x+W) \cap S_U = \emptyset$. So, we obtain $(x+W) \cap C = \emptyset$ since $C \subset S_U$ by definition of C.

Finally, it remains to prove that $(x - W) \cap C = \emptyset$ too. Otherwise, there is some $u \in (x - W) \cap C$, or equivalently $x \in u + W$ and $u \in C$. On one hand the inclusion $x \in u + W$ entails $(u + W) \cap C \neq \emptyset$ since x has been fixed in C, and on the other hand the inclusion $u \in C$ implies by the claim above the contradiction $(u + W) \cap C = \emptyset$. $\square$

PROPOSITION 2.243. *Let X be a finite-dimensional normed space and S be a nonempty closed subset. Then the set*

$$\{x \in S : T^B(S;x) \cap -T^B(S;x) = \{0\}\}$$

is countable.

PROOF. For the countable family $(S_i)_{i \in \mathbb{N} \cup \{0\}}$ furnished by Proposition 2.241 above with $K(x) = X$ for all x, we have $T^B(S_i;x) = X \cap T^B(S_i;x) = \{0\}$ for every $x \in S_i$, which ensures that any point in S_i is isolated in S_i according to Proposition 2.239. Therefore, by Lemma 2.244 below each S_i is a countable set, justifying the countability of the set in the proposition. $\square$

LEMMA 2.244. *Let T be a topological space with a countable basis of open sets (which is the case of any separable metrizable space). For any subset S of T, the set of points of S which are isolated in S is a countable set.*

PROOF. Denote by S_0 the set of points of S which are isolated in S. Suppose that S_0 is nonempty and fix a countable basis $(U_n)_{n \in \mathbb{N}}$ of open sets of T. For each point $x \in S_0$ choose some $p(x) \in \mathbb{N}$ such that $S \cap U_{p(x)} = \{x\}$. The mapping $p : S_0 \to \mathbb{N}$ being obviously one-to-one, the set S_0 is countable. $\square$

2.7.2. Lipschitz surfaces. In the same spirit of the previous subsection only concerned with the set of points $x \in S$ with $X \cap T^B(S;x) = \{0\}$, we will consider, for a finite-dimensional vector space F, the structure of the set of points $x \in S$ such that $F \cap T^B(S;x)$ is the null space. The structure will be related to Lipschitz surfaces. In order to define such surfaces, let us recall the concept of topological vector complements. For two algebraic complement vector subspaces E and F of a vector space X, so $X = E \oplus F$, we denote as usual by $\pi^{E,F} : X \to E$ and $\pi^{F,E} : X \to F$ the linear mappings such that

$$(2.88) \quad x = \pi^{E,F}(x) + \pi^{F,E}(x) \quad \text{with } \pi^{E,F}(x) \in E, \ \pi^{F,E}(x) \in F \quad \text{for all } x \in X.$$

The mapping $\pi^{E,F}$ is generally called the *projector mapping onto E parallel to F*. When in addition X is a normed space and the mappings $\pi^{E,F}$ and $\pi^{F,E}$ are *continuous*, we say that E and F are *topological complement vector subspaces of X* or are *topologically complemented in X*. We recall that any finite-dimensional vector subspace of the normed space X has a topological complement vector subspace in X. When X is a Banach space, it is known that two vector subspaces of X are topologically complemented if and only if they are closed and algebraically complemented. It is also known that, in the Banach space X, any closed vector

2.7. BASIC TANGENTIAL PROPERTIES THROUGH MEASURE THEORY

subspace E which is *finite codimensional* (that is, the quotient space X/E is finite-dimensional) has a topological vector complement. We refer, for example, to [**174**] for the results recalled above on complement vector subspaces.

DEFINITION 2.245. Let X be a normed space and F be a nonzero closed vector subspace of X. A nonempty set C of X is called a *Lipschitz F-surface* when there exist a topological complement vector subspace E of F in X and a Lipschitz mapping $f : E \to F$ such that $C = \{x + f(x) : x \in E\}$.

When $\dim F = p \geq 1$, one says that C is a *p-codimensional Lipschitz surface* or a *Lipschitz surface of codimension p* of X; a Lipschitz surface of codimension 1 is called a *Lipschitz hypersurface* of X.

When X is finite-dimensional, it is readily seen that any Lipschitz F-surface with $F = X$ is a singleton set of X.

It is worth pointing out diverse basic properties of Lipschitz surfaces in the next two propositions.

PROPOSITION 2.246. *Let X be a normed space, F be a vector subspace of X of dimension $p > 1$ and F_0 be a vector subspace of F of dimension m with $p > m \geq 1$. Then any Lipschitz F-surface of X is contained in a Lipschitz F_0-surface.*

In particular, any Lipschitz surface of X of codimension $p > m \geq 1$ is a subset of a Lipschitz surface of codimension m.

PROOF. Let C be a Lipschitz F-surface of X. Choose, according to the above definition, a topological vector complement E of F in X (so $X = E \oplus F$) and a Lipschitz mapping $f : E \to F$ such that $C = \{x + f(x) : x \in E\}$; note the continuity of the linear mappings $\pi^{E,F} : X \to E$ and $\pi^{F,E} : X \to F$. Take a vector complement G of F_0 in F, so $F = G \oplus F_0$, $\dim G = p - m$, and the well-defined linear mappings $\pi^{G,F_0} : F \to G$ and $\pi^{F_0,G} : F \to F_0$ are continuous since F is finite-dimensional. We observe that, for the topological sum $E_0 := E \oplus G$, we have $X = E_0 \oplus F_0$ and the latter sum is topological since any $x \in X$ can be written as

$$x = \pi^{E,F}(x) + \pi^{F,E}(x) = \left(\pi^{E,F}(x) + \pi^{G,F_0}(\pi^{F,E}(x))\right) + \pi^{F_0,G}(\pi^{F,E}(x))$$

with $x \mapsto \pi^{E_0,F_0}(x) = \pi^{E,F}(x) + \pi^{G,F_0}(\pi^{F,E}(x))$ continuous, and hence also $x \mapsto x - \pi^{E_0,F_0} = \pi^{F_0,E_0}(x)$. Considering the Lipschitz mapping $f_0 : E_0 \to F_0$ defined by $f_0(u) := (\pi^{F_0,E_0} \circ f \circ \pi^{E,F})(u)$ for all $u \in E_0$, the set $S := \{u + f_0(u) : u \in E_0\}$ is a Lipschitz F_0-surface (hence an m-codimensional Lipschitz surface) of X. We claim that $C \subset S$. Indeed, fix any $x \in E$. Since $f(x) \in F = G \oplus F_0$, we can write $f(x) = f_G(x) + f_{F_0}(x)$ with $f_G(x) \in G$ and $f_{F_0}(x) \in F_0$ and hence $\pi^{F_0,E_0}(f(x)) = f_{F_0}(x)$. For $u(x) := x + f_G(x) \in E_0$, we have $\pi^{E,F}(u(x)) = x$ since $x \in E$ and $f_G(x) \in G \subset F$. Then we see that

$$x + f(x) = (x + f_G(x)) + f_{F_0}(x) = u(x) + (\pi^{F_0,E_0} \circ f \circ \pi^{E,F})(u(x)) = u(x) + f_0(u(x)),$$

which implies the desired inclusion $C \subset S$. This inclusion justifies that C is contained in a Lipschitz F_0-surface, so the first assertion is established. The second assertion is a direct consequence of the first. □

PROPOSITION 2.247. *Let X be a normed space, F be a nonzero closed vector subspace of X which has a topological vector complement in X, and C be a subset of X. The following assertions are equivalent:*
(a) *the set C is a Lipschitz F-surface;*

(b) there exists a topological vector complement E of F in X such that the restriction $\pi^{E,F}|_C : C \to E$ of $\pi^{E,F}$ to C is a bijection and its inverse is Lipschitz;
(c) for any topological vector complement E of F in X, the mapping $\pi^{E,F}|_C : C \to E$ is a bijection and its inverse is Lipschitz;
(d) for any topological vector complement E of F in X, there exists a Lipschitz mapping $f : E \to F$ such that $C = \{x + f(x) : x \in E\}$.

PROOF. Suppose that (a) holds, so there exists a topological vector complement E of F in X and a Lipschitz mapping $f : E \to F$ such that $C = \{x + f(x) : x \in E\}$. Obviously $\pi^{E,F}|_C$ is a bijection and its inverse is Lipschitz since $(\pi^{E,F}|_C)^{-1}(u) := u + f(u)$ for all $u \in E$. So (a) implies (b).

Suppose that there exits a topological vector complement E_0 of F in X fulfilling (b), that is, $\pi^{E_0,F}|_C : C \to E_0$ is a bijection and its inverse is Lipschitz. Take any topological vector complement E of F in X. Write any $x \in X$ as $x = \pi^{E_0,F}(x) + \pi^{F,E_0}(x)$. Then taking $x \in E_0$ we have

$$\pi^{E_0,F}|_E \left(\pi^{E,F}|_{E_0}(x) \right) = \pi^{E_0,F}\left(\pi^{E,F}(x)\right) = \pi^{E_0,F}\left(x - \pi^{F,E}(x)\right) = x$$

(the latter equality being due to the fact that $x \in E_0$ and $-\pi^{F,E}(x) \in F$), and similarly $\pi^{E,F}|_{E_0}\left(\pi^{E_0,F}|_E(y)\right) = y$ for all $y \in E$; so $\pi^{E,F}|_{E_0} : E_0 \to E$ is a linear isomorphism with $\left(\pi^{E,F}|_{E_0}\right)^{-1} = \pi^{E_0,F}|_E$. On the other hand, for any $u \in C$, we can write

$$\pi^{E,F}|_{E_0}\left(\pi^{E_0,F}|_C(u)\right) = \pi^{E,F}\left(\pi^{E_0,F}(u)\right) = \pi^{E,F}\left(u - \pi^{F,E_0}(u)\right)$$
$$= \pi^{E,F}(u) = \pi^{E,F}|_C(u),$$

so the equality $\pi^{E,F}|_C = \left(\pi^{E,F}|_{E_0}\right) \circ \left(\pi^{E_0,F}|_C\right)$ says that $\pi^{E,F}|_C : C \to E$ is a bijection and its inverse $\left(\pi^{E,F}|_C\right)^{-1} = \left(\pi^{E_0,F}|_C\right)^{-1} \circ \left(\pi^{E_0,F}|_E\right)$ is Lipschitz thanks to the above Lipschitz property of $\left(\pi^{E_0,F}|_C\right)^{-1}$. The properties in (c) are then justified.

Suppose that (c) holds and take any topological vector complement E of F in X. By (c) we know that $\pi^{E,F}|_C : C \to E$ is bijective and $q_C := (\pi^{E,F}|_C)^{-1} : E \to C$ is Lipschitz; further, for any $u \in E$,

$$\pi^{E,F}(q_C(u) - u) = \pi^{E,F}(q_C(u)) - \pi^{E,F}(u) = \pi^{E,F}|_C(q_C(u)) - u = u - u = 0,$$

so $q_C(u) - u \in F$ for all $u \in E$. Then, we can define the mapping $f : E \to F$ with $f(u) := q_C(u) - u$ for all $u \in E$, and this mapping f is Lipschitz with $C = \{q_C(u) : u \in E\} = \{u + f(u) : u \in E\}$. It ensues that the set C is a Lipschitz F-surface, which ensures (d).

Finally, the assertion (d) trivially implies (a). □

Now we establish a first property concerning the set of points x where the intersection of $T^B(S;x)$ with a fixed dimensional vector subspace is reduced to zero.

PROPOSITION 2.248. *Let S be a nonempty subset of a normed space X and F be a nonzero finite-dimensional vector subspace of X. Then the set*

$$C := \{x \in S : F \cap T^B(S;x) = \{0\}\}$$

can be covered by countably many Lipschitz F-surfaces.

PROOF. The vector subspace F being finite-dimensional, we can choose a topological complement vector space E so that $X = E \oplus F$, which allows us to write any $x \in X$ in one way as $x = x_E + x_F$ with $x_E \in E$, $x_F \in F$. Note the continuity of the mappings $\pi^{E,F} : X \to E$ and $\pi^{F,E} : X \to F$ with $\pi^{E,F}(x) = x_E$ and $\pi^{F,E}(x) = x_F$ for all $x \in X$. For each $k \in \mathbb{N}$, put

$$C_k := \{x \in C : (x' \in S \text{ and } \|x'_F - x_F\| \le 1/k) \Rightarrow \|x'_F - x_F\| \le k\,\|x'_E - x_E\|\}.$$

Obviously, $\bigcup_k C_k \subset C$. In fact the equality holds. Indeed, suppose the contrary, that is, there is some $x \in C$ with $x \notin \bigcup_k C_k$. Then, for each $k \in \mathbb{N}$, there exists $x^k \in S$ with $\|x^k_F - x_F\| \le 1/k$ and $\|x^k_F - x_F\| > k\,\|x^k_E - x_E\|$. For $t_k = \|x^k_F - x_F\|$, $e_k := t_k^{-1}(x^k_E - x_E)$ and $h_k := t_k^{-1}(x^k_F - x_F)$, we note that $t_k > 0$ with $t_k \downarrow 0$, $e_k \to 0$ since $\|e_k\| < 1/k$ by what precedes, and we also note that $x + t_k(e_k + h_k) = x^k \in S$. Since F is finite-dimensional and $\|h_k\| = 1$ with $h_k \in F$, some subsequence $(h_{s(k)})_k$ converges as $k \to \infty$ to some $h \in F$ with $\|h\| = 1$. Consequently, $(e_{s(k)} + h_{s(k)})_k$ converges to h too. This combined with the inclusion $x + t_{s(k)}(e_{s(k)} + h_{s(k)}) \in S$ guarantees that $h \in T^B(S; x)$, which contradicts the equality $F \cap T^B(S; x) = \{0\}$ since $h \in F$ with $\|h\| = 1$. So, the equality $C = \bigcup_k C_k$ holds true.

For each $k \in \mathbb{N}$, consider a cover of the finite-dimensional space F by a sequence $(B_{k,n})_n$ of subsets of F with diam $B_{k,n} < 1/k$, and put

$$C_{k,n} := \{x \in C_k : \pi^{F,E}(x) \in B_{k,n}\},$$

so $C = \bigcup_{k,n \in \mathbb{N}} C_{k,n}$. Observe, for any pair $x', x'' \in C_{k,n}$, that the definition of C_k yields $\|x'_F - x''_F\| \le k\,\|x'_E - x''_E\|$. Putting $P_{k,n} := \pi^{E,F}(C_{k,n})$, the latter inequality guarantees the existence of a mapping $f_{k,n} : P_{k,n} \to F$ which is Lipschitz on $P_{k,n}$ with k as a Lipschitz constant and such that

$$C_{k,n} = \{u + f_{k,n}(u) : u \in P_{k,n}\}.$$

According to Proposition 2.79, we can extend $f_{k,n}$ to a Lipschitz mapping, still denoted by $f_{n,k}$, on the whole subspace E. This finishes the proof. □

2.7.3. Metrics on the set of vector subspaces. Let $\text{Svect}(X)$ denote the set of all nonzero vector subspaces of a non-null normed space X. Recall that $\mathbb{S}_X$ denotes the unit sphere of X, that is,

$$\mathbb{S}_X := \{u \in X : \|u\| = 1\}.$$

For $F, G \in \text{Svect}(X)$ we define

(2.89) $\quad d_{\text{Svect}}(F, G) := \max\{\text{exc}(F \cap \mathbb{S}_X, G \cap \mathbb{S}_X), \text{exc}(G \cap \mathbb{S}_X, F \cap \mathbb{S}_X)\},$

that is, $d_{\text{Svect}}(F, G) = \text{haus}(F \cap \mathbb{S}_X, G \cap \mathbb{S}_X)$, and we define also

(2.90) $\quad \widehat{d}_{\text{Svect}}(F, G) = \max\{\text{exc}(F \cap \mathbb{S}_X, G), \text{exc}(G \cap \mathbb{S}_X, F)\}.$

We recall (see Section 1.8) that $\text{exc}(C, D) := \sup_{x \in C} d(x, D)$ and $\text{haus}(C, D)$ denote respectively the Hausdorff-Pompeiu excess of C over D and the Hausdorff-Pompeiu semidistance between two subsets C and D of X. For all nonzero closed vector subspaces F, G of X, we obviously have

$$\widehat{d}_{\text{Svect}}(F, G) \le 1,$$

and the inequalities $d(x, G) \le d(x, G \cap \mathbb{S}_X) \le 2\,d(x, G)$ for all $x \in F \cap \mathbb{S}_X$ hold true by Lemma 1.105(b), as well as the inequalities

$$\text{exc}(F \cap \mathbb{S}_X, G) \le \text{exc}(F \cap \mathbb{S}_X, G \cap \mathbb{S}_X) \le 2\,\text{exc}(F \cap \mathbb{S}_X, G)$$

along with

(2.91) $$\widehat{d}_{\text{Svect}}(F,G) \leq d_{\text{Svect}}(F,G) \leq 2\widehat{d}_{\text{Svect}}(F,G).$$

The function d_{Svect} is easily seen to be a distance over $\text{Svect}(X)$, while the function $\widehat{d}_{\text{Svect}}$ does not fulfill in general the triangle inequality as seen in Exercise 1.109 in the previous chapter. However, (2.91) tells us that they generate the same topology on $\text{Svect}(X)$. Further, as we will see, it is often more convenient to work with the pseudo-distance $\widehat{d}_{\text{Svect}}$.

When the closed vector subspaces F, G of X are not reduced to zero, the inequality $\sup_{x \in F \cap \mathbb{S}_X} d(x, G) \leq \sup_{x \in F \cap \mathbb{B}_X} d(x, G)$ is evident and the convexity of the function $d(\cdot, G)$ guarantees also the converse inequality (since the supremum of a real-valued convex function on the convex hull of a set $C \subset X$ coincides with the supremum of the function over C). So the equalities

(2.92) $$\widehat{d}_{\text{Svect}}(F,G) = \max\{\text{exc}(F \cap \mathbb{B}_X, G), \text{exc}(G \cap \mathbb{B}_X, F)\} =: \widehat{\text{haus}}_{\rho_0}(F,G),$$

for $\rho_0 := 1$, hold true. Consequently, by Subsection 1.9.3 in the previous chapter, the topology associated with the distance d_{Svect}, or equivalently with the pseudo-distance $\widehat{d}_{\text{Svect}}$, on the collection of nonzero closed vector subspaces of X coincides on that collection with the topology induced by the Attouch-Wets topology. We recall that this is due to the equality $\widehat{\text{haus}}_\rho(F,G) = \rho \widehat{\text{haus}}_1(F,G)$ for every $\rho > 0$, according to the fact that

$$\sup_{x \in F \cap \rho \mathbb{B}_X} d(x,G) = \sup_{x \in F \cap \mathbb{B}_X} d(\rho x, G) = \rho \sup_{x \in F \cap \mathbb{B}_X} d(x,G).$$

One of the advantages to often deal with the pseudo-distance $\widehat{d}_{\text{Svect}}(F,G)$ (instead of $d_{\text{Svect}}(F,G)$) appears in the duality equality established below with $\widehat{d}_{\text{Svect}}(F^\perp, G^\perp)$, where $F^\perp$ denotes the vector space of $x^* \in X^*$ which are null on the vector subspace F of X, that is, $F^\perp := \{x^* \in X^* : \langle x^*, x \rangle = 0, \forall x \in F\}$. In fact, we will prove a more general duality result related to closed convex cones P, Q. The duality result requires the following lemma.

LEMMA 2.249. *Let Q be a nonempty convex cone of a normed space $(X, \|\cdot\|)$. Then, for any $x \in X$ and $x^* \in X^*$, one has*

$$d(x,Q) = \max_{u^* \in Q^\circ \cap \mathbb{B}_{X^*}} \langle u^*, x \rangle \quad \text{and} \quad d(x^*, Q^\circ) = \sup_{u \in (\text{cl } Q) \cap \mathbb{B}_X} \langle x^*, u \rangle.$$

PROOF. For any $u^* \in Q^\circ \cap \mathbb{B}_{X^*}$, we have, for all $y \in Q$,

$$\langle u^*, x \rangle \leq \langle u^*, x - y \rangle \leq \|x - y\|,$$

hence $\langle u^*, x \rangle \leq d(x, Q)$. On the other, d_Q is sublinear and Lipschitz, so we can choose $u_0^* \in \partial d_Q(x)$, which is equivalent to $\langle u_0^*, x \rangle = d_Q(x)$ and $\langle u_0^*, \cdot \rangle \leq d_Q(\cdot)$ according to Proposition 2.65, thus in particular we have $\langle u_0^*, x \rangle = d_Q(x)$ and $u_0^* \in Q^\circ \cap \mathbb{B}_{X^*}$. The equality concerning $d(x, Q)$ is then justified.

Concerning $d(x^*, Q^\circ)$, we observe as above that, for any $u \in \mathbb{B}_X \cap \text{cl } Q$, we have, for all $y^* \in Q^\circ$,

$$\langle x^*, u \rangle \leq \langle x^* - y^*, u \rangle \leq \|x^* - y^*\|,$$

hence $\langle x^*, u \rangle \leq d(x^*, Q^\circ)$. We then deduce $\sup_{u \in \mathbb{B}_X \cap \text{cl } Q} \langle x^*, u \rangle \leq d(x^*, Q^\circ)$. To prove the converse inequality of the latter, we may suppose $d(x^*, Q^\circ) > 0$. Fix any real $0 < r < d(x^*, Q^\circ)$, so $(x^* + r\mathbb{B}_{X^*}) \cap Q^\circ = \emptyset$. The Hahn-Banach separation theorem applied to the weak star closed convex set Q° and the weak star compact convex set $x^* + r\mathbb{B}_{X^*}$ yields a real β and an element $u_0 \in X$ with $\|u_0\| = 1$ such that

(2.93) $\qquad \langle y^*, u_0 \rangle < \beta \ \forall y^* \in Q^\circ, \quad \text{and} \quad \langle x^* + rb^*, u_0 \rangle > \beta \ \forall b^* \in \mathbb{B}_{X^*}.$

Since Q° is a cone containing zero, the first of the latter inequalities gives $\beta > 0$ as well as $\langle z^*, u_0 \rangle \le 0$ for all $z^* \in Q^\circ$, or equivalently $u_0 \in \operatorname{cl} Q$; so, the second inequality in (2.93) yields

$$\sup_{u \in \mathbb{B}_X \cap \operatorname{cl} Q} \langle x^*, u \rangle \ge \langle x^*, u_0 \rangle \ge \beta + r > r.$$

It ensues that $d(x^*, Q^\circ) \le \sup_{u \in \mathbb{B}_X \cap \operatorname{cl} Q} \langle x^*, u \rangle$ as desired, and this finishes the proof. $\square$

THEOREM 2.250 (duality for truncated Hausdorff pseudo-distance of cones). Let $(X, \|\cdot\|)$ be a normed space, P, Q be two closed convex cones of X, and F, G be two closed vector subspaces of X. Then, one has:

$$\operatorname{exc}(P \cap \mathbb{B}_X, Q) = \operatorname{exc}(Q^\circ \cap \mathbb{B}_{X^*}, P^\circ), \quad \widehat{\operatorname{haus}}_1(P, Q) = \widehat{\operatorname{haus}}_1(P^\circ, Q^\circ),$$

$$\widehat{\operatorname{haus}}_1(F, G) = \widehat{\operatorname{haus}}_1(F^\perp, G^\perp),$$

and if in addition F and G are both non-null

$$\widehat{d}_{\operatorname{Svect}}(F, G) = \widehat{d}_{\operatorname{Svect}}(F^\perp, G^\perp).$$

PROOF. Lemma 2.249 above allows us to write

$$\operatorname{exc}(P \cap \mathbb{B}_X, Q) = \sup_{x \in P \cap \mathbb{B}_X} d(x, Q) = \sup_{x \in P \cap \mathbb{B}_X} \sup_{x^* \in Q^\circ \cap \mathbb{B}_{X^*}} \langle x^*, x \rangle$$

$$= \sup_{x^* \in Q^\circ \cap \mathbb{B}_{X^*}} \sup_{x \in P \cap \mathbb{B}_X} \langle x^*, x \rangle$$

$$= \sup_{x^* \in Q^\circ \cap \mathbb{B}_{X^*}} d(x^*, P^\circ) = \operatorname{exc}(Q^\circ \cap \mathbb{B}_{X^*}, P^\circ).$$

So, the first equality of the theorem is justified, and the two others involving $\widehat{\operatorname{haus}}_1(\cdot, \cdot)$ follow from that equality. Finally, (2.92) gives the equality in the statement related to $\widehat{d}_{\operatorname{Svect}}(\cdot, \cdot)$. $\square$

We also have for closed subspaces in a Hilbert space:

THEOREM 2.251 (truncated Hausdorff pseudo-distance of vector subspaces via orthogonal projection). Let F and G be two closed subspaces of a Hilbert space H. Then one has

$$\widehat{\operatorname{haus}}_1(F, G) = \|P_F - P_G\|,$$

where P_F denotes the orthogonal projection on F and $\|P_F - P_G\|$ is the usual norm of the continuous linear mapping $P_F - P_G$.

In particular, if in addition both F and G are non-null in H, then $\widehat{d}_{\operatorname{Svect}}(F, G) = \|P_F - P_G\|$.

PROOF. Fix any $x \in \mathbb{B}_H$ and note by the equality

$$(P_F - P_G)x = P_F \circ (I - P_G)(x) - (I - P_F) \circ P_G(x)$$

(where I denotes the identity on H) that

$$\|(P_F - P_G)x\|^2 = \|P_F \circ (I - P_G)(x)\|^2 + \|(I - P_F) \circ P_G(x)\|^2,$$

and hence (since $P_F + P_{F^\perp} = I$)

(2.94) $\quad \|(P_F - P_G)x\|^2 = \|(I - P_{F^\perp}) \circ P_{G^\perp}(x)\|^2 + \|(I - P_F) \circ P_G(x)\|^2.$

We derive that

$$\|(P_F - P_G)x\|^2 = d(P_{G^\perp}(x), F^\perp)^2 + d(P_G(x), F)^2$$
$$\leq \widehat{\mathrm{haus}}_1(F^\perp, G^\perp)^2 \|P_{G^\perp}(x)\|^2 + \widehat{\mathrm{haus}}_1(F, G)^2 \|P_G(x)\|^2$$
$$= \widehat{\mathrm{haus}}_1(F, G)^2 \|x\|^2,$$

where the latter equality is due to Theorem 2.250 and to the equality $\|x\|^2 = \|P_G(x)\|^2 + \|P_{G^\perp}(x)\|^2$. This being true for every $x \in \mathbb{B}_X$, it follows that

$$\|P_F - P_G\| \leq \widehat{\mathrm{haus}}_1(F, G).$$

To obtain the converse inequality, we observe from (2.94) that for any $x \in G \cap \mathbb{B}_X$ we have $\|P_F - P_G\| \geq \|x - P_F(x)\| = d(x, F)$, so $\|P_F - P_G\| \geq \widehat{\mathrm{exc}}_1(G, F)$. Permuting F and G we also obtain $\|P_G - P_F\| \geq \widehat{\mathrm{exc}}_1(F, G)$. It results that

$$\|P_F - P_G\| \geq \max\{\widehat{\mathrm{exc}}_1(F, G), \widehat{\mathrm{exc}}_1(G, F)\} = \widehat{\mathrm{haus}}_1(F, G),$$

which finishes the proof. $\square$

Our next task is to investigate some topological properties of the *set* $\mathrm{Svect}_p(X)$ *of p-dimensional vector subspaces* of X, where p is an integer with $p \leq \dim X$.

LEMMA 2.252. *Let* $(X, \|\cdot\|)$ *be a normed space and let* $p \in \mathbb{N}$ *with* $p \leq \dim X$. *Let also* F *be a p-dimensional vector subspace of* X *and* $v_1, \cdots, v_p$ *be a basis of* F. *Then, for each real* $\varepsilon > 0$, *there exists a real* $\delta > 0$ *such that, for any vectors* $w_k \in B_X(v_k, \delta)$ *for* $k = 1, \cdots, p$, *the vector space* G *spanned by* $w_1, \cdots, w_p$ *is p-dimensional and satisfies the inequality* $\widehat{d}_{\mathrm{Svect}}(G, F) < \varepsilon$.

PROOF. For the norm on $\mathbb{R}^p$ given, for any $\zeta := (\zeta_1, \cdots, \zeta_p) \in \mathbb{R}^p$, by $\|\zeta_1 v_1 + \cdots + \zeta_p v_p\|$ and for the usual sum norm $\|\zeta\|_1 := |\zeta_1| + \cdots + |\zeta_p|$, there exists (by their equivalence) a real constant $\alpha > 0$ such that $\|\zeta_1 v_1 + \cdots + \zeta_p v_p\| \geq 2\alpha \|\zeta\|_1$ for all $\zeta \in \mathbb{R}^p$. Then, for any $w_k \in X$ with $\|w_k - v_k\| < \alpha$ for $k = 1, \cdots, p$, we have for all $\zeta \in \mathbb{R}^p$

$$\left\| \sum_{k=1}^p \zeta_k w_k \right\| \geq \left\| \sum_{k=1}^p \zeta_k v_k \right\| - \left\| \sum_{k=1}^p \zeta_k (w_k - v_k) \right\|$$

(2.95) $\qquad\qquad\qquad \geq 2\alpha \|\zeta\|_1 - \alpha \|\zeta\|_1 = \alpha \|\zeta\|_1.$

Fix any real $\varepsilon > 0$ and choose a positive real $\delta < \min\{\alpha, \varepsilon\alpha\}$. Fix also any $w_k \in X$ with $\|w_k - v_k\| < \delta$ for $k = 1, \cdots, p$ and denote by G the vector space spanned by $w_1, \cdots, w_p$. By (2.95), for all $\zeta \in \mathbb{R}^p$, we have $\|\sum_{k=1}^p \zeta_k w_k\| \geq \alpha \|\zeta\|_1$ and this guarantees the linear independence of the vectors $w_1, \cdots, w_p$, so the vector space G is p-dimensional. Further, taking any $w \in G$ with $\|w\| = 1$ and choosing $\zeta \in \mathbb{R}^p$ such that $w = \zeta_1 w_1 + \cdots + \zeta_p w_p$, the inequality (2.95) again gives $\|\zeta\|_1 \leq 1/\alpha$, so, for $v := \zeta_1 v_1 + \cdots + \zeta_p v_p \in F$ we have

$$\|w - v\| \leq \sum_{k=1}^p |\zeta_k| \|w_k - v_k\| \leq \delta/\alpha,$$

hence $\sup_{w \in G \cap \mathbb{S}_X} d(w, F) \leq \delta/\alpha < \varepsilon$.

Similarly, taking any $v \in F$ with $\|v\| = 1$ and choosing $\xi \in \mathbb{R}^p$ such that $v = \xi_1 v_1 + \cdots + \xi_p v_p$, the choice of α assures us that $\|\xi\|_1 \leq 1/(2\alpha)$, hence putting $w := \xi_1 w_1 + \cdots + \xi_p w_p$ we see that

$$\|v - w\| \leq \sum_{k=1}^{p} |\xi_k| \|v_k - w_k\| \leq \delta/\alpha,$$

hence $\sup_{v \in F \cap \mathbb{S}_X} d(v, G) \leq \delta/\alpha < \varepsilon$. We then obtain $\widehat{d}_{\text{Svect}}(G, F) < \varepsilon$, which concludes the proof. □

With the help of the above lemma we can study the separability of $\text{Svect}(X)$ and the compactness of $\text{Svect}_p(X)$.

PROPOSITION 2.253. *Let $(X, \|\cdot\|)$ be a normed space and let $p \in \mathbb{N}$ with $p \leq \dim X$. The following hold:*
(a) *If X is separable, then the metric space $(\text{Svect}_p(X), d_{\text{Svect}})$ is separable too.*
(b) *If X is finite-dimensional, then the metric space $(\text{Svect}_p(X), d_{\text{Svect}})$ is compact.*

PROOF. (a) Take a countable dense subset Q of X and note that the set $\mathcal{F}$ of all p-dimensional subspaces of X which have a basis with vectors from Q is countable. Fix any $F \in \text{Svect}_p(X)$ with $v_1, \cdots, v_p$ as a basis and fix also any real $\varepsilon > 0$. For the real $\delta > 0$ given by Lemma 2.252 choose, according to the density of Q, vectors $w_k \in Q$ with $\|w_k - v_k\| < \delta$ for $k = 1, \cdots, p$. For the vector space G spanned by $w_1, \cdots, w_p$, Lemma 2.252 says that $G \in \mathcal{F}$ and $\widehat{d}_{\text{Svect}}(G, F) < \varepsilon$. This combined with (2.91) justifies the assertion (a).
(b) Without loss of generality, we may suppose that the norm $\|\cdot\|$ is associated with an inner product $\langle \cdot, \cdot \rangle$ on X. Let $(F_n)_{n \in \mathbb{N}}$ be a sequence in $\text{Svect}_p(X)$. For each $n \in \mathbb{N}$ choose an orthogonal basis $v_{1,n}, \cdots, v_{p,n}$ of F_n with $\|v_{k,n}\| = 1$ for $k = 1, \cdots, p$. Since X is finite-dimensional, extracting a subsequence we may suppose that $(v_{k,n})_n$ converges to some vector $v_k \in X$ with $\|v_k\| = 1$ and $\langle v_k, v_j \rangle = 0$ for each $k = 1, \cdots, p$ and each $j = 1, \cdots, p$ with $j \neq k$. Then, the vectors $v_1, \cdots, v_p$ form an orthogonal basis (of unit vectors) of the vector subspace F that they span. Fix any real $\varepsilon > 0$. Choosing $\delta > 0$ given by Lemma 2.252 and choosing also $n_0 \in \mathbb{N}$ such that $\|v_{k,n} - v_k\| < \delta$ for all $k = 1, \cdots, p$ and $n \geq n_0$, we obtain $\widehat{d}_{\text{Svect}}(F_n, F) < \varepsilon$, which justifies the convergence of $(F_n)_n$ to F with respect to $\widehat{d}_{\text{Svect}}$, and so the compactness of $(\text{Svect}_p(X), d_{\text{Svect}})$. This finishes the proof of the proposition. □

2.7.4. Points of nullity of trace of B-tangent cone on subspace. With the help of the two previous subsections, we are now in a position to prove the following negligibility type of the set of points of nullity of the intersection of the B-tangent cone with a p-dimensional vector subspace of a separable normed space.

THEOREM 2.254 (points of nullity of trace of B-tangent cone on subspace). *Let S be a nonempty subset of a separable normed space X and let p be a positive integer with $p < \dim X$. Let C be the set of points $x \in S$ for which there exists a p-dimensional vector subspace (depending on x) whose intersection with $T^B(S; x)$ is $\{0\}$. Then the set C can be covered by countably many Lipschitz surfaces of codimension p.*

PROOF. Take from (a) of the above proposition a countable family $(F_n)_{n \in \mathbb{N}}$ which is dense in $\mathrm{Svect}_p(X)$ with respect to d_{Svect}. For each $F \in \mathrm{Svect}_p(X)$, putting
$$C_F := \{x \in S : F \cap T^B(S; x) = \{0\}\} = \{x \in S : F \cap \mathbb{S}_X \cap T^B(S; x) = \emptyset\},$$
we claim that $\bigcup_{F \in \mathrm{Svect}_p(X)} C_F = \bigcup_{n \in \mathbb{N}} C_{F_n}$. Indeed, fix any $F \in \mathrm{Svect}_p(X)$ and any $x \in C_F$. By definition of C_F and by compactness of $F \cap \mathbb{S}_X$, there exits some real $\varepsilon > 0$ such that
$$d(v, T^B(S; x)) > 2\varepsilon \quad \text{for all } v \in F \cap \mathbb{S}_X.$$
By density of $(F_n)_n$, choose some $n_0 \in \mathbb{N}$ satisfying $d_{\mathrm{Svect}}(F_{n_0}, F) < \varepsilon$. For each $u \in F_{n_0} \cap \mathbb{S}_X$ we can then choose some $\xi(u) \in F \cap \mathbb{S}_X$ such that $\|u - \xi(u)\| < \varepsilon$, which entails
$$d(u, T^B(S; x)) \geq d(\xi(u), T^B(S; x)) - \|u - \xi(u)\| > 2\varepsilon - \varepsilon = \varepsilon.$$
This guarantees that $F_{n_0} \cap \mathbb{S}_X \cap T^B(S; x) = \emptyset$, or equivalently $x \in C_{F_{n_0}}$, and this justifies the claim.

We then have $C = \bigcup_{n \in \mathbb{N}} C_{F_n}$, and then the statement of the proposition follows from Proposition 2.248. □

2.7.5. Tangential properties through Hausdorff measure. In the case where X is finite-dimensional, more can be said through the Hausdorff measure. Before stating and proving a first proposition relative to this case, we need a lemma concerning the Bouligand-Peano tangent cone at Lebesgue density point.

Let λ denote the *outer Lebesgue measure* on $\mathbb{R}^N$ given, for every set $A \subset \mathbb{R}^N$, by $\lambda(A)$ equal to the infimum of $\sum_{n \in \mathbb{N}} \mathrm{meas}_N(P_n)$ over the sequences $(P_n)_n$ of N-rectangles in $\mathbb{R}^N$ with $A \subset \bigcup_{n \in \mathbb{N}} P_n$. Above $\mathrm{meas}_N(P_n)$ denotes the N-dimensional measure of the N-rectangle P_n. Given a set S of $\mathbb{R}^N$, a point $\overline{x} \in S$ is called a λ-*density point* or an N-*dimensional outer Lebesgue density point* of S whenever
$$\lim_{r \downarrow 0} \frac{\lambda(S \cap B(\overline{x}, r))}{\lambda(B(\overline{x}, r))} = 1.$$
It is known (see Theorem D.1 in Appendix) that λ-almost every point of S is a λ-density point of S.

LEMMA 2.255. *Let S be a subset of a finite-dimensional Euclidean space X and let $\overline{x} \in S$ be a density point of S with respect to the outer Lebesgue measure on X (relative to a fixed orthonormal basis of X). The following hold:*
(a) *The Bouligand-Peano tangent cone $T^B(S; \overline{x})$ of S at $\overline{x}$ is the whole space X.*
(b) *Let a real $0 < \varepsilon < 1$. There exists some real $r > 0$ such that with any $u \in B(\overline{x}, r)$ one can associate some $p(u) \in S$ so that*
$$\|p(u) - \overline{x}\| \leq \|u - \overline{x}\| \quad \text{and} \quad \|p(u) - u\| \leq \varepsilon \|u - \overline{x}\|.$$

PROOF. Let λ denote the outer Lebesgue measure on X (relative to a fixed orthonormal basis of X) and let $\overline{x} \in S$ be a λ-density point of S.
(a) Suppose that $T^B(S; \overline{x}) \neq X$ and take a unit vector $v \in X$ with $v \notin T^B(S; \overline{x})$. This means that there exits some positive real $\eta < 1$ such that
$$(\overline{x} +]0, \eta] B(v, \eta)) \cap S = \emptyset.$$
Fix any positive real $r < \eta$ and put $B_r := B(\overline{x}, r)$, $S_r := S \cap B_r$ and
$$\Delta_r := B_r \cap (\overline{x} +]0, \eta] B(v, \eta)).$$

Note that the ratio $c := \lambda(\Delta_r)/\lambda(B_r)$ is positive and independent of r (for $r \leq \eta$). Since $\Delta_r \subset B_r \setminus S_r$, we have
$$\lambda(S_r) = \lambda(B_r) - \lambda(B_r \setminus S_r) \leq \lambda(B_r) - \lambda(\Delta_r),$$
which yields
$$\frac{\lambda(S_r)}{\lambda(B_r)} \leq 1 - \frac{\lambda(\Delta_r)}{\lambda(B_r)} = 1 - c$$
and this contradicts that $\bar{x}$ is a λ-density point of S. The assertion (a) is then established.

(b) Let $0 < \varepsilon < 1$. Suppose that the assertion in (b) does not hold, that is, for any real $r > 0$ there exists some $u_r \in B(\bar{x}, r)$ such that
$$S \cap B[u_r, \varepsilon\|u_r - \bar{x}\|] \cap B[\bar{x}, \|u_r - \bar{x}\|] = \emptyset.$$
For $t(r) := \|u_r - \bar{x}\|$ we see that $t(r) > 0$ and $t(r) \downarrow 0$ as $r \downarrow 0$. Put $z_r := u_r + (\varepsilon/2)(\bar{x} - u_r)$, $D_{t(r)} := B(z_r, (\varepsilon/2)t(r))$, and as above put also $B_{t(r)} := B(\bar{x}, t(r))$ and $S_{t(r)} := S \cap B_{t(r)}$. Since
$$D_{t(r)} = B\left(z_r, \frac{1}{2}\varepsilon\|u_r - \bar{x}\|\right) \subset B[u_r, \varepsilon\|u_r - \bar{x}\|] \cap B[\bar{x}, \|u_r - \bar{x}\|]$$
(as easily seen), we have $D_{t(r)} \cap S = \emptyset$, that is, $D_{t(r)} \subset B_{t(r)} \setminus S_{t(r)}$. Consequently,
$$\lambda(S_{t(r)}) = \lambda(B_{t(r)}) - \lambda(B_{t(r)} \setminus S_{t(r)}) \leq \lambda(B_{t(r)}) - \lambda(D_{t(r)}),$$
which gives
$$\frac{\lambda(S_{t(r)})}{\lambda(B_{t(r)})} \leq 1 - \frac{\lambda(D_{t(r)})}{\lambda(B_{t(r)})} = 1 - \frac{\varepsilon^N}{2^N},$$
where N denotes the dimension of X. This is in contradiction with the λ-density point property of $\bar{x}$ in S and concludes the proof of the lemma. $\square$

We can now state the first result which complements, in the finite-dimensional setting, the result of Proposition 2.248. The statement involves the Hausdorff measure $\mathcal{H}_k$ recalled in Section D in Appendix).

PROPOSITION 2.256. Let S be a nonempty subset of a finite-dimensional Euclidean space X with dimension N and let F be a vector subspace of dimension p with $0 < p < N$. Then the set
$$C := \{x \in S : F \cap T^B(S; x) = \{0\}\}$$
is the union of a sequence of sets $(C_i)_{i \in \mathbb{N}}$ with $\mathcal{H}_{N-p}(C_i) < +\infty$ such that, for $\mathcal{H}_{N-p}$-almost every $x \in C_i$, there exists a surjective linear mapping $\Lambda : X \to F$ (depending on x) such that
$$\operatorname{Ker} \Lambda = T^B(C_i; x) = T^B(S; x),$$
thus $T^B(S; x)$ is an $(N - p)$-dimensional vector subspace of X.

Furthermore, each set C_i is contained in a Lipschitz F-surface (hence of p-codimension).

PROOF. Let E be the orthogonal complement subspace of F in X, so we can write each $x \in X$ as $x = x_E + x_F$ with $x_E \in E$ and $x_F \in F$. For each $k \in \mathbb{N}$, we modify slightly the basic idea of the proof of Proposition 2.248 in setting
$$C_k := \{x \in C : (x' \in S \text{ and } \|x' - x\| \leq 1/k) \Rightarrow \|x'_F - x_F\| \leq k \|x'_E - x_E\|\}.$$

Obviously, $\bigcup_k C_k \subset C$. We claim that the equality holds. Indeed suppose that there is some $x \in C$ with $x \notin \bigcup_k C_k$, hence, for each $k \in \mathbb{N}$, there exists $x^k \in S \cap B[x, 1/k]$ with $\|x_F^k - x_F\| > k \|x_E^k - x_E\|$. For $t_k := \|x^k - x\|$, we note that $t_k > 0$ with $t_k \downarrow 0$. Putting $h^k := t_k^{-1}(x_F^k - x_F)$ and $e^k := t_k^{-1}(x_E^k - x_E)$ we see that $e^k \to 0$ since $\|e^k\| \leq 1/k$ and that $x + t_k(h^k + e^k) = x^k \in S$ for all $k \in \mathbb{N}$. Since $\|h^k\| \leq 1$, extracting a subsequence if necessary we may suppose that $(h^k)_k$ converges to some $h \in F$. This vector $h \in T^B(S; x)$ and the equality $\|h^k + e^k\| = 1$ ensures $\|h\| = 1$. So h is a nonzero vector in $F \cap T^B(S; x)$, which is in contradiction with the relation $F \cap T^B(S; x) = \{0\}$ since $x \in C$. This justifies the claim.

Since X is finite-dimensional, we can choose, for each $k \in \mathbb{N}$, a sequence of sets $(C_{k,n})_n$ with diam $C_{k,n} \leq 1/k$ and such that $C_k = \bigcup_{n \in \mathbb{N}} C_{k,n}$. Observe, for any pair $x', x'' \in C_{k,n}$, that the definition of C_k yields $\|x'_F - x''_F\| \leq k \|x'_E - x''_E\|$. Denoting by $P_{k,n}$ the orthogonal projection of $C_{k,n}$ on E, the latter inequality guarantees the existence of a mapping $f_{k,n} : P_{k,n} \to F$ which is Lipschitz on $P_{k,n}$ with k as a Lipschitz constant and such that

$$C_{k,n} = \{u + f_{k,n}(u) : u \in P_{k,n}\}.$$

According to Proposition 2.79, we can extend $f_{k,n}$ to a Lipschitz mapping, still denoted by $f_{k,n}$, on the whole subspace E with values into F and with Lipschitz constant pk. The Lipschitz property of $f_{k,n}$ and the boundedness of the set $P_{k,n}$ ensure that $\mathcal{H}_{N-p}(C_{k,n}) < +\infty$.

Let λ_E denote the $(N - p)$-dimensional outer Lebesgue measure on the Euclidean space E (relative to a fixed orthonormal basis of E). Fix any pair $k, n \in \mathbb{N}$ such that $\lambda_E(P_{k,n}) > 0$ and denote by $\widehat{P}_{k,n}$ the set of points of $P_{k,n}$ which are both points of λ_E-density of $P_{k,n}$ and of Fréchet differentiability of the Lipschitz mapping $f_{k,n}$, both concepts of outer density point and Fréchet differentiability being taken relative to the Euclidean subspace E. The set $P_{k,n} \setminus \widehat{P}_{k,n}$ being of measure zero for the $(N - p)$-dimensional Lebesgue measure on E and $f_{k,n}$ being Lipschitz on $P_{n,k}$, we see that

$$\mathcal{H}_{N-p}(\{u + f_{k,n}(u) : u \in P_{k,n} \setminus \widehat{P}_{k,n}\}) = 0.$$

It remains to show the tangential properties for the set

$$G_{k,n} := \{u + f_{k,n}(u) : u \in \widehat{P}_{k,n}\}$$

(which is the image of the graph of the restriction of $f_{k,n}$ to $\widehat{P}_{k,n}$ under the bijective linear mapping $(u, v) \mapsto u + v$ from $E \times F$ into X). Fix any $\overline{x} \in G_{k,n}$, so $\overline{x}_F = f_{k,n}(\overline{x}_E)$. Denote by $\zeta = Df_{k,n}(\overline{x}_E)$ the Fréchet derivative of $f_{k,n}$ at $\overline{x}_E$ and define the linear mapping $\Lambda : X \to F$ by $\Lambda(h) := h_F - \zeta(h_E)$ for all $h \in X$. Denote by $B_E(u, r)$ the open ball in the vector space E with respect to the induced norm, for $u \in E$ and $r > 0$.

Since $\overline{x}_E$ is a λ_E-density point of $P_{k,n}$, the assertion (a) in Lemma 2.255 says that E coincides with the Bouligand-Peano tangent cone of $P_{k,n}$ at $\overline{x}_E$ relative to the vector space E. So, for any nonzero fixed vector $u \in E$, there are sequences $(u_m)_m$ in E converging to u and $(t_m)_m$ tending to 0 with $t_m > 0$ such that, for all $m \in \mathbb{N}$, we have $\overline{x}_E + t_m u_m \in P_{k,n}$, hence

$$(\overline{x}_E + t_m u_m) + f_{k,n}(\overline{x}_E + t_m u_m) \in C_{k,n}.$$

The Fréchet differentiability of $f_{k,n}$ at $\overline{x}_E$ furnishes a sequence of vectors $(v_m)_m$ in F converging to 0 in F with $f_{k,n}(\overline{x}_E + t_m u_m) = f_{k,n}(\overline{x}_E) + t_m(\zeta(u_m) + v_m)$, thus

$$\bigl(\overline{x}_E + f_{k,n}(\overline{x}_E)\bigr) + t_m\bigl(u_m + \zeta(u_m) + v_m\bigr) \in C_{k,n},$$

which implies $u + \zeta(u) \in T^B(C_{k,n}; \overline{x})$. This easily entails the inclusion $\operatorname{Ker}(\Lambda) \subset T^B(C_{k,n}; \overline{x})$.

Now fix any $0 < \varepsilon < 1$. By (b) of Lemma 2.255 there exists some $r_0 > 0$ such that with any $u \in B_E(\overline{x}_E, r_0)$ we can associate $b(u) \in P_{k,n}$ satisfying

(2.96) $\qquad \|b(u) - \overline{x}_E\| \le \|u - \overline{x}_E\| \quad \text{and} \quad \|b(u) - u\| \le \varepsilon\|u - \overline{x}_E\|.$

Put, for any $u \in B_E(\overline{x}_E, r_0)$,

$$\Delta_{k,n}(u) := f_{k,n}(b(u)) - \overline{x}_F - \zeta(b(u) - \overline{x}_E)$$

and note that, for any $y \in F$,

$$y - \overline{x}_F - \zeta(u - \overline{x}_E) = \Delta_{k,n}(u) + (y - f_{k,n}(b(u))) + \zeta(b(u) - u).$$

Consider any $x \in S$ with $\|x - \overline{x}\| < \min\{1/(4pk^2), r_0\}$. Then $\|b(x_E) - x_E\| \le \varepsilon \|x_E - \overline{x}_E\| \le 1/(2k)$ and

$$\|f_{k,n}(b(x_E)) - \overline{x}_F\| = \|f_{k,n}(b(x_E)) - f_{k,n}(\overline{x}_E)\|$$
$$\le pk \|b(x_E) - \overline{x}_E\| \le pk \|x_E - \overline{x}_E\| \le 1/(4k),$$

thus $\|f_{k,n}(b(x_E)) - x_F\| \le 1/(2k)$, so $\|b(x_E) + f_{k,n}(b(x_E)) - x\| \le 1/k$. Noting that $b(x_E) + f_{k,n}(b(x_E))$ belongs to $C_{k,n} \subset C_k$, the definition of C_k ensures that $\|x_F - f_{k,n}(b(x_E))\| \le k \|x_E - b(x_E)\|$, so taking $y = x_F$ and $u = x_E$ in the latter equality above involving y and $\Delta_{k,n}(u)$, we see that

$$\|x_F - \overline{x}_F - \zeta(x_E - \overline{x}_E)\| \le \|\Delta_{k,n}(x_E)\| + (k + \|\zeta\|)\|x_E - b(x_E)\|.$$

For x_E close enough to $\overline{x}_E$, say $\|x_E - \overline{x}_E\| \le \eta_k$ for some positive real $\eta_k < r_0$, the Fréchet differentiability of $f_{k,n}$ at $\overline{x}_E$ and the inequality $\|b(x_E) - \overline{x}_E\| \le \|x_E - \overline{x}_E\|$ (according to the definition of $b(\cdot)$) giving

$$\|\Delta_{k,n}(x_E)\| \le \varepsilon \|b(x_E) - \overline{x}_E\| \le \varepsilon \|x_E - \overline{x}_E\|,$$

we deduce by the second inequality in (2.96)

$$\|\Lambda(x - \overline{x})\| = \|x_F - \overline{x}_F - \zeta(x_E - \overline{x}_E)\| \le \varepsilon(1 + k + \|\zeta\|)\|x_E - \overline{x}_E\|$$

for all $x \in S \cap B_X(\overline{x}, \rho_k)$ where $\rho_k := \min\{\eta_k, 1/(4pk^2)\}$. This easily entails $T^B(S; \overline{x}) \subset \operatorname{Ker}(\Lambda)$, hence

$$\operatorname{Ker}(\Lambda) \subset T^B(C_{k,n}; \overline{x}) \subset T^B(S; \overline{x}) \subset \operatorname{Ker}(\Lambda).$$

The equality between all the latter members holds true. Further, for $L := \operatorname{Ker}\Lambda$, the equality $L = T^B(S; \overline{x})$ gives $F \cap L = \{0\}$, hence $\dim F + \dim L = \dim(F \oplus L) \le N$, so $\dim L \le N - p$. On the other hand, $\dim L = \dim \operatorname{Ker}\Lambda \ge N - p$ since Λ is a linear mapping from X into F with $\dim F = p$. Consequently, $\dim L = N - p$ and the proof is finished by using a bijection between $\mathbb{N}$ and $\mathbb{N} \times \mathbb{N}$ allowing us to write C_i in place of $C_{k,n}$. $\square$

THEOREM 2.257 (F. Roger). *Let S be a nonempty subset of a finite dimensional Euclidean space X with dimension N, let p be an integer with $0 < p < N$. Let C denote the set of $x \in S$ for which there exists a p-dimensional vector subspace of X (depending on x) whose intersection with $T^B(S; x)$ is $\{0\}$. Then the set C is the union of a sequence of sets $(C_i)_{i \in \mathbb{N}}$ with $\mathcal{H}_{N-p}(C_i) < +\infty$ such that,* for $\mathcal{H}_{N-p}$-almost every $x \in C_i$, the Bouligand-Peano tangent cone $T^B(S;x)$ is an $(N-p)$-dimensional vector subspace of X.

Further, each set C_i is contained in a Lipschitz surface of codimension p.

PROOF. As in the proof of Theorem 2.254 take from (a) of Proposition 2.253 a countable family $(F_n)_{n\in\mathbb{N}}$ which is dense in $\mathrm{Svect}_p(X)$ with respect to d_{Svect}. For each $F \in \mathrm{Svect}_p(X)$, put
$$C_F := \{x \in S : F \cap T^B(S;x) = \{0\}\} = \{x \in S : F \cap \mathbb{S}_X \cap T^B(S;x) = \emptyset\}.$$
We have seen in the proof of Theorem 2.254 that $C = \bigcup_{F \in \mathrm{Svect}_p(X)} C_F = \bigcup_{n \in \mathbb{N}} C_{F_n}$. The equality $C = \bigcup_{n \in \mathbb{N}} C_{F_n}$ combined with Proposition 2.256 justifies the statement of the theorem. □

The next proposition deals with the case where F is a half-line instead of a vector space. It will be also at the heart of the proof of Theorem 2.260 below. The statement of the proposition requires to introduce some concepts. Let x^* be a nonzero continuous linear functional on a normed space X. We denote by $\mathrm{Hypl}(x^*)$ the vector hyperplane
$$\mathrm{Hypl}(x^*) := \{h \in X : \langle x^*, h \rangle = 0\}$$
and similarly we define the closed and open half-spaces
$$\mathrm{Hypl}^{\leq}(x^*) := \{h \in X : \langle x^*, h \rangle \leq 0\} \text{ and } \mathrm{Hypl}^{<}(x^*) := \{h \in X : \langle x^*, h \rangle < 0\}.$$
The closed and open half-spaces $\mathrm{Hypl}^{\geq}(x^*)$ and $\mathrm{Hypl}^{>}(x^*)$ are defined analogously. If, for a subset S of X and $x \in S$, we have
$$\mathrm{Hypl}(x^*) \subset T^B(S;x) \subset \mathrm{Hypl}^{\leq}(x^*),$$
we say that $\mathrm{Hypl}(x^*)$ is a *one-sided (or extreme) tangent hyperplane* of S at x.

PROPOSITION 2.258. Let S be a nonempty subset of a finite-dimensional Euclidean space X with dimension N and let v be a nonzero vector of X. Then the set $C := \{x \in S : v \notin T^B(S;x)\}$ is the union of a sequence of sets $(C_i)_{i \in \mathbb{N}}$ with $\mathcal{H}_{N-1}(C_i) < +\infty$ such that for $\mathcal{H}_{N-1}$-almost every $x \in C_i$ there exists a nonzero $x^* \in X$ (depending on x) with $\langle x^*, v \rangle > 0$ such that
$$\{h \in X : \langle x^*, h \rangle = 0\} \subset T^B(C_i;x)$$
and
$$T^B(S;x) \subset \{h \in X : \langle x^*, h \rangle \leq 0\};$$
hence in particular
$$\mathrm{Hypl}(x^*) \subset T^B(S;x) \subset \mathrm{Hypl}^{\leq}(x^*) \quad \text{and} \quad v \notin \mathrm{Hypl}^{\leq}(x^*);$$
so at $\mathcal{H}_{N-1}$-almost every point of C the set S has a one-sided tangent hyperplane which does not contain v.

Further, each set C_i is included in a Lipschitz hypersurface.

PROOF. Without loss of generality, we may suppose that $\|v\| = 1$. Put $e_N := v$ and choose $e_1, \cdots, e_{N-1}$ in X such that $e_1, \cdots, e_N$ form an orthonormal basis of X. Write each $x \in X$ as $x = x_E + x_N e_N$ with $x_N \in \mathbb{R}$ and $x_E \in E$, where E is the vector subspace spanned by $e_1, \cdots, e_{N-1}$. Modifying slightly the idea of the proofs of Propositions 2.248 and 2.256, for each $k \in \mathbb{N}$, put

(2.97) $C_k := \{x \in C : (x' \in S \text{ and } \|x' - x\| \leq 1/k) \Rightarrow x'_N - x_N \leq k \|x'_E - x_E\|\}.$

Obviously, $\bigcup_k C_k \subset C$. We claim that the equality holds. Indeed suppose, by contradiction, that there is some $x \in C$ with $x \notin \bigcup_k C_k$, hence for each $k \in \mathbb{N}$ there exists $x^k \in S \cap B[x, 1/k]$ with $x_N^k - x_N > k \|x_E^k - x_E\|$. For $t_k := \|x^k - x\|$ and $h^k := t_k^{-1}(x^k - x)$, we note that $t_k > 0$ with $t_k \downarrow 0$ and $x + t_k h^k = x^k \in S$, and (since $\|h^k\| = 1$ and X is finite-dimensional) extracting a subsequence we may suppose that $(h^k)_k$ converges as $k \to \infty$ to some $h \in X$ with $\|h\| = 1$, and obviously $h \in T^B(S; x)$. Since

$$\|h_E^k\| = \frac{\|x_E^k - x_E\|}{\|x^k - x\|} \leq \frac{\|x_E^k - x_E\|}{x_N^k - x_N} < \frac{1}{k},$$

we have $h_E^k \to 0$ as $k \to \infty$, hence $h_E = 0$. Further,

$$1 \leq \left(\frac{x_N^k - x_N}{\|x^k - x\|}\right)^{-1} \leq 1 + \frac{\|x_E^k - x_E\|}{x_N^k - x_N}$$

thus $h_N^k = (x_N^k - x_N)/\|x^k - x\| \to 1$ as $k \to \infty$, which entails $h_N = 1$. This yields $h = v$ hence $v \in T^B(S; x)$, which is in contradiction with the relation $v \notin T^B(S; x)$ since $x \in C$. The claim is then established.

Since X is finite-dimensional, for any $k \in \mathbb{N}$, we can choose a sequence of sets $(C_{k,n})_n$ with diam $C_{k,n} \leq 1/k$ and such that $C_k = \bigcup_{n \in \mathbb{N}} C_{k,n}$. Observe, for any pair $x', x'' \in C_{k,n}$, that the definition of C_k yields $|x_N' - x_N''| \leq k \|x_E' - x_E''\|$. Denoting by $P_{k,n}$ the orthogonal projection of $C_{k,n}$ on E, the latter inequality guarantees the existence of a function $f_{k,n} : P_{k,n} \to \mathbb{R}$ which is Lipschitz on $P_{k,n}$ with k as a Lipschitz constant and such that

$$C_{k,n} = \{u + f_{k,n}(u)e_N : u \in P_{k,n}\}.$$

According to Proposition 2.79, we can extend $f_{k,n}$ to a k-Lipschitz function, still denoted by $f_{k,n}$, on the whole vector subspace E. The Lipschitz property of $f_{k,n}$ and the boundedness of the set $P_{k,n}$ ensure that $\mathcal{H}_{N-1}(C_{k,n}) < +\infty$.

Denote by λ_E the outer $(N-1)$-Lebesgue measure on the Euclidean subspace E (relative to the orthonormal basis $e_1, \cdots, e_{N-1}$ of E). Fix any pair $k, n \in \mathbb{N}$ such that $\lambda_E(P_{k,n}) > 0$ and denote by $\widehat{P}_{k,n}$ the set of points of $P_{k,n}$ which are both points of λ_E-density of $P_{k,n}$ and of Fréchet differentiability of $f_{k,n}$, both concepts of outer density point and Fréchet differentiability being taken relative to the Euclidean subspace E. The set $P_{k,n} \setminus \widehat{P}_{k,n}$ being of measure zero for the $(N-1)$-dimensional Lebesgue measure on E and $f_{k,n}$ being Lipschitz continuous, we see that

$$\mathcal{H}_{N-1}(\{u + f_{k,n}(u)e_N : u \in P_{k,n} \setminus \widehat{P}_{k,n}\}) = 0.$$

It remains to show the tangential properties for the set $G_{k,n} := \{u + f_{k,n}(u)e_N : u \in \widehat{P}_{k,n}\}$ (which is the image of the graph of the restriction of $f_{k,n}$ to $\widehat{P}_{k,n}$ under the bijective linear mapping $(u, r) \mapsto u + re_N$ from $E \times \mathbb{R}$ into X). Fix any $\bar{x} \in G_{k,n}$, so $\bar{x}_N = f_{k,n}(\bar{x}_E)$. Denote by $\zeta = Df_{k,n}(\bar{x}_E)$ the Fréchet derivative of $f_{k,n}$ at $\bar{x}_E$ and define $x^* \in X$ by $\langle x^*, h \rangle = h_N - \zeta(h_E)$ for all $h \in X$. Denote by $B_E(u, r)$ the open ball in the vector space E with respect to the induced norm, for any $u \in E$ and any real $r > 0$.

Since $\bar{x}_E$ is a λ_E-density point of $P_{k,n}$, it follows from the assertion (a) of Lemma 2.255 that E coincides with the Bouligand-Peano tangent cone of $P_{k,n}$ at $\bar{x}_E$ relative to the vector space E. So, for any fixed nonzero $u \in E$, there are sequences $(u_m)_m$ in E converging to u and $(t_m)_m$ tending to 0 with $t_m > 0$ such that, for all $m \in \mathbb{N}$, we have $\overline{x}_E + t_m u_m \in P_{k,n}$ hence

$$(\overline{x}_E + t_m u_m) + f_{k,n}(\overline{x}_E + t_m u_m)e_N \in C_{k,n}.$$

The Fréchet differentiability of $f_{k,n}$ at $\overline{x}_E$ furnishes a sequence of reals $(\alpha_m)_m$ converging to 0 with $f_{k,n}(\overline{x}_E + t_m u_m) = f_{k,n}(\overline{x}_E) + t_m(\zeta(u_m) + \alpha_m)$, thus

$$(\overline{x}_E + f_{k,n}(\overline{x}_E)e_N) + t_m\big(u_m + (\zeta(u_m) + \alpha_m)e_N\big) \in C_{k,n},$$

which implies $u + \zeta(u)e_N \in T^B(C_{k,n}; \overline{x})$. The inclusion $\{h \in X : \langle x^*, h \rangle = 0\} \subset T^B(C_{k,n}; \overline{x})$ is then justified.

Now fix any $0 < \varepsilon < 1$. By the assertion (b) of Lemma 2.255, there exists some $r_0 > 0$ such that with any $u \in B_E(\overline{x}_E, r_0)$ we can associate some $p(u) \in P_{k,n}$ such that

$$\|p(u) - \overline{x}_E\| \leq \|u - \overline{x}_E\| \quad \text{and} \quad \|p(u) - u\| \leq \varepsilon \|u - \overline{x}_E\|.$$

For any $u \in B_E(\overline{x}_E, r_0)$, put $\Delta_{k,n}(u) := f_{k,n}(p(u)) - \overline{x}_N - \zeta(p(u) - \overline{x}_E)$ and note that, for any $\rho \in \mathbb{R}$,

$$\rho - \overline{x}_N - \zeta(u - \overline{x}_E) = \Delta_{k,n}(u) + (\rho - f_{k,n}(p(u))) + \zeta(p(u) - u).$$

Consider any $x \in S$ with $\|x - \overline{x}\| \leq \min\{1/(4k^2), r_0\}$. Then $\|p(x_E) - x_E\| \leq \varepsilon \|x_E - \overline{x}_E\| \leq 1/(2k)$ and

$$|f_{k,n}(p(x_E)) - \overline{x}_N| = |f_{k,n}(p(x_E)) - f_{k,n}(\overline{x}_E)|$$
$$\leq k\|p(x_E) - \overline{x}_E\| \leq k\|x_E - \overline{x}_E\| \leq 1/(4k),$$

thus $|f_{k,n}(p(x_E)) - x_N| \leq 1/(2k)$, so $\|p(x_E) + f_{k,n}(p(x_E))e_N - x\| \leq 1/k$. Noting that $p(x_E) + f_{k,n}(p(x_E))e_N$ belongs to $C_{k,n} \subset C_k$, the definition of C_k entails that $x_N - f_{k,n}(p(x_E)) \leq k\|x_E - p(x_E)\|$, so taking $\rho = x_N$ and $u = x_E$ in the latter equality above involving ρ and $\Delta_{k,n}(u)$, we see that

$$x_N - \overline{x}_N - \zeta(x_E - \overline{x}_E) \leq |\Delta_{k,n}(x_E)| + (k + \|\zeta\|)\|x_E - p(x_E)\|.$$

For x_E close enough to $\overline{x}_E$, say $\|x_E - \overline{x}_E\| \leq \eta_k$ for some positive real $\eta_k < r_0$, the Fréchet differentiability of $f_{k,n}$ at $\overline{x}_E$ and the inequality $\|p(x_E) - \overline{x}_E\| \leq \|x_E - \overline{x}_E\|$ (according to the definition of $p(\cdot)$) giving

$$|\Delta_{k,n}(x_E)| \leq \varepsilon \|p(x_E) - \overline{x}_E\| \leq \varepsilon \|x_E - \overline{x}_E\|,$$

we deduce by the inequality $\|p(x_E) - x_E\| \leq \varepsilon \|x_E - \overline{x}_E\|$ that

$$\langle x^*, x - \overline{x} \rangle = x_N - \overline{x}_N - \zeta(x_E - \overline{x}_E) \leq \varepsilon(1 + k + \|\zeta\|)\|x_E - \overline{x}_E\|$$

for all $x \in S \cap B_X(\overline{x}, \rho_k)$ where $\rho_k := \min\{\eta_k, 1/(4k^2)\}$. From this it is easily seen that $T^B(S; \overline{x}) \subset \{h \in X : \langle x^*, h \rangle \leq 0\}$. On the other hand, from the definition of x^* we have $\langle x^*, e_N \rangle = 1$. The proof is then finished by using a bijection between $\mathbb{N}$ and $\mathbb{N} \times \mathbb{N}$ allowing us to write C_i in place of $C_{k,n}$. $\square$

In addition to Proposition 2.258, the proof of Theorem 2.260 will also use the following lemma.

LEMMA 2.259. *Let K be a cone of a normed space X and let nonzero $x_1^*, x_2^* \in X^*$ be such that, for $i = 1, 2$,*

$$\mathrm{Hypl}(x_i^*) \subset K \subset \mathrm{Hypl}^{\leq}(x_i^*)$$

and such that $\mathrm{Hypl}(x_1^*) \neq K$.

Then $x_2^* = r\,x_1^*$ for some real $r > 0$, hence
$$\text{Hypl}(x_1^*) = \text{Hypl}(x_2^*) \quad \text{and} \quad \text{Hypl}^{\leq}(x_1^*) = \text{Hypl}^{\leq}(x_2^*).$$

PROOF. If $\langle x_1^*, h \rangle = 0$, then $\pm h \in \text{Hypl}(x_1^*) \subset K \subset \text{Hypl}^{\leq}(x_2^*)$ hence $\langle x_2^*, \pm h \rangle \leq 0$, that is, $\langle x_2^*, h \rangle = 0$. The inclusion of the kernel of x_1^* into the kernel of x_2^* is known (and this is easily seen) to ensure that $x_2^* = r\,x_1^*$ for some $r \in \mathbb{R}$, and $r \neq 0$ since x_2^* is nonzero by assumption. Choose $\overline{h} \in K \setminus \text{Hypl}(x_1^*)$ according to the assumption $K \neq \text{Hypl}(x_1^*)$ and $\text{Hypl}(x_1^*) \subset K$. The choice of $\overline{h}$ and the inclusions $K \subset \text{Hypl}^{\leq}(x_i^*)$, for $i = 1, 2$, entail $\langle x_1^*, \overline{h} \rangle < 0$ and $\langle x_2^*, \overline{h} \rangle \leq 0$. From the latter inequalities and from the equality $\langle x_2^*, \overline{h} \rangle = r\,\langle x_1^*, \overline{h} \rangle$ we see that $r > 0$. This justifies the first equality $x_2^* = rx_1^*$ with $r > 0$ of the lemma and the two other equalities are direct consequences. □

We can now state and establish a theorem concerning the richness of points where the Bouligand-Peano tangent cone is either a vector hyperplane or a half-space of the space.

THEOREM 2.260 (F. Roger). *Let S be a subset of a finite-dimensional Euclidean space X with dimension N and C be the set of $x \in S$ such that $T^B(S; x) \neq X$. Then the following properties hold.*
(a) *The set C is the union of a countable family of subsets of X with finite $\mathcal{H}_{N-1}$-Hausdorff measures and included in Lipschitz hypersurfaces.*
(b) *At $\mathcal{H}_{N-1}$-almost every point $x \in C$, either $T^B(S; x)$ is a vector hyperplane or $T^B(S; x)$ is a half-space.*

PROOF. Let $(v_k)_{k \in \mathbb{N}}$ be a sequence which is dense in the unit sphere of X and, for each k, put $C_k := \{x \in S : v_k \notin T^B(S; x)\}$. For any $x \in C$ there exists some unit vector $\overline{v} \notin T^B(S; x)$, so, as a neighborhood of $\overline{v}$ in the unit sphere, the intersection of $X \setminus T^B(S; x)$ with the unit sphere contains some v_j, hence $x \in C_j$. This means $C \subset \bigcup_k C_k$ and this inclusion is an equality since the reverse inclusion is obvious. By Proposition 2.258 the set S has a one-sided tangent hyperplane at $\mathcal{H}_{N-1}$-almost every point of C_k, hence, for some set $P \subset C$ with $\mathcal{H}_{N-1}(P) = 0$, the set S has a one-sided tangent hyperplane at every point of $C \setminus P$. The proposition also ensures that the set (C_k and also) C is the union of a countable family of subsets of X which have finite $\mathcal{H}_{N-1}$-measure and are included in Lipschitz hypersurfaces.

Now let Q be the set of $x \in C$ such that both following properties (i) and (ii) hold:

: (i) the set S has a one-sided tangent hyperplane at x different from the B-tangent cone $T^B(S; x)$, that is, there exists a nonzero $\zeta(x) \in X^*$ with
$$\text{Hypl}(\zeta(x)) \subset T^B(S; x) \subset \text{Hypl}^{\leq}(\zeta(x)) \quad \text{and} \quad \text{Hypl}(\zeta(x)) \neq T^B(S; x);$$
: (ii) $T^B(S; x)$ is not a half-space.

Consider, for each integer k, the set
$$Q_k := \{x \in Q : v_k \in \text{Hypl}^{\leq}(\zeta(x)),\ v_k \notin T^B(S; x)\}$$
$$= \{x \in Q : v_k \in \text{Hypl}^{<}(\zeta(x)),\ v_k \notin T^B(S; x)\}.$$

Take any $x \in Q$. The properties (i) and (ii) guarantee that the open cone
$$\text{Hypl}^{<}(\zeta(x)) \cap (X \setminus T^B(S; x))$$

is nonempty, thus its intersection with the unit sphere of X contains some v_k since it is a nonempty open set of the unit sphere with respect to the induced topology. This means $x \in Q_k$, hence $Q = \bigcup_k Q_k$. The latter equality and Proposition 2.258 give some set $G \subset Q$ with $\mathcal{H}_{N-1}(G) = 0$ such that, for each integer k, we can find, for any $x \in Q_k \setminus G$ (if any), some nonzero $u^*(x) \in X^*$ satisfying

(2.98) $\qquad \mathrm{Hypl}(u^*(x)) \subset T^B(S;x) \subset \mathrm{Hypl}^{\leq}(u^*(x))$ and $\langle u^*(x), v_k \rangle > 0$.

Let us show that $Q_k \setminus G = \emptyset$. Suppose the contrary and take $x \in Q_k \setminus G$. By Lemma 2.259 there exists a real $r > 0$ such that $\zeta(x) = r u^*(x)$. So we obtain by definition of Q_k

$$\langle r u^*(x), v_k \rangle < 0, \quad \text{which contradicts the inequality in (2.98).}$$

Consequently, one has $Q \setminus G = \emptyset$, which gives $Q \subset G$ hence $\mathcal{H}_{N-1}(Q) = 0$.

Finally, $\mathcal{H}_{N-1}(P \cup Q) = 0$ and for any $x \in C \setminus (P \cup Q)$ either $T^B(S;x)$ is a vector hyperplane or a half-space. $\square$

It is worth pointing out that for some non tangentially regular sets S, all truncations $U \cap (x + T^C(S;x))$, where U is a neighborhood of x, may be far to S. Indeed for the simple set $S := \{(x',r) \in \mathbb{R}^2 : r = |x'|\}$ we have $T^C(S;(0,0)) = \{(0,0)\}$, whereas $T^B(S;(0;0)) = S$. So, besides the above tools there is also need for other concepts in nonsmooth analysis. This is the object of the next chapter.

2.8. Further results

2.8.1. Intersection of C-normal cones. For convex sets S_1, S_2 it is clear from Theorem 2.2(d) that the normal cone of S_2 at x is included in the one of S_1 whenever $x \in S_1 \subset S_2$. More generally, for nonconvex sets S_1, S_2 and a point $x \in S_1 \subset S_2$, we know that $T^B(S_1;x) \subset T^B(S_2;x)$, and hence $N^C(S_2;x) \subset N^C(S_1;x)$ when S_1 and S_2 are both tangentially regular at x. Such a property is not valid with the Clarke normal cone for general sets. Indeed, we have already seen in Remark 2.4 that the Clarke tangent cone is not isotone, and hence the Clarke normal cone is not antitone, that is, the inclusion $N^C(S_2;x) \subset N^C(S_1;x)$ may fail when $x \in S_1 \subset S_2$. In fact, for $x \in (\mathrm{bdry}\, S_1) \cap (\mathrm{bdry}\, S_2)$ and $S_1 \subset S_2$ we may even have $N^C(S_1;x) \cap N^C(S_2;x) = \{0\}$ while both cones are nonzero, as illustrated with the sets

$$S_1 := \{(r,s) \in \mathbb{R}^2 : s \geq 0\} \quad \text{and} \quad S_2 := \{(r,s) \in \mathbb{R}^2 : s \geq -\sqrt{|r|}\},$$

and the point $x := (0,0)$; indeed

$$T^C(S_1;(0,0)) = \mathbb{R} \times [0,+\infty[\quad \text{and} \quad T^C(S_2;(0,0)) = \{0\} \times \mathbb{R},$$

which is equivalent to

$$N^C(S_1;(0,0)) = \{0\} \times]-\infty, 0] \quad \text{and} \quad N^C(S_2;(0,0)) = \mathbb{R} \times \{0\}.$$

Nevertheless, besides the situation of tangentially regular sets, the next proposition provides a general condition under which there exists a common nonzero element in an intersection $\bigcap_{i \in I} N^C(S_i;x)$.

PROPOSITION 2.261. *Let $(S_i)_{i \in I}$ be a family of subsets of a normed space X such that either $S_i \subset S_j$ or $S_j \subset S_i$ for any $i, j \in I$. Let $\bar{x} \in \bigcap_{i \in I} S_i$ for which there* is an open convex set K in X with $0 \in \text{bdry}\, K$ such that $(\bar{x} + K) \cap S_i = \emptyset$ for all $i \in I$ $\left(\text{which entails } \bar{x} \in \bigcap_{i \in I} \text{bdry}\, S_i\right)$. Then one has

$$\bigcap_{i \in I} N^C(S_i; \bar{x}) \neq \{0\}.$$

EXERCISE 2.262. Suppose the contrary of the conclusion of the proposition, that is, $\bigcap_{i \in I} N^C(S_i; \bar{x}) = \{0\}$.

(a) Argue through (1.30) that $\overline{\text{co}}\left(\bigcup_{i \in I} T^C(S_i; \bar{x})\right) = X$, and then there exists some

$$h \in K \cap \text{co}\left(\bigcup_{i \in I} T^C(S_i; \bar{x})\right).$$

(b) Argue that there are $m \in \mathbb{N}$ and $i_k \in I$ as well as $h_k \in T^C(S_{i_k}; \bar{x})$, for $k = 1, \cdots, m$, such that

$$h = h_1 + \cdots + h_m \quad \text{and} \quad S_{i_1} \subset S_{i_2} \subset \cdots \subset S_{i_m}.$$

(c) Putting $t_n = 1/n$, by definition of the Clarke tangent cone there is a sequence $(h_1^n)_n$ converging to h_1 such that $\bar{x} + t_n h_1^n \in S_{i_1} \subset S_{i_2}$ for all $n \in \mathbb{N}$. Similarly, there is a sequence $(h_2^n)_n$ converging to h_2 such that $(\bar{x} + t_n h_1^n) + t_n h_2^n \in S_{i_2}$ for all $n \in \mathbb{N}$. At the last step, there is $(h_m^n)_n$ converging to h_m such that $\bar{x} + t_n(h_1^n + h_2^n + \cdots + h_m^n) \in S_{i_m}$ for all $n \in \mathbb{N}$.

Since $h_1^n + \cdots + h_m^n \to h \in K$ as $n \to \infty$ and since K is an open convex set with $0 \in \text{bdry}\, K$, argue that $\bar{x} + t_n(h_1^n + \cdots + h_m^n) \in (\bar{x} + K) \cap S_{i_m}$ for large n and conclude by contradiction. □

In (c) above we notice that the crucial feature is that we can find $h_2^n \to h_2$ with $x_n + t_n h_2^n \in S_{i_2}$ whenever $S_{i_2} \ni x_n \to \bar{x}$ and $t_n \downarrow 0$ with $(t_n^{-1}(x_n - \bar{x}))_n$ convergent in X. Such vectors can be useful in some nonsmooth problems. Given a subset S of a normed space X and $x \in S$, a vector $h \in X$ belongs to the *Michel-Penot tangent cone* $T^{MP}(S; x)$ of S at x provided for any $S \ni x_n \to x$ and $t_n \downarrow 0$ with $(t_n^{-1}(x_n - x))_n$ convergent in X, there exists a sequence $h_n \to h$ such that $x_n + t_n h_n \in S$ for all $n \in \mathbb{N}$. We denote by $N^{MP}(S; x)$ the polar cone and we call it the *Michel-Penot normal cone* of S at x. As usual, these tangent and normal cones are empty by convention for $x \notin S$. Obviously $T^C(S; x) \subset T^{MP}(S; x)$, and hence $N^{MP}(S; x) \subset N^C(S; x)$; both inclusions are equalities when S is convex.

EXERCISE 2.263. Show that the cone $T^{MP}(S; x)$ is convex and check that Proposition 2.261, under the same assumptions, still holds with $N^{MP}(S_i; \bar{x})$ in place of $N^C(S_i; \bar{x})$. □

For an open set U of a normed space X and a function $f : U \to \mathbb{R} \cup \{-\infty, +\infty\}$ finite at $x \in U$, the *Michel-Penot subdifferential* $\partial_{MP} f(x)$ of f at x is defined by

$$\partial_{MP} f(x) := \{x^* \in X^* : (x^*, -1) \in N^{MP}(\text{epi}\, f; (x, f(x)))\}.$$

Clearly, $\partial_{MP} f(x) \subset \partial_C f(x)$.

EXERCISE 2.264. If f is Lipschitz continuous near x, show that for the support function of $\partial_{MP}f(x)$ the following equality holds

$$\sigma(h, \partial_{MP}f(x)) = \sup_{v \in X} \limsup_{t \downarrow 0} t^{-1}(f(x+tv+th) - f(x+tv)).$$

Derive that $\partial_{MP}f(x) = \{D_G f(x)\}$ whenever f is Lipschitz continuous near x and Gâteaux differentiable at x. □

We notice that the second property of Exercise 2.264 tells us in particular that the multimapping $\partial_{MP}f$ of a locally Lipschitz function f does not enjoy (contrarily to the Clarke subdifferential) the upper w^*-semicontinuity.

For a family of functions, the following common subgradient property holds.

PROPOSITION 2.265. Let $(f_i)_{i \in I}$ be a family of functions from an open set U into $\mathbb{R} \cup \{-\infty, +\infty\}$ such that either $f_i \leq f_j$ or $f_j \leq f_i$ for any $i, j \in I$. Let $\bar{x} \in U$ such that all $f_i(\bar{x})$ coincide and are finite, that is, there exists some $r \in \mathbb{R}$ with $r = f_i(\bar{x})$ for all $i \in I$. Assume that there are a vector $h \in X$ and a function $\varphi : U \to \mathbb{R} \cup \{-\infty, +\infty\}$ with $\varphi(\bar{x}) = r$ such that $\varphi \leq f_i$ for all $i \in I$ and such that $\underline{d}_H^+ \varphi(\bar{x}; h) > -\infty$. Then, either $\bigcap_{i \in I} \partial_{MP} f_i(\bar{x}) \neq \emptyset$ or there is some nonzero $x^* \in X^*$ such that $(x^*, 0) \in \bigcap_{i \in I} N^{MP}(\text{epi } f_i; (\bar{x}, f_i(\bar{x})))$.

EXERCISE 2.266. Prove the proposition. (Hint: apply Exercise 2.263 with $S_i := \text{epi } f_i$). □

The following corollary with $\varphi := f_1$ follows directly.

COROLLARY 2.267. Let $(f_n)_{n \in \mathbb{N}}$ be a sequence of locally Lipschitz continuous functions from an open set U of a normed space X into $\mathbb{R}$ with $f_n \leq f_{n+1}$ for all $n \in \mathbb{N}$. Let $\bar{x} \in U$ with $f_n(\bar{x}) = f_{n+1}(\bar{x})$ for all $n \in \mathbb{N}$. Then one has

$$\emptyset \neq \bigcap_{n \in \mathbb{N}} \partial_{MP} f_i(\bar{x}) \subset \bigcap_{n \in \mathbb{N}} \partial_C f_i(\bar{x}).$$

2.8.2. Subset of a Cartesian product. Given two normed spaces X and Y we know by Chapter 1 that any subset of $X \times Y$ can be seen as the graph of a multimapping $M : X \rightrightarrows Y$. For a pair $(\bar{x}, \bar{y}) \in \text{gph } M$ we know that a pair $(\bar{v}, \bar{w}) \in I(\text{gph } M; (\bar{x}, \bar{y}))$ (the interior tangent cone of gph M at $(\bar{x}, \bar{y})$) if and only if there are a real $\varepsilon > 0$, a neighborhood $U \times U'$ of $(\bar{x}, \bar{y})$ and a neighborhood $V \times W$ of $(\bar{v}, \bar{w})$ in $X \times Y$ such that

$$(U \times U') \cap \text{gph } M +]0, \varepsilon[(V \times W) \subset \text{gph } M,$$

or equivalently for all $t \in]0, \varepsilon[$, $v \in V$ and $(x, y) \in U \times U'$ with $y \in M(x)$,

(2.99) $\quad y + tW \subset M(x + tv), \quad$ that is, $\quad W \subset t^{-1}(M(x+tv) - y).$

The left one of the two latter inclusions clearly entails that int $M(x + tv) \neq \emptyset$, which is very restrictive for a multimapping; think to the case of a (single-valued) mapping. Instead of the interior tangent cone of the graph of a multimapping, a more suitable notion must be considered. For simplicity of notation, we will write $I(M; (x, y))$, $T^C(M; (x, y)), \cdots$, in place of $I(\text{gph } M; (x, y))$, $T^C(\text{gph } M; (x, y)), \cdots$.

Modifying slightly the right inclusion in (2.99), we then define the *quasi interiorly tangent cone* $Q(M; (\bar{x}, \bar{y}))$ of the multimapping M at $(\bar{x}, \bar{y})$ as follows. A pair $(\bar{v}, \bar{w}) \in Q(M; (\bar{x}, \bar{y}))$ provided that for each neighborhood W of $\bar{w}$, there exist

2.8. FURTHER RESULTS

a real $\varepsilon > 0$, a neighborhood U of $\bar{x}$ in X, a neighborhood U' of $\bar{y}$ in Y and a neighborhood V of $\bar{v}$ in X such that, for all $(x,y) \in U \times U'$ with $y \in M(x)$, $t \in]0, \varepsilon[$ and $v \in V$,

(2.100) $\big((x,y) + t(\{v\} \times W)\big) \cap \operatorname{gph} M \neq \emptyset$, that is, $W \cap t^{-1}(M(x+tv) - y) \neq \emptyset$.

This clearly means (see Definition 1.2)

$$(\bar{v}, \bar{w}) \in Q\big(M; (\bar{x}, \bar{y})\big) \Leftrightarrow \bar{w} \in \liminf_{\substack{v \to \bar{v} \\ (x,y) \to_M (\bar{x}, \bar{y}) \\ t \downarrow 0}} t^{-1}(M(x+tv) - y),$$

where $(x,y) \to_M (\bar{x}, \bar{y})$ stands for $(x,y) \to (\bar{x}, \bar{y})$ with $(x,y) \in \operatorname{gph} M$. As usual, by convention $Q(M; (\bar{x}, \bar{y})) = \emptyset$ whenever $\bar{y} \notin M(\bar{x})$.

In terms of sequences, by the first sequential characterization in Proposition 1.14(a) we see that $(\bar{v}, \bar{w}) \in Q(M; (\bar{x}, \bar{y}))$ if and only if, for any sequence $(x_n, y_n)_n$ in $\operatorname{gph} M$ converging to $(\bar{x}, \bar{y})$, any sequence $(t_n)_n$ in $]0, +\infty[$ tending to 0 and any sequence $(v_n)_n$ in X converging to $\bar{v}$ there is a sequence $(w_n)_n$ in Y converging to $\bar{w}$ such that

$$y_n + t_n w_n \in M(x_n + t_n v_n) \quad \text{for all } n \in \mathbb{N}.$$

The second sequential characterization in Proposition 1.14(a) allows also to say that $(\bar{v}, \bar{w}) \in Q(M; (\bar{x}, \bar{y}))$ if and only if, for any sequence $(x_n, y_n)_n$ in $\operatorname{gph} M$ converging to $(\bar{x}, \bar{y})$, any sequence $(t_n)_n$ in $]0, +\infty[$ tending to 0 and any sequence $(v_n)_n$ in X converging to $\bar{v}$ there exist subsequences $(x_{s(n)}, y_{s(n)})_n$, $(t_{s(n)})_n$, $(v_{s(n)})_n$ and a sequence $(w_n)_n$ in Y converging to $\bar{w}$ such that

$$y_{s(n)} + t_{s(n)} w_n \in M(x_{s(n)} + t_{s(n)} v_{s(n)}) \quad \text{for all } n \in \mathbb{N}.$$

Since everything will be clear in its context, for convenience we will identify in this subsection the set $Q(M; (x,y))$ with the multimapping from X into Y whose it is the graph, so $Q(M; (x,y))(v)$ will denote the set of all $w \in Y$ with $(v,w) \in Q(M; (x,y))$. The notation $T^C(M; (x,y))(v)$ will have a similar interpretation.

Obviously the following inclusions hold true:

$$I\big(M; (\bar{x}, \bar{y})\big) \subset Q\big(M; (\bar{x}, \bar{y})\big) \subset T^C\big(M; (\bar{x}, \bar{y})\big).$$

The next proposition shows that the quasi interiorly tangent cone enjoys desired properties similar to those of the interiorly tangent cone.

PROPOSITION 2.268. *Let $M: X \rightrightarrows Y$ be a multimapping between two normed spaces X, Y and let $\bar{x} \in X$ and $\bar{y} \in Y$ with $\bar{y} \in M(\bar{x})$. The following hold:*
(a) *One has the inclusion*

$$Q\big(M; (\bar{x}, \bar{y})\big) + T^C\big(M; (\bar{x}, \bar{y})\big) \subset Q\big(M; (\bar{x}, \bar{y})\big),$$

and hence $Q(M; (\bar{x}, \bar{y}))$ is a convex cone in $X \times Y$.
(b) *For any $v \in X$ the set $Q(M; (\bar{x}, \bar{y}))(v)$ is closed and convex in Y.*
(c) *Whenever $Q(M; (\bar{x}, \bar{y}))$ is nonempty, one has the equality*

$$T^C\big(M; (\bar{x}, \bar{y})\big) = \operatorname{cl}_{X \times Y} Q\big(M; (\bar{x}, \bar{y})\big).$$

EXERCISE 2.269. Prove the properties of the proposition. (Hint: follow the proof of similar properties with $I(S; \cdot)$ in Theorem 2.13). □

A particular situation when the quasi interiorly tangent cone of the multi-mapping M coincides with its Clarke tangent cone arises under the Aubin-Lipschitz property of M. This property (which is a suitable extension of Lipschitz behavior of single valued mapping) will be largely explored in Chapter 7. At this stage, let us say that the multimapping M fulfills the *Aubin-Lipschitz* (or *pseudo Lipschitz*) *property at* $\bar{x}$ *for* $\bar{y} \in M(\bar{x})$ provided there are a real $\gamma \geq 0$ and neighborhoods U and W of $\bar{x}$ and $\bar{y}$ respectively such that

$$M(x) \cap W \subset M(x') + \gamma \|x - x'\| \mathbb{B}_Y \quad \text{for all } x, x' \in U.$$

PROPOSITION 2.270. *Let* $M : X \rightrightarrows Y$ *be a multimapping between two normed spaces* X, Y *which satisfies the Aubin-Lipschitz property at* $\bar{x}$ *for* $\bar{y} \in M(\bar{x})$. *Then for every* $v \in X$ *one has the equality*

$$Q(M; (\bar{x}, \bar{y}))(v) = T^C(M; (\bar{x}, \bar{y}))(v).$$

EXERCISE 2.271. Prove the proposition by using the sequential characterizations of the quasi interiorly tangent cone. □

The goal of the next exercise is to give a condition under which $I(M; (\bar{x}, \bar{y}))(\bar{v})$ coincides with the interior of $Q(M; (\bar{x}, \bar{y}))(\bar{v})$. Of course, such an equality with the non-emptiness of the former set suggests the non-emptiness of interiors of images. The condition below is a kind of uniform property of interior.

EXERCISE 2.272. Let $M : X \rightrightarrows Y$ be a multimapping between two normed spaces and $\bar{x} \in X$. Assume that there is an open convex set K with $0 \in \operatorname{cl} K$ such that $M(x) + K \subset M(x)$ for all x near $\bar{x}$. Fix any $\bar{v} \in X$.
(a) Argue that $I(M; (\bar{x}, \bar{y}))(\bar{v}) \subset \operatorname{int}_Y\Big((Q(M; (\bar{x}, \bar{y})))(\bar{v})\Big)$.
(b) Fix $\bar{w} \in \operatorname{int}(Q(M; (\bar{x}, \bar{y})))(\bar{v})$ (supposing the latter set is nonempty) and fix also a convex neighborhood Ω of zero in X with $\bar{w} + \Omega \subset Q(M; (\bar{x}, \bar{y}))(\bar{v})$. Let U_0 be an open neighborhood of $\bar{x}$ such that $M(x) + K \subset M(x)$ for all $x \in U_0$. Choose $a \in (\bar{w} + \Omega) \cap (\bar{w} - K)$ and put $c := 2^{-1}(a + \bar{w})$ and $d := 2^{-1}(a + c)$.
(b.1) Argue that $c \in \bar{w} + \Omega$ and then $d \in \bar{w} + \Omega$.
(b.2) Writing $a = 2^{-1}(a + a) \in 2^{-1}(a + \bar{w} - K) = c - 2^{-1}K$, deduce that $d \in c - \frac{1}{4}K =: W_1$.
(c) Argue that there are a neighborhood $U \subset U_0$ of $\bar{x}$, a neighborhood U' of $\bar{y}$, a neighborhood V of $\bar{v}$, and a real $\varepsilon \in \,]0,1[$ such that

$$\big((x, y) + t(\{v\} \times W_1)\big) \cap \operatorname{gph} M \neq \emptyset,$$

for all $(x, y) \in (U \times U') \cap \operatorname{gph} M$, $v \in V$ and $t \in \,]0, \varepsilon[$.
(d) Writing $\bar{w} = 2c - a \in 2c - \bar{w} + K$ we see that $\bar{w} \in c + \frac{1}{2}K =: W$, and hence W is a neighborhood of $\bar{w}$. Take any $(x, y) \in (U \times U') \cap \operatorname{gph} M$, any $t \in \,]0, \varepsilon[$ and any $(v, w) \in V \times W$.
(d.1) Argue that there is some $w_1 \in W_1$ such that $y + tw_1 \in M(x + tv)$.
(d.2) Writing $y + tw = y + tw_1 + t(w - w_1)$ argue that $y + tw \in M(x + tv)$, and then $(\bar{v}, \bar{w}) \in I(M; (\bar{x}, \bar{y}))$.
(e) Deduce that $I(M; (\bar{x}, \bar{y}))(\bar{v}) = \operatorname{int}_Y\Big(Q(M; (\bar{x}, \bar{y}))(\bar{v})\Big)$. □

2.8.3. Compactly epi-Lipschitz sets.
In Theorem 2.235 we saw that
$$\operatorname*{Lim\,inf}_{S \ni x \to \bar{x}} T^B(S; x) \subset T^C(S; \bar{x}),$$

for any closed set S of a Banach space and any $\bar{x} \in S$, and we also saw in that theorem and Proposition 2.238 that the inclusion is an equality whenever either X is finite-dimensional or S satisfies the interior tangent property at $\bar{x}$. We are then interested in a class of sets including both previous ones and for which the above equality still holds true for any set in the class.

We know by definition that a subset S of X has the interior tangent property at $\bar{x} \in S$ if there exists some vector $h \in I(S; \bar{x})$ (the interior tangent cone of S at $\bar{x}$), that is, there exist a real $\varepsilon > 0$, a neighborhood U of $\bar{x}$ and a neighborhood V of zero such that $S \cap U +]0, \varepsilon[(h + V) \subset S$, or equivalently

$$S \cap U + tV \subset S + t\{-h\} \quad \text{for all } t \in]0, \varepsilon[.$$

Noticing that $\{-h\}$ is obviously a compact set yields to the following definition.

DEFINITION 2.273. A subset S of a normed space X is said to satisfy the *compact tangent property/condition* at point $\bar{x} \in S$ provided there are a nonempty compact set K in X, a real $\varepsilon > 0$, a neighborhood U of $\bar{x}$ in X and a neighborhood V of zero in X such that

$$S \cap U + tV \subset S + tK \quad \text{for all } t \in]0, \varepsilon[.$$

When $\bar{x} \in \text{bdry } S$ and S is closed near $\bar{x}$ and has the compact tangent property at $\bar{x}$, one says the set S is *compactly epi-Lipschitz* at $\bar{x}$ (following in this way the characterization of epi-Lipschitz sets at boundary points in Theorem 2.20).

REMARK 2.274. (a) By what precedes any set at $\bar{x}$ with the interior tangent property satisfies the compact tangent property at $\bar{x}$.
(b) If X is finite-dimensional, taking $V = K = \mathbb{B}_X$, we see that any subset of X has the compact tangent property at any of its points. This says in particular that the class of sets with the interior tangent property is strictly included in that with the compact tangent property.
(c) If X is infinite-dimensional, then no compact set in X has the compact tangent property at some point.
(d) Let X be an infinite-dimensional normed space. Since singletons of X are not compactly epi-Lipschitz by (c), we see that, under the inclusions $\bar{x} \in S_1 \subset S_2$ and the compact tangent property of S_2 at $\bar{x}$, the set S_1 may fail to have the compact tangent property at $\bar{x}$. □

THEOREM 2.275 (J.M. Borwein and H. Strójwas). Let S be a subset of a Banach space X which is closed near $\bar{x} \in S$ and satisfies the compact tangent property at $\bar{x}$. Then one has the equality

$$\liminf_{S \ni x \to \bar{x}} T^B(S; x) = T^C(S; \bar{x}).$$

The proof is established in the following exercise.

EXERCISE 2.276. For $x \in S$, let us denote by $\mathcal{T}(x)$ the cone defined by $h \in \mathcal{T}(x)$ provided that for each neighborhood V of zero in X there are a nonempty compact set K in X with $K \subset V$, a neighborhood U of x and a real $\varepsilon > 0$ such that, for all $t \in]0, \varepsilon[$

(2.101) $$S \cap U + th \subset S + tK.$$

(a) Prove that $\mathcal{T}(\bar{x}) \subset \liminf_{S \ni x \to \bar{x}} T^B(S; x)$ (Hint: Fix any $h \in \mathcal{T}(\bar{x})$ and any closed neighborhood V of zero. Choose a nonempty compact set $K \subset V$, a neighborhood

U of $\bar{x}$ and a real $\varepsilon > 0$ such that (2.101) is satisfied. Let $(t_n)_n$ in $]0, \varepsilon[$ tending to 0. For any fixed $u \in S \cap U$ choose $v_n(u) \in K$ such that $h \in t_n^{-1}(S - u) + v_n(u)$ and deduce that, for some cluster point $v(u)$ of $(v_n(u))_n$, one has $h \in T^B(S; u) + v(u) \subset T^B(S; u) + V$. Deduce that $h \in \underset{S \ni u \to \bar{x}}{\text{Lim inf}}\, T^B(S; u) \subset T^C(S; \bar{x})$.)

(b) Now let $h \in T^C(S; \bar{x})$ and let W be a neighborhood of zero. Argue the existence of a compact set $K \subset W$, and ε, U and V satisfying Definition 2.273.

(b.1) Argue the existence of $\varepsilon' \in]0, \varepsilon[$, a neighborhood $U' \subset U$ of $\bar{x}$ and a neighborhood $V' \subset V$ of zero such that for every $t \in]0, \varepsilon'[$

$$U' + t(h - V') \subset U \quad \text{and} \quad h \in t^{-1}(S - u) + V' \quad \text{for all } u \in U' \cap S.$$

(b.2) Fix $u \in U' \cap S$ and $t \in]0, \varepsilon'[$, and choose some $v \in V'$ with $u + t(h - v) \in S$. Argue that

$$u + th = u + t(h - v) + tv \in S + tK.$$

(b.3) Deduce that $h \in \mathcal{T}(\bar{x})$.

(b.4) Conclude that both members in the theorem coincide with $\mathcal{T}(\bar{x})$. □

We know that $\text{int}\, T^C(S; \bar{x}) \neq \emptyset$ whenever the set S has the interior tangent property at $\bar{x} \in S$. Our next aim is to establish the converse for sets satisfying the compact tangent property. Let us begin with the following lemma.

LEMMA 2.277. *Let S be a subset of a normed space X and let K be a compact set included in $T^C(S; \bar{x})$, where $\bar{x} \in S$. Then for any neighborhood V of zero in X there are a real $\varepsilon > 0$ and a neighborhood U of $\bar{x}$ in X such that*

$$S \cap U + tK \subset S + tV \quad \text{for all } t \in]0, \varepsilon[.$$

EXERCISE 2.278. Prove the lemma. □

LEMMA 2.279. *Let S be a subset of a normed space which has the compact tangent property at $\bar{x} \in S$ and let $h \in \text{int}\, T^C(S; \bar{x})$. Then there are a compact set $K \subset \mathbb{B}_X$ and reals $\delta, \varepsilon, r > 0$ such that for all $t \in]0, \varepsilon[$*

$$(\bar{x} + \delta \mathbb{B}_X) \cap S + t(h + r\mathbb{B}_X) \subset S + t(r/2)K.$$

The proof of the lemma is the purpose of the following exercise.

EXERCISE 2.280. Fix $s \in]0, 1[$ with $h + s\mathbb{B}_X \subset T^C(S; \bar{x})$.

(a) Argue that there are a compact set $K \subset s\mathbb{B}_X$ and reals $\delta_1, \varepsilon_1 > 0$, $r \in]0, 1[$ such that

(2.102) $\quad (\bar{x} + \delta_1 \mathbb{B}_X) \cap S + t(r\mathbb{B}_X) \subset S + tK \quad \text{for all } t \in]0, \varepsilon_1[,$

and then choose positive reals $\delta_2 < \delta_1$ and $\varepsilon_2 < \varepsilon_1$ such that

(2.103) $\quad (\bar{x} + \delta_2 \mathbb{B}_X) + t(h + K + (r^2/2)\mathbb{B}_X) \subset \bar{x} + \delta_1 \mathbb{B}_X \quad \text{for all } t \in]0, \varepsilon_2[.$

Since $h + K \subset T^C(S; \bar{x})$, by Lemma 2.277 take positive reals $\delta_3 < \delta_2$ and $\varepsilon_3 < \varepsilon_2$ such that

(2.104) $\quad (\bar{x} + \delta_3 \mathbb{B}_X) \cap S + t(h + K) \subset S + t(r^2/2)\mathbb{B}_X \quad \text{for all } t \in]0, \varepsilon_3[,$

and then choose positive reals $\delta < \delta_3$ and $\varepsilon < \varepsilon_3$ such that

(2.105) $\quad (\bar{x} + \delta \mathbb{B}_X) + t(r\mathbb{B}_X - K) \subset \bar{x} + \delta_3 \mathbb{B}_X \quad \text{for all } t \in]0, \varepsilon[.$

(b) Check by (2.102), (2.105), (2.104), (2.103) and (2.102) again, that for all $t \in \,]0, \varepsilon]$
$$(\bar{x} + \delta \mathbb{B}_X) \cap S + t(h + r\mathbb{B}_X) \subset (\bar{x} + \delta_3 \mathbb{B}_X) \cap S + t(h + K)$$
$$\subset (\bar{x} + \delta_1 \mathbb{B}_X) \cap S + t(r^2/2) \mathbb{B}_X \subset S + t(r/2)K,$$
which justifies the lemma. □

PROPOSITION 2.281. *Let S be a subset of a Banach space X which is closed near $\bar{x} \in S$. The following assertions are equivalent:*
(a) *the set S has the compact tangent property at $\bar{x}$ and $\mathrm{int}\, T^C(S; \bar{x}) \neq \emptyset$;*
(b) *the set S satisfies the interior tangent property at $\bar{x}$.*

PROOF. The assertion (b) obviously entails (a). Let us prove the converse. We may suppose that S is closed. By the assumption (a) fix some $h \in \mathrm{int}\, T^C(S; \bar{x})$ and, recalling that the closed convex hull of a compact set in a Banach space is compact, choose a compact convex set $K \subset \mathbb{B}_X$ and reals $\delta, \varepsilon, r > 0$ such that the property in Lemma 2.279 is fulfilled. Let $\delta_1 > 0$ and $\varepsilon_1 \in \,]0, \varepsilon[$ such that
$$\delta_1 + 2\varepsilon_1 (\|h\| + 2^{-1}r) < \delta.$$
We claim that
$$(2.106) \qquad (\bar{x} + \delta_1 \mathbb{B}_X) \cap S + \,]0, \varepsilon_1[\,(h + 2^{-1}r\mathbb{B}_X) \subset S.$$
Suppose the contrary, that is, there are
$$x_0 \in (\bar{x} + \delta_1 \mathbb{B}_X) \cap S, \quad \varepsilon_0 \in \,]0, \varepsilon_1[, \quad h_0 \in h + 2^{-1}r\mathbb{B}_X$$
satisfying $x_0 + \varepsilon_0 h_0 \notin S$. Then, there is some positive real $\eta < 2^{-1}\varepsilon_0 r$ such that
$$(2.107) \qquad (x_0 + \varepsilon_0 h_0 + \eta \mathbb{B}_X) \cap S = \emptyset,$$
which entails
$$(2.108) \qquad \eta < \|x_0 + \varepsilon_0 h_0 - x_0\| = \varepsilon_0 \|h_0\|.$$
Since the set $T := \{\tau \in [0, \varepsilon_0] : x_0 + \tau h_0 \in S + \eta K\}$ is nonempty (by the above choice of $K, \delta, \varepsilon, r$ through Lemma 2.279) and closed (by the closedness of $S + \eta K$), we can define $\varepsilon_2 := \max T$. The inclusion $K \subset \mathbb{B}_X$ and (2.107) ensure that $0 < \varepsilon_2 < \varepsilon_0$. On the other hand, by definition of ε_2 there is some $v \in K$ such that $x_2 := x_0 + \varepsilon_2 h_0 - \eta v \in S$. Putting $h_2 := h_0 + 2^{-1}rv$, we have $x_2 \in \bar{x} + \delta \mathbb{B}_X$ and $h_2 \in h + r\mathbb{B}_X$, which yields, by the choice of $K, \delta, \varepsilon, r$ satisfying Lemma 2.279, that for every $t \in \,]0, \varepsilon[$
$$x_2 + th_2 \subset S + 2^{-1}trK.$$
It results that, for every $t \in \,]0, 2r^{-1}\eta[$, writing $x_0 + (\varepsilon_2 + t)h_0 = x_2 + th_2 + (\eta - 2^{-1}tr)v$, we see that
$$x_0 + (\varepsilon_2 + t)h_0 \in S + 2^{-1}trK + (\eta - 2^{-1}tr)v \subset S + \eta K,$$
which contradicts the definition of ε_2. Consequently, (2.106) holds true, and hence the set S has the interior tangent property at $\bar{x}$. □

In the finite-dimensional setting we have the following direct corollary.

COROLLARY 2.282 (R. T. Rockafellar). *Let X be a finite-dimensional normed space and S be a subset which is closed near $\bar{x} \in S$. The set S has the interior tangent property at $\bar{x}$ if and only if $\mathrm{int}\, T^C(S; \bar{x}) \neq \emptyset$.*

The next exercise provides some characterizations of convex sets which have the compact tangent property.

EXERCISE 2.283. Let X be a normed space.
(a) If $S \subset X$ has the compact tangent property at $\bar{x} \in S$, prove that there is some compact set K in X such that $0 \in \text{int}(S + K)$.
(b) Show that a convex set S in X has the compact tangent property at any of its points if and only if there is a compact set K in X such that $0 \in \text{int}(S + K)$. (Hint: For the implication $\Leftarrow$, choose a convex neighborhood V of zero such that $2V \subset S + K$ and fixing any $\bar{x} \in S$, prove that for every $t \in\,]0, 1[$, the inclusion $S \cap (\bar{x} + V) + tV \subset S + t(\bar{x} + K)$ holds true.)
(c) Deduce that a nonempty convex set S has the compact tangent property at any of its points if and only if S satisfies the compact tangent property at some of its points.
(d) Prove that a closed convex set S in X has the compact tangent property if and only if there is a finite subset F in X such that $0 \in \text{int}(S + \text{co}\,F)$. (Hint: For the implication $\Rightarrow$ choose by (b) a compact set K with $0 \in \text{int}(S + K)$. Then take a closed ball V centered at zero with $V \subset S + K$ and a finite set $F \subset K$ such that $K \subset F + 2^{-1}V$. With $P := \text{co}\,F$, show by induction that

$$2^{-1}V \subset (2^{-1} + \cdots + 2^{-n})S + (2^{-1} + \cdots + 2^n)P + 2^{-n-1}V,$$

and then deduce that $2^{-1}V \subset \text{cl}(S + P) = S + P$).
(e) Argue by (a) that any convex set containing a set which has the compact tangent property at some of its points, satisfies the compact tangent property at all its points. $\square$

2.9. Comments

For optimization problems and nonsmooth mechanics, a unilateral variational analysis of *first-order variations* of nonsmooth functions via the concept of subdifferential essentially began with the works of J. J. Moreau and R. T. Rockafellar for nonsmooth convex functions. Earlier works of S. Mandelbrodt [**686**] and W. Fenchel [**404, 405**] started the extension of Legendre transform [**653**] to the situation of nonsmooth convex functions by introducing the conjugate of such functions. Fenchel [**405**] also studied a duality theory for optimization problems governed by convex functions defined on convex subsets in $\mathbb{R}^n$. To set this duality theory, given a convex set C in $\mathbb{R}^n$ and a convex function $f : C \to \mathbb{R}$, Fenchel defined the conjugate of the pair (C, f) as the pair (Γ, ϕ), where Γ is the set of $\xi \in \mathbb{R}^n$ such that $x \mapsto \langle \xi, x \rangle - f(x)$ is bounded from above on C and $\phi : \Gamma \to \mathbb{R}$ is given by $\phi(\xi) := \sup_{x \in C} (\langle \xi, x \rangle - f(x))$ for all $\xi \in \Gamma$; see more in the comments in Section 3.24 of Chapter 3. To offer a desirable flexibility to that concept, J. J. Moreau and R. T. Rockafellar independently in the beginning of 1960's dealt with extended real-valued functions, and for such a convex function $f : X \to \mathbb{R} \cup \{-\infty, +\infty\}$ they defined its conjugate $f^* : X^* \to \mathbb{R} \cup \{-\infty, +\infty\}$ as $f^*(x^*) := \sup_{x \in X} (\langle x^*, x \rangle - f(x))$ for all x^* in the topological dual X^* of the locally convex space X. Moreau [**738**] and Rockafellar [**844**] introduced the *subdifferential* $\partial f(x)$ of the convex function f at a point $x \in X$ (where $|f(x)| < +\infty$) as

$$\partial f(x) = \{x^* \in X^* : \langle x^*, h \rangle \le f(x + h) - f(x), \forall h \in X\}.$$

It is worth noticing that x is a minimizer of f on X if and only if $0 \in \partial f(x)$, while with $X = \mathbb{R}^n$ the concept of Schwartz's distribution is inadequate for stating an optimality condition for f at x. We also note that the inclusion $x^* \in \partial f(x)$ is linked with the conjugate f^*, in the sense that it holds if and only if $f^*(x^*) + f(x) \le \langle x^*, x \rangle$.

A rich calculus of conjugates and subdifferentials of convex functions as well as the duality theory with such functions will be developed in Chapter 3.

Another crucial step in the unilateral analysis of first-order variations of functions via a concept of subdifferential was reached in the 1973 doctoral thesis of F. H. Clarke [**232**] with nondifferentiable nonconvex functions. In this thesis and in [**233**], Clarke introduced for a locally Lipschitz function $f : \mathbb{R}^n \to \mathbb{R}$ its generalized gradient at a point $x \in \mathbb{R}^n$ as

$$\partial f(x) = \operatorname{co}\left\{\lim_{n \to +\infty} \nabla f(x_n) : \operatorname{Dom} \nabla f \ni x_n \to x\right\}.$$

Therein, he showed that the support function of $\partial f(x)$ is equal to $f^\circ(x;\cdot)$, where

(2.109) $$f^\circ(x;h) = \limsup_{t \downarrow 0, u \to x} t^{-1}[f(u+th) - f(u)]$$

and he established for locally Lipschitz functions a vast calculus for generalized gradients for finite sum, composition with $\mathcal{C}^1$ mappings, supremum of finitely as well as infinitely many locally Lipschitz functions. Clarke also introduced in the thesis his tangent cone and normal cone of a set $S \subset \mathbb{R}^n$ at $x \in S$ under the form

$$T^C(S;x) = \{h \in \mathbb{R}^n : d_S^\circ(x;h) = 0\} \quad \text{and} \quad N^C(S;x) = \operatorname{cl}(\mathbb{R}_+ \partial_C d_S(x))$$

and he showed that $N^C(S;x)$ coincides with the (negative) polar cone of $T^C(S;x)$. Then, given an extended real-valued function $f : \mathbb{R}^n \to \mathbb{R} \cup \{+\infty\}$, Clarke showed in [**232**] that the above concept of generalized gradient can be extended to such a function f at $x \in \operatorname{dom} f$ by setting

$$\partial f(x) = \{\zeta \in \mathbb{R}^n : (\zeta, -1) \in N^C(\operatorname{epi} f; (x, f(x)))\}.$$

Through those concepts, Clarke [**232**] (see also [**234, 235, 236, 237, 238, 241, 242**]) provided diverse applications and showed in particular how they allowed to obtain optimality conditions for nonsmooth nonconvex problems in calculus of variations and optimal control. The two other basic subdifferentials for nonsmooth nonconvex functions with applications to those problems and with rich calculus, say the Mordukhovich limiting subdifferential and the Ioffe approximate subdifferential, will be extensively developed in Chapters 4 and 5 respectively.

Regarding the situation of a Banach space X, after introducing the generalized directional derivative $f^\circ(x;\cdot)$ of a locally Lipschitz function $f : X \to \mathbb{R}$ by the equality (2.109) and the Clarke subdifferential, under the name of generalized gradient, of such a function as the subset

$$\partial f(x) = \{x^* \in X^* : \langle x^*, \cdot \rangle \le f^\circ(x;\cdot)\},$$

F. H. Clarke [**233, 239**] defined (following his thesis [**232**]), for a set S with $x \in S$, the Clarke tangent and normal cones $T^C(S;x)$ and $N^C(S;x)$ in the forms

$$T^C(S;x) = \{h \in X : d_S^\circ(x;h) = 0\} \quad \text{and} \quad N^C(S;x) = \operatorname{cl}_{w(X^*,X)}(\mathbb{R}_+ \partial_C d_S(x)).$$

Then, F. H. Clarke [**239**] showed with the above definitions that the Clarke tangent and normal cones are mutually polar, proved calculus rules for the sum of two locally Lipschitz functions, and established necessary optimality conditions for optimal control problems (see [**232**]) and mathematical programming problems (see [**239**]). The necessary optimality conditions for optimal control problems in the very large setting in [**232**] have been proved therein for the first time with general nonsmooth nonconvex data; that made clear the efficiency and interest of the above concepts.

J.-B. Hiriart-Urruty [**481, 484**] provided in his 1979 paper [**481, 484**] (submitted: May, 1977) the general fundamental sequential description (a_1) in Theorem 2.2 for the Clarke tangent cone; the form (a_2) in Theorem 2.2 has been used as tangent cone in L. Thibault [**906, 908**]. For a vector $h \in X$, the equivalences between (a_1), (a_2) and the equality $d_S^\circ(x; h) = 0$ (the original definition by Clarke of C-tangent vector) are due to Hiriart-Urruty [**484**]. The other fundamental description in terms of neighborhoods (for general topological vector spaces) in Definition 2.1 was given by R. T. Rockafellar in his 1980 paper [**856**] (submitted: March, 1978). The fact that the C-tangent cone $T^C(S; x)$ can be expressed as the inner limit (or limit inferior) of the multimapping $(t, u) \mapsto t^{-1}(S - u)$ as $t \downarrow 0$ and $S \ni u \to x$, was also noticed for the first time by Rockafellar [**856**].

The arguments in terms of neighborhoods in the proof of Theorem 2.2 follow R. T. Rockafellar [**856**]. The equivalence between (a) and (b) in Proposition 2.8 is due to J.-B. Hiriart-Urruty [**485**, Theorem 1] as well as the main ideas of the proof given for this equivalence. For Proposition 2.10 relative to the Clarke tangent cone of the boundary of a set, we refer to M.-O. Czarnecki and L. Thibault [**305**]; the method therein uses main ideas of M. Quincampoix [**825**] for a previous similar result relative to the Bouligand-Peano tangent cone.

The vectors in the interior tangent cone $I(S; x)$ appeared for the definition of epi-Lipschitz sets in the fundamental paper [**856**, p. 267] of R. T. Rockafellar as a brother of the *hypertangent cone*. In [**856**, (5.7)] Rockafellar defined a vector h in X to be hypertangent to a subset S at $\bar{x} \in S$ if there exist a neighborhood U of $\bar{x}$ and a real $\varepsilon > 0$ such that

(2.110) $\qquad x +]0, \varepsilon[h \subset S \quad \text{for all } x \in U \cap S.$

With the use of those types of vectors (interiorly tangent and hypertangent) R. T. Rockafellar proved, in the same paper and in [**854**], Theorem 2.13 and Theorem 2.20 of characterisation of epi-Lipschitz sets as well as Propositions 2.29 and 2.33; working with epigraphs of locally Lipschitz functions, Proposition 2.29 had been previously noticed by J.-B. Hiriart-Urruty [**484**, the end of page 7]. Epi-Lipschitz sets as defined in Definition 2.19 were also utilized in the 1967 book [**762**, p. 14-15] for existence and properties of trace functions for Sobolev spaces with an *open connected set Ω in $\mathbb{R}^N$ with Lipschitz boundary* in the sense in [**762**, p. 14]. Concerning Theorem 2.20 see also L. Cafarelli [**191**, Proof of Theorem 2]. Regarding the interior tangent property, the notion of sets with the *segment property/condition* was utilized in the geometry of types of open sets Ω involved for embedding theorems for Sobolev spaces. The open set Ω in $\mathbb{R}^N$ is defined to satisfy the *segment property/condition* if for every $\bar{x} \in \text{bdry}\,\Omega$ (see, for example, the 1975 book [**2**, p. 54] of R. A. Adams and also the second edition [**3**, p. 82] of R. A. Adams and J. J. F. Fournier) there are a unit vector $v \in \mathbb{R}^N$ and reals $\varepsilon, \eta > 0$ such that $x +]0, \varepsilon[v \subset \Omega$ for all $x \in \overline{\Omega} \cap B(\bar{x}, \eta)$, so we see that unit hypertangent vectors (as in (2.110)) coincide with such vectors. Clearly, the translation of the interior tangent condition in (2.3) corresponds to the latter condition with the open set V in place of the singleton point v. A result similar to Proposition 2.16 with the Bouligand-Peano tangent cone was first established by M. Quicampoix [**825**, Theorem 2.2]. Remark 2.17(b) was noticed by J.-B. Hiriart-Urruty [**484**, Proof of Proposition 5] with the help of subdifferential and directional calculus applied to the equality $d_S = \max\{d_{S^-}, d_{S^+}\}$, where $S := \{g(\cdot) = 0\}$, $S^- := \{g(\cdot) \leq 0\}$, $S^+ := \{g(\cdot) \geq 0\}$;

the statement in terms of Clarke normal cone in [**484**, Proposition 5] is in fact equivalent to the equality in Remark 2.17 here.

In infinite dimensional normed spaces, each one of the three basic types of differentiability (Gâteaux, Hadamard, Fréchet) is known to be an efficient tool for diverse problems in mathematical analysis. The three are also fundamental in the development of various variational properties in several sections in the book. The Gâteaux derivative was involved by R. Gâteaux in [**431**, p. 326] (see also [**433**, p. 11])[1] with the name of *"variation première"* (English: *"first variation"*) and with the notation $\delta U(z, \cdot)$ for a functional $U : C([0,1], \mathbb{R}) \to \mathbb{R}$ at a function $z \in C([0,1], \mathbb{R})$, where $\delta U(z, \cdot)$ is required to be a continuous linear functional from $C([0,1], \mathbb{R})$ into $\mathbb{R}$ such that for each $h \in C([0,1], \mathbb{R})$ the function $t \mapsto U(z + th)$ is derivable at 0 with $\delta U(z, h)$ as derivate at 0. The adaptation (as it is usual and done in the text) to the *Gâteaux differentiability* for normed spaces is then direct. Given a real-valued function f of two real variables, it appears from his 1923 note [**455**] that, for J. Hadamard the differential of f at (x_0, y_0) in the form $df(x_0, y_0) = p\,dx + q\,dy$ must mean that, for any mapping of one real variable $t \mapsto g(t) = (x(t), y(t))$ with $x(\cdot)$ and $y(\cdot)$ derivable at t_0 with $(x(t_0), y(t_0)) = (x_0, y_0)$, one has

$$\frac{d(f \circ g)}{dt}(t_0) = p\frac{dx}{dt}(t_0) + q\frac{dy}{dt}(t_0).$$

Translating this with today notations, it is clear that the function f is differentiable at (x_0, y_0) in the above Hadamard sense provided that, for any mapping of one real variable $t \mapsto \varphi(t) = (\varphi_1(t), \varphi_2(t)) \in \mathbb{R}^2$ defined near 0 with $\varphi(0) = (x_0, y_0)$ and with φ_i derivable at 0, the function $f \circ \varphi$ is derivable at 0. This feature clearly entails that $p = \frac{\partial f}{\partial x}(x_0, y_0)$ and $q = \frac{\partial f}{\partial y}(x_0, y_0)$ (but the converse fails in the sense that the existence of both partial derivatives at (x_0, y_0) does not imply the differentiability of f at (x_0, y_0) in the above Hadamard sense). According to this, the *Hadamard differentiability* at $x_0 \in X$ of a mapping F between two normed spaces X, Y is defined by some authors by the existence of some continuous linear mapping $A : X \to Y$ such that $F \circ \varphi$ is derivable at 0 with $(F \circ \varphi)'(0) = A(\varphi'(0))$ for any one real variable mapping $\varphi :\,]-r, r[\,\to X$ derivable at 0 and satisfying $\varphi(0) = x_0$. This is known by Proposition 2.38 to be equivalent to the definition of Hadamard differentiability taken in the manuscript. The concept of differentiability at x_0 of the mapping F as the existence of a continuous linear mapping[2] $A : X \to Y$ such that

$$\frac{1}{\|h\|}\|F(x_0 + h) - F(x_0) - A(h)\| \to 0 \quad \text{as } \|h\| \to 0,$$

was probably first considered in infinite dimensions in this form[3] by M. Fréchet in [**420**, p. 846-847] and [**425**, p. 139] with $Y = \mathbb{R}$ and $X = C([a, b], \mathbb{R})$ endowed with the norm of uniform convergence, and in [**426**, p. 308] for X, Y general normed spaces (called *"espaces ($\mathcal{D}$) vectoriels"* in [**426**, p. 304]). This is usually called

---

[1]Gâteaux died in the Front in the First World War at the age of 25 on October 3, 1913. The publication [**433**] and the previous one [**432**] correspond to diverse drafts of works left by Gâteaux before his departure to the Front. Those two publications were edited by P. Levy with various added comments.

[2]Notice that a continuous linear mapping A is just called "linear mapping" by Fréchet who defined the linearity property by requiring that A be additive and continuous.

[3]The term *"écart"* (*"gap"*) between $x_0 + h$ and x_0 with the notation $\Delta(h)$ was employed by Fréchet in [**420**] instead of $\|h\|$.

the *Fréchet differentiability*. As said in [**421**, p. 1051], the main motivation in [**420**] was the case where X is a vector space of *"elements of any nature, lines, functions, etc"*. Fréchet devoted a large treatment of this notion to functionals of Calculus of variations in pages 139-161 of his 1914 paper [**425**] and to differential calculus in [**423, 424**]. A very detailed study of this concept of *"differential"* in analysis was also carried out by Fréchet in [**426**]; he also treated therein, for the differential in his sense, the change of variables, the situation when X is the product of finitely many vector spaces, and applications to integral functionals. High order differentiability definitions and properties of integral functionals depending on several parameters were provided by Fréchet in [**426**]. Another related form for functions of n real variables (and equivalent for such functions) was employed earlier in his 1893 monograph [**903**, p. 133] by O. Stolz for a function f of several real variables, and this form was utilized some years later by J. Pierpont in his 1905 book [**802**, p. 268]. Fréchet [**423**, p. 388] cited this paper of Stolz and this book of Pierpont. The form used by Stolz [**903**] amounts to saying that a real-valued function f of real variables $x_1, \cdots, x_n$ is differentiable at a point $(z_1, \cdots, z_n)$ if it admits partial derivatives at this point and if, for each $i \in \{1, \cdots, n\}$ there is a function $(x_1, \cdots, x_n) \mapsto \varepsilon_i(x_1, \cdots, x_n)$ tending to 0 as $x_i \to 0$ for all i, such that for $h_1, \cdots, h_n$ near 0 in $\mathbb{R}$

$$f(z_1 + h_1, \cdots, z_n + h_n) = f(z_1, \cdots, z_n) + \sum_{i=1}^{n} h_i \frac{\partial f}{\partial x_i}(z) + \sum_{i=1}^{n} h_i \varepsilon_i(h_1, \cdots, h_n).$$

W.-H. Young presented and developed in his 1910 book [**978**] (as pointed out by M. Fréchet [**423**, p. 388]) all the advantages offered for functions of several real variables by the differential in the sense considered by Stolz in [**903**]. Much earlier to the above authors, as pointed out by S. Dolecki and G. H. Greco [**342, 343**] there were works by G. Peano related to differentiability of functions of several variables. Denoting by $u \times v$ the scalar product of two vectors in the plane/space and putting $\mathrm{mod}(u) := \sqrt{u \times u}$, G. Peano[4] declared in his 1887 manuscript [**782**, p. 131] that a function f of two/three real variables is differentiable at x_0 provided that there exists a vector v such that

$$f(x_0 + h) - f(x_0) = v \times (h + \varepsilon(h)) \quad \text{with } \varepsilon(h) \to 0 \text{ as } h \to 0.$$

In the later 1908 manuscript [**784**, p. 334] Peano employed instead the form

$$\frac{f(x_0 + h) - f(x_0) - v \times h}{\mathrm{mod}(h)} \to 0 \quad \text{as } h \to 0,$$

and such a vector v is called therein the *"derivata"* of f at x_0. By means of this, optimality conditions are stated in the two aforementioned Peano manuscripts in presence or not of a constraint set. If f is a complex-valued function of a complex variable, Peano [**784**, p. 330] employed in the latter above property the term $g(h)$, with a linear mapping g, in place of $v \times h$.

Theorem 2.47 and its proof are taken from L. Zajíček [**988**, p. 158] (see also [**990**]). The expressions in Theorem 2.72(b) appeared in R. T. Rockafellar [**856**] as extensions of $f^\circ(x; \cdot)$ for non locally Lipschitz functions. Propositions 2.74 and 2.76 are due to F. H. Clarke [**239**] and the example in Remark 2.75 of a set whose graph

---

[4]We learned Peano's contributions to differentiability and tangency in his books [**782, 784**] from the papers [**342, 343**] by S. Dolecki and G. H. Greco giving tribute to Peano for his works on those subjects.

of Clarke normal cone is not closed is due to R. T. Rockafellar [**854**]. The result in Proposition 2.78 was established in M.-O. Czarnecki and A.N. Gudovich [**304**] with a different proof, the proof given here follows M.-O. Czarnecki and L. Thibault [**305**]; for many other results in this line and applications of such representations of epi-Lipschitz sets to constrained equilibria of multimappings and to differential inclusions we refer to [**113, 279, 281, 282, 301, 302, 303, 637**] and references therein.

The classical theorem on differentiability of locally Lipschitz continuous functions defined on $\mathbb{R}^N$ was given, as well-known, by H. Rademacher [**829**]. Several simple proofs of this theorem are actually known; see, for example, D. N. Bessis and F. H. Clarke [**101**], J. P. R. Christensen [**228**], A. Nekvinda and L. Zajíček [**763**]. The proof given in the treatise follows the one of Proposition 6.41 in the book [**89**] by Y. Benyamini and J. Lindenstrauss. Several extensions of Rademacher theorem for locally Lipschitz continuous functions defined on infinite-dimensional Banach spaces were obtained in the last quarter of the twentieth century and the subject is still an active field of investigations. For certain Banach spaces X, the differentiability of locally Lipschitz continuous functions outside a set which is null in some measure sense was established by N. Aronszajn [**23**], J. P. R. Christensen [**227, 228**], R. R. Phelps [**800**]. Concerning the dense property of points of differentiability, D. Preiss [**819**] established another deep result in proving that any locally Lipschitz continuous function on a Banach space X is Fréchet (resp. Gâteaux) differentiable on a dense set in X provided that X can be renormed by an equivalent norm which is Fréchet (resp. Gâteaux) differentiable off zero. Other results can be found here in Chapters 12 and 13.

Proposition 2.79 of extension of real-valued Lipschitz functions was apparently independently established by E. J. McShane and by H. Whitney. Such a formula was offered by H. Whitney in his 1934 paper [**972**] (submitted: December 1932). In a footnote in the first page (that is, p. 63) of [**972**], for a continuous function $f : S \to \mathbb{R}$ on a compact set S in $\mathbb{R}^n$ (hence uniformly continuous on S), Whitney defined a continuous extension $F : \mathbb{R}^n \to \mathbb{R}$ by the formula

$$F(x) = \sup_{u \in S} \big(f(u) - h(d(x,y))\big) \quad \text{for all } x \in \mathbb{R}^n,$$

where $h : [0, +\infty[\to [0, +\infty[$ is any continuous nondecreasing function satisfying $h(0) = 0$; for example, h can be taken to be the modulus of continuity of f over S. Under the γ-Lipschitz property of f on S, the expression of $\varphi(x) = \sup_{u \in S} \big(f(u) - \gamma d(x,u)\big)$ as stated in Proposition 2.79 in the book was offered by E. J. McShane in his 1934 paper [**703**] (submitted: June 1934). In fact, assuming that S is a subset of a metric space E and $f : S \to \mathbb{R}$ is γ-Lipschitz on S, McShane [**703**, Theorem 1] used the formula to obtain an extension of f preserving the same γ-Lipschitz condition on the whole space E. Concerning the expression of $\Phi(x) = \inf_{u \in S} \big(f(u) + \gamma d(x,u)\big)$ in the same Proposition 2.79, it results from the opposite of φ in changing f in $-f$. It is worth mentioning that, using the expressions of both φ and Φ and assuming that S is a compact set in the Euclidean space $\mathbb{R}^n$, G. Aronsson studied in his 1966 paper [**22**, p. 552] the differentiability of any γ-Lipschitz extension of f at any point $x \in \mathbb{R}^n \setminus S$ where $\varphi(x) = \Phi(x)$. Various properties concerning the Clarke subdifferential of Φ can be found in the paper [**489**] by J.-B. Hiriart-Urruty.

Proposition 2.80(a) expressing the distance from a convex set by means of its support function is in general established by using the Hahn-Banach theorem or the infimal convolution (see, for example, Moreau [**746**, p. 11]). The equality in (c) of Proposition 2.80 was established by L. Hörmander [**502**, p. 185-186] by another approach.

Theorem 2.83 on the gradient representation of the Clarke subdifferential is due F. H. Clarke [**232, 233**]. The vector subspace property of the Clarke tangent cone of the graph of locally Lipschitz mapping in Proposition 2.89 was observed by R. T. Rockafellar in his 1979 paper [**854**]; the proof given here is different. Directionally Lipschitz functions in Definition 2.91 were introduced by R. T. Rockafellar [**855**] and Propositions 2.92 and 2.94 were demonstrated therein. The equality in (a) of Proposition 2.95 is due to J.-B. Hiriart-Urruty [**484**]. For $\bar{x} \in S$, $T^C(S;\bar{x})$ and $N^C(S;\bar{x})$ defined as in the book by

$$T^C(S;\bar{x}) = \liminf_{S \ni x \to \bar{x}, t \downarrow 0} \frac{1}{t}(S - x) \quad \text{and} \quad N^C(S;\bar{x}) = \left(T^C(S;\bar{x})\right)^o,$$

the main ideas of arguments for the equality

$$N^C(S;\bar{x}) = \mathrm{cl}_{w^*}\left(\partial_C d_S(\bar{x})\right)$$

in (c) of Proposition 2.95 are implicitely contained in the papers [**484**] of Hiriart-Urruty and [**854, 855**] of Rockafellar. Regarding this latter equality we mention that, for the use of a tight concept of generalized equilibria in nonconvex economy, B. Cornet and M.-O. Czarnecki defined, for S closed near $\bar{x} \in \mathrm{bdry}\, S$, another smaller normal cone as

$$\mathbb{R}_+ \limsup_{S \not\ni x \to \bar{x}} \partial_C d_S(x);$$

this cone was utilized for studies of marginal pricing rules in economics by J.-M. Bonnisseau and B. Cornet [**114**], by J.-M. Bonnisseau, B. Cornet and M.-O. Czarnecki [**115**] and by many other authors.

Theorem 2.98 was first established by F. H. Clarke [**232, 233**] when all the functions are locally Lipschitz. The case of the sum of two functions with merely one Lipschitz (and even directionally Lipschitz plus a qualification condition) is due to R. T. Rockafellar [**855**]. The proof of Theorem 2.98 given in the text follows a method employed by A. D. Ioffe [**514**] for the approximate subdifferential of the sum. This method of A. D. Ioffe has been also used by A. Jourani and L. Thibault [**587**]; see also J.-P. Penot [**795**]. The Moreau-Rockafellar theorem (see Theorem 2.105) has been first established in locally convex space by J. J. Moreau [**738**] in 1963 when both convex functions f and g are proper lower semicontinuous and g is continuous at some point of $\mathrm{dom}\, f \cap \mathrm{dom}\, g$; the general form in Theorem 2.105 removing the lower semicontinuity of both functions was demonstrated by R. T. Rockafellar in his 1966 paper [**848**]. Here Theorem 2.105 is obtained as consequence of Lemmas 2.100 and 2.101 in the book.

As already said in the preface, instead of the classical tangent line (resp. plane) to a curve (resp. surface), the concept of tangent at a point x of a set $S \subset \mathbb{R}^n$ that G. Bouligand considered in his 1928 paper [**157**, p. 29] corresponds to the set $x + \limsup_{t \downarrow 0} t^{-1}(S - x)$. The *"set of demi-tangents"* is used as terminology by Bouligand in [**157**], but later he utilized the name *"contingent"* (see [**163, 164**]). In addition to [**157**], Bouligand devoted a large series of papers illustrating the interest and the efficiency of this tangent set [**158, 159, 160, 161, 162, 163**].

Diverse works in the 1930's followed Bouligand's idea in the use of that concept; we cite, for example, the works of L. Chamard [**217**], G. Durand [**367, 368**], J. Mirguet [**714**], G. Rabaté [**826**], F. Roger [**867**]. In an earlier 1887 monograph [**782**, p. 305] G. Peano considered another notion of tangent set given by the set of y such that $\lim_{t\downarrow 0} d(y - x, \frac{1}{t}(S-x)) = 0$, which corresponds to $x + \liminf_{t\downarrow 0} t^{-1}(S-x)$ according to Proposition 1.12(a). By means of this set, Peano proved in [**782**] necessary optimality conditions for a solution x_0 of the problem of maximization of a differentiable function f (of two or three real variables) subject to the constraint set S, which we can write in the today usual form:

(2.111) \qquad Maximize $f(x)$ subject to $x \in S$.

In his other 1908 manuscript [**784**, p. 331], Peano denoted by $\mathrm{Tang}(S, x)$ the set of y such that 0 is a limit point of the mapping $t \mapsto d(y - x, \frac{1}{t}(S-x))$ as $t \downarrow 0$; this clearly coincides by Proposition 1.12(b) to $x + \limsup_{t\downarrow 0} t^{-1}(S-x)$, that is, to $x + T^B(S; x)$. By means of $\mathrm{Tang}(S, \cdot)$, Peano stated in page 335 of [**784**] sharper necessary optimality conditions for a solution x_0 of (2.111), as $Df(x_0) \times (x - x_0) \leq 0$ for all $x \in \mathrm{Tang}(S, x_0)$, where $u \times v$ (as said above) was used by Peano to denote the Euclidean inner product and where $Df(x_0)$ indicates the vector "*derivata de functione f in puncto x_0*" (in Peano's terminology in [**784**, p. 334]). The cone $x + T^B(S; x)$ was also considered by F. Severi in his 1930 paper [**886**].

Proposition 2.115 was observed by E. Polovinkin. Proposition 2.117 is due to M. Quincampoix (see [**825**, Corollary 2.4]) and Remark 2.118(a) is Theorem 2.2 in [**825**]. The directional derivative $f^B(x; \cdot) = \underline{d}^+_H f(x; \cdot)$ has been introduced in the study of optimization problems by J.-P. Penot [**785**] and its properties are thoroughly studied therein. The (Clarke) directional subregularity of a locally function f at a point x has been introduced by F. H. Clarke [**232**] as the equality in (b_2) of Proposition 2.126. The equality in (c) of that Proposition 2.126, for the Clarke subdifferential of $f_1 + \cdots + f_m$, was established by Clarke in [**232**] under the local Lipschitz property of all the functions $f_1, \cdots, f_m$. The extension to the case where merely $f_2, \cdots, f_m$ are locally Lipschitz is due to R. T. Rockafellar [**855**]; therein the result is even proved under (the tangential regularity of all the functions) and the directionally Lipschitz property of $f_2, \cdots, f_m$ plus a qualification condition.

The proof of the famous Graves theorem (see [**448**]), Theorem 2.139, and the proof of Corollary 2.140 reproduce A. D. Ioffe's arguments for Theorem 1.12 and Theorem 1.17 in his book [**526**]. The equality $T^B(G^{-1}(0); \bar{x}) = DG(\bar{x})^{-1}(0)$ in Corollary 2.142 appeared in the 1934 paper [**682**] of L. A. Lyusternik where is shown the stronger equality $\liminf_{t\downarrow 0} t^{-1}(G^{-1}(0) - \bar{x}) = DG(\bar{x})^{-1}(0)$ under the same assumption of surjectivity of the strict Fréchet derivative of G at $\bar{x}$. Proposition 2.144(b) was noticed and established by A. Jourani and M. Sene [**580**] with a different proof. The results in Proposition 2.146 and Corollary 2.147 were observed and established in J. V. Burke, M. C. Ferris and M. Qian [**182**, pages 203–206]; the case of a multimapping can be found in M. Bounkhel and L. Thibault [**171**]. Example 2.149 as well as its proof are taken from J. M. Borwein and M. Fabian [**127**].

The signed distance function from a set has been introduced in Optimization by J.-B. Hiriart-Urruty in his 1979 paper [**484**]; we refer also for its use in diverse contexts to [**230, 324, 325**] and references therein. Proposition 2.152 and its proof are taken from J.-B. Hiriart-Urruty [**484**, Theorem 3].

Theorem 2.154 and Theorem 2.155 have been established by B. Cornet and M.-O. Czarnecki in finite dimensions; as stated and proved they are taken from M.-O. Czarnecki and L. Thibault [**305**]. For similar and other results and for applications to equilibria, we refer to J.-M. Bonnisseau and B. Cornet [**113**], B. Cornet [**279**], B. Cornet and M.-O. Czarnecki [**282**], A. Cwiszewski and W. Kryszewski [**301, 302**], M.-O. Czarnecki [**303**], M.-O. Czarnecki and A. N. Gudovich [**304**].

The results concerning the local Lipschitz property of real-valued continuous convex functions are quite classical. Concerning Theorems 2.157, 2.158, 2.161 and Corollaries 2.159 and 2.164, their versions for functions defined on open convex sets of the finite-dimensional space $\mathbb{R}^m$ can be found in R. T. Rockafellar [**852**], and we refer to [**852**] for historical comments. Here we followed the approach in M. Jouak and L. Thibault [**563**]. The case of convex mappings with values in a normed space Y ordered with a normal convex cone is less classical. It was known from M. Valadier [**945, 946**] that a convex mapping upper bounded around a point $\bar{x}$ by a constant vector of Y is continuous at $\bar{x}$. The equivalence between (b) and (c) in Theorem 2.174 was first observed and proved by J. M. Borwein [**121**]. Using that equivalence, Proposition 2.172 and Theorems 2.173 and 2.177 were established in M. Jouak and L. Thibault [**563**]; see also [**118, 119**] for similar vector results. The expression of the signed distance function sgd_S in Lemma 2.165 and its convexity whenever S is convex appeared in J.-B. Hiriart-Urruty's papers [**484**, Proposition 1] and [**485**, Proposition 4], and the proof of (a) in Lemma 2.165 is taken from [**484**]. The concavity on a convex set S of the distance function $d(\cdot, X \setminus S)$ from its complement $X \setminus S$ in Lemma 2.165 appeared previously in N. Bourbaki [**174**, p. 150, Exercise 18]. Lemma 2.175 and Proposition 2.176 as well as their main arguments are taken from A. Carioli and L. Vesely [**206**].

The mean value theorem in Theorem 2.180 was first observed by G. Lebourg [**650**]. The characterization of strict Fréchet differentiability via C-subdifferential in (b) of Proposition 2.181 was noticed by J. M. Borwein [**124**, Proposition 3.1]. The class of compactly Lipschitzian mappings was introduced by L. Thibault [**907, 908, 909**]. The statements and proofs of Propositions 2.184 and 2.185 as well as Theorem 2.186 are taken from L. Thibault [**918**]. The proofs of Proposition 2.187 and Theorem 2.188 make use of arguments developed in A. Jourani and L. Thibault [**582**]. For the result on composition of convex vector-valued mappings in Theorem 2.189 we followed C. Combari, M. Laghdir and L. Thibault [**269, 271**]. Theorem 2.192 on the extension of Lipschitz mappings was first given in 1934 by M. D. Kirszbraun [**605**] when both spaces X and Y are finite-dimensional Euclidean spaces; the theorem was rediscovered by F. A. Valentine in his 1945 paper [**952**] (there was also a two-dimensional 1943 version [**951**] of Valentine). The proofs in [**605, 952**] used Helly's theorem on intersection of convex sets. A version between general Hilbert spaces appeared in the 1970 paper [**713**] by G. Minty. In the book [**865**, Theorem 9.58] by R. T. Rockafellar and R. J-B. Wets one can find a proof for X and Y finite dimensional involving the Zorn lemma and the result in Lemma 2.193 with a finite set I demonstrated via the minimization of a suitable differentiable function. The proof in [**865**] is still valid when X and Y are Hilbert spaces according to the weak compactness of closed balls in such spaces. The proof of Theorem 2.192 in the manuscript combines the application of the subdifferential of maximum of finitely many functions with main arguments from Proposition 1.13 and Theorem 1.12 in the book [**89**] of Y. Benyamini and J. Lindenstrauss. A proof of Theorem 2.192 with Lemma 2.193, but without involving the Zorn lemma can be found in the 2004 lectures [**475**, p. 7] by J. Heinonen. Several assertions in Proposition 2.194 were first observed by J.-B. Hiriart Urruty [**482**]. Theorem 2.198 was established in an equivalent form by F. H. Clarke [**243**, Theorem 2.8.2] and another similar result can also be found in [**243**, Theorem 2.8.6]. The equality $f'(x;h) = \max\{\langle Dg_t(x), h\rangle : t \in T(x)\}$ seems to have been first established in 1966 by Danskin [**315**] when T is compact and both $g(\cdot,\cdot)$ and $D_1 g(\cdot,\cdot)$ are continuous on $U \times T$. Corollary 2.202 is an adaptation of the statement and arguments of Theorem 2.11 in M. Valadier [**945**]; see also [**243**, Corollary 2, p. 100]. Theorem 2.204 is due to Valadier and the proof in the book follows Valadier's arguments in [**944, 945**]. A Version in locally convex spaces will be given and proved in Theorem 3.41 in the next chapter. Corollary 2.205 was established by M. Volle [**961**].

Lemma 2.213 was first observed by F. H. Clarke [**232, 233**]. The result in Proposition 2.216 was established by E. Ernst and M. Volle in their 2016 paper [**388**] through arguments of duality theory with nonconvex functions. The particular case of Corollary 2.218 was previously given and proved by T. Champion in his 2004 article [**218**]; see also the papers [**69**] by L. Ban and W. Song and [**387**] by E. Ernst and M. Volle. The variational principle in Theorem 2.221 was discovered by I. Ekeland [**381**] (see also Ekeland's survey [**381**]); the proof given here follows that in J.-P. Aubin and I. Ekeland [**43**]. Theorem 2.225 and Theorem 2.227 are due to E. Bishop and R. R. Phelps [**103, 104**]. The use of Ekeland variational principle in the proof (given in the text) of necessary optimality conditions with Lagrange multipliers in Theorem 2.228 is due F. H. Clarke [**239**]; the proof in the text is exactly that in [**239**]. Necessary optimality conditions, with equality constraints in a less general form (nonsmooth inequality constraints and smooth equality constraints), were previously established with Clarke subdifferentials through a different approach by J.-B. Hiriart-Urruty [**482**] (involving the calculus of the normal cone of the feasible set defined by the constraints).

The inclusion in Theorem 2.235 was first observed and proved by B. Cornet [**277**] in finite dimensions and the equality in the same context was proved by B. Cornet [**278**] and J.-P. Penot [**786**]. The inclusion in its full generality of Banach space as stated in Theorem 2.235 as well as the characterization of Clarke tangent cone in Proposition 2.234 are due to J. S. Treiman [**934**]; see Theorems 2.1 and 3.1 in [**934**]. The proofs given here for Proposition 2.234 and Theorem 2.235 follow Penot's approach in [**787**]. Theorem 2.232 is due to J. Daneš [**306**] and is generally called Daneš' drop theorem; the proof which is given is taken from Penot [**787**] and we refer to [**307**] for several related results. It is worth pointing out that the assumption $r < d(b, S)$ in Theorem 2.232 cannot be relaxed in $S \cap B[b, r] = \emptyset$. Indeed, let us say that the norm $\|\cdot\|$ of a Banach space X has the property ($\mathcal{D}$) provided that for every closed subset S with $S \cap \mathbb{B}_{\|\cdot\|} = \emptyset$ there exists some $c \in S$ such that $S \cap \mathrm{co}\left(\{c\} \cup \mathbb{B}_{\|\cdot\|}\right) = \{c\}$. S. Rolewicz [**871**, Theorem 5] proved in 1987 that any Banach space $(X, \|\cdot\|)$ with the property ($\mathcal{D}$) is reflexive. Soon later, V. Montesinos [**717**, Theorem 3] showed that the norm $\|\cdot\|$ of a Banach space $(X, \|\cdot\|)$ has the property ($\mathcal{D}$) if and only if the space X is reflexive and the norm $\|\cdot\|$ is sequentially Kadec-Klee. One can also find in [**717**] diverse other results related to the separable reduction of property ($\mathcal{D}$), the heredity of property ($\mathcal{D}$) for induced norms, quotient norms and the canonical norm of $L_X^p(T, \mathcal{T}, \mu)$. For other inclusions in the line of the one in Theorem 2.235 we refer to J.-P. Aubin and H. Frankowska [**44**] and to A. Jourani and T. Zakaryan [**592**]. Example 2.236 showing that the inclusion of the Clarke tangent cone into the limit inferior of the Bouligand-Peano tangent cone generally fails in infinite dimensions was constructed by J. S. Treiman [**934**]. Proposition 2.238 was also first observed in [**934**, Theorem 3.3].

The use in infinite dimensions of sets which can be covered by countably many Lipschitz surfaces apparently began with the two 1978 papers by L. Zajíček [**984, 985**]. In finite dimensions, such sets were considered already in 1928 by W. H. Young [**980**] in $\mathbb{R}^2$ in another form (and called therein "*ensemble ridé*"), and in 1946 by P. Erdös [**386**]. Definition 2.245 follows the definition of *Lipschitz F-surface* as stated by Zajíček in [**994**, Definition 3.1], and Proposition 2.247 and its proof reproduces Lemma 3.2 in the same paper [**994**] by Zajíček. Proposition 2.248 follows A. Nekvinda and L. Zajíček [**764**, lemma 2.10].

For closed subspaces E, F of a normed space X, as written in the historical comments concerning Chapter 1 the pseudo-distance $\widehat{d}_{\mathrm{Svect}}(E, F)$, called the geometric opening, seems to have been first introduced in 1948 by M. G. Krein, M. A. Krasnoselskiĭ and D. P. Milman [**618**] and in another earlier equivalent form by M. G. Krein and M. A. Krasnoselskiĭ in their 1947 paper [**617**]. It was also considered later by I. C. Gohberg and M. G. Krein [**441**] and by T. Kato [**598**]. The distance $d_{\mathrm{Svect}}(E, F)$ has been proposed by I. C. Gohberg and A. S. Markus [**442**]; a different but equivalent distance to d_{Svect} was introduced earlier to [**442**] by J. D. Newburgh [**765**]. We also refer the interested reader to E. R. Berkson [**92**], H. O. Cordes and J. P. Labrousse [**276**], J. D. Newburgh [**765**] for other results and other distances. The equality $\widehat{\mathrm{haus}}_1(P, Q) = \widehat{\mathrm{haus}}_1(P^\circ, Q^\circ)$ in Theorem 2.250 is due to D. W. Walkup and R. J.-B. Wets [**968**] and extended to convex functions by G. Beer [**81**]; we refer also to J.-P. Penot [**789**]. The particular equality $\widehat{d}_{\mathrm{Svect}}(E, F) = \widehat{d}_{\mathrm{Svect}}(E^\perp, F^\perp)$ was already known by M. G. Krein and M. A. Krasnoselskiĭ [**617**] in the Hilbert setting, and by M. G. Krein, M. A. Krasnoselskiĭ and D. P. Milman [**618**] in the case of a Banach space. Theorem 2.251 can be found in [**9, 599, 778**].

Proposition 2.241 and Proposition 2.243 appeared in F. Roger [**867**] (see Theorem I therein). Theorem 2.254 was established in the finite-dimensional framework by F. Roger with a different proof in "Théorème II (1^{iere} *partie*)" of [**867**]. The statement and proof of Lemma 2.252 which contributes to the proof (given in the book) of the latter theorem are taken from L. Zajíček [**994**, Lemma 2.4]; the lemma can also be seen from the earlier result by A. Largillier [**645**, Theorem 2.2] in complex Banach spaces. For (a) in Proposition 2.253 we followed L. Zajíček [**996**, Lemma 2.1]. Theorems 2.257 and 2.260 are also due to F. Roger; they correspond respectively to "Théorème II (2^{ieme} *partie, cas particulier*)"and "Théorème II (*énoncé général*)"in [**867**]. The approach presented here through Proposition 2.258 follows that in $\mathbb{R}^2$ of Saks' book [**876**]. A version in $\mathbb{R}^2$ of Theorem 2.260 was also stated by A. Kolmogoroff and J. Verčenko [**611, 612**] independently of [**867**]; we also refer to A. S. Besicovitch [**99**], U. S. Haslam-Jones [**471, 472**], S. Saks [**875**] for related properties.

The concept of quasi interiorly tangent cone of a multimapping has been introduced by L. Thibault [**911**] and all the results in Subsection 2.8.2 are taken from that paper. As seen in Subsection 2.8.3 the notion of compactly epi-Lipschitz sets, introduced by J. M. Borwein and H. Strowjas [**142**, Definition 2.1], leads to various extensions of properties of sets with the interior tangent property and of epi-Lipschitz sets. All the results in that subsection were obtained by Borwein and Strowjas [**142**].

The result in Corollary 2.267 was first observed and proved by A. S. Lewis and R. Lucchetti [**658**] with three functions and all the functions locally Lipschitz; the approach therein is based on the following nonlinear Fenchel-type duality established by A. S. Lewis and D. A. Ralph [**660**, Theorem 2]: Let Q be a nonempty convex compact set in $\mathbb{R}^n$ and let $f, g : \mathbb{R}^n \to \mathbb{R} \cup \{+\infty\}$ be two proper convex lower semicontinuous functions with $Q \supset (\mathrm{dom}\, f) \cup (\mathrm{dom}\, g)$. Let $h : \mathbb{R}^n \to \mathbb{R}$ be a function which is Lipschitz on an open set containing Q and such that $-g(x) \leq h(x) \leq f(x)$ for all $x \in Q$. Then there exist $c \in Q$ and $c^* \in \partial_C h(c)$ such that

(2.112) $$f^*(c^*) + g^*(-c^*) \leq 0,$$

where f^* and g^* are the Legendre-Fenchel conjugates of f and g respectively (see Definition 2.68 for the definition of Legendre-Fenchel conjugate).

As stated in a normed space and with any non-increasing finite sequence, the result in Corollary 2.267 follows from Theorem 4.2 by J. M. Borwein and S. Fitzpatrick [**130**]. We must emphasize that [**130**, Theorem 4.2] stated in $\mathbb{R}^n$ still holds true in any normed space X, as noticed in [**128**, page 10] since [**128**, Lemma 3.1] is valid in any normed space. In fact, [**130**, Theorem 4.2] shows in a normed space X, for functions $f_1 \geq \cdots f_m \geq h \geq g_1 \geq \cdots \geq g_n$ with h Lipschitz, f_i and $-g_j$ lower semicontinuous, and satisfying $f_i(\bar{x}) = h(\bar{x}) = g_j(\bar{x})$ for all i, j, that there is some $x^* \in \partial_C h(\bar{x})$ such that $x^* \in \partial_C f_i(\bar{x})$ and $x^* \in -\partial_C(-g_j)(\bar{x})$ for all i, j. The more general result in Proposition 2.261 as well its proof, its consequence in Proposition 2.265 and its version with the Michel-Penot normal cone, are all taken from J. Benoist and J.-B. Hiriart-Urruty [**88**].

CHAPTER 3

Convexity and duality in locally convex spaces

Many examples of convex functions have been given in Subsection 2.2.4 and many properties of such functions in normed spaces have been established in that subsection and in Section 2.3. The main aim of this chapter is to develop the convex duality theory and to derive basic formulas for Legendre-Fenchel conjugates and subdifferentials of convex functions.

3.1. Convex functions on topological vector spaces

In addition to the characterization of the subdifferential in Proposition 2.69 in the previous chapter, we will see that, for a large class of convex functions, every property of a function f in the class corresponds to some property of its subdifferential, and corresponds also to some property of its Legendre-Fenchel conjugate f^*. This will then offer a dual way to study several properties of f. In fact, in many situations the framework in which one has to work with convex functions is that of locally convex space; it is for example the case of lower $w(X^*, X)$-semicontinuous convex functions defined on the topological dual X^* of a Banach space X. Further, taking a proper lower semicontinuous convex function $f : X \to \mathbb{R} \cup \{+\infty\}$ on a Banach space X, we will see later that $w(X^*, X)$ is one of the right locally convex topologies θ on X^* such that the Legendre-Fenchel conjugate of $f^* : X^* \to \mathbb{R} \cup \{+\infty\}$, with X^* endowed with the θ-topology, coincides with f.

3.1.1. Subdifferential and directional derivatives of convex functions on topological vector spaces. Let us consider a topological vector space (X, τ_X) that we will also denote by X when there is no risk of ambiguity. Like in the case of a normed space, the duality pairing related to X and to its topological dual X^* is the bilinear functional $\langle \cdot, \cdot \rangle_{X^*, X} : X^* \times X \to \mathbb{R}$ defined by $\langle x^*, x \rangle_{X^*, X} = x^*(x)$, the value of x^* at x. If there is no risk, we will simply write as done in the previous chapters $\langle \cdot, \cdot \rangle$.

Convex functions are already defined in the previous chapter. Obviously, any positively linear combination of convex functions (that is, $\sum_{i=1}^{m} r_i f_i$ with $r_i > 0$ and f_i convex) is convex and the composition of a convex function with an inner affine mapping is convex. The case of composition with an inner vector valued convex mapping has already been considered for normed spaces in the previous chapter, and it will be analyzed later in this chapter in the context of more general spaces. For any convex function $\varphi : X \times Y \to \mathbb{R} \cup \{-\infty, +\infty\}$, the convexity of the function f defined on X by $f(x) = \inf_{y \in Y} \varphi(x, y)$ has been seen in the previous chapter (see Proposition 2.54). We also easily notice two additional useful properties:

PROPOSITION 3.1. Let X, Y be two topological vector spaces and $\varphi : X \times Y \to \mathbb{R} \cup \{-\infty, +\infty\}$ be an extended real-valued function. Let f be the function defined by $f(x) := \inf_{y \in Y} \varphi(x, y)$ for all $x \in X$ and let $\pi_X : X \times Y \to X$ the projector on X defined by $\pi_X(x, y) := x$.

(a) Recalling that the epigraph strict of f is defined as
$$\text{epi}_s f := \{(x, r) \in X \times \mathbb{R} : f(x) < r\},$$
one has
$$\text{epi}_s f = \pi_X(\text{epi}_s \varphi),$$
so one recovers the convexity of f whenever φ is convex.

(b) If $\pi_X(\text{epi}\,\varphi)$ is closed, then the usual epigraph satisfies the equality
$$\text{epi}\, f = \pi_X(\text{epi}\,\varphi),$$
and the infimum defining $f(x)$ is attained whenever $f(x)$ is finite.

Let us turn to the subdifferential of convex functions in topological vector spaces. In the case of a convex function f on a normed space, we have seen in the preceding chapter that its Clarke subdifferential at a point x with $|f(x)| < +\infty$ coincides with the set of slopes of continuous affine functions minorizing f and equal to f at x. In this chapter we consider this set of slopes for any function defined on a topological vector space.

DEFINITION 3.2. Given a nonempty subset U of the topological vector space X and a function $f : U \to \mathbb{R} \cup \{-\infty, +\infty\}$ on U, we define its *Moreau-Rockafellar subdifferential* by

(3.1) $\partial_{MR} f(x) := \{x^* \in X^* : \langle x^*, u - x \rangle \leq f(u) - f(x),\ \forall u \in U\}$ if $|f(x)| < +\infty$,

and $\partial_{MR} f(x) := \emptyset$ if $f(x)$ is not finite. Each $x^* \in \partial_{MR} f(x)$ is a *Moreau-Rockafellar subgradient of f at x*. If U and f are convex, we will write $\partial f(x)$ in place of $\partial_{MR} f(x)$, as is usual in the literature. Under the convexity of U and f, we simply say that $\partial f(x)$ is *the subdifferential of f at x* and any element of $\partial f(x)$ is called a *subgradient of f at x*.

Evidently
$$\partial_{MR}(rf)(x) = r\partial_{MR} f(x) \quad \text{for every real } r > 0.$$

It is also clear that the effective domain $\text{dom}\, f = \{x \in U : f(x) < +\infty\}$ is convex whenever U is a convex set and f is convex on U. Nevertheless, even if U and f are convex, the domain $\text{Dom}\,\partial f$ of the subdifferential multimapping ∂f may fail to be convex as will be seen in Example 3.193.

Another quite obvious property is the following.

PROPOSITION 3.3. Let $f : U \to \mathbb{R} \cup \{-\infty, +\infty\}$ be a function on a nonempty convex subset U of a topological vector space X and let C be a nonempty convex set of U. If $\partial_{MR}(x) \neq \emptyset$ for every $x \in C$, then f is convex on C.

PROOF. Assume that $\partial_{MR} f(u) \neq \emptyset$ for all $u \in C$. First, we note that f is finite on C by the above definition. Fix any $x, y \in C$ and $t \in]0, 1[$, and put $z_t := (1 - t)x + ty$. For each $t \in]0, 1[$ choose $z_t^* \in \partial_{MR} f(z_t)$. By definition of $\partial_{MR} f(z_t)$ we can write

$$f(z_t) = (1-t)f(z_t) + tf(z_t) \leq (1-t)\big(f(x) - \langle z_t^*, x - z_t \rangle\big) + t\big(f(y) - \langle z_t^*, y - z_t \rangle\big),$$

which is equivalent to $f(z_t) \leq (1-t)f(x) + tf(y)$. This justifies the convexity of f on C. □

In the situation of convex functions, it is clear that $\partial f(x)$ extends to topological vector spaces the definition given in (2.21) for convex functions on normed spaces. The normal cone of a convex set as seen in the previous chapter is thus also extended to topological vector spaces as follows.

DEFINITION 3.4. If C is a nonempty convex set in the topological vector space X, its *normal cone* $N(C; x)$ at $x \in X$ is defined by $N(C; x) = \partial \Psi_C(x)$, which means that for $x \in C$

(3.2) $\qquad N(C; x) = \{x^* \in X^* : \langle x^*, u - x \rangle \leq 0, \; \forall u \in C\},$

and $N(C; x) = \emptyset$ for $x \notin C$.

The following easy property makes clear that subgradients can also be recovered from normals.

PROPOSITION 3.5. *Let X be a topological vector space and $f : X \to \mathbb{R} \cup \{-\infty, +\infty\}$ be a convex function. Then one has*
$$\partial f(x) = \{x^* \in X^* : (x^*, -1) \in N(\mathrm{epi}\, f; (x, f(x)))\}$$
at any $x \in X$ where f is finite.

Given a nonempty convex set U in X and a function (resp. convex function) $f : U \to \mathbb{R} \cup \{-\infty, +\infty\}$, its Moreau-Rockafellar subdifferential $\partial_{MR} f(x)$ (resp. $\partial f(x)$) at $x \in U$ coincides with the Moreau-Rockafellar subdifferential at x of its extension f_0 given by $f_0(u) = f(u)$ for all $u \in U$ and $f_0(u) = +\infty$ for all $u \in X \setminus U$. Accordingly, we will mainly work in this chapter with extended real-valued functions defined on the whole topological vector space X.

Assume that f is convex on X and fix $x \in X$ such that f is finite at x. It is clear that
$$x^* \in \partial f(x) \Rightarrow \langle x^*, \cdot \rangle \leq d_H^+(x; \cdot),$$
where $d_H^+(x; \cdot)$ is the Hadamard lower directional derivative of f at x as defined in (2.41), that is,
$$d_H^+(x; h) = \liminf_{t \downarrow 0, v \to h} t^{-1}\big(f(x + tv) - f(x)\big) \quad \text{for all } h \in X.$$
On the other hand, recalling that the (standard) directional derivative
$$f'(x; h) := \lim_{t \downarrow 0} t^{-1}\big(f(x + th) - f(x)\big)$$
of the convex function f at x satisfies the equality $f'(x; h) = \inf_{t>0} t^{-1}[f(x+th) - f(x)]$ (see (2.20)) we also see that

(3.3) $\qquad x^* \in \partial f(x) \Leftrightarrow \langle x^*, \cdot \rangle \leq f'(x; \cdot).$

Noting that $f'(x; \cdot)$ is convex (by the convexity of f) and $f'(x; 0) = 0$, it follows that

(3.4) $\qquad \partial f(x) = \partial f'(x; \cdot)(0).$

Further, taking any reals $r > f'(x; h)$ and $s > f'(x; -h)$, we see that $r + s \geq f'(x; h + (-h)) = 0$, or equivalently $-s \leq r$. We deduce that $-f'(x; -h) \leq f'(x; h)$ for all $h \in X$.

Since $\underline{d}_H^+(x;\cdot) \le f'(x;\cdot)$, it results also for the convex function f that

(3.5) $\qquad x^* \in \partial f(x) \Leftrightarrow \langle x^*, \cdot \rangle \le \underline{d}_H^+(x;\cdot).$

Obviously, $\underline{d}_H^+(x;\cdot)$ is positively homogeneous and lower semicontinuous (but not necessarily null at zero). Under the convexity of f it is also convex. Indeed, let $h, h' \in X$ be such that $\underline{d}_H^+(x;h) < +\infty$ and $\underline{d}_H^+(x;h') < +\infty$. Choose nets $(s_j, h_j)_{j \in J}$ converging to $(0, h)$ with $s_j > 0$ and $(t_j, h'_j)_{j \in J}$ converging to $(0, h')$ with $t_j > 0$ such that

$$\underline{d}_H^+(x;h) = \lim_{j \in J} s_j^{-1}\big(f(x+s_j h_j) - f(x)\big), \quad \underline{d}_H^+(x;h') = \lim_{j \in J} t_j^{-1}\big(f(x+t_j h'_j) - f(x)\big)$$

(it is known, and easily seen by taking the Cartesian product in a suitable way, that both nets can be indexed by the same directed set J). For some $j_0 \in J$ we may write for every $j \succeq j_0$

$$s_j^{-1}\big(f(x+s_j h_j) - f(x)\big) + t_j^{-1}\big(f(x+t_j h'_j) - f(x)\big)$$
$$= \frac{s_j + t_j}{s_j t_j}\left(\frac{t_j}{s_j + t_j}f(x+s_j h_j) + \frac{s_j}{s_j + t_j}f(x+t_j h'_j) - f(x)\right),$$

which by the convexity of f gives with $r_j := s_j t_j / (s_j + t_j)$

$$\frac{f(x+s_j h_j) - f(x)}{s_j} + \frac{f(x+t_j h'_j) - f(x)}{t_j} \ge \frac{f(x+r_j(h_j + h'_j)) - f(x)}{r_j}.$$

Since $r_j \downarrow 0$, it ensues that $\underline{d}_H^+(x;h) + \underline{d}_H^+(x;h') \ge \underline{d}_H^+(x;h+h')$, confirming the convexity of $\underline{d}_H^+(x;\cdot)$. We summarize all the above features in the proposition:

PROPOSITION 3.6. *Let X be a topological vector space and $f : X \to \mathbb{R} \cup \{-\infty, +\infty\}$ be a convex function and $x \in X$ be a point where f is finite. The following hold.*
(a) *The directional derivative $f'(x;\cdot)$ is a positively homogeneous convex function with $f'(x;0) = 0$ satisfying $\partial f(x) = \partial f'(x;\cdot)(0)$, that is,*

$$\partial f(x) = \{x^* \in X^* : \langle x^*, h \rangle \le f'(x,h), \forall h \in X\};$$

further $-f'(x;-h) \le f'(x;h)$ for all $h \in X$.
(b) *The lower directional H-derivative $\underline{d}_H^+(x;\cdot)$ is lower semicontinuous, convex and positively homogeneous (not necessarily null at 0), and*

$$\partial f(x) = \{x^* \in X^* : \langle x^*, h \rangle \le \underline{d}_H^+(x;h), \forall h \in X\}.$$

Note that we will also see in Corollary 3.79 that, under the convexity of f, the subdifferential $\partial f(x) \ne \emptyset$ at $x \in X$ with $|f(x)| < +\infty$ if and only if $\underline{d}_H^+(x;0) = 0$.

The next proposition establishes two other properties of the directional derivative.

PROPOSITION 3.7. *Let X be a topological vector space, $f : X \to \mathbb{R} \cup \{-\infty, +\infty\}$ be a convex function and $x \in X$ be a point where f is finite. The following hold.*
(a) *One has the equality*

$$\mathrm{dom}\, f'(x;\cdot) = \mathbb{R}_+(\mathrm{dom}\, f - x).$$

(b) *One also has*

$$\mathbb{R}_+\big(\mathrm{epi}\, f - (x, f(x))\big) \subset \mathrm{epi}\, f'(x;\cdot) \subset \mathrm{cl}\big(\mathbb{R}_+\big(\mathrm{epi}\, f - (x, f(x))\big)\big).$$

PROOF. Recall first the equality $f'(x; h) = \inf_{t>0} t^{-1}(f(x+th) - f(x))$.

(a) From the previous equality it is clear that the right side in (a) is included in the left one. To prove the converse inclusion, fix any $h \in \text{dom } f'(x; \cdot)$. There exists a real $r > \inf_{t>0} t^{-1}(f(x+th) - f(x))$, or equivalently $f(x+th) < f(x) + tr$ for some real $t > 0$, which gives $h \in t^{-1}(\text{dom } f - x)$. It follows that $h \in \mathbb{R}_+(\text{dom } f - x)$. This finishes the justification of the assertion (a).

(b) Take any (h, r) in $]0, +\infty[(\text{epi } f - (x, f(x)))$. There exists a real $t > 0$ such that
$$(x, f(x)) + t(h, r) \in \text{epi } f, \quad \text{hence } t^{-1}(f(x+th) - f(x)) \leq r,$$
which implies $f'(x; h) \leq r$. This and the equality $f'(x; 0) = 0$ ensure that the left inclusion in (b) holds true. To show the other inclusion, fix any (h, r) in epi $f'(x; \cdot)$. Then $r \geq \inf_{t>0} t^{-1}(f(x+th) - f(x))$, so for each $n \in \mathbb{N}$ there some real $t_n > 0$ such that
$$r + (1/n) \geq t_n^{-1}(f(x+t_n h) - f(x)),$$
or equivalently $(h, r+(1/n)) \in t_n^{-1}(\text{epi } f-(x, f(x)))$. This entails that (h, r) belongs to $\text{cl}(\mathbb{R}_+(\text{epi } f - (x, f(x))))$, and the proof of the proposition is finished. □

Let X be a topological vector space, $f : X \to \mathbb{R} \cup \{-\infty, +\infty\}$ be a convex function and $x \in X$ be a point where f is finite. By (b) in Proposition 3.7 above we have that
$$\text{cl}(\text{epi } f'(x; \cdot)) = \text{cl}(\mathbb{R}_+(\text{epi } f - (x, f(x)))).$$
On the other hand, as for (2.39) and (2.38) we can easily see that
$$\text{epi } \underline{d}_H^+ f(x; \cdot) = \text{cl}(\mathbb{R}_+(\text{epi } f - (x, f(x)))).$$
It ensues that
$$\text{epi } \underline{d}_H^+ f(x; \cdot) = \text{cl}(\text{epi } f'(x; \cdot)),$$
which combined with what precedes gives that $\underline{d}_H^+ f(x; \cdot)$ is the lower semicontinuous hull of $f'(x; \cdot)$. We state this as follows.

PROPOSITION 3.8. Let $f : X \to \mathbb{R} \cup \{-\infty, +\infty\}$ be an extended real-valued convex function on a topological vector space X. Then for any $x \in X$ where f is finite one has
$$\underline{d}_H^+ f(x; h) = \liminf_{h' \to h} f'(x; h')$$
for every $h \in X$.

In addition to Definition 3.2 we also extend, with $\varepsilon \geq 0$, the definition of ε-subdifferential in (2.64) of a convex function $f : X \to \mathbb{R} \cup \{-\infty, +\infty\}$ on the topological vector space X by
$$\partial_\varepsilon f(x) := \{x^* \in X^* : \langle x^*, u - x \rangle \leq f(u) - f(x) + \varepsilon, \ \forall u \in X\} \quad \text{if } |f(x)| < +\infty,$$
and $\partial_\varepsilon f(x) := \emptyset$ if $f(x)$ is not finite. It is clear that
(3.6) $$\partial f(x) = \bigcap_{\varepsilon > 0} \partial_\varepsilon f(x).$$

Further, with $|f(x)| < +\infty$ and with the directional ε-derivative defined in (2.65) as
$$f'_\varepsilon(x; h) := \inf_{t>0} \frac{f(x+th) - f(x) + \varepsilon}{t},$$

the properties established in the framework of normed spaces just after (2.65) in the previous chapter are valid in the framework of topological vector spaces. We state them as follows:

PROPOSITION 3.9. *Let X be a topological vector space and $f : X \to \mathbb{R} \cup \{-\infty, +\infty\}$ be a convex function. Let a real $\varepsilon \geq 0$ and $x \in X$ with $|f(x)| < +\infty$. The following hold:*
(a) *The directional ε-derivative $f'_\varepsilon(x; \cdot)$ is a positively homogeneous convex function with $f'_\varepsilon(x; 0) = 0$;*
(b) *for every $h \in X$ one has $\lim_{\varepsilon \downarrow 0} f'_\varepsilon(x; h) = f'(x; h)$;*
(c) *the directional ε-derivative characterizes the ε-subdifferential in the form*
$$x^* \in \partial_\varepsilon f(x) \Leftrightarrow \langle x^*, \cdot \rangle \leq f'_\varepsilon(x; \cdot), \text{ or equivalently } \partial_\varepsilon f(x) = \partial f'_\varepsilon(x; \cdot)(0).$$

From the very definition of $\partial f(x)$ and $\partial_\varepsilon f(x)$ we also readily see the following:

PROPOSITION 3.10. *Let X be a topological vector space, $f, g : X \to \mathbb{R} \cup \{+\infty\}$ be convex functions. For any $x, a \in X$, any reals $\varepsilon \geq 0$ and $r > 0$*
$$\partial_\varepsilon \big(f(\cdot + a)\big)(x) = \partial_\varepsilon f(x + a), \quad \partial_\varepsilon (rf)(x) = r \partial_{\varepsilon/r} f(x),$$
and for any reals $\varepsilon_1, \varepsilon_2 \geq 0$ with $\varepsilon_1 + \varepsilon_2 \leq \varepsilon$
$$\partial_{\varepsilon_1} f(x) + \partial_{\varepsilon_2} g(x) \subset \partial_\varepsilon (f + g)(x).$$
If X is assumed to be Hausdorff locally convex and if $A : E \to X$ is a continuous linear mapping from a Hausdorff locally convex space E into X, then for any $u \in E$
$$A^* \big(\partial_\varepsilon f(Au)\big) \subset \partial_\varepsilon (f \circ A)(u).$$

For sums of separate convex functions the following equalities are also directly seen.

PROPOSITION 3.11. *Let X_k, $k = 1, \cdots, m$ be topological vector spaces and $f_k : X_k \to \mathbb{R} \cup \{+\infty\}$ be convex functions. For $\varphi(x_1, \cdots, x_m) := f_1(x_1) + \cdots + f_m(x_m)$ and any real $\varepsilon \geq 0$ one has*
$$\partial \varphi(x_1, \cdots, x_m) = \partial f_1(x_1) \times \cdots \times \partial f_m(x_m)$$
along with
$$\partial_\varepsilon \varphi(x) = \bigcup_{\varepsilon_i \geq 0, \varepsilon_1 + \cdots + \varepsilon_m = \varepsilon} \partial_{\varepsilon_1} f_1(x_1) \times \cdots \times \partial_{\varepsilon_m} f_m(x_m).$$

The case of a sum with a continuous linear functional as well as the case of a positively homogeneous convex function presents particular formulas. Recall that for a subset C of a topological vector space X its support function is defined on the topological dual X^* by
$$\sigma_C(x^*) = \sigma(x^*, C) := \sup_{x \in C} \langle x^*, x \rangle \quad \text{for all } x^* \in X^*.$$

PROPOSITION 3.12. *Let X be a topological vector space, $f : X \to \mathbb{R} \cup \{-\infty, +\infty\}$ be a convex function and $\varepsilon \geq 0$.*
(a) *For any real β and any continuous linear function a^* on X one has*
$$\partial_\varepsilon (a^* + \beta + f)(x) = a^* + \partial_\varepsilon f(x) \quad \text{for all } x \in X.$$
(b) *If the convex function f is positively homogeneous with $f(0) = 0$, then one has*
$$\partial_\varepsilon f(0) = \partial f(0)$$

along with
$$\partial f(x) = \{x^* \in \partial f(0) : \langle x^*, x \rangle = f(x)\} \quad \text{for all } x \in X$$
and
$$\partial_\varepsilon f(x) = \{x^* \in \partial f(0) : \langle x^*, x \rangle \geq f(x) - \varepsilon\} \quad \text{for all } x \in X \text{ with } f(x) > -\infty.$$

(c) If X is a Hausdorff locally convex space and C is a nonempty subset of X, then for the support function $\sigma_C := \sigma(\cdot, C)$ and endowing X^* with the $w(X^*, X)$ topology one has
$$\partial \sigma_C(0) = \overline{\operatorname{co}}\, C$$
along with for all $x^* \in X^*$
$$\partial \sigma_C(x^*) = \{x \in \overline{\operatorname{co}}\, C : \langle x^*, x \rangle = \sigma_C(x^*)\}$$
and
$$\partial_\varepsilon \sigma_C(x^*) = \{x \in \overline{\operatorname{co}}\, C : \langle x^*, x \rangle \geq \sigma_C(x^*) - \varepsilon\},$$
where the subdifferential and ε-subdifferential of σ_C at any point in X^* are then taken in X.

Similarly, for any nonempty subset D of X^*, one has
$$\partial \sigma_D(0) = \overline{\operatorname{co}}^*\, D$$
along with for all $x \in X$
$$\partial \sigma_D(x) = \{x^* \in \overline{\operatorname{co}}^*\, D : \langle x^*, x \rangle = \sigma_D(x)\}$$
and
$$\partial_\varepsilon \sigma_D(x) = \{x^* \in \overline{\operatorname{co}}^*\, D : \langle x^*, x \rangle \geq \sigma_D(x) - \varepsilon\}.$$

PROOF. (a) We may suppose that f is finite at x. Then $x^* \in \partial_\varepsilon(a^* + \beta + f)(x)$ means that, for all $u \in X$, one has $\langle x^*, u - x \rangle \leq f(u) + a^*(u) - f(x) - a^*(x) + \varepsilon$, or equivalently $\langle x^* - a^*, u - x \rangle \leq f(u) - f(x) + \varepsilon$. The latter inequality for all $u \in X$ translates that $x^* - a^* \in \partial_\varepsilon f(x)$, so the equality in (a) is justified.

(b) Assume in addition that f is positively homogeneous with $f(0) = 0$. Let $x \in X$ with $|f(x)| < +\infty$. Taking $x^* \in \partial_\varepsilon f(x)$ (if any), we have for every $u \in X$ and every real $t > 0$
$$t^{-1}\langle x^*, u \rangle = \langle x^*, t^{-1} u \rangle \leq f(x + t^{-1} u) - f(x) + \varepsilon \leq t^{-1} f(u) + \varepsilon,$$
hence $\langle x^*, u \rangle \leq f(u) + t\varepsilon$. Passing to the limit as $]0, +\infty[\ni t \to 0$ gives $\langle x^*, u \rangle \leq f(u)$ for all $u \in X$, that is, $x^* \in \partial f(0)$. It results in particular that $\partial_\varepsilon f(0) \subset \partial f(0)$, which confirms the first equality in (b) since the converse inclusion is evident.

If $f(x)$ is not finite (resp. $f(x) = +\infty$), the second (resp. third) equality in (b) is obvious since both members are empty in this case. So, suppose that $f(x)$ is finite. Let $x^* \in \partial f(0)$ with $\langle x^*, x \rangle \geq f(x) - \varepsilon$. Then for any $y \in X$ we have from the inclusion $x^* \in \partial f(0)$ that $\langle x^*, y \rangle \leq f(y)$, which combined with the inequality $\langle x^*, x \rangle \geq f(x) - \varepsilon$ gives $\langle x^*, y - x \rangle \leq f(y) - f(x) + \varepsilon$ for all $y \in X$. It follows that $x^* \in \partial_\varepsilon f(x)$.

Conversely, let $x^* \in \partial_\varepsilon f(x)$. We already know by what precedes that $x^* \in \partial f(0)$. We also have by definition of $\partial_\varepsilon f(x)$
$$-\langle x^*, x \rangle = \langle x^*, 0 - x \rangle \leq f(0) - f(x) + \varepsilon = -f(x) + \varepsilon,$$
or equivalently $\langle x^*, x \rangle \geq f(x) - \varepsilon$. The second equality in the assertion (b) is then justified, and the third equality in (b) follows directly from the second.

(c) Endow the topological dual X^* of the Hausdorff locally convex space X with the topology $w(X^*, X)$, so the topological dual of X^* is identified with X. Since $C \neq \emptyset$, we have $\sigma_C(0) = 0$, then $x \in \partial \sigma_C(0)$ means that

$$\langle u^*, x \rangle \leq \sigma_C(u^*) = \sigma_{\overline{co}\, C}(u^*) \quad \text{for all } u^* \in X,$$

which in turn is equivalent to $x \in \overline{co}\, C$ since the latter set is a $w(X, X^*)$-closed convex set. This proves the first equality in (c). The second and third equalities are then consequences of the second equality in (b). The other equalities with σ_D follow in a similar way. $\square$

EXAMPLE 3.13. Let $(X, \|\cdot\|)$ be a normed space. Taking $D = \mathbb{B}_{X^*}$ in Proposition 3.12, we obtain

$$\partial \|\cdot\|(0) = \mathbb{B}_{X^*} \quad \text{and} \quad \partial_\varepsilon \|\cdot\|(x) = \{x^* \in \mathbb{B}_{X^*} : \langle x^*, x \rangle \geq \|x\| - \varepsilon\}$$

for any real $\varepsilon \geq 0$ and any $x \in X$. The equality $\partial \|\cdot\|(0) = \mathbb{B}_{X^*}$, already seen in Example 2.66(a), then appears as a particular case of the preceding general formula for the ε-subdifferential of norms. $\square$

3.1.2. Topological and Lipschitz properties of convex functions on topological vector spaces.

To express the Lipschitz property of continuous convex functions on a locally convex space X, we need to record that the seminorm $\mathfrak{p}_V$ associated with a symmetric convex neighborhood V of zero in X is defined as the Minkowski gauge function of V. Given a vector space E, the *Minkowski gauge function* $j_C : E \to [0, +\infty]$ of any convex set C of E containing zero is defined by

$$(3.7) \qquad j_C(x) := \inf\{t > 0 : t^{-1}x \in C\} \quad \text{for all } x \in E.$$

The gauge function j_C is known and easily seen to be convex, positively homogeneous with $j_C(0) = 0$. When additionally C is symmetric and $]0, +\infty[\, C = E$, the function j_C is finite-valued and clearly a seminorm on E. The gauge function of a symmetric convex neighborhood V of zero of X is then a seminorm, and it is (known and) not difficult to see that it is further continuous; it will be denoted, as it is usual, by $\mathfrak{p}_V$ instead of j_V.

As a preparatory step to the Lipschitz continuity in a locally convex space, we start with the continuity in a general topological vector space. Following the above definition of $\mathfrak{p}_V$ for a symmetric convex neighborhood V of a locally convex space, given a general topological vector space X and any *balanced neighborhood* V of zero in X it will be of interest in the next proposition to put

$$(3.8) \qquad \mathfrak{q}_V(x) = \inf\{t > 0 : t^{-1}x \in V\}.$$

Clearly, this function $\mathfrak{q}_V$ is even, finite-valued and positively homogeneous with $\mathfrak{q}_V(X) \subset [0, +\infty[$ and $\mathfrak{q}_V(0) = 0$. Further, it is not difficult to see that $\mathfrak{q}_V$ is continuous at 0 and that for $t > 0$ one has the implication $\mathfrak{q}_V(x) < t \Rightarrow t^{-1}x \in V$ (and when V is additionally closed, the equivalence $\mathfrak{q}_V(x) \leq t \Leftrightarrow t^{-1}x \in V$ holds true). We can then prove the following general continuity property at a point for a convex function on a topological vector space.

PROPOSITION 3.14. *Let U be a nonempty open convex set of a topological vector space X and $(f_i)_{i \in I}$ be a family of real-valued convex functions $f_i : U \to \mathbb{R}$. Assume that there exists a balanced neighborhood V of zero with $\bar{x} + V + V \subset U$ and a real $\beta \geq 0$ with $f_i(x) - f_i(\bar{x}) \leq \beta$ for all $i \in I$ and $x \in \bar{x} + V + V$. Then*

for $\mathfrak{q}_V$ defined as above (that is, $\mathfrak{q}_V(x) := \inf\{t > 0 : t^{-1}x \in V\}$) one has for each $i \in I$
$$|f_i(x) - f_i(y)| \leq 2\beta\,\mathfrak{q}_V(x - y) \quad \text{for all } x, y \in \overline{x} + V;$$
in particular the family $(f_i)_{i \in I}$ is equicontinuous on $\overline{x} + \operatorname{int} V$.

PROOF. We follow the proof of Lemma 2.156. Fix any $i \in I$ and put $g_i(x) := f_i(x + \overline{x}) - f_i(\overline{x})$ for all $x \in -\overline{x} + U$, so $g_i(x) \leq \beta$ for all $x \in W := V + V$. For any $x \in W$ we also see from the convexity of g_i that $g_i(x) - g_i(0) \geq g_i(0) - g_i(-x)$ thus $g_i(x) \geq -g_i(-x) \geq -\beta$ since $g_i(0) = 0$. It ensues that $|g_i(x)| \leq \beta$ for all $x \in W$. Now fix arbitrary $x, y \in V$ and take any real $s > \mathfrak{q}_V(x - y)$. Putting $z_s := y + s^{-1}(y - x)$ we have $z_s \in W$ since $s^{-1}(y - x) \in V$, and we also have $y = (1 - t)x + tz_s$, where $t = (1 + s)^{-1}s < 1$. By the convexity of g_i again we see that

$$(3.9) \qquad g_i(y) \leq (1 - t)g_i(x) + tg_i(z_s) = g_i(x) + t\big(g_i(z_s) - g_i(x)\big),$$

hence we obtain that $g_i(y) - g_i(x) \leq 2\beta t \leq 2\beta s$. We deduce that $g_i(y) - g_i(x) \leq 2\beta\mathfrak{q}_V(x - y)$, which gives by symmetry $|g_i(y) - g_i(x)| \leq 2\beta\,\mathfrak{q}_V(y - x)$. $\square$

As in the situation of normed spaces we have the following corollary.

COROLLARY 3.15. *Let U be a nonempty open convex set of a topological vector space X and $f : U \to \mathbb{R} \cup \{+\infty\}$ be a convex function. The following assertions are equivalent.*
(a) *The function f is bounded from above over some nonempty open subset of U;*
(b) $\operatorname{int}(\operatorname{dom} f) \neq \emptyset$ *and f is continuous on $\operatorname{int}(\operatorname{dom} f)$.*

Proposition 3.14 (resp. its corollary 3.15) assumes that the function is bounded from above on a neighborhood of the point of interest (resp. on a nonempty open subset of U). In the case when the convex function is bounded from above on the entire space X, a stronger property holds true as the following proposition says.

PROPOSITION 3.16. *Let X be a vector space and $f : X \to \mathbb{R} \cup \{-\infty, +\infty\}$ be a convex function.*
(a) *If f is bounded from above on the entire space X, then f is constant on X.*
(b) *If $f \not\equiv -\infty$ and is majorized on X by an affine function $\ell(\cdot) + c$, where $\ell : X \to \mathbb{R}$ is linear and $c \in \mathbb{R}$, then f is affine on X; more precisely, $f = \ell(\cdot) + \beta$ for some constant $\beta \in \mathbb{R}$.*

PROOF. (a) If $f \equiv -\infty$, there is nothing to do. Suppose that $f \not\equiv -\infty$, so $f(\overline{x}) > -\infty$ for some $\overline{x}$, hence $f(\overline{x})$ is finite by the boundedness assumption. Put $g(\cdot) := f(\cdot + \overline{x}) - f(\overline{x})$. Noting that $g(0) = 0$ and g is bounded from above by some real β, we see that for every $x \in X$ and for every $t \in\,]0, 1[$

$$g(x) \leq tg\left(\frac{1}{t}x\right) \leq t\beta,$$

hence $g(x) \leq 0$. On the other hand, for every $x \in X$ the latter inequality yields $g(-x) \leq 0$, hence for any real $r > g(x)$ we have

$$0 = g(0) = g\left(\frac{1}{2}x + \frac{1}{2}(-x)\right) \leq \frac{1}{2}r + \frac{1}{2}g(-x) \leq \frac{1}{2}r,$$

which gives $0 \leq g(x)$. It results that $g(x) = 0$ for all $x \in X$, which justifies that f is constant on X.

(b) Considering $f_0 = f - \ell(\cdot)$, the convex function f_0 is bounded from above (by c) on the entire space X with $f_0 \not\equiv -\infty$. By what precedes in (a) it results that there is a constant $\beta \in \mathbb{R}$ such that $f_0 = \beta$, that is, $f = \ell(\cdot) + \beta$. □

Now let us look at the context of a locally convex space X in considering a nonempty open convex set U in this locally convex space X and a real-valued convex function $f : U \to \mathbb{R}$. Fix $\bar{x} \in U$ and let V be a symmetric convex neighborhood of zero such that $\bar{x} + 2V \subset U$. If $f(\cdot) - f(\bar{x})$ is bounded from above by some real $\beta > 0$ on $\bar{x} + 2V$, then (since $\mathfrak{p}_V = \mathfrak{q}_V$ because V is balanced) Proposition 3.14 guarantees that

$$|f(x) - f(y)| \leq 2\beta\,\mathfrak{p}_V(x-y) \quad \text{for all } x, y \in \bar{x} + V,$$

which translates that f is Lipschitz on $\bar{x} + V$ relative to the seminorm $\mathfrak{p}_V$.

DEFINITION 3.17. Recall in the context of the locally convex space X that a function $g : W \to \mathbb{R}$ on an open set W of X is *Lipschitz* (or *Lipschitz continuous*) on an open set $W_0 \subset W$ if there are a real $\gamma \geq 0$ and a continuous seminorm $\mathfrak{p}$ on X such that

$$|g(x) - g(y)| \leq \gamma\,\mathfrak{p}(x-y) \quad \text{for all } x, y \in W_0;$$

in such a case, one also says that g is Lipschitz (or Lipschitz continuous) on W_0 relative to the continuous seminorm $\mathfrak{p}$. Since $\gamma\mathfrak{p}$ is a continuous seminorm, it is clear that g is Lipschitz on W_0 if and only if there is a continuous seminorm $\mathfrak{p}_0$ on X such that

$$|g(x) - g(y)| \leq \mathfrak{p}_0(x-y) \quad \text{for all } x, y \in W_0.$$

The function g is said to be *locally Lipschitz (or locally Lipschitz continuous)* near $\bar{x} \in W$ when there are an open neighborhood $W_0 \subset W$ of $\bar{x}$ and a continuous seminorm relative to which g is Lipschitz on W_0. If g is Lipschitz near each point of an open set $W' \subset W$, one says that f is locally Lipshcitz (or locally Lipschitz continuous) on W'.

The above Lipschitz property of g yields the following extension of Theorem 2.158 with exactly the same proof.

THEOREM 3.18 (Lipschitz property of convex functions on locally convex spaces). Let U be a nonempty open convex set of a locally convex space X and $f : U \to \mathbb{R} \cup \{+\infty\}$ be a convex function. (I) Given a point $\bar{x} \in U$ the following are equivalent.
(α) The function f is bounded from above near $\bar{x}$;
(β) the function f is finite at $\bar{x}$ and continuous at this point;
(γ) the function f is finite at $\bar{x}$ and Lipschitz near $\bar{x}$.
(II) One also has the equivalence of the following assertions.
(a) The function f is bounded from above over some nonempty open subset of U;
(b) the function f is finite at some point in U and continuous at this point;
(c) $\text{int}\,(\text{dom}\,f) \neq \emptyset$ and f is continuous on $\text{int}\,(\text{dom}\,f)$;
(d) $\text{int}\,(\text{dom}\,f) \neq \emptyset$ and f is locally Lipschitz on $\text{int}\,(\text{dom}\,f)$.

Recalling that a *Fréchet space* X is a Hausdorff locally convex space whose topology is metrizable by a distance d such that (X, d) is a complete metric space, it is clear that Lemma 2.160 is still true with the same proof for such a space X. So, Corollary 2.162 is still valid with appropriate changes of arguments. We formulate these features in the following theorem.

THEOREM 3.19 (Lipschitz property of convex functions on Fréchet spaces). *Let U be an open convex set of a Fréchet space X and $f : U \to \mathbb{R} \cup \{+\infty\}$ be a proper lower semicontinuous convex function.*
(A) *The equality* $\operatorname{core}(\operatorname{dom} f) = \operatorname{int}(\operatorname{dom} f)$ *holds.*
(B) *Given $\bar{x} \in U$ the following assertions are equivalent:*
(a) *the function f is finite at $\bar{x}$ and locally Lipschitz continuous near $\bar{x}$ relative to a continuous seminorm on X (in the sense of Definition 3.17);*
(b) *the function f is finite at $\bar{x}$ and continuous at $\bar{x}$;*
(c) *the interiority property $\bar{x} \in \operatorname{int}(\operatorname{dom} f)$ is satisfied;*
(d) *the inclusion $\bar{x} \in \operatorname{core}(\operatorname{dom} f)$ holds.*
(C) *If $\operatorname{core}(\operatorname{dom} f) = \operatorname{int}(\operatorname{dom} f)$ is nonempty, f is continuous therein.*

A relative continuity property also holds for a proper lower semicontinuous convex function at relative boundary points of the (effective) domain in general topological vector spaces.

PROPOSITION 3.20. *Let U be a nonempty open convex set of a Hausdorff topological vector space X and $f : U \to \mathbb{R} \cup \{+\infty\}$ be a lower semicontinuous convex function.*
(a) *The restriction of f to any line segment $[a,b]$ with $a,b \in \operatorname{dom} f$ is continuous on $[a,b]$.*
(b) *More generally, given the convex hull $S := \operatorname{co}\{a_0, a_1, \cdots, a_m\}$ of $m+1$ affinely independent points in $\operatorname{dom} f$, the restriction of f to S is continuous.*

PROOF. We already know by Corollary 2.159 that at any point in $\operatorname{rint} S$ the function f is continuous relative to S. Let any $\bar{x}$ in the relative boundary of S. The set S can be decomposed as a finite union $S = \bigcup_{i \in I} S_i$, where each $S_i := \operatorname{co}\{\bar{x}, a_{1,i}, \cdots, a_{m_i,i}\}$ is the convex hull of affinely independent points $\bar{x}, a_{1,i}, \cdots, a_{m_i,i}$ including the point $\bar{x}$ of interest. The continuity of f at $\bar{x}$ relative to S is then reduced to the continuity of f at $\bar{x}$ relative to S_i for each $i \in I$. So fix any $i \in I$. Any point $x \in S_i$ admits a unique representation $x = \lambda_0(x)\bar{x} + \lambda_1(x)a_{1,i} + \cdots + \lambda_{m_i}(x)a_{m_i,i}$ with $\lambda_0(x) + \cdots + \lambda_{m_i}(x) = 1$ and $\lambda_k(x) \geq 0$ for all $k = 0, \cdots, m_i$. Noting that $x - \bar{x} = \lambda_1(x)(a_{1,i} - \bar{x}) + \cdots + \lambda_{m_i}(a_{m_i,i} - \bar{x})$ and keeping in mind that the vectors $a_{1,i} - \bar{x}, \cdots, a_{m_i,i} - \bar{x}$ are linearly independent, we see that $S_i \ni x \to \bar{x}$ if and only if $\lambda_1(x) \to 0, \cdots, \lambda_{m_i,i}(x) \to 0$, so in such a case $\lambda_0(x) \to 1$. Further, the convexity of f ensures that for each $x \in S_i$

$$f(x) \leq \lambda_0(x)f(\bar{x}) + \lambda_1(x)f(a_{1,i}) + \cdots + \lambda_{m_i,i}(x)f(a_{m_i,i}),$$

hence $\limsup_{S_i \ni x \to \bar{x}} f(x) \leq f(\bar{x})$. This combined with the lower semicontinuity of f implies that $f(\bar{x}) = \lim_{S_i \ni x \to \bar{x}} f(x)$, which means that f is continuous at $\bar{x}$ relative to S_i, so the assertion (b) is established. The other assertion (a) clearly follows from the assertion (b). □

For one real variable convex function we derive the following corollary.

COROLLARY 3.21. *Let $f : \mathbb{R} \to \mathbb{R} \cup \{+\infty\}$ be a proper lower semicontinuous convex function of one real variable. Then the restriction of f to $\operatorname{cl}(\operatorname{dom} f)$ is continuous as function with values in $\mathbb{R} \cup \{-\infty, +\infty\}$.*

PROOF. We may suppose that $\operatorname{dom} f$ is not a singleton. By convexity $\operatorname{dom} f$ is an interval I in $\mathbb{R}$ whose left and right endpoints are $a \in [-\infty, +\infty[$ and $b \in$

$]-\infty, +\infty]$ with $a < b$. By Proposition 3.20(a) we already know that f is continuous on $]a, b[$. If $a \in I$ (resp. $b \in I$), the restriction $f_{|_I}$ of f to I is continuous at a (resp. at b) by Proposition 3.20(a) again. If $a \in \mathbb{R} \setminus I$ (resp. $b \in \mathbb{R} \setminus I$), $f(a) = +\infty$ (resp. $f(b) = +\infty$), hence by lower semicontinuity $\lim_{x \downarrow a} f(x) = +\infty$ (resp. $\lim_{x \uparrow b} f(x) = +\infty$), which tells us that $f_{|_I}$ is continuous at a (resp. at b). □

Now we prove that the finiteness of a convex function at some point of the interior of its effective domain entails its properness.

PROPOSITION 3.22. *Let X be a vector space and $f : X \to \mathbb{R} \cup \{-\infty, +\infty\}$ be a convex function. If f is finite at some point $\overline{x} \in \mathrm{core}\,(\mathrm{dom}\, f)$, then f does not take the value $-\infty$, that is, f is proper.*

PROOF. Assume that f is finite at a point $\overline{x} \in \mathrm{core}\,(\mathrm{dom}\, f)$. Suppose that $f(u) = -\infty$ for some u. Since $\overline{x} \in \mathrm{core}\,(\mathrm{dom}\, f)$, there is some $v \in \mathrm{dom}\, f$ such that $\overline{x} \in]u, v[$, that is, $\overline{x} = su + tv$ with $s, t > 0$ satisfying $s + t = 1$. Then $f(\overline{x}) \leq sf(u) + tf(v) = -\infty$, which contradicts the finiteness of $f(\overline{x})$. □

As shown next, properness of lower semicontinuous convex functions amounts to finiteness at some point.

PROPOSITION 3.23. *Let X be a topological vector space and $f : X \to \mathbb{R} \cup \{-\infty, +\infty\}$ be a lower semicontinuous convex function. If f is finite at some point, then it does not take the value $-\infty$, that is, it is proper.*

PROOF. Suppose that there are $\overline{x}, \overline{u} \in X$ with $|f(\overline{x})| < +\infty$ and $f(\overline{u}) = -\infty$. If $f(x) = -\infty$ for all $x \in]\overline{u}, \overline{x}[$, then by lower semicontinuity $f(\overline{x}) \leq \liminf_{\substack{x \in]\overline{u},\overline{x}[\\ x \to \overline{x}}} f(x) = -\infty$, which is a contradiction. Otherwise, if f is finite at some point $a \in]\overline{u}, \overline{x}[$, then by convexity, for some $t \in]0, 1[$ we have $f(a) \leq tf(\overline{u}) + (1-t)f(\overline{x}) = -\infty$, which is also a contradiction. □

COROLLARY 3.24. *Let X be a topological vector space and $f : X \to \mathbb{R} \cup \{-\infty, +\infty\}$ be a lower semicontinuous positively homogeneous convex function. Then f is proper if and only if $f(0) = 0$.*

PROOF. If $f(0) = 0$, Proposition 3.23 above directly yields that f is proper. Conversely, suppose that f is proper. Then by positive homogeneity we have either $f(0) = +\infty$ or $f(0) = 0$. On the other hand, choosing $a \in X$ with $f(a) \in \mathbb{R}$, we also have
$$0 = \lim_{t \downarrow 0} tf(a) = \lim_{t \downarrow 0} f(ta) \geq f(0),$$
where the inequality is due to the lower semicontinuity of f at 0. It follows that $f(0) = 0$, which finishes the proof. □

3.1.3. Lipschitz property of convex functions under growth conditions. In this subsection we study other forms of Lipschitz property of convex functions under certain growth conditions. The related results will utilize the following lemma with two well-known inequalities which have their own interests.

LEMMA 3.25. *Let $a, b \in [0, +\infty[$ and $s \in]0, +\infty[$.*
(a) *If $s \geq 1$ one has*
$$(a+b)^s \leq 2^{s-1}(a^s + b^s).$$

(b) If $s \in \,]0,1[$, then one has
$$(a+b)^s \leq 2^s (a^s + b^s).$$

PROOF. (a) Assume $s \geq 1$. The function from $[0, +\infty[$ into $\mathbb{R}$ given by $t \mapsto t^s$ being convex, we have
$$\left(\frac{a+b}{2}\right)^s \leq \frac{1}{2}a^s + \frac{1}{2}b^s, \text{ or equivalently } (a+b)^s \leq 2^{s-1}(a^s + b^s).$$

(b) By continuity it suffices to prove the inequality in (b) when $s \in \,]0,1[$ is a rational number, say $s = p/q$ with $1 \leq p < q$ in $\mathbb{N}$. Indeed, by (a) and by the inequality $(\alpha+\beta)^{1/q} \leq \alpha^{1/q} + \beta^{1/q}$ for $\alpha, \beta \in [0, +\infty[$ (due to the evident inequality $\alpha + \beta \leq (\alpha^{1/q} + \beta^{1/q})^q$), we can write
$$(a+b)^{p/q} \leq \left(2^{p-1}(a^p + b^p)\right)^{1/q} = 2^{\frac{p-1}{q}}(a^p + b^p)^{1/q} \leq 2^{p/q}\left(a^{p/q} + b^{p/q}\right),$$
which gives the desired inequality. □

PROPOSITION 3.26. Let $f : X \to \mathbb{R}$ be a convex function on a normed space $(X, \|\cdot\|)$ satisfying the growth condition $f(x) \leq \alpha \|x\|^s + \beta$ for all $x \in X$, where α, β, s are 3 real constants with $s \geq 1$, $\alpha > 0$ and $\beta \in \mathbb{R}$. Then one has for all $x, y \in X$
$$|f(x) - f(y)| \leq \left(\alpha_s \|x\|^{s-1} + \alpha_s \|y\|^{s-1} + \beta_s\right) \|x - y\|,$$
where $\alpha_s := \alpha(1 + 2^{3s-2})$ and $\beta_s := 2^s \alpha + 2\beta - 2f(0)$.

PROOF. For $\mu := \beta - f(0)$ and the convex function g with $g(x) := f(x) - f(0)$ we have $g(x) \leq \alpha \|x\|^s + \mu$ and $g(0) = 0$, hence
$$g(x) = g(x) - g(0) \geq g(0) - g(-x) = -g(-x) \geq -\alpha \|x\|^s - \mu.$$

Let $x \neq y$ in X, $v := (1 + \|x\| + \|y\|) \frac{y-x}{\|y-x\|}$ and $t := \frac{\|y-x\|}{1+\|x\|+\|y\|}$. Since $0 < t < 1$ the convexity of g gives $t^{-1}\bigl(g(x + tv) - g(x)\bigr) \leq g(x + v) - g(x)$, which means with $u := (y-x)/\|y-x\|$
$$g(y) - g(x) \leq \frac{\|y-x\|}{1 + \|x\| + \|y\|}\bigl(g(x + (1 + \|x\| + \|y\|)u) - g(x)\bigr).$$

Noting that $\|u\| = 1$ we deduce
$$g(y) - g(x) \leq \|y - x\| \frac{\alpha(1 + 2\|x\| + \|y\|)^s + \mu + \alpha\|x\|^s + \mu}{1 + \|x\| + \|y\|}$$
$$\leq \|y - x\| \left[\frac{\alpha(1 + 2\|x\| + \|y\|)}{1 + \|x\| + \|y\|}(1 + 2\|x\| + \|y\|)^{s-1} + 2\mu + \alpha\|x\|^{s-1}\right]$$
$$\leq \|y - x\|\left[2\alpha\bigl(1 + 2(\|x\| + \|y\|)\bigr)^{s-1} + 2\mu + \alpha\|x\|^{s-1}\right].$$

Writing by Lemma 3.25
$$2\alpha(1 + 2(\|x\| + \|y\|))^{s-1} + 2\mu + \alpha\|x\|^{s-1}$$
$$\leq \alpha 2^s\bigl(1 + 2^{s-1}(\|x\| + \|y\|)^{s-1}\bigr) + \alpha\|x\|^{s-1} + 2\mu$$
$$\leq \alpha 2^s + \alpha 2^{2s-1} 2^{s-1}\bigl(\|x\|^{s-1} + \|y\|^{s-1}\bigr) + \alpha\|x\|^{s-1} + 2\mu$$

and setting $\alpha_s := \alpha(1 + 2^{3s-2})$ and $\beta_s := 2^s \alpha + 2\mu = 2^s \alpha + 2\beta - 2f(0)$, it results that
$$f(y) - f(x) = g(y) - g(x) \leq \bigl(\alpha_s \|x\|^{s-1} + \alpha_s \|y\|^{s-1} + \beta_s\bigr)\|x - y\|.$$

By symmetry the same inequality holds for $f(x) - f(y)$, hence
$$|f(x) - f(y)| \leq (\alpha_s \|x\|^{s-1} + \alpha_s \|y\|^{s-1} + \beta_s) \|x - y\|.$$
This inequality being also true for $x = y$, the proof is finished. □

A similar Lipschitz property holds true, under growth conditions, for functions of m variables which are separately, for each component, either convex or concave. Given vector spaces $X_1, \cdots, X_m$, a function $f : X_1 \times \cdots \times X_m \to \mathbb{R} \cup \{-\infty, +\infty\}$ is called *either convex or concave with respect to each space X_i* if for each $i = 1, \cdots, m$, either $f(x_1, \cdots, x_{i-1}, \cdot, x_{i+1}, \cdots, x_m)$ is convex on X_i for every $x \in X$ or $f(x_1, \cdots, x_{i-1}, \cdot, x_{i+1}, \cdots, x_m)$ is concave on X_i for every $x \in X$, where X is the Cartesian product $X := X_1 \times \cdots \times X_m$. Functions which are convex with respect to each space X_i are also called *separately convex*.

PROPOSITION 3.27. *Let $(X_i, \|\cdot\|_i)$, $i = 1, \cdots, m$, be normed spaces and $f : X \to \mathbb{R}$ be a function on $X = X_1 \times \cdots \times X_m$. Let $\|\cdot\|$ be anyone of the three canonical product norms $\|\cdot\|_1$, $\|\cdot\|_2$ and $\|\cdot\|_\infty$ on X. Assume that f is either convex or concave with respect to each space X_i, $i = 1, \cdots, m$. Assume also that f satisfies the growth condition $|f(x)| \leq \alpha \|x\|^s + \beta$ for all $x \in X$, where α, β, s are 3 real constants with $s \geq 1$, $\alpha > 0$ and $\beta \in \mathbb{R}$. Then one has for all $x, y \in X$*
$$|f(x) - f(y)| \leq (\alpha_s \|x\|^{s-1} + \alpha_s \|y\|^{s-1} + \beta_s) \|x - y\|,$$
where $\alpha_s := m\alpha 2^{s-1}(1 + 2^{2s-1})$ and $\beta_s := m(2^s \alpha + 2\beta)$.

PROOF. Fix any $x \neq y$ in X and define $f_1 : X_1 \to \mathbb{R}$ by $f_1(z) = f(z, x_2, \cdots, x_m)$ for all $z \in X_1$ if f is convex with respect to X_1 (resp $f_1(z) = -f(z, x_2, \cdots, x_m)$ for all $z \in X_1$ if f is concave with respect to X_1). Analogously to the proof of Proposition 3.26 put
$$u := \frac{y_1 - x_1}{\|y - x\|}, \quad v := (1 + \|x\| + \|y\|)u, \quad t := \frac{\|y - x\|}{1 + \|x\| + \|y\|},$$
and note that $\|u\| \leq 1$ and $0 < t < 1$. By convexity of f_1 we have
$$t^{-1}(f_1(x_1 + tv) - f_1(x_1)) \leq f_1(x_1 + v) - f_1(x_1),$$
or equivalently
$$f_1(y_1) - f_1(x_1) \leq \frac{\|y - x\|}{1 + \|x\| + \|y\|} (f_1(x_1 + (1 + \|x\| + \|y\|)u) - f_1(x_1)),$$
which gives
$$f_1(y_1) - f_1(x_1)$$
$$\leq \|y - x\| \frac{|f(x_1 + (1 + \|x\| + \|y\|)u, x_2, \cdots, x_m)| + |f(x_1, x_2, \cdots, x_m)|}{1 + \|x\| + \|y\|}$$
$$\leq \|y - x\| \frac{\alpha(1 + 2\|x\| + 2\|y\|)^s + \beta + \alpha(\|x\| + \|y\|)^s + \beta}{1 + \|x\| + \|y\|}.$$
From this we obtain as in the proof of Proposition 3.26 that
$$f_1(y_1) - f_1(x_1) \leq \|y - x\|(a\|x\|^{s-1} + a\|y\|^{s-1} + b),$$
with $a := \alpha 2^{s-1}(1 + 2^{2s-1})$ and $b := 2^s \alpha + 2\beta$.

Since $-1 < -t$ we also have
$$\frac{f_1(y_1 - v) - f_1(y_1)}{-1} \leq \frac{f_1(y_1 - tv) - f_1(y_1)}{-t},$$

which is equivalent to
$$f_1(x_1) - f_1(y_1) \le t\big(f_1(y_1 - v) - f_1(y_1)\big),$$
hence
$$f_1(x_1) - f_1(y_1)$$
$$\le \|y - x\| \frac{|f(y_1 - (1 + \|x\| + \|y\|)u, x_2, \cdots, x_m)| + |f(y_1, x_2, \cdots, x_m)|}{1 + \|x\| + \|y\|}$$
$$\le \|y - x\| \frac{\alpha(1 + 2\|x\| + 2\|y\|)^s + \beta + \alpha(\|x\| + \|y\|)^s + \beta}{1 + \|x\| + \|y\|}.$$

As above, this ensures that
$$f_1(x_1) - f_1(y_1) \le \|y - x\|\big(a\|x\|^{s-1} + a\|y\|^{s-1} + b\big),$$
which combined with the similar above inequality for $f_1(x_1) - f_1(y_1)$ yields
$$|f(x_1, x_2, \cdots, x_m) - f_1(y_1, x_2, \cdots, x_m)| \le \big(a\|x\|^{s-1} + a\|y\|^{s-1} + b\big)\|x - y\|.$$
The real $|f(y_1, \cdots, y_{i-1}, x_i, x_{i+1}, \cdots, x_m) - f(y_1, \cdots, y_{i-1}, y_i, x_{i+1}, \cdots, x_m)|$ is majorized by the same right-hand member of the latter inequality. Then writing
$$f(x) - f(y) = \sum_{i=1}^{m} f(y_1, \cdots, y_{i-1}, x_i, x_{i+1}, \cdots, x_m) - f(y_1, \cdots, y_{i-1}, y_i, x_{i+1}, \cdots, x_m)$$
it results that
$$|f(x) - f(y)| \le m\big(a\|x\|^{s-1} + a\|y\|^{s-1} + b\big)\|x - y\|.$$
This finishes the proof. $\square$

3.1.4. Coercive convex functions. For a proper convex function we saw in Theorem 3.18 that its boundedness from above near a point entails its Lipschitz continuity near that point. Before considering the coercivity of a convex function, let us examine the situation when the convex function is bounded from below near a point. The related proposition will prepare the result on coercivity properties in Proposition 3.29.

PROPOSITION 3.28. *Let $(X, \|\cdot\|)$ be a normed space and $f : X \to \mathbb{R} \cup \{-\infty, +\infty\}$ be a convex function which is bounded from below near a point in $\operatorname{dom} f$, which holds in particular whenever f is lower semicontinuous at a point where it is finite. Then f is bounded from below over any bounded set in X.*

PROOF. Denote $x_0 \in \operatorname{dom} f$ a point near which f is bounded from below and notice that $f(x_0)$ is finite. Choose reals $r, \mu > 0$ such that $f(u) \ge -\mu$ for all $u \in B[x_0, r]$. Consider any real $\rho > r$ and take any $x \in B[x_0, \rho]$ with $\|x - x_0\| > r$. Denote z the point in $[x_0, x]$ such that $\|z - x_0\| = r$, that is,
$$z = \frac{\|z - x\|}{\|x - x_0\|} x_0 + \frac{r}{\|x - x_0\|} x \quad \text{and} \quad r + \|z - x\| = \|x - x_0\|.$$
This yields by convexity of f
$$-\mu \le f(z) \le \frac{\|z - x\|}{\|x - x_0\|} f(x_0) + \frac{r}{\|x - x_0\|} f(x) \le \frac{\|z - x\|}{\|x - x_0\|} (f(x_0))^+ + \frac{r}{\|x - x_0\|} f(x),$$
which gives
$$-\mu\|x - x_0\| \le \|z - x\|(f(x_0))^+ + rf(x), \text{ hence } -\mu\rho \le \rho(f(x_0))^+ + rf(x).$$

Putting $\mu_0 := \rho(\mu + (f(x_0))^+)/r$ and noting that $\mu_0 \geq \mu$ since $\rho/r \geq 1$, we obtain $-\mu_0 \leq f(x)$ for all $x \in B[x_0, \rho]$. □

Let us turn now to the condition $\lim_{\|x\| \to +\infty} f(x) = +\infty$, usually called the *coercivity condition* in the literature.

PROPOSITION 3.29. *Let $(X, \|\cdot\|)$ be a normed space and $f : X \to \mathbb{R} \cup \{+\infty\}$ be a proper convex function which is bounded from below near a point in $\operatorname{dom} f$, which holds in particular whenever f is lower semicontinuous at point where it is finite. The following are equivalent.*
(a) *The function f is coercive in the sense that* $\lim_{\|x\| \to +\infty} f(x) = +\infty$;
(b) *there exists two reals $\alpha > 0$ and $\beta \in \mathbb{R}$ such that $f(\cdot) \geq \alpha \|\cdot\| + \beta$;*
(c) *the inequality* $\liminf_{\|x\| \to +\infty} \frac{f(x)}{\|x\|} > 0$ *holds;*
(d) *all the sublevel sets $\{f(\cdot) \leq r\}$ with $r \in \mathbb{R}$ are bounded.*

PROOF. To show (a)⇒(b) it is easily seen (with $f(\cdot + x_0) - f(x_0)$ in place of f for $x_0 \in \operatorname{dom} f$) that we may and do suppose that $f(0) = 0$. By the coercivity condition choose a real $\mu > 0$ such that for any $x \in X$ with $\|x\| \geq \mu$ we have $f(x) \geq 1$, so by convexity of f with $f(0) = 0$

$$1 \leq f\left(\frac{\mu}{\|x\|} x\right) \leq \frac{\mu}{\|x\|} f(x), \text{ hence } f(x) \geq \mu^{-1} \|x\|.$$

On the other hand, by Proposition 3.28 above f is bounded from below on $\mu\mathbb{B}$, thus there is a real $\gamma \geq 0$ such that $f(x) \geq -\gamma$ for all $x \in \mu\mathbb{B}$. Putting $\alpha := \mu^{-1}$ and $\beta := -\alpha\mu - \gamma$ it ensues that $f(x) \geq \alpha\|x\| + \beta$ for all $x \in X$, which justifies the implication (a)⇒(b).

The other implications (b)⇒(c)⇒(a) and (a)⇔(d) are straightforward. □

3.1.5. Subdifferentiability and topological properties of subdifferential of convex functions.
By the previous chapter we know that the subdifferential of a convex function on a normed space is nonempty at any point of continuity. The next theorem shows that this feature still holds in topological vector spaces.

THEOREM 3.30 (subdifferentiability of convex functions at continuity points). *Let X be a topological vector space and $f : X \to \mathbb{R} \cup \{+\infty\}$ be a convex function which is finite at $\bar{x} \in X$ and continuous at $\bar{x}$.*
(a) *One has $\partial f(\bar{x}) \neq \emptyset$ and*

$$f'(\bar{x}; h) = \max\{\langle x^*, h \rangle : x^* \in \partial f(\bar{x})\} \quad \text{for all } h \in X.$$

(b) *If in addition the space X is Hausdorff and locally convex, then $\partial f(\bar{x})$ is $w(X^*, X)$-compact.*

PROOF. Take a real β and a symmetric neighborhood V of zero in X such that $f(\bar{x} + h) \leq \beta$ for all $h \in V$. Putting $\mu := \beta - f(\bar{x})$ we see that, for every $h \in V$

$$-\mu \leq -f(\bar{x} - h) + f(\bar{x}) \leq f'(\bar{x}; h) \leq f(\bar{x} + h) - f(\bar{x}) \leq \mu,$$

so the positively homogeneous function $f'(\bar{x}; \cdot)$ is finite-valued on V, and hence also on X. Further, $f'(\bar{x}; \cdot)$ is bounded from above on V (by μ), so by Proposition 3.14 it is continuous at zero. Therefore, the convex positively homogeneous function $f'(\bar{x}; \cdot)$ is finite and continuous on X. Fix any $\bar{h} \in X$. By the analytic Hahn-Banach theorem there exists some linear functional $\ell(\cdot) \leq f'(\bar{x}; \cdot)$ with $\ell(\bar{h}) = f'(\bar{x}; \bar{h})$;

further ℓ is continuous by the continuity of $f'(\overline{x};\cdot)$. We then deduce that $\ell \in \partial f(\overline{x})$ and $f'(\overline{x};\overline{h}) = \max\{\langle x^*, \overline{h}\rangle : x^* \in X\}$ by the construction of ℓ. This justifies (a), and (b) follows from the fact that $|\langle x^*, h\rangle| \leq \mu$ for all $x^* \in \partial f(\overline{x})$ and all $h \in V$ according to the above inequalities $-\mu \leq f'(\overline{x}; h) \leq \mu$ for all $h \in V$. □

Given a real-valued continuous convex function f on an open convex set U of a Hausdorff locally convex space X, the assertion (a) in the above theorem ensures, for each $h \in X$, that $x \mapsto \sigma(h, \partial f(x))$ is upper semicontinuous on U. Then Theorem 1.48 combined with the assertion (b) in the above theorem yields the following upper semicontinuity property:

PROPOSITION 3.31. *Let U be an open convex set of a Hausdorff locally convex space (X, τ_X) and $f : U \to \mathbb{R}$ be a continuous convex function. Then the multimapping $\partial f : U \rightrightarrows (X^*, w(X^*, X))$ is τ_X-to-weak* upper semicontinuous.*

In the context of normed spaces two other results of outer semicontinuity can be established.

PROPOSITION 3.32. *Let X be a normed space and $f : X \to \mathbb{R} \cup \{+\infty\}$ be a convex function.*
(a) *If f is lower semicontinuous at $\overline{x} \in \mathrm{dom}\, f$, then for each real $\varepsilon \geq 0$ the multimapping $\partial_\varepsilon f$ is $\|\cdot\|$-to-$\|\cdot\|_*$ outer semicontinuous at $\overline{x}$.*
(b) *If f is lower semicontinuous, then for every real $\varepsilon \geq 0$ the graph $\mathrm{gph}\, \partial_\varepsilon f$ of its ε-subdifferential is both $\|\cdot\| \times \|\cdot\|_*$ closed and sequentially $\|\cdot\| \times$ weak* closed (resp. sequentially weak $\times \|\cdot\|_*$ closed).*
(c) *More generally, if f is lower semicontinuous at $\overline{x} \in \mathrm{dom}\, f$, given any real $\varepsilon \geq 0$ and any net $(x_j, x_j^*)_{j \in J}$ in $\mathrm{gph}\, \partial_\varepsilon f$ converging $\|\cdot\| \times$ weak* (resp. weak $\times \|\cdot\|_*$) to $(\overline{x}, x^*)$ with $(x_j^*)_{j \in J}$ (resp. $(x_j)_{j \in J}$) bounded in norm, one has $x^* \in \partial_\varepsilon f(\overline{x})$.*

PROOF. To prove (a), suppose that f is lower semicontinuous at $\overline{x} \in \mathrm{dom}\, f$ and fix any real $\varepsilon \geq 0$. Take any $x^* \in \underset{x \to \overline{x}}{\mathrm{Lim\,sup}}\, \partial_\varepsilon f(x)$. By Proposition 1.14 there exist sequences $(x_n)_n$ converging strongly to $\overline{x}$ and $(x_n^*)_n$ in X^* converging strongly to x^* with $x_n^* \in \partial_\varepsilon f(x_n)$ for all $n \in \mathbb{N}$. Taking any $y \in X$, we have for every $n \in \mathbb{N}$ that $\langle x_n^*, y - x_n\rangle \leq f(y) - f(x_n) + \varepsilon$, which yields by the lower semicontinuity of f at $\overline{x}$ that $\langle x^*, y - \overline{x}\rangle \leq f(y) - f(\overline{x}) + \varepsilon$. This being true for all $y \in X$ we obtain that $x^* \in \partial_\varepsilon f(\overline{x})$, which justifies the outer semicontinuity of $\partial_\varepsilon f$ at $\overline{x}$.

The assertions (b) and (c) are obtained with similar arguments. □

We provide now a lower semicontinuous convex function whose graph of subdifferential is not topologically $\|\cdot\| \times$ weak* closed.

EXAMPLE 3.33 (J. Borwein, S. Fitzpatrick and R. Girgensohn). Let H be an infinite dimensional separable Hilbert space and $(e_k)_{k \in \mathbb{N}}$ be a Hilbert basis with $\|e_k\| = 1$ (for example, $H = \ell_\mathbb{R}^2(\mathbb{N})$). Let P denote the set of all prime numbers $p \geq 2$ and $\mathbb{N}_2 := \{k \in \mathbb{N} : k \geq 2\}$. For each $p \in P$ and each $k \in \mathbb{N}_2$ let

$$\zeta_{p;k} := \frac{1}{p}(e_p + e_{p^k}) \quad \text{and} \quad \xi_{p,k} := e_p + (p-1)e_{p^k}.$$

For $p_1, p_2 \in P$ with $p_1 \neq p_2$, we note that $p_1^{k_1} \neq p_2^{k_2}$ for all $k_1, k_2 \in \mathbb{N}_2$. From this we obtain

$$\langle \xi_{p_1, k_1}, \zeta_{p_2, k_2}\rangle = \begin{cases} 0 & \text{if } p_1 \neq p_2 \\ 1/p_1 & \text{if } p_1 = p_2,\, k_1 \neq k_2 \\ 1 & \text{if } p_1 = p_2,\, k_1 = k_2. \end{cases}$$

Consider the lower semicontinuous convex function f defined on H by

$$f(x) := \sup\left(\langle e_1, x\rangle + 1, \sup_{(p,k)\in P\times \mathbb{N}_2} \langle \xi_{p,k}, x\rangle\right) \quad \text{for all } x \in H.$$

Since $f(0) = 1$ and $f(-e_1) = 0$, the function f is proper and $0 \notin \partial f(0)$ since $f(-e_1) < f(0)$. On the other hand, for each $(p,k) \in P \times \mathbb{N}_2$ we have $f(\zeta_{p,k}) = 1$, and for every $x \in H$ we also have $f(x) \geq \langle \xi_{p,k}, x\rangle$, hence

$$f(x) - f(\zeta_{p,k}) = f(x) - 1 \geq \langle \xi_{p,k}, x\rangle - 1 = \langle \xi_{p,k}, x - \zeta_{p,k}\rangle.$$

It ensues that $\xi_{p,k} \in \partial f(\zeta_{p,k})$ for every $(p,k) \in P \times \mathbb{N}_2$.

Now fix any real $\varepsilon > 0$ and any finite set $G \subset H$. Since $\|\zeta_{p,k}\| = 2/p$ for all $p \in P$ and $e_n \to 0$ weakly as $n \to +\infty$, there exists some $p_0 \in P$ such that $\langle e_{p_0}, v\rangle < \varepsilon/2$ for all $v \in G$ and $\|\zeta_{p_0,k}\| < \varepsilon$ for all $k \in \mathbb{N}_2$. Since $e_{p_0^k} \to 0$ weakly as $k \to \infty$, we can choose some $k_0 \in \mathbb{N}_2$ such that

$$\langle e_{p_0^{k_0}}, v\rangle < \frac{\varepsilon}{2(p_0 - 1)} \quad \text{for all } v \in G.$$

We deduce that $\|\zeta_{p_0,k_0}\| < \varepsilon$ and that for each $v \in G$

$$\langle \xi_{p_0,k_0}, v\rangle = \langle e_{p_0}, v\rangle + (p_0 - 1)\langle e_{p_0^{k_0}}, v\rangle < \varepsilon.$$

This tells us that $(0,0)$ belongs to the $\|\cdot\| \times$ weak closure of $\operatorname{gph}\partial f$ whereas $(0,0) \notin \operatorname{gph}\partial f$ (as seen above). It results that the graph $\operatorname{gph}\partial f$ of ∂f is not topologically $\|\cdot\| \times$ weak closed in $H \times H$. $\square$

Given a nonempty convex set S of a normed space X, it is clear that its indicator function Ψ_S is lower semicontinuous at each point in $\operatorname{dom}\Psi_S = S$, hence by Proposition 3.32 above the multimapping $N(S;\cdot)$ is outer semicontinuous at each point in S. The following corollary is then derived.

COROLLARY 3.34. *Let S be a nonempty convex set of a normed space X. Then the multimapping $N(S;\cdot)$ is $\|\cdot\|$-to-$\|\cdot\|_*$ outer semicontinuous at each point $\bar{x} \in S$, that is, for any sequences $(x_n)_n$ in S converging strongly to $\bar{x}$ and $(x_n^*)_n$ in X^* converging strongly to x^* with $x_n^* \in N(S;x_n)$ for all $n \in \mathbb{N}$, one has $x^* \in N(S;\bar{x})$.*

If in addition S is closed, then the graph of $N(S;\cdot)$ is both $\|\cdot\| \times \|\cdot\|_*$ closed and sequentially $\|\cdot\| \times$ weak* closed. In fact, under the closedness of S, given any net $(x_j, x_j^*)_{j\in J}$ in $\operatorname{gph} N(S;\cdot)$ converging $\|\cdot\| \times$ weak* (resp. weak $\times \|\cdot\|_*$) to $(\bar{x}, x^*)$ with $(x_j^*)_{j\in J}$ (resp. $(x_j)_{j\in J}$) bounded in norm, one has $x^* \in N(S;\bar{x})$.

For proper lower semicontinuous convex functions f the distance $d(0, \partial f(\cdot))$ (known as a specific type of *metric slope* for f) fulfills the lower semicontinuity property. Recall that $d(0, \partial f(x)) = \inf_{x^* \in \partial f(x)} \|x^*\|$, so $d(0, \partial f(x)) = +\infty$ whenever $\partial f(x)$ is empty.

PROPOSITION 3.35. *Let $f : X \to \mathbb{R} \cup \{+\infty\}$ be a proper lower semicontinuous convex function on a normed space X. Then the function $x \mapsto d(0, \partial f(x))$ is lower semicontinuous.*

PROOF. Fix any $\bar{x} \in X$ and put $\lambda := \liminf_{x \to \bar{x}} d(0, \partial f(x))$. We may suppose $\lambda < +\infty$ since otherwise the desired inequality $d(0, \partial f(\bar{x})) \leq \lambda$ is obvious. Choose a sequence $(x_n)_n$ in X converging to $\bar{x}$ such that

$$\lim_{n\to\infty} d(0, \partial f(x_n)) = \lambda.$$

There exists $N \in \mathbb{N}$ such that for each $n \geq N$ we have $d(0, \partial f(x_n)) < +\infty$, so we can choose $x_n^* \in \partial f(x_n)$ satisfying $\|x_n^*\| \leq d(0, \partial f(x_n)) + (1/n)$. The sequence $(x_n^*)_{n \geq N}$ is bounded in X^*, hence it admits a subnet $(x_{s(j)}^*)_{j \in J}$ weakly* converging in X^* to some x^*. Proposition 3.32(c) yields that $x^* \in \partial f(\bar{x})$, hence by weak* semicontinuity of the dual norm

$$d(0, \partial f(\bar{x})) \leq \|x^*\| \leq \liminf_{j \in J} d(0, \partial f(x_{s(j)})) = \lambda = \liminf_{x \to \bar{x}} d(0, \partial f(x)),$$

which justifies the desired lower semicontinuity property. $\square$

The next proposition establishes a boundedness property of subdifferential under a Lipschitz-like property.

PROPOSITION 3.36. Let X be a topological vector space and $f : X \to \mathbb{R} \cup \{+\infty\}$ be a convex function which is finite at $\bar{x} \in X$. Assume that the are balanced neighborhoods V, W of zero in X such that

$$|f(x) - f(y)| \leq \mathfrak{q}_W(x - y) \quad \text{for all } x, y \in \bar{x} + V + V.$$

(a) For any real $\varepsilon > 0$ and any $x \in \bar{x} + V$ one has

$$\partial_\varepsilon f(x) \subset \{x^* \in X^* : \langle x^*, h \rangle \leq \mathfrak{q}_W(h) + \varepsilon, \forall h \in V\}.$$

(b) For $x \in \bar{x} + V$ and $x^* \in \partial f(x)$ one has

$$\langle x^*, h \rangle \leq \mathfrak{q}_W(h) \quad \text{for all } h \in X.$$

PROOF. (a) Considering any $x^* \in \partial_\varepsilon f(x)$ with $x \in \bar{x} + V$, it suffices to write for every $h \in V$

$$\langle x^*, h \rangle \leq f(x + h) - f(x) + \varepsilon \leq \mathfrak{q}_W(h) + \varepsilon.$$

(b) Taking $\varepsilon = 0$ in (a), we obtain $\langle x^*, h \rangle \leq \mathfrak{q}_W(h)$ for every $h \in V$, so by positive homogeneity of $\mathfrak{q}_W$ it results that $\langle x^*, h \rangle \leq \mathfrak{q}_W(h)$ for all $h \in X$. $\square$

The assertion (b) in Theorem 3.30 above is in fact a particular case of the following result.

PROPOSITION 3.37. Let X be a topological vector space and $f : X \to \mathbb{R} \cup \{+\infty\}$ be a convex function which is finite at $\bar{x} \in X$ and continuous at $\bar{x}$.
(a) There exist a balanced neighborhood W of zero in X, a neighborhood W_0 of zero in X and a neighborhood U of $\bar{x}$ such that for each real $\varepsilon \geq 0$

$$\bigcup_{x \in U} \partial_\varepsilon f(x) \subset \{x^* \in X^* : \langle x^*, h \rangle \leq \mathfrak{q}_W(h) + \varepsilon, \forall h \in W_0\},$$

so the set $\bigcup_{x \in U} \partial_\varepsilon f(x)$ is equicontinuous in X^*.
(b) If X is locally convex, there exist a continuous seminorm $\mathfrak{p}$ on X, a neighborhood U of $\bar{x}$ and a neighborhood W_0 of zero in X such that for each real $\varepsilon \geq 0$

$$\bigcup_{x \in U} \partial_\varepsilon f(x) \subset \{x^* \in X^* : \langle x^*, h \rangle \leq \mathfrak{p}(h) + \varepsilon, \forall h \in W_0\};$$

in particular there is a neighborhood V of zero in X (depending on ε) such that

$$\bigcup_{x \in U} \partial_\varepsilon f(x) \subset V^\circ,$$

so $\bigcup_{x \in U} \partial_\varepsilon f(x)$ is relatively $w(X^*, X)$-compact if the locally convex space X is Hausdorff.

PROOF. By Proposition 3.14 there are two balanced neighborhoods W, W' of zero in X such that
$$|f(x) - f(y)| \leq \mathfrak{q}_W(x - y) \quad \text{for all } x, y \in \bar{x} + W' + W'.$$
So, the first half of the assertion (a) follows from Proposition 3.36.

Let us prove the equicontinuity of $\bigcup_{x \in U} \partial_\varepsilon f(x)$ in X^*. Take any real $\eta > 0$. Choose $s \in]0,1]$ such that $s\varepsilon < \eta/2$ and choose also (by the continuity of $\mathfrak{q}_W$ at zero) a balanced neighborhood $W_0' \subset W'$ of zero such that $\mathfrak{q}_W(h) \leq \eta/2$ for all $h \in W_0'$. Then for any $x^* \in \bigcup_{x \in U} \partial_\varepsilon f(x)$ we have for every $h' \in W_0 := sW_0'$
$$\langle x^*, h' \rangle \leq \mathfrak{q}_W(h') + s\varepsilon \leq (s\eta/2) + s\varepsilon < \eta.$$
This justifies the equicontinuity of $\bigcup_{x \in U} \partial_\varepsilon f(x)$ in X^*.

(b) It is easy to derive the assertion (b) from (a). □

Keeping in mind that a convex function is continuous near $\bar{x}$ whenever it is continuous at $\bar{x}$ and that $\partial_\varepsilon f(x)$ is weak*-closed in X^*, the following corollary is directly derived.

COROLLARY 3.38. *Let X be a Hausdorff locally convex space and $f : X \to \mathbb{R} \cup \{+\infty\}$ be a convex function which is finite at $\bar{x} \in X$ and continuous at $\bar{x}$. Then there exists an open neighborhood U of $\bar{x}$ such that $\partial_\varepsilon f(x)$ is $w(X^*, X)$-compact for every $x \in U$ and every real $\varepsilon \geq 0$.*

In Proposition 2.211 it has been shown, for any convex function f on a finite dimensional space, that $\partial f(x)$ is unbounded at any $x \in \text{bdry}(\text{dom } f)$ with $\partial f(x) \neq \emptyset$. The next proposition establishes in infinite dimensions the property of non-boundedness of subdifferential of any convex function f at subdifferentiability points of the boundary of int (dom f).

PROPOSITION 3.39. *Let X be a Hausdorff locally convex space and $f : X \to \mathbb{R} \cup \{+\infty\}$ be a convex function with int (dom f) $\neq \emptyset$. Then for any $x \notin \text{int}(\text{dom } f)$ either $\partial f(x) = \emptyset$ or $\partial f(x)$ is not w^*-bounded.*

PROOF. Take (if there exists) any $\bar{x} \in \text{Dom } \partial f$ with $\bar{x} \notin \text{int}(\text{dom } f)$. The latter property furnishes by the geometric Hahn-Banach separation theorem some non zero $u^* \in X^*$ such that $\langle u^*, x \rangle \leq \langle u^*, \bar{x} \rangle$ for all $x \in \text{int}(\text{dom } f)$, hence also for all $x \in \text{dom } f$. So, choosing $\bar{x}^* \in \partial f(\bar{x})$ it ensues that for every real $t \geq 0$
$$\langle \bar{x}^* + tu^*, x - \bar{x} \rangle \leq f(x) - f(\bar{x}) \quad \text{for all } x \in \text{dom } f,$$
hence $\bar{x}^* + \mathbb{R}_+ u^* \subset \partial f(\bar{x})$. This tells us that $\partial f(\bar{x})$ is not w^*-bounded, which finishes the proof. □

3.1.6. Subdifferential of suprema of infinitely many convex functions in locally convex spaces. Theorem 2.204 provided, in the setting of normed spaces, a general rule for the subdifferential of the supremum of infinitely many convex functions under the continuity property of the supremum at the reference point. In this subsection we show how that result can be extended to locally convex spaces through the Lipschitz properties in the previous subsection. The proof is an adaptation of that of Theorem 2.204.

LEMMA 3.40. Let T be any nonempty set and X be a locally convex space. Let $g : X \times T \to \mathbb{R} \cup \{+\infty\}$ be a function such that $g_t := g(\cdot, t)$ is convex for every $t \in T$ and let $f(x) := \sup_{t \in T} g(x, t)$ for all $x \in X$. Let $\bar{x}$ be a point such that the convex function f is finite at $\bar{x}$ and bounded from above near $\bar{x}$. Then for any real $0 < \eta < 1$ there exists a neighborhood U_η of $\bar{x}$ such that for every $x \in U_\eta$ and every $t \in T_\eta$ one has
$$\partial g_t(x) \subset \partial_\eta g_t(\bar{x}) \cap \partial_{2\eta} f(\bar{x}),$$
where $T_\eta(\bar{x}) := \{t \in T : g_t(\bar{x}) \geq f(\bar{x}) - \eta\}$.

PROOF. Let W be a convex neighborhood of zero such that f is bounded from above on $\bar{x} + 2W$. Fix any real $0 < \eta < 1$. The family $\{g_t - g_t(\bar{x}) : t \in T_\eta\}$ is equibounded from above on $\bar{x} + 2W$, where $T_\eta := T_\eta(\bar{x})$. By Proposition 3.14 there exist a real $\gamma > 1$ and a symmetric convex neighborhood $V \subset W$ of zero (with γ and V depending on η) such that for each $t \in T_\eta$
$$|g_t(x') - g_t(x)| \leq \gamma \mathfrak{p}_V(x' - x) \quad \text{for all } x, x' \in \bar{x} + V,$$
where $\mathfrak{p}_V$ is the Minkowski gauge function of V. Put $U_\eta := \bar{x} + \frac{\eta}{2\gamma} V$ and take any $t \in T_\eta$ and any $x^* \in \partial g_t(x)$. Then, for every $y \in X$ one obtains by Proposition 3.36(b)
$$\langle x^*, y - \bar{x} \rangle = \langle x^*, y - x \rangle + \langle x^*, x - \bar{x} \rangle \leq g_t(y) - g_t(x) + \gamma \mathfrak{p}_V(x - \bar{x})$$
$$\leq g_t(y) - g_t(\bar{x}) + \eta.$$
From the latter inequality we also have by definition of $T_\eta(\bar{x})$
$$\langle x^*, y - \bar{x} \rangle \leq f(y) - f(\bar{x}) + 2\eta \quad \text{for all } y \in X,$$
so the proof is finished. □

THEOREM 3.41 (M. Valadier's theorem for subdifferential of suprema of infinitely many convex functions II). Let T be any nonempty set and X be a Hausdorff locally convex space. Let $g : X \times T \to \mathbb{R} \cup \{+\infty\}$ be a function such that $g_t := g(\cdot, t)$ is convex for every $t \in T$ and let $f(x) := \sup_{t \in T} g(x, t)$ for all $x \in X$. Let $\bar{x}$ be a point such that the convex function f is finite at $\bar{x}$ and bounded from above near $\bar{x}$. Let $T_\eta(\bar{x}) := \{t \in T : g(\bar{x}, t) \geq f(\bar{x}) - \eta\}$. Then the equality
$$\partial f(\bar{x}) = \bigcap_{\eta > 0, V \in \mathcal{N}(\bar{x})} \overline{\operatorname{co}}^* \left(\bigcup_{t \in T_\eta(\bar{x}), x \in V} \partial g_t(x) \right)$$
holds true, where $\mathcal{N}(\bar{x})$ denotes the class of neighborhoods of $\bar{x}$ in X.

PROOF. Let us start with the inclusion $\subset$. Fixing any $h \in X$, any real $\varepsilon > 0$, any real $\eta > 0$ and any neighborhood V of $\bar{x}$ such that f is bounded from above on V, it is enough to find $\tau \in T_\eta(\bar{x})$ and $x \in V$ such that $g'_\tau(x; h) \geq f'(\bar{x}; h) - \varepsilon$. Note that $f'(\bar{x}; h)$ is finite by the continuity of the convex function f at $\bar{x}$ (see Proposition 3.14). Choose a real $s \in]0, 1]$ with $\bar{x} + sh \in V$ and such that

(3.10) $\qquad (2s)^{-1} \big(f(\bar{x} + 2sh) - f(\bar{x}) \big) \leq f'(\bar{x}; h) + (\eta/4).$

Set $x := \bar{x} + sh$ and take $\tau \in T$ such that

(3.11) $\qquad f(x) - g_\tau(x) \leq \min\{\eta/4, s\varepsilon\}.$

Writing $f(x) - f(\bar{x}) = (f(\bar{x} + sh) - f(\bar{x})) \geq sf'(\bar{x}; h)$, it follows that

(3.12) $\qquad g_\tau(x) \geq f(x) - (\eta/4) \geq sf'(\bar{x}; h) + f(\bar{x}) - (\eta/4).$

Further, for $u := \bar{x} + 2sh$ we notice from (3.10) (since $g_\tau(u) \leq f(u)$) that
$$g_\tau(u) \leq f(\bar{x}) + 2sf'(\bar{x}; h) + (s\eta/2) \leq f(\bar{x}) + 2sf'(\bar{x}; h) + \eta/2.$$

Combining the latter inequality and (3.12) with the (Jensen) inequality $2g_\tau(x) \leq g_\tau(\bar{x}) + g_\tau(u)$ we obtain $g_\tau(\bar{x}) \geq f(\bar{x}) - \eta$, that is, $\tau \in T_\eta(\bar{x})$. On the other hand, we have
$$g'_\tau(x; h) \geq (-s)^{-1}(g_\tau(x - sh) - g_\tau(x)) = s^{-1}(g_\tau(x) - g_\tau(\bar{x}))$$
$$\geq s^{-1}(g_\tau(x) - f(\bar{x})) \geq s^{-1}(f(x) - f(\bar{x})) - \varepsilon,$$

where the latter inequality is due to (3.11). It results that $g'_\tau(x; h) \geq f'(\bar{x}; h) - \varepsilon$ as desired.

Let us show the converse inclusion. Fix any real $0 < \eta < 1$. By Lemma 3.40 there is a neighborhood U_η of $\bar{x}$ such that
$$Q_\eta := \bigcup \{\partial g_t(x) : t \in T_\eta, x \in U_\eta\} \subset \partial_{2\eta} f(\bar{x}),$$

hence $\overline{co}^* Q_\eta \subset \partial_{2\eta} f(\bar{x})$. Therefore, $\bigcap_{\eta>0} \overline{co}^* Q_\eta \subset \partial f(\bar{x})$, and this clearly entails the desired inclusion $\supset$ of the theorem. The proof is then complete. $\square$

The next corollary from Theorem 3.41 is similar to the result in the normed setting in Corollary 2.205.

COROLLARY 3.42. *Under the hypotheses and notation in Theorem 3.41 one also has the equality*
$$\partial f(\bar{x}) = \bigcap_{\eta>0} \overline{co}^* \left(\bigcup_{t \in T_\eta(\bar{x})} \partial_\eta g_t(\bar{x}) \right).$$

PROOF. As in Corollary 2.205 put $Q_\eta := \bigcup_{t \in T_\eta(\bar{x})} \partial_\eta g_t(\bar{x})$.

To prove first the inclusion $\supset$, take any real $\eta > 0$ and any $x^* \in Q_\eta$. Let $\tau \in T_\eta(\bar{x})$ be such that $x^* \in \partial g_\tau(\bar{x})$. For every $y \in X$ we can write
$$\langle x^*, y - \bar{x} \rangle \leq g_\tau(y) - g_\tau(\bar{x}) + \eta \leq f(y) - f(\bar{x}) + 2\eta,$$

which gives $x^* \in \partial_{2\eta} f(\bar{x})$. It ensues that $Q_\eta \subset \partial_{2\eta} f(\bar{x})$, thus $\overline{co}^*(Q_\eta) \subset \partial_{2\eta} f(\bar{x})$. This clearly entails the desired inclusion $\supset$ of the corollary.

Now let us show the converse inclusion $\subset$. Fix any real $0 < \eta < 1$. Let U_η be the neighborhood of $\bar{x}$ given by Lemma 3.40. Take any $t \in T_\eta(\bar{x})$, any $x \in U_\eta$ and any $x^* \in \partial g_t(x)$. Lemma 3.40 tells us that $x^* \in \partial_\eta g_t(\bar{x})$, hence $x^* \in Q_\eta$. From this and Theorem 3.41 we see that the left-hand member in the corollary is included in $\bigcap_{\eta>0} \overline{co}^*(Q_\eta)$, which finishes the proof. $\square$

We turn now to descriptions by means of exact subgradients of $g(\cdot, t)$ at the reference point $\bar{x}$ under certain compact-like conditions of the index set T.

3.1. CONVEX FUNCTIONS ON TOPOLOGICAL VECTOR SPACES

PROPOSITION 3.43. *Let T be a topological space and U be an open convex set of a Hausdorff locally convex space X. Let $g : U \times T \to \mathbb{R} \cup \{+\infty\}$ be a function such that $g_t := g(\cdot, t)$ is convex for every $t \in T$ and let $f : U \to \mathbb{R} \cup \{+\infty\}$ be defined by*

$$f(x) = \sup_{t \in T} g(x, t) \quad \text{for all } x \in U.$$

Let $\overline{x} \in \operatorname{core}(\operatorname{dom} f)$ be a point such that $g(\cdot, t)$ is continuous at $\overline{x}$ for every $t \in T$. Assume that $g(x, \cdot)$ is upper semicontinuous on T for every $x \in \operatorname{dom} f$ and that any minimizing sequence of the supremum $\sup_{t \in T} g(\overline{x}, t)$ admits a convergent subnet. With $T(\overline{x}) := \{t \in T : g(\overline{x}, t) = f(\overline{x})\}$ one has the equality

$$\partial f(\overline{x}) = \overline{\operatorname{co}}^* \left(\bigcup_{t \in T(\overline{x})} \partial g_t(\overline{x}) \right).$$

PROOF. Denote by Q the right-hand side of the equality to prove. The inclusion of $Q \subset \partial f(\overline{x})$ being easily verified, let us prove the converse inclusion. Fix any $h \in X$. By the assumption $\overline{x} \in \operatorname{core} \operatorname{dom} f$ take a real $\alpha \in \,]0, 1]$ such that f is finite on $\overline{x} + [0, \alpha]h \subset U$, and put $v := \alpha h$. Let $(r_n)_n$ be a sequence in $]0, \alpha]$ tending to 0. For each $n \in \mathbb{N}$ choose $t_n \in T$ such that

$$f(\overline{x} + r_n v) \leq g_{t_n}(\overline{x} + r_n v) + r_n^2,$$

so writing

$$(1 - r_n) f(\overline{x}) \geq (1 - r_n) g_{t_n}(\overline{x}) \geq g_{t_n}(\overline{x} + r_n v) - r_n g_{t_n}(\overline{x} + v)$$
$$\geq f(\overline{x} + r_n v) - r_n^2 - r_n g_{t_n}(\overline{x} + v) \geq f(\overline{x} + r_n v) - r_n^2 - r_n f(\overline{x} + v).$$

Since f is lower semicontinuous at $\overline{x}$, it ensues that $g_{t_n}(\overline{x}) \to f(\overline{x})$ as $n \to \infty$. Then $(t_n)_n$ is a maximizing sequence for the supremum $\sup_{\tau \in T} g_\tau(\overline{x})$, hence by assumption it admits a subnet $(t_{s(j)})_{j \in J}$ which converges to some $t \in T$. Put $\tau_j := t_{s(j)}$ for each $j \in J$. By upper semicontinuity of $g(\overline{x}, \cdot)$ we have

$$f(\overline{x}) = \lim_{j \in J} g(\overline{x}, \tau_j) \leq g(\overline{x}, t) = g_t(\overline{x}),$$

so $f(\overline{x}) = g_t(\overline{x})$ and $t \in T(\overline{x})$. Fix any real $r \in \,]0, \alpha]$ and choose an integer $j_0 \in J$ such that $\rho_j := r_{s(j)} < r$ for every $j \in J$ with $j \geq j_0$. For any $j \in J$ with $j \geq j_0$

$$r^{-1}\left(g_{\tau_j}(\overline{x} + rh) - g_{\tau_j}(\overline{x})\right) \geq \rho_j^{-1}\left(g_{\tau_j}(\overline{x} + \rho_j h) - g_{\tau_j}(\overline{x})\right)$$
$$\geq \rho_j^{-1}\left(f(\overline{x} + \rho_j h) - f(\overline{x})\right) - \rho_j,$$

hence taking the upper limit on $j \in J$ we get by upper semicontinuity of $g(\overline{x}+rh, \cdot)$ (since $\overline{x} + rh \in \operatorname{dom} f$) that

$$r^{-1}\left(g_t(\overline{x} + rh) - g_t(\overline{x})\right) \geq f'(\overline{x}; h) \geq \sigma\left(h, \partial f(\overline{x})\right).$$

This being true for every $r \in \,]0, \alpha]$ we deduce that

$$\sigma(h, Q) \geq \sigma\left(h, \partial g_t(\overline{x})\right) = g_t'(\overline{x}; h) \geq \sigma\left(h, \partial f(\overline{x})\right).$$

The inequality $\sigma(h, Q) \geq \sigma\left(h, \partial f(\overline{x})\right)$ being true for every $h \in X$, it results that $\partial f(\overline{x}) \subset Q$, which finishes the proof of the proposition. □

In the case of compactness of the index set T we have the following theorem.

THEOREM 3.44 (A.D. Ioffe and V.M. Tikhomirov). *Let T be a Hausdorff topological compact space and U be an open convex set of a Hausdorff locally convex space X. Let $g : U \times T \to \mathbb{R} \cup \{+\infty\}$ be a function such that $g_t := g(\cdot, t)$ is convex for every $t \in T$ and let $f : U \to \mathbb{R} \cup \{+\infty\}$ be defined by*

$$f(x) = \sup_{t \in T} g(x,t) \quad \text{for all } x \in U.$$

Let $\bar{x} \in \operatorname{dom} f$ be such that $g(\cdot, t)$ is continuous at $\bar{x}$ for every $t \in T$. Assume that $g(x, \cdot)$ is upper semicontinuous on T for each $x \in \operatorname{dom} f$. With $T(\bar{x}) := \{t \in T : g(\bar{x}, t) = f(\bar{x})\}$ one has the equality

$$\partial f(\bar{x}) = \overline{\operatorname{co}}^* \left(\bigcup_{t \in T(\bar{x})} \partial g_t(\bar{x}) \right).$$

PROOF. Consider a real $\rho > f(\bar{x})$. Notice that $g(\bar{x}, t) < \rho$ for every $t \in T$. To prove that $\bar{x} \in \operatorname{core}(\operatorname{dom} f)$, fix any $h \in X$. For each $t \in T$ there is by continuity of $g(\cdot, t)$ at $\bar{x}$ some real $r_t > 0$ with $\bar{x} + r_t h \in U$ such that $g(\bar{x} + r_t h, t) < \rho$, and by upper semicontinuity of $g(\bar{x} + r_t h, \cdot)$ the set $V_t := \{\tau \in T : g(\bar{x} + r_t h, \tau) < \rho\}$ is open and clearly contains t. By compactness of T there are $t_1, \cdots, t_m$ in T such that $T = V_{t_1} \cup \cdots \cup V_{t_m}$. Denote $r := \min\{r_{t_1}, \cdots, r_{t_m}\} > 0$. For each $t \in T$ there is $k \in \{1, \cdots, m\}$ such that $t \in V_{t_k}$, that is, $g(\bar{x} + r_{t_k} h, t) < \rho$, which combined with the inequality $g(\bar{x}, t) < \rho$ and the convexity of $g(\cdot, t)$ ensures that $g(\bar{x} + rh, t) < \rho$ since $0 < r \le r_{t_k}$. This being true for every $t \in T$ we deduce that $f(\bar{x} + rh) \le \rho$, hence $f(\bar{x} + \eta h) \le \rho$ for all $\eta \in [0, r]$ since $f(\bar{x}) < \rho$ and f is convex. Therefore, we obtain $\bar{x} + [0, r]h \subset \operatorname{dom} f$, hence $\bar{x} \in \operatorname{core}(\operatorname{dom} f)$. The result of the theorem then follows from Proposition 3.43. $\square$

3.1.7. Subdifferential properties of one variable convex functions.
Convex functions of one real variable present certain subdifferential special features. Consider a convex function $f : \mathbb{R} \to \mathbb{R} \cup \{+\infty\}$ of one real variable and $x \in \operatorname{dom} f$. From the non-decreasing property on $\mathbb{R} \setminus \{0\}$ of the function $t \mapsto \frac{f(x+t) - f(x)}{t}$ (see (2.19)), we see that the right and left derivates of f at x

$$f'_+(x) := D^+ f(x) := \lim_{t \downarrow 0} \frac{f(x+t) - f(x)}{t} = \inf_{t > 0} \frac{f(x+t) - f(x)}{t}$$

and

$$f'_-(x) := D^- f(x) := \lim_{t \uparrow 0} \frac{f(x+t) - f(x)}{t} = \sup_{t < 0} \frac{f(x+t) - f(x)}{t}$$

exist in $\mathbb{R} \cup \{-\infty, +\infty\}$ and $f'_-(x) \le f'_+(x)$. More generally, for any $u \in \operatorname{dom} f$ with $u < x$ (if there are), one has

$$f'_+(u) \le \frac{f(x) - f(u)}{x - u} = \frac{f(u) - f(x)}{u - x} \le f'_-(x),$$

and hence f'_- and f'_+ are non-decreasing on $\operatorname{dom} f$ and

$$f'_-(u) \le f'_+(u) \le f'_-(x) \le f'_+(x).$$

If f'_+ is continuous at $x \in \operatorname{int}(\operatorname{dom} f)$, the latter inequalities give

$$f'_+(x) = \lim_{u \uparrow x} f'_+(u) \le f'_-(x) \le f'_+(x),$$

hence $f'_+(x) = f'_-(x)$, or equivalently f is differentiable at x. If $J := \text{int}(\text{dom } f) \neq \emptyset$, then by local Lipschitz property of f on J the non-decreasing function f'_+ is finite on J, and hence there is a countable set $C \subset J$ such that f'_+ is continuous at any point in $J \setminus C$, hence f is differentiable at every $x \in J \setminus C$.

We summarize those properties as follows.

PROPOSITION 3.45. Let $f : \mathbb{R} \to \mathbb{R} \cup \{+\infty\}$ be a proper convex function of one real variable.
(a) The left and right derivates f'_- and f'_+ are non-decreasing functions from the interval $\text{dom } f$ into $\mathbb{R} \cup \{-\infty, +\infty\}$ and for $u, x \in \text{dom } f$ with $u < x$ one has
(3.13) $$f'_-(u) \leq f'_+(u) \leq f'_-(x) \leq f'_+(x).$$
(b) The set of points in the interval $\text{dom } f$ where f is not differentiable, is at most countable.

The property (b) fails for convex functions of several variables, as the following example shows.

EXAMPLE 3.46. The classical convex function $f : \mathbb{R}^2 \to \mathbb{R}$ defined by $f(x, y) = |x|$ for all $(x, y) \in \mathbb{R}^2$ is not differentiable at any point of the continuum subset $\{0\} \times \mathbb{R}$ of its domain. □

Making $t \downarrow 0$ and $r \to 1$ in the equality
$$\frac{f(x + tr) - f(x)}{t} = r \frac{f(x + tr) - f(x)}{tr},$$
we see that
$$\underline{d}f(x; 1) = f'(x; 1) = f'_+(x).$$
Similarly,
$$\underline{d}f(x; -1) = f'(x; -1) = -f'_-(x).$$
Then, identifying the dual of $\mathbb{R}$ with $\mathbb{R}$ (as usual) any subgradient of f is a real number and
$$r \in \partial f(x) \iff ru \leq f'(x; u) \, \forall u \in \mathbb{R} \iff f'_-(x) \leq r \leq f'_+(x).$$
We translate the latter property as a proposition.

PROPOSITION 3.47. Let $f : \mathbb{R} \to \mathbb{R} \cup \{+\infty\}$ be a convex function of one real variable. For any $x \in \text{dom } f$ one has
$$\partial f(x) = \mathbb{R} \cap [f'_-(x), f'_+(x)].$$

The next proposition examines the diverse situations for the point $x \in \text{dom } f$.

PROPOSITION 3.48. Let $f : \mathbb{R} \to \mathbb{R} \cup \{+\infty\}$ be a proper convex function of one real variable and let $x \in \text{dom } f$.
(A) If $\text{dom } f = \{x\}$, then $\partial f(x) = \mathbb{R}$.
(B) Assume that $\text{int}(\text{dom } f) \neq \emptyset$.
    (a) If $x \in \text{int}(\text{dom } f)$, then $f'_+(x)$ and $f'_-(x)$ are finite and
$$\partial f(x) = [f'_-(x), f'_+(x)].$$
    (b) If x is the right endpoint of the interval $\text{dom } f$, then $f'_+(x) = +\infty$, $f'_-(x) \in]-\infty, +\infty]$,
$$\partial f(x) = \emptyset \text{ if } f'_-(x) = +\infty \quad \text{and} \quad \partial f(x) = [f'_-(x), +\infty[\text{ if } f'_-(x) \in \mathbb{R}.$$

(c) If x is the left endpoint of dom f, then $f'_-(x) = -\infty$, $f'_+(x) \in [-\infty, +\infty[$, $\partial f(x) = \emptyset$ if $f'_+(x) = -\infty$ and $\partial f(x) =]-\infty, f'_+(x)]$ if $f'_+(x) \in \mathbb{R}$.

PROOF. It suffices to prove (B). If $x \in \operatorname{int}(\operatorname{dom} f)$, the left and right derivates of f at x are finite since f is locally Lipschitz continuous on $\operatorname{int}(\operatorname{dom} f)$ by Corollary 2.159, and the equality $\partial f(x) = [f'_-(x), f'_+(x)]$ holds true by Proposition 3.47 above. The assertion (a) in (B) is then justified. Suppose now that x is the right endpoint of the interval dom f. Then $f(x+t) = +\infty$ for every real $t > 0$, so $f'_+(x) = +\infty$. Further, fixing $\tau > 0$ with $x - \tau \in \operatorname{dom} f$, we have for all $t \in [-\tau, 0[$

$$\frac{f(x-\tau) - f(x)}{-\tau} \leq \frac{f(x+t) - f(x)}{t},$$

hence $f'_-(x) > -\infty$. The equalities in (b) concerning $\partial f(x)$ follow from Proposition 3.47. Then (b) is justified, while (c) is obtained in a similar way. □

The next proposition shows that lower semicontinuous convex functions of one real variable are *subdifferentially determined*. Extension to Banach spaces will be established in Theorem 3.204.

PROPOSITION 3.49. *Let $f, g : \mathbb{R} \to \mathbb{R} \cup \{+\infty\}$ be two proper lower semicontinuous convex functions of one real variable such that $\partial f(x) \subset \partial g(x)$ for all $x \in \mathbb{R}$. Then f and g are equal up to an additive real constant, that is, there exists a real constant C such that $f = g + C$.*

PROOF. If dom f is a singleton dom $f = \{\bar{x}\}$, then by Proposition 3.48 we have $\mathbb{R} = \partial f(\bar{x}) \subset \partial g(\bar{x})$, hence Proposition 3.48 again yields dom $g = \{\bar{x}\}$, which in turn entails that f and g are equal up to an additive constant. Now suppose that $\operatorname{int}(\operatorname{dom} f) \neq \emptyset$. By convexity dom f and dom g are intervals I_f and I_g in $\mathbb{R}$. Since f is continuous on $\operatorname{int}(\operatorname{dom} f)$ (see Corollary 2.159) we have by Theorem 3.30(a)

$$\operatorname{int}(\operatorname{dom} f) = \operatorname{int}(\operatorname{Dom} \partial f) \subset \operatorname{int}(\operatorname{Dom} \partial g) = \operatorname{int}(\operatorname{dom} g),$$

so dom $g \neq \emptyset$. Denote by a and b the left and right endpoints of I_f in $\mathbb{R} \cup \{-\infty, +\infty\}$. Since f and g are locally Lipschitz on $]a, b[$ (see Corollary 2.159), by Rademacher theorem there exists a Lebesgue negligible subset N in $]a, b[$ such that f and g are derivable at the points in the set $]a, b[\setminus N$ with $f' = g'$ on this set, and for all $x, y \in]a, b[$

$$f(x) - f(y) = \int_y^x f'(t)\,dt = \int_y^x g'(t)\,dt = g(x) - g(y).$$

The latter equality means that for some constant C, $f(x) = g(x) + C$ for all $x \in]a, b[$, and by Corollary 3.21 we still have

(3.14) $\qquad f(x) = g(x) + C \quad \text{for all } x \in [a, b] \cap \mathbb{R}.$

This tells us in particular that $I_f \subset I_g$.

If $b = +\infty$, it is clear that b coincides with the endpoint of I_g since $I_f \subset I_g$.

If $b \in \mathbb{R} \setminus I_f$, we have $f(b) = +\infty$, thus the lower semicontinuity of f gives $\lim_{x \uparrow b} f(x) = +\infty$, which by (3.14) implies that $\lim_{x \uparrow b} g(x) = +\infty$. It ensues by the continuity of the restriction of g to I_g (see Corollary 3.21) that $b \notin I_g$, thus b is also the right endpoint of I_g.

Suppose now that $b \in I_f$. If $\partial f(b) = \emptyset$, then by (b) in Proposition 3.48(B) $f'_-(b) = +\infty$, hence $g'_-(b) = +\infty$ by (3.14), which ensures that b is the right endpoint of I_g. If $\partial f(b) \neq \emptyset$, by (b) in Proposition 3.48(B) again we have $f'_-(b) \in \mathbb{R}$ and $\partial f(b) = [f'_-(b), +\infty[$. It ensues that $[f'_-(b), +\infty[\subset \partial g(b)$, thus by the assertions in Proposition 3.48 the element b is the right endpoint of I_g.

Similarly, the element a coincides with the left extremity of I_g. Putting altogether we obtain that $I_f = I_g$, so by (3.14) it results that $f(x) = g(x) + C$ for all $x \in \mathbb{R}$. □

Given a proper lower semicontinuous convex function $f : \mathbb{R} \to \mathbb{R} \cup \{+\infty\}$ and $x \in]a, b] \subset \mathrm{dom}\, f$, the inequality (3.13) and the non-decreasing property of f'_- and f'_+ entail that
$$f'_-(x) \geq \lim_{u\uparrow x} f'_+(u) \geq \lim_{u\uparrow x} f'_-(u).$$

On the other hand, for any $v \in]a, x[$ by the continuity of the restriction of f to $]a, x]$ (see Corollary 3.21)
$$\frac{f(v) - f(x)}{v - x} = \lim_{u\uparrow x} \frac{f(v) - f(u)}{v - u} \leq \lim_{u\uparrow x} f'_-(u),$$
which ensures that
$$f'_-(x) = \lim_{v\uparrow x} \frac{f(v) - f(x)}{v - x} \leq \lim_{u\uparrow x} f'_-(u).$$
It results that $f'_-(x) = \lim_{u\uparrow x} f'_-(u) = \lim_{u\uparrow x} f'_+(u)$, and similar equalities also hold true for $f'_+(x)$ if $x \in [a, b[\subset \mathrm{dom}\, f$. We have then proved the following proposition.

PROPOSITION 3.50. *Let $f : \mathbb{R} \to \mathbb{R} \cup \{+\infty\}$ be a proper lower semicontinuous convex function. If f is finite on $]a, b]$ (resp. on $[a, b[$), then*
$$f'_-(x) = \lim_{u\uparrow x} f'_-(u) = \lim_{u\uparrow x} f'_+(u) \quad \text{for any } x \in]a, b]$$
(resp. $f'_+(x) = \lim_{u\downarrow x} f'_+(u) = \lim_{u\downarrow x} f'_-(u) \quad \text{for any } x \in [a, b[$).

Now, given a convex function $f : X \to \mathbb{R} \cup \{+\infty\}$ on a topological vector space X, for any $x, y \in X$, $x^* \in \partial f(x)$ and $y^* \in \partial f(y)$, by definition
$$\langle x^*, x - y \rangle \geq f(x) - f(y) \quad \text{and} \quad \langle -y^*, x - y \rangle \geq f(y) - f(x),$$
hence $\langle x^* - y^*, x - y \rangle \geq 0$ since $f(x)$ and $f(y)$ are finite. This property has been already seen in Proposition 2.182 in the setting of normed spaces as the monotonicity property. The definition below translates its extension to the context of topological vector spaces. It also defines the concept of maximal monotonicity.

DEFINITION 3.51. *Let X be a topological vector space and $M : X \rightrightarrows X^*$ be a multimapping from X into its topological dual. One says that M is monotone whenever*
$$\langle x^* - y^*, x - y \rangle \geq 0$$
for all $x, y \in \mathrm{Dom}\, M$, $x^* \in M(x)$ and $y^* \in M(y)$.

If further in the class of monotone multimappings from X into X^* ordered by the inclusion $\subset$ of graphs, M is a maximal element, one says that M is a maximal monotone multimapping. This amounts to saying that any monotone multimapping from X into X^* whose graph contains $\mathrm{gph}\, M$ coincides with M.

Clearly, a monotone multimapping $M : X \rightrightarrows X^*$ is maximal if and only if for $(y, y^*) \in X \times X^*$ the inequality

$$\langle y^* - x^*, y - x \rangle \geq 0 \quad \text{for all } (x, x^*) \in \operatorname{gph} M$$

entails that $y^* \in M(y)$.

When $X = \mathbb{R}$ it is clear that M is monotone if and only if for any $s, t \in \operatorname{Dom} M$ with $s < t$ one has

(3.15) $\qquad s^* \leq t^* \quad \text{for all} \quad s^* \in M(s), t^* \in M(t).$

We show next via Proposition 3.48 and Proposition 3.50 that the subdifferential of a proper lower semicontinuous convex function from $\mathbb{R}$ to $\mathbb{R} \cup \{+\infty\}$ is a maximal monotone multimapping. The result will be generalized to the Banach setting in Theorem 3.206. Recall first the following: Given $\alpha < \beta$ in $\mathbb{R}$ and a continuous function $\varphi : [\alpha, \beta] \to \mathbb{R}$ whose left (resp. right) derivate $\varphi'_-(t)$ exists and is finite for every $t \in]\alpha, \beta[$ (resp. $\varphi'_+(t)$ exists and is finite for every $t \in]\alpha, \beta[$), then there are $\gamma_1, \gamma_2 \in]\alpha, \beta[$ such that

(3.16) $\qquad \varphi'_-(\gamma_1) \leq \dfrac{\varphi(\beta) - \varphi(\alpha)}{\beta - \alpha} \leq \varphi'_-(\gamma_2)$

$\qquad \text{(resp. } \varphi'_+(\gamma_1) \leq \dfrac{\varphi(\beta) - \varphi(\alpha)}{\beta - \alpha} \leq \varphi'_+(\gamma_2)\text{)};$

thus this function φ is Lipschitz continuous on $[\alpha, \beta]$ if and only if φ'_- (resp. φ'_+) is bounded on $]\alpha, \beta[$.

PROPOSITION 3.52. *Let $f : \mathbb{R} \to \mathbb{R} \cup \{+\infty\}$ be a proper lower semicontinuous convex function of one real variable. The subdifferential multimapping $\partial f : \mathbb{R} \rightrightarrows \mathbb{R}$ is maximal monotone.*

PROOF. By Proposition 3.48 $\operatorname{dom} f$ is a nonempty interval I, and clearly we may suppose $\operatorname{int} I \neq \emptyset$. Denote by a and b the left and right endpoints of I in $\mathbb{R} \cup \{-\infty, +\infty\}$. Let $M : \mathbb{R} \rightrightarrows \mathbb{R}$ be a monotone multimapping with $\operatorname{gph} M \supset \operatorname{gph} \partial f$ and let $(t, t^*) \in \operatorname{gph} M$.

Suppose that $a < t < b$. For $s \in]t, b[$ we have by Proposition 3.48(B) $f'_+(s) \in \partial f(s)$, which by (3.15) gives $f'_+(s) \geq t^*$. This ensures $f'_+(t) = \lim_{s \downarrow t} f'_+(s) \geq t^*$ by Proposition 3.50. Similarly, for $s \in]a, t[$ we have $f'_-(s) \leq t^*$, hence $f'_-(t) = \lim_{s \uparrow t} f'_-(s) \leq t^*$. It results by Proposition 3.48(B) that $t^* \in \partial f(t)$, thus $M(t) = \partial f(t)$.

Suppose that $t = b \in I$ and $f'_-(b) < +\infty$. Then $f'_-(b)$ is finite and taking a sequence $(s_n)_n$ in $]a, b[$ tending to b, we obtain $f'_-(s_n) \to f'_-(b)$ by Proposition 3.50. Since $f'_-(s_n) \in \partial f(s_n)$ by Proposition 3.48(B), we also have $f'_-(s_n) \leq t^*$ by (3.15), which gives $t^* \geq f'_-(t)$, so $t^* \in \partial f(t)$ by Proposition 3.48(B-b). This ensures that $M(t) = \partial f(t)$.

Suppose that $t = b \in I$ and $f'_-(b) = +\infty$. Taking a sequence $(s_n)_n$ in $]a, b[$ with $s_n \uparrow b$, we have $f'_-(s_n) \to +\infty$ by Proposition 3.50. This, combined with the inclusion $f'_-(s_n) \in \partial f(s_n)$ and with the property (3.15), furnishes the contradiction $t^* \geq +\infty$.

Suppose that $t = b \notin I$. Then by lower semicontinuity $f(s) \to +\infty$ as $s \to b$. Since f is continuous on $]a, b[$ and f'_- is non-decreasing on $]a, b[$, by (3.16) there is a sequence $s_n \uparrow b$ in $]a,b[$ such that $f'_-(s_n) \to +\infty$. Since $f'_-(s_n) \in \partial f(s_n)$ (see Proposition 3.48(B)), this gives by (3.15) the contradiction $t^* \geq +\infty$.

Suppose that $t > b$. Then $b \in \mathbb{R}$, so using (b) in Proposition 3.48, noting that $\partial f(b) = [f'_-(b), +\infty[$ if $b \in I$ with $f'_-(b) < +\infty$ and reasoning as in the two latter cases we can find a sequence $(s_n, s_n^*)_n$ in $\mathrm{gph}\, \partial f$ with $s_n^* \to +\infty$ and with $s_n \to b$ and $s_n \leq b < t$. This entails by (3.15) the contradiction $t^* \geq +\infty$.

Similar properties being true for the left endpoint a, it results that $\mathrm{gph}\, M = \mathrm{gph}\, \partial f$, which confirms the maximal monotonicity of ∂f. □

3.2. Subdifferentiability of convex functions in finite dimensions

Theorem 3.30 established the non-vacuity of the subdifferential of a convex function at continuity points while in Proposition 3.48 a complete description is obtained for subdifferentials on convex functions of one real variable. This section deals with the subdifferentiability of convex functions defined on finite-dimensional spaces.

3.2.1. Subdifferentiability via the relative interior. In finite dimensions, our aim in this subsection is to establish the subdifferentiability of a convex function under a basic relative interior condition. Let us first recall (as in Appendix B) that a subset F of a vector space X is *affine* if by definition, for any $x_1, \cdots, x_k$ in F and any reals $\alpha_1, \cdots, \alpha_k$ with $\alpha_1 + \cdots + \alpha_k = 1$ one has

$$\alpha_1 x_1 + \cdots + \alpha_k x_k \in F,$$

otherwise stated, every *affine combination* of F remains in F. When F is nonempty, taking any $u \in F$ we see that $F - u$ is a vector subspace of X. Consequently, any nonempty affine subset F of X is of the form $u + E$, where $u \in X$ and E is a vector subspace of X; such a vector space E is unique and its dimension is called the dimension of F, which is generally stated as $\dim F := \dim E$. Given any subset S of X, the intersection of all affine subsets of X containing S is an affine set, and it is the smallest affine subset of X containing S. It is called the *affine hull* of S and denoted by $\mathrm{aff}\, S$. When S is nonempty, it is not difficult to see that $\mathrm{aff}\, S$ coincides with the set of all affine combinations of elements in S, which means that $x \in \mathrm{aff}\, S$ if and only if there are $k \in \mathbb{N}$, $x_1, \cdots, x_k \in S$ and $\alpha_1, \cdots, \alpha_k \in \mathbb{R}$ such that $x = \alpha_1 x_1 + \cdots + \alpha_k x_k$.

Assume now that X is a finite-dimensional normed space and let C be a convex set in X. We recall that the *relative interior* $\mathrm{rint}\, C$ of C is the interior of C relative the affine hull $\mathrm{aff}\, C$ endowed with the induced topology, otherwise stated,

$$\mathrm{rint}\, C = \mathrm{int}_{\mathrm{aff}\, C}(C).$$

For any affine subset F of X, we note by its closedness that

$$\mathrm{rint}\, F = F = \mathrm{cl}\, F.$$

We also know (see Proposition B.7 in Appendix) that $\mathrm{rint}\, C \neq \emptyset$ whenever the convex set C in the finite-dimensional space X is nonempty. By means of this we can prove the following subdifferentiability property.

PROPOSITION 3.53. *Let X be a finite-dimensional normed space and $f : X \to \mathbb{R} \cup \{-\infty, +\infty\}$ be a convex function which is finite at some point in $\mathrm{rint}\, (\mathrm{dom}\, f)$. Then one has*

$$\partial f(x) \neq \emptyset \quad \text{for every } x \in \mathrm{rint}\, (\mathrm{dom}\, f).$$

PROOF. Fix any point $\bar{x} \in \operatorname{rint}(\operatorname{dom} f)$. Without loss of generality we may and do suppose that $\bar{x} = 0$. The affine hull of $\operatorname{dom} f - \bar{x}$ coincides with the vector subspace $E := \mathbb{R}(\operatorname{dom} f)$ and $\operatorname{rint}(\operatorname{dom} f) = \operatorname{int}_E(\operatorname{dom} f)$. Denoting by $f_0 : E \to \mathbb{R} \cup \{+\infty\}$ the restriction of f on E we have $0 \in \operatorname{int}_E(\operatorname{dom} f_0)$ and f_0 is finite at some point in $\operatorname{int}_E(\operatorname{dom} f_0)$. It ensues by Proposition 3.22 that f_0 does not take the value $-\infty$. The proper convex function f_0 is then (Lipschitz) continuous near 0 by Corollary 2.159, hence $\partial f_0(0) \neq \emptyset$ (see, for example, Theorem 3.30(a)). Choose some linear mapping $\ell_0 : E \to \mathbb{R}$ with $\ell_0 \in \partial f_0(0)$. Taking a linear extension $\ell : X \to \mathbb{R}$ of ℓ_0, we obtain that $\ell(x) \leq f(x) - f(0)$ for all $x \in X$, since $f(x) = +\infty$ for all $x \in X \setminus E$. It results that $\ell \in \partial f(0)$, justifying that $\partial f(0) \neq \emptyset$. □

As a direct corollary we have:

COROLLARY 3.54. *Let X be a finite-dimensional normed space and $f : X \to \mathbb{R} \cup \{+\infty\}$ be a proper convex function. Then $\operatorname{rint}(\operatorname{dom} f) \neq \emptyset$ and*

$$\partial f(x) \neq \emptyset \quad \text{for every } x \in \operatorname{rint}(\operatorname{dom} f),$$

so in particular $\operatorname{Dom}(\partial f) \neq \emptyset$.

3.2.2. Subdifferentiability of polyhedral convex functions. In addition to the previous subsection, another important case of subdifferentiability is that of polyhedral convex functions. Let X be a finite-dimensional normed space. Recall that a *polyhedral convex set* C in X is an intersection $\bigcap_{i \in I} \mathcal{H}_i$ of finitely many closed (affine) half-spaces; one also says that C is a *convex polyhedron*. Such a set C is evidently closed. Taking the index set $I = \emptyset$, we see that the global space X is the greatest polyhedral convex set in X. Every vector hyperplane in X being clearly polyhedral convex, so is every vector subspace. From this we see that every affine set in X (including X and $\emptyset$) is polyhedral convex. From the definition it is also clear that C is polyhedral convex if and only if there are $a_i^* \in X^*$, $\beta_i \in \mathbb{R}$, $i = 1, \cdots, k$ such that

$$(3.17) \qquad C = \{x \in X : \langle a_i^*, x \rangle \leq \beta_i, \; \forall i = 1, \cdots, k\};$$

this is known in the literature as a *half-space representation*. Such a convex set C is evidently closed. The Minkowski-Weyl theorem for convex polyhedra (see Theorem B.19 in Appendix) says that a nonempty set C in X is polyhedral convex if and only if there are points $a_1, \cdots, a_p$ in X and directions associated with $a_{p+1}, \cdots, a_q$ in X such that $x \in C$ provided there are $t_1, \cdots, t_p \geq 0$ with $t_1 + \cdots + t_p = 1$ and $t_{p+1}, \cdots, t_q \geq 0$ such that

$$(3.18) \qquad x = t_1 a_1 + \cdots + t_p a_p + t_{p+1} a_{p+1} + \cdots + t_q a_q;$$

this is often called a *vertex/ray representation*.

PROPOSITION 3.55. *Let $A : X \to Y$ be an affine mapping between finite-dimensional normed spaces and let C and D be polyhedral convex sets in X and Y respectively. Then $A(C)$ and $A^{-1}(D)$ are polyhedral convex in Y and X respectively.*

PROOF. The polyhedral convexity of $A(C)$ easily follows from (3.18) whereas that of $A^{-1}(D)$ is a consequence of (3.17). □

3.2. SUBDIFFERENTIABILITY OF CONVEX FUNCTIONS IN FINITE DIMENSIONS

Since the polyhedral convexity is obviously preserved by finite Cartesian product, the preceding proposition ensures that the sum of finitely many polyhedral convex sets is polyhedral convex. This and the evident case (by (3.17)) of intersection yield the following proposition.

PROPOSITION 3.56. *The sum and intersection of finitely many polyhedral convex sets of a finite dimensional normed space are polyhedral convex.*

Now we recall that a function $f : X \to \mathbb{R} \cup \{-\infty, +\infty\}$ is *polyhedral convex* if its epigraph is polyhedral convex in $X \times \mathbb{R}$. Therefore, by Proposition 3.23 any polyhedral convex function which is finite at some point does not take the value $-\infty$, that is, it is proper. The particular functions $-\omega_X$ and ω_X (with $\omega_X \equiv +\infty$) are the smallest and greatest polyhedral convex functions on X.

Noticing that $\mathrm{epi}\,(f \circ A) = \widehat{A}^{-1}(\mathrm{epi}\,f)$, where $\widehat{A}(x,t) = (A(x),t)$ for every $(x,t) \in X \times \mathbb{R}$, and writing for each $r \in \mathbb{R}$

$$\{f(\cdot) \leq r\} = \pi_X\big((\mathrm{epi}\,f) \cap (X \times]-\infty, r])\big),$$

the very definition of polyhedral convex functions and Proposition 3.55 entail:

PROPOSITION 3.57. *Let X, Y be two finite-dimensional normed spaces.*
(a) *The composition of any polyhedral convex function on Y with an affine mapping from X into Y is polyhedral convex.*
(b) *The sublevel sets of any polyhedral convex function on X are polyhedral convex.*

Given a polyhedral convex function $\varphi : X \times Y \to \mathbb{R} \cup \{-\infty, +\infty\}$ on the Cartesian product of two finite-dimensional normed spaces, by Proposition 3.55 the set $\pi_{X \times \mathbb{R}}(\mathrm{epi}\,\varphi)$ is polyhedral convex, hence closed. Consequently, by Proposition 3.1 this set concides with the epigraph of the function f defined on X by $f(x) = \inf_{y \in Y} \varphi(x,y)$, so f is polyhedral convex. Moreover, such a function f is proper if and only if it finite at some point according to the above comment using Proposition 3.23. We state those features in the proposition:

PROPOSITION 3.58. *Let X, Y be two finite-dimensional normed spaces and $\varphi : X \times Y \to \mathbb{R} \cup \{-\infty, +\infty\}$ be a polyhedral convex function. Then the function f defined on X by $f(x) := \inf_{y \in Y} \varphi(x,y)$ for all $x \in X$ is polyhedral convex. The function f is in addition proper if and only if it is finite at some point. Further, the infimum in the definition of $f(x)$ is attained for each x where $f(x)$ is finite.*

Given the polyhedral convex function $\varphi : X \times Y \to \mathbb{R} \cup \{-\infty, +\infty\}$, we also note that for each $b \in Y$ we have

$$\mathrm{epi}\,\varphi(\cdot, b) = \pi_{X \times \mathbb{R}}\big((\mathrm{epi}\,\varphi) \cap (X \times \{b\} \times \mathbb{R})\big),$$

thus the function $\varphi(\cdot, b)$ is polyhedral convex.

According to the definition of polyhedral convex functions (as functions whose epigraphs are convex polyhedra), the two basic types of functions, *affine-max* $x \mapsto \zeta(x) := \max_{i \in I}\big(\langle a_i^*, x\rangle + \beta_i\big)$ (with a finite set I) and indicators Ψ_C of polyhedral convex sets C, are clearly other examples of polyhedral convex functions. Further, since $\mathrm{epi}\,(\zeta + \Psi_C) = (\mathrm{epi}\,\zeta) \cap (C \times \mathbb{R})$, the function $\zeta + \Psi_C$ is polyhedral convex. On the other hand, using the definition we see that the epigraph of any polyhedral convex function on X different from $-\omega_X$ is the intersection of finitely many half-spaces which are either vertical or epigraphs of affine functions. All together justify the following proposition.

PROPOSITION 3.59. Given a finite-dimensional normed space X, every polyhedral convex function f on X different from $-\omega_X$ is of the form

$$x \mapsto \max_{i \in I} \left(\langle a_i^*, x \rangle + \beta_i \right) + \Psi_C(x),$$

where C is a polyhedral convex set in X, I is a finite index set, $a_i^* \in X^*$ and $\beta_i \in \mathbb{R}$.

Recalling the equalities $\Psi_{C_1} + \cdots + \Psi_{C_m} = \Psi_C$ with $C := \bigcap_{k=1}^{m} C_k$ and

$$\sup_{i_1 \in I_1} r_{i_1} + \cdots + \sup_{i_m \in I_m} r_{i_m} = \sup_{(i_1, \cdots, i_m) \in I_1 \times \cdots I_m} (r_{i_1} + \cdots + r_{i_m}),$$

the following corollary is direct.

COROLLARY 3.60. *Every positively linear combination of polyhedral convex functions on a finite-dimensional normed space X is polyhedral convex.*

In the representation in Proposition 3.59 and with $\zeta(\cdot) := \max_{i \in I}\left(\langle a_i^*, \cdot \rangle + \beta_i \right)$, for any $x \in C$ taking $i_x \in I$ with $\zeta(x) = \langle a_{i_x}^*, x \rangle + \beta_{i_x}$, we see that $x^* := a_{i_x}^* \in \partial \zeta(x)$, which directly gives that $x^* \in \partial(\zeta + \Psi_C)(x)$ according to the inequality defining a subgradient of a convex function. We translate this in the corollary:

COROLLARY 3.61. *If a polyhedral convex function f on a finite-dimensional normed space X is finite at some point, then $\partial f(x) \neq \emptyset$ at every $x \in \operatorname{dom} f$.*

The next corollary is concerned with the support function of a polyhedral convex set.

COROLLARY 3.62. *Let X be a finite-dimensional normed space and C be a polyhedral convex set in X. The support function σ_C of C is polyhedral convex.*

PROOF. We may suppose $C \neq \emptyset$, otherwise the result is evident. By the Minkowski-Weyl theorem recalled above there exist points $a_1, \cdots, a_p \in X$ and directions associated to $a_{p+1}, \cdots, a_q$ such that

$$C = \left\{ \sum_{i=1}^{p} t_i a_i + \sum_{i=p+1}^{q} t_i a_i : t_i \geq 0 \,\forall i = 1, \cdots, q, \sum_{i=1}^{p} t_i = 1 \right\}.$$

Then taking any $x^* \in X^*$ we derive that

$$\sigma_C(x^*) = \sup \left\{ \sum_{i=1}^{p} t_i \langle x^*, a_i \rangle : t_i \geq 0, \sum_{i=1}^{p} t_i = 1 \right\}$$

$$+ \sup \left\{ \sum_{i=p+1}^{q} t_i \langle x^*, a_i \rangle : t_i \geq 0 \,\forall i = p+1, \cdots, q \right\}$$

$$= \max_{1 \leq i \leq p} \langle x^*, a_i \rangle + \Psi_D(x^*),$$

where $D := \{x^* \in X^* : \langle x^*, a_i \rangle \leq 0 \,\forall i = p+1, \cdots, q\}$. This justifies the polyhedral convexity of the function σ_C according to Proposition 3.59. $\square$

The above corollary and the polyhedral convexity of sublevel sets of polyhedral convex sets (see Proposition 3.57) give:

COROLLARY 3.63. *Let X be a finite-dimensional normed space. The polar of any polyhedral convex set in X is polyhedral convex in X^*.*

3.3. Conjugates in the locally convex setting

From now on in this chapter we will mainly deal with locally convex spaces and all *locally convex spaces will be assumed to be Hausdorff* (which guarantees for any such non-null space the topological dual is non-null).

3.3.1. General properties and examples of Legendre-Fenchel conjugate.
Given a nonempty subset U of a topological vector space X and a function $f : U \to \mathbb{R} \cup \{-\infty, +\infty\}$, we extend Definition 2.68 by defining its *Legendre-Fenchel conjugate* $f^* : X^* \to \mathbb{R} \cup \{-\infty, +\infty\}$ by

$$f^*(x^*) = \sup_{x \in U} \left(\langle x^*, x \rangle - f(x)\right) \quad \text{for all } x^* \in X^*.$$

Considering the extension $f_0 : X \to \mathbb{R} \cup \{-\infty, +\infty\}$ given by $f_0(x) = f(x)$ if $x \in U$ and $f_0(x) = +\infty$ if $x \in X \setminus U$, it is clear that the Legendre-Fenchel conjugate of f coincides with the Legendre-Fenchel conjugate of f_0, that is,

$$f^*(x^*) = f_0^*(x^*) = \sup_{x \in X} \left(\langle x^*, x \rangle - f_0(x)\right) \quad \text{for all } x^* \in X^*.$$

Sometimes it will also be convenient to say, for short, conjugate or conjugate function instead of Legendre-Fenchel conjugate.

We readily derive the following first properties.

PROPOSITION 3.64. *Let $f, g : X \to \mathbb{R} \cup \{-\infty, +\infty\}$ be two functions on a Hausdorff locally convex space X. The following properties hold.*
(a) *The conjugate function f^* is convex and lower $w(X^*, X)$-semicontinuous.*
(b) $f^*(x^*) = \sup_{x \in \text{dom } f} \left(\langle x^*, x \rangle - f(x)\right)$, *with the usual convention that the supremum is $-\infty$ when $\text{dom } f = \emptyset$.*
(c) $\inf_X f = -f^*(0)$, *and hence f is proper and bounded from below if and only if f^* is finite at zero.*
(d) *One always have the Fenchel inequality*

$$\langle x^*, x \rangle - f(x) \leq f^*(x^*).$$

(e) $f \leq g \Rightarrow g^* \leq f^*$.
(f) *If $f^* \equiv +\infty$, then there is no continuous affine function minorizing f; the equality $f^* \equiv +\infty$ holds in particular whenever $f(x_0) = -\infty$ for some $x_0 \in X$.*
(g) f^* *takes the value $-\infty$ somewhere if and only if $f \equiv +\infty$; in this case $f^* \equiv -\infty$.*
(h) f^* *is proper if and only if $f \not\equiv +\infty$ and f is minorized by a continuous affine function.*
(i) *The inclusion $\bar{x}^* \in \partial_{MR} f(\bar{x})$ holds if and only if $f(\bar{x})$ is finite and $\langle \bar{x}^*, \bar{x} \rangle = f(\bar{x}) + f^*(\bar{x}^*)$; further*

$$\bar{x}^* \in \partial_{MR} f(\bar{x}) \Rightarrow \bar{x} \in \partial f^*(\bar{x}^*).$$

(j) *More generally, under the convexity of f one has for any real $\varepsilon \geq 0$ that $\bar{x}^* \in \partial_\varepsilon f(\bar{x})$ if and only if $f(\bar{x})$ is finite and $\langle \bar{x}^*, \bar{x} \rangle + \varepsilon \geq f(\bar{x}) + f^*(\bar{x}^*)$.*

A list of Legendre-Fenchel conjugates can be addressed. Let X be a Hausdorff locally convex space. For $f, f_i : X \to \mathbb{R} \cup \{-\infty, +\infty\}$ with $i \in I$, and for $a \in X$, $a^* \in X^*$, $\alpha \in \mathbb{R}$, the following properties can be easily derived from the definition of Legendre-Fenchel conjugate (arguments are provided for some of them):

- For pointwise infimum and supremum of many functions $(f_i)_{i \in I}$ from X into $\mathbb{R} \cup \{-\infty, +\infty\}$ one has

$$(3.19) \qquad \left(\inf_{i \in I} f_i\right)^* = \sup_{i \in I} f_i^* \quad \text{and} \quad \left(\sup_{i \in I} f_i\right)^* \leq \inf_{i \in I} f_i^*.$$

The functions $f_1 := \Psi_{[-1,0]}$ and $f_2 := \Psi_{[0,1]}$ on $\mathbb{R}$ allow us to see that the above inequality fails to be an equality.

- For $\omega_X : X \to \mathbb{R} \cup \{-\infty, +\infty\}$ defined by $\omega_X(x) = +\infty$ for all $x \in X$ one has

$$(\omega_X)^* = -\omega_{X^*} \quad \text{and} \quad (-\omega_X)^* = \omega_{X^*}.$$

- Concerning the parity of functions one has

$$(3.20) \qquad f \text{ even} \implies f^* \text{ even}.$$

It suffices to write

$$f^*(-x^*) = \sup_{x \in X}[\langle -x^*, x \rangle - f(x)] = \sup_{x \in X}[\langle x^*, -x \rangle - f(-x)] = f^*(x^*).$$

- For $p, q \in\,]1, +\infty[$ with $1/p + 1/q = 1$ one has with the absolute value function $|\cdot|$ on $\mathbb{R}$

$$(3.21) \qquad \left(\frac{1}{p}|\cdot|^p\right)^* = \frac{1}{q}|\cdot|^q,$$

and for the function $\varphi : \mathbb{R} \to \mathbb{R} \cup \{+\infty\}$ defined by $\varphi(t) = \begin{cases} \frac{1}{p}t^p & \text{if } t \geq 0 \\ +\infty & \text{if } t < 0, \end{cases}$ one has $\varphi^*(s) = \frac{1}{q}(s^+)^q$ for all $s \in \mathbb{R}$, where $s^+ := \max\{s, 0\}$ and q is as above.

- For $\theta : \mathbb{R} \to \mathbb{R} \cup \{-\infty, +\infty\}$ even on $\mathbb{R}$, nondecreasing on $[0, +\infty[$ and with $\theta(0) = 0$, the conjugate function θ^* is convex, even on $\mathbb{R}$, nondecreasing on $[0, +\infty[$ and $\theta^*(0) = 0$, and in fact

$$(3.22) \qquad \theta^*(s) = \theta^*(|s|) = \sup_{r \geq 0}[r|s| - \theta(r)] \quad \text{for all } s \in \mathbb{R}.$$

Since the function $s \mapsto r|s| - \theta(r)$ is nondecreasing on $[0, +\infty[$ for each $r \geq 0$, all the stated properties will follow once (3.22) is established. The equality $\theta^*(s) = \theta^*(|s|)$ comes from (3.20). Further, noting that

$$\sup_{t \geq 0}[-t|s| - \theta(-t)] = \sup_{t \geq 0}[-t|s| - \theta(t)] = -\theta(0)$$

because $t \mapsto -t|s| - \theta(t)$ is nonincreasing on $[0, +\infty[$, we see that

$$\theta^*(|s|) = \max\left\{\sup_{r \geq 0}[-r|s| - \theta(-r)], \sup_{r \geq 0}[r|s| - \theta(r)]\right\} = \sup_{r \geq 0}[r|s| - \theta(r)].$$

- For a subset $S \subset X$

$$(3.23) \qquad (\Psi_S)^*(x^*) = \sigma(x^*, S) = \sigma_S(x^*),$$

where we recall that both $\sigma(\cdot, S)$ and σ_S denote the support function of S defined by $\sigma(x^*, S) := \sup_{x \in S}\langle x^*, x \rangle$ (with the convention that this supremum is $+\infty$ when

- $S = \emptyset$).
- For any function $f : X \to \mathbb{R} \cup \{-\infty, +\infty\}$ one has

(3.24) $\qquad f^*(x^*) = \sigma_{\mathrm{epi}\, f}(x^*, -1) \quad \text{for all } x^* \in X^*.$

This directly follows from the definitions of the conjugate function and the support function.

- If X is a finite-dimensional normed space and $f : X \to \mathbb{R} \cup \{-\infty, +\infty\}$ is polyhedral convex, then

(3.25) $\qquad$ its conjugate f^* is also polyhedral convex.

This is a direct consequence of (3.24) above and of Corollary 3.62.

- If the subset S of X contains zero, then $(\Psi_S)^*$ coincides also with the gauge function j_{S° of the polar set S° of S (see (3.7) and (1.27) for definitions), that is,

(3.26) $\qquad (\Psi_S)^*(x^*) = \sigma(x^*, S) = j_{S^\circ}(x^*) \quad \text{for all } x^* \in X^*.$

Indeed, we have for every $x^* \in X^*$

$$j_{S^\circ}(x^*) = \inf\{t > 0 : t^{-1}x^* \in S^\circ\} = \inf\{t > 0 : \sigma(t^{-1}x^*, S) \le 1\}$$
$$= \inf\{t > 0 : \sigma(x^*, S) \le t\} = \sigma(x^*, S),$$

where the latter equality is due to the fact that $\sigma(\cdot, S) \ge 0$ because $0 \in S$. This combined with (3.23) justifies the desired equalities.

- For $g(x) = f(x) - \langle a^*, x \rangle$

(3.27) $\qquad g^*(x^*) = f^*(x^* + a^*).$

- For $g(x) = f(x) - \alpha$

(3.28) $\qquad g^*(x^*) = f^*(x^*) + \alpha.$

- For $g(x) = f(x - a)$

(3.29) $\qquad g^*(x^*) = f^*(x^*) + \langle x^*, a \rangle.$

- For $g(x) = f(x - a) - \langle a^*, x \rangle - \alpha$

(3.30) $\qquad g^*(x^*) = f^*(x^* + a^*) + \langle x^*, a \rangle + \alpha + \langle a^*, a \rangle.$

- For any real constant $\lambda \ne 0$ and $g(x) = f(\lambda x)$

(3.31) $\qquad g^*(x^*) = f^*\left(\frac{1}{\lambda}x^*\right).$

- For any real constant $\lambda > 0$ and $g(x) = \lambda f(x)$

(3.32) $\qquad g^*(x^*) = \lambda f^*\left(\frac{1}{\lambda}x^*\right).$

- For the indicator function $\Psi_{\{a\}}$ of the singleton $\{a\}$

(3.33) $\qquad \Psi_{\{a\}}{}^*(x^*) = \langle x^*, a \rangle.$

- For $g(x) = \langle a^*, x \rangle$

(3.34) $\qquad g^*(x^*) = \Psi_{\{a^*\}}(x^*).$

- For the indicator function Ψ_K of a nonempty convex cone $K \subset X$, recalling that K° coincides with the negative polar cone of K in the topological dual space of X one has

(3.35) $\qquad (\Psi_K)^* = \Psi_{K^\circ}.$

- Given Hausdorff locally convex spaces X_i and $f_i : X_i \to \mathbb{R} \cup \{-\infty, +\infty\}$, $i = 1, \cdots, m$, for $X = X_1 \times \cdots \times X_m$ and $f(x) = f_1(x_1) + \cdots + f_m(x_m)$ well defined on X one has

(3.36) $\quad f^*(x^*) = f_1^*(x_1^*) + \cdots + f_m^*(x_m^*)\quad$ for all $x^* \in X^* = X_1^* \times \cdots \times X_m^*$.

- Assuming that $(X, \|\cdot\|_X)$ is a normed space, denoting here and below in such a case by $\|\cdot\|_{X^*}$ the usual dual norm on X^* of the norm $\|\cdot\|_X$, one has for any real constant $\lambda > 0$

(3.37) $\quad (\Psi_{\lambda \mathbb{B}_X})^* = \lambda \|\cdot\|_{X^*}\quad$ and $\quad (\lambda \|\cdot\|_X)^* = \Psi_{\lambda \mathbb{B}_{X^*}}$,

where we recall that $\mathbb{B}_X$ and $\mathbb{B}_{X^*}$ are the closed unit balls of $(X, \|\cdot\|_X)$ and $(X^*, \|\cdot\|_{X^*})$ centered at the origins of X and X^*, respectively.

- For an even function $\theta : \mathbb{R} \to \mathbb{R} \cup \{-\infty, +\infty\}$ which is nondecreasing on $[0, +\infty[$ with $\theta(0) = 0$ one has for a normed space $(X, \|\cdot\|_X)$

(3.38) $\quad (\theta \circ \|\cdot\|_X)^* = \theta^* \circ \|\cdot\|_{X^*}$.

Indeed with X non-null and $x^* \in X^*$ we can write by (3.22)

$$\begin{aligned}(\theta \circ \|\cdot\|_X)^*(x^*) &= \sup_{x \in X}[\langle x^*, x\rangle - \theta(\|x\|_X)]\\ &= \sup_{r \in [0,+\infty[} \sup_{u \in \mathbb{S}_X}[\langle x^*, ru\rangle - \theta(r)]\\ &= \sup_{r \in [0,+\infty[}[r\|x^*\|_{X^*} - \theta(r)]\\ &= \theta^*(\|x^*\|_{X^*}).\end{aligned}$$

- For $p, q \in \,]1, +\infty[$ with $1/p + 1/q = 1$ particularizing (3.38) with $\theta := (1/p)|\cdot|^p$ and using (3.21) we obtain for a normed space $(X, \|\cdot\|_X)$

(3.39) $\quad \left(\dfrac{1}{p}\|\cdot\|_X^p\right)^* = \dfrac{1}{q}\|\cdot\|_{X^*}^q.$

- For $p, q \in \,]1, +\infty[$ with $1/p + 1/q = 1$ and for any real constant $\lambda > 0$ the latter equality combined with (3.32) furnishes for a normed space $(X, \|\cdot\|_X)$

(3.40) $\quad \left(\dfrac{1}{p\lambda}\|\cdot\|_X^p\right)^* = \dfrac{\lambda^{q-1}}{q}\|\cdot\|_{X^*}^q,$

so in particular

(3.41) $\quad \left(\dfrac{1}{2\lambda}\|\cdot\|_X^2\right)^* = \dfrac{\lambda}{2}\|\cdot\|_{X^*}^2.$

- For $\varphi : \mathbb{R}^m \to \mathbb{R}$ defined by $\varphi(x) = \sum_{i=1}^{m} \exp(x_i)$, using (3.36) one has with the convention $0 \log(0) = 0$

(3.42) $\quad \varphi^*(y) = \begin{cases} \sum_{i=1}^{m}(y_i \log(y_i) - y_i) & \text{if } y \in (\mathbb{R}_+)^m \\ +\infty & \text{otherwise.} \end{cases}$

- For φ defined on $\mathbb{R}$ by $\varphi(t) = \begin{cases} \log(t) & \text{if } t > 0 \\ +\infty & \text{if } t \le 0, \end{cases}$ one has $\varphi^* \equiv +\infty$.

- For φ defined on $\mathbb{R}$ by $\varphi(t) = \begin{cases} -\log t & \text{if } t > 0 \\ +\infty & \text{if } t \le 0, \end{cases}$ one has
$$\varphi^*(s) = \begin{cases} -1 - \log(-s) & \text{if } s < 0 \\ +\infty & \text{if } s \ge 0. \end{cases}$$

- For $\varphi : \mathbb{R} \to \mathbb{R}$ with $\varphi(t) = t^+$ one has $\varphi^* = \Psi_{[0,1]}$.
- For $\varphi : \mathbb{R} \to \mathbb{R}$ with $\varphi(t) = \sqrt{1+t^2}$ for all $t \in \mathbb{R}$ one has
$$\varphi^*(s) = \begin{cases} -\sqrt{1-s^2} & \text{if } |t| \le 1 \\ +\infty & \text{if } |t| > 1. \end{cases}$$

- In addition to the pointwise infimum in (3.19), given a continuous linear mapping $A : X \to Y$ between two locally convex spaces and a function $f : X \to \mathbb{R} \cup \{-\infty, +\infty\}$, it can be of interest to consider the conjugate of the function $h : X \to \mathbb{R} \cup \{-\infty, +\infty\}$ defined by
$$h(y) = \inf\{f(x) : x \in A^{-1}(y)\} \quad \text{for all } y \in Y.$$

Fix any $y^* \in Y^*$ and note that

$$\begin{aligned} h^*(y^*) &= \sup_{y \in Y} \left(\langle y^*, y \rangle - h(y) \right) = \sup_{y \in Y} \sup_{x \in A^{-1}(y)} \left(\langle y^*, y \rangle - f(x) \right) \\ &= \sup_{(x,y) \in \text{gph } A} \left(\langle y^*, y \rangle - f(x) \right) = \sup_{x \in X} \sup_{y \in \{A(x)\}} \left(\langle y^*, y \rangle - f(x) \right) \\ &= \sup_{x \in X} \left(\langle y^*, Ax \rangle - f(x) \right) = \sup_{x \in X} \left(\langle A^*y^*, x \rangle - f(x) \right) = f^*(A^*y^*). \end{aligned}$$

This tells us that
$$h^* = f^* \circ A^*.$$

- Another function of great interest defined via an infimum is the distance function d_S from a subset S in a normed space $(X, \|\cdot\|_X)$ given by $d_S(x) = \inf_{y \in S} \|x - y\|$. Noticing that

$$\begin{aligned} (d_S)^*(x^*) &= \sup_{x \in X} \left(\langle x^*, x \rangle - \inf_{y \in S} \|x - y\|_X \right) = \sup_{y \in S} \sup_{x \in X} \left(\langle x^*, x \rangle - \|x - y\|_X \right) \\ &= \sup_{y \in S} [\langle x^*, y \rangle + \sup_{x \in X} (\langle x^*, x - y \rangle - \|x - y\|_X)] \\ &= \sup_{y \in S} \langle x^*, y \rangle + \sup_{u \in X} (\langle x^*, u \rangle - \|u\|_X) \end{aligned}$$

we see by (3.37) that

(3.43) $$(d_S)^*(x^*) = (\Psi_S)^*(x^*) + \Psi_{\mathbb{B}_{X^*}}(x^*).$$

- Writing $d_S(x) = \inf_{y \in X} (\|x - y\|_X + \Psi_S(y))$ reveals that the distance function is a particular case of a large class of functions known in the literature as infimal convolution functions. Given a vector space X and two functions $f, g : X \to \mathbb{R} \cup \{+\infty\}$, the extended real

(3.44) $$(f \square g)(x) := \inf_{y \in X} (f(x - y) + g(y))$$

is well-defined in $\mathbb{R} \cup \{-\infty, +\infty\}$ for all $x \in X$. The function $f \square g : X \to \mathbb{R} \cup \{-\infty, +\infty\}$ is called the *infimal convolution (or inf-convolution) of functions f and g*. It is also clear that

$$(3.45) \qquad (f \square g)(x) := \inf_{u+v=x} (f(u) + g(v)),$$

where the infimum is taken over all pairs $(u, v) \in X \times X$ such that $u + v = x$. Further, when the functions f and g are convex, $f \square g$ is also convex according to Proposition 2.54, otherwise stated

$$(3.46) \qquad (f \text{ and } g \text{ convex}) \Rightarrow (f \square g \text{ convex}).$$

REMARK 3.65. More generally, for functions f, g taking values in $\mathbb{R} \cup \{-\infty, +\infty\}$, one extends the definition of $f \square g : X \to \mathbb{R} \cup \{-\infty, +\infty\}$ by setting

$$(f \square g)(x) := \inf_{y \in X} (f(x-y) \dotplus g(y)) \quad \text{for all } x \in X,$$

where $\dotplus$ is the *upper extended addition* in $\mathbb{R} \cup \{-\infty, +\infty\}$ with

$$(-\infty) \dotplus (+\infty) = +\infty \quad \text{and} \quad (+\infty) \dotplus (-\infty) = +\infty.$$

Doing so, one can see that the infimal convolution is associative, that is, for $f_i : X \to \mathbb{R} \cup \{-\infty, +\infty\}$, with $i = 1, 2, 3$ one has

$$(f_1 \square f_2) \square f_3 = f_1 \square (f_2 \square f_3),$$

since one easily verifies the equality

$$((f_1 \square f_2) \square f_3)(x) = \inf_{x_1 + x_2 + x_3 = x} (f_1(x_1) + f_2(x_2) + f_3(x_3))$$

and the similar one for $(f_1 \square (f_2 \square f_3))(x)$. So, the infimal convolution operation on functions from X into $\mathbb{R} \cup \{-\infty, +\infty\}$ is associative, commutative, and admits the indicator function $\Psi_{\{0\}}$ of the singleton set $\{0\}$ as identity element.

According to the associativity property, the convolution function

$$f_1 \square \cdots \square f_m : X \to \mathbb{R} \cup \{-\infty, +\infty\}$$

can be defined by induction as $(f_1 \square \cdots \square f_{m-1}) \square f_m$. When all the functions f_i take their values in $\mathbb{R} \cup \{+\infty\}$, for example, it is not difficult to see by induction that for every $x \in X$

$$(f_1 \square \cdots \square f_m)(x) = \inf\{f_1(x_1) + \cdots + f_m(x_m) : (x_1, \cdots, x_m) \in X^m, x_1 + \cdots + x_m = x\}.$$

Then for $\bar{x} \in X$ where $(f_1 \square \cdots \square + f_m)(\bar{x})$ is finite, ones says that the infimal convolution is *exact at $\bar{x}$* or *attained for $\bar{x}$* if there exist $\bar{x}_1, \cdots, \bar{x}_m$ in X such that

$$(f_1 \square \cdots \square + f_m)(\bar{x}) = f_1(\bar{x}_1) + \cdots + f_m(\bar{x}_m).$$

In the case of two extended real-valued functions f, g on X, the exactness or attainment property of the infimal convolution $f \square g$ amounts to saying that there is some $\bar{y} \in X$ such that

$$(f \square g)(\bar{x}) = f(\bar{x} - \bar{y}) + g(\bar{y}),$$

as easily seen. $\square$

We noticed above in (3.46) the stability of convexity under infimal convolution. Using Proposition 3.57(a) and Corollary 3.60, and applying Proposition 3.58 we see that the following similar stability also holds true for the polyhedral convexity.

PROPOSITION 3.66. Given a finite-dimensional normed space X, any infimal convolution of finitely many polyhedral convex functions on X is polyhedral convex.

Recall that the infimal convolution is said to be *exact at a point* $\bar{x} \in X$ if there exists some $\bar{y} \in X$ such that

(3.47) $$(f\square g)(\bar{x}) = f(\bar{x} - \bar{y}) + g(\bar{y}).$$

When this property holds at every point $\bar{x} \in X$ (resp. $\bar{x} \in S \subset X$), we will say that the infimal convolution is *exact* (resp. *exact on S*). The infimal convolution will be largely studied in other chapters and it will appear as a basic function. At this stage, assuming that X is a locally convex space the previous arguments also furnish the following formula for the conjugate of infimal convolution:

(3.48) $$(f\square g)^* = f^* + g^*.$$

We can also provide a first fundamental case where the infimal convolution is exact and lower semicontinuous. It is related to the compactness assumption of the sublevels $\{f \leq r\} := \{x \in X : f(x) \leq r\}$ of one function.

PROPOSITION 3.67. *Let (X, τ_X) be a Hausdorff topological vector space and $f, g : X \to \mathbb{R} \cup \{+\infty\}$ be extended real-valued functions. Assume that all the sublevels $\{f \leq r\}$ (with $r \in \mathbb{R}$) of f are τ_X-compact and that g is bounded from below and lower τ_X-semicontinuous. Then the function $f\square g$ is lower τ_X-semicontinuous, and the infimal convolution is exact at any point in X.*

If in addition the functions f, g are proper, then $f\square g$ is also proper.

PROOF. We note first that $\varphi := f\square g$ is bounded from below and that f is lower semicontinuous. If either $f \equiv +\infty$ or $g \equiv +\infty$, the first result is obvious. So, suppose that both f, g are proper. For each $x \in X$ we easily see that all the sublevels of the function $f(\cdot) + g(x - \cdot)$ are compact, hence $f\square g$ is exact at each $x \in \operatorname{dom} f + \operatorname{dom} g$, which also says that $f\square g$ is proper. If $x \notin \operatorname{dom} f + \operatorname{dom} g$, then $f(\cdot) + g(x - \cdot) \equiv +\infty$ and $f\square g$ is still exact at x. Thus, the infimal convolution $f\square g$ is exact at any point in X.

Now fix any $\bar{x} \in X$ and take any real $r < \varphi(\bar{x})$. Set $s := \inf_X g$ and note that $s \in \mathbb{R}$. The set $C := \{x \in X : f(x) \leq r - s\}$ is compact. This leads us to define φ_1, φ_2 on X by putting for every $x \in X$

$$\varphi_1(x) = \inf_{y \in C}\left(f(y) + g(x - y)\right) \quad \text{and} \quad \varphi_2(x) = \inf_{y \notin C}\left(f(y) + g(x - y)\right),$$

so $\varphi(\cdot) = \min\{\varphi_1(\cdot), \varphi_2(\cdot)\}$. It is clear that $\varphi_2(x) > r$ for all $x \in X$. On the other hand, using the lower semicontinuity of the function $(x, y) \mapsto f(y) + g(x - y)$ on $X \times C$ and Berge Theorem (see Theorem 1.43(b)), we see that φ_1 is lower semicontinuous on X. Since $\varphi_1(\bar{x}) > r$ (because $\varphi_1(\bar{x}) \geq \varphi(\bar{x})$), there is a neighborhood V of $\bar{x}$ such that $\varphi_1(x) > r$ for all $x \in V$. It results that $\varphi(x) > r$ for all $x \in V$, which justifies the lower semicontinuity of φ at $\bar{x}$ and finishes the proof. $\square$

The next example shows that the infimal convolution of lower semicontinuous functions may fail to be lower semicontinuous, even though the data functions are in addition convex. Before stating the example, we notice that it is easily seen that given two sets S_1, S_2 in a vector space X, the infimal convolution $\Psi_{S_1} \square \Psi_{S_2}$ coincides with the indicator function of the Minkowski sum $S_1 + S_2$, that is,

(3.49) $$\Psi_{S_1} \square \Psi_{S_2} = \Psi_{S_1 + S_2}.$$

EXAMPLE 3.68. In $\mathbb{R}^2$ denote $S_1 := \{(x,r) \in \mathbb{R}\times\mathbb{R} : r \geq \exp(x)\}$ and $S_2 := \mathbb{R}\times\{0\}$. The indicator functions Ψ_{S_i}, $i = 1, 2$, are convex and lower semicontinuous, but $\Psi_{S_1} \square \Psi_{S_2} = \Psi_{S_1+S_2}$ is not lower semicontinuous since the set $S_1+S_2 = \mathbb{R}\times]0,+\infty[$ is not closed in $\mathbb{R} \times \mathbb{R}$. To see the equality $S_1 + S_2 = \mathbb{R}\times]0,+\infty[$, it suffices to note that the inclusion $\supset$ is due to the fact any $(u,v) \in \mathbb{R}\times]0,+\infty[$ can be written as

$$(u,v) = (\log(v), v) + (u - \log(v), 0)$$

and that the converse inclusion $\subset$ is evident. □

It is worth noticing the following description of Moreau-Rockafellar subgradients (see Definition 3.2) of the infimal convolution under attainment.

PROPOSITION 3.69. Let $f_i : X \to \mathbb{R} \cup \{+\infty\}$, $i = 1,\cdots,m$, be functions on a topological vector space X such that the infimal convolution $f_1 \square \cdots \square f_m$ is finite at $\overline{x} \in X$.
(a) If the infimal convolution is attained at $\overline{x}$, then for any $\overline{x}_1, \cdots, \overline{x}_m$ with

$$(f_1 \square \cdots \square f_m)(\overline{x}) = f_1(\overline{x}_1) + \cdots + f_m(\overline{x}_m)$$

one has

$$\partial_{MR}(f_1 \square \cdots \square f_m)(\overline{x}) = \partial_{MR} f_1(\overline{x}_1) \cap \cdots \cap \partial_{MR} f_m(\overline{x}_m).$$

(b) Assume that the functions f_i are convex. Then, given any real $\eta > 0$ and any $\overline{x}_1, \cdots, \overline{x}_m$ in X with

$$f_1(\overline{x}_1) + \cdots + f_m(\overline{x}_m) \leq (f_1 \square \cdots \square f_m)(\overline{x})) + \eta,$$

one has for any real $\varepsilon \geq 0$

$$\partial_\varepsilon (f_1 \square \cdots \square f_m)(\overline{x}) \subset \partial_{\varepsilon+\eta} f_1(\overline{x}_1) \cap \cdots \cap \partial_{\varepsilon+\eta} f_m(\overline{x}_m).$$

PROOF. (a) It suffices to prove the result for $m = 2$. So, put $f = f_1$ and $g = f_2$. Fix any y with $(f \square g)(\overline{x}) = f(y) + g(\overline{x} - y)$. Take any $x^* \in \partial_{MR}(f \square g)(\overline{x})$. Then for any $u \in X$ we have

$$\langle x^*, u - \overline{x}\rangle \leq \inf_{z \in X} \big(f(z) + g(u-z)\big) - f(y) - g(\overline{x}-y)$$
$$\leq f(y) + g(u-y) - f(y) - g(\overline{x}-y) = g(u-y) - g(\overline{x}-y),$$

hence $\langle x^*, x-(\overline{x}-y)\rangle \leq g(x)-g(\overline{x}-y)$ for every $x \in X$. Therefore, $x^* \in \partial_{MR} g(\overline{x}-y)$ and similarly $x^* \in \partial_{MR} f(y)$, so the inclusion $\subset$ is justified.

Conversely, take any $x^* \in \partial_{MR} f(y) \cap \partial_{MR} g(\overline{x} - y)$. Given any $x \in X$ we have for every $z \in X$

$$\langle x^*, z-y\rangle \leq f(z) - f(y) \text{ and } \langle x^*, x-z-(\overline{x}-y)\rangle \leq g(x-z) - g(\overline{x}-y),$$

hence adding both inequalities gives for every $z \in X$

$$\langle x^*, x - \overline{x}\rangle \leq f(z) + g(x-z) - f(y) - g(\overline{x}-y),$$

otherwise stated $\langle x^*, x - \overline{x}\rangle \leq (f \square g)(x) - (f \square g)(\overline{x})$. This being true for every $x \in X$, we conclude that $x^* \in \partial_{MR}(f \square g)(\overline{x})$. The assertion (a) is then proved.
(b) The proof of (b) is similar and omitted. □

Considering now f as any function from a topological vector space (X, τ_X) into $\mathbb{R} \cup \{-\infty, +\infty\}$, we denote by cl f (or $\text{cl}_{\tau_X} f$ when the topology τ_X needs to be precised) the pointwise supremum of all lower semicontinuous functions from X into $\mathbb{R} \cup \{-\infty, +\infty\}$ majorized by f. It is an exercise to see that

(3.50) $\quad (\text{cl } f)(x) = \liminf_{u \to x} f(u) \quad \text{and} \quad \text{epi}(\text{cl } f) = \text{cl}_{X\times\mathbb{R}}(\text{epi } f),$

where we recall that the notation $\operatorname{cl}_{X\times\mathbb{R}}$ means the topological closure in $X\times\mathbb{R}$. Accordingly, the function $\operatorname{cl}_{\tau_X} f$ is called the τ_X-*closure hull* (τ_X-*lower semicontinuous hull, or τ_X-closure*) of f.

It is known and easily seen that for any $\bar{x} \in X$
$$\liminf_{x \to \bar{x}} f(x) = \min\{f(\bar{x}), \liminf_{\substack{x \to \bar{x} \\ x \neq \bar{x}}} f(x)\}.$$

The simple function $f : X \to \mathbb{R}$ defined by $f(x) = 1$ if $x \neq \bar{x}$ and $f(\bar{x}) = 0$ shows that one may have $\liminf_{x \to \bar{x}} f(x) < \liminf_{\substack{x \to \bar{x} \\ x \neq \bar{x}}} f(x)$. For the convex function $\Psi_{\{\bar{x}\}}$ we also see that $\liminf_{x \to \bar{x}} \Psi_{\{\bar{x}\}}(x) < \liminf_{\substack{x \to \bar{x} \\ x \neq \bar{x}}} \Psi_{\{\bar{x}\}}(x)$. Despite these two examples, the next proposition shows that equality holds whenever f is a convex function whose effective domain contains at least a point different from $\bar{x}$.

PROPOSITION 3.70. *Let X be a topological vector space and $f : X \to \mathbb{R} \cup \{-\infty, +\infty\}$ be a convex function.*
(a) *If $\operatorname{dom} f$ contains a point different from $\bar{x}$, then*
$$\liminf_{x \to \bar{x}} f(x) = \liminf_{\substack{x \to \bar{x} \\ x \neq \bar{x}}} f(x).$$

(b) *If $\operatorname{dom} f$ contains more than one point, then for every $x \in X$*
$$(\operatorname{cl} f)(x) = \liminf_{\substack{u \to x \\ u \neq x}} f(u).$$

PROOF. The assertion (b) being a direct consequence of (a), it suffices to show

(3.51) $$\liminf_{\substack{x \to \bar{x} \\ x \neq \bar{x}}} f(x) \leq f(\bar{x})$$

under the assumption in (a). We may suppose $f(\bar{x}) < +\infty$. Choose $a \in \operatorname{dom} f$ with $a \neq \bar{x}$. If $f(a) = -\infty$, we have $f(x) = -\infty$ for all $x \in]a, \bar{x}[$, so (3.51) holds. Suppose now that $f(a)$ is finite. Writing for every $t \in]0, 1[$
$$f(ta + (1-t)\bar{x}) \leq tf(a) + (1-t)f(\bar{x}),$$
and taking the limit inferior as $]0, 1[\ni t \to 0$, we see that (3.51) holds again. □

Following the above scheme, given a function f, we also notice that the supremum of all convex (resp. lower semicontinuous convex) functions majorized by f is a convex (resp. lower semicontinuous convex) function majorized by f. Then the greatest convex (resp. greatest lower semicontinuous convex) function majorized by f exists and it coincides with the latter supremum function. The convex hull and closed convex hull of a function are then defined as follows.

DEFINITION 3.71. Given a vector space X (resp. a Hausdorff locally convex space (X, τ_X)) and a function $f : X \to \mathbb{R} \cup \{-\infty, +\infty\}$, the *convex hull* $\operatorname{co} f$ (resp. *closed convex hull* $\overline{\operatorname{co}} f$) of f is defined as the pointwise supremum of all convex (resp. lower τ_X-semicontinuous convex) functions majorized by f; so $\operatorname{co} f$ (resp. $\overline{\operatorname{co}} f$) is the greatest convex (resp. the greatest lower τ_X-semicontinuous convex) function majorized by f.

Obviously, one has

(3.52) $$\overline{\operatorname{co}} f \leq \operatorname{co} f \leq f \quad \text{and} \quad \overline{\operatorname{co}} f \leq \operatorname{cl} f \leq f.$$

PROPOSITION 3.72. Let X be a vector space and $f : X \to \mathbb{R} \cup \{-\infty, +\infty\}$ be a function on X.
(a) The inclusion $\operatorname{co}(\operatorname{epi} f) \subset \operatorname{epi}(\operatorname{co} f)$ holds.
(b) For every $x \in X$ one has the equalities

$$(\operatorname{co} f)(x) = \inf\{r \in \mathbb{R} : (x, r) \in \operatorname{co}(\operatorname{epi} f)\}$$

$$= \inf\left\{\sum_{i=1}^m \lambda_i f(x_i) : m \in \mathbb{N}, \lambda_i > 0, \sum_{i=1}^m \lambda_i = 1, \sum_{i=1}^m \lambda_i x_i = x\right\},$$

where the convention $(+\infty) + (-\infty) = +\infty$ is taken.

PROOF. (a) By definition one has $\operatorname{co} f \leq f$, hence $\operatorname{epi} f \subset \operatorname{epi}(\operatorname{co} f)$. This and the convexity of the set $\operatorname{epi}(\operatorname{co} f)$ yields the desired inclusion $\operatorname{co}(\operatorname{epi} f) \subset \operatorname{epi}(\operatorname{co} f)$.
(b) Denote by $\varphi(x)$ (resp. $\Phi(x)$) the infimum in (b) involving $\operatorname{co}(\operatorname{epi} f)$ (resp. $f(x_i)$). Let $g := \operatorname{co} f : X \to \mathbb{R} \cup \{-\infty, +\infty\}$ be the greatest convex function majorized by the function f.

Let us prove first $g \leq \Phi$ and $g \leq \varphi$. Fix any $x \in X$. Since $\operatorname{co}(\operatorname{epi} f) \subset \operatorname{epi} g$ by (a), it ensues that

$$g(x) = \inf\{r \in \mathbb{R} : (x, r) \in \operatorname{epi} g\} \leq \inf\{r \in \mathbb{R} : (x, r) \in \operatorname{co}(\operatorname{epi} f)\} = \varphi(x).$$

Concerning $\Phi(x)$, we may suppose that $\Phi(x) < +\infty$. Take any nonempty finite subset I of $\mathbb{N}$, any I-uple of reals $(\lambda_i)_{i \in I}$ with $\lambda_i > 0$ and $\sum_{i \in I} \lambda_i = 1$, any I-uple $(x_i)_{i \in I}$ with $\sum_{i \in I} \lambda_i x_i = x$. If $f(x_{i_0}) = +\infty$ for some $i_0 \in I$, clearly $g(x) \leq \sum_{i \in I} \lambda_i f(x_i)$. If $f(x_i) < +\infty$ for all $i \in I$, by convexity of g and by the inequality $g \leq f$ we have

$$g(x) \leq \sum_{i \in I} \lambda_i g(x_i) \leq \sum_{i \in I} \lambda_i f(x_i).$$

So, in any case $g(x) \leq \sum_{i \in I} \lambda_i f(x_i)$, and taking appropriate infimum gives $g(x) \leq \Phi(x)$. Thus both inequalities $g \leq \Phi$ and $g \leq \varphi$ are justified.

We verity now that the functions Φ and φ are convex. Fix any $x, y \in X$ and any reals $\alpha, \beta \in {]}0, +\infty{[}$ with $\alpha + \beta = 1$. Concerning φ, take any reals r, s with (x, r) and (y, s) in $\operatorname{co}(\operatorname{epi} f)$. Then $(\alpha x + \beta y, \alpha r + \beta s) \in \operatorname{co}(\operatorname{epi} f)$, hence $\varphi(\alpha x + \beta y) \leq \alpha r + \beta s$. Taking the infimum over $r \in \mathbb{R}$ with $(x, r) \in \operatorname{co}(\operatorname{epi} f)$ and then the infimum over $s \in \mathbb{R}$ with $(y, s) \in \operatorname{co}(\operatorname{epi} f)$ gives $\varphi(\alpha x + \beta y) \leq \alpha \varphi(x) + \beta \varphi(y)$, so φ is convex.

For Φ, take any finite uples $(\lambda_i)_{i \in I}$ with $\lambda_i > 0$ and $\sum_{i \in I} \lambda_i = 1$ and $(x_i)_{i \in I}$ with $\sum_{i \in I} \lambda_i x_i = x$, and take also any finite uples $(\mu_j)_{j \in J}$ with $\mu_j > 0$ and $\sum_{j \in J} \mu_j = 1$ and $(y_j)_{j \in J}$ with $\sum_{j \in J} \mu_j y_j = y$. Put $K := (\{1\} \times I) \cup (\{2\} \times J)$ and define

$$(t_k, z_k) = \begin{cases} (\alpha \lambda_i, x_i) & \text{if } k = (1, i) \\ (\beta \mu_j, y_j) & \text{if } k = (2, j). \end{cases}$$

We note that $t_k > 0$, $\sum_{k \in K} t_k = 1$ and $\sum_{k \in K} t_k z_k = \alpha x + \beta y$, hence, if all $f(x_i) < +\infty$ and all $f(y_j) < +\infty$, we have

$$\Phi(\alpha x + \beta y) \leq \sum_{k \in K} t_k f(z_k) = \alpha \sum_{i \in I} f(x_i) + \beta \sum_{j \in J} f(y_j).$$

Then $\Phi(\alpha x + \beta y) \leq \alpha \sum_{i \in I} f(x_i) + \beta \sum_{j \in J} f(y_j)$ in any case, since the inequality is evident if $f(x_{i_0}) = +\infty$ for some $i_0 \in I$ or $f(y_{j_0}) = +\infty$ for some $j_0 \in J$. Taking appropriately the infimum over $(\lambda_i)_{i \in I}$ and $(x_i)_{i \in I}$, and then over $(\mu_j)_{j \in J}$ and $(y_j)_{j \in J}$ yields
$$\Phi(\alpha x + \beta y) \leq \alpha \Phi(x) + \beta \Phi(y),$$
which justifies the convexity of Φ.

The functions Φ and φ being convex and majorized by f, it results from the definition of g that $\Phi \leq g$ and $\varphi \leq g$. Combining this with the previous inequalities $g \leq \Phi$ and $g \leq \varphi$, we conclude that $\Phi = g$ and $\varphi = g$. □

The inclusion in the assertion (a) of Proposition 3.72 may be strict as shown in the following example.

EXAMPLE 3.73. Consider the function $f : \mathbb{R} \to \mathbb{R}$ defined by $f(0) = 1$ and $f(x) = |x|$ if $x \neq 0$, we see that $\operatorname{co} f = |\cdot|$ on the entire real line $\mathbb{R}$. Then $\operatorname{epi}(\operatorname{co} f) = \operatorname{epi}(|\cdot|)$ whereas
$$\operatorname{co}(\operatorname{epi} f) = \big(\operatorname{epi}(|\cdot|)\big) \setminus \{(0,0)\},$$
so $\operatorname{co}(\operatorname{epi} f) \neq \operatorname{epi}(\operatorname{co} f)$. □

PROPOSITION 3.74. Let X be a Hausdorff locally convex space and $f : X \to \mathbb{R} \cup \{-\infty, +\infty\}$ be a function on X. Then one has

(3.53) $$\operatorname{epi}(\overline{\operatorname{co}} f) = \overline{\operatorname{co}}(\operatorname{epi} f) = \operatorname{cl}\big(\operatorname{epi}(\operatorname{co} f)\big).$$

PROOF. From the inequalities $\overline{\operatorname{co}} f \leq \operatorname{co} f \leq f$ we have

$\operatorname{epi} f \subset \operatorname{epi}(\operatorname{co} f) \subset \operatorname{epi}(\overline{\operatorname{co}} f)$, hence $\overline{\operatorname{co}}(\operatorname{epi} f) \subset \operatorname{cl}\big(\operatorname{epi}(\operatorname{co} f)\big) \subset \operatorname{epi}(\overline{\operatorname{co}} f)$,

the latter inclusion being due to the closedness property of the set $\operatorname{epi}(\overline{\operatorname{co}} f)$.

It remains to prove that $\operatorname{epi}(\overline{\operatorname{co}} f) \subset \overline{\operatorname{co}}(\operatorname{epi} f)$. To do so, let us first verify that $\overline{\operatorname{co}}(\operatorname{epi} f)$ is the epigraph of a function. Take any $(x, r) \in \operatorname{co}(\operatorname{epi} f)$ and any real $s \geq r$. If $(x, r) \in \operatorname{epi} f$, then $f(x) \leq r \leq s$, so $(x, s) \in \operatorname{epi} f$. Suppose $(x, r) \notin \operatorname{epi} f$. There are an integer $m \geq 2$, an m-uple $(\lambda_i)_{i=1}^m$ in $]0, +\infty[$ with $\sum_{i=1}^m \lambda_i = 1$, an m-uple $(x_i, r_i)_{i=1}^m$ in $\operatorname{epi} f$ with $\sum_{i=1}^m \lambda_i (x_i, r_i) = (x, r)$. Put $s_m := (s - r + \lambda_m r_m)/\lambda_m$ and $s_i := r_i$ for $i = 1, \cdots, m-1$. Clearly $s_i \geq f(x_i)$ for every $i = 1, \cdots, m-1$, and $s_m \geq r_m \geq f(x_m)$ since $s - r \geq 0$. Further, we note that $(x, s) = \sum_{i=1}^m \lambda_i (x_i, s_i)$, hence $(x, s) \in \operatorname{co}(\operatorname{epi} f)$. Therefore, if $(x, r) \in \operatorname{co}(\operatorname{epi} f)$, then $(x, s) \in \operatorname{co}(\operatorname{epi} f)$ for every real $s \geq r$. Consider now $(x, r) \in \overline{\operatorname{co}}(\operatorname{epi} f)$ and any real $s > r$. Choose a sequence $(x_n, r_n)_n$ in $\operatorname{co}(\operatorname{epi} f)$ converging to (x, r) and a sequence of reals $(s_n)_n$ tending to s. For n large enough we have $s_n > r_n$, hence by what precedes $(x_n, s_n) \in \operatorname{co}(\operatorname{epi} f)$, which gives $(x, s) \in \overline{\operatorname{co}}(\operatorname{epi} f)$. We deduce that $(x, r) \in \overline{\operatorname{co}}(\operatorname{epi} f)$ entails that $(x, s) \in \overline{\operatorname{co}}(\operatorname{epi} f)$ for every real $s \geq r$. This property yields that the closed set $\overline{\operatorname{co}}(\operatorname{epi} f)$ is the epigraph of the function $g : X \to \mathbb{R} \cup \{-\infty, +\infty\}$ defined by

$$g(x) = \inf\{r \in \mathbb{R} : (x, r) \in \overline{\operatorname{co}}(\operatorname{epi} f)\},$$

with the usual fact that the infimum of the empty set is $+\infty$. Since $\operatorname{epi} f \subset \overline{\operatorname{co}}(\operatorname{epi} f) = \operatorname{epi} g$, the lower semicontinuous convex function g satisfies the inequality

$g \leq f$, thus $g \leq \overline{\text{co}} f$ (by definition of $\overline{\text{co}} f$). It results that
$$\text{epi}\,(\overline{\text{co}}\, f) \subset \text{epi}\, g = \overline{\text{co}}(\text{epi}\, f),$$
which finishes the proof of the proposition. □

Regarding the Legendre-Fenchel conjugate of convex hulls of functions we have the following equalities.

PROPOSITION 3.75. *Let (X, τ_X) be a Hausdorff locally convex space and $f : X \to \mathbb{R} \cup \{-\infty, +\infty\}$ be any extended real-valued function. Then one has*
$$(\text{co}\, f)^* = (\overline{\text{co}}\, f)^* = (\text{cl}\, f)^* = f^*.$$

PROOF. Notice first the following equivalences
$$(x^*, s) \in \text{epi}\, f^* \Leftrightarrow s \geq \langle x^*, x \rangle - f(x),\ \forall x \in X$$
$$\Leftrightarrow \langle x^*, \cdot \rangle - s \leq f(\cdot)$$
$$\Leftrightarrow \langle x^*, \cdot \rangle - s \leq (\overline{\text{co}}\, f)(\cdot)$$
$$\Leftrightarrow s \geq \langle x^*, x \rangle - (\overline{\text{co}}\, f)(x),\ \forall x \in X$$
$$\Leftrightarrow (x^*, s) \in \text{epi}\,(\overline{\text{co}}\, f)^*.$$
It ensues that $(\overline{\text{co}}\, f)^* = f^*$. This equality and the inequalities in (3.52) yield the desired equalities of the proposition. □

The next proposition establishes another useful result concerning now the subdifferential of the closed convex hull.

PROPOSITION 3.76. *Let X be a Hausdorff locally convex space and $f : X \to \mathbb{R} \cup \{-\infty, +\infty\}$ be a function which is finite at $\overline{x}$. If $\partial_{MR} f(\overline{x}) \neq \emptyset$, that is, there exists some $\overline{x}^* \in X^*$ such that*
$$\langle \overline{x}^*, \cdot \rangle - f(\cdot) \leq \langle \overline{x}^*, \overline{x} \rangle - f(\overline{x}),$$
then $(\overline{\text{co}}\, f)(\overline{x}) = (\text{co}\, f)(\overline{x}) = f(\overline{x})$ and
$$\partial(\text{co}\, f)(\overline{x}) = \partial(\overline{\text{co}}\, f)(\overline{x}) = \partial_{MR} f(\overline{x}) := \{x^* \in X^* : \langle x^*, \cdot \rangle - f(\cdot) \leq \langle x^*, \overline{x} \rangle - f(\overline{x})\}.$$

PROOF. By assumption the continuous convex function $g := \langle \overline{x}^*, \cdot - \overline{x} \rangle + f(\overline{x})$ is majorized by f, hence $g \leq \overline{\text{co}} f \leq \text{co}\, f \leq f$. In particular, we deduce that
$$f(\overline{x}) = g(\overline{x}) \leq (\overline{\text{co}}\, f)(\overline{x}) \leq (\text{co}\, f)(\overline{x}) \leq f(\overline{x}),$$
thus $(\overline{\text{co}}\, f)(\overline{x}) = (\text{co}\, f)(\overline{x}) = f(\overline{x})$.

Now, putting $Q := \{x^* \in X^* : \langle x^*, \cdot \rangle - f(\cdot) \leq \langle x^*, \overline{x} \rangle - f(\overline{x})\}$ we know (see Definition 3.2) that $\partial_{MR} f(\overline{x}) = Q$. For any $x^* \in \partial(\text{co}\, f)(\overline{x})$, the above equality $(\text{co}\, f)(\overline{x}) = f(\overline{x})$ yields for all $x \in X$
$$\langle x^*, x - \overline{x} \rangle \leq (\text{co}\, f)(x) - (\text{co}\, f)(\overline{x}) \leq f(x) - f(\overline{x}),$$
so $x^* \in Q$, and hence $\partial(\text{co}\, f)(\overline{x}) \subset Q$. Similarly, $\partial(\overline{\text{co}}\, f)(\overline{x}) \subset \partial(\text{co}\, f)(\overline{x})$. It remains to prove that $Q \subset \partial(\overline{\text{co}}\, f)(\overline{x})$. Fix any $x^* \in Q$. The reasoning at the beginning of the proof with x^* in place of $\overline{x}^*$ gives for all $x \in X$
$$\langle x^*, x - \overline{x} \rangle + f(\overline{x}) \leq (\overline{\text{co}}\, f)(x), \text{ or equivalently } \langle x^*, x - \overline{x} \rangle \leq (\overline{\text{co}}\, f)(x) - (\overline{\text{co}}\, f)(\overline{x}).$$
This means that $x^* \in \partial(\overline{\text{co}}\, f)(\overline{x})$, and the proof is finished. □

3.3.2. Pointwise supremum of continuous affine functions.
Proposition 3.64(h) characterizes the properness of f^* via the existence of a continuous affine function minorizing f. The existence of such an affine function is given below. More generally, the proposition characterizes proper functions which are pointwise supremum of continuous affine functions.

PROPOSITION 3.77. *Let (X, τ_X) be a Hausdorff locally convex space and $f : X \to \mathbb{R} \cup \{-\infty, +\infty\}$ be a proper lower τ_X-semicontinuous convex function. The following hold.*
(a) *The function f is the pointwise supremum of all τ_X-continuous affine functions majorized by f.*
(b) *The function f is the pointwise supremum of all finite-valued τ_X-continuous convex functions majorized by f.*
(c) *The conjugate function f^* is a proper lower w^*-semicontinuous convex function.*

PROOF. Fix any $b \in X$ and any real number $\beta < f(b)$ (such numbers exist since $f(b) > -\infty$). Since $(b, \beta) \notin \operatorname{epi} f$ and $\operatorname{epi} f$ is closed and convex, there exists by the Hahn-Banach separation theorem some pair $(b^*, s) \in X^* \times \mathbb{R}$ and some $\gamma \in \mathbb{R}$ such that for all $(x, r) \in \operatorname{epi} f$

$$\langle b^*, x \rangle + sr < \gamma < \langle b^*, b \rangle + s\beta. \tag{3.54}$$

Choosing $b_0 \in \operatorname{dom} f$ (since f is proper), we obtain with $x = b_0$ that for any real $\rho > \max\{f(b_0), 0\}$ we have $s < \rho^{-1}(\gamma - \langle b^*, b_0 \rangle)$, and hence making $\rho \to +\infty$ we get $s \leq 0$. We have to analyze two cases:

Case 1: $s < 0$.
In this case for $a^* := -s^{-1} b^*$ and $\alpha := -s^{-1} \gamma$, we deduce from (3.54) that for all $(x, r) \in \operatorname{epi} f$

$$\langle a^*, x \rangle - r < \alpha < \langle a^*, b \rangle - \beta, \tag{3.55}$$

which entails in particular

$$\beta < \langle a^*, b \rangle - \alpha \quad \text{and} \quad \langle a^*, x \rangle - \alpha \leq f(x) \text{ for all } x \in X. \tag{3.56}$$

Case 2: $s = 0$.
In this second case (3.54) is reduced to

$$\langle b^*, x \rangle < \gamma < \langle b^*, b \rangle \quad \text{for all } x \in \operatorname{dom} f. \tag{3.57}$$

Choose some real $\beta_0 < f(b_0)$, some $(b_0^*, s_0) \in X^* \times \mathbb{R}$ and some $\gamma_0 \in \mathbb{R}$ such that, like for (3.54), we have for all $(x, r) \in \operatorname{epi} f$

$$\langle b_0^*, x \rangle + s_0 r < \gamma_0 < \langle b_0^*, b_0 \rangle + s_0 \beta_0.$$

Taking $(x, r) = (b_0, f(b_0))$ we see that $s_0 < 0$, and hence according to the first case there is some $(a_0^*, \alpha_0) \in X^* \times \mathbb{R}$ such that

$$\langle a_0^*, x \rangle - \alpha_0 \leq f(x) \quad \text{for all } x \in X.$$

Combining this with the left inequality in (3.57) we deduce for any $x \in X$ and any real $t > 0$

$$\langle a_0^*, x \rangle - \alpha_0 + t(\langle b^*, x \rangle - \gamma) \leq f(x). \tag{3.58}$$

Since $\langle b^*, b \rangle - \gamma > 0$ according to the right inequality in (3.57), we may choose some real $t_0 > 0$ such that

$$\langle a_0^*, b \rangle - \alpha_0 + t_0(\langle b^*, b \rangle - \gamma) > \beta.$$

This inequality and (3.58) entail, for $a^* := a_0^* + t_0 b^*$ and $\alpha := \alpha_0 + t_0 \gamma$,
$$\beta < \langle a^*, b \rangle - \alpha \quad \text{and} \quad \langle a^*, x \rangle - \alpha \leq f(x) \text{ for all } x \in X.$$

Consequently, in any case for any real $\beta < f(b)$ we obtain the continuous affine function $\langle a^*, \cdot \rangle - \alpha$ satisfying (3.56). This implies that $f(b) = \sup\{\varphi(b) : \varphi \in \mathcal{A}(f)\}$, where $\mathcal{A}(f)$ denotes the set of all continuous affine functions majorized by f. This is the assertion (a).

The assertion (b) is a direct consequence of (a).

Finally, (c) follows from the above assertion (a) and from the assertions (g) and (h) in Proposition 3.64. □

COROLLARY 3.78. *Let X be a Hausdorff locally convex space and let $f : X \to \{-\infty, +\infty\}$ be a proper lower semicontinuous positively homogeneous convex function. Then f is the pointwise supremum of all continuous linear functionals majorized by f.*

PROOF. By Corollary 3.24 we know that $f(0) = 0$ and by Proposition 3.77(a) there is a (nonempty) family $(\langle a_i^*, \cdot \rangle + \beta_i)_{i \in I}$ of continuous affine functions such that $f(x) = \sup_{i \in I}(\langle a_i^*, x \rangle + \beta_i)$. Clearly $\sup_{i \in I} \beta_i = f(0) = 0$, so $\beta_i \leq 0$ for all $i \in I$, which implies that $f(x) \leq \sup_{i \in I} \langle a_i^*, x \rangle$ for all $x \in X$. On the other hand, let any $x \in \operatorname{dom} f$ and any $i \in I$. For every $n \in \mathbb{N}$ we have $\langle a_i^*, nx \rangle + \beta_i \leq f(nx)$, or equivalently, $\langle a_i^*, x \rangle + (\beta_i/n) \leq f(x)$, hence $\langle a_i^*, x \rangle \leq f(x)$. We then conclude that $f(x) = \sup_{i \in I} \langle a_i^*, x \rangle$ for all $x \in X$. □

COROLLARY 3.79. *Let X be a Hausdorff locally convex space and let $f : X \to \mathbb{R} \cup \{-\infty, +\infty\}$ be a convex function which is finite at $\overline{x} \in X$. Then $\partial f(\overline{x}) \neq \emptyset$ if and only if $\underline{d}_H^+ f(\overline{x}; 0) = 0$; in such a case $\underline{d}_H^+ f(\overline{x}; \cdot)$ coincides with the support function relative to X of the set $\partial f(\overline{x})$.*

PROOF. By (3.5) we know that $x^* \in \partial f(\overline{x})$ if and only if $\langle x^*, \cdot \rangle \leq \underline{d}_H^+ f(\overline{x}; \cdot)$. If $\partial f(\overline{x}) \neq \emptyset$, we then see that $0 \leq \underline{d}_H^+ f(\overline{x}; 0)$, hence $\underline{d}_H^+ f(\overline{x}; 0) = 0$ since $\underline{d}_H^+ f(\overline{x}; 0) \leq 0$ by its very definition. Conversely, suppose that $\underline{d}_H^+ f(\overline{x}; 0) = 0$. By Proposition 3.23 the positively homogeneous convex function $\underline{d}_H^+ f(\overline{x}; \cdot)$ is proper, hence Corollary 3.78 above implies that the set $\partial f(\overline{x})$ is nonempty and its support function coincides with $\underline{d}_H^+ f(\overline{x}; \cdot)$. □

REMARK 3.80. Under notation of Corollary 3.79, it is worth noticing that, when $\underline{d}_H^+ f(\overline{x}; 0) \neq 0$, one has $\underline{d}_H^+ f(\overline{x}; 0) = -\infty$, hence $\underline{d}_H^+ f(\overline{x}; h) \in \{-\infty, +\infty\}$ since Proposition 3.23 says that any lower semicontinuous convex function $\varphi : X \to \mathbb{R} \cup \{-\infty, +\infty\}$ finite at some point does not take the value $-\infty$. □

Given an extended real-valued function f on an open set of a normed space X which is finite at $\overline{x}$, we saw in Theorem 2.72 that the Rockafellar directional derivative $f^\uparrow(\overline{x}; \cdot)$ is lower semicontinuous, convex and positively homogeneous with $f^\uparrow(\overline{x}; 0) \in \{0, -\infty\}$, and $x^* \in \partial_C f(\overline{x})$ if and only if $\langle x^*, h \rangle \leq f^\uparrow(\overline{x}; h)$ for all $h \in X$. Then repeating the arguments in Corollary 3.79 we obtain:

COROLLARY 3.81. *Let $f : U \to \mathbb{R} \cup \{-\infty, +\infty\}$ be an extended real-valued function on an open set U of a normed space X and let $\overline{x} \in U$ be a point where f is finite. Then $\partial_C f(\overline{x}) \neq \emptyset$ if and only if $f^\uparrow(\overline{x}; 0) = 0$; in such a case $f^\uparrow(\overline{x}; \cdot)$ coincides with the support function relative to X of the set $\partial_C f(\overline{x})$, and $f^\uparrow(\overline{x}; \cdot) > -\infty$ along with $\partial_C f(\overline{x}) = \partial f^\uparrow(\overline{x}; \cdot)(0)$.*

REMARK 3.82. Similarly to Remark 3.80, if $f^\uparrow(\bar{x};0) \neq 0$, then the inclusion $f^\uparrow(\bar{x};\cdot) \in \{-\infty, +\infty\}$ holds along with the equality along with $f^\uparrow(\bar{x};0) = -\infty$. □

We also have the following corollary providing two useful characterizations of support functions.

COROLLARY 3.83. *Let X be a Hausdorff locally convex space and let $\varphi : X^* \to \mathbb{R} \cup \{-\infty, +\infty\}$ be an extended real-valued function on the topological dual X^*. The following are equivalent.*
(a) *The function φ is the support function of a nonempty closed convex set in X;*
(b) *the function φ is positively homogeneous, convex, lower $w(X^*, X)$ semicontinuous and $\varphi(0) = 0$;*
(c) *the function φ is proper, positively homogeneous, convex and lower $w(X^*, X)$ semicontinuous.*

PROOF. Endow X^* with the $w(X^*, X)$ topology, so the topological dual of $(X^*, w(X^*, X))$ is identified with X.

The implication (a)⇒(b) is evident and the implication (b)⇒(c) follows from Corollary 3.24. Now suppose that φ is proper, positively homogeneous and $w(X^*, X)$ lower semicontinuous. Denote by C the set of $x \in X$ such that $\langle \cdot, x \rangle$ is majorized by φ. The set C is closed and convex, and Corollary 3.78 entails that the set C is nonempty and φ coincides with its support function. □

3.3.3. Biconjugate and Fenchel-Moreau theorem.

Endowing X^* with the topology $\beta(X^*, X)$ of uniform convergence on bounded subsets of a Hausdorff locally convex space (X, τ_X), the topological dual of $(X^*, \beta(X^*, X))$ is generally denoted by X^{**} and called the topological bidual of (X, τ_X). Given a function $f : X \to \mathbb{R} \cup \{-\infty, +\infty\}$, the Legendre-Fenchel conjugate over X^{**} of the function $f^* : X^* \to \mathbb{R} \cup \{-\infty, +\infty\}$ will be called the *Legendre-Fenchel biconjugate relative to X^{**} of f*, and it will be denoted by f^{**}. This can be translated by $f^{**} : X^{**} \to \mathbb{R} \cup \{-\infty, +\infty\}$ with

$$(3.59) \qquad f^{**}(x^{**}) = \sup_{x^* \in X^*} \left(\langle x^{**}, x^* \rangle_{X^{**}, X^*} - f^*(x^*) \right) \quad \text{for all } x^{**} \in X^{**}.$$

Now, equip X^* with any locally convex topology for which the topological dual of X^* is identified with X through the mapping assigning to each $x \in X$ the continuous linear function $x^* \mapsto \langle x^*, x \rangle$ on X^*; for example, the weak* topology $w(X^*, X)$ is known to enjoy such a property and it is the weakest locally convex topology on X^* with the property. Doing so, we then consider the Legendre-Fenchel conjugate of f^* relative to X, which is given over X by

$$(3.60) \qquad x \mapsto \sup_{x^* \in X^*} \left(\langle x^*, x \rangle_{X^*, X} - f^*(x^*) \right);$$

this is the biconjugate relative to X of f. Invoking the mapping $j : X \to X^{**}$ with $j(x) := \langle \cdot, x \rangle \in X^{**}$, so $\langle j(x), x^* \rangle_{X^{**}, X^*} = \langle x^*, x \rangle_{X^*, X}$, we see that the function in (3.60) coincides with the restriction to X of the function f^{**} in (3.59). This means that the biconjugate relative to X is the restriction to X of the biconjugate relative to X^{**}.

The above analysis also reveals that given a convex function $\varphi : X^* \to \mathbb{R} \cup \{-\infty, +\infty\}$ which is finite at $x^* \in X^*$ we have to distinguish at the point x^* its subdifferential in X^{**} defined as

$$(3.61) \qquad \partial_{X^{**}} \varphi(x^*) := \{ x^{**} \in X^{**} : \langle x^{**}, u^* - x^* \rangle \leq \varphi(u^*) - \varphi(x^*), \forall u^* \in X^* \}$$

from its subdifferential in X defined as

(3.62) $\quad \partial_X \varphi(x^*) := \{x \in X : \langle u^* - x^*, x \rangle \leq \varphi(u^*) - \varphi(x^*), \forall u^* \in X^*\}.$

When φ is not finite at x^*, both $\partial_{X^{**}}\varphi(x^*)$ and $\partial_X\varphi(x^*)$ are defined as the empty set. Through the usual identification we see that

(3.63) $\quad \partial_X \varphi(x^*) = X \cap \partial_{X^{**}} \varphi(x^*) \quad \text{for all } x^* \in X^*.$

The ε-subdifferential of φ at x^* in X^{**} and in X are defined similarly. Sometimes, when there is no ambiguity on the topology taken on X^* we will omit the subscript to merely write $\partial \varphi(x^*)$.

In the particular case when $(X, \|\cdot\|)$ is a normed space, we know that $\beta(X^*, X)$ is the topology over X^* associated with the dual norm $\|\cdot\|_*$, given by $\|x^*\|_* := \sup_{\|x\|\leq 1} |\langle x^*, x\rangle|$. So, the biconjugate in (3.59) and the subdifferential in (3.61) are relative to the usual topological bidual normed space X^{**} of X.

Concerning the biconjugate relative to X, we first note that the use of the Fenchel inequality $\langle x^*, x \rangle - f(x) \leq f^*(x^*)$ in (3.60) yields that, for any function $f : X \to \mathbb{R} \cup \{-\infty, +\infty\}$, one always has

(3.64) $\quad f^{**}(x) \leq f(x) \quad \text{for all } x \in X.$

The next theorem shows that the equality holds whenever f is proper, lower semicontinuous and convex.

THEOREM 3.84 (W. Fenchel; J.J. Moreau: theorem for biconjugate). Let (X, τ_X) be a Hausdorff locally convex space and $f : X \to \mathbb{R} \cup \{-\infty, +\infty\}$ be an extended real-valued function. Then the following hold.
(a) The biconjugate relative to X of f is the pointwise supremum of all τ_X-continuous affine functions on X majorized by f.
(b) The biconjugate relative to X of f is also the pointwise supremum of all proper lower τ_X-semicontinuous convex functions on X majorized by f.
(c) If f is proper, lower τ_X-semicontinuous and convex, then f coincides with its biconjugate relative to X, that is,

$$f^{**}(x) = f(x) \quad \text{for all } x \in X.$$

PROOF. (a) Considering any continuous affine function φ on X with $\varphi \leq f$ (if any), when we take twice the conjugate we obtain that $\varphi^{**}(x) \leq f^{**}(x)$ for all $x \in X$. By (3.28), (3.34) and (3.33) we have $\varphi^{**}(x) = \varphi(x)$ for all $x \in X$. We deduce that $\varphi(x) \leq f^{**}(x)$ for all $x \in X$. This along with the equality

(3.65) $\quad f^{**}(x) = \sup_{x^* \in \operatorname{dom} f^*} \left(\langle x^*, x \rangle - f^*(x^*) \right) \quad \text{for all } x \in X,$

ensures the assertion (a) whenever f^* is proper.

In the alternative case where f^* is not proper, then according to assertion (g) in Proposition 3.64 either $f^* \equiv -\infty$ or $f^* \equiv +\infty$. If $f^* \equiv -\infty$ (resp. $f^* \equiv +\infty$), the assertion (a) follows from (3.65) and (g) (resp. (f)) in Proposition 3.64.
(b) and (c). Each one of assertions (b) and (c) follows from the above assertion (a) and from Proposition 3.77(a). $\square$

Taking Proposition 3.64(j) into account, we directly derive from Theorem 3.84(c) the following duality property for ε-subdifferential.

COROLLARY 3.85. Let $f : X \to \mathbb{R} \cup \{+\infty\}$ be a proper lower semicontinuous convex function on a Hausdorff locally convex space X and let a real $\varepsilon \geq 0$. Then one has
$$x^* \in \partial_\varepsilon f(x) \Leftrightarrow x \in \partial_\varepsilon f^*(x^*).$$

Through Theorem 3.84 we can also show some features of the biconjugate relative to X^{**}.

COROLLARY 3.86. Let X be a Hausdorff locally convex space and $f : X \to \mathbb{R} \cup \{-\infty, +\infty\}$ be an extended real-valued function. The following hold.
(a) The biconjugate f^{**} relative to X^{**} of f is the pointwise supremum on X^{**} of all $w(X^{**}, X^*)$-continuous affine functions on X^{**} majorized by f on X.
(b) The biconjugate f^{**} relative to X^{**} of f is the pointwise supremum on X^{**} of all finite-valued $w(X^{**}, X^*)$-continuous convex functions on X^{**} majorized by f on X.
(c) The biconjugate f^{**} relative to X^{**} of f is the pointwise supremum on X^{**} of all proper lower $w(X^{**}, X^*)$-semicontinuous convex functions on X^{**} majorized by f on X.
(d) The biconjugate f^{**} relative to X of f is the largest lower $w(X^{**}, X^*)$ semicontinuous convex function on X^{**} majorized by f on X.
(e) If f is finite-valued, convex and continuous, then its biconjugate f^{**} relative to X^{**} is the largest lower $w(X^{**}, X^*)$-semicontinuous convex extension of f to X^{**}.

PROOF. (a) Let $g : X^{**} \to \mathbb{R} \cup \{-\infty, +\infty\}$ be defined by $g(x) = f(x)$ for all $x \in X$ and $g(x^{**}) = +\infty$ for all $x^{**} \in X^{**} \setminus X$. Endowing X^{**} with the $w(X^{**}, X^*)$-topology and X^* with the $w(X^*, X^{**})$-topology we obtain, according to Theorem 3.84(a), that the biconjugate of g relative to X^{**} is the pointwise supremum of all $w(X^{**}, X^*)$-continuous affine functions on X^{**} majorized by g on X^{**}, or equivalently majorized by f on X according to the definition of g. Observe that $g^* : X^* \to \mathbb{R} \cup \{-\infty, +\infty\}$ is equal to f^* by the definition of g again and note that the topological duals of $(X^*, \beta(X^*, X))$ and $(X^*, w(X^*, X^{**}))$ both coincide with X^{**}. Then endowing X^* with the $w(X^*, X^{**})$-topology and taking the conjugates of $g^* : (X^*, w(X^*, X^{**})) \to \mathbb{R} \cup \{-\infty, +\infty\}$ and $f^* : (X^*, w(X^*, X^{**})) \to \mathbb{R} \cup \{-\infty, +\infty\}$, we obtain that $f^{**} : X^{**} \to \mathbb{R} \cup \{-\infty, +\infty\}$ coincides with the biconjugate of g relative to X^{**}, and hence is the pointwise supremum of all $w(X^{**}, X^*)$-continuous affine functions on X^{**} majorized by g on X^{**}, or equivalently majorized by f on X.
(b) and (c). Denoting by $s_i(\cdot)$ the pointwise supremum in (i), for $i = a, b, c$, we obviously have $s_a \leq s_b \leq s_c$. Further, any proper lower $w(X^{**}, X^*)$-semicontinuous convex function φ on X^{**} with $\varphi(x) \leq f(x)$ for all $x \in X$ is, by Proposition 3.77(a), the pointwise supremum of a collection of $w(X^{**}, X^*)$-continuous affine functions majorized by φ on X^{**}, and hence majorized by f on X. So, $s_c \leq s_a$, and thus (b) and (c) hold true according to (a).
(d) The assertion (d) follows from (c).
(e) The assertion (e) follows from (d) above and from Theorem 3.84(c). □

The biconjugate of a function is connected to its closed convex hull as follows.

PROPOSITION 3.87. Let X be a Hausdorff locally convex space and $f : X \to \mathbb{R} \cup \{-\infty, +\infty\}$ be any extended real-valued function with $\operatorname{dom} f \neq \emptyset$.
(a) If f is bounded from below by a continuous affine function (or equivalently, $\overline{\operatorname{co}} f$

is proper), then $\overline{\mathrm{co}}\, f$ coincides with the biconjugate of f relative to X, that is,
$$f^{**}(x) = (\overline{\mathrm{co}}\, f)(x) \quad \text{for all } x \in X.$$
(b) If there is no continuous affine function minorizing f (or equivalently, $\overline{\mathrm{co}}\, f$ is not proper), then $f^{**}(x^{**}) = -\infty$ for all $x^{**} \in X^{**}$.

PROOF. (a) If $\overline{\mathrm{co}}\, f$ is proper, then $\overline{\mathrm{co}}\, f$ is a proper lower semicontinuous convex function, so by Theorem 3.84(c) and Proposition 3.75
$$(\overline{\mathrm{co}}\, f)(x) = (\overline{\mathrm{co}}\, f)^{**}(x) = f^{**}(x) \quad \text{for all } x \in X.$$
(b) Now suppose that $\overline{\mathrm{co}}\, f$ is not proper. Noting by the inequality $\overline{\mathrm{co}}\, f \leq f$ that $(\overline{\mathrm{co}}\, f)(a) < +\infty$ for some $a \in X$ (keep in mind that $\mathrm{dom}\, f \neq \emptyset$ by assumption), we deduce from Proposition 3.23 that $\overline{\mathrm{co}}\, f(a) = -\infty$. Consequently, $(\overline{\mathrm{co}}\, f)^* \equiv +\infty$, hence $f^* \equiv +\infty$ according to Proposition 3.75. This entails that $f^{**}(x^{**}) = -\infty$ for all $x^{**} \in X^{**}$. $\square$

We provide now a first context where the biconjugate relative to X of a convex function coincides with its closure.

PROPOSITION 3.88. Let X be a Hausdorff locally convex space and $f : X \to \mathbb{R} \cup \{-\infty, +\infty\}$ be a convex function.
(a) If f is lower semicontinuous at a point $\bar{x} \in \mathrm{dom}\, f$, then $f^{**}(\bar{x}) = f(\bar{x})$.
(b) If f is lower semicontinuous at some point $\bar{x} \in X$ with $|f(\bar{x})| < +\infty$, then the biconjugate of f relative to X coincides with its closure, that is,
$$f^{**}(x) = (\mathrm{cl}\, f)(x) \quad \text{for all } x \in X;$$
furthermore, $\mathrm{cl}\, f$ is proper.

PROOF. The function f being convex, we have $\overline{\mathrm{co}}\, f = \mathrm{cl}\, f$, hence in particular $(\overline{\mathrm{co}}\, f)(\bar{x}) = (\mathrm{cl}\, f)(\bar{x}) = f(\bar{x})$, according to the lower semicontinuity of f at $\bar{x}$ (in both cases (a) and (b)).
(b) Assume that $|f(\bar{x})| < +\infty$. Then by Proposition 3.23 the function $\overline{\mathrm{co}}\, f$ is proper. From Proposition 3.87(a) we also have
$$f^{**}(x) = (\overline{\mathrm{co}}\, f)(x) = (\mathrm{cl}\, f)(x) \quad \text{for all } x \in X,$$
so (b) is justified.
(a) Now assume that $f(\bar{x}) < +\infty$. If $f(\bar{x})$ is finite, the equality $f^{**}(\bar{x}) = f(\bar{x})$ follows from (b) above; and if otherwise $f(\bar{x}) = -\infty$, then the inequality $f^{**}(\bar{x}) \leq f(\bar{x})$ gives $f^{**}(\bar{x}) = f(\bar{x})$. This completes the proof. $\square$

We also give another situation when the biconjugate of a convex function coincides with its closure hull.

PROPOSITION 3.89. Let X be a Hausdorff ocally convex space and $f : X \to \mathbb{R} \cup \{-\infty, +\infty\}$ be a convex function minorized by a finite-valued lower semicontinuous function. Then the biconjugate of f relative to X coincides with $\mathrm{cl}\, f$, that is,
$$f^{**}(x) = \liminf_{u \to x} f(u) \quad \text{for all } x \in X.$$

PROOF. Since f is minorized by a finite-valued lower semicontinuous function, its closure $\mathrm{cl}\, f$ takes its values in $]-\infty, +\infty]$, and hence by Proposition 3.23 either $\mathrm{cl}\, f \equiv +\infty$ or $\mathrm{cl}\, f$ is a proper lower semicontinuous convex function. In the second case, by Theorem 3.84(c)

(3.66) $\qquad (\mathrm{cl}\, f)(x) = (\mathrm{cl}\, f)^{**}(x) \quad \text{for all } x \in X,$

and this equality still holds when $\operatorname{cl} f \equiv +\infty$ since $(\operatorname{cl} f)^* \equiv -\infty$ in this other case. Furthermore, the equality $(\operatorname{cl} f)^* = f^*$ in Proposition 3.75 yields $(\operatorname{cl} f)^{**} = f^{**}$. Putting this and the equality $\operatorname{cl} f(x) = \liminf_{u \to x} f(u)$ into (3.66) we obtain the desired result. □

In finite dimensions the following corollary holds.

COROLLARY 3.90. *Assume that the Hausdorff locally convex space X is finite-dimensional and f is a convex function on X with values in $\mathbb{R} \cup \{+\infty\}$. Then one has $f^{**} = \operatorname{cl} f$, that is,*

$$f^{**}(x) = \liminf_{u \to x} f(u) \quad \text{for all } x \in X.$$

PROOF. We may suppose $\operatorname{dom} f \neq \emptyset$ and $0 \in \operatorname{dom} f$. Denote $E := \operatorname{Vect}(\operatorname{dom} f)$ the vector subspace spanned by $\operatorname{dom} f$ and $f_{|E}$ the restriction of f to E. Choosing $x_0 \in \operatorname{rint}(\operatorname{dom} f) = \operatorname{int}_E(\operatorname{dom} f)$, the convex function $f_{|E}$ is continuous at x_0, hence $\partial(f_{|E})(x_0) \neq \emptyset$. From this, it is easily seen that f is minorized on X by an affine function, thus the corollary follows from Proposition 3.89 above. □

REMARK 3.91. If the Hausdorff locally convex space X is infinite-dimensional with X^* different from the algebraic dual of X, the above corollary fails. Indeed, let ℓ be a non-continuous linear functional on X and C be a closed convex subset with $\operatorname{int} C \neq \emptyset$ and $C \neq X$. The convex function $f := \ell + \Psi_C$ takes its values in $\mathbb{R} \cup \{+\infty\}$. Taking the biconjugate f^{**} relative to X, by Theorem 3.84(a) $f^{**} \equiv -\infty$ (since there is no continuous affine function minorizing f), while (as easily seen) $(\operatorname{cl} f)(x) = -\infty$ for all $x \in C$ and $(\operatorname{cl} f)(x) = +\infty$ for all $x \in X \setminus C$. □

As a second corollary from Proposition 3.89, we can derive a general result describing the Legendre-Fenchel conjugate of a sum of functions by means of the data functions. The result shows that the conjugate of a sum coincides with the closure of the infimal convolution of the conjugates.

COROLLARY 3.92. *Let X be a Hausdorff locally convex space and $f, g : X \to \mathbb{R} \cup \{+\infty\}$ be two proper lower semicontinuous convex functions with $\operatorname{dom} f \cap \operatorname{dom} g \neq \emptyset$. Then one has the equality*

$$(f + g)^* = \operatorname{cl}_{w^*}(f^* \square g^*),$$

where the closure is taken with respect to the $w(X^*, X)$ topology (as indicated with cl_{w^*}).

PROOF. Note that $f + g$ is a proper lower semicontinuous convex function, thus f^*, g^* and $(f + g)^*$ are proper lower w^*-semicontinuous convex functions (see Proposition 3.77(c)). Endow X^* with the $w(X^*, X)$ topology, so X^{**} coincides with X and f, g coincide with their biconjugates according to Fenchel-Moreau theorem (see Theorem 3.84(c)). Then the equality (3.48) yields

(3.67) $\qquad f + g = (f^* \square g^*)^*, \quad \text{hence } (f + g)^* = (f^* \square g^*)^{**}.$

On the other hand, considering any $x^* \in X^*$ we have for all $y^* \in X^*$ and $x \in X$

$$f^*(x^* - y^*) + g^*(y^*) \geq \langle x^* - y^*, x \rangle - f(x) + \langle y^*, x \rangle - g(x) = \langle x^*, x \rangle - (f + g)(x),$$

so taking the infimum over $y^* \in X^*$ and then the supremum over $x \in X$ we obtain $(f^* \square g^*)(x^*) \geq (f + g)^*(x^*)$ for every $x^* \in X^*$. This entails in particular that the convex function $(f^* \square g^*)$ is bounded from below by a continuous affine function $\langle \cdot, a \rangle + \beta$, since there is such an affine function minorizing the proper w^*-semicontinuous convex function $(f+g)^*$ by Proposition 3.77(a). This and Proposition 3.89 tells us that $(f^* \Box g^*)^{**} = \mathrm{cl}_{w^*}(f^* \Box g^*)$, which by (3.67) confirms the equality of the corollary. □

The equality in Corollary 3.92 may fail without involving the closure operation in the right-hand side, as illustrated in the following example.

EXAMPLE 3.93. Coming back to the closed convex subsets $S_1 := \mathrm{epi}\,(\exp)$ and $S_2 := \mathbb{R} \times \{0\}$ of $\mathbb{R}^2$ in Example 3.68 and considering their support functions $f := \sigma(\cdot, S_1)$ and $g := \sigma(\cdot, S_2)$, we see that

$$f^* \Box g^* = \Psi_{S_1} \Box \Psi_{S_2} = \Psi_{S_1 + S_2}.$$

Recalling that $S_1 + S_2$ is not closed in $\mathbb{R}^2$ (since $S_1 + S_2 = \mathbb{R} \times]0, +\infty[$ as seen in Example 3.68), the infimal convolution function $f^* \Box g^*$ is not closed, hence it clearly fails to coincide with $(f+g)^*$. □

Through Proposition 3.88 we can establish a basic property of non vacuity of ε-subdifferential of convex functions with $\varepsilon > 0$.

PROPOSITION 3.94. *Let X be a Hausdorff locally convex space and $f : X \to \mathbb{R} \cup \{-\infty, +\infty\}$ be a convex function which is finite at $\overline{x} \in X$. Then f is lower semicontinuous at $\overline{x}$ if and only if $\partial_\varepsilon f(\overline{x}) \neq \emptyset$ for every real $\varepsilon > 0$.*

PROOF. Suppose first that for each fixed $\varepsilon > 0$ one has $\partial_\varepsilon f(\overline{x}) \neq \emptyset$, and choose $\overline{x}^* \in \partial_\varepsilon f(\overline{x})$. This yields for all $x \in X$

$$f(\overline{x}) \leq f(x) + \langle \overline{x}^*, \overline{x} - x \rangle + \varepsilon,$$

which ensures that $f(\overline{x}) \leq \varepsilon + \liminf_{x \to \overline{x}} f(x)$. This being true for all $\varepsilon > 0$, it ensues that $f(\overline{x}) \leq \liminf_{x \to \overline{x}} f(x)$, that is, f is lower semicontinuous at $\overline{x}$.

Conversely, suppose that f is lower semicontinuous at $\overline{x}$ and fix any real $\varepsilon > 0$. By Proposition 3.23 we know that the lower semicontinuous convex function f is proper, hence Proposition 3.88 gives $f(\overline{x}) = (f^*)^*(\overline{x})$, so by definition of conjugate there is some $\overline{x}^* \in X^*$ such that $f(\overline{x}) - \varepsilon < \langle \overline{x}^*, \overline{x} \rangle - f^*(\overline{x}^*)$. This gives $\partial_\varepsilon f(\overline{x}) \neq \emptyset$ by Proposition 3.64(j), and the proof is finished. □

We saw in (3.19) that the conjugate of the supremum of a family of functions is not greater than the infimum of the family of conjugates. Proposition 3.89 allows us to provide a quite general situation when we can express the conjugate of the supremum of a family of functions in terms of the infimum of the family of conjugates.

PROPOSITION 3.95. *Let X be a Hausdorff locally convex space and $(f_i)_{j \in J}$ be a nondecreasing net of proper lower semicontinuous convex functions on X with $\sup_{j \in J} f_j$ finite at some point in X. Then one has*

$$\left(\sup_{j \in J} f_j\right)^* = \mathrm{cl}_{w^*}\left(\inf_{j \in J} f_j^*\right).$$

PROOF. Endow X^* with the $w(X^*, X)$ topology, so the conjugate of any function defined on X^* is taken relative to X as well as any biconjugate. By the equality in (3.19) and by the Fenchel-Moreau Theorem 3.84 we have $(\inf_{j \in J} f_j^*)^* = \sup_{j \in J}(f_j^{**}) = \sup_{j \in J} f_j$, which implies that $(\inf_{j \in J} f_j^*)^{**} = (\sup_{j \in J} f_j)^*$. On the other hand, denote $f := \sup_{j \in J} f_j$ and notice by Proposition 3.77(c) that f^* is proper (since f is proper, lower semicontinuous and convex) and that $f^* \leq \inf_{j \in J} f_j^*$ by the inequality in (3.19). Notice also that the net $(f_j^*)_{j \in J}$ is nonincreasing, so the function $\inf_{j \in J} f_j^*$ is convex as the pointwise limit of convex functions. The function f^* being proper, convex and lower w^*-semicontinuous, by Proposition 3.77 it is minorized by a $w(X^*, X)$-continuous affine function $\langle \cdot, a \rangle + \beta$, so the convex function $\inf_{j \in J} f_j^*$ is also minorized by $\langle \cdot, a \rangle + \beta$. Proposition 3.89 tells us that $(\inf_{j \in J} f_j^*)^{**} = \mathrm{cl}_{w^*}(\inf_{j \in J} f_j^*)$. This combined with what precedes entails the desired equality $(\sup_{j \in J} f_j)^* = \mathrm{cl}_{w^*}(\inf_{j \in J} f_j^*)$. □

3.3.4. Dual conditions for coercivity of convex functions. Proposition 3.29 provides diverse primal criteria for the coercivity of convex functions. The next proposition furnishes a dual criterion by means of the Legendre-Fenchel conjugate.

PROPOSITION 3.96. *Let $(X, \|\cdot\|)$ be a normed space and $f : X \to \mathbb{R} \cup \{-\infty, +\infty\}$ be a convex function which is bounded from below near a point in $\mathrm{dom}\, f$, which holds in particular whenever f is lower semicontinuous at a point where it is finite. Then f is coercive $\left(\text{that is, } \lim_{\|x\| \to +\infty} f(x) = +\infty\right)$ if and only if the conjugate f^* is continuous at the origin of X^*.*

PROOF. By Proposition 3.29 we know that f is coercive if and only if there are reals $\alpha > 0$ and $\beta \in \mathbb{R}$ such that $f \geq \alpha \|\cdot\| + \beta$. Noting that $f \geq f^{**}$ on X, the latter is equivalent to $f^* \leq (\alpha \|\cdot\| + \beta)^*$, which itself is equivalent to $f^* \leq \Psi_{\alpha \mathbb{B}_{X^*}} - \beta$ according to (3.28) and (3.37). Consequently, f is coercive if and only if there are reals $\alpha > 0$ and $\beta \in \mathbb{R}$ such that f^* is bounded from above on $\alpha \mathbb{B}_{X^*}$ by $-\beta$, which means by Theorem 3.18(I) that f^* is $\|\cdot\|_*$-continuous at the origin of X^* (since f^* does not take the value $-\infty$ because f is finite at some point in X). □

The next proposition establishes a dual criterion in terms of conjugate for the condition $\lim_{\|x\| \to +\infty} \frac{f(x)}{\|x\|} = +\infty$ clearly stronger than the previous coercivity condition $\lim_{\|x\| \to +\infty} f(x) = +\infty$.

PROPOSITION 3.97. *Let $(X, \|\cdot\|)$ be a normed space and $f : X \to \mathbb{R} \cup \{+\infty\}$ be a proper lower semicontinuous convex function.*
(a) *One has $\lim_{\|x\| \to +\infty} \frac{f(x)}{\|x\|} = +\infty$ if and only if f^* is bounded on bounded sets of the normed space X^*.*
(b) *The function f is bounded on bounded sets of X if and only if the condition $\lim_{\|x^*\|_* \to +\infty} \frac{f^*(x^*)}{\|x^*\|_*} = +\infty$ is satisfied.*

PROOF. (a) Assume that $\lim_{\|x\| \to +\infty} \frac{f(x)}{\|x\|} = +\infty$. Take any real $\alpha > 0$. Then there is a real $\mu > 0$ such that $f(x) \geq \alpha \|x\|$ for every $x \in X$ with $\|x\| \geq \mu$. By Proposition 3.28 we also know that f is bounded from below on $\mu \mathbb{B}_X$, so there is a real $\gamma > 0$ such that $f(x) \geq -\gamma$ for all $x \in \mu \mathbb{B}_X$. Putting $\beta := -\gamma - \alpha \mu$ we have for any $x \in X$ with $\|x\| \leq \mu$

$$\alpha \|x\| + \beta \leq \alpha \mu + \beta = -\gamma \leq f(x).$$

Since $\beta < 0$, we also have $\alpha \|x\| + \beta \leq \alpha \|x\| \leq f(x)$ for every $x \in \mu \mathbb{B}_X$. Therefore, $\alpha \|x\| + \beta \leq f(x)$ for all $x \in X$, then $f^* \leq \Psi_{\alpha \mathbb{B}_{X^*}} - \beta$ by (3.28) and (3.37), which entails that f^* is bounded from above on $\alpha\mathbb{B}_{X^*}$. Further, as a proper weak* lower semicontinuous convex function, f^* is minorized by a weak* continuous affine function $x^* \mapsto \langle x^*, a\rangle + \rho$ with $a \in X$ and $\rho \in \mathbb{R}$ (see Proposition 3.77(a)), hence f^* is bounded from below on $\alpha\mathbb{B}_{X^*}$. Altogether, for any real $\alpha > 0$ the conjugate f^* is bounded on $\alpha\mathbb{B}_{X^*}$, which justifies the implication $\Longrightarrow$ in (a).

Conversely, assume that f^* is bounded on any bounded set in X^*. Take any real $\mu > 0$. There exists some real $\beta > 0$ such that $f^*(x^*) \leq \beta$ for all $x^* \in \mu\mathbb{B}_{X^*}$, or equivalently $f^* \leq \Psi_{\mu\mathbb{B}_{X^*}} + \beta$. Since $f = f^{**}$ on X, we see that $f \geq \mu\|\cdot\| - \beta$ on X. This ensures that $\liminf_{\|x\|\to+\infty} \frac{f(x)}{\|x\|} \geq \mu$. This inequality being true for every real $\mu > 0$, we deduce that $\liminf_{\|x\|\to+\infty} \frac{f(x)}{\|x\|} = +\infty$, or equivalently $\lim_{\|x\|\to+\infty} \frac{f(x)}{\|x\|} = +\infty$.

(b) By (a) the condition with the limit in (b) holds if and only if the biconjugate f^{**} relative to X^{**} is bounded on any ball $\mu\mathbb{B}_{X^{**}}$. Since f^{**} is minorized by a continuous affine function $x^{**} \mapsto \langle x^{**}, a^*\rangle + \rho$, the latter boundedness property is equivalent to $f^{**}(x^{**}) \leq \beta$ for all $x \in \mu\mathbb{B}_{X^{**}}$ for some real β, that is, $\mu\mathbb{B}_{X^{**}} \subset \{f^{**} \leq \beta\}$. Since $\mu\mathbb{B}_X$ is $w(X^{**}, X^*)$-dense in $\mu\mathbb{B}_{X^{**}}$ by Goldstine theorem (recalled in Theorem C.7 in Appendix) and $\{f^{**}(\cdot) \leq \beta\}$ is $w(X^{**}, X^*)$-closed by $w(X^{**}, X^*)$-lower semicontinuity of f^{**}, we obtain that the latter inclusion is equivalent to $\mu\mathbb{B}_X \subset \{f^{**}(\cdot) \leq \beta\}$, which itself is equivalent to $\mu\mathbb{B}_X \subset \{f(\cdot) \leq \beta\}$ since f^{**} coincides with f on X. It results that the condition with limit in (b) holds if and only if f is bounded from above on bounded sets of X, hence if and only if f is bounded on bounded sets of X (since f is minorized by a continuous affine function). $\square$

3.3.5. Global Lipschitz property of conjugate functions.
Given a convex function on a Hausdorff locally convex space, Proposition 3.14 and Theorem 3.18 establishes its Lipschitz property near a point whenever it is bounded from above near that point. In this subsection we study the relationship between the global Lipschitz property of the function (resp. conjugate) and the boundedness of the domain of the conjugate (resp. function).

For a nonempty bounded symmetric convex set B of a Hausdorff locally convex space X (resp. of its topological dual X^*) we will denote as usual by p_B its support function defined on X^* (resp. on X) by

$$p_B(x^*) := \sigma_B(x^*) = \sup_{u \in B} \langle x^*, u\rangle \quad (\text{resp. } p_B(x) := \sigma_B(x) = \sup_{u^* \in B} \langle u^*, x\rangle.$$

PROPOSITION 3.98. *Let X be a Hausdorff locally convex space and $f : X \to \mathbb{R} \cup \{+\infty\}$ be a proper function.*
(a) *If $\mathrm{dom}\, f \subset B$ for some bounded symmetric convex set B in X and f is bounded from below on B, then the conjugate f^* is Lipschitz on X^* with*

$$|f^*(x^*) - f^*(y^*)| \leq p_B(x^* - y^*) \quad \text{for all } x^*, y^* \in X^*.$$

(b) *If f is Lipschitz on X, in the sense that for some nonempty w^*-bounded symmetric w^*-closed convex set B in X^**

$$|f(x) - f(y)| \leq p_B(x - y) \quad \text{for all } x, y \in X,$$

then one has $\mathrm{dom}\, f^* \subset B$.

PROOF. (a) Assume that $\mathrm{dom}\, f \subset B$. For $x \in \mathrm{dom}\, f \subset B$, consider the function $\varphi_x(\cdot) := \langle \cdot, x\rangle - f(x)$ on X^*. Given $x \in \mathrm{dom}\, f \subset B$ we see that for any

$x^*, y^* \in X^*$ we have
$$\varphi_x(x^*) \leq \varphi_x(y^*) + \sigma_B(x^* - y^*),$$
so using the equality $f^*(x^*) = \sup_{x \in B} \varphi_x(x^*)$ (due to the definition of f^* and the inclusion dom $f \subset B$) we deduce that $f^*(x^*) \leq f^*(y^*) + \sigma_B(x^* - y^*)$. Further, it is clear from the assumptions that f^* is finite on X^*, hence the latter inequality gives
$$|f^*(x^*) - f^*(y^*)| \leq p_B(x^* - y^*) \quad \text{for all } x^*, y^* \in X^*.$$

(b) Assume now that for some nonempty w^*-bounded symmetric w^*-closed convex set B in X^*
$$|f(x) - f(y)| \leq p_B(x - y) \quad \text{for all } x, y \in X.$$
Then for any $x \in X$ (keeping in mind that $p_B = \sigma_B$) we have for every $x^* \in X^*$
$$\langle x^*, x \rangle - f(x) \geq \langle x^*, x \rangle - \sigma_B(x) - f(0).$$
If $x^* \notin B$ there is some $u \in X$ with $\langle x^*, u \rangle > \sigma_B(u)$. Then for any real $t > 0$ writing
$$\langle x^*, tu \rangle - f(tu) \geq t(\langle x^*, u \rangle - \sigma_B(u)) - f(0),$$
we see that $f^*(x^*) \geq t(\langle x^*, u \rangle - \sigma_B(u)) - f(0)$ for every real $t > 0$, hence $f^*(x^*) = +\infty$. This entails that dom $f^* \subset B$. □

In the case of a normed space $(X, \|\cdot\|)$, taking $B = \mathbb{B}_X$ and $B = \mathbb{B}_{X^*}$ in (a) and (b) respectively in Proposition 3.98 yields the following first corollary.

COROLLARY 3.99. *Let $(X, \|\cdot\|)$ be a normed space and $f : X \to \mathbb{R} \cup \{+\infty\}$ be a proper function.*
(a) *If for some real $\gamma > 0$ the function f is bounded from below on $\gamma \mathbb{B}_X$ and dom $f \subset \gamma \mathbb{B}_X$, the conjugate f^* is finite and γ-Lipschitz on X^*.*
(b) *If the function f is finite and γ-Lipschitz on X, then dom $f^* \subset \gamma \mathbb{B}_{X^*}$.*

Now recall that for a proper lower semicontinuous convex function $f : X \to \mathbb{R} \cup \{+\infty\}$ on the normed space X, one has $f = f^{**}_{|X}$ (see Theorem 3.84) and that f (resp. f^*) is minorized (see Proposition 3.77(a)) by a continuous affine function $\langle a^*, \cdot \rangle + \beta$ (resp. $\langle a^{**}, \cdot \rangle + \beta'$). The next corollary then follows from the preceding one.

COROLLARY 3.100. *Let $(X, \|\cdot\|)$ be a normed space and $f : X \to \mathbb{R}$ be a proper lower semicontinuous convex function on X, and let a real $\gamma > 0$.*
(a) *One has dom $f \subset \gamma \mathbb{B}_X$ if and only if f^* is finite and γ-Lipschitz on X^*.*
(b) *The function f is finite and γ-Lipschitz on X if and only dom $f^* \subset \gamma \mathbb{B}_{X^*}$.*

3.4. B-differentiability and continuity of subdifferential

This section is essentially devoted to study for a bornology $\mathcal{B}$ (see below) diverse links between the $\mathcal{B}$-differentiability of a convex function and its subdifferential and ε-subdifferential.

3.4.1. $\mathcal{B}$-differentiability: Definition and intrinsic characterizations for convex functions.
The characterization of Hadamard/Fréchet differentiability in Proposition 2.37 leads to consider for a topological vector space X a family $\mathcal{B}$ of nonempty symmetric bounded subsets of X such that $\bigcup_{B \in \mathcal{B}} B = X$. Such a family is called a (symmetric) *bornology* of X. Taking Proposition 2.37 into account, one extends the concept of differentiabilty as follows.

DEFINITION 3.101. Let U be an open subset of a topological vector space X and let $G : U \to Y$ be a mapping from U into a Hausdorff topological vector space Y. The mapping G is said to be $\mathcal{B}$-*differentiable at a point* $\overline{x} \in U$ provided there exists a continuous linear mapping $A : X \to Y$ such that for each $B \in \mathcal{B}$
$$t^{-1}\big(G(\overline{x} + th) - G(\overline{x}) - tA(h)\big) \to 0 \quad \text{as } t \downarrow 0$$
uniformly with respect to $h \in B$.

Similarly one says that G is *strictly* $\mathcal{B}$-*differentiable at* $\overline{x}$ if the exists a continuous linear mapping $A : X \to Y$ such that for each $B \in \mathcal{B}$
$$t^{-1}\big(G(x + th) - G(x) - tA(h)\big) \to 0 \quad \text{as } t \downarrow 0 \text{ and } x \to \overline{x}$$
uniformly with respect to $h \in B$.

For a nonempty subset $C \subset U$, one says that G is *uniformly* $\mathcal{B}$-*differentiable on* C if there is a mapping $\Lambda : C \to L(X, Y)$ such that for each $B \in \mathcal{B}$
$$t^{-1}\big(G(x + th) - G(x) - t\Lambda(x)(h)\big) \to 0 \quad \text{as } t \downarrow 0$$
uniformly with respect to $(x, h) \in C \times B$.

The above concepts are extended to a function $f : X \to \mathbb{R} \cup \{+\infty\}$, $\overline{x} \in \text{dom } f$ and $C \subset \text{dom } f$.

When G is $\mathcal{B}$-differentiable at $\overline{x}$, it is easy via the Hausdorff property of Y to see that the above mapping $A \in L(X, Y)$ is unique; it is called the $\mathcal{B}$-*derivative of* G *at* $\overline{x}$ and it is denoted by $D_\mathcal{B} G(\overline{x})$ or $DG(\overline{x})$ if there is no risk of confusion. Using the symmetry of $B \in \mathcal{B}$, it is not difficult to check that in the above definition of (strict, uniform) $\mathcal{B}$-differentiability one can equivalently require $t \to 0$ in place of $t \downarrow 0$.

It is also clear that the strict $\mathcal{B}$-differentiability entails the $\mathcal{B}$-differentiability. If $\mathcal{B}'$ is another symmetric bornology in X with $\mathcal{B} \subset \mathcal{B}'$, it is evident that the $\mathcal{B}'$-differentiability of G at $\overline{x}$ implies its $\mathcal{B}$-differentiability at $\overline{x}$. When X, Y are normed spaces and $\mathcal{B}$ is the class of nonempty symmetric bounded (resp. compact, or finite) subsets of X, we see by Proposition 2.37 that the (strict) $\mathcal{B}$-differentiability corresponds to the (strict) Fréchet (resp. Hadamard, or Gâteaux) differentiability. It is usual to continue to say that G is Hadamard (resp. Gâteaux) differentiable at $\overline{x}$ when $\mathcal{B}$ is the class of nonempty symmetric compact (resp. finite) subsets of X.

We are mainly concerned in this section with the $\mathcal{B}$-differentiability of convex functions. Let us start with the following simple property in infinite dimensions which complements the finite dimensional result of Proposition 2.211.

PROPOSITION 3.102. Let $f : U \to \mathbb{R}$ be a real-valued convex function on an open convex set U of a Hausdorff topological vector space X. If f is Gâteaux differentiable at a point $\overline{x} \in U$, then $\partial f(\overline{x})$ is a singleton and $\partial f(\overline{x}) = \{Df(\overline{x})\}$.

Conversely, if f is continuous at $\overline{x}$ and $\partial f(\overline{x})$ is a singleton $\partial f(\overline{x}) = \{\overline{x}^*\}$, then f is Gâteaux differentiable at $\overline{x}$ with $Df(\overline{x}) = \overline{x}^*$.

3.4. ẞ-DIFFERENTIABILITY AND CONTINUITY OF SUBDIFFERENTIAL

PROOF. Assume that f is Gâteaux differentiable at $\bar{x}$. Taking any $x \in U$ and putting $h := x - \bar{x}$, we have for every $t \in]0, 1]$

$$t^{-1}\bigl(f(\bar{x} + th) - f(\bar{x})\bigr) \leq f(\bar{x} + h) - f(\bar{x}),$$

hence passing to the limit as $t \downarrow 0$ we obtain $\langle Df(\bar{x}), h \rangle \leq f(\bar{x} + h) - f(\bar{x})$. The latter inequality means that $\langle Df(\bar{x}), x - \bar{x} \rangle \leq f(x) - f(\bar{x})$ for all $x \in U$, that is, $Df(\bar{x}) \in \partial f(\bar{x})$. This says in particular that $\partial f(\bar{x}) \neq \emptyset$. So, take any $x^* \in \partial f(\bar{x})$ and fix any $h \in X$. Choosing a real $r > 0$ such that $\bar{x}+]0, r]h \subset U$ we can write for every $t \in]0, r]$

$$\langle x^*, h \rangle \leq t^{-1}\bigl(f(\bar{x} + th) - f(\bar{x})\bigr),$$

hence passing to limit as $t \downarrow 0$ gives $\langle x^*, h \rangle \leq \langle Df(\bar{x}), h \rangle$. This being true for every $h \in X$, it follows that $x^* = Df(\bar{x})$. Altogether justifies the first assertion of the proposition.

Conversely, suppose that f is continuous at $\bar{x}$ and $\partial f(\bar{x}) = \{\bar{x}^*\}$. The continuity of f at $\bar{x}$ entails by Theorem 3.30(a) that for all $h \in X$

$$\lim_{t \downarrow 0} t^{-1}\bigl(f(\bar{x} + th) - f(\bar{x})\bigr) = \max\{\langle x^*, h \rangle : x^* \in \partial f(x)\} = \langle \bar{x}^*, h \rangle,$$

which justifies that f is (by definition) Gâteaux differentiable at $\bar{x}$ along with the equality $Df(\bar{x}) = \bar{x}^*$. □

The converse property in the above proposition is established under the continuity requirement. In fact, the converse fails in general as illustrated by the function seen in Example 2.212.

Taking into account Proposition 3.102, the following question arises: Does the Gâteaux differentiability at a point $\bar{x}$ of a real-valued convex function entail its continuity at $\bar{x}$?. We give below a simple example providing a negative answer.

EXAMPLE 3.103. Consider an infinite dimensional normed space X and a non-continuous linear functional $\zeta : X \to \mathbb{R}$. Define the convex function $f : X \to \mathbb{R}$ by $f(x) := \bigl(\zeta(x)\bigr)^2$ for all $x \in X$. This function is clearly Gâteaux differentiable at 0 (with $D_G f(0) = 0$) while it is not continuous at 0. The function f is even in addition lower semicontinuous at 0 (but not near 0). □

Despite the preceding comments and example, for a convex function $f : X \to \mathbb{R} \cup \{+\infty\}$ on a Banach space X the next proposition shows that its Gâteaux differentiability at a point $\bar{x}$ entails its continuity at $\bar{x}$ whenever it is lower semicontinuous near $\bar{x}$.

PROPOSITION 3.104. Let $f : U \to \mathbb{R} \cup \{+\infty\}$ be a lower semicontinuous convex function on an open convex set U of a Banach space X. Assume that f is Gâteaux differentiable at a point $\bar{x} \in \mathrm{dom}\, f$. Then f is continuous at $\bar{x}$.

PROOF. The Gâteaux differentiability of f at $\bar{x}$ ensures by definition that $\bar{x} \in \mathrm{core}(\mathrm{dom}\, f)$. Consequently, the continuity of f at $\bar{x}$ easily follows from Corollary 2.162. □

REMARK 3.105. It is worth pointing out that, by Example 3.103, the lower semicontinuity of f *near* the point $\bar{x}$ in the above proposition cannot be replaced by its lower semicontinuity merely at $\bar{x}$. □

The next proposition extends Proposition 2.178 to the context of topological vector spaces.

PROPOSITION 3.106. *Let U be an open convex set of a Hausdorff topological vector space X and $f : U \to \mathbb{R}$ be a real-valued continuous convex function. Let $\mathcal{B}$ be a symmetric bornology in X. Then the function f is $\mathcal{B}$-differentiable at a point $\bar{x} \in U$ if and only if it is strictly $\mathcal{B}$-differentiable at $\bar{x}$.*

PROOF. Assume that f is $\mathcal{B}$-differentiable at $\bar{x}$. Using the function $f - Df(\bar{x})(\cdot)$ we may suppose that $Df(\bar{x}) = 0$ in X^*. By Proposition 3.14 there is a function $q : X \to [0, +\infty[$ with $q(0) = 0$ continuous at 0 and a neighborhood W of zero in X with $\bar{x} + W + W + W \subset U$ such that $|f(x) - f(y)| \le q(x-y)$ for all elements $x, y \in \bar{x} + W + W + W$. Fix any $B \in \mathcal{B}$. Take any real $\varepsilon > 0$ and choose a real $r > 0$ with $\bar{x} + rB \subset W$ such that

$$\eta(r) := \sup_{v \in B} \left| \frac{f(\bar{x} + rv) - f(\bar{x})}{r} \right| \le \varepsilon/3.$$

Choose a neighborhood $V \subset W$ of zero in X such that $q(u) \le (\varepsilon r)/3$ for all $u \in V + V$. Choose also a real $\delta \in]0, r[$ such that $[-\delta, \delta]B \subset V$. Take any $x \in \bar{x} + V$ and any $t \in]0, \delta[$. Take also any $h \in B$. We note first that

$$t^{-1}\big(f(x+th) - f(x)\big) = (-t)^{-1}\big(f(x + th - th) - f(x+th)\big)$$
$$\le r^{-1}\big(f(x + th + rh) - f(x+th)\big).$$

Considering the equality

$$r^{-1}\big(f(x+th+rh) - f(x+th)\big)$$
$$= r^{-1}\Big(\big(f(x+th+rh) - f(\bar{x}+rh)\big) + \big(f(\bar{x}+rh) - f(\bar{x})\big) + \big(f(\bar{x}) - f(x+th)\big)\Big),$$

we see that

$$t^{-1}\big(f(x+th) - f(x)\big) \le r^{-1}\big(q(x - \bar{x} + th) + r\eta(r) + q(x - \bar{x} + th)\big) \le \varepsilon,$$

where the inclusion $x - \bar{x} + th \in V + V$ is used in the latter inequality.

Analogously, we have

$$t^{-1}\big(f(x) - f(x+th)\big) = (-t)^{-1}\big(f(x + (-t)(-h)) - f(x)\big)$$
$$\le r^{-1}\big(f(x + r(-h)) - f(x)\big),$$

which gives

$$t^{-1}\big(f(x) - f(x+th)\big)$$
$$\le r^{-1}\Big(\big(f(x - rh) - f(\bar{x} - rh)\big) + \big(f(\bar{x} - rh) - f(\bar{x})\big) + \big(f(\bar{x}) - f(x)\big)\Big)$$
$$\le r^{-1}\big(q(x - \bar{x}) + r\eta(r) + q(x - \bar{x})\big) \le \varepsilon.$$

It follows that $t^{-1}|f(x+th) - f(x)| \le \varepsilon$. Consequently, for all $t \in]0, \delta[$ and $x \in \bar{x} + V$

$$\sup_{h \in B} \big|t^{-1}\big(f(x+th) - f(x)\big)\big| \le \varepsilon,$$

which translates the strict $\mathcal{B}$-differentiability of f at $\bar{x}$. The proof is then finished. □

We establish now an intrinsic characterization of $\mathcal{B}$-differentiability of convex functions.

3.4. $\mathcal{B}$-DIFFERENTIABILITY AND CONTINUITY OF SUBDIFFERENTIAL

PROPOSITION 3.107. *Let U be an open convex set of a Hausdorff topological vector space X and $f : U \to \mathbb{R}$ be a real-valued continuous convex function. Let $\mathcal{B}$ be a symmetric bornology in X. The following hold.*
(a) *The function f is $\mathcal{B}$-differentiable (or equivalently strictly $\mathcal{B}$-differentiable) at a point $\bar{x} \in U$ if and only if for each $B \in \mathcal{B}$*

$$t^{-1}\bigl(f(\bar{x} + th) + f(\bar{x} - th) - 2f(\bar{x})\bigr) \to 0 \quad \text{as } t \downarrow 0$$

uniformly with respect to $h \in B$.
(b) *The function f is uniformly $\mathcal{B}$-differentiable on a nonempty subset $C \subset U$ if and only if for each $B \in \mathcal{B}$*

$$t^{-1}\bigl(f(x + th) + f(x - th) - 2f(x)\bigr) \to 0 \quad \text{as } t \downarrow 0$$

uniformly with respect to $(x, h) \in C \times B$.

PROOF. We already saw in Proposition 3.106 that the $\mathcal{B}$-differentiability of f at $\bar{x}$ is equivalent to its strict $\mathcal{B}$-differentiability at $\bar{x}$. So it suffices to establish (b) since the characterization of $\mathcal{B}$-differentiability in (a) follows from (b) with $C = \{\bar{x}\}$.

Let us start the proof of (b) with the implication $\Longrightarrow$. Suppose that f is uniformly $\mathcal{B}$-differentiable on C. Fix any $B \in \mathcal{B}$. There is a function $\eta :]0, +\infty[\times C \times X \to \mathbb{R}$ with $\sup_{(u,v) \in C \times B} |\eta(t, u, v)| \to 0$ as $t \downarrow 0$ such that for some $r > 0$ one has $C+]0, r[B \subset U$ and

$$f(x + th) - f(x) = t\langle Df(x), h\rangle + t\eta(t, x, h) \quad \text{for all } t \in]0, r[, (x, h) \in C \times B.$$

This entails that for all $t \in]0, r[$ and $(x, h) \in C \times B$

$$t^{-1}|f(x+th)+f(x-th)-2f(x)| = |\eta(t, x, h)+\eta(t, x, -h)| \le 2 \sup_{(u,v)\in C\times B} |\eta(t, u, v)|,$$

which justifies the implication $\Longrightarrow$.

Conversely, suppose that the property in (b) is satisfied. Fix any real $\varepsilon > 0$. By continuity of the convex function f on U choose for each $x \in U$ some $\zeta(x) \in \partial f(x) \ne \emptyset$ (see Theorem 3.30(a)). Fix any $B \in \mathcal{B}$ and note that $\lim_{t \downarrow 0} t^{-1}\rho(t) = 0$, where for some $r > 0$

$$\rho(t) := \sup_{(u,v)\in C\times B} |f(u + tv) + f(u - tv) - 2f(u)| \quad \text{for all } t \in]0, r[.$$

By this there exist a real $\delta > 0$ such that $C+]0, \delta[B \subset U$ and $t^{-1}\rho(t) \le \varepsilon$ for all $t \in]0, \delta[$. For any $t \in]0, \delta[$ and any $x \in C$ applying twice the inclusion $\zeta(x) \in \partial f(x)$ yields for all $h \in B$

$$\rho(t) \ge f(x + th) + f(x - th) - 2f(x) \ge f(x + th) - f(x) - t\langle \zeta(x), h\rangle \ge 0,$$

which gives

$$0 \le \frac{f(x + th) - f(x) - t\langle \zeta(x), h\rangle}{t} \le t^{-1}\rho(t) \le \varepsilon.$$

This entails that

$$\sup_{(x,h)\in C\times B} \left|\frac{f(x + th) - f(x) - t\langle \zeta(x), h\rangle}{t}\right| \to 0 \quad \text{as } t \downarrow 0,$$

which justifies the uniform $\mathcal{B}$-differentiability of f over C. □

Concerning the Fréchet derivative of the Legendre-Fenchel conjugate f^* of a proper lower semicontinuous function $f : X \to \mathbb{R} \cup \{+\infty\}$ on a Banach space X, we will see in Theorem 4.58 and Corollary 4.60 in the next chapter that $D_F f^*(\bar{x}^*) \in X$ whenever f^* is Fréchet differentiable at the point $\bar{x}^*$, that is, there is some $\bar{x} \in X$ such that

$$\langle D_F f^*(\bar{x}^*), v^* \rangle = \langle v^*, \bar{x} \rangle \quad \text{for all } v^* \in X^*.$$

3.4.2. $\mathcal{B}$-differentiability of convex functions and continuous selections of subdifferentials.

With the (symmetric) bornology $\mathcal{B}$ on X we associate on X^* the topology of uniform convergence over the sets $B \in \mathcal{B}$. This topology coincides with the topology generated by the family of seminorms $(p_B)_{B \in \mathcal{B}}$ on X^* defined by

(3.68) $$p_B(x^*) := \sup_{x \in B} |\langle x^*, x \rangle| \quad \text{for all } x^* \in X^*;$$

we also note that $p_B(x^*) = \sup_{x \in B} \langle x^*, x \rangle$, since each $B \in \mathcal{B}$ is symmetric. With this topology at hands, we can characterize the $\mathcal{B}$-differentiability of convex functions via the continuity of selections of their subdifferentials.

Given a multimapping $M : T \rightrightarrows Y$ between two sets and given a subset $U \subset T$, we recall that a mapping $\zeta : U \to Y$ is a *selection of the multimapping* M provided that $\zeta(t) \in M(t)$ for every $t \in U$.

THEOREM 3.108 (differentiability of convex functions via continuous selection of subdifferential). *Let U be an open convex set of a Hausdorff topological vector space (X, τ) and $f : U \to \mathbb{R}$ be a real-valued continuous convex function. Let $\mathcal{B}$ be a symmetric bornology in X and $t_\mathcal{B}$ the topology on X^* of uniform convergence over the sets in $\mathcal{B}$. Given $\bar{x} \in U$ the following assertions are equivalent:*
(a) *the function f is $\mathcal{B}$-differentiable, or equivalently strictly $\mathcal{B}$-differentiable, at $\bar{x}$;*
(b) *any selection $\zeta : U \to X^*$ of ∂f is τ-to-$t_\mathcal{B}$ continuous at $\bar{x}$;*
(c) *there exists a selection $\zeta : U \to X^*$ of ∂f which is τ-to-$t_\mathcal{B}$ continuous at $\bar{x}$.*

PROOF. The implication (b)$\Rightarrow$(c) being obvious, let us prove (c)$\Rightarrow$(a). Let $\zeta(\cdot)$ be given by (c) and let any $B \in \mathcal{B}$. Take any real $\varepsilon > 0$. By continuity of $\zeta(\cdot)$ choose some neighborhood V of zero in X such $\bar{x} + V \subset U$ and $\sup_{v \in B} |\langle \zeta(x) - \zeta(\bar{x}), v \rangle| \leq \varepsilon$ for all $x \in \bar{x} + V$. Choose a real $\delta > 0$ such that $[-\delta, \delta]B \subset V$. Fixing $t \in]0, \delta]$ we obtain for every $h \in B$

$$\langle \zeta(\bar{x} + th), -th \rangle \leq f(\bar{x}) - f(\bar{x} + th),$$

hence

$$f(\bar{x} + th) - f(\bar{x}) - t\langle \zeta(\bar{x}), h \rangle \leq t \langle \zeta(\bar{x} + th) - \zeta(\bar{x}), h \rangle \leq t\varepsilon,$$

which in turn gives by the inclusion $\zeta(\bar{x}) \in \partial f(\bar{x})$

$$0 \leq t^{-1} \big(f(\bar{x} + th) - f(\bar{x}) - t\langle \zeta(\bar{x}), h \rangle \big) \leq \varepsilon.$$

From this we deduce that for every $t \in]0, \delta[$

$$\sup_{h \in B} \big| t^{-1} \big(f(\bar{x} + th) - f(\bar{x}) - t\langle \zeta(\bar{x}), h \rangle \big) \big| \leq \varepsilon,$$

so f is $\mathcal{B}$-differentiable at $\bar{x}$ with $\zeta(\bar{x})$ as its $\mathcal{B}$-derivative.

3.4. $\mathcal{B}$-DIFFERENTIABILITY AND CONTINUITY OF SUBDIFFERENTIAL

Now let us show (a)$\Rightarrow$(b). Assume that f is $\mathcal{B}$-differentiable at $\bar{x}$, so $\partial f(\bar{x})$ is a singleton by Proposition 3.102, hence $D_\mathcal{B} f(\bar{x}) = \zeta(\bar{x})$. Fix any $B \in \mathcal{B}$. We note that $\lim_{t \downarrow 0} t^{-1} \eta(t) = 0$, where for some $\delta > 0$

$$\eta(t) := \sup_{v \in B} |f(\bar{x} + tv) - f(\bar{x}) - t\langle \zeta(\bar{x}), v \rangle| \text{ for all } t \in]0, \delta].$$

Let V be a neighborhood of zero in X such that $\bar{x} + V + V \subset U$. Fix any real $\varepsilon > 0$ and set $\varepsilon' := \varepsilon/3$. Choose a real $r \in]0, \delta[$ such that $r^{-1} \eta(r) \leq \varepsilon'$ and $rB \subset V$. Take any $x \in \bar{x} + V$ and any $h \in B$. By definition of $\eta(\cdot)$ we have

$$\langle \zeta(x), \bar{x} + rh - x \rangle \leq f(\bar{x} + rh) - f(x)$$
$$\leq f(\bar{x}) + r\langle \zeta(\bar{x}), h \rangle + \eta(r) - f(x),$$

which implies that

$$r\langle \zeta(x) - \zeta(\bar{x}), h \rangle \leq \langle \zeta(x), x - \bar{x} \rangle + f(\bar{x}) - f(x) + \eta(r)$$
$$\leq f(2x - \bar{x}) - f(x) + f(\bar{x}) - f(x) + \eta(r).$$

It ensues that

$$\langle \zeta(x) - \zeta(\bar{x}), h \rangle \leq r^{-1} |f(2x - \bar{x}) - f(x)| + r^{-1} |f(\bar{x}) - f(x)| + r^{-1} \eta(r),$$

hence by the symmetry of B

$$\sup_{h \in B} |\langle \zeta(x) - \zeta(\bar{x}), h \rangle| \leq r^{-1} |f(2x - \bar{x}) - f(x)| + r^{-1} |f(\bar{x}) - f(x)| + \varepsilon'.$$

By continuity of f at $\bar{x}$, choose a neighborhood $V_0 \subset V$ of zero such that $r^{-1} |f(u) - f(v)| \leq \varepsilon'$ for all $u, v \in \bar{x} + V_0 + V_0$. It follows that

$$\sup_{h \in B} |\langle \zeta(x) - \zeta(\bar{x}), h \rangle| \leq \varepsilon \quad \text{for all } x \in \bar{x} + V_0.$$

This being true for any $B \in \mathcal{B}$ and any real $\varepsilon > 0$, we conclude that $\zeta(\cdot)$ is τ-to-$t_\mathcal{B}$ continuous at $\bar{x}$. $\square$

Two corollaries directly follow.

COROLLARY 3.109. Let U be an open convex set of a Hausdorff topological vector space (X, τ) and $f : U \to \mathbb{R}$ be a real-valued continuous convex function. Given $\bar{x} \in U$ the following assertions are equivalent:
(a) the function f is Gâteaux differentiable, or equivalently strictly Gâteaux differentiable, at $\bar{x}$;
(b) any selection $\zeta : U \to X^*$ of ∂f is τ-to-w^* continuous at $\bar{x}$;
(c) there exists a selection $\zeta : U \to X^*$ of ∂f which is τ-to-w^* continuous at $\bar{x}$.

COROLLARY 3.110. Let U be an open convex set of a normed space X and $f : U \to \mathbb{R}$ be a real-valued continuous convex function. Given $\bar{x} \in U$ the following assertions are equivalent:
(a) the function f is Fréchet differentiable, or equivalently strictly Fréchet differentiable, at $\bar{x}$;
(b) any selection $\zeta : U \to X^*$ of ∂f is $\|\cdot\|$-to-$\|\cdot\|_*$ continuous at $\bar{x}$;
(c) there exists a selection $\zeta : U \to X^*$ of ∂f which is $\|\cdot\|$-to-$\|\cdot\|_*$ continuous at $\bar{x}$.

3.4.3. $\mathcal{B}$-differentiability of convex functions and continuity of their subdifferentials.
In Proposition 2.181 we characterized the strict Fréchet differentiability of locally Lipschitz functions by means of certain continuity properties of their Clarke subdifferential. Our next goal is to go further with convex functions in characterizing the $\mathcal{B}$-differentiability of such a function via a continuity property of its subdifferential or its ε-subdifferential multimapping. We begin with the following result which has its own interest.

PROPOSITION 3.111. *Let X be a topological vector space and $f : X \to \mathbb{R} \cup \{+\infty\}$ be a convex function which is finite at $\bar{x}$ and continuous at $\bar{x}$. Then for every real $\varepsilon > 0$ there exist a real $\varepsilon' \in \,]0, \varepsilon]$ and a neighborhood U of $\bar{x}$ such that*
$$\partial_{\varepsilon'} f(x) \subset \partial_\varepsilon f(\bar{x}) \quad \text{for all } x \in U.$$

PROOF. Fix any real $\varepsilon > 0$ and put $\varepsilon' := \varepsilon/5$. Choose a neighborhood V of zero in X such that $|f(\bar{x} + w) - f(\bar{x})| \leq \varepsilon'$ for all $w \in V + V$. Take any $x \in U := \bar{x} + V$ and any $x^* \in \partial_{\varepsilon'} f(x)$. Put $v := x - \bar{x} \in V$ and take any $y \in \operatorname{dom} f$. Writing
$$\langle x^*, y - \bar{x} \rangle = \langle x^*, v \rangle + \langle x^*, y - \bar{x} - v \rangle$$
$$\leq f(\bar{x} + v + v) - f(\bar{x} + v) + \varepsilon' + f(y) - f(\bar{x} + v) + \varepsilon',$$
we obtain that
$$\langle x^*, y - \bar{x} \rangle \leq f(y) - f(\bar{x}) + 2(f(\bar{x}) - f(\bar{x} + v)) + (f(\bar{x} + v + v) - f(\bar{x})) + 2\varepsilon'$$
$$\leq f(y) - f(\bar{x}) + 2\varepsilon' + \varepsilon' + 2\varepsilon',$$
hence $\langle x^*, y - \bar{x} \rangle \leq f(y) - f(\bar{x}) + \varepsilon$. This justifies the inclusion $x^* \in \partial_\varepsilon f(\bar{x})$ and finishes the proof. $\square$

The following corollary follows directly from the previous proposition.

COROLLARY 3.112. *Let X be a topological vector space and $f : X \to \mathbb{R} \cup \{+\infty\}$ be a convex function which is finite at $\bar{x}$ and continuous at $\bar{x}$. Then for every real $\varepsilon > 0$ there exists a neighborhood U of $\bar{x}$ such that*
$$\partial f(x) \subset \partial_\varepsilon f(\bar{x}) \quad \text{for all } x \in U.$$

The next proposition provides a first sufficient condition for continuity of a convex function by means of ε-subdifferential. Other conditions will be given later.

PROPOSITION 3.113. *Let X be a Hausdorff topological vector space and $f : X \to \mathbb{R} \cup \{+\infty\}$ be a convex function. Assume that there exist a real $\varepsilon > 0$, an open set U in X and a selection $\zeta : U \to X^*$ of $\partial_\varepsilon f$ on U such that the family $(\zeta(u))_{u \in U}$ is equicontinuous in X^*. Then f is continuous on U.*

PROOF. Fix any $\bar{x} \in U$. Let any real $\eta > 0$. Choose a real $r \in \,]0, 1]$ with $r\varepsilon \leq \eta/2$ and a symmetric neighborhood V of zero in X such that $\bar{x} + V \subset U$ and $|\langle \zeta(u), v \rangle| \leq \eta/2$ for all $u \in U$ and $v \in V$. Take any $v \in V$. We have
$$\langle \zeta(\bar{x} + v), -v \rangle \leq f(\bar{x}) - f(\bar{x} + v) + \varepsilon,$$
and hence
$$f(\bar{x} + v) - f(\bar{x}) \leq \langle \zeta(\bar{x} + v), v \rangle + \varepsilon \leq \varepsilon + \eta/2.$$
It follows by the convexity of f that
$$f(\bar{x} + rv) - f(\bar{x}) \leq r\bigl(f(\bar{x} + v) - f(\bar{x})\bigr) \leq r\varepsilon + (r\eta)/2 \leq \eta.$$

Since $-v \in V$, we also have $f(\bar{x} - rv) - f(\bar{x}) \leq \eta$, thus by the convexity of f again
$$f(\bar{x} + rv) - f(\bar{x}) \geq -(f(\bar{x} - rv) - f(\bar{x})) \geq -\eta.$$
Therefore, $|f(\bar{x} + rv) - f(\bar{x})| \leq \eta$ for all $v \in V$, which entails the continuity of f at $\bar{x}$ and finishes the proof. □

As a corollary we have the following characterization of continuity.

COROLLARY 3.114. *Let X be a Hausdorff locally convex space and $f : X \to \mathbb{R} \cup \{+\infty\}$ be a convex function. Then f is continuous at a point $\bar{x} \in \mathrm{dom}\, f$ if and only if there exist a real $\varepsilon > 0$, a neighborhood U of $\bar{x}$ and a selection $\zeta : U \to X^*$ of the multimapping $\partial_\varepsilon f$ such that the family $(\zeta(u))_{u \in U}$ is equicontinuous in X^*.*

PROOF. The implication $\Longleftarrow$ follows directly from Proposition 3.113 above. To prove the converse implication, suppose that f is continuous at $\bar{x}$. Using both Proposition 3.37 and Theorem 3.30 we obtain a neighborhood U of $\bar{x}$ such that $\bigcup_{u \in U} \partial f(u)$ is equicontinuous and $\partial f(u) \neq \emptyset$ for every $u \in U$. Then we can choose a mapping $\zeta : U \to X^*$ with $\zeta(u) \in \partial f(u)$ for every $u \in U$, and the family $(\zeta(u))_{u \in U}$ is equicontinuous by what precedes. Taking any real $\varepsilon > 0$ the mapping $\zeta(\cdot)$ is also a selection of the multimapping $\partial_\varepsilon f$. This justifies the implication $\Longrightarrow$. □

Under $\mathcal{B}$-differentiability, the ε-subdifferential of a convex function is contained in a suitable enlargement of the subdifferential as shown below. Before stating the proposition, recall (see the very beginning of Chapter 1) that, for a function $\varphi : X \to \mathbb{R} \cup \{-\infty, +\infty\}$ and $r \in \mathbb{R} \cup \{-\infty, +\infty\}$, we denote
$$\{\varphi \leq r\} := \{x \in X : \varphi(x) \leq r\}.$$
When it needs to be emphasized that φ is a function, we write $\{\varphi(\cdot) \leq r\}$ in place of $\{\varphi \leq r\}$.

PROPOSITION 3.115. *Let X be a Hausdorff topological vector space, $\mathcal{B}$ be a symmetric bornology on X and $f : X \to \mathbb{R} \cup \{+\infty\}$ be a convex function which is $\mathcal{B}$-differentiable at $\bar{x} \in \mathrm{dom}\, f$. Then for every $B \in \mathcal{B}$, every real $r > 0$ and every $\varepsilon \in]0, r[$ one has*
$$\partial_\varepsilon f(\bar{x}) \subset \partial f(\bar{x}) + r\{p_B(\cdot) \leq 1\},$$
where we recall that $\{p_B(\cdot) \leq 1\} := \{x^* \in X^* : p_B(x^*) \leq 1\}$ and $p_B(x^*) := \sup_{v \in B} |\langle x^*, v \rangle|$ (see (3.111)).

PROOF. Denoting $\bar{x}^* := Df(\bar{x})$, by Proposition 3.102 we have $\partial f(\bar{x}) = \{\bar{x}^*\}$. Fix any $B \in \mathcal{B}$, any real $r > 0$ and any $\varepsilon \in]0, r[$. By $\mathcal{B}$-differentiability there exists a real $t > 0$ such that
$$\left|t^{-1}\left(f(\bar{x} + tv) - f(\bar{x})\right) - \langle \bar{x}^*, v \rangle\right| \leq r - \varepsilon \quad \text{for all } v \in B.$$
It ensues that for every $x^* \in \partial_\varepsilon f(\bar{x})$ and every $v \in B$
$$\langle x^*, v \rangle \leq t^{-1}\left(f(\bar{x} + tv) - f(\bar{x})\right) + \varepsilon \leq \langle \bar{x}^*, v \rangle + r,$$
which, combined with the analogous inequality with $-v$ in place of v, yields the inequality $|\langle x^* - \bar{x}^*, v \rangle| \leq r$, so $p_B(x^* - \bar{x}^*) \leq r$ as desired. □

Combining this proposition with Proposition 3.111 gives:

COROLLARY 3.116. Let X be a Hausdorff topological vector space, $\mathcal{B}$ be a symmetric bornology on X and $f : X \to \mathbb{R} \cup \{+\infty\}$ be a convex function which is continuous at $\bar{x} \in \operatorname{dom} f$ and $\mathcal{B}$-differentiable at $\bar{x}$. Then for every $B \in \mathcal{B}$ and every real $r > 0$ there exists an $\varepsilon_0 \in]0, r[$ and a neighborhood U of $\bar{x}$ in X such that for all $\varepsilon \in [0, \varepsilon_0]$ and all $x \in U$

$$\partial_\varepsilon f(x) \subset \partial f(\bar{x}) + r\{p_B(\cdot) \leq 1\}.$$

The following lemma prepares the next theorem.

LEMMA 3.117. Let X be a Hausdorff topological vector space and $f : X \to \mathbb{R} \cup \{+\infty\}$ be a convex function. Assume that f is subdifferentiable at $\bar{x} \in \operatorname{dom} f$ (that is, $\partial f(\bar{x}) \neq \emptyset$) and that for each $B \in \mathcal{B}$ and each real $r > 0$ there exists a neighborhood U of $\bar{x}$ such that

$$\partial f(\bar{x})(v) \subset \partial f(x)(v) + [-r, +r] \quad \text{for all } x \in U, v \in B,$$

where $\partial f(\bar{x})(v) := \{x^*(v) : x^* \in \partial f(\bar{x})\}$. Then the function f is $\mathcal{B}$-differentiable at the point $\bar{x}$.

PROOF. Choose some $\bar{x}^* \in \partial f(\bar{x})$. Fix any $B \in \mathcal{B}$ and any real $r > 0$ and let U be a neighborhood of $\bar{x}$ given by the assumption. Let a real $\tau > 0$ such that $\bar{x} + [0, \tau]B \subset U$. Consider any $t \in]0, \tau]$ and any $v \in B$, and choose $x^* \in \partial f(\bar{x} + tv)$ such that

$$|\langle \bar{x}^*, v\rangle - \langle x^*, v\rangle| \leq r.$$

Then, writing

$$t^{-1}\big(f(\bar{x} + tv) - f(\bar{x})\big) = -t^{-1}\big(f(\bar{x} + tv - tv) - f(\bar{x} + tv)\big) \leq \langle x^*, v\rangle,$$

we deduce that

$$t^{-1}\big(f(\bar{x} + tv) - f(\bar{x})\big) - \langle \bar{x}^*, v\rangle \leq \langle x^* - \bar{x}^*, v\rangle \leq r.$$

Consequently, we obtain for all $v \in B$ and all $t \in]0, \tau]$

$$0 \leq t^{-1}\big(f(\bar{x} + tv) - f(\bar{x})\big) - \langle \bar{x}^*, v\rangle \leq r,$$

which says that f is $\mathcal{B}$-differentiable at $\bar{x}$. □

We are now ready to characterize the $\mathcal{B}$-differentiability of a convex function by a certain continuity of its subdifferential multimapping.

THEOREM 3.118 (E. Asplund and R.T. Rockafellar). Let X be a Hausdorff topological vector space and $f : X \to \mathbb{R} \cup \{+\infty\}$ be a convex function which is continuous at $\bar{x} \in \operatorname{dom} f$, and let $\mathcal{B}$ be a symmetric bornology on X. The following assertions are equivalent:
(a) the function f is $\mathcal{B}$-differentiable at $\bar{x}$;
(b) for every $B \in \mathcal{B}$ and every real $r > 0$ there exists a neighborhood U of $\bar{x}$ such that for all $x \in U$

$$\partial f(\bar{x}) \subset \partial f(x) + r\{p_B(\cdot) \leq 1\} \quad \text{and} \quad \partial f(x) \subset \partial f(\bar{x}) + r\{p_B(\cdot) \leq 1\};$$

(c) for every $B \in \mathcal{B}$ and every real $r > 0$ there exist a real $\varepsilon_0 > 0$ and a neighborhood U of $\bar{x}$ such that for all $\varepsilon \in [0, \varepsilon_0]$ and all $x \in U$

$$\partial f(\bar{x}) \subset \partial_\varepsilon f(x) + r\{p_B(\cdot) \leq 1\} \quad \text{and} \quad \partial_\varepsilon f(x) \subset \partial f(\bar{x}) + r\{p_B(\cdot) \leq 1\};$$

3.4. ℬ-DIFFERENTIABILITY AND CONTINUITY OF SUBDIFFERENTIAL

(d) for every $B \in \mathcal{B}$ and every real $r > 0$ there exists a neighborhood U of $\bar{x}$ such that for all $x \in U$
$$\partial f(\bar{x}) \subset \partial f(x) + r\{p_B(\cdot) \leq 1\};$$

(e) for every $B \in \mathcal{B}$ and every real $r > 0$ there exists a neighborhood U of $\bar{x}$ such that for all $x \in U$ and all $v \in B$
$$\partial f(\bar{x})(v) \subset \partial f(x)(v) + [-r, r].$$

If the Hausdorff topological vector space X is locally convex, one can add to the above list of equivalences anyone of (f) and (g):

(f) for every $B \in \mathcal{B}$ and every real $r > 0$ there exist a real $\varepsilon_0 > 0$ and a neighborhood U of $\bar{x}$ such that for all $\varepsilon \in]0, \varepsilon_0]$ and all $x \in U$
$$\partial f(\bar{x}) \subset \partial_\varepsilon f(x) + r\{p_B(\cdot) \leq 1\} \quad \text{and} \quad \partial_\varepsilon f(x) \subset \partial f(\bar{x}) + r\{p_B(\cdot) \leq 1\}.$$

(g) for every $B \in \mathcal{B}$ and every real $r > 0$ there exist a real $\varepsilon_0 > 0$ and a neighborhood U of $\bar{x}$ such that for all $\varepsilon \in]0, \varepsilon_0]$ and all $x \in U$
$$\partial f(\bar{x}) \subset \partial_\varepsilon f(x) + r\{p_B(\cdot) \leq 1\}.$$

PROOF. By the continuity of f at $\bar{x}$ we note first that $\partial f(\bar{x}) \neq \emptyset$ according to Theorem 3.30(a).

Then the implication (e)⇒(a) follows from Lemma 3.117, while the implications (c)⇒(d)⇒(e) are obvious.

Now suppose that (a) holds, that is, f is ℬ-differentiable at $\bar{x}$. Take any real $r > 0$ and any $B \in \mathcal{B}$. By Corollary 3.116 there are a real $\varepsilon_0 > 0$ and a neighborhood U of $\bar{x}$ in X such that for all $\varepsilon \in [0, \varepsilon_0]$ and all $x \in U$
$$\partial_\varepsilon f(x) \subset \partial f(\bar{x}) + r\{p_B(\cdot) \leq 1\},$$

which in turn implies that
$$\partial f(\bar{x}) \subset \partial_\varepsilon f(x) + r\{p_B(\cdot) \leq 1\},$$

since $\partial f(\bar{x})$ is a singleton and $\{p_B(\cdot) \leq 1\}$ is a symmetric set. This justifies the implication (a)⇒(c), so the assertions (a), (c), (d), (e) are equivalent.

Finally, since the implications (c)⇒(b)⇒(d) are evident, we conclude that all the assertions (a)-(e) are pairwise equivalent.

Concerning (f) and (g) we note first that the implications (c)⇒(f)⇒(g) are obvious. To prove (g)⇒(d), assume now that X is locally convex and that (g) is satisfied. Take any $B \in \mathcal{B}$ and any real $r > 0$. By (g) there exist a real $\varepsilon_0 > 0$ and a neighborhood U of $\bar{x}$ such that for all $\varepsilon \in]0, \varepsilon_0]$ and all $x \in U$

(3.69)
$$\partial f(\bar{x}) \subset \partial_\varepsilon f(x) + r\{p_B(\cdot) \leq 1\}.$$

By Corollary 3.38 striking the neighborhood U of $\bar{x}$ we may suppose that $\partial f(x)$ is $w(X^*, X)$-compact for all $x \in U$. Further, the set $\{p_B(\cdot) \leq 1\}$ is $w(X^*, X)$-closed by the lower $w(X^*, X)$-semicontinuity of p_B due to the definition $p_B(x^*) = \sup_{u \in B} |\langle x^*, u \rangle|$. Therefore, it is easily seen that for each $x \in U$

(3.70)
$$\bigcap_{\varepsilon \in]0, \varepsilon_0]} (\partial_\varepsilon f(x) + r\{p_B(\cdot) \leq 1\}) = \partial f(x) + r\{p_B(\cdot) \leq 1\}.$$

Indeed, taking any x^* in the left-hand member and taking $(\varepsilon_n)_n$ in $]0, \varepsilon_0]$ tending to 0 there exists for each n some $y_n^* \in \partial_{\varepsilon_n} f(x)$ such that $x^* - y_n^* \in r\{p_B(\cdot) \leq 1\}$. Taking a subnet if necessary, we may suppose that $(y_n^*)_n$ $w(X^*, X)$-converges to

some y^*. Using the definition of $\partial_{\varepsilon_n} f(x)$ we see that $y^* \in \partial f(x)$. Further, the $w(X^*, X)$-closedness of $\{p_B(\cdot) \le 1\}$ gives
$$x^* - y^* \in r\{p_B(\cdot) \le 1\}, \quad \text{hence } x^* \in \partial f(x) + r\{p_B(\cdot) \le 1\}.$$
This justifies the equality (3.70). From that equality (3.70) and the inclusion in (3.69) we obtain that for all $x \in U$
$$\partial f(\bar{x}) \subset \partial f(x) + r\{p_B(\cdot) \le 1\}.$$
This tells us that (d) holds true as desired, so the proof is finished. $\square$

The two following corollaries in the normed space setting directly follow.

COROLLARY 3.119 (E. Asplund and R.T. Rockafellar). *Let $(X, \|\cdot\|)$ be a normed space and $f : X \to \mathbb{R} \cup \{+\infty\}$ be a convex function which is continuous at $\bar{x} \in \mathrm{dom}\, f$. Then the following are equivalent:*
(a) *the function f is Fréchet differentiable at $\bar{x}$;*
(b) *for every real $r > 0$ there exists a neighborhood U of $\bar{x}$ such that for all $x \in U$*
$$\partial f(\bar{x}) \subset \partial f(x) + r\mathbb{B}_{X^*} \quad \text{and} \quad \partial f(x) \subset \partial f(\bar{x}) + r\mathbb{B}_{X^*},$$
otherwise stated, the multimapping ∂f is Hausdorff continuous at $\bar{x}$, that is, with respect to the Hausdorff-Pompeiu distance on subsets of X^*;
(c) *for every real $r > 0$ there exist a real $\varepsilon_0 > 0$ and a neighborhood U of $\bar{x}$ such that for all $\varepsilon \in [0, \varepsilon_0]$ and all $x \in U$*
$$\partial f(\bar{x}) \subset \partial_\varepsilon f(x) + r\mathbb{B}_{X^*} \quad \text{and} \quad \partial_\varepsilon f(x) \subset \partial f(\bar{x}) + r\mathbb{B}_{X^*};$$
(d) *for every real $r > 0$ there exist a real $\varepsilon_0 > 0$ and a neighborhood U of $\bar{x}$ such that for all $\varepsilon \in \,]0, \varepsilon_0]$ and all $x \in U$*
$$\partial f(\bar{x}) \subset \partial_\varepsilon f(x) + r\mathbb{B}_{X^*} \quad \text{and} \quad \partial_\varepsilon f(x) \subset \partial f(\bar{x}) + r\mathbb{B}_{X^*};$$
(e) *for every real $r > 0$ there exists a neighborhood U of $\bar{x}$ such that for all $x \in U$*
$$\partial f(\bar{x}) \subset \partial f(x) + r\mathbb{B}_{X^*},$$
that is, the multimapping ∂f is Hausdorff lower semicontinuous at $\bar{x}$;
(f) *for every real $r > 0$ there exist a real $\varepsilon_0 > 0$ and a neighborhood U of $\bar{x}$ such that for all $\varepsilon \in \,]0, \varepsilon_0]$ and all $x \in U$*
$$\partial f(\bar{x}) \subset \partial_\varepsilon f(x) + r\mathbb{B}_{X^*}.$$

The second corollary provides, in addition to Proposition 2.211 and Proposition 3.102, other subdifferential conditions for the Gâteaux differentiability of convex functions.

COROLLARY 3.120 (E. Asplund and R.T. Rockafellar). *Let X be a Hausdorff locally convex space and $f : X \to \mathbb{R} \cup \{+\infty\}$ be a convex function which is continuous at $\bar{x} \in \mathrm{dom}\, f$. Then the following are equivalent:*
(a) *the function f is Gâteaux differentiable at $\bar{x}$;*
(b) *for every $w(X^*, X)$ neighborhood W of zero in X^* there exists a neighborhood U of $\bar{x}$ such that for all $x \in U$*
$$\partial f(\bar{x}) \subset \partial f(x) + W \quad \text{and} \quad \partial f(x) \subset \partial f(\bar{x}) + W;$$
(c) *for every $w(X^*, X)$ neighborhood W of zero in X^* there exist a real $\varepsilon_0 > 0$ and a neighborhood U of $\bar{x}$ such that for all $\varepsilon \in [0, \varepsilon_0]$ and all $x \in U$*
$$\partial f(\bar{x}) \subset \partial_\varepsilon f(x) + W \quad \text{and} \quad \partial_\varepsilon f(x) \subset \partial f(\bar{x}) + W;$$

3.4. $\mathcal{B}$-DIFFERENTIABILITY AND CONTINUITY OF SUBDIFFERENTIAL

(d) for every $w(X^*, X)$ neighborhood W of zero in X^* there exist a real $\varepsilon_0 > 0$ and a neighborhood U of $\overline{x}$ such that for all $\varepsilon \in \,]0, \varepsilon_0]$ and all $x \in U$

$$\partial f(\overline{x}) \subset \partial_\varepsilon f(x) + W \quad \text{and} \quad \partial_\varepsilon f(x) \subset \partial f(\overline{x}) + W;$$

(e) for every $w(X^*, X)$ neighborhood W of zero in X^* there exists a neighborhood U of $\overline{x}$ such that for all $x \in U$

$$\partial f(\overline{x}) \subset \partial f(x) + W;$$

(f) for every $w(X^*, X)$ neighborhood W of zero in X^* there exist a real $\varepsilon_0 > 0$ and a neighborhood U of $\overline{x}$ such that for all $\varepsilon \in \,]0, \varepsilon_0]$ and all $x \in U$

$$\partial f(\overline{x}) \subset \partial_\varepsilon f(x) + W.$$

We also have another corollary.

COROLLARY 3.121. *Let X be a Hausdorff locally convex space and $f : X \to \mathbb{R} \cup \{+\infty\}$ be a convex function which is continuous at $\overline{x} \in \mathrm{dom}\, f$, and let $\mathcal{B}$ be a symmetric bornology on X. Let $\overline{x}^* \in X^*$. The following assertions are equivalent:*
(a) *the function f is $\mathcal{B}$-differentiable at $\overline{x}$ with $Df(\overline{x}) = \overline{x}^*$;*
(b) *for every $B \in \mathcal{B}$ and every real $r > 0$ there exists a neighborhood U of $\overline{x}$ such that for all $x \in U$*

$$\{\overline{x}^*\} \subset \partial f(x) + r\{p_B(\cdot) \leq 1\} \quad \text{and} \quad \partial f(x) \subset \overline{x}^* + r\{p_B(\cdot) \leq 1\};$$

(c) *for every $B \in \mathcal{B}$ and every real $r > 0$ there exists a neighborhood U of $\overline{x}$ such that for all $x \in U$*

$$\partial f(x) \subset \overline{x}^* + r\{p_B(\cdot) \leq 1\};$$

(d) *for every $B \in \mathcal{B}$ and every real $r > 0$ there exist a real $\varepsilon_0 > 0$ and a neighborhood U of $\overline{x}$ such that for all $\varepsilon \in \,]0, \varepsilon_0]$ (resp. all $\varepsilon \in [0, \varepsilon_0]$) and all $x \in U$*

$$\{\overline{x}^*\} \subset \partial_\varepsilon f(x) + r\{p_B(\cdot) \leq 1\} \quad \text{and} \quad \partial_\varepsilon f(x) \subset \overline{x}^* + r\{p_B(\cdot) \leq 1\};$$

(e) *for every $B \in \mathcal{B}$ and every real $r > 0$ there exist a real $\varepsilon_0 > 0$ and a neighborhood U of $\overline{x}$ such that for all $\varepsilon \in \,]0, \varepsilon_0]$ (resp. all $\varepsilon \in [0, \varepsilon_0]$) and all $x \in U$*

$$\partial_\varepsilon f(x) \subset \overline{x}^* + r\{p_B(\cdot) \leq 1\}.$$

PROOF. We note first that (b)$\Leftrightarrow$(c) (resp. (d)$\Leftrightarrow$(e)) since the left-hand inclusion in (b) (resp. (d)) is a consequence of the right-hand one. We also note by Theorem 3.118 that (a)$\Rightarrow$(b) and (a)$\Rightarrow$(d).

It remains to prove (c)$\Rightarrow$(a) and (e)$\Rightarrow$(a). Let us justify the second implication relative to $]0, \varepsilon_0]$, the proof of the others being similar. Suppose that (e) holds. By the inclusion $\partial f(\overline{x}) \subset \partial_\varepsilon f(\overline{x})$ this entails in particular $\partial f(\overline{x}) \subset \overline{x}^* + r\{p(\cdot) \leq 1\}$ for every $B \in \mathcal{B}$ and every $r > 0$. Since $\partial f(\overline{x}) \neq \emptyset$ by continuity of f at $\overline{x}$ (see Theorem 3.30(a)), choosing any $x^* \in \partial f(\overline{x})$, taking any $v \in X$ and considering some $B \in \mathcal{B}$ with $B \ni v$ it ensues that $\langle x^*, v\rangle \leq \langle \overline{x}^*, v\rangle + r$ for all $r > 0$, hence $\langle x^*, v\rangle \leq \langle \overline{x}^*, v\rangle$, which in turn entails that $x^* = \overline{x}^*$. Therefore $\partial f(\overline{x}) = \{\overline{x}^*\}$.

Fix any $B \in \mathcal{B}$ and any real $r > 0$. Clearly, the inclusion in (e) furnishes a real $\varepsilon_0 > 0$ and a neighborhood U of $\overline{x}$ such that for all $\varepsilon \in \,]0, \varepsilon_0]$ and all $x \in U$

$$\{\overline{x}^*\} \subset \partial_\varepsilon f(x) + r\{p_B(\cdot) \leq 1\} \quad \text{and} \quad \partial_\varepsilon f(x) \subset \overline{x}^* + r\{p_B(\cdot) \leq 1\},$$

which means

$$\partial f(\overline{x}) \subset \partial_\varepsilon f(x) + r\{p_B(\cdot) \leq 1\} \quad \text{and} \quad \partial_\varepsilon f(x) \subset \partial f(\overline{x}) + r\{p_B(\cdot) \leq 1\}.$$

By Theorem 3.118 and the equality $\partial f(\bar{x}) = \{\bar{x}^*\}$ the assertion (a) follows, and the proof is finished. □

3.4.4. Lipschitz continuity of ε-subdifferential. Denoting by Cont f the sets of points where a function $f : X \to \mathbb{R} \cup \{+\infty\}$ is finite and continuous, the next theorem proves that the multimapping

$$]0, +\infty[\times \operatorname{Cont} f \ni (\varepsilon, x) \mapsto \partial_\varepsilon f(x)$$

is locally Lipschitz in a certain sense. This shows in particular that the inclusions with $\varepsilon \in]0, \varepsilon_0]$

$$\partial_\varepsilon f(\bar{x}) \subset \partial_\varepsilon f(x) + r\{p_B(\cdot) \le 1\} \quad \text{and} \quad \partial_\varepsilon f(x) \subset \partial_\varepsilon f(\bar{x}) + r\{p_B(\cdot) \le 1\}$$

may be used neither in place of the inclusions in (c) of Theorem 3.118 nor in place of the inclusions in (b) of Corollary 3.121.

THEOREM 3.122 (Lipschitz continuity of ε-subdifferential). *Let X be a topological vector space and $f : X \to \mathbb{R} \cup \{+\infty\}$ be a convex function with $\operatorname{Cont} f \ne \emptyset$. Let $\varepsilon_0 \in]0, +\infty[$ and $\bar{x} \in \operatorname{Cont} f$. Then for any nonempty bounded subset $B \subset X$ there exist three reals $\alpha, \beta, \gamma > 0$ with $\alpha < \varepsilon_0 < \beta$, a neighborhood U of $\bar{x}$ and a closed balanced neighborhood W of zero in X such that for all $\varepsilon, \varepsilon' \in]\alpha, \beta[$ and all $x, x' \in U$*

$$\partial_\varepsilon f(x) \subset \partial_{\varepsilon'} f(x') + \gamma \left(|\varepsilon - \varepsilon'| + \mathfrak{q}_W(x - x') \right) \{p_B(\cdot) \le 1\}.$$

PROOF. Fix two reals α, β with $0 < \alpha < \varepsilon_0 < \beta$ and $\alpha^{-1}(\beta - \alpha) < 1/4$. Let B be any nonempty bounded subset of X. By Proposition 3.37(a) there is a neighborhood V_0 of zero in X such that $\bigcup_{x \in \bar{x} + V_0} \partial_\beta f(x)$ is equicontinuous in X^*. Then we can choose a neighborhood $V_1 \subset V_0$ of zero in X such that $(x^* - y^*)(V_1) \subset [-\varepsilon/2, \varepsilon/2]$ for all $x^*, y^* \in \partial_\beta f(\bar{x} + V_0)$, so choosing a real $s > 0$ such that $B \subset sV_1$ we see that

(3.71) $\quad (x^* - y^*)(B) \subset [-\varepsilon s/2, \varepsilon s/2] \quad$ for all $x^*, y^* \in \partial_\beta f(\bar{x} + V_0)$.

We also know by Proposition 3.14 that there is a closed balanced neighborhood W of zero and a balanced neighborhood $V_2 \subset V_1$ of zero in X such that for all $y, y' \in V_2$ and for all $x^* \in \partial_\beta f(\bar{x} + V_0)$

(3.72) $|f(\bar{x}+y)-f(\bar{x}+y')| \le (2s)^{-1}\mathfrak{q}_W(y-y')$ and $|\langle x^*, y-y'\rangle| \le (2s)^{-1}\mathfrak{q}_W(y-y')$.

Choose another balanced neighborhood $V \subset V_2$ of zero in X such that

$$(\alpha s)^{-1}\mathfrak{q}_W(y - y') < 1/4 \quad \text{for all } y, y' \in V.$$

Fix two arbitrary points $v, v' \in V$ and two arbitrary reals $\varepsilon, \varepsilon' \in]\alpha, \beta[$ and put

(3.73) $\qquad t := (1/\varepsilon')\big(|\varepsilon - \varepsilon'| + s^{-1}\mathfrak{q}_W(v - v')\big).$

We note that $0 < t < 1/2$, so setting $\varepsilon'_1 := t\varepsilon'$ and $\varepsilon'_2 := (1+t)\varepsilon'$ we obtain for any $y \in V$

$$t\partial_{\varepsilon'_1} f(\bar{x} + y) + (1-t)\partial_{\varepsilon'_2} f(\bar{x} + y) \subset \partial_{\varepsilon'} f(\bar{x} + y),$$

which ensures that

$$\partial_{\varepsilon'_2} f(\bar{x} + y) \subset \partial_{\varepsilon'} f(\bar{x} + y) + (1-t)^{-1}t\big(\partial_{\varepsilon'} f(\bar{x} + y) - \partial_{\varepsilon'_1} f(\bar{x} + y)\big).$$

Since $\varepsilon'_1 < \varepsilon'$, we deduce that for every $y \in V$

(3.74) $\qquad \partial_{\varepsilon'_2} f(\bar{x} + y) \subset \partial_{\varepsilon'} f(\bar{x} + y) + (1-t)^{-1}t\big(\partial_{\varepsilon'} f(\bar{x} + y) - \partial_{\varepsilon'} f(\bar{x} + y)\big).$

Take now any $u^* \in \partial_\varepsilon f(\bar{x}+v)$. By the very definition of ε-subdifferential we have for every $u \in X$

$$\begin{aligned}\langle u^*, u-\bar{x}-v'\rangle &= \langle u^*, u-\bar{x}-v\rangle + \langle u^*, v-v'\rangle\\ &\leq f(u)-f(\bar{x}+v)+\varepsilon+\langle u^*, v-v'\rangle\\ &= \bigl(f(u)-f(\bar{x}+v')\bigr)+\bigl(f(\bar{x}+v')-f(\bar{x}+v)\bigr)+\langle u^*, v-v'\rangle+\varepsilon,\end{aligned}$$

which gives by (3.72)

$$\begin{aligned}\langle u^*, u-\bar{x}-v'\rangle &\leq f(u)-f(\bar{x}+v')+s^{-1}\mathfrak{q}_W(v-v')+\varepsilon\\ &\leq f(u)-f(\bar{x}+v')+\bigl(s^{-1}\mathfrak{q}_W(v-v')+|\varepsilon-\varepsilon'|+\varepsilon\bigr).\end{aligned}$$

Consequently, invoking the definition of t in (3.73) we get that $u^* \in \partial_{\varepsilon'_2} f(\bar{x}+v')$, thus by (3.74)

$$u^* \in \partial_{\varepsilon'} f(\bar{x}+v') + (1-t)t^{-1}\bigl(\partial_{\varepsilon'} f(\bar{x}+v') - \partial_{\varepsilon'} f(\bar{x}+v')\bigr).$$

This tells us that there are v^*, x_1^*, x_2^* all three in $\partial_{\varepsilon'} f(\bar{x}+v')$ such that $u^* = v^* + (1-t)^{-1}t(x_1^* - x_2^*)$. Using (3.71) and the inequality $(1-t)^{-1} < 2$ (keep in mind that $0 < t < 1/2$), it ensues that for every $b \in B$

$$|\langle u^* - v^*, b\rangle| = (1-t)^{-1}t|\langle x_1^* - x_2^*, b\rangle| \leq 2^{-1}(1-t)^{-1}st \leq st.$$

Putting $\gamma := \alpha^{-1}(1+s)$ and noticing that

$$st = (1/\varepsilon')\bigl(s|\varepsilon - \varepsilon'| + \mathfrak{q}_B(v-v')\bigr) \leq \gamma\bigl(|\varepsilon - \varepsilon'| + \mathfrak{q}_B(v-v')\bigr),$$

it follows that for every $b \in B$

$$|\langle u^* - v^*, b\rangle| \leq \gamma\bigl(|\varepsilon - \varepsilon'| + \mathfrak{q}_B(v-v')\bigr),$$

or equivalently $p_B(u^* - v^*) \leq \gamma\bigl(|\varepsilon - \varepsilon'| + \mathfrak{q}_B(v-v')\bigr)$. It results that

$$u^* \in v^* + \gamma\bigl(|\varepsilon - \varepsilon'| + \mathfrak{q}_B(v-v')\bigr)\{p_B(\cdot) \leq 1\},$$

which yields

$$\partial_\varepsilon f(\bar{x}+v) \subset \partial_{\varepsilon'} f(\bar{x}+v') + \gamma\bigl(|\varepsilon - \varepsilon'| + \mathfrak{q}_B(v-v')\bigr)\{p_B(\cdot) \leq 1\}$$

and finishes the proof of the theorem. □

The following corollary translates the above theorem in the case of a normed space X; it is obtained with B as the closed unit ball in X.

COROLLARY 3.123. *Let X be a normed space and $f : X \to \mathbb{R} \cup \{+\infty\}$ be a convex function with* Cont $f \neq \emptyset$. *Then the multimapping* $(\varepsilon, x) \mapsto \partial_\varepsilon f(x)$ *is locally Lipschitz on* $]0, +\infty[\times$ Cont f *with respect to the Hausdorff distance on subsets of* X^*, *that is, for any $\varepsilon_0 \in]0, +\infty[$ and any $\bar{x} \in$ Cont f there are a real $\gamma > 0$, an open interval $I \subset]0, +\infty[$ containing ε_0 and a neighborhood $U \subset$ Cont f of $\bar{x}$ in X such that*

$$\partial_\varepsilon f(x) \subset \partial_{\varepsilon'} f(x') + \gamma\bigl(|\varepsilon - \varepsilon'| + \|x - x'\|\bigr)\mathbb{B}_{X^*}$$

for all $\varepsilon, \varepsilon' \in I$ and $x, x' \in U$.

3.5. Asymptotic functions and cones

Given a convex function $f : X \to \mathbb{R} \cup \{-\infty, +\infty\}$, Corollary 3.79 established that the Hadamard directional (lower) derivative $\underline{d}_H^+(\bar{x}; \cdot)$ is the support function of $\partial f(\bar{x})$ whenever $\partial f(\bar{x}) \neq \emptyset$. Assuming that the convex function is proper and lower semicontinuous, for any real $\varepsilon > 0$ we will show similarly that the support function of $\partial_\varepsilon f(\bar{x})$ coincides with the ε-directional derivative $f'_\varepsilon(\bar{x}; \cdot)$. We will obtain that as an application of the concept of asymptotic function that we now define and develop.

3.5.1. Definitions and general properties.
Let $f : X \to \mathbb{R} \cup \{-\infty, +\infty\}$ be a function on a topological vector space X. At any $x \in X$ where f is finite, consider similarly to the definition of $\underline{d}_H^+ f(x; v)$ the expression $\liminf_{\substack{t \to +\infty \\ u \to v}} t^{-1}(f(x + tu) - f(x))$.

Clearly, the latter limit inferior coincides with $\liminf_{\substack{t \to +\infty \\ u \to v}} t^{-1} f(x + tu)$. Further, writing

$$t^{-1} f(x + tu) = t^{-1} f(t(u + t^{-1} x)) \quad \text{and} \quad t^{-1} f(tw) = t^{-1} f(x + t(w - t^{-1} x)),$$

we see that respectively

$$\liminf_{\substack{t \to +\infty \\ u \to v}} t^{-1} f(x + tu) \geq \liminf_{\substack{t \to +\infty \\ w \to v}} t^{-1} f(tw) \quad \text{and} \quad \liminf_{\substack{t \to +\infty \\ w \to v}} t^{-1} f(tw) \geq \liminf_{\substack{t \to +\infty \\ u \to v}} t^{-1} f(x + tu).$$

This gives $\liminf_{\substack{t \to +\infty \\ u \to v}} t^{-1} f(x + tu) = \liminf_{\substack{t \to +\infty \\ u \to v}} t^{-1} f(tu)$ and leads to the following definition.

DEFINITION 3.124. Let X be a topological vector space and $f : X \to \mathbb{R} \cup \{-\infty, +\infty\}$ be an extended real-valued function. The *asymptotic function* of f (also called *horizon function* or *recession function* of f) is the function $f^\infty : X \to \mathbb{R} \cup \{-\infty, +\infty\}$ defined by

$$f^\infty(v) := \liminf_{\substack{t \to +\infty \\ u \to v}} t^{-1} f(tu) = \liminf_{\substack{r \downarrow 0 \\ u \to v}} r f(r^{-1} u);$$

(one also finds in the literature the notation $(f0^+)(v)$). Further, for any $x \in X$ with $|f(x)| < +\infty$ one also has

$$f^\infty(v) = \liminf_{\substack{t \to +\infty \\ u \to v}} t^{-1} \big(f(x + tu) - f(x)\big) = \liminf_{\substack{r \downarrow 0 \\ u \to v}} r \big(f(x + r^{-1} u) - f(x)\big).$$

The following properties follow directly from the definition.

PROPOSITION 3.125. Let $f, g : X \to \mathbb{R} \cup \{-\infty, +\infty\}$ be functions from a topological vector space X.
(a) The function f^∞ is positively homogeneous and lower semicontinuous.
(b) If f is positively homogeneous, then $f^\infty = \operatorname{cl} f$.
(c) $f \leq g \Rightarrow f^\infty \leq g^\infty$.
(d) The inequality $f^\infty + g^\infty \leq (f + g)^\infty$ holds whenever the sums are well-defined; and the inequality is an equality if additionally g is continuous and positively homogeneous.

Let S be a subset of X. Since we have $\Psi_S(X) \subset \{0, +\infty\}$, the (lower semicontinuous positively homogeneous) function $(\Psi_S)^\infty$ takes also values in $\{0, +\infty\}$, hence it is the indicator function of a closed cone S^∞. Further, from the very definition of asymptotic function, $(\Psi_S)^\infty(v) = 0$ if and only if there are nets of positive reals $(t_j)_{j \in J}$ tending to $+\infty$ and $(v_j)_{j \in J}$ converging to v such that $\Psi_S(t_j v_j) = 0$, that is, $v_j \in t_j^{-1} S$ for all $j \in J$. This means that $S^\infty = \mathrm{Lim\,sup}_{t \to +\infty}(t^{-1} S)$. It is then also clear by Proposition 1.16(b) that for any $x \in S$ one also has $S^\infty = \mathrm{Lim\,sup}_{t \to +\infty}(t^{-1}(S - x))$.

DEFINITION 3.126. Let S be a subset of a topological vector space X. The closed cone
$$S^\infty := \mathrm{Lim\,sup}_{t \to +\infty} \left(\frac{1}{t} S \right)$$
is called the *asymptotic cone* of S or the *cone of asymptotic directions* of S; it is the closed cone satisfying the equality $(\Psi_S)^\infty = \Psi_{S^\infty}$. When S is nonempty, one also has for any $x \in S$
$$S^\infty = \mathrm{Lim\,sup}_{t \to +\infty} \left(\frac{1}{t}(S - x) \right).$$
In some situations it can be more convenient to write S_∞ in place of S^∞; this is in particular the case when a superscript is already present. (The cone S^∞ is also called in the literature the *recession cone* of S and denoted by $0^+ S$).

PROPOSITION 3.127. *Let S, Q be subsets of a topological vector space X and R be a subset of another topological vector space Y.*

(a) *The equalities* $(\mathrm{cl}\, S)^\infty = S^\infty$ *and* $S^\infty = \bigcap_{r > 0} \mathrm{cl} \left(\bigcup_{t \geq r} \frac{1}{t} S \right)$ *hold, and if* $S \neq \emptyset$ *one also has for any* $x \in S$
$$S^\infty = \bigcap_{r > 0} \mathrm{cl} \left(\bigcup_{t \geq r} \frac{1}{t}(S - x) \right).$$

(b) *If S is convex, then S^∞ is a closed convex cone; further, for any $x \in S$ (if $S \neq \emptyset$)*
$$S^\infty = \bigcap_{t > 0} \mathrm{cl} \left(\frac{1}{t}(S - x) \right) = \mathrm{Lim}_{t \to +\infty} \left(\frac{1}{t}(S - x) \right),$$
where $\mathrm{Lim}\, t^{-1}(S - x)$ means that the semilimits inferior and superior coincide.

(c) *If S is a cone, then $S^\infty = \mathrm{cl}\, S$.*

(d) $Q \subset S \Rightarrow Q^\infty \subset S^\infty$.

(e) $(S \times R)^\infty \subset S^\infty \times R^\infty$ *and this inclusion is an equality whenever S is convex.*

(f) *If Q is nonempty and bounded, then*
$$Q^\infty = \{0\} \quad \text{and} \quad (S + Q)^\infty = S^\infty.$$

(g) *The equality* $(S \cup Q)^\infty = S^\infty \cup Q^\infty$ *holds; if Q is nonempty and bounded and S is nonempty, then* $(S \cup Q)^\infty = S^\infty$.

(h) *If X is Hausdorff and finite-dimensional, then Q is bounded if and only if $Q^\infty = \{0\}$.*

PROOF. The assertions (a), (c), (d) follow directly from the definition of S^∞. Concerning (f), assuming that Q is nonempty and bounded, Proposition 1.16(b) tells us that, for any $v \in S^\infty$ there are $t_j \to +\infty$ with $t_j > 0$ and $v_j \in Q$ with $(1/t_j) v_j \to v$, so $v = 0$ since $1/t_j \to 0$ and $(v_j)_j$ is bounded. This confirms the first equality in (f), and the other equality in (f) is obtained similarly.

Concerning the first equality in (g), we may suppose that $S \cup Q \neq \emptyset$, since otherwise it is obvious. Let any $u \in (S \cup Q)^\infty$. By definition there are a net $(x_j)_{j \in J}$ in $S \cup Q$ and a net $(t_j)_{j \in J}$ in $]0, +\infty[$ tending to 0 such that $t_j x_j \to u$. Clearly, there is a subnet $(x_{s(i)})_{i \in I}$ contained in one of the two sets, say S. This and the convergence $t_{s(i)} x_{s(i)} \to u$ say that $u \in S^\infty$. It ensues that $(S \cup Q)^\infty \subset S^\infty \cup Q^\infty$, and this inclusion is an equality since the converse inclusion is a direct consequence of (d). If Q is nonempty and bounded and S is nonempty, then $(S \cup Q)^\infty = S^\infty \cup \{0\}$ (according to $Q^\infty = \{0\}$ from (f)), so $(S \cup Q)^\infty = S^\infty$ since $0 \in S^\infty$ due the nonemptiness of S.

Assume now that X is Hausdorff and finite-dimensional and that Q is unbounded. Choose a norm $\|\cdot\|$ on X whose associated topology coincides with the initial Hausdorff vector topology of X. There is a sequence $(v_n)_n$ in Q with $\|v_n\| \to +\infty$, so with $t_n = \|v_n\|$ we see that $t_n \to +\infty$ and a subsequence of $t_n^{-1} v_n$ converges to some $u \in S^\infty$ with $\|u\| = 1$, which justifies (h) according to the first equality in (f).

Let us argue (b). The case $S = \emptyset$ is trivial since $S^\infty = \emptyset$ in this case. Assume that S is a nonempty convex set and fix any $x \in S$. By Definition 3.126 we have
$$S^\infty = \bigcap_{r>0} \mathrm{cl}\left(\bigcup_{t \geq r} t^{-1}(S-x)\right) \quad \text{(see Proposition 1.7)}.$$
Noting from the inclusion $0 \in S - x$ and the convexity of $S - x$ that $t^{-1}(S-x) \subset r^{-1}(S-x)$ for any reals $t \geq r > 0$, we deduce that $S^\infty = \bigcap_{r>0} \mathrm{cl}\left(r^{-1}(S-x)\right)$. This confirms the convexity of S^∞ as well as the first equality in (b). From the equality $\limsup_{r \to +\infty} r^{-1}(S-x) = \bigcap_{r>0} \mathrm{cl}\left(r^{-1}(S-x)\right)$ it is also clear that the latter outer semilimit (or semilimit superior) coincides with the corresponding inner semilimit (or semilimit inferior).

Finally, let us consider (e). The inclusion in (e) can be seen directly or with the use of Proposition 1.5(a). The latter inclusion being trivially an equality if anyone of the two sets is empty, suppose that S is convex and both sets are nonempty. Taking $(x, y) \in S \times R$ the assertion (b) tells us that the inner and outer semilimits (or semilimits inferior and superior) of $t^{-1}(S-x)$ as $t \to +\infty$ coincide, hence the equality in (e) can be deduced by applying Proposition 1.5(a) to the outer semilimit of $t^{-1}\big((S-x) \times (R-y)\big)$ as $t \to +\infty$. $\square$

The asymptotic cone of a closed convex set enjoys simple useful descriptions.

PROPOSITION 3.128. *Let S be a nonempty closed convex set of a topological vector space X. Then one has*
$$S^\infty = \bigcap_{t>0} \frac{1}{t}(S-x) = \{u \in X : x + \mathbb{R}_+ u \subset S\} \quad \text{for every } x \in S,$$
along with the equalities
$$S^\infty = \{u \in X : S + \mathbb{R}_+ u \subset S\} = \{u \in X : u + S \subset S\}.$$

PROOF. Since the convex set S is closed, the equality between the first two members of the first chain follows from the first equality in (b) of Proposition 3.127; further, the third member of this first chain is just a translation of the second member there. On the other hand, the above equality $S^\infty = \{u \in X : x + \mathbb{R}_+ u \subset S\}$ for every $x \in S$ clearly gives the equality between S^∞ and the second member in the second chain. Now take any u in the third member of the second chain, that is,

$u + S \subset S$. This entails that $2u + S = u + (u + S) \subset u + (S) \subset S$, so by induction we have $nu + S \subset S$ for every $n \in \mathbb{N}$. Considering any real $t \geq 0$ and choosing $k \in \mathbb{N}$ with $k \geq t$, we obtain by convexity of S

$$tu + S = \frac{t}{k}(ku + S) + \left(1 - \frac{t}{k}\right) S \subset \frac{t}{k} S + \left(1 - \frac{t}{k}\right) S = S,$$

which says that $S + \mathbb{R}_+ u \subset S$, and hence the third member of the second chain is included in the second member of this chain. This finishes the proof. □

As a first corollary we have:

COROLLARY 3.129. *For every nonempty closed convex set S of a Hausdorff locally convex space X, one has*

$$S^\infty = (\mathrm{dom}\,\sigma_S)^\circ,$$

where we recall that $\sigma_S := \sigma(\cdot, S)$ is the support function of S and where the polar is taken in X for X^* endowed with the $w(X^*, X)$ topology.

PROOF. We note first that $\mathrm{dom}\,\sigma_S$ is a convex cone in X.

Consider any $u \in S^\infty$. Take any $x^* \in \mathrm{dom}\,\sigma_S$ and put $\beta := \sigma_S(x^*) \in \mathbb{R}$. Choosing $\bar{x} \in S$, by Proposition 3.128 for every real $t > 0$ we have $\bar{x} + tu \in S$, hence $\langle x^*, \bar{x} + tu \rangle \leq \beta$, which in turn gives $\langle x^*, u \rangle \leq t^{-1}(\beta - \langle x^*, \bar{x} \rangle)$. Taking the limit as $t \to +\infty$ entails $\langle x^*, u \rangle \leq 0$. This being true for any $x^* \in \mathrm{dom}\,\sigma_S$, it follows that $u \in (\mathrm{dom}\,\sigma_S)^\circ$.

Conversely, endow X^* with the $w(X^*, X)$ topology (so its topological dual is identified with X) and consider any $u \in (\mathrm{dom}\,\sigma_S)^\circ$. Take any $x^* \in X^*$ and suppose for a moment that $\sigma_S(x^*) < +\infty$, that is, $x^* \in \mathrm{dom}\,\sigma_S$. Then, $\langle x^*, u \rangle \leq 0$, thus

$$\sigma(x^*, u + S) = \langle x^*, u \rangle + \sigma(x^*, S) \leq \sigma(x^*, S).$$

The latter inequality being obvious if $\sigma(x^*, S) = +\infty$, it results that $\sigma(x^*, u+S) \leq \sigma(x^*S)$ for all $x^* \in X^*$, which implies that $u + S \subset S$ since S is a closed convex set. This says that $u \in S^\infty$ according to Proposition 3.128 again. This finishes the proof of the corollary. □

From the above proposition we can also derive the expression of the asymptotic cone of the polar S° of a closed convex set S containing zero in terms of the support function of the set S. Because of the superscript in S°, according to our convention in Definition 3.126 we will denote the asymptotic cone by $(S^\circ)_\infty$ instead of $(S^\circ)^\infty$.

COROLLARY 3.130. *Let X be a Hausdorff locally convex set and S be a nonempty set in X. Then one has the equality*

$$(S^\circ)_\infty = \{x^* \in X^* : \sigma(x^*, S) \leq 0\}.$$

If in addition $0 \in \mathrm{cl}\,S$, one also has

$$(S^\circ)_\infty = \{x^* \in X^* : \sigma(x^*, S) = 0\}.$$

PROOF. Endow X^* with the $w(X^*, X)$-topology and note that S° is a $w(X^*, X)$ closed convex set in X^* containing zero.

By Proposition 3.128 we have $(S^\circ)_\infty + S^\circ \subset S^\circ$, hence in particular $(S^\circ)_\infty \subset S^\circ$. Consider any $x^* \in (S^\circ)_\infty$. Noting for any $n \in \mathbb{N}$ that $nx^* \in (S^\circ)_\infty$, it ensues that $nx^* \in S^\circ$, so $\sigma(nx^*, S) \leq 1$, which clearly implies $\sigma(x^*, S) \leq 1/n$. This being true for any $n \in \mathbb{N}$, it follows that $\sigma(x^*, S) \leq 0$.

Conversely, suppose that $\sigma(x^*, S) \leq 0$. Then for any $u^* \in S^\circ$ we have
$$\sigma(x^* + u^*, S) \leq \sigma(x^*, S) + \sigma(u^*, S) \leq 1,$$
hence $x^* + u^* \in S^\circ$. Consequently, $x^* + S^\circ \subset S^\circ$, thus Proposition 3.128 again tells us that $x^* \in (S^\circ)_\infty$.

The first equality is then justified and the second follows from the first and the inequality $\sigma(\cdot, S) \geq 0$ due to the assumption $0 \in \mathrm{cl} S$. □

While the inclusion $\left(\bigcap_{i \in I} S_i\right)^\infty \subset \bigcap_{i \in I} S_i^\infty$ always holds, the following third corollary proves the equality for closed convex sets with nonempty intersection.

COROLLARY 3.131. *Let $(S_i)_{i \in I}$ be a family of closed convex subsets of a topological vector space X such that $\bigcap_{i \in I} S_i \neq \emptyset$. Then one has*
$$\left(\bigcap_{i \in I} S_i\right)^\infty = \bigcap_{i \in I} S_i^\infty.$$

PROOF. The left member is evidently included in the right. Take any u in the right member. For any $i \in I$, by Proposition 3.128 we have $u + S_i \subset S_i$, hence taking the intersection gives $u + \bigcap_{i \in I} S_i \subset \bigcap_{i \in I} S_i$. Proposition 3.128 again tells us that $u \in \left(\bigcap_{i \in I} S_i\right)^\infty$. This justifies the desired equality since the second member is always included in the first as said above. □

Concerning the case of two nonclosed convex sets we also have:

COROLLARY 3.132. *Let S_1, S_2 be two convex sets of a topological vector space X such that $S_1 \cap \mathrm{int}\, S_2 \neq \emptyset$. Then, one has*
$$(S_1 \cap S_2)^\infty = S_1^\infty \cap S_2^\infty.$$

PROOF. We know that $\mathrm{cl}\,(S_1 \cap S_2) = (\mathrm{cl}\, S_1) \cap (\mathrm{cl}\, S_2)$ (see, Proposition B.6 in Appendix). Recalling by Proposition 3.127(a) that $S^\infty = (\mathrm{cl}\, S)^\infty$, it ensues from Corollary 3.131 that with $C_i = \mathrm{cl}\, S_i$ one has
$$(S_1 \cap S_2)^\infty = \bigl(\mathrm{cl}\,(S_1 \cap S_2)\bigr)^\infty = (C_1 \cap C_2)^\infty = C_1^\infty \cap C_2^\infty.$$
Applying again the equality $(\mathrm{cl}\, S)^\infty = S^\infty$, the desired equality $(S_1 \cap S_2)^\infty = S_1^\infty \cap S_2^\infty$ follows. □

The next proposition shows, for a convex function f, the equality between $(\partial f(\overline{x}))^\infty$ and $N(\mathrm{dom}\, f; \overline{x})$ whenever $\partial f(\overline{x}) \neq \emptyset$.

PROPOSITION 3.133 (Asymptotic cone of the subdifferential of a convex function). *Let X be a Hausdorff locally convex space and $f : X \to \mathbb{R} \cup \{+\infty\}$ be a convex function. If f is finite at $\overline{x} \in X$, then for any real $\varepsilon \geq 0$ with $\partial_\varepsilon f(\overline{x}) \neq \emptyset$ one has*
$$(\partial_\varepsilon f(\overline{x}))^\infty = N(\mathrm{dom}\, f; \overline{x}),$$
where the asymptotic cone is taken in X^* endowed with any topology compatible with the duality between X and X^*.

3.5. ASYMPTOTIC FUNCTIONS AND CONES

PROOF. Fix any $\varepsilon \geq 0$ with $\partial_\varepsilon f(\overline{x}) \neq \emptyset$ and choose $\overline{x}^* \in \partial_\varepsilon f(\overline{x})$. For any real $t > 0$ it is readily checked that $\overline{x}^* + tN(\operatorname{dom} f; \overline{x}) \subset \partial_\varepsilon f(\overline{x})$, or equivalently $N(\operatorname{dom} f; \overline{x}) \subset \frac{1}{t}(\partial_\varepsilon f(\overline{x}) - \overline{x}^*)$. This combined with the first equality in Proposition 3.127(b) yields that $N(\operatorname{dom} f; \overline{x}) \subset (\partial_\varepsilon f(\overline{x}))^\infty$. Conversely, endow X^* with any topology compatible with the duality between X and X^* and take any $u^* \in (\partial_\varepsilon f(\overline{x}))^\infty$. By the second equality in Proposition 3.127(b) for each $t > 0$ there exists $u_t^* \in t^{-1}(\partial_\varepsilon f(\overline{x}) - \overline{x}^*)$ such that $u_t^* \to u^*$ as $t \to +\infty$. Then $\overline{x}^* + tu_t^* \in \partial_\varepsilon f(\overline{x})$, hence for every $x \in \operatorname{dom} f$ we have

$$\langle u_t^* + t^{-1}\overline{x}^*, x - \overline{x} \rangle \leq t^{-1}(f(x) - f(\overline{x} + \varepsilon)),$$

which entails $\langle u^*, x - \overline{x} \rangle \leq 0$. Consequently, $u^* \in N(\operatorname{dom} f; \overline{x})$, which finishes the proof. □

The following corollary of Proposition 3.133 directly follows from Proposition 3.94 above.

COROLLARY 3.134. *Let X be a Hausdorff locally convex space and $f : X \to \mathbb{R} \cup \{+\infty\}$ be a proper lower semicontinuous convex function. If f is finite at $\overline{x} \in X$, then for any real $\varepsilon > 0$ one has*

$$(\partial_\varepsilon f(\overline{x}))^\infty = N(\operatorname{dom} f; \overline{x}),$$

where the asymptotic cone is taken in X^* endowed with any topology compatible with the duality between X and X^*.

When S is a closed set with $0 \notin S$, its asymptotic cone yields the following remarkable description of its closed generated cone.

PROPOSITION 3.135. *Let S be a closed set of a topological vector space X with $0 \notin S$. Then one has*

$$\operatorname{cl}(\mathbb{R}_+ S) = (\mathbb{R}_+ S) \cup S^\infty.$$

PROOF. By the equality $S^\infty = \bigcap_{r > 0} \operatorname{cl}\left(\bigcup_{t \geq r} \frac{1}{t} S\right)$ in Proposition 3.127(a) we have $S^\infty \subset \operatorname{cl}(\mathbb{R}_+ S)$, and hence the inclusion $\supset$ of the proposition clearly holds. To show the converse inclusion, we may suppose that $S \neq \emptyset$. Take any nonzero $h \in \operatorname{cl}(\mathbb{R}_+ S)$ and write $h = \lim_{j \in J} t_j x_j$, where $(t_j)_j$ and $(x_j)_j$ are nets in $]0, +\infty[$ and S respectively. There is a subnet of $(t_j)_j$, that we do not relabel, tending to some $t \in [0, +\infty]$. We first observe that $t < +\infty$, since otherwise we would have $x_j \to 0$, and this would contradict the assumption 0 is not in the closed set S. In the case when $t > 0$ we have $x_j \to t^{-1} h$, so $t^{-1} h \in S$, hence $h \in tS \subset \mathbb{R}_+ S$. If $t = 0$, then $\tau_j := 1/t_j \to +\infty$ and

$$h = \lim_j \frac{1}{\tau_j} x_j \in \operatorname{Lim\,sup}_{\tau \to +\infty} \frac{1}{\tau} S = S^\infty.$$

Consequently, in any case we obtain $h \in (\mathbb{R}_+ S) \cup S^\infty$, which justifies the desired inclusion $\subset$ of the proposition. □

Above we arrived to the asymptotic cone of a set through the asymptotic function of the indicator function. Conversely, the asymptotic function of a function f is connected to the asymptotic cone of the epigraph of f as follows.

PROPOSITION 3.136. Let $f : X \to \mathbb{R} \cup \{-\infty, +\infty\}$ be a function on a topological vector space X. One has the equality
$$\operatorname{epi}(f^\infty) = (\operatorname{epi} f)^\infty.$$

PROOF. Fix first any $(v, \alpha) \in (\operatorname{epi} f)^\infty$. There are a net $(t_j)_{j \in J}$ in $]0, +\infty[$ tending to $+\infty$ and a net $(v_j, \alpha_j)_j$ converging to (v, α) such that $t_j(v_j, \alpha_j) \in \operatorname{epi} f$ for all $j \in J$ (see Proposition 1.16(b)). Then $f(t_j v_j) \leq t_j \alpha_j$, or equivalently $t_j^{-1} f(t_j v_j) \leq \alpha_j$, hence $f^\infty(v) \leq \alpha$ according to Definition 3.124, which justifies the inclusion $(\operatorname{epi} f)^\infty \subset \operatorname{epi}(f^\infty)$.

Conversely, let any $(v, \alpha) \in \operatorname{epi}(f^\infty)$, that is, $f^\infty(v) \leq \alpha$. Take any real $\beta > \alpha$. Choose nets $(t_j)_{j \in J}$ in $]0, +\infty[$ tending to $+\infty$ and $(v_j)_{j \in J}$ converging to v such that $\beta > f^\infty(v) = \lim_{j \in J} t_j^{-1} f(t_j v_j)$. There is some $j_0 \in J$ such that for each $j \succeq j_0$ we have $t_j^{-1} f(t_j v_j) < \beta$, hence $t_j(v_j, \beta) \in \operatorname{epi} f$. It ensues that $(v, \beta) \in (\operatorname{epi} f)^\infty$ according to Proposition 1.16(b). The latter inclusion being true for every real $\beta > \alpha$, the closedness of $(\operatorname{epi} f)^\infty$ ensures that $(v, \alpha) \in (\operatorname{epi} f)^\infty$, which confirms the converse of the inclusion justified above. $\square$

3.5.2. Asymptotic functions under convexity. We study now diverse additional properties of f^∞ under the convexity of f.

PROPOSITION 3.137. Let $f, g : X \to \mathbb{R} \cup \{-\infty, +\infty\}$ be functions on a topological vector space X, let $b \in X$ and let $A : E \to X$ be a continuous linear mapping from a topological vector space E into X.
(a) If f is convex, then f^∞ is a lower semicontinuous positively homogeneous convex function.
(b) If f is proper, convex and lower semicontinuous, then for any $x \in \operatorname{dom} f$ and any $v \in X$
$$f^\infty(v) = \lim_{t \to +\infty} t^{-1}[f(x + tv) - f(x)] = \sup_{t > 0} t^{-1}[f(x + tv) - f(x)],$$
or equivalently
$$f^\infty(v) = \lim_{r \downarrow 0} r[f(x + r^{-1} v) - f(x)] = \sup_{r > 0} r[f(x + r^{-1} v) - f(x)].$$
(c) If f and g are proper, convex and lower semicontinuous with $\operatorname{dom} f \cap \operatorname{dom} g \neq \emptyset$, then
$$(f + g)^\infty = f^\infty + g^\infty.$$
(d) If f is proper, convex and lower semicontinuous with $A^{-1}(b - \operatorname{dom} f) \neq \emptyset$, then for $\varphi : E \to \mathbb{R} \cup \{+\infty\}$ defined by $\varphi(u) = f(A(u) + b)$ for all $u \in E$, one has
$$\varphi^\infty = f^\infty \circ A.$$

PROOF. Assertion (a) follows from Proposition 3.136 and Proposition 3.127(b).
(b) Assume now that f is proper, convex and lower semicontinuous, and fix any $x \in \operatorname{dom} f$. We note first that the nondecreasing property of the function $t \mapsto t^{-1}[f(x+tv) - f(x)]$ guarantees the existence of $\lim_{t \to +\infty} t^{-1}[f(x+tv) - f(x)]$ as well as its equality with $\sup_{t > 0} t^{-1}[f(x+tv) - f(x)]$. On the other hand, by Proposition 3.136 we see that $(v, \alpha) \in \operatorname{epi}(f^\infty)$ if and only if $(v, \alpha) \in (\operatorname{epi} f)^\infty$, which in turn (since $\operatorname{epi} f$ is closed and convex) is equivalent to (v, α) belongs to $\bigcap_{t > 0} t^{-1}(\operatorname{epi} f - (x, f(x)))$. The latter means that $t^{-1}[f(x + tv) - f(x)] \leq \alpha$ for all $t > 0$, otherwise stated

$\sup_{t>0} t^{-1}[f(x+tv) - f(x)] \leq \alpha$. It results that $f^\infty(v) = \sup_{t>0} t^{-1}[f(x+tv) - f(x)]$ for all $v \in X$ as desired.

(c)-(d) Finally, properties (c) and (d) directly follow from (b). □

REMARK 3.138. We can also provide an analytic proof for Proposition 3.137 above. Again, the nondecreasing property of the function $t \mapsto t^{-1}[f(x+tv) - f(x)]$ guarantees the existence of $\zeta(v) := \lim_{t\to+\infty} t^{-1}[f(x+tv) - f(x)]$ as well as the equality $\zeta(v) = \sup_{t>0} t^{-1}[f(x+tv) - f(x)]$.

The inequality $f^\infty(v) \leq \zeta(v)$ being obvious from the definition of $f^\infty(v)$, we have to justify the converse inequality. For any real $t > 0$ the lower semicontinuity and convexity of f ensure that

$$f(x+tv) \leq \liminf_{\substack{\theta \to +\infty \\ u \to v}} f\left(\left(1 - \frac{t}{\theta}\right)x + \frac{t}{\theta}(\theta u)\right) \leq \liminf_{\substack{\theta \to +\infty \\ u \to v}} \left(\left(1 - \frac{t}{\theta}\right)f(x) + \frac{t}{\theta}f(\theta u)\right),$$

which yields

$$f(x+tv) \leq f(x) + t \liminf_{\substack{\theta \to +\infty \\ u \to v}} \left(\theta^{-1} f(\theta u)\right).$$

The latter inequality is equivalent to $t^{-1}[f(x+tv) - f(x)] \leq f^\infty(v)$ for every $t > 0$, thus $\zeta(v) \leq f^\infty(v)$. This finishes the proof. □

PROPOSITION 3.139. Let X be a Hausdorff locally convex space and $f : X \to \mathbb{R} \cup \{+\infty\}$ be a proper lower semicontinuous convex function. The asymptotic function of f coincides with the support function relative to X of $\operatorname{dom} f^*$, that is,

$$f^\infty(v) = \sigma(v, \operatorname{dom} f^*) \quad \text{for all } v \in X.$$

PROOF. Fix any $\bar{x} \in \operatorname{dom} f$, so $f^\infty(u) = \sup_{r>0} r[f(\bar{x} + r^{-1}u) - f(\bar{x})]$ for all $u \in X$ by Proposition 3.137(b) above. We note through (3.30), (3.31) and (3.32) that the proper lower semicontinuous convex function $u \mapsto r[f(\bar{x} + r^{-1}u) - f(\bar{x})]$ is the conjugate function relative to X of the function on X^* given by $x^* \mapsto r[f(\bar{x}) + f^*(x^*) - \langle x^*, \bar{x}\rangle]$, so putting $g(x^*) := f(\bar{x}) + f^*(x^*) - \langle x^*, \bar{x}\rangle$ we have by what precedes that for any $u \in X$

$$f^\infty(u) = \sup_{r>0}(rg)^*(u) = (\inf_{r>0} rg)^*(u),$$

where we use (3.19) for the second equality on the right. By the Fenchel inequality we see that $g(x^*) \geq 0$ for all $x^* \in X^*$, hence $\inf_{r>0} rg(x^*) = +\infty$ if $x^* \notin \operatorname{dom} f^*$ and $\inf_{r>0} rg(x^*) = 0$ if $x^* \in \operatorname{dom} f^*$ (keep in mind that f^* is proper by Proposition 3.77). This means that $\inf_{r>0} rg$ coincides with the indicator function of $\operatorname{dom} f^*$, hence f^∞ coincides with the conjugate relative to X of this indicator function, that is, with the support function relative to X of $\operatorname{dom} f^*$. □

For a lower semicontinuous function $f : X \to \mathbb{R} \cup \{+\infty\}$, extending naturally to $[0,+\infty[\times X$ the function $(r, u) \mapsto rf(r^{-1}u)$ by taking $\liminf_{\substack{r \downarrow 0 \\ u' \to u}} rf(r^{-1}u') = f^\infty(u)$ as value at $(0, u)$, we see that such an extension is lower semicontinuous on $[0, +\infty[\times X$ and, by Proposition 3.137(b), it takes its values in $\mathbb{R} \cup \{+\infty\}$ whenever in addition f is proper and convex. We state this result in the following proposition.

PROPOSITION 3.140. Let X be a locally convex space and $f : X \to \mathbb{R} \cup \{+\infty\}$ be a proper lower semicontinuous convex function. The extended real-valued function φ defined on $[0,+\infty[\times X$ by

$$\varphi(r,u) := rf(r^{-1}u) \text{ if } (r,u) \in]0,+\infty[\times X \text{ and } \varphi(r,u) := f^\infty(u) \text{ if } r = 0,$$

takes its values in $\mathbb{R} \cup \{+\infty\}$ and is lower semicontinuous relative to $[0,+\infty[\times X$.

We now prove the lower semicontinuity of $f'_\varepsilon(\bar{x};\cdot)$.

THEOREM 3.141 (lower semicontinuity of ε-directional derivative). Let X be a Hausdorff locally convex space and $f : X \to \mathbb{R} \cup \{+\infty\}$ be a proper lower semicontinuous convex function. For any real $\varepsilon > 0$ and any point $\bar{x} \in \operatorname{dom} f$ the function $f'_\varepsilon(\bar{x};\cdot)$ is lower semicontinuous, and it is the support function of $\partial_\varepsilon f(\bar{x})$.

PROOF. Applying with the function $x \mapsto f(\bar{x}+x) - f(\bar{x}) + \varepsilon$ Proposition 3.140 and using Proposition 3.137(b), we obtain that the function φ on $[0,+\infty[\times X$ with

$$\varphi(r,u) = r[f(\bar{x}+r^{-1}u) - f(\bar{x}) + \varepsilon] \text{ if } r > 0 \text{ and } \varphi(0,u) = \lim_{r \downarrow 0} r[f(\bar{x}+r^{-1}u) - f(\bar{x}) + \varepsilon]$$

is well-defined, takes its values in $\mathbb{R} \cup \{+\infty\}$ and is lower semicontinuous. On the other hand,

$$\varphi(0,u) = \lim_{r \downarrow 0} r[f(\bar{x}+r^{-1}u) - f(\bar{x}) + \varepsilon] \geq \inf_{r > 0} r[f(\bar{x}+r^{-1}u) - f(\bar{x}) + \varepsilon],$$

hence

(3.75) $$\inf_{r \in [0,+\infty[} \varphi(r,u) = \inf_{r > 0} \varphi(r,u) = f'_\varepsilon(\bar{x};u),$$

where the equality on the right follows from the definition of $f'_\varepsilon(\bar{x};u)$.

Now, since the function $x \mapsto g(x) := f(\bar{x}+x) - f(\bar{x}) + \varepsilon$ is proper, lower semicontinuous and convex with $g(0) = \varepsilon > 0$, there exist by Proposition 3.77(a) some $a^* \in X^*$ and some real $\beta > 0$ such that $\langle a^*, \cdot \rangle + \beta \leq g$. Then for all $u \in X$ and $r \in]0,+\infty[$ we have

$$\langle a^*, u \rangle + r\beta = r\langle a^*, r^{-1}u \rangle + r\beta \leq rg(r^{-1}u) = \varphi(r,u),$$

and we note from this and (3.75) that $f'_\varepsilon(\bar{x};u) > -\infty$. Fix any $\bar{u} \in X$ and fix also any real $\rho < f'_\varepsilon(\bar{x};\bar{u})$, and take a real ρ_0 strictly between the two latter values. Choose a neighborhood U_1 of $\bar{u}$ over which $\langle a^*, \cdot \rangle$ is bounded from below. Since $\beta > 0$ there is some real $r_0 > 0$ such that for all $r > r_0$ and $u \in U_1$ we have $\langle a^*, u \rangle + r\beta > \rho_0$, and hence $\varphi(r,u) > \rho_0$. Since $[0,r_0]$ is compact and the function φ is lower semicontinuous on $[0,r_0] \times X$, Berge theorem (see Theorem 1.43(b)) tells us that the function $\zeta : X \to \mathbb{R} \cup \{+\infty\}$, defined by $\zeta(u) := \inf_{r \in [0,r_0]} \varphi(r,u)$ for all $u \in X$, is lower semicontinuous. Since $\rho_0 < \inf_{r \geq 0} \varphi(r,\bar{u}) \leq \zeta(\bar{u})$, there is some neighborhood $U_0 \subset U_1$ of $\bar{u}$ such that $\rho_0 < \zeta(u)$ for all $u \in U_0$. It ensues that $\rho < \inf_{r \geq 0} \varphi(r,u) = f'_\varepsilon(\bar{x};u)$ for all $u \in U_0$, so the function $f'_\varepsilon(\bar{x};\cdot)$ is lower semicontinuous at $\bar{u}$.

In addition to the above lower semicontinuity, we know by Proposition 3.9(a) that $f'_\varepsilon(\bar{x};\cdot)$ is a positively homogeneous convex function with $f'_\varepsilon(\bar{x};0) = 0$, hence $f'_\varepsilon(\bar{x};\cdot)$ is the supremum of all continuous linear functions that it majorizes (see Corollaries 3.78 and 3.24). This combined with the equivalence $x^* \in \partial_\varepsilon f(\bar{x}) \Leftrightarrow \langle x^*, \cdot \rangle \leq f'_\varepsilon(\bar{x};\cdot)$ (see Proposition 3.9(c)) tells us that $f'_\varepsilon(\bar{x};\cdot)$ is the support function of $\partial_\varepsilon f(\bar{x})$. The proof is then complete. $\square$

3.6. Brønsted-Rockafellar theorem

Let a real $\varepsilon > 0$ and $f : X \to \mathbb{R} \cup \{+\infty\}$ be a proper lower semicontinuous convex function on a locally convex space X. By Proposition 3.94 we know that $\partial f_\varepsilon(\bar{x}) \neq \emptyset$ at any $\bar{x} \in \operatorname{dom} f$. In this section, when X is a Banach space, we show that ε-subgradient $\bar{x}^* \in \partial_\varepsilon f(\bar{x})$ is $\sqrt{\varepsilon}$-close to a true subgradient of f at a point $\sqrt{\varepsilon}$-close to $\bar{x}$. Recalling that $\|\cdot\|_*$ denotes the dual norm of $\|\cdot\|$, the exact statement is the following so-called Brønsted-Rockafellar theorem:

THEOREM 3.142. [A. Brønsted and R.T. Rockafellar] Let $(X, \|\cdot\|)$ be a Banach space, $f : X \to \mathbb{R} \cup \{-\infty, +\infty\}$ be a proper lower semicontinuous convex function and $g : X^* \to \mathbb{R} \cup \{+\infty\}$ be a proper lower $w(X^*, X)$-semicontinuous convex function. Let two reals $\varepsilon > 0$, $\lambda > 0$.
(a) If $\bar{x}^* \in \partial_\varepsilon f(\bar{x})$ with $\varepsilon > 0$, then there exist $z \in \operatorname{dom} f$ and $z^* \in \partial f(z)$ such that
$$\|z - \bar{x}\| \leq \lambda, \ \|z^* - \bar{x}^*\|_* \leq \frac{\varepsilon}{\lambda}, \ |f(z) - f(\bar{x})| \leq \varepsilon + \lambda\|\bar{x}^*\|_*.$$
(b) If $\bar{x} \in X$ satisfies $\bar{x} \in \partial_\varepsilon g(\bar{x}^*)$, then there exist $z^* \in \operatorname{dom} g$ and $z \in X$ with $z \in \partial g(z^*)$ such that
$$\|z^* - \bar{x}^*\|_* \leq \lambda, \ \|z - \bar{x}\| \leq \frac{\varepsilon}{\lambda}, \ |g^*(z) - g^*(\bar{x})| \leq \varepsilon + \frac{\varepsilon}{\lambda}\|\bar{x}^*\|_*.$$

The theorem will be a consequence of a more general version.

THEOREM 3.143. [J.M. Borwein's version of Brønsted-Rockafellar theorem] Let $f : X \to \mathbb{R} \cup \{-\infty, +\infty\}$ be a proper lower semicontinuous convex function on a Banach space $(X, \|\cdot\|)$. If $\bar{x}^* \in \partial_\varepsilon f(\bar{x})$ with $\varepsilon > 0$, then for any reals $\lambda > 0$ and $\beta \geq 0$ there exist $z \in \operatorname{dom} f$, $z^* \in \partial f(z)$ and $\gamma \in [-1, 1]$ such that
  (a) $\|z - \bar{x}\| + \beta|\langle \bar{x}^*, z - \bar{x}\rangle| \leq \lambda$;
  (b) $\left\|z^* - \left(1 + \frac{\varepsilon\beta\gamma}{\lambda}\right)\bar{x}^*\right\| \leq \varepsilon/\lambda$;
  (c) $|f(z) - \langle \bar{x}^*, z - \bar{x}\rangle - f(\bar{x})| \leq \varepsilon$ and $|f(z) - f(\bar{x})| \leq \varepsilon + \lambda\bar{\beta}$ with $\bar{\beta}$ defined by $\bar{\beta} = \begin{cases} 1/\beta & \text{if } \beta > 0 \\ \|\bar{x}^*\| & \text{if } \beta = 0 \end{cases}$;
  (d) $|\langle z^* - \bar{x}^*, z - \bar{x}\rangle| \leq \varepsilon$ and $|f(z) - \langle z^*, z - \bar{x}\rangle - f(\bar{x})| \leq 2\varepsilon$;
  (e) $z^* \in \partial_{2\varepsilon} f(\bar{x})$;
  (f) $|\langle z^* - \bar{x}^*, x\rangle| \leq \frac{\varepsilon}{\lambda}(\|x\| + \beta|\langle \bar{x}^*, x\rangle|)$ for all $x \in X$.

PROOF. Fix arbitrary elements $\varepsilon > 0$, $\lambda > 0$, $\beta \geq 0$ and $(\bar{x}, \bar{x}^*) \in X \times X^*$ with $\bar{x}^* \in \partial_\varepsilon f(\bar{x})$. Define $g := f - \langle \bar{x}^*, \cdot \rangle$. We notice that $g(\bar{x})$ is finite and
$$(3.76) \qquad g(\bar{x}) \leq \inf_X g + \varepsilon.$$
Endow X with the equivalent norm N defined by $N(x) := \|x\| + \beta|\langle \bar{x}^*, x\rangle|$ for all $x \in X$, and denote by N_* the associated dual norm. We note that
$$\|x\| \leq N(x) \leq (1 + \beta\|\bar{x}^*\|_*)\|x\|, \text{ hence } N_*(x^*) \leq \|x^*\|_* \leq (1 + \beta\|\bar{x}^*\|_*)N_*(x^*).$$
We claim that the proof is finished if we show the following:
(3.77) there exist $z \in \operatorname{dom} g = \operatorname{dom} f$ and $y^* \in \partial g(z)$ and $\gamma \in [-1, 1]$ such that:
  - (π_1) $N(z - \bar{x}) \leq \lambda$;
  - (π_2) $\left\|y^* - \frac{\varepsilon\beta\gamma}{\lambda}\bar{x}^*\right\|_* \leq \frac{\varepsilon}{\lambda}$ and $N_*(y^*) \leq \varepsilon/\lambda$;

- (π_3) $|g(z) - g(\bar{x})| \leq \varepsilon$;
- (π_4) $y^* \in \partial_{2\varepsilon} g(\bar{x})$.

Indeed, (π_1) translates the inequality (a) and entails that

$$(3.78) \qquad |\langle \bar{x}^*, z - \bar{x} \rangle| \leq \lambda \bar{\beta}.$$

Put $z^* := y^* + \bar{x}^*$ and observe that $z^* \in \partial f(z)$. Observe also that the inequalities in (π_2) translate the inequality in (b) and the inequality $N_*(z^* - \bar{x}^*) \leq \varepsilon/\lambda$. It ensues that for every $u \in X$

$$|\langle z^* - \bar{x}^*, u \rangle| \leq N_*(z^* - \bar{x}^*) N(u) \leq \frac{\varepsilon}{\lambda}(\|u\| + \beta |\langle \bar{x}^*, u \rangle|)$$

and that

$$(3.79) \qquad |\langle z^* - \bar{x}^*, z - \bar{x} \rangle| \leq N_*(z^* - \bar{x}^*) N(z - \bar{x}) \leq \varepsilon,$$

so (f) and the first inequality in (d) hold. Further, (π_3) is equivalent to $|f(z) - \langle \bar{x}^*, z - \bar{x} \rangle - f(\bar{x})| \leq \varepsilon$, which leads on the one hand by (3.78) to

$$|f(z) - f(\bar{x})| \leq \varepsilon + |\langle \bar{x}^*, z - \bar{x} \rangle| \leq \varepsilon + \lambda \bar{\beta},$$

and on the other hand by (3.79) to

$$|f(z) - \langle z^*, z - \bar{x} \rangle - f(\bar{x})| \leq 2\varepsilon.$$

Therefore, (c) holds as well as the second inequality in (d). Finally, it is easily seen that (e) follows from (π_4) and the definition of g.

Let us show the properties (π_1)-(π_4). Endowing X with the equivalent norm N and applying the Ekeland variational principle to the function g with $a := \bar{x}$ (see Theorem 2.221) we obtain a point $b \in X$ such that

$$N(b - \bar{x}) \leq \lambda, \ g(b) + \frac{\varepsilon}{\lambda} N(b - \bar{x}) \leq g(\bar{x}),$$

$$g(b) < g(x) + \frac{\varepsilon}{\lambda} N(x - b) \quad \text{for all } x \in X \setminus \{\bar{x}\}.$$

It results that $0 \in \partial(g + h)(b)$, where $h := \frac{\varepsilon}{\lambda} N(\cdot - b)$. The Moreau-Rockafellar subdifferential sum rule in Theorem 2.105 (due to the continuity of h) furnishes $b_1^* \in \partial g(b)$ and $b_2^* \in \partial h(b)$ such that $b_1^* + b_2^* = 0$. On the one hand, we note that $\partial h(b) = \{u^* \in X^* : N_*(u^*) \leq \frac{\varepsilon}{\lambda}\}$, so $N_*(b_1^*) = N_*(b_2^*) \leq \varepsilon/\lambda$. On the other hand, since $h = \frac{\varepsilon}{\lambda}(\|\cdot\| + \beta |\langle \bar{x}^*, \cdot \rangle|)$, applying Theorem 2.105 again and Theorem 2.135 we obtain some $\mu \in [-1, 1]$ such that

$$b_1^* - \frac{\varepsilon \beta \mu}{\lambda} \bar{x}^* \in \frac{\varepsilon}{\lambda} \mathbb{B}_{\|\cdot\|_*}, \text{ that is, } \left\|b_1^* - \frac{\varepsilon \beta \mu}{\lambda} \bar{x}^*\right\|_* \leq \frac{\varepsilon}{\lambda}.$$

It remains to examine two cases:

(I) $g(\bar{x}) - \varepsilon \leq g(b)$ or (II) $g(b) < g(\bar{x}) - \varepsilon$.

Case (I). In this first case we have $0 \leq g(\bar{x}) - g(b) \leq \varepsilon$, where the first inequality is due to the fact $g(b) + \frac{\varepsilon}{\lambda} N(b - \bar{x}) \leq g(\bar{x})$. Choosing $z := b$ and $y^* = b_1^*$, we also have $y^* \in \partial g(z)$ and the properties (π_1), (π_2), (π_3) are immediate. Concerning (π_4)

it suffices to see that for any $y \in X$

$$\begin{aligned}\langle y^*, y - \overline{x}\rangle &= \langle y^*, y - z\rangle + \langle y^*, z - \overline{x}\rangle \\ &\leq g(y) - g(z) + N_*(y^*)N(z - \overline{x}) \\ &\leq g(y) - g(\overline{x}) + g(\overline{x}) - g(z) + \varepsilon \\ &\leq g(y) - g(\overline{x}) + 2\varepsilon,\end{aligned}$$

which says that $y^* \in \partial_{2\varepsilon} g(\overline{x})$.

Case (II). In this second case we have $g(b) < g(\overline{x}) - \varepsilon < g(\overline{x})$. Define on $[0,1]$ the function φ (depending on $\overline{x}$ and b) by $\varphi(t) := g(t\overline{x} + (1-t)b)$ for all $t \in [0,1]$. This function remains convex and is finite on $[0,1]$ since $\{b, \overline{x}\} \subset \mathrm{dom}\, g$. Since φ is also lower semicontinuous on $[0,1]$, it is continuous therein (see Proposition 3.20). Therefore, there exists $t_0 \in]0,1[$ such that $\varphi(t_0) = g(\overline{x}) - \varepsilon$. Choosing $z := t_0\overline{x} + (1-t_0)b$, we have $g(z) = g(\overline{x}) - \varepsilon$, and hence $|g(z) - g(\overline{x})| = \varepsilon$ and $N(z - \overline{x}) \leq \lambda$ since

$$N(z - \overline{x}) = (1 - t_0)N(b - \overline{x}) \leq N(b - \overline{x}) \leq \lambda.$$

Moreover, (3.76) ensures $0 \leq g(y) - g(\overline{x}) + \varepsilon = g(y) - g(z)$ for all $y \in X$, that is, $0 \in \partial g(z)$. In this second case (II) the choice of $y^* = 0$, $\gamma = 0$ and z as defined above via t_0 fulfills the properties (π_1), (π_2), (π_3), and the property (π_4) follows from (3.76).

In both cases (I) and (II) we found z, y, γ satisfying (3.77) and the properties (π_1)-(π_4). This finishes the proof of the theorem. □

PROOF OF THEOREM 3.142. To derive (a) in Theorem 3.142 from Theorem 3.143, it suffices in the latter to take $\beta = 0$ and invoke (a), (b), (c). Concerning (b) in Theorem 3.142, we note by Theorem 3.84(c) (applied to g with X^* endowed with the $w(X^*, X)$ topology) that there exists a proper lower semicontinuous convex function $f : X \to \mathbb{R} \cup \{+\infty\}$ (in fact, the conjugate of g relative to X) such that $f^* = g$. Then the inclusion assumption in (b) of Theorem 3.142 means $\overline{x}^* \in \partial_\varepsilon f(\overline{x})$, so applying (a) in Theorem 3.142 to f with $\varepsilon > 0$ and with $\lambda' := \varepsilon/\lambda$ in place of λ yields $z \in X$ and $z^* \in X^*$ fulfilling the properties in (b). □

The following example shows that Theorems 3.142 and 3.143 may fail in Fréchet spaces in place of Banach spaces.

EXAMPLE 3.144. [A. Brønsted and R.T. Rockafellar] By [610] let X be a Fréchet space which has a closed convex subset S containing neither support points nor entire lines, and for which there is a nonzero $u \in X$ such that $\mathbb{R}_+ u \subset S$. The latter inclusion says that $0 \in S$ and $u \in S^\infty$ (the asymptotic cone of S) by a result in Proposition 3.128 on asymptotic cones of convex sets. Consider the extended real-valued function f defined on X by

$$f(x) = \inf\{t \in \mathbb{R} : x + tu \in S\} \quad \text{for all } x \in X,$$

and note that f does not take the value $-\infty$ since S does not contain any entire line. Note also that $f(0)$ is finite since $\mathbb{R}_+ u \subset S$, so f is a proper function. Writing

$$f(x) = \inf_{t \in \mathbb{R}} \left(t + \Psi_S(x + tu)\right)$$

we see that the function f is convex (see Proposition 2.54). By the latter equality we also observe that for any $x \in X$ and any $s \in \mathbb{R}$

$$f(x+su) = \inf_{t \in \mathbb{R}} \left(t + \Psi_S(x+(t+s)u)\right) = -s + \inf_{t \in \mathbb{R}} \left((s+t) + \Psi_S(x+(s+t)u)\right) = -s + f(x),$$

then $\{f(\cdot) \leq s\} = -su + \{f(\cdot) \leq 0\}$. To see from this that the convex function f is also lower semicontinuous, it suffices to show that $\{f(\cdot) \leq 0\} = S$. First the equivalence $x \in S \Leftrightarrow x + 0u \in S$ ensures that $S \subset \{f(\cdot) \leq 0\}$. Now, suppose that $f(x) \leq 0$, so $t_0 := -f(x)$ is a non-negative real since f does not take the value $-\infty$. By the closedness of S and the definition of f we have $x - t_0 u \in S$, which in turn gives $x - t_0 u + \mathbb{R}_+ u \subset S$ by Proposition 3.128. It ensues in particular that $x \in S$. This combined with what precedes justifies the desired equality $\{f(\cdot) \leq 0\} = S$, hence the proper convex function f is lower semicontinuous.

Now, we claim that $\partial f(x) = \emptyset$ for all $x \in X$. Suppose there is some $x \in X$ with $\partial f(x) \neq \emptyset$ and take $x^* \in \partial f(x)$. For every $s \in \mathbb{R}$ using the above equality $f(x + su) = -s + f(x)$ along with the fact $f(x) \in \mathbb{R}$, we see that

$$s \langle x^*, u \rangle = \langle x^*, x + su - x \rangle \leq f(x + su) - f(x) = -s.$$

The inequality $s \langle x^*, u \rangle \leq -s$ for every $s \in \mathbb{R}$ gives $\langle x^*, u \rangle = -1$. Therefore, $x^* \neq 0$ and for every $y \in S$

$$\langle x^*, y - (x + f(x)u) \rangle = \langle x^*, y - x \rangle + f(x) \leq f(y) \leq 0,$$

so $x^* \in N(S; x + f(x)u)$. This is a contradiction since $N(S; z) = \{0\}$ for every $z \in S$ according to the fact that S has no support points.

In conclusion, the function f is a proper lower semicontinuous convex function with $\mathrm{Dom}\,\partial f = \emptyset$, which illustrates that Theorems 3.142 and 3.143 are not valid for Fréchet spaces in place of Banach spaces. $\square$

Taking $\beta = 1$ and $\lambda = \sqrt{\varepsilon}$ in Theorem 3.143 yields the following corollary.

COROLLARY 3.145. Let $f : X \to \mathbb{R} \cup \{-\infty, +\infty\}$ be a proper lower semicontinuous convex function on a Banach space $(X, \|\cdot\|)$. If $\overline{x}^* \in \partial_\varepsilon f(\overline{x})$ with $\varepsilon > 0$, then there exist $z \in \mathrm{dom}\, f$, $z^* \in \partial f(z)$ and $\gamma \in [-1, 1]$ such that
  (a) $\|z - \overline{x}\| + |\langle \overline{x}^*, z - \overline{x} \rangle| \leq \sqrt{\varepsilon}$;
  (b) $\|z^* - (1 + \gamma \sqrt{\varepsilon})\overline{x}^*\| \leq \sqrt{\varepsilon}$;
  (c) $|f(z) - \langle \overline{x}^*, z - \overline{x} \rangle - f(\overline{x})| \leq \varepsilon$ and $|f(z) - f(\overline{x})| \leq \varepsilon + \sqrt{\varepsilon}$;
  (d) $|\langle z^* - \overline{x}^*, z - \overline{x} \rangle| \leq \varepsilon$ and $|f(z) - \langle z^*, z - \overline{x} \rangle - f(\overline{x})| \leq 2\varepsilon$;
  (e) $z^* \in \partial_{2\varepsilon} f(\overline{x})$;
  (f) $|\langle z^* - \overline{x}^*, x \rangle| \leq \sqrt{\varepsilon}(\|x\| + |\langle \overline{x}^*, x \rangle|)$ for all $x \in X$.

3.7. Duality for the sum with a linear function

Proposition 2.69 and Proposition 3.64(i) revealed how the Legendre-Fenchel conjugate of a convex function is connected with its subdifferential. Thanks to Theorem 3.84, when f is in addition proper and lower semicontinuous more can be said in the following assertions. The equalities in (a) have been already noticed in (3.63) and the others are easily verified.

PROPOSITION 3.146. Let X be a Hausdorff locally convex space and $f : X \to]-\infty, +\infty]$ be a proper convex function. The following hold with $\partial f^*(\cdot) := \partial_{X^{**}} f^*(\cdot)$.
(a) For any real $\varepsilon > 0$ one has the equalities

$$\partial f^*(\cdot) \cap X = (\partial f)^{-1}(\cdot) \quad \text{and} \quad \partial_\varepsilon f^*(\cdot) \cap X = (\partial_\varepsilon f)^{-1}$$

3.7. DUALITY FOR THE SUM WITH A LINEAR FUNCTION

whenever f is in addition lower semicontinuous.

(b) One also has

$$\inf_X f = -f^*(0) \quad \text{and more generally} \quad \inf_{x \in X}[f(x) - \langle \bar{x}^*, x \rangle] = -f^*(\bar{x}^*),$$

and if f is additionally lower semicontinuous

$$\operatorname{Argmin} f = \partial f^*(0) \cap X \text{ and } \operatorname{Argmin}(f - \langle \bar{x}^*, \cdot \rangle) = \partial f^*(\bar{x}^*) \cap X.$$

Let (X, τ_X) be a Hausdorff locally convex space. If we denote by $\Gamma_0(X, \tau_X)$ the set all proper lower τ_X-semicontinuous convex functions from X into $\mathbb{R} \cup \{-\infty, +\infty\}$ and by $\Gamma(X, \tau_X)$ the set $\Gamma(X, \tau_X) := \Gamma_0(X, \tau_X) \cup \{-\omega_X, +\omega_X\}$, it is clear that $\Gamma_0(X, \tau_X) = \Gamma_0(X, w(X, X^*))$ and $\Gamma(X, \tau_X) = \Gamma(X, w(X, X^*))$. Then we see that the Legendre-Fenchel (conjugate) transform assigns to each function $f \in \Gamma_0(X, w(X, X^*))$ (resp. $f \in \Gamma(X, w(X, X^*)))$ a function $f^* \in \Gamma_0(X^*, w(X^*, X))$ (resp. $f^* \in \Gamma(X^*, w(X^*, X)))$. Further, endowing X^* with its weak* topology $w(X^*, X)$ (so, the topological dual of $(X^*, w(X^*, X))$ is identified with X), we also see by Theorem 3.84(c) that, for $f \in \Gamma_0(X, w(X, X^*))$ (resp. $f \in \Gamma(X, w(X, X^*)))$ the Legendre-Fenchel biconjugate $f^{**} = (f^*)^*$ relative to X coincides with f. This is summarized in the following duality diagrams

$$\begin{array}{ccccc} \Gamma_0(X, w(X, X^*)) & \xrightarrow{*} & \Gamma_0(X^*, w(X^*, X)) & \xrightarrow{*} & \Gamma_0(X, w(X, X^*)) \\ f & \mapsto & f^* & \mapsto & f^{**} = f \end{array}$$

$$\begin{array}{ccccc} \Gamma(X, w(X, X^*)) & \xrightarrow{*} & \Gamma(X^*, w(X^*, X)) & \xrightarrow{*} & \Gamma(X, w(X, X^*)) \\ f & \mapsto & f^* & \mapsto & f^{**} = f \end{array}$$

We recall that the above notation ω_X stands for the function on X identically equals to $+\infty$.

Let us put in light another duality scheme. Let $f \in \Gamma_0(X, \tau_X)$ and $\bar{x}^* \in X^*$. Proposition 3.146(b) tells us that

$$\inf_{x \in X}[f(x) + \langle \bar{x}^*, x \rangle] = -f^*(-\bar{x}^*) = - \inf_{x^* \in X^*}[f^*(-x^*) + \Psi_{\{\bar{x}^*\}}(x^*)].$$

Noting that the conjugate of the function $g := \langle \bar{x}^*, \cdot \rangle$ is $g^* = \Psi_{\{\bar{x}^*\}}$ (see (3.34)), we can write with the above particular function g the duality equality

(3.80) $$\inf_{x \in X}[f(x) + g(x)] = - \inf_{x^* \in X^*}[f^*(-x^*) + g^*(x^*)].$$

Further, in the particular above context with $f, g \in \Gamma_0(X)$, endowing X^* with its weak* topology and putting $f_1(x^*) := f^*(-x^*)$ and $g_1(x^*) := g^*(x^*)$, with the problem $\inf_{x^* \in X^*}[f^*(-x^*) + g^*(x^*)]$ is associated through the above process the problem: $-\inf_{x \in X}[f_1^*(-x) + g_1^*(x)]$. Since $g_1^*(x) = g(x)$ and $f_1^*(x) = f(-x)$ (see (3.31)), the above process then associates with $-\inf_{x^* \in X^*}[f^*(-x^*) + g^*(x^*)]$ nothing else but the starting problem $\inf_{x \in X}[f(x) + g(x)]$.

Consequently, the optimization problem

$$\text{Maximize} \quad -f^*(-x^*) - g^*(x^*) \quad \text{over } x^* \in X^*$$

is then revealed as a dual problem of the original optimization problem

$$\text{Minimize} \quad f(x) + g(x) \quad \text{over } x \in X.$$

Such a duality can also be seen via a point of view of perturbation. Considering a perturbation of the above function $g = \langle \bar{x}^*, \cdot \rangle$ in the form $x \mapsto g(x+y)$ where $y \in Y := X$ is the perturbation parameter, one has to consider for each $y \in Y = X$

$$\inf_{x \in X} \varphi(x, y), \quad \text{where } \varphi(x, y) := f(x) + g(x+y),$$

so the original problem $\inf_{x \in X}[f(x) + g(x)]$ corresponds to $\inf_{x \in X} \varphi(x, 0)$. The Legendre-Fenchel conjugate of $\varphi(\cdot, \cdot)$ can be computed as follows:

$$\varphi^*(x^*, y^*) = \sup_{x \in X, y \in X} [\langle x^*, x \rangle + \langle y^*, y \rangle - f(x) - \langle \bar{x}^*, x \rangle - \langle \bar{x}^*, y \rangle]$$

$$= \sup_{x \in X}[\langle x^* - \bar{x}^*, x \rangle - f(x)] + \sup_{y \in X}[\langle y^* - \bar{x}^*, y \rangle]$$

(3.81)
$$= f^*(x^* - \bar{x}^*) + \Psi_{\{\bar{x}^*\}}(y^*).$$

We then see that

$$f^*(-\bar{x}^*) = \inf_{y^* \in X^*} \varphi^*(0, y^*), \text{ or equivalently } -f^*(-\bar{x}^*) = \sup_{y^* \in X^*} -\varphi^*(0, y^*),$$

so Proposition 3.146(b) gives with the particular function $g = \langle \bar{x}^*, \cdot \rangle$

$$\inf_{x \in X} \varphi(x, 0) = \sup_{y^* \in X^*} -\varphi^*(0, y^*),$$

and further, $\inf_{x \in X} \varphi(x, 0) = \inf_{x \in X}[f(x) + g(x)]$ as seen above by definition of φ, and $\sup_{y^* \in X^*} -\varphi^*(0, y^*) = \sup_{y^* \in X^*} [-f^*(-y^*) - g^*(y^*)]$ by (3.81). We then retrieve the previous duality equality (3.80) via the other duality equality

(3.82)
$$\inf_{x \in X} \varphi(x, 0) = \sup_{y^* \in Y^*} -\varphi^*(0, y^*)$$

obtained with the particular function $g = \langle \bar{x}^*, \cdot \rangle$ and with $Y := X$. Such a duality will be seen in the next section as a particular case of a general duality scheme involving perturbations of original problems.

3.8. Duality in convex optimization

Let us continue with the above approach of perturbations. We start by illustrating the point of view of perturbations with the general model problem of optimization considered in Section 2.5.4 and that we denote by $(\mathcal{P}_{\text{model}})$ or $(\mathcal{P})$ for short:

$$(\mathcal{P}) \begin{cases} \text{Minimize} & f_0(x) \\ \text{subject to the constraints} & f_i(x) \leq 0 \ \forall i \in I, \ f_i(x) = 0 \ \forall i \in J, \ x \in C. \end{cases}$$

Here, $I := \{1, \cdots, m\}$ and $J := \{m+1, \cdots, m+n\}$ are finite sets, $f, f_i : X \to \mathbb{R}$ are real-valued functions, and C is a nonempty subset of a locally convex space X. For convenience of notation we index the functions in the inequality constraints as $f_1, \cdots, f_m$ and in the equality constraints as $f_{m+1}, \cdots, f_{m+n}$ instead of $f_1, \cdots, f_p$ and $f_{p+1}, \cdots, f_{p+q}$ as it was the case in Section 2.5.4. We reserve p, q to denote appropriate functions which will appear in the development below.

The extended real

$$\inf\{f_0(x) : f_i(x) \leq 0 \ \forall i \in I, \ f_i(x) = 0 \ \forall i \in J, \ x \in C\}$$

is called the *value of the optimization problem* $(\mathcal{P})$. It is denoted by $\inf(\mathcal{P})$ or $\text{val}(\mathcal{P})$.

3.8. DUALITY IN CONVEX OPTIMIZATION

The problem $(\mathcal{P})$ is said to be a *convex minimization problem* when the functions $f_0, f_1, \cdots, f_m$ are convex, the functions $f_{m+1}, \cdots, f_{m+n}$ are affine, and the set C is convex.

Denote the set of *feasible points* (assumed to be nonempty) by C_0, that is,
$$C_0 := \{x \in C : f_i(x) \leq 0 \; \forall i \in I, f_i(x) = 0 \; \forall i \in J\}.$$
Considering the function $\overline{f} : X \to \mathbb{R} \cup \{+\infty\}$ defined by $\overline{f}(x) = f_0(x)$ if $x \in C_0$ and $\overline{f}(x) = +\infty$ otherwise, it is easily verified that $\operatorname{val}(\mathcal{P}) = \inf_{x \in X} \overline{f}(x)$ and that a point $\overline{x} \in X$ is a solution of the problem $(\mathcal{P})$ if and only if the equality $\overline{f}(\overline{x}) = \inf_{x \in X} \overline{f}(x)$ is satisfied. So, the problem $(\mathcal{P})$ is equivalent to the unconstrained minimization problem

Minimize $\quad \overline{f}(x) \quad$ over x in the whole space X.

Furthermore,
$$\operatorname{val}(\mathcal{P}) = \inf_{x \in X} \overline{f}(x) = -(\overline{f})^*(0),$$
where $(\overline{f})^*$ is the Legendre-Fenchel conjugate of $\overline{f}$. When it is needed to emphasize the variable with respect to which the minimization problem is considered, it will be convenient to write x as subscript, that is,

Minimize$_x \quad \overline{f}(x) \quad$ over x in the whole space X.

For theoretical and numerical reasons, in presence of perturbation parameters of data, one is generally led to study the stability (resp. sensibility) of the problem, that is, generalized semicontinuity (resp. subdifferentiability) of the value of the problem with respect to the perturbations. In the case of the above problem, for instance, it is natural for $Y := \mathbb{R}^I \times \mathbb{R}^J$ to consider the perturbed problem $(\mathcal{P}_y)$ (with $y \in Y$)
$$\begin{cases} \text{Minimize}_x \quad f_0(x) \\ \text{subject to the constraints} \quad f_i(x) + y_i \leq 0 \; \forall i \in I, \; f_i(x) + y_i = 0 \; \forall i \in J, \; x \in C. \end{cases}$$
One immediately observes that the original problem $(\mathcal{P})$ is just the problem $(\mathcal{P}_0)$ for $y = 0$ and that the above perturbation form takes into account eventual errors in the constraint $(f_1(x), \cdots, f_{m+n}(x)) \in \mathbb{R}_+^m \times \{0_{\mathbb{R}^n}\}$. If we define the function $\varphi : X \times Y \to \mathbb{R} \cup \{-\infty, +\infty\}$ by
$$\varphi(x,y) = \begin{cases} f_0(x) & \text{if } f_i(x) + y_i \leq 0 \; \forall i \in I, \; f_i(x) + y_i = 0 \; \forall i \in J, \; x \in C \\ +\infty & \text{otherwise}, \end{cases}$$
then $(\mathcal{P}_y)$ takes the following form:

$(\mathcal{P}_y) \quad$ Minimize $\quad \varphi(x,y) \quad$ over x in the whole space X.

Consider the function $p : Y \to \mathbb{R} \cup \{-\infty, +\infty\}$ defined by

(3.83) $$p(y) := \inf_{x \in X} \varphi(x,y).$$

For each $y \in Y$, we see that $p(y)$ is the value of the perturbed problem $(\mathcal{P}_y)$. The function p is called the *perturbation function* of the value of $(\mathcal{P})$, or simply the *value function* of $(\mathcal{P})$ or the *performance function* of $(\mathcal{P})$. Some authors also say the *marginal function*.

The *stability* of $(\mathcal{P})$ is in general related to semicontinuity properties of the function p. In the same way, the *sensibility* is related to subdifferentiability properties of p.

We have already noticed that the problem $(\mathcal{P})$ corresponds to the problem $(\mathcal{P}_0)$ for $y = 0$, that is, we may write $(\mathcal{P})$ in the form

$$(\mathcal{P}) \quad \text{Minimize} \quad \varphi(x, 0) \quad \text{over } x \text{ in the whole space } X.$$

Furthermore, coming back to the above function f, forgetting the perturbation y and taking the conjugacy transform merely with respect to the variable x, we have also already mentioned the first equality below

$$(3.84) \qquad \text{val}(\mathcal{P}) = \inf_{x \in X} (\overline{f})(x) = -(\overline{f})^*(0)$$

(see Proposition 3.146(b) for the second equality). Since $\overline{f}(x) = \varphi(x, 0)$, invoking now the conjugacy transform with respect to the pair of variables (x, y), we can write for any $y^* \in Y^*$

$$\begin{aligned}
(\overline{f})^*(0) &= \sup_{x \in X} [\langle 0, x \rangle - \overline{f}(x)] \\
&= \sup_{x \in X} [\langle 0, x \rangle + \langle y^*, 0 \rangle - \varphi(x, 0)] \\
&\leq \sup_{x \in X, y \in Y} [\langle 0, x \rangle + \langle y^*, y \rangle - \varphi(x, y)] \\
&= \varphi^*(0, y^*),
\end{aligned}$$

and hence $-(\overline{f})^*(0) \geq \sup_{y^* \in Y^*} -\varphi^*(0, y^*)$. Taking (3.84) and the above form of $(\mathcal{P})$ into account, it ensues that

$$(3.85) \qquad \inf_{x \in X} \varphi(x, 0) \geq \sup_{y^* \in Y^*} -\varphi^*(0, y^*).$$

This inequality leads us to study the maximization problem associated with the concave function $y^* \mapsto -\varphi^*(0, y^*)$, that is,

$$\text{Maximize} \quad -\varphi^*(0, y^*) \quad \text{over } y^* \text{ in the whole space } Y^*.$$

Unlike (3.82) with the particular function $g = \langle \overline{x}^*, \cdot \rangle$, note that the equality in place of the inequality in (3.85) will require certain conditions on the data.

More generally, consider two Hausdorff locally convex spaces X and Y with topological dual spaces X^* and Y^*, and consider an extended real-valued function $\varphi : X \times Y \to \mathbb{R} \cup \{-\infty, +\infty\}$. Denote by $(\mathcal{P})$ the *primal optimization problem*

$$(\mathcal{P}) \quad \text{Minimize} \quad \varphi(x, 0) \quad \text{over } x \text{ in the whole space } X$$

and denote by $\text{Sol}(\mathcal{P})$ its set of solutions. According to the previous analysis concerning the model problem with inequality/equality constraints and according also to what has been seen for (3.80) and (3.82) in a particular case, the problem

$$(\mathcal{D}(\mathcal{P})) \quad \text{Maximize} \quad -\varphi^*(0, y^*) \quad \text{over } y^* \text{ in the whole space } Y^*$$

is called the *dual problem* of the minimization problem $(\mathcal{P})$. This is the *Rockafellar dual problem* of $(\mathcal{P})$. Sometimes, it will be convenient to write $(\mathcal{D}_{Y^*}(\mathcal{P}))$ when we need to emphasize the dual space Y^* over which the dual problem is taken, and to just write $(\mathcal{D})$ when there is no risk of confusion. We will denote by $\text{Sol}(\mathcal{D})$ the set of solutions of the dual problem.

Obviously, the problem $\mathcal{P}$ depends only on the function $x \mapsto \varphi(x, 0)$, but the function $(x, y) \mapsto \varphi(x, y)$ involves the perturbation. Thus, the function φ defines the way the problem is perturbed.

Extending the particular case (3.83), the *performance/value* (or *marginal*) *function* associated with the problem $(\mathcal{P})$ is the function $p : Y \to \mathbb{R} \cup \{-\infty, +\infty\}$ defined by
$$p(y) := \inf_{x \in X} \varphi(x, y).$$

PROPOSITION 3.147. *With above notation, one always has*
$$\text{val}\,(\mathcal{D}(\mathcal{P})) \leq \text{val}\,(\mathcal{P}).$$

PROOF. According to the Fenchel inequality, for all $x \in X$ and $y^* \in Y^*$ we have
$$\langle (0, y^*), (x, 0) \rangle - \varphi(x, 0) \leq \varphi^*(0, y^*),$$
which means
$$-\varphi^*(0, y^*) \leq \varphi(x, 0).$$
This inequality being true for all $x \in X$ and $y^* \in Y^*$, we obtain
$$\sup_{y^* \in Y^*} -\varphi^*(0, y^*) \leq \inf_{x \in X} \varphi(x, 0),$$
which is the inequality in the proposition. □

Before passing to the dual problem of a maximization problem, let us mention that the excess of val $(\mathcal{P})$ over val $(\mathcal{D}(\mathcal{P}))$ is generally called the *duality gap* of $(\mathcal{P})$. The duality gap can be analyzed through the performance function. First, we note that the definition of the latter function gives
$$p(0) = \text{val}\,(\mathcal{P}).$$

Let us now compute the Legendre-Fenchel biconjugate of p relative to Y. At a first step for any $y^* \in Y^*$ we have
$$p^*(y^*) = \sup_{y \in Y} [\langle y^*, y \rangle - p(y)]$$
$$= \sup_{y \in Y} [\langle y^*, y \rangle - \inf_{x \in X} \varphi(x, y)]$$
$$= \sup_{y \in Y} \left(\sup_{x \in X} [\langle y^*, y \rangle - \varphi(x, y)] \right),$$
from which we see that
$$p^*(y^*) = \sup_{x \in X, y \in Y} [\langle 0, x \rangle + \langle y^*, y \rangle - \varphi(x, y)]$$
$$= \varphi^*(0, y^*).$$

At a second step, from the above equality we derive for every $y \in Y$
$$p^{**}(y) = \sup_{y^* \in Y^*} [\langle y^*, y \rangle - \varphi^*(0, y^*)],$$
and hence $p^{**}(0) = \sup_{y^* \in Y^*} -\varphi^*(0, y^*)$, that is,
$$p^{**}(0) = \text{val}\,(\mathcal{D}(\mathcal{P})).$$

Consequently, whenever either the performance function p is a proper lower semi-continuous convex function or satisfies $p \equiv +\infty$ or $p \equiv -\infty$, there is no duality gap according to Theorem 3.84(c).

Let us define now the dual problem of a maximization problem. Analogously to the above study, any maximization problem over a Hausdorff locally convex space V can be seen in the form

$$(\mathcal{Q}) \quad \text{Maximize} \quad \gamma(0, v) \quad \text{over } v \text{ in the whole space } V,$$

where $\gamma : U \times V \to \mathbb{R} \cup \{-\infty, +\infty\}$ is a bivariate function and U is another Hausdorff locally convex space representing the space of perturbations. With the opposite function $\Phi := -\gamma$ (which is convex whenever γ is concave), the problem $(\mathcal{Q})$ takes the form

$$(\mathcal{Q}) \quad \text{Maximize} \quad -\Phi(0, v) \quad \text{over } v \text{ in the whole space } V.$$

Proposition 3.147 ensures that $\sup_{u^* \in U^*} -\Phi^*(u^*, 0) \leq \inf_{v \in V} \Phi(0, v)$, or equivalently

$$\sup_{v \in V} -\Phi(0, v) \leq \inf_{u^* \in U^*} \Phi^*(u^*, 0),$$

which leads to define the dual problem $(\mathcal{D}(\mathcal{Q}))$ or $(\mathcal{D}_{U^*}(\mathcal{Q}))$ as

$$(\mathcal{D}(\mathcal{Q})) \quad \text{Minimize} \quad \Phi^*(u^*, 0) \quad \text{over } u^* \text{ in the whole space } U^*.$$

With the problem (Q) is associated the value function $u \mapsto \sup_{v \in V} -\Phi(u, v)$ which is the opposite of the (inf-)performance function $u \mapsto \inf_{v \in V} \Phi(u, v)$.

Now let us come back to the initial Hausdorff locally convex spaces X and Y and let us endow their topological dual spaces X^* and Y^* with any locally convex topologies compatible with the duality pairings between X and X^* and between Y and Y^*. The topological dual spaces of X^* and Y^* with respect to these topologies are then identified with X and Y respectively. So, applying the above form of $(\mathcal{D}(\mathcal{Q}))$ with $Q = \mathcal{D}(\mathcal{P})$, we arrive to the dual problem $(\mathcal{D}(\mathcal{D}(\mathcal{P})))$ or $(\mathcal{D}_X(\mathcal{D}(\mathcal{P})))$ as the problem

(3.86) $\quad (\mathcal{D}(\mathcal{D}(\mathcal{P}))) \quad \text{Minimize} \quad \varphi^{**}(x, 0) \quad \text{over } x \text{ in the whole space } X.$

Continuing with X^* and Y^* equipped with topologies compatible with the respective dualities, the Fenchel-Moreau theorem (see Theorem 3.84(c)) tells us that, whenever φ is a proper lower semicontinuous convex function on $X \times Y$, the problem $(\mathcal{D}(\mathcal{D}(\mathcal{P})))$ is nothing else but $(\mathcal{P})$, that is, the dual problem of the dual problem of $(\mathcal{P})$ is $(\mathcal{P})$. Otherwise stated, the bidual problem of $(\mathcal{P})$ coincides with the primal problem $(\mathcal{P})$ whenever φ is a proper lower semicontinuous convex function on $X \times Y$.

Our aim now is to investigate conditions under which the inequality in Proposition 3.147 is an equality, that is, there is no duality gap between $(\mathcal{P})$ and $(\mathcal{D})$, where $(\mathcal{D})$ serves to abbreviate $(\mathcal{D}(\mathcal{P}))$ as already done in diverse preceding places. In regards to that we denote by q the (inf-)performance function associated with $(\mathcal{D})$, that is, according to the previous analysis, the function $q : X^* \to \mathbb{R} \cup \{-\infty, +\infty\}$ defined by

(3.87) $\quad q(x^*) = \inf_{y^* \in Y^*} \varphi^*(x^*, y^*) \quad \text{for all } x^* \in X^*.$

Clearly, we have

(3.88) $\quad \text{val}(\mathcal{D}) = -q(0).$

We begin with the following proposition concerning the performance/value function and its Legendre-Fenchel conjugate. Its first assertions (a) and (b) just restate the result of the computation of $p^*(y^*)$ and $p^{**}(0)$ done in the above analysis of the duality gap of the problem $(\mathcal{P})$.

PROPOSITION 3.148. With notation above, the following properties hold:
(a) For all $y^* \in Y^*$ one has
$$p^*(y^*) = \varphi^*(0, y^*).$$
(b) The values of the primal and dual problems satisfy the equalities
$$\operatorname{val}(\mathcal{P}) = p(0) \quad \text{and} \quad \operatorname{val}(\mathcal{D}) = p^{**}(0),$$
so there is no duality gap if and only if $p(0) = p^{**}(0)$.
(c) An element $\bar{y}^* \in Y^*$ is a solution of $(\mathcal{D})$ if and only if
$$\langle \bar{y}^*, y \rangle - p(y) \leq -p^{**}(0) \quad \text{for all } y \in Y.$$
(d) The function p is convex whenever the perturbation function φ is convex.

PROOF. Assertions (a) and (b) are already shown (as said above) and the assertion (d) is a consequence of Proposition 2.54. Let us prove (c). According to (a) an element $\bar{y}^* \in Y^*$ is a solution of $(\mathcal{D})$ if and only if $p^*(\bar{y}^*) = \inf_{y^* \in Y^*} p^*(y^*)$, and hence it suffices to note that

$$p^*(\bar{y}^*) = \inf_{y^* \in Y^*} p^*(y^*) \Leftrightarrow p^*(\bar{y}^*) = -p^{**}(0) \Leftrightarrow \langle \bar{y}^*, y \rangle - p(y) \leq -p^{**}(0) \ \forall y \in Y,$$

where the first equivalence uses the equality $\inf_Y g = -g^*(0)$ for any function $g: Y \to \mathbb{R} \cup \{-\infty, +\infty\}$. □

THEOREM 3.149 (duality result I: case $\varphi(\cdot, \cdot)$). With notation above, the following hold.
(a) If $p(0) = -\infty$, then there is no duality gap, more precisely,
$$\operatorname{val}(\mathcal{D}) = \operatorname{val}(\mathcal{P}) = -\infty,$$
and every $y^* \in Y^*$ is a solution of $(\mathcal{D})$.
(b) If $p(0)$ is finite and there exists some $\bar{y}^* \in Y^*$ such that

(3.89) $$p(0) = \inf_{y \in Y} [p(y) - \langle \bar{y}^*, y \rangle], \text{ that is, } \bar{y}^* \in \partial_{MR} p(0),$$

then one has:
(b$_1$) there is no duality gap, and in fact
$$\operatorname{val}(\mathcal{D}) = \operatorname{val}(\mathcal{P}) \in \mathbb{R}$$
(the most interesting case);
(b$_2$) the dual problem $(\mathcal{D})$ admits at least one solution;
(b$_3$) the set of solutions of $(\mathcal{D})$ coincides with the set of elements $\bar{y}^* \in Y^*$ satisfying (3.89);
(b$_4$) the set of solutions of $(\mathcal{D})$ is a weakly* compact convex set in Y^* whenever in addition the function p is bounded from above on a neighborhood of zero.

PROOF. (a) If $p(0) = -\infty$, then
$$\operatorname{val}(\mathcal{P}) = \inf_{x \in X} \varphi(x,0) = p(0) = -\infty.$$
Using the inequality $\operatorname{val}(\mathcal{D}) \leq \operatorname{val}(\mathcal{P})$ in Proposition 3.147 we obtain
$$\operatorname{val}(\mathcal{D}) = \operatorname{val}(\mathcal{P}) = -\infty.$$
Further, the latter equality $\operatorname{val}(\mathcal{D}) = -\infty$ tells us that $\sup_{y^* \in Y^*} -\varphi^*(0, y^*) = -\infty$, hence every $y^* \in Y^*$ realizes the supremum of the function $-\varphi^*(0, \cdot)$, that is, it is a solution of $(\mathcal{D})$.

(b) The assertion (b_2) follows from (b_3). So, fix any $\overline{y}^*$ satisfying (3.89). Since $p(0)$ is finite by assumption, we have
$$\operatorname{val}(\mathcal{P}) = p(0) \in \mathbb{R},$$
and according to (3.89)
$$p^*(\overline{y}^*) = \langle \overline{y}^*, 0 \rangle - p(0) = -\operatorname{val}(\mathcal{P}).$$
Using the inequality in Proposition 3.147 again we obtain
$$(3.90) \qquad \operatorname{val}(\mathcal{D}) \leq \operatorname{val}(\mathcal{P}) = -p^*(\overline{y}^*) = -\varphi^*(0, \overline{y}^*),$$
the latter equality being due to Proposition 3.148. Since, by definition, $\operatorname{val}(\mathcal{D}) = \sup_{y^* \in Y^*} -\varphi^*(0, y^*)$, we can conclude according to (3.90), that $\overline{y}^*$ is a solution of $(\mathcal{D})$ and the values of $(\mathcal{D})$ and $(\mathcal{P})$ coincide. This confirms (b_3) and (b_1). Further, the above equality between the values of $(\mathcal{D})$ and $(\mathcal{P})$ can be rewritten as $p(0) = p^{**}(0)$, so (b_3) is true by Proposition 3.148(c).

Now suppose in addition that p is bounded from above near 0, that is, there exists a real $\mu > 0$ and a symmetric neighborhood W of zero in Y such that $p(y) \leq \mu$ for all $y \in W$. Therefore, for every $y \in W$ we see that $|\langle y^*, y \rangle| \leq \mu + |p(0)|$ for all $y^* \in \operatorname{Sol}(\mathcal{D})$, so the weakly* closed set $\operatorname{Sol}(\mathcal{D})$ is weakly* bounded, and hence weakly* compact. $\square$

If the function p is convex and $p(0)$ is finite, (3.89) means $\overline{y}^* \in \partial p(0)$. This entails the following first corollary of Theorem 3.149.

COROLLARY 3.150. Assume with the above notation that p is convex and $\partial p(0) \neq \emptyset$. Then $\operatorname{val}(\mathcal{P}) = \operatorname{val}(\mathcal{D}) \in \mathbb{R}$ (so there is no duality gap) and one has
$$\operatorname{Sol}(\mathcal{D}) = \partial p(0).$$

Keeping in mind that $(\mathcal{P})$ is the dual of $(\mathcal{D})$ when φ is proper, lower semicontinuous and convex, we directly obtain the dualization form from Corollary 3.150.

COROLLARY 3.151. Assume with the above notation that φ is proper, lower semicontinuous and convex, and that both $\partial p(0)$ and $\partial q(0)$ are nonempty. Then $\operatorname{val}(\mathcal{P}) = \operatorname{val}(\mathcal{D}) \in \mathbb{R}$ (so there is no duality gap) and one has
$$\operatorname{Sol}(\mathcal{P}) = \partial q(0) \quad \text{and} \quad \operatorname{Sol}(\mathcal{D}) = \partial p(0).$$

Under the convexity of φ, one can also use the subdifferentials of φ and φ^* to provide a characterization of pairs of solutions of primal and dual optimization problems, that is, pairs $(\overline{x}, \overline{y}^*)$ such that $\overline{x}$ is a solution of problem $(\mathcal{P})$ and $\overline{y}^*$ a solution of problem $(\mathcal{D})$.

3.8. DUALITY IN CONVEX OPTIMIZATION

PROPOSITION 3.152. *Assume that the function $\varphi : X \times Y \to \mathbb{R} \cup \{-\infty, +\infty\}$ is proper, lower semicontinuous and convex. Then one has the equivalences:*

$$\left(\begin{array}{l} \overline{x} \in \mathrm{Sol}(\mathcal{P}) \\ \overline{y}^* \in \mathrm{Sol}(\mathcal{D}) \\ \mathrm{val}(\mathcal{P}) = \mathrm{val}(\mathcal{D}) \in \mathbb{R} \end{array} \right) \Leftrightarrow (0, \overline{y}^*) \in \partial \varphi(\overline{x}, 0) \Leftrightarrow (\overline{x}, 0) \in \partial \varphi^*(0, \overline{y}^*).$$

PROOF. The equivalence between the inclusions $(0, \overline{y}^*) \in \partial \varphi(\overline{x}, 0)$ and $(\overline{x}, 0) \in \partial \varphi^*(0, \overline{y}^*)$ follows (with X^* and Y^* endowed with the weak* topologies) from the equality $\varphi^{**} = \varphi$ due to the assumption that φ is proper lower semicontinuous and convex (see Theorem 3.84(c)). On the other hand, these inclusions mean that $\varphi(\overline{x}, 0)$ and $\varphi^*(0, \overline{y}^*)$ are finite and

$$\langle \overline{y}^*, y \rangle - \varphi(x, y) \leq -\varphi(\overline{x}, 0) \text{ and } \langle x^*, \overline{x} \rangle - \varphi^*(x^*, y^*) \leq -\varphi^*(0, \overline{y}^*)$$

for all $x \in X$, $y \in Y$, $x^* \in X^*$ and $y^* \in Y^*$. Taking $y = 0$ in the former inequality and $x^* = 0$ in the latter yields $\varphi(\overline{x}, 0) \leq \varphi(\cdot, 0)$ and $-\varphi^*(0, \cdot) \leq -\varphi^*(0, \overline{y}^*)$ respectively, that is, $\overline{x} \in \mathrm{Sol}(\mathcal{P})$ and $\overline{y}^* \in \mathrm{Sol}(\mathcal{D})$. Further, by Proposition 3.64(i) the inclusion $(0, \overline{y}^*) \in \partial \varphi(\overline{x}, 0)$ means that

$$\langle 0, \overline{x} \rangle + \langle \overline{y}^*, 0 \rangle = \varphi(\overline{x}, 0) + \varphi^*(0, \overline{y}^*),$$

or equivalently $\varphi(\overline{x}, 0) = -\varphi^*(0, \overline{y})$.

Conversely, suppose that the properties related to the values and sets of solutions of $(\mathcal{P})$ and $(\mathcal{D})$ hold. The assumption of equality $\mathrm{val}(\mathcal{P}) = \mathrm{val}(\mathcal{D})$ along with the finiteness of this common value implies that $\varphi(\overline{x}, 0) = -\varphi^*(0, \overline{y}^*)$, or equivalently

$$\langle 0, \overline{x} \rangle + \langle \overline{y}^*, 0 \rangle = \varphi(\overline{x}, \overline{x}) + \varphi^*(0, \overline{y}^*).$$

We deduce by the convexity of φ that $(0, \overline{y}^*) \in \partial \varphi(\overline{x}, 0)$ according to Proposition 3.64(i) again. □

Let us come back to the model problem with constraints associated with f_i, $i \in I \cup J$. Putting $Y := \mathbb{R}^I \times \mathbb{R}^J$, $Y_+ := \mathbb{R}^I_+ \times \{0_{\mathbb{R}^J}\}$ and

$$G(x) := (f_1(x), \cdots, f_m(x), f_{m+1}(x), \cdots, f_{m+n}(x)),$$

we see that for all $x \in X$ and $y \in Y$
(3.91)
$$\overline{f}(x) = (f_0 + \Psi_C)(x) + \Psi_{-Y_+}(G(x)) \quad \text{and} \quad \varphi(x, y) = (f_0 + \Psi_C)(x) + \Psi_{-Y^*}(G(x) + y).$$

Let us set this in a more general context. Let Y_+ be a convex cone in Y containing zero and $\leq_{Y_+}$ be the associated preorder defined in Y by

$$y \leq_{Y_+} y' \Leftrightarrow y' - y \in Y_+.$$

Let ∞_Y be an abstract greatest element adjoined to Y. With any mapping $G : X \to Y \cup \{+\infty_Y\}$ we associate its effective domain $\mathrm{dom}\, G := \{x \in X : G(x) \in Y\}$ and its epigraph $\mathrm{epi}\, G$ as the subset of $X \times Y$ given by

$$\mathrm{epi}\, G := \{(x, y) \in X \times Y : G(x) \leq_{Y_+} y\}.$$

Let us also recall that a function $g : Y \to \mathbb{R} \cup \{+\infty\}$ is said to be *non Y_+-decreasing* over a nonempty set $S \subset Y$ whenever $g(y) \leq g(y')$ for all $y, y' \in S$ with $y \leq_{Y_+} y'$. When $S = Y$, we simply say that g is non Y_+-decreasing.

EXAMPLE 3.153. (a) Any $y^* \in Y_+^*$ is clearly non Y_+-decreasing.
(b) The indicator function Ψ_{-Y_+} of the opposite of the cone Y_+ is non Y_+-decreasing. Indeed, let $y, y' \in Y$ with $y \leq_{Y_+} y'$, that is, $y \in y' - Y_+$. If $y' \notin -Y_+$, one has $\Psi_{-Y_+}(y') = +\infty$, and hence the inequality $\Psi_{-Y_+}(y) \leq \Psi_{-Y_+}(y')$ is trivial. If $y' \in -Y_+$, then $y \in y' - Y_+ \subset -Y_+ - Y_+ \subset -Y_+$, so $\Psi_{-Y_+}(y) = 0 = \Psi_{-Y+}(y')$. In any case, we obtain $\Psi_{-Y_+}(y) \leq \Psi_{-Y+}(y')$, which justifies the non Y_+-decreasing property of Ψ_{-Y_+}. □

Given a mapping $G: X \to Y \cup \{+\infty_Y\}$ and two functions $f: X \to \mathbb{R} \cup \{+\infty\}$ and $g: Y \to \mathbb{R} \cup \{+\infty\}$, consider the optimization problem

(3.92)   ($\mathcal{P}$)   Minimize $f(x) + g(G(x))$ over x in the whole space X,

where by convention we define

(3.93) $$g(+\infty_Y) = +\infty.$$

This convention gives for the null functional 0_{Y^*} (that is, the zero in Y^*)

(3.94) $$0_{Y^*} \circ G = \Psi_{\mathrm{dom}\, G}.$$

Taking (3.91) into account, we naturally associate with ($\mathcal{P}$) the perturbation function $\varphi: X \times Y \to \mathbb{R} \cup \{-\infty, +\infty\}$ defined by

$$\varphi(x, y) = f(x) + g(G(x) + y),$$

and hence the dual problem is

($\mathcal{D}$)   Maximize $-\varphi^*(0, y^*)$ over y^* in the whole space Y^*.

Let us compute $\varphi^*(0, y^*)$ by writing

$$\varphi^*(0, y^*) = \sup_{x \in X, y \in Y} [\langle y^*, y \rangle - (f(x) + g(G(x) + y))]$$

$$= \sup_{x \in \mathrm{dom}\, f \cap \mathrm{dom}\, G} \left[-f(x) + \sup_{y \in Y} (\langle y^*, y \rangle - g(G(x) + y)) \right]$$

$$= \sup_{x \in \mathrm{dom}\, f \cap \mathrm{dom}\, G} \left[-f(x) - \langle y^*, G(x) \rangle + \sup_{y \in Y} (\langle y^*, G(x) + y \rangle - g(G(x) + y)) \right]$$

$$= \sup_{x \in \mathrm{dom}\, f \cap \mathrm{dom}\, G} \left[-f(x) - \langle y^*, G(x) \rangle + \sup_{z \in Y} (\langle y^*, z \rangle - g(z)) \right],$$

which gives

$$\varphi^*(0, y^*) = \sup_{x \in \mathrm{dom}\, f \cap \mathrm{dom}\, G} [-f(x) - \langle y^*, G(x) \rangle + g^*(y^*)]$$

$$= g^*(y^*) + \sup_{x \in \mathrm{dom}\, f \cap \mathrm{dom}\, G} [\langle 0, x \rangle - (f + y^* \circ G)(x)]$$

(3.95) $$= g^*(y^*) + (f + y^* \circ G)^*(0),$$

where the latter equality is due to the fact that $(f + y^* \circ G)(x) = +\infty$ for every $x \notin \mathrm{dom}\, f \cap \mathrm{dom}\, G$ since by our convention (3.93)

(3.96)
$(y^* \circ G)(x) = \langle y^*, G(x) \rangle$ if $G(x) \in Y$ and $(y^* \circ G)(x) = +\infty$ if $G(x) = +\infty_Y$.

The dual problem then takes the form

(3.97) ($\mathcal{D}$) Maximize $-[g^*(y^*) + (f + y^* \circ G)^*(0)]$ over y^* in the whole space Y^*, with $y^* \circ G$ defined as in (3.96).

Before continuing the analysis of the dual problem we emphasize a particular face of the convention (3.93).

REMARK 3.154. With G and Y^+ as above, when g is a constant real-valued function $g \equiv \lambda$ with $\lambda \in \mathbb{R}$, the convention (3.93) yields to define the function $\lambda G : Y \to \mathbb{R} \cup \{+\infty\}$ by

$$(\lambda G)(x) = \lambda G(x) \text{ if } x \in \operatorname{dom} G \quad \text{and} \quad (\lambda G)(x) = +\infty \text{ if } x \notin \operatorname{dom} G,$$

so in particular (as already seen in (3.94)) in a more general setting)

(3.98) $$0 G = \Psi_{\operatorname{dom} G},$$

that is, $0G$ equals to the indicator function of $\operatorname{dom} G$. □

The next lemma will allow us to write the dual problem $(\mathcal{D})$ in a form involving the *positive dual cone* Y_+^* of Y_+ defined by

$$Y_+^* := \{y^* \in Y^* : \langle y^*, y \rangle \geq 0, \ \forall y \in Y_+\}.$$

Clearly, Y_+^* is a weakly* closed convex cone in Y^* and $Y_+^* = (-Y_+)^\circ$ the negative polar cone of $-Y_+$.

LEMMA 3.155. If $g : Y \to \mathbb{R} \cup \{+\infty\}$ is proper and non Y_+-decreasing, then

$$g^*(y^*) = +\infty \quad \text{for all } y^* \notin Y_+^*.$$

PROOF. Indeed, fix any $y^* \in Y^* \setminus Y_+^*$. There exists some $y_0 \in -Y_+$ with $\langle y^*, y_0 \rangle > 0$. Choose some point $b \in Y$ with $g(b)$ finite. Then, for every real $t > 0$ we have

$$g^*(y^*) = \sup_{y \in Y} [\langle y^*, y \rangle - g(y)] \geq \langle y^*, b + ty_0 \rangle - g(b + ty_0),$$

which combined with the inequality $g(b + ty_0) \leq g(b)$ (keep in mind that g is non Y_+-decreasing) entails that

$$g^*(y^*) \geq t \langle y^*, y_0 \rangle + \langle y^*, b \rangle - g(b).$$

The latter inequality being true for every real $t > 0$, it results that $g^*(y^*) = +\infty$ (since $\langle y^*, y_0 \rangle > 0$). □

According to the above lemma, when g is proper and non Y_+-decreasing, the dual problem may also be written in the form

(3.99) $(\mathcal{D})$ $\begin{cases} \text{Maximize} & -[g^*(y^*) + (f + y^* \circ G)^*(0)] \\ \text{subject to the constraint} & y^* \in Y_+^*. \end{cases}$

For the performance function $p : Y \to \mathbb{R} \cup \{-\infty, +\infty\}$ of the problem $(\mathcal{P})$ we have

(3.100) $$p(y) = \inf_{x \in X} [f(x) + g(G(x) + y)] \quad \text{for all } y \in Y,$$

where we recall that $g(+\infty_Y) = +\infty$ by convention. We then see that $y \in \operatorname{dom} p$ if and only if there exists some $x \in \operatorname{dom} f \cap \operatorname{dom} G$ such that $G(x) + y \in \operatorname{dom} g$, that is, $y \in \operatorname{dom} g - G(\operatorname{dom} f \cap \operatorname{dom} G)$. Consequently,

(3.101) $$\operatorname{dom} p = \operatorname{dom} g - G(\operatorname{dom} f \cap \operatorname{dom} G).$$

Before establishing a duality theorem for problem $(\mathcal{P})$, we recall (see (2.60)) that a mapping $G : X \to Y \cup \{+\infty_Y\}$ is Y_+-convex (or convex relative to the convex cone Y_+) provided that for all $t \in \,]0, 1[$, $x, x' \in X$

$$G(tx + (1-t)x') \leq_{Y_+} tG(x) + (1-t)G(x').$$

It is clear that $\dom G$ is convex whenever the mapping G is Y_+-convex. Further, one can verify that G is Y_+-convex if and only if $\epi G$ is a convex set in $X \times Y$.

EXAMPLE 3.156. Let $f_i : X \to \mathbb{R}$ with $i = 1, \cdots, m+n$, $I := \{1, \cdots, m\}$ and $J := \{m+1, \cdots, m+n\}$. Consider $Y := \mathbb{R}^I \times \mathbb{R}^J$ and the mapping $G : X \to Y$ defined by
$$G(x) := (f_1(x), \cdots, f_m(x), f_{m+1}(x), \cdots, f_{m+n}(x)) \quad \text{for all } x \in X,$$
and consider also the convex cone $Y_+ := \mathbb{R}^I_+ \times \{0_{\mathbb{R}^J}\}$ which contains the origin of Y. It is easily seen that the mapping G is Y_+-convex if and only if each function f_i, with $i \in I$, is convex and each function f_i, with $i \in J$, is affine. □

Let us state some properties which are easily verified.

LEMMA 3.157. Assume that $G : X \to Y \cup \{+\infty_Y\}$ is Y_+-convex and $g : Y \to \mathbb{R} \cup \{+\infty\}$ is convex. The following hold.
(a) If g is non Y_+-decreasing on $G(\dom G) + Y_+$, then the function $g \circ G$ is convex.
(b) If g is non Y_+-decreasing, then for any convex subset $S \subset \dom G$ the set $\dom g - G(S)$ and the sets $\{g \leq r\} - G(S)$ are convex for all $r \in \mathbb{R}$, where we recall that $\{f \leq r\} := \{x \in X : f(x) \leq r\}$.

If f, g and G are convex and if g is non Y_+-decreasing, then by Lemma 3.157 above the function
$$(x, y) \mapsto \varphi(x, y) = f(x) + g(G(x) + y)$$
is convex, since the mapping $G_0 : X \times Y \to Y \cup \{+\infty_Y\}$ defined by $G_0(x, y) = G(x) + y$ is Y_+-convex over $X \times Y$; this tells us that the performance/value function p is convex.

The next theorem provides a verifiable basic condition under which there is no duality gap. It involves the *Mackey topology* $\mu(Y, Y^*)$ of the Hausdorff locally convex space (Y, τ_Y), that is, (as recalled in Appendix) the greatest locally convex topology θ on Y for which the topological dual space of (Y, θ) coincides with Y^*, where Y^* denotes the topological dual of (Y, τ_Y). This topology $\mu(Y, Y^*)$ is known to be generated by the family of seminorms $(p_K)_K$ with the nonempty $w(Y^*, Y)$-compact sets K in Y^*, where

(3.102) $$p_K(y) = \sup_{y^* \in K} |\langle y^*, y \rangle| \quad \text{for all } y \in Y.$$

Similarly, the *Mackey topology* $\mu(Y^*, Y)$ of the topological dual space Y^* is the greatest locally convex topology θ on Y^* for which the topological dual space of (Y^*, θ) is identified with Y through the duality pairing. This topology is generated by the family of seminorms $(q_K)_K$ with the nonempty $w(Y, Y^*)$-compact sets K in Y, where

(3.103) $$q_K(y^*) = \sup_{y \in K} |\langle y^*, y \rangle| \quad \text{for all } y^* \in Y^*.$$

In general, one writes p_K in place of q_K when there is no risk of ambiguity. Any vector topology θ on Y (resp. on Y^*) for which the topological dual of (Y, θ) (resp. of (Y^*, θ)) is Y^* (resp. coincides with Y through the canonical pairing) is said to be *compatible* with the duality between Y and Y^*.

Now recall the equality (see (3.94))
$$0_{Y^*} \circ G = \Psi_{\dom G}$$

which is implicit in the next theorem along with in Corollaries 3.159 and 3.160.

THEOREM 3.158 (duality result II: case $\mathbf{f} + \mathbf{g} \circ \mathbf{G}$). Let X and Y be two Hausdorff locally convex spaces. Let $G : X \to Y \cup \{+\infty_Y\}$ be a Y_+-convex mapping, $f : X \to \mathbb{R} \cup \{+\infty\}$ be a convex function, and $g : Y \to \mathbb{R} \cup \{+\infty\}$ be a convex function which is non Y_+-decreasing. Then the duality result

$$\inf_{x \in X} [f(x) + g(G(x))] = \max_{y^* \in Y^*} [-g^*(y^*) - (f + y^* \circ G)^*(0)]$$
$$= \max_{y^* \in Y^*_+} [-g^*(y^*) - (f + y^* \circ G)^*(0)]$$

holds under the following condition:
For the vector subspace

$$Z := \text{span}[\text{dom } g - G(\text{dom } f \cap \text{dom } G)]$$

endowed with the topology induced by the Mackey topology of Y, there exists some number $r \in \mathbb{R}$ such that

$$0 \in \text{int}_Z (\{g \leq r\} - G(\{f \leq r\} \cap \text{dom } G)).$$

Further, if in addition $\text{val}(\mathcal{P})$ is finite, then $\text{Sol}(\mathcal{D}) = \partial p(0)$.

PROOF. If $p(0) = -\infty$, the result is a direct consequence of the assertion (a) in Theorem 3.149. So, suppose $p(0) > -\infty$ and for

$$Z := \text{span}[\text{dom } g - G(\text{dom } f \cap \text{dom } G)],$$

choose by assumption a number $r \in \mathbb{R}$ and a neighborhood W of zero in Z such that

(3.104) $$W \subset \{g \leq r\} - G(\{f \leq r\} \cap \text{dom } G).$$

Denote by $p_0 : Z \to \mathbb{R} \cup \{-\infty, +\infty\}$ the restriction of p on Z. Observe that p_0 is convex and fix any $y \in W$. According to (3.104) there are some $y_r \in \{g \leq r\}$ and $x_r \in \{f \leq r\} \cap \text{dom } G$ such that $y = y_r - G(x_r)$, that is, $G(x_r) + y = y_r$. Then

$$p_0(y) = p(y) \leq f(x_r) + g(G(x_r) + y) = f(x_r) + g(y_r) \leq 2r,$$

which entails that p_0 is bounded from above on the neighborhood W of zero in Z. In particular, $p_0(0) < +\infty$, and hence $p_0(0)$ is finite because $p(0) > -\infty$ (by our assumption on p) and $p_0(0) = p(0)$ (since p_0 is the restriction of p on Z). The convex function p_0 being finite at zero and bounded from above on a neighborhood of zero, it is continuous at zero by Theorem 3.18. This and Theorem 3.30 ensure that $\partial p_0(0) \neq \emptyset$. Fix any $z^* \in \partial p_0(0)$. The linear functional z^* being continuous on Z, by the Hahn-Banach theorem it may be extended to a linear functional ζ on Y which is continuous with respect to the Mackey topology $\mu(Y, Y^*)$. But the topological dual of $(Y, \mu(Y, Y^*))$ is Y^*, and hence $\zeta \in Y^*$. Further, by (3.101) we have $\text{dom } p \subset Z$, which entails $p(y) = +\infty$ for all $y \in Y \setminus Z$. Combining this with the inclusion $z^* \in \partial p_0(0)$, we see that

$$\langle \zeta, y - 0 \rangle_{Y,Y^*} + p(0) \leq p(y) \quad \text{for all } y \in Y,$$

or equivalently $\zeta \in \partial p(0)$. This says that the condition (3.89) is fulfilled with $\bar{y}^* := \zeta$. The desired first equality of the theorem then follows from the assertion (b_1) in (b) of Theorem 3.149.

Concerning the second equality in the theorem, observe that for any $y \in -Y_+$ we have

$$p(y) = \inf_{x \in \text{dom } G}[f(x) + g(G(x) + y)] \leq \inf_{x \in \text{dom } G}[f(x) + g(G(x))] = p(0),$$

the inequality being due to the nondecreasing property of g. So, for $a^* := \zeta \in \partial p(0)$ we have

$$\langle a^*, y \rangle \leq p(y) - p(0) \leq 0,$$

which means $a^* \in Y_+^*$. The second equality in the theorem is then justified.

Finally, assume in addition that val$(\mathcal{D})$ is finite, that is, $p(0) \in \mathbb{R}$. Noting that p is convex (by the convexity of φ) the equality Sol$(\mathcal{D}) = \partial p(0)$ follows from Lemma 3.148. The proof is then complete. $\square$

Let us derive from the previous theorem some useful corollaries. The first one considers two easily verifiable conditions.

COROLLARY 3.159. *Let X and Y be two Hausdorff locally convex spaces. Let $G : X \to Y \cup \{+\infty_Y\}$ be a Y_+-convex mapping, $f : X \to \mathbb{R} \cup \{+\infty\}$ be a convex function, and $g : Y \to \mathbb{R} \cup \{+\infty\}$ be a convex function which is non Y_+-decreasing. Then the duality result*

$$\inf_{x \in X}[f(x) + g(G(x))] = \max_{y^* \in Y^*}[-g^*(y^*) - (f + y^* \circ G)^*(0)]$$

$$= \max_{y^* \in Y_+^*}[-g^*(y^*) - (f + y^* \circ G)^*(0)]$$

holds under either hypothesis (a) or hypothesis (b):
(a) *There exists some point $\bar{x} \in \text{dom } f \cap \text{dom } G$ such that g is finite and continuous at $G(\bar{x})$ with respect to a locally convex topology θ compatible with the duality between Y and Y^*;*
(b) *the space Y is a normed space and for the vector subspace*

$$Z := \text{span}[\text{dom } g - G(\text{dom } f \cap \text{dom } G)]$$

endowed with the norm induced by the one of Y, there exists some number $r \in \mathbb{R}$ such that

$$0 \in \text{int}_Z(\{g \leq r\} - G(\{f \leq r\} \cap \text{dom } G)).$$

PROOF. (b) Under the hypothesis (b), the corollary directly follows from Theorem 3.158.
(a) Under the hypothesis (a), there exists some θ-neighborhood V of zero in Y such that $G(\bar{x}) + V \subset \text{dom } g$, and hence (since $\bar{x} \in \text{dom } f \cap \text{dom } G$)

$$V = (G(\bar{x}) + V) - G(\bar{x}) \subset \text{dom } g - G(\text{dom } f \cap \text{dom } G).$$

This implies that $Z = Y$. Further, for the real number $r := 1 + \max\{g(G(\bar{x})), f(\bar{x})\}$, there exists a θ-neighborhood W of zero in Y such that

$$g(G(\bar{x}) + y) \leq r \quad \text{for all } y \in W.$$

Combining this with the inclusion $\bar{x} \in \{f \leq r\} \cap \text{dom } G$, for every $y \in W$ we can write

$$y = (G(\bar{x}) + y) - G(\bar{x}) \in \{g \leq r\} - G(\{f \leq r\} \cap \text{dom } G),$$

and hence

$$W \subset \{g \leq r\} - G(\{f \leq r\} \cap \text{dom } G).$$

Since W is also a $\mu(Y, Y^*)$-neighborhood of zero (because $0 \subset \mu(Y, Y^*)$), the result follows again from Theorem 3.158. □

The second corollary is related to the situation when Y is finite-dimensional.

COROLLARY 3.160. *Let X be a Hausdorff locally convex space and Y be a finite-dimensional normed space. Let $G : X \to Y \cup \{+\infty_Y\}$ be a Y_+-convex mapping, $f : X \to \mathbb{R} \cup \{+\infty\}$ be a convex function, and $g : Y \to \mathbb{R} \cup \{+\infty\}$ be a convex function which is non Y_+-decreasing. Assume that*

$$0 \in \mathrm{rint}\,[\mathrm{dom}\,g - G(\mathrm{dom}\,f \cap \mathrm{dom}\,G)].$$

Then one has

$$\inf_{x \in X}[f(x) + g(G(x))] = \max_{y^* \in Y^*}[-g^*(y^*) - (f + y^* \circ G)^*(0)]$$
$$= \max_{y^* \in Y_+^*}[-g^*(y^*) - (f + y^* \circ G)^*(0)].$$

PROOF. Setting $Z := \mathrm{span}[\mathrm{dom}\,g - G(\mathrm{dom}\,f \cap \mathrm{dom}\,G)]$, we have by assumption $0 \in \mathrm{int}_Z[\mathrm{dom}\,g - G(\mathrm{dom}\,f \cap \mathrm{dom}\,G)]$. Denote by m the dimension of the vector space Z. Suppose for a moment that $m \geq 1$ and fix a basis $\{e_1, \cdots, e_m\}$ of Z. Then for each $j = 1, \cdots, m$ there exist $\alpha_j \in \,]0, +\infty[$, $x_j, x'_j \in \mathrm{dom}\,f \cap \mathrm{dom}\,G$ and $y_j, y'_j \in \mathrm{dom}\,g$ such that

(3.105) $\qquad \alpha_j e_j = y_j - G(x_j) \quad \text{and} \quad -\alpha_j e_j = y'_j - G(x'_j).$

The system $\{\alpha_1 e_1, \cdots, \alpha_m e_m\}$ being a basis of Z, it is easily seen that the set $\mathrm{co}\,\{\alpha_1 e_1, \cdots, \alpha_m e_m, -\alpha_1 e_1; \cdots, -\alpha_m e_m\}$ is a neighborhood of zero in Z, and hence

(3.106) $\qquad 0 \in \mathrm{int}_Z\big(\mathrm{co}\,\{\alpha_1 e_1, \cdots, \alpha_m e_m, -\alpha_1 e_1; \cdots, -\alpha_m e_m\}\big).$

If we put $r := \max_{1 \leq j \leq m}\big(\max\{g(y_j), g(y'_j), f(x_j), f(x'_j)\}\big)$, the equalities in (3.105) imply that $\alpha_j e_j$ and $-\alpha_j e_j$ belong to $\{g \leq r\} - G(\{f \leq r\} \cap \mathrm{dom}\,G)$, and hence according to the convexity of the latter set (see (b) in Lemma 3.157)

$$\mathrm{co}\,\{\alpha_1 e_1, \cdots, \alpha_m e_m, -\alpha_1 e_1; \cdots, -\alpha_m e_m\} \subset \{g \leq r\} - G(\{f \leq r\} \cap \mathrm{dom}\,G).$$

Taking (3.106) into account we obtain

(3.107) $\qquad 0 \in \mathrm{int}_Z\big(\{g \leq r\} - G(\{f \leq r\} \cap \mathrm{dom}\,G)\big).$

In the remaining case when $m = 0$, we have $Z = \{0\}$ and $\mathrm{dom}\,g - G(\mathrm{dom}\,f \cap \mathrm{dom}\,G) \neq \emptyset$ (by the inclusion $0 \in \mathrm{int}_Z\big(\mathrm{dom}\,g - G(\mathrm{dom}\,f \cap \mathrm{dom}\,G))\big)$, so noticing that

$$\mathrm{dom}\,g - G(\mathrm{dom}\,f \cap \mathrm{dom}\,G) = \bigcup_{r \in \mathbb{R}} \big(\{g \leq r\} - G(\{f \leq r\} \cap \mathrm{dom}\,G)\big),$$

it ensues that there exists some $r \in \mathbb{R}$ such that $\{g \leq r\} - G(\{f \leq r\} \cap \mathrm{dom}\,G) \neq \emptyset$. We derive that

$$\{g \leq r\} - G(\{f \leq r\} \cap \mathrm{dom}\,G) = \{0\} = Z,$$

and hence (3.107) still holds true.

The equalities in the corollary then follow from Theorem 3.158. (Observe also that for $m = 0$, the equalities can be checked by direct computation). □

The third corollary assumes that G is a continuous linear mapping.

COROLLARY 3.161 (duality result III: Rockafellar's version). Let X and Y be Hausdorff locally convex spaces, $f : X \to \mathbb{R} \cup \{+\infty\}$ and $g : Y \to \mathbb{R} \cup \{+\infty\}$ be convex functions, and $A : X \to Y$ be a continuous linear mapping. Then the duality result

$$\inf_{x \in X} [f(x) + g(Ax)] = \max_{y^* \in Y^*} -[g^*(y^*) + f^*(-A^*y^*)]$$

holds in each one of the cases (a), (b), (c), (d) below:

(a) For the vector subspace $Z := \operatorname{span}[\operatorname{dom} g - A(\operatorname{dom} f)]$ endowed with the topology induced by the Mackey topology of Y, there exists some number $r \in \mathbb{R}$ such that

$$0 \in \operatorname{int}_Z(\{g \leq r\} - A(\{f \leq r\}));$$

(b) there exists some point $\bar{x} \in \operatorname{dom} f$ such that g is finite and continuous at $A\bar{x}$ with respect to a locally convex topology θ on Y compatible with the duality between Y and Y^*;

(c) the space Y is a normed space and for $Z := \operatorname{span}[\operatorname{dom} g - A(\operatorname{dom} f)]$ endowed with the norm induced by the one of Y, there exists some number $r \in \mathbb{R}$ such that

$$0 \in \operatorname{int}_Z[\{g \leq r\} - A(\{f \leq r\})];$$

(d) the space Y is a finite-dimensional normed space and

$$0 \in \operatorname{rint}[\operatorname{dom} g - A(\operatorname{dom} f)].$$

PROOF. Consider the convex cone $Y_+ = \{0_Y\}$. It is readily seen that A is Y_+-convex and that g is non Y_+-decreasing. Therefore, we can apply Theorem 3.158 in case (a), Corollary 3.159 in cases (b) and (c), and Corollary 3.160 in case (d) to obtain

$$\inf_{x \in X} [f(x) + g(Ax)] = \max_{y^* \in Y^*} -[g^*(y^*) + (f + y^* \circ A)^*(0)].$$

Further, by (3.27) we have

$$(f + y^* \circ A)^*(0) = f^*(-y^* \circ A) = f^*(-A^*y^*),$$

hence we can conclude that the desired equality of the corollary holds true. □

With A as the identity we obtain:

COROLLARY 3.162 (Fenchel's version). Let X be a Hausdorff locally convex space, $f, g : X \to \mathbb{R} \cup \{+\infty\}$ be convex functions. Then the duality result

$$\inf_{x \in X} [f(x) + g(x)] = \max_{x^* \in X^*} -[g^*(x^*) + f^*(-x^*)]$$

holds in each one of the cases (a), (b), (c), (d) below:

(a) For the vector subspace $Z := \operatorname{span}[\operatorname{dom} g - \operatorname{dom} f)]$ endowed with the topology induced by the Mackey topology of X, there exists some number $r \in \mathbb{R}$ such that

$$0 \in \operatorname{int}_Z(\{g \leq r\} - \{f \leq r\});$$

(b) there exists some point $\bar{x} \in \operatorname{dom} f$ such that g is finite and continuous at $\bar{x}$ with respect to a locally convex topology θ on X compatible with the duality between X and X^*;

(c) the space X is a normed space and for $Z := \operatorname{span}[\operatorname{dom} g - \operatorname{dom} f]$ endowed with the norm induced by the one of X, there exists some number $r \in \mathbb{R}$ such that

$$0 \in \operatorname{int}_Z[\{g \leq r\} - \{f \leq r\}];$$

(d) the space Y is a finite-dimensional normed space and
$$0 \in \text{rint}\,[\text{dom}\,g - \text{dom}\,f].$$

3.9. Duality, infsup property, Lagrange multipliers

Consider the (primal) minimization problem

$$(\mathcal{P}) \begin{cases} \text{Minimize} & f(x) \\ \text{subject to the constraints} & G(x) \leq_{Y_+} 0 \text{ and } x \in C, \end{cases}$$

with a mapping $G : X \to Y$, a function $f : X \to \mathbb{R} \cup \{+\infty\}$ and a set $C \subset X$. Assume that the convex cone Y_+ contains zero and assume also that set of feasible points $C \cap G^{-1}(-Y_+)$ is nonempty.

(I) Another form of the dual problem.
Using the indicator functions Ψ_C and Ψ_{-Y_+} we can write $(\mathcal{P})$ in the form

(3.108)
$$(\mathcal{P}) \quad \text{Minimize} \quad (f + \Psi_C)(x) + \Psi_{-Y_+}(G(x)) \quad \text{over } x \text{ in the whole space } X.$$

Keep also in mind by Example 3.153(b) that the convex function Ψ_{-Y_+} is non Y_+-decreasing.

Taking the latter property of Ψ_{-Y_+} into account, (3.99) may be used to obtain the following form of the dual problem

$$(\mathcal{D}) \begin{cases} \text{Maximize}_{y^*} & -[(\Psi_{-Y_+})^*(y^*) + (f + \Psi_C + y^* \circ G)^*(0)] \\ \text{subject to the constraint} & y^* \in Y_+^*. \end{cases}$$

Since $-Y_+^*$ is a convex cone, we have $(\Psi_{-Y_+})^* = \Psi_{(-Y_+)^\circ}$ (see (3.35)), and hence $(\Psi_{-Y_+})^* = \Psi_{Y_+^*}$ because $(-Y_+)^\circ = Y_+^*$. Then, for any $y^* \in Y_+^*$

$$(\Psi_{-Y_+})^*(y^*) + (f + \Psi_C + y^* \circ G)^*(0) = (f + \Psi_C + y^* \circ G)^*(0)$$
$$= -\inf_{x \in X} \left(f(x) + \Psi_C(x) + y^* \circ G(x)\right),$$

since the equality $h^*(0) = -\inf_{x \in X} h(x)$ always holds for any extended real-valued function h on X. This yields the following form for the dual problem

(3.109) $(\mathcal{D}) \begin{cases} \text{Maximize}_{y^*} & \inf_{x \in X} \left(f(x) + \langle y^*, G(x) \rangle + \Psi_C(x)\right) \\ \text{subject to the constraint} & y^* \in Y_+^*. \end{cases}$

Furthermore, one may apply Theorem 3.158 or its Corollary 3.160 to obtain sufficient conditions guaranteeing that there is no duality gap. For instance, if Y is a finite-dimensional normed space, the condition

$$0 \in \text{rint}\,[Y_+ + G(C \cap \text{dom}\,f)]$$

ensures that there is no duality gap.

(II) The value of the primal problem in an inf sup form.
Assume in this part (II) that Y_+ is *closed* and that X, Y are still Hausdorff locally convex spaces. The above form of $(\mathcal{D})$ tells us that

(3.110) $$\text{val}\,(\mathcal{D}) = \sup_{y^* \in Y_+^*} \inf_{x \in C} \left(f(x) + \langle y^*, G(x) \rangle\right).$$

Observe that for any feasible point x of $(\mathcal{P})$ we have $-G(x) \in Y_+$, and hence for any $y^* \in Y_+^*$ we have $\langle y^*, -G(x) \rangle \geq 0$, that is, $\langle y^*, G(x) \rangle \leq 0$. Consequently, for any feasible point x of $(\mathcal{P})$ we obtain $\sup_{y^* \in Y_+^*} \langle y^*, G(x) \rangle = 0$, because $0 \in Y_+^*$. Then, for any feasible point x of $(\mathcal{P})$ which is in $\operatorname{dom} f$

$$\sup_{y^* \in Y_+^*} \big(f(x) + \langle y^*, G(x) \rangle \big) = f(x) + \sup_{y^* \in Y_+^*} \langle y^*, G(x) \rangle = f(x),$$

and hence $\sup_{y^* \in Y_+^*} \big(f(x) + \langle y^*, G(x) \rangle \big) = f(x)$ for every feasible point x of $(\mathcal{P})$ since the equality is obviously true when $x \notin \operatorname{dom} f$. Therefore, we have

$$(3.111) \qquad \inf_{\substack{x \in C \\ G(x) \leq_{Y_+} 0}} f(x) = \inf_{\substack{x \in C \\ G(x) \leq_{Y_+} 0}} \sup_{y^* \in Y_+^*} \big(f(x) + \langle y^*, G(x) \rangle \big).$$

On the other hand, if $G(x) \notin -Y_+$, the closedness property of the convex cone $-Y_+$ allows us to apply the Hahn-Banach separation theorem to obtain some $b^* \in Y^*$ such that

$$\langle b^*, -G(x) \rangle < 0 \leq \langle b^*, y \rangle \quad \text{for all } y \in Y_+,$$

and hence $b^* \in Y_+^*$ and $\langle b^*, G(x) \rangle > 0$. In the case $-G(x) \notin Y_+$ and $x \in \operatorname{dom} f$ we then have

$$\sup_{y^* \in Y_+^*} \big(f(x) + \langle y^*, G(x) \rangle \big) = f(x) + \sup_{y^* \in Y_+^*} \langle y^*, G(x) \rangle$$

$$\geq f(x) + \sup_{t \in \,]0, +\infty[} t \langle b^*, G(x) \rangle$$

$$= f(x) + \langle b^*, G(x) \rangle \sup_{t \in \,]0, +\infty[} t$$

$$= +\infty,$$

which gives

$$\sup_{y^* \in Y_+^*} \big(f(x) + \langle y^*, G(x) \rangle \big) = +\infty.$$

Then for any $x \in X$ with $-G(x) \notin Y_+$ the last equality holds since it is obvious for $x \notin \operatorname{dom} f$. This combined with (3.111) yields

$$\operatorname{val}(\mathcal{P}) = \inf_{x \in C} \sup_{y^* \in Y_+^*} \big(f(x) + \langle y^*, G(x) \rangle \big).$$

From the latter equality and from (3.110) we derive that there is no duality gap if and only if

$$\sup_{y^* \in Y_+^*} \inf_{x \in C} \big(f(x) + \langle y^*, G(x) \rangle \big) = \inf_{x \in C} \sup_{y^* \in Y_+^*} \big(f(x) + \langle y^*, G(x) \rangle \big),$$

which is also equivalent to

$$\sup_{y^* \in Y_+^*} \inf_{x \in X} \big(f(x) + \langle y^*, G(x) \rangle + \Psi_C(x) \big) = \inf_{x \in X} \sup_{y^* \in Y_+^*} \big(f(x) + \langle y^*, G(x) \rangle + \Psi_C(x) \big).$$

(III) Lagrange multipliers.
Let us come back for a moment to the model problem and suppose that X is a Banach space, the functions $f_0, \cdots, f_{m+n}$ are continuously differentiable and the set C is closed. Suppose also that $\bar{x}$ is a solution of $(\mathcal{P})$ for which there

exists (see, for example, Corollary 2.230) a normalized Lagrange multiplier uple $\bar{\lambda} = (\bar{\lambda}_1, \cdots, \bar{\lambda}_{m+n})$ in $Y := \mathbb{R}^I \times \mathbb{R}^J$, that is, one has as optimality conditions

$$0 \in Df_0(\bar{x}) + \sum_{i \in I \cup J} \bar{\lambda}_i Df_i(\bar{x}) + N^C(C; \bar{x})$$

and each $\bar{\lambda}_i$ with $i \in I$ satisfies $\bar{\lambda}_i \geq 0$ along with $\bar{\lambda}_i f_i(\bar{x}) = 0$. Remembering in this setting that $G(x) = (f_1(x), \cdots, f_{m+n}(x))$, the two latter conditions mean that $\bar{\lambda} \in N(Y_+; G(\bar{x}))$, where $Y_+ := \mathbb{R}_+^I \times \{0_{\mathbb{R}^J}\}$. Considering the function $\mathcal{L}$ defined on $\mathbb{R}^I \times \mathbb{R}^J$ by $\mathcal{L}(x, \lambda) = +\infty$ if $(x, \lambda) \in (X \setminus C) \times (Y^* \setminus Y_+^*)$ and

$$\mathcal{L}(x, \lambda) = f_0(x) + \langle \lambda, G(x) \rangle + \Psi_C(x) - \Psi_{Y_+}(\lambda) \quad \text{if } (x, \lambda) \in (X \times Y_+^*) \cup (C \times Y^*).$$

The optimality conditions can be rewritten as

$$0 \in \partial \mathcal{L}(\cdot, \bar{\lambda})(\bar{x}) \quad \text{and} \quad 0 \in \partial(-\mathcal{L}(\bar{x}, \cdot))(\bar{\lambda}).$$

This makes clear the connection between the above optimality conditions and the above function $\mathcal{L}$ called accordingly the *Lagrangian function* of the model problem.

Concerning the more general primal problem (3.108), taking into account the above process and the form (3.109) of its dual problem $(\mathcal{D})$ its *Lagrangian function* $\mathcal{L} : X \times Y^* \to \mathbb{R} \cup \{-\infty, +\infty\}$ is defined by $\mathcal{L}(x, y^*) = +\infty$ if $(x, y^*) \in (X \setminus C) \times (Y^* \setminus Y_+^*)$ and

$$\mathcal{L}(x, \lambda) = f(x) + \langle y^*, G(x) \rangle + \Psi_C(x) - \Psi_{Y_+}(y^*) \quad \text{if } (x, \lambda) \in (X \times Y_+^*) \cup (C \times Y^*).$$

Putting $\mathcal{L}_0(x, y^*) := f(x) + \langle y^*, G(x) \rangle + \Psi_C(x)$, the form of $(\mathcal{D})$ in (3.109) becomes

$$(\mathcal{D}) \begin{cases} \text{Maximize} \quad \inf_{x \in X} \mathcal{L}_0(x, y^*) = \inf_{x \in X} \big(f(x) + \langle y^*, G(x) \rangle + \Psi_C(x) \big) \\ \text{subject to the constraint} \quad y^* \in Y_+^*, \end{cases}$$

or equivalently

$$(\mathcal{D}) \quad \text{Maximize} \quad \inf_{x \in X} \mathcal{L}(x, y^*) \quad \text{over } y^* \text{ in the whole space } Y^*.$$

Consider now the problem in (3.92) assuming that $G(X) \subset Y$ and $f(X) \subset \mathbb{R} \cup \{+\infty\}$. Since $(f + y^* \circ G)^*(0) = -\inf_{x \in X}(f(x) + \langle y^*, G(x) \rangle)$ its dual problem in (3.97) can be rewritten as

$$(\mathcal{D}) \quad \text{Maximize} \quad \inf_{x \in X} \mathcal{L}(x, y^*) \quad \text{over } y^* \text{ in the whole space } Y^*,$$

where $\mathcal{L} : X \times Y^* \to \mathbb{R} \cup \{-\infty, +\infty\}$ is the corresponding *Lagrangian function* defined by $\mathcal{L}(x, y^*) = +\infty$ if $(x, y^*) \in (X \setminus \mathrm{dom}\, f) \times (Y^* \setminus \mathrm{dom}\, g^*)$ and

(3.112)
$$\mathcal{L}(x, y^*) = f(x) + \langle y^*, G(x) \rangle - g^*(y^*) \quad \text{if } (x, y^*) \in (X \times \mathrm{dom}\, g^*) \cup (\mathrm{dom}\, f \times Y^*).$$

3.10. Linear optimization problem

We proceed in this section to showing how the above analysis applies to linear optimization problems.

Let P and Q be two *closed convex cones* in Hausdorff locally convex spaces X and Y respectively (generally, the cone P represents a cone of type X_+ and Q a cone of type $-Y_+$). Let also $c^* \in X^*$, $b \in Y$ and a continuous linear mapping $A : X \to Y$.

Consider the primal problem

$$(\mathcal{P}) \begin{cases} \text{Minimize}_x & \langle c^*, x \rangle \\ \text{subject to the constraints} & Ax - b \in Q \text{ and } x \in P. \end{cases}$$

Such a problem is called a *linear minimization problem*, because of the special forms of the constraints and the objective function.

It will be shown that the dual problem takes a form similar to the primal one. If we put $Y_+ := -Q$, we may write $(\mathcal{P})$ in the form

$$(\mathcal{P}) \begin{cases} \text{Minimize}_x & \langle c^*, x \rangle \\ \text{subject to the constraints} & Ax - b \leq_{Y_+} 0 \text{ and } x \in P. \end{cases}$$

The form (3.109) leads us, for $z^* \in Y^*$, to make the computation

$$\inf_{x \in X} \left(\langle c^*, x \rangle + \langle z^*, Ax - b \rangle + \Psi_P(x) \right) = \inf_{x \in P} \left(\langle c^*, x \rangle + \langle A^* z^*, x \rangle - \langle z^*, b \rangle \right)$$

$$= -\langle z^*, b \rangle + \inf_{x \in P} \left(\langle c^*, x \rangle + \langle A^* z^*, x \rangle \right).$$

Using the Hahn-Banach separation theorem, we can see that

$$\inf_{x \in P} \langle A^* z^* + c^*, x \rangle = \begin{cases} -\infty & \text{if } A^* z^* + c^* \notin -P^\circ \\ 0 & \text{if } A^* z^* + c^* \in -P^\circ. \end{cases}$$

Consequently, according to (3.109), the dual problem $(\mathcal{D})$ of $(\mathcal{P})$ takes the form

$$(\mathcal{D}) \begin{cases} \text{Maximize}_{z^*} & -\langle z^*, b \rangle \\ \text{subject to the constraints} & A^* z^* + c^* \in -P^\circ \text{ and } z^* \in Y_+^*. \end{cases}$$

Observing now that $Y_+^* = Q^\circ$, changing z^* into $-y^*$ yields the following form for the dual problem

$$(\mathcal{D}) \begin{cases} \text{Maximize}_{y^*} & \langle y^*, b \rangle \\ \text{subject to the constraints} & A^* y^* - c^* \in P^\circ \text{ and } y^* \in -Q^\circ, \end{cases}$$

which is a linear maximization problem. This is the general scheme of primal and dual linear optimization problems.

Let us proceed now to the translation in the context of finite-dimensional Euclidean spaces. Assume that $X = \mathbb{R}^n$ and $Y = \mathbb{R}^m$ endowed with their usual inner products and let $c \in \mathbb{R}^n$, $b \in \mathbb{R}^m$ and A be a matrix of order $m \times n$. Take $P = \mathbb{R}^n_+$ and $Q = -\mathbb{R}^m_+$.

With respect to the usual inner products, the dual spaces of $\mathbb{R}^n$ and $\mathbb{R}^m$ coincide with $\mathbb{R}^n$ and $\mathbb{R}^m$ respectively. The primal and dual linear problems take the forms

$$(\mathcal{P}) \begin{cases} \text{Minimize}_x & \langle c, x \rangle \\ \text{subject to the constraints} & Ax - b \leq_{\mathbb{R}^m_+} 0 \text{ and } 0 \leq_{\mathbb{R}^n_+} x \end{cases}$$

and

$$(\mathcal{D}) \begin{cases} \text{Maximize}_y & \langle b, y \rangle \\ \text{subject to the constraints} & A^t y - c \leq_{\mathbb{R}^n_+} 0 \text{ and } y \leq_{\mathbb{R}^m_+} 0, \end{cases}$$

where A^t denotes the transpose matrix of A.

In the setting of above Euclidean spaces, if $(\mathcal{P})$ has at least one feasible point, then one has

$$\text{val}\,(\mathcal{D}) = \text{val}(\mathcal{P}).$$

Indeed, in the case of such a linear primal problem we notice by Corollary 3.63 and Proposition 3.58 that the performance function p is polyhedral convex with $p(0) < +\infty$, so if $p(0) \in \mathbb{R}$ then
$$\partial p(0) \neq \emptyset$$
according to Corollary 3.61, and hence Theorem 3.149 allows to conclude. Further, when val($\mathcal{P}$) is finite, the same Theorem 3.149 also ensures in addition that both the primal and dual problems admit solutions.

3.11. Sum and chain rules of subdifferential under convexity

With the functions f_1 and f_2 in Example 2.99 we saw with $\bar{x} = 0$ that $\partial(f_1 + f_2)(\bar{x}) \not\subset \partial f_1(\bar{x}) + \partial f_2(\bar{x})$ while the functions f_1 and f_2 are convex. In this section we show how the above duality scheme yields conditions under which sum and chain rules of subdifferential for convex functions are valid. We mention first that any condition ensuring the non-vacuity of the subdifferential $\partial p(0)$ of the performance function is called a *qualification condition*.

We start with the following lemma.

LEMMA 3.163. *Let Y be a locally convex space and Y_+ be a convex cone in Y containing zero. Let $g : Y \to \mathbb{R} \cup \{+\infty\}$ be convex function which is non Y_+-decreasing. Then the inclusion $\partial g(Y) \subset Y_+^*$ holds.*

PROOF. Fix any $y \in \mathrm{Dom}\,\partial g$ and any $y^* \in \partial g(y)$. Then for any $h \in Y_+$ one has
$$\langle y^*, -h \rangle \leq g(y - h) - g(y) \leq 0,$$
which gives $\langle y^*, h \rangle \geq 0$. We conclude that $y^* \in Y_+^*$. □

In all this section we use again the equality (see (3.94))
$$0_{Y^*} \circ G = \Psi_{\mathrm{dom}\,G}.$$

PROPOSITION 3.164. *Let X and Y be Hausdorff locally convex spaces and Y_+ be a convex cone in Y containing zero. Let $G : X \to Y \cup \{+\infty_Y\}$ be a Y_+-convex mapping, $f : X \to \mathbb{R} \cup \{+\infty\}$ be a convex function and $g : Y \to \mathbb{R} \cup \{+\infty\}$ be a convex function which is non Y_+-decreasing. Let*
$$Z := \mathrm{span}\,[\mathrm{dom}\,g - G(\mathrm{dom}\,f \cap \mathrm{dom}\,G)]$$
be endowed with the topology induced by the Mackey topology of Y and let $a^* \in X^*$. Assume that there exists some real number $r > 0$ such that
$$0 \in \mathrm{int}_Z\,[\{g \leq r\} - G(\{f \leq r\} \cap \mathrm{dom}\,G \cap \{\langle a^*, \cdot \rangle \geq -r\})].$$
Then the following hold.
(α) One has the equalities
$$(f + g \circ G)^*(a^*) = \min_{y^* \in Y^*}\,[g^*(y^*) + (f + y^* \circ G)^*(a^*)]$$
$$= \min_{y^* \in Y_+^*}\,[g^*(y^*) + (f + y^* \circ G)^*(a^*)].$$

(β) If in addition $a^* \in \partial(f + g \circ G)(a)$, then there exists some $y^* \in \partial g(G(a)) \subset Y_+^*$ such that $a^* \in \partial(f + y^* \circ G)(a)$.

PROOF. (α) Let Z and $r > 0$ be as given in the statement. Put $f_0 := f - \langle a^*, \cdot \rangle$ and note that $\operatorname{dom} f_0 = \operatorname{dom} f$. For any $x \in \{f \leq r\} \cap \{\langle a^*, \cdot \rangle \geq -r\}$, we have
$$f_0(x) = f(x) - \langle a^*, x \rangle \leq 2r.$$
It ensues by the assumption that $0 \in \operatorname{int}_Z[\{g \leq 2r\} - G(\{f_0 \leq 2r\} \cap \operatorname{dom} G)]$. The assumption in Theorem 3.158 is satisfied for f_0, and then applying this theorem we obtain that
$$\inf_{x \in X} [f(x) + g(G(x)) - \langle a^*, x \rangle] = \max_{y^* \in Y^*} -[g^*(y^*) + (f + y^* \circ G - \langle a^*, \cdot \rangle)^*(0)]$$
$$= \max_{y^* \in Y^*_+} -[g^*(y^*) + (f + y^* \circ G - \langle a^*, \cdot \rangle)^*(0)].$$

Since we know by (3.27) that
$$(f + y^* \circ G - \langle a^*, \cdot \rangle)^*(0) = (f + y^* \circ G)^*(a^*),$$
it results that
$$(f + g \circ G)^*(a^*) = \sup_{x \in X} [\langle a^*, x \rangle - (f(x) + g(G(x)))]$$
$$= \min_{y^* \in Y^*} [g^*(y^*) + (f + y^* \circ G)^*(a^*)]$$
$$= \min_{y^* \in Y^*_+} [g^*(y^*) + (f + y^* \circ G)^*(a^*)].$$

(β) Assume in addition that $a^* \in \partial(f + g \circ G)(a)$ and choose by (α) some $y^*_a \in Y^*_+$ such that
$$\inf_{x \in X} [f(x) - \langle a^*, x \rangle + g(G(x))] = -g^*(y^*_a) - (f + y^*_a \circ G)^*(a^*).$$
Using this and the equality
$$\inf_{x \in X} [f(x) - \langle a^*, x \rangle + g(G(x))] = f(a) - \langle a^*, a \rangle + g(G(a)),$$
because $0 \in \partial(f + g \circ G - \langle a^*, \cdot \rangle)(a)$ (since $a^* \in \partial(f + g \circ G)(a)$), we obtain that
$$f(a) - \langle a^*, a \rangle + g(G(a)) = -g^*(y^*_a) - (f + y^*_a \circ G)^*(a^*),$$
and hence
$$[(f + y^*_a \circ G)(a) + (f + y^*_a \circ G)^*(a^*) - \langle a^*, a \rangle] + [g(G(a)) + g^*(y^*_a) - \langle y^*_a, G(a) \rangle] = 0.$$
By the Fenchel inequality $\langle u^*, u \rangle - h(u) \leq h^*(u^*)$, each above bracket [ ] expression is non-negative, thus they are both zero, that is,
$$g(G(a)) + g^*(y^*_a) - \langle y^*_a, G(a) \rangle = 0$$
and
$$(f + y^*_a \circ G)(a) + (f + y^*_a \circ G)^*(a^*) - \langle a^*, a \rangle = 0.$$
The characterization of subgradient of convex function in terms of conjugate (see Proposition 3.64(i)) yields
$$y^*_a \in \partial g(G(a)) \quad \text{and} \quad a^* \in \partial(f + y^*_a \circ G)(a).$$
This finishes the proof of the proposition since we know that $\partial g(G(a)) \subset Y^*_+$ by Lemma 3.163. $\square$

Through the above proposition we can prove the following theorem with useful formulas for the Legendre-Fenchel conjugate and subdifferential of $f + g \circ G$.

THEOREM 3.165 (basic theorem for conjugate and subdifferential of $f + g \circ G$: convex functions with g fully non-decreasing). Let X and Y be Hausdorff locally convex spaces and Y_+ be a convex cone in Y containing zero. Let $G : X \to Y \cup \{+\infty_Y\}$ be a Y_+-convex mapping, $f : X \to \mathbb{R} \cup \{+\infty\}$ be a convex function and $g : Y \to \mathbb{R} \cup \{+\infty\}$ be a convex function which is non Y_+-decreasing. Then for any $x^* \in X^*$ and any $x \in X$ one has both the equalities

$$(f + g \circ G)^*(x^*) = \min_{y^* \in Y^*} [g^*(y^*) + (f + y^* \circ G)^*(x^*)]$$
$$= \min_{y^* \in Y^*_+} [g^*(y^*) + (f + y^* \circ G)^*(x^*)]$$

and the equality

$$\partial(f + g \circ G)(x) = \bigcup_{y^* \in \partial g(G(x))} \partial(f + y^* \circ G)(x)$$

under anyone of the following conditions:
(a) With $Z := \operatorname{span}[\operatorname{dom} g - G(\operatorname{dom} f \cap \operatorname{dom} G)]$ endowed with the topology induced by the Mackey topology of Y, for each $w(X, X^*)$-neighborhood V of zero in X there is some real $r > 0$ such that

$$0 \in \operatorname{int}_Z [\{g \leq r\} - G(\{f \leq r\} \cap \operatorname{dom} G \cap rV)];$$

(b) there exists some point $\bar{x} \in \operatorname{dom} f \cap \operatorname{dom} G$ such that g is finite and continuous at $G(\bar{x})$ with respect to a locally convex topology θ compatible with the duality between the spaces Y and Y^*;
(c) the space Y is a normed space and, for $Z := \operatorname{span}[\operatorname{dom} g - G(\operatorname{dom} f \cap \operatorname{dom} G)]$ endowed with the norm induced by the one of Y, there exists some real number $r > 0$ such that

$$0 \in \operatorname{int}_Z [\{g \leq r\} - G(\{f \leq r\} \cap \operatorname{dom} G \cap r\mathbb{B}_X)];$$

(d) the space Y is a finite-dimensional normed space and

$$0 \in \operatorname{rint}[\operatorname{dom} g - G(\operatorname{dom} f \cap \operatorname{dom} G)].$$

PROOF. Let $Z := \operatorname{span}[\operatorname{dom} g - G(\operatorname{dom} f \cap \operatorname{dom} G)]$ be endowed with the topology induced on Z by the Mackey topology $\mu(Y, Y^*)$ of Y.
(I) Assume that condition (a) holds. First, fix any $x^* \in X^*$ and consider the $w(X, X^*)$-neighborhood $V := \{\langle x^*, \cdot \rangle \geq -1\}$ of zero in X. Choose by assumption a real $r > 0$ such that

$$0 \in \operatorname{int}_Z [\{g \leq r\} - G(\{f \leq r\} \cap \operatorname{dom} G \cap rV)].$$

Since $rV = \{rx : \langle x^*, x \rangle \geq -1\} = \{u \in X : \langle x^*, u \rangle \geq -r\}$, we see that

$$0 \in \operatorname{int}_Z [\{g \leq r\} - G(\{f \leq r\} \cap \operatorname{dom} G \cap \{\langle x^*, \cdot \rangle \geq -r\})].$$

We can apply Proposition 3.164 to obtain the equalities concerning $(f + g \circ G)^*(x^*)$. Further, Proposition 3.164 also tells us that x^* belongs to $\bigcup_{y^* \in \partial g(G(x))} \partial(f + y^* \circ G)(x)$ whenever it belongs to $\partial(f + g \circ G)(x)$. This justifies the inclusion

$$\partial(f + g \circ G)(x) \subset \bigcup_{y^* \in \partial g(G(x))} \partial(f + y^* \circ G)(x),$$

and the converse inclusion is a consequence of Lemma 3.163 above and Lemma 3.166 below. This finishes the proof under condition (a).

(II) It remains to show that condition (a) is implied by anyone of conditions (b), (c) and (d). Let V be any convex $w(X, X^*)$-neighborhood of zero in X.

Suppose first that condition (b) is satisfied. Choose a 0-neighborhood W of zero in Y such that $g(w + G(\bar{x})) \leq 1 + g(G(\bar{x}))$ for all $w \in W$, and choose also a positive real $r > \max\{1 + g(G(\bar{x})), f(\bar{x})\}$ such that $\bar{x} \in rV$. Then

$$W = (G(\bar{x}) + W) - G(\bar{x}) \subset \{g \leq r\} - G(\{f \leq r\} \cap \operatorname{dom} G \cap rV),$$

which ensures on the one hand that $Z := \operatorname{span}[\operatorname{dom} g - G(\operatorname{dom} f \cap \operatorname{dom} G)] = Y$, and on the other hand that

$$0 \in \operatorname{int}_Z[\{g \leq r\} - G(\{f \leq r\} \cap \operatorname{dom} G \cap rV)],$$

where $Z = Y$ is endowed with the Mackey topology $\mu(Y, Y^*)$ since W is a $\mu(Y, Y^*)$-neighborhood of zero. This means that condition (b) implies condition (a).

Under condition (c), choosing a real $\alpha > 0$ such that $\alpha \mathbb{B}_X \subset V$ and putting $s := \max\{r, r/\alpha\}$ we see from condition (c) that

$$0 \in \operatorname{int}_Z[\{g \leq s\} - G(\{f \leq s\} \cap \operatorname{dom} G \cap sV)],$$

so condition (a) holds true.

Finally, suppose that condition (d) is satisfied, so

$$0 \in \operatorname{int}_Z[\operatorname{dom} g - G(\operatorname{dom} f \cap \operatorname{dom} G)].$$

Denote by m the dimension of the vector space Z. Suppose in a first case that $m \geq 1$ and, as in the proof of Corollary 3.160, fix a basis $\{e_1, \cdots, e_m\}$ of Z. Then for each $j = 1, \cdots, m$ there exist $\alpha_j \in]0, +\infty[$, $x_j, x'_j \in \operatorname{dom} f \cap \operatorname{dom} G$ and $y_j, y'_j \in \operatorname{dom} g$ such that

(3.113) $\qquad \alpha_j e_j = y_j - G(x_j) \quad \text{and} \quad -\alpha_j e_j = y'_j - G(x'_j).$

Choose a real $s > 0$ such that $x_j \in sV$ and $x'_j \in sV$ for all $j = 1, \cdots, m$. Since the system $\{\alpha_1 e_1, \cdots, \alpha_m e_m\}$ is a basis of Z, the set

$$\operatorname{co}\{\alpha_1 e_1, \cdots, \alpha_m e_m, -\alpha_1 e_1, \cdots, -\alpha_m e_m\}$$

is a neighborhood of zero in Z, or equivalently

(3.114) $\qquad 0 \in \operatorname{int}_Z(\operatorname{co}\{\alpha_1 e_1, \cdots, \alpha_m e_m, -\alpha_1 e_1, \cdots, -\alpha_m e_m\}).$

If we take a non-negative real $s' \geq \max\limits_{1 \leq j \leq m}(\max\{|g(y_j)|, g(y'_j), f(x_j), f(x'_j)\})$ and put $r := s + s'$, the equalities in (3.113) imply that $\alpha_j e_j$ and $-\alpha_j e_j$ belong to $\{g \leq r\} - G(\{f \leq r\} \cap \operatorname{dom} G \cap rV)$, and hence according to the convexity of the latter set (see (b) in Lemma 3.157)

$$\operatorname{co}\{\alpha_1 e_1, \cdots, \alpha_m e_m, -\alpha_1 e_1, \cdots, -\alpha_m e_m\} \subset \{g \leq r\} - G(\{f \leq r\} \cap \operatorname{dom} G \cap rV).$$

Using (3.114) we obtain

(3.115) $\qquad 0 \in \operatorname{int}_Z(\{g \leq r\} - G(\{f \leq r\} \cap \operatorname{dom} G \cap rV)).$

In the remaining case when $m = 0$, we have $Z = \{0\}$ and $\operatorname{dom} g - G(\operatorname{dom} f \cap \operatorname{dom} G) \neq \emptyset$ (by the inclusion $0 \in \operatorname{int}_Z(\operatorname{dom} g - G(\operatorname{dom} f \cap \operatorname{dom} G))$). The equality

$$\operatorname{dom} g - G(\operatorname{dom} f \cap \operatorname{dom} G) = \bigcup_{r \in]0, +\infty[} (\{g \leq r\} - G(\{f \leq r\} \cap \operatorname{dom} G \cap rV))$$

then tells us that there exists some $r > 0$ such that $\{g \leq r\} - G(\{f \leq r\} \cap \operatorname{dom} G) \neq \emptyset$. Therefore, we have

$$\{g \leq r\} - G(\{f \leq r\} \cap \operatorname{dom} G \cap rV) = \{0\} = Z,$$

and hence (3.115) still holds true, which finishes the proof of the theorem. □

Before continuing the study of subdifferential rules for convex functions, it is worth noticing that Example 2.107 in the the previous chapter tells us that the inclusion of the left-hand side into the right-hand one in Theorem 3.165 fails in general without condition.

LEMMA 3.166. *Let X and Y be locally convex spaces and Y_+ be a convex cone in Y containing zero. Let $G : X \to Y \cup \{+\infty_Y\}$ be a Y_+-convex mapping, $f : X \to \mathbb{R} \cup \{+\infty\}$ be a convex function and $g : Y \to \mathbb{R} \cup \{+\infty\}$ be a convex function which is non Y_+-decreasing on $G(\operatorname{dom} G) + Y_+$. Then for any $x \in X$ one has the inclusion*

$$\bigcup_{y^* \in \partial g(G(x))} \partial (f + y^* \circ G)(x) \subset \partial (f + g \circ G)(x).$$

PROOF. Notice first by Lemma 3.157(a) that $f + g \circ G$ is convex. There is nothing to prove if the left-hand side member is empty. So, take any element x^* in this member. There exists $y^* \in \partial g(G(x))$ such that for all $u \in \operatorname{dom}(f + g \circ G)$

$$\langle x^*, u - x \rangle \leq f(u) + \langle y^*, G(u) \rangle - f(x) - \langle y^*, G(x) \rangle$$

(note that $f(x) + y^* \circ G(x)$ is finite). Since $y^* \in \partial g(G(x))$, we have $g \circ G(x) \in \mathbb{R}$, and for all $y \in Y$

$$\langle y^*, y - G(x) \rangle \leq g(y) - g(G(x)).$$

Combining this with the previous inequality yields for all $u \in \operatorname{dom}(f + g \circ G)$

$$\langle x^*, u - x \rangle \leq \big(f(u) + g(G(u))\big) - \big(f(x) + g(G(x))\big),$$

which means that $x^* \in \partial (f + g \circ G)(x)$. This justifies the inclusion of the lemma. □

Taking $f = 0$ yields the following first corollary.

COROLLARY 3.167. *Let X and Y be Hausdorff locally convex spaces and Y_+ be a convex cone in Y containing zero. Let $G : X \to Y \cup \{+\infty_Y\}$ be a Y_+-convex mapping and $g : Y \to \mathbb{R} \cup \{+\infty\}$ be a convex function which is non Y_+-decreasing. Then for any $x^* \in X^*$ and $x \in X$ one has the equalities*

$$(g \circ G)^*(x^*) = \min_{y^* \in Y^*} [g^*(y^*) + (y^* \circ G)^*(x^*)] = \min_{y^* \in Y^*_+} [g^*(y^*) + (y^* \circ G)^*(x^*)]$$

as well as the equality

$$\partial (g \circ G)(x) = \bigcup_{y^* \in \partial g(G(x))} \partial (y^* \circ G)(x)$$

under anyone of conditions (a), (b), (c) and (d) below:
(a) *With $Z := \operatorname{span}[\operatorname{dom} g - G(\operatorname{dom} G)]$ endowed with the topology induced by the Mackey topology of Y, for each $w(X, X^*)$-neighborhood V of zero in X there is some real $r > 0$ such that*

$$0 \in \operatorname{int}_Z [\{g \leq r\} - G(\operatorname{dom} G \cap rV)];$$

(b) *there exists some point $\bar{x} \in \operatorname{dom} G$ such that g is finite and continuous at $G(\bar{x})$ with respect to a locally convex topology compatible with the duality between the*

spaces Y and Y^*;
(c) the space Y is a normed space and, for $Z := \operatorname{span}[\operatorname{dom} g - G(\operatorname{dom} G)]$ endowed with the norm induced by the one of Y, there exists some real number $r > 0$ such that
$$0 \in \operatorname{int}_Z [\{g \leq r\} - G(\operatorname{dom} G \cap r\mathbb{B}_X)];$$
(d) the space Y is a finite-dimensional normed space and
$$0 \in \operatorname{rint}[\operatorname{dom} g - G(\operatorname{dom} G)].$$

A second corollary of Theorem 3.165 corresponds to the case when G is a continuous linear mapping A.

COROLLARY 3.168. *Let X, Y be Hausdorff locally convex spaces, $f : X \to \mathbb{R} \cup \{+\infty\}$ and $g : Y \to \mathbb{R} \cup \{+\infty\}$ be convex functions and $A : X \to Y$ be a continuous linear mapping. Then for any $x^* \in X^*$ and any $x \in X$ one has both equalities*
$$(f + g \circ A)^*(x^*) = \min_{y^* \in Y^*} [g^*(y^*) + f^*(x^* - A^*y^*)]$$
and
$$\partial(f + g \circ A)(x) = \partial f(x) + A^*(\partial g(G(x)))$$
in each one of the following cases (a), (b), (c) and (d):
(a) *With $Z := \operatorname{span}[\operatorname{dom} g - A(\operatorname{dom} f)]$ endowed with the topology induced by the Mackey topology of Y, for each $w(X, X^*)$-neighborhood V of zero in X there is some real $r > 0$ such that*
$$0 \in \operatorname{int}_Z [\{g \leq r\} - A(\{f \leq r\} \cap rV)];$$
(b) *there exists some $\bar{x} \in \operatorname{dom} f$ such that g is finite and continuous at $A\bar{x}$ with respect to a locally convex topology θ compatible with the duality between the spaces Y and Y^*;*
(c) *the space Y is a normed space and for $Z := \operatorname{span}[\operatorname{dom} g - A(\operatorname{dom} f)]$ there exists some real number $r > 0$ such that*
$$0 \in \operatorname{int}_Z [\{g \leq r\} - A(\{f \leq r\} \cap r\mathbb{B}_X)];$$
(d) *the space Y is a finite-dimensional normed space and*
$$0 \in \operatorname{rint}[\operatorname{dom} g - A(\operatorname{dom} f)].$$

PROOF. Apply Theorem 3.165 with $Y_+ := \{0_Y\}$, and hence $Y_+^* = Y^*$. Observe also that for each $y^* \in Y^*$ we have $y^* \circ A \in X^*$, and hence from (3.27)
$$(f + y^* \circ A)^*(x^*) = f^*(x^* - y^* \circ A) = f^*(x^* - A^*y^*),$$
and from the very definition (see (3.1))
$$\partial(f + y^* \circ A)(x) = y^* \circ A + \partial f(x) = A^*y^* + \partial f(x).$$
Noticing that
$$\bigcup_{y^* \in \partial g(Ax)} \{A^*y^*\} = A^*(\partial g(Ax)),$$
it results that
$$\bigcup_{y^* \in \partial g(Ax)} \partial(f + y^* \circ A)(x) = \bigcup_{y^* \in \partial g(Ax)} (A^*y^* + \partial f(x)) = \partial f(x) + \bigcup_{y^* \in \partial g(Ax)} \{A^*y^*\}$$
$$= \partial f(x) + A^*(\partial g(Ax)).$$
All together justify the corollary according to Theorem 3.165. □

3.11. SUM AND CHAIN RULES OF SUBDIFFERENTIAL UNDER CONVEXITY

REMARK 3.169. It is worth noticing that the condition (d) in Corollary 3.168 is equivalent to $(\mathrm{rint}\,(\mathrm{dom}\,)) \cap A(\mathrm{rint}\,(\mathrm{dom}\,f))) \neq \emptyset$ according to Proposition B.15 and Corollary B.17 in Appendix. □

Taking A as the identity in Corollary 3.168 furnishes a third corollary.

COROLLARY 3.170 (conjugate and subdifferential of sum of convex functions in locally convex space). Let X be a Hausdorff locally convex space, $f, g : X \to \mathbb{R} \cup \{+\infty\}$ be convex functions. Then for any $x^* \in X^*$ and $x \in X$ one has

$$(f+g)^*(x^*) = \min_{y^* \in X^*}[f^*(x^* - y^*) + g^*(y^*)] = (f^* \Box g^*)(x^*)$$

as well as

$$\partial(f+g)(x) = \partial f(x) + \partial g(x)$$

in each one of the following cases (a), (b), (c) and (d):
(a) With $Z := \mathrm{span}\,[\mathrm{dom}\,g - \mathrm{dom}\,f)]$ endowed with the topology induced by the Mackey topology of X, for each $w(X, X^*)$-neighborhood V of zero in X there is some real $r > 0$ such that

$$0 \in \mathrm{int}_Z[\{g \le r\} - (\{f \le r\} \cap rV)];$$

(b) **(Moreau-Rockafellar continuity condition):** one of the functions is finite and continuous at a point where the other function is finite;
(c) the space X is a normed space and for $Z := \mathrm{span}\,[\mathrm{dom}\,g - \mathrm{dom}\,f]$ there exists some real number $r > 0$ such that

$$0 \in \mathrm{int}_Z\,[\{g \le r\} - (\{f \le r\} \cap r\mathbb{B}_X)];$$

(d) **(Rockafellar relative interior condition):** the space X is a finite-dimensional normed space and

$$0 \in \mathrm{rint}\,[\mathrm{dom}\,g - \mathrm{dom}\,f].$$

Taking $f = 0$ in Corollary 3.168 and noting that $f^* = \Psi_{\{0\}}$ we obtain the following fourth corollary with chain rule.

COROLLARY 3.171 (conjugate and subdifferential chain rule for convex functions in locally convex space). Let X, Y be Hausdorff locally convex spaces, $g : Y \to \mathbb{R} \cup \{+\infty\}$ be a convex function and $A : X \to Y$ be a continuous linear mapping. Then for any $x^* \in X^*$ and $x \in X$ one has

$$(g \circ A)^*(x^*) = g^*(A^*y^*) \quad \text{if } x^* = A^*y^* \quad \text{and} \quad (g \circ A)^*(x^*) = +\infty \quad \text{if } x^* \notin A^*(Y^*)$$

as well as

$$\partial(g \circ A)(x) = A^*(\partial g(Ax))$$

in each one of the following cases (a), (b), (c) and (d):
(a) With $Z := \mathrm{span}\,[\mathrm{dom}\,g - A(X)]$ endowed with the topology induced by the Mackey topology of Y, for each $w(X, X^*)$-neighborhood V of zero in X there is some real $r > 0$ such that

$$0 \in \mathrm{int}_Z\,[\{g \le r\} - A(rV)];$$

(b) there exists some $\bar{x} \in X$ such that g is finite and continuous at $A\bar{x}$ with respect to a locally convex topology θ compatible with the duality between Y and Y^*;

(c) the space Y is a normed space and for $Z := \operatorname{span}[\operatorname{dom} g - A(\operatorname{dom} f)]$ there exists some real number $r > 0$ such that
$$0 \in \operatorname{int}_Z [\{g \leq r\} - A(r\mathbb{B}_X)];$$
(d) the space Y is a normed finite-dimensional space and
$$0 \in \operatorname{rint}[\operatorname{dom} g - A(X)].$$

Before turning to a general sandwich theorem, it is worth noticing that Example 2.107 in the the previous chapter tells us that, in all the above subdifferential rules in this section, the inclusion of the left-hand side into the right-hand one fails in general without any condition.

THEOREM 3.172 (sandwich theorem). *Let X, Y be Hausdorff locally convex spaces, $f : X \to \mathbb{R} \cup \{+\infty\}$ and $g : Y \to \mathbb{R} \cup \{+\infty\}$ be convex functions and $A : X \to Y$ be a continuous linear mapping such that $f \geq -g \circ A$. Then there exist $b^* \in Y^*$ and $\beta \in \mathbb{R}$ with*
$$f \geq \langle b^* \circ A, \cdot \rangle + \beta \geq -g \circ A,$$
in each one of the following cases (a), (b), (c) and (d):
(a) With $Z := \operatorname{span}[\operatorname{dom} g - A(\operatorname{dom} f)]$ endowed with the topology induced by the Mackey topology of Y, for each $w(X, X^*)$-neighborhood V of zero in X there is some real $r > 0$ such that
$$0 \in \operatorname{int}_Z[\{g \leq r\} - A(\{f \leq r\} \cap rV)];$$
(b) there exists some $\bar{x} \in \operatorname{dom} f$ such that g is finite and continuous at $A\bar{x}$ with respect to a locally convex topology θ compatible with the duality between the spaces Y and Y^*;
(c) the space Y is a normed space and for $Z := \operatorname{span}[\operatorname{dom} g - A(\operatorname{dom} f)]$ there exists some real number $r > 0$ such that
$$0 \in \operatorname{int}_Z[\{g \leq r\} - A(\{f \leq r\} \cap r\mathbb{B}_X)];$$
(d) the space Y is a finite-dimensional normed space and
$$0 \in \operatorname{rint}[\operatorname{dom} g - A(\operatorname{dom} f)].$$

PROOF. Assume anyone of assumptions (a)-(d). Taking $x^* = 0$ in Corollary 3.168 furnishes some $b^* \in Y^*$ such that
$$0 \leq \inf_{x \in X}(f + g \circ A) = \sup_{y^* \in Y^*} \left(-f^*(A^*y^*) - g^*(-y^*)\right) = -f^*(A^*b^*) - g^*(-b^*),$$
and $g^*(-b^*)$ is finite since the inequality $0 \leq \inf_{x \in X}(f + g \circ A)$ combining with anyone of (a)-(d) ensures that $\inf_{x \in X}(f + g \circ A)$ is finite. This gives $-f^*(A^*b^*) \geq g^*(-b^*) = \beta$ with $\beta \in \mathbb{R}$. By definition of Legendre-Fenchel conjugate this can be translated as
$$\inf_{x \in X}\left(f(x) - \langle A^*b^*, x \rangle\right) \geq \sup_{y \in Y}\left(\langle(-b^*, y\rangle - g(y)\right) = \beta,$$
which yields for all $x \in X$ and all $y \in Y$
$$f(x) \geq \langle b^* \circ A, x \rangle + \beta \quad \text{and} \quad -g(y) \leq \langle b^*, y \rangle + \beta.$$
Finally, for any $x \in X$ the two latter inequalities and the choice $y = Ax$ guarantee that
$$f(x) \geq \langle b^* \circ A, x \rangle + \beta \quad \text{and} \quad -g(Ax) \leq \langle b^* \circ A, x \rangle + \beta,$$

and the proof is complete. □

EXAMPLE 3.173. Define the (lower semicontinuous) convex functions $f, g : \mathbb{R} \to \mathbb{R} \cup \{+\infty\}$ by

$$f(x) = \begin{cases} \sqrt{x} & \text{if } x \geq 0 \\ +\infty & \text{if } x < 0 \end{cases} \quad \text{and} \quad g(x) = \begin{cases} \sqrt{-x} & \text{if } x \leq 0 \\ +\infty & \text{if } x > 0. \end{cases}$$

Clearly, $-g \leq f$ but there is no affine function $\varphi : \mathbb{R} \to \mathbb{R}$ satisfying $-g \leq \varphi \leq f$. Notice that the qualification condition $0 \in \text{rint}(\text{dom } g - \text{dom } f)$ is violated since $\text{dom } g - \text{dom } f =]-\infty, 0]$ here. □

3.12. Application to chain rule for C-subgradients and C-normals

Proposition 2.141 established an inclusion formula for the C-tangent cone of inverse image under a strictly differentiable mapping with surjective derivative between Banach spaces. This section applies Corollary 3.171 via that inclusion to obtain the corresponding C-normal cone inclusion as well as a chain rule for C-subdifferential in a non-Lipschitz framework.

THEOREM 3.174 (C-subdifferential chain rule for non-Lipschitz functions). Let X, Y be Banach spaces and $G : X \to Y$ be a mapping which is strictly Fréchet differentiable at $\bar{x} \in X$ with $A := DG(\bar{x})$ surjective. Let $\bar{y} := G(\bar{x})$.
(a) For any subset S of Y with $\bar{y} \in S$ one has

$$N^C\left(G^{-1}(S); \bar{x}\right) \subset A^*\left(N^C(S; \bar{y})\right);$$

if in addition S is tangentially regular at $\bar{y}$, then the above inclusion is an equality.
(b) For any function $g : Y \to \mathbb{R} \cup \{+\infty\}$ which is finite at $\bar{y}$ one has

$$\partial_C(g \circ G)(\bar{x}) \subset A^*(\partial_C g(\bar{y}));$$

the latter inclusion is an equality whenever the epigraph $\text{epi } g$ is tangentially regular at $(\bar{y}, g(\bar{y}))$.

PROOF. (a) By Proposition 2.141 we know that

$$A^{-1}\left(T^C(S; \bar{y})\right) \subset T^C(G^{-1}(S); \bar{x}),$$

with equality if the set S is tangentially regular at $\bar{y}$. Denoting by T_S and $T_{G^{-1}(S)}$ the C-tangent cones of S and $G^{-1}(S)$ at $\bar{y}$ and $\bar{x}$ respectively, the above tangential inclusion yields

$$\Psi_{T_{G^{-1}(S)}} \leq \Psi_{A^{-1}(T_S)} = \Psi_{T_S} \circ A \quad \text{with } \Psi_{T_{G^{-1}(S)}}(0) = 0 = (\Psi_{T_S} \circ A)(0).$$

By convexity of the functions $\Psi_{T_{G^{-1}(S)}}$ and $\Psi_{T_S} \circ A$ we deduce that

$$N^C\left(G^{-1}(S); \bar{x}\right) = \partial \Psi_{T_{G^{-1}(S)}}(0) \subset \partial(\Psi_{T_S} \circ A)(0).$$

On the other hand, by the surjectivity of A we have $\text{span}[\text{dom } \Psi_{T_S} - A(X)] = Y$, and by the Banach-Schauder open mapping theorem (see Theorem C.3 in Appendix) there exists some real $r > 0$ such that $\mathbb{B}_Y \subset A(r\mathbb{B}_X)$. This ensures that

$$\mathbb{B}_Y \subset \{\Psi_{T_S} \leq r\} - A(r\mathbb{B}_X), \text{ hence } 0 \in \text{int}_Y[\{\Psi_{T_S} \leq r\} - A(r\mathbb{B}_X)].$$

Corollary 3.171 yields

$$\partial(\Psi_{T_S} \circ A)(0) = A^*\left(\partial \Psi_{T_S}(A0)\right) = A^*\left(\partial \Psi_{T_S}(0)\right) = A^*\left(N^C(S; \bar{y})\right).$$

We deduce that $N^C\left(G^{-1}(S); \bar{x}\right) \subset A^*\left(N^C(S; \bar{y})\right)$, with equality if the set S is tangentially regular at $\bar{y}$.

(b) Fix any $x^* \in \partial_C(g \circ G)(\bar{x})$. Let $\widehat{G}: X \times \mathbb{R} \to Y \times \mathbb{R}$ be defined by $\widehat{G}(x,r) := (G(x), r)$, so for $\bar{r} := g(\bar{y})$ we see that $\widehat{G}$ is strictly differentiable at $(\bar{x}, \bar{r})$ with $D\widehat{G}(\bar{x}, \bar{r})(h, r) = (Ah, r) =: \widehat{A}(h, r)$. Since $\operatorname{epi}(g \circ G) = \widehat{G}^{-1}(\operatorname{epi} g)$ and $\widehat{A}$ is surjective, the normal inclusion (resp. equality if $\operatorname{epi} g$ is tangentially regular at $(\bar{y}, g(\bar{y}))$) in (a) yields (resp. is equivalent to) $(x^*, -1) \in \widehat{A}^*\left(N^C(\operatorname{epi} g; (\bar{y}, \bar{r}))\right)$. This means that there exists $(y^*, -s) \in N^C(\operatorname{epi} g; (\bar{y}, \bar{r}))$ such that

$$(x^*, -1) = \widehat{A}^*(y^*, -s) = (y^* \circ A, -s),$$

so $s = 1$, $y^* \in \partial_C g(\bar{y})$ and $x^* = y^* \circ A$. This justifies the desired inclusion $\partial_C(g \circ G)(\bar{x}) \subset A^*(\partial_C g(\bar{y}))$ (resp. the desired corresponding equality if additionally $\operatorname{epi} g$ is tangentially regular at $(\bar{y}, g(\bar{y}))$). The proof is finished. □

3.13. Calculus rules for normals and tangents to convex sets

In the previous section calculus rules for subdifferential of convex functions in Section 3.11 have been applied to sets of the form $G^{-1}(S)$, where the mapping G is required to be strictly Fréchet differentiable at the point of interest. The first part of this section is devoted to applications of Section 3.11 to more general calculus rules of normals to convex sets of the form $C \cap A^{-1}(S)$, where A is a continuous linear mapping.

First, notice that the arguments in the normed setting for Theorem 2.2(d) and Proposition 2.113(e) guarantee also that for a convex set S of a Hausdorff locally convex space X and for $x \in S$, one has

$$\underset{t \downarrow 0,\, u \xrightarrow{S} x}{\operatorname{Lim\,inf}} \frac{1}{t}(S - u) = \operatorname{cl}(\mathbb{R}_+(S - x)) = \underset{t \downarrow 0}{\operatorname{Lim\,sup}} \frac{1}{t}(S - x).$$

DEFINITION 3.175. Given a convex set S of a Hausdorff locally convex space X and $x \in S$, the common closed convex cone in the above chain of equalities is denoted $T(S; x)$ and called the *tangent cone* of the convex set S at the point x. For $x \notin S$, by convention one puts $T(S; x) = \emptyset$.

Recalling that the normal cone of the convex set S at $x \in S$ (see Definition 3.4) satisfies

$$N(S; x) = \{x^* \in X^* : \langle x^*, y - x \rangle \le 0,\, \forall y \in S\},$$

the equality $T(S; x) = \operatorname{cl}(\mathbb{R}_+(C - x))$ yields that $N(S; x)$ coincides with the negative polar cone in X^* of $T(S; x)$, and that $T(S; x)$ in turn coincides with the negative polar cone in X of $N(S; x)$. Otherwise stated,

(3.116) $\qquad N(S; x) = (T(S; x))^\circ \quad \text{and} \quad T(S; x) = {}^\circ(N(S; x)).$

PROPOSITION 3.176. Let X, Y be Hausdorff locally convex spaces, C and S be convex sets in X and Y respectively, and $A: X \to Y$ be a continuous linear mapping. Then for every $x \in X$ the equalities

$$N(C \cap A^{-1}(S); x) = N(C; x) + A^*(N(S; Ax))$$

and

$$T(C \cap A^{-1}(S); x) = T(C; x) \cap A^{-1}(T(S; Ax))$$

hold in each one of the following cases (a), (b), (c) and (d):
(a) With $Z := \operatorname{span}[S - A(C)]$ endowed with the topology induced by the Mackey

3.13. CALCULUS RULES FOR NORMALS AND TANGENTS TO CONVEX SETS

topology of X, for each $w(X, X^*)$-neighborhood V of zero in X there is some real $r > 0$ such that
$$0 \in \operatorname{int}_Z[S - A(C \cap rV)];$$
(b) the condition $C \cap A^{-1}(\operatorname{int}_\theta S) \neq \emptyset$ is fulfilled, where θ is any locally convex topology on Y compatible with the duality between Y and Y^*;
(c) the space X is a normed space and for $Z := \operatorname{span}[S - A(C)]$ there exists some real number $r > 0$ such that
$$0 \in \operatorname{int}_Z[S - A(C \cap r\mathbb{B}_X)];$$
(d) the space X is a finite-dimensional normed space and
$$0 \in \operatorname{rint}[S - A(C)].$$

PROOF. Observe first that both left-hand and right-hand sides of the desired equalities are empty whenever $Ax \notin S$. Putting $Q := C \cap A^{-1}(S)$, we note that the indicator function Ψ_Q of Q is given by
$$\Psi_Q = \Psi_C + \Psi_S \circ A.$$
So, applying applying Corollary 3.168 justifies the conclusion related to normal cones.

Recall that endowing X^* with its $w(X^*, X)$ topology, for closed convex cones P_1, P_2 in X^* one has $°(P_1 + P_2) = (°P_1) \cap (°P_2)$ by taking polar in X (see (1.32)). Recall also that for a closed convex cone K in Y one has $°(A^*(K°)) = A^{-1}(K)$ (see (1.33)). Therefore, the conclusion for tangent cones follows from the previous one for normal cones. □

The next two corollaries directly follow.

COROLLARY 3.177. *Let X be a Hausdorff locally convex space and S_1, S_2 be convex sets in X. Then one has for every $x \in X$*
$$N(S_1 \cap S_2; x) = N(S_1; x) + N(S_2; x) \quad \text{and} \quad T(S_1 \cap S_2; x) = T(S_1; x) \cap T(S_2; x)$$
in each one of the following cases (a), (b), (c) and (d):
(a) With $Z := \operatorname{span}[S_1 - S_2]$ endowed with the topology induced by the Mackey topology of X, for each $w(X, X^*)$-neighborhood V of zero in X there is some real $r > 0$ such that
$$0 \in \operatorname{int}_Z[S_1 - (S_2 \cap rV)];$$
(b) the condition $S_1 \cap \operatorname{int}_\theta S_2 \neq \emptyset$ is fulfilled, where θ is any locally convex topology on X compatible with the duality between X and X^*;
(c) the space X is a normed space and for $Z := \operatorname{span}[S_1 - S_2]$ there exists some real number $r > 0$ such that
$$0 \in \operatorname{int}_Z[S_1 - (S_2 \cap r\mathbb{B}_X)];$$
(d) the space X is a finite-dimensional normed space and
$$0 \in \operatorname{rint}[S_1 - S_2].$$

COROLLARY 3.178. *Let X, Y be Hausdorff locally convex spaces, S be a convex set in Y, and $A : X \to Y$ be a continuous linear mapping. Then one has for every $x \in X$*
$$N(A^{-1}(S); x) = A^*(N(S; Ax)) \quad \text{and} \quad T(A^{-1}(S); x) = A^{-1}(T(S; Ax)).$$

in each one of the following cases (a), (b), (c) and (d):
(a) With $Z := \text{span}\,[S - A(X)]$ endowed with the topology induced by the Mackey topology of X, for each $w(X, X^*)$-neighborhood V of zero in X there is some real $r > 0$ such that
$$0 \in \text{int}_Z[S - A(rV)];$$
(b) the condition $A^{-1}\,(\text{int}_\theta S) \neq \emptyset$ is fulfilled, where θ is any locally convex topology on Y compatible with the duality between Y and Y^*;
(c) the space X is a normed space and for $Z := \text{span}\,[S - A(X)]$ there exists some real number $r > 0$ such that
$$0 \in \text{int}_Z\,[S - A(r\mathbb{B}_X)];$$
(d) the space X is a finite-dimensional normed space and
$$0 \in \text{rint}\,[S - A(X)].$$

We turn now to normals to the direct images of convex sets under continuous linear mappings.

PROPOSITION 3.179. *Let X, Y be Hausdorff locally convex spaces, $A : X \to Y$ be a continuous linear mapping between X and Y, and C be a convex set in X. Then, given any $y \in A(C)$ one has for every $x \in C \cap A^{-1}(y)$*
$$N(A(C); y) = (A^*)^{-1}\bigl(N(C; x)\bigr) \quad \text{and} \quad T(A(C); y) = \text{cl}_Y\bigl(A(T(C; x))\bigr).$$

PROOF. Fix $x \in C \cap A^{-1}(y)$. For $y^* \in Y^*$ one has $y^* \in N(A(C); y)$ if and only if for any $u \in C$
$$\langle y^*, Au - Ax \rangle \leq 0, \quad \text{or equivalently} \quad \langle A^*y^*, u - x \rangle \leq 0.$$
This means that $y^* \in N(A(C); y)$ if and only if $A^*(y^*) \in N(C; x)$, which confirms the first equality of the proposition.

Concerning the second equality with tangent cones, note first that given a convex cone K in X, for the (negative) polar cone of $A(K)$ one has the equality $(A(K))^\circ = (A^*)^{-1}(K^\circ)$ (see (1.33)). So, for X^* endowed with the $w(X^*, X)$ topology, taking the negative polar in X gives ${}^\circ\bigl((A(K))^\circ\bigr) = {}^\circ\bigl((A^*)^{-1}(K^\circ)\bigr)$, which is equivalent to ${}^\circ\bigl((A^*)^{-1}(K^\circ)\bigr) = \text{cl}_X(A(K))$. Then, taking $K = T(C; x)$ in the latter equality and using the above equality related to normal cones, we obtain that $T(A(C); y) = \text{cl}_Y\bigl(A(T(C, x))\bigr)$ since ${}^\circ\bigl(N(A(C); y)\bigr) = T(A(C); y)$. $\square$

Normals to Minkowski sum of finitely many convex sets are described as follows.

COROLLARY 3.180. *Let $S_1, \cdots, S_m$ be convex sets of a Hausdorff locally convex space X and let $S := S_1 + \cdots + S_m$ be the Minkowski sum of these sets. Then, given any $x \in S$ one has for every representation $x = x_1 + \cdots + x_m$, with $x_i \in S_i$,*
$$N(S; x) = N(S_1; x_1) \cap \cdots \cap N(S_m; x_m)$$
and
$$T(S; x) = \text{cl}_X\bigl(T(S_1, x_1) + \cdots + T(S_m; x_m)\bigr).$$

PROOF. Consider $C := S_1 \times \cdots \times S_m$ and $A : X \times \cdots \times X \to X$ defined by
$$A(u_1, \cdots, u_m) = u_1 + \cdots + u_m \quad \text{for every } (u_1, \cdots, u_m) \in X \times \cdots \times X,$$
and note that $S = A(S_1 \times \cdots \times S_m)$. Since $A^* : X^* \to X^* \times \cdots \times X^*$ is given by
$$A^*(u^*) = (u^*, \cdots, u^*) \quad \text{for every } u^* \in X^*,$$
the corollary follows from Proposition 3.179. $\square$

Given two convex sets S_1, S_2 of a Hausdorff locally convex space X and $\bar{x} \in S_1 \cap S_2$, it is clear that $\mathbb{R}_+(S_1 - S_2) = X$ entails that $N(S_1 - S_2; 0) = \{0\}$. The latter equality is equivalent to $N(S_1; \bar{x}) \cap (-N(S_2; \bar{x})) = \{0\}$ according to Corollary 3.180 above. On the other hand, if X is a finite-dimensional normed space the equality $N(S_1 - S_2; 0) = \{0\}$ is equivalent to $\mathbb{R}_+(S_1 - S_2) = \{0\}$ by Lemma 3.182 below. We then have the second corollary:

COROLLARY 3.181. *Let S_1, S_2 be two convex sets of a finite-dimensional normed space X and let $\bar{x} \in S_1 \cap S_2$. Then one has the equivalence*

$$\mathbb{R}_+(S_1 - S_2) = X \iff N(S_1; \bar{x}) \cap (-N(S_2; \bar{x})) = \{0\}.$$

LEMMA 3.182. *Let C be a convex set of a finite dimensional normed space X and let $\bar{x} \in C$. The following equivalences hold:*

$$\bar{x} \in \text{int}_X C \iff \mathbb{R}_+(C - \bar{x}) = X \iff N(C; \bar{x}) = \{0\}.$$

PROOF. The implications $\bar{x} \in \text{int}_X C \Rightarrow \mathbb{R}_+(C - \bar{x}) = X \Rightarrow N(C; \bar{x}) = \{0\}$ are obvious. It remains to show that $N(C; \bar{x}) = \{0\}$ entails $\bar{x} \in \text{int}_X C$. Assume that $\bar{x} \notin \text{int}_X C$.

Case 1: $\text{int}_X C \neq \emptyset$. By separation theorem there is a nonzero $x^* \in X^*$ such that $\langle x^*, x \rangle < \langle x^*, \bar{x} \rangle$ for all $x \in \text{int}_X C$, hence $\langle x^*, x - \bar{x} \rangle \leq 0$ for all $x \in C \subset \text{cl}(\text{int}_X C)$, since C is a convex set with nonempty interior (see Proposition B.5 in Appendix).

Case 2: $\text{int}_X C = \emptyset$. Corollary B.8 in Appendix says that $E := \text{span}(C - \bar{x}) \neq X$. Since the vector subspace E is closed, there is a nonzero $x^* \in X^*$ which is null on E, hence $\langle x^*, x - \bar{x} \rangle = 0$ for all $x \in C$. This ensures that $x^* \in N(C; \bar{x})$.

In any case, we obtain that $N(C; \bar{x}) \neq \{0\}$. This justifies that $N(C; \bar{x}) = \{0\}$ entails $\bar{x} \in \text{int}_X C$, so the proof of the lemma is finished. □

THEOREM 3.183 (subdifferential characterization of differentiability of convex functions in finite dimensions). *Let $f : X \to \mathbb{R} \cup \{+\infty\}$ be a proper convex function on a finite dimensional normed space X and let $\bar{x} \in X$ where f is finite.*
(I) *One has $\bar{x} \in \text{bdry}(\text{dom } f)$ if and only if the subdifferential $\partial f(\bar{x})$ is either empty or unbounded.*
(II) *The following assertions are equivalent:*
(a) *one has $\partial f(\bar{x})$ is a singleton;*
(b) *the function f is strictly Hadamard (or equivalently strictly Fréchet) differentiable at $\bar{x}$;*
(c) *the function f is Gâteaux differentiable at $\bar{x}$.*

PROOF. Denote $C := \text{dom } f$.
(I) Assume that $\bar{x} \in \text{bdry } C$. By Lemma 3.182 the nonempty cone $N(C; \bar{x})$ is non-null, hence unbounded. On the other hand, from the equality $f = f + \Psi_C$ we have $\partial f(\bar{x}) + N(C; \bar{x}) \subset \partial f(\bar{x})$, so $\partial f(\bar{x})$ is unbounded if it is nonempty. This justifies the implication $\Rightarrow$.

If $\bar{x} \notin \text{bdry } C$, then $\bar{x} \in \text{int } C$, which ensues that f is Lipschitz continuous near $\bar{x}$ (see Corollary 2.159), thus $\partial f(\bar{x})$ is nonempty and bounded. The equivalence in (I) is established.

(II) The implications (b)$\Rightarrow$(c) and (c)$\Rightarrow$(a) being evident, it remains to show the implication (a)$\Rightarrow$(b). Assume that $\partial f(\bar{x})$ is a singleton. By (I) we have $\bar{x} \in \text{int } C$, which as above entails that f is Lipschitz continuous near $\bar{x}$. Therefore, by

Propositions 3.102 and 3.106 the function f is strictly Gâteaux differentiable at $\bar{x}$, hence strictly Hadamard differentiable at $\bar{x}$ since it is Lipschitz continuous near $\bar{x}$. This translates (b) and finishes the proof. □

The next example illustrates that the equivalence between (a) and (b) in (II) of the theorem fails in infinite dimensions.

EXAMPLE 3.184. Let C be the closed convex set of the infinite-dimensional Banach space X as in Example 2.210, with $0 \in \mathrm{bdry}\, C$ and $N(C; 0) = \{0\}$. Consider the convex function f given by the indicator function of C, so $\partial f(0) = \{0\}$ since $\partial f(0) = N(C; 0)$. At the boundary point $0 \in \mathrm{bdry}\,(\mathrm{dom}\, f)$ of $\mathrm{dom}\, f$ the subdifferential $\partial f(0) = \{0\}$, which says that f satisfies the property (a) in the theorem while neither (b) nor (c) holds. □

3.14. Chain rule with partially nondecreasing functions

Let us come back to the function $f + g \circ G$, where $f : X \to \mathbb{R} \cup \{+\infty\}$ and $g : Y \to \mathbb{R} \cup \{+\infty\}$ are convex functions on Hausdorff locally convex spaces X and Y respectively, $G : X \to Y \cup \{+\infty_Y\}$ is a Y_+-convex mapping. We focus now on the case when g is non Y_+-decreasing merely on the convex set $G(\mathrm{dom}\, G) + Y_+$. We will work with the function

$$(x, u) \mapsto f(x) + g(u) + \Psi_{-Y_+}(G(x) - u)$$

on $X \times Y$. Define $f_0 : X \times Y \to \mathbb{R} \cup \{+\infty\}$ and $G_0 : X \times Y \to Y \cup \{+\infty_Y\}$ by

$$f_0(x, u) = f(x) + g(u) \quad G_0(x, u) = G(x) - u \quad \text{for all } (x, u) \in X \times Y.$$

Setting $g_0 := \Psi_{-Y_+}$, we see that

$$f(x) + g(u) + \Psi_{-Y_+}(G(x) - u) = (f_0 + g_0 \circ G_0)(x, u) \quad \text{for all } (x, u) \in X \times Y.$$

Noticing that $g_0 = \Psi_{-Y_+}$ is non Y_+-decreasing on the whole space we can prove Proposition 3.185. In this proposition and the rest of this section the equality (see (3.94))

$$0_{Y^*} \circ G = \Psi_{\mathrm{dom}\, G}$$

will be employed (as in the previous sections).

PROPOSITION 3.185 (conjugate and subdifferential of $f + g \circ G$: convex functions with g partially non-decreasing). Let X and Y be Hausdorff locally convex spaces and Y_+ be a convex cone in Y containing zero. Let $G : X \to Y \cup \{+\infty_Y\}$ be a Y_+-convex mapping, $f : X \to \mathbb{R} \cup \{+\infty\}$ be a convex function and $g : Y \to \mathbb{R} \cup \{+\infty\}$ be a convex function which is non Y_+-decreasing on $G(\mathrm{dom}\, G) + Y_+$. Then for any $x^* \in X^*$ and $x \in X$ both the equality

$$(f + g \circ G)^*(x^*) = \min_{y^* \in Y_+^*} [g^*(y^*) + (f + y^* \circ G)^*(x^*)].$$

and the equality

$$\partial(f + g \circ G)(x) = \bigcup_{y^* \in Y_+^* \cap \partial g(G(x))} \partial(f + y^* \circ G)(x)$$

hold under anyone of conditions (a), (b), (c) and (d) below:
(a) With $Z := \mathrm{span}\,[\mathrm{dom}\, g - G(\mathrm{dom}\, f \cap \mathrm{dom}\, G) - Y_+]$ endowed with the topology

induced by the Mackey topology of Y, for each $w(X, X^*)$-neighborhood V of zero in X there is some real $r > 0$ such that
$$0 \in \text{int}_Z\, [\{g \leq r\} - G(\{f \leq r\} \cap \text{dom}\, G \cap rV) - Y_+];$$
(b) there exists some point $\bar{x} \in \text{dom}\, f \cap \text{dom}\, G$ such that g is finite and continuous at $G(\bar{x})$ with respect to a locally convex topology compatible with the duality between Y and Y^*;
(c) the space Y is a normed space and, for $Z := \text{span}\, [\text{dom}\, g - G(\text{dom}\, f \cap \text{dom}\, G) - Y_+]$ endowed with the norm induced by the one of Y, there exists some real number $r > 0$ such that
$$0 \in \text{int}_Z\, [\{g \leq r\} - G(\{f \leq r\} \cap \text{dom}\, G \cap r\mathbb{B}_X) - Y_+];$$
(d) the space Y is a finite-dimensional normed space and
$$0 \in \text{rint}\, [\text{dom}\, g - G(\text{dom}\, f \cap \text{dom}\, G) - Y_+].$$

The proof of the proposition is the object of the following exercise.

EXERCISE 3.186. We keep notation f_0, g_0 and G_0 preceding the statement of the proposition.
(α) Verify that $\text{dom}\, f_0 = (\text{dom}\, f) \times (\text{dom}\, g)$ and
$$G_0(\text{dom}\, f_0 \cap \text{dom}\, G_0 \cap (S \times S')) = G(\text{dom}\, f \cap \text{dom}\, G \cap S) - (\text{dom}\, g \cap S').$$
(β) For $x^* \in X^*$ and $u^*, y^* \in Y^*$ show that
$$(f_0 + y^* \circ G_0)^*(x^*, u^*) = f^*(x^*) + (g + y^* \circ G)^*(y^* + u^*)$$
and
$$(f + g \circ G)^*(x^*) = (f_0 + g_0 \circ G_0)^*(x^*, 0).$$
(γ) Show that $x^* \in \partial(f + g \circ G)(x)$ if and only if $(x^*, 0) \in \partial(f_0 + g_0 \circ G_0)(x, G(x))$.
(δ) Derive the conclusions of the proposition from Proposition 3.164. □

COROLLARY 3.187. Let X and Y be Hausdorff locally convex spaces and Y_+ be a convex cone in Y containing zero. Let $G : X \to Y \cup \{+\infty_Y\}$ be a Y_+-convex mapping and $g : Y \to \mathbb{R} \cup \{+\infty\}$ be a convex function which is non Y_+-decreasing on $G(\text{dom}\, G) + Y_+$. Then for any $x^* \in X^*$ and $x \in X$ both the equality
$$(g \circ G)^*(x^*) = \min_{y^* \in Y_+^*}\, [g^*(y^*) + (y^* \circ G)^*(x^*)]$$
and the equality
$$\partial(g \circ G)(x) = \bigcup_{y^* \in Y_+^* \cap \partial g(G(x))} \partial(y^* \circ G)(x)$$
hold under anyone of conditions (a), (b), (c) and (d) below:
(a) With $Z := \text{span}\, [\text{dom}\, g - G(\text{dom}\, G) - Y_+]$ endowed with the topology induced by the Mackey topology of Y, for each $w(X, X^*)$-neighborhood V of zero in X there is some real $r > 0$ such that
$$0 \in \text{int}_Z\, [\{g \leq r\} - G(\text{dom}\, G \cap rV) - Y_+];$$
(b) there exists some point $\bar{x} \in \text{dom}\, G$ such that g is finite and continuous at $G(\bar{x})$ with respect to a locally convex topology compatible with the duality between Y and Y^*;
(c) the space Y is a normed space and, for $Z := \text{span}\, [\text{dom}\, g - G(\text{dom}\, G) - Y_+]$

endowed with the norm induced by the one of Y, there exists some real number $r > 0$ such that
$$0 \in \mathrm{int}_Z\, [\{g \leq r\} - G(\mathrm{dom}\, G \cap r\mathbb{B}_X) - Y_+];$$
(d) the space Y is a finite-dimensional normed space and
$$0 \in \mathrm{rint}\, [\mathrm{dom}\, g - G(\mathrm{dom}\, G) - Y_+].$$

EXERCISE 3.188. Prove the above corollary.

A particular situation with $Y = \mathbb{R}$ is often of interest.

COROLLARY 3.189. Let X be a Hausdorff locally convex space and let $G : X \to \mathbb{R} \cup \{+\infty\}$ be a convex function. Let $g : \mathbb{R} \to \mathbb{R} \cup \{+\infty\}$ be a convex function which is nondecreasing on $G(\mathrm{dom}\, G) + \mathbb{R}_+$ and derivable at $\bar{y} := G(\bar{x})$ for a point $\bar{x} \in \mathrm{dom}\, G$. Then one has
$$\partial(g \circ G)(\bar{x}) = g'(\bar{y})\partial G(\bar{x}) \text{ if } g'(\bar{y}) > 0 \text{ and } \partial(g \circ G)(\bar{x}) = N(\mathrm{dom}\, G; \bar{y}) \text{ if } g'(\bar{y}) = 0.$$

PROOF. By assumption the function g is continuous at $G(\bar{x})$, hence the condition (d) in Corollary 3.187 is satisfied. Consequently, with $\bar{y}^* : \mathbb{R} \to \mathbb{R}$ defined by $\bar{y}^*(y) := g'(\bar{y})y$ we obtain
$$\partial(g \circ G)(\bar{x}) = \partial(\bar{y}^* \circ G)(\bar{x}).$$
On the other hand, for $g'(\bar{y}) = 0$ we have $\partial(\bar{y}^* \circ G)(\bar{x}) = \partial \Psi_{\mathrm{dom}\, G}(\bar{x})$, while for $g'(\bar{y}) > 0$ we have
$$\partial(\bar{y}^* \circ G)(\bar{x}) = \partial(g'(\bar{y})G)(\bar{x}) = g'(\bar{y})\partial G(\bar{x}).$$
The corollary is then justified. $\square$

3.15. Extended rules for subdifferential of maximum of finitely many convex functions

In this section we use the above extended chain rules to establish, in the situation of convex functions, extended rules for the subdifferential of the maximum of finitely many convex functions. These extended rules will generalize, under the convexity of the data functions, the rules obtained in Proposition 2.190.

PROPOSITION 3.190. Let X be a Hausdorff locally convex space and
$$f(x) := \max_{i \in I} f_i(x) \quad \text{for all } i \in I,$$
where I is a nonempty finite set and each $f_i : X \to \mathbb{R} \cup \{+\infty\}$ is a convex function. Then for any $\bar{x} \in \mathrm{dom}\, f$ one has
$$\partial f(\bar{x}) = \bigcup \left\{ \partial \left(\sum_{i \in I} \lambda_i f_i \right)(\bar{x}) : \lambda_i \geq 0, \sum_{i \in I} \lambda_i = 1, \lambda_i = 0\ \forall i \in I \setminus I(\bar{x}) \right\}$$
$$= \bigcup \left\{ \partial \left(\sum_{i \in I(\bar{x}), \lambda_i > 0} \lambda_i f_i + \Psi_{\mathrm{dom}\, f} \right)(\bar{x}) : \lambda_i \geq 0, \sum_{i \in I(\bar{x})} \lambda_i = 1 \right\}$$
$$= \bigcup \left\{ \partial \left(\sum_{i \in I(\bar{x})} \lambda_i f_i + \Psi_{\mathrm{dom}\, f} \right)(\bar{x}) : \lambda_i \geq 0, \sum_{i \in I(\bar{x})} \lambda_i = 1 \right\},$$
where (we recall that) $\lambda_i f_i = \Psi_{\mathrm{dom}\, f_i}$ if $\lambda_i = 0$ and where $I(\bar{x}) := \{i \in I : f_i(\bar{x}) = f(\bar{x})\}$.

PROOF. Consider $Y := \mathbb{R}^I$ and $Y_+ := \mathbb{R}^I_+$ and define $g : Y = \mathbb{R}^I \to \mathbb{R}$ by $g(y) = \max_{i \in I} y_i$ for all $y = (y_i)_{i \in I} \in Y$ and $G : X \to Y \cup \{+\infty_Y\}$ by $G(x) = (f_i(x))_{i \in I}$ if $f_i(x)$ is finite for every $i \in I$ and $G(x) = +\infty_Y$ otherwise. It is clear with our convention (3.93) that $f = g \circ G$. By continuity of g the condition (d) in Corollary 3.187 is satisfied. Further, putting $\Lambda := \{\lambda = (\lambda_i)_{i \in I} : \lambda_i \geq 0, \sum_{i \in I} \lambda_i = 1\}$, for each $\bar{y} = (\bar{y}_i)_{i \in I}$ we know that

$$\partial g(\bar{y}) = \{L_\lambda : \lambda \in \Lambda, \lambda_i = 0 \text{ if } \bar{y}_i < g(\bar{y})\},$$

where the linear functional $L_\lambda : Y \to \mathbb{R}$ is defined by $L_\lambda(y) = \sum_{i \in I} \lambda_i y_i$ for all $y \in Y$. Taking $\bar{y} := G(\bar{x})$ and noting with our convention (3.93) that $L_\lambda \circ G = \sum_{i \in I} \lambda_i f_i$, the first equality in the proposition follows from Corollary 3.187 under condition (d). Concerning the second equality, it is due to the fact that for each $\lambda \in \Lambda$ satisfying $\lambda_i = 0$ for all $i \notin I(\bar{x})$, with $J_\lambda := I \setminus \{i \in I(\bar{x}) : \lambda_i > 0\}$,

$$\sum_{i \in I} \lambda_i f_i = \sum_{\substack{i \in I(\bar{x}) \\ \lambda_i > 0}} \lambda_i f_i + \sum_{i \in J_\lambda} \Psi_{\mathrm{dom}\, f_i} = \sum_{\substack{i \in I(\bar{x}) \\ \lambda_i > 0}} \lambda_i f_i + \Psi_{\mathrm{dom}\, f},$$

as easily verified. Finally, since

$$\sum_{\substack{i \in I(\bar{x}) \\ \lambda_i > 0}} \lambda_i f_i + \Psi_{\mathrm{dom}\, f} = \sum_{\substack{i \in I(\bar{x}) \\ \lambda_i > 0}} \lambda_i (f_i + \Psi_{\mathrm{dom}\, f}) = \sum_{i \in I(\bar{x})} \lambda_i (f_i + \Psi_{\mathrm{dom}\, f})$$

for each $\lambda \in \Lambda$, the third equality follows from the second. $\square$

The next corollaries concern situations where qualification conditions are present. The first corollary uses the Moreau-Rockafellar condition.

COROLLARY 3.191. Let f_i and f be as in Proposition 3.190 and let $\bar{x} \in \mathrm{dom}\, f$.
(a) If there is $i_0 \in I(\bar{x})$ such that all the functions f_i with $i \in I(\bar{x}) \setminus \{i_0\}$ are continuous at a same point in $\mathrm{dom}\, f$, then

$$\partial f(\bar{x}) = \mathrm{co}\left(\bigcup_{i \in I(\bar{x}) \setminus \{i_0\}} \partial f_i(\bar{x}) \bigcup \partial(f_{i_0} + \Psi_{\mathrm{dom}\, f})(\bar{x})\right) + N(\mathrm{dom}\, f; \bar{x}).$$

(b) If all the functions f_i with $i \in I$ except perhaps one are continuous at a same point in $\mathrm{dom}\, f$, then

$$\partial f(\bar{x}) = \mathrm{co}\left(\bigcup_{i \in I(\bar{x})} \partial f_i(\bar{x})\right) + \sum_{i \in I} N(\mathrm{dom}\, f_i; \bar{x}).$$

PROOF. (a) Take any $x^* \in \partial f(\bar{x})$ if any. By Proposition 3.190 above there is $(\lambda_i)_{i \in I}$ in $\mathbb{R}^I_+$ with $\lambda_i = 0$ for every $i \in I \setminus I(\bar{x})$ and $\sum_{i \in I} \lambda_i = 1$ such that $x^* \in \partial\left(\sum_{i \in I} \lambda_i f_i\right)(\bar{x})$. Putting $J := \{i \in I(\bar{x}) : \lambda_i > 0\}$ and noticing that

$$\sum_{i \in I} \lambda_i f_i = \sum_{i \in J} \lambda_i f_i + \sum_{i \in I \setminus J} \Psi_{\mathrm{dom}\, f_i} = \sum_{i \in J} \lambda_i f_i + \Psi_{\mathrm{dom}\, f}$$

we see that

$$
\text{(3.117)} \qquad \sum_{i \in I} \lambda_i f_i = \begin{cases} \sum_{i \in J \setminus \{i_0\}} \lambda_i f_i + \lambda_{i_0}(f_{i_0} + \Psi_{\text{dom } f}) & \text{if } \lambda_{i_0} > 0 \\ \sum_{i \in J \setminus \{i_0\}} \lambda_i f_i + \Psi_{\text{dom } f} & \text{if } \lambda_{i_0} = 0. \end{cases}
$$

If $\lambda_{i_0} > 0$, each function $\lambda_i f_i$ with $i \in J \setminus \{i_0\}$ is finite and continuous at a same point in $\text{dom}(f_{i_0} + \Psi_{\text{dom } f}) = \text{dom } f$ by the assumption in (a), then by (3.117) and Corollary 3.170 (under the Moreau-Rockafellar condition) for sum rule it ensues that

$$\partial \left(\sum_{i \in I} \lambda_i f_i \right)(\bar{x}) = \sum_{i \in J \setminus \{i_0\}} \lambda_i \partial f_i(\bar{x}) + \lambda_{i_0} \partial(f_{i_0} + \Psi_{\text{dom } f})(\bar{x}).$$

If $\lambda_{i_0} = 0$, then (3.117) and the continuity of each function $\lambda_i f_i$ with $i \in J \setminus \{i_0\}$ at a same point of $\text{dom } f$ again yield

$$\partial \left(\sum_{i \in I} \lambda_i f_i \right)(\bar{x}) = \sum_{i \in J \setminus \{i_0\}} \lambda_i \partial f_i(\bar{x}) + \partial \Psi_{\text{dom } f}(\bar{x}).$$

In either the case $\lambda_{i_0} > 0$ or $\lambda_{i_0} = 0$ we obtain with $D := \text{dom } f$ that

$$x^* \in \partial \left(\sum_{i \in I} \lambda_i f_i \right)(\bar{x}) \subset \text{co} \left(\bigcup_{i \in I(\bar{x}) \setminus \{i_0\}} \partial f_i(\bar{x}) \bigcup \partial(f_{i_0} + \Psi_D)(\bar{x}) \right) + N(D; \bar{x}),$$

which justifies the inclusion

$$\partial f(\bar{x}) \subset \text{co} \left(\bigcup_{i \in I(\bar{x}) \setminus \{i_0\}} \partial f_i(\bar{x}) \bigcup \partial(f_{i_0} + \Psi_{\text{dom } f})(\bar{x}) \right) + N(\text{dom } f; \bar{x}).$$

The converse inclusion being easily verified, the equality in (a) is proved.

(b) We note first that all the functions $\Psi_{\text{dom } f_i}$ except perhaps one are finite and continuous at a same point in $\text{dom } f$, hence $N(\text{dom } f; \bar{x}) = \sum_{i \in I} N(\text{dom } f_i; \bar{x})$. Let $k \in I$ be such that all the functions f_i with $i \in I \setminus \{k\}$ are continuous at a same point in $\text{dom } f$. If $k \notin I(\bar{x})$, the equality in (b) then clearly follows from (a). Suppose that $k \in I(\bar{x})$. Putting $i_0 := k$ we see that $f_{i_0} + \Psi_{\text{dom } f} = f_{i_0} + \sum_{i \in I \setminus \{i_0\}} \Psi_{\text{dom } f_i}$ and each function $\Psi_{\text{dom } f_i}$ with $i \in I \setminus \{i_0\}$ is finite and continuous at a same point in $\text{dom } f_{i_0}$. Therefore, with $D_i := \text{dom } f_i$

$$\partial \left(f_{i_0} + \sum_{i \in I \setminus \{i_0\}} \Psi_{D_i} \right)(\bar{x}) = \partial f_{i_0}(\bar{x}) + \sum_{i \in I \setminus \{i_0\}} \partial \Psi_{D_i}(\bar{x}) = \partial f_{i_0}(\bar{x}) + \sum_{i \in I} \partial \Psi_{D_i}(\bar{x}),$$

where the second equality is due to the fact that $\partial f_{i_0}(\bar{x}) = \partial f_{i_0}(\bar{x}) + \partial \Psi_{D_{i_0}}(\bar{x})$ (as easily checked). This combined with the above equality $N(\text{dom } f; \bar{x}) = \sum_{i \in I} N(D_i; \bar{x})$ and with the equality $f_{i_0} + \Psi_{\text{dom } f} = f_{i_0} + \sum_{i \in I \setminus \{i_0\}} \Psi_{\text{dom } f_i}$ entails by (a) again the equality in (b). $\square$

REMARK 3.192. Since $\text{dom } f = \bigcap_{i \in I} \text{dom } f_i$, as already seen in Remark 2.106 the condition that each function f_i with $i \in I(\bar{x}) \setminus \{i_0\}$ (resp. $i \in I$) is finite and continuous at a same point in dom f is equivalent to requiring that each such function f_i is finite and continuous at some point $a_i \in \text{dom } f$. □

Before stating the next corollary we take advantage of Corollary 3.191(b) to give the example (promised just after Proposition 3.5) of a proper lower semicontinuous convex function f for which $\text{Dom } \partial f$ is nonconvex.

EXAMPLE 3.193. Consider the continuous convex function $f_1 : \mathbb{R}^2 \to \mathbb{R}$ with $f_1(x, y) = |y|$ for all $(x, y) \in \mathbb{R}^2$ and the proper lower semicontinuous convex function $f_2 : \mathbb{R}^2 \to \mathbb{R} \cup \{+\infty\}$ with

$$f_2(x, y) = \begin{cases} 1 - \sqrt{x} & \text{if } x \geq 0 \\ +\infty & \text{if } x < 0. \end{cases}$$

Let $f : \mathbb{R}^2 \to \mathbb{R} \cup \{+\infty\}$ be the proper lower semicontinuous convex function defined by

$$f(x, y) = \max\{f_1(x, y), f_2(x, y)\} \quad \text{for all } (x, y) \in \mathbb{R}^2.$$

It is clear that $\partial f(x, y) = \emptyset$ if $x < 0$ since $f(x, y) = +\infty$ in this case. By 3.191(b) we see that $\partial f(x, y) = \emptyset$ if $(x, y) \in \{0\} \times]-1, 1[$, and that $\partial f(x, y) \neq \emptyset$ in all other cases. Therefore,

$$\text{Dom } \partial f = ([0, +\infty[\times \mathbb{R}) \setminus (\{0\} \times]-1, 1[),$$

so the domain of the subdifferential multimapping ∂f is nonconvex. □

The second corollary directly follows from Corollary 3.194(b).

COROLLARY 3.194. *Let f_i and f be as in Proposition 3.190 and let $\bar{x} \in \text{dom } f$. If all the functions f_i are continuous at $\bar{x}$, then*

$$\partial f(\bar{x}) = \text{co}\left(\bigcup_{i \in I(\bar{x})} \partial f_i(\bar{x})\right).$$

The third corollary considers the Rockafellar relative interior condition in finite dimensions.

COROLLARY 3.195. *Let f_i and f be as in Proposition 3.190 and let $\bar{x} \in \text{dom } f$. If X is finite-dimensional and $\bigcap_{i \in I} \text{rint}(\text{dom } f_i) \neq \emptyset$, then*

$$\partial f(\bar{x}) = \text{co}\left(\bigcup_{i \in I(\bar{x})} \partial f_i(\bar{x})\right) + \sum_{i \in I} N(\text{dom } f_i; \bar{x}).$$

PROOF. As in the proof of (a) of the previous corollary, take any $x^* \in \partial f(\bar{x})$ if any. By Proposition 3.190 above there is $(\lambda_i)_{i \in I}$ in $\mathbb{R}_+^I$ with $\lambda_i = 0$ for every $i \in I \setminus I(\bar{x})$ and $\sum_{i \in I} \lambda_i = 1$ such that $x^* \in \partial\left(\sum_{i \in I} \lambda_i f_i\right)(\bar{x})$. Denoting $J := \{i \in I(\bar{x}) : \lambda_i > 0\}$ and observing that

$$\sum_{i \in I} \lambda_i f_i = \sum_{i \in J} \lambda_i f_i + \sum_{i \in I \setminus J} \Psi_{\text{dom } f_i}$$

we deduce by Corollary 3.170(d) that

$$\partial\left(\sum_{i \in I} \lambda_i f_i\right)(\bar{x}) = \sum_{i \in J} \lambda_i \partial f_i(\bar{x}) + \sum_{i \in I \setminus J} \partial \Psi_{\text{dom } f_i}(\bar{x}),$$

and hence
$$\partial\left(\sum_{i\in I}\lambda_i f_i\right)(\overline{x}) \subset \operatorname{co}\left(\bigcup_{i\in I(\overline{x})} \partial f_i(\overline{x})\right) + \sum_{i\in I} N(\operatorname{dom} f_i; \overline{x}).$$

The inclusion $\partial f(\overline{x}) \subset \operatorname{co}\left(\bigcup_{i\in I(\overline{x})} \partial f_i(\overline{x})\right) + \sum_{i\in I} N(\operatorname{dom} f_i; \overline{x})$ then holds true, and this inclusion is in fact an equality since the converse is easily verified. □

3.16. Limiting formulas for subdifferential of convex functions

Under constraint qualification conditions for convex functions f, g and a linear mapping A, we obtained formulas in Section 3.11 for the description of any subgradient of $f + g \circ A$ at $\overline{x}$ in terms of subgradients of the functions f and g at $\overline{x}$ and $A\overline{x}$ respectively. Removing the constraint qualification conditions, we will establish in this section, in the Banach space framework, similar limiting descriptions. We will also deal with the composition $g \circ G$, where in place of the linear mapping A we consider a general convex vector-valued mapping G.

3.16.1. Limiting sum/chain rule for subdifferential of convex functions.
In order to begin with the basic case $f + g \circ A$, we prove first two lemmas.

LEMMA 3.196. *Let $(X, \|\cdot\|_X)$ and $(Y, \|\cdot\|_Y)$ be normed spaces, $A: X \to Y$ be a continuous linear mapping, C be a nonempty subset of X and D be a subset of Y with $A(C) \subset D$. Let $\phi, \varphi: C \to \mathbb{R} \cup \{+\infty\}$ be two functions bounded from below on C and $g: D \to \mathbb{R} \cup \{+\infty\}$ be a function bounded from below on D with $\operatorname{dom}\phi \cap \operatorname{dom}\varphi \cap A^{-1}(\operatorname{dom} g) \neq \emptyset$. Assume that ϕ and g are weakly semicontinuous relative to C and D respectively, and that the set $\{x \in C : \varphi(x) \leq \rho\}$ is weakly compact in X for every real ρ. Then, with*
$$\Delta(u, x, y) := \frac{1}{2}(\|x - u\|_X^2 + \|y - Au\|_Y^2 + \|y - Ax\|_Y^2)$$
and with
$$w_r(u, x, y) := \phi(u) + \varphi(x) + g(y) + r\Delta(u, x, y)$$
for all $u, x \in X$ and $y \in Y$, one has
$$\inf_{u\in C, x\in C, y\in D} w_r(u, x, y) \xrightarrow[r\to+\infty]{} \inf_{x\in C}[\phi(x) + \varphi(x) + g(Ax)].$$
Furthermore, for any $(u_r, x_r, y_r) \in C \times C \times D$ satisfying
$$w_r(u_r, x_r, y_r) - \inf_{u\in C, x\in C, y\in D} w_r(u, x, y) \xrightarrow[r\to+\infty]{} 0,$$
one has
$$r\Delta(u_r, x_r, y_r) \xrightarrow[r\to+\infty]{} 0.$$

PROOF. Put $I := \inf_{x\in C}[\phi(x) + \varphi(x) + g(Ax)]$ and for each real $r > 0$ put also $I(r) := \inf_{u\in C, x\in C, y\in D} w_r(u, x, y)$. Obviously $I(r)$ and I are finite and $I(r) \leq I$ for all $r > 0$.

Let any $\varepsilon_r > 0$ with $\varepsilon_r \to 0$ as $r \to +\infty$, so $(\varepsilon_r)_{r\geq \overline{r}}$ is bounded for some real $\overline{r} > 0$. For each $r > 0$ take $u_r, x_r \in C$ and $y_r \in D$ such that $w_r(u_r, x_r, y_r) \leq I(r) + \varepsilon_r$. It ensues that

(3.118) $$\phi(u_r) + \varphi(x_r) + g(y_r) \leq w_r(u_r, x_r, y_r) \leq I(r) + \varepsilon_r.$$

3.16. LIMITING FORMULAS FOR SUBDIFFERENTIAL OF CONVEX FUNCTIONS

Since the functions ϕ, φ, g are bounded from below and $I(r) \leq I$, it is clear from the above inequality on the right that $\|x_r - u_r\|_X \to 0$, $\|y_r - Au_r\|_Y \to 0$ and $\|y_r - Ax_r\|_Y \to 0$ as $r \to +\infty$.

Take any sequence $(r_k)_k$ in $]0, +\infty[$ converging to $+\infty$. Putting $\mu := I + \sup_{0 < r \leq \bar{r}} \varepsilon_r - \inf_C \phi - \inf_D g$, for k large enough the sequence $(x_{r_k})_k$ belongs to the w-compact set $\{x \in C : \varphi(x) \leq \mu\}$, and hence it admits a subnet $(x_{r_{s(j)}})_{j \in J}$ weakly converging to a point $c \in C$. Putting $u'_j := u_{r_{s(j)}}$, $x'_j := x_{r_{s(j)}}$ and $y'_j := y_{r_{s(j)}}$ for every $j \in J$, the subnets $(u'_j)_j$ and $(y'_j)_j$ weakly converge to c and Ac respectively (since $\|x'_j - u'_j\|_X \to 0$ and $\|y'_j - Ax'_j\|_Y \to 0$). This and the weak semicontinuity of the functions ϕ, φ, g along with (3.118) entail with $\zeta_r := w_r(u_r, x_r, y_r)$ and with $\alpha := \phi(c) + \varphi(c) + g(Ac)$ that

$$I \leq \alpha \leq \liminf_{j \in J} [\phi(u'_j) + \varphi(x'_j) + g(y'_j)] \leq \liminf_{j \in J} \zeta_{r_{s(j)}} \leq \liminf_{j \in J} I(r_{s(j)}) \leq I.$$

Then we easily see that $I(r) \to I$ and $\xi_r := \phi(u_r) + \varphi(x_r) + g(y_r) \to I$ along with $\zeta_r \to I$ as $r \to +\infty$, hence also $r\Delta(u_r, x_r, y_r) = \zeta_r - \xi_r \to 0$ as $r \to +\infty$.

Finally, let any (u_r, x_r, y_r) in $C \times C \times D$ such that $w_r(u_r, x_r, y_r) - I(r) \to 0$ as $r \to +\infty$. Putting $\varepsilon_r := (1/r) + |w_r(u_r, x_r, y_r) - I(r)| > 0$ we have $w_r(u_r, x_r, y_r) \leq I(r) + \varepsilon_r$ with $\varepsilon_r \to 0$ as $r \to +\infty$, thus from what precedes we deduce that $r\Delta(u_r, x_r, y_r) \to 0$ as $r \to +\infty$. $\square$

REMARK 3.197. Of course, in the proof of Lemma 3.196 from the sequence $(x_{r_k})_k$ in the weak compact set $\{x \in C : \varphi(x) \leq \mu\}$, by the Eberlein-Šmulian theorem, a weak convergent subsequence can be extracted instead of the weak convergent subnet. $\square$

Given a function $\varphi : X \to \mathbb{R} \cup \{+\infty\}$ on a normed space X, we say that φ is *boundedly weakly inf-compact* provided that for each closed ball B in X and each real ρ the set $B \cap \{x \in X : \varphi(x) \leq \rho\}$ is weakly compact in X. If X is a reflexive Banach space, it is clear that any lower semicontinuous convex function on X is boundedly weakly inf-compact.

LEMMA 3.198. Let $(X, \|\cdot\|_X)$ and $(Y, \|\cdot\|_Y)$ be Banach spaces, $A : X \to Y$ be a continuous linear mapping, $\phi, \varphi : X \to \mathbb{R} \cup \{+\infty\}$ and $g : Y \to \mathbb{R} \cup \{+\infty\}$ be lower semicontinuous convex functions. Let $\bar{x} \in X$ be such that $\phi(\bar{x})$, $\varphi(\bar{x})$ and $g(A\bar{x})$ are finite. Assume that φ is boundedly weakly inf-compact and that $\bar{x}$ is a minimizer of $\phi + \varphi + g \circ A$ over X. Then for any real $\varepsilon > 0$ there are $(u_\varepsilon, u_\varepsilon^*)$ in $\text{gph}\,\partial\phi$, $(x_\varepsilon, x_\varepsilon^*)$ in $\text{gph}\,\partial\varphi$ and $(y_\varepsilon, y_\varepsilon^*)$ in $\text{gph}\,\partial g$ such that

$$\|u_\varepsilon - \bar{x}\|_X \leq \varepsilon, \ |\phi(u_\varepsilon) - \phi(\bar{x})| \leq \varepsilon, \ \|x_\varepsilon - \bar{x}\|_X \leq \varepsilon, \ |\varphi(x_\varepsilon) - \varphi(\bar{x})| \leq \varepsilon,$$
$$\|y_\varepsilon - A\bar{x}\|_Y \leq \varepsilon, \ |g(y_\varepsilon) - g(A\bar{x})| \leq \varepsilon, \ \|u_\varepsilon^* + x_\varepsilon^* + y_\varepsilon^* \circ A\| \leq \varepsilon,$$
$$\max\{\|u_\varepsilon^*\|, \|x_\varepsilon^*\|, \|y_\varepsilon^*\|\} \cdot (\|u_\varepsilon - x_\varepsilon\| + \|y_\varepsilon - Au_\varepsilon\| + \|y_\varepsilon - Ax_\varepsilon\|) \leq \varepsilon.$$

PROOF. Fix $\varepsilon > 0$ and keep notation in the statement of Lemma 3.196. Applying this lemma with the closed convex sets $C := \bar{x} + (1 + \|A\|)^{-1}\varepsilon \mathbb{B}_X$ and $D := \bar{y} + \varepsilon \mathbb{B}_Y$, where $\bar{y} := A\bar{x}$, we obtain

$$\inf_{u \in C, x \in C, y \in D} w_r(u, x, y) \xrightarrow[r \to +\infty]{} \phi(\bar{x}) + \varphi(\bar{x}) + g(\bar{y}),$$

and hence for each integer $n \geq 1$ there exists a real $r_n > n$ such that

$$(3.119) \qquad \phi(\bar{x}) + \varphi(\bar{x}) + g(\bar{y}) < \inf_{u \in C, x \in C, y \in D} w_{r_n}(u, x, y) + \frac{\varepsilon^2}{n^2}.$$

Since $\Delta(\bar{x},\bar{x},A\bar{x}) = 0$, the latter inequality can be written in the form

(3.120) $$w_{r_n}(\bar{x},\bar{x},\bar{y}) < \inf\{w_{r_n}(u,x,y) : u,x \in C, y \in D\} + \frac{\varepsilon^2}{n^2}.$$

Endowing $X \times X \times Y$ with the sum norm, the Ekeland variational principle in the complete metric space $C \times C \times D$ furnishes $u_n \in C$, $x_n \in C$ and $y_n \in D$ such that

(3.121) $$w_{r_n}(u_n,x_n,y_n) \leq w_{r_n}(\bar{x},\bar{x},\bar{y}), \quad \|u_n - \bar{x}\| + \|x_n - \bar{x}\| + \|y_n - \bar{y}\| \leq \frac{\varepsilon}{n},$$

and (u_n, x_n, y_n) is a minimizer over $C \times C \times D$ of the function

$$(u,x,y) \mapsto \phi(u) + \varphi(x) + g(y) + r_n\Delta(u,x,y)$$
$$+ \frac{\varepsilon}{n}(\|u - \bar{x}\| + \|x - \bar{x}\| + \|y - \bar{y}\|),$$

so for n large enough $(0,0,0)$ belongs to the subdifferential of this function at the point (u_n,x_n,y_n) which is in the interior of $C \times C \times D$ for large n. For the function $\Phi : X \times X \times Y \to \mathbb{R} \cup \{+\infty\}$ defined by

$$\Phi(u,x,y) := \phi(u) + \varphi(x) + g(y) \quad \text{for all } (u,x,y) \in X \times X \times Y,$$

we know that $\partial\Phi(u,x,y) = \partial\phi(u) \times \partial\varphi(x) \times \partial g(y)$ (see Proposition 3.11). So, the continuity of the convex function Δ allows us to apply the Moreau-Rockafellar theorem 2.105 or Corollary 3.170 to obtain $u_n^* \in \partial\phi(u_n)$, $x_n^* \in \partial\varphi(x_n)$ and $y_n^* \in \partial g(y_n)$ such that

$$-(u_n^*, x_n^*, y_n^*) \in r_n\partial\Delta(u_n,x_n,y_n) + \frac{\varepsilon}{n}\mathbb{B}.$$

Observing that for any $(u^*, x^*, y^*) \in \partial\Delta(u,x,y)$ we have $u^* + x^* + y^* \circ A = 0$, it ensues that

$$\|u_n^* + x_n^* + y_n^* \circ A\| \leq (2 + \|A\|)\frac{\varepsilon}{n}.$$

On the other hand, from (3.119) and (3.121) we see that

(3.122) $$w_{r_n}(u_n,x_n,y_n) \xrightarrow[n\to\infty]{} \phi(\bar{x}) + \varphi(\bar{x}) + g(\bar{y}),$$

and by definition of the function w_r we also have

$$\phi(u_n) + \varphi(x_n) + g(y_n) \leq w_{r_n}(u_n,x_n,y_n).$$

Since $u_n \to \bar{x}$, $x_n \to \bar{x}$ and $y_n \to \bar{y}$ as $n \to \infty$, using the lower semicontinuity of ϕ, φ and g we derive from the latter inequality and from (3.122) that

$$\phi(u_n) + \varphi(x_n) + g(y_n) \xrightarrow[n\to\infty]{} \phi(\bar{x}) + \varphi(\bar{x}) + g(\bar{y}).$$

This and the lower semicontinuity of ϕ, φ, g once again imply that $\phi(u_n) \to \phi(\bar{x})$, $\varphi(x_n) \to \varphi(\bar{x})$ and $g(y_n) \to g(\bar{y})$ as $n \to \infty$. Furthermore, by (3.120) and (3.121) we have

$$w_{r_n}(u_n,x_n,y_n) - \inf_{u \in C, x \in C, y \in D} w_{r_n}(u,x,y) \xrightarrow[n\to\infty]{} 0,$$

thus Lemma 3.196 ensures that

(3.123) $$r_n(\|u_n - x_n\|^2 + \|y_n - Au_n\|^2 + \|y_n - Ax_n\|^2) \xrightarrow[n\to\infty]{} 0.$$

Now fix any $(u^*, x^*, y^*) \in \partial\Delta(u, x, y)$. For any $v \in X$ with $\|v\| = 1$ and any real $t > 0$

$$\langle u^*, tv\rangle \leq \frac{1}{2}(\|u - x + tv\|^2 - \|u - x\|^2) + \frac{1}{2}(\|y - Au - tAv\|^2 - \|y - Au\|^2)$$

$$= \frac{1}{2}(\|u - x + tv\| - \|u - x\|)(\|u - x + tv\| + \|u - x\|)$$

$$+ \frac{1}{2}(\|y - Au - tAv\| - \|y - Au\|)(\|y - Au - tAv\| + \|y - Au\|),$$

which gives

$$\langle u^*, tv\rangle \leq \frac{1}{2}t(\|u - x + tv\| + \|u - x\|) + \frac{1}{2}t\|A\|(\|y - Au - tAv\| + \|y - Au\|),$$

so dividing by t and then taking the limit as $t \downarrow 0$ furnishes with $K := 1 + \|A\|$

(3.124) $\quad \langle u^*, v\rangle \leq \|u - x\| + \|A\| \|y - Au\|$, so $\|u^*\| \leq K(\|u - x\| + \|y - Au\| + \|y - Ax\|)$.

Similarly, we obtain

$$\|x^*\| \leq \|u - x\| + \|A\| \|y - Ax\| \quad \text{and} \quad \|y^*\| \leq \|y - Au\| + \|y - Ax\|,$$

which yields

$$\|x^*\| \leq K(\|u-x\|+\|y-Au\|+\|y-Ax\|) \text{ and } \|y^*\| \leq K(\|u-x\|+\|y-Au\|+\|y-Ax\|).$$

Using the latter inequalities and the second one in (3.124) gives

$$\max\{\|u_n^*\|, \|x_n^*\|, \|y_n^*\|\}(\|u_n - x_n\| + \|y_n - Au_n\| + \|y_n - Ax_n\|) \xrightarrow[n\to\infty]{} 0.$$

It suffices to choose some integer n sufficiently large to complete the proof. $\square$

We can now prove limiting formulas for subgradients of $f + g \circ A$. In the statement of the theorem, as usual, $\lim_{w^*}$ and $\lim_{\|\cdot\|_*}$ denote respectively the weak* limit and strong limit in a normed dual space.

THEOREM 3.199 (limiting sum/chain subdifferential rule for convex functions). Let $(X, \|\cdot\|_X)$ and $(Y, \|\cdot\|_Y)$ be two Banach spaces, $A: X \to Y$ be a continuous linear mapping, and $f: X \to \mathbb{R} \cup \{+\infty\}$ and $g: Y \to \mathbb{R} \cup \{+\infty\}$ be two lower semicontinuous convex functions. Let $\bar{x} \in X$ with $f(\bar{x}) < +\infty$ and $g(A\bar{x}) < +\infty$. Then, $x^* \in \partial(f + g \circ A)(\bar{x})$ if and only if there are nets (resp. sequences if either f is boundedly weakly inf-compact or X is reflexive) $(x_j)_j$ in X, $(y_j)_j$ in Y, $(x_j^*)_j$ in X^* and $(y_j^*)_j$ in Y^* with $x_j^* \in \partial f(x_j)$ and $y_j^* \in \partial g(y_j)$ such that

(3.125) $\qquad x^* = \lim_{w^*}(x_j^* + y_j^* \circ A)$ $\quad$ (resp. $x^* = \lim_{\|\cdot\|_*}(x_j^* + y_j^* \circ A)$)

and such that anyone of the following conditions holds:

(i) $(x_j, f(x_j)) \to (\bar{x}, f(\bar{x}))$, $(y_j, g(y_j)) \to (A\bar{x}, g(A\bar{x}))$, $\langle x_j^*, x_j - \bar{x}\rangle \to 0$, $\langle y_j^*, y_j - A\bar{x}\rangle \to 0$ and $(\|x_j^*\| + \|y_j^*\|)\|y_j - Ax_j\|_Y \to 0$;

(ii) $(x_j, f(x_j)) \to (\bar{x}, f(\bar{x}))$, $(y_j, g(y_j)) \to (A\bar{x}, g(A\bar{x}))$, $\langle x_j^*, x_j - \bar{x}\rangle \to 0$, $\langle y_j^*, y_j - A\bar{x}\rangle \to 0$;

(iii) $x_j \to \bar{x}$, $y_j \to A\bar{x}$, $f(x_j) - \langle x_j^*, x_j - \bar{x}\rangle \to f(\bar{x})$ and $g(y_j) - \langle y_j^*, y_j - A\bar{x}\rangle \to g(A\bar{x})$;

(iv) $x_j \to \bar{x}$, $y_j \to A\bar{x}$, $\langle x_j^*, x_j - \bar{x}\rangle \to 0$ and $\langle y_j^*, y_j - A\bar{x}\rangle \to 0$;

(v) $x_j \to \bar{x}$, $y_j \to A\bar{x}$, $\liminf \langle x_j^*, \bar{x} - x_j\rangle \geq 0$ and $\liminf \langle y_j^*, A\bar{x} - y_j\rangle \geq 0$.

PROOF. We denote by (o) the assertion $x^* \in \partial(f + g \circ A)(\overline{x})$ and by (i),$\cdots$,(v) the above condition (i),$\cdots$,(v) with the additional equality in (3.125). The proof will be established by showing that all the assertions (o) to (v) are equivalent.

(o)$\Rightarrow$(i). Let $x^* \in \partial(f + g \circ A)(\overline{x})$, so $\overline{x}$ is a minimizer over X of the function $(f - x^*) + g \circ A$.

If f is boundedly weakly inf-compact or X is reflexive, we easily see that $(f - x^*)$ is also boundedly weakly inf-compact, so we will apply in this case Lemma 3.198 with the null function in the place of ϕ and with $f - x^*$ in the place of φ. In this case we set $J := \mathbb{N}$ and write $j = n \in \mathbb{N}$.

If f is not boundedly weakly inf-compact, we denote by $\mathcal{F}$ the collection of finite-dimensional vector subspaces of X containing $\overline{x}$ and J will be the directed set $\mathbb{N} \times \mathcal{F}$ with $(n, L) \preceq (n', L')$ if and only if $n \leq n'$ and $L \subset L'$. Noting that $\overline{x}$ is a minimizer of $(f - x^*) + \Psi_L + g \circ A$ and that Ψ_L is boundedly inf-compact, we will apply in this case Lemma 3.198 with $f - x^*$ in place of ϕ and with Ψ_L in place of φ. We will write $j = (n, L) \in J$.

From Lemma 3.198 there exist (x_j, x_j^*) in gph ∂f, (y_j, y_j^*) in gph ∂g and (z_j, z_j^*) in gph $\partial \Psi_L$ (resp. $z_j \in X$ and $z_j^* = 0$ if f is boundedly weakly inf-compact) with

(3.126) $$\|x_j - \overline{x}\| \leq \frac{1}{n}, \ \|y_j - \overline{y}\| \leq \frac{1}{n}, \ \|z_j - \overline{x}\| \leq \frac{1}{n},$$

(3.127) $$|f(x_j) - f(\overline{x})| \leq \frac{1}{n}, \ |g(y_j) - g(\overline{y})| \leq \frac{1}{n},$$

(3.128) $$\|x_j^* + z_j^* + y_j^* \circ A - x^*\| \leq \frac{1}{n} \ (\text{resp. } \|x_j^* + y_j^* \circ A - x^*\| \leq \frac{1}{n}),$$

(3.129) $\max\{\|x_j^* - x^*\|, \|z_j^*\|, \|y_j^*\|\} \cdot (\|x_j - z_j\| + \|y_j - Az_j\| + \|y_j - Ax_j\|) \leq \frac{1}{n}.$

The latter inequality and (3.126) easily imply that

$$(\|x_j^*\| + \|y_j^*\|)\|y_j - Ax_j\| \to 0.$$

Since $\partial \Psi_L(v) = L^{\perp} := \{v^* \in X^* : \langle v^*, w \rangle = 0, \forall w \in L\}$ for any $v \in L$, it follows that

$$x_j^* + y_j^* \circ A \in x^* - z_j^* + \frac{1}{n}\mathbb{B}_{X^*} \subset x^* + L^{\perp} + \frac{1}{n}\mathbb{B}_{X^*} \ (\text{resp. } x_j^* + y_j^* \circ A \in x^* + \frac{1}{n}\mathbb{B}_{X^*}),$$

and hence

$$x^* = \lim_{w^*}(x_j^* + y_j^* \circ A) \quad (\text{resp. } x^* = \lim_{\|\cdot\|}(x_j^* + y_j^* \circ A)).$$

Moreover, for some $b_j^* \in \mathbb{B}_{X^*}$, since $z_j \in L$ and $z_j^* \in L^{\perp}$ (resp. since $z_j^* = 0$) one has

$$\langle x_j^* + y_j^* \circ A, x_j - \overline{x} \rangle = \left\langle x^* + \frac{1}{n}b_j^*, x_j - \overline{x} \right\rangle - \langle z_j^*, x_j - \overline{x} \rangle$$

$$= \left\langle x^* + \frac{1}{n}b_j^*, x_j - \overline{x} \right\rangle - \langle z_j^*, z_j - \overline{x} \rangle + \langle z_j^*, z_j - x_j \rangle$$

$$= \left\langle x^* + \frac{1}{n}b_j^*, x_j - \overline{x} \right\rangle - \langle z_j^*, z_j - x_j \rangle,$$

and hence by (3.126) and (3.129)

(3.130) $$\langle x_j^* + y_j^* \circ A, x_j - \overline{x} \rangle \to 0.$$

Note by the inequality of subgradient of convex functions that
$$\langle x_j^*, \bar{x} - x_j \rangle \leq f(\bar{x}) - f(x_j) \quad \text{and} \quad \langle y_j^*, \bar{y} - y_j \rangle \leq g(\bar{y}) - g(y_j),$$
thus taking into account (3.126) we obtain

(3.131) $\quad \limsup \langle x_j^*, \bar{x} - x_j \rangle \leq 0 \quad \text{and} \quad \limsup \langle y_j^*, \bar{y} - y_j \rangle \leq 0.$

Further, noticing that $\limsup \langle y_j^*, \bar{y} - Ax_j \rangle = \limsup \langle y_j^*, \bar{y} - y_j \rangle$ since $\|y_j^*\| \, \|y_j - Ax_j\| \to 0$ by (3.129), we derive from (3.130)
$$0 \leq \liminf \langle x_j^*, \bar{x} - x_j \rangle + \limsup \langle y_j^*, \bar{y} - y_j \rangle,$$
and using this and both inequalities in (3.131) we get
$$0 \leq \liminf \langle x_j^*, \bar{x} - x_j \rangle \leq \limsup \langle x_j^*, \bar{x} - x_j \rangle \leq 0,$$
which ensures that
$$\lim \langle x_j^*, \bar{x} - x_j \rangle = 0.$$
In the same way
$$\lim \langle y_j^*, \bar{y} - y_j \rangle = 0,$$
and this completes the proof of the implication (o)$\Rightarrow$(i).

(i)$\Rightarrow$(ii) and (ii)$\Rightarrow$(iii) are trivial.

(iii)$\Rightarrow$(o). Suppose that $x^* = \lim_{w^*}(x_j^* + y_j^* \circ A)$ with $x_j^* \in \partial f(x_j)$ and $y_j^* \in \partial g(y_j)$ satisfying (iii). Then for every $x \in X$ we have

$$\langle x_j^* + y_j^* \circ A, x - \bar{x} \rangle$$
$$= (\langle x_j^*, x - x_j \rangle + \langle x_j^*, x_j - \bar{x} \rangle) + (\langle y_j^*, Ax - y_j \rangle + \langle y_j^*, y_j - A\bar{x} \rangle)$$
(3.132) $\quad \leq [f(x) - (f(x_j) - \langle x_j^*, x_j - \bar{x} \rangle)] + [g \circ A(x) - (g(y_j) - \langle y_j^*, y_j - A\bar{x} \rangle)],$

and passing to the limit we obtain
$$\langle x^*, x - \bar{x} \rangle \leq [f(x) - f(\bar{x})] + [g \circ A(x) - g \circ A(\bar{x})].$$
The latter being true for all $x \in X$, it ensues that $x^* \in \partial(f + g \circ A)(\bar{x})$, which corresponds to (o).

(i)$\Rightarrow$(iv) and (iv)$\Rightarrow$(v) are obvious.

(v)$\Rightarrow$(o). Assuming (v), for every $x \in X$ we have as in (3.132)
$$\langle x_j^*, \bar{x} - x_j \rangle + \langle y_j^*, A\bar{x} - y_j \rangle + f(x_j) + g(y_j) \leq f(x) + g \circ A(x) - \langle x_j^* + y_j^* \circ A, x - \bar{x} \rangle,$$
and hence
$$\liminf \langle x_j^*, \bar{x} - x_j \rangle + \liminf \langle y_j^*, A\bar{x} - y_j \rangle + \liminf f(x_j) + \liminf g(y_j)$$
$$\leq f(x) + g \circ A(x) - \langle x^*, x - \bar{x} \rangle.$$
According to the assumptions
$$0 \leq \liminf \langle x_j^*, \bar{x} - x_j \rangle \quad \text{and} \quad 0 \leq \liminf \langle y_j^*, A\bar{x} - y_j \rangle$$
and to the lower semicontinuity of f and g it results that
$$f(\bar{x}) + g(A\bar{x}) \leq f(x) + g(Ax) - \langle x^*, x - \bar{x} \rangle.$$
This ensures that $x^* \in \partial(f + g \circ A)(\bar{x})$, that is, (o) holds. The proof of the theorem is then complete. $\square$

The first corollary is obtained with f as the null function.

COROLLARY 3.200 (limiting chain rule for a convex function and a linear mapping: Banach space case). Let $(X, \|\cdot\|_X)$ and $(Y, \|\cdot\|_Y)$ be two Banach spaces, $A : X \to Y$ be a continuous linear mapping, and and $g : Y \to \mathbb{R} \cup \{+\infty\}$ be a lower semicontinuous convex function. Let $\bar{x} \in X$ with $g(A\bar{x}) < +\infty$. Then, $x^* \in \partial(g \circ A)(\bar{x})$ if and only if there are nets (resp. sequences if X is reflexive) $(y_j)_j$ in Y and $(y_j^*)_j$ in Y^* with $y_j^* \in \partial g(y_j)$ such that

$$(3.133) \qquad x^* = \lim_{w^*} y_j^* \circ A \quad (\text{resp. } x^* = \lim_{\|\cdot\|_*} y_j^* \circ A)$$

and such that anyone of the following conditions holds:

(i) $(y_j, g(y_j)) \to (A\bar{x}, g(A\bar{x}))$, $\langle y_j^*, y_j - A\bar{x} \rangle \to 0$ and $\|y_j^*\| \|y_j - Ax_j\|_Y \to 0$;
(ii) $(y_j, g(y_j)) \to (A\bar{x}, g(A\bar{x}))$, $\langle y_j^*, y_j - A\bar{x} \rangle \to 0$;
(iii) $y_j \to A\bar{x}$ and $g(y_j) - \langle y_j^*, y_j - A\bar{x} \rangle \to g(A\bar{x})$;
(iv) $y_j \to A\bar{x}$ and $\langle y_j^*, y_j - A\bar{x} \rangle \to 0$;
(v) $y_j \to A\bar{x}$ and $\liminf \langle y_j^*, y_j - A\bar{x} \rangle \geq 0$.

The second corollary is derived from Theorem 3.199 with A as the identity mapping on X.

COROLLARY 3.201 (limiting sum rule for convex functions). Let $(X, \|\cdot\|_X)$ and $(Y, \|\cdot\|_Y)$ be two Banach spaces, and $f : X \to \mathbb{R} \cup \{+\infty\}$ and $g : Y \to \mathbb{R} \cup \{+\infty\}$ be two lower semicontinuous convex functions. Let $\bar{x} \in X$ with $f(\bar{x}) < +\infty$ and $g(\bar{x}) < +\infty$. Then, $x^* \in \partial(f + g)(\bar{x})$ if and only if there are nets (resp. sequences if either f is boundedly weakly inf-compact or X is reflexive) $(x_j)_j$ and $(y_j)_j$ in X, $(x_j^*)_j$ and $(y_j^*)_j$ in X^* with $x_j^* \in \partial f(x_j)$ and $y_j^* \in \partial g(y_j)$ such that

$$(3.134) \qquad x^* = \lim_{w^*} (x_j^* + y_j^*) \quad (\text{resp. } x^* = \lim_{\|\cdot\|_*} (x_j^* + y_j^*))$$

and such that anyone of the following conditions holds:

(i) $(x_j, f(x_j)) \to (\bar{x}, f(\bar{x}))$, $(y_j, g(y_j)) \to (\bar{x}, g(\bar{x}))$, $\langle x_j^*, x_j - \bar{x} \rangle \to 0$, $\langle y_j^*, y_j - \bar{x} \rangle \to 0$ and $(\|x_j^*\| + \|y_j^*\|)\|y_j - x_j\|_Y \to 0$;
(ii) $(x_j, f(x_j)) \to (\bar{x}, f(\bar{x}))$, $(y_j, g(y_j)) \to (\bar{x}, g(\bar{x}))$, $\langle x_j^*, x_j - \bar{x} \rangle \to 0$, $\langle y_j^*, y_j - \bar{x} \rangle \to 0$;
(iii) $x_j \to \bar{x}$, $y_j \to \bar{x}$, $f(x_j) - \langle x_j^*, x_j - \bar{x} \rangle \to f(\bar{x})$ and $g(y_j) - \langle y_j^*, y_j - \bar{x} \rangle \to g(\bar{x})$;
(iv) $x_j \to \bar{x}$, $y_j \to \bar{x}$, $\langle x_j^*, x_j - \bar{x} \rangle \to 0$ and $\langle y_j^*, y_j - \bar{x} \rangle \to 0$;
(v) $x_j \to \bar{x}$, $y_j \to \bar{x}$, $\liminf \langle x_j^*, x_j - \bar{x} \rangle \geq 0$ and $\liminf \langle y_j^*, y_j - \bar{x} \rangle \geq 0$.

EXERCISE 3.202. Let $(X, \|\cdot\|_X)$ and $(Y, \|\cdot\|_Y)$ be two Banach spaces, $A : X \to Y$ be a continuous linear mapping, and $f : X \to \mathbb{R} \cup \{+\infty\}$ and $g : Y \to \mathbb{R} \cup \{+\infty\}$ be two lower semicontinuous convex functions. Let $\bar{x} \in X$ with $f(\bar{x}) < +\infty$ and $g(A\bar{x}) < +\infty$. The aim of the exercise is to prove the equality

$$\partial(f + g \circ A)(\bar{x}) = \bigcap_{\varepsilon > 0} \operatorname{cl}_{w^*} \left(\partial_\varepsilon f(\bar{x}) + A^*(\partial_\varepsilon g(A\bar{x})) \right).$$

(a) Show that the set on the right-hand side of the equality is included in the one on the left-hand side.
(b) Fix any $x^* \in \partial(f + g \circ A)(\bar{x})$ and take any real $\varepsilon > 0$. Consider the nets in (iii) in Theorem 3.199.
(b$_1$) Argue that there exists some $j_\varepsilon \in J$ such that for all $j \succeq j_\varepsilon$

$$f(x_j) - \langle x_j^*, x_j - \bar{x} \rangle \geq f(\bar{x}) - \varepsilon \quad \text{and} \quad g(y_j) - \langle y_j^*, y_j - A\bar{x} \rangle \geq g(A\bar{x}) - \varepsilon.$$

3.16. LIMITING FORMULAS FOR SUBDIFFERENTIAL OF CONVEX FUNCTIONS

(b$_2$) Deduce for every $j \succeq j_\varepsilon$ and every $x \in X$ that
$$\langle x_j^*, x - \bar{x} \rangle \leq f(x) - f(\bar{x}) + \varepsilon,$$
and hence $x_j^* \in \partial_\varepsilon f(\bar{x})$.

(b$_3$) Argue similarly that $y_j^* \in \partial_\varepsilon g(A\bar{x})$ for every $j \succeq j_\varepsilon$.

(b$_4$) Deduce that $x^* \in \mathrm{cl}_{w^*}\left(\partial_\varepsilon f(\bar{x}) + A^*(\partial_\varepsilon g(A\bar{x}))\right)$ and conclude. □

3.16.2. Limiting rules for subdifferential of composition with inner vector-valued convex mapping. We apply the previous formulas to study the composition with an inner vector-valued convex mapping. Before stating the theorem of this section recall (see (3.94)) that with G as defined in the theorem one has
$$0_{Y^*} \circ G = \Psi_{\mathrm{dom}\, G}.$$

THEOREM 3.203 (limiting chain rule for a convex function and a convex vector mapping). *Let X and Y be two normed spaces and Y_+ be a convex cone of Y. Assume that $G : X \to Y \cup \{+\infty_Y\}$ is a convex mapping with closed epigraph and that $g : Y \to \mathbb{R} \cup \{+\infty\}$ is a proper lower semicontinuous convex function which is nondecreasing over $G(\mathrm{dom}\, G) + Y_+$. Let $\bar{x} \in \mathrm{dom}\,(g \circ G)$.*

(a) *If X and Y are reflexive Banach spaces, then $x^* \in \partial(g \circ G)(\bar{x})$ if and only if there exist sequences $x_n \to \bar{x}$ in X, $y_n \to G(\bar{x})$ in Y, $x_n^* \to x^*$ in X^*, $e_n^* \to 0$ in Y^*, and $y_n^* \in Y_+^*$ such that*
$$y_n^* + e_n^* \in \partial g(y_n) \quad \text{and} \quad x_n^* \in \partial(y_n^* \circ G)(x_n)$$
and such that anyone of (i$_a$) and (ii$_a$) holds:

(i$_a$) $g(y_n) \to g(G(\bar{x}))$, $\langle y_n^*, y_n - G(\bar{x})\rangle \to 0$, $\langle y_n^*, G(x_n) - G(\bar{x})\rangle \to 0$, and $\|y_n^*\|\|y_n - y_n'\| \to 0$ for some $y_n' \in G(x_n) + Y_+$ with $y_n' \to G(\bar{x})$ and $\langle y_n^*, y_n'\rangle = \langle y_n^*, G(x_n)\rangle$;

(ii$_a$) $g(y_n) - \langle y_n^*, y_n - G(\bar{x})\rangle \to g(G(\bar{x}))$ and $\langle y_n^*, G(x_n) - G(\bar{x})\rangle \to 0$.

(b) *If X and Y are general Banach spaces, then $x^* \in \partial(g \circ G)(\bar{x})$ if and only if there exist nets $x_n \to \bar{x}$ in X, $y_n \to G(\bar{x})$ in Y, $x_n^* \xrightarrow{w^*} x^*$ in X^*, $e_n^* \xrightarrow{w^*} 0$ in Y^*, and $y_n^* \in Y_+^*$ such that*
$$y_n^* + e_n^* \in \partial g(y_n) \quad \text{and} \quad x_n^* \in \partial(y_n^* \circ G)(x_n)$$
and such that anyone of (i$_b$) and (ii$_b$) holds:

(i$_b$) $g(y_n) \to g(G(\bar{x}))$, $\langle y_n^* + e_n^*, y_n - G(\bar{x})\rangle \to 0$,
$\langle y_n^*, G(x_n) - G(\bar{x})\rangle - \langle x_n^*, x_n - \bar{x}\rangle \to 0$, $(\|y_n^*\| + \|e_n^*\|)\|y_n - y_n'\| \to 0$ for some $y_n' \in G(x_n) + Y_+$ with $y_n' \to G(\bar{x})$ and $\langle y_n^*, y_n'\rangle = \langle y_n^*, G(x_n)\rangle$;

(ii$_b$) $g(y_n) - \langle y_n^* + e_n^*, y_n - G(\bar{x})\rangle \to g(G(\bar{x}))$ and
$\langle y_n^*, G(x_n) - G(\bar{x})\rangle - \langle x_n^*, x_n - \bar{x}\rangle \to 0$.

(c) *If X is a general Banach space and Y is finite-dimensional, then $x^* \in \partial(g \circ G)(\bar{x})$ if and only if there exist nets $x_n \to \bar{x}$ in X, $y_n \to G(\bar{x})$ in Y, $x_n^* \xrightarrow{w^*} x^*$ in X^*, $e_n^* \to 0$ in Y^*, and $y_n^* \in Y_+^*$ such that*
$$y_n^* + e_n^* \in \partial g(y_n) \quad \text{and} \quad x_n^* \in \partial(y_n^* \circ G)(x_n)$$
and such that anyone of (i$_c$) and (ii$_c$) holds:

(i$_c$) $g(y_n) \to g(G(\bar{x}))$, $\langle y_n^*, y_n - G(\bar{x})\rangle \to 0$, $\langle y_n^*, G(x_n) - G(\bar{x})\rangle \to 0$, $\langle x_n^*, x_n - \bar{x}\rangle \to 0$, $\|y_n^*\|\|y_n - y_n'\| \to 0$ for some $y_n' \in G(x_n) + Y_+$ with $y_n' \to G(\bar{x})$ and $\langle y_n^*, y_n'\rangle = \langle y_n^*, G(x_n)\rangle$;

(ii$_c$) $g(y_n) - \langle y_n^*, y_n - G(\bar{x})\rangle \to g(G(\bar{x}))$, $\langle y_n^*, G(x_n) - G(\bar{x})\rangle \to 0$, and $\langle x_n^*, x_n - \bar{x}\rangle \to 0$.

PROOF. Set $\bar{y} := G(\bar{x})$ and, as usual, set also $g_1(x,y) := g(y)$ and $g_2(x,y) := \Psi_{\text{epi}\,G}(x,y)$ for all $x \in X$ and $y \in Y$. It is easily checked that $x^* \in \partial(g \circ G)(\bar{x})$ if and only if $(x^*, 0) \in \partial(g_1 + g_2)(\bar{x}, \bar{y})$. By Corollary 3.201, this amounts to saying, whenever both Banach spaces X and Y are reflexive (resp. either X or Y is nonreflexive), that there exist sequences (resp. nets)

$$(3.135) \qquad (0, y_n^* + e_n^*) + (x_n^*, -y_n^*) \to (x^*, 0) \quad \text{strongly (resp. weakly-*)}$$

with

$$y_n^* + e_n^* \in \partial g(y_n),\ (x_n^*, -y_n^*) \in \partial \Psi_{\text{epi}\,G}(x_n, y_n'),$$

$$(3.136) \qquad y_n \to \bar{y},\ (x_n, y_n') \to (\bar{x}, \bar{y}),$$

$$(3.137) \qquad g(y_n) \to g(\bar{y}),\ \langle y_n^* + e_n^*, y_n - \bar{y}\rangle \to 0,$$

$$(3.138) \qquad \langle y_n^*, y_n' - \bar{y}\rangle - \langle x_n^*, x_n - \bar{x}\rangle \to 0$$

$$(3.139) \qquad (\|y_n^* + e_n^*\| + \|x_n^*\| + \|y_n^*\|)\|y_n - y_n'\| \to 0.$$

We observe that (3.135) is equivalent to $x_n^* \to x^*$ and $e_n^* \to 0$ strongly (resp. weakly-*) and that the inclusion $(x_n^*, -y_n^*) \in \partial \Psi_{\text{epi}\,G}(x_n, y_n')$ means that

$$(3.140) \qquad \langle x_n^*, x - x_n\rangle - \langle y_n^*, y - y_n'\rangle \leq 0 \quad \text{for all } (x,y) \in \text{epi}\,G.$$

On one hand, for any $y' \in Y_+$, taking in the latter inequality $x = x_n$ and $y = y_n' + y'$ yields $\langle y_n^*, y'\rangle \geq 0$ thus $y_n^* \in Y_+^*$. On the other hand, taking $x = x_n$ and $y = G(x_n)$ in (3.140) we obtain $\langle y_n^*, y_n' - G(x_n)\rangle \leq 0$, and hence $\langle y_n^*, y_n' - G(x_n)\rangle = 0$ since $y_n^* \in Y_+^*$ and $y_n' - G(x_n) \in Y_+$. Consequently, (3.140) can be rewritten as

$$\langle x_n^*, x - x_n\rangle \leq \langle y_n^*, y\rangle - \langle y_n^*, G(x_n)\rangle \quad \text{for all } (x,y) \in \text{epi}\,G,$$

which, according to the inclusion $y_n^* \in Y_+^*$, is equivalent to

$$\langle x_n^*, x - x_n\rangle \leq y_n^* \circ G(x) - y_n^* \circ G(x_n) \quad \text{for all } x \in \text{dom}\,G.$$

The function $y_n^* \circ G$ being convex (since $y_n^* \in Y_+^*$), the inequality (3.140) is then equivalent to $x_n^* \in \partial(y_n^* \circ G)(x_n)$.

(a) Suppose first that X and Y are reflexive Banach spaces, so the properties above hold true for sequences $(x_n)_n$, $(x_n^*)_n$ etc. Since $x_n^* \overset{\|\ \|}{\to} x^*$, we observe that (3.138) is equivalent to $\langle y_n^*, y_n' - \bar{y}\rangle \to 0$, which is equivalent to $\langle y_n^*, G(x_n) - \bar{y}\rangle \to 0$ thanks to the equality $\langle y_n^*, y_n' - G(x_n)\rangle = 0$. Further, since $e_n^* \overset{\|\ \|}{\to} 0$, the second convergence in (3.137) is equivalent to $\langle y_n^*, y_n - \bar{y}\rangle \to 0$, and the convergence in (3.139) is equivalent to $\|y_n^*\|\|y_n - y_n'\| \to 0$. So, the conditions in (a) with (i$_a$) hold true and (i$_a$) obviously entails (ii$_a$).

On the other hand, taking sequences as given by (a) with (ii$_a$), we have, for each $x \in \text{dom}\,G$,

$$\langle x_n^*, x - x_n\rangle$$
$$\leq y_n^*(G(x)) - y_n^*(G(x_n))$$
$$= \langle y_n^* + e_n^*, G(x) - y_n\rangle - \langle e_n^*, G(x) - y_n\rangle + \langle y_n^*, y_n - \bar{y}\rangle - \langle y_n^*, G(x_n) - \bar{y}\rangle$$
$$\leq g(G(x)) - \{g(y_n) - \langle y_n^*, y_n - \bar{y}\rangle\} - \langle e_n^*, G(x) - y_n\rangle - \langle y_n^*, G(x_n) - \bar{y}\rangle,$$

and passing to the limit gives $\langle x^*, x - \bar{x}\rangle \leq g(G(x)) - g(G(\bar{x}))$, and hence $x^* \in \partial(g \circ G)(\bar{x})$. The equivalences in the assertion (a) are then established.

(b) Suppose that X and Y are general Banach spaces. Consider the nets obtained in the part preceding the arguments of (a). From (3.139) we deduce $(\|y_n^*\| + \|e_n^*\|)\|y_n' - y_n\| \to 0$. Further, since $\langle y_n^*, y_n' - G(x_n)\rangle = 0$, it results from (3.138) that $\langle y_n^*, G(x_n) - G(\bar{x})\rangle - \langle x_n^*, x_n - \bar{x}\rangle \to 0$. All the properties in (b) with (i_b) are then justified and the ones concerning (ii_b) readily follow.

Now let us show that (b) with (ii_b) implies $x^* \in \partial(g \circ G)(\bar{x})$. Considering the nets as given by (b) with (ii_b), we have, for each $x \in \mathrm{dom}\, G$,

$$\langle x_n^*, x - \bar{x}\rangle = \langle x_n^*, x - x_n\rangle + \langle x_n^*, x_n - \bar{x}\rangle$$
$$\leq y_n^* \circ G(x) - y_n^* \circ G(x_n) + \langle x_n^*, x_n - \bar{x}\rangle$$
$$= \langle y_n^* + e_n^*, G(x) - y_n\rangle - \langle e_n^*, G(x) - \bar{y}\rangle + \langle y_n^* + e_n^*, y_n - \bar{y}\rangle$$
$$- \langle y_n^*, G(x_n) - \bar{y}\rangle + \langle x_n^*, x_n - \bar{x}\rangle$$

which gives

$$\langle x_n^*, x - \bar{x}\rangle \leq g(G(x)) - \{g(y_n) - \langle y_n^* + e_n^*, y_n - \bar{y}\rangle\} - \langle e_n^*, G(x) - \bar{y}\rangle$$
$$- \{\langle y_n^*, G(x_n) - \bar{y}\rangle - \langle x_n^*, x_n - \bar{x}\rangle\},$$

and hence $\langle x^*, x - \bar{x}\rangle \leq g(G(x)) - g(G(\bar{x}))$. Therefore, $x^* \in \partial(g \circ G)(\bar{x})$, so the equivalences in the assertion (b) are then proved.

(c) Consider the nets given by (b) with (i_b). From the finite-dimensional property of Y the net $(e_n^*)_n$ strongly converges to zero. The convergence $\langle y_n^* + e_n^*, y_n - \bar{y}\rangle \to 0$ from (i_b) then ensures $\langle y_n^*, y_n - \bar{y}\rangle \to 0$. Combining this with the convergence $\|y_n^*\|\|y_n' - y_n\| \to 0$ and with the equality $\langle y_n^*, y_n'\rangle = \langle y_n^*, G(x_n)\rangle$ from (i_b), we obtain $\langle y_n^*, G(x_n) - \bar{y}\rangle \to 0$. It results from this and from the convergence $\langle y_n^*, G(x_n) - \bar{y}\rangle - \langle x_n^*, x_n - \bar{x}\rangle \to 0$ that $\langle x_n^*, x_n - \bar{x}\rangle \to 0$. All the properties in (i_c) are then deduced from the ones in (i_b), and evidently (ii_c) follows from (i_c).

Finally, as in (a) above, one also shows that (c) with (ii_c) entails that $x^* \in \partial(g \circ G)(\bar{x})$. $\square$

3.17. Subdifferential determination and maximal monotonicity for convex functions on Banach spaces

This section is devoted to the extension to Banach spaces of the subdifferential determination and maximal monotonicity property established for one real variable convex functions in Proposition 3.49 and Proposition 3.52 respectively.

THEOREM 3.204 (J.J. Moreau; R.T. Rockafellar: subdifferential determination of convex functions). Let X be a Banach space and $f, g : X \to \mathbb{R} \cup \{+\infty\}$ be two proper lower semicontinuous convex functions. The following are equivalent:

(a) $\partial f(x) \subset \partial g(x)$ holds for all $x \in X$;
(b) there exists a real constant C such that

$$f(x) = g(x) + C \quad \text{for all } x \in X.$$

PROOF. Only the implication (a)$\Rightarrow$(b) needs to be proved. Fix a point $a \in \mathrm{Dom}\,\partial f$ (recall by Proposition 2.223 that $\mathrm{Dom}\,\partial f \neq \emptyset$), so $f(a)$ and $g(a)$ are finite. We may suppose that $a = 0$. For each $b \in X$ consider the continuous linear mapping $A_b : \mathbb{R} \to X$ defined by $A_b(t) = tb$ for all $t \in \mathbb{R}$. Note that the lower semicontinuous convex functions $f \circ A_b$ and $g \circ A_b$ are finite at 0, and hence proper.

We begin by showing that $\partial(f \circ A_b)(t) \subset \partial(g \circ A_b)(t)$ for all $t \in \mathbb{R}$. Fix any (t, t^*) in $\mathrm{gph}\,\partial(f \circ A_b)$. By (iv) in Corollary 3.200 we have $t^* = \lim x_j^* \circ A_b$ with

$$x_j^* \in \partial f(x_j), \quad \langle x_j^*, x_j - A_b t \rangle \to 0 \quad \text{and} \quad x_j \to A_b t = tb.$$

The inclusion assumption entails that $x_j^* \in \partial g(x_j)$, and hence by (iv) in Corollary 3.200 once again $t^* \in \partial(g \circ A_b)(t)$. Therefore, the desired inclusion $\partial(f \circ A_b)(t) \subset \partial(g \circ A_b)(t)$ holds for all $t \in \mathbb{R}$.

By Proposition 3.49 we obtain $f(0) + g(b) = g(0) + f(b)$. This being true for every $b \in X$, putting $C := f(0) - g(0)$ we deduce with $b = x$ that $f(x) = g(x) + C$ for all $x \in X$, which finishes the proof. $\square$

REMARK 3.205. Example 3.144 provides a Fréchet space X and a proper lower semicontinuous convex function $f : X \to \mathbb{R} \cup \{+\infty\}$ such that $\partial f(x) = \emptyset$ for all $x \in X$. This ensures that Theorem 3.204 fails in the setting of Fréchet spaces. $\square$

We demonstrate next the maximal monotonicity of subdifferential of proper lower semicontinuous convex functions on Banach spaces with two different proofs. The first one uses the limiting chain rule with some arguments similar to the proof of Theorem 3.204 above.

THEOREM 3.206 (maximal monotonicity of subdifferential of a convex function). Let X be a Banach space and $f : X \to \mathbb{R} \cup \{+\infty\}$ be a proper lower semicontinuous convex function. Then the subdifferential multimapping $\partial f : X \rightrightarrows X^*$ is maximal monotone.

PROOF. Suppose that there exists $(a, a^*) \in X \times X^*$ with $a^* \notin \partial f(a)$ and $\langle x^* - a^*, x - a \rangle \geq 0$ for all $(x, x^*) \in \mathrm{gph}\,\partial f$. We may suppose that $a = 0$ and $a^* = 0$. Since $0 \notin \partial f(0)$, there exists $b \in X$ with $f(b) < f(0)$. Define $A : \mathbb{R} \to X$ by $A(t) = tb$ and $f_0 : \mathbb{R} \to \mathbb{R} \cup \{+\infty\}$ by $f_0(t) = f(At)$ for all $t \in \mathbb{R}$. Then $f_0(1) < f_0(0)$, hence $0 \notin \partial f_0(0)$. Fix any $(t, t^*) \in \mathrm{gph}\,\partial f_0$. By (iv) in Corollary 3.200 there exist nets $x_j \to At = tb$, $x_j^* \in \partial f(x_j)$ with $x_j^* \circ A \to t^*$ and $\langle x_j^*, x_j - At \rangle \to 0$, hence $\langle x_j^*, x_j \rangle \to \langle t^*, t \rangle$. Since by assumption $\langle x^*, x \rangle \geq 0$ for all $(x, x^*) \in \mathrm{gph}\,\partial f$, it ensues that $\langle x_j^*, x_j \rangle \geq 0$, thus $t^* t = \langle t^*, t \rangle \geq 0$. This contradicts the maximal monotonicity of ∂f_0 due to Proposition 3.52, hence it follows that ∂f is maximal monotone. $\square$

We can also provide another proof based on the use of the basic kernel function $(1/2)\|\cdot\|^2$ and the Brønsted-Rockafellar theorem for ε-subdifferential.

SECOND PROOF OF THEOREM 3.206. Let $(a, a^*) \in X \times X^*$ with $\langle x^* - a^*, x - a \rangle \geq 0$ for all $(x, x^*) \in \mathrm{gph}\,\partial f$. Define $\varphi := g + k$, where $g := f(\cdot + a)$ and $k := (1/2)\|\cdot\|^2$. By the Moreau-Rockafellar condition of continuity of k (see Corollary 3.170) we have $\partial \varphi(x) = \partial f(x + a) + \partial k(x)$ for all $x \in X$. Since g is a proper lower semicontinuous convex function, we know that g^* is proper (see Proposition 3.77(c)), hence there is some $b^* \in X^*$ such that $g^*(b^*)$ is finite. We also note that for any $x \in X$

$$\langle a^*, x \rangle - \varphi(x) = \langle a^*, x \rangle - g(x) - k(x)$$
$$\leq \langle a^*, x \rangle - \langle b^*, x \rangle + g^*(b^*) - k(x) = g^*(b^*) + \langle a^* - b^*, x \rangle - k(x)$$
$$\leq g^*(b^*) + k^*(a^* - b^*) = g^*(b^*) + \frac{1}{2}\|a^* - b^*\|_*^2,$$

where the latter equality is due to (3.39). Fixing any $n \in \mathbb{N}$, it ensues that $\varphi^*(a^*)$ is finite according to the finiteness of $g^*(b^*)$, hence there is some $a_n \in X$ such that

$$(3.141) \qquad \varphi^*(a^*) - \frac{1}{n^2} \leq \langle a^*, a_n \rangle - \varphi(a_n), \quad \text{that is, } a^* \in \partial_{1/n^2}\varphi(a_n),$$

where the latter inclusion is given by Proposition 3.64(j). The Brønsted-Rockafellar theorem (see Theorem 3.142) furnishes $(u_n, u_n^*) \in \operatorname{gph} \partial \varphi$ such that

$$(3.142) \qquad \|u_n - a_n\| \leq 1/n \quad \text{and} \quad \|u_n^* - a^*\|_* \leq 1/n.$$

Since $\partial \varphi(u_n) = \partial f(u_n + a) + \partial k(u_n)$, there is $v_n^* \in \partial k(u_n)$ such that $u_n^* - v_n^* \in \partial f(u_n + a)$. This entails that

$$\langle u_n^* - v_n^* - a^*, u_n + a - a \rangle \geq 0, \quad \text{that is, } \langle v_n^*, u_n \rangle \leq \langle u_n^* - a^*, u_n \rangle.$$

Using this and the equality $\langle v_n^*, u_n \rangle = \|u_n\|^2$ (due to the inclusion $v_n^* \in \partial k(u_n)$) (see (2.25)) we see from the second inequality in (3.142) that $\|u_n\|^2 \leq \|u_n\|/n$. Then $\|u_n\| \leq 1/n$, which combined with the first inequality in (3.142) gives $\|a_n\| \leq 2/n$, so $a_n \to 0$. Taking the lower semicontinuity of φ into account, it results from the inequality in (3.141) as $n \to \infty$ that $\varphi(0) + \varphi^*(a^*) \leq 0$, hence $a^* \in \partial \varphi(0)$ (see Proposition 3.64(j)). We derive that $a^* \in \partial f(a) + \partial k(0)$, so $a^* \in \partial f(a)$ since $\partial k(0) = \{0\}$ (see (2.25)). This confirms the maximal monotonicity of the multimapping ∂f. □

3.18. Normals to convex sublevel sets

C-normals to sublevel sets of functions (on a normed space) which are locally Lipschitz continuous at $\bar{x}$ have been studied in Proposition 2.196 and the case of convex functions continuous at $\bar{x}$ is a particular case (in the normed space setting). In this section we are interested in the situation where no continuity property is required for the convex functions. We begin with the locally convex framework under Slater qualification condition.

3.18.1. Normals to convex sublevels under Slater condition in locally convex spaces.

PROPOSITION 3.207. Let X be a Hausdorff locally convex space and $(f_i)_{i \in I}$ be a finite family of convex functions from X into $\mathbb{R} \cup \{+\infty\}$. Let $(r_i)_{i \in I}$ be a family in $\mathbb{R}$ and let $S := \{x \in X : f_i(x) \leq r_i, \forall i \in I\}$. Assume that the Slater qualification condition $\bigcap_{i \in I} \{f_i < r_i\} \neq \emptyset$ holds. Then for any $\bar{x} \in S$ one has

$$N(S; \bar{x}) = \bigcup_{\lambda \in \mathbb{R}_+^I} \partial \left(\sum_{i \in I(\bar{x}, \lambda)} \lambda_i f_i + \sum_{i \in I \setminus I(\bar{x}, \lambda)} \Psi_{\operatorname{dom} f_i} \right)(\bar{x}),$$

where $I(\bar{x}, \lambda) := \{i \in I(\bar{x}) : \lambda_i > 0\}$ with $I(\bar{x}) := \{i \in I : f_i(\bar{x}) = r_i\}$.

PROOF. We follow the main lines of the proof of Proposition 3.190. Let $Y := \mathbb{R}^I$ and $Y_+ := \mathbb{R}_+^I$. Consider $g := \Psi_{-Y_+}$ from Y into $\mathbb{R} \cup \{+\infty\}$ and define $G : X \to Y \cup \{+\infty_Y\}$ by $G(x) = (f_i(x) - r_i)_{i \in I}$ if $f_i(x)$ is finite for every $i \in I$ and $G(x) = +\infty_Y$ otherwise. It is clear that g and G are convex and that g is Y_+-nondecreasing on Y. We also notice with our convention (3.93) that $\Psi_S = g \circ G$. Choosing $a \in \bigcap_{i \in I} \{f_i < r_i\} \neq \emptyset$ we see that $G(a) \in \operatorname{int}(\operatorname{dom} g)$, so the condition (d) in Corollary 3.187 is satisfied. Further, putting $\bar{y} = G(\bar{x})$ it is clear that

$$\partial g(\bar{y}) = \{L_\lambda : \lambda \in \mathbb{R}_+^I, \lambda_i = 0 \text{ if } i \notin I(\bar{x}))\},$$

where the linear functional $L_\lambda : Y \to \mathbb{R}$ is defined by $L_\lambda(y) = \sum_{i \in I} \lambda_i y_i$ for all $y \in Y$. Observing with our convention (3.93) that $L_\lambda \circ G = \sum_{i \in I} \lambda_i f_i$ and $0 f_i = \operatorname{dom} f_i$, the equality in the proposition follows from Corollary 3.187 under condition (d). □

As a first corollary we directly have:

COROLLARY 3.208. *Let X be a Hausdorff locally convex space and $(f_i)_{i \in I}$ be a finite family of convex functions from X into $\mathbb{R} \cup \{+\infty\}$. Let $(r_i)_{i \in I}$ be a family in $\mathbb{R}$ and let $S := \{x \in X : f_i(x) \leq r_i, \forall i \in I\}$. Assume that the Slater qualification condition $\bigcap_{i \in I}\{f_i < r_i\} \neq \emptyset$ holds. If all the functions f_i are continuous at $\bar{x} \in S$, then one has*

$$N(S; \bar{x}) = \sum_{i \in I(\bar{x})} \mathbb{R}_+ \partial f_i(\bar{x}).$$

The next two corollaries are concerned with one function.

COROLLARY 3.209. *Let X be a Hausdorff locally convex space and $f : X \to \mathbb{R} \cup \{+\infty\}$ be a convex function. Let $r \in \mathbb{R}$ and let $S := \{x \in X : f(x) \leq r\}$. Assume that the Slater qualification condition $\{f < r\} \neq \emptyset$ holds. The for any $\bar{x} \in S$ one has*

$$N(S; \bar{x}) = \begin{cases} \mathbb{R}_+ \partial f(\bar{x}) \cup N(\operatorname{dom} f; \bar{x}) & \text{if } f(\bar{x}) = r \\ N(\operatorname{dom} f; \bar{x}) & \text{if } f(\bar{x}) < r. \end{cases}$$

PROOF. If $f(\bar{x}) < r$, then $I(\bar{x}) = \emptyset$, so the result is clearly a consequence of Proposition 3.207 above. Suppose now that $f(\bar{x}) = r$, that is, $I(\bar{x})$ is nonempty and singleton. Proposition 3.207 again tells us that $N(S; \bar{x}) = \bigcup_{\lambda \in \mathbb{R}_+} \partial(\lambda f)(\bar{x})$. Recalling that $0f = \Psi_{\operatorname{dom} f}$ and $\partial(\lambda f)(\bar{x}) = \lambda \partial f(\bar{x})$ for every real $\lambda > 0$, the equality of the corollary follows. □

Assuming that f is continuous at $\bar{x}$ we directly obtain:

COROLLARY 3.210. *Let X be a Hausdorff locally convex space and $f : X \to \mathbb{R}$ be a convex function which is continuous at $\bar{x} \in \operatorname{dom} f$. Assume that the Slater condition $\{f < f(\bar{x})\} \neq \emptyset$ holds. Then for $S := \{f \leq f(\bar{x})\}$ one has*

$$N(S; \bar{x}) = \mathbb{R}_+ \partial f(\bar{x}).$$

The next proposition closes this subsection with vector convex sublevels in the form $\{x \in X : F(x) \leq_{Y_+} \bar{y}\}$, where Y_+ is a convex cone Y_+ of a Hausdorff locally convex space and $F : X \to Y \cup \{+\infty_Y\}$ is a convex mapping from a Hausdorff locally convex space X into $Y \cup \{+\infty_Y\}$. In its statement as seen in (3.94))

$$0_{Y^*} \circ G = \Psi_{\operatorname{dom} G}$$

for any mapping $G : X \to Y \cup \{+\infty_Y\}$.

PROPOSITION 3.211. *Let X and Y be Hausdorff locally convex spaces and Y_+ be a convex cone in Y containing zero. Let $F : X \to Y \cup \{+\infty_Y\}$ be a Y_+-convex mapping, $\bar{y} \in Y$ and $\bar{x} \in X$ with $F(\bar{x}) \leq_{Y_+} \bar{y}$. Then for the sublevel set $S := \{x \in X : F(x) \leq_{Y_+} \bar{y}\}$ one has the equality*

$$N(S; \bar{x}) = \bigcup_{y^* \in Y_+^*, \langle y^*, F(\bar{x}) - \bar{y} \rangle = 0} \partial(y^* \circ F)(\bar{x})$$

under anyone of conditions (a), (b), (c) and (d) below:
(a) With $Z := \text{span}\,[Y_+ + F(\text{dom}\, G) - \bar{y}]$ endowed with the topology induced by the Mackey topology of Y, for each $w(X, X^*)$-neighborhood V of zero in X there is some real $\rho > 0$ such that
$$0 \in \text{int}_Z\,[Y_+ + F(\text{dom}\, F \cap \rho V) - \bar{y}];$$
(b) the Slater qualification condition is satisfied, that is, there some $x_0 \in \text{dom}\, F$ such that the inclusion $F(x_0) \in \bar{y} - \text{int}_Y Y_+$ is satisfied, where the interior $\text{int}_Y Y_+$ is taken with respect to the Mackey topology of Y;
(c) the space Y is a normed space and, for $Z := \text{span}\,[Y_+ + F(\text{dom}\, F) - \bar{y}]$ endowed with the norm induced by the one of Y, there exists some real number $\rho > 0$ such that
$$0 \in \text{int}_Z\,[Y_+ + F(\text{dom}\, F \cap \rho \mathbb{B}_X) - \bar{y}];$$
(d) the space Y is a finite-dimensional normed space and
$$0 \in \text{rint}\,[Y_+ + F(\text{dom}\, F) - \bar{y}].$$

PROOF. Put $g := \Psi_{-Y_+}$ and $G := F - \bar{y}$. We note that $\Psi_S = g \circ G$, so $N(S; \bar{x}) = \partial(g \circ G)(\bar{x})$. By Example 3.153 the function g is Y_+-nondecreasing and by Proposition 2.65 we have $\partial g(G(\bar{x})) = \{y^* \in Y_+^* : \langle y^*, G(\bar{x})\rangle = 0\}$. We also notice that the Slater condition $F(x_0) \in \bar{y} - \text{int}_Y Y_+$ assures us that $G(x_0) \in -\text{int}_Y Y_+$, thus $g = \Psi_{-Y_+}$ is finite and continuous at $G(x_0) \in Y$ with respect to the Mackey topology. The proposition then follows from Corollary 3.153. □

3.18.2. Horizon subgradients of convex functions.
In Proposition 3.207 above and its corollary the Slater qualification condition has been assumed. When no qualification condition is required, certain general descriptions of the normal cone of the sublevel set of a convex function will involve the horizon subdifferential of the convex function.

DEFINITION 3.212. Given a locally convex space X and a convex function $f : X \to \mathbb{R} \cup \{+\infty\}$, the *horizon subdifferential* (or *singular subdifferential*) of f at $\bar{x} \in \text{dom}\, f$ is defined by
$$\partial^\infty f(\bar{x}) := \{x^* \in X^* : (x^*, 0) \in N(\text{epi}\, f; (\bar{x}, f(\bar{x})))\}.$$
The elements of $\partial^\infty f(x)$ are called *horizon* (or *singular*) *subgradients* of f at $\bar{x}$. When f is not finite at $\bar{x}$ one defines $\partial^\infty f(\bar{x}) = \emptyset$.

If X is a normed space, it is obvious that $\partial^\infty f(\bar{x}) = \partial_C^\infty f(\bar{x})$ for any convex function f. Horizon subgradients of convex functions enjoy in Banach spaces another property even more remarkable. Indeed, when X is a Banach space and the convex function f is lower semicontinuous, any horizon subgradient can be recovered from true subgradients at nearby points. This is confirmed by one of the results of the next proposition which shows in such a framework that
$$\partial^\infty f(\bar{x}) = {}^{\text{seq}} \operatorname*{Lim\,sup}_{x \to_f \bar{x},\, \mu \downarrow 0} \mu\, \partial f(x),$$
that is, $x^* \in \partial^\infty f(\bar{x})$ if and only if there exist sequences $(x_n)_n$ in X, $(x_n^*)_n$ in X^*, $(\mu_n)_n$ in $]0, +\infty[$ such that $\mu_n \downarrow 0$, $x_n \to \bar{x}$, $f(x_n) \to f(\bar{x})$, $x_n^* \in \partial f(x_n)$ and $\mu_n x_n^* \xrightarrow{w^*} x^*$. Given a multimapping $M : T \rightrightarrows X^*$ from a metric space T into the topological dual space X^*, we recall (see Section 1.2) that ${}^{\text{seq}} \operatorname*{Lim\,sup}_{T_0 \ni t \to \bar{t}} M(t)$

denotes the *sequential* Lim sup with X^* endowed with the w^*-topology whereas $\underset{T_0 \ni t \to \bar{t}}{\text{Lim sup}} M(t)$ stands for the usual Lim sup with X^* endowed with the topology associated with its dual norm. With these notations we have:

THEOREM 3.213 (limiting descriptions of horizon subgradients of convex functions via true subgradients). *Let X be a Banach space and let $f : X \to \mathbb{R} \cup \{+\infty\}$ be a proper lower semicontinuous convex function. Then one has, for every $\bar{x} \in \operatorname{dom} f$,*

$$\partial^\infty f(\bar{x}) = N(\operatorname{dom} f; \bar{x}) = {}^{\text{seq}} \underset{x \overset{f}{\to} \bar{x}, \mu \downarrow 0}{\text{Lim sup}} \mu \, \partial f(x)$$

$$= \underset{x \overset{f}{\to} \bar{x}, \mu \downarrow 0}{\text{Lim sup}} \mu \, \partial f(x) = \underset{x \to \bar{x}, \mu \downarrow 0}{\text{Lim sup}} \mu \, \partial f(x) = {}^{\text{seq}} \underset{x \to \bar{x}, \mu \downarrow 0}{\text{Lim sup}} \mu \, \partial f(x).$$

Since $0 \in \partial^\infty f(\bar{x})$, the same equalities also hold with $0 \le \mu \to 0$ in place of $\mu \downarrow 0$.

PROOF. Consider any $\bar{x} \in \operatorname{dom} f$. The inclusions of the fourth member into the fifth and the fifth into the sixth are obvious. Let us show that the sixth member is included into the second, that is,

(3.143) $$\qquad {}^{\text{seq}} \underset{x \to \bar{x}, \mu \downarrow 0}{\text{Lim sup}} \mu \, \partial f(x) \subset N(\operatorname{dom} f; \bar{x}).$$

Let $x^* \in {}^{\text{seq}} \underset{x \to \bar{x}, \mu \downarrow 0}{\text{Lim sup}} \mu \, \partial f(x)$. By definition, there exist $(x_n)_n$ in X, $(x_n^*)_n$ in X^* and $(\mu_n) \downarrow 0$ such that $x_n \to \bar{x}, * x_n^* \in \partial f(x_n)$ and $\mu_n x_n^* \overset{w^*}{\to} x^*$. Take a real $\varepsilon > 0$ and by the lower semicontinuity of f choose some integer n_0 such that $f(\bar{x}) - \varepsilon \le f(x_n)$ for all $n \ge n_0$. Fixing any $x \in \operatorname{dom} f$, the subdifferential inequality implies that, for every $n \ge n_0$,

(3.144) $$\langle \mu_n x_n^*, x - x_n \rangle \le \mu_n \left(f(x) - f(x_n) \right) \le \mu_n (f(x) - f(\bar{x}) + \varepsilon).$$

Passing to the limit as $n \to +\infty$ in both extreme members of (3.144), we obtain $\langle x^*, x - \bar{x} \rangle \le 0$ because $x_n \to \bar{x}$, $\mu_n x_n^* \overset{w^*}{\to} x^*$ and $\mu_n \to 0$ as $n \to +\infty$. Since this is true for every $x \in \operatorname{dom} f$, we derive that $x^* \in N(\operatorname{dom} f; \bar{x})$, which shows the inclusion (3.143).

To prove the inclusion of the second member into the fourth, fix any $x^* \in N(\operatorname{dom} f; \bar{x})$. For each real $\eta > 0$, we know from the lower semicontinuity of the convex function f (see Proposition 3.94) that $\partial_{\eta^2} f(\bar{x}) \ne \emptyset$, so we choose some $y_\eta^* \in \partial_{\eta^2} f(\bar{x})$. Putting $\mu(\eta) := \eta/(1 + \|y_\eta^*\|)$, it is clear from the definitions of ε-subdifferential and normal cone that $z_\eta^* := y_\eta^* + \mu(\eta)^{-1} x^* \in \partial_{\eta^2} f(\bar{x})$. Since X is a Banach space, the Borwein version of Brønsted-Rockafellar theorem (see Theorem 3.143) yields some $x_\eta \in X$ and $x_\eta^* \in \partial f(x_\eta)$ such that

$$\|x_\eta - \bar{x}\| \le \eta, \ |f(x_\eta) - f(\bar{x})| \le \eta(\eta + 1), \ \|x_\eta^* - z_\eta^*\| \le \eta(1 + \|z_\eta^*\|).$$

Observing that the latter inequality entails

$$\|\mu(\eta) x_\eta^* - \mu(\eta) y_\eta^* - x^*\| \le \eta \bigl(\mu(\eta) + \mu(\eta) \|y_\eta^*\| + \|x^*\| \bigr)$$

and noting that $\mu(\eta) \|y_\eta^*\| \le \eta \to 0$ as $\eta \downarrow 0$, we see that $\mu(\eta) x_\eta^* \to x^*$ with $\mu(\eta) \downarrow 0$ as $\eta \downarrow 0$. This justifies the desired inclusion $N(\operatorname{dom} f; \bar{x}) \subset \underset{x \overset{f}{\to} \bar{x}, \mu \downarrow 0}{\text{Lim sup}} \mu \, \partial f(x)$. So, we have established the equality between the second, fourth, fifth and sixth members. Further, the fourth member is obviously included into the third and the third is also obviously included into the sixth, so we can add the third member into the preceding list of equalities.

Finally, the equality $N(\mathrm{dom}\, f; \bar{x}) = \{x^* \in X^* : (x^*, 0) \in N(\mathrm{epi}\, f; (\bar{x}, f(\bar{x})))\}$ follows easily from the definition of normal cone to a convex set, so the proof of the theorem is finished. $\square$

3.18.3. Limiting formulas for normals to convex sublevels: Reflexive Banach space case.
Given a lower semicontinuous convex function f, we establish various limiting descriptions of the normal cone at $\bar{x} \in S$ of the sublevel set

$$S = \{f \le r\} := \{x \in X : f(x) \le r\}.$$

The formulas make use of no qualification condition. We start with the framework of reflexive Banach spaces. To avoid risk of confusion recall that our notation $\mu \downarrow 0$ means $\mu \to 0$ with $\mu > 0$.

THEOREM 3.214 (A. Cabot and L. Thibault). Let X be a reflexive Banach space and let $f : X \to \mathbb{R} \cup \{+\infty\}$ be a proper lower semicontinuous convex function. Let $r \in \mathbb{R}$ be such that the sublevel set $S = \{f \le r\}$ is nonempty. For any $\bar{x} \in S$ one has

$$(3.145) \quad N(S; \bar{x}) = \operatorname*{Lim\,sup}_{\substack{x \to \bar{x},\, \mu \ge 0 \\ \mu(f(x) - r) \to 0}} \partial(\mu f)(x)$$

$$(3.146) \quad = \operatorname*{Lim\,sup}_{\substack{x \to \bar{x},\, \mu \ge 0 \\ \mu(f(x) - r) \to 0}} \mu \partial f(x)$$

$$(3.147) \quad = \begin{cases} \operatorname*{Lim\,sup}_{x \to \bar{x}} \mathbb{R}_+ \partial f(x) & \text{if } f(\bar{x}) = r \\ \operatorname*{Lim\,sup}_{x \to \bar{x},\, \mu \downarrow 0} \mu \partial f(x) = N(\mathrm{dom}\, f; \bar{x}) & \text{if } f(\bar{x}) < r. \end{cases}$$

(Concerning the second member of the first equality we recall that $0f = \Psi_{\mathrm{dom}\, f}$.)

PROOF. Let us start with the inclusion

$$(3.148) \quad N(S; \bar{x}) \subset \operatorname*{Lim\,sup}_{\substack{x \to \bar{x},\, \mu \ge 0 \\ \mu(f(x) - r) \to 0}} \partial(\mu f)(x).$$

Define the lower semicontinuous convex function $G : X \to \mathbb{R} \cup \{+\infty\}$ by $G(x) = f(x) - r$. Observe that $S = \{x \in X : G(x) \in -\mathbb{R}_+\}$ and hence $\Psi_S = \Psi_{-\mathbb{R}_+} \circ G$. This implies that

$$N(S; \bar{x}) = \partial \Psi_S(\bar{x}) = \partial \left(\Psi_{-\mathbb{R}_+} \circ G \right)(\bar{x}).$$

The function $\Psi_{-\mathbb{R}_+}$ is lower semicontinuous, convex and nondecreasing. Let us fix $x^* \in N(S; \bar{x})$ and apply Theorem 3.203(a) with $Y = \mathbb{R}$, $Y_+ = \mathbb{R}_+$, $g = \Psi_{-\mathbb{R}_+}$ and the function G defined above. We obtain the existence of sequences $(x_n)_n$ in X, $(y_n)_n$ in $\mathbb{R}$, $(x_n^*)_n$ in X^*, $(e_n^*)_n$ in $\mathbb{R}$ and $(y_n^*)_n$ in $\mathbb{R}_+$ such that $x_n \to \bar{x}$, $y_n \to f(\bar{x}) - r$, $x_n^* \to x^*$, $e_n^* \to 0$ and

$$(3.149) \quad y_n^* + e_n^* \in N(-\mathbb{R}_+; y_n),$$

$$(3.150) \quad x_n^* \in \partial(y_n^* f)(x_n)$$

$$(3.151) \quad y_n^* (f(x_n) - f(\bar{x})) \to 0 \quad \text{as } n \to +\infty.$$

Let us suppose that $f(\bar{x}) < r$. Since $y_n \to f(\bar{x}) - r$ as $n \to +\infty$, we have $y_n < 0$ for n large enough. It ensues that $N(-\mathbb{R}_+; y_n) = \{0\}$ and the inclusion (3.149) then implies that

(3.152) $$y_n^* = -e_n^* \to 0 \quad \text{as } n \to +\infty, \quad \text{if } f(\bar{x}) < r.$$

Now observe that
$$y_n^*(f(x_n) - r) = y_n^*(f(x_n) - f(\bar{x})) + y_n^*(f(\bar{x}) - r).$$

By (3.151) and (3.152), we immediately deduce that

(3.153) $$y_n^*(f(x_n) - r) \to 0 \quad \text{as } n \to +\infty.$$

If $f(\bar{x}) = r$, the property (3.153) still holds true as a direct consequence of (3.151). Finally, we have built sequences $x_n \to \bar{x}$, $x_n^* \to x^*$, $(y_n^*)_n$ in $\mathbb{R}_+$ satisfying (3.150) and (3.153), which clearly shows that

$$x^* \in \operatorname*{Lim\,sup}_{\substack{x \to \bar{x},\, \mu \geq 0 \\ \mu(f(x) - r) \to 0}} \partial(\mu f)(x).$$

The inclusion (3.148) is proved.

Let us now establish the inclusion

(3.154) $$\operatorname*{Lim\,sup}_{\substack{x \to \bar{x},\, \mu \geq 0 \\ \mu(f(x) - r) \to 0}} \partial(\mu f)(x) \subset \operatorname*{Lim\,sup}_{\substack{x \to \bar{x},\, \mu \geq 0 \\ \mu(f(x) - r) \to 0}} \mu \partial f(x).$$

Let $x^* \in \operatorname*{Lim\,sup}_{\substack{x \to \bar{x},\, \mu \geq 0 \\ \mu(f(x) - r) \to 0}} \partial(\mu f)(x)$. By definition, there exist sequences $x_n \to \bar{x}$, $x_n^* \to x^*$, $(\mu_n)_n$ in $\mathbb{R}_+$ satisfying $x_n^* \in \partial(\mu_n f)(x_n)$ and $\mu_n(f(x_n) - r) \to 0$. Suppose first that there exists a subsequence of $(\mu_n)_n$, still denoted by $(\mu_n)_n$, such that $\mu_n > 0$ for every $n \in \mathbb{N}$. We then have $x_n^* \in \mu_n \partial f(x_n)$, and hence

(3.155) $$x^* \in \operatorname*{Lim\,sup}_{\substack{x \to \bar{x},\, \mu \geq 0 \\ \mu(f(x) - r) \to 0}} \mu \partial f(x).$$

Now suppose that $\mu_n = 0$ for n large enough. Recalling that $0f = \Psi_{\operatorname{dom} f}$, we obtain $x_n^* \in N(\operatorname{dom} f; x_n)$ for n large enough, say $n \geq n_0$. Since the multimapping $N(\operatorname{dom} f; \cdot)$ is outer semicontinuous at $\bar{x}$ with respect to the norms in X and X^* (see Corollary 3.34), we deduce at the limit as $n \to +\infty$ that $x^* \in N(\operatorname{dom} f; \bar{x})$. By Theorem 3.213, we have $N(\operatorname{dom} f; \bar{x}) = \operatorname*{Lim\,sup}_{\substack{x \to \bar{x},\, \mu \downarrow 0 \\ f}} \mu \partial f(x)$, hence

$$x^* \in \operatorname*{Lim\,sup}_{\substack{x \to \bar{x},\, \mu \downarrow 0 \\ f(x) \to f(\bar{x})}} \mu \partial f(x) \subset \operatorname*{Lim\,sup}_{\substack{x \to \bar{x},\, \mu \geq 0 \\ \mu(f(x) - r) \to 0}} \mu \partial f(x).$$

Therefore, the inclusion (3.155) is satisfied in both cases, which proves (3.154).

Let us now show the inclusion

(3.156) $$\operatorname*{Lim\,sup}_{\substack{x \to \bar{x},\, \mu \geq 0 \\ \mu(f(x) - r) \to 0}} \mu \partial f(x) \subset N(S; \bar{x}).$$

Let $x^* \in \operatorname*{Lim\,sup}_{\substack{x \to \bar{x},\, \mu \geq 0 \\ \mu(f(x) - r) \to 0}} \mu \partial f(x)$. By definition, there exist sequences $(x_n)_n$ in X, $(x_n^*)_n$ in X^* and $(\mu_n)_n$ in $\mathbb{R}_+$ such that $x_n^* \in \partial f(x_n)$, $x_n \to \bar{x}$, $\mu_n x_n^* \to x^*$ and

$\mu_n(f(x_n) - r) \to 0$ as $n \to +\infty$. Let us fix any $x \in S$. The subdifferential inequality gives
$$\langle x_n^*, x - x_n \rangle \le f(x) - f(x_n) \le r - f(x_n).$$
Multiplying by μ_n, we find $\langle \mu_n x_n^*, x - x_n \rangle \le \mu_n(r - f(x_n))$. Taking the limit as $n \to +\infty$, we obtain that $\langle x^*, x - \bar{x} \rangle \le 0$ because $x_n \to \bar{x}$, $\mu_n x_n^* \to x^*$ and $\mu_n(r - f(x_n)) \to 0$ as $n \to +\infty$. Since this is true for every $x \in S$, we conclude that $x^* \in N(S; \bar{x})$, which justifies the inclusion (3.156). By combining inclusions (3.148), (3.154) and (3.156), we obtain formulas (3.145) and (3.146).

To prove the formula (3.147), let us establish the following lemma.

LEMMA 3.215. Assume that there exist sequences $(x_n)_n$ in X, $(x_n^*)_n$ in X^* and $(\mu_n)_n$ in $\mathbb{R}_+$ such that $x_n^* \in \partial f(x_n)$, $x_n \to \bar{x}$ and $\mu_n x_n^* \to x^*$ as $n \to +\infty$. Then we have
$$\lim_{n \to +\infty} \mu_n (f(x_n) - f(\bar{x})) = 0.$$

PROOF OF LEMMA 3.215. Since $x_n^* \in \partial f(x_n)$, the subdifferential inequality gives
$$f(x_n) - f(\bar{x}) \le \langle x_n^*, x_n - \bar{x} \rangle.$$
Multiplying by μ_n and taking the upper limit as $n \to +\infty$, we obtain that
$$\limsup_{n \to +\infty} \mu_n (f(x_n) - f(\bar{x})) \le 0, \tag{3.157}$$
because $x_n \to \bar{x}$ and $\mu_n x_n^* \to x^*$. To complete the proof of the lemma, it remains to show that
$$\liminf_{n \to +\infty} \mu_n (f(x_n) - f(\bar{x})) \ge 0. \tag{3.158}$$
First suppose that the sequence $(\mu_n)_n$ is bounded from above, say by $M > 0$. We then have
$$\mu_n (f(x_n) - f(\bar{x})) \ge -M (f(x_n) - f(\bar{x}))^-,$$
where we recall that $\alpha^- := \max\{0, -\alpha\}$ denotes the negative part of the extended real α. The lower semicontinuity of f implies that $\liminf_{n \to +\infty} f(x_n) \ge f(\bar{x})$, hence $\lim_{n \to +\infty} (f(x_n) - f(\bar{x}))^- = 0$. Taking the lower limit as $n \to +\infty$ in the above inequality, we obtain (3.158).

Now suppose that the sequence $(\mu_n)_n$ is not bounded from above. There exists a subsequence of $(\mu_n)_n$, that we do not relabel, such that $\lim_{n \to +\infty} \mu_n = +\infty$. Observe that $x_n^* = (\mu_n x_n^*)/\mu_n \to 0$ because $\mu_n x_n^* \to x^*$ as $n \to +\infty$. Since $x_n \to \bar{x}$ as $n \to +\infty$ and $x_n^* \in \partial f(x_n)$, we deduce that $0 \in \partial f(\bar{x})$, due to the $\|\cdot\| \times \|\cdot\|_*$ closedness of the graph of the multimapping ∂f (see Proposition 3.32). Therefore $\bar{x}$ is a minimizer of the convex function f over X. This implies that $f(x_n) \ge f(\bar{x})$ for every $n \in \mathbb{N}$, hence the formula (3.158) is still satisfied in this case. By combining inequalities (3.157) and (3.158), it ensues that $\lim_{n \to +\infty} \mu_n (f(x_n) - f(\bar{x})) = 0$, which finishes the proof of the lemma. $\square$

We can now continue the proof of the theorem in proving the formula (3.147). First suppose that $f(\bar{x}) = r$. Lemma 3.215 tells us that
$$\operatorname{Lim\,sup}_{x \to \bar{x}} \mathbb{R}_+ \partial f(x) \subset \operatorname{Lim\,sup}_{\begin{cases} x \to \bar{x},\, \mu \ge 0 \\ \mu(f(x) - f(\bar{x})) \to 0 \end{cases}} \mu \partial f(x)$$
and since the reverse inclusion is obviously true, both members are equal. Then, it suffices to use the formula (3.146).

Now suppose that $f(\bar{x}) < r$. Observe that

$$N(S;\bar{x}) = \operatorname*{Lim\,sup}_{\substack{x \to \bar{x},\, \mu \geq 0 \\ \mu(f(x)-r) \to 0}} \mu \partial f(x) \quad \text{in view of formula (3.146)}$$

$$\subset \operatorname*{Lim\,sup}_{\substack{x \to \bar{x},\, \mu \geq 0 \\ \mu(f(x)-r) \to 0 \\ \mu(f(x) - f(\bar{x})) \to 0}} \mu \partial f(x) \quad \text{from Lemma 3.215}$$

$$\subset \operatorname*{Lim\,sup}_{x \to \bar{x},\, \mu \downarrow 0} \mu \partial f(x) \quad \text{because } f(\bar{x}) \neq r$$

$$= N(\operatorname{dom} f; \bar{x}) \quad \text{from Theorem 3.213,}$$

note that in the latter inclusion above we also use the obvious inclusion $0 \in \operatorname*{Lim\,sup}_{x \to \bar{x},\, \mu \downarrow 0} \mu \partial f(x)$ (keep in mind that $\operatorname{Dom} \partial f$ is dense in $\operatorname{dom} f$ by Proposition 2.223).
Since $S \subset \operatorname{dom} f$, the inclusion $N(\operatorname{dom} f; \bar{x}) \subset N(S;\bar{x})$ is obvious, so we obtain

$$N(S;\bar{x}) = \operatorname*{Lim\,sup}_{x \to \bar{x},\, \mu \downarrow 0} \mu \partial f(x) = N(\operatorname{dom} f; \bar{x}),$$

which concludes the proof of the theorem. □

COROLLARY 3.216. *Let X be a reflexive Banach space and let $f : X \to \mathbb{R} \cup \{+\infty\}$ be a lower semicontinuous convex function. For $\bar{x} \in \operatorname{dom} f$ and for the sublevel set $S = \{f \leq f(\bar{x})\}$ one has*

(a) $N(S;\bar{x}) = \mathbb{R}_+ \partial f(\bar{x}) \cup N(\operatorname{dom} f; \bar{x}) \cup \operatorname*{Lim\,sup}_{x \to \bar{x},\, \mu \to +\infty} \mu \partial f(x)$.

Assuming in addition the Slater condition $\{f < f(\bar{x})\} \neq \emptyset$, one obtains

(b) $N(S;\bar{x}) = \mathbb{R}_+ \partial f(\bar{x}) \cup N(\operatorname{dom} f; \bar{x})$.

(c) *If additionally $\partial f(\bar{x}) \neq \emptyset$, then $N(S;\bar{x}) = \operatorname{cl}(\mathbb{R}_+ \partial f(\bar{x}))$.*

PROOF. (a) In view of the formula (3.147), it suffices to prove the equality

(3.159) $\operatorname*{Lim\,sup}_{x \to \bar{x}} \mathbb{R}_+ \partial f(x) = \mathbb{R}_+ \partial f(\bar{x}) \cup N(\operatorname{dom} f; \bar{x}) \cup \operatorname*{Lim\,sup}_{x \to \bar{x},\, \mu \to +\infty} \mu \partial f(x)$.

Let $x^* \in \operatorname*{Lim\,sup}_{x \to \bar{x}} \mathbb{R}_+ \partial f(x)$. By definition, there exist a sequence $(x_n)_n$ in X, a sequence $(x_n^*)_n$ in X^* and a sequence $(\mu_n)_n$ in $\mathbb{R}_+$ such that $x_n \to \bar{x}$, $x_n^* \to x^*$ as $n \to +\infty$ and $x_n^* \in \mu_n \partial f(x_n)$. There exists a subsequence of $(\mu_n)_n$, that we do not relabel, such that $\lim_{n \to +\infty} \mu_n = \bar{\mu} \in \mathbb{R}_+ \cup \{+\infty\}$. If $\bar{\mu} \in \mathbb{R}_+ \setminus \{0\}$, we deduce from the closedness of the graph of the multimapping ∂f that $x^*/\bar{\mu} \in \partial f(\bar{x})$, hence $x^* \in \mathbb{R}_+ \partial f(\bar{x})$. Now suppose that $\lim_{n \to +\infty} \mu_n = 0$. We then obtain that $x^* \in \operatorname*{Lim\,sup}_{x \to \bar{x},\, 0 \geq \mu \to 0} \mu \partial f(x)$, and therefore $x^* \in N(\operatorname{dom} f; \bar{x})$ according to Theorem 3.213. Finally, if $\lim_{n \to +\infty} \mu_n = +\infty$, we find $x^* \in \operatorname*{Lim\,sup}_{x \to \bar{x},\, \mu \to +\infty} \mu \partial f(x)$. The above arguments show that

$$x^* \in \mathbb{R}_+ \partial f(\bar{x}) \cup N(\operatorname{dom} f; \bar{x}) \cup \operatorname*{Lim\,sup}_{x \to \bar{x},\, \mu \to +\infty} \mu \partial f(x),$$

hence the first inclusion in the formula (3.159) is proved. Since the reverse inclusion is obvious, the proof of (a) is complete.

(b) Let us now assume the Slater condition, thus implying that $0 \notin \partial f(\bar{x})$. By the closedness of the graph of ∂f we immediately obtain that the set $\operatorname*{Lim\,sup}_{x \to \bar{x},\, \mu \to +\infty} \mu \partial f(x)$ is empty. Formula (b) then follows from (a).

(c) Now assume in addition that $\partial f(\bar{x}) \neq \emptyset$, so Proposition 3.133 tells us that we have $N(\operatorname{dom} f; \bar{x}) = (\partial f(\bar{x}))^\infty$. Recalling that $\operatorname{cl}(\mathbb{R}_+ C) = \mathbb{R}_+ C \cup C^\infty$ for any nonempty closed convex set $C \subset X$ such that $0 \notin C$ (see Proposition 3.135) and applying this fact with $C = \partial \varphi(\bar{x})$, we obtain (c) from (b). $\square$

REMARK 3.217. (a) Since $N(\operatorname{dom} f; \bar{x}) = \{0\}$ if $\bar{x} \in \operatorname{int}(\operatorname{dom} f)$, from (b) we retrieve the equality $N(S; \bar{x}) = \mathbb{R}_+ \partial f(\bar{x})$ of Corollary 3.210 whenever the Slater condition holds and the convex function f is continuous at $\bar{x}$.
(b) We note that formula (b) is a particular case of Corollary 3.209. $\square$

3.18.4. Limiting formulas for normals to convex sublevels: General Banach space case.
In the framework of general Banach spaces we obtain the following result similar to Theorem 3.214 but here the upper limit has to be taken with respect to the weak star topology of X^* and an additional condition involving the bracket $\langle \cdot, x - \bar{x} \rangle$ needs to be required.

THEOREM 3.218 (limiting descriptions of normals of convex sublevels: general Banach space). Let X be a Banach space and let $f : X \to \mathbb{R} \cup \{+\infty\}$ be a lower semicontinuous convex function. Let $r \in \mathbb{R}$ be such that the sublevel set $S = \{f \leq r\}$ is nonempty and let $\bar{x} \in S$. Then one has

$$N(S; \bar{x}) = {}^{w^*}\!\operatorname{Lim\,sup}_{\substack{x \to \bar{x}, \mu \geq 0 \\ \mu(f(x) - r) \to 0 \\ \langle \cdot, x - \bar{x} \rangle \to 0}} \partial(\mu f)(x) = {}^{w^*}\!\operatorname{Lim\,sup}_{\substack{x \to \bar{x}, \mu \geq 0 \\ \mu(f(x) - r) \to 0 \\ \mu \langle \cdot, x - \bar{x} \rangle \to 0}} \mu \partial f(x),$$

where the second (resp. third) member stands for the set of weak star limits $w^*\lim x_i^*$ (resp. $w^*\lim \mu_i u_i^*$) such that $\mu_i \geq 0$, $x_i^* \in \partial(\mu_i f)(x_i)$ (resp. $u_i^* \in \partial f(x_i)$), the net $(x_i)_i$ strongly converges to $\bar{x}$, $\mu_i(f(x_i) - r) \to 0$ and $\langle x_i^*, x_i - \bar{x} \rangle \to 0$ (resp. $\mu_i \langle u_i^*, x_i - \bar{x} \rangle \to 0$).
Further, if $f(\bar{x}) < r$ one also has

$$(3.160) \qquad N(S; \bar{x}) = {}^{w^*}\!\operatorname{Lim\,sup}_{\substack{x \to \bar{x}, \mu \downarrow 0 \\ \mu \langle \cdot, x - \bar{x} \rangle \to 0}} \mu \partial f(x) = N(\operatorname{dom} f; \bar{x}).$$

PROOF. First we note that in the proof of Theorem 3.214, according to Theorem 3.203(c), $(x_n)_n$, $(y_n)_n$, $(x_n^*)_n$, $(y_n^*)_n$, and (e_n^*) have to be taken as *nets* instead of sequences, the strong convergence $x_n^* \to x^*$ has to be replaced by the weak-star convergence $x_n^* \xrightarrow{w^*} x^*$, and we have in addition $\langle x_n^*, x_n - \bar{x} \rangle \to 0$. With those elements at hands, one can show with arguments in the line of Theorem 3.214 that

$$N(S; \bar{x}) \subset {}^{w^*}\!\operatorname{Lim\,sup}_{\substack{x \to \bar{x}, \mu \geq 0 \\ \mu(f(x) - r) \to 0 \\ \langle \cdot, x - \bar{x} \rangle \to 0}} \partial(\mu f)(x),$$

where the second member denotes the set of weak star limits of nets $(z_n^*)_n$ for which there are a net $(x_n)_n$ converging to $\bar{x}$ and a net $(\mu_n)_n$ in $\mathbb{R}_+$ such that $z_n^* \in \partial(\mu_n f)(x_n)$, $\mu_n(f(x_n) - r) \to 0$ and $\langle z_n^*, x_n - \bar{x} \rangle \to 0$. Let us show that the latter second member is included in the same upper limit but with $\mu \partial f(x)$ in place of $\partial(\mu f)(x)$. Let x^* in the upper limit with $\partial(\mu f)(x)$. By definition, there exist nets $x_n \to \bar{x}$, $x_n^* \xrightarrow{w^*} x^*$, $(\mu_n) \subset \mathbb{R}_+$ satisfying $x_n^* \in \partial(\mu_n f)(x_n)$, $\mu_n(f(x_n) - r) \to 0$ and $\langle x_n^*, x_n - \bar{x} \rangle \to 0$. First suppose that there exists a subnet of (μ_n), that we do not relabel, such that $\mu_n > 0$ for every n. We then have $x_n^* \in \mu_n \partial f(x_n)$, and hence

(3.161) $$x^* \in \underset{\substack{x \to \bar{x}, \mu \geq 0 \\ \mu(f(x) - r) \to 0 \\ \mu\langle \cdot, x - \bar{x}\rangle \to 0}}{w^* \operatorname{Lim sup}} \mu \partial f(x).$$

Now suppose, for some element n_0 of the set of indices, that $\mu_n = 0$ for all $n \succeq n_0$. Recalling that $0f = \Psi_{\operatorname{dom} f}$, we obtain $x_n^* \in N(\operatorname{dom} f; x_n)$ for n large enough. Writing, for every $x \in \operatorname{dom} f$,

$$\langle x_n^*, x - \bar{x}\rangle = \langle x_n^*, x - x_n\rangle + \langle x_n^*, x_n - \bar{x}\rangle \leq \langle x_n^*, x_n - \bar{x}\rangle$$

and taking the convergence $\langle x_n^*, x_n - \bar{x}\rangle \to 0$ into account, we see at the limit on n that $x^* \in N(\operatorname{dom} f; \bar{x})$. By Theorem 3.213 we have $N(\operatorname{dom} f; \bar{x}) = \underset{x \to \bar{x}, \mu \downarrow 0}{\operatorname{Lim sup}} \mu \partial f(x)$,

hence

$$x^* \in \underset{\substack{x \to \bar{x}, \mu \downarrow 0 \\ f(x) \to f(\bar{x})}}{\operatorname{Lim sup}} \mu \partial f(x) \subset \underset{\substack{x \to \bar{x}, \mu \geq 0 \\ \mu(f(x) - r) \to 0 \\ \mu\langle \cdot, x - \bar{x}\rangle \to 0}}{w^* \operatorname{Lim sup}} \mu \partial f(x).$$

Therefore the inclusion (3.161) is satisfied in both cases, which proves the desired inclusion.

Now let us show the inclusion

(3.162) $$\underset{\substack{x \to \bar{x}, \mu \geq 0 \\ \mu(f(x) - r) \to 0 \\ \mu\langle \cdot, x - \bar{x}\rangle \to 0}}{w^* \operatorname{Lim sup}} \mu \partial f(x) \subset N(S; \bar{x}).$$

Fix x^* in the first member. By definition, there are nets $(x_n)_n$ in X, $(x_n^*)_n$ in X^* and $(\mu_n)_n$ in $\mathbb{R}_+$ such that $x_n^* \in \partial f(x_n)$, $x_n \to \bar{x}$, $\mu_n x_n^* \overset{w^*}{\to} x^*$, $\mu_n(f(x_n) - r) \to 0$, and $\mu_n\langle x_n^*, x_n - \bar{x}\rangle \to 0$. For any fixed element $x \in S$, the subdifferential inequality gives

$$\langle \mu_n x_n^*, x - x_n\rangle \leq \mu_n\bigl(f(x) - f(x_n)\bigr) \leq \mu_n(r - f(x_n)),$$

thus $\langle x^*, x - \bar{x}\rangle \leq 0$ because $\mu_n x_n^* \overset{w^*}{\to} x^*$, $\mu_n\langle x_n^*, x_n - \bar{x}\rangle \to 0$, and $\mu_n(r - f(x_n)) \to 0$. We deduce that $x^* \in N(S; \bar{x})$, which shows the inclusion (3.162) and finishes the proof of the first two equalities of the theorem.

Finally, assume that $f(\bar{x}) < r$. Under this additional hypothesis, for the nets at the very beginning of the proof of the theorem, we obtain as in (3.152) that $y_n^* \to 0$ with $y_n^* \geq 0$, or equivalently $\mu_n \to 0$ with $\mu_n \geq 0$ since $\mu_n = y_n^*$. It ensues that

$$N(S; \bar{x}) \subset \underset{\substack{x \to \bar{x}, 0 \geq \mu \to 0 \\ \mu\langle \cdot, x - \bar{x}\rangle \to 0}}{w^* \operatorname{Lim sup}} \partial(\mu f)(x) \subset \underset{\substack{x \to \bar{x}, \mu \downarrow 0 \\ \mu\langle \cdot, x - \bar{x}\rangle \to 0}}{w^* \operatorname{Lim sup}} \partial(\mu f)(x),$$

where the latter equality is due to the fact that zero belongs to the Lim sup with $\mu \downarrow 0$ as easily seen since $\operatorname{Dom} \partial f$ is dense in $\operatorname{dom} f$ by Proposition 2.223. By arguing as above we deduce that

(3.163) $$N(S; \bar{x}) \subset \underset{\substack{x \to \bar{x}, \mu \downarrow 0 \\ \mu\langle \cdot, x - \bar{x}\rangle \to 0}}{w^* \operatorname{Lim sup}} \mu \partial f(x).$$

Let us now show the inclusion

(3.164) $$\underset{\begin{cases} x \to \overline{x},\, \mu \downarrow 0 \\ \mu\langle \cdot,\, x - \overline{x}\rangle \to 0 \end{cases}}{w^* \text{-} \operatorname{Lim\,sup}} \mu \, \partial f(x) \subset N(\operatorname{dom} f; \overline{x}).$$

Fix any x^* in the left member. By definition there exist nets $(\mu_n)_n$ in $]0, +\infty[$ with $\mu_n \to 0$, $(x_n)_n$ in X converging strongly to $\overline{x}$, $(u_n^*)_n$ in X^* with $u_n^* \in \partial f(x_n)$ such that $\mu_n u_n^* \overset{w^*}{\to} x^*$ and $\mu_n \langle u_n^*, x_n - \overline{x}\rangle \to 0$. Consider any $x \in \operatorname{dom} f$ and any real $\varepsilon > 0$. From the lower semicontinuity of f, choose some index element n_0 such that for any $n \succeq n_0$ we have $f(x_n) \geq f(\overline{x}) - \varepsilon$, where $\preceq$ denotes the directed preorder of the set of elements n. Then, for every $n \succeq n_0$ we have

$$\langle u_n^*, x - x_n\rangle \leq f(x) - f(x_n) \leq f(x) - f(\overline{x}) + \varepsilon,$$

hence

$$\langle \mu_n u_n^*, x - \overline{x}\rangle = \langle \mu_n u_n^*, x - x_n\rangle + \langle \mu_n u_n^*, x_n - \overline{x}\rangle$$
$$\leq \mu_n(f(x) - f(\overline{x}) + \varepsilon) + \mu_n \langle u_n^*, x_n - \overline{x}\rangle,$$

so passing to the limit gives $\langle x^*, x - \overline{x}\rangle \leq 0$. Since this is true for every $x \in \operatorname{dom} f$, we obtain $x^* \in N(\operatorname{dom} f; \overline{x})$, which confirms the inclusion (3.164). Finally, since $\overline{x} \in S \subset \operatorname{dom} f$, it is easily seen that

(3.165) $$N(\operatorname{dom} f; \overline{x}) \subset N(S; \overline{x}).$$

By combining the inclusions (3.163), (3.164) and (3.165), we find the equalities in (3.160). The proof is then complete. $\square$

3.18.5. Limiting formulas for normals to intersection of finitely many sublevels. In this section, we consider a finite family of functions $(f_i)_{i \in I}$ from X into $\mathbb{R} \cup \{+\infty\}$, which are assumed to be lower semicontinuous and convex. Consider $(r_i)_{i \in I}$ in $\mathbb{R}$ and the set

$$S = \{x \in X : f_i(x) \leq r_i, \forall i \in I\} = \bigcap_{i \in I} \{f_i \leq r_i\}.$$

Given $\overline{x} \in S$, our aim is to give a sequential formula for the normal cone $N(S; \overline{x})$, without requiring any qualification condition. The key ingredients in that direction are Corollary 3.201 and Theorem 3.203. We focus our attention to the case where the Banach space X is reflexive; of course adaptations to the case of a nonreflexive Banach space can be realized as in Subsection 3.18.4. First, a suitable application of Theorem 3.203 leads to the following result.

THEOREM 3.219 (limiting descriptions of normals of intersections of convex sublevels). Let X be a reflexive Banach space, I be a nonempty finite set and $f_i : X \to \mathbb{R} \cup \{+\infty\}$ be a lower semicontinuous convex function, for each $i \in I$. Let $(r_i)_{i \in I}$ in $\mathbb{R}$ and let the set $S = \bigcap_{i \in I} \{f_i \leq r_i\}$. For any $\overline{x} \in S$ one has

(3.166) $$N(S; \overline{x}) = \underset{\begin{cases} x \to \overline{x},\, \mu_i \geq 0 \\ \sum_{i \in I} \mu_i (f_i(x) - r_i) \to 0 \end{cases}}{\operatorname{Lim\,sup}} \partial\left[\sum_{i \in I} \mu_i f_i\right](x).$$

PROOF. Let us start with the inclusion $\subset$. Analogously to the proof of Theorem 3.214, consider $Y := \mathbb{R}^I$ and its convex cone $Y_+ := (\mathbb{R}_+)^I$ and define the function $G : X \to Y \cup \{+\infty_Y\}$ by

$$G(x) = \begin{cases} (f_i(x) - r_i)_{i \in I} & \text{if } x \in \bigcap_{i \in I} \text{dom } f_i \\ +\infty_Y & \text{otherwise.} \end{cases}$$

The space $Y = \mathbb{R}^I$ is endowed with the preorder $\leq_{Y_+}$ associated with the convex cone $Y_+ = (\mathbb{R}_+)^I$, that is, $y \leq_{Y_+} y' \Leftrightarrow y' - y \in Y_+$. The abstract maximal element $\{+\infty_Y\}$ is adjoined to Y. The convexity of each function f_i implies the convexity of G in the vector sense. The closedness of epi G is easily obtained from the closedness of the epigraph epi f_i for every $i \in I$. Now note that $S = \{x \in X : G(x) \in -Y^+\}$, and hence $\Psi_S = \Psi_{-Y_+} \circ G$. This implies that

$$N(S; \overline{x}) = \partial \Psi_S(\overline{x}) = \partial \left(\Psi_{-Y_+} \circ G \right)(\overline{x}).$$

The function Ψ_{-Y_+} is lower semicontinuous, convex and nondecreasing with respect to the preorder $\leq_{Y_+}$. Let us fix $x^* \in N(S; \overline{x})$ and apply Theorem 3.203(a) with $g = \Psi_{-Y_+}$ and the mapping G defined above. We obtain the existence of sequences $(x_n)_n$ in X, $(y_n)_n$ in $Y = \mathbb{R}^I$, $(x_n^*)_n$ in X^*, $(e_n^*)_n$ in Y^* and $(y_n^*)_n$ in Y_+^* such that $x_n \to \overline{x}$, $y_n \to G(\overline{x})$, $x_n^* \to x^*$, $e_n^* \to 0$ and

$$y_n^* + e_n^* \in N((-\mathbb{R}_+)^I; y_n), \quad x_n^* \in \partial(y_n^* \circ G)(x_n) \quad \text{and} \quad \langle y_n^*, G(x_n) - G(\overline{x}) \rangle \to 0.$$

Let us denote respectively by $(y_{i,n})_{i \in I}$ and by $(y_{i,n}^*)_{i \in I}$ and $(e_{i,n}^*)_{i \in I}$ the coordinates of the vector y_n in the canonical basis of $\mathbb{R}^I$ and of the linear functionals y_n^* and e_n^* in the canonical basis of $Y^* = (\mathbb{R}^I)^*$. Since $y_n^* \circ G = \sum_{i \in I} y_{i,n}^* (f_i - r_i)$ according to (3.93), we obtain

$$(3.167) \qquad x_n^* \in \partial \left[\sum_{i \in I} y_{i,n}^* f_i \right](x_n)$$

and

$$(3.168) \qquad \sum_{i \in I} y_{i,n}^* (f_i(x_n) - f_i(\overline{x})) \to 0 \quad \text{as } n \to +\infty.$$

Let us denote by $I(\overline{x})$ the set of (so-called) *active indices*:

$$I(\overline{x}) = \{i \in I : f_i(\overline{x}) = r_i\}.$$

Let us fix any $i \notin I(\overline{x})$. Since $y_{i,n} \to f_i(\overline{x}) - r_i$ as $n \to +\infty$ and since $f_i(\overline{x}) < r_i$, we have $y_{i,n} < 0$ for n large enough. It ensues that $N(-\mathbb{R}_+; y_{i,n}) = \{0\}$ and formula $y_{i,n}^* + e_{i,n}^* \in N(-\mathbb{R}_+; y_n)$ then implies that $y_{i,n}^* = -e_{i,n}^* \to 0$ as $n \to +\infty$. Thus, we have proved that

$$(3.169) \qquad \forall i \notin I(\overline{x}), \quad y_{i,n}^* \to 0 \quad \text{as } n \to +\infty.$$

Now observe that

$$\sum_{i \in I} y_{i,n}^* (f_i(x_n) - r_i) = \sum_{i \in I} y_{i,n}^* (f_i(x_n) - f_i(\overline{x})) + \sum_{i \in I} y_{i,n}^* (f_i(\overline{x}) - r_i)$$

$$(3.170) \qquad\qquad = \sum_{i \in I} y_{i,n}^* (f_i(x_n) - f_i(\overline{x})) + \sum_{i \notin I(\overline{x})} y_{i,n}^* (f_i(\overline{x}) - r_i).$$

By (3.168), (3.169) and (3.170), we immediately deduce that

(3.171) $$\sum_{i \in I} y^*_{i,n}(f_i(x_n) - r_i) \to 0 \text{ as } n \to +\infty.$$

Finally, we have built sequences $x_n \to \overline{x}$, $x^*_n \to x^*$, $(y^*_{i,n})_n$ in $\mathbb{R}_+$ satisfying (3.167) and (3.171), which clearly shows that

(3.172) $$x^* \in \underset{\substack{x \to \overline{x},\ \mu_i \geq 0 \\ \sum_{i \in I} \mu_i (f_i(x) - r_i) \to 0}}{\operatorname{Lim\,sup}} \partial \left[\sum_{i \in I} \mu_i f_i\right](x).$$

Conversely, take any x^* satisfying (3.172) and let us prove that $x^* \in N(S; \overline{x})$. By definition, there exist sequences $x_n \to \overline{x}$, $x^*_n \to x^*$, $(\mu_{i,n})_n$ in $\mathbb{R}_+$ for each $i \in I$, such that

$$x^*_n \in \partial \left[\sum_{i \in I} \mu_{i,n} f_i\right](x_n) \quad \text{and} \quad \sum_{i \in I} \mu_{i,n}(f_i(x_n) - r_i) \to 0 \text{ as } n \to +\infty.$$

Let us fix any $x \in S$. By using the first relation above and the convexity of the function $\sum_{i \in I} \mu_{i,n} f_i$, we find (since $x \in S$)

$$\langle x^*_n, x - x_n \rangle \leq \sum_{i \in I} \mu_{i,n}(f_i(x) - f_i(x_n)) \leq \sum_{i \in I} \mu_{i,n}(r_i - f_i(x_n)).$$

Recalling that $x_n \to \overline{x}$, $x^*_n \to x^*$ and $\sum_{i \in I}^m \mu_{i,n}(r_i - f_i(x_n)) \to 0$ as $n \to +\infty$, and taking the limit as $n \to +\infty$ in the above inequality, we obtain that $\langle x^*, x - \overline{x} \rangle \leq 0$. Since this is true for every $x \in S$, we conclude that $x^* \in N(S; \overline{x})$. □

REMARK 3.220. The condition $\sum_{i \in I} \mu_{i,n}(f_i(x_n) - r_i) \to 0$ as $n \to +\infty$ can be interpreted as a *relaxed complementary slackness condition*. □

THEOREM 3.221. Let X be a reflexive Banach space and let $(f_i)_{i \in I}$ be a finite family of lower semicontinuous convex functions from X into $\mathbb{R} \cup \{+\infty\}$. Let $S = \bigcap_{i=1}^m \{f_i \leq r_i\}$, where $r_1, \cdots, r_m \in \mathbb{R}$. For any $\overline{x} \in S$ one has

(3.173) $$N(S; \overline{x}) \subset \underset{\substack{x \to \overline{x},\ \mu_i \geq 0 \\ \mu_i \to 0 \text{ if } i \notin I(\overline{x})}}{\operatorname{Lim\,sup}} \partial \left[\sum_{i \in I} \mu_i f_i\right](x),$$

with $I(\overline{x}) = \{i \in I : f_i(\overline{x}) = r_i\}$. If in addition the following Slater qualification condition is satisfied

(3.174) $$\bigcap_{i \in I} \{f_i < r_i\} \neq \emptyset,$$

then the above inclusion holds as an equality.

PROOF. The inclusion (3.173) is obtained as a by-product of the proof of Theorem 3.219, see the formula (3.169) in particular. The proof of the reverse inclusion is based on the following lemma.

LEMMA 3.222. Assume that there exist sequences $(x_n)_n$ in X, $(x^*_n)_n$ in X^* and $(\mu_{i,n})_n$ in $\mathbb{R}_+$, for each $i \in I$, such that $x_n \to \overline{x}$, $x^*_n \to x^*$, $\mu_{i,n} \to 0$ if $i \notin I(\overline{x})$ and $x^*_n \in \partial \left[\sum_{i \in I} \mu_{i,n} f_i\right](x_n)$. Assume also that the Slater condition (3.174) holds. Then the following properties are satisfied

(a) There exists $M > 0$ such that $\sum_{i \in I} \mu_{i,n} \leq M$ for every $n \in \mathbb{N}$.

(b) $\lim_{n\to+\infty} \sum_{i\in I} \mu_{i,n}(f_i(x_n) - f_i(\bar{x})) = 0$.

PROOF OF LEMMA 3.222. (a) Let us argue by contradiction and suppose that the sequence $(M_n)_n$ defined by $M_n = \sum_{i\in I} \mu_{i,n}$ is not bounded. There exists a subsequence of $(M_n)_n$, that we do not relabel, such that $\lim_{n\to+\infty} M_n = +\infty$. For each $i \in I$, consider the sequence $(\nu_{i,n})$ defined by $\nu_{i,n} = \mu_{i,n}/M_n$. Each sequence $(\nu_{i,n})$ satisfies $\nu_{i,n} \in [0,1]$ for every $n \in \mathbb{N}$. Thus, we can extract in $\mathbb{R}^I$ a subsequence of $((\nu_{i,n})_{i\in I})_n$, that we do not relabel, such that $\lim_{n\to+\infty} \nu_{i,n} = \bar{\nu}_i \in [0,1]$ for each $i \in I$. The real numbers $\bar{\nu}_i$ satisfy $\sum_{i\in I} \bar{\nu}_i = 1$ and $\bar{\nu}_i = 0$ if $i \notin I(\bar{x})$ because $\mu_{i,n} \to 0$ in this case. Let us now fix $x \in \bigcap_{i\in I}\{f_i < r_i\}$, which is nonempty by assumption. Observing that $x_n^*/M_n \in \partial\left[\sum_{i\in I} \nu_{i,n} f_i\right](x_n)$, the subdifferential inequality yields

$$(3.175) \qquad \sum_{i=1}^{m} \nu_{i,n} f_i(x) \geq \sum_{i\in I} \nu_{i,n} f_i(x_n) + \langle x_n^*/M_n, x - x_n\rangle.$$

Since $x_n \to \bar{x}$, $x_n^* \to x^*$ and $M_n \to +\infty$ as $n \to +\infty$, we immediately obtain

$$(3.176) \qquad \lim_{n\to+\infty} \langle x_n^*/M_n, x - x_n\rangle = 0.$$

On the other hand, we have

$$(3.177) \qquad \lim_{n\to+\infty} \sum_{i=1}^{m} \nu_{i,n} f_i(x) = \sum_{i=1}^{k} \bar{\nu}_i f_i(x),$$

because $\lim_{n\to+\infty} \nu_{i,n} = \bar{\nu}_i$. Finally, the lower semicontinuity of each function f_i implies that

$$\liminf_{n\to+\infty} \nu_{i,n} f_i(x_n) \geq \bar{\nu}_i f_i(\bar{x}),$$

hence

$$(3.178) \qquad \liminf_{n\to+\infty} \sum_{i=1}^{m} \nu_{i,n} f_i(x_n) \geq \sum_{i\in I} \liminf_{n\to+\infty} \nu_{i,n} f_i(x_n) \geq \sum_{i\in I} \bar{\nu}_i f_i(\bar{x}).$$

By combining (3.175), (3.176), (3.177) and (3.178), we find

$$\sum_{i\in I} \bar{\nu}_i f_i(x) \geq \sum_{i\in I} \bar{\nu}_i f_i(\bar{x}).$$

Recalling that $\bar{\nu}_i = 0$ if $i \notin I(\bar{x})$, it ensues that

$$(3.179) \qquad \sum_{i\in I(\bar{x})} \bar{\nu}_i f_i(x) \geq \sum_{i\in I(\bar{x})} \bar{\nu}_i f_i(\bar{x}) = \sum_{i\in I(\bar{x})} \bar{\nu}_i r_i.$$

On the other hand, since $x \in \bigcap_{i\in I}\{f_i < r_i\}$ and since $\bar{\nu}_i > 0$ for at least one $i \in I(\bar{x})$, we have

$$\sum_{i\in I(\bar{x})} \bar{\nu}_i f_i(x) < \sum_{i\in I(\bar{x})} \bar{\nu}_i r_i,$$

which yields a contradiction with (3.179).

(b) From (a) there exists a real $M > 0$ such that $\sum_{i\in I} \mu_{i,n} \leq M$ for every $n \in \mathbb{N}$. We then have

$$\mu_{i,n}\left(f_i(x_n) - f_i(\bar{x})\right) \geq -M\left(f_i(x_n) - f_i(\bar{x})\right)^-,$$

where we recall the notation $\alpha^- := \max\{-\alpha, 0\}$. The lower semicontinuity of f_i implies that $\liminf_{n\to+\infty} f_i(x_n) \geq f_i(\overline{x})$, hence $\lim_{n\to+\infty}(f_i(x_n) - f_i(\overline{x}))^- = 0$. Taking the lower limit as $n \to +\infty$ in the above inequality, we obtain the inequality $\liminf_{n\to+\infty} \mu_{i,n}(f_i(x_n) - f_i(\overline{x})) \geq 0$. It ensues that

$$(3.180) \quad \liminf_{n\to+\infty} \sum_{i\in I} [\mu_{i,n}(f_i(x_n) - f_i(\overline{x}))] \geq \sum_{i\in I} \liminf_{n\to+\infty} [\mu_{i,n}(f_i(x_n) - f_i(\overline{x}))] \geq 0.$$

Using the assumption $x_n^* \in \partial\left[\sum_{i\in I} \mu_{i,n} f_i\right](x_n)$, the subdifferential inequality gives

$$\sum_{i\in I}[\mu_{i,n}(f_i(x_n) - f_i(\overline{x}))] \leq \langle x_n^*, x_n - \overline{x}\rangle.$$

Since $\lim_{n\to+\infty}\langle x_n^*, x_n - \overline{x}\rangle = 0$, we deduce that

$$(3.181) \quad \limsup_{n\to+\infty} \sum_{i\in I}[\mu_{i,n}(f_i(x_n) - f_i(\overline{x}))] \leq 0.$$

Combining inequalities (3.180) and (3.181) yields

$$\lim_{n\to+\infty} \sum_{i\in I}^{m} [\mu_{i,n}(f_i(x_n) - f_i(\overline{x}))] = 0,$$

which finishes the proof the lemma. $\square$

Let us come back to the proof of Theorem 3.221. Lemma 3.222 tells us that

$$\operatorname{Lim\,sup}_{\left\{\begin{array}{l} x \to \overline{x},\, \mu_i \geq 0 \\ \mu_i \to 0 \text{ if } i \notin I(\overline{x}) \end{array}\right.} \partial\left[\sum_{i\in I} \mu_i f_i\right](x) \subset \operatorname{Lim\,sup}_{\left\{\begin{array}{l} x \to \overline{x},\, \mu_i \geq 0 \\ \mu_i \to 0 \text{ if } i \notin I(\overline{x}) \\ \sum_{i\in I} \mu_i(f_i(x) - f_i(\overline{x})) \to 0 \end{array}\right.} \partial\left[\sum_{i\in I} \mu_i f_i\right](x)$$

$$\subset \operatorname{Lim\,sup}_{\left\{\begin{array}{l} x \to \overline{x},\, \mu_i \geq 0 \\ \sum_{i\in I} \mu_i(f_i(x) - r_i) \to 0 \end{array}\right.} \partial\left[\sum_{i\in I} \mu_i f_i\right](x)$$

$$= N(S; \overline{x}) \quad \text{in view of Theorem 3.219.}$$

The proof of the theorem is then complete. $\square$

REMARK 3.223. It is worth showing how the above theorem allows us to recover, in the context of Banach spaces, the formula in Corollary 3.208 for the normal cone $N(S; \overline{x})$ under both the continuity of the functions f_i, $i \in I$, and the Slater condition (3.174):

If, with notation in Theorem 3.221, the Slater condition (3.174) is satisfied and if all the functions f_i are continuous at $\overline{x} \in S$, then

$$N_S(\overline{x}) = \sum_{i\in I(\overline{x})} \mathbb{R}_+ \partial f_i(\overline{x}),$$

with the convention that the sum over an empty set of indices equals zero.

Indeed, from Theorem 3.221 we know that

$$N(S; \overline{x}) = \operatorname{Lim\,sup}_{\left\{\begin{array}{l} x \to \overline{x},\, \mu_i \geq 0 \\ \mu_i \to 0 \text{ if } i \notin I(\overline{x}) \end{array}\right.} \partial\left[\sum_{i\in I} \mu_i f_i\right](x).$$

On the other hand, by continuity of f_i there is an open neighborhood U of $\bar{x}$ such that all functions f_i are bounded on U. Further, the Moreau-Rockafellar condition being satisfied, we have for every $x \in X$

$$\partial \left[\sum_{i \in I} \mu_i f_i \right](x) = \sum_{i \in I} \mu_i \, \partial f_i(x).$$

Hence, it suffices to show that

(3.182) $\displaystyle \operatorname*{Lim\,sup}_{\begin{cases} x \to \bar{x},\, \mu_i \geq 0 \\ \mu_i \to 0 \text{ if } i \notin I(\bar{x}) \end{cases}} \sum_{i \in I} \mu_i \, \partial f_i(x) = \sum_{i \in I(\bar{x})} \mathbb{R}_+ \, \partial f_i(\bar{x}).$

Let $x^* \in \operatorname*{Lim\,sup}_{\begin{cases} x \to \bar{x},\, \mu_i \geq 0 \\ \mu_i \to 0 \text{ if } i \notin I(\bar{x}) \end{cases}} \sum_{i \in I} \mu_i \, \partial f_i(x)$. By definition, there exist sequences $(x_n)_n$ in U, $(x_{i,n}^*)_n$ in X^*, $(\mu_{i,n})_n$ in $\mathbb{R}_+$ such that $x_{i,n}^* \in \partial f_i(x_n)$, $x_n \xrightarrow[n \to +\infty]{} \bar{x}$, $\sum_{i \in I} \mu_{i,n} x_{i,n}^* \xrightarrow[n \to +\infty]{} x^*$ and $\mu_{i,n} \xrightarrow[n \to +\infty]{} 0$ if $i \notin I(\bar{x})$. According to the boundedness of each function f_i on U, the sequence $(x_{i,n}^*)$ is bounded, hence has a convergent subsequence with respect to the weak star topology of X^*. On the other hand, Lemma 3.222 says that each sequence $(\mu_{i,n})$ is bounded. Therefore, by extracting iteratively subsequences, we can build an increasing map $s : \mathbb{N} \to \mathbb{N}$ such that

(3.183) $\quad \forall i \in I, \quad x_{i,s(n)}^* \xrightarrow[n \to +\infty]{w^*} x_i^* \quad \text{and} \quad \mu_{i,s(n)} \xrightarrow[n \to +\infty]{} \bar{\mu}_i,$

for some $x_1^*, \cdots, x_m^* \in X^*$ and $\bar{\mu}_1, \cdots, \bar{\mu}_k \in \mathbb{R}_+$. From this and the sequential $\| \, \| \times w^*$-closedness property of the graph of each multimapping ∂f_i, we immediately obtain that $x_i^* \in \partial f_i(\bar{x})$. Since $\sum_{i \in I} \mu_{i,n} x_{i,n}^* \xrightarrow[n \to +\infty]{} x^*$, we deduce from (3.183) that $\sum_{i \in I} \bar{\mu}_i x_i^* = x^*$. Recalling that $\bar{\mu}_i = 0$ for $i \notin I(\bar{x})$, this implies that $x^* = \sum_{i \in I(\bar{x})} \bar{\mu}_i x_i^* \in \sum_{i \in I(\bar{x})} \mathbb{R}_+ \, \partial f_i(\bar{x})$. Therefore the inclusion

$$\operatorname*{Lim\,sup}_{\begin{cases} x \to \bar{x},\, \mu_i \geq 0 \\ \mu_i \to 0 \text{ if } i \notin I(\bar{x}) \end{cases}} \sum_{i \in I} \mu_i \, \partial f_i(x) \subset \sum_{i \in I(\bar{x})} \mathbb{R}_+ \, \partial f_i(\bar{x})$$

is proved and since the reverse inclusion is immediate, the equality (3.182) is demonstrated. $\square$

In Theorem 3.219 and Theorem 3.221 the normal cone $N(S; \bar{x})$ is described through the subdifferential of positively linear combinations of the functions f_i, $i = 1, \cdots, k$. The next theorem provides an additional description via the separate subdifferentials of the functions f_i. For such a description we use the set

$$\operatorname*{Lim\,sup}_{\begin{cases} x_i \to \bar{x},\, \mu_i \geq 0 \\ \mu_i \, (f_i(x_i) - r_i) \to 0 \\ \mu_i \langle x_i^*, x_i - \bar{x} \rangle \to 0 \end{cases}} \sum_{i \in I} \mu_i \, \partial f_i(x_i)$$

defined as the set of limits $\lim_{n \to +\infty} \sum_{i \in I} \mu_{i,n} x_{i,n}^*$ such that $\mu_{i,n} \geq 0$, $x_{i,n} \xrightarrow[n \to +\infty]{} \bar{x}$, $x_{i,n}^* \in \partial f_i(x_{i,n})$, $\mu_{i,n}(f_i(x_{i,n}) - r_i) \xrightarrow[n \to +\infty]{} 0$ and $\mu_{i,n} \langle x_{i,n}^*, x_{i,n} - \bar{x} \rangle \xrightarrow[n \to +\infty]{} 0$ for $i \in I$. In the same vein, we denote by

$$\operatorname*{Lim\,sup}_{\begin{cases} x_i \to \bar{x},\, \mu_i \geq 0 \\ \mu_i \, (f_i(x_i) - r_i - \langle x_i^*, x_i - \bar{x} \rangle) \to 0 \end{cases}} \sum_{i \in I} \mu_i \, \partial f_i(x_i),$$

and respectively

$$\operatorname*{Lim\,sup}_{\left\{\begin{array}{l}x_i \to \overline{x},\, \mu_i \geq 0 \\ \sum_{i \in I} \mu_i \left(f_i(x_i) - r_i - \langle x_i^*, x_i - \overline{x}\rangle\right) \to 0\end{array}\right.} \sum_{i \in I} \mu_i \, \partial f_i(x_i)$$

the sets, where the last conditions of the above definition are replaced respectively by

$$\mu_{i,n} \left(f(x_{i,n}) - r_i - \langle x_{i,n}^*, x_{i,n} - \overline{x}\rangle\right) \xrightarrow[n \to +\infty]{} 0 \text{ for } i \in I,$$

and

$$\sum_{i \in I} \mu_{i,n} \left(f_i(x_{i,n}) - r_i - \langle x_{i,n}^*, x_{i,n} - \overline{x}\rangle\right) \xrightarrow[n \to +\infty]{} 0.$$

By combining Corollary 3.201 and Theorem 3.214, we show in the next theorem that the normal cone $N(S; \overline{x})$ to the set $S = \bigcap_{i \in I}\{f_i \leq r_i\}$ coincides with the above upper limits.

THEOREM 3.224. *Let X be a reflexive Banach space and let $(f_i)_{i \in I}$ be a finite family of lower semicontinuous convex functions from X into $\mathbb{R} \cup \{+\infty\}$. Let $S = \bigcap_{i \in I}\{f_i \leq r_i\}$, where $r_i \in \mathbb{R}$ for each $i \in I$. For any $\overline{x} \in S$, one has*

$$N(S; \overline{x}) = \operatorname*{Lim\,sup}_{\left\{\begin{array}{l}x_i \to \overline{x},\, \mu_i \geq 0 \\ \mu_i(f_i(x_i) - r_i) \to 0 \\ \mu_i \langle x_i^*, x_i - \overline{x}\rangle \to 0\end{array}\right.} \sum_{i \in I} \mu_i \, \partial f_i(x_i)$$

$$= \operatorname*{Lim\,sup}_{\left\{\begin{array}{l}x_i \to \overline{x},\, \mu_i \geq 0 \\ \mu_i(f_i(x_i) - r_i - \langle x_i^*, x_i - \overline{x}\rangle) \to 0\end{array}\right.} \sum_{i \in I} \mu_i \, \partial f_i(x_i)$$

$$= \operatorname*{Lim\,sup}_{\left\{\begin{array}{l}x_i \to \overline{x},\, \mu_i \geq 0 \\ \sum_{i \in I} \mu_i (f_i(x_i) - r_i - \langle x_i^*, x_i - \overline{x}\rangle) \to 0\end{array}\right.} \sum_{i \in I} \mu_i \, \partial f_i(x_i).$$

PROOF. Let $x^* \in N(S; \overline{x})$. First observe that $\Psi_S = \sum_{i \in I} \Psi_{\{f_i \leq r_i\}}$, hence

$$N_S(\overline{x}) = \partial \Psi_S(\overline{x}) = \partial \left(\sum_{i \in I} \Psi_{\{f_i \leq r_i\}}\right)(\overline{x}).$$

From Corollary 3.201, for each $i \in I$ there exist sequences $(u_{i,n})_n$ in X, $(u_{i,n}^*)_n$ in X^* such that $u_{i,n}^* \in N(\{f_i \leq r_i\}; u_{i,n})$ and

(3.184) $\quad u_{i,n} \xrightarrow[n \to +\infty]{} \overline{x}, \quad \sum_{i \in I} u_{i,n}^* \xrightarrow[n \to +\infty]{} x^* \quad \text{and} \quad \langle u_{i,n}^*, u_{i,n} - \overline{x}\rangle \xrightarrow[n \to +\infty]{} 0.$

By Theorem 3.214, we have

$$N(\{f_i \leq r_i\}; u_{i,n}) = \operatorname*{lim\,sup}_{\left\{\begin{array}{l}x \to u_{i,n},\, \nu \geq 0 \\ \nu(f_i(x) - f_i) \to 0\end{array}\right.} \nu \, \partial f_i(x).$$

Hence there exist sequences $(\nu_{i,n,k})_k$ in $\mathbb{R}$, $(v_{i,n,k})_k$ in X and $(v_{i,n,k}^*)_k$ in X^* such that

(3.185) $\quad \nu_{i,n,k} \geq 0, \quad v_{i,n,k}^* \in \partial f_i(v_{i,n,k}),$

(3.186)

$\nu_{i,n,k} \xrightarrow[k \to +\infty]{} u_{i,n}, \quad \nu_{i,n,k} v_{i,n,k}^* \xrightarrow[k \to +\infty]{} u_{i,n}^*, \quad \nu_{i,n,k}(f_i(v_{i,n,k}) - r_i) \xrightarrow[k \to +\infty]{} 0.$

From (3.186), there exists an increasing function $s : \mathbb{N} \to \mathbb{N}$ such that for every $i \in I$ and every $n \in \mathbb{N}$,

$$(3.187) \qquad \|v_{i,n,s(n)} - u_{i,n}\| \leq \frac{1}{n}, \quad |\langle u_{i,n}^*, v_{i,n,s(n)} - u_{i,n}\rangle| \leq \frac{1}{n},$$

$$(3.188) \quad \|\nu_{i,n,s(n)} v_{i,n,s(n)}^* - u_{i,n}^*\| \leq \frac{1}{n} \quad \text{and} \quad |\nu_{i,n,s(n)}(f_i(v_{i,n,s(n)}) - r_i)| \leq \frac{1}{n}.$$

Let us then set $\mu_{i,n} := \nu_{i,n,s(n)}$, $x_{i,n} := v_{i,n,s(n)}$ and $x_{i,n}^* := v_{i,n,s(n)}^*$. In view of (3.185), we have $\mu_{i,n} \geq 0$ and $x_{i,n}^* \in \partial f_i(x_{i,n})$. By using (3.184), (3.187) and (3.188), and by denoting by card(I) the cardinal (that is, the number of elements) of I we find

$$\|x_{i,n} - \overline{x}\| \leq \|x_{i,n} - u_{i,n}\| + \|u_{i,n} - \overline{x}\|$$
$$\leq \frac{1}{n} + \|u_{i,n} - \overline{x}\| \xrightarrow[n \to +\infty]{} 0,$$

$$\left\|\sum_{i \in I} \mu_{i,n} x_{i,n}^* - x^*\right\| \leq \left\|\sum_{i \in I} \mu_{i,n} x_{i,n}^* - \sum_{i \in I} u_{i,n}^*\right\| + \left\|\sum_{i \in I}^m u_{i,n}^* - x^*\right\|$$
$$\leq \sum_{i \in I} \|\mu_{i,n} x_{i,n}^* - u_{i,n}^*\| + \left\|\sum_{i \in I} u_{i,n}^* - x^*\right\|$$
$$\leq \frac{\text{card}(I)}{n} + \left\|\sum_{i \in I} u_{i,n}^* - x^*\right\| \xrightarrow[n \to +\infty]{} 0,$$

$$|\langle \mu_{i,n} x_{i,n}^*, x_{i,n} - \overline{x}\rangle| \leq |\langle \mu_{i,n} x_{i,n}^* - u_{i,n}^*, x_{i,n} - u_{i,n}\rangle| +$$
$$|\langle \mu_{i,n} x_{i,n}^* - u_{i,n}^*, u_{i,n} - \overline{x}\rangle| +$$
$$|\langle u_{i,n}^*, x_{i,n} - u_{i,n}\rangle| + |\langle u_{i,n}^*, u_{i,n} - \overline{x}\rangle|$$
$$\leq \|\mu_{i,n} x_{i,n}^* - u_{i,n}^*\|\|x_{i,n} - u_{i,n}\| +$$
$$\|\mu_{i,n} x_{i,n}^* - u_{i,n}^*\|\|u_{i,n} - \overline{x}\| +$$
$$|\langle u_{i,n}^*, x_{i,n} - u_{i,n}\rangle| + |\langle u_{i,n}^*, u_{i,n} - \overline{x}\rangle|$$
$$\leq \frac{1}{n^2} + \frac{1}{n}\|u_{i,n} - \overline{x}\| + \frac{1}{n} + |\langle u_{i,n}^*, u_{i,n} - \overline{x}\rangle| \xrightarrow[n \to +\infty]{} 0,$$

and

$$|\mu_{i,n}(f_i(x_{i,n}) - r_i)| \leq \frac{1}{n}.$$

Finally, for each $i \in I$ we have built sequences $(x_{i,n})_n$ in X, $(x_{i,n}^*)_n$ in X^*, $(\mu_{i,n})_n$ in $\mathbb{R}_+$ such that $x_{i,n}^* \in \partial f_i(x_{i,n})$ with $x_{i,n} \xrightarrow[n \to +\infty]{} \overline{x}$, $\sum_{i \in I} \mu_{i,n} x_{i,n}^* \xrightarrow[n \to +\infty]{} x^*$, $\mu_{i,n}\langle x_{i,n}^*, x_{i,n} - \overline{x}\rangle \xrightarrow[n \to +\infty]{} 0$ and $\mu_{i,n}(f_i(x_{i,n}) - r_i) \xrightarrow[n \to +\infty]{} 0$. Hence we obtain that $x^* \in \limsup_{\begin{cases} x_i \to \overline{x}, \mu_i \geq 0 \\ \mu_i(f_i(x_i) - r_i) \to 0 \\ \mu_i\langle x_i^*, x_i - \overline{x}\rangle \to 0 \end{cases}} \sum_{i \in I} \mu_i \partial f_i(x_i)$, which proves the inclusion

$$N(S; \overline{x}) \subset \operatorname{Lim\,sup}_{\begin{cases} x_i \to \overline{x}, \mu_i \geq 0 \\ \mu_i(f(x_i) - r_i) \to 0 \\ \mu_i\langle x_i^*, x_i - \overline{x}\rangle \to 0 \end{cases}} \sum_{i \in I} \mu_i \partial f_i(x_i) =: Q.$$

The inclusions

$$Q \subset \operatorname*{Lim\,sup}_{\left\{\begin{array}{l}x_i \to \overline{x},\, \mu_i \geq 0 \\ \mu_i\,(f_i(x_i) - r_i - \langle x_i^*, x_i - \overline{x}\rangle) \to 0\end{array}\right.} \sum_{i \in I} \mu_i\, \partial f_i(x_i)$$

$$\subset \operatorname*{Lim\,sup}_{\left\{\begin{array}{l}x_i \to \overline{x},\, \mu_i \geq 0 \\ \sum_{i \in I} \mu_i\,(f_i(x_i) - r_i - \langle x_i^*, x_i - \overline{x}\rangle) \to 0\end{array}\right.} \sum_{i \in I} \mu_i\, \partial f_i(x_i)$$

being obvious, it remains to prove that

$$\operatorname*{Lim\,sup}_{\left\{\begin{array}{l}x_i \to \overline{x},\, \mu_i \geq 0 \\ \sum_{i \in I} \mu_i\,(f_i(x_i) - \lambda_i - \langle x_i^*, x_i - \overline{x}\rangle) \to 0\end{array}\right.} \sum_{i \in I} \mu_i\, \partial f_i(x_i) \subset N(S; \overline{x}).$$

For that purpose, let us suppose that $x^* \in X^*$ is such that $\sum_{i \in I}^m \mu_{i,n}\, x_{i,n}^* \xrightarrow[n \to +\infty]{} x^*$ with $\mu_{i,n} \geq 0$, $x_{i,n}^* \in \partial f_i(x_{i,n})$, $x_{i,n} \xrightarrow[n \to +\infty]{} \overline{x}$ and

(3.189) $$\sum_{i \in I} \mu_{i,n}\,\big(f_i(x_{i,n}) - \lambda_i - \langle x_{i,n}^*, x_{i,n} - \overline{x}\rangle\big) \xrightarrow[n \to +\infty]{} 0.$$

Let us fix any $x \in S$. Since $x_{i,n}^* \in \partial f_i(x_{i,n})$ and since f_i is convex, we have

$$\langle x_{i,n}^*, x - x_{i,n}\rangle \leq f_i(x) - f_i(x_{i,n})$$
$$\leq r_i - f_i(x_{i,n}) \quad \text{because } x \in S.$$

This can be rewritten as

$$\langle x_{i,n}^*, x - \overline{x}\rangle \leq r_i - f_i(x_{i,n}) + \langle x_{i,n}^*, x_{i,n} - \overline{x}\rangle.$$

After multiplication by $\mu_{i,n}$ and summation over $i \in I$, we find

$$\left\langle \sum_{i \in I} \mu_{i,n}\, x_{i,n}^*, x - \overline{x}\right\rangle \leq \sum_{i \in I} \mu_{i,n}\,\big((\lambda_i - f_i(x_{i,n})) + \langle x_{i,n}^*, x_{i,n} - \overline{x}\rangle\big).$$

Recalling the formula (3.189) and the fact that $\sum_{i \in I} \mu_{i,n}\, x_{i,n}^* \xrightarrow[n \to +\infty]{} x^*$, we obtain at the limit as $n \to +\infty$ that $\langle x^*, x - \overline{x}\rangle \leq 0$. Since this is true for every $x \in S$, we conclude that $x^* \in N(S; \overline{x})$. The proof is complete. □

3.18.6. Limiting formulas for normals to vector convex sublevels. We consider now the situation of sublevels of vector-valued convex mappings with closed epigraphs. We recall (see (3.94)) that with the mapping F in the two theorems below

$$0_{Y^*} \circ F = \Psi_{\operatorname{dom} F}.$$

THEOREM 3.225 (limiting descriptions of normals of vector convex sublevels: reflexive Banach space). *Let X and Y be reflexive Banach spaces and Y_+ be a closed convex cone of Y. Let $F : X \to Y \cup \{+\infty_Y\}$ be a convex mapping whose epigraph is closed. Let $\overline{x} \in \operatorname{dom} F$ and $S = \{x \in X : F(x) \leq_{Y_+} F(\overline{x})\}$. Then for $\overline{y} := F(\overline{x})$ one has*

$$N(S; \overline{x}) = \operatorname*{Lim\,sup}_{\left\{\begin{array}{l}x \to \overline{x},\, y^* \in Y_+^* \\ \langle y^*, F(x) - \overline{y}\rangle \to 0\end{array}\right.} \partial(y^* \circ F)(x),$$

where the right-hand member stands for the set of strong limits $\lim x_n^*$ of sequences $(x_n^*)_n$ in X^* such that $x_n^* \in \partial(y_n^* \circ F)(x_n)$, $y_n^* \in Y_+^*$, the sequence $(x_n)_n$ strongly converges to $\overline{x}$ and $\langle y_n^*, F(x_n) - \overline{y}\rangle \to 0$.

PROOF. To prove that the left member is included in the right, define the convex mapping $G : X \to Y \cup \{+\infty_Y\}$ by $G(x) = F(x) - \overline{y}$ and note that one has $S = \{x \in X : G(x) \in -Y_+\}$, and hence $\Psi_S = \Psi_{-Y_+} \circ G$. This entails that

$$N(S; \overline{x}) = \partial \Psi_S(\overline{x}) = \partial \left(\Psi_{-Y_+} \circ G \right)(\overline{x}).$$

The function Ψ_{-Y_+} is lower semicontinuous, convex and Y_+-nondecreasing. Let us fix $x^* \in N(S; \overline{x})$ and apply Theorem 3.203(a) with $g = \Psi_{-Y_+}$ and the mapping G defined above. This furnishes the existence of sequences $(x_n)_n$ in X, $(y_n)_n$ in Y, $(x_n^*)_n$ in X^*, $(e_n^*)_n$ in Y^* and $(y_n^*)_n$ in Y_+^* such that $x_n \to \overline{x}$, $y_n \to F(\overline{x}) - \overline{y}$, $x_n^* \to x^*$, $e_n^* \to 0$ and

$$y_n^* + e_n^* \in N(-Y_+; y_n),$$

$$x_n^* \in \partial(y_n^* \circ F)(x_n)$$

$$\langle y_n^*, F(x_n) - \overline{y} \rangle \to 0 \quad \text{as } n \to +\infty.$$

From this we obtain the desired inclusion of the left member in the statement into the right, and the converse inclusion can be directly shown. $\square$

THEOREM 3.226 (limiting descriptions of normals of vector convex sublevels: general Banach space). Let X and Y be Banach spaces and Y_+ be a closed convex cone of Y. Let $F : X \to Y \cup \{+\infty_Y\}$ be a convex mapping whose epigraph is closed. Let $\overline{x} \in \operatorname{dom} F$ and $S = \{x \in X : F(x) \leq F(\overline{x})\}$. Then for $\overline{y} := F(\overline{x})$ one has

$$N(S; \overline{x}) = {}^{w^*}\!\!\!\operatorname*{Lim\,sup}_{\substack{x \to \overline{x}, y^* \in Y_+^* \\ \langle \cdot, x - \overline{x} \rangle - \langle y^*, F(x) - \overline{y} \rangle \to 0}} \partial(y^* \circ F)(x),$$

where the right-hand member stands for the set of weak star limits ${}^{w^*}\!\lim x_i^*$ of nets $(x_i^*)_i$ in X^* such that $x_i^* \in \partial(y_i^* \circ F)(x_i)$, $y_i^* \in Y_+^*$, the net $(x_i)_i$ strongly converges to $\overline{x}$ and $\langle x_i^*, x_i - \overline{x} \rangle - \langle y_i^*, F(x_i) - \overline{y} \rangle \to 0$.

PROOF. Take any x^* in the left-hand member. Following the proof of Theorem 3.225, according to Theorem 3.203(b), $(x_n)_n$, $(y_n)_n$, $(x_n^*)_n$, $(y_n^*)_n$, and (e_n^*) have to be taken as *nets* instead of sequences, the strong convergence $x_n^* \to x^*$ has to be replaced by the weak-star convergence $x_n^* \overset{w^*}{\to} x^*$, and we have in addition $\langle x_n^*, x_n - \overline{x} \rangle - \langle y_n^*, F(x_n) - \overline{y} \rangle \to 0$. Then, one obtains as in the proof of Theorem 3.214 that

$$N(S; \overline{x}) \subset {}^{w^*}\!\!\!\operatorname*{Lim\,sup}_{\substack{x \to \overline{x}, y^* \in Y_+^* \\ \langle \cdot, x - \overline{x} \rangle - \langle y^*, F(x) - \overline{y} \rangle \to 0}} \partial(y^* \circ F)(x),$$

where the right member is as described in the statement. The converse inclusion is easy to verify. $\square$

Without any qualification condition descriptions of normals to convex sublevels by means of ε-subgradients at the reference point $\overline{x}$ instead of nearby points will be also established in Corollary 3.297.

3.19. Continuity of conjugate functions, weak compactness of sublevels, minimum attainment

In this section we focus our attention on three properties for a convex function $f : X \to \mathbb{R} \cup \{+\infty\}$:
- the continuity of the conjugate f^* at x^* with respect to the Mackey topology;
- the weak compactness of sublevels $\{f - \langle x^*, \cdot \rangle \leq r\}$ for $r \in \mathbb{R}$;
- the attainment of the minimum of $f - \langle x^*, \cdot \rangle$.

3.19.1. Continuity of conjugate functions and weak compactness of sublevels.
Given a nonempty subset V of a locally convex space X, it will be convenient in this subsection to denote by $\sigma_V : X^* \to \mathbb{R} \cup \{+\infty\}$, instead of $\sigma(\cdot, V)$ its support function, that is,

$$\sigma_V(x^*) := \sup_{x \in V} \langle x^*, x \rangle.$$

LEMMA 3.227. Let X be a locally convex space and $f : X \to \mathbb{R} \cup \{-\infty, +\infty\}$ be a function.
(a) If f is bounded from above on a subset V of X by a real μ, then

$$\{f^* \leq r\} \subset \{\sigma_V \leq r + \mu\} \quad \text{for any } r \in \mathbb{R}.$$

(b) If f is finite-valued on X and if V is a subset of X with $0 \in V$, then

$$\left(x' - x \in V \Rightarrow f(x') - f(x) \leq 1\right) \implies \left(\mathrm{dom}\, f^* \subset \{\sigma_V \leq 1\}\right).$$

EXERCISE 3.228 (Proof of Lemma 3.227). (a) Show directly (a) from the definition of conjugate function of f.
(b) Assume that f is finite-valued and $f(x') - f(x) \leq 1$ whenever $x' - x \in V$. Fix any $x^* \in X^*$ with $\sigma_V(x^*) > 1$ and choose $v \in V$ such that $\langle x^*, v \rangle > 1$.
(b$_1$) For each $n \in \mathbb{N}$ argue that $f(nv) \leq f(0) + n$ and $f^*(x^*) \geq -f(0) + n(\langle x^*, v \rangle - 1)$.
(b$_2$) Derive that $x^* \notin \mathrm{dom}\, f^*$ and conclude. $\square$

Given a vector topology θ on the locally convex space X and a function $f : X \to \mathbb{R} \cup \{-\infty, +\infty\}$ which is bounded from above on a θ-neighborhood of $\bar{x} \in X$, the function $x \mapsto f_0(x) = f(x + \bar{x})$ is bounded from above on a symmetric θ-neighborhood V_0 of zero by some real $\mu > 0$. Then for each real r, Lemma 3.227(a) entails that for $s := (|r| + \mu)^{-1}$

$$\{f^* - \langle \cdot, \bar{x} \rangle \leq r\} = \{f_0^* \leq r\} \subset \{\sigma_{V_0} \leq r + \mu\} \subset \{\sigma_{V_0} \leq 1/s\} = (sV_0)^\circ.$$

This justifies the following proposition.

PROPOSITION 3.229 (J.J. Moreau). Let $f : X \to \mathbb{R} \cup \{-\infty, +\infty\}$ be a function on a Hausdorff locally convex space (X, τ_X) and $\bar{x} \in X$.
(a) If f is bounded from above on a θ-neighborhood of $\bar{x}$, where θ is a vector topology on X (not necessarily compatible with the duality between X and X^*), then for each $r \in \mathbb{R}$ there is a θ-neighborhood V of zero such that $\{f^* - \langle \cdot, \bar{x} \rangle \leq r\} \subset V^\circ$.
(b) If f is bounded from above on a τ_X-neighborhood of $\bar{x}$, then for each $r \in \mathbb{R}$ the set $\{f^* - \langle \cdot, \bar{x} \rangle \leq r\}$ is $w(X^*, X)$-compact.

Consider now the topological dual $(E^*, \|\cdot\|_{E^*})$ of a normed space $(E, \|\cdot\|_E)$ and denote by θ the topology on E^* associated with the dual norm $\|\cdot\|_{E^*}$. If for the locally convex space (X, τ_X) we take $(E^*, w(E^*, E))$ (that is, E^* endowed with the topology $w(E^*, E)$), then the topological dual of (X, τ_X) coincides with

E. Further, for any θ-neighborhood V of zero in $X = E^*$ there is a real $\alpha > 0$ such that $\alpha^{-1}\mathbb{B}_{E^*} \subset V$, hence taking the polar with X equipped with the τ_X-topology, we see that $V^\circ \subset \alpha \mathbb{B}_E$.

COROLLARY 3.230 (R.T. Rockafellar). *Let $(E, \|\cdot\|_E)$ be a normed space and $\|\cdot\|_{E^*}$ be the dual norm on the topological dual E^*. If a function $f : E^* \to \mathbb{R} \cup \{-\infty, +\infty\}$ is bounded from above on a $\|\cdot\|_{E^*}$-neighborhood of a point $\bar{x}^* \in E^*$, then for each $r \in \mathbb{R}$ the set $\{f^* - \langle \bar{x}^*, \cdot \rangle \leq r\}$ is $\|\cdot\|_E$-bounded in E, where the conjugate f^* of f is taken in E.*

COROLLARY 3.231. *Let $f : X \to \mathbb{R} \cup \{-\infty, +\infty\}$ be a function on a Hausdorff locally convex space (X, τ_X) and $\bar{x} \in X$. If f is bounded from above on a $w(X, X^*)$-neighborhood of $\bar{x}$, then $\mathrm{span}\,(\mathrm{dom}\,f^*)$ is finite-dimensional and for each $r \in \mathbb{R}$ the set $\{f^* - \langle \cdot, \bar{x}\rangle \leq r\}$ is bounded relative to the space $\mathrm{span}\,(\mathrm{dom}\,f^*)$.*

EXERCISE 3.232 (Proof of Corollary 3.231). Suppose that f is bounded from above on a $w(X, X^*)$-neighborhood of $\bar{x}$, so as above one may and do suppose that $\bar{x} = 0$. If $f^* \equiv +\infty$ there is nothing to prove, so suppose also that $f^* \not\equiv +\infty$.
(α) Argue that there are $r_0 \in \mathbb{R}$ with $r_0 > \inf_{X^*} f^*$ and $u_1^*, \cdots, u_m^* \in X^*$ such that $\{f^* \leq r_0\} \subset \mathrm{co}\,\{u_1^*, \cdots, u_m^*\}$, and derive a finite-dimensional subspace L of X^* such that $\{f^* \leq r_0\} \subset L$. (Use Lemma 3.227).
(β) Argue that $\{f^* \leq r\} \subset L$ for all reals $r \leq r_0$.
(γ) Let any real $r > r_0$. By the inequality $r_0 > \inf_{X^*} f^*$ fix some $x_0^* \in X^*$ with $s_0 := f^*(x_0^*) < r_0$ and let $\lambda \in {]}0, 1[$ with $r_0 = (1-\lambda)s_0 + \lambda r$.
(γ_1) Argue that $x_0^* \in L$ and that $x^* \in \{f^* \leq r\} \Rightarrow (1-\lambda)x_0^* + \lambda x^* \in \{f^* \leq r_0\}$.
(γ_2) Deduce that $\{f^* \leq r\} \subset L$.
(γ_3) Derive that $\mathrm{dom}\,f^* \subset L$ and use (I-a) in the proposition to conclude for each $r \in \mathbb{R}$ that $\{f^* \leq r\}$ is bounded in L. □

The next proposition states a remarkable characterization of uniform continuity, with respect to a vector topology θ, of a function $f \in \Gamma(X, w(X, X^*))$ via the domain of its conjugate f^*.

PROPOSITION 3.233. *Let (X, τ_X) be a Hausdorff locally convex space and θ be a vector topology on X (not necessarily compatible with the duality between X and X^*). Let $f : X \to \mathbb{R} \cup \{-\infty, +\infty\}$ be a proper lower τ_X-semicontinuous convex function. Then, f is finite and uniformly θ-continuous on X if and only if there is some θ-neighborhood V of zero in X such that $\mathrm{dom}\,f^* \subset V^\circ$.*

PROOF. The implication $\Rightarrow$ follows from Lemma 3.227(b). For the converse implication, we note by assumption that there is a balanced θ-neighborhood V of zero in X such that $\mathrm{dom}\,f^* \subset V^\circ$. Consider any real $\varepsilon > 0$ and take any x_1, x_2 in X with $x_1 - x_2 \in \varepsilon V$. Then, for every $u^* \in \mathrm{dom}\,f^* \subset V^\circ$ we have $\langle u^*, x_1 - x_2 \rangle \leq \varepsilon$, hence
$$\langle u^*, x_1 \rangle - f^*(u^*) \leq \langle u^*, x_2 \rangle - f^*(u^*) + \varepsilon,$$
which gives by taking the supremum in both members over $u^* \in \mathrm{dom}\,f^*$ that $f^{**}(x_1) \leq f^{**}(x_2) + \varepsilon$. The function f being proper, convex and lower $w(X, X^*)$-semicontinuous, we have $f^{**} = f$ on X, thus $f(x_1) \leq f(x_2) + \varepsilon$ for all $x_1, x_2 \in X$ with $x_1 - x_2 \in \varepsilon V$. Fixing $a \in X$ with $f(a) \in \mathbb{R}$ we see that f is finite on $a + \varepsilon V$, hence by induction f is finite on $a + n\varepsilon V$ for all $n \in \mathbb{N}$. Since $X = \bigcup_{n \in \mathbb{N}} n\varepsilon V$, the

function f is finite-valued on X. We then conclude that f is finite-valued on X and uniformly θ-continuous on X. □

If X is a normed space, it is easy to see that the polar of $\mathbb{B}_X$ (resp. of $\mathbb{B}_{X^*}$ relative to X) is equal to $\mathbb{B}_{X^*}$ (resp. $\mathbb{B}_X$). The following corollary of the above proposition then follows.

COROLLARY 3.234 (boundedness of domain of conjugate). Let $(X, \|\cdot\|_X)$ be a normed space and f be a proper convex extended real-valued function from X (resp. X^*) into $\mathbb{R} \cup \{-\infty, +\infty\}$ which is lower $\|\cdot\|_X$-semicontinuous (resp. $w(X^*, X)$-semicontinuous). Let f^* denote the conjugate of f (resp. the conjugate of f relative to X). Then, f is finite and uniformly continuous on X (resp. on X^*) with respect to the norm if and only if the set $\mathrm{dom}\, f^*$ is bounded in X^* (resp. in X) with respect to the dual norm on X^* (resp. with respect to the norm on X).

Our next step is to obtain a characterization of θ-continuity at a point of a function $f \in \Gamma(X, w(X, X^*))$ via sublevel sets of the conjugate f^* in the line of the property in Lemma 3.227(a). We begin with the following two lemmas. Recall that, for a subset C of a vector space, j_C denotes its Minkowski gauge function.

LEMMA 3.235. Let $\varphi : Y \to \mathbb{R} \cup \{+\infty\}$ be a lower semicontinuous convex function on a Hausdorff locally convex space Y with $\varphi(0) = 0$ and let $C := \{\varphi \leq r\}$ with $r \in]0, +\infty[$. Let $v \in Y$ be such that $C \cap \mathbb{R}_+ v = [0, \tau v]$ for some real $\tau > 0$. Then one has
$$r\, j_C(tv) \geq \varphi(tv) \quad \text{for all } t \in [0, \tau]$$
$$r\, j_C(tv) \leq \varphi(tv) \quad \text{for all } t > \tau,$$

PROOF. Clearly, C is a $w(Y, Y^*)$-closed convex set containing zero and $C \cap \mathbb{R}_+ u = [0, u]$ for $u := \tau v$. Without loss of generality, we may suppose $v \neq 0$, so $j_C(u) = 1$. Since $u \in C$ and φ is convex with $\varphi(0) = 0$, for any $t \in [0, 1]$ we have
$$r\, j_C(tu) = tr\, j_C(u) = tr \geq t\varphi(u) \geq \varphi(tu).$$
It remains to show that $r\, j_C(tu) \leq \varphi(tu)$ for all $t > 1$. If $\varphi(tu) = +\infty$ for all $t > 1$ there is noting to prove. Suppose that there is some real $s > 1$ with $\varphi(su)$ finite. The restriction to $[0, su]$ of the lower semicontinuous convex function φ is finite-valued and continuous (see Proposition 3.20). We then have $\varphi(u) = r$ since otherwise the inequality $\varphi(u) < r$ would furnish some $s_0 \in]1, s[$ with $\varphi(s_0 u) < r$, which would give $s_0 u \in C$, and this is impossible because $C \cap \mathbb{R}_+ u = [0, u]$. Since $\varphi(0) = 0$, it results that for every real $t > 1$
$$\varphi(tu) \geq t\varphi(u) = tr = rt\, j_C(u) = r\, j_C(tu),$$
which clearly finishes the proof of the lemma. □

LEMMA 3.236. Let $\varphi : Y \to \mathbb{R} \cup \{+\infty\}$ be a lower semicontinuous convex function on a Hausdorff locally convex space Y with $\varphi(0) = 0$ and let $C := \{\varphi \leq r\}$ with $r \in]0, +\infty[$. Let a nonzero $v \in Y$ be such that $C \cap \mathbb{R}_+ v = \{0\}$. Then $\varphi(tv) = +\infty$ for every real $t > 0$.

PROOF. Suppose that $\varphi(t_0 v) < +\infty$ for some real $t_0 > 0$. Since $C \cap \mathbb{R}_+ v = \{0\}$, we have $\varphi(t_0 v) > r$. Further, the restriction to $[0, t_0]$ of the lower semicontinuous convex function $t \mapsto \varphi(tv)$ is finite-valued and continuous (see Proposition 3.20), so by the intermediate value theorem there is some $t_1 \in]0, t_0[$ such that $\varphi(t_1 v) = r$. This contradicts the equality $C \cap \mathbb{R}_+ v = \{0\}$ and finishes the proof of the lemma. □

THEOREM 3.237 (J.J. Moreau: continuity of $f \in \Gamma_0(X)$ via f^*). Let (X, τ_X) be a Hausdorff locally convex space, X^* be its topological dual and $f : X \to \mathbb{R} \cup \{+\infty\}$ be a proper lower τ_X-semicontinuous convex function. Let also θ be a vector topology on X (not necessarily compatible with the duality between X and X^*). For $\bar{x} \in X$ the following assertions are equivalent:
(a) the function f is finite at $\bar{x}$ and θ-continuous at $\bar{x}$;
(b) for every $r \in \mathbb{R}$ there exists a θ-neighborhood V of zero in X such that $\{f^* - \langle \cdot, \bar{x} \rangle \leq r\} \subset V^\circ$;
(c) there exists some real $r > \inf_{X^*} f^*$ and a θ-neighborhood V of zero in X such that $\{f^* - \langle \cdot, \bar{x} \rangle \leq r\} \subset V^\circ$.

PROOF. Since $f^* - \langle \cdot, \bar{x} \rangle = g^*$, where $g(x) = f(x + \bar{x})$, we may suppose that $\bar{x} = 0$. The implication (a) $\Rightarrow$ (b) follows from Lemma 3.227(a) and the implication (b) $\Rightarrow$ (c) is trivial. Suppose that (c) holds, that is, suppose that there are a real $r > \inf_{X^*} f^*$ and a balanced θ-neighborhood V of zero in X such that $\{f^* \leq r\} \subset V^\circ$. Since f^* is lower $w(X^*, X)$-semicontinuous, convex and proper (see Proposition 3.77(c)), there exist $\beta \in \mathbb{R}$ and $a \in X$ such that $\langle \cdot, a \rangle + \beta \leq f^*$ according to Proposition 3.77(a). Choose a real $\alpha > 0$ such that $\alpha a \in V$. Then for each $x^* \in \{f^* \leq r\}$ we have

$$f^*(x^*) \geq \beta + \langle x^*, a \rangle = \beta + \alpha^{-1} \langle x^*, \alpha a \rangle \geq \beta - \alpha^{-1},$$

so f^* is bounded from below on $\{f^* \leq r\}$, and hence bounded from below on X^*. This combined with the properness of f^* implies $\inf_{X^*} f^* \in \mathbb{R}$.

Put $\gamma := \inf_{X^*} f^* = -f(0)$ and (since $r > \inf_{X^*} f^*$) choose $\bar{x}^* \in X^*$ such that

$$\lambda := f^*(\bar{x}^*) \in [\gamma, r[.$$

It ensues that $\bar{x}^* \in \{f^* \leq r\} \subset V^\circ$, that is, $\langle \bar{x}^*, x \rangle \leq 1$ for all $x \in V$, thus $x \mapsto \langle \bar{x}^*, x \rangle$ is θ-continuous on X. To obtain the θ-continuity of f at zero it then suffices to show the θ-continuity at zero of the function $g := f - \langle \bar{x}^*, \cdot \rangle + \lambda$. We know (see (3.30)) that $g^*(x^*) = f^*(x^* + \bar{x}^*) - \lambda$ for all $x^* \in X^*$, hence $g^*(0) = 0$ along with

(3.190) $$\inf_{X^*} g^* = \inf_{X^*} f^* - \lambda = \gamma - \lambda \leq 0.$$

Furthermore,

(3.191)
$$C := \{g^* \leq r - \lambda\} = \{f^* \leq r\} - \bar{x}^* \subset \{f^* \leq r\} - \{f^* \leq r\} \subset V^\circ + V^\circ = 2V^\circ.$$

Fix any nonzero $u^* \in X^*$ and note by (3.191) that $C \cap \mathbb{R}_+ u^* = [0, \tau u^*]$ for some real $\tau \geq 0$. Suppose first that $\tau > 0$, so for any real $t > \tau$ by Lemma 3.235 we have $(r - \lambda) j_C(tu^*) \leq g^*(tu^*)$, hence

$$(r - \lambda) j_C(tu^*) + \gamma - r \leq g^*(tu^*).$$

Further, for any $t \in [0, \tau]$ noting that $tu^* \in C$ (since $[0, \tau u^*] = C \cap \mathbb{R}_+ u^*$) we see from the latter inequality and from (3.190) that

$$(r - \lambda) j_C(tu^*) + \gamma - r \leq r - \lambda + \gamma - r = \gamma - \lambda \leq \inf_{X^*} g^* \leq g^*(tu^*).$$

It results that $(r - \lambda) j_C + \gamma - r \leq g^*$ on $\mathbb{R}_+ u^*$. If $\tau = 0$, then Lemma 3.236 says that $g^*(tu^*) = +\infty$ for all $t > 0$. Thus, in any case we have that $(r - \lambda) j_C + \gamma - r \leq g^*$ on $\mathbb{R}_+ u^*$, and this clearly entails that $(r - \lambda) j_C + \gamma - r \leq g^*$ on the whole space X^*. It follows that $g \leq ((r-\lambda)j_C)^* - \gamma + r$ because the $w(X, X^*)$-lower semicontinuous proper convex function g is the conjugate of g^* on X (see Theorem 3.84(c)). Using (3.26) and 3.84(c) again, we also see with $s := r - \lambda > 0$ that $(sj_C)^*(x) = j_C^*(s^{-1}x) = \Psi_{C^\circ}(s^{-1}x)$, so $(sj_C)^* = \Psi_{sC^\circ}$. This and the latter inequality related to g yield that g is bounded from above on sC° by $r - \gamma$. Since $C \subset 2V^\circ = (2^{-1}V)^\circ$, we also have $(s/2)V \subset sC^\circ$, thus g is bounded from above on the θ-neighborhood $(s/2)V$ of zero. This guarantees (by Proposition 3.14) that the convex function g is θ-continuous at zero and finishes the proof. □

From the definition of the Mackey topology of a topological dual space as recalled in (3.103) we see that a weakly compact set in a (Hausdorff) locally convex space X is a weakly closed set in X included in the polar (relative to X) of a neighborhood of zero in X^* with respect to the Mackey topology of X^*. So, reminding that the biconjugate relative to X of a proper lower semicontinuous convex function f coincides with f and applying Theorem 3.237 with f^* in place of f, we obtain the following corollary.

COROLLARY 3.238 (J.J. Moreau: weak compactness of sublevels). Let (X, τ_X) be a Hausdorff locally convex space, X^* be its topological dual and $f : X \to \mathbb{R} \cup \{+\infty\}$ be a proper lower τ_X-semicontinuous convex function. For $\bar{x}^* \in X^*$ the following assertions are equivalent:
(a) the conjugate function f^* is finite at $\bar{x}^*$ and continuous at $\bar{x}^*$ with respect to the Mackey topology on X^*;
(b) for every $r \in \mathbb{R}$ the set $\{f - \langle \bar{x}^*, \cdot \rangle \leq r\}$ is $w(X, X^*)$-compact;
(c) there exists some real $r > \inf_X f$ such that the set $\{f - \langle \bar{x}^*, \cdot \rangle \leq r\}$ is $w(X, X^*)$-compact.

Given a Hausdorff locally convex space (X, τ_X) and its topological dual X^*, let us associate with each nonempty symmetric convex τ_X-bounded set B in X (resp. $w(X^*, X)$-bounded set B in X^*) the seminorm p_B on X^* (resp. on X) defined by

$$p_B(x^*) := \sup_{x \in B} \langle x^*, x \rangle \quad \text{for all } x^* \in X^*$$

$$(\text{resp. } p_B(x) := \sup_{x^* \in B} \langle x^*, x \rangle \quad \text{for all } x \in X).$$

The locally convex topology on X^* (resp. on X) generated by this family of seminorms, with nonempty symmetric convex τ_X-bounded (resp. $w(X^*, X)$-bounded) sets B in X (resp. in X^*), is known in the literature (as recalled in Appendix) as the bounded topology on X^* (resp. on X). This topology is usually denoted by $\beta(X^*, X)$ (resp. $\beta(X, X^*)$) (and when X is a normed space and τ_X is the topology associated with the norm, we note that $\beta(X^*, X)$ (resp. $\beta(X, X^*)$) coincides with the topology on X^* associated with the dual norm (resp. on X associated with the primal norm). From the very definition of $\beta(X^*, X)$ (resp. $\beta(X, X^*)$) we see that for any $\beta(X^*, X)$-neighborhood (resp. $\beta(X, X^*)$-neighborhood) V of zero in X^* (resp. in X) the polar set V° in X (resp. in X^*) is τ_X-bounded in X (resp. $w(X^*, X)$-bounded in X^*). Applying Theorem 3.237 with primal space $(X^*, w(X^*, X))$, with the function $g := f^* \in \Gamma_0(X^*, w(X^*, X))$ and with θ as the topology $\beta(X^*, X)$, we obtain the following equivalences.

THEOREM 3.239 (R.T. Rockafellar: general boundedness of sublevels). Let (X, τ_X) be a Hausdorff locally convex space, X^* be its topological dual and $f : X \to \mathbb{R} \cup \{+\infty\}$ be a proper lower τ_X-semicontinuous convex function. Let $\beta(X^*, X)$ the

bounded topology on X^* recalled above. For $\bar{x}^* \in X^*$ the following assertions are equivalent:
(a) the function f^* is finite at $\bar{x}^*$ and $\beta(X^*, X)$-continuous at $\bar{x}^*$;
(b) for every $r \in \mathbb{R}$ the set $\{f - \langle \bar{x}^*, \cdot \rangle \leq r\}$ is τ_X-bounded in X;
(c) there exists some real $r > \inf_X f$ such that the set $\{f - \langle \bar{x}^*, \cdot \rangle \leq r\}$ is τ_X-bounded in X.

REMARK 3.240. Let any function $f : X \to \mathbb{R} \cup \{-\infty, +\infty\}$ on a locally convex space X and let $\bar{x} \in X$. In addition to the above theorem and to Corollary 3.230, a direct consequence of Lemma 3.227 also tells us that the sublevels $\{f^* - \langle \cdot, \bar{x} \rangle \leq r\}$ (with $r \in \mathbb{R}$) are $\beta(X^*, X)$ bounded whenever the function f is bounded from above on a $\beta(X, X^*)$-neighborhood of point $\bar{x}$. $\square$

In the important setting of normed spaces the above result becomes:

COROLLARY 3.241 (R.T. Rockafellar: boundedness of sublevels in normed spaces). Let $(X, \|\cdot\|_X)$ be a normed space, $(X^*, \|\cdot\|_{X^*})$ be the dual normed space and $f : X \to \mathbb{R} \cup \{+\infty\}$ be a proper lower semicontinuous convex function. For $\bar{x}^* \in X^*$ the following assertions are equivalent:
(a) the function f^* is finite at $\bar{x}^*$ and $\|\cdot\|_{X^*}$-continuous at $\bar{x}^*$;
(b) for every $r \in \mathbb{R}$ the set $\{f - \langle \bar{x}^*, \cdot \rangle \leq r\}$ is $\|\cdot\|_X$-bounded in X;
(c) there exists some real $r > \inf_X f$ such that the set $\{f - \langle \bar{x}^*, \cdot \rangle \leq r\}$ is $\|\cdot\|_X$-bounded in X.

Theorem 3.237 also admits as corollary the following basic result in normed spaces.

COROLLARY 3.242 (J.J. Moreau: weak-star compactness of sublevels in normed spaces). Let $(X, \|\cdot\|)$ be a normed space and $f : X \to \mathbb{R} \cup \{+\infty\}$ be a proper lower semicontinuous convex function. For $\bar{x} \in X$ the following assertions are equivalent:
(a) the function f is finite at $\bar{x}$ and $\|\cdot\|$-continuous at $\bar{x}$;
(b) for every $r \in \mathbb{R}$ the set $\{f^* - \langle \cdot, \bar{x} \rangle \leq r\}$ is $w(X^*, X)$-compact in X^*;
(c) there exists some real $r > \inf_{X^*} f^*$ such that the set $\{f^* - \langle \cdot, \bar{x} \rangle \leq r\}$ is $w(X^*, X)$-compact in X^*.

3.19.2. Attainment of the minimum of $f - \langle x^*, \cdot \rangle$. Given a proper lower semicontinuous convex function $f : X \to \mathbb{R} \cup \{+\infty\}$, the weak compactness of $\{f - \langle x^*, \cdot \rangle \leq r\}$ for all $r \in \mathbb{R}$ obviously entails that the function $f - \langle x^*, \cdot \rangle$ attains its minimum on X. This latter attainment property is equivalent to say that $x^* \in \partial f(x)$ for some $x \in X$, or equivalently $x^* \in \text{Dom} \, \partial f^*$ for $\partial f^* : X^* \rightrightarrows X$ with X^* equipped with the $w(X^*, X)$-topology. Further, by Corollary 3.238, for a nonempty open set $U \subset X^*$ with respect to the Mackey topology $\mu(X^*, X)$ in X^*, the sublevels of the function $f - \langle x^*, \cdot \rangle$ are weakly compact in X for every $x^* \in U$ if and only if f^* is continuous on U with respect to the Mackey topology on X^*, and anyone of these two properties implies that $U \subset \text{Dom} \, \partial f^*$. If in addition X is a reflexive Banach space, the Mackey topology on X^* coincides with the dual norm topology on X^* (as known and easily seen), so X^* endowed with this topology is a complete space and for the nonempty open set U

$(f^* \text{ finite and continuous on } U) \Leftrightarrow (U \subset \text{dom} f^*) \Leftrightarrow (U \subset \text{Dom} \, \partial f^*)$.

Consequently, if X is a reflexive Banach space we see, for a nonempty open set $U \subset X^*$ with respect to the Mackey topology, that the following three properties

are equivalent:
- the function $f - \langle x^*, \cdot \rangle$ attains its minimum on X for every $x^* \in U$;
- the function f^* is finite on U and continuous on U with respect to the Mackey topology on X^*;
- the inclusion $U \subset \operatorname{Dom} \partial f^*$ holds for $\partial f^* : X^* \rightrightarrows X$.

Our next goal is to showing the equivalence of these three properties in any Banach space (without requiring the reflexivity). We will even study the equivalences in the framework of a complete Hausdorff locally convex space.

We recall by Corollary 3.85 that, if the function $f : X \to \mathbb{R} \cup \{+\infty\}$ is proper, lower semicontinuous and convex, then

(3.192) $\qquad (\partial f)^{-1}(x^*) = \partial f^*(x^*) \quad \text{for all } x^* \in X^*,$

where ∂f^* is considered as a multimapping from X^* into X, or equivalently $\partial f^*(x^*)$ is taken for X^* endowed with the weak* topology, so its topological dual is X.

The statement of Theorem 3.245 will involve the following class of functions.

DEFINITION 3.243. Consider a set $K \subset X^*$ with $0 \in K$. We define $\mathcal{E}(K)$ as the class of all functions $\varphi : X \to \mathbb{R} \cup \{+\infty\}$ such that
  (i) $\varphi^* : X^* \to \mathbb{R} \cup \{+\infty\}$ is proper with $\varphi^*(0) = 0$;
  (ii) $K_\infty \subset \operatorname{dom} \varphi^* \subset K$, where we recall that K_∞ denotes the asymptotic cone of K (see Definition 3.126);
  (iii) for each $x^* \in K \setminus K_\infty$, $\sup\{\varphi^*(\eta x^*) : \eta > 0, \eta x^* \in \operatorname{dom} \varphi^*\} = +\infty$;
  (iv) for each $x^* \in \operatorname{dom} \varphi^*$, $(\partial_{MR}\varphi)^{-1}(x^*) \neq \emptyset$.

Concerning the above condition (iv) it is worth noting that given $x^* \in X^*$ one has

(3.193) $\qquad (\partial_{MR}\varphi)^{-1}(x^*) \neq \emptyset \Leftrightarrow \varphi - \langle x^*, \cdot \rangle \text{ attains its minimum on } X.$

EXAMPLE 3.244. A first example is furnished with $K = X^*$. Indeed, in such a case taking for φ the indicator function $\Psi_{\{0\}}$ of the singleton $\{0\}$ in X, the function φ^* is the null linear functional defined on X. Then, concerning Definition 3.243 conditions (i) is obvious and $K_\infty = \operatorname{dom} \varphi^* = K$ since $K_\infty = X^*$, so condition (ii) is satisfied. Finally, condition (iv) is clear and the equality $K \setminus K_\infty = \emptyset$ ensures that condition (iii) is also satisfied. We then conclude that $\mathcal{E}(X^*) \neq \emptyset$. □

Other basic examples of subsets $K \subset X^*$ with $\mathcal{E}(K) \neq \emptyset$ will be given later. Before that, we illustrate with Theorem 3.245 below how the nonemptiness of $\mathcal{E}(K)$ yields to various first compactness properties of sublevels.

THEOREM 3.245 (relative w-compactness of sublevels). Let X be a complete Hausdorff locally convex space. Let $K \subset X^*$ with $0 \in K$, $\mathbb{R}_+ K = X^*$ and $\mathcal{E}(K) \neq \emptyset$, and let $f : X \to \mathbb{R} \cup \{+\infty\}$ be a function such that $(\partial_{MR}f)^{-1}(x^*) \neq \emptyset$ for all $x^* \in K$. Then for every $r \in \mathbb{R}$ and every $\varphi \in \mathcal{E}(K)$ the sets

$$\operatorname{Ep}_f^\varphi(r) := \{(x, \alpha) \in X \times \mathbb{R} : (x, \alpha) \in \operatorname{epi} f + \operatorname{epi} \varphi \text{ and } \alpha \leq r\},$$
$$S_{f \Box \varphi}(r) := \{(x, \alpha) \in X \times \mathbb{R} : f \Box \varphi(x) \leq \alpha \leq r\},$$
$$S_f(r) := \{(x, \alpha) \in X \times \mathbb{R} : f(x) \leq \alpha \leq r\},$$
$$\{f \leq r\} := \{x \in X : f(x) \leq r\}$$

are all relatively weakly compact.

PROOF. Fix any $\varphi \in \mathcal{E}(K)$. By our assumption $(\partial_{MR}f)^{-1}(0) \neq \emptyset$ we see that $\inf_X f$ is finite (see (3.193)) and that choosing $x_0 \in \partial f^*(0) \neq \emptyset$ (see Proposition 3.64(i)) and setting $\beta_0 := -\inf_X f \in \mathbb{R}$ give $f^* \geq \langle \cdot, x_0 \rangle + \beta_0$ on X^*. Putting $\alpha_0 := 1 - \inf_X f$, for the function $g := f(\cdot + x_0) + \alpha_0$ we have $\inf_X g = 1$ and $g^* = f^* - \langle \cdot, x_0 \rangle - \alpha_0 \geq \beta_0 - \alpha_0$. Further, it is not difficult to see that

$$\mathrm{Ep}_g^\varphi(r) = (-x_0, \alpha_0) + \mathrm{Ep}_f^\varphi(r - \alpha_0), \quad S_{g \square \varphi}(r) = (-x_0, \alpha_0) + S_{f \square \varphi}(r - \alpha_0)$$

$$S_g(r) = (-x_0, \alpha_0) + S_f(r - \alpha_0), \quad \{g \leq r\} = -x_0 + \{f \leq r - \alpha_0\}.$$

Accordingly, we may and do suppose that $\inf_X f = 1$ and $\inf_{X^*} f^* > -\infty$. Then we consider $E := \mathrm{epi}\, f + \mathrm{epi}\, \varphi$ and (since $\inf_X \varphi = 0$ by Definition 3.243) we can define $T : E \to X \times \mathbb{R}$ by $T(x, \alpha) = \alpha^{-1}(x, -1)$. It is clear that T is a homeomorphism from E to $T(E)$ relative to the topologies induced by the weak topology.

Let us prove that $A := T(E) \cup \{(0, 0)\}$ is relatively weakly compact. Fix any arbitrary $(x^*, \rho) \in K \times \mathbb{R}$ and consider two cases.

Case 1: $x^* \in K_\infty$ and $f^*(\eta x^*) + \varphi^*(\eta x^*) \leq \eta \rho$ for all $\eta > 0$.

In this case for any real $\eta > 0$, and for any $a, b \in X$

$$\langle x^*, a \rangle - \eta^{-1} f(a) + \langle x^*, b \rangle - \eta^{-1} \varphi(b) \leq \rho.$$

Consequently, for every $(x, \nu) = (a + b, \alpha + \beta) \in E$ with $(a, \alpha) \in \mathrm{epi}\, f$ and $(b, \beta) \in \mathrm{epi}\, \varphi$ and for every $\eta > 0$

$$\langle (x^*, \rho), T(x, \nu) \rangle = (\alpha + \beta)^{-1} (\langle x^*, a \rangle + \langle x^*, b \rangle - \rho)$$
$$\leq (\alpha + \beta)^{-1} \left(\eta^{-1} f(a) + \eta^{-1} \varphi(b) \right) \leq \eta^{-1}.$$

Since the real $\eta > 0$ is arbitrary, we get $\langle (x^*, \rho), T(x, \nu) \rangle \leq 0$, and in addition

$$\langle (x^*, \rho), (0, 0) \rangle = 0,$$

hence it ensues that (x^*, ρ) attains its maximum over A at $(0, 0)$.

Case 2: Either (i) $x^* \in K_\infty$ and $f^*(\eta x^*) + \varphi^*(\eta x^*) > \eta \rho$ for some $\eta > 0$, or (ii) $x^* \in K \setminus K_\infty$.

In either case (i) or case (ii) we consider the proper lower semicontinuous convex function h on $\mathbb{R}$ defined by $h(t) = f^*(tx^*) + \varphi^*(tx^*) - \rho t$ for any $t \in \mathbb{R}$; this is a continuous function over its effective domain (see Corollary 3.21). Furthermore, in both cases $h(0) = f^*(0) + \varphi^*(0) = -1 < 0$ and $\sup_{\substack{t \in \mathrm{dom}\, h \\ t > 0}} h(t) > 0$. Indeed, in case (i) the latter inequality is evident, and in case (ii) noting that the assumption $(\partial_{MR}f)^{-1}(x^*) \neq \emptyset$ for all $x^* \in K$ implies that f^* is finite on $K \supset \mathrm{dom}\, \varphi^*$ (see (ii) in Definition 3.243), we see that

$$\{t > 0 : t \in \mathrm{dom}\, h\} = \{t > 0 : tx^* \in \mathrm{dom}\, \varphi^*\},$$

and this implies that

$$\sup_{\substack{t \in \mathrm{dom}\, h \\ t > 0}} h(t) \geq \inf_{X^*} f^* + \sup_{\substack{tx^* \in \mathrm{dom}\, \varphi^* \\ t > 0}} \varphi^*(tx^*) - t_0 \max\{0, \rho\} = +\infty,$$

where $t_0 := \sup\{t > 0 : tx^* \in \mathrm{dom}\, \varphi^*\} < +\infty$ because $x^* \notin K_\infty$ and where we use (iii) in Definition 3.243. Then, by the intermediate value theorem there exists a real $\delta > 0$ such that

$$f^*(\delta x^*) + \varphi^*(\delta x^*) = \delta \rho.$$

Since $\delta x^* \in \operatorname{dom} \varphi^*$ and $\delta x^* \in K$ (because $\operatorname{dom} \varphi^* \subset K$), there exist by Proposition 3.64(i) and the hypothesis relative to φ and f some $a_0, b_0 \in X$ such that $\varphi^*(\delta x^*) = \langle \delta x^*, b_0 \rangle - \varphi(b_0)$ and $f^*(\delta x^*) = \langle \delta x^*, a_0 \rangle - f(a_0)$, and this implies that

$$\langle x^*, a_0 \rangle + \langle x^*, b_0 \rangle - \delta^{-1}(f(a_0) + \varphi(b_0)) = \rho,$$

and for all $a, b \in X$

$$\langle x^*, a \rangle + \langle x^*, b \rangle - \delta^{-1}(f(a) + \varphi(b)) \leq \rho.$$

Then, for every $(x, \nu) = (a + b, \alpha + \beta) \in E$ with some $(a, \alpha) \in \operatorname{epi} f$ and $(b, \beta) \in \operatorname{epi} \varphi$, we have

$$\langle (x^*, \rho), T(x, \nu) \rangle = (\alpha + \beta)^{-1}(\langle x^*, a \rangle + \langle x^*, b \rangle - \rho)$$
$$\leq (\alpha + \beta)^{-1}(\delta^{-1}f(a) + \delta^{-1}\varphi(b)) \leq \delta^{-1},$$

$\langle (x^*, \rho), T(a_0 + b_0, f(a_0) + \varphi(b_0)) \rangle = \delta^{-1}$ and $\langle (x^*, \rho), (0, 0) \rangle = 0$, and all together entail that (x^*, ρ) attains its maximum over A at $T(a_0 + b_0, f(a_0) + \varphi(b_0))$.

Since $\mathbb{R}_+(K \times \mathbb{R}) = X^* \times \mathbb{R}$, we conclude that every $(x^*, \rho) \in X^* \times \mathbb{R}$ attains its maximum over A. Then, since the locally convex space X is complete, by James' theorem (see Theorem C.10 in Appendix) A is relatively weakly compact.

Fix any real $r \geq 1$ and consider any net $(x_i, \nu_i)_{i \in I}$ in $\operatorname{Ep}_f^\varphi(r)$. Since $1 \leq \nu_i \leq r$, we may suppose that $(\nu_i)_i$ converges to some $\nu \in [1, r]$. Since $z_i := \nu_i^{-1}(x_i, -1)$ is in $T(E)$, there is a subnet $(z_{s(j)})_{j \in J}$ converging weakly to some (u, λ) in $X \times \mathbb{R}$, so $(x_{s(j)})_{j \in J}$ converges weakly to νu in X. Then $(x_{s(j)}, \nu_{s(j)})_{j \in J}$ converges weakly, which justifies that $\operatorname{Ep}_f^\varphi(r)$ is relatively weakly compact.

Now noting that $(x, \nu) \in S_{f \square \varphi}(r)$ entails that $(x, \nu + 1) \in \operatorname{Ep}_f^\varphi(r + 1)$, we see that $S_{f \square \varphi}(r) \subset (0, -1) + \operatorname{Ep}_f^\varphi(r + 1)$, hence $S_{f \square \varphi}(r)$ is relatively weakly compact.

In order to prove the desired property of $S_f(r)$, consider $x_0 \in \operatorname{dom} \varphi$ and take the function $\tilde{f}(\cdot) := f(\cdot + x_0)$. Since $\tilde{f} \square \varphi(x) \leq f(x) + \varphi(x_0)$, we get that $S_f(r) \subset (0, -\varphi(x_0)) + S_{\tilde{f} \square \varphi}(r + \varphi(x_0))$, which proves that $S_f(r)$ is relatively weakly compact.

Finally, the relative weak compactness of the sublevel set $\{f \leq r\}$ follows from the inclusion $\{f \leq r\} \times \{r\} \subset S_f(r)$. □

The next lemmas provide various conditions for the non-emptiness of $\mathcal{E}(K)$. The first lemma considers the case of a strict sublevel of a finite-valued lower w^*-semicontinuous convex function on X^*.

LEMMA 3.246. *Let X be a Hausdorff locally convex space and let $\beta \in \mathbb{R} \cup \{+\infty\}$. Let $g : X^* \to \mathbb{R}$ be a lower weak* semicontinuous convex function such that $0 \in \{g < \beta\} := \{x^* \in X^* : g(x^*) < \beta\}$ and*

$$\partial g(x^*) \neq \emptyset, \ \forall x^* \in \{g < \beta\}.$$

Then $\mathcal{E}(\{g < \beta\}) \neq \emptyset$; in fact, there exists a proper lower semicontinuous convex function $\varphi \in \mathcal{E}(\{g < \beta\})$ such that φ^* is bounded from below on X^*.

PROOF. By Example 3.244 we may and do suppose that $\beta \in \mathbb{R}$. Consider a non-decreasing convex function $\rho : \mathbb{R} \to]0, +\infty]$ which on $]-\infty, \beta[$ is finite, injective and continuously differentiable with $\rho'(u) > 0$ for all $u \in]-\infty, \beta[$, and which satisfies $\lim_{u \uparrow \beta} \rho(u) = +\infty$; for example one can consider ρ defined by $\rho(u) = +\infty$ for

all $u \in [\beta, +\infty[$ and
$$\rho(u) = \frac{1}{\beta - u} \quad \text{for all } u \in\,]-\infty, \beta[.$$

Now defining $h(x^*) = \rho(g(x^*)) - \rho(g(0))$ for all $x^* \in X^*$, it is not difficult to see that h is a proper lower weak* semicontinuous convex function and $\operatorname{dom} h = \{g < \beta\}$. Then by Theorem 3.84 (applied to the space X^* equipped with the $w(X^*, X)$ topology) the proper lower semicontinuous convex function $\varphi := h^*$ from X into $\mathbb{R} \cup \{+\infty\}$ satisfies the equality $h = \varphi^*$. Further, $\operatorname{dom} \varphi^* = \operatorname{dom} h = \{g < \beta\}$ ensuring in particular (ii) in Definition 3.243, and on the other hand we have $\varphi^*(0) = 0$ and $\varphi^*(X^*) \subset \mathbb{R} \cup \{+\infty\}$, that is, (i) in the same definition. We claim that φ^* also satisfies the following properties (iii) and (iv) in Definition 3.243:

(iii) $\sup_{\eta x^* \in \operatorname{dom} \varphi^*,\, \eta > 0} \varphi^*(\eta x^*) = +\infty$ for all $x^* \in \{g < \beta\} \setminus (\{g < \beta\})_\infty$;

(iv) $(\partial \varphi)^{-1}(x^*) \neq \emptyset$ for all $x^* \in \operatorname{dom} \varphi^*$.

Indeed, pick $x^* \in \{g < \beta\} \setminus (\{g < \beta\})_\infty$. Then $0 \in \{g < \beta\}$ and $x^* \notin (\{g < \beta\})_\infty$, so for some real $\lambda > 0$ we have $g(\lambda x^*) \geq \beta$. Therefore, by the continuity on $\mathbb{R}$ of the function $s \mapsto g(sx^*)$, there is some $\eta_0 > 0$ such that $g(\eta_0 x^*) = \beta$. Putting $\eta_1 := \inf\{\eta > 0 : g(\eta x^*) = \beta\}$, we see by the continuity of the function $s \mapsto g(sx^*)$ that η_1 is a real number with $0 < \eta_1 \leq \eta_0$ such that $g(\eta_1 x^*) = \beta$ and $\lim_{\eta \uparrow \eta_1} g(\eta x^*) = \beta$. Then $\eta x^* \in \operatorname{dom} h$ for every $\eta \in\,]0, \eta_1[$ and $\lim_{\eta \uparrow \eta_1} h(\eta x^*) = +\infty$. This ensures that (iii) holds.

By the chain rule in Corollary 3.187 we also have that
$$\partial \varphi^*(x^*) = \rho'(g(x^*)) \partial g(x^*), \quad \text{for all } x^* \text{ with } \rho'(g(x^*)) > 0,$$
then (iv) holds by (3.192), so $\mathcal{E}(\{f < \beta\}) \neq \emptyset$. Finally, since $\rho \geq 0$ we have that $\varphi^* = h$ is bounded from below. $\square$

The second lemma is related to the polar in X^* of a circled weakly compact set. Given a nonempty subset K of a locally convex space X, in addition to the concept of polar of K (already used in the previous chapters), we define its strict polar as the set

(3.194) $\qquad K^{\circ,s} := \{x^* \in X^* : \langle x^*, x \rangle < 1,\ \forall x \in K\}$.

LEMMA 3.247. *Let X be a Hausdorff locally convex space and K be a nonempty circled convex subset in X with $\mathbb{R}_+ K^\circ = X^*$.*
(a) If K is relatively weakly compact, then $\mathcal{E}(K^\circ) \neq \emptyset$; in fact under the relative weak compactness of K there exists a proper convex function $\varphi \in \mathcal{E}(K^\circ)$ with $\varphi^* \geq 0$ and whose all sublevel sets are weakly compact in the space X.
(b) If in addition the space X is complete, then $\mathcal{E}(K^\circ) \neq \emptyset$ if and only if K is relatively weakly compact.

PROOF. (a) Suppose that K is relatively weakly compact. Consider $\zeta : \mathbb{R} \to \mathbb{R} \cup \{+\infty\}$ with $\zeta(t) = \tan(t)$ for $t \in [0, \pi/2[$ and $\zeta(t) = +\infty$ otherwise, and by means of the support function σ_K of K define $h : X^* \to \mathbb{R} \cup \{+\infty\}$ by
$$h(x^*) := \zeta\left(\sigma_K(x^*) \frac{\pi}{2}\right) + \delta_{K^{\circ,s}}(x^*) \quad \text{for all } x^* \in X^*.$$
We notice that h is convex and continuous on $K^{\circ,s}$ with respect to the Mackey topology $\mu(X^*, X)$, and so lower weak* semicontinuous on X^*. Then, we define

$\varphi : X \to \mathbb{R} \cup \{+\infty\}$ by
$$\varphi(x) = h^*(x) = \sup_{x^* \in X^*} \left(\langle x^*, x \rangle - h(x^*) \right) \quad \text{for all } x \in X,$$
and we know by the Fenchel-Moreau theorem (see Theorem 3.84(c)) that its conjugate function satisfies $\varphi^* = h$. Moreover, $\varphi^*(0) = 0$ and:
(α) We have by Corollary 3.130
$$(K^\circ)_\infty = \{x^* \in X^* : \sigma_K(x^*) \le 0\} = \{x^* \in X^* : \sigma_K(x^*) = 0\}$$
$$\subset \mathrm{dom}\, \varphi^* = \{x^* \in X^* : \sigma_K(x^*) < 1\} = K^{\circ,s};$$
(β) For any $x^* \in K^\circ \setminus (K^\circ)_\infty$, or equivalently $0 < \sigma_K(x^*) \le 1$, since $\tan(t) \to +\infty$ as $t \uparrow \pi/2$ we see that
$$\sup_{\substack{\eta x^* \in \mathrm{dom}\, \varphi^* \\ \eta > 0}} \varphi^*(\eta x^*) = +\infty;$$
(γ) We already noticed that $\varphi^* = h$ is Mackey continuous at every point of $K^{\circ,s} = \mathrm{dom}\, \varphi^*$, hence endowing X^* with the Mackey topology we have for each $x^* \in \mathrm{dom}\, \varphi^*$ that $X \supset \partial \varphi^*(x^*) \ne \emptyset$ by Theorem 3.30(a). This ensures that $(\partial \varphi)^{-1}(x^*) \ne \emptyset$ for all $x^* \in \mathrm{dom}\, \varphi^*$.

From (α), (β) and (γ) we derive that $\varphi \in \mathcal{E}(K^\circ)$. On the other hand, since φ^* is Mackey continuous at each point of $K^{\circ,s}$ (as said above), Moreau's result in Corollary 3.238 guarantees that for each $x^* \in K^{\circ,s}$ all the sublevel sets of $\varphi - \langle x^*, \cdot \rangle$ are weakly compact, hence in particular (for $x^* = 0$) all the sublevels of φ are weakly compact. Finally, since by its definition $h \ge 0$, we have $\varphi^* \ge 0$, so the proof of the assertion (a) is complete.
(b) Suppose that X is complete and $\mathcal{E}(K^\circ) \ne \emptyset$. Then, the support function $f := \sigma_{K^\circ}$ of K° relative to X satisfies the properties: $0 \in (\partial f)^{-1}(x^*)$ for every $x^* \in K^\circ$, and $\mathbb{R}_+ K^\circ = X^*$ (by our assumptions). Then, by Theorem 3.245 we have that $\{f \le 1\}$ is relatively weakly compact, and since $K \subset \{f \le 1\}$ we get the relative weak compactness of K. This and the above assertion (a) justify the equivalence in (b). □

The third lemma studies the situation of another set constructed from a finite-valued lower w^*-semicontinuous convex function on X^*.

LEMMA 3.248. *Let X be a Hausdorff locally convex space and let $\alpha \in \mathbb{R} \cup \{-\infty\}$ and $\beta \in \mathbb{R} \cup \{+\infty\}$. Let $g : X^* \to \mathbb{R}$ be a lower weak* semicontinuous convex function such that*
$$0 \in \{\alpha < g < \beta\} := \{x^* : \alpha < g(x^*) < \beta\}, \text{ and}$$
$$\partial g(x^*) \ne \emptyset, \forall x^* \in \{\alpha < g < \beta\}.$$
Then, there exists a weak* neighborhood of zero $U \subset X^*$ such that
$$\mathcal{E}(U \cap \{\alpha < g < \beta\}) \ne \emptyset;$$
in fact there exists $\varphi \in \mathcal{E}(U \cap \{\alpha < g < \beta\})$ such that φ^* is bounded from below on X^*

PROOF. Since $g(0) > \alpha$ and g is lower w^*-semicontinuous at 0, we may choose a w^*-closed convex circled w^*-neighborhood U of zero in X^* such that $g(x^*) > \alpha$ for all $x^* \in U$. Noting that the polar $K := U^\circ$ of U in X is $w(X, X^*)$-compact and that U is equal to the polar of K in X^*, by Lemma 3.247 there exists a proper convex function $\varphi_1 \in \mathcal{E}(U)$ with weakly compact sublevels such that φ_1^* is bounded from below. Now, define the function $\tilde{g}$ on X^* by $\tilde{g}(x^*) := \max\{g(x^*), \alpha\}$ for all $x^* \in X^*$. Then, it is clear that $\tilde{g}$ is a lower weak* semicontinuous convex function, $0 \in \{\tilde{g} < \beta\}$ and for any $u^* \in \{\tilde{g} < \beta\}$ (directly from the definition of the subdifferential)

$$\partial \tilde{g}(u^*) \supset \partial g(u^*) \text{ if } g(u^*) > \alpha \quad \text{and} \quad \partial \tilde{g}(u^*) \ni 0 \text{ if } g(u^*) \leq \alpha,$$

where the second inclusion is due to the fact that, under the condition $g(u^*) \leq \alpha$, one has $\alpha \in \mathbb{R}$ along with $\tilde{g}(u^*) = \alpha$ and $\tilde{g}(\cdot) \geq \alpha$. Consequently, by Lemma 3.246 there exists a proper lower semicontinuous convex function $\varphi_2 \in \mathcal{E}(\{\tilde{g} < \beta\})$ such that φ_2^* is bounded from below. Further, φ_2 is bounded from below on X since $(\partial \varphi_2)^{-1}(0) \neq \emptyset$ by (i) and (iv) in Definition 3.243. The convex function $\varphi := \varphi_1 \square \varphi_2$ is then proper and lower semicontinuous (see Proposition 3.67 applied with the $w(X, X^*)$ topology on X). Let us verify that $\varphi \in \mathcal{E}(U \cap \{\alpha < g < \beta\})$. Indeed

(i) $\varphi^* = \varphi_1^* + \varphi_2^*$ (see (3.48)), consequently φ^* is proper and $\varphi^*(0) = 0$.
(ii) Since $\operatorname{dom} \varphi^* = \operatorname{dom} \varphi_1^* \cap \operatorname{dom} \varphi_2^*$ and $U \cap \{\alpha < g < \beta\} = U \cap \{\tilde{g} < \beta\}$ we can write

$$(U \cap \{\alpha < g < \beta\})_\infty = (U \cap \{\tilde{g} < \beta\})_\infty \subset U_\infty \cap (\{\tilde{g} < \beta\})_\infty$$
$$\subset \operatorname{dom} \varphi_1^* \cap \operatorname{dom} \varphi_2^* = \operatorname{dom} \varphi^* = \operatorname{dom} \varphi_1^* \cap \operatorname{dom} \varphi_2^*$$
$$\subset U \cap \{\tilde{g} < \beta\} = U \cap \{\alpha < g < \beta\},$$

where the inclusions in the two latter lines are due to the property (iii) in Definition 3.243.
(iii) Put $C^1 := U$ and $C^2 := \{\tilde{g} < \beta\}$ and note that $U \cap \{\alpha < g < \beta\} = C^1 \cap C^2$. Since the convex sets C^1, C^2 satisfy $C^2 \cap \operatorname{int}_{w^*} C^1 \neq \emptyset$ (0 belongs to this intersection), equipping X^* with the $w(X^*, X)$ topology Corollary 3.132 tells us that

(3.195) $$(C^1 \cap C^2)_\infty = C^1_\infty \cap C^2_\infty.$$

Fix $x^* \in (C^1 \cap C^2) \setminus (C^1 \cap C^2)_\infty$. Define $I := \{\eta > 0 : \eta x^* \in \operatorname{dom} \varphi^*\}$ and $I_i := \{\eta > 0 : \eta x^* \in \operatorname{dom} \varphi_i^*\}$, $i = 1, 2$, and note that I_i is an interval and $I = I_1 \cap I_2$. Since by (3.195)

$$(C^1 \cap C^2) \setminus (C^1 \cap C^2)_\infty = ((C^1 \cap C^2) \setminus C^1_\infty) \cup ((C^1 \cap C^2) \setminus C^2_\infty),$$

for at least one $i \in \{1, 2\}$ one has $x^* \in (C^1 \cap C^2) \setminus C^i_\infty$, and for this i we have by (iii) in Definition 3.243

(3.196) $$\sup\{\varphi_i^*(\eta x^*) : \eta > 0, \eta x^* \in \operatorname{dom} \varphi_i^*\} = \sup\{\varphi_i^*(\eta x^*) : \eta \in I_i\} = +\infty,$$

which ensures in particular for this i that the interval I_i is of the form $I_i =]0, \lambda_i[$ for some real $\lambda_i > 0$ as easily seen from the inclusion $x^* \notin C^i_\infty$ and the continuity of the function $t \mapsto \varphi_i^*(tx^*)$ relative to its domain with 0 in this domain. Further, if $x^* \in C^i_\infty$, then for every real $r > 0$ we have $rx^* \in C^i_\infty \subset \operatorname{dom} \varphi_i^*$ by (ii) in Definition 3.243, and hence $I_i :=]0, +\infty[$. We have obtained:
- In the case $x^* \notin C^1_\infty$ and $x^* \notin C^2_\infty$, we have $I_1 =]0, \lambda_1[$ and $I_2 =]0, \lambda_2[$ with $\lambda_i \in]0, +\infty[$.
- In the case $x^* \in C^1_\infty$ and $x^* \notin C^2_\infty$, we have $I_1 =]0, +\infty[$ and $I_2 =]0, \lambda_2[$ with $\lambda_2 \in]0, +\infty[$.
- In the case $x^* \in C^2_\infty$ and $x^* \notin C^1_\infty$, we have $I_2 =]0, +\infty[$ and $I_1 =]0, \lambda_1[$ with $\lambda_1 \in]0, +\infty[$.

In any case, we get that $I = I_1 \cap I_2$ is a nonempty interval in the form $I =]0, \lambda[$ with λ equal to the minimum of the right endpoints of I_1 and I_2. This combined with the continuity of the convex functions $t \mapsto \varphi_i^*(tx^*)$ on their domains entail that
$$\sup_{\eta \in I} \varphi^*(\eta x^*) = \sup_{\eta \in]0,\lambda[} \left(\varphi_1^*(\eta x^*) + \varphi_2^*(\eta x^*)\right) = +\infty,$$
where in the latter equality we have used that $\sup_{\eta \in]0,\lambda[} \varphi_i^*(\eta x^*) = +\infty$ for a certain $i \in \{1,2\}$ and $t \mapsto \varphi_j^*(tx^*)$ is bounded from below for the other $j \in \{1,2\}$ different from i. Therefore, the property $\sup\{\varphi^*(\eta x^*) : \eta > 0, \eta x^* \in \operatorname{dom} \varphi^*\}$ for any $x^* \in (U \cap \{\alpha < g < \beta\}) \setminus (U \cap \{\alpha < g < \beta\})_\infty$ is established.

(iv) Take $x^* \in \operatorname{dom} \varphi^*$. Since $(\partial \varphi_1)^{-1}(x^*) \neq \emptyset$ and $(\partial \varphi_2)^{-1}(x^*) \neq \emptyset$ by (iv) in Definition 3.243 (because $\operatorname{dom} \varphi^* = \operatorname{dom} \varphi_1^* \cap \operatorname{dom} \varphi_2^*$), endowing X^* with the $w(X^*, X)$-topology we have that
$$X \supset \partial \varphi^*(x^*) \supset \partial \varphi_1^*(x^*) + \partial \varphi_2^*(x^*) \neq \emptyset,$$
so $(\partial \varphi)^{-1}(x^*) \neq \emptyset$ by (3.192).

Altogether justify that $\varphi \in \mathcal{E}(U \cap \{\alpha < g < \beta\})$ and the proof is finished. $\square$

We recall (see Definition 2.55) that the core of a set A in a vector space X, denoted by $\operatorname{core} A$, is defined by
$$\operatorname{core} A := \{a \in A : \forall x \in X, \exists \delta > 0, \forall \lambda \in [-\delta, \delta], a + \lambda x \in A\}.$$

THEOREM 3.249 (relative w-compactness of sublevels of linear perturbations: I). Let X be a complete Hausdorff locally convex space and $\alpha \in \mathbb{R} \cup \{-\infty\}$ and $\beta \in \mathbb{R} \cup \{+\infty\}$. Let $f : X \to \mathbb{R} \cup \{+\infty\}$ be a function such that the core of the set $\{\alpha < f^* < \beta\}$ is nonempty and
$$(\partial_{MR} f)^{-1}(x^*) \neq \emptyset, \ \forall x^* \in \{\alpha < f^* < \beta\}.$$
Then for every $x^* \in \operatorname{core}(\{\alpha < f^* < \beta\})$ and every $r \in \mathbb{R}$ the sets
$$S_{f-x^*}(r) := \{(x, \nu) \in X \times \mathbb{R} : f(x) - \langle x^*, x \rangle \leq \nu \leq r\},$$
$$\{f - \langle x^*, \cdot \rangle \leq r\} := \{x \in X : f(x) - \langle x^*, x \rangle \leq r\}$$
are relatively weakly compact.

PROOF. Fix any $u^* \in \operatorname{core}(\{\alpha < f^* < \beta\})$. Note that for $h := f - \langle u^*, \cdot \rangle$
$$-u^* + \{y^* : \alpha < f^*(y^*) < \beta\} = \{y^* - u^* : \alpha < f^*(y^*) < \beta\}$$
$$= \{x^* : \alpha < f^*(u^* + x^*) < \beta\}$$
$$= \{x^* : \alpha < h^*(x^*) < \beta\},$$
which implies in particular that $0 \in \operatorname{core}(\{\alpha < h^* < \beta\})$. On the other hand, we have (as easily seen)
$$x^* \in \partial_{MR} f(x) \Leftrightarrow x^* - u^* \in \partial_{MR} h(x),$$
so $(\partial_{MR} f)^{-1}(x^*) = (\partial_{MR} h)^{-1}(x^* - u^*)$. It ensues that for $\alpha < h^*(x^*) < \beta$, we have $\alpha < f^*(u^* + x^*) < \beta$, hence $(\partial_{MR} f)^{-1}(u^* + x^*) \neq \emptyset$, or equivalently $(\partial_{MR} h)^{-1}(x^*) \neq \emptyset$. Therefore, the assumptions of Lemma 3.248 are satisfied with $g = h^*$, thus there exists a weak*-neighborhood U of zero in X^* such that for $K := U \cap \{\alpha < h^* < \beta\}$ the set $\mathcal{E}(K)$ is nonempty. Further, we have $0 \in K$ and
$$\mathbb{R}_+ K = \mathbb{R}_+(U \cap \{\alpha < h^* < \beta\}) = X^*,$$

where the latter equality is due to the inclusion $0 \in \operatorname{core}(\{\alpha < h^* < \beta\})$. Since $(\partial_{MR} h)^{-1}(x^*) \neq \emptyset$ for every $x^* \in K \subset \{\alpha < h^* < \beta\}$, we can apply Theorem 3.245 to obtain that $S_h(r)$ and $\{h \leq r\}$ are relatively weakly compact. We conclude that $S_{f-u^*}(r)$ and $\{f - \langle u^*, \cdot \rangle \leq r\}$ are relatively weakly compact. □

We use the above theorem to show a weak compactness property for a pair of closed bounded convex sets.

COROLLARY 3.250. *Let X be a complete Hausdorff locally convex space and let A and B be nonempty bounded closed convex sets of X such that $0 \notin \operatorname{cl}(A - B)$. If every $x^* \in X^*$ with*

(3.197) $$\sup_{x \in B} \langle x^*, x \rangle < \inf_{x \in A} \langle x^*, x \rangle$$

attains its infimum on A and its supremum on B, then both A and B are weakly compact.

PROOF. Consider the set $C := \operatorname{cl}(B - A)$, and the indicator function $f := \Psi_C$ of C, so $f^* = \sigma_C$.

We claim that the core of the set $\{f^* < 0\} = \{-\infty < f^* < 0\}$ is nonempty and $(\partial f)^{-1}(x^*) \neq \emptyset$ for all $x^* \in \{f^* < 0\}$. On the one hand, to see that the core of the set $\{f^* < 0\}$ is nonempty, by the assumption $0 \notin \operatorname{cl}(A - B)$ choose some $x_0^* \in X^*$ such that $\sigma_C(x_0^*) < 0$. Take any $x^* \in X^*$. The boundedness of C ensures that on $\mathbb{R}$ the convex function h defined for all $t \in \mathbb{R}$ by $h(t) := \sigma_C(x_0^* + tx^*)$ is finite, hence continuous. Then, by the inequality $h(0) < 0$ there is a real $\delta > 0$ such that $\sigma_C(x_0^* + tx^*) = h(t) < 0$ for all $t \in [-\delta, \delta]$, which means that $x_0^* + [-\delta, \delta]x^* \subset \{f^* < 0\}$. This says that x_0^* belongs to the core of $\{f^* < 0\}$. On the other hand, the set $\{f^* < 0\}$ is just the set of all points $x^* \in X^*$ such that (3.197) holds, and we notice by (3.192) and Proposition 3.12(c) that

$$(\partial f)^{-1}(x^*) = \partial f^*(x^*) = \{x \in C : \sigma_C(x^*) = \langle x^*, x \rangle\}.$$

Hence, by the attainment assumption related to (3.197) the set $(\partial f)^{-1}(x^*)$ is nonempty for all $x^* \in \{f^* < 0\}$. The claim is then justified.

Thus, Theorem 3.249 allows us to conclude that for all $x^* \in \operatorname{core}(\{f^* < 0\})$ and $r \in \mathbb{R}$ the set $\{f - \langle x^*, \cdot \rangle \leq r\}$ is relatively weakly compact. In particular, fixing $x_0^* \in \operatorname{core}(\{f^* < 0\})$ and putting $r_0 := \sigma_C(-x_0^*)$ we have that $C \subset \{f - \langle x_0^*, \cdot \rangle \leq r_0\}$. Therefore, the sets A and B are relatively weakly compact. □

Now, besides Corollary 3.238, we establish another relative compactness result of sublevels of $f - \langle x^*, \cdot \rangle$ by means of attainment of its infimum. The proof uses Theorem 3.245 above.

THEOREM 3.251 (relative w-compactness of sublevels of linear perturbations: II). *Let X be a complete Hausdorff locally convex space. Let U be a nonempty open set in X^* with respect to the Mackey topology and let $f : X \to \mathbb{R} \cup \{+\infty\}$ be a function such that $f - \langle x^*, \cdot \rangle$ attains its minimum for every $x^* \in U$. Then for every $r \in \mathbb{R}$ and every $x^* \in U$ the sets*

$$S_{f-x^*}(r) = \{(x, \alpha) \in X \times \mathbb{R} : f(x) - \langle x^*, x \rangle \leq \alpha \leq r\},$$
$$\{f - \langle x^*, \cdot \rangle \leq r\} = \{x \in X : f(x) - \langle x^*, x \rangle \leq r\}$$

are relatively weakly compact. In particular, f^* is Mackey continuous at some point of its domain.

PROOF. Considering $x_0^* \in U$ and the function $f - \langle x_0^*, \cdot \rangle$ in place of f, we may suppose that $0 \in U$. From the definition of the Mackey topology, there exists a nonempty convex circled weakly compact subset K of X such that $K^\circ \subset U$, thus $f - \langle x^*, \cdot \rangle$ attains its minimum for every $x^* \in K^\circ$, which is also equivalent to $(\partial_{MR} f)^{-1}(x^*) \neq \emptyset$ for all $x^* \in K^\circ$. Noting also that $\mathbb{R}_+ K^\circ = X^*$, by Lemma 3.247 we can take $\varphi \in \mathcal{E}(K^\circ)$, then by Theorem 3.245 the sets $S_f(r)$ and $\{f \leq r\}$ are relatively weakly compact. Finally, by Moreau's result in Corollary 3.238, f^* is Mackey continuous at zero as desired. $\square$

The following corollary directly follows from the above theorem.

COROLLARY 3.252. Let X be a complete Hausdorff locally convex space and $f : X \to \mathbb{R} \cup \{+\infty\}$ be a lower weakly semicontinuous function. If $f - \langle x^*, \cdot \rangle$ attains its minimum for every $x^* \in X^*$, then the sublevel set $\{f \leq r\}$ is weakly compact for any $r \in \mathbb{R}$.

From Theorem 3.251 we can also deduce the next two corollaries.

COROLLARY 3.253. Let X be a complete Hausdorff locally convex space. Let D be a nonempty weakly compact subset of X with $0 \notin D$. If A is a bounded subset of X such that every $x^* \in X^*$ with $\inf_{x \in D} \langle x^*, x \rangle > 0$ attains its supremum on A, then A is relatively weakly compact in X.

PROOF. Consider the set $U := \{x^* \in X^* : \inf_{x \in D} \langle x^*, x \rangle > 0\}$. Due to the fact that D is weakly compact, the set U is an open set in X^* with respect to the Mackey topology. Then, applying Theorem 3.251 with the indicator function $f := \Psi_A$, we get that for every $r \in \mathbb{R}$ and every $x^* \in U$ the set $\{f - \langle x^*, \cdot \rangle \leq r\}$ is relatively weakly compact. In particular, fixing a point $x_0^* \in U$ and putting $r_0 := \sigma_A(-x_0^*)$, we have that $A \subset \{f - \langle x_0^*, \cdot \rangle \leq r_0\}$, which implies that A is relatively weakly compact. $\square$

COROLLARY 3.254. Let X be a complete Hausdorff locally convex space. Let $f : X \to \mathbb{R} \cup \{+\infty\}$ be a proper lower semicontinuous convex function and let U be a nonempty open set in X^* with respect to the Mackey topology. Then the following statements are equivalent:
(a) f^* is Mackey continuous at some point in $\operatorname{dom} f^*$ and $U \subset \operatorname{dom} f^*$;
(b) for every $x^* \in U$ all the sublevels $\{f - \langle x^*, \cdot \rangle \leq r\}$ are weakly compact;
(c) for every $x^* \in U$ the function $f - \langle x^*, \cdot \rangle$ attains its minimum, or equivalently the set $(\partial_{MR} f)^{-1}(x^*) \neq \emptyset$.

PROOF. Noting that (a) is equivalent to saying that f^* is Mackey continuous on the open set U, we see that the equivalence between (a) and (b) follows from Moreau's result in Corollary 3.238.

On the other hand, the implication (b) $\Rightarrow$ (c) is straightforward. Finally, (c) $\Rightarrow$ (b) is given by Theorem 3.251. $\square$

EXAMPLE 3.255. Consider any nonempty weakly compact circled convex set K of a Banach space X such that $K^\circ \neq X^*$. The support function $f := \sigma_{K^\circ}$ satisfies the assumption of Theorem 3.251 with $U := K^{\circ,s}$. However, it does not satisfy the assumption of Corollary 3.252, because the effective domain of the subdifferential of $f^* = \Psi_{K^\circ}$ is not the whole space X^*. $\square$

EXAMPLE 3.256. Consider a non-reflexive Banach space X. Define $f := \|\cdot\|$ on X and note that its conjugate function is given by $f^* = \Psi_{\mathbb{B}_{X^*}}$, where we recall that $\mathbb{B}_{X^*}$ denotes the closed unit ball in X^* centered at zero. Moreover, for every $x^* \in \mathbb{B}_{X^*}$ one has that $f^*(x^*) = \langle x^*, 0 \rangle - f(0)$, hence $f - x^*$ attains its minimum. Nevertheless, the set $\mathbb{B}_X = \{x \in X : f(x) \le 1\}$ is not weakly compact. This example shows that the assumption $\mathcal{E}(\mathbb{B}_{X^*}) \ne \emptyset$ is crucial in Theorem 3.245. □

Example 3.255 shows that Theorem 3.251 cannot be extended to the case when U therein is the open/closed unit ball in the dual space of a Banach space. However, we can establish the following characterization of semi-reflexivity of a Hausdorff locally convex space in terms of the non-emptiness of $\mathcal{E}(B^\circ)$ for some family of bounded sets B in X. Given a Hausdorff locally convex space (X, τ_X) we recall that the strong topology $\beta(X^*, X)$ on X^* is the topology generated by the uniform convergence over bounded sets of X. The bidual of X, denoted by X^{**}, is the topological dual of $(X^*, \beta(X^*, X))$. The Hausdorff locally convex space (X, τ_X) is called *semi-reflexive* if the canonical embedding (or evaluation mapping) $X \ni x \to \langle \cdot, x \rangle \in X^{**}$ is surjective (see Theorem C.9 in Appendix for more details). In contrast, (X, τ_X) is called *reflexive* if the canonical embedding is a homeomorphism from (X, τ_X) onto $(X^{**}, \beta(X^{**}, X^*))$, where $\beta(X^{**}, X^*)$ is the topology on X^{**} of uniform convergence over bounded sets in $(X^*, \beta(X^*, X))$. It is worth mentioning that every semi-reflexive normed space is a reflexive Banach space (see this property in Appendix C).

COROLLARY 3.257. *A complete Hausdorff locally convex vector space (X, τ_X) is semi-reflexive if and only if $\mathcal{E}(B^\circ) \ne \emptyset$ for every nonempty bounded circled convex set B in X.*

PROOF. We know that a Hausdorff locally convex space X is semi-reflexive if and only if every bounded subset of X is relatively weakly compact (see Theorem C.9 in Appendix).

First, assume that X is semi-reflexive and fix any nonempty bounded circled convex set B in X, so B is relatively weakly compact. Moreover $\mathbb{R}_+ B^\circ = X^*$. Indeed, taking any $x^* \in X^*$, by the boundedness of B there exists some $\lambda > 0$ such that
$$\lambda B \subset V := \{x^*\}^\circ = \{x \in X : \langle x^*, x \rangle \le 1\}.$$
Then $\mathbb{R}_+ B^\circ \supset \lambda^{-1} B^\circ \supset V^\circ \ni x^*$, which confirms by the arbitrariness of x^* that $\mathbb{R}_+ B^\circ = X^*$ (this can also be seen from the fact that B° is a $\beta(X^*, X)$ neighborhood of zero in X^*). Then, by Lemma 3.247 the set $\mathcal{E}(B^\circ)$ is nonempty.

Conversely, take a nonempty bounded circled convex set B of X with $\mathcal{E}(B^\circ) \ne \emptyset$. By Lemma 3.247 the set B is relatively weakly compact. Since the boundedness is preserved under circled convex hull, we deduce under the property of the corollary that every bounded subset of X is relatively compact, so X is semi-reflexive. □

3.20. Subdifferential of distance functions from convex sets

Given a nonempty convex set C in a normed space $(X, \|\cdot\|)$, we already saw in Proposition 2.54 that the distance function d_C is convex. We also proved in Proposition 2.97 that for any $x \in C$

(3.198) $\qquad \partial d_C(x) = N(C; x) \cap \mathbb{B}_{X^*} \quad \text{and} \quad N(C; x) = \mathbb{R}_+ \partial d_C(x).$

In this section we will establish some additional properties.

Let us begin with the situation when $\operatorname{Proj}_C(x) \neq \emptyset$, which holds in particular whenever X is a reflexive Banach space and the convex set C is closed. In fact, the next proposition covers this situation as well as the one where there is no nearest point. A similar result for nonconvex sets will be provided in a more general context in Proposition 4.32.

PROPOSITION 3.258. Let C be a nonempty convex set of a normed vector space $(X, \|\cdot\|)$ and let $\bar{x} \in X$. For any $x^* \in \partial d_C(\bar{x})$ the following hold.
(a) For any sequence $(y_n)_n$ in C with $\|\bar{x} - y_n\| \to d_C(\bar{x})$ one has
$$\langle x^*, \bar{x} - y_n \rangle \to d_C(\bar{x}) \quad \text{as } n \to \infty.$$
(b) If in addition $\operatorname{Proj}_C(\bar{x}) \neq \emptyset$, then for any $\bar{y} \in \operatorname{Proj}_C(\bar{x})$
$$\langle x^*, \bar{x} - \bar{y} \rangle = d_C(\bar{x}).$$

PROOF. The assertion (b) clearly follows from (a). Take any $x^* \in \partial d_C(\bar{x})$ and any sequence $(y_n)_n$ in C such that $\|\bar{x} - y_n\| \to d_C(\bar{x})$ as $n \to \infty$. By property of subdifferential of convex functions we have for each $n \in \mathbb{N}$
$$\langle x^*, y_n - \bar{x} \rangle \leq d_C(y_n) - d_C(\bar{x}),$$
or equivalently $d_C(\bar{x}) \leq \langle x^*, \bar{x} - y_n \rangle$. On the other hand, the inclusion $x^* \in \partial d_C(\bar{x})$ ensures that $\|x^*\| \leq 1$ according to the Lipschitz property of d_C with Lipschitz constant 1, and hence $\langle x^*, \bar{x} - y_n \rangle \leq \|\bar{x} - y_n\|$. All together, we obtain
$$d_C(\bar{x}) \leq \langle x^*, \bar{x} - y_n \rangle \leq \|\bar{x} - y_n\|,$$
which confirms that $\langle x^*, \bar{x} - y_n \rangle \to d_C(\bar{x})$. □

When the convex set C is closed and the point $\bar{x}$ is outside C, the subdifferential $\partial d_C(\bar{x})$ can be expressed by means of normals to a certain enlargement of C. Let us first consider some features for enlargements of sets.

DEFINITION 3.259. For any $r \geq 0$ we define the *closed r-enlargement* of a set S of the normed space X as
$$\operatorname{Enl}_r S := \{u \in X : d_S(u) \leq r\}.$$
We also define the set of points at exact r-distance to S as
$$D_r(S) := \{u \in X : d_S(u) = r\}.$$

It is worth noticing that $\operatorname{Enl}_r S$ is convex whenever the set S is convex.

LEMMA 3.260. Let S be a subset of a normed space $(X, \|\cdot\|)$. Then, for any real $r \geq 0$ and any $u \in X$ such that $d(u, S) \geq r$, one has
$$d(u, S) = r + d(u, \operatorname{Enl}_r S) = r + d(u, D_r(S)).$$

PROOF. Put $E_r(S) := \operatorname{Enl}_r S$. Fix any $u \in X$ with $d(u, S) \geq r$. For all $y \in E_r(S)$, we have
$$d(u, S) \leq d(y, S) + \|u - y\| \leq r + \|u - y\|,$$
and taking the infimum over $y \in E_r(S)$ gives
$$d(u, S) \leq r + d(u, E_r(S)) \leq r + d(u, D_r(S)).$$

It remains to show that the last member of the latter inequality is less or equal to the first. Fix any $y \in S$ and consider the real-valued continuous function h

defined on $[0, +\infty[$ by $h(t) := d\big(tu + (1-t)y, S\big)$. Observing that $h(0) = 0$ (because $y \in S$) and $h(1) \geq r$, we may apply the classical intermediate value theorem to find some $t_0 \in [0, 1]$ such that $h(t_0) = r$. Putting $z := t_0 u + (1 - t_0)y$, we have $d(z, S) = r$, or equivalently $z \in D_r(S)$, and further

$$\|u - y\| = \|u - z\| + \|z - y\|.$$

Consequently, because $y \in S$ we obtain

$$\|u - y\| \geq \|u - z\| + d(z, S) = \|u - z\| + r,$$

and since $z \in D_r(S)$, it follows that

$$\|u - y\| \geq d(u, D_r(S)) + r.$$

This yields $d(u, S) \geq d(u, D_r(S)) + r$ which completes the proof of the lemma. $\square$

REMARK 3.261. It is worth observing that the left equality of the lemma can be obtained without using the intermediate value theorem. Indeed, let $u \in X$ with $d(u, S) \geq r$, and suppose without loss of generality $d(u, S) > 0$ (otherwise, the equality is obvious). Fix any $y \in S$. Putting $z := y + r\frac{u-y}{\|u-y\|}$ we see that $z \in [y, u]$ (since $r \leq \|u - y\|$) and that $z \in E_r(S)$ since $d(z, S) \leq \|z - y\| = r$. Then

$$\|u - y\| = \|u - z\| + \|z - y\| = \|u - z\| + r \geq d(u, E_r(S)) + r,$$

so $d(u, S) \geq r + d(u, E_r(S))$. The converse inequality is obtained as in the above proof of the lemma. $\square$

We can now state and prove the result linking subgradients of the distance function from a convex set and normals to a suitable enlargement of the set.

PROPOSITION 3.262. Let C be a nonempty convex set in a normed vector space $(X, \|\cdot\|)$ and $\overline{x} \notin \operatorname{cl} C$. For $r := d_C(\overline{x}) > 0$ and for the closed r-enlargement $\operatorname{Enl}_r C$ of C one has

$$\partial d_C(\overline{x}) = N(\operatorname{Enl}_r C; \overline{x}) \cap \{x^* \in X^* : \|x^*\| = 1\}.$$

PROOF. Put $E_r(C) := \operatorname{Enl}_r C$. To show the inclusion of the left-hand side into the right one, fix any $x^* \in \partial d_C(\overline{x})$. We have

(3.199) $\qquad \langle x^*, x - \overline{x} \rangle \leq d_C(x) - d_C(\overline{x}) \quad$ for all $x \in X$.

Since $d_C(x) - d_C(\overline{x}) \leq 0$ for all $x \in E_r(C)$, it ensues that $\langle x^*, x - \overline{x} \rangle \leq 0$ for all $x \in E_r(C)$, which tells us that $x^* \in N(E_r(C); \overline{x})$.

Let us show that $\|x^*\| = 1$. The function d_C being Lipschitz with constant one, it is clear that $\|x^*\| \leq 1$. Consider a sequence $(y_n)_n$ in C with $\|\overline{x} - y_n\|$ converging to $d_C(\overline{x})$ as $n \to \infty$. Proposition 3.258 gives that $\langle x^*, v_n/\|v_n\| \rangle \to 1$ as $n \to \infty$ for the vectors $v_n := \overline{x} - y_n$. This justifies that $\|x^*\| = 1$, and finishes the proof of the inclusion of the left-hand side into the right.

Now we proceed to prove the converse inclusion. Fix any $x^* \in N(E_r(C); \overline{x})$ with $\|x^*\| = 1$. First, notice that $x^* \in \partial \operatorname{dist}(\cdot, E_r(C))(\overline{x})$ by (3.198) recalled above, thus

$$\langle x^*, x - \overline{x} \rangle \leq \operatorname{dist}(x, E_r(C)) - \operatorname{dist}(\overline{x}, E_r(C)) \quad \text{for all } x \in X.$$

This and Lemma 3.260 above give

(3.200) $\qquad \langle x^*, x - \overline{x} \rangle \leq d_C(x) - d_C(\overline{x}) \quad$ for all $x \in X \setminus E_r(C)$.

Fix now any $x \in E_r(C)$ and put $t_x := d_C(\overline{x}) - d_C(x) \geq 0$. By the equality $\|x^*\| = 1$, choose a sequence of unit vectors $(u_n)_n$ in X such that $\langle x^*, u_n \rangle \to 1$. For each $n \in \mathbb{N}$, writing

$$d_C(x + t_x u_n) \leq d_C(x) + t_x = d_C(\overline{x}) = r,$$

we see that $x + t_x u_n \in E_r(C)$, and hence $\langle x^*, x + t_x u_n - \overline{x} \rangle \leq 0$. From this we derive that for every $n \in \mathbb{N}$

$$\langle x^*, x - \overline{x} \rangle = \langle x^*, x + t_x u_n - \overline{x} \rangle - \langle x^*, t_x u_n \rangle \leq \langle x^*, u_n \rangle (d_C(x) - d_C(\overline{x})).$$

This combined with (3.200) yields that $\langle x^*, x - \overline{x} \rangle \leq d_C(x) - d_C(\overline{x})$ for all $x \in X$. It results that $x^* \in \partial d_C(\overline{x})$, which finishes the proof of the proposition. □

3.21. Moreau envelope, strongly convex functions

Given a lower semicontinuous convex function, we will see in this section that its Moreau envelope furnishes a remarkable way to approximate the function by continuously differentiable convex functions. We will also show how it allows us to describe many important variational properties.

3.21.1. Moreau envelope.
We already saw various features concerning conjugate and subgradients of infimal convolution in the equality (3.48) and in Proposition 3.69. We will study now the specific infimal convolution of a function with the kernel $\frac{1}{2\lambda}\|\cdot\|^2$.

DEFINITION 3.263. Let $(X, \|\cdot\|)$ be a normed space and $f : X \to \mathbb{R} \cup \{+\infty\}$ be a proper function. For any real $\lambda > 0$ the *Moreau envelope with index λ* of the function f is the function $e_\lambda f$ defined as the infimal convolution $e_\lambda f := f \square \frac{1}{2\lambda}\|\cdot\|^2$, that is,

$$e_\lambda f(x) = \inf_{y \in X} \left(f(y) + \frac{1}{2\lambda}\|x - y\|^2 \right) \quad \text{for all } x \in X.$$

The *proximal multimapping with index λ* of the function f is defined by

$$\text{Prox}_\lambda f(x) := \left\{ y \in X : f(y) + \frac{1}{2\lambda}\|x - y\|^2 = e_\lambda f(x) \right\} \quad \text{for all } x \in X.$$

When $\text{Prox}_\lambda f(x)$ is a singleton, we will denote by $P_\lambda f(x)$ its unique element. If $\text{Prox}_\lambda f(x)$ is a singleton for any $x \in X$, the mapping $P_\lambda f : X \to X$ is called the *proximal mapping with index λ* of the function f.

When $\lambda = 1$, we will write $(\text{Prox } f)(x)$ and $(Pf)(x)$ instead of $(\text{Prox}_\lambda f)(x)$ and $(P_\lambda f)(x)$. Noting that with $\lambda > 0$

$$\inf_{y \in X} \left(f(y) + \frac{1}{2\lambda}\|x - y\|^2 \right) = \frac{1}{\lambda} \inf_{y \in X} \left(\lambda f(y) + \frac{1}{2}\|x - y\|^2 \right),$$

we see that $(e_\lambda f)(x) = \lambda^{-1}(e_1(\lambda f))(x)$ and

(3.201) $\quad (\text{Prox}_\lambda f)(x) = (\text{Prox}(\lambda f))(x)$ and $(P_\lambda f)(x) = (P(\lambda f))(x)$.

Clearly, if f is the indicator function Ψ_S of a nonempty set S in X, then

(3.202) $\quad e_\lambda \Psi_S(x) = \frac{1}{2\lambda} d_S^2(x)$ and $\text{Prox}_\lambda \Psi_S(x) = \text{Proj}_S(x)$.

EXAMPLE 3.264. In addition to the above example in (3.202) consider a Hilbert space H, a closed set S in H and a real $\lambda > 0$. Given $a \in H$ and $\beta \in \mathbb{R}$, let $f : H \to \mathbb{R} \cup \{+\infty\}$ be defined by
$$f(\cdot) = \langle a, \cdot \rangle + \beta + \Psi_S(\cdot).$$
For any $x \in H$ we notice that for every $y \in H$
$$\frac{1}{2\lambda}\|x - y\|^2 + f(y) = \frac{1}{2\lambda}\left(\|x - y\|^2 + 2\langle \lambda a, y\rangle + \Psi_S(y)\right) + \beta$$
$$= \frac{1}{2\lambda}\left(\|(x - y) - \lambda a\|^2 + \Psi_S(y) + 2\lambda\langle a, x\rangle - \lambda^2\|a\|^2\right) + \beta,$$
which gives
$$\inf_{y \in H}\left(\frac{1}{2\lambda}\|x - y\|^2 + f(y)\right) = \langle a, x\rangle - \frac{\lambda}{2}\|a\|^2 + \beta + \frac{1}{2\lambda}\inf_{y \in S}\|(x - \lambda a) - y\|^2.$$
It ensues that
$$e_\lambda f(x) = \langle a, x\rangle - \frac{\lambda}{2}\|a\|^2 + \beta + \frac{1}{2\lambda}d_S^2(x - \lambda a) \quad \text{and} \quad \operatorname{Prox}_\lambda f(x) = \operatorname{Proj}_S(x - \lambda a).$$

In the particular case when $S = H$ we obtain for the function $g := \langle a, \cdot\rangle + \beta$ that
$$e_\lambda g(x) = \langle a, x\rangle - \frac{\lambda}{2}\|a\|^2 + \beta \quad \text{and} \quad \operatorname{Prox}_\lambda g(x) = x - \lambda a.$$

We also note that taking f above with $\beta = 0$ and a as the zero vector in H, that is, $f = \Psi_S$, we retrieve that $e_\lambda \Psi_S(x) = \frac{1}{2\lambda}d_S^2(x)$ and $\operatorname{Prox}_\lambda \Psi_S(x) = \operatorname{Proj}_S(x)$ as in (3.202). $\square$

We start now the study of basic features of Moreau envelope with its Lipschitz behavior and with certain convergence properties.

THEOREM 3.265 (Lipschitz property and convergence of Moreau envelope). Let $(X, \|\cdot\|)$ be a normed space and $f : X \to \mathbb{R} \cup \{+\infty\}$ be a proper function for which there exist reals $\alpha \geq 0$, $\beta \geq 0$ and $\gamma \in \mathbb{R}$ such that
$$f(x) \geq -\alpha\|x\|^2 - \beta\|x\| + \gamma \quad \text{for all } x \in X,$$
that is, f is bounded from below by a negative quadratic function. Then the following hold.
(a) For any real $\lambda > 0$ one has the inclusion $\operatorname{Argmin}_X f \subset \operatorname{Argmin}_X(e_\lambda f)$ along with the inequality $e_\lambda f(x) \leq f(x)$ for every $x \in X$ and the equality
$$\inf_{x \in X} f(x) = \inf_{x \in X}(e_\lambda f)(x).$$
(b) For any real $\lambda \in {]}0, \frac{1}{2\alpha}[$ the Moreau envelope $e_\lambda f$ is finite-valued on X and it is Lipschitz continuous on each ball $r\mathbb{B}_X$ for some real Lipschitz constant $L \geq r/\lambda$. Above by convention $1/(2\alpha) = +\infty$ for $\alpha = 0$.
(c) Given $x \in X$ with $\sup_{\lambda \in {]}0, 1/(2\alpha)[} e_\lambda f(x) < +\infty$ and given $(\theta(\lambda))_{0 < \lambda < 1/(2\alpha)}$ with $\theta(\lambda) > 0$ and $\theta(\lambda) \downarrow 0$ as $\lambda \downarrow 0$, then for any family $(y_\lambda)_{0 < \lambda < 1/(2\alpha)}$ with
$$f(y_\lambda) + \frac{1}{2\lambda}\|x - y_\lambda\|^2 \leq e_\lambda f(x) + \theta(\lambda)$$
one has
$$y_\lambda \to x \quad \text{and} \quad f(y_\lambda) \to \liminf_{u \to x} f(u) \quad \text{as } \lambda \downarrow 0.$$

(d) The family $(e_\lambda f)_\lambda$ pointwise converges as $\lambda \downarrow 0$ to the lower semicontinuous hull of the function f.
(e) If in addition f is lower semicontinuous, then the family $(e_\lambda f)_\lambda$ pointwise converges to f as $\lambda \downarrow 0$; further, for any nonincreasing net $(\lambda_j)_{j \in J}$ in $]0, \frac{1}{2\alpha}[$ tending to 0 and any net $(x_j)_{j \in J}$ in X converging strongly to $\bar{x} \in X$ one has
$$f(\bar{x}) \leq \liminf_{j \in J} e_{\lambda_j} f(x_j).$$

PROOF. (a) On one hand, the inequality $f(x) \geq e_\lambda f(x)$ for every $x \in X$ is obvious (since $f(x) = f(x) + \frac{1}{2\lambda}\|x - x\|^2$) and it clearly ensures that $\inf_X f \geq \inf_X e_\lambda f$. On the other hand, noting for every $y \in X$ that $f(y) + \frac{1}{2\lambda}\|x - y\|^2 \geq f(y)$ we see that
$$(e_\lambda f)(x) = \inf_{y \in X}\left(f(y) + \frac{1}{2\lambda}\|x - y\|^2\right) \geq \inf_{y \in X} f(y) = \inf_X f,$$
hence $\inf_{x \in X}(e_\lambda f)(x) \geq \inf_X f$. This and the previous converse inequality justify the equality in (a).

Regarding the inclusion in (a), it suffices to observe that if $u \in \mathrm{Argmin}_X f$ one has by the preceding equality that
$$f(u) = \inf_X f = \inf_X(e_\lambda f) \leq (e_\lambda f)(u) \leq f(u),$$
so $(e_\lambda f)(u) = \inf_X(e_\lambda f)$, hence $u \in \mathrm{Argmin}_X(e_\lambda f)$.

(b) Fix any real $\lambda \in]0, \frac{1}{2\alpha}[$. Take $x \in X$. For any $y \in X$ we have by the minorization by the negative quadratic function that
$$f(y) + \frac{1}{2\lambda}\|x - y\|^2 \geq -\alpha\|y\|^2 - \beta\|y\| + \gamma + \frac{1}{2\lambda}\|x\|^2 + \frac{1}{2\lambda}\|y\|^2 - \frac{1}{\lambda}\|x\|\|y\|$$
$$(3.203) \qquad \geq \left(\frac{1}{2\lambda} - \alpha\right)\|y\|^2 - \left(\beta + \frac{1}{\lambda}\|x\|\right)\|y\| + \gamma =: \zeta(\|y\|),$$

which ensures that the function $f + \frac{1}{2\lambda}\|x - \cdot\|^2$ is bounded from below on the space X, since the continuous function $\zeta : [0, +\infty[\to \mathbb{R}$ is bounded from below on $[0, +\infty[$ because $\zeta(t) \to +\infty$ as $t \to +\infty$. Therefore, the infimum of the function $y \mapsto f(y) + \frac{1}{2\lambda}\|x - y\|^2$ over X is finite according to the properness of f, thus the Moreau envelope $e_\lambda f$ is finite-valued on X.

Now fix a point $y_0 \in X$ where f is finite and a real $r > 0$. For $\varepsilon > 0$, $x \in r\mathbb{B}_X$, and $z \in X$ such that $e_\lambda f(x) + \varepsilon \geq f(z) + \frac{1}{2\lambda}\|x - z\|^2$ we have
$$f(y_0) + \frac{1}{\lambda}(r^2 + \|y_0\|^2) + \varepsilon \geq f(y_0) + \frac{1}{2\lambda}\|x - y_0\|^2 + \varepsilon$$
$$\geq e_\lambda f(x) + \varepsilon \geq f(z) + \frac{1}{2\lambda}\|x - z\|^2.$$

On the other hand, by (3.203)
$$f(z) + \frac{1}{2\lambda}\|x - z\|^2 \geq \left(\frac{1}{2\lambda} - \alpha\right)\|z\|^2 - \left(\beta + \frac{r}{\lambda}\right)\|z\| + \gamma,$$

hence
$$f(y_0) + \frac{1}{\lambda}(r^2 + \|y_0\|^2) + \varepsilon - \gamma \geq \left(\frac{1}{2\lambda} - \alpha\right)\|z\|^2 - \left(\beta + \frac{r}{\lambda}\right)\|z\|.$$

Consequently, there is a real $\eta > 0$ (depending only on λ, r, ε and y_0) such that $\|z\| \leq \eta$, so
$$e_\lambda f(x) = \inf_{z \in \eta \mathbb{B}_X} \left(f(z) + \frac{1}{2\lambda}\|x - z\|^2 \right) \quad \text{for all } x \in r\mathbb{B}_X.$$

Noting that for each $z \in \eta \mathbb{B}_X$ the function $\frac{1}{2\lambda}\|\cdot - z\|^2$ is Lipschitz continuous on $r\mathbb{B}_X$ with $\frac{1}{\lambda}(r + \eta)$ as Lipschitz constant therein, the latter equality for $e_\lambda f(x)$ guarantees that
$$|e_\lambda f(x) - e_\lambda f(x')| \leq \frac{1}{\lambda}(r + \eta)\|x - x'\|.$$

This confirms the desired Lipschitz property of $e_\lambda f$ on bounded sets.

(c) Let x, $(\theta(\lambda))_\lambda$ and $(y_\lambda)_\lambda$ be as in the statement of (c) (notice that such a family $(y_\lambda)_\lambda$ exists since $e_\lambda f(x)$ is finite by (b) for every $\lambda \in \,]0, \frac{1}{2\alpha}[$). Put $g(u) := \sup_{\lambda \in \,]0, \frac{1}{2\alpha}[} e_\lambda f(u)$ for every $u \in X$ and note that g is lower semicontinuous since each $e_\lambda f(\cdot)$ is continuous by (b). The family $(e_\lambda f(x))_\lambda$ being clearly nonincreasing in λ, (as already noticed above) it converges to $\sup_{\lambda \in \,]0, \frac{1}{2\alpha}[} e_\lambda f(x) = g(x)$ as $\lambda \downarrow 0$. For each $\lambda \in \,]0, \frac{1}{2\alpha}[$ we can write by the inequality assumption on y_λ that

(3.204) $$\frac{1}{2\lambda}\|x - y_\lambda\|^2 - \alpha\|y_\lambda\|^2 - \beta\|y_\lambda\| + \gamma \leq g(x) + \theta(\lambda),$$

hence
$$\left(\frac{1}{2\lambda} - \alpha \right)\|y_\lambda\|^2 - \frac{1}{\lambda}\|x\|\|y_\lambda\| - \beta\|y_\lambda\| + \frac{1}{2\lambda}\|x\|^2 \leq g(x) + \theta(\lambda) - \gamma,$$

which entails that

(3.205) $$\frac{1}{\lambda}\|y_\lambda\|\left[\left(\frac{1}{2} - \alpha\lambda\right)\|y_\lambda\| - \|x\| - \beta\lambda\right] \leq g(x) + \theta(\lambda) - \gamma.$$

It follows that there are real constants $\lambda_0 \in \,]0, \frac{1}{2\alpha}[$ and $r > 0$ such that
$$\|y_\lambda\| \leq r \quad \text{for all } \lambda \in \,]0, \lambda_0[,$$
since otherwise there would exist a sequence $(\lambda_n)_n$ in $]0, \frac{1}{2\alpha}[$ with $\lambda_n \downarrow 0$ such that $\|y_{\lambda_n}\| \to +\infty$, which would contradict (3.205). This and (3.204) ensure that for every $\lambda \in \,]0, \lambda_0[$
$$\|x - y_\lambda\|^2 \leq 2\lambda\bigl(g(x) + \theta(\lambda) + \alpha r^2 + \beta r - \gamma\bigr),$$
which in turn entails that $y_\lambda \to x$ as $\lambda \downarrow 0$.

Further, since $f(y_\lambda) \leq g(x) + \theta(\lambda)$ for every $\lambda \in \,]0, \frac{1}{2\alpha}[$ according to the assumption on y_λ, we obtain that

(3.206) $$\liminf_{u \to x} f(u) \leq \liminf_{\lambda \downarrow 0} f(y_\lambda) \leq \limsup_{\lambda \downarrow 0} f(y_\lambda) \leq g(x).$$

On the other hand, $f(\cdot) \geq g(\cdot)$ since $e_\lambda f(\cdot) \leq f(\cdot)$. Then recalling that g is lower semicontinuous it ensues that $\liminf_{u \to x} f(u) \geq \liminf_{u \to x} g(u) = g(x)$, which combined with (3.206) ensures that $\lim_{\lambda \downarrow 0} f(y_\lambda) = \liminf_{u \to x} f(u)$.

(d) Denote $\overline{f}$ the lower semicontinuous hull of the function f and note that the family $(e_\lambda f)_\lambda$ being clearly nonincreasing in λ, it converges pointwise to $g(\cdot) := \sup_{\lambda \in \,]0, \frac{1}{2\alpha}[} e_\lambda f(\cdot)$ as $\lambda \downarrow 0$. It remains to show that $g = \overline{f}$. Given any $x \in X$ where g

is finite, for each $\lambda \in \,]0, \frac{1}{2\alpha}[$ we know by (b) that $e_\lambda f(x)$ is finite, so we can choose $y_\lambda \in X$ such that

$$e_\lambda f(x) \leq f(y_\lambda) + \frac{1}{2\lambda}\|x - y_\lambda\|^2 \leq e_\lambda f(x) + \lambda,$$

which entails by (c) that $\overline{f}(x) \leq g(x)$ if $g(x)$ is finite. The inequality $\overline{f}(x) \leq g(x)$ is also obvious if $g(x) = +\infty$, hence it follows that $\overline{f}(\cdot) \leq g(\cdot)$ on X. On the other hand, $f(\cdot) \geq g(\cdot)$ since $e_\lambda f(\cdot) \leq f(\cdot)$. Recalling that g is lowersemicontinuous and that $\overline{f}$ is the greatest lower semicontinuous function which minorizes f, it ensues that $\overline{f}(\cdot) \geq g(\cdot)$. This combined with what precedes confirms that $g(\cdot) = \overline{f}(\cdot)$ on X as desired.

(e) The convergence property in (e) is a direct consequence of (d). Regarding the lower limit property, consider nets $(\lambda_j)_{j \in J}$ and $(x_j)_{j \in J}$ satisfying the assumptions in (e). Then for each $i \in J$ by continuity of $x \mapsto e_{\lambda_i} f(x)$ we have

$$e_{\lambda_i} f(\overline{x}) = \liminf_{j \in J} e_{\lambda_i} f(x_j) \leq \liminf_{j \in J} e_{\lambda_j} f(x_j),$$

where the latter inequality is due to the nonincreasing property of the function $\lambda \mapsto e_\lambda f(x)$ on $]0, \frac{1}{2\alpha}[$. Since $f(\overline{x}) = \lim_{i \in J} e_{\lambda_i} f(\overline{x})$ by the above convergence property, we deduce that $f(\overline{x}) \leq \liminf_{j \in J} e_{\lambda_j} f(x_j)$. This finishes the proof of the theorem. $\square$

REMARK 3.266. Let (X, d) be a metric space (instead of a normed space) and $f : X \to \mathbb{R} \cup \{+\infty\}$ be a proper function such that for some $x_0 \in X$ and some reals $\alpha \geq 0$, $\beta \geq 0$ and $\gamma \in \mathbb{R}$

$$f(\cdot) \geq -\alpha d^2(\cdot, x_0) - \beta d(\cdot, x_0) + \gamma \quad \text{on } X.$$

In this context, defining the Moreau envelope with index $\lambda > 0$ as

$$e_\lambda f(x) := \inf_{y \in X} \left(f(y) + \frac{1}{2\lambda} d^2(x, y) \right) \quad \text{for all } x \in X,$$

it is easily seen that the properties in Theorem 3.265 still hold true with almost the same arguments. $\square$

When the function f is convex, the following additional properties hold. The properties in the assertion (c) are related to strictly convex norms. It is known and easily seen that $\|\cdot\|^2$ is a strictly convex function on the vector space X whenever the norm $\|\cdot\|$ of X is strictly convex in the sense that $\|(1-t)x + ty\| < 1$ for all $x, y \in \mathbb{S}_X$ with $x \neq y$ and $t \in \,]0, 1[$ (see Subsection 18.1.1 for more). Then, when the function f is convex and the norm $\|\cdot\|$ is *strictly convex*, the function $y \mapsto f(y) + \frac{1}{2\lambda}\|x - y\|^2$ is strictly convex on X, then the set $\text{Prox}_\lambda f(x)$ is either empty or a singleton. Recall also that the norm $\|\cdot\|$ of a normed space $(X, \|\cdot\|)$ has the (sequential) *Kadec-Klee property* provided that $\|x_n - x\| \to 0$ whenever $(x_n)_n$ converges weakly to x and $\|x_n\| \to \|x\|$ as $n \to \infty$; diverse examples of such norms are given in Subsection 18.1.2.

THEOREM 3.267 (differentiability of Moreau envelope under convexity). Let $(X, \|\cdot\|)$ be a normed space and $f : X \to \mathbb{R} \cup \{+\infty\}$ be a proper lower semicontinuous convex function.
(a) For any real $\lambda > 0$ the Moreau envelope $e_\lambda f$ is finite-valued, convex on X and

Lipschitz on bounded sets, and for every $x \in X$ with $\operatorname{Prox}_\lambda f(x) \neq \emptyset$
$$\partial(e_\lambda f)(x) = \partial f(y) \cap \partial\left(\frac{1}{2\lambda}\|\cdot\|^2\right)(x-y) \quad \text{for all } y \in \operatorname{Prox}_\lambda f(x).$$

(b) The family $(e_\lambda f)_{\lambda>0}$ converges pointwise to f as $\lambda \downarrow 0$, and for any nonincreasing net $(\lambda_j)_{j \in J}$ in $]0, +\infty[$ tending to 0 and any net $(x_j)_{j \in J}$ in X converging weakly to $\overline{x} \in X$ one has
$$f(\overline{x}) \leq \liminf_{j \in J} e_{\lambda_j} f(x_j).$$

(c) If X is a reflexive Banach space, then one has for each real $\lambda > 0$
$$\operatorname{Argmin}_X f = \operatorname{Argmin}_X (e_\lambda f),$$
and for each $x \in X$
$$\operatorname{Prox}_\lambda f(x) \neq \emptyset,$$
along with
$$d\bigl(0, \partial(e_\lambda f)(x)\bigr) \leq \sup_{y^* \in \partial(e_\lambda f)(x)} \|y^*\| \leq d(0, \partial f(x)).$$

(d) If X is a reflexive Banach space and its norm $\|\cdot\|$ is strictly convex, then for any real $\lambda > 0$ the proximal mapping $P_\lambda f$ is well defined and norm-to-weak continuous on X and for any $x \in \operatorname{dom} f$
$$P_\lambda f(x) \to x \quad \text{and} \quad f\bigl(P_\lambda f(x)\bigr) \to f(x) \quad \text{as } \lambda \downarrow 0;$$
if in addition the norm $\|\cdot\|$ is (sequentially) Kadec-Klee, then $P_\lambda f$ is norm-to-norm continuous on X.

(e) If X is a reflexive Banach space and its norm $\|\cdot\|$ is both strictly convex and Gâteaux differentiable off zero, then for any real $\lambda > 0$ the Moreau envelope $e_\lambda f$ is Gâteaux differentiable on X, the proximal mapping $P_\lambda f$ is well defined and norm-to-weak continuous on X, and for every $x \in X$
$$\{D(e_\lambda f)(x)\} = \partial f\bigl(P_\lambda f(x)\bigr) \cap \{\lambda^{-1} J(x - P_\lambda f(x))\},$$
where $J := D(\tfrac{1}{2}\|\cdot\|^2)$ is the Gâteaux derivative of $\tfrac{1}{2}\|\cdot\|^2$.

(f) If X is a reflexive Banach space and its norm is strictly convex, (sequentially) Kadec-Klee and Fréchet differentiable off zero, then the Moreau envelope $e_\lambda f$ is of class C^1 on X.

PROOF. (a) By Proposition 3.77 we know that f is bounded from below by a continuous affine function, so Theorem 3.265(a) tells us that for each real $\lambda > 0$, the function $e_\lambda f$ is finite-valued and Lipschitz on bounded sets. The convexity of $e_\lambda f$ and the equality in (a) follow from Proposition 3.69.

(b) The convergence property follows from the same property in (e) in Theorem 3.265. To justify the other property with lower limit, we follow the arguments for the similar property in (e) in Theorem 3.265. Fix nets $(\lambda_j)_{j \in J}$ and $(x_j)_{j \in J}$ satisfying the assumptions in (b). Then for each $i \in J$ by weak lower semicontinuity of (the continuous convex function) $x \mapsto e_{\lambda_i} f(x)$ we have
$$e_{\lambda_i} f(\overline{x}) \leq \liminf_{j \in J} e_{\lambda_i} f(x_j) \leq \liminf_{j \in J} e_{\lambda_j} f(x_j),$$
where the latter inequality is due to the nonincreasing property of the function $\lambda \mapsto e_\lambda f(x)$ on $]0, \tfrac{1}{2\alpha}[$. Since $f(\overline{x}) = \lim_{i \in J} e_{\lambda_i} f(\overline{x})$, we obtain that $f(\overline{x}) \leq \liminf_{j \in J} e_{\lambda_j} f(x_j)$.

(c) Assume that X is reflexive. Given $x \in X$, using the boundedness from below of f by a continuous affine function, we see that the weakly lower semicontinuous function $\varphi := f(\cdot) + \frac{1}{2\lambda}\|x - \cdot\|^2$ satisfies $\varphi(y) \to +\infty$ as $\|y\| \to +\infty$, and hence it attains its infimum on X by weak compactness of closed balls in the reflexive space X. This says that $\text{Prox}_\lambda f(x) \neq \emptyset$.

Now take any $u \in \text{Argmin}_X(e_\lambda f)$, which is equivalent to $0 \in \partial(e_\lambda f)(u)$ by convexity of $e_\lambda f$. Choosing $v \in \text{Prox}_\lambda f(u)$, by the equality in (a) we have $0 \in \partial f(v) \cap \partial(\frac{1}{2\lambda}\|\cdot\|^2)(u-v)$. The inclusion $0 \in \partial(\frac{1}{2\lambda}\|\cdot\|^2)(u-v)$ tells us that $u - v$ is a minimizer of $\|\cdot\|^2$, hence $v = u$. Therefore, $0 \in \partial f(u)$, so $u \in \text{Argmin}_X f$ by convexity of f. It ensues that $\text{Argmin}_X(e_\lambda f) \subset \text{Argmin}_X f$, and this inclusion is an equality since the converse has been established in Theorem 3.265(a).

Let us prove the inequalities in (c). Fix any $y^* \in \partial(e_\lambda f)(x)$ and $y \in \text{Prox}_\lambda f(x)$. By (a) we have $y^* \in \partial f(y) \cap \partial(\frac{1}{2\lambda}\|\cdot - y\|^2)(x)$. By the description of $\partial(\frac{1}{2}\|\cdot\|^2)$ in (2.25), we have

$$\|y^*\| = \frac{1}{\lambda}\|x - y\| \quad \text{and} \quad \langle y^*, x - y\rangle = \frac{1}{\lambda}\|x - y\|^2.$$

From this and from the monotonicity of ∂f we derive that for any $x^* \in \partial f(x)$ (if this set is nonempty)

$$\|x^*\|\,\|x - y\| \geq \langle x^*, x - y\rangle \geq \langle y^*, x - y\rangle = \|y^*\|\,\|x - y\|,$$

hence $\|x^*\| \geq \|y^*\|$ if $y \neq x$. Further, if $y = x$, we have $y^* \in \partial(\frac{1}{2\lambda}\|\cdot\|^2)(0)$, hence $y^* = 0$. It follows that in any case $\|x^*\| \geq \|y^*\| \geq d(0, \partial(e_\lambda f)(x))$. This being true for every $x^* \in \partial f(x)$, it ensues that

$$d(0, \partial f(x)) \geq \sup_{y^* \in \partial(e_\lambda f)(x)} \|y^*\| \geq d(0, \partial(e_\lambda f)(x)),$$

and this furnishes the inequalities in (c) whenever $\partial f(x) \neq \emptyset$. We also notice that the latter inequalities are evident if $\partial f(x) = \emptyset$, since in this situation we know that we have $d(0, \partial f(x)) = +\infty$.

(d) Under the strict convexity of the norm $\|\cdot\|$ of the reflexive Banach space X the function $\varphi := f(\cdot) + \frac{1}{2\lambda}\|x - \cdot\|^2$ in the proof of (a) is then strictly convex, so its minimizer is unique, which means that $P_\lambda f(x)$ exists for every $x \in X$. The convergence properties in (d) directly follow from Theorem 3.265(c) since $\sup_{\lambda > 0} e_\lambda f(x) \leq f(x)$.

To show the norm-to-weak continuity of $P_\lambda f$ consider any $x \in X$ and any net $(x_i)_{i \in I}$ in X converging strongly to x. Without loss of generality we may and do suppose that $(x_i)_i$ is bounded. Let us put $y_i := P_\lambda f(x_i)$ and let us choose $y_i^* \in \lambda \partial(e_\lambda f)(x_i)$ (since $\partial(e_\lambda f)(x_i) \neq \emptyset$ by continuity of the convex function $e_\lambda f$). By (a) we have $y_i^* \in \partial(\frac{1}{2}\|\cdot\|^2)(x_i - y_i)$, so $\langle y_i^*, x_i - y_i\rangle = \|x_i - y_i\|^2$ and $\|y_i^*\| = \|x_i - y_i\|$ (see (2.25)). Since $\lambda^{-1} y_i^* \in \partial(e_\lambda f)(x_i)$ and $e_\lambda f$ is Lipschitz on bounded sets, the net $(y_i^*)_i$ is bounded as well as the net $(y_i)_i$. Take any subnet (that we do not relabel) converging weakly to some $y \in X$. Writing

$$e_\lambda f(x_i) = f(y_i) + \frac{1}{2\lambda}\|x_i - y_i\|^2$$

and using the continuity of $e_\lambda f$ and the weak lower semicontinuity of f and $\|\cdot\|^2$ we obtain

(3.207) $\quad e_\lambda f(x) \geq f(y) + \frac{1}{2\lambda} \limsup_i \|x - y_i\|^2 \geq f(y) + \frac{1}{2\lambda}\|x - y\|^2,$

which ensures that $y = P_\lambda f(x)$. Then any weakly convergent subnet of the bounded net $(y_i)_i$ in the reflexive Banach space X converges weakly to $P_\lambda f(x)$, hence all the net $(y_i)_i$ converges weakly to $P_\lambda f(x)$. This guarantees that $P_\lambda f$ is norm-to-weak continuous.

Assume that the norm $\|\cdot\|$ is in addition (sequentially) Kadec-Klee. To prove the norm-to-norm continuity of $P_\lambda f$ take any $x \in X$ and any sequence $(x_n)_n$ in X with $\|x_n - x\| \to 0$ as $n \to \infty$. Put $y_n := P_\lambda f(x_n)$. The precedent arguments guarantee that the sequence $(y_n)_n$ converges weakly to $y := P_\lambda f(x)$. Further, from the left inequality in (3.207) we see that

$$f(y) + \frac{1}{2\lambda}\|x - y\|^2 = e_\lambda f(x) \geq f(y) + \frac{1}{2\lambda}\limsup_{n\to\infty} \|x - y_n\|^2.$$

Noting that $f(y)$ is finite since $e_\lambda f(x)$ is finite, we deduce that $\limsup_{n\to\infty} \|x - y_n\|^2 \leq \|x - y\|^2$, which combined with the weak lower semicontinuity of $\|\cdot\|^2$ entails that $\|x - y_n\|^2 \to \|x - y\|^2$. Since $(x - y_n)_n$ converges weakly to $x - y$, the (sequential) Kadec-Klee property of $\|\cdot\|$ assures us that $\|y_n - y\| \to 0$. This confirms the norm-to-norm continuity of $P_\lambda f$.

(e) Assume that X is reflexive and $\|\cdot\|$ is both strictly convex and Gâteaux differentiable off zero. We already know by (d) that $P_\lambda f$ is well defined and norm-to-weak continuous on X. The equality in (a) can then be rewritten in the form

$$\partial(e_\lambda f)(x) = \partial f\big(P_\lambda f(x)\big) \cap \{\lambda^{-1} J\big(x - P_\lambda f(x)\big)\}.$$

The set $\partial(e_\lambda f)(x)$ (which is nonempty by continuity and convexity of $e_\lambda f$) is then a singleton for every $x \in X$, thus the continuous convex function $e_\lambda f$ is Gâteaux differentiable on X (see Proposition 3.102) and the formula for its derivative follows from the above equality as well.

(f) Under the assumptions of (f) the Moreau envelope $e_\lambda f$ is Gâteaux differentiable with $D(e_\lambda f)(x) = \lambda^{-1} J\big(x - P_\lambda f(x)\big)$ by (e), the mapping $P_\lambda f$ is norm-to-norm continuous by (d) and the duality mapping $J := D(\frac{1}{2}\|\cdot\|^2)$ is norm-to-norm* continuous by Fréchet differentiability of $\|\cdot\|^2$ (see Theorem 2.179). Consequently, the Gâteaux derivative $D(e_\lambda f)$ is norm-to-norm* continuous, which means that $e_\lambda f$ is of class $\mathcal{C}^1$, so the proof is complete. $\square$

Now let us turn to the setting of Hilbert spaces.

THEOREM 3.268 (contraction of proximal mapping under convexity). Let $(H, \|\cdot\|)$ be a Hilbert space and $f : H \to \mathbb{R} \cup \{+\infty\}$ be a proper lower semicontinuous convex function. The following hold.

(a) The Moreau envelope $e_\lambda f$ is continuously differentiable on X and for any $x \in X$

$$\operatorname{Prox}_\lambda f(x) = \{P_\lambda f(x)\} \text{ and } \{D(e_\lambda f)(x)\} = \partial f\big(P_\lambda f(x)\big) \cap \{\lambda^{-1}\big(x - P_\lambda f(x)\big)\},$$

hence (with I denoting the identity mapping on H)

$$P_\lambda f(x) = (I + \lambda \partial f)^{-1}(x).$$

(b) For any $x, y \in H$

$$\langle x - y, P_\lambda f(x) - P_\lambda f(y)\rangle \geq \|P_\lambda f(x) - P_\lambda f(y)\|^2,$$

so in particular the mapping $P_\lambda f$ is Lipschitz on the space H with Lipschitz constant 1, that is,

$$\|P_\lambda f(x) - P_\lambda f(y)\| \leq \|x - y\| \quad \text{for all } x, y \in H.$$

Further, for $Q_\lambda f(x) := x - P_\lambda f(x)$ one also has for all $x, y \in H$

$$\langle x - y, Q_\lambda f(x) - Q_\lambda f(y)\rangle \geq \|Q_\lambda f(x) - Q_\lambda f(y)\|^2 \text{ and } \|Q_\lambda f(x) - Q_\lambda f(y)\| \leq \|x - y\|,$$

so $D(e_\lambda f)$ is Lipschitz on H with $1/\lambda$ as Lipschitz constant.

(c) Denoting $C := \operatorname{cl}(\operatorname{dom} f)$ one has for any $x \in H$
$$P_\lambda f(x) \to \operatorname{proj}_C(x) \quad \text{as } \lambda \downarrow 0.$$

PROOF. First we notice by Theorem 3.267(e) that $e_\lambda f$ is Gâteaux differentiable and that the equality
$$\{D(e_\lambda f)(x)\} = \partial f(P_\lambda f(x)) \cap \{\lambda^{-1}(x - P_\lambda f(x))\}$$
holds true since J is the identity mapping according to the Hilbert structure of $(H, \|\cdot\|)$. The convexity of the function f and the inclusion $\lambda^{-1}(x - P_\lambda f(x)) \in \partial f(P_\lambda f(x))$ (due to latter equality) also yield for any $x, y \in X$
$$\langle \lambda^{-1}(x - P_\lambda f(x)) - \lambda^{-1}(y - P_\lambda f(y)), P_\lambda f(x) - P_\lambda f(y) \rangle \geq 0,$$
or equivalently
$$\langle x - y, P_\lambda f(x) - P_\lambda f(y) \rangle \geq \|P_\lambda f(x) - P_\lambda f(y)\|^2.$$
From this it is also clear that $P_\lambda f$ is a contraction on H. The latter inequality also yields with simple calculations to the inequalities concerning $Q_\lambda f$, and the contraction property of $Q_\lambda f$ combined with the above expression of $D(e_\lambda f)(x)$ ensures that $e_\lambda f$ is Fréchet differentiable on H with $D(e_\lambda f)$ Lipschitz on H with $1/\lambda$ as Lipschitz constant. The assertions (a) and (b) are then established.

Regarding (c), Theorem 3.267(d) entails for $x \in \operatorname{dom} f$ that $P_\lambda f(x) \to x$ as $\lambda \downarrow 0$. If $x \in C := \operatorname{cl}(\operatorname{dom} f)$, for every $\varepsilon > 0$ we can choose $x_\varepsilon \in \operatorname{dom} f$ with $\|x_\varepsilon - x\| < \varepsilon$. Since by what precedes $P_\lambda f(x_\varepsilon) \to x_\varepsilon$ as $\lambda \downarrow 0$, we can also choose $\lambda_\varepsilon > 0$ such that $\|P_\lambda f(x_\varepsilon) - x_\varepsilon\| < \varepsilon$ for all $\lambda \in]0, \lambda_\varepsilon]$. Then for every $\lambda \in]0, \lambda_\varepsilon]$ writing
$$\|P_\lambda f(x) - x\| \leq \|P_\lambda f(x) - P_\lambda f(x_\varepsilon)\| + \|P_\lambda f(x_\varepsilon) - x_\varepsilon\| + \|x_\varepsilon - x\|,$$
we see by the above contraction property of $P_\lambda f$ that
$$\|P_\lambda f(x) - x\| \leq 2\|x_\varepsilon - x\| + \|P_\lambda f(x_\varepsilon) - x_\varepsilon\| < 3\varepsilon,$$
which gives that $P_\lambda f(x) \to x$ as $\lambda \downarrow 0$. Now suppose that $x \notin C$ and choose $b \in H$ and $\gamma \in \mathbb{R}$ such that $f(\cdot) \geq \langle b, \cdot \rangle + \gamma$. Consider any sequence $(\lambda_n)_n$ in $]0, +\infty[$ with $\lambda_n \downarrow 0$ and put $y_n := P_{\lambda_n} f(x)$. Taking some $z \in C$ and noting by the contraction property of $P_{\lambda_n} f(\cdot)$ that
$$\|y_n\| \leq \|P_{\lambda_n} f(z)\| + \|P_{\lambda_n} f(x) - P_{\lambda_n} f(z)\| \leq \|P_{\lambda_n} f(z)\| + \|x - z\|,$$
we see that the sequence $(y_n)_n$ is bounded since $(P_{\lambda_n} f(z))_n$ is convergent by what precedes. Consider any weakly convergent subsequence that we do not relabel and let y be its weak limit. The inclusion $\lambda_n^{-1}(x - y_n) \in \partial f(y_n)$ (due to (a)) tells us that $y_n \in \operatorname{Dom} \partial f$, hence $y \in C$. Given any $u \in \operatorname{dom} f$ we also have
$$f(u) \geq f(y_n) + \langle \lambda_n^{-1}(x - y_n), u - y_n \rangle,$$
which gives
$$(3.208) \qquad \lambda_n f(u) \geq -\lambda_n \|b\| \|y_n\| + \lambda_n \gamma + \langle x - y_n, u \rangle - \langle x, y_n \rangle + \|y_n\|^2,$$
so the weak lower semicontinuity of $\|\cdot\|^2$ furnishes as $n \to \infty$
$$0 \geq \langle x - y, u \rangle - \langle x, y \rangle + \limsup_{n \to \infty} \|y_n\|^2$$
$$\geq \langle x - y, u \rangle - \langle x, y \rangle + \|y\|^2 = \langle x - y, u - y \rangle,$$

and the latter inequalities still hold for any $u \in C$. The inequality $\langle x - y, u - y \rangle \leq 0$ for every $u \in C$ implies $y = \operatorname{proj}_C(x)$. Further, taking $u = y$ in the inequality involving $\limsup_{n \to \infty} \|y_n\|^2$ we also have

$$0 \geq \langle x - y, y \rangle - \langle x, y \rangle + \limsup_{n \to \infty} \|y_n\|^2,$$

or equivalently $\limsup_{n \to \infty} \|y_n\|^2 \leq \|y\|^2$, which yields $\|y_n\|^2 \to \|y\|^2$ according to the weak lower semicontinuity of $\|\cdot\|^2$. This and the weak convergence of $(y_n)_n$ to y entail that $y_n \to y = \operatorname{proj}_C(x)$ strongly. It follows that $y_\lambda \to \operatorname{proj}_C(x)$ strongly as $\lambda \downarrow 0$, which finishes the proof of the theorem. $\square$

We recall the notation $(Pf)(x)$ for $(P_\lambda f)(x)$ when $\lambda = 1$.

THEOREM 3.269 (J.J. Moreau: proximal mappings of pair of conjugates). Let $f : H \to \mathbb{R} \cup \{+\infty\}$ be a proper lower semicontinuous convex function on a Hilbert space H and let $x, u, v \in H$. Then the assertions (a) and (b) are equivalent:
(a) $\quad x = u + v$ and $f(u) + f^*(v) = \langle u, v \rangle$;
(b) $\quad u = (Pf)(x)$ and $v = (Pf^*)(x)$.

PROOF. Assume that (a) holds, so $f(u)$ and $f^*(v)$ are finite. For any $z \in X$ we have

$$\langle u, v \rangle - f(u) = f^*(v) \geq \langle z, v \rangle - f(z),$$

so putting $\varphi(z) := \frac{1}{2}\|x - z\|^2 + f(z)$ we obtain for any $z \in H$

$$\varphi(z) - \varphi(u) = \frac{1}{2}\|u + v - z\|^2 + f(z) - \frac{1}{2}\|v\|^2 - f(u)$$
$$\geq \frac{1}{2}\|(u - z) + v\|^2 - \frac{1}{2}\|v\|^2 + \langle z - u, v \rangle$$
$$= \frac{1}{2}\|u - z\|^2.$$

This tells us that u is the minimizer of φ, that is, $u = (Pf)(x)$. Since $f = (f^*)^*$ according to the Fenchel-Moreau theorem (see Theorem 3.84), by symmetry we also have $v = (Pf^*)(x)$, then the implication (a)$\Rightarrow$(b) is established.

Conversely, assume that (b) is satisfied and put $v' := x - u$, so $x = u + v'$. Let us prove that $f(u) + f^*(v') = \langle u, v' \rangle$. Take any $z \in H$ and any $t \in \,]0, 1[$ and note that the equality $u = (Pf)(x)$ assures us that

$$\frac{1}{2}\|x - tz - (1-t)u\|^2 + f(tz + (1-t)u) \geq \frac{1}{2}\|x - u\|^2 + f(u),$$

which in turn by Jensen inequality implies that

$$\frac{1}{2}\|x - tz - (1-t)u\|^2 + tf(z) + (1-t)f(u) \geq \frac{1}{2}\|x - u\|^2 + f(u),$$

or equivalently

$$\frac{1}{2}\|v'\|^2 - \frac{1}{2}\|v' - t(z-u)\|^2 + f(u) \leq tf(z) + (1-t)f(u).$$

This inequality can be rewritten after computation as

$$-\frac{1}{2}t^2\|z - u\|^2 + t\langle z - u, v' \rangle \leq tf(z) - tf(u),$$

which yields after dividing by $t > 0$

$$\langle z, v' \rangle - f(z) \le \langle u, v' \rangle - f(u) + \frac{1}{2} t \|z - u\|^2.$$

Making $t \downarrow 0$ we obtain for every $z \in H$

$$\langle z, v' \rangle - f(z) \le \langle u, v' \rangle - f(u),$$

hence $f^*(v') = \langle u, v' \rangle - f(u)$, or equivalently $f(u) + f^*(v') = \langle u, v' \rangle$ as desired. This combined with the equality $u + v' = x$ (by definition of v') entails by the above implication (a)$\Rightarrow$(b) that $v' = (Pf^*)(x)$, thus $v' = v$ by the assumption (b). It results that $x = u + v$ and $f(u) + f^*(v) = \langle u, v \rangle$, which corresponds to (a). The proof is finished. □

Regarding the Moreau envelope $(e_1 f)(\cdot)$ (that is, $(e_\lambda f)(\cdot)$ with $\lambda = 1$), we have the following corollary.

COROLLARY 3.270. *Let $f : H \to \mathbb{R} \cup \{+\infty\}$ be a proper lower semicontinuous convex function on a Hilbert space H. Then for any $x \in H$ one has*

$$(e_1 f)(x) + (e_1 f^*)(x) = \frac{1}{2}\|x\|^2.$$

PROOF. Fix $x \in X$ and denote $u := (Pf)(x)$ and $v := (Pf^*)(x)$. By Theorem 3.269 we have $u + v = x$ and $f(u) + f^*(v) = \langle u, v \rangle$. These equalities yield

$$(e_1 f)(x) + (e_1 f^*)(x) = f(u) + \frac{1}{2}\|x - u\|^2 + f^*(v) + \frac{1}{2}\|x - v\|^2$$
$$= \langle u, v \rangle + \frac{1}{2}\|v\|^2 + \frac{1}{2}\|u\|^2 = \frac{1}{2}\|u + v\|^2,$$

from which follows the desired equality $(e_1 f)(x) + (e_1 f^*)(x) = \frac{1}{2}\|x\|^2$. □

The next corollary of Theorem 3.269 is concerned with the analog of the equality $(Pf)(x) + (Pf^*)(x) = x$, when the proximal mapping with any index $\lambda > 0$ is considered.

COROLLARY 3.271. *Let $f : H \to \mathbb{R} \cup \{+\infty\}$ be a proper lower semicontinuous convex function on a Hilbert space H. Then for any real $\lambda > 0$ one has*

$$\left(P_\lambda f \right)(x) + \lambda \left(P_{\frac{1}{\lambda}} f^* \right)\left(\frac{1}{\lambda} x \right) = x.$$

PROOF. By (3.201) we know that $(P_\lambda f)(x) = (P(\lambda f))(x)$ and we also know by (3.32) that $(\lambda f)^*(\cdot) = \lambda f^*(\frac{1}{\lambda} \cdot)$. On the other hand, for $v_\lambda := (P(\lambda f)^*)(x)$ we also have

$$(\lambda f)^*(v_\lambda) + \frac{1}{2}\|x - v_\lambda\|^2 = \inf_{y \in H} \left((\lambda f)^*(y) + \frac{1}{2}\|x - y\|^2 \right),$$

which gives

$$f^*\left(\frac{1}{\lambda} v_\lambda\right) + \frac{1}{2\lambda}\|x - v_\lambda\|^2 = \inf_{y \in H} \left(f^*\left(\frac{1}{\lambda} y\right) + \frac{1}{2\lambda}\|x - y\|^2 \right)$$
$$= \inf_{y \in H} \left(f^*\left(\frac{1}{\lambda} y\right) + \frac{\lambda}{2}\left\|\frac{1}{\lambda} x - \frac{1}{\lambda} y\right\|^2 \right)$$
$$= \inf_{y' \in H} \left(f^*(y') + \frac{1}{2\lambda^{-1}}\left\|\frac{1}{\lambda} x - y'\right\|^2 \right),$$

or equivalently

$$f^*\left(\frac{1}{\lambda}v_\lambda\right) + \frac{1}{2\lambda^{-1}}\left\|\frac{1}{\lambda}x - \frac{1}{\lambda}v_\lambda\right\|^2 = \inf_{y'\in H}\left(f^*(y') + \frac{1}{2\lambda^{-1}}\left\|\frac{1}{\lambda}x - y'\right\|^2\right).$$

It follows that

$$\lambda^{-1}v_\lambda = (P_{\lambda^{-1}}f^*)(\lambda^{-1}x), \quad \text{or equivalently } v_\lambda = \lambda(P_{\lambda^{-1}}f^*)(\lambda^{-1}x).$$

Consequently, the equality $x = (P(\lambda f))(x) + (P(\lambda f)^*)(x)$ furnished by Theorem 3.269 becomes

$$x = \left(P_\lambda f\right)(x) + \lambda\left(P_{\frac{1}{\lambda}}f^*\right)(\lambda^{-1}x)$$

as stated in the corollary. □

3.21.2. Strongly convex functions. In this subsection we will characterize Moreau envelopes with index $\lambda > 0$ as functions which are conjugates of σ-strongly convex functions with $\sigma = \lambda$.

DEFINITION 3.272. Given a convex set C in a normed space $(X, \|\cdot\|)$, an extended real-valued function $f : C \to \mathbb{R} \cup \{+\infty\}$ is called *strongly convex* on C provided that there exists a real $\sigma > 0$ such that for all $x, y \in C$ and $t \in]0, 1[$

$$f((1-t)x + ty) \leq (1-t)f(x) + tf(y) - \frac{1}{2}\sigma t(1-t)\|x-y\|^2.$$

One also says that f is σ-strongly convex on C, or strongly convex on C with constant σ. When C is the whole space X, one just says that f is σ-strongly convex.

Clearly, any strongly convex function f is strictly convex on its effective domain dom f, but the function $x \mapsto x^4$ on $\mathbb{R}$ shows that the converse fails.

Strongly convex functions are of great interest mainly in Hilbert spaces. Recalling by (2.17) that the norm of a Hilbert space H satisfies, for all $x, y \in H$ and $t \in]0, 1[$, the equality

$$\|(1-t)x + ty\|^2 = (1-t)\|x\|^2 + t\|y\|^2 + t(1-t)\|x-y\|^2,$$

we see that the σ-strong convexity of f on the convex set C of H amounts to the convexity on C of the function $f - \frac{\sigma}{2}\|\cdot\|^2$. We state this as follows:

PROPOSITION 3.273. Let H be a Hilbert space whose associated norm is $\|\cdot\|$, let $f : C \to \mathbb{R} \cup \{+\infty\}$ be a proper function on a convex set C of H and let $\sigma \in]0, +\infty[$. Then f is σ-strongly convex on C if and only if $f - \frac{\sigma}{2}\|\cdot\|^2$ is convex on C.

The following lemma will be used in the proof of the theorem of characterization of functions which are Moreau envelopes. It has its own interest. Recall that a multimapping $M : U \rightrightarrows X^*$ from a subset U of a normed space X into its dual normed space X^* is σ-*strongly co-monotone* on U for a real $\sigma > 0$ provided that for all (x, x^*) and (y, y^*) in gph M one has

$$\langle x^* - y^*, x - y\rangle \geq \sigma\|x^* - y^*\|^2.$$

LEMMA 3.274. Let U be a nonempty open convex set of a Hilbert space H, let $\sigma > 0$ be a positive real and let $f : U \to \mathbb{R}$ be a Gâteaux differentiable function on U. Consider the following assertions.

(a) The gradient mapping $\nabla f : U \to H$ is σ-strongly co-monotone on U.
(b) The gradient mapping $\nabla f : U \to H$ is $1/\sigma$-Lipschitz continuous on U.
(c) For all $x, y \in U$ one has

$$|f(y) - f(x) - \langle \nabla f(x), y - x \rangle| \leq \frac{1}{2\sigma}\|y - x\|^2.$$

Then one has (a) $\Rightarrow$ (b) $\Rightarrow$ (c).

PROOF. (a) $\Rightarrow$ (b) Take any $x, y \in U$. By (a) we have

$$\langle \nabla f(x) - \nabla f(y), x - y \rangle \geq \sigma \|\nabla f(x) - \nabla f(y)\|^2,$$

which ensures that

$$\|\nabla f(x) - \nabla f(y)\| \, \|x - y\| \geq \sigma \|\nabla f(x) - \nabla f(y)\|^2,$$

hence we obtain $\|\nabla f(x) - \nabla f(y)\| \leq (1/\sigma)\|x - y\|$, which translates the $1/\sigma$-Lipschitz continuity of ∇f on U.
(b) $\Rightarrow$ (c) Take any $x, y \in U$ and by (a) write

$$f(y) = f(x) + \int_0^1 \langle \nabla f(x + t(y - x)), y - x \rangle \, dt$$

$$= f(x) + \int_0^1 \langle \nabla f(x + t(y - x)) - \nabla f(x), y - x \rangle \, dt + \langle \nabla f(x), y - x \rangle$$

$$\leq f(x) + \frac{1}{\sigma}\|y - x\|^2 \int_0^1 t \, dt + \langle \nabla f(x), y - x \rangle,$$

which translates the inequality

$$f(y) \leq f(x) + \frac{1}{2\sigma}\|y - x\|^2 + \langle \nabla f(x), y - x \rangle.$$

Changing in $-f$ and combining the new inequality with the previous gives the desired inequality in (c). $\square$

We will also see in Theorem 10.37 that the $1/\sigma$-Lipschitz property of ∇f on the open convex set U of the Hilbert space H is equivalent to the property

$$|\langle \nabla f(x) - \nabla f(y), x - y \rangle| \leq (1/\sigma)\|x - y\|^2 \text{ for all } x, y \in U;$$

other equivalences to the same property will be also shown in this Theorem 10.37.

THEOREM 3.275 (conjugate under strong convexity and characterization of Moreau envelopes). Let $f : H \to \mathbb{R} \cup \{+\infty\}$ be a proper lower semicontinuous convex function on a Hilbert space H with inner product $\langle \cdot, \cdot \rangle$ and associated norm $\|\cdot\|$. Given a real $\sigma > 0$ the following are equivalent:
(a) the function f is σ-strongly convex;
(b) there exists a proper lower semicontinuous convex function $g : H \to \mathbb{R} \cup \{+\infty\}$ such that $f^* = e_\sigma g$, that is, f^* is the Moreau envelope with index σ of a proper lower semicontinuous convex function g;
(c) the conjugate function f^* is Gâteaux differentiable on H and its gradient is σ-strongly co-monotone on H, that is,

$$\langle \nabla f^*(v_1) - \nabla f^*(v_2), v_1 - v_2 \rangle \geq \sigma \|\nabla f^*(v_1) - \nabla f^*(v_2)\|^2 \quad \text{for all } v_1, v_2 \in H;$$

(d) the conjugate function f^* is Gâteaux differentiable on H and its gradient ∇f^* is $(1/\sigma)$-Lipschitz on H;

(e) the conjugate function f^* is Gâteaux differentiable on H and

$$|f^*(v') - f^*(v) - \langle \nabla f^*(v), v' - v \rangle| \leq \frac{1}{2\sigma} \|v' - v\|^2 \quad \text{for all } v, v' \in H;$$

(f) the conjugate function f^* is Gâteaux differentiable on H and

$$f^*(v') \leq f^*(v) + \langle \nabla f^*(v), v' - v \rangle + \frac{1}{2\sigma} \|v' - v\|^2 \quad \text{for all } v, v' \in H.$$

PROOF. (a)$\Leftrightarrow$(b) By Proposition 3.273 we know that (a) holds if and only if there exists a proper lower semicontinuous function $\zeta : H \to \mathbb{R} \cup \{+\infty\}$ such that $\zeta = f - \frac{\sigma}{2} \|\cdot\|^2$, or equivalently $f = \zeta + \frac{\sigma}{2} \|\cdot\|^2$. This is equivalent in turn, by the Moreau-Rockafellar condition in Corollary 3.170, to $f^* = \zeta^* \Box \frac{1}{2\sigma} \|\cdot\|^2$ (since $\frac{\sigma}{2} \|\cdot\|^2$ is continuous and its conjugate is $\frac{1}{2\sigma} \|\cdot\|^2$ (see (3.41))), so (a)$\Leftrightarrow$(b) is established.

(a)$\Rightarrow$(c) Suppose that (a) is satisfied. Since (a)$\Rightarrow$(b), Theorem 3.267(b) tells us that f^* is Gâteaux differentiable on H. On the other hand, by Proposition 3.273 there is a convex function $\zeta : H \to \mathbb{R} \cup \{+\infty\}$ such that $f = \zeta + \frac{\sigma}{2} \|\cdot\|^2$, hence for each $x \in H$ we have $\partial f(x) = \partial \zeta(x) + \sigma x$ by the Moreau-Rockafellar result in Corollary 3.170. Then for (x_i, v_i) in gph ∂f, with $i = 1, 2$, we can choose $z_i \in \partial \zeta(x_i)$ such that $v_i = z_i + \sigma x_i$, thus

$$\langle v_1 - v_2, x_1 - x_2 \rangle = \langle z_1 - z_2, x_1 - x_2 \rangle + \sigma \|x_1 - x_2\|^2 \geq \sigma \|x_1 - x_2\|^2.$$

Recalling that $v \in \partial f(x)$ is equivalent to $x \in \partial f^*(v)$ (since f is a proper lower semicontinuous convex function), we have for any $v_i \in H$ that $v_i \in \partial f(x_i)$ for $x_i := \nabla f^*(v_i)$, hence

$$\langle v_1 - v_2, \nabla f^*(v_1) - \nabla f^*(v_2) \rangle \geq \sigma \|\nabla f^*(v_1) - \nabla f^*(v_2)\|^2,$$

which translates (c).

(c)$\Rightarrow$(d) This implication follows from Lemma 3.274.

(d)$\Rightarrow$(e) This implication also follows from Lemma 3.274.

(e)$\Rightarrow$(f) This implication is evident.

(f)$\Rightarrow$(a) According to (f) we have for every $v' \in H$

$$f^*(v') = \inf_{v \in H} \left(f^*(v) + \langle \nabla f^*(v), v' - v \rangle + \frac{1}{2\sigma} \|v' - v\|^2 \right),$$

so (f being the conjugate of f^* by the Fenchel-Moreau theorem (see Theorem 3.84(c))) we can write for any $x \in H$

$$f(x) = \sup_{v' \in H} \left(\langle v', x \rangle - f^*(v') \right)$$

$$= \sup_{v, v' \in H} \left(\langle v', x \rangle - f^*(v) - \langle v' - v, \nabla f^*(v) \rangle - \frac{1}{2\sigma} \|v' - v\|^2 \right)$$

$$= \sup_{v, u \in H} \left(\langle v + u, x \rangle - f^*(v) - \langle u, \nabla f^*(v) \rangle - \frac{1}{2\sigma} \|u\|^2 \right).$$

We deduce that for every $x \in H$

$$f(x) = \sup_{v \in H} \left(\langle v, x \rangle - f^*(v) + \sup_{u \in H} \left(\langle u, x - \nabla f^*(v) \rangle - \frac{1}{2\sigma} \|u\|^2 \right) \right)$$

$$= \sup_{v \in H} \left(\langle v, x \rangle - f^*(v) + \frac{\sigma}{2} \|x - \nabla f^*(v)\|^2 \right),$$

so we obtain, with the convex function $\zeta : H \to \mathbb{R} \cup \{+\infty\}$ given by
$$\zeta(x) := \sup_{v \in H} \left(\langle v - \sigma \nabla f^*(v), x \rangle + \frac{\sigma}{2} \|\nabla f^*(v)\|^2 - f^*(v) \right),$$
the equality $f(x) = \zeta(x) + \frac{\sigma}{2}\|x\|^2$. Proposition 3.273 tells us that f is σ-strongly convex, hence the proof of the theorem is finished. □

Let $f : U \to \mathbb{R} \cup \{+\infty\}$ be a proper function on an open convex set U of a Hilbert space H. If f is σ-strongly convex on U, then for all pairs (u_i, v_i) in gph ∂f with $u_i \in U$, $i = 1, 2$, one has

(3.209) $$\langle v_1 - v_2, u_1 - u_2 \rangle \geq \sigma \|u_1 - u_2\|^2,$$

that is, the multimapping ∂f is σ-strongly monotone on U. Indeed, by σ-strong convexity of f on U there is a convex function $\zeta : U \to \mathbb{R} \cup \{+\infty\}$ such that $f = \zeta + \frac{\sigma}{2}\|\cdot\|^2$, so $\partial f(u) = \sigma u + \partial \zeta(u)$ for every $u \in U$ by the Moreau-Rockafellar theorem. Take any $(u_i, v_i) \in$ gph ∂f, $i = 1, 2$. We can choose $y_i \in \partial \zeta(u_i)$ such that $v_i = \sigma u_i + y_i$, and hence
$$\langle v_1 - v_2, u_1 - u_2 \rangle = \sigma \|u_1 - u_2\|^2 + \langle y_1 - y_2, u_1 - u_2 \rangle \geq \sigma \|u_1 - u_2\|^2.$$
In fact, we will see in Corollary 6.70 that the σ-strong monotonicity of the Clarke subdifferential $\partial_C f$ on U is a characterization of the σ-strong convexity of proper lower semicontinuous functions $f : U \to \mathbb{R} \cup \{+\infty\}$.

The equivalence between (a), (b), (c) (d), (e) and (f) in the above theorem can be rewritten in the following form of characterizations of Moreau envelopes.

COROLLARY 3.276 (characterizations of Moreau envelopes). Let $\varphi : H \to \mathbb{R} \cup \{+\infty\}$ be a proper lower semicontinuous convex function on a Hilbert space H and let $\lambda \in]0, +\infty[$. The following are equivalent:
(a) the function φ is the Moreau envelope with index λ of a proper lower semicontinuous convex function on H;
(b) the conjugate function φ^* is λ-strongly convex on H;
(c) the function φ is Gâteaux differentiable on H and its gradient $\nabla\varphi$ is Lipschitz on the whole space H with Lipschitz constant $1/\lambda$;
(d) the function φ is Gâteaux differentiable on H and
$$|\varphi(x') - \varphi(x) - \langle \nabla\varphi(x), x' - x \rangle| \leq \frac{1}{2\lambda} \|x' - x\|^2 \quad \text{for all } x, x' \in H;$$
(e) the function φ is Gâteaux differentiable on H and
$$\varphi(x') \leq \varphi(x) + \langle \nabla\varphi(x), x' - x \rangle + \frac{1}{2\lambda} \|x' - x\|^2 \quad \text{for all } x, x' \in H;$$
(f) the function φ is Gâteaux differentiable on H and its gradient $\nabla\varphi$ is λ-strongly co-monotone on H, that is, for all $x_1, x_2 \in U$
$$\langle \nabla\varphi(x_1) - \nabla\varphi(x_2), x_1 - x_2 \rangle \geq \lambda \|\nabla\varphi(x_1) - \nabla\varphi(x_2)\|^2.$$

We establish now characterizations of mappings which are proximal mappings.

THEOREM 3.277 (characterizations of proximal mappings). Let $\Pi : H \to H$ be a mapping from a Hilbert space H into itself. The following are equivalent:
(a) the mapping Π is a proximal mapping with index $\lambda = 1$, that is, there is a proper lower semicontinuous convex function $f : H \to \mathbb{R} \cup \{+\infty\}$ with $(Pf)(\cdot) = \Pi(\cdot)$;
(b) there exists a continuous convex function $\varphi : H \to \mathbb{R}$ Gâteaux differentiable on

H with $\nabla\varphi(x) = \Pi(x)$ for every $x \in H$ and such that the function $\frac{1}{2}\|\cdot\|^2 - \varphi$ is convex;
(c) there exist proper lower semicontinuous convex functions $\varphi, \varrho : H \to \mathbb{R} \cup \{+\infty\}$ such that $\varphi + \varrho = \frac{1}{2}\|\cdot\|^2$ and such that $\Pi(x) \in \partial\varphi(x)$ for all $x \in H$;
(d) the mapping Π is a contraction and there exists a proper lower semicontinuous convex function $\varphi : H \to \mathbb{R} \cup \{+\infty\}$ such that $\Pi(x) \in \partial\varphi(x)$ for all $x \in H$.

PROOF. (a)$\Rightarrow$(b) Assume that (a) holds and put $g := f^*$. By Theorem 3.267 the Moreau envelope (with index $\lambda = 1$) $\varphi := e_1 g$ of g is a continuous convex function which is Gâteaux differentiable on H and for each $x \in H$ one has the equality $x - (Pg)(x) = \nabla\varphi(x)$. This and Theorem 3.269 give $(Pf)(x) = \nabla\varphi(x)$ for every $x \in H$, hence $\Pi(x) = \nabla\varphi(x)$ for every $x \in H$. On the other hand, we know by Corollary 3.270 that $(e_1 f)(\cdot) + \varphi(\cdot) = \frac{1}{2}\|\cdot\|^2$, so the implication (a)$\Rightarrow$(b) is justified.
(b)$\Rightarrow$(c) This implication is trivial.
(c)$\Rightarrow$(a) By continuity of finite-valued lower semicontinuous convex functions on Banach spaces (see Theorem 3.19) and by the assumption (c) there is a continuous convex function $\varrho : H \to \mathbb{R}$ such that $\varphi = \frac{1}{2}\|\cdot\|^2 - \varrho$, and by Proposition 3.77(a) there are families $(a_i)_{i \in I}$ in H and $(\beta_i)_{i \in I}$ in $\mathbb{R}$ such that $\varrho(\cdot) = \sup_{i \in I}(\langle a_i, \cdot \rangle + \beta_i)$. Then
$$\varphi(x) = \inf_{i \in I}\left(\frac{1}{2}\|x\|^2 - \langle a_i, x\rangle - \beta_i\right) \quad \text{for all } x \in H,$$
which by (3.19) and (3.30) gives for every $v \in H$
$$\varphi^*(v) = \sup_{i \in I}\left(\frac{1}{2}\|v + a_i\|^2 + \beta_i\right)$$
$$= \frac{1}{2}\|v\|^2 + \sup_{i \in I}\left(\langle a_i, v\rangle + \frac{1}{2}\|a_i\|^2 + \beta_i\right).$$
This says by Proposition 3.273 that φ^* is σ-strongly convex with $\sigma = 1$, hence it ensues by the equivalence (a)$\Leftrightarrow$(b) in Theorem 3.275 that there is a proper lower semicontinuous convex function $g : H \to \mathbb{R} \cup \{+\infty\}$ such that $\varphi = e_\sigma g$ with $\sigma = 1$. The function φ is then Gâteaux differentiable (see Theorem 3.267(c)) and $\nabla\varphi = \Pi$ since by assumption $\Pi(x) \in \partial\varphi(x)$. Denoting $f := g^*$ one obtains from Theorem 3.267(c) again and from Theorem 3.269 that for every $x \in H$
$$\Pi(x) = \nabla(e_\sigma g)(x) = x - (Pg)(x) = (Pf)(x),$$
so $\Pi(\cdot) = (Pf)(\cdot)$, which translates (a).
(a)$\Rightarrow$(d) Under (a) there is a lower semicontinuous convex function $f : H \to \mathbb{R} \cup \{+\infty\}$ such that $\Pi(\cdot) = (Pf)(\cdot)$, so Π is a contraction by Theorem 3.267(c). Further, by Theorem 3.269 one has for every $x \in H$
$$\Pi(x) = (Pf)(x) = x - (Pf^*)(x) = \nabla(e_1 f^*)(x).$$
The implication (a)$\Rightarrow$(d) is then justified.
(d)$\Rightarrow$(a) Assume (d). The lower semicontinuous convex function φ is finite on H (since $\operatorname{Dom}\partial\varphi = H$), hence continuous on H (see Theorem 3.19). Then Corollary 3.110 tells us that φ is Fréchet differentiable on H with $\nabla\varphi(\cdot) = \Pi(\cdot)$, so $\nabla\varphi(\cdot)$ is a contraction. By Corollary 3.276 there is a proper lower semicontinuous convex function $g : H \to \mathbb{R} \cup \{+\infty\}$ such that $\varphi = e_\sigma g$ with $\sigma = 1$, hence
$$\nabla\varphi(x) = x - (Pg)(x) = (Pg^*)(x) \quad \text{for all } x \in H$$

(by Theorems 3.267(c) and 3.269). The proof is then complete. □

EXAMPLE 3.278. As example of application, we show that any convex combinations of proximal mappings (with index $\lambda = 1$) is a proximal mapping. Indeed, let H be a Hilbert space and $\Pi_i : H \to H$ be proximal mappings on H with $i = 1, \cdots, m$. Let $\Pi := t_1 \Pi_1 + \cdots + t_m \Pi_m$. By the characterization (c) in Theorem 3.277 there exists, for each $i = 1, \cdots, m$, a pair (φ_i, ϱ_i) of lower semicontinuous convex functions from H into $\mathbb{R}$ such that $\varphi_i + \varrho_i = \frac{1}{2} \|\cdot\|^2$ and $\Pi_i(x) \in \partial \varphi_i(x)$ for every $x \in X$. Then putting

$$\varphi := t_1 \varphi_1 + \cdots + t_m \varphi_m \quad \text{and} \quad \varrho := t_1 \varrho_1 + \cdots + t_m \varrho_m,$$

the functions φ, ϱ are clearly lower semicontinuous and convex with $\varphi + \varrho = \frac{1}{2} \|\cdot\|^2$, and for every $x \in H$

$$\Pi(x) \in t_1 \partial \varphi_1(x) + \cdots + t_m \partial \varphi_m(x) \subset \partial \varphi(x).$$

The characterization (c) in Theorem 3.277 once again says that Π is a proximal mapping. □

3.22. Gâteaux differentiability at subdifferentiability points

In the previous section we saw how the global Lipschitz property of the Gâteaux derivative of a convex function on a Hilbert space is connected with the strong convexity of its Legendre-Fenchel conjugate. Now, we study a similar connection for the Gâteaux differentiability over the interior of the domain of a proper lower semicontinuous convex function on a reflexive Banach space.

THEOREM 3.279 (dual characterization of uniqueness of subgradient). Let $f : X \to \mathbb{R} \cup \{+\infty\}$ be a proper lower semicontinuous convex function on a reflexive Banach space X. The following are equivalent:
(a) the subdifferential $\partial f(x)$ is a singleton for any $x \in \mathrm{Dom}\,\partial f$;
(b) the conjugate function f^* is strictly convex on any convex subset of $\mathrm{Dom}\,\partial f^*$.

PROOF. First, observe that $(\partial f)^{-1} = \partial f^*$ since X is reflexive and f is a proper lower semicontinuous convex function.
(a)⇒(b) Suppose that f^* is not strictly convex on some convex subset of $\mathrm{Dom}\,\partial f^*$. This means that there are two distinct points x_0^*, x_1^* in X^* such that $[x_0^*, x_1^*] \subset \mathrm{Dom}\,\partial f^*$ along with $f(x_t^*) = (1-t)f(x_0^*) + tf(x_1^*)$ for some $t \in]0,1[$, where $x_t^* := (1-t)x_0^* + tx_1^*$. Choose $u \in \partial f^*(x_t^*)$. Then $f(u) + f^*(x_t^*) = \langle x_t^*, u \rangle$ according to Proposition 3.64(i), which combined with the preceding equality yields

$$(1-t)[f(u) + f^*(x_0^*) - \langle x_0^*, u \rangle] + t[f(u) + f^*(x_1^*) - \langle x_1^*, u \rangle] = 0.$$

Since $t \in]0,1[$ and each bracket $[\,]\geq 0$ (by the Fenchel inequality), we deduce that each bracket $[\,]$ is null, so $x_0^* \in \partial f(u)$ and $x_1^* \in \partial f(u)$, and this says that $\partial f(u)$ is not a singleton. This ensures that (a)⇒(b) holds true.
(b)⇒(a) Now suppose for some $u \in \mathrm{Dom}\,\partial f$ that $\partial f(u)$ contains two distinct elements x_0^*, x_1^*. For each $t \in [0,1]$ put $x_t^* := (1-t)x_0^* + tx_1^*$ and notice (by convexity of $\partial f(u)$) that $x_t^* \in \partial f(u)$ for every $t \in [0,1]$. Taking any $t \in]0,1[$ and using Proposition 3.64(i) we can write

$$f^*(x_t^*) = \langle x^*, u \rangle - f(u) = (1-t)[\langle x_0^*, u \rangle - f(u)] + t[\langle x_1^*, u \rangle - f(u)]$$
$$= (1-t)f^*(x_0^*) + tf^*(x_1^*),$$

which entails that f^* is not strictly convex on the convex subset $[x_0^*, x_1^*]$ of $\operatorname{Dom} \partial f^*$, since $[x_0^*, x_1^*] \subset \partial f(u) \subset \operatorname{Dom} \partial f^*$. $\square$

COROLLARY 3.280 (dual characterization of G-differentiability at subdifferentiability points). Let X be a Banach space and $f : X \to \mathbb{R} \cup \{+\infty\}$ be a proper lower semicontinuous convex function. Assume that either X is finite dimensional or X is reflexive and $\operatorname{int}(\operatorname{dom} f) \neq \emptyset$. Then the following are equivalent:
(a) the function f is Gâteaux differentiable on $\operatorname{int}(\operatorname{dom} f) \neq \emptyset$ and $\partial f(x) = \emptyset$ for every $x \in \operatorname{bdry}(\operatorname{dom} f)$;
(b) the conjugate function f^* is strictly convex on every convex subset of $\operatorname{Dom}(\partial f^*)$.

PROOF. It suffices to invoke Proposition 3.281 below ensuring that (a) in the corollary is equivalent to (a) in Theorem 3.279. $\square$

PROPOSITION 3.281. Let X be a Banach space and $f : X \to \mathbb{R} \cup \{+\infty\}$ be a proper convex function. Assume that either X is finite dimensional or f is lower semicontinuous with $\operatorname{int}(\operatorname{dom} f) \neq \emptyset$. Then the following are equivalent:
(a) the set $\partial f(x)$ is a singleton for every $x \in \operatorname{Dom} \partial f$;
(b) the function f is strictly Hadamard differentiable on $\operatorname{int}(\operatorname{dom} f) \neq \emptyset$ and $\partial f(x) = \emptyset$ for all $x \in \operatorname{bdry}(\operatorname{dom} f)$;
(c) the function f is Gâteaux differentiable on $\operatorname{int}(\operatorname{dom} f) \neq \emptyset$ and $\partial f(x) = \emptyset$ for all $x \in \operatorname{bdry}(\operatorname{dom} f)$.

PROOF. The implications (b)$\Rightarrow$(c) and (c)$\Rightarrow$(a) being evident, we have to show (a)$\Rightarrow$(b). Under the assumptions we know by Propositions 2.211 and 3.39 that at any $x \in \operatorname{bdry}(\operatorname{dom} f)$ the set $\partial f(x)$ is either empty or unbounded, and we also know by Corollary 3.54 and by Theorems 3.19 and 3.30 that $\operatorname{Dom}(\partial f) \neq \emptyset$. Then the property (a) is equivalent (under our assumptions) to saying that $\operatorname{int}(\operatorname{dom} f) = \operatorname{Dom}(\partial f) \neq \emptyset$ along with $\partial f(x)$ is a singleton for every $x \in \operatorname{int}(\operatorname{dom} f)$. Since (again under our assumptions) f is continuous on $\operatorname{int}(\operatorname{dom} f)$ (see Corollary 2.159 and Theorem 3.19), by Proposition 3.102 the property (a) ensures that f is Gâteaux differentiable on $\operatorname{int}(\operatorname{dom} f) \neq \emptyset$ and $\partial f(x) = \emptyset$ for all $x \in \operatorname{bdry}(\operatorname{dom} f)$, which is the property (c). $\square$

3.23. Further results

In this section we develop, on the one hand, additional duality results with partial conjugate, and on the other hand additional sum and chain rules in terms of ε-subdifferentials of the data convex functions.

3.23.1. Duality with partial conjugate.
For the problem

$$\text{Minimize} f(x) + g(G(x)) \quad \text{over } x \text{ in the whole space } X$$

with the assumption $G(X) \subset Y$, we have seen in (3.112) that $\mathcal{L}(x, y^*) = f(x) + \langle y^*, G(x) \rangle - g^*(y^*)$ for $(x, y^*) \in X \times (\operatorname{dom} g^*) \cup (\operatorname{dom} f) \times Y^*$, where $\mathcal{L}$ is the Lagrangian function. On the other hand, considering the perturbation function $(x, y) \mapsto \varphi(x, y) = f(x) + g(G(x) + y)$ of the problem, we note that for any $x \in \operatorname{dom} f$ and $y^* \in Y^*$

$$f(x) + \langle y^*, G(x) \rangle - g^*(y^*) = f(x) + \langle y^*, G(x) \rangle - \sup_{y \in Y}[\langle y^*, G(x) + y \rangle - g(G(x) + y)]$$

$$= -\sup_{y \in Y}[\langle y^*, y \rangle - f(x) - g(G(x) + y)] = -(\varphi_x)^*(y^*),$$

where φ_x denotes the function $y \mapsto \varphi(x,y)$ on Y. This gives for $x \in \operatorname{dom} f$ and $y^* \in Y^*$ that
$$\mathcal{L}(x,y^*) = -(\varphi_x)^*(y^*) = \inf_{y \in Y}[\varphi(x,y) - \langle y^*, y \rangle],$$
and we observe that the equalities still hold if $x \notin \operatorname{dom} f$.

Accordingly, given a general function $\varphi : X \times Y \to \mathbb{R} \cup \{-\infty, +\infty\}$ and the associated problem

$$(\mathcal{P}_{\varphi(\cdot,0)}) \quad \text{Minimize} \quad \varphi(x,0) \quad \text{over } x \text{ in the whole space } X,$$

we extend to the latter problem the concept of Lagrangian by defining its *Lagrangian function* $\mathcal{L} : X \times Y^* \to \mathbb{R} \cup \{-\infty, +\infty\}$ by

$$\mathcal{L}(x,y^*) := -(\varphi_x)^*(y^*) = \inf_{y \in Y}[\varphi(x,y) - \langle y^*, y \rangle] \quad \text{for all } (x,y^*) \in X \times Y^*.$$

Since

$$-\varphi^*(0,y^*) = -\sup_{(x,y) \in X \times Y}[\langle 0, x \rangle + \langle y^*, y \rangle - \varphi(x,y)]$$
$$= -\sup_{x \in X} \sup_{y \in Y}[\langle y^*, y \rangle - \varphi_x(y)]$$
$$= -\sup_{x \in X}(\varphi_x)^*(y^*) = \inf_{x \in X} -(\varphi_x)^*(y^*),$$

the dual problem of $(\mathcal{P}_{\varphi(\cdot,0)})$ can be written as

$$(\mathcal{D}(\mathcal{P}_{\varphi(\cdot,0)})) \quad \text{Maximize} \quad \inf_{x \in X} \mathcal{L}(x,y^*) \quad \text{over } y^* \text{ in the whole space } Y^*.$$

We also note from the definition that $\mathcal{L}(x,\cdot)$ is concave and $w(Y,Y^*)$ upper semicontinuous, and that

(3.210) $$\varphi(x,y) = \sup_{y^* \in Y^*} [\mathcal{L}(x,y^*) + \langle y^*, y \rangle]$$

whenever φ is proper and $\varphi(x,\cdot)$ is lower semicontinuous and convex for every $x \in X$. Indeed, in such a case if either $\varphi(x,\cdot) \equiv +\infty$ or $\varphi(x,\cdot)$ is finite at some point, (taking the biconjugate relative to Y) we have $\varphi_x(\cdot) = ((\varphi_x)^*)^*$, and hence from $-(\varphi_x)^*(y^*) = \mathcal{L}(x,y^*)$ we obtain $\varphi_x(y) = \sup_{y^* \in Y^*}[\langle y^*, y \rangle + \mathcal{L}(x,y^*)]$ by the Fenchel-Moreau theorem (see Theorem 3.84).

PROPOSITION 3.282. Assume that φ is proper and $\varphi(x,\cdot)$ is convex and lower semicontinuous for each $x \in X$. The following hold:
(a) The function $\mathcal{L}(\cdot,y^*)$ is convex on X for every $y^* \in Y^*$ if and only if the function φ is convex on $X \times Y$.
(b) If φ is convex on $X \times Y$, then one has
$$(x^*, y^*) \in \partial \varphi(x,y) \Leftrightarrow \big(x^* \in \partial \mathcal{L}(\cdot, y)(x) \text{ and } y^* \in \partial(-\mathcal{L})(x, \cdot)(y) \big).$$

EXERCISE 3.283 (Proof of Proposition 3.282). (α) Show the implication $\Rightarrow$ in (a) from (3.210) and deduce the equivalence in (a).
(β) Assume that φ is convex on $X \times Y$.
(β_1) Argue that
$$y^* \in \partial \varphi(x,\cdot)(y) \Leftrightarrow y \in \partial(-\mathcal{L})(x,\cdot)(y^*) \Leftrightarrow \varphi(x,y) - \langle y^*, y \rangle = \mathcal{L}(x,y^*) \in \mathbb{R}.$$
(β_2) Let $(x^*, y^*) \in \partial \varphi(x,y)$. Show that $y \in \partial(-\mathcal{L})(x,\cdot)(y^*)$ and $x^* \in \partial \mathcal{L}(\cdot, y^*)(x)$.
(β_3) Show the equivalence in (b). □

3.23.2. Calculus for ε-subdifferential of convex functions.
This subsection develops sum and chain rules for the ε-subdifferential under the qualification conditions used in Theorem 3.165. Recall first (see (3.94)) that for a convex cone Y_+ of a Hausdorff locally convex space Y and a mapping $G : X \to Y \cup \{+\infty_Y\}$ on a Hausdorff locally convex space X

$$0_{Y^*} \circ G = \Psi_{\operatorname{dom} G}.$$

This is considered in all this subsection.

THEOREM 3.284 (basic theorem for ε-subdifferential of $f + g \circ G$: convex functions under qualification conditions). Let X and Y be locally convex spaces and Y_+ be a convex cone in Y containing zero. Let $G : X \to Y \cup \{+\infty_Y\}$ be a Y_+-convex mapping, $f : X \to \mathbb{R} \cup \{+\infty\}$ be a convex function and $g : Y \to \mathbb{R} \cup \{+\infty\}$ be a convex function which is non Y_+-decreasing. Then for any real $\varepsilon \geq 0$ and any $x \in X$ one has the equality

$$\partial_\varepsilon (f + g \circ G)(x) = \bigcup_{\varepsilon_i \geq 0, \varepsilon_1 + \varepsilon_2 = \varepsilon, y^* \in Y_+^* \cap \partial_{\varepsilon_2} g(G(x))} \partial_{\varepsilon_1} (f + y^* \circ G)(x)$$

under anyone of the following conditions:
(a) With $Z := \operatorname{span} [\operatorname{dom} g - G(\operatorname{dom} f \cap \operatorname{dom} G)]$ endowed with the topology induced by the Mackey topology of Y, for each $w(X, X^*)$-neighborhood V of zero in X there is some real $r > 0$ such that

$$0 \in \operatorname{int}_Z [\{g \leq r\} - G(\{f \leq r\} \cap \operatorname{dom} G \cap rV)];$$

(b) there exists some point $\bar{x} \in \operatorname{dom} f \cap \operatorname{dom} G$ such that g is finite and continuous at $G(\bar{x})$ with respect to a locally convex topology compatible with the duality between Y and Y^*;
(c) the space Y is a normed space and, for $Z := \operatorname{span} [\operatorname{dom} g - G(\operatorname{dom} f \cap \operatorname{dom} G)]$ endowed with the norm induced by the one of Y, there exists some real number $r > 0$ such that

$$0 \in \operatorname{int}_Z [\{g \leq r\} - G(\{f \leq r\} \cap \operatorname{dom} G \cap r\mathbb{B}_X)];$$

(d) the space Y is a finite-dimensional normed space and

$$0 \in \operatorname{rint} [\operatorname{dom} g - G(\operatorname{dom} f \cap \operatorname{dom} G)].$$

PROOF. We will prove only the inclusion $\subset$, the other being easily verified. Take $a^* \in \partial_\varepsilon (f + g \circ G)(a)$. Under anyone of assumptions (a),$\cdots$,(d) there exists, by Theorem 3.165, some $y_a^* \in Y_+^*$ such that

$$\inf_{x \in X} [f(x) - \langle a^*, x \rangle + g(G(x))] = -g^*(y_a^*) - (f + y_a^* \circ G)^*(a^*).$$

According to this and the inequality

$$\varepsilon + \inf_{x \in X} [f(x) - \langle a^*, x \rangle + g(G(x))] \geq f(a) - \langle a^*, a \rangle + g(G(a))$$

(due to the inclusion $a^* \in \partial_\varepsilon (f + g \circ G)(a)$) it ensues that

$$\varepsilon \geq [(f + y_a^* \circ G)(a) + (f + y_a^* \circ G)^*(a^*) - \langle a^*, a \rangle] + [g(G(a)) + g^*(y_a^*) - \langle y_a^*, G(a) \rangle].$$

By the Fenchel inequality $\langle u^*, u \rangle - h(u) \leq h^*(u^*)$, each above bracket [ ] expression is non-negative, thus they are $\varepsilon_1 \geq 0$, $\varepsilon_2 \geq 0$ with $\varepsilon_1 + \varepsilon_2 = \varepsilon$ such that

$$(f + y_a^* \circ G)(a) + (f + y_a^* \circ G)^*(a^*) - \langle a^*, a \rangle \leq \varepsilon_1$$

and
$$g(G(a)) + g^*(y_a^*) - \langle y_a^*, G(a)\rangle \leq \varepsilon_2.$$
The characterization of ε-subgradient of convex function in terms of conjugate (see Proposition 3.64(j)) ensures that
$$y_a^* \in \partial_{\varepsilon_2} g(G(a)) \quad \text{and} \quad a^* \in \partial_{\varepsilon_1}(f + y_a^* \circ G)(a).$$
This finishes the proof. □

For $Y_+ = \{0\}$ we already saw that $Y_+^* = Y^*$. It is then easy to deduce the following corollary when G is a continuous linear mapping.

COROLLARY 3.285. *Let X, Y be locally convex spaces, $f : X \to \mathbb{R} \cup \{+\infty\}$ and $g : Y \to \mathbb{R} \cup \{+\infty\}$ be convex functions and $A : X \to Y$ be a continuous linear mapping. Then for any $\varepsilon \geq 0$ and any $x \in X$ one has*
$$\partial_\varepsilon (f + g \circ A)(x) = \bigcup_{\varepsilon_i \geq 0, \varepsilon_1 + \varepsilon_2 = \varepsilon} \left(\partial_{\varepsilon_1} f(x) + A^*(\partial_{\varepsilon_2} g(G(x)))\right)$$
in each one of the following cases (a), (b), (c) and (d):
(a) *With $Z := \operatorname{span}[\operatorname{dom} g - A(\operatorname{dom} f)]$ endowed with the topology induced by the Mackey topology of Y, for each $w(X, X^*)$-neighborhood V of zero in X there is some real $r > 0$ such that*
$$0 \in \operatorname{int}_Z [\{g \leq r\} - A(\{f \leq r\} \cap rV)];$$
(b) *there exists some $\bar{x} \in \operatorname{dom} f$ such that g is finite and continuous at $A\bar{x}$ with respect to a locally convex topology compatible with the duality between Y and Y^*;*
(c) *the space Y is a normed space and for $Z := \operatorname{span}[\operatorname{dom} g - A(\operatorname{dom} f)]$ there exists some real number $r > 0$ such that*
$$0 \in \operatorname{int}_Z [\{g \leq r\} - A(\{f \leq r\} \cap r\mathbb{B}_X)];$$
(d) *the space Y is a finite-dimensional normed space and*
$$0 \in \operatorname{rint}[\operatorname{dom} g - A(\operatorname{dom} f)].$$

PROOF. Apply Theorem 3.165 with $Y_+ := \{0_Y\}$, and hence $Y_+^* = Y^*$. Observe also that for each $y^* \in Y^*$ we have $y^* \circ A \in X^*$, and hence (see Proposition 3.12(a))
$$\partial_{\varepsilon_1}(f + y^* \circ A)(x) = y^* \circ A + \partial_{\varepsilon_1} f(x) = A^* y^* + \partial_{\varepsilon_1} f(x).$$
Noticing that
$$\bigcup_{y^* \in \partial_{\varepsilon_2} g(Ax)} \{A^* y^*\} = A^*(\partial_{\varepsilon_2} g(Ax)),$$
it results that
$$\bigcup_{y^* \in \partial_{\varepsilon_2} g(Ax)} \partial_{\varepsilon_1}(f + y^* \circ A)(x) = \bigcup_{y^* \in \partial_{\varepsilon_2} g(Ax)} (A^* y^* + \partial_{\varepsilon_1} f(x))$$
$$= \partial_{\varepsilon_1} f(x) + \bigcup_{y^* \in \partial_{\varepsilon_2} g(Ax)} \{A^* y^*\}$$
$$= \partial_{\varepsilon_1} f(x) + A^*(\partial_{\varepsilon_2} g(Ax)).$$
All together justify the corollary according to Theorem 3.284. □

Taking A as the identity in the previous corollary furnishes a second corollary.

COROLLARY 3.286 (ε-subdifferential of sum of convex functions under qualification conditions). Let X be a locally convex space, $f, g : X \to \mathbb{R} \cup \{+\infty\}$ be convex functions. Then for any real $\varepsilon \geq 0$ and $x \in X$ one has

$$\partial_\varepsilon (f+g)(x) = \bigcap_{\varepsilon_i \geq 0, \varepsilon_1 + \varepsilon_2 = \varepsilon} (\partial_{\varepsilon_1} f(x) + \partial_{\varepsilon_2} g(x))$$

in each one of the following cases (a), (b), (c) and (d):
(a) With $Z := \operatorname{span}[\operatorname{dom} g - \operatorname{dom} f)]$ endowed with the topology induced by the Mackey topology of X, for each $w(X, X^*)$-neighborhood V of zero in X there is some real $r > 0$ such that

$$0 \in \operatorname{int}_Z [\{g \leq r\} - (\{f \leq r\} \cap rV)];$$

(b) (**Moreau-Rockafellar continuity condition**): one of the functions is finite and continuous at a point where the other function is finite;
(c) the space X is a normed space and for $Z := \operatorname{span}[\operatorname{dom} g - \operatorname{dom} f]$ there exists some real number $r > 0$ such that

$$0 \in \operatorname{int}_Z [\{g \leq r\} - (\{f \leq r\} \cap r\mathbb{B}_X)];$$

(d) (**Rockafellar relative interior condition**): the space X is a finite-dimensional normed space and

$$0 \in \operatorname{rint}[\operatorname{dom} g - \operatorname{dom} f].$$

Taking $f = 0$ in Corollary 3.285 we obtain the following fourth corollary with chain rule.

COROLLARY 3.287 (ε-subdifferential chain rule for convex functions under qualification conditions). Let X, Y be locally convex spaces, $g : Y \to \mathbb{R} \cup \{+\infty\}$ be a convex function and $A : X \to Y$ be a continuous linear mapping. Then for any real $\varepsilon \geq 0$ and $x \in X$ one has

$$\partial_\varepsilon (g \circ A)(x) = A^*(\partial_\varepsilon g(G(x)))$$

in each one of the following cases (a), (b), (c) and (d):
(a) With $Z := \operatorname{span}[\operatorname{dom} g - A(X)]$ endowed with the topology induced by the Mackey topology of Y, for each $w(X, X^*)$-neighborhood V of zero in X there is some real $r > 0$ such that

$$0 \in \operatorname{int}_Z [\{g \leq r\} - A(rV)];$$

(b) there exists some $\overline{x} \in X$ such that g is finite and continuous at $A\overline{x}$ with respect to a locally convex topology compatible with the duality between Y and Y^*;
(c) the space Y is a normed space and for $Z := \operatorname{span}[\operatorname{dom} g - A(X)]$ there exists some real number $r > 0$ such that

$$0 \in \operatorname{int}_Z [\{g \leq r\} - A(r\mathbb{B}_X)];$$

(d) the space Y is a finite-dimensional normed space and

$$0 \in \operatorname{rint}[\operatorname{dom} g - A(X)].$$

3.23.3. Extended calculus for ε-subdifferential of convex functions.

The calculus rules in the previous subsection requires qualification conditions. For Banach spaces X, Y, a continuous linear mapping $A : X \to Y$ and proper lower semicontinuous convex functions $f : X \to \mathbb{R} \cup \{+\infty\}$ and $g : Y \to \mathbb{R} \cup \{+\infty\}$, without any qualification condition Exercise 3.202 says that

$$\partial(f + g \circ A)(\bar{x}) = \bigcap_{\varepsilon > 0} \mathrm{cl}_{w^*} \left(\partial_\varepsilon f(\bar{x}) + A^*(\partial_\varepsilon g(A\bar{x})) \right).$$

In this subsection we will remove the completeness of the spaces and establish the same equality for general locally convex spaces X, Y. We will even obtain a similar formula for ε-subdifferential of $f + g \circ A$ in the locally convex space setting.

The following lemma will be one of the keys of the approach that we follow in this subsection to derive ε-subdifferential rules.

LEMMA 3.288. *Let (Z, θ) be a topological vector space and $\varphi : Z \to \mathbb{R} \cup \{-\infty, +\infty\}$ be a convex function. For any real $r > \inf_Z \varphi$ one has*

$$\{(\mathrm{cl}_\theta \varphi)(\cdot) \leq r\} = \mathrm{cl}_\theta \{\varphi(\cdot) < r\} = \mathrm{cl}_\theta \{\varphi(\cdot) \leq r\}.$$

PROOF. Fix any real $r > \inf_Y \varphi$. Remembering that $\mathrm{cl}_\theta \varphi \leq \varphi$, we see that $\{\varphi \leq r\} \subset \{\mathrm{cl}_\theta \varphi \leq r\}$, thus

$$\mathrm{cl}_\theta \{\varphi < r\} \subset \mathrm{cl}_\theta \{\varphi \leq r\} \subset \{\mathrm{cl}_\theta \varphi \leq r\}.$$

It remains to show that the third latter set is included in the first. Let any $z \in \{\mathrm{cl}_\theta \varphi \leq r\}$. Choose some $a \in Z$ such that $\varphi(a) < r$ (according to the inequality $r > \inf_Y \varphi$). The convexity of $\mathrm{cl}_\theta \varphi$ yields for any $t \in {]0, 1[}$

$$(\mathrm{cl}_\theta \varphi)(z_t) \leq (1 - t)(\mathrm{cl}_\theta \varphi)(z) + t(\mathrm{cl}_\theta \varphi)(a) < r,$$

where $z_t := (1-t)z + ta$. Take any neighborhood W of z and choose a neighborhood V of zero with $z + V + V \subset W$. Since $z_t \to z$ as $t \downarrow 0$ there is some $s \in {]0, 1[}$ such that $b := z_s \in z + V$. Then we have $\liminf_{u \to b} \varphi(u) = (\mathrm{cl}_\theta \varphi)(b) < r$, hence there is some $u_0 \in b + V$ such that $\varphi(u_0) < r$. It ensues that $u_0 \in \{\varphi < r\} \cap W$, which finishes the proof. $\square$

REMARK 3.289. In fact, the above proof shows that the equality in the lemma holds true whenever the function φ satisfies the weaker *quasi-convexity* condition $\varphi((1 - t)z_1 + tz_2) \leq \max\{\varphi(z_1), \varphi(z_2)\}$ for all $z_1, z_2 \in Z$ and $t \in [0, 1]$. $\square$

Remember that, under anyone of qualification conditions (a),$\cdots$,(d) in Corollary 3.167, one has $(g \circ G)^*(x^*) = \inf_{y^* \in Y^*} [g^*(y^*) + (y^* \circ G)^*(x^*)]$. Conversely, without any qualification condition the next lemma shows that the conjugate relative to X of the function in x^* of the right-hand side coincides with $g \circ G$ under certain lower semicontinuity properties. For all the development below in this subsection we record that $g \circ G$ and $(y^* \circ G)$ are defined with $(g \circ G)(x) := +\infty$ and $(y^* \circ G)(x) := +\infty$ for every $x \notin \mathrm{dom}\, G$, so

$$0_{Y^*} \circ G = \Psi_{\mathrm{dom}\, G}$$

as already said in (3.94).

LEMMA 3.290. *Let X and Y be two locally convex spaces and Y_+ be a convex cone in Y containing zero. Let $g : Y \to \mathbb{R} \cup \{+\infty\}$ be a proper lower semicontinuous convex function which is non Y_+-decreasing and let $G : X \to Y \cup \{+\infty_Y\}$ be a proper Y_+-convex mapping which is Y_+^*-scalarly lower semicontinuous in the sense* that $y^* \circ G$ is lower semicontinuous for every $y^* \in Y_+^*$. The function $\varphi : X^* \to \mathbb{R} \cup \{-\infty, +\infty\}$ defined by

$$\varphi(x^*) = \inf_{y^* \in \operatorname{dom} g^*} [g^*(y^*) + (y^* \circ G)^*(x^*)]$$

is convex and $\varphi^*(x) = (g \circ G)(x)$ for all $x \in X$.

PROOF. Recall first (see Lemma 3.155) that $\operatorname{dom} g^* \subset Y_+^*$, so for each $y^* \in \operatorname{dom} g^*$ the function $y^* \circ G$ is proper, lower semicontinuous and convex.
Since the function $(x^*, y^*) \mapsto (y^* \circ G)^*(x^*) = \sup_{u \in \operatorname{dom} G} [\langle x^*, u \rangle - \langle y^*, G(u) \rangle]$ is convex, so is the function φ by Proposition 2.54. Concerning the conjugate of φ it suffices to write for any $x \in X$

$$\varphi^*(x) = \sup_{x^* \in X^*} \sup_{y^* \in \operatorname{dom} g^*} [\langle x^*, x \rangle - g^*(y^*) - (y^* \circ G)^*(x^*)]$$
$$= \sup_{y^* \in \operatorname{dom} g^*} \sup_{x^* \in X^*} [-g^*(y^*) + \langle x^*, x \rangle - (y^* \circ G)^*(x^*)]$$
$$= \sup_{y^* \in \operatorname{dom} g^*} [-g^*(y^*) + (y^* \circ G)^{**}(x)],$$

which by the Fenchel-Moreau theorem (see Theorem 3.84(c)) yields

$$\varphi^*(x) = \sup_{y^* \in \operatorname{dom} g^*} [-g^*(y^*) + (y^* \circ G)(x)].$$

It ensues that for $x \notin \operatorname{dom} G$ one has $\varphi^*(x) = +\infty = (g \circ G)(x)$, and for $x \in \operatorname{dom} G$ one has

$$\varphi^*(x) = \sup_{y^* \in \operatorname{dom} g^*} [-g^*(y^*) + \langle y^*, G(x) \rangle] = g^{**}(G(x)) = g(G(x)),$$

since $g^{**}(y) = g(y)$ for all $y \in Y$ by Theorem 3.84 again. This confirms the equality of the lemma. □

EXERCISE 3.291. Let X and Y be two locally convex spaces and Y_+ be a convex cone in Y containing zero. Let $g : Y \to \mathbb{R} \cup \{+\infty\}$ be a proper lower semicontinuous convex function which is non Y_+-decreasing and let $G : X \to Y \cup \{+\infty_Y\}$ be a proper Y_+-convex mapping which is Y_+^*-scalarly lower semicontinuous. Let $x \in \operatorname{dom}(g \circ G)$ and $\varepsilon > 0$, and let $\Lambda(\varepsilon) := \{(\varepsilon_1, \varepsilon_2) : \varepsilon_1 \geq 0, \varepsilon_2 \geq 0, \varepsilon_1 + \varepsilon_2 = \varepsilon\}$. For $\varepsilon_1, \varepsilon_2 \geq 0$ put $Q_{\varepsilon_1, \varepsilon_2} := \bigcup_{y^* \in \partial_{\varepsilon_1} g(G(x))} \partial_{\varepsilon_2} (y^* \circ G)(x)$.

(a) Justify that $Q_{\varepsilon_1, \varepsilon_2} \subset \partial_\varepsilon (g \circ G)(x)$ for $(\varepsilon_1, \varepsilon_2) \in \Lambda(\varepsilon)$ and derive that

$$\operatorname{cl}_{w^*} \left(\bigcup_{(\varepsilon_1, \varepsilon_2) \in \Lambda(\varepsilon)} Q_{\varepsilon_1, \varepsilon_2} \right) \subset \partial_\varepsilon (g \circ G)(x).$$

(b) For $\varepsilon_1 > 0$ and $\varepsilon_2 > 0$ show that $Q_{\varepsilon_1, \varepsilon_2} \neq \emptyset$ (Use that $y^* \circ G$ is proper, lower semicontinuous and convex for $y^* \in \partial_{\varepsilon_1} g(G(x)) \subset \operatorname{dom} g^* \subset Y_+^*$). □

THEOREM 3.292 (ε-subdifferential chain rule: convex functions without qualification condition). Let X and Y be two locally convex spaces and Y_+ be a convex cone in Y containing zero. Let $g : Y \to \mathbb{R} \cup \{+\infty\}$ be a proper lower semicontinuous convex function which is non Y_+-decreasing and let $G : X \to$

$Y \cup \{+\infty_Y\}$ be a proper Y_+-convex mapping which is Y_+^*-scalarly lower semi-continuous. Then for any real $\varepsilon > 0$ and any $x \in X$ one has with $\Lambda(\varepsilon) := \{(\varepsilon_1, \varepsilon_2) : \varepsilon_1 \geq 0, \varepsilon_2 \geq 0, \varepsilon_1 + \varepsilon_2 = \varepsilon\}$

$$\partial_\varepsilon (g \circ G)(x) = \mathrm{cl}_{w^*} \left(\bigcup_{(\varepsilon_1, \varepsilon_2) \in \Lambda(\varepsilon)} \bigcup_{y^* \in \partial_{\varepsilon_1} g(G(x))} \partial_{\varepsilon_2} (y^* \circ G)(x) \right).$$

PROOF. Take any $\bar{x} \in X$ and note that the equality at $x = \bar{x}$ is trivial if $\bar{x} \notin \mathrm{dom}\,(g \circ G)$ since both sides are empty in this case. Suppose that $\bar{x} \in \mathrm{dom}\,(g \circ G)$. By Exercise 3.291 the right-hand side of the desired equality of the theorem is non empty and included in the left-hand side.

Let us show the converse inclusion. Considering the function f_0 and the mapping G_0 with $g_0(y) := g(y + G(\bar{x})) - g(G(\bar{x}))$ and $G_0(x) := G(x + \bar{x}) - G(\bar{x})$, we may and do suppose that $\bar{x} = 0$, $G(\bar{x}) = 0$ and $g(0) = 0$. Considering the function φ in Lemma 3.290 and taking its conjugate relative to X we have $\varphi^* = g \circ G$, so in particular $\inf_{X^*} \varphi = -\varphi^*(0) = -(g \circ G)(0) = 0$. The convex function φ is then bounded from below, so endowing X^* with its $w(X^*, X)$ topology Proposition 3.89 tells us that $\varphi^{**} = \mathrm{cl}_{w^*} \varphi$. Therefore, the equality $\varphi^* = g \circ G$ gives $(g \circ G)^* = \varphi^{**} = \mathrm{cl}_{w^*} \varphi$. It ensues (remembering $\bar{x} = 0$) that

$$\partial_\varepsilon (g \circ G)(\bar{x}) = \{(g \circ G)^* \leq \varepsilon\} = \{\mathrm{cl}_{w^*} \varphi \leq \varepsilon\}.$$

Further, since $\inf_{X^*} \varphi = 0 < \varepsilon$, we obtain $\{\mathrm{cl}_{w^*} \varphi \leq \varepsilon\} = \mathrm{cl}_{w^*} \{\varphi < \varepsilon\}$ by Lemma 3.288. Fix any $x^* \in \{\varphi < \varepsilon\}$. Keeping in mind that $g(0) = 0$ and $(g \circ G)(0) = 0$, we note (from the definition of conjugate) that $g^* \geq 0$ and $(g \circ G)^* \geq 0$. Using this we see from the inequality $\varphi^*(x^*) < \varepsilon$ and from the definition of φ in Lemma 3.290 that there are $y^* \in \mathrm{dom}\,g^*$, $\varepsilon_1 \geq 0$ and $\varepsilon_2 \geq 0$ with $\varepsilon_1 + \varepsilon_2 = \varepsilon$ such that $g^*(y^*) \leq \varepsilon_1$ and $(y^* \circ G)^*(x^*) \leq \varepsilon_2$, hence $y^* \in \partial_{\varepsilon_1} g(0)$ and $x^* \in \partial(y^* \circ G)(0)$. This and the w^*-closedness of the right-hand side of the inclusion in the theorem guarantee that the desired converse inclusion holds true. The proof is then finished. $\square$

Before deriving a first corollary let us justify the following simple lemma.

LEMMA 3.293. Let X, Y be two locally convex spaces and $f : X \to \mathbb{R} \cup \{+\infty\}$ and $g : Y \to \mathbb{R} \cup \{+\infty\}$ be two convex functions. For the convex function $\varphi : X \times Y \to \mathbb{R} \cup \{+\infty\}$ defined by $\varphi(x, y) = f(x) + g(y)$ one has for any real $\varepsilon \geq 0$ and any $(x, y) \in X \times Y$

$$\partial_\varepsilon \varphi(x, y) = \bigcup_{(\varepsilon_1, \varepsilon_2) \in \Lambda(\varepsilon)} \left(\partial_{\varepsilon_1} f(x) \times \partial_{\varepsilon_2} g(y) \right),$$

where $\Lambda(\varepsilon) := \{(\varepsilon_1, \varepsilon_2) : \varepsilon_1 \geq 0, \varepsilon_2 \geq 0, \varepsilon_1 + \varepsilon_2 = \varepsilon\}$.

PROOF. Fix any real $\varepsilon \geq 0$, any $\bar{x} \in \mathrm{dom}\,f$ and any $\bar{y} \in \mathrm{dom}\,g$. Putting $\varphi(x, y) := f(x) + g(y)$ we know (see (3.36)) that $\varphi^*(x^*, y^*) = f^*(x^*) + g^*(y^*)$, so it ensues that $(x^*, y^*) \in \partial \varphi(\bar{x}, \bar{y})$ if and only if

$$[f(\bar{x}) + f^*(x^*) - \langle x^*, \bar{x} \rangle] + [g(\bar{y}) + g^*(y^*) - \langle y^*, \bar{y} \rangle] \leq \varepsilon.$$

Since each expression between the brackets $[\,]$ is non-negative, the latter inequality amounts to saying that there are $\varepsilon_1 \geq 0$ and $\varepsilon_2 \geq 0$ with $\varepsilon_1 + \varepsilon_2 = \varepsilon$ such that the expression between the first (resp. second) bracket is not greater than ε_1 (resp.

ε_2), or equivalently $x^* \in \partial_{\varepsilon_1} f(\overline{x})$ and $y^* \in \partial_{\varepsilon_2} g(\overline{y})$. Otherwise stated, we have

$$\partial_\varepsilon \varphi(\overline{x}, \overline{y}) = \bigcup_{(\varepsilon_1, \varepsilon_2) \in \Lambda(\varepsilon)} (\partial_{\varepsilon_1} f(\overline{x}) \times \partial_{\varepsilon_2} g(\overline{y})),$$

which is the desired equality. □

COROLLARY 3.294 (ε-subdifferential of $f + g \circ G$: convex functions without qualification condition). Let X and Y be two locally convex spaces and Y_+ be a convex cone in Y containing zero. Let $g : Y \to \mathbb{R} \cup \{+\infty\}$ be a proper lower semicontinuous convex function which is non Y_+-decreasing and let $G : X \to Y \cup \{+\infty_Y\}$ be a proper Y_+-convex mapping which is Y_+^*-scalarly lower semicontinuous. Let also $f : X \to \mathbb{R} \cup \{+\infty\}$ be a proper lower semicontinuous convex function. Then for any real $\varepsilon > 0$ and any $x \in X$ one has

$$\partial_\varepsilon (f + g \circ G)(x) = \mathrm{cl}_{w^*} \left(\bigcup_{(\varepsilon_1, \varepsilon_2, \varepsilon_3) \in \Lambda(\varepsilon)} \bigcup_{y^* \in \partial_{\varepsilon_2} g(G(x))} (\partial_{\varepsilon_1} f(x) + \partial_{\varepsilon_3} (y^* \circ G)(x)) \right),$$

where $\Lambda(\varepsilon) := \{(\varepsilon_1, \varepsilon_2, \varepsilon_3) : \varepsilon_i \geq 0, \varepsilon_1 + \varepsilon_2 + \varepsilon_3 = \varepsilon\}$.

PROOF. Consider the convex cone $\{0\} \times Y_+$ in $X \times Y$ and the proper convex mapping $G_0 : X \to (X \times Y) \cup \{+\infty_{X \times Y}\}$ defined by $G_0(x) = (x, G(x))$ for all $x \in X$. Note that for any (x^*, y^*) in the positive dual cone in $X^* \times Y^*$, that is, $x^* \in X^*$ and $y^* \in Y_+^*$, the function $x \mapsto \langle x^*, x \rangle + \langle y^*, G(x) \rangle$ is lower semicontinuous, hence the mapping G_0 is $X^* \times Y_+^*$-scalarly lower semicontinuous. The proper lower semicontinuous convex function $g_0 : X \times Y \to \mathbb{R} \cup \{+\infty\}$ defined by $g_0(x, y) := f(x) + g(y)$ is obviously non $\{0\} \times Y_+$-decreasing, and by Lemma 3.293 we know that $(x^*, y^*) \in \partial_{\varepsilon_1} g_0(x, y)$ if and only there are $\varepsilon_1', \varepsilon_1'' \geq 0$ with $\varepsilon_1' + \varepsilon_1'' = \varepsilon_1$ such that $x^* \in \partial_{\varepsilon_1'} f(x)$ and $y^* \in \partial_{\varepsilon_1''}(y)$. We note also that $((x^*, y^*) \circ G_0)(x) = (x^* + y^* \circ G)(x)$ and $\partial_{\varepsilon_2}(x^* + y^* \circ G)(x) = x^* + \partial_{\varepsilon_2}(y^* \circ G)(x)$. It then follows from Theorem 3.292 that

$$\partial_\varepsilon (f + g \circ G)(x) = \mathrm{cl}_{w^*} [\bigcup_{\substack{\varepsilon_1 + \varepsilon_2 = \varepsilon \\ \varepsilon_1 \geq 0, \varepsilon_2 \geq 0}} \bigcup_{\substack{\varepsilon_1' \geq 0, \varepsilon_1'' \geq 0, \varepsilon_1' + \varepsilon_1'' = \varepsilon_1 \\ x^* \in \partial_{\varepsilon_1'} f(x), y^* \in \partial_{\varepsilon_1''} g(G(x))}} (x^* + \partial_{\varepsilon_2}(y^* \circ G)(x))]$$

$$= \mathrm{cl}_{w^*} [\bigcup_{\substack{\varepsilon_1' + \varepsilon_1'' + \varepsilon_2 = \varepsilon \\ \varepsilon_1' \geq 0, \varepsilon_1'' \geq 0, \varepsilon_2 \geq 0}} \bigcup_{y^* \in \partial_{\varepsilon_1''} g(G(x))} (\partial_{\varepsilon_1'} f(x) + \partial_{\varepsilon_2}(y^* \circ G)(x))],$$

and this finishes the proof. □

A second corollary describes the exact subdifferential $\partial (f + g \circ G)(x)$ in terms of ε-subdifferentials of the data.

COROLLARY 3.295 (subdifferential of $f + g \circ G$ via ε-subdifferentials: convex functions without qualification condition). Let X and Y be two locally convex spaces and Y_+ be a convex cone in Y containing zero. Let $g : Y \to \mathbb{R} \cup \{+\infty\}$ be a proper lower semicontinuous convex function which is non Y_+-decreasing and let $G : X \to Y \cup \{+\infty_Y\}$ be a proper Y_+-convex mapping which is Y_+^*-scalarly lower semicontinuous. Let also $f : X \to \mathbb{R} \cup \{+\infty\}$ be a proper lower semicontinuous convex function. Then for any $x \in X$ one has

$$\partial(f + g \circ G)(x) = \bigcap_{\varepsilon > 0} \mathrm{cl}_{w^*} \left(\bigcup_{y^* \in \partial_\varepsilon g(G(x))} \left(\partial_\varepsilon f(x) + \partial_\varepsilon(y^* \circ G)(x) \right) \right).$$

PROOF. Trivially, we have for $\varepsilon_1' + \varepsilon"_1 + \varepsilon_2 = \varepsilon$ and $\varepsilon_1', \varepsilon"_1, \varepsilon_2 \geq 0$

$$\bigcup_{y^* \in \partial_{\varepsilon"_1} g(G(x))} \left(\partial_{\varepsilon_1'} f(x) + \partial_{\varepsilon_2}(y^* \circ G)(x) \right) \subset \bigcup_{y^* \in \partial_\varepsilon g(G(x))} \left(\partial_\varepsilon f(x) + \partial_\varepsilon(y^* \circ G)(x) \right),$$

hence by Corollary 3.294 the left-hand side of the equality of the corollary is included in the right-hand side. The converse equality is also valid since from the very definition of ε-subdifferential one easily sees for any $\varepsilon \geq 0$ and $y^* \in \partial_\varepsilon g(G(x))$ that $\partial_\varepsilon f(x) + \partial_\varepsilon(y^* \circ G)(x) \subset \partial_{3\varepsilon}(f + g \circ G)(x)$. □

Taking f as the null function in the above corollary gives:

COROLLARY 3.296. Let X and Y be two locally convex spaces and Y_+ be a convex cone in Y containing zero. Let $g : Y \to \mathbb{R} \cup \{+\infty\}$ be a proper lower semicontinuous convex function which is non Y_+-decreasing and let $G : X \to Y \cup \{+\infty_Y\}$ be a proper Y_+-convex mapping which is Y_+^*-scalarly lower semicontinuous. Then for any $x \in X$ one has

$$\partial(g \circ G)(x) = \bigcap_{\varepsilon > 0} \mathrm{cl}_{w^*} \left(\bigcup_{y^* \in \partial_\varepsilon g(G(x))} \partial_\varepsilon(y^* \circ G)(x) \right).$$

A fourth corollary expresses normals to sublevels in terms of ε-subgradients at reference points without any qualification condition. It complements the results in Section 3.18 where nearby points are involved.

COROLLARY 3.297. Let X and Y be two locally convex spaces and Y_+ be a closed convex cone in Y containing zero. Let $G : X \to Y \cup \{+\infty_Y\}$ be a proper Y_+-convex mapping which is Y_+^*-scalarly lower semicontinuous, let $\overline{y} \in Y$ and let $\overline{x} \in S := \{x \in X : G(x) \leq_{Y_+} \overline{y}\}$. Then one has

$$N(S; \overline{x}) = \bigcap_{\varepsilon > 0} \mathrm{cl}_{w^*} \left(\bigcup_{y^* \in P_\varepsilon} \partial_\varepsilon(y^* \circ G)(\overline{x}) \right),$$

where $P_\varepsilon := \{y^* \in Y_+^* : -\varepsilon \leq \langle y^*, G(\overline{x}) - \overline{y} \rangle\}$ and where we recall that $0_{Y^*} \circ G = \Psi_{\mathrm{dom}\, G}$.

If in addition $G(\overline{x}) = \overline{y}$, one also has

$$N(S; \overline{x}) = \bigcap_{\varepsilon > 0} \mathrm{cl}_{w^*} \left(\bigcup_{y^* \in Y_+^*} \partial_\varepsilon(y^* \circ G)(\overline{x}) \right).$$

PROOF. Considering the mapping $G(\cdot) - \overline{y}$ we may and do suppose that $\overline{y} = 0$. Putting $g := \Psi_{-Y_+}$ we see that

$$\Psi_S = g \circ G, \quad \text{so } N(S; \overline{x}) = \partial(g \circ G)(\overline{x}).$$

The function Ψ_{-Y_+} is positively homogeneous, convex and null at zero, then for each real $\varepsilon > 0$ we have by Proposition 3.12(b)

$$\partial_\varepsilon g(G(\overline{x})) = \{y^* \in Y^* : y^* \in \partial \Psi_{-Y_+}(0), \langle y^*, G(\overline{x}) \rangle \geq -\varepsilon\},$$

so $\partial_\varepsilon g(G(\bar{x})) = P_\varepsilon := \{y^* \in Y_+^* : -\varepsilon \leq \langle y^*, G(\bar{x})\rangle\}$. Further, the function g is lower semicontinuous by the closedness of Y^+ and it is non Y_+-decreasing according to Example 3.153(b). Therefore, the first equality of the corollary follows from Corollary 3.298.

Assume in addition that $G(\bar{x}) = 0$. By the first equality of the corollary we clearly have $N(S; \bar{x}) \subset \bigcap_{\varepsilon > 0} \mathrm{cl}_{w^*} Q_\varepsilon$, where

$$Q_\varepsilon := \bigcup_{y^* \in Y_+^*} \partial_\varepsilon(y^* \circ G)(\bar{x}).$$

Now take any $y^* \in Y_+^*$ and any $x^* \in \partial_\varepsilon(y^* \circ G)(\bar{x})$. For any $x \in S$ we have the inequality $\langle y^*, G(x)\rangle \leq 0$, which in turn yields

$$\langle x^*, x - \bar{x}\rangle \leq y^* \circ G(x) - y^* \circ G(\bar{x}) + \varepsilon \leq \varepsilon,$$

so we easily derive from this that given any $x \in S$ we have $\langle x^*, x - \bar{x}\rangle \leq \varepsilon$ for every $x^* \in \mathrm{cl}_{w^*} Q_\varepsilon$. Consequently, for any $x^* \in \bigcap_{\varepsilon > 0} \mathrm{cl}_{w^*} Q_\varepsilon$ we obtain $\langle x^*, x - \bar{x}\rangle \leq 0$ for all $x \in S$, hence $x^* \in N(S; \bar{x})$. This finishes the proof of the corollary. □

A fifth corollary corresponds to the chain rule with a continuous linear mapping.

COROLLARY 3.298. *Let $A : X \to Y$ be a continuous linear mapping between two locally convex spaces X and Y and $g : Y \to \mathbb{R} \cup \{+\infty\}$ be a proper lower semicontinuous convex function. The equality*

$$\partial_\varepsilon(g \circ A)(x) = \mathrm{cl}_{w^*} [A^*(\partial_\varepsilon g(Ax))]$$

holds for any real $\varepsilon > 0$ and any $x \in X$.

PROOF. Since $\partial_\varepsilon(y^* \circ A)(\bar{x}) = \{y^* \circ A\} = \{A^* y^*\}$, the result of the corollary follows by applying Corollary 3.296 with the linear mapping A for G and $\{0\}$ as the convex cone Y_+. □

The following sixth corollary directly follows from the previous corollary.

COROLLARY 3.299 (J.-B. Hiriart-Urruty and R.R. Phelps: subdifferential chain rule). *Let $A : X \to Y$ be a continuous linear mapping between two locally convex spaces X and Y and $g : Y \to \mathbb{R} \cup \{+\infty\}$ be a proper lower semicontinuous convex function. For any $x \in X$ one has*

$$\partial(g \circ A)(x) = \bigcap_{\varepsilon > 0} \mathrm{cl}_{w^*} [A^*(\partial_\varepsilon g(Ax))].$$

REMARK 3.300. It is worth pointing out that a direct simple proof can also be provided for the above corollary. Let $\bar{x} \in A^{-1}(\mathrm{dom}\, g)$ and $\bar{y} := A\bar{x}$. Take any $\varepsilon > 0$. Denoting $\varphi := g \circ A$ we see for any $v \in X$ that $\varphi'(\bar{x}; v) = g'(\bar{y}; Av)$. Then, since $g'_\varepsilon(\bar{y}; \cdot)$ is the support function of $\partial_\varepsilon g(\bar{x})$ (see Theorem 3.141), using (1.26) it follows that

$$\varphi'(\bar{x}; v) \leq g'_\varepsilon(A\bar{x}; Av) = \sigma\left(Av, \partial_\varepsilon g(\bar{y})\right) = \sigma\left(v, A^*(\partial_\varepsilon g(\bar{y}))\right).$$

By (1.16) it ensues for any $x^* \in \partial \varphi(\bar{x})$ that $x^* \in \mathrm{cl}_{w^*}[A^*(\partial_\varepsilon g(\bar{x})]$. This justifies the inclusion of the left-hand side into the right-hand one. The converse inclusion is easily verified. □

Another corollary is concerned with the ε-subdifferential of the sum of finitely many proper lower semicontinuous convex functions.

3.23. FURTHER RESULTS

COROLLARY 3.301. *Let X be a locally convex space and $f, g : X \to \mathbb{R} \cup \{+\infty\}$ be two proper lower semicontinuous convex functions. For any real $\varepsilon > 0$ and $x \in X$ one has with $\Lambda(\varepsilon) := \{(\varepsilon_1, \varepsilon_2) : \varepsilon_1 \geq 0, \varepsilon_2 \geq 0, \varepsilon_1 + \varepsilon_2 = \varepsilon\}$*

$$\partial_\varepsilon (f+g)(x) = \mathrm{cl}_{w^*} \left(\bigcup_{(\varepsilon_1, \varepsilon_2) \in \Lambda(\varepsilon)} (\partial_{\varepsilon_1} f(x) + \partial_{\varepsilon_2} g(x)) \right).$$

PROOF. Take the convex cone $X_+ := \{0\}$ and G as the identity mapping Id_X on X, we see that Id_X is X_+-convex and X^*-scalarly lower semicontinuous. Further, the function g is non X_+-decreasing. We deduce from Corollary 3.294 that with $\Lambda_3(\varepsilon) := \{(\varepsilon_1, \varepsilon_2, \varepsilon_3) : \varepsilon_i \geq 0, \varepsilon_1 + \varepsilon_2 + \varepsilon_3 = \varepsilon\}$ and $\Lambda_2(\varepsilon) := \{(\varepsilon_1, \varepsilon_2) : \varepsilon_i \geq 0, \varepsilon_1 + \varepsilon_2 = \varepsilon\}$

$$\partial_\varepsilon (f+g)(x) = \mathrm{cl}_{w^*} \left(\bigcup_{(\varepsilon_1, \varepsilon_2, \varepsilon_3) \in \Lambda_3(\varepsilon)} \bigcup_{y^* \in \partial_{\varepsilon_2} g(G(x))} (\partial_{\varepsilon_1} f(x) + y^*) \right)$$

$$= \mathrm{cl}_{w^*} \left(\bigcup_{(\varepsilon_1, \varepsilon_2) \in \Lambda_2(\varepsilon)} (\partial_{\varepsilon_1} f(x) + \partial_{\varepsilon_2} g(x)) \right),$$

which confirms the equality of the corollary. □

Given $\varepsilon_1, \varepsilon_2 \geq 0$ with $\varepsilon_1 + \varepsilon_2 = \varepsilon$, we notice that

$$\partial_{\varepsilon_1} f(x) + \partial_{\varepsilon_2} f(x) \subset \partial_\varepsilon f(x) + \partial_\varepsilon f(x) \subset \partial_{\varepsilon_1'} f(x) + \partial_{\varepsilon_2'} f(x),$$

where $\varepsilon_1' = \varepsilon_2' = \varepsilon$ and $\varepsilon_1' + \varepsilon_2' = 2\varepsilon$. Then, the following formula for the exact subdifferential of a sum is directly derived from Corollary 3.301.

COROLLARY 3.302 (J.-B. Hiriart-Urruty and R.R. Phelps: subdifferential sum rule). *Let X be a locally convex space and $f, g : X \to \mathbb{R} \cup \{+\infty\}$ be two proper lower semicontinuous convex functions. For any $x \in X$ one has*

$$\partial (f+g)(x) = \bigcap_{\varepsilon > 0} \mathrm{cl}_{w^*} [\partial_\varepsilon f(x) + \partial_\varepsilon g(x)].$$

REMARK 3.303. Like in Remark 3.300 a direct simple proof can be given for the above corollary. Let $\overline{x} \in \mathrm{dom}\, f \cap \mathrm{dom}\, g$ and take any $\varepsilon > 0$. Denoting $\varphi := f + g$ and remembering the expression of $\varphi'(\overline{x}; \cdot)$ as a limit, we see for any $v \in X$ that $\varphi'(\overline{x}; v) = f'(\overline{x}; v) + g'(\overline{x}; v)$. Then, using the definition of $f'_\varepsilon(\overline{x}; \cdot)$ as an infimum, we get $\varphi'(\overline{x}; v) \leq f'_\varepsilon(\overline{x}; v) + g'_\varepsilon(\overline{x}; v)$. Since $f'_\varepsilon(\overline{x}; \cdot)$ is the support function of $\partial_\varepsilon f(\overline{x})$ (see Theorem 3.141), it ensues from (1.26) that

$$\varphi'(\overline{x}; v) \leq \sigma(v; \partial_\varepsilon f(\overline{x})) + \sigma(v; \partial_\varepsilon g(\overline{x})) = \sigma(v, \partial_\varepsilon f(\overline{x}) + \partial_\varepsilon g(\overline{x})).$$

By (1.16) we deduce for any $x^* \in \partial \varphi(\overline{x})$ that $x^* \in \mathrm{cl}_{w^*} [\partial_\varepsilon f(\overline{x}) + \partial_\varepsilon g(\overline{x})]$. This justifies the inclusion of the left-hand side into the right-hand. The converse inclusion is easily verified. □

Corollary 3.194 covers the case of supremum of finitely many convex functions which are continuous at a reference point $\overline{x}$. As above the use of ε-subdifferential allows us in the following proposition to complement this result (as well as the others in Section 3.15) with the situation when the continuity assumption is not fulfilled but merely the lower semicontinuity is present.

PROPOSITION 3.304 (A. Brønsted: Subdifferential maximum rule). Let X be a locally convex space and $f_i : X \to \mathbb{R} \cup \{+\infty\}$, $i = 1, \cdots, m$, be proper lower semicontinuous convex functions. For any $x \in X$ with $f_1(x) = \cdots = f_m(x)$ one has

$$\partial(\max\{f_1, \cdots, f_m\})(x) = \bigcap_{\varepsilon > 0} \overline{co}^*[\partial_\varepsilon f_1(x) \cup \cdots \cup \partial_\varepsilon f_m(x)],$$

where the maximum is taken in $\mathbb{R} \cup \{-\infty, +\infty\}$.

PROOF. Fix any $\overline{x} \in X$ with $f_1(\overline{x}) = \cdots = f_m(\overline{x})$. We may suppose that the value $f_1(\overline{x}) = \cdots = f_m(\overline{x})$ is finite, since otherwise the equality is trivial at $\overline{x}$. Put $I := \{1, \cdots, m\}$ and $f(x) := \max\{f_1(x), \cdots, f_m(x)\}$ for all $x \in X$. For every $\varepsilon > 0$ and every $i \in I$ the inequality $f_i \leq f$ along with the equality $f_i(\overline{x}) = f(\overline{x})$ yield $\partial_\varepsilon f_i(\overline{x}) \subset \partial_\varepsilon f(\overline{x})$, thus $\overline{co}^* \left(\bigcup_{i \in I} \partial_\varepsilon f_i(\overline{x}) \right) \subset \partial_\varepsilon f(\overline{x})$, which clearly implies the inclusion of the right-hand side in the proposition into the left-hand one. To prove the converse inclusion, we proceed like in Remark 3.303. Note first that, for every $v \in X$ and every real $t > 0$

$$t^{-1}[f(\overline{x} + tv) - f(\overline{x})] = \max_{i \in I} t^{-1}[f_i(\overline{x} + tv) - f(\overline{x})] = \max_{i \in I} t^{-1}[f_i(\overline{x} + tv) - f_i(\overline{x})],$$

which entails as $t \downarrow 0$ that $f'(\overline{x}; v) = \max_{i \in I} f_i'(\overline{x}; v)$. Then for any real $\varepsilon > 0$, using the inequality $f_i'(\overline{x}; \cdot) \leq (f_i)_\varepsilon'(\overline{x}; \cdot)$ and remembering by Theorem 3.141 that the directional ε-derivative $(f_i)_\varepsilon'(\overline{x}; \cdot)$ is the support function of $\partial_\varepsilon f_i(\overline{x})$, it follows that

$$f'(\overline{x}; v) \leq \max_{i \in I}(f_i)_\varepsilon'(\overline{x}; v) = \max_{i \in I} \sigma(v, \partial_\varepsilon f_i(\overline{x})).$$

For any $x^* \in \partial f(\overline{x})$ it ensues that $\langle x^*, v \rangle \leq \sigma \left(v, \bigcup_{i \in I} \partial_\varepsilon f_i(\overline{x}) \right)$ for every $v \in X$, and hence $x^* \in \overline{co}^* \left(\bigcup_{i \in I} \partial_\varepsilon f_i(\overline{x}) \right)$. From this we obtain the inclusion $\subset$ of the proposition, so the proof is finished. $\square$

REMARK 3.305. Fix $\overline{x} \in X$ with $f_1(\overline{x}) = \cdots = f_m(\overline{x})$ finite. On $Y := \mathbb{R}^m$ consider the convex cone $Y_+ := \mathbb{R}_+^m$ and define the Y_+-convex mapping $G : X \to Y \cup \{+\infty_Y\}$ by $G(x) = (f_1(x), \cdots, f_m(x))$ if $x \in \text{dom}\, f_1 \cap \cdots \cap \text{dom}\, f_m$ and $G(x) = +\infty_Y$ otherwise. Consider also the continuous convex function $g : Y \to Y \cup \{+\infty\}$ defined by $g(y) := \max\{y_1, \cdots, y_m\}$ for all $y = (y_1, \cdots, y_m) \in Y = \mathbb{R}^m$. We note that g is non Y_+-decreasing and we recall that $g(+\infty_Y) = +\infty$ by convention. Putting $f(x) := \max_{1 \leq i \leq m} f_i(x)$ we see that $f = g \circ G$. Therefore, using Theorem 3.292 and Corollary 3.301 one can provide another proof for Proposition 3.304. However, the complete computation and arguments make this second proof much longer, so we kept the previous direct proof. $\square$

3.23.4. ε-Subdifferential determination of convex functions and cyclic monotonicity. In Theorem 3.204 we saw that two proper lower semicontinuous convex functions on a Banach space are equal up to an additive constant if and only if they have the same subdifferential at every point in the space. In this subsection we remove the completeness of the space by using the ε-subdifferential. The proof is quite similar to the one of Theorem 3.204.

THEOREM 3.306 (ε-subdifferential determination of convex functions). Let X be a locally convex space and $f, g : X \to \mathbb{R} \cup \{+\infty\}$ be two proper lower semicontinuous convex functions. The following are equivalent:

(a) there is a real $\alpha > 0$ such that $\partial_\varepsilon f(x) \subset \partial_\varepsilon g(x)$ holds for all $\varepsilon \in \,]0,\alpha[$ and all $x \in X$;
(b) the inclusion $\partial_\varepsilon f(x) \subset \partial_\varepsilon g(x)$ holds for every $\varepsilon > 0$ and every $x \in X$;
(c) there exists a real constant C such that
$$f(x) = g(x) + C \quad \text{for all } x \in X.$$

PROOF. The implications (c)$\Rightarrow$(b)$\Rightarrow$(a) are evident. It remains to prove the implication (a)$\Rightarrow$(c). Fix a point $a \in \operatorname{dom} f$ and for any $\varepsilon \in \,]0,\alpha[$ note by Proposition 3.94 that $\partial_\varepsilon f(a) \neq \emptyset$, and hence $g(a)$ is also finite by (a). We may suppose that $a = 0$. For each $b \in X$ consider the continuous linear mapping $A_b : \mathbb{R} \to X$ defined by $A_b(t) = tb$ for all $t \in \mathbb{R}$ and note that the lower semicontinuous convex functions $f \circ A_b$ and $g \circ A_b$ are finite at 0, and hence proper.

Fix any $(t,t^*) \in \operatorname{gph} \partial(f \circ A_b)$. By Corollary 3.299 we have that
$$t^* \in \bigcap_{0<\varepsilon<\alpha} \operatorname{cl}_{w^*}\bigl(A_b^*(\partial_\varepsilon f(A_b t))\bigr).$$

This and (a) imply that
$$t^* \in \bigcap_{0<\varepsilon<\alpha} \operatorname{cl}_{w^*}\bigl(A_b^*(\partial_\varepsilon g(A_b t))\bigr) = \bigcap_{\varepsilon>0} \operatorname{cl}_{w^*}\bigl(A_b^*(\partial_\varepsilon g(A_b t))\bigr),$$

so by Corollary 3.299 one again we obtain $t^* \in \partial(g \circ A_b)(t)$. This translates that the inclusion $\partial(f \circ A_b)(t) \subset \partial(g \circ A_b)(t)$ holds for all $t \in \mathbb{R}$. Applying Proposition 3.49 we have $f(0) + g(b) = g(0) + f(b)$. This being true for every $b \in X$, setting $C := f(0) - g(0)$ yields (with $x = b$) the desired equality $f(x) = g(x) + C$ for all $x \in X$. $\square$

We apply the above theorem to derive the maximality of the cyclically monotone family $(\partial_\varepsilon f)_{\varepsilon \geq 0}$.

DEFINITION 3.307. Given an interval I in $\mathbb{R}$ whose left endpoint is $0 \in I$, a family $(T_\varepsilon)_{\varepsilon \in I}$ of multimappings from a locally convex space X into its topological dual X^* is said to be a *cyclically monotone family* provided that for any integer $m \geq 1$, $\varepsilon_i \in I$, $(x_i, x_i^*) \in X \times X^*$ with $x_i^* \in T_{\varepsilon_i}(x_i)$, $i = 0, 1, \cdots, m$, one has
$$\sum_{i=0}^{m-1} \langle x_i^*, x_i - x_{i+1}\rangle + \langle x_m^*, x_m - x_0\rangle \geq -\sum_{i=0}^{m} \varepsilon_i.$$

When I is the singleton $\{0\}$, one just says that the *multimapping* $T = T_0$ *is cyclically monotone*.

The cyclically monotone family $(T_\varepsilon)_{\varepsilon \in I}$ is called *maximal* if for any cyclically monotone family $(S_\varepsilon)_{\varepsilon \in I}$ of multimappings from X into X^* such that $T_\varepsilon(x) \subset S_\varepsilon(x)$ for all $\varepsilon \in I$ and $x \in X$, one has $S_\varepsilon(x) = T_\varepsilon(x)$ for all $\varepsilon \in I$ and $x \in X$. If $I = \{0\}$, the multimapping $T = T_0$ is then said in this case to be a *maximal cyclically monotone multimaping*.

It is easily seen that the family $(\partial_\varepsilon f)_{\varepsilon \in I}$, where f is a proper convex function, is cyclically monotone. In fact we will see next that any maximal cyclically monotone family $(T_\varepsilon)_{\varepsilon \in I}$ is of the form $(\partial_\varepsilon f)_{\varepsilon \in I}$ for some proper lower semicontinuous convex function $f : X \to \mathbb{R} \cup \{+\infty\}$.

THEOREM 3.308 (characterization of maximal cyclically monotone family of multimappings). *Let X be a locally convex space and $(T_\varepsilon)_{\varepsilon \in I}$ be a cyclically monotone family of multimappings from X into X^*, where I is an interval in $\mathbb{R}$ with left endpoint $0 \in I$ and with $\operatorname{int} I \neq \emptyset$. Then the family $(T_\varepsilon)_{\varepsilon \in I}$ is maximal cyclically monotone if and only if there exists a proper lower semicontinuous convex function $f : X \to \mathbb{R} \cup \{+\infty\}$ such that $T_\varepsilon = \partial_\varepsilon f$ for all $\varepsilon \in I$.*

Further, the function f is unique up to an additive real constant.

The proof involves the following lemma.

LEMMA 3.309. *Let X be a locally convex space and $(T_\varepsilon)_{\varepsilon \in I}$ be any family of multimappings from X into X^*, where I is any interval in $\mathbb{R}$ with left endpoint $0 \in I$. The following are equivalent:*

(a) *the family is cyclically monotone;*
(b) *there exists a proper lower semicontinuous convex function $\varphi : X \to \mathbb{R} \cup \{+\infty\}$ such that $T_\varepsilon(x) \subset \partial_\varepsilon \varphi(x)$ for all $\varepsilon \in I$ and all $x \in X$.*

PROOF. We only need to prove (a)$\Rightarrow$(b). Clearly we may suppose $\operatorname{Dom} T_{\varepsilon_0} \neq \emptyset$ for some $\varepsilon_0 \in I$, so we choose $x_0 \in \operatorname{Dom} T_{\varepsilon_0}$ and $x_0^* \in T_{\varepsilon_0}(x_0)$. For each $u \in X$ define

$$\varphi(u) := \sup\left\{ \sum_{i=0}^{m-1} \langle x_i^*, x_{i+1} - x_i \rangle + \langle x_m^*, u - x_m \rangle - \sum_{i=0}^{m} \varepsilon_i : \varepsilon_i \in I, (x_i, x_i^*) \in \operatorname{gph} T_{\varepsilon_i}, i = 1, \cdots, m, m \in \mathbb{N} \right\}.$$

Clearly, φ is a lower semicontinuous convex function with $\varphi(X) \subset \mathbb{R} \cup \{+\infty\}$. Further, by the cyclic monotonicity of $(T_\varepsilon)_{\varepsilon \in I}$ we have $\varphi(x_0) \leq 0$, so φ is proper. Fix any $\varepsilon \in I$ and any $x \in X$.

Suppose that $T_\varepsilon(x) \neq \emptyset$ and take any $x^* \in T_\varepsilon(x)$. For any fixed $u \in X$ and for $\gamma := \langle x^*, u - x \rangle - \varepsilon$, considering any family $(x_i, x_i^*)_{i=1}^m$ with $(x_i, x_i^*) \in \operatorname{gph} T_{\varepsilon_i}$ and putting $x_{m+1} := x$, $x_{m+1}^* := x^*$, $\varepsilon_{m+1} = \varepsilon$ we obtain by definition of $\varphi(u)$

$$\varphi(u) \geq \sum_{i=0}^{m} \langle x_i^*, x_{i+1} - x_i \rangle + \langle x_{m+1}^*, u - x_{m+1} \rangle - \sum_{i=0}^{m+1} \varepsilon_i$$

$$= \sum_{i=0}^{m-1} \langle x_i^*, x_{i+1} - x_i \rangle + \langle x_m^*, x - x_m \rangle - \sum_{i=0}^{m} \varepsilon_i + \gamma,$$

so taking the supremum over families $(x_i, x_i^*)_{i=1}^m$ with $(x_i, x_i^*) \in \operatorname{gph} T_{\varepsilon_i}$ and $m \in \mathbb{N}$ yields

$$\varphi(u) \geq \varphi(x) + \gamma, \text{ that is } \varphi(u) \geq \varphi(x) + \langle x^*, u - x \rangle - \varepsilon.$$

The latter inequality being true for every $u \in X$, it ensues that $x^* \in \partial_\varepsilon \varphi(x)$. We then conclude that $T_\varepsilon(x) \subset \partial_\varepsilon \varphi(x)$ for all $\varepsilon \in I$ and $x \in X$. $\square$

PROOF OF THEOREM 3.308. If the family $(T_\varepsilon)_{\varepsilon \in I}$ is maximal cyclically monotone, the conclusion follows from Lemma 3.309 above, where equality holds by the maximality assumption of the family $(T_\varepsilon)_{\varepsilon \in I}$.

For the converse implication, it suffices to show that, given a proper lower semicontinuous convex function $f : X \to \mathbb{R} \cup \{+\infty\}$, the family $(\partial_\varepsilon f)_{\varepsilon \in I}$ is maximal cyclically monotone. We already know that the latter family is cyclically monotone.

Let us prove its maximality. Consider any cyclically monotone family $(S_\varepsilon)_{\varepsilon \in I}$ of multimappings from X into X^* satisfying
$$\partial_\varepsilon f(x) \subset S_\varepsilon(x) \quad \text{for all } \varepsilon \in I \text{ and } x \in X.$$
By Lemma 3.309 there is a proper lower semicontinuous convex function $g : X \to \mathbb{R} \cup \{+\infty\}$ such that $S_\varepsilon(x) \subset \partial_\varepsilon g(x)$ for all $\varepsilon \in I$ and $x \in X$. Choosing $\alpha > 0$ with $]0, \alpha[\subset I$ (keep in mind that $\text{int } I \neq \emptyset$), it ensues that
$$\partial_\varepsilon f(x) \subset \partial_\varepsilon g(x) \quad \text{for all } \varepsilon \in]0, \alpha[,\ x \in X.$$
By virtue of Theorem 3.306 there is a real constant C such that $f = g + C$ on X, hence for all $\varepsilon \in I$ and $x \in X$ we have
$$\partial_\varepsilon f(x) \subset S_\varepsilon(x) \subset \partial_\varepsilon g(x) = \partial_\varepsilon f(x).$$
We deduce that $S_\varepsilon(x) = \partial_\varepsilon f(x)$ for all $\varepsilon \in I$ and $x \in X$, which justifies the desired maximality of the family $(\partial_\varepsilon f)_{\varepsilon \in I}$. Finally, the uniqueness of f up to an additive constant follows from Theorem 3.306. □

In the framework of Banach spaces we obtain the following similar result.

THEOREM 3.310 (R.T. Rockafellar: maximal cyclic monotonicity of subdifferential). *A multimapping $T : X \rightrightarrows X^*$ from a Banach space X into X^* is maximal cyclically monotone if and only if there exists a proper lower semicontinuous convex function $f : X \to \mathbb{R} \cup \{+\infty\}$ such that $T = \partial f$.*

This function is unique up to an additive real constant.

PROOF. We argue as in Theorem 3.308. The implication $\Rightarrow$ follows from Lemma 3.309 with $I := \{0\}$ and $T_0 := T$, where equality holds by the maximality assumption of T.

For the converse implication $\Leftarrow$, fix a proper lower semicontinuous convex function $f : X \to \mathbb{R} \cup \{+\infty\}$. It is clear that ∂f is cyclically monotone. It remains to show its maximality. Take any monotone multimapping $S : X \rightrightarrows X^*$ with $\partial f(x) \subset S(x)$ for all $x \in X$. By Lemma 3.309 there is a proper lower semicontinuous convex function $g : X \to \mathbb{R} \cup \{+\infty\}$ such that $S(x) \subset \partial g(x)$ for all $x \in X$. By Theorem 3.204 there is a real constant C such that $f = g + C$ on X, hence we have for all $x \in X$
$$\partial f(x) \subset S(x) \subset \partial g(x) = \partial f(x).$$
We deduce that $S(x) = \partial f(x)$ for all $x \in X$, which justifies the desired maximality of the cyclic monotone multimapping ∂f. Finally, the uniqueness of f up to an additive constant is a consequence of Theorem 3.204. □

3.23.5. Limiting subdifferential chain rule for convex functions on locally convex spaces. In Subsection 3.16.1 we developed limiting calculus rules for $\partial(g \circ A)$ where $A : X \to Y$ is a continuous linear mapping between two Banach spaces X and Y. By restricting ourselves to formulas abandoning (or weakening) the condition $\|y_j^*\| \|y_j - Ax_j\| \to 0$ in Corollary 3.200, we will work in this subsection with the more general case when X is a locally convex space.

Let $A : X \to Y$ be a continuous linear mapping between Hausdorff locally convex spaces X, Y and let $g : Y \to \mathbb{R} \cup \{+\infty\}$ be a proper lower semicontinuous convex function. Consider any locally convex topology θ on X^* for which the topological dual of (X^*, θ) is identified with X, that is, θ is between the weak* topology and the Mackey topology of X^*. By the convexity (for $\eta > 0$) of the set $A^*(\partial_\eta f(x))$ we have $\text{cl}_{w^*}(A^*(\partial_\eta g(x))) = \text{cl}_\theta(A^*(\partial_\eta g(A x)))$, so Corollary 3.200 gives for any $x \in X$

$$\partial(g \circ A)(x) = \bigcap_{\eta > 0} \text{cl}_\theta\left(A^*(\partial_\eta g(A x))\right).$$

THEOREM 3.311 (limiting chain rule for a convex function and a linear mapping: locally convex space case). Let X be a Hausdorff locally convex space, $(Y, \|\cdot\|)$ be a Banach space and θ be any locally convex topology on X^* with $(X^*, \theta)^* \equiv X$. Let $A : X \to Y$ be a continuous linear mapping and and $g : Y \to \mathbb{R} \cup \{+\infty\}$ be a lower semicontinuous convex function. Let $\bar{x} \in X$ with $g(A\bar{x}) < +\infty$. Then, $x^* \in \partial(g \circ A)(\bar{x})$ if and only if there are nets (resp. sequences if X is a reflexive Banach space) $(y_j)_j$ in Y and $(y_j^*)_j$ in Y^* with $y_j^* \in \partial g(y_j)$ such that

(3.211) $\qquad x^* = \lim_\theta y_j^* \circ A \quad (\text{resp. } x^* = \lim_{\|\cdot\|_*} y_j^* \circ A)$

and such that anyone of the following conditions holds:

(i) $y_j \to A\bar{x}$, $\langle y_j^*, y_j - A\bar{x}\rangle \to 0$ and $\text{dist}(y_j, A(X))\|y_j^*\|_* \to 0$;
(ii) $y_j \to A\bar{x}$, $\langle y_j^*, y_j\rangle \to \langle x^*, \bar{x}\rangle$ and $\text{dist}(y_j, A(X))\|y_j^*\|_* \to 0$;
(iii) $y_j \to A\bar{x}$ and $\langle y_j^*, y_j\rangle \to \langle x^*, \bar{x}\rangle$;
(iv) $y_j \to A\bar{x}$ and $\langle y_j^*, y_j - A\bar{x}\rangle \to 0$;
(v) $y_j \to A\bar{x}$ and $\liminf \langle y_j^*, A\bar{x} - y_j\rangle \geq 0$;
(vi) $y_j \to A\bar{x}$, $g(y_j) \to g(A\bar{x})$ and $\langle y_j^*, y_j\rangle \to \langle x^*, \bar{x}\rangle$.

PROOF. Denoting $(k)_0$ the condition (k) plus (3.211), it is easy to see that $(i)_0 \Rightarrow (ii)_0 \Rightarrow (iii)_0$ and $(iii)_0 \Rightarrow (iv)_0 \Rightarrow (v)_0$ along with $(iv)_0 \Leftrightarrow (vi)_0$. Further, as in the proof of Theorem 3.199, it is not difficult to prove that $(v)_0$ entails $x^* \in \partial(g \circ A)(\bar{x})$. Assume now $x^* \in \partial(g \circ A)(\bar{x})$. Let $\varepsilon > 0$ and let V be any θ-open symmetric convex neighborhood of zero in X^* and $p = p_V$ be the Minkowski gauge of V. To show that $(i)_0$ holds, it suffices to find $v \in Y$ and $v^* \in \partial g(v)$ such that

(3.212) $\qquad \|v - A\bar{x}\| < \varepsilon, \ |\langle v^*, v - A\bar{x}\rangle| < \varepsilon,$

(3.213) $\qquad \text{dist}(v, A(X))\|v^*\|_* < \varepsilon,$

and

(3.214) $\qquad A^* v^* - x^* \in V, \text{ or equivalently } p(A^* v^* - x^*) < 1.$

Denote $U := \{u \in X : \langle \cdot, u\rangle \leq p(\cdot)\}$. For any $y^* \in Y^*$, we note that

$$\sup_{u \in U} |\langle y^*, Au\rangle| = \sup_{u \in U} |\langle A^* y^*, u\rangle| \leq p(A^* y^*),$$

so $A(U)$ is bounded in Y by the Banach-Steinhauss theorem. Put

$$M := \sup_{u \in U} \|Au\| < +\infty.$$

Choose a real $\eta > 0$ such that

$$2\sqrt{\eta}(1 + \sqrt{\eta}) < \varepsilon \text{ and } (2M + 2p(x^*) + 1)\sqrt{\eta} < 1.$$

By the equality preceding the statement of the theorem, there exists some $z^* \in \partial_\eta g(A\bar{x})$ such that $p(A^* z^* - x^*) < 1/2$. Consider the equivalent norm $\|\|\cdot\|\|$ on Y defined by

$$\|\|y\|\| := \|y\| + |\langle z^*, y\rangle| + (1 + \|z^*\|)\text{dist}(y, A(X)) \quad \text{for all } y \in Y,$$

and let $\|\cdot\|_*$ the dual norm on Y^* of $\|\cdot\|$. Observe that $\|z^*\|_* \leq 1$ since $|\langle z^*, y\rangle| \leq \|y\|$ for all $y \in Y$. By the Brønsted-Rockafellar theorem (see Theorem 3.142), there are $v \in X$ and $v^* \in \partial g(v)$ such that

$$\|v - A\overline{x}\| < \sqrt{\eta} \quad \text{and} \quad \|v^* - z^*\|_* < \sqrt{\eta},$$

so $\|v^*\|_* \leq \|v^* - z^*\|_* + \|z^*\|_* \leq \sqrt{\eta} + 1$. We proceed to show that v and v^* have the desired properties. First, we notice that $\|v - A\overline{x}\| \leq \|v - A\overline{x}\| \leq \sqrt{\eta} < \varepsilon$ and

$$|\langle v^*, v - A\overline{x}\rangle| \leq |\langle v^* - z^*, v - A\overline{x}\rangle| + |\langle z^*, v - A\overline{x}\rangle|$$
$$\leq \|v^* - z^*\|_* \|v - A\overline{x}\| + \|v - A\overline{x}\|$$
$$\leq \eta + \sqrt{\eta} < \varepsilon,$$

which justifies (3.212). Regarding (3.214), choose by the Hahn-Banach theorem some $u \in X$ such that

$$\langle \cdot, u\rangle \leq p(\cdot) \text{ on } X^* \text{ and } \langle A^*v^* - A^*z^*, u\rangle = p(A^*v^* - A^*z^*),$$

so in particular $u \in U$. Observe by the inequality $p(A^*z^* - x^*) < 1/2$ that

$$\|Au\| = \|Au\| + |\langle z^*, Au\rangle| + 0 = \|Au\| + |\langle A^*z^*, u\rangle|$$
$$\leq \|Au\| + p(A^*z^*) < \|Au\| + p(x^*) + 1/2 \leq M + p(x^*) + 1/2.$$

It follows that

$$p(A^*v^* - A^*z^*) = \langle A^*v^* - A^*z^*, u\rangle = \langle v^* - z^*, Au\rangle$$
$$\leq \|Au\|\sqrt{\eta} < (M + p(x^*) + 1/2)\sqrt{\eta} < 1/2.$$

Therefore, we obtain that

$$p(A^*v^* - x^*) \leq p(A^*v^* - A^*z^*) + p(A^*z^* - x^*) < 1/2 + 1/2 = 1,$$

which gives (3.214). We now pass to (3.213). Since $\|y\| \leq 2(1 + \|z^*\|_*)\|y\|$ for all $y \in Y$, we can write

$$\|v^*\|_* \leq 2(1 + \|z^*\|_*)\|v^*\|_* \leq 2(1 + \|z^*\|_*)(1 + \sqrt{\eta}).$$

This and the equality $\operatorname{dist}(v, A(X)) = \operatorname{dist}(v - A\overline{x}, A(X))$ yield

$$\operatorname{dist}(v, A(X))\|v^*\|_* \leq \operatorname{dist}(v - A\overline{x}, A(X))2(1 + \|z^*\|_*)(1 + \sqrt{\eta})$$
$$= 2(1 + \|z^*\|_*)\operatorname{dist}(v - A\overline{x}, A(X))(1 + \sqrt{\eta})$$
$$\leq 2\|v - A\overline{x}\|(1 + \sqrt{\eta}) \leq 2\sqrt{\eta}(1 + \sqrt{\eta}) < \varepsilon,$$

where the inequality

$$(1 + \|z^*\|_*)\operatorname{dist}(v - A\overline{x}, A(X)) \leq \|v - A\overline{x}\|$$

(involved above) is due to the definition of $\|\cdot\|$. The property (3.213) is then established and the proof of the theorem is finished. □

3.24. Comments

As mentioned by E. F. Beckenbach [80, p. 441], the inequality translating the convexity of a function attracted the attention of some authors already in the last quarter of the 19th century. Consider a function $f :]a, b[\to \mathbb{R}$ of one real variable. In 1889 M. O. Hölder established that, if f is of class $\mathcal{C}^2$ with $f'' \geq 0$, the inequality

(3.215) $$f\left(\sum_{i=1}^{k} \lambda_i x_i\right) \leq \sum_{i=1}^{k} \lambda_i f(x_i)$$

holds for all $\lambda_i \geq 0$ satisfying $\sum_{i=1}^{k} \lambda_i = 1$. J. Hadamard [452] proved in 1893 that, if f admits an increasing derivative, the inequality

$$f\left(\frac{x+y}{2}\right) < \frac{1}{y-x} \int_x^y f(t)\, dt$$

holds true for all $a < x < y < b$. In addition to Beckenbach's paper [80], D. S. Mitrinović and I. B. Lacković revealed in their 1985 article [715, p. 229] that, earlier to the previous authors, Ch. Hermite wrote in a 1883 short note the following: "Soit $f(x)$ une fonction qui varie toujours dans le même sens de $x = a$, à $x = b$. On aura les relations

$$(b-a)f\left(\frac{a+b}{2}\right) < \int_a^b f(x)\, dx < (b-a)\frac{f(a)+f(b)}{2}$$

ou bien

$$(b-a)f\left(\frac{a+b}{2}\right) > \int_a^b f(x)\, dx > (b-a)\frac{f(a)+f(b)}{2}$$

suivant que la courbe $y = f(x)$ tourne sa convexité ou sa concavité vers l'axe des abscisses".[1] Around the period of 1885's, other classes of functions for which were also obtained inequalities of convexity type, appeared in some articles. We refer for example to [453, 454, 401]. The interest of the study of functions satisfying such a kind of inequality apparently began in 1905 with J. L. W. V. Jensen [559, 560]. In this paper, Jensen declared the function f to be *convex* provided that the inequality

(3.216) $$f\left(\frac{x+y}{2}\right) \leq \frac{f(x)+f(y)}{2},$$

is satisfied; nowadays it is said that f is *midpoint convex*. By means of midpoint convex functions, Jensen established in [560] diverse new inequalities and recovered a lot of old basic ones, in particular (3.215) under the additional continuity of f (see [560, inequality (5), p. 180]). In a second step, Jensen studied for themselves midpoint convex functions, putting in light various properties like: continuity under boundedness from above, existence of left-hand and right-hand derivates under the same additional hypothesis of boundedness from above, etc. The great importance of convex functions was then revealed by Jensen, and his name is generally associated in the literature with the definition of convex functions. It is also worth mentioning that, at the end of page 191 of [560], Jensen wrote: "la notion (de fonctions convexes) devra trouver sa place dans les expositions élémentaires de la théorie des fonctions réelles".[2]

---

[1]Traduction: "Let $f(x)$ be a function which always varies in the same sense from $x = a$ to $x = b$. One will have the relations, either

$$(b-a)f\left(\frac{a+b}{2}\right) < \int_a^b f(x)\, dx < (b-a)\frac{f(a)+f(b)}{2}$$

or

$$(b-a)f\left(\frac{a+b}{2}\right) > \int_a^b f(x)\, dx > (b-a)\frac{f(a)+f(b)}{2},$$

depending on whether the curve $y = f(x)$ turns its convexity or its concavity towards the axis of abscissa."

[2]Traduction: "the notion (of convex functions) should be treated in elementary presentations of the theory of real-valued functions"

Since the 1960s diverse new concepts have been introduced for nondifferentiable convex functions of vector variables. Actually, many books are completely (or for a large part) devoted to the theory of finite/infinite dimensional convex analysis through the basic tools of conjugate functions and subdifferentials. We quote among them [**35, 43, 55, 72, 78, 134, 146, 214, 384, 494, 529, 649, 744, 801, 852, 853, 865, 1000**]. The present chapter of the manuscript contains many results which did not appear in any previous book: It is the case of Sections 3.16, 3.18, 3.19. Further, the development and presentation that we followed in several places are different from those of the predecessors of the book: in regards to this, we refer, for example, to the approaches that we followed for the subdifferential determination and maximal monotonicity of the subdifferential etc.

We continue the comments with a historical view of the analytical development of conjugates in convex analysis. The notion of conjugate is at the origin of the theory of convex analysis with extended real-valued functions, that is, with values in $\mathbb{R} \cup \{-\infty, +\infty\}$.

Let us first examine as in [**924**] three earlier papers of S. Mandelbrojt, of W. Fenchel, and of Z. Birnbaum and W. Orlicz. Considering a function f of one real variable and using notation $\overline{\text{borne}}$ to mean the supremum, S. Mandelbrojt stated in his 1939 article [**686**] the following theorem:

Theorem (Mandelbrojt) Toute fonction $f(x)$ de la forme

(3.217) $$f(x) = \overline{\underset{-\infty < t < +\infty}{\text{borne}}}\, [xt - \varphi(t)],$$

où $\varphi(t)$ est une fonction quelconque, définie sur la droite entière $(-\infty < t < +\infty)$, est une fonction convexe. Réciproquement: quelle que soit la fonction convexe $f(x)$, il existe une fonction $\varphi(t)$ telle que la relation (3.217) ait lieu; on peut, en particulier, poser

(3.218) $$\varphi(t) = \overline{\underset{-\infty < x < +\infty}{\text{borne}}}\, [xt - f(x)].$$

Before beginning with the proof, Mandelbrojt added: "À chaque fonction convexe $f(x)$ correspond une autre fonction convexe, $\varphi(t)$ qu'on peut appeler la fonction convexe *associée* à $f(x)$ et qui est liée à celle-ci par la double relation (3.217) et (3.218)".

Here are the English translations.

Theorem A (Mandelbrojt) Every function $f(x)$ in the form

$$f(x) = \overline{\underset{-\infty < t < +\infty}{\text{borne}}}\, [xt - \varphi(t)],$$

where $\varphi(t)$ is any function, defined on the whole real line $(-\infty < t < +\infty)$, is a convex function. Conversely: For any convex function $f(x)$, there exists a function $\varphi(t)$ such that the relation (3.217) holds; one can, in particular, choose

$$\varphi(t) = \overline{\underset{-\infty < x < +\infty}{\text{borne}}}\, [xt - f(x)].$$

"With any convex function $f(x)$ is associated another convex function, $\varphi(t)$ which can be called the convex function *associated* with $f(x)$ and which is related to the latter by both relations (3.217) and (3.218)".

At the end of this 1939 paper [**686**] Mandelbrojt wrote: "We believe that this result is new, but the actual events do not allow us to be sure". In fact, Z. Birnbaum and W. Orlicz had considered earlier in 1931 a transformation in this vein. Let $M : \mathbb{R} \to \mathbb{R}$ be a function such that

(i) $M(0) = 0$, $M(u) > 0$ if $u > 0$, $M(-u) = M(u)$;
(ii) there are $\alpha > 0$, $\beta > 0$ satisfying $M(u) > \beta$ for all $u > \alpha$;
(iii) $\dfrac{M(u)}{|u|} \underset{|u|\to 0}{\longrightarrow} 0$ and $\dfrac{M(u)}{|u|} \underset{|u|\to +\infty}{\longrightarrow} +\infty$.

To every such function M (called an N'-function in [**102**, p. 8, Definition 5]) Z. Birnbaum and W. Orlicz associated in their 1931 paper [**102**] (see page 8 therein) a function N defined on $\mathbb{R}$ by

$$N(v) := \max_{u \geq 0}[uv - M(u)] \quad \text{if } v \geq 0 \quad \text{and} \quad N(v) = N(|v|) \quad \text{if } v < 0.$$

Birnbaum and Orlicz showed (see pages 9 and 10 in [**102**]) that the associated function N is an N'-function which is convex. We must notice that the duality in Mandelbrojt's theorem above does not appear.

Let us come back for a moment to Theorem A in analyzing Mandelbrojt's arguments. Let $f(x)$ convex be given. With $\varphi(t)$ defined as in (3.218) above, to prove for all t, the inequality

$$\overline{\text{borne}}_{-\infty < t < +\infty} [xt - \varphi(t)] \geq f(x),$$

Mandelbrojt used the right side derivative $f'_+(x)$ to write for every x

$$\overline{\text{borne}}_{-\infty < t < +\infty} \left\{ xt - \overline{\text{borne}}_{-\infty < \tau < +\infty}[t\tau - f(\tau)] \right\} \geq xf'_+(x) - \overline{\text{borne}}_{-\infty < \tau < +\infty}[\tau f'_+(x) - f(\tau)].$$

This right side derivative is also used to write in a second step, for all τ

$$\tau f'_+(x) - f(\tau) \leq x f'_+(x) - f(x), \text{ hence } \overline{\text{borne}}_{-\infty < \tau < +\infty}[\tau f'_+(x) - f(\tau)] = x f'_+(x) - f(x).$$

Then, the conclusion in [**686**] is

$$\overline{\text{borne}}_{-\infty < t < +\infty} \left\{ xt - \overline{\text{borne}}_{-\infty < \tau < +\infty}[t\tau - f(\tau)] \right\} \geq x f'_+(x) - x f'_+(x) + f(x) = f(x).$$

It is clear through the arguments reproduced above that Mandelbrojt implicitly assumed that the convex function f is finite on the whole real line. However, the function φ is finite-valued on the interval of all reals t such that the function $\tau \mapsto \tau t - f(\tau)$ is bounded from above, and this interval may fail to be the whole real line. Consequently, there is no complete duality between f and φ.

Taking this fact into account W. Fenchel dealt in the 1949 paper [**404**] with convex functions defined on convex subsets of $\mathbb{R}^n$. As in [**404**] denoting $\Sigma x\xi = \sum_{i=1}^{n} x_i \xi_i$ for $x, \xi \in \mathbb{R}^n$, and saying interior point (resp. boundary point) of a convex set C of $\mathbb{R}^n$ instead of relative interior (resp. relative boundary) point of C, the formulation word for word of Fenchel's result is the following.

Theorem B (Fenchel [**404**]) Let C be a convex point set in $\mathbb{R}^n$ and $f(x)$ a function defined in C convex and semicontinuous from below and such that $\lim_{C \ni x \to u} f(x) = \infty$ for each boundary point u of C which does not belong to C. Then there exists one and only one point set Γ in $\mathbb{R}^n$ and one and only one function $\phi(\xi)$ defined in Γ with exactly the same properties as C and $f(x)$ such that

(3.219) $$\Sigma x\xi \le f(x) + \phi(\xi),$$

where to every interior point x of C there corresponds at least one point ξ of Γ for which equality holds.

In the same way C, $f(x)$ correspond to Γ, $\phi(\xi)$.

In the proof in [**404**] the set Γ is defined as the set

(3.220) $$\Gamma := \{\xi \in \mathbb{R}^n : x \mapsto \Sigma x\xi - f(x) \text{ is bounded from above in } C\},$$

and ϕ in turn is defined by

(3.221) $$\phi(\xi) = \sup_{x \in C} \left(\Sigma x\xi - f(x)\right) \quad \text{for all } \xi \in \Gamma.$$

One of the main tools in the development of the proof is the (finite-dimensional) property that at each relative interior point of C there is a support hyperplane of the hypersurface determined by f on C.

The pair (Γ, ϕ) is called in [**404**] the conjugate of (C, f). In the Lectures [**405**, p. 109] Fenchel also showed:[3]

Theorem C (Fenchel [**405**]) Let (C_i, f_i), $i = 1, 2$, be as (C, f) in Theorem B and let (Γ_i, ϕ_i) be the conjugate of (C_i, f_i). Assume that

$$\text{rint } C_1 \cap \text{rint } C_2 \ne \emptyset \quad \text{and} \quad \text{rint } \Gamma_1 \cap -\text{rint } \Gamma_2 \ne \emptyset.$$

Then the equality

$$\min_{x \in C_1 \cap C_2} (f_1(x) + f_2(x)) = \max_{\xi \in \Gamma_1 \cap -\Gamma_2} \left(-\phi_1(\xi) - \phi_2(-\xi)\right)$$

holds true.

Theorem B of Fenchel establishes a complete duality in the sense that (C, f) is also the conjugate of (Γ, φ). However, the statement is not so amenable. On the one hand, the property $\lim_{C \ni x \to u} f(x) = \infty$ at every relative boundary point u of C which does not belong to C is required in the involved class of concerned lower semi-continuous convex functions. So, for functions constructed from finitely/infinitely many others, working both with the latter and with the set where the new function is defined is of course not amenable.

In the early 1960s J. J. Moreau and R. T. Rockafellar realized independently that the right setting which one must work with is the one of functions taking on values in the extended real line $\overline{\mathbb{R}} := \mathbb{R} \cup \{-\infty, +\infty\}$. Given a locally convex space X, a function $f : X \to \mathbb{R} \cup \{-\infty, +\infty\}$ is then declared in this context to be convex whenever the classical inequality holds true with the extended addition $\dotplus$ (given by $(-\infty)\dotplus(+\infty) := +\infty$) and the real scalar in $]0, 1[$, or equivalently the epigraph of f is a convex set in $X \times \mathbb{R}$. Following Fenchel and Mandelbrojt, the *conjugate* function of any extended real-valued function $f : X \to \mathbb{R} \cup \{-\infty, +\infty\}$, is

---

[3]The corresponding result in [**405**] is stated as the maximization of $g(x) - f(x)$, with g concave and f convex. In Theorem C, we have translated the result with two convex functions.

defined by Moreau [**733**] and Rockafellar [**844**] as the extended real-valued function $f^* : X^* \to \mathbb{R} \cup \{-\infty, +\infty\}$ with

$$f^*(x^*) := \sup_{x \in X} \left(\langle x^*, x \rangle - f(x)\right) \quad \text{for all } x^* \in X^*.$$

This point of view of extended real-valued functions offers the flexibility to deal with addition, infimum or supremum of finitely/infinitely many functions, as well as with composition and integral.

Concerning the aforementioned results by Fenchel and working with extended real-valued functions, Theorem B has been extended by Moreau to infinite dimensions, and the extension of Theorem C by Rockafellar provided a much more general context allowing the presence of inequality constraints, which contributes to derive the linear duality theorem very easily.

Because of the need for Mechanics of Continua (see [**743**, Section 6]) Moreau's main concern was with infinite-dimensional spaces. He considered the set $\Gamma_0(X)$ of all lower semicontinuous convex functions from X into $\mathbb{R} \cup \{-\infty, +\infty\}$ which are proper, that is, $f(X) \subset \mathbb{R} \cup \{+\infty\}$ and $f \not\equiv +\infty$ and the set $\Gamma(X) = \Gamma_0(X) \cup \{-\omega_X, +\omega_X\}$, where we recall that ω_X denotes the function identically equal to $+\infty$ on X. In this context, Moreau established in his 1962 paper[**733**] the following conjugate duality result, where the conjugate of f^* is taken with X^* endowed with the topology $w(X^*, X)$, so the topological dual of X^* coincides with X.

Theorem D (Moreau [**733**]) Let X be a (Hausdorff) locally convex space. Then any function $f \in \Gamma(X)$ coincides with its biconjugate $f^{**} = (f^*)^*$.

Further, for any $f \in \Gamma_0(X)$ one has $f^* \in \Gamma_0(X^*)$, where X^* is equipped with the $w(X^*, X)$ topology.

In his 1963 PhD thesis [**844**], Rockafellar proved a general duality result, reported also in his paper [**845**], that we can translate as follows:

Theorem E (Rockafellar [**844, 845**]) Let X and Y be finite-dimensional partially-ordered vector spaces in which the non-negative convex cones X_+ and Y_+ are polyhedral. Let A be a linear mapping from X into Y, and let $f_1 : X \to \mathbb{R} \cup \{+\infty\}$ and $f_2 : Y \to \mathbb{R} \cup \{+\infty\}$ be proper convex functions. Assume that there exists at least one $\bar{x} \in \mathrm{rint}\,(\mathrm{dom}\,f_1)$ such that $\bar{x} \geq_{X_+} 0$ and $A\bar{x} \geq_{Y_+} \bar{y}$ for some $\bar{y} \in \mathrm{rint}\,(\mathrm{dom}\,f_2)$. Then the following equality

$$\inf\{f_1(x) + f_2(y) : x \geq_{X_+} 0, Ax \geq_{Y_+} y\}$$
$$= \max\{-f_1^*(x^*) - f_2^*(-y^*) : y^* \geq_{Y_+^*} 0, A^*y^* \leq_{X_+^*} x^*\}$$

holds where $A^* : Y^* \to X^*$ denotes the adjoint of A and X_+^*, Y_+^* are the non-negative dual cones with

$$X_+^* := \{x^* \in X^* : \langle x^*, x \rangle \geq 0, \; \forall x \in X_+\};$$

above the infimum is taken with respect to (x, y) and the maximum with respect to (x^*, y^*).

Rockafellar also derived the linear duality theorem as well as the duality result

(3.222) $$\inf_{x \in X} \left(f_1(x) + f_2(Ax)\right) = \max_{y^* \in Y^*} \left(-f_1^*(-A^*y^*) - f_2^*(y^*)\right)$$

under the assumption $\mathrm{rint}\,(\mathrm{dom}\,f_1) \cap A(\mathrm{rint}\,(\mathrm{dom}\,f_2)) \neq \emptyset$; see also other comments below on this duality equality.

Theorem D of Moreau and Theorem E of Rockafellar (see (3.222)) obviously encompass Fenchel's results in Theorem B and Theorem C respectively.

Connections are made in [**404, 849, 865**] between the above concept of conjugate and the transform by A.-M. Legendre in [**653**], known as Legendre transform. This was born from Legendre's investigations on the (PDE) partial differential equation (whose earlier studies of a particular case by G. Monge motivated Legendre, as written in [**653**, p. 309])

$$A(p,q)\frac{\partial^2 u}{\partial x^2}(x,y) + B(p,q)\frac{\partial^2 u}{\partial x \partial y}(x,y) + C(p,q)\frac{\partial^2 u}{\partial y^2}(x,y) = 0,$$

where (as in [**653**]) $p := \frac{\partial u}{\partial x}(x,y)$ and $q := \frac{\partial u}{\partial y}(x,y)$, and the coefficients A, B, C depend merely on p, q (see [**653**, p. 314]). The first step in [**653**, Section II] was to transform the above equation in a suitable way into another one with p, q as variables. Assuming ∇u is a diffeomorphism from $\mathbb{R}^2$ onto $\mathbb{R}^2$, Legendre's idea is to write $\nabla u(x(p,q), y(p,q)) = (p,q)$, or equivalently $(x(p,q), y(p,q)) = (\nabla u)^{-1}(p,q)$. Then x, y, u can be considered as functions of (p, q). Noting that the regularity of u ensures that $\frac{\partial p}{\partial y} = \frac{\partial q}{\partial x}$, calculation gives $\frac{\partial x}{\partial q}(p,q) = \frac{\partial y}{\partial p}(p,q)$, so $x(p,q)\,dp + y(p,q)\,dq$ is an exact differential form on $\mathbb{R}^2$. Then let $\mathbb{R}^2 \ni (p,q) \mapsto \omega(p,q) \in \mathbb{R}$ be such that $d\omega = x\,dp + y\,dq$. One sees that

$$x(p,q) = \frac{\partial \omega}{\partial p}(p,q), \quad y(p,q) = \frac{\partial \omega}{\partial q}(p,q), \quad d(px + yq - \omega) = du,$$

so a suitable choice of constant allows one to obtain the function ω such that $p\,x(p,q) + q\,y(p,q) - \omega(p,q) = u(p,q)$ for all $(p,q) \in \mathbb{R}^2$; I followed Legendre's notation in denoting by $\omega(\cdot)$ the function determined in this way. Altogether yields the dual pair of equalities

$$p = \frac{\partial u}{\partial x}(x,y), \quad q = \frac{\partial u}{\partial y}(x,y), \quad u(x,y) + \omega(p,q) = px + qy,$$

and

$$x = \frac{\partial \omega}{\partial p}(p,q), \quad y = \frac{\partial \omega}{\partial q}(p,q), \quad u(x,y) + \omega(p,q) = px + qy.$$

Further, ω is of class C^2 and after calculation Legendre obtained the new (PDE)

$$A(p,q)\frac{\partial^2 \omega}{\partial q^2}(p,q) - B(p,q)\frac{\partial^2 \omega}{\partial p \partial q}(p,q) + C(p,q)\frac{\partial^2 \omega}{\partial p^2}(p,q) = 0,$$

and he noticed that the new (PDE) is of the same type but much easier since it does not contain first order partial derivatives. In fact, before the development reproduced above Legendre utilized the method to study the famous equation with $A(p,q) = 1+q^2$, $B(p,q) = 2pq$ and $C(p,q) = 1+p^2$ (related to Monge's works). In [**653**] Legendre also applied his transform to various classes of (PDE).

Let us also emphasize another viewpoint from the book of R. Courant and D. Hilbert [**296**] concerning Legendre's transform. In the first step of the process, the idea was to establish a correspondence between a smooth surface (regular in a certain sense) and the family of its tangent planes. Of course, the main point is to recover the surface knowing the family of tangent planes. Let us follow the presentation in [**296**, Chapter I, Section 6]. Consider the surface (S) which is the graph of a function $x \mapsto u(x)$, where $x := (x_1, x_2) \in \mathbb{R}^2$; suppose that u is smooth on $\mathbb{R}^2$ (say C^1 at this stage). This surface can be seen also as the envelope of its tangent planes. Denote by $(\bar{x}, \bar{z}) \in \mathbb{R}^2 \times \mathbb{R}$ the generic point of a nonvertical plane whose equation is

$$(P_{\xi,\omega}) \quad \bar{z} - \xi_1 \bar{x}_1 - \xi_2 \bar{x}_2 + \omega = 0, \quad \text{or equivalently} \quad \bar{z} - \langle \xi, \bar{x} \rangle + \omega = 0;$$

the pair $(\xi, \omega) \in \mathbb{R}^2 \times \mathbb{R}$ is called in [296] the *plane coordinates* of $(P_{\xi,\omega})$.

Given $x \in \mathbb{R}^2$ one knows that, with $(\bar{x}, \bar{z}) \in \mathbb{R}^2 \times \mathbb{R}$ as generic point, the equation of the plane tangent to (S) at the point $(x, u(x)) \in S$ has as equation

$$\bar{z} - u(x) - \langle \bar{x} - x, \nabla u(x) \rangle = 0,$$

so its plane coordinate (ξ, ω) satisfies

$$\xi = \nabla u(x) \quad \text{and} \quad \omega = \langle x, \nabla u(x) \rangle - u(x).$$

The surface (S) is determined also by the family of its tangent planes, that is, the family $\left((P_{\xi,\omega(\xi)})\right)_\xi$, where $\omega(\xi)$ needs to be such that $(P_{\xi,\omega(\xi)})$ is tangent to (S). Given $\xi \in \mathbb{R}^2$, if x is the element in $\mathbb{R}^2$ such that $(P_{\xi,\omega(\xi)})$ is tangent to (S) at $(x, u(x))$, then calculating x in terms of ξ (when it is possible) by means of the equality $\xi = \nabla u(x)$ furnishes the value in terms of ξ of

$$\omega(\xi) = \langle \xi, x \rangle - u(x)$$

as a function of ξ. This is then feasible with $\xi \mapsto \omega(\xi)$ as a $\mathcal{C}^1$ mapping provided that ∇u is a diffeomorphism from $\mathbb{R}^2$ onto itself. So, under this additional assumption on ∇u, with the surface (S) is associated the surface (Σ) which is the graph of $\xi \mapsto \omega(\xi)$, where $(\xi, \omega(\xi))_{\xi \in \mathbb{R}^2}$ are as above the plane coordinates of planes tangent to (S).

Writing $\omega(\xi) = \langle \xi, x(\xi) \rangle - u(x(\xi)) = \xi_1 x_1(\xi) + \xi_2 x_2(\xi) - u(x_1(\xi), x_2(\xi))$ it results also that

$$\frac{\partial \omega}{\partial \xi_1}(\xi) = x_1(\xi) + \xi_1 \frac{\partial x_1}{\partial \xi_1}(\xi) + \xi_2 \frac{\partial x_2}{\partial \xi_1}(\xi) - \frac{\partial u}{\partial x_1}(x(\xi)) \frac{\partial x_1}{\partial \xi_1}(\xi) - \frac{\partial u}{\partial x_2}(x(\xi)) \frac{\partial x_2}{\partial \xi_1}(\xi) = x_1(\xi),$$

and, in the same way, $\frac{\partial \omega}{\partial \xi_2}(\xi) = x_2(\xi)$. Further, by what precedes one has the equality $u(x) = \langle x, \xi \rangle - \omega(\xi)$. Consequently, one sees that $\nabla \omega$ exists and is a diffeomorphism from $\mathbb{R}^2$ onto $\mathbb{R}^2$ and that (S) is in turn the surface associated with (Σ) through the same above process, in the sense that (S) is the surface described in $\mathbb{R}^2 \times \mathbb{R}$ by the *plane coordinates* of planes tangent to (Σ). This duality is confirmed also by the following dual equalities obtained above:

$$\omega(\xi) + u(x) = \langle \xi, x \rangle, \quad \xi = \nabla u(x), \quad x = \nabla \omega(\xi).$$

Of course, the above analysis is still valid with $\mathbb{R}^n \times \mathbb{R}$ in place of $\mathbb{R}^2 \times \mathbb{R}$.

When the function u is in addition convex, it is clearly seen in the two above presentations that ω is the conjugate function of u as it was pointed out by Fenchel [404, 405].

Fenchel [404, the bottom of page 73] and Moreau [744, Chapter 14, 1st ed.; Section 14.4, 2nd ed.] relied also the conjugate transform to the famous 1912 inequality of W. H. Young[4] (see (1) in [979]).

---

[4] With Young's notation in [979] the inequality can be stated in the form: If U is an increasing function of $u \geq 0$ with $U(0) = 0$, then with $v \geq 0$,

$$uv \leq \int_0^u U(t)\, dt + \int_0^v U^{-1}(t)\, dt.$$

It was established in [979] in view of applications to Fourier series.

It is also worth pointing out that W. L. Jones in his 1960 dissertation [**561**] adapted to infinite-dimensional spaces the approach by Fenchel in associating to a pair (C, f) the conjugate pair (Γ, ϕ), where as above $\Gamma := \{x^* \in X^* : \sup_{x \in C} (\langle x^*, x \rangle - f(x)) < +\infty\}$ and $\phi : \Gamma \to \mathbb{R}$ is defined by $\varphi(x^*) := \sup_{x \in \Gamma} (\langle x^*, x \rangle - f(x))$ for all $x^* \in \Gamma$. Independently of J. J. Moreau (as written in the introduction of [**179**, p. 3]) A. Brønsted studied, in locally convex spaces, conjugates of functions in his 1964 paper [**179**]. Brønsted followed, as done Jones, in [**179**] the definition of Fenchel via a pair (Γ, ϕ) as above; he improved in Section 3 therein various results of W. L. Jones and established other results in that section and in Sections 4 and 5.

All the above comments on the paper by Legendre and on the papers by Mandelbrojt and by Fenchel reproduce the corresponding part in [**924**]. Let us pass now to the sources of other concepts and of the results developed in the present chapter. The history of Legendre-Fenchel conjugate being largely developed in the preceding paragraphs, we begin with the concept of subdifferential.

For a nonempty set $S \subset \mathbb{R}^N$, a hyperplane $\{\langle \zeta, \cdot \rangle = \alpha\}$ is said to be a *supporting flat* of S by W. Fenchel [**405**, p. 45] (which we call an *extended supporting hyperplane* of S) if $\sup_{x \in S}\langle \zeta, x \rangle = \alpha$; whenever the supremum is attained, the hyperplane is a support hyperplane in the usual standard sense. Given a convex function $f : C \to \mathbb{R}$ with closed epigraph in $\mathbb{R}^N \times \mathbb{R}$ and its conjugate $\varphi : \Gamma \to \mathbb{R}$ (in the Fenchel sense in (3.220) and (3.221) above), and given elements $x_0 \in C$ and $\xi_0 \in \Gamma$, W. Fenchel considered in [**405**, p. 102] the extended supporting hyperplane

$$\{(x, r) \in \mathbb{R}^N \times \mathbb{R} : \langle \xi_0, x \rangle - r = \varphi(\xi_0)\}$$

of $\operatorname{epi}\varphi$, following in this way the tradition in geometry. Fenchel then denoted by $C(\xi_0)$ the projection of this hyperplane on $\mathbb{R}^N$ and he noticed that

(3.223) $\qquad x \in C(\xi_0) \Leftrightarrow \langle \xi_0, x \rangle - f(x) = \varphi(\xi_0).$

Dually, taking the projection $\Gamma(x_0)$ of the analogous hyperplane with $(x_0, f(x_0))$ in place of $(\xi_0, \varphi(\xi_0))$, Fenchel arrived to

(3.224) $\qquad \xi \in \Gamma(x_0) \Leftrightarrow \langle \xi, x_0 \rangle - \varphi(\xi) = f(x_0).$

From (3.223) and (3.224) and from the fact that (C, f) is equal to the conjugate of (Γ, φ), it is clear that $C(\xi_0)$ and $\Gamma(x_0)$ coincide with the subdifferentials of the convex functions φ and f at ξ_0 and x_0 respectively, that is,

$$C(\xi_0) = \{x \in \mathbb{R}^N : \langle \xi - \xi_0, x \rangle \leq \varphi(\xi) - \varphi(\xi_0), \forall \xi \in \Gamma\}$$

$$\Gamma(x_0) = \{\xi \in \mathbb{R}^n : \langle \xi, x - x_0 \rangle \leq f(x) - f(x_0), \forall x \in C\}.$$

With the set $\Gamma(x_0)$ at hands, Fenchel undertook the computation of the conjugate in his above sense of the directional derivative $f'(x_0; \cdot)$. Assume that $\Gamma(x_0) \neq \emptyset$. At a first step, Fenchel established that the conjugate of the function

$$(x_0 + \mathbb{R}_+(C - x_0)) \ni x \mapsto f(x_0) + f'(x_0; x - x_0)$$

is the function

$$\Gamma(x_0) \ni \xi \mapsto \langle \xi, x_0 \rangle - f(x_0),$$

and from this he achieved his objective in deriving that the conjugate in his sense of $(\mathbb{R}_+(C - x_0), f'(x_0; \cdot))$ is $(\Gamma(x_0), 0)$, that is, the function null on $\Gamma(x_0)$. Neither calculus rules nor other use of $\Gamma(x_0)$ (resp. $C(\xi_0)$) were provided by Fenchel. Nowadays, $\Gamma(x_0)$ is called the subdifferential of f at x_0 and it is denoted by $\partial f(x_0)$. The fundamental role of this concept in convex analysis was revealed independently by J. J. Moreau [**738**] and R. T. Rockafellar [**844**]. They showed that the concept is, besides that of conjugate, at the heart of the theory of non-differentiable convex functions. For such functions, Moreau and Rockafellar developed a rich subdifferential calculus and showed how this concept of subdifferential can be used as an efficient tool to replace the usual derivative in the study of problems with non-differentiable data in analysis, optimization, mechanics etc.

Proposition 3.14 is a classical result in the normed setting; as stated we involved the use of the function $\mathfrak{q}_V$ as in [**562**, Proposition 2.5]. The equicontinuity property in Proposition 3.37 has been shown in 1964 by J. J. Moreau [**740**]; the proof in the manuscript of this property and the others in Proposition 3.37 follow the approach in [**912**, Proposition 4.2]. Proposition 3.20(b) and its proof are taken from the book of R. T. Rockafellar and R. J-B. Wets [**865**] (see Theorem 2.35 in [**865**]). Example 3.33 and its arguments are due to J. M. Borwein, S. Fitzpatrick and R. Girgensohn [**132**, example 1]. The lower semicontinuity of $d(0, \partial f(\cdot))$ in Proposition 3.35 was observed by A. D. Ioffe and Y. Sekiguchi in their paper [**528**, Proposition 7] as a consequence of the concept of perfect regularity for convex multimappings developed there. Theorem 3.30 and Proposition 3.31 were noticed by J. J. Moreau in his 1963/1965 papers [**737, 742**]. Proposition 3.50 is classical; our arguments follow those in Theorem 24.1 in the book [**852**] of R. T. Rockafellar. Proposition 3.47 was noticed in the same book [**852**, p. 229].

Our definition of polyhedral convex functions in Subsection 3.2.2 is that used by R. T. Rockafellar [**852**, p. 172]. Corollary 3.60 and Corollary 3.61 were proved in Theorem 19.4 and Theorem 23.10 respectively of [**852**]. Concerning conjugate and polarity, in [**852**] Rockafellar established first in [**852**, Theorem 19.2] the preservation of polyhedral convexity of functions under the Legendre-Fenchel transform and then derived the result of preservation for sets under polarity (see Corollary 19.2.1 in [**852**]). Here we took the reverse path in deducing in (3.25) the polyhedral convexity of f^* from that of polar sets in Corollary 3.62.

Theorem 3.41 is the complete version in locally convex spaces as established by M. Valadier [**944, 945**]; as precised before its statement, it is the extension of the version in normed spaces in Theorem 2.204. Corollary 3.42 is the adaptation to locally convex spaces of the version for normed spaces in Corollary 2.205; as said in the section of comments in the previous chapter the result in normed spaces of Corollary 2.205 was given by M. Volle [**961**]. Theorem 3.44 was proved by A. D. Ioffe and V. M. Tikhomirov [**529**] and its proof follows [**529**]; the arguments of Proposition 3.43 are contained for a large part in [**529**]. Similar statements with compactness of the index set T and with exact subgradients of the data functions g_t at the reference point $\bar{x}$ were previously obtained in [**315, 944, 945**] under different continuity/differentiability conditions for the data functions. We also refer to the papers [**288, 289**] by R. Correa, A. Hantoute, to [**463**] by A. Hantoute and M. A. López and to [**464**] by A. Hantoute, M. A. López and C. Zalinescu, where one can find diverse other formulas and many references for subgradients of suprema of infinitely many convex functions.

The attainment result for infimal convolution of functions in Proposition 3.67 is taken from Moreau [**744**, Proposition 4.4] (see also Moreau [**735, 737**]). For the history of infimal convolution of functions we refer to Section 4.8 in the next chapter. Proposition 3.70 corresponds to Remark 5.3 in Moreau [**744**]. The results in Proposition 3.77 with extended real-valued proper lower semicontinuous functions were obtained in similar forms in finite dimensions in 1949 by W. Fenchel [**404, 405**] and in infinite dimensions by J. J. Moreau [**733**] in 1962. In locally convex spaces, earlier to [**733**], the situation of such functions which are additionally positively homogeneous was studied in 1955 by L. Hörmander [**502**, Théorème 5] and the result in Corollary 3.83 was given therein. The result in Corollary 3.81 characterizing the nonemptiness of the Clarke subdifferential $\partial_C f(\overline{x})$ by means of the equality $f^\uparrow(\overline{x}; 0) = 0$ (via the Rockafellar directional derivative $f^\uparrow(\overline{x}; \cdot)$) was noticed by R. T. Rockafellar [**856**]. Theorem 3.84 was first established, as mentioned above through Theorem B and (3.221), in 1949 by W. Fenchel [**404**] in finite dimensions in the above setting of a convex function $f : C \to \mathbb{R}$ defined on a convex subset C of $\mathbb{R}^n$, lower semicontinuous relative to C and such that

$$(3.225) \qquad \lim_{C \ni u \to x} f(u) = +\infty \quad \text{for every } x \in (\operatorname{cl} C) \setminus C.$$

The version of Theorem 3.84 with the more flexible situation of extended real-valued functions was achieved in 1962 by J. J. Moreau [**733**] in the setting of locally convex spaces; see also the 1964 paper by A. Brønsted [**179**] in the framework (as done Fenchel in finite dimensions) of a convex function $f : C \to \mathbb{R}$ with a convex set C of a locally convex space. Moreau's proof is different from the one by Fenchel, and it corresponds for a very large part to the proof that we followed in the manuscript. Both Corollary 3.90 and Remark 3.91 are taken from Remark 5.2 in Moreau [**744**]. The statement of Proposition 3.94 was observed and demonstrated by E. Asplund and R. T. Rockafellar [**30**, p. 452, l. 24-29]. The result in Proposition 3.95 has been mentioned by A. Hantoute and M. A. Lopez in the proof of Lemma 2 in their paper [**463**].

Proposition 3.96 was probably first given in an analogous form by J. J. Moreau in [**740**] and [**744**, Chapter 8] and by R. T. Rockafellar [**847**]. While the condition $\lim_{\|x\| \to +\infty} = +\infty$ is known in the literature as the *coercivity condition* for the function f, it is worth noticing that the condition $\lim_{\|x\| \to +\infty} \frac{f(x)}{\|x\|} = +\infty$ is called the *strong coercivity condition* by some authors. Those conditions are also useful for the study of some properties of Young integrands, see for example A. Fougères's paper [**416**, Théorème 2.6] and references therein.

The Fréchet differentiability version in normed spaces of Theorem 3.108 (that is, Corollary 3.110) is contained in E. Asplund [**28**, Lemma 7]. Theorem 3.118 and Corollary 3.121 are due to E. Asplund and R. T. Rockafellar, see Theorem 3 and Corollary 2 in their 1969 paper [**30**].

Asymptotic cones were developed for convex sets in finite-dimensional spaces by J. J. Stoker in his 1940 paper [**904**] under the name of *characteristic cones*. As said in the line -2 of page 165 in [**904**], the idea of the *characteristic cone* for convex sets was contained in the Steinitz 1913 paper [**902**]. Asymptotic cones of convex sets in infinite-dimensional spaces were later studied in the 1962 paper [**226**] of G. Choquet. Asymptotic functions for convex functions in infinite dimensions were defined and fully developed in the 1966 paper [**847**] of R. T. Rockafellar (see the line -3 in page 51 of [**847**] for the definition). A large part of Section 3.5 follows that paper [**847**] of Rockafellar. Proposition 3.137 is contained in [**847**, Corollary 3C] and Proposition 3.139 corresponds to [**847**, Corollary 3D]. The result in Proposition 3.140 is [**847**, Theorem 3F]. Proposition 3.135 is due to G. Choquet [**226**]. Proposition 3.128 and

Corollary 3.129 are taken from Rockafellar [**847**, Theorem 2A]. We also refer to the book of A. Auslender and M. Teboulle [**45**], and to J.-P. Dedieu [**319**] and D. T. Luc [**678**] for further results. Second-order asymptotic cones and functions can be found in the paper [**459**].

The statement in Theorem 3.142 first appeared in 1965 in the lemma in page 608 of the 1965 paper of A. Brønsted and R. T. Rockafellar [**181**, p. 608]. It was recognized as a fundamental result in variational analysis, and it is cited in the literature as Brønsted-Rockafellar theorem. The proof of Brønsted and Rockafellar [p. 608][**181**] used some arguments similar to those of E. Bishop and R. R. Phelps [**104**] for support hyperplanes. Theorem 3.143 is a substantially more general version (of Theorem 3.142) published in 1982 by J. M. Borwein in [**120**, Theorem 1]. The proof given for Theorem 3.143 follows that in [**120**]. Example 3.144 was constructed by Brønsted and Rockafellar [**181**, p. 610] by means of a convex set provided by V. Klee [**610**] satisfying the properties involved in [**181**] and in Example 3.144 in the book.

The history of duality in optimization is largely commented in the book [**865**, p. 531] to which we refer the reader. As said by R. T. Rockafellar and R. J-B. Wets in [**865**] the origin of such a duality is probably the 1928 paper [**963**] by J. von Neumann on game theory. Another crucial step was concerned with the duality theory for linear optimization problems. D. Gale, H. W. Kuhn and A. W. Tucker in their paper [**430**] of 1951 formulated the dual problem in linear programming and proved the linear duality theorem. In [**316**, p. 44] G. B. Dantzig reported a discussion with J. Von Neumann in October 1947 at the Institute for Advanced Study at Princeton where he presented to J. von Neumann an Air Force problem and a linear programming approach. Through Dantzig's account in [**316**] of von Neumann's mathematical analysis during the discussion, the analogy was clear for von Neumann between linear programming and the theory of zero-sum two-person game. In the paper [**964**] published one month later by von Neumann, as analyzed by T. H. Kjeldsen [**606**], given $a \in \mathbb{R}^n$, $b \in \mathbb{R}^m$ and an $m \times n$-matrix A, it was shown that if the maximization problem

$$(3.226) \qquad \text{Maximize}_x \ \langle a, x \rangle \quad \text{subject to } Ax \leq b,\ x \geq 0,$$

has a solution, then there exists $\overline{y} \in \mathbb{R}^m$ satisfying ${}^t A\overline{y} \geq a$ such that $\langle b, \overline{y} \rangle \leq \langle b, y \rangle$ for all $y \geq 0$ with ${}^t Ay \geq a$. This is clearly related to the dual of the latter linear maximization problem (3.226). The fundamental step of the extension of the duality theory to convex optimization problems occurred principally with works by Fenchel and by Rockafellar. Considering, as in Theorem C above of Fenchel, two convex sets $C_1, C_2 \subset \mathbb{R}^n$ with $C_1 \cap C_2 \neq \emptyset$ and two convex functions $f_i : C_i \to \mathbb{R}$, $i = 1, 2$, which are lower semicontinuous relative to C_i and satisfy the above property (3.225), W. Fenchel associated to the primal problem

$$\text{Minimize } f_1(x) + f_2(x) \quad \text{over } x \in C_1 \cap C_2,$$

the other problem

$$\text{Maximize } -\phi_1(y) - \phi_2(-y) \quad \text{over } y \in \Gamma_1 \cap -\Gamma_2,$$

with the assumption $\Gamma_1 \cap \Gamma_2 \neq \emptyset$, where Γ_1, Γ_2 are defined as in (3.220). Fenchel stated in (2) of [**405**, p. 106], in addition to Theorem C, that the value of the two above problems coincide. Unfortunately, that equality does not hold because of a small error related to the conjugate of the sum. According to R. T. Rockafellar

[850, p. 182, line 11] the error was first discovered by J. Stoer; Rockafellar provided in [850, p. 182-183] two examples contradicting the above duality equality, showing in this way that (at the opposite of finite-dimensional linear programming) a qualification condition must be required for duality in convex programming even in finite dimensions. In the same paper [850] Rockafellar considered the general case of a duality scheme in the locally convex setting. Given two Hausdorff locally convex spaces X, Y, a continuous linear mapping $A : X \to Y$ and two proper convex functions $f_1 : X \to \mathbb{R} \cup \{-\infty, +\infty\}$ and $f_2 : Y \to \mathbb{R} \cup \{-\infty, +\infty\}$ with $C := \operatorname{dom} f_1$ and $D := \operatorname{dom} f_2$, Rockafellar [850] associated to the primal problem

$$\text{Minimize}_x \ f_1(x) + f_2(Ax) \quad \text{subject to } x \in C, \ Ax \in D,$$

the dual problem

$$\text{Maximize}_{y^*} \ -f_1^*(-A^*y^*) - f_2^*(y^*) \quad \text{subject to } y^* \in D_0, \ A^*y^* \in -C_0,$$

where $C_0 := \operatorname{dom} f_1^*$ and $D_0 := \operatorname{dom} f_2^*$, and where we recall that A^* is the adjoint of A. In fact as done in Rockafellar [850], it is equivalent to study the problems

$$(P) \ \text{Minimize}_x \ f_1(x) + f_2(Ax) \ \text{over } x \in X,$$

$$(P^*) \ \text{Maximize}_{y^*} \ -f_1^*(-A^*y^*) - f_2^*(y^*) \ \text{over } y^* \in Y^*.$$

The approach in [850] is based on the analysis of the performance function

(3.227) $$Y \ni y \mapsto p(y) := \inf_{x \in X} \left(f_1(x) + g(y + Ax) \right)$$

with the perturbation variable y. Rockafellar's Theorem 3 in [850] showed the duality equality $\inf(P) = \max(P^*)$ whenever (P) is *"stably set"* in the sense of [850], a condition which can easily be seen to be equivalent to

$$\partial p(0) \neq \emptyset \quad \text{when } \inf(P) \text{ is finite}.$$

In Rockafellar's paper [850] Theorem 1 proved that (P) is "stably set" whenever

$$0 \in A(\operatorname{dom} f_1) - \operatorname{cont} f_2$$

(remind that $\operatorname{cont} f_2$ denotes the set of points where f_2 is finite and continuous), while Theorem 2 in the same Rockafellar's paper [850] showed that (P) is "stably set" provided that X, Y are finite-dimensional and

(3.228) $$0 \in \operatorname{rint} \left(A(\operatorname{dom} f_1) - \operatorname{dom} f_2 \right) = A(\operatorname{rint} (\operatorname{dom} f_1)) - \operatorname{rint}(\operatorname{dom} f_2).$$

In fact, the latter finite-dimensional statement already appeared (see (3.227) above) in the paper [845] of 1964 where Rockafellar derived in a quite easy way (shorter than arguments in pages 113-115 in Fenchel's lectures [405]) the dual linear problem and the linear duality theorem of D. Gale, H. W. Kuhn and A. W. Tucker [430]. While the approach in [850] is based on the subdifferentiability at 0 of the performance function p in (3.227) above, Rockafellar deduced in [845] under (3.228) the above finite-dimensional convex duality from Theorem 2 in [845] which says: Given a proper convex function $h : E \to \mathbb{R} \cup \{+\infty\}$ on a finite-dimensional vector space E and a polyhedral convex cone $K \subset E$ with $K \cap \operatorname{rint} (\operatorname{dom} h) \neq \emptyset$, then

$$\inf\{h(x) : x \in K\} = \max\{-h^*(x^*) : x^* \in -K^\circ\}.$$

In [852] Rockafellar introduced, with finite-dimensional vector spaces X and Y, a much more general perturbation approach with a function $X \times Y \ni \varphi(x, y) \in \mathbb{R} \cup \{-\infty, +\infty\}$, where y is a parameter and

$$\text{Minimize } \varphi(x, 0) \quad \text{over } x \in X$$

corresponds to the primal minimization problem, and he continued this approach with Banach spaces X and Y in [853]. This is the approach that we have followed in Section 3.8. For the dual of the minimization problem of $f(x)+g(G(x))$ we followed the paper [269] which adapted Rockafellar's arguments in [850, 852, 853] with a convex vector valued mapping $G : X \to Y \cup \{+\infty_Y\}$ in place of a continuous linear mapping $A : X \to Y$. Assertion (b) and assertion (d) in Corollary 3.161 correspond to the combination of Theorem 3 in Rockafellar's paper [850] with Theorem 1 and Theorem 2 respectively in the same paper [850].

Concerning Section 3.11, Theorem 3.165 and Corollary 3.167 are taken from the articles [269, 271] by C. Combari, M. Laghdir and L. Thibault; similar results and others can also be found in D. Azé paper [54] and C. Zalinescu's book [1000]. The assertion (b) in Corollary 3.170 have been first proved in 1963 by J. J. Moreau (under the additional condition of lower semicontinuity of the functions) with the use of the infimal convolution of the conjugate functions f^*, g^* (which written as $y^* \mapsto \inf_{x^* \in X^*} \left(f^*(x^*) + g^*(y^* - x^*) \right)$ appears as a performance function with y^* as parameter). The result has been extended by R. T. Rockafellar in his 1966 paper [848], in any topological vector space without the semicontinuity condition: if we consider $x = 0$, $f(0) = g(0) = 0$ and take $x^* \in \partial(f + g)(0)$ the method in [848] consists in applying the Hahn-Banach separation theorem to the epigraph of $f - \langle x^*, \cdot \rangle$ and the hypograph of $-g$. Assertion (d) in Corollary 3.170 and assertion (d) in Corollary 3.171 have been first established in Rockafellar's dissertation of 1963 [844]. For other convex duality and subdifferential sum/chain rules with convex functions in Fréchet spaces as well as for CS-closed or lower CS-closed convex functions, we refer to the paper [866] by B. Rodriguez and S. Simons and the paper [16] by C. Amara and M. Ciligot-Travain, and to the books by S. Simons [890] and by C. Zalinescu [1000]; see also references therein. Proposition 3.179 and Corollary 3.180 were fisrt noticed by R. T. Rockafellar and R.J-B. Wets in Theorem 6.43 and Exercise 6.44 of their book [865]; nonconvex similar results, as in [865], will be established in Proposition 4.115 and Corollary 4.116. Example 3.193 is taken from J. M. Borwein and J. D. Vanderwerff's book [146, Exercise 4.1.18].

General subdifferential sum/chain rules for convex functions without any qualification condition started with the 1993 paper [496] of J.-B. Hiriart-Urruty and R. R. Phelps where formulas in Corollary 3.299 and Corollary 3.302 were established for the first time. These formulas for $\partial(g \circ A)(\bar{x})$ and $\partial(f + g)(\bar{x})$ involve ε-subdifferentials of f, g of the data at the same reference point $\bar{x}$. Soon later, H. Attouch, J.-B. Baillon and M. Théra obtained with their 1994 paper [33] in the Hilbert framework the formula in Corollary 3.201 with sequences and condition (v) for the sum, that is, for $\partial(f + g)(\bar{x})$. The formula in Corollary 3.201 with sequences and condition (iv) also appeared in [33] with Hilbert spaces. While the approach in [496] deals with expressions for Legendre-Fenchel conjugates of some convex functions, the sequential result is obtained in [33] as a consequence of general results established therein for variational sums of maximal monotone operators in Hilbert spaces. The formulas in Theorem 3.199 in the general Banach setting began on the one hand with the 1995 paper [914] by L. Thibault (presented at the "Seventh French-German Conference on Optimization in Dijon", June 27-July 2, 1994), and on the other hand with the two other papers [916, 917] of L. Thibault. The papers [792] by J.-P. Penot, [575] by A. Jourani, [593] by F. Jules, [594] by F. Jules and M. Lassonde, [**413**] by S. Fitzpatrick and S. Simons, [**287**] by R. Correa, A. Hantoute and A. Jourani were also devoted to sequential formulas for $\partial(f+g)$ and $\partial(g \circ A)$ under the convexity of f and g. The approach that we followed for the proof of Theorem 3.199 is exactly the one used by L. Thibault in [**921**, Theorem 1]. For limiting formulas for the continuous sum

$$x \mapsto \int_T f(x,t)\,d\mu(t)$$

on a Banach space X with convex functions $f(\cdot, t)$ we refer the reader to the 2006 paper [**522**] of A. D. Ioffe and also to the 2008 paper [**675**] by O. Lopez and L. Thibault.

The subdifferential determination Theorem 3.204 was established in the general framework of Banach spaces by R. T. Rockafellar in his 1966 and 1970 papers [**846, 851**]. It was previously shown in Hilbert space by J. J. Moreau in Proposition 8.b of his 1965 paper [**741**], where the use of the function $(1/2)\|\cdot\|^2$ has been initiated in the study of such properties of subdifferentials of convex functions; the arguments in [**741**] are also valid in the context of reflexive Banach spaces (as noticed by Moreau [**741**, p. 87, 1967 version] or [**741**, Remark 10.5, 2003 version]). The proof in the manuscript of Theorem 3.204 through the limiting chain rule is due to L. Thibault (see [**921**, Proposition 3]). Other proofs of Theorem 3.204 can be found in [**435, 548, 614, 741, 851, 926, 927, 1015**]. Theorem 3.206 on maximal monotonicity of subdifferentials of proper lower semicontinuous convex functions on Banach spaces is due to R. T. Rockafellar with his 1966 and 1970 papers [**846, 851**]. Such a property was previously noticed and established in Theorem 2 of the 1964 paper [**712**] by G. J. Minty for real-valued continuous convex functions on general topological vector spaces. It was proved by J. J. Moreau for lower semicontinuous convex functions on Hilbert spaces in Proposition 12.b in his 1965 paper [**741**] through the use of the proximal mapping prox_f associated to the kernel $(1/2)\|\cdot\|^2$ (see [**741**]). The proof of Theorem 3.206 in the manuscript follows the limiting chain rule approach of L. Thibault in [**921**, Proposition 4]. The second proof, that we gave, reproduces the one by M. Marques-Alves and B. F. Svaiter [**694**, Theorem 2.1] with certain simplifications by S. Simons [**891**, Theorem 2.1].

The formula $N(\{f \leq r\}; \bar{x}) = \mathbb{R}_+ \partial f(\bar{x})$ with $f(\bar{x}) = r$, under the Slater condition $\{f < r\} \neq \emptyset$, for the normal cone of the sublevel set of a continuous convex function f is classical; see the previous chapter. Under the Slater condition again and in the finite dimensional setting, R. T. Rockafellar showed in Theorem 23.7 of his 1970 book [**852**] that

$$N(\{f \leq r\}; \bar{x}) = \text{cl}\left(\mathbb{R}_+ \partial f(\bar{x})\right)$$

for any convex function $f : \mathbb{R}^n \to \mathbb{R} \cup \{+\infty\}$. The general extension in Theorem 3.214 is due to A. Cabot and L. Thibault [**190**, Theorem 3.1]. This and all the other results in Section 3.18 as well as all the proofs are taken from that paper [**190**] by Cabot and Thibault.

Theorem 3.237 is due to J. J. Moreau [**740**] and its proof follows the one of Moreau [**744**, Proposition 8.2]. Theorem 3.239 and Corollary 3.241 correspond to R. T. Rockafellar's result in [**847**, Theorem 7A(a)], and the results in Corollaries 3.238 and 3.242 appeared in Moreau [**740**].

The statement in Corollary 3.252 has been established for Banach spaces by J. Saint-Raymond in Theorem 2.4 of his paper [**874**] of 2013; W. B. Moors provided in his 2017 paper [**718**] a short and much simpler proof in the same setting of Banach space. Theorem 3.245 corresponds to Theorem 3.2 in the paper [**797**] by P. Pérez-Aros and L. Thibault; the proof in the manuscript reproduces the one in [**797**] which itself uses some ideas in Moors' paper [**718**]. The result in Corollary 3.250 was first proved by B. Cascales, J. Orihuela and A. Pérez [**207**, Theorem 2] in 2017 for Banach spaces whole dual ball is weak* convex block compact (see the definition in [**207**]), and it has been established for any Banach space in the 2017 paper of W. B. Moors [**719**, Theorem 1]. The extension presented in the book to complete locally convex spaces is due to P. Pérez-Aros and L. Thibault [**797**, Corollary 3.8]. The result in Corollary 3.253 is due to J. Orihuela [**777**] in the setting of Banach spaces. In the book we reproduced its extension to locally convex spaces by Pérez-Aros and Thibault [**797**, Corollary 3.10]. The interest for such a result probably began with the 2002 paper [**323**] by F. Delbaen and the 2006 paper [**567**] by E. Jouini, W. Schachermayer and N. Touzi, both related to mathematical economics and finance. Let $(\Omega, \mathcal{F}, P)$ be probability space and let us recall (see [**323**, Definition 3]) that a function $\rho : L^\infty(\Omega, \mathcal{F}, P) \to \mathbb{R}$ is a *coherent risk measure* if it is sublinear with $\rho(X) \leq 0$ if $X \geq 0$ and with

$$\rho(X + c) = \rho(X) - c$$

for any real constant c. The investigation of characterizations of coherent risk measures ρ led Delbaen to study in [**323**, Theorem 8], through the famous James compactness theorem, the attainment of the supremum

$$\sup\{E(-XY) : Y \in \mathcal{Y}\} = \sup\{E(-XY) - \Psi_\mathcal{Y}(Y) : Y \in L^1(\Omega, \mathcal{F}, P)\}$$

for every $X \in L^\infty(\Omega, \mathcal{F}, \mathbb{R})$, where $\mathcal{Y}$ is a certain $w(L^1, L^\infty)$-closed convex set in $L^1(\Omega, \mathcal{F}, P)$. Considering instead a *monetary utility function* $U : L^\infty(\Omega, \mathcal{F}, P) \to \mathbb{R}$, or equivalently a *convex coherent risk measure* $\rho := -U$ (that is, the sublinearity of ρ is replaced by its convexity), Jouini, Schachermayer and Touzi employed in [**567**, Theorem 5.2] an extension (in separable spaces) of James theorem by J. Orihuela for the attainment of the corresponding supremum

$$\sup\{E(-XY) - V(Y) : Y \in L^1(\Omega, \mathcal{F}, P)\}$$

for every $X \in L^\infty(\Omega, \mathcal{F}, P)$, where $V : L^1(\Omega, \mathcal{F}, P) \to \mathbb{R}_+ \cup \{+\infty\}$ is a certain $w(L^1, L^\infty)$-lower semicontinuous convex function. Here $E(\cdot)$ denotes the *expectation* with respect to the probability P.

The proof of the principal inclusion

$$N(\text{Enl}_r C; \bar{x}) \cap \mathbb{S}_{X^*} \subset \partial d_C(\bar{x})$$

for $\bar{x}$ outside the closure of the convex set C in Proposition 3.262 is an adaptation to convex sets of A.Y Kruger's arguments in [**619**] for Fréchet subgradients and normals.

The Moreau envelope $e_\lambda f$ and proximal mapping $\text{Prox}_\lambda f$, with $\lambda = 1$ in Definition 3.263, were introduced by J. J. Moreau in his papers [**735**, **741**] and studied there mainly for extended real-valued convex functions f on Hilbert spaces. The parametrized form with any real $\lambda > 0$ was probably considered for the first time by H. Brézis [**176**] for the study of existence and properties of the absolutely continuous solution u from $[0, T]$ into a Hilbert space H for the differential evolution inclusion

$$\frac{du}{dt} + \partial \phi(u(t)) \ni g \quad \text{with } u(0) = u_0 \in \text{cl}(\text{dom}\,\phi),$$

where $g \in L^2([0,T], H)$. Above the solution is considered in the strong sense (that is, it is required to be absolutely continuous) under the initial condition with data u_0 in the form $u_0 \in \text{cl}(\text{dom}\,\phi)$ instead of the classical one $u_0 \in \text{dom}\,\phi$. The approach in [**176**] is based on the properties and convergence of the family $(u_\lambda)_{\lambda>0}$ of solutions of the differential evolution equations

$$\frac{du_\lambda}{dt} + \nabla(e_\lambda \phi)(u_\lambda(t)) = g \quad \text{with } u_\lambda(0) = 0.$$

One of first extensive presentations of Moreau envelope was given in Sections 2.7 and 3.4 of H. Attouch's book [**32**]. Note that the Moreau envelope $e_\lambda f$ was called in [**32**] *Moreau-Yosida approximation* of index λ of f because of the fact that, when f is a proper lower semicontinuous convex function on a Hilbert space, $\nabla(e_\lambda f)$ coincides with the *Yosida approximate* of index λ of the maximal monotone operator ∂f. Since the functions $e_\lambda f$ are still of great interest for diverse fundamental classes of nonconvex functions (as will be seen in other chapters), we followed the terminology "Moreau envelope" utilized by R. T. Rockafellar and R. J-B. Wets in their book [**865**]. Regarding the results in Section 3.21 of our book, the Lipschitz continuity property in (b) in Theorem 3.265 was already clear in Moreau's paper [**741**, Propositions 5.b and 7.d] for convex functions f on Hilbert spaces, and the pointwise convergence property of $(e_\lambda f)_\lambda$ to f in (e) of the same Theorem 3.265 probably first appeared in Brézis' paper [**176**, Proposition 10] (see also Brézis's book [**177**]) for convex functions f on Hilbert spaces. The assertion of Gâteaux differentiability of the Moreau envelope in (e) in Theorem 3.267 was established by Moreau [**741**, Propositions 5.b and 7.d] for convex functions f on Hilbert spaces. The inequality

$$f(\bar{x}) \leq \liminf_{j \in J} e_{\lambda_j} f(x_j)$$

in (b) in the same Theorem 3.267 has been noticed by P. Pérez-Aros and E. Vilches [**798**, Proposition 2.2(d)]; for this inequality as well as for the similar inequality in (e) in Theorem 3.265 the argumants in the book are the analogs of those in [**798**] for sequences. For the proof of the inequalities in the assertion (c) of Theorem 3.267 we followed the approach by L. Thibault and D. Zagrodny [**929**, Proposition 2.1]. The arguments for the convergence property

$$P_\lambda f(x) \to \text{proj}\,_C(x) \quad \text{as } \lambda \downarrow 0$$

(with $C := \text{cl}(\text{dom}\,f)$) in (c) in Theorem 3.268 follow the main ideas in the proof of assertion (ii) in Proposition 17.2.2 of the book of H. Attouch, G. Buttazzo and G. Michaille [**35**]. Theorem 3.269 was proved by Moreau [**741**, Proposition 4.a], and the arguments in the book follow those in [**741**]. The equivalences with $\sigma = 1$ between (a), (b) and (c) in Theorem 3.275 (resp. in Corollary 3.276) appeared in equivalent forms in Moreau's paper [**741**, Proposition 9.b]. The other equivalent assertion (f) in Theorem 3.275 (resp. in Corollary 3.276) is taken from Proposition 12.60(d) in the book of R. T. Rockafellar and R. J-B. Wets [**865**]. The proof in the book for the principal implication (f)$\Rightarrow$(a) in Theorem 3.275 follows the arguments by Rockafellar and Wets [**865**, Proposition 12.60]; the argument of the proof of the implication (b) $\Rightarrow$ (c) in Lemma 3.274 is also taken from [**865**]. The equivalences in Theorem 3.277 along with the arguments for the main implication (c)$\Rightarrow$(a) therein are taken from Propositions 9.b and 10.b of Moreau [**741**]. Example 3.278 corresponds to Proposition 9.d in Moreau [**741**]. Adaptations of Moreau envelopes with other referential kernels can be found in A. Fougères and A. Truffert [**417**].

The statement of Theorem 3.279 appeared for the first time in an equivalent form in the proof of Theorem 26.3 of R. T. Rockafellar's book [**852**]. The proof of Theorem 3.279 in the book follows the arguments in the book of Rockafellar and Wets [**865**, Theorem 11.13]. The equivalence between (a) and (b) in Corollary 3.280 was discovered by Rockafellar in his book [**852**, Theorem 26.3], and it also appeared in the book of Rockafellar and Wets [**865**, Theorem 11.13]. The functions satisfying the property (a) (resp. (b)) in Corollary 3.280 are called *essentially smooth* (resp. *essentially strictly convex*) in page 251 (resp. page 253) of Rockafellar's book [**852**] while the name *almost differentiable* (resp. *almost strictly convex*) is attributed in Rockafellar and Wets [**865**, Theorem 11/13].

As already said above, subdifferential sum/chain rules for convex functions without qualification conditions were initiated by the 1993 paper [**496**] of J.-B. Hiriart-Urruty and R. R. Phelps. However, subdifferential calculus for convex functions without qualification condition was already begun much earlier with the 1972 paper [**180**] by A. Brønsted; but therein merely the statement of Proposition 3.304 for the maximum of two convex functions was established. In their paper [**496**] Hiriart-Urruty and Phelps provided many more results on sum, composition, infimal convolution etc. Corollary 3.302 corresponds to [**496**, Theorem 2.1], and Corollary 3.299 to [**496**, Theorem 3.1]. Reducing to $\overline{x} = 0$ along with $f(0) = g(0) = \varepsilon$ with $\varepsilon > 0$ being fixed, and denoting by f_0 and g_0 the support functions of $\partial_\varepsilon f(0)$ and $\partial_\varepsilon g(0)$ respectively, the method in [**180**] to show the statement of Proposition 3.304, with two functions f, g, consisted in proving the left equality of the following ones

$$\partial(\max\{f_0, g_0\})(0) = \overline{\mathrm{co}}^*[\partial f_0(0) \cup \partial g_0(0)] = \overline{\mathrm{co}}^*[\partial_\varepsilon f(0) \cup \partial_\varepsilon g(0)],$$

and in showing in a second step that

$$\partial(\max\{f, g\})(0) \subset \partial(\max\{f_0, g_0\})(0)$$

by applying the Hahn-Banach separation theorem between the epigraph of f and an appropriate disjoint line segment in $X \times \mathbb{R}$. The methods developed in [**496**] are more direct and more intuitive: For the principal inclusion

$$\partial(f + g)(\overline{x}) \subset \mathrm{cl}_{w^*}[\partial_\varepsilon f(\overline{x}) + \partial_\varepsilon g(\overline{x})]$$

(with $\varepsilon > 0$) in the statement of Corollary 3.302, reducing to $\overline{x}^* = 0$ in the left-hand side, or equivalently, to the inequality

$$(f + g)^*(0) + (f + g)(\overline{x}) \leq 0,$$

and noticing that $(f + g)^* = \mathrm{cl}_{w^*}(f^* \square g^*)$, the method in [**496**] consisted in seeing by Lemma 3.288 that

$$0 \in \{x^* \in X^* : (f^* \square g^*)(x^*) \leq r + \varepsilon/2\},$$

where $r := -(f + g)(\overline{x})$; then writing

$$(f^* \square g^*)(x^*) - r = \inf_{u^* + v^* = x^*} \left([f^*(u^*) - \langle u^*, \overline{x} + f(\overline{x})\rangle] + [g^*(v^*) - \langle v^*, \overline{x} + g(\overline{x})\rangle]\right) \leq \varepsilon/2,$$

the authors in [**496**] derived the existence of $u^*, v^* \in X^*$ with $u^* + v^* = x^*$ such that the sum of the above brakets $[\ ] + [\ ] < \varepsilon$. Since the expressions between brakets are non-negative, it is deduced in [**496**] that $u^* \in \partial_\varepsilon f(\overline{x})$ and $v^* \in \partial_\varepsilon g(\overline{x})$. The approach that we used for ε-subdifferential calculus in Subsection 3.23.3 is in some sense in the spirit of the latter arguments; it mainly followed the ideas in the 1995 paper [**495**] of J.-B. Hiriart-Urruty, M. Moussaoui, A. Seeger and M. Volle. The proof in the book for Theorem 3.292 reproduced the one of [**495**, Theorem 8.1]. Corollary 3.298 corresponds to [**495**, Theorem 7.1]. Corollary 3.297 complements the sequential descriptions in Subsection 3.18.6; other results for normals to convex sublevels via ε-subgradients can be found in the 2017 paper by A. Hantoute and A. Svensson [**465**].

Theorems 3.306 and 3.308 reproduce the statements and proofs of S. Marcellin and L. Thibault [**688**, Theorems 1.2 and 2.1]; see also [**272**, Theorem 1.2] for another result similar to Theorem 3.306 where the function g in the right-hand side is not assumed to be convex. Lemma 3.309 is the adaptation of the same result proved by R. T. Rockafellar [**846**] with $\varepsilon = 0$; in [**846**] it is proved that any cyclically monotone multimapping $M : X \rightrightarrows X^*$ is pointwise included in the subdifferential of some proper lower semicontinuous function from X into $\mathbb{R} \cup \{+\infty\}$. Theorem 3.310 is due to Rockafellar [**846**].

For a general Hausdorff locally convex space X, a continuous linear mapping $A : X \to Y$ from X into a Banach space $(X, \|\cdot\|)$ and a proper lower semicontinuous convex function $g : Y \to \mathbb{R} \cup \{-\infty, +\infty\}$, the proof of the limiting chain rule in Theorem 3.311 is mainly based on the proof of S. Fitzpatrick and S. Simons [**413**, Theorem 3.3]. The descriptions (i) and (ii) with the use of

$$\mathrm{dist}(x_j, A(X))\|x_j^*\|_* \to 0$$

in Theorem 3.311 are taken from [**413**, Theorem 3.3].

CHAPTER 4

Mordukhovich limiting normal cone and subdifferential

Besides the nonsmooth variational tools in Chapter 2, the basic concepts of Mordukhovich normal cone and subdifferential will be central in the text. These concepts are defined through the crucial notions of Fréchet subgradients and normals. Given a function $f : U \to \mathbb{R} \cup \{-\infty, +\infty\}$ from an open set U of a normed space $(X, \|\cdot\|)$ and a point $x \in U$ with $|f(x)| < +\infty$, clearly the function f is Fréchet differentiable at x if and only if for every real $\varepsilon > 0$ there exists a neighborhood $V \subset U$ of x such that

$$-\varepsilon \|x' - x\| \leq f(x') - f(x) - \langle x^*, x' - x \rangle \leq \varepsilon \|x' - x\| \quad \text{for all } x' \in V,$$

with $x^* = Df(x)$. Considering only the left inequality leads to Definition 4.1 below.

4.1. Fréchet normal and subgradient

According to the above comments we devote this section to the study of Fréchet normals and subgradients.

4.1.1. Definitions and first properties. We begin by translating (as said above) the aforementioned left inequality in the following definition.

DEFINITION 4.1. Let U be a neighborhood of a point x in the normed space $(X, \|\cdot\|)$ and $f : U \to \mathbb{R} \cup \{-\infty, +\infty\}$ be an extended real-valued function which is finite at this point $x \in U$. A continuous linear functional $x^* \in X^*$ is called a *Fréchet subgradient* or *F-subgradient* of f at x provided that for any real $\varepsilon > 0$ there exists a neighborhood $V \subset U$ of x such that

(4.1) $\qquad \langle x^*, x' - x \rangle \leq f(x') - f(x) + \varepsilon \|x' - x\| \quad \text{for all } x' \in V,$

or otherwise stated

$$\liminf_{x' \to x} \frac{f(x') - f(x) - \langle x^*, x' - x \rangle}{\|x' - x\|} \geq 0.$$

The set $\partial_F f(x)$ of all Fréchet subgradients of f at x is the *Fréchet subdifferential* or *F-subdifferential* of f at x. If f is not finite at x, one sets $\partial_F f(x) = \emptyset$. When $\partial_F f(x) \neq \emptyset$ one says that f is *Fréchet subdifferentiable* at x.

For a subset S of X, the Fréchet subdifferential of the indicator function Ψ_S at x is called the *Fréchet normal cone* of S at x and it is denoted by $N^F(S; x)$, that is, $N^F(S; x) := \partial_F \Psi_S(x)$. Any element $x^* \in N^F(S; x)$ is called a *Fréchet normal (functional)* or *F-normal* of (or to) S at x. Obviously, for $x \notin S$ one has $N^F(S; x) = \emptyset$, and for $x \in S$ one has $x^* \in N^F(S; x)$ if and only if for any real $\varepsilon > 0$ there exists some neighborhood V of x such that

(4.2) $\qquad \langle x^*, x' - x \rangle \leq \varepsilon \|x' - x\| \quad \text{for all } x' \in S \cap V,$

that is,
$$\limsup_{S \ni x' \to x} \frac{\langle x^*, x' - x \rangle}{\|x' - x\|} \leq 0.$$

Given the neighborhood U of x in the normed space X and $f : U \to \mathbb{R} \cup \{-\infty, +\infty\}$, for the extension $\overline{f}$ of f as the function from X into $\mathbb{R} \cup \{-\infty, +\infty\}$ defined by $\overline{f}(u) = f(u)$ if $u \in U$ and $\overline{f}(u) = +\infty$ otherwise, it is clear that $\partial_F f(x) = \partial_F \overline{f}(x)$. So, as for the study of Clarke subdifferential, we will often suppose (without loss of generality) that the extended real-valued function f is defined *on the whole space X*.

Obviously, from (4.1) we have:

PROPOSITION 4.2. Let U be a nonempty open set of a normed space $(X, \|\cdot\|)$ and $f : U \to \mathbb{R} \cup \{-\infty, +\infty\}$ be an extended real-valued function. Let also S be a nonempty subset of X.
(a) Using any equivalent norm on X the above definition of $\partial_F f(x)$ (hence also of $N^F(S; x)$) remains unaltered.
(b) If $g : U \to \mathbb{R} \cup \{-\infty, +\infty\}$ is another function satisfying $g(x) = f(x)$ and $g(u) \leq f(u)$ for u near x, then $\partial_F g(x) \subset \partial_F f(x)$.

Before proceeding with examples of expressions of F-subdifferentials, let us relate the F-subdifferential of a function to the F-subdifferential of its lower semicontinuous hull, and also to the normal cone to its epigraph. Given an open set U of a topological space and a function $f : U \to \mathbb{R} \cup \{-\infty, +\infty\}$, recall first that the *lower semicontinuous hull* cl f of f is the greatest lower semicontinuous function on U majorized by f. The epigraph of cl f coincides with the closure in $U \times \mathbb{R}$ of the epigraph of f and (see (3.50))

$$(\operatorname{cl} f)(x) = \liminf_{u \to x} f(u) \quad \text{for all } x \in U.$$

PROPOSITION 4.3. Let $(X, \|\cdot\|)$ be a normed space and $f : X \to \mathbb{R} \cup \{-\infty, +\infty\}$ be a function on X.
(a) If f is Fréchet subdifferentiable at x, that is, $\partial_F f(x) \neq \emptyset$, then f is lower semicontinuous at x.
(b) If f is lower semicontinuous at x with $|f(x)| < +\infty$, then the Fréchet subdifferentials of f and of its lower semicontinuous hull at x coincide, that is,
$$\partial_F f(x) = \partial_F (\operatorname{cl} f)(x).$$

PROOF. (a) The non-emptiness of $\partial_F f(x)$ says that f is finite at x and fixing $x^* \in \partial_F f(x)$ and any $\varepsilon > 0$ the above definition yields some neighborhood U of x such that

(4.3) $\qquad f(x) + \langle x^*, u - x \rangle - \varepsilon \|u - x\| \leq f(u) \quad \text{for all } u \in U.$

Therefore, we obtain $f(x) \leq \liminf_{u \to x} f(u)$, which means that f is lower semicontinuous at x.
(b) Set $f_0 := \operatorname{cl} f$. Assume that f is lower semicontinuous at $x \in X$ and $|f(x)| < +\infty$, hence $f_0(x) = f(x)$. Take any $x^* \in \partial_F f(x)$ (if any) and any real $\varepsilon > 0$. There is a neighborhood U of x such that the inequality (4.3) holds. The left-hand side of this inequality being in u a lower semicontinuous function we deduce for all $u \in U$
$$f_0(x) + \langle x^*, u - x \rangle - \varepsilon \|u - x\| \leq f_0(u),$$

thus $x^* \in \partial_F f_0(x)$ and $\partial_F f(x) \subset \partial_F f_0(x)$. By Proposition 4.2(b) the converse inclusion is trivial since $f_0 \leq f$ and $f_0(x) = f(x)$ as seen above. □

The way that the F-subdifferential of a function is related to the epigraph is provided in the next proposition; the formula is similar with what we already saw for the connection between C-subdifferential and C-normal cone to epigraph. In the same time, we will also prove another analytical description of the F-subdifferential of a function f via Fréchet derivatives of functions which are below f; this furnishes another justification of the terminology of "Fréchet subdifferential".

PROPOSITION 4.4. Let $(X, \|\cdot\|)$ be a normed space. For any function $f : X \to \mathbb{R} \cup \{-\infty, +\infty\}$ which is finite at $x \in X$, the following are equivalent:
(a) $x^* \in \partial_F f(x)$;
(b) $(x^*, -1) \in N^F(\text{epi } f; (x, f(x)))$;
(c) there exists a neighborhood U of x and function $\varphi : U \to \mathbb{R}$ Fréchet differentiable at x and such that $x^* = D\varphi(x)$, $\varphi(x) = f(x)$ and $\varphi(x') \leq f(x')$ for all $x' \in U$.

PROOF. Suppose first that x^* satisfies the property of (a). Fix any $\varepsilon > 0$ and take a neighborhood U of x such that

$$\langle x^*, x' - x \rangle \leq f(x') - f(x) + \varepsilon \|x' - x\| \quad \text{for all } x' \in U.$$

This obviously implies, for all $(x', r) \in (U \times \mathbb{R}) \cap \text{epi } f$, that

$$\langle x^*, x' - x \rangle - 1(r - f(x)) \leq \varepsilon(\|x' - x\| + |r - f(x)|),$$

which ensures that $(x^*, -1) \in N^F(\text{epi } f; (x, f(x)))$, that is, (b) holds true.

Suppose now that the property of (a) does not hold for x^*. There exists some $\varepsilon > 0$ and a sequence $x_n \to x$ such that

$$f(x_n) < f(x) + \langle x^*, x_n - x \rangle - \varepsilon \|x_n - x\| \quad \text{for all } n \in \mathbb{N},$$

hence in particular $\|x_n - x\| > 0$. For $r_n := f(x) + \langle x^*, x_n - x \rangle - \varepsilon \|x_n - x\|$, we then have $(x_n, r_n) \in \text{epi } f$ and obviously $r_n \to f(x)$. Observing for all $n \in \mathbb{N}$ that (with the sum norm on $X \times \mathbb{R}$)

$$\frac{\langle x^*, x_n - x \rangle - (r_n - f(x))}{\|(x_n, r_n) - (x, f(x))\|} = \frac{\varepsilon \|x_n - x\|}{\|(x_n - x, \langle x^*, x_n - x \rangle - \varepsilon \|x_n - x\|)\|}$$

$$\geq \frac{\varepsilon}{1 + \|x^*\| + \varepsilon},$$

we deduce that $(x^*, -1) \notin N^F(\text{epi } f; (x, f(x)))$. We have thus proved the equivalence between (a) and (b).

Since (c) readily implies (a), let us suppose (a) and prove (c). Let x^* satisfying the property (a). This property gives a neighborhood U of x over which f is bounded from below. Define $\varphi : U \to \mathbb{R}$ with $\varphi(x') := \min\{f(x'), f(x) + \langle x^*, x' - x \rangle\}$ for all $x' \in U$. Of course, $\varphi(x) = f(x)$ and $\varphi(x') \leq f(x')$ for all $x' \in U$. Further, for each $\varepsilon > 0$ there exists by (a) some neighborhood $V \subset U$ of x such that for all $x' \in V$ we have $f(x') - f(x) - \langle x^*, x' - x \rangle \geq -\varepsilon \|x' - x\|$, hence

$$0 \geq \varphi(x') - \varphi(x) - \langle x^*, x' - x \rangle \geq -\varepsilon \|x' - x\|.$$

This entails the Fréchet differentiability of φ at x with $D\varphi(x) = x^*$, justifying that (a) and (c) are equivalent. □

EXAMPLE 4.5. With Proposition 4.4(c) at hand, the equalities below are directly seen for the functions in Example 2.36.
(a) $\partial_F(|\cdot|)(0) = [-1, 1]$ and $\partial_F(-|\cdot|)(0) = \emptyset$.
(b) $\partial(\sqrt{|\cdot|})(0) = \mathbb{R}$ and $\partial(-\sqrt{|\cdot|})(0) = \emptyset$.
(c) For the functions $f, g : \mathbb{R} \to \mathbb{R} \cup \{+\infty\}$ defined by

$$f(r) := \begin{cases} \sqrt{r} & \text{if } r \geq 0 \\ +\infty & \text{if } r < 0 \end{cases} \qquad g(r) := \begin{cases} -\sqrt{r} & \text{if } r \geq 0 \\ +\infty & \text{if } r < 0, \end{cases}$$

we have $\partial_F f(0) = \mathbb{R}$ and $\partial_F g(0) = \emptyset$.
(d) Considering the function $f : \mathbb{R} \to \mathbb{R}$ defined by

$$f(r) := \begin{cases} \sqrt{r} & \text{if } r \geq 0 \\ 0 & \text{if } r < 0, \end{cases}$$

we have $\partial_F f(0) = [0, +\infty[$ and $\partial_F(-f)(0) = \emptyset$. □

In the next example we focus on the F-subdifferential of various types of positively homogeneous functions.

EXAMPLE 4.6. (a) If $\varphi : X \to \mathbb{R} \cup \{-\infty, +\infty\}$ is a positively homogeneous function on a normed space X with $\varphi(0) = 0$, then

$$\partial_F \varphi(0) = \{x^* \in X^* : \langle x^*, x \rangle \leq \varphi(x), \forall x \in X\}.$$

Indeed, let any $x^* \in \partial_F \varphi(0)$ and any $x \in X$. Take any $\varepsilon > 0$ and choose $\delta > 0$ such that $\langle x^*, u \rangle \leq \varphi(u) + \varepsilon \|u\|$ for all $u \in B[0, \delta]$. Choosing $\delta_0 \in]0, \delta]$ such that $\|tx\| \leq \delta$ for all $t \in [0, \delta_0]$, we see that for every $t \in]0, \delta_0]$ one has $t\langle x^*, x \rangle \leq t\varphi(x) + t\varepsilon \|x\|$, or equivalently $\langle x^*, x \rangle \leq \varphi(x) + \varepsilon \|x\|$. This being true for every $\varepsilon > 0$, it results that $x^* \in \partial_F \varphi(0)$ implies that $\langle x^*, x \rangle \leq \varphi(x)$ for all $x \in X$, so the desired equality is justified since the converse implication is trivial.
(b) The vacuity of $\partial_F(-|\cdot|)(0)$ in Example 4.5(a) above is in fact a particular case of a general property. Let a normed space X and a function $\varphi : X \to \mathbb{R} \cup \{-\infty, +\infty\}$ with $\varphi(0) = 0$ and such that there is $v \in X$ with $\varphi(v) < 0$ and such that $\varphi(tv) = |t|\varphi(v)$. Then

$$\partial_F \varphi(0) = \emptyset.$$

Indeed, suppose that there exists $x^* \in \partial_F \varphi(0)$. Then for any $t \in \mathbb{R}$ we would have by (a) that $t\langle x^*, v \rangle \leq \varphi(tv) = |t|\varphi(v)$. If $\langle x^*v \rangle$ is either positive or negative, we would get a contradiction by making either $t \to +\infty$ or $t \to -\infty$. If $\langle x^*, v \rangle = 0$, the inequality $0 \leq \varphi(v)$ (with $t = 1$) would again give a contradiction.
(c) Consider the function $f : \mathbb{R}^2 \to \mathbb{R}$ in Example 2.191 defined by $f(x, y) = |\alpha|x| + \beta y|$ with $\alpha, \beta \in \mathbb{R}$. If $\alpha = 0$, we see that $f(x, y) = |\beta| |y|$, so $\partial_F f(0, 0) = \{0\} \times [-|\beta|, |\beta|]$; and similarly if $\beta = 0$, we see that $\partial_F f(0, 0) = [-|\alpha|, |\alpha|] \times \{0\}$. Further, the case when $\alpha < 0$ is reduced to the one when $\alpha > 0$ since $|\alpha|x| + \beta y| = |(-\alpha)|x| - \beta y|$. We then focus on the case when $\alpha > 0$ and $\beta \neq 0$.

Let any $(a, b) \in \partial_F f(0, 0)$. By (a) above we know that $ax + by \leq |\alpha|x| + \beta y|$ for all $(x, y) \in \mathbb{R}^2$. With $y = -(\alpha/\beta)|x|$ we note that $ax - (\alpha/\beta)b|x| \leq 0$ for all $x \in \mathbb{R}$. Taking $x > 0$ (resp. $x < 0$) in the latter inequality yields $a \leq \alpha(b/\beta)$ (resp. $a \geq -\alpha(b/\beta)$), hence $|a| \leq \alpha(b/\beta)$ and $b/\beta \geq 0$. Further, making $x = 0$ in the first inequality gives $by \leq |\beta| |y|$ for all $y \in \mathbb{R}$, thus $|b/\beta| \leq 1$. It ensues that

$$\partial_F f(0, 0) \subset \left\{ (a, b) \in \mathbb{R}^2 : |a| \leq \frac{\alpha b}{\beta}, \frac{b}{\beta} \leq 1 \right\}.$$

Now fix any (a, b) in the right-hand side of the latter inclusion. Fix any $(x, y) \in \mathbb{R}^2$. If $\alpha|x| + \beta y \geq 0$, we have

$$ax + by \leq |a|\,|x| + by \leq \frac{b}{\beta}\alpha|x| + by = \frac{b}{\beta}(\alpha|x| + \beta y),$$

hence

$$ax + by \leq \alpha|x| + \beta y = |\alpha|x| + \beta y|.$$

If $\alpha|x| + \beta y < 0$, this inequality and the inequality $b/\beta \geq 0$ (due to the fact that $|a| \leq \alpha(b/\beta)$ with $\alpha > 0$) ensure that

$$ax + by = ax + \frac{b}{\beta}\beta y \leq ax + \frac{b}{\beta}(-\alpha|x|) \leq |a|\,|x| - \frac{\alpha b}{\beta}|x|,$$

thus

$$ax + by \leq \left(|a| - \frac{\alpha b}{\beta}\right)|x| \leq 0 \leq |\alpha|x| + \beta y|.$$

So, we obtain $ax + by \leq f(x, y)$ for all $(x, y) \in \mathbb{R}^2$, which means by (a) above again that $(a, b) \in \partial_F f(0, 0)$. We have then established that for $\alpha > 0$ and $\beta \neq 0$

$$\partial_F f(0,0) = \left\{(a, b) \in \mathbb{R}^2 : |a| \leq \frac{\alpha b}{\beta}, \ \frac{b}{\beta} \leq 1\right\}.$$

(d) With f as in (c) and $(\alpha, \beta) \neq (0, 0)$, the function $-f$ satisfies the assumptions in (b) with $v = (1, 0)$ if $\alpha \neq 0$ (resp. $v = (0, 1)$ if $\beta \neq 0$), hence it follows that $\partial_F(-f)(0, 0) = \emptyset$. □

Now we start with some properties of Fréchet normals. The first properties are given in the next proposition. They follow directly from (4.2) and Proposition 4.3).

PROPOSITION 4.7. *For any subset S of the normed space X and $x \in S$, one has*

(4.4) $$N^F(S; x) = N^F(\operatorname{cl} S; x),$$

and for every neighborhood U of x

$$N^F(S; x) = N^F(S \cap U; x).$$

If S' is a subset of S with $x \in S'$, then $N^F(S; x) \subset N^F(S'; x)$; in particular

$$N^F(S; x) \subset N^F(\operatorname{bdry} S; x) \quad \text{for all } x \in \operatorname{bdry} S.$$

If the set S is convex, it is easy to see that the inequality (4.2) holds for all $x' \in S$, and hence $N^F(S; x)$ and $\partial_F f(x)$ coincide respectively with the standard normal cone and subdifferential for the convex set S and a convex function f in Definition 2.64.

Fix now any $x^* \in N^F(S; x)$ and $h \in T^B(S; x)$. Consider any real $\varepsilon > 0$ and take by the definition above some neighborhood V of x such that (4.2) holds. By the inclusion $h \in T^B(S; x)$ there exist sequences $t_n \downarrow 0$ and $h_n \to h$ such that $x + t_n h_n \in S$ hence $x + t_n h_n \in S \cap V$ for n large enough, which ensures by (4.2) that $\langle x^*, h_n \rangle \leq \varepsilon \|h_n\|$. We deduce that $\langle x^*, h \rangle \leq 0$. Those features and others are stated in the next proposition. The assertion (b) is very useful. It establishes the $\|\cdot\|_*$-closedness of the Fréchet subdifferential $\partial_F f(x)$.

PROPOSITION 4.8. Let S be a subset of a normed space X with $x \in S$, U be an open set of X with $x \in U$, and $f : U \to \mathbb{R} \cup \{-\infty, +\infty\}$ be a function which is finite at x.
(a) The following inclusions always hold
$$N^F(S;x) \subset \left(T^B(S;x)\right)^\circ \subset N^C(S;x)$$
and
$$\partial_F f(x) \subset \{x^* \in X^* : \langle x^*, h \rangle \leq f^B(x;h) \; \forall h \in X\} \subset \partial_C f(x).$$
(b) The sets $N^F(S;x)$ and $\partial_F f(x)$ are convex and strongly closed in X^*.
(c) If S is convex (resp. U is a convex set and f is a convex function on U), then $N^F(S;x)$ (resp. $\partial_F f(x)$) coincides with the standard normal cone (resp. subdifferential) for convex sets (resp. convex functions) (see Definition 2.64), that is,
$$N^F(S;x) = \{x^* \in X^* : \langle x^*, x' - x \rangle \leq 0 \; \forall x' \in S\} =: N(S;x)$$
and
$$\partial_F f(x) = \{x^* \in X^* : \langle x^*, x' - x \rangle \leq f(x') - f(x) \; \forall x' \in U\} =: \partial f(x).$$
(d) If X is finite-dimensional, then
$$N^F(S;x) = \left(T^B(S;x)\right)^\circ \text{ and } \partial_F f(x) = \{x^* \in X^* : \langle x^*, h \rangle \leq f^B(x;h) \; \forall h \in X\}.$$

PROOF. The assertion (c) has been established above as well as the first inclusion in (a), that is, $N^F(S;x) \subset \left(T^B(S;x)\right)^\circ$. The second inclusion $\left(T^B(S;x)\right)^\circ \subset N^C(S;x)$ is known (see Proposition 2.113). The subdifferential inclusions in (a) then follow according to Proposition 4.4(b) and to (2.39).

Concerning (b) we first note that the convexity property follows readily from the definition. Consider now any sequence $(x_n^*)_n$ of $N^F(S;x)$ converging strongly to some $x^* \in X^*$ and fix any real $\varepsilon > 0$. Choose an integer k such that $\|x_k^* - x^*\| \leq \varepsilon/2$. The inclusion $x_k^* \in N^F(S;x)$ yields some positive real $\delta < \varepsilon/2$ such that for all $x' \in B(x,\delta) \cap S$
$$\langle x_k^*, x' - x \rangle \leq (\varepsilon/2)\|x' - x\|,$$
thus
$$\langle x^*, x' - x \rangle \leq \|x_k^* - x^*\|_* \|x' - x\| + \langle x_k^*, x' - x \rangle \leq \varepsilon \|x' - x\|.$$
This means $x^* \in N^F(S;x)$, hence $N^F(S;x)$ is strongly closed in X^*, which obviously entails that $\partial_F f(x)$ is strongly closed as well according to Proposition 4.4(b).

Assuming that X is finite-dimensional, to obtain (d) it remains to show the inclusion $\left(T^B(S;x)\right)^\circ \subset N^F(S;x)$. Take $x^* \notin N^F(S;x)$. By definition, there exist some real $\varepsilon > 0$ and some sequence $(x_n)_n$ in S converging to x such that $\langle x^*, x_n - x \rangle > \varepsilon \|x_n - x\|$, for all $n \in \mathbb{N}$. For $h_n := (x_n - x)/\|x_n - x\|$, a subsequence of $(h_n)_n$ converges to some $h \in X$ by the compactness of $\mathbb{B}_X$. Then, on the one hand $h \in T^B(S;x)$ since $x + t_n h_n = x_n \in S$ for $t_n := \|x_n - x\|$, and on the other hand $\langle x^*, h \rangle \geq \varepsilon$. Consequently, $x^* \notin \left(T^B(S;x)\right)^\circ$ as desired. □

REMARK 4.9. Denoting by $L^\perp$ the orthogonal in X^* of a vector subspace L of X, the equality $\partial_F \Psi_{x+L}(x) = L^\perp$ is easily seen to be a direct consequence of the second equality in the above assertion (c). It will be used in several places later in this section. □

COROLLARY 4.10. Let S be a subset of a normed space X with $x \in S$. The regularity property $N^C(S;x) = N^F(S;x)$ of the Clarke normal cone at x entails the tangential regularity $T^C(S;x) = T^B(S;x)$ of S at x.

4.1. FRÉCHET NORMAL AND SUBGRADIENT

PROOF. The assumption $N^C(S;x) = N^F(S;x)$ combined with Proposition 4.8(a) ensures that $\left(T^B(S;x)\right)^\circ = N^C(S;x)$. It results that

$$T^B(S;x) \subset \left(N^C(S;x)\right)^\circ = T^C(S;x),$$

so the equality $T^B(S;x) = T^C(S;x)$ holds, since the other inclusion $T^C(S;x) \subset T^B(S;x)$ is always true. $\square$

Figure 4.1 shows (with the same set S in Figure 2.1) the Fréchet normal cone of S at x_i, for $i \in \{1, 7, 11\}$, with the representation of $x_i + N^F(S;x_i)$. At the point x_i, for the other indices i, we have

$$N^F(S;x_2) =]-\infty, 0] \times \{0\}, \quad N^F(S;x_3) = N^F(S;x_4) = \{(0,0)\},$$

$$N^F(S;x_5) = N^F(S;x_6) = \{(v_1, v_2) \in \mathbb{R}^2 : v_1 \geq 0 \text{ and } v_2 = -v_1\},$$

$$N^F(S;x_8) = \{(v_1, v_2) \in \mathbb{R}^2 : v_1 \leq 0, v_2 \leq v_1\},$$

$$N^F(S;x_9) = N^F(S;x_{10}) = \{(0,0)\}, \quad N^F(S;x_{12}) = N^F(S;x_{13}) = \mathbb{R} \times \{0\}.$$

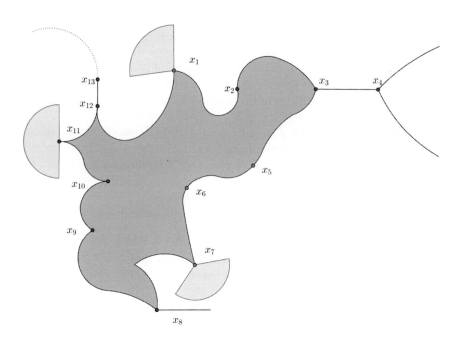

FIGURE 4.1. Fréchet normal cones.

It is worth pointing out that the expressions of Fréchet subdifferentials of the functions in Example 4.5 above can also been obtained with normal cones in Figure 4.1 through the equivalence $x^* \in \partial_F f(x) \Leftrightarrow (x^*, -1) \in N^F\bigl(\text{epi } f; (x, f(x))\bigr)$.

In addition to properties of the F-subdifferential in Propositions 4.2 and 4.3, we continue the study of F-subdifferential. The next proposition considers the Fermat rule with F-subdifferential and some other important properties.

PROPOSITION 4.11. Let X be a normed space, $f, g : X \to \mathbb{R} \cup \{-\infty, +\infty\}$ be two functions, and let $x \in X$ be a point where both functions are finite. Let Z be another normed space and $G : Z \to X$ be a mapping which is Fréchet differentiable at a point z with $G(z) = x$.
(a) If f and g coincide near x, then $\partial_F f(x) = \partial_F g(x)$.
(b) If x is a local minimum point for f, then $0 \in \partial_F f(x)$.
(c) If f is locally Lipschitz continuous near x with $\gamma \geq 0$ as a Lipschitz constant, then $\partial_F f(x) \subset \gamma \mathbb{B}_{X^*}$.
(d) For any real $r > 0$ the equality $\partial_F(rf)(x) = r \partial_F f(x)$ is fulfilled.
(e) The inclusions

$$\partial_F f(x) + \partial_F g(x) \subset \partial_F(f+g)(x) \text{ and } DG(z)^*\big(\partial_F g(G(z))\big) \subset \partial_F(g \circ G)(z)$$

always hold.
(f) If f is finite near x and Fréchet differentiable at x, then

$$\partial_F f(x) = \{Df(x)\} \quad \text{and} \quad \partial_F(f+g)(x) = Df(x) + \partial_F g(x).$$

(g) If f is finite near x and Gâteaux differentiable at x, then

$$\partial_F f(x) \subset \{D_G f(x)\}.$$

PROOF. Assertions (a)-(d) as well as the first inclusion in (e) follow readily from the definition of Fréchet subgradient. Concerning the second inclusion in (e), fixing any $x^* \in \partial_F g(x)$ there exists by Proposition 4.4(c) a neighborhood V of $x = G(z)$ and a function $\varphi : V \to \mathbb{R}$ differentiable at x with $x^* = D\varphi(x)$, $\varphi(x) = g(x)$ and $\varphi(x') \leq g(x')$ for all $x' \in V$. Choosing, by continuity of G at z, a neighborhood U of z such that $G(U) \subset V$ and putting $\varphi_0(z') := \varphi \circ G(z')$ for all $z' \in U$ it is readily seen that φ_0 fulfills the conditions in Proposition 4.4(c) for the function $g \circ G$ with $D\varphi_0(z) = x^* \circ DG(z)$, so $x^* \circ DG(z) \in \partial(g \circ G)(z)$ as desired.

To justify (g), assume that f is finite near x and Gâteaux differentiable at x. Consider an arbitrary $x^* \in \partial_F f(x)$ (if any) and fix any $h \in X$. Take any $\varepsilon > 0$. There exists a neighborhood V of 0 in X such that f is finite on $x + V$ and for all $v \in V$

$$\langle x^*, v \rangle \leq f(x+v) - f(x) + \varepsilon \|v\|.$$

There exists also a real $\delta > 0$ such that $]0, \delta[h \subset V$ and for all $t \in]0, \delta[$

$$|f(x+th) - f(x) - t\langle D_G f(x), h \rangle| \leq t\varepsilon \|h\|.$$

It results that $\langle x^* - D_G f(x), h \rangle \leq 2\varepsilon \|h\|$, which entails that $\langle x^* - D_G f(x), h \rangle \leq 0$. This being true for all $h \in X$, it ensues that $x^* = D_G f(x)$. We conclude that $\partial_F f(x) \subset \{D_G f(x)\}$.

Let us turn now to (f). Suppose that f is Fréchet differentiable at x. Obviously, the definition of Fréchet subgradient entails that $Df(x) \in \partial_F f(x)$. This and the above inclusion $\partial_F f(x) \subset \{D_G f(x)\}$ prove the first part of (f).

Finally, by the latter equality and (e) we have $-Df(x) + \partial_F(f+g)(x) \subset \partial_F g(x)$, which combined with (e) again ensures the desired equality

$$\partial_F(f+g)(x) = Df(x) + \partial_F g(x).$$

$\square$

REMARK 4.12. The equality in (d) of Proposition 4.11 above may fail for $r = 0$. Indeed, if we take $r = 0$ and $f = -|\cdot|$ on $\mathbb{R}$, we have that $r\partial f(0) = \emptyset$ while $\partial(rf)(0) = \{0\}$. $\square$

Let us characterize the Fréchet differentiability of f in terms of subdifferentiability.

PROPOSITION 4.13. *Let X be a normed space, $f : X \to \mathbb{R} \cup \{-\infty, +\infty\}$ be an extended real-valued function, and $x \in X$ be a point where f is finite. The following hold:*
(a) *The function f is Fréchet differentiable at x if and only if both functions f and $-f$ are Fréchet subdifferentiable at x. In such a case*
$$\{Df(x)\} = \partial_F f(x) = -\partial_F(-f)(x).$$
(b) *If $\partial_F f(x) \neq \emptyset$, then $-\partial_F(-f)(x) \subset \partial_F f(x)$.*
(c) *If f is convex and continuous at x, then $\partial_F(-f)(x) \neq \emptyset$ if and only if f is Fréchet differentiable at x; in this case $\partial_F(-f)(x) = \{-D_F f(x)\}$.*

PROOF. (a) Suppose that $\partial_F f(x) \neq \emptyset$ and $\partial_F(-f)(x) \neq \emptyset$. Take $u^* \in \partial_F f(x)$ and $v^* \in \partial_F(-f)(x)$. Fix any $\varepsilon > 0$ and choose a neighborhood U of x such that for all $x' \in U$
$$\langle u^*, x' - x \rangle \leq f(x') - f(x) + \varepsilon \|x' - x\| \quad \text{and} \quad \langle v^*, x' - x \rangle \leq -f(x') + f(x) + \varepsilon \|x' - x\|.$$
This ensures that f is finite on U and
$$\langle u^* + v^*, x' - x \rangle \leq 2\varepsilon \|x' - x\| \quad \text{for all } x' \in U,$$
hence $\|u^* + v^*\| \leq 2\varepsilon$. So, $v^* = -u^*$ and this equality gives
$$|f(x') - f(x) - \langle u^*, x' - x \rangle| \leq \varepsilon \|x' - x\| \quad \text{for all } x' \in U,$$
which guarantees the Fréchet differentiability of f at x.

The reverse implication is obvious and the equality of the proposition follows from the arguments above.

(b) If $\partial_F(-f)(x)$ is empty, the inclusion of (b) is trivial, and otherwise the inclusion follows immediately from (a).

(c) Finally, under the assumptions in (c) we have $\partial_F f(x) = \partial f(x) \neq \emptyset$ (see Theorem 3.30), so (c) is a direct consequence of (a). □

The Gâteaux and Fréchet differentiabilities of Moreau envelopes of convex functions have been obtained in Theorems 3.267 and 3.268, and of course the differentiability of the square distance function to convex sets is a particular case. Here we show how the above proposition can also be applied, for example, to derive the Fréchet differentiability of the square distance function d_S^2 of a closed convex set S. Before proceeding to the application, we need to fix notation. For any subset S of a normed space $(X, \|\cdot\|)$ and $u \in X$, we recall (see (2.69)) that $\text{Proj}_S(u)$ denotes the (possibly empty) set of all *nearest (or closest) points of u in S*, that is,
$$\text{Proj}_S(u) := \{y \in S : \|u - y\| = d(u, S)\}.$$
When the set $\text{Proj}_S(u)$ is a singleton, $P_S(u)$ will denote its unique point.

COROLLARY 4.14. *Let $(X, \|\cdot\|)$ be a reflexive Banach space whose norm $\|\cdot\|$ is Fréchet differentiable off zero, and let S be a nonempty closed convex set of X. Then the function d_S^2 is Fréchet differentiable on X and, for every $x \in X$,*
$$D(d_S^2)(x) = D(\|\cdot - z\|^2)(x) \quad \text{for all } z \in \text{Proj}_S(x),$$
where the distance is taken through the given differentiable norm $\|\cdot\|$.

In the particular case where $(X, \|\cdot\|)$ is a Hilbert space, then
$$\nabla(d_S^2)(x) = 2(x - P_S(x)) \quad \text{for all } x \in X.$$

PROOF. Fix any $x \in X$. Since d_S^2 is a continuous convex function, to obtain its Fréchet differentiability at x, it suffices by Proposition 4.13(c) to show that $\partial_F(-d_S^2)(x) \neq \emptyset$. Choose a real $r > d_S(x)$ and put $S_r := S \cap B[x, r]$. By the weak compactness of the bounded closed convex set S_r of X and the weak lower semicontinuity of $\|\cdot\|$, there exists some point $z \in S_r$ such that $\|x-z\| = \inf_{y \in S_r} \|x - y\|$, and it is immediate to see that $z \in \operatorname{Proj}_S(x)$. So, putting $\varphi(u) := \|u - z\|^2 - d_S^2(u)$ for all $u \in X$, the point x is a global minimizer of φ, and this entails according to (b) and (f) in Proposition 4.11 and to the Fréchet differentiability of $\|\cdot - z\|^2$ on X that $0 \in \partial_F\varphi(x) = D(\|\cdot - z\|^2)(x) + \partial_F(-d_S^2)(x)$, thus in particular $\partial_F(-d_S^2)(x) \neq \emptyset$. This finishes the proof, since in the Hilbert setting $\nabla(\|\cdot - z\|^2)(x) = 2(x - z)$ and $z = P_S(x)$. $\square$

Before considering F-subdifferential chain rules, let us recall the concept of one-sided Lipschitz mappings.

DEFINITION 4.15. For two metric spaces X, Y and an open set U in X, a mapping $G : U \to Y$ is *one-sided γ-Lipschitz* at $\overline{x} \in U$ if there is a neighborhood $U_0 \subset U$ such that

(4.5) $$d\big(G(x), G(\overline{x})\big) \leq \gamma d(x, \overline{x}) \quad \text{for all } x \in U_0.$$

One also says that G is one-sided γ-Lipschitz at $\overline{x}$ relative to such a set U_0.

Clearly, the mapping G is one-sided Lipschitz at the point $\overline{x}$ whenever it is Fréchet differentiable at that point.

PROPOSITION 4.16. Let U be an open set of a normed space X and let $G : U \to Y$ be a mapping from U into a normed space Y which is one-sided Lipschitz at a point $\overline{x} \in U$. Let $g : W \to \mathbb{R} \cup \{-\infty, +\infty\}$ be an extended real-valued function on an open set $W \supset G(U)$ and let $\overline{y} := G(\overline{x})$.
(a) One has the inclusion
$$\bigcup_{y^* \in \partial_F g(\overline{y})} \partial_F(y^* \circ G)(\overline{x}) \subset \partial_F(g \circ G)(\overline{x}).$$

(b) If g is F-differentiable at $\overline{y}$, then
$$\partial_F(g \circ G)(\overline{x}) = \partial_F(g'(\overline{y}) \circ G)(\overline{x}).$$

PROOF. We may assume that W is the whole space Y. Denote by $\gamma > 0$ a one-sided Lipschitz constant of G at $\overline{x}$.
(a) We may suppose that the left hand-side in (a) is nonempty. Fix any $y^* \in \partial_F g(\overline{y})$ such that $\partial_F(y^* \circ G)(\overline{x}) \neq \emptyset$ and take any x^* in this set. We note that $g(\overline{y})$ is finite. Fix any $\varepsilon > 0$. Choose a neighborhood V of $\overline{y}$ such that for $\varepsilon' := \varepsilon/2$
$$\langle y^*, y - \overline{y} \rangle \leq g(y) - g(\overline{y}) + (\varepsilon'/\gamma)\|y - \overline{y}\| \quad \text{for all } y \in V,$$
and choose a neighborhood $U_0 \subset U$ of $\overline{x}$ relative to which G is one-sided γ-Lipschitz along with $G(U_0) \subset V$. Shrinking U_0 if necessary, we may also suppose that for every $x \in U_0$
$$\langle x^*, x - \overline{x} \rangle \leq y^* \circ G(x) - y^* \circ G(\overline{x}) + \varepsilon'\|x - \overline{x}\|,$$

which entails that
$$\langle x^*, x - \bar{x}\rangle \leq g(G(x)) - g(G(\bar{x})) + \varepsilon\|x - \bar{x}\|.$$
This justifies (a).

(b) Assume now g is F-differentiable at $\bar{y}$. Put $\bar{y}^* := g'(\bar{y})$ and fix any $\varepsilon > 0$. Choose a neighborhood V of $\bar{y}$ such that
$$|g(y) - g(\bar{y}) - \langle \bar{y}^*, y - \bar{y}\rangle| \leq (\varepsilon/\gamma)\|y - \bar{y}\| \quad \text{for all } y \in V.$$
Choose a neighborhood $U_0 \subset U$ relative to which G is one-sided γ-Lipschitz at $\bar{x}$, such that $G(U_0) \subset V$. Setting $\rho := g \circ G - \bar{y}^* \circ G$, we see that $|\rho(x) - \rho(\bar{x})| \leq \varepsilon\|x - \bar{x}\|$ for all $x \in U_0$, hence ρ is F-differentiable at $\bar{x}$ with $D\rho(\bar{x}) = 0$. Using this and the equality $g \circ G = \bar{y}^* \circ G + \rho$, it results by Proposition 4.11(f) that $\partial_F(g \circ G)(\bar{x}) = \partial_F(\bar{y}^* \circ G)(\bar{x})$. □

COROLLARY 4.17. *Let U be an open set of a normed space X and let $f : U \to \mathbb{R} \cup \{-\infty, +\infty\}$ be a function which is finite at $\bar{x} \in U$ and one-sided Lipschitz at $\bar{x}$. Let $\varphi : I \to \mathbb{R} \cup \{-\infty, +\infty\}$ be an extended real-valued function on an interval $I \supset f(U)$ which is differentiable at $\bar{t} := f(\bar{x})$. Then for the function $\varphi \circ f$ well-defined near $\bar{x}$ the equality*
$$\partial_F(\varphi \circ f)(\bar{x}) = \partial_F(rf)(\bar{x})$$
holds, where r denotes the derivate $r := \varphi'(\bar{t})$.

If in addition $\varphi'(\bar{t}) > 0$, one has
$$\partial_F(\varphi \circ f)(\bar{x}) = \varphi'(\bar{t})\partial_F f(\bar{x}).$$

PROOF. First, we notice for the derivative $D\varphi(\bar{t})$ that $D\varphi(\bar{t}) \circ f(x) = \varphi'(\bar{t})f(x)$ for x near $\bar{x}$. Further, if the real $\varphi'(\bar{t}) > 0$, we also have $\partial_F(\varphi'(\bar{t})f)(\bar{x}) = \varphi'(\bar{t})\partial_F f(\bar{x})$. The corollary then follows from Proposition 4.16(b). □

The Fréchet subdifferential of a difference of two functions is related to the Minkowski difference (see 2.2.11) of the Fréchet subdifferentials of the functions as follows.

PROPOSITION 4.18. *Let U be an open set of a normed space X and $f, g : U \to \mathbb{R} \cup \{-\infty, +\infty\}$ be two functions such that $f - g$ is well-defined on U and finite at a point $x \in U$. Then one has*
$$\partial_F(f - g)(x) \subset \partial_F f(x) \stackrel{*}{-} \partial_F g(x) := \bigcap_{y^* \in \partial_F g(x)} (\partial_F f(x) - y^*).$$

PROOF. We may suppose that both $\partial_F g(x)$ and the left-hand side are nonempty. Let any $x^* \in \partial_F(f - g)(x)$ and any $y^* \in \partial_F g(x)$. Take any $\varepsilon > 0$ and choose a neighborhood $U_0 \subset U$ of x such that for all $u \in U_0$ one has $\langle y^*, u - x\rangle \leq g(u) - g(x) + (\varepsilon/2)\|u - x\|$ and
$$\langle x^*, u - x\rangle \leq f(u) - g(u) - f(x) + g(x) + (\varepsilon/2)\|u - x\|.$$
Both inequalities ensure that there is a neighborhood $U_0' \subset U_0$ on which g and $-g$ are greater than $-\infty$, so g is finite on U_0'. Then, for each $u \in U_0'$ adding the inequalities give $\langle x^* + y^*, u - x\rangle \leq f(u) - f(x) + \varepsilon\|u - x\|$, which entails that $x^* + y^* \in \partial_F f(x)$. □

Concerning the case of the (Cartesian) product of finitely many spaces, we have the following:

PROPOSITION 4.19. Let $X, Y, X_1, \cdots, X_m$ be normed vector spaces, $g : Y \to \mathbb{R} \cup \{-\infty, +\infty\}$ be a function on Y, and, for each $i = 1, \cdots, m$, let $f_i : X_i \to \mathbb{R} \cup \{-\infty, +\infty\}$ be a function on X_i. Let also $G : X \to Y$ be a mapping. Then the following properties hold.
(a) For $\varphi(u_1, \cdots, u_m) := f_1(u_1) + \cdots + f_m(u_m)$ (assumed to be) well-defined on an open set U of $X_1 \times \cdots \times X_m$ and for any point $(x_1, \cdots, x_m) \in U$, one has

$$\partial_F \varphi(x_1, \cdots, x_m) = \partial_F f_1(x_1) \times \cdots \times \partial_F f_m(x_m).$$

(b) For $h : X \times Y \to \mathbb{R} \cup \{-\infty, +\infty\}$ with $h(x', y') = g(y')$ for all $(x', y') \in X \times Y$, one has for any $x \in X$ and $y \in Y$,

$$\partial_F h(x, y) = \{0\} \times \partial_F g(y).$$

(c) Assume that G is Fréchet differentiable at $x \in X$. Then one has $(x^*, -y^*) \in N^F(\mathrm{gph}\, G; (x, G(x)))$ if and only if

$$x^* = y^* \circ DG(x) =: DG(x)^*(y^*).$$

PROOF. The assertion (a) follows easily from the definition of Fréchet subgradient and the assertion (b) is a particular case of (a).

To verify (c) fix $(x^*, -y^*) \in N^F(\mathrm{gph}\, G; (x, G(x)))$ and take any real $\varepsilon > 0$. Choose neighborhoods U and V of x and $G(x)$ such that for all $x' \in U$ and $y' \in V$

$$\langle x^*, x' - x \rangle - \langle y^*, y' - G(x) \rangle \leq \varepsilon(\|x' - x\| + \|y' - G(x)\|).$$

According to the Fréchet differentiability of G at x, shrinking U if necessary we may suppose that $G(U) \subset V$ and that

$$\|G(x') - G(x) - DG(x)(x' - x)\| \leq \varepsilon \|x' - x\| \quad \text{for all } x' \in U.$$

Then, for any $x' \in U$, taking $y' := G(x')$ we obtain

$$\langle x^* - y^* \circ DG(x), x' - x \rangle \leq \varepsilon(1 + \|DG(x)\| + \varepsilon + \|y^*\|)\|x' - x\|,$$

thus $\|x^* - y^* \circ DG(x)\| \leq \varepsilon(1 + \|DG(x)\| + \varepsilon + \|y^*\|)$. Since the latter inequality holds true for any real $\varepsilon > 0$, we deduce $x^* = y^* \circ DG(x) = DG(x)^*(y^*)$, establishing the implication $\Rightarrow$ of (c). The reverse implication is similar and easier. $\square$

The following easy inclusions provide estimates for F-subdifferentials of infimum and supremum of many functions.

PROPOSITION 4.20. Let U be an open set of a normed space X and $(f_t)_{t \in T}$ be a family of functions from U into $\mathbb{R} \cup \{-\infty, +\infty\}$. Let $\varphi(u) := \inf_{t \in T} f_t(u)$ and $\Phi(u) := \sup_{t \in T} f_t(u)$ for all $u \in U$ and let $x \in U$ where $\varphi(x)$ (resp. $\Phi(x)$) is finite. Let $T_\varphi(x) := \{t \in T : f_t(x) = \varphi(x)\}$ (resp. $T_\Phi(x) := \{t \in T : f_t(x) = \Phi(x)\}$). One has

$$\partial_F \varphi(x) \subset \bigcap_{t \in T_\varphi(x)} \partial_F f_t(x) \quad \text{and} \quad \bigcup_{t \in T_\Phi(x)} \partial_F f_t(x) \subset \partial_F \Phi(x).$$

PROOF. Clearly, the first (resp. second) inclusion holds if $T_\varphi(x)$ (resp. $T_\Phi(x)$) is empty. Otherwise, we note that $f_t(x)$ is finite at any $t \in T_\varphi(x)$ (resp. any $t \in T_\Phi(x)$). Then the result follows from the inequality $\varphi(\cdot) \leq f_t(\cdot)$ and the equality $\varphi(x) = f_t(x)$ for all $t \in T_\varphi(x)$ (resp. from the inequality $f_t(\cdot) \leq \Phi(\cdot)$ and the equality $f_t(x) = \Phi(x)$ for all $t \in T_\Phi(x)$). $\square$

Fréchet normals to epigraphs have particular properties. The assertion (e) below says in particular that $N^F(\text{epi } f;(x,f(x)))$ (through which we gave in Proposition 4.4 a description of $\partial_F f(x)$ with certain normal vectors) can be recovered from $\partial_F f(x)$ whenever f is Fréchet subdifferentiable at x.

PROPOSITION 4.21. *Let $(X, \|\cdot\|)$ be a normed space, $f : X \to \mathbb{R} \cup \{-\infty, +\infty\}$ be an extended real-valued function, and let $x \in X$ be a point with $|f(x)| < +\infty$. The following hold:*
(a) *For any $(x^*, -r^*) \in N^F(\text{epi } f;(x,r))$ one has $r^* \geq 0$.*
(b) *If $(x^*, -r^*) \in N^F(\text{epi } f;(x,r))$ with $r^* > 0$, then $r = f(x)$.*
(c) *If $(x^*, 0) \in N^F(\text{epi } f;(x,r))$, then one also has $(x^*, 0) \in N^F(\text{epi } f;(x,f(x)))$.*
(d) *If $(x^*, -r^*) \in N^F(\text{epi } f;(x,r))$ with $r > f(x)$, then $r^* = 0$.*
(e) *If f is Fréchet subdifferentiable at x (that is, $\partial_F f(x) \neq \emptyset$), then*

$$N^F(\text{epi } f;(x,f(x))) = \text{cl}_{\|\cdot\|_*}\big(\mathbb{R}_+(\partial_F f(x) \times \{-1\})\big),$$

where $\text{cl}_{\|\cdot\|_*}$ denotes the closure in $X^* \times \mathbb{R}$ with respect to the dual norm.

PROOF. (a) Take any $(x^*, -r^*) \in N^F(\text{epi } f;(x,r))$, hence $r \geq f(x)$. Fix any real $\varepsilon > 0$ and take some $\delta > 0$ such that

$$\langle x^*, x' - x \rangle - r^*(r' - f(x)) \leq \varepsilon(\|x' - x\| + |r' - f(x)|)$$

for all $(x', r') \in \text{epi } f$ with $\|x' - x\| \leq \delta$ and $|r' - r| \leq \delta$. Taking $x' = x$ and $r' = r + \delta$ yields $-r^*(\delta + r - f(x)) \leq \varepsilon(\delta + r - f(x))$, that is, $r^* \geq -\varepsilon$. This being true for any $\varepsilon > 0$, we obtain $r^* \geq 0$.
(b) Observe first that $r \geq f(x)$. Fix a positive real $\varepsilon < r^*$ and choose $\delta > 0$ such that

(4.6) $$\langle x^*, x' - x \rangle - r^*(r' - r) \leq \varepsilon(\|x' - x\| + |r' - r|)$$

for all $(x', r') \in \text{epi } f$ with $\|x' - x\| \leq \delta$ and $|r' - r| \leq \delta$. For the nonnegative real $\lambda = \min\{\delta, r - f(x)\}$ taking $x' = x$ and $r' = r - \lambda$, we have $(x', r') \in \text{epi } f$ with $|r' - r| = \lambda \leq \delta$, hence (4.6) gives $\lambda(r^* - \varepsilon) \leq 0$. The latter inequality entails that $\lambda \leq 0$, thus $r = f(x)$.
(c) As above, observe that $r \geq f(x)$. Consider any $\varepsilon > 0$ and choose $\delta > 0$ such that

$$\langle x^*, x' - x \rangle \leq \varepsilon(\|x' - x\| + |r' - r|)$$

for all $(x', r') \in \text{epi } f$ with $\|x' - x\| \leq \delta$ and $|r' - r| \leq \delta$. Fix any $(x', r') \in \text{epi } f$ with $\|x' - x\| \leq \delta$ and $|r' - f(x)| \leq \delta$. For $r'' := r' + r - f(x)$, we have $(x', r'') \in \text{epi } f$ (because $r - f(x) \geq 0$) and $|r'' - r| = |r' - f(x)| \leq \delta$, and hence we obtain

$$\langle x^*, x' - x \rangle \leq \varepsilon(\|x' - x\| + |r'' - r|) = \varepsilon(\|x' - x\| + |r' - f(x)|).$$

This guarantees the inclusion $(x^*, 0) \in N^F(\text{epi } f;(x,f(x)))$.
(d) The assertion (d) follows directly from (a) and (b).
(e) For the equality in (e), the inclusion of the second member into the first one follows from the geometric description of $\partial_F f(x)$ in Proposition 4.4(b) and the strong closedness property of the cone $N(f;x) := N^F(\text{epi } f;(x,f(x)))$ (see (b) in Proposition 4.8). Assume now $\partial_F f(x) \neq \emptyset$ and fix $u^* \in \partial_F f(x)$. Take any $(x^*, -r^*)$ in $N(f;x)$ and note that $r^* \geq 0$ by (a). For $0 < s < 1$ and $x_s^* = (1-s)x^* + su^*$, we have $y_s^* := (x_s^*, -r^*(1-s) - s)$ in $N(f;x)$ according to the convexity of $N(f;x)$ (see Proposition 4.8(b)), thus $y_s^* \in \mathbb{R}_+(\partial_F f(x) \times \{-1\})$ since $r^*(1-s) + s > 0$. The proof is then finished since $y_s^* \to (x^*, -r^*)$ as $s \downarrow 0$. $\square$

REMARK 4.22. The above assertions (c) and (d) ensure in particular that the inclusion
$$N^F(\text{epi } f; (x,r)) \subset N^F(\text{epi } f; (x, f(x)))$$
always holds true for every $r \in \mathbb{R}$. □

PROPOSITION 4.23. Let U be an open set of the normed space X and $f : U \to \mathbb{R} \cup \{-\infty, +\infty\}$ be a function which is finite near a point $x \in U$ and Lipschitz continuous there with $\gamma \geq 0$ as a Lipschitz constant.
(a) One has
$$(x^*, -r) \in N^C(\text{epi } f; (x, f(x))) \Rightarrow \|x^*\| \leq r\gamma,$$
hence in particular
$$(x^*, 0) \in N^C(\text{epi } f; (x, f(x))) \Rightarrow x^* = 0.$$
(b) If $\partial_F f(x) \neq \emptyset$, the F-normal cone to the epigraph of f at x can be described by the F-subdifferential as follows
$$N^F(\text{epi } f; (x, f(x))) = \mathbb{R}_+(\partial_F f(x) \times \{-1\}).$$

PROOF. (a) Put $N(f; x) = N^F(\text{epi } f; (x, f(x)))$. Fix any $(x^*, -r) \in N^F(f; x)$. Take any $\varepsilon > 0$. There exists a neighborhood W of $(x, f(x))$ such that for any $(u, s) \in W \cap \text{epi } f$
$$\langle x^*, u - x \rangle - r(s - f(x)) \leq \varepsilon(\|u - x\| + |s - f(x)|).$$
Choose a neighborhood U of x such that f is γ-Lipschitz on U and $(u, f(u))$ belongs to W for all $u \in U$. Then for each $u \in U$ we have
$$\langle x^*, u - x \rangle \leq r(f(u) - f(x)) + \varepsilon(\|u - x\| + |f(u) - f(x)|),$$
hence
$$\langle x^*, u - x \rangle \leq (r\gamma + \varepsilon(1+\gamma))\|u - x\|,$$
which entails that $\|x^*\| \leq (r\gamma + \varepsilon(1+\gamma))$. This being true for every $\varepsilon > 0$, it results that $\|x^*\| \leq r\gamma$, which justifies the assertion (a).
(b) Denote $Q := \partial_F f(x) \times \{-1\}$ and note that $(0,0) \in \mathbb{R}_+ Q$ by the non-vacuity of $\partial_F f(x)$. From the geometric description of F-subgradient in Proposition 4.4(b) the set $\mathbb{R}_+ Q$ is clearly included in the first member of (b). Conversely, fix any $(x^*, -r) \in N(f; x)$. If $r > 0$, we can write by definition of F-subgradient
$$(x^*, -r) = r(r^{-1}x^*, -1) \in r(\partial_F f(x) \times \{-1\}) \subset \mathbb{R}_+(\partial_F f(x) \times \{-1\}) = \mathbb{R}_+ Q.$$
If $r = 0$, by (a) we have $x^* = 0$, so $(x^*, -r) = (0, 0) \in \mathbb{R}_+ Q$. So, in any case we have $(x^*, -r) \in \mathbb{R}_+ Q$. It ensues that $N(f; x) \subset \mathbb{R}_+ Q$, hence we conclude that $N(f; x) = \mathbb{R}_+ Q$. □

In addition to the analytical definition and to the analytical characterization of the Fréchet subdifferential in Proposition 4.4, the next proposition provides other useful descriptions.

Let us first prove a lemma which has its own interest.

LEMMA 4.24. Let $\sigma : [0, +\infty[\to [0, +\infty[$ be a function such that
$$\sigma(0) = \sigma'_+(0) = 0 \quad \text{and} \quad \sigma(t) \leq \alpha t + \beta \quad \text{for all } t \geq 0,$$
for two real constants α, β with $\alpha \geq 0$. Then there exists an even $\mathcal{C}^1$ function $\theta : \mathbb{R} \to [0, +\infty[$ nondecreasing on $[0, +\infty[$ such that
$$\theta(0) = \theta'(0) = 0 \quad \text{and} \quad \theta(t) > \sigma(t) \quad \text{for all } t > 0.$$

PROOF. Putting $\rho(0) := 0$ and $\rho(t) := \sigma(t)/t$ for $t > 0$, the function ρ is continuous at 0. For $\rho_0(t) := \sup_{0 \leq s \leq t} \rho(s)$, the function ρ_0 is obviously nondecreasing and it is easily seen from the assumptions that it is finite on $[0, +\infty[$, continuous at 0 with $\rho_0(0) = \rho(0) = 0$. For each $t \geq 0$, $I(t) := \int_t^{2t} \rho_0(s)\,ds$ is well defined and I is locally absolutely continuous on $[0, +\infty[$. Setting $\rho_1(0) := 0$ and $\rho_1(t) := I(t)/t$ for $t > 0$, we see that $\rho_0(t) \leq \rho_1(t) \leq \rho_0(2t)$, hence ρ_1 is continuous at 0. The function ρ_1 is also locally absolutely continuous on $]0, +\infty[$ and for almost every $t \in]0, +\infty[$

$$\rho_1'(t) = t^{-1}[2\rho_0(2t) - \rho_0(t)] - t^{-2}I(t)$$
$$\geq t^{-1}[2\rho_0(2t) - \rho_0(t)] - t^{-1}\rho_0(2t) = t^{-1}[\rho_0(2t) - \rho_0(t)] \geq 0.$$

This guarantees that ρ_1 is nondecreasing on $[0, +\infty[$. For $\theta_0(t) := \int_t^{2t} \rho_1(s)\,ds$, the function θ_0 is derivable on $[0, +\infty[$ with $\theta_0'(t) = 2\rho_1(2t) - \rho_1(t)$, ensuring that θ_0 is nondecreasing and of class $\mathcal{C}^1$ on $[0, +\infty[$ with $\theta_0(0) = \theta_0'(0) = 0$. Further, for $\rho_2(0) := 0$ and $\rho_2(t) := t^{-1}\theta_0(t) = t^{-1}\int_t^{2t} \rho_1(s)\,ds$, repeating the above arguments yields $\rho_2(t) \geq \rho_1(t)$, which entails $\theta_0(t) \geq \sigma(t)$ for all $t \geq 0$. Putting, for all $t \geq 0$, $\theta(t) := t^2 + \theta_0(t)$ and $\theta(-t) := \theta(t)$, the function θ fulfills the desired properties. □

The description in (a) below says that the function φ in Proposition 4.4(c) with $\varphi(x) = f(x)$ may be required near the reference point x to be *strictly below* (and not merely below) f outside x.

PROPOSITION 4.25. *Let $(X, \|\cdot\|)$ be a normed space and $f : X \to \mathbb{R} \cup \{+\infty\}$ be a function which is finite at $x \in X$.*
(a) *An element $x^* \in \partial_F f(x)$ if and only if there exist an open convex neighborhood U of x and a function $\varphi : U \to \mathbb{R}$ which is Fréchet differentiable at x with $D\varphi(x) = x^*$ and such that*

$$\varphi(x) = f(x) \quad \text{and} \quad \varphi(u) < f(u) \quad \text{for all } u \in U \text{ with } u \neq x.$$

(b) *Assume that the norm $\|\cdot\|$ is Fréchet differentiable off zero. Then in the description in (a) the function φ can be required in addition to be $\mathcal{C}^1$ on U; further it can be taken $U = X$ whenever f is bounded from below over the whole space X.*

PROOF. The implication $\Leftarrow$ in (a) follows directly from Proposition 4.4(c).

Now fix $x^* \in \partial_F f(x)$ and take an open set $U \ni x$ over which f is bounded from below by some real γ (resp. $U = X$ if f is bounded from below on X). Putting for $t \geq 0$

$$\sigma(t) = \sup\{f(x) - f(u) + \langle x^*, u - x \rangle : u \in U, \|u - x\| \leq t\},$$

we have $\sigma(t) \geq 0$ and $\sigma(t) \leq \alpha t + \beta$ for $\alpha = \|x^*\|$ and $\beta = f(x) - \gamma$. Fix any $\varepsilon > 0$ and by the inclusion $x^* \in \partial_F f(x)$ choose some ball $B(x, \delta) \subset U$ such that $f(x) - f(u) + \langle x^*, u - x \rangle \leq \varepsilon \|u - x\|$ for all $u \in B(x, \delta)$. Take any positive real $t < \delta$. Then for all $u \in U$ with $\|u - x\| \leq t$ we get

$$f(x) - f(u) + \langle x^*, u - x \rangle \leq \varepsilon \|u - x\| \leq \varepsilon t,$$

thus $\sigma(t) \leq \varepsilon t$. Combining this with the obvious relations $\sigma(0) = 0$ and $\sigma(t) \geq 0$ for all $t \geq 0$, we see that σ has a right derivative $\sigma_+'(0) = 0$ at 0.

We may apply the above lemma and let θ be the function given by the lemma. The function φ on U with

$$\varphi(u) := -\theta(\|u - x\|) + f(x) + \langle x^*, u - x \rangle \quad \text{for all } u \in U,$$

is easily seen to meet the required properties in (a), and in (b) also when the additional assumptions in (b) hold (using under (b) the $\mathcal{C}^1$ property on $X \setminus \{0\}$ of $\|\cdot\|$ according to Theorem 2.179). □

4.1.2. Fréchet subgradients of distance functions. We already saw in Subsection 3.3.1 that the distance function is a particular case of the infimal convolution of two functions. While the Moreau-Rockafellar subdifferential is described in Proposition 3.69, the present subsection will study the Fréchet subdifferential of the infimal convolution; such functions are known to be of great interest in various fields of analysis. First, let us put as a definition the situation considered in (3.44) for two functions and in Remark 3.65 when more than two functions are involved.

DEFINITION 4.26. For functions $f_i : X \to \mathbb{R} \cup \{+\infty\}$ with $i = 1, \cdots, n$, the *infimal convolution* or *inf-convolution* $f_1 \square \cdots \square f_n$ (also called *epi-sum*) of $f_1, \cdots, f_n$ is the function from $X \to \mathbb{R} \cup \{-\infty, +\infty\}$ defined by

$$(f_1 \square \cdots \square f_n)(x) := \inf_{\sum_{i=1}^n x_i = x} \big(f_1(x_1) + \cdots + f_n(x_n)\big) \quad \text{for all } x \in X,$$

the infimum being taken over the set of all n-tuples $(x_1, \cdots, x_n) \in X^n$ such that $\sum_{i=1}^n x_i = x$. One says that this infimal convolution is *exact* at a point $\overline{x} \in X$ or *attained* for $\overline{x}$ if there exist $\overline{x}_1, \cdots, \overline{x}_n \in X$ such that the equality

$$(f_1 \square \cdots \square f_n)(\overline{x}) = f_1(\overline{x}_1) + \cdots + f_n(\overline{x}_n)$$

holds true.

According to Remark 3.65 the use of the upper extended addition $\dot{+}$ allows to consider $(f_1 \square f_2) \square f_3$ and $f_1 \square (f_2 \square f_3)$, and to see that both values $((f_1 \square f_2) \square f_3)(x)$ and $(f_1 \square (f_2 \square f_3))(x)$ coincide with

$$\inf_{x_1 + x_2 + x_3 = x} \big(f_1(x_1) + f_2(x_2) + f_3(x_3)\big).$$

This means that with $f_1 \square f_2 \square f_3$ as given in Definition 4.26 one has

$$(f_1 \square f_2) \square f_3 = f_1 \square (f_2 \square f_3) = f_1 \square f_2 \square f_3.$$

In the case of two functions, as already seen in (3.44), $f_1 \square f_2$ can also be written in the form

$$(f_1 \square f_2)(x) = \inf_{y \in X} \big(f_1(y) + f_2(x - y)\big) \quad \text{for all } x \in X.$$

A general inclusion (resp. inequality) holds for the Fréchet subdifferential (resp. the Bouligand directional derivative) at points where the infimal convolution is attained. This provides, for nonconvex functions, a result in the line of the one for convex functions in Proposition 3.69.

PROPOSITION 4.27. Let $f_i : X \to \mathbb{R} \cup \{+\infty\}$, with $i = 1, \cdots, n$, be n functions on the normed space X. Assume that the infimal convolution $f_1 \square \cdots \square f_n$ is finite at $\overline{x} \in X$ and attained at $(\overline{x}_1, \cdots, \overline{x}_n)$ for $\overline{x}$, that is, $\overline{x}_1 + \cdots + \overline{x}_n = \overline{x}$ and

$$(f_1 \square \cdots \square f_n)(\overline{x}) = f_1(\overline{x}_1) + \cdots + f_n(\overline{x}_n).$$

Then

$$\partial_F (f_1 \square \cdots \square f_n)(\overline{x}) \subset \bigcap_{i=1}^n \partial_F f_i(\overline{x}_i)$$

and for any $h \in X$
$$(f_1 \square \cdots \square f_n)^B(\overline{x}; h) \leq \min_{1 \leq i \leq n} f_i^B(\overline{x}_i; h).$$

PROOF. We observe that each f_i is finite at $\overline{x}_i$. Put $f := f_1 \square \cdots \square f_n$ and fix $x^* \in \partial_F f(\overline{x})$. Consider any $\varepsilon > 0$ and choose $\delta > 0$ such that for any $x \in B(\overline{x}, \delta)$ we have
$$\langle x^*, x - \overline{x} \rangle \leq f(x) - f(\overline{x}) + \varepsilon \|x - \overline{x}\|.$$
Fix any $k \in \{1, \cdots, n\}$ and any $u \in B(\overline{x}_k, \delta)$. Since
$$\overline{x} + u - \overline{x}_k \in B(\overline{x}, \delta) \quad \text{and} \quad \overline{x} + u - \overline{x}_k = \overline{u}_1 + \cdots + \overline{u}_n$$
for $\overline{u}_i = \overline{x}_i$ if $i \neq k$ and $\overline{u}_k = u$, taking $x = \overline{x} + u - \overline{x}_k$ in the above inequality yields
$$\begin{aligned}\langle x^*, u - \overline{x}_k \rangle &= \langle x^*, (\overline{x} + u - \overline{x}_k) - \overline{x} \rangle \\ &\leq f(\overline{x} + u - \overline{x}_k) - f_1(\overline{x}_1) - \cdots - f_n(\overline{x}_n) + \varepsilon \|u - \overline{x}_k\| \\ &\leq f_1(\overline{u}_1) + \cdots + f_n(\overline{u}_n) - f_1(\overline{x}_1) - \cdots - f_n(\overline{x}_n) + \varepsilon \|u - \overline{x}_k\| \\ &= f_k(u) - f_k(\overline{x}_k) + \varepsilon \|u - \overline{x}_k\|.\end{aligned}$$
This ensures $x^* \in \partial_F f_k(\overline{x}_k)$ and the desired inclusion.

The proof concerning the Bouligand directional derivative is similar. $\square$

Assuming as above that the inf-convolution $\varphi := f_1 \square \cdots \square f_n$ is finite at $\overline{x}$ and that $\varphi(\overline{x}) = f_1(\overline{x}_1) + \cdots + f_n(\overline{x}_n)$ with $\overline{x} = \overline{x}_1 + \cdots + \overline{x}_n$, we notice that, for any $h \in X$
$$\begin{aligned}-\varphi(\overline{x} + h) + \varphi(\overline{x}) &\geq -f_1(\overline{x}_1 + h)) - f_2(\overline{x}_2) - \cdots - f_n(\overline{x}_n) + f_1(\overline{x}_1) + \cdots + f_n(\overline{x}_n) \\ &= -f_1(\overline{x}_1 + h) + f_1(\overline{x}_1)\end{aligned}$$
and from this we easily see that
$$\partial_F(-f_1)(\overline{x}_1) \subset \partial_F(-\varphi)(\overline{x}) \quad \text{and} \quad (-f_1)^B(\overline{x}_1; h) \leq (-\varphi)^B(\overline{x}; h).$$
We have then proved the following proposition.

PROPOSITION 4.28. *Let $f_1, \cdots, f_n : X \to \mathbb{R} \cup \{+\infty\}$ be functions on the normed space X and let $\overline{x}$ be a point at which $f_1 \square \cdots \square f_n$ is finite. Assume that there exist $\overline{x}_1, \cdots, \overline{x}_n \in X$ such that $(f_1 \square \cdots \square f_n)(\overline{x}) = f_1(\overline{x}_1) + \cdots + f_n(\overline{x}_n)$. Then one has, for $i = 1, \cdots, n$,*
$$\partial_F(-f_i)(\overline{x}_i) \subset \partial_F\bigl(-(f_1 \square \cdots \square f_n)\bigr)(\overline{x}),$$
$$(-f_i)^B(\overline{x}_i; h) \leq \bigl(-(f_1 \square \cdots \square f_n)\bigr)^B(\overline{x}; h) \quad \text{for all } h \in X.$$

When one performs the inf-convolution of a function f with a sufficiently differentiable kernel function g, the differentiability of $f \square g$ can be obtained as follows.

PROPOSITION 4.29. *Let $f, g : X \to \mathbb{R} \cup \{+\infty\}$ be two functions on the normed space X and let $\overline{x}$ be a point at which $f \square g$ is finite. Assume that there exists an open neighborhood U of $\overline{x}$ and a selection $\sigma : U \to X$ of the multimapping $x \mapsto \operatorname{Argmin}_X\bigl(f(\cdot) + g(x - \cdot)\bigr)$ such that σ is continuous at $\overline{x}$ and g is strictly Fréchet differentiable at $\overline{x} - \sigma(\overline{x})$. Then $f \square g$ is Fréchet differentiable at $\overline{x}$ and*
$$D(f \square g)(\overline{x}) = Dg\bigl(\overline{x} - \sigma(\overline{x})\bigr).$$

PROOF. By Proposition 4.28 above, for small h, we have
$$(f\square g)(\overline{x}+h) - (f\square g)(\overline{x}) \leq Dg(\overline{x}-\sigma(\overline{x}))(h) + o(\|h\|).$$
On the other hand, by the strict Fréchet differentiability of g at $\overline{x}-\sigma(\overline{x})$ and by the continuity of σ at $\overline{x}$ we have, for h sufficiently small,
$$(f\square g)(\overline{x}+h) - (f\square g)(\overline{x})$$
$$= f(\sigma(\overline{x}+h)) + g(\overline{x}+h-\sigma(\overline{x}+h)) - (f\square g)(\overline{x})$$
$$= f(\sigma(\overline{x}+h)) + g(\overline{x}-\sigma(\overline{x}+h)) + Dg(\overline{x}-\sigma(\overline{x}))(h) + o(\|h\|) - (f\square g)(\overline{x}).$$
Since $f(\sigma(\overline{x}+h)) + g(\overline{x}-\sigma(\overline{x}+h)) - (f\square g)(\overline{x}) \geq 0$, we derive
$$(f\square g)(\overline{x}+h) - (f\square g)(\overline{x}) \geq Dg(\overline{x}-\sigma(\overline{x}))(h) + o(\|h\|).$$
Combining this with the first inequality above we deduce the Fréchet differentiability of $f\square g$ at $\overline{x}$ as well as the formula of the proposition. $\square$

Various other properties of the inf-convolution with kernel functions will be developed in Chapter 14. We study now how the Fréchet normal cone is related to the distance function. The results will be used later in particular, to establish some criteria for the prox-regularity of a set in Chapter 15. Let us start with the point of reference in the set.

PROPOSITION 4.30. *Let S be a set of the normed space $(X, \|\cdot\|)$ and $x \in S$. Then*
$$\partial_F d_S(x) = N^F(S;x) \cap \mathbb{B}_{X^*} \quad \text{and} \quad N^F(S;x) = \mathbb{R}_+ \partial_F d_S(x).$$

PROOF. Let $x^* \in \partial_F d_S(x)$. Since, for all $u \in X$
$$d_S(u) = \inf_{y \in S}\|u-y\| = \inf\{\Psi_S(u_1) + \|u_2\| : u_1, u_2 \in X, u_1+u_2 = x\} = (\Psi_S \square \|\cdot\|)(u)$$
and since this infimal convolution is attained at $(x,0)$ for x because $d_S(x) = 0 = \Psi_S(x) + \|0\|$, Proposition 4.27 above says that
$$\partial_F d_S(x) \subset \partial_F \Psi_S(x) \cap \partial\|\cdot\|(0) = N^F(S;x) \cap \mathbb{B}_{X^*}.$$
(Note that the inclusion in $N^F(S;x) \cap \mathbb{B}_{X^*}$ follows also directly from the analytic definition of the Fréchet subdifferential with the function d_S and from (c) in Proposition 4.11).

Let us show the reverse inclusion. Fix any $x^* \in N^F(S;x)$ with $\|x^*\| \leq 1$ and fix any $\varepsilon > 0$. For $\varepsilon' := \varepsilon/4$ there exists some real $\delta > 0$ such that

(4.7) $$\langle x^*, x'-x \rangle \leq \varepsilon'\|x'-x\| \quad \text{for all } x' \in S \cap B(x,\delta).$$

Putting $\gamma := \min\{1, \varepsilon/4, \delta/3\}$, for any fixed element $z \in B(x,\gamma)$, we can choose $y_z \in S$ satisfying

(4.8) $$\|y_z - z\| \leq d_S(z) + \gamma\|z-x\|.$$

Then $y_z \in B(x,\delta)$, because according to the latter inequality we have

(4.9) $$\|y_z - x\| \leq \|y_z - z\| + \|z - x\| \leq 3\|z-x\| < \delta,$$

and hence this and (4.7) yield
$$\langle x^*, z-x \rangle = \langle x^*, y_z - x\rangle + \langle x^*, z - y_z\rangle$$
$$\leq \varepsilon'\|y_z - x\| + \|y_z - z\|$$
$$\leq 3\varepsilon'\|z-x\| + d_S(z) + \gamma\|z-x\|,$$

the latter inequality being due to the second inequality in (4.9) and to (4.8). So, for all $z \in B(x, \gamma)$
$$\langle x^*, z - x \rangle \leq d_S(z) - d_S(x) + \varepsilon \|z - x\|,$$
which says that $x^* \in \partial_F d_S(x)$, so the first equality of the proposition is established. The second concerning normal cones is a direct consequence of the first one. □

Now we examine the case of points of reference outside the closure of the set.

For any set S of the normed space X and any $r \geq 0$, recall (see Definition 3.259) that the closed r-enlargement of S is given by
$$\mathrm{Enl}_r S := \{u \in X : d_S(u) \leq r\}.$$
In Lemma 3.260 for any $u \in X$ such that $d_S(u) \geq r$ we proved that
$$d(u, S) = r + d(u, \mathrm{Enl}_r S) = r + d(u, D_r(S)),$$
where we recall that $D_r(S) := \{x \in X : d_S(x) = r\}$.

The result concerning the Fréchet subdifferential of the distance function at a point outside the closure of S is the following. It is an extension to nonconvex sets of the previous result in Proposition 3.262 for convex sets.

PROPOSITION 4.31. *Let S be a subset of a normed space $(X, \|\cdot\|)$ and $x \notin \mathrm{cl}\, S$. For $r := d_S(x) > 0$ and for the closed r-enlargement $\mathrm{Enl}_r S$ of S one has*
$$\partial_F d_S(x) = N^F(\mathrm{Enl}_r S; x) \cap \{x^* \in X^* : \|x^*\| = 1\}.$$

PROOF. Put $E_r(S) := \mathrm{Enl}_r S$. We begin by showing the inclusion of the left-hand member into the right-hand one. Fix $x^* \in \partial_F d_S(x)$ and fix also any $\varepsilon > 0$. By the analytical description of Fréchet subgradient there exists $\delta > 0$ such that

(4.10) $\quad \langle x^*, x' - x \rangle \leq d_S(x') - d_S(x) + \varepsilon \|x' - x\| \quad$ for all $x' \in B[x, \delta]$.

Since $d_S(x') - d_S(x) \leq 0$ for all $x' \in E_r(S)$, we obtain
$$\langle x^*, x' - x \rangle \leq \varepsilon \|x' - x\| \quad \text{for all } x' \in E_r(S) \cap B[x, \delta],$$
which ensures that $x^* \in N^F(E_r(S); x)$.

Now, we show that $\|x^*\| = 1$. Put $t := \min\{1, \varepsilon, \delta/(1 + d_S(x))\}$ and choose $x_t \in S$ such that
$$\|x_t - x\| \leq d_S(x) + t^2.$$
For $x' := x + t(x_t - x)$, since
$$\|x' - x\| \leq t\|x_t - x\| \leq t d_S(x) + t^3 \leq t(1 + d_S(x)) \leq \delta,$$
we get by (4.10)
$$\langle x^*, x' - x \rangle \leq \|x' - x_t\| - \|x - x_t\| + t^2 + \varepsilon t \|x - x_t\|$$
$$= (1 - t)\|x - x_t\| - \|x - x_t\| + t^2 + \varepsilon t \|x - x_t\|$$
$$= -t\|x - x_t\| + t^2 + \varepsilon t \|x - x_t\|.$$
Thus, taking the definition of x' into account we see that
$$\langle x^*, x_t - x \rangle \leq -\|x - x_t\| + t + \varepsilon \|x - x_t\|$$
$$\leq -\|x - x_t\| + \varepsilon(1 + \|x - x_t\|),$$
and hence
$$\frac{\langle x^*, x - x_t \rangle}{\|x - x_t\|} \geq 1 - \varepsilon \left(1 + \frac{1}{\|x - x_t\|}\right) \geq 1 - \varepsilon \left(1 + \frac{1}{d_S(x)}\right).$$

This ensures that $\|x^*\| \geq 1$, thus $\|x^*\| = 1$ since $x^* \in \partial_F d_S(x) \subset \mathbb{B}_{X^*}$ according to the 1-Lipschitz property of d_S (see (c) of Proposition 4.11). The first desired inclusion of the proposition is established.

Now we proceed to prove that the right-hand member of the equality of the proposition is included in the left-hand. Fix any $x^* \in N^F(E_r(S); x)$ with $\|x^*\| = 1$ and fix any $\varepsilon > 0$. On the one hand, notice that $x^* \in \partial_F \mathrm{dist}\,(\cdot, E_r(S))(x)$ by Proposition 4.30. So, there exists $\delta_1 > 0$ such that

$$\langle x^*, x' - x \rangle \leq \mathrm{dist}\,(x', E_r(S)) - \mathrm{dist}\,(x, E_r(S)) + \varepsilon \|x' - x\| \quad \text{for all } x' \in B[x, \delta_1].$$

This and Lemma 3.260 above give

$$(4.11) \quad \langle x^*, x' - x \rangle \leq d_S(x') - d_S(x) + \varepsilon \|x' - x\| \quad \text{for all } x' \in B[x, \delta_1] \setminus E_r(S).$$

On the other hand, since $x^* \in N^F(E_r(S); x)$, there exists $\delta_2 > 0$ such that

$$(4.12) \quad \langle x^*, x' - x \rangle \leq \frac{\varepsilon}{2} \|x' - x\| \quad \text{for all } x' \in E_r(S) \cap B[x, \delta_2].$$

Fix now a positive real $\delta_3 < \delta_2/2$ and $x' \in E_r(S) \cap B[x, \delta_3]$, and put $t_{x'} := d_S(x) - d_S(x') \geq 0$. Take any real $\lambda > 0$. Since $\|x^*\| = 1$, we can choose $u \in X$, with $\|u\| = 1$, such that $\langle x^*, u \rangle > 1 - \lambda$. Then $x' + t_{x'} u \in E_r(S) \cap B[x, \delta_2]$, because

$$d_S(x' + t_{x'} u) \leq d_S(x') + t_{x'} = d_S(x) = r,$$

and

$$\|x' + t_{x'} u - x\| \leq \|x' - x\| + t_{x'} \leq 2\|x' - x\| \leq 2\delta_3 \leq \delta_2.$$

This and (4.12) give

$$\langle x^*, x' + t_{x'} u - x \rangle \leq \frac{\varepsilon}{2} \|x' + t_{x'} u - x\| \leq \varepsilon \|x' - x\|,$$

which ensures that

$$\begin{aligned}\langle x^*, x' - x \rangle &= \langle x^*, x' + t_{x'} u - x \rangle - \langle x^*, t_{x'} u \rangle \\ &\leq \varepsilon \|x' - x\| - t_{x'}(1 - \lambda) \\ &= \varepsilon \|x' - x\| + (1 - \lambda)(d_S(x') - d_S(x)).\end{aligned}$$

This being true for every real $\lambda > 0$, we deduce that

$$\langle x^*, x' - x \rangle \leq \varepsilon \|x' - x\| + d_S(x') - d_S(x).$$

Putting $\delta := \min\{\delta_1, \delta_3\}$ and combining the latter inequality with (4.11) we obtain

$$\langle x^*, x' - x \rangle \leq d_S(x') - d_S(x) + \varepsilon \|x' - x\| \quad \text{for all } x' \in B[x, \delta].$$

So, $x^* \in \partial_F d_S(x)$ and the proof is complete. $\square$

Given a nonempty convex set C of a normed space $(X, \|\cdot\|)$ and $\bar{x} \in X$, we already observed in Proposition 3.258 that $\langle x^*, \bar{x} - y_n \rangle \to d_C(\bar{x})$ for every minimizing sequence $(y_n)_n$ in C for the distance $d_C(\bar{x})$, that is, $\|\bar{x} - y_n\| \to d_C(\bar{x})$. In addition to Proposition 4.30 and Proposition 4.31, the following extension of the above property to nonconvex sets is sometimes of great interest.

PROPOSITION 4.32. *Let S be a nonempty subset of a normed space X and $x^* \in \partial_F d_S(\bar{x})$ with $\bar{x} \in X$.*
(a) *One has*

$$d_S(\bar{x}) = \lim_{y \in S,\, \|y - \bar{x}\| \to d_S(\bar{x})} \langle x^*, \bar{x} - y \rangle,$$

that is, for any sequence $(y_n)_n$ in S with $\|\bar{x} - y_n\| \to d_S(\bar{x})$ as $n \to \infty$

$$\langle x^*, \bar{x} - y_n \rangle \to d_S(\bar{x}) \quad \text{as } n \to \infty.$$

(b) In particular, for any $\bar{y} \in S$ with $\|\bar{x} - \bar{y}\| = d_S(\bar{x})$ (if any) one has

$$\langle x^*, \bar{x} - \bar{y} \rangle = d_S(\bar{x}).$$

PROOF. It is enough to show (a) since (b) follows from (a). The assertion (a) being obvious if $\bar{x} \in \operatorname{cl} S$, assume that $\bar{x} \in X \setminus \operatorname{cl} S$ with $\partial_F d_S(\bar{x}) \neq \emptyset$ and let $x^* \in \partial_F d_S(\bar{x})$. We know by Proposition 4.31 that $\|x^*\|_* = 1$. Take any sequence $y_n \in S$ with $\|y_n - \bar{x}\| \to d_S(\bar{x})$ and $\|y_n - \bar{x}\| - d_S(\bar{x}) < 1/2$. Put $t_n := (1/n) + \left(\|\bar{x} - y_n\| - d_S(\bar{x})\right)^{1/2}$. Since $t_n(y_n - \bar{x}) \to 0$, from the inclusion $x^* \in \partial_F d_S(\bar{x})$ we can see that

$$\liminf_n \{t_n^{-1}[d_S(\bar{x} + t_n(y_n - \bar{x})) - d_S(\bar{x})] - \langle x^*, y_n - \bar{x} \rangle\} \geq 0.$$

Further, we also have for n large enough

$$d_S(\bar{x} + t_n(y_n - \bar{x})) - d_S(\bar{x})$$
$$\leq \|\bar{x} + t_n(y_n - \bar{x}) - y_n\| - d_S(\bar{x})$$
$$\leq \|\bar{x} + t_n(y_n - \bar{x}) - y_n\| - \|\bar{x} - y_n\| + \left(\|\bar{x} - y_n\| - d_S(\bar{x})\right)$$
$$= -t_n\|\bar{x} - y_n\| + \left(\|\bar{x} - y_n\| - d_S(\bar{x})\right)$$

(the equality being due to the inclusion $t_n \in]0, 1[$), and hence

$$\liminf_n \{-\|\bar{x} - y_n\| - \langle x^*, y_n - \bar{x} \rangle + t_n\} \geq 0,$$

which gives

$$d_S(\bar{x}) = \lim_n \|\bar{x} - y_n\| \leq \liminf_n \langle x^*, \bar{x} - y_n \rangle.$$

On the other hand, using the inequality $\|x^*\| \leq 1$, we see that

$$d_S(\bar{x}) = \lim_n \|\bar{x} - y_n\| \geq \limsup_n \langle x^*, \bar{x} - y_n \rangle.$$

Combining these last two inequalities gives that

$$d_S(\bar{x}) = \lim_n \langle x^*, \bar{x} - y_n \rangle,$$

which justifies the equality in the assertion (a). $\square$

4.2. Separable reduction principle for F-subdifferentiability

In the theory of Geometry of Banach Spaces, one is often interested in knowing if a property is satisfied on the entire Banach space whenever it is fulfilled on every closed separable vector subspace. In such a case the verification of the property is then reduced to its study on closed separable vector subspaces; one then says that the property is *separably determined*. This section is devoted to the separable determination of a suitable Fréchet subdifferentiability and other similar properties.

4.2.1. Preparatory lemmas. The proof of the theorem concerning the separable determination of F-subdifferentiability needs three lemmas. The first lemma is concerned with the characterization of the existence of subgradients with specific properties for convex functions.

LEMMA 4.33. *Let X, Y, Z be three Banach spaces, $A : X \to Y$ and $T : Z \to Y$ be two continuous linear mappings, and $\varphi : Y \to \mathbb{R} \cup \{+\infty\}$ be a convex function which is finite at zero. Given reals $c \geq 0$, $\varepsilon > 0$ and $\rho \geq 0$, the following two properties are equivalent:*
(a) *there exist $\varepsilon' \in]0, \varepsilon[$ and $w \in Z$ with $\|w\| = 1$ such that*
$$\varphi(y) \geq \varphi(0) - c\|y - Ax - Tz\| - \varepsilon'\|x\| - \varepsilon'\|z\| - \rho\|z - w\| + \rho$$
for all $x \in X$, $y \in Y$, and $z \in Z$;
(b) *there is some $y^* \in \partial\varphi(0)$ such that $\|y^*\| \leq c$, $\|A^*y^*\| < \varepsilon$, and $\big|\|T^*y^*\| - \rho\big| < \varepsilon$.*

PROOF. Suppose that (a) holds. Then, for all $(x, y, z) \in X \times Y \times Z$, we have
$$\varphi_1(x, y, z) := \varphi(y) + c\|y - Ax - Tz\| + \varepsilon'\|x\| + \varepsilon'\|z\| + \rho\|z - w\| - \rho \geq \varphi_1(0, 0, 0),$$
and hence $(0, 0, 0) \in \partial\varphi_1(0, 0, 0)$. All the convex functions in the latter sum being continuous except φ, by the Moreau-Rockafellar theorem (see Theorem 2.105) we easily obtain some $y^* \in \partial\varphi(0)$, $v^* \in Y^*$, $x^* \in X^*$, and $z^*, w^* \in Z^*$ such that $\|v^*\| \leq c$, $\|x^*\| \leq \varepsilon'$, $\|z^*\| \leq \varepsilon'$, and $\langle w^*, -w \rangle = \|w^*\| = \rho$ and
$$(0, 0, 0) = (0, y^*, 0) + (-A^*v^*, v^*, -T^*v^*) + (x^*, 0, 0) + (0, 0, z^*) + (0, 0, w^*),$$
or equivalently $y^* = -v^*$, $A^*y^* = -x^*$, $T^*y^* + w^* = -z^*$. It results that $\|-y^*\| \leq c$, $\|A^*y^*\| \leq \varepsilon' < \varepsilon$, and
$$\big|\|T^*y^*\| - \rho\big| = \big|\|T^*y^*\| - \| - w^*\|\big| \leq \|T^*y^* + w^*\| \leq \varepsilon' < \varepsilon,$$
so the assertion (b) holds true.

Conversely suppose that (b) is satisfied. Choose some real ε' such that $\|A^*y^*\| < \varepsilon' < \varepsilon$ and $\big|\|T^*y^*\| - \rho\big| < \varepsilon' < \varepsilon$. By the Bishop-Phelps theorem (see Theorem 2.225) we also choose some $z^* \in Z^*$ such that $\|z^*\| = 1$ and $\big\|\|T^*y^*\|z^* - T^*y^*\big\| < \varepsilon' - \big|\|T^*y^*\| - \rho\big|$ and such that z^* attains its norm on the unit sphere of Z. Then let $w \in Z$ with $\|w\| = 1$ and $\langle z^*, w \rangle = \|z^*\|$. Put $w^* := \rho z^*$ and note that $\langle w^*, w \rangle = \|w^*\| = \rho$ and
$$\big\| w^* - \|T^*y^*\|z^* \big\| = \big|\rho - \|T^*y^*\|\big|, \quad \text{thus} \quad \|w^* - T^*y^*\| < \varepsilon'.$$
For all $x \in X$, $y \in Y$ and $z \in Z$, writing
$$\varphi(y) \geq \varphi(0) + \langle y^*, y \rangle$$
$$= \varphi(0) + \langle y^*, y - Ax - Tz \rangle + \langle A^*y^*, x \rangle$$
$$+ \langle T^*y^* - w^*, z \rangle + \langle w^*, z - w \rangle + \langle w^*, w \rangle,$$
it then follows that
$$\varphi(y) \geq \varphi(0) - c\|y - Ax - Tz\| - \varepsilon'\|x\| - \varepsilon'\|z\| - \rho\|z - w\| + \rho,$$
which is the inequality in (a). □

Before stating the next lemmas let us fix some notation. Given reals $c \geq 0$, $\varepsilon > 0$ and $\rho \geq 0$, for the Banach spaces X, Y, Z, for the continuous linear mappings

$A : X \to Y$ and $T : Z \to Y$ as in the above lemma, and for $w \in Z$ with $\|w\| = 1$, we put (taking the above lemma into account)

(4.13) $\quad g_{(c,\varepsilon,\rho),w}(x,v,z) := c\|v - Ax - Tz\| + \varepsilon\|x\| + \varepsilon\|z\| + \rho\|z - w\| - \rho,$

and we note that $g_{(c,\varepsilon,\rho),w}$ is a continuous convex function which is null at $(0,0,0)$.

We also denote:
- by Δ the set of all non-increasing sequences $\delta = (\delta_n) \in\,]0,+\infty[^\mathbb{N}$, that is, $\delta_n \geq \delta_{n+1}$;
- by Λ the set of all sequences $\lambda = (\lambda_n) \in \mathbb{R}_+^\mathbb{N}$ with $\sum_{n=1}^\infty \lambda_n = 1$ and with finite support, that is, the set $\{n \in \mathbb{N} : \lambda_n \neq 0\}$ is finite;
- by Υ the set of all sequences $\nu = (\nu_n) \in \mathbb{N}^\mathbb{N}$ such that the set $\{n \in \mathbb{N} : \nu_n \neq 1\}$ is finite;
- by $\mathcal{H}(\nu,\delta)$ the set of all sequences $h = (h_n) \in Y^\mathbb{N}$ such that $\|h_n\| < \delta_{\nu_n}$, where $\nu = (\nu_n)$ and $\delta = (\delta_n)$ are given in Υ and Δ respectively.

In the rest of this subsection we adopt the *convention* $\lambda_n f(u) = 0$ when $f(u) = +\infty$ and the real $\lambda_n = 0$.

LEMMA 4.34. *For any function* $f : Y \to \mathbb{R} \cup \{+\infty\}$ *which is finite at* $y \in Y$ *and any sequence* $\delta = (\delta_n)$ *in* Δ, *the function* $\varphi_\delta : Y \to \mathbb{R} \cup \{-\infty, +\infty\}$ *with*

$$\varphi_\delta(v) := \inf \left\{ \sum_{n=1}^\infty \lambda_n \left(f(y + h_n) + \frac{1}{\nu_n}\|h_n\| \right) - f(y) : \right.$$
$$\left. (\lambda_n) \in \Lambda, (\nu_n) \in \Upsilon, (h_n) \in \mathcal{H}(\nu,\delta), \sum_{n=1}^\infty \lambda_n h_n = v \right\},$$

is convex.

PROOF. Let $0 < t < 1$ and $(v', r'), (v'', r'')$ be two pairs in $Y \times \mathbb{R}$ with $\varphi_\delta(v') < r'$ and $\varphi_\delta(v'') < r''$. From the definition of φ_δ there are $(\lambda'_n), (\lambda''_n)$ in Λ, $(\nu'_n), (\nu''_n)$ in Υ, $(h'_n), (h''_n)$ in $\mathcal{H}(\nu,\delta)$ such that

$$\sum_{n=1}^\infty \lambda'_n h'_n = v', \ r' > \sum_{n=1}^\infty \lambda'_n \left(f(y + h'_n) + \frac{1}{\nu'_n}\|h'_n\| \right) - f(y),$$

and similar properties for (v'', r''). Fix some integer n_0 such that $\lambda'_n = \lambda''_n = 0$ for all $n > n_0$, and put

$$\begin{array}{llll} \lambda_n = t\lambda'_n, & \nu_n = \nu'_n, & h_n = h'_n, & \text{if } 1 \leq n \leq n_0, \\ \lambda_n = (1-t)\lambda''_{n-n_0}, & \nu_n = \nu''_{n-n_0}, & h_n = h''_{n-n_0}, & \text{if } n_0 < n \leq 2n_0, \\ \lambda_n = 0, & \nu_n = 1, & h_n = 0, & \text{if } n > 2n_0. \end{array}$$

Obviously $(\lambda_n) \in \Lambda$, $(\nu_n) \in \Upsilon$, and $\|h_n\| < \delta_{\nu_n}$ for all $n \in \mathbb{N}$, so it ensues that $(h_n) \in \mathcal{H}(\nu,\delta)$; further, we also see that $\sum_{n=1}^\infty \lambda_n h_n = tv' + (1-t)v''$. We then deduce

that
$$tr' + (1-t)r'' + f(y)$$
$$> \sum_{n=1}^{n_0} \left(t\lambda'_n \left(f(y + h'_n) + \frac{1}{\nu'_n} \|h'_n\| \right) + (1-t)\lambda''_n \left(f(y + h''_n) + \frac{1}{\nu''_n} \|h''_n\| \right) \right)$$
$$= \sum_{n=1}^{n_0} \lambda_n \left(f(y + h_n) + \frac{1}{\nu_n} \|h_n\| \right) + \sum_{n=1}^{n_0} \lambda_{n+n_0} \left(f(y + h_{n+n_0}) + \frac{1}{\nu_{n+n_0}} \|h_{n+n_0}\| \right)$$
$$= \sum_{n=1}^{\infty} \lambda_n \left(f(y + h_n) + \frac{1}{\nu_n} \|h_n\| \right) \geq \varphi_\delta(tv' + (1-t)v'') + f(y),$$

which says that φ_δ is convex as desired. $\square$

The third lemma provides a characterization of the existence of Fréchet subgradients with specific properties similar to those obtained in Lemma 4.33.

LEMMA 4.35. *Let* $f : Y \to \mathbb{R} \cup \{+\infty\}$ *be a function which is finite at* $y \in Y$ *and let reals* $c \geq 0$, $\varepsilon > 0$ *and* $\rho \geq 0$. *The following assertions are equivalent:*
(a) *there exist some* $\varepsilon' \in]0, \varepsilon[$, *some* $w \in Z$ *with* $\|w\| = 1$, *and some sequence* $\delta = (\delta_n)$ *in* $\mathbb{Q}^{\mathbb{N}} \cap \Delta$ *such that*
$$\sum_{n=1}^{\infty} \lambda_n \left(f(y + h_n) + \frac{1}{\nu_n} \|h_n\| \right) + g_{(c, \varepsilon', \rho), w}\left(x, \sum_{n=1}^{\infty} \lambda_n h_n, z \right) \geq f(y),$$
for all $x \in X$, $z \in Z$, $(\lambda_n) \in \Lambda$, $(\nu_n) \in \Upsilon$, *and* $(h_n) \in \mathcal{H}(\nu, \delta)$;
(b) *there exists some* $y^* \in \partial_F f(y)$ *such that* $\|y^*\| \leq c$, $\|A^*y^*\| < \varepsilon$, *and* $\big|\|T^*y^*\| - \rho\big| < \varepsilon$.

PROOF. Suppose that (a) is satisfied and, for the sequence $\delta \in \Delta$ in (a), consider the function φ_δ in Lemma 4.34. Put $\varphi(v) = \varphi_\delta(v)$ if $\|v\| \leq \delta_1$ and $\varphi(v) = +\infty$ otherwise. Obviously, from the definition of φ_δ and from the inequality in (a) we have that
$$(4.14) \qquad \varphi(v) \geq -g_{(c, \varepsilon', \rho), w}(0, v, 0) > -\infty \quad \text{for all } v \in Y,$$
and from Lemma 4.34 the function φ is convex. Further, it is also easily seen from the definition of φ_δ that $\varphi(0) \leq 0$. On the other hand, the definitions of φ_δ and $g_{(c, \varepsilon', \rho), w}$ and the left inequality in (4.14) ensure that
$$\varphi(0) = \varphi(0) + g_{(c, \varepsilon', \rho), w}(0, 0, 0) \geq 0, \text{ so } \varphi(0) = 0.$$
By (a) of the present lemma the condition (a) in Lemma 4.33 is satisfied. Lemma 4.33 then yields some $y^* \in \partial \varphi(0)$ such that $\|y^*\| \leq c$, $\|A^*y^*\| < \varepsilon$, and $\big|\|T^*y^*\| - \rho\big| < \varepsilon$. Take any real $\eta > 0$ and choose some $p \in \mathbb{N}$ with $1/p < \eta$. Fix any $v \in Y$ with $\|v\| < \delta_p$. Put $\lambda_p := 1$, $h_p := v$, and $\nu_p = p$, and for every $n \in \mathbb{N}$ with $n \neq p$, put $\lambda_n := 0$, $h_n := 0$, and $\nu_n := 1$. We see that $(\lambda_n) \in \Lambda$, $\nu = (\nu_n)$ is in Υ, and $(h_n) \in \mathcal{H}(\nu, \delta)$, so it follows from the definition of φ_δ and the inclusion $y^* \in \partial \varphi(0)$ that
$$f(y + v) + \eta\|v\| - f(y) \geq f(y + v) + \frac{1}{\nu_p}\|v\| - f(y) \geq \varphi(v) \geq \varphi(0) + \langle y^*, v \rangle = \langle y^*, v \rangle,$$
which means that $y^* \in \partial_F f(y)$, thus (b) is fulfilled.

Now suppose that (b) is satisfied and take some $y^* \in \partial_F f(y)$ with the properties in (b). For each integer $n \in \mathbb{N}$ we choose some positive rational number $\delta_n \in \mathbb{Q}$ such

that $f(y+v) + \frac{1}{n}\|v\| - f(y) \geq \langle y^*, v \rangle$ for all $v \in Y$ with $\|v\| \leq \delta_n$. We may suppose that the sequence $\delta := (\delta_n)$ is nonincreasing, so (δ_n) is in $\mathbb{Q}^{\mathbb{N}} \cap \Delta$. Consider again the convex function φ_δ in Lemma 4.34 associated with that sequence δ and define $\varphi(v) = \varphi_\delta(v)$ if $\|v\| \leq \delta_1$ and $\varphi(v) = +\infty$ otherwise. According to the definition of φ_δ we have $\varphi(0) \leq 0$ as above. Fix any $v \in Y$ with $\|v\| \leq \delta_1$ and take any sequences (λ_n) in Λ, $\nu = (\nu_n)$ in Υ, and (h_n) in $\mathcal{H}(\nu, \delta)$ with $\sum_{n=1}^{\infty} \lambda_n h_n = v$. Observing that
$f(y + h_n) + \frac{1}{\nu_n}\|h_n\| - f(y) \geq \langle y^*, h_n \rangle$ for all n because $\|h_n\| < \delta_{\nu_n}$ according to the definition of $\mathcal{H}(\nu, \delta)$, it ensues that

$$\sum_{n=1}^{\infty} \lambda_n \left(f(y + h_n) + \frac{1}{\nu_n}\|h_n\| \right) - f(y) \geq \sum_{n=1}^{\infty} \lambda_n \langle y^*, h_n \rangle = \langle y^*, v \rangle.$$

Then it results from the definition of φ_δ that, for all $v \in Y$ with $\|v\| \leq \delta_1$, we have $\varphi(v) \geq \langle y^*, v \rangle \geq \varphi(0) + \langle y^*, v \rangle$, and the left of the two latter inequalities tells us in particular that φ does not take on the value $-\infty$ and that $\varphi(0) = 0$ as well. Consequently, $y^* \in \partial \varphi(0)$ and with $c \geq 0$, $\varepsilon > 0$ and $\rho \geq 0$ as in the statement of (b) we can apply Lemma 4.33 to obtain the assertion (a) of the lemma. $\square$

4.2.2. Separable reduction of Fréchet subdifferentiability. We can now state and prove the separable reduction property of Fréchet subdifferentiability. The theorem even establishes the reduction of existence of Fréchet subgradient with some additional conditions. As stated the theorem will be used below and later.

THEOREM 4.36 (M. Fabian and A.D. Ioffe: separable reduction of Fréchet subdifferentiability plus estimates). Let X, Y, Z be Banach spaces, $A : X \to Y$ and $T : Z \to Y$ be continuous linear mappings, and $f : Y \to \mathbb{R} \cup \{+\infty\}$ be a proper function. Let also L be a separable subspace of Y. Then given a countable set $\mathcal{D}$ of triples (c, ε, ρ) with $c \geq 0$, $\varepsilon > 0$ and $\rho \geq 0$, there exist closed separable subspaces $L_0 \subset Y$ with $L_0 \supset L$, $X_0 \subset X$ and $Z_0 \subset Z$ such that $A(X_0) \subset L_0$, $T(Z_0) \subset L_0$, and such that for any $(c, \varepsilon, \rho) \in \mathcal{D}$ and any $y \in L_0 \cap \mathrm{dom}\, f$,

(4.15) $\qquad \{y^* \in \partial_F f(y) : \|y^*\| \leq c, \|A^*y^*\| < \varepsilon, |\|T^*y^*\| - \rho| < \varepsilon\} \neq \emptyset$

whenever

(4.16) $\qquad \{l^* \in \partial_F f_0(y) : \|l^*\| \leq c, \|A_0^* l^*\| < \varepsilon, |\|T_0^* l^*\| - \rho| < \varepsilon\} \neq \emptyset,$

where $f_0 := f_{|L_0}$ denotes the restriction of f to L_0, and $A_0 := A_{|X_0} : X_0 \to L_0$ and $T_0 := T_{|Z_0} : Z_0 \to L_0$ are defined by $A_0(x) = A(x)$ for all $x \in X_0$ and $T_0(z) = T(z)$ for all $z \in Z_0$.

Further, in the case when $Y_1 \times \cdots \times Y_m$ (resp. Y^m) is considered in place of Y, then the closed separable subspace containing L can be chosen in the form $L_{0,1} \times \cdots \times L_{0,m}$ with $L_{0,i} \subset Y_i$ (resp. in the form $L_0 \times \cdots \times L_0$ with $L_0 \subset Y$).

PROOF. **Step 1.** Consider the new set $\mathcal{D}$ (that we do not relabel) by adding into $\mathcal{D}$ any triple (c, ε', ρ) with $\varepsilon' \in \mathbb{Q}_> := \mathbb{Q} \cap]0, +\infty[$ and such that there exists some $\varepsilon > \varepsilon'$ for which $(c, \varepsilon, \rho) \in \mathcal{D}$. Obviously this new set $\mathcal{D}$ is still denumerable. For any $y \in Y$, $d := (c, \varepsilon, \rho) \in \mathcal{D}$, $w \in Z$ with $\|w\| = 1$, $\lambda \in \Lambda$, $\nu \in \Upsilon$, and $\delta \in \Delta$,

define
(4.17)
$$I(y,d,w,\lambda,\nu,\delta) := \inf\left\{\sum_{n\geq 1}\lambda_n\left(f(y+h_n)+\frac{1}{\nu_n}\|h_n\|\right)+g_{d,w}\left(x,\sum_{n\geq 1}\lambda_n h_n, z\right):\right.$$
$$(h_n)\in\mathcal{H}(\nu,\delta),\ x\in X,\ z\in Z\bigg\},$$

where $g_{d,w}(\cdot,\cdot,\cdot)$ is defined as in (4.13). Consider any $r\in\mathbb{Q}_>$. If $I(y,d,w,\lambda,\nu,\delta) > -\infty$, we choose some $(\widehat{h},\widehat{x},\widehat{z})$ in $\mathcal{H}(\nu,\delta)\times X\times Z$ such that
(4.18)
$$I(y,d,w,\lambda,\nu,\delta) > \sum_{n=1}^{\infty}\lambda_n\left(f\left(y+\widehat{h}_n\right)+\frac{1}{\nu_n}\left\|\widehat{h}_n\right\|\right)+g_{d,w}\left(\widehat{x},\sum_{n=1}^{\infty}\lambda_n\widehat{h}_n,\widehat{z}\right) - r.$$

On the contrary, if $I(y,d,w,\lambda,\nu,\delta) = -\infty$, we choose some $(\widehat{h},\widehat{x},\widehat{z})$ in $\mathcal{H}(\nu,\delta)\times X\times Z$ such that

(4.19) $$-\frac{1}{r}\sum_{n=1}^{\infty}\lambda_n\left(f\left(y+\widehat{h}_n\right)+\frac{1}{\nu_n}\left\|\widehat{h}_n\right\|\right)+g_{d,w}\left(\widehat{x},\sum_{n=1}^{\infty}\lambda_n\widehat{h}_n,\widehat{z}\right).$$

Note that the three so chosen elements
$$\widehat{h}(y,d,w,\lambda,\nu,r,\delta),\ \widehat{x}(y,d,w,\lambda,\nu,r,\delta),\ \widehat{z}(y,d,w,\lambda,\nu,r,\delta)$$
depend on the uple $(y,d,w,\lambda,\nu,r,\delta)$.

Step 2. We will construct a sequence of triples $(E_k,F_k,G_k)_{k\in\mathbb{N}}$, where (E_k), (F_k), (G_k) are increasing sequences of closed separable subspaces of X, Y, and Z respectively. The construction will proceed in such a way that, at each step, $A(E_k)$ and $T(G_k)$ are subsets of F_k, and at each step we will choose a dense countable subset $C(F_k)$ of F_k and a dense countable subset $C(G_k)$ of the unit sphere of G_k.

Step 3. Put $E_0 := \{0\}$, $F_0 := L$, and $G_0 := \{0\}$. Suppose that E_i, F_i and G_i as well as $C(F_i)$ and $C(G_i)$ have been defined for $i = 1,\cdots,k-1$. Take any $y\in C(F_{k-1})$, any $d\in\mathcal{D}$, any $w\in C(G_{k-1})$, any $\lambda = (\lambda_n)\in\mathbb{Q}^{\mathbb{N}}\cap\Lambda$, any $\nu = (\nu_n)\in\Upsilon$, any $r\in\mathbb{Q}_>$, and any $\delta = (\delta_n)\in\mathbb{Q}^{\mathbb{N}}\cap\Delta$. Note that we may take only countably many distinct $\widehat{x}(y,d,w,\lambda,\nu,r,\delta)$, $\widehat{z}(y,d,w,\lambda,\nu,r,\delta)$, and $\widehat{h}(y,d,w,\lambda,\nu,r,\delta)$. Indeed, the uples $(y,d,w,\lambda,\nu,r,\delta)$ are chosen from the countable set $C(F_{k-1})\times\mathcal{D}\times C(G_{k-1})\times(\mathbb{Q}^{\mathbb{N}}\cap\Lambda)\times\Upsilon\times\mathbb{Q}_>\times(\mathbb{Q}^{\mathbb{N}}\cap\Delta)$. Further, for any fixed y,d,w,λ,ν,r there is some $\overline{n}\in\mathbb{N}$ such that $\lambda_n = 0$ for all $n > \overline{n}$. Obviously, the choice of h_n for such n has no effect on the value of the function whose infimum is taken in (4.17), and thus it can be required that $\widehat{x}(y,d,w,\lambda,\nu,r,\delta) = \widehat{x}(y,d,w,\lambda,\nu,r,\delta')$, $\widehat{z}(y,d,w,\lambda,\nu,r,\delta) = \widehat{z}(y,d,w,\lambda,\nu,r,\delta')$, and $\widehat{h}(y,d,w,\lambda,\nu,r,\delta) = \widehat{h}(y,d,w,\lambda,\nu,r,\delta')$ whenever $\delta,\delta'\in\mathbb{Q}^{\mathbb{N}}\cap\Delta$ satisfy $\delta_{\nu_1} = \delta'_{\nu_1},\cdots,\delta_{\nu_{\overline{n}}} = \delta'_{\nu_{\overline{n}}}$.

We define E_k (resp. G_k) as the closure of the vector space spanned by the union of E_{k-1} (resp. G_{k-1}) and the countable set of all elements $\widehat{x}(y,d,w,\lambda,\nu,r,\delta)$ (resp. $\widehat{z}(y,d,w,\lambda,\nu,r,\delta)$) with $y\in C(E_{k-1})$, $d\in\mathcal{D}$, $w\in C(G_{k-1})$, $\lambda\in\mathbb{Q}^{\mathbb{N}}\cap\Lambda$, $\nu\in\Upsilon$, $r\in\mathbb{Q}_>$, $\delta\in\mathbb{Q}^{\mathbb{N}}\cap\Delta$. Similarly, we define F_k as the closure of the vector space spanned by the union of $F_{k-1}\cup A(E_k)\cup T(G_k)$ and the countable sets $\{\widehat{h}(y,d,w,\lambda,\nu,r,\delta)_n : n\in\mathbb{N}\}$ with the parameters $y,d,w,\lambda,\nu,r,\delta$ describing the same countable sets as above. Evidently, E_k, F_k and G_k are separable. We take countable sets $C(F_k)$ dense in F_k and $C(G_k)$ dense in the unit sphere of G_k, with $C(F_k)\supset C(F_{k-1})$ and $C(G_k)\supset C(G_{k-1})$, and then continue the process.

Define the vector spaces E, F and G as the closures of $\bigcup_{k\in\mathbb{N}} E_k$, $\bigcup_{k\in\mathbb{N}} F_k$, and $\bigcup_{k\in\mathbb{N}} G_k$, so $\widehat{x}(y,d,w,\lambda,\nu,r,\delta) \in E$, $\widehat{z}(y,d,w,\lambda,\nu,r,\delta) \in G$ for all parameters y, d, w, λ, ν, r, δ.

Step 4. To show that $X_0 := E$, $L_0 := F$ and $Z_0 := G$ are the desired spaces, let us fix any $y \in F$ and any $(c,\varepsilon,\rho) \in \mathcal{D}$, and let us prove that the set in (4.15) is nonempty whenever the one in (4.16) is nonempty. Suppose that there is an element $l^* \in F^*$ such that

(4.20) $\qquad l^* \in \partial_F f_0(y), \quad \|l^*\| \le c, \quad \|A_0^* l^*\| < \varepsilon, \quad |\|T_0^* l^*\| - \rho| < \varepsilon,$

where f_0, A_0 and T_0 denote the restrictions of f, A and T to E, F and G respectively. Lemma 4.35 applied to f_0, A_0 and T_0 yields some $\varepsilon' \in \mathbb{Q}_>$ with $\varepsilon' < \varepsilon$, some $w \in G$ with $\|w\| = 1$, and some $\delta = (\delta_n)$ in $\mathbb{Q} \cap \Delta$ such that the inequality in (a) of that lemma is satisfied for all $x \in E$, $z \in G$, $\lambda \in \Lambda$, $\nu = (\nu_n) \in \Upsilon$, and $(h_n) \in \mathcal{H}(\nu,\delta)$ with $h_n \in F$ for all $n \in \mathbb{N}$. It remains to prove that the condition $h_n \in F$ can be dropped and the inequality in (a) of Lemma 4.35 is still true for all $x \in X$ and all $z \in Z$. Indeed, with this at hand, Lemma 4.35 again will conclude the proof of the theorem.

For this purpose, fix any $x \in X$, any $z \in Z$, any $(\lambda_n) \in \Lambda$, any $\nu = (\nu_n) \in \Upsilon$, and any $(h_n) \in \mathcal{H}(\nu,\delta)$ with $h_n \in Y$. Suppose first that $(\lambda_n) \in \mathbb{Q}^\mathbb{N} \cap \Lambda$. Noting that $N := \{n \in \mathbb{N} : \lambda_n > 0\}$ is finite, take any $r \in \mathbb{Q}_>$ such that $\|h_n\| < \delta_{\nu_n} - 2r$ for every $n \in N$ and such that

(4.21) $\qquad\qquad -1/r < f(y) - r - cr - \rho r.$

Take $k \in \mathbb{N}$ such that $\text{dist}(y, F_k) < r$ and $\text{dist}(w, \mathbb{S}_{G_k}) < r$, where $\mathbb{S}_{G_k}$ denotes the closed unit sphere of G_k centered at zero. Take also some $y' \in C(F_k)$ (thus $y' \in F$ since $C(F_k) \subset F_k$) and some $w' \in C(G_k)$ such that $\|y' - y\| < r$ and $\|w' - w\| < r$. For each $n \in N$ put $h'_n := h_n + (y - y')$, so $y + h_n = y' + h'_n$, $\|h_n\| > \|h'_n\| - r$ and $\|h'_n\| < \delta_{\nu_n} - r$. Put also $h'_n = 0$ for all $n \in \mathbb{N} \setminus N$. Consider the triple $d := (c, \varepsilon', \rho) \in \mathcal{D}$ and observe from the definition of $g_{d,w}$ in (4.13) that

$$\sum_{n=1}^{\infty} \lambda_n \left(f(y + h_n) + \frac{1}{\nu_n} \|h_n\| \right) + g_{d,w}\left(x, \sum_{n=1}^{\infty} \lambda_n h_n, z\right)$$

(4.22) $\displaystyle \ge \sum_{n=1}^{\infty} \lambda_n \left(f(y' + h'_n) + \frac{1}{\nu_n} \|h'_n\| \right) + g_{d,w'}\left(x, \sum_{n=1}^{\infty} \lambda_n h'_n, z\right) - r - cr - \rho r.$

Let $\delta' = (\delta'_n) \in \mathbb{Q}^\mathbb{N} \cap \Delta$ be such that $\delta'_m < \delta_m$ for all $m \in \mathbb{N}$ and that $\delta'_{\nu_n} = \delta_{\nu_n} - r$ for all $n \in N$, so $(h'_n) \in \mathcal{H}(\nu, \delta')$. Denote $x' := \widehat{x}(y', d, w', \lambda, \nu, r, \delta')$, $z' := \widehat{z}(y', d, w', \lambda, \nu, r, \delta')$ and $\overline{h}' := \widehat{h}(y', d, w', \lambda, \nu, r, \delta')$, where

$$\widehat{x}(y', d, w', \lambda, \nu, r, \delta'), \ \widehat{z}(y', d, w', \lambda, \nu, r, \delta') \text{ and } \widehat{h}(y', d, w', \lambda, \nu, r, \delta')$$

are the elements obtained in Step 1. We observe, by the construction of E_k and E, G_k and G, F_k and F, that $x' \in E$, $z' \in G$, and $\overline{h}'_n \in F$ for all $n \in \mathbb{N}$.

Taking into account the relationship between (h_n) and (h'_n) above, put on the one hand $\overline{h}_n := \overline{h}'_n + (y' - y)$ for every $n \in N$, so $y + \overline{h}_n = y' + \overline{h}'_n$, $\|\overline{h}'_n\| > \|\overline{h}_n\| - r$, and $\|\overline{h}_n\| < \delta_{\nu_n}$. On the other hand, for every $n \in \mathbb{N} \setminus N$ put $\overline{h}_n = 0$. We then see that $(\overline{h}_n) \in \mathcal{H}(\nu, \delta)$ and also $\overline{h}_n \in F$ for all $n \in \mathbb{N}$ (since $y, y', \overline{h}'_n \in F$). As for

(4.22), from the definition of $g_{d,w'}$ we have

$$\sum_{n=1}^{\infty} \lambda_n \left(f(y' + \overline{h}_n') + \frac{1}{\nu_n} \|\overline{h}_n'\| \right) + g_{d,w'} \left(x', \sum_{n=1}^{\infty} \lambda_n \overline{h}_n', z' \right)$$

$$\geq \sum_{n=1}^{\infty} \lambda_n \left(f\left(y + \overline{h}_n\right) + \frac{1}{\nu_n} \|\overline{h}_n\| \right) + g_{d,w} \left(x', \sum_{n=1}^{\infty} \lambda_n \overline{h}_n, z \right) - r - cr - \rho r$$

(4.23) $\quad \geq f(y) - r - cr - \rho r,$

where the last inequality is due to the property in Lemma 4.35 obtained in the beginning of this Step 4 for the spaces E, F, G.

If $I(y', d, w', \lambda, \nu, \delta') = -\infty$, then we would have from (4.19)

$$-\frac{1}{r} > \sum_{n=1}^{\infty} \lambda_n \left(f\left(y' + \overline{h}_n'\right) + \frac{1}{\nu_n} \|\overline{h}_n'\| \right) + g_{d,w'} \left(x', \sum_{n=1}^{\infty} \lambda_n \overline{h}_n', z' \right),$$

which would contradict (4.23) and (4.21). Thus, the inequality $I(y', d, w', \lambda, \nu, \delta') > -\infty$ holds true, so (4.18) entails

$$\sum_{n=1}^{\infty} \lambda_n \left(f(y' + h_n') + \frac{1}{\nu_n} \|h_n'\| \right) + g_{d,w'} \left(x, \sum_{n=1}^{\infty} \lambda_n h_n', z \right)$$

$$> \sum_{n=1}^{\infty} \lambda_n \left(f\left(y' + \overline{h}_n'\right) + \frac{1}{\nu_n} \|\overline{h}_n'\| \right) + g_{d,w'} \left(x', \sum_{n=1}^{\infty} \lambda_n \overline{h}_n', z' \right) - r.$$

Combining the latter inequality and the inequalities (4.22) and (4.23), and making $r \downarrow 0$, we easily obtain

$$\sum_{n=1}^{\infty} \lambda_n \left(f(y + h_n) + \frac{1}{\nu_n} \|h_n\| \right) + g_{d,w} \left(x, \sum_{n=1}^{\infty} \lambda_n h_n, z \right) \geq f(y).$$

The latter inequality being true for all $\lambda \in \mathbb{Q}^{\mathbb{N}} \cap \Lambda$, the continuity of $g_{d,w}$ ensures that it holds true for all $\lambda \in \Lambda$. Lemma 4.35 allows us to confirm the first part of the theorem.

Regarding the case when $Y_1 \times \cdots \times Y_m$ (resp. Y^m, that is, $Y_i = Y$ for $i = 1, \cdots, m$) is considered in place of Y, put $F_0 := L$ and denote first by F_k' the closure of the vector space spanned by the union of $F_{k-1} \cup A(E_k) \cup T(G_k)$ and the countable sets $\{\widehat{h}(y, d, w, \lambda, \nu, r, \delta)_n : n \in \mathbb{N}\}$ with the parameters $y, d, w, \lambda, \nu, r, \delta$ describing the same countable sets as in Step 3, and denote also by $F_{k,1}, \cdots, F_{k,m}$ the projection of F_k' on $Y_1, \cdots, Y_m$ respectively. Defining F_k as $F_k := F_{k,1} \times \cdots \times F_{k,m}$ (resp. $F_k := V_k \times \cdots \times V_k$, where V_k is the closed vector subspace spanned by $\bigcup_{1 \leq i \leq m} F_{k,i}$) and noting that the closure F of $\bigcup_{k \in \mathbb{N}} F_k$ is a cartesian product, it is enough to proceed as above in Step 4 (observing in the case Y^m that the Cartesian product is in the form $L_0 \times \cdots \times L_0$). □

THEOREM 4.37 (M. Fabian and A.D. Ioffe: separable reduction of inclusion of zero in a sum of Fréchet subdifferentials). Let X be a Banach space and $f_1, \cdots, f_m$ be functions from X into $\mathbb{R} \cup \{+\infty\}$ which are locally bounded from below. Let also reals $r > 0$, $s \geq 0$, and let L be a separable subspace of X. Then there exists a closed separable subspace L_0 containing L such that, for any $x_1, \cdots, x_m$ in

$L_0 \cap \mathrm{dom}\, f$, the inclusion

$$0 \in \big(\partial_F f_1(x_1) \setminus s\mathbb{B}_{X^*}\big) + \partial_F f_2(x_2) + \cdots + \partial_F f_m(x_m) + B_{X^*}(0,r)$$

$$\big(\text{resp.} \quad 0 \in \partial_F f_1(x_1) + \partial_F f_2(x_2) + \cdots + \partial_F f_m(x_m) + B_{X^*}(0,r)\big)$$

is satisfied whenever

$$0 \in \big(\partial_F f_{0,1}(x_1) \setminus s\mathbb{B}_{L_0^*}\big) + \partial_F f_{0,2}(x_2) + \cdots + \partial_F f_{0,m}(x_m) + B_{L_0^*}(0,r)$$

(resp. if and only if

$$0 \in \partial_F f_{0,1}(x_1) + \partial_F f_{0,2}(x_2) + \cdots + \partial_F f_{0,m}(x_m) + B_{L_0^*}(0,r)\big),$$

where $f_{0,i} := f_i|_{L_0}$ denotes the restriction of f_i to L_0.

PROOF. Put $Y := X$, define $A : X \to Y^m$ by $A(x) := (x, \cdots, x)$, and set $f(y_1, \cdots, y_m) = f_1(y_1) + \cdots + f_m(y_m)$ for all $(y_1, \cdots, y_m) \in Y^m$. Put also $Z = X$ and define $T : Z \to Y^m$ as $T(x) := (x, 0, \cdots, 0)$ (resp. as the null mapping). For $y^* = (x_1^*, \cdots, x_m^*)$, we have $A^* y^* = x_1^* + \cdots + x_m^*$ and $T^* y^* = x_1^*$ (resp. $T^* = 0$). Consider the class $\mathcal{D}$ of triples $(c, \varepsilon, \rho) \in \mathbb{Q}^3$ with $c \geq 0$, $\varepsilon > 0$, and $\rho \geq 0$. Consider $X_0 \subset X$, $L_0 \times \cdots \times L_0 \subset Y^m$ and $Z_0 \subset Z$ given by Theorem 4.36, and denote by f_0 the restriction of f to $L_0 \times \cdots \times L_0$, by A_0 the restriction of A to X_0, and by T_0 the restriction of T to Z_0. Suppose that the second inclusion involving s (resp. without s) in the statement of the theorem is satisfied, so there are $l_i^* \in \partial_F f_{0,i}(x_i)$ such $\|l_1^* + \cdots + l_m^*\| < r$ and $\|l_1^*\| > s$ (resp. with no restriction on l_1^*). In the case involving s choose a real σ such that $s < \sigma < \|l_1^*\| < \sigma + 2r$ and put $\varepsilon := r$ and $\rho := \sigma + r$ (resp. in the case where s is not involved, put $\varepsilon := r$ and $\rho := 0$). Then for $l^* := (l_1^*, \cdots, l_m^*)$, we have $l^* \in \partial_F f_0(x_1, \cdots, x_m)$ and $\|A_0^* l^*\| < \varepsilon$, along with $|\|T_0^* l^*\| - \rho| < \varepsilon$ since $\sigma < \|l_1^*\| < \sigma + 2r$ (resp. since $T_0^* = 0$). Putting $c := \|l^*\|$, Theorem 4.36 gives $(x_1^*, \cdots, x_m^*) \in \partial_F f(x_1, \cdots, x_m)$ such that $\|A^*(x_1^*, \cdots, x_m^*)\| < \varepsilon = r$ and $|\|T^*(x_1^*, \cdots, x_m^*)\| - \rho| < \varepsilon = r$ (resp. $\|0x^*\| < \varepsilon$), hence $\|x_1^* + \cdots + x_m^*\| < r$ and $\|x_1^*\| > \sigma > s$ (resp. with no restriction on x_1^*). Consequently, as desired

$$0 \in \big(\partial_F f_1(x_1) \setminus s\mathbb{B}_{X^*}\big) + \partial_F f_2(x_2) + \cdots + \partial_F f_m(x_m) + B_{X^*}(0,r)$$

$$\big(\text{resp.} \quad 0 \in \partial_F f_1(x_1) + \partial_F f_2(x_2) + \cdots + \partial_F f_m(x_m) + B_{X^*}(0,r)\big).$$

Finally, if $0 \in \partial_F f_1(x_1) + \partial_F f_2(x_2) + \cdots + \partial_F f_m(x_m) + B_{X^*}(0,r)$, there are $x_i^* \in \partial_F f_i(x_i)$ with $\|x_1^* + \cdots + x_m^*\| < r$. Denoting by $x_{0,i}^*$ the restriction of x_i^* to L_0 we know that $\|x_{0,1}^* + \cdots + x_{0,m}^*\| < r$ (according to the definition of the norm of a linear functional), and further it is easy to see that $x_{0,i}^* \in \partial_F f_{0,i}(x_i)$ for all $i = 1, \cdots, m$. This completes the proof. $\square$

4.3. Fuzzy calculus rules for Fréchet subdifferentials

In addition to the above separable reduction of Fréchet subdifferentiability, the fuzzy sum rules for Fréchet subdifferentials will require smooth variational principles.

4.3.1. Borwein-Preiss variational principle.

Let f be an extended real-valued lower semicontinuous function over a Banach space $(X, \|\cdot\|)$ which is bounded from below (and which does not attain its infimum). As we saw in Chapter 2, the Ekeland variational principle provides a point b which is a minimizer of the perturbed function $f + \varepsilon \|\cdot - b\|$. If the function f is (real-valued and) differentiable, the perturbed function $f + \varepsilon \|\cdot - b\|$ is not differentiable at the point b, since a norm is never differentiable at the origin. So, it is natural, when $(X, \|\cdot\|)$ is a Hilbert space, to look for a perturbation in the form $f + \varepsilon \|\cdot - b\|^2$. Our goal is more generally devoted to the study of the context where X can be endowed with an equivalent norm which is differentiable out of the origin, that is, differentiable at any nonzero point of X. In such a space, we will obtain a perturbed function in the form $f + \varepsilon \Delta$ with a smooth function Δ that we express in terms of the renorm of X. This will allow us through such a variational principle to show for appropriate Banach spaces the richness of the set $\mathrm{Dom}\, \partial_F f$ in $\mathrm{dom}\, f$ as well as estimates of $\partial_F(f+g)$ in terms of $\partial_F f$ and $\partial_F g$.

We will in fact focus the study to the investigation of a strong minimizer for the perturbed function.

DEFINITION 4.38. Let (E, d) be a metric space and $g : E \to \mathbb{R} \cup \{+\infty\}$ (resp. $g : E \to \mathbb{R} \cup \{-\infty\}$) be a function which is bounded from below (resp. from above). One says that a point $b \in E$ is a *strong minimizer* (resp. *strong maximizer*) of g over E when $g(b) = \inf_E g$ (resp. $g(b) = \sup_E g$) and any minimizing sequence $(x_n)_n$ of g, that is, $g(x_n) \to \inf_E g$, (resp. maximizing sequence) converges to b with respect to the distance d. Such a point b is obviously unique whenever it exists. One also says that the infimum (resp. supremum) of f over E is *strongly attained* at $b \in E$.

If in addition f is lower (resp. upper) semicontinuous, it is clear that the infimum (resp. supremum) of f over E is strongly attained if and only if every minimizing (resp. maximizing) sequence converges in E.

Notice also that every strong minimizer is obviously a strict minimizer, that is, $g(b) < g(x)$ for all $x \neq b$ in E. The converse fails as shown by the simple example given by the function $g : \mathbb{R} \to \mathbb{R}$ with $g(x) = x^2 e^x$.

Assume that the function g is finite at some point of E and bounded from below (resp. from above). For each real $\varepsilon > 0$, put

$$A_\varepsilon(g) := \left\{ x \in E : g(x) \leq \varepsilon + \inf_E g \right\} \quad \left(\text{resp. } A_\varepsilon(g) := \left\{ x \in E : g(x) \geq -\varepsilon + \sup_E g \right\}\right).$$

If the metric space (E, d) is supposed to be *complete* and if g is lower (resp. upper) semicontinuous on E, then g admits a strong minimizer (resp. maximizer) point if and only if

$$\inf_{\varepsilon > 0} \left(\mathrm{diam}\, A_\varepsilon(g) \right) = 0,$$

or equivalently there exits a sequence of positive numbers $(\varepsilon_n)_n$ such that

(4.24) $$\mathrm{diam}\, A_{\varepsilon_n}(g) \to 0 \text{ as } n \to \infty.$$

THEOREM 4.39 (a general metric variational principle). Let (E, d) be a complete metric space and $f : E \to \mathbb{R} \cup \{+\infty\}$ be a proper lower semicontinuous function which is bounded from below on E. Let $a \in E$ and $\varepsilon > 0$ be such that

$$f(a) < \inf_E f + \varepsilon.$$

4.3. FUZZY CALCULUS RULES FOR FRÉCHET SUBDIFFERENTIALS

Let also a sequence of functions $\rho_n : E \times E \to [0, +\infty]$ satisfying:
(i) $\rho_n(x, x) = 0$ for every $x \in E$;
(ii) for any $x' \in E$ and any $n \in \mathbb{N}$ the function $\rho_n(\cdot, x')$ is lower semicontinuous;
(iii) for all $x_n \in E$ and $E_n \subset E$ (nonempty),

$$\left(\sup_{x \in E_n} \rho_n(x, x_n) \leq 1, \forall n \right) \Rightarrow \operatorname{diam} E_n \xrightarrow[n \to \infty]{} 0.$$

Then for any nonincreasing sequence $(\mu_n)_{n \in \mathbb{N}}$ of the open interval $]0, 1[$, there exists a sequence $(b_n)_{n \in \mathbb{N}}$ in E with $b_1 = a$ converging to some point $b \in E$ and such that, for $\rho_\infty(x) := \sum_{n=1}^{\infty} \mu_n \rho_n(x, b_n)$ one has
(a) $\rho_n(b, b_n) < 1$ and $f(b_n) \leq f(a)$, for all $n \in \mathbb{N}$;
(b) $f(b) + \varepsilon \rho_\infty(b) \leq f(a)$;
(c) the point b is a strong minimizer of $f + \varepsilon \rho_\infty$ over E.

In particular

$$f(b) + \varepsilon \rho_\infty(b) < f(x) + \varepsilon \rho_\infty(x) \quad \text{for all } x \neq b \text{ in } E.$$

PROOF. Put $b_1 = a$ and $f_1 = f$. We can by induction define f_{n+1}, choose b_{n+1} and consider E_n satisfying

(4.25) $$f_{n+1}(x) := f_n(x) + \varepsilon \mu_n \rho_n(x, b_n)$$

(4.26) $$f_{n+1}(b_{n+1}) \leq \frac{\mu_{n+1}}{2} f_n(b_n) + \left(1 - \frac{\mu_{n+1}}{2}\right) \inf_E f_{n+1} \leq f_n(b_n)$$

(4.27) $$E_n := \left\{ x \in E : f_{n+1}(x) \leq f_{n+1}(b_{n+1}) + \frac{1}{2} \mu_n \varepsilon \right\}.$$

To argue (4.26), note that $\inf_E f_{n+1} \leq f_{n+1}(b_n) = f_n(b_n)$. If the latter inequality is strict, then $\inf_E f_{n+1} < \frac{\mu_{n+1}}{2} f_n(b_n) + (1 - \frac{\mu_{n+1}}{2}) \inf_E f_{n+1}$ and this ensures the existence of b_{n+1}; if else $\inf_E f_{n+1} = f_n(b_n)$, it suffices to take $b_{n+1} = b_n$.

From (4.26) we derive that $f(b_n) \leq f_n(b_n) \leq f_1(b_1) = f(a)$, so the inequality $f(b_n) \leq f(a)$ is justified.

Observe that E_n is nonempty (since $b_{n+1} \in E_n$) and closed according to the lower semicontinuity of f_{n+1}, and observe also that $f_n \leq f_{n+1}$. Since $0 < \mu_n < 1$, it follows from (4.26) and from the assumption $f(a) < \inf_E f + \varepsilon$ that

(4.28) $$f_{n+1}(b_{n+1}) - \inf_E f_{n+1} \leq \frac{1}{2} \mu_{n+1} \left[f_n(b_n) - \inf_E f_{n+1} \right]$$
(4.29) $$\leq f_n(b_n) - \inf_E f_n \leq \cdots \leq f_1(b_1) - \inf_E f_1 < \varepsilon.$$

Claim 1: $E_{n+1} \subset E_n$.
Fix $x \in E_{n+1}$. The inequality $\mu_{n+1} \leq \mu_n$ and the inequality (4.26) entail

$$f_{n+1}(x) \leq f_{n+2}(x) \leq f_{n+2}(b_{n+2}) + \frac{1}{2} \mu_{n+1} \varepsilon \leq f_{n+1}(b_{n+1}) + \frac{1}{2} \mu_n \varepsilon,$$

thus $x \in E_n$ as desired.

Claim 2: $\operatorname{diam} E_n \to 0$.

Observe that

(4.30)
$$f_n(b_n) - \inf_E f_n \leq \frac{1}{2}\mu_n\left[f_{n-1}(b_{n-1}) - \inf_E f_n\right] \leq \frac{1}{2}\mu_n\left[f_{n-1}(b_{n-1}) - \inf_E f_{n-1}\right] < \frac{1}{2}\mu_n\varepsilon,$$

where the first and third inequalities are due to (4.28) and (4.29) respectively, and the second one to the nondecreasing property of the sequence $(f_n)_n$. Fix now any $x \in E_n$ and write by (4.27) and (4.25)

(4.31) $\quad \varepsilon\mu_n\rho_n(x, b_n) \leq f_{n+1}(b_{n+1}) - f_n(x) + \frac{1}{2}\mu_n\varepsilon \leq f_{n+1}(b_{n+1}) - \inf_E f_n + \frac{1}{2}\mu_n\varepsilon.$

Since $f_{n+1}(b_{n+1}) \leq f_n(b_n)$ according to (4.26), we deduce from (4.30) and (4.31) that

(4.32) $\qquad \varepsilon\mu_n\rho_n(x, b_n) < \varepsilon\mu_n, \quad \text{hence } \rho_n(x, b_n) < 1.$

This ensures by the assumption (iii) that $\operatorname{diam} E_n \to 0$, which proves the claim.

Note that Claim 2 yields the existence of some $b \in E$ such that $\bigcap_{n\in\mathbb{N}} E_n = \{b\}$. Further, on the one hand $b_n \to b$ since b and b_n belong to E_n, and on the other hand the same inclusion $b \in E_n$ also entails $\rho_n(b, b_n) < 1$ according to (4.32), which is the assertion (a).

Claim 3: $f(b) + \varepsilon\rho_\infty(b) \leq f_n(b_n)$ for all $n \in \mathbb{N}$.

Observe that, for $Y_n := \{x \in E : f_{n+1}(x) \leq f_{n+1}(b_{n+1})\}$, the sequence $(Y_n)_{n\in\mathbb{N}}$ is a nonincreasing sequence of nonempty closed sets with $Y_n \subset E_n$. Consequently, $\bigcap_{n\in\mathbb{N}} Y_n = \{b\}$, and hence for any integer $k \geq n+1$, it results that

$$f_k(b) \leq f_k(b_k) \leq f_n(b_n) \leq f_1(b_1) = f(a),$$

where the first inequality is due to the inclusion $b \in Y_{k-1}$ and the second and third inequalities are due to (4.26). Noticing from (4.25) that $f_k(b) \to f(b) + \varepsilon\rho_\infty(b)$ as $k \to \infty$, we deduce that

$$f(b) + \varepsilon\rho_\infty(b) \leq f_n(b_n) \leq f(a),$$

which translates the claim and the assertion (b) as well.

Claim 4: The point b is a strong minimizer of $g := f + \varepsilon\rho_\infty$.

Observe first that the function g is lower semicontinuous and bounded from below and that the assertion (b) says in particular that g is proper. Take any $x \in A_{\frac{1}{2}\mu_n\varepsilon}(g)$, that is, $g(x) \leq \inf_E g + \frac{1}{2}\mu_n\varepsilon$. Then

$$f_{n+1}(x) \leq g(x) \leq g(b) + \frac{1}{2}\mu_n\varepsilon \leq f_{n+1}(b_{n+1}) + \frac{1}{2}\mu_n\varepsilon,$$

where the first inequality is due to the definitions of f_{n+1} and g, and the last inequality is due to Claim 3. This means that $A_{\frac{1}{2}\mu_n\varepsilon}(g) \subset E_n$, thus $\operatorname{diam} A_{\frac{1}{2}\mu_n\varepsilon}(g) \to 0$ as $n \to \infty$. Consequently, by (4.24) the point b is a strong minimum over E of g, which is Claim 4 and completes the proof of the theorem. □

We observe in the next corollary that the Ekeland variational principle can be seen as a consequence of the above theorem, but with the additional property that the point b is a strong minimizer of the perturbed function.

COROLLARY 4.40 (Ekeland variational principle with strong minimizer). In Theorem 2.221 on Ekeland variational principle, the pair of assertions (b) and (c) can be replaced by the following pair (b') and (c'):
(b') $f(b) + \frac{\varepsilon}{2\lambda}d(a,b) \leq f(a)$;
(c') the point b is a strong minimizer over E of the function $f + \frac{\varepsilon}{\lambda}d(\cdot, b)$.

PROOF. We may apply Theorem 4.39 with
$$\rho_n(x,y) := \frac{n}{\lambda}d(x,y) \quad \text{and} \quad \mu_n := \frac{1}{n2^n}.$$
Observe that, for b_n and b given by the theorem, we have for $x \neq b$
$$f(b) < f(x) + \varepsilon\bigl(\rho_\infty(x) - \rho_\infty(b)\bigr)$$
$$= f(x) + \frac{\varepsilon}{\lambda}\sum_{k=1}^{\infty}\frac{1}{2^k}\bigl(d(x,b_k) - d(b,b_k)\bigr)$$
(4.33)
$$\leq f(x) + \frac{\varepsilon}{\lambda}d(x,b).$$

This is the assertion (c) in the Ekeland variational principle, and assertions (a) and (b') follow directly from conclusions (a) and (b) in Theorem 4.39.

It remains to show that b is a strong minimizer of the function $f + \frac{\varepsilon}{\lambda}d(\cdot, b)$. Let $(x_n)_n$ be a minimizing sequence of the latter function. Since $f(b) = \inf_{x \in E}\bigl(f(x) + \frac{\varepsilon}{\lambda}d(x,b)\bigr)$ by the latter inequality above, we have
$$f(x_n) + \frac{\varepsilon}{\lambda}d(x_n, b) \to f(b),$$
which implies $f(x_n) + \varepsilon\rho_\infty(x_n) \to f(b) + \varepsilon\rho_\infty(b)$ according to inequalities (4.33). The assertion (c) in Theorem 4.39 then yields $x_n \to b$, hence b is a strong minimizer of the function $f + \frac{\varepsilon}{\lambda}d(\cdot, b)$ over E. $\square$

We come now to the Borwein-Preiss variational principle which is another useful variational principle that we will also derive from Theorem 4.39.

THEOREM 4.41 (Borwein-Preiss variational principle with strong minimizer). Let $(X, \|\cdot\|)$ be a Banach space and $f : X \to \mathbb{R} \cup \{+\infty\}$ be a proper lower semicontinuous function which is bounded from below on X. Let two reals $p \geq 1$ and $\lambda > 0$ and let $(\alpha_n)_{n \in \mathbb{N}}$ be a nonincreasing sequence in $]0,1[$ with $\sum_{n=1}^{\infty}\alpha_n = 1$.
Let $a \in X$ and $\varepsilon > 0$ with
$$f(a) < \inf_X f + \varepsilon.$$
Then there exists a sequence $(b_n)_{n \in \mathbb{N}}$ with $b_1 = a$ converging to some point $b \in X$ and such that, for $\Delta_p(x) := \sum_{n=1}^{\infty}\alpha_n\|x - b_n\|^p$ one has
(a) for all $n \in \mathbb{N}$ the inequalities $f(b_n) \leq f(a)$ and $\|b - b_n\| < \lambda$, hence in particular $\|b - a\| < \lambda$;
(b) $f(b) + \frac{\varepsilon}{\lambda^p}\Delta_p(b) \leq f(a)$;
(c) the point b is a strong minimizer over X of the function $f + \frac{\varepsilon}{\lambda^p}\Delta_p$.

In particular, one has
$$f(b) + \frac{\varepsilon}{\lambda^p}\Delta_p(b) < f(x) + \frac{\varepsilon}{\lambda^p}\Delta_p(x) \quad \text{for all } x \neq b \text{ in } X.$$

PROOF. It suffices to apply Theorem 4.39 with
$$\rho_n(x,y) := \frac{n}{\lambda^p}\|x-y\|^p \quad \text{and} \quad \mu_n := \frac{\alpha_n}{n}.$$
□

Let us examine the case when X is a Hilbert space H.

THEOREM 4.42 (Borwein-Preiss variational principle in Hilbert space). Let H be a Hilbert space and $\|\cdot\|$ be the norm associated with the inner product, and let $f : H \to \mathbb{R} \cup \{+\infty\}$ be a proper lower semicontinuous function which is bounded from below on H.

Let $a \in H$ and $\varepsilon > 0$ be such that
$$f(a) < \inf_H f + \varepsilon.$$
Then for every real $\lambda > 0$ there exist two elements b and q in H such that
(a) $\|b-a\| < \lambda$ and $\|b-q\| < \lambda$;
(b) $f(b) + \frac{\varepsilon}{2\lambda^2}\|b-a\|^2 \leq f(a)$;
(c) the point b is a strong minimizer of the function $f + \frac{\varepsilon}{\lambda^2}\|\cdot-q\|^2$, hence in particular
$$f(b) + \frac{\varepsilon}{\lambda^2}\|b-q\|^2 < f(x) + \frac{\varepsilon}{\lambda^2}\|x-q\|^2 \quad \text{for all } x \neq b \text{ in } H;$$
(d) for the concave quadratic function $\varphi(x) := -(\varepsilon/\lambda^2)(\|x-q\|^2 - \|b-q\|^2) + f(b)$ and for $\zeta := -2(\varepsilon/\lambda^2)(b-q)$ one has $\zeta = \nabla\varphi(b)$, $\varphi(b) = f(b)$, and
$$\varphi(x) < f(x) \quad \text{for all } x \neq b \text{ in } H,$$
and the latter inequality gives in particular
$$\langle \zeta, x-b \rangle \leq f(x) - f(b) + \frac{\varepsilon}{\lambda^2}\|x-b\|^2 \quad \text{for all } x \in H.$$

PROOF. Applying Theorem 4.41 with $p = 2$ and $\alpha_n := 1/2^n$ we immediately obtain the first inequality of (a) and the property (b) of the theorem from the assertions (a) and (b) in Theorem 4.41. Note that the sequence $(b_n)_n$ furnished by Theorem 4.41 is bounded. So, putting $q := \sum_{n=1}^{\infty} \frac{1}{2^n} b_n$, we have for every $x \in H$

$$\Delta_2(x) = \sum_{n=1}^{\infty} \frac{1}{2^n}\|x-b_n\|^2$$
$$= \sum_{n=1}^{\infty} \frac{1}{2^n}\|x\|^2 - 2\left\langle x, \sum_{n=1}^{\infty} \frac{1}{2^n} b_n \right\rangle + \sum_{n=1}^{\infty} \frac{1}{2^n}\|b_n\|^2$$
$$= \left(\|x\|^2 - 2\langle x,q \rangle + \|q\|^2\right) - \left(\|q\|^2 - \sum_{n=1}^{\infty} \frac{1}{2^n}\|b_n\|^2\right),$$

hence $\Delta_2(x) = \|x-q\|^2 + \gamma$ for some constant γ (independent of x and n). Consequently, the property (c) follows from the assertion (c) in Theorem 4.41. Let us verify the second inequality of (a). From the assertion (a) in Theorem 4.41 we have $\|b-b_n\| < \lambda$ for all $n \in \mathbb{N}$, thus by definition of q we get

$$\|b-q\| \leq \sum_{n=1}^{\infty} \frac{1}{2^n}\|b-b_n\| < \lambda,$$

hence the desired inequality.

Finally, the concave quadratic function φ satisfies obviously by (c) the inequality $\varphi(x) < f(x)$ for all $x \neq b$ and the second inequality in (d) follows from the latter inequality combined with the equality
$$\varphi(x) = \langle \zeta, x - b \rangle - (\varepsilon/\lambda^2)\|x - b\|^2 + f(b).$$
$\square$

The next lemma establishes some properties of the function Δ_p involved in the Borwein-Preiss principle. In particular, it is shown that this function is differentiable whenever the norm is differentiable outside of zero, making in this way the perturbed function $f + \frac{\varepsilon}{\lambda^p}\Delta_p$ differentiable whenever $p > 1$ and f is itself differentiable.

LEMMA 4.43. *Let $(X, \|\cdot\|)$ be a normed space and $(b_n)_{n\in\mathbb{N}}$ be a bounded sequence in X, and let $\sum \alpha_n$ be a convergent series of nonnegative real numbers. Then, for any real $p \geq 1$ the function Δ_p defined by*
$$\Delta_p(x) := \sum_{n=1}^{\infty} \alpha_n \|x - b_n\|^p$$
is finite, convex, and continuous on X.

If in addition $p > 1$ and the norm $\|\cdot\|$ is Fréchet (resp. Gâteaux) differentiable outside the origin, then Δ_p is Fréchet (resp. Gâteaux) differentiable on the whole space X. Further, for any upper bound $\gamma(x)$ of the sequence $(\|x - b_n\|)_{n\in\mathbb{N}}$ one has
$$\|D\Delta_p(x)\|_* \leq p\big(\gamma(x)\big)^{p-1}\sum_{n=1}^{\infty}\alpha_n,$$
hence in particular $\|D\Delta_p(\cdot)\|_*$ is bounded over any bounded set of X.

PROOF. The function Δ_p is obviously convex. Further, for any fixed $a \in X$ it is easily seen that there exists some constant $\sigma > 0$ such that $\alpha_n\|x - b_n\| \leq \sigma\alpha_n$ for all $x \in B(a, 1)$ and all $n \in \mathbb{N}$. This combined with the convergence of the series $\sum \alpha_n$ entails that Δ_p is continuous on $B(a, 1)$, so it is continuous on X.

Now suppose that $p > 1$ and the norm $\|\cdot\|$ is Fréchet (resp. Gâteaux) differentiable outside zero. Take $B = \mathbb{B}_X$ if $\|\ \|$ is Fréchet differentiable and B as any singleton of $\mathbb{B}_X$ if $\|\cdot\|$ is Gâteaux differentiable. Noting for $t > 0$ that $t^{-1}(\|th\|^p - \|0\|^p) = t^{p-1}\|h\|^p$, we see that the function $\|\cdot\|^p$ is Fréchet differentiable at the origin, hence it is Fréchet (resp. Gâteaux) differentiable on the whole space X. Then, so is each function $g_n := \alpha_n\|\cdot - b_n\|^p$. Putting $\Lambda(x) = D\|\cdot\|(x)$ if $x \neq 0$ and taking as $\Lambda(0)$ the null linear functional on X, we have for any $x \in X$
$$Dg_n(x) = p\alpha_n\|x - b_n\|^{p-1}\Lambda(x - b_n).$$
Fix $x \in X$ and $\varepsilon > 0$. Denote by $\gamma(x)$ an upper bound of the sequence $(\|x-b_n\|)_{n\in\mathbb{N}}$. Since $\|Dg_n(x)\|_* \leq p\alpha_n\big(\gamma(x)\big)^{p-1}$, the series $\sum Dg_n(x)$ converges in X^*, hence we can put $x^* := \sum_{n=1}^{\infty} Dg_n(x)$. Taking $h \in B$ we then obtain for any $n \in \mathbb{N}$ and $t \in]0, 1]$
$$t^{-1}\big|\|x - b_n + th\|^p - \|x - b_n\|^p\big|$$
$$= t^{-1}\big|\|x - b_n + th\| - \|x - b_n\|\big|\big(p(\theta\|x - b_n + th\| + (1 - \theta)\|x - b_n\|)\big)^{p-1}$$
$$\leq t^{-1}\|th\|p\big(1 + \gamma(x)\big)^{p-1} \leq p\big(1 + \gamma(x)\big)^{p-1},$$

where the equality holds for some $\theta \in [0,1]$, hence
$$t^{-1}|g_n(x+th) - g_n(x) - t\langle Dg_n(x), h\rangle| \leq p\alpha_n\Big((\gamma(x))^{p-1} + (1+\gamma(x))^{p-1}\Big).$$

We can then choose an integer N such that, for all $h \in B$, $t \in]0,1]$, and $n \geq N$
$$\sum_{n=1+N}^{\infty} t^{-1}|g_n(x+th) - g_n(x) - t\langle Dg_n(x), h\rangle| \leq \varepsilon/2.$$

But the differentiability of each g_n at x ensures the existence of some positive $\delta < 1$ such that, for all $h \in B$ and $t \in]0,\delta]$
$$\sum_{n=1}^{N} t^{-1}|g_n(x+th) - g_n(x) - t\langle Dg_n(x), h\rangle| \leq \varepsilon/2.$$

It then follows that, for all $h \in B$ and $t \in]0,\delta]$
$$t^{-1}|\Delta_p(x+th) - \Delta_p(x) - t\langle x^*, h\rangle| \leq \varepsilon,$$

thus Δ_p is Fréchet (resp. Gâteaux) differentiable at x with $D\Delta_p(x) = x^*$, and this equality ensures
$$\|D\Delta_p(x)\|_* \leq p(\gamma(x))^{p-1}\sum_{n=1}^{\infty} \alpha_n.$$

$\square$

As a direct consequence of the above lemma and Theorem 4.41 we have:

THEOREM 4.44 (a variational principle under differentiable norm). Assume that the norm $\|\cdot\|$ is Fréchet differentiable (resp. $\mathcal{C}^1$, $\mathcal{C}^{1,\sigma}$) outside zero. Then, putting $\varphi(x) := -(\varepsilon/\lambda^p)(\Delta_p(x) - \Delta_p(b)) + f(b)$ and $\zeta := D\varphi(b)$, one has in addition to the conclusions of Theorem 4.41
$$\|\zeta\| \leq p\varepsilon/\lambda, \quad \varphi(b) = f(b) \quad \text{and} \quad \varphi(x) < f(x) \quad \text{for all } x \in X \setminus \{b\}.$$

4.3.2. Fuzzy sum rule for Fréchet subdifferential under Fréchet differentiable renorm. To reach a sum rule for the Fréchet subdifferential we need the next lemma which also has its own interest. Recall first that, for a nonempty set C of the normed space X, the modulus of uniform continuity relative to C of a function $f: C \to \mathbb{R}$ is defined by
$$\omega_f(t) := \sup\{|f(u) - f(v)| : u, v \in C, \|u - v\| \leq t\} \quad \text{for all } t \geq 0,$$

so f is uniformly continuous on C (relative to the induced metric on C) if and only if $\omega_f(t) \to 0$ as $t \downarrow 0$. Recall also that, given a Hausdorff locally convex space (X, τ) with topological dual X^*, a Hausdorff locally convex topology θ on X is compatible with the duality between X and X^* provided that the topological dual of (X, θ) coincides with X^*.

LEMMA 4.45. Let $(X, \|\cdot\|)$ be a normed space, C be a nonempty subset of X, and $f_i : C \to \mathbb{R} \cup \{+\infty\}$, $i = 1, \cdots, n$ be n functions on C which are simultaneously finite at a certain point of C. Then, for any real $p > 0$
$$\lim_{r \to +\infty} \inf_{u \in C^n} \left(\sum_{i=1}^n f_i(u_i) + \sum_{1 \leq i < j \leq n} r\|u_i - u_j\|^p\right) = \inf_{x \in C} \sum_{i=1}^n f_i(x)$$

under anyone of conditions (a) and (b):
(a) for some Hausdorff locally convex topology θ on X compatible with the duality between X and X^*, the functions f_i are lower θ-semicontinuous and bounded from below on C and one of them, say f_n, has all its sublevel sets in C θ-compact, that is, $\{x \in C : f_n(x) \leq \rho\}$ is θ-compact for every real ρ;
(b) the functions f_i are bounded from below on C and all of them, except perhaps f_n, are finite on C and uniformly continuous on C (relative to the induced metric on C).

PROOF. Put $I := \inf_{x \in C} \sum_{i=1}^{n} f_i(x)$ and put also, for each real $r > 0$,

$$I(r) := \inf_{u \in C^n} \left(\sum_{i=1}^{n} f_i(u_i) + \sum_{1 \leq i < j \leq n} r \|u_i - u_j\|^p \right).$$

Obviously $I(r)$ and I are finite and $I(r) \leq I$ for all $r > 0$.

For each $r > 0$ take $(u_{1,r}, \cdots, u_{n,r}) \in C^n$ such that

(4.34) $$\sum_{i=1}^{n} f_i(u_{i,r}) + \sum_{1 \leq i < j \leq n} r \|u_{i,r} - u_{j,r}\|^p \leq I(r) + (1/r).$$

In both cases of the lemma, the boundedness from below of the functions f_i on C ensures, for each pair (i,j) with $i < j$, that $\|u_{i,r} - u_{j,r}\| \to 0$ as $r \to +\infty$.

If $f_1, \cdots, f_{n-1}$ are uniformly continuous on C, denoting by $\omega_i(\cdot)$ the modulus of uniform continuity relative to C of f_i (for $i = 1, \cdots, n-1$) we have

$$I(r) \geq \sum_{i=1}^{n} f_i(u_{i,r}) + \sum_{1 \leq i < j \leq n} r \|u_{i,r} - u_{j,r}\|^p - (1/r)$$

$$\geq \sum_{i=1}^{n} f_i(u_{n,r}) - \sum_{i=1}^{n-1} \omega_i(\|u_{i,r} - u_{n,r}\|) + \sum_{1 \leq i < j \leq n} r \|u_{i,r} - u_{j,r}\|^p - (1/r)$$

$$\geq I - \sum_{i=1}^{n-1} \omega_i(\|u_{i,r} - u_{n,r}\|) - (1/r).$$

This entails $I - \sum_{i=1}^{n-1} \omega_i(\|u_{i,r} - u_{n,r}\|) - (1/r) \leq I(r) \leq I$, thus $I(r) \to I$ as $r \to +\infty$.

Consider now the case (a) where the sublevel sets in C of f_n are θ-compact. Take any sequence $(r_k)_k$ in $]0, +\infty[$ converging to $+\infty$ and put $u'_{i,k} := u_{i,r_k}$. We note that the sequence $(u'_{n,k})_k$ belongs to the θ-compact set $\{x \in C : f_n(x) \leq s\}$, where $s := I + 1 - \sum_{i=1}^{n-1} \inf_C f_i$. Some subnet (not relabeled) of $(u'_{n,k})_k$ θ-converges to some point $c \in C$, and for each $i = 1, \cdots, n-1$ the sequence $(u'_{i,k})_k$ θ-converges to c as well, since $\|u'_{i,k} - u'_{n,k}\| \to 0$. The lower θ-semicontinuity property of $f_1, \cdots, f_n$ on C yields

$$\sum_{i=1}^{n} f_i(c) \leq \liminf_k \sum_{i=1}^{n} f_i(u'_{i,k}) \leq \liminf_k I(r_k),$$

the right inequality being due to (4.34). This combined with the inequality $I(r) \leq I$ entails that $I(r_k) \to I$, completing the proof. □

PROPOSITION 4.46. Let X be a Banach space admitting an equivalent norm $\|\cdot\|$ which is Fréchet differentiable off zero and let $f_i : X \to \mathbb{R} \cup \{-\infty, +\infty\}$, $i = 1, \cdots, m$ be functions which are finite at $\bar{x} \in X$. Assume that $\bar{x}$ is a local minimizer of $f_1 + \cdots + f_m$. The following hold:

(a) If f_1 is lower semicontinuous near $\bar{x}$ and $f_2, \cdots, f_m$ are finite near $\bar{x}$ and uniformly continuous near $\bar{x}$, then for any real $\varepsilon > 0$ there exist $x_1, \cdots, x_m \in B(\bar{x}, \varepsilon)$ (all depending on x^* and ε) with $|f_1(x_1) - f_1(\bar{x})| < \varepsilon$ such that
$$0 \in \partial_F f_1(x_1) + \cdots + \partial_F f_m(x_m) + \varepsilon \mathbb{B}_{X^*}.$$

(b) If all the functions $f_1, \cdots, f_m$ are lower semicontinuous near $\bar{x}$, then for any real $\varepsilon > 0$, and any finite-dimensional vector subspace L of X there exist $x_1, \cdots, x_m \in B(\bar{x}, \varepsilon)$ (all depending on x^*, ε, and L) with $|f_i(x_i) - f_i(\bar{x})| < \varepsilon$ for $i = 1, \cdots, m$, such that
$$0 \in \partial_F f_1(x_1) + \cdots + \partial_F f_m(x_m) + \varepsilon \mathbb{B}_{X^*} + L^\perp,$$
where $L^\perp := \{u^* \in X^* : \langle u^*, u \rangle = 0, \forall u \in L\}$.

PROOF. Endow X with the Fréchet differentiable norm $\|\cdot\|$. Choose $\delta > 0$ such that $\bar{x}$ is a minimizer of $f_1 + \cdots + f_m$ on the set $C := B[\bar{x}, \delta]$. We may suppose that $f_1(\bar{x}) + \cdots + f_m(\bar{x}) = 0$.

(a) We shrike δ such that f_m is lower semicontinuous on C and $f_2, \cdots, f_m$ are finite on C and uniformly continuous on C. Endow $X^m = X \times \cdots \times X$ with the Fréchet differentiable (off zero) norm given by $\|x\| = \left(\|x_1\|^2 + \cdots + \|x_m\|^2\right)^{1/2}$ and fix a real sequence $(r_k)_k$ with $r_k > 0$ and $r_k \to +\infty$ as $k \to +\infty$. According to the uniform continuity of $f_2, \cdots, f_m$ on C, the above lemma says that, for $h(x) := f_1(x_1) + \cdots + f_m(x_m)$,

$$\eta'_k := \inf_{x \in C^m} \left(h(x) + \frac{r_k}{2} \sum_{1 \le i < j \le m} \|x_i - x_j\|^2 \right) \xrightarrow[k \to \infty]{} \inf_C (f_1 + \cdots + f_m) = 0.$$

Further, for $h_0(x) := h(x) + \frac{r_k}{2} \sum_{1 \le i < j \le m} \|x_i - x_j\|^2$, we have $\eta'_k \le h_0(\bar{x}, \cdots, \bar{x}) = 0$. We may then define $\eta_k := \sqrt{-\eta'_k + (1/k)}$ and we see that
$$h_0(\bar{x}, \cdots, \bar{x}) < \inf_{C^m} h_0 + \eta_k^2$$
since $h_0(\bar{x}, \cdots, \bar{x}) = 0 < \eta'_k + (-\eta'_k + (1/k))$. Thus the Borwein-Preiss principle applied to the function $h_0 + \Psi_{C^m}$ (see Theorem 4.41) with $p = 2$, $\varepsilon = \eta_k^2$ and $\lambda = \sqrt{\eta_k}$ yields a sequence $(b^{n,k})_n$ in C^m converging to some $b^k = (b_1^k, \cdots, b_m^k) \in C^m$ with $b^{n,k} = (b_1^{n,k}, \cdots, b_m^{n,k}) \in C^m$, and such that $\|(b_1^k, \cdots, b_m^k) - (\bar{x}, \cdots, \bar{x})\| < \sqrt{\eta_k}$ and, for
$$h_1(x) := h_0(x) + \eta_k \Delta_1(x_1) + \cdots + \eta_k \Delta_m(x_m),$$
the function h_1 attains its minimum on C^m at $(b_1^k, \cdots, b_m^k)$; here with a fixed sequence $(\alpha_n)_n$ as in the statement of Theorem 4.41, the functions $\Delta_1, \cdots, \Delta_m$ are defined by
$$\Delta_i(x_i) = \sum_{n=1}^\infty \alpha_n \|x_i - b_i^{n,k}\|^2, \text{ for } i = 1, \cdots, m.$$

For k large enough, we have $b_i^k \in B(\bar{x}, \delta)$, hence $b_1^k, \cdots, b_m^k$ are local minimizers of the functions
$$x_1 \mapsto h_1(x_1, b_2^k, \cdots, b_m^k), \cdots, x_m \mapsto h_1(b_1^k, \cdots, b_{m-1}^k, x_m)$$

respectively. So, denoting by $J(u)$ the Fréchet derivative of $\frac{1}{2}\|\cdot\|^2$ at u we obtain that, for each $i \in \{1, \cdots, m\}$

$$-r_k \sum_{j=1, j\neq i}^{j=m} J(b_i^k - b_j^k) - \eta_k D\Delta_i(b_i^k) \in \partial_F f_i(b_i^k).$$

Adding together the latter inclusions (with $i = 1, \cdots, m$) and taking into account the boundedness of $D\Delta_i$ over any bounded set (see Lemma 4.43) yield

$$0 \in \partial_F f_1(b_1^k) + \cdots + \partial_F f_m(b_m^k) + \sigma_k \mathbb{B}_{X^*}$$

for some positive real $\sigma_k \to 0$ as $k \to \infty$.

To conclude the assertion (a) let us show the proximity between $f_i(b_i^k)$ and $f_i(\bar{x})$. We proceed in such a way that the arguments are also valid for the proximity of $f_i(x_i)$ and $f_i(\bar{x})$ in the case (b). We know that the point $(b_1^k, \cdots, b_m^k)$ above given by the Borwein-Preiss principle satisfies

$$h_0(b_1^k, \cdots, b_m^k) + \eta_k \Delta_1(b_1^k) + \cdots + \eta_k \Delta_m(b_m^k) \leq h_0(\bar{x}, \cdots, \bar{x}) = 0,$$

and this ensures

$$f_1(b_1^k) + \cdots + f_m(b_m^k) \leq 0 = f_1(\bar{x}) + \cdots + f_m(\bar{x}).$$

The latter inequality combined with the lower semicontinuity property of $f_1, \cdots, f_m$ at $\bar{x}$ easily entails that $f_i(b_i^k) \to f_i(\bar{x})$ for all $i = 1, \cdots, m$. The assertion (a) is then established.

(b) Concerning (b), we may suppose that all functions $f_1, \cdots, f_m$ are lower semicontinuous on C. Fix any finite-dimensional space L of X. Obviously, $\bar{x}$ is a minimizer on C of $f_1 + \cdots + f_m + f_{m+1}$, where $f_{m+1} := \Psi_{\bar{x}+L}$. Fix $r_k > 0$ with $r_k \to +\infty$. The sublevel sets of f_{m+1} on C being evidently compact, we may apply again the above lemma to obtain that

$$\eta'_k := \inf_{x \in C^{m+1}} \left(\sum_{i=1}^{m+1} f_i(x_i) + \frac{r_k}{2} \sum_{1 \leq i < j \leq m+1} \|x_i - x_j\|^2 \right) \xrightarrow[k \to \infty]{} \inf_C (f_1 + \cdots + f_{m+1}) = 0$$

and $\eta'_k \leq 0$. Proceeding as above yields some $x_i \in B(\bar{x}, \varepsilon)$ with $|f_i(x_i) - f_i(\bar{x})| < \varepsilon$ and some $x_i^* \in \partial_F f_i(x_i)$ such that $0 \in x_1^* + \cdots + x_m^* + x_{m+1}^* + \varepsilon \mathbb{B}_{X^*}$. Since $\partial_F f_{m+1}(x_{m+1}) = L^\perp$, it results that $x_i \in B(\bar{x}, \varepsilon)$, $|f_i(x_i) - f_i(\bar{x})| < \varepsilon$, for $i \in \{1, \cdots, m\}$, and

$$0 \in \partial_F f_1(x_1) + \cdots + \partial_F f_m(x_m) + \varepsilon \mathbb{B}_{X^*} + L^\perp,$$

which completes the proof. $\square$

The sum rule in the theorem below for a Fréchet subgradient of the sum of finitely many functions at a point $\bar{x}$ involves Fréchet subgradients of the data functions not at the reference point $\bar{x}$ but at nearby points. Such a rule is termed *fuzzy sum rule* in the literature. We prove first the result in the setting of Banach space with a Fréchet differentiable renorm.

THEOREM 4.47 (fuzzy sum rule for Fréchet subdifferential under differentiable renorm). Let X be a Banach space admitting an equivalent norm $\|\cdot\|$ which is Fréchet differentiable off zero and let $f_i : X \to \mathbb{R} \cup \{-\infty, +\infty\}$, $i = 1, \cdots, m$ be functions which are finite at $\bar{x} \in X$. The following hold:
(a) If f_1 is lower semicontinuous near $\bar{x}$ and $f_2, \cdots, f_m$ are finite near $\bar{x}$ and

uniformly continuous near $\bar{x}$, then for any $x^* \in \partial_F(f_1 + \cdots + f_m)(\bar{x})$ and any real $\varepsilon > 0$ there exist $x_1, \cdots, x_m \in B(\bar{x}, \varepsilon)$ (all depending on x^* and ε) with $|f_1(x_1) - f_1(\bar{x})| < \varepsilon$ such that

$$x^* \in \partial_F f_1(x_1) + \cdots + \partial_F f_m(x_m) + \varepsilon \mathbb{B}_{X^*}.$$

(b) If all the functions $f_1, \cdots, f_m$ are lower semicontinuous near $\bar{x}$, then for any $x^* \in \partial_F(f_1 + \cdots + f_m)(\bar{x})$, any real $\varepsilon > 0$, and any finite-dimensional vector subspace L of X there exist $x_1, \cdots, x_m \in B(\bar{x}, \varepsilon)$ (all depending on x^*, ε, and L) with $|f_i(x_i) - f_i(\bar{x})| < \varepsilon$ for $i = 1, \cdots, m$, such that

$$x^* \in \partial_F f_1(x_1) + \cdots + \partial_F f_m(x_m) + \varepsilon \mathbb{B}_{X^*} + L^\perp,$$

where $L^\perp := \{u^* \in X^* : \langle u^*, u \rangle = 0, \forall u \in L\}$.

PROOF. Fix any real $\varepsilon > 0$. We know by Proposition 4.25 that the inclusion $x^* \in \partial_F(f_1 + \cdots + f_m)(\bar{x})$ means that there exists an open neighborhood U of $\bar{x}$ and a $\mathcal{C}^1$ function $\varphi : U \to \mathbb{R}$ such that $D\varphi(\bar{x}) = x^*$ along with

$$\varphi(\bar{x}) = f_1(\bar{x}) + \cdots + f_m(\bar{x}) \quad \text{and} \quad \varphi(u) \leq f_1(u) + \cdots + f_m(u) \; \forall u \in U.$$

We may suppose $\|D\varphi(u) - D\varphi(\bar{x})\| \leq \varepsilon/2$ for all $u \in U$. Putting $f_{m+1} := -\varphi$, the point $\bar{x}$ is a local minimizer of $f_1 + \cdots + f_{m+1}$, so applying the above proposition gives in the case of assertion (a) some real $\delta \in]0, \varepsilon/2[$ and some $x_i \in B(\bar{x}, \delta)$ with $|f_i(x_i) - f_i(\bar{x})| < \delta$, for $i \in \{1, \cdots, m+1\}$, and such that

$$0 \in \partial_F f_1(x_1) + \cdots + \partial_F f_m(x_m) + \partial_F f_{m+1}(x_{m+1}) + (\varepsilon/2)\mathbb{B}_{X^*}.$$

Using the equality $\partial_F f_{m+1}(x_{m+1}) = \{-D\varphi(x_{m+1})\}$, we easily derive the inclusion in the assertion (a). The arguments for (b) are similar. □

4.3.3. Applications to convex functions and Asplund spaces. From the above fuzzy sum rule, one can derive the Ekeland-Lebourg theorem concerning the sets of points of Fréchet differentiability of continuous convex functions. Let us prove first the following lemma which particularizes in normed spaces the features in Proposition 3.107 and also provides some new properties. Before stating the lemma, let us recall that a mapping $g : U \to Y$ from an open set U of a normed space X into a normed space Y is said to be *uniformly Fréchet differentiable* on a subset $C \subset U$ if it is Fréchet differentiable at each point of C and

$$\lim_{t \to 0} t^{-1}\big(g(x + th) - g(x) - t\, Dg(x)(h)\big) = 0$$

uniformly with respect to $(x, h) \in C \times \mathbb{B}_X$.

LEMMA 4.48. *Let U be an open convex set of a normed space X and $f : U \to \mathbb{R}$ be a continuous convex function. The following hold.*
(a) *The function f is Gâteaux differentiable at $\bar{x} \in U$ (or equivalently strictly Gâteaux differentiable at $\bar{x}$) if and only if for each $h \in X$*

$$\lim_{t \downarrow 0} t^{-1}\big(f(\bar{x} + th) + f(\bar{x} - th) - 2f(\bar{x})\big) = 0,$$

or equivalently

$$\limsup_{t \downarrow 0} t^{-1}\big(f(\bar{x} + th) + f(\bar{x} - th) - 2f(\bar{x})\big) \leq 0.$$

(b) The function f is Fréchet differentiable at $\bar x \in U$ (or equivalently strictly Fréchet differentiable at $\bar x$) if and only if in (a) the limit is uniform with respect to $h \in \mathbb{B}_X$ (or with respect to $h \in X$ with $\|h\| = 1$), or equivalently

$$\limsup_{h \to 0} \frac{f(\bar x + h) + f(\bar x - h) - 2f(\bar x)}{\|h\|} \leq 0.$$

Similarly f is uniformly Fréchet differentiable on a set $C \subset U$ if and only if the limit in (a) exists uniformly with respect to $(\bar x, h) \in C \times \mathbb{B}_X$.
(c) The set of points where f is Fréchet differentiable is a (maybe empty) G_δ set in U, that is, a countable intersection of open sets of X included in U.
(d) If in addition X is separable, then the set of points of Gâteaux differentiability of f is also a G_δ-set in U.

PROOF. Proposition 2.178 tells us that the Gâteaux (resp. Fréchet) differentiability of the continuous convex function f at $\bar x$ is equivalent to its strict Gâteaux (resp. strict Fréchet) differentiability at $\bar x$.

We also note that $d_D^+ f(\bar x; h) := \lim_{t \downarrow 0} t^{-1} \big(f(\bar x + th) - f(\bar x)\big)$ exists and $d_D^+ f(\bar x; \cdot)$ is finite, continuous and sublinear according to the convexity and continuity of f.
(a) By what precedes, the nullity property of the limit in (a) entails that the equality $d_D^+ f(\bar x; -h) = -d_D^+ f(\bar x; h)$ holds, and this combined with the previous remark ensures that $d_D^+ f(\bar x; \cdot)$ is a continuous linear functional. This proves the implication $\Leftarrow$ in (a), so the equivalence holds true since the converse implication is obvious. The second equivalence between the nullity of the limit and the nonpositivity of the upper limit (or limit superior) follows from the inequality $f(\bar x + v) + f(\bar x - v) - 2f(\bar x) \geq 0$ (for $\bar x \pm v \in U$) due to the convexity of f.
(b) To obtain the equivalence for the uniform Fréchet differentiability on C, it is enough to show the implication $\Leftarrow$. Suppose that the condition of the limit in (a) holds uniformly with respect to $\bar x \in C$ and with respect to $h \in \mathbb{B}_X$, so in particular f is Gâteaux differentiable by (a) at each $\bar x \in C$ with Gâteaux derivative $A := D_G f(\bar x)$. Then, by the convexity of f there is some $\delta > 0$ such that, for all $t \in]0, \delta[$, $\bar x \in C$ and $h \in \mathbb{B}_X$,

$$t^{-1}\big(f(\bar x + th) - f(\bar x) - tA(h)\big) \geq 0 \quad \text{and} \quad t^{-1}\big(f(\bar x - th) - f(\bar x) + tA(h)\big) \geq 0.$$

This combined with the equality

$$t^{-1}\big(f(\bar x + th) + f(\bar x - th) - 2f(\bar x)\big)$$
$$= t^{-1}\big(f(\bar x + th) - f(\bar x) - tA(h)\big) + t^{-1}\big(f(\bar x - th) - f(\bar x) + tA(h)\big)$$

entails that $\lim_{t \downarrow 0} t^{-1}\big(f(\bar x + th) - f(\bar x) - tA(h)\big) = 0$ uniformly with respect to $\bar x \in C$ and with respect to $h \in \mathbb{B}_X$, so f is uniformly Fréchet differentiable on C. The other equivalence for the Fréchet differentiability at $\bar x$ is a consequence of the previous one with C as the singleton $C = \{\bar x\}$.
(c) From the local Lipschitz property of f, it is easily seen, for each $n \in \mathbb{N}$ that the set

$$U_n := \{x \in U : \exists r \in]0, 1[, \, x + r\mathbb{B}_X \subset U, \, \sup_{h \in \mathbb{B}_X}\big(f(x + rh) + f(x - rh) - 2f(x)\big) < r/n\}$$

is open, and the non-negativity and nondecreasing property of $t \mapsto t^{-1}\big(f(x+th) + f(x-th) - 2f(x)\big)$ with $t > 0$ ensures that $U_n = \{x \in U : \exists r \in]0, 1[, \, x + r\mathbb{B}_X \subset U, 0 \leq \sup_{h \in \mathbb{B}_X}\big(f(x+th) + f(x-th) - 2f(x)\big) < t/n, \, \forall t \in]0, r]\}$. From this and (b) we see that the set where f is Fréchet differentiable coincides with $\bigcap_{n \in \mathbb{N}} U_n$.

(d) Finally, suppose that X is separable. Taking a dense sequence $(h_n)_{n \in \mathbb{N}}$ in X and putting

$$U_{n,m} := \{x \in U : \exists r \in \,]0,1[,\, x \pm rh_m \subset U,\, f(x+rh_m) + f(x-rh_m) - 2f(x) < r/n\},$$

each set $U_{n,m}$ is open, and the set where f is Gâteaux differentiable is easily seen (as above) to coincide with $\bigcap_{n,m \in \mathbb{N}} U_{n,m}$. □

THEOREM 4.49 (I. Ekeland and G. Lebourg). *Let X be a Banach space admitting an equivalent norm which is Fréchet differentiable off zero. Then any continuous convex function from an open convex set U of X into $\mathbb{R}$ is Fréchet differentiable on a dense G_δ subset of U.*

PROOF. Let $f : U \to \mathbb{R}$ be a continuous convex function and fix any $x \in U$ and any $\varepsilon > 0$. Then $0 \in \partial_F((-f) + f)(x)$, so Theorem 4.47(a) furnishes some $u, v \in U \cap B(x, \varepsilon)$ such that $0 \in \partial_F(-f)(u) + \partial f(v) + \varepsilon \mathbb{B}_{X^*}$, hence $\partial_F(-f)(u) \neq \emptyset$. Since $\partial_F f(u) = \partial f(u) \neq \emptyset$ by the convexity and continuity of f, it results that f is Fréchet differentiable at u according to Proposition 4.13(a). Consequently, the set of points in U where f is Fréchet differentiable is dense in U and it is G_δ by Lemma 4.48(c) above. □

The latter theorem was first proved by E. Asplund in the context of any Banach space X admitting an equivalent norm, such that the corresponding dual norm on X^* is locally uniformly convex. So, the following definition is usually and reasonably accepted in the literature.

DEFINITION 4.50. A Banach space X is called an *Asplund space* provided that every real-valued continuous convex function on an open convex set U of X is Fréchet differentiable on a dense G_δ subset of U, or equivalently on a dense subset of U according to Lemma 4.48(c).

The Ekeland-Lebourg theorem can then be reformulated by saying that every Banach space, with an equivalent norm which is Fréchet differentiable off zero, is Asplund. In particular, every reflexive Banach space is Asplund since any such space admits an equivalent Fréchet differentiable norm off zero (see Theorem C.11 in Appendix). The Lebesgue and Sobolev spaces $L^p(\Omega, \mathbb{R})$ and $W^{m,p}(\Omega, \mathbb{R})$ with $p \in \,]1, \infty[$ (and Ω open in $\mathbb{R}^n$) are examples of such spaces. On the contrary, these spaces with either $p = 1$ or $p = \infty$ as well as the Banach space of continuous functions $\mathcal{C}([0,1], \mathbb{R})$ are not Asplund spaces (see, for example, the comments after Corollary C.13 in Appendix). It is also known (see Theorem C.12 in Appendix) that any Banach space whose dual is separable admits an equivalent Fréchet differentiable (off zero) renorm, so it is Asplund.

In fact, the Fréchet differentiable renorming of separable subspaces characterizes Asplund spaces as follows (see Theorem C.12 in Appendix):

THEOREM 4.51 (characterizations of Asplund space via separable subspaces). *For a Banach space X, the following are equivalent:*
(a) *the space X is Asplund;*
(b) *the topological dual of any separable vector subspace of X is separable;*
(c) *any separable vector subspace of X admits an equivalent norm which is Fréchet differentiable off zero.*

4.3.4. Fuzzy sum rule for Fréchet subdifferential in Asplund space.

We now show how the separable reduction in the previous section allows us to obtain in the framework of a general Asplund space conclusions similar to those in Theorem 4.47.

THEOREM 4.52 (fuzzy sum rule for Fréchet subdifferential in Asplund space). Let X be an Asplund space and let $f_i : X \to \mathbb{R} \cup \{-\infty, +\infty\}$, $i = 1, \cdots, m$, be functions which are finite at $\bar{x} \in X$. Then the assertions (a) and (b) in Theorem 4.47 still hold true.

PROOF. Without loss of generality we may suppose both in (a) and (b) that the functions f_i are locally bounded from below.
(a) Let $\varepsilon > 0$ and $x^* \in \partial_F (f_1 + \cdots + f_m)(\bar{x})$ or equivalently $0 \in \partial_F (f_1 + \cdots + f_m + f_{m+1})(\bar{x})$, where $f_{m+1} := \langle -x^*, \cdot \rangle$. Fix a closed separable vector subspace L of the space X with $\bar{x} \in L$ and by Theorem 4.37 take a closed separable vector subspace $L_0 \supset L$ such that for all $u_1, \cdots, u_m, u_{m+1}$ in L_0, the inclusion

$$0 \in \partial_F f_1(u_1) + \cdots + \partial_F f_m(u_m) + \partial_F f_{m+1}(u_{m+1}) + \varepsilon \mathbb{B}_{X^*}$$

is satisfied if and only if

$$0 \in \partial_F f_{0,1}(u_1) + \cdots + \partial_F f_{0,m}(u_m) + \partial_F f_{0,m+1}(u_{m+1}) + \varepsilon \mathbb{B}_{L_0^*},$$

where $f_{0,i}$ denotes the restriction of f_i to L_0. On the other hand, note that $0 \in \partial_F (f_{0,1} + \cdots + f_{0,m} + f_{0,m+1})(\bar{x})$ according to the definition of Fréchet subdifferential and to the inclusion $0 \in \partial_F (f_1 + \cdots + f_m + f_{m+1})(\bar{x})$. Since L_0 admits by Theorem 4.51(c) an equivalent norm which is Fréchet differentiable off zero, Theorem 4.47 yields some $x_1, \cdots, x_m, x_{m+1}$ in L_0 with $|f_{0,1}(x_1) - f(\bar{x})| < \varepsilon$ and $\|x_i - \bar{x}\| < \varepsilon$ for $i = 1, \cdots, m + 1$ such that

$$0 \in \partial_F f_{0,1}(x_1) + \cdots + \partial_F f_{0,m}(x_m) + \partial_F f_{0,m+1}(x_{m+1}) + \varepsilon \mathbb{B}_{L_0^*}.$$

It results by what precedes that

$$0 \in \partial_F f_1(x_1) + \cdots + \partial_F f_m(x_m) + \partial_F f_{m+1}(x_{m+1}) + \varepsilon \mathbb{B}_{X^*},$$

which means that $|f(x_1) - f(\bar{x})| < \varepsilon$ and $\|x_i - \bar{x}\| < \varepsilon$ for $i = 1, \cdots, m$, and

$$x^* \in \partial_F f_1(x_1) + \cdots + \partial_F f_m(x_m) + \varepsilon \mathbb{B}_{X^*}.$$

This finishes the proof of (a).
(b) Fix any finite-dimensional vector subspace L of X and note that one has $x^* \in \partial_F (f_1 + \cdots + f_m + \Psi_{\bar{x}+L})(\bar{x})$. Put $f_{m+1} := \langle -x^*, \cdot \rangle$ and $f_{m+2} := \Psi_{\bar{x}+L}$, and write the above inclusion as

$$0 \in \partial_F (f_1 + \cdots + f_m + f_{m+1} + f_{m+2})(\bar{x}).$$

Theorem 4.37 gives us a closed separable vector subspace L_0 containing $\{\bar{x}\} \cup L$ such that for all $u_1, \cdots, u_m, u_{m+1}, u_{m+2}$ in L_0, the inclusion

$$0 \in \partial_F f_1(u_1) + \cdots + \partial_F f_m(u_m) + \partial_F f_{m+1}(u_{m+1}) + \partial_F f_{m+2}(u_{m+2}) + \varepsilon \mathbb{B}_{X^*}$$

is satisfied if and only if

$$0 \in \partial_F f_{0,1}(u_1) + \cdots + \partial_F f_{0,m}(u_m) + \partial_F f_{0,m+1}(u_{m+1}) + \partial_F f_{0,m+2}(u_{m+2}) + \varepsilon \mathbb{B}_{L_0^*},$$

where $f_{0,i}$ denotes as above the restriction of f_i to L_0. As $0 \in \partial_F \left(\sum_{i=1}^{m+2} f_{0,i} \right)(\bar{x})$, by Theorem 4.47(b) one easily finds $x_i \in L_0$ with $\|x_i - \bar{x}\| < \varepsilon$ and $|f_{0,i}(x_i) - f_{0,i}(\bar{x})| < \varepsilon$

($i=1,\cdots,m+2$) such that
$$0 \in \partial_F f_{0,1}(x_1) + \cdots + \partial_F f_{0,m}(x_m) + \partial_F f_{0,m+1}(x_{m+1}) + \partial_F f_{0,m+2}(x_{m+2}) + \varepsilon \mathbb{B}_{L_0^*}$$
since, denoting by $L^{\perp_{L_0^*}}$ the orthogonal of L in the topological dual space L_0^* of L_0, one notices that
$$\partial_F f_{0,m+2}(x_{m+2}) + L^{\perp_{L_0^*}} = \partial_F f_{0,m+2}(x_{m+2})$$
according to the equality $\partial_F f_{0,m+2}(x_{m+2}) = L^{\perp_{L_0^*}}$. Thus, by the above equivalence one obtains
$$0 \in \partial_F f_1(x_1) + \cdots + \partial_F f_m(x_m) + \partial_F f_{m+1}(x_{m+1}) + \partial_F f_{m+2}(x_{m+2}) + \varepsilon \mathbb{B}_{X^*}.$$
It ensues that $\|x_i - \overline{x}\| < \varepsilon$, $|f_i(x_i) - f_i(\overline{x})| < \varepsilon$, $i = 1, \cdots, m$, and
$$x^* \in \partial_F f_1(x_1) + \cdots + \partial_F f_m(x_m) + \varepsilon \mathbb{B}_{X^*} + L^\perp,$$
which finishes the proof. □

A density property of Fréchet subdifferentiability points can be deduced from the previous theorem. Such a result will be established in Proposition 6.50 for a large class of subdifferentials.

COROLLARY 4.53 (density of Fréchet subdifferentiability points). Let X be an Asplund space and $f : X \to \mathbb{R} \cup \{+\infty\}$ be a lower semicontinuous function which is finite at some point. Then $\mathrm{Dom}\, \partial_F f$ is graphically dense in $\mathrm{dom}\, f$, that is, for any $x \in \mathrm{dom}\, f$ there is a sequence $(x_n)_n$ in $\mathrm{Dom}\, \partial_F f$ such that $(x_n, f(x_n)) \to (x, f(x))$.

PROOF. Fix any $x \in \mathrm{dom}\, f$ and observe that x is a minimizer point of the function $f + \Psi_{\{x\}}$, hence $0 \in \partial_F(f + \Psi_{\{x\}})(x)$. For each real $\varepsilon > 0$, the assertion (b) in the above theorem furnishes some point $x' \in \mathrm{Dom}\, \partial_F f$ such that $\|x' - x\| < \varepsilon$ and $|f(x') - f(x)| < \varepsilon$. □

The property in the above corollary in fact characterizes the Asplund property of the Banach space as justified below:

COROLLARY 4.54. Let X be a Banach space. The following assertions are equivalent.
(a) The Banach space X is an Asplund space;
(b) the fuzzy calculus (a) in Theorem 4.52 holds for $m = 2$ with $f_1 : X \to \mathbb{R} \cup \{+\infty\}$ lower semicontinuous and $f_2 : X \to \mathbb{R}$ locally Lipschitz continuous;
(c) the fuzzy calculus (b) in Theorem 4.52 holds for $m = 2$ with $f_1, f_2 : X \to \mathbb{R} \cup \{+\infty\}$ lower semicontinuous;
(d) for any lower semicontinuous function $f : X \to \mathbb{R} \cup \{+\infty\}$, the effective domain $\mathrm{Dom}\, \partial_F f$ of $\partial_F f$ is graphically dense in the effective domain $\mathrm{dom}\, f$ of f, that is, for any $x \in \mathrm{dom}\, f$ there exists a sequence $(x_n)_n$ in $\mathrm{Dom}\, \partial_F f$ such that $(x_n, f(x_n))_n$ converges to $(x, f(x))$;
(e) for any nonempty open set U of X and any locally Lipschitz continuous function $f : U \to \mathbb{R}$, the effective domain $\mathrm{Dom}\, \partial_F f$ is dense in U;
(f) for any locally Lipschitz continuous function $f : X \to \mathbb{R}$, the effective domain $\mathrm{Dom}\, \partial_F f$ is dense in X.

PROOF. The implications (a) $\Rightarrow$ (b) and (a) $\Rightarrow$ (c) follow from Theorem 4.52 and the implication (c) $\Rightarrow$ (d) is shown in the proof of the above corollary. Let us prove (b) $\Rightarrow$ (d). Fix any $\overline{x} \in \mathrm{dom}\, f$ and any $\varepsilon > 0$. By lower semicontinuity

of f choose some real $r > 0$ such that $f(\bar{x}) - \varepsilon < f(x)$ for all $x \in B[\bar{x}, r]$, hence $f(\bar{x}) \leq \inf_{B[\bar{x},r]} f + \varepsilon$. Taking $0 < \lambda < \min\{\varepsilon, r/2\}$ the Ekeland variational principle in Theorem 2.221 gives some $u \in B[\bar{x}, \lambda]$ with $f(u) \leq f(\bar{x}) \leq f(u) + \varepsilon$ and such that
$$f(u) \leq f(x) + (\varepsilon/\lambda)\|x - u\| \quad \text{for all } x \in B[\bar{x}, r].$$
Since $\|u - \bar{x}\| < r$, the point u is a local minimizer of $f + (\varepsilon/\lambda)\|\cdot - u\|$. Consequently $0 \in \partial_F(f + (\varepsilon/\lambda)\|\cdot - u\|)(u)$, which by (b) gives some $v \in X$ with $\partial_F f(v) \neq \emptyset$, $\|v - u\| < \varepsilon$ and $|f(v) - f(u)| < \varepsilon$, hence $\|v - \bar{x}\| < 2\varepsilon$ and $|f(v) - f(\bar{x})| < 2\varepsilon$. This translates the desired implication (b) $\Rightarrow$ (d).

Now suppose (d) and let $f : U \to \mathbb{R}$ be a locally Lipschitz function on an open set U of X. Fix $\bar{x} \in U$ and choose $r > 0$ with $B[\bar{x}, r] \subset U$. Define the lower semicontinuous function $\overline{f} : X \to \mathbb{R} \cup \{+\infty\}$ by $\overline{f}(x) = f(x)$ if $x \in B[\bar{x}, r]$ and $\overline{f}(x) = +\infty$ if $x \in X \setminus B[\bar{x}, r]$. For any real $\varepsilon > 0$, choosing $0 < \varepsilon' < \min\{\varepsilon, r\}$, the property (d) applied to $\overline{f}$ yields some $u \in B(\bar{x}, \varepsilon')$ with $\partial_F \overline{f}(u) \neq \emptyset$, so $u \in U$ with $\|u - \bar{x}\| < \varepsilon$ and $\partial_F f(u) \neq \emptyset$. This justifies the implication (d) $\Rightarrow$ (e).

The implication (e) $\Rightarrow$ (f) is obvious. To see that (f) implies (e), consider any locally Lipschitz function $f : U \to \mathbb{R}$ on an open set U of X and any $\bar{x} \in U$. Take also any $\varepsilon > 0$. Let $U_0 \subset U$ be an open neighborhood of $\bar{x}$ over which f is Lipschitz. We know (see Proposition 2.79) that $f|_{U_0}$ (the restriction of f to U_0) can be extended to X to a Lipschitz function $\overline{f}$. By (f) there is some $u \in U_0 \cap B(\bar{x}, \varepsilon)$ such that $u \in \text{Dom}\, \partial_F \overline{f}$, thus $u \in \text{Dom}\, \partial_F f$ since $u \in U_0$. This translates the implication (f) $\Rightarrow$ (e).

It remains to show the implication (e) $\Rightarrow$ (a). Suppose (e) and fix any continuous convex function $f : U \to \mathbb{R}$ on an open convex set U of X. By (e) there exists a dense set D of U such that $\partial_F(-f)(x) \neq \emptyset$ for all $x \in D$. But for the continuous convex function f, we have $\partial_F f(x) = \partial f(x) \neq \emptyset$ for all $x \in U$ (see Proposition 2.63 and (d) in Theorem 2.70). Proposition 4.13 tells us that f is Fréchet differentiable at each point of the dense set D, which entails the Asplund property of the Banach space X. This finishes the proof. $\square$

The next example illustrates that the lower semicontinuity requirement in Corollary 4.53 and Corollary 4.54(d) cannot be removed. In fact, it provides a non lower semicontinuous function of one real variable whose Fréchet subdifferential is empty at any point.

EXAMPLE 4.55 (A.N. Singh). Clearly, every $x \in {]0, 1[}$ admits a unique representation as
$$x = \sum_{k=1}^{\infty} \frac{a_k(x)}{3^k} \quad \text{with } a_k(x) \in \{0, 1, 2\} \text{ and infinitely many } a_k(x) \neq 2.$$
For each $x \in {]0, 1[}$ put, via the above representation,
$$f(x) := -\sum_{k=1}^{\infty} \frac{b_k(x)}{2^k}, \quad \text{where } b_k(x) := \begin{cases} 1 & \text{if } a_k(x) = 2 \\ 0 & \text{otherwise.} \end{cases}$$
Consider $u := \frac{\alpha_1}{3} + \cdots + \frac{\alpha_{p-1}}{3^{p-1}} + \frac{1}{3^p}$ with $\alpha_1, \cdots, \alpha_{p-1} \in \{0, 1, 2\}$, and for each $n \in \mathbb{N}$ set
$$u_n := \frac{\alpha_1}{3} + \cdots + \frac{\alpha_{p-1}}{3^{p-1}} + \frac{0}{3^p} + \frac{2}{3^{p+1}} + \cdots + \frac{2}{3^{p+n}}.$$

One has $f(u) = -\frac{\beta_1}{2} - \cdots - \frac{\beta_{p-1}}{2^{p-1}}$ while

$$f(u_n) = -\frac{\beta_1}{2} - \cdots - \frac{\beta_{p-1}}{2^{p-1}} - \left(\frac{1}{2^{p+1}} + \cdots + \frac{1}{2^{p+n}}\right),$$

where, for each $k \in \{1, \cdots, p-1\}$, (as above) $\beta_k = 1$ if $\alpha_k = 2$ and $\beta_k = 0$ otherwise. It is readily seen that $u_n \to u$ and $f(u_n) \to f(u) - 1/2^p$ as $n \to \infty$. Since $u_n < u$, it follows that f is not lower semicontinuous at u. This also easily yields that f is lower semicontinuous near no point in $]0,1[$.

Now we claim that at any $x \in]0,1[$ the right-hand lower Dini derivate of f is $-\infty$, so $\partial_F f(x) = \emptyset$. Indeed, fix any $x \in]0,1[$ and take its above representation $x = \sum_{k=1}^{\infty} \frac{a_k(x)}{3^k}$. There is an increasing sequence $(s(n))_{n \in \mathbb{N}}$ of integers in $\mathbb{N}$ such that $a_{s(n)}(x) \in \{0,1\}$. For each $n \in \mathbb{N}$ putting

$$x_n := \frac{a_1(x)}{3} + \cdots + \frac{a_{s(n)-1}(x)}{3^{s(n)-1}} + \frac{2}{3^{s(n)}} + \sum_{k=s(n)+1}^{\infty} \frac{a_k(x)}{3^k},$$

we see that

$$x_n - x = \frac{2 - a_{s(n)}(x)}{3^{s(n)}} \quad \text{and} \quad f(x_n) - f(x) = -\frac{1}{2^{s(n)}}$$

along with $2 - a_{s(n)} \in \{1,2\}$ since $a_{s(n)} \in \{0,1\}$. It ensues that as $n \to \infty$

$$x_n \downarrow x \quad \text{and} \quad \frac{f(x_n) - f(x)}{x_n - x} = -\frac{3^{s(n)}}{2^{s(n)}(2 - a_{s(n)})} \to -\infty,$$

which confirms that the right-hand lower Dini derivate of f at x is $-\infty$. $\square$

4.3.5. Fuzzy chain rule for Fréchet subdifferential. A fuzzy chain rule for lower semicontinuous or Lipschitz continuous functions can be derived from the theorem on fuzzy sum rule. Let us state first a lemma.

LEMMA 4.56. *Let X and Y be normed spaces and $G : X \to Y$ be a mapping which is one-sided Lipschitz at a point $\bar{x} \in X$. Then*

$$(x^*, -y^*) \in N^F(\operatorname{gph} G; (\bar{x}, G(\bar{x}))) \Leftrightarrow x^* \in \partial_F(y^* \circ G)(\bar{x}).$$

PROOF. Let $\delta_0 > 0$ be such that G is one-sided γ_G-Lipschitz at $\bar{x}$ relative to $B(\bar{x}, \delta_0)$. Fix $(x^*, -y^*) \in N^F(\operatorname{gph} G; (\bar{x}, G(\bar{x})))$. For any $\varepsilon > 0$ there exists some $0 < \delta' < \delta_0$ such that for all $(x, y) \in B((\bar{x}, G(\bar{x})), \delta')$ with $y = G(x)$ we have

$$\langle x^*, x - \bar{x}\rangle - \langle y^*, y - G(\bar{x})\rangle \leq \varepsilon(\|x - \bar{x}\| + \|y - G(\bar{x})\|).$$

Choose a positive real $\delta < \delta'$ such that $\|G(x) - G(\bar{x})\| < \delta'$ for all $x \in B(\bar{x}, \delta)$. Then for any $x \in B(\bar{x}, \delta)$ we obtain

$$\langle x^*, x - \bar{x}\rangle \leq y^* \circ G(x) - y^* \circ G(\bar{x}) + \varepsilon(1 + \gamma_G)\|x - \bar{x}\|,$$

which translates that $x^* \in \partial_F(y^* \circ G)(\bar{x})$, hence the implication $\Rightarrow$ of the lemma holds. The proof of the converse implication is similar and easier. $\square$

THEOREM 4.57 (fuzzy chain rule for Fréchet subdifferential). *Let X and Y be Asplund spaces, G be a mapping from X into Y, and $g : Y \to \mathbb{R} \cup \{+\infty\}$ be a function which is finite at $G(\bar{x})$, where $\bar{x} \in X$.*
(a) *If g is uniformly continuous near $G(\bar{x})$ and the graph of G is closed near*

$(\bar{x}, G(\bar{x}))$, then for any $x^* \in \partial_F(g \circ G)(\bar{x})$ and $\varepsilon > 0$ there exist $x \in B(\bar{x}, \varepsilon)$, $y \in B(G(\bar{x}), \varepsilon)$, $(u^*, -y^*) \in N^F(\operatorname{gph} G; (x, G(x)))$ such that

$$x^* \in u^* + \varepsilon \mathbb{B}_{X^*} \quad \text{and} \quad y^* \in \partial_F g(y) + \varepsilon \mathbb{B}_{Y^*}.$$

(b) If g is uniformly continuous near $G(\bar{x})$ and G is Lipschitz continuous near $\bar{x}$, then for any $x^* \in \partial_F(g \circ G)(\bar{x})$ and $\varepsilon > 0$ there exist $x \in B(\bar{x}, \varepsilon)$, $y \in B(G(\bar{x}), \varepsilon)$, and $y^* \in \partial_F g(y)$ such that

$$x^* \in \partial_F(y^* \circ G)(x) + \varepsilon \mathbb{B}_{X^*}.$$

(c) If g is lower semicontinuous near $G(\bar{x})$ and the graph of G is closed near $(\bar{x}, G(\bar{x}))$, then for any $x^* \in \partial_F(g \circ G)(\bar{x})$, any $\varepsilon > 0$, any finite-dimensional subspace L of X, and any finite-dimensional subspace Q of Y, there exist $x \in B(\bar{x}, \varepsilon)$, $y \in B(G(\bar{x}), \varepsilon)$ with $|g(y) - g(G(\bar{x}))| < \varepsilon$, $(u^*, -y^*) \in N^F(\operatorname{gph} G; (x, G(x)))$ such that

$$x^* \in u^* + \varepsilon \mathbb{B}_{X^*} + L^\perp \quad \text{and} \quad y^* \in \partial_F g(y) + \varepsilon \mathbb{B}_{Y^*} + Q^\perp.$$

(d) If g is lower semicontinuous near $G(\bar{x})$ and G is Lipschitz continuous near $\bar{x}$, then for any $x^* \in \partial_F(g \circ G)(\bar{x})$, any $\varepsilon > 0$, any finite-dimensional subspace L of X, and any finite-dimensional subspace Q of Y, there exist $x \in B(\bar{x}, \varepsilon)$, $y \in B(G(\bar{x}), \varepsilon)$ with $|g(y) - g(G(\bar{x}))| < \varepsilon$, and $y^* \in \partial_F g(y) + \varepsilon \mathbb{B}_{Y^*} + Q^\perp$ such that

$$x^* \in \partial_F(y^* \circ G)(x) + \varepsilon \mathbb{B}_{X^*} + L^\perp.$$

PROOF. (a) Consider any $x^* \in \partial_F(g \circ G)(\bar{x})$ and put $f(x, y) := g(y)$ and $h(x, y) := \Psi_{\operatorname{gph} G}(x, y)$ for all $(x, y) \in X \times Y$. It is not difficult to see that $(x^*, 0) \in \partial_F(f + h)(\bar{x}, G(\bar{x}))$. Since f is uniformly continuous near $(\bar{x}, G(\bar{x}))$ and gph G is closed near $(\bar{x}, G(\bar{x}))$, Theorem 4.52 above combined with the assertion (b) of Proposition 4.19 says that for any $\varepsilon > 0$ there exist

$$(u, y) \in B\bigl((\bar{x}, G(\bar{x})), \varepsilon\bigr), \ (x, G(x)) \in B\bigl((\bar{x}, G(\bar{x})), \varepsilon\bigr), \ (0, v^*) \in \partial_F f(u, y),$$

and

$$(u^*, -y^*) \in N^F\bigl(\operatorname{gph} G, (x, G(x))\bigr)$$

such that

(4.35) $(x^*, 0) \in (0, v^*) + (u^*, -y^*) + \varepsilon \mathbb{B}$, hence $x^* \in u^* + \varepsilon \mathbb{B}$ and $y^* \in v^* + \varepsilon \mathbb{B}$.

Consequently, $y^* \in \partial_F g(y) + \varepsilon \mathbb{B}_{Y^*}$ since $v^* \in \partial_F g(y)$. All the properties in (a) are reached.

(b) Since G is Lipschitz continuous on some ball $B(\bar{x}, \delta)$, requiring the point x in (a) to be also in $B(\bar{x}, \delta)$ one observes first by the above lemma that $(u^*, -y^*) \in N^F(\operatorname{gph} G; (x, G(x)))$ means $u^* \in \partial_F(y^* \circ G)(x)$. So $x^* \in \partial_F(y^* \circ G)(x) + \varepsilon \mathbb{B}_{X^*}$ with $y^* \in \partial_F g(y) + \varepsilon \mathbb{B}_{Y^*}$, hence $y^* = y_1^* + \varepsilon b^*$ for some $y_1^* \in \partial_F g(y)$ and $b^* \in \mathbb{B}_{Y^*}$. Denoting by γ a Lipschitz constant of G over $B(\bar{x}, \varepsilon)$ (shrinking ε if necessary), the fuzzy sum rule applied to $y^* \circ G = y_1^* \circ G + \varepsilon b^* \circ G$ along with the γ-Lipschitz property of G ensures that, for some $x' \in B(\bar{x}, \varepsilon)$, one has $x^* \in \partial_F(y_1^* \circ G)(x') + (2 + \gamma) \varepsilon \mathbb{B}_{X^*}$.

(c) It is enough to repeat the proof of (a) with $\varepsilon \mathbb{B}_{X^*} + L^\perp$ and $\varepsilon \mathbb{B}_{Y^*} + Q^\perp$ in place of $\varepsilon \mathbb{B}_{X^*}$ and $\varepsilon \mathbb{B}_{Y^*}$ respectively in (4.35). Since g here is not continuous at $G(\bar{x})$, we must add in the application of the fuzzy sum rule that $|f(u, y) - f(\bar{x}, G(\bar{x}))| < \varepsilon$, that is, $|g(y) - g(G(\bar{x}))| < \varepsilon$.

(d) The arguments are similar to those for (b). □

4.3.6. Stegall variational principle, Fréchet derivative of conjugate function. Before passing to the next section, let us establish another variational principle with linear perturbation. We need first a characterization of Fréchet differentiability of the Legendre-Fenchel conjugate f^* of a function f. Given a Hausdorff locally convex space (X, τ_X) with topological dual X^* and a function $f : X \to \mathbb{R} \cup \{-\infty, +\infty\}$, recall (see Definition 2.68) that the Legendre-Fenchel conjugate $f^* : X^* \to \mathbb{R} \cup \{-\infty, +\infty\}$ is defined by

(4.36) $$f^*(x^*) := \sup_{x \in X} \left(\langle x^*, x \rangle - f(x)\right) \quad \text{for all } x^* \in X^*.$$

If in place of (X, τ) we consider the space $(X^*, w(X^*, X))$ (so, the topological dual of $(X^* w(X^*, X))$ is identified with X through the usual duality pairing), then (as seen in the previous chapter) for $g : X^* \to \mathbb{R} \cup \{-\infty, +\infty\}$ defined on X^*, its conjugate function in the pairing from X^* to X is given by $g^* : X \to \mathbb{R} \cup \{-\infty, +\infty\}$ with

$$g^*(x) = \sup_{x^* \in X^*} \left(\langle x^*, x \rangle - g(x^*)\right) \quad \text{for all } x \in X.$$

Clearly f^* is convex and $w(X^*, X)$ lower semicontinuous and g^* is convex and $w(X, X^*)$ lower semicontinuous. Further, f^* (resp. g^*) takes values in $\mathbb{R} \cup \{+\infty\}$ whenever f (resp. g) is proper.

THEOREM 4.58 (F-derivative of conjugate function). *Let $(X, \|\cdot\|)$ be a Banach space, and $f : X \to \mathbb{R} \cup \{+\infty\}$ and $g : X^* \to \mathbb{R} \cup \{+\infty\}$ be both norm lower semicontinuous. Assume that f^* is finite at $u^* \in X^*$ and g^* is finite at $\overline{x} \in X$, where g^* defined on X is the conjugate of g in the pairing from X^* to X as above.*
(a) *If the function f^* is Fréchet differentiable at u^*, then the function $\langle u^*, \cdot \rangle - f$ has a strong maximizer $\overline{u} \in X$ and $Df^*(u^*) = \langle \cdot, \overline{u} \rangle$.*
Conversely, if the function $\langle u^*, \cdot \rangle - f$ has a strong maximizer $\overline{u} \in X$ and $\mathrm{dom}\, f$ is bounded in X, then f^* is Fréchet differentiable at u^* and $Df^*(u^*) = \langle \cdot, \overline{u} \rangle$.
(b) *If the function g^* is Fréchet differentiable at $\overline{x}$, then the function $\langle \cdot, \overline{x} \rangle - g$ has a strong maximizer $x^* \in X^*$ and $Dg^*(\overline{x}) = x^*$.*
Conversely, if the function $\langle \cdot, \overline{x} \rangle - g$ has a strong maximizer $x^* \in X^*$ and $\mathrm{dom}\, g$ is bounded in X^*, then the function g^* is Fréchet differentiable at $\overline{x}$ and $Dg^*(\overline{x}) = x^*$.

PROOF. Of course, we may suppose $X \neq \{0\}$.
Suppose first that g^* is Fréchet differentiable at $\overline{x}$. Consider for each $n \in \mathbb{N}$ the nonempty $\|\cdot\|_*$-closed set

$$A_n := \{y^* \in X^* : \langle y^*, \overline{x} \rangle - g(y^*) \geq g^*(\overline{x}) - \varepsilon_n^2\}$$

where $\varepsilon_n := 1/n$, and put $\delta_n := \min\{n, \mathrm{diam}\, A_n\}$. Choose $x_n^*, y_n^* \in A_n$ such that $\|x_n^* - y_n^*\|_* \geq \frac{2\delta_n}{3}$, hence there is some $u_n \in \mathbb{B}_X$ satisfying $\langle x_n^* - y_n^*, u_n \rangle \geq \delta_n/2$. It ensues by definition of conjugate and by definition of A_n that

$$g^*(\overline{x} + \varepsilon_n u_n) + g^*(\overline{x} - \varepsilon_n u_n) - 2g^*(\overline{x}) \geq \left(\langle x_n^*, \overline{x} + \varepsilon_n u_n \rangle - g(x_n^*) - g^*(\overline{x})\right)$$
$$+ \left(\langle y_n^*, \overline{x} - \varepsilon_n u_n \rangle - g(y_n^*) - g^*(\overline{x})\right)$$
$$\geq -2\varepsilon_n^2 + \varepsilon_n \langle x_n^* - y_n^*, u_n \rangle,$$

which yields

$$\delta_n \leq 2\langle x_n^* - y_n^*, u_n \rangle \leq 4\varepsilon_n + 2\frac{g^*(\overline{x} + \varepsilon_n u_n) + g^*(\overline{x} - \varepsilon_n u_n) - 2g^*(\overline{x})}{\varepsilon_n}.$$

We then obtain that $\delta_n \to 0$ according to the Fréchet differentiability of g^* at $\bar{x}$ (see Lemma 4.48). It ensues that diam $A_n \to 0$, so by (4.24) the upper $\|\cdot\|_*$-semicontinuous function $\langle \cdot, \bar{x}\rangle - g$ admits a strong maximizer $x^* \in X^*$.

Further, the equality $g^*(\bar{x}) = \langle x^*, \bar{x}\rangle - g(x^*)$ and the definition of conjugate function give

$$0 = \lim_{h \to 0} \frac{g^*(\bar{x}+h) - g^*(\bar{x}) - \langle Dg^*(\bar{x}), h\rangle}{\|h\|}$$

$$\geq \limsup_{h \to 0} \frac{\langle x^*, \bar{x}+h\rangle - g(x^*) - \langle x^*, \bar{x}\rangle + g(x^*) - \langle Dg^*(\bar{x}), h\rangle}{\|h\|}$$

$$= \inf_{\eta > 0} \sup_{0 < \|h\| \leq \eta} \frac{\langle x^* - Dg^*(\bar{x}), h\rangle}{\|h\|} = \inf_{\eta > 0} \|x^* - Dg^*(\bar{x})\| = \|x^* - Dg^*(\bar{x})\|,$$

so $x^* = Dg^*(\bar{x})$. The implication $\Rightarrow$ in (b) is proved.

Suppose now that $\langle \cdot, \bar{x}\rangle - g$ has a strong maximizer $x^* \in X^*$ and $\operatorname{dom} g$ is bounded in X^*. Since

(4.37) $\quad g^*(x) = \sup_{v^* \in \operatorname{dom} g} \left(\langle v^*, x\rangle - g(v^*)\right) \quad$ for all $x \in X$,

we easily see from the boundedness of $\operatorname{dom} g$ that g^* is finite and Lipschitz on X. Consider any sequence of nonzero vectors $(h_n)_n$ in X converging to 0 and for each $n \in \mathbb{N}$, with $\varepsilon_n := 1/n$ choose by (4.37) $x_n^*, y_n^* \in \operatorname{dom} g$ such that

$$\langle x_n^*, \bar{x}+h_n\rangle - g(x_n^*) > g^*(\bar{x}+h_n) + \varepsilon_n \|h_n\|, \quad \langle y_n^*, \bar{x}-h_n\rangle - g(y_n^*) > g^*(\bar{x}-h_n) + \varepsilon_n \|h_n\|.$$

The latter inequalities and the definition of conjugate function give

$$g^*(\bar{x}) \geq \langle x_n^*, \bar{x}\rangle - g(x_n^*) > g^*(\bar{x}+h_n) - \langle x_n^*, h_n\rangle + \varepsilon_n \|h_n\|,$$
$$g^*(\bar{x}) \geq \langle y_n^*, \bar{x}\rangle - g(y_n^*) > g^*(\bar{x}-h_n) + \langle y_n^*, h_n\rangle + \varepsilon_n \|h_n\|,$$

and we note that the sequences $(x_n^*)_n$ and $(y_n^*)_n$ are bounded in X^* thanks to the boundedness of $\operatorname{dom} g$. We derive that $\langle x_n^*, \bar{x}\rangle - g(x_n^*) \to g^*(\bar{x})$ and $\langle y_n^*, \bar{x}\rangle - g(y_n^*) \to g^*(\bar{x})$ as $n \to \infty$. This says that both sequences $(x_n^*)_n$ and $(y_n^*)_n$ are maximizing sequences of the function $\langle \cdot, \bar{x}\rangle - g$ which by assumption has a strong maximizer. It follows that $\|x_n^* - y_n^*\| \to 0$ as $n \to \infty$. So, writing

$$g^*(\bar{x}+h_n) + g^*(\bar{x}-h_n) - 2g^*(\bar{x})$$
$$< \left(\langle x_n^*, \bar{x}+h_n\rangle - g(x_n^*) - \varepsilon_n\|h_n\| - g^*(\bar{x})\right)$$
$$+ \left(\langle y_n^*, \bar{x}-h_n\rangle - g(y_n^*) - \varepsilon_n\|h_n\| - g^*(\bar{x})\right)$$
$$\leq \langle x_n^* - y_n^*, h_n\rangle - 2\varepsilon_n\|h_n\| \leq \|h_n\|(\|x_n^* - y_n^*\| + 2\varepsilon_n),$$

we see that $\limsup_{n \to \infty} \dfrac{g^*(\bar{x}+h_n) + g^*(\bar{x}-h_n) - 2g^*(\bar{x})}{\|h_n\|} \leq 0$, hence Lemma 4.48(b) tells us that g^* is Fréchet differentiable at $\bar{x}$. As in the proof of the previous implication we can show that $Dg^*(\bar{x}) = x^*$. This finishes the proof of (b) and of the theorem since the arguments for (a) are similar. $\square$

REMARK 4.59. The existence of a strong maximizer of $\langle u^*, \cdot\rangle - f$ is not sufficient to guarantee the Fréchet differentiability of f^* at u^*. The continuous function $f: \mathbb{R} \to \mathbb{R}$, with $f(x) = \min\{1, \frac{1}{2}x^2\}$, and the functional $u^* = 0$ are such that

$u^*(\cdot) - f(\cdot)$ admits a strong maximizer (at 0). However, with $f_1(x) = 1$ and $f_2(x) = \frac{1}{2}x^2$, we have for all $y \in \mathbb{R}$

$$f^*(y) = \max\{f_1^*(y), f_2^*(y)\} = \max\left\{-1 + \Psi_{\{0\}}(y), \frac{1}{2}y^2\right\} = \Psi_{\{0\}}(y),$$

hence we see that the conjugate function f^* is not differentiable at $u^* = 0$; note that $\operatorname{dom} f$ is unbounded. □

The following direct corollary of Theorem 4.58 must be emphasized.

COROLLARY 4.60. *Let $(X, \|\cdot\|)$ be a Banach space and $f : X \to \mathbb{R} \cup \{+\infty\}$ be a lower $\|\cdot\|$-semicontinuous function. If the Legendre-Fenchel conjugate function f^* is Fréchet differentiable at a point $u^* \in X^*$, then $Df^*(u^*) \in X$.*

Let us define now a variant of Asplund spaces.

DEFINITION 4.61. *Given a Banach space X, one says analogously to Definition 4.50 that X^* is weak* Asplund provided that every weak* lower semicontinuous convex function from X^* into $\mathbb{R}$ is Fréchet differentiable on a dense G_δ subset of the space $(X^*, \|\cdot\|_*)$.*

REMARK 4.62. It is worth noticing that, for a normed space X, any weak* lower semicontinuous convex function $g : X^* \to \mathbb{R} \cup \{+\infty\}$ is $\|\cdot\|_{X^*}$-continuous on $\operatorname{int}_{\|\cdot\|_{X^*}}(\operatorname{dom} g)$ (which coincides here with the set $\operatorname{core}(\operatorname{dom} g)$). Indeed, since the weak* topology $w(X^*, X)$ is included in the topology associated with the norm $\|\cdot\|_{X^*}$, the convex function g is lower $\|\cdot\|_{X^*}$-semicontinuous on the Banach space $(X^*, \|\cdot\|_{X^*})$. Then, Corollary 2.162 confirms the above $\|\cdot\|_{X^*}$-continuity property of the function g on $\operatorname{int}_{\|\cdot\|_{X^*}}(\operatorname{dom} g)$ (as well as the coincidence of this set with $\operatorname{core}(\operatorname{dom} g)$). □

THEOREM 4.63 (C. Stegall variational principle). *Let $(X, \|\cdot\|)$ be a Banach space.*
(a) *Assume that X is an Asplund space, S is a nonempty bounded $\|\cdot\|_*$-closed subset of X^* and $g : X^* \to \mathbb{R} \cup \{+\infty\}$ is a lower $\|\cdot\|_*$-semicontinuous function which is bounded from below on S and proper on S. Then the set of $v \in X$ such that $g + \langle \cdot, v \rangle$ attains a strong minimum on S is a dense G_δ set in $(X, \|\cdot\|)$.*
(b) *Assume that X^* is a weak* Asplund space, S is a nonempty bounded $\|\cdot\|$-closed subset of X and $f : X \to \mathbb{R} \cup \{+\infty\}$ is a lower $\|\cdot\|$-semicontinuous function which is bounded from below on S and proper on S. Then the set of $v^* \in X^*$ such that $f + \langle v^*, \cdot \rangle$ attains a strong minimum on S is a dense G_δ set of $(X^*, \|\cdot\|_*)$.*

PROOF. (a) Obviously $\varphi := g + \Psi_S$ is proper and lower $\|\cdot\|_*$-semicontinuous on X^*. Further the convex function φ^* (conjugate of φ in the duality pairing from X^* to X) is easily seen to be finite and Lipschitz on X (thanks to the boundedness of S), hence it is Fréchet differentiable on a dense G_δ set in $(X, \|\cdot\|)$. By (b) in Theorem 4.58 this amounts to saying that the set of $v \in X$ where $\langle \cdot, -v \rangle - \varphi$ attains a strong maximum on X^*, or equivalently $\langle \cdot, v \rangle + g$ attains a strong minimum on S, is a dense G_δ set in $(X, \|\cdot\|)$.
(b) As in (a) consider the proper lower $\|\cdot\|$-semicontinuous function $\varphi := f + \Psi_S$ on X. From the definition of conjugate we see that φ^* is weak* lower semicontinuous on X^* and we also see that φ^* is finite on X^* (by the boundedness of S). Then, φ^* is Fréchet differentiable on a set containing a dense G_δ set of $(X^*, \|\cdot\|_*)$. Consequently, (a) in Theorem 4.58 yields that the set of $v^* \in X^*$ where $\langle -v^*, \cdot \rangle - \varphi$ attains a

strong maximum on X, or equivalently $\langle v^*, \cdot \rangle + f$ attains a strong minimum on S, is a dense G_δ set of $(X^*, \|\cdot\|_*)$. □

4.4. Mordukhovich limiting subdifferential in Asplund space

When the Banach space X is Asplund, Corollary 4.53 ensures that the set $\{u \in X : \partial_F f(u) \neq \emptyset\}$ is graphically dense in the set where the lower semicontinuous function f is finite. On the other hand, Theorem 4.52 says that the Fréchet subdifferential enjoys on the Asplund space X only fuzzy calculus rules, that is, for any function $g : X \to \mathbb{R} \cup \{+\infty\}$ which is Lipschitz continuous near x, for $x^* \in \partial_F(f+g)(x)$ and $\varepsilon > 0$, there exist $x', x'' \in X$ such that

(4.38) $$x^* \in \partial_F f(x') + \partial_F g(x'') + \varepsilon \mathbb{B}_{X^*}$$

and $\|x' - x\| + |f(x') - f(x)| < \varepsilon$. So, in Asplund spaces a limiting process is needed to obtain calculus rules only with the reference point; the need of such a process is also confirmed by the fact that, in the set S of Figure 4.1, the geometry of S near x_i, for $i \in \{3, 4\}$, is not enough translated by the Fréchet normal cone $N^F(S; x_i)$. The limit process will be carried out merely with the weak star limit of sequences $(x_n^*)_n$ with $x_n^* \in \partial_F f(x_n)$ since (see Theorem C.15 in Appendix) Asplund spaces have (like reflexive Banach spaces) the following property:

(4.39) $\begin{cases} \text{any bounded sequence of the topological dual of an Asplund space} \\ \text{has a weak star convergent subsequence.} \end{cases}$

4.4.1. Definitions, properties, calculus. The above analysis leads to the definition:

DEFINITION 4.64. Let $(X, \|\cdot\|)$ be an Asplund space, U be a neighborhood of a point x in X, and $f : U \to \{-\infty, +\infty\}$ be an extended real-valued function which is finite at $x \in U$. A continuous linear functional $x^* \in X^*$ is a *Mordukhovich limiting subgradient* (or *limiting subgradient*, or *L-subgradient*) of f at x if there exists a sequence $((x_n, f(x_n)))_n$ converging to $(x, f(x))$ and a sequence $(x_n^*)_n$ converging weakly star to x^* such that $x_n^* \in \partial_F f(x_n)$. The set $\partial_L f(x)$ of all limiting subgradients of f at x is the *Mordukhovich limiting subdifferential* (or *limiting subdifferential*, or *L-subdifferential*) of f at x, otherwise stated (with notation in (1.6))

$$\partial_L f(x) = {}^{\text{seq}}\!\operatorname*{Lim\,sup}_{u \to_f x} \partial_F f(u).$$

As usual, if $f(x)$ is not finite, one puts $\partial_L f(x) = \emptyset$.

Given the neighborhood U of x and the function $f : U \to \mathbb{R} \cup \{-\infty, +\infty\}$, clearly $\partial_L f(x) = \partial_L \overline{f}(x)$ where $\overline{f}$ is the natural extension of f as a function on X, that is, $\overline{f}(u) = f(u)$ if $u \in U$ and $\overline{f}(u) = +\infty$ otherwise. So, very often (as with the Clarke subdifferential) when dealing with $\partial_L f$, the function f will be supposed to be defined on the whole space X.

According to (1.6) we see, for $|f(x)| < +\infty$ and for X^* endowed with the $w(X^*, X)$-topology, that $\partial_L f(x)$ is the sequential outer w^*-limit (or sequential w^*-limit superior) of $\partial_F f(u)$ as $u \to_f x$.

DEFINITION 4.65. When f is the indicator function $\Psi_S : X \to \mathbb{R} \cup \{-\infty, +\infty\}$ of a set S in the Asplund space X, $\partial_L \Psi_S(x)$ is called the *Mordukhovich limiting* normal cone (or *limiting normal cone*, or *L-normal cone*) $N^L(S;x)$ of S at x. This means through Definition 4.64 that, with $x \in S$, by (1.6)

(4.40) $$N^L(S;x) = {}^{\text{seq}}\underset{S \ni u \to x}{\text{Lim sup}}\, N^F(S;u),$$

that is,

$${}^{\text{seq}}\underset{S \ni u \to x}{\text{Lim sup}}\, N^F(S;u) := \left\{ \begin{array}{l} x^* \in X^* : \exists \text{ sequences } S \ni x_n \to x, x_n^* \xrightarrow{w^*} x^*, \\ \text{with } x_n^* \in N^F(S;x_n) \end{array} \right\},$$

and that $N^L(S;x) = \emptyset$ if $x \notin S$. Any element $x^* \in N^L(S;x)$ is called a *limiting normal* or (*L-normal*).

EXAMPLE 4.66. Taking Example 4.5 into account, we directly obtain the following equalities for the functions in Example 2.36.
(a) $\partial_L(|\cdot|)(0) = [-1,1]$ and $\partial_L(-|\cdot|)(0) = \{-1,1\}$.
(b) $\partial_L(\sqrt{|\cdot|})(0) = \mathbb{R}$ and $\partial_L(-\sqrt{|\cdot|})(0) = \emptyset$.
(c) For the functions $f, g : \mathbb{R} \to \mathbb{R} \cup \{+\infty\}$ defined by

$$f(r) := \begin{cases} \sqrt{r} & \text{if } r \geq 0 \\ +\infty & \text{if } r < 0 \end{cases} \qquad g(r) := \begin{cases} -\sqrt{r} & \text{if } r \geq 0 \\ +\infty & \text{if } r < 0, \end{cases}$$

we have $\partial_L f(0) = \mathbb{R}$ and $\partial_L g(0) = \emptyset$.
(d) Considering the function $f : \mathbb{R} \to \mathbb{R}$ defined by

$$f(r) := \begin{cases} \sqrt{r} & \text{if } r \geq 0 \\ 0 & \text{if } r < 0, \end{cases}$$

we have $\partial_L f(0) = [0, +\infty[$ and $\partial_L(-f)(0) = \{0\}$.
(e) Let $f : \mathbb{R} \to \mathbb{R}$ be the Lipschitz function in Example 2.84 defined by $f(0) = 0$ and $f(r) = r^2 \sin(1/r)$ for all $r \neq 0$. From the equalities in that Example 2.84 we directly derive that

$$\partial_L f(0) = [-1, 1] = \partial_C f(0).$$

$\square$

For the F-normal cone and L-normal cone we directly have:

PROPOSITION 4.67. *Let $A : X \to Y$ be an isomorphism between normed spaces X, Y and let S be a subset of Y. Let $b \in Y$ and $x \in A^{-1}(b + S)$.*
(a) *Ones has*

$$N^F(A^{-1}(b+S);x) = A^*(N^F(S;A(x)-b)).$$

(b) *If X and Y are Asplund spaces, then*

$$N^L(A^{-1}(b+S);x) = A^*(N^L(S;A(x)-b)).$$

The following provides some first properties for functions. Assertions (a), (b), (c), (d) of the proposition directly follow from the above definitions and from the corresponding properties for the Fréchet subdifferential. Extensions to any normed vector space will be established in Proposition 4.106 through Fréchet ε-subgradients.

PROPOSITION 4.68. *Let X be an Asplund space, S be a subset of X with $x \in S$, and $f, g : X \to \mathbb{R} \cup \{-\infty, +\infty\}$ be two functions which are finite at x.*
(a) *The equality $N^L(S;x) = N^L(S \cap U;x)$ holds for any neighborhood U of x. If f and g coincide near x, then $\partial_L f(x) = \partial_L g(x)$.*
(b) *$x^* \in \partial_L f(x)$ if and only if $(x^*, -1) \in N^L(\text{epi } f; (x, f(x)))$.*

(c) For any real $r > 0$ the equality $\partial_L(rf)(x) = r\partial_L f(x)$ holds.
(d) If f is finite and Lipschitz continuous near x with $\gamma \geq 0$ as a Lipschitz constant, then $\partial_L f(x) \subset \gamma \mathbb{B}_{X^*}$.
(e) The inclusions

$$N^L(S;x) \subset \mathbb{R}_+ \partial_C d_S(x) \subset N^C(S;x) \quad \text{and} \quad \partial_L f(x) \subset \partial_C f(x)$$

always hold.
(f) If f is finite near x and strictly Fréchet differentiable at x, then

$$\partial_L f(x) = \partial_C f(x) = \{Df(x)\}.$$

PROOF. As said above only (e) and (f) need to be proved. Fix any $x^* \in N^L(S;x)$ and choose sequences $(x_n)_n$ in S converging to x and $(x_n^*)_n$ converging weak star to x^* with $x_n^* \in N^F(S;x_n)$. Fix some real $\gamma \geq \|x_n^*\|$ for all $n \in \mathbb{N}$. By Proposition 4.30, for each $n \in \mathbb{N}$ we have $x_n^* \in \gamma \partial_F d_S(x_n)$ hence $x_n^* \in \gamma \partial_C d_S(x_n)$ according to the inclusion of the Fréchet subdifferential into the Clarke one (see Proposition 4.8). So, the inclusion $x^* \in \gamma \partial_C d_S(x)$ holds (see Proposition 2.74) by the w^*-closedness at x of the Clarke subdifferential of the Lipschitz function d_C. This proves the inclusion $N^L(S;x) \subset \mathbb{R}_+ \partial_C d_S(x)$ and the right member of this inclusion is known to be contained in $N^C(S;x)$ (see Proposition 2.95). Then, the inclusion of the subdifferentials follows from this and (b), which proves (e).

Finally, if f is strictly Fréchet differentiable at x, then the equality $\partial_L f(x) = \{Df(x)\}$ follows from (e) and the equalities between $\{Df(x)\}$ and each one of the sets $\partial_C f(x)$ and $\partial_F f(x)$ (see (e) of Proposition 2.50 and (f) of Proposition 4.11). □

REMARK 4.69. (a) Unlike the Clarke subdifferential, the equality $\partial_L(-f)(x) = -\partial_L f(x)$ can fail for locally Lipschitz continuous functions. The simple example $f = |\cdot|$ in $\mathbb{R}$ confirms this fact since $\partial_L f(0) = [-1,1]$, while $\partial_L(-f)(0) = \{-1,1\}$ as seen in Example 4.66(a).
(b) Given a finite sequence $f_1 \leq f_2 \leq \cdots \leq f_m$ of functions from a normed space X into $\mathbb{R} \cup \{+\infty\}$ with $f_1(\overline{x}) = f_m(\overline{x}) < +\infty$, it is obvious that

$$\partial_F f_1(\overline{x}) \subset \cdots \subset \partial_F f_m(\overline{x});$$

further, we have seen in Corollary 2.267 that $\bigcap_{k=1}^{m} \partial_C f_k(\overline{x}) \neq \emptyset$ whenever f_1 is Lipschitz near $\overline{x}$. None of the latter properties holds with the limiting subdifferential, even in $X = \mathbb{R}$; it is enough to consider $\overline{x} := 0$ and the functions $f_1, f_2 : \mathbb{R} \to \mathbb{R}$ defined by $f_1(x) := -|x|$ and $f_2(x) := 0$ for all $x \in \mathbb{R}$, and to notice that $\partial_L f_1(0) = \{-1,1\}$ and $\partial_L f_2(0) = \{0\}$. □

The L-subdifferential of the function $-|\cdot|$ at 0 is in fact a particular case of the following result.

PROPOSITION 4.70. Let U be an open set of an Asplund space X and $f : U \to \mathbb{R}$ be a continuous concave function. Then for any $\overline{x} \in U$ one has

$$\partial_L f(\overline{x}) = {}^{\text{seq}}\!\!\!\limsup_{\text{Dom } Df \ni x \to \overline{x}} \{Df(x)\}.$$

PROOF. The function $-f$ being convex and continuous, it ensues that

$$\partial_F(-f)(x) = \partial(-f)(x) \neq \emptyset \quad \text{for all } x \in U.$$

Consequently, Proposition 4.13(c) ensures that $\partial_F f(x) \neq \emptyset$ if and only if f is Fréchet differentiable at x. This combined with the definition of $\partial f(\overline{x})$ justifies the equality of the proposition. $\square$

Figure 4.2 illustrates (with the same set S in Figure 2.1) the Mordukhovich limiting normal cone of S at x_i, for $i \in \{1, 7, 11\}$, with the representation of $x_i + N^F(S; x_i)$. At the point x_i, for the other indices i, we have

$$N^L(S; x_2) =]-\infty, 0] \times \{0\},$$

$$N^L(S; x_3) = (\{0\} \times \mathbb{R}) \cup ([0, +\infty[\times \{0\}) \cup \{(v_1, v_2) \in \mathbb{R}^2 : v_1 \geq 0 \text{ and } v_2 = v_1\},$$

$$N^L(S; x_4) = (\{0\} \times \mathbb{R}) \cup \{(v_1, v_2) \in \mathbb{R}^2 : v_2 = \pm v_1\},$$

$$N^L(S; x_5) = N^L(S; x_6) = \{(v_1, v_2) \in \mathbb{R}^2 : v_1 \geq 0 \text{ and } v_2 = -v_1\},$$

$$N^L(S; x_8) = \{(v_1, v_2) \in \mathbb{R}^2 : v_1 \leq 0, v_2 \leq v_1\} \cup (\{0\} \times [0, +\infty[) \cup ([0, +\infty[\times \{0\}),$$

$$N^L(S; x_9) = \left\{(v_1, v_2) \in \mathbb{R}^2 : v_1 \leq 0, v_2 = \pm \frac{4}{3} v_1\right\}, \quad N^L(S; x_{10}) = \{0\} \times \mathbb{R},$$

$$N^L(S; x_{12}) = \mathbb{R} \times \{0\}, \quad N^L(S; x_{13}) = \mathbb{R}^2.$$

We then see that the set S is L-normally regular in the sense of Definition 4.77 at all the points x_i except for $i = 3, 4, 8, 9, 10, 13$.

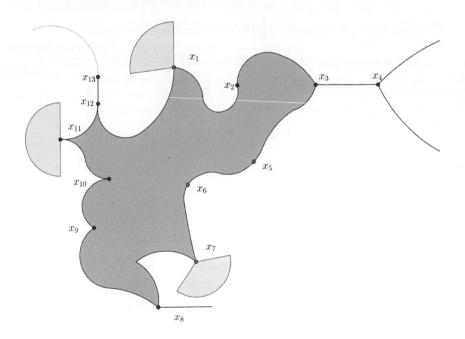

FIGURE 4.2. Mordukhovich limiting normal cones.

Two additional properties concerning the closure and the boundary of a set are stated next.

PROPOSITION 4.71. Let S be a nonempty set of the normed space X.
(a) For any $x \in S$ one has
$$N^L(S;x) \subset N^L(\operatorname{cl} S;x).$$
(b) For any $x \in \operatorname{bdry} S$ one has
$$N^L(S;x) \subset N^L(\operatorname{bdry} S;x).$$

PROOF. The property (a) follows directly from the definition of limiting normal and from the equality $N^F(S;u) = N^F(\operatorname{cl} S;u)$ for $u \in S$ in Proposition 4.7. Concerning (b), fix any $x \in \operatorname{bdry} S$ and any nonzero $x^* \in N^L(S;x)$. There exist sequences $(x_n)_n$ in S converging to x and $(x_n^*)_n$ in X^* converging weakly star to x^* with $x_n^* \in N^F(S;x_n)$ for all $n \in \mathbb{N}$. Since $x^* \neq 0$, one has $x_n^* \neq 0$ for large n, say for $n \geq n_0$. It results that, for each integer $n \geq n_0$, one has $x_n \in \operatorname{bdry} S$, hence $x_n^* \in N^F(\operatorname{bdry} S;x_n)$ according to Proposition 4.7 again. We derive that $x^* \in N^L(\operatorname{bdry} S;x)$ as desired. □

REMARK 4.72. Unlike the Fréchet normal cone (see Proposition 4.7), the converse inclusion of (a) in the above proposition fails in general for the Mordukhovich limiting normal cone. Indeed, for the set S of $\mathbb{R}$ given by
$$S := \{0\} \cup \left(\bigcup_{n \in \mathbb{N}} \right]\frac{1}{2n+1}, \frac{1}{2n}\left[\right),$$
we have $N^F(S;0) = N^L(S;0) =]-\infty,0]$ while $N^L(\operatorname{cl} S;0) = \mathbb{R}$.

The converse inclusion of (b) in Proposition 4.7 also fails in general, since with $S = [0,1]$ we have $N^L(S;1) = [0,+\infty[$ whereas $N^L(\operatorname{bdry} S;1) = \mathbb{R}$. □

4.4.2. Calculus rules. The next proposition presents some first subdifferential rules. The results will be extended to general normed vector spaces in Proposition 4.111.

PROPOSITION 4.73. Let $X, Y, X_1, \cdots, X_m$ be Asplund spaces, $g : Y \to \mathbb{R} \cup \{-\infty, +\infty\}$ be a function on Y, and for each $i = 1, \cdots, m$ let $f_i : X_i \to \mathbb{R} \cup \{-\infty, +\infty\}$ be a function and S_i be a subset of X_i. Let also $G : X \to Y$ be a mapping. Then the following hold:
(a) If $\varphi(u_1, \cdots, u_m) := f_1(u_1) + \cdots + f_m(u_m)$ is well defined on an open set U of $X_1 \times \cdots \times X_m$ and finite at $(x_1, \cdots, x_m) \in U$, and if either f_i is continuous at x_i relative to $f_i^{-1}(\mathbb{R})$ for $i = 2, \cdots, m$ or each f_i is lower semicontinuous at x_i relative to $f_i^{-1}(\mathbb{R})$ for $i = 1, \cdots, m$, then one has
$$\partial_L \varphi(x_1, \cdots, x_m) = \partial_L f_1(x_1) \times \cdots \times \partial_L f_m(x_m);$$
consequently, with $x_i \in S_i$ for $i = 1, \cdots, m$
$$N^L\big(S_1 \times \cdots \times S_m; (x_1, \cdots, x_m)\big) = N^L(S_1;x_1) \times \cdots \times N^L(S_m;x_m).$$
(b) For $h : X \times Y \to \mathbb{R} \cup \{+\infty\}$ with $h(x',y') = g(y')$ for all $(x',y') \in X \times Y$, one has at any $x \in X$ and $y \in Y$,
$$\partial_L h(x,y) = \{0\} \times \partial_L g(y).$$
(c) Assume that G is strictly Fréchet differentiable at $x \in X$. Then one has the equality $N^L\big(\operatorname{gph} G; (x, G(x))\big) = N^F\big(\operatorname{gph} G; (x, G(x))\big)$, and further
$$(x^*, -y^*) \in N^L\big(\operatorname{gph} G; (x, G(x))\big) \text{ if and only if } x^* = y^* \circ DG(x) = DG(x)^*(y^*).$$

PROOF. Without loss of generality, we may suppose $m = 2$. Concerning the first equality of (a), the inclusion of the right member into the left one follows easily from the definition of the Mordukhovich limiting subdifferential and from Proposition 4.19. Fix now any $(x_1^*, x_2^*) \in \partial_L \varphi(x_1, x_2)$. There exist sequences $(x_{1,n}, x_{2,n})_n$ converging to (x_1, x_2) with $f_1(x_{1,n}) + f_2(x_{2,n}) \to f_1(x_1) + f_2(x_2)$ and $(x_{1,n}^*, x_{2,n}^*)_n$ w^*-converging to (x_1^*, x_2^*) with $(x_{1,n}^*, x_{2,n}^*) \in \partial_F \varphi(x_{1,n}, x_{2,n})$. For $i = 1, 2$, Proposition 4.19 again ensures that $x_{i,n}^* \in \partial_F f_i(x_{i,n})$, and from either the continuity of f_2 relative to $f_2^{-1}(\mathbb{R})$ or the lower semicontinuity of f_1 relative to $f_1^{-1}(\mathbb{R})$ and of f_2 relative to $f_2^{-1}(\mathbb{R})$ it is easily seen that $f_i(x_{i,n}) \to f_i(x_i)$ for $i = 1, 2$. Therefore, $x_i^* \in \partial_L f_i(x_i)$, establishing the first equality in (a).

Taking $f_i = \Psi_{S_i}$ and observing that each Ψ_{S_i} is continuous relative to S_i, the second equality in (a) follows directly from the previous one. Similarly, one deduces the equality in (b).

Let us prove the assertion (c). Fix $(x^*, -y^*) \in N^L\big(\operatorname{gph} G; (x, G(x))\big)$ and fix also any $\varepsilon > 0$. By the strict Fréchet differentiability of G at x choose some open neighborhood U of x over which G is γ-Lipschitz and such that

$$\|G(x') - G(x'') - DG(x)(x' - x'')\| \leq \varepsilon \|x' - x''\| \quad \text{for all } x', x'' \in U.$$

Take now sequences $x_n \to x$ with $x_n \in U$ and $(x_n^*, -y_n^*) \to (x^*, -y^*)$ weakly star in $X^* \times Y^*$ with $(x_n^*, -y_n^*) \in N^F\big(\operatorname{gph} G; (x_n, G(x_n))\big)$ and choose some real $\beta \geq 0$ such that $\|y_n^*\| \leq \beta$ for all $n \in \mathbb{N}$. Then for each $n \in \mathbb{N}$ there is some neighborhood $U_n \subset U$ of x_n such that, for every $x' \in U_n$

$$\langle x_n^*, x' - x_n \rangle - \langle y_n^*, G(x') - G(x_n) \rangle \leq \varepsilon(\|x' - x_n\| + \|G(x') - G(x_n)\|),$$

hence

$$\langle x_n^* - y_n^* \circ DG(x), x' - x_n \rangle \leq \varepsilon(1 + \beta + \gamma)\|x' - x_n\|,$$

which ensures that $\|x_n^* - y_n^* \circ DG(x)\| \leq \varepsilon(1+\beta+\gamma)$. This entails by the weak star lower semicontinuity of the norm in X^* that $\|x^* - y^* \circ DG(x)\| \leq \varepsilon(1+\beta+\gamma)$. This being true for every $\varepsilon > 0$ we obtain $x^* = y^* \circ DG(x) = DG(x)^*(y^*)$. Combining this with (c) of Proposition 4.19 yields, for $S := \operatorname{gph} G$ and $z := (x, G(x))$,

$$(x^*, -y^*) \in N^L(S; z) \Rightarrow x^* = y^* \circ DG(x) \Rightarrow (x^*, -y^*) \in N^F(S; z)$$
$$\Rightarrow (x^*, -y^*) \in N^L(S; z),$$

and this guarantees the assertion (c) and finishes the proof. □

In the finite-dimensional setting, using the classical diagonal process one sees that the graph of the multimapping $N^L(S; \cdot)$ is closed. The result fails in infinite dimensions as stated in the following proposition.

PROPOSITION 4.74. (a) If X is finite-dimensional, then for any set $S \subset X$ the graph of the multimapping $N^L(S; \cdot)$ is closed.
(b) In infinite dimensions, even for $X = \ell^2(\mathbb{N})$ there are closed sets $S \subset \ell^2(\mathbb{N})$ and $x \in \operatorname{bdry} S$ such that $N^L(S; x)$ is not norm closed; consequently the multimapping $N^L(S; \cdot)$ is not norm-to-norm closed at x, hence neither sequentially norm-to-weak closed at x.

PROOF. The assertion (a) is justified above and (b) follows from the next Example 4.75. □

EXAMPLE 4.75 (S. Fitzpatrick's example). Let $(e_k)_{k\geq 1}$ be the standard orthonormal basis of $\ell^2(\mathbb{N})$ and let $e_k^* := e_k$ for all $k \geq 1$. Put
$$S := \{r(e_1 - je_j) + s(je_1 - e_k) : k > j > 1, r, s \geq 0\} \cup \{re_1 : r \geq 0\}$$
and note that S is norm closed. We are going to show that $N^L(S;0)$ is not norm closed.

Claim 1. $e_1^* + j^{-1}e_j^* + je_k^* \in N^F(S; k^{-1}(je_1 - e_k))$ and $e_1^* + j^{-1}e_j^* \in N^L(S;0)$ for $1 < j < k$.

Fix integers $k > j > 1$. Observe first that $\|re_1 - k^{-1}(je_1 - e_k)\|^2 = (r - k^{-1}j)^2 + k^{-2} \geq k^{-2}$, hence the ball $B(k^{-1}(je_1 - e_k), k^{-2})$ contains no point of the form re_1. For $r > 0$ we have
$$\langle e_1^* + j^{-1}e_j^* + je_k^*, r(e_1 - je_j) + s(je_1 - e_k) - k^{-1}(je_1 - e_k)\rangle$$
$$= (r - r) + (sj - sj) + (k^{-1}j - k^{-1}j) = 0,$$
and this combined with the previous observation yields
$$e_1^* + j^{-1}e_j^* + je_k^* \in N^F(S; k^{-1}(je_1 - e_k)),$$
which confirms the first inclusion of the claim.

For each integer $j > 1$, since $(k^{-1}(je_1 - e_k))_{k>j}$ converges strongly to 0 and $(e_1^* + j^{-1}e_j^* + je_k^*)_{k>j}$ converges weakly to $e_1^* + j^{-1}e_j^*$ as $k \to \infty$, we have the second inclusion $e_1^* + j^{-1}e_j^* \in N^L(S;0)$.

Claim 2. $e_1^* \notin N^L(S;0)$.

We proceed by contradiction in supposing that $e_1^* \in N^L(S;0)$. By the definition of $N^L(S;0)$ there exist a sequence $(x_n)_n$ in S norm converging to 0 and a sequence $(x_n^*)_n$ converging weakly to e_1^* with $x_n^* \in N^F(S;x_n)$.

If for infinitely many n the point x_n is of the form $x_n = r_n e_1$, say for $n \in N$ with $N \subset \mathbb{N}$ infinite, then for each $n \in N$ we have $x_n + re_1 \in S$ for all $r > 0$, so
$$0 \geq \limsup_{r\downarrow 0}\langle x_n^*, \frac{x_n + re_1 - x_n}{\|x_n + re_1 - x_n\|}\rangle = \langle x_n^*, e_1\rangle,$$
hence making $N \ni n \to \infty$ yields the contradiction $0 \geq 1$.

Then for all n large enough $x_n = r_n(e_1 - j_n e_{j_n}) + s_n(j_n e_1 - e_{k_n})$ with $1 < j_n < k_n$, thus $x_n + s(j_n e_1 - e_{k_n}) \in S$ for all $s > 0$, and consequently
$$0 \geq \limsup_{s\downarrow 0}\langle x_n^*, \frac{x_n + s(j_n e_1 - e_{k_n}) - x_n}{\|x_n + s(j_n e_1 - e_{k_n}) - x_n\|}\rangle = \langle x_n^*, \frac{j_n e_1 - e_{k_n}}{\sqrt{1 + j_n^2}}\rangle.$$
This entails, for some constant $K > 0$ (independent of n) and for n large enough,
$$\langle x_n^*, e_1\rangle \leq j_n^{-1}\langle x_n^*, e_{k_n}\rangle \leq j_n^{-1}\|x_n^*\| \leq j_n^{-1}K.$$
Since the first member tends to 1 as $n \to \infty$ the set $\{j_n : n \in \mathbb{N}\}$ is bounded in $\mathbb{N}$, that is, finite, hence there exists some integer $j_0 > 1$ and an infinite subset $N \subset \mathbb{N}$ such that $j_n = j_0$ for all $n \in N$.

Similarly, for n large enough we have $x_n + r(e_1 - j_n e_{j_n}) \in S$ for all $r > 0$, hence
$$0 \geq \limsup_{r\downarrow 0}\left\langle x_n^*, \frac{x_n + r(e_1 - j_n e_{j_n}) - x_n}{\|x_n + r(e_1 - j_n e_{j_n}) - x_n\|}\right\rangle = \left\langle x_n^*, \frac{e_1 - j_n e_{j_n}}{\sqrt{1 + j_n^2}}\right\rangle,$$
hence $\langle x_n^*, e_1\rangle \leq j_n\langle x_n^*, e_{j_n}\rangle$. Then for large $n \in N$ we obtain $\langle x_n^*, e_1\rangle \leq j_0\langle x_n^*, e_{j_0}\rangle$, and taking the limit as $N \ni n \to \infty$ gives the contradiction $1 \leq 0$.

In conclusion, we have from Claims 1, 2 that $e_1^* + j^{-1}e_j^* \in N^L(S;0)$ with $(e_1^* + j^{-1}e_j^*)_{j\geq 2}$ norm converging to $e_1^* \notin N^L(S;0)$, which justifies that $N^L(S;0)$ is not norm closed. $\square$

Let us turn now to some crucial additional properties of the Mordukhovich limiting subdifferential in the setting of Asplund space.

We have already observed (for example, with $f = -|\cdot|$) that the Fréchet subdifferential of a locally Lipschitz continuous function can be empty at some points. Such a situation does not occur with the Mordukhovich limiting subdifferential on Asplund space.

PROPOSITION 4.76. *Let X be an Asplund space and f be a function which is finite and Lipschitz continuous near $x \in X$. Then $\partial_L f(x) \neq \emptyset$.*

PROOF. By the density result of $\mathrm{Dom}\,\partial_F f$ in Corollary 4.53 there exist sequences $(x_n)_n$ in X with $x_n \to x$ and $(x_n^*)_n$ in X^* with $x_n^* \in \partial_F f(x_n)$. Denoting by γ a Lipschitz constant of f near x, we have $\|x_n^*\| \leq \gamma$ for n large enough. By the Asplund property of X, the sequence $(x_n^*)_n$ has a convergent subsequence converging weak star to some x^* (see (4.39)). Evidently, $x^* \in \partial_L f(x)$. $\square$

The second result in the Asplund setting concerns the sum rule. It provides, for the Mordukhovich limiting subdifferential of the sum of functions, a point-based rule involving the subdifferentials of the data functions at the reference point. A case for the equality in the sum rule formula will need the concept of L-subdifferential regularity.

DEFINITION 4.77. A function $f : X \to \mathbb{R} \cup \{+\infty\}$ is *L-subdifferentially regular* at a point $x \in \mathrm{dom}\, f$ if the Fréchet and Mordukhovich limiting subdifferentials of f at x coincide, that is, $\partial_F f(x) = \partial_L f(x)$. Specifying the case where f is the indicator function of a set S with $x \in S$, one says that the set S is *L-normally regular* at x when $N^F(S;x) = N^L(S;x)$.

THEOREM 4.78 (sum rule for L-subdifferential). *Let X be an Asplund space and $f_1, \cdots, f_m : X \to \mathbb{R} \cup \{+\infty\}$ be functions which are finite at $x \in X$ and lower semicontinuous near x. If all but one of these functions are Lipschitz continuous near x, then one has*

(4.41) $$\partial_L(f_1 + \cdots + f_m)(x) \subset \partial_L f_1(x) + \cdots + \partial_L f_m(x).$$

The inclusion is an equality whenever in addition either all the functions are L-subdifferentially regular at x (and in this case, so is the sum function), or all the functions but one of them are strictly Fréchet differentiable at the point x.

PROOF. It is enough to consider that $m = 2$ as well as $f_1 =: f$ is lower semicontinuous near x and $f_2 =: g$ is Lipschitz continuous near x. Fix $x^* \in \partial_L(f+g)(x)$ and take sequences $x_n \to x$ with $f(x_n) + g(x_n) \to f(x) + g(x)$, $x_n^* \xrightarrow{w^*} x^*$ with $x_n^* \in \partial_F(f+g)(x_n)$. By Theorem 4.52 on fuzzy sum rule there are $u_n, v_n \in B(x, 1/n)$ with $|f(u_n) - f(x_n)| < 1/n$, $u_n^* \in \partial_F f(u_n)$, $v_n^* \in \partial_F g(v_n)$ with $\|x^* - (u_n^* + v_n^*)\| < 1/n$. The sequence $(v_n^*)_n$ is bounded by the Lipschitz property of g near x and thus (see (4.39)) according to the Asplund property of X a subsequence (that we do not relabel) of $(v_n^*)_n$ converges weakly star to some v^*. By definition $v^* \in \partial_L g(x)$. Further, the continuity of g at x entails that $f(u_n) \to f(x)$ and since $u_n^* \to x^* - v^*$ weakly star in X^* we obtain that $x^* - v^* \in \partial_L f(x)$, which yields $x^* \in \partial_L f(x) + \partial_L g(x)$. This shows the desired inclusion of the theorem.

If in addition g is strictly Fréchet differentiable at x, we can write by the inclusion above $\partial_L f(x) \subset \partial_L(f+g)(x) - Dg(x)$, hence $\partial_L f(x) + Dg(x) \subset \partial_L(f+g)(x)$. This ensures the desired equality. From the general inclusions $\partial_F f(x) + \partial_F g(x) \subset \partial_F(f+g)(x) \subset \partial_L(f+g)(x)$ (see Proposition 4.11), the equality also holds whenever f and g are L-subdifferentially regular at x; the L-subdifferential regularity of $f+g$ at x follows also from the same above general inclusions. $\square$

The equality $\partial_L(f+g)(x) = \partial_L f(x) + Dg(x)$ in the above theorem, under the strict Fréchet differentiability of g, will be extended to any normed space in Proposition 4.106(f).

EXAMPLE 4.79. The converse inclusion in (4.41) fails in general. The same simple functions $f_1 := |\cdot|$ and $f_2 := -|\cdot|$ on $\mathbb{R}$ in Example 2.102 furnish $\partial_L(f_1 + f_2)(0) = \{0\}$, while $\partial_L f_1(0) = [-1, 1]$ and $\partial_L f_2(0) = \{-1, 1\}$. $\square$

Conditions ensuring the formula $N^L(S_1 \cap \cdots \cap S_m; x) \subset N^L(S_1; x) + \cdots + N^L(S_m; x)$ will be provided in Corollaries 7.93 and 7.94.

The next theorem deals with the chain rule for $g \circ G$ when the mapping G is compactly Lipschitzian.

THEOREM 4.80 (chain rule for L-subdifferential). Let $G : X \to Y$ be a mapping between Asplund spaces X and Y, and $g : Y \to \mathbb{R}$ be a function which is Lipschitz continuous near $y = G(x)$.
(a) If G is compactly Lipschitzian at x, then $g \circ G$ is Lipschitz continuous near x and
$$\partial_L(g \circ G)(x) \subset \bigcup_{y^* \in \partial_L g(y)} \partial_L(y^* \circ G)(x).$$
(b) If G is strictly Fréchet differentiable at x, then one has
$$\partial_L(g \circ G)(x) \subset DG(x)^*\big(\partial_L g(y)\big), \tag{4.42}$$
where $DG(x)^*$ denotes the adjoint of $DG(x)$ and
$$DG(x)^*(\partial_L g(y)) := \{DG(x)^*(y^*) : y^* \in \partial_L g(y)\}.$$

PROOF. (a) Let U be an open neighborhood of x such that G is γ_G-Lipschitz continuous on U and $V \supset G(U)$ be an open neighborhood of $G(x)$ over which g is γ_g-Lipschitz continuous. The function $g \circ G$ is $\gamma_g \gamma_G$-Lipschitz continuous on U. Fix any $x^* \in \partial_L(g \circ G)(x)$. Taking positive reals $\varepsilon_n \downarrow 0$, the definition of $\partial_L(g \circ G)(x)$ and Theorem 4.57 give sequences $(x_n)_n$ of U converging to x, $(x_n^*)_n$ in X^* converging weakly star to x^*, $(y_n)_n$ of V converging to $y := G(x)$, and $(y_n^*)_n$ in Y^* such that
$$y_n^* \in \partial_F g(y_n) \quad \text{and} \quad x_n^* \in \partial_F(y_n^* \circ G)(x_n) + \varepsilon_n \mathbb{B},$$
thus for each n we have $x_n^* \in u_n^* + \varepsilon_n \mathbb{B}$ for some $u_n^* \in \partial_F(y_n^* \circ G)(x_n)$. Since $\|y_n^*\| \le \gamma_g$, by the Asplund property of Y (see (4.39)) a subsequence of $(y_n^*)_n$ (that we do not relabel) converges weakly star to some $y^* \in Y^*$. Obviously, we have $y^* \in \partial_L g(y)$. Further, by the fuzzy sum rule in Theorem 4.52 for each $n \in \mathbb{N}$ there are x_n' and x_n'' in U with $\|x_n' - x_n\| < \varepsilon_n$ and $\|x_n'' - x_n\| < \varepsilon_n$, and $v_n^* \in \partial_F\big((y_n^* - y^*) \circ G\big)(x_n'')$ such that
$$u_n^* - v_n^* \in \partial_F(y^* \circ G)(x_n') + \varepsilon_n \mathbb{B}_{X^*}, \quad \text{hence} \quad x_n^* - v_n^* + 2\varepsilon_n b_n^* \in \partial_F(y^* \circ G)(x_n')$$

for some $b_n^* \in \mathbb{B}_{X^*}$. The assertion (c) of Proposition 2.187 implies that the sequence $(v_n^*)_n$ converges weakly star to zero in X^*. Consequently, the last inclusion above guarantees that $x^* \in \partial_L(y^* \circ G)(x)$, establishing the desired inclusion of (a) in the theorem.

(b) If G is strictly Fréchet differentiable at x, then for any $y^* \in \partial_L g(G(x))$, the function $y^* \circ G$ is strictly Fréchet differentiable at x, which yields $\partial_L(y^* \circ G)(x) = \{D(y^* \circ G)(x)\}$, that is, $\partial_L(y^* \circ G)(x) = \{DG(x)^*(y^*)\}$. The inclusion in the above assertion (a) then yields

$$\partial_L(g \circ G)(x) \subset \{DG(x)^*(y^*) : y^* \in \partial_L g(G(x))\} = DG(x)^*(\partial_L g(G(x))),$$

which finishes the proof of the theorem. □

The case when the strict differentiability is assumed for the function g, and not for the mapping G as in (b) above, will be treated in the setting of any normed space in Corollary 4.109.

At this stage, as a corollary of Theorem 4.80 we can deduce an estimate rule for the limiting subdifferential of the pointwise maximum of finitely many functions.

COROLLARY 4.81. *Let U be an open convex set of an Asplund space X and $(f_i)_{i \in I}$ be a finite family of locally Lipschitz continuous real-valued functions defined on U, and let $f(x) = \max_{i \in I} f_i(x)$. For any $x \in U$ and for $I(x) := \{i \in I : f_i(x) = f(x)\}$, one has*

$$\partial_L f(x) \subset \bigcup_{\lambda \in \Lambda} \partial_L \left(\sum_{i \in I(x)} \lambda_i f_i \right)(x),$$

where $\Lambda := \left\{ \lambda \in \mathbb{R}^{I(x)} : \lambda_i \geq 0, \sum_{i \in I(x)} \lambda_i = 1 \right\}$.

PROOF. Without loss of generality me may suppose that $I := \{1, \cdots, n\}$. Then put $Y := \mathbb{R}^n$ and $g(y) = \max_{i \in I} y_i$. Consider also the mapping $G : U \to Y$ with $G(u) := (f_1(u), \cdots, f_n(u))$ for all $u \in U$. This mapping G is Lipschitz continuous near x, thus compactly Lipschitzian at x since Y is finite-dimensional. By Proposition 2.190 we know that $\partial_L g(G(x)) = \partial_C g(G(x))$ coincides with the set of $\lambda \in \mathbb{R}^n$ with $\lambda_i \geq 0$ and such that $\sum_{i \in I} \lambda_i = 1$ and $\lambda_i = 0$ for all $i \notin I(x)$. So, the desired inclusion follows from Theorem 4.80. □

Unlike the C-subdifferential, the case of the minimum of finitely many functions cannot be reduced to the situation of the maximum through the opposite. The result which is valid in this case for L-subdifferential is the following. It does not require any chain rule.

PROPOSITION 4.82. *Let U be an open set of an Asplund space X and $(f_i)_{i \in I}$ be a finite family of functions from U into $\mathbb{R} \cup \{-\infty, +\infty\}$, and let $f(x) = \min_{i \in I} f_i(x)$. For any $x \in U$ where f is finite and where the functions f_i ($i \in I$) are lower semicontinuous, one has*

$$\partial_L f(x) \subset \bigcup_{i \in I(x)} \partial_L f_i(x),$$

where $I(x) := \{i \in I : f_i(x) = f(x)\}$.

PROOF. Fix any $x^* \in \partial_L f(x)$ (that we suppose nonempty). There are sequences $(x_n, f(x_n))_n$ converging to $(x, f(x))$ with $f(x_n)$ finite and $(x_n^*)_n$ converging weakly* to x^* with $x_n^* \in \partial_F f(x_n)$. For each $n \in \mathbb{N}$ choose $i_n \in I$ with $f_{i_n}(x_n) = f(x_n)$. Since I is finite, there are some $j \in I$ and some increasing function $s: \mathbb{N} \to \mathbb{N}$ such that $i_{s(n)} = j$ for all $n \in \mathbb{N}$. Then $f_j(x_{s(n)}) = f(x_{s(n)})$ for all $n \in \mathbb{N}$, hence in particular $\lim_{n\to\infty} f_j(x_{s(n)}) = f(x) \leq f_j(x)$. This and the lower semicontinuity of f_j at x imply that $f_j(x_{s(n)}) \to f_j(x)$ as $n \to \infty$ and $f_j(x) = f(x)$, that is, $j \in I(x)$. Since $x_{s(n)}^* \in \partial_F f_j(x_{s(n)})$ for all $n \in \mathbb{N}$, we derive that $x^* \in \partial_L f_j(x)$, and this translates the inclusion of the proposition. □

EXAMPLE 4.83. For $\alpha, \beta \in \mathbb{R}$ consider the function $f: \mathbb{R}^2 \to \mathbb{R}$ in Example 2.191 defined by $f(x, y) := |\alpha|x| + \beta y|$ and put $g(x, y) := \alpha|x| + \beta y$. Let us compute the L-subdifferential of f and $-f$ at $(0, 0)$. If $\alpha = 0$, we have $f(x, y) = |\beta||y|$ and $-f(x, y) = -|\beta||y|$, so $\partial_L f(0, 0) = \{0\} \times [-|\beta|, |\beta|]$, and for $-f$ we have $\partial_L(-f)(0, 0) = \{(0, 0)\}$ if $\beta = 0$ and $\partial_L(-f)(0, 0) = \{0\} \times \{-\beta, \beta\}$ if $\beta \neq 0$ according to Example 4.66(a). Similarly, if $\beta = 0$, the expressions $f(x, y) = |\alpha||x|$ and $-f(x, y) = -|\alpha||x|$ yield as above $\partial_L f(0, 0) = [-|\alpha|, |\alpha|] \times \{0\}$ as well as $\partial_L(-f)(0, 0) = \{(0, 0)\}$ if $\alpha = 0$ and $\partial_L(-f)(0, 0) = \{-\alpha, \alpha\} \times \{0\}$ if $\alpha \neq 0$. Noting that $|\alpha|x| + \beta y| = |(-\alpha)|x| - \beta y|$, it remains to study the situation when $\alpha > 0$ and $\beta \neq 0$.

(a) Theorem 4.80(a) says that $\partial_L f(0, 0) \subset \bigcup_{r \in [-1, 1]} \partial_L(rg)(0, 0)$. The function g is convex and $\partial_L g(0, 0) = [-\alpha, \alpha] \times \{\beta\}$. By Proposition 4.73 and Example 4.66(a) we also have $\partial_L(-g)(0, 0) = \{-\alpha, \alpha\} \times \{-\beta\}$. It ensues that

$$\partial_L f(0,0) \subset \left(\bigcup_{r \in [0,1]} ([-\alpha r, \alpha r] \times \{\beta r\}) \right) \cup \left(\bigcup_{r \in]0,1]} (\{-\alpha r, \alpha r\} \times \{-\beta r\}) \right).$$

In fact, the inclusion is an equality. To see that, consider the F-subdifferential of f at (x, y) by studying the following cases apart:

- Case 1: $\alpha|x| + \beta y > 0$: For (u, v) sufficiently close to (x, y) we have $f(u, v) = \alpha|u| + \beta v$, hence

$$\partial_F f(x, y) = \begin{cases} [-\alpha, \alpha] \times \{\beta\} & \text{if } x = 0 \\ \{-\alpha\} \times \{\beta\} & \text{if } x < 0 \\ \{\alpha\} \times \{\beta\} & \text{if } x > 0. \end{cases}$$

- Case 2: $\alpha|x| + \beta y < 0$: For (u, v) sufficiently close to (x, y) we have $f(u, v) = -\alpha|u| - \beta v$, hence

$$\partial_F f(x, y) = \begin{cases} \emptyset \times \{-\beta\} = \emptyset & \text{if } x = 0 \\ \{\alpha\} \times \{-\beta\} & \text{if } x < 0 \\ \{-\alpha\} \times \{-\beta\} & \text{if } x > 0, \end{cases}$$

where we have used Example 4.5(a) for the equality to the empty set.

- Case 3: $\alpha|x| + \beta y = 0$:

If $x = 0$, then $y = 0$, so Example 4.6(c) tells us that $\partial_F f(x, y) = \{(a, b) : |a| \leq \frac{\alpha b}{\beta}, \frac{b}{\beta} \leq 1\}$, which gives with $r := b/\beta \in [0, 1]$ that

$$\partial_F f(x, y) = \bigcup_{r \in [0,1]} ([-\alpha r, \alpha r] \times \{\beta r\}).$$

If $x > 0$, then for (u, v) close enough to (x, y) we have $f(u, v) = |\alpha u + \beta v|$, so taking the convexity of f near (x, y) into account, Theorem 2.135 gives $\partial_F f(x, y) = [-1, 1](\alpha, \beta)$.

Similarly, if $x < 0$, then for (u, v) close enough to (x, y) we have $f(u, v) = |-\alpha u + \beta v|$, which gives $\partial_F f(x, y) = [-1, 1](-\alpha, \beta)$.

We summarize the case 3 when $\alpha|x| + \beta y = 0$ (as for the two previous cases 1 and 2) in the form

$$\partial_F f(x, y) = \begin{cases} \bigcup_{r \in [0,1]} ([-\alpha r, \alpha r] \times \{\beta r\}) & \text{if } x = 0 \\ \{(-\alpha r, \beta r) : r \in [-1, 1]\} & \text{if } x < 0 \\ \{(\alpha r, \beta r) : r \in [-1, 1]\} & \text{if } x > 0. \end{cases}$$

In conclusion, taking $\operatorname*{Lim\,sup}_{(x,y) \to (0,0)} \partial_F f(x, y)$ confirms that for $\alpha > 0$ and $\beta \neq 0$

$$\partial_L f(0, 0) = \left(\bigcup_{r \in [0,1]} ([-\alpha r, \alpha r] \times \{\beta r\}) \right) \cup \left(\bigcup_{r \in]0,1]} (\{-\alpha r, \alpha r\} \times \{-\beta r\}) \right).$$

(b) Defining $\varphi(x, y) := -f(x, y) = -|\alpha|x| + \beta y|$ and noting that

$$\varphi(x, y) = \min\{g(x, y), -g(x, y)\} \quad \text{with} \quad g(x, y) := \alpha|x| + \beta y,$$

Proposition 4.82 tells us that

$$\partial_L \varphi(0, 0) \subset ([-\alpha, \alpha] \times \{\beta\}) \cup (\{-\alpha, \alpha\} \times \{-\beta\}).$$

Here also the inclusion is in fact an equality. To verify this, let us proceed as in (a) with the F-subdifferentials of φ.

- Case 1: $\alpha|x| + \beta y > 0$: For (u, v) sufficiently close to (x, y) we have $\varphi(u, v) = -\alpha|u| - \beta v$, hence

$$\partial_F f(x, y) = \begin{cases} \emptyset \times \{-\beta\} = \emptyset & \text{if } x = 0 \\ \{\alpha\} \times \{-\beta\} & \text{if } x < 0 \\ \{-\alpha\} \times \{-\beta\} & \text{if } x > 0. \end{cases}$$

- Case 2: $\alpha|x| + \beta y < 0$: For (u, v) sufficiently close to (x, y) we have $\varphi(u, v) = \alpha|u| + \beta v$, hence

$$\partial_F f(x, y) = \begin{cases} [-\alpha, \alpha] \times \{\beta\} & \text{if } x = 0 \\ \{-\alpha\} \times \{\beta\} & \text{if } x < 0 \\ \{\alpha\} \times \{\beta\} & \text{if } x > 0. \end{cases}$$

At this stage, we see from the two above cases that

$$([-\alpha, \alpha] \times \{\beta\}) \cup (\{-\alpha, \alpha\} \times \{-\beta\}) \subset \operatorname*{Lim\,sup}_{(x,y) \to (0,0)} \partial_F f(x, y).$$

This and the corresponding above converse inclusion confirms for $\alpha > 0$ and $\beta \neq 0$ the equality

$$\partial_L \varphi(0, 0) = ([-\alpha, \alpha] \times \{\beta\}) \cup (\{-\alpha, \alpha\} \times \{-\beta\}).$$

□

A mean-value inequality can also be derived from Theorems 4.78 and 4.80 for locally Lipschitz continuous functions.

THEOREM 4.84 (mean value inequality with L/F-subdifferential: Lipschitz functions). Let U be an open convex set of an Asplund space X and $f : U \to \mathbb{R}$ be a locally Lipschitz continuous real-valued function. Then for any $a, b \in U$ with $a \neq b$ the following hold.
(a) There exist some $c \in [a, b[:= \{a + t(b - a) : t \in [0, 1[\}$ (resp. $c \in]a, b[$) and $c^* \in \partial_L f(c)$ (resp. $c^* \in \partial_L f(c) \cup (- \partial_L(-f)(c))$) such that
$$f(b) - f(a) \leq \langle c^*, b - a \rangle \quad (\text{resp. } f(b) - f(a) = \langle c^*, b - a \rangle).$$
(b) For any $\varepsilon > 0$ there are $c \in U$ with $d(c, [a, b]) < \varepsilon$ and $c^* \in \partial_F f(c)$ such that
$$f(b) - f(a) \leq \langle c^*, b - a \rangle + \varepsilon.$$

PROOF. Fix $a, b \in U$ with $a \neq b$ and fix also an open interval $I \supset [0, 1]$ with $a + t(b - a) \in U$ for all $t \in I$. Put
$$\varphi(t) := f(a + t(b - a)) + t(f(a) - f(b)) \quad \text{for all } t \in I.$$
According to the equality $\varphi(0) = \varphi(1)$, the continuous function φ attains its minimum on $[0, 1]$ at some point $s \in [0, 1[$. In consequence, $0 \in \partial_F(\varphi + \Psi_{[0,1]})(s)$.
(a) The limiting subdifferential sum rule in Theorem 4.78 says that $0 \in \partial_L \varphi(s) + N^L([0, 1]; s)$ since φ is Lipschitz continuous near s. For $c := a + s(b-a)$ by Theorems 4.78 and 4.80 we obtain some $c^* \in \partial_L f(c)$ such that
$$\langle c^*, b - a \rangle + f(a) - f(b) \in -N^L([0, 1]; s).$$
Using the convexity of $[0, 1]$ it is easily seen that $N^L([0, 1]; s) \subset]-\infty, 0]$, thus
$$f(b) - f(a) \leq \langle c^*, b - a \rangle.$$
Similarly, at some point $r \in]0, 1[$ the function φ attains a minimum or a maximum on $]0, 1[$, hence $0 \in \partial_L \varphi(r) \cup \partial_L(-\varphi)(r)$. For $e := a + r(b - a)$, Theorem 4.80 furnishes some $e^* \in \partial_L f(e) \cup (- \partial_L(-f)(e))$ such that $f(b) - f(a) = \langle e^*, b - a \rangle$. This finishes the proof of (a).
(b) Fix $\varepsilon > 0$. Taking any positive real $\varepsilon' < \varepsilon/(2 + 2\|b - a\|)$ with $s - \varepsilon' \in I$ and $s + \varepsilon' < 1$, the fuzzy sum rule in Theorem 4.52 yields $s', s'' \in]s - \varepsilon', s + \varepsilon'[$ and $\zeta \in \partial_F \varphi(s')$ such that $\zeta \in -N^F([0, 1]; s'') + [-\varepsilon', +\varepsilon']$. By Theorem 4.57 there are some $c \in U \cap B(a + s'(b-a), \varepsilon')$ and $c^* \in \partial_F f(c)$ satisfying $|\zeta - \langle c^*, b - a \rangle - f(a) + f(b)| \leq \varepsilon'$. Since $N^F([0, 1]; s'') \subset]-\infty, 0]$ (because $s'' \in [0, 1[$), we deduce
$$f(b) - f(a) \leq \langle c^*, b - a \rangle + \varepsilon,$$
and from the inclusion $a + s(b - a) \in [a, b]$ it is easily seen that $d(c, [a, b]) \leq \varepsilon'(1 + \|b - a\|) < \varepsilon$. □

4.4.3. L-Subdifferential of distance function. This subsection examines the Mordukhovich limiting subdifferential of the distance function d_S. We begin with the case of points inside the set S.

THEOREM 4.85 (L-subdifferential of distance function at inside point). Assume that $(X, \|\cdot\|)$ is an Asplund space and S is a subset of X which is closed near $x \in S$. Then the following hold:
(a) $\partial_L d_S(x) = {}^{\text{seq}} \underset{S \ni y \to x}{\text{Lim sup}} \, \partial_F d_S(y)$.
(b) The Mordukhovich limiting subdifferential of the distance function d_S is related to the limiting normal cone as follows:
(4.43) $\quad \partial_L d_S(x) \subset N^L(S; x) \cap \mathbb{B}_{X^*}$ and $N^L(S; x) = \mathbb{R}_+ \partial_L d_S(x)$.

(c) If X is finite-dimensional, ones has
$$\partial_L d_S(x) = N^L(S; x) \cap \mathbb{B}_{X^*},$$
that is, the inclusion in (b) is an equality.

The proof of the theorem is based on the following lemmas which have their own interests and which will also be used in other places in the manuscript.

LEMMA 4.86. Let $(X, \|\cdot\|)$ be a Banach space and $f : X \to \mathbb{R} \cup \{+\infty\}$ be a function which is finite at $\bar{x}$ and lower semicontinuous on a closed ball $B[\bar{x}, \delta]$. Let $x^* \in X^*$ and $\varepsilon, \lambda, \eta > 0$ with $\lambda < \delta$ such that
$$\langle x^*, x - \bar{x} \rangle \leq f(x) - f(\bar{x}) + \varepsilon \|x - \bar{x}\| + \eta \quad \forall x \in B[\bar{x}, \delta].$$
Then there exist $\bar{u} \in B[\bar{x}, \lambda]$ with $f(\bar{u}) \leq f(\bar{x})$ such that
$$\langle x^*, x - \bar{u} \rangle \leq f(x) - f(\bar{u}) + (\varepsilon + \lambda^{-1}\eta)\|x - \bar{u}\| \quad \forall x \in B[\bar{x}, \delta],$$
hence in particular
$$x^* \in \partial_F \big(f + (\varepsilon + \lambda^{-1}\eta)\|\cdot - \bar{u}\|\big)(\bar{u}).$$

PROOF. Putting $\varphi(x) := f(x) - \langle x^*, x \rangle + \varepsilon \|x - \bar{x}\|$ and $E := B[\bar{x}, \delta]$, the inequality of the assumption means $\varphi(\bar{x}) \leq \inf_E \varphi + \eta$, hence the Ekeland principle in Theorem 2.221 furnishes some $\bar{u} \in E$ with $\|\bar{u} - \bar{x}\| \leq \lambda$ such that for all $x \in E$
$$f(\bar{u}) - \langle x^*, \bar{u} \rangle + \varepsilon \|\bar{u} - \bar{x}\| \leq f(x) - \langle x^*, x \rangle + \varepsilon \|x - \bar{x}\| + \lambda^{-1}\eta \|x - \bar{u}\|,$$
thus (since $\varepsilon \|x - \bar{x}\| - \varepsilon \|\bar{u} - \bar{x}\| \leq \varepsilon \|x - \bar{u}\|$) the latter inequality entails
$$\langle x^*, x - \bar{u} \rangle \leq f(x) - f(\bar{u}) + (\varepsilon + \lambda^{-1}\eta)\|x - \bar{u}\|.$$
Further, observing that $\|\bar{u} - \bar{x}\| \leq \lambda < \delta$, the set E is a neighborhood of $\bar{u}$, so the latter inequality guarantees that $x^* \in \partial_F \big(f + (\varepsilon + \lambda^{-1}\eta)\|\cdot - \bar{u}\|\big)(\bar{u})$. □

LEMMA 4.87. Let $(X, \|\cdot\|)$ be a Banach space and S be a closed set of X. Let $\alpha, \eta > 0$ with $\alpha < \eta/2$, $u \in X$ and $u^* \in X^*$ be such that, for all $u' \in B[u, \alpha]$,
$$\langle u^*, u' - u \rangle \leq d_S(u') - d_S(u) + \eta \|u' - u\|.$$
Then, for any $y \in S$ with $\|u - y\| \leq \alpha^2 + d_S(u)$, there is some $z \in S \cap B[y, \alpha/2]$ such that
$$\langle u^*, z' - z \rangle \leq (\eta + 2\alpha)\|z' - z\| \quad \text{for all } z' \in S \cap B[y, \alpha],$$
hence in particular $u^* \in \partial_F \big(\Psi_S + (\eta + 2\alpha)\|\cdot - z\|\big)(z)$.

PROOF. Fix any $y \in S$ with $\|y - u\| \leq d_S(u) + \alpha^2$. For all $y' \in B[y, \alpha]$ we have
$$\langle u^*, y' - y \rangle \leq d_S(u + y' - y) - d_S(u) + \eta \|y' - y\|$$
$$\leq d_S(y') + \|u - y\| - d_S(u) + \eta \|y' - y\|,$$
and hence $\langle u^*, y' - y \rangle \leq d_S(y') + \eta \|y' - y\| + \alpha^2$, thus
$$\langle u^*, y' - y \rangle \leq \Psi_S(y') - \Psi_S(y) + \eta \|y' - y\| + \alpha^2.$$
Applying Lemma 4.86 above with $\lambda := \alpha/2$ we obtain some $z \in S$ with $\|z - y\| \leq \alpha/2$ such that
$$\langle u^*, z' - z \rangle \leq (\eta + 2\alpha)\|z' - z\| \quad \text{for all } z' \in S \cap B[y, \alpha]$$
and $u^* \in \partial_F \big(\Psi_S + (\eta + 2\alpha)\|\cdot - z\|\big)(z)$. □

The third lemma is related to the approximation of Fréchet subgradient of the distance function d_S at point outside S by Fréchet subgradient of d_S at point inside the set S.

LEMMA 4.88. Let S be a closed subset of an Asplund space X, let $u \in X$ and let $u^* \in \partial_F d_S(u)$. Then for every $\varepsilon > 0$, there exist $x \in S$ and $x^* \in \partial_F d_S(x)$ such that
$$\|x - u\| \leq \varepsilon + d_S(u) \quad \text{and} \quad \|x^* - u^*\| \leq \varepsilon.$$

PROOF. Fix $\varepsilon > 0$ and choose $\eta > 0$ with $\eta < \min\{1, \varepsilon/6\}$. By the analytic description of the Fréchet subdifferential, we may choose some positive number $\alpha < \eta/2$ such that for all $u' \in B[u, \alpha]$
$$\langle u^*, u' - u \rangle \leq d_S(u') - d_S(u) + \eta \|u' - u\|.$$

Fix some $y \in S$ satisfying

(4.44) $$\|u - y\| \leq \alpha^2 + d_S(u).$$

Applying Lemma 4.87 we obtain some $z \in S$ with $\|z - y\| \leq \alpha/2$ such that,

(4.45) $$u^* \in \partial_F\big(\Psi_S(\cdot) + (\eta + 2\alpha)\|\cdot - z\|\big)(z) \subset \partial_F(\Psi_S + 2\eta\|\cdot - z\|)(z),$$

the second inclusion being due to the inequality $\eta + 2\alpha \leq 2\eta$. According to the fuzzy sum rule in Theorem 4.52, the latter entails the existence of some $x \in S$ with $\|x - z\| < \eta$ and such that
$$u^* \in \partial_F \Psi_S(x) + 3\eta \mathbb{B}_{X^*} = N^F(S; x) + 3\eta \mathbb{B}_{X^*}.$$

Then we may choose some $e^* \in \mathbb{B}_{X^*}$ satisfying
$$v^* := u^* + 3\eta e^* \in N^F(S; x).$$

Since $u^* \in \partial_F d_S(u)$, we have $\|u^*\| \leq 1$ (see (d) of Proposition 4.68), thus $\|v^*\| \leq 1 + 3\eta$. Applying Proposition 4.30 we get
$$x^* := (1 + 3\eta)^{-1} v^* \in \partial_F d_S(x).$$

Further, since $\|u^* - e^*\| \leq 2$ we have
$$\|x^* - u^*\| = \frac{3\eta}{1 + 3\eta} \|u^* - e^*\| \leq \frac{6\eta}{1 + 3\eta} < \varepsilon.$$

On the other hand, by (4.44)
$$\|x - u\| \leq \|x - z\| + \|z - y\| + \|y - u\| \leq \eta + \alpha + \alpha^2 + d_S(u),$$
which implies
$$\|x - u\| < \varepsilon + d_S(u).$$
This finishes the proof. □

PROOF OF THEOREM 4.85. Without loss of generality we may suppose that the set S is closed.

(a) According to the definition of the Mordukhovich limiting subdifferential, the right member in (a) is included in the left one. To show the converse inclusion fix any $x^* \in \partial_L d_S(x)$. There exist sequences $(u_n)_n$ in X and $(u_n^*)_n$ in X^*, with $u_n^* \in \partial_F d_S(u_n)$, such that
$$\|\cdot\| - \lim_{n \to \infty} u_n = x \quad \text{and} \quad w^* - \lim_{n \to \infty} u_n^* = x^*.$$

Applying Lemma 4.88 above for each $n \in \mathbb{N}$ and for $\varepsilon_n = 1/n$, we obtain a sequence $(x_n)_n$ in S, and a sequence $(x_n^*)_n$ in X^* with $x_n^* \in \partial_F d_S(x_n)$ such that for all $n \in \mathbb{N}$

$$\|x_n - u_n\| \leq \varepsilon_n + d_S(u_n) \quad \text{and} \quad \|x_n^* - u_n^*\| \leq \varepsilon_n.$$

We deduce that $\|\cdot\| - \lim_{n\to\infty} x_n = x$ and $w^* - \lim_{n\to\infty} x_n^* = x^*$, which proves the assertion (a) of the theorem.

(b) The inclusion $\partial_L d_S(x) \subset N^L(S;x) \cap \mathbb{B}_{X^*}$ follows from the equality in (a) and the equality $\partial_F d_S(u) = N^F(S;u) \cap \mathbb{B}_{X^*}$ for all $u \in S$, and it obviously implies that $\mathbb{R}_+ \partial_L d_S(x) \subset N^L(S;x)$. It remains to show the converse of the latter inclusion for the proof of (b). Fix any element $x^* \in N^L(S;x)$. There exist a sequence $(x_n)_n$ in S $\|\cdot\|$-converging to x, and a sequence $(x_n^*)_n$ in X^* converging weakly star to x^* with $x_n^* \in N^F(S;x_n)$. Let $\gamma > 0$ such that $\|x_n^*\| \leq \gamma$ for all $n \in \mathbb{N}$. By Proposition 4.30 one has $\gamma^{-1} x_n^* \in \partial_F d_S(x_n)$, and hence $\gamma^{-1} x^* \in \partial_L d_S(x)$. We conclude that $x^* \in \mathbb{R}_+ \partial_L d_S(x)$.

(c) Suppose now that X is finite-dimensional and fix $x^* \in N^L(S;x) \cap \mathbb{B}_{X^*}$. Take $(x_n)_n$ in S converging to x and $(x_n^*)_n$ in X^* converging to x^* with $x_n^* \in N^F(S;x_n)$. For each integer k there exists some integer $s(k) > k$ such that $\|x_{s(k)}^*\| \leq 1 + (1/k)$, hence $(1 + (1/k))^{-1} x_{s(k)}^* \in \partial_F d_S(x_{s(k)})$ by Proposition 4.30 again. This guarantees that $x^* \in \partial_L d_S(x)$ and finishes the proof. $\square$

The assertions (b) and (c) in the above theorem fail if the set S is not closed near $x \in S$. Indeed, considering in $\mathbb{R}$ the set

$$S := \{0\} \cup \left(\bigcup_{n \in \mathbb{N}} \left]\frac{1}{2n+1}, \frac{1}{2n}\right[\right),$$

we saw in Remark 4.72 that $N^L(S;0) =]-\infty, 0]$ and $N^L(\overline{S};0) = \mathbb{R}$, where $\overline{S} := \operatorname{cl} S$. Then, while $N^L(S;0) =]-\infty, 0]$, we notice that

$$\mathbb{R}_+ \partial_L d_S(0) = \mathbb{R}_+ \partial_L d_{\overline{S}}(0) = N^L(\overline{S};0) = \mathbb{R},$$

where the first equality is due to the well-known obvious equality $d_S = d_{\overline{S}}$ and the second is due to (b) in the above theorem.

The next example also shows that there are closed sets in infinite dimensions for which the inclusion in the assertion (b) of Theorem 4.85 is strict, or in other words the equality in the assertion (c) in that theorem fails for closed sets in infinite dimensions.

EXAMPLE 4.89 (G.E. Ivanov and L. Thibault: a closed set S in infinite dimensions with $\partial_L d_S(x) \neq N^L(S;x) \cap \mathbb{B}$). Let $H := \ell_\mathbb{R}^2(\mathbb{N} \cup \{0\})$ be the usual Hilbert space of functions $x : \mathbb{N} \cup \{0\} \to \mathbb{R}$ such that $\sum_{n=0}^{\infty} x_n^2 < +\infty$ endowed with its usual inner product. Consider the following sets in H:

$$S_k = \{x \in H : x_0 + x_k \leq 0\}, \quad k \geq 1,$$

$$S_0 = \{x \in H : x_0 = 0\}, \quad S = \bigcup_{k=0}^{\infty} S_k.$$

Define the mapping $\gamma : H \to \mathbb{R}$ by

(4.46) $$\gamma(x) := x_0 + \inf_{n \geq 1} x_n \quad \text{for all } x \in H,$$

4.4. MORDUKOVICH LIMITING SUBDIFFERENTIAL

and note that for $x, y \in H$ we have

$$x_0 + x_n \leq y_0 + y_n + |x_0 - y_0| + |x_n - y_n| \leq y_0 + y_n + \sqrt{2}\|x - y\|_2 \quad \forall n \in \mathbb{N},$$

so taking the infimum over $n \in \mathbb{N}$ gives $\gamma(x) \leq \gamma(y) + \sqrt{2}\|x-y\|_2$, hence γ is Lipschitz continuous.

- We claim that the set S is closed. Indeed, fix any $x \in H \setminus S$. Then $x_0 \neq 0$ and $x_0 + x_k > 0$ for every $k \geq 1$. Passing to the limit in the latter inequality and using the relation $\lim_k x_k = 0$ we obtain $x_0 \geq 0$, and hence $x_0 > 0$. Using once more the relation $\lim_k x_k = 0$, we can find some $k_0 \geq 1$ such that $x_0 + x_k > \frac{x_0}{2}$ for all $k > k_0$. Consequently, $\gamma(x) \geq \min\left\{\frac{x_0}{2}, \min_{k \leq k_0}(x_0 + x_k)\right\} > 0$. By continuity of $\gamma : H \to \mathbb{R}$ we have $\gamma(x') > 0$ for any x' in some neighborhood U of x in H. Thus, for any $x' \in U$, it ensues that, on the one hand $x'_0 = x'_0 + \lim_k x'_k \geq \gamma(x') > 0$, and on the other hand, $x'_0 + x'_k \geq \gamma(x') > 0$ for every $k \geq 1$. This means that $U \cap S = \emptyset$, and hence S is closed, which justifies the claim.

- Let $e_0, e_1, \ldots$ be the standard basis of H, that is, $e_k(n) = 1$ if $n = k$ and $e_k(n) = 0$ if $n \neq k$. Let $e_0^*, e_1^*, \ldots$ be the standard basis of H^*, that is, $\langle e_k^*, x \rangle = x_k$ for any $x \in H$.

Observe that $e_0^* + e_k^* \in N^F\left(S; \frac{e_0 - e_k}{k}\right)$ for any $k \geq 1$, as each $x \in S$ sufficiently close to $\frac{e_0 - e_k}{k}$ belongs to S_k. Since the sequence $\left(\frac{e_0 - e_k}{k}\right)_k$ converges to 0 and the sequence $(e_0^* + e_k^*)_k$ converges weakly star to e_0^*, we obtain the inclusion $e_0^* \in N^L(S; 0)$, hence $e_0^* \in N^L(S; 0) \cap \mathbb{B}_{H^*}$.

- Now observe that for each $x \in H$

$$d_S(x) = \inf_{k \geq 0} d_{S_k}(x),$$

(4.47)
$$d_{S_0}(x) = |x_0|,$$

(4.48)
$$d_{S_k}(x) = \begin{cases} \frac{1}{\sqrt{2}}(x_0 + x_k) & \text{if } x_0 + x_k \geq 0, \\ 0 & \text{if } x_0 + x_k < 0, \end{cases} \quad k \geq 1.$$

Let us show that

(4.49)
$$d_S(x) = \max\left\{\frac{1}{\sqrt{2}}\gamma(x), 0\right\} \quad \forall x \in H.$$

Fix any $x \in H$. If $\gamma(x) < 0$, then by (4.46) there exists $k \geq 1$ such that $x_0 + x_k < 0$, so $x \in S_k \subset S$. Thus, (4.49) holds true if $\gamma(x) < 0$. Now suppose that $\gamma(x) \geq 0$. According to (4.46) we have $x_0 + x_k \geq 0$ for all $k \geq 1$, and hence $x_0 \geq 0$ (after passing to the limit as $k \to +\infty$). By (4.46)–(4.48) we see that

$$\inf_{k \geq 1} d_{S_k}(x) = \frac{1}{\sqrt{2}}\gamma(x), \quad d_{S_0}(x) = |x_0| = x_0 = x_0 + \lim_k x_k \geq \gamma(x) \geq \frac{1}{\sqrt{2}}\gamma(x).$$

So, $d_S(x) = \inf_{k \geq 0} d_{S_k}(x) = \inf_{k \geq 1} d_{S_k}(x) = \frac{1}{\sqrt{2}}\gamma(x)$, and (4.49) holds true again.

- For each $r \in \mathbb{R}$ define $j_r : H \to H$ by $j_r(x) = (x_0 + r, x_1, x_2, \cdots)$, that is, $\langle e_0^*, j_r(x) \rangle = x_0 + r$ and $\langle e_k^*, j_r(x) \rangle = x_k$ for every $k \in \mathbb{N}$. For every $x \in H$ it follows by (4.46) that

$$|\gamma(x) - \gamma(j_r(x))| \leq |r|,$$

hence by (4.49)
$$|d_S(x) - d_S(j_r(x))| \leq \frac{1}{\sqrt{2}}|r|.$$

For $x \in H$ and $\zeta \in \partial_F d_S(x)$ with $\zeta = \sum_{k=0}^{\infty} \zeta_k e_k^*$, for any $\varepsilon > 0$ we obtain by definition of Fréchet subdifferential that, for any $r \in \mathbb{R}$ with absolute value small enough
$$r\zeta_0 = \langle \zeta, j_r(x) - x \rangle \leq d_S(j_r(x)) - d_S(x) + \varepsilon \|j_r(x) - x\|_2$$
$$\leq \frac{1}{\sqrt{2}}|r| + \varepsilon|r| = \left(\frac{1}{\sqrt{2}} + \varepsilon\right)|r|,$$

hence $|\zeta_0| \leq \varepsilon + (1/\sqrt{2})$. Therefore, if $\zeta \in \partial_L d_S(0)$, then $|\zeta_0| \leq \frac{1}{\sqrt{2}}$. So, $e_0^* \notin \partial_L d_S(0)$ while $e_0^* \in N^L(S;0) \cap \mathbb{B}_{H^*}$ as seen above.
• we conclude that $\partial_L d_S(0) \neq N^L(S;0) \cap \mathbb{B}_{H^*}$. □

Theorem 4.85 allows us to characterize the boundary points of closed sets of finite-dimensional spaces.

COROLLARY 4.90. *Let X be a finite-dimensional normed vector space, S be a closed set of X, and $x \in S$. Then the following are equivalent.*
(a) *The point $x \in \operatorname{bdry} S$;*
(b) *there exists some $x^* \in \partial_L d_S(x)$ with $\|x^*\| = 1$;*
(c) $N^L(S;x) \neq \{0\}$.

PROOF. If $x \notin \operatorname{bdry} S$, then $x \in \operatorname{int} S$, hence $N^L(S;x) = \{0\}$, proving the implication $(c) \Rightarrow (a)$.

Suppose that $x \in \operatorname{bdry} S$. There exists a sequence of points $(x_n)_n$ in $X \setminus S$ converging to x. For each n the set $\partial_L d_S(x_n)$ is nonempty, hence by the definition of the Mordukhovich limiting subdifferential, there are some $u_n \in B(x_n, 1/n) \setminus S$ and some $u_n^* \in \partial_F d_S(u_n)$. We know that $\|u_n^*\| = 1$, thus a subsequence of $(u_n^*)_n$ converges to some u^* with $\|u^*\| = 1$. Of course, we have $u^* \in \partial_L d_S(x)$, hence the implication $(a) \Rightarrow (b)$ holds.

Finally, the implication $(b) \Rightarrow (c)$ follows from the inclusion of $\partial_L d_S(x)$ into $N^L(S;x)$ in Theorem 4.85. □

4.4.4. Mordukhovich limiting subdifferential in normed space. Lemma 4.88 is at the heart of the proof of the equality
$$(4.50) \qquad N^L(S;x) = \mathbb{R}_+ \partial_L d_S(x)$$
established in Theorem 4.85 for a closed set S of an Asplund space X with $x \in S$. The fuzzy sum rule is crucial for dealing with the Fréchet subdifferential of the sum function in the last member of (4.45) in Lemma 4.88. So, to extend formula 4.50 to arbitrary (non-Asplund) Banach space, the form of the sum function in the last member of (4.45) leads to introduce an approximate Fréchet subdifferential as follows.

DEFINITION 4.91. *Let U be a neighborhood of a point x in a normed space $(X, \|\cdot\|)$ and $f : U \to \mathbb{R} \cup \{-\infty, +\infty\}$ be a function which is finite at x. For any real $\varepsilon \geq 0$ one defines the Fréchet ε-subdifferential $\partial_{F,\varepsilon} f(x)$ of f at x as*
$$(4.51) \qquad \partial_{F,\varepsilon} f(x) = \partial_F \big(f + \varepsilon \| \cdot - x \| \big)(x),$$

where the function $f + \varepsilon\|\cdot - x\|$ is defined on the set U; each element in $\partial_{F,\varepsilon} f(x)$ is called a Fréchet ε-subgradient of f at x. So, $x^* \in \partial_{F,\varepsilon} f(x)$ if and only if for any real $\varepsilon' > \varepsilon$ there exists a neighborhood $V \subset U$ of x such that

$$\langle x^*, u - x \rangle \leq f(u) - f(x) + \varepsilon'\|u - x\| \quad \text{for all } u \in V.$$

One also puts $\partial_{F,\varepsilon} f(x) = \emptyset$ whenever $|f(x)| = +\infty$. Obviously, for the natural extension $\overline{f}$ of f considered in Definition 4.64, with $\overline{f}(u) = f(u)$ if $u \in U$ and $\overline{f}(u) = +\infty$ if $u \in X \setminus U$, one has $\partial_{F,\varepsilon} f(x) = \partial_{F,\varepsilon} \overline{f}(x)$.

For a set S of X with $x \in S$, the set $N^{F,\varepsilon}(S;x) := \partial_{F,\varepsilon} \Psi_S(x)$ (where $\Psi_S : X \to \mathbb{R} \cup \{-\infty, +\infty\}$ denotes the indicator function) is the set of Fréchet ε-normals of S at the point x. One also defines $N^{F,\varepsilon}(S;x) = \emptyset$ if $x \notin S$.

Of course, $\partial_{F,\varepsilon} f(x)$ coincides with $\partial_F f(x)$ for $\varepsilon = 0$, and $\partial_F f(x) \subset \partial_{F,\varepsilon} f(x)$ for all $\varepsilon \geq 0$. It is also worth pointing out that unlike the Fréchet subdifferential, for $\varepsilon > 0$ the set $\partial_{F,\varepsilon} f(x)$ evidently depends on the given norm $\|\cdot\|$ on X. However if $\|\cdot\|'$ is another norm on X with $\|\cdot\|' \leq \beta\|\cdot\|$ for some real constant $\beta > 0$, then evidently

$$(4.52) \qquad \partial'_{F,\varepsilon} f(x) \subset \partial_{F,\beta\varepsilon} f(x) \quad \text{for all } x \in X,$$

where $\partial'_{F,\varepsilon} f$ denotes the Fréchet ε-subdifferential with respect to the norm $\|\cdot\|'$.

If $\partial_{F,\varepsilon} f(x) \neq \emptyset$, then as a direct consequence of (4.51) and Corollary 4.3 we have the following proposition.

PROPOSITION 4.92. *Let U be an open neighborhood of x in a normed space X and $f : U \to \mathbb{R} \cup \{-\infty, +\infty\}$ be a function which is finite at x with $\partial_{F,\varepsilon} f(x) \neq \emptyset$ for some $\varepsilon \geq 0$. Then f is lower semicontinuous at x and $\partial_{F,\varepsilon} f(x)$ coincides with the Fréchet ε-subdifferential at x of the lower semicontinuous hull of f.*

The next proposition provides some other elementary properties of $\partial_{F,\varepsilon} f(x)$.

PROPOSITION 4.93. *Let $(X, \|\cdot\|)$ be a normed vector space, $f, g : X \to \mathbb{R} \cup \{-\infty, +\infty\}$ be two functions which are finite at $x \in X$, and S be a subset of X with $x \in S$.*
(a) *If f and g coincide near x, then $\partial_{F,\varepsilon} f(x) = \partial_{F,\varepsilon} g(x)$ for all $\varepsilon \geq 0$.*
(b) *For $r > 0$ one has*

$$\partial_{F,\varepsilon}(rf)(x) = r\partial_{F,\varepsilon/r} f(x) \quad \text{and} \quad rN^{F,\varepsilon}(S;x) = N^{F,r\varepsilon}(S;x).$$

(c) *If f is Lipschitz continuous near x with $\gamma \geq 0$ as a Lipschitz constant, then $\partial_{F,\varepsilon} f(x) \subset (\gamma + \varepsilon)\mathbb{B}_{X^*}$.*
(d) *For $X \times \mathbb{R}$ endowed with the sum norm $\|(x,r)\| := \|x\| + |r|$ one has for any real $\varepsilon \geq 0$*

$$x^* \in \partial_{F,\varepsilon} f(x) \Rightarrow (x^*, -1) \in N^{F,\varepsilon}(\operatorname{epi} f; (x, f(x))).$$

Conversely, for any $\varepsilon \in [0, 1[$, one has

$$(x^*, -1) \in N^{F,\varepsilon}(\operatorname{epi} f; (x, f(x))) \Rightarrow x^* \in \partial_{F,\varepsilon'} f(x) \quad \text{for } \varepsilon' := \frac{\varepsilon(1 + \|x^*\|_*)}{1 - \varepsilon}.$$

PROOF. Since by definition $\partial_{F,\varepsilon} f(x) = \partial_F(f + \varepsilon\|\cdot - x\|)$, assertions (a), (b), and (c) follow directly from the corresponding assertions in Proposition 4.11.

Concerning the implications in the assertion (d), they are obtained as in the proof of Proposition 4.4. □

Second components of Fréchet ε-normals to epigraphs have properties similar to those in Proposition 4.21 for Fréchet normals.

PROPOSITION 4.94. *Let $(X, \|\cdot\|)$ be a normed space and $f : X \to \mathbb{R} \cup \{-\infty, +\infty\}$ be an extended real-valued function. Endow $X \times \mathbb{R}$ with the sum norm.*
(a) *If $\varphi : X \times \mathbb{R} \to \mathbb{R}$ is a function such that $\varphi(x, \cdot)$ is Lipschitz continuous near $r \in \mathbb{R}$ with $\gamma \geq 0$ as a Lipschitz constant, then for any $\varepsilon \geq 0$*

$$\left\{ \begin{array}{c} (x^*, -r^*) \in \partial_{F,\varepsilon}(\Psi_{\mathrm{epi}\,f} + \varphi)(x, r) \\ \text{and } r^* > \varepsilon + \gamma \end{array} \right\} \Rightarrow r = f(x).$$

In particular

$$\left\{ \begin{array}{c} (x^*, -r^*) \in N^{F,\varepsilon}\big(\mathrm{epi}\,f; (x, r)\big) \\ \text{and } r^* > \varepsilon \end{array} \right\} \Rightarrow r = f(x).$$

(b) *If f is finite at x, then*

$$(x^*, -r^*) \in N^{F,\varepsilon}\big(\mathrm{epi}\,f; (x, f(x))\big) \Rightarrow r^* \geq -\varepsilon.$$

(c) *If $(x^*, -r^*) \in N^{F,\varepsilon}\big(\mathrm{epi}\,f; (x, r)\big)$ with $r > f(x)$, then $-\varepsilon \leq r^* \leq \varepsilon$.*

PROOF. Fix $(x^*, -r^*) \in \partial_{F,\varepsilon}(\Psi_{\mathrm{epi}\,f} + \varphi)(x, r)$ with $r^* > \varepsilon + \gamma$. By the definition of ε-Fréchet subdifferential, for any real $\varepsilon' > \varepsilon$ there exists some real $\delta > 0$ such that

$$\langle x^*, u - x \rangle - r^*(s - r) \leq \varphi(u, s) - \varphi(x, r) + \varepsilon' \|(u, s) - (x, r)\|$$

for all $(u, s) \in \mathrm{epi}\,f \cap B((x, r), \delta)$, and such that $\varphi(x, \cdot)$ is γ-Lipschitz on $]r - \delta, r + \delta[$. Since $(x, r) \in \mathrm{epi}\,f$, we have $r \geq f(x)$. Suppose that $r > f(x)$. We can then choose some real s such that $f(x) < s < r$ and $(x, s) \in B((x, r), \delta)$. Taking (x, s) in place of (u, s) in the above inequality yields

$$-r^*(s - r) \leq (\varepsilon' + \gamma)(r - s), \quad \text{hence} \quad r^* \leq \varepsilon' + \gamma.$$

Consequently $r^* \leq \varepsilon + \gamma$, which contradicts the assumption $r^* > \varepsilon + \gamma$. This justifies the equality $r = f(x)$, proving (a).

Concerning (b) fix $(x^*, -r^*) \in N^{F,\varepsilon}\big(\mathrm{epi}\,f; (x, f(x))\big)$. Take any $\varepsilon' > \varepsilon$ and choose some $\delta > 0$ such that

$$\langle x^*, x' - x \rangle - r^*(r' - f(x)) \leq \varepsilon'(\|x' - x\| + |r' - f(x)|)$$

for all $(x', r') \in \mathrm{epi}\,f$ with $\|x' - x\| \leq \delta$ and $|r' - f(x)| \leq \delta$. For $x' = x$ and $r' = f(x) + \delta$ we obtain $-r^*\delta \leq \varepsilon'\delta$, which easily gives $r^* \geq -\varepsilon$.

Under the assumptions of (c), one has $r - f(x) > 0$, and for any $\varepsilon' > \varepsilon$ there exists some $0 < \delta < r - f(x)$ such that

$$\langle x^*, x' - x \rangle - r^*(r' - r) \leq \varepsilon'(\|x' - x\| + |r' - r|)$$

for all $(x', r') \in \mathrm{epi}\,f$ with $\|x' - x\| \leq \delta$ and $|r' - r| \leq \delta$. For $r' \in [r - \delta, r + \delta]$, we observe that $(x, r') \in \mathrm{epi}\,f$, then $-r^*(r' - r) \leq \varepsilon'|r' - r|$. This being true for all $r' \in [r - \delta, r + \delta]$, it follows that $|r^*| \leq \varepsilon$, finishing the proof of the proposition. $\square$

In addition to the characterization via Fréchet subdifferential in Proposition 4.13(a), we establish another characterization via Fréchet ε-subdifferential.

4.4. MORDUKOVICH LIMITING SUBDIFFERENTIAL

PROPOSITION 4.95. *Let $(X, \|\cdot\|)$ be a normed space and $f : X \to \mathbb{R} \cup \{-\infty, +\infty\}$ be a function which is finite at $\bar{x} \in X$. Then f is Fréchet differentiable at $\bar{x}$ if and only if for every real $\varepsilon > 0$ both sets $\partial_{F,\varepsilon} f(\bar{x})$ and $\partial_{F,\varepsilon}(-f)(\bar{x})$ are nonempty.*

PROOF. It suffices to prove the implication $\Leftarrow$. Fix any $n \in \mathbb{N}$ and choose $u_n^* \in \partial_{F,\varepsilon_n} f(\bar{x})$ and $v_n^* \in -\partial_{F,\varepsilon_n}(-f)(\bar{x})$ where $\varepsilon_n := 1/(2n)$. There exists a neighborhood V_n of $\bar{x}$ such that for all $x \in V_n$

(4.53) $$\langle u_n^*, x - \bar{x}\rangle \leq f(x) - f(\bar{x}) + \|x - \bar{x}\|/n$$

and

(4.54) $$\langle -v_n^*, x - \bar{x}\rangle \leq -f(x) + f(\bar{x}) + \|x - \bar{x}\|/n.$$

Take any integer $m \geq n$ and any $x \in V_n \cap V_m$, and note that $f(x)$ is finite. Writing (4.53) for m and combining with (4.54) for n give

$$\langle u_m^* - v_n^*, x - \bar{x}\rangle \leq 2\|x - \bar{x}\|/n,$$

which implies that

(4.55) $$\|u_m^* - v_n^*\|_* \leq 2/n.$$

Similarly, writing (4.54) for m and combining with (4.53) for n give

$$\langle u_n^* - v_m^*, x - \bar{x}\rangle \leq 2\|x - \bar{x}\|/n,$$

which implies that

(4.56) $$\|u_n^* - v_m^*\|_* \leq 2/n.$$

From (4.55) and (4.56) it is readily seen that both sequences $(u^*)_n$ and $(v^*)_n$ are Cauchy sequences, and according to the completeness of $(X^*, \|\cdot\|_*)$ and to (4.55) they strongly converge to a same element $x^* \in X^*$.

To prove that x^* is the Fréchet derivative of f at $\bar{x}$ consider any real $\varepsilon > 0$ and fix an integer $k \geq 3/\varepsilon$. For any $x \in V_k$ we have by (4.53) and (4.54)

(4.57) $$\langle u_k^*, x - \bar{x}\rangle - \|x - \bar{x}\|/k \leq f(x) - f(\bar{x}) \leq \langle v_k^*, x - \bar{x}\rangle + \|x - \bar{x}\|/k.$$

On the other hand, in (4.55) and (4.56) taking $n = k$ and making $m \to +\infty$ yield

$$\|u_k^* - x^*\|_* \leq 2/k \quad \text{and} \quad \|v_k^* - x^*\|_* \leq 2/k.$$

Combining the latter inequalities with (4.57) we obtain for any $x \in V_k$

$$\langle x^*, x - \bar{x}\rangle - 3\|x - \bar{x}\|/k \leq f(x) - f(\bar{x}) \leq \langle x^*, x - \bar{x}\rangle + 3\|x - \bar{x}\|/k,$$

so keeping in mind that $k \geq 3/\varepsilon$ it follows that for any $x \in V_k$

$$|f(x) - f(\bar{x}) - \langle x^*, x - \bar{x}\rangle| \leq \varepsilon \|x - \bar{x}\|.$$

We then conclude that f is Fréchet differentiable at $\bar{x}$ (with x^* as F-derivative). $\square$

Let us now establish the result preparing the extension of the definition of $\partial_L f(x)$ to the context of any Banach space.

PROPOSITION 4.96. *Let X be an Asplund space and $f : X \to \mathbb{R} \cup \{-\infty, +\infty\}$ be a function which is finite at $x \in X$ and lower semicontinuous near x. Then the following hold:*
(a) *For any $\varepsilon' > \varepsilon \geq 0$ and any $x^* \in \partial_{F,\varepsilon} f(x)$ there are $u \in X$ with $\|u - x\| \leq \varepsilon$*

and $|f(u) - f(x)| \leq \varepsilon$, and $u^* \in \partial_F f(u)$ with $\|u^* - x^*\| \leq \varepsilon'$.
(b) One has the equality
$$\partial_L f(x) = {}^{\text{seq}}\operatorname*{Lim\,sup}_{\varepsilon \downarrow 0, u \to_f x} \partial_{F,\varepsilon} f(u).$$

PROOF. (a) The assertion (a) obviously holds for $\varepsilon = 0$, since in this case it is enough to take $u = x$ and $u^* = x^*$. For $\varepsilon > 0$ using the Lipschitz property of the function $\varepsilon \| \cdot - x \|$ and applying to $f + \varepsilon \| \cdot - x \|$ the fuzzy sum rule in Theorem 4.52 yields easily the desired elements u and u^*.
(b) The assertion (b) readily follows from the assertion (a). □

The assertion (b) above allows us to extend the definition of $\partial_L f(x)$ to any normed space as follows.

DEFINITION 4.97. Let U be a neighborhood of a point x in a normed space X and $f : U \to \mathbb{R} \cup \{-\infty, +\infty\}$ be a function which is finite at $x \in U$ and lower semicontinuous near x. Taking into account the assertion (b) of the proposition above one defines the *Mordukhovich (extended) limiting subdifferential* of f at x as the set of elements $x^* \in X^*$ for which there exist sequences $(x_n)_n$ in X with $(x_n, f(x_n))_n$ converging to $(x, f(x))$, $(\varepsilon_n)_n$ of positive numbers tending to 0, and $(x_n^*)_n$ in X^* converging weakly star to x^* with $x_n^* \in \partial_{F,\varepsilon_n} f(x_n)$ for all $n \in \mathbb{N}$. As for Asplund spaces, we denote this set by $\partial_L f(x)$ according to Proposition 4.96.

So, the *Mordukhovich (extended) limiting normal cone*, that we still denote by $N^L(S;x)$, of a set $S \subset X$ closed near $x \in S$, is defined as $\partial_L \Psi_S(x)$, where we recall that $\Psi_S : X \to \mathbb{R} \cup \{-\infty, +\infty\}$ is the indicator function of S.

By (4.52) we see that the above definitions do not depend on equivalent norms on X. Further, extending f with $+\infty$ outside U, we may, as above, essentially restrict ourselves to extended real-valued functions defined on the whole normed space X.

First, we show that the definition of the Mordukhovich (extended) limiting subdifferential leads to the following extension of Theorem 4.85.

THEOREM 4.98 (*L*-subdifferential of distance function via *F*-subgradients at inside points). Let $(X, \|\cdot\|)$ be any Banach space and S be a subset of X which is closed near $x \in S$. Then the following hold:
(a) $\partial_L d_S(x) = {}^{\text{seq}}\operatorname*{Lim\,sup}_{\varepsilon \downarrow 0, S \ni y \to x} \partial_{F,\varepsilon} d_S(y).$
(b) The (extended) limiting subdifferential of the distance function d_S is related to the (extended) limiting normal cone as follows:

(4.58) $\qquad \partial_L d_S(x) \subset N^L(S;x) \cap \mathbb{B}_{X^*}$ and $N^L(S;x) = \mathbb{R}_+ \partial_L d_S(x).$

The following lemma, similar to Lemma 4.88 but with its own right, will be central in the proof of the theorem.

LEMMA 4.99. Let S be a closed subset of a Banach space X, let $u \in X$ and let $u^* \in \partial_{F,\varepsilon} d_S(u)$ with $\varepsilon > 0$. Then for every $\sigma > 0$, there exist $x \in S$ such that for $\varepsilon' = \varepsilon(1 + \sigma)$
$$\|x - u\| \leq \sigma + d_S(u) \quad \text{and} \quad u^* \in (1 + 2\varepsilon + 2\varepsilon\sigma)\partial_{F,\varepsilon'} d_S(x).$$

PROOF. Fix $\sigma > 0$ and put $\eta = \varepsilon + (\varepsilon\sigma/2)$. Choose a positive number $\alpha < \min\{1, \varepsilon\sigma/4, \sigma/2\}$ such that for all $u' \in B[u, \alpha]$
$$\langle u^*, u' - u \rangle \leq d_S(u') - d_S(u) + \eta \|u' - u\|.$$

This tells us first that $\|u^*\|_* \le 1+\eta$. Fix some $y \in S$ satistying
$$\|u-y\| \le \alpha^2 + d_S(u).$$
Noticing that $\alpha < \eta/2$, we may apply Lemma 4.87 to get some $z \in S \cap B[y,\alpha]$ with $\|z-y\| \le \alpha/2$ such that for all $z' \in S \cap B[y,\alpha]$
$$\langle u^*, z'-z\rangle \le (\eta+2\alpha)\|z'-z\|,$$
hence for all $z' \in S \cap B[z,\alpha/2] \subset S \cap B[y,\alpha]$ we have $\langle u^*, z'-z\rangle \le (\eta+2\alpha)\|z'-z\|$. This means that the point z is a minimizer over $C := S \cap B[z,\alpha/2]$ of the function $\langle -u^*, \cdot\rangle + (\eta+2\alpha)\|\cdot - z\|$ which is $(1+2\varepsilon+2\varepsilon\sigma)$-Lipschitzian on C. We obtain by the penalization result in Lemma 2.213 that for all $z' \in B[z,\alpha/2]$
$$\langle u^*, z'-z\rangle \le (1+2\varepsilon+2\varepsilon\sigma)d_C(z') - (1+2\varepsilon+2\varepsilon\sigma)d_C(z) + (\eta+2\alpha)\|z'-z\|,$$
which ensures for all $z' \in B[z,\alpha/4]$ that
$$\langle u^*, z'-z\rangle \le (1+2\varepsilon+2\varepsilon\sigma)d_S(z') - (1+2\varepsilon+2\varepsilon\sigma)d_S(z) + (\eta+2\alpha)\|z'-z\|,$$
since $d_C(z') = d_S(z')$ by Lemma 2.219. This says, for $\varepsilon' := \varepsilon + \varepsilon\sigma > \eta + 2\alpha$, that $u^* \in (1+2\varepsilon+2\varepsilon\sigma)\partial_{F,\varepsilon'}d_S(z)$ by Proposition 4.93(b) since $1+2\varepsilon+2\varepsilon\sigma > 1$. Further, $z \in S$ and
$$\|z-u\| \le \|z-y\| + \|y-u\| \le \alpha + \alpha^2 + d_S(u) \le \sigma + d_S(u).$$
Taking $x = z$ completes the proof. □

PROOF OF THEOREM 4.98. The assertion (a) follows (as for (a) of Theorem 4.85) from the lemma above.

Concerning (b), it is first easy to see that, for any $y^* \in \partial_{F,\varepsilon} d_S(y)$ with $y \in S$ we have $y^* \in \partial_{F,\varepsilon}\Psi_S(y)$ and $\|y^*\|_* \le 1+\varepsilon$. This combined with the assertion (a) above ensures the first inclusion of (b). Consider now any $x^* \in N^L(S;x)$ and take $\varepsilon_n \downarrow 0$, $S \ni x_n \to x$, and $x_n^* \to x^*$ weakly star with $x_n^* \in \partial_{F,\varepsilon_n}\Psi_S(x_n)$. Choosing $\gamma > 0$ with $\|x_n^*\|_* \le \gamma$ for all n, it is not difficult to see (like in the last part of the proof of Lemma 4.99) that $x_n^* \in \partial_{F,\varepsilon_n}(\gamma+1)d_S(x_n)$. This ensures the desired second inclusion in (b) and completes the proof. □

From Theorem 4.98(a) we derive the following result which will be crucial, for example, in Proposition 4.114 as well as in other places of the text.

PROPOSITION 4.100. Let $(X, \|\cdot\|)$ be a Banach space and S be a subset of X which is closed near $\bar{x} \in S$. For any lower semicontinuous function $\varphi : U \to \mathbb{R} \cup \{+\infty\}$ defined on an open neighborhood U of $\bar{x}$ such that
$$d_S(x) \le \varphi(x) \quad \text{for all } x \in U \text{ and } d_S(x) = \varphi(x) \quad \text{for all } x \in U \cap S,$$
one has the inclusion
$$\partial_L d_S(\bar{x}) \subset \partial_L \varphi(\bar{x}).$$
In particular, for any real $\gamma \ge 1$
$$\partial_L d_S(\bar{x}) \subset \gamma \partial_L d_S(\bar{x}).$$

PROOF. From the assumption we see that $\partial_{F,\varepsilon}d_S(x) \subset \partial_{F,\varepsilon}\varphi(x)$ for all $x \in U \cap S$. It results by Theorem 4.98(a) that

$$\partial_L d_S(\overline{x}) = {}^{\text{seq}}\text{Lim sup}_{\varepsilon \downarrow 0, S \ni x \to \overline{x}} \partial_{F,\varepsilon}d_S(x) \subset {}^{\text{seq}}\text{Lim sup}_{\varepsilon \downarrow 0, S \ni x \to \overline{x}} \partial_{F,\varepsilon}\varphi(x)$$
$$\subset {}^{\text{seq}}\text{Lim sup}_{\varepsilon \downarrow 0, x \to_\varphi \overline{x}} \partial_{F,\varepsilon}\varphi(x) \subset \partial_L \varphi(\overline{x}),$$

where the second inclusion is due to the fact that $\varphi(x) = 0 = \varphi(\overline{x})$ for all $x \in U \cap S$. This justifies the desired inclusion in (a).

Concerning the second inclusion, it suffices to take $\varphi = \gamma d_S$ and to note via Proposition 4.93(b) that $\partial_L \varphi(\overline{x}) = \gamma \partial d_S(\overline{x})$. $\square$

At points inside or outside the set S, there is a link (as seen above for exact Fréchet subgradients) between Fréchet ε-subgradients of d_S and Fréchet ε-normals.

PROPOSITION 4.101. Let S be a subset of the normed space $(X, \|\cdot\|)$ and $x \in X$. Let also $\varepsilon \geq 0$, $\theta \geq 0$, $0 \leq \theta' \leq 1$. The following hold:
(a) For $x \in S$ one has

$$\partial_{F,\varepsilon}d_S(x) \subset N^{F,\varepsilon}(S;x) \cap (1+\varepsilon)\mathbb{B}_{X^*} \text{ and } N^{F,\varepsilon}(S;x) \cap (1+\theta)\mathbb{B}_{X^*} \subset \partial_{F,\theta+2\varepsilon}d_S(x).$$

(b) For $x \notin \text{cl}\, S$, putting $E := \{u \in X : d_S(u) \leq d_S(x)\}$ one has

$$\partial_{F,\varepsilon}d_S(x) \subset N^{F,\varepsilon}(E;x) \cap \{x^* \in X^* : 1 - \varepsilon \leq \|x^*\| \leq 1 + \varepsilon\}$$

and $N^{F,\varepsilon}(E;x) \cap \{x^* \in X^* : 1 - \theta' \leq \|x^*\| \leq 1 + \theta\} \subset \partial_{F,\theta''+2\varepsilon}d_S(x)$,

where $\theta'' := \max\{\theta, \theta'\}$.

PROOF. To prove (a) we adapt the proof of Proposition 4.30. First, fix any $x^* \in \partial_{F,\varepsilon}d_S(x)$. For any $\eta > 0$, by definition there is $\delta > 0$ such that, for all $x' \in B(x, \delta)$ we have

$$\langle x^*, x' - x \rangle \leq d_S(x') - d_S(x) + (\varepsilon + \eta)\|x' - x\|,$$

hence $\langle x^*, x' - x \rangle \leq (1 + \varepsilon + \eta)\|x' - x\|$. This being true for all $x' \in B(x, \delta)$ and all $\eta > 0$, it ensues that $\|x^*\| \leq 1 + \varepsilon$. On the other hand, taking any $x' \in S \cap B(x, \delta)$ in the above inequality gives $\langle x^*, x' - x \rangle \leq (\varepsilon + \eta)\|x' - x\|$, hence $x^* \in N^{F,\varepsilon}(S;x)$. The inclusion $\partial_{F,\varepsilon}d_S(x) \subset N^{F,\varepsilon}(S;x) \cap (1 + \varepsilon)\mathbb{B}_{X^*}$ is established.

To prove the second inclusion in (a), fix any $x^* \in N^{F,\varepsilon}(S;x)$ with $\|x^*\| \leq 1 + \theta$ and take any $\eta > 0$. There exists some real $\delta > 0$ such that

(4.59) $\langle x^*, x' - x \rangle \leq (\varepsilon + \eta)\|x' - x\|$ for all $x' \in S \cap B(x, 2\delta)$.

With $C := S \cap B(x, 2\delta)$, by Lipschitz property of $u \mapsto \langle -x^*, u - x \rangle + (\varepsilon + \eta)\|u - x\|$ we derive (see Lemma 2.213) that, for all $x' \in B(x, 2\delta)$

$$\langle x^*, x' - x \rangle \leq (1 + \theta + \varepsilon + \eta)d_C(x') - (1 + \theta + \varepsilon + \eta)d_C(x) + (\varepsilon + \eta)\|x' - x\|$$
$$\leq d_C(x') - d_C(x) + (\theta + 2\varepsilon + 2\eta)\|x' - x\|.$$

For any $x' \in B(x, \delta)$, using by Lemma 2.219 the equality $d_C(x') = d_S(x')$ it follows that

$$\langle x^*, x' - x \rangle \leq d_S(x') - d_S(x) + (\theta + 2\varepsilon + 2\eta)\|x' - x\|,$$

which justifies that $x^* \in \partial_{F,\theta+2\varepsilon}d_S(x)$, and finishes the proof of (a).

Similar adaptations of the proof of Proposition 4.31 yield (b). $\square$

Let us continue the study of subdifferentials of the distance function d_S at $x \notin S$. We consider now the case where there is at least one nearest point of x in S. When the set $\text{Proj}\,_S(u)$ (of nearest points of u in S) is a singleton, we recall the convention to denote by $P_S(u)$ its unique point.

We consider first the case of a general normed space and relates in such a context the Fréchet subdifferential of the distance function of a closed set and the set of nearest points.

PROPOSITION 4.102. *Let S be a subset of a normed space $(X, \|\cdot\|)$ and $x \notin \text{cl}\, S$. If $\text{Proj}\,_S(x) \neq \emptyset$, then one has*

$$\partial_F d_S(x) \subset \bigcap_{y \in \text{Proj}\,_S(x)} N^F(S; y) \cap \{x^* \in X^* : \|x^*\| = 1\}.$$

PROOF. Fix any element $x^* \in \partial_F d_S(x)$. One knows by Proposition 4.31 that $\|x^*\| = 1$. Consider any $y \in \text{Proj}\,_S(x) \neq \emptyset$ and take any $\varepsilon > 0$. Choose some $\delta > 0$ such that

$$\langle x^*, u - x \rangle \leq d_S(u) - d_S(x) + \varepsilon \|u - x\| \quad \text{for any } u \in B(x, \delta).$$

Hence for each $y' \in B(y, \delta) \cap S$, one obtains

$$\begin{aligned}
\langle x^*, y' - y \rangle &= \langle x^*, (y' - y + x) - x \rangle \\
&\leq d_S(y' - y + x) - \|x - y\| + \varepsilon \|y' - y\| \\
&\leq \varepsilon \|y' - y\|,
\end{aligned}$$

where the last inequality is due to the fact that

$$d_S(y' - y + x) - \|x - y\| \leq \|(y' - y + x) - y'\| - \|x - y\| = 0$$

since $y' \in S$. This ensures the desired inclusion of the proposition. $\square$

REMARK 4.103. The inclusion in the above proposition may be strict even for convex sets in $\mathbb{R}^2$. Indeed, for $S := \mathbb{R}_+ \times \mathbb{R}_+$ and $x = (-1, -1)$, since $P_S(x) = (0, 0)$ we have

$$\partial_F d_S(x) = \{x/\sqrt{2}\} \quad \text{by Corollary 4.14, while}$$

$$N^F(S; (0,0)) \cap \{u \in \mathbb{R}^2 : \|u\| = 1\} = \{(s, t) \in \mathbb{R}^2 : s \leq 0, t \leq 0, s^2 + t^2 = 1\},$$

which confirms that the inclusion in the above proposition may be strict. $\square$

The next proposition deals in the Hilbert setting with Fréchet subdifferentials and lower Dini directional derivatives of distance functions in the special cases of weakly closed sets and ball-compact sets (we recall that a set in a normed space is ball-compact if its intersection with any closed ball is strongly compact). Before stating the proposition let us show first that the two latter concepts are different. To do so, consider an infinite-dimensional Hilbert space H and note first that H and $\mathbb{B}_H$ are weakly closed and not ball-compact. On the other hand, take for each $n \in \mathbb{N}$ a vector subspace H_n with dimension n and take $x_{n,1}, \cdots, x_{n,i_n}$ in H_n with $\|x_{n,k}\| = n$ such that the compact set $\{x \in H_n : \|x\| = n\}$ is covered by the balls $B(x_{n,i}, 1/n)$ with $i \in I_n := \{1, \cdots, i_n\}$. The set $S := \{x_{n,i} : n \in \mathbb{N}, i \in I_n\}$ is ball-compact since its intersection with any ball is a finite set. However, S is not weakly closed. Indeed, take any weak neighborhood V of zero and choose $\varepsilon > 0$ and $u_1, \cdots, u_k$ in the unit sphere $\mathbb{S}_H$ such that

$$\{x \in H : \langle u_j, x \rangle < \varepsilon, j = 1, \cdots, k\} \subset V.$$

Fix an integer $n > \max\{k, 1/\varepsilon\}$. Since $k < n = \dim H_n$ there is some nonzero $y \in H_n$ with $\langle u_j, y \rangle = 0$ for all $j \in \{1, \cdots, k\}$. Put $h := ny/\|y\| \in H_n$ and choose some $i \in \{1, \cdots, i_n\}$ such that $h \in B(x_{n,i}, 1/n)$. Then for each $j \in \{1, \cdots, k\}$ we have

$$\langle u_j, x_{n,i} \rangle = \langle u_j, x_{n,i} - h \rangle \leq \|x_{n,i} - h\| < 1/n < \varepsilon,$$

so $S \cap V \neq \emptyset$. Consequently, $0 \in \mathrm{cl}_w S$, hence S is not weakly closed since $0 \notin S$.

PROPOSITION 4.104. Let S be a nonempty set in a Hilbert space H which is either ball-compact or weakly closed and let $x \in H \setminus S$.
(a) For any $h \in H$ one has

$$(d_S)^B(x; h) = \min_{y \in \mathrm{Proj}_S(x)} \frac{\langle x - y, h \rangle}{d_S(x)}.$$

(b) If $\partial_F d_S(x) \neq \emptyset$, then $P_S(x)$ exists (that is, $\mathrm{Proj}_S(x)$ is a singleton) and

$$\partial_F d_S(x) = \{(x - P_S(x))/d_S(x)\}.$$

(c) If the Hilbert space H is finite-dimensional, then $\partial_F d_S(x) \neq \emptyset$ if and only if $\mathrm{Proj}_S(x)$ is a singleton.

PROOF. (a) It is readily seen that $\mathrm{Proj}_S(u) \neq \emptyset$ for all $u \in H$ if S is ball-compact and this non-emptiness property still holds true, if S is weakly closed, by the weak lower semicontinuity of the norm $\|\cdot\|$. We first observe that for any $y \in \mathrm{Proj}_S(x)$ the inequality $(d_S)^B(x; h) \leq \frac{\langle x-y,h \rangle}{d_S(x)}$ easily follows from Proposition 4.27 (note also that the latter inequality can be directly verified). Consider now a sequence $(t_n)_n$ of positive reals converging to 0 and such that $(d_S)^B(x; h) = \lim_n t_n^{-1}(d_S(x + t_n h) - d_S(x))$. For each n choose some $y_n \in \mathrm{Proj}_S(x + t_n h)$ and note that $\|x + t_n h - y_n\| = d_S(x + t_n h) \to d_S(x)$, so $\|x - y_n\| \to d_S(x)$ as $n \to \infty$. By the boundedness of $(y_n)_n$ we may suppose (extracting a subsequence) that $(y_n)_n$ converges strongly (resp. weakly) to some y if S is ball-compact (resp. weakly closed). The continuity (resp. weak lower semicontinuity) of $\|\cdot\|$ and the strong (resp. weak) closedness property of S easily yield that $y \in \mathrm{Proj}_S(x)$. Further, observing that $d_S(x) \leq \|x - y_n\|$, we also have

$$t_n^{-1}(d_S(x + t_n h) - d_S(x)) \geq t_n^{-1}(\|x + t_n h - y_n\| - \|x - y_n\|)$$
$$= \frac{2\langle x - y_n, h \rangle + t_n \|h\|^2}{\|x + t_n h - y_n\| + \|x - y_n\|},$$

and hence

$$(d_S)^B(x; h) \geq \langle x - y, h \rangle / d_S(x).$$

Therefore, the desired equality for $(d_S)^B(x; h)$ follows.
(b) As above, we have $\mathrm{Proj}_S(x) \neq \emptyset$. Take any $y \in \mathrm{Proj}_S(x)$. Since $d_S(x) = (\Psi_S \square \|\cdot\|)(x)$, Proposition 4.27 guarantees that $\partial_F d_S(x) \subset \{\frac{x-y}{\|x-y\|}\}$ since $\|\cdot\|$ is differentiable at $x - y \neq 0$. The assumption of nonemptiness of $\partial_F d_S(x)$ then implies that $\partial_F d_S(x)$ and $\mathrm{Proj}_S(x)$ are singleton sets with $\mathrm{Proj}_S(x) = \{P_S(x)\}$, and that $\partial_F d_S(x) = \{(x - P_S(x))/d_S(x)\}$, which translates the assertion (b).
(c) Assume that H is finite-dimensional and $\mathrm{Proj}_S(x)$ is a singleton. The assertion (a) ensures that $\partial_F d_S(x) = \{(x - P_S(x))/d_S(x)\}$ according to (d) of Proposition 4.8, which entails that $\partial_F d_S(x)$ is nonempty. $\square$

The link between $\partial_L d_S(x)$ and $\operatorname{Proj}_S(x)$ will be established in Proposition 5.69 for a weakly closed subset S of a Hilbert space and $x \notin S$.

When $\operatorname{Proj}_S(x) = \emptyset$, the inclusion in Proposition 4.102 does not furnish any information since its right-hand member is the whole space X^*. In such a case, the nonempty set
$$\operatorname{Proj}_{S,\varepsilon}(x) := \{y \in S : \|y - x\| \leq d_S(x) + \varepsilon\}$$
of ε-nearest points of x in S can be used. As shown below, the set of approximate nearest points then provides another situation, in addition to Theorem 4.98, where Fréchet ε-subdifferentials are useful to obtain an upper estimate of the Fréchet subdifferential of the distance function d_S at $x \notin S$ when nearest points in S of x are missing.

PROPOSITION 4.105. *Let S be a closed subset of a Banach space X and $x \notin S$. Then*
$$\partial_F d_S(x) \subset \bigcap_{\varepsilon > 0} \bigcup_{y \in \operatorname{Proj}_{S,\varepsilon}(x)} N^{F,\varepsilon}(S;y) \cap \{x^* \in X^* : \|x^*\| = 1\}.$$

PROOF. Fix any $x^* \in \partial_F d_S(x)$ and any $\varepsilon > 0$. Consider any $0 < \varepsilon' < \varepsilon/2$ and choose $\delta > 0$ such that
$$\langle x^*, u - x \rangle \leq d_S(u) - d_S(x) + \varepsilon' \|u - x\| \quad \text{for all } u \in B[x, \delta].$$
Take any $0 < \eta < \min\{1, \varepsilon', \delta\}$ and choose $y \in S$ with $\|y - x\| < d_S(x) + \eta^2$. For any $u \in S \cap B[y, \delta]$, writing $u - y = (u - y + x) - x$ and noting that $u - y + x \in B[x, \delta]$ it ensues that
$$\langle x^*, u - y \rangle \leq d_S(u - y + x) - \|x - y\| + \eta^2 + \varepsilon' \|u - y\| \leq \varepsilon' \|u - y\| + \eta^2,$$
where the latter inequality is due to the inclusion $u \in S$. Applying Lemma 4.86 with $f := \Psi_S$ and $\lambda := \eta$, we obtain some $\bar{u} \in S$ with $\|\bar{u} - y\| \leq \eta$ such that
$$x^* \in \partial_F(\Psi_S + (\varepsilon' + \eta)\|\cdot - \bar{u}\|)(\bar{u}), \text{ hence } x^* \in N^{F,\varepsilon}(S;\bar{u}).$$
Further, since
$$\|\bar{u} - x\| \leq \|\bar{u} - y\| + \|y - x\| \leq \eta + d_S(x) + \eta^2 \leq d_S(x) + \varepsilon,$$
we also have $\bar{u} \in \operatorname{Proj}_{S,\varepsilon}(x)$, which concludes the proof. □

The next proposition provides some additional properties of the Mordukhovich limiting subdifferential in the context of any normed space. It extends Proposition 4.68 to general normed spaces. Assertions (a), (b), (c) directly follow from Definition 4.97 and from the corresponding properties for the Fréchet ε-subdifferential in Proposition 4.93.

PROPOSITION 4.106. *let X be a normed space, S be a subset of X closed near $x \in S$, and $f, g : X \to \mathbb{R} \cup \{-\infty, +\infty\}$ be two functions which are finite at x and lower semicontinuous near x.*
(a) *The equality $N^L(S;x) = N^L(S \cap U;x)$ holds for any neighborhood U of x. If f and g coincide near x, then $\partial_L f(x) = \partial_L g(x)$.*
(b) *$x^* \in \partial_L f(x)$ if and only if $(x^*, -1) \in N^L\big(\operatorname{epi} f; (x, f(x))\big)$.*
(c) *For any real $r > 0$ the equality $\partial_L(rf)(x) = r\partial_L f(x)$ holds.*
(d) *If f is finite and Lipschitz continuous near x with $\gamma \geq 0$ as a Lipschitz constant,*

then $\partial_L f(x) \subset \partial_C f(x) \subset \gamma \mathbb{B}_{X^*}$.
(e) If f is convex, then
$$\partial_L f(x) = \{x^* \in X^* : \langle x^*, x' - x \rangle \leq f(x') - f(x) \ \forall x' \in X\}.$$
(f) If g is finite near x and strictly Fréchet differentiable at x, then
$$\partial_L g(x) = \{Dg(x)\} \quad \text{and} \quad \partial_L(f+g)(x) = \partial_L f(x) + Dg(x).$$
(g) If X is a Banach space, the following inclusions hold:
$$N^L(S;x) \subset \mathbb{R}_+ \partial_C d_S(x) \subset N^C(S;x) \quad \text{and} \quad \partial_L f(x) \subset \partial_C f(x).$$

PROOF. As said above only $(d) - (g)$ need to be proved.
(d) Assume that f is Lipschitz near x. Noting that
$$\partial_{F,\varepsilon} f(u) = \partial_F(f + \varepsilon \|\cdot - u\|)(u) \subset \partial_C(f + \varepsilon \|\cdot - u\|)(u) \subset \partial_C f(u) + \varepsilon \mathbb{B}_{X^*}$$
for u close enough to x, the inclusion $\partial_L f(x) \subset \partial_C f(x)$ follows from the $\|\cdot\|$-to- $w(X^*, X)$ closedness property of $\partial_C f$ at x (see Proposition 2.74). The other inclusion $\partial_C f(x) \subset \gamma \mathbb{B}_{X^*}$ is already known.
(e) Denote (as usual) by $\partial f(x)$ the right member of (e). For any $\varepsilon > 0$ and any u where f is finite we have by Propositions 2.63 and 4.8 and by Theorem 2.98
$$\partial_{F,\varepsilon} f(u) = \partial_F(f + \varepsilon \|\cdot - u\|)(u) \subset \partial f(u) + \varepsilon \mathbb{B}_{X^*}.$$
Fix an arbitrary $x^* \in \partial_L f(x)$ and take sequences $\varepsilon_n \downarrow 0$, $(x_n, f(x_n)) \to (x, f(x))$, and $x_n^* \to x^*$ weak star with $x_n^* \in \partial_{F,\varepsilon_n} f(x_n)$. Choose by what precedes $e_n^* \in \varepsilon_n \mathbb{B}_{X^*}$ such that $x_n^* + e_n^* \in \partial f(x_n)$. Using the definition of ∂f it is easily seen that $x^* \in \partial f(x)$, thus $\partial_L f(x) \subset \partial f(x)$. This and the obvious inclusions
$$\partial f(x) \subset \partial_F f(x) \subset \partial_L f(x)$$
justify the desired equality in (e).
(f) Fix any $x^* \in \partial_L(f+g)(x)$. By the strict Fréchet differentiability of g at x choose some open neighborhood U of x over which g is finite and a function $\varepsilon : U \times U \to [0, +\infty[$ with $\varepsilon(x', x'') \to 0$ as $(x', x'') \to (x, x)$ such that
$$|g(x') - g(x'') - Dg(x)(x' - x'')| \leq \|x' - x''\| \varepsilon(x', x'') \quad \text{for all } x', x'' \in U.$$
For each $n \in \mathbb{N}$ choose $\eta_n \in]0, 1/n[$ with $B(x, 2\eta_n) \subset U$ such that $\varepsilon(x', x'') < 1/n$ for all $x', x'' \in B(x, 2\eta_n)$. Take by definition of the (extended) limiting subdifferential, sequences $\varepsilon_n \to 0$ with $\varepsilon_n > 0$, $x_n \to x$ with $x_n \in U$ and $(f+g)(x_n) \to (f+g)(x)$, and $x_n^* \to x^*$ weakly star in X^* with $x_n^* \in \partial_{F,\varepsilon_n}(f+g)(x_n)$. Extracting a subsequence if necessary, we may suppose that for every $n \in \mathbb{N}$ one has $\|x_n - x\| < \eta_n$. For each $n \in \mathbb{N}$ there is $\delta_n \in]0, \eta_n[$ such that
$$\langle x_n^*, x' - x_n \rangle \leq (f+g)(x') - (f+g)(x_n) + 2\varepsilon_n \|x' - x_n\| \quad \text{for all } x' \in B(x_n, \delta_n).$$
For each $n \in \mathbb{N}$, noting that $\varepsilon(x', x_n) \leq 1/n$ for all $x' \in B(x_n, \delta_n)$ and putting $\varepsilon'_n := 2\varepsilon_n + (1/n)$, it follows that
$$\langle x_n^* - Dg(x), x' - x_n \rangle \leq f(x') - f(x_n) + \varepsilon'_n \|x' - x_n\| \quad \text{for all } x' \in B(x_n, \delta_n),$$
which ensures that $x_n^* - Dg(x) \in \partial_{F,\varepsilon'_n} f(x_n)$. This and the property $f(x_n) \to f(x)$ entail that $x^* - Dg(x) \in \partial_L f(x)$, which guarantees the inclusion $\partial_L(f+g)(x) \subset \partial_L f(x) + Dg(x)$. The converse inclusion being deduced from the previous one applied to the equality $f = (f+g) + (-g)$, the second equality in (f) is proved.

Concerning the first equality in (f), it is a consequence of the second one and of (e), both applied with f as the null function.

(g) From (d) above and from (b) in Theorem 4.98 we directly deduce the inclusion $N^L(S;x) \subset \mathbb{R}_+ \partial_C d_S(x)$ and the right-hand member of this inclusion is known to be contained in $N^C(S;x)$ (see Proposition 2.95). Then, regarding the situation of functions, the general inclusion $\partial_L f(x) \subset \partial_C f(x)$ follows, with f lower semicontinuous near x, according to (b), establishing (g). □

REMARK 4.107. (a) In the setting of a Banach space the equality concerning $\partial_L f(x)$ in assertion (e) of the above theorem can also be seen through the last inclusion in the assertion (g) in accordance with the inclusions

$$\{x^* \in X^* : \langle x^*, x' - x \rangle \leq f(x') - f(x), \forall x' \in X\} \subset \partial_F f(x) \subset \partial_L f(x) \subset \partial_C f(x).$$

(b) Even in $\mathbb{R}$ we have already seen in Remark 4.69 that the equality $\partial_L(-f)(x) = -\partial_L f(x)$ may fail for locally Lipschitz continuous functions. □

Proceeding as in Corollary 2.104 and Proposition 2.109, we obtain the following corollaries from Proposition 4.106(f).

COROLLARY 4.108. Let $f, g : U \to \mathbb{R}$ be two extended real-valued functions on an open set U of a normed space X. Assume that $g = f + \rho$, where $\rho : U \to \mathbb{R}$ is strictly Fréchet differentiable at a point $\bar{x} \in U$ with $D\rho(\bar{x}) = 0$. Then one has $\partial_L g(\bar{x}) = \partial_L f(\bar{x})$.

COROLLARY 4.109. Let $G : U \to Y$ be a mapping from an open set U of a normed space X into a normed space Y and let $g : V \to \mathbb{R}$ be a function from an open set V of Y containing $G(U)$. Assume that G is Lipschitz continuous near $\bar{x} \in U$ and g is strictly Fréchet differentiable at $\bar{y} := G(\bar{x})$, and let $\bar{y}^* := Dg(\bar{y})$. Then one has

$$\partial_L(g \circ G)(\bar{x}) = \partial_L(\bar{y}^* \circ G)(\bar{x}).$$

In particular, given a real-valued function $f : U \to \mathbb{R}$ which is Lipschitz continuous near $\bar{x} \in U$ and a function $\varphi : I \to \mathbb{R}$ on an open interval $I \supset f(U)$ which is strictly differentiable at $f(\bar{x})$, then with $r := \varphi'(f(\bar{x}))$ one has

$$\partial_L(\varphi \circ f)(\bar{x}) = \partial_L(rf)(\bar{x}).$$

We immediately apply the above corollary in the following example concerning the product of finitely many functions and the quotient.

EXAMPLE 4.110. Let $f, g : U \to \mathbb{R}$ be two locally Lipschitz functions on an open set U of a normed space X. Considering as in Example 2.110 the mapping $G : U \to \mathbb{R}^2$ and the functions $p : \mathbb{R}^2 \to \mathbb{R}$ and $q : D_q \to \mathbb{R}$ defined by $G(x) = (f(x), g(x))$ for all $x \in U$, $p(r,s) = rs$ for all $(r,s) \in \mathbb{R}^2$, and $q(r,s) = r/s$ for all $(r,s) \in D_q := \{(\rho, \sigma) \in \mathbb{R}^2 : \sigma \neq 0\}$, the application of the above corollary with $p \circ G$ and $q \circ G$ gives respectively (a) and (b) below:
(a) For every $x \in U$ one has

$$\partial_L(fg)(x) = \partial_L\big(g(x)f + f(x)g\big)(x) \subset \partial_L\big(g(x)f\big)(x) + \partial_L\big(f(x)g\big)(x).$$

(b) For every $x \in U$ with $g(x) \neq 0$ one has

$$\partial_L\left(\frac{f}{g}\right)(x) = \frac{1}{(g(x))^2} \partial_L\big(g(x)f - f(x)g\big)(x).$$

□

The next proposition presents some other subdifferential rules for $\partial_L f(x)$. It extends Proposition 4.73 to general normed vector spaces.

PROPOSITION 4.111. Let $X, Y, X_1, \cdots, X_m$ be normed vector spaces, $g : Y \to \mathbb{R} \cup \{-\infty, +\infty\}$ be a function on Y, and $f_i : X_i \to \mathbb{R} \cup \{-\infty, +\infty\}$ be a function on X_i, for $i = 1, \cdots, m$. Let also $G : X \to Y$ be a mapping. Then, with $X_1 \times \cdots \times X_m$ endowed with the sum norm $\|(u_1, \cdots, u_m)\| := \|u_1\| + \cdots + \|u_m\|$ and $X \times Y$ also endowed with the sum norm, the following hold:

(a) For $\varphi(u_1, \cdots, u_m) := f_1(u_1) + \cdots + f_m(u_m)$ well defined on an open set U of the product space $X_1 \times \cdots \times X_m$, one has for any real $\varepsilon \geq 0$ and any $(x_1, \cdots, x_m) \in U$

$$\partial_{F,\varepsilon} \varphi(x_1, \cdots, x_m) = \partial_{F,\varepsilon} f_1(x_1) \times \cdots \times \partial_{F,\varepsilon} f_m(x_m),$$

hence in particular, for any set $S_i \subset X_i$ and $x_i \in S_i$, for $i = 1, \cdots, m$,

$$N^{F,\varepsilon}(S_1 \times \cdots \times S_m; (x_1, \cdots, x_m)) = N^{F,\varepsilon}(S_1; x_1) \times \cdots \times N^{F,\varepsilon}(S_m; x_m).$$

(b) If $\varphi(u_1, \cdots, u_m) := f_1(u_1) + \cdots + f_m(u_m)$ is well defined on an open set U of $X_1 \times \cdots \times X_m$ containing $(x_1, \cdots, x_m)$, and if each f_i is lower semicontinuous near x_i, then one has

$$\partial_L \varphi(x_1, \cdots, x_m) = \partial_L f_1(x_1) \times \cdots \times \partial_L f_m(x_m);$$

consequently, if S_i is a subset of X_i which is closed near $x_i \in S_i$,

$$N^L(S_1 \times \cdots \times S_m; (x_1, \cdots, x_m)) = N^L(S_1; x_1) \times \cdots \times N^L(S_m, x_m).$$

(c) For $h : X \times Y \to \mathbb{R} \cup \{-\infty, +\infty\}$ with $h(x', y') = g(y')$ for all $(x', y') \in X \times Y$ one has at any $(x, y) \in X \times Y$,

$$\partial_{F,\varepsilon} h(x, y) = (\varepsilon \mathbb{B}_{X^*}) \times \partial_{F,\varepsilon} g(y) \quad \text{and} \quad \partial_L h(x, y) = \{0\} \times \partial_L g(y),$$

where in the second equality g is assumed to be lower semicontinuous near y.

(d) Assume that G is strictly Fréchet differentiable at $x \in X$. Then one has

$$N^L(\operatorname{gph} G; (x, G(x))) = N^F(\operatorname{gph} G; (x, G(x))),$$

and further $(x^*, -y^*) \in N^L(\operatorname{gph} G; x, G(x))$ if and only if $y^* = DG(x)^*(x^*)$, where $DG(x)^*$ denotes the adjoint linear mapping of $DG(x)$.

PROOF. The equalities in the assertion (a) are direct consequences of the definitions of Fréchet ε-subdifferential. The first equality in the assertion (b) follows from (a) and the definition of (extended) limiting subdifferential as in Proposition 4.73, and the second equality in (b) is obtained from the first with $f_i = \Psi_{S_i}$. The assertion (c) is a particular case of (b).

Let us prove the assertion (d). Fix $(x^*, -y^*) \in N^L(\operatorname{gph} G; (x, G(x)))$ and fix also any $\varepsilon > 0$. By the strict Fréchet differentiability of G at x choose some open neighborhood U of x over which G is γ-Lipschitz and such that

$$\|G(x') - G(x'') - DG(x)(x' - x'')\| \leq \varepsilon \|x' - x''\| \quad \text{for all } x', x'' \in U.$$

Take now by definition of the (extended) limiting normal cone, sequences of positive numbers $\varepsilon_n \to 0$, $x_n \to x$ with $x_n \in U$, and $(x_n^*, -y_n^*) \to (x^*, -y^*)$ weakly star in $X^* \times Y^*$ with $(x_n^*, -y_n^*) \in N^{F,\varepsilon_n}(\operatorname{gph} G; (x_n, G(x_n)))$ and choose some real $\beta \geq 0$ such that $\|y_n^*\| \leq \beta$ for all $n \in \mathbb{N}$. Then for each $n \in \mathbb{N}$, putting $\varepsilon_n' := 2\varepsilon_n$ there is some neighborhood $U_n \subset U$ of x_n such that

$$\langle x_n^*, x' - x_n \rangle - \langle y_n^*, G(x') - G(x_n) \rangle \leq \varepsilon_n'(\|x' - x_n\| + \|G(x') - G(x_n)\|) \,\forall x' \in U_n,$$

hence for n large enough

$$\langle x_n^* - y_n^* \circ DG(x), x' - x_n \rangle \leq \varepsilon(1 + \beta + \gamma)\|x' - x_n\| \,\forall x' \in U_n,$$

which ensures that $\|x_n^* - y_n^* \circ DG(x)\| \leq \varepsilon(1+\beta+\gamma)$. This entails by the weak star lower semicontinuity of the norm in X^* that $\|x^* - y^* \circ DG(x)\| \leq \varepsilon(1+\beta+\gamma)$. This being true for every $\varepsilon > 0$ we obtain $x^* = y^* \circ DG(x) = DG(x)^*(y^*)$. This combined with Proposition 4.19 completes the proof of (d). □

We explore now some estimates for Fréchet normals to intersection, inverse and direct images. Let us start with intersection and inverse images.

PROPOSITION 4.112. Let X and Y be two normed spaces, S be a subset of Y, and $G : X \to Y$ be a mapping which is Fréchet differentiable at a point $x \in G^{-1}(S)$. Let also $S_1, \cdots, S_m$ be subsets of X with $x \in S_i$ for all $i = 1, \cdots, m$.
(a) For any $\varepsilon_1, \cdots, \varepsilon_m \geq 0$ and $\varepsilon = \varepsilon_1 + \cdots + \varepsilon_m$, one has
$$N^{F,\varepsilon_1}(S_1; x) + \cdots + N^{F,\varepsilon_m}(S_m; x) \subset N^{F,\varepsilon}(S_1 \cap \cdots \cap S_m; x).$$
(b) For any fixed real $\varepsilon \geq 0$ and for $\varepsilon' := \varepsilon \|DG(x)\|$, one has
$$\{y^* \circ DG(x) : y^* \in N^{F,\varepsilon}(S; G(x))\} \subset N^{F,\varepsilon'}(G^{-1}(S); x).$$
In particular
$$DG(x)^* \big(N^F(S; G(x))\big) \subset N^F(G^{-1}(S); x).$$

PROOF. We prove only (b). Fix any $\varepsilon \geq 0$ and any $y^* \in N^{F,\varepsilon}(S; G(x))$. Take any $\eta > 0$ and choose a neighborhood V of $G(x)$ such that
$$\langle y^*, y - G(x) \rangle \leq (\varepsilon + \eta)\|y - G(x)\| \quad \text{for all } y \in C \cap V.$$
By continuity and differentiability of G at x consider a neighborhood U of x such that $G(U) \subset V$ and $\|G(x') - G(x) - DG(x)(x' - x)\| \leq \eta\|x' - x\|$ for all $x' \in U$. Then, for any $x' \in U \cap G^{-1}(S)$ we have
$$\langle y^*, G(x') - G(x) \rangle \leq (\varepsilon + \eta)\|G(x') - G(x)\|,$$
hence
$$\langle y^* \circ DG(x), x' - x \rangle \leq \big(\eta\|y^*\| + (\varepsilon + \eta)(\|DG(x)\| + \eta)\big)\|x' - x\|.$$
This translates for $\varepsilon' := \varepsilon\|DG(x)\|$ that $y^* \circ DG(x) \in N^{F,\varepsilon'}(G^{-1}(S); x)$ as required.

For the second inclusion, it suffices to take $\varepsilon = 0$ (note that (e) in Proposition 4.11 could be used). □

Under a metric inequality, the second inclusion in (b) of the proposition above is an equality as established in Proposition 4.113 below. Such types of inequality are already used in Proposition 2.138 and they will be studied in detail in Chapter 7. In preparation for the proof of the proposition, let us recall that, for any continuous linear mapping $A : X \to Y$ between Banach spaces X and Y with $A(X)$ closed in Y, one has (see Theorem C.5 in Appendix) the equality

(4.60) $$A^*(Y^*) = \big(\text{Ker}(A)\big)^\perp.$$

Consider now a surjective continuous linear mapping $A : X \to Y$ from a normed space X onto a Banach space Y. Put $Q := \{v^* \in Y^* : \|v^* \circ A\| \leq 1\}$. Take any $y \in Y$ and choose $x \in X$ such that $y = Ax$, so for every $v^* \in Q$ we have
$$|\langle v^*, y \rangle| = |\langle v^* \circ A, x \rangle| \leq \|v^* \circ A\| \|x\| \leq \|x\|,$$
thus $\sup_{v^* \in Q} |v^*(y)| < +\infty$ for any $y \in Y$. The uniform boundedness principle (known also as the Banach-Steinhauss theorem) tells us that Q is norm-bounded in Y^*,

that is, there is some real $\beta > 0$ such that $\|v^*\| \leq \beta$ for all $v^* \in Q$. For any $y^* \in Y^*$, observing for every $\varepsilon > 0$ that $y^*/(\varepsilon + \|y^* \circ A\|) \in Q$, we derive that $\|y^*\| \leq \beta(\varepsilon + \|y^* \circ A\|)$. It follows that

(4.61) $$\|y^*\| \leq \beta\|y^* \circ A\| \quad \text{for all } y^* \in Y^*.$$

Such an inequality will also been seen in Theorem C.6 in Appendix as a characterization of the surjectivity of A under the completeness of both spaces.

PROPOSITION 4.113. *Let X and Y be Banach spaces, S be a subset of Y, $G : X \to Y$ be a mapping, and $\bar{x} \in G^{-1}(S)$. Let also $g : Y \to \mathbb{R} \cup \{-\infty, +\infty\}$ be a function which is finite at $G(\bar{x})$. Assume that G is Fréchet differentiable at $\bar{x}$ with $DG(\bar{x})$ surjective and that there exist some real $\gamma \geq 0$ and some neighborhoods U of $\bar{x}$ and V of $\bar{y} := G(\bar{x})$ such that the metric inequality*

(4.62) $$d(x, G^{-1}(y)) \leq \gamma \|G(x) - y\| \quad \text{for all } x \in U, y \in V \cap S$$

holds. Then one has the inclusion

$$N^F(G^{-1}(S); \bar{x}) \cap \mathbb{B}_{X^*} \subset DG(\bar{x})^* \big(N^F(S; G(\bar{x})) \cap \beta \mathbb{B}_{Y^*}\big),$$

where β is any constant satisfying (4.61) for the surjective mapping $DG(\bar{x})$, and

$$N^F(G^{-1}(S); \bar{x}) = DG(\bar{x})^* \big(N^F(S; G(\bar{x}))\big);$$

if the inequality (4.62) is satisfied for all $y \in V$ instead of $y \in V \cap S$, then

$$\partial_F(g \circ G)(\bar{x}) = DG(\bar{x})^* \big(\partial_F g(G(\bar{x}))\big).$$

PROOF. Let U and γ as given by the assumptions. Denote by A the Fréchet derivative of G at $\bar{x}$, that is, $A := DG(\bar{x})$. Fix $x^* \in N^F(G^{-1}(S); \bar{x}) \cap \mathbb{B}_{X^*}$, so that $x^* \in \partial_F d(\cdot, G^{-1}(S))(\bar{x})$ according to Proposition 4.30. We proceed to show first that $x^* \in (\text{Ker}(A))^\perp$, that is, x^* is null over Ker A. Fix any nonzero v in Ker A and any $\varepsilon > 0$, and choose some $\delta > 0$ such that $B(\bar{x}, \delta) \subset U$ and $\langle x^*, x - \bar{x}\rangle \leq \varepsilon\|x - \bar{x}\|$ for all $x \in G^{-1}(S) \cap B(\bar{x}, \delta)$. For any $t > 0$ such that $\bar{x} + tv \in U$, the metric inequality assumption furnishes some $x_t \in G^{-1}(\bar{y})$ such that

$$\|(\bar{x} + tv) - x_t\| \leq 2\gamma\|G(\bar{x} + tv) - G(\bar{x})\| = t\zeta(t),$$

where $\zeta(t) \downarrow 0$ as $t \downarrow 0$ since $DG(\bar{x})(v) = 0$. This entails $t^{-1}(x_t - \bar{x}) \to v$ as $t \downarrow 0$, so in particular $x_t \neq \bar{x}$ for small $t > 0$ and $G^{-1}(S) \ni x_t \to \bar{x}$ as $t \downarrow 0$. Therefore, for $t > 0$ small enough, we have

$$\langle x^*, x_t - \bar{x}\rangle \leq \varepsilon\|x_t - \bar{x}\|, \text{ hence } \langle x^*, t^{-1}(x_t - \bar{x})\rangle \leq \varepsilon\|t^{-1}(x_t - \bar{x})\|,$$

and making $t \downarrow 0$ gives $\langle x^*, v\rangle \leq \varepsilon\|v\|$, which ensures $\langle x^*, v\rangle = 0$ as desired. By (4.60) above we deduce that $x^* \in A^*(Y^*)$, hence there is some $y^* \in Y^*$ such that $x^* = y^* \circ A$. Further, for the real constant $\beta > 0$ (independent of x^*, y^*) given by (4.61) we have $\|y^*\| \leq \beta \|x^*\|$.

Let us show that $y^* \in N^F(S; \bar{y})$. Fix any $\varepsilon > 0$. Choose some $\delta > 0$ such that $B(\bar{x}, \delta) \subset U$ and $\langle x^*, x - \bar{x}\rangle \leq \varepsilon\|x - \bar{x}\|$ for all $x \in G^{-1}(S) \cap B(\bar{x}, \delta)$ as above and such that

$$\|G(x) - G(\bar{x}) - A(x - \bar{x})\| \leq \varepsilon\|x - \bar{x}\| \quad \text{for all } x \in B(\bar{x}, \delta)$$

as well. Choose also $\eta > 0$ with $B(G(\bar{x}), \eta) \subset V$ and such that $2\gamma\eta < \delta$. Fix any $y \in S \cap B(\bar{y}, \eta)$ with $y \neq \bar{y}$ and, by the metric inequality assumption, choose $u_y \in G^{-1}(y)$ such that $\|u_y - \bar{x}\| < 2\gamma\|y - G(\bar{x})\| < \delta$. Then we have

$$\langle y^*, A(u_y - \bar{x})\rangle = \langle x^*, u_y - \bar{x}\rangle \leq \varepsilon\|u_y - \bar{x}\|,$$

thus
$$\langle y^*, y - \bar{y}\rangle = \langle y^*, G(u_y) - G(\bar{x})\rangle \leq \varepsilon(1+\beta)\|u_y - \bar{x}\| \leq 2\gamma\varepsilon(1+\beta)\|y - \bar{y}\|,$$
which means that $y^* \in N^F(S; \bar{y})$, hence $y^* \in N^F(S; \bar{y}) \cap \beta \mathbb{B}_{Y^*}$.

The latter relation entails the inclusion $N^F(G^{-1}(S); \bar{x}) \subset A^*(N^F(S; G(\bar{x})))$, and this inclusion holds as an equality since the converse is always true.

Concerning the equality of the statement relative to the Fréchet subdifferentials, considering the mapping $\widehat{G} : X \times \mathbb{R} \to Y \times \mathbb{R}$ with $\widehat{G}(x, r) := (G(x), r)$, it is clear that the above result can be applied with the set epi g in $X \times \mathbb{R}$. From this, the desired equality is easily obtained. □

A similar metric inequality yields also estimates for the limiting normal cone to inverse images.

PROPOSITION 4.114. *Let X and Y be Asplund spaces, S be a subset of Y, and $G : X \to Y$ be a mapping from X into Y. Let $\bar{x} \in G^{-1}(S)$ and $\bar{y} := G(\bar{x})$. Assume that, for some real constant $\gamma \geq 0$, the metric inequality*
$$d(x, G^{-1}(S)) \leq \gamma d(G(x), S) \quad \text{for all } x \text{ near } \bar{x}$$
is satisfied.

(a) *If G is compactly Lipschitzian at $\bar{x}$, then*
$$\partial_L d_{G^{-1}(S)}(\bar{x}) \subset \bigcup_{y^* \in \partial_L d_S(\bar{y})} \gamma \partial_L(y^* \circ G)(\bar{x}).$$

(b) *If G is strictly Fréchet differentiable at $\bar{x}$, then*
$$\partial_L d_{G^{-1}(S)}(\bar{x}) \subset DG(\bar{x})^*(\gamma \partial_L d_S(\bar{y}))$$
and if S is closed near $\bar{y}$,
$$N^L(G^{-1}(S); \bar{x}) \subset DG(\bar{x})^*(N^L(S; \bar{y})).$$

PROOF. Put $C := G^{-1}(S)$. By the metric inequality assumption, fix an open neighborhood U of $\bar{x}$ such that $d_C(x') \leq \gamma(d_S \circ G)(x')$ for all $x' \in U$ and note that $d_C(x') = \gamma(d_S \circ G)(x')$ for all $x' \in C$. By Proposition 4.100 we know that $\partial_L d_C(\bar{x}) \subset \gamma \partial_L(d_S \circ G)(\bar{x})$. This and Theorem 4.80 ensure both inclusions in (a) and (b) related to the distance functions. The inclusion in (b) concerning the normal cones follows from the corresponding one with the distance functions and from Theorem 4.85(b). □

Under similar metric inequalities various other estimates for limiting normal cones will be established in Chapter 7. For direct images, in the line of convex situation in Proposition 3.179, we have:

PROPOSITION 4.115. *Let C be a subset of the normed space X with $y \in S := G(C)$, where G is a mapping from X into a normed space Y. The following hold:*
(a) *Let $x \in G^{-1}(y) \cap C$ be a point where G is Fréchet differentiable. For any fixed real $\varepsilon \geq 0$ and for $\varepsilon' := \varepsilon \|DG(x)\|$ one has*
$$N^{F,\varepsilon}(S; y) \subset \{y^* \in Y^* : y^* \circ DG(x) \in N^{F,\varepsilon'}(C; x)\}.$$
(b) *If Δ_G denotes the set of points in $G^{-1}(y) \cap C$ where G is Fréchet differentiable, then one has*

$$N^F(S;y) \subset \bigcap_{x \in \Delta_G} (DG(x)^*)^{-1}(N^F(C;x)).$$

(c) If there exists some neighborhood V of y such that $G^{-1}(V) \cap C \subset K$ for some compact set K of X, C is closed and G is $\mathcal{C}^1$ on an open set U containing $\mathrm{cl}\bigl(G^{-1}(V) \cap C\bigr)$, then

$$N^L(S;y) \subset \bigcup_{x \in G^{-1}(y) \cap C} \{y^* \in Y^* : y^* \circ DG(x) \in N^L(C;x)\}.$$

PROOF. Let $x \in G^{-1}(y) \cap C$ be a point where G is Fréchet differentiable. Fix any $y^* \in N^{F,\varepsilon}(S;y)$ and fix also any $\eta > 0$. Choose a neighborhood V of y such that

$$\langle y^*, y' - y \rangle \leq (\varepsilon + \eta)\|y' - y\| \quad \text{for all } y' \in G(C) \cap V.$$

Let U be a neighborhood of x such that $G(U) \subset V$ and

$$\|G(x') - G(x) - DG(x)(x' - x)\| \leq \eta \|x' - x\| \quad \text{for all } x' \in U.$$

Then for any $x' \in U \cap C$ we have

$$\begin{aligned}\langle y^*, G(x') - G(x) \rangle &\leq (\varepsilon + \eta)\|G(x') - G(x)\| \\ &\leq (\varepsilon + \eta)\bigl(\|DG(x)(x' - x)\| + \eta\|x' - x\|\bigr) \\ &\leq (\varepsilon + \eta)(\eta + \|DG(x)\|)\|x' - x\|,\end{aligned}$$

so that

$$\langle y^* \circ DG(x), x' - x \rangle \leq \bigl(\eta\|y^*\| + (\varepsilon + \eta)(\eta + \|DG(x)\|)\bigr)\|x' - x\|.$$

This ensures the desired inclusion $y^* \circ DG(x) \in N^{F,\varepsilon'}(C;x)$ in (a). The assertion in (b) follows directly from (a).

Now let us prove (c). Let V and U be given by the assumption in (c). Observe that S is closed near y. Fix $y^* \in N^L(S;y)$ and choose sequences $(\varepsilon_n)_n$ tending to 0, $(y_n)_n$ in $S \cap V$ converging to y and $(y_n^*)_n$ converging weakly star to y^* in Y^* with $y_n^* \in N^{F,\varepsilon_n}(S;y_n)$. For each n put $\varepsilon_n' := \varepsilon_n \|DG(x_n)\|$ and take some $x_n \in G^{-1}(y_n) \cap C$. Since $x_n \in K$, taking a subsequence if necessary, we may then suppose that $(x_n)_n$ converges to some $x \in \mathrm{cl}(G^{-1}(V) \cap C) \subset U$. The continuity of G on U gives $y = G(x)$, thus $x \in G^{-1}(y) \cap C$ according to the closedness of C. Further, for n large enough by (a) we have $y_n^* \circ DG(x_n) \in N^{F,\varepsilon_n'}(C;x_n)$, so by the $\mathcal{C}^1$ property of DG on U we obtain $y^* \circ DG(x) \in N^L(C;x)$, justifying (c). $\square$

Fréchet and limiting normals to the sum of finitely many sets can be derived:

COROLLARY 4.116. *Let $S_1, \cdots, S_m$ be nonempty subsets of the normed space X and $S := S_1 + \cdots + S_m$.*
(a) *For any $x \in S$ one has*

$$N^F(S;x) \subset \bigcap_{x_1 + \cdots + x_m = x, x_i \in S_i} \bigl(N^F(S_1;x_1) \cap \cdots \cap N^F(S_m;x_m)\bigr).$$

(b) *Assume that each set S_i is closed. If there is a neighborhood V of $x \in S$ such that the set of $(x_1, \cdots, x_m) \in S_1 \times \cdots \times S_m$ satisfying $x_1 + \cdots + x_m \in V$ is included in a compact set of X, then*

$$N^L(S;x) \subset \bigcup_{x_1 + \cdots + x_m = x, x_i \in S_i} \bigl(N^L(S_1;x_1) \cap \cdots \cap N^L(S_m;x_m)\bigr).$$

PROOF. Consider the continuous linear mapping $G : X \times \cdots \times X \to X$ with $G(u_1, \cdots, u_m) = u_1 + \cdots + u_m$ and observe that $S = G(C)$ where $C := S_1 \times \cdots \times S_m$. Fix $x \in S$ and note that
$$G^{-1}(x) \cap C = \{(x_1, \cdots, x_m) \in C : x_1 + \cdots + x_m = x\}.$$
For any $u = (u_1, \cdots, u_m) \in S_1 \times \cdots \times S_m = C$ write by Proposition 4.19
$$N^F(C; u) = N^F(S_1; u_1) \times \cdots \times N^F(S_m; u_m)$$
and the similar equality for the limiting normal cone, when S_i is closed, according to Proposition 4.111. For any $y^* \in X^*$ observe that the inclusion $y^* \circ G \in (X \times \cdots \times X)^*$ and the identification of the latter topological dual space with $X^* \times \cdots \times X^*$ ensures the identification of $y^* \circ G$ with $(y^*, \cdots, y^*)$. By those observations, $y^* \circ G \in N^F(C; u)$ means $y^* \in N^F(S_1; u_1) \cap \cdots \cap N^F(S_m; u_m)$, and the same meaning also holds for $N^L(C; u)$. Consequently the desired properties (a) and (b) of the corollary are obtained by applying Proposition 4.115. $\square$

4.5. Representation of C-subdifferential via limiting subgradients

We have already seen in any Banach space X that $\partial_L f(x) \subset \partial_C f(x)$ (see Propositions 4.68(c) and 4.106(g)). When X is an Asplund space and f is lower semicontinuous near x, a complete description of $\partial_C f(x)$ in terms of $\partial_L f(x)$ and of horizontal limiting subgradients is available.

4.5.1. Horizon L-subgradient and representation of C-subdifferential.

We start with the introduction of horizon limiting subgradient.

DEFINITION 4.117. Let X be an Asplund space, U be a neighborhood of a point $x \in X$, and $f : U \to \mathbb{R} \cup \{-\infty, +\infty\}$ be a function which is finite at $x \in X$. One says that $x^* \in X^*$ is a *horizon limiting subgradient* (or *singular limiting subgradient*) of f at x provided that $(x^*, 0) \in N^L(\text{epi } f; (x, f(x)))$. The set of all horizon limiting subgradients of f at x is the *horizon limiting subdifferential* of f at x and it is denoted by $\partial_L^\infty f(x)$. One also says *horizon L-subgradient* and *horizon L-subdifferential*. Similarly, the horizon F-subgradient is defined by $\partial_F^\infty f(x) = \{x^* \in X^* : (x^*, 0) \in N^F(\text{epi } f; (x, f(x)))\}$ and any $x^* \in \partial_F^\infty f(x)$ is a horizon F-subgradient. If $|f(x)| = +\infty$, one puts $\partial_F^\infty f(x) = \partial_L^\infty f(x) = \emptyset$.

Obviously $\partial_L^\infty f(x)$ is a cone in X^* containing zero whenever $f(x)$ is finite, and $\partial_L^\infty \Psi_S(x) = N^L(S; x)$ for any subset S of X with $x \in S$. When f is Lipschitz continuous near x, for all $h \in X$, we know by Theorem 2.70 that $(h, f^\circ(x; h)) \in T^C(\text{epi } f; (x, f(x)))$, hence for any $x^* \in \partial_L^\infty f(x)$ we have $\langle x^*, h \rangle \leq 0$ since $(x^*, 0)$ belongs to $N^C(\text{epi } f; (x, f(x)))$. In consequence, $\partial_L^\infty f(x) = \{0\}$ whenever f is Lipschitz continuous near x. Using this and the equivalence $x^* \in \partial_L f(x)$ if and only if $(x^*, -1) \in N^L(\text{epi } f; (x, f(x)))$ in Proposition 4.68(b), we obtain as in (2.32) and in Proposition 2.87 the following descriptions of L-normal cone to epigraphs. Note that the description in (b) below utilizes also the non-vacuity of $\partial_L f(x)$ when f is finite and Lipschitz near x.

PROPOSITION 4.118. *Let $f : U \to \mathbb{R}\{-\infty, +\infty\}$ be a function on an open set U of an Asplund space X.*
(a) *One has the equality*
$$N^L(\text{epi } f; (x, f(x))) = \mathbb{R}_+ \big(\partial_L f(x) \times \{-1\} \cup \partial_L^\infty f(x) \times \{0\}\big).$$

(b) If f is finite and Lipschitz continuous near $x \in U$, then
$$N^L(\operatorname{epi} f; (x, f(x))) = \mathbb{R}_+ (\partial_L f(x) \times \{-1\}).$$

Before establishing the description of $\partial_C f(x)$ let us state the following lemma.

LEMMA 4.119. *Let C, C' be two subsets of a vector space X. Then*
$$\operatorname{co}(C + C') = \operatorname{co} C + \operatorname{co} C'.$$

PROOF. The inclusion of the left member into the second one is obvious. Let arbitrary $x \in \operatorname{co} C + \operatorname{co} C'$. Then $x = \sum_{k \in K} t_k x_k + \sum_{k' \in K'} t'_{k'} x'_{k'}$, where K and K' are finite sets, $x_k \in C$, $x'_{k'} \in C'$, $t_k, t'_{k'} \geq 0$ with $\sum_{k \in K} t_k = 1$ and $\sum_{k' \in K'} t'_{k'} = 1$. So, $x = \sum_{(k,k') \in K \times K'} t_k t'_{k'} (x_k + x'_{k'})$ with $\sum_{(k,k') \in K \times K'} t_k t'_{k'} = 1$. This means that $x \in \operatorname{co}(C + C')$, proving the second desired inclusion. □

THEOREM 4.120 (*C-normal cone via L-normals*). *Let X be an Asplund space, S be a subset of X which is closed near $x \in S$, and $f : X \to \mathbb{R} \cup \{+\infty\}$ be a function which is finite at x and lower semicontinuous near x. The following hold:*
(a) $N^C(S; x) = \overline{\operatorname{co}}^* (N^L(S; x))$.
(b) $\partial_C f(x) = \overline{\operatorname{co}}^* (\partial_L f(x) + \partial_L^\infty f(x))$.
(c) *If in addition f is Lipschitz continuous near x, then $\partial_C f(x) = \overline{\operatorname{co}}^* (\partial_L f(x))$.*

PROOF. We start with the proof of (c). Fix any $h \in X$ and choose sequences $(x_n)_n$ of X converging to x and $(t_n)_n$ of positive numbers tending to 0 such that $f^\circ(x; h) = \lim_{n \to \infty} t_n^{-1} (f(x_n + t_n h) - f(x_n))$. By the mean value inequality in the assertion (b) of Theorem 4.84, for each integer n large enough there are $c_n \in [x_n, x_n + t_n h] + (1/n)\mathbb{B}_X$ and $c_n^* \in \partial_F f(c_n)$ such that
$$t_n^{-1} (f(x_n + t_n h) - f(x_n)) \leq \langle c_n^*, h \rangle + (1/n).$$
The boundedness of $(c_n^*)_n$ and the Asplund property of X (see (4.39)) give a subsequence of $(c_n^*)_n$ converging weak star to some c^*. Then $c^* \in \partial_L f(x)$ and
$$f^\circ(x; h) \leq \langle c^*, h \rangle \leq \sigma(h, \partial_L f(x)),$$
where we recall that $\sigma(\cdot, \partial_L f(x))$ denotes the support function of $\partial_L f(x)$. This means that $\partial_C f(x) \subset \overline{\operatorname{co}}^* (\partial_L f(x))$, thus the desired equality in (c) is established according to the inclusion $\partial_L f(x) \subset \partial_C f(x)$ (see (g) in Proposition 4.106) and to the convexity and w^*-closedness of $\partial_C f(x)$ by (d) in Theorem 2.70.

Applying (c) to the distance function d_S yields $\partial_C d_S(x) = \overline{\operatorname{co}}^* (\partial_L d_S(x))$, hence the descriptions of $N^C(S; x)$ and $N^L(S; x)$ in terms of d_S (see Proposition 2.95 and Theorem 4.85) imply that $N^C(S; x) = \overline{\operatorname{co}}^* (N^L(S; x))$, which is the equality in (a).

Now let us turn to (b). We first observe that, for $u^* \in \partial_L f(x)$ and $v^* \in \partial_L^\infty f(x)$ we have both $(u^*, -1)$ and $(v^*, 0)$ in $N^L(\operatorname{epi} f; (x, f(x))) \subset N^C(\operatorname{epi} f; (x, f(x)))$, thus $(u^* + v^*, -1) \in N^C(\operatorname{epi} f; (x, f(x)))$ since the latter set is a convex cone. This entails the inclusion of the second member of (b) into the first one.

It remains to show the converse inclusion of (b). Take first
$$(u^*, -1) \in \operatorname{co}(N^L(\operatorname{epi} f; (x, f(x)))).$$
There exist a finite set K, $(x_k^*, -r_k) \in N^L(\operatorname{epi} f; (x, f(x)))$, and $t_k \geq 0$ with $\sum_{k \in K} t_k = 1$ such that $(u^*, -1) = \sum_{k \in K} t_k (x_k^*, -r_k)$. By (a) in Proposition 4.21 we

know that $r_k \geq 0$, hence for $K_1 := \{k \in K : r_k > 0\}$ and $K_0 := \{k \in K : r_k = 0\}$ we have $\sum_{k \in K_1} t_k r_k = 1$ and

$$(u^*, -1) = \sum_{k \in K_1} t_k r_k (x_k^*/r_k, -1) + \sum_{k \in K_0} t_k (x_k^*, 0).$$

Putting $u_k^* := x_k^*/r_k$ and $v_k^* := 0$ for all $k \in K_1$, and $v_k^* := x_k^*$ for $k \in K_0$, we obtain

$$u^* = \sum_{k \in K_1} t_k r_k u_k^* + \sum_{k \in K} t_k v_k^* \in \operatorname{co} \partial_L f(x) + \operatorname{co} \partial_L^\infty f(x),$$

or equivalently $u^* \in \operatorname{co}(\partial_L f(x) + \partial_L^\infty f(x))$ by Lemma 4.119 above. Take now an arbitrary $x^* \in \partial_C f(x)$. By (a) we have $(x^*, -1) \in \overline{\operatorname{co}}^*(N^L(\operatorname{epi} f; (x, f(x)))$, thus it is easily seen that $(x^*, -1)$ is the weak star limit of a net $(x_i^*, -1)_{i \in I}$ of elements of co $N^L(\operatorname{epi} f; (x, f(x)))$. By what precedes, each $x_i^* \in \operatorname{co}(\partial_L f(x) + \partial_L^\infty f(x))$, hence $x^* \in \overline{\operatorname{co}}^*(\partial_L f(x) + \partial_L^\infty f(x))$, completing the proof of the theorem. □

REMARK 4.121. The assertion (a) (resp. (b)) in the above theorem fails if the set S (resp. function f) is not closed (resp. lower semicontinuous) near x. Indeed, with the set S of $\mathbb{R}$ in Remark 4.72 given by

$$S := \{0\} \cup \left(\bigcup_{n \in \mathbb{N}} \left]\frac{1}{2n+1}, \frac{1}{2n}\right[\right),$$

we see that $N^L(S; 0) =]-\infty, 0]$, while $N^C(S; 0) = \mathbb{R}$. □

Taking f as the distance function in Theorem 4.120(c) we obtain the following corollary.

COROLLARY 4.122. Let S be a subset of an Asplund space X. Then, for all $x \in X$ one has

$$\partial_C d_S(x) = \overline{\operatorname{co}}^*(\partial_L d_S(x)).$$

For a ball-compact set S in a Hilbert space H and $x \notin S$ we have seen in Proposition 2.207(a) that $\partial_C d_S(x) = \frac{1}{d_S(x)} \overline{\operatorname{co}}(x - \operatorname{Proj}_S(x))$. Below with the use of Corollary 4.122 we will show that this result also holds true for weakly closed sets in H. We need to establish first the following lemma which will also be useful in other places in the text.

LEMMA 4.123. Let S be a nonempty set of a normed space $(X, \|\cdot\|)$ and $u \in X$. Assume that $y \in \operatorname{Proj}_S(u)$. Then one has

$$y \in \operatorname{Proj}_S(y + t(u - y)) \quad \text{for all } t \in [0, 1].$$

If $(X, \|\cdot\|)$ is a Hilbert space, one even has

$$y = P_S(y + t(u - y)) \quad \text{for all } t \in [0, 1[.$$

PROOF. The result being obviously true with $t = 0$, fix any $t \in]0, 1]$ and put $z_t := y + t(u - y)$. Considering any $x \in S$, we have

$$\|z_t - y\| = t\|u - y\| \leq t\|u - x\| = \|tu - tx\|,$$

and since $tu = z_t - (1-t)y$ we obtain

$$\|z_t - y\| \leq \|z_t - (1-t)y - tx\|, \quad \text{hence } \|z_t - y\| \leq (1-t)\|z_t - y\| + t\|z_t - x\|.$$

The latter inequality is equivalent to $t\|z_t-y\| \leq t\|z_t-x\|$, that is, $\|z_t-y\| \leq \|z_t-x\|$. This means that $y \in \operatorname{Proj}_S(z_t)$.

Assume now that $(X, \|\cdot\|)$ is a Hilbert space and let $\langle \cdot, \cdot \rangle$ be the inner product on X generating the norm $\|\cdot\|$. We may suppose that S is not a singleton. Fix any $t \in [0,1[$. For any $x \in S$ with $x \neq y$, we have

$$\begin{aligned}\|y+t(u-y)-x\|^2 &= t^2\|u-y\|^2 + \|y-x\|^2 + 2t\langle u-y, y-x\rangle \\ &= t^2\|u-y\|^2 + (1-t)\|y-x\|^2 \\ &\quad + t(\|y-x\|^2 + 2\langle u-y, y-x\rangle + \|u-y\|^2) - t\|u-y\|^2 \\ &= t^2\|u-y\|^2 + (1-t)\|y-x\|^2 + t(\|u-x\|^2 - \|u-y\|^2),\end{aligned}$$

which, combined with the inequalities $(1-t)\|y-x\|^2 > 0$ and $\|u-x\| \geq \|u-y\|$, yields

$$\|y+t(u-y)-x\|^2 > t^2\|u-y\|^2 = \|y+t(u-y)-y\|^2.$$

Consequently, y is the unique nearest point of $y + t(u-y)$ in S, that is, $y = P_S(y+t(u-y))$. $\square$

REMARK 4.124. The result in the above lemma related to Hilbert spaces will be extended in Lemma 18.8 to strictly convex normed spaces. $\square$

PROPOSITION 4.125. Let S be a nonempty weakly closed set in a Hilbert space H. For any $x \in H \setminus S$ one has the equalities

$$\partial_C d_S(x) = \frac{1}{d_S(x)}\overline{\operatorname{co}}(x - \operatorname{Proj}_S(x)) \text{ and } (d_S)^\circ(x;h) = \max_{y \in \operatorname{Proj}_S(x)} \frac{\langle x-y, h\rangle}{d_S(x)} \; \forall h \in H.$$

PROOF. It suffices to prove the first equality, since the second follows from the first. Using the equality $\partial_L d_S(x) = {}^{\text{seq}}\operatorname{Lim\,sup}_{u\to x} \partial_F d_S(u)$ we easily deduce from Proposition 4.104(b) that $\partial_L d_S(x) \subset \frac{1}{d_S(x)}(x - \operatorname{Proj}_S(x))$. This implies the inclusion of the left member into the right according to the equality $\partial_C d_S(x) = \overline{\operatorname{co}}(\partial_L d_S(x))$ in Corollary 4.122.

On the other hand, take an arbitrary element $y \in \operatorname{Proj}_S(x)$ and note by Lemma 4.123 that $y = P_S(x_t)$ for any $t \in \,]0,1[$, where $x_t := x + t(y-x)$. By Proposition 4.104(a) it ensues that for any $t \in \,]0,1[$

(4.63) $$(x_t - y)/d_S(x_t) \in \partial_C d_S(x_t).$$

Since $(x_t - y)/d_S(x_t)$ tends to $(x-y)/d_S(x)$ as $t \downarrow 0$, it results by the norm-to-weak outer semicontinuity of $\partial_C d_S$ that $(x - \operatorname{Proj}_S(x))/d_S(x)$ is included into $\partial_C d_S(x)$, which in turn trivially entails the inclusion of $\frac{1}{d_S(x)}\overline{\operatorname{co}}(x - \operatorname{Proj}_S(x))$ into $\partial_C d_S(x)$. This finishes the proof. $\square$

REMARK 4.126. Assume that the Hilbert space H is finite-dimensional and that S is a nonempty closed subset. Let $x \in H \setminus S$ and let any $y \in \operatorname{Proj}_S(x)$. For any $t \in \,]0,1[$ and for $x_t := x + t(y-x)$, as above we have $y = P_S(x_t)$, so Proposition 4.104(a) guarantees the inclusion $(x_t - y)/d_S(x_t) \in \partial_F d_S(x_t)$ since H is finite-dimensional. It ensues that $(x - \operatorname{Proj}_S(x))/d_S(x)$ is included in $\partial_L d_S(x)$. Since the converse inclusion has been seen in the beginning of the proof of the above proposition, the equality

$$\partial_L d_S(x) = \frac{1}{d_S(x)}(x - \operatorname{Proj}_S(x))$$

4.5.2. Analytic description of horizon limiting subgradient. Horizon limiting subgradients can also be described in an analytic way. In order to establish that description let us prove an approximation result for horizon F-subgradient.

PROPOSITION 4.127. *Let X be an Asplund space and $f : X \to \mathbb{R} \cup \{+\infty\}$ be a function which is finite at $\bar{x}$ and lower semicontinuous near $\bar{x}$. Let $x^* \in X^*$ with $(x^*, 0) \in N^F\bigl(\operatorname{epi} f; (\bar{x}, f(\bar{x}))\bigr)$. Then for any real $\varepsilon > 0$ there exist $u \in \operatorname{dom} f$ with $\|(u, f(u)) - (\bar{x}, f(\bar{x}))\| \leq \varepsilon$ and $(u^*, -s) \in N^F\bigl(\operatorname{epi} f; (u, f(u))\bigr)$ with $0 < s \leq \varepsilon$ and $\|u^* - x^*\| \leq \varepsilon$.*

PROOF. We may suppose that f is lower semicontinuous on X and that $\bar{x} = 0$, $f(\bar{x}) = 0$, and $\|x^*\| = 1$. Endow $X \times \mathbb{R}$ with the max norm. Fix $\varepsilon \in {]0, 1/2[}$ and choose some real $0 < \eta(\varepsilon) < \varepsilon$ such that $f(x) \geq -\varepsilon$ for all $x \in \eta(\varepsilon)\mathbb{B}_X$ and

(4.64) $\qquad \langle x^*, x \rangle < \varepsilon(\|x\| + |\rho|) \quad$ for all $(x, \rho) \in (\operatorname{epi} f) \cap \eta(\varepsilon)\mathbb{B}_{X \times \mathbb{R}} \setminus \{(0, 0)\}$.

Choose also some $\delta(\varepsilon) > 0$ such that $(1 - \varepsilon)\delta(\varepsilon) < \varepsilon\eta(\varepsilon)$ and note that $\delta(\varepsilon) < \eta(\varepsilon)$. Consider the closed convex cone

(4.65) $\qquad\qquad\qquad K_\varepsilon := \{x \in X : \langle x^*, x \rangle \geq \varepsilon\|x\|\}$

and observe that

(4.66) $\qquad\qquad\qquad f(x) \geq 0 \quad$ for all $x \in K_\varepsilon \cap \delta(\varepsilon)\mathbb{B}_X$.

Indeed, if $f(x) < 0$ for some $x \in K_\varepsilon \cap \delta(\varepsilon)\mathbb{B}_X$, we would have $(x, 0) \in (\operatorname{epi} f) \cap \delta(\varepsilon)\mathbb{B}_{X \times \mathbb{R}} \setminus \{(0, 0)\}$, thus (4.64) would give $\langle x^*, x \rangle < \varepsilon\|x\|$, which is in contradiction with the inclusion $x \in K_\varepsilon$. Further, taking any $x \notin K_\varepsilon$ and any $y \in K_{2\varepsilon}$, we have on the one hand

$$\|x - y\| \geq \langle x^*, y - x \rangle = \langle x^*, y \rangle - \langle x^*, x \rangle \geq 2\varepsilon\|y\| - \varepsilon\|x\|,$$

and on the other hand $2\varepsilon\|x - y\| \geq 2\varepsilon\|x\| - 2\varepsilon\|y\|$, so adding with the latter inequality yields $(1 + 2\varepsilon)\|x - y\| \geq \varepsilon\|x\|$. Consequently, we derive that

(4.67) $\qquad\qquad\qquad d(x, K_{2\varepsilon}) \geq \dfrac{\varepsilon}{1 + 2\varepsilon}\|x\| \quad$ for all $x \notin K_\varepsilon$.

Now, for each real $r > 0$, consider the function $g_{\varepsilon, r}$ on X given by

$$g_{\varepsilon, r}(x) := \varepsilon f(x) + \frac{1}{r} d(x, K_{2\varepsilon}) - \langle x^*, x \rangle + 2\varepsilon\|x\| \quad \text{for all } x \in X,$$

and note that $g_{\varepsilon, r}$ is lower semicontinuous and bounded from below on $\delta(\varepsilon)\mathbb{B}_X$. Applying the form of Ekeland variational principle in Corollary 2.222 furnishes some $x_{\varepsilon, r} \in \delta(\varepsilon)\mathbb{B}_X$ such that

$$g_{\varepsilon, r}(x_{\varepsilon, r}) \leq g_{\varepsilon, r}(x) + r\|x - x_{\varepsilon, r}\| \quad \text{for all } x \in \delta(\varepsilon)\mathbb{B}_X.$$

With $x = 0$ we obtain $g_{\varepsilon, r}(x_{\varepsilon, r}) \leq r\|x_{\varepsilon, r}\|$, or equivalently

$$\varepsilon f(x_{\varepsilon, r}) + \frac{1}{r} d(x_{\varepsilon, r}, K_{2\varepsilon}) - \langle x^*, x_{\varepsilon, r} \rangle + 2\varepsilon\|x_{\varepsilon, r}\| \leq r\|x_{\varepsilon, r}\|,$$

which entails $d(x_{\varepsilon, r}, K_{2\varepsilon}) \to 0$ as $r \downarrow 0$ (since f is bounded from below on $\delta(\varepsilon)\mathbb{B}$).

If there is some $r(\varepsilon) \in {]0, \varepsilon[}$ such that $x_{\varepsilon, r(\varepsilon)} \in K_\varepsilon$, then putting $x_\varepsilon := x_{\varepsilon, r(\varepsilon)}$ we have $\langle x^*, x_\varepsilon \rangle \geq \varepsilon\|x_\varepsilon\|$ by definition of K_ε. In the case $f(x_\varepsilon) \leq \eta(\varepsilon)$, by (4.66) the point $(x_\varepsilon, f(x_\varepsilon))$ is in $(\operatorname{epi} f) \cap \eta(\varepsilon)\mathbb{B}_{X\times\mathbb{R}}$, thus by (4.64) we obtain $\varepsilon f(x_\varepsilon) + \varepsilon\|x_\varepsilon\| - \langle x^*, x_\varepsilon \rangle \geq 0$; and in the case $f(x_\varepsilon) > \eta(\varepsilon)$ we can write

$$\langle x^*, x_\varepsilon \rangle \leq \|x_\varepsilon\| = \varepsilon\|x_\varepsilon\| + (1-\varepsilon)\|x_\varepsilon\| \leq \varepsilon\|x_\varepsilon\| + (1-\varepsilon)\delta(\varepsilon)$$
$$\leq \varepsilon\|x_\varepsilon\| + \varepsilon\eta(\varepsilon) \leq \varepsilon\|x_\varepsilon\| + \varepsilon f(x_\varepsilon).$$

So in any case, $\varepsilon f(x_\varepsilon) + \varepsilon\|x_\varepsilon\| - \langle x^*, x_\varepsilon \rangle \geq 0$, hence

$$g_{\varepsilon,r(\varepsilon)}(x_{\varepsilon,r(\varepsilon)}) \geq \varepsilon\|x_\varepsilon\| + \frac{1}{r(\varepsilon)}d(x_\varepsilon, K_{2\varepsilon}) \geq \varepsilon\|x_\varepsilon\|.$$

From this and the above inequality $g_{\varepsilon,r(\varepsilon)} \leq r(\varepsilon)\|x_{\varepsilon,r(\varepsilon)}\|$, it results that $\varepsilon\|x_\varepsilon\| \leq r(\varepsilon)\|x_\varepsilon\|$, hence $x_\varepsilon = 0 \in B(0, \delta(\varepsilon))$.

If $x_{\varepsilon,r} \notin K_\varepsilon$ for all $r \in]0, \varepsilon[$, then by (4.67)

$$\frac{\varepsilon}{1+2\varepsilon}\|x_{\varepsilon,r}\| \leq d(x_{\varepsilon,r}, K_{2\varepsilon}) \quad \text{for all } r \in]0, \varepsilon[,$$

and since $d(x_{\varepsilon,r}, K_{2\varepsilon}) \to 0$ as $r \downarrow 0$, we can choose some $r(\varepsilon) < \varepsilon$ such that $x_\varepsilon := x_{\varepsilon,r(\varepsilon)} \in B(0, \delta(\varepsilon))$.

In any situation, we get some $0 < r(\varepsilon) < \varepsilon$ such that $x_\varepsilon := x_{\varepsilon,r(\varepsilon)} \in B(0,\delta(\varepsilon))$ is a minimizer on this open ball of the function $g_{\varepsilon,r(\varepsilon)} + r(\varepsilon)\|\cdot - x_\varepsilon\|$, so $0 \in \partial_F(g_{\varepsilon,r(\varepsilon)} + r(\varepsilon)\|\cdot - x_\varepsilon\|)(x_\varepsilon)$. Theorem 4.52 gives some $u_\varepsilon, v_\varepsilon \in B(x_\varepsilon, \varepsilon)$ and $u_\varepsilon^* \in \partial_F f(u_\varepsilon)$, $v_\varepsilon^* \in \partial d_{K_{2\varepsilon}}(v_\varepsilon)$ such that

$$\|\varepsilon u_\varepsilon^* + (1/r(\varepsilon))v_\varepsilon^* - x^*\| \leq 3\varepsilon + r(\varepsilon).$$

By Lemma 4.88 there are $w_\varepsilon \in K_{2\varepsilon}$ with $\|w_\varepsilon - v_\varepsilon\| \leq d(v_\varepsilon, K_{2\varepsilon}) + \varepsilon$ and $w_\varepsilon^* \in \partial d_{K_{2\varepsilon}}(w_\varepsilon)$ with $\|w_\varepsilon^* - v_\varepsilon^*\| \leq (r(\varepsilon))^2$. The inclusion $w_\varepsilon^* \in \partial d_{K_{2\varepsilon}}(w_\varepsilon)$ tells us that $\|w_\varepsilon^*\| \leq 1$ and by Lemma 4.128 below there is $\sigma_\varepsilon \geq 0$ such that $w_\varepsilon^* = \sigma_\varepsilon(-x^* - 2\varepsilon b_\varepsilon^*)$ with $b_\varepsilon^* \in \mathbb{B}_{X^*}$, since $\partial d_{K_{2\varepsilon}}(w_\varepsilon) \subset (K_{2\varepsilon})^\circ$. Further,

$$\|\varepsilon u_\varepsilon^* + r(\varepsilon)^{-1} w_\varepsilon^* - x^*\| \leq 3\varepsilon + r(\varepsilon) + r(\varepsilon) = 3\varepsilon + 2r(\varepsilon),$$

which implies

$$\|\varepsilon u_\varepsilon^* - (1 + r(\varepsilon)^{-1}\sigma_\varepsilon)x^*\| \leq 2r(\varepsilon)^{-1}\sigma_\varepsilon\varepsilon + 3\varepsilon + 2r(\varepsilon).$$

Finally, setting $s_\varepsilon := \varepsilon/(1 + r(\varepsilon)^{-1}\sigma_\varepsilon)$ it ensues that

$$\|s_\varepsilon u_\varepsilon^* - x^*\| \leq \frac{2r(\varepsilon)^{-1}\sigma_\varepsilon}{1 + r(\varepsilon)^{-1}\sigma_\varepsilon}\varepsilon + \frac{3}{1 + r(\varepsilon)^{-1}\sigma_\varepsilon}(\varepsilon + r(\varepsilon))$$

with the second member of the inequality as well as s_ε tending to 0 as $\varepsilon \downarrow 0$. This finishes the proof of the proposition. $\square$

LEMMA 4.128. *Let a normed space $(X, \|\cdot\|)$, a nonzero $x^* \in X^*$, and a real $\varepsilon \geq 0$. For the convex cone $K := \{x \in X : \langle x^*, x\rangle \geq \varepsilon\|x\|\}$ the negative polar K° of K in X^* is given by*

$$K^\circ = \operatorname{cl}_{w^*}\big(\mathbb{R}_+(-x^* + \varepsilon\mathbb{B}_{X^*})\big).$$

Further, if $\varepsilon < \|x^*\|$ then

$$K^\circ = \mathbb{R}_+(-x^* + \varepsilon\mathbb{B}_{X^*}).$$

PROOF. Observe that, endowing X^* with the weak* topology, an element $x \in X$ belongs to the negative polar in X of the weak* closed convex set $\operatorname{cl}_{w^*}(\mathbb{R}_+(-x^* + \varepsilon\mathbb{B}_{X^*}))$ if and only if $\varepsilon\langle u^*, x\rangle - \langle x^*, x\rangle \leq 0$ for all $u^* \in \mathbb{B}_{X^*}$, otherwise stated

$\varepsilon\|x\| - \langle x^*, x\rangle \leq 0$, which is equivalent to $x \in K$. This justifies the first equality of the lemma.

The second equality follows from the fact that the convex set $\mathbb{R}_+(-x^* + \varepsilon \mathbb{B}_{X^*})$ is weak* closed whenever $\varepsilon < \|x^*\|$ (as easily seen) since under this assumption $-x^* + \varepsilon \mathbb{B}_{X^*}$ is a weak* compact convex set of X^* which does not contain zero. □

REMARK 4.129. Writing $K = \{x \in X : g(x) \leq 0\}$ with $g(x) := \varepsilon\|x\| - \langle x^*, x\rangle$, in the case $\varepsilon < \|x^*\|$ Proposition 2.196 can also be applied to obtain the second equality in the lemma. □

THEOREM 4.130 (analytic description of horizon L-subgradients). Let X be an Asplund space and $f : X \to \mathbb{R} \cup \{+\infty\}$ be a function which is finite at x and lower semicontinuous near x. Then

$$\partial_L^\infty f(x) = {}^{\text{seq}}\limsup_{t\downarrow 0, u\to_f x} t\partial_F f(u),$$

that is, $x^* \in \partial_L^\infty f(x)$ if and only if there are sequences $((x_n, f(x_n)))_n$ converging to $(x, f(x))$, $(t_n)_n$ in $]0, +\infty[$ tending to 0 and $x_n^* \in \partial_F f(x_n)$ such that $(t_n x_n^*)_n$ converges weakly* to x^*.

PROOF. It is not difficult to see that $(x^*, 0) \in N^L(\text{epi } f; (x, f(x)))$ whenever the property in the statement of the theorem is satisfied. Conversely, suppose that $(x^*, 0) \in N^L(\text{epi } f; (x, f(x)))$. Then there is a sequence $(u_n, \rho_n)_n$ in epi f converging to $(x, f(x))$ and a sequence $(u_n^*, -r_n)$ converging weakly* to $(x^*, 0)$ with $(u_n^*, -r_n) \in N^F(\text{epi } f; (u_n, \rho_n))$. If for some subsequence of $(r_n)_n$ (that we do not relabel) we have $r_n > 0$ for all n, then $\rho_n = f(u_n)$ (see Proposition 4.21(b)), and putting $x_n^* := r_n^{-1} u_n^*$ and $t_n := r_n$ we obtain that $t_n \downarrow 0$, $(t^n x_n^*)_n$ converges weakly* to x^* with $x_n^* \in \partial_F f(x_n)$, where $x_n := u_n$ and $(x_n, f(x_n))_n$ converges to $(x, f(x))$.

Suppose that $r_n = 0$ for all n large enough, say $n \geq n_0$. By Proposition 4.21(c) we know that $(u_n^*, 0) \in N^F(\text{epi } f; (u_n, f(u_n)))$, and $f(u_n) \to f(x)$ since $f(u_n) \leq \rho_n$ and f is lower semicontinuous at x. Further, by the above proposition, for each integer $n \geq n_0$, we can choose some real $0 < t_n < 1/n$, $x_n \in X$ with $\|x_n - u_n\| < 1/n$ and $|f(x_n) - f(u_n)| < 1/n$ and $v_n^* \in X^*$ with $\|v_n^* - u_n^*\| < 1/n$ and $(v_n^*, -t_n) \in N^F(\text{epi } f; (x_n, f(x_n)))$. Setting $x_n^* := t_n^{-1} v_n^*$, we have also in this second case that $(t_n x_n^*)_n$ converges weakly* to x^* with $t_n \downarrow 0$ and $x_n^* \in \partial_F f(x_n)$, as well as $x_n \to x$ and $f(x_n) \to f(x)$. The proof of the theorem is then complete. □

4.6. Proximal normal cone and subdifferential

One of the main objectives of this section is to describe another limiting process leading, in Hilbert spaces, to the Mordukhovich limiting subdifferential.

4.6.1. Definition and properties of proximal subgradient. In the context of Hilbert space, the natural concept of perpendicularity or proximality is of great interest.

DEFINITION 4.131. Let H be a Hilbert space with the inner product $\langle \cdot, \cdot \rangle$ and let $\|\cdot\|$ be the norm associated with the inner product. Let S be a subset of H and $x \in S$. A vector $v \in H$ is said to be *a proximal normal vector of* (or *a perpendicular vector to*) S at x provided there exists some real $r > 0$ such that

(4.68) $$x \in \text{Proj}_S(x + rv).$$

The set of all such vectors v is obviously a cone. It is called the *cone of proximal normal vectors of S at x* and denoted by $N^P(S;x)$.

For a neighborhood U of a point x in H and a function $f: U \to \mathbb{R} \cup \{-\infty, +\infty\}$ which is finite at x, the *proximal subdifferential* $\partial_P f(x)$ of f at x is then

$$\partial_P f(x) := \{v \in H : (v, -1) \in N^P(\operatorname{epi} f; (x, f(x)))\},$$

where $H \times \mathbb{R}$ is endowed with the usual Hilbert product structure. Any element in $\partial_P f(x)$ is called a *proximal subgradient* of f at x. If f is not finite at x, one puts $\partial_P f(x) = \emptyset$.

In the Hilbert setting above, we observe that a unit vector v (that is, $\|v\| = 1$) satisfies (4.68) if and only if $S \cap B_H(x + rv, r) = \emptyset$. One then says that the unit proximal normal v at x is *realized by the r-ball $B_H(x+rv, r)$*. Of course, any r'-ball $B_H(x+r'v, r')$, with $0 < r' < r$, realizes also the unit proximal normal vector v at x. We also note that, for $s > 0$,

$$x \in \operatorname{Proj}_S(x + sv) \Leftrightarrow s^2 \|v\|^2 \leq \|x' - x\|^2 - 2s\langle v, x' - x\rangle + s^2\|v\|^2 \quad \forall x' \in S$$

(4.69)
$$\Leftrightarrow \langle v, x' - x\rangle \leq \frac{1}{2s}\|x' - x\|^2 \quad \forall x' \in S.$$

PROPOSITION 4.132. *Let S be a subset of a Hilbert space H and $x \in S$. Then, $v \in N^P(S;x)$ if and only if anyone of the following properties (a), (b), (c) holds:*
(a) *there exists some real $s > 0$ such that $x = P_S(x + sv)$;*
(b) *there exists a real $\sigma \geq 0$ such that*

$$\langle v, x' - x\rangle \leq \sigma\|x' - x\|^2 \quad \text{for all } x' \in S;$$

(c) *there are a neighborhood U of x and a real $\sigma \geq 0$ such that*

$$\langle v, x' - x\rangle \leq \sigma\|x' - x\|^2 \quad \text{for all } x' \in S \cap U.$$

PROOF. Denote by (p) the property $v \in N^P(S;x)$. This property (p) is equivalent to (a) by Lemma 4.123. The equivalence between (p) and (b) follows from (4.69), and the assertion (b) obviously entails (c). Fix any $v \in H$ satisfying (c), hence consider two real numbers $\delta > 0$ and $\sigma \geq 0$ such that

$$\langle v, x' - x\rangle \leq \sigma\|x' - x\|^2 \quad \text{for all } x' \in S \cap B(x, \delta).$$

If $x' \in S \setminus B(x, \delta)$, then $\|x' - x\| \geq \delta$ which implies

$$\langle v, x' - x\rangle \leq \|v\|\,\|x' - x\| \leq \frac{\|v\|}{\delta}\delta\|x' - x\| \leq \frac{\|v\|}{\delta}\|x' - x\|^2.$$

So, putting $\sigma' := \max\{\sigma, \frac{\|v\|}{\delta}\}$ we obtain

$$\langle v, x' - x\rangle \leq \sigma'\|x' - x\|^2 \quad \text{for all } x' \in S,$$

which is the assertion (b). $\square$

Definition 4.131 is also relevant in the context of *uniformly convex Banach space*. This will be discussed in Chapter 18. In the next definition, we use the local variational characterization of the assertion (c) in Proposition 4.132 in the context of a general normed space.

DEFINITION 4.133. Let $(X, \|\cdot\|)$ be a normed space and S be a subset of X, and let $x \in S$. A continuous linear functional $x^* \in X^*$ is called a *variational proximal* *normal functional of* the set S at x when there exist a real constant $\sigma \geq 0$ and a neighborhood U of x such that

(4.70) $$\langle x^*, x' - x \rangle \leq \sigma \|x' - x\|^2 \quad \text{for all } x' \in U \cap S.$$

The set of such functionals (which is obviously a cone) is the *functional variational proximal normal cone* $N^{VP}(S;x)$ of S at x.

Let U be a neighborhood of a point x in X and $f: U \to \mathbb{R} \cup \{-\infty, +\infty\}$ be an extended real-valued function which is finite at $x \in U$. Its *functional variational proximal subdifferential* $\partial_{VP} f(x)$ is then defined by

$$\partial_{VP} f(x) := \{x^* \in X^* : (x^*, -1) \in N^{VP}(\operatorname{epi} f; (x, f(x)))\}.$$

Any continuous linear functional in $\partial_{VP} f(x)$ is called a *functional variational proximal subgradient* of f at x. If f is not finite at x, one puts $\partial_{VP} f(x) = \emptyset$. When there is no risk of ambiguity, we will only say variational proximal subdifferential (resp. subgradient, normal).

Clearly the definitions of $N^{VP}(S;x)$ and $\partial_{VP} f(x)$ do not change if we take any other equivalent norm on X. Further, the definition of $N^{VP}(S;\cdot)$ guarantees:

PROPOSITION 4.134. *Given a subset S of the normed space X and $x \in S$, one has*

(4.71) $$N^{VP}(S;x) = N^{VP}(\operatorname{cl} S; x),$$

and for every neighborhood U of x

$$N^{VP}(S;x) = N^{VP}(S \cap U; x).$$

If S' is a subset of S with $x \in S'$, then $N^{VP}(S;x) \subset N^{VP}(S';x)$; in particular

$$N^{VP}(S;x) \subset N^{VP}(\operatorname{bdry} S; x) \quad \text{for all } x \in \operatorname{bdry} S.$$

The following description of variational proximal normals to inverse image under an isomorphism is a direct consequence of the above definition.

PROPOSITION 4.135. *Let $A: X \to Y$ be an isomorphism between normed spaces X, Y, let S be a subset of Y and let $b \in Y$. For any $x \in A^{-1}(b+S)$ one has*

$$N^{VP}(A^{-1}(b+S); x) = A^*(N^{VP}(S; A(x) - b)).$$

As in the previous chapters, extending f in $\overline{f}$ to X with $+\infty$ outside U, one has $\partial_{VP} f(x) = \partial_{VP} \overline{f}(x)$ (resp. $\partial_P f(x) = \partial_P \overline{f}(x)$ if X is Hilbert).

PROPOSITION 4.136. *Let X be a normed space, S be a subset of X, and $f: X \to \mathbb{R} \cup \{-\infty, +\infty\}$ be a function on X. The following hold for all $x \in X$:*
(a) *The sets $N^{VP}(S;x)$ and $\partial_{VP} f(x)$ are convex in X^*.*
(b) *One has the inclusions*

$$N^{VP}(S;x) \subset N^F(S;x) \subset N^L(S;x) \quad \text{and} \quad \partial_{VP} f(x) \subset \partial_F f(x) \subset \partial_L f(x);$$

hence in particular $\partial_{VP} f(x) \subset \gamma \mathbb{B}_{X^*}$ whenever f is γ-Lipschitz continuous near x.
(c) *If S (resp. f) is convex, then $N^{VP}(S;x)$ (resp. $\partial_{VP} f(x)$) coincides with the standard normal cone of S (resp. the standard subdifferential of f) at x, otherwise stated*

$$N^{VP}(S;x) = \{x^* \in X^* : \langle x^*, x' - x \rangle \leq 0, \ \forall x' \in S\} \quad \text{if } x \in S$$
$$(\text{resp. } \partial_{VP} f(x) = \{x^* \in X^* : \langle x^*, x' - x \rangle \leq f(x') - f(x), \ \forall x' \in X\}).$$

(d) When X is a Hilbert space, then (through the Riesz identification)
$$N^P(S;x) = N^{VP}(S;x) \quad \text{and} \quad \partial_P f(x) = \partial_{VP} f(x).$$

PROOF. Assertions (a) and (b) follow from the above definition and from the variational description of Fréchet normal along with the definition of Fréchet subdifferential; the inclusion related to the Lipschitz property of f is a consequence of (b) and Proposition 4.11(c). Assuming that the set S is convex and setting $\mathcal{N} := \{x^* \in X^* : \langle x^*, x' - x \rangle \leq 0, \forall x' \in S\}$, we know that $N^F(S;x) = \mathcal{N}$ by Proposition 4.8(c). This combined with the obvious inclusions $\mathcal{N} \subset N^{VP}(S;x) \subset N^F(S;x)$ yields the first desired equality of (c), which in turn entails the second.

Concerning the assertion (d), it suffices, for $x \in S$, to apply the assertion (c) of Proposition 4.132. □

We now show that variational proximal subgradients also enjoy variational analytic descriptions.

PROPOSITION 4.137. Let S be a subset of the normed space $(X, \|\cdot\|)$ and let $f : X \to \mathbb{R} \cup \{-\infty, +\infty\}$ be an extended real-valued function. Given a point x where f is finite, a continuous linear functional $x^* \in \partial_{VP} f(x)$ if and only if there are a real constant $\sigma \geq 0$ and a neighborhood U of x such that

(4.72) $\qquad \langle x^*, x' - x \rangle \leq f(x') - f(x) + \sigma \|x' - x\|^2 \quad \text{for all } x' \in U.$

PROOF. The proof of the equivalence is quite similar to the proof for the equivalence between assertions (b) and (c) in Proposition 4.132. Let $x^* \in X^*$ satisfying the property of the proposition, that is, there exist a pair of real numbers $\sigma \geq 0$ and $\delta > 0$ such that
$$\langle x^*, x' - x \rangle \leq f(x') - f(x) + \sigma \|x' - x\|^2 \quad \text{for all } x' \in B(x, \delta).$$
Then, for all $(x', r) \in (B(x, \delta) \times \mathbb{R}) \cap \text{epi } f$
$$\langle x^*, x' - x \rangle - (r - f(x)) \leq \sigma \|x' - x\|^2 \leq \sigma \left(\|x' - x\|^2 + |r - f(x)|^2 \right),$$
which implies that $(x^*, -1) \in N^{VP}\bigl(\text{epi}; (x, f(x))\bigr)$, hence $x^* \in \partial_{VP} f(x)$.

Suppose now that $x^* \in X^*$ does not satisfy the property of the proposition. Putting $\sigma_n := n$ and $\delta_n := 1/n$, for each $n \in \mathbb{N}$ there is some $x_n \in B(x, \delta_n)$ such that
$$f(x_n) < f(x) + \langle x^*, x_n - x \rangle - \sigma_n \|x_n - x\|^2.$$
For $r_n := f(x) + \langle x^*, x_n - x \rangle - \sigma_n \|x_n - x\|^2$, we then have $(x_n, r_n) \in \text{epi } f$ and $r_n \to f(x)$ because of the inequality $\sigma_n \|x_n - x\|^2 < 1/n$. Observe for all $n \in \mathbb{N}$ that (with the norm $\|(x, r)\| = (\|x\|^2 + |r|^2)^{1/2}$ on $X \times \mathbb{R}$)
$$\frac{\langle x^*, x_n - x \rangle - (r_n - f(x))}{\|(x_n, r_n) - (x, f(x))\|^2} = \frac{\sigma_n \|x_n - x\|^2}{\|(x_n - x, \langle x^*, x_n - x \rangle - \sigma_n \|x_n - x\|^2)\|^2}$$
$$\geq \frac{\sigma_n}{1 + 2\|x^*\|^2 + 2\sigma_n^2 \|x_n - x\|^2} \geq \frac{\sigma_n}{3 + 2\|x^*\|^2},$$
where the last inequality is due to the relation $\sigma_n \|x_n - x\| \leq 1$. Since $\sigma_n \to +\infty$, we deduce that $(x^*, -1) \notin N^{VP}\bigl(\text{epi}; (x, f(x))\bigr)$, and the characterization of the proposition is established. □

EXAMPLE 4.138. (a) It is worth emphasizing that the sets $N^{VP}(S;x)$ and $\partial_{VP}f(x)$ may be nonclosed (with respect to the norm). Consider the function $f : \mathbb{R} \to \mathbb{R} \cup \{+\infty\}$ with $f(x) = -x^{1+\alpha}$ if $x \geq 0$ and $f(x) = +\infty$ if $x < 0$. For $\alpha \in {]}0,1[$, it is easily seen from Proposition 4.137(a) above that $\partial_P f(0) = {]}-\infty, 0[$. Denoting by S the epigraph of f, we also see through Proposition 4.137(a) and Definition 4.133 that

$$N^P(S;(0,0)) = (]-\infty, 0[\times]-\infty, 0]) \cup \{(0,0)\}.$$

Since $\partial_F f(x)$ and $N^F(S;x)$ are always norm closed according to Proposition 4.7, the function f and the set S above provide examples of a function of one real variable and of a set in $\mathbb{R}^2$ where the inclusions $\partial_P f(x) \subset \partial_F f(x)$ and $N^P(S;x) \subset N^F(S;x)$ in Proposition 4.136(b) are strict.
(b) We may even have, for a $\mathcal{C}^1$ function, $\partial_{VP} f(\bar{x}) \neq \{Df(\bar{x})\}$, that is $\partial_P f(\bar{x}) = \emptyset$. For $\alpha \in {]}0,1[$ as above, such a function is given by $f : \mathbb{R} \to \mathbb{R}$ with $f(x) = |x|^{1+\alpha}$ if $x < 0$ and $f(x) = -|x|^{1+\alpha}$ if $x \geq 0$ since $\partial_P f(0) = \emptyset$ as easily checked via (a) of the above proposition. □

Nevertheless, we will see later that, in the Hilbert setting, the proximal normal may be used in place of the Fréchet normal cone in the limiting process leading to the Mordukhovich limiting normal cone.

The assertion (a) in the following proposition has been already noticed after Definition 4.133. The other assertion (b) is a direct consequence of Proposition 4.137(a) above.

PROPOSITION 4.139. Let S be a set in a normed space $(X, \|\cdot\|)$ and $f : X \to \mathbb{R} \cup \{-\infty, +\infty\}$ be an extended real-valued function.
(a) Using any equivalent norm on X the definitions of $\partial_{VP}f(x)$ and $N^{VP}(S;x)$) remains unaltered.
(b) If $g : X \to \mathbb{R} \cup \{-\infty, +\infty\}$ is another function satisfying $g \leq f$ and $g(x) = f(x)$, then $\partial_{VP} g(x) \subset \partial_{VP} f(x)$.

The statement of the following proposition is similar to its Fréchet counterpart in Proposition 4.3. Using Proposition 4.137(a) the proof proceeds as the one of Proposition 4.3.

PROPOSITION 4.140. Let $(X, \|\cdot\|)$ be a normed space and $f : X \to \mathbb{R} \cup \{-\infty, +\infty\}$ be an extended real-valued function. If $\partial_{VP} f(x) \neq \emptyset$, then the following hold:
(a) the function f is lower semicontinuous at x;
(b) one has the equality

$$\partial_{VP} f(x) = \partial_{VP}(\operatorname{cl} f)(x).$$

We observed in Example 4.138(b) that $\partial_{VP} f(x)$ may be empty for certain $\mathcal{C}^1$ functions f. The next proposition proves, among diverse properties, that $\partial_{VP} f(x)$ is a singleton whenever f is differentiable near x with Df one-sided Lipschitz at x, in particular whenever f is $\mathcal{C}^{1,1}$ near x. Recall that a mapping $G : U \to Y$ from an open set U of a normed space X into a normed space Y is said to be $\mathcal{C}^{1,1}$ on U when it is Fréchet differentiable on U and the derivative DG is locally Lipschitz continuous on U. Let us establish first a lemma which has its own interest.

LEMMA 4.141. Let X be a normed space and $f, g : U \to \mathbb{R} \cup \{-\infty, +\infty\}$ be two extended real-valued functions. Let $x \in U$ with $|f(x)| < +\infty$ be such that there

exist a neighborhood $U_0 \subset U$ and a real $\gamma > 0$ such that
$$g(u) - \gamma\|u - x\|^2 \le f(u) \le g(u) + \gamma\|u - x\|^2 \quad \text{for all } u \in U_0.$$
Then $\partial_{VP} f(x) = \partial_{VP} g(x)$.

PROOF. First, note that $g(x) = f(x)$. Take any $x^* \in \partial_{VP} f(x)$ (if any). There exist a real $\sigma \ge 0$ and a convex neighborhood $V \subset U_0$ of x such that for all $u \in V$ one has
$$\langle x^*, u - x \rangle \le f(u) - f(x) + \sigma\|u - x\|^2 \le g(u) - g(x) + (\gamma + \sigma)\|u - x\|^2,$$
so $x^* \in \partial_{VP} g(x)$. It results that $\partial_{VP} f(x) \subset \partial_{VP} g(x)$, and hence $\partial_{VP} f(x) = \partial_{VP} g(x)$ since the converse inclusion also holds in a similar way. $\square$

PROPOSITION 4.142. *Let X be a normed space, $f, g : X \to \mathbb{R} \cup \{-\infty, +\infty\}$ be two functions, and let $x \in X$ be a point where both functions are finite. Let Z be another normed space and $G : Z \to X$ be a mapping which is Fréchet differentiable near a point z with $G(z) = x$ and such that DG is one-sided Lipschitz at z (which holds in particular if G is $\mathcal{C}^{1,1}$ near z).*
(a) *If f and g coincide near x, then $\partial_{VP} f(x) = \partial_{VP} g(x)$.*
(b) *If x is a local minimum point for f, then $0 \in \partial_{VP} f(x)$.*
(c) *If f is locally Lipschitz continuous near x with $\gamma \ge 0$ as a Lipschitz constant, then $\partial_{VP} f(x) \subset \gamma \mathbb{B}_{X^*}$.*
(d) *For any real $r > 0$ the equality $\partial_{VP}(rf)(x) = r\partial_{VP} f(x)$ is fulfilled.*
(e) *The inclusions*
$$\partial_{VP} f(x) + \partial_{VP} g(x) \subset \partial_{VP}(f + g)(x) \text{ and } DG(z)^*\big(\partial_{VP} g(G(z))\big) \subset \partial_{VP}(g \circ G)(z)$$
always hold.
(f) *If f is finite and Fréchet differentiable near x with Df one-sided Lipschitz at x (which holds in particular if f is $\mathcal{C}^{1,1}$ near x), then*
$$\partial_{VP} f(x) = \{Df(x)\} \quad \text{and} \quad \partial_{VP}(f + g)(x) = Df(x) + \partial_{VP} g(x).$$
(g) *If f is finite near x and Gâteaux differentiable at x, then*
$$\partial_{VP} f(x) \subset \{D_G f(x)\}.$$

PROOF. Assertions (a)-(d) are evident, the first inclusion in (e) is straightforward and the inclusion in (g) follows from Proposition 4.11(g) since $\partial_{VP} f(x) \subset \partial_F f(x)$. Let us start with the first equality of (f). Fix a real $\gamma > 0$ and an open convex neighborhood U of x over which f is finite and differentiable and such that $|\langle Df(u) - Df(x), u - x \rangle| \le \gamma\|u - x\|^2$ for all $u \in U$. For any $u \in U$, defining $g(u) := f(x) + \langle Df(x), u - x \rangle$, we see that
$$|f(u) - g(u)| = \left| \int_0^1 \langle Df(x + t(u - x)) - Df(x), u - x \rangle \, dt \right| \le (\gamma/2)\|u - x\|^2,$$
so $\partial_{VP} g(x) = \partial_{VP} f(x) = \{Df(x)\}$ by Lemma 4.141.

The first inclusion in (e) and the first equality of (f) ensure that $Df(x) + \partial_{VP} g(x) \subset \partial_{VP}(f + g)(x)$. Writing $g = (f + g) - f$ and using the one-sided Lipschitz property of $D(-f)$ at x yield according to the last reasoning above that $D(-f)(x) + \partial_{VP}(f + g)(x) \subset \partial_{VP} g(x)$. All together guarantee the desired equality $\partial_{VP}(f + g)(x) = Df(x) + \partial_{VP} g(x)$.

It remains to prove the second inclusion in (e). Fix any $x^* \in \partial_{VP} g(x)$ (if any), so there are two reals $\sigma > 0$ and $\delta > 0$ such that
$$\langle x^*, u - x \rangle \leq g(u) - g(x) + \sigma \|u - x\|^2 \quad \text{for all } u \in B(x, \delta).$$
Choose a real $\gamma > 0$ and an open ball $W = B(z, \eta)$ in Z such that $G(W) \subset B(z, \delta)$ along with $\|DG(y) - DG(z)\| \leq \gamma \|y - z\|$ for all $y \in W$. Put $A := DG(z)$ and $\beta := \gamma \eta + \|A\|$. Take any $y \in W$ and note that
$$\langle x^*, G(y) - G(z) \rangle \leq g(G(y)) - g(G(z)) + \sigma \|G(y) - G(z)\|^2 =: \zeta(y).$$
Further, writing
$$G(y) - G(z) = A(y - z) + \int_0^1 (DG(z + t(y - z)) - DG(z))(y - z)\, dt$$
we see that for some constant $\beta \geq 0$ we have $\|G(y) - G(z)\| \leq \beta \|y - z\|$ for all $y \in W$. Altogether we deduce that
$$\langle x^* \circ A, y - z \rangle \leq \zeta(y) - \int_0^1 \langle x^*, (DG(z + t(y - z)) - DG(z))(y - z) \rangle\, dt$$
$$\leq g(G(y)) - g(G(z)) + \alpha \|y - z\|^2,$$
where $\alpha := \sigma \beta^2 + \gamma \|x^*\|/2$. It follows that $x^* \circ A \in \partial_{VP}(g \circ G)(z)$, which finishes the proof. $\square$

The next proposition examines in its first part the case when both $\partial_{VP} f(x)$ and $\partial_{VP}(-f)(x)$ are nonempty.

PROPOSITION 4.143. *Let X be a normed space, $f : X \to \mathbb{R} \cup \{-\infty, +\infty\}$ be an extended real-valued function, and $x \in X$ be a point where f is finite.*
(a) If $\partial_{VP} f(x) \neq \emptyset$ and $\partial_{VP}(-f)(x) \neq \emptyset$, then f is Fréchet differentiable at x and
$$\{Df(x)\} = \partial_{VP} f(x) = -\partial_{VP}(-f)(x).$$
(b) If $\partial_{VP} f(x) \neq \emptyset$, then $-\partial_{VP}(-f)(x) \subset \partial_{VP} f(x)$

PROOF. (a) Assuming that both variational proximal subdifferentials at x of f and $-f$ are nonempty, then so are the Fréchet subdifferentials, and hence by Proposition 4.13, the function f is Fréchet differentiable at x with $\partial_F f(x)$ and $-\partial_F(-f)(x)$ equal to $\{Df(x)\}$. This and the non-vacuity assumption of $\partial_{VP} f(x)$ and $\partial_{VP}(-f)(x)$ yield the equalities of (a) since the variational proximal subdifferential is included in the Fréchet one.
(b) Now suppose that $\partial_{VP} f(x) \neq \emptyset$. If $\partial_{VP}(-f)(x) = \emptyset$, there is nothing to prove, and otherwise the inclusion in (b) follows from (a). $\square$

We establish now certain chain rule properties for the variational proximal subdifferential.

PROPOSITION 4.144. *Let U be an open set of a normed space X and let $G : U \to Y$ be a mapping from U into a normed space Y which is one-sided Lipschitz at a point $\bar{x} \in U$. Let $g : W \to \mathbb{R} \cup \{-\infty, +\infty\}$ be an extended real-valued function on an open set $W \supset G(U)$ and let $\bar{y} := G(\bar{x})$.*
(a) One has the inclusion
$$\bigcup_{y^* \in \partial_{VP} g(\bar{y})} \partial_{VP}(y^* \circ G)(\bar{x}) \subset \partial_{VP}(g \circ G)(\bar{x}).$$

(b) If in addition g is differentiable near $\bar{y}$ with Dg one-sided Lipschitz at $\bar{y}$ (which holds in particular if g is $\mathcal{C}^{1,1}$ at $\bar{y}$), then
$$\partial_{VP}(g \circ G)(\bar{x}) = \partial_{VP}(g'(\bar{y}) \circ G)(\bar{x}).$$

PROOF. Denote by $\gamma > 0$ a one-sided Lipschitz constant of G at $\bar{x}$ and $U_0 \subset U$ be an open neighborhood of $\bar{x}$ such that $\|G(x) - G(\bar{x})\| \leq \gamma \|x - \bar{x}\|$ for all $x \in U$.
(a) We may suppose that the left hand-side in (a) is nonempty. Fix any $y^* \in \partial_P g(\bar{y})$ such that $\partial_P(y^* \circ G)(\bar{x}) \neq \emptyset$ and take any $x^* \in \partial_P(y^* \circ G)(\bar{x})$. We note that $g(\bar{y})$ is finite. Choose a real $\sigma > 0$ and a ball $V = B(\bar{y}, \delta)$ such that
$$\langle y^*, y - \bar{y} \rangle \leq g(y) - g(\bar{y}) + \sigma \|y - \bar{y}\|^2 \quad \text{for all } y \in V,$$
and choose another real $\sigma' > 0$ and a ball $U_1 = B(\bar{x}, \eta) \subset U_0$ with $G(U_1) \subset V$ and such that for every $x \in U_1$
$$\langle x^*, x - \bar{x} \rangle \leq y^* \circ G(x) - y^* \circ G(\bar{x}) + \sigma' \|x - \bar{x}\|^2.$$
Putting $\alpha := \sigma' + \sigma \gamma^2$ we obtain for all $x \in U_1$
$$\langle x^*, x - \bar{x} \rangle \leq g(G(x)) - g(G(\bar{x})) + \alpha \|x - \bar{x}\|^2,$$
where $\alpha := \sigma' + \sigma \gamma^2$. This justifies (a).
(b) Assume now g is differentiable near $\bar{y}$ with Dg one-sided Lipschitz at $\bar{y}$ and put $\bar{y}^* := g'(\bar{y})$, so there are a real $\beta > 0$ and a ball $V = B(G(\bar{x}), \eta)$ such that $|Dg(y) - Dg(G(\bar{x})| \leq \beta \|y - G(\bar{x})\|$ for all $y \in V$. Choose a neighborhood $U_3 \subset U_0$ such $G(U_3) \subset V$. Set
$$f_1(x) := g \circ G(x) - g \circ G(\bar{x}) \text{ and } f_2(x) := \bar{y}^* \circ G(x) - \bar{y}^* \circ G(\bar{x}) \text{ for all } x \in U.$$
For any $x \in U_3$ we have
$$|f_1(x) - f_2(x)| = \left| \int_0^1 \langle Dg(G(\bar{x}) + t(G(x) - G(\bar{x}))) - Dg(G(\bar{x})), G(x) - G(\bar{x}) \rangle \, dt \right|$$
$$\leq \frac{\beta}{2} \|G(x) - G(\bar{x})\|^2 \leq \frac{\gamma^2 \beta}{2} \|x - \bar{x}\|^2,$$
so Lemma 4.141 tells us that $\partial_{VP} f_1(\bar{x}) = \partial_{VP} f_2(\bar{x})$, which is equivalent to $\partial_{VP}(g \circ G)(\bar{x}) = \partial_{VP}(\bar{y}^* \circ G)(\bar{x})$. The proof is then complete. □

COROLLARY 4.145. Let U be an open set of a normed space X and let $f : U \to \mathbb{R} \cup \{-\infty, +\infty\}$ be a function which is finite at $\bar{x} \in U$ and one-sided Lipschitz at $\bar{x}$. Let $\varphi : I \to \mathbb{R} \cup \{-\infty, +\infty\}$ be an extended real-valued function on an interval $I \supset f(U)$ which is differentiable near $\bar{t} := f(\bar{x})$ with $D\varphi$ one-sided Lipschitz at $\bar{t}$. Then for the function $\varphi \circ f$ well-defined near $\bar{x}$ the equality
$$\partial_{VP}(\varphi \circ f)(\bar{x}) = \partial_{VP}(rf)(\bar{x})$$
holds, where r denotes the derivate $r := \varphi'(\bar{t})$.
If in addition $\varphi'(\bar{t}) > 0$, one has
$$\partial_{VP}(\varphi \circ f)(\bar{x}) = \varphi'(\bar{t}) \partial_{VP} f(\bar{x}).$$

PROOF. First, we notice for the derivative $D\varphi(\bar{t})$ that $D\varphi(\bar{t}) \circ f(x) = \varphi'(\bar{t}) f(x)$ for x near $\bar{x}$. Further, if the real $\varphi'(\bar{t}) > 0$, we also have $\partial_{VP}(\varphi'(\bar{t}) f)(\bar{x}) = \varphi'(\bar{t}) \partial_{VP} f(\bar{x})$. The corollary then follows from Proposition 4.144(b). □

As for the Fréchet subdifferential, concerning the case of the (Cartesian) product of finitely many spaces, we have the following:

PROPOSITION 4.146. Let $X, Y, X_1, \cdots, X_m$ be normed vector spaces, $g : Y \to \mathbb{R} \cup \{-\infty, +\infty\}$ be a function on Y, and, for each $i = 1, \cdots, m$ let $f_i : X_i \to \mathbb{R} \cup \{-\infty, +\infty\}$ be a function on X_i. Let also $G : X \to Y$ be a mapping from X into Y. Then the following properties hold.

(a) For $\varphi(u_1, \cdots, u_m) := f_1(u_1) + \cdots + f_m(u_m)$ (assumed to be) well defined on an open set U of $X_1 \times \cdots \times X_m$ and for any point $(x_1, \cdots, x_m) \in U$, one has

$$\partial_{VP}\varphi(x_1, \cdots, x_m) = \partial_{VP}f(x_1) \times \cdots \times \partial_{VP}f_m(x_m).$$

(b) For $h : X \times Y \to \mathbb{R} \cup \{-\infty, +\infty\}$ with $h(x', y') = g(y')$ for all $(x', y') \in X \times Y$, one has at any $x \in X$ and $y \in Y$,

$$\partial_{VP}h(x, y) = \{0\} \times \partial_{VP}g(y).$$

(c) Assume that G is differentiable near x with DG one-sided Lipschitz at $x \in X$ (which holds in particular if f is $\mathcal{C}^{1,1}$ near x). Then, one has

$$(x^*, -y^*) \in N^{VP}(\operatorname{gph} G; (x, G(x))) \Leftrightarrow x^* = y^* \circ DG(x) =: DG(x)^*(y^*).$$

PROOF. The assertion (a) follows easily from the local characterization of variational proximal subgradient in Proposition 4.137(a) and the assertion (b) is a particular case of (a).

To verify (c), we first note that, for any $(x^*, -y^*) \in N^{VP}(\operatorname{gph} G; (x, G(x)))$, we have $x^* = DG(x)^*(y^*)$ by (c) in Proposition 4.19 and by the inclusion of the variational proximal normal cone into the Fréchet one. Conversely, suppose that $x^* = DG(x)^*(y^*)$ and let a real $\gamma \geq 0$ and an open convex neighborhood U of x such that $\|DG(u) - DG(x)\| \leq \gamma \|u - x\|$ for all $u \in U$. Then for all $(x', y') \in (U \times X) \cap \operatorname{gph} G$, it follows that

$$\langle x^*, x' - x \rangle - \langle y^*, y' - G(x) \rangle = \langle -y^*, G(x') - G(x) - DG(x)(x' - x) \rangle$$

$$= \left\langle -y^*, \int_0^1 (DG(x + t(x' - x)) - DG(x))(x' - x) dt \right\rangle$$

$$\leq \gamma \|y^*\|_* \|x' - x\|^2 \int_0^1 t \, dt = (1/2)\gamma \|y^*\|_* \|x' - x\|^2,$$

so the desired inclusion $(x^*, -y^*) \in N^{VP}(\operatorname{gph} G; (x, G(x)))$ holds true. □

As Fréchet normals, variational proximal normals to epigraphs have particular properties. The assertion (e) below reveals that $N^{VP}(\operatorname{epi} f; (x, f(x)))$ (through which $\partial_{VP}f(x)$ is defined above) is included in the closure of the cone generated by $\partial_{VP}f(x) \times \{-1\}$ whenever $\partial_{VP}f(x) \neq \emptyset$.

PROPOSITION 4.147. Let $(X, \|\cdot\|)$ be a normed space, $f : X \to \mathbb{R} \cup \{-\infty, +\infty\}$ be an extended real-valued function, and $x \in X$ be a point where f is finite. The following hold:

(a) For any $(x^*, -r^*) \in N^{VP}(\operatorname{epi} f; (x, f(x)))$ one has $r^* \geq 0$.
(b) If $(x^*, -r^*) \in N^{VP}(\operatorname{epi} f; (x, r))$ with $r^* > 0$, then $r = f(x)$.
(c) If $(x^*, 0) \in N^{VP}(\operatorname{epi} f; (x, r))$, then one also has $(x^*, 0) \in N^{VP}(\operatorname{epi} f; (x, f(x)))$.
(d) If $(x^*, -r^*) \in N^{VP}(\operatorname{epi} f; (x, r))$ with $r > f(x)$, then $r^* = 0$.
(e) If $\partial_{VP}f(x) \neq \emptyset$, then

$$\mathbb{R}_+(\partial_{VP}f(x) \times \{-1\}) \subset N^{VP}(\operatorname{epi} f; (x, f(x))) \subset \operatorname{cl}_{\|\cdot\|_*}(\mathbb{R}_+(\partial_{VP}f(x) \times \{-1\})),$$

where $\operatorname{cl}_{\|\cdot\|_*}$ denotes the closure in $X^* \times \mathbb{R}$ with respect to the dual norm.

PROOF. The assertions (a), (b) and (d) follow from the corresponding ones in Proposition 4.21 and from the inclusion $N^{VP}(\cdot;\cdot) \subset N^F(\cdot;\cdot)$.
(c) Fix $(x^*, 0) \in N^{VP}(\text{epi } f; (x, r))$ and observe first that $r \geq f(x)$. Choose $\sigma \geq 0$ and $\delta > 0$ such that
$$\langle x^*, x' - x \rangle \leq \sigma(\|x' - x\|^2 + |r' - r|^2)$$
for all $(x', r') \in \text{epi } f$ with $\|x' - x\| \leq \delta$ and $|r' - r| \leq \delta$. Fix any $(x', r') \in \text{epi } f$ with $\|x' - x\| \leq \delta$ and $|r' - f(x)| \leq \delta$. For $r'' := r' + r - f(x)$ we have $(x', r'') \in \text{epi } f$ (because $r - f(x) \geq 0$) and $|r'' - r| = |r' - f(x)| \leq \delta$, and hence we obtain
$$\langle x^*, x' - x \rangle \leq \sigma(\|x' - x\|^2 + |r'' - r|^2) = \sigma(\|x' - x\|^2 + |r' - f(x)|^2).$$
This guarantees the inclusion $(x^*, 0) \in N^{VP}(\text{epi } f; (x, f(x)))$.
(e) Concerning (e), the first inclusion follows from the definition of $\partial_{VP} f(x)$. Assume now $\partial_{VP} f(x) \neq \emptyset$ and fix $u^* \in \partial_{VP} f(x)$. Take any $(x^*, -r^*)$ in $N(f; x) := N^{VP}(\text{epi } f; (x, f(x)))$ and note that $r^* \geq 0$ by (a). For $0 < s < 1$ and $x_s^* = (1-s)x^* + su^*$, we have $y_s^* := (x_s^*, -r^*(1-s) - s)$ in $N(f; x)$ according to the convexity of $N(f; x)$ (see Proposition 4.136(b)), thus $y_s^* \in \mathbb{R}_+\big(\partial_{VP} f(x) \times \{-1\}\big)$ since $r^*(1-s) + s > 0$. The proof is then finished since $y_s^* \to (x^*, -r^*)$ as $s \downarrow 0$. □

REMARK 4.148. The above assertions (c) and (d) ensure in particular that the inclusion
$$N^{VP}(\text{epi }; (x, r)) \subset N^{VP}(\text{epi } f; (x, f(x)))$$
always holds true for every $r \in \mathbb{R}$. □

Proceeding as in Proposition 4.23 we see that VP-normals to epigraphs enjoy the following properties.

PROPOSITION 4.149. Let U be an open set of the normed space X and $f : U \to \mathbb{R} \cup \{-\infty, +\infty\}$ be a function which is finite near a point $x \in U$ and Lipschitz continuous there with $\gamma \geq 0$ as a Lipschitz constant.
(a) One has
$$(x^*, -r) \in N^{VP}(\text{epi } f; (x, f(x))) \Rightarrow \|x^*\| \leq r\gamma,$$
hence in particular
$$(x^*, 0) \in N^{VP}(\text{epi } f; (x, f(x))) \Rightarrow x^* = 0.$$
(b) If $\partial_{VP} f(x) \neq \emptyset$, the VP-normal cone to the epigraph of f at x can be described by the VP-subdifferential as follows
$$N^{VP}(\text{epi } f; (x, f(x))) = \mathbb{R}_+\big(\partial_{VP} f(x) \times \{-1\}\big).$$

In some situations it can arise that one needs to utilize a global inequality on the space X characterizing variational proximal subgradients.

PROPOSITION 4.150. Let $(X, \|\cdot\|)$ be a normed space and $f : X \to \mathbb{R} \cup \{+\infty\}$ be a function which is finite at $\bar{x} \in X$ and minorized by a quadratic function of the norm, that is, there are some reals $\alpha > 0$, $\beta > 0$ and $\gamma \in \mathbb{R}$ such that
$$f(x') \geq -\alpha \|x'\|^2 - \beta \|x'\| + \gamma \quad \text{for all } x' \in X.$$
(a) Let reals $r > 0$, $s > 0$, $\delta > 0$, $\sigma > 0$ and a subset
$$W \subset \big(B_X(\bar{x}, 2\delta) \cap \{f(\cdot) \leq f(\bar{x}) + s\}\big) \times \mu B_{X^*}$$
such that for every $(x, x^*) \in W$
$$\langle x^*, x' - x \rangle \leq f(x') - f(x) + \sigma \|x' - x\|^2 \quad \text{for all } x' \in B(\bar{x}, 2\delta).$$

Then there exists a real $\sigma_0 > \sigma$ such that for every $(x, x^*) \in W$ with $x \in B(\bar{x}, \delta)$ one has

$$\langle x^*, x' - x \rangle < f(x') - f(x) + \sigma_0 \|x' - x\|^2 \quad \text{for all } x' \in X \setminus \{x\}.$$

(b) In particular, for any $\bar{x}^* \in \partial_{VP} f(\bar{x})$ there exists a real $\sigma_0 > 0$ such that

$$\langle \bar{x}^*, x' - \bar{x} \rangle < f(x') - f(\bar{x}) + \sigma_0 \|x' - \bar{x}\|^2 \quad \text{for all } x' \in X \setminus \{\bar{x}\}.$$

(c) Given a subset S of X, $\bar{x} \in S$ and $\bar{x}^* \in N^{VP}(S; \bar{x})$, there exists some real $\sigma_0 > 0$ such that

$$\langle \bar{x}^*, x' - \bar{x} \rangle < \sigma_0 \|x' - \bar{x}\|^2 \quad \text{for all } x' \in S \setminus \{\bar{x}\}.$$

PROOF. The assertion (c) follows from (b), and (b) is a consequence of (a) with $W = \{(\bar{x}, \bar{x}^*)\}$. Let us prove (a). Fix any $(x, x^*) \in W$ with $\|x - \bar{x}\| < \delta$. Putting $\lambda := \|\bar{x}\| + \delta$ we note by the quadratic minorization that for all $x' \in X$

$$f(x') \geq -\alpha \|x' - x\|^2 - (2\alpha \|x\| + \beta)\|x' - x\| + (\gamma - \alpha \|x\|^2 - \beta \|x\|)$$
$$\geq -\alpha \|x' - x\|^2 - (2\alpha\lambda + \beta)\|x' - x\| + (\gamma - \alpha\lambda^2 - \beta\lambda).$$

On the other hand, for any real $\rho > 0$ we have for all $x' \in X$

$$f(x) + \langle x^*, x' - x \rangle - \rho\|x' - x\|^2 \leq f(\bar{x}) + s + \mu\|x' - x\| - \rho\|x' - x\|^2.$$

Choose a real $\rho_0 > \sigma$ (independent of (x, x^*)) such that for all reals $t \geq \delta$

$$-\rho_0 t^2 + \mu t + f(\bar{x}) + s \leq -\alpha t^2 - (2\alpha\lambda + \beta)t + (\gamma - \alpha\lambda^2 - \beta\lambda).$$

From the above inequalities and from the conditions assumed for the set W it results that for all $x' \in X \setminus \{x\}$ (by considering the two cases $\|x' - x\| \geq \delta$ or not)

$$f(x') \geq f(x) + \langle x^*, x' - x \rangle - \rho_0 \|x' - x\|^2,$$

and hence taking a real $\sigma_0 > \rho_0$ we derive that

$$f(x') > f(x) + \langle x^*, x' - x \rangle - \sigma_0 \|x' - x\|^2.$$

$\square$

4.6.2. Proximal subgradients of distance functions. As for the Fréchet subdifferential, the following inclusion holds for the variational proximal subdifferential of an infimal convolution at points where the latter is attained.

PROPOSITION 4.151. *Let $f_i : X \to \mathbb{R} \cup \{+\infty\}$, with $i = 1, \cdots, n$, be n functions on the normed space X such that the infimal convolution $f_1 \square \cdots \square f_n$ is finite at $\bar{x} \in X$ and attained at $(\bar{x}_1, \cdots, \bar{x}_n)$ for $\bar{x}$, that is, $\bar{x}_1 + \cdots + \bar{x}_n = \bar{x}$ and*

$$(f_1 \square \cdots \square f_n)(\bar{x}) = f_1(\bar{x}_1) + \cdots + f_n(\bar{x}_n).$$

Then, one has the inclusion

$$\partial_{VP}(f_1 \square \cdots \square f_n)(\bar{x}) \subset \bigcap_{i=1}^{n} \partial_{VP} f_i(\bar{x}_i).$$

PROOF. The proof is similar to that for the Fréchet subdifferential. The finiteness of the inf-convolution at $\bar{x}$ implies that each f_i is finite at $\bar{x}_i$. Put $f := f_1 \square \cdots \square f_n$ and fix $x^* \in \partial_{VP} f(\bar{x})$. There exist $\sigma \geq 0$ and $\delta > 0$ such that for any $x \in B(\bar{x}, \delta)$

$$\langle x^*, x - \bar{x} \rangle \leq f(x) - f(\bar{x}) + \sigma\|x - \bar{x}\|^2.$$

Fix any $k \in \{1, \cdots, n\}$ and any $u \in B(\overline{x}_k, \delta)$. Since
$$\overline{x} + u - \overline{x}_k \in B(\overline{x}, \delta) \quad \text{and} \quad \overline{x} + u - \overline{x}_k = \overline{u}_1 + \cdots + \overline{u}_n$$
for $\overline{u}_i = \overline{x}_i$ if $i \neq k$ and $\overline{u}_k = u$, taking $x = \overline{x} + u - \overline{x}_k$ in the above inequality yields
$$\begin{aligned} \langle x^*, u - \overline{x}_k \rangle &= \langle x^*, (\overline{x} + u - \overline{x}_k) - \overline{x} \rangle \\ &\leq f(\overline{x} + u - \overline{x}_k) - f_1(\overline{x}_1) - \cdots - f_n(\overline{x}_n) + \sigma \|u - \overline{x}_k\|^2 \\ &\leq f_1(\overline{u}_1) + \cdots + f_n(\overline{u}_n) - f_1(\overline{x}_1) - \cdots - f_n(\overline{x}_n) + \sigma \|u - \overline{x}_k\|^2 \\ &= f_k(u) - f_k(\overline{x}_k) + \sigma \|u - \overline{x}_k\|^2. \end{aligned}$$

This ensures $x^* \in \partial_{VP} f_k(\overline{x}_k)$ and the desired inclusion. □

Concerning the variational proximal subdifferential of the opposite of the inf-convolution, the arguments for Proposition 4.28 justify also the following.

PROPOSITION 4.152. *Let $f_1, \cdots, f_n : X \to \mathbb{R} \cup \{+\infty\}$ be functions on the normed space X and let $\overline{x}$ be a point at which $f_1 \square \cdots \square f_n$ is finite. Assume that there exist $\overline{x}_i \in X$, with $i = 1, \cdots, n$, such that $(f_1 \square \cdots \square f_n)(\overline{x}) = f_1(\overline{x}_1) + \cdots + f_n(\overline{x}_n)$. Then one has, for $i = 1, \cdots, n$,*
$$\partial_{VP}(-f_i)(\overline{x}_i) \subset \partial_{VP}\big(-(f_1 \square \cdots \square f_n)\big)(\overline{x}).$$

Now let us study the relationship between the variational proximal normal cone and the variational proximal subdifferential of the distance function d_S at points in the set S.

PROPOSITION 4.153. *Let S be a set of the normed space $(X, \|\cdot\|)$ and $x \in S$. Then*
$$\partial_{VP} d_S(x) = N^{VP}(S; x) \cap \mathbb{B}_{X^*} \quad \text{and} \quad N^{VP}(S; x) = \mathbb{R}_+ \partial_{VP} d_S(x).$$

PROOF. We begin by showing the inclusion $\partial_{VP} d_S(x) \subset N^{VP}(S; x)$. Let $x^* \in \partial_{VP} d_S(x)$. We first observe that $\|x^*\| \leq 1$ since $\partial_{VP} d_V(x) \subset \partial_F d_C(x)$. Further, by the analytic description of variational proximal subgradient in Proposition 4.137(a), there exist a real $\sigma \geq 0$ and a neighborhood U of x such that
$$\langle x^*, x' - x \rangle \leq d_S(x') - d_S(x) + \sigma \|x' - x\|^2 \quad \text{for all } x' \in U,$$
and hence in particular for all $x' \in S \cap U$ we have $\langle x^*, x' - x \rangle \leq \sigma \|x' - x\|^2$. This ensures that $x^* \in N^{VP}(S; x)$, so $\partial_{VP} d_S(x) \subset N^{VP}(S; x) \cap \mathbb{B}_{X^*}$. (Note that the inclusion can also be justified through the above proposition as in Proposition 4.30).

Let us show the reverse inclusion. Fix any $x^* \in N^{VP}(S; x)$ with $\|x^*\| \leq 1$. Take two real numbers $\sigma \geq 0$ and $\delta > 0$ such that

(4.73) $$\langle x^*, x' - x \rangle \leq \sigma \|x' - x\|^2 \quad \text{for all } x' \in S \cap B(x, \delta).$$

Fix $\gamma := \min\{1, \delta/3\}$ and fix also any $z \in B(x, \gamma)$. Choose $y_z \in S$ such that

(4.74) $$\|y_z - z\| \leq d_S(z) + \|z - x\|^2.$$

Then $y_z \in B(x, \delta)$, because according to the latter inequality we have

(4.75) $$\|y_z - x\| \leq \|y_z - z\| + \|z - x\| \leq 3\|z - x\| < 3\gamma \leq \delta,$$

and hence this last inequality along with (4.73) yield
$$\begin{aligned}\langle x^*, z - x\rangle &= \langle x^*, y_z - x\rangle + \langle x^*, z - y_z\rangle \\ &\leq \sigma\|y_z - x\|^2 + \|y_z - z\| \\ &\leq 9\sigma\|z - x\|^2 + d_S(z) + \|z - x\|^2,\end{aligned}$$
the last inequality being due to the second inequality in (4.75) and to (4.74). So, for all $z \in B(x, \gamma)$
$$\langle x^*, z - x\rangle \leq d_S(z) - d_S(x) + (9\sigma + 1)\|z - x\|^2,$$
which says that $x^* \in \partial_{VP} d_S(x)$, and finishes the proof of the first equality of the proposition. The second equality follows directly from the first. $\square$

Let us consider now the case of points outside the closure of the set S.

PROPOSITION 4.154. *Let S be a subset of a normed space $(X, \|\cdot\|)$ and $x \notin \mathrm{cl}\,S$. For $r := d_S(x) > 0$ and for the closed r-enlargement $\mathrm{Enl}_r\, S$ of S one has*
$$\partial_{VP} d_S(x) = N^{VP}(\mathrm{Enl}_r\, S; x) \cap \{x^* \in X^* : \|x^*\| = 1\}.$$

PROOF. Put $E_r(S) := \mathrm{Enl}_r\, S$. We begin by showing the inclusion of the first member into the second one. Fix $x^* \in \partial_{VP} d_S(x)$. There exists two real numbers $\sigma \geq 0$ and $\delta > 0$ such that
$$\langle x^*, x' - x\rangle \leq d_S(x') - d_S(x) + \sigma\|x' - x\|^2 \quad \text{for all } x' \in B(x, \delta).$$
Since $d_S(x') - d_S(x) \leq 0$ for all $x' \in E_r(S)$, we obtain
$$\langle x^*, x' - x\rangle \leq \sigma\|x' - x\|^2 \quad \text{for all } x' \in E_r(S) \cap B(x, \delta),$$
which means that $x^* \in N^{VP}(E_r(S); x)$. Further, since $\partial_{VP} d_S(x) \subset \partial_F d_S(x)$, Proposition 4.31 ensures that $\|x^*\| = 1$.

We now turn to the proof of the reverse inclusion. Fix any $x^* \in N^{VP}(E_r(S); x)$ with $\|x^*\| = 1$ and fix also any $\eta \in]0, 1[$. By Proposition 4.153, we have $x^* \in \partial_{VP} \mathrm{dist}\,(\cdot, E_r(S))(x)$, and hence there exist $\sigma_1 \geq 0$ and $\delta_1 > 0$ such that
$$\langle x^*, x' - x\rangle \leq \mathrm{dist}\,(x', E_r(S)) - \mathrm{dist}\,(x, E_r(S)) + \sigma_1\|x' - x\|^2 \quad \text{for all } x' \in B[x, \delta_1].$$
By Lemma 3.260 one obtains
$$(4.76) \quad \langle x^*, x' - x\rangle \leq d_S(x') - d_S(x) + \sigma_1\|x' - x\|^2 \text{ for all } x' \in B[x, \delta_1] \setminus E_r(S).$$
On the other hand, since $x^* \in N^{VP}(E_r(S); x)$, there exist a nonnegative $\sigma \geq \sigma_1$ and $\delta_2 > 0$ such that
$$(4.77) \quad \langle x^*, x' - x\rangle \leq \sigma\|x' - x\|^2 \quad \text{for all } x' \in E_r(S) \cap B[x, \delta_2].$$
Fix now $\delta := \min\{\delta_1, \delta_2/2\}$ and $x' \in E_r(S) \cap B[x, \delta]$, and put $t_{x'} := d_S(x) - d_S(x') \geq 0$. Take any real $\lambda > 0$. Since $\|x^*\| = 1$, we can choose $u \in X$, with $\|u\| = 1$, such that $\langle x^*, u\rangle > 1 - \lambda$. Then $x' + t_{x'}u \in E_r(S) \cap B[x, \delta_2]$, because as in the proof of Proposition 4.31
$$d_S(x' + t_{x'}u) \leq d_S(x') + t_{x'} = d_S(x) = r,$$
and
$$\|x' + t_{x'}u - x\| \leq \|x' - x\| + t_{x'} \leq 2\|x' - x\| \leq 2\delta \leq \delta_2.$$

This and the inequality $\|x' + t_{x'}u - x\| \leq 2\|x' - x\|$ yield through (4.77)

$$\begin{aligned}\langle x^*, x' - x\rangle &= \langle x^*, x' + t_{x'}u - x\rangle - \langle x^*, t_{x'}u\rangle \\ &\leq \sigma\|x' + t_{x'}u - x\|^2 - t_{x'}(1-\lambda) \\ &\leq 4\sigma\|x' - x\|^2 + (1-\lambda)\left(d_S(x') - d_S(x)\right).\end{aligned}$$

Taking the limit as $\lambda \downarrow 0$, we deduce that

$$\langle x^*, x' - x\rangle \leq 4\sigma\|x' - x\|^2 + d_S(x') - d_S(x).$$

Combining the latter inequality with (4.76) we obtain

$$\langle x^*, x' - x\rangle \leq d_S(x') - d_S(x) + 4\sigma\|x' - x\|^2 \quad \text{for all } x' \in B[x, \delta].$$

Consequently, $x^* \in \partial_{VP} d_S(x)$ and the proof is complete. $\square$

4.6.3. Proximal fuzzy calculus and proximal representation of the limiting subdifferential. In the context of Hilbert space, fuzzy calculus rules are valid for the proximal subdifferential. Let us start with the fuzzy sum rule.

THEOREM 4.155 (proximal fuzzy sum rule). *Let H be a Hilbert space and let $f_1, \cdots, f_m : H \to \mathbb{R} \cup \{-\infty, +\infty\}$ be functions which are finite at $\bar{x} \in X$. The following hold:*

(a) *If f_1 is lower semicontinuous near $\bar{x}$ and $f_2, \cdots, f_m$ are uniformly continuous near $\bar{x}$, then for any $v \in \partial_P(f_1 + \cdots + f_m)(\bar{x})$ and any real $\varepsilon > 0$ there exist $x_1, \cdots, x_m \in B(\bar{x}, \varepsilon)$ (all depending on v and ε) with $|f_1(x) - f_1(\bar{x})| < \varepsilon$ such that*

$$v \in \partial_P f_1(x_1) + \cdots + \partial_P f_m(x_m) + \varepsilon \mathbb{B}_H.$$

(b) *If all the functions $f_1, \cdots, f_m$ are lower semicontinuous near $\bar{x}$, then for any $v \in \partial_P(f_1 + \cdots + f_m)(\bar{x})$, any real $\varepsilon > 0$, and any finite-dimensional subspace L of H, there exist $x_1, \cdots, x_m \in B(\bar{x}, \varepsilon)$ (all depending on v, ε, and L) with $|f_i(x_i) - f_i(\bar{x})| < \varepsilon$, such that*

$$v \in \partial_P f_1(x_1) + \cdots + \partial_P f_m(x_m) + \varepsilon \mathbb{B}_H + L^\perp.$$

PROOF. We may consider $m = 2$, $f_1 =: f$ and $f_2 =: g$. Fix $v \in \partial_P(f+g)(\bar{x})$ and choose $\delta > 0$ such that on the set $C := B[\bar{x}, \delta]$ the functions f, g are bounded from below. The method below is an adaptation of the one in Proposition 4.46.
(a) Shrinking δ if necessary, we may suppose that f is lower semicontinuous on C and g is uniformly continuous on C, and that for some real $\sigma \geq 0$ we have

$$0 \leq \left(f(x) - f(\bar{x}) - \langle v, x - \bar{x}\rangle\right) + \left(g(x) - g(\bar{x}) + \frac{\sigma}{2}\|x - \bar{x}\|^2\right) =: \varphi(x) \quad \text{for all } x \in C.$$

Endow $H \times H$ with the natural inner product and the associated norm, and fix a real sequence $(r_k)_k$ with $r_k > 0$ and $r_k \to +\infty$ as $k \to +\infty$. Put

$$h(x, y) := \left(f(x) - f(\bar{x}) - \langle v, x - \bar{x}\rangle\right) + \left(g(y) - g(\bar{x}) + \frac{\sigma}{2}\|y - \bar{x}\|^2\right).$$

According to the uniform continuity property of $g + \frac{\sigma}{2}\|\cdot - \bar{x}\|^2$ on C, Lemma 4.45 says that

$$\eta'_k := \inf_{(x,y) \in C \times C}\left(h(x, y) + \frac{r_k^2}{2}\|x - y\|^2\right) \to \inf_C \varphi = 0 \quad \text{as } k \to \infty,$$

and obviously, for $h_0(x, y) := h(x, y) + \frac{r_k^2}{2}\|x - y\|^2$ we have $\eta'_k \leq h_0(\bar{x}, \bar{x}) = 0$. Putting $\eta_k := \sqrt{-\eta'_k + (1/k)}$ and writing the inequality $0 < \eta'_k + (-\eta'_k + (1/k))$ as

$h_0(\bar{x}, \bar{x}) < \inf_{(x,y) \in C \times C} h_0(x, y) + \eta_k^2$, the Borwein-Preiss principle in Hilbert space (see Theorem 4.42) applied with $\varepsilon = \eta_k^2$ and $\lambda = \eta_k \sqrt{2}$ yields (b'_k, b''_k) and (q'_k, q''_k) in $C \times C$ such that

$$\|(b'_k, b''_k) - (q'_k, q''_k)\| < \eta_k \sqrt{2}, \ \|(q'_k, q''_k) - (\bar{x}, \bar{x})\| < \eta_k \sqrt{2},$$

and for

$$h_1(x, y) := h(x, y) + \frac{r_k^2}{2}\|x - y\|^2 + \frac{1}{2}\|x - q'_k\|^2 + \frac{1}{2}\|y - q''_k\|^2$$

the function h_1 attains its minimum on $C \times C$ at (b'_k, b''_k). For k large enough $b'_k, b''_k \in B(\bar{x}, \delta)$, hence b'_k and b''_k are local minimum of $h_1(\cdot, b''_k)$ and $h_1(b'_k, \cdot)$ respectively. So, by (b) of Proposition 4.142 we obtain

$$v - r_k(b'_k - b''_k) - (b'_k - q'_k) \in \partial_P f(b'_k)$$
$$r_k(b'_k - b''_k) - \sigma(b''_k - \bar{x}) - (b''_k - q''_k) \in \partial_P g(b''_k).$$

Adding both inclusions gives

$$v \in \partial_P f(b'_k) + \partial_P g(b''_k) + \sigma_k \mathbb{B}_H$$

for some positive $\sigma_k \to 0$.

To conclude the assertion (a) let us show the proximity between $f(b'_k)$ and $f(\bar{x})$. We know that the point (b'_k, b''_k) above given by the Borwein-Preiss principle satisfies

$$h_0(b'_k, b''_k) + \frac{1}{4}\|(b'_k, b''_k) - (\bar{x}, \bar{x})\|^2 \leq h_0(\bar{x}, \bar{x}) = 0,$$

hence in particular

$$f(b'_k) + g(b''_k) \leq f(\bar{x}) + g(\bar{x}) + \langle v, b'_k - \bar{x}\rangle - \frac{\sigma}{2}\|b''_k - \bar{x}\|^2.$$

The latter inequality combined with the lower semicontinuity property of f and g at $\bar{x}$ easily entails that $f(b'_k) \to f(\bar{x})$ and $g(b''_k) \to g(\bar{x})$. The assertion (a) is then established.

(b) Concerning the assertion (b) it suffices to follow the proof of the assertion (b) in Proposition 4.46 with appropriate adaptations as above. □

In the context of Hilbert space, three first crucial properties concerning the density of $\text{Dom}\,\partial_P f$ and the approximation of Fréchet subgradients by proximal subgradients can be derived from the above theorem.

THEOREM 4.156 (L-subdifferential as outer limit of P-subdifferential). *Let H be a Hilbert space, S be a subset of H, and $f : U \to \mathbb{R} \cup \{+\infty\}$ be a proper lower semicontinuous function on an open set U of H.*
(a) The set $\text{Dom}\,\partial_P f$ is graphically dense in $\text{dom}\,f$, that is, for any $x \in \text{dom}\,f$ there is a sequence $(x_n)_n$ in $\text{Dom}\,\partial_P f$ such that $(x_n, f(x_n)) \to (x, f(x))$.
(b) For any $x \in U$ and $v \in \partial_F f(x)$, and for any $\varepsilon > 0$ there are $x' \in U$ and $v' \in H$ such that

$$v' \in \partial_P f(x') \ \text{ with } \|v' - v\| < \varepsilon, \ \|x' - x\| < \varepsilon, \ \text{and } |f(x') - f(x)| < \varepsilon.$$

(c) The limiting subdifferential $\partial_L f(x)$ coincides with the norm-to-weak sequential outer limit of the proximal subdifferential $\partial_P f(u)$ as $(u, f(u)) \to (x, f(x))$, that is,

$$\partial_L f(x) = {}^{\text{seq}}\!\operatorname*{Lim\,sup}_{u \to_f x} \partial_P f(u).$$

(d) If S is closed near $x \in S$, then $N^L(S;x) = {}^{\text{seq}} \operatorname*{Lim\,sup}_{S \ni u \to x} N^P(S;u)$.

(e) For any $x \in \operatorname{dom} f$, the equality $\partial^\infty_L f(x) = {}^{\text{seq}} \operatorname*{Lim\,sup}_{t \downarrow 0, u \to_f x} t \partial_P f(u)$ holds.

PROOF. Without loss of generality we may suppose that the function f is defined and lower semicontinuous on H and that the set S is closed.

(a) Fix any $x \in \operatorname{dom} f$ and observe as in Corollary 4.53 that x is a minimum point of $f + \Psi_{\{x\}}$ hence $0 \in \partial_P(f + \Psi_{\{x\}})(x)$. For each $\varepsilon > 0$, the assertion (a) in the theorem above gives some $x' \in \operatorname{Dom} \partial_P f$ with $\|x' - x\| < \varepsilon$ and $|f(x') - f(x)| < \varepsilon$.

(b) Take $v \in \partial_F f(x)$ and $\varepsilon > 0$. It is easily seen from the definition of Fréchet subgradient that the point x is a local minimum point of $f - \langle v, \cdot \rangle + \varepsilon \| \cdot - x \|$, hence $0 \in \partial_P(f - \langle v, \cdot \rangle + \varepsilon \| \cdot - x \|)(x)$ or equivalently $v \in \partial_P(f + \varepsilon \| \cdot - x \|)(x)$ according to Proposition 4.142(b). The assertion (a) of Theorem 4.155 then yields (via the inclusion $\partial_P(\| \cdot - x \|)(H) \subset \mathbb{B}_H$) some $(x', v') \in \operatorname{gph} \partial_P f$ such that $\|x' - x\| < \varepsilon$, $|f(x') - f(x)| < \varepsilon$, $\|v' - v\| < 2\varepsilon$.

(c) The assertion (c) follows directly from (b) and the definition of Mordukhovich limiting subdifferential.

(d)-(e) The property (d) is a direct consequence of (c) with $f := \Psi_S$ and (e) follows from (b) and from Theorem 4.130. □

The following corollary directly follows from (c) in the above theorem and from Theorem 4.120(c).

COROLLARY 4.157. Let H be a Hilbert space and $f : U \to \mathbb{R}$ be a function on an open set U of H which is Lipschitz near $\bar{x} \in U$. Then one has

$$\partial_C f(\bar{x}) = \overline{\operatorname{co}}^* \left({}^{\text{seq}} \operatorname*{Lim\,sup}_{x \to \bar{x}} \partial_P f(x) \right).$$

Any L-subgradient of distance function d_S at $x \in S$ is the weak limit of P-subgradients of d_S at points inside S.

PROPOSITION 4.158. Let H be a Hilbert space and S be a subset of H which closed near $x \in S$. Then one has

$$\partial_L d_S(\bar{x}) = {}^{\text{seq}} \operatorname*{Lim\,sup}_{S \ni u \to x} \partial_P d_S(u).$$

PROOF. We may suppose that S is closed. We know (see Theorem 4.85(a)) that

(4.78) $$\partial_L d_S(x) = {}^{\text{seq}} \operatorname*{Lim\,sup}_{S \ni v \to x} \partial_F d_S(v).$$

Further, Theorem 4.156(b) ensures that for any $\eta > 0$ and $v^* \in N^F(S;v)$ there exists some $w \in S \cap B(v, \eta)$ and $w^* \in N^P(S;w)$ with $\|w^* - v^*\| < \eta$. The equality $\partial_P d_S(v) = N^P(S;v) \cap \mathbb{B}$ for $v \in S$ in Proposition 4.153 and the analogous one for the Fréchet normal cone easily yield that for any $\eta > 0$, $v \in S$, and $v^* \in \partial_F d_S(v)$, there exists some $w \in S \cap B(v, \eta)$ and $w^* \in \partial_P d_S(w)$ with $\|w^* - v^*\| < \eta$. Combining this with the equality (4.78) we obtain that

$$\partial_L d_S(x) = {}^{\text{seq}} \operatorname*{Lim\,sup}_{S \ni v \to u} \partial_P d_S(v),$$

which is the desired equality. □

4.6. PROXIMAL NORMAL CONE AND SUBDIFFERENTIAL

A proximal fuzzy chain rule for Lipschitz continuous or lower semicontinuous functions can be derived from the theorem on fuzzy sum rule like for the Fréchet subdifferential. Let us establish first a lemma similar to Lemma 4.56.

LEMMA 4.159. *Let X and Y be normed spaces, $G : X \to Y$ be a mapping which is Lipschitz continuous near $\bar{x}$. Then*
$$(x^*, -y^*) \in N^{VP}(\operatorname{gph} G; (\bar{x}, G(\bar{x}))) \Leftrightarrow x^* \in \partial_{VP}(y^* \circ G)(\bar{x}).$$

PROOF. Consider reals $\delta_0 > 0$ and $\gamma_G \geq 0$ such that G is γ_G-Lipschitz continuous on $B(\bar{x}, \delta_0)$. Fix $(x^*, -y^*) \in N^{VP}(\operatorname{gph} G; (\bar{x}, G(\bar{x})))$. There exist some real $\sigma \geq 0$ and some real $0 < \delta' < \delta_0$ such that for all $(x, y) \in B((\bar{x}, G(\bar{x})), \delta')$ with $y = G(x)$ we have
$$\langle x^*, x - \bar{x}\rangle - \langle y^*, y - G(\bar{x})\rangle \leq \sigma(\|x - \bar{x}\|^2 + \|y - G(\bar{x})\|^2).$$
Choose a positive $\delta < \delta'$ such that $\|G(x) - G(\bar{x})\| < \delta'$ for all $x \in B(\bar{x}, \delta)$. Then for any $x \in B(\bar{x}, \delta)$ we obtain
$$\langle x^*, x - \bar{x}\rangle \leq y^* \circ G(x) - y^* \circ G(\bar{x}) + \sigma(1 + \gamma_G^2)\|x - \bar{x}\|^2,$$
which translates that $x^* \in \partial_{VP}(y^* \circ G)(\bar{x})$, hence the implication $\Rightarrow$ of the lemma. The converse implication is similar and easier. □

THEOREM 4.160 (proximal fuzzy chain rule). *Let H and H' be Hilbert spaces, G be a mapping from H into H', and $g : H' \to \mathbb{R} \cup \{+\infty\}$ be a function which is finite at $G(\bar{x})$, where $\bar{x} \in H$.*

(a) *If g is uniformly continuous near $G(\bar{x})$ and the graph of G is closed near $(\bar{x}, G(\bar{x}))$, then for any $v \in \partial_P(g \circ G)(\bar{x})$ and $\varepsilon > 0$ there exist $x \in B(\bar{x}, \varepsilon)$, $y \in B(G(\bar{x}), \varepsilon)$, $(u, -w) \in N^P(\operatorname{gph} G; (x, G(x)))$ such that*
$$v \in u + \varepsilon \mathbb{B}_H \quad \text{and} \quad w \in \partial_P g(y) + \varepsilon \mathbb{B}_{H'}.$$

(b) *If g is uniformly continuous near $G(\bar{x})$ and G is Lipschitz continuous near $\bar{x}$, then for any $v \in \partial_P(g \circ G)(\bar{x})$ and $\varepsilon > 0$ there exist $x \in B(\bar{x}, \varepsilon)$, $y \in B(G(\bar{x}), \varepsilon)$, and $w \in \partial_P g(y)$ such that*
$$v \in \partial_P(\langle w, G(\cdot)\rangle)(x) + \varepsilon \mathbb{B}_H.$$

(c) *If g is lower semicontinuous near $G(\bar{x})$ and the graph of G is closed near $(\bar{x}, G(\bar{x}))$, then for any $v \in \partial_P(g \circ G)(\bar{x})$, any $\varepsilon > 0$, any finite-dimensional subspace L of H, and any finite-dimensional subspace Q of H', there exist $x \in B(\bar{x}, \varepsilon)$, $y \in B(G(\bar{x}), \varepsilon)$ with $|g(y) - g(G(\bar{x}))| < \varepsilon$, $(u, -w) \in N^P(\operatorname{gph} G; (x, G(x)))$ such that*
$$v \in u + \varepsilon \mathbb{B}_H + L^\perp \quad \text{and} \quad w \in \partial_P g(y) + \varepsilon \mathbb{B}_{H'} + Q^\perp.$$

(d) *If g is lower semicontinuous near $G(\bar{x})$ and G is Lipschitz continuous near $\bar{x}$, then for any $v \in \partial_P(g \circ G)(\bar{x})$, any $\varepsilon > 0$, any finite-dimensional subspace L of H, and any finite-dimensional subspace Q of H', there exist $x \in B(\bar{x}, \varepsilon)$, $y \in B(G(\bar{x}), \varepsilon)$ with $|g(y) - g(G(\bar{x}))| < \varepsilon$, and $w \in \partial_P g(y) + \varepsilon \mathbb{B}_{H'} + Q^\perp$ such that*
$$v \in \partial_P(\langle w, G(\cdot)\rangle)(x) + \varepsilon \mathbb{B}_H + L^\perp.$$

PROOF. The proof is similar to that of Theorem 4.57.
(a) Consider any $v \in \partial_P(g \circ G)(\bar{x})$ and put $f(x, y) := g(y)$ and $h(x, y) := \Psi_{\operatorname{gph} G}(x, y)$ for all $(x, y) \in H \times H'$. It is not difficult to see that $(v, 0) \in \partial_P(f + h)(\bar{x}, G(\bar{x}))$.

Since f is uniformly continuous near $(\bar{x}, G(\bar{x}))$ and $\mathrm{gph}\, G$ is closed near $(\bar{x}, G(\bar{x}))$, the proximal fuzzy sum rule theorem above ensues that, for any $\varepsilon > 0$, there exist

$$(x', y) \in B((\bar{x}, G(\bar{x})), \varepsilon),\ (x, G(x)) \in B((\bar{x}, G(\bar{x})), \varepsilon),\ (0, w') \in \partial_P f(x', y),$$

and

$$(u, -w) \in N^P\bigl(\mathrm{gph}\, G; (x, G(x))\bigr)$$

such that

(4.79) $(v, 0) \in (0, w') + (u, -w) + \varepsilon \mathbb{B}$,   hence $v \in u + \varepsilon \mathbb{B}$ and $w \in w' + \varepsilon \mathbb{B}$.

Consequently, $w \in \partial_P g(y) + \varepsilon \mathbb{B}_{H'}$ since $w' \in \partial_P g(y)$. All the properties in (a) are reached.

(b) Since G is Lipschitz continuous on some ball $B(\bar{x}, \delta)$, shrinking $\varepsilon < \delta$ one observes first by the above lemma that $(u, -w) \in N^P\bigl(\mathrm{gph}\, G; (x, G(x))\bigr)$ means $u \in \partial_P(\langle w, G(\cdot) \rangle)(x)$. So $v \in \partial_P(\langle w, G(\cdot) \rangle)(x) + \varepsilon \mathbb{B}_H$ with $w \in \partial_P g(y) + \varepsilon \mathbb{B}_{H'}$, hence $w = w_1 + \varepsilon b$ for some $w_1 \in \partial_P g(y)$ and $b \in \mathbb{B}_{H'}$. Denoting by γ a Lipschitz constant of G over $B(\bar{x}, \varepsilon)$, the fuzzy sum rule applied to $\langle w, G(\cdot) \rangle = \langle w_1, G(\cdot) \rangle + \varepsilon \langle b, G(\cdot) \rangle$ along with the γ-Lipschitz property of G ensures that, for some $x' \in B(\bar{x}, \varepsilon)$, one has $v \in \partial_P(\langle w_1, G(\cdot) \rangle)(x') + (2 + \gamma) \varepsilon \mathbb{B}_H$.

(c) It is enough to repeat the proof of (a) with $\varepsilon \mathbb{B}_H + L^\perp$ and $\varepsilon \mathbb{B}_{H'} + Q^\perp$ in place of $\varepsilon \mathbb{B}_H$ and $\varepsilon \mathbb{B}_{H'}$ respectively in (4.79). Since g here is not continuous at $G(\bar{x})$, we must add in the application of the fuzzy sum rule that $|f(x', y) - f(\bar{x}, G(\bar{x}))| < \varepsilon$, that is, $|g(y) - g(G(\bar{x}))| < \varepsilon$.

(d) The arguments are similar to those for (b). $\square$

With the above fuzzy chain rules at hand, following the arguments in the proof of Theorem 4.84(b) we obtain:

THEOREM 4.161 (mean value inequality for Lipschitz functions with proximal subdifferentials). *Let U be an open convex set of a Hilbert space H and $f : U \to \mathbb{R}$ be a locally Lipschitz continuous real-valued function. Let $a, b \in U$ with $a \neq b$. Then, for any $\varepsilon > 0$ there are $c \in U$ with $d(c, [a, b]) < \varepsilon$ and $c^* \in \partial_P f(c)$ such that*

$$f(b) - f(a) \leq \langle c^*, b - a \rangle + \varepsilon.$$

4.7. Further results

4.7.1. F-normal cone to graphs of multimappings. Given a multimapping $M : X \rightrightarrows Y$ between two normed spaces (resp. Banach spaces) X and Y and a pair $(\bar{x}, \bar{y})$ in its graph, we know by Proposition 4.30 (resp. Theorem 4.98(b)) that the Fréchet (resp. limiting) normal cone to $\mathrm{gph}\, M$ at $(\bar{x}, \bar{y})$ coincides with the cone generated by the Fréchet (resp. limiting) subdifferential at $(\bar{x}, \bar{y})$ of the distance function $d(\cdot, \mathrm{gph}\, M)$. As will be seen next, this normal cone also coincides with the cone generated by the corresponding subdifferential of the other natural distance function to images Δ_M defined by

(4.80) $\Delta_M(x, y) = d(y, M(x))$ for all $(x, y) \in X \times Y$.

Here we keep the usual convention $d(y, M(x)) = +\infty$ when $M(x) = \emptyset$.

PROPOSITION 4.162. *Let $M : X \rightrightarrows Y$ be a multimapping between two normed spaces X, Y and let $(\bar{x}, \bar{y}) \in \mathrm{gph}\, M$. Then one has*

$$\partial_F \Delta_M(\bar{x}, \bar{y}) = N^F\bigl(\mathrm{gph}\, M; (\bar{x}, \bar{y})\bigr) \cap (X^* \times \mathbb{B}_{Y^*})$$

and
$$N^F\bigl(\mathrm{gph}\, M;(\overline{x},\overline{y})\bigr) = \mathbb{R}_+ \partial_F \Delta_M(\overline{x},\overline{y}).$$

PROOF. It suffices to prove the first equality since it clearly entails the second. Fix any (x^*, y^*) in $\partial_F \Delta_M(\overline{x},\overline{y})$ and take any $\varepsilon > 0$. There exists some $\eta > 0$ such that for all $x \in B[\overline{x}, \eta]$ and all $y \in B[\overline{y}, \eta]$ one has
$$\langle x^*, x - \overline{x}\rangle + \langle y^*, y - \overline{y}\rangle \le d(y, M(x)) - d(\overline{y}, M(\overline{x})) + \varepsilon(\|x - \overline{x}\| + \|y - \overline{y}\|).$$
Taking $x = \overline{x}$ we see that for all $y \in B[\overline{y}, \eta]$
$$\langle y^*, y - \overline{y}\rangle \le d(y, M(\overline{x})) - d(\overline{y}, M(\overline{x})) + \varepsilon\|y - \overline{y}\|.$$
This says that $y^* \in \partial_F d_{M(\overline{x})}(\overline{y})$, and hence $\|y^*\| \le 1$ according to Proposition 4.30. Further, taking $x \in B[\overline{x}, \eta]$ and $y \in B[\overline{y}, \eta]$ with (x, y) in gph M we obtain
$$\langle x^*, x - \overline{x}\rangle + \langle y^*, y - \overline{y}\rangle \le \varepsilon(\|x - \overline{x}\| + \|y - \overline{y}\|).$$
The inclusion of $\partial_F \Delta_M(\overline{x}, \overline{y})$ into $X^* \times \mathbb{B}_{Y^*}$ is then justified.

Let us show the converse inclusion. Fix any (x^*, y^*) in $N^F\bigl(\mathrm{gph}\, M; (\overline{x}, \overline{y})\bigr)$ with $\|y^*\| < 1$ and take any $\varepsilon > 0$ with $\varepsilon + \|y^*\| < 1$. There exists some real $\eta' > 0$ such that for all $(u, v) \in \mathrm{gph}\, M$ with $\|u - \overline{x}\| \le 3\eta'$ and $\|v - \overline{y}\| \le 3\eta'$ one has
$$(4.81) \qquad \langle x^*, u - \overline{x}\rangle + \langle y^*, v - \overline{y}\rangle \le \varepsilon(\|u - \overline{x}\| + \|v - \overline{y}\|).$$
Fix a real $\eta > 0$ such that $(2 + \|x^*\|)\eta < \eta'$. Take any $x \in B[\overline{x}, \eta]$ and any $y \in B[\overline{y}, \eta]$. If $d(y, M(x)) > \eta'$, then
$$\langle x^*, x - \overline{x}\rangle + \langle y^*, y - \overline{y}\rangle \le (\|x^*\| + 1)\eta < d(y, M(x)),$$
which entails
$$(4.82)\ \langle x^*, x - \overline{x}\rangle + \langle y^*, y - \overline{y}\rangle \le d(y, M(x)) - d(\overline{y}, M(\overline{x})) + \varepsilon(\|x - \overline{x}\| + \|y - \overline{y}\|).$$
Suppose now that $d(y, M(x)) \le \eta'$. Choose $y' \in M(x)$ such that
$$(4.83) \qquad \|y' - y\| \le d(y, M(x)) + \varepsilon(\|x - \overline{x}\| + \|y - \overline{y}\|),$$
and note that
$$\|y' - \overline{y}\| \le \|y - \overline{y}\| + \|y - y'\| \le \eta' + 3\eta < 3\eta'.$$
This and (4.81) ensure that
$$\langle x^*, x - \overline{x}\rangle + \langle y^*, y - \overline{y}\rangle = \langle x^*, x - \overline{x}\rangle + \langle y^*, y' - \overline{y}\rangle + \langle y^*, y - y'\rangle$$
$$\le \varepsilon(\|x - \overline{x}\| + \|y' - \overline{y}\|) + \|y^*\|\|y - y'\|$$
$$\le \varepsilon(\|x - \overline{x}\| + \|y - \overline{y}\|) + (\varepsilon + \|y^*\|)\|y - y'\|.$$
Since $(\varepsilon + \|y^*\|)\|y - y'\| \le \|y - y'\|$, using (4.83) we deduce that
$$\langle x^*, x - \overline{x}\rangle + \langle y^*, y - \overline{y}\rangle \le 2\varepsilon(\|x - \overline{x}\| + \|y - \overline{y}\|) + d(y, M(x)) - d(\overline{y}, M(\overline{x})).$$
This inequality and (4.82) imply that $(x^*, y^*) \in \partial_F \Delta_M(\overline{x}, \overline{y})$. Since the Fréchet subdifferential of any function is strongly closed according to Proposition 4.8(b), it results that $(x^*, y^*) \in \partial_F \Delta_M(\overline{x}, \overline{y})$ for every $(x^*, y^*) \in N^F\bigl(\mathrm{gph}\, M; (\overline{x}, \overline{y})\bigr)$ with $\|y^*\| \le 1$. This finishes the proof of the proposition. $\square$

PROPOSITION 4.163. *Let $M : X \rightrightarrows Y$ be a multimapping with closed graph between two Banach spaces X, Y and let $(\overline{x}, \overline{y}) \in \mathrm{gph}\, M$. Then one has*
$$N^L\bigl(\mathrm{gph}\, M; (\overline{x}, \overline{y})\bigr) = \mathbb{R}_+ \partial_L \Delta_M(\overline{x}, \overline{y}).$$

EXERCISE 4.164. Prove the proposition by adapting arguments of the proof of Theorem 4.98. □

4.7.2. L-subdifferential versus C-subdifferential in the real line. This section is mainly devoted to some comparisons between L-subdifferential and C-subdifferential of functions of one real variable. We will begin by showing that these two subdifferentials of a Lipschitz function of one real variable coincide except at countable points. We need first a lemma.

LEMMA 4.165. Let $f : I \to \mathbb{R}$ be a locally Lipschitz function on an open interval I of $\mathbb{R}$. If $0 \in \partial_C f(\overline{x}) \setminus \partial_L f(\overline{x})$, then there exists a real $\delta > 0$ with $]\overline{x} - \delta, \overline{x} + \delta[\subset I$ and such that f is increasing on $]\overline{x} - \delta, \overline{x}[$ and f decreasing on $]\overline{x}, \overline{x} + \delta[$.

PROOF. Without loss of generality we may suppose that $I = \mathbb{R}$. Assume by contradiction that for every real $\eta > 0$ there exist x, x' in $]\overline{x}, \overline{x} + \eta[$ such that
$$\overline{x} < x < x' < \overline{x} + \eta \quad \text{and} \quad f(x') \geq f(x).$$
Since $0 \notin \partial_L f(\overline{x})$ and $0 \in \text{co}\,\partial_L f(\overline{x}) = \partial_C f(\overline{x})$, there is some real $\xi < 0$ with $\xi \in \partial_L f(\overline{x})$. Therefore, for the open neighborhood $]\overline{x} - \eta, x[$ of $\overline{x}$ there exists some $z \in]\overline{x} - \eta, x[$ and some $\zeta \in \partial_F f(z)$ with $\zeta < 0$. From the very definition of $\partial_F f(z)$ there is some $z' \in]\overline{x} - \eta, z[$ such that $f(z') > f(z)$, hence we obtain
$$\overline{x} - \eta < z' < z < x < x' < \overline{x} + \eta, \quad f(z') > f(z) \quad \text{and} \quad f(x') \geq f(x).$$
It ensues that f attains its minimum on $[z', x']$ at some point $u \in]z', x'[$, hence $0 \in \partial_F f(u)$ and $\|u - \overline{x}\| < \eta$. This being true for any real $\eta > 0$, it follows that $0 \in \partial_L f(\overline{x})$, which is in contradiction with the assumption $0 \notin \partial_L f(\overline{x})$. This ensures that there is some real $\eta > 0$ such that f is increasing on $]\overline{x} - \eta, \overline{x}[$.

Similarly, there exists some real $\eta' > 0$ such that f is decreasing on $]\overline{x}, \overline{x} + \eta'[$. □

THEOREM 4.166 (G. Katriel). Let I be an open interval of $\mathbb{R}$ and $f : I \to \mathbb{R}$ be a locally Lipschitz function. The following hold.
(a) For any $\alpha \in \mathbb{R}$ the set $\{x \in I : \alpha \in \partial_C f(x) \setminus \partial_L f(x)\}$ is a set whose each point (if any) is isolated in it.
(b) The set $\{x \in I : \partial_C f(x) \neq \partial_L f(x)\}$ is countable.

PROOF. (a) Fix any $\alpha \in \mathbb{R}$ such that the set
$$P_\alpha := \{x \in I : \alpha \in \partial_C f(x) \setminus \partial_L f(x)\}$$
is nonempty. Fix any $\overline{x} \in P_\alpha$. Putting $g(x) := f(x) - \alpha x$ for all $x \in I$, we see that $0 \in \partial_C g(\overline{x}) \setminus \partial_L g(\overline{x})$. Take a real $\delta > 0$ as given by Lemma 4.165 above. This lemma assures us that there is no point $x \neq \overline{x}$ in $P_\alpha \cap]\overline{x} - \delta, \overline{x} + \delta[$, hence the point $\overline{x}$ is isolated in P_α as desired.
(b) To prove the assertion (b) it suffices to show that
$$D := \{x \in I : \partial_C f(x) \neq \partial_L f(x)\} = \bigcup_{q \in \mathbb{Q}} P_q.$$
Take any $\overline{x} \notin \bigcup_{q \in \mathbb{Q}} P_q$, so $\partial_C f(\overline{x}) \cap \mathbb{Q} = \partial_L f(\overline{x}) \cap \mathbb{Q}$. If $\partial_C f(\overline{x})$ is a singleton, then by the inclusion $\emptyset \neq \partial_L f(\overline{x}) \subset \partial_C f(\overline{x})$ we trivially have $\partial_C f(\overline{x}) = \partial_L f(\overline{x})$, so $\overline{x} \notin D$. Otherwise, $\partial_C f(\overline{x})$ is a nonempty closed interval which is not a singleton, hence
$$\partial_C f(\overline{x})) = \text{cl}\,(\partial_C f(\overline{x}) \cap \mathbb{Q}) = \text{cl}\,(\partial_L f(\overline{x}) \cap \mathbb{Q}) \subset \partial_L f(\overline{x})$$

since the set $\partial_L f(\overline{x})$ is closed. It ensues again that $\overline{x} \notin D$. This and the obvious inclusion $\bigcup_{q \in \mathbb{Q}} P_q \subset D$ entail that $D = \bigcup_{q \in \mathbb{Q}} P_q$. Since each set P_q is countable by (a), it results that D is countable. The proof is finished. □

The above theorem tells us that the set of points where the C-subdifferential and the L-subdifferential of a locally Lipschitz function on an open interval of $\mathbb{R}$ differ is very small. Further, for any locally Lipschitz function f near x which is L-subdifferentially regular at x, it is clear that $\partial_L f(x) = \partial_C f(x)$. The following exercise shows that this coincidence also holds true for one-variable locally Lipschitz functions which are differentiable near x.

EXERCISE 4.167. Let I be an open interval of $\mathbb{R}$ and $f : I \to \mathbb{R}$ be a locally Lipschitz function which is differentiable on I. Then for every $x \in I$ one has $\partial_L f(x) = \partial_C f(x)$. (Hint: Fix any $x \in I$. Argue that $\partial_C f(x) = [\liminf_{u \to x} f'(u), \limsup_{u \to x} f'(u)]$ and, for any λ in the interior of the latter interval (if this interior is nonempty), argue that there are two sequences $(u_n)_n$ and $(v_n)_n$ in I tending to x such that $f'(u_n) < \lambda < f'(v_n)$ for all $n \in \mathbb{N}$; by Darboux intermediate theorem for f' consider, for each $n \in \mathbb{N}$ some w_n between u_n and v_n satisfying $f'(w_n) = \lambda$ and conclude that $\lambda \in \partial_L f(x)$). □

4.8. Comments

The limiting normal cone $N^L(S; x)$ of a closed S in $\mathbb{R}^N$ endowed with its usual Euclidean norm was first defined for $x \in S$ (under the name of *cone conjugated to S at x* and with notation $K_S(x)$) by B. S. Mordukhovich in relation (4.2) in his 1976 paper [720] in the form

$$N^L(S; x) = \bigcap_{\eta > 0} \mathrm{cl} \left(\bigcup_{u \in B(x, \eta)} \mathbb{R}_+ (u - \mathrm{Proj}_S(u)) \right).$$

This can be rewritten as $N^L(S; x) = \limsup_{u \to x} \mathbb{R}_+(u - \mathrm{Proj}_S(u))$ according to Proposition 1.7 in the present manuscript. For a lower semicontinuous function $f : \mathbb{R}^N \to \mathbb{R} \cup \{+\infty\}$ Mordukhovich then defined in relation (4.3) in [720] the limiting subdifferential of f at a point $x \in \mathbb{R}^N$ where f is finite as

$$\partial_L f(x) = \{\xi \in \mathbb{R}^N : (\xi, -1) \in N^L(\mathrm{epi}\, f; (x, f(x)))\};$$

in [720] this set was denoted by $Df(x)$ and called the *D-derivative of f at x*. The terminology of *limiting normal cone* and *limiting subdifferential* is nowadays very usual. The limiting normal cone was introduced as above by Mordukhovich in [720] as a by-product of his method employed there in the proofs of Theorems 3.1 and 5.1 to obtain optimality conditions for constrained optimal control problems by suitable metric approximations of such problems through smooth unconstrained problems. Via those concepts, necessary optimality conditions proved in [720] are very robust and sharp. The method also allowed Mordukhovich to derive in his 1980 paper [721], for finite-dimensional minimization problems with inequality/equality constraints and set-constraint, necessary optimality conditions in terms of limiting subdifferential and normal cone. By the method of approximations, Mordukhovich [722] also established in finite dimensions the formula

$$N^L(S_1 \cap S_2; x) \subset N^L(S_1; x) + N^L(S_2; x)$$

provided that for the closed sets S_1, S_2 the only pair of vectors $\xi_i \in N^L(S_i; x)$ with $\xi_1 + \xi_2 = 0$ is $\xi_1 = \xi_2 = 0$. It is worth mentioning that the formula can also be derived from other methods by A. D. Ioffe [**514**] and R. T. Rockafellar [**860**]. Various finite-dimensional calculus rules can be found in the books of Mordukhovich [**723**] and of Rockafellar and Wets [**865**], and the main rules are established in the present chapter and others will be proved in Subsection 7.7.4 of Chapter 7. The development of the theory of limiting normal/subgradient in infinite dimensions began with the paper [**635**] by A. Y. Kruger and B. S. Mordukhovich and the paper [**620**] by A.Y. Kruger mainly in the framework of Banach spaces admitting an equivalent Fréchet differentiable renorm off zero. The next step of the theory in Asplund spaces, as treated in the book, was mainly and deeply carried out in 1996 by B. S. Mordukhovich and Y. Shao [**730**] on the basis of the 1989 deep paper by M. Fabian [**391**] on approximate sum rule and density property for Fréchet subgradients (F-subgradients, for short) of lower semicontinuous functions in Asplund spaces. The name of *Fréchet subgradient* is due to the fact that the inequality in its definition (see Definition 4.1) corresponds to one of the two inequalities resulting from the decoupling of the fundamental type of variation utilized by M. Fréchet to define the differentiability in infinite dimensions.

A limiting subgradient of a lower semicontinuous function was defined in the three aforementioned papers [**635, 620, 730**] as a weak-star graphical limit, in a certain sense, of F-subgradients (or approximate F-subgradients). The F-subgradient, as in Definition 4.1, was probably first considered for optimization theory by M. S. Bazaraa, J. J. Goode and M. Z. Nashed in their 1974 paper [**73**]. The full use of F-subgradients and approximate F-subgradients in the generation of a robust limiting subdifferential in Banach spaces started with the 1980 paper by A. Y. Kruger and B. S. Mordukhovich [**635**] and the 1981 paper by A. Y. Kruger [**620**]. A certain approximate F-subgradient, called local ε-support, was employed in 1976 by I. Ekeland and G. Lebourg [**383**, Definition 1.1] to establish the generic Fréchet differentiability of continuous convex functions on Banach spaces admitting Fréchet differentiable bump function. Notice that F-subgradients were also utilized independently under notation $D^-u(x)$ by M. Crandall and P.-L. Lions in their 1983 article [**298**, p. 16] (submitted in 1981) related to viscosity subsolutions of Hamilton-Jacobi partial differential equations (see also [**297**]). For the proof of the equivalence (a) $\Leftrightarrow$ (b) between F-subgradient and F-normal to epigraph in Proposition 4.4 we followed the arguments in the proof of Theorem 1.86 in Mordukhovich's book [**725**]. Example 4.6(c) is inspired by a similar one in [**725**, p. 92]. The equality in (e) of Proposition 4.21 was apparently first established in J.-P. Penot's book [**795**, Corollary 4.17]. Lemma 4.24 and its proof are taken from R. T. Rockafellar and R. J-B. Wets [**865**, Proof of Theorem 6.11]. Proposition 4.25 in its form is also taken from [**865**, Proposition 8.5].

The infimal convolution (inf-convolution, for short) was probably first considered in W. Fenchel's 1953 lectures [**405**]. Given two convex functions $f_i : C_i \to \mathbb{R}$ with epi f_i closed in $\mathbb{R}^N \times \mathbb{R}$ and their conjugates $\varphi_i : \Gamma_i \to \mathbb{R}$ in Fenchel's sense (see Section 3.24 of comments), $i = 1, 2$, the investigation in [**405**, p. 96] of the expression of the conjugate of $f_1 + f_2 : C_1 \cap C_2 \to \mathbb{R}$ by means of φ_i led Fenchel to obtain there under mild conditions that the conjugate of $f_1 + f_2$ is given by the function defined in $\Gamma_1 + \Gamma_2$ by

$$\xi \mapsto \inf_{\xi_i \in \Gamma_i, \xi_1 + \xi_2 = \xi} \left(\varphi_1(\xi_1) + \varphi_2(\xi_2) \right) \quad \text{for all } \xi \in \Gamma_1 + \Gamma_2.$$

This function clearly coincides on $\Gamma_1 + \Gamma_2$ with the infimal convolution of the extensions of φ_1 and φ_2 to $\mathbb{R}^N$ by $+\infty$ outside Γ_1 and Γ_2 respectively. The infimal convolution was extensively developed by J. J. Moreau [**735, 737, 745**]. Other large developments under convexity can be found in J. J. Moreau's lectures [**744**] and R. T. Rockafellar's book [**852**]. In accordance with the problem, coming from optimal allocation of resources in economics (see [**596, 597**]), which consists in determining the maximum of $F(x) = F_1(x_1) + \cdots + F_n(x_n)$ over the reals $x_i \in \mathbb{R}_+$ satisfying $x_1 + \cdots + x_n = x$, R. Bellman and W. Karush also considered in their 1961 paper [**83**] the function on $\mathbb{R}_+^N$

$$x \mapsto \max_{0 \leq y \leq x} \big(f_1(y) + f_2(x-y)\big) = \max_{x_i \geq 0, x_1 + x_2 = x} \big(f_1(x_1) + f_2(x_2)\big),$$

where f_1, f_2 are two real-valued functions on $\mathbb{R}_+^N$ and $\mathbb{R}_+ := [0, +\infty[$. Bellman and Karush used in [**84, 85**] the name *maximum convolution* and the notation $f_1 \oplus f_2$. This function $f_1 \oplus f_2$ clearly coincides with the opposite of the inf-convolution $g_1 \square g_2$ where g_i is the extension to $\mathbb{R}^N$ of the opposite of f_i defined by $g_i(x) = +\infty$ for x outside $\mathbb{R}_+^N$. The results in Propositions 4.27 on F-subdifferential and F-differentiability of infimal convolution are quite direct and the arguments are the same with convex data. Other results related to subgradients of infimal convolution functions can be found in [**257, 260, 264, 266, 267, 534, 540, 541, 542, 543, 590, 759, 760, 770, 974**]. Proposition 4.30 was first obtained by A. Y. Kruger [**619**]. The formula in Proposition 4.31 for the distance function at outset point is also due to Kruger; the proof presented here follows, as in [**172**], Kruger's arguments in [**619**] with certain complements. The property (a) in Proposition 4.32, for any minimizing sequence of the distance function and any of its F-subgradient, is taken from J. M. Borwein and S. Fitzpatrick [**128**, Proposition 1.4] as well as the main ideas of the proof. This Proposition 4.32 is an extension of the similar result in Proposition 3.258 for convex sets. Diverse other results on F-subgradients and approximate F-subgradients can be found in a series of Kruger's papers [**624, 625, 626, 627, 628**].

The first result related to the separable reduction of Fréchet subdifferentiability has been established in 1985 by M. Fabian and N. V. Zhivkov [**400**] with the proof that a Banach space X is Asplund if and only if, for any proper lower semicontinuous function $f : X \to \mathbb{R} \cup \{+\infty\}$ and for any real $\varepsilon > 0$, the set $\text{Dom}\, \partial_{F,\varepsilon} f$ is dense in $\text{dom}\, f$. Four years later, Fabian [**391**] proved, with the help of the Borwein-Preiss variational principle, the result for $\varepsilon = 0$, that is, for Fréchet subdifferentials. For the proof it has been shown in [**391**] that, given a function $f : X \to \mathbb{R} \cup \{+\infty\}$ and an open set U with $U \cap \text{dom}\, f \neq \emptyset$, there is a closed separable subspace $X_0 \subset X$ such that the Fréchet subdifferential of f is nonempty at some point in $U \cap X_0$ whenever it is so for the restriction $f_{|X_0}$ of f to X_0. From the latter result Theorem 4.52 was derived in [**391**]. Another work related to the separable reduction of Fréchet subdifferentiability but with a certain additional property has been done by M. Fabian and B. S. Mordukhovich [**397**] (see also [**396**]). A less long proof of the separable reduction for the sum of a convex function and a nonconvex one has been given by J.-P. Penot [**794**]. Soon after, A. D. Ioffe established Theorem 4.36 without the space Z and the linear mapping T. The general case of Theorem 4.36 as stated in the text has been given and established by M. Fabian and A. D. Ioffe [**394**] in a very elegant way. Section 4.2 reproduces entirely the statements and proofs of Fabian and Ioffe [**394**]. We took from [**394**] all the results and the proofs in: Lemmas 4.33, 4.34, 4.35 and Theorems 4.36, 4.37. We also refer to M. Fabian and A. D. Ioffe [**395**] for other results related to the separable reduction for Fréchet subdifferentials. Previously to Fabian's article [**391**] fuzzy sum rules for the Fréchet (resp. Hadamard) ε-subdifferential, with $\varepsilon > 0$, was established by Ioffe in his 1983 article [**511**] for Banach spaces with Fréchet (resp. Gâteaux) differentiable renorms. The arguments in [**511**] required the restriction $\varepsilon > 0$, since those arguments employed the Ekeland variational principle instead of the Borwein-Preiss variational principle which was not yet established at that moment.

The Borwein-Preiss variational principle in Theorem 4.42 is in the line of the Ekeland variational principle, but the perturbation there is smooth whenever the norm of the underlying space is itself smooth. This is of great interest in many situations as in the case mentioned above and in diverse other cases in this chapter. Theorem 4.39 corresponds to an adaptation to metric spaces by P. D. Loewen and X. Wang [**674**, Theorem 2.2] of the principle established in [**140**, Theorem 2.6] by J. M. Borwein and D. Preiss in the context of normed spaces. The proof of Theorem 4.42 in the book reproduces that of Loewen and Wang [**674**, Theorem 2.2], which itself is largely inspired by the proof of Borwein and Preiss [**140**, Theorem 2.6]. The statement of Theorem 4.41 is exactly that of the principle of Borwein and Preiss [**140**, Theorem 2.6] while Theorem 4.42 corresponds to Remark 2.7(c) in the same paper [**140**] by Borwein and Preiss. Lemma 4.43 was noticed in [**140**, Proposition 2.4].

For a sum of functions $f_1, \cdots, f_n$ the *decoupling penalization form*

$$(4.84) \qquad \sum_{i=1}^{n} f_i(u_i) + \sum_{1 \leq i < j \leq n}^{n} r\|u_i - u_j\|^p$$

in Lemma 4.45 was introduced in diverse papers for different purposes. In his 1984 paper [**514**, p. 400] (submitted in 1981), given $0 \in \partial_F(f+g)(\overline{x})$ for two lower semicontinuous functions f, g on the usual Euclidean space $\mathbb{R}^N$ and fixing $\delta > 0$, A. D. Ioffe used the function in u, v, x in $\mathbb{R}^N$ given by

$$p_r(u, v, x) := f(u) + g(v) + (r/2)(\|u - x\|^2 + \|v - x\|^2) + \delta\|x - \overline{x}\|$$

and taking a suitable $0 < \varepsilon < \delta$ and a minimizer (u_r, v_r, x_r) of p_r over $B_r \times B_r \times B_r$, with $B_r := B[\overline{x}, r]$, Ioffe showed that $u_r \to_f \overline{x}$, $v_r \to_g \overline{x}$ and that $(\partial_F f(u_r) + \partial_F g(v_r)) \cap \delta \mathbb{B}_{\mathbb{R}^N} \neq \emptyset$ via the smoothness of $\|\cdot\|^2$, obtaining in this way a fuzzy sum rule. In fact, the Hadamard subdifferential is used in [**514**] but it coincides with the F-subdifferential in finite dimensions as shown in Proposition 5.5(b) in Chapter 5. The penalization form (4.84) was also used in the context of Hamilton-Jacobi equations by M. Crandall, H. Ishii and P.-L. Lions in Lemma 3.1 and Proposition 3.7 of their 1992 paper [**297**].

The name of Asplund space in Definition 4.50 is traditionally used in the literature because of the fundamental 1968 paper [**28**] by E. Asplund. In page 31 of that paper [**28**] Asplund defined a Banach space X to be a *strong differentiability space* provided that every real-valued continuous convex function on an open convex set U in X is Fréchet differentiable on a dense G_δ set in U. Then Asplund proved in Theorem 1 therein that, if X is a Banach space admitting an equivalent norm whose dual norm in X^* is locally uniformly convex, then X is a strong differentiability space. Given an open convex set U of any normed space and a continuous convex function $f: U \to \mathbb{R}$, Asplund also showed in Lemma 6 in the same paper

[**28**] that whenever the set Q of Fréchet differentiability of f is dense in U, then Q is a dense G_δ set in U. In the infinite dimensional setting, previously to [**28**], for a continuous convex function f on a Banach space X, it was essentially shown by S. Mazur [**700**, p. 76-77] in 1933 that f is Gâteaux differentiable on a dense G_δ set of X whenever X is separable, and J. Lindenstrauss [**664**] proved in 1963 the Fréchet differentiability of f on a dense (hence dense G_δ) set of X whenever X is a reflexive Banach space. The reflexive Banach space was also assumed in [**664**] to admit an equivalent Fréchet differentiable norm, which was much later known to be always true. The density result of Lindenstrauss in [**664**] appeared as a byproduct of his general study for which Banach spaces X, Y the norm attaining operators from X into Y are dense in the space of all continuous linear operators from X into Y, which corresponds to a study of a vector version of the famous dual Bishop-Phelps theorem where $Y = \mathbb{R}$ (see Theorem 2.225). Theorem 4.49 on the Asplund property of any Banach space admitting an equivalent Fréchet differentiable (off zero) renorm (even with Banach spaces admitting F-differentiable bump functions) is due to I. Ekeland and G. Lebourg who established the property in their 1976 paper [**383**, Theorem 2.6]. The Ekeland variational principle and the notion of ε-support linear functional (giving rise to some Fréchet ε-subgradient) are the basic tools in [**383**].

Theorem 4.52 and Corollary 4.53 are due to M. Fabian [**391**]. Example 4.55 is taken from the 1945 paper [**892**] by A. N. Singh. In fact, the function f in Example 4.55 is the opposite of the function given by Singh [**892**] which is continuous on the right on $]0, 1[$ and whose right upper Dini derivate is $+\infty$ everywhere on $]0, 1[$; a previous example of such a function was given earlier in 1934 by S. Mazurkiewicz [**702**]. (We also mention that inspired by the development in [**892**], W. Sierpinski and A. N. Singh [**889**] provided in 1949 a nondecreasing function on $]0, 1[$, continuous on the right, discontinuous on a dense set and such that the right lower Dini derivate is null everywhere on $]0, 1[$.) Theorem 4.58 is essentially contained in Proposition 1 of Asplund [**28**], which itself is (as said in [**29**, p. 37]) a careful reformulation of certain older statements by V. L. Šmulian in his 1941 paper [**896**] (see also Šmulian's series of papers on the differentiability of norms and the structure of the unit sphere: [**893, 894, 895, 896, 897**]). The case when the functions f and g in Theorem 4.58 are assumed to be convex was also considered by Asplund and Rockafellar in Theorem 1 and Corollary 5 in their paper [**30**]. The property in Corollary 4.60 that the Fréchet derivative $Df^*(u^*)$ (when it exists) belongs to X for the conjugate f^* of any lower $\|\cdot\|$-semicontinuous function f on a Banach space X was noticed in [**28**, Corollary 1]. The proof in the book of Theorem 4.58 follows that of Lemma 4.1 in the 2001 paper [**674**] of P. D. Loewen and X. Wang. The results in Theorem 4.60 are due to C. Stegall and were proved in his 1978 paper [**900**]; the proofs given here are those of Loewen and Wang for Theorems 4.2 and 4.4 in their paper [**674**].

The sequential forms of Definitions 4.64 and 4.65 of Mordukhovich subdifferential and normal in the context of Asplund spaces were employed in the 1996 paper [**730**] by B. S. Mordukhovich and Y. Shao (see also [**725**]). Proposition 4.70 is taken from Ioffe [**514**, Proposition 7]. Example 4.75 is due to S. Fitzpatrick who communicated it to Mordukhovich; as stated and developed, it is taken from [**725**, Example 1.7]. The sum rule in Theorem 4.78 in the Asplund space setting was established by Mordukhovich and Shao [**730**, Theorem 4.1] even for more general functions (see also Mordokhovich's book [**725**]); the result was previously obtained in finite dimensions by B. S. Mordukhovich (see also the proof by A. D. Ioffe [**514**] and the paper of R. T. Rockafellar [**860**] where it is implicitely contained). In reflexive spaces or Banach spaces with a F-differentiable renorm a similar sum rule was provided by Kruger and Mordukhovich [**635**] in 1980 and by Kruger [**620**] in 1981 respectively. In Mordukhovich and Shao's paper [**730, 725**] the sum rule was proved via the application of an extremality principle of local separation type. Theorem 4.80 followed the arguments by A. Jourani and L. Thibault [**582, 583**]. Proposition 4.82 is taken from Ioffe [**514**, Theorem 7], from Mordukhovich and Shao [**730**, Theorem 4.6] and from Proposition 1.113 in Mordukhovich's book [**725**]. The development of Example 4.83 follows Example 2.49 in the same book [**725**] by Mordukhovich. Theorem 4.85(b) appeared in Thibault [**913**] and the assertion (a) is taken from D. Aussel, A. Daniilidis and L. Thibault [**49**]. Lemmas 4.86, 4.87 and 4.88 follow [**49**] and [**913**]. Example 4.89 and its development are taken from G.E. Ivanov and L. Thibault [**541**].

Propositions 4.93 and 4.94 were essentially shown by Kruger [**619**]. Proposition 4.95 and its proof correspond to Theorem 1.3 in the paper [**383**] by I. Ekeland and G. Lebourg. Theorem 4.98 is similar to Theorem 4.85; see [**913**] for its assertion (b). Proposition 4.101 was mainly proved in Kruger [**619**]. Proposition 4.104 is essentially contained in Example 8.53 in the book [**865**] of R. T. Rockafellar and R. J-B. Wets. Proposition 4.105 was demonstrated by B. S. Mordukhovich and N. M. Nam [**727**, Theorem 3.6] (and this result of [**727**] was reproduced in [**725**, Theorem 1.103]). Corollary 4.116 corresponds for a very large part to Exercise 6.44 in Rockafellar and Wets [**865**].

The formula in (a) of Theorem 4.120 started with the equality

$$(4.85) \qquad N^C(S;x) = \overline{\operatorname{co}}\left(\operatorname*{Lim\,sup}_{\operatorname{Dom} P_S \ni u \to x} \mathbb{R}_+(u - P_S(u))\right)$$

proved by F. H. Clarke in his 1973 dissertation thesis [**232**] and in his 1975 paper [**233**, Proposition 3.2] (submitted in January 1974); here S is any closed set in $\mathbb{R}^N$ equipped with its usual Euclidean structure and $x \in S$. For a lower semicontinuous function $f : \mathbb{R}^N \to \mathbb{R} \cup \{+\infty\}$, R. T. Rockafellar then showed in his 1981 paper [**858**, Theorem 1] that

$$(4.86) \qquad \partial_C f(x) = \overline{\operatorname{co}}\left(\operatorname*{Lim\,sup}_{u \to_f x} \partial_P f(u) + \operatorname*{Lim\,sup}_{t \downarrow 0, u \to_f x} t \partial_P f(u)\right),$$

and the equality

$$(4.87) \qquad \operatorname*{Lim\,sup}_{t \downarrow 0, u \to_f x} t \partial_P f(u)) = \left\{\zeta \in \mathbb{R}^N : (\zeta, 0) \in \operatorname*{Lim\,sup}_{\operatorname{epi} f \ni (u,r) \to (x, f(x))} N_P(\operatorname{epi} f; (u, r))\right\}$$

can be seen from the arguments in [**858**, p. 427-428] and this is clearly said with (1.12) in [**860**, p. 667] and extensively employed there. The equality (4.85) was extended in 1986 by J. M. Borwein and H. Strowjas [**143**, Theorem 3.1] (submitted in November 1984) for reflexive spaces endowed with Kadec-Klee norms; two different simplified proofs were provided one year later by P. D. Loewen [**668**] in Hilbert space and by J. M. Borwein and J.R. Giles [**133**, Theorem 4] for any closed set S in a Banach space X equipped with a uniformly Gâteaux differentiable norm such that $\operatorname{Dom}\operatorname{Proj}_S$ is dense in X. Previously to [**143**], for a closed set S in a Banach space X admitting an equivalent F-differentiable norm and for a lower semicontinuous function $f : X \to \mathbb{R} \cup \{+\infty\}$ J. S. Treiman established in his 1983 dissertation thesis [**933**] (see also [**937**]) that

$$N^C(S;x) = \overline{\text{co}}^* \left({}^{\text{seq}} \underset{\varepsilon \downarrow 0, u \to x}{\text{Lim sup}} N^{F,\varepsilon}(S;u) \right)$$

and

$$\partial_C f(x) = \overline{\text{co}}^* \left({}^{\text{seq}} \underset{\varepsilon \downarrow 0, u \to_f x}{\text{Lim sup}} \partial_{F,\varepsilon} f(u) + {}^{\text{seq}} \underset{\varepsilon, t \downarrow 0, u \to_f x}{\text{Lim sup}} t \partial_{F,\varepsilon} f(u) \right).$$

The equality for $N^C(S;x)$ was obtained later for reflexive Banach spaces in [**143**, Corollary 3.2] with $N^F(S,u)$ in place of $N^{F,\varepsilon}(S;u)$, and in a Banach space admitting an equivalent F-differentiable norm the other equality for $\partial_C f(x)$ with $\varepsilon = 0$ can be seen as a consequence of Theorems 3 and 4 in the 1990 paper [**518**] (submitted in January 1989) by A. D. Ioffe. The proof of Theorem 4.120 follows the arguments in Mordukhovich's book [**725**] and in Modukhovich and Shao article [**730**, Theorem 8.11]. The arguments in the proofs of Proposition 4.127 and Theorem 4.130 reproduce those by A. D. Ioffe [**518**, p. 190-191]. Example 4.138 is taken from the book [**865**] of Rockafellar and Wets. The arguments for (a) in Proposition 4.150 are those in [**865**]. Proposition 4.154 was first shown by F. H. Clarke, R. J. Stern and P. R. Wolenski [**250**] in Hilbert spaces; the proof in the manuscript for normed vector spaces is an adaptation by M. Bounkhel and L. Thibault in [**172**, Theorem 4.3] of the method of A. Y. Kruger [**619**] for the Fréchet subdifferential of the distance function at outset point. Theorem 4.160 is taken from the book [**247**] by F. H. Clarke, Yu. S. Ledyaev, R. J. Stern and P. R. Wolenski.

Propositions 4.162 and 4.163 are taken from Thibault's articles [**922**, Proposition 4.1] and [**913**] respectively. Theorem 4.166 is due to G. Katriel; its proof and the one of Lemma 4.165 correspond Katriel's proof of Theorem 1 in his 1995 paper [**600**]. In fact, Katriel [**600**, Theorem 1] established the result for any lower semicontinuous function on any open interval of $\mathbb{R}$. It is worth mentioning that the situation of functions of several variables is completely different. Given any real $\varepsilon > 0$, Katriel [**600**, p. 591-593] constructed an example of a locally Lipschitz function on $\mathbb{R}^2$ such that the set of points in $\mathbb{R}^2$ where the limiting and Clarke subdifferential coincide has Lebesgue measure less that ε. The statement and development in Exercise 4.167 reproduce Theorem 1 by J. M. Borwein and X. Wang in [**147**].

CHAPTER 5

Ioffe approximate subdifferential

As we saw in Chapter 4, full calculus for the Mordukhovich limiting subdifferential generally requires the Asplund property of the underlying spaces. The present chapter develops the concept of Ioffe approximate subdifferential $\partial_A f(x)$ in the context of a general Banach space X. We will also see that $\partial_A f(x) \subset \partial_C f(x)$; further, on the one hand we will have $\partial_L f(x) \subset \partial_A f(x)$ under the Asplund property of X, and on the other hand it will be proved that $\mathrm{cl}_{w(X^*,X)} \partial_A f(x) = \mathrm{cl}_{w(X^*,X)} \partial_L f(x)$ whenever f is lower semicontinuous near x and X admits an equivalent norm which is Fréchet differentiable off zero. The Hadamard directional derivative and subgradients will play crucial roles in the development of properties of Ioffe approximate subdifferential.

5.1. Hadamard subgradient

5.1.1. General properties. Let $f : U \to \mathbb{R} \cup \{-\infty, +\infty\}$ be a function from an open set U of a normed space X which is finite at $x \in U$ and let $h \in X$. Recalling the expressions

$$\underline{d}_H^+ f(x; h) := \liminf_{v \to h, t \downarrow 0} t^{-1}\bigl(f(x+tv) - f(x)\bigr)$$

and $\quad \underline{d}_D^+ f(x; h) := \liminf_{t \downarrow 0} t^{-1}\bigl(f(x+th) - f(x)\bigr)$

of the *Hadamard* and *Dini* (right-hand lower) *directional derivatives*, we see that we always have (as already pointed out in (2.43))

$$\underline{d}_H^+ f(x; \cdot) \leq \underline{d}_D^+ f(x; \cdot), \quad \text{hence in particular} \quad \underline{d}_H^+ f(x; 0) \leq 0$$

(where $\underline{d}_H^+ f(x; 0)$ may be equal to $-\infty$).

In addition to the second above inequality, the right-hand lower Hadamard directional derivative in the direction zero allows us to characterize the lower one-sided Lipschitz property. One-sided Lipschitz mappings have been defined in (4.5). In the case of (extended) real-valued functions, the unilateral concept of lower (resp. upper) one-sided Lipschitz property is of great interest.

DEFINITION 5.1. A function $f : U \to \mathbb{R} \cup \{-\infty, +\infty\}$ from an open set U of a normed space X is said to be *lower one-sided Lipschitz* at a point $\bar{x}$ in U, where f is finite, provided that there exist some real $\gamma \geq 0$ and some neighborhood $U_0 \subset U$ of $\bar{x}$ such that

$$f(x) - f(\bar{x}) \geq -\gamma \|x - \bar{x}\| \quad \text{for all } x \in U_0.$$

PROPOSITION 5.2. Let $f : U \to \mathbb{R} \cup \{-\infty, +\infty\}$ be a function from an open set U of a normed space X and $\bar{x} \in U$ with $|f(\bar{x})| < +\infty$. The following assertions are equivalent:
(a) $\underline{d}_H^+ f(\bar{x}; 0) = 0$;
(b) $\underline{d}_H^+ f(\bar{x}; 0) \geq 0$;

(c) $\underline{d}_H^+ f(\overline{x}; 0)$ is finite;
(d) $\underline{d}_H^+ f(\overline{x}; 0) > -\infty$;
(e) $\liminf_{x \to \overline{x}} \frac{f(x) - f(\overline{x})}{\|x - \overline{x}\|} > -\infty$, that is, f is lower one-sided Lipschitz at $\overline{x}$.

In particular, anyone of conditions (a)-(e) entails that f is lower semicontinuous at $\overline{x}$.

PROOF. Since the inequality $\underline{d}_H^+ f(\overline{x}; 0) \leq 0$ is always true (as observed above), we see that (a) and (b) are equivalent, and also (c) and (d) are equivalent. On the other hand, the positive homogeneity of $\underline{d}_H^+ f(\overline{x}; \cdot)$ ensures the equivalence between (a) and (c), so the assertions (a)-(d) are pairwise equivalent.

(e) $\Rightarrow$ (b). Suppose (e) and take a real $\gamma \geq 0$ and a neighborhood $V \subset U$ of $\overline{x}$ such that $f(x) - f(\overline{x}) \geq -\gamma \|x - \overline{x}\|$ for all $x \in V$. Then for any real $t > 0$ and any $h \in X$ with $\overline{x} + th \in V$ we have

$$t^{-1}\big(f(\overline{x} + th) - f(\overline{x})\big) \geq -\gamma \|h\|, \quad \text{so} \quad \liminf_{t \downarrow 0, h \to 0} t^{-1}\big(f(\overline{x} + th) - f(\overline{x})\big) \geq 0,$$

which is the property (b).

(b) $\Rightarrow$ (e). Suppose (b) and choose a real number $r > 0$ such that, for all $t \in]0, r]$ and $h \in B[0, r]$, one has

$$t^{-1}\big(f(\overline{x} + th) - f(\overline{x})\big) \geq -1, \quad \text{that is,} \quad f(\overline{x} + th) - f(\overline{x}) \geq -t.$$

For any $x \in B(\overline{x}, r^2)$ with $x \neq \overline{x}$, taking $t := r^{-1}\|x - \overline{x}\|$ and $h := r(x - \overline{x})/\|x - \overline{x}\|$ yields $f(x) - f(\overline{x}) \geq -r^{-1}\|x - \overline{x}\|$, which gives (e).

Finally, the lower semicontinuity property of f at $\overline{x}$ is clearly implied by (e). $\square$

With the Hadamard directional derivative is associated the Hadamard subgradients and the Hadamard subdifferential as follows:

DEFINITION 5.3. Let $f : U \to \mathbb{R} \cup \{-\infty, +\infty\}$ be a function from an open set U of a normed space X and $x \in U$ with $|f(x)| < +\infty$. A continuous linear functional $x^* \in X^*$ is a *Hadamard subgradient*, or *H-subgradient*, of f at x if

$$\langle x^*, h \rangle \leq \underline{d}_H^+ f(x; h) \quad \text{for all } h \in X.$$

The set $\partial_H f(x)$ of all Hadamard subgradients of f at x is called the *Hadamard subdifferential*, or *H-subdifferential*, of f at x.

Similarly, for any real $\varepsilon \geq 0$ the *Hadamard ε-subdifferential* of f at x is defined by

$$\partial_{H,\varepsilon} f(x) := \{x^* \in X^* : \langle x^*, h \rangle \leq \underline{d}_H^+ f(x; h) + \varepsilon \|h\|, \forall h \in X\}.$$

We see that $\partial_{H,\varepsilon} f(x)$ coincides with $\partial_H f(x)$ for $\varepsilon = 0$.

If $|f(x)| = +\infty$, one puts by convention $\partial_H f(x) = \partial_{H,\varepsilon} f(x) = \emptyset$. When $\partial_H f(x) \neq \emptyset$ one says that f is *Hadamard subdifferentiable*, or *H-subdifferentiable*, at the point x.

As in the previous chapters, for $\overline{f} : X \to \mathbb{R} \cup \{-\infty, +\infty\}$ with $\overline{f}(u) = f(u)$ if u is in the open set U and $\overline{f}(u) = +\infty$ otherwise, one has $\underline{d}_H^+ f(x; \cdot) = \underline{d}_H^+ \overline{f}(x; \cdot)$ and $\partial_{H,\varepsilon} f(x) = \partial_{H,\varepsilon} \overline{f}(x)$, for $\varepsilon \geq 0$ and $x \in U$. So, we will indifferently consider either functions $f : U \to \mathbb{R} \cup \{-\infty, +\infty\}$ or $f : X \to \mathbb{R} \cup \{-\infty, +\infty\}$. Similarly, when operations, like sum $f + g$, composition $g \circ G$ etc, are considered, we will merely assume that the functions $f + g$, $g \circ G$, are well defined near x to work with their Hadamard directional derivatives and subdifferentials at x.

5.1. HADAMARD SUBGRADIENT

With $g = f + \varepsilon \|\cdot - x\|$ and $\varepsilon \geq 0$ the equality $\underline{d}_H^+ f(x; h) + \varepsilon \|h\| = \underline{d}_H^+ g(x; h)$ clearly holds true, and it guarantees that

(5.1) $$\partial_{H,\varepsilon} f(x) = \partial_H (f + \varepsilon \|\cdot - x\|)(x).$$

As a first consequence of the definition of Hadamard subdifferential, Proposition 5.2 directly entails the following:

PROPOSITION 5.4. *Let $f : U \to \mathbb{R} \cup \{-\infty, +\infty\}$ be a function on an open set in a normed space X and $x \in U$ with $|f(x)| < +\infty$. Then f is lower semicontinuous at x whenever it is Hadamard subdifferentiable at x, or more generally whenever $\partial_{H,\varepsilon} f(x) \neq \emptyset$ for some $\varepsilon \geq 0$.*

For a function $f : X \to \mathbb{R} \cup \{-\infty, +\infty\}$, the next proposition shows in particular that the Fréchet subdifferential $\partial_F f(x)$ and the Hadamard subdifferential $\partial_H f(x)$ coincide whenever $\mathrm{dom}\, f := \{u \in X : f(u) < +\infty\}$ is contained in a finite-dimensional affine subspace.

PROPOSITION 5.5. *Let $f : U \to \mathbb{R} \cup \{-\infty, +\infty\}$ be a function from an open set U of a normed space X and let $x \in X$ with $|f(x)| < +\infty$. For any real $\varepsilon \geq 0$, the following hold:*
(a) $\partial_{F,\varepsilon} f(x) \subset \partial_{H,\varepsilon} f(x).$
(b) *The inclusion is an equality whenever, for some neighborhood V of x in X, the set $V \cap \mathrm{dom}\, f$ is contained in a finite-dimensional affine subspace of X (which holds in particular whenever X is finite-dimensional).*
(c) *Let E be a finite-dimensional vector subspace of X, let $r \in \mathbb{R}$ and let*

$$(x^*, -r^*) \in \partial_{H,\varepsilon}(\Psi_{(\mathrm{epi}\, f) \cap ((x,r) + E \times \mathbb{R})})(x, r).$$

Then, one has $r = f(x)$ whenever $r^* > \varepsilon$.

PROOF. The assertion (a) is easily seen from the definitions of Fréchet and Hadamard subdifferentials.
(b) Denote by X_0 a finite-dimensional vector subspace of X such that $x + X_0$ contains $V \cap \mathrm{dom}\, f$. Suppose that there is some $x^* \in \partial_{H,\varepsilon} f(x)$ which is not in $\partial_{F,\varepsilon} f(x)$. Then there is some real $\eta > 0$ and some sequence $(x_n)_n$ in $U \cap V$ converging to x with

$$f(x_n) - f(x) - \langle x^*, x_n - x \rangle < -(\varepsilon + \eta) \|x_n - x\| \quad \text{for all } n \in \mathbb{N}.$$

Clearly, for each $n \in \mathbb{N}$, we have $x_n \in \mathrm{dom}\, f$ and $x_n \neq x$, so setting $h_n := (x_n - x)/\|x_n - x\|$, we see that $h_n \in X_0$. Consequently, a subsequence (that we do not relabel) of $(h_n)_n$ converge to some $h \in X$ with $\|h\| = 1$. Then, for each $n \in \mathbb{N}$, setting $t_n := \|x_n - x\|$ the previous inequality gives

$$t_n^{-1}(f(x + t_n h_n) - f(x)) < \langle x^*, h_n \rangle - (\varepsilon + \eta) = \langle x^*, h_n \rangle - (\varepsilon + \eta)\|h\|.$$

Since $t_n \downarrow 0$, it results that $\underline{d}_H^+ f(x; h) < \langle x^*, h \rangle - \varepsilon \|h\|$, which entails the contradiction $x^* \notin \partial_{H,\varepsilon} f(x)$.
(c) Concerning (c), extending f to X with $f(u) = +\infty$ if $u \in X \setminus U$, we first note that $(\mathrm{epi}\, f) \cap ((x, r) + E \times \mathbb{R}) = \mathrm{epi}\, g$, where $g : X \to \mathbb{R} \cup \{-\infty, +\infty\}$ is defined by $g(u) = f(u)$ if $u \in x + E$ and $g(u) = +\infty$ otherwise. This and (b) say that $(x^*, -r^*) \in \partial_{F,\varepsilon} \Psi_{\mathrm{epi}\, g}(x, r)$, so (c) follows from Proposition 4.94(b). □

From the above proposition we derive the following corollary which will be used in various places below.

COROLLARY 5.6. Let $f : X \to \mathbb{R} \cup \{-\infty, +\infty\}$ be a function on a normed space X and let $x \in X$ with $|f(x)| < +\infty$. Let E be a finite-dimensional vector subspace of X and let two reals $\varepsilon \geq 0$ and $\eta > 0$.
(a) If $x^* \in \partial_{H,\varepsilon} f(x)$, then the point x is a local minimizer on the constraint set $x + E$ of the function
$$f(\cdot) - \langle x^*, \cdot - x \rangle + (\varepsilon + \eta) \| \cdot - x \|.$$
(b) If $x^* \in \partial_{H,\varepsilon}(f + \Psi_{x+E})(x)$ and f is Lipschitz near x with $\gamma \geq 0$ as a Lipschitz constant, then x is a local unconstrained minimizer of the function
$$f(\cdot) - \langle x^*, \cdot - x \rangle + (\varepsilon + \eta) \| \cdot - x \| + (\varepsilon + \eta + \gamma + \|x^*\|) d(\cdot, x + E).$$

PROOF. (a) From the inclusion $x^* \in \partial_{H,\varepsilon} f(x)$ we note that there is some real $\delta_0 > 0$ such that $f(x') > -\infty$ for all $x' \in B(x, \delta_0)$, so the function $f + \Psi_E(\cdot - x)$ is well defined on $B(x, \delta_0)$. From the definition of $\underline{d}_H^+ f(x; \cdot)$ we see that $x^* \in \partial_{H,\varepsilon}(f + \Psi_E(\cdot - x))(x)$, so $x^* \in \partial_{F,\varepsilon}(f + \Psi_E(\cdot - x))(x)$ by (b) in the above proposition. By the definition of Fréchet ε-subdifferential, for any real $\eta > 0$ there is a positive real $\delta < \delta_0$ such that
$$f(x) - \langle x^*, x \rangle \leq f(x') - \langle x^*, x' \rangle + \Psi_E(x' - x) + (\varepsilon + \eta) \|x' - x\|, \ \forall x' \in B(x, \delta).$$
This translates (a) of the proposition.
(b) Concerning (b), we know by (a) that x is a local minimizer of $\varphi := f - \langle x^*, \cdot - x \rangle + (\varepsilon + \eta) \| \cdot - x \|$ on $x + E$. Since φ is κ-Lipschitz near x with $\kappa := \varepsilon + \eta + \gamma + \|x^*\|$, the point x is an unconstrained local minimizer of $\varphi + \kappa d(\cdot, x + E)$ (see Lemma 2.213 and Lemma 2.219). $\square$

The next proposition establishes various elementary properties of Hadamard subdifferential.

PROPOSITION 5.7. Let X be a normed space and let $\varepsilon \in [0, +\infty[$. Let $f, g : X \to \mathbb{R} \cup \{-\infty, +\infty\}$ be two functions and let $x \in X$ be a point where both functions are finite. Let also Z be another normed space and $G : Z \to X$ be a mapping which is Hadamard differentiable at a point z with $G(z) = x$.
(a) If f and g coincide near x, then $\partial_{H,\varepsilon} f(x) = \partial_{H,\varepsilon} g(x)$.
(b) If x is a local minimizer of f, then $0 \in \partial_H f(x)$.
(c) If f is locally Lipschitz continuous near x with $\gamma \geq 0$ as a Lipschitz constant, then $\partial_{H,\varepsilon} f(x) \subset (\gamma + \varepsilon) \mathbb{B}_{X^*}$.
(d) For any real $r > 0$ the equality $\partial_{H,\varepsilon}(rf)(x) = r \partial_{H, r^{-1}\varepsilon} f(x)$ is fulfilled.
(e) If f is proper and convex on an open convex neighborhood U of x, then
$$\partial_H f(x) = \{x^* \in X^* : \langle x^*, u - x \rangle \leq f(u) - f(x), \ \forall u \in U\}.$$
(f) If $f + g$ is well defined near x, one has
$$\partial_H f(x) + \partial_H g(x) \subset \partial_H(f + g)(x),$$
and concerning $g \circ G$ the inclusion
$$DG(x)^* (\partial_H g(G(z))) \subset \partial_H(g \circ G)(z)$$
holds, and this latter inclusion is an equality whenever G is a $\mathcal{C}^1$-diffeomorphism near z.
(g) If f is finite near x and Hadamard differentiable at x, then
$$\partial_H f(x) = \{Df(x)\} \quad \text{and} \quad \partial_H(f + g)(x) = Df(x) + \partial_H g(x).$$

PROOF. Assertions (a) – (e) as well as the first inclusion in (f) follow readily from the definition. Concerning the second inclusion in (f), fix any $x^* \in \partial_H g(x)$ and any $h \in Z$, and set $A := DG(z)$. From the Hadamard differentiability of G at z there exists a mapping $e :]0,+\infty[\times Z \to X$ with $\lim_{t \downarrow 0, v \to h} e(t,v) = 0$ such that $G(z+tv) = x + t(Av + e(t,v))$. Consider a sequence $(h_n)_n$ in Z converging to h and a sequence $(t_n)_n$ in $]0,+\infty[$ tending to 0 such that

$$\underline{d}_H^+(g \circ G)(z;h) = \lim_{n \to \infty} t_n^{-1}\big(g \circ G(z + t_n h_n) - g \circ G(z)\big).$$

It ensues that

$$\langle x^*, Ah \rangle \leq \liminf_{n \to \infty} t_n^{-1}\big(g(x + t_n(Ah_n + e(t_n, h_n))) - g(x)\big)$$
$$= \lim_{n \to \infty} t_n^{-1}\big(g \circ G(z + t_n h_n) - g \circ G(z)\big),$$

or equivalently $\langle x^* \circ A, h \rangle \leq \underline{d}_H^+(g \circ G)(z;h)$. This means that $x^* \circ A \in \partial_H(g \circ G)(z)$, so $DG(x)^*\big(\partial_H g(G(z))\big) \subset \partial_H(g \circ G)(z)$ as desired. Further, if G is a C^1-diffeomorphism near z, considering $(g \circ G) \circ G^{-1}$ near x we obtain the converse inclusion by applying the previous one.

Let us now turn to (g). Suppose that f is Hadamard differentiable at x. Then for any $h \in X$,

$$\langle Df(x), h \rangle = \lim_{t \downarrow 0, h' \to h} t^{-1}\big(f(x+th') - f(x)\big) = \underline{d}_H^+ f(x;h),$$

and this is easily seen to imply that $\partial_H f(x) = \{Df(x)\}$, proving the first part of the assertion (g).

Finally, by the latter equality and (f) we have $-Df(x) + \partial_H(f+g)(x) \subset \partial_H g(x)$, which combined with (f) again ensures the desired equality

$$\partial_H(f+g)(x) = Df(x) + \partial_H g(x).$$
$\square$

In addition to the above chain rule for $g \circ G$ in (f), assuming now that G is one-sided Lipschitz at $\overline{x}$ and g is F-differentiable at $\overline{y} := G(\overline{x})$, it is easily seen for every $h \in X$ that $\underline{d}_H^+(g \circ G)(\overline{x}; h) = \underline{d}_H^+(\overline{y}^* \circ G)(\overline{x}; h)$, where $\overline{y}^* := g'(\overline{y})$. This justifies the following proposition.

PROPOSITION 5.8. Let $G : U \to Y$ be a mapping from an open set U of the normed space X into a normed space Y and let $g : V \to \mathbb{R} \cup \{-\infty, +\infty\}$ be a function from an open set V of Y containing $G(U)$. Assume that G is one-sided Lipschitz at $\overline{x} \in U$ and g is Fréchet differentiable at $\overline{y} := G(\overline{x})$. Then $g \circ G$ is one-sided Lipschitz at $\overline{x}$ and

$$\partial_H(g \circ G)(\overline{x}) = \partial_H\big(g'(\overline{y}) \circ G\big)(\overline{x}).$$

As a direct corollary we have:

COROLLARY 5.9. Let U be an open set of a normed space X and let $f : U \to \mathbb{R} \cup \{-\infty, +\infty\}$ be a function which is finite at $\overline{x} \in U$ and one-sided Lipschitz at $\overline{x}$. Let $\varphi : I \to \mathbb{R} \cup \{-\infty, +\infty\}$ be an extended real-valued function on an interval $I \supset f(U)$ which is differentiable at $\overline{t} := f(\overline{x})$. Then for the function $\varphi \circ f$ well-defined near $\overline{x}$ the equality

$$\partial_H(\varphi \circ f)(\overline{x}) = \partial_H(rf)(\overline{x})$$

holds, where r denotes the derivate $r := \varphi'(\bar{t})$.

If in addition $\varphi'(\bar{t}) > 0$, one has

$$\partial_H(\varphi \circ f)(\bar{x}) = \varphi'(\bar{t})\partial_H f(\bar{x}).$$

Proposition 5.8 provides a formula by means of $\partial_H(\bar{y}^* \circ G)(\bar{x})$ with $\bar{y}^* := g'(\bar{y})$. It is also of interest to notice, like Lemma 4.56 for F-subgradients, how a set in the form $\partial_H(y^* \circ G)(\bar{x})$ is connected with the B-tangent cone of the graph of G.

PROPOSITION 5.10. *Let $G : X \to Y$ be a mapping between normed spaces X, Y and let $\bar{x} \in X$, $x^* \in X^*$ and $y^* \in Y^*$.*
(a) *One has the implication*

$$x^* \in \partial_H(y^* \circ G)(\bar{x}) \implies \langle x^*, h \rangle \leq \langle y^*, v \rangle \ \forall (h, v) \in T^B\big(\operatorname{gph} G; (\bar{x}, G(\bar{x}))\big).$$

(b) *If G is compactly Lipschitzian at $\bar{x}$, then the above implication is an equivalence.*

PROOF. (a) Suppose that $x^* \in \partial_H(y^* \circ G)(\bar{x})$. Take any (h, v) in the tangent cone $T^B\big(\operatorname{gph} G; (\bar{x}, G(\bar{x}))\big)$. There exist sequences $(t_n)_n$ tending to 0 with $t_n > 0$ and $(h_n, v_n)_n$ converging to (h, v) with $(\bar{x} + t_n h_n, G(\bar{x}) + t_n v_n)$ in $\operatorname{gph} G$, which means $v_n = t_n^{-1}\big(G(\bar{x} + t_n h_n) - G(\bar{x})\big)$. By the inclusion $x^* \in \partial_H(y^* \circ G)(\bar{x})$ it ensues that

$$\langle x^*, h \rangle \leq \underline{d}_H^+(y^* \circ G)(\bar{x}; h) \leq \lim_{n \to \infty} t_n^{-1}\big((y^* \circ G)(\bar{x} + t_n h_n) - (y^* \circ G)(\bar{x})\big)$$

$$= \Big\langle y^*, \lim_{n \to \infty} t_n^{-1}\big(G(\bar{x} + t_n h_n) - G(\bar{x})\big) \Big\rangle,$$

which gives $\langle x^*, h \rangle \leq \langle y^*, v \rangle$.
(b) Assume that G is compactly Lipschitzian at $\bar{x}$ and suppose that the property involving the B-tangent cone holds. Fix any $h \in X$ and choose sequences $(t_n)_n$ tending to 0 with $t_n > 0$ and $(h_n)_n$ converging to h such that

$$\underline{d}_H^+(y^* \circ G)(\bar{x}; h) = \lim_{n \to \infty} t_n^{-1}\big((y^* \circ G)(\bar{x} + t_n h_n) - (y^* \circ G)(\bar{x})\big).$$

By the compactly Lipschitzian property of the mapping G at $\bar{x}$ we may suppose that $t_n^{-1}\big(G(\bar{x} + t_n h_n) - G(\bar{x})\big)$ converges to some vector $v \in Y$. We deduce that (h, v) belongs to $T^B\big(\operatorname{gph} G; (\bar{x}, G(\bar{x}))\big)$, which entails that

$$\langle x^*, h \rangle \leq \langle y^*, v \rangle = \underline{d}_H^+(y^* \circ G)(\bar{x}; h).$$

The inequality $\langle x^*, h \rangle \leq \underline{d}_H^+(y^* \circ G)(\bar{x}; h)$, being true for every $h \in X$, we conclude that $x^* \in \partial_H(y^* \circ G)(\bar{x})$. □

Let us characterize the Hadamard differentiability of f in terms of subdifferentiability. The statement and arguments are similar to those of Proposition 4.13 related to the F-subdifferential.

PROPOSITION 5.11. *Let X be a normed space, $f : X \to \mathbb{R} \cup \{-\infty, +\infty\}$ be an extended real-valued function, and $x \in X$ be a point where f is finite. The following hold:*
(a) *The function f is Hadamard differentiable at x if and only if both functions f and $-f$ are Hadamard subdifferentiable at x. In such a case, one has*

$$\{Df(x)\} = \partial_H f(x) = -\partial_H(-f)(x).$$

(b) *If $\partial_H f(x) \neq \emptyset$, then one has $-\partial_H(-f)(x) \subset \partial_H f(x)$.*

PROOF. (a) Suppose that $\partial_H f(x) \neq \emptyset$ and $\partial_H(-f)(x) \neq \emptyset$. Take $u^* \in \partial_H f(x)$ and $v^* \in \partial_H(-f)(x)$, and fix any $h \in X$. There exist functions ε_i with $\lim_{t\downarrow 0, v \to h} \varepsilon_i(t, v) = 0$ such that, for all $t > 0$ small enough and v near h,

$$\langle u^*, v\rangle \leq t^{-1}\big(f(x+tv)-f(x)\big)+\varepsilon_1(t,v), \quad \langle v^*, v\rangle \leq t^{-1}\big(-f(x+tv))+f(x)\big)+\varepsilon_2(t,v).$$

This ensures that $f(x + tv)$ is finite for all $t > 0$ small enough and v near h, and for such t and v

$$\langle u^* + v^*, v\rangle \leq \varepsilon_1(t,v) + \varepsilon_2(t,v),$$

hence $\langle u^* + v^*, h\rangle \leq 0$ for every $h \in X$. So, $v^* = -u^*$ and this equality gives, for all $t > 0$ small enough and v near h,

$$-\varepsilon_1(t,v) \leq t^{-1}\big(f(x+tv) - f(x) - t\langle u^*, v\rangle\big) \leq \varepsilon_2(t,v),$$

which guarantees the Hadamard differentiability of f at x.

The reverse implication is obvious and the equality of the proposition follows from the above arguments.

(b) Assume that $\partial_H f(x) \neq \emptyset$. If $\partial_H(-f)(x)$ is empty, the inclusion of (b) is trivial, and otherwise the inclusion follows immediately from (a). □

5.1.2. Hadamard subdifferential of sums of functions. We start the study of the Hadamard subdifferential of sums of functions with the case of a sum with separate variables.

PROPOSITION 5.12. *Let $X, Y, X_1, \cdots, X_m$ be normed vector spaces, $g : Y \to \mathbb{R} \cup \{-\infty, +\infty\}$ be a function on Y, and, for each $i = 1, \cdots, m$, let $f_i : X_i \to \mathbb{R} \cup \{-\infty, +\infty\}$ be a function on X_i. Then the following properties hold.*
(a) *Given $\varepsilon \geq 0$ and finite-dimensional vector subspaces E_i of X_i, endowing $X_1 \times \cdots \times X_m$ with the sum norm, for $\varphi(u_1, \cdots, u_m) := f_1(u_1)+\cdots+f_m(u_m)$ (assumed to be) well defined on an open set U of $X_1 \times \cdots \times X_m$ and for any $x = (x_1, \cdots, x_m) \in U$ one has with $E := E_1 \times \cdots \times E_m$*

$$\partial_{H,\varepsilon}(\varphi + \Psi_{x+E})(x) = \partial_{H,\varepsilon}(f_1 + \Psi_{x_1+E_1})(x_1) \times \cdots \times \partial_{H,\varepsilon}(f_m + \Psi_{x_m+E_m})(x_m).$$

(b) *For $h : X \times Y \to \mathbb{R} \cup \{-\infty, +\infty\}$ with $h(x', y') = g(y')$ for all $(x', y') \in X \times Y$, one has for any $x \in X$ and $y \in Y$, and for any finite-dimensional vector subspaces E and F of X and Y respectively,*

$$\partial_{H,\varepsilon}(h + \Psi_{(x,y)+E\times F})(x, y) = (\varepsilon \mathbb{B}_{X^*} + E^\perp) \times \partial_{H,\varepsilon}(g + \Psi_{y+F})(y),$$

where $X \times Y$ is endowed with the sum norm.

PROOF. We may suppose that φ is finite at x. Putting $\varphi_0 = \varphi + \Psi_{x+E_1 \times \cdots \times E_m}$ we note that

$$\varphi_0(u_1, \cdots, u_m) = (f_1 + \Psi_{x_1+E_1})(u_1) + \cdots + (f_m + \Psi_{x_m+E_m})(u_m),$$

so (a) follows from Proposition 5.5(b) and Proposition 4.19.

Concerning (b) it suffices to apply (a) and to note by (5.1) and by Moreau-Rockafellar theorem (see Theorem 2.105) that $\partial_{H,\varepsilon}\Psi_{y+F}(y) = \varepsilon \mathbb{B}_{X^*} + F^\perp$. □

Hadamard subdifferential also enjoys a rich fuzzy calculus under Gâteaux differentiable renorm. Let us begin with the following proposition concerning the local minimization of a sum of functions. Taking a Gâteaux (hence Hadamard) differentiable norm (off zero), its proof follows from a direct adaptation of that of Proposition 4.46. Indeed, it suffices to use, for the functions Δ_i in the proof of Proposition

4.46, the properties in Lemma 4.43 giving their Hadamard differentiability and the boundedness over any bounded set of their Hadamard derivatives.

PROPOSITION 5.13. Let X be a Banach space admitting an equivalent norm $\|\cdot\|$ which is Gâteaux differentiable off zero and let $f_i : X \to \mathbb{R} \cup \{-\infty, +\infty\}$, $i = 1, \cdots, m$, be functions which are finite at $\bar{x} \in X$. Assume that $\bar{x} \in X$ is a local minimizer of $f_1 + \cdots + f_m$. The following hold:
(a) If f_1 is lower semicontinuous near $\bar{x}$ and $f_2, \cdots, f_m$ are finite near $\bar{x}$ and uniformly continuous near $\bar{x}$, then for any real $\varepsilon > 0$ there exist $x_1, \cdots, x_m \in B(\bar{x}, \varepsilon)$ (all depending on x^* and ε) with $|f_1(x_1) - f_1(\bar{x})| < \varepsilon$ such that
$$0 \in \partial_H f_1(x_1) + \cdots + \partial_H f_m(x_m) + \varepsilon \mathbb{B}_{X^*}.$$
(b) If all the functions $f_1, \cdots, f_m$ are lower semicontinuous near $\bar{x}$, then for any real $\varepsilon > 0$, and any finite-dimensional vector subspace E of X there exist $x_1, \cdots, x_m \in B(\bar{x}, \varepsilon)$ (all depending on x^*, ε, and E) with $|f_i(x_i) - f_i(\bar{x})| < \varepsilon$ for $i = 1, \cdots, m$, such that
$$0 \in \partial_H f_1(x_1) + \cdots + \partial_H f_m(x_m) + \varepsilon \mathbb{B}_{X^*} + E^\perp,$$
where we recall that $E^\perp := \{u^* \in X^* : \langle u^*, u \rangle = 0, \forall u \in E\}$.

Let X be a Banach space admitting a Gâteaux differentiable renorm. If $f : X \to \mathbb{R} \cup \{+\infty\}$ is semicontinuous on an open set $U \subset X$ and finite at $\bar{x} \in U$, noting that $\bar{x}$ is minimizer of $f + \Psi_{\{\bar{x}\}}$, for any real $\varepsilon > 0$ the above proposition provides some $x \in B(\bar{x}, \varepsilon) \cap U$ such that $|f(x) - f(\bar{x})| < \varepsilon$ and $\partial_H f(x) \neq \emptyset$. Otherwise stated, we have the following corollary.

COROLLARY 5.14. Let X be a Banach space admitting a Gâteaux differentiable (off zero) equivalent renorm and let $f : X \to \mathbb{R} \cup \{+\infty\}$ be a proper lower semicontinuous function. Then $\mathrm{Dom}\,\partial_H f$ is sequentially graphically dense in $\mathrm{dom}\, f$.

We use now Proposition 5.13 to establish fuzzy calculus rules for the Hadamard subdifferential of a sum of functions under the assumption of a Gâteaux differentiable renorm of the underlying space.

THEOREM 5.15 (fuzzy sum rule for H-subdifferential under differentiable renorm). Let X be a Banach space admitting an equivalent norm $\|\cdot\|$ which is Gâteaux differentiable off zero and let $f_i : X \to \mathbb{R} \cup \{-\infty, +\infty\}$, $i = 1, \cdots, m$, be functions which are finite at $\bar{x} \in X$ and lower semicontinuous near $\bar{x}$. Then for any $x^* \in \partial_H(f_1 + \cdots + f_m)(\bar{x})$, any real $\varepsilon > 0$, and any finite-dimensional vector subspace E of X, there exist $x_1, \cdots, x_m \in B(\bar{x}, \varepsilon)$ (all depending on x^*, ε, and E) with $|f_i(x_i) - f_i(\bar{x})| < \varepsilon$ for $i = 1, \cdots, m$, such that
$$x^* \in \partial_H f_1(x_1) + \cdots + \partial_H f_m(x_m) + \varepsilon \mathbb{B}_{X^*} + E^\perp.$$

PROOF. Fix any real $\varepsilon > 0$ and any finite-dimensional subspace E of X, and put $\varepsilon' := \varepsilon/2$. With $f_{m+1}(x) := -\langle x^*, x \rangle$, $f_{m+2}(x) := \varepsilon' \|x - \bar{x}\|$ and $f_{m+3}(x) := \Psi_{\bar{x}+E}(x)$, Corollary 5.6(a) says that $\bar{x}$ is a local minimizer of $f_1 + \cdots + f_{m+3}$. Proposition 5.13(b) then furnishes $x_i \in X$ and $x_i^* \in \partial_H f(x_i)$, with $\|x_i - \bar{x}\| < \varepsilon'$ and $|f_i(x_i) - f_i(\bar{x})| < \varepsilon'$, such that $0 \in \partial_H f_1(x_1) + \cdots + \partial_H f_{m+3}(x_{m+3}) + \varepsilon' \mathbb{B}_{X^*} + E^\perp$. Noting that $\partial_H f_{m+1}(x_{m+1}) = \{-x^*\}$, $\partial_H f_{m+2}(x_{m+2}) \subset \varepsilon' \mathbb{B}_{X^*}$ and $\partial_H \Psi_{\bar{x}+E}(\bar{x}) = E^\perp$, we obtain
$$x^* \in \partial_H f_1(x_1) + \cdots + \partial_H f_m(x_m) + 2\varepsilon' \mathbb{B}_{X^*} + E^\perp,$$
which is the desired inclusion. □

Similarly to Theorem 4.57, using Proposition 5.10(b) we have the following fuzzy chain rule.

THEOREM 5.16 (chain rule for H-subdifferential under differentiable renorm). Let X and Y be Banach spaces admitting Gâteaux differentiable (off zero) equivalent norms, G be a mapping from X into Y, and $g : Y \to \mathbb{R} \cup \{+\infty\}$ be a function which is finite at $G(\overline{x})$, where $\overline{x} \in X$.
(a) If g is lower semicontinuous near $G(\overline{x})$ and the graph of G is closed near $(\overline{x}, G(\overline{x}))$, then for any $x^* \in \partial_H(g \circ G)(\overline{x})$, any $\varepsilon > 0$, any finite-dimensional subspace E of X, and any finite-dimensional subspace Q of Y, there exist $x \in B(\overline{x}, \varepsilon)$, $y \in B(G(\overline{x}), \varepsilon)$ with $|g(y) - g(G(\overline{x}))| < \varepsilon$, $(u^*, -v^*) \in \left(T^B\bigl(\text{gph}\, G; (x, G(x))\bigr)\right)^\circ$ such that
$$x^* \in u^* + \varepsilon \mathbb{B}_{X^*} + E^\perp \quad \text{and} \quad v^* \in \partial_H g(y) + \varepsilon \mathbb{B}_{Y^*} + Q^\perp.$$
(b) If g is lower semicontinuous near $G(\overline{x})$ and G is compactly Lipschitzian at $\overline{x}$, then for any $x^* \in \partial_H(g \circ G)(\overline{x})$, any $\varepsilon > 0$, any finite-dimensional subspace E of X, and any finite-dimensional subspace Q of Y, there exist $x \in B(\overline{x}, \varepsilon)$, $y \in B(G(\overline{x}), \varepsilon)$ with $|g(y) - g(G(\overline{x}))| < \varepsilon$, and $y^* \in \partial_H g(y) + \varepsilon \mathbb{B}_{Y^*} + Q^\perp$ such that
$$x^* \in \partial_H(y^* \circ G)(x) + \varepsilon \mathbb{B}_{X^*} + E^\perp.$$

5.2. Ioffe A-subdifferential on separable Banach spaces

We start now with the Ioffe approximate subdifferential of functions defined on separable Banach spaces. The main motivation to study first the situation of such spaces is the sequential description provided in Theorem 5.18(b). Such a sequential consideration is particularly useful when one has to work with integral spaces like separable $L^1_X(T, \mu)$ spaces, where X is a separable Banach space and μ is a σ-finite measure on T equipped with a σ-field countably generated in a suitable way. For such $L^1_X(T, \mu)$ spaces, the use of sequences in any analysis is known to be very crucial. It is also worth noting that *general $L^1_X(T, \mu)$ spaces are not Asplund*.

5.2.1. Definition for Lipschitz functions and comparisons.

DEFINITION 5.17. Let U be an open subset of a *separable Banach space* X and let $f : U \to \mathbb{R} \cup \{-\infty, +\infty\}$ be a function which is finite and Lipschitz near a point $\overline{x} \in U$. The *Ioffe approximate subdifferential* or *A-subdifferential* $\partial_A f(\overline{x})$ is defined by
$$\partial_A f(\overline{x}) = \limsup_{x \to \overline{x}} \partial_H f(x),$$
where the topological Lim sup is taken with respect to the weak* topology in X^* (and the norm topology in X).

If $\gamma \geq 0$ denotes a Lipschitz constant of f on open neighborhood $V \subset U$ of $\overline{x}$, we note that $\partial_H f(x) \subset \gamma \mathbb{B}_{X^*}$ for all $x \in V$, and that the topology induced on $\mathbb{B}_{X^*}$ by $w(X^*, X)$ is metrizable according to the separable property of X. Recall also that any separable Banach space admits an equivalent norm which is Gâteaux differentiable off zero.

THEOREM 5.18 (sequential description and compactness of A-subdifferential for Lipschitz functions on separable Banach spaces). Let U be an open set of a separable Banach space $(X, \|\cdot\|)$ and $f : U \to \mathbb{R} \cup \{-\infty, +\infty\}$ be finite and γ-Lipschitz near $\overline{x} \in U$. The following hold:

(a) The set $\partial_A f(\overline{x})$ is nonempty and weak* compact in X^*, and
$$\partial_A f(\overline{x}) \subset \partial_C f(\overline{x}) \subset \gamma \mathbb{B}_{X^*}.$$

(b) The inclusion $x^* \in \partial_A f(\overline{x})$ holds if and only if there are sequences $(x_n)_n$ in U converging to $\overline{x}$ and $(x_n^*)_n$ in X^* converging weakly* to x^* such that $x_n^* \in \partial_H f(x_n)$ for all $n \in \mathbb{N}$; otherwise stated
$$\partial_A f(\overline{x}) = {}^{\text{seq}}\limsup_{x \to \overline{x}} \partial_H f(x).$$

PROOF. We may suppose that f is finite and γ-Lipschitz on U. Clearly, the inclusion $\partial_A f(\overline{x}) \subset \gamma \mathbb{B}_{X^*}$ holds true.

(b) We note first that (b) follows from the inclusion $\partial_H f(x) \subset \gamma \mathbb{B}_{X^*}$ for x near $\overline{x}$ and (thanks to the separable property of X) from the metrizability of $\gamma \mathbb{B}_{X^*}$ endowed with the topology induced by $w(X^*, X)$ (see Proposition 1.14(b)).

(a) By definition $\partial_A f(\overline{x})$ is w^*-closed and included in the w^*-compact set $\gamma \mathbb{B}_{X^*}$, so $\partial_A f(\overline{x})$ is w^*-compact. Further, take a sequence $(u_n)_n$ in U converging to $\overline{x}$ with $\partial_H f(u_n) \neq \emptyset$ (see Corollary 5.14) and choose $u_n^* \in \partial_H f(u_n)$ for any $n \in \mathbb{N}$. By the sequential w^*-compactness of $\gamma \mathbb{B}_{X^*}$ we may suppose, without loss of generality that $(u_n^*)_n$ converges weakly* to some $u^* \in X^*$, so $u^* \in \partial_A f(\overline{x})$, justifying the non-vacuity of $\partial_A f(\overline{x})$.

Finally, by (b) we have
$$\partial_A f(\overline{x}) = {}^{\text{seq}}\limsup_{x \to \overline{x}} \partial_H f(x) \subset {}^{\text{seq}}\limsup_{x \to \overline{x}} \partial_C f(x) \subset \partial_C f(\overline{x}),$$
where the latter inclusion is due to the outer semicontinuity on U of $\partial_C f$ (see Proposition 2.74(a)). $\square$

As preparatory steps for the next proposition where X is Asplund, we need the following lemmas.

LEMMA 5.19. *Let U be an open set of an Asplund space X and $f : U \to \mathbb{R}$ be a Lipschitz function with $\gamma \geq 0$ as a Lipschitz constant. Let $x \in U$ and $x^* \in \partial_H f(x)$. Then for any weak* neighborhood V of zero in X^* and any real $\delta > 0$ there are $u \in U$ with $\|u - x\| < \delta$ and $u^* \in \partial_F f(u)$ with $u^* - x^* \in V$.*

PROOF. Fix any real $\delta > 0$ and any weak* neighborhood V of zero in X^*. Take a real $\varepsilon > 0$ and a finite-dimensional subspace E in X such that $2\varepsilon \mathbb{B}_{X^*} + E^\perp \subset V$. By Corollary 5.6 we know that x is a local minimizer on the constraint set $x + E$ of the function
$$x' \mapsto \varphi(x') := f(x') - \langle x^*, x' - x \rangle + \varepsilon \|x' - x\|.$$
Since $\|x^*\| \leq \gamma$ (see Proposition 5.7(c)), by Lemma 2.213 the point x is also a local unconstrained minimizer of the function
$$x' \mapsto \varphi(x') + (1 + 2\gamma) d_{x+E}(x') = \varphi(x') + (1 + 2\gamma) d_E(x' - x),$$
hence $0 \in \partial_F(\varphi + (1 + 2\gamma) d_E(\cdot - x))(x)$. The space X being an Asplund space, by Theorem 4.47 there are $u \in U$ with $\|u - x\| < \delta$ and $u^* \in \partial_F f(u)$ such that $x^* \in u^* + 2\varepsilon \mathbb{B}_{X^*} + E^\perp$, hence $x^* \in u^* + V$ as desired. $\square$

Recalling that closed balls of the topological dual of any Asplund space are sequentially w^*-compact, we derive the second lemma as follows.

LEMMA 5.20. *Let U be an open set of an Asplund space X and $f : U \to \mathbb{R}$ be a Lipschitz function. Then, for any $\bar{x} \in U$ one has*

$$\operatorname{Lim\,sup}_{x \to \bar{x}} \partial_H f(x) = \operatorname{cl}_{w^*}\left(\operatorname{^{seq}Lim\,sup}_{x \to \bar{x}} \partial_H f(x)\right) = \operatorname{cl}_{w^*}\left(\partial_L f(\bar{x})\right).$$

PROOF. Denoting by γ a Lipschitz constant of f on the open set U of $\bar{x}$, according to the inclusion $\partial_H f(x) \subset \gamma \mathbb{B}_{X^*}$, for all $x \in U$, and to the sequential w^*-compactness of $\gamma \mathbb{B}_{X^*}$ (recalled above), the first equality between the first and the second member in the lemma follows from Proposition 1.17.

Since $\partial_F f(x) \subset \partial_H f(x)$, the third member is evidently included into the second. Conversely, fix any $x^* \in {}^{\text{seq}}\operatorname{Lim\,sup}_{x \to \bar{x}} \partial_H f(x)$ and fix also any w^*-neighborhood V of zero in X^*. Choose a real $\varepsilon > 0$ and a finite-dimensional vector subspace E of X such that $2\varepsilon \mathbb{B}_{X^*} + E^\perp \subset V$. By definition of sequential outer limit, take a sequence $(x_n)_n$ in U converging to $\bar{x}$ and a sequence $(x_n^*)_n$ in X^* converging weakly* to x^* with $x_n^* \in \partial_H f(x_n)$ for all $n \in \mathbb{N}$. By Lemma 5.19 above there are $u_n \in U$ with $\|u_n - x_n\| \leq 1/n$ and $u_n^* \in \partial_F f(u_n)$ such that $x_n^* \in u_n^* + 2\varepsilon \mathbb{B}_{X^*} + E^\perp$. By the weak* sequential compactness of $\gamma \mathbb{B}_{X^*}$ take, from the sequence $(u_n^*)_n$ (which is in $\gamma \mathbb{B}_{X^*}$), a subsequence converging weakly* to some $u^* \in X^*$. Noting that $u_n \to \bar{x}$, it ensues that $u^* \in \partial_L f(\bar{x})$ and $x^* \in u^* + 2\varepsilon \mathbb{B}_{X^*} + E^\perp$ (where the latter inclusion is due to the weak* closedness of $2\varepsilon \mathbb{B}_{X^*} + E^\perp$). This yields $x^* \in \partial_L f(\bar{x}) + V$. This being true for any w^*-neighborhood V of zero in X^*, it ensues that $x^* \in \operatorname{cl}_{w^*}\left(\partial_L f(\bar{x})\right)$, which justifies the inclusion of the second member into the third and finishes the proof of the lemma. □

With the help of the latter lemma we can consider the situation when the separable Banach space is Asplund. In such spaces the approximate subdifferential and the limiting one coincide for Lipschitz functions. Recalling that a separable Banach space is Asplund if and only if its topological dual is strongly separable, we state the result in the following form.

PROPOSITION 5.21. *Let U be an open set of a Banach space X whose topological dual X^* is (strongly) separable. For any function $f : U \to \mathbb{R} \cup \{-\infty, +\infty\}$ which is finite and Lipschitz near $\bar{x} \in U$, one has*

$$\partial_A f(\bar{x}) = \partial_L f(\bar{x}).$$

PROOF. By Definition 5.17 the first member of Lemma 5.20 coincides with $\partial_A f(\bar{x})$. Concerning the third member of the same lemma, denote by γ a Lipschitz constant of f near $\bar{x}$, so for $\partial_F f(x) \subset \gamma \mathbb{B}_{X^*}$ for all x near $\bar{x}$. Since the topology induced on $\gamma \mathbb{B}_{X^*}$ by $w(X^*, X)$ is metrizable (according to the separability of X), Proposition 1.14(b) entails that, for X^* endowed with the $w(X^*, X)$ topology the sequential outer limit $^{\text{seq}}\operatorname{Lim\,sup}_{x \to \bar{x}} \partial_F f(x)$ coincides with its topological outer limit, hence it is w^*-closed. This ensures that $\partial_L f(\bar{x})$ coincides with the second member of Lemma 5.20, so the equality in the proposition follows from the equality between the first and third members in Lemma 5.20. □

The calculus rule for the A-subdifferential of the sum of Lipschitz functions on separable Banach spaces is quite direct.

PROPOSITION 5.22. *Let U be an open set of a separable Banach space $(X, \|\cdot\|)$ and $f, g : U \to \mathbb{R} \cup \{+\infty\}$ be finite and Lipschitz near $\bar{x} \in U$. Then one has*

$$\partial_A (f + g)(\bar{x}) \subset \partial_A f(\bar{x}) + \partial_A g(\bar{x}).$$

PROOF. We may suppose that f, g are finite and Lipschitz on U. Let $x^* \in \partial_A(f+g)(\bar{x})$. Fix any w^*-closed neighborhood V of zero in $(X^*, w(X^*, X))$. By Theorem 5.18(b) there are a sequence $(x_n)_n$ in U converging to $\bar{x}$ and a sequence $(x_n^*)_n$ in X^* converging weakly* to x^* with $x_n^* \in \partial_H(f+g)(x_n)$ for all $n \in \mathbb{N}$. By Theorem 5.15, for each $n \in \mathbb{N}$, there are $u_n, v_n \in U$ with $u_n, v_n \in (x_n + (1/n)\mathbb{B}_X)$ and $u_n^* \in \partial_H f(u_n)$, $v_n^* \in \partial_H g(v_n)$ such that $x_n^* \in u_n^* + v_n^* + V$. Since $(u_n^*, v_n^*)_n$ is bounded, we can choose a weak* limit (u^*, v^*) of a subsequence of $(u_n^*, v_n^*)_n$, thus $u^* \in \partial_A f(\bar{x})$, $v^* \in \partial_A g(\bar{x})$, and $x^* \in u^* + v^* + V$. This ensures that $x^* \in \partial_A f(\bar{x}) + \partial_A g(\bar{x}) + V$, which guarantees that $x^* \in \partial_A f(\bar{x}) + \partial_A g(\bar{x})$ since $\partial_A f(\bar{x}) + \partial_A g(\bar{x})$ is weakly* closed (keeping in mind that $\partial_A f(\bar{x})$ is w^*-compact by Theorem 5.18(a)). $\square$

The cases of multiplication with a positive real and of composition with a $\mathcal{C}^1$-diffeomorphism is more direct in the sense that they follow directly from (d) and (f) in Proposition 5.7.

PROPOSITION 5.23. Let X, Y be separable Banach spaces, $G : X \to Y$, $g : Y \to \mathbb{R} \cup \{-\infty, +\infty\}$, and $f : X \to \mathbb{R} \cup \{-\infty, +\infty\}$.
(a) If f is finite at $\bar{x} \in X$ and Lipschitz near $\bar{x}$, then for any real $r > 0$ one has

$$\partial_A(rf)(\bar{x}) = r\partial_A f(\bar{x}).$$

(b) If G is a $\mathcal{C}^1$-diffeomorphism near $\bar{x} \in X$ and g is finite at $\bar{y} := G(\bar{x})$ and Lipschitz near $\bar{y}$, then

$$\partial_A(g \circ G)(\bar{x}) = DG(\bar{x})^* \big(\partial_A g(\bar{y}) \big).$$

More general chain rules will be established later.

5.2.2. A-normal cone in separable Banach spaces.
Given a continuous convex function $f : U \to \mathbb{R}$ on an open convex set U of a separable Banach space X, we easily see from Theorem 5.18(b) that $\partial_A f(x) = \partial f(x)$ according to the usual definition of $\partial f(x)$ for convex functions (see Definition 2.64). We state this property in the proposition:

PROPOSITION 5.24. Let X be a separable Banach space and $f : U \to \mathbb{R}$ be a continuous convex function on an open convex set U of X. Then, one has $\partial_A f(x) = \partial f(x)$ for all $x \in U$.

This proposition combined with Proposition 2.97 guarantees that, for any nonempty convex set S of the separable Banach space X and for $\bar{x} \in S$, the set $\mathbb{R}_+ \partial_A d_S(\bar{x})$ coincides with the normal cone of the convex set S at $\bar{x}$ in the usual sense for convex sets.

In the context of an Asplund space X, we know by Theorem 4.85(b) that the limiting normal cone $N^L(S; \bar{x})$ of a closed S at $\bar{x} \in S$ can be defined equivalently either as $^{\text{seq}}\mathop{\text{Lim sup}}\limits_{x \to \bar{x}} \partial_F \Psi_S(x)$ or $\mathbb{R}_+\left(^{\text{seq}}\mathop{\text{Lim sup}}\limits_{x \to \bar{x}} \partial_F d_S(x)\right)$.

The situation with Hadamard subdifferential in place of Fréchet one is very different in infinite-dimensional spaces. Considering in a separable Banach space X the closed set S in Example 2.236 we saw that

$$T^C(S; 0) = \mathbb{R}_+ b \quad \text{and} \quad T^B\left(S; \frac{1}{m}a\right) = \{0\},$$

where a and b are nonzero. We deduce that $\partial_H \Psi_S(\frac{1}{m}a) = X^*$ for all $m \in \mathbb{N}$, hence $\underset{x \to 0}{\operatorname{Lim\,sup}}\, \partial_H \Psi_S(x) = X^*$. Then, $\underset{x \to 0}{\operatorname{Lim\,sup}}\, \partial_H \Psi_S(x) = X^*$ is strictly bigger than the Clarke normal cone $N^C(S;0) = \{x^* \in X^* : \langle x^*, b \rangle \leq 0\}$. This example shows that, in an infinite-dimensional separable Banach space X, the outer limit $\underset{x \to \bar{x}}{\operatorname{Lim\,sup}}\, \partial_H \Psi_S(x)$ is *too large* to be used to define the approximate normal cone of a set S of X at a point $\bar{x} \in S$. According to this and to the situations examined above either with the convexity of S or with the limiting normal cone, the approximate normal cone is defined through the distance function as follows.

DEFINITION 5.25. Let X be a separable Banach space and let $\|\cdot\|$ be an equivalent norm on X which is Gâteaux differentiable off zero. Given a subset S of X and a point $\bar{x} \in S$, its *Ioffe approximate normal cone* or *A-normal cone* $N^A(S;\bar{x})$ at $\bar{x}$ is defined by

$$N^A(S;\bar{x}) = \mathbb{R}_+ \partial_A d_S(\bar{x}).$$

For $\bar{x} \notin S$, as usual we put $N^A(S;\bar{x}) = \emptyset$.

By the equality $d_{\operatorname{cl} S}(\cdot) = d_S(\cdot)$ we obviously have

$$N^A(S;x) = N^A(\operatorname{cl} S; x) \quad \text{for all } x \in S.$$

In order to establish in Proposition 5.29 below that the above concept is invariant under Gâteaux differentiable equivalent norms of X, we need various preparatory results. The first one in the next lemma relies Hadamard subgradients of the distance function at points outside a set to Hadamard subgradients at points inside the set. It is in the line of Lemma 4.88.

LEMMA 5.26. *Let $(X, \|\cdot\|)$ be a Banach space whose norm $\|\cdot\|$ is Gâteaux differentiable off zero and let S be a closed subset of X. Let $u \notin S$ and $u^* \in \partial_H d_S(u)$. Then for any real $\varepsilon > 0$ and any nonzero finite-dimensional vector subspace F of X, there are $\bar{x} \in S$ and $x^* \in \partial_H d_S(\bar{x})$ such that*

$$\|\bar{x} - u\| \leq d_S(u) + \varepsilon \quad \text{and} \quad x^* - u^* \in \varepsilon \mathbb{B}_{X^*} + F^\perp.$$

PROOF. We may suppose $\varepsilon < 1$. Put $\mu := 8(\|u\| + 2 + 2d_S(u))^2$ and choose a real $\delta > 0$ with $\delta(1 + \mu) < \varepsilon/6$. By Corollary 5.6(a) there is a positive real $\eta < \min\{\delta/4, d_S(u)\}$ such that

$$d_S(u) \leq d_S(u+h) - \langle u^*, h \rangle + \frac{\delta}{4}\|h\| \quad \text{for all } h \in F \cap \eta \mathbb{B}_X = \eta \mathbb{B}_F,$$

where $\mathbb{B}_F := F \cap \mathbb{B}_X$. Setting

$$\varphi(x,h) := \|x - (u+h)\| - \langle u^*, h \rangle + \frac{\delta}{4}\|h\| \quad \text{for all } x \in X, h \in X,$$

we see that

$$\inf_{x \in S, h \in \eta \mathbb{B}_F} \varphi(x,h) = d_S(u) = \inf_{x \in S} \varphi(x,0).$$

Note also that, for $S_0 := S \cap B[u, 1 + d_S(u)]$,

$$\inf_{x \in S_0, h \in \eta \mathbb{B}_F} \varphi(x,h) \geq \inf_{x \in S, h \in \eta \mathbb{B}_F} \varphi(x,h) = d_S(u) = d_{S_0}(u)$$

$$= \inf_{x \in S_0} \varphi(x,0) \geq \inf_{x \in S_0, h \in \eta \mathbb{B}_F} \varphi(x,h),$$

so $\inf_{S_0 \times \eta \mathbb{B}_F} \varphi = d_S(u) = d_{S_0}(u) = \inf_{x \in S_0} \varphi(x, 0)$. Choose a point $a \in S_0$ such that

$$\varphi(a, 0) < \eta^3 + \inf_{x \in S_0} \varphi(x, 0) = \eta^3 + \inf_{x \in S_0, h \in \eta \mathbb{B}_F} \varphi(x, h).$$

Endowing $X \times X$ with the norm defined by $\|(x, h)\| = (\|x\|^2 + \|h\|^2)^{1/2}$ for $(x, h) \in X \times X$ and taking $\lambda := \eta$, the Borwein-Preiss variational principle applied to $\varphi + \Psi_{S_0 \times \eta \mathbb{B}_F}$ with $p = 2$ (see Theorem 4.41) yields with $\alpha_n := 1/2^n$ a sequence $(x_n, h_n)_n$ in $S_0 \times \eta \mathbb{B}_F$, with $(x_1, h_1) = (a, 0)$, converging to some $(\overline{x}, \overline{h})$ in $S_0 \times \eta \mathbb{B}_F$ and such that, for

$$\Delta_1(x) := \sum_{n=1}^{\infty} \alpha_n \|x - x_n\|^2, \quad \Delta_2(h) := \sum_{n=1}^{\infty} \alpha_n \|h - h_n\|^2, \quad \Delta(x, h) = \Delta_1(x) + \Delta_2(h),$$

one has $\|\overline{x} - x_n\|^2 + \|\overline{h} - h_n\|^2 < \lambda^2$ for all $n \in \mathbb{N}$, and also

$$\varphi(\overline{x}, \overline{h}) + \eta \Delta(\overline{x}, \overline{h}) \leq \varphi(x, h) + \eta \Delta(x, h) \quad \text{for all } (x, h) \in S_0 \times \eta \mathbb{B}_F.$$

We have $\|\overline{h}\| \leq \eta$ and we note that

$$\|\overline{x} - u\| \leq \|\overline{x} - a\| + \|a - u\| = \|\overline{x} - a\| + \varphi(a, 0) \leq \lambda + \eta^3 + d_S(u) < 1 + d_S(u),$$

where the equality $\|a - u\| = \varphi(a, 0)$ is due to the definition of φ. Note also by the inclusion $(x_n, h_n) \in B[u, 1 + d_S(u)] \times \eta \mathbb{B}_F$ and by the definition of μ that $\mu \geq \sup_{S_0 \times \eta \mathbb{B}_F} \Delta$. It ensues that

$$\eta \mu \geq \sup_{S_0 \times \eta \mathbb{B}_F} \eta \Delta = \sup_{S_0 \times \eta \mathbb{B}_F} \eta \Delta + \inf_{S_0 \times \eta \mathbb{B}_F} \varphi - d_S(u)$$

$$\geq \inf_{S_0 \times \eta \mathbb{B}_F} (\varphi + \eta \Delta) - d_S(u) \geq \varphi(\overline{x}, \overline{h}) - d_S(u),$$

which entails (using the inequality $\|u^*\| \leq 1$) that

$$\eta \mu \geq \|\overline{x} - u - \overline{h}\| - \langle u^*, \overline{h} \rangle + \frac{\delta}{4} \|\overline{h}\| - d_S(u) \geq \|\overline{x} - u\| - 2\|\overline{h}\| - d_S(u),$$

so $\|\overline{x} - u\| \leq d_S(u) + 2\eta + \eta \mu < d_S(u) + \varepsilon < 1 + d_S(u)$.

Now, observe that, for any $x \in B[u, 1 + d_S(u)]$, Lemma 4.43 ensures that

$$\|D_H \Delta_1(x)\| \leq 2 \sup_{n \in \mathbb{N}} \|x - x_n\| \leq 4(1 + d_S(u)) \leq \mu,$$

hence the function $\varphi(\cdot, h) + \eta \Delta_1(\cdot)$ is γ-Lipschitz on $B[u, 1 + d_S(u)]$ with $\gamma := 1 + \eta \mu$. Combining this with the above inequality $\|\overline{x} - u\| < 1 + d_S(u)$, Lemma 2.215 and Lemma 2.219 imply that, for some open neighborhood U_0 of $\overline{x}$, the pair $(\overline{x}, \overline{h})$ is a minimizer on $U_0 \times X$ of the function

$$(x, h) \mapsto \varphi(x, h) + \eta \Delta_1(x) + \eta \Delta_2(h) + \gamma d_S(x) + \Psi_{\eta \mathbb{B}_F}(h),$$

thus on the one hand (with $h = \overline{h}$)

$$\gamma \big(d_S(x) - d_S(\overline{x})\big) \geq \eta \big(-\Delta_1(x) + \Delta_1(\overline{x})\big) - \|x - (u + \overline{h})\| + \|\overline{x} - (u + \overline{h})\| \quad \text{for all } x \text{ near } \overline{x}.$$

Since $d_S(u + \overline{h}) \geq d_S(u) - \|\overline{h}\| \geq d_S(u) - \eta > 0$, we deduce that $\|\overline{x} - (u + \overline{h})\| > 0$, thus the norm $\|\cdot\|$ is Gâteaux (hence Hadamard) differentiable at $\overline{x} - (u + \overline{h})$. Setting $y^* := -D_H \|\cdot\|(\overline{x} - u - \overline{h})$ and keeping in mind that Δ_1 is Hadamard differentiable (see Lemma 4.43), it results that

$$v^* := y^* - \eta D_H \Delta_1(\overline{x}) \in \gamma \partial_H d_S(\overline{x}) = (1 + \eta \mu) \partial_H d_S(\overline{x}).$$

On the other hand, (fixing x as $\bar{x}$) we also see that for all $h \in X$
$$\frac{\delta}{4}(\|h\| - \|\bar{h}\|) + \Psi_{\eta \mathbb{B}_F}(h) - \Psi_{\eta \mathbb{B}_F}(\bar{h})$$
$$\geq \eta\big(-\Delta_2(h) + \Delta_2(\bar{h})\big) - \|-h + \bar{x} - u\| + \|-\bar{h} + \bar{x} - u\| + \langle u^*, h - \bar{h}\rangle,$$
which gives according to the Hadamard differentiability of Δ_2 and to the Moreau-Rockafellar theorem (see Theorem 2.105)
$$u^* - y^* - \eta\, D_H \Delta_2(\bar{h}) \in \partial\left(\frac{\delta}{4}\|\cdot\| + \Psi_{\eta \mathbb{B}_F}\right)(\bar{h}) = \frac{\delta}{4}\partial\|\cdot\|(\bar{h}) + \partial\Psi_{\eta \mathbb{B}_F}(\bar{h}) \subset \frac{\delta}{4}\mathbb{B}_{X^*} + F^\perp,$$
where the inclusion $\partial\Psi_{\eta\mathbb{B}_F}(\bar{h}) \subset F^\perp$ is due to the inequality $\|\bar{h}\| \leq \eta$. Writing (by Lemma 4.43 again)
$$\|D_H \Delta_2(\bar{h})\| \leq 2\sup_{n \in \mathbb{N}} \|\bar{h} - h_n\| \leq 4\eta \leq \mu,$$
we derive that (keeping in mind that $\eta\mu < \delta\mu < \varepsilon/6$)
$$u^* - v^* \in \eta\, D_H \Delta_1(\bar{x}) + \eta\, D_H \Delta_2(\bar{h}) + \frac{\delta}{4}\mathbb{B}_{X^*} + F^\perp \subset \left(\frac{\varepsilon}{3} + \frac{\varepsilon}{24}\right)\mathbb{B}_{X^*} + F^\perp,$$
which entails that
$$u^* - v^* \in \frac{\varepsilon}{2}\mathbb{B}_{X^*} + F^\perp.$$
Putting $\sigma := \eta\mu$ and $x^* := (1+\sigma)^{-1} v^*$ we have $x^* \in \partial_H d_S(\bar{x})$ and $\sigma < \varepsilon/6$ (since $\eta(1+\mu) < \varepsilon/6$), and recalling that $\|u^*\| \leq 1$ we see that
$$u^* - x^* = \frac{1}{1+\sigma}(u^* - v^*) + \frac{\sigma}{1+\sigma}u^* \in \frac{\varepsilon}{2(1+\sigma)}\mathbb{B}_{X^*} + F^\perp + \sigma\mathbb{B}_{X^*} \subset \varepsilon\mathbb{B}_{X^*} + F^\perp.$$
This finishes the proof of the lemma. □

Let us now state the following proposition whose proof derives directly from the above lemma and from the fact that $d_{\operatorname{cl} S}(\cdot) = d_S(\cdot)$. Besides its obvious own interest, it will serve to argue the next lemma from which will follow our desired invariance of $N^A(S;\cdot)$ under equivalent Gâteaux differentiable norms.

PROPOSITION 5.27. *Let X be a separable Banach space endowed with a norm $\|\cdot\|$ which is Gâteaux differentiable off zero, let S be a subset of X and let $\bar{x} \in S$. Then, one has*
$$\partial_A d_S(\bar{x}) = \operatorname*{Lim\,sup}_{\overline{S} \ni x \to \bar{x}} \partial_H d_S(x),$$
where $\overline{S} := \operatorname{cl} S$.

LEMMA 5.28. *Let X be a separable Banach space endowed with a norm $\|\cdot\|$ which is Gâteaux differentiable off zero, and let S be a subset of X. Let U be an open set of X with $U \cap S \neq \emptyset$ and $\varphi : U \to \mathbb{R}$ be a locally Lipschitz function such that $d_S(x) \leq \varphi(x)$ for all $x \in U$ and $\varphi(x) = 0$ for all $x \in U \cap S$. Then for any $x \in U \cap S$, one has $\partial_A d_S(x) \subset \partial_A \varphi(x)$.*

PROOF. Fix any $x \in U \cap S$. By continuity of φ we have $\varphi(u) = 0$ for all $u \in U \cap \overline{S}$, where $\overline{S} := \operatorname{cl} S$. From this and the inequality assumption we see that $\partial_H d_S(u) \subset \partial_H \varphi(u)$, for all $u \in U \cap \overline{S}$, so
$$\operatorname*{Lim\,sup}_{\overline{S} \ni u \to x} \partial_H d_S(u) \subset \operatorname*{Lim\,sup}_{\overline{S} \ni u \to x} \partial_H \varphi(u) \subset \operatorname*{Lim\,sup}_{u \to x} \partial_H \varphi(u).$$
This combined with the previous proposition entails the desired inclusion. □

Thanks to the latter lemma we can now state and prove our desired proposition stating that the definition $N^A(S;x) = \mathbb{R}_+ \partial_A d_S(x)$ does not depend on equivalent Gâteaux differentiable norms. More generally, we will see in the next section that it does not depend on equivalent norms of X either (removing in this way the Gâteaux differentiability restriction).

PROPOSITION 5.29. Let X be a separable Banach space endowed with a Gâteaux differentiable (off zero) norm $\|\cdot\|$, let S be a subset S of X and let $\overline{x} \in S$. Then for any other equivalent Gâteaux differentiable (off zero) norm $\|\|\cdot\|\|$ on X, one has

$$\mathbb{R}_+ \partial_A d_{S,\|\cdot\|}(\overline{x}) = \mathbb{R}_+ \partial_A d_{S,\|\|\cdot\|\|}(\overline{x}),$$

that is, the definition of $N^A(\cdot;\cdot)$ in Definition 5.25 does not depend on equivalent Gâteaux differentiable norms.

PROOF. Let $\|\|\cdot\|\|$ be another equivalent Gâteaux differentiable (off zero) norm on X. Fix two reals $\alpha > 0$ and $\beta > 0$ such that $\alpha\|\|\cdot\|\| \leq \|\cdot\| \leq \beta\|\|\cdot\|\|$. Then

$$\alpha\, d_{S,\|\|\cdot\|\|}(x) \leq d_{S,\|\cdot\|}(x) \leq \beta\, d_{S,\|\|\cdot\|\|}(x) \quad \text{for all } x \in X.$$

This combined with Lemma 5.28 tells us that

$$\partial_A d_{S,\|\cdot\|}(\overline{x}) \subset \alpha^{-1} \partial_A d_{S,\|\|\cdot\|\|}(\overline{x}) \text{ and } \partial_A d_{S,\|\|\cdot\|\|}(\overline{x}) \subset \beta\, \partial_A d_{S,\|\cdot\|}(\overline{x}).$$

The conclusion of the proposition directly follows. □

Although in any infinite separable Banach space $\operatorname{Lim\,sup}_{S \ni x \to \overline{x}} \partial_H \Psi_S(x)$ is strictly bigger than $N^A(S;\overline{x})$ for some closed sets as seen in the analysis preceding Definition 5.25, the situation is different in finite-dimensional spaces.

PROPOSITION 5.30. If X is a finite-dimensional Euclidean space, then

$$N^A(S;\overline{x}) = \operatorname{Lim\,sup}_{\overline{S} \ni x \to \overline{x}} \partial_H \Psi_{\overline{S}}(x),$$

for any set S of X and $\overline{x} \in S$, where $\overline{S} := \operatorname{cl} S$.

PROOF. Fix $\overline{x} \in S$. Without loss of generality we may assume that S is closed. For each $x \in S$, it is easy to see that $\partial_H d_S(x) \subset \partial_H \Psi_S(x)$ (according to the definition of the Hadamard subdifferential) and this gives

$$\operatorname{Lim\,sup}_{S \ni x \to \overline{x}} \partial_H d_S(x) \subset \operatorname{Lim\,sup}_{x \to \overline{x}} \partial_H \Psi_S(x).$$

The right-hand side being clearly a cone, it results by Proposition 5.27 that

$$N^A(S;\overline{x}) = \mathbb{R}_+ \operatorname{Lim\,sup}_{S \ni x \to \overline{x}} \partial_H d_S(x) \subset \operatorname{Lim\,sup}_{x \to \overline{x}} \partial_H \Psi_S(x).$$

Take now any $x^* \in \operatorname{Lim\,sup}_{x \to \overline{x}} \partial_H \Psi_S(x)$ and any real $\varepsilon > 0$. The space X being finite-dimensional, there are sequences $(x_n)_n$ in S converging to $\overline{x}$ and $(x_n^*)_n$ in $X^* = X$ converging to x^* such that $x_n^* \in \partial_H \Psi_S(x_n)$ with $\|x_n^*\| \leq \|x^*\| + \varepsilon =: r$. Using again that X is finite-dimensional, Proposition 5.5 says that Fréchet and Hadamard subdifferentials coincide, so Proposition 4.30 guarantees that $x_n^* \in r\partial_F d_S(x_n) = r\partial_H d_S(x_n)$. We derive that $x^* \in r \operatorname{Lim\,sup}_{S \ni x \to \overline{x}} \partial_H d_S(x) \subset N^A(S;\overline{x})$. This completes the proof. □

Many other properties of the approximate normal cone will be established in Section 5.4 in the context of general Banach spaces. Through such properties the approximate subdifferential of general functions will also be defined and studied. First, we need to develop a large part of the theory in general Banach spaces for the class of locally Lipschitz functions.

5.3. Ioffe A-subdifferential of Lipschitz functions on Banach spaces

Since the Hadamard subdifferentiability of a Lipschitz function over a general Banach space is not guaranteed at enough points, taking Corollary 5.14 and the previous section into account we will invoke in such a framework the restrictions to separable Banach subspaces.

5.3.1. Definition, properties and sum. Given a Banach space X, we will denote by $\mathcal{S}(X)$ the collection of all *closed separable vector subspaces* of X. Similarly, $\mathcal{F}(X)$ will be the collection of all *finite-dimensional vector subspaces* of X. Sometimes, we will write $\mathcal{S}$ and $\mathcal{F}$ when there is no risk of confusion.

DEFINITION 5.31. Let U be an open set of a Banach space X and $f : U \to \mathbb{R} \cup \{-\infty, +\infty\}$ be a function which is finite near $\bar{x} \in U$ and locally Lipschitz near $\bar{x}$. The *Ioffe approximate subdifferential* or *A-subdifferential* $\partial_A f(\bar{x})$ of f at $\bar{x}$ is defined by

$$\bigcap_{E \in \mathcal{S}(X)} \operatorname*{Lim\,sup}_{x \to \bar{x}} \partial_H (f + \Psi_{x+E})(x).$$

Here, as in Definition 5.17, the topological Lim sup is taken with respect to the weak* topology on X^* (and the norm topology on X).

Since $\partial_H f(x) \subset \partial_H (f + \Psi_{x+E})(x)$, it is clear that the latter definition coincides with Definition 5.17 whenever the Banach space X is separable.

On the other hand, given any $E \in \mathcal{S}(X)$, if two functions f and g coincide near $\bar{x}$ and are Lipschitz near $\bar{x}$, obviously $f + \Psi_{x+E}$ and $g + \Psi_{x+E}$ coincide near $\bar{x}$, so $\partial_H (f + \Psi_{x+E})(x) = \partial_H (g + \Psi_{x+E})(x)$ for all x near $\bar{x}$. Consequently, we can say that $\partial_A f(\bar{x})$ is a local concept, or more precisely:

PROPOSITION 5.32. Let X be a Banach space and $f, g : X \to \mathbb{R} \cup \{-\infty, +\infty\}$ be two functions which coincide near $\bar{x} \in X$ and are Lipschitz (and finite) near $\bar{x}$. Then one has the equality

$$\partial_A f(\bar{x}) = \partial_A g(\bar{x}).$$

As for Proposition 5.23, the next proposition follows directly from (d) and (f) in Proposition 5.7.

PROPOSITION 5.33. Let X, Y be Banach spaces, $G : X \to Y$, $g : Y \to \mathbb{R} \cup \{-\infty, +\infty\}$, and $f : X \to \mathbb{R} \cup \{-\infty, +\infty\}$.
(a) If f is finite at $\bar{x} \in X$ and Lipschitz near $\bar{x}$, then for any real $r > 0$ one has

$$\partial_A (rf)(\bar{x}) = r\partial_A f(\bar{x}).$$

(b) If G is a $\mathcal{C}^1$-diffeomorphism near $\bar{x} \in X$ and g is finite at $\bar{y} := G(\bar{x})$ and Lipschitz near $\bar{y}$, then

$$\partial_A (g \circ G)(\bar{x}) = DG(\bar{x})^* \big(\partial_A g(\bar{y})\big).$$

As for C-subdifferential, under the Lipschitz property of f, we will show in the next theorem that the multimapping $\partial_A f$ takes on w^*-compact values and is outer semicontinuous.

Let us note first the following lemma.

LEMMA 5.34. *Let $f : U \to \mathbb{R} \cup \{+\infty\}$ be a function from an open set U of a normed space X which is finite at $x \in U$, and let E be a closed vector subspace of X. Let $g : E \to \mathbb{R} \cup \{+\infty\}$ be defined by $g(v) := f(x+v)$ for all $v \in E$. The following hold:*

(a) *One has for any $h \in X$,*

$$\underline{d}_H^+(f + \Psi_{x+E})(x;h) = +\infty \text{ if } h \notin E \text{ and } \underline{d}_H^+(f + \Psi_{x+E})(x;h) = \underline{d}_H^+ g(0;h) \text{ if } h \in E,$$

which ensures, for $x^* \in X^*$ and $v \in E$ with $x+v \in \operatorname{dom} f$, the equivalence

$$x^* \in \partial_H(f + \Psi_{x+v+E})(x+v) \iff x^*_{|E} \in \partial_H g(v).$$

(b) *If in addition f is finite on U and locally Lipschitz therein,*

$$\underline{d}_H^+(f + \Psi_{x+E})(x;h) = \underline{d}_H^+ f(x;h) \text{ for all } h \in E,$$

so in particular, for $x^* \in X^*$, one has the equivalence

$$x^* \in \partial_H(f + \Psi_{x+E})(x) \iff \langle x^*, h \rangle \le \underline{d}_H^+ f(x;h), \forall h \in E.$$

PROOF. Since $X \setminus E$ is open and $\Psi_{x+E}(x+tv) = +\infty$ for any $v \notin E$ and any real $t > 0$, it is clear that $\underline{d}_H^+(f + \Psi_{x+E})(x;h) = +\infty$ if $h \notin E$, and for all $h \in E$

$$\underline{d}_H^+(f + \Psi_{x+E})(x;h) = \liminf_{E \ni v \to h, t \downarrow 0} t^{-1}\big(f(x+tv) - f(x)\big) = \underline{d}_H^+ g(0;h).$$

The assertion (a) is then proved.

Suppose now f is finite and locally Lipschitz on U. Fix any $h \in E$. Then, according to what precedes

$$\begin{aligned} \underline{d}_H^+(f + \Psi_{x+E})(x;h) &= \liminf_{E \ni v \to h, t \downarrow 0} t^{-1}\big(f(x+tv) - f(x)\big) \\ &= \liminf_{t \downarrow 0} t^{-1}\big(f(x+th) - f(x)\big) \\ &= \underline{d}_H^+ f(x;h), \end{aligned}$$

where both last equalities are due to the Lipschitz property of f. $\square$

THEOREM 5.35 (outer semicontinuity and w^*-compactness of A-subdifferential of Lipschitz function). *Let X be a Banach space, U be an open neighborhood of $\overline{x} \in X$, and $f : U \to \mathbb{R}$ be a function which is Lipschitz on U with $\gamma \ge 0$ as a Lipschitz constant. The following properties hold.*

(a) *The multimapping $\partial_A f(\cdot)$ from U into $(X^*, w(X^*, X))$ is both outer semicontinuous and upper semicontinuous on U and*

$$\partial_A f(\overline{x}) = \operatorname*{Lim\,sup}_{x \to \overline{x}} \partial_A f(x).$$

(b) *The inclusion $\partial_H f(\overline{x}) \subset \partial_A f(\overline{x})$ holds, so in particular $D_H f(\overline{x}) \in \partial_A f(\overline{x})$ whenever f is in addition Hadamard differentiable at $\overline{x}$.*

(c) *The A-subdifferential $\partial_A f(\overline{x})$ is nonempty, w^*-compact, and*

$$\partial_A f(\overline{x}) \subset \partial_C f(\overline{x}) \subset \gamma \mathbb{B}_{X^*};$$

in particular $\partial_A f(\overline{x}) = \{Df(\overline{x})\}$ whenever f is strictly Hadamard differentiable at the point $\overline{x}$.

PROOF. Taking any $E \in \mathcal{S}(X)$, for any $u \in U$ and $u^* \in \partial_H(f + \Psi_{u+E})(u)$, we have by Lemma 5.34 above
$$\langle u^*, h \rangle \leq \underline{d}_H^+ f(u; h) \leq \gamma \|h\|, \text{ for all } h \in E.$$
It easily ensues for any $x \in U$ that $\|x^*\| \leq \gamma$ for every $x^* \in \partial_A f(x)$, otherwise stated, $\partial_A f(x) \subset \gamma \mathbb{B}_{X^*}$ for any $x \in U$.
(a) Take any $y \in U$. Given any $E \in \mathcal{S}(X)$ it is clear that

$$\operatorname*{Lim\,sup}_{x \to y} \partial_A f(x) = \operatorname*{Lim\,sup}_{x \to y} \left(\bigcap_{L \in \mathcal{S}(X)} \operatorname*{Lim\,sup}_{u \to x} \partial_H (f + \Psi_{u+L})(u) \right)$$
$$\subset \operatorname*{Lim\,sup}_{x \to y} \left(\operatorname*{Lim\,sup}_{u \to x} \partial_H (f + \Psi_{u+E})(u) \right)$$
$$\subset \operatorname*{Lim\,sup}_{x \to y} \partial_H (f + \Psi_{x+E})(x),$$

where the latter inclusion is due to Proposition 1.11. Thus, taking the intersection over $E \in \mathcal{S}(X)$ gives $\operatorname*{Lim\,sup}_{x \to y} \partial_A f(x) \subset \partial_A f(y)$. Since the converse inclusion is evident, we obtain the outer semicontinuity property of $\partial_A f$ on U. Further, for each $x \in U$ we have that $\partial_A f(x)$ is w^*-closed by Definition 5.31 and $\partial_A f(x) \subset \gamma \mathbb{B}_{X^*}$ as seen above. According to the w^*-compactness of $\mathbb{B}_{X^*}$, the multimapping $\partial_A f$ is then $\| \cdot \|$-to-w^* upper semicontinuous on U (see Proposition 1.39).
(b) For any $E \in \mathcal{S}(X)$ it is clear that $\partial_H f(\bar{x}) \subset \partial_H (f + \Psi_{\bar{x}+E})(\bar{x})$, so

$$\partial_H f(\bar{x}) \subset \bigcap_{E \in \mathcal{S}(X)} \partial_H (f + \Psi_{\bar{x}+E})(\bar{x}) \subset \bigcap_{E \in \mathcal{S}(X)} \operatorname*{Lim\,sup}_{x \to \bar{x}} \partial_H (f + \Psi_{x+E})(x).$$

This yields the first inclusion in (b) and the second is direct.
(c) Take any $E \in \mathcal{S}(X)$ with $\bar{x} \in E$ and denote by f_E the restriction of f to E. By Corollary 5.14, for each $n \in \mathbb{N}$, there is $x_{n,E} \in U \cap \operatorname{Dom} \partial_H f_E$ with $\|x_{n,E} - \bar{x}\| < 1/n$, so we can choose $u^*_{n,E} \in \partial_H f_E(x_{n,E})$. For any $n \in \mathbb{N}$, we clearly have $\|u^*_{n,E}\|_{E^*} \leq \gamma$, so extending $u^*_{n,E}$ to $x^*_{n,E} \in X^*$ with the same norm, obviously

$$x^*_{n,E} \in \partial_H (f + \Psi_{x_{n,E}+E})(x_{n,E}) \quad \text{and} \quad \|x^*_{n,E}\|_{X^*} \leq \gamma.$$

Endowing $\mathcal{S}(X)$ with the directed order $\subset$ and denoting by x^* a w^*-limit point of the net $(x^*_{n,E})_{(n,E) \in \mathbb{N} \times \mathcal{S}(X)}$ it is easily seen that $x^* \in \partial_A f(\bar{x})$. This justifies the non-vacuity of $\partial_A f(\bar{x})$ and its w^*-compactness follows from its w^*-closedness (by definition) and from its inclusion in $\gamma \mathbb{B}_{X^*}$.

Since we know that the inclusion $\partial_C f(\bar{x}) \subset \gamma \mathbb{B}_{X^*}$ is also valid (see Theorem 2.70(d)), it remains to show that $\partial_A f(\bar{x}) \subset \partial_C f(\bar{x})$. Fix any $x^* \in \partial_A f(\bar{x})$ and fix also any $E \in \mathcal{S}(X)$. There are nets $(x_j)_{j \in J}$ converging to $\bar{x}$ and $(x^*_j)_{j \in J}$ converging weakly* to x^* such that, for each $j \in J$ we have $x^*_j \in \partial_H(f + \Psi_{x_j+E})(x_j)$, hence for every $h \in E$
$$\langle x^*_j, h \rangle \leq \underline{d}_H^+ f(x_j; h) \leq f^\circ(x_j; h).$$
For each $h \in E$, taking the limit superior with $j \in J$ gives by the upper semicontinuity of $f^\circ(\cdot; h)$ (see Proposition 2.74(a))
$$\langle x^*, h \rangle \leq \limsup_{j \in J} f^\circ(x_j; h) \leq f^\circ(\bar{x}; h).$$
It ensues that $\langle x^*, h \rangle \leq f^\circ(\bar{x}; h)$ for all $h \in X$, that is, $x^* \in \partial_C f(\bar{x})$, justifying the inclusion $\partial_A f(\bar{x}) \subset \partial_C f(\bar{x})$.

Assume that f is strictly Hadamard differentiable at $\bar{x}$. Then $\partial_C f(\bar{x}) = \{Df(\bar{x})\}$ by Proposition 2.76, so the inclusions $Df(\bar{x}) \in \partial_A f(\bar{x}) \subset \partial_C f(\bar{x})$, from (b) and what precedes, guarantee that $\partial_A f(\bar{x}) = \{Df(\bar{x})\}$. This finishes the proof. □

The above assertions (b) and (c) tell us that $\partial_H f(\bar{x}) \subset \partial_A f(\bar{x}) \subset \partial_C f(\bar{x})$. Under the convexity of f we know that the first and third members coincide with the usual subdifferential of f (see Proposition 5.7(e) and Proposition 2.63). We then have the following corollary (which can also be proved easily in a direct way).

COROLLARY 5.36. *Let X be a Banach space and $f : U \to \mathbb{R}$ be a real-valued continuous convex function on an open convex set U of X. Then, for any $\bar{x} \in U$*

$$\partial_A f(\bar{x}) = \partial f(\bar{x}) := \{x^* \in X^* : \langle x^*, x - \bar{x} \rangle \le f(x) - f(\bar{x}), \forall x \in U\}.$$

Let us say that a collection $\mathcal{L}$ of vector subspaces of a Banach space X is *A-admissible* if each finite-dimensional vector subspace is included in some $L \in \mathcal{L}$ and, for any $L_1, L_2 \in \mathcal{L}$ there is some $L_3 \in \mathcal{L}$ containing $L_1 \cup L_2$. Consequently, the pair $(\mathcal{L}, \subset)$ is directed, and for any cofinal family $(L_j)_{j \in J}$ of $\mathcal{L}$ (see Section A in Appendix), it is easily seen that, for any net $(e_j^*)_{j \in J}$ with $e_j^* \in L_j^{\perp}$ we have that $(e_j^*)_j$ w^*-converges to 0 in X^*; recall that $(L_j)_{j \in J}$ (with $L_j \in \mathcal{L}$) is cofinal in the directed pair $(\mathcal{L}, \subset)$ whenever each set of $\mathcal{L}$ is contained in some L_j (see Section A in Appendix).

Consider now a subset Q of X^*. Given any $M \in \mathcal{L}$, any $V \in \mathcal{N}_U(\bar{x})$ (the collection of neighborhoods of $\bar{x}$ included in U), and any w^*-neighborhood W of zero in X^*, observing that $\partial_{H,\varepsilon}(f + \Psi_{x+L})(x) \subset \partial_{H,\varepsilon}(f + \Psi_{x+M})(x)$ whenever $L \supset M$ (according to the obvious inequality $\underline{d}_H^+(f + \Psi_{x+L})(x;\cdot) \le \underline{d}_H^+(f + \Psi_{x+M})(x;\cdot)$ in such a case), we see that

$$\bigcup_{x \in V} \partial_{H,\varepsilon}(f + \Psi_{x+M})(x) \cap Q = \bigcup_{x \in V, L \in \mathcal{L}, L \supset M} \partial_{H,\varepsilon}(f + \Psi_{x+L})(x) \cap Q.$$

Taking the closures of both members and then the intersection over $V \in \mathcal{N}_U(\bar{x})$, over $\varepsilon > 0$ (resp. with $\varepsilon = 0$) and over $M \in \mathcal{L}$, by Proposition 1.7 we obtain

$$\bigcap_{L \in \mathcal{L}} \operatorname{Lim\,sup}_{x \to \bar{x}, \varepsilon \downarrow 0} \left(\partial_{H,\varepsilon}(f + \Psi_{x+L})(x) \cap Q \right) = \operatorname{Lim\,sup}_{L \in \mathcal{L}, x \to \bar{x}, \varepsilon \downarrow 0} \left(\partial_{H,\varepsilon}(f + \Psi_{x+L})(x) \cap Q \right),$$

(resp. the same equality with $\varepsilon = 0$).

We have established the following result.

LEMMA 5.37. *Let U be an open set of a Banach space X and $f : U \to \mathbb{R} \cup \{-\infty, +\infty\}$ be finite at $\bar{x} \in U$ and locally Lipschitz near $\bar{x}$. Then, for any subset Q of X^* and for any A-admissible collection $\mathcal{L}$ of vector subspaces of X, one has*

$$\bigcap_{L \in \mathcal{L}} \operatorname{Lim\,sup}_{x \to \bar{x}} \left(\partial_H(f + \Psi_{x+L})(x) \cap Q \right) = \operatorname{Lim\,sup}_{L \in \mathcal{L}, x \to \bar{x}} \left(\partial_H(f + \Psi_{x+L})(x) \cap Q \right)$$

and

$$\bigcap_{L \in \mathcal{L}} \operatorname{Lim\,sup}_{x \to \bar{x}, \varepsilon \downarrow 0} \left(\partial_{H,\varepsilon}(f + \Psi_{x+L})(x) \cap Q \right) = \operatorname{Lim\,sup}_{L \in \mathcal{L}, x \to \bar{x}, \varepsilon \downarrow 0} \left(\partial_{H,\varepsilon}(f + \Psi_{x+L})(x) \cap Q \right),$$

otherwise stated, x^* belongs to the first member of the first equality if and only if there are nets $(x_j)_{j \in J}$ in U, $(x_j^*)_{j \in J}$ in X^* and a cofinal family $(L_j)_{j \in J}$ in $\mathcal{L}$ such that $(x_j)_j$ converges to $\bar{x}$, $(x_j^*)_j$ converges weakly to x^* and*

$$x_j^* \in \partial_H(f + \Psi_{x_j + L_j})(x_j) \cap Q \quad \text{for all } j \in J;$$

a similar description holds true with the equality involving ε.

Our next aim is to show that the four members in the above lemma coincide when $Q = X^*$ and coincide also with those associated with $Q = \gamma' \mathbb{B}_{X^*}$, where γ' is greater than a Lipschitz constant of f near $\bar{x}$. To do so, we will use some other lemmas.

LEMMA 5.38. Let X be a Banach space and L be a closed vector subspace admitting an equivalent norm (to the induced norm) which is Gâteaux differentiable off zero. Let $f_i : X \to \mathbb{R} \cup \{+\infty\}$, $i = 1, \cdots, m$, be lower semicontinuous functions. Then for any $x^* \in \partial_H(f_1 + \cdots + f_m + \Psi_{x+L})(x)$, any real $\varepsilon > 0$, and any finite-dimensional vector subspace F of X, there exist $x_1, \cdots, x_m \in x+L$ with $\|x_i - x\| < \varepsilon$ and $|f_i(x_i) - f_i(x)| < \varepsilon$ for $i = 1, \cdots, m$, such that

$$x^* \in \partial_H(f_1 + \Psi_{x_1+L})(x_1) + \cdots + \partial_H(f_m + \Psi_{x_m+L})(x_m) + \varepsilon \mathbb{B}_{X^*} + F^\perp.$$

PROOF. Take any vector basis of F (assumed to be nonzero) and denote by F_0 the vector space generated by the elements in this basis outside L. Since F_0 is finite-dimensional, $E := L \oplus F_0$ is a closed vector subspace of X whose induced norm admits an equivalent Gâteaux differentiable norm (off zero), and clearly E contains F.

Putting $\zeta^* := x^*_{|E}$ (the restriction of x^* to E) and considering $g_i : L \to \mathbb{R} \cup \{+\infty\}$ defined by $g_i(v) = f_i(x+v) + \Psi_L(v)$ for all $v \in E$ with $i = 1, \cdots, m$, we have $\zeta^* \in \partial_H(g_1 + \cdots + g_m)(0)$. Denoting by $\|\cdot\|_E$ the induced norm on E, by Theorem 5.15 there are $v_i \in L$ with $\|v_i\|_E < \varepsilon$ and $|g_i(v_i) - g_i(0)| < \varepsilon$ (which argues that $v_i \in L$), $\zeta_i^* \in \partial_H g_i(v_i)$ and $q^* \in \mathbb{B}_{E^*}$ such that $\zeta_1^* + \cdots + \zeta_m^* - \zeta^* - \varepsilon q^*$ is null on F. Extend q^* to $b^* \in X^*$ with $\|b^*\|_{X^*} = \|q^*\|_{E^*} \leq 1$ (so $b^* \in \mathbb{B}_{X^*}$) and extend also ζ_i^* to a continuous linear functional x_i^* on X. Setting $x_i = x + v_i$, by Lemma 5.34(a) it is clear that

$$x_i^* \in \partial_H(f_i + \Psi_{x+L})(x_i) = \partial_H(f_i + \Psi_{x_i+L})(x_i)$$

(the equality being due to the fact that $x_i + L = x + L$). Finally, it is also clear that $x_1^* + \cdots + x_m^* - x^* - \varepsilon b^*$ is null on F, and hence belongs to $F^\perp$. The lemma is then demonstrated. □

The next lemma deals with the case $\mathcal{L} = \mathcal{F}$.

LEMMA 5.39. Let X be a Banach space and $f : X \to \mathbb{R} \cup \{+\infty\}$ be a function which is finite at $\bar{x}$ and Lipschitz near $\bar{x}$. Then, one has

$$\bigcap_{F \in \mathcal{F}(X)} \operatorname{Lim\,sup}_{x \to \bar{x}} \partial_H(f + \Psi_{x+F})(x) = \bigcap_{F \in \mathcal{F}(X)} \operatorname{Lim\,sup}_{x \to \bar{x}, \varepsilon \downarrow 0} \partial_{H,\varepsilon}(f + \Psi_{x+F})(x).$$

PROOF. It is enough to prove the inclusion of the second member into the first. Fix any real $\varepsilon > 0$ and also $x^* \in \partial_{H,\varepsilon}(f + \Psi_{x+F})(x)$, with $x \in U$ and $F \in \mathcal{F}$. Then $x^* \in \partial_H(f + \varepsilon\|\cdot - x\| + \Psi_{x+F})(x)$, so the above lemma and the Moreau-Rockafellar theorem (see Theorem 2.105) yield $u \in B(x, \varepsilon) \cap (x+F)$ such that

$$x^* \in \partial(f + \Psi_{u+F})(u) + 2\varepsilon \mathbb{B}_{X^*} + F^\perp.$$

From this and Lemma 5.37 the result easily follows. □

LEMMA 5.40. Let U be an open set of a Banach space X and $f : U \to \mathbb{R}$ be a function which is Lipschitz on U with $\gamma \geq 0$ as a Lipschitz constant. Then, for any

A-admissible collection $\mathcal{L}$ of vector subspaces of X, the following hold:

(a) For any $x \in U$, any vector subspace L of X and any real $\varepsilon \geq 0$, one has
$$\partial_{H,\varepsilon}(f + \Psi_{x+L})(x) = \partial_{H,\varepsilon}(f + \Psi_{x+L})(x) \cap (\gamma + \varepsilon)\mathbb{B}_{X^*} + L^\perp.$$

(b) Consequently, for $\bar{x} \in U$ one has the following equalities
$$\bigcap_{L \in \mathcal{L}} \operatorname{Lim\,sup}_{x \to \bar{x}} \partial_H(f + \Psi_{x+L})(x) = \bigcap_{L \in \mathcal{L}} \operatorname{Lim\,sup}_{x \to \bar{x}} \left(\partial_H(f + \Psi_{x+L})(x) \cap \gamma \mathbb{B}_{X^*} \right)$$

and
$$\bigcap_{L \in \mathcal{L}} \operatorname{Lim\,sup}_{x \to \bar{x}, \varepsilon \downarrow 0} \partial_{H,\varepsilon}(f + \Psi_{x+L})(x) = \bigcap_{L \in \mathcal{L}} \operatorname{Lim\,sup}_{x \to \bar{x}, \varepsilon \downarrow 0} \left(\partial_{H,\varepsilon}(f + \Psi_{x+L})(x) \cap (\gamma + \varepsilon)\mathbb{B}_{X^*} \right).$$

PROOF. (a) The right-hand side being evidently included in the other, fix any $x^* \in \partial_{H,\varepsilon}(f + \Psi_{x+L})(x)$. Then $\|x^*_{|L}\|_{L^*} \leq \gamma + \varepsilon$, so $x^*_{|L}$ can be extended to an element u^* in X^* with $\|u^*\|_{X^*} \leq \gamma + \varepsilon$, and $u^* \in \partial_{H,\varepsilon}(f + \Psi_{x+L})(x) \cap (\gamma + \varepsilon)\mathbb{B}_{X^*}$ according to Lemma 5.34(b). Since $x^* - u^*$ is null on L, or equivalently belongs to $L^\perp$, the equality in (a) follows.

(b) We will prove the result involving ε, the other one following in the same way. Obviously, only the inclusion of the first member into the second one needs to be shown. Fix any x^* in the first member. Take any $L \in \mathcal{L}$. By Lemma 5.37 there are a cofinal family $(L_j)_{j \in J}$ of $\mathcal{L}$ and nets $(\varepsilon_j)_{j \in J}$ tending to 0 with $\varepsilon_j > 0$, $(x_j)_{j \in J}$ in U converging to $\bar{x}$ and $(x^*_j)_{j \in J}$ converging weakly* to x^* such that $x^*_j \in \partial_{H,\varepsilon_j}(f + \Psi_{x_j + L_j})(x_j)$. By (a), for each $j \in J$, there exists $e^*_j \in L^\perp_j$ such that $x^*_j + e^*_j \in \partial_{H,\varepsilon_j}(f + \Psi_{x_j + L_j})(x_j) \cap (\gamma + \varepsilon_j)\mathbb{B}_{X^*}$. Since $(e^*_j)_j$ w^*-converges to 0, the desired inclusion is obtained. □

LEMMA 5.41. *Let L be any vector subspace of the Banach space X, K be any w^*-compact set of X^*, and let $M : T \rightrightarrows (X^*, w(X^*, X))$ be a multimapping from a topological space T into X^* such that $M(T) \subset K$. Then for X^* endowed with the $w(X^*, X)$ topology and for any $\bar{t} \in \operatorname{cl}_T(T_0)$ (with $T_0 \subset T$) one has*

(5.2)
$$\bigcap_{F \in \mathcal{F}(X), F \subset L} \operatorname{Lim\,sup}_{t \to T_0 \bar{t}} (M(t) + F^\perp) = \left(\operatorname{Lim\,sup}_{t \to T_0 \bar{t}} (M(t)) \right) + L^\perp = \operatorname{Lim\,sup}_{t \to T_0 \bar{t}} (M(t) + L^\perp)$$

hence in particular
$$\bigcap_{F \in \mathcal{F}(X), F \subset L} (K + F^\perp) = K + L^\perp.$$

PROOF. The equality related to $K + L^\perp$ directly follows from (5.2) with $M(t) = K$ for all $t \in T$. Further, we note that the middle-hand side of (5.2) is included in the right-hand one which itself is included in the left-hand. It remains to show that this left-hand side is included in the middle. Denote $\mathcal{F}_L := \{F \in \mathcal{F}(X) : F \subset L\}$. Fix any x^* in the left-hand side of (5.2). Then for each $F \in \mathcal{F}_L$ there are a net $(t_{F,i})_{i \in I_F}$ in T_0 converging to $\bar{t}$, a net $(\zeta_{F,i})_{i \in I_F}$ in X^* with $\zeta_{F,i} \in M(t_{F,i})$ and a net $(e_{F,i})_{i \in I_F}$ with $e_{F,i} \in F^\perp$ such that
$$x^* = \lim_{i \in I_F} (\zeta_{F,i} + e_{F,i})$$
with respect to weak* topology in X^*. Set $J := \Pi_{F \in \mathcal{F}_L} I_F$ and endow $\mathcal{F}_L \times J = \mathcal{F}_L \times \Pi_{F \in \mathcal{F}_L} I_F$ with the product preorder, where the preorder of $\mathcal{F}_L$ is the inclusion

C. This means that, for $E, G \in \mathcal{F}_L$ and for $j = (j_F)_{F \in \mathcal{F}_L}$ and $\gamma = (\gamma_F)_{F \in \mathcal{F}_L}$ in $J = \Pi_{F \in \mathcal{F}_L} I_F$ the relation $(E, j) \preceq (G, \gamma)$ is equivalent to $E \subset G$ and $j_F \preceq_F \gamma_F$ for every $F \in \mathcal{F}_L$. For each $(F, j) \in \mathcal{F}_L \times J$ put

$$\tau_{F,j} := t_{F, j_F}, \quad \xi_{F,j} := \zeta_{F, j_F} \quad \text{and} \quad \ell_{F,j} := e_{F, j_F}.$$

Then we know (see Proposition A.2 in Appendix) that the net $(\tau_{F,j})_{(F,j) \in \mathcal{F}_L \times J}$ converges to $\bar{t}$. Further, by the inclusion of $M(\tau_{F,j})$ in the weak* compact set K a (non relabeled) subnet of $(\xi_{F,j})_{(F,j) \in \mathcal{F}_L \times J}$ weakly* converges to some $\xi \in X^*$, so the corresponding subnet of $(\ell_{F,j})_{(F,j) \in \mathcal{F}_L \times J}$ weakly* converges also to some $\ell \in X^*$ with $x^* = \xi + \ell$. It is clear that $\ell \in L^\perp$, hence x^* belongs to the middle-hand side of (5.2) as desired. This finishes the proof of the lemma. □

With the above lemmas at hands we can now state and prove the theorem providing the desired equalities concerning the approximate subdifferential of a locally Lipschitz function.

THEOREM 5.42 (description of A-subdifferential of Lipschitz functions with Hadamard ε-subdifferential). Let X be a Banach space and $\mathcal{L}$ be an A-admissible collection of closed vector subspaces admitting each one an equivalent renorm (to the induced norm) which is Gâteaux differentiable off zero. Let also $f : U \to \mathbb{R}$ be a Lipschitz function on an open neighborhood U of $\bar{x} \in X$ with $\gamma \geq 0$ as a Lipschitz constant. Then, the following equalities hold:

$$\partial_A f(\bar{x}) = \bigcap_{L \in \mathcal{L}} \operatorname*{Lim\,sup}_{x \to \bar{x}} \partial_H (f + \Psi_{x+L})(x) = \bigcap_{L \in \mathcal{L}} \operatorname*{Lim\,sup}_{x \to \bar{x}} \left(\partial_H (f + \Psi_{x+L})(x) \cap \gamma \mathbb{B}_{X^*} \right)$$

$$= \bigcap_{L \in \mathcal{L}} \operatorname*{Lim\,sup}_{x \to \bar{x}, \varepsilon \downarrow 0} \left(\partial_{H, \varepsilon}(f + \Psi_{x+L})(x) \cap (\gamma + \varepsilon) \mathbb{B}_{X^*} \right)$$

$$= \bigcap_{L \in \mathcal{L}} \operatorname*{Lim\,sup}_{x \to \bar{x}, \varepsilon \downarrow 0} \partial_{H, \varepsilon}(f + \Psi_{x+L})(x).$$

PROOF. In their order in the theorem, denote the above members by G, $G_\mathcal{L}$, $G_\mathcal{L}^\gamma$, $G_{\mathcal{L},0}^\gamma$ and $G_{\mathcal{L},0}$. By Lemma 5.40(b) we know that $G_\mathcal{L} = G_\mathcal{L}^\gamma$ and $G_{\mathcal{L},0} = G_{\mathcal{L},0}^\gamma$. By definition of $G_\mathcal{L}$ and $G_{\mathcal{L},0}$ we also have $G_\mathcal{L} \subset G_{\mathcal{L},0} \subset G_{\mathcal{F},0}$, and $G_{\mathcal{F},0} = G_\mathcal{F}$ by Lemma 5.39, where $\mathcal{F} := \mathcal{F}(X)$. Since $G = G_\mathcal{S}$ by definition of $\partial_A f$, we only have to show that $G_\mathcal{F} \subset G_\mathcal{L}$.

First, fix any $F \in \mathcal{F}$. For every $L \in \mathcal{L}$ with $L \supset F$, the equality $f + \Psi_{x+F} = f + \Psi_{x+F} + \Psi_{x+L}$ and Lemma 5.38 (with $f_1 := f$ and $f_2 := \Psi_{x+F}$) ensure that

$$\partial_H(f + \Psi_{x+F})(x) \subset \operatorname*{Lim\,sup}_{u \to x} \left(\partial_H(f + \Psi_{x+L})(u) + F^\perp \right)$$

$$\subset \operatorname*{Lim\,sup}_{u \to x} \left(\partial_H(f + \Psi_{u+L})(u) + F^\perp \right)$$

$$= \operatorname*{Lim\,sup}_{u \to x} \left(\partial_H(f + \Psi_{u+L})(u) \cap \gamma \mathbb{B}_{X^*} + F^\perp \right),$$

where the latter equality is due to Lemma 5.40(a) and to the equality $L^\perp + F^\perp = F^\perp$. This and Proposition 1.11 yield

$$G_{\mathcal{F}} \subset \bigcap_{F \in \mathcal{F}} \operatorname*{Lim\,sup}_{x \to \overline{x}} \left(\bigcap_{L \in \mathcal{L}, L \supset F} \operatorname*{Lim\,sup}_{u \to x} \left(\partial_H(f + \Psi_{u+L})(u) \cap \gamma \mathbb{B}_{X^*} + F^\perp \right) \right)$$

$$\subset \bigcap_{F \in \mathcal{F}} \bigcap_{L \in \mathcal{L}, L \supset F} \operatorname*{Lim\,sup}_{u \to \overline{x}} \left(\partial_H(f + \Psi_{u+L})(u) \cap \gamma \mathbb{B}_{X^*} + F^\perp \right),$$

so using the w^*-compactness of $\gamma \mathbb{B}_{X^*}$ and applying Lemma 5.41 it results that

$$G_{\mathcal{F}} \subset \bigcap_{L \in \mathcal{L}} \operatorname*{Lim\,sup}_{u \to \overline{x}} \left(\partial_H(f + \Psi_{u+L})(u) \cap \gamma \mathbb{B}_{X^*} + L^\perp \right).$$

Since $\partial_H(f + \Psi_{u+L})(u) + L^\perp = \partial_H(f + \Psi_{u+L})(u)$, we deduce that

$$G_{\mathcal{F}} \subset \bigcap_{L \in \mathcal{L}} \operatorname*{Lim\,sup}_{u \to \overline{x}} \partial_H(f + \Psi_{u+L})(u) = G_{\mathcal{L}},$$

which finishes the proof. □

COROLLARY 5.43. *Let X be a Banach space admitting an equivalent Gâteaux differentiable (off zero) renorm and let $f : U \to \mathbb{R}$ be a Lipschitz function on an open neighborhood U of $\overline{x} \in X$ with $\gamma \geq 0$ as a Lipschitz constant. Then one has*

$$\partial_A f(\overline{x}) = \operatorname*{Lim\,sup}_{x \to \overline{x}} \partial_H f(x) = \operatorname*{Lim\,sup}_{x \to \overline{x}} \left(\partial_H f(x) \cap \gamma \mathbb{B}_{X^*} \right)$$

$$= \operatorname*{Lim\,sup}_{x \to \overline{x}, \varepsilon \downarrow 0} \left(\partial_{H,\varepsilon} f(x) \cap (\gamma + \varepsilon) \mathbb{B}_{X^*} \right) = \operatorname*{Lim\,sup}_{x \to \overline{x}, \varepsilon \downarrow 0} \partial_{H,\varepsilon} f(x).$$

PROOF. Under the assumptions, the class $\mathcal{L} := \{X\}$ formed by the unique space X is obviously A-admissible and X admits an equivalent Gâteaux differentiable renorm. It suffices to apply Theorem 5.42 with the A-admissible class $\mathcal{L} := \{X\}$. □

In addition to Theorem 5.35(c) the next theorem shows that $\partial_C f(\overline{x})$ is the weak* closed convex hull of $\partial_A f(\overline{x})$ whenever f is Lipschitz near $\overline{x}$. For a locally Lipschitz function g on a separable Banach space E we will see in Theorem 13.41 that g is Hadamard differentiable (or equivalently Gâteaux differentiable) on a specific dense set Δ_g and that the equality for the Clarke directional derivative in Theorem 2.83 is extended as

(5.3) $\quad g^\circ(x; h) = \max \left\{ \left\langle \lim_{n \to \infty} D_H g(x_n), h \right\rangle : \Delta_g \ni x_n \to x \right\}$ for all $x, h \in E$.

THEOREM 5.44 (C-subdifferential of Lipschitz function as w^*-closed convex hull of A-subdifferential). *Let X be a Banach space and $f : X \to \mathbb{R} \cup \{-\infty, +\infty\}$ be finite at $\overline{x} \in X$ and Lipschitz near $\overline{x}$. Then $\partial_A f(\overline{x})$ is a nonempty w^*-compact set in X^* and*

$$\partial_C f(\overline{x}) = \overline{\operatorname{co}}^* \left(\partial_A f(\overline{x}) \right).$$

PROOF. Without loss of generality we may suppose that f is locally Lipschitz on X. Fix any nonzero $h \in X$. Take any $x^* \in \partial_A f(\overline{x})$ and choose nets $(x_j)_{j \in J}$ converging to $\overline{x}$ and $(x_j^*)_{j \in J}$ converging weakly* to x^* such that $x_j^* \in \partial_H(f + \Psi_{x_j + \mathbb{R}h})(x_j)$. Then $\langle x_j^*, h \rangle \leq \underline{d}_H^+ f(x_j; h) \leq f^\circ(x_j; h)$, hence

$$\langle x^*, h \rangle \leq \limsup_{j \in J} f^\circ(x_j; h) \leq f^\circ(\overline{x}; h).$$

This says that

(5.4)
$$\sup_{x^* \in \partial_A f(\bar{x})} \langle x^*, h \rangle \leq f^\circ(\bar{x}; h).$$

Concerning the converse inequality, consider sequences $(t_n)_n$ in $]0, +\infty[$ tending to 0 and $(x_n)_n$ converging to $\bar{x}$ such that

(5.5)
$$f^\circ(\bar{x}; h) = \lim_{n \to \infty} t_n^{-1} \big(f(x_n + t_n h) - f(x_n) \big),$$

and denote by E_0 the closed vector subspace generated by $\{\bar{x}, h\} \cup \{x_n : n \in \mathbb{N}\}$. Take any $E \in \mathcal{S}_0 := \{L \in \mathcal{S}(X) : L \supset E_0\}$. Let f_E be the restriction of f to the separable Banach space E. From (5.5) it is clear that $f^\circ(\bar{x}; h) = f_E^\circ(\bar{x}; h) =: \ell$. By (5.3), for each $n \in \mathbb{N}$, there are $u_{n,E} \in B_E(\bar{x}, 1/n) \cap \Delta_{f_E}$ and $u_{n,E}^* := D_H f_E(u_n)$ with $|u_{n,E}^*(h) - \ell| < 1/n$. Extend $u_{n,E}^*$ to a continuous linear functional $x_{n,E}^*$ to the whole space X with $\|x_{n,E}^*\| = \|u_{n,E}^*\| \leq \gamma$. Clearly, $x_{n,E}^* \in \partial_H(f + \Psi_{u_{n,E}+E})(u_{n,E})$, and the nets $(u_{n,E})_{(n,E) \in \mathbb{N} \times \mathcal{S}_0}$ and $(\langle x_{n,E}^*, h \rangle)_{(n,E) \in \mathbb{N} \times \mathcal{S}_0}$ converges to $\bar{x}$ and ℓ respectively. Denoting by x^* the limit of a w^*-convergent subnet of $(x_{n,E}^*)_{(n,E) \in \mathbb{N} \times \mathcal{S}_0}$, it results that $x^* \in \partial_A f(\bar{x})$ and $\ell = \langle x^*, h \rangle$. Combining the latter equality with (5.4) yields $f^\circ(\bar{x}; h) = \max\{\langle x^*, h \rangle : x^* \in \partial_A f(\bar{x})\}$ for all $h \in X$. This justifies that $\partial_C f(\bar{x}) = \overline{\text{co}}^*(\partial_A f(\bar{x}))$, and finally $\partial_A f(\bar{x})$ is nonempty and w^*-compact as already seen in Theorem 5.35(c). □

Before establishing a first sum rule let us define the A-subdifferential regularity of a function.

DEFINITION 5.45. Let U be an open set of a Banach space X and $f : U \to \mathbb{R} \cup \{-\infty, +\infty\}$ be finite at $\bar{x} \in U$ and Lipschitz continuous near $\bar{x}$. The function f is called A-subdifferentially regular at $\bar{x}$ provided $\partial_A f(\bar{x}) = \partial_H f(\bar{x})$.

THEOREM 5.46 (A-subdifferential of a sum of Lipschitz functions). Let U be an open set of a Banach space X and $f, g : U \to \mathbb{R}$ be two locally Lipschitz functions. Then, for any $\bar{x} \in U$ one has

$$\partial_A(f + g)(\bar{x}) \subset \partial_A f(\bar{x}) + \partial_A g(\bar{x}).$$

The inclusion is an equality whenever both functions f and g are in addition A-subdifferentially regular at $\bar{x}$.

PROOF. We may suppose that both f and g are Lipschitz on U. Let $\gamma \geq 0$ be a common Lipschitz constant and let $x^* \in \partial_A(f + g)(\bar{x})$. By Theorem 5.42 and Lemma 5.38 there are a cofinal family $(F_j)_{j \in J}$ of $\mathcal{F}$ and nets $(u_j)_{j \in J}$, $(v_j)_{j \in J}$ in U both converging to $\bar{x}$, $(u_j^*)_{j \in J}$ and $(v_j^*)_{j \in J}$ in X^* such that $u_j^* \in \partial_H(f + \Psi_{u_j + F_j})(u_j)$, $v_j^* \in \partial_H(g + \Psi_{v_j + F_j})(v_j)$ and $(u_j^* + v_j^*)_{j \in J}$ converges weakly* to x^*. By Lemma 5.40(a) for each $j \in J$ there are

$$\zeta_j^* \in \partial_H(f + \Psi_{u_j + F_j}) \cap \gamma \mathbb{B}_{X^*}, \quad \xi_j^* \in \partial_H(g + \Psi_{v_j + F_j}) \cap \gamma \mathbb{B}_{X^*},$$

$e_j^* \in F_j^\perp$ and $\ell_j^* \in F_j^\perp$ such that $u_j^* = \zeta_j^* + e_j^*$ and $v_j^* = \xi_j^* + \ell_j^*$. We know that $(e_j^*)_j$ and $(\ell_j^*)_j$ converge weakly* to zero, and by w^*-compactness of $\gamma \mathbb{B}_{X^*}$ we may suppose that $(\zeta_j^*)_j$ and $(\xi_j^*)_j$ converge weakly* in X^* to u^* and v^* respectively. It results that $u^* \in \partial_A f(\bar{x})$, $v^* \in \partial_A g(\bar{x})$ and $x^* = u^* + v^*$, proving the desired inclusion.

Under the A-subdifferential regularity assumption, by Proposition 5.7(f) we obtain
$$\partial_H f(\bar{x}) + \partial_H g(\bar{x}) \subset \partial_H (f+g)(\bar{x}) \subset \partial_A (f+g)(\bar{x})$$
$$\subset \partial_A f(\bar{x}) + \partial_A g(\bar{x}) = \partial_H f(\bar{x}) + \partial_H g(\bar{x}),$$
which translates the desired equality as well as the A-subdifferential regularity of $f+g$ at $\bar{x}$. □

Concerning the case of sum with separate variables a similar conclusion in equality form is valid. Taking the admissible family of $X_1 \times \cdots \times X_m$ constituted by finite-dimensional vector subspaces in the form $E_1 \times \cdots \times E_m$, and using Theorem 5.42, Proposition 5.12 and Proposition 1.5(b), we directly derive the following result.

PROPOSITION 5.47. Let $X, Y, X_1, \cdots, X_m$ be Banach spaces, $g : Y \to \mathbb{R}$ be a locally Lipschitz function on Y, and for each $i = 1, \cdots, m$, let $f_i : X_i \to \mathbb{R}$ be a locally Lipschitz function on X_i. Then the following properties hold.
(a) For $\varphi(u_1, \cdots, u_m) := f_1(u_1) + \cdots + f_m(u_m)$ and for any $x = (x_1, \cdots, x_m) \in X_1 \times \cdots \times X_m$, one has
$$\partial_A \varphi(x) = \partial_A f_1(x_1) \times \cdots \times \partial_A f_m(x_m).$$
(b) For $h : X \times Y \to \mathbb{R}$ with $h(x', y') = g(y')$ for all $(x', y') \in X \times Y$, one has for any $x \in X$ and $y \in Y$,
$$\partial_A h(x, y) = \{0\} \times \partial_A g(y).$$

5.4. A-normal cone and A-subdifferential of general functions

Taking into account the comments preceding Definition 5.25, we continue to use the distance function to define the approximate normal cone in general Banach spaces.

5.4.1. A-normal cone in general Banach spaces.
The definition below repeats Definition 5.25 in the framework of general Banach spaces.

DEFINITION 5.48. let S be a subset of a general Banach space $(X, \|\cdot\|)$. The *Ioffe approximate normal cone* or *A-normal cone* $N^A(S; \bar{x})$ of S at $\bar{x} \in S$ is defined by
$$N^A(S; \bar{x}) = \mathbb{R}_+ \partial_A d_S(\bar{x}).$$
If $\bar{x} \notin S$, one sets $N^A(S; \bar{x}) = \emptyset$.

The function d_S being Lipschitz continuous, the above definition and Theorem 5.35(c) directly gives the following:

PROPOSITION 5.49. Let S be a nonempty subset of a Banach space $(X, \|\cdot\|)$ and $\bar{x} \in S$. Then the following inclusions hold
$$N^A(S; \bar{x}) \subset \mathbb{R}_+ \partial_C d_S(\bar{x}) \subset N^C(S; \bar{x}).$$

As in the case of separable Banach spaces, from Definition 5.48 it is clear that
$$N^A(S; \bar{x}) = N^A(\operatorname{cl} S; \bar{x}) \quad \text{for any } \bar{x} \in S.$$

Recalling (by Lemma 2.219) that, given $\bar{x} \in S$ we have
$$d(x, S) = d\big(x, S \cap B(\bar{x}, 2r)\big) \quad \text{for all } x \in B(\bar{x}, r),$$
we also see by Proposition 5.32 that the approximate normal cone is a local concept:

PROPOSITION 5.50. *Let S be a subset of a Banach space X and let $\bar{x} \in S$. Then, for any neighborhood U of $\bar{x}$ one has*
$$N^A(S;\bar{x}) = N^A(S \cap U; \bar{x}).$$

When S is convex, the following directly follows from Proposition 2.97 and Corollary 5.36.

PROPOSITION 5.51. *Let S be a convex set of a Banach space X. For any $\bar{x} \in S$, one has*
$$N^A(S;\bar{x}) = \mathbb{R}_+ \partial d_S(\bar{x}) = N(S;\bar{x}) := \{x^* \in X^* : \langle x^*, x - \bar{x} \rangle \leq 0, \forall x \in S\}.$$

Through the above proposition we see in particular that, for convex sets, Definition 5.48 does not depend on equivalent norms. The fact that it is so for any set, will be established in Theorem 5.57. Some intermediate results are necessary to achieve this theorem. Let us begin with the following one.

LEMMA 5.52. *Let X be a Banach space and $f_1, \cdots, f_m : X \to \mathbb{R} \cup \{+\infty\}$ be lower semicontinuous functions. Let also E be a closed vector subspace whose induced norm admits an equivalent Gâteaux differentiable renorm (off zero) and let $x^* \in \partial(f_1 + \cdots + f_m + \Psi_{x+E})(x)$. Then, for any real $\varepsilon > 0$ and any finite-dimensional vector subspace F of E, there are $x_i \in B(x, \varepsilon)$ with $|f_i(x_i) - f_i(x)| < \varepsilon$ such that*
$$x^* \in \partial_H(f_1 + \Psi_{x+E})(x_1) + \cdots + \partial_H(f_m + \Psi_{x+E})(x_m) + \varepsilon \mathbb{B}_{X^*} + F^\perp.$$

PROOF. Consider $\overline{f}_i : E \to \mathbb{R} \cup \{+\infty\}$ defined by $\overline{f}_i(v) = f_i(x+v)$ for all $v \in E$. Denoting by $x^*_{|E}$ the restriction of x^* to E, by Lemma 5.34(a) we have that $x^*_{|E} \in \partial(\overline{f}_1 + \cdots + \overline{f}_m)(0)$. By Theorem 5.15 there are $v_i \in E \cap B(0, \varepsilon)$ with $|\overline{f}_i(v_i) - \overline{f}_i(0)| < \varepsilon$, $v_i^* \in E^*$ with $v_i^* \in \partial_H \overline{f}_i(v_i)$, $b_0^* \in \mathbb{B}_{E^*}$, and $e_0^* \in E^*$ null on F such that
$$x^*_{|E} = v_1^* + \cdots + v_m^* + \varepsilon b_0^* + e_0^*.$$
Define x_i^* and b^* as continuous linear functionals on X extending, with the same norms, v_i^* and b_0^* respectively. Then $\zeta^* := x^* - x_1^* - \cdots - x_m^* - \varepsilon b^* \in F^\perp$, and Lemma 5.34(a) again ensures that $x_i^* \in \partial_H(f_i + \Psi_{x+E})(x_i)$, where $x_i := x + v_i$. The proof is then complete. □

Given a vector subspace E of a normed space X and a mapping g from an open set U of X into a normed space Y, we say that the mapping g is Fréchet *differentiable at a point* $a \in U$ *along* E whenever there is a continuous linear mapping $\Lambda : E \to Y$ such that for all $h \in E$ with $a + h \in U$ we have

(5.6) $\qquad g(a+h) = g(a) + \Lambda(h) + \|h\|\varepsilon(h), \quad \text{with } \varepsilon(h) \to 0_Y \text{ as } h \to 0_E.$

This amounts to saying that the mapping $h \mapsto g(a+h)$ restricted to the open set $E \cap (-a + U)$ in E is Fréchet differentiable at 0_E. The Gâteaux (resp. Hadamard) differentiability of g at a along E is defined in a similar way.

LEMMA 5.53. *Let F be a finite-dimensional vector subspace of a normed space $(X, \|\cdot\|)$. Then for each real $\delta > 0$ there is an equivalent norm $\|\|\cdot\|\|$ on X which is differentiable along F at every nonzero point in X and such that*
$$\big| \|x\| - \|\|x\|\| \big| \leq \delta \|x\| \quad \text{for all } x \in X.$$

PROOF. We may suppose $0 < \delta < 1$. Fix an inner product $(\cdot|\cdot)_F$ on F and denote by $A : F \to F^*$ the linear mapping defined by $A(u) := (u|\cdot)_F$ for every $u \in F$, so $\langle v, A(u)\rangle_{F,F^*} = (u|v)_F$. Relative to the norm induced by $\|\cdot\|$ on F and to the associated dual norm on F^*, according to the isomorphism property of A, we have
$$(1/\|A^{-1}\|)\|u\| \leq \|A(u)\| \leq \|A\|\,\|u\| \quad \text{for all } u \in F.$$
The function $q : F \to \mathbb{R}$ defined by $q(u) := \langle u, A(u)\rangle_{F,F^*}$ for all $u \in F$, is then a positively definite quadratic form on F. For each real $t > 0$, define the function $\varphi_t : X \to [0, +\infty[$ by
$$\varphi_t(x) := \min_{u \in F} \left(\|x - u\|^2 + tq(u) \right) \quad \text{for all } x \in X.$$
It is not difficult to verify that $\sqrt{\varphi_t(\cdot)}$ is a norm on X. So, we proceed to prove that, for t large enough, this norm enjoys the required properties.

Fix any $x \in X$. Since the function $g : F \to \mathbb{R}$, with $g(u) = \|x - u\|^2 + tq(u)$ for all $u \in F$, is strictly convex on F along with $g(u) \to +\infty$ as $\|u\| \to +\infty$, it has a unique minimizer π_x on the finite-dimensional space F. It ensues that
$$0 \in \partial g(\pi_x) \subset 2tA(\pi_x) + 2\|x - \pi_x\|\mathbb{B}_{F^*},$$
and hence $t\|A(\pi_x)\| \leq \|x - \pi_x\| \leq \|x\| + \|\pi_x\|$. Putting $\gamma := 1/\|A^{-1}\|$ and taking $t > 1/\gamma$, we obtain
$$\gamma t \|\pi_x\| \leq \|x\| + \|\pi_x\|, \quad \text{hence } \|\pi_x\| \leq (\gamma t - 1)^{-1}\|x\|.$$
It then follows from the definition of φ_t that
$$\|x\|^2 \geq \varphi_t(x) \geq \|x - \pi_x\|^2 \geq \|x\|^2 - 2\|x\|\,\|\pi_x\|$$
$$\geq \|x\|^2 - 2(\gamma t - 1)^{-1}\|x\|^2 = \left(1 - 2(\gamma t - 1)^{-1}\right)\|x\|^2,$$
so fixing $t > 1/\gamma$ and such that $2(\gamma t - 1)^{-1} < \delta^2$, we see that
$$\left| \|x\| - \sqrt{\varphi_t(x)} \right| \leq \delta\|x\| \quad \text{for all } x \in X.$$

Now fix any $x \in X$ and let us show that the restriction function Φ_t of $\varphi_t(x + \cdot)$ to the vector space F is differentiable at 0_F. This function Φ_t being convex and continuous, $\partial \Phi_t(0_F) \neq \emptyset$. On the other hand, for any $h \in F$
$$\Phi_t(h) = \min_{u \in F}\left(\|x + h - u\|^2 + tq(u)\right) = \min_{v \in F}\left(\|x - v\|^2 + tq(v + h)\right)$$
$$\leq \|x - \pi_x\|^2 + tq(\pi_x + h) = \|x - \pi_x\|^2 + tq(\pi_x) + t\langle h, 2A(\pi_x)\rangle_{F,F^*} + tq(h)$$
$$= \varphi_t(x) + t\langle h, 2A(\pi_x)\rangle_{F,F^*} + tq(h) = \Phi_t(0_F) + t\langle h, 2A(\pi_x)\rangle_{F,F^*} + tq(h).$$

Since $q(h)/\|h\| \to 0$ as $h \to 0_F$, we derive that $-2tA(\pi_x) \in \partial_H(-\Phi_t)(0_F)$. This combined with the non-vacuity of $\partial \Phi_t(0_F)$ guarantees (see Proposition 5.11) the differentiability of Φ_t at 0_F, or equivalently the differentiability of φ_t at x along F. This justifies the differentiability of $\sqrt{\varphi_t(\cdot)}$ along F at any nonzero $x \in X$, so the proof is finished. $\square$

LEMMA 5.54. *Let $(X, \|\cdot\|)$ be a Banach space, S be a nonempty closed subset of X, and F be a finite-dimensional vector subspace of X. Let $u \notin S$ and $u^* \in \partial_H(d_S + \Psi_{u+F})(u)$. Then, for any real $\varepsilon > 0$, there are $\bar{x} \in S$ and $x^* \in \partial_{H,\varepsilon}(d_S + \Psi_{\bar{x}+F})(\bar{x})$ such that*
$$\|\bar{x} - u\| \leq d_S(u) + \varepsilon \quad \text{and} \quad x^* - u^* \in \varepsilon \mathbb{B}_{X^*} + F^\perp.$$

5.4. A-NORMAL CONE AND A-SUBDIFFERENTIAL OF GENERAL FUNCTIONS

PROOF. We may suppose $\varepsilon < 1$. By Corollary 5.6 there is a positive real $\delta < \min\{\varepsilon/2, d_S(u)\}$ such that

$$d_S(u) \leq d_S(u+h) - \langle u^*, h \rangle + \frac{\varepsilon}{3}\|h\| \quad \text{for all } h \in F \cap \delta\mathbb{B}_X = \delta\mathbb{B}_F,$$

where $\mathbb{B}_F := F \cap \mathbb{B}_X$. As in Lemma 5.26, setting

$$\varphi(x,h) := \|x - (u+h)\| - \langle u^*, h \rangle + \frac{\varepsilon}{3}\|h\| \quad \text{for all } x \in X, h \in X,$$

we see that

(5.7) $$\inf_{x \in S, h \in \delta\mathbb{B}_F} \varphi(x,h) = d_S(u) = \inf_{x \in S} \varphi(x,0).$$

Fix a positive real $\eta < \delta/2$ with $2\eta\|u^*\| < \varepsilon/3$. Choose a point $a \in S$ such that

(5.8) $$\varphi(a,0) = \|a - u\| \leq (\eta^2/3) + d_S(u).$$

Let $\|\!\|\cdot\|\!\|$ be an equivalent norm on X, given by the above lemma, which is differentiable at any nonzero point of X along F and such that

(5.9) $$\big|\|\!\|x\|\!\| - \|x\|\big| \leq \frac{\eta^2}{3(\|a-u\| + 2\delta)}\|x\| \quad \text{for all } x \in X.$$

Define

$$\varphi_0(x,h) := \|\!\|x - (u+h)\|\!\| - \langle u^*, h \rangle + \frac{\varepsilon}{3}\|h\| \quad \text{for all } x, h \in X.$$

Then, for any $x \in X$ with $\|x - a\| \leq \delta$ and any $h \in X$ with $\|h\| \leq \delta$, noting that

$$\|x - u - h\| \leq \|x - a\| + \|a - u\| + \|h\| \leq \|a - u\| + 2\delta,$$

we deduce that

$$|\varphi_0(x,h) - \varphi(x,h)| = \big|\|\!\|x - (u+h)\|\!\| - \|x - (u+h)\|\big|$$

(5.10) $$\leq \frac{\eta^2}{3(\|a-u\| + 2\delta)}\|x - u - h\| \leq \frac{1}{3}\eta^2.$$

It results that

$$\inf\{\varphi_0(x,h) : h \in \delta\mathbb{B}_F, x \in S \cap B_X[a,\delta]\}$$
$$\geq \inf\{\varphi(x,h) : h \in \delta\mathbb{B}_F, x \in S \cap B_X[a,\delta]\} - \eta^2/3$$
$$\geq \inf\{\varphi(x,h) : h \in \delta\mathbb{B}_F, x \in S\} - \eta^2/3,$$

and recalling by (5.7) that the latter infimum is equal to $d_S(u)$, we obtain by (5.8) and then by (5.10) that

$$\inf\{\varphi_0(x,h) : h \in \delta\mathbb{B}_F, x \in S \cap B_X[a,\delta]\} \geq d_S(u) - \frac{1}{3}\eta^2$$
$$\geq \varphi(a,0) - \frac{2}{3}\eta^2 \geq \varphi_0(a,0) - \eta^2.$$

By the Ekeland variational principle there are $\overline{h} \in F$ with $\|\overline{h}\| \leq \eta < \delta$ and $\overline{x} \in X$ with $\|\overline{x} - a\| \leq \eta < \delta$ such that

$$\varphi_0(\overline{x}, \overline{h}) \leq \varphi_0(x,h) + \eta(\|x - \overline{x}\| + \|h - \overline{h}\|) \quad \text{for all } x \in S \cap B_X[a,\delta], h \in B_F[0,\delta].$$

The function $\varphi_0(\cdot, h) + \eta(\|\cdot - \overline{x}\| + \|h - \overline{h}\|)$ being $(1 + 2\eta)$-Lipschitz with respect to the norm $\|\cdot\|$ according to (5.9), it follows (see Lemma 2.219 and Lemma 2.215) that for all x in X near $\overline{x}$ and $h \in F$ with $\|h - \overline{h}\| < \delta - \eta$

(5.11) $$\varphi_0(\overline{x}, \overline{h}) \leq \varphi_0(x,h) + (1 + 2\eta)d_S(x) + \eta(\|x - \overline{x}\| + \|h - \overline{h}\|).$$

On the other hand, note that
$$\|\bar{x} - u - \bar{h}\| \geq \|a - u\| - \|a - \bar{x}\| - \|\bar{h}\| \geq d_S(u) - 2\eta > 0,$$
hence $\|\cdot\|$ is differentiable at $\bar{x} - u - \bar{h} \neq 0$ along F. Let $z^* \in X^*$ be such that the restriction of $-z^*$ to F coincides with the derivative of $\|\cdot\|$ at $\bar{x} - u - \bar{h}$ along the space F. From (5.11) we deduce that for all $v \in F$ and $h \in F$
$$0 \leq \langle -z^*, v - h \rangle - \langle u^*, h \rangle + (1 + 2\eta)\underline{d}_H^+(d_S)(\bar{x}; v) + \eta\|v\| + \left(\eta + \frac{\varepsilon}{3}\right)\|h\|.$$
With $h = 0$ we obtain for all $v \in F$,
$$\langle z^*, v \rangle \leq (1 + 2\eta)\underline{d}_H^+(d_S)(\bar{x}; v) + \eta\|v\|,$$
and taking $v = 0$ gives for all $h \in F$
$$\langle u^* - z^*, h \rangle \leq \left(\eta + \frac{\varepsilon}{3}\right)\|h\|,$$
which entails that $u^* - z^* \in (\eta + (\varepsilon/3))\mathbb{B}_{X^*} + F^\perp$. Putting $x^* := (1 + 2\eta)^{-1}z^*$ we observe that for all $v \in F$,
$$\langle x^*, v \rangle \leq \underline{d}_H^+(d_S)(\bar{x}; v) + \frac{\eta}{1 + 2\eta}\|v\| \leq \underline{d}_H^+(d_S + \Psi_{\bar{x}+F})(\bar{x}; v) + \eta\|v\|,$$
so $x^* \in \partial_{H,\varepsilon}(d_S + \Psi_{\bar{x}+F})(\bar{x})$. We also observe that
$$u^* - x^* = \frac{1}{1 + 2\eta}(u^* - z^*) + \frac{2\eta}{1 + 2\eta}u^* \in \frac{\eta + (\varepsilon/3)}{1 + 2\eta}\mathbb{B}_{X^*} + \frac{\varepsilon}{3}\mathbb{B}_{X^*} + F^\perp \subset \varepsilon\mathbb{B}_{X^*} + F^\perp.$$
The proof is then complete. $\square$

From the latter lemma we can obtain an important description of the approximate subdifferential of the distance function at points inside the set.

PROPOSITION 5.55. *Let $(X, \|\cdot\|)$ be a Banach space, let S be a subset of X and let $\bar{x} \in S$. Then, one has*
$$\partial_A d_S(\bar{x}) = \bigcap_{F \in \mathcal{F}(X)} \operatorname*{Lim\,sup}_{\overline{S} \ni x \to \bar{x}, \varepsilon \downarrow 0} \left(\partial_{H,\varepsilon}(d_S + \Psi_{x+F})(x) \cap (1 + \varepsilon)\mathbb{B}_{X^*}\right)$$
$$= \bigcap_{F \in \mathcal{F}(X)} \operatorname*{Lim\,sup}_{\overline{S} \ni x \to \bar{x}, \varepsilon \downarrow 0} \partial_{H,\varepsilon}(d_S + \Psi_{x+F})(x),$$
where $\overline{S} := \operatorname{cl} S$.

PROOF. We may suppose that S is closed. Lemma 5.54 guarantees the equality
$$\bigcap_{F \in \mathcal{F}} \operatorname*{Lim\,sup}_{x \to \bar{x}, \varepsilon \downarrow 0} \partial_{H,\varepsilon}(d_S + \Psi_{x+F})(x) = \bigcap_{F \in \mathcal{F}} \operatorname*{Lim\,sup}_{\overline{S} \ni x \to \bar{x}, \varepsilon \downarrow 0} \partial_{H,\varepsilon}(d_S + \Psi_{x+F})(x),$$
whose right-hand side coincides with $\bigcap_{F \in \mathcal{F}} \operatorname*{Lim\,sup}_{\overline{S} \ni x \to \bar{x}, \varepsilon \downarrow 0} \partial_{H,\varepsilon}(d_S + \Psi_{x+F})(x) \cap (1+\varepsilon)\mathbb{B}_{X^*}$
according to Lemma 5.40(a), while the left-hand side is equal to $\partial_A d_S(\bar{x})$ by Theorem 5.42. The desired equalities then follow. $\square$

The next lemma is an easy extension of Lemma 5.28.

LEMMA 5.56. Let $(X, \|\cdot\|)$ be a Banach space and let S be a subset of X. Let U be an open set of X with $U \cap S \neq \emptyset$ and $\varphi : U \to \mathbb{R}$ be a locally Lipschitz function such that $d_S(x) \leq \varphi(x)$ for all $x \in U$ and $\varphi(x) = 0$ for all $x \in U \cap S$. Then for any $\bar{x} \in U \cap S$, one has $\partial_A d_S(\bar{x}) \subset \partial_A \varphi(\bar{x})$.

In particular, for any real $r > 0$ and $\bar{x} \in S$ one has

$$[0, r] \partial_A d_S(\bar{x}) = r \partial_A d_S(\bar{x}).$$

PROOF. The arguments for the justification of the inclusion are similar to those in Lemma 5.28. Fix any $\bar{x} \in U \cap S$. Note by continuity that $\varphi(x) = 0$ for all $x \in U \cap \overline{S}$, where $\overline{S} := \operatorname{cl} S$. From this and the inequality assumption we see that $\partial_{H,\varepsilon}(d_S + \Psi_{\bar{x}+F})(x) \subset \partial_{H,\varepsilon}(\varphi + \Psi_{\bar{x}+F})(x)$, for all $x \in U \cap \overline{S}$ and $\varepsilon > 0$. This and Proposition 5.55 ensure that

$$\partial_A d_S(\bar{x}) = \bigcap_{F \in \mathcal{F}} \operatorname{Lim\,sup}_{\overline{S} \ni x \to \bar{x}, \varepsilon \downarrow 0} \partial_{H,\varepsilon}(d_S + \Psi_{\bar{x}+F})(x)$$

$$\subset \bigcap_{F \in \mathcal{F}} \operatorname{Lim\,sup}_{\overline{S} \ni x \to \bar{x}, \varepsilon \downarrow 0} \partial_{H,\varepsilon}(\varphi + \Psi_{\bar{x}+F})(x) \subset \bigcap_{F \in \mathcal{F}} \operatorname{Lim\,sup}_{x \to \bar{x}, \varepsilon \downarrow 0} \partial_{H,\varepsilon}(\varphi + \Psi_{\bar{x}+F})(x).$$

This combined with Theorem 5.42 entails the inclusion of the lemma. The equality $[0, r] \partial_A d_S(\bar{x}) = r \partial_A d_S(\bar{x})$ is a direct consequence of the above inclusion and the equality $\partial_A(\rho f)(\bar{x}) = \rho \partial_A f(\bar{x})$ for $\rho \geq 0$ and f locally Lipschitz near $\bar{x}$ (see Proposition 5.33(a)). □

We are now in a position to prove the invariance of the approximate normal cone with respect to equivalent norms.

THEOREM 5.57 (invariance of A-normal cone with respect to equivalent renorm). Let $(X, \|\cdot\|)$ be a Banach space, let S be a subset of X and let $\bar{x} \in S$. Then for any other equivalent norm $\|\cdot\|$, one has

$$\mathbb{R}_+ \partial_A d_{S, \|\cdot\|}(\bar{x}) = \mathbb{R}_+ \partial_A d_{S, \|\cdot\|}(\bar{x}),$$

that is, the definition of $N^A(\cdot; \cdot)$ in Definition 5.48 does not depend on equivalent norms.

PROOF. Let $\|\cdot\|$ be another equivalent norm on X. Fix two reals $\alpha > 0$ and $\beta > 0$ such that $\alpha \|\cdot\| \leq \|\cdot\| \leq \beta \|\cdot\|$. Then

$$\alpha \, d_{S, \|\cdot\|}(x) \leq d_{S, \|\cdot\|}(x) \leq \beta \, d_{S, \|\cdot\|}(x) \quad \text{for all } x \in X.$$

Through this we obtain by Lemma 5.56 that

$$\partial_A d_{S, \|\cdot\|}(\bar{x}) \subset \alpha^{-1} \partial_A d_{S, \|\cdot\|}(\bar{x}) \text{ and } \partial_A d_{S, \|\cdot\|}(\bar{x}) \subset \beta \, \partial_A d_{S, \|\cdot\|}(\bar{x}),$$

so the conclusion of the lemma is derived. □

PROPOSITION 5.58. Let $S_1, \cdots, S_m$ be subsets of Banach spaces $X_1, \cdots, X_m$, and let $\bar{x}_i \in S_i$ for each $i = 1, \cdots, m$. Then with $\bar{x} := (\bar{x}_1, \cdots, \bar{x}_m)$ one has

$$N^A(S_1 \times \cdots \times S_m; \bar{x}) = N^A(S_1; \bar{x}_1) \times \cdots \times N^A(S_m; \bar{x}_i).$$

PROOF. Put $S := S_1 \times \cdots \times S_m$ and endow $X := X_1 \times \cdots \times X_m$ with the sum norm. Then for any $x = (x_1, \cdots, x_m) \in X$, we have $d_S(x) = d_{S_1}(x_1) + \cdots + d_{S_m}(x_m)$. Proposition 5.47 guarantees that

$$\partial_A d_S(\bar{x}) = \partial_A d_{S_1}(\bar{x}_1) \times \cdots \times \partial_A d_{S_m}(\bar{x}_m),$$

and from this equality, using Theorem 5.57 and the equality in Lemma 5.56, the desired equality of the proposition is easily deduced. □

5.4.2. A-subdifferential of general functions. We begin this subsection by relying the A-subdifferential of a locally Lipschitz function to the A-normal cone of its epigraph. Let us first establish a lemma on the distance function of the epigraph of such a function.

LEMMA 5.59. *Let $(X, \|\cdot\|_X)$ and $(Y, \|\cdot\|_Y)$ be two normed spaces, and let Q be a nonempty set of X.*
(a) *If a function $f : Q \to \mathbb{R}$ is Lipschitz on Q with $\gamma > 0$ as a Lipschitz constant, then considering the norm $\|\cdot\|$ on $X \times \mathbb{R}$ given by $\|(x,r)\| := \gamma\|x\|_X + |r|$, one has*
$$d_{\|\cdot\|}((x,r), \operatorname{epi} f) = (f(x) - r)^+ \quad \text{for all } (x,r) \in Q \times \mathbb{R},$$
and this distance is achieved at $(x, f(x))$ whenever $(x,r) \notin \operatorname{epi} f$ with $x \in Q$.
(b) *If a mapping $G : Q \to Y$ is γ-Lipschitz on Q with $\gamma > 0$, then with respect to the norm $\|\cdot\|$ on $X \times Y$ defined by $\|(x,y)\| := \gamma\|x\|_X + \|y\|_Y$ one has*
$$d_{\|\cdot\|}((x,y), \operatorname{gph} G) = \|y - G(x)\|_Y \quad \text{for all } (x,y) \in Q \times Y.$$

PROOF. (a) Let $(x,r) \in Q \times \mathbb{R}$. The equality being obvious for $(x,r) \in \operatorname{epi} f$, suppose that $r < f(x)$. For any $(u,s) \in \operatorname{epi} f$, noting that $u \in Q$ and putting $\sigma := s - f(u) \geq 0$, we can write
$$\gamma\|u - x\| + |s - r| = \gamma\|u - x\| + |f(u) + \sigma - r| \geq |f(x) + \sigma - r| \geq f(x) - r,$$
so $\gamma\|u - x\| + |s - r| \geq (f(x) - r)^+$ with equality for $(u,s) = (x, f(x))$. This justifies the assertion (a).
(b) For any $(u,v) \in \operatorname{gph} G$, we have $u \in Q$ and
$$\gamma\|u - x\|_X + \|v - y\| = \gamma\|u - x\|_X + \|G(u) - y\|_Y \geq \|G(x) - y\|_Y$$
with equality for $(u,v) = (x, G(x))$. This translates (b). □

PROPOSITION 5.60. *Let X be a Banach space and $f : X \to \mathbb{R} \cup \{-\infty, +\infty\}$ be a function which finite at $\bar{x} \in X$ and Lipschitz near $\bar{x}$. The following equality holds*
$$\partial_A f(\bar{x}) = \{x^* \in X^* : (x^*, -1) \in N^A(\operatorname{epi} f; (\bar{x}, f(\bar{x})))\}.$$

PROOF. Put $S := \operatorname{epi} f$. According to Proposition 2.79, Proposition 5.32 and Proposition 5.50, we may suppose that f is Lipschitz on the whole space X with a real $\gamma > 0$ as a Lipschitz constant.

Endow $X \times \mathbb{R}$ with the norm $\|\cdot\|$ defined by $\|(x,r)\| := \gamma\|x\| + |r|$. Fix any separable Banach subspace E of X and any $(u, u^*) \in X \times X^*$ with $u^* \in \partial_H(f + \Psi_{u+E})(u)$. For any $h \in E$ and $s \in \mathbb{R}$ there exist a function ε on $]0, +\infty[$ with $\varepsilon(t) \to 0$ as $t \downarrow 0$ such that for every $t > 0$,
$$\langle u^*, th \rangle \leq f(u + th) - f(u) + t\varepsilon(t) = (f(u+th) - f(u) - ts) + ts + t\varepsilon(t),$$
which by Lemma 5.59(a) yields
$$\langle u^*, h \rangle - s \leq t^{-1} d_S(u + th, f(u) + ts) + \varepsilon(t).$$
It ensues that $(u^*, -1) \in \partial_H(d_S + \Psi_{(u,f(u))+E\times\mathbb{R}})(u, f(u))$. For any $x^* \in \partial_A f(\bar{x})$, we then derive by Theorem 5.57 that
$$(x^*, -1) \in \partial_A d_S(\bar{x}, f(\bar{x})) \subset N^A(S; (\bar{x}, f(\bar{x}))).$$

Now fix $x^* \in X^*$ satisfying $(x^*, -1) \in N^A(S; (\bar{x}, f(\bar{x})))$, so by Theorem 5.57 again there is some real $\lambda > 0$ with $(x^*, -1) \in \lambda \partial_A d_S(\bar{x}, f(\bar{x}))$. Take any finite-dimensional vector subspace F of X. By Proposition 5.55 there are nets $(\varepsilon_j)_j$ in

$]0,+\infty[$, $(x_j^*, r_j)_j$ in S and $(x_j^*, -r_j^*)_j$ in $X^* \times \mathbb{R}$ such that $\varepsilon_j \downarrow 0$, $x_j \to \bar{x}$, $r_j \to f(\bar{x})$, $r_j^* \to 1$ with $r_j^* > 0$, $x_j^* \to x^*$ weakly*, all with j in a directed set J and such that

$$(x_j^*, -r_j^*) \in \partial_{H,\varepsilon_j/2}(\lambda d_S + \Psi_{(x_j, r_j)+F\times\mathbb{R}})(x_j, r_j), \quad \text{for all } j \in J.$$

By Proposition 5.5(c) we may suppose that $r_j = f(x_j)$ for all $j \in J$. For each $j \in J$, Proposition 5.5(b) tells us that there is some real $\delta_j > 0$ such that, for all $r \in]f(x_j) - \delta_j, f(x_j) + \delta_j[$ and $x \in B(x_j, \delta_j)$ with $x \in x_j + F$,

$$\langle x_j^*, x - x_j \rangle - r_j^*(r - f(x_j)) \leq \lambda d_S(x, r) + \varepsilon_j(\|x - x_j\| + |r - f(x_j)|).$$

Putting $u_j^* := x_j^*/r_j^*$, $\varepsilon_j' := (\varepsilon_j/r_j^*)(1+\gamma)$ and choosing $\delta_j' \in]0, \delta_j[$ such that $f(x) \in]f(x_j) - \delta_j, f(x_j) + \delta_j[$ for all $x \in B(x_j, \delta_j')$, it ensues that, for all $x \in B(x_j, \delta_j')$ with $x \in x_j + F$, we have

$$\langle u_j^*, x - x_j \rangle \leq f(x) - f(x_j) + (\varepsilon_j/r_j^*)(\|x - x_j\| + |f(x) - f(x_j)|)$$
$$\leq f(x) - f(x_j) + \varepsilon_j' \|x - x_j\|.$$

For each $j \in J$, this gives that $u_j^* \in \partial_{H,\varepsilon_j'}(f + \Psi_{x_j+F})(x_j)$, hence $x^* \in \limsup_{x \to \bar{x}, \varepsilon \downarrow 0} \partial_{H,\varepsilon}(f + \Psi_{x+F})(x)$. This being true for any $F \in \mathcal{F}(X)$, it results by Theorem 5.42 that $x^* \in \partial_A f(\bar{x})$, and the proof is complete. $\square$

The latter proposition allows us to extend the definition of A-subdifferential to general functions as follows.

DEFINITION 5.61. Let X be a Banach space and $f : X \to \{-\infty, +\infty\}$ be a function which is finite at $\bar{x} \in X$. The *Ioffe approximate subdifferential* or *A-subdifferential* $\partial_A f(\bar{x})$ of f at $\bar{x}$ is defined by

$$\partial_A f(\bar{x}) = \{x^* \in X^* : (x^*, -1) \in N^A(\operatorname{epi} f; (\bar{x}, f(\bar{x})))\}.$$

This yields in particular that $\partial_A f(\bar{x}) = \emptyset$ whenever f is not finite at $\bar{x}$.

We first show that the A-subdifferential of the indicator function of a set coincides with the A-normal cone of the set.

PROPOSITION 5.62. For any subset S of a Banach space X and $\bar{x} \in S$, one has

$$\partial_A \Psi_S(\bar{x}) = N^A(S; \bar{x}).$$

PROOF. Since $\operatorname{epi} \Psi_S = S \times [0, +\infty[$, Proposition 5.58 relative to the A-normal cone of a Cartesian product and Proposition 5.51 (applied to the convex set $[0, +\infty[$ in $\mathbb{R}$) guarantee that

$$N^A(\operatorname{epi} \Psi_S; (\bar{x}, 0)) = N^A(S; \bar{x}) \times]-\infty, 0].$$

Therefore, $(x^*, -1) \in N^A(\operatorname{epi} \Psi_S; (\bar{x}, 0))$ if and only $x^* \in N^A(S; \bar{x})$, or equivalently $x^* \in \partial_A \Psi_S(\bar{x}) \Leftrightarrow x^* \in N^A(S; \bar{x})$, which translates the desired equality of the proposition. $\square$

If two functions f_1, f_2 are finite at $\bar{x}$ and coincide on a neighborhood V of $\bar{x}$, then it is evident that $\operatorname{epi} f_1$ and $\operatorname{epi} f_2$ coincide on the neighborhood $V \times \mathbb{R}$ of $(\bar{x}, r)$, where $r := f_1(\bar{x}) = f_2(\bar{x})$. Then, the assertion (a) below follows from Proposition 5.50, while (b) is easily seen from Proposition 5.51.

PROPOSITION 5.63. Let X be a Banach space.
(a) If $f_1, f_2 : X \to \mathbb{R} \cup \{-\infty, +\infty\}$ are finite at $\bar{x} \in X$ and coincide near $\bar{x}$, then $\partial_A f_1(\bar{x}) = \partial_A f_2(\bar{x})$.
(b) If $f : U \to \mathbb{R} \cup \{+\infty\}$ is a convex function on an open convex set U of X, then for any $x \in U$,
$$\partial_A f(x) = \partial f(x) := \{x^* \in X^* : \langle x^*, u - x \rangle \leq f(u) - f(x), \forall u \in U\}.$$

Unlike the Lipschitz setting in Proposition 5.33, the result in the framework of non-Lipschitz functions for the A-subdifferential with the multiplication by a positive real requires less direct arguments.

PROPOSITION 5.64. For any real $r > 0$ and any function $f : X \to \mathbb{R} \cup \{-\infty, +\infty\}$ on a Banach space X, one has
$$\partial_A (rf)(x) = r \partial_A f(x) \quad \text{for all } x \in X.$$

PROOF. Fix any real $r > 0$. Endow $X \times \mathbb{R}$ with the usual sum norm $\|\cdot\|$ and denote by d_S the distance function from a set $S \subset X \times \mathbb{R}$ with respect to that norm. Consider also the norm $\|\cdot\|$ on $X \times \mathbb{R}$ defined by $\|(u, \sigma)\| := \|u\| + r|\sigma|$, and denote by $d_S^{\|\cdot\|}$ the distance function from S relative to the norm $\|\cdot\|$. For any $(u, \sigma) \in X \times \mathbb{R}$, we note that
$$d_{\mathrm{epi}\,(rf)}(u, \sigma) = \inf\{\|(u, \sigma) - (v, \theta)\| : (v, \theta) \in \mathrm{epi}\,(rf)\}$$
$$= \inf\{\|(u, \sigma) - (v, r(r^{-1}\theta))\| : (v, r^{-1}\theta) \in \mathrm{epi}\,f\}$$
$$= \inf\{\|(u, \sigma) - (v, r\mu)\| : (v, \mu) \in \mathrm{epi}\,f\}$$
$$= \inf\{\|u - v\| + r|r^{-1}\sigma - \mu| : (v, \mu) \in \mathrm{epi}\,f\} = d_{\mathrm{epi}\,f}^{\|\cdot\|}(u, r^{-1}\sigma).$$
Consequently, by Proposition 5.33 we have with $\lambda > 0$
$$(x^*, -1) \in \lambda \partial_A d_{\mathrm{epi}\,(rf)}(x, rf(x))$$
$$\Leftrightarrow (x^*, -1) \in \lambda \partial_A d_{\mathrm{epi}\,f}^{\|\cdot\|}(x, f(x)) \circ (\mathrm{Id}_X, r^{-1}\mathrm{Id}_{\mathbb{R}})$$
$$\Leftrightarrow (x^*, -1) = \lambda(\zeta^*, -r^{-1}\alpha) \text{ with } (\zeta^*, -\alpha) \in \partial_A d_{\mathrm{epi}\,f}^{\|\cdot\|}(x, f(x))$$
$$\Leftrightarrow (x^*, -1) = \lambda\alpha(\xi^*, -r^{-1}) \text{ with } (\xi^*, -1) \in \mathbb{R}_+ \partial_A d_{\mathrm{epi}\,f}^{\|\cdot\|}(x, f(x)).$$
The latter is equivalent to $\lambda\alpha = r$ and $x^* = \lambda\alpha\xi^* = r\xi^*$ with $\xi^* \in \partial_A f(x)$ (see Theorem 5.57). This means that $x^* \in \partial_A (rf)(x)$ if and only if $x^* \in r\partial_A f(x)$, which finishes the proof. $\square$

Like C-normal and L-normal cones, the A-normal cone of an epigraph enjoys particular properties.

PROPOSITION 5.65. Let X be a Banach space and $f : X \to \mathbb{R} \cup \{+\infty\}$ be a function which is finite at $\bar{x}$ and lower semicontinuous near $\bar{x}$. Let $(x^*, s) \in N^A(\mathrm{epi}\,f; (\bar{x}, f(\bar{x})))$. The following hold:
(a) One has $s \leq 0$.
(b) If in addition f is γ-Lipschitz near $\bar{x}$, then $s\gamma \leq -\|x^*\|$.

PROOF. We may suppose that f is lower semicontinuous on X. Put $S := \mathrm{epi}\,f$.
(a) For any $(u, r) \in S$ and any real $t > 0$, noting that $(u, r + t) \in S$ we see that $t^{-1}d_S(u, r + t) = 0$, so $\underline{d}_H^+(d_S)((u, r); (0, 1)) = 0$. Given any $\varepsilon > 0$ and any finite-dimensional vector subspace E of X, it results that, for any $(u^*, \sigma) \in$

$\partial_{H,\varepsilon}(d_S + \Psi_{(u,r)+E\times\mathbb{R}})(u,r)$ we have $\langle u^*, 0\rangle + \sigma \leq \varepsilon$, that is, $\sigma \leq \varepsilon$. This combined with Proposition 5.55 yields $s \leq 0$, proving (a).
(b) Suppose that f is γ-Lipschitz on $B(\bar{x}, 2\delta)$. Fix any $h \in X$. For any $(u,r) \in S$ with $u \in B(\bar{x}, \delta)$ and any $t > 0$ with $t\|h\| < \delta$, we can write
$$f(u+th) \leq f(u) + t\gamma\|h\| \leq r + t\gamma\|h\|,$$
thus $(u+th, r+t\gamma\|h\|) \in S$. This gives
$$d_S(u+th, r+t\gamma\|h\|) = 0, \text{ hence } \underline{d}_H^+(d_S)\big((u,r);(h,\gamma\|h\|)\big) = 0.$$
Taking any $\varepsilon > 0$ and any finite-dimensional vector subspace E of X with $h \in E$, it ensues that, for any $(u^*, \sigma) \in \partial_{H,\varepsilon}(d_S + \Psi_{(u,r)+E\times\mathbb{R}})(u,r)$ we obtain
$$\langle u^*, h\rangle + \sigma\gamma\|h\| \leq \varepsilon(1+\gamma)\|h\|.$$
This combined with Proposition 5.55 entails that $\langle x^*, h\rangle + s\gamma\|h\| \leq 0$. Since h is arbitrary in X, we conclude that $\|x^*\| + s\gamma \leq 0$ as desired. □

As the subdifferentials in the previous chapters, the A-subdifferential fulfills the Fermat optimality condition as established in the assertion (b) below.

PROPOSITION 5.66. *Let X be a Banach space and $f: U \to \mathbb{R} \cup \{-\infty, +\infty\}$ be an extended real-valued function on an open set U of X.*
(a) *For all $x \in U$ the inclusions*
$$\partial_F f(x) \subset \partial_A f(x) \subset \partial_C f(x)$$
always hold.
(b) *If $\bar{x} \in U$ is a local minimizer of f with $|f(\bar{x})| < +\infty$, then $0 \in \partial_A f(\bar{x})$.*

PROOF. (a) Fix any $x \in U$. The three sets in (a) are empty if $|f(x)| = +\infty$. Suppose that $f(x)$ is finite and note that $\partial_A f(x) \subset \partial_C f(x)$ by Proposition 5.49. Fix any $x^* \in \partial_F f(x)$ (if any). Putting $y := (x, f(x))$ we know by Proposition 4.4(b) that $(x^*, -1) \in N^F(\text{epi } f; y)$. Since $N^F(\text{epi } f; y) \cap \mathbb{B}_{X^* \times \mathbb{R}}$ coincides with $\partial_F d_{\text{epi } f}(y)$ by Proposition 4.30, we have $(x^*, -1) \in \beta \partial_F d_{\text{epi } f}(y)$, where $\beta := \|(x^*, -1)\|$. Using the property that $\partial_H g(u) \subset \partial_A g(u)$ whenever g is Lipschitz near u (see Theorem 5.35(b)), it ensues that $(x^*, -1) \in \beta \partial_A d_{\text{epi } f}(y)$. This gives $x^* \in \partial_A f(x)$, which translates the first inclusion in (a).
(b) The assertion (b) clearly follows from the first inclusion in (a). □

We consider now functions φ of type $\varphi(x,y) = g(y)$.

PROPOSITION 5.67. *Let $g: V \to \mathbb{R}\{-\infty, +\infty\}$ be an extended real-valued function on an open set V of a Banach space Y. Let X be a Banach space and let $\varphi: X \times V \to \mathbb{R} \cup \{-\infty, +\infty\}$ be defined by*
$$\varphi(x,y) := g(y) \quad \text{for all } (x,y) \in X \times V.$$
Then for any $x \in X$ and $y \in V$ one has
$$\partial_A \varphi(x,y) = \{0\} \times \partial_A g(y).$$

PROOF. Fix any $x \in X$ and $y \in V$. The result being obvious if $|g(y)| = +\infty$, we suppose that g is finite at y. Noting that $\text{epi } \varphi = X \times \text{epi } g$, Proposition 5.58 tells us that $(x^*, y^*, -1)$ belongs to $N^A(\text{epi } \varphi; (x, y, \varphi(x,y)))$ if and only if $x^* \in N^A(X; x)$ and $(y^*, -1)$ is in $N^A(\text{epi } g; (y, g(y)))$, which in turn means that $x^* = 0$ and $y^* \in \partial_A g(y)$. This justifies the equality of the proposition. □

The next theorem complements Proposition 5.21. It concerns the situation of reflexive spaces and Banach spaces with a Fréchet differentiable renorm.

THEOREM 5.68 (A-subdifferential versus L-subdifferential in reflexive spaces or F-smooth spaces). *Let S be a subset of a Banach space which is closed near $\bar{x} \in S$ and let $f : U \to \mathbb{R} \cup \{-\infty, +\infty\}$ be a function on an open set U of X which is finite at $\bar{x} \in U$ and lower semicontinuous near $\bar{x}$. The following hold:*
(a) *If X is reflexive, then*
$$\partial_A f(\bar{x}) = \partial_L f(\bar{x}) \quad \text{and} \quad N^A(S; \bar{x}) = N^L(S; \bar{x}).$$
(b) *If X admits an equivalent norm which is Fréchet differentiable off zero, then*
$$\mathrm{cl}_{w^*}\big(\partial_A f(\bar{x})\big) = \mathrm{cl}_{w^*}\big(\partial_L f(\bar{x})\big) \quad \text{and} \quad \mathrm{cl}_{w^*}\big(N^A(S; \bar{x})\big) = \mathrm{cl}_{w^*}\big(N^L(S; \bar{x})\big).$$

PROOF. Assume first that f is Lipschitz on U with $\gamma \geq 0$ as a Lipschitz constant. Under the assumption on the space X in either (a) or (b), X is Asplund and admits a Gâteaux differentiable (off zero) renorm, hence Lemma 5.20 and Corollary 5.43 tell us that

$$(5.12) \quad \partial_A f(\bar{x}) = \mathop{\mathrm{Lim\,sup}}_{x \to \bar{x}} \partial_H f(x) = \mathrm{cl}_{w^*}\Big(\mathop{\mathrm{seq\,Lim\,sup}}_{x \to \bar{x}} \partial_H f(x)\Big) = \mathrm{cl}_{w^*}\big(\partial_L f(\bar{x})\big).$$

(a) To prove the assertion (a) we deal first with the case when f is Lipschitz on U with constant $\gamma \geq 0$. Since $\partial_F f(x) \subset \gamma \mathbb{B}_{X^*}$, for all $x \in U$, and since the space X^* is reflexive, Proposition 1.18 ensures that $\partial_L f(\bar{x})$ is w^*-closed, so from (5.12) we obtain that $\partial_A f(\bar{x}) = \partial_L f(\bar{x})$.

Using the definition of $N^A(\cdot; \cdot)$ and the known equality $N^L(S; \bar{x}) = \mathbb{R}_+ \partial_L d_S(\bar{x})$, we deduce through the above equality applied to the Lipschitz function $f = d_S$ that

$$N^A(S; \bar{x}) = \mathbb{R}_+ \partial_A d_S(\bar{x}) = \mathbb{R}_+ \partial_L d_S(\bar{x}) = N^L(S; \bar{x}),$$

which furnishes the equality in (a) for the normal cones.

If f is lower semicontinuous near $\bar{x}$, applying the above equality between the normal cones with the set epi f, the equality $\partial_A f(\bar{x}) = \partial_L f(\bar{x})$ directly follows.
(b) From (5.12) applied with the function $f = d_S$, we have

$$\partial_A d_S(\bar{x}) = \mathrm{cl}_{w^*}\big(\partial_L d_S(\bar{x})\big).$$

This entails that $\partial_A d_S(\bar{x}) \subset \mathrm{cl}_{w^*}\big(N^L(S; \bar{x})\big)$ and $\partial_L d_S(\bar{x}) \subset N^A(S; \bar{x})$, which easily gives

$$\mathrm{cl}_{w^*}\big(N^A(S; \bar{x})\big) = \mathrm{cl}_{w^*}\big(N^L(S; \bar{x})\big).$$

Assume now that f is lower semicontinuous near $\bar{x}$ and take any $x^* \in \partial_A f(\bar{x})$, so $(x^*, -1) \in N^A\big(\mathrm{epi}\, f; (\bar{x}, f(\bar{x}))\big)$. By the latter equality above applied with $S := \mathrm{epi}\, f$ there is some net $(u_j^*, -r_j)_{j \in J}$ converging weakly* to $(x^*, -1)$ such that $r_j > 0$ and

$$(u_j^*, -r_j) \in N^L\big(\mathrm{epi}\, f; (\bar{x}, f(\bar{x}))\big) \quad \text{for all } j \in J.$$

Putting $x_j^* := r_j^{-1} u_j^*$, it results that $x_j^* \to x^*$ weakly* and

$$(x_j^*, -1) \in N^L\big(\mathrm{epi}\, f; (\bar{x}, f(\bar{x}))\big), \quad \text{that is, } x_j^* \in \partial_L f(\bar{x}).$$

This means that $x^* \in \mathrm{cl}_{w^*}\big(\partial_L f(\bar{x})\big)$. From this we easily see that

$$\mathrm{cl}_{w^*}\big(\partial_A f(\bar{x})\big) \subset \mathrm{cl}_{w^*}\big(\partial_L f(\bar{x})\big),$$

and this inclusion is in fact an equality since the converse inclusion is evident. The proof of the theorem is then finished. □

The above theorem allows us to extend the equality obtained in Remark 4.126 for closed subsets of finite-dimensional Euclidean spaces to weakly closed sets (and ball-compact sets) of Hilbert spaces.

PROPOSITION 5.69. *Let S be a nonempty set in a Hilbert space H which is either ball-compact or weakly closed. Then for any $\bar{x} \in H \setminus S$ one has*

$$\partial_A d_S(\bar{x}) = \partial_L d_S(\bar{x}) = \frac{1}{d_S(\bar{x})}(\bar{x} - \operatorname{Proj}_S(\bar{x})).$$

PROOF. We already know by Theorem 5.68 that $\partial_A d_S(\bar{x}) = \partial_L d_S(\bar{x})$. On the other hand, using the equality $\partial_L d_S(\bar{x}) = {}^{\text{seq}}\operatorname{Lim\,sup}_{x \to \bar{x}} \partial_F d_S(x)$ it easily follows from Proposition 4.104(b) that $\partial_L d_S(\bar{x}) \subset \frac{1}{d_S(\bar{x})}(\bar{x} - \operatorname{Proj}_S(\bar{x}))$.

Regarding the converse inclusion, take an arbitrary element $y \in \operatorname{Proj}_S(\bar{x})$ and (as in the proof of Proposition 4.125) note by Lemma 4.123 that $y = P_S(x_t)$ for any $t \in {]}0,1[$, where $x_t := \bar{x} + t(y - \bar{x})$. Note also that $(x_t - y)/d_S(x_t)$ tends to $(\bar{x} - y)/d_S(\bar{x})$ as $t \downarrow 0$. By Proposition 4.104(a) it ensues that

$$(x_t - y)/d_S(x_t) \in \partial_H d_S(x_t),$$

so $(\bar{x}-y)/d_S(\bar{x})$ lies in $\partial_A d_S(\bar{x})$ by Corollary 5.43. This ensures the desired inclusion of $(\bar{x} - \operatorname{Proj}_S(\bar{x}))/d_S(\bar{x})$ into $\partial_A d_S(\bar{x})$, which finishes the proof. □

5.4.3. Chain rule for A-subdifferential and mean value inequality.

Before proving the chain rule for A-subdifferential of the composition of a Lipschitz function with a compactly Lipschitzian mapping we need to establish a lemma.

LEMMA 5.70. *Let X, Y be Banach spaces, U be an open set of X, and $\bar{x} \in U$. Let $G : U \to Y$ be a mapping which is Lipschitz on U and $g : Y \to \mathbb{R} \cup \{-\infty, +\infty\}$ be a function which is Lipschitz near $\bar{y} := G(\bar{x})$. Then for each $x^* \in \partial_A(g \circ G)(\bar{x})$ there is some $y^* \in \partial_A g(\bar{y})$ such that*

$$(x^*, -y^*) \in N^A\big(\operatorname{gph} G; (\bar{x}, \bar{y})\big) = \mathbb{R}_+ \partial_A d_{\operatorname{gph} G}(\bar{x}, \bar{y}).$$

PROOF. Let V be an open neighborhood of $\bar{y}$ over which g is κ-Lipschitz. Shrinking U if necessary we may suppose that $G(U) \subset V$. Let γ be a Lipschitz constant of G and $\|\cdot\|$ be the norm on $X \times Y$ given by $\|(x,y)\| := \gamma \|x\|_X + \|y\|_Y$. Denoting by d_G the distance function from the set $\operatorname{gph} G$ with respect to the norm $\|\cdot\|$, we know by Lemma 5.59(b) that $d_G(x, y) = \|y - G(x)\|$ for all $(x,y) \in U \times Y$. For all $x \in U$ and $y \in V$, it ensues that

$$g \circ G(x) \leq g(y) + \kappa \|y - G(x)\| = g(y) + \kappa d_G(x, y) =: f(x, y).$$

Then, for any closed separable vector subspace $E \times L$ of $X \times Y$ and any $(h, v) \in E \times L$, we deduce that, for every $x \in U$ and $t > 0$ sufficiently small

$$t^{-1}\big(g \circ G(x + th) - g \circ G(x)\big) \leq t^{-1}\big(f(x + th, G(x) + tv) - f(x, G(x))\big),$$

which entails that, for all $(h, v) \in X \times Y$,

$$\underline{d}_H^+(g \circ G + \Psi_{x+E})(x; h) \leq \underline{d}_H^+(f + \Psi_{(x,G(x))+E \times L})\big((x, G(x)); (h, v)\big).$$

It follows that

$$\partial_H(g \circ G + \Psi_{x+E})(x) \times \{0\} \subset \partial_H(f + \Psi_{(x,G(x))+E \times L})(x, G(x)),$$

which ensures that

$$\partial_A(g \circ G)(\overline{x}) \times \{0\} \subset \bigcap_{E \in \mathcal{S}(X), L \in \mathcal{S}(Y)} \limsup_{x \to \overline{x}} \partial_H(f + \Psi_{(x,G(x))+E \times L})(x, G(x))$$

$$\subset \bigcap_{E \in \mathcal{S}(X), L \in \mathcal{S}(Y)} \limsup_{(x,y) \to (\overline{x},\overline{y})} \partial_H(f + \Psi_{(x,y)+E \times L})(x, y) = \partial_A f(\overline{x}, \overline{y}),$$

where the latter equality is due to Theorem 5.42. Since $\partial_A f(\overline{x}, \overline{y}) \subset \{0\} \times \partial_A g(\overline{y}) + \kappa \partial_A d_G(\overline{x}, \overline{y})$ according to Theorem 5.46, to Proposition 5.47, and to Proposition 5.33, given $x^* \in \partial_A(g \circ G)(\overline{x})$ we obtain some $y^* \in \partial_A g(\overline{y})$ such that, as desired (according to Theorem 5.57),

$$(x^*, -y^*) \in \kappa \partial_A d_G(\overline{x}, \overline{y}) \subset N^A(\mathrm{gph}\, G; (\overline{x}, \overline{y})),$$

which justifies the lemma. □

THEOREM 5.71 (chain rule for A-subdifferential). Let X, Y be Banach spaces and U be an open set of X. Let $G : U \to Y$ be a mapping which is compactly Lipschitzian at $\overline{x} \in U$ and $g : Y \to \mathbb{R} \cup \{-\infty, +\infty\}$ be a function which is finite and Lipschitz near $\overline{y} := G(\overline{x})$. Then $g \circ G$ is Lipschitz near $\overline{x}$ and

$$\partial_A(g \circ G)(\overline{x}) \subset \bigcup_{y^* \in \partial_A g(\overline{y})} \partial_A(y^* \circ G)(\overline{x}).$$

PROOF. We may suppose that G and $g \circ G$ are Lipschitz on U and the neighborhood W of $\overline{x}$ in Definition 2.183 coincides with U. By Theorem 2.186(c) we may also suppose that the multimapping $K : X \rightrightarrows X$ in Definition 2.183 (taking nonempty compact values) is upper semicontinuous on the whole space X. Endow $X \times Y$ with the norm $\|\cdot\|$ defined by $\|(x, y)\| := \gamma \|x\| + \|y\|$, where $\gamma > 0$ is a Lipschitz constant of G on U. Denote by d_G the distance function from the set $\mathrm{gph}\, G$ relative to the norm $\|\cdot\|$. Fix any $x^* \in \partial_A(g \circ G)(\overline{x})$, so there is by the above lemma some real $\lambda > 0$ and $y^* \in \partial_A g(\overline{x})$ such that $(x^*, -y^*) \in \lambda \partial_A d_G(\overline{x}, \overline{y})$. We can write $(x^*, -y^*) = \lambda(u^*, -v^*)$ with (see Theorem 5.42)

$$(u^*, -v^*) \in \bigcap_{F \times E \in \mathcal{F}(X) \times \mathcal{S}(Y)} \limsup_{(x,y) \to (\overline{x},\overline{y})} \partial_H(d_G + \Psi_{(x,y)+F \times E})(x, y).$$

Fix any $F \in \mathcal{F}(X)$. The set $F \cap \mathbb{B}_X$ being compact, its range $K(F \cap \mathbb{B}_X)$ by the upper semicontinuous multimapping K is strongly compact in Y (see Theorem 1.35), so the closed vector subspace E of Y spanned by the set $K(F \cap \mathbb{B}_X)$ is separable, thus $E \in \mathcal{S}(Y)$. By Theorem 5.42 choose nets $(x_j, y_j)_{j \in J}$ converging to $(\overline{x}, \overline{y})$ and $(u_j^*, -v_j^*)_{j \in J}$ converging weakly* to $(u^*, -v^*)$ in $X^* \times Y^*$ with

$$(u_j^*, -v_j^*) \in \partial_H(d_G + \Psi_{(x_j,y_j)+F \times E})(x_j, y_j) \cap \mathbb{B}_{\|\cdot\|_*}, \quad \forall j \in J.$$

Fix $j \in J$ and $h \in F \cap \mathbb{B}_X$, and put for all $t > 0$ such that $x_j + th \in U$,

$$v(t) := t^{-1}\big(G(x_j + th) - G(x_j)\big) + r(t, x_j; h) b(t, x_j; h) \in K(h),$$

where $b(t, x_j; h) \in \mathbb{B}_Y$. Choose a sequence $(t_n)_n$ in $]0, 1]$ tending to 0 such that

$$d_H^+(v_j^* \circ G)(x_j; h) = \lim_{n \to \infty} t_n^{-1} \langle v_j^*, G(x_j + t_n h) - G(x_j) \rangle.$$

Putting $v_n := v(t_n)$, we may suppose that $(v_n)_n$ converges to some $v \in K(h) \subset E$. There is a sequence $(s_n)_n$ in $]0, +\infty[$ tending to 0 such that, with $\eta_n := \|v_n - v\|$,

we have for n large enough
$$\langle u_j^*, t_n h\rangle - \langle v_j^*, t_n v\rangle \le d_G(x_j + t_n h, y_j + t_n v) - d_G(x_j, y_j) + t_n s_n$$
$$\le d_G\big(x_j + t_n h, y_j + G(x_j + t_n h) - G(x_j)\big) - d_G(x_j, y_j) + t_n\big(s_n + \eta_n + r(t_n, x_j; h)\big)$$
$$= \|y_j - G(x_j)\| - \|y_j - G(x_j)\| + t_n\big(s_n + \eta_n + r(t_n, x_j; h)\big),$$
where the latter equality is due to Lemma 5.59. It ensues that
$$\langle u_j^*, h\rangle \le \langle v_j^*, v\rangle + s_n + \eta_n + r(t_n, x_j; h)$$
$$= t_n^{-1}\langle v_j^*, G(x_j + t_n h) - G(x_j)\rangle + \langle v_j^*, v - v_n\rangle + \langle v_j^*, w_n\rangle + s_n + \eta_n + r(t_n, x_j; h),$$
where $w_n := r(t_n, x_j; h) b(t_n, x_j; h)$. We deduce that $\langle u_j^*, h\rangle \le \underline{d}_H^+(v_j^* \circ G)(x_j; h)$ for all $h \in F$, hence
$$u_j^* \in \partial_H\big(v_j^* \circ G + \Psi_{x_j + F}\big)(x_j).$$
The space F being finite-dimensional, the point x_j (see Corollary 5.6) is, for any real $\varepsilon > 0$, a local minimizer on the constraint set $x_j + F$ of the function
$$\varphi_j := v_j^* \circ G - \langle u_j^*, \cdot\rangle + \varepsilon \|\cdot - x_j\|,$$
thus x_j is an unconstrained local minimizer of the function $\varphi_j + \beta d(\cdot, x_j + F)$ (see Lemma 2.213), where $\beta := \gamma \|v_j^*\| + \|u_j^*\| + \varepsilon$. The A-subdifferential formula for a sum of locally Lipschitz functions (see Theorem 5.46) yields
$$u_j^* \in \partial_A(v_j^* \circ G)(x_j) + \varepsilon \mathbb{B}_{X^*} + F^\perp,$$
which gives (by Theorem 5.46)
$$u_j^* \in \partial_A(v^* \circ G)(x_j) + \partial_A\big((v_j^* - v^*) \circ G\big)(x_j) + \varepsilon \mathbb{B}_{X^*} + F^\perp.$$
Each one of the three first sets in the right-hand side being included in a w^*-compact set, Theorem 5.35 and Proposition 2.187 imply that $u^* \in \partial_A(v^* \circ G)(\overline{x}) + F^\perp$. From this we see by the w^*-closedness of $\partial_A(v^* \circ G)(\overline{x})$ due to Definition 5.31 that $u^* \in \partial_A(v^* \circ G)(\overline{x})$, so $x^* \in \partial_A(y^* \circ G)(\overline{x})$ by Proposition 5.33(a). This finishes the proof. $\square$

The first corollary of Theorem 5.71 is related to the situation when the mapping G is strictly H-differentiable.

COROLLARY 5.72. *Let X, Y be Banach spaces and U be an open set of X. Let $G : U \to Y$ be a mapping which is strictly Hadamard differentiable at $\overline{x} \in U$ and $g : Y \to \mathbb{R} \cup \{-\infty, +\infty\}$ be a function which is finite and Lipschitz near $\overline{y} := G(\overline{x})$. Then, one has*
$$(5.13) \qquad \partial_A(g \circ G)(\overline{x}) \subset DG(\overline{x})^*\big(\partial_A g(\overline{y})\big),$$
where $DG(\overline{x})^*$ denotes the adjoint of $DG(\overline{x})$ and
$$DG(\overline{x})^*(\partial_A g(\overline{y})) := \{DG(\overline{x})^*(y^*) : y^* \in \partial_A g(\overline{y})\}.$$

PROOF. For any $y^* \in \partial_A g(G(\overline{x}))$, the function $y^* \circ G$ is strictly Hadamard differentiable at $\overline{x}$, which yields $\partial_A(y^* \circ G)(\overline{x}) = \{D(y^* \circ G)(\overline{x})\}$ (see Theorem 5.35), that is, $\partial_A(y^* \circ G)(\overline{x}) = \{DG(x)^*(y^*)\}$. Theorem 5.71 then yields
$$\partial_A(g \circ G)(\overline{x}) \subset \{DG(x)^*(y^*) : y^* \in \partial_A g(G(\overline{x}))\} = DG(\overline{x})^*\big(\partial_A g(G(\overline{x}))\big).$$
$\square$

A second corollary concerns an estimate rule for the approximate subdifferential of the pointwise maximum of finitely many functions. It is in the line of Corollary 4.81.

COROLLARY 5.73. *Let U be an open convex set of a Banach space X and $(f_i)_{i \in I}$ be a finite family of locally Lipschitz real-valued functions defined on U, and let $f(x) = \max_{i \in I} f_i(x)$. For any $\bar{x} \in U$ and for $I(\bar{x}) := \{i \in I : f_i(\bar{x}) = f(\bar{x})\}$, one has*

$$\partial_A f(\bar{x}) \subset \bigcup_{\lambda \in \Lambda} \partial_A \left(\sum_{i \in I(\bar{x})} \lambda_i f_i \right)(\bar{x}),$$

where $\Lambda := \{\lambda \in \mathbb{R}^{I(\bar{x})} : \lambda_i \geq 0, \sum_{i \in I(\bar{x})} \lambda_i = 1\}$.

PROOF. Without loss of generality me may suppose that $I := \{1, \cdots, n\}$. Set $Y := \mathbb{R}^n$ and $g(y) = \max_{i \in I} y_i$. The mapping $G : U \to Y$, defined by $G(x) := (f_1(x), \cdots, f_n(x))$ for all $x \in U$, is locally Lipschitz continuous near $\bar{x}$, thus compactly Lipschitzian at $\bar{x}$ since Y is finite-dimensional. By Proposition 2.190 and Proposition 5.63, $\partial_A g(G(\bar{x})) = \partial_C g(G(\bar{x}))$ coincides with the set of $\lambda \in \mathbb{R}^n$ with $\lambda_i \geq 0$ and such that $\sum_{i \in I} \lambda_i = 1$ and $\lambda_i = 0$ for all $i \notin I(\bar{x})$. The inclusion of the corollary then follows from Theorem 5.71. □

We derive now a mean-value theorem with A-subgradients for locally Lipschitz functions. It is similar to Theorem 4.84 related to L-subgradients.

THEOREM 5.74 (mean value inequality with A-subdifferential for Lipschitz functions). *Let U be an open convex set of a Banach space X and $f : U \to \mathbb{R}$ be a locally Lipschitz real-valued function. Then for any $a, b \in U$ with $a \neq b$, there exist some $c \in [a, b[:= \{a + t(b - a) : t \in [0, 1[\}$ (resp. $c \in]a, b[$) and $c^* \in \partial_A f(c)$ (resp. $c^* \in \partial_A f(c) \cup (-\partial_A(-f)(c))$) such that*

$$f(b) - f(a) \leq \langle c^*, b - a \rangle \quad (\text{resp. } f(b) - f(a) = \langle c^*, b - a \rangle).$$

PROOF. Fix $a, b \in U$ with $a \neq b$ and fix also an open interval $I \supset [0, 1]$ with $a + t(b - a) \in U$ for all $t \in I$. Put

$$\varphi(t) := f(a + t(b - a)) + t(f(a) - f(b)) \quad \text{for all } t \in I.$$

This function φ is easily seen to be Lipschitz on an open set in $\mathbb{R}$ containing $[0, 1]$ with some Lipschitz constant κ. By the equality $\varphi(0) = \varphi(1)$, this function φ attains its minimum on $[0, 1]$ at some point $s \in [0, 1[$, hence s is an unconstrained local minimizer of $\varphi + \kappa d(\cdot, [0, 1])$. In consequence, $0 \in \partial_A(\varphi + \kappa d(\cdot, [0, 1]))(s)$. The A-subdifferential sum rule in Theorem 5.46 tells us that

$$0 \in \partial_A \varphi(s) + \kappa d(\cdot, [0, 1])(s) \subset \partial_A \varphi(s) + N([0, 1]; s).$$

Putting $c := a + s(b - a)$, Theorem 5.46 and Corollary 5.72 give some $c^* \in \partial_A f(c)$ such that

$$\langle c^*, b - a \rangle + f(a) - f(b) \in -N([0, 1]; s).$$

Using the convexity of $[0, 1]$ we see that $N([0, 1]; s) \subset]-\infty, 0]$, thus

$$f(b) - f(a) \leq \langle c^*, b - a \rangle.$$

Similarly, at some point $r \in]0, 1[$ the function φ attains either a minimum or a maximum on $]0, 1[$, hence $0 \in \partial_A \varphi(r) \cup \partial_A(-\varphi)(r)$. With $e := a + r(b - a)$, Corollary

5.72 furnishes some $e^* \in \partial_A f(e) \cup (-\partial_A(-f)(e))$ such that $f(b) - f(a) = \langle e^*, b-a \rangle$. This finishes the proof. □

5.4.4. Representation of C-subdifferential with A-subgradients.

We have already established in any Banach space X that $\partial_A f(\bar{x}) \subset \partial_C f(\bar{x})$ (see Proposition 5.66(a)). When f is lower semicontinuous near $\bar{x}$, we show how the the above mean value inequality allows us to provide below a complete description of $\partial_C f(\bar{x})$ in terms of $\partial_A f(\bar{x})$ and horizon A-subgradients.

We begin with an adaptation of Definitions 2.86 and 4.117.

DEFINITION 5.75. Let X be a Banach space, U be a neighborhood of a point $x \in X$, and $f : U \to \mathbb{R} \cup \{-\infty, +\infty\}$ be a function which is finite at x. One says that $x^* \in X^*$ is a *horizon A-subgradient* (or *singular A-subgradient*) of f at x if $(x^*, 0) \in N^A(\text{epi } f; (x, f(x)))$. The set of all horizon A-subgradients of f at x is the *horizon A-subdifferential* of f at x and it is denoted by $\partial_A^\infty f(x)$. If $|f(x)| = +\infty$, one puts $\partial_A^\infty f(x) = \emptyset$.

The set $\partial_A^\infty f(x)$ is evidently a cone in X^* containing zero and according to Proposition 5.58, $\partial_A^\infty \Psi_S(x) = N^A(S; x)$ for any subset S of X with $x \in S$. When f is Lipschitz continuous near x, Proposition 5.65(b) tells us that $\partial_A^\infty f(x) = \{0\}$.

The statement and proof of the following representation of Clarke subdifferential with approximate subgradients are quite similar to those of Theorem 4.120.

THEOREM 5.76 (representation of C-subdifferential with A-subgradients). Let X be a Banach space, S be a subset of X which is closed near $\bar{x} \in S$, and $f : X \to \mathbb{R} \cup \{+\infty\}$ be a function which is finite at $\bar{x}$ and lower semicontinuous near $\bar{x}$. The following hold:
(a) $N^C(S; \bar{x}) = \overline{\text{co}}^*(N^A(S; \bar{x}))$.
(b) $\partial_C f(\bar{x}) = \overline{\text{co}}^*(\partial_A f(\bar{x}) + \partial_A^\infty f(\bar{x}))$.
(c) If f is Lipschitz continuous near $\bar{x}$, then $\partial_C f(\bar{x}) = \overline{\text{co}}^*(\partial_A f(\bar{x}))$.

PROOF. Let us begin with (c). Fix any $h \in X$ and choose sequences $(t_n)_n$ in $]0, +\infty[$ tending to 0 and $(x_n)_n$ in X converging to $\bar{x}$ such that $f^\circ(\bar{x}; h) = \lim_{n\to\infty} t_n^{-1}(f(x_n + t_n h) - f(x_n))$. By the mean value inequality in Theorem 5.74, for each integer n large enough there are $c_n \in [x_n, x_n + t_n h]$ and $c_n^* \in \partial_A f(c_n)$ such that
$$t_n^{-1}(f(x_n + t_n h) - f(x_n)) \le \langle c_n^*, h \rangle.$$
Considering a weak* limit point c^* of the bounded sequence $(c_n^*)_n$, we have $c^* \in \partial_A f(\bar{x})$ (by the outer semicontinuity of $\partial_A f$ in Theorem 5.35) and
$$f^\circ(\bar{x}; h) \le \langle c^*, h \rangle \le \sigma(h, \partial_A f(\bar{x})),$$
where as usual $\sigma(\cdot, \partial_A f(\bar{x}))$ is the support function of $\partial_A f(\bar{x})$. This ensures that $\partial_C f(\bar{x}) \subset \overline{\text{co}}^*(\partial_A f(\bar{x}))$, which entails the desired equality in (c) according to the inclusion $\partial_A f(\bar{x}) \subset \partial_C f(\bar{x})$ (see (a) in Proposition 5.66) and to the convexity and w^*-closedness of $\partial_C f(\bar{x})$ by (d) in Theorem 2.70.

Applying (c) to the distance function d_S yields $\partial_C d_S(\bar{x}) = \overline{\text{co}}^*(\partial_A d_S(\bar{x}))$. Then the descriptions of $N^C(S; \bar{x})$ and $N^A(S; x)$ in terms of d_S (see Proposition 2.95 and Definition 5.48) imply that $N^C(S; \bar{x}) = \overline{\text{co}}^*(N^A(S; \bar{x}))$, which is the equality in (a).

Now let us turn to (b). We note that, for $u^* \in \partial_A f(\bar{x})$ and $v^* \in \partial_A^\infty f(\bar{x})$ we have both $(u^*, -1)$ and $(v^*, 0)$ in $N^A(\text{epi } f; (\bar{x}, f(\bar{x}))) \subset N^C(\text{epi } f; (\bar{x}, f(\bar{x})))$, thus

$(u^* + v^*, -1) \in N^C(\operatorname{epi} f; (\overline{x}, f(\overline{x})))$ since the latter set is a convex cone. This justifies the inclusion of the right-hand side of (b) into the left one.

Let us prove the converse inclusion of (b). Take first

$$(u^*, -1) \in \operatorname{co}\left(N^A(\operatorname{epi} f; (\overline{x}, f(\overline{x})))\right).$$

There exist a finite set K, $(x_k^*, -r_k) \in N^A(\operatorname{epi} f; (\overline{x}, f(\overline{x})))$, and $t_k \geq 0$ with $\sum_{k \in K} t_k = 1$ such that $(u^*, -1) = \sum_{k \in K} t_k(x_k^*, -r_k)$. By Proposition 5.65 we know that $r_k \geq 0$, hence for $K_1 := \{k \in K : r_k > 0\}$ and $K_0 := \{k \in K : r_k = 0\}$ we have $\sum_{k \in K_1} t_k r_k = 1$ and

$$(u^*, -1) = \sum_{k \in K_1} t_k r_k (x_k^*/r_k, -1) + \sum_{k \in K_0} t_k(x_k^*, 0).$$

Putting $u_k^* := x_k^*/r_k$ and $v_k^* := 0$ for all $k \in K_1$, and $v_k^* := x_k^*$ for $k \in K_0$, we obtain

$$u^* = \sum_{k \in K_1} t_k r_k u_k^* + \sum_{k \in K} t_k v_k^* \in \operatorname{co} \partial_A f(\overline{x}) + \operatorname{co} \partial_A^\infty f(\overline{x}),$$

or equivalently $u^* \in \operatorname{co}\left(\partial_A f(\overline{x}) + \partial_A^\infty f(\overline{x})\right)$ by Lemma 4.119. Take now an arbitrary $x^* \in \partial_C f(\overline{x})$. By (a) we have $(x^*, -1) \in \overline{\operatorname{co}}^*\left(N^A(\operatorname{epi} f; (\overline{x}, f(\overline{x})))\right)$, thus it is easily seen that $(x^*, -1)$ is the weak star limit of a net $(x_i^*, -1)_{i \in I}$ of elements of $\operatorname{co} N^A(\operatorname{epi} f; (\overline{x}, f(\overline{x})))$. By what precedes, each $x_i^* \in \operatorname{co}\left(\partial_A f(\overline{x}) + \partial_A^\infty f(\overline{x})\right)$, hence $x^* \in \overline{\operatorname{co}}^*\left(\partial_A f(\overline{x}) + \partial_A^\infty f(\overline{x})\right)$, completing the proof of the theorem. □

Note that the example in Remark 4.121 tells us that the assertion (a) (resp. (b)) in the above theorem fails if the set S (resp. the function f) is not closed (resp. not lower semicontinuous) near $\overline{x}$.

For a locally Lipschitz function f, recalling that $\partial_C f(\overline{x})$ is a singleton if and only if f is strictly Hadamard differentiable at $\overline{x}$ (see Proposition 2.76), and keeping in mind that $\partial_A f(\overline{x}) \neq \emptyset$, we directly derive from Theorem 5.76(c) the following corollary.

COROLLARY 5.77. *Let X be a Banach space and $f : X \to \mathbb{R} \cup \{-\infty, +\infty\}$ be finite at $\overline{x} \in X$ and Lipschitz near $\overline{x}$. Then $\partial_A f(\overline{x})$ is a singleton if and only if f is strictly Hadamard differentiable at $\overline{x}$.*

5.4.5. Extended A-subdifferential sum rule. This subsection is devoted to extend the A-subdifferential sum rule of two Lipschitz functions to the sum where one of the two functions is not required to be Lipschitz.

THEOREM 5.78 (extended sum rule for A-subdifferential). *Let X be a Banach space and $f_1, \cdots, f_m : X \to \mathbb{R} \cup \{+\infty\}$ be functions which are finite at $\overline{x} \in X$. If all but one of these functions are Lipschitz near $\overline{x}$, then one has*

(5.14) $$\partial_A(f_1 + \cdots + f_m)(\overline{x}) \subset \partial_A f_1(\overline{x}) + \cdots + \partial_A f_m(\overline{x}).$$

The inclusion is an equality whenever in addition the subdifferential regularity $\partial_A f_k(\overline{x}) = \partial_F f_k(\overline{x})$ is satisfied for $k = 1, \cdots, m$ (and in this case, the latter subdifferential regularity also holds at $\overline{x}$ for the sum function).

If all the functions $f_2, \cdots, f_m$ are strictly Fréchet differentiable at the point $\overline{x}$, then the equality

$$\partial_A(f_1 + f_2 + \cdots + f_m)(\overline{x}) = \partial_A f_1(\overline{x}) + D_F f_2(\overline{x}) + \cdots + D_F f_m(\overline{x})$$

holds true.

5.4. A-NORMAL CONE AND A-SUBDIFFERENTIAL OF GENERAL FUNCTIONS

The proof of the theorem will proceed with several lemmas. It is sufficient to work with $m = 2$ along with $f := f_1$ is real-valued and γ-*Lipschitz on* X (see Proposition 2.79) and $g = f_2 : X \to \mathbb{R} \cup \{+\infty\}$. Denote by $\|\cdot\|_\gamma$ and $\|\cdot\|$ the norms on $X \times \mathbb{R}$ and $X \times \mathbb{R} \times \mathbb{R}$ defined by

$$\|(x,r)\|_\gamma := \gamma\|x\| + |r| \quad \text{and} \quad \|(x,r,s)\| := \gamma\|x\| + |r| + |s|,$$

and denote also by $d_{\|\cdot\|_\gamma}(\cdot, C')$ and $d_{\|\cdot\|}(\cdot, C'')$ the distance functions from sets $C' \subset X \times \mathbb{R}$ and $C'' \subset X \times \mathbb{R} \times \mathbb{R}$. As in the proof for the C-subdifferential of a sum in Theorem 2.98, consider (with a slight modification of notation) the sets

$$S := \{(u, \alpha, \beta) \in X \times \mathbb{R} \times \mathbb{R} : f(u) + g(u) \le \alpha + \beta\},$$

$$S' := \{(u, \alpha, \beta) \in X \times \mathbb{R} \times \mathbb{R} : f(u) \le \alpha, g(u) \le \beta\}.$$

By definition of S it appears that $(u, \alpha, \beta) \in S$ if and only if $(u, \alpha + \theta, \beta - \theta) \in S$ for every $\theta \in \mathbb{R}$. From this we easily see that, relative to any norm on $X \times \mathbb{R} \times \mathbb{R}$, the equality

(5.15) $$\operatorname{dist}((x,r,s), S) = \operatorname{dist}((x, r+\theta, s-\theta), S)$$

holds true for all $\theta \in \mathbb{R}$; further, the obvious inclusion $S' \subset S$ ensures that

$$\operatorname{dist}((x,r,s), S) \le \operatorname{dist}((x,r,s), S') \quad \text{for all } (x,r,s) \in X \times \mathbb{R} \times \mathbb{R}.$$

LEMMA 5.79. *Under the above assumptions and notations, for all* (x,r,s) *in* $X \times \mathbb{R} \times \mathbb{R}$ *one has*

$$d_{\|\cdot\|}((x,r,s), S) \le 2\Big(d_{\|\cdot\|_\gamma}((x,r), \operatorname{epi} f) + d_{\|\cdot\|_\gamma}((x,s), \operatorname{epi} g)\Big).$$

PROOF. Take for a moment any $(x,s) \in \operatorname{epi} g$. For any real $r \le f(x)$, the Lipschitz property of f ensures that, for every $(u, \alpha, \beta) \in S'$, putting $\sigma := \alpha - f(u) \ge 0$ we have

$$|\alpha - r| + |\beta - s| + \gamma\|u - x\| = |f(u) + \sigma - r| + |\beta - s| + \gamma\|u - x\|$$
$$\ge |f(x) + \sigma - r| + |\beta - s|$$
$$\ge |f(x) + \sigma - r| \ge f(x) + \sigma - r \ge f(x) - r,$$

which gives

$$|\alpha - r| + |\beta - s| + \gamma\|u - x\| \ge \big(f(x) - r\big)^+,$$

with equality for $(u, \alpha, \beta) = (x, f(x), s) \in S'$. It ensues that for every real $r \le f(x)$

$$d_{\|\cdot\|}((x,r,s), S) \le d_{\|\cdot\|}((x,r,s), S') = \big(f(x) - r\big)^+ = d_{\|\cdot\|_\gamma}((x,r), \operatorname{epi} f),$$

where the second equality is due to Lemma 5.59(a). It follows that $d_{\|\cdot\|}((x,r,s); S) \le d_{\|\cdot\|_\gamma}((x,r), \operatorname{epi} f)$ for any $r \in \mathbb{R}$ since $d_{\|\cdot\|}((x,r,s); S) = 0$ for every real $r > f(x)$. This says in particular that, for every $(x,s) \in \operatorname{epi} g$ and every $r \in \mathbb{R}$

$$d_{\|\cdot\|_\gamma}((x,r), \operatorname{epi} f) - d_{\|\cdot\|}((x,r,s), S) \ge 0,$$

and evidently the function $(x', r', s') \mapsto d_{\|\cdot\|_\gamma}(x', r') - d_{\|\cdot\|}((x', r', s'), S')$ is 2-Lipschitz with respect to the norm $\|\cdot\|$ on $X \times \mathbb{R} \times \mathbb{R}$. Lemma 2.213 entails that, for all $(x,r,s) \in X \times \mathbb{R} \times \mathbb{R}$

$$d_{\|\cdot\|_\gamma}((x,r), \operatorname{epi} f) - d_{\|\cdot\|}((x,r,s), S') + 2d_{\|\cdot\|_\gamma}((x,s), \operatorname{epi} g) \ge 0,$$

and from this we obtain

$$d_{\|\cdot\|}((x,r,s), S) \le 2\Big(d_{\|\cdot\|_\gamma}((x,r), \operatorname{epi} f) + d_{\|\cdot\|_\gamma}((x,s), \operatorname{epi} g)\Big),$$

which is the desired inequality of the lemma. □

LEMMA 5.80. Under the above assumptions and notations, one has
$$(x^*, -r^*) \in \partial_A d_{\|\cdot\|_\gamma}(\cdot, \operatorname{epi}(f+g))(\overline{x}, f(\overline{x}) + g(\overline{x}))$$
$$\Longrightarrow (x^*, -r^*, -r^*) \in \partial_A d_{\|\cdot\|}(\cdot, S)(\overline{x}, f(\overline{x}), g(\overline{x})).$$

PROOF. First, we observe that
$$d_{\|\cdot\|}((x, r, s), S)$$
$$= \inf\{\gamma \|x' - x\| + |r' - r| + |s' - s| : x' \in X, r', s' \in \mathbb{R}, r' + s' \geq f(x') + g(x')\}$$
$$\geq \inf\{\gamma \|x' - x\| + |(r' + s') - (r + s)| : x' \in X, r', s' \in \mathbb{R}, r' + s' \geq f(x') + g(x')\}$$
$$= d_{\|\cdot\|_\gamma}((x, r + s), \operatorname{epi}(f + g)).$$

Put $\varphi(x, r, s) := d_{\|\cdot\|}((x, r, s), S)$ and $\varphi_\gamma(x, t) := d_{\|\cdot\|_\gamma}((x, t), \operatorname{epi}(f + g))$. Taking $x \in X$ and $r, s \in \mathbb{R}$ with $r + s \geq f(x) + g(x)$, and taking also $F \in \mathcal{F}(X)$ we deduce that, for any $(h, \rho, \sigma) \in F \times \mathbb{R} \times \mathbb{R}$ and $\varepsilon > 0$,
$$\underline{d}_H^+ \varphi_\gamma((x, r + s); (h, \rho + \sigma)) + \varepsilon \|(h, \rho + \sigma)\|_\gamma$$
$$\leq \underline{d}_H^+ \varphi((x, r, s); (h, \rho, \sigma)) + \varepsilon \|(h, \rho, \sigma)\|.$$

Then, for $(u^*, -\rho^*) \in \partial_{H,\varepsilon}(\varphi_\gamma + \Psi_{(x, r+s) + F \times \mathbb{R}})(x, r+s)$ we obtain that one has relative to the norm $\|\cdot\|$
$$(u^*, -\rho^*, -\rho^*) \in \partial_{H,\varepsilon}(\varphi + \Psi_{(x, r, s) + F \times \mathbb{R} \times \mathbb{R}})(x, r, s).$$

Using this and using the fact that any $(u_j, t_j)_j$ in $\operatorname{epi}(f+g)$ converging to $(\overline{x}, f(\overline{x}) + g(\overline{x}))$ can be written as $(u_j, r_j + s_j)$ with $r_j := f(u_j) \to f(\overline{x})$ (by continuity of f) and $s_j := t_j - r_j \to g(\overline{x})$, it is not difficult to see through Proposition 5.55 that the implication of the lemma holds true. □

LEMMA 5.81. Under the above assumptions and notations, for each
$$(x^*, -r^*, -s^*) \in \partial_A d_{\|\cdot\|}(\cdot, S)(\overline{x}, f(\overline{x}), g(\overline{x}))$$
there exist x_1^*, x_2^* in X^* such that $x^* = x_1^* + x_2^*$ and
$$(x_1^*, -r^*) \in 2\partial_A d_{\|\cdot\|_\gamma}(\cdot, \operatorname{epi} f)(\overline{x}, f(\overline{x})), \ (x_2^*, -s^*) \in 2\partial_A d_{\|\cdot\|_\gamma}(\cdot, \operatorname{epi} g)(\overline{x}, g(\overline{x})).$$

PROOF. Put for all $(x, r, s) \in X \times \mathbb{R} \times \mathbb{R}$
$$\varphi(x, r, s) := d_{\|\cdot\|}((x, r, s), S), \Phi_\gamma(x, r, s) := 2d_{\|\cdot\|_\gamma}((x, r), \operatorname{epi} f) + 2d_{\|\cdot\|_\gamma}((x, s), \operatorname{epi} g)$$
and note by Lemma 5.79 that
$$\varphi(x, r, s) \leq \Phi_\gamma(x, r, s).$$

Take any finite-dimensional vector subspace F of X. By Proposition 5.55 there are (with a directed index set J) nets $(\varepsilon_j)_j$ in $]0, +\infty[$ tending to 0, $(x_j, r_j, s_j)_j$ in S converging to $(\overline{x}, f(\overline{x}), g(\overline{x}))$ and $(x_j^*, -r_j^*, -s_j^*)_j$ in $X^* \times \mathbb{R} \times \mathbb{R}$ converging weakly* to $(x^*, -r^*, -s^*)$ and such that, for all $j \in J$

(5.16) $\quad (x_j^*, -r_j^*, -s_j^*) \in \partial_{H, \varepsilon_j}(\varphi + \Psi_{(x_j, r_j, s_j) + F \times \mathbb{R} \times \mathbb{R}})(x_j, r_j, s_j).$

Putting $r_j' := f(x_j)$ and $s_j' := r_j + s_j - f(x_j)$, we see that $(x_j, r_j') \in \operatorname{epi} f, (x_j, s_j') \in \operatorname{epi} g$. Further, given $(h, \rho, \sigma) \in F \times \mathbb{R} \times \mathbb{R}$, by (5.15) we have, for every $j \in J$
$$\varphi(x_j + th, r_j + t\rho, s_j + t\sigma) = \varphi(x_j + th, r_j' + t\rho, s_j' + t\sigma)$$

5.4. A-NORMAL CONE AND A-SUBDIFFERENTIAL OF GENERAL FUNCTIONS

along with
$$\varphi(x_j, r'_j, s'_j) = \varphi(x_j, r_j, s_j) = 0.$$
It ensues that, for all $(h, \rho, \sigma) \in F \times \mathbb{R} \times \mathbb{R}$
$$\underline{d}_H \varphi\big((x_j, r_j, s_j); (h, \rho, \sigma)\big) + \varepsilon_j \|(h, \rho, \sigma)\|$$
$$= \underline{d}_H \varphi\big((x_j, r'_j, s'_j); (h, \rho, \sigma)\big) + \varepsilon_j \|(h, \rho, \sigma)\|$$
$$\leq \underline{d}_H \Phi_\gamma \big((x_j, r'_j, s'_j); (h, \rho, \sigma)\big) + \varepsilon_j \|(h, \rho, \sigma)\|,$$
which yields that one has relative to the norm $\|\cdot\|$ on $X \times \mathbb{R} \times \mathbb{R}$
$$\partial_{H,\varepsilon_j}(\varphi + \Psi_{(x_j, r_j, s_j) + F \times \mathbb{R} \times \mathbb{R}})(x_j, r_j, s_j) \subset \partial_{H,\varepsilon_j}(\Phi_\gamma + \Psi_{(x_j, r'_j, s'_j) + F \times \mathbb{R} \times \mathbb{R}})(x_j, r'_j, s'_j).$$
This combined with Theorem 5.42 tells us that
$$(x^*, -r^*, -s^*) \in \partial_A \Phi_\gamma \big(\overline{x}, f(\overline{x}), g(\overline{x})\big).$$
On the other hand, Theorem 5.46 guarantees that by
$$\partial_A \Phi_\gamma \big(\overline{x}, f(\overline{x}), g(\overline{x})\big) \subset 2\partial_A \Phi_{\gamma,1}\big(\overline{x}, f(\overline{x}), g(\overline{x})\big) + 2\partial_A \Phi_{\gamma,2}\big(\overline{x}, f(\overline{x}), g(\overline{x})\big),$$
where $\Phi_{\gamma,1}(x, r, s) := 2d_{\|\cdot\|_\gamma}(\cdot, \operatorname{epi} f)(x, r)$ and $\Phi_{\gamma,2}(x, r, s) := 2d_{\|\cdot\|_\gamma}(\cdot, \operatorname{epi} g)(x, s)$. Using Proposition 5.47(b) we easily derive $x_1^*, x_2^* \in X^*$ such that $x^* = x_1^* + x_2^*$ along with
$$(x_1^*, -r^*) \in 2\partial_A d_{\|\cdot\|_\gamma}(\cdot, \operatorname{epi} f)\big(\overline{x}, f(\overline{x})\big) \text{ and } (x_2^*, -s^*) \in 2\partial_A d_{\|\cdot\|_\gamma}(\cdot, \operatorname{epi} g)\big(\overline{x}, g(\overline{x})\big).$$
This translates the conclusion of the lemma. □

THE END OF THE PROOF OF THEOREM 5.78. Let $x^* \in \partial_A(f + g)(\overline{x})$, that is, $(x^*, -1) \in \lambda \partial_A d_{\|\cdot\|_\gamma}(\cdot, \operatorname{epi}(f + g))\big(\overline{x}, f(\overline{x}) + g(\overline{x})\big)$ for some real $\lambda > 0$. By Lemma 5.80 we have
$$(x^*, -1, -1) \in \lambda \partial_A d_{\|\cdot\|}(\cdot, S)\big(\overline{x}, f(\overline{x}), g(\overline{x})\big),$$
so Lemma 5.81 furnishes $x_1^*, x_2^* \in X^*$ such that $x^* = x_1^* + x_2^*$ with
$$(x_1^*, -1) \in 2\lambda \partial_A d_{\|\cdot\|_\gamma}(\cdot, \operatorname{epi} f)\big(\overline{x}, f(\overline{x})\big), (x_2^*, -1) \in 2\lambda \partial_A d_{\|\cdot\|_\gamma}(\cdot, \operatorname{epi} g)\big(\overline{x}, g(\overline{x})\big).$$
The two latter inclusions entail that $x_1^* \in \partial_A f(\overline{x})$ and $x_2^* \in \partial_A g(\overline{x})$, so the inclusion $\partial_A(f + g)(\overline{x}) \subset \partial_A f(\overline{x}) + \partial_A g(\overline{x})$ is justified.

If in addition the A-subdifferentials of f, g at $\overline{x}$ and their F-subdifferentials at $\overline{x}$ coincide, it ensues from the sum rule for F-subgradients (see Proposition 4.11) that
$$\partial_A(f + g)(\overline{x}) \subset \partial_A f(\overline{x}) + \partial_A g(\overline{x}) = \partial_F f(\overline{x}) + \partial_F g(\overline{x}) \subset \partial_F(f + g)(\overline{x}).$$
Since the last member is included in the first according to Proposition 5.66(a), we deduce that all the members coincide, which provides the desired equality of the theorem as well as the coincidence between $\partial_A(f + g)(\overline{x})$ and $\partial_F(f + g)(\overline{x})$.

If f is strictly Fréchet differentiable at $\overline{x}$, then $\partial_A f(\overline{x}) = D_F f(\overline{x})$ by Theorem 5.35(c), so by the above first inclusion
$$\partial_A(f + g)(\overline{x}) \subset D_F f(\overline{x}) + \partial_A g(\overline{x}) \text{ and } \partial_A g(\overline{x}) \subset D_F(-f)(\overline{x}) + \partial_A(f + g)(\overline{x}),$$
which readily gives $\partial_A(f + g)(\overline{x}) = D_F f(\overline{x}) + \partial_A g(\overline{x})$. The proof of the theorem is now complete. □

Proceeding as in Corollary 2.104 and Proposition 2.109, we obtain the following corollaries from Proposition 4.106(f).

COROLLARY 5.82. Let $f, g : U \to \mathbb{R}$ be two extended real-valued functions on an open set U of a Banach space X. Assume that $g = f + \rho$, where $\rho : U \to \mathbb{R}$ is strictly Fréchet differentiable at a point $\bar{x} \in U$ with $D\rho(\bar{x}) = 0$. Then one has $\partial_A g(\bar{x}) = \partial_A f(\bar{x})$.

COROLLARY 5.83. Let $G : U \to Y$ be a mapping from an open set U of a Banach space X into a normed space Y and let $g : V \to \mathbb{R}$ be a function from an open set V of Y containing $G(U)$. Assume that G is Lipschitz continuous near $\bar{x} \in U$ and g is strictly Fréchet differentiable at $\bar{y} := G(\bar{x})$, and let $\bar{y}^* := Dg(\bar{y})$. Then one has

$$\partial_A(g \circ G)(\bar{x}) = \partial_A(\bar{y}^* \circ G)(\bar{x}).$$

In particular, given a real-valued function $f : U \to \mathbb{R}$ which is Lipschitz continuous near $\bar{x} \in U$ and a function $\varphi : I \to \mathbb{R}$ on an open interval $I \supset f(U)$ which is strictly differentiable at $f(\bar{x})$, then with $r := \varphi'(f(\bar{x}))$ one has

$$\partial_A(\varphi \circ f)(\bar{x}) = \partial_A(rf)(\bar{x}).$$

The above corollary applies to the product of finitely many functions and the quotient.

EXAMPLE 5.84. Let $f, g : U \to \mathbb{R}$ be two locally Lipschitz functions on an open set U of a Banach space X. Considering as in Example 2.110 the mapping $G : U \to \mathbb{R}^2$ and the functions $p : \mathbb{R}^2 \to \mathbb{R}$ and $q : D_q \to \mathbb{R}$ defined by $G(x) = (f(x), g(x))$ for all $x \in U$, $p(r, s) = rs$ for all $(r, s) \in \mathbb{R}^2$, and $q(r, s) = r/s$ for all $(r, s) \in D_q := \{(\rho, \sigma) \in \mathbb{R}^2 : \sigma \neq 0\}$, the application of the above corollary with $p \circ G$ and $q \circ G$ gives respectively (a) and (b) below:

(a) For every $x \in U$ one has

$$\partial_A(fg)(x) = \partial_A\big(g(x)f + f(x)g\big)(x) \subset \partial_A\big(g(x)f\big)(x) + \partial_A\big(f(x)g\big)(x).$$

(b) For every $x \in U$ with $g(x) \neq 0$ one has

$$\partial_A\left(\frac{f}{g}\right)(x) = \frac{1}{(g(x))^2} \partial_A\big(g(x)f - f(x)g\big)(x).$$

$\square$

5.5. Further results

5.5.1. A-normals to compactly epi-Lipschitz sets. Given a set S of a normed space which has the interior tangent property at $\bar{x} \in S$ with respect to a nonzero vector $\bar{h}$, we saw in Proposition 2.34 that there exist a neighborhood U of $\bar{x}$ and a real $\gamma > 0$ such that

$$\|x^*\| \leq \gamma |\langle x^*, \bar{h} \rangle| \quad \text{for all } x \in U \cap S \text{ and } x^* \in N^C(S; x).$$

We show in this subsection that a similar feature holds true, in Banach spaces, for A-normals to sets satisfying the compact tangent property. Recall (see Definition 2.273) that a set S in a normed space X has the compact tangent property at $\bar{x} \in S$ provided that there exist a compact set K in X and two reals $r > 0, \gamma > 0$ such that

$$S \cap B(\bar{x}, r) + t\gamma \mathbb{B}_X \subset S - tK \quad \text{for all } t \in \,]0, r[.$$

5.5. FURTHER RESULTS

PROPOSITION 5.85. *Let S be a nonempty closed set in a Banach space X which has the compact tangent property at $\bar{x} \in S$ for some compact set K in X. The following properties hold.*
(a) *There exists a neighborhood U of $\bar{x} \in S$ such that for all $x \in U \cap S$ and $x^* \in N^A(S;x)$*
$$\inf_{v \in K} \langle x^*, v \rangle + \|x^*\| \leq 0.$$
(b) *For any net $(x_j)_{j \in J}$ in S converging to $\bar{x}$ and any bounded net $(x_j^*)_{j \in J}$ in X^* with $x_j^* \in N^A(S;x_j)$ for all $j \in J$ one has the equivalence*
$$x_j^* \to 0 \text{ weakly}^* \iff x_j^* \to 0 \text{ strongly}.$$

PROOF. It suffices to show (a) since (b) easily follows from (a). By the compact tangent property of S at $\bar{x}$ choose real numbers $\gamma > 0$ and $r > 0$ such that for all $t \in \,]0, r[$ the inclusion

(5.17) $$S \cap B(\bar{x}, 2r) + t\gamma \mathbb{B}_X \subset S - tK$$

holds. Let $x \in S \cap B(\bar{x}, r)$ and $x^* \in \partial_A d_S(x)$. Fix any real $\delta > 0$ and choose some finite subset $K_\delta \subset K$ such that $K \subset K_\delta + \delta \mathbb{B}_X$. Take any $u \in S \cap B(x, r)$ and $b \in \gamma \mathbb{B}_X$. Take also a sequence of reals $(t_n)_n$ tending to 0 with $0 < t_n < r$. For each integer n there exists by (5.17) some $v_n \in K$ (depending on u and b) with $u + t_n(b + v_n) \in S$. Choose some $v_n^\delta \in K_\delta$ such that $\|v_n - v_n^\delta\| \leq \delta$. Taking a subsequence if necessary, we may suppose that $(v_n^\delta)_{n \in \mathbb{N}}$ converges to some $v^\delta \in K_\delta$. Put $L_b := \mathrm{span}[\{b\} \cup K_\delta]$ (the vector space spanned by $\{b\} \cup K_\delta$) and for any $\varepsilon > 0$, consider any $u^* \in \partial_{H,\varepsilon}(d_S + \Psi_{u+L_b})(u)$. On one hand, we have

$$\langle u^*, b + v^\delta \rangle \leq d_H^-(d_S)(u; b + v^\delta) + \varepsilon \|b + v^\delta\|$$
(5.18)
$$\leq d_H^-(d_S)(u; b + v^\delta) + \varepsilon \zeta,$$

where $\zeta := \gamma + \max_{y \in K} \|y\|$. On the other hand, choosing $b_n^\delta \in \mathbb{B}_X$ such that $v_n = v_n^\delta + \delta b_n^\delta$, we have

$$t_n^{-1}[d_S(u + t_n(b + v_n^\delta)) - d_S(u)] \leq t_n^{-1}[d_S(u + t_n(b + v_n^\delta + \delta b_n^\delta)) - d_S(u)] + t_n^{-1}\|t_n \delta b_n^\delta\|$$
$$= \delta \|b_n^\delta\| \leq \delta,$$

and hence
$$d_H^+(d_S)(u; b + v^\delta) \leq \delta.$$

Using (5.18) we obtain

(5.19) $$\langle u^*, b + v^\delta \rangle \leq \delta + \varepsilon \zeta.$$

By Proposition 5.55 we know that $x^* \in \limsup_{S \ni u \to x, \varepsilon \downarrow 0} \partial_{H,\varepsilon}(d_S + \Psi_{u+L_b})(u)$. Therefore there exist nets $u_{j,\varepsilon} \to x$ with $u_{j,\varepsilon} \in S \cap B(x,r)$ and $u_{j,\varepsilon}^* \overset{w^*}{\to} x^*$ with $u_{j,\varepsilon}^* \in \partial_{H,\varepsilon}(d_S + \Psi_{u_{j,\varepsilon}+L_b})(u_{j,\varepsilon})$, the nets being indexed by both j and ε, where we recall that $\overset{w^*}{\to}$ denotes the weak-star convergence in X^*. Since $x \in B(\bar{x}, r)$, we may suppose that $u_{j,\varepsilon} \in S \cap B(\bar{x}, r)$. By (5.19) there exists some $v_{j,\varepsilon}^\delta \in K_\delta$ such that $\langle u_{j,\varepsilon}^*, b + v_{j,\varepsilon}^\delta \rangle \leq \delta + \varepsilon \zeta$. Since K_δ is a finite set, we easily obtain some $w \in K_\delta$ for which $\langle x^*, b + w \rangle \leq \delta$, and hence

$$\langle x^*, b \rangle + \inf_{v \in K} \langle x^*, v \rangle \leq \delta.$$

Taking the supremum over $b \in \gamma \mathbb{B}_X$ and making $\delta \downarrow 0$, it results that for any $x^* \in \partial_A d_S(x)$
$$\gamma \|x^*\| + \inf_{v \in K} \langle x^*, v \rangle \le 0,$$
and this inequality still obviously holds for all $x^* \in \mathbb{R}_+ \partial_A d_S(x) = N^A(S; x)$. This translates the property (a) and finishes the proof. $\square$

5.5.2. Subdifferentially pathological Lipschitz functions.

A locally Lipschitz function $f : U \to \mathbb{R}$ from an open set U of a Banach space X into $\mathbb{R}$ is declared to be *subdifferentially pathological* with respect to a subdifferential ∂ when $\partial f(x)$ is a big set for any point $x \in U$. If S is a subset of $]0,1[$ such that both sets $I \cap S$ and $I \cap \complement_{]0,1[} S$ have positive Lebesgue measure for any interval $I :=]r,s[$ with $0 < r < s < 1$, then putting

$$(5.20) \qquad f(x) := \int_0^x \mathbf{1}_S(t)\,dt \quad \text{for all } x \in]0,1[,$$

the Lipschitz function $f :]0,1[\to \mathbb{R}$ is subdifferentially pathological with respect to the Clarke subdifferential since $\partial_C f(x) = [0,1]$ for all $x \in]0,1[$. The latter equality can be seen from the gradient representation (see Theorem 2.83)

$$\partial_C f(x) = \operatorname{co}(\{\lim_{n \to \infty} \nabla f(x_n) : D_f \ni x_n \to x\}),$$

where D_f is any subset of $\operatorname{dom} \nabla f$ with full Lebesgue measure in $]0,1[$.

The aim of this subsection is to study, via a Baire category approach, such a pathology with respect to subdifferentials $\partial_A, \partial_L, \partial_C$.

Let U be an open set of a Banach space X and $T : U \rightrightarrows X^*$ be a multimapping which takes nonempty convex w^*-compact values and which is upper semicontinuous with respect to the topology induced by the norm on U and to the weak* topology on X^*. By Proposition 1.86 we know that T is locally bounded, that is, for each $x \in U$ there is a real $\beta > 0$ and a neighborhood $V \subset U$ of x such that $T(V) \subset \beta \mathbb{B}_{X^*}$. We denote by $\mathcal{R}_T(U)$ the set of real-valued locally Lipschitz functions $f : U \to \mathbb{R}$ such that

$$\partial_C f(x) \subset T(x) \quad \text{for all } x \in U;$$

if T is a constant multimapping with $T(x) = K$ for all $x \in U$, we will write $\mathcal{R}_K(U)$. When $\mathcal{R}_T(U)$ is nonempty, it will be endowed with the distance

$$d(f,g) := \inf\{1, \sup_{x \in U} |f(x) - g(x)|\} \quad \text{for all } f,g \in \mathcal{R}_T(U).$$

Assume that $\mathcal{R}_T(u) \ne \emptyset$ and let us show that $(\mathcal{R}_T(U), d)$ is complete. Let $(f_n)_n$ be a Cauchy sequence in $\mathcal{R}_T(U)$. Clearly, for every $x \in U$, the real sequence $(f_n(x))_n$ converges to some $f(x)$. Fixing any $x \in U$ and taking, by the local boundedness of T, a real $\beta > 0$ and a convex neighborhood $V \subset U$ of x with $T(V) \subset \beta \mathbb{B}_{X^*}$, we have for all $u, v \in V$ (according to the mean value equality in Theorem 2.180)

$$|f(u) - f(v)| = \lim_{n \to \infty} |f_n(u) - f_n(v)| \le \beta \|u - v\|,$$

so f is locally Lipschitz on U.

Now fix any $x \in U$, any $h \in X$ and any $\varepsilon > 0$. By the upper semicontinuity of the support function $u \mapsto \sigma(h, T(u))$ (see Theorem 1.48(a)), choose a convex neighborhood $U_0 \subset U$ of x such that $\sigma(h, T(u)) < \sigma(h, T(x)) + \varepsilon$ for all $u \in U_0$. Taking $\delta > 0$ satisfying $B(x, \delta) +]0, \delta[h \subset U_0$, for each $u \in B(x, \delta)$ and $t \in]0, \delta[$

there exists by the mean value theorem some $u_t \in [u, u+th]$ and $u_t^* \in \partial_C f(u_t)$ such that
$$t^{-1}(f(u+th) - f(u)) = \langle u_t^*, h \rangle \leq \sigma(h, T(u_t)) < \sigma(h, T(x)) + \varepsilon.$$
This justifies that $\sigma(h, \partial_C f(x)) = f^\circ(x; h) \leq \sigma(h, T(x))$, hence $\partial_C f(x) \subset T(x)$, that is, $f \in \mathcal{R}_T(U)$. Since it is easily seen that $d(f_n, f) \to 0$ as $n \to \infty$, it results that the metric space $(\mathcal{R}_T(U), d)$ is *complete*.

The result on Baire category property of subdifferentially pathological Lipschitz functions requires some preparatory lemmas.

LEMMA 5.86. *Let U be an open set of a separable Banach space $(X, \|\cdot\|)$ and $f : U \to \mathbb{R}$ a locally Lipschitz function. Then there exists a countable subset D of $\mathrm{gph}\,\partial_H f$ such that*
$$\mathrm{gph}\,\partial_A f = \mathrm{cl}_{U \times X^*} D,$$
where the closure is taken with U endowed with the topology induced by $\|\cdot\|$ and X^* endowed with the weak* topology.

PROOF. Denote by τ the topology induced on $U \times X^*$ by the product topology of the norm topology of X and the weak* topology of X^*. For each integer $n \in \mathbb{N}$ set $U_n := \{x \in U : \partial_H f(x) \subset n\mathbb{B}_{X^*}\}$ and note that $U \times X^* = \bigcup_{n \in \mathbb{N}} U_n \times n\mathbb{B}_{X^*}$ according to the local Lipschitz property of f. By the w^*-compactness and metrizability of $\mathbb{B}_{X^*}$, for each $n \in \mathbb{N}$ we see that $U_n \times n\mathbb{B}_{X^*}$ is τ-separable, and hence $(\mathrm{gph}\,\partial_H f) \cap (U_n \times n\mathbb{B}_{X^*})$ is also separable. Then, for each $n \in \mathbb{N}$, we can choose a countable set D_n such that
$$D_n \subset (\mathrm{gph}\,\partial_H f) \cap (U_n \times n\mathbb{B}_{X^*}) \quad \text{and} \quad (\mathrm{gph}\,\partial_H f) \cap (U_n \times n\mathbb{B}_{X^*}) \subset \mathrm{cl}_\tau D_n.$$
For the countable set $D := \bigcup_{n \in \mathbb{N}} D_n$ we obtain
$$D \subset \mathrm{gph}\,\partial_H f = \bigcup_{n \in \mathbb{N}} (\mathrm{gph}\,\partial_H f) \cap (U_n \times n\mathbb{B}_{X^*}) \subset \bigcup_{n \in \mathbb{N}} \mathrm{cl}_\tau D_n \subset \mathrm{cl}_\tau D.$$
From the equality $\mathrm{gph}\,\partial_A f = \mathrm{cl}_\tau(\mathrm{gph}\,\partial_H f)$ (see Definition 5.17), we derive that $\mathrm{gph}\,\partial_A f = \mathrm{cl}_\tau D$. □

LEMMA 5.87. *Let $(X, \|\cdot\|)$ be a Banach space whose norm $\|\cdot\|$ is Gâteaux differentiable off zero, let $f : X \to \mathbb{R} \cup \{+\infty\}$ be a proper lower semicontinuous function and let Y be a finite-dimensional vector subspace of X. Let $\bar{x} \in X$ and $\varepsilon, \delta > 0$ be such that*

  (i) *f is bounded from below on $\bar{x} + \delta\mathbb{B}_X$;*
  (ii) *$f(u) - f(\bar{x}) > -\varepsilon\delta$ for all $u \in \bar{x} + \delta\mathbb{B}_Y$.*

Then there exists some $\bar{u} \in X$ with $\|\bar{u} - \bar{x}\| < \delta$ and some $u^* \in \partial_H f(\bar{u})$ with $\|u^*_{|Y}\| < 2\varepsilon$.

The lemma is proved in the following exercise.

EXERCISE 5.88. Let $\kappa > 0$ be such that
$$\inf\{f(u) + \kappa d^2(u, \bar{x} + Y) : u \in \bar{x} + \delta\mathbb{B}_X\} > f(\bar{x}) - \varepsilon\delta.$$
(a) Argue that $d^2_{\bar{x}+Y}$ is Gâteaux differentiable and $D_G(d^2_{\bar{x}+Y})(u)_{|Y} = 0$ for all $u \in X$.
(b) Putting $\varphi := f + \kappa d^2_{\bar{x}+Y} + \Psi_{\bar{x}+\delta\mathbb{B}_X}$, argue (through the Ekeland variational

principle and Proposition 5.13(a)) the existence of some $\bar{u} \in X$ with $\|\bar{u} - \bar{x}\| < \delta$ and some $\zeta^* \in X^*$ with $\|\zeta^*\| < 2\varepsilon$ such that $-\zeta^* \in \partial_H \varphi(\bar{u})$.
(c) Put $u^* := -\zeta^* - \kappa D_G d_{\bar{x}+Y}^2(\bar{u})$ and derive that $u^* \in \partial_H f(\bar{u})$ and $\|u^*_{|Y}\| < 2\varepsilon$.

LEMMA 5.89. *Let U be an open set of a separable Banach space X and $T : U \rightrightarrows X^*$ be a multimapping which is weak* upper semicontinuous with nonempty weak* compact convex values, and such that $\mathcal{R}_T(U)$ is nonempty. Then for each $g \in \mathcal{R}_T(U)$, the set*
$$\{f \in \mathcal{R}_T(U) : \partial_A g(x) \subset \partial_A f(x), \, \forall x \in U\}$$
contains a G_δ dense set in $(\mathcal{R}_T(U), d)$.

The proof of the lemma is the purpose of the following exercise.

EXERCISE 5.90. Let $(X_n)_n$ be an increasing sequence of finite-dimensional vector subspaces of X such that $X = \mathrm{cl}(\bigcup_{n \in \mathbb{N}} X_n)$ and, for each $(x, x^*) \in \mathrm{gph}\, \partial_H g$ and $n \in \mathbb{N}$ let
$$Q_{(x,x^*,n)} := \{f \in \mathcal{R}_T(U) : \exists y \in B_U(x, 1/n), y^* \in \partial_H f(y) \text{ with } \|(x^* - y^*)_{|X_n}\| \leq 4/n\}.$$
We consider three parts.
(I) Fix $(x, x^*) \in \mathrm{gph}\, \partial_H g$ and $n \in \mathbb{N}$. The objective in this part (I) is to show that $\mathrm{int}\, Q_{(x,x^*,n)}$ is dense in $(\mathcal{R}_T(U), d)$.
   Let $f_0 \in \mathcal{R}_T(U)$ and $\varepsilon \in]0, 1[$. For every $u \in U$ put
$$f_1(u) := g(u) + f_0(x) - g(x) - \varepsilon/3, \quad f_2(u) := \min\{f_0(u), f_1(u)\}$$
and $f_3(u) := \max\{f_2(u), f_0(u) - 2\varepsilon/3\}$.
(a) Check that $f_1, f_2, f_3 \in \mathcal{R}_T(U)$, $f_2 \leq f_0$, $f_0 - 2\varepsilon/3 \leq f_3 \leq f_0$ and $d(f_3, f_0) < \varepsilon$.
(b) Check also that
$$f_0(x) - \frac{2\varepsilon}{3} < f_3(x) = f_2(x) = f_1(x) = f_0(x) - \frac{\varepsilon}{3} < f_0(x),$$
and deduce an open neighborhood $U_0 \subset U$ of x such that $f_1 = f_2 = f_3$ on U_0.
(c) From the compactness of $\mathbb{S}_{X_n}$, the inclusion $x^* \in \partial_H g(x)$ and the local Lipschitz property of g, derive a real $\delta \in]0, 1/n[$ such that $B[x, \delta] \subset U_0$, the function g is Lipschitz on $B[x, \delta]$ and
$$(g - x^*)(x + th) - (g - x^*)(x) > \frac{-t}{n} \geq \frac{-\delta}{n}, \quad \forall t \in]0, \delta], h \in \mathbb{S}_{X_n}.$$
(d) Now fix any $r \in]0, \delta/(2n)[$ and any $f \in B(f_3, r)$. Check that
$$(f - x^*)(u) - (f - x^*)(x) > -\frac{2\delta}{n} \quad \text{for all } u \in \delta \mathbb{B}_{X_n},$$
and by Lemma 5.87 take $y \in B(x, \delta)$ and $v^* \in \partial_H(f - x^*)(y)$ with $\|v^*_{|X_n}\| < 4/n$.
   For $y^* := v^* + x^*$ we see that $y^* \in \partial_H f(y)$ and $\|(y^* - x^*)_{|X_n}\| < 4/n$, so $f \in Q_{(x,x^*,n)}$. This means that $B(f_3, r) \subset Q_{(x,x^*,n)}$, and hence $f_3 \in B(f_0, \varepsilon) \cap \mathrm{int}\, Q_{(x,x^*,n)}$.
   This justifies that $\mathrm{int}\, Q_{(x,x^*,n)}$ is dense in $(\mathcal{R}_T(U), d)$ as desired.
(II) Fix $(x, x^*) \in \mathrm{gph}\, \partial_H g$ and take any $f \in Q_{(x,x^*)} := \bigcap_{n \in \mathbb{N}} Q_{(x,x^*,n)}$. For each $n \in \mathbb{N}$ choose $x_n \in B_U(x, 1/n)$ and $x_n^* \in \partial_H f(x_n)$ with $\|x_{n|X_n}^* - x_{|X_n}^*\| \leq 4/n$.
(a) Argue that $(x_n^*)_n$ converges pointwise to x^* on $\bigcup_{n \in \mathbb{N}} X_n$ (keep in mind that the sequence $(X_n)_n$ is increasing).
(b) Argue that $(x_n^*)_n$ is bounded (by the local Lipschitz property of f), and hence

$(x_n^*)_n$ converges weakly* to x^* in X^*.
(c) Derive that $x^* \in \partial_A f(x)$.
(III) By Lemma 5.86 take a countable set $D \subset \operatorname{gph} \partial_H g$ such that $\operatorname{gph} \partial_A g = \operatorname{cl}_\tau D$, where τ is the topology induced on $U \times X^*$ by the the product topology of the norm topology on X and the weak* topology on X^*. Let $Q := \bigcap_{(x,x^*) \in D} Q_{(x,x^*)}$.

(a) For any $f \in Q$, argue by (II) that $D \subset \operatorname{gph} \partial_A f$ and that
$$\operatorname{gph} \partial_A g = \operatorname{cl}_\tau D \subset \operatorname{gph} \partial_A f$$
(keep in mind by Theorem 5.35(a) that $\operatorname{gph} \partial_A f$ is τ-closed in $U \times X^*$).
(b) Conclude that
$$Q \subset \{f \in \mathcal{R}_T(U) : \partial_A g(x) \subset \partial_A f(x), \forall x \in U\}$$
and that Q contains by (I) a G_δ dense set in $(\mathcal{R}_T(U), d)$, which justifies the above lemma. □

Given a metric space E, a normed space X and a locally bounded multimapping $\Phi : E \rightrightarrows X^*$ with nonempty values, recall by Proposition 1.88 that there exists a unique smallest weak* upper semicontinuous multimapping $\Gamma : E \rightrightarrows X^*$ with w^*-compact values (resp. with w^*-compact convex values) such that $\Gamma(x) \supset \Phi(x)$ for all $x \in U$.

PROPOSITION 5.91. Let U be an open set of a separable Banach space X and $(g_n)_n$ be a sequence of locally equi-Lipschitz functions from U into $\mathbb{R}$.
(a) If we denote by M (resp. T) the smallest upper semicontinuous multimapping from U into $(X^*, w(X^*, X))$ with w^*-compact values (resp. w^*-compact convex values) and such that the image of x contains $\bigcup_{n \in \mathbb{N}} \partial_A g_n(x)$ for all $x \in U$, then $\mathcal{R}_T(U) \neq \emptyset$ and the set
$$\{f \in \mathcal{R}_T(U) : M(x) \subset \partial_A f(x) \subset \partial_C f(x) = T(x), \forall x \in U\}$$
contains a G_δ dense set in $(\mathcal{R}_T(U), d)$.
(b) Similarly, if we denote by T the smallest upper semicontinuous multimapping from U into $(X^*, w(X^*, X))$ with w^*-compact convex values and such that the image of x contains $\bigcup_{n \in \mathbb{N}} \partial_C g_n(x)$ for all $x \in U$, then $\mathcal{R}_T(U) \neq \emptyset$ and the set
$$\{f \in \mathcal{R}_T(U) : \partial_C f(x) = T(x), \forall x \in U\}$$
contains a G_δ dense set in $(\mathcal{R}_T(U), d)$.

PROOF. Clearly $\mathcal{R}_T(U) \neq \emptyset$ in both cases.
(a) For each $n \in \mathbb{N}$ Lemma 5.89 says that $Q_n := \{f \in \mathcal{R}_T(U) : \partial_A g_n(x) \subset \partial_A f(x), \forall x \in U\}$ contains a G_δ dense set in $(\mathcal{R}_T(U), d)$, hence $Q := \bigcap_{n \in \mathbb{N}} Q_n$ also contains a G_δ dense set in $(\mathcal{R}_T(U), d)$. Fixing any $x \in U$ and any $f \in Q$, it is clear that $\bigcup_{n \in \mathbb{N}} \partial_A g_n(x) \subset \partial_A f(x)$, which entails
$$T(x) \subset \partial_C f(x) \quad \text{and} \quad M(x) \subset \partial_A f(x) \subset \partial_C f(x) \subset T(x),$$
where the inclusion $\partial_C f(x) \subset T(x)$ is due to the definition of $\mathcal{R}_T(U)$. We conclude that
$$M(x) \subset \partial_A f(x) \subset \partial_C f(x) = T(x).$$
(b) For each $n \in \mathbb{N}$, from Lemma 5.89 and from the equality $\partial_C g_n(x) = \overline{\operatorname{co}}^*(\partial_A g_n(x))$ we see that the set $\{f \in \mathcal{R}_T(U) : \partial_C g_n(x) \subset \partial_C f(x), \forall x \in U\}$ contains a G_δ dense set in $(\mathcal{R}_T(U), d)$. Then it suffices to proceed as in (a). □

We can now derive the G_δ dense property of subdifferentially pathological Lipschitz functions with respect to $\partial_A, \partial_L, \partial_C$.

PROPOSITION 5.92. *Let U be an open subset of a separable Banach space X. The following hold:*
(a) *Given any bounded sequence $(a_n^*)_n$ in X^* and $\Xi := \{a_n^* : n \in \mathbb{N}\}$, then $(\mathcal{R}_{\overline{co}^*(\Xi)}(U) \neq \emptyset$ and) the set of Lipschitz functions $f : U \to \mathbb{R}$ such that $\mathrm{cl}_{w^*}(\Xi) \subset \partial_A f(x) \subset \overline{co}^*(\Xi)$ for all $x \in U$, contains a G_δ dense set in $(\mathcal{R}_{\overline{co}^*(\Xi)}(U), d)$.*
(b) *For any nonempty w^*-compact convex subset K in X^*, (one has $\mathcal{R}_K(U) \neq \emptyset$ and) the set of Lipschitz functions $f : U \to \mathbb{R}$ such that $\partial_C f(x) = K$ for all $x \in U$, contains a G_δ dense set in $(\mathcal{R}_K(U), d)$.*
(c) *If X^* is strongly separable (that is, X is a separable Asplund space), then the assertion (a) holds true with ∂_L in place of ∂_A.*

PROOF. The assertion (a) follows directly from (a) in the above proposition with $g_n := \langle a_n^*, \cdot \rangle$.

To prove (b) take a sequence $(a_n^*)_n$ which is w^*-dense in K, so $K = \overline{co}^*(\{a_n^* : n \in \mathbb{N}\})$. It suffices to apply (b) in the above proposition with $g_n := \langle a_n^*, \cdot \rangle$.

Finally, when X^* is strongly separable, using in such a case (by Proposition 5.21) the equality between limiting and approximate subdifferentials of locally Lipschitz functions on X, we see that (c) is a consequence of (a). $\square$

EXERCISE 5.93. Let $g_1, \cdots, g_m$ be a finite family of real-valued locally Lipschitz functions from an open set U of a separable Banach space X and let $T : U \rightrightarrows X$ be defined by
$$T(x) := co\left(\bigcup_{k=1}^m \partial_C g(x)\right) \quad \text{for all } x \in U.$$
Show that the set $\{f \in \mathcal{R}_T(U) : \partial_C f(x) = T(x), \forall x \in U\}$ contains a G_δ dense set in $(\mathcal{R}_T(U), d)$. (Hint: argue that T is weak* upper semicontinuous and apply Proposition 5.91(b)).

The next exercise characterizes multimappings which are C-subdifferentials of locally Lipschitz functions of one real variable. Given a dense set D in an open set U of a topological space T and a function φ defined on D with values in $\mathbb{R} \cup \{-\infty, +\infty\}$ the functions defined on U by

(5.21) $\qquad x \mapsto \liminf_{D \ni u \to x} \varphi(u) \quad \text{and} \quad x \mapsto \limsup_{D \ni u \to x} \varphi(u) \quad \text{for all } x \in U,$

are easily seen to be lower semicontinuous and upper semicontinuous respectively. If U is an open interval of $\mathbb{R}$ and φ is defined on the whole interval U, denoting by λ the one-dimensional Lebesgue measure, one says that φ is λ-*robustly lower* (resp. *upper*) *semicontinuous* at $x \in U$ when, for any λ-null set N in U,

$$\varphi(x) = \liminf_{N \not\ni u \to x} \varphi(u) \quad (\text{resp. } \varphi(x) = \limsup_{N \not\ni u \to x} \varphi(u)).$$

If this holds for all $x \in U$, we say that φ is lower (resp. upper) λ-robustly semicontinuous on U. In such a case φ is in particular lower (resp. upper) semicontinuous on U according to the above observation.

EXERCISE 5.94. Let U be an open interval in $\mathbb{R}$ and $\alpha, \beta : I \to \mathbb{R}$ be two locally bounded functions such that $\alpha \leq \beta$. The aim of the exercise is to show that the following assertions are equivalent:

(i) α is λ-robustly lower semicontinuous and β is λ-robustly upper semicontinuous;
(ii) there exists a locally Lipschitz function $f : U \to \mathbb{R}$ such that
$$\partial_C f(x) = [\alpha(x), \beta(x)] \quad \text{for all } x \in U;$$
(iii) for $T : U \rightrightarrows \mathbb{R}$ defined by $T(x) := [\alpha(x), \beta(x)]$ for all $x \in U$, the set $\mathcal{R}_T(U)$ is nonempty and the set $\{f \in \mathcal{R}_T(U) : \partial_C f(x) = [\alpha(x), \beta(x)], \forall x \in U\}$ contains a G_δ dense set in $(\mathcal{R}_T(U), d)$.

(I) Suppose that (i) holds true.
(a) Justify that α and β are locally λ-integrable on U, so fixing $x_0 \in U$ one can define the locally Lipschitz functions $g_1, g_2 : U \to \mathbb{R}$ by
$$g_1(x) := \int_{x_0}^{x} \alpha(u)\, du \quad \text{and} \quad g_2(x) := \int_{x_0}^{x} \beta(u)\, du \quad \text{for all } x \in U.$$
(b) Argue that there exists a λ-null set $N \subset U$ such that at each point $u \in U \setminus N$ both functions g_1 and g_2 are derivable with $g'_1(u) = \alpha(u)$ and $g'_2(u) = \beta(u)$. Argue also (via the gradient representation in Theorem 2.83) that, for every $x \in U$
$$\partial_C g_1(x) = \left[\liminf_{N \not\ni u \to x} g'_1(u), \limsup_{N \not\ni u \to x} g'_1(u)\right] = \left[\alpha(x), \limsup_{N \not\ni u \to x} g'_1(u)\right]$$
$$\partial_C g_2(x) = \left[\liminf_{N \not\ni u \to x} g'_2(u), \limsup_{N \not\ni u \to x} g'_2(u)\right] = \left[\liminf_{N \not\ni u \to x} g'_2(u), \beta(x)\right].$$
(c) Argue that, for every $x \in U$
$$\limsup_{N \not\ni u \to x} g'_1(u) \leq \beta(x) \quad \text{and} \quad \liminf_{N \not\ni u \to x} g'_2(u) \geq \alpha(x).$$
(d) Derive that $\operatorname{co}(\partial_C g_1(x) \cup \partial_C g_2(x)) = [\alpha(x), \beta(x)] =: T(x)$ for all $x \in U$, and deduce from Exercise 5.93 that the implication (i)$\Rightarrow$(ii) holds true.
(II) Argue via the gradient representation in Theorem 2.83 that (ii)$\Rightarrow$(i) and conclude that the three assertions (i), (ii), (iii) are equivalent.

5.6. Comments

The approximate subdifferential that we presented in this chapter, was developed by A. D. Ioffe in a series of six papers [509, 511, 514, 515, 517, 525]. The 1981 mimeographed publication [509] contains the bases of the theory. It was published as a preprint in "Cahiers CEREMADE de l'Université Paris IX Dauphine". The finite-dimensional setting was completely analyzed in the paper [514] published in 1984 (and submitted in 1981). In this paper [514], given an extended real-valued function $f : \mathbb{R}^N \to \mathbb{R} \cup \{+\infty\}$ on $\mathbb{R}^N$ endowed with its natural inner product, its approximate subdifferential at a point $\bar{x} \in \operatorname{dom} f$ is defined as
$$\partial_a f(\bar{x}) := \operatorname*{Lim\,sup}_{x \to_f \bar{x}} \partial_H f(x).$$
Through this formula and the penalisation decoupling technique in (4.84), for functions f, g defined on $\mathbb{R}^N$, the key sum rule
(5.22) $$\partial_a(f + g)(\bar{x}) \subset \partial_a f(\bar{x}) + \partial_a g(\bar{x})$$
is established under a certain qualification condition (fulfilled in particular in the basic situation where one of the functions is Lipschitz near $\bar{x}$). This yields on one

hand a corresponding chain rule along with various applications and on the other the key normal intersection rule

$$(5.23) \qquad N^a(S_1 \cap S_2; \bar{x}) \subset N^a(S_1; \bar{x}) + N^a(S_2; \bar{x}),$$

where S_1, S_2 are subsets in $\mathbb{R}^N$ containing $\bar{x}$ and satisfying suitable qualification conditions at $\bar{x}$ and where $N^a(S_i; \bar{x}) := \partial_a \Psi_{S_i}(\bar{x})$. It was also noticed in [**514**, p. 389 and p. 384 (Theorem 1)] that $\partial_a f(\bar{x})$ coincides in finite dimensions with the Mordukhovich limiting subdifferential $\partial_L f(\bar{x})$ (introduced in [**720**] via the normal cone $\underset{u \to \bar{x}}{\operatorname{Lim\,sup}} \mathbb{R}_+ (u - \operatorname{Proj}_{\overline{S}}(u))$). Previously, for a closed set S in $\mathbb{R}^N$ and $\bar{x} \in S$ it was shown in [**509**, p. 26-28] by Ioffe that $N^a(S; \bar{x})$ coincides in finite dimensions with the latter form $\underset{u \to \bar{x}}{\operatorname{Lim\,sup}} \mathbb{R}_+ (u - \operatorname{Proj}_{\overline{S}}(u))$ of the limiting normal cone. Considering a Banach space X admitting an equivalent renorm $\|\cdot\|$ which is nowhere Gâteaux differentiable and recalling (see Proposition 5.11(a)) that a continuous convex function $f : X \to \mathbb{R}$ is Gâteaux differentiable at x if and only if $\partial_H(-f)(x) \neq \emptyset$, one sees with $\varphi := -\|\cdot\|$ that $\partial_H \varphi(x) = \emptyset$ for every $x \in X$. The situation of some Banach spaces with diverse locally Lipschitz continuous functions whose H-subdifferentials are empty everywhere, led Ioffe to use certain finite dimensional reduction processes. Given a Hausdorff locally convex space X and a function $f : X \to \mathbb{R} \cup \{+\infty\}$, Ioffe in [**509, 515**] extends the above finite-dimensional definition for $\bar{x} \in \operatorname{dom} f$ as well as the key sum and chain rules with

$$(5.24) \qquad \partial_a f(\bar{x}) = \bigcap_{L \in \mathcal{F}(X)} \underset{x \to_f \bar{x}}{\operatorname{Lim\,sup}} \partial_H (f + \Psi_{x+L})(x);$$

the notation $\partial_A f$ is employed in [**515**] but we keep the notation $\partial_a f$ for adjustment with the style in the book and below. In (5.24) by $\mathcal{F}(X)$ we denote (as done in the previous sections of the chapter) the class of all finite dimensional vector subspaces of the vector space X. Ioffe observed that, in any infinite-dimensional Banach space (separable or not), there are examples of closed sets S and points $\bar{x} \in S$ such that $\partial_a \Psi_S(\bar{x})$ is strictly bigger than both $\operatorname{cl}_{w^*} (\mathbb{R}_+ \partial_a d_S(\bar{x}))$ and $\mathbb{R}_+ \partial_a d_S(\bar{x})$. The two latter sets were considered by Ioffe [**517**] as normal cones, called there the G-normal cone and the nucleus of the G-normal cone. Under certain conditions, it was shown in [**517**] that each of these two normal cones fulfills the key property that the normal cone of an intersection $S_1 \cap S_2$ is included in the sum of normal cones of S_1 and S_2 at the same point. For a function f on a Banach space X, the subdifferential associated with the G-normal cone (resp. associated with its nucleus) is defined in [**509, 517**] (resp. in [**517**]) as the set of $x^* \in X^*$ such that $(x^*, -1)$ belongs to the corresponding normal cone of the epigraph of f. These G-normal cone and G-subdifferential as well as their nuclei are shown in [**517**] to satisfy (under some conditions) the normal intersection rule of type (5.23) and the subdifferential sum rule of type (5.22). It was also proved in [**517**] the important property that the Clarke normal cone is equal to the weak* closed convex hull of the G-normal cone as well as of its nucleus. In the development of this chapter, given a general Banach space $(X, \|\cdot\|)$ we have retained $\mathbb{R}_+ \partial_a d_S(\bar{x})$ as the definition of the *Ioffe approximate normal cone* with the notation $N_A(S; \bar{x})$; the Ioffe approximate subdifferential $\partial_A f(\bar{x})$ for an extended real-valued function f is then defined as the set of $x^* \in X^*$ such that $(x^*, -1)$ is contained in the Ioffe approximate normal cone of the epigraph of f at $(\bar{x}, f(\bar{x}))$. These normal cone $N^A(S; \cdot)$ and subdifferential

$\partial_A f$ are precisely those used by Ioffe at least since his 2000 survey paper [**521**] (see also [**520, 525, 526**]), but with notations $N^G(S;\cdot)$ and $\partial_G f$ instead.

The concept of Hadamard subdifferential is not due to J. Hadamard. The terminology of Hadamard subdifferential in Definition 5.3 (like the one of Fréchet subdifferential in Definition 4.1) is utilized to emphasize the feature that $\underline{d}_H^+ f(x;h)$ corresponds to the right-hand (unilateral) inequality coming from the definition of Hadamard differentiability of real-valued functions. This concept of Hadamard subdifferential has been introduced by J.-P. Penot in [**785**], and Proposition 5.7 was established therein. The arguments for Lemma 5.20 are contained in Ioffe [**518**, Lemma 4], in the paper [**730**, Theorem 9.2] by B. S. Mordukhovich and Y. Shao, in Mordukhovich's book [**725**, Theorem 3.59], and in the paper [**525**, Theorem 5.4] by Ioffe. Proposition 5.21 is a particular case of (9.8) in Theorem 9.2 by Mordukhovich and Shao [**730**]. Lemma 5.26 is taken from Ioffe [**525**, Theorem 7.10].

Lemmas 5.39 and 5.41 reproduce Lemmas 5.2 and 5.3 in Ioffe's paper [**515**] while the proof of Theorem 5.42 followed the arguments of Ioffe for the proof of Theorem 5.1 in the same paper [**515**]. Theorem 5.44 first appeared in [**515**, (3.3) in Proposition 3.3] with a different proof.

Lemma 5.53 and its proof correspond to Lemma 2.3 in Ioffe's paper [**517**] (see also [**509**, p. 37-39]) while the statement and proof of Lemma 5.54 are taken from the proof of Proposition 2.4 in the same Ioffe paper [**517**] (see also [**509**, p. 39-42]). Proposition 5.55 reproduces Proposition 2.4 in [**517**] and Theorem 5.57 is the adaptation of Ioffe's arguments in [**517**, p. 8] (see also [**509**, p. 43]). The assertions (a) and (b) in Lemma 5.59 were noticed by Ioffe in the proofs of Lemma 5.2 and Proposition 3.7 respectively in [**517**]. The statement of Proposition 5.65 is contained in the statement of [**517**, Proposition 3.5]. The assertion (a) in Theorem 5.68 can be seen as a particular case of (9.10) in [**730**] by Mordukhovich and Shao. Lemma 5.70 is due to Ioffe [**517**]. Theorem 5.71 was proved by A. Jourani and L. Thibault [**582, 583**]; a similar result was shown earlier by Ioffe [**517**, Theorem 7.8] under the assumption that G admits at $\bar{x}$ a strict prederivative with norm compact values. Theorem 5.76 is contained in Ioffe [**517**, Proposition 3.4 and Theorem 7.2(d)]. The statement and proof of Theorem 5.78 for the A-subdifferential of sums followed the papers [**517**, Theorems 5.6 and 7.4] and [**525**, p. 106] (see also Ioffe's previous preprint [**509**, p. 64-67]).

Proposition 5.85 was established in a slightly similar form by A. Jourani and L. Thibault [**585**, Lemma 2.3]. Results of this type with Fréchet normals probably began with P. D. Loewen [**670**]. The proof given in the manuscript for Proposition 5.85 follows the arguments by R. Correa, P. Gajardo and L. Thibault [**286**, Proposition 3.6]. The pathological function $f(x) = \int_0^x \mathbf{1}_S(t)\,dt$ is due to R. T. Rockafellar [**859**]. Except this, the whole development in Subsection 5.5.2 is taken from J. M. Borwein, W. B. Moors and X. Wang [**139**]. Previously to [**139**], for every convex set K in $\mathbb{R}^n$, E. Jouini [**566**] constructed a Lipschitz function $f: \mathbb{R}^n \to \mathbb{R}$ such that $\partial_C f(x) = K$ for all $x \in \mathbb{R}^n$; such a function for a polytope K in $\mathbb{R}^n$ was constructed earlier by Jouini [**565**]. For other results on pathological Lipschitz functions with big subdifferentials at any point, we refer to Borwein, Moors and Wang [**138, 139**] and to Borwein and Wang [**148**].

CHAPTER 6

Sequential mean value inequalities

6.1. Mean value inequalities with Dini derivatives

This section is devoted to various mean value inequalities by means of one-sided Dini derivatives and to their consequences.

6.1.1. Mean value inequalities with lower/upper Dini directional derivatives. In this first subsection we start with mean value inequalities in terms of (right-hand) lower and upper Dini directional derivatives. The concept of Dini directional derivative has been already partly considered in (2.42) just after we introduced the notion of Bouligand directional derivative in Definition 2.119 in Chapter 2. Here the concept is defined and involved in a general setting.

DEFINITION 6.1. Let U be a set of a normed space $(X, \|\cdot\|)$ and $f : U \to \mathbb{R} \cup \{-\infty, +\infty\}$ be a function which is finite at $x \in U$. For any $h \in X$ such that $]0, \tau]h \subset (U - x)$ for some $\tau > 0$, the *(right-hand) lower and upper Dini directional derivatives* $\underline{d}_D^+ f(x; h)$ and $\overline{d}_D^+ f(x; h)$ of f at x in the direction h is defined by

$$\underline{d}_D^+ f(x; h) := \liminf_{t \downarrow 0} t^{-1}[f(x+th) - f(x)], \quad \overline{d}_D^+ f(x; h) := \limsup_{t \downarrow 0} t^{-1}[f(x+th) - f(x)].$$

So, $\underline{d}_D^+ f(x; \cdot)$ and $\overline{d}_D^+ (x; \cdot)$ are defined on the whole space X whenever U is a neighborhood of the point x.

When $X = \mathbb{R}$ and U is an open interval I of $\mathbb{R}$, it will be sometimes convenient, for $r \in I$, to denote as usual $\underline{D}^+ f(r)$ and $\overline{D}^+ f(r)$ in place of $\underline{d}_D^+ f(r; 1)$ and $\overline{d}_D^+ f(r; 1)$, and to consider also the other similar concepts $\underline{D}^- f(r)$ and $\overline{D}^- f(r)$; that is,

$$\underline{D}^+ f(r) := \liminf_{s \downarrow r} \frac{f(s) - f(r)}{s - r} \quad \text{and} \quad \overline{D}^+ f(r) := \limsup_{s \downarrow r} \frac{f(s) - f(r)}{s - r}$$

$$\underline{D}^- f(r) := \liminf_{s \uparrow r} \frac{f(s) - f(r)}{s - r} \quad \text{and} \quad \overline{D}^- f(r) := \limsup_{s \uparrow r} \frac{f(s) - f(r)}{s - r}.$$

They are the *unilateral* (or *one-sided*) *Dini semiderivates* (also called *extreme unilateral Dini derivates*).

Assuming that U is a neighborhood of x, we see that $\underline{d}_D^+ f(x; \cdot)$ and $\overline{d}_D^+ f(x; \cdot)$ are positively homogeneous, that is, for all reals $r > 0$

$$\underline{d}_D^+ f(x; rh) = r\underline{d}_D^+ f(x; h) \quad \text{and} \quad \overline{d}_D^+ f(x; rh) = r\overline{d}_D^+ f(x; h),$$

and of course $\underline{d}_D^+ f(x; \cdot) \leq \overline{d}_D^+ f(x; \cdot)$. Further,

$$f^B(x; \cdot) = \underline{d}_H^+ f(x; \cdot) \leq \underline{d}_D^+ (x; \cdot)$$

and the latter inequality is an equality whenever the function f is Lipschitz near the point x. When the duality between $\underline{d}_D^+ f(x;\cdot)$ and $\overline{d}_D^+ f(x;\cdot)$ does not need to be involved in an analysis, it is also usual to denote by $f^D(x;\cdot)$ the (right hand) lower Dini directional derivative in place of $\underline{d}_D^+ f(x;\cdot)$; this mimics the symbol $f^B(x;\cdot)$ for the Bouligand directional derivative.

THEOREM 6.2 (mean value inequality with Dini directional derivatives). Let a, b be two distinct points of a normed space $(X, \|\cdot\|)$.
(a) If $f : [a,b] \to \mathbb{R} \cup \{+\infty\}$ is an extended real-valued function which is finite at a and lower semicontinuous relative to $[a,b]$, then, for any real $r \leq f(b)$, there exists some $c \in [a, b[\cap \operatorname{dom} f$ (depending on r) such that

(6.1) $$r - f(a) \leq \underline{d}_D^+ f(c; b - a)$$

and

$$\|b - a\|(f(c) - f(a)) \leq \|c - a\|(r - f(a)).$$

(b) If in addition f is continuous at the point a relative to $[a,b]$, then there is some c in the open interval $]a, b[$ such that (6.1) holds.
(c) If $f : [a,b] \to \mathbb{R}$ is a real-valued continuous function, then there exist some $c_1, c_2 \in]a, b[$ such that

$$f(b) - f(a) \leq \underline{d}_D^+ f(c_1; b - a) \quad \text{and} \quad \overline{d}_D^+ f(c_2; b - a) \leq f(b) - f(a).$$

PROOF. (a) Consider the lower semicontinuous function g on $[a,b]$ defined by $g(x) = f(x)$ if $x \neq b$ and $g(b) = r$. For each $t \in [0,1]$, put as usual

$$\varphi(t) := g(a + t(b-a)) + t(g(a) - g(b)).$$

The function φ is lower semicontinuous relative to $[0, 1]$ and $\varphi(0) = \varphi(1) = f(a)$, then there is some $\tau \in [0, 1[$ such that the minimum of φ on $[0, 1]$ is attained at τ (and $\varphi(\tau)$ is finite since $\varphi(\tau) \leq f(a)$). This obviously yields $\underline{d}_D^+ \varphi(\tau; 1) \geq 0$, hence, for $c := a + \tau(b - a) \in [a, b[$, we have that $f(c)$ is finite and

$$\underline{d}_D^+ g(c; b - a) + g(a) - g(b) \geq 0, \quad \text{or equivalently} \quad \underline{d}_D^+ g(c; b - a) \geq r - f(a),$$

which translates the desired inequality (6.1) since $\underline{d}_D^+ g(c; b - a) = \underline{d}_D^+ f(c; b - a)$ according to the inclusion $c \in [a, b[$.

On the other hand, the inequality $\varphi(\tau) \leq \varphi(0)$ means $f(c) - f(a) \leq \tau(r - f(a))$. Further, keeping in mind that $c = a + \tau(b - a)$ we see that $\|c - a\| = \tau\|b - a\|$, which combined with the latter inequality gives the second inequality in (a).
(b) Assume in addition that f is continuous at a relative to $[a,b]$ and put $\zeta(t) := g(a + t(b - a))$ for all $t \in [0, 1]$, with g as above. Suppose by contradiction that

(6.2) $$\underline{d}_D^+ \zeta(t; 1) < \zeta(1) - \zeta(0) =: \delta \quad \text{for all } t \in]0, 1[\cap \operatorname{dom} \zeta.$$

From the upper semicontinuity of ζ at 0 choose some $t_0 \in]0, 1[$ such that ζ is finite on $[0, t_0]$. Taking the finiteness of $\zeta(t_0)$ and (6.2) into account, we see by the mean value inequality above (applied with the function ζ on $[t_0, \tau]$) that $\zeta(\tau)$ is finite for all $\tau \in]t_0, 1]$, and hence ζ is finite on $[0, 1]$. The inequality $\underline{d}_D^+ \zeta(t_0; 1) < \delta$ and the definition of $\underline{d}_D^+ \zeta(t_0; 1)$ as a limit inferior furnishes a real $s \in]0, 1 - t_0[$ such that $\zeta(t_0 + s) - \zeta(t_0) < \delta s$. Further, for each positive real $\sigma < \min\{t_0, 1 - t_0 - s\}$, from (6.2) and the mean value inequality above we have $\zeta(t_0) - \zeta(\sigma) \leq \delta(t_0 - \sigma)$ and $\zeta(1 - \sigma) - \zeta(t_0 + s) \leq \delta(1 - \sigma - t_0 - s)$, so $\zeta(t_0) - \zeta(0) \leq \delta t_0$ and $\zeta(1) - \zeta(t_0 + s) \leq$

6.1. MEAN VALUE INEQUALITIES WITH DINI DERIVATIVES

$\delta(1-t_0-s)$ thanks to the upper semicontinuity of ζ at 0 and its lower semicontinuity at 1. We deduce

$$\delta = \zeta(1) - \zeta(0) = \big(\zeta(1) - \zeta(t_0 + s)\big) + \big(\zeta(t_0 + s) - \zeta(t_0)\big) + \big(\zeta(t_0) - \zeta(0)\big)$$
$$< \delta(1 - t_0 - s) + \delta s + \delta t_0 = \delta,$$

which gives a contradiction.

(c) The first inequality of (c) is a consequence of (a) and the second inequality follows from the first applied to $-f$ since $\underline{d}_D^+(-f)(x;h) = -\overline{d}_D^+ f(x;h)$. □

Through the above theorem, characterizations of Lipschitz and convex properties can be obtained in terms of lower Dini directional derivative.

COROLLARY 6.3. *Let U be an open convex set of a normed space X and $f : U \to \mathbb{R} \cup \{+\infty\}$ be a proper lower semicontinuous function. For $\gamma \geq 0$, the function f is Lipschitz continuous on U with γ as a Lipschitz constant if and only if*

$$\underline{d}_D^+ f(x;h) \leq \gamma \|h\| \quad \text{for all } x \in U \cap \mathrm{dom}\, f \text{ and } h \in X.$$

PROOF. The Lipschitz property obviously implies the inequality of the corollary. Let us prove the reverse implication. Fix any $x \in U$ where f is finite and take any $y \in U$ with $y \neq x$. Then, for any real $r \leq f(y)$, the above theorem gives some $c \in [x,y[\cap \mathrm{dom}\, f$ such that $r - f(x) \leq \underline{d}_D^+ f(c; y - x)$, hence $r - f(x) \leq \gamma \|y - x\|$. This entails $f(y) - f(x) \leq \gamma \|y - x\|$. Therefore, $f(y)$ is finite, that is, f is finite on U and, for all $x, y \in U$, we have $f(y) - f(x) \leq \gamma \|y - x\|$, which is equivalent to $|f(y) - f(x)| \leq \gamma \|y - x\|$. □

COROLLARY 6.4. *Let U be a convex set of a normed space X and $f : U \to \mathbb{R} \cup \{+\infty\}$ be a proper lower semicontinuous function (with respect to the induced topology). Then f is convex on U if and only if*

$$\underline{d}_D^+ f(x; y - x) + \underline{d}_D^+ f(y; x - y) \leq 0$$

for all $x, y \in U \cap \mathrm{dom}\, f$ where the left-hand side is well-defined.

PROOF. We need only to prove the convexity of f under the assumption of the inequality. Fix any two points $x, y \in U \cap \mathrm{dom}\, f$ where f is finite with $y \neq x$. Consider any $t \in]0,1[$ and put $z := x + t(y - x)$. For any real $r \leq f(z)$, choose by the above theorem $c_1 \in [x, z[\cap \mathrm{dom}\, f$ and $c_2 \in [y, z[\cap \mathrm{dom}\, f$ such that

$$r - f(x) \leq t\,\underline{d}_D^+ f(c_1; y - x) \quad \text{and} \quad r - f(y) \leq (1-t)\,\underline{d}_D^+ f(c_2; x - y).$$

Multiplying the first inequality by $(1-t)$ and the second by t produces after addition

$$r - (1-t)f(x) - tf(y) \leq t(1-t)[\underline{d}_D^+ f(c_1; y - x) + \underline{d}_D^+ f(c_2; x - y)]$$
$$= t(1-t) \frac{\|y - x\|}{\|c_1 - c_2\|} [\underline{d}_D^+ f(c_1; c_2 - c_1) + \underline{d}_D^+ f(c_2; c_1 - c_2)] \leq 0,$$

which means that f is convex on U. □

Classical analysis says that a real-valued derivable function of one real variable with nonpositive derivate on an interval is nonincreasing on this interval. The next corollary generalizes this result to the nonincreasing property of a lower semicontinuous function with respect to a cone. A function $f : C \to \mathbb{R} \cup \{+\infty\}$ is said to be *K-nonincreasing on C*, for a cone K of X, whenever $f(y) \leq f(x)$ for all $x, y \in C$ with $y - x \in K$.

COROLLARY 6.5. *Let U be an open convex set of a normed space X, $f : U \to \mathbb{R} \cup \{+\infty\}$ be a proper lower semicontinuous function, and K be a nonempty cone of X. The function f is K-nonincreasing on U if and only if $\underline{d}_D^+ f(x; h) \leq 0$ for all $x \in U \cap \mathrm{dom}\, f$ and $h \in K$.*

PROOF. To show the implication $\Rightarrow$, fix any $x \in U \cap \mathrm{dom}\, f$ and $h \in K$. Choosing $\delta > 0$ such that $x +]0, \delta[h \subset U$, we observe that
$$t^{-1}\big(f(x+th) - f(x)\big) \leq 0 \quad \text{for all } t \in]0, \delta[,$$
which entails the inequality $\underline{d}_D^+ f(x; h) \leq 0$.

Let us prove the reverse implication. Fix $x, y \in U$ with $x \neq y$, $f(x) < +\infty$ and $y - x \in K$. From Theorem 6.2, for any real $r \leq f(y)$, there exists some $c \in [x, y[$ with $c \in U \cap \mathrm{dom}\, f$ such that $r - f(x) \leq \underline{d}_D^+ f(c; y - x) \leq 0$, hence $f(y) \leq f(x)$. So f is K-nonincreasing on U. $\square$

6.1.2. Sub-sup regularity and saddle functions. Recall, for a locally Lipschitz function $f : U \to \mathbb{R}$ on an open set U of the normed space X, that $f^B(u; h) = \underline{d}_D^+ f(u; h)$ for all $u \in U$ and $h \in X$. We then see that the Clarke directional subregularity of such a function f at $x \in U$ (in the sense of Definition 2.127 via (b_1) in Proposition 2.126) entails that $\underline{d}_D^+ f(\cdot; h)$ is upper semicontinuous at x since $f^\circ(\cdot; h)$ is upper semicontinuous according to Proposition 2.74(a). In fact, the above Dini mean value inequality allows us to show in (b) below that the latter upper semicontinuity of the (right-hand) lower Dini directional derivative characterizes the (Clarke) directional subregularity.

PROPOSITION 6.6. *Let U be an open set of the normed space X and $f : U \to \mathbb{R}$ be a real-valued locally Lipschitz function. The following hold:*
(a) *For all $x \in U$ and $h \in X$,*
$$f^\circ(x; h) = \limsup_{u \to x} \underline{d}_D^+ f(u; h) = \limsup_{u \to x} \overline{d}_D^+ f(u; h).$$
(b) *The function f is (Clarke) subregular at $x \in U$ in a direction h if and only if the right-hand lower (upper) Dini directional derivative $\underline{d}_D^+ f(\cdot; h)$ ($\overline{d}_D^+ f(\cdot; h)$) is upper-semicontinuous at x.*

PROOF. (a) Fix $h \in X$. For any $u \in U$, we obviously have
$$\underline{d}_D^+ f(u; h) \leq \overline{d}_D^+ f(u; h) \leq f^\circ(u; h),$$
so $\limsup_{u \to x} \underline{d}_D^+ f(u; h) \leq \limsup_{u \to x} \overline{d}_D^+ f(u; h) \leq f^\circ(x; h)$ according to the upper semicontinuity of $f^\circ(\cdot; h)$ established in Proposition 2.74(a). To prove that the third member is not less than the first, we may suppose $h \neq 0$. Choose a real $\eta > 0$ such that $U_0 + [0, \eta]h \subset U$, where $U_0 := B(x, \eta)$. For any $y \in U_0$ and $t \in]0, \eta]$, by Theorem 6.2 there is some $\theta(y, t) \in [0, t]$ such that
$$t^{-1}\big(f(y+th) - f(y)\big) \leq \underline{d}_D^+ f(y + \theta(y, t) h; h).$$
Since $\theta(y, t) \downarrow 0$ as $y \to x$ and $t \downarrow 0$, passing to the limit superior as $y \to x$ and $t \downarrow 0$ gives as desired
$$f^\circ(x; h) \leq \limsup_{u \to x} \underline{d}_D^+ f(u; h).$$
(b) The assertion (b) follows from (a) and the comments preceding the statement of the proposition. $\square$

Taking the previous proposition into account, we introduce a concept of separate directional regularity.

DEFINITION 6.7. Let U and V be two open sets of normed spaces X and Y respectively. A locally Lipschitz function $f : U \times V \to \mathbb{R}$ is said to be *sub-sup regular* at $(x, y) \in U \times V$ in the direction $(h, k) \in X \times Y$ provided that $\underline{d}_D^+ f(\cdot, \cdot; h, 0)$ is upper semicontinuous at (x, y) and $\overline{d}_D^+ f(\cdot, \cdot; 0, k)$ is lower semicontinuous at (x, y). When the properties hold true for all directions $(h, k) \in X \times Y$ one says that f is *directionally sub-sup regular* at (x, y).

Similarly, one says that f is *sub-sub regular* at (x, y) in the direction (h, k) whenever both $\underline{d}_D^+ f(\cdot, \cdot; h, 0)$ and $\underline{d}_D^+ f(\cdot, \cdot; 0, k)$ are upper semicontinuous at (x, y).

REMARK 6.8. The above definition of sub-sup regularity amounts to saying that the locally Lipschitz functions f and $-f$ are (Clarke) subregular at (x, y) in the directions $(h, 0)$ and $(0, k)$ respectively. □

Continuous saddle functions are particular examples of directionally sub-sup regular functions, as it will be seen below. Given two convex sets U and V of vector spaces X and Y respectively, a function $f : U \times V \to \mathbb{R}$ is a *saddle (or convex-concave) function* whenever, for each $(x, y) \in U \times V$, the function $f(\cdot, y)$ is convex and the function $f(x, \cdot)$ is concave (that is, its opposite is convex). When $f(\cdot, y)$ and $f(x, \cdot)$ are convex for all $(x, y) \in U \times V$, the function f is called *separately convex (or biconvex)*. Before proving the sub-sup (resp. sub-sub) regularity of continuous saddle (resp. separately convex) functions, let us establish their local Lipschitz property. We even prove the local equi-Lipschitz property of equi-continuous families of such functions.

PROPOSITION 6.9. Let U and V be two open convex sets of normed spaces X and Y respectively and let $(f_i)_{i \in I}$ be a family of functions from $U \times V$ into $\mathbb{R}$. Assume that either all the functions f_i are saddle or all the functions f_i are separately convex. Then the following are equivalent:
(a) the family $(f_i)_{i \in I}$ is locally equi-Lipschitz on $U \times V$;
(b) the family $(f_i)_{i \in I}$ is equi-continuous on $U \times V$;
(c) For each $(\overline{x}, \overline{y}) \in U \times V$ the family $\big(f_i - f_i(\overline{x}, \overline{y})\big)_{i \in I}$ is locally equi-bounded around $(\overline{x}, \overline{y})$.

PROOF. The implications (a) ⇒ (b) ⇒ (c) being obvious, we have only to prove (c) ⇒ (a). So, suppose (c) and fix $\overline{x} \in U$, $\overline{y} \in V$, open convex neighborhoods $U_0 \subset U$ and $V_0 \subset V$ of $\overline{x}$ and $\overline{y}$ respectively along with a real $\beta \geq 0$ such that $|f_i(x, y) - f_i(\overline{x}, \overline{y})| \leq \beta$ for all $x \in U_0$, $y \in V_0$ and $i \in I$. By Lemma 2.156 the families of functions $\big(f_i(\cdot, y) - f_i(\overline{x}, \overline{y})\big)_{i \in I, y \in V_0}$ and $\big(f_i(x, \cdot) - f_i(\overline{x}, \overline{y})\big)_{i \in I, x \in U_0}$ are equi-Lipschitz on open neighborhoods $U_0' \subset U_0$ and $V_0' \subset V_0$ of $\overline{x}$ and $\overline{y}$ with a common Lipschitz constant γ. Then, for each $i \in I$, we have, for all $(x, y) \in U_0' \times V_0'$ and $(x', y') \in U_0' \times V_0'$,

$$|f_i(x, y) - f_i(x', y')| \leq |f_i(x, y) - f_i(x', y)| + |f_i(x', y) - f_i(x', y')|$$
$$\leq \gamma\big(\|x - x'\| + \|y - y'\|\big),$$

which translates the assertion (a) of equi-Lipschitz property of the family $(f_i)_{i \in I}$ near the point $(\overline{x}, \overline{y})$. □

PROPOSITION 6.10. Let U and V be two open convex sets of normed spaces X and Y respectively and let $f : U \times V \to \mathbb{R}$ be a real-valued function.
(a) If f is a continuous saddle function, then f is directionally sub-sup regular at any point in $U \times V$.
(b) If f is continuous and separately convex, then f is directionally sub-sub regular at any point in $U \times V$.

PROOF. Observe first by the above proposition that f is locally Lipschitz on $U \times V$. Fix $(h, k) \in X \times V$ and $(\overline{x}, \overline{y}) \in U \times V$. Choose open convex neighborhoods $U_0 \subset U$ and $V_0 \subset V$ of $\overline{x}$ and $\overline{y}$ respectively and a real $\eta > 0$ such that

$$U_0 \times V_0 + [0, \eta[\, \text{co}\, \{(h, k), (h, 0), (0, k)\} \subset U \times V.$$

(a) Under the condition (a), for any $(x, y) \in U_0 \times V_0$, from the convexity of $f(\cdot, y)$ and the concavity of $f(x, \cdot)$ we see that

$$\underline{d}_D^+ f(x, y; h, 0) = f'(x, y; h, 0) = \inf_{t \in]0, \eta[} t^{-1}[f(x + th, y) - f(x, y)]$$

and

$$\underline{d}_D^+ f(x, y; 0, k) = f'(x, y; 0, k) = \sup_{t \in]0, \eta[} t^{-1}[f(x, y + tk) - f(x, y)].$$

Then $\underline{d}_D^+ f(\cdot, \cdot; h, 0)$ is upper semicontinuous and $\underline{d}_D^+ f(\cdot, \cdot; 0, k)$ is lower semicontinuous at $(\overline{x}, \overline{y})$, or in other words, f is directionally sub-sup regular at $(\overline{x}, \overline{y})$ according to Proposition 6.6.
(b) Analogously, under the condition (b), for any $(x, y) \in U_0 \times V_0$, from the convexity of both $f(\cdot, y)$ and $f(x, \cdot)$ it results that

$$\underline{d}_D^+ f(x, y; h, 0) = f'(x, y; h, 0) = \inf_{t \in]0, \eta[} t^{-1}[f(x + th, y) - f(x, y)]$$

and

$$\underline{d}_D^+ f(x, y; 0, k) = f'(x, y; 0, k) = \inf_{t \in]0, \eta[} t^{-1}[f(x, y + tk) - f(x, y)].$$

Consequently, both functions $\underline{d}_D^+ f(\cdot, \cdot; h, 0)$ and $\underline{d}_D^+ f(\cdot, \cdot; 0, k)$ are upper semicontinuous, which translates the directional sub-sub regularity of f at $(\overline{x}, \overline{y})$ thanks to Proposition 6.6 again. □

The next proposition provides a useful expression for the *Clarke directional derivative* of sub-sup regular locally Lipschitz functions.

PROPOSITION 6.11. Let U and V be two open convex sets of normed spaces X and Y respectively and $f : U \times V \to \mathbb{R}$ be a locally Lipschitz function. Then f is sub-sup regular at $(x, y) \in U \times V$ in a direction $(h, -k)$ if and only if the directional derivatives $f'(x, y; h, 0)$ and $f'(x, y; 0, -k)$ exist and

$$f^\circ(x, y; h, k) = f'(x, y; h, 0) - f'(x, y; 0, -k).$$

PROOF. Recalling from Definition 2.127 that a locally Lipschitz function φ is C-subregular at a point a in a direction w if and only if $\varphi'(a; w)$ exists and $\varphi^\circ(a; w) = \varphi'(a, w)$, we see that the implication $\Leftarrow$ holds true.
Conversely, suppose that f is sub-sup regular at (x, y) in the direction $(h, -k)$. By the assertion (b_2) of Proposition 2.126, $f'(x, y; h, 0)$ and $f'(x, y; 0, -k)$ exist,

which yields

$$f^{\circ}(x,y;h,k) \geq \limsup_{t\downarrow 0} t^{-1}[f((x,y-tk)+t(h,k)) - f(x,y-tk)]$$
$$= \limsup_{t\downarrow 0} \left(t^{-1}[f(x+th,y) - f(x,y)] - t^{-1}[f(x,y-tk) - f(x,y)]\right)$$
$$= f'(x,y;h,0) - f'(x,y;0,-k).$$

To prove the reverse inequality, fix $\eta > 0$ such that

$$U_0 \times V_0 + [0,\eta[\,\mathrm{co}\,\{(h,-k),(h,0),(0,-k)\} \subset U \times V,$$

where $U_0 := B(x,\eta)$ and $V_0 := B(y,\eta)$. For any $(u,v) \in U_0 \times V_0$ and $t \in\,]0,\eta[$, we then observe that

$$f(u+th, v+tk) - f(u,v)$$
$$= [f(u+th,v) - f(u,v)] - [f(u+th, v+tk-tk) - f(u+th, v+tk)]$$
$$= \int_0^t \underline{d}_D^+ f(u+sh,v;h,0)\,ds - \int_0^t \overline{d}_D^+ f(u+th, v+tk-sk; 0, -k)\,ds.$$

Then, for $\varepsilon > 0$, invoking (by the sub-sup regularity of f) the upper and lower semicontinuity of $\underline{d}_D^+ f(\cdot,\cdot;h,0)$ and $\overline{d}_D^+ f(\cdot,\cdot;0,-k)$ respectively, we deduce

$$f(u+th,v+tk) - f(u,v) \leq \int_0^t [f'(x,y;h,0) + \varepsilon]\,ds - \int_0^t [f'(x,y;0,-k) - \varepsilon]\,ds$$

for $t > 0$ sufficiently small and (u,v) sufficiently close to (x,y). Consequently,

$$f^{\circ}(x,y;h,k) \leq f'(x,y;h,0) - f'(x,y;0,-k) + 2\varepsilon,$$

and this finishes the proof. □

The classical example of the function $f : \mathbb{R}^2 \to \mathbb{R}$, with $f(x,y) = xy/(x^2+y^2)$ if $(x,y) \neq (0,0)$ and $f(0,0) = 0$, makes clear that the existence of partial directional derivatives does not guarantee the existence of joint directional derivatives; here $f'(0,0;1,0)$ and $f'(0,0;0,1)$ exist while $f'(0,0;1,1)$ does not exist. The next proposition says that the result does hold whenever sub-sup regularity is satisfied.

PROPOSITION 6.12. *Let U and V be two open sets of normed spaces X and Y respectively and $f : U \times V \to \mathbb{R}$ be a locally Lipschitz function which is sub-sup regular at $(x,y) \in U \times V$ in the direction (h,k). Then the directional derivative $f'(x,y;h,k)$ exists and*

$$f'(x,y;h,k) = f'(x,y;h,0) + f'(x,y;0,k).$$

PROOF. By the previous proposition we know that the directional derivatives $f'(x,y;h,0)$ and $f'(x,y;0,k)$ exist. Fix a real number $\eta > 0$ such that we have $(x,y) + [0,\eta]\,\mathrm{co}\{(h,k),(h,0),(0,k)\} \subset U \times V$. For any fixed real $t \in\,]0,\eta[$, we can write

$$f(x+th, y+tk) - f(x,y)$$
$$= [f(x+th, y+tk) - f(x, y+tk)] + [f(x, y+tk) - f(x,y)]$$
$$= \int_0^t \underline{d}_D^+ f(x+sh, y+tk; h, 0)\,ds + [f(x, y+tk) - f(x,y)].$$

Then for each real $\varepsilon > 0$, the latter equality allows us to deduce, from the upper semicontinuity of $\underline{d}_D^+ f(\cdot,\cdot;h,0)$ at (x,y) and from the existence and definition of $f'(x,y;0,k)$ as a limit, that

$$f(x+th, y+tk) - f(x,y) \leq t[\underline{d}_D^+ f(x,y;h,0) + \varepsilon] + t[f'(x,y;0,k) + \varepsilon]$$

for $t > 0$ small enough, or equivalently

$$t^{-1}[f(x+th, y+tk) - f(x,y)] \leq f'(x,y;h,0) + f'(x,y;0,k) + 2\varepsilon.$$

Analogously, using the lower semicontinuity of $\overline{d}_D^+ f(\cdot,\cdot;0,k)$ and the existence of $f'(x,y;h,0)$ we can obtain

$$t^{-1}[f(x+th, y+tk) - f(x,y)] \geq f'(x,y;h,0) + f'(x,y;0,k) - 2\varepsilon,$$

which completes the proof. □

Through the above directional results, we can investigate the Clarke subdifferential of separately directionally regular Lipschitz functions.

PROPOSITION 6.13. *Let U and V be two open convex sets of normed spaces X and Y respectively and $f : U \times V \to \mathbb{R}$ be a locally Lipschitz function. Assume that f is subregular at $(x,y) \in U \times V$ in the first direction; that is, $\underline{d}_D^+ f(\cdot,\cdot;h,0)$ is upper semicontinuous at (x,y) for every $h \in X$. Then*

$$(x^*, y^*) \in \partial_C f(x,y) \Rightarrow x^* \in \partial_C f(\cdot, y)(x).$$

PROOF. From Proposition 6.6 and from (b_2) in Proposition 2.126 we have, for each $h \in X$,

$$g^\circ(x;h) \leq f^\circ(x,y;h,0) = f'(x,y;h,0) = g'(x;h), \quad \text{where } g := f(\cdot, y).$$

So, for $(x^*, y^*) \in \partial_C f(x,y)$ we obtain

$$\langle x^*, h \rangle = \langle (x^*, y^*), (h, 0) \rangle \leq f^\circ(x,y;h,0) = g'(x;h) = g^\circ(x;h) \quad \text{for all } h \in X,$$

or equivalently, $x^* \in \partial_C g(x) = \partial_C f(\cdot, y)(x)$. □

COROLLARY 6.14. *Let U and V be open convex subsets of normed spaces X and Y, $f : U \times V \to \mathbb{R}$ be a real-valued function, and $(x,y) \in U \times V$. The inclusion*

$$\partial_C f(x,y) \subset \partial_C f(\cdot, y)(x) \times \partial_C f(x, \cdot)(y)$$

holds under anyone of the following conditions (a) and (b):
(a) the function f is locally Lipschitz and directionally sub-sub regular at (x,y);
(b) the function f is continuous and separately convex.

PROOF. Under the assumption (a) the result follows directly from the previous proposition. The case of (b) is a consequence of (a) and Proposition 6.10(b) along with Proposition 6.9. □

REMARK 6.15. The converse inclusion of the above proposition fails in general even for jointly convex function. Indeed, consider two differentiable convex functions $\varphi, \psi : \mathbb{R} \times \mathbb{R} \to \mathbb{R}$ with different partial derivatives $D_i\varphi(\overline{x},\overline{y}) \neq D_i\psi(\overline{x},\overline{y})$, for $i = 1, 2$, (for example $\varphi(x,y) = x$, $\psi(x,y) = y$, $\overline{x} = \overline{y} = 1$). Then, for $f(x,y) := \max\{\varphi(x,y), \psi(x,y)\}$, one has

$$\partial f(\overline{x},\overline{y}) = \mathrm{co}\{\nabla\varphi(\overline{x},\overline{y}), \nabla\psi(\overline{x},\overline{y})\} \quad \text{if } \varphi(\overline{x},\overline{y}) = \psi(\overline{x},\overline{y}),$$

which cannot be expressed as a cartesian product of two sets of $\mathbb{R}$. □

We establish now the description of the Clarke subdifferential of continuous saddle function by means of partial subdifferentials.

THEOREM 6.16 (*C*-subdifferential of saddle functions). *Let U and V be open convex subsets of normed spaces X and Y, $f : U \times V \to \mathbb{R}$ be a real-valued function, and $(x, y) \in U \times V$. The equality*
$$\partial_C f(x, y) = \partial_C f(\cdot, y)(x) \times \partial_C f(x, \cdot)(y)$$
holds under anyone of the following conditions (a) *and* (b):
(a) *the function f is locally Lipschitz and directionally sub-sup regular at (x, y);*
(b) *the function f is a continuous saddle function.*

PROOF. (a) Denote $\varphi := f(\cdot, y)$ and $\psi := f(x, \cdot)$ and fix any $h \in X$ and $k \in Y$. By the upper semicontinuity of $\underline{d}_D^+ f(\cdot, y; h, 0) = \underline{d}_D^+ \varphi(\cdot; h)$ it follows from Proposition 6.6 and Proposition 2.126(b_2) that the directional derivatives $f'(x, y; h, 0)$ and $\varphi'(x; h)$ exist and
$$f'(x, y; h, 0) = \varphi'(x; h) = \varphi^\circ(x; h) = \sup_{x^* \in \partial_C \varphi(x)} \langle x^*, h \rangle.$$

Analogously, we have
$$-f'(x, y; 0, -k) = -\psi'(y; -k) = (-\psi)^\circ(y; -k) = \psi^\circ(y; k) = \sup_{y^* \in \partial_C \psi(y)} \langle y^*, k \rangle.$$

Combining the previous equalities with Proposition 6.11 yields
$$f^\circ(x, y; h, k) = \sup_{x^* \in \partial_C \varphi(x)} \langle x^*, h \rangle + \sup_{y^* \in \partial_C \psi(y)} \langle y^*, k \rangle$$
$$= \sup\{\langle x^*, h \rangle + \langle y^*, k \rangle : (x^*, y^*) \in \partial_C \varphi(x) \times \partial_C \psi(y)\},$$
which leads to the desired equality of the theorem.
(b) The case (b) follows directly from (a) above and Proposition 6.10(a) along with Proposition 6.9. □

6.1.3. Extended gradient representations of subdifferentials. Given a locally Lipschitz function $f : \mathbb{R}^m \to \mathbb{R}$, Theorem 2.83 established the gradient representation
$$\partial_C f(\bar{x}) = \text{co}\left\{\lim_{n \to \infty} \nabla f(x_n) : \text{Dom}\, \nabla f \ni x_n \to \bar{x}\right\}$$
for the *C*-subdifferential. This subsection is devoted to extensions of the result to extended real-valued functions on normed spaces which are either convex or directionally Lipschitz and subregular. Let us begin with a lemma. Given a function $f : U \to \mathbb{R} \cup \{-\infty, +\infty\}$ on an open set U of a normed space X, it will be convenient to denote, for $\alpha > 0$ and $x \in U$ with $|f(x)| < +\infty$,

(6.3) $\qquad B_f(x, \alpha) := \{u \in X : \|u - x\| < \alpha \text{ and } |f(u) - f(x)| < \alpha\}.$

The lemma will involve, as it appears in Theorem 2.72(b), the expression of the Rockafellar directional derivative $f^\uparrow(\bar{x}; \cdot)$ of the function f at $\bar{x}$ when it is lower semicontinuous at $\bar{x}$. The subregularity at any point u near which f is locally Lipschitz will be used (like in the previous subsection) as characterized by the equality $f^\circ(u; \cdot) = f'(u; \cdot)$ in Proposition 2.126(b_2), so in such a case

(6.4) $\qquad f'(u; h) = \max\{\langle x^*, h \rangle : x^* \in \partial_C f(u)\} \quad \text{for every } h \in X$

according to Theorem 2.70(d). The notation $w^* \lim_j x_j^*$ will be employed to emphasize (when needed) the weak* limit of $(x_j^*)_j$ in X^*. For a subset $S \subset X$, we recall that a set $\mathcal{D} \subset S$ is f-*graphically dense* in S if for any $x \in S$ there exists a sequence $(x_n)_n$ in $\mathcal{D}$ with $x_n \to_f x$, that is, $x_n \to x$ along with $f(x_n) \to f(x)$.

LEMMA 6.17. *Let* $f : U \to \mathbb{R} \cup \{+\infty\}$ *be a function on an open set* U *of a normed space* X *and let* $\bar{x} \in \operatorname{dom} f$ *be a point where* f *is lower semicontinuous and directionally Lipschitz and for which there is an open neighborhood* $U_0 \subset U$ *of* $\bar{x}$ *such that* f *is locally Lipschitz on* $U_0 \cap \operatorname{int} \operatorname{dom} f$. *Let* $\mathcal{D} \subset U_0 \cap \operatorname{int} \operatorname{dom} f$ *be a subset which is* f-*graphically dense in* $U_0 \cap \operatorname{dom} f$ *and at each point of which* f *is subregular. Then for the support function* $\sigma(\cdot, Q)$ *of the subset*

$$Q := \overline{\operatorname{co}}^* \left\{ w^* \lim_j \zeta_j : \zeta_j \in \partial_C f(x_j), \mathcal{D} \ni x_j \to_f \bar{x} \right\}$$

of X *one has*

$$\sigma(h, Q) = \limsup_{\mathcal{D} \ni x \to_f \bar{x}} f'(x; h) \quad \text{for every } h \in \operatorname{int} \operatorname{dom} f^\uparrow(\bar{x} \cdot),$$

where the set whose w^*-*closed convex hull is taken, is the set of* w^*-*limits of convergent nets (resp. convergent sequences if* $\mathbb{B}_{X^*}$ *is sequentially* w^*-*compact)* $w^* \lim_j D_G f(x_j)$ *with* $\mathcal{D} \ni x_j \to_f \bar{x}$. *Above we use for the Rockafellar directional derivative* $f^\uparrow(\bar{x}; \cdot)$ *the expression in Theorem 2.72(b).*

PROOF. We may extend f to X by setting $f(x) = +\infty$ for all $x \in X \setminus U$. Fix any $h \in \operatorname{int} \operatorname{dom} f^\uparrow(\bar{x}; \cdot)$. If $Q \neq \emptyset$, according to the Lipschitz property and subregularity of f at any point $u \in \mathcal{D}$ we have by (6.4)

$$\sigma(h, Q) = \sup \left\{ \left\langle w^* \lim_j \zeta_j, h \right\rangle : \zeta_j \in \partial_C f(x_j), \mathcal{D} \ni x_j \to_f \bar{x} \right\}$$

$$\leq \sup \left\{ \lim_j \langle \zeta_j, h \rangle : \zeta_j \in \partial_C f(x_j), \mathcal{D} \ni x_j \to_f \bar{x} \right\}$$

$$\leq \sup \left\{ \lim_j f'(x_j; , h) : \mathcal{D} \ni x_j \to_f \bar{x} \right\}$$

$$= \limsup_{\mathcal{D} \ni x \to_f \bar{x}} f'(x; h),$$

so $\sigma(h, Q) \leq \limsup_{\mathcal{D} \ni x \to_f \bar{x}} f'(x; h)$, and the inequality is still valid if $Q = \emptyset$ (since $\sigma(h, Q) = -\infty$ if $Q = \emptyset$).

Let us prove the converse inequality $\limsup_{\mathcal{D} \ni x \to_f \bar{x}} f'(x; h) \leq \sigma(h, Q)$. We may and do suppose $\limsup_{\mathcal{D} \ni x \to_f \bar{x}} f'(x; h) > -\infty$. The function f being directionally Lipschitz at $\bar{x}$ in the direction h according to Proposition 2.168, we can choose (by Proposition 2.90(d)) reals $\alpha > 0$ and $\beta > 0$ such that for all $x \in B_f(\bar{x}, \alpha)$, $h' \in B(h, \alpha)$ and $t \in \,]0, \alpha[$

(6.5) $$t^{-1}\big(f(x + th') - f(x)\big) \leq \beta,$$

where $B_f(\bar{x}, \alpha)$ is as defined in (6.3). By the assumption of Lipschitz property and subregularity of f at any point in $\mathcal{D}$ we can choose for each $x \in \mathcal{D}$ some $\xi(x) \in \partial_C f(x)$ such that $f'(x; h) = \langle \zeta(x), h \rangle$ according to (6.4). Choose a net (resp. sequence) $(x_j)_j$ in $\mathcal{D} \cap B_f(\bar{x}, \alpha)$ such that

$$x_j \to_f \bar{x} \quad \text{and} \quad \lim_j f'(x_j; h) = \limsup_{\mathcal{D} \ni x \to_f \bar{x}} f'(x; h).$$

Then, $\lim_j f'(x_j; h) > -\infty$, hence there are a real $\gamma > 0$ and an element j_0 such that $f'(x_j; h) > -\gamma$ for all $j \geq j_0$. Fix for a moment any $j \geq j_0$ and any $v \in B(0,1)$. By (6.5) and the equality $f^\circ(u; \cdot) = f'(u; \cdot)$ for every $u \in \mathcal{D}$ (due to the subregularity of f at any point in $\mathcal{D}$)

$$\langle \xi(x_j), h + \alpha v \rangle \leq f'(x_j; h + \alpha v) \leq \beta,$$

which entails

$$\langle \xi(x_j), \alpha v \rangle \leq \beta - \langle \xi(x_j), h \rangle = \beta - f'(x_j; h) \leq \beta + \gamma,$$

and hence

$$\langle \xi(x_j), v \rangle \leq \alpha^{-1}(\beta + \gamma) =: \rho.$$

It ensues that $\|\xi(x_j)\| \leq \rho$. Then, $(\xi(x_j))_j$ admits a subnet (resp. subsequence), that we do not relabel, w^*-converging to some $x^* \in X^*$, so we have $x^* = \,^{w^*}\!\lim_j \xi(x_j)$ in Q, and clearly

$$\limsup_{\mathcal{D} \ni x \to_f \bar{x}} f'(x; h) = \lim_j f'(x_j; h) = \lim_j \langle \xi(x_j), h \rangle = \langle x^*, h \rangle \leq \sigma(h, Q).$$

This finishes the proof of the lemma. □

The first representation is concerned with the extended representation of C-subdifferential of directionally Lipschitz functions by means of subgradients on particular sets.

THEOREM 6.18 (subgradient $\mathcal{D}$-representation of C-subdifferential of directionally Lipschitz functions). *Let X be a normed space and $f : U \to \mathbb{R} \cup \{+\infty\}$ be a function on an open set U of X. Let $\bar{x} \in U$ be a point where f is finite and directionally Lipschitz. Assume that there is an open neighborhood $U_0 \subset U$ of $\bar{x}$ over which f is lower semicontinuous and such that f is locally Lipschitz on $U_0 \cap \text{int dom } f$ and subregular at each point in $U_0 \cap \text{int dom } f$. Then for any set $\mathcal{D} \subset U_0 \cap \text{int dom } f$ which is f-graphically dense in $U_0 \cap \text{dom } f$, one has*

$$\partial_C f(\bar{x}) = \overline{\text{co}}^* \left\{ {}^{w^*}\!\lim_j \zeta_j : \zeta_j \in \partial_C f(x_j), \mathcal{D} \ni x_j \to_f \bar{x} \right\} + N(\text{dom } f^\uparrow(\bar{x}; \cdot); 0)$$

$$= \bigcap_{\eta > 0} \overline{\text{co}}^* \left(\partial_C f(\mathcal{D} \cap B_f(\bar{x}, \eta)) \right) + N(\text{dom } f^\uparrow(\bar{x}; \cdot); 0),$$

where the first set whose w^*-closed convex hull is taken, is the set of w^*-limits of convergent nets (resp. convergent sequences if $\mathbb{B}_{X^*}$ is sequentially w^*-compact) ${}^{w^*}\!\lim_j \zeta_j$ with $\mathcal{D} \ni x_j \to_f \bar{x}$.

PROOF. Extend f to the entire space X by setting $f(x) = +\infty$ for all $x \in X \setminus U$. We note by the directional Lipschitz property of f at $\bar{x}$ that any neighborhood of $\bar{x}$ meets $\text{int dom } f$. Denote $Q := \overline{\text{co}}^* L$, where

$$L := \left\{ {}^{w^*}\!\lim_j \zeta_j : \zeta_j \in \partial_C f(x_j), \mathcal{D} \ni x_j \to_f \bar{x} \right\}.$$

It is clear that L is included in $\overline{\text{co}}^* \left(\partial_C(\mathcal{D} \cap B_f(\bar{x}, \eta)) \right)$ for every $\eta > 0$, thus Q is included in $\bigcap_{\eta > 0} \overline{\text{co}}^* \left(\partial_C(\mathcal{D} \cap B_f(\bar{x}, \eta)) \right)$.

(a) Let us show the inclusion $Q + N(\text{dom } f^\uparrow(\bar{x}; \cdot); 0) \subset \partial_C f(\bar{x})$. We may suppose

$Q \neq \emptyset$, otherwise the inclusion is evident. Recalling that $f^\uparrow(\bar{x}; 0) \in \{0, -\infty\}$ (see Theorem 2.72(c)), we have $f^\uparrow(\bar{x}; 0) \leq 0$, hence
$$0 \in \text{dom}\, f^\uparrow(\bar{x}; \cdot) = \{v \in X : f^\uparrow(\bar{x}; v) < +\infty\}.$$
Take any $u^* = {}^{w^*}\lim_j \zeta_j$ with $\zeta_j \in \partial_C f(x_j)$ and $\mathcal{D} \ni x_j \to_f \bar{x}$. By the graphical $\|\cdot\|$-to-w^* outer semicontinuity of the multimapping $\partial_C f$ at $\bar{x}$ due to the directionally Lipschitz property of f at this point (as seen in Proposition 2.94), it follows that $u^* \in \partial_C f(\bar{x})$. It ensues that $L \subset \partial_C f(\bar{x})$, and hence $Q \subset \partial_C f(\bar{x})$ since $\partial_C f(\bar{x})$ is a w^*-closed convex set in X^*. So,

(6.6) $\qquad \sigma(h, Q) \leq f^\uparrow(\bar{x}; h) \quad \text{for all } h \in X,$

and the inequality still holds if the set Q is empty. For any $x^* \in Q$ and any $y^* \in N(\text{dom}\, f^\uparrow(\bar{x}; \cdot); 0)$, we deduce for every $h \in \text{dom}\, f^\uparrow(\bar{x}; \cdot)$

$$\langle x^*, h\rangle \leq f^\uparrow(\bar{x}; h) \quad \text{and} \quad \langle y^*, h\rangle \leq 0,$$

hence $\langle x^* + y^*, h\rangle \leq f^\uparrow(\bar{x}; h)$. Consequently, $x^* + y^* \in \partial_C f(\bar{x})$, which justifies the desired inclusion $Q + N(\text{dom}\, f^\uparrow(\bar{x}; \cdot); 0) \subset \partial_C f(\bar{x})$.

(b) Let us show the more involved inclusion $\partial_C f(\bar{x}) \subset Q + N(\text{dom}\, f^\uparrow(\bar{x}; \cdot); 0)$.

(I) First, we claim that

$$f^\uparrow(\bar{x}; h) \leq \limsup_{\mathcal{D} \ni x \to_f \bar{x}} f'(x; h) \quad \text{for any } h \in \text{int}\,\text{dom}\, f^\uparrow(\bar{x}; \cdot).$$

If $0 \in \text{int}\,\text{dom}\, f^\uparrow(\bar{x}; \cdot)$, the inequality is obvious for $h = 0$ since $f^\uparrow(\bar{x}; 0) \in \{-\infty, 0\}$. Fix any nonzero $h \in \text{int}\,\text{dom}\, f^\uparrow(\bar{x}; \cdot)$ with $\limsup_{\mathcal{D} \ni x \to_f \bar{x}} f'(x; h) < +\infty$ and fix any real $\gamma > \limsup_{\mathcal{D} \ni x \to_f \bar{x}} f'(x; h)$. There exists a real $\eta > 0$ such that $B(\bar{x}, \eta) \subset U_0$ and such that

$$f'(u; h) < \gamma \quad \text{for every } u \in \mathcal{D} \cap B_f(\bar{x}, \eta).$$

Then for every $x \in B_f(\bar{x}, \eta)$ we have

(6.7) $\qquad \limsup_{\mathcal{D} \ni u \to_f x} f'(u; h) \leq \gamma.$

By lower semicontinuity of f at $\bar{x}$ there is a real $\delta \in {]0, \eta[}$ such that

(6.8) $\qquad f(\bar{x}) - \eta < f(x) \quad \text{for every } x \in B(\bar{x}, \delta).$

There are also $\alpha_0 \in {]0, \|h\|[}$ and $\beta \in {]0, +\infty[}$ such that

(6.9) $\qquad B_f(\bar{x}, \alpha_0) + [0, \alpha_0] B(h, \alpha_0) \subset U_0$

and

(6.10) $\quad t^{-1}\big(f(x + th') - f(x)\big) \leq \beta \quad \text{for all } x \in B_f(\bar{x}, \alpha_0), h' \in B(h, \alpha_0), t \in {]0, \alpha_0]},$

where such α_0 and β are due to the directional Lipschitz property of f at $\bar{x}$ in the direction h since $h \in \text{int}\,\text{dom}\, f^\uparrow(\bar{x}; \cdot)$ (see Proposition 2.168). Choose a real α such that

$$0 < \alpha < \min\{1, \alpha_0, (1 + \beta)^{-1}\eta, (2 + \|h\|)^{-1}\delta\}.$$

Fix any $x \in B_f(\bar{x}, \alpha)$, any $t \in {]0, \alpha[}$ and any $h' \in B(h, \alpha)$, so $h' \neq 0$. Since by assumption f is locally Lipschitz on $U_0 \cap \text{int}\,\text{dom}\, f$, we can choose for each $u \in U_0 \cap \text{int}\,\text{dom}\, f$ some $\xi(u) \in \partial_C f(u)$. On the other hand, by (6.10) we have for any $r \in {]0, \alpha]}$

$$f(x + rh') \leq f(x) + r\beta,$$

which combined with the lower semicontinuity of f at x gives $f(x+rh') \to f(x)$ as $r \downarrow 0$. It ensues that the restriction of $r \mapsto f(x+rh')$ to $[0,\alpha]$ is continuous since $x+\,]0,\alpha]h' \subset U_0 \cap \mathrm{int}\,\mathrm{dom}\,f$ by (6.9) and (6.10). Writing
$$t^{-1}\bigl(f(x+th')-f(x)\bigr) = t^{-1}\bigl((-f)(x)-(-f)(x+th')\bigr),$$
the mean value inequality in Theorem 6.2(c) furnishes some $s \in \,]0,t[$ such that

(6.11) $\quad t^{-1}\bigl(f(x+th')-f(x)\bigr) \leq \underline{d}_D^+(-f)(x+sh';-h') = -\overline{d}_D^+ f(x+sh';-h').$

Further, observing that
$$\|x+sh'-\overline{x}\| \leq \|x-\overline{x}\| + s(\|h\|+1) < \alpha(2+\|h\|) < \delta < \eta,$$
by (6.10) again we see that $x+sh' \in B(\overline{x},\eta) \cap \mathrm{int}\,(\mathrm{dom}\,f)$. Then, by our assumption, f is Lipschitz near $x+sh'$ and f is subregular at this point $x+sh'$, hence $\overline{d}_D^+ f(\cdot;-h')$ is upper semicontinuous at $x+sh'$ (see Proposition 6.6), which gives
$$\overline{d}_D^+ f(x+sh';-h') = \limsup_{x' \to x+sh'} \overline{d}_D^+ f(x';-h') \geq \limsup_{D \ni x' \to_f x+sh'} f'(x';-h'),$$
and hence, by the fact that $f'(u;-h') = f^\circ(u;-h') \geq \langle \xi(u),-h' \rangle$ for any $u \in D$, we obtain
$$\overline{d}_D^+ f(x+sh';-h') \geq \limsup_{D \ni x' \to_f x+sh'} \langle \xi(x'),-h' \rangle.$$

On the other hand, since $\|x+sh'-\overline{x}\| < \delta$, we have $f(\overline{x})-\eta < f(x+sh')$ by (6.8). We also note by (6.10), by the inclusion $x \in B_f(\overline{x},\alpha)$ and by the choice of α that
$$f(x+sh') \leq f(x) + s\beta \leq f(\overline{x}) + \alpha + \alpha\beta < f(\overline{x}) + \eta.$$
We deduce that $|f(x+sh')-f(\overline{x})| < \eta$, hence $x+sh' \in B_f(\overline{x},\eta)$. Using the equality $f^\circ(u;\cdot) = f'(u;\cdot)$ for any $u \in U_0 \cap \mathrm{int}\,(\mathrm{dom}\,f)$, we also have by (6.11) that

$$\begin{aligned} t^{-1}\bigl(f(x+th')-f(x)\bigr) &\leq -\limsup_{D \ni x' \to_f x+sh'} \langle \xi(x'),-h' \rangle \\ &= \liminf_{D \ni x' \to_f x+sh'} \langle \xi(x'),h' \rangle \\ &\leq \liminf_{D \ni x' \to_f x+sh'} f'(x';h'), \end{aligned}$$

which in turn by (6.7) and the above inclusion $x+sh' \in B_f(\overline{x},\eta)$ gives
$$t^{-1}\bigl(f(x+th')-f(x)\bigr) \leq \gamma.$$

From this we see that
$$f^\uparrow(\overline{x};h) \leq \limsup_{D \ni u \to_f \overline{x}} f'(u;h)$$
which confirms the claim.

(II) To prove the inclusion $\partial_C f(\overline{x}) \subset Q + N(\mathrm{dom}\,f^\uparrow(\overline{x};\cdot);0)$, we may and do suppose that $\partial_C f(\overline{x}) \neq \emptyset$, so $f^\uparrow(\overline{x};0) = 0$ and $f^\uparrow(\overline{x};\cdot) > -\infty$ by Corollary 3.81. Put $P := \{0\} \cup \mathrm{int}\,\mathrm{dom}\,f^\uparrow(\overline{x};\cdot)$. By Lemma 6.17 the inequality in (I) above can be translated as
$$f^\uparrow(\overline{x};h) \leq \sigma(h,Q) \quad \text{for all } h \in \mathrm{int}\,\mathrm{dom}\,f^\uparrow(\overline{x};\cdot),$$
which entails in particular that $Q \neq \emptyset$ since $f^\uparrow(\overline{x};h) > -\infty$ as said above; then we have
$$f^\uparrow(\overline{x};h) \leq \varphi(h) := \sigma(h,Q) + \Psi_P(h) \quad \text{for all } h \in X,$$
and both functions $\sigma(\cdot,Q)$ and Ψ_P are null at 0 and take values in $\mathbb{R} \cup \{+\infty\}$. Since $f^\uparrow(\overline{x};0) = 0 = \varphi(0)$ and that the convex function Ψ_P is continuous on

int dom $f^\uparrow(\bar{x};\cdot) \neq \emptyset$ and finite there and since int dom $f^\uparrow(\bar{x};\cdot) \subset \operatorname{dom}\sigma(\cdot,Q)$ by (6.6), we deduce via the Moreau-Rockafellar theorem (see Theorem 2.105) that

$$\partial_C f(\bar{x}) = \partial f^\uparrow(\bar{x};\cdot)(0) \subset \partial\varphi(0) = \partial\sigma(\cdot,Q)(0) + N(P;0) = Q + N(\operatorname{dom} f^\uparrow(\bar{x};\cdot);0).$$

This is the desired inclusion, which combined with (a) guarantees the equality

$$\partial_C f(\bar{x}) = Q + N(\operatorname{dom} f^\uparrow(\bar{x};\cdot);0).$$

(c) Denoting $Q_0 := \bigcap_{\eta>0} \overline{\operatorname{co}}^*\big(\partial_C f(\mathcal{D} \cap B_f(\bar{x},\eta))\big)$, it remains to prove that $Q_0 + N(\operatorname{dom} f^\uparrow(\bar{x};\cdot);0) \subset \partial_C f(\bar{x})$ since we already noticed that $Q \subset Q_0$. We may and do suppose $Q_0 \neq \emptyset$, so in particular $\sigma(0,Q_0) = 0$. For each $h \in X$, keeping in mind (6.4) for any $u \in \mathcal{D}$, we note that for every $\eta > 0$

$$\sigma(h,Q_0) \leq \sup_{u \in \mathcal{D} \cap B_f(\bar{x},\eta)} \sigma(h,\partial_C f(u)) = \sup_{u \in \mathcal{D} \cap B_f(\bar{x},\eta)} f'(u;h) \quad \text{for all } \eta > 0,$$

which gives

$$\sigma(h,Q_0) \leq \inf_{\eta>0} \sup_{u \in \mathcal{D} \cap B_f(\bar{x},\eta)} f'(u,h) = \limsup_{\mathcal{D} \ni u \to_f \bar{x}} f'(u;h).$$

Combining this with (6.6) and Lemma 6.17 it follows for every $h \in \operatorname{int} \operatorname{dom} f^\uparrow(\bar{x};\cdot)$ that $\sigma(h,Q_0) \leq f^\uparrow(\bar{x};h)$, and hence by the non-vacuity assumption of Q_0

$$(6.12) \qquad -\infty < \sigma(h,Q_0) \leq f^\uparrow(\bar{x};h).$$

For any $h \in \operatorname{int} \operatorname{dom} f^\uparrow(\bar{x};\cdot)$ we also see from the directional Lipschitz property of f at $\bar{x}$ in the direction h that $f^\uparrow(\bar{x};\cdot)$ is bounded from above near h. Therefore, both functions $f^\uparrow(\bar{x};\cdot)$ and $\sigma(\cdot,Q_0)$ are finite on $\operatorname{int} \operatorname{dom} f^\uparrow(\bar{x};\cdot)$ and continuous there, and $f^\uparrow(\bar{x};0) = 0$ by Remark 3.82. Choose $h_0 \in \operatorname{int} \operatorname{dom} f^\uparrow(\bar{x};\cdot)$. For any $h \in \operatorname{dom} f^\uparrow(\bar{x};\cdot)$ we know that the restriction of $f^\uparrow(\bar{x};\cdot)$ to $[h,h_0]$ is continuous (since $f^\uparrow(\bar{x};\cdot)$ is convex and lower semicontinuous), thus by the lower semicontinuity of $\sigma(\cdot,Q_0)$ we obtain from (6.12) that $\sigma(h,Q_0) \leq f^\uparrow(\bar{x};h)$. This yields with $P_0 = \operatorname{dom} f^\uparrow(\bar{x};\cdot)$ that

$$\sigma(h,Q_0) + \Psi_{P_0}(h) \leq f^\uparrow(\bar{x};h) \quad \text{for every } h \in X,$$

which in turn justifies that

$$Q_0 + N(P_0;0) \subset \partial\big(\sigma(\cdot,Q_0) + \Psi_{P_0}\big)(0) \subset \partial f^\uparrow(\bar{x};\cdot)(0) = \partial_C f(\bar{x}).$$

This finishes the proof of the theorem. $\square$

The next theorem is related to the basic situation of differentiability on a sufficiently rich set.

THEOREM 6.19 (gradient $\mathcal{D}$-representation of C-subdifferential of directionally Lipschitz functions). Let X be a normed space and $f : U \to \mathbb{R} \cup \{+\infty\}$ be a function on an open set U of X. Let $\bar{x} \in U$ be a point where f is finite and directionally Lipschitz. Assume that there is an open neighborhood $U_0 \subset U$ of $\bar{x}$ over which f is lower semicontinuous and such that f is locally Lipschitz on $U_0 \cap \operatorname{int} \operatorname{dom} f$ and subregular at each point in $U_0 \cap \operatorname{int} \operatorname{dom} f$. Then for any set $\mathcal{D} \subset U_0 \cap \operatorname{int} \operatorname{dom} f$

which is f-graphically dense in $U_0 \cap \operatorname{dom} f$ and on which f is Gâteaux differentiable, one has

$$\partial_C f(\bar{x}) = \overline{\operatorname{co}}^* \left\{ \underset{j}{w^* \lim} D_G f(x_j) : \mathcal{D} \ni x_j \to_f \bar{x} \right\} + N(\operatorname{dom} f^\uparrow(\bar{x}; \cdot); 0)$$

$$= \bigcap_{\eta > 0} \overline{\operatorname{co}}^* \left(D_G f(\mathcal{D} \cap B_f(\bar{x}, \eta)) \right) + N(\operatorname{dom} f^\uparrow(\bar{x}; \cdot); 0),$$

where the first set whose w^*-closed convex hull is taken, is the set of w^*-limits of convergent nets (resp. convergent sequences if $\mathbb{B}_{X^*}$ is sequentially w^*-compact) $\underset{j}{w^* \lim} D_G f(x_j)$ with $\mathcal{D} \ni x_j \to_f \bar{x}$.

PROOF. At each point $u \in \mathcal{D} \subset U_0 \cap \operatorname{int} \operatorname{dom} f$ the function f, is by assumption, locally Lipschitz and subregular, so

$$f^\uparrow(u; \cdot) = f^\circ(u; \cdot) = f'(u; \cdot) = \langle D_G f(u), \cdot \rangle,$$

hence $\partial_C f(u) = \{D_G f(u)\}$. Therefore, all the assumptions are fulfilled to apply Theorem 6.18. □

We state now two corollaries related to locally Lipschitz subregular functions. The first is a direct consequence of Theorem 6.18.

COROLLARY 6.20 (subgradient $\mathcal{D}$-representation of C-subdifferential of locally Lipschitz functions). Let X be a normed space and $f : U \to \mathbb{R}$ be a function on an open set U of X which is locally Lipschitz and subregular on U. Then for any set $\mathcal{D} \subset U$ which is dense in U, one has for any $x \in U$

$$\partial_C f(x) = \overline{\operatorname{co}}^* \left\{ \underset{j}{w^* \lim} \zeta_j : \zeta_j \in \partial_C f(x_j), \mathcal{D} \ni x_j \to x \right\} = \bigcap_{\eta > 0} \overline{\operatorname{co}}^* \left(D_G f(\mathcal{D} \cap B(x, \eta)) \right),$$

where the set whose w^*-closed convex hull is taken, is the set of w^*-limits of convergent nets (resp. convergent sequences if $\mathbb{B}_{X^*}$ is sequentially w^*-compact) $\underset{j}{w^* \lim} \zeta_j$ with $\mathcal{D} \ni x_j \to x$.

The second corollary follows from Theorem 6.19.

COROLLARY 6.21 (gradient $\mathcal{D}$-representation of C-subdifferential of locally Lipschitz functions). Let X be a normed space and $f : U \to \mathbb{R}$ be a function on an open set U of X which is locally Lipschitz and subregular on U. Then for any set $\mathcal{D} \subset U$ which is dense in U and on which f is Gâteaux differentiable, one has for any $x \in U$

$$\partial_C f(x) = \overline{\operatorname{co}}^* \left\{ \underset{j}{w^* \lim} D_G f(x_j) : \mathcal{D} \ni x_j \to x \right\} = \bigcap_{\eta > 0} \overline{\operatorname{co}}^* \left(D_G f(\mathcal{D} \cap B(x, \eta)) \right),$$

where the first set whose w^*-closed convex hull is taken, is the set of w^*-limits of convergent nets (resp. convergent sequences if $\mathbb{B}_{X^*}$ is sequentially w^*-compact) $\underset{j}{w^* \lim} D_G f(x_j)$ with $\mathcal{D} \ni x_j \to x$.

We turn now to the situation of convex functions.

THEOREM 6.22 (subgradient $\mathcal{D}$-representation of subdifferential of convex functions). Let U be an convex open set of a normed space X and $f : U \to \mathbb{R} \cup \{+\infty\}$ be a lower semicontinuous convex function which is continuous at some point in

$\text{dom}\, f$. Let $\mathcal{D} \subset \text{int dom}\, f$ be a set which is dense in $\text{int dom}\, f$. Then for any $x \in \text{dom}\, f$ one has the equalities

$$\partial f(x) = \overline{\text{co}}^* \left\{ w^* \lim_j \zeta_j : \zeta_j \in \partial f(x_j), \mathcal{D} \ni x_j \to_f x \right\} + N(\text{dom}\, f; x)$$

$$= \overline{\text{co}}^* \left\{ w^* \lim_j \zeta_j : \zeta_j \in \partial f(x_j), \mathcal{D} \ni x_j \to x, (\|\zeta_j\|)_j \text{ bounded} \right\} + N(\text{dom}\, f; x)$$

$$= \bigcap_{\eta > 0} \overline{\text{co}}^* \left(\partial f(\mathcal{D} \cap B_f(\overline{x}, \eta)) \right) + N(\text{dom}\, f; x),$$

where the first two sets whose w^*-closed convex hulls are taken, are sets with w^*-limits of convergent nets (resp. convergent sequences if $\mathbb{B}_{X^*}$ is sequentially w^*-compact) $w^* \lim_j \zeta_j$ with $\mathcal{D} \ni x_j \to_f \overline{x}$ in the first set, and $\mathcal{D} \ni x_j \to \overline{x}$ in the second.

PROOF. Fix any $x \in \text{dom}\, f$ and note that $\text{int dom}\, f \neq \emptyset$. Note also by Proposition 2.167 that f is directionally Lipschitz at x. Denote

$$L := \left\{ w^* \lim_j \zeta_j : \zeta_j \in \partial f(x_j), \mathcal{D} \ni x_j \to_f x \right\}$$

and

$$L_b := \left\{ w^* \lim_j \zeta_j : \zeta_j \in \partial f(x_j), \mathcal{D} \ni x_j \to x, (\|\zeta_j\|)_j \text{ bounded} \right\}.$$

By continuity of f on $\text{int dom}\, f$ (see Theorem 2.158), the set $\mathcal{D}$ is f-graphically dense in $\text{int dom}\, f$. Further, fixing $a \in \text{int dom}\, f$, for any $u \in \text{dom}\, f$ with $u \neq a$ the restriction of f to $[u, a]$ is continuous (see Proposition 3.20) and $]u, a] \subset \text{int dom}\, f$ (see Proposition B.5(b) in Appendix). Then, $\text{int dom}\, f$ is f-graphically dense in $\text{dom}\, f$, hence $\mathcal{D}$ is also f-graphically dense in $\text{dom}\, f$.

Moreover, $\text{int dom}\, f^\uparrow(x; \cdot) =]0, +\infty[(\text{int dom}\, f - x)$ by Propositions 2.167 and 2.168, hence we have

$$x^* \in N(\text{dom}\, f^\uparrow(x; \cdot); 0) \Leftrightarrow \langle x^*, y - x \rangle \leq 0 \,\forall y \in \text{int dom}\, f \Leftrightarrow x^* \in N(\text{dom}\, f; x).$$

Then we can apply Theorem 6.19 to obtain the equality

$$\partial f(x) = \overline{\text{co}}^* \left\{ w^* \lim_j \zeta_j : \zeta_j \in \partial f(x_j), \mathcal{D} \ni x_j \to_f x \right\} + N(\text{dom}\, f; x).$$

Now (if any) take any $u^* = w^* \lim_j \zeta_j$ with $\zeta_j \in \partial f(x_j)$ and $\mathcal{D} \ni x_j \to_f x$. By directional Lipschitz property of f at x, choose $h \in X$ and reals $\alpha > 0$ and $\beta > 0$ such that for any $x' \in B_f(x, \alpha)$, any $h' \in B(h, \alpha)$ and any $t \in]0, \alpha]$

(6.13) $$t^{-1}\left(f(x' + th') - f(x') \right) \leq \beta.$$

Choose j_0 such that for all $j \geq j_0$

$$x_j \in B_f(x, \alpha) \quad \text{and} \quad \langle \zeta_j, h \rangle \geq \langle u^*, h \rangle - 1 =: \gamma.$$

For any $j \geq j_0$ we see by (6.13) that for every $v \in B(0, 1)$

$$\langle \zeta_j, h + \alpha v \rangle \leq \beta, \text{ hence } \langle \zeta_j, v \rangle \leq \alpha^{-1}(\beta - \gamma).$$

It ensues that $(\|\zeta_j\|)_{j \geq j_0}$ is bounded. We deduce that $L \subset L_b$, so

(6.14) $$\partial f(x) = \overline{\text{co}}^*(L) + N(\text{dom}\, f; x) \subset \overline{\text{co}}^*(L_b) + N(\text{dom}\, f; x).$$

Conversely, (if any) consider any $u^* = \underset{j}{w^* \lim} \zeta_j$ with $\zeta_j \in \partial f(x_j)$ and $x_j \to x$, and such that $(\|\zeta_j\|)_j$ is bounded. From the boundedness of $(\|\zeta_j\|)_j$ we easily see that $\langle \zeta_j, x_j \rangle \to \langle u^*, x \rangle$. For any $u \in U$ and any j we have
$$\langle \zeta_j, u - x_j \rangle \leq f(u) - f(x_j),$$
which gives by the lower semicontinuity of f at x
$$\langle u^*, u - x \rangle \leq f(u) - f(x).$$
This being true for every $u \in U$, it results that $u^* \in \partial f(x)$, so $L_b \subset \partial f(x)$. As above we derive that $\overline{\text{co}}^*(L_b) + N(\text{dom } f; x) \subset \partial f(x)$, which combined with (6.14) finishes the proof. □

The next theorem particularizes the previous one in the fundamental case when the convex function is Gâteaux differentiable at each point in $\mathcal{D}$.

THEOREM 6.23 (gradient $\mathcal{D}$-representation of subdifferential of convex functions). Let U be an open convex set of a normed space X and $f : U \to \mathbb{R} \cup \{+\infty\}$ be a lower semicontinuous convex function which is continuous at some point in dom f. Let $\mathcal{D} \subset \text{int dom } f$ be a set which is dense in int dom f and such that f is Gâteaux differentiable at each point in $\mathcal{D}$. Then for any $x \in \text{dom } f$ one has

$$\partial f(x) = \overline{\text{co}}^* \left\{ \underset{j}{w^* \lim} D_G f(x_j) : \mathcal{D} \ni x_j \to_f x \right\} + N(\text{dom } f; x)$$

$$= \overline{\text{co}}^* \left\{ \underset{j}{w^* \lim} D_G f(x_j) : \mathcal{D} \ni x_j \to x, (\|D_G(x_j)\|)_j \text{ bounded} \right\} + N(\text{dom } f; x)$$

$$= \bigcap_{\eta > 0} \overline{\text{co}}^* \left(D_G f(\mathcal{D} \cap B_f(\overline{x}, \eta)) \right) + N(\text{dom } f; x)$$

where the first two sets whose w^*-closed convex hull are taken, are sets with w^*-limits of convergent nets (resp. convergent sequences if $\mathbb{B}_{X^*}$ is sequentially w^*-compact) $\underset{j}{w^* \lim} D_G f(x_j)$ with $\mathcal{D} \ni x_j \to_f \overline{x}$ in the first set, and $\mathcal{D} \ni x_j \to \overline{x}$ in the second.

Recalling (see Corollary 2.163) that any lower semicontinuous convex function $f : U \to \mathbb{R} \cup \{+\infty\}$ on an open convex set U of a Banach space is continuous on int dom f if the latter set is nonempty, we directly derive the following corollary.

COROLLARY 6.24. Let U be an open convex set of a Banach space X and $f : U \to \mathbb{R} \cup \{+\infty\}$ be a lower semicontinuous convex function with int dom $f \neq \emptyset$. Let $\mathcal{D} \subset \text{int dom } f$ be a set which is dense in int dom f and such that f is Gâteaux differentiable at each point in $\mathcal{D}$. Then for any $x \in \text{dom } f$ one has

$$\partial f(x) = \overline{\text{co}}^* \left\{ \underset{j}{w^* \lim} D_G f(x_j) : \mathcal{D} \ni x_j \to_f x \right\} + N(\text{dom } f; x)$$

$$= \overline{\text{co}}^* \left\{ \underset{j}{w^* \lim} D_G f(x_j) : \mathcal{D} \ni x_j \to x, (\|D_G(x_j)\|)_j \text{ bounded} \right\} + N(\text{dom } f; x)$$

$$= \bigcap_{\eta > 0} \overline{\text{co}}^* \left(D_G f(\mathcal{D} \cap B_f(\overline{x}, \eta)) \right) + N(\text{dom } f; x)$$

where the first two sets whose w^*-closed convex hulls are taken, are sets of w^*-limits of convergent nets (resp. convergent sequences if $\mathbb{B}_{X^*}$ is sequentially w^*-compact) $\underset{j}{w^* \lim} D_G f(x_j)$ with $\mathcal{D} \ni x_j \to_f \overline{x}$ in the first set, and $\mathcal{D} \ni x_j \to \overline{x}$ in the second.

The case of continuous convex functions is also of interest and the next two corresponding corollaries directly follow from Theorem 6.22 for the first and from Theorem 6.23 for the second.

COROLLARY 6.25. *Let U be an open convex set of a normed space X and $f : U \to \mathbb{R}$ be a continuous convex function. Let $\mathcal{D} \subset U$ be a subset which is dense in U. Then for any $x \in U$ one has*

$$\partial f(x) = \overline{co}^* \left\{ w^* \lim_j \zeta_j : \zeta_j \in \partial f(x_j), \mathcal{D} \ni x_j \to x \right\} = \bigcap_{\eta > 0} \overline{co}^* \left(\partial f(\mathcal{D} \cap B(\overline{x}, \eta)) \right)$$

where the first set whose w^*-closed convex hull is taken, is the set of w^*-limits of convergent nets (resp. convergent sequences if $\mathbb{B}_{X^*}$ is sequentially w^*-compact) $w^* \lim_j \zeta_j$ with $\mathcal{D} \ni x_j \to \overline{x}$.

COROLLARY 6.26. *Let U be an open convex set of a normed space X and $f : U \to \mathbb{R}$ be a continuous convex function. Let $\mathcal{D} \subset U$ be a subset which is dense in U and such that f is Gâteaux differentiable at each point in $\mathcal{D}$. Then for any $x \in U$ one has*

$$\partial f(x) = \overline{co}^* \left\{ w^* \lim_j D_G f(x_j) : \mathcal{D} \ni x_j \to x \right\} = \bigcap_{\eta > 0} \overline{co}^* \left(D_G f(\mathcal{D} \cap B(\overline{x}, \eta)) \right)$$

where the first set whose w^*-closed convex hull is taken, is the set of w^*-limits of convergent nets (resp. convergent sequences if $\mathbb{B}_{X^*}$ is sequentially w^*-compact) $w^* \lim_j D_G f(x_j)$ with $\mathcal{D} \ni x_j \to \overline{x}$.

6.1.4. Conditions for monotonicity and other properties via Dini semiderivates. The mean value inequality in Theorem 6.2 and its consequences in both previous subsections, as for example the nondecreasing property in Corollary 6.5, required a certain semicontinuity/continuity condition. In the case of functions of one real variable, other important properties and mean value inequalities can be established by means of Dini semiderivates even when semicontinuity properties fail. Let us begin with some results for the nondecreasing property.

PROPOSITION 6.27. *Let I be an open interval of $\mathbb{R}$ and $f : I \to \mathbb{R}$ be a real-valued function satisfying the following conditions (i) and (ii):*
(i) *for every $r \in I$*

$$\liminf_{t > r, t \to r} f(t) \leq f(r) \leq \liminf_{t < r, t \to r} f(t);$$

(ii) *the set $\{f(t) : t \in I, \underline{D}^+ f(t) \geq 0\}$ does not contain any nonempty open interval. Then the function f is nonincreasing on I.*

PROOF. Let us proceed by contradiction and suppose that there are $r, s \in I$ with $r < s$ and $f(r) < f(s)$. Denoting by S the set of $t \in I$ such that $\underline{D}^+ f(t) \geq 0$, the assumption (ii) gives some $\sigma \notin f(S)$ such that $f(r) < \sigma < f(s)$. Denote by θ the upper bound of the reals $t \in [r, s]$ such that $f(t) \leq \sigma$. Since $f(s) > \sigma$, we have $\liminf_{t \uparrow s} f(t) > \sigma$ by the assumption (i), so there exists some $\varepsilon > 0$ such that $f(t) > \sigma$ for all $t \in [s - \varepsilon, s]$, which entails $\theta \leq s - \varepsilon$, hence $\theta < s$.

Let us show that $f(\theta) = \sigma$. If $f(\theta) < \sigma$, by the first inequality of the assumption (i) we would have $\liminf f(t) < \sigma$ as $t \to \theta$ with $t > \theta$, which would produce, according to the above inequality $\theta < s$, some $t_0 \in]\theta, s]$ with $f(t_0) < \sigma$, contradicting the definition of θ. Consequently, we have $f(\theta) \geq \sigma$. Further, by the definition of θ there exists a sequence $(t_n)_n$ in $[r,\theta]$ tending to θ with $f(t_n) \leq \sigma$. This combined with the second inequality in the assumption (i) implies $f(\theta) \leq \liminf_{n\to\infty} f(t_n) \leq \sigma$. So $f(\theta) = \sigma$, which guarantees $\theta \notin S$ since $\sigma \notin f(S)$.

On the other hand, the definition of θ ensures for all $t \in]\theta, s]$ that $f(t) > \sigma$, thus
$$\frac{f(t) - f(\theta)}{t - \theta} = \frac{f(t) - \sigma}{t - \theta} > 0,$$
which yields $\underline{D}^+ f(\theta) \geq 0$ and contradicts the above fact that $\theta \notin S$. We then conclude that f is nonincreasing on I. □

Diverse consequences of Proposition 6.27 are obtained in its Corollaries 6.28-6.31.

COROLLARY 6.28. *Let I be an open interval of $\mathbb{R}$ and $f : I \to \mathbb{R}$ be a real-valued function satisfying the following conditions (i) and (ii):*
(i) *for every $r \in I$*
$$\liminf_{t>r,t\to r} f(t) \leq f(r) \leq \liminf_{t<r,t\to r} f(t);$$
(ii) *the inequality $\underline{D}^+ f(t) \leq 0$ holds for all $t \in I$ except for those of a countable subset.*

Then the function f is nonincreasing on I.

PROOF. Fixing any real $\varepsilon > 0$ and putting $g(t) := f(t) - \varepsilon t$, we see that the assumption (i) is fulfilled for g, and further $\underline{D}^+ g(t) \leq -\varepsilon < 0$ for all $t \in I$ except for those in a countable subset S of I. Of course, $g(S)$ is countable hence $\{g(t) : t \in I, \underline{D}^+ g(t) \geq 0\}$ contains no nonempty open intervals. The above proposition guarantees that g is nonincreasing, so for $s, t \in I$ with $s \leq t$ we have $f(t) - \varepsilon t \leq f(s) - \varepsilon s$ for all $\varepsilon > 0$ hence $f(t) \leq f(s)$, which means that f is nonincreasing. □

COROLLARY 6.29. *Let I be an open interval of $\mathbb{R}$ and $f : I \to \mathbb{R}$ be a real-valued function such that for every $r \in I$*
$$\liminf_{t>r,t\to r} f(t) \leq f(r) \leq \liminf_{t<r,t\to r} f(t).$$
Then, for $q(f; r, s) := (f(s) - f(r))/(s - r)$, one has
$$\sup\{q(f; r, s) : r, s \in I, r \neq s\} = \sup_{t \in I} \underline{D}^+ f(t) = \sup_{t \in I} \overline{D}^+ f(t).$$

PROOF. Put $\beta := \sup_{t \in I} \underline{D}^+ f(t)$. Since $\underline{D}^+ f(t) \leq \overline{D}^+ f(t)$, it is easily seen that
$$\beta \leq \sup_{t \in I} \overline{D}^+ f(t) \leq \sup\{q(f; r, s) : r, s \in I, r \neq s\}.$$

Thus it suffices to suppose $\beta < +\infty$ and to show that $\sup\{q(f; r, s) : r, s \in I, r \neq s\} \leq \beta$. Fix any real $\alpha > \beta$ and note that, for $g(t) = f(t) - \alpha t$, we have $\underline{D}^+ g(t) = \underline{D}^+ f(t) - \alpha$, so $\underline{D}^+ g(t) < 0$ for all $t \in I$. For any $r, s \in I$ with $r < s$, the above corollary ensures $f(s) - f(r) \leq \alpha(s - r)$, hence $q(f; r, s) \leq \beta$. Consequently, $\sup\{q(f; r, s) : r, s \in I, r \neq s\} \leq \beta$ as desired. □

COROLLARY 6.30. Let I be an open interval of $\mathbb{R}$ and $f : I \to \mathbb{R}$ be a continuous real-valued function. For $q(f;r,s) := (f(s) - f(r))/(s-r)$, the following equalities hold:

$$\sup\{q(f;r,s) : r,s \in I, r \neq s\} = \sup_{t \in I} \underline{D}^+ f(t) = \sup_{t \in I} \underline{D}^- f(t)$$
$$= \sup_{t \in I} \overline{D}^+ f(t) = \sup_{t \in I} \overline{D}^- f(t)$$

and

$$\inf\{q(f;r,s) : r,s \in I, r \neq s\} = \inf_{t \in I} \underline{D}^+ f(t) = \inf_{t \in I} \underline{D}^- f(t)$$
$$= \inf_{t \in I} \overline{D}^+ f(t) = \inf_{t \in I} \overline{D}^- f(t).$$

PROOF. We know by the above corollary that the first two members and the fourth of the first chain are equal. Consider the open interval $J := -I$ and the continuous function $g : J \to \mathbb{R}$ defined by $g(t) := -f(-t)$ for all $t \in J$. It is easily seen that $\underline{D}^+ g(t) = \underline{D}^- f(-t)$ and $\overline{D}^+ g(t) = \overline{D}^- f(-t)$, hence the previous equalities give

$$\sup\{q(g;r,s) : r,s \in J, r \neq s\} = \sup_{t \in J} \underline{D}^- f(-t) = \sup_{t \in J} \overline{D}^- f(-t).$$

Observing, for $r,s \in J$ with $r \neq s$, that $q(g;r,s) = q(f;-r,-s)$ we deduce that

$$\sup\{q(f;r,s) : r,s \in I, r \neq s\} = \sup\{q(g;r,s) : r,s \in J, r \neq s\}$$
$$= \sup_{t \in I} \underline{D}^- f(t) = \sup_{t \in I} \overline{D}^- f(t),$$

which completes the proof of the first chain of equalities.

The second chain of equalities is obtained in substituting $-f$ for f. □

The next corollary is a direct consequence of the latter corollary.

COROLLARY 6.31. Let I be an open interval of $\mathbb{R}$ and $f : I \to \mathbb{R}$ be a continuous real-valued function. If anyone of the four Dini unilateral semiderivates of f is continuous at $r \in I$, then so are the three others, and all four Dini unilateral semiderivates at r are equal, so f is derivable at r.

Lemma 2.244 in Chapter 2 established, in a topological space T with a countable basis of neighborhoods, for any subset $S \subset T$, the countability of the set of points of S which are isolated in S. To prepare the next proposition, the following lemma provides the same result for the right-hand side topology. For $S \subset \mathbb{R}$, recall that $r \in S$ is isolated in S on the right (resp. on the left) hand side if $S \cap [r, r+\varepsilon[= \{r\}$ (resp. $S \cap]r - \varepsilon, r] = \{r\}$) for some real $\varepsilon > 0$.

LEMMA 6.32. Let S be a nonempty subset of $\mathbb{R}$. Then the set of points of S which are isolated in S on the right (resp. on the left) is countable.

PROOF. Denote by S_0 the set of points of S which are isolated on the right in S. Putting $E_n := \{r \in S : [r, r + (1/n)[\cap S = \{r\}\}$, for each $n \in \mathbb{N}$, we have $S_0 = \bigcup_{n \in \mathbb{N}} E_n$. Further, for each integer $k \in \mathbb{Z}$, the interval $[k/n, (k+1)/n[$ contains at most one point of E_n. Then, the equality $E_n = \bigcup_{k \in \mathbb{Z}} E_n \cap [k/n, (k+1)/n[$ ensures that E_n is countable, and hence the set S_0 is countable. □

Of course $\underline{d}_D^+ f(r;1) = \underline{D}^+ f(r) \leq \overline{D}^+ f(r) = -\underline{d}_D^+(-f)(r;1)$ for all r where f is finite. The assertion (b) below says that the inequality $\underline{d}_D^+ f(r;1) = \underline{D}^+ f(r) \leq \overline{D}^- f(r) = -\underline{d}_D^+ f(r;-1)$ also holds for all r where f is finite except those of a countable subset.

PROPOSITION 6.33. *Let $f : \mathbb{R} \to \mathbb{R} \cup \{-\infty, +\infty\}$ be an extended real-valued function on $\mathbb{R}$. Each of the following sets is countable:*
(a) *the set of elements $r \in \mathbb{R}$ such that*

$$\liminf_{t \to r} f(t) < \liminf_{t > r, t \to r} f(t) \text{ or } \limsup_{t \to r} f(t) > \limsup_{t > r, t \to r} f(t);$$

(b) *the set of elements $r \in \mathbb{R}$ such that $|f(r)| < +\infty$ and*

$$\overline{D}^- f(r) < \underline{D}^+ f(r)) \quad (\text{or equivalently } - \underline{d}_D^+ f(r;-1) < \underline{d}_D^+ f(r;1)).$$

PROOF. (a) For each rational number $q \in \mathbb{Q}$ denote by S_q the set of $r \in \mathbb{R}$ such that

$$\liminf_{t \to r} f(t) < q < \liminf_{t > r, t \to r} f(t).$$

Fix any $q \in \mathbb{Q}$ such that S_q is nonempty. For each $r \in S_q$, the second inequality yields some $\varepsilon > 0$ such that $f(t) > q$ for all $t \in {]r, r+\varepsilon[}$. Suppose that $]r, r+\varepsilon[\cap S_q \neq \emptyset$ and fix some s in this intersection. Then

$$\liminf_{t \to s} f(t) < q < \liminf_{t > s, t \to s} f(t),$$

hence there exists some sequence $(t_n)_n$ tending to s with $t_n \leq s$ and $f(t_n) < q$. Taking some integer N such that $t_N \in {]r, r+\varepsilon[}$ we see that $f(t_N) < q$, which contradicts the above inequality $f(t) > q$ for all $t \in {]r, r+\varepsilon[}$. Then we have $S_q \cap {]r, r+\varepsilon[} = \emptyset$, and this means that any point of S_q is isolated in S_q with respect to the topology on the right-hand side of $\mathbb{R}$, and hence S_q is a countable set according to the above lemma. Consequently, the set of $r \in \mathbb{R}$ satisfying the first inequality in the statement (a) is countable since it coincides with $\bigcup_{q \in \mathbb{Q}} S_q$.
(b) Denote by S the set defined by (b). For each $m \in \mathbb{Z}$ and $n \in \mathbb{N}$, denote by $S_{m,n}$ the set of $r \in \mathbb{R}$ where f is finite and such that $\overline{D}^- f(r) < m/n < \underline{D}^+ f(r)$. Fix any $m \in \mathbb{Z}$ and $n \in \mathbb{N}$ such that $S_{m,n} \neq \emptyset$. Putting $f_{m,n}(t) := f(t) - mt/n$ and fixing $r \in S_{m,n}$, we see that $\overline{D}^- f_{m,n}(r) < 0 < \underline{D}^+ f_{m,n}(r)$. From this we easily see that r is a strict local minimizer of $f_{m,n}$, which guarantees that the set $S_{m,n}$ is countable according to Proposition 2.51 in Chapter 2. We then conclude that the set in (b) is countable as the union of the sets $S_{m,n}$. □

6.1.5. Mean value inequality for images of sets and Denjoy-Young-Saks theorem.
The next properties are related to measure theory. Our first objective is to obtain, for a subset S of an interval $I \subset \mathbb{R}$ and a function $f : I \to \mathbb{R}$, integral mean value inequalities in the form of estimates of the Lebesgue measure of $f(S)$ in terms of integrals of the absolute value of unilateral Dini semiderivates of f along S. The second objective is to derive the famous Denjoy-Young-Saks theorem on unilateral semiderivates.

For a real-valued function f, in addition to the above inequality in Proposition 6.33(b) saying that

$$\underline{D}^+ f(r) = \underline{d}_D^+ f(r;1) \leq -\underline{d}_D^+ f(r;-1) = \overline{D}^- f(r)$$

for all r except those of a countable set, we establish first in Proposition 6.35 below that those unilateral Dini semiderivates $\underline{d}_D^+ f(r;1)$ and $-\underline{d}_D^+ f(r;-1)$ in fact coincide for almost all r whenever $\underline{d}_D^+ f(\cdot;1)$ takes on merely finite values. The proof of the proposition utilizes the following lemma relative to Bouligand-Peano tangent cones of subsets of $\mathbb{R}^2$.

LEMMA 6.34. *Let $f : I \to \mathbb{R}$ be a real-valued function defined on an open interval I of $\mathbb{R}$ and let $r \in I$.*
(a) *One has*
$$T^B\bigl(\mathrm{gph}\, f; (r, f(r))\bigr) \setminus (\{0\} \times]-\infty, 0[) \subset \mathrm{epi}\, \underline{d}_D^+ f(r;\cdot).$$
(b) *If for some $\alpha \in \mathbb{R}$*
$$\{(u,v) \in \mathbb{R} \times \mathbb{R} : v = \alpha u\} \subset T^B\bigl(\mathrm{gph}\, f; (r, f(r))\bigr)$$
and
$$T^B\bigl(\mathrm{gph}\, f; (r, f(r))\bigr) \setminus (\{0\} \times]-\infty, 0[) \subset \{(u,v) \in \mathbb{R} \times \mathbb{R} : v \geq \alpha u\},$$
then
$$\alpha = \underline{D}^+ f(r) = \underline{d}_D^+ f(r;1) = -\underline{d}_D^+ f(r;-1) = \overline{D}^- f(r)$$
and f is lower semicontinuous at r.

PROOF. For convenience, let us use the symbol $f^D(r;\cdot)$ in place of $\underline{d}_D^+ f(r;\cdot)$.
(a) Fix $(u,v) \in T^B\bigl(\mathrm{gph}\, f;(r,f(r))\bigr)$. Suppose first $u \neq 0$. By the definition of the Bouligand-Peano tangent cone there are sequences $(\tau_n)_n$ tending to 0 with $\tau_n > 0$ and $(u_n,v_n)_n$ converging to (u,v) such that $(r,f(r)) + \tau_n(u_n,v_n) \in \mathrm{gph}\, f$, or equivalently $v_n = \tau_n^{-1}[f(r + \tau_n u_n) - f(r)]$.

If $u > 0$, then for n large enough $u_n > 0$ and for $t_n := \tau_n u_n$ we have $t_n > 0$ with $t_n \to 0$ and $t_n^{-1}[f(r + t_n) - f(r)] \to u^{-1} v$, so $u^{-1} v \geq f^D(r;1)$ according to the definition of $f^D(r;1)$ as a lower limit; the latter means $v \geq f^D(r;u)$ since $u > 0$.

Similarly for $u < 0$, we see that $t_n := -\tau_n u_n > 0$ for large n, and $t_n \to 0$ along with $t_n^{-1}[f(r - t_n) - f(r)] \to -u^{-1} v$, so $-u^{-1} v \geq f^D(r;-1)$ or equivalently $v \geq -u f^D(r;-1) = f^D(r;u)$ since $-u > 0$.

Consequently, we have $v \geq f^D(r;u)$.

Suppose now $u = 0$ and $v \geq 0$. In this case we still have $v \geq 0 \geq f^D(r;0) = f^D(r;u)$.

(b) By the first inclusion of the assumption in (b) we have $(1,\alpha)$ and $(-1,-\alpha)$ in the Bouligand-Peano tangent cone of $\mathrm{gph}\, f$ at $(r, f(r))$, so by (a) above

(6.15) $$\alpha \geq f^D(r;1) \quad \text{and} \quad -\alpha \geq f^D(r;-1).$$

By the definition of Bouligand-Peano tangent cone, choose sequences $(t_n)_n$ and $(\tau_n)_n$ in $]0, +\infty[$ tending to 0, $(u_n, v_n)_n$ and $(\rho_n, \sigma_n)_n$ both converging to $(1, \alpha)$ such that
$$v_n = t_n^{-1}[f(r + t_n u_n) - f(r)] \quad \text{and} \quad -\sigma_n = \tau_n^{-1}[f(r - \tau_n \rho_n) - f(r)],$$
so for $s_n := t_n u_n$ and $\zeta_n := \tau_n \rho_n$ we have $s_n \to 0$, $\zeta_n \to 0$, and $s_n, \zeta_n \in]0, +\infty[$ for large n. Since for n large enough
$$u_n^{-1} v_n = s_n^{-1}[f(r + s_n) - f(r)] \quad \text{and} \quad -\rho_n^{-1} \sigma_n = \zeta_n^{-1}[f(r - \zeta_n) - f(r)],$$
we deduce that the functions $t \mapsto t^{-1}[f(r+t) - f(r)]$ and $t \mapsto t^{-1}[f(r-t) - f(r)]$ have cluster points in $\mathbb{R}$ as $t \downarrow 0$. So, consider any cluster points ρ_+ and ρ_- in $\mathbb{R}$ of the latter functions as $t \downarrow 0$. It is easily seen that $(1, \rho_+)$ and $(-1, \rho_-)$ belong to

$T^B\big(\mathrm{gph}\, f; (r, f(r))\big)$, hence $\rho_+ \geq \alpha$ and $\rho_- \geq -\alpha$ according to the second inclusion in the assumption (b). Consequently $f^D(r; 1) \geq \alpha$ and $f^D(r; -1) \geq -\alpha$, which combined with (6.15) gives $f^D(r; 1) = \alpha = -f^D(r; -1)$.

Finally, from the finiteness of $f^D(r; 1)$ and $f^D(r; -1)$ we easily obtain that $\liminf_{t \downarrow 0} f(r+t) \geq f(r)$ and $\liminf_{t \downarrow 0} f(r-t) \geq f(r)$, thus f is lower semicontinuous at r and the proof is complete. □

PROPOSITION 6.35. *Let S be a nonempty set of an open interval I of $\mathbb{R}$ and $f : I \to \mathbb{R}$ be a real-valued function. If $\underline{D}^+ f(t)$ (resp. $\overline{D}^- f(t)$) is finite at every $t \in S$ except for those of a countable subset, then there exists a Lebesgue negligible subset S_0 of S such that*

$$\underline{d}_D^+ f(t; -1) = -\underline{d}_D^+ f(t; 1) \text{ or equivalently } \underline{D}^+ f(t) = \overline{D}^- f(t), \quad \text{for all } t \in S \setminus S_0,$$

and $\mathcal{H}\big(\mathrm{gph}\, f \cap (S_0 \times \mathbb{R})\big) = 0$, where $\mathcal{H}$ denotes the one-dimensional Hausdorff outer measure in $\mathbb{R}^2$.

Similarly, if $\underline{D}^- f(t)$ (resp. $\overline{D}^+ f(t)$) is finite at every $t \in S$ except for those of a countable subset, then there exists a Lebesgue negligible subset $S_0 \subset S$ such that

$$\underline{D}^- f(t) = \overline{D}^+ f(t) \quad \text{for all } t \in S \setminus S_0.$$

PROOF. As in the proof of the previous lemma we use the symbol $f^D(r; \cdot)$ in place of $\underline{d}_D^+ f(r; \cdot)$. Without loss of generality, we may suppose that $\underline{D}^+ f(r) = f^D(r; 1)$ is finite for all $r \in S$. So, for each $r \in S$, setting $T(r) := T^B\big(\mathrm{gph}\, f; (r, f(r))\big)$, on the one hand $(1, f^D(r; 1)) \in T(r)$ as easily seen, and on the other hand $T(r)$ is not the whole space $\mathbb{R}^2$ since $(1, f^D(r; 1) - \varepsilon)$ (with $\varepsilon > 0$) does not belong to $T(r)$ as it follows from (a) in Lemma 6.34. Then, by Theorem 2.260 in Chapter 2 there exists a set $S_0 \subset S$ such that $\mathcal{H}(\mathrm{gph}\, f \cap (S_0 \times \mathbb{R})) = 0$ (which entails $\lambda(S_0) = 0$) and such that, for any fixed $r \in S \setminus S_0$, the Bouligand-Peano tangent cone of $\mathrm{gph}\, f$ at $(r, f(r))$ is either a vector line or a half-plane.

Suppose first that $T(r)$ is a vector line. From the inclusion of $(1, f^D(r; 1)) \in T(r)$ we see that $T(r)$ is not $\{0\} \times \mathbb{R}$, hence there exists $\alpha \in \mathbb{R}$ such that $T(r) = \{(u, v) \in \mathbb{R}^2 : v = \alpha u\}$, so (b) of Lemma 6.34 gives $\alpha = \underline{D}^+ f(r) = \overline{D}^- f(r)$.

Suppose now that $T(r)$ is a half-plane. If the boundary of $T(r)$ is vertical, that is, $\{0\} \times \mathbb{R}$, then the half-plane $T(r)$ is equal to $\{(u, v) \in \mathbb{R}^2 : u \geq 0\}$ since $(1, f^D(r; 1)) \in T(r)$. Thus, (a) in Lemma 6.34 yields $\{(u, v) \in \mathbb{R}^2 : u > 0\} \subset \mathrm{epi}\, f^D(r; \cdot)$ and this furnishes the contradiction $(1, f^D(r; 1) - \varepsilon) \in \mathrm{epi}\, f^D(r; \cdot)$ for $\varepsilon > 0$.

Consequently, the boundary of the half-plane $T(r)$ is not vertical. So either

$$T(r) = \{(u, v) \in \mathbb{R}^2 : v \geq \alpha u\} \quad \text{for some } \alpha \in \mathbb{R}$$

or $\quad T(r) = \{(u, v) \in \mathbb{R}^2 : v \leq \beta u\} \quad \text{for some } \beta \in \mathbb{R}.$

If the latter equality (with β) is fulfilled, we deduce through (a) in Lemma 6.34 that $v \geq f^D(r; 1)$ for all $v \leq \beta$, which is impossible since $f^D(r; 1)$ is finite. Therefore, $T(r) = \{(u, v) \in \mathbb{R}^2 : v \geq \alpha u\}$ for some $\alpha \in \mathbb{R}$ and (b) in Lemma 6.34 entails $\alpha = f^D(r; 1) = -f^D(r; -1)$.

The case concerning the finiteness of $\underline{D}^- f(\cdot)$ (resp. $\overline{D}^+ f(\cdot)$) is reduced to the previous one through the function $g : J \to \mathbb{R}$ with $J := \{-r : r \in I\}$ and $g(t) := -f(-t)$ for every $t \in J$, since $\underline{D}^- f(r) = \underline{D}^+ g(-r)$ and $\overline{D}^- f(r) = \overline{D}^+ g(-r)$ for every $r \in I$. The proof of the proposition is complete. □

COROLLARY 6.36. Let $f : I \to \mathbb{R}$ be a real-valued function defined on an open interval I of $\mathbb{R}$ and S be a nonempty set of I. Under anyone of the conditions (a), (a'), (b), (b') below, the function f is derivable at almost all points in S.
(a) For all $t \in S$, the unilateral semiderivates $\underline{D}^+ f(t)$ and $\overline{D}^+ f(t)$ are finite;
(a') for all $t \in S$, $\underline{D}^- f(t)$ and $\overline{D}^- f(t)$ are finite;
(b) for all $t \in S$, $\underline{D}^+ f(t) = \underline{D}^- f(t)$ and the common value is finite;
(b') for all $t \in S$, $\overline{D}^- f(t) = \overline{D}^+ f(t)$ and the common value is finite.

PROOF. (a)-(a'). Suppose that $\underline{D}^+ f(t)$ and $\overline{D}^+ f(t) = -\underline{D}^+(-f)(t)$ are finite for every $t \in S$. By Proposition 6.35 there is a subset $S_0 \subset S$ with $\mathcal{H}(\text{gph } f \cap (S_0 \times \mathbb{R})) = 0$ such that for every $t \in S \setminus S_0$

$$\underline{D}^+ f(t) = \overline{D}^- f(t) \quad \text{and} \quad \underline{D}^+(-f)(t) = \overline{D}^-(-f)(t),$$

the latter equality being equivalent to $\overline{D}^+ f(t) = \underline{D}^- f(t)$. Since

$$\underline{D}^+ f(t) \leq \overline{D}^+ f(t) \quad \text{and} \quad \underline{D}^- f(t) \leq \overline{D}^- f(t),$$

we obtain the equality between the four unilateral semiderivates $\underline{D}^+ f(t)$, $\overline{D}^+ f(t)$, $\underline{D}^- f(t)$ and $\overline{D}^- f(t)$. This means that f is derivable at any $t \in S \setminus S_0$.

The proof of (a') is similar.

(b)-(b'). Suppose now that the condition (b) holds. As above, the finiteness of the unilateral semiderivates in (b) allows us to apply Proposition 6.35 to obtain a subset $S_0 \subset S$ with $\mathcal{H}(\text{gph } f \cap (S_0 \times \mathbb{R})) = 0$ such that for all $t \in S \setminus S_0$

$$\underline{D}^+ f(t) = \overline{D}^- f(t) \quad \text{and} \quad \underline{D}^- f(t) = \overline{D}^+ f(t).$$

Combining the latter equalities with the equality in (b) we see that the four Dini unilateral semiderivates coincide at every point in $S \setminus S_0$.

The proof of (b') is similar. □

The following lemma which has its own interest prepares the next theorem concerning estimates of the outer Lebesgue measure of $f(S)$ in terms of Dini unilateral semiderivates.

LEMMA 6.37. Let S be a nonempty set of $\mathbb{R}$ and $f : \mathbb{R} \to \mathbb{R}$ be a real-valued function such that for some real $K \geq 0$ either the condition (a) or (b) is fulfilled:
(a) $-K \leq \underline{D}^+ f(t)$ and $\overline{D}^- f(t) \leq K$ for all $t \in S$;
(b) $-K \leq \underline{D}^- f(t)$ and $\overline{D}^+ f(t) \leq K$ for all $t \in S$.
Then $\lambda(f(S)) \leq K\lambda(S)$, where λ denotes the outer Lebesgue measure and $K\lambda(S) = 0$ by convention if $K = 0$ and $\lambda(S) = +\infty$.

PROOF. It suffices to prove the lemma under the assumption (a). Considering the increasing sequence $(S \cap [-k, k])_k$ if necessary, we may suppose that $\lambda(S) < +\infty$. Fix any real $\varepsilon > 0$ and, for each $n \in \mathbb{N}$, denote by S_n the set of $r \in S$ such that $-(K + \varepsilon)|t - r| \leq f(t) - f(r)$ for all $t \in {]}r - 1/n, r + 1/n{[}$. Obviously we have $S_n \subset S_{n+1}$ and both inequalities in the assumption ensure that $S = \bigcup_{n \in \mathbb{N}} S_n$.

Now fix any $n \in \mathbb{N}$ and consider a sequence of open intervals $(I_{m,n})_m$ with diam $(I_{m,n}) < 1/n$ and such that $S_n \subset \bigcup_{m \in \mathbb{N}} I_{m,n}$ and $\sum_{m \in \mathbb{N}} \lambda(I_{m,n}) \leq \lambda(S_n) + \varepsilon$. For each $m \in \mathbb{N}$ and each pair of elements $r, s \in S_n \cap I_{m,n}$ we have by definition of the set S_n

$$|f(s) - f(r)| \leq (K + \varepsilon)|s - r| \leq (K + \varepsilon)\lambda(I_{m,n}),$$

6.1. MEAN VALUE INEQUALITIES WITH DINI DERIVATIVES

hence $\lambda(f(S_n \cap I_{m,n})) \leq (K+\varepsilon)\lambda(I_{m,n})$. Combining this with the above properties of the sequence $(I_{m,n})_m$ gives

$$\lambda(f(S_n)) \leq \sum_{m \in \mathbb{N}} \lambda(f(S_n \cap I_{m,n})) \leq (K+\varepsilon) \sum_{m \in \mathbb{N}} \lambda(I_{m,n}) \leq (K+\varepsilon)\bigl(\lambda(S_n)+\varepsilon\bigr).$$

From the increasing property of the sequence $(S_n)_n$ we obtain $\lambda(f(S)) \leq (K+\varepsilon)\bigl(\lambda(S)+\varepsilon\bigr)$, thus $\lambda(f(S)) \leq K\lambda(S)$ as desired. $\square$

Before stating the theorem we need the following observation. Suppose that the real-valued function $f : I \to \mathbb{R}$ defined on an open interval I is Borel measurable. Fix a sequence $(r_n, f(r_n))_{n \in \mathbb{N}}$ which is dense in gph f and note that $(r_n)_{n \in \mathbb{N}}$ is itself dense in I. Taking any real α and setting, for $m, n, p \in \mathbb{N}$,

$$T_{m,n,p} := \{t \in I : t \notin\,]r_n - 1/m, r_n[\text{ or } f(r_n) - f(t) \geq (\alpha + 1/p)(r_n - t)\}$$

it is not difficult to see that

$$\{t \in I : \underline{D}^+ f(t) \geq \alpha\} = \bigcap_{p \in \mathbb{N}} \bigcup_{m \in \mathbb{N}} \bigcap_{n \in \mathbb{N}} T_{m,n,p},$$

hence $\underline{D}^+ f(\cdot)$ is a Borel function. Similarly, the function $\overline{D}^+ f(\cdot)$ is Borel measurable.

THEOREM 6.38 (Integral mean value inequality with Dini semiderivates for images of sets). *Let $f : I \to \mathbb{R}$ be a real-valued function defined on an interval I of $\mathbb{R}$ and S be a nonempty subset of $\mathbb{R}$.*
(a) *If, for some real $K \geq 0$, one has $|\underline{D}^+ f(t)| \leq K$ (resp. $|\overline{D}^+ f(t)| \leq K$) for all $t \in S$ except for those of a countable subset of S, then $\lambda(f(S)) \leq K\lambda(S)$, where λ denotes the outer Lebesgue measure and $K\lambda(S) = 0$ by convention if $K = 0$ and $\lambda(S) = +\infty$.*
(b) *If there exists a non-negative real-valued Lebesgue measurable function ρ on I such that $|\underline{D}^+ f(t)| \leq \rho(t)$ (resp. $|\overline{D}^+ f(t)| \leq \rho(t)$) for all $t \in S$ except for those of a countable subset of S and if the set S is Lebesgue measurable, then*

$$\lambda(f(S)) \leq \int_S \rho(t)\, dt.$$

(c) *If $\underline{D}^+ f(t)$ (resp. $\overline{D}^+ f(t)$) is finite for all $t \in S$ except for those of a countable subset of S and if the set S and the function f are Borel measurable, then*

$$\lambda(f(S)) \leq \int_S |\underline{D}^+ f(t)|\, dt \quad (\text{resp. } \lambda(f(S)) \leq \int_S |\overline{D}^+ f(t)|\, dt).$$

PROOF. (a) It suffices to prove (a) under the assumption on $\underline{D}^+ f(\cdot)$. Without loss of generality, we may suppose that the inequality in the assumption holds for all $t \in S$. Denote by S_0 the set of elements $t \in S$ such that $\underline{D}^+ f(t) \neq \overline{D}^- f(t)$. By Proposition 6.35 $\mathcal{H}((\text{gph } f) \cap (S_0 \times \mathbb{R})) = 0$, where $\mathcal{H}$ denotes the one-dimensional Hausdorff measure, hence $\lambda(f(S_0)) = 0$. Since $|\underline{D}^+ f(t)| = |\overline{D}^- f(t)| \leq K$ for all $t \in S \setminus S_0$, Lemma 6.37 above says that $\lambda(f(S \setminus S_0)) \leq K\lambda(S \setminus S_0)$, thus we obtain $\lambda(f(S)) \leq K\lambda(S)$.
(b) Assume that $|\underline{D}^+ f(t)| \leq \rho(t)$ (resp. $|\overline{D}^+ f(t)| \leq \rho(t)$) for all $t \in S$ except for those of a countable subset of S and that the set S is Lebesgue measurable. We may

suppose that S is bounded. For each real $\varepsilon > 0$ denote by S_n the set of elements $t \in S$ such that $(n-1)\varepsilon \le \rho(t) < n\varepsilon$. By the first assertion above we have

$$\lambda(f(S)) \le \sum_{n \in \mathbb{N}} \lambda(f(S_n)) \le \sum_{n \in \mathbb{N}} n\varepsilon \lambda(S_n) \le \int_S \rho(t)\,dt + \varepsilon \lambda(S),$$

so that $\lambda(f(S)) \le \int_S \rho(t)\,dt$.

(c) Under the Borel measurability of the function f, we have seen above that the Dini semiderivate $\underline{D}^+ f(\cdot)$ (resp. $\overline{D}^+ f(\cdot)$) is Borel measurable. So, denoting by A the countable subset of S where $\underline{D}^+ f(\cdot)$ (resp. $\overline{D}^+ f(\cdot)$) is not finite, it suffices to apply (b) with the function ρ defined by $\rho(t) = |\underline{D}^+ f(\cdot)|$ (resp. $\rho(t) = |\overline{D}^+ f(t)|$) for $t \in I \setminus A$ and $\rho(t) = 0$ for $t \in A$. $\square$

The following corollary concerns the case where the boundedness of $|\underline{D}^+ f(\cdot)|$ in the above theorem is replaced by merely the finiteness of $\underline{D}^+ f(\cdot)$.

COROLLARY 6.39. *Let T be a nonempty set of $\mathbb{R}$ and $f : \mathbb{R} \to \mathbb{R}$ be a real-valued function. Assume that $\underline{D}^+ f(t)$ (resp. $\overline{D}^+ f(t)$) is finite for all $t \in T$ except for those of a countable subset of T. Then $\lambda(f(S)) = 0$ for any set $S \subset T$ with $\lambda(S) = 0$, where λ denotes the outer Lebesgue measure.*

PROOF. We prove the corollary only for the case concerning $\underline{D}^+ f(\cdot)$. We may suppose that $\underline{D}^+ f(t)$ is finite for every $t \in T$. Let S be a subset of T with $\lambda(S) = 0$. For each $n \in \mathbb{N}$, denote by S_n the set of $t \in S$ such that $|\underline{D}^+ f(t)| \le n$. By the first assertion of the above theorem, we have $\lambda(f(S_n)) \le n\lambda(S_n) = 0$. This combined with the equality $S = \cup_{n \in \mathbb{N}} S_n$ guarantees that $\lambda(f(S)) = 0$. $\square$

The next proposition studies the situation where a Dini unilateral semiderivate is infinite. The following lemma prepares the proposition.

LEMMA 6.40. *Let S be a subset of a Euclidean space X with dimension 2 and let e_1 be a unit vector of X. Let C be the set of points in S where $\mathbb{R}e_1$ is an extreme tangent vector line to S. Let e_2 be a unit vector orthogonal to e_2. Then the orthogonal projection of C on $\mathbb{R}e_2$ is one-dimensional Lebesgue negligible.*

PROOF. Putting $C^{\le} := \{x \in C : T^B(S;x) \subset \{ue_1 + ve_2 : v \le 0\}\}$ and $C^{\ge} := \{x \in C : T^B(S;x) \subset \{ue_1 + ve_2 : v \ge 0\}\}$ we see, by the assumption, that $C = C^{\le} \cup C^{\ge}$ along with $e_2 \notin T^B(S;x)$ for all $x \in C^{\le}$, and $-e_2 \notin T^B(S;x)$ for all $x \in C^{\ge}$. By Proposition 2.258 there exist a sequence $(P_n)_n$ of subsets of $\mathbb{R}$ and a sequence $(f_n)_n$ of Lipschitz continuous functions from $\mathbb{R}$ into $\mathbb{R}$ such that, for $G_n := \{ue_1 + ve_2 : u \in P_n, v = f_n(u)\}$, we have $C^{\le} = \bigcup_n G_n$. Fix n and denote by Q_n be the Lebesgue negligible set of $\mathbb{R}$ where f_n is not derivable (according to the Rademacher theorem). Let Δ_n be the set of elements in $P_n \setminus Q_n$ which are isolated either on the right or on the left in $P_n \setminus Q_n$. Lemma 6.32 tells us that Δ_n is countable, hence $Q'_n := Q_n \cup \Delta_n$ is negligible in $\mathbb{R}$. For each fixed element $r \in P_n \setminus Q'_n$, we can choose some sequences $(t_k)_k$ and $(\tau_k)_k$ in $]0, +\infty[$ tending to 0 with $r + t_k \in P_n$ and $r - \tau_k \in P_n$ for all $k \in \mathbb{N}$. For $v_k := t_k^{-1}[f_n(r+t_k) - f_n(r)]$ and $w_k := -\tau_k^{-1}[f_n(r-\tau_k) - f_n(r)]$ we have, for all k,

$$(re_1 + f_n(r)e_2) + t_k(e_1 + v_k e_2) \in G_n \text{ and } (re_1 + f_n(r)e_2) + \tau_k(-e_1 - w_k e_2) \in G_n$$

hence $\pm(e_1 + f'_n(r)e_2) \in T^B(G_n; (re_1 + f_n(r)e_2))$ since both sequences $(v_k)_k$ and $(w_k)_k$ converge to the usual derivate $f'_n(r)$. Since $G_n \subset S$, it results that $\pm(e_1 +$

$f'_n(r)e_2$) belong to $T^B(S; re_1 + f_n(r)e_2)$. The latter inclusion and the definition of $C^\leq$ entail $\pm f'_n(r) \leq 0$, that is, $f'_n(r) = 0$ for all $r \in P_n \setminus Q'_n$. Denoting by A and A_n the orthogonal projections on $\mathbb{R}e_2$ of $C^\geq$ and G_n respectively, we obviously have $A = \bigcup_n A_n$ and $A_n = \{f_n(r)e_2 : r \in P_n\}$, and $A_n = A'_n \cup A''_n$, where $A'_n := \{f_n(r)e_2 : r \in P_n \setminus Q'_n\}$ and $A''_n := \{f_n(r)e_2 : r \in P_n \cap Q'_n\}$. The set A'_n has Lebesgue measure zero in $\mathbb{R}e_2$ by Lemma 6.37, and the set A''_n has also Lebesgue measure zero in $\mathbb{R}e_2$ since $P_n \cap Q'_n$ is negligible in $\mathbb{R}$ and f_n is Lipschitz continuous on $\mathbb{R}$. Consequently, the set $A_n = A'_n \cup A''_n$ is Lebesgue negligible in $\mathbb{R}e_2$ hence so is the set A.

Similarly, the orthogonal projection B of $C^\geq$ on $\mathbb{R}e_2$ is Lebesgue negligible in $\mathbb{R}e_2$. We then conclude that the orthogonal projection of C on $\mathbb{R}e_2$ is Lebesgue negligible in $\mathbb{R}e_2$. □

The situation where a Dini unilateral semiderivate is infinite is a particular case of:

PROPOSITION 6.41. *Let I be an interval of $\mathbb{R}$ and $f : I \to \mathbb{R}$ be a real-valued function. The set of $r \in I$ such that*

$$\lim_{t \downarrow 0} |t^{-1}[f(r+t) - f(r)]| = +\infty \quad \left(\text{resp. } \lim_{t \uparrow 0} |t^{-1}[f(r+t) - f(r)]| = +\infty\right)$$

is Lebesgue negligible.

PROOF. Denote by Q the set of $r \in I$ such that $\lim_{t \downarrow 0} |t^{-1}[f(r+t) - f(r)]| = +\infty$ and put $S := \operatorname{gph} f$ and $C := S \cap (Q \times \mathbb{R})$. We claim, for each fixed $r \in Q$, that $T(r) \subset]-\infty, 0] \times \mathbb{R}$, where $T(r) := T^B(S; (r, f(r)))$. Suppose by contradiction that some $(u, v) \in T(r)$ with $u > 0$ and $v \in \mathbb{R}$. By the sequential characterization of Bouligand-Peano tangent cone, there are sequences $(t_n)_n$ in $]0, +\infty[$ tending to 0 and $(u_n, v_n)_n$ in $\mathbb{R}^2$ converging to (u, v) such that $v_n = t_n^{-1}[f(r + t_n u_n) - f(r)]$. We may suppose $u_n > 0$ for all n, so $|u_n^{-1} v_n| = |(t_n u_n)^{-1}[f(r + t_n u_n) - f(r)]|$ and by definition of Q we have $|u_n^{-1} v_n| \to +\infty$ as $n \to \infty$, which is a contradiction since $|u_n^{-1} v_n| \to |u^{-1} v|$.

The claim above ensures in particular $T(r) \neq \mathbb{R}^2$ for all $r \in Q$, hence denoting by $\mathcal{H}$ the one-dimensional Hausdorff measure in $\mathbb{R}^2$, by Theorem 2.260 there is a subset $C_0 \subset C$ with $\mathcal{H}(C_0) = 0$ such that for every $(r, f(r)) \in C \setminus C_0$ the tangent cone $T(r)$ is either a vector line or a half-plane. This and the inclusion $T(r) \subset]-\infty, 0] \times \mathbb{R}$ entail, for each $(r, f(r)) \in C \setminus C_0$, that either $T(r) = \{0\} \times \mathbb{R}$ or $T(r) =]-\infty, 0] \times \mathbb{R}$, hence in particular the set S has $\{0\} \times \mathbb{R}$ as an extreme tangent vector line at any $(r, f(r))$ in $C \setminus C_0$. By Lemma 6.40 the orthogonal projection on $\mathbb{R} \times \{0\}$ of $C \setminus C_0$ is Lebesgue negligible in $\mathbb{R} \times \{0\}$. On the other hand, the orthogonal projection on $\mathbb{R} \times \{0\}$ of C_0 is also Lebesgue negligible in $\mathbb{R} \times \{0\}$ since $\mathcal{H}(C_0) = 0$. Consequently, the orthogonal projection of C on $\mathbb{R} \times \{0\}$ is Lebesgue negligible in $\mathbb{R} \times \{0\}$, thus Q is Lebesgue negligible in $\mathbb{R}$ since the orthogonal projection of C on $\mathbb{R} \times \{0\}$ is $Q \times \{0\}$. The proof is then complete. □

From Proposition 6.35, Corollary 6.36 and Proposition 6.41 the following famous Denjoy-Young-Saks theorem directly follows. See the section of comments at the end of the chapter for the contribution of each of those three authors to the theorem.

THEOREM 6.42 (Denjoy-Young-Saks theorem on unilateral semiderivates). *Let I be an open interval of $\mathbb{R}$ and f be a real-valued function. Then for Lebesgue*

almost every $t \in I$, one of the following properties is fulfilled:
(a) $\overline{D}^+ f(t) = \overline{D}^- f(t) = +\infty$ and $\underline{D}^+ f(t) = \underline{D}^- f(t) = -\infty$;
(b) $\overline{D}^+ f(t) = +\infty$, $\underline{D}^- f(t) = -\infty$, and $\underline{D}^+ f(t) = \overline{D}^- f(t)$ with this common value finite;
(c) $\overline{D}^- f(t) = +\infty$, $\underline{D}^+ f(t) = -\infty$, and $\underline{D}^- f(t) = \overline{D}^+ f(t)$ with this common value finite;
(d) the four unilateral semiderivates of f at t are finite and equal, that is, f is derivable at t.

REMARK 6.43. Each one of the above properties may even arise for almost every $t \in I$.
(a) Let A_{-1} and A_1 be two disjoint countable dense subsets of $\mathbb{R}$, e.g., $A_1 := \mathbb{Q}$ and $A_{-1} := \sqrt{2} + \mathbb{Q}$. Define $f : \mathbb{R} \to \mathbb{R}$ by $f(t) = -1$ if $t \in A_{-1}$, $f(t) = 1$ if $t \in A_1$ and $f(t) = 0$ otherwise. Then, for every $t \notin (A_{-1} \cup A_1)$ we have

$$\overline{D}^+ f(t) = \overline{D}^- f(t) = +\infty \quad \text{and} \quad \underline{D}^+ f(t) = \underline{D}^- f(t) = -\infty,$$

and of course $A_{-1} \cup A_1$ is Lebesgue negligible.
(b) For f defined on $\mathbb{R}$ by $f(t) = 1$ if $t \in \mathbb{Q}$ and $f(t) = 0$ otherwise, we have for every $t \notin \mathbb{Q}$

$$\overline{D}^+ f(t) = +\infty, \; \underline{D}^- f(t) = -\infty, \text{ and } \underline{D}^+ f(t) = \overline{D}^- f(t) = 0.$$

(c) Taking now $f(t) = -1$ if $t \in \mathbb{Q}$ and $f(t) = 0$ if $t \in \mathbb{R} \setminus \mathbb{Q}$, we obtain for every $t \notin \mathbb{Q}$

$$\overline{D}^- f(t) = +\infty, \; \underline{D}^+ f(t) = -\infty, \text{ and } \underline{D}^- f(t) = \overline{D}^+ f(t) = 0.$$

(d) Finally, concerning (d) it suffices to take any function f derivable on the whole set $\mathbb{R}$. □

6.1.6. Mean value inequality with Dini subgradients.
Theorem 6.38 on the inequality $\lambda(f(S)) \leq \int_S |\underline{D}^+ f(t)| \, dt$ (when $\underline{D}^+ f$ is finite on S) as well as the Denjoy-Saks theorem being obtained, our next aim is to take advantage of some above results (namely Lemma 6.37, Propositions 6.35 and 6.41, and Theorem 6.38) to establish a mean value inequality in integral form for Borelian functions defined on normed spaces in terms of Dini subgradients.

DEFINITION 6.44. Let U be an open set of a normed space X with $x \in U$ and $f : U \to \mathbb{R} \cup \{-\infty, +\infty\}$ be an extended real-valued function which is finite at x. By means of the (right-hand) lower Dini directional derivative $\underline{d}_D^+ f(x; \cdot)$, we define the *Dini subdifferential* of f at x by

$$\partial_D f(x) := \{x^* \in X^* : \langle x^*, h \rangle \leq \underline{d}_D^+ f(x; h) \text{ for all } h \in X\}.$$

Each element in $\partial_D f(x)$ is called a *Dini subgradient* of f at x.

It is easily seen that

$$\partial_D f(x) := \{x^* \in X^* : -\underline{d}_D^+ f(x; -h) \leq \langle x^*, h \rangle \leq \underline{d}_D^+ f(x; h) \text{ for all } h \in X\}.$$

So, if X coincides with the real line, that is, $X = \mathbb{R}$, then for all $r \in U$

$$\partial_D f(r) = \{\alpha \in \mathbb{R} : -\underline{d}_D^+ f(r; -1) \leq \alpha \leq \underline{d}_D^+ f(r; 1)\}$$
(6.16)
$$= \{\alpha \in \mathbb{R} : \overline{D}^- f(r) \leq \alpha \leq \underline{D}^+ f(r)\}.$$

Before starting with the study of mean value inequalities with Dini subgradients, let us note the following result which is in the same line with Propositions 4.13 and 5.11.

PROPOSITION 6.45. *Let U be an open set of a normed space X with $x \in U$ and $f : U \to \mathbb{R} \cup \{-\infty, +\infty\}$ be an extended real-valued function which is finite at x. Then f is Gâteaux differentiable at x if and only if both sets $\partial_D f(x)$ and $\partial_D(-f)(x)$ are nonempty.*

PROOF. Only the implication $\Leftarrow$ needs to be proved. Take any $u^* \in \partial_D f(x)$ and $v^* \in \partial_D(-f)(x)$. Fixing any $h \in X$ and any $\varepsilon > 0$ there exists a real $\delta > 0$ such that for all $0 < t < \delta$ we have $x \pm th \in U$ and

$$-\varepsilon + \langle u^*, v \rangle \leq t^{-1}[f(x+tv) - f(x)] \quad \text{and} \quad -\varepsilon + \langle v^*, v \rangle \leq t^{-1}[-f(x+tv) + f(x)],$$

for $v = \pm h$. On the one hand, this ensures that $f(x \pm th)$ is finite, so adding the two inequalities with $v = h$ yields $\langle u^* + v^*, h \rangle \leq 2\varepsilon$ and hence $\langle u^* + v^*, h \rangle \leq 0$ by making $\varepsilon \downarrow 0$. The latter inequality being true for all $h \in X$ we deduce that $u^* + v^* = 0$. Consequently, the previous inequalities can be rewritten as

$$-\varepsilon \leq t^{-1}[f(x+tv) - f(x)] - \langle u^*, v \rangle \leq \varepsilon \quad \text{for } v = \pm h \text{ and } 0 < t < \delta,$$

or equivalently

$$-\varepsilon \leq t^{-1}[f(x+th) - f(x)] - \langle u^*, h \rangle \leq \varepsilon \quad \text{for } 0 < |t| < \delta.$$

This justifies the Gâteaux differentiability of f at x. □

REMARK 6.46. Besides the above proposition we must emphasize that the condition requiring that $d_D^+ f(x; \cdot)$ is linear and continuous is not sufficient to guarantee the Gâteaux differentiability of f at x. Indeed, for the function $f : \mathbb{R} \to \mathbb{R}$ with $f(x) = 1$ if $x \in \mathbb{Q}$ and $f(x) = 0$ otherwise, we have for each $x \in \mathbb{Q}$

$$d_D^+ f(x; \cdot) \equiv 0 \text{ on } \mathbb{R}, \quad \text{while} \quad d_D^+(-f)(x; \cdot) \equiv -\infty \text{ on } \mathbb{R} \setminus \{0\}.$$

□

Now, in preparation for the mean value inequality in terms of Dini subgradients let us establish the following lemma.

LEMMA 6.47. *Let I be an open interval of $\mathbb{R}$ and $f : I \to \mathbb{R}$ be a Borel measurable function. Let S be a Borelian subset of I such that $\partial_D f(t) \cup \partial_D(-f)(t) \neq \emptyset$ for all $t \in S$ except for those of a countable subset. Then the function φ given, for each $t \in I$, by*

$$\varphi(t) := \inf\{|\alpha| : \alpha \in \partial_D f(t) \cup \partial_D(-f)(t)\}$$

is Borel measurable, finite-valued on S outside a countable set, and

$$\lambda(f(S)) \leq \int_S \varphi(t)\, dt,$$

where λ denotes the outer Lebesgue measure.

PROOF. We may suppose that $\partial_D f(t) \cup \partial_D(-f)(t) \neq \emptyset$ for all $t \in S$. The function φ is then finite on S. For each $n \in \mathbb{N}$, consider the sets

$$T_n^+ := \{t \in S : \partial_D f(t) \cap]-n, n[\neq \emptyset\} \quad \text{and} \quad T_n^- := \{t \in S : \partial_D(-f)(t) \cap]-n, n[\neq \emptyset\}.$$

These sets and the function φ are Borel measurable according to (6.16) and to the Borel measurability of the unilateral semiderivates of f and $-f$ involved in the description of $\partial_D f$ and $\partial_D(-f)$ in (6.16). Since $-n < \underline{D}^+ f(t)$ and $\overline{D}^- f(t) < n$ for every $t \in T_n^+$, Lemma 6.37 tells us that $\lambda(f(T_n^+ \cap T)) \leq n\,\lambda(T_n^+ \cap T)$ for every Borel subset T of S; similarly we also have $\lambda(f(T_n^- \cap T)) \leq n\,\lambda(T_n^- \cap T)$. Combining both latter inequalities with the assumption $\partial_D f(t) \cup \partial_D(-f)(t) \neq \emptyset$ for every $t \in I$, we see that

(6.17) $\quad \lambda(f(T)) = 0 \quad$ for any Borelian set $T \subset S$ with $\lambda(T) = 0$.

Consider now the Borelian sets
$$T' := \{t \in S : \partial_D f(t) \neq \emptyset\} \quad \text{and} \quad T'' := \{t \in S : \partial_D(-f)(t) \neq \emptyset\}.$$
The definition of T' and (6.16) ensure that $T' = T_1' \cup T_2'$ where
$$T_1' := \{t \in T' : -\infty < \underline{D}^+ f(t) < +\infty\}$$
and $\quad T_2' := \{t \in T' : \underline{D}^+ f(t) = \overline{D}^+ f(t) = +\infty\}.$
Proposition 6.35 gives a Borelian negligible subset N of T_1' such that
$$\partial_D f(t) = \{\underline{D}^+ f(t)\} = \{\overline{D}^- f(t)\} \quad \text{for all } t \in T_1' \setminus N,$$
hence for any Borelian subset T of $T_1' \setminus N$ we have by Theorem 6.38
$$\lambda(f(T)) \leq \int_T |\partial_D f(t)|\, dt,$$
where $|\partial_D f(t)|$ is taken as the absolute value of the single element of $\partial_D f(t)$ for $t \in T_1' \setminus N$. Further, we also have $\lambda(f(N)) = 0$ according to (6.17). The relation (6.17) also ensures $\lambda(f(T_2')) = 0$ since $\lambda(T_2') = 0$ by Proposition 6.41. This combined with the latter inequality yields
$$\lambda(f(T)) \leq \int_T |\partial_D f(t)|\, dt \quad \text{for any Borelian set } T \subset T',$$
where $|\partial_D f(t)|$ is taken as 0 for any $t \in N$. Similarly, the same inequality holds true with $\partial_D(-f)$ in place of $\partial_D f$ for any Borelian set $T \subset T''$.

Now put
$$S' := \{t \in S : \varphi(t) = \inf\{|\alpha| : \alpha \in \partial_D f(t)\}\}$$
and
$$S'' := \{t \in S : \varphi(t) < \inf\{|\alpha| : \alpha \in \partial_D f(t)\}\}.$$
As above we see that the sets S' and S'' are Borelian, and we also see through the definition of φ that
$$S'' \subset \{t \in S : \varphi(t) = \inf\{|\alpha| : \alpha \in \partial_D(-f)(t)\}\}.$$
Further, since all the values of φ are finite, we have $S' \subset T'$ and $S'' \subset T''$. According to what precedes, we deduce
$$\lambda(f(S')) \leq \int_{S'} |\partial_D f(t)|\, dt = \int_{S'} \varphi(t)\, dt$$
and $\quad \lambda(f(S'')) \leq \int_{S''} |\partial_D(-f)(t)|\, dt = \int_{S''} \varphi(t)\, dt,$
which entails, according to the equalities $S = S' \cup S''$ and $S' \cap S'' = \emptyset$,
$$\lambda(f(S)) \leq \int_S \varphi(t)\, dt.$$
The proof of the lemma is then finished. $\square$

With the above lemma at hands, we can establish a mean value inequality in terms of Dini subgradients of Borelian functions. Here unlike Theorem 6.2, the function f is not required to be lower semicontinuous. However, for each x, an assumption of Dini subdifferentiability of either f or $-f$ is crucial. So, the next theorem and Theorem 6.2 are different in their statements.

THEOREM 6.48 (mean value inequality with Dini subgradients). Let U be an open convex set of a normed space X and $f : U \to \mathbb{R}$ be a real-valued Borel measurable function such that
$$\partial_D f(x) \cup \partial_D(-f)(x) \neq \emptyset$$
for all $x \in U$ except for those of a countable subset. Let
$$\Phi(x) = \inf\{\|x^*\| : x^* \in \partial_D f(x) \cup \partial_D(-f)(x)\} \quad \text{for all } x \in U.$$
Then denoting by λ the one-dimensional outer Lebesgue measure one has, for all $a, b \in U$,
$$\lambda\big(f([a,b])\big) \leq \|b - a\| \int_0^1 \Phi(a + t(b-a))\, dt.$$
If, in addition to the above assumptions, f is continuous relative to $[a, b]$, then
$$f(b) - f(a) \leq \|b - a\| \int_0^1 \Phi(a + t(b-a))\, dt.$$

PROOF. Fix $a, b \in U$ with $a \neq b$ and let I be an open interval of $\mathbb{R}$ containing $[0, 1]$ such that $a + t(b - a) \in U$ for all $t \in I$. Consider $g : I \to \mathbb{R}$ defined by $g(t) := f(a + t(b-a))$ for all $t \in I$. The function g is Borel measurable and, for any $x^* \in \partial_D f(a + t(b-a))$, it is asily seen that $\langle x^*, b - a \rangle \in \partial_D g(t)$. Then, for all $t \in I$ except for those of a countable subset, we have $\partial_D g(t) \cup \partial_D(-g)(t) \neq \emptyset$ and
$$\inf\{|\alpha| : \alpha \in \partial_D g(t) \cup \partial_D(-g)(t)\} \leq \|b - a\| \Phi(a + t(b-a)).$$
Therefore, Lemma 6.47 above yields
$$\lambda\big(f([a,b])\big) = \lambda\big(g([0,1])\big) \leq \|b - a\| \int_0^1 \Phi(a + t(b-a))\, dt,$$
as desired.

Finally, under the additional continuity of f relative to $[a, b]$, the interval delimited by $f(a)$ and $f(b)$ is included in $f([a, b])$, and hence $f(b) - f(a) \leq \lambda(f([a,b]))$. The proof is then finished. $\square$

Through Lemma 6.47 a condition for the Lipschitz behavior can be given as follows:

PROPOSITION 6.49. Let U be an open convex set of a normed space X and $f : U \to \mathbb{R}$ be a real-valued function for which there exists some real $\gamma \geq 0$ such that $\partial_D f(x) \cap \gamma \mathbb{B}_{X^*} \neq \emptyset$ for all $x \in U$. Then, for all $a, b \in U$, one has
$$\lambda\big(f([a,b])\big) \leq \gamma \|b - a\|,$$
where λ denotes the one-dimensional outer Lebesgue measure.

If in addition the restriction of f to any line segment of U is upper semicontinuous, then f is Lipschitz continuous on U with γ as a Lipschitz constant.

PROOF. Fix $a, b \in U$ with $a \neq b$ and consider the open interval I of $\mathbb{R}$ and the function g defined in the proof of the latter theorem above. Taking $t \in I$ and $x^* \in \partial_D f(a+t(b-a))$ we have $\langle x^*, b-a \rangle$ in $\partial_D g(t)$, and the nonemptiness of $\partial_D g(t)$ easily implies the lower semicontinuity of g on I. Then g is Borel measurable and for all $t \in I$
$$\inf\{|\alpha| : \alpha \in \partial_D g(t) \cup \partial_D(-g)(t)\} \leq \gamma \|b-a\|,$$
so Lemma 6.47 gives
$$\lambda\big(f([a,b])\big) = \lambda\big(g([0,1])\big) \leq \gamma \|b-a\|.$$

Now assume in addition that f is upper semicontinuous on every line segment of U. Then for any $a, b \in U$ the function g above is continuous $[0,1]$, thus the interval with extremities $g(0) = f(a)$ and $g(1) = f(b)$ is included in $g([0,1])$. Therefore, $|g(1) - g(0)| \leq \lambda\big(g([0,1])\big)$ and this combined with what precedes ensures
$$|f(b) - f(a)| \leq \gamma \|b-a\|,$$
justifying the Lipschitz property of f over U. □

6.2. Zagrodny mean value inequality

The previous section produced a mean value inequality for a lower semicontinuous function in terms of lower Dini directional derivative. In Theorems 4.84 and 2.180 we established mean value inequalities for a *locally Lipschitz continuous* function $f : X \to \mathbb{R}$ in the form
$$f(b) - f(a) \leq \langle c^*, b-a \rangle,$$
where c^* is some element of the subdifferential of f at some $c \in]a, b[$ for the Mordukhovich limiting subdifferential if X is an Asplund space, and the Clarke subdifferential if X is a normed space (with the bonus of equality in place of the inequality in the case of the Clarke subdifferential). Our objective here is to explore the area of mean value inequalities and various other properties for extended real-valued lower semicontinuous functions through subdifferentials of these functions. Note that a mean value inequality has been proved for a non-Lipschitz function $f : U \to \mathbb{R}$ in Theorem 6.48 with Dini subgradients but there the function f has been required to satisfy the subdifferentiability assumption
$$\partial_D f(x) \cup \partial_D(-f)(x) \neq \emptyset$$
for all x in the open set U except for a countable set of points of U. Such a subdifferentiability assumption will be avoided in the present section.

In order to obtain and state the results in a unified way for several subdifferentials, it is very convenient to deal with any subdifferential fulfilling some suitable properties.

Let U be a nonempty open set of a normed space X and $\mathcal{F}(U)$ be a class of functions from U into $\mathbb{R} \cup \{-\infty, +\infty\}$ which contains the restrictions to U of continuous convex functions on X and is stable by addition with these functions. Given a subdifferential for functions in $\mathcal{F}(U)$, which is in particular an operator ∂ from $\mathcal{F}(U) \times U$ into subsets of the topological dual space X^* (assigning to any every pair $(f, x) \in \mathcal{F}(U) \times U$ a set $\partial f(x) \subset X^*$), consider the following fundamental properties:

Prop.1: $\partial f(x) = \emptyset$ if $|f(x)| = +\infty$ and $0 \in \partial f(x)$ whenever $x \in U$ is a local minimum point of f with $|f(x)| < +\infty$;

Prop.2: $\partial f(x) = \partial g(x)$ whenever f and g coincide on a neighborhood of x;

Prop.3: if f is finite at $x \in U$ and the restriction $f_{|V}$ of f to a convex neighborhood $V \subset U$ of x is lower semicontinuous and convex, then $\partial f(x)$ coincides with the standard subdifferential of the convex function $f_{|V}$ at x, that is,
$$\partial f(x) = \{x^* \in X^* : \langle x^*, u - x \rangle \le f(u) - f(x) \ \forall u \in V\};$$

Prop.4: for $f \in \mathcal{F}(U)$ lower semicontinuous near x and for the restriction g to U of a finite-valued, convex, and continuous function on X, if x is a local minimum point for $f+g$, then for any real $\varepsilon > 0$ there are $x', x'' \in U \cap B(x, \varepsilon)$ with $|f(x') - f(x)| < \varepsilon$ and such that
$$0 \in \partial f(x') + \partial g(x'') + \varepsilon \mathbb{B}_{X^*}.$$

When $\mathcal{F}(U)$ is the class of all extended real-valued functions on U, we will just say a subdifferential on U with properties **Prop.1-Prop.4**. If $f : V \to \mathbb{R} \cup \{-\infty, +\infty\}$ is a function defined on a neighborhood $V \subset U$ of x such that $\overline{f} \in \mathcal{F}(U)$, where $\overline{f}$ is the extension of f to U with $\overline{f}(x') = +\infty$ for $x' \in U \setminus V$, we will set $\partial f(x) := \partial \overline{f}(x)$.

If ∂ is a subdifferential on X and S is a subset of X, we will write $N(S; x)$ in place of $\partial \Psi_S(x)$ and we will call $N(S; x)$ the normal cone of S at x associated with the subdifferential ∂. When the subdifferential ∂ needs to be emphasized, we will write $N^{\partial}(S; x)$.

All the subdifferentials studied in the previous chapters fulfill properties **Prop.1-Prop.4** according to the calculus rules established for those subdifferentials. More precisely those properties hold:
- for the standard subdifferential for convex functions on the class $\mathcal{F}_{\text{conv}}(X)$ of all extended real-valued convex functions on a normed space X, see Theorem 2.105;
- for the Clarke subdifferential on the class of all extended real-valued functions on the open set U of a normed space X, see Theorem 2.98;
- for the Ioffe approximate subdifferential on the class of all extended real-valued functions on the open set U of a Banach space X, see Theorem 5.78;
- for the Fréchet subdifferential on the class of all extended real-valued functions on the open set U of an Asplund space X, see Theorem 4.52;
- for the Mordukhovich limiting subdifferential on the class of all extended real-valued functions on the open set U of an Asplund space X, see Theorem 4.78;
- for the proximal subdifferential on the class of all extended real-valued functions on the open set U of a Hilbert space X, see Theorem 4.155(a);
- for the Hadamard/Bouligand (resp. Dini) subdifferential on a Banach space admitting a Gâteaux differentiable equivalent norm, see Proposition 5.13.

Another interesting class will be involved later in the development of Rolle-type theorems. Let $\mathcal{F}(U)$ be the class of functions of the form $f = h + g$ where $h : U \to \mathbb{R}$ is Gâteaux differentiable on the open set U of a normed space X and g is the restriction to U of a real-valued continuous convex function on X. For each $x \in U$, the directional derivative $f'(x; \cdot)$ exits and $f'(x; \cdot) = D_G h(x)(\cdot) + g'(x; \cdot)$, hence $f'(x; \cdot)$ is positively homogeneous, continuous and convex on X. Denote by

$\partial f(x)$ the subdifferential of $f'(x;\cdot)$ at the origin in the sense of convex analysis. The class $\mathcal{F}(U)$ contains the restrictions to U of real-valued continuous convex functions and is stable by addition with those functions. It is also easily seen that the operator ∂ fulfills Properties **Prop.1-Prop.4** for $\mathcal{F}(U)$.

We are now in a position to examine various properties related to subdifferentials of functions.

6.2.1. Density properties for subdifferentials.
Before establishing the Zagrodny mean value inequality in the next subsection, let us show some density results for subdfifferentials.

PROPOSITION 6.50. *Let U be an open set of a Banach space $(X, \|\cdot\|)$ and ∂ be a subdifferential on a class $\mathcal{F}(U)$ satisfying* **Prop.1-Prop.4**. *Then for any proper lower semicontinuous function $f : U \to \mathbb{R} \cup \{+\infty\}$ in $\mathcal{F}(U)$, the effective domain $\mathrm{Dom}\,\partial f$ is graphically dense in the effective domain $\mathrm{dom}\,f$ of f, that is, for any $x \in \mathrm{dom}\,f$ there exists a sequence $(x_n)_n$ in $\mathrm{Dom}\,\partial f$ such that $(x_n, f(x_n))_n$ converges to $(x, f(x))$.*

PROOF. We proceed as in the proof of the implication $(a) \Rightarrow (d)$ in Corollary 4.54. Fix any $\overline{x} \in \mathrm{dom}\,f$, any $\varepsilon > 0$. From the lower semicontinuity of f choose some $r > 0$ such that $B[\overline{x}, r] \subset U$ and $f(\overline{x}) - \varepsilon < f(x)$ for all $x \in B[\overline{x}, r]$, hence $f(\overline{x}) \leq \inf_{B[\overline{x},r]} f + \varepsilon$. Taking $0 < \lambda < \min\{\varepsilon, r/2\}$ the Ekeland variational principle (see Theorem 2.221) gives some $u \in B[\overline{x}, \lambda]$ with $f(u) \leq f(\overline{x}) \leq f(u) + \varepsilon$ and such that
$$f(u) \leq f(x) + (\varepsilon/\lambda)\|x - u\| \quad \text{for all } x \in B[\overline{x}, r].$$
Since $\|u - \overline{x}\| < r$, the point u is a local minimizer of $f + (\varepsilon/\lambda)\|\cdot - u\|$. Consequently, properties **Prop.4** and **Prop.3** gives some $v \in X$ with $\partial f(v) \neq \emptyset$, $\|v - u\| < \varepsilon$ and $|f(v) - f(u)| < \varepsilon$, hence $\|v - \overline{x}\| < 2\varepsilon$ and $|f(v) - f(\overline{x})| < 2\varepsilon$. This finishes the proof. □

The second proposition concerns the density of the range set of the subdifferential.

PROPOSITION 6.51. *Let $(X, \|\cdot\|)$ be a Banach space, ∂ be a subdifferential on a class $\mathcal{F}(X)$ satisfying* **Prop.1-Prop.4**, *and f be an extended real-valued proper lower semicontinuous function in $\mathcal{F}(X)$. Assume the following coercivity-type condition holds: there exists a lower semicontinuous function $\theta : [0, +\infty[\to \mathbb{R} \cup \{+\infty\}$ with $\theta(t) \to +\infty$ as $t \to +\infty$ and such that $f(x) \geq \|x\|\theta(\|x\|)$ for all $x \in X$. Then the range $\partial f(X)$ is dense in $(X^*, \|\cdot\|_*)$.*

PROOF. Fix any $x^* \in X^*$ and any $\varepsilon > 0$. From the equality $\lim_{t \to +\infty} \theta(t) = +\infty$ choose some real $r > 0$ such that $\theta(t) - \|x^*\|_* \geq 1$ for all $t > r$, and from the lower semicontinuity of θ choose some real γ such that $t \mapsto (\theta(t) - \|x^*\|_*)t$ is bounded from below by γ on the compact interval $[0, r]$. So, for all $x \in X$ we have
$$f(x) - \langle x^*, x \rangle \geq \big(\theta(\|x\|) - \|x^*\|_*\big)\|x\| \geq \min\{r, \gamma\},$$
hence the lower semicontinuous function $f - \langle x^*, \cdot \rangle$ is bounded from below on X. Taking $a \in \mathrm{dom}\,f$, the form of the Ekeland variational principle in Corollary 2.222 gives some $b \in X$ which is a minimizer on X of $f - \langle x^*, \cdot \rangle + \varepsilon\|\cdot - b\|$, so from properties **Prop.4** and **Prop.3** there exists some $u \in X$ and $u^* \in \partial f(u)$ with $u^* - x^* \in 2\varepsilon \mathbb{B}_{X^*}$. This translates the density of $\partial f(X)$ in $(X^*, \|\cdot\|_*)$. □

6.2.2. Zagrodny mean value theorem. We state and prove now the Zagrodny mean value inequality theorem.

THEOREM 6.52 (D. Zagrodny sequential mean value inequality). *Let $(X, \|\cdot\|)$ be a Banach space, U be an open convex set of X and ∂ be a subdifferential on U satisfying* **Prop.1-Prop.4**. *Let $f : U \to \mathbb{R} \cup \{+\infty\}$ be a function which is lower semicontinuous on U and let $a, b \in U$ with $a \neq b$ and f finite at a. Then for any real $r \leq f(b)$ there exist $c \in [a, b[\cap \mathrm{dom}\, f$, $c_n \to_f c$ with $c_n \in U \cap \mathrm{Dom}\, \partial f$ and $c_n^* \in \partial f(c_n)$ (resp. $c \in [a, b[\cap \mathrm{dom}\, f$ and $c_n \to_f c$ with $c_n \in U \cap \mathrm{dom}\, f$) such that:*

(i) $\quad r - f(a) \leq \liminf\limits_{n \to \infty} \langle c_n^*, b - a \rangle$ (resp. $r - f(a) \leq \liminf\limits_{n \to \infty} \underline{d}_H^+ f(c_n; b - a)$);

(ii) *for every $y \in c + [0, +\infty[(b - a)$ one has*

$$\|y - c\| \frac{r - f(a)}{\|b - a\|} \leq \liminf_{n \to \infty} \langle c_n^*, y - c_n \rangle \quad (\text{resp. } \|y - c\| \frac{r - f(a)}{\|b - a\|} \leq \liminf_{n \to \infty} \underline{d}_H^+ f(c_n; y - c_n));$$

(iii) $\quad \|b - a\|(f(c) - f(a)) \leq \|c - a\|(r - f(a))$;

(iv) $\|c_n^*\| d(c_n, [a, b]) \to 0$ as $n \to \infty$ (resp. $\left(\sup_{\|h\| \leq 1}(-\underline{d}_H^+ f(c_n; h))^+\right) d(c_n, [a, b]) \to 0$ as $n \to \infty$, where $t^+ := \max\{t, 0\}$ for $t \in \mathbb{R} \cup \{-\infty, +\infty\}$).

PROOF. Take an open convex set $U_0 \supset [a, b]$ with $\mathrm{cl}\, U_0 \subset U$. Choose $z^* \in X^*$ such that $\langle z^*, b - a \rangle = \|b - a\|$ and put $\overline{f}(b) = r$ and $\overline{f}(x) = f(x)$ for all $x \neq b$ in U. The function $\overline{f}$ is still lower semicontinuous. The lower semicontinuous function φ, defined by

$$\varphi(x) = \overline{f}(x) - \frac{r - f(a)}{\|b - a\|} \langle z^*, x - a \rangle \quad \text{for all } x \in U,$$

satisfies the equality $\varphi(a) = \varphi(b)$, and hence attains its minimum on $[a, b]$ at some point $c \in [a, b[$. The functions φ and f are then finite at c. Writing $\varphi(c) \leq \varphi(b)$ easily gives (iii).

Now choose a real $\rho > 0$ such that φ is bounded from below over $V := [a, b] + \rho \mathbb{B}_X$ by some real γ and $V \subset U_0$. Put $\varphi_V(x) = \varphi(x)$ if $x \in V$ and $\varphi_V(x) = +\infty$ if $x \in U \setminus V$. Take for each integer $n \geq 1$ a real $\rho_n \in {]0, \rho[}$ such that $\varphi(x) \geq \varphi(c) - 1/n^2$ for all $x \in [a, b] + \rho_n \mathbb{B}_X$ and choose a real $t_n \geq n$ such that $\gamma + t_n \rho_n \geq \varphi(c) - 1/n^2$. Then one has

$$\varphi(c) \leq \inf_{x \in \mathrm{cl}\, U_0} \left(\varphi_V(x) + t_n d_{[a,b]}(x) \right) + \frac{1}{n^2}.$$

So, by the Ekeland variational principle (see Theorem 2.221) applied on the (complete metric) space $\mathrm{cl}\, U_0$ to the lower semicontinuous function $F_n := \varphi_V + t_n d_{[a,b]}$ with $\varepsilon = 1/n^2$ and $\lambda = 1/n$, there exists some $u_n \in \mathrm{cl}\, U_0$ such that

(6.18)
$$\|u_n - c\| \leq 1/n, \quad \varphi_V(u_n) \leq \varphi_V(u_n) + t_n d_{[a,b]}(u_n) = F_n(u_n) \leq F_n(c) = \varphi(c),$$

and $\quad F_n(u_n) \leq F_n(x) + \frac{1}{n} \|x - u_n\| \quad$ for all $x \in \mathrm{cl}\, U_0$.

We may suppose that $u_n \in \mathrm{int}\, V$ for all n. The lower semicontinuity of φ at $c \neq b$ combined with (6.18) entails that

$$(u_n, f(u_n)) \to (c, f(c)) \quad \text{and} \quad t_n d_{[a,b]}(u_n) \to 0 \text{ as } n \to \infty.$$

Without loss of generality, we may suppose $f(u_n)$ is finite for all n.

Let us establish first the properties relative to the lower Hadamard directional derivative $\underline{d}_H^+ f(\cdot; \cdot)$. For each integer n since u_n is a minimizer on $\mathrm{int}\, V$ of $\varphi + $

$t_n d_{[a,b]} + \frac{1}{n}\|\cdot - u_n\|$, we easily see, for all $h \in X$, that $\underline{d}_H^+ \varphi(u_n; h) + t_n d'_{[a,b]}(u_n; h) + \frac{1}{n}\|h\| \geq 0$, which (through the relation $u_n \neq b$) is equivalent to

(6.19) $\quad \underline{d}_H^+ f(u_n; h) - \dfrac{r - f(a)}{\|b - a\|}\langle z^*, h\rangle + t_n d'_{[a,b]}(u_n; h) + \dfrac{1}{n}\|h\| \geq 0.$

Choosing $q_n \in [a, b]$ satisfying $\|u_n - q_n\| = d_{[a,b]}(u_n)$, we have $\|u_n - q_n\| \leq \|u_n - c\|$ hence $q_n \to c \in [a, b[$, so, without loss of generality, we may suppose that $q_n \neq b$ for all n. Further, the convexity of $d_{[a,b]}$ ensures that $d'_{[a,b]}(u_n; \cdot)$ is sublinear hence

$$d'_{[a,b]}(u_n; b - q_n) \leq d'_{[a,b]}(u_n; b - u_n) + d'_{[a,b]}(u_n; u_n - q_n)$$
$$\leq d_{[a,b]}(b) - d_{[a,b]}(u_n) + \|u_n - q_n\|$$
$$= -d_{[a,b]}(u_n) + d_{[a,b]}(u_n) = 0.$$

From this, we obtain on the one hand $d'_{[a,b]}(u_n; b - a) \leq 0$ (since $b - a = \lambda(b - q_n)$ for some $\lambda > 0$) and using this and (6.19) with $h = b - a$ gives $r - f(a) \leq \liminf_{n \to \infty} \underline{d}_H^+ f(u_n; b - a)$, which is the part of (i) involving the lower Hadamard directional derivative. On the other hand, for any $y \in c + [0, +\infty[(b - a)$, from the previous inequality $d'_{[a,b]}(u_n; b - a) \leq 0$ we see that $d'_{[a,b]}(u_n; y - c) \leq 0$, thus

$$d'_{[a,b]}(u_n; y - u_n) \leq d'_{[a,b]}(u_n; y - c) + d'_{[a,b]}(u_n; c - u_n)$$
$$\leq d_{[a,b]}(c) - d_{[a,b]}(u_n) = -d_{[a,b]}(u_n) \leq 0.$$

Taking $h = y - u_n$ in (6.19) then produces

$$\|y - c\| \dfrac{r - f(a)}{\|b - a\|} \leq \liminf_{n \to \infty} \underline{d}_H^+ f(u_n; y - u_n)$$

since $\langle z^*, y - u_n \rangle \to \langle z^*, y - c \rangle = \|y - c\|$.

The inequality $d'_{[a,b]}(u_n; h) \leq \|h\|$ and (6.19) guarantee that

$$\sup_{\|h\|\leq 1} (-\underline{d}_H^+ f(u_n; h))^+ \leq (\beta + t_n)$$

for some constant $\beta \geq 0$ independent of n, and this justifies that

$$\left(\sup_{\|h\|\leq 1}(-\underline{d}_H^+ f(u_n; h))^+\right) d_{[a,b]}(u_n) \to 0 \quad \text{as } n \to \infty.$$

All the properties relative to the lower Hadamard directional derivative are then established with $c_n := u_n$.

It remains to prove the properties relative to the subdifferential. The minimizer property of $u_n \in \text{int } V$ for the function F_n allows us to apply the property **Prop. 4** of the subdifferential ∂ with $\varepsilon = 1/(nt_n)$ for $\overline{f}$ in place of f and $-\dfrac{r - f(a)}{\|b-a\|}\langle z^*, \cdot - a\rangle + t_n d_{[a,b]} + \frac{1}{n}\|\cdot - u_n\|$ in place of the continuous convex function g. Using the property **Prop. 3**, we then find $b_n^* \in \mathbb{B}_{X^*}$, $v_n^* \in \partial d_{[a,b]}(v_n)$ with $\|v_n - u_n\| < \frac{1}{nt_n}$ and $v_n \in U$, $c_n^* \in \partial \overline{f}(c_n)$ with $\|c_n - u_n\| < \frac{1}{nt_n}$, $c_n \in U$ and $|\overline{f}(c_n) - \overline{f}(u_n)| < \frac{1}{nt_n}$ such that

$$-c_n^* + \dfrac{r - f(a)}{\|b - a\|} z^* = t_n v_n^* + \dfrac{2}{n} b_n^*.$$

We observe that $\|v_n^*\| \leq 1$, and since $(u_n)_n$ and $(c_n)_n$ tend to $c \neq b$ as $n \to \infty$, for large n, say $n \geq n_0$, we have $\overline{f}(u_n) = f(u_n)$, $\overline{f}(c_n) = f(c_n)$, and $c_n^* \in \partial f(c_n)$ (see the property **Prop.2**). Put $x_n^* := c_n^* - \frac{r - f(a)}{\|b - a\|} z^*$. The convexity of $d_{[a,b]}$ gives

$$\langle v_n^*, b - v_n \rangle \leq d_{[a,b]}(b) - d_{[a,b]}(v_n) = -d_{[a,b]}(v_n) \leq 0.$$

Choosing $p_n \in [a, b]$ with $\|v_n - p_n\| = d_{[a,b]}(v_n)$ we have $p_n \to c$, so we may suppose $p_n \neq b$ for all $n \geq n_0$. Writing

$$\langle v_n^*, b - p_n \rangle = \langle v_n^*, b - v_n \rangle + \langle v_n^*, v_n - p_n \rangle \leq d_{[a,b]}(b) - d_{[a,b]}(v_n) + \|v_n^*\| \|v_n - p_n\|$$
$$\leq -d_{[a,b]}(v_n) + d_{[a,b]}(v_n) = 0,$$

we derive that $\langle t_n v_n^*, b - a \rangle \leq 0$. Note also by Proposition 4.102 that $v_n^* \in N([a, b]; p_n)$. Fix any $y \in c + [0, +\infty[(b - a)$, that is, $y - c = \mu(b - a)$ with $\mu \geq 0$. It ensues that for every $n \geq n_0$

$$\langle t_n v_n^*, y - p_n \rangle = \langle t_n v_n^*, y - c \rangle + t_n \langle v_n^*, c - p_n \rangle$$
$$\leq \langle t_n v_n^*, y - c \rangle = \mu \langle t_n v_n^*, b - a \rangle \leq 0.$$

We also observe from the inequalities $\langle t_n v_n^*, b - a \rangle \leq 0$ and $\|x_n^* + t_n v_n^*\| \leq 2/n$ that $\liminf_{n \to \infty} \langle x_n^*, b - a \rangle \geq 0$. Further, on the one hand we have $(c_n, f(c_n))_n$ tends to $(c, f(c))$ and $t_n d_{[a,b]}(c_n) \to 0$ because

$$t_n d_{[a,b]}(c_n) \leq t_n d_{[a,b]}(u_n) + t_n \|c_n - u_n\| \leq t_n d_{[a,b]}(u_n) + 1/n,$$

hence $\|x_n^*\| d_{[a,b]}(c_n) \to 0$ since $\|x_n^*\| \leq t_n + (2/n)$. On the other hand, writing

$$t_n \|p_n - v_n\| = t_n d_{[a,b]}(v_n) \leq t_n d_{[a,b]}(u_n) + t_n \|u_n - v_n\| \leq t_n d_{[a,b]}(u_n) + 1/n$$

yields $t_n \|p_n - v_n\| \to 0$, and hence $t_n \|p_n - c_n\| \to 0$ since

$$t_n \|p_n - c_n\| \leq t_n \|p_n - v_n\| + t_n \|v_n - c_n\| \leq t_n \|p_n - v_n\| + 2/n.$$

Using the inequality $\|x_n^* + t_n v_n^*\| \leq 2/n$ and observing that

$$\langle x_n^*, y - c_n \rangle = \langle x_n^* + t_n v_n^*, y - c_n \rangle - \langle t_n v_n^*, y - p_n \rangle - \langle t_n v_n^*, p_n - c_n \rangle$$
$$\geq \langle x_n^* + t_n v_n^*, y - c_n \rangle - \langle t_n v_n^*, p_n - c_n \rangle,$$

we deduce that $\liminf_{n \to \infty} \langle x_n^*, y - c_n \rangle \geq 0$.

Since $x_n^* = c_n^* - \frac{r - f(a)}{\|b - a\|} z^*$, we obtain $\langle c_n^*, b - a \rangle = \langle x_n^*, b - a \rangle + r - f(a)$ and

$$\liminf_{n \to \infty} \langle c_n^*, y - c_n \rangle = \liminf_{n \to \infty} \langle x_n^*, y - c_n \rangle + \lim_{n \to \infty} \frac{r - f(a)}{\|b - a\|} \langle z^*, y - c_n \rangle$$
$$\geq \frac{r - f(a)}{\|b - a\|} \langle z^*, y - c \rangle = \frac{r - f(a)}{\|b - a\|} \|y - c\|,$$

as well as $\|c_n^*\| d_{[a,b]}(c_n) \to 0$ as $n \to \infty$. All the properties of the theorem relative to the subdifferential ∂ then follow. □

REMARK 6.53. The above proof makes clear that the same sequence $(c_n)_n$ may be taken for both statements with subdifferential and lower Hadamard directional derivative whenever the subdifferential ∂ enjoys, in place of **Prop.4**, the exact sum subdifferential property: $\partial(f + g)(x) \subset \partial f(x) + \partial g(x)$ for all continuous convex function g. □

From the property (i) in Theorem 6.52 we directly deduce:

COROLLARY 6.54. *Let $(X, \|\cdot\|)$ be a Banach space, U be an open convex set of X and ∂ be a subdifferential on U satisfying* **Prop.1-Prop.4**. *Let $f : U \to \mathbb{R} \cup \{+\infty\}$ be a function which is lower semicontinuous on U and finite at $a \in U$. Then, for any $b \in U$ and any $\varepsilon > 0$, one has*

$$|f(b) - f(a)| \leq \|b - a\| \sup\{\|c^*\| : c^* \in \partial f(c), \, c \in [a, b] + \varepsilon \mathbb{B}_X\},$$

as well as

$$|f(b) - f(a)| \leq \sup\{|\underline{d}_H^+ f(c; b - a)| : c \in [a, b] + \varepsilon \mathbb{B}_X\}.$$

PROOF. Fix any $\varepsilon > 0$ and consider $\beta \in [0, +\infty]$ as the second member of the first (resp. second) inequality in the corollary. From the property (i) in Theorem 6.52 we have $r - f(a) \leq \beta$ for any real $r \leq f(b)$, so $f(b) - f(a) \leq \beta$. If $f(b)$ is finite, we can permute a, b to obtain $f(a) - f(b) \leq \beta$, hence $|f(b) - f(a)| \leq \beta$. In the case where $f(b) = +\infty$, then $f(b) - f(a) \geq 0$, thus $|f(b) - f(a)| = f(b) - f(a)$. In any case, we see that $|f(b) - f(a)| \leq \beta$. □

6.2.3. Subdifferential and tangential characterizations of Lipschitz properties.
From the above mean value inequalities, we first derive tangential and subdifferential characterizations of Lipschitz property.

THEOREM 6.55 (subdifferential and tangential characterizations of Lipschitz property). *Let X be a Banach space, U be an open convex set of X and ∂ be a subdifferential on U satisfying* **Prop.1-Prop.4**. *Let $f : U \to \mathbb{R} \cup \{+\infty\}$ be a function which is finite at some point in U and let $\gamma \geq 0$. Consider the following assertions:*
(a) *f is Lipschitz continuous on U with γ as a Lipschitz constant therein;*
(b) *f is lower semicontinuous on U and $\partial f(x) \subset \gamma \mathbb{B}_{X^*}$ for all $x \in U$;*
(c) *f is lower semicontinuous on U and $\underline{d}_H^+ f(x; h) \leq \gamma \|h\|$ for all $x \in U \cap \mathrm{dom}\, f$ and $h \in X$.*

Then (a) $\Leftrightarrow$ (c) and (b) $\Rightarrow$ (a); also (a) $\Leftrightarrow$ (b) whenever ∂f is included in the Clarke subdifferential.

PROOF. Assume (c) (resp. (b)). Fix any $x \in U$ where f is finite and take any $y \in U$ with $y \neq x$ and any real $r \leq f(y)$. Choose according to the compactness of $[x, y]$ some $\eta > 0$ such that $V := [x, y] + \eta \mathbb{B}_X \subset U$ and consider the lower semicontinuous function $\overline{f}$ defined on U by $\overline{f}(u) := f(u)$ if $u \in V$ and $\overline{f}(u) := +\infty$ otherwise. The Zagrodny mean value inequality applied to the function $\overline{f}$ furnishes $c_n \to_f c \in [x, y[$ with $c_n \in [x, y] + B(0, \eta) \subset U$ and $c_n^* \in \partial f(c_n)$ such that

$$r - f(x) \leq \underline{d}_H^+ f(c_n; y - x) \leq \gamma\|y - x\| \quad (\text{resp. } r - f(x) \leq \liminf_{n \to \infty} \langle c_n^*, y - x \rangle \leq \gamma\|y - x\|).$$

Consequently, on the one hand f is finite at y and hence on the whole set U, and on the other hand, for all $x, y \in U$, we obtain $f(y) - f(x) \leq \gamma\|y - x\|$, thus changing x and y yields $|f(y) - f(x)| \leq \gamma\|y - x\|$. This justifies the implications (c) $\Rightarrow$ (a) and (b) $\Rightarrow$ (a).

The implication (a) $\Rightarrow$ (c) is obvious and we know that $\partial_C f(x) \subset \gamma \mathbb{B}_{X^*}$ for all $x \in U$ whenever f is γ-Lipschitz continuous on the open set U (see Theorem 2.70(d)). The proof is then finished. □

The first corollary of the above theorem concerns the Lipschitz property on bounded sets.

COROLLARY 6.56. Let $f : X \to \mathbb{R} \cup \{+\infty\}$ be a proper lower semicontinuous function on a Banach space X and let ∂ be a subdifferential on X satisfying **Prop.1-Prop.4** and included in the Clarke subdifferential. The following are equivalent:
(a) the function f is Lipschitz on any bounded set;
(b) the multimapping $\partial_C f$ maps any bounded set on a bounded set;
(c) for any open bounded set U there is a real $\gamma_U \geq 0$ such that $\underline{d}_H^+ f(x; h) \leq \gamma_U \|h\|$ for all $x \in U \cap \operatorname{dom} f$ and $h \in X$.

If in addition f is convex, anyone of properties (d) and (e) below may be added to the list of equivalences:
(d) the function f is bounded on any bounded set;
(e) the function f is bounded from above on any bounded set.

PROOF. The equivalences between (a), (b) and (c) follow from Corollary 6.55. The implications (a)$\Rightarrow$(d) and (d)$\Rightarrow$(e) are evident. The final implication (e)$\Rightarrow$(a) results from Lemma 2.156. □

As a direct consequence of Theorem 6.55 with $\gamma = 0$ we obtain the following second corollary giving subdifferential and tangential characterizations of constant functions.

COROLLARY 6.57. Let U be an open convex set of a Banach space X, ∂ be a subdifferential on U satisfying **Prop.1-Prop.4**, and $f : U \to \mathbb{R} \cup \{+\infty\}$ be a proper function. Consider the following assertions:
(a) The function f is constant on U;
(b) the function f is lower semicontinuous on U and $\partial f(x) \subset \{0\}$ for all $x \in U$;
(c) the function f is lower semicontinuous on U and $\underline{d}_H^+ f(x; h) \leq 0$ for all $x \in U \cap \operatorname{dom} f$ and $h \in X$.
Then (a) $\Leftrightarrow$ (c) and (b) $\Rightarrow$ (a); also (a) $\Leftrightarrow$ (b) whenever ∂f is included in the Clarke subdifferential.

We establish now tangential and subdifferential characterizations of directional Lipschitz property. We prove first the following lemma.

LEMMA 6.58. Let $(X, \|\cdot\|)$ be a normed space, $f : X \to \mathbb{R} \cup \{+\infty\}$ be a lower semicontinuous function which is directionally Lipschitz at $\overline{x} \in X$ in a direction $\overline{h}$ with constants $\beta \in \mathbb{R}$, $\varepsilon > 0$, $\delta > 0$ satisfying

(6.20) $$t^{-1}[f(x + th) - f(x)] \leq \beta,$$

for all $t \in]0, \delta]$, $h \in B(\overline{h}, \delta)$, $x \in B(\overline{x}, \delta) \cap \operatorname{dom} f$ with $|f(x) - f(\overline{x})| < \varepsilon$.
The following properties hold:
(a) For any $x \in B(\overline{x}, \delta) \cap \operatorname{dom} f$ with $|f(x) - f(\overline{x})| < \varepsilon$ one has

(6.21) $$\sup_{b \in \delta \mathbb{B}_X} \underline{d}_H^+ f(x, \overline{h} + b) \leq \beta.$$

(b) If the subdifferential ∂f, for some subdifferential ∂ on X satisfying properties **Prop.1-Prop.4**, is included in the Clarke one, then

(6.22) $\langle x^*, \overline{h} \rangle + \delta \|x^*\| \leq \beta \quad \forall x^* \in \partial f(x), \forall x \in B(\overline{x}, \delta)$ with $|f(x) - f(\overline{x})| < \varepsilon$.

PROOF. Let $\beta \in \mathbb{R}$, $\varepsilon > 0$, $\delta > 0$ satisfying (6.20). Fix first any $x \in B(\overline{x}, \delta) \cap \operatorname{dom} f$ with $|f(x) - f(\overline{x})| < \varepsilon$. Obviously, $\underline{d}_H^+ f(x; h) \leq \beta$ for all $h \in B(\overline{h}, \delta)$ and the inequality still holds for all $h \in B[\overline{h}, \delta]$ by the lower semicontinuity property of $\underline{d}_H^+ f(x; \cdot)$. The assertion (a) is then established.

Now fix any $x \in B(\bar{x}, \delta) \cap \operatorname{Dom} \partial f$ with $|f(x) - f(\bar{x})| < \varepsilon$, and $x^* \in \partial f(x)$. Fix also any $b \in X$ with $\|b\| < 1$ and put $h = \bar{h} + \delta b$. It is easily seen that $(h, \beta) \in T^C(\operatorname{epi} f; (x, f(x)))$. Since $(x^*, -1) \in N^C(\operatorname{epi} f; (x, f(x)))$, we deduce $\langle x^*, \bar{h} + \delta b \rangle - \beta \leq 0$. Consequently $\langle x^*, \bar{h} \rangle + \delta \|x^*\| \leq \beta$, so the proof is complete. $\square$

THEOREM 6.59 (subdifferential and tangential characterizations of directionally Lipschitz functions). Let X be a Banach space, $f : X \to \mathbb{R} \cup \{+\infty\}$ be a lower semicontinuous function, $\bar{x} \in \operatorname{dom} f$, and $\bar{h} \in X$. Let ∂ be a subdifferential on X satisfying **Prop.1-Prop.4** and such that ∂f is included in the Clarke subdifferential of f. Then the following are equivalent:
(a) the function f is directionally Lipschitz at $\bar{x}$ in the direction $\bar{h}$ with constants $\beta, \varepsilon', \delta'$ satisfying (6.20);
(b) there are constants $\varepsilon > 0$ and $\delta > 0$ such that (6.21) is fulfilled with $\beta, \varepsilon, \delta$;
(c) there are constants $\varepsilon > 0$ and $\delta > 0$ such that (6.22) is fulfilled with $\beta, \varepsilon, \delta$.

PROOF. The above lemma guarantees the implications $(a) \Rightarrow (b)$ and $(a) \Rightarrow (c)$.
Let us prove $(c) \Rightarrow (a)$ (resp. $(b) \Rightarrow (a)$). Without loss of generality we may suppose that $\varepsilon < 1$ and $\delta < \varepsilon$, and that they are such that

$$f(\bar{x}) < f(x) + \varepsilon \quad \text{for all } x \in B(\bar{x}, \delta).$$

Take $\varepsilon' > 0$ such that $\varepsilon'(\|\bar{h}\| + |\beta| + 2) < \delta$, and $\delta' \in]0, \varepsilon'[$. Fix any $x \in B(\bar{x}, \delta')$ with $|f(x) - f(\bar{x})| < \varepsilon'$, any nonzero $h \in B(\bar{h}, \delta')$, and any $t \in]0, \varepsilon'[$. Consider $f(x + th)$ and suppose $f(x + th) > f(x) + t\beta$. We may choose and fix a real $\mu \in]0, 1[$ such that $f(x + th) > f(x) + t(\beta + \mu)$. Set $r := f(x) + t(\beta + \mu)$. Apply the Zagrodny mean value theorem to get some $c_n \to_f c \in [x, x + th[$ such that (iii) of that theorem is fulfilled and

$$r - f(x) \leq \liminf_{n \to \infty} \langle c_n^*, th \rangle$$

for some sequence $c_n^* \in \partial f(c_n)$

$$(\text{resp. } r - f(x) \leq \liminf_{n \to \infty} d_H^+ f(c_n; th)).$$

Then, for n large enough, $\|c_n - \bar{x}\| \leq \|c_n - c\| + \varepsilon'(\|\bar{h}\| + \delta') + \delta' < \delta$.

CASE I: $c = x$.
Then we have $f(c_n) \to f(x)$ and for n large enough

$$|f(c_n) - f(\bar{x})| \leq |f(c_n) - f(x)| + |f(x) - f(\bar{x})| \leq |f(c_n) - f(x)| + \varepsilon' < \varepsilon$$

CASE II: $c \neq x$.
Then the property (iii) of the Zagrodny mean value theorem gives

$$t\|h\|(f(c) - f(x)) \leq \|c - x\|(r - f(x)).$$

We rewrite the latter inequality as

$$t\|h\|(f(c) - f(\bar{x})) \leq \|c - x\|(r - f(x)) + t\|h\|(f(x) - f(\bar{x}))$$
$$\leq \|c - x\|(r - f(x)) + t\|h\|\varepsilon'$$
$$\leq t\|h\|(r - f(x) + \varepsilon'),$$

from where

$$f(c) - f(\bar{x}) \leq r - f(x) + \varepsilon' = t(\beta + \mu) + \varepsilon'$$
$$\leq t(|\beta| + \mu) + \varepsilon' \leq \varepsilon'(|\beta| + \mu + 1) \leq \varepsilon'(|\beta| + 2).$$

Therefore, for sufficiently large n, we can write
$$f(c_n) - f(\overline{x}) = f(c_n) - f(c) + f(c) - f(\overline{x}) \leq |f(c_n) - f(c)| + \varepsilon'(|\beta| + 2) < \varepsilon.$$
Further, for n large enough, the points $c_n \in B(\overline{x}, \delta)$ and hence the choice of δ guarantees $f(\overline{x}) - f(c_n) < \varepsilon$. Therefore, we obtain $|f(c_n) - f(\overline{x})| < \varepsilon$.

In both cases I and II we have for n large enough, $c_n \in B(\overline{x}, \delta)$ and $|f(c_n) - f(\overline{x})| < \varepsilon$, then (6.22) gives $\langle c_n^*, h \rangle \leq \beta$ (resp. (6.21) gives $\underline{d}_H^+ f(c_n; h) \leq \beta$). So,
$$r - f(x) \leq \liminf_{n \to \infty} \langle c_n^*, th \rangle \leq t\beta, \quad \left(\text{resp. } r - f(x) \leq \liminf_{n \to \infty} \underline{d}_H^+ f(c_n; th) \leq t\beta \right),$$
that is, $t(\beta + \mu) \leq t\beta$. This means $\mu \leq 0$, which is a contradiction, since $\mu > 0$. We conclude $f(x + th) \leq f(x) + t\beta$, and further, if $0 \in B(\overline{h}, \delta)$, the inequality still holds for $h = 0$ by the lower semicontinuity property of f. This finishes the proof of the theorem. □

The following first corollary provides characterizations of sets with the interior tangent condition through properties of tangent and normal cones. Before stating the corollary, recall that the lower Hadamard directional derivative coincides with the Bouligand directional derivative.

COROLLARY 6.60 (tangential and normal characterizations of epi-Lipschitz sets). Let S be a subset of a Banach space X which is closed near $\overline{x} \in S$ and let $\overline{h} \in X$. Let ∂ be a subdifferential on X satisfying **Prop.1-Prop.4** and such that the ∂-normal set $N^\partial(S; \cdot) := \partial \Psi_S(\cdot)$ is included in the Clarke normal cone of S. Then the following properties are equivalent:
(a) the set S has the interior tangent property at $\overline{x}$ in the direction $\overline{h}$;
(b) there exist some $\delta > 0$ and some neighborhood U of $\overline{x}$ such that
$$\overline{h} + \delta \mathbb{B}_X \subset T^C(S; x) \quad \text{for all } x \in U \cap S;$$
(c) there exist some $\delta > 0$ and some neighborhood U of $\overline{x}$ such that
$$\overline{h} + \delta \mathbb{B}_X \subset T^B(S; x) \quad \text{for all } x \in U \cap S;$$
(d) there exist some $\delta > 0$ and some neighborhood U of $\overline{x}$ such that
$$\langle x^*, \overline{h} \rangle + \delta \|x^*\| \leq 0 \quad \text{for all } x \in U \cap S \text{ and } x^* \in N^\partial(S; x);$$
(e) there exists some $\delta > 0$ and some neighborhood U of $\overline{x}$ such that
$$\overline{h} + \delta \mathbb{B}_X \subset \left(N^\partial(S; x) \right)^\circ \quad \text{for all } x \in U \cap S,$$
where $\left(N^\partial(S; x) \right)^\circ$ denotes the negative polar cone of the set $N^\partial(S; x)$.

PROOF. Without loss of generality, we may suppose that S is closed.
(a) $\Rightarrow$ (b). If S has the interior tangent property at $\overline{x}$ in the direction $\overline{h}$, by definition of such a property we have some $\delta > 0$ and some open neighborhood U of $\overline{x}$ such that
$$S \cap U +]0, \delta[B(\overline{h}, \delta) \subset S.$$
This immediately guarantees that $B(\overline{h}, \delta) \subset T^C(S; x)$ for all $x \in U \cap S$, which ensures (b).
(b) $\Rightarrow$ (c) This implication is obvious since $T^C(S; x) \subset T^B(S; x)$.
(c) $\Rightarrow$ (a) Recall that a set S has the interior tangent property at a point $\overline{x} \in S$ in a direction $\overline{h}$ if and only if its indicator function Ψ_S is directionally Lipschitz at $\overline{x}$ in the direction $\overline{h}$ and recall also that the Bouligand directional derivative

of Ψ_S is the indicator of the Bouligand tangent cone. So, the desired implication directly follows from the characterization (b) in Theorem 6.59 above applied with the indicator function Ψ_S in place of the function f.

(a) $\Leftrightarrow$ (d) Obviously (d) implies (a) by the characterization (c) in Theorem 6.59 applied again with Ψ_S in place of the function f.

Suppose now (a) holds. The same characterization in terms of the Clarke subdifferential of Ψ_S furnishes some $\beta \in \mathbb{R}$, $\delta > 0$, and some neighborhood U of $\bar{x}$ such that $\langle x^*, \overline{h}\rangle + \delta\|x^*\| \leq \beta$ for all $x^* \in N^C(S;x)$ with $x \in U \cap S$. Fix any $x \in U \cap S$ and any $x^* \in N^C(S;x)$. Since $N^C(S;x)$ is a cone, for any real $t > 0$ we have $t^{-1}x^* \in N^C(S;x)$ hence

$$\langle t^{-1}x^*, \overline{h}\rangle + \delta\|t^{-1}x^*\| \leq \beta, \text{ that is, } \langle x^*, \overline{h}\rangle + \delta\|x^*\| \leq t\beta.$$

Taking the limit as $t \downarrow 0$, we obtain $\langle x^*, \overline{h}\rangle + \delta\|x^*\| \leq 0$. Since $N^\partial(S;x) \subset N^C(S;x)$, the latter inequality holds in particular for all $x^* \in N^\partial(S;x)$, showing that (a) entails (d).

(e) $\Rightarrow$ (a) Suppose (e) and take any $x \in U \cap S$, $x^* \in N^\partial(S;x)$. For all $b \in \mathbb{B}_X$ we have $\overline{h} + \delta b \in \left(N^\partial(S;x)\right)^\circ$, then $\langle x^*, \overline{h} + \delta b\rangle \leq 0$. Consequently,

$$\langle x^*, \overline{h}\rangle + \delta\|x^*\| = \sup_{b \in \mathbb{B}_X} \langle x^*, \overline{h} + \delta b\rangle \leq 0,$$

that is, (d) holds. So, the desired implication (e) $\Rightarrow$ (a) is true since (d) and (a) are equivalent.

(a) $\Rightarrow$ (e) Assume (a) and choose some $\delta > 0$ and some neighborhood U of $\bar{x}$ satisfying (b) according to the equivalence above between (a) and (b). Fix any $x \in U \cap S$. We have $\overline{h} + \delta \mathbb{B}_X \subset T^C(S;x)$. Further, the inclusion $N^\partial(S;x) \subset N^C(S;x)$ implies

$$T^C(S;x) = \left(N^C(S;x)\right)^\circ \subset \left(N^\partial(S;x)\right)^\circ.$$

We can conclude that $\overline{h} + \delta\mathbb{B}_X \subset \left(N^\partial(S;x)\right)^\circ$, which finishes the proof of the corollary. $\square$

As a second direct corollary with $\overline{h} = 0$ we derive, in addition to Theorem 6.55, the following characterization of the Lipschitz property of a function near a fixed point. Here instead of requiring that the image under ∂f of a whole neighborhood of $\bar{x}$ be included in a closed ball centered at zero, the property is assumed merely for a graphical neighborhood of $\bar{x}$.

COROLLARY 6.61. Let U be an open set of a Banach space X. Let $f : U \to \mathbb{R} \cup \{+\infty\}$ be a function which is finite at $\bar{x} \in U$ and lower semicontinuous near $\bar{x}$, and let $\gamma \geq 0$. Let ∂ be a subdifferential on U satisfying **Prop.1-Prop.4** and such that ∂f is included in the Clarke subdifferential of f. Then the following properties are pairwise equivalent:

(a) The function f is γ-Lipschitz continuous near $\bar{x}$;

(b) there exists $\delta > 0$ such that for all $x \in B(\bar{x}, \delta)$ with $|f(x) - f(\bar{x})| < \delta$ one has

$$\sup_{b \in \mathbb{B}_X} d_H^+ f(x;b) \leq \gamma;$$

(c) there exists $\delta > 0$ such that for all $x \in B(\bar{x}, \delta) \cap \text{Dom}\,\partial f$ with $|f(x) - f(\bar{x})| < \delta$ one has

$$\partial f(x) \subset \gamma \mathbb{B}_{X^*}.$$

The above characterization involves the subdifferentials at nearby points of $\bar{x}$. In the finite-dimensional setting, a useful characterization of the Lipschitz continuity of f at $\bar{x}$ through the Clarke subdifferential of f at merely the reference point $\bar{x}$ can be deduced.

COROLLARY 6.62. Assume that the normed space X is finite-dimensional. Let $f : U \to \mathbb{R} \cup \{+\infty\}$ be a function on an open set U of X which is finite at $\bar{x} \in U$ and lower semicontinuous near $\bar{x}$. Then the following properties are pairwise equivalent:
(a) The function f is Lipschitz continuous near $\bar{x}$;
(b) $\partial_C f(\bar{x})$ is nonempty and bounded;
(c) $\partial_L^\infty f(\bar{x}) = \{0\}$.

PROOF. The assertion (a) obviously implies (b), and the equality $\partial_C f(\bar{x}) = \overline{\mathrm{co}}(\partial_L f(\bar{x}) + \partial_L^\infty f(\bar{x}))$ (see Theorem 4.120) ensures that (b) entails (c) since $\partial_L^\infty f(\bar{x})$ is a cone of X^* containing zero. To show that (c) implies (a), we proceed by contradiction. Suppose that f is not Lipschitz continuous near $\bar{x}$. Applying the assertion (c) of Corollary 6.61 with the Fréchet subdifferential, we obtain a sequence $(x_n)_n$ converging to $\bar{x}$ with $(f(x_n))_n$ tending to $f(\bar{x})$ and a sequence $(x_n^*)_n$ with $\|x_n^*\| \to +\infty$ as $n \to \infty$ and $x_n^* \in \partial_F f(x_n)$ for all $n \in \mathbb{N}$. We may suppose $\|x_n^*\| \neq 0$ for all n. The inclusion $x_n^* \in \partial_F f(x_n)$ means $(x_n^*, -1) \in N^F(\mathrm{epi}\, f; (x_n, f(x_n)))$, or equivalently
$$(\|x_n^*\|^{-1} x_n^*, -\|x_n^*\|^{-1}) \in N^F(\mathrm{epi}\, f; (x_n, f(x_n))).$$
Extracting a subsequence of $(\|x_n^*\|^{-1} x_n^*)_n$ converging to a vector u^* in X with $\|u^*\| = 1$, the latter inclusion guarantees that $(u^*, 0) \in N^L(\mathrm{epi}\, f; (\bar{x}, f(\bar{x})))$, that is, $u^* \in \partial_L^\infty f(\bar{x})$. This justifies the implication (c) $\Rightarrow$ (a) and finishes the proof. $\square$

EXAMPLE 6.63. The characterization (b) of the latter corollary does not hold with the Mordukhovich limiting subdifferential $\partial_L f(\bar{x})$ in place of the Clarke subdifferential $\partial_C f(\bar{x})$. Indeed the simple continuous function $f : \mathbb{R} \to \mathbb{R}$, with $f(x) = 0$ if $x \leq 0$ and $f(x) = -\sqrt{x}$ otherwise, is not Lipschitz continuous near $\bar{x} := 0$ but $\partial_L f(\bar{x}) = \{0\}$. Observe also that $\partial_C f(\bar{x}) = \partial_L^\infty f(\bar{x}) =]-\infty, 0]$. $\square$

REMARK 6.64. If X is an Asplund space and f is lower semicontinuous near $\bar{x}$, then the assertion (a) of Corollary 6.62 (that is, the Lipschitz continuity of f near $\bar{x}$) is equivalent to anyone of the following conditions (b') and (c') in place of (b) and (c) respectively:
(b') The epigraph of f is compactly epi-Lipschitzian at $(\bar{x}, f(\bar{x}))$ and $\partial_C f(\bar{x})$ is nonempty and bounded;
(c') The epigraph of f is compactly epi-Lipschitzian at $(\bar{x}, f(\bar{x}))$ and $\partial_L^\infty f(\bar{x}) = \{0\}$.

For the implication (b') $\Rightarrow$ (c') nothing has to be added and regarding (a) $\Rightarrow$ (b') we only need to keep in mind that the Lipschitz continuity of f near $\bar{x}$ guarantees that epi f is epi-Lipschitz at $(\bar{x}, f(\bar{x}))$, hence compactly epi-Lipschitzian at that point. Now suppose that (a) does not hold. We keep the sequences $(x_n)_n$ in X and $(x_n^*)_n$ in X^* as generated in the proof of Corollary 6.62. For $u_n^* := \|x_n^*\|^{-1} x_n^*$, the Asplund property of X allows us to extract from $(u_n^*)_n$ a subsequence (that we do not relabel) converging weakly star to some $u^* \in X^*$, so $(u^*, 0) \in N^L(\mathrm{epi}\, f; (\bar{x}, f(\bar{x})))$, that is, $u^* \in \partial_L^\infty f(\bar{x})$. From the compactly epi-Lipschitzian property of f (see Proposition 5.85), we have $u^* \neq 0$, hence the condition (c') does not hold. This justifies the implication (c') $\Rightarrow$ (a). $\square$

The following non-boundedness property is directly derived from Corollary 6.62. It is an extension of the main implication $\Rightarrow$ in Proposition 2.211 to nonconvex functions.

COROLLARY 6.65. *Let $f: U \to \mathbb{R} \cup \{+\infty\}$ be a lower semicontinuous function on an open set U of a finite-dimensional normed space X. Then at any point $\bar{x} \in \mathrm{bdry}(\mathrm{dom}\, f)$ either $\partial_C f(\bar{x})$ is empty or $\partial_C f(\bar{x})$ is unbounded.*

The next corollary is related to subdifferential conditions for the strict F-differentiability in finite dimensions. It can be seen as a nonconvex result partially in the line of Proposition 3.281.

COROLLARY 6.66. *Assume that the normed space X is finite-dimensional. Let $f: U \to \mathbb{R} \cup \{+\infty\}$ be a function on an open set U of X which is finite at $\bar{x} \in U$ and lower semicontinuous near $\bar{x}$. Then the following properties are pairwise equivalent:*
(a) *The function f is strictly Hadamard (or equivalently strictly Fréchet) differentiable $\bar{x}$;*
(b) *$\partial_C f(\bar{x})$ is a singleton;*
(c) *$\partial_L f(\bar{x})$ is a singleton and $\partial_L^\infty f(\bar{x}) = \{0\}$.*

PROOF. The implication (a)$\Rightarrow$(b) follows from the assertion (e) in Proposition 2.50(e), saying that the C-subdifferential is reduced to the gradient at any point of strict H-differentiability. The equivalence (b)$\Leftrightarrow$(c) is due to the formula

$$\partial_C f(\bar{x}) = \overline{\mathrm{co}}\big(\partial_L f(\bar{x}) + \partial_L^\infty f(\bar{x})\big)$$

(see Theorem 4.120) and the cone property of $\partial_L^\infty f(\bar{x})$). Finally, (b) entails that f is Lipschitz near $\bar{x}$ by Corollary 6.62, hence the implication (b)$\Rightarrow$(a) results from Proposition 2.76 characterizing the strict H-differentiability at a point $\bar{x}$ where f is Lipschitz around as the singleton property of $\partial_C f(\bar{x})$. $\square$

Notice that Example 6.63 tells us that the condition $\partial_L^\infty f(\bar{x}) = \{0\}$ cannot be removed from (c) in Corollary 6.66.

6.2.4. Subdifferential and tangential characterizations of monotonicity and convexity. Classical analysis says that a real-valued derivable function of one real variable with nonpositive derivative on an interval is nonincreasing on this interval. The next proposition generalizes this result to the nonincreasing property of a lower semicontinuous function with respect to a cone. Given a subset C of a vector space X, a function $f: C \to \mathbb{R} \cup \{+\infty\}$ is said to be K-nonincreasing on C, for a cone K of X, whenever $f(y) \leq f(x)$ for all $x, y \in C$ with $y - x \in K$.

PROPOSITION 6.67. *Let U be an open convex set of a Banach space X and ∂ be a subdifferential on X satisfying **Prop.1-Prop.4**. Let $f: U \to \mathbb{R} \cup \{+\infty\}$ be a proper lower semicontinuous function and K be a nonempty cone of X. Consider the assertions:*
(a) *The function f is $\overline{\mathrm{co}}\, K$-nonincreasing on U;*
(b) *the function f is K-nonincreasing on U;*
(c) *$\underline{d}_H^+ f(x; h) \leq 0$ for all $x \in U \cap \mathrm{dom}\, f$ and $h \in K$;*
(d) *$\partial f(x) \subset K^\circ$ for all $x \in U$.*

Then (b) $\Leftrightarrow$ (c) $\Leftrightarrow$ (a) $\Leftarrow$ (d). Further, the four assertions are pairwise equivalent whenever ∂f is included in the Clarke subdifferential of f.

PROOF. The implication (a) ⇒ (b) is obvious. To see that (b) entails (c), fix any $x \in U \cap \operatorname{dom} f$ and $h \in K$. Choosing $\delta > 0$ such that $x+]0,\delta[h \subset U$, we observe that
$$t^{-1}[f(x+th) - f(x)] \leq 0 \quad \text{for all } t \in\,]0,\delta[,$$
thus in particular $\underline{d}_H^+ f(x;h) \leq 0$.

Let us show (c) ⇒ (b) (resp. (d) ⇒ (a)). Fix $x, y \in U$ with $x \neq y$, $f(x) < +\infty$ and $y - x \in K$ (resp. $y - x \in \overline{\operatorname{co}}\, K$). For any real $r \leq f(y)$ the Zagrodny mean value inequality gives $c_n \to_f c \in [x, y]$ with $c_n \in U \cap \operatorname{dom} f$ (resp. $c_n \to_f c \in [x, y[$ with $c_n \in U \cap \operatorname{dom} f$ and $c_n^* \in \partial f(c_n)$) such that
$$r - f(x) \leq \liminf_{n \to \infty} \underline{d}_H^+ f(c_n; y - x) \leq 0 \quad (\text{resp. } r - f(x) \leq \liminf_{n \to \infty} \langle c_n^*, y - x \rangle \leq 0),$$
which obviously entails $f(y) \leq f(x)$.

Now we prove that (b) entails the inclusion $\partial_C f(U) \subset (\overline{\operatorname{co}}\, K)^\circ$. Assume (b) and take $x \in U \cap \operatorname{Dom} \partial_C f$, $x^* \in \partial_C f(x)$, and $h \in K$. First, let us show that $(h,0) \in T^C(\operatorname{epi} f; (x, f(x)))$. Take any sequences $(t_n)_n$ tending to 0 with $t_n > 0$ and $(x_n, r_n)_n$ in $\operatorname{epi} f$ converging to $(x, f(x))$. Choosing an integer N such that $x_n + t_n h \in U$ for all $n \geq N$, we obtain by the K-nonincreasing property of f on U
$$f(x_n + t_n h) \leq f(x_n) \leq r_n,$$
which gives $(x_n, r_n) + t_n(h, 0) \in \operatorname{epi} f$ for all $n \geq N$. This justifies the inclusion $(h, 0) \in T^C(\operatorname{epi} f; (x, f(x)))$. Since $x^* \in \partial_C f(x)$ means
$$(x^*, -1) \in (T^C(\operatorname{epi} f; (x, f(x))))^\circ,$$
we deduce $\langle x^*, h \rangle \leq 0$. Consequently, $x^* \in K^\circ = (\overline{\operatorname{co}}\, K)^\circ$, as desired.

Denoting by (d') the property (d) with ∂_C in place of ∂, we have proved (d) ⇒ (a), (b) ⇔ (c) and (a) ⇒ (b) ⇒ (d') ⇒ (a). So (a) – (c) are equivalent, and if, in addition $\partial \subset \partial_C$, those properties are also equivalent to (d) since in that case by what precedes (a) ⇒ (d') ⇒ (d) ⇒ (a). This finishes the proof. □

We establish now some very useful characterizations of convexity for lower semicontinuous functions defined on Banach spaces.

THEOREM 6.68 (subdifferential and tangential characterizations of convex functions). Let U be an open convex set of a Banach space X and ∂ be a subdifferential on U satisfying properties **Prop.1-Prop.4**. Let $f : U \to \mathbb{R} \cup \{+\infty\}$ be a proper lower semicontinuous function. The following assertions are equivalent:
(a) the function f is convex;
(b) for any $x \in U \cap \operatorname{dom} f$ one has
$$\underline{d}_H^+ f(x; y - x) \leq f(y) - f(x) \quad \text{for all } y \in U;$$
(c) for any $x, y \in U \cap \operatorname{dom} f$, $\underline{d}_H^+ f(x; y - x) + \underline{d}_H^+ f(y; x - y)$ is well defined and
$$\underline{d}_H^+ f(x; y - x) + \underline{d}_H^+ f(y; x - y) \leq 0;$$
(d) for any $y \in U$, $x \in U \cap \operatorname{Dom} \partial f$ and $x^* \in \partial f(x)$
$$\langle x^*, y - x \rangle \leq f(y) - f(x);$$
(e) the operator ∂f is monotone, that is, for all $x, y \in U \cap \operatorname{Dom} \partial f$, $x^* \in \partial f(x)$, and $y^* \in \partial f(y)$, one has
$$\langle x^* - y^*, x - y \rangle \geq 0.$$

PROOF. (a) ⇒ (d) and (a) ⇒ (b). Assume that f is convex. On one hand, from property **Prop.3** we know that ∂f coincides with the subdifferential of f in the standard sense for convex functions, which justifies (a) ⇒ (d).

On the other hand, for any $x, y \in U \cap \operatorname{dom} f$ the convexity of f easily gives by the slope inequality (or as easily seen) $t^{-1}[f(x+t(y-x))-f(x)] \leq f(y)-f(x)$ for all $t \in]0,1]$. We deduce $\underline{d}_H^+ f(x; y-x) \leq f(y) - f(x)$, hence we have proved the desired implications (a) ⇒ (d) and (a) ⇒ (b).

(b) ⇒ (c). For $x, y \in U \cap \operatorname{dom} f$, we have by (b) the inequalities $\underline{d}_H^+ f(x; y-x) \leq f(y) - f(x)$ and $\underline{d}_H^+ f(y; x-y) \leq f(x) - f(y)$, and adding both inequalities gives

$$\underline{d}_H^+ f(x; y-x) + \underline{d}_H^+ f(y; x-y) \leq 0,$$

where the left member is well defined thanks to both former inequalities. So, the implication (b) ⇒ (c) is established.

(d) ⇒ (e). This inequality is obvious.

(e) ⇒ (a) and (d) ⇒ (a). Fix as above $x, y \in U \cap \operatorname{dom} f$ with $x \neq y$ and $s, t \in]0, 1[$. Put $z := sx + ty$ and take any reals $\rho < r < f(z)$. Applying the Zagrodny mean value theorem on $[x, z[$ to the function f we obtain some $c \in [x, z[\cap \operatorname{dom} f$, $c_n \to_f c$ as $n \to \infty$ with $c_n \in U$, and $c_n^* \in \partial f(c_n)$ (resp. some $c \in [x, z[\cap \operatorname{dom} f$ and $c_n \to_f c$ with $c_n \in U \cap \operatorname{dom} f$) such that for all n

$$\left\langle c_n^*, \frac{y - c_n}{\|y - c_n\|} \right\rangle > \frac{\rho - f(x)}{\|z - x\|} = \frac{\rho - f(x)}{t\|x - y\|}$$

$$\left(\text{resp. } \underline{d}_H^+ f\left(c_n; \frac{y - c_n}{\|y - c_n\|}\right) > \frac{\rho - f(x)}{\|z - x\|} = \frac{\rho - f(x)}{t\|x - y\|}\right),$$

according to (ii) of the Zagrodny mean value theorem. Choose $\alpha \in [0, 1[$ such that $z = \alpha c + (1 - \alpha)y$ and put $z_n := \alpha c_n + (1 - \alpha)y$. Since $z_n \to z$ and f is lower semicontinuous at z we may suppose that $\rho < r < f(z_n)$ for all n. For each n we apply again the Zagrodny mean value theorem on $[y, z_n[$ to obtain some $e_n \in [y, z_n[\cap \operatorname{dom} f$, $e_{n,k} \to_f e_n$ as $k \to \infty$ with $e_{n,k} \in U$, and $e_{n,k}^* \in \partial f(e_{n,k})$ (resp. $e_n \in [y, z_n[\cap \operatorname{dom} f$, $e_{n,k} \to_f e_n$ as $k \to \infty$ with $e_{n,k} \in \operatorname{dom} f$) such that

$$\left\langle e_{n,k}^*, \frac{c_n - e_{n,k}}{\|c_n - e_{n,k}\|} \right\rangle > \frac{\rho - f(y)}{\|z_n - y\|} = \frac{\rho - f(y)}{s_n \|x - y\|} = \frac{t\rho - tf(y)}{s_n t\|x - y\|}$$

$$\left(\text{resp. } \underline{d}_H^+ f\left(e_{n,k}; \frac{c_n - e_{n,k}}{\|c_n - e_{n,k}\|}\right) > \frac{\rho - f(y)}{\|z_n - y\|} = \frac{\rho - f(y)}{s_n \|x - y\|} = \frac{t\rho - tf(y)}{s_n t\|x - y\|}\right),$$

where s_n is chosen so that $s_n \|x - y\| = \|z_n - y\|$, hence $s_n \to s$ as $n \to \infty$. Noting that

$$\left\langle c_n^*, \frac{e_n - c_n}{\|e_n - c_n\|} \right\rangle = \left\langle c_n^*, \frac{y - c_n}{\|y - c_n\|} \right\rangle > \frac{\rho - f(x)}{t\|x - y\|}$$

$$\left(\text{resp. } \underline{d}_H^+ f\left(c_n; \frac{e_n - c_n}{\|e_n - c_n\|}\right) = \underline{d}_H^+ f\left(c_n; \frac{y - c_n}{\|y - c_n\|}\right) > \frac{\rho - f(x)}{t\|x - y\|}\right),$$

we can choose for each n some $\sigma(n) \geq n$ such that for $d_n := e_{n, \sigma(n)}$ and $d_n^* := e_{n, \sigma(n)}^*$

$$\left\langle c_n^*, \frac{d_n - c_n}{\|d_n - c_n\|} \right\rangle > \frac{\rho - f(x)}{t\|x - y\|} = \frac{s_n \rho - s_n f(x)}{s_n t\|x - y\|}$$

$\Big($resp. by the lower semicontinuiy of $\underline{d}_H^+ f(c_n;\cdot)\Big)$

$$\underline{d}_H^+ f\left(c_n; \frac{d_n - c_n}{\|d_n - c_n\|}\right) > \frac{\rho - f(x)}{t\|x - y\|} = \frac{s_n\rho - s_n f(x)}{s_n t\|x - y\|}\Big).$$

Since we also have

$$\left\langle d_n^*, \frac{c_n - d_n}{\|d_n - c_n\|}\right\rangle > \frac{t\rho - tf(y)}{s_n t\|x - y\|} \quad \left(\text{resp. } \underline{d}_H^+ f(d_n; \frac{c_n - d_n}{\|d_n - c_n\|}) > \frac{t\rho - tf(y)}{s_n t\|x - y\|}\right),$$

we deduce according to the assumption in (e) (resp. (c)) that

$$0 \geq \left\langle c_n^* - d_n^*, \frac{d_n - c_n}{\|d_n - c_n\|}\right\rangle > \frac{(s_n + t)\rho - s_n f(x) - tf(y)}{s_n t\|x - y\|},$$

$\Big($resp.

$$0 \geq \frac{1}{\|d_n - c_n\|}\big(\underline{d}_H^+ f(c_n; d_n - c_n) + \underline{d}_H^+ f(d_n; c_n - d_n)\big) > \frac{(s_n + t)\rho - s_n f(x) - tf(y)}{s_n t\|x - y\|}\Big),$$

which ensures $0 > (s_n + t)\rho - s_n f(x) - tf(y)$. Since $s_n \to s$ as $n \to \infty$ and ρ is arbitrarily less than $f(z)$, we conclude that $f(z) \leq sf(x) + tf(y)$. This finishes the proof. $\square$

Through the choice of f as the indicator function Ψ_S of a closed set we can establish the following characterizations of convexity of sets.

COROLLARY 6.69 (tangential and normal characterizations of convex sets). Let S be a closed set of a Banach space X and let ∂ be a subdifferential on X satisfying properties **Prop.1-Prop.4**. Let $N^\partial(S;\cdot) := \partial \Psi_S(\cdot)$ be the associated ∂-normal set. The following properties are pairwise equivalent:
(a) The set S is convex;
(b) the inclusion $S \subset x + T^C(S; x)$ holds for all $x \in S$;
(c) the inclusion $S \subset x + T^B(S, x)$ holds for all $x \in S$;
(d) for the normal set $N^\partial(S;\cdot) = \partial \Psi_S$ and for all $x, y \in S$, $x^* \in N^\partial(S; x)$, $y^* \in N^\partial(S; y)$ one has

$$\langle x^* - y^*, x - y\rangle \geq 0;$$

(e) the set S satisfies the inclusion

$$S \subset \{u \in X : \langle x^*, u - x\rangle \leq 0, \forall (x, x^*) \in \mathrm{gph}\, N^\partial(S;\cdot)\}.$$

PROOF. (a) $\Rightarrow$ (b) If S is convex, for any $x \in S$ we know that $\mathrm{cl}\big(\mathbb{R}_+(S - x)\big) = T^C(S; x)$, so the inclusion $S - x \subset T^C(S; x)$ holds.
(b) $\Rightarrow$ (c) This implication is obvious since we always have $T^C(S; x) \subset T^B(S; x)$ for all $x \in S$.
(c) $\Rightarrow$ (a) Suppose (c) is fulfilled. Then for all $x, y \in S$ we have $y - x \in T^B(S; x)$ and $x - y \in T^B(S; y)$, or equivalently $\Psi_{T^B(S;x)}(y - x) + \Psi_{T^B(S;y)}(x - y) \leq 0$. Since $\Psi_{T^B(S;x)}(\cdot) = (\Psi_S)^B(x;\cdot) = \underline{d}_H^+(\Psi_S)(x;\cdot)$, the latter inequality means

$$\underline{d}_H^+(\Psi_S)(x; y - x) + \underline{d}_H^+(\Psi_S)(y, x - y) \leq 0 \quad \text{for all } x, y \in \mathrm{dom}\, \Psi_S.$$

Theorem 6.68 then says that the lower semicontinuous function Ψ_S is convex, that is, the set S is convex.
(a) $\Leftrightarrow$ (d) This equivalence follows directly from the equivalence between (a) and (e) in Theorem 6.68 by taking for f the indicator function of S.
(d) $\Rightarrow$ (e) Suppose (d) holds and fix $x \in S$. Since (d) is equivalent to (a), by

property **Prop.3** we have $N^\partial(S;x) = \{x^* \in X^* : \langle x^*, y - x \rangle \leq 0 \ \forall y \in S\}$. Thus $S \subset \{u \in X : \langle x^*, u - x \rangle \leq 0, \forall (x, x^*) \in \mathrm{gph}\, N^\partial(S;\cdot)\}$, which is the property (e).
(e) $\Rightarrow$ (d) Suppose (e) and take $x, y \in S$, and $x^* \in N^\partial(S;x)$, $y^* \in N^\partial(S;y)$. By (e) we have $\langle x^*, y - x \rangle \leq 0$ and $\langle y^*, x - y \rangle \leq 0$. Consequently, $\langle x^* - y^*, x - y \rangle \geq 0$, which finishes the proof of the corollary. $\square$

Given a real $\sigma > 0$ and a proper function $f : U \to \mathbb{R} \cup \{+\infty\}$ on an open convex set U of a Hilbert space H it is clear that for any $x \in \mathrm{dom}\, f$ we have for every $h \in H$

$$d_H^+\left(f - \frac{\sigma}{2}\|\cdot\|^2\right)(x; h) = d_H^+ f(x; h) - \sigma \langle x, h \rangle \text{ and } \partial\left(f - \frac{\sigma}{2}\|\cdot\|^2\right)(x) = -\sigma x + \partial f(x),$$

where ∂f is either $\partial_C f$, or $\partial_L f$, or $\partial_A f$, or ∂_F, or ∂_P. The following corollary then follows directly from Theorem 6.68 and from Proposition 3.273 for strongly convex functions.

COROLLARY 6.70 (subdifferential and tangential characterizations of strongly convex functions). Let $f : U \to \mathbb{R} \cup \{+\infty\}$ be a proper lower semicontinuous function on a nonempty convex set U of a Hilbert space H and let ∂f be either the C-subdifferential, or L-subdifferential, or A-subdifferential, or F-subdifferential, or P-subdifferential. Given a real $\sigma > 0$ the following are equivalent:
(a) the function f is σ-strongly convex on U;
(b) for any $y \in U$ and any $x \in \mathrm{dom}\, f$ one has

$$d_H^+ f(x; y - x) \leq f(y) - f(x) - \frac{\sigma}{2}\|y - x\|^2;$$

(c) for any $x, y \in \mathrm{dom}\, f$ one has

$$d_H^+ f(x; y - x) + d_H^+ f(y; x - y) \leq -\sigma\|x - y\|^2;$$

(d) for any $y \in U$, any $x \in \mathrm{dom}\, \partial f$ and any $v \in \partial f(x)$ one has

$$\langle v, y - x \rangle \leq f(y) - f(x) - \frac{\sigma}{2}\|y - x\|^2;$$

(e) for any (x, v) and (x', v') in $\mathrm{gph}\, \partial f$ one has

$$\langle v - v', x - x' \rangle \geq \sigma\|x - x'\|^2,$$

that is, the subdifferential multimapping ∂f is σ-strongly monotone on U.

6.3. Approximate and sequential Rolle-type theorems

Let ∂ be a subdifferential (in the sense of the previous subsection) on a class $\mathcal{F}(U)$ (resp. on U), where U is an open set of the normed space X. Given a subset $V \supset U$ and a function $f : V \to \mathbb{R} \cup \{-\infty, +\infty\}$ with its restriction $f_{|U}$ in $\mathcal{F}(U)$, we will write $\partial f(x)$ in place of $\partial f_{|U}(x)$ for all $x \in U$.

First, we observe that, for finite-dimensional vector spaces, the classical Rolle theorem can be easily extended as follows.

THEOREM 6.71 (Rolle theorem with several variables). Let X be a finite-dimensional vector space, U be a nonempty open bounded set of X and $f : \mathrm{cl}\, U \to \mathbb{R}$ be a continuous function such that f is constant on the boundary $\mathrm{bdry}\, U$ of U. Let ∂ be a subdifferential with properties **Prop.1**-**Prop.4** on a class $\mathcal{F}(U)$ containing the restrictions to U of both functions f and $-f$. Then there exists some $z \in U$ such that $0 \in \partial f(z) \cup \partial(-f)(z)$.

PROOF. The proof follows the usual argument. By compactness of cl U, the continuous function f attains on cl U its minimum m_1 at z_1 and its maximum m_2 at z_2. If $m_1 = m_2$, then f is constant on cl U, so $0 \in \partial f(z)$ for every $z \in U$. Suppose $m_1 < m_2$. Since cl $U = U \cup \text{bdry}\, U$ and f is constant on bdry U, it follows that at least one of z_1, z_2 is in U. It results by **Prop.1** that $0 \in \partial f(z_1)$ or $0 \in \partial(-f)(z_2)$. □

Such a result fails in infinite dimensions even for smooth functions as shown in the following example.

EXAMPLE 6.72 (S.A. Shkarin). Consider $X = L^2([0,1])$ the Lebesgue space of classes of squared integrable functions on $[0,1]$ endowed with its usual inner product $\langle x, y \rangle := \int_0^1 x(t) y(t) \, dt$ and the associated norm $\|\cdot\|$. Let $A : X \to X$ be the continuous linear operator defined by $A(x)(t) := tx(t)$ for all $x \in X$ and let $\varphi(t) = t(1-t)$ for all $t \in [0, 1]$. For each $x \in X$, put
$$f(x) = (1 - \|x\|^2) g(x),$$
where $g(x) = \langle A(x), x \rangle + 2\langle \varphi, x \rangle + c$ and c is some real constant that will be determined later. Obviously, the function f is infinitely Fréchet differentiable on X and $f(x) = 0$ for all $x \in X$ with $\|x\| = 1$.

The first step is to show that it is possible to choose c in such a way that $g(x) > 0$ for all $x \in X$. Fix any $x \in X$ with $\|x\| = 1$ and any $\lambda \in \mathbb{R}$, and note that
$$g(\lambda x) = \lambda^2 \langle A(x), x \rangle + 2\lambda \langle \varphi, x \rangle + c.$$
Requiring $c > 0$ to be such that $\varphi(t) < \sqrt{ct}$ for all $t \in [0, 1]$ except for finitely many $t \in [0, 1]$, we have (according to the continuity of both functions)
$$\langle \varphi, x \rangle^2 < c \left(\int_0^1 |x(t)| \sqrt{t} \, dt \right)^2, \text{ hence } \langle \varphi, x \rangle^2 < c \int_0^1 |x(t)|^2 t \, dt,$$
so
$$\inf_{\lambda \in \mathbb{R}} g(\lambda x) = c - \frac{\langle \varphi, x \rangle^2}{\langle A(x), x \rangle} > c - c \left(\int_0^1 |x(t)|^2 t \, dt \right) \left(\int_0^1 |x(t)|^2 t \, dt \right)^{-1} \geq 0.$$
This yields $g(x) > 0$ for all $x \in X$. Observing that a real $t \in \,]0, 1]$ satisfies $\varphi(t) < \sqrt{ct}$ if and only if $t(1-t)^2 < c$, it is easily seen that the best choice for c is $\max_{0 < t \leq 1} t(1-t)^2 = \frac{4}{27}$ which is attained at $t = 1/3$. We make this choice $c := \frac{4}{27}$ and note that $\varphi(t) < \sqrt{ct}$ for all $t \in \,]0, 1] \setminus \{1/3\}$.

The second step is to show that $\nabla f(x) \neq 0$ whenever $\|x\| < 1$. Note first that, for all $x \in X$,
$$\nabla f(x) = 2(1 - \|x\|^2)(A(x) + \varphi) - 2\big(\langle A(x), x \rangle + 2\langle \varphi, x \rangle + c\big) x.$$
Therefore, if $\nabla f(u) = 0$ for some $u \in X$ with $\|u\| < 1$, we see that $A(u) + \varphi = \mu u$ for the real $\mu = \frac{g(u)}{1 - \|u\|^2} > 0$. Then $u(t) = \frac{\varphi(t)}{\mu - t}$ for almost all $t \in [0, 1]$, which ensures

(6.23) $\qquad \mu \geq 1 \quad$ since $u \in L^2[0, 1]$.

Using the expression above of $\nabla f(x)$, we also have
$$(1 - \|u\|^2)(A(u) + \varphi) = \big(\langle A(u), u \rangle + 2\langle \varphi, u \rangle + c \big) u,$$
so
$$\mu \left(1 - \int_0^1 \frac{\varphi(t)^2}{(\mu - t)^2} \, dt \right) u = \left(\int_0^1 \left[\frac{\varphi(t)^2 t}{(\mu - t)^2} + 2 \frac{\varphi(t)^2}{\mu - t} \right] dt + c \right) u.$$

Noting that $u \neq 0$ because $\nabla f(0) = 2\varphi \neq 0$, the latter is equivalent to

$$\mu = \int_0^1 \frac{\varphi(t)^2 \mu}{(\mu - t)^2} \, dt + \int_0^1 \left[\frac{\varphi(t)^2 t}{(\mu - t)^2} + 2 \frac{\varphi(t)^2}{\mu - t} \right] dt + c,$$

which gives

$$\mu = \zeta(\mu) + c \quad \text{where} \quad \zeta(\mu) := \int_0^1 \frac{t^2(1-t)^2(3\mu - t)}{(\mu - t)^2} \, dt.$$

Since the function ζ is nonincreasing on $[1, +\infty[$, we have $\sup_{r \geq 1} \zeta(r) = \zeta(1) = 3/4$. It follows that $\mu = \zeta(\mu) + c \leq \frac{3}{4} + \frac{4}{27}$ hence $\mu < 1$ which contradicts the observation $\mu \geq 1$ in (6.23). The contradiction so obtained says that the gradient of the corresponding smooth function f is not zero at any point of the open unit ball. $\square$

Despite the above counter example in infinite Hilbert space, we show below that *approximate* Rolle theorems are valid in infinite-dimensional Banach spaces even for nondifferentiable real-valued functions.

THEOREM 6.73 (approximate Rolle-type theorem). *Let $(X, \|\cdot\|)$ be a Banach space and U be a nonempty open set of X with $U \neq X$ and let $a \in U$ and $r := d(a, X \setminus U)$. Let also $f : \operatorname{cl} U \to \mathbb{R}$ be a bounded continuous function such that $f(\operatorname{bdry} U) \subset [\gamma - \varepsilon, \gamma + \varepsilon]$ for some constants $\gamma \in \mathbb{R}$ and $\varepsilon > 0$ and let ∂ be a subdifferential on a class $\mathcal{F}(U)$ satisfying* **Prop.1-Prop.4** *and containing the restrictions to U of both functions f and $-f$. Then, one has*

$$\inf\{\|z^*\| : z^* \in \partial f(z) \cup \partial(-f)(z), \ z \in U\} \leq 2\varepsilon/r.$$

The proof will use both lemmas below.

LEMMA 6.74. *Let U be a nonempty open set of a Banach space $(X, \|\cdot\|)$ with $U \neq X$ and $f : \operatorname{cl} U \to \mathbb{R}$ be a function which is lower semicontinuous, bounded from below and such that $\inf_{x \in \operatorname{cl} U} f(x) < \inf_{x \in \operatorname{bdry} U} f(x)$. Let also ∂ be a subdifferential on a class $\mathcal{F}(U)$ satisfying* **Prop.1-Prop.4** *and containing the restriction of f to U. Then for any real $\eta > 0$ there exist some point $z \in U$ which is a local minimizer of $f + \eta \|\cdot - z\|$ on U and some $z' \in U$ and $z^* \in \partial f(z')$ such that $\|z^*\| < \eta$.*

PROOF. Let $\eta > 0$. Consider the real number $\delta := \inf_{x \in \operatorname{bdry} U} f(x) - \inf_{x \in \operatorname{cl} U} f(x) > 0$ and choose some $a \in \operatorname{cl} U$ such that

(6.24) $$f(a) < \delta + \inf_{x \in \operatorname{cl} U} f(x).$$

By definition of δ, this is equivalent to $f(a) < \inf_{x \in \operatorname{bdry} U} f(x)$, hence $a \in U$. For each real $\sigma \in \,]0, \eta[$, the Ekeland variational principle (see Theorem 2.221) gives some $b \in \operatorname{cl} U$ such that

$$f(b) + \sigma\|b - a\| \leq f(a), \quad \|b - a\| \leq \delta/\sigma$$

and $\quad f(b) \leq f(x) + \sigma\|x - b\| \quad$ for all $x \in \operatorname{cl} U$.

If $b = a$, then $b \in U$; and if $b \neq a$ we have $f(a) \geq f(b) + \sigma\|b - a\| > f(b)$, so (6.24) yields

$$f(b) < \delta + \inf_{x \in \operatorname{cl} U} f(x) = \inf_{x \in \operatorname{bdry} U} f(x),$$

thus $b \in U$. In any case $b \in U$, and hence b is a local minimizer of $f + \sigma \|\cdot - b\|$ on U, thus also of $f + \eta \|\cdot - b\|$. So, $z := b$ fulfills the first conclusion of the lemma.

Furthermore, since $b \in U$ is a local minimizer of $f + \sigma \| \cdot - b \|$ on U, according to Property **Prop.4** of the subdifferential ∂, there are $z', z'' \in U$ such that

$$0 \in \partial f(z') + \sigma \partial \| \cdot - b \| (z'') + \frac{1}{2}(\eta - \sigma) \mathbb{B}_{X^*}.$$

Consequently, by Property **Prop.3** there is some $z^* \in \partial f(z')$ such that $\|z^*\| < \eta$, which translates the second conclusion of the lemma. □

LEMMA 6.75. *Let U be a nonempty open set of a Banach space $(X, \| \cdot \|)$ with $U \neq X$ and let $a \in U$ and $r := d(a, X \setminus U)$. Let $f : \operatorname{cl} U \to \mathbb{R}$ be a lower semicontinuous function such that $f(\operatorname{cl} U) \subset [-\varepsilon, \varepsilon]$ for some real $\varepsilon > 0$. Then for any real $\eta > 0$, there exist some Lipschitz convex function h with 1 as a Lipschitz constant, some positive real $\lambda < (2\varepsilon/r) + \eta$, and some $z \in U$ which is a local minimizer of $f + \lambda h$ on U.*

PROOF. Let $\eta > 0$. For any real $\sigma > 0$, consider the lower semicontinuous function $F : \operatorname{cl} U \to \mathbb{R}$ defined by $F(x) := f(x) + r^{-1}(2\varepsilon + \sigma)\|x - a\|$ for all $x \in \operatorname{cl} U$ and note that F is bounded from below on $\operatorname{cl} U$. For each $x \in \operatorname{bdry} U$, we have $\|x - a\| \geq r$, thus $F(x) \geq f(x) + (2\varepsilon + \sigma) \geq f(a) + \sigma = F(a) + \sigma$, where the second inequality is due to the fact that $f(x) - f(a) \geq -2\varepsilon$. This entails that $\inf_{x \in \operatorname{cl} U} F(x) < \inf_{x \in \operatorname{bdry} U} F(x)$, so the first conclusion of Lemma 6.74 above gives some $u \in U$ which is a local minimizer of $F + \sigma \| \cdot - u \|$ on U. Consequently, it suffices to choose the positive real σ to satisfy $\sigma(1 + r^{-1}) < \eta$ in order to get the desired property. □

PROOF OF THEOREM 6.73. Fix any real $\eta > 0$.

If $\inf_{x \in \operatorname{cl} U} f(x) < \inf_{x \in \operatorname{bdry} U} f(x)$, the second conclusion of Lemma 6.74 provides some $z_1 \in U$ and $z_1^* \in \partial f(z_1)$ with $\|z_1^*\| < \eta$; and if $\sup_{x \in \operatorname{bdry} U} f(x) < \sup_{x \in \operatorname{cl} U} f(x)$, then we can apply the latter conclusion to the lower semicontinuous function $(-f)$ to get some $z_2 \in U$ and $z_2^* \in \partial(-f)(z_2)$ with $\|z_2^*\| < \eta$.

If $\inf_{x \in \operatorname{cl} U} f(x) = \inf_{x \in \operatorname{bdry} U} f(x)$ and $\sup_{x \in \operatorname{cl} U} f(x) = \sup_{x \in \operatorname{bdry} U} f(x)$, then, for $g(x) = f(x) - \gamma$, the assumption $g(\operatorname{bdry} U) \subset [-\varepsilon, +\varepsilon]$ guarantees that $g(\operatorname{cl} U) \subset [-\varepsilon, \varepsilon]$. Lemma 6.75 above gives some Lipschitz convex function h with 1 as a Lipschitz constant, some positive real $\lambda < (2\varepsilon/r) + \eta$, and some $z \in U$ which is a local minimizer of $g + \lambda h$ on U. From Properties **Prop.3** and **Prop.4** of the subdifferential we obtain some $z_3 \in U$ and $z_3^* \in \partial f(z_3)$ with $\|z_3^*\| < \eta + 2r^{-1}\varepsilon$.

In any case, we obtain some $z \in U$ and $z^* \in \partial f(z) \cup \partial(-f)(z)$ with $\|z^*\| < \eta + 2r^{-1}\varepsilon$, thus

$$\inf\{\|z^*\| : z^* \in \partial f(z) \cup \partial(-f)(z),\ z \in U\} \leq 2\varepsilon/r,$$

which finishes the proof of the theorem. □

A first direct consequence is the following sequential Rolle-type theorem:

THEOREM 6.76 (sequential Rolle-type theorem). *Let $(X, \| \cdot \|)$ be a Banach space and U be a nonempty open set of X with $U \neq X$. Let also $f : \operatorname{cl} U \to \mathbb{R}$ be a bounded continuous function which is constant on the boundary of U, and let ∂*

be a subdifferential on a class $\mathcal{F}(U)$ satisfying **Prop.1-Prop.4** and containing the restrictions to U of both f and $-f$. Then, one has
$$\inf\{\|z^*\| : z^* \in \partial f(z) \cup \partial(-f)(z),\ z \in U\} = 0,$$
or equivalently there are sequences $(z_n)_n$ in U and $(z_n^*)_n$ in X^* with $z_n^* \to 0$ strongly as $n \to \infty$ and such that either $z_n^* \in \partial f(z_n)$ for all $n \in \mathbb{N}$ or $z_n^* \in \partial(-f)(z_n)$ for all integers $n \in \mathbb{N}$.

Now considering a real-valued continuous bounded function f on the whole space X, that is, $\sup_{x \in X} |f(x)| \leq \varepsilon$ for some real $\varepsilon > 0$, we obtain from Theorem 6.73 that, for every real $r > 0$,
$$\inf\{\|z^*\| : z^* \in \partial f(z) \cup \partial(-f)(z),\ z \in U_r\} \leq 2\varepsilon/r,$$
where $U_r := B(0, r)$. So, for every real $r > 0$,
$$\inf\{\|z^*\| : z^* \in \partial f(z) \cup \partial(-f)(z),\ z \in X\} \leq 2\varepsilon/r,$$
hence
$$\inf\{\|z^*\| : z^* \in \partial f(z) \cup \partial(-f)(z),\ z \in X\} = 0.$$
We state this in the following corollary:

COROLLARY 6.77. *Let $(X, \|\cdot\|)$ be a Banach space and $f : X \to \mathbb{R}$ be a function which is continuous and bounded on X. Let ∂ be a subdifferential on a class $\mathcal{F}(X)$ satisfying* **Prop.1-Prop.4** *and containing both functions f and $-f$. Then, one has*
$$\inf\{\|z^*\| : z^* \in \partial f(z) \cup \partial(-f)(z),\ z \in X\} = 0,$$
or equivalently there are sequences $(z_n)_n$ in X and $(z_n^*)_n$ in X^* with $z_n^* \to 0$ strongly as $n \to \infty$ and such that either $z_n^* \in \partial f(z_n)$ for all $n \in \mathbb{N}$ or $z_n^* \in \partial(-f)(z_n)$ for all $n \in \mathbb{N}$.

We have already observed (see Proposition 4.13(b)) that $-\partial_F(-f)(x) \subset \partial_F f(x)$ whenever $\partial_F f(x) \neq \emptyset$ and that the same property holds true with the variational proximal subdifferential (see Proposition 4.143(b)); it is also easy to see that the property holds true as well for the Hadamard subdifferential and the Dini subdifferential. Considering the general case when $-\partial(-f)(x) \subset \partial f(x)$ whenever $\partial f(x) \neq \emptyset$ produces the following third corollary.

COROLLARY 6.78. *Let $(X, \|\cdot\|)$ be a Banach space and U be a nonempty open set of X with $U \neq X$. Let also $f : \operatorname{cl} U \to \mathbb{R}$ be a bounded continuous function which is constant on the boundary of U. The following hold:*
(a) *If ∂ is a subdifferential on a class $\mathcal{F}(U)$ satisfying* **Prop.1-Prop.4** *and containing both $f|_U$ and $-f|_U$ and such that $-\partial(-\varphi)(x) \subset \partial\varphi(x)$ whenever $\partial\varphi(x) \neq \emptyset$ with $\varphi \in \mathcal{F}(U)$, and if f is ∂-subdifferentiable on U (that is, $\partial f(x) \neq \emptyset$ for all $x \in U$), then*
$$\inf\{\|z^*\| : z^* \in \partial f(U)\} = 0.$$
(b) *If X is an Asplund space and f is Fréchet subdifferentiable on U, then $\inf\{\|z^*\| : z^* \in \partial_F f(U)\} = 0$.*
(c) *If X is a Hilbert space and f is subdifferentiable on U with respect to the proximal subdifferential, then $\inf\{\|z^*\| : z^* \in \partial_P f(U)\} = 0$.*
(d) *If X has a Gâteaux differentiable renorm off zero and if f is ∂_H-subdifferentiable (resp. ∂_D-subdifferentiable) on U, then $\inf\{\|z^*\| : z^* \in \partial_H f(U)\} = 0$ (resp.*

$\inf\{\|z^*\| : z^* \in \partial_D f(U)\} = 0)$.
(e) If f is Gâteaux differentiable on U, then $\inf\{\|D_G f(z)\| : z \in U\} = 0$.
(f) If f is locally Lipschitz on U, then $\inf\{\|z^*\| : z^* \in \partial_C f(U)\} = 0$.

PROOF. Under the assumptions of (a), for any $z \in U$ we have $\partial f(z) \neq \emptyset$ hence, according to the assumptions, $-\partial(-f)(z) \subset \partial f(z)$; the same inclusion also holds under the assumption of (f). This yields that $\{\|z^*\| : z^* \in \partial f(U) \cup \partial(-f)(U)\}$ coincides with $\{\|z^*\| : z^* \in \partial f(U)\}$, and this combined with Theorem 6.76 produces the conclusions of (a) and (f).

The assertions (b), (c) and (d) follow from (a) and from the comments preceding the statement of the corollary.

It remains to prove (e). Suppose that f is Gâteaux differentiable on U and consider for $\mathcal{F}(U)$ the class of functions $h + g$ where $h : U \to \mathbb{R}$ is Gâteaux differentiable on U and $g : U \to \mathbb{R}$ is the restriction to U of a finite-valued continuous convex function on X. Note that $\mathcal{F}(U)$ contains f and $-f$ as well as all the restrictions to U of finite-valued continuous convex functions. On this class $\mathcal{F}(U)$ we know (see the examples of subdifferentials in the beginning of the previous subsection) that the operator ∂ given by $\partial f(x) = D_G h(x) + \partial_{\text{conv}} g(x)$ for all $x \in U$ is well defined and is a subdifferential (here, to avoid confusion, the symbol $\partial_{\text{conv}} g(x)$ is used to denote the standard convex subdifferential of the convex function g). Since $\partial(-f)(x) = -D_G f(x) = -\partial f(x)$ for all $x \in U$, we can apply Theorem 6.76 to obtain the desired property. □

REMARK 6.79. Of course, using Theorem 6.73 in place of Theorem 6.76 in the arguments of the proof of the latter corollary produces the assertions (a)-(f) in ε-forms whose statements are left to the reader. □

6.4. Multidirectional mean value inequalities

For a real $\delta > 0$ and a set S of a normed space X, in addition to notation $\text{Enl}_\delta S$ in Definition 3.259 it will be convenient in this section (as it will be clear in a moment) to denote also by $B_\delta[S]$ the closed δ-enlargement of S, that is,

$$B_\delta[S] := \{x \in X : d_S(x) \leq \delta\}.$$

DEFINITION 6.80. Let S be a nonempty set of a normed space $(X, \|\cdot\|)$ and $f : X \to \mathbb{R} \cup \{-\infty, +\infty\}$ be an extended real-valued function. We define

$$\linf_S f := \lim_{\delta \downarrow 0} \inf_{x \in B_\delta[S]} f(x) = \sup_{\delta > 0} \inf_{x \in B_\delta[S]} f(x)$$

which is the limit as $\delta \downarrow 0$ of the infimum of f over the δ-enlargement of S; of course, $\linf_S f$ is invariant with respect to any norm equivalent to $\|\cdot\|$. Sometimes, it may be convenient to denote $\linf_{x \in S} f(x)$ in place of $\linf_S f$.

As first properties of $\linf_S f$ we have:

PROPOSITION 6.81. Let f, g be two extended real-valued functions on a normed space $(X, \|\cdot\|)$ and S, S' be subsets of X. The following properties hold.
(a) $\linf_S f \leq \inf_S f$;
(b) $(S \subset S' \Rightarrow \linf_{S'} f \leq \linf_S f)$ and $(f \leq g$ on $B_\delta[S]$ for some $\delta \Rightarrow \linf_S f \leq \linf_S g)$;
(c) $\linf_S (r + f) = r + \linf_S f$ for any constant $r \in \mathbb{R}$, $\linf_S (rf) = r\linf_S f$ for any real constant $r > 0$, and $\linf_S f + \linf_S g \leq \linf_S (f + g)$ whenever the sums are well

defined;
(d) for $S = \{\bar{x}\}$, one has $\operatorname{linf}_S f = \liminf_{x \to \bar{x}} f(x)$;
(e) $d_S(x_n) \to 0 \Rightarrow \operatorname{linf}_S f \leq \liminf_{n \to \infty} f(x_n)$;
(f) there exists a sequence $(x_n)_n$ in X with $d_S(x_n) \to 0$ such that $\operatorname{linf}_S f = \lim_{n \to \infty} f(x_n)$.

PROOF. The assertions (a)-(e) are direct consequences of the definition. Let us check (f). Put $\lambda := \operatorname{linf}_S f$ and $\delta_n := 1/n$. Suppose $\lambda > -\infty$. For each $n \in \mathbb{N}$ the inequality $\lambda \geq \inf_{B_{\delta_n}[S]} f$ yields some $x_n \in B_{\delta_n}[S]$ such that $\inf_{B_{\delta_n}[S]} f \leq f(x_n) \leq \lambda + \delta_n$, thus $\lambda = \lim_{n \to \infty} f(x_n)$ and $d_S(x_n) \to 0$. If $\lambda = -\infty$, for each $n \in \mathbb{N}$, we have $\inf_{B_{\delta_n}[S]} f = -\infty$ so that there exists $x_n \in B_{\delta_n}[S]$ such that $f(x_n) < -\delta_n^{-1}$, and this also guarantees that $\lambda = \lim_{n \to \infty} f(x_n)$ and $d_S(x_n) \to 0$. □

Regarding the converse inequality of $\operatorname{linf}_S f \leq \inf_S f$, it is worth observing the following:

PROPOSITION 6.82. *Let $(X, \|\cdot\|)$ be a normed space and f be an extended real-valued function on X. Under anyone of conditions (a), (b) and (c) below one has the equality $\operatorname{linf}_S f = \inf_S f$.*
(a) *The function f is real-valued and uniformly continuous on $B_{\delta_0}[S]$ for some $\delta_0 > 0$.*
(b) *For some topology τ weaker than the norm topology, the set S is τ-compact and the function f is τ-lower semicontinuous at each point of S.*
(c) *For some topology τ weaker than the norm topology, the set S is τ-closed and the function f is τ-inf-compact relative to $B_{\delta_0}[S]$ for some $\delta_0 > 0$.*

PROOF. It is enough to prove $\inf_S f \leq \operatorname{linf}_S f$.
(a) Fix any $\varepsilon > 0$ and choose $0 < \delta' < \delta_0$ such that $|f(u) - f(u')| \leq \varepsilon$ for all $u, u' \in B_{\delta_0}[S]$ with $\|u - u'\| \leq \delta'$. Let $0 < \delta < \delta'$. Taking any $u \in B_\delta[S]$ there exists $x \in S$ with $\|x - u\| < \delta'$, which gives

$$f(u) \geq f(x) - \varepsilon \geq \inf_S f - \varepsilon, \quad \text{hence} \quad \inf_{B_\delta[S]} f \geq \inf_S f - \varepsilon.$$

The latter entails $\operatorname{linf}_S f \geq \inf_S f - \varepsilon$, hence $\operatorname{linf}_S f \geq \inf_S f$.
(b) and (c). Suppose that $\operatorname{linf}_S f < +\infty$ and fix any real $r < \operatorname{linf}_S f$. According to the definition of $\operatorname{linf}_S f$, for each $n \in \mathbb{N}$, we can choose $x_n \in S$ and $v_n \in (1/n)\mathbb{B}_X$ such that $f(x_n + v_n) < r$. By the assumption (b) or (c) we can take a subnet of $(x_n + v_n)_n$ τ-converging to some $\bar{x} \in S$. From the τ-lower semicontinuity of f at $\bar{x}$ we obtain

$$\inf_{x \in S} f(x) \leq f(\bar{x}) \leq r,$$

and we conclude that $\inf_S f \leq \operatorname{linf}_S f$. □

In Theorem 6.52 we established a mean value inequality related to $r - f(a)$ for any real $r \leq f(b)$. In this section we are interested in a similar inequality related to $r - f(a)$ for any real $r \leq \operatorname{linf}_S f$, that is, we are concerned with a closed convex subset S of X in place of the point b.

In the remainder of this section, X will be a Banach space and $\mathcal{F}(X)$ and $\mathcal{F}(X \times \mathbb{R})$ will be two classes on which is (simultaneously) generated a subdifferential ∂ in the sense of Section 6.2. *Any result involving ∂f will implicitly assume either $f \in \mathcal{F}(X)$ or $f \in \mathcal{F}(X \times \mathbb{R})$*. In addition to Properties **Prop.1-Prop.4** we also require that X admits an equivalent norm $\|\cdot\|$ such that with respect to $\|\cdot\|$ and to the norm $(\|x\|^2 + |r|^2)^{1/2}$ on $X \times \mathbb{R}$ the following properties **Prop.4'** and **Prop.5** hold:

Prop.4': for $f \in \mathcal{F}(X)$ (resp. $f \in \mathcal{F}(X \times \mathbb{R})$) lower semicontinuous near x in X (resp. in $X \times \mathbb{R}$) one has

$$\partial(f + \zeta)(x) \subset \partial f(x) + \partial \zeta(x)$$

whenever ζ is a function on X (resp. on $X \times \mathbb{R}$) of the form $\zeta(x) := K d_S^2(x) + \sum_{n \in \mathbb{N}} \alpha_n \|x - a_n\|^2$, where S is a closed convex set of X (resp. of $X \times \mathbb{R}$), K is a real $K \geq 0$, the sequence $(a_n)_n$ is a convergent sequence of X (resp. of $X \times \mathbb{R}$), and $\Sigma \alpha_n$ is a convergent series of reals $\alpha_n \geq 0$; so, it is implicitely assumed that such functions ζ belong to $\mathcal{F}(X)$ (resp. $\mathcal{F}(X \times \mathbb{R})$).

With respect to $\mathcal{F}(X \times \mathbb{R})$ the subdifferential ∂ will be assumed to satisfy also the property:

Prop.5: for $f \in \mathcal{F}(X)$, $\alpha \in \mathbb{R}$ and $\varphi(x,t) = f(x) + \alpha t$ for $(x,t) \in X \times \mathbb{R}$, the function $\varphi \in \mathcal{F}(X \times \mathbb{R})$ and $\partial \varphi(x,t) \subset \partial f(x) \times \{\alpha\}$.

- Properties **Prop.4'** and **Prop.5** are obviously fulfilled whenever $\mathcal{F}(X)$ and $\mathcal{F}(X \times \mathbb{R})$ are the classes of convex functions and ∂ is the standard subdifferential for convex functions.
- The properties are also obviously fulfilled for the classes of all functions on X and on $X \times \mathbb{R}$ when ∂ is either the Clarke subdifferential and X is a normed space (see Proposition 2.108 and Theorem 2.98), or the Mordukhovich limiting subdifferential and X is an Asplund space (see Proposition 4.73 and Theorem 4.78), or the Ioffe approximate subdifferential and X is a Banach space (see Proposition 5.67 and Theorem 5.78).
- The properties are fulfilled for the classes of all functions on X and $X \times \mathbb{R}$ whenever X is a reflexive Banach space and ∂ is the Fréchet subdifferential ∂_F.

Indeed, the reflexive Banach space X admits an equivalent norm $\|\cdot\|$ with $\|\cdot\|^2$ Fréchet differentiable on X (see Theorem C.11 in Appendix) and so is the corresponding norm on $X \times \mathbb{R}$ as constructed above. Then for any closed convex set S of X (resp. $X \times \mathbb{R}$) the function d_S^2 is Fréchet differentiable (see Corollary 4.14) hence so are functions of the form ζ on X (resp. on $X \times \mathbb{R}$) in **Prop.4'**. Then for any such a function ζ, we have $\partial_F (f + \zeta)(x) = \partial_F f(x) + D\zeta(x)$ for all x, where $D\zeta(x)$ denotes the Fréchet derivative of ζ at x; this translates Property **Prop.4'**. Regarding Property **Prop.5** it suffices to apply Proposition 4.19.
- If $(X, \|\cdot\|)$ is a Hilbert space and ∂ is the proximal subdifferential ∂_P, Corollary 4.14 says that $\nabla d_S^2(x) = 2(x - P_S(x))$, so functions of type ζ above are $\mathcal{C}^{1,1}$ and Property **Prop.4'** is fulfilled according to Proposition 4.142(b); so both Properties **Prop.4'** and **Prop.5** are satisfied according to Proposition 4.146. $\square$

In preparation for the mean value inequalities that we intend to achieve (related to a point a and a closed convex set S), we start with several lemmas.

LEMMA 6.83. *Let $f : X \to \mathbb{R} \cup \{+\infty\}$ be a lower semicontinuous function defined on the Banach space X and S be a closed subset of X. Let $\varepsilon > 0$, $\lambda > 0$ and $\bar{y} \in S$ be such that*

(i) $B[\bar{y}, \lambda] \subset S$ *and* (ii) $f(\bar{y}) < \inf_S f + \varepsilon.$

Then there exist some $\bar{x}, \bar{u} \in B(\bar{y}, \lambda)$ such that
(a) $f(\bar{x}) \leq f(\bar{y}) < \inf_S f + \varepsilon$ and $f(\bar{u}) \leq f(\bar{y}) < \inf_S f + \varepsilon$;
(b)
$$0 \leq \underline{d}_H^+ f(\bar{x}; h) + \frac{2\varepsilon}{\lambda}\|h\| \text{ for all } h \in X \quad \text{and} \quad 0 \in \partial f(\bar{u}) + \frac{2\varepsilon}{\lambda}\mathbb{B}_{X^*}.$$

PROOF. The Ekeland (resp. Borwein-Preiss) variational principle applied with the function $f + \Psi_S$, see Theorems 2.221 and 4.41, furnishes some $\bar{x} \in S$ such that
$$\|\bar{x} - \bar{y}\| \leq \lambda/2 \text{ (resp. } \|\bar{x} - \bar{y}\| < \lambda), \quad f(\bar{x}) \leq f(\bar{y}) < \inf_S f + \varepsilon,$$

(6.25) $$f(\bar{x}) \leq f(x) + \frac{\varepsilon}{\lambda}\varphi(x) \quad \text{for all } x \in S,$$

where $\varphi(\cdot) := 2\|\cdot - \bar{x}\|$ (resp. $\varphi(\cdot) := \frac{1}{\lambda}\Delta_2(\cdot) := \frac{1}{\lambda}\sum_{n=1}^{\infty} \alpha_n \|\cdot - b_n\|^2$ with $\alpha_n \geq 0$ and $\sum_{n=1}^{\infty} \alpha_n = 1$, $(b_n)_n$ is a sequence of X converging to $\bar{x}$ with $\|b_n - \bar{x}\| < \lambda$ for all $n \in \mathbb{N}$). For $\lambda' := \lambda - \|\bar{x} - \bar{y}\| > 0$, in both cases we have $B(\bar{x}, \lambda') \subset B(\bar{y}, \lambda)$, hence $B(\bar{x}, \lambda') \subset S$. The inequality (6.25) then ensures that $\bar{x}$ is a local minimizer of the function $f + \frac{\varepsilon}{\lambda}\varphi$.

Let us establish the inequality concerning the lower Hadamard directional derivative. Fix any $h \in X$ and choose some $\eta > 0$ such that
$$\bar{x} + [0, \eta]B(h, \eta) \subset B(\bar{x}, \lambda'),$$

so $\bar{x} + [0, \eta]B(h, \eta) \subset S$, which gives that for all $0 < t < \eta$ and $h' \in B(h, \eta)$, we have by (6.25)
$$f(\bar{x}) \leq f(\bar{x} + th') + \frac{2\varepsilon}{\lambda}\|\bar{x} + th' - \bar{x}\|, \text{ that is, } \frac{f(\bar{x} + th') - f(\bar{x})}{t} + \frac{2\varepsilon}{\lambda}\|h'\| \geq 0.$$

Consequently,
$$0 \leq \liminf_{t \downarrow 0, h' \to h} \left(\frac{f(\bar{x} + th') - f(\bar{x})}{t} + \frac{2\varepsilon}{\lambda}\|h'\| \right)$$
$$= \liminf_{t \downarrow 0, h' \to h} \frac{f(\bar{x} + th') - f(\bar{x})}{t} + \frac{2\varepsilon}{\lambda}\|h\|,$$

that is, $0 \leq \underline{d}_H^+ f(\bar{x}; h) + \frac{2\varepsilon}{\lambda}\|h\|$ as desired.

To prove the subdifferential inclusion, observe first that using arguments similar to those in Lemma 4.43 one easily sees that the directional derivative of the continuous convex function Δ_2 satisfies $(\Delta_2)'(x; h) \leq 2(\sum_{n=1}^{\infty} \alpha_n \|x - b_n\|)\|h\|$ for all $x, h \in X$. So, for each $h \in \mathbb{B}_X$ the inequalities $\|\bar{x} - b_n\| < \lambda$ yield $(\Delta_2)'(\bar{x}; h) \leq 2\lambda$, and hence $\partial \Delta_2(\bar{x}) \subset 2\lambda \mathbb{B}_{X^*}$, or equivalently $\partial \varphi(\bar{x}) \subset 2\mathbb{B}_{X^*}$. The point $\bar{x}$ being a local minimizer of $f + (\varepsilon/\lambda)\varphi$, we conclude by Property **Prop.4'** that $0 \in \partial f(\bar{x}) + \frac{2\varepsilon}{\lambda}\mathbb{B}_{X^*}$. □

LEMMA 6.84. *Let $f : X \to \mathbb{R} \cup \{+\infty\}$ be a lower semicontinuous function defined on the Banach space X and S be a closed convex subset of X. Let $\varepsilon > 0$ and $\bar{y} \in S$ be such that*
$$f(\bar{y}) < \inf_S f + \varepsilon.$$

Then, for every $\lambda > 0$ such that f is bounded from below on $B[\bar{y}, \lambda]$, there exist $\bar{x}, \bar{u} \in B(\bar{y}, \lambda)$ and some real $k > 0$ such that

(a) $f(\overline{x}) < f(\overline{y}) + \varepsilon$ and $f(\overline{u}) < f(\overline{y}) + \varepsilon$;

(b)
$$0 \le \underline{d}_H^+ f(\overline{x}; h) + k(d_S^2)'(\overline{x}; h) + \frac{2\varepsilon}{\lambda}\|h\| \quad \text{for all } h \in X;$$

(c) $0 \in \partial f(\overline{u}) + k\partial d_S^2(\overline{u}) + \frac{2\varepsilon}{\lambda}\mathbb{B}_{X^*}$.

PROOF. Fix any $\lambda > 0$ such that f is bounded from below on $B[\overline{y}, \lambda]$. According to the assumption $f(\overline{y}) < \inf_S f + \varepsilon$ fix any positive real $\eta < \varepsilon$ such that $f(\overline{y}) - \eta < \inf_S f$, so the definition of $\inf_S f$ yields some real $\delta > 0$ such that

(6.26) $$f(\overline{y}) - \eta < \inf_{B_\delta[S]} f.$$

Consider the closed set
$$C := B_\delta[S] \cup B[\overline{y}, \lambda].$$
The function f is bounded from below over C since it is bounded from below over $B[\overline{y}, \lambda]$ by assumption, and f is bounded from below over $B_\delta[S]$ as well according to (6.26). We can then choose some real $k > 0$ such that
$$f(\overline{y}) < \inf_C f + k\delta^2 + \eta.$$

Now set $g := f + kd_S^2$. We show that $g(\overline{y}) = f(\overline{y}) < \inf_C g + \varepsilon$. If $x \in C \setminus B_\delta[S]$, then $\delta < d_S(x)$, and hence
$$f(\overline{y}) < f(x) + k\delta^2 + \eta \le f(x) + kd_S^2(x) + \eta = g(x) + \eta.$$
If $x \in B_\delta[S]$, we have by (6.26)
$$f(\overline{y}) < f(x) + \eta \le f(x) + kd_S^2(x) + \eta = g(x) + \eta.$$
It follows that
$$g(\overline{y}) = f(\overline{y}) \le \inf_C g + \eta < \inf_C g + \varepsilon.$$
Lemma 6.83 above then furnishes some points $\overline{x}, \overline{u} \in B(\overline{y}, \lambda)$ with
$$\max\{g(\overline{x}), g(\overline{u})\} \le g(\overline{y}) < \inf_C g + \varepsilon$$
(thus satisfying (a)) and such that
$$0 \le \underline{d}_H^+ g(\overline{x}; h) + \frac{2\varepsilon}{\lambda}\|h\| \quad \text{for all } h \in X,$$
$$0 \in \partial g(\overline{u}) + \frac{2\varepsilon}{\lambda}\mathbb{B}_{X^*}, \quad \text{hence} \quad 0 \in \partial f(\overline{u}) + k\partial d_S^2(\overline{u}) + \frac{2\varepsilon}{\lambda}\mathbb{B}_{X^*},$$
and the latter inclusion, which is due to **Prop.4'** and **Prop.3**, translates the assertion (c). Further, since the convex function d_S^2 is Lipschitz continuous near $\overline{x}$, it is easily seen that
$$(d_S^2)'(\overline{x}; h) = \lim_{t \downarrow 0} \frac{d_S^2(\overline{x} + th) - d_S^2(\overline{x})}{t} = \lim_{t \downarrow 0, h' \to h} \frac{d_S^2(\overline{x} + th') - d_S^2(\overline{x})}{t},$$
and this entails that
$$\liminf_{t \downarrow 0, h' \to h} \frac{g(\overline{x} + th') - g(\overline{x})}{t} = \liminf_{t \downarrow 0, h' \to h} \frac{f(\overline{x} + th') - f(\overline{x})}{t} + k\,(d_S^2)'(\overline{x}; h),$$
or equivalently
$$\underline{d}_H^+ g(\overline{x}; h) = \underline{d}_H^+ f(\overline{x}; h) + k(d_S^2)'(\overline{x}; h).$$
This and the above inequality related to $\underline{d}_H^+ g(\overline{x}; h)$ justify the assertion (b). □

For a point $a \in X$ and a convex subset S of X we denote by $[a, S]$ the convex hull of $\{a\} \cup S$, that is,

$$[a, S] := \{(1-t)a + ty : t \in [0, 1] \text{ and } y \in S\}.$$

When the convex set S is in addition closed and bounded, $\operatorname{co}(\{a\} \cup S)$ is closed, hence in this case $[a, S]$ coincides with the closed drop $D(a, S) := \overline{\operatorname{co}}(\{a\} \cup S)$ introduced before the Daneš' drop theorem (see Theorem 2.232).

LEMMA 6.85. *Let S be a nonempty closed convex set of the Banach space X and let $a, \overline{x} \in X$ be such that*

$$d_D(\overline{x}) < d_S(\overline{x}),$$

where $D := [a, S]$. Then, for any real $\delta \geq 0$ one has

$$(d_{B_\delta[D]}^2)'(\overline{x}; d - a) \leq 0 \quad \text{for all } d \in D.$$

PROOF. Fix any real $\delta \geq 0$. Obviously, for every $d \in X$,

$$(d_{B_\delta[D]}^2)'(\overline{x}; d - a) = 2 d_{B_\delta[D]}(\overline{x}) (d_{B_\delta[D]})'(\overline{x}; d - a),$$

so $(d_{B_\delta[D]}^2)'(\overline{x}; d - a) = 0$ whenever $\overline{x} \in B_\delta[D]$.

Suppose that $\overline{x} \notin B_\delta[D]$. We know by Lemma 3.260 that for all u in the open set $X \setminus B_\delta[D]$

$$d_D(u) = d_{B_\delta[D]}(u) + \delta,$$

so $d_D'(\overline{x}; h) = (d_{B_\delta[D]})'(\overline{x}; h)$ for all $h \in X$. Now consider a sequence $(x_n)_n$ in D such that for all $n \in \mathbb{N}$

(6.27) $$d_D(\overline{x}) \leq \|\overline{x} - x_n\| < d_D(\overline{x}) + \frac{1}{n^2}.$$

Then, for each $n \in \mathbb{N}$ we have for all $d \in X$

$$(d_{B_\delta[D]}^2)'(\overline{x}; d - x_n) = 2 d_{B_\delta[D]}(\overline{x})(d_{B_\delta[D]})'(\overline{x}; d - x_n) = 2 d_{B_\delta[D]}(\overline{x}) d_D'(\overline{x}; d - x_n)$$
$$\leq 2 d_{B_\delta[D]}(\overline{x})\big(d_D'(\overline{x}; d - \overline{x}) + \|\overline{x} - x_n\|\big),$$

where the inequality is due to the Lipschitz property of $d_D'(\overline{x}; \cdot)$ with 1 as Lipschitz constant. It follows, for each $n \in \mathbb{N}$, according to the convexity of d_D that, for all $d \in D$,

$$(d_{B_\delta[D]}^2)'(\overline{x}; d - x_n) \leq 2 d_{B_\delta[D]}(\overline{x})\big(d_D(d) - d_D(\overline{x}) + \|\overline{x} - x_n\|\big)$$

(6.28) $$= 2 d_{B_\delta[D]}(\overline{x})\big(-d_D(\overline{x}) + \|\overline{x} - x_n\|\big) \leq \frac{2}{n^2} d_{B_\delta[D]}(\overline{x}),$$

the last inequality being due to (6.27).

Now from the assumption $d_D(\overline{x}) < d_S(\overline{x})$ choose a real $\alpha > 0$ such that $d_D(\overline{x}) + \alpha < d_S(\overline{x})$. Fix any real $0 < \varepsilon < \alpha$ and put $\gamma := \alpha - \varepsilon > 0$. For every $d' \in D$ satisfying $\|d' - \overline{x}\| < d_D(\overline{x}) + \varepsilon$, we have

$$d_S(d') \geq d_S(\overline{x}) - \|d' - \overline{x}\| > \alpha + d_D(\overline{x}) - \|d' - \overline{x}\|, \quad \text{hence} \quad d_S(d') > \gamma.$$

Choosing by (6.27) an integer N such that $\|x_n - \overline{x}\| < d_D(\overline{x}) + \varepsilon$ for all $n \geq N$, we deduce, for each $n \geq N$, that $d_S(x_n) > \gamma$. So, fixing any $n \geq N$, we see that $x_n \notin S$, hence there exists some $t_n \in [0, +\infty[$ and $y_n \in S$ such that

$$x_n = \frac{1}{1 + t_n} a + \frac{t_n}{1 + t_n} y_n,$$

which entails $x_n - a = t_n(y_n - x_n)$ and
$$\|x_n - a\| = t_n\|x_n - y_n\| \geq t_n d_S(x_n) \geq t_n \gamma,$$
so $t_n \leq \gamma^{-1}\|x_n - a\|$. Fix any $d \in D$. The sublinearity of $(d^2_{B_\delta[D]})'(\overline{x}; \cdot)$ gives
$$(d^2_{B_\delta[D]})'(\overline{x}; d - a) \leq (d^2_{B_\delta[D]})'(\overline{x}; d - x_n) + (d^2_{B_\delta[D]})'(\overline{x}; x_n - a)$$
$$= (d^2_{B_\delta[D]})'(\overline{x}; d - x_n) + t_n(d^2_{B_\delta[D]})'(\overline{x}; y_n - x_n)$$
and (since $y_n \in D$) the inequality (6.28) yields
$$(d^2_{B_\delta[D]})'(\overline{x}; d - a) \leq \frac{2}{n^2} d_{B_\delta[D]}(\overline{x})(1 + t_n) \leq \frac{2}{n^2} d_{B_\delta[D]}(\overline{x})(1 + \gamma^{-1}\|x_n - a\|).$$
The sequence $(x_n)_n$ being bounded according to (6.27), we deduce that
$$(d^2_{B_\delta[D]})'(\overline{x}; d - a) \leq 0$$
as desired. □

LEMMA 6.86. *Let S be a nonempty closed convex set of the Banach space X, $a \in X$, $D := [a, S]$ and $f : X \to \mathbb{R} \cup \{+\infty\}$ be a lower semicontinuous function which is finite at some point of D. Let $\delta > 0$ be such that f is bounded from below on $B_\delta[D]$ and let $(y_n)_n$ be a sequence in X such that*
(i) $d_D(y_n) \to 0$ and $f(y_n) \to \liminf_D f$ as $n \to \infty$;
(ii) $y_n \notin B_\delta[S]$ for all $n \in \mathbb{N}$.

Then for any sequence $(\varepsilon_n)_n$ in $]0, +\infty[$ tending to 0, there exist a subsequence $(y_{s(n)})_n$ and sequences $(c_n)_n$ and $(c'_n)_n$ in $\text{dom } f$ and a sequence $(c^*_n)_n$ with $c^*_n \in \partial f(c'_n)$ such that
(a) $\|c_n - y_{s(n)}\| \to 0$, $f(c_n) \to \liminf_D f$ and $f(c'_n) \to \liminf_D f$ as $n \to \infty$;
(b) *for all $n \in \mathbb{N}$ and $d \in D$,*
$$d^+_H f(c_n; d - c_n) \geq -\varepsilon_n\|d - c_n\| \quad \text{and} \quad \langle c^*_n, d - c'_n \rangle \geq -\varepsilon_n\|d - c'_n\|;$$
(c) *for all $n \in \mathbb{N}$ and $d \in D$,*
$$d^+_H f(c_n; d - a) \geq -\varepsilon_n\|d - a\| \quad \text{and} \quad \langle c^*_n, d - a \rangle \geq -\varepsilon_n\|d - a\|.$$

PROOF. Let $\delta > 0$ and $(y_n)_n$ be given as in the lemma. Consider any sequence $(\varepsilon_n)_n$ in $]0, +\infty[$ tending to 0. The assumption of boundedness from below of f on $B_\delta[D]$ ensuring that $\liminf_D f$ is finite, by definition of $\liminf_D f$ we can find a decreasing sequence $(\delta_n)_n$ in $]0, \delta[$ with $\delta_n < 1/n$ and such that
$$\liminf_D f - \frac{\varepsilon_n^2}{2} < \inf_{B_{2\delta_n}[D]} f, \quad \text{hence} \quad \liminf_D f + \frac{\varepsilon_n^2}{2} < \inf_{B_{\delta_n}[D]} f + \varepsilon_n^2.$$
From the assumption (i) choose a subsequence $(y_{s(n)})_n$ such that for all n
$$d_D(y_{s(n)}) < \delta_n \quad \text{and} \quad f(y_{s(n)}) < \liminf_D f + (\varepsilon_n^2/2),$$
hence
$$f(y_{s(n)}) < \inf_{B_{\delta_n}[D]} f + \varepsilon_n^2.$$
Choose an integer N such that, for all $n \geq N$, we have $d_D(y_{s(n)}) < \delta/2$ and $\varepsilon_n < \delta/4$. For each integer $n \geq N$, it ensues $B[y_{s(n)}, 2\varepsilon_n] \subset B_\delta[D]$, and hence f is bounded from below on $B[y_{s(n)}, 2\varepsilon_n]$. Applying Lemma 6.84 we obtain sequences $(c_n)_n$ and $(c'_n)_n$ in $B(y_{s(n)}, 2\varepsilon_n)$, and a sequence $(k_n)_n$ in $]0, +\infty[$ such that
(α) $f(c_n) < f(y_{s(n)}) + \varepsilon_n^2$ and $f(c'_n) < f(y_{s(n)}) + \varepsilon_n^2$;

(β) $0 \leq \underline{d}_H^+ f(c_n; h) + k_n (d^2_{B_{\delta_n}[D]})'(c_n; h) + \varepsilon_n \|h\|$ for all $h \in X$;

(γ) $0 \in \partial f(c'_n) + k_n \partial d^2_{B_{\delta_n}[D]}(c'_n) + \varepsilon_n \mathbb{B}_{X^*}$,

and the latter inclusion furnishes $c_n^* \in \partial f(c'_n)$ such that

(6.29) $\qquad -c_n^* \in k_n \partial d^2_{B_{\delta_n}[D]}(c'_n) + \varepsilon_n \mathbb{B}_{X^*}.$

Obviously $\|c_n - y_{s(n)}\| \to 0$ and $\|c'_n - y_{s(n)}\| \to 0$ as $n \to \infty$ and, since

$$\max\{d_D(c_n), d_D(c'_n)\} \leq \delta_n + 2\varepsilon_n =: \eta_n,$$

we have

$$\inf_{B_{\eta_n}[D]} f \leq f(c_n) \leq \inf_{B_{\delta_n}[D]} f + 2\varepsilon_n^2 \quad \text{and} \quad \inf_{B_{\eta_n}[D]} f \leq f(c'_n) \leq \inf_{B_{\delta_n}[D]} f + 2\varepsilon_n^2,$$

and those inequalities imply $f(c_n) \to \liminf_D f$ and $f(c'_n) \to \liminf_D f$ as $n \to \infty$. Further, for $z_n := c_n$ (resp. $z_n := c'_n$) we have for all $d \in D$

$$(d^2_{B_{\delta_n}[D]})'(z_n; d - z_n) \leq d^2_{B_{\delta_n}[D]}(d) - d^2_{B_{\delta_n}[D]}(z_n) = -d^2_{B_{\delta_n}[D]}(z_n) \leq 0,$$

which, through (β) and the inclusion (6.29), entails

$$\underline{d}_H^+ f(c_n; d - c_n) \geq -\varepsilon_n \|d - c_n\| \quad \text{and} \quad \langle c_n^*, d - c'_n \rangle \geq -\varepsilon_n \|d - c'_n\|.$$

To prove the remaining assertion (c) we show that $d_D(z_n) < d_S(z_n)$, which will allow us to apply Lemma 6.85. Since $\|z_n - y_{s(n)}\| \to 0$, taking another integer greater than N if necessary, we may suppose that $\|z_n - y_{s(n)}\| < \delta/2$ for all integers $n \geq N$. Noting that $d_S(y_{s(n)}) > \delta$ because $y_{s(n)} \notin B_\delta[S]$ by assumption, we obtain, for all $n \geq N$,

$$d_S(z_n) \geq d_S(y_{s(n)}) - \|z_n - y_{s(n)}\| > \delta - (\delta/2),$$

hence $d_S(z_n) > \delta/2$. On the other hand, the convergences $\|z_n - y_{s(n)}\| \to 0$ and $d_D(y_{s(n)}) \to 0$ imply $d_D(z_n) \to 0$, and taking again another integer greater than N we may suppose $d_D(z_n) < \delta/2$ for all $n \geq N$. So, for each $n \geq N$, we have $d_D(z_n) < d_S(z_n)$ as desired, and Lemma 6.85 says that $(d^2_{B_{\delta_n}[D]})'(z_n; d - a) \leq 0$. This combined with ($\beta$) and the inclusion (6.29) justifies the inequalities in (c) and finishes the proof of the lemma. $\square$

LEMMA 6.87. *Let S be a nonempty closed convex set of the Banach space X, $a \in X$, $D := [a, S]$ and $f : X \to \mathbb{R} \cup \{+\infty\}$ be a lower semicontinuous function which is finite at some point of D. Assume that f is bounded from below on $B_{\delta_0}[D]$ for some $\delta_0 > 0$ and assume that there exists $\bar{y} \in D \setminus S$ such that $f(\bar{y}) \leq \liminf_S f$.*

Then, for any sequence of positive reals $(\varepsilon_n)_n$ tending to 0, there exist sequences $(c_n)_n$ and $(c'_n)_n$ in $(X \setminus B_\delta[S]) \cap \mathrm{dom}\, f$ for some $\delta > 0$ and a sequence $(c_n^*)_n$ with $c_n^* \in \partial f(c'_n)$ such that
(a) $d_D(c_n) \to 0$, $d_D(c'_n) \to 0$, $f(c_n) \to \liminf_D f$, and $f(c'_n) \to \liminf_D f$ as $n \to \infty$;
(b) *for all $n \in \mathbb{N}$ and $d \in D$,*

$$\underline{d}_H^+ f(c_n; d - c_n) \geq -\varepsilon_n \|d - c_n\| \quad \text{and} \quad \langle c_n^*, d - c'_n \rangle \geq -\varepsilon_n \|d - c'_n\|;$$

(c) *for all $n \in \mathbb{N}$ and $d \in D$,*

$$\underline{d}_H^+ f(c_n; d - a) \geq -\varepsilon_n \|d - a\| \quad \text{and} \quad \langle c_n^*, d - a \rangle \geq -\varepsilon_n \|d - a\|.$$

If in addition $f(\bar{y}) = \liminf_D f$, then the sequences $(c_n)_n$ and $(c'_n)_n$ can be taken such that

$$(c_n, f(c_n)) \to (\bar{y}, f(\bar{y})) \quad \text{and} \quad (c'_n, f(c'_n)) \to (\bar{y}, f(\bar{y})) \quad \text{as } n \to \infty.$$

PROOF. We note that $f(\bar{y}) \geq \liminf_D f$ according to the inclusion $\bar{y} \in D$. First, suppose $f(\bar{y}) = \liminf_D f$. Since $\bar{y}$ is not in the closed set S, there exists some real $0 < \delta < \delta_0$ such that $\bar{y} \notin B_\delta[S]$, so the constant sequence $(y_n)_n$ with $y_n := \bar{y}$ fulfills the assumptions (i) and (ii) of Lemma 6.86 which then furnishes the desired sequences $(c_n)_n$, $(c'_n)_n$ and $(c^*_n)_n$ with the properties (a)-(c) of the lemma. Further, in this case where $f(\bar{y}) = \liminf_D f$, for $z_n := c_n$ or $z_n := c'_n$, the convergences in (a) of Lemma 6.86 mean $(z_n, f(z_n)) \to (\bar{y}, f(\bar{y}))$, which is the additional property at the end of the statement of the lemma.

Now suppose $f(\bar{y}) > \liminf_D f$, hence $\liminf_D f < \liminf_S f$ since $f(\bar{y}) \leq \liminf_S f$ by assumption. From Proposition 6.82 there exists a sequence $(y_n)_n$ such that $d_D(y_n) \to 0$ and $f(y_n) \to \liminf_D f$ as $n \to \infty$. On the other hand, from the inequality $\liminf_S f > \liminf_D f$ (observed above) there exists some real $0 < \delta < \delta_0/2$ such that
$$\liminf_S f \geq \inf_{B_{2\delta}[S]} f > \liminf_D f$$
and since $f(y_n) \to \liminf_D f$, for large n we have $f(y_n) < \inf_{B_{2\delta}[S]} f$, hence $y_n \notin B_{2\delta}[S]$. Lemma 6.86 again can be applied to obtain the results of (a)-(c). Moreover, since $\|c_n - y_{s(n)}\| \to 0$ and $\|c'_n - y_{s(n)}\| \to 0$, we see, for large n, that $c_n \notin B_\delta[S]$ and $c'_n \notin B_\delta[S]$, which finishes the proof of the lemma. □

We are now in a position to state and prove the muldirectional mean value inequalities. The first theorem is concerned with the general case of unbounded convex sets. Its statement is the Aussel-Corvellec-Lassonde version of the Clarke-Ledyaev multidirectional mean value inequality.

THEOREM 6.88 (multidirectional mean value inequality for unbounded convex set). Let S be a nonempty closed convex set of the Banach space X, $a \in X$, $D := [a, S]$ and $f : X \to \mathbb{R} \cup \{+\infty\}$ be a lower semicontinuous function which is finite at a. Assume that f is bounded from below on $B_{\delta_0}[D]$ for some $\delta_0 > 0$. Then, for any real $r \leq \liminf_S f$ and any sequence of positive reals $(\varepsilon_n)_n$ tending to 0, there exist sequences $(t_n)_n$ and $(s_n)_n$ in $[0,1]$, $(c_n)_n$ and $(c'_n)_n$ in $\operatorname{dom} f$ with $(c_n, t_n)_n$ and $(c'_n, s_n)_n$ outside $B_\delta[S \times \{1\}]$ for some $\delta > 0$, and a sequence $(c^*_n)_n$ with $c^*_n \in \partial f(c_n)$ such that

(a) for $\widehat{D} := \operatorname{co}(\{(a, 0)\} \cup S \times \{1\})$, $d_{\widehat{D}}(c_n, t_n) \to 0$ and $d_{\widehat{D}}(c'_n, s_n) \to 0$, hence in particular $d_D(c_n) \to 0$ and $d_D(c'_n) \to 0$ as $n \to \infty$; also

$$\liminf_D f \leq \liminf_{n \to \infty} f(c_n) \leq \liminf_D f + |r - f(a)| \text{ and } \liminf_D f \leq \liminf_{n \to \infty} f(c'_n) \leq \liminf_D f + |r - f(a)|;$$

(b) for all $n \in \mathbb{N}$, $t \in [0, 1]$ and $d \in a + t(S - a)$,
$$\underline{d}^+_H f(c_n; d - c_n) \geq (t - t_n)(r - f(a)) - \varepsilon_n(\|d - c_n\| + |t - t_n|),$$
hence in particular $\underline{d}^+_H f(c_n; d - c_n) \geq -|r - f(a)| - \varepsilon_n \|d - c_n\| - \varepsilon'_n$, for a sequence of positive reals $(\varepsilon'_n)_n$ with $\liminf_{n \to \infty} \varepsilon'_n = 0$; and similarly
$$\langle c^*_n, d - c'_n \rangle \geq (t - s_n)(r - f(a)) - \varepsilon_n(\|d - c'_n\| + |t - s_n|),$$
hence in particular $\langle c^*_n, d - c'_n \rangle \geq -|r - f(a)| - \varepsilon_n \|d - c'_n\| - \varepsilon''_n$, for a sequence of positive reals ε''_n with $\liminf_{n \to \infty} \varepsilon''_n = 0$;

(c) for all $n \in \mathbb{N}$ and $y \in S$,
$$\underline{d}^+_H f(c_n; y - a) \geq r - f(a) - \varepsilon_n \|y - a\| - \varepsilon_n \text{ and } \langle c^*_n, y - a \rangle \geq r - f(a) - \varepsilon_n \|y - a\| - \varepsilon_n.$$

If in addition $f(z) - \tau(r - f(a)) = \text{linf}_{(x,t)\in\widehat{D}} (f(x) - t(r - f(a)))$ for some $(z, \tau) \in \widehat{D} \setminus S \times \{1\}$, then the sequences $(c_n)_n$ and $(c'_n)_n$ can be taken such that
$$(c_n, f(c_n)) \to (z, f(z)) \quad \text{and} \quad (c'_n, f(c'_n)) \to (z, f(z)) \quad \text{as } n \to \infty.$$

PROOF. Consider the lower semicontinuous function $\varphi : X \times \mathbb{R} \to \mathbb{R} \cup \{+\infty\}$ defined by
$$\varphi(x, t) = f(x) - t(r - f(a)) \quad \text{for all } (x, t) \in X \times \mathbb{R}$$
and put $\widehat{a} := (a, 0)$, $\widehat{S} := S \times \{1\}$, and $\widehat{D} := [\widehat{a}, \widehat{S}] \subset X \times \mathbb{R}$. Consider the norm $\|(x, t)\| := (\|x\|^2 + t^2)^{1/2}$ on $X \times \mathbb{R}$. We claim that
(6.30)
$$\text{linf}_{\widehat{D}} \varphi \leq \text{linf}_D f \text{ if } r \geq f(a) \quad \text{and} \quad \text{linf}_{\widehat{D}} \varphi \leq \text{linf}_D f - (r - f(a)) \text{ if } r < f(a).$$
Fixing any $\delta > 0$ and any $x \in B_\delta[D]$, we can choose some $t \in [0, 1]$ such that $(x, t) \in B_\delta[\widehat{D}]$. Then, if $r - f(a) \geq 0$ we have $\inf_{B_\delta[\widehat{D}]} \varphi \leq \varphi(x, t) \leq f(x)$, thus $\inf_{B_\delta[\widehat{D}]} \varphi \leq \inf_{B_\delta[D]} f$, which justifies the inequality of the claim. If $r - f(a) < 0$, then $\inf_{B_\delta[\widehat{D}]} \varphi \leq \varphi(x, t) \leq f(x) - (r - f(a))$, so
$$\inf_{B_\delta[\widehat{D}]} \varphi \leq \inf_{B_\delta[D]} f - (r - f(a)),$$
which ensures also the claim.

Now we show that $\varphi(\widehat{a}) = f(a) \leq \text{linf}_{\widehat{S}} \varphi$. Let $\varepsilon > 0$. Since $r \leq \text{linf}_S f = \sup_{\delta>0} \inf_{B_\delta[S]} f$, there exists $\delta > 0$ such that $r - \varepsilon < \inf_{B_\delta[S]} f$, or equivalently
$$f(a) - \varepsilon < \inf_{B_\delta[S]} f - (r - f(a)).$$
Choose a positive real $\eta < \delta$ such that for all $t \in [1 - \eta, 1 + \eta]$
$$f(a) - \varepsilon < \inf_{B_\delta[S]} f - t(r - f(a)).$$
Then, for all $(x, t) \in B_\eta[\widehat{S}] \subset B_\eta[S] \times [1 - \eta, 1 + \eta]$,
$$f(a) - \varepsilon < f(x) - t(r - f(a)) = \varphi(x, t),$$
which ensures that $f(a) - \varepsilon \leq \inf_{(x,t)\in B_\eta[\widehat{S}]} \varphi(x, t) \leq \text{linf}_{\widehat{S}} \varphi$, so
$$\varphi(\widehat{a}) = f(a) \leq \text{linf}_{\widehat{S}} \varphi.$$

On the other hand, the function $t \mapsto -t(r - f(a))$ being bounded on $[-\eta, 1 + \eta]$, the function φ is bounded from below on $B_\eta[\widehat{D}] \subset B_\eta[D] \times [-\eta, 1 + \eta]$. Noting also that $\widehat{a} \in \widehat{D} \setminus \widehat{S}$, we can apply Lemma 6.87 with $\widehat{a}$ in place of $\overline{y}$ to obtain sequences $(c_n, t_n)_n$ and $(c'_n, s_n)_n$ in $\text{dom}\,\varphi$ and outside $B_\delta[\widehat{D}]$ for some $\delta > 0$, and $(c^*_n, r_n) \in \partial\varphi(c'_n, s_n)$ such that

(α) $d_{\widehat{D}}(c_n, t_n) \to 0$, $d_{\widehat{D}}(c'_n, s_n) \to 0$, $\varphi(c_n, t_n) \to \text{linf}_{\widehat{D}} \varphi$ and $\varphi(c'_n, s_n) \to \text{linf}_{\widehat{D}} \varphi$;

(β) for all $n \in \mathbb{N}$ and $\widehat{d} \in \widehat{D}$
$$\underline{d}^+_H \varphi((c_n, t_n); \widehat{d} - (c_n, t_n)) \geq -\varepsilon_n \|\widehat{d} - (c_n, t_n)\|$$
and $\langle (c^*_n, r_n), \widehat{d} - (c'_n, s_n) \rangle \geq -\varepsilon_n \|\widehat{d} - (c'_n, s_n)\|$;

(γ) for all $n \in \mathbb{N}$ and $\widehat{d} \in \widehat{D}$
$$\underline{d}^+_H \varphi((c_n, t_n); \widehat{d} - \widehat{a}) \geq -\varepsilon_n \|\widehat{d} - \widehat{a}\| \quad \text{and} \quad \langle (c^*_n, r_n), \widehat{d} - \widehat{a} \rangle \geq -\varepsilon_n \|\widehat{d} - \widehat{a}\|.$$

Fix any $y \in S$. The inclusion $(y, 1) \in \widehat{D}$ gives through the first inequality in (γ)

$$\underline{d}_H^+ f(c_n; y - a) - (r - f(a)) = \underline{d}_H^+ \varphi((c_n, t_n); (y - a, 1)) = \underline{d}_H^+ \varphi((c_n, t_n); (y, 1) - \widehat{a})$$
$$\geq -\varepsilon_n \|(y - a, 1)\| \geq -\varepsilon_n \|y - a\| - \varepsilon_n,$$

which justifies the first inequality in (c) of the theorem. Since $\partial \varphi(c_n', s_n) \subset \partial f(c_n') \times \{-(r - f(a))\}$ according to property **Prop.5**, we have $r_n = -(r - f(a))$ and the second inequality in (γ) entails for $\widehat{d} := (y, 1)$

$$\langle c_n^*, y - a \rangle - (r - f(a)) = \langle (c_n^*, r_n), (y - a, 1) \rangle = \langle (c_n^*, r_n), (y, 1) - \widehat{a} \rangle$$
$$\geq -\varepsilon_n \|(y - a, 1)\| \geq -\varepsilon_n \|y - a\| - \varepsilon_n,$$

and this gives the second inequality in (c) of the theorem.

Now fixing any $t \in [0, 1]$ and any $d \in a + t(S - a)$ we see that $(d, t) \in \widehat{D}$, and the first inequality in (β) yields

$$\underline{d}_H^+ f(c_n; d - c_n) - (t - t_n)(r - f(a)) = \underline{d}_H^+ \varphi((c_n, t_n); (d - c_n, t - t_n))$$
$$= \underline{d}_H^+ \varphi((c_n, t_n); (d, t) - (c_n, t_n))$$
$$\geq -\varepsilon_n(\|d - c_n\| + |t - t_n|).$$

Denoting by t_0 a cluster point of $(t_n)_n$, the latter inequality also gives

$$\underline{d}_H^+ f(c_n; d - c_n)$$
$$\geq (t - t_0)(r - f(a)) - \varepsilon_n \|d - c_n\| - \varepsilon_n |t - t_0| - |t_0 - t_n|(|r - f(a)| + \varepsilon_n)$$
$$\geq -|r - f(a)| - \varepsilon_n \|d - c_n\| - \varepsilon_n',$$

where $\varepsilon_n' := \varepsilon_n + |t_0 - t_n|(|r - f(a)| + \varepsilon_n)$ and $\liminf_{n \to \infty} \varepsilon_n' = 0$. The first two inequalities in (b) of the theorem are then justified and the other inequalities in (b) are obtained similarly.

Let us show, for $(z_n, \tau_n) := (c_n, t_n)$ (resp. $(z_n, \tau_n) := (c_n', s_n)$), that $d_D(z_n) \to 0$. Since the convex set $D \times [0, 1]$ contains both the point $\widehat{a} = (a, 0)$ and the set $\widehat{S} = S \times \{1\}$, the inclusion $\widehat{D} \subset D \times [0, 1]$ holds true, so for each $(d, t) \in \widehat{D}$ we have $d \in D$, hence

$$d_D(z_n) \leq \|z_n - d\| \leq \|(z_n, \tau_n) - (d, t)\|.$$

It follows that $d_D(z_n) \leq d_{\widehat{D}}(z_n, \tau_n)$, thus $d_D(z_n) \to 0$. Further, denoting by τ a cluster point of $(\tau_n)_n$, we see from (α) that there exists some subsequence of $(f(z_n))_n$ which tends to $\lambda := \liminf_{\widehat{D}} \varphi + \tau(r - f(a))$, which combined with (6.30) gives $\lambda \leq \liminf_D f + |r - f(a)|$. This justifies the property concerning $\liminf_{n \to \infty} f(z_n)$.

Finally, under the assumption related to (z, τ), that is, $\varphi(z, \tau) = \liminf_{\widehat{D}} \varphi$ for some $(z, \tau) \in \widehat{D} \setminus \widehat{S}$, the conclusion of Lemma 6.87 related to $\overline{y}$, with (z, τ) in place of $\overline{y}$, says that the above sequence $(z_n, \tau_n)_n$ can be taken so that $(z_n, \tau_n) \to_\varphi (z, \tau)$ thus $z_n \to_f z$. The proof of the theorem is finished. $\square$

We turn now to the case when the closed convex S is bounded.

THEOREM 6.89 (multidirectional mean value inequality for bounded convex set). Let S be a nonempty bounded closed convex set of the Banach space X, $a \in X$, $D := [a, S]$ and $f : X \to \mathbb{R} \cup \{+\infty\}$ be a lower semicontinuous function which is finite at a. Assume that f is bounded from below on $B_{\delta_0}[D]$ for some $\delta_0 > 0$. Then, for any real $r \leq \liminf_S f$ and any sequence of positive reals $(\varepsilon_n)_n$ tending to 0, there exist sequences $(t_n)_n$ and $(s_n)_n$ in $[0, 1]$ tending to $t_0 \in [0, 1[$

and $s_0 \in [0, 1[$ respectively, $(c_n)_n$ and $(c'_n)_n$ in dom f with $(c_n, t_n)_n$ and $(c'_n, s_n)_n$ outside $B_\delta[S \times \{1\}]$ for some $\delta > 0$, and a sequence $(c_n^*)_n$ with $c_n^* \in \partial f(c'_n)$) such that
(a) $d(c_n, a + t_0(S - a)) \to 0$ and $d(c'_n, a + s_0(S - a)) \to 0$ as $n \to \infty$;
(b) for all $t \in [0, 1]$ and $d \in a + t(S - a)$,

$$\liminf_{n \to \infty} \inf_{d \in a+t(S-a)} \underline{d}_H^+ f(c_n; d - c_n) \geq (t - t_0)(r - f(a))$$

$$\liminf_{n \to \infty} \inf_{d \in a+t(S-a)} \langle c_n^*, d - c'_n \rangle \geq (t - s_0)(r - f(a));$$

(c) for all $n \in \mathbb{N}$

$$\liminf_{n \to \infty} \inf_{y \in S} \underline{d}_H^+ f(c_n; y - a) \geq r - f(a) \text{ and } \liminf_{n \to \infty} \inf_{y \in S} \langle c_n^*, y - a \rangle \geq r - f(a).$$

If in addition $f(z) - \tau(r - f(a)) = \text{linf}_{(x,t) \in \widehat{D}} (f(x) - \tau(r - f(a)))$ for some $(z, \tau) \in \widehat{D} \setminus S \times \{1\}$, then the sequences $(c_n)_n$ and $(c'_n)_n$ can be taken such that

$$(c_n, f(c_n)) \to (z, f(z)) \text{ and } (c'_n, f(c'_n)) \to (z, f(z)) \text{ as } n \to \infty.$$

PROOF. Take the sequences provided by Theorem 6.88 and choose cluster points t_0 and s_0 of the sequences $(t_n)_n$ and $(s_n)_n$ respectively. The set S being bounded, the set D is bounded and the sequences $(c_n)_n$ and $(c'_n)_n$ as well. So, the inequalities in (b) and (c) follow from (b) and (c) in Theorem 6.88. Now let us put $\tau_0 := t_0$ (resp. $\tau_0 := s_0$) and $(z_n, \tau_n) := (c_n, t_n)$ (resp. $(z_n, \tau_n) := (c'_n, s_n)$) and let us show that $d_{a+\tau_0(S-a)}(z_n) \to 0$. Since $d_{\widehat{D}}(z_n, \tau_n) \to 0$ for

$$\widehat{D} := \text{co}(\{(a, 0) \cup S \times \{1\}) = \{(a + t(y - a), t) : y \in S, t \in [0, 1]\},$$

there exist $y_n \in S$ and $\theta_n \in [0, 1]$ such that $\|z_n - (a + \theta_n(y_n - a))\| + |\tau_n - \theta_n| \to 0$. Then a subsequence $(\theta_{\sigma(n)})_n$ tends to τ_0 and $\|z_{\sigma(n)} - (a + \tau_0(y_{\sigma(n)} - a))\| \to 0$ because $\|(\theta_{\sigma(n)} - \tau_0)(y_{\sigma(n)} - a)\| \to 0$ thanks to the boundedness of S. Therefore, $d_{a+\tau_0(S-a)}(z_{\sigma(n)}) \to 0$. It remains to see that $\tau_0 < 1$. If $\tau_0 = 1$, then putting $\tau_0 = 1$ in the last convergence above gives $d_S(z_{\sigma(n)}) \to 0$, hence $d((z_{\sigma(n)}, \tau_{\sigma(n)}), S \times \{1\}) \to 0$, which is in contradiction with the fact that Theorem 6.88 says that, for some $\delta > 0$, all the points $(z_n, \tau_n) \notin B_\delta[S \times \{1\}]$. This finishes the proof of the theorem. $\square$

In the final case when the convex set S is compact, the statement takes the following form.

THEOREM 6.90 (multidirectional mean value inequality for compact convex set). Let S be a nonempty compact convex set of the Banach space X, $a \in X$, $D := [a, S]$ and $f : X \to \mathbb{R} \cup \{+\infty\}$ be a lower semicontinuous function which is finite at a. Then, for any real $r \leq \min_S f$, there exist $t_0 \in [0, 1[$ and $s_0 \in [0, 1[$ respectively, $c \in a + t_0(S - a)$ and $c' \in a + s_0(S - a)$, $(c_n)_n$ and $(c'_n)_n$ in dom f, and a sequence $(c_n^*)_n$ with $c_n^* \in \partial f(c'_n)$) such that
(a) $(c_n, f(c_n)) \to (c, f(c))$ and $(c'_n, f(c'_n)) \to (c', f(c'))$ as $n \to \infty$;
(b) for all $t \in [0, 1]$,

$$\liminf_{n \to \infty} \inf_{d \in a+t(S-a)} \underline{d}_H^+ f(c_n; d - c_n) \geq (t - t_0)(r - f(a))$$

$$\liminf_{n \to \infty} \inf_{d \in a+t(S-a)} \langle x_n^*, d - c'_n \rangle \geq (t - s_0)(r - f(a));$$

(c) for all $n \in \mathbb{N}$,
$$\liminf_{n\to\infty} \inf_{y\in S} d_H^+ f(c_n; y-a) \geq r - f(a) \text{ and } \liminf_{n\to\infty} \inf_{y\in S} \langle c_n^*, y-a\rangle \geq r - f(a).$$

PROOF. The set D is obviously compact. According to Proposition 6.82, $\text{linf}_S f = \inf_S f$ and $\text{linf}_{\widehat{D}} \varphi = \inf_{\widehat{D}} \varphi$ for $\widehat{D} := \text{co}(\{(a,0)\} \cup S \times \{1\})$ and $\varphi : X \times \mathbb{R} \to \mathbb{R} \cup \{+\infty\}$ with $\varphi(x,t) := f(x) - t(r - f(a))$ as in the proof of Theorem 6.89. Since $r \leq \inf_S f$, we have, for all $x \in S$,
$$\varphi(a,0) = f(a) \leq f(a) - (r - f(x)) = \varphi(x,1),$$
hence there exists some $(z,\tau) \in \widehat{D} \setminus S \times \{1\}$ such that $\min_{\widehat{D}} \varphi = \varphi(z,\tau)$. So the theorem follows from the last conclusion in the statement of Theorem 6.89. □

6.5. Comments

Mean value inequalities with Dini directional derivatives are classical. For an equivalent form of the inequality in (a) of Theorem 6.2 we refer, for example, to J.-B. Hiriart Urruty [487]. The arguments for the form (b) of the mean value inequality in Theorem 6.2 followed L. Zajíček [995]. Characterizations of convexity similar to Corollary 6.4 were given by D. T. Luc and S. Swaminathan [680]. Separate regular locally Lipschitz functions were studied by R. Correa and L. Thibault [295]. Except Proposition 6.9 appeared in M. Jouak and L. Thibaults's paper [562], all the results in Subsection 6.1.2 are taken from the paper [295] by Correa and Thibault.

Given a lower semicontinuous convex function $f : \mathbb{R}^m \to \mathbb{R} \cup \{+\infty\}$ with $\text{int dom } f \neq \emptyset$, it is shown in R. T. Rockafellar's book [852, Theorem 25.6] that for any $x \in \text{dom } f$
$$\partial f(x) = \overline{\text{co}} \left\{ \lim_j \nabla f(x_j) : \text{Dom } \nabla f \ni x_j \to x \right\} + N(\text{dom } f; x).$$

Rockafellar's proof in [852] is based on V. L. Klee's representation theorem of any closed convex set in $\mathbb{R}^m$ by means of its extreme points and extreme directions. Theorem 6.23 is an extension of the above result to infinite dimensions; the arguments for its proof are different from those in [852]. The particular case of Theorem 6.19 with locally Lipschitz functions was established by R. Correa and A. Jofre [290, Theorems 2.3 and 4.2]. The main ideas of the proof of Theorem 6.23 followed those of the paper [564] of A. Jofre and L. Thibault which contains a less detailed proof in Proposition 3.4 therein for a similar result. Corollary 6.24 corresponds to [564, Corollary 3.5].

Theorem 6.42 is known in the literature as *the Denjoy-Young-Saks theorem*. Theorem 6.42 was stated and established for the first time by A. Denjoy in his 1915 very long paper [327] of 136 pages containing also various other deep results, applications, examples and counter-examples. The theorem was announced by Denjoy in a previous note [326]. The paper [327] was the first of a series of long papers [327, 328, 329, 330] of Denjoy developing his work on *the derivation and its inverse calculus*.[1] Denjoy's papers [326, 327] studied the case of continuous functions. Among several extensions to arbitrary functions of one-real variable we cite G. C. Young [977], F. Riesz [836], S. Saks [876]. G. C. Young's 1915 paper proved Theorem 6.42 for measurable functions; the measurability assumption was

---

[1]Denjoy wrote in a footnote in [327, p. 105]: "This is the first part of a much more extensive work *on the derivation and its inverse calculus*".

removed by S. Saks to establish the theorem for any real-valued function in his book [**876**]. Some generalizations to differential coefficients of higher orders were obtained by A. Denjoy [**331**], F. Roger [**868**], J. Marcinkiewicz and A. Zygmund [**692**], J. Marcinkiewicz [**691**]. Our development for the proof of Denjoy-Young-Saks Theorem 6.42 follows S. Saks' book [**876**].

Theorem 6.48 is due to G. Godefroy [**443**]; other previous less general statements in this line with the non-emptiness of $\partial_D f(x)$ in place of $\partial_D f(x) \cup \partial_D(-f)(x)$ can be found in D. Azagra [**50**]. The proofs of Theorem 6.48 and Lemma 6.47 are taken from Godefroy's paper [**443**]. The main part of Proposition 6.49 is also in [**50, 443**].

The Zagrodny mean value theorem (Theorem 6.52) is due to D. Zagrodny [**981**]. It has been established in [**981**] for the Clarke subdifferential, but as shown in [**915**] slight modifications of arguments in [**981**] yield to the case of a general abstract subdifferential; various comments and comparisons can also be found in Zagrodny's papers [**982, 983**]. The presentation in the manuscript uses ideas from [**981, 915**] and proves Theorem 6.52 both for a general subdifferential with properties **Prop.1-Prop.2** and for the Bouligand directional derivative. The inequality in (ii) of Theorem 6.52 is stated in [**981, 915**] for $y = b$, and its form with any $y \in c + [0, +\infty[(b - a)$ is taken from H. V. Ngai and J.-P. Penot [**769**]. The idea of the introduction of an abstract subdifferential satisfying Properties **Prop.1-Prop.4** started with R. Correa, A. Jofre and L. Thibault [**292, 293**] with the aim of getting a unified proof of several results as those in Theorem 6.68 and Corollary 6.69 in the book as well as many others. A. Ioffe [**514, 516**] considered previously an abstract subdifferential for another purpose: he proved in [**514**, p. 414, Theorem 9] and [**516**, p. 590, Remark 2] that the approximate subdifferential is the smallest abstract subdifferential with full calculus which is upper semicontinuous for real-valued locally Lipschitz continuous functions. The subdifferential characterization given by the equivalence (a) ⇔ (b) in Theorem 6.55 was first established by J. S. Treiman [**935**] for the Clarke subdifferential through the techniques of E. Bishop and R. R. Phelps [**103, 104**]; the equivalence in this context is also a direct consequence of the mean value inequality of D. Zagrodny [**981**]. The equivalence was obtained later by R. Redheffer and W. Walter [**832**] in $\mathbb{R}^n$ with the proximal subdifferential and by V. Weckesser [**970**] with the Fréchet subdifferential in a Radon-Nikodým Banach space X admitting an equivalent norm which is differentiable off zero. P. D. Loewen [**672**] showed that the equivalence in terms of Fréchet subdifferential still holds in any Banach space admitting an equivalent Fréchet differentiable norm off zero. The equivalence for the proximal subdifferential in Hilbert space was proved by F. H. Clarke, R. J. Stern and P. R. Wolenski [**249**]. All those results have been unified with the general context of abstract subdifferential by L. Thibault and D. Zagrodny [**926**], see also R. Correa, P. Gajardo and L. Thibault [**283**]. The equivalence (a) ⇔ (c) (in terms of the Bouligand directional derivative) in Theorem 6.55 seems to be new as formulated in such a generality. The first subdifferential characterization of directionally Lipschitz functions was shown with the Clarke subdifferential by J. Treiman [**935**]. The abstract subdifferential approach for the equivalence (a) ⇔ (c) in Theorem 6.59 follows L. Thibault and N. Zlateva [**930**]; the equivalence (a) ⇔ (b) in the same Theorem 6.59 seems to be new. The characterization (c) of the interior tangent property for sets in Corollary 6.60 is probably new; see also R. Correa, P. Gajardo and L. Thibault [**285**] for (d) in the same Corollary 6.60. The equivalence (a) ⇔ (b) in Corollary 6.62 is due to R. T. Rockafellar [**856**]. The equivalence (a)⇔(c) appeared in R. T. Rockafellar and R. J-B. Wets' book [**865**, Theorem 9.13(b)]; see also B. Mordukhovich's book [**725**, Theorem 3.52(d)] for the same characterization $\partial_L^\infty(\overline{x}) = \{0\}$ in the Banach setting under a certain compactness property of epi f at $(\overline{x}, f(\overline{x}))$. The equivalence (a)⇔(c) in Corollary 6.66 was given in Rockafellar and Wets' book [**865**, Theorem 9.18(c)]. The subdifferential characterization of the nondecreasing property in Proposition 6.67 was first obtained by F. H. Clarke, R. J. Stern and P. R. Wolenski [**249**] for the proximal subdifferential on Hilbert space. The general subdifferential statement of Proposition 6.67 comes from R. Correa, P. Gajardo and L. Thibault [**283**]; see also B. Mordukhovich [**725**, Theorem 3.55] for the Fréchet subdifferential on Asplund space. The subdifferential characterization of convex functions in Theorem 6.68 was first observed and proved by R. A. Poliquin [**804**] in $\mathbb{R}^n$. The proof in [**804**] is based on a quadratic conjugate notion as well as on proximal subgradients. The general subdifferential version of Theorem 6.68 in Banach spaces was obtained by R. Correa, A. Jofre and L. Thibault [**292, 293**] through the Zagrodny mean value theorem; see also [**291**] for simple arguments in reflexive Banach spaces. The proof presented here uses ideas and techniques from those papers and from H.-V. Ngai and J.-P. Penot [**769**]. Corollary 6.69(c) seems to be new. In addition to the Hadamard directional derivative characterizations in the text of the above properties, characterizations with Dini directional derivatives can be found in Dinh The Luc and Swaminathan [**680**] and F. H. Clarke, R. J. Stern and P. R. Wolenski [**249**].

The failure of Rolle's theorem in infinite dimensions was first established by S. A. Shkarin who proved this failure in [**888**] for infinite-dimensional superreflexive spaces and for non-reflexive spaces with smooth norms. The counter-example 6.72 related to Rolle theorem comes from Shkarin's paper [**888**]; our presentation follows D. Azé and J.-B. Hiriart-Urruty [**58**]. Other concrete counter-examples can be found in [**406, 408**]. A general characterization of infinite-dimensional Banach spaces where Rolle's theorem fails is the following theorem by D. Azagra and M. Jiménez-Sevilla in [**53**, Theorem 1.1, Corollary 1.2]: An infinite-dimensional Banach space X has a $\mathcal{C}^p$ smooth bump function if and only if there exists a bounded contractible open subset U of X and a continuous function $f : \mathrm{cl}\, U \to \mathbb{R}$ such that f is $\mathcal{C}^p$ smooth in U, zero on the boundary bdry U, and $Df(x) \neq 0$ for all $x \in U$.

The approximate Rolle-like theorems provided by Theorem 6.73 and Theorem 6.76 are adaptations of results from D. Azagra and R. Deville [**51**] and from D. Azagra, J. Ferrera and F. López-Mesas [**52**].

Multidirectional mean value inequalities began with two fundamental papers by F. H. Clarke and Yu. S. Ledyaev [**245, 246**]. The first result proved by Clarke and Ledyaev [**245**, Theorem 2.1] in 1994 can be stated as follows:

THEOREM 6.91 (F.H. Clarke and Yu.S. Ledyaev). *Let C_1, C_2 be two nonempty closed bounded convex sets in a Banach space X and let $f : X \to \mathbb{R}$ be a function continuously differentiable on an open set containing $\mathrm{co}(C_1 \cup C_2)$. Assume that either C_1 or C_2 is compact. Then for each $\varepsilon > 0$ there exists a point $c \in \mathrm{co}(C_1 \cup C_2)$ such that*

$$\inf_{C_2} f - \sup_{C_1} f < \langle Df(c), y - x \rangle - \varepsilon$$

for all $x \in C_1$ and $y \in C_2$.

The second multidirectional mean value inequality by Clarke and Ledyaev [**246**, Theorem 2.1] was published in the same 1994 year. We state it, with the concept $\operatorname{linf}_S f$ in Definition 6.80, in the following form:

THEOREM 6.92 (F.H. Clarke and Yu.S. Ledyaev). *Let S be a nonempty closed bounded convex set of a Hilbert space H and $f : H \to \mathbb{R} \cup \{+\infty\}$ be a lower semicontinuous function. Let $a \in H$ be a point where f is finite and such that f is bounded from below on $[a, S] + \delta_0 \mathbb{B}_H$ for some real $\delta_0 > 0$. Then for any real $r < \operatorname{linf}_S f$ and any real $\varepsilon > 0$ there exits $z \in [a, S] + \varepsilon \mathbb{B}_H$ and $\zeta \in \partial_P f(z)$ such that*
$$r - f(a) < \langle \zeta, y - a \rangle \quad \text{for all } y \in S$$
along with $f(z) < \inf_{[a,S]} f + |r| + \varepsilon$.

Theorem 6.92 was independently extended by D. Aussel, J.-N. Corvellec and M. Lassonde [**48**] to Banach spaces, by M. Ivanov and N. Zlateva [**544, 546**] to certain nonconvex sets, by Q. J. Zhu [**1013**] to smooth Banach spaces. A further version has been established in 2018 by M. Hamamdjiev and M. Ivanov [**462**]. Various equivalent conditions for multidirectional mean value inequalities were provided by A. D. Ioffe [**519**], M. Lassonde [**646**], Q. J. Zhu [**1014**]. A different type of multidirectional mean value inequality was given by D. T. Luc [**679**]. The approach that we adopted in the book is that of Aussel, Corvellec and Lassonde [**48**]: it offers a great generality in Banach spaces with efficient extended multidirectional mean value inequalities. All the results in the concerned Section 6.4 in their subdifferential form along with their main arguments are taken from that paper [**48**] by Aussel, Corvellec and Lassonde.

CHAPTER 7

Metric regularity

Let $M : X \rightrightarrows Y$ be a multimapping between two metric spaces. For a given $\bar{y} \in Y$ and the inclusion (or generalized equation)
$$M(x) \ni \bar{y} \quad \text{with unknown } x,$$
the *metric regularity* of M is concerned with estimates of the distance of a point x from the solution set $M^{-1}(\bar{y})$ of the generalized equation, that is, estimates of $d(x, M^{-1}(\bar{y}))$. It is also concerned, for perturbations y of $\bar{y}$ in some subset of Y, with estimates of $d(x, M^{-1}(y))$, that is, the distance of the point x from the solution set of the perturbed generalized equation with y in place of $\bar{y}$. We will see that a linear estimation of $d(x, M^{-1}(y))$ in terms of $d(y, M(x))$ is connected with a certain Lipschitz behavior of the inverse multimapping M^{-1} and that this Lipschitz behavior is reduced to the usual Lipschitz property whenever M^{-1} is a (single-valued) mapping.

Lipschitz properties of functions and (single-valued) mappings played crucial roles in the development of subdifferential calculus rules in Chapters 2, 4 and 5. The present chapter studies the metric regularity of a multimapping as well as the Lipschitz property with which the metric regularity is connected through the inverse multimapping. The metric regularity will be a key tool for estimates of coderivatives, for calculus of tangent and normal cones to constraint sets, and for optimality conditions as well. It will also be fundamental in other chapters for some properties related to the subsmoothness of sets, prox-regularity of sets etc.

7.1. Aubin-Lipschitz property and metric regularity

In addition to the Lipschitz continuity of a multimapping with respect to the Hausdorff-Pompeiu distance, as we will see below there is also need for another Lipschitz-like property, known as Aubin property in the literature. It will be shown that this Lipschitz-like property for a multimapping characterizes the metric regularity of its inverse.

DEFINITION 7.1. Let $M : X \rightrightarrows Y$ be a multimapping between two metric spaces and P be a subset of X, and let $\bar{x} \in P$ and $\bar{y} \in M(\bar{x})$. One says that M has the *Aubin-Lipschitz* (or *pseudo-Lipschitz*, or *truncated Lipschitz-like*) *property relative to P at $\bar{x}$ for $\bar{y}$* if there are a real constant $\gamma \geq 0$ and neighborhoods U of $\bar{x}$ and V of $\bar{y}$ such that
$$M(x') \cap V \subset \bigcup_{y \in M(x)} B[y, \gamma d(x', x)] \quad \text{for all } x' \in U \text{ and } x \in U \cap P,$$
or equivalently, when Y is a normed space,

(7.1) $\qquad M(x') \cap V \subset M(x) + \gamma \|x' - x\| \mathbb{B}_Y \quad \text{for all } x' \in U \text{ and } x \in U \cap P.$

Such a real γ is called an *Aubin-Lipschitz constant relative to P at $\bar{x}$ for $\bar{y}$* (over the neighborhood U for the neighborhood V). The *Aubin-Lipschitz rate* (or *modulus*) $\mathrm{lip}_P[M](\bar{x}|\bar{y})$ of M *relative to P at $\bar{x}$ for $\bar{y}$* is defined as the infimum of all constants $\gamma \geq 0$ for which there are neighborhoods U and V such that the latter inclusion holds.

When P coincides with the whole space X, that is, there are a real constant $\gamma \geq 0$ and neighborhoods U of $\bar{x}$ and V of $\bar{y}$ such that

(7.2) $$M(x') \cap V \subset \bigcup_{y \in M(x)} B[y, \gamma d(x', x)] \quad \text{for all } x, x' \in U,$$

one just says that the multimapping M has the *Aubin-Lipschitz* (or *pseudo-Lipschitz*, or *truncated Lipschitz-like*) *property at $\bar{x}$ for $\bar{y}$* (over the neighborhood U for the neighborhood V); the above rate is written as $\mathrm{lip}\,[M](\bar{x}|\bar{y})$.

REMARK 7.2. (a) Since $M(\bar{x}) \cap V \neq \emptyset$, taking $x' = \bar{x}$ in the inclusion of the definition we see that $M(x) \neq \emptyset$ for all $x \in U \cap P$.
(b) In fact, taking $x' = \bar{x}$ and noting that $\bar{y} \in M(\bar{y}) \cap V$, we see that for each $x \in U \cap P$ there is some $\zeta(x) \in M(x)$ with $d(\zeta(x), \bar{y}) \leq \gamma d(x, \bar{x})$. From this it is clear that for any neighborhood V' of $\bar{y}$ there is a neighborhood U' of $\bar{x}$ such that

$$M(x) \cap V' \neq \emptyset \quad \text{for all } x \in U'.$$

(c) In some cases it can be useful to note that the property in the definition can be written for all $x' \in U$ and $x \in U \cap P$ as

$$\bigl(y' \in M(x') \cap V\bigr) \implies \bigl(B[y', \gamma d(x', x)] \cap M(x) \neq \emptyset\bigr).$$

□

Besides the case $P = X$, another case of great interest is when $P = \{\bar{x}\}$.

DEFINITION 7.3. When in the latter definition the set P is reduced to $\bar{x}$, that is $P = \{\bar{x}\}$, the related property becomes

$$M(x') \cap V \subset \bigcup_{y \in M(\bar{x})} B[y, \gamma d(x', \bar{x})] \quad \text{for all } x' \in U,$$

and M is said to be *calm* (or *one-sided Aubin-Lipschitz*) *at $\bar{x}$ for $\bar{y}$*; such a real $\gamma \geq 0$ is a constant of calmness (or one-sided Aubin-Lipschitz property) of M at $\bar{x}$ for $\bar{y}$ (over the neighborhood U for the neighborhood V). The rate $\mathrm{lip}_P[M](\bar{x}|\bar{y})$ with $P = \{\bar{x}\}$ is then denoted as $\mathrm{calm}\,[M](\bar{x}|\bar{y})$ and it is called the *rate* (or *modulus*) *of calmness* (or *of one-sided Aubin-Lipschitz property*) of M at $\bar{x}$ for $\bar{y}$.

REMARK 7.4. When M is defined only on a neighborhood U_0 of $\bar{x}$, one defines the above concepts for M as the corresponding ones of its natural extension obtained by setting $M(x) := \emptyset$ for all $x \in X \setminus U_0$. □

In Definition 7.1, shrinking the neighborhood U for which the inclusion $x \in U \cap P$ is involved, the point x' can be allowed to move in the whole space X as established in the following proposition:

PROPOSITION 7.5. *Let $M : X \rightrightarrows Y$ be a multimapping between two metric spaces and P be a subset of X. Let $\bar{x}$ be a point of P and $\bar{y} \in M(\bar{x})$. The multimapping M has the Aubin-Lipschitz property relative to P at $\bar{x}$ for $\bar{y}$ with*

$\gamma \geq 0$ as an Aubin-Lipschitz constant if and only if there are neighborhoods U' of $\bar{x}$ and V' of $\bar{y}$ such that

$$M(x') \cap V' \subset \bigcup_{y \in M(x)} B[y, \gamma d(x', x)] \quad \text{for all } x' \in X \text{ and } x \in U' \cap P.$$

PROOF. The implication $\Leftarrow$ being obvious, let us prove the converse. Let $\gamma \geq 0$ and U, V be given by Definition 7.1, and let $\delta, \eta > 0$ be such that $B[\bar{x}, \delta] \subset U$ and $B[\bar{y}, \eta] \subset V$. Choose $0 < \delta' < \delta$ and $0 < \eta' < \eta$ such that $2\gamma\delta' + \eta' < \gamma\delta$, and put $U' := B[\bar{x}, \delta']$ and $V' := B[\bar{y}, \eta']$. Fix any $x \in U' \cap P$. From Definition 7.1 we have

$$M(x') \cap V' \subset \bigcup_{y \in M(x)} B[y, \gamma d(x', x)] \quad \text{for all } x' \in B[\bar{x}, \delta];$$

note that this entails in particular the existence of some $v \in M(x)$ such that $\bar{y} \in B[v, \gamma d(\bar{x}, x)]$ and hence $B[\bar{y}, \eta'] \subset B[v, \gamma\delta' + \eta']$, and note also that $\gamma\delta' + \eta' \leq \gamma\delta - \gamma\delta'$. For any fixed $x' \in X \setminus B[\bar{x}, \delta]$, the latter inequality combined with the inequality $d(x', x) > \delta - \delta'$ yields $\gamma\delta' + \eta' \leq \gamma\delta - \gamma\delta' < \gamma d(x', x)$, which entails $V' = B[\bar{y}, \eta'] \subset B[v, \gamma d(x', x)]$. It ensues that $M(x') \cap V' \subset \bigcup_{y \in M(x)} B[y, \gamma d(x', x)]$ for all $x' \in X$, which finishes the proof. $\square$

In the particular case where $P = \{\bar{x}\}$ the above proposition tells us that M is calm at $\bar{x}$ for $\bar{y}$ with γ as a constant of calmness if and only if there exists a neighborhood V' of $\bar{y}$ such that

$$M(x') \cap V' \subset \bigcup_{y \in M(\bar{x})} B[y, \gamma d(x', \bar{x})] \quad \text{for all } x' \in X,$$

which ensures

(7.3) $\quad d(y', M(\bar{x})) \leq \gamma d(x', \bar{x}) \quad \text{for all } x' \in X, y' \in M(x') \cap V'.$

Further, the latter inequality property with the constant γ entails the former inclusion form with any constant $\gamma' > \gamma$; indeed, taking any $x' \neq \bar{x}$ the inequality entails, for every $y' \in M(x') \cap V'$, that $d(y', M(\bar{x})) < \gamma' d(x', \bar{x})$, which gives some $y \in M(\bar{x})$ with $y' \in B(y, \gamma' d(x', \bar{x}))$, hence for every $x' \in X$ we derive that $M(x') \cap V' \subset \bigcup_{y \in M(\bar{x})} B[y, \gamma d(x', \bar{x})]$. For the particular multimapping $\Phi_f : \mathbb{R} \to X$ with $\Phi_f(t) := \{x \in X : f(x) \leq t\}$, where $f : X \to \mathbb{R} \cup \{+\infty\}$, and for a point $\bar{x} \in X$ where f is finite, the fulfillment of the inequality (7.3) for $\Phi := \Phi_f$ at $\bar{t} := f(\bar{x})$ for $\bar{x}$, with $\gamma \geq 0$ and $V' = B[\bar{x}, \varepsilon]$ (where $\varepsilon > 0$), signifies that $d(x, \{f(\cdot) \leq \bar{t}\}) \leq \gamma|t - \bar{t}|$ for all $t \in \mathbb{R}$ and $x \in B[\bar{x}, \varepsilon]$ with $f(x) \leq t$. So taking $x \in B[\bar{x}, \varepsilon]$ with $f(\bar{x}) \leq f(x) < +\infty$, we obtain with $t = f(x)$ that $d(x, \{f(\cdot) \leq \bar{t}\}) \leq \gamma|f(x) - f(\bar{x})| = \gamma(f(x) - f(\bar{x}))$; on the other hand if $f(x) \leq f(\bar{x}) = \bar{t}$, then $d(x, \{f(\cdot) \leq \bar{t}\}) = 0$. It ensues that

$$d(x, \{f(\cdot) \leq \bar{t}\}) \leq \gamma(f(x) - f(\bar{x}))^+ \quad \text{for all } x \in B[\bar{x}, \varepsilon].$$

Conversely, suppose that the latter inequality holds and fix any $t \in \mathbb{R}$ and $x \in \Phi_f(t)$, that is, $f(x) \leq t$. If $\bar{t} = f(\bar{x}) \leq f(x)$, then $(f(x) - f(\bar{x}))^+ = f(x) - f(\bar{x}) \leq t - \bar{t}$, and hence $d(x, \{f(\cdot) \leq \bar{t}\}) \leq \gamma|t - \bar{t}|$. If $f(x) < \bar{t}$, then $d(x, \{f(\cdot) \leq \bar{t}\}) = 0$. It results that $d(x, \{f(\cdot) \leq \bar{t}\}) \leq \gamma|t - \bar{t}|$ in any case.

We can then state:

PROPOSITION 7.6. For an extended real-valued function $f : X \to \mathbb{R} \cup \{+\infty\}$ and $\bar{x} \in \operatorname{dom} f$, if the sublevel multimapping $t \mapsto \{f(\cdot) \leq t\}$ from $\mathbb{R}$ to the metric space X is calm at $\bar{t} := f(\bar{x})$ for $\bar{x}$ with $\gamma \geq 0$ as a real constant of calmness, then the function f satisfies the (local) *error bound property*

$$d(x, \{f(\cdot) \leq \bar{t}\}) \leq \gamma \big(f(x) - f(\bar{x})\big)^+ \quad \text{for all } x \text{ near } \bar{x}.$$

Conversely, the above local error bound property entails the calmness of $t \mapsto \{f(\cdot) \leq t\}$ at $\bar{t} = f(\bar{x})$ for $\bar{x}$ with any $\gamma' > \gamma$ as a constant of calmness.

The Aubin-Lipschitz property can also be characterized as a Lipschitz behavior of the function $d(y, M(\cdot))$ as follows:

PROPOSITION 7.7. Let $M : X \rightrightarrows Y$ be a multimapping between two metric spaces and $(\bar{x}, \bar{y}) \in \operatorname{gph} M$. For $\gamma > 0$, consider the assertions:
(a) the multimapping M satisfies the Aubin-Lipschitz property at $\bar{x}$ for $\bar{y}$ with γ as a constant;
(b) there are neighborhoods U and V of $\bar{x}$ and $\bar{y}$ such that for every $y \in V$ the function $d(y, M(\cdot))$ is γ-Lipschitz on U;
(c) there are neighborhoods U and V of $\bar{x}$ and $\bar{y}$ such that for any $y, y' \in V$ and $x, x' \in U$

$$|d(y', M(x')) - d(y, M(x))| \leq d(y', y) + \gamma d(x', x).$$

Then (a) $\Rightarrow$ (b) $\Leftrightarrow$ (c); moreover, the assertion (b) with γ implies (a) with any $\gamma' > \gamma$ in place of γ.

The three assertions with the same constant γ are equivalent provided any point near $\bar{y}$ has (at least) a nearest point in $M(x)$ for x near $\bar{x}$, which is the case when Y is a finite-dimensional normed space and M takes on closed values near $\bar{x}$.

PROOF. Evidently (b) and (c) are equivalent. Suppose that (b) is fulfilled with γ, U and V (resp. suppose also that any point in V has a nearest point in $M(x)$). Fix any $\gamma' > \gamma$ and take any $x, x' \in U$ with $x' \neq x$. For any $y' \in M(x') \cap V$, the γ-Lipschitz property assumption gives $d(y', M(x)) \leq \gamma d(x, x')$. From the inequality $\gamma < \gamma'$ (resp. from the nearest point property of $M(x)$), there exists some $y \in M(x)$ with $d(y', y) < \gamma' d(x', x)$ (resp. $d(y', y) \leq \gamma d(x', x)$), so $M(x') \cap V \subset \bigcup_{y \in M(x)} B[y, \gamma' d(x', x)]$ (resp. the same inclusion holds with γ in place of γ'), which translates (a) with γ' (resp. with γ).

Now let us suppose that (a) holds and let U and V be the neighborhoods given by Definition 7.1. Take a real $\varepsilon > 0$ such that $B[\bar{y}, 3\varepsilon] \subset V$ and choose $\delta > 0$ such that $\delta \gamma < \varepsilon/2$ and $U' := B[\bar{x}, \delta] \subset U$. Fix any $y \in V' := B[\bar{y}, \varepsilon/2]$ and $x, x' \in U'$. Take any $y' \in M(x') \cap B[\bar{y}, 3\varepsilon]$ and note that $d(y', M(x)) \leq \gamma d(x', x)$ according to the Aubin-Lipschitz property. By the same property again we see that $d(\bar{y}, M(x')) \leq \gamma d(x', \bar{x})$, so $\bar{\delta} := d(\bar{y}, M(x')) < \varepsilon/2$, and hence $B[\bar{y}, \bar{\delta} + 2\varepsilon] \subset B[\bar{y}, 3\varepsilon]$.

From the inequality $d(y', M(x)) \leq \gamma d(x', x)$ we obtain $d(y, M(x)) \leq d(y, y') + \gamma d(x', x)$. This being true for all $y' \in M(x') \cap B[\bar{y}, 3\varepsilon]$, it results that

$$d(y, M(x)) \leq d\big(y, M(x') \cap B[\bar{y}, 3\varepsilon]\big) + \gamma d(x', x)$$
$$\leq d\big(y, M(x') \cap B[\bar{y}, \bar{\delta} + 2\varepsilon]\big) + \gamma d(x', x) = d(y, M(x')) + \gamma d(x', x),$$

where the equality is due to Lemma 2.219. So, (a) $\Rightarrow$ (b) holds true and the proof is finished. $\square$

REMARK 7.8. Obviously, the additional nearest point assumption in the proposition is fulfilled for closed convex sets $M(x)$ in Hilbert space. We will see later in Chapter 15 on *prox-regular sets* that the property is also satisfied for the general class of such sets. □

For the multimapping $M : X \rightrightarrows Y$ between two metric spaces, let be given $\bar x$ and $\bar y$ with $\bar y \in M(\bar x)$, that is, $\bar x$ is a given solution of the inclusion $M(x) \ni \bar y$ with unknown x. One is often interested to know whether for y near $\bar y$ the inclusion $M(x) \ni y$ (with unknown x) has a solution near $\bar x$, or more precisely, whether a solution can be guaranteed with an *error bound of linear rate* of $d(y, \bar y)$ as y moves near $\bar y$. It is also crucial, as mentioned in the very beginning of this chapter, given $\bar y$ to look for estimates to the solution set $M^{-1}(\bar y)$. Both questions naturally lead to introduce the following definition.

DEFINITION 7.9. Let $M : X \rightrightarrows Y$ be a multimapping between two metric spaces X and Y. Let Q be a subset of Y and let $(\bar x, \bar y) \in \operatorname{gph} M$ with $\bar y \in Q$. One says that M is *metrically regular relative to Q at $\bar x$ for $\bar y$* if there exist a real $\gamma \geq 0$ and neighborhoods U and V of $\bar x$ and $\bar y$ respectively such that

(7.4) $\qquad d(x, M^{-1}(y)) \leq \gamma d(y, M(x)) \quad \text{for all } x \in U \text{ and } y \in V \cap Q;$

such a real γ is a metric regularity constant of M relative to Q at $\bar x$ for $\bar y$ (over the neighborhood U and for the neighborhood V). The *rate* (also called *modulus*) *of metric regularity* $\operatorname{reg}_Q[M](\bar x|\bar y)$ of M *relative to Q at $\bar x$ for $\bar y$* is defined as the infimum of all $\gamma \geq 0$ for which there are neighborhoods U and V of $\bar x$ and $\bar y$ such that the latter inequality is fulfilled. If M is a (single-valued) mapping, one only says that it is metrically regular relative to Q at $\bar x$ and one writes $\operatorname{reg}_Q[M](\bar x)$.

When in the above inequality U and V are the whole spaces X and Y respectively, one says that M is *globally metrically regular relative to Q*.

The first case when $Q = Y$ is fundamental.

DEFINITION 7.10. When Q is the whole space Y, one omits it in the above concept and one only says that M is *metrically regular at $\bar x$ for $\bar y$*, which is translated by the existence of a real $\gamma \geq 0$ and neighborhoods U and V of $\bar x$ and $\bar y$ respectively such that

(7.5) $\qquad d(x, M^{-1}(y)) \leq \gamma d(y, M(x)) \quad \text{for all } x \in U \text{ and } y \in V,$

so γ is a constant of metric regularity of M over the neighborhood U for the neighborhood V. The above rate (with $Q = Y$) in this case is written as $\operatorname{reg}[M](\bar x|\bar y)$ and it is the *rate* (or *modulus*) *of metric regularity of M at $\bar x$ for $\bar y$*. When in addition to (7.5) there are some real $r > 0$ such that $M^{-1}(y) \cap B(\bar x, r)$ is a singleton for all $y \in V$, one says that M is *strongly metrically regular at $\bar x$ for $\bar y$*.

REMARK 7.11. It is readily seen that the above inequality in (7.4) holds true whenever $d(y, M(x)) = 0$ and $M(x)$ is closed; so

$$\operatorname{reg}_Q[M](\bar x|\bar y) = \limsup_{y \notin M(x), x \to \bar x, Q \ni y \to \bar y} \frac{d(x, M^{-1}(y))}{d(y, M(x))}$$

provided that M takes on closed values near $\bar x$. □

The second case $Q = \{\bar y\}$ is also of great interest.

DEFINITION 7.12. When $Q = \{\bar{y}\}$ in what precedes, one again omits Q to say that M is *metrically subregular at $\bar{x}$ for $\bar{y}$*, which then means that there exist a real $\gamma \geq 0$ and a neighborhood U of $\bar{x}$ such that

(7.6) $$d(x, M^{-1}(\bar{y})) \leq \gamma d(\bar{y}, M(x)) \quad \text{for all } x \in U.$$

The rate $\operatorname{reg}_Q[M](\bar{x}|\bar{y})$ is then denoted by $\operatorname{subreg}[M](\bar{x}|\bar{y})$ and it is called the *rate (or modulus) of metric subregularity of M at $\bar{x}$ for $\bar{y}$*. It is then the infimum of all reals $\gamma \geq 0$ for which there is a neighborhood U of $\bar{x}$ such that the latter inequality is satisfied. When in addition to (7.5) $M^{-1}(\bar{y}) \cap U = \{\bar{x}\}$, one says that M is *strongly metrically subregular at $\bar{x}$ for $\bar{y}$*.

As shown in the next theorem, the above constant of metric regularity is linked with the constant of openness with linear rate.

DEFINITION 7.13. Let $M : X \rightrightarrows Y$ be a multimapping between two metric spaces X and Y. Let Q be a subset of Y and let $(\bar{x}, \bar{y}) \in \operatorname{gph} M$ with $\bar{y} \in Q$. One says that M is *open with a linear rate relative to Q at $\bar{x}$ for $\bar{y}$* if there exist $c > 0$, $r > 0$ and neighborhoods U and V of $\bar{x}$ and $\bar{y}$ respectively such that, for all $(x, y) \in \operatorname{gph} M$ with $x \in U$ and $y \in V$ and for all $t \in]0, 1]$, one has

$$B[y, ctr] \cap Q \subset M(B[x, tr]);$$

in such case one says that c is a *constant of linear openness of M at $\bar{x}$ for $\bar{y}$* (over the neighborhood U for the neighborhood V). The *rate* (also called *constant*) $\operatorname{ope}_Q[M](\bar{x}|\bar{y})$ of linear openness of M relative to Q at $\bar{x}$ for $\bar{y}$ is defined as the supremum of all such coefficients $c > 0$, with the convention that the supremum is zero if there is no coefficient $c > 0$. If M is a (single-valued) mapping, it is enough to say that it is open with a linear rate relative to Q at $\bar{x}$ and one writes $\operatorname{ope}_Q[M](\bar{x})$. If Q is the whole space Y, it will be omitted, that is, one will say that M is *open with a linear rate at $\bar{x}$ for $\bar{y}$* and the corresponding rate will be denoted as $\operatorname{ope}[M](\bar{x}|\bar{y})$.

When for a coefficient $c > 0$ the above inclusion holds with $U = X$ and $V = Y$, for all reals $r > 0$, the multimapping M is said to be *globally open with a linear rate relative to Q* and c is called a *coefficient of global openness with linear rate*.

It is worth observing that, for $Q = Y$, the inclusion above translates a full openness with linear rate since it becomes, for any $(x, y) \in \operatorname{gph} M$ with $x \in U$ and $y \in V$,

$$B[y, ctr] \subset M(B[x, tr]) \quad \text{for all } t \in [0, 1].$$

Unlike the above case, if Q is reduced to the singleton $\{\bar{y}\}$, the openness relative to $Q = \{\bar{y}\}$ is not a true openness since it means that there exist reals $c \geq 0$ and $r > 0$ along with neighborhoods U of $\bar{x}$ and V of $\bar{y}$ such that for any $x \in U$ and $t \in]0, 1]$

$$\left(\bar{y} \in \bigcup_{y \in M(x) \cap V} B[y, ctr]\right) \implies \left(\bar{y} \in M(B[x, tr])\right).$$

The proof of the theorem linking the various above constants uses the two following lemmas which will also be involved in some other results.

LEMMA 7.14. Let $M : X \rightrightarrows Y$ be a multimapping between metric spaces, U be a nonempty subset of X and Q, T, V be subsets of Y with $Q \cap T \cap V \neq \emptyset$. Consider, for $\gamma > 0$, the assertions:

(a) $d(x, M^{-1}(y)) \le \gamma d(y, M(x) \cap T)$ for all $x \in U$ and $y \in T \cap Q \cap V$;

(b) $M^{-1}(y) \cap U \subset \bigcup_{x' \in M^{-1}(y')} B[x', \gamma d(y, y')]$ for all $y \in T$ and $y' \in T \cap Q \cap V$.

The assertion (b) entails (a); and if (a) is fulfilled with γ and either the sets $M^{-1}(y)$ are closed in X for all $y \in T \cap Q \cap V$ or the sets $M(x) \cap T$ are closed in Y for all $x \in U$, then (b) holds true with any $\gamma' > \gamma$ in place of γ.

Furthermore, with $V = T$ the assertions (b'), (c) and (c') are pairwise equivalent:

(b') $M^{-1}(y) \cap U \subset \bigcup_{x' \in M^{-1}(y')} B[x', \gamma d(y, y')]$ for all $y \in T$ and $y' \in T \cap Q$;

(c) $B[y, \gamma^{-1}s] \cap Q \cap T \subset M(B[x, s])$ for all $x \in U$, $y \in M(x) \cap T$ and $s \ge 0$;

(c') $B[y, \gamma^{-1}s] \cap Q \cap T \subset M(B[x, s])$ for all $x \in U$, $y \in M(x) \cap T$ and all reals s with $0 \le s \le \gamma \operatorname{diam} T$.

PROOF. Suppose that (b) is true. Fix any $x \in U$ and $y \in T \cap Q \cap V$, and take any $z \in M(x) \cap T$. The first and last inclusions mean $x \in M^{-1}(z) \cap U$ and $z \in T$, hence by (b) there is some $u \in M^{-1}(y)$ such that $x \in B[u, \gamma d(z, y)]$ or equivalently $d(x, u) \le \gamma d(z, y)$, which ensures that $d(x, M^{-1}(y)) \le \gamma d(y, z)$. It results that $d(x, M^{-1}(y)) \le \gamma d(y, M(x) \cap T)$, which proves (b) $\Rightarrow$ (a).

Now suppose that (a) holds as well as the closedness property related to M^{-1} (resp. M), and fix any $\gamma' > \gamma$. Fix also any $y \in T$ and $y' \in T \cap Q \cap V$, and take any $x \in M^{-1}(y) \cap U$. By (a) we have the inequality $d(x, M^{-1}(y')) \le \gamma d(y', M(x) \cap T)$. Since $\gamma' > \gamma$ and $M^{-1}(y')$ (resp. $M(x) \cap T$) is closed, there is some $x' \in M^{-1}(y')$ such that $d(x, x') \le \gamma' d(y', M(x) \cap T)$, and hence $d(x, x') \le \gamma' d(y', y)$ since $y \in M(x) \cap T$. It ensues that $x \in B[x', \gamma' d(y, y')]$, which justifies the desired implication.

The implication (c) $\Rightarrow$ (c') is obvious. Let us show that (b') $\Rightarrow$ (c). Suppose (b') and fix any $x \in U$, any $y \in M(x) \cap T$ and any real $s \ge 0$. Take any $y' \in B[y, \gamma^{-1}s] \cap Q \cap T$, which means $\gamma d(y, y') \le s$ and $y' \in Q \cap T$. Since $x \in M^{-1}(y) \cap U$ and $y \in T$, from (b') there exists $x' \in M^{-1}(y')$ such that $d(x, x') \le \gamma d(y, y') \le s$, hence $y' \in M(B[x, s])$, which translates the property (c).

Finally, suppose (c') holds with U and T, and take any $y \in T$, any $y' \in T \cap Q$, and any $x \in M^{-1}(y) \cap U$. Setting $s := \gamma d(y, y')$ and noting that $y' \in B[y, \gamma^{-1}s] \cap Q \cap T$ with $s \le \gamma \operatorname{diam} T$, the assertion (c') furnishes some $x' \in B[x, s]$ with $y' \in M(x')$, or equivalently $x \in B[x', \gamma d(y, y')]$ and $x' \in M^{-1}(y')$. This justifies the implication (c') $\Rightarrow$ (b'), and the proof is complete. □

REMARK 7.15. It is clear from the above proof that (a) and (b) are equivalent whenever every $x \in U$ has a nearest point in $M^{-1}(y)$ for any $y \in T \cap Q \cap V$, which is satisfied in particular whenever X is a finite-dimensional normed space and M^{-1} takes on closed values on $T \cap Q \cap V$. □

LEMMA 7.16. *Let $M : X \rightrightarrows Y$ be a multimapping between metric spaces, Q be a subset of Y containing $\bar{y}$ and U be a subset of X such that for some reals $\gamma > 0$ and $\eta > 0$*

$$M^{-1}(y) \cap U \subset \bigcup_{x' \in M^{-1}(y')} B[x', \gamma d(y, y')]$$

for all $y \in B[\bar{y}, \eta]$ and $y' \in Q \cap B[\bar{y}, \eta]$. Then

$$B[y, \gamma^{-1}s] \cap Q \subset M(B[y, s])$$

for all $x \in U$, $y \in M(x) \cap B[\bar{y}, \eta/2]$ and $s \in [0, \gamma\eta/2]$.

PROOF. Under the assumptions of the lemma, the property of (b') with $T = V = B[\bar{y}, \eta]$ in Lemma 7.14 is satisfied. The implication (b') $\Rightarrow$ (c) with $T = V = B[\bar{y}, \eta]$ in that lemma entails that $B[y, \gamma^{-1}s] \cap Q \cap B[\bar{y}, \eta] \subset M(B[y, s])$ for all $x \in U$, $y \in M(x) \cap B[\bar{y}, \eta]$, $s \geq 0$. Putting $r := \gamma\eta/2$ we have $B[y, \gamma^{-1}r] \subset B[\bar{y}, \eta]$ for all $y \in B[\bar{y}, \eta/2]$. It ensues that $B[y, \gamma^{-1}s] \cap Q \subset M(B[y, s])$ for all $x \in U$, $y \in M(x) \cap B[\bar{y}, \eta/2]$, and $0 \leq s \leq r$. $\square$

THEOREM 7.17 (metric regularity and openness). Let $M : X \rightrightarrows Y$ be a multimapping from a metric space X into a metric space Y. Let Q be a subset of Y and let $(\bar{x}, \bar{y}) \in \text{gph}\, M$ with $\bar{y} \in Q$. Assume that either M takes on closed values near $\bar{x}$ (that is, $M(x)$ is closed for all x near $\bar{x}$) or the inverse multimapping M^{-1} takes on closed values near $\bar{y}$. Then the following equalities hold true:

$$\text{reg}_Q[M](\bar{x}|\bar{y}) = \text{lip}_Q[M^{-1}](\bar{y}|\bar{x}) = 1/\text{ope}_Q[M](\bar{x}|\bar{y}),$$

with the convention $1/0 = +\infty$; and concerning the two last constants, M is in fact open with a linear rate at $\bar{x}$ for $\bar{y}$ with $c > 0$ as a coefficient of openness with linear rate if and only if M^{-1} has the Aubin-Lipschitz property at $\bar{y}$ for $\bar{x}$ with $1/c$ as an Aubin-Lipschitz constant.

Consequently, the following assertions are equivalent:
(a) the multimapping M is metrically regular relative to Q at $\bar{x}$ for $\bar{y}$;
(b) the multimapping M is open relative to Q with a linear rate at $\bar{x}$ for $\bar{y}$;
(c) the inverse multimapping M^{-1} satisfies the Aubin-Lipschitz property relative to Q at $\bar{y}$ for $\bar{x}$.

PROOF. It is enough to show the equalities of the theorem. Take any real $\gamma \geq 0$ (if any) such that the inequality in Definition 7.9 holds for some neighborhoods U and V of $\bar{x}$ and $\bar{y}$ respectively. We may suppose that $M(x)$ is closed for all $x \in U$ (resp. $M^{-1}(y)$ is closed for all $y \in V$). For any real $\gamma' > \gamma$, the above lemma with $T = Y$ tells us that the property in Definition 7.1 is satisfied with U, V and γ' in place of γ, so the inequality $\text{reg}_Q[M](\bar{x}|\bar{y}) \geq \text{lip}_Q[M^{-1}](\bar{y}|\bar{x})$ holds true.

Let us prove the converse inequality. For the multimapping M^{-1}, take any real $\gamma \geq 0$ (if any) and neighborhoods V and U of $\bar{y}$ and $\bar{x}$ respectively such that the inclusion in Proposition 7.5 is satisfied (with Q in place of P). Then the above lemma with $T = Y$ ensures that the property in Definition 7.1 holds true with U, V and γ, and from this we obtain the desired converse inequality.

Let $c > 0$ (if any), $r > 0$, U and V be neighborhoods of $\bar{x}$ and $\bar{y}$ satisfying the property in Definition 7.13. Shrinking the neighborhood V we may suppose that $\text{diam}\, V \leq cr$, so the property of (c') with $T = V$ in Lemma 7.14 is fulfilled. The implication (c') $\Rightarrow$ (b') with $T = V$ in that lemma tells us that M^{-1} is Aubin-Lipschitz relative to Q at $\bar{y}$ for $\bar{x}$ with c^{-1} as an Aubin-Lipschitz constant. This justifies the inequality $\text{Lip}_Q[M^{-1}](\bar{y}|\bar{x}) \leq 1/\text{ope}_Q[M](\bar{x}|\bar{y})$.

Conversely, suppose that M^{-1} is Aubin-Lipschitz relative to Q at $\bar{y}$ for $\bar{x}$ with a real $\gamma > 0$ and neighborhoods V and U of $\bar{y}$ and $\bar{x}$ satisfying the property in Definition 7.1. Take $\eta > 0$ with $B[\bar{y}, \eta] \subset V$. For $r := \gamma\eta/2$ Lemma 7.16 tells us that $B[y, \gamma^{-1}s] \cap Q \subset M(B[y, s])$ for all $x \in U$, $y \in M(x) \cap B[\bar{y}, \eta/2]$, and $0 \leq s \leq r$. This means that $\gamma^{-1} \leq \text{ope}\,[M](\bar{x}|\bar{y})$, so $\text{Lip}_Q[M^{-1}](\bar{y}|\bar{x}) \geq 1/\text{ope}_Q[M](\bar{x}|\bar{y})$, and the proof is finished. $\square$

Taking $U = X$ and $T = V = Y$ in Lemma 7.14 yields similar global results:

7.1. AUBIN-LIPSCHITZ PROPERTY AND METRIC REGULARITY

PROPOSITION 7.18. *Let $M : X \rightrightarrows Y$ be a multimapping with closed graph from a metric space X into a metric space Y, and let Q be a nonempty subset of Y. For $\gamma > 0$, consider the following assertions:*

(a) $d(x, M^{-1}(y)) \leq \gamma d(y, M(x))$ for all $x \in X$ and $y \in Q$;
(b) *the inverse multimapping M^{-1} satisfies the following Lipschitz condition:*

$$M^{-1}(y') \subset \bigcup_{x \in M^{-1}(y)} B[x, \gamma d(y', y)] \quad \text{for all } y \in Y, y' \in Q;$$

(c) *the inclusion $B[y, \gamma^{-1}s] \cap Q \subset M(B[x, s])$ holds for all $(x, y) \in \operatorname{gph} M$ and $s > 0$.*

Then (c) $\Leftrightarrow$ (b) $\Rightarrow$ (a), and on the other hand (a) with γ implies (b) with any real $\gamma' > \gamma$.

Furthermore, the three conditions (a), (b) and (c) are equivalent whenever X is a finite-dimensional normed space.

Consider a multimapping M between two normed spaces X and Y and points $\bar{x} \in X$ and $\bar{y} \in M(\bar{x})$. Assume that $\bar{x}$ is a minimizer of a κ-Lipschitz function $f : X \to \mathbb{R}$ over the constraint set $M^{-1}(\bar{y})$. Then (as seen in Lemma 2.213) the point $\bar{x}$ is a minimizer of the function $x \mapsto f(x) + \kappa d(x, M^{-1}(\bar{y}))$ over X. If M is metrically regular at $\bar{x}$ for $\bar{y}$, then $\bar{x}$ is a local minimizer of the function $x \mapsto f(x) + \kappa \gamma d(\bar{y}, M(x))$. However, the function $x \mapsto d(\bar{y}, M(x))$ may fail to be lower semicontinuous, and calculus rules for Fréchet or limiting subdifferentials require such a lower semicontinuity. It is then useful to introduce the concept of graphical metric regularity.

DEFINITION 7.19. *Let $M : X \rightrightarrows Y$ be a multimapping and $(\bar{x}, \bar{y}) \in \operatorname{gph} M$. Let also Q be a subset of Y with $\bar{y} \in Q$. We will say that M is* graphically metrically regular relative to Q at $\bar{x}$ for $\bar{y}$ *provided there exist some real $\gamma \geq 0$ and neighborhoods U and V of $\bar{x}$ and $\bar{y}$ respectively such that*

(7.7) $\qquad d(x, M^{-1}(y)) \leq \gamma d((x, y), \operatorname{gph} M) \quad \text{for all } x \in U, y \in V \cap Q.$

When $Q = Y$, this means that there are a real $\gamma \geq 0$ and neighborhoods U and V of $\bar{x}$ and $\bar{y}$ respectively such that

$$d(x, M^{-1}(y)) \leq \gamma d((x, y), \operatorname{gph} M) \quad \text{for all } x \in U, y \in V,$$

and one only says that M is graphically metrically regular at $\bar{x}$ for $\bar{y}$. *Analogously, the case $Q = \{\bar{y}\}$ amounts to saying that there are a real $\gamma \geq 0$ and a neighborhood U of $\bar{x}$ such that*

$$d(x, M^{-1}(\bar{y})) \leq \gamma d((x, \bar{y}), \operatorname{gph} M) \quad \text{for all } x \in U,$$

so in such a case one just says that the multimapping M is graphically metrically subregular at $\bar{x}$ for $\bar{y}$.

Fortunately, both concepts of metric regularity and graphical metric regularity coincide in quite a large generality:

PROPOSITION 7.20. *Let $M : X \rightrightarrows Y$ be a multimapping between two metric spaces and $(\bar{x}, \bar{y}) \in \operatorname{gph} M$. Let Q be a subset of Y with $\bar{y} \in Q$. Then the following hold:*

(a) *The multimapping M is metrically regular relative to Q at $\bar{x}$ for $\bar{y}$ with $\gamma > 0$ as a metric regularity constant if and only if there are neighborhoods U and V of $\bar{x}$ and $\bar{y}$ such that*

$$d(x, M^{-1}(y)) \leq d_\gamma((x, y), \operatorname{gph} M) \quad \text{for all } x \in U, y \in V \cap Q,$$

where d_γ denotes the distance on $X \times Y$ with $d_\gamma((u,v),(u',v')) := d(u,u') + \gamma d(v,v')$.

(b) The multimapping M is metrically regular relative to Q at $\bar{x}$ for $\bar{y}$ if and only if it is graphically metrically regular relative to Q at $\bar{x}$ for $\bar{y}$.

PROOF. Since $d_\gamma((x,y),\operatorname{gph} M) \leq \gamma d(y, M(x))$ as easily seen, the implication $\Longleftarrow$ in (a) is obvious. Let us show the converse implication. Let $\gamma > 0$, $U := B[\bar{x}, 2\delta]$ and $V := B[\bar{y}, 2\delta]$ (with $\delta > 0$) given by the definition of metric regularity of M at $\bar{x}$ for $\bar{y}$. According to the definition of the distance d_γ choose a real $\eta > 0$ such that $B_{d_\gamma}[(\bar{x}, \bar{y}), 2\eta] \subset U \times V$. Choose also a neighborhood $U' \subset U$ of $\bar{x}$ and $V' \subset V$ of $\bar{y}$ such that $U' \times V' \subset B_{d_\gamma}[(\bar{x}, \bar{y}), \eta]$. We know (see Lemma 2.219) that

$$d_\gamma((x,y),\operatorname{gph} M) = d_\gamma((x,y),(\operatorname{gph} M) \cap B_{d_\gamma}[(\bar{x},\bar{y}), 2\eta])$$

for all $x \in U'$ and $y \in V'$. Further, for any fixed $x \in U'$ and $y \in V' \cap Q$, we have, for all $(u,v) \in (\operatorname{gph} M) \cap B_{d_\gamma}[(\bar{x},\bar{y}), 2\eta]$, according to the inequality of metric regularity

$$d(x, M^{-1}(y)) \leq d(x,u) + d(u, M^{-1}(y)) \leq d(x,u) + \gamma d(y, M(u))$$
$$\leq d(x,u) + \gamma d(y,v) = d_\gamma((x,y),(u,v)).$$

The latter inequality yields $d(x, M^{-1}(y)) \leq d_\gamma\big((x,y),(\operatorname{gph} M) \cap B_{d_\gamma}[(\bar{x},\bar{y}), 2\eta]\big)$, or equivalently (according to what precedes) $d(x, M^{-1}(y)) \leq d_\gamma((x,y), \operatorname{gph} M)$, which translates the desired property in (a). This finishes the proof since the assertion (b) follows from (a). $\square$

The next proposition provides another characterization of metric regularity.

PROPOSITION 7.21. Let $M : X \rightrightarrows Y$ be a multimapping between two metric spaces and $(\bar{x}, \bar{y}) \in \operatorname{gph} M$. Let Q be a subset of Y with $\bar{y} \in Q$. The multimapping M is metrically regular relative to Q at $\bar{x}$ for $\bar{y}$ with $\gamma \geq 0$ as a metric regularity constant if and only if there exists some real $\delta > 0$ such that

(7.8) $$d(x, M^{-1}(y)) \leq \gamma d(y, M(x))$$

for all $x \in B[\bar{x}, \delta]$, $y \in B[\bar{y}, \delta] \cap Q$ with $\gamma d(y, M(x)) < \delta$.

PROOF. The implication $\Longrightarrow$ is obvious. Let $\delta > 0$ satisfying the property of the proposition, and choose a real $0 < \eta < (1+\gamma)^{-1}\delta$. Fix any $x \in B[\bar{x}, \eta]$ and $y \in B[\bar{y}, \eta] \cap Q$, and let us show that $d(x, M^{-1}(y)) \leq \gamma d(y, M(x))$. The required inequality being true if $\gamma d(y, M(x)) < \delta$ according to the choice of δ, we may suppose that $\gamma d(y, M(x)) \geq \delta$. Noting that $\gamma d(y, M(\bar{x})) \leq \gamma d(y, \bar{y}) < \delta$, the property related to the choice of δ yields

$$d(x, M^{-1}(y)) \leq d(\bar{x}, M^{-1}(y)) + d(x, \bar{x}) \leq \gamma d(y, M(\bar{x})) + d(x, \bar{x})$$
$$\leq \gamma d(y, \bar{y}) + d(x, \bar{x}) < \delta \leq \gamma d(y, M(x)).$$

The multimapping M is then metrically regular at $\bar{x}$ for $\bar{y}$ with γ as a metric regularity constant. $\square$

The metric regularity of a multimapping can be reduced to that of a (single valued) mapping as follows:

PROPOSITION 7.22. Let $M : X \rightrightarrows Y$ be a multimapping between two metric spaces and $(\bar{x}, \bar{y}) \in \operatorname{gph} M$ and let Q be a subset of Y with $\bar{y} \in Q$. Let $\Pi_{M,Y} : \operatorname{gph} M \to Y$ be the mapping on $\operatorname{gph} M$ defined by $\Pi_{M,Y}(x,y) := y$ and consider the

7.1. AUBIN-LIPSCHITZ PROPERTY AND METRIC REGULARITY

distance on gph M given by $d\big((x,y),(x',y')\big) := d(x,x') + d(y,y')$. The following assertions hold.

(a) One has the equality

$$\mathrm{reg}_{\,Q}[\Pi_{M,Y}](\overline{x}|\overline{y}) = 1 + \mathrm{reg}_{\,Q}[M](\overline{x}|\overline{y}).$$

(b) The multimapping M is metrically regular relative to Q at $\overline{x}$ for $\overline{y}$ if and only if the (single valued) mapping $\Pi_{M,Y}$ is metrically regular relative to Q at $\overline{x}$.

PROOF. The assertion (b) being a direct consequence of (a), we have to prove (a). Suppose that $\mathrm{reg}_Q[M](\overline{x}|\overline{y}) < +\infty$ and take any real $\gamma > \mathrm{reg}_Q[M](\overline{x}|\overline{y})$. There are neighborhoods U and V of $\overline{x}$ and $\overline{y}$ such that $d(x', M^{-1}(y')) \leq \gamma d(y', M(x'))$ for all $x' \in U$ and $y' \in V \cap Q$. Fix any $(x,y) \in (U \times V) \cap \mathrm{gph}\, M$ and $v \in V \cap Q$. Take any real $r > \gamma d(v,y)$. Since

$$d(x, M^{-1}(v)) \leq \gamma d(v, M(x)) \leq \gamma d(v,y) < r,$$

we can choose some $u \in M^{-1}(v)$ such that $d(x,u) < r$, hence

$$d\big((x,y), \Pi_{M,Y}^{-1}(v)\big) \leq d\big((x,y),(u,v)\big) = d(x,u) + d(y,v) < r + d(y,v).$$

This gives $d\big((x,y), \Pi_{M,Y}^{-1}(v)\big) \leq (1+\gamma)d(v,y)$, which translates the metric regularity relative to Q of $\Pi_{M,Y}$ with $1+\gamma$ as a constant. It ensues that $1+\gamma \geq \mathrm{reg}_Q[\Pi_{M,Y}](\overline{x},\overline{y})$ and hence $1 + \mathrm{reg}_Q[M](\overline{x}|\overline{y}) \geq \mathrm{reg}_Q[\Pi_{M,Y}](\overline{x},\overline{y})$.

Let us show the converse inequality. Suppose that $\mathrm{reg}_Q[\Pi_{M,Y}](\overline{x},\overline{y}) < +\infty$ and consider any real $\gamma > 0$ satisfying $1 + \gamma > \mathrm{reg}_Q[\Pi_{M,Y}](\overline{x},\overline{y})$. There is a real $\eta > 0$ such that

$$d\big((x,y), \Pi_{M,Y}^{-1}(v)\big) \leq (1+\gamma)d(v,y)$$

for all $(x,y) \in (U \times V) \cap \mathrm{gph}\, M$ and $v \in V \cap Q$, where $U := B[\overline{x}, \eta]$ and $V := B[\overline{y}, \eta]$. Choose a real $0 < \delta < \tfrac{1}{2}\min\{\eta, \gamma\eta\}$. Fix any $x \in B[\overline{x}, \delta]$ and $y \in B[\overline{y}, \delta] \cap Q$ with $\gamma d(y, M(x)) < \delta$. Take any real r with $\gamma d(y, M(x)) < r < \delta$. Choose some $y' \in M(x)$ such that $\gamma d(y,y') < r$, so $d(y,y') < \gamma^{-1}r < \gamma^{-1}\delta < \eta/2$ hence $d(y', \overline{y}) < \eta$. By what precedes we have

$$d\big((x,y'), \Pi_{M,Y}^{-1}(y)\big) \leq (1+\gamma)d(y',y) < \gamma^{-1}(1+\gamma)r.$$

Choose $(u,z) \in \Pi_{M,Y}^{-1}(y)$ (that is, $z = y$ and $u \in M^{-1}(y)$) such that

$$d(y',z) + d(x,u) = d\big((x,y'),(u,z)\big) < \gamma^{-1}(1+\gamma)r,$$

which entails $d(y',y) + d(x, M^{-1}(y)) < \gamma^{-1}(1+\gamma)r$, and hence $d(y, M(x)) + d(x, M^{-1}(y)) < \gamma^{-1}(1+\gamma)r$. The latter inequality being true for all r with $\gamma d(y, M(x)) < r < \delta$, it results that $d(y, M(x)) + d(x, M^{-1}(y)) \leq (1+\gamma)d(y, M(x))$, or equivalently $d(x, M^{-1}(y)) \leq \gamma d(y, M(x))$. Proposition 7.21 above says that M is metrically regular relative to Q at $\overline{x}$ for $\overline{y}$ with γ as a metrically regular constant, so $\mathrm{reg}_Q[M](\overline{x}|\overline{y}) \leq \gamma$, hence $\mathrm{reg}_Q[M](\overline{x}|\overline{y}) \leq -1 + \mathrm{reg}_Q[\Pi_{M,Y}](\overline{x},\overline{y})$. The desired inequality then holds true and the proof is complete. □

Regarding the rate of linear openness of M, it is of interest to show that it can also be given as the limit inferior of some ratio of openness coefficient. For a multimapping M between topological spaces X and Y, recall that $(x,y) \xrightarrow[M]{} (\overline{x}, \overline{y})$ means $(x,y) \to (\overline{x}, \overline{y})$ with $(x,y) \in \mathrm{gph}\, M$.

PROPOSITION 7.23. Let $M: X \rightrightarrows Y$ be a multimapping between metric spaces X and Y and $(\overline{x}, \overline{y}) \in \operatorname{gph} M$. Let also Q be a subset of Y with $\overline{y} \in Q$. Then

$$\operatorname{ope}_Q[M](\overline{x}|\overline{y}) = \liminf_{(x,y) \xrightarrow[M]{} (\overline{x},\overline{y}),\, s\downarrow 0} \left(\frac{1}{s} \sup\{r \geq 0: B[y,r] \cap Q \subset M(B[x,s])\} \right).$$

PROOF. Denote the second member of the theorem by σ and set $\theta(s,x,y) := \sup\{r \geq 0: B[y,r] \cap Q \subset M(B[x,s])\}$. Fix any real $c > 0$ (if any) for which there are $r, \delta > 0$ such that $B[y, ctr] \cap Q \subset M(B[x, tr])$ for all $(x,y) \in \operatorname{gph} M$ with $x \in B[\overline{x}, \delta]$ and $y \in B[\overline{y}, \delta]$ and for all $t \in {]0,1]}$. Putting $\delta_0 := \min\{r, \delta\}$, for all $s \in {]0, \delta_0]}$ and $(x,y) \in \operatorname{gph} M$ with $x \in B[\overline{x}, \delta_0]$ and $y \in B[\overline{y}, \delta_0]$, we have $B[y, cs] \cap Q \subset M(B[x, s])$, which means $cs \leq \theta(s,x,y)$, or equivalently $c \leq \frac{1}{s}\theta(s,x,y)$. It follows that $c \leq \liminf_{(x,y) \xrightarrow[M]{} (\overline{x},\overline{y}),\, s\downarrow 0} \frac{1}{s}\theta(s,x,y)$, so $\operatorname{ope}_Q[M](\overline{x}|\overline{y}) \leq \sigma$.

Now fix any $0 < c < \sigma$ (if any). By definition of limit inferior there exists some real $\delta > 0$ such that for every $t \in {]0, \delta]}$ and every $(x,y) \in \operatorname{gph} M$ with $x \in B[\overline{x}, \delta]$ and $y \in B[\overline{y}, \delta]$ we have $c < \frac{1}{t}\theta(t,x,y)$, thus (by definition of $\theta(t,x,y)$) there is some real $r(t,x,y) > ct$ such that $B[y, r(t,x,y)] \cap Q \subset M(B[x,t])$, which entails that $B[y, ct] \cap Q \subset M(B[x,t])$ since $ct < r(t,x,y)$. The latter inclusion ensures that $c \leq \operatorname{ope}_Q[M](\overline{x}|\overline{y})$, hence $\sigma \leq \operatorname{ope}_Q[M](\overline{x}|\overline{y})$, so the required equality of the proposition holds true. $\square$

7.2. Openness and metric regularity of convex multimappings: Robinson-Ursescu theorem

This section is concerned with the openness or metric regularity of multimappings with convex graph. Given two normed spaces X and Y, the graph of a multimapping $M: X \rightrightarrows Y$ is easily seen to be convex (resp. a convex cone) if and only if for all $x, x' \in X$ and $t, t' \in [0,1]$ with $t + t' = 1$ (resp. $t, t' \in [0, +\infty[$)

$$tM(x) + t'M(x') \subset M(tx + t'x').$$

For a multimapping with convex graph it then results that $M(C)$ is convex for any convex set C in X, and if in addition $0 \in M(0)$, then $tM(x) \subset M(tx)$ for all $t \in [0,1]$ and $x \in X$. A multimapping whose graph is convex is often called a *convex multimapping*.

Recall that $\mathbb{U}_X$ denotes the open unit ball of X, that is, $\mathbb{U}_X := B_X(0,1)$.

LEMMA 7.24. Let $M: X \rightrightarrows Y$ be a multimapping between a Banach space X and a normed space Y for which $0 \in M(0)$. Assume that the graph $\operatorname{gph} M$ of M is closed and convex and that $r\mathbb{U}_Y \subset \operatorname{cl}_Y(M(\mathbb{B}_X))$ for some real $r > 0$. Then one has $r\mathbb{U}_Y \subset M(\mathbb{B}_X)$.

PROOF. Fix $v \in r\mathbb{U}_Y$. Choose $0 < t < 1$ such that $\|v\| < rt$ and put $\theta_t := \min\{t, 1-t\}$. Observe that, for any $k \in \mathbb{N}$ and any real $s > 0$, the assumption of the lemma implies that $rs^k \mathbb{U}_Y \subset \operatorname{cl}_Y(s^k M(\mathbb{B}_X))$, and hence, for each $\varepsilon > 0$ and each $w \in rs^k \mathbb{U}_Y$ there exists some $u \in X$ and $\pi(u) \in M(u)$ such that

$$\|u\| \leq 1 \quad \text{and} \quad \|w - s^k \pi(u)\| < \varepsilon.$$

Since $\|v\| < rt$, with $\varepsilon = r\theta_t^2$ we obtain some $u_1 \in X$ and $\pi(u_1) \in M(u_1)$ such that

$$\|u_1\| \leq 1 \quad \text{and} \quad \|y - t\pi(u_1)\| < r\theta_t^2.$$

Taking the latter inequality into account and choosing $\varepsilon = r\theta_t^3$ we find $u_2 \in X$ and $\pi(u_2) \in M(u_2)$ such that

$$\|u_2\| \leq 1 \quad \text{and} \quad \|y - t\pi(u_1) - \theta_t^2 \pi(u_2)\| < r\theta_t^3.$$

We then construct by induction sequences $(u_n)_{n\geq 1}$ in X and $(\pi(u_n))_{n\geq 1}$ in Y with $\pi(u_n) \in M(u_n)$ for all $n \geq 1$ and such that

$$\|u_n\| \leq 1 \ \forall\, n \geq 1 \ \text{and} \ \|y - \big(t\pi(u_1) + \theta_t^2 \pi(u_2) + \cdots + \theta_t^n \pi(u_n)\big)\| < r\theta_t^{n+1} \ \forall\, n \geq 2.$$

Putting $x_1 = tu_1$ and $x_n := tu_1 + \theta_t^2 u_2 + \cdots + \theta_t^n u_n$ for $n \geq 2$, the first inequality tells us that $(x_n)_{n \geq 1}$ is a Cauchy sequence in the Banach space X, so it converges to some $x \in X$. The second inequality yields, with $y_1 = t\pi(u_1)$ and $y_n := t\pi(u_1) + \theta_t^2 \pi(u_2) + \cdots + \theta_t^n \pi(u_n)$ for $n \geq 2$, that $y_n \to y$ as $n \to \infty$. Noting that

$$t + \theta_t^2 + \cdots + \theta_t^n < t + \frac{\theta_t^2}{1 - \theta_t} \leq t + \frac{\theta_t^2}{1-t} \leq t + \frac{(1-t)^2}{1-t} = 1$$

(where the second and third inequalities are due to the properties $\theta_t \leq t$ and $\theta_t \leq 1-t$ respectively), from the convexity of gph M and the inclusion $0 \in M(0)$ it directly follows that $y_n \in M(x_n)$, and hence $y \in M(x)$ by the closedness of gph M. Further, writing $\|x_n\| \leq t + \theta_t^2 + \cdots + \theta_t^n < 1$ we also see that $\|x\| \leq 1$. The inclusion $r\mathbb{U}_Y \subset M(\mathbb{B}_X)$ is then justified. □

REMARK 7.25. From the above lemma it is worth noticing the following feature. Let $A : X \to Y$ be a continuous surjective linear mapping between two Banach spaces. By the Banach-Schauder open mapping theorem (see Theorem C.3 in Appendix) we know that there is a real constant $c > 0$ such that $c\mathbb{B}_Y \subset A(\mathbb{B}_X)$. Then, for any $0 < c' < c$ and any continuous linear mapping $A' : X \to Y$ with $\|A' - A\| \leq c - c'$ one has

$$c'\mathbb{U}_Y \subset A'(\mathbb{B}_X).$$

Indeed, writing

$$c'\mathbb{B}_Y + (c - c')\mathbb{B}_Y \subset A(\mathbb{B}_X) \subset A'(\mathbb{B}_X) + (A - A')(\mathbb{B}_X) \subset A'(\mathbb{B}_X) + (c - c')\mathbb{B}_Y,$$

and taking support functions we see that $c'\mathbb{B}_Y \subset \mathrm{cl}_Y\big(A'(\mathbb{B}_X)\big)$, so Lemma 7.24 tells us that $c'\mathbb{U}_Y \subset A'(\mathbb{B}_X)$. □

Before stating the theorem recall (see Definition 2.55) that the *core* of a set in a vector space X is the set

(7.9) $\qquad \mathrm{core}\, S := \{x \in X : \forall y \in X,\ \exists r > 0,\ x + ty \in S, \forall t \in [-r, r]\}.$

Clearly, $x \in \mathrm{core}\, S$ if and only if for any $y \in X$ there is some $r > 0$ such that $x + ty \in S$ for all $t \in [0, r]$. Obviously, when S is convex, one has $\mathrm{core}\, S = \{x \in S : \forall y \in X, \exists r > 0,\ x + ry \in S\}$, otherwise stated, an element $x \in S$ belongs to the core of the convex set S if and only if $\mathbb{R}_+(S - x) = X$, where $\mathbb{R}_+ := [0, +\infty[$.

THEOREM 7.26 (S.M. Robinson; C. Ursescu). *Let X and Y be two Banach spaces and $M : X \rightrightarrows Y$ be a multimapping whose graph is closed and convex and let $(\bar{x}, \bar{y}) \in \mathrm{gph}\, M$. Then M is open with a linear rate (or equivalently metrically regular) at $\bar{x}$ for $\bar{y}$ if and only if $\bar{y} \in \mathrm{core}\, M(X)$.*

In fact, under the latter inclusion there exists some real $c > 0$ such that

$$d(x, M^{-1}(y)) \leq (c - \|y - \bar{y}\|)^{-1}(1 + \|x - \bar{x}\|)d(y, M(x)) \ \text{for all}\ x \in X, y \in B(\bar{y}, c).$$

PROOF. Considering $M_0(x) := M(x + \bar{x}) - \bar{y}$ we may suppose that $\bar{x} = 0$ and $\bar{y} = 0$. The inclusion $0 \in \operatorname{core} M(X)$ is clearly a necessary condition. Conversely, assuming that inclusion we easily see that $Y = \bigcup_{n \in \mathbb{N}} nC_n$ with $C_n := M(n\mathbb{B}_X)$, so

$$Y = \bigcup_{n \in \mathbb{N}} n(C_n \cap -C_n) \subset \bigcup_{n \in \mathbb{N}} n \overline{(C_n \cap -C_n)},$$

where the bar denotes the closure. The Baire category theorem and the symmetry of the convex sets $\overline{C_n \cap -C_n}$ give some $n \in \mathbb{N}$ for which $0 \in \operatorname{int}_Y \overline{(C_n \cap -C_n)}$, hence $0 \in \operatorname{int}_Y \overline{(C_n)}$. Consequently, we can find $\rho > 0$ such that $\rho \mathbb{U}_Y \subset \overline{M(n\mathbb{B}_X)}$. Since $n^{-1} M(n\mathbb{B}_X) \subset M(\mathbb{B}_X)$, it ensues that $n^{-1} \rho \mathbb{U}_Y \subset \overline{M(\mathbb{B}_X)}$. Taking any fixed positive real $c < \rho/(2n)$ Lemma 7.24 ensures that $c\mathbb{B}_Y \subset M(\mathbb{B}_X)$. It results for every $t \in [0,1]$

$$tc\mathbb{B}_Y = (1-t)0 + t(c\mathbb{B}_Y) \subset (1-t)M(0) + tM(\mathbb{B}_X) \subset M(t\mathbb{B}_X),$$

and this tranlates the linear openness of M at $\bar{x} = 0$ for $\bar{y} = 0$.

Regarding the inequality in the theorem, it follows from the next lemma. $\square$

LEMMA 7.27. *Let X and Y be two normed spaces and $M : X \rightrightarrows Y$ be a multimapping whose graph is convex and let $(\bar{x}, \bar{y}) \in \operatorname{gph} M$. Assume that for some real $c > 0$*

$$\bar{y} + c\mathbb{U}_Y \subset M(\bar{x} + \mathbb{B}_X).$$

Then, for all $x \in X$ and $y \in \bar{y} + c\mathbb{U}_Y$ one has

$$d(x, M^{-1}(y)) \leq (c - \|y - \bar{y}\|)^{-1}(1 + \|x - \bar{x}\|)d(y, M(x)).$$

PROOF. Fix any $x \in \operatorname{Dom} M$ and $y \in \bar{y} + c\mathbb{U}_Y$, and fix also any real $\varepsilon > 0$. Take any $z \in M(x)$ and define

$$s := c - \|y - \bar{y}\| \quad \text{and} \quad y' := y + s \frac{y - z}{\varepsilon + \|y - z\|}.$$

Since $\|y' - \bar{y}\| < \|y - \bar{y}\| + s = c$, there exists some $x' \in \bar{x} + \mathbb{B}_X$ such that $y' \in M(x')$. Observing that $y = (1-t)z + ty'$ with $t := (\varepsilon + s + \|y - z\|)^{-1}(\varepsilon + \|y - z\|)$, we obtain

$$y \in (1-t)M(x) + tM(x') \subset M\big((1-t)x + tx'\big),$$

hence

$$d(x, M^{-1}(y)) \leq \|x - (1-t)x - tx'\| = t\|x - x'\| \leq t(\|x - \bar{x}\| + 1).$$

Noting that $t \leq (s+\varepsilon)^{-1}(\|y - z\| + \varepsilon)$, we deduce that $d(x, M^{-1}(y)) \leq s^{-1}(\|x - \bar{x}\| + 1)\|y - z\|$. This being true for arbitrary element $z \in M(x)$, we conclude that $d(x, M^{-1}(y)) \leq (c - \|y - \bar{y}\|)^{-1}(\|x - \bar{x}\| + 1)d(y, M(x))$. $\square$

7.3. Criteria and estimates of rates of openness and metric regularity of multimappings

We proceed now to the determination/estimation of the rates of openness and metric regularity of multimappings. We start with linear mappings. Before stating the theorem on the rate of openness of a linear mapping $A : X \to Y$, we observe for $\bar{x} \in X$ and $\bar{y} := A(\bar{x})$ that $B[\bar{y}, ct] \subset A(B[\bar{x}, t])$ (with $t > 0$) if and only if $c\mathbb{B}_Y \subset A(\mathbb{B}_X)$, so the rate of openness $\operatorname{ope}[A](x)$ of A at $x \in X$ does not depend on x, and in fact

(7.10) $\qquad \operatorname{ope}[A](x) = \sup\{c > 0 : c\mathbb{B}_Y \subset A(\mathbb{B}_X)\} =: \operatorname{ope}[A].$

7.3. CRITERIA FOR METRIC REGULARITY

THEOREM 7.28 (rate of openness of continuous linear mappings). *For any continuous linear mapping A from a Banach space X into a normed space Y, one has*

$$\operatorname{ope}[A] = \inf\{\|A^*y^*\| : \|y^*\| = 1\}.$$

PROOF. If Y is reduced to zero, the result is obvious since $\inf \emptyset = +\infty$. Suppose that Y is nonzero.

Denote by μ the second member of the equality of the theorem. Fix $c > 0$ (if any) with $c\mathbb{B}_Y \subset A(\mathbb{B}_X)$ and take any $y^* \in Y^*$ with $\|y^*\| = 1$. For every $b \in \mathbb{B}_Y$ we can choose some $u \in \mathbb{B}_X$ such that $cb = A(u)$, which yields $c\langle y^*, b\rangle = \langle A^*y^*, u\rangle \leq \|A^*y^*\|$, and hence $c = c\|y^*\| \leq \|A^*y^*\|$. This ensures that $\operatorname{ope}[A] \leq \mu$.

Conversely, suppose $\mu > 0$ and consider any positive real $c < \mu$. For all $y^* \in Y^*$ we have $\mu\|y^*\| \leq \|A^*y^*\| = \|y^* \circ A\|$, which entails that $\sigma(y^*, \mu\mathbb{B}_Y) \leq \sigma(y^*, A(\mathbb{B}_X))$ (recall $\sigma(\cdot, \cdot)$ denotes the support function), hence $\mu(\mathbb{B}_Y) \subset \operatorname{cl}_Y(A(\mathbb{B}_X))$. Lemma 7.24 tells us that $\mu\mathbb{U}_Y \subset A(\mathbb{B}_X)$, which ensures the inclusion $c\mathbb{B}_Y \subset A(\mathbb{B}_X)$. It results that $\mu \leq \operatorname{ope}[A]$, and the proof is finished. $\square$

Observe that for the linear mapping $A : X \to Y$ and for $x \in X$ we have

$$x^* = A^*y^* \Leftrightarrow \langle x^*, u\rangle - \langle y^*, Au\rangle \leq 0 \ \forall u \in X \Leftrightarrow \langle x^*, u-x\rangle - \langle y^*, Au-Ax\rangle \leq 0 \ \forall u \in X,$$

otherwise stated $x^* = A^*y^* \Leftrightarrow (x^*, -y^*) \in N(\operatorname{gph} A; (x, Ax))$, where $N(\cdot; \cdot)$ is the standard normal cone of the convex set (in fact, vector subspace) $\operatorname{gph} A$. Accordingly, considering a subdifferential ∂ and a multimapping $M : X \rightrightarrows Y$ between two normed spaces X and Y, we define its *∂-coderivative* at $(x, y) \in \operatorname{gph} M$ as the multimapping $D^*_\partial M(x, y) : Y^* \rightrightarrows X^*$ given for every $y^* \in Y^*$ by

(7.11) $$D^*_\partial M(x, y)(y^*) = \{x^* \in X^* : (x^*, -y^*) \in N^\partial(\operatorname{gph} M; (x, y))\},$$

where $N^\partial(\operatorname{gph} M; (x, y))$ denotes the normal cone associated to the subdifferential operator ∂, that is, $N^\partial(\operatorname{gph} M; \cdot) = \partial \Psi_{\operatorname{gph} M}(\cdot)$. The above theorem then leads to associate with M and the point $(x, y) \in \operatorname{gph} M$ the dual constant

(7.12) $$|D^*_\partial M(x,y)|_{\inf} = \inf\{\|x^*\| : x^* \in D^*_\partial M(x,y)(y^*), \|y^*\| = 1\}$$

relative to the ∂-coderivative of M at (x, y), where by convention the infimum is $+\infty$ if the set on the right-hand side is empty. When ∂ is the Fréchet, proximal, limiting, approximate, or Clarke subdifferential, we will write $D^*_F M$, $D^*_P M$, $D^*_L M$, $D^*_A M$, $D^*_C M$. As shown in the following theorem, the dual constant provides suitable estimates for linear rates of openness of multimappings.

THEOREM 7.29 (coderivative estimate of linear rate of openness). *Let $M : X \rightrightarrows Y$ be a multimapping between two normed spaces and $(\bar{x}, \bar{y}) \in \operatorname{gph} M$.*
(a) *The following inequality holds true*

$$\operatorname{ope}[M](\bar{x}|\bar{y}) \leq \liminf_{(x,y) \xrightarrow{M} (\bar{x}, \bar{y})} |D^*_F M(x, y)|_{\inf}.$$

(b) *Assume that X and Y are Banach spaces. If $\operatorname{gph} M$ is closed near $(\bar{x}, \bar{y})$ and ∂ is a subdifferential on $X \times Y$ satisfying the properties **Prop.1-Prop.4** in the previous chapter, then*

$$\liminf_{(x,y) \xrightarrow{M} (\bar{x}, \bar{y})} |D^*_\partial M(\bar{x}, \bar{y})|_{\inf} \leq \operatorname{ope}[M](\bar{x}|\bar{y});$$

so M is metrically regular at $\bar{x}$ for $\bar{y}$ whenever $\liminf\limits_{(x,y)\xrightarrow{M}(\bar{x},\bar{y})}|D^*_\partial M(x,y)|_{\inf} > 0$.

(c) If X and Y are Asplund spaces and gph M is closed near $(\bar{x},\bar{y})$, then
$$\text{ope}\,[M](\bar{x}|\bar{y}) = \liminf\limits_{(x,y)\xrightarrow{M}(\bar{x},\bar{y})}|D^*_F M(x,y)|_{\inf};$$
thus M is metrically regular at $\bar{x}$ for $\bar{y}$ if and only if $\liminf\limits_{(x,y)\xrightarrow{M}(\bar{x},\bar{y})}|D^*_F M(x,y)|_{\inf} > 0$.

(d) If both spaces X and Y are finite-dimensional, then
$$\text{ope}\,[M](\bar{x}|\bar{y}) = |D^*_L M(\bar{x},\bar{y})|_{\inf} = \min\{\|x^*\| : x^* \in D^*_L M(x,y)(y^*),\; \|y^*\|=1\},$$
where by convention the minimum is $+\infty$ whenever the set over which it is taken is empty.

PROOF. It suffices to deal with the case when Y is not reduced to zero.
(a) We may suppose that $\text{ope}\,[M](\bar{x},\bar{y}) > 0$ and the right member of the inequality in (a) is finite. Fix any two real numbers c and β such that $0 < c < \text{ope}\,[M](\bar{x}|\bar{y})$ and $\beta > \liminf\limits_{(x,y)\xrightarrow{M}(\bar{x},\bar{y})}|D^*_F M(x,y)|_{\inf}$. There exists a real number $\delta > 0$, a pair $(x,y) \in$ gph M with $\|x-\bar{x}\| \leq \delta$ and $\|y-\bar{y}\| \leq \delta$, and a pair $(x^*,-y^*) \in N^F(\text{gph}\,M;(x,y))$ with $\|y^*\| = 1$ and $\|x^*\| < \beta$, such that $B[y,cr] \subset M\big(B[x,r]\big)$ for all $r \in [0,\delta]$. Put $c' := \max\{c,1\}$ and $\beta' := (\beta - \|x^*\|)/2$. From the definition of the Fréchet normal cone there is some $0 < r < \delta$ such that
$$\langle(x^*,-y^*),(u,v)-(x,y)\rangle \leq (\beta'/c')\|(u,v)-(x,y)\| \leq \beta' r$$
for all $(u,v) \in (\text{gph}\,M)\cap B[(x,y),c'r]$, where the norm $\|\cdot\|$ on $X\times Y$ is the sup norm. Choose some $z \in cr\mathbb{B}_Y$ such that $\langle -y^*,z\rangle > cr - \beta' r$ and put $y' := y+z$. Then $y' \in B[y,cr] \subset M\big(B[x,r]\big)$, so there is some $x' \in B[x,r]$ such that $y' \in M(x')$. This entails $(x',y') \in (\text{gph}\,M)\cap B[(x,y),c'r]$, hence $\langle(x^*,-y^*),(x',y')-(x,y)\rangle \leq \beta' r$. On the other hand,
$$\langle(x^*,-y^*),(x',y')-(x,y)\rangle = \langle x^*, x'-x\rangle + \langle -y^*, z\rangle > -r\|x^*\| + cr - \beta' r.$$
The latter inequality combined with the previous inequality yields $c < \|x^*\| + 2\beta' = \beta$, and this implies (a).

(b) We may suppose that gph M is closed and $\text{ope}\,[M](\bar{x}|\bar{y})$ is finite. Fix any $c,c' \in \mathbb{R}$ with $\text{ope}\,[M](\bar{x}|\bar{y}) < c < c'$ and fix also a positive real $\gamma < 1/c'$. For any $\delta > 0$ there exist a positive real $r < \frac{1}{2}\min\{\gamma\delta,\delta\}$ (depending on δ), $(x,y) \in$ gph M with $\|x-\bar{x}\| \leq \delta/2$ and $\|y-\bar{y}\| \leq \delta/2$, and $z \in B[y,cr]$ such that $M^{-1}(z)\cap B[x,r] = \emptyset$.

Step I For the continuous function $f_1 : X\times Y \to \mathbb{R}$ defined by $f_1(u,v) := \|v-z\|$, we have $f_1(u,v) > 0$ for all $(u,v) \in$ gph M with $u \in B[x,r]$, and also $f_1(x,y) \leq cr$. Let f_3 be the indicator function of gph M. For any $\mu \in\,]c,c'[$, considering the norm $\|(u,v)\|_\gamma := \max\{\|u\|,\gamma\|v\|\}$ on $X\times Y$ and applying the Ekeland variational principle (see Theorem 2.221) furnish some minimizer (x',y') of the function $f_1 + f_2 + f_3$, where $f_2(u,v) := \mu\|(u,v)-(x',y')\|_\gamma$ for all $(u,v) \in X\times Y$, in such a way that the minimizer (x',y') satisfies $\|(x',y')-(x,y)\|_\gamma < cr/\mu < r$ and $\|y'-z\| \leq \|y-z\|$. Keeping in mind that the dual norm of $\|\cdot\|_\gamma$ is $\|u^*\| + \gamma^{-1}\|v^*\|$ for all $(u^*,v^*) \in X^*\times Y^*$, we see that $\|u^*\| + \gamma^{-1}\|v^*\| \leq \mu$ for all $(u^*,v^*) \in \partial f_2(u,v)$ with $(u,v) \in X\times Y$.

Step II Note that $\gamma\mu < \gamma c' < 1$ and choose a positive real $\varepsilon < \min\{\frac{1-\gamma\mu}{\gamma+1}, \frac{c'-\mu}{c'+1}\}$. By properties **Prop.1-Prop.4** choose $(x_i,y_i) \in X\times Y$ and $(u_i^*,v_i^*) \in \partial f_i(x_i,y_i)$

such that $f_1(x_1, y_1) > 0$, $(x_3, y_3) \in \operatorname{gph} M$ with $\|x_3 - x'\| \leq \delta/2 - r$ and $\|y_3 - y'\| \leq \delta/2 - \gamma^{-1}r$, and $\|u_1^* + u_2^* + u_3^*\| + \|v_1^* + v_2^* + v_3^*\| < \varepsilon$. Since $u_1^* = 0$ according to the definition of f_1, it results that $\|u_2^* + u_3^*\| < \varepsilon$ and hence $\|u_3^*\| \leq \|u_2^*\| + \varepsilon$. Noting that $\|v_1^*\| = 1$ (since $\|y_1 - z\| = f_1(x_1, y_1) > 0$) and using the inequality $\|u_2^*\| + \gamma^{-1}\|v_2^*\| \leq \mu$ (observed above), we also have

$$\|v_3^*\| \geq \|v_1^*\| - \|v_2^*\| - \|v_1^* + v_2^* + v_3^*\| \geq 1 - \gamma(\mu - \|u_2^*\|) - \varepsilon > 0.$$

It ensues that on the one hand

$$\|x_3 - \bar{x}\| \leq \|x_3 - x'\| + \|x' - x\| + \|x - \bar{x}\| \leq \frac{\delta}{2} - r + r\frac{c}{\mu} + \frac{\delta}{2} < \delta,$$

$$\|y_3 - \bar{y}\| \leq \|y_3 - y'\| + \|y' - y\| + \|y - \bar{y}\| \leq \frac{\delta}{2} - \gamma^{-1}r + \gamma^{-1}r\frac{c}{\mu} + \frac{\delta}{2} < \delta,$$

and on the other hand, putting $x^* := u_3^*/\|v_3^*\|$ and $y^* := -v_3^*/\|v_3^*\|$, we see that $x^* \in D_\partial^* M(x_3, y_3)(y^*)$ along with $\|y_3^*\| = 1$ and

$$\|x^*\| \leq \frac{\|u_2^*\| + \varepsilon}{1 - \gamma(\mu - \|u_2^*\|) - \varepsilon} \leq \frac{\mu + \varepsilon}{1 - \varepsilon} < c',$$

where the second inequality is due to the inclusion $\|u_2^*\| \in [0, \mu]$ and to the decreasing property on $[0, \mu]$ of the function

$$t \mapsto \frac{1 + \gamma t - \gamma\mu - \varepsilon}{t + \varepsilon} = \gamma + \frac{1 - \varepsilon(\gamma + 1) - \gamma\mu}{t + \varepsilon}$$

(since $1 - \varepsilon(\gamma + 1) - \gamma\mu > 0$). We conclude that $\lambda(\bar{x}, \bar{y}) \leq c'$, where $\lambda(\bar{x}, \bar{y}) := \liminf_{(x,y) \xrightarrow{M} (\bar{x}, \bar{y})} |D_\partial^* M(x, y)|_{\inf}$, and hence $\lambda(\bar{x}, \bar{y}) \leq \operatorname{ope}[M](\bar{x}, \bar{y})$, which is the desired inequality in (b). The condition ensuring the metric regularity in (b) then follows from Theorem 7.17.
(c) The assertion (c) follows from (a) and (b).
(d) The equality between $\liminf_{(x,y) \xrightarrow{M} (\bar{x}, \bar{y})} |D_F^* M(x, y)|_{\inf}$ and $|D_L^* M(\bar{x}, \bar{y})|_{\inf}$ as well as the equality between the second and third members in (d) are easily seen from the finite dimension of both X and Y. So, using (c) we see that both equalities in (d) hold true. □

The definition of linear openness of the multimapping $M : X \rightrightarrows Y$ (between normed spaces X and Y) at $\bar{x}$ for $\bar{y} \in M(\bar{x})$ with a constant $c > 0$ can be rewritten (see Definition 7.13) as the existence of $r > 0$ and neighborhoods U and V of $\bar{x}$ and $\bar{y}$ such that

$$c\mathbb{B}_Y \subset \frac{1}{t}\left(M(x + t\mathbb{B}_X) - y\right) \quad \text{for all } 0 < t \leq r, x \in U, y \in V, \text{ with } y \in M(x).$$

Defining the *Frankowska constant* $\operatorname{fran}[M](\bar{x}|\bar{y})$ of M at $\bar{x}$ for $\bar{y} \in M(\bar{x})$ as the supremum of all reals $\rho > 0$ for which there are neighborhoods U and V of $\bar{x}$ and $\bar{y}$ respectively such that, for all $x \in U$ and $y \in V$ with $y \in M(x)$,

(7.13) $$\rho \mathbb{B}_Y \subset \operatorname*{Lim\,sup}_{t \downarrow 0} \frac{1}{t}(M(x + t\mathbb{B}_X) - y),$$

it is then clear that

(7.14) $$\operatorname{ope}[M](\bar{x}|\bar{y}) \leq \operatorname{fran}[M](\bar{x}|\bar{y}).$$

The next theorem shows that equality holds in the context of Banach spaces. We prove first a lemma.

LEMMA 7.30. *Let $M : X \rightrightarrows Y$ be a multimapping between two Banach spaces whose graph is closed near $(\overline{x}, \overline{y}) \in \operatorname{gph} M$. Let $0 \leq \eta < \rho$ be two reals for which there are neighborhoods U and V of $\overline{x}$ and $\overline{y}$ respectively such that, for all $x \in U$ and $y \in V$ with $y \in M(x)$,*

$$\rho \mathbb{B}_Y \subset \operatorname{Lim\,sup}_{t \downarrow 0} \frac{1}{t}(M(x + t\mathbb{B}_X) - y) + \eta \mathbb{B}_Y.$$

Then $\rho - \eta \leq \operatorname{ope}[M](\overline{x}|\overline{y})$.

PROOF. We may suppose that Y is not zero and that $\operatorname{gph} M$ is closed and $\operatorname{ope}[M](\overline{x}|\overline{y})$ is finite. Fix any $c, c' \in \mathbb{R}$ with $\operatorname{ope}[M](\overline{x}|\overline{y}) < c < c'$ and fix also $\delta_0 > 0$ such that

$$\rho \mathbb{B}_Y \subset \operatorname{Lim\,sup}_{t \downarrow 0} \frac{1}{t}(M(x + t\mathbb{B}_X) - y) + \eta \mathbb{B}_Y, \forall x \in B(\overline{x}, \delta_0), y \in B(\overline{y}, \delta_0) \text{ with } y \in M(x).$$

Fix a positive real $\gamma < \min\{1/(\rho + \eta), 1/c'\}$ and define $\|(\zeta, \xi)\|_\gamma := \max\{\|\zeta\|, \gamma\|\xi\|\}$ for all (ζ, ξ) in $X \times Y$. For any $0 < \delta < \delta_0$ there exist a positive real $r < \frac{1}{2}\min\{\gamma\delta, \delta\}$, a pair $(x, y) \in \operatorname{gph} M$ with $\|x - \overline{x}\| < \delta/2$ and $\|y - \overline{y}\| < \delta/2$, and $z \in B[y, cr]$ such that $M^{-1}(z) \cap B[x, r] = \emptyset$. Let $f_1 : X \times Y \to \mathbb{R}$ with $f_1(\zeta, \xi) = \|\xi - z\|$ and let $f_3 := \Psi_{\operatorname{gph} M}$. Taking any $\mu \in]c, c'[$ the arguments and notations of Step I in the proof of (b) in Theorem 7.29 furnish some minimizer $(x', y') \in \operatorname{gph} M$ of the function $f_1 + f_2 + f_3$, where $f_2(\zeta, \xi) := \mu\|(\zeta, \xi) - (x', y')\|_\gamma$ for all $(\zeta, \xi) \in X \times Y$, in such a way that the minimizer (x', y') satisfies $\|(x', y') - (x, y)\|_\gamma < r$, so $\|x' - x\| < r$, $\|x' - \overline{x}\| < \delta_0$, $\|y' - \overline{y}\| < \delta_0$. For any $v \in \mathbb{B}_Y$ there exist $w \in \mathbb{B}_Y$ and sequences $(u_n)_n$ in $\mathbb{B}_X$, $(t_n)_n$ in $]0, +\infty[$ with $t_n \downarrow 0$, and $(e_n)_n$ in Y with $e_n \to 0$ such that $y' + t_n(\rho v + \eta w + e_n) \in M(x' + t_n u_n)$ for all n. Then, for all n

$$\left(\|y' - z + t_n(\rho v + \eta w + e_n)\| - \|y' - z\|\right) + \mu\|(t_n u_n, t_n(\rho v + \eta w + e_n))\|_\gamma \geq 0,$$

which yields with $z' := y' - z \neq 0$

$$\mu = \mu \max\{1, \gamma(\rho + \eta)\} \geq \mu \lim_{n \to \infty} \frac{1}{t_n} \max\{t_n \|u_n\|, t_n \gamma \|\rho v + \eta w + e_n\|\}$$

$$= \mu \lim_{n \to \infty} \frac{1}{t_n} \|(t_n u_n, t_n(\rho v + \eta w + e_n))\|_\gamma$$

$$\geq \lim_{n \to \infty} \frac{1}{t_n}(\|z'\| - \|z' + t_n(\rho v + \eta w + e_n)\|).$$

Observing that $\|z' + t_n(\rho v + \eta w + e_n)\| \leq \|z' + t_n \rho v\| + t_n \|e_n\| + t_n \eta$, it ensues that $\mu \geq -\eta + \sup_{v \in \mathbb{B}_Y} \varphi(v)$, where $\varphi(v) := \lim_{n \to \infty} \frac{1}{t_n}(\|z'\| - \|z' + t_n \rho v\|)$. Since $z' \neq 0$, one can check with $v_0 := -z'/\|z'\| \in \mathbb{B}_Y$ that $\varphi(v_0) = \rho$, so $\sup_{v \in \mathbb{B}_Y} \varphi(v) = \rho$ thanks to the inequality $\varphi(v) \leq \rho$ for all $v \in \mathbb{B}_Y$. Consequently, $\mu \geq \rho - \eta$, hence the desired inequality $\rho - \eta \leq \operatorname{ope}[M](\overline{x}|\overline{y})$ follows. □

Since we already observed in (7.14) that $\operatorname{ope}[M](\overline{x}|\overline{y}) \leq \operatorname{fran}[M](\overline{x}|\overline{y})$ and since taking, in the above lemma with $\eta = 0$, the supremum over $\rho > 0$ satisfying (7.13) gives the reverse inequality, the following theorem is established.

THEOREM 7.31 (H. Frankowska: determination of openness rate). Let $M : X \rightrightarrows Y$ be a multimapping between two Banach spaces whose graph is closed near $(\bar{x}, \bar{y}) \in \operatorname{gph} M$. Then, the Frankowska constant and the linear rate of openness of M at $\bar{x}$ for $\bar{y}$ coincide, that is,
$$\operatorname{fran}[M](\bar{x}|\bar{y}) = \operatorname{ope}[M](\bar{x}|\bar{y}).$$

For $(x, y) \in \operatorname{gph} M$ we denote by $T^B M(x, y) : X \rightrightarrows Y$ the multimapping whose graph is the Bouligand-Peano tangent cone to the graph of M at (x, y), that is,
$$\operatorname{gph} T^B M(x, y) = T^B (\operatorname{gph} M; (x, y)).$$
Then, another consequence of Lemma 7.30 is the following.

THEOREM 7.32 (J.-P. Aubin: tangential condition for openness). Let $M : X \rightrightarrows Y$ be a multimapping between two Banach spaces whose graph is closed near $(\bar{x}, \bar{y}) \in \operatorname{gph} M$. If there exist reals $c > \eta \geq 0$ and neighborhoods U and V of $\bar{x}$ and $\bar{y}$ respectively such that
$$c \mathbb{B}_Y \subset T^B M(x, y)(\mathbb{B}_X) + \eta \mathbb{B}_Y \quad \text{for all } x \in U, y \in V, \text{ with } y \in M(x),$$
then M is linearly open, or equivalently metrically regular, at $\bar{x}$ for $\bar{y}$; further for any $0 < \gamma < c - \eta$, the reals γ and $1/\gamma$ are constants of linear openness and metric regularity respectively.

PROOF. Fix any $x \in U$ and $y \in V$ with $y \in M(x)$. Take any $v \in \mathbb{B}_Y$ and choose by the assumption some $u \in \mathbb{B}_X$ and $w \in \mathbb{B}_Y$ such that $(u, cv - \eta w) \in T^B(\operatorname{gph} M; (x, y))$. There are sequences $(t_n)_n$ in $]0, +\infty[$ tending to 0, $(u_n)_n$ and $(z_n)_n$ converging to u and $cv - \eta w$ in X and Y respectively satisfying $y + t_n z_n \in M(x + t_n u_n) \subset M(x + t_n \mathbb{B}_X)$, hence $z_n \in t_n^{-1}(M(x + t_n \mathbb{B}_X) - y)$ for all n. Consequently, $cv - \eta w \in \operatorname{Lim\,sup}_{t \downarrow 0} \frac{1}{t}(M(x + t\mathbb{B}_X) - y)$, hence $c\mathbb{B}_Y \subset \operatorname{Lim\,sup}_{t \downarrow 0} \frac{1}{t}(M(x + t\mathbb{B}_X) - y) + \eta \mathbb{B}_Y$, which implies $0 < c - \eta \leq \operatorname{ope}[M](\bar{x}|\bar{y})$ according to Lemma 7.30. □

Coming back to coderivative conditions for metric regularity, the assertions (c) and (d) below translate in a more amenable way the characterization of Theorem 7.29(c) in the context of Asplund spaces.

THEOREM 7.33 (F-coderivative criteria for metric regularity). Let $M : X \rightrightarrows Y$ be a multimapping whose graph is closed near $(\bar{x}, \bar{y}) \in \operatorname{gph} M$. Assume that X and Y are Asplund spaces. The following are equivalent:
(a) the multimapping M is metrically regular at $\bar{x}$ for $\bar{y}$;
(b) $\liminf_{(x,y) \xrightarrow{M} (\bar{x},\bar{y})} |D_F^* M(x, y)|_{\inf} > 0$;
(c) there exists some real $c > 0$ and some neighborhoods U of $\bar{x}$ and V of $\bar{y}$ such that $\|x^*\| \geq c \|y^*\|$ for all $x^* \in D_F^* M(x, y)(y^*)$ with $(x, y) \in \operatorname{gph} M$, $x \in U$ and $y \in V$.

If the Asplund space Y is not reduced to zero, the following property can be added to the list of equivalences:
(d) there exists some real $c > 0$ and some neighborhoods U of $\bar{x}$ and V of $\bar{y}$ such that $\|x^*\| \geq c$ for all $x^* \in D_F^* M(x, y)(y^*)$ with $\|y^*\| = 1$ and $(x, y) \in \operatorname{gph} M$, $x \in U$ and $y \in V$.

PROOF. We may suppose that Y is not reduced to zero. Under the assumption of Asplund property of X and Y the equivalence between (a) and (b) follows from (c) in Theorem 7.29. Obviously, the assertion (b) is equivalent to (d), and (d) is equivalent to (c). □

Since the Ioffe approximate subdifferential possesses full calculus on general Banach spaces, we also have the following theorem. Given a subset S of a Banach space which is closed near $\bar{u}$, recall that S is A-normally regular near a point $\bar{u} \in S$ provided there is an open neighborhood U of $\bar{u}$ such that for all $u \in S \cap U$ one has $N^A(S; u) = N^F(S; u)$.

THEOREM 7.34 (metric regularity through A-coderivative). Let X and Y be two Banach spaces and $M : X \rightrightarrows Y$ be a multimapping whose graph is closed near $(\bar{x}, \bar{y}) \in \operatorname{gph} M$.
(a) The multimapping M is metrically regular at $\bar{x}$ for $\bar{y}$ whenever there exists some real $c > 0$ and some neighborhoods U of $\bar{x}$ and V of $\bar{y}$ such that $\|x^*\| \geq c$ for all $x^* \in D_A^* M(x,y)(y^*)$ with $\|y^*\| = 1$ and $(x, y) \in \operatorname{gph} M$, $x \in U$ and $y \in V$.
(b) Under the A-normal regularity of the set $\operatorname{gph} M$ near $(\bar{x}, \bar{y})$, the statement in (a) becomes an equivalence, and in fact
$$\operatorname{ope}[M](\bar{x}|\bar{y}) = \liminf_{(x,y) \xrightarrow{M} (\bar{x},\bar{y})} |D_A^* M(x,y)|_{\inf}.$$

PROOF. The assertion (a) follows from (b) in Theorem 7.29 and the assertion (b) is a consequence of both (a) and (b) in Theorem 7.29. □

Given a subdifferential ∂ on $X \times Y$, one defines the *kernel* of the ∂-coderivative of $M : X \rightrightarrows Y$ at $(x, y) \in \operatorname{gph} M$ as
$$\operatorname{Ker} D_\partial^* M(x,y) := \{y^* \in Y^* : 0 \in D_\partial^* M(x,y)(y^*)\}.$$
In the finite-dimensional setting, (d) in Theorem 7.29 directly justifies the following criterion involving the kernel of the Mordukhovich limiting coderivative.

THEOREM 7.35 (B.S. Mordukhovich: criterion for metric regularity). Let X and Y be two finite-dimensional normed spaces and $M : X \rightrightarrows Y$ be a multimapping with closed graph near $(\bar{x}, \bar{y}) \in \operatorname{gph} M$. Then M is metrically regular at $\bar{x}$ for $\bar{y}$ if and only if $\operatorname{Ker} D_L^* M(\bar{x}, \bar{y}) = \{0\}$.

Except Theorem 7.35 in finite dimensions, the above sufficient conditions involve the coderivative of M at points near the reference point $(\bar{x}, \bar{y})$. A compactness concept will allow us to get, in the infinite-dimensional setting, sufficient conditions via the coderivative at only the reference point.

DEFINITION 7.36. Let $M : X \rightrightarrows Y$ be a multimapping between two Banach spaces and $(\bar{x}, \bar{y}) \in \operatorname{gph} M$, and let ∂ be a subdifferential on $X \times Y$. We say that M is ∂-*coderivatively compact* or ∂-*codirectionally compact* (resp. *sequentially ∂-coderivatively* or *sequentially ∂-codirectionally compact*) at $\bar{x}$ for $\bar{y}$ whenever for any net (resp. sequence) $(x_n, y_n, x_n^*, y_n^*)_n$ with $x_n^* \in D_\partial^* M(x_n, y_n)(y_n^*)$, $x_n^* \to 0$ strongly and $y_n^* \to 0$ weakly*, one has $y_n^* \to 0$ strongly. For the subdifferential ∂_A, ∂_F or ∂_L we simply say A-coderivatively, F-coderivatively or L-coderivatively.

Obviously, the coderivative compactness entails the sequential one. Further, M is clearly coderivatively compact at any point of its domain whenever the space Y is finite-dimensional.

7.3. CRITERIA FOR METRIC REGULARITY

The necessary condition for metric regularity in Theorem 7.29(c) is amenable essentially in Asplund spaces. Indeed, the condition is in terms of Fréchet normal cone and this normal cone enjoys fuzzy calculus merely in Asplund spaces. So, in preparation for the next characterization with the approximate normal cone at the reference point under coderivative compactness, the following lemma is needed.

LEMMA 7.37. *Let X and Y be two Banach spaces and $M : X \rightrightarrows Y$ be a multimapping whose graph is closed near $(\bar{x}, \bar{y}) \in \operatorname{gph} M$. If M is not metrically regular at $\bar{x}$ for $\bar{y}$, then there exist sequences $\sigma_n \to 0$ with $\sigma_n > 0$, $x_n \to \bar{x}$, $y_n \to \bar{y}$ and $z_n \to \bar{y}$ such that for every integer n the following properties (i) and (ii) hold:*
(i) $y_n \notin M(x_n)$ and $z_n \in M(x_n)$;
(ii) *for all $(x, y) \in \operatorname{gph} M$*
$$\|z_n - y_n\| \le \|y - y_n\| + \sigma_n(\|x - x_n\| + \|y - z_n\|).$$

PROOF. We may suppose that $\operatorname{gph} M$ is closed. Since M is not metrically regular at $\bar{x}$ for $\bar{y}$, by Proposition 7.21 for each integer $n \in \mathbb{N}$ there exist $x'_n \in B(\bar{x}, 1/n)$ and $y_n \in B(\bar{y}, 1/n)$ such that $y_n \notin M(x'_n)$ and
$$d(y_n, M(x'_n)) < 1/n \quad \text{and} \quad d(x'_n, M^{-1}(y_n)) > n \, d(y_n, M(x'_n)).$$
Fix any $n \in \mathbb{N}$. By the above inequalities we can choose $y'_n \in M(x'_n)$ satisfying

(7.15) $$\|y'_n - y_n\| < \min\left\{\frac{1}{n}, \frac{1}{n}d(x'_n, M^{-1}(y_n))\right\}.$$

Note that $\varepsilon_n := \|y'_n - y_n\| > 0$ (since $y_n \notin M(x'_n)$) and that, for the continuous function f_n on $X \times Y$ defined by $f_n(x, y) := \|y - y_n\|$, we clearly have
$$f_n(x'_n, y'_n) \le \inf_{\operatorname{gph} M} f_n + \varepsilon_n.$$
Putting $\lambda_n := \min\{n\varepsilon_n, \sqrt{\varepsilon_n}\}$, the Ekeland variational principle provides for each $n \in \mathbb{N}$ some (x_n, z_n) in $\operatorname{gph} M$ such that $\|x_n - x'_n\| + \|z_n - y'_n\| \le \lambda_n$ and
$$f_n(x_n, z_n) \le f_n(x, y) + \sigma_n(\|x - x_n\| + \|y - z_n\|) \quad \text{for all } (x, y) \in \operatorname{gph} M,$$
where $\sigma_n := \varepsilon_n / \lambda_n = \max\{1/n, \sqrt{\varepsilon_n}\}$. Writing by (7.15)
$$\|x_n - x'_n\| \le \lambda_n \le n\varepsilon_n < d(x'_n, M^{-1}(y_n)),$$
we see that $y_n \notin M(x_n)$. Further, evidently $\sigma_n \to 0$ and $y_n \to \bar{y}$, and since $\lambda_n \to 0$ we also have $x_n \to \bar{x}$ and $z_n \to \bar{y}$. The proof is then finished. □

THEOREM 7.38 (metric regularity under coderivative compactness). *Let X and Y be two Banach spaces (resp. Asplund spaces) and $M : X \rightrightarrows Y$ be a multimapping whose graph is closed near $(\bar{x}, \bar{y}) \in \operatorname{gph} M$. If M is sequentially A-coderivatively (resp. sequentially F-coderivatively) compact at $\bar{x}$ for $\bar{y}$ and if the condition $\operatorname{Ker} D^*_A M(\bar{x}, \bar{y}) = \{0\}$ (resp. $\operatorname{Ker} D^*_L M(\bar{x}, \bar{y}) = \{0\}$) is satisfied, then M is metrically regular at $\bar{x}$ for $\bar{y}$.*

PROOF. Consider first the case with the ∂_A-coderivative. Let us proceed by contradiction in supposing that M is not metrically regular at $\bar{x}$ for $\bar{y}$. Taking the sequences given by the above lemma, by Lemma 2.213 (of penalization) and by calculus rules (keep in mind that $y_n - z_n \neq 0$, so $\partial\|\cdot\|(z_n - y_n) \subset \mathbb{S}_{Y^*}$) we have
$$(0, 0) \in \{0\} \times \mathbb{S}_{Y^*} + \sigma_n(\mathbb{B}_{X^*} \times \mathbb{B}_{Y^*}) + (1 + \sigma_n)\partial_A d_{\operatorname{gph} M}(x_n, z_n).$$

Then, there exist $(x_n^*, -z_n^*) \in (1+\sigma_n)\partial_A d_{\text{gph } M}(x_n, z_n)$ with $\|x_n^*\| \leq \sigma_n$ and $y_n^* \in Y^*$ with $\|y_n^*\| = 1$ such that $\|y_n^* - z_n^*\| \leq \sigma_n$. Let $(z_{s(j)}^*)_j$ be any weakly* convergent subnet of the bounded sequence $(z_n^*)_n$ and let z^* be its weak* limit. Since $\|y_n^* - z_n^*\| \to 0$, the subnet $(y_{s(j)}^*)_j$ converges weakly* to some y^*. Using the inclusion $(\tau_n x_n^*, -\tau_n z_n^*) \in \partial_A d_{\text{gph } M}(x_n, z_n)$ with $\tau_n := (1+\sigma_n)^{-1}$, by the weak* outer semicontinuity property of approximate subdifferential of locally Lipschitz functions (see Theorem 5.35(a)), we obtain

$$(0, -z^*) \in \partial_A d_{\text{gph } M}(\bar{x}, \bar{y}), \text{ thus } z^* \in \text{Ker } D_A^* M(\bar{x}, \bar{y}).$$

The kernel assumption yields $z^* = 0$. It results that 0 is the weak* limit of every weakly* convergent subnet of the bounded sequence $(z_n^*)_n$, hence all the sequence $(z_n^*)_n$ converges weakly* to 0. The sequential ∂_A-coderivative compactness tells us that the sequence $(z_n^*)_n$ converges strongly to 0, which is a contradiction since $\|y_n^*\| = 1$ and $\|y_n^* - z_n^*\| \leq \sigma_n$ with $\sigma_n \to 0$ as $n \to \infty$. The proof with the ∂_A-coderivative is complete.

Concerning the case with the sequential ∂_F-coderivative compactness and Asplund spaces X and Y, taking Theorem 7.29 into account, it suffices to show that

$$\liminf_{(x,y) \xrightarrow{M} (\bar{x},\bar{y})} |D_F^* M(x,y)|_{\inf} > 0.$$

Suppose that the inequality is not true, that is, the latter limit inferior is null. Then there exists a sequence $(x_n, y_n, x_n^*, y_n^*)_n$ with $(x_n, y_n) \in B((\bar{x}, \bar{y}), 1/n) \cap \text{gph } M$, $\|y_n^*\| = 1$, $\|x_n^*\| \leq 1/n$, and $x_n^* \in D_F^* M(x_n, y_n)(y_n^*)$. The latter yields $(x_n^*, -y_n^*) \in N^F(\text{gph } M; (x_n, y_n))$.

By the weak* sequential compactness of $\mathbb{B}_{Y^*}$ since Y is an Asplund space (see Theorem C.15 in Appendix), let y^* be the w^*-limit of a subsequence $(y_{s(n)}^*)_n$. It ensues that $(0, -y^*) \in N^L(\text{gph } M; (\bar{x}, \bar{y}))$, hence $0 \in D_L^* M(\bar{x}, \bar{y})(y^*)$. The nullity assumption of the kernel of $D_L^* M(\bar{x}, \bar{y})$ ensures the equality $y^* = 0$. The sequential ∂_F-coderivative compactness implies that $\|y_{s(n)}^*\| \to 0$, and this contradicts the equality $\|y_{s(n)}^*\| = 1$ for all n and finishes the proof. $\square$

The Fréchet coderivative compactness is also necessary for the metric regularity in Asplund spaces:

PROPOSITION 7.39. *Let X and Y be Asplund spaces and $M : X \rightrightarrows Y$ be a multimapping whose graph is closed near $(\bar{x}, \bar{y}) \in \text{gph } M$. If M is metrically regular at $\bar{x}$ for $\bar{y}$, then M is sequentially ∂_F-coderivatively compact at $\bar{x}$ for $\bar{y}$.*

PROOF. Suppose that M is metrically regular at $\bar{x}$ for $\bar{y}$ and let any sequence $(x_n, y_n)_n$ in gph M and $(x_n^*, y_n^*)_n$ with $x_n^* \in D_F^* M(x_n, y_n)(y_n^*)$, $x_n^* \to 0$ strongly and $y_n^* \to 0$ weakly star. From Theorem 7.33(c) there exists some real $c > 0$ such that $\|x_n^*\| \geq c\|y_n^*\|$ for all n large enough. Since $\|x_n^*\| \to 0$, it results that $\|y_n^*\| \to 0$, and this confirms the sequential ∂_F-coderivative compactness of M at $\bar{x}$ for $\bar{y}$. $\square$

In addition to the above coderivative conditions (which involve elements of dual spaces) and to the Aubin tangential condition in Theorem 7.32, another primal condition can be obtained through Theorem 7.29.

THEOREM 7.40 (J.-P. Aubin: tangential condition with closed convex hull). *Let X and Y be two Asplund spaces and $M : X \rightrightarrows Y$ be a multimapping whose*

graph is closed near $(\bar{x},\bar{y})\in\operatorname{gph} M$. Assume that there is a real $c>0$ such that
$$c\mathbb{B}_Y \subset \overline{\operatorname{co}}\bigl(T^B M(x,y)(\mathbb{B}_X)\bigr) \quad \text{for all } (x,y)\in \operatorname{gph} M \text{ near } (\bar{x},\bar{y}).$$
Then M is metrically regular at $\bar{x}$ for $\bar{y}$ with the real $\gamma := 1/c$ as a constant of metric regularity.

PROOF. Let U and V be neighborhoods of $\bar{x}$ and $\bar{y}$ respectively such that the inclusion in the statement holds for all $x \in U$ and $y \in V$ with $y \in M(x)$. Fix any pairs (x,y) as above and (x^*,y^*) with $x^* \in D_F^* M(x,y)(y^*)$. Take any $v \in \mathbb{B}_Y$ and choose (according to the assumption) a sequence $(v_n)_n$ converging to v such that $cv_n \in \operatorname{co}\bigl(T^B M(x,y)(\mathbb{B}_X)\bigr)$. We then have $cv_n = \sum_{k \in K_n} t_{n,k} v_{n,k}$, where K_n is a finite set, $t_{n,k} \geq 0$, $\sum_k t_{n,k} = 1$, and $(u_{n,k},v_{n,k})$ in $T^B(\operatorname{gph} M;(x,y))$ with $u_{n,k} \in \mathbb{B}_X$. Since the Fréchet normal cone is contained in the polar of the Bouligand-Peano tangent cone, we obtain $\sum_k t_{n,k}\langle x^*,u_{n,k}\rangle - t_{n,k}\langle y^*,v_{n,k}\rangle \leq 0$, so $-\|x^*\| - c\langle y^*,v_n\rangle \leq 0$. This ensures that $-c\langle y^*,v\rangle \leq \|x^*\|$, thus $c\|y^*\| \leq \|x^*\|$. The assertion (c) of Theorem 7.29 guarantees the desired metric regularity of M with constant $1/c$. $\square$

7.4. Metrically regular transversality of system of sets

Let us consider now subsets $S_1,\cdots,S_m$ of a normed space X and $\bar{x} \in S_1 \cap \cdots \cap S_m$, and let us associate the multimapping $M_{S_1,\cdots,S_m} : X \rightrightarrows X^m$ given by
$$M_{S_1,\cdots,S_m}(x) := -(x,\cdots,x) + S_1 \times \cdots \times S_m \quad \text{for all } x \in X.$$
For any $u = (u_1,\cdots,u_m) \in X^m$, observe that $x' \in M_{S_1,\cdots,S_m}^{-1}(u)$ means $x' \in \bigcap_{i=1}^m (S_i - u_i)$, so
$$d\bigl(x, M_{S_1,\cdots,S_m}^{-1}(u)\bigr) = d\Bigl(x, \bigcap_{i=1}^m (S_i - u_i)\Bigr);$$
endowing X^m with the sum norm, we also note that
$$\begin{aligned} d\bigl((u_1,\cdots,u_m), M_{S_1,\cdots,S_m}(x)\bigr) &= d\bigl((u_1,\cdots,u_m),(S_1-x)\times\cdots\times(S_m-x)\bigr) \\ &= d\bigl((x+u_1,\cdots,x+u_m), S_1\times\cdots\times S_m\bigr) \\ &= d(x+u_1,S_1) + \cdots + d(x+u_m,S_m), \end{aligned}$$
where the last equality is due to the choice of the sum norm on X^m. Consequently, with the above multimapping $M_{S_1,\cdots,S_m}$, with $\bar{y} := (0,\cdots,0) \in M_{S_1,\cdots,S_m}(\bar{x})$ or equivalently $\bar{x} \in S_1 \cap \cdots \cap S_m$, and with a subset Q_0 of X^m containing $(0,\cdots,0)$ in place of the set Q in Definition 7.9, the inequality (7.4) means, for the real $\gamma \geq 0$ and the neighborhoods $U = B[\bar{x},\delta]$ and $V = B[\bar{x},\delta]\times\cdots\times B[\bar{x},\delta]$ (with some real $\delta > 0$) in (7.4), that

$$(7.16) \qquad d\Bigl(x, \bigcap_{i=1}^m (S_i - u_i)\Bigr) \leq \gamma\bigl(d(x+u_1,S_1) + \cdots + d(x+u_m,S_m)\bigr)$$

for all $x \in \bar{x} + \delta \mathbb{B}_X$ and $u_1,\cdots,u_m \in \delta \mathbb{B}_X$ with $(u_1,\cdots,u_m) \in Q_0$. The latter inequality is obviously symmetric with respect to the sets $S_1,\cdots,S_m$ whenever the set Q_0 itself is symmetric in the sense that $(u_1,\cdots,u_m) \in Q_0 \Leftrightarrow (u_{\pi(1)},\cdots,u_{\pi(m)}) \in Q_0$ for every permutation π over $\{1,\cdots,m\}$.

DEFINITION 7.41. Let $S_1, \cdots, S_m$ be subsets of a normed space X and let $\bar{x} \in S_1 \cap \cdots \cap S_m$. When the inequality (7.16) is fulfilled, we say that the sets $S_1, \cdots, S_m$ are *metrically regularly transversal at $\bar{x}$ relative to the subset Q_0 of X^m* containing $(0, \cdots, 0)$. The infimum of all reals $\gamma \geq 0$ for which there is $\delta > 0$ such that (7.16) is satisfied for all $x \in B[\bar{x}, \delta]$ and $u_i \in \delta\mathbb{B}_X$ with $(u_1, \cdots, u_m) \in Q_0$ is called the *rate (or modulus) of metrically regular transversality* of the sets $S_1, \cdots, S_m$ at $\bar{x}$ relative to Q_0. It coincides with the metric regularity rate of $M_{S_1, \cdots, S_m}$ at $\bar{x}$ for $\bar{y} = (0, \cdots, 0)$ relative to Q_0.

When $Q_0 = X^m$, one just says that the intersection of sets is *metrically regularly transversal at $\bar{x}$*, and the inequality becomes

$$(7.17) \qquad d\left(x, \bigcap_{i=1}^{m}(S_i - u_i)\right) \leq \gamma\bigl(d(x+u_1, S_1), \cdots, d(x+u_m, S_m)\bigr)$$

for all $x \in \bar{x} + \delta\mathbb{B}_X$ and $u_1, \cdots, u_m \in \delta\mathbb{B}_X$. The infimum of reals $\gamma \geq 0$ for which there is some $\delta > 0$ fulfilling the latter property is the *rate (or modulus) of metrically regular transversality* of the sets $S_1, \cdots, S_m$ at $\bar{x}$; we denote it by $\text{reg}_\cap [S_1, \cdots, S_m](\bar{x})$.

DEFINITION 7.42. According to the definition of metric subregularity of a multimapping and to the above analysis, when $Q_0 = \{(0, \cdots, 0)\}$, the sets $S_1, \cdots, S_m$ are said to be *metrically subregularly transversal at $\bar{x}$* and this is translated by the existence of a real $\gamma \geq 0$ and a neighborhood U of $\bar{x}$ such that

$$d\left(x, \bigcap_{i=1}^{m} S_i\right) \leq \gamma\bigl(d(x, S_1) + \cdots + d(x, S_m)\bigr)$$

for all $x \in U$. The infimum of reals $\gamma \geq 0$ for which there is some neighborhood U of $\bar{x}$ such that the latter inequality holds for all $x \in U$ is the *rate (or modulus) of metrically subregular tranversality* of the sets $S_1, \cdots, S_m$ at $\bar{x}$; we denote it by $\text{subreg}_\cap [S_1, \cdots, S_m](\bar{x})$.

In finite-dimensional differential geometry, the transversality of two $\mathcal{C}^1$ submanifolds M_1, M_2 of $\mathbb{R}^n$ at a point $m_0 \in M_1 \cap M_2$ is defined either in the tangential form $T(M_1; m_0) - T(M_2; m_0) = \mathbb{R}^n$ or in the equivalent normal form $N(M_1; m_0) \cap (-N(M_2; m_0)) = \{0\}$. We also know that under such a transversality, $M_1 \cap M_2$ is a submanifold at m_0 and the equalities

$$T(M_1 \cap M_2; m_0) = T(M_1; m_0) \cap T(M_2; m_0)$$

and

$$N(M_1 \cap M_2; m_0) = N(M_1; m_0) + N(M_2; m_0).$$

hold true. We will see in Corollary 7.51 that for two closed subsets S_1, S_2 in $\mathbb{R}^n$ and $\bar{x} \in S_1 \cap S_2$, the condition $N^L(S_1; \bar{x}) \cap (-N^L(S_2; \bar{x})) = \{0\}$ is equivalent to the metric condition (7.17). So, the two $\mathcal{C}^1$-submanifolds M_1, M_2 in $\mathbb{R}^n$ are transversal at m_0 (in the sense of differential geometry) if and only if they are metrically regularly transversal in the sense of Definition 7.41. On the other hand, Corollary 7.93 and Proposition 7.97 will say that the metrically subregular transversality furnishes the right calculus for the normal and tangent cones to the intersection $S_1 \cap S_2$ at $\bar{x}$. All together, the above terminology "metric transversality" is then utilized to translate into a metric context the term "transversality" of differential geometry.

Coming back to (7.16) and taking $m = 2$ one obtains

(7.18) $$d(x, (S_1 - u_1) \cap (S_2 - u_2)) \leq \gamma(d(x + u_1, S_1) + d(x + u_2, S_2))$$

for all $x \in \bar{x} + \delta\mathbb{B}_X$ and $u_1, u_2 \in \delta\mathbb{B}_X$ with $(u_1, u_2) \in Q_0$.

For Q_0 in the form $Q_0 = \{0\} \times Q'_0$, where $0 \in Q'_0 \subset X$, the latter inequality ensures

(7.19) $$d(x, S_1 \cap (S_2 - u)) \leq \gamma(d(x, S_1) + d(x + u, S_2))$$

for all $x \in \bar{x} + \delta\mathbb{B}_X$ and $u \in (\delta\mathbb{B}_X) \cap Q'_0$.

On the other hand, taking $\delta' := \delta/2$ we obtain from the latter inequality, for all $x \in \bar{x} + \delta'\mathbb{B}_X$ and $u_1, u_2 \in (\delta'\mathbb{B}_X) \cap Q'_0$, that

$$d(x + u_1, S_1 \cap (S_2 + u_1 - u_2)) \leq \gamma(d(x + u_1, S_1) + d(x + u_2, S_2)),$$

or equivalently

$$d(x, (S_1 - u_1) \cap (S_2 - u_2)) \leq \gamma(d(x + u_1, S_1) + d(x + u_2, S_2)).$$

So, in particular, the two sets S_1 and S_2 are in metrically regular transversality at $\bar{x}$ with γ as a constant of metrically regular transversality if and only if, for some $\delta > 0$, the inequality (7.19) holds true for all $x \in \bar{x} + \mathbb{B}_X$ and $u \in \delta\mathbb{B}_X$.

Now let $M : X \rightrightarrows Y$ be a multimapping between two vector spaces and $(\bar{x}, \bar{y}) \in \text{gph } M$, and let C and Q be subsets of Y with $\bar{y} \in C \cap Q$. Applying (7.19) with $X \times Y$ in place of X, $S_1 = \text{gph } M$, $S_2 = X \times C$, and $Q'_0 = \{0\} \times (\bar{y} - Q)$ as the subset of $X \times Y$ containing the origin, we obtain reals $\gamma \geq 0$ and $\delta > 0$ such that

$$d((x, y), X \times (C - z) \cap \text{gph } M) \leq \gamma[d(y + z, C) + d((x, y), \text{gph } M)]$$

for all $x \in \bar{x} + \delta\mathbb{B}_X$, $y \in \bar{y} + \delta\mathbb{B}_Y$ and $z \in (\delta\mathbb{B}_Y) \cap (\bar{y} - Q)$.

Since $u \in M^{-1}(C - z)$ whenever $(u, c - z) \in X \times (C - z) \cap \text{gph } M$ for some $c \in C$, we see that (for $X \times Y$ endowed with the sum norm)

$$d((x, y), X \times (C - z) \cap \text{gph } M)$$
$$= \inf\{d(x, x') + d(y, y') : (x', y') \in X \times Y, y' \in (C - z) \cap M(x')\}$$
$$\geq \inf\{d(x, x') + d(y, y') : x' \in M^{-1}(C - z), y' \in Y\}$$
$$= \inf\{d(x, x') : x' \in M^{-1}(C - z)\} + \inf\{d(y, y') : y' \in Y\}$$
$$= d(x, M^{-1}(C - z)).$$

So, when the above inequality related to $X \times (C - z)$ and $\text{gph } M$ is satisfied, we also have

(7.20) $$d(x, M^{-1}(C - z)) \leq \gamma[d(y + z, C) + d((x, y), \text{gph } M)]$$

for all $x \in \bar{x} + \delta\mathbb{B}_X$, $y \in \bar{y} + \delta\mathbb{B}_Y$ and $z \in (\delta\mathbb{B}_Y) \cap (\bar{y} - Q)$. When the latter inequality is fulfilled, the multimapping M is said to be *metrically regular at $\bar{x}$ for the point $\bar{y}$ and the subset C of Y relative to Q*.

If we apply (7.20) with $C = \{\bar{y}\}$ and take $z = \bar{y} - y$ with $y \in Q$, we obtain that, for all $x \in \bar{x} + \delta\mathbb{B}_X$ and $y \in (\bar{y} + \delta\mathbb{B}_Y) \cap Q$,

$$d(x, M^{-1}(y)) \leq \gamma d((x, y), \text{gph } M),$$

which is the graphical metric regularity relative to Q in the form (7.7). So, this and all the analysis above can be summarized in the following theorem.

THEOREM 7.43 (equivalence of various metric regularities). All the forms (7.4)-(7.20) of relative metric regularity are equivalent in the sense that each one can be deduced from any other one.

In particular, for $Q = X$ (resp. $Q = \{\bar{x}\}$), $Q_0 = X^m$ (resp. $Q_0 = \{(0,\cdots,0)\}$), and $Q_0' = X$ (resp. $Q_0' = \{0\} \times \{0\}$) all the corresponding forms of metric regularity (resp. metric subregularity) and metrically regular transversality (resp. metrically subregular transversality) are equivalent in the sense that each one can be deduced from any other one.

In addition to the above theorem, the next result shows how the metrically regular transversality of finitely many sets can be directly reduced to that of two sets.

PROPOSITION 7.44. With the normed space X and above notations, let $C := S_1 \times \cdots \times S_m$ and let D denote the diagonal of X^m, that is, $D := \{(x,\cdots,x) : x \in X\}$. Then the sets $S_1,\cdots,S_m$ are metrically regularly transversal at $\bar{x} \in \bigcap_{i=1}^{m} S_i$ if and only if the two sets C and D of the normed space X^m are metrically regularly transversal at $(\bar{x},\cdots,\bar{x})$.

PROOF. Equip X^m with the sum norm. Consider $x \in X$ and $u_i, v_i \in X$, for $i = 1,\cdots,m$. We observe that

$$(v_1,\cdots,v_m) \in (C - (u_1,\cdots,u_m)) \cap D \Leftrightarrow v_1 = \cdots = v_m \in \bigcap_{i=1}^{m}(S_i - u_i),$$

hence

$$d((x,\cdots,x),(C-(u_1,\cdots,u_m)) \cap D) = \inf_{v \in \bigcap_{i=1}^{m}(S_i - u_i)} (\|x-v\| + \cdots + \|x-v\|)$$

$$= m\, d\left(x, \bigcap_{i=1}^{m}(S_i - u_i)\right).$$

Consequently,

$$d((x,\cdots,x),(C-(u_1,\cdots,u_m)) \cap D)$$
$$\leq \gamma[d((x,\cdots,x), S - (u_1,\cdots,u_m)) + d((x,\cdots,x), D)]$$

means

$$m\, d\left(x, \bigcap_{i=1}^{m}(S_i - u_i)\right) \leq \gamma d((x,\cdots,x),(S_1 - u_1) \times \cdots \times (S_m - u_m)),$$

or equivalently

$$d\left(x, \bigcap_{i=1}^{m}(S_i - u_i)\right) \leq (\gamma/m)[d(x, S_1 - u_1) + \cdots + d(x, S_m - u_m)].$$

From the latter inequality we see first that the metrically regular transversality of C and D at $(\bar{x},\cdots,\bar{x})$ entails the metrically regular transversality of $S_1,\cdots,S_m$ at $\bar{x}$.

Conversely, suppose that $S_1,\cdots,S_m$ are metrically regularly transversal at $\bar{x}$. Then there exit neighborhoods W of $\bar{x}$ and U of zero along with a real $\beta \geq 0$ such that, for all $x \in W$ and $u_i \in U$, $i = 1, \cdots, m$,

$$d\left(x, \bigcap_{i=1}^{m}(S_i - u_i)\right) \leq \beta[d(x, S_1 - u_1) + \cdots + d(x, S_m - u_m)].$$

So, fixing $u_1, \cdots, u_m \in U$, this means, according to what precedes, that for all $x \in W$,

$$0 \leq \beta\, d((x, \cdots, x), C - (u_1, \cdots, u_m)) - m\, d((x, \cdots, x), (C - (u_1, \cdots, u_m)) \cap D),$$

or equivalently, for all $(x_1, \cdots, x_m) \in D \cap (W \times \cdots \times W)$,

$$0 \leq \beta\, d((x_1, \cdots, x_m), C - (u_1, \cdots, u_m)) - m\, d((x_1, \cdots, x_m), (C - (u_1, \cdots, u_m)) \cap D).$$

Since the expression on the right-hand side is a Lipschitz function of $(x_1, \cdots, x_m)$ with $\beta + m$ as a Lipschitz constant, it follows (according to Lemma 2.213) that for every $\chi \in (x_1, \cdots, x_m) \in W \times \cdots \times W$

$$0 \leq \beta\, d(\chi, C - (u_1, \cdots, u_m)) - m\, d(\chi, (C - (u_1, \cdots, u_m)) \cap D)$$
$$+ (\beta + m) d(\chi, D \cap (W \times \cdots \times W)).$$

By Lemma 2.219, there is a neighborhood $W_0 \subset W$ of $\bar{x}$ such that

$$d((x_1, \cdots, x_m), D \cap (W \times \cdots \times W)) = d((x_1, \cdots, x_m), D) \quad \text{for all } x_i \in W_0.$$

It results that, for all $x_i \in W_0$ and $u_i \in U$, $i = 1, \cdots, m$, the inequality

$$m\, d((x_1, \cdots, x_m), (C - (u_1, \cdots, u_m)) \cap D)$$
$$\leq (\beta + m)[d((x_1, \cdots, x_m), C - (u_1, \cdots, u_m)) + d((x_1, \cdots, x_m), D)]$$

is fulfilled, which translates the metrically regular transversality of the two sets C and D at $(\bar{x}, \cdots, \bar{x})$ according to the characterization form (7.19) of metrically regular transversality of two sets. This finishes the proof. □

Let us come back to the multimapping $M_{S_1, \cdots, S_m}(x) := (S_1 - x) \times \cdots \times (S_m - x)$ from X into X^m associated with the sets $S_1, \cdots, S_m$. Let $\bar{x} \in S_1 \cap \cdots \cap S_m$ and $\bar{y} = (0, \cdots, 0)$. For $x_i \in S_i$ or equivalently $(x_1 - x, \cdots, x_m - x) \in M(x)$, where $M := M_{S_1, \cdots, S_m}$, the inclusion

$$B[x_1 - x, r] \times \cdots \times B[x_m - x, r] \subset M(x + s\mathbb{B}_X)$$

signifies

$$B[x_1 - x, r] \times \cdots \times B[x_m - x, r] \subset \bigcup_{\zeta \in \mathbb{B}_X}(S_1 \times \cdots \times S_m - (x, \cdots, x) - s(\zeta, \cdots, \zeta)),$$

which is equivalent to

$$\left(\bigcap_{i=1}^{m}(S_i - x_i - rb_i)\right) \cap s\mathbb{B}_X \neq \emptyset \quad \text{for all } b_i \in \mathbb{B}_X.$$

Consequently, Theorem 7.17 and Proposition 7.23 justify the following theorem:

THEOREM 7.45 (inverse of rate of metrically regular transversality). Let be given subsets $S_1, \cdots, S_m$ of a normed space X which are closed near a point $\bar{x}$ in the intersection $S_1 \cap \cdots \cap S_m$. The following assertions hold.
(a) The inverse of the rate of metrically regular transversality of the sets $S_1, \cdots, S_m$

at $\bar{x}$, that is, $\big(\mathrm{reg}_\cap [S_1, \cdots, S_m](\bar{x})\big)^{-1}$, coincides with the supremum of all reals $c \geq 0$ for which there exist $r, \delta > 0$ such that, for all $t \in [0, 1]$ and $x_i \in S_i \cap B[\bar{x}, \delta]$,

$$\left(\bigcap_{i=1}^m (S_i - x_i - ctrb_i)\right) \cap tr\mathbb{B}_X \neq \emptyset, \quad \forall b_i \in \mathbb{B}_X.$$

Furthermore, one has the other equality

$$\big(\mathrm{reg}_\cap [S_1, \cdots, S_m](\bar{x})\big)^{-1} = \liminf_{s \downarrow 0; x_i \xrightarrow{S_i} \bar{x}} \frac{1}{s}\theta(s, x_1, \cdots, x_m),$$

where $\theta(s, x_1, \cdots, x_m)$ is the supremum of all reals $r \geq 0$ such that

$$\left(\bigcap_{i=1}^m (S_i - x_i - rb_i)\right) \cap s\mathbb{B}_X \neq \emptyset \quad \text{for all } b_i \in \mathbb{B}_X.$$

(b) The sets $S_1, \cdots, S_m$ are metrically regularly transversal at $\bar{x}$ if and only if there are reals $c, r, \delta > 0$ such that for any $t \in [0, 1]$ and $x_i \in S_i \cap B[\bar{x}, \delta]$, $i = 1, \cdots, m$,

$$\left(\bigcap_{i=1}^m (S_i - x_i - ctr\, b_i)\right) \cap tr\mathbb{B}_X \neq \emptyset \quad \text{for all } b_i \in \mathbb{B}_X.$$

The function $s \mapsto \theta(s, x_1, \cdots, x_m)$ in the above theorem is obviously *nondecreasing*. Further, for any real $0 \leq r < \theta(s, x_1, \cdots, x_m)$, we obviously have

$$r\mathbb{B}_X \subset S_i - x_i + s\mathbb{B}_X \quad \text{for all } i = 1, \cdots, m.$$

From this we can see that the function θ enjoys the following other properties:

PROPOSITION 7.46. *The following hold:*
(a) *If* $0 \notin \bigcap_{i=1}^m \mathrm{int}(\mathrm{cl}\, S_i - x_i)$, *then* $\theta(s, x_1, \cdots, x_m) \to 0$ *as* $s \downarrow 0$.
(b) *If the sets* $S_1, \cdots, S_m$ *are closed, then the assertion* (a) *is an equivalence.*

PROOF. For simplicity, put $\theta(s) := \theta(s, x_1, \cdots, x_m)$.
(a) Suppose that $\theta(s) \not\to 0$ as $s \downarrow 0$. There exists an $\varepsilon > 0$ and a sequence $s_n \to 0$ with $\theta(s_n) > \varepsilon$, and hence by what precedes the proposition we have $\varepsilon\mathbb{B}_X \subset S_i - x_i + s_n\mathbb{B}_X$. Then $\varepsilon\mathbb{B}_X \subset \mathrm{cl}\, S_i - x_i$, thus $0 \in \bigcap_{i=1}^m \mathrm{int}\,(\mathrm{cl}\, S_i - x_i)$.

(b) Suppose the closedness of S_i and $0 \in \bigcap_{i=1}^m (\mathrm{int}\, S_i - x_i)$. There is $\varepsilon > 0$ such that, for each i we have $\varepsilon\mathbb{B}_X \subset S_i - x_i$, hence, for all $b_i \in \mathbb{B}_X$, the inclusion $0 \in S_i - x_i - \varepsilon b_i$ holds true. Thus, for every $s \geq 0$, we obtain $0 \in \left(\bigcap_{i=1}^m (S_i - x_i - \varepsilon b_i)\right) \cap s\mathbb{B}_X$, and this ensures that $\theta(s) \geq \varepsilon$. It results that $\theta(s) \not\to 0$ as $s \downarrow 0$. $\square$

In the case of two sets it is obvious that

$$(S_1 - x_1 - rb_1) \cap (S_2 - x_2 - rb_2) \cap (s + r)\mathbb{B}_X \neq \emptyset \quad \text{for all } b_1, b_2 \in \mathbb{B}_X$$

whenever

$$rb_1 - rb_2 \in (S_1 - x_1) \cap s\mathbb{B}_X - (S_2 - x_2) \cap s\mathbb{B}_X \quad \text{for all } b_1, b_2 \in \mathbb{B}_X,$$

and the latter is equivalent to

$$2r\mathbb{B}_X \subset (S_1 - x_1) \cap s\mathbb{B}_X - (S_2 - x_2) \cap s\mathbb{B}_X.$$

Then, besides the above function θ consider the supremum $w(s, x_1, x_2)$ of all reals $r \geq 0$ such that

$$r\mathbb{B}_X \subset (S_1 - x_1) \cap s\mathbb{B}_X - (S_2 - x_2) \cap s\mathbb{B}_X.$$

Like θ the function $s \mapsto w(s, x_1, x_2)$ is nondecreasing. On the other hand, the assertion (a) in the next lemma says in particular that $w(s, x_1, x_2) \to 0$ as $s \downarrow 0$ (while $\theta(s, x_1, x_2)$ may fail to tend to 0 as $s \downarrow 0$ according to Proposition 7.46 above).

LEMMA 7.47. *With notation above, one has the following inequalities:*
(a) $w(s, x_1, x_2) \leq 2\min\{s, \theta(s + w(s, x_1, x_2)/2, x_1, x_2)\}$;
(b) $w(s, x_1, x_2) \leq 2\theta(2s, x_1, x_2)$;
(c) $2\theta(s, x_1, x_2) \leq w(s + \theta(s, x_1, x_2), x_1, x_2)$;
(d) $2\min\{s, \theta(s, x_1, x_2)\} \leq w(2s, x_1, x_2)$.

PROOF. The inequality $w(s, x_1, x_2) \leq 2s$ follows directly from the definition of $w(\cdot)$. Further, for any $r \geq 0$ in the set over which the supremum in the definition of $w(s, x_1, x_2)$ is taken, the above analysis tells us that

$$(S_1 - x_1 - (r/2)b_1) \cap (S_2 - x_2 - (r/2)b_2) \cap (s + r/2)\mathbb{B}_X \neq \emptyset \quad \text{for all } b_1, b_2 \in \mathbb{B}_X,$$

hence $r/2 \leq \theta(s + r/2, x_1, x_2) \leq \theta(s + w(s, x_1, x_2)/2, x_1, x_2)$ according to the nondecreasing property of $\theta(\cdot, x_1, x_2)$. Consequently,

$$w(s, x_1, x_2) \leq 2\theta(s + w(s, x_1, x_2)/2, x_1, x_2),$$

and hence the inequality in (a) is proved.

Concerning (b) it suffices, according to (a) and the nondecreasing property of $\theta(\cdot, x_1, x_2)$, to write

$$w(s, x_1, x_2) \leq 2\theta(s + w(s, x_1, x_2)/2, x_1, x_2) \leq 2\theta(2s, x_1, x_2).$$

It remains to prove (c) and (d). We may suppose that $\theta(s, x_1, x_2) \neq 0$. Consider any real $0 \leq r < \theta(s, x_1, x_2)$ and consider also any $b \in \mathbb{B}_X$. Then

$$(S_1 - x_1 - rb) \cap (S_2 - x_2 + rb) \cap s\mathbb{B}_X \neq \emptyset,$$

so $2rb \in (S_1 - x_1) \cap (s + r)\mathbb{B}_X - (S_2 - x_2) \cap (s + r)\mathbb{B}_X$. This yields

$$2r\mathbb{B}_X \subset (S_1 - x_1) \cap (s + r)\mathbb{B}_X - (S_2 - x_2) \cap (s + r)\mathbb{B}_X,$$

which entails $2r \leq w(s + r, x_1, x_2)$. The latter inequality and the nondecreasing property of $w(\cdot, x_1, x_2)$ give on the one hand

$$2r \leq w(s + \theta(s, x_1, x_2), x_1, x_2), \quad \text{hence } 2\theta(s, x_1, x_2) \leq w(s + \theta(s, x_1, x_2), x_1, x_2),$$

and this ensures in particular $2\theta(s, x_1, x_2) \leq w(2s, x_1, x_2)$ whenever $\theta(s, x_1, x_2) \leq s$. If $\theta(s, x_1, x_2) > s$, we can choose $r = s$ in what precedes, and the above inequality $2r \leq w(s + r, x_1, x_2)$ becomes $2s \leq w(2s, x_1, x_2)$. Consequently, the inequality $2\min\{s, \theta(s, x_1, x_2)\} \leq w(2s, x_1, x_2)$ holds true, so both assertions (c) and (d) are established. $\square$

The above lemma guarantees that $\liminf\limits_{s \downarrow 0; x_i \xrightarrow{S_i} \overline{x}} \frac{1}{s}\theta(s, x_1, x_2) > 0$ if and only if $\liminf\limits_{s \downarrow 0; x_i \xrightarrow{S_i} \overline{x}} \frac{1}{s}w(s, x_1, x_2) > 0$. This and Theorem 7.45 justify the following theorem:

THEOREM 7.48 (A.Y. Kruger: metric regular transversality theorem for two sets). Let S_1 and S_2 be two sets of a normed space X which are closed near $\bar{x} \in S_1 \cap S_2$. The following assertions are equivalent:
(a) the sets S_1 and S_2 are metrically regularly transversal at $\bar{x}$;
(b) $\liminf\limits_{s \downarrow 0; x_i \underset{S_i}{\to} \bar{x}} \frac{1}{s} w(s, x_1, x_2) > 0$;
(c) there are reals $c, r, \delta > 0$ such that, for all $t \in [0, 1]$,
$$crt\mathbb{B}_X \subset (S_1 - x_1) \cap rt\mathbb{B}_X - (S_2 - x_2) \cap rt\mathbb{B}_X \quad \text{for all } x_i \in S_i \cap B[\bar{x}, \delta].$$

Now let us proceed to dual criteria for m sets $S_1, \cdots, S_m$ of a normed space X and $\bar{x}$ in the intersection. Consider the associated multimapping $M := M_{S_1, \cdots, S_m}$ as defined above with $M(x) = (S_1 - x) \times \cdots \times (S_m - x)$ for all $x \in X$. For the continuous linear mapping A from $X \times X_1 \times \cdots \times X_m$ into $X \times X_1 \times \cdots \times X_m$ with $A(u, u_1, \cdots, u_m) := (u, u_1 + u, \cdots, u_m + u)$, where $X_1 = \cdots = X_m = X$, the graph of M can be written as
$$\text{gph } M = \{(x, x_1 - x, \cdots, x_m - x) : x \in X, (x_1, \cdots, x_m) \in S_1 \times \cdots \times S_m\}$$
$$= A^{-1}(X \times S_1 \times \cdots \times S_m).$$
Further, A is bijective and A^{-1} is continuous with
$$A^{-1}(v, v_1, \cdots, v_m) = (v, v_1 - v, \cdots, v_m - v)$$
for all $(v, v_1, \cdots, v_m) \in X \times X_1 \times \cdots \times X_m$. Noting for the indicator function of gph M that $\Psi_{\text{gph } M} = \Psi_{X \times S_1 \times \cdots \times S_m} \circ A$ and applying subdifferential chain rules of previous chapters, we see that, for the Fréchet, Mordukhovich limiting, or Ioffe approximate normal cone N, we have, for $(x, x_1 - x, \cdots, x_m - x) \in \text{gph } M$,
$$(x^*, x_1^*, \cdots, x_m^*) \in N\big(\text{gph } M; (x, x_1 - x, \cdots, x_m - x)\big)$$
if and only if
$$(x^*, x_1^*, \cdots, x_m^*) \circ A^{-1} \in N\big(X \times S_1 \times \cdots \times S_m; (x, x_1, \cdots, x_m)\big),$$
or equivalently
$$(x^*, x_1^*, \cdots, x_m^*) \circ A^{-1} \in \{0\} \times N(S_1; x_i) \times \cdots \times N(S_m; x_m).$$
Observing, for all $(v, v_1, \cdots, v_m) \in X \times X_1 \times \cdots \times X_m$, that
$$\langle (x^*, x_1^*, \cdots, x_m^*), A^{-1}(v, v_1, \cdots, v_m)\rangle = \langle x^*, v\rangle + \langle x_1^*, v_1 - v\rangle + \cdots + \langle x_m^*, v_m - v\rangle$$
$$= \langle x^* - x_1^* - \cdots - x_m^*, v\rangle + \langle x_1^*, v_1\rangle + \cdots + \langle x_m^*, v_m\rangle$$
$$= \langle (x^* - x_1^* - \cdots - x_m^*, x_1^*, \cdots, x_m^*), (v, v_1, \cdots, v_m)\rangle,$$
we obtain $(x^*, x_1^*, \cdots, x_m^*) \circ A^{-1} = (x^* - x_1^* - \cdots - x_m^*, x_1^*, \cdots, x_m^*)$. Consequently,
$$(x^*, x_1^*, \cdots, x_m^*) \in N\big(\text{gph } M; (x, x_1 - x, \cdots, x_m - x)\big)$$
if and only if
$$x^* = x_1^* + \cdots + x_m^*, \ x_1^* \in N(S_1; x_1), \cdots, x_m^* \in N(S_m; x_m).$$
For the the Fréchet, Mordukhovich limiting, or Ioffe approximate $D^*M(x, y)$, it ensues that
(7.21)
$$D^*M(x, x_1 - x, \cdots, x_m - x)(x_1^*, \cdots, x_m^*) = \{-(x_1^* + \cdots + x_m^*)\} \text{ if } x_i^* \in -N(S_i : x_i),$$
and $D^*M(x, x_1 - x, \cdots, x_m - x) = \emptyset$ otherwise.

From Theorem 7.33 applied with $Y := X^m$ and $\bar{y} := (0, \cdots, 0)$ we then have:

7.4. METRICALLY REGULAR TRANSVERSALITY

THEOREM 7.49 (criteria for metrically regular transversality with F-normals).
Let $S_1, \cdots, S_m$ be subsets of an Asplund space X which are closed near a point $\bar{x}$ of their intersection. The following are equivalent:
(a) the sets $S_1, \cdots, S_m$ are metrically regularly transversal at $\bar{x}$;
(b)
$$\liminf_{\substack{x_i \to \bar{x} \\ S_i}} \left(\inf \left\{ \left\| \sum_{i=1}^m x_i^* \right\| : x_i^* \in N^F(S_i; x_i), \sum_{i=1}^m \|x_i^*\| = 1 \right\} \right) > 0;$$
(c) there exists some real $c > 0$ and some neighborhood U of the point $\bar{x}$ such that $\left\| \sum_{i=1}^m x_i^* \right\| \geq c$ for all $x_i \in S_i \cap U$ and $x_i^* \in N^F(S_i; x_i)$ with $\sum_{i=1}^m \|x_i^*\| = 1$;
(d) there exists some real $c > 0$ and some neighborhood U of the point $\bar{x}$ such that $\left\| \sum_{i=1}^m x_i^* \right\| \geq c \sum_{i=1}^m \|x_i^*\|$ for all $x_i \in S_i \cap U$ and $x_i^* \in N^F(S_i; x_i)$.

Similarly, the description (7.21) for the Ioffe approximate coderivative and Theorem 7.34 plus Theorem 7.49 yield:

THEOREM 7.50 (conditions for metrically regular transversality with A-normals). Let $S_1, \cdots, S_m$ be subsets of a Banach space X which are closed near a point $\bar{x}$ of their intersection. If there exists some real $c > 0$ and some neighborhood U of the point $\bar{x}$ such that $\left\| \sum_{i=1}^m x_i^* \right\| \geq c \sum_{i=1}^m \|x_i^*\|$ for all $x_i \in S_i \cap U$ and $x_i^* \in N^A(S_i; x_i)$, then the sets $S_1, \cdots, S_m$ are metrically regularly transversal at $\bar{x}$.
The converse also holds provided the sets S_i are A-normally regular near $\bar{x}$, that is $N^A(S_i; x) = N^F(S_i; x)$ for $x \in S$ near $\bar{x}$.

In the finite-dimensional setting, we obtain the following result which approves of the terminology "metrically regular transversality" by analogy to the notion of "transversality of submanifolds" in differential geometry:

COROLLARY 7.51. Let $S_1, \cdots, S_m$ be subsets of a finite-dimensional normed space X which are closed near a point $\bar{x}$ of their intersection. Then, the sets $S_1, \cdots, S_m$ are metrically regularly transversal at $\bar{x}$ if and only if
$$\left(x_i^* \in N^L(S_i; \bar{x}), i = 1, \cdots, m, \text{ and } \sum_{i=1}^m x_i^* = 0 \right) \implies x_1^* = \cdots = x_m^* = 0,$$
that is, if and only if these sets are L-normally transversal at $\bar{x}$ according to Definition 7.52 below.

DEFINITION 7.52. Continuing to follow the terminology in differential geometry, given a subdifferential ∂ and sets $S_1, \cdots, S_m$ of a normed space X, we say that these sets are ∂-*normally transversal* (or ∂-*normally transverse*) *at a point* $\bar{x}$ in their intersection provided that
$$\left(x_i^* \in N^\partial(S_i; \bar{x}), i = 1, \cdots, m, \text{ and } \sum_{i=1}^m x_i^* = 0 \right) \implies x_1^* = \cdots = x_m^* = 0.$$
When ∂ is the A-subdifferential (resp. F-subdifferential, L-subdifferential, C-subdifferential), we simply say that the sets are A-normally (resp. F-normally, L-normally, C-normally) transversal (or transverse) at $\bar{x}$.

Taking (7.21) into account, the definition of coderivative compactness for multimappings leads to introduce the following concept of normal compactness for finite systems of sets.

DEFINITION 7.53. Let be given a subdifferential ∂ and sets $S_1, \cdots, S_m$ of a Banach space X which are closed near a point $\bar{x}$ of their intersection. We say that the system of sets $S_1, \cdots, S_m$ is ∂-normally (resp. sequentially ∂-normally) compact at $\bar{x}$ for the intersection operation when for any nets (resp. sequences) $(x_{i,n}, x_{i,n}^*)_n$ in $X \times X^*$ with $x_{i,n}^* \in N^\partial(S_i; x_{i,n})$ and $x_{i,n} \to \bar{x}$, $i = 1, \cdots, m$, one has
$$\left(\left\| \sum_{i=1}^m x_{i,n}^* \right\| \to 0 \text{ and } (x_{1,n}^*, \cdots, x_{m,n}^*) \to (0, \cdots, 0) \text{ weakly}^* \right)$$
$$\implies \|x_{i,n}^*\| \to 0, \ i = 1, \cdots, m.$$

REMARK 7.54. It is clear that the above definition is of interest when the system contains at least two sets, otherwise for a system with a single set the property trivially holds. □

Of course, for any subdifferential ∂ for which the description (7.21) holds true, the multimapping $M = M_{S_1, \cdots, S_m}$ is (sequentially) ∂-coderivatively compact at $\bar{x}$ for $(0, \cdots, 0)$ if and only if the system of sets $S_1, \cdots, S_m$ is (sequentially) ∂-normally compact at $\bar{x}$. On the other hand, it is also clear that the ∂-normal transversality property
$$\left(x_i^* \in N^\partial(S_i; \bar{x}), i = 1, \cdots, m, \text{ and } x_1^* + \cdots + x_m^* = 0 \right) \implies x_1^* = \cdots = x_m^* = 0$$
in Definition 7.52, is equivalent to $\operatorname{Ker} D_\partial^* M(\bar{x}, (0, \cdots, 0)) = \{(0, \cdots, 0)\}$ according to (7.21), where $M := M_{S_1, \cdots, S_m}$. So, from Theorem 7.38 it results:

THEOREM 7.55 (metrically regular transversality under sequentially transversal condition). Let $S_1, \cdots, S_m$ be subsets of a Banach space (resp. Asplund space) X which are closed near a point $\bar{x}$ of their intersection and which are A-normally (resp. L-normally) transversal at $\bar{x}$, that is,
$$\left(x_i^* \in N^A(S_i; \bar{x}) \text{ and } x_1^* + \cdots + x_m^* = 0 \right) \implies x_1^* = \cdots = x_m^* = 0.$$
If the system of sets $S_1, \cdots, S_m$ is sequentially A-normally (resp. sequentially L-normally) compact at $\bar{x}$ for the intersection operation, then these sets are metrically regularly transversal at $\bar{x}$.

The coderivative compactness of $M_{S_1, \cdots, S_m}$ can also be obtained through the concept of normal compactness of sets.

DEFINITION 7.56. Given a subdifferential ∂ on a normed space X, one says that a subset S of X is ∂-normally compact (resp. sequentially ∂-normally compact) at $\bar{x} \in S$ provided that, for any net (resp. sequence) $(x_n)_n$ in S converging to $\bar{x}$ and any net (resp. sequence) $(x_n^*)_n$ converging weakly star to 0 with $x_n^* \in N^\partial(S; x_n)$, one has $\|x_n^*\| \to 0$. For the subdifferential ∂_A, ∂_F or ∂_L we simply say A-normally, F-normally or L-normally compact, and the same for the sequential normal compactness.

By Proposition 2.34 we see that a set S of a Banach space is A-normally compact at $\bar{x} \in S$ whenever it satisfies the interior tangent condition at $\bar{x}$. More generally, the following proposition shows the similar sequential result for sets with the compact tangent property. It readily follows from Proposition 5.85.

PROPOSITION 7.57. Let S be a subset of a Banach space X which is closed near a point $\bar{x} \in S$ and has the compact tangent property at $\bar{x}$. Then the set S is sequentially A-normally compact at $\bar{x}$.

Coming back to the description (7.21) for the Ioffe approximate (resp. Fréchet) coderivative, we see that $M_{S_1,\cdots,S_m}$ is (sequentially) A-coderivatively (resp. (sequentially) F-coderivatively) compact at $\bar{x}$ for $(0,\cdots,0)$ whenever all but at most one of the sets S_i are (sequentially) A-normally (resp. (sequentially) F-normally) compact. From this, Theorem 7.38, and the description (7.21) it results the following:

THEOREM 7.58 (metrically regular transversality under normal compactness). Let $S_1,\cdots,S_m$ be subsets of a Banach (resp. Asplund) space X which are closed near a point $\bar{x}$ in their intersection and which are A-normally (resp. F-normally) transversal at $\bar{x}$, that is, for $x_i^* \in N^A(S_i;\bar{x})$ (resp. $x_i^* \in N^F(S_i;\bar{x})$) $i = 1,\cdots,m$,
$$x_i^* + \cdots + x_m^* = 0 \implies x_1^* = \cdots = x_m^* = 0.$$
If all but at most one of the sets are sequentially A-normally (resp. sequentially F-normally) compact at $\bar{x}$, then the sets $S_1,\cdots,S_m$ are metrically regularly transversal at $\bar{x}$.

In the case of intersection of two closed convex sets, a global metric inequality holds true.

THEOREM 7.59 (metrically regular transversality from convexity). Let S_1, S_2 be two closed convex sets of a normed space $(X, \|\cdot\|)$ and $\bar{x} \in S_1 \cap S_2$.
(a) Assume that there is a real $r > 0$ such that
$$r\mathbb{U}_X \subset (S_1 - \bar{x}) \cap \mathbb{B}_X - (S_2 - \bar{x}) \cap \mathbb{B}_X.$$
Then, with $c := r/(1+r)$ one has, for all $x \in X$ and $y_1, y_2 \in X$ with $\|y_1\| + \|y_2\| < c$,
$$d\big(x, (S_1-y_1)\cap(S_2-y_2)\big) \leq (c-\|y_1\|-\|y_2\|)^{-1}(1+\|x-\bar{x}\|)\big(d(x, S_1-y_1) + d(x, S_2-y_2)\big).$$
(b) If $(X, \|\cdot\|)$ is a Banach space and $0 \in \text{core}(S_1 - S_2)$, then for some real $c > 0$ the inequality in (a) holds true.

PROOF. As we will see below, we derive (b) from (a).
(a) Consider the multimapping $M : X \rightrightarrows X \times X$ with $M(x) := (S_1 - x) \times (S_2 - x)$ for all $x \in X$. We note that $\bar{y} = (0,0) \in M(\bar{x})$ and we easily see that the graph of M is closed and convex. Endow $X \times X$ with the sum norm $\|(x, x')\|_1 := \|x\| + \|x'\|$ and denote by $\mathbb{U}_{1,X\times X}$ the open unit ball of $X \times X$ relative to this norm. Assume the condition in (a) is fulfilled. We claim that
$$\frac{r}{1+r}\mathbb{U}_{1,X\times X} \subset M(\bar{x} + \mathbb{B}_X).$$
Indeed, take any $(u_1, u_2) \in \mathbb{U}_{1,X\times X}$. We have $r\|u_1 - u_2\| < r$, hence by the assumption in (a) there are $x_i \in S_i$ (with $i = 1, 2$) such that
$$r(u_1 - u_2) = (x_1 - \bar{x}) - (x_2 - \bar{x}), \quad \|x_1 - \bar{x}\| \leq 1, \|x_2 - \bar{x}\| \leq 1.$$
With $z := x_1 - \bar{x} - ru_1 = x_2 - \bar{x} - ru_2$ and $z' := (1+r)^{-1}z$ we obtain
$$r(u_1, u_2) = (x_1 - \bar{x} - z, x_2 - \bar{x} - z),$$

which is equivalent to
$$\frac{r}{1+r}(u_1, u_2) = \left(\frac{1}{1+r}(x_1 - \bar{x}) - z', \frac{1}{1+r}(x_2 - \bar{x}) - z'\right).$$
Since 0 belongs to $S_1 - \bar{x}$ and $S_2 - \bar{x}$, it ensues that
$$\frac{r}{1+r}(u_1, u_2) \in (S_1 - \bar{x} - z', S_2 - \bar{x} - z') = M(\bar{x} + z').$$
This and the inequality $\|z'\| \leq 1$ confirms the claim $\frac{r}{1+r}\mathbb{U}_{1,X\times X} \subset M(\bar{x} + \mathbb{B}_X)$. Putting $c := r/(1+r)$ Lemma 7.27 yields, for all $x \in X$ and $y \in c\mathbb{U}_{1,X\times X}$
$$d(x, M^{-1}(y)) \leq (c - \|y_1\| - \|y_2\|)^{-1}(1 + \|x - \bar{x}\|)d(y, M(x)),$$
or equivalently for all $x \in X$ and $y_1, y_2 \in X$ with $\|y_1\| + \|y_2\| < c$,
$$d\big(x, (S_1-y_1)\cap(S_2-y_2)\big) \leq (c-\|y_1\|-\|y_2\|)^{-1}(1+\|x-\bar{x}\|)\big(d(x, S_1-y_1)+d(x, S_2-y_2)\big).$$
(b) Define now M as the multimapping from $X \times X$ into X given by $M(x,y) = x-y$ if $(x,y) \in S_1 \times S_2$ and $M(x,y) = \emptyset$ otherwise and endow $X \times X$ with the max norm. The graph of M is clearly closed and convex and by the assumption in (b) we have $(0,0) \in \operatorname{core} M(X)$. Since $0 \in M(\bar{x}, \bar{x})$, the Robinson-Ursescu theorem (see Theorem 7.26) furnishes some real $r > 0$ such that
$$r\mathbb{B}_X \subset M\big((\bar{x} + \mathbb{B}_X) \times (\bar{x} + \mathbb{B}_X)\big) = S_1 \cap (\bar{x} + \mathbb{B}_X) - S_2 \cap (\bar{x} + \mathbb{B}_X),$$
and this is equivalent to
$$r\mathbb{B}_X \subset (S_1 - \bar{x}) \cap \mathbb{B}_X - (S_2 - \bar{x}) \cap \mathbb{B}_X.$$
The conclusion in (b) then follows from (a). $\square$

Constraints or *feasible sets* in either optimization problems or many other problems, including many equality-inequality constraints, often arise in the representation form $g^{-1}(D)$, where $g : X \to Y$ is a mapping between two normed spaces and D is a subset of Y. Observing that $x \in g^{-1}(D)$ is equivalent to $0 \in g(x) - D$ leads to appeal to the multimapping $M_g = M_{g,D} : X \rightrightarrows Y$ defined by $M_g(x) := g(x) - D$. Since
$$d(x, M_g^{-1}(y)) = d(x, g^{-1}(y+D)) \quad \text{and} \quad d(y, M_g(x)) = d(g(x), y+D),$$
saying, for $\bar{x} \in g^{-1}(D)$, that M_g is metrically regular $\bar{x}$ for $0 \in M_g(\bar{x})$ relative to a set $Q \ni 0$ in the image space Y amounts to saying that there are a real $\gamma \geq 0$ and neighborhoods U of $\bar{x}$ in X and V of 0 in Y such that
$$d(x, g^{-1}(y+D)) \leq \gamma d(g(x), y+D) \quad \text{for all } x \in U, \, y \in V \cap Q.$$
It is worth noticing that the case $D = \{g(\bar{x})\}$ means that the mapping g is metrically regular at $\bar{x}$. As it appears several times above, the cases $Q = X$ and $Q = \{0\}$ are of great interest too. The case $Q = X$ leads to a real $\gamma \geq 0$ and neighborhoods U of $\bar{x}$ in X and V of 0 in Y such that

(7.22) $\qquad d(x, g^{-1}(y+D)) \leq \gamma d(g(x), y+D) \quad \text{for all } x \in U, \, y \in V;$

we then say that *the mapping g is metrically regular at $\bar{x}$ with respect to the subset D of the image space Y*. When the role of $g^{-1}(D)$ as a set-constraint needs to be emphasized, we also say that *the constraint/feasible set-representative $g^{-1}(S)$ or the representative inverse image of the set S by g is metrically regular at $\bar{x}$*. By analogy to the transversality of sets (see Definition 7.42) and to the terminology in differential geometry, we will also say that *the mapping g is at $\bar{x}$ metrically* *regularly transversal to the set* D. The infimum, denoted by reg.$_{,D}[g](\bar{x})$ of reals $\gamma > 0$ satisfying (7.22) is *the rate (or modulus) of metric regularity of g with respect to the subset D of the image space Y (or of the constraint/feasible set-representative $g^{-1}(D)$).*

If $Q = \{0\}$, that is, there are a real $\gamma \geq 0$ and a neighborhood U of $\bar{x}$ in X such that

(7.23) $\qquad d(x, g^{-1}(D)) \leq \gamma d(g(x), D) \quad \text{for all } x \in U,$

we say that *the mapping g is metrically subregular at $\bar{x}$ with respect to the subset D of the image space Y, or the constraint/feasible set-representative $g^{-1}(D)$ is metrically subregular at $\bar{x}$, or the mapping g is at $\bar{x}$ metrically subregularly transversal to the set D.* The infimum, denoted by subreg.$_{,D}[g](\bar{x})$, of such reals $\gamma > 0$ is *the rate (or modulus) of metric subregularity of g with respect to the subset D of the image space Y (or of the constraint/feasible set-representative $g^{-1}(D)$).*

Similarly, when the constraint or feasible set is in the form $C \cap g^{-1}(D)$ with $C \subset X$, for $M(x) = M_{g,C,D}$ defined by $M(x) = g(x) - D$ if $x \in C$ and $M(x) = \emptyset$ if $x \in X \setminus C$, the metric regularity of M at $\bar{x} \in C \cap g^{-1}(D)$ for $0 \in M(\bar{x})$ relative to a set $Q \ni 0$ in Y means that there are a real $\gamma_0 \geq 0$ and neighborhoods U_0 of $\bar{x}$ in X and V of 0 in Y such that

$$d(x, C \cap g^{-1}(y+D)) \leq \gamma_0 d(g(x), y+D) \quad \text{for all } x \in U_0 \cap C,\ y \in V \cap Q.$$

If in addition g is locally Lipschitz, then (by Lemmas 2.213 and 2.219) the metric regularity relative to Q becomes equivalent to the existence of a real $\gamma \geq 0$ and neighborhoods U and V of $\bar{x}$ and 0 respectively such that

$$d(x, C \cap g^{-1}(y+D)) \leq \gamma \Big(d(x, C) + d(g(x), y+D)\Big) \quad \text{for all } x \in U,\ y \in V \cap Q.$$

The case $Q = X$ leads to a real $\gamma \geq 0$ and neighborhoods U of $\bar{x}$ in X and V of 0 in Y such that

(7.24) $\quad d(x, C \cap g^{-1}(y+D)) \leq \gamma \Big(d(x, C) + d(g(x), y+D)\Big) \quad \text{for all } x \in U,\ y \in V;$

we then say as above that *the mapping g is metrically regular at $\bar{x}$ with respect to the subset C of the domain space X and the subset D of the image space Y (or the contraint/feasible set-representative $C \cap g^{-1}(D)$ is metrically regular at $\bar{x}$.*

If $Q = \{0\}$, that is, there are a real $\gamma \geq 0$ and a neighborhood U of $\bar{x}$ in X such that

(7.25) $\qquad d(x, C \cap g^{-1}(D)) \leq \gamma \Big(d(x, C) + d(g(x), D)\Big) \quad \text{for all } x \in U,$

we say that *the mapping g is metrically subregular at $\bar{x}$ with respect to the subset C of the domain space X and the subset D of the image space Y (or the contraint/feasible set-representative $C \cap g^{-1}(D)$ is metrically subregular at $\bar{x}$.*

All the above conditions are *qualification conditions* of the constraint in the sense (as we will see later) that they allow us to compute tangent and normal cones to $g^{-1}(D)$ (resp. $C \cap g^{-1}(D)$) and they guarantee the existence of Lagrange multipliers for optimization problem in the form of minimization of a real-valued function $f(x)$ subject to the constraint $g(x) \in D$ (resp. subject to the constraints $x \in C$ and $g(x) \in D$) when f and g are locally Lipschitz around a minimizer $\bar{x}$.

7.5. Metric regularity of convex feasible sets, Hoffman inequality

We consider in this section the case when the mapping g involved in the end of the previous section is convex with respect to a convex cone; more general nonconvex cases will be studied in the next section. We will utilize concepts and notations in Subsection 2.3.3 and Section 3.8 related to vector valued convex mappings with values in $Y \cup \{+\infty_Y\}$, where Y is a normed vector space endowed with a closed convex cone K and where $+\infty_Y$ is an abstract greatest element adjoined to Y with respect to the preorder $\leq_Y$ associated with the convex cone K.

PROPOSITION 7.60. *Let X and Y be Banach spaces and K be a closed convex cone of Y. Let C be a closed convex set of X and $g : X \to Y \cup \{+\infty_K\}$ be a K-convex mapping whose K-epigraph is closed in $X \times Y$. Let $\bar{x} \in X$ and $\bar{y} \in Y$ with $\bar{x} \in C \cap g^{-1}(-K + \bar{y})$. Then the inclusion*

$$0 \in \mathrm{core}\big((g(C) \cap Y) - \bar{y} + K \big)$$

is a necessary and sufficient condition for the existence of a real $c > 0$ such that, for all $x \in C \cap \mathrm{dom}\, g$ and $y \in B(\bar{y}, c)$

$$d\big(x, g^{-1}(-K + y)\big) \leq (c - \|y - \bar{y}\|)^{-1}(1 + \|x - \bar{x}\|) d(g(x) - y, -K).$$

If in addition g is Lipschitz near $\bar{x}$ (which is the case whenever g is continuous at $\bar{x}$ and K is $\|\cdot\|$-normal), then the above inclusion is also equivalent to the existence of a real $\gamma \geq 0$ and neighborhoods U and V of $\bar{x}$ and $\bar{y}$ respectively such that

$$d\big(x, g^{-1}(-K + y)\big) \leq \gamma \big(d(x, C) + d(g(x) - y, -K)\big) \quad \text{for all } x \in U, y \in V.$$

PROOF. Considering $g_0(x) = g(x) - \bar{y}$ we may and do suppose $\bar{y} = 0$. Define the multimapping $M : X \rightrightarrows Y$ by $M(x) := g(x) + K$ if $x \in C \cap \mathrm{dom}\, g$ and $M(x) = \emptyset$ otherwise. Clearly

$$\mathrm{gph}\, M = \{(x, y) \in X \times Y : x \in C, (x, y) \in \mathrm{epi}_K g\},$$

so $\mathrm{gph}\, M$ is closed and convex and $0 \in \mathrm{core}\, M(X)$. The Robinson-Ursescu theorem (see Theorem 7.26) tells us that the inclusion $0 \in \mathrm{core}\, M(X)$ is a necessary and sufficient condition for the existence of some $c > 0$ such that

$$d(x, M^{-1}(y)) \leq (c - \|y - \bar{y}\|)^{-1}(1 + \|x - \bar{x}\|) d(y, M(x))$$

for all $x \in X$ and $y \in B(\bar{y}, c)$. This clearly yields the first inequality in the theorem.

From this first inequality we easily obtain a real $\gamma \geq 0$ and neighborhoods U of $\bar{x}$ and V of $\bar{y}$ such that $d\big(x, g^{-1}(-K+y)\big) \leq \gamma d(g(x) - y, -K)$ for all $x \in U \cap C \cap \mathrm{dom}\, g$ and $y \in V$. So, if in addition g is Lipschitz near $\bar{x}$ (which holds under the continuity of g at $\bar{x}$ and the normality of K according to Proposition 2.174), shrinking U we may suppose that $U \subset \mathrm{dom}\, g$. The analysis ahead the statement of the theorem tells us that shrinking again U and V as well, and taking in place of the real γ a suitable number greater than that real, the second inequality in the theorem holds true. On the other hand, this second inequality entails the metric regularity of M at $\bar{x}$ for $\bar{y}$, and hence its linear openness, which in turn implies the inclusion condition in the theorem. The second equivalence is then justified. □

When the K-convex mapping is locally Lipschitz, a similar result holds with $K - D$ in place of K, where D is a convex set.

PROPOSITION 7.61. Let X and Y be Banach spaces and K be a closed convex cone of Y. Let C and D be closed convex sets of X and Y respectively, $g : X \to Y$ be a K-convex continuous mapping. Let $\overline{x} \in C$ and $\overline{y} \in Y$ with $g(\overline{x}) \in D - K + \overline{y}$. Assume that $K - D$ is closed. Then the inclusion

$$0 \in \mathrm{core}\big(g(C) + K - D - \overline{y}\big)$$

is a necessary and sufficient condition for the existence of a real $c > 0$ such that, for all $x \in C$ and $y \in B(\overline{y}, c)$

$$d\big(x, g^{-1}(D - K + y)\big) \leq (c - \|y - \overline{y}\|)^{-1}(1 + \|x - \overline{x}\|) d(g(x) - y, D - K).$$

If in addition g is locally Lipschitz (which is the case whenever K is $\|\cdot\|$-normal), the above inclusion is also equivalent to the existence of a real $\gamma \geq 0$ and neighborhoods U of $\overline{x}$ in X and V of $\overline{y}$ in Y such that

$$d\big(x, g^{-1}(D - K + y)\big) \leq \gamma\big(d(x, C) + d(g(x) - y, D - K)\big) \quad \text{for all } x \in U, y \in V.$$

PROOF. Putting $M(x) = g(x) - \overline{y} + K - D$ if $x \in C$ and $M(x) = \emptyset$ if $x \notin C$, we see that $0 \in M(\overline{x})$. Further, from the equality

$$\mathrm{gph}\, M = \{(x, y) \in X \times Y : x \in C, y \in g(x) - \overline{y} + K - D\},$$

the closedness of $\mathrm{gph}\, M$ is clear and using the same equality it is not difficult to check its convexity. So, it suffices to argue as in the proof of Proposition 7.60 above. □

The case when the cone K has the form $K = K_1 \times K_2 \subset Y_1 \times Y_2$ with $\mathrm{int}_{Y_1} K_1 \neq \emptyset$ provides, under the K-convexity of g, another equivalent condition for the metric regularity.

PROPOSITION 7.62. Let X, Y_1, Y_2 be Banach spaces and K_1 and K_2 be closed convex cones of Y_1 and Y_2 respectively. Let C be a closed convex set of X and $g_i : X \to Y_i$ be a K_i-convex continuous mapping with $i = 1, 2$. Let $\overline{x} \in C$ with $g_1(\overline{x}) \leq_{K_1} 0$ and $g_2(\overline{x}) \leq_{K_2} 0$. Assume that $\mathrm{int}_{Y_1} K_1 \neq \emptyset$. Then, with $g(x) := (g_1(x), g_2(x))$ one has

$$(0, 0) \in \mathrm{core}\,\big(g(C) + K_1 \times K_2\big),$$

if and only if both conditions (i) and (ii) are satisfistied:
(i) There exists some $x_0 \in C$ such that $g_2(x_0) \leq_{K_2} 0$ and $g_1(x_0) \in -\mathrm{int}_{Y_1} K_1$;
(ii) $0 \in \mathrm{core}(g_2(C) + K_2)$.

Further, conditions (i) and (ii) are also equivalent to the existence of a real $c > 0$ such that, for all $x \in C$ and $y = (y_1, y_2) \in B_{Y_1}(0, c) \times B_{Y_2}(0, c)$

$$d\big(x, g_1^{-1}(-K_1 + y_1) \cap g_2^{-1}(-K_2 + y_2)\big)$$
$$\leq (c - \|y\|)^{-1}(1 + \|x - \overline{x}\|)\big(d(g_1(x) - y_1, -K_1) + d(g_2(x) - y_2, -K_2)\big),$$

where $\|y\| = \max\{\|y_1\|, \|y_2\|\}$.

PROOF. Suppose that $(0, 0) \in \mathrm{core}\,\big(g(C) + K_1 \times K_2\big)$. This inclusion clearly yields $0 \in \mathrm{core}_{Y_2}\big(g_2(C) + K_2\big)$, that is, condition (ii). Fixing some $k_1 \in \mathrm{int}_{Y_1} K_1$ the same inclusion ensures that $-t(g_1(\overline{x}) + k_1, g_2(\overline{x})) \in (g_1(u), g_2(u)) + K_1 \times K_2$ for some real $t > 0$ and some $u \in C$. It results that

$$0 = \frac{t}{1+t}\big(g_1(\overline{x}) + k_1\big) + \frac{1}{1+t}\big(-t(g_1(\overline{x}) + k_1)\big)$$

$$\in \frac{t}{1+t}g_1(\overline{x}) + \mathrm{int}_{Y_1} K_1 + \frac{1}{1+t}g_1(u) + K_1$$

$$\subset g_1\left(\frac{t}{1+t}\overline{x} + \frac{1}{1+t}u\right) + \mathrm{int}_{Y_1} K_1,$$

where the latter inclusion is due to the equality $K_1 + \mathrm{int}_{Y_1} K_1 + K_1 = \mathrm{int}_{Y_1} K_1$ and to the K_1-convexity of g_1. On the other hand,

$$0 = \frac{t}{1+t}g_2(\overline{x}) + \frac{1}{1+t}\big(-t(g_2(\overline{x}))\big) \in \frac{t}{1+t}g_2(\overline{x}) + \frac{1}{1+t}g_2(u) + K_2$$

$$\subset g_2\left(\frac{t}{1+t}\overline{x} + \frac{1}{1+t}u\right) + K_2.$$

So with $x_0 := (1+t)^{-1}(t\overline{x}+u) \in C$ we obtain $g_1(x_0) \in -\mathrm{int}_{Y_1} K_1$ and $g_2(x_0) \in -K_2$, which gives condition (i).

Suppose now that both conditions (i) and (ii) are satisfied. From the inclusion $g_1(x_0) \in -\mathrm{int}_{Y_1} K_1$ and the continuity of g_1 there are reals $s > 0$ and $0 < t < 1$ such that $s\mathbb{B}_{Y_1} + g_1(x_0 + t\mathbb{B}_X) \subset -\mathrm{int}_{Y_1} K_1$. On the other hand, the inclusion $0 \in \mathrm{core}_{Y_2}\big(g_2(C)+K_2\big)$ and Theorem 7.26 furnish a real $r \in \,]0, s[$ such that $r\theta \mathbb{B}_{Y_2} \subset g_2\big(C \cap (x_0 + \theta \mathbb{B}_X)\big) + K_2$ for all $\theta \in [0,1]$. Take any $v_1 \in \mathbb{B}_{Y_1}$ and $v_2 \in \mathbb{B}_{Y_2}$. By what precedes there exists some $b \in \mathbb{B}_X$ such that $x_0 + tb \in C$ and $rtv_2 \in g_2(x_0+tb)+K_2$. Since $g_1(x_0+tb) \in -\mathrm{int}_{Y_1} K_1$, with $u := x_0+tb \in C$ we have $0 \in g_1(u) + K_1$ and $rtv_2 \in g_2(u) + K_2$, or equivalently $(0, rtv_2) \in g(u) + K_1 \times K_2$. Further, our choice of s guarantees that $-rv_1 + g_1(x_0) \in -K_1$, that is, $rv_1 \in g_1(x_0) + K_1$. This entails $(rv_1, 0) \in g(x_0) + K_1 \times K_2$ since $0 \in g_2(x_0) + K_2$ by condition (i). It follows that

$$\frac{rt}{1+t}(v_1, v_2) = \frac{1}{1+t}(0, rtv_2) + \frac{t}{1+t}(rv_1, 0)$$

$$\in \frac{1}{1+t}g(u) + K_1 \times K_2 + \frac{t}{1+t}g(x_0) + K_1 \times K_2,$$

which implies according to the convexity of g

$$\frac{rt}{1+t}(v_1, v_2) \in g\left(\frac{t}{1+t}u + \frac{1}{1+t}x_0\right) + K_1 \times K_2 \subset g(C) + K_1 \times K_2.$$

This tells us that $(0,0) \in \mathrm{core}\,\big(g(C) + K_1 \times K_2\big)$.

Finally, the equivalence with the existence of a real $c > 0$ for which the inequality of the theorem is satisfied follows from Proposition 7.61. □

As a direct corollary with $K_2 = \{0\}$ we have:

COROLLARY 7.63. *Let X, Y_1, Y_2 be Banach spaces and K be a closed convex cone of Y_1 with $\mathrm{int}_{Y_1} K \neq \emptyset$. Let C be a closed convex set of X and $g : X \to Y_1$ be a K-convex locally Lipschitz continuous mapping. Let $b \in Y_2$ and $A : X \to Y_2$ be a continuous linear mapping. Let $\overline{x} \in C$ with $g(\overline{x}) \leq_{K_1} 0$ and $A(\overline{x}) = b$. Then, the constraint or feasible set-representative*

$$\{x \in X : x \in C, g(x) \leq_{K_1} 0, A(x) - b = 0\}$$

is metrically regular at $\overline{x}$ for $(g(\overline{x}), A(\overline{x}) - b)$ if and only if both conditions (i) and (ii) are satisfied:

(i) There exists some $x_0 \in C$ such that $A(x_0) = b$ and $g(x_0) \in -\mathrm{int}_{Y_1} K$;
(ii) $0 \in \mathrm{core}(A(C) - b)$.

Further, in such a case there exists a real $c > 0$ such that, for all $x \in C$ and $y = (y_1, y_2) \in B_{Y_1}(0, c) \times B_{Y_2}(0, c)$

$$d(x, g^{-1}(-K_1 + y_1) \cap A^{-1}(b + y_2))$$
$$\leq (c - \|y\|)^{-1}(1 + \|x - \bar{x}\|)(d(g_1(x) - y_1, -K) + \|A(x) - b - y_2\|),$$

where $\|y\| = \max\{\|y_1\|, \|y_2\|\}$.

The next theorem provides an estimate for the distance to the set of solutions of a linear system defined by a vector equality and by finitely many inequalities. Recall first that, given a continuous linear mapping $A : X \to Y$ between two Banach spaces X, Y, its *range* $A(X)$ is closed in X if and only if the set $A^*(Y^*)$ (that is, the range of the adjoint) is norm closed in X^* (see the equivalence (a)$\Leftrightarrow$(b) in Theorem C.5 in Appendix). By the equivalence (a) $\Leftrightarrow$(d) in the same Theorem C.5 in Appendix, the closedness of $A(X)$ in Y is also equivalent to the property that $A^*(Y^*)$ coincides with the orthogonal in X^* of the kernel of A, so in such a case $A^*(Y^*)$ is $w(X^*, X)$ closed in X^*. The proof of the theorem will also use the following lemma.

LEMMA 7.64. *Let E be a Hausdorff locally convex space. Then for any closed vector subspace E_0 and for any finite dimensional vector subspace F of E the vector subspace $E_0 + F$ is closed in E.*

PROOF. Let us proceed by induction on the dimension of F. Suppose first that $\dim F = 1$, so $F = \mathbb{R}v$ for some non zero $v \in E$. If $v \in E_0$, then $E_0 + \mathbb{R}v = E_0$ and we are done. Suppose $v \notin E_0$, so by the Hahn-Banach theorem there is some $\zeta \in E^*$ such that $\langle \zeta, v \rangle = 1$ and $\langle \zeta, u \rangle = 0$ for all $u \in E_0$. Take any net $(x_j)_{j \in J}$ in $E_0 + \mathbb{R}v$ converging to some $x \in E$. For each $j \in J$ there are $u_j \in E_0$ and $r_j \in \mathbb{R}$ such that $x_j = u_j + r_j v$, hence $r_j = \langle \zeta, x_j \rangle$. Then the net $(r_j)_{j \in J}$ tends in $\mathbb{R}$ to $r = \langle \zeta, x \rangle$, which in turn entails that $(u_j)_{j \in J}$ converges in E to $u := x - rv$, and $u \in E_0$ by closedness of E_0. It results from the equality $x = u + rv$ that $x \in E_0 + \mathbb{R}v$, hence $E_0 + \mathbb{R}v$ is closed in E.

Suppose that the property holds for any finite dimensional vector subspace of dimension less or equal to $n - 1$ with $n \geq 2$ and take any vector subspace F whose dimension is n. Let $\{v_1, \cdots, v_{n-1}, v_n\}$ be a vector basis of F and let $F_{n-1} := \mathrm{span}\{v_1, \cdots, v_{n-1}\}$. The vector subspace $E_{n-1} = E_0 + F_{n-1}$ is closed, hence by what precedes $E_0 + F = E_{n-1} + \mathbb{R}v_n$ is closed in E as desired. $\square$

THEOREM 7.65 (Hoffman-Ioffe metric inequality in Banach spaces). *Let $A : X \to Y$ be a continuous linear mapping between Banach spaces X, Y such that $A(X)$ is closed and let $a_i^* : X \to \mathbb{R}$, $i = 1, \cdots, n$, be continuous linear functionals. Let $b \in Y$ and $\beta_i \in \mathbb{R}$, and let*

$$S := \{x \in X : Ax = b, \langle a_i^*, x \rangle \leq \beta_i, i = 1, \cdots, n\}.$$

If $S \neq \emptyset$, then there exists a real $\gamma \geq 0$, depending only on A and a_i^* (but neither on b nor on β_i), such that for all $x \in X$

$$d(x, S) \leq \gamma \left(\|Ax - b\| + \sum_{i=1}^{n} (\langle a_i^*, x \rangle - \beta_i)^+ \right),$$

where $r^+ := \max\{r, 0\}$ as usual.

PROOF. Consider first the case when all second members b and β_i, for all i, are null. Denote by S_0 the solution set of the system in this case and let us proceed by induction on the number of inequalities. According to Corollary 7.63 we note that the result is clear if there is no inequality, since in this case $S_0 = A_0^{-1}(0)$, where $A_0 : X \to A(X)$ denotes the surjective continuous linear mapping between the Banach spaces X and $Z := A(X)$ defined by $A_0(x) := A(x)$ for all $x \in X$. Suppose that, with second members $b = 0$ and $\beta_i = 0$, the result holds when there are at most $n-1$ inequalities and let us prove it with n inequalities.

In a first step, suppose that the convex set $G := A_0^*(Z^*) + \mathrm{co}\{a_1^*, \cdots, a_n^*\}$ does not contain 0. Since $A_0^*(Z^*)$ is w^*-closed in X^* according to the equivalence (a)$\Leftrightarrow$(d) of Theorem C.5 in Appendix (recalled before the statement of Lemma 7.64), the convex set G is w^*-closed in X^* by w^*-compactness of $\mathrm{co}\{a_1^*, \cdots, a_n^*\}$. Then, by the separation theorem there are some $v \in X$ and some real α such that, for all $z^* \in Z^*$ and $\lambda_i \geq 0$ with $\lambda_1 + \cdots + \lambda_n = 1$

$$\langle A_0^* z^*, v \rangle + \lambda_1 \langle a_1^*, v \rangle + \cdots + \lambda_n \langle a_n^*, v \rangle < \alpha < \langle 0, v \rangle = 0.$$

For each $i = 1, \cdots, n$, taking $z^* = 0$, $\lambda_i = 1$ and $\lambda_j = 0$ for $j \neq i$, gives $\langle a_i^*, v \rangle < 0$. Taking $\lambda_2 = \cdots = \lambda_n = 0$ and $\lambda_1 = 1$, we obtain $\langle z^*, A_0 v \rangle \leq \langle a_1^*, -v \rangle$ for all $z^* \in Z^*$, so $A_0 v = 0$. Since $S_0 = \{x \in X : A_0 x = 0, \langle a_i^*, x \rangle \leq 0, i = 1, \cdots, n\}$, Corollary 7.63, applied with $\bar{x} = 0$, $C = X$, $y = 0$, easily furnishes some real $\kappa > 0$ such that for all $x \in \mathbb{B}$,

$$d(x, S_0) \leq \kappa \left(\|A_0 x\| + \sum_{i=1}^n \langle a_i^*, x \rangle^+ \right).$$

Since S_0 is a cone and $A_0 x = A x$, we derive, for all $x \in X$

$$d(x, S_0) \leq \kappa \left(\|Ax\| + \sum_{i=1}^n \langle a_i^*, x \rangle^+ \right).$$

Now suppose that $0 \in G$. There exists some a_i^*, say a_n^*, which can be written in the form

(7.26) $$a_n^* = -\sum_{i=1}^{n-1} \alpha_i a_i^* - A_0^* v^*,$$

with $v^* \in Z^*$ and $\alpha_i \geq 0$, $i = 1, \cdots, n-1$. It ensues that $\langle a_n^*, x \rangle = 0$ for all $x \in S_0$. Consider the continuous linear mapping $\Lambda : X \to Y \times \mathbb{R}$ defined by $\Lambda x := (Ax, a_n^*(x))$ for all $x \in X$ and note (as easily checked) that $\Lambda^* : Y^* \times \mathbb{R} \to X^*$ is given $\Lambda^*(y^*, r) = A^* y^* + r a_n^*$ for all (y^*, r) in $Y^* \times \mathbb{R}$. It ensues that $\Lambda^*(Y^* \times \mathbb{R}) = A^*(Y^*) + \mathbb{R} a_n^*$. Since $A^*(Y^*)$ is norm closed in X^* by Theorem C.5 in Appendix (recalled above) and by the closedness assumption of $A(X)$, Lemma 7.64 says that $\Lambda^*(Y^* \times \mathbb{R})$ is norm closed in X^*. The equivalence (a)$\Leftrightarrow$(b) in Theorem C.5 again gives that $\Lambda(X)$ is closed in $Y \times \mathbb{R}$. It is also clear that

$$S_0 = \{x \in X : \Lambda x = 0, \langle a_i^*, x \rangle \leq 0, i = 1, \cdots, n-1\}.$$

By the induction assumption there is some real $\mu > 0$ such that

$$d(x, S_0) \leq \mu \left(\|\Lambda x\| + \sum_{i=1}^{n-1} \langle a_i^*, x \rangle^+ \right) \leq \mu \left(\|Ax\| + |\langle a_n^*, x \rangle| + \sum_{i=1}^{n-1} \langle a_i^*, x \rangle^+ \right).$$

7.5. METRIC REGULARITY OF FEASIBLE SETS

On the other hand, (recalling that $r^- = \max\{0, -r\}$) (7.26) gives for all $x \in X$,

$$\langle a_n^*, x\rangle^- = \max\left\{0, \sum_{i=1}^{n-1}\alpha_i\langle a_i^*, x\rangle + \langle v^*, Ax\rangle\right\} \leq \sum_{i=1}^{n-1}\alpha_i\langle a_i^*, x\rangle^+ + \|v^*\|\,\|Ax\|,$$

so $|\langle a_n^*, x\rangle| \leq \langle a_n^*, x\rangle^+ + \sum_{i=1}^{n-1}\alpha_i\langle a_i^*, x\rangle^+ + \|v^*\|\,\|Ax\|$. Combining this with what precedes furnishes a real $\kappa' > 0$ such that for all $x \in X$,

$$d(x, S_0) \leq \kappa'\left(\|Ax\| + \sum_{i=1}^{n}\langle a_i^*, x\rangle^+\right).$$

The reals κ, κ' being obtained through the form of the set S_0, we note that they depend only on A and a_i^* (and neither on b nor on β_i).

For any nonempty subset J of $\{1, \cdots, n\}$ set

$$S_{0,J} := \{y \in X : Ay = 0, \langle a_i^*, y\rangle \leq 0, \forall i \in J\}$$

and $S_{0,J} = \{y \in X : Ay = 0\}$ for $J = \emptyset$. By what precedes there is a real constant $\gamma > 0$ (independent of J, b, β_i) such that for any subset J (either nonempty or not) of $\{1, \cdots, n\}$

$$d(x, S_{0,J}) \leq \gamma\left(\|Ax\| + \sum_{i \in J}\langle a_i^*, x\rangle^+\right) \quad \text{for all } x \in X.$$

Consider now the general situation when b, β_i are not necessarily null. Fix any $x \notin S$ (the result being obvious if $x \in S$). Since the set S is nonempty, one of the following possibilities occur: $d(x, S) = d(x, \{u : Au = b\})$ or there are some nonempty set $J_x \subset \{1, \cdots, n\}$ and some $\bar{u} \in X$ (depending also on x) such that $A\bar{u} = b$, $\langle a_i^*, \bar{u}\rangle = \beta_i$ for all $i \in J_x$ and

$$d(x, S) = d(x, \{u : Au = b, \langle a_i^*, u\rangle \leq \beta_i, \forall i \in J_x\}).$$

Under the existence of such nonempty set J_x and $\bar{u}$ we can write by the above analysis

$$d(x, S) = d(x, \{u \in X : A(u - \bar{u}) = 0, \langle a_i^*, u - \bar{u}\rangle \leq 0, \forall i \in J_x\})$$

$$= d(x - \bar{u}, S_{0,J}) \leq \gamma\left(\|A(x - \bar{u})\| + \sum_{i \in J_x}\langle a_i^*, x - \bar{u}\rangle^+\right),$$

which yields

$$d(x, S) \leq \gamma\left(\|A(x) - b\| + \sum_{i \in J_x}(\langle a_i^*, x\rangle - \beta_i)^+\right)$$

$$\leq \gamma\left(\|A(x) - b\| + \sum_{i=1}^{n}(\langle a_i^*, x\rangle - \beta_i)^+\right).$$

If $d(x, S) = d(x, \{u : Au = b\})$, similar arguments with $J = \emptyset$ give the same latter inequality. The proof is then complete. □

If X is finite-dimensional, the following corollary directly follows.

COROLLARY 7.66 (Hoffman metric inequality in finite dimensions). *Let X be a finite-dimensional normed space, $a_i^* : X \to \mathbb{R}$, $i = 1, \cdots, n$, be linear functionals, and $A : X \to Y$ be a linear mapping from X into a Banach space Y. Let $b \in Y$ and $\beta_i \in \mathbb{R}$, and let*

$$S := \{x \in X : Ax = b, \langle a_i^*, x \rangle \leq \beta_i, i = 1, \cdots, n\}.$$

If $S \neq \emptyset$, then there is some real $\gamma \geq 0$ such that, for all $x \in X$

$$d(x, S) \leq \gamma \left(\|Ax - b\| + \sum_{i=1}^{n} (\langle a_i^*, x \rangle - \beta_i)^+ \right).$$

7.6. Metric regularity and Lipschitz additive perturbation

We turn now to the preservation of linear openness (or equivalently metric regularity) under suitable Lipschitz additive perturbation.

THEOREM 7.67 (persistence of metric regularity under Lipschitz perturbation). *Let M and Φ be multimappings from a complete metric space X into a Banach space Y, let $\varphi : X \to Y$ be a single valued mapping, and let real numbers $c > \lambda \geq 0$. Let $(\bar{x}, \bar{y}) \in \operatorname{gph} M$.*
(a) Assume that there exists a real $\rho > 0$ such that φ is λ-Lipschitz on $B[\bar{x}, \rho]$ and such that for $U := B[\bar{x}, \rho]$, $V := B[\bar{y}, \rho]$, the multimapping M^{-1} satisfies the Aubin-Lipschitz property in (7.2) with constant c^{-1} over V for U, that is, for all $y \in V$ and $y' \in V$

$$M^{-1}(y) \cap U \subset \bigcup_{x' \in M^{-1}(y')} B[x', c^{-1} d(y, y')].$$

Let $\eta := \min\{r, r/(2\lambda)\}$ where

$$r := \left(1 + \frac{\max\{1, c^{-1}\}}{1 - c^{-1}\lambda} \right)^{-1} \frac{\rho}{5 + \lambda}.$$

Then the multimapping $(M + \varphi)^{-1}$ satisfies the Aubin-Lipschitz property with constant $(c - \lambda)^{-1}$ over $B[\bar{y} + \varphi(\bar{x}), r/2]$ for $B[\bar{x}, \eta]$, that is,

$$(M + \varphi)^{-1}(q) \cap B[\bar{x}, \eta] \subset \bigcup_{x' \in (M+\varphi)^{-1}(q')} B[x', (c - \lambda)^{-1} d(q, q')]$$

for all $q \in B[\bar{q}, r/2]$ and all $q' \in B[\bar{q}, r/2]$, where $\bar{q} := \bar{y} + \varphi(\bar{x})$.
In particular, if M is open with a linear rate at $\bar{x}$ for $\bar{y} \in M(\bar{x})$ with c as a constant of openness and if φ is λ-Lipschitz near $\bar{x}$, then the multimapping $M + \varphi$ is open with linear rate at $\bar{x}$ for $\bar{y} + \varphi(\bar{x})$ with $c - \lambda$ as a coefficient of openness.
(b) If M is globally open with a linear rate with c as a constant of openness and if Φ is globally λ-Lipschitz in the sense

$$\Phi(x') \subset \Phi(x) + \lambda \, d(x', x) \mathbb{B}_Y \quad \text{for all } x, x' \in X,$$

then the set valued mapping $M + \Phi$ is globally open with a linear rate and with $c - \lambda$ as a coefficient of openness.

PROOF. Let $\alpha := \lambda/c$ and consider Φ as equal to φ in the case (a). In the case (a) let $\rho, r, \eta > 0$ as given. Fix any pair $(x, y) \in \operatorname{gph} M$ with $x \in B[\bar{x}, \eta]$ and $y \in B[\bar{y}, r]$, put $z := \varphi(x)$ and $\bar{z} := \varphi(\bar{x})$, and consider any $v \in B[\bar{y} + \bar{z}, 3r]$.

7.6. PERTURBATION OF METRIC REGULARITY

In the case (b) fix instead any $(x,y) \in \operatorname{gph} M$ without any other restriction and fix any $z \in \Phi(x)$ and any $\bar{z} \in \Phi(\bar{x})$ with $\|\bar{z} - z\| \leq \lambda d(x, \bar{x})$. Observe also that M^{-1} is globally c^{-1}-Lipschitz on Y in the sense of (b) of Proposition 7.18.

Let $y_0 = y$, $x_0 = x$, $z_0 = z$, and $w_0 = y + z - v$. Put $y_1 = y_0 - w_0$. In the case (a) we observe that $\|w_0\| \leq r(4 + \lambda)$ and $\|y_1 - \bar{y}\| = \|v - z - \bar{y}\| \leq r(3 + \lambda)$. Therefore, in the case (a) and (b) as well, we can choose $x_1 \in M^{-1}(y_1)$ with $d(x_1, x_0) \leq c^{-1}\|y_1 - y_0\| = c^{-1}\|w_0\|$; since in the case (a) the latter inequality implies $d(x_1, \bar{x}) \leq r(1 + c^{-1}(4 + \lambda))$, it follows that $z_0 \in \Phi(x_1) + \lambda d(x_0, x_1)\mathbb{B}_Y$ in that case, and hence in anyone of cases (a) and (b) we can choose $z_1 \in \Phi(x_1)$ with $\|z_1 - z_0\| \leq \lambda d(x_1, x_0)$ and we put $w_1 = z_1 - z_0$. Suppose that we have constructed y_k, x_k, z_k, and w_k, for $k = 1, \cdots, n$ such that

$$(\mathcal{P}_k) \begin{cases} y_k = y_{k-1} - w_{k-1}, \ x_k \in M^{-1}(y_k) \text{ with } d(x_k, x_{k-1}) \leq c^{-1}\|w_{k-1}\|, \\ z_k \in \Phi(x_k) \text{ with } \|z_k - z_{k-1}\| \leq \lambda d(x_k, x_{k-1}), \\ w_k = z_k - z_{k-1}. \end{cases}$$

Then, for any such fixed k, we see that $\|w_k\| \leq \lambda d(x_k, x_{k-1}) \leq \alpha\|w_{k-1}\|$, thus $\|w_k\| \leq \alpha^k\|w_0\|$. It ensues that $d(x_k, x_{k-1}) \leq \alpha^{k-1}\|w_0\|/c$, and hence $d(x_k, x_0) \leq \frac{1}{1-\alpha}\frac{\|w_0\|}{c}$. Note also that $\|z_k - z_{k-1}\| = \|w_k\| \leq \alpha^k\|w_0\|$. In the case (a) we have in addition from what precedes $d(x_k, \bar{x}) \leq r(1 + \frac{4+\lambda}{c(1-\alpha)})$ and

$$\|y_k - y_0\| = \left\|\sum_{i=0}^{k-1} w_i\right\| \leq \frac{1}{1-\alpha}\|w_0\| \leq \frac{r(4+\lambda)}{1-\alpha},$$

which entails $\|y_k - \bar{y}\| \leq r(1 + \frac{4+\lambda}{1-\alpha})$; also $\|z_k - z_{k-1}\| \leq \alpha^k\|w_0\| \leq \alpha^k r(4+\lambda)$, which gives $\|z_k - z_0\| \leq \frac{r(4+\lambda)}{1-\alpha}$, hence $\|z_k - \bar{z}\| \leq r(1 + \frac{4+\lambda}{1-\alpha})$. Setting $y_{n+1} = y_n - w_n$ in any case, we see in the case (a) that $\|w_n\| \leq \alpha^n\|w_0\| \leq r(4+\lambda)$, hence

$$\|y_{n+1} - \bar{y}\| \leq \|y_n - \bar{y}\| + \|w_n\| \leq r\left(5 + \lambda + \frac{4+\lambda}{1-\alpha}\right).$$

In the case (a) or (b) the Aubin-Lipschitz or (see Proposition 7.18) the global Lipschitz property of M^{-1} allows us to choose some $x_{n+1} \in M^{-1}(y_{n+1})$ with $d(x_{n+1}, x_n) \leq c^{-1}d(y_n, y_{n+1}) = c^{-1}\|w_n\| \leq c^{-1}\alpha^n\|w_0\|$. In the case (a) this ensures that $d(x_{n+1}, x_n) \leq c^{-1}\alpha^n\|w_0\| \leq \frac{r(4+\lambda)}{c}\alpha^n$, which implies $d(x_{n+1}, x_0) \leq \frac{r(4+\lambda)}{c}\frac{1}{1-\alpha}$, and hence $d(x_{n+1}, \bar{x}) \leq r(1 + \frac{4+\lambda}{c(1-\alpha)})$. Similarly, the Lipschitz property of Φ gives us some $z_{n+1} \in \Phi(x_{n+1})$ with $\|z_{n+1} - z_n\| \leq \lambda d(x_{n+1}, x_n)$, so putting $w_{n+1} = z_{n+1} - z_n$, we see that y_{n+1}, x_{n+1}, z_{n+1} and w_{n+1} can be chosen to satisfy the same above conditions. This justifies by induction the existence of a sequence $(y_n, x_n, z_n, w_n)_n$ such that $(\mathcal{P}_n)$ holds true for all $n \in \mathbb{N}$.

From what precedes it appears that $d(x_n, x_{n-1}) \leq \alpha^{n-1}\|w_0\|/c$ and $\|w_n\| = \|z_n - z_{n-1}\| \leq \alpha^n\|w_0\|$. Therefore, $(x_n)_n$ and $(z_n)_n$ are Cauchy sequences which then converge to some u and ζ respectively and $\zeta \in \Phi(u)$, and also $\|w_n\| \to 0$. From the equalities $y_k = y_{k-1} - w_{k-1}$ and $w_k = z_k - z_{k-1}$, $k = 1, \cdots, n$, it is not difficult to see that $y_n + z_n = v + w_n$, so $y_n \to v - \zeta$ and $v - \zeta \in M(u)$. It results that $v \in \Gamma(u) := M(u) + \Phi(u)$. From the above inequality $d(x_n, x_0) \leq \frac{1}{1-\alpha}\frac{\|w_0\|}{c}$ and the definition of w_0 we also have

(7.27) $$d(u, x) \leq \frac{1}{c(1-\alpha)}\|v - (y+z)\| = \frac{1}{c-\lambda}\|v - (y+z)\|.$$

From the latter inequality we readily deduce in the case (b), for any $y \in M(x)$ and $z \in \Phi(x)$, and for any real $t > 0$, that the inclusion $v \in B[y + z, (c - \lambda)t]$ guarantees that $u \in B[x,t] \cap \Gamma^{-1}(v)$, otherwise stated $B[y+z, (c-\lambda)t] \subset \Gamma(B[x,t])$. This establishes (b).

In the case (a), fix any $v \in B[\bar{y} + \bar{z}, r/2]$, any $q \in B[\bar{y} + \bar{z}, r/2]$ and any $x \in \Gamma^{-1}(q) \cap B[\bar{x}, \eta]$, so $q = y + z$ with $y \in M(x)$ and $z = \varphi(x)$. Keeping in mind that $\bar{z} = \varphi(\bar{x})$ we have

$$\|y - \bar{y}\| \le \|q - \bar{y} - \bar{z}\| + \|\varphi(\bar{x}) - \varphi(x)\| \le (r/2) + \lambda\|x - \bar{x}\| \le r.$$

Then by what precedes there exists $u \in \Gamma^{-1}(v)$ such that (7.27) holds true, so $x \in B[u, (c - \lambda)^{-1} d(q, v)]$. It results that for all $q, v \in B[\bar{y} + \bar{z}, r/2]$

$$\Gamma^{-1}(q) \cap B[\bar{x}, \eta] \subset \bigcup_{u \in \Gamma^{-1}(v)} B[u, (c - \lambda)^{-1} d(q, v)],$$

which justifies the desired Aubin-Lipschitz property of $(M + \varphi)^{-1}$ with constant $(c - \lambda)^{-1}$ over $B[\bar{y} + \bar{z}, r/2]$ for $B[\bar{x}, \eta]$.

Finally, the linear openness with constant $c - \lambda$ of $M + \varphi$ at $\bar{x}$ for $\bar{y} + \bar{z}$ follows from its above Aubin-Lipschitz property and from Lemma 7.16. □

COROLLARY 7.68. *Let M be a multimapping from a complete metric space X into a Banach space Y whose graph is closed near a point $(\bar{x}, \bar{y}) \in \mathrm{gph}\, M$ and let $\varphi : X \to Y$ be a single valued mapping which is locally Lipschitz near $\bar{x}$.*
(a) *For $\bar{z} := \bar{y} + \varphi(\bar{x})$, one has the inequality*

$$\mathrm{ope}\,[M + \varphi](\bar{x}|\bar{z}) \ge \mathrm{ope}\,[M](\bar{x}|\bar{y}) - \mathrm{Lip}\,[\varphi](\bar{x}),$$

and if $\mathrm{Lip}[\varphi](\bar{x}) \le \bigl(\mathrm{reg}\,[M](\bar{x}|\bar{y})\bigr)^{-1}$ (with the convention $r^{-1} = 0$ if $r = +\infty$ and $r^{-1} = +\infty$ if $r = 0$), one also has

$$\mathrm{reg}\,[M + \varphi](\bar{x}|\bar{z}) \le \bigl((\mathrm{reg}\,[M](\bar{x}|\bar{y}))^{-1} - \mathrm{lip}\,[\varphi](\bar{x})\bigr)^{-1}.$$

(b) *If $\mathrm{lip}[\varphi](\bar{x}) = 0$, then*

$$\mathrm{ope}\,[M + \varphi](\bar{x}|\bar{z}) = \mathrm{ope}\,[M](\bar{x}|\bar{y}) \quad \text{and} \quad \mathrm{reg}\,[M + \varphi](\bar{x}|\bar{z}) = \mathrm{reg}\,[M](\bar{x}|\bar{y}).$$

PROOF. (a) The first inequality is obviously true when its right member is nonpositive; if $\mathrm{lip}[\varphi](\bar{x}) < \mathrm{ope}[M](\bar{x}|\bar{y})$ the inequality directly follows from Theorem 7.67 above. The second inequality results from the preceding one since the rate of openness of a multimapping coincides with the inverse of its rate of metric regularity (see Theorem 7.17).
(b) Regarding (b) we have by (a) the inequality $\mathrm{ope}\,[M + \varphi](\bar{x}|\bar{z}) \ge \mathrm{ope}\,[M](\bar{x}|\bar{y})$, which also implies the converse inequality through the equality $M = (M+\varphi)+(-\varphi)$ and the observation $\mathrm{lip}[-\varphi](\bar{x}) = \mathrm{lip}\,[\varphi](\bar{x}) = 0$. □

REMARK 7.69. It is worth observing that such a perturbation result fails for the metric subregularity. Indeed, consider the mapping $\varphi : \mathbb{R} \to \mathbb{R}$ with $\varphi(x) = -x^2$ and note that $\mathrm{lip}[\varphi](0) = 0$. Let also the multimapping $M : \mathbb{R} \rightrightarrows \mathbb{R}$ with $M(x) = \emptyset$ if $x < 0$, $M(0) = [0, +\infty[$ and $M(x) = \{0\}$ if $x > 0$. Obviously $M^{-1} = M$, hence $M^{-1}(y) \subset M^{-1}(0)$, and this says in particular that M^{-1} is calm at 0 for 0, that is, M^{-1} has the Aubin-Lipschitz property at 0 for 0 relative to $Q := \{0\}$. Consequently M is metrically subregular at 0 for 0. On the other hand, $(M+\varphi)^{-1}$ is single-valued with $(M+\varphi)^{-1}(y) = 0$ if $y > 0$ and $(M+\varphi)^{-1}(y) = \sqrt{|y|}$ if $y \le 0$.

We then see that $(M + \varphi)^{-1}$ is not calm at 0 for 0, or equivalently $M + \varphi$ is not metrically subregular at 0 for 0. □

The next basic metric regularity result concerns the statement of Proposition 7.61 when strict differentiability is required for the mapping g in place of convexity.

THEOREM 7.70 (S.M. Robinson: theorem for metric regularity of constraint systems). *Let X and Y be Banach spaces and D be a closed convex set of Y. Let C be a closed convex set of X and $g : X \to Y$ be a mapping which is strictly Fréchet differentiable at $\bar{x} \in C \cap g^{-1}(D)$. Let $\bar{y} := g(\bar{x})$ and $A := Dg(\bar{x})$. The following hold:*
(a) *For the multimappings $M, M_0 : X \rightrightarrows Y$, defined for $x \notin C$ by $M(x) = M_0(x) = \emptyset$ and for $x \in C$ by*

$$M(x) = g(x) - D \quad \text{and} \quad M_0(x) = A(x - \bar{x}) + \bar{y} - D,$$

one has $\operatorname{reg}[M](\bar{x}|0) = \operatorname{reg}[M_0](\bar{x}|0)$.
(b) *Consequently, the inclusion*

$$0 \in \operatorname{core}\big(A(C - \bar{x}) - (D - \bar{y}) \big)$$

is a necessary and sufficient condition for the metric regularity of the multimapping M at $\bar{x}$ for 0, or equivalently for the metric regularity qualification of the constraint $C \cap g^{-1}(D)$ at $\bar{x}$, that is, the existence of a real $\gamma \geq 0$ and neighborhoods U of $\bar{x}$ in X and V of 0 in Y such that

$$d\big(x, g^{-1}(D + y)\big) \leq \gamma\big(d(x, C) + d(g(x) - y, D)\big) \quad \text{for all } x \in U, y \in V.$$

PROOF. Both multimappings M and M_0 clearly have closed graphs near $(\bar{x}, 0)$. With $\varphi(x) := g(x) - \bar{y} - A(x - \bar{x})$, we also observe that on the one hand $\operatorname{lip}[\varphi](\bar{x}) = 0$, and on the other hand $M(x) = M_0(x) + \varphi(x)$ for all $x \in X$. Then, Corollary 7.68(b) tells us that $\operatorname{reg}[M](\bar{x}|0) = \operatorname{reg}[M_0](\bar{x}|0)$.

Further, the mapping $g_0 : X \to Y$ with $g_0(x) = A(x - \bar{x})$ is affine, hence K-convex with $K = \{0\}$. So, from Proposition 7.61 applied with g_0 and $K = \{0\}$ the inclusion in (b) is a necessary and sufficient condition for the metric regularity, at $\bar{x}$ for 0, of the multimapping M_0, hence of M as well. Since g is Lipschitz near $\bar{x}$, the latter amounts by (7.24) to the existence of γ, U and V satisfying the inequality in the theorem. □

By Remark 7.25, given a mapping $g : X \to Y$ between two Banach spaces of class $\mathcal{C}^1$ on a neighborhood W of $\bar{x}$ and such that $Dg(\bar{x})$ is surjective, there is an open neighborhood $U \subset W$ of $\bar{x}$ and a real $r > 0$ such that for all $x \in U$ the inclusion $r\mathbb{B}_Y \subset Dg(x)(\mathbb{B}_X)$ holds, so $Dg(x)$ is surjective. We derive a similar local property from the above theorem for the core-inclusion in its assertion (b) whenever the mapping g is of class $\mathcal{C}^1$.

PROPOSITION 7.71. *Let X and Y be Banach spaces and D be a closed convex set of Y. Let C be a closed convex set of X and $g : X \to Y$ be a mapping which is continuously differentiable near $\bar{x} \in C \cap g^{-1}(D)$. Assume that*

$$0 \in \operatorname{core}\big(Dg(\bar{x})(C - \bar{x}) - (D - g(\bar{x})) \big).$$

Then there exists an open neighborhood U of $\bar{x}$ over which g is continuously differentiable and such that for every $x \in U \cap C \cap g^{-1}(D)$

$$0 \in \operatorname{core}\big(Dg(x)(C - x) - (D - g(x)) \big).$$

PROOF. By Theorem 7.70(b) there exist a real $\gamma \geq 0$, a neighborhood V of zero in Y and an open neighborhood U of $\bar{x}$ over which g is of class $\mathcal{C}^1$ and such that
$$d(u, g^{-1}(D+v)) \leq \gamma(d(u,C) + d(g(u)-v, D)) \quad \text{for all } u \in U, v \in V.$$
Fixing any $x \in U$, the same assertion (b) in Theorem 7.70 (in fact, the implication $\Leftarrow$ therein) entails that
$$0 \in \mathrm{core}\big(Dg(x)(C-x) - (D - g(x))\big),$$
as desired. $\square$

When $D = -K \times \{0\}$ with a convex cone K with nonempty interior the condition in Theorem 7.70 takes the following form:

PROPOSITION 7.72. Let X, Y_1, Y_2 be Banach spaces and K be a closed convex cone of Y_1 with $\mathrm{int}_{Y_1} K \neq \emptyset$, let $g : X \to Y_1 \times Y_2$ be a mapping with $g(x) = (g_1(x), g_2(x)) \in Y_1 \times Y_2$, and let $\bar{x} \in X$ with $g_1(\bar{x}) \in -K$ and $g_2(\bar{x}) = 0$. Assume that g is strictly Fréchet differentiable at $\bar{x}$. Then the constraint $g^{-1}(-K \times \{0\})$ is metrically regular at $\bar{x}$ if and only if the following constraint qualification condition is satisfied:

(CQC) $\quad \begin{cases} Dg_2(\bar{x}) \text{ is surjective and there is some } \bar{v} \in X \text{ with} \\ Dg_2(\bar{x})(\bar{v}) = 0 \text{ and } Dg_1(\bar{x})(\bar{v}) \in -\mathrm{int}_{Y_1} K. \end{cases}$

PROOF. Put $D := -K \times \{0\}$, $A = Dg(\bar{x})$ with $Av := (A_1 v, A_2 v)$. By Theorem 7.70 we know that the metric regularity of the constraint $g^{-1}(-K \times \{0\})$ is equivalent to the inclusion $0 \in \mathrm{core}(A(X) + K \times \{0\})$, or equivalently $Y = A(X) + K \times \{0\}$ since $A(X) + K \times \{0\}$ is a cone.

The equality $Y = A(X) + K \times \{0\}$ clearly yields $Y_2 = A_2(X)$ and fixing $e_1 \in \mathrm{int}\, K$ we can choose $\bar{v} \in X$ such that $(-e_1, 0) \in A(\bar{v}) + K \times \{0\}$. This ensures that $A_2 \bar{v} = 0$ and
$$A_1(\bar{v}) \in -e_1 - K \subset -(\mathrm{int}\, K) - K = -\mathrm{int}\, K,$$
so (CQC) is fulfilled.

Conversely, suppose (CQC) is satisfied. Take any $(y_1, y_2) \in Y_1 \times Y_2$. From the surjectivity of A_2 choose $u \in X$ such that $y_2 = A_2(u)$. Rewriting the inclusion $A_1(\bar{v}) \in -\mathrm{int}\, K$ as $0 \in \mathrm{int}(A_1(\bar{v}) + K)$ we can choose a real $t > 0$ such that $t^{-1}(y_1 - A_1(u)) \in A_1(\bar{v}) + K$, hence $y_1 \in A_1(u + t\bar{v}) + K$. Using this and the equality $A_2(\bar{v}) = 0$ it results that $(y_1, y_2) \in A(u + t\bar{v}) + K \times \{0\}$, so $Y_1 \times Y_2 = A(X) + K \times \{0\}$. This completes the proof. $\square$

COROLLARY 7.73. Let $\bar{x}$ be a point in the constraint set
$$\{x \in X : g_1(x) \leq 0, \cdots, g_m(x) \leq 0, g(x) = 0\},$$
where $g : X \to Y$ is a mapping from the Banach space X into a Banach space Y and $g_1, \cdots, g_m : X \to \mathbb{R}$ are real-valued functions defined on X. Assume that g and $g_1, \cdots, g_m$ are strictly Fréchet differentiable at $\bar{x}$ and denote by $I(\bar{x}) := \{i \in \{1, \cdots, m\} : g_i(\bar{x}) = 0\}$ the set of indices of active inequality constraints. Then the constraint is metrically regular at $\bar{x}$ if and only if the following Mangasarian-Fromovitz constraint qualification condition is satisfied:

(MFCQC) $\quad \begin{cases} Dg(\bar{x}) \text{ is surjective and there is some } \bar{v} \in X \text{ with} \\ Dg(\bar{x})(\bar{v}) = 0 \text{ and } Dg_i(\bar{x})(\bar{v}) < 0, \forall i \in I(\bar{x}). \end{cases}$

PROOF. Put $I := I(\bar{x})$, $Y_1 := \mathbb{R}^I$, $K_1 := (\mathbb{R}_+)^I$ (where $\mathbb{R}_+ := [0, +\infty[$), and define $g_I : X \to \mathbb{R}^I$ by $g_I(x) = (g_i(x))_{i \in I}$ for all $x \in X$. There exists a neighborhood U of $\bar{x}$ such that the intersection of the constraint set of the theorem coincides with

$$U \cap \{x \in X : g_i(x) \leq 0, \forall i \in I, g(x) = 0\} = U \cap G^{-1}(-K_1 \times \{0\}),$$

where $G(x) := (g_I(x), g(x)) \in Y_1 \times Y$. Since U is a neighborhood of $\bar{x}$, the metric regularity qualification at $\bar{x}$ of the constraint set in the theorem is equivalent to the metric regularity of the constraint set $G^{-1}(-K_1 \times \{0\})$. The corollary then follows from Proposition 7.72. □

Applying now Theorem 7.70 with $D = \{g(\bar{x})\}$ and $C = X$ yields:

THEOREM 7.74 (Lyusternik-Graves theorem). Let $g : X \to Y$ be a mapping between two Banach spaces which is strictly Fréchet differentiable at $\bar{x} \in X$. Then

$$\operatorname{reg}[g](\bar{x}) = \operatorname{reg}[Dg(\bar{x})] \quad \text{and} \quad \operatorname{ope}[g](\bar{x}) = \operatorname{ope}[Dg(\bar{x})],$$

so g is metrically regular (or equivalently open with a linear rate) at $\bar{x}$ if and only if the derivative $Dg(\bar{x})$ is surjective.

The following corollary is another basic metric regularity inequality. It is a direct consequence of Theorem 7.74 and Lemma 7.76 below.

COROLLARY 7.75 (metric regularity of inverse image under surjectivity of strict derivative). Let $g : X \to Y$ be a mapping between two Banach spaces which is strictly Fréchet differentiable at a point $\bar{x}$ with $Dg(\bar{x})$ surjective. Let D be any subset of Y containing $g(\bar{x})$. Then there exist a real $\gamma \geq 0$, a neighborhood U of $\bar{x}$ in X and a neighborhood V of zero in Y such that

$$d\left(x, g^{-1}(y+S)\right) \leq \gamma \, d(g(x) - y, S) = \gamma \, d(g(x), y + S) \quad \text{for all } x \in U, y \in V.$$

LEMMA 7.76. Let $g : X \to Y$ be a mapping between normed spaces which is continuous at $\bar{x} \in X$. Then the mapping g is metrically regular at $\bar{x}$ if and only if it is metrically regular at $\bar{x}$ with respect to any subset S of the image space Y with $g(\bar{x}) \in S$.

PROOF. The implication $\Leftarrow$ is obvious by taking S as the singleton $\{g(\bar{x})\}$. Suppose now that g is metrically regular at $\bar{x}$. There are two reals $\gamma > 0$ and $\varepsilon > 0$ such that $d(x, g^{-1}(w)) \leq \gamma \|g(x) - w\|$ for all $x \in B(\bar{x}, \varepsilon)$ and $w \in B(\bar{y}, 3\varepsilon)$, where $\bar{y} := g(\bar{x})$. By continuity of g at $\bar{x}$ choose a real $\delta \in \,]0, \varepsilon[$ such that for every $x \in B(\bar{x}, \delta)$ one has $g(x) \in B(\bar{y}, \varepsilon/2)$. Fix any $x \in B(\bar{x}, \delta)$ and any $y \in B_Y(0, \varepsilon/2)$. For every $w \in S \cap B(\bar{y}, 2\varepsilon)$ we have

$$d(x, g^{-1}(y+S)) \leq d(x, g^{-1}(y+w)) \leq \gamma \|g(x) - y - w\|,$$

hence $d(x, g^{-1}(y+S)) \leq \gamma \, d(g(x) - y, S \cap B(\bar{y}, 2\varepsilon))$. We deduce that

$$d(x, g^{-1}(y+S)) \leq \gamma \, d(g(x) - y, S) = \gamma \, d(g(x), y + S)$$

since $d(g(x) - y, S \cap B(\bar{y}, 2\varepsilon)) = d(g(x) - y, S)$ according to Lemma 2.219. This translates the desired implication $\Rightarrow$. □

7.7. Optimality conditions and calculus of tangent and normal cones under metric subregularity

This section is devoted to illustrating how the metric subregularity can serve as a constraint qualification condition to calculate/estimate tangent and normal cones, and to establish the existence of Lagrange multipliers for constraint optimization problems.

7.7.1. Optimality conditions under metric subregularity or other conditions.
A general optimality condition with Lagrange multipliers has been established in Theorem 2.228 with the C-subdifferential for optimization problems with finitely many inequality/equality constraints with locally Lipschitz functions. This subsection considers the more general situation of possible vector inequality/equality constraints. We start with the existence of Lagrange multipliers for a constraint set $C \cap g^{-1}(D)$, where $g : X \to Y$ is a mapping which is Lipschitz near $\bar{x} \in C \cap g^{-1}(D)$, and C and D are closed subsets of the Banach spaces X and Y respectively. With such a constraint we associated in Section 7.4 the multimapping $M_{g,C,D} : X \rightrightarrows Y$ defined by $M_{g,C,D}(x) = g(x) - D$ if $x \in C$ and $M_{g,C,D}(x) = \emptyset$ otherwise. We saw that the metric subregularity of that multimapping at $\bar{x}$ for $0 \in M_{g,C,D}(\bar{x})$ means that there are a real constant $\gamma \geq 0$ and a neighborhood U of $\bar{x}$ such that

(7.28) $\quad d(x, C \cap g^{-1}(D)) \leq \gamma\big(d(x, C) + d(g(x), D)\big) \quad \text{for all } x \in U.$

In such a case we said in (7.25) of Section 7.4 that the *metric subregularity constraint qualification condition* is satisfied for the constraint $C \cap g^{-1}(D)$ at $\bar{x}$.

THEOREM 7.77 (Lagrange multipliers). Let $g : X \to Y$ be a mapping between two Banach spaces X, Y, let C and D be two closed subsets of X and Y respectively, and let $f : X \to \mathbb{R}$ be a locally Lipschitz function. Let $\bar{x} \in C \cap g^{-1}(D)$ be a local solution of the optimization problem

$$\text{Minimize} \quad f(x) \quad \text{subject to } x \in C \cap g^{-1}(D).$$

Assume that g is compactly Lipschitzian at $\bar{x}$ and the metric subregularity qualification condition is satisfied for the constraint $C \cap g^{-1}(D)$ at $\bar{x}$. Then the following necessary optimality conditions with Lagrange multipliers hold:
(a) There exist $\lambda^* \in N^A(D; g(\bar{x}))$ and a real $k \geq 0$ such that

$$0 \in \partial_A\big(f + \lambda^* \circ g + k d_C\big)(\bar{x}), \quad \text{so in particular } 0 \in \partial_A f(\bar{x}) + \partial_A(\lambda^* \circ g)(\bar{x}) + N^A(C; \bar{x}).$$

(b) If X and Y are Asplund spaces, there exist $\lambda^* \in N^L(D; g(\bar{x}))$ and a real $k \geq 0$ such that

$$0 \in \partial_L\big(f + \lambda^* \circ g + k d_C\big)(\bar{x}), \quad \text{so in particular } 0 \in \partial_L f(\bar{x}) + \partial_L(\lambda^* \circ g)(\bar{x}) + N^L(C; \bar{x}).$$

PROOF. Put $S := C \cap g^{-1}(D)$ and denote by γ_0 a local Lipschitz constant of f near $\bar{x}$. Lemma 2.213 tells us that $\bar{x}$ is a local minimizer of $f + \gamma_0 d_S$, that is, $f(\bar{x}) \leq f(x) + \gamma_0 d_S(x)$ for all x near $\bar{x}$. From the metric subregularity qualification of the constraint $C \cap g^{-1}(D)$ at $\bar{x}$ (see (7.28)) there is a real constant $\gamma \geq 0$ such that $d(x, S) \leq \gamma\big(d_C(x) + d_D(g(x))\big)$ for x near $\bar{x}$. It results that $f(\bar{x}) \leq f(x) + \gamma_0 \gamma\big(d_C(x) + d_D(g(x))\big)$ for all x near $\bar{x}$. Put $k := \gamma_0 \gamma$ and consider the mapping $\Phi : X \to \mathbb{R} \times Y \times \mathbb{R}$ with $\Phi(x) := (f(x), g(x), k d_C(x))$, and the function $\varphi : \mathbb{R} \times Y \times \mathbb{R} \to \mathbb{R}$ with $\varphi(r, y, t) := r + k d_D(y) + t$ for all $(r, y, t) \in \mathbb{R} \times Y \times \mathbb{R}$. By what precedes the point $\bar{x}$ is a local minimizer of the function $\varphi \circ \Phi$, thus

$0 \in \partial_A(\varphi \circ \Phi)(\bar{x})$ (resp. $0 \in \partial_L(\varphi \circ \Phi)(\bar{x})$). Since the function φ is Lipschitz and the mapping Φ is compactly Lipschitzian at $\bar{x}$, Theorem 5.71 (resp. Theorem 4.80) and easy computation furnish some $\lambda^* \in k\partial_A d_D(g(\bar{x})) \subset N^A(D; g(\bar{x}))$ (resp. $\lambda^* \in k\partial_L d_D(g(\bar{x})) \subset N^L(D; g(\bar{x}))$) such that

$$0 \in \partial_A(f + \lambda^* \circ g + kd_C)(\bar{x})$$

(resp. $0 \in \partial_L(f + \lambda^* \circ g + kd_C)(\bar{x})$). From the latter the second inclusion in (a) (resp. (b)) follows. $\square$

If K is a closed convex cone of a normed space Z, we know by the sublinearity of Ψ_{-K} (see (2.26)) that $z^* \in N(-K; \bar{z})$ if and only if $z^* \in Z_+^*$ and $\langle z^*, \bar{z} \rangle = 0$, where $Z_+^* := \{z^* \in Z^* : \langle z^*, z \rangle \geq 0, \forall z \in K\}$. So, using Theorem 7.72 we directly derive from Theorem 7.77 a first corollary:

COROLLARY 7.78. Let X, Y_1, Y_2 be Banach spaces, let K be a closed convex cone of Y_1 with $\text{int}_{Y_1} K \neq \emptyset$, and let $f : X \to \mathbb{R}$, $g_i : X \to Y_i$, $i = 1, 2$, be mappings which are strictly Fréchet differentiable at $\bar{x} \in X$ with $g_1(\bar{x}) \in -K$ and $g_2(\bar{x}) = 0$. Assume that the following constraint qualification condition is satisfied:

(CQC) $\begin{cases} Dg_2(\bar{x}) \text{ is surjective and there is some } \bar{v} \in X \text{ with} \\ Dg_2(\bar{x})(\bar{v}) = 0 \text{ and } Dg_1(\bar{x})(\bar{v}) \in -\text{int } K. \end{cases}$

If $\bar{x}$ is a local solution of the optimization problem

Minimize $f(x)$ subject to $g_1(x) \in -K$, $g_2(x) = 0$,

then there exist Lagrange multipliers $\lambda_2^* \in Y_2^*$ and $\lambda_1^* \in (Y_1)_+^*$ with $\langle \lambda_1^*, g_1(\bar{x}) \rangle = 0$ such that

$$0 = Df(\bar{x}) + \lambda_1^* \circ Dg_1(\bar{x}) + \lambda_2^* \circ Dg_2(\bar{x}).$$

The second corollary of Theorem 7.77 (which complements Corollary 2.231) is the Mangasarian-Fromovitz necessary optimality condition in presence of finitely many inequality constraints but also with the possibility of infinitely many equality constraints. It uses the above description of $N(-K; \bar{z})$ with $K := \mathbb{R}_+^m \times \{0_Y\}$ along with Corollary 7.73.

COROLLARY 7.79 (optimality condition under Mangasarian-Fromovitz qualification:2). Let X and Y be Banach spaces and $f, g_1, \cdots, g_m : X \to \mathbb{R}$ and $g : X \to Y$ be mappings which are strictly Fréchet differentiable $\bar{x}$ with $g(\bar{x}) = 0$ and $g_i(\bar{x}) \leq 0$, $i = 1, \cdots, m$. Assume that the following Mangasarian-Fromovitz constraint qualification condition is satisfied

(MFCQC) $\begin{cases} Dg(\bar{x}) \text{ is surjective and there is some } \bar{v} \in X \text{ with} \\ Dg(\bar{x})(\bar{v}) = 0 \text{ and } Dg_i(\bar{x})(\bar{v}) < 0, \forall i \in I(\bar{x}), \end{cases}$

where $I(\bar{x}) := \{i \in \{1, \cdots, m\} : g_i(\bar{x}) = 0\}$ denotes the set of indices of active inequality constraints.

If $\bar{x}$ is a local solution of the optimization problem

Minimize $f(x)$ subject to $g_1(x) \leq 0, \cdots, g_m(x) \leq 0, g(x) = 0$,

then there exist Lagrange multipliers $\lambda^* \in Y^*$ and $\lambda_i \in [0, +\infty[$ with $\lambda_i g_i(\bar{x}) = 0$, $i = 1, \cdots, m$, such that

$$0 = Df(\bar{x}) + \lambda_1 Dg_1(\bar{x}) + \cdots + \lambda_m Dg_m(\bar{x}) + \lambda^* \circ Dg(\bar{x}).$$

7.7.2. Estimates of coderivatives under metric subregularity or other conditions; regularity of nonsmooth constraints.
In addition to Theorem 7.77 the metric subregularity also allows us to obtain estimates of the coderivative of the multimapping $M_{g,C,D}$.

PROPOSITION 7.80. *Let C and D be closed subsets of Banach spaces X and Y respectively and let $g : X \to Y$ be a mapping which is compactly Lipschitzian at $\bar{x} \in C$. For $\bar{y} \in g(\bar{x}) - D$ and for the associated multimapping $M := M_{g,C,D}$ from X into Y with*

$$M(x) = g(x) - D \quad \text{if } x \in C \quad \text{and} \quad M(x) = \emptyset \quad \text{otherwise,}$$

the following estimate holds for the approximate (resp. limiting, if X and Y are Asplund spaces) coderivative:

$$\begin{cases} D_A^* M(\bar{x}, \bar{y})(y^*) \subset \bigcup_{k \geq 0} \partial_A(y^* \circ g + kd_C)(\bar{x}) & \text{if } y^* \in N^A(D; g(\bar{x}) - \bar{y}) \\ D_A^* M(\bar{x}, \bar{x})(y^*) = \emptyset & \text{if } y^* \notin N^A(D; g(\bar{x}) - \bar{y}); \end{cases}$$

(resp. if X and Y are Asplund spaces

$$\begin{cases} D_L^* M(\bar{x}, \bar{y})(y^*) \subset \bigcup_{k \geq 0} \partial_L(y^* \circ g + kd_C)(\bar{x}) & \text{if } y^* \in N^L(D; g(\bar{x}) - \bar{y}) \\ D_L^* M(\bar{x}, \bar{y})(y^*) = \emptyset & \text{if } y^* \notin N^L(D; g(\bar{x}) - \bar{y}) \,). \end{cases}$$

PROOF. Note that for any $x \in C$ and any $y \in Y$

$$d\big((x,y), \operatorname{gph} M\big) \leq d(y, M(x)) = d(g(x) - y, D),$$

so $f(\bar{x}, \bar{y}) = 0 \leq f(x, y)$, where $f(x, y) := d(g(x) - y, D) - d\big((x,y), \operatorname{gph} M\big)$. Using the local Lipschitz property of f near $(\bar{x}, \bar{y})$, by Lemma 2.213 there exists some real $\gamma \geq 0$ such that

$$d\big((x,y), \operatorname{gph} M\big) \leq d(g(x) - y, D) + \gamma d(x, C) \quad \text{for all } (x, y) \text{ near } (\bar{x}, \bar{y}).$$

Define the mapping $\Phi : X \times Y \to \mathbb{R} \times Y$ by $\Phi(x, y) = (\gamma d_C(x), g(x) - y)$ and the function $\varphi : \mathbb{R} \times Y \to \mathbb{R}$ by $\varphi(r, v) = r + d_D(v)$. We see that $d\big((x,y), \operatorname{gph} M\big) \leq \varphi \circ \Phi(x, y)$ for all (x, y) near $(\bar{x}, \bar{y})$ and the inequality is an equality for all $(x, y) \in \operatorname{gph} M$. It ensues (see Lemma 5.56 (resp. Proposition 4.100(a))) that $\partial d_{\operatorname{gph} M}(\bar{x}, \bar{y}) \subset \partial(\varphi \circ \Phi)(\bar{x}, \bar{y})$ where ∂ stands for ∂_A (resp. ∂_L if X and Y are Asplund spaces). Denoting by $N(\cdot, \cdot)$ the associated normal cone and taking $(x^*, -y^*) \in N\big(\operatorname{gph} M; (\bar{x}, \bar{y})\big)$ there exists some real $\beta \geq 0$ such that $(x^*, -y^*) \in \beta \partial d_{\operatorname{gph} M}(\bar{x}, \bar{y})$ (see Definition 5.48 (resp. Theorem 4.85(b)), so $(x^*, -y^*) \in \beta \partial(\varphi \circ \Phi)(\bar{x}, \bar{y})$. Since the function φ is Lipschitz on $Y \times \mathbb{R}$ and the mapping Φ is easily seen to be compactly Lipschitzian at $(\bar{x}, \bar{y})$, by Theorem 5.71 (resp. Theorem 4.80) there exists some $\zeta^* \in \partial \varphi(0, g(\bar{x}) - \bar{y})$ such that $(x^*, -y^*) \in \beta(\zeta^* \circ \Phi)(\bar{x}, \bar{y})$. Clearly $\zeta^* = (1, v^*) \in \mathbb{R} \times Y^*$ with $v^* \in \partial d_D(g(\bar{x}) - \bar{y}) \subset N(D; g(\bar{x}) - \bar{y})$, and $(\zeta^* \circ \Phi)(x, y) = \gamma d_C(x) + (v^* \circ g)(x) - \langle v^*, y \rangle$ for all $(x, y) \in X \times Y$. It follows by Proposition 5.47 (resp. Proposition 4.73(a)) that $x^* \in \beta \partial(\gamma d_C + v^* \circ g)(\bar{x})$ and $y^* = \beta v^*$, hence $y^* \in N(D; g(\bar{x}) - \bar{y})$ and $x^* \in \partial(y^* \circ g + \gamma \beta d_C)(\bar{x})$. This means that for any $x^* \in D^* M(\bar{x}, \bar{y})(y^*)$ we have $y^* \in N(D; g(\bar{x}) - \bar{y})$ and there exists some real $k \geq 0$ such that $x^* \in \partial(y^* \circ g + kd_C)(\bar{x})$, which finishes the proof. $\square$

Whenever g is strictly Fréchet differentiable and C and D are normally regular at the appropriate points, the above inclusions are equalities as justified in the next proposition.

7.7. OPTIMALITY AND CALCULUS OF TANGENT AND NORMAL CONES

PROPOSITION 7.81. Let C and D be closed subsets of normed spaces X and Y respectively, let $g : X \to Y$ be a mapping from X into Y, and let $\bar{x} \in C$ and $\bar{y} \in g(\bar{x}) - D$. If g is Fréchet differentiable at $\bar{x}$, then for the associated multimapping $M := M_{g,C,D}$ from X into Y with

$$M(x) = g(x) - D \quad \text{if } x \in C \quad \text{and} \quad M(x) = \emptyset \quad \text{otherwise,}$$

one has

$$\{(x^*, y^*) : y^* \in N^F(D, g(\bar{x}) - \bar{y})), x^* \in y^* \circ Dg(\bar{x}) + N^F(C; \bar{x})\} \subset \mathrm{gph}\, D_F^* M(\bar{x}, \bar{y}).$$

If g is strictly Fréchet differentiable at $\bar{x}$ with X, Y Banach spaces and if the sets C and D are A-normally (resp. L-normally) regular at $\bar{x}$ and $g(\bar{x}) - \bar{y}$ respectively, then the approximate (resp. limiting, if X and Y are Asplund spaces) coderivative of M at $(\bar{x}, \bar{y})$ coincides with the Fréchet coderivative $D_F^* M(\bar{x}, \bar{y})$ and

$$D_A^* M(\bar{x}, \bar{y})(y^*) = \begin{cases} y^* \circ Dg(\bar{x}) + N^A(C; \bar{x}) & \text{if } y^* \in N^A(D; g(\bar{x}) - \bar{y}) \\ \emptyset & \text{if } y^* \notin N^A(D; g(\bar{x}) - \bar{y}) \end{cases}$$

(resp. if X and Y are Asplund spaces, the similar equality holds with $D_L^* M(\cdot, \cdot)$ and $N^L(\cdot; \cdot)$ in place of $D_A^* M(\cdot, \cdot)$ and $N^A(\cdot; \cdot)$).

PROOF. Fix any $u^* \in N^F(C; \bar{x})$ and $y^* \in N^F(D; g(\bar{x}) - \bar{y})$, and set $x^* := u^* + y^* \circ Dg(\bar{x})$. Write

$$\langle x^*, x - \bar{x} \rangle - \langle y^*, y - \bar{y} \rangle$$
$$= \langle u^*, x - \bar{x} \rangle - \langle y^*, -Dg(\bar{x})(x - \bar{x}) + g(x) - g(\bar{x}) \rangle + \langle y^*, g(x) - y - g(\bar{x}) + \bar{y} \rangle.$$

Taking any $\varepsilon > 0$ and using the Fréchet differentiability of g at $\bar{x}$ we then see that, for all x near $\bar{x}$ and y near $\bar{y}$ with $(x, y) \in \mathrm{gph}\, M$ or equivalently $x \in C$ and $g(x) - y \in D$, we have

$$\langle x^*, x - \bar{x} \rangle - \langle y^*, y - \bar{y} \rangle \leq \varepsilon \|x - \bar{x}\| + \varepsilon \|y^*\| \, \|x - \bar{x}\| + \varepsilon \|(g(x) - g(\bar{x})) - (y - \bar{y})\|$$
$$\leq \varepsilon(1 + \|y^*\| + \gamma)\|x - \bar{x}\| + \varepsilon \|y - \bar{y}\|,$$

where γ denotes a Lipschitz constant of g near $\bar{x}$. It ensues that the pair $(x^*, -y^*)$ belongs to $N^F(\mathrm{gph}\, M; (\bar{x}, \bar{y}))$, so $x^* \in D_F^* M(\bar{x}, \bar{y})(y^*)$, which justifies the inclusion in the first conclusion of the proposition.

Recalling that strict Fréchet differentiability at a point guarantees compact Lipschitz property at the point, the second conclusion easily follows from the above inclusion and from Proposition 7.80. □

Taking Proposition 7.80 into account, a natural condition on the data g, C, D to guarantee the nullity of the kernel of the coderivative is the following one. Given the mapping g and the closed sets C and D as above and taking ∂ as the approximate (resp. limiting) subdifferential, we say that the triple (g, C, D) has the *nullity kernel ∂-normal property* at $(\bar{x}, g(\bar{x}) - \bar{y}) \in C \times D$ provided

(7.29)
$$\Big(y^* \in N^\partial(D; g(\bar{x}) - \bar{y}) \text{ and } 0 \in \partial(y^* \circ g + kd_C)(\bar{x}) \text{ for some } k \geq 0 \Big) \Rightarrow y^* = 0;$$

we will write (NKNP) for short. It is clear from Proposition 7.80 that this condition entails the nullity of the kernel of the corresponding coderivative of $M := M_{g,C,D}$, that is,

$$(\mathrm{NKNP}) \Rightarrow \Big(\mathrm{Ker}\, D_\partial^* M(\bar{x}, g(\bar{x}) - \bar{y}) = \{0\} \Big).$$

When $\partial = \partial_A$ (resp. $\partial = \partial_L$) we simply say the nullity kernel A-normal (resp. L-normal) property.

It is also natural to consider for the triple (g, C, D) the following *sequential nullity ∂-normal compactness property (SNNCP)* at $(\bar{x}, g(\bar{x}) - \bar{y})$:

(7.30)
$$\begin{pmatrix} C \ni x_n \to \bar{x}, g(x_n) - D \ni y_n \to 0, y_n^* \in N^\partial(D; g(x_n) - y_n), \\ x_n^* \in \partial(y_n^* \circ g)(x_n) + N^\partial(C; x_n), x_n^* \xrightarrow{\|\cdot\|} 0, y_n^* \xrightarrow{w^*} 0 \end{pmatrix} \Rightarrow \|y_n^*\| \to 0.$$

To see, for $\partial = \partial_A$ and X, Y Banach spaces or $\partial = \partial_L$ and X, Y Asplund spaces, that the latter condition implies the sequential coderivative compactness of $M := M_{g,C,D}$ at $\bar{x}$ for $\bar{y}$, take sequences $\operatorname{gph} M \ni (x_n, y_n) \to (\bar{x}, \bar{y})$, $x_n^* \in D_\partial^* M(x_n, y_n)(y_n^*)$ with $x_n^* \xrightarrow{\|\cdot\|} 0$ and $y_n^* \xrightarrow{w^*} 0$. The inclusion $\operatorname{gph} M \ni (x_n, y_n)$ means $x_n \in C$ and $g(x_n) - y_n \in D$ while the other inclusion $x_n^* \in D_\partial^* M(x_n, y_n)(y_n^*)$ entails by Proposition 7.80 that $x_n^* \in \partial(y_n^* \circ g)(x_n) + N^\partial(C; x_n)$. Then (SNNCP) yields $\|y_n^*\| \to 0$, so the desired sequential coderivative compactness of M holds true as desired. If $\partial = \partial_A$ (resp. $\partial = \partial_L$) we say the sequential nullity A-normal (resp. L-normal) compactness property.

It is clear that the property (SNNCP) is fulfilled whenever the set D is sequentially ∂-normally compact at $g(\bar{x}) - \bar{y}$.

If g is the identity on the space X and $\bar{x} \in C \cap D$, the above property can be reformulated as

(7.31)
$$\begin{pmatrix} C \ni u_n \to \bar{x}, u_n^* \in N^\partial(C; u_n), D \ni v_n \to \bar{x}, v_n^* \in N^\partial(D; v_n), \\ u_n^* + v_n^* \xrightarrow{\|\cdot\|} 0, v_n^* \xrightarrow{w^*} 0 \end{pmatrix} \Rightarrow \|v_n^*\| \to 0.$$

In such case, we say that the pair of sets (C, D) is *sequentially ∂-normally mixed compact* at $\bar{x}$.

All the above features and Theorem 7.38 guarantee the following theorem via an analysis similar to (7.24).

THEOREM 7.82 (metric regularity for nonsmooth constraint systems: General case). *Let C and D be closed subsets of Banach spaces X and Y respectively, let $g : X \to Y$ be a mapping which is compactly Lipschitzian at $\bar{x} \in C$, and let $\bar{y} \in g(\bar{x}) - D$. Assume that (NKNP) and (SNNCP) are satisfied for the triple (g, C, D) at $(\bar{x}, g(\bar{x}) - \bar{y})$ with $\partial = \partial_A$ (resp. with $\partial = \partial_L$ and X, Y Asplund spaces); the second condition (SNNCP) being in particular satisfied if D is sequentially ∂-normally compact at $g(\bar{x}) - \bar{y}$. Then, there exist a real $\gamma \geq 0$ and neighborhoods U of $\bar{x}$ in X and V of $\bar{y}$ in Y such that for all $x \in U$ and $y \in V$ the following metric inequality holds true:*

$$d(x, C \cap g^{-1}(y + D)) \leq \gamma(d(x, C) + d(g(x) - y, D)).$$

REMARK 7.83. With Theorem 7.82 at hand, it is worth noticing that, under the $\mathcal{C}^1$ property of g near $\bar{x}$, condition (CQC) in Corollary 7.78 is satisfied if and only if both (NKNP) and (SNKNP) are fulfilled with $C = X$ and $D = -K \times \{0\}$. To argue that, keep notation in Corollary 7.78 and put $M(x) := g(x) + K \times \{0\}$ for all $x \in X$, where $g(x) = (g_1(x), g_2(x)) \in Y$ with $Y = Y_1 \times Y_2$. We know by Proposition 7.72 that (CQC) implies that the multimapping M is metrically regular at $\bar{x}$ for 0, which yields by Theorem 7.29(a) the existence of a real $c > 0$ such that $\|x^*\| \geq c\|y^*\|$

for all $x^* \in D_F^* M(x,y)(y^*)$ and $(x,y) \in \operatorname{gph} M$ near $(\bar{x}, 0)$. The latter means by Proposition 7.81 that $\|x^*\| \geq c\|y^*\|$ for all (x^*, y^*) such that $x^* = y^* \circ Dg(x)$ with $y^* \in N^A(-K \times \{0\}; g(x) - y)$ and (x,y) near $(\bar{x}, 0)$ with $g(x) - y \in -K \times \{0\}$. This clearly entails both (NKNP) and (SNKNP) at $(\bar{x}, g(\bar{x})) \in X \times (-K \times \{0\})$ for the triple $(g, X, -K \times \{0\})$, and both those properties imply the metric regularity of M at $\bar{x}$ for 0 by Theorem 7.82. Finally, this metric regularity of M at $\bar{x}$ for 0 entails condition (CQC) by Proposition 7.72. The aforementioned equivalence is then established. □

When the space Y is finite-dimensional, the following corollary of Theorem 7.82 directly follows.

COROLLARY 7.84 (metric regularity for nonsmooth constraint systems: finite-dimensional image space). Let X be a Banach space and Y be a finite-dimensional normed space. Let C and D be closed subsets of X and Y respectively, let $g : X \to Y$ be a mapping which is Lipschitz near $\bar{x} \in C$, and let $\bar{y} \in g(\bar{x}) - D$. Assume that (NKNP) is satisfied for the triple (g, C, D) at $(\bar{x}, g(\bar{x}) - \bar{y})$ with $\partial = \partial_A$ (resp. with $\partial = \partial_L$ and X Asplund space), that is,

$$\Big(y^* \in N^\partial(D; g(\bar{x}) - \bar{y}) \text{ and } 0 \in \partial(y^* \circ g + kd_C)(\bar{x}) \text{ for some } k \geq 0 \Big) \Rightarrow y^* = 0.$$

Then, there exist a real $\gamma \geq 0$ and neighborhoods U of $\bar{x}$ in X and V of $\bar{y}$ in Y such that for all $x \in U$ and $y \in V$ the following metric inequality holds true:

$$d\big(x, C \cap g^{-1}(y + D)\big) \leq \gamma\big(d(x, C) + d(g(x) - y, D)\big).$$

7.7.3. General optimality conditions. In absence of qualification condition, necessary optimality conditions can be obtained in terms of general Lagrange multipliers with a multiplier $\lambda_0 \geq 0$ assigned to the objective of the minimization problem as follows.

THEOREM 7.85 (general Lagrange multipliers under SNNCP). Let $g : X \to Y$ be a mapping between two Banach spaces X, Y, let C and D be two closed subsets of X and Y respectively, and let $f : X \to \mathbb{R}$ be a locally Lipschitz function. Let $\bar{x} \in C \cap g^{-1}(D)$ be a local solution of the optimization problem

$$\text{Minimize} \quad f(x) \quad \text{subject to } x \in C \cap g^{-1}(D).$$

Assume that g is compactly Lipschitzian at $\bar{x}$. Then the following Lagrange necessary optimality conditions hold:
(a) If (SNNCP) is satisfied at $(\bar{x}, g(\bar{x}))$ with ∂_A, there exist a real $\lambda_0 \geq 0$ and a functional $\lambda^* \in N^A(D; g(\bar{x}))$ with $(\lambda_0, \lambda^*) \neq 0$ along with a real $k \geq 0$ such that

$$0 \in \partial_A\big(\lambda_0 f + \lambda^* \circ g + kd_C\big)(\bar{x}),$$

so in particular

$$0 \in \lambda_0 \partial_A f(\bar{x}) + \partial_A(\lambda^* \circ g)(\bar{x}) + N^A(C; \bar{x}).$$

(b) If X and Y are Asplund spaces and (SNNCP) is satisfied at $(\bar{x}, g(\bar{x}))$ with ∂_L, there exist a real $\lambda_0 \geq 0$ and a functional $\lambda^* \in N^L(D; g(\bar{x}))$ with $(\lambda_0, \lambda^*) \neq 0$ along with a real $k \geq 0$ such that

$$0 \in \partial_L\big(\lambda_0 f + \lambda^* \circ g + kd_C\big)(\bar{x}),$$

so in particular

$$0 \in \lambda_0 \partial_L f(\bar{x}) + \partial_L(\lambda^* \circ g)(\bar{x}) + N^L(C; \bar{x}).$$

PROOF. We provide the proof in the context of ∂_A-subdifferential. If in addition (NKNP) is satisfied, then by Theorem 7.82 the metric regularity (hence subregularity) qualification condition is fulfilled; so the conclusion of the theorem follows from Theorem 7.77 with $\lambda_0 = 1$.

If (NKNP) is not satisfied at $(\bar{x}, g(\bar{x}))$, the conclusion of the theorem directly results with $\lambda_0 = 0$ and $\lambda^* \neq 0$ from the definition of (NKNP). □

Since (SNNCP) holds true whenever the space Y is finite-dimensional, the next two corollaries follow.

COROLLARY 7.86 (general Lagrange multipliers: finite-dimensional image space case). Let X be a Banach space and Y be a finite-dimensional normed space. Let $f : X \to \mathbb{R}$ and $g : X \to Y$ be locally Lipschitz and let C and D be two closed subsets of X and Y respectively. Let $\bar{x} \in C \cap g^{-1}(D)$ be a local solution of the optimization problem

$$\text{Minimize} \quad f(x) \quad \text{subject to } x \in C \cap g^{-1}(D).$$

Then the following necessary optimality conditions hold:
(a) There exist a real $\lambda_0 \geq 0$ and a functional $\lambda^* \in N^A(D; g(\bar{x}))$ with $(\lambda_0, \lambda^*) \neq 0$ along with a real $k \geq 0$ such that

$$0 \in \partial_A(\lambda_0 f + \lambda^* \circ g + k d_C)(\bar{x}),$$

so in particular

$$0 \in \lambda_0 \partial_A f(\bar{x}) + \partial_A(\lambda^* \circ g)(\bar{x}) + N^A(C; \bar{x}).$$

(b) If X is an Asplund space, there exist a real $\lambda_0 \geq 0$ and a functional $\lambda^* \in N^L(D; g(\bar{x}))$ with $(\lambda_0, \lambda^*) \neq 0$ along with a real $k \geq 0$ such that

$$0 \in \partial_L(\lambda_0 f + \lambda^* \circ g + k d_C)(\bar{x}),$$

so in particular

$$0 \in \lambda_0 \partial_L f(\bar{x}) + \partial_L(\lambda^* \circ g)(\bar{x}) + N^L(C; \bar{x}).$$

The second corollary is in the line of Theorem 2.228 established with the C-subdifferential.

COROLLARY 7.87 (general Lagrange multipliers: finitely many inequality and equality constraints). Let X be a Banach space (resp. an Asplund space), S a closed subset of X and $f_i : X \to \mathbb{R}$ a locally Lipschitz function with $i = 0, 1, \cdots, p+q$. Let $\bar{x}$ be a local solution of the optimization problem

$$\begin{cases} \text{Minimize} \quad f_0(x) \\ \text{subject to } f_1(x) \leq 0, \cdots, f_p(x) \leq 0, f_{p+1}(x) = 0, \cdots, f_{p+q}(x) = 0, x \in C. \end{cases}$$

Then the following necessary optimality condition hold. There exist reals $\lambda_0, \cdots, \lambda_{p+q}$ not all null with $\lambda_i \geq 0$ for $0 \leq i \leq p$ and $\lambda_i f_i(\bar{x}) = 0$ for $1 \leq i \leq p$ and a real $k \geq 0$ such that

$$0 \in \partial_A(\lambda_0 f_0 + \lambda_1 f_1 + \cdots + \lambda_{p+q} f_{p+q} + k d_C)(\bar{x}),$$

so in particular

$$0 \in \lambda_0 \partial_A f_0(\bar{x}) + \partial_A(\lambda_1 f_1)(\bar{x}) + \cdots + \partial_A(\lambda_{p+q} f_{p+q})(\bar{x}) + N^A(C; \bar{x})$$

(resp. the same inclusions with ∂_L in place of ∂_A).

7.7.4. Normal/tangent cone calculus and chain rule.
Diverse results for calculus of tangent and C-normal cones have been established in Subsection 2.2.13 and Subsection 2.4.1. Proposition 7.80 leads to think that the metric subregularity should help in the calculus of normal and tangent cones in other situations. This is confirmed as follows.

THEOREM 7.88 (A/L-normal cone under metric subregularity). *Let $g : X \to Y$ be a mapping between two Banach spaces X and Y, let C be a subset of X closed near $\bar{x} \in C$, and let D be a subset of Y closed near $g(\bar{x}) \in D$. If g is compactly Lipschitzian at $\bar{x}$ and if the metric subregularity qualification condition (7.28) is satisfied for $C \cap g^{-1}(D)$ at $\bar{x}$, then*

$$N^A(C \cap g^{-1}(D); \bar{x}) \subset \bigcup_{k \geq 0, y^* \in N^A(D; g(\bar{x}))} \partial_A(y^* \circ g + k d_C)(\bar{x}).$$

If X and Y are Asplund spaces, the same inclusion holds true with the limiting normal cone and subdifferential.

PROOF. Put $S := C \cap g^{-1}(D)$. The metric subregularity condition furnishes a real $\gamma \geq 0$ and neighborhoods U of $\bar{x}$ such that

$$d(x, S) \leq \gamma\bigl(d(x, C) + d(g(x), D)\bigr) \quad \text{for all } x \in U.$$

Define $\varphi(r, y) = r + \gamma d_D(y)$ for all $(r, y) \in \mathbb{R} \times Y$ and $\Phi(x) := (\gamma d_C(x), g(x)) \in \mathbb{R} \times Y$ for all $x \in X$. The latter inequality becomes $d_S(x) \leq \varphi \circ \Phi(x)$ for all $x \in U$, with equality for all $x \in S$. Denoting by ∂ the approximate subdifferential (resp. limiting subdifferential if X and Y are Asplund), Lemma 5.56 (resp. Proposition 4.100) tells us that $\partial d_S(\bar{x}) \subset \partial(\varphi \circ \Phi)(\bar{x})$. Since φ is locally Lipschitz and Φ is compactly Lipschitzian at $\bar{x}$, given $x^* \in \partial d_S(\bar{x})$ we derive the existence of some $y^* \in \gamma \partial d_D(g(\bar{x}))$ such that $x^* \in \partial(y^* \circ g + \gamma d_C)(\bar{x})$. This justifies the inclusion of the proposition since $N^\partial(S; \bar{x}) = \mathbb{R}_+ \partial d_S(\bar{x})$. $\square$

A first useful and basic corollary considers the situation when g is strictly differentiable with surjective derivative. The corollary directly follows from the above theorem and corollary 7.75.

COROLLARY 7.89 (A/L-normal cone of inverse image under surjective strict derivative). *Let $g : X \to Y$ be a mapping between Banach spaces which is strictly Fréchet differentiable at $\bar{x} \in X$ with $DG(\bar{x})$ surjective. If D is a subset of Y closed near $g(\bar{x}) \in D$, then*

$$N^A\bigl(g^{-1}(D); \bar{x}\bigr) \subset Dg(\bar{x})^* \bigl(N^A(D; g(\bar{x}))\bigr);$$

the inclusion is an equality whenever additionally D is A-normally regular at $g(\bar{x})$.

If X and Y are Asplund spaces, the same inclusion (resp. equality under L-regularity of D at $g(\bar{x})$) holds true with the limiting normal cone.

Through the above corollary one derives, as already done in the proof of the assertion (b) in Theorem 3.174, the following other corollary.

COROLLARY 7.90 (A/L-chain rule with surjective strict derivative). *Let $g : X \to Y$ be a mapping between Banach spaces which is strictly Fréchet differentiable at $\bar{x} \in X$ with $DG(\bar{x})$ surjective and let $f : Y \to \mathbb{R} \cup \{+\infty\}$ a function which is lower semicontinuous near $g(\bar{x})$. Then*

$$\partial_A(f \circ g)(\bar{x}) \subset Dg(\bar{x})^* \bigl(\partial_A f(g(\bar{x}))\bigr).$$

The same inclusion is also valid with limiting subdifferential when X and Y are Asplund spaces.

If in Theorem 7.88 the mapping g is additionally strictly Hadamard differentiable, the arguments therein show that

$$\partial_A d_S(\bar{x}) \subset \gamma Dg(\bar{x})^* \left(\partial_A d_D(g(\bar{x}))\right) + \gamma \partial_A d_C(\bar{x}),$$

and hence

$$\partial_A d_S(\bar{x}) \subset \gamma Dg(\bar{x})^* \left(\partial_C d_D(g(\bar{x}))\right) + \gamma \partial_C d_C(\bar{x}).$$

Since the sum of the two sets in the right-hand side is clearly convex and weakly* compact, we obtain the following corollary.

COROLLARY 7.91. *Let $g : X \to Y$ be a mapping between two Banach spaces X and Y, let C be subset of X closed near $\bar{x} \in C$, and let D be a subset of Y closed near $g(\bar{x}) \in D$. If g is strictly Hadamard differentiable at $\bar{x}$ and if the metric subregularity qualification condition (7.28) is satisfied for $C \cap g^{-1}(D)$ at $\bar{x}$ with a real constant $\gamma > 0$, then for $S := C \cap g^{-1}(D)$ one has*

$$\partial_C d_S(\bar{x}) \subset \gamma Dg(\bar{x})^* \left(\partial_C d_D(g(\bar{x}))\right) + \gamma \partial_C d_C(\bar{x}).$$

Since the inclusion $N^F(C;\bar{x}) + Dg(\bar{x})^*\left(N^F(D;g(\bar{x}))\right) \subset N^F\left(C \cap g^{-1}(D);\bar{x}\right)$ is easily verified when g is Fréchet differentiable at $\bar{x}$, the following result follows according to Theorem 7.88.

PROPOSITION 7.92. *Let $g : X \to Y$ be a mapping between two normed spaces X and Y, let C be a subset of X closed near $\bar{x} \in C$, and let D be a subset of Y closed near $g(\bar{x}) \in D$. If g is Fréchet differentiable at $\bar{x}$, then*

$$N^F(C;\bar{x}) + Dg(\bar{x})^*\left(N^F(D;g(\bar{x}))\right) \subset N^F\left(C \cap g^{-1}(D);\bar{x}\right).$$

If g is strictly Fréchet differentiable at $\bar{x}$ with X, Y Banach spaces (resp. Asplund spaces), if the metric subregularity qualification condition (7.28) is satisfied for $C \cap g^{-1}(D)$ at $\bar{x}$, and if the sets C and D are A-normally (resp. L-normally) regular at $\bar{x}$ and $g(\bar{x})$ respectively, then $C \cap g^{-1}(D)$ is A-normally (resp. L-normally) regular at $\bar{x}$ and

$$N^A\left(C \cap g^{-1}(D); \bar{x}\right) = N^A(C;\bar{x}) + Dg(\bar{x})^*\left(N^A(D;g(\bar{x}))\right)$$

(resp. the same equality holds true with the limiting normal cone and subdifferential).

Taking as g the identity mapping from X into itself in Theorem 7.88 and Proposition 7.92, we can derive the following fundamental estimates for normal cones of intersections of sets. The special case of intersection of convex sets will be considered in Corollary 7.110.

COROLLARY 7.93. *Let X be a normed space and C, D be two subsets which are closed near $\bar{x} \in C \cap D$.*
(a) *The following inclusion holds*

$$N^F(C;\bar{x}) + N^F(D;\bar{x}) \subset N^F(C \cap D; \bar{x}).$$

(b) *If the sets C and D are metrically subregularly transversal at $\bar{x}$ and if X, Y are Banach spaces (resp. Asplund spaces), then*

$$N^A(C \cap D; \bar{x}) \subset N^A(C;\bar{x}) + N^A(D;\bar{x})$$

(resp. the same inclusion holds with $N^L(\cdot;\cdot)$ in place of $N^A(\cdot;\cdot)$). If in addition both sets C and D are A-normally (resp. L-normally) regular at $\bar{x}$, then the above inclusion is an equality (resp. an equality in terms of $N^L(\cdot;\cdot)$).

Under the tangential condition $I(C;\bar{x}) \cap T^C(D;\bar{x}) \neq \emptyset$ we saw in Theorem 2.13(d) that the inclusion $N^C(C \cap D; \bar{x}) \subset N^C(C;\bar{x}) + N^C(D;\bar{x})$ holds true. Under the same tangential condition, the next corollary shows among other things the similar inclusion with the A-normal cone. Before stating the corollary, we notice that for two subsets C and D of a normed space X and $\bar{x} \in C \cap D$, the C-tangential transversality at $\bar{x}$ implies their C-normal transversality at $\bar{x}$, that is,

$$(7.32) \quad (T^C(C;\bar{x}) - T^C(D;\bar{x}) = X) \Longrightarrow (N^C(C;\bar{x}) \cap (-N^C(D;\bar{x})) = \{0\}).$$

Indeed, take any $x^* \in N^C(C;\bar{x}) \cap (-N^C(D;\bar{x}))$. For every $v \in X$ there exist $v_C \in T^C(C;\bar{x})$ and $v_D \in T^C(D;\bar{x})$ such that $v = v_C - v_D$, so

$$\langle x^*, v \rangle = \langle x^* v_C \rangle + \langle -x^*, v_D \rangle \leq 0.$$

This being true for all $v \in X$, it ensues that $x^* = 0$ as desired.

COROLLARY 7.94. *Let C and D be two subsets of a Banach space X which are closed near $\bar{x} \in C \cap D$. The inclusion*

$$N^A(C \cap D; \bar{x}) \subset N^A(C;\bar{x}) + N^A(D;\bar{x})$$

(resp. the same with $N^L(\cdot;\cdot)$) holds under anyone of the following conditions (a), (b), (c), (d) and (e):
(a) The pair of sets (C, D) is sequentially A-normally (resp. F-normally) mixed compact at $\bar{x}$ (see (7.31)) and the sets C, D are A-normally (resp. L-normally if X is an Asplund space) transversal at $\bar{x}$, that is, $N^A(C;\bar{x}) \cap (-N^A(D;\bar{x})) = \{0\}$ (resp. the same with $N^L(\cdot,\cdot)$ in place of $N^A(\cdot;\cdot)$).
(b) One of the sets C, D is sequentially A-normally (resp. F-normally) compact at $\bar{x}$ and the sets C, D are A-normally (resp. L-normally if X is an Asplund space) transversal at $\bar{x}$.
(c) $T^C(C;\bar{x}) \cap I(D;\bar{x}) \neq \emptyset$.
(d) The space X is finite-dimensional and the sets C, D are L-normally transversal at $\bar{x}$, that is, $N^L(C;\bar{x}) \cap (-N^L(D;\bar{x})) = \{0\}$.
(e) The space X is finite-dimensional and the sets C, D are C-tangentially transversal at $\bar{x}$, that is, $T^C(C;\bar{x}) - T^C(D;\bar{x}) = X$.

PROOF. The C-tangential condition in (e) entails that the equality $N^A(C;\bar{x}) \cap (-N^A(D;\bar{x})) = \{0\}$ holds according to (7.32). Consequently, the inclusion in the corollary, under anyone of the four assumptions (a), (b), (d) and (e), follows from Corollary 7.93 and Theorem 7.82.

Now assume the condition in (c). On the one hand, the set D satisfies the compact tangent condition, hence by Proposition 2.34 we see that it is A-normally compact. On the other hand, there exists by (c) some $u \in T^C(C;\bar{x})$ and a neighborhood V of zero in X such that $u + V \subset I(D;\bar{x})$, so

$$V = (u + V) - u \subset T^C(D;\bar{x}) - T^C(C;\bar{x}).$$

We deduce that $X = T^C(C;\bar{x}) - T^C(D;\bar{x})$, thus $N^A(C;\bar{x}) \cap (-N^A(D;\bar{x})) = \{0\}$ by (7.32). The conditions in (b) are then satisfied, hence by what precedes we obtain the desired inclusion in the corollary. □

Under the convexity of the sets C, D in Proposition 7.92, the easier-to-verify Robinson qualification condition also guarantees the equality for the normal cone to $C \cap g^{-1}(D)$. The corresponding result concerning the subdifferential of $f_1 + f_2 \circ g$ will be developed in Corollary 11.9 with a strictly differentiable mapping $g : X \to Y$ and convex functions $f_1 : X \to \mathbb{R} \cup \{+\infty\}$ and $f_2 : Y \to \mathbb{R} \cup \{+\infty\}$.

THEOREM 7.95 (normal cone under Robinson condition). *Let X, Y be Banach spaces and D be a closed convex subset of Y. Let C be a closed convex subset of X and $g : X \to Y$ be a mapping which is strictly Fréchet differentiable at $\bar{x} \in C \cap g^{-1}(D)$. Assume that, for $\bar{y} := g(\bar{x})$ the following Robinson qualification condition holds:*
$$0 \in \mathrm{core}\big(Dg(\bar{x})(C - \bar{x}) - (D - \bar{y})\big), \text{ that is, } \mathbb{R}_+\big(Dg(\bar{x})(C - \bar{x}) - (D - \bar{y})\big) = Y.$$
Then, for $S := C \cap g^{-1}(D)$ one has
$$N^F(S; \bar{x}) = N^L(S; \bar{x}) = N^A(S; \bar{x}) = N^C(S; \bar{x}) = N(C; \bar{x}) + Dg(\bar{x})^*\big(N(D; \bar{y})\big).$$

PROOF. Put $A : Dg(\bar{x})$. We know by Theorem 7.70 that the Robinson qualification condition ensures that $C \cap g^{-1}(D)$ enjoys the metric regularity condition, hence Proposition 7.92 tells us that

(7.33) $\qquad N^F(S; \bar{x}) = N^A(S; \bar{x}) = N(C; \bar{x}) + A^*\big(N(D; \bar{y})\big),$

and the first equality also ensures that the three members coincide with $N^L(S; \bar{x})$ (since $N^F(Q; \cdot) \subset N^L(Q; \cdot) \subset N^A(Q; \cdot)$ for any set Q). We now proceed to showing that $N(C; \bar{x}) + A^*\big(N(D; \bar{y})\big)$ is weakly* closed in X^*. Consider the multimapping $M : X \rightrightarrows Y$ defined by $M(x) := A(x) - (D - \bar{y})$ if $x \in C - \bar{x}$ and $M(x) = \emptyset$ otherwise. The graph of M is clearly closed and convex and $0 \in \mathrm{core}\, M(X)$ according to the Robinson qualification condition again. Since $0 \in M(0)$, it follows by Robinson-Ursescu theorem (see Theorem 7.26) that there exists some real $r > 0$ such that $r \mathbb{B}_Y \subset M(\mathbb{B}_X)$, or equivalently
$$r \mathbb{B}_Y \subset -A\big((C - \bar{x}) \cap \mathbb{B}_X\big) + (D - \bar{y}).$$
Let $(x_j^*)_{j \in J}$ be a bounded net in $N(C; \bar{x}) + A^*\big(N(D; \bar{y})\big)$ converging weakly* to some x^* in X^*, and let a real $\gamma \geq 0$ such that $\|x_j^*\| \leq \gamma$ for all $j \in J$. For each $j \in J$ there are $u_j^* \in N(C; \bar{x})$ and $v_j^* \in N(D; \bar{y})$ such that $x_j^* = u_j^* + v_j^* \circ A$. We claim that the net $(v_j^*)_{j \in J}$ is bounded. Indeed, fix any $b \in \mathbb{B}_Y$ and choose $d \in D$ and $c \in C$ with $c - \bar{x} \in \mathbb{B}_X$ such that $rb = -A(c - \bar{x}) + (d - \bar{y})$, so
$$\langle v_j^*, rb \rangle = \langle v_j^*, d - \bar{y} \rangle - \langle v_j^* \circ A, c - \bar{x} \rangle \leq \langle u_j^*, c - \bar{x} \rangle + \langle -x_j^*, c - \bar{x} \rangle \leq \|x_j^*\| \leq \gamma,$$
hence $\|v_j^*\| \leq \gamma/r$, which justifies the claim. By the boundedness of $(x_j^*)_{j \in J}$ we derive that the net $(u_j^*)_{j \in J}$ is also bounded. Extracting a subnet from $(u_j^*, v_j^*)_{j \in J}$ if necessary, we may suppose that $(u_j^*)_{j \in J}$ and $(v^*)_{j \in J}$ converge weakly* to some u^* and v^* in X^* and Y^* respectively. Consequently, $(v_j^* \circ A)_{j \in J}$ converges to $v^* \circ A$ in X^*, so $x^* = u^* + v^* \circ A$ along with $u^* \in N(C; \bar{x})$ and $v^* \in N(D; \bar{y})$ according to the weak* closedness of $N(C; \bar{x})$ and $N(D; \bar{y})$. Consequently, $x^* \in N(C; \bar{x}) + A^*\big(N(D; \bar{y})\big)$ and the Krein-Šmulian theorem (see Theorem C.4 in Appendix) tells us that the convex set $N(C; \bar{x}) + A^*\big(N(D; \bar{y})\big)$ enjoys the weak* closedness property in X^*, as desired.

Combining the latter weak* closedness property with (7.33) as well as with the equality $N^C(S; \bar{y}) = \overline{\mathrm{co}}^*\big(N^A(S; \bar{y})\big)$ (see Theorem 5.76), we conclude that $N^C(S; \bar{x}) = N(C; \bar{x}) + A^*\big(N(D; \bar{y})\big)$. $\square$

In the important case where the constraints set is in the form in Corollary 7.73, Theorem 7.70, condition (MFCQC) in Corollary 7.73 and Theorem 7.95 directly ensure the following basic theorem.

THEOREM 7.96 (normal cone under Mangasarian-Fromovitz condition). Let $\bar{x}$ be a point in the constraint set
$$S := \{x \in X : g_1(x) \leq 0, \cdots, g_m(x) \leq 0, g(x) = 0\},$$
where $g : X \to Y$ is a mapping between two Banach spaces X, Y and $g_1, \cdots, g_m : X \to \mathbb{R}$ are real-valued functions defined on X. Assume that g and $g_1, \cdots, g_m$ are strictly Fréchet differentiable at $\bar{x}$ and denote by $I(\bar{x}) := \{i \in \{1, \cdots, m\} : g_i(\bar{x}) = 0\}$. If Mangasarian-Fromovitz condition (MFCQC) in Corollary 7.73 is satisfied at $\bar{x}$, then

$$N^C(S; \bar{x}) = N^A(S; \bar{x}) = \left\{ \sum_{i \in I(\bar{x})} \lambda_i Dg_i(\bar{x}) : \lambda_i \geq 0, i \in I(\bar{x}) \right\} + \{y^* \circ Dg(\bar{x}) : y^* \in Y^*\}.$$

If in addition X and Y are Asplund spaces, then the three above members coincide with $N^L(S; \bar{x})$.

In addition to the tangential calculus in Chapter 2, other tangent cone formulas can be established with the help of metric subregularity. Given an open set U of a normed space X, for a mapping $g : U \to Y$ from U into a normed space Y which is compactly Lipschitzian at $\bar{x} \in U$ and for
$$q_g(t, x; h) := t^{-1}\big(g(x + th) - g(x)\big),$$
by Theorem 2.186 the set
$$D_g(\bar{x}; h) := \left\{ \lim_{n \to \infty} q_g(t_n, x_n; h) : t_n \downarrow 0, x_n \to \bar{x} \right\}$$
is nonempty and included in the compact set $K(h)$ involving in Definition 2.183 of compactly Lipschitzian mappings. Furthermore, for any sequences $t_n \downarrow 0$, $x_n \to \bar{x}$ and $h_n \to h$ (with $x_n, x_n + t_n h_n \in U$), by Theorem 2.186 again the sequence $(q_g(t_n, x_n; h_n))_n$ contains a subsequence converging to an element in $D_g(\bar{x}; h)$; from this it is easily seen that $D_g(\bar{x}; h)$ is compact. By Lipschitz property of g around $\bar{x}$ it is also clear that
$$\lim_{n \to \infty} q_g(t_n, x_n; h_n) = \lim_{n \to \infty} q_g(t_n, x_n; h)$$
whenever anyone of the limits exists.

Besides the description of tangent cones with $D_g(\bar{x}; \cdot)$ in the next proposition, it is worth observing the following properties:
$$D_g(\bar{x}; \lambda h) = \lambda D_g(\bar{x}; h) \quad \text{for all } \lambda \geq 0, h \in X;$$
$$D_g(\bar{x}; h + h') \subset D_g(\bar{x}; h) + D_g(\bar{x}; h') \quad \text{for all } h, h' \in X.$$
A multimapping $M : X \rightrightarrows Y$ satisfying such properties, that is,

(7.34) $\qquad M(h + h') \subset M(h) + M(h')$ and $M(\lambda h) = \lambda M(h)$ for $\lambda \geq 0$

is generally called a *fan*.

Weakening the sequential characterization in Theorem 2.186 of compactly Lipschitzian mapping in keeping all x_n as $\bar{x}$, we say that the mapping g is *one-sided compactly Lipschitzian* at $\bar{x}$ provided that, given any $h \in X$, for every sequence

$t_n \downarrow 0$ and every sequence $h_n \to h$ with $\bar{x}+t_n h_n \in U$, the sequence $\left(q_g(t_n,\bar{x};h_n)\right)_n$ has a limit point. For such a mapping the set

$$D_{g,\bar{x}}(\bar{x};h) := \left\{\lim_{n\to\infty} q_g(t_n,\bar{x};h) : t_n \downarrow 0, h_n \to h\right\}$$

is nonempty. Clearly, any mapping which is either compactly Lipschitzian at $\bar{x}$ or Hadamard differentiable at $\bar{x}$, is one-sided compactly Lipschitzian at $\bar{x}$.

PROPOSITION 7.97. Let $g : X \to Y$ be a mapping between normed spaces X, Y, let C be a subset of X closed near $\bar{x} \in C$ and D be a subset of Y closed near $g(\bar{x}) \in D$.
(a) If g is one-sided compactly Lipschitzian at $\bar{x}$, then

$$T^B(C \cap g^{-1}(D); \bar{x}) \subset T^B(C; \bar{x}) \cap \{h \in X : D_{g,\bar{x}}(\bar{x};h) \cap T^B(D; g(\bar{x})) \neq \emptyset\}.$$

(b) If g is compactly Lipschitzian at $\bar{x}$ and if the metric subregularity condition (7.28) is satisfied at $\bar{x}$, then

$$T^C(C; \bar{x}) \cap \{h \in X : D_g(\bar{x};h) \subset T^C(D; g(\bar{x}))\} \subset T^C(C \cap g^{-1}(D); \bar{x}).$$

(c) If g is strictly Hadamard differentiable at $\bar{x}$, if the metric subregularity condition (7.28) is satisfied at $\bar{x}$, and if the sets C and D are (Clarke) tangentially regular at $\bar{x}$ and $g(\bar{x})$ respectively, then the set $C \cap g^{-1}(D)$ is (Clarke) tangentially regular at $\bar{x}$ and

$$T^C(C \cap g^{-1}(D); \bar{x}) = T^C(C; \bar{x}) \cap Dg(\bar{x})^{-1}(T^C(D : g(\bar{x}))).$$

PROOF. Put $S := C \cap g^{-1}(D)$.
(a) Fix $h \in T^B(S; \bar{x})$. There exist sequences $t_n \downarrow 0$ and $h_n \to h$ such that, for every $n \in \mathbb{N}$ we have $\bar{x} + t_n h_n \in S$, or equivalently $\bar{x} + t_n h_n \in C$ and $g(\bar{x} + t_n h_n) \in D$. The first inclusion $\bar{x} + t_n h_n \in C$ for all n tells us that $h \in T^B(C; \bar{x})$. Setting $v_n := q_g(t_n, \bar{x}; h_n)$, the second inclusion $g(\bar{x} + t_n h_n) \in D$ can be written as $g(\bar{x}) + t_n v_n \in D$. Choosing a limit point v of the sequence $(v_n)_n$ we obtain $v \in D_{g,\bar{x}}(\bar{x};h)$ and $v \in T^B(D; g(\bar{x}))$, and this justifies the inclusion in the assertion (a).
(b) The metric subregularity condition furnishes a real $\gamma > 0$ such that, for all x near $\bar{x}$

$$d(x, S) \leq \gamma\big(d(x, C) + d(g(x), D)\big).$$

Fix any $h \in X$ and by Proposition 2.95(a) choose sequences $t_n \downarrow 0$ and $x_n \to \bar{x}$ with $x_n \in S$ such that $d_S^\circ(\bar{x}; h) = \lim_{n\to\infty} t_n^{-1} d_S(x_n + t_n h)$. Setting $v_n := q_g(t_n, x_n; h)$, we can write

$$\gamma^{-1} d_S(x_n + t_n h) \leq d_C(x_n + t_n h) + d_D(g(x_n + t_n h)) = d_C(x_n + t_n h) + d_D(g(x_n)) + t_n v_n).$$

Let $v^h \in D_g(\bar{x}; h)$ be the limit of a subsequence $(v_{s(n)})_n$. Since $x_n \in C$ and $g(x_n) \in D$, it ensues that

$$\gamma^{-1} d_S^\circ(\bar{x}; h) \leq \limsup_{n\to\infty} t_n^{-1} d_C(x_{s(n)} + t_{s(n)} h) + \limsup_{n\to\infty} t_n^{-1} d_D\big(g(x_{s(n)}) + t_{s(n)} v_{s(n)}\big)$$

$$\leq d_C^\circ(\bar{x}; h) + d_D^\circ(g(\bar{x}); v^h),$$

which ensures that

$$\gamma^{-1} d_S^\circ(\bar{x}; h) \leq d_C^\circ(\bar{x}; h) + \sup\{d_D^\circ(g(\bar{x}); v) : v \in D_g(\bar{x}; h)\}.$$

Recalling that $h \in T^C(S; \bar{x})$ if and only if $d_S^\circ(\bar{x}; h) \leq 0$ (see Proposition 2.95(b)), the inclusion in (b) follows.

(c) Since the strict Hadamard differentiability of g at $\bar{x}$ entails its compact Lipschitz property of $\bar{x}$, we easily deduce (c) from (a) and (b). □

Let us consider the fundamental case of the set S in Theorem 7.96. For $Y_0 := \mathbb{R}^{I(\bar{x})} \times \{0_Y\}$, for $g_0(x) := (g_i(x))_{i \in I(\bar{x})}, g(\bar{x}))$ and for $D_0 := \mathbb{R}_+^{I(\bar{x})} \times \{0_Y\}$, it is clear that, for some neighborhood U of $\bar{x}$, the set $S \cap U$ coincides with the intersection with U of the set $g_0^{-1}(D_0)$. So applying the assertion (c) in Theorem 7.97 with D_0, g_0 and $C = X$ gives the following basic theorem on tangent cones of sets.

THEOREM 7.98 (tangent cone under Mangasarian-Fromovitz condition). Let $\bar{x}$ be a point in the constraint set

$$S := \{x \in X : g_1(x) \leq 0, \cdots, g_n(x) \leq 0, g(x) = 0\},$$

where $g : X \to Y$ is a mapping between two Banach spaces X, Y and $g_1, \cdots, g_m : X \to \mathbb{R}$ are real-valued functions defined on X. Assume that g and $g_1, \cdots, g_n$ are strictly Fréchet differentiable at $\bar{x}$ and denote by $I(\bar{x}) := \{i \in \{1, \cdots, n\} : g_i(\bar{x}) = 0\}$. If the Mangasarian-Fromovitz condition (MFCQC) in Corollary 7.73 is satisfied at $\bar{x}$, then

$$T^B(S; \bar{x}) = T^C(S; \bar{x}) = \{h \in X : Dg_i(\bar{x})h \leq 0, \forall i \in I(\bar{x}), Dg(\bar{x})h = 0\}.$$

We establish now some calculus for the tangential derivative of the multi-mapping $M_{g,C,D}$.

PROPOSITION 7.99. Let $g : X \to Y$ be a mapping between two normed spaces X, Y and let C and D be two subsets of X and Y respectively. For the multimapping $M := M_{g,C,D}$ with $M(x) = g(x) - D$ if $x \in C$ and $M(x) = \emptyset$ otherwise, and for $\bar{x} \in C$ and $\bar{y} \in M(\bar{x}) = g(\bar{x}) - D$, the following hold:
(a) If g is one-sided compactly Lipschitzian at $\bar{x}$, then for every $h \in T^B(C; \bar{x})$, one has

$$T^B M(\bar{x}, \bar{y})(h) \subset D_{g,\bar{x}}(\bar{x}; h) - T^B(D; g(\bar{x}) - \bar{y}),$$

and $T^B M(\bar{x}, \bar{y})(h) = \emptyset$ whenever $h \notin T^B(C; \bar{x})$.
(b) If g is Hadamard differentiable at $\bar{x}$, then for every $h \in T^B(C; \bar{x})$

$$\{v \in Y : (h, Dg(\bar{x})h - v) \in T^B(C \times D; (\bar{x}, g(\bar{x}) - \bar{y}))\} \subset T^B M(\bar{x}, \bar{y})(h).$$

PROOF. (a) Fix $v \in T^B M(\bar{x}, \bar{y})(h)$ (if any). There exist sequences $t_n \to 0$ with $t_n > 0$ and $(h_n, v_n) \to (h, v)$ such that $\bar{y} + t_n v_n \in M(\bar{x} + t_n h_n)$, or equivalently $\bar{x} + t_n h_n \in C$ and $g(\bar{x} + t_n h_n) - \bar{y} - t_n v_n \in D$. Putting $q_n := q_g(t_n, \bar{x}; h_n)$, the second latter inclusion means $g(\bar{x}) - \bar{y} + t_n(q_n - v_n) \in D$. Choosing a limit point $q \in D_{g,\bar{x}}(\bar{x}; h)$, we derive that $q - v \in T^B(D; g(\bar{x}) - \bar{y})$. So, we obtain $h \in T^B(C; \bar{x})$ and $v \in D_{g,\bar{x}}(\bar{x}; h) - T^B(D; g(\bar{x}) - \bar{y})$, which proves all the assertion (a).
(b) Now assume that g is Hadamard differentiable at $\bar{x}$ and note that the first member is empty whenever $h \notin T^B(C; \bar{x})$. So, fix $h \in T^B(C; \bar{x})$ and $v \in X$ such that

$$(h, Dg(\bar{x})h - v) \in T^B(C \times D; (\bar{x}, g(\bar{x}) - \bar{y})).$$

There exist sequences $t_n \downarrow 0$, $h_n \to h$ and $v_n \to v$ such that, for every $n \in \mathbb{N}$ we have $\bar{x} + t_n h_n \in C$ and $g(\bar{x}) - \bar{y} + t_n(Dg(\bar{x})h_n - v_n) \in D$. By the Hadamard differentiability of g at $\bar{x}$ there is a sequence $e_n \to 0$ in Y such that $g(\bar{x}) + t_n Dg(\bar{x})h_n = g(\bar{x} + t_n h_n) + t_n e_n$, thus

$g(\bar{x} + t_n h_n) - \bar{y} + t_n(-v_n + e_n) \in D$, that is, $\bar{y} + t_n(v_n - e_n) \in g(\bar{x} + t_n h_n) - D$,

with $\bar{x} + t_n h_n \in C$. This means that
$$(\bar{x}, \bar{y}) + t_n(h_n, v_n - e_n) \in \text{gph}\, M \quad \text{for all } n \in \mathbb{N},$$
which ensures that $v \in T^B M(\bar{x}, \bar{y})(h)$, and justifies the assertion (b). □

At this stage, taking the assertion (b) of the above proposition into account and recalling that the Bouligand-Peano tangent cone and the Clarke tangent cone are given by
$$T^B(S; x) = \limsup_{t \downarrow 0} \frac{1}{t}(S - x) \quad \text{and} \quad T^C(S; x) = \liminf_{t \downarrow 0;\, S \ni u \to x} \frac{1}{t}(S - u),$$
we introduce the adjacent tangent cone as follows. Given a set S of a normed space X and a point $x \in S$ the *adjacent tangent cone* $T^\flat(S; x)$ of S at x is defined by

(7.35) $$T^\flat(S; x) := \liminf_{t \downarrow 0} \frac{1}{t}(S - x).$$

So, a vector $h \in T^\flat(C; x)$ if and only if for any sequence $(t_n)_n$ tending to 0 with $t_n > 0$ there exists a sequence $h_n \to h$ such that $x + t_n h_n \in C$ for all $n \in \mathbb{N}$. It is clear that
$$T^C(S; x) \subset T^\flat(S; x) \subset T^B(S; x).$$
Given another set S' of another normed space X' and given $x' \in S'$, it is also easily seen that
$$T^\flat(S; x) \times T^B(S; x') \subset T^B(S \times S'; (x, x')).$$
According to this inclusion, the assertion (b) of the next proposition follows directly from Proposition 7.99(b).

PROPOSITION 7.100. *Let $g : X \to Y$ be a mapping between two normed spaces X, Y and let C and D be two subsets of X and Y respectively. For the multimapping $M := M_{g,C,D}$ with $M(x) = g(x) - D$ if $x \in C$ and $M(x) = \emptyset$ otherwise, and for $\bar{y} \in M(\bar{x}) = g(\bar{x}) - D$, the following hold:*
(a) *If g is compactly Lipschitzian at $\bar{x}$, then for every $h \in T^\flat(C; \bar{x})$, one has*
$$D_{g,\bar{x}}(\bar{x}; h) - T^\flat(D; g(\bar{x}) - \bar{y}) \subset T^B M(\bar{x}, \bar{y})(h),$$
and the equality holds whenever, for C at $\bar{x}$ as well as for D at $g(\bar{x}) - \bar{y}$, the adjacent and Bouligand-Peano tangent cones coincide.
(b) *If g is Hadamard differentiable at $\bar{x}$, then for every $h \in T^B(C; \bar{x})$*
$$Dg(\bar{x})h - T^\flat(D; g(\bar{x}) - \bar{y}) \subset T^B M(\bar{x}, \bar{y})(h)$$
(while $T^B M(\bar{x}, \bar{y}) = \emptyset$ for $h \notin T^B(C; \bar{x})$), and similarly for every $h \in T^\flat(D; g(\bar{x}) - \bar{y})$
$$Dg(\bar{x})h - T^B(D; g(\bar{x}) - \bar{y}) \subset T^B M(\bar{x}, \bar{y})(h).$$

PROOF. As said above, (b) follows from Proposition 7.99(b). To prove (a), fix any $h \in T^\flat(C; \bar{x})$ and $q \in D_{g,\bar{x}}(\bar{x}; h)$, and fix also any $v \in T^\flat(D; g(\bar{x}) - \bar{y})$. Let $t_n \downarrow 0$ be a sequence such that $q = \lim_{n \to \infty} q_g(t_n, \bar{x}; h)$. There exist sequences $h_n \to h$ and $v_n \to v$ such that $\bar{x} + t_n h_n \in C$ and $g(\bar{x}) - \bar{y} + t_n v_n \in D$. Putting $q_n := q_g(t_n, \bar{x}; h_n)$ we note that $q_n \to q$ and, for every n,
$$g(\bar{x} + t_n h_n) - t_n q_n - \bar{y} + t_n v_n \in D, \text{ that is, } \bar{y} + t_n(q_n - v_n) \in g(\bar{x} + t_n h_n) - D$$
along with $\bar{x} + t_n h_n \in C$. Then $\bar{y} + t_n(q_n - v_n) \in M(\bar{x} + t_n h_n)$ with $h_n \to h$ and $q_n - v_n \to q - v$, which guarantees that $q - v \in T^B M(\bar{x}, \bar{y})(h)$. This proves the desired inclusion $D_{g,\bar{x}}(\bar{x}; h) - T^\flat(D; g(\bar{x}) - \bar{y}) \subset T^B M(\bar{x}, \bar{y})(h)$ for all $h \in$

7.7. OPTIMALITY AND CALCULUS OF TANGENT AND NORMAL CONES

$T^b(C;\bar{x})$. The equality under the additional assumption in (a) directly follows from Proposition 7.99(a). □

Using the above proposition and Theorem 7.32 of Aubin condition for openness with the multimapping $M_{g,C,D}$ we arrive at the following metric regularity inequality.

PROPOSITION 7.101. *Let $g : X \to Y$ be a mapping between two Banach spaces X, Y, let C a subset of X closed near $\bar{x} \in C$ and let D be a subset of Y closed near $g(\bar{x}) - \bar{y} \in D$. Assume that g is Lipschitz near $\bar{x}$ and Hadamard differentiable near $\bar{x}$ and that there exist two reals $c > 0$ and $\eta \in [0, c[$ and two neighborhoods U of $\bar{x}$ and V of $\bar{y}$ such that, for all $x \in U \cap C$ and $y \in V \cap (g(x) - D)$,*

$$\mathbb{B}_Y \subset Dg(x)\left(T^B(C;x) \cap c\mathbb{B}_X\right) - T^b(D; g(x) - y) + \eta \mathbb{B}_Y.$$

Then, the multimapping $M_{g,C,D}$ is metrically regular at $\bar{x}$ for $\bar{y}$, otherwise stated, there exist a real $\gamma \geq 0$ and neighborhoods U_0 of $\bar{x}$ and V_0 of $\bar{y}$ such that

$$d\big(x, C \cap g^{-1}(y + D)\big) \leq \gamma \Big(d(x, C) + d\big(g(x) - y, D\big)\Big) \quad \text{for all } x \in U_0, \ y \in V_0.$$

Particularizing with $\bar{y} = 0$ and $\eta = 0$ allows us to prove a C-normal cone formula of great importance.

PROPOSITION 7.102. *Let $g : X \to Y$ be a mapping between two Banach spaces X, Y, let C a subset of X closed near $\bar{x} \in C$ and let D be a subset of Y closed near $g(\bar{x}) \in D$. Assume that g is continuously differentiable near $\bar{x}$ and that there exist a real $\sigma > 0$ and two neighborhoods U of $\bar{x}$ and V of $g(\bar{x})$ such that, for all $x \in U \cap C$ and $y \in V \cap D$,*

$$\sigma \mathbb{B}_Y \subset Dg(x)\left(T^C(C;x) \cap \mathbb{B}_X\right) - T^C(D; y).$$

Then one has with $A := Dg(\bar{x})$

$$N^C\left(C \cap g^{-1}(D); \bar{x}\right) \subset N^C(C; \bar{x}) + A^* \left(N^C(D; g(\bar{x}))\right),$$

and the inclusion is an equality whenever the sets C and D are tangentially regular at $\bar{x}$ and $g(\bar{x})$ respectively.

PROOF. Proposition 7.101 and Proposition 7.97 ensure that

$$T^C(C;\bar{x}) \cap A^{-1}\left(T^C(D; g(\bar{x}))\right) \subset T^C\left((C \cap g^{-1}(D)); \bar{x}\right)$$

with the inclusion replaced by the equality if in addition the sets C and D are tangentially regular at $\bar{x}$ and $g(\bar{x})$ respectively. Therefore, putting $T_C := T^C(C;\bar{x})$ and $T_D := T^C(D; g(\bar{x}))$ and taking the negative polar in the above inclusion, we see that

$$N^C\left(C \cap g^{-1}(D); \bar{x}\right) \subset N(T_C \cap A^{-1}(T_D); 0)$$

with equality under the additional tangential regularity. Since the assumption of the theorem entails that $0 \in \operatorname{core}\left(A(T_C) - T_D\right)$, Theorem 7.95 gives

$$N(T_C \cap A^{-1}(T_D); 0) = N(T_C; 0) + A^* \left(N(T_D; 0)\right).$$

This and the equalities $N^C(C;\bar{x}) = N(T_C; 0)$ and $N^C(D; g(\bar{x})) = N(T_D; 0)$ finish the proof. □

Taking g as the identity over the space X yields the following corollary.

COROLLARY 7.103. Let C and D be two subsets of a Banach space X both closed near a point $\bar{x} \in C \cap D$. Assume that there exist a real $\sigma > 0$ and a neighborhood U of $\bar{x}$ such that for all $x \in U \cap C$ and $y \in U \cap D$

$$\sigma \mathbb{B}_X \subset T^C(C;x) \cap \mathbb{B}_X - T^C(D;y) \cap \mathbb{B}_X.$$

Then one has

$$N^C(C \cap D; \bar{x}) \subset N^C(C; \bar{x}) + N^C(D; \bar{x}),$$

and the inclusion is an equality whenever the sets C and D are in addition tangentially regular at $\bar{x}$.

7.8. More on subdifferential calculus for convex functions

In addition to Section 3.11, this section provides, in the context of Banach spaces, formulas for sum and chain rules of convex functions under various verifiable conditions.

THEOREM 7.104 (conjugate and subdifferential of $f + g \circ G$ under convexity in Banach spaces). Let X and Y be Banach spaces and Y_+ be a convex cone in Y containing zero. Let $G : X \to Y \cup \{+\infty_Y\}$ be a Y_+-convex mapping with closed epigraph and $f : X \to \mathbb{R} \cup \{+\infty\}$ and $g : Y \to \mathbb{R} \cup \{-\infty, +\infty\}$ be lower semicontinuous convex functions.
(a) If g is non Y_+-decreasing and if $\mathbb{R}_+[\operatorname{dom} g - G(\operatorname{dom} f \cap \operatorname{dom} G)]$ is a closed vector subspace of Y, then for any $x^* \in X^*$ and $x \in X$ one has the equalities

$$(f + g \circ G)^*(x^*) = \min_{y^* \in Y^*} [g^*(y^*) + (f + y^* \circ G)^*(x^*)]$$
$$= \min_{y^* \in Y_+^*} [g^*(y^*) + (f + y^* \circ G)^*(x^*)]$$

and the equality

$$\partial(f + g \circ G)(x) = \bigcup_{y^* \in \partial g(G(x))} \partial(f + y^* \circ G)(x).$$

(b) If g is non Y_+-decreasing on $G(\operatorname{dom} G) + Y_+$ and if $\mathbb{R}_+[\operatorname{dom} g - G(\operatorname{dom} f \cap \operatorname{dom} G) - Y_+]$ is a closed vector subspace of Y, then for any $x^* \in X^*$ and $x \in X$ one has the equality

$$(f + g \circ G)^*(x^*) = \min_{y^* \in Y_+^*} [g^*(y^*) + (f + y^* \circ G)^*(x^*)]$$

and the equality

$$\partial(f + g \circ G)(x) = \bigcup_{y^* \in Y_+^* \cap \partial g(G(x))} \partial(f + y^* \circ G)(x).$$

PROOF. The assertion (a) follows from (b) since $\operatorname{dom} g - Y_+ = \operatorname{dom} g$ if g is non Y_+-decreasing on X. To prove (b), we may suppose that $0 \in \operatorname{dom} g - G(\operatorname{dom} f \cap \operatorname{dom} G)$ since otherwise $\operatorname{dom}(f + g \circ G) = \emptyset$ and then both equalities are trivial. Therefore, setting $C := \operatorname{dom} g - G(\operatorname{dom} f \cap \operatorname{dom} G) - Y_+$ we see that, related to the closed vector subspace $Z := \mathbb{R}_+C$, we have $0 \in \operatorname{int}_Z C$. Let $E := X \times Y \times \mathbb{R} \times \mathbb{R}$ and $E_Z := X \times Z \times \mathbb{R} \times \mathbb{R}$ and, with $P := (x, y, r, s)$, let

$$S_f := \{P \in E : x \in \operatorname{dom} G, f(x) \leq r\} \quad \text{and} \quad S_g := \{P' \in E : g(y') \leq s'\}.$$

Let the multimapping $M : E \times E \to E_Z$ from the Banach space $E \times E$ into the Banach space E_Z be defined by

$$M(P, P') = \{x - x'\} \times (G(x) - y' + Y_+) \times \{r - r'\} \times \{s - s'\} \quad \text{if } (P, P') \in S_f \times S_g$$

and $M(P, P') = \emptyset$ otherwise. It is easily seen that $\text{gph } M$ is a closed convex set in $E \times E \times E_Z$. Now choose a symmetric neighborhood W of zero in Z such that $W \subset C$. We claim that $\Omega := X \times W \times \mathbb{R} \times \mathbb{R} \subset M(E \times E)$. Indeed, let any $(u, w, \rho, \sigma) \in \Omega$. By the inclusion $W \subset C$ there exist $x_w \in \text{dom } f \cap \text{dom } G$, $q_w \in \text{dom } g$ and $y_w \in Y_+$ such that $-w = q_w - G(x_w) - y_w$. Writing

$$(u, w, \rho, \sigma) = (x_w - (x_w - u), G(x_w) - q_w + y_w, f(x_w) - (f(x_w) - \rho), g(q_w) - (g(q_w) - \sigma))$$

we see that $(u, w, \rho, \sigma) \in M(E \times E)$, which justifies the claim. Fixing $\overline{x} \in \text{dom } f \cap \text{dom } G$ with $\overline{y} := G(\overline{x}) \in \text{dom } g$ and putting $\overline{P} := (\overline{x}, \overline{y}, f(\overline{x}), g(\overline{y}))$, by Robinson-Ursescu theorem (see Theorem 7.26) we have $0 \in \text{int}_Z M\big((\overline{P} + \mathbb{B}_E) \times (\overline{P} + \mathbb{B}_E)\big)$. This gives a real $\lambda > 0$ such that $\lambda \mathbb{B}_X \times \lambda \mathbb{B}_Z \times [-\lambda, +\lambda] \times [-\lambda, +\lambda]$ is included in the set of uples $(x - x', v - y', r - r', s - s') \in E_Z$ such that $x, x' \in \overline{x} + \mathbb{B}_X$, $y, y' \in \overline{y} + \mathbb{B}_Y$, $r, r' \in [f(\overline{x}) - 1, f(\overline{x}) + 1]$, $s, s' \in [g(\overline{y}) - 1, g(\overline{y}) + 1]$, $x \in \text{dom } G$, $f(x) \leq r$, $g(y') \leq s'$ and $v \in G(x) + Y_+$. Putting $\mu := \max\{1 + |f(\overline{x})|, 1 + |g(\overline{y})|, 1 + \|\overline{x}\|\}$ we derive that $\lambda \mathbb{B}_Z \subset G(\{f \leq \mu\} \cap \text{dom } G \cap \mu \mathbb{B}_X) + Y_+ - \{g \leq \mu\}$. The equalities in the present theorem are then consequences of Theorem 3.165 since the hypothesis (c) therein is satisfied. $\square$

The following corollaries directly follow.

COROLLARY 7.105. *Let X and Y be Banach spaces and $A : X \to Y$ be a continuous linear mapping. Let $f : X \to \mathbb{R} \cup \{+\infty\}$ and $g : Y \to \mathbb{R} \cup \{+\infty\}$ be lower semicontinuous convex functions. Assume that $\mathbb{R}_+[\text{dom } g - A(\text{dom } f)]$ is a closed vector subspace of Y. Then for any $x^* \in X^*$ and $x \in X$ one has the equality*

$$(f + g \circ A)^*(x^*) = \min_{y^* \in Y^*} [g^*(y^*) + f^*(x^* - A^*y^*)]$$

and the equality

$$\partial(f + g \circ A)(x) = \partial f(x) + A^*(\partial g(Ax)).$$

COROLLARY 7.106. *Let X and Y be Banach spaces and $A : X \to Y$ be a continuous linear mapping. Let $f : X \to \mathbb{R} \cup \{+\infty\}$ and $g : Y \to \mathbb{R} \cup \{+\infty\}$ be lower semicontinuous convex functions. Assume that $0 \in \text{core}[\text{dom } g - A(\text{dom } f)]$. Then for any $x^* \in X^*$ and $x \in X$ one has the equality*

$$(f + g \circ A)^*(x^*) = \min_{y^* \in Y^*} [g^*(y^*) + f^*(x^* - A^*y^*)]$$

and the equality

$$\partial(f + g \circ A)(x) = \partial f(x) + A^*(\partial g(Ax)).$$

COROLLARY 7.107 (Attouch-Brezis). *Let X and Y be Banach spaces and let $f, g : X \to \mathbb{R} \cup \{+\infty\}$ be lower semicontinuous convex functions. Assume that $\mathbb{R}_+[\text{dom } f - \text{dom } g]$ is a closed vector subspace of Y. Then for any $x^* \in X^*$ and $x \in X$ one has both the equalities*

$$(f + g)^*(x^*) = \min_{y^* \in X^*} [f^*(x^* - y^*) + g^*(y*)] = (f^* \square g^*)(x^*)$$

and the equality

$$\partial(f + g)(x) = \partial f(x) + \partial g(x).$$

COROLLARY 7.108 (Robinson-Rockafellar). Let X be a Banach space and let $f, g : X \to \mathbb{R} \cup \{+\infty\}$ be lower semicontinuous convex functions. Assume that $0 \in \operatorname{core}[\operatorname{dom} g - \operatorname{dom} f]$. Then for any $x^* \in X^*$ and $x \in X$ one has the equalities
$$(f+g)^*(x^*) = \min_{y^* \in X^*} [f^*(x^* - y^*) + g^*(y^*)] = (f^* \square g^*)(x^*)$$
and the equality
$$\partial(f+g)(x) = \partial f(x) + \partial g(x).$$

Taking as f and g indicator functions of convex sets in Corollary 7.105 and Corollary 7.106 yields the following corollary. The result under the assumption (7.37) below of the corollary can also be seen from Theorem 7.95; here the Krein-Šmulian theorem is not utilized at the opposite of 7.95.

COROLLARY 7.109. Let X, Y be Banach spaces and $A : X \to Y$ be a continuous linear mapping. Let C and D be nonempty closed convex sets of X and Y respectively. For any $x \in C \cap A^{-1}(D)$ one has the equality
$$N(C \cap A^{-1}(D); x) = N(C; x) + A^*(N(D; Ax))$$
under anyone of qualification conditions (7.36) or (7.37)

(7.36) $\qquad \mathbb{R}_+[D - A(C)]$ is a closed vector subspace of Y,

(7.37) $\qquad\qquad 0 \in \operatorname{core}[D - A(C)].$

The fundamental case of intersection of closed convex sets corresponds to A as the identity mapping.

COROLLARY 7.110. Let C and D be nonempty closed convex sets of a Banach space X. For any $x \in C \cap D$ one has the equality
$$N(C \cap D; x) = N(C; x) + N(D; x)$$
under anyone of qualification conditions (7.38) or (7.39)

(7.38) $\qquad \mathbb{R}_+[D - C]$ is a closed vector subspace of Y,

(7.39) $\qquad\qquad 0 \in \operatorname{core}[D - C].$

7.9. Further results

This section is devoted to metric regularity/subregularity properties of multimappings of two types: multimappings whose graphs are polyhedral convex and multimappings which are subdifferentials.

7.9.1. Metric subregularity of polyhedral multimappings.

Recall that a closed *polyhedral convex* set of a Euclidean space is a finite intersection of closed half-spaces. The next proposition shows that multimappings with closed polyhedral convex graphs are upper-Lipschitz semicontinuous.

DEFINITION 7.111. A multimapping $M : X \rightrightarrows Y$ between two normed spaces is called *upper-Lipschitz semicontinuous* at a point $\bar{x} \in \operatorname{Dom} M$ provided there is a real $\gamma \geq 0$ and a neighborhood U of $\bar{x}$ such that
$$M(x) \subset M(\bar{x}) + \gamma \|x - \bar{x}\| \mathbb{B}_Y \quad \text{for all } x \in U.$$
In such a case, one says that M is upper-Lipschitz semicontinuous at $\bar{x}$ with constant γ on U.

REMARK 7.112. Obviously, such a multimapping is Hausdorff upper semicontinuous at $\bar{x}$ (see Definition 1.74). □

PROPOSITION 7.113. Let X, Y be Banach spaces and $M : X \rightrightarrows Y$ be a multimapping whose graph is a finite intersection of closed half-spaces in $X \times Y$, that is, there are $a_i^* \in X^*$, $b_i^* \in Y^*$, $\beta_i \in \mathbb{R}$, $i = 1, \cdots, n$ such that

$$\operatorname{gph} M = \{(x, y) \in X \times Y : \langle a_i^*, x \rangle + \langle b_i^*, y \rangle \leq \beta_i, \forall i = 1, \cdots, n\}.$$

Then, there is a real $\gamma \geq 0$ such that, for any $\bar{x} \in \operatorname{Dom} M$

$$M(x) \subset M(\bar{x}) + \gamma \|x - \bar{x}\| \, \mathbb{B}_Y \quad \text{for all } x \in X,$$

hence in particular M is upper-Lipschitz semicontinuous at any point in $\operatorname{Dom} M$.

PROOF. Let a_i^*, b_i^*, β_i as in the statement and let $\bar{x} \in \operatorname{Dom} M$. Since

$$M(\bar{x}) = \{y \in Y : \langle b_i^*, y \rangle \leq \beta_i - \langle a_i^*, \bar{x} \rangle, \forall i = 1, \cdots, n\},$$

Hoffman-Ioffe Theorem 7.65 furnishes a real $\gamma_0 \geq 0$ (depending only on $b_1^*, \cdots, b_n^*$) which allows us to write, for all $x \in \operatorname{Dom} M$ and $y \in M(x)$

$$d(y, M(\bar{x})) \leq \gamma_0 \sum_{i=1}^{n} (\langle b_i^*, y \rangle + \langle a_i^*, \bar{x} \rangle - \beta_i)^+$$

$$\leq \gamma_0 \sum_{i=1}^{n} \left((\langle a_i^*, \bar{x} - x \rangle)^+ + (\langle a_i^*, x \rangle + \langle b_i^*, y \rangle - \beta_i)^+ \right)$$

$$= \gamma_0 \sum_{i=1}^{n} (\langle a_i^*, \bar{x} - x \rangle)^+ \leq \gamma_0 \left(\sum_{i=1}^{n} \|a_i^*\| \right) \|x - \bar{x}\|.$$

Taking any real $\gamma > \gamma_0 \sum_{i=1}^{n} \|a_i^*\|$ it ensues that, for all $x \in X$

$$M(x) \subset M(\bar{x}) + \gamma \|x - \bar{x}\| \, \mathbb{B}_Y.$$

□

PROPOSITION 7.114. Let $M : \mathbb{R}^q \rightrightarrows \mathbb{R}^p$ be a multimapping whose graph is a finite union of closed polyhedral convex sets in $\mathbb{R}^q \times \mathbb{R}^p$. Then, for any $(\bar{x}, \bar{y}) \in \operatorname{gph} M$ the multimapping M is metrically subregular at $\bar{x}$ for $\bar{y}$.

EXERCISE 7.115 (Proof of Proposition 7.114). Let $M, \bar{x}, \bar{y}$ as in the statement of the proposition.
(a) Argue that there are multimappings $M_1, \cdots, M_\nu : \mathbb{R}^q \rightrightarrows \mathbb{R}^p$ with closed polyhedral convex graphs such that $M(x) = \bigcup_{j=1}^{\nu} M_j(x)$ for all $x \in \mathbb{R}^q$.
(b) Let $J = \{1, \cdots, \nu\}$ and $J_0 := \{j \in J : \bar{y} \in M_j(\bar{x})\}$. For each $j \in J_0$, argue by Proposition 7.113 that there is $\gamma_j \in [0, +\infty[$ such that for all $y \in \mathbb{R}^p$

$$M_j^{-1}(y) \subset M_j^{-1}(\bar{y}) + \gamma_j \|y - \bar{y}\| \, \mathbb{B}_X,$$

and derive that $d(x, M^{-1}(\bar{y})) \leq d(x, M_j^{-1}(\bar{y})) \leq \gamma_j \, d(\bar{y}, M_j(x))$ for every $x \in \mathbb{R}^q$.
(c) Let $j \in J \setminus J_0$.
(c$_1$) Argue that $\operatorname{gph} M_j = \operatorname{gph} G_{1,j} \cap \cdots \cap \operatorname{gph} G_{n_j, j}$, where

$$\operatorname{gph} G_{i,j} = \{(x, y) \in \mathbb{R}^q \times \mathbb{R}^p : \langle a_{i,j}^*, x \rangle + \langle b_{i,j}^*, y \rangle \leq \beta_{i,j}\}$$

with $a_{i,j}^* \in \mathbb{R}^q$, $b_{i,j}^* \in \mathbb{R}^p$ and $\beta_{i,j} \in \mathbb{R}$ (with $i = 1, \cdots, n_j$).
(c$_2$) Taking $k \in \{1, \cdots, n_j\}$ with $\bar{y} \notin G_{k,j}(\bar{x})$, argue that there exists some $\delta_k > 0$ such that $d(\bar{y}, G_{k,j}(x)) \geq \delta_k$ for every $x \in B(\bar{x}, \delta_k)$, and deduce that for every $x \in B(\bar{x}, \delta_k)$

$$d(x, M^{-1}(\bar{y})) \leq \|x - \bar{x}\| \leq \delta_k \leq d(\bar{y}, G_{k,j}(x)) \leq d(\bar{y}, M_j(x)).$$

(d) Derive with $\gamma := \max\{1, \max_{j \in J_0} \gamma_j\}$ that there exists some $\delta > 0$ such that, for each $j \in J$ one has $d(x, M^{-1}(\bar{y})) \leq \gamma d(\bar{y}, M_j(x))$ for all $x \in B(\bar{x}, \delta)$, and conclude that the statement of Proposition 7.114 holds true. □

7.9.2. Metric regularity/subregularity of subdifferential and growth conditions. In addition to the above concepts of metric regularity/subregularity, it can arise in some situations that the θ-order metric regularity/subregularity must be used. Such a concept will allow us below to establish some subdifferential characterizations of some growth conditions of functions.

DEFINITION 7.116. Let X, Y be two metric spaces. Given a real $\theta > 0$ we say that a multimapping $M : X \rightrightarrows Y$ is *θ-order metrically regular* (resp. *subregular*) with constant $\gamma \in [0, +\infty[$ at $\bar{x}$ for $\bar{y} \in M(\bar{x})$ provided that in place of (7.5) (resp. (7.6)) one can find neighborhoods U and V of $\bar{x}$ and $\bar{y}$ (resp. a neighborhood U of $\bar{x}$) such that, for all $(x, y) \in U \times V$ (resp. $x \in U$)

$$d(x, M^{-1}(y)) \leq \gamma \big(d(y, M(x))\big)^\theta \text{ (resp. } d(x, M^{-1}(\bar{y})) \leq \gamma \big(d(\bar{y}, M(x))\big)^\theta \text{)}.$$

The *θ-order strong metric regularity* (resp. *subregularity*) of M at $\bar{x}$ for $\bar{y}$ holds when in addition there is a neighborhood U' of $\bar{x}$ such that $M^{-1}(y) \cap U'$ is a singleton for all $y \in V$ (resp. when in addition $M^{-1}(\bar{y}) \cap U$ is a singleton).

REMARK 7.117. The inequality of θ-order metric regularity ensures that, for all $y \in V$ one has $d(\bar{x}, M^{-1}(y)) \leq \gamma d(y, M(\bar{x}))^\theta$, which implies that for each neighborhood U' of $\bar{x}$ there is a neighborhood V' of $\bar{y}$ such that $M^{-1}(y) \cap U' \neq \emptyset$ for all $y \in V'$. Consequently, it is easy to see that the definition of θ-order strong metric regularity with constant γ can be stated in anyone of the two forms:
(a) there are a real $r > 0$ and neighborhoods U of $\bar{x}$ and V of $\bar{y}$ such that for all $x \in U$ and $y \in V$,

$$d(x, M^{-1}(y)) \leq \gamma d(y, M(x))^\theta$$

and $M^{-1}(y) \cap B(\bar{x}, r)$ is a singleton;
(b) there are neighborhoods U of $\bar{x}$ and V of $\bar{y}$ such that for all $x \in U$ and $y \in V$,

$$d(x, M^{-1}(y)) \leq \gamma d(y, M(x))^\theta$$

and $M^{-1}(y) \cap U$ is a singleton. □

Analogously to the metric regularity, the Aubin-Lipschitz property can be adapted to the case of θ-order as follows.

DEFINITION 7.118. Let $\theta \in]0, +\infty[$ and let $M : X \rightrightarrows Y$ be a multimapping between two metric spaces. Let also $(\bar{x}, \bar{y}) \in \operatorname{gph} M$. Extending Definition 7.1 we say that M satisfies the *θ-order Aubin-Hölder* (or, *pseudo-Hölder, truncated Hölder-like*) *property* at $\bar{x}$ for $\bar{y}$ if there are a real $\gamma \geq 0$ and neighborhoods U and V of $\bar{x}$ and $\bar{y}$ respectively such that

$$M(x') \cap V \subset \bigcup_{y \in M(x)} B[y, \gamma \, d(x', x)^\theta] \quad \text{for all } x, x' \in U.$$

Such a real $\gamma \geq 0$ is called a *θ-order Aubin-Hölder* (or, *pseudo-Hölder, truncated Hölder-like*) *constant* of M (over U for V). When we have instead

$$M(x') \cap V \subset \bigcup_{y \in M(\overline{x})} B[y, \gamma\, d(x', \overline{x})^\theta] \quad \text{for all } x' \in U,$$

the multimapping is said to be *θ-order calm* (or *one-sided Aubin-Hölder*) at $\overline{x}$ for $\overline{y}$; the real $\gamma \geq 0$ is then a *θ-order calmness* (or *one-sided Aubin-Hölder*) *constant* of M (over U). The associated *θ-order rate* (or *modulus*) is defined like in Definition 7.1 and Definition 7.3.

REMARK 7.119. In both Definition 7.116 and Definition 7.118 instead of the function $[0, +\infty[\ni t \mapsto t^\theta$ a general appropriate function $w : [0, +\infty[\to [0, +\infty[$ could be involved with the use of $\gamma w(\|x' - x\|)$ in place of $\gamma \|x' - x\|^\theta$. The same remark is also valid for the study below of stable minimizers and (lower) growth conditions. However the practical interest of a (lower) growth condition of a function f around a point $\overline{x}$ in a normed space remains clearly in the form

$$f(x) \geq f(\overline{x}) + \langle \overline{x}^*, x - \overline{x} \rangle + \kappa \|x - \overline{x}\|^s$$

for x near $\overline{x}$. This justifies our choice of θ-order form in Definition 7.116 and Definition 7.118. □

The particular case when $\theta > 1$ in Definition 7.116 (as well as in Definition 7.118, see the proof below) furnishes a certain property of constant value.

PROPOSITION 7.120. *Let X, Y be two normed vector spaces and let a real number $\theta > 1$. Let $g : X \to Y$ be a mapping and $M : X \rightrightarrows Y$ be a multimapping whose inverse takes on closed values and let $(\overline{x}, \overline{y}) \in \operatorname{gph} M$. The following properties hold.*
(a) *If there are some open connected set U in X and some real $\gamma \geq 0$ such that*

$$\|g(x') - g(x)\| \leq \gamma \|x' - x\|^\theta \quad \text{for all } x, x' \in U,$$

then g is constant on U.
(b) *If there are a real $\gamma \geq 0$ and neighborhoods U of $\overline{x}$ in X and V of $\overline{y}$ in Y such that*

$$d(x, M^{-1}(y)) \leq \gamma\, d(y, M(x))^\theta \quad \text{for all } x \in U, y \in V,$$

then there are some neighborhoods $U_0 \subset U$ of $\overline{x}$ and $V_0 \subset V$ of $\overline{y}$ such that

$$M^{-1}(y) \cap U_0 = M^{-1}(y') \cap U_0 \quad \text{for all } y, y' \in V_0,$$

that is, the multimapping $y \mapsto M^{-1}(y) \cap U_0$ is constant on V_0.

PROOF. (a) It is enough to see that the inequality assumption in (a) entails that the Fréchet derivative of g exists and is null at every point in the open connected set U.
(b) First we note by the assumption in (b) that $\Gamma(y) := M^{-1}(y) \neq \emptyset$ for all $y \in V$. Fixing a real $\gamma_0 > \gamma$ we also derive from the assumption in (b) that

$$\Gamma(y') \cap U \subset \Gamma(y) + \gamma_0 \|y' - y\|^\theta \mathbb{B}_X \quad \text{for all } y, y' \in V.$$

From this (as in the proof of Proposition 7.7) it follows that there exist a neighborhood $U_0 \subset U$ of $\overline{x}$ and an open convex neighborhood $V_0 \subset V$ of $\overline{y}$ such that for each $x \in U_0$

$$d(x, \Gamma(y')) \leq d(x, \Gamma(y)) + \gamma_0 \|y' - y\|^\theta \quad \text{for all } y, y' \in V_0.$$

Taking any $y' \in V_0$ the latter inequality and the property (a) imply that, for every $y \in V_0$ and every $x \in U_0 \cap \Gamma(y)$

$$d(x, \Gamma(y')) = d(x, \Gamma(y)) = 0, \text{ so } x \in \Gamma(y').$$

This says that $\Gamma(y) \cap U_0 \subset \Gamma(y') \cap U_0$, hence by symmetry $\Gamma(y) \cap U_0 = \Gamma(y') \cap U_0$ for all $y, y' \in V_0$. $\square$

Like in Proposition 7.5 the point x' in Definition 7.118 can be allowed to move in the whole space X as stated next.

PROPOSITION 7.121. Let $M : X \rightrightarrows Y$ be a multimapping between two metric spaces and let $(\bar{x}, \bar{y}) \in \mathrm{gph}\, M$. Let reals $\theta > 0$ and $\gamma > 0$.
(a) The multimapping M has the θ-order Aubin-Hölder property at $\bar{x}$ for $\bar{y}$ with γ as constant if and only if there are neighborhoods U' of $\bar{x}$ and V' of $\bar{y}$ such that

$$M(x') \cap V' \subset \bigcup_{y \in M(x)} B[y, \gamma\, d(x', x)^\theta] \quad \text{for all } x' \in X \text{ and } x \in U'.$$

(b) The multimapping M has the θ-order calmness (or one-sided Aubin-Hölder) property at $\bar{x}$ for $\bar{y}$ with γ as constant if and only if there is a neighborhood V' of $\bar{y}$ such that

$$M(x') \cap V' \subset \bigcup_{y \in M(\bar{x})} B[y, \gamma\, d(x', \bar{x})^\theta] \quad \text{for all } x' \in X.$$

EXERCISE 7.122 (Proof of the proposition). Let P be a subset of X containing the point $\bar{x}$. Suppose that there are $\delta > 0$ and $\eta > 0$ such that

$$M(x') \cap B[\bar{y}, \eta] \subset \bigcup_{y \in M(x)} B[y, \gamma\, d(x', x)^\theta] \quad \text{for all } x \in B[\bar{x}, \delta] \cap P \text{ and } x' \in B[\bar{x}, \delta].$$

Choose $0 < \delta' < \delta$ and $0 < \eta' < \eta$ such that $\gamma(\delta - \delta')^\theta - \gamma(\delta')^\theta - \eta' > 0$, and let $U' := B[\bar{x}, \delta']$ and $V' := B[\bar{y}, \eta']$. Let any $x \in U' \cap P$.
(a) Argue the existence of some $v \in M(x)$ such that $d(\bar{y}, v) \leq \gamma\, d(\bar{x}, x)^\theta$, and hence $B[\bar{y}, \eta'] \subset B[v, \gamma(\delta')^\theta + \eta']$.
(b) Let any $x' \in X \setminus B[\bar{x}, \delta]$. Argue that

$$\gamma(\delta')^\theta + \eta' \leq \gamma(\delta - \delta')^\theta < \gamma\, d(x', x)^\theta,$$

and hence $V' = B[\bar{y}, \eta'] \subset B[v, \gamma\, d(x', x)^\theta]$.
(c) Derive that $M(x') \cap V' \subset \bigcup_{y \in M(x)} B[y, \gamma\, d(x', x)^\theta]$ for all $x' \in X$.
(d) Conclude that both equivalences of the proposition hold true. $\square$

In order to obtain the equivalence between θ-order Aubin-Hölder property and θ-order metric regularity (as already seen in Theorem 7.17 when $\theta = 1$), let us state the following lemma.

LEMMA 7.123. Let $M : X \rightrightarrows Y$ be a multimapping between metric spaces, U be a nonempty subset of X and Q, V be subsets of Y with $Q \cap V \neq \emptyset$. Consider, for a real $\gamma \geq 0$, the assertions:
(a) $d(x, M^{-1}(y)) \leq \gamma\, d(y, M(x))^\theta$ for all $x \in U$ and $y \in Q \cap V$;
(b) $M^{-1}(y) \cap U \subset \bigcup_{x' \in M^{-1}(y')} B[x', \gamma\, d(y, y')^\theta]$ for all $y \in Y$ and $y' \in Q \cap V$.

The assertion (b) entails (a); and if (a) is fulfilled with γ and either the sets $M^{-1}(y)$ are closed in X for all $y \in Q \cap V$ or the sets $M(x)$ are closed in Y for all $x \in U$, then (b) holds true with any $\gamma' > \gamma$ in place of γ.

If X (resp. Y) is finite-dimensional and the sets $M^{-1}(y)$ (resp. $M(x)$) are closed for all $y \in Q \cap V$ (resp. for all $x \in U$), then (a) and (b) are equivalent.

EXERCISE 7.124. Prove the lemma by adapting the proof of (b)⇒(a) and the proof of ((a)+closedness hypothesis)⇒(b) with γ' in Lemma 7.14. □

The above lemma (with $Q = Y$ and then $Q = \{\bar{x}\}$) combined with Proposition 7.121 directly yields the following properties.

PROPOSITION 7.125. Let $M : X \rightrightarrows Y$ be a multimapping between metric spaces, let $(\bar{x}, \bar{y}) \in \operatorname{gph} M$, and let a real $\theta > 0$.
(a) Assume that either M takes on closed values near $\bar{x}$ or M^{-1} takes on closed values near $\bar{y}$. Then M is θ-order metrically regular at $\bar{x}$ for $\bar{y}$ with constant $\gamma \geq 0$ whenever the inverse multimapping M^{-1} satisfies the θ-order Aubin-Hölder property at $\bar{y}$ for $\bar{x}$ with constant γ.

Conversely, if M is θ-order metrically regular at $\bar{x}$ for $\bar{y}$ with constant $\gamma \geq 0$, then for any real $\gamma' > \gamma$ the inverse multimapping M^{-1} satisfies the θ-order Aubin-Hölder property at $\bar{y}$ for $\bar{x}$ with constant γ'.
(b) Assume that either M takes on closed values near $\bar{x}$ or $M^{-1}(\bar{y})$ is closed. Then M is θ-order metrically subregular at $\bar{x}$ for $\bar{y}$ with constant $\gamma \geq 0$ whenever the inverse multimapping M^{-1} satisfies the θ-order calmness (or one-sided Aubin-Hölder) property at $\bar{y}$ for $\bar{x}$ with constant γ.

Conversely, if M is θ-order metrically subregular at $\bar{x}$ for $\bar{y}$ with constant $\gamma \geq 0$, then for any real $\gamma' > \gamma$ the inverse multimapping M^{-1} satisfies the θ-order calmness (or one-sided Aubin-Hölder) property at $\bar{y}$ for $\bar{x}$ with constant γ'.

We turn now to characterizing strong metric regularity/subregularity through a selection property. Such characterizations are of interest in various situations.

PROPOSITION 7.126. Let a real $\theta > 0$ and let $M : X \rightrightarrows Y$ be a multimapping between two normed spaces. Let also $(\bar{x}, \bar{y}) \in \operatorname{gph} M$ and let a real $\gamma \geq 0$.
(I) The following assertions are equivalent:
(a) the multimapping M is θ-order strongly metrically regular at $\bar{x}$ for $\bar{y}$ with constant γ;
(b) there are a neighborhood U of $\bar{x}$, a neighborhood V of $\bar{y}$ and a mapping $\zeta : V \to X$ such that $\zeta(y) \in M^{-1}(y)$ for all $y \in V$ and
$$\|x - \zeta(y)\| \leq \gamma d(y, M(x))^\theta \quad \text{for all } x \in U, y \in V;$$
(c) there are a neighborhood U of $\bar{x}$, a neighborhood V of $\bar{y}$ and a mapping $\zeta : V \to X$ continuous at $\bar{y}$ with $\zeta(\bar{y}) = \bar{x}$ such that $\zeta(y) \in M^{-1}(y)$ for all $y \in V$ and
$$\|x - \zeta(y)\| \leq \gamma d(y, M(x))^\theta \quad \text{for all } x \in U, y \in V;$$
(d) there are a neighborhood U of $\bar{x}$, a neighborhood V of $\bar{y}$ and a mapping $\zeta : V \to X$ such that $M^{-1}(y) \cap U = \{\zeta(y)\}$ for all $y \in V$ and
$$\|x - \zeta(y)\| \leq \gamma d(y, M(x))^\theta \quad \text{for all } x \in U, y \in V;$$
(e) there are a neighborhood U of $\bar{x}$, a neighborhood V of $\bar{y}$ and a mapping $\zeta : V \to X$ such that $M^{-1}(y) \cap U = \{\zeta(y)\}$ for all $y \in V$ and
$$\|\zeta(y') - \zeta(y)\| \leq \gamma \|y' - y\|^\theta \quad \text{for all } y, y' \in V.$$

(II) The multimapping M is θ-order strongly metrically subregular at $\bar{x}$ for $\bar{y}$ with constant $\gamma \geq 0$ if and only if there is a neighborhood U of $\bar{x}$ such that
$$\|x - \bar{x}\| \leq \gamma \, d\big(\bar{y}, M(x)\big)^{\theta} \quad \text{for all } x \in U.$$

PROOF. We only prove (I) and leave (II) to the reader with similar arguments.
(b)$\Rightarrow$(c). Let U, V and ζ be given by (b). Taking $x = \bar{x}$ in the inequality in (b) we see that $\zeta(y) \to \bar{x}$ as $y \to \bar{y}$, and taking $x = \bar{x}$ and $y = \bar{y}$ in the same inequality in (b) gives $\zeta(\bar{y}) = \bar{x}$. Thus, (b)$\Rightarrow$(c) holds true.
(c)$\Rightarrow$(d) Let U, V and ζ be given by (c). From the inequality in (c) we have $\|u - \zeta(y)\| = 0$ for all $y \in V$ and $u \in M^{-1}(y) \cap U$, hence $M^{-1}(y) \cap U \subset \{\zeta(y)\}$ for all $y \in V$. On the other hand, since ζ is continuous at $\bar{y}$ with $\zeta(\bar{y}) = \bar{x}$, we may choose a neighborhood $V_0 \subset V$ of $\bar{y}$ with $\zeta(V_0) \subset U$. It then follows that $M^{-1}(y) \cap U = \{\zeta(y)\}$ for all $y \in V_0$. This finishes the justification of (c)$\Rightarrow$(d).
(d)$\Rightarrow$(e). Let U and V given by (d). For any $y, y' \in V$, since $\zeta(y') \in U$ we may take $x = \zeta(y')$ in the inequality in (d) to obtain
$$\|\zeta(y') - \zeta(y)\| \leq \gamma \, d\big(y, M(\zeta(y'))\big)^{\theta} \leq \gamma \|y - y'\|^{\theta}.$$
This translates (e).
(e)$\Rightarrow$(a). Let U, V and ζ be given by (e). For any $y, y' \in V$, using the equality $M^{-1}(y) \cap U = \{\zeta(y)\}$ along with the inequality in (e) we obtain
$$M^{-1}(y) \cap U \subset M^{-1}(y') + \gamma \|y - y'\|^{\theta} \, \mathbb{B}_X,$$
so Proposition 7.125 tells us that M is θ-order metrically regular at $\bar{x}$ for $\bar{y}$ with constant γ. This combined with Remark 7.117 justifies the implication (e)$\Rightarrow$(a).
The last implication (a)$\Rightarrow$(b) is the subject of the next exercise. $\square$

EXERCISE 7.127 (Proof of (a)$\Rightarrow$(b) in Proposition 7.126). Suppose (a) and let reals $r > 0$ and $\delta > 0$ be such that r, $U := B(\bar{x}, \delta)$ and $V := B(\bar{y}, \delta)$ satisfy Definition 7.116 via Remark 7.117, that is, for all $x \in U$ and $y \in V$
$$d\big(x, M^{-1}(y)\big) \leq \gamma \, d\big(y, M(x)\big)^{\theta} \quad \text{and} \quad M^{-1}(y) \cap B(\bar{x}, r) \text{ is a singleton.}$$
Let any real $\gamma_0 > \gamma$.
(α) Argue that for each $y \in B(\bar{y}, \delta)$
$$d\big(\bar{x}, M^{-1}(y)\big) \leq \gamma \|y - \bar{y}\|^{\theta},$$
and derive the existence of a mapping $\zeta : B(\bar{y}, \delta) \to X$ such that $\zeta(\bar{y}) = \bar{x}$ and for all $y \in B(\bar{y}, \delta)$
$$\zeta(y) \in M^{-1}(y) \quad \text{and} \quad \|\zeta(y) - \bar{x}\| \leq \gamma_0 \|y - \bar{y}\|^{\theta}.$$
(β) Argue that for every $y \in Y$ with $\|y - \bar{y}\| < \min\{\delta, (r/\gamma_0)^{1/\theta}\}$
$$M^{-1}(y) \cap B(\bar{x}, r) = \{\zeta(y)\}.$$
(γ) Let $\eta := \min\{\delta, r\}$ and $\rho := \min\{\delta, \big(\eta/(3\gamma_0)\big)^{1/\theta}\}$. For any $y \in B(\bar{y}, \rho)$, argue that $\|\zeta(y) - \bar{x}\| < \eta/3$ and derive that, for any $x \in B(\bar{x}, \eta/3)$ one has
$$d\big(x, M^{-1}(y)\big) = d\big(x, M^{-1}(y) \cap B(x, 2\eta/3)\big) = \|x - \zeta(y)\|.$$
(δ) With $\delta_0 := \min\{\rho, \eta/3\}$ derive that (b) in the statement of the proposition holds true with $U := B(\bar{x}, \delta_0)$ and $V := B(\bar{y}, \delta_0)$. $\square$

Our aim now is to study how the above metric concepts of regularity and subregularity of a subdifferential multimapping of a function are linked to various growth conditions of the function. The analysis is intended to embrace the three main subdifferentials with full sum rule: $\partial_C, \partial_L, \partial_A$. Then, in all the remainder of this subsection, given an appropriate normed space X, we assume that ∂ is a subdifferential on the class of extended real-valued functions on X such that:
- $\partial f(x) = \emptyset$ if $|f(x)| = +\infty$, and $0 \in \partial f(x)$ whenever x is a local minimizer of f with $|f(x)| < +\infty$;
- when f is convex, $\partial f(x)$ coincides with the standard subdifferential of convex functions;
- the full sum rule $\partial (f+g)(x) \subset \partial f(x) + \partial g(x)$ holds whenever f is lower semicontinuous near x and g is finite and Lipschitz continuous near $x \in X$;
- $\partial f(x) \subset \partial_C f(x)$ for all x.

We start our analysis with two preliminary lemmas.

LEMMA 7.128. *For any reals $s \geq 1$ and $0 < \alpha < 1$, one has*
$$s\alpha^{s-1}(1-\alpha) < 1.$$

PROOF. With $g(t) := t^s$ for all $t \in [0,1]$ it is enough to note that for some $\tau \in]\alpha, 1[$
$$1 > 1^s - \alpha^s = (1-\alpha)g'(\tau) = s(1-\alpha)\tau^{s-1} \geq s(1-\alpha)\alpha^{s-1}.$$
$\square$

LEMMA 7.129. *Let $\alpha \in]0,1[$, $s \in]1, +\infty[$ and $\gamma \in]0, +\infty[$. Let $c : \mathbb{R} \to \mathbb{R}$ be defined for all $t \in \mathbb{R}$ by*
$$c(t) := s(1-\alpha)\alpha^{s-1}|t| + t - s(1-\alpha)\alpha^{s-1}\gamma^{1-s}.$$
Then one has $c(t) \geq t$ for all $t \leq -\gamma^{1-s}$, and $c(t) < t$ for all $t \in]-\gamma^{1-s}, \gamma^{1-s}[$. Further, the continuous piecewise linear function c is increasing on $\mathbb{R}$ and
$$c(]-\infty, \gamma^{1-s}[) \subset]-\infty, \gamma^{1-s}[\quad \text{and} \quad c(]-\infty, 0]) \subset]-\infty, 0[.$$

PROOF. If $t \leq -\gamma^{1-s}$, then $|t| = -t$, so
$$c(t) = t + s(1-\alpha)\alpha^{s-1}(-t - \gamma^{1-s}) \geq t,$$
which is the first inequality of the lemma. Concerning the second inequality, for any $t \in]-\gamma^{1-s}, \gamma^{1-s}[$ we have with $A := s(1-\alpha)\alpha^{s-1}$
$$c(t) = A(|t| - \gamma^{1-s}) + t < t.$$
On the other hand, noting that
$$c(t) = (1-A)t - \gamma^{1-s}A \text{ if } t \leq 0, \quad \text{and} \quad c(t) = (1+A)t - \gamma^{1-s}A \text{ if } t \geq 0,$$
we see that the function c is increasing on $\mathbb{R}$ since $1 - A > 0$ by Lemma 7.128. Finally, the increasing property and the equalities $c(\gamma^{1-s}) = \gamma^{1-s}$ and $c(0) = -s(1-\alpha)\alpha^{s-1}\gamma^{1-s}$ justify the inclusions of the lemma. $\square$

We can now begin the study of the link between metric subregularity of ∂f and certain growth conditions for the function f. For convenience we will write $1 + 1/\theta$ instead of $1 + (1/\theta)$.

LEMMA 7.130. Let $f : X \to \mathbb{R} \cup \{+\infty\}$ be a proper lower semicontinuous function on a Banach space X and let a real $\theta > 0$. Let $(\bar{x}, \bar{x}^*) \in \operatorname{gph} \partial f$ and $Q \subset (\partial f)^{-1}(\bar{x}^*)$ with $\bar{x} \in Q$. Assume that there are real numbers $\gamma > 0$, $\delta > 0$ such that

(7.40) $$d(x, Q) \leq \gamma d(\bar{x}^*, \partial f(x))^\theta \quad \text{for all } x \in B(\bar{x}, \delta).$$

Assume also that for $s := 1 + 1/\theta$, there exist some $r > 0$ and $\sigma \in]-\infty, \gamma^{-1/\theta}[$ such that

(7.41) $$f(x) \geq f(\bar{x}) + \langle \bar{x}^*, x - \bar{x} \rangle - \frac{\sigma}{s} d(x, Q)^s \quad \text{for all } x \in B[\bar{x}, r].$$

Then, given any $\alpha \in]0, 1[$ and any $n \in \mathbb{N}$ one has

$$f(x) \geq f(\bar{x}) + \langle \bar{x}^*, x - \bar{x} \rangle - \frac{c_n(\sigma)}{s} d(x, Q)^s \quad \text{for all } x \in B\left[\bar{x}, \frac{\min\{r, \delta\}}{(2 - \alpha)^n}\right],$$

where $c_n := c \circ \cdots \circ c$ is the n-times composition of the function $c : \mathbb{R} \to \mathbb{R}$ in Lemma 7.129.

EXERCISE 7.131 (Proof of the above lemma). Consider the function $\eta :]0, +\infty[\to]0, +\infty[$ with $\eta(t) := (1/(2 - \alpha)) \min\{t, \delta\}$ for all $t \in]0, +\infty[$.
(I) Suppose that $\sigma \leq -\gamma^{1-s}$. Noting by Lemma 7.129 that $c(\sigma) \geq \sigma$, derive from (7.41) that

(7.42) $$f(x) \geq f(\bar{x}) + \langle \bar{x}^*, x - \bar{x} \rangle - \frac{c(\sigma)}{s} d(x, Q)^s, \quad \text{for all } x \in B[\bar{x}, \eta(r)].$$

(II) Let $g(x) := f(x) - \langle \bar{x}^*, x - \bar{x} \rangle + \frac{\sigma}{s} d(x, Q)^s$ for all $x \in X$, and consider now the case when $\sigma \in]-\gamma^{1-s}, \gamma^{1-s}[$.
Suppose that there is $x_0 \in B[\bar{x}, \eta(r)]$ such that

$$f(x_0) < f(\bar{x}) + \langle \bar{x}^*, x_0 - \bar{x} \rangle - \frac{c(\sigma)}{s} d(x_0, Q)^s.$$

(a) Argue that

$$\inf_{B[\bar{x}, r]} g \geq f(\bar{x}) > g(x_0) + \frac{c(\sigma) - \sigma}{s} d(x_0, Q)^s,$$

and hence (since $c(\sigma) < \sigma$ by Lemma 7.129) there is some $\sigma_0 \in]c(\sigma), \sigma[$ such that

$$\inf_{B[\bar{x}, r]} g + \frac{\sigma - \sigma_0}{s} d(x_0, Q)^s > g(x_0).$$

(b) Derive that $d(x_0, Q) > 0$ and that by the Ekeland variational principle there is $u_0 \in B[\bar{x}, r]$ such that

$$\|u_0 - x_0\| \leq (1 - \alpha) d(x_0, Q)$$

and

$$g(u_0) \leq g(x) + \frac{\sigma - \sigma_0}{s(1 - \alpha)} d(x_0, Q)^{s-1} \|x - u_0\|, \forall x \in B[\bar{x}, r].$$

(c) Noting that

$$\|u_0 - \bar{x}\| \leq \|u_0 - x_0\| + \|x_0 - \bar{x}\| \leq (1 - \alpha) d(x_0, Q) + \|x_0 - \bar{x}\|$$
$$\leq (2 - \alpha) \|x_0 - \bar{x}\| \leq (2 - \alpha) \eta(r) < r,$$

and using the chain rule in Theorem 2.188 for the Clarke subdifferential and the inclusion $\partial_C d(\cdot, Q) \subset \mathbb{B}_{X^*}$ deduce that

$$0 \in \partial f(u_0) - \bar{x}^* + \left(|\sigma|\, d(u_0, Q)^{s-1} + \frac{\sigma - \sigma_0}{s(1-\alpha)} d(x_0, Q)^{s-1}\right) \mathbb{B}_{X^*},$$

and hence

$$d(\bar{x}^*, \partial f(u_0)) \leq |\sigma|\, d(u_0, Q)^{s-1} + \frac{\sigma - \sigma_0}{s(1-\alpha)} d(x_0, Q)^{s-1}.$$

(d) Observing that

$d(x_0, Q) \leq \|u_0 - x_0\| + d(u_0, Q) \leq (1-\alpha)d(x_0, Q) + d(u_0, Q)$, so $\alpha\, d(x_0, Q) \leq d(u_0, Q)$,

deduce from (c) that

$$d(\bar{x}^*, \partial f(u_0)) \leq \left(|\sigma| + \frac{\sigma - \sigma_0}{s(1-\alpha)\alpha^{s-1}}\right) d(u_0, Q)^{s-1}.$$

(e) Combining this with the assumption (7.40) derive the contradiction $\sigma_0 \leq c(\sigma)$ with the choice of σ_0.

(III) Deduce that for each real $\sigma < \gamma^{1-s}$ the inequality (7.42) holds true for all $x \in B[\bar{x}, \eta(r)]$.

(IV) Fix any real $\sigma < \gamma^{1-s}$. Since $c(\sigma) < \gamma^{1-s}$ (because $c(\sigma) \leq \sigma$ by Lemma 7.129), argue that one may apply the conclusion in (III) with $c(\sigma)$ in place of σ and $\eta(r)$ in place of r to obtain with $c_2 := c \circ c$ and $\eta_2 := \eta \circ \eta$ that

$$f(x) \geq f(\bar{x}) + \langle \bar{x}^*, x - \bar{x}\rangle - \frac{c_2(\sigma)}{s} d(x, Q)^s, \quad \text{for all } x \in B[\bar{x}, \eta_2(r)].$$

Conclude by induction and by justifying that $\eta_n(\sigma) = (1/(2-\alpha)^n)\min\{r, \sigma\}$. $\square$

In the next lemma, we denote by $E(\tau)$ the relative integer part of $\tau \in \mathbb{R}$, that is, $E(\tau) \in \mathbb{Z}$ and $E(\tau) \leq \tau < E(\tau) + 1$.

LEMMA 7.132 (J. Yao, X.Y. Zheng and J. Zhu). *Let reals $\alpha \in\,]0,1[$ and $\theta > 0$, and let*

$$\lambda_\alpha := 1 - (1 + 1/\theta)(1-\alpha)\alpha^{1/\theta} \quad \text{and} \quad \Lambda_\alpha := 1 + (1 + 1/\theta)(1-\alpha)\alpha^{1/\theta}.$$

Let $f : X \to \mathbb{R} \cup \{+\infty\}$ be a proper lower semicontinuous function on a Banach space X and $r \in\,]0, +\infty]$. Let $(\bar{x}, \bar{x}^*) \in \operatorname{gph} \partial f$ and $Q \subset (\partial f)^{-1}(\bar{x}^*)$ with $\bar{x} \in Q$. Assume that there are real numbers $\gamma > 0$ and $\delta > 0$ such that

$$d(x, Q) \leq \gamma\, d(\bar{x}^*, \partial f(x))^\theta \quad \text{for all } x \in B(\bar{x}, \delta).$$

Assume also that there exist some $r > 0$ and $\sigma \in [0, \gamma^{-1/\theta}[$ such that

$$f(x) \geq f(\bar{x}) + \langle \bar{x}^*, x - \bar{x}\rangle - \frac{\sigma}{s} d(x, Q)^s \quad \text{for all } x \in B[\bar{x}, r],$$

where $s := 1 + 1/\theta$. Let

$$p := E\left(1 - \frac{\log(1 - \sigma\gamma^{1/\theta})}{\log \Lambda_\alpha}\right), \quad L := (\gamma^{-1/\theta} - \sigma)\Lambda_\alpha^p - \gamma^{-1/\theta},$$

$$\kappa_n := (L - \gamma^{-1/\theta})\lambda_\alpha^{n-1} + \gamma^{-1/\theta}.$$

The following hold:
(a) $p \geq 1$, $0 < L < \gamma^{-1/\theta}$, $\kappa_n \to \gamma^{-1/\theta}$ as $n \to \infty$;
(b) for each integer $n \in \mathbb{N}$ and $r_n := \left(1/(2-\alpha)^{p+n-1}\right)\min\{r,\delta\}$ one has

$$f(x) - \langle \bar{x}^*, x \rangle \geq f(\bar{x}) - \langle \bar{x}^*, \bar{x} \rangle + \frac{\kappa_n}{s} d(x,Q)^s \quad \text{for all } x \in B[\bar{x}, r_n].$$

PROOF. (a) Obviously $\Lambda_\alpha > 1$ and $0 < 1 - \sigma\gamma^{1/\theta} \leq 1$, hence $\log \Lambda_\alpha > 0$ and $\log(1 - \sigma\gamma^{1/\theta}) \leq 0$. We derive that

$$1 - \frac{\log(1-\sigma\gamma^{1/\theta})}{\log \Lambda_\alpha} \geq 1, \quad \text{and} \quad p \geq 1.$$

Furthermore, since $0 < (1+1/\theta)(1-\alpha)\alpha^{1/\theta} < 1$ by Lemma 7.128, we have $0 < \lambda_\alpha < 1$, hence $\kappa_n \to \gamma^{-1/\theta}$. By the definition of p we also have

$$-\frac{\log(1-\sigma\gamma^{1/\theta})}{\log \Lambda_\alpha} < p \leq 1 - \frac{\log(1-\sigma\gamma^{1/\theta})}{\log \Lambda_\alpha},$$

or equivalently

$$\log\left(\frac{1}{1-\sigma\gamma^{1/\theta}}\right) < \log(\Lambda_\alpha^p) \leq \log\left(\frac{\Lambda_\alpha}{1-\sigma\gamma^{1/\theta}}\right),$$

which means

(7.43) $$\frac{1}{1-\sigma\gamma^{1/\theta}} < \Lambda_\alpha^p \leq \frac{\Lambda_\alpha}{1-\sigma\gamma^{1/\theta}}.$$

Noting that $\gamma^{1/\theta}L = (1-\sigma\gamma^{1/\theta})\Lambda_\alpha^p - 1$ we obtain from the second inequality in (7.43) that

$$0 < \gamma^{1/\theta}L \leq \Lambda_\alpha - 1 = (1+1/\theta)(1-\alpha)\alpha^{1/\theta} < 1,$$

so $0 < L < \gamma^{-1/\theta}$. This finishes the justification of the properties in (a).
The proof of (b) is the subject of the next exercise. □

EXERCISE 7.133 (Proof of (b) in Lemma 7.132). Define $\varphi, \psi : \mathbb{R} \to \mathbb{R}$ by putting for all $t \in \mathbb{R}$

$$\varphi(t) := \Lambda_\alpha(t - \gamma^{-1/\theta}) + \gamma^{-1/\theta} \quad \text{and} \quad \psi(t) := \lambda_\alpha(t + \gamma^{-1/\theta}) - \gamma^{-1/\theta},$$

so with $s := 1 + 1/\theta$ the function c in Lemma 7.129 can be rewritten as

(7.44) $\quad c(t) = \varphi(t)$ if $t \in [0, \gamma^{-1/\theta}[$ and $c(t) = \psi(t)$ if $t \in]-\infty, 0]$.

Put $c_0(t) = \varphi_0(t) = \psi_0(t) := t$ for all $t \in \mathbb{R}$ and for each $n \in \mathbb{N}$ set $\varphi_n := \varphi \circ \cdots \circ \varphi$, and define ψ_n and c_n as similar compositions of the functions ψ and c n-times.
(a) Check that for each $n \in \mathbb{N} \cup \{0\}$ one has for all $t \in \mathbb{R}$

$$\varphi_n(t) = \Lambda_\alpha^n(t - \gamma^{-1/\theta}) + \gamma^{-1/\theta} \quad \text{and} \quad \psi_n(t) = \lambda_\alpha^n(t + \gamma^{-1/\theta}) - \gamma^{-1/\theta}.$$

(b) For each $n \in \mathbb{N}$ check by the inclusion $c(]-\infty, 0]) \subset]-\infty, 0[$ in Lemma 7.129 that $c_n(t) = \psi_n(t)$ for all $t \leq 0$.
(c) Argue by (7.43) that for all $k = 0, 1, \cdots, p-1$

$$\Lambda_\alpha^k \leq \Lambda_\alpha^{p-1} \leq \frac{1}{1-\sigma\gamma^{1/\theta}} < \Lambda_\alpha^p.$$

(d) Using the inequalities $0 \leq \sigma < \gamma^{-1/\theta}$ deduce from (c) that for all $k = 0, 1, \cdots, p-1$

$$\varphi_p(\sigma) < 0 \leq \varphi_k(\sigma) < \gamma^{-1/\theta}.$$

(e) Deduce by (7.44) that $c_p(\sigma) = \varphi_p(\sigma) < 0$, and by Lemma 7.130 that for each $n \in \mathbb{N}$

(7.45) $\quad f(x) \geq f(\overline{x}) + \langle \overline{x}^*, x - \overline{x} \rangle - \dfrac{c_{n-1+p}(\sigma)}{s} d(x, Q)^s, \quad \forall x \in B[\overline{x}, \eta_{n-1+p}(r)],$

where $s := 1 + 1/\theta$.

(f) Verify that $\kappa_1 = L = -\varphi_p(\sigma) = -c_p(\sigma)$ and derive from (b) and from the definition of κ_n and ψ_{n-1} that for each $n \in \mathbb{N}$

$$c_{n-1+p}(\sigma) = \psi_{n-1}(c_p(\sigma)) = \psi_{n-1}(-L) = -\kappa_n.$$

(g) From the latter equality in (f) and from (7.45) conclude that (b) in Lemma 7.132 holds. □

We are now able to state a first lower growth condition of f around $\overline{x}$ under a θ-order subregularity-like property of ∂f.

PROPOSITION 7.134. *Let $f : X \to \mathbb{R} \cup \{+\infty\}$ be a proper lower semicontinuous function on a Banach space X and let $\theta \in {]0, +\infty[}$ and $s := 1 + 1/\theta$. Let $(\overline{x}, \overline{x}^*) \in \mathrm{gph}\, \partial f$ and $Q \subset (\partial f)^{-1}(\overline{x}^*)$ with $\overline{x} \in Q$. Assume that there are $\sigma \in [0, \gamma^{-1/\theta}[$ and $r > 0$ such that*

$$f(x) \geq f(\overline{x}) + \langle \overline{x}^*, x - \overline{x} \rangle - \dfrac{\sigma}{s} d(x, Q)^s \quad \text{for all } x \in B_X[\overline{x}, r].$$

Assume also that there are real numbers $\gamma > 0$ and $\delta > 0$ such that

$$d(x, Q) \leq \gamma\, d(\overline{x}^*, \partial f(x))^\theta \quad \text{for all } x \in B(\overline{x}, \delta).$$

Then for every real $\kappa \in {]0, \gamma^{-1/\theta}[}$ there is some $r_0 > 0$ (depending only on $\kappa, \gamma, \sigma, \theta$) such that

$$f(x) \geq f(\overline{x}) + \langle \overline{x}^*, x - \overline{x} \rangle + \dfrac{\kappa}{s} d(x, Q)^s \quad \text{for all } x \in B(\overline{x}, r_0).$$

PROOF. With notation in Lemma 7.132 it is enough to fix any $\alpha \in {]0, 1[}$ and to choose some positive integer n such that $\kappa < \kappa_n < \gamma^{-1/\theta}$, where κ_n is given by Lemma 7.132. □

As a partial converse of the above proposition we have the following result.

PROPOSITION 7.135. *Let $f : X \to \mathbb{R} \cup \{+\infty\}$ be a proper lower semicontinuous function on a Banach space X and let $\theta \in {]0, +\infty[}$ and $s := 1 + 1/\theta$. Let $(\overline{x}, \overline{x}^*) \in \mathrm{gph}\, \partial f$ and $Q \subset (\partial f)^{-1}(\overline{x}^*)$ with $\overline{x} \in Q$. Assume that there are real numbers $\kappa > 0$ and $\eta > 0$ such that*

$$f(x) \geq f(\overline{x}) + \langle \overline{x}^*, x - \overline{x} \rangle + \dfrac{\kappa}{s} d(x, Q)^s \quad \text{for all } x \in B(\overline{x}, \eta),$$

and assume also that for some $\sigma \in [0, \kappa[$ one has, for all $u \in Q \cap B(\overline{x}, \eta)$ and $(x, x^*) \in \mathrm{gph}\, \partial f$ with $x \in B(\overline{x}, \eta)$ and $x^* \in B(\overline{x}^*, \eta)$,

$$f(u) \geq f(x) + \langle x^*, u - x \rangle - \dfrac{\sigma}{s} d(x, Q)^s.$$

Then for $\gamma := s^\theta (\kappa - \sigma)^{-\theta}$ and $r := \min \left\{ \dfrac{\eta}{2}, \left(\dfrac{\eta s}{2(\kappa - \sigma)} \right)^\theta \right\}$ one has

$$d(x, Q) \leq \gamma\, d(\overline{x}^*, \partial f(x))^\theta \quad \text{for all } x \in B(\overline{x}, r).$$

The proof of the proposition is the subject of the following exercise.

EXERCISE 7.136 (Proof of Proposition 7.134). With the assumptions of the proposition let
$$r := \min\left\{\frac{\eta}{2}, \left(\frac{\eta s}{2(\kappa - \sigma)}\right)^\theta\right\}.$$
Fix any $x \in B(\bar{x}, r)$ and note that $d(x, Q) \leq \|x - \bar{x}\| < r$.
(I) If $d(\bar{x}^*, \partial f(x)) \geq \eta/2$, we have
$$d(\bar{x}^*, \partial f(x)) \geq \eta/2 \geq \frac{\kappa - \sigma}{s} r^{1/\theta} \geq \frac{\kappa - \sigma}{s} d(x, Q)^{1/\theta}.$$
(II) Suppose that $d(\bar{x}^*, \partial f(x)) < \eta/2$ and take any $x^* \in \partial f(x) \cap B(\bar{x}^*, \eta/2)$. Consider any $u \in Q$ with $\|u - x\| \leq \|x - \bar{x}\|$, so $\|u - \bar{x}\| < \eta$. Since $\bar{x}^* \in \partial f(u)$, we have by the second assumption of inequality
$$f(\bar{x}) \geq f(u) + \langle \bar{x}^*, \bar{x} - u\rangle - \frac{\sigma}{s} d(u, Q)^{1+1/\theta} = f(u) + \langle \bar{x}^*, \bar{x} - u\rangle,$$
which yields by the second assumption of inequality again
$$\|x^* - \bar{x}^*\| \, \|x - u\|$$
$$\geq \langle x^* - \bar{x}^*, x - u\rangle = \langle x^*, x - u\rangle + \langle \bar{x}^*, u - \bar{x}\rangle - \langle \bar{x}^*, x - \bar{x}\rangle$$
$$\geq f(x) - f(u) - \frac{\sigma}{s} d(x, Q)^{1+1/\theta} + f(u) - f(\bar{x}) - \langle \bar{x}^*, x - \bar{x}\rangle$$
$$= f(x) - f(\bar{x}) - \langle \bar{x}^*, x - \bar{x}\rangle - \frac{\sigma}{s} d(x, Q)^{1+1/\theta}$$
$$\geq \frac{\kappa - \sigma}{s} d(x, Q)^{1+1/\theta},$$
where the latter inequality is due to the first assumption of inequality.
(a) Deduce that $\|x^* - \bar{x}^*\| \geq \frac{\kappa - \sigma}{s} d(x, Q)^{1/\theta}$ for all $x^* \in \partial f(x) \cap B(\bar{x}^*, \eta/2)$.
(b) Keeping in mind that $d(\bar{x}^*, \partial f(x)) < \eta/2$ deduce that
$$d(\bar{x}^*, \partial f(x)) \geq \frac{\kappa - \sigma}{s} d(x, Q)^{1/\theta}.$$
(III) Conclude that $d(x, Q) \leq \gamma d(\bar{x}^*, \partial f(x))^\theta$ for all $x \in B(\bar{x}, r)$. $\square$

With $Q = (\partial f)^{-1}(\bar{x})$ in both Proposition 7.134 and Proposition 7.135 we obtain the following first corollary.

COROLLARY 7.137. Let $f : X \to \mathbb{R} \cup \{+\infty\}$ be a proper lower semicontinuous function on a Banach space X and let $\theta \in \,]0, +\infty[$ and $s := 1 + 1/\theta$. Let $(\bar{x}, \bar{x}^*) \in \mathrm{gph}\,\partial f$.
(a) If the multimapping ∂f is θ-order metrically subregular with constant $\gamma > 0$ at $\bar{x}$ for $\bar{x}^*$ and if there are $\sigma \in [0, \gamma^{-1/\theta}[$ and $r > 0$ such that
$$f(x) \geq f(\bar{x}) + \langle \bar{x}^*, x - \bar{x}\rangle - \frac{\sigma}{s} d\big(x, (\partial f)^{-1}(\bar{x}^*)\big)^s \quad \text{for all } x \in B_X[\bar{x}, r],$$
then for every real $\kappa \in \,]0, \gamma^{-1/\theta}[$ there is some real $\eta > 0$ such that
$$f(x) \geq f(\bar{x}) + \langle \bar{x}^*, x - \bar{x}\rangle + \frac{\kappa}{s} d\big(x, (\partial f)^{-1}(\bar{x}^*)\big)^s \quad \text{for all } x \in B_X(\bar{x}, \eta).$$
(b) If there are reals $\kappa > 0$ and $\eta > 0$ such that
$$f(x) \geq f(\bar{x}) - \langle \bar{x}^*, x - \bar{x}\rangle + \frac{\kappa}{s} d\big(x, (\partial f)^{-1}(\bar{x}^*)\big)^s \quad \text{for all } x \in B_X(\bar{x}, \eta)$$

and if there is some $\sigma \in [0, \kappa[$ such that, for all $u \in B_X(\bar{x}, \eta) \cap (\partial f)^{-1}(\bar{x}^*)$ and $(x, x^*) \in \mathrm{gph}\, \partial f$ with $x \in B_X(\bar{x}, \eta)$ and $x^* \in B_{X^*}(\bar{x}^*, \eta)$

$$f(u) \geq f(x) + \langle x^*, u - x \rangle - \frac{\sigma}{s} d\big(x, (\partial f)^{-1}(x^*)\big)^s,$$

then for $\gamma := s^\theta (\kappa - \sigma)^{-\theta}$ the multimapping ∂f is θ-order metrically subregular with constant γ at $\bar{x}$ for $\bar{x}^*$.

Noticing that $\bar{x}^* \in \partial f(\bar{x})$ whenever $\bar{x}$ is a local minimizer of $f - \langle \bar{x}^*, \cdot \rangle$, we derive the following second corollary.

COROLLARY 7.138. *Let $f : X \to \mathbb{R} \cup \{+\infty\}$ be a proper lower semicontinuous function on a Banach space X and let $\theta \in]0, +\infty[$ and $s := 1 + 1/\theta$. Let $\bar{x}^* \in X^*$ and let $\bar{x} \in \mathrm{dom}\, f$ be a local minimizer of $f - \langle \bar{x}^*, \cdot \rangle$. The following hold:*
(a) *If ∂f is θ-order metrically subregular with constant $\gamma > 0$ at $\bar{x}$ for $\bar{x}^*$, then for every $\kappa \in]0, \gamma^{-1/\theta}[$ there exists $\eta > 0$ such that*

$$(7.46) \qquad f(x) \geq f(\bar{x}) + \langle \bar{x}^*, x - \bar{x} \rangle + \frac{\kappa}{s} d\big(x, (\partial f)^{-1}(\bar{x}^*)\big)^s, \ \forall x \in B(\bar{x}, \eta).$$

(b) *Conversely, given reals $\kappa > 0$ and $\sigma \in [0, \kappa[$, the multimapping ∂f is θ-order metrically subregular with constant $s^\theta (\kappa - \sigma)^{-\theta}$ at $\bar{x}$ for $\bar{x}^*$ provided that (7.46) is satisfied with the constant κ on some ball $B(\bar{x}, \eta)$ and provided that one has for all $u \in B(\bar{x}, \eta) \cap (\partial f)^{-1}(\bar{x}^*)$ and $(x, x^*) \in \mathrm{gph}\, \partial f$ with $x \in B(\bar{x}, \eta)$ and $x^* \in B_{X^*}(\bar{x}^*, \eta)$*

$$f(u) \geq f(x) + \langle x^*, u - x \rangle - \frac{\sigma}{s} d\big(x, (\partial f)^{-1}(\bar{x}^*)\big)^s.$$

In the case when f is convex, observing that $\bar{x}^* \in \partial f(\bar{x})$ if and only if $\bar{x}$ is a global minimizer of $f - \langle \bar{x}^*, \cdot \rangle$, we see that the following statement directly derives from Corollary 7.138 with $\sigma = 0$ therein.

COROLLARY 7.139. *Let $f : X \to \mathbb{R} \cup \{+\infty\}$ be a proper lower semicontinuous convex function on a Banach space X and let $\theta \in]0, +\infty[$. Let $(\bar{x}, \bar{x}^*) \in \mathrm{gph}\, f$. Then the multimapping ∂f is θ-order metrically subregular at $\bar{x}$ for $\bar{x}^*$ if and only if (7.46) holds with $s := 1 + 1/\theta$ and some reals $\kappa > 0$ and $\eta > 0$.*

Before stating other corollaries, we introduce the concept of stable minimizers.

DEFINITION 7.140. *Let $f : X \to \mathbb{R} \cup \{+\infty\}$ be a function on a normed space X and $\bar{x} \in \mathrm{dom}\, f$, and let $s \in [1, +\infty[$.*
(a) *One says that $\bar{x}$ is an s-order local minimizer of f with constant $\kappa > 0$ provided there is a real $\eta > 0$ such that one has the following s-order growth condition:*

$$f(x) \geq f(\bar{x}) + \frac{\kappa}{s} \|x - \bar{x}\|^s \quad \text{for all } x \in B_X(\bar{x}, \eta);$$

when $s = 1$ it is usual to call $\bar{x}$ a sharp local minimizer of f.
(b) *When for the constant $\kappa > 0$ there are reals $\delta > 0$, $\eta > 0$ and a mapping $\zeta : B_{X^*}(0, \delta) \to B_X(\bar{x}, \eta)$ with $\zeta(0) = \bar{x}$ such that one has the reinforced property*

$$f(x) - \langle u^*, x \rangle \geq f(\zeta(u^*)) - \langle u^*, \zeta(u^*) \rangle + \frac{\kappa}{s} \|x - \zeta(u^*)\|^s,$$

for all $u^* \in B_{X^*}(0, \delta)$ and $x \in B_X(\bar{x}, \eta)$, one says that $\bar{x}$ is a stable s-order local minimizer of f with constant $\kappa > 0$. This property is also called uniform s-order growth condition (for small linear perturbations).

REMARK 7.141. (a) Of course, any s-order local minimizer is a (usual) local minimizer.

Taking $u^* = 0$ in the inequality in (b) above yields for every $x \in B(\overline{x}, \eta)$

(7.47) $$f(x) \geq f(\overline{x}) + \kappa \|x - \overline{x}\|^s \geq f(\overline{x}),$$

hence any stable s-order local minimizer with constant $\kappa > 0$ is an s-order local minimizer with the same constant κ.

On the other hand, for any $u^* \in B_{X^*}(0, \delta)$, since $f(\overline{x}) \leq f(\zeta(u^*))$ by (7.47), taking $x = \overline{x}$ in the inequality in (b) above we obtain

$$\frac{\kappa}{s} \|\overline{x} - \zeta(u^*)\|^s \leq f(\overline{x}) - f(\zeta(u^*)) + \langle u^*, \zeta(u^*) - \overline{x} \rangle$$
$$\leq \langle u^*, \zeta(u^*) - \overline{x} \rangle \leq \|u^*\| \|\zeta(u^*) - \overline{x}\|,$$

so $s^{-1}\kappa \|\overline{x} - \zeta(u^*)\|^{s-1} \leq \|u^*\|$. For $s > 1$ the mapping ζ in (b) above is then continuous at 0 in X^*.

(b) For any $u^* \in B_{X^*}(0, \delta)$, we also see that $\zeta(u^*)$ is the unique element in $B_X(\overline{x}, \eta)$ satisfying the condition in (b), and further

$$\underset{B(\overline{x},\eta)}{\operatorname{Argmin}}(f - \langle u^*, \cdot \rangle) = \{\zeta(u^*)\}.$$

(c) s-order local minimizers are sensitive with respect to small perturbations. For $f(x, y) := (|x| + |y|)^2$ for all $(x, y) \in \mathbb{R}^2$, the point $(0, 0)$ is a second-order minimizer of f, while no second-order minimizer exists for the linearly perturbed function $(x, y) \mapsto f(x, y) + a(x + y)$ with nonzero real a. Clearly, such an instability does not hold for stable s-order local minimizers. □

Sharp minimizers of convex functions enjoy particular properties.

PROPOSITION 7.142. Let $f : X \to \mathbb{R} \cup \{+\infty\}$ be a function on a normed space X and let $\overline{x} \in \operatorname{dom} f$.
(I) If $\overline{x}$ is a sharp (that is, 1st-order) local minimizer for f with constant $\kappa > 0$, then one has

$$\kappa \mathbb{B}_{X^*} \subset \partial f(\overline{x}).$$

(II) If f is convex, then the following assertions are equivalent:
(a) the point $\overline{x}$ is a sharp local minimizer of f with constant $\kappa > 0$;
(b) the inclusion $\kappa \mathbb{B}_{X^*} \subset \partial f(\overline{x})$ holds;
(c) for any $u^* \in \kappa \mathbb{B}_{X^*}$ one has

$$(\kappa - \|u^*\|)\|x - \overline{x}\| \leq f(x) - f(\overline{x}) - \langle u^*, x - \overline{x} \rangle \quad \text{for all } x \in X;$$

(d) for any $u^* \in \kappa \mathbb{U}_{X^*}$ one has

$$\underset{X}{\operatorname{Argmin}}(f - \langle u^*, \cdot \rangle) = \{\overline{x}\};$$

(e) for any $u^* \in \kappa \mathbb{B}_{X^*}$ one has

$$f^*(u^*) = \langle u^*, \overline{x} \rangle - f(\overline{x}).$$

The point $\overline{x}$ is then a sharp local minimizer of the convex function f if and only if for any $s \in [1, +\infty[$ there are reals $\kappa > 0$, $\delta > 0$ and $\eta > 0$ such that

$$f(x) - \langle u^*, x \rangle \geq f(\overline{x}) - \langle u^*, \overline{x} \rangle + \frac{\kappa}{s} \|x - \overline{x}\|^s$$

for all $x \in B(\overline{x}, \eta)$ and $u^* \in \delta \mathbb{U}_{X^*}$.

EXERCISE 7.143 (Proof of Proposition 7.142). Consider notation in the above proposition.
(I) Using the equality $\kappa \|x - \bar{x}\| = \sup_{u^* \in \kappa \mathbb{B}_{X^*}} \langle u^*, x - \bar{x} \rangle$, derive the assertion (I) in the proposition.
(II) Assume that f is convex.
(α) Show by the convexity of f that the implication in (I) is an equivalence, so (a)$\Leftrightarrow$(b) is true.
(β) Argue (through the inclusion $u^* + (\kappa - \|u^*\|)\mathbb{B}_{X^*} \subset \kappa \mathbb{B}_{X^*}$ for every $u^* \in \kappa \mathbb{B}_{X^*}$) that

$\kappa \mathbb{B}_{X^*} \subset \partial f(\bar{x})$
$\Leftrightarrow (\kappa - \|u^*\|)\mathbb{B}_{X^*} \subset \partial (f - \langle u^*, \cdot \rangle)(\bar{x}), \ \forall u^* \in \kappa \mathbb{B}_{X^*}$
$\Leftrightarrow f(x) - \langle u^*, x \rangle \geq f(\bar{x}) - \langle u^*, \bar{x} \rangle + (\kappa - \|u^*\|)\|x - \bar{x}\|, \ \forall u^* \in \kappa \mathbb{B}_{X^*}, \forall x \in X$
$\Rightarrow \underset{X}{\operatorname{Argmin}}(f - \langle u^*, \cdot \rangle) = \{\bar{x}\}, \ \forall u^* \in \kappa \mathbb{U}_{X^*},$

and derive that (b) $\Leftrightarrow$ (c) $\Rightarrow$ (d).
(γ) Argue that (d) $\Rightarrow$ (b) and that (e) $\Leftrightarrow$ (b), and conclude with the equivalences (a) $\Leftrightarrow \cdots \Leftrightarrow$ (e).
(δ) Finally from (c) derive the equivalence involving the real $s \geq 1$. □

The rest of the analysis is now mainly focused on s-order local minimizers with $s > 1$. With such a real $s > 1$ we observe first a Hölder continuity property for the mapping ζ involved in the definition of stable s-order local minimizer.

PROPOSITION 7.144. Let $f : X \to \mathbb{R} \cup \{+\infty\}$ be a function on a normed space X and $\bar{x} \in \operatorname{dom} f$ be a stable s-order local minimizer for f with a real $s > 1$. Let $\kappa > 0$, $\delta > 0$, $\eta > 0$ and $\zeta : B_{X^*}(0, \delta) \to B_X(\bar{x}, \eta)$ be as in Definition 7.140(b). Then the mapping ζ is Hölder on $B_{X^*}(0, \delta)$, more precisely

$$\|\zeta(v^*) - \zeta(u^*)\| \leq (2\kappa/s)^{-1/(s-1)} \|v^* - u^*\|^{1/(s-1)}, \ \forall u^*, v^* \in B_{X^*}(0, \delta).$$

The proof of the proposition is established in the following exercise.

EXERCISE 7.145. (a) Argue first that f is finite at $\zeta(u^*)$ for every $u^* \in B_{X^*}(0, \delta)$.
(b) Derive that $(2\kappa/s) \|\zeta(v^*) - \zeta(u^*)\|^s \leq \langle v^* - u^*, \zeta(v^*) - \zeta(u^*) \rangle$ and conclude with the Hölder inequality property. □

Using Proposition 7.126(II), another corollary of Proposition 7.134 and Proposition 7.135 is obtained with $Q = \{\bar{x}\}$ in those propositions.

COROLLARY 7.146. Let $f : X \to \mathbb{R} \cup \{+\infty\}$ be a proper lower semicontinuous function on a Banach space X and let $\theta \in]0, +\infty[$ and $s := 1 + 1/\theta$. Let $(\bar{x}, \bar{x}^*) \in \operatorname{gph} \partial f$.
(a) If the multimapping ∂f is θ-order strongly metrically subregular with constant $\gamma > 0$ at $\bar{x}$ for $\bar{x}^*$ and if there are $\sigma \in [0, \gamma^{-1/\theta}[$ and $r > 0$ such that

$$f(x) \geq f(\bar{x}) + \langle \bar{x}^*, x - \bar{x} \rangle - \frac{\sigma}{s} \|x - \bar{x}\|^s \quad \text{for all } x \in B_X[\bar{x}, r],$$

then for every real $\kappa \in]0, \gamma^{-1/\theta}[$ the point $\bar{x}$ is an s-order local minimizer of $f - \langle \bar{x}^*, \cdot \rangle$ with constant κ.
(b) If for some real $\kappa > 0$ the point $\bar{x}$ is an s-order local minimizer of $f - \langle \bar{x}^*, \cdot \rangle$

with constant κ and if there is some $\sigma \in [0, \kappa[$ such that, for every $(x, x^*) \in \operatorname{gph} \partial f$ with $x \in B(\bar{x}, \eta)$ and $x^* \in B_{X^*}(\bar{x}, \eta)$

$$f(\bar{x}) \geq f(x) + \langle x^*, \bar{x} - x \rangle - \frac{\sigma}{s} \|x - \bar{x}\|^s,$$

then for $\gamma := s^\theta (\kappa - \sigma)^{-\theta}$ the multimapping ∂f is θ-order strongly metrically subregular with constant γ at $\bar{x}$ for $\bar{x}^*$.

By the inclusion $\bar{x}^* \in \partial f(\bar{x})$ whenever $\bar{x}$ is a local minimizer of $f - \langle \bar{x}^*, \cdot \rangle$, we derive with $\sigma = 0$ in Corollary 7.146 the following corollary.

COROLLARY 7.147. *Let $f : X \to \mathbb{R} \cup \{+\infty\}$ be a proper lower semicontinuous function on a Banach space X and let $\theta \in]0, +\infty[$ and $s := 1 + 1/\theta$. Let $\bar{x}^* \in X^*$ and let $\bar{x} \in \operatorname{dom} f$ be a local minimizer of $f - \langle \bar{x}^*, \cdot \rangle$. If ∂f is θ-order strongly metrically subregular with constant $\gamma > 0$ at $\bar{x}$ for $\bar{x}^*$, then for every $\kappa \in]0, \gamma^{-1/\theta}[$ the point $\bar{x}$ is an s-order local minimizer of $f - \langle \bar{x}^*, \cdot \rangle$ with constant κ.*

The next corollary involves the concepts of prox-regularity and subdifferential continuity of a function. Many other properties of such functions will be developed later in Chapter 11.

DEFINITION 7.148. *Let $f : U \to \mathbb{R} \cup \{+\infty\}$ be a function on an open set U of a normed space X, let ∂ be a subdifferential on U and let $(\bar{x}, \bar{x}^*) \in \operatorname{gph} \partial f$ and $s \in]1, +\infty[$.*
(a) *Given a real $\sigma \geq 0$ the function f is said to be s-order σ-prox-regular relative to the ∂-subdifferential at $\bar{x}$ for $\bar{x}^*$ provided that there exists a real $\eta > 0$ such that, for any $(x, x^*) \in \operatorname{gph} \partial f$ with $x \in B_X(\bar{x}, \eta)$ and $|f(x) - f(\bar{x})| < \eta$ and with $x^* \in B_{X^*}(\bar{x}^*, \eta)$, one has*

$$\langle x^*, u - x \rangle \leq f(u) - f(x) + \frac{\sigma}{s} \|u - x\|^s \quad \text{for all } u \in B_X(\bar{x}, \eta).$$

When $s = 2$ one just says that f is σ-prox-regular relative to ∂ at $\bar{x}$ for $\bar{x}^*$.
(b) *The function f is said to be ∂-subdifferentially continuous at $\bar{x}$ for $\bar{x}^*$ if the function $(x, x^*) \mapsto f(x)$ is continuous at $(\bar{x}, \bar{x}^*)$ relative to $\operatorname{gph} \partial f$, that is, $f(x) \to f(\bar{x})$ as $\operatorname{gph} \partial f \ni (x, x^*) \to (\bar{x}, \bar{x}^*)$.*

When f is ∂-subdifferentially continuous at $\bar{x}$ for $\bar{x}^* \in \partial f(\bar{x})$, then f is s-order σ-prox-regular at $\bar{x}$ for $\bar{x}^*$ if and only if there exists a real $\eta > 0$ such that, for every $(x, x^*) \in \operatorname{gph} \partial f$ with $x \in B_X(\bar{x}, \eta)$ and $x^* \in B_{X^*}(\bar{x}^*, \eta)$, one has

$$\langle x^*, u - x \rangle \leq f(u) - f(x) + \frac{\sigma}{s} \|u - x\|^s, \ \forall u \in B_X(\bar{x}, \eta).$$

Obviously the (usual) continuity of a function at a ∂-subdifferentiability point entails its ∂-subdifferential continuity at that point. Any convex function is also subdifferentially continuous at any point of the domain of its subdifferential. Indeed, consider any convex function $f : U \to \mathbb{R} \cup \{+\infty\}$ and any $(\bar{x}, \bar{x}^*) \in \operatorname{gph} \partial f$. Writing for every $(x, x^*) \in \operatorname{gph} \partial f$

$$f(\bar{x}) + \langle \bar{x}^*, x - \bar{x} \rangle \leq f(x) \leq f(\bar{x}) - \langle x^*, \bar{x} - x \rangle,$$

we see that $f(x) \to f(\bar{x})$ as $\operatorname{gph} \partial f \ni (x, x^*) \to (\bar{x}, \bar{x}^*)$, hence f is subdifferentially continuous at $\bar{x}$ for $\bar{x}^*$. Clearly, the convex function f is also s-order σ-prox-regular relative to ∂ at $\bar{x}$ for $\bar{x}^*$ for any reals $s \geq 1$ and $\sigma \geq 0$. A lot of other examples will be provided in Chapter 11.

With Definition 7.148 at hands we then directly derive the following assertions from Corollary 7.146.

COROLLARY 7.149. *Let $f : X \to \mathbb{R} \cup \{+\infty\}$ be a proper lower semicontinuous function defined on a Banach space X and let $\theta \in]0, +\infty[$ and $s := 1 + 1/\theta$. Let $(\bar{x}, \bar{x}^*) \in \mathrm{gph}\, \partial f$. The following assertions hold.*
(a) *If the multimapping ∂f is θ-order strongly metrically subregular with constant $\gamma > 0$ at $\bar{x}$ for $\bar{x}^*$ and if there is some $\sigma \in [0, \gamma^{-1/\theta}[$ such that at $\bar{x}$ for $\bar{x}^*$, relative to ∂, the function f is s-order σ-prox-regular and subdifferentially continuous, then for every real $\kappa \in]0, \gamma^{-1/\theta}[$ the point $\bar{x}$ is an s-order local minimizer of the function $f - \langle \bar{x}^*, \cdot \rangle$ with constant κ.*
(b) *If for some real $\kappa > 0$ the point $\bar{x}$ is an s-order local minimizer of $f - \langle \bar{x}^*, \cdot \rangle$ with constant κ and if there is some $\sigma \in [0, \kappa[$ such that at $\bar{x}$ for $\bar{x}^*$, relative to ∂, the function f is s-order σ-prox-regular and subdifferentially continuous, then for $\gamma := s^\theta (\kappa - \sigma)^{-\theta}$ the multimapping ∂f is θ-order strongly metrically subregular with constant γ at $\bar{x}$ for $\bar{x}^*$.*

In the case when f is convex the following statement directly derives from Corollary 7.146 again.

COROLLARY 7.150. *Let $f : X \to \mathbb{R} \cup \{+\infty\}$ be a proper lower semicontinuous convex function on a Banach space X and $(\bar{x}, \bar{x}^*) \in \mathrm{gph}\, f$. Let $\theta \in]0, +\infty[$ and $s := 1 + 1/\theta$. Then the multimapping ∂f is θ-order strongly metrically subregular at $\bar{x}$ for $\bar{x}^*$ if and only if the point $\bar{x}$ is an s-order local minimizer of $f - \langle \bar{x}^*, \cdot \rangle$.*

The next corollary considers the case when $\bar{x}$ is a local minimizer of the function f itself.

COROLLARY 7.151. *Let $f : X \to \mathbb{R} \cup \{+\infty\}$ be a proper lower semicontinuous function on a Banach space X and let $\theta \in]0, +\infty[$ and $s := 1 + 1/\theta$. Let $\bar{x} \in \mathrm{dom}\, f$ be a local minimizer of f. The following hold:*
(a) *If ∂f is θ-order strongly metrically subregular with constant $\gamma > 0$ at $\bar{x}$ for 0, then for every $\kappa \in]0, \gamma^{-1/\theta}[$ the point $\bar{x}$ is an s-order local minimizer of f with constant κ.*
(b) *If the proper lower semicontinuous function f is convex, then as a partial converse, ∂f is θ-order strongly metrically subregular with constant $s^\theta \kappa^{-\theta}$ at $\bar{x}$ for 0 whenever $\bar{x}$ is an s-order local minimizer of f with constant κ.*
(c) *If the proper lower semicontinuous function f is convex, then ∂f is θ-order strongly metrically subregular at $\bar{x}$ for 0 if and only if $\bar{x}$ is an s-order local minimizer of f.*

REMARK 7.152. Without additional assumption on f, the converse of (a) in the above corollary fails in general. For the lower semicontinuous function $f : \mathbb{R} \to \mathbb{R} \cup \{+\infty\}$ with

$$f(0) = 0, \ f(x) = (1/2)x^2 - x^2 \sin(1/x) \text{ if } x > 0, \ f(x) = +\infty \text{ if } x < 0,$$

it is easily verified that the point $\bar{x} := 0$ is a second-order minimizer of f. Nevertheless, there is no $\eta > 0$ for which $(\partial_L f)^{-1}(0) \cap B(0, \eta)$ is a singleton since 0 is not an isolated point of $(\partial_L f)^{-1}(0)$ (as it can be easily checked), so $\partial_L f$ is not strongly metrically subregular at $\bar{x}$ for 0. □

The next proposition carries out the situation when ∂f enjoys the stronger property of strong metric regularity. Let us establish first a lemma.

LEMMA 7.153 (J. Yao, X.Y. Zheng and J. Zhu). Let $f : X \to \mathbb{R} \cup \{+\infty\}$ be a proper lower semicontinuous function on a Banach space X and let $\theta \in \,]0,+\infty[$ and $s := 1 + 1/\theta$. Let $(\bar{x}, \bar{x}^*) \in \operatorname{gph} \partial f$ and let $\mathcal{Q} : X^* \rightrightarrows X$ be a multimapping such that $\bar{x} \in \mathcal{Q}(\bar{x}^*)$ and $\mathcal{Q}(x^*) \subset (\partial f)^{-1}(x^*)$ for all $x^* \in X^*$. Let a real $\gamma > 0$ be such that there is $\delta > 0$ for which

(7.48) $\quad d(x, \mathcal{Q}(x^*)) \leq \gamma d(x^*, \partial f(x))^\theta \quad$ for all $x \in B_X(\bar{x}, \delta), x^* \in B_{X^*}(\bar{x}^*, \delta)$.

Assume also that there is $\sigma \in [0, \gamma^{-1/\theta}[$ such that, for every $x' \in B_X(\bar{x}, \delta)$ and every $(x^*, x) \in \operatorname{gph} \mathcal{Q}$ with $x^* \in B_{X^*}(\bar{x}^*, \delta)$ and $x \in B_X(\bar{x}, \delta)$

(7.49) $\quad \langle x^*, x' - x \rangle \leq f(x') - f(x) + \dfrac{\sigma}{s} d(x', \mathcal{Q}(x^*))^s.$

Then for any $\kappa \in \,]0, \gamma^{-1/\theta}[$ there are reals $\delta_0 > 0$ and $r_0 > 0$ such that for every $x \in B_X(\bar{x}, r_0)$ and every $(u^*, u) \in \operatorname{gph} \mathcal{Q}$ with $u^* \in B_{X^*}(\bar{x}^*, \delta_0)$, $u \in B_X(\bar{x}, r_0)$ one has

$$f(x) \geq f(u) + \langle u^*, x - u \rangle + \dfrac{\kappa}{s} d(x, \mathcal{Q}(u^*))^s.$$

PROOF. Put $s := 1 + 1/\theta$ and fix any $\kappa \in \,]0, \gamma^{-1/\theta}[$. Note that (7.48) gives for all $x^* \in B(\bar{x}^*, \delta)$

$$d(\bar{x}, \mathcal{Q}(x^*)) \leq \gamma d(x^*, \partial f(\bar{x})) \leq \gamma \|x^* - \bar{x}^*\|^\theta,$$

which clearly yields

(7.50) $\quad\quad\quad\quad\quad\quad d(\bar{x}, \mathcal{Q}(x^*)) \to 0 \quad$ as $x^* \to \bar{x}^*.$

Choose some positive integer N such that $\kappa < \kappa_N < \gamma^{-1/\theta}$, where κ_n is given by the expression in Lemma 7.132 depending only on $\alpha, \gamma, \sigma, \theta$ (with $\kappa_n \to \gamma^{-1/\theta}$ as $n \to \infty$). Let $r := r_N := \dfrac{\delta/2}{(2-\alpha)^{p+N-1}}$, where p is also given by the expression in Lemma 7.132 depending only on $\alpha, \gamma, \sigma, \theta$. By (7.50) choose some $\delta_0 \in \,]0, \delta[$ such that $\mathcal{Q}(x^*) \cap B(\bar{x}, r) \neq \emptyset$ for all $x^* \in B(\bar{x}^*, \delta_0)$. Fix any $u^* \in B(\bar{x}^*, \delta_0)$ and any $u \in \mathcal{Q}(u^*) \cap B(\bar{x}, r/2)$. From (7.48) we have

$$d(x, \mathcal{Q}(u^*)) \leq \gamma d(u^*, \partial f(x))^\theta \quad \text{for all } x \in B(u, \delta/2),$$

and from (7.49) we also have

$$f(x) \geq f(u) + \langle u^*, x - u \rangle - \dfrac{\sigma}{s} d(x, \mathcal{Q}(u^*))^s \quad \text{for all } x \in B_X[u, \delta/2].$$

This allows us to apply Proposition 7.134 with (u, u^*) and $\mathcal{Q}(u^*)$ in place of $(\bar{x}, \bar{x}^*)$ and $\mathcal{Q}$ respectively, to obtain

$$f(x) \geq f(u) + \langle u^*, x - u \rangle + \dfrac{\kappa_N}{s} d(x, \mathcal{Q}(u^*))^s \quad \text{for all } x \in B(u, r),$$

which entails that

$$f(x) \geq f(u) + \langle u^*, x - u \rangle + \dfrac{\kappa}{s} d(x, \mathcal{Q}(u^*))^s \quad \text{for all } x \in B(\bar{x}, r/2).$$

This finishes the proof. □

The first consequence of Lemma 7.153 is directly obtained with the choice $\mathcal{Q}(x^*) = (\partial f)^{-1}(x^*)$ for all $x^* \in X^*$.

PROPOSITION 7.154. Let $f : X \to \mathbb{R} \cup \{+\infty\}$ be a proper lower semicontinuous function on a Banach space X and let $\theta \in {]}0, +\infty[$ and $s := 1 + 1/\theta$. Let $(\bar{x}, \bar{x}^*) \in \mathrm{gph}\, \partial f$ be such that the multimapping ∂f is θ-order metrically regular at $\bar{x}$ for $\bar{x}^*$ with constant $\gamma > 0$. Assume also that there is $\sigma \in [0, \gamma^{-1/\theta}[$ and $\delta > 0$ such that, for every $x' \in B_X(\bar{x}, \delta)$ and every $(x, x^*) \in \mathrm{gph}\, \partial f$ with $x \in B_X(\bar{x}, \delta)$ and $x^* \in B_{X^*}(\bar{x}^*, \delta)$

$$\langle x^*, x' - x \rangle \leq f(x') - f(x) + \frac{\sigma}{s} d\big(x', (\partial f)^{-1}(x^*)\big)^s.$$

Then for any $\kappa \in {]}0, \gamma^{-1/\theta}[$ there are reals $\delta_0 > 0$ and $r_0 > 0$ such that for every $x \in B_X(\bar{x}, r_0)$ and every $(u, u^*) \in \mathrm{gph}\, \partial f$ with $u^* \in B_{X^*}(\bar{x}^*, \delta_0)$, $u \in B_X(\bar{x}, r_0)$ one has

$$f(x) \geq f(u) + \langle u^*, x - u \rangle + \frac{\kappa}{s} d\big(x, (\partial f)^{-1}(u^*)\big)^s.$$

The next proposition which will be derived concerns the situation when f is prox-regular and ∂f is strongly metrically regular.

PROPOSITION 7.155. Let $f : X \to \mathbb{R} \cup \{+\infty\}$ be a proper lower semicontinuous function on a Banach space X and let $\theta \in {]}0, +\infty[$ and $s := 1 + 1/\theta$. Let $(\bar{x}, \bar{x}^*) \in \mathrm{gph}\, \partial f$. Assume that ∂f is θ-order strongly metrically regular at $\bar{x}$ for $\bar{x}^*$ with constant $\gamma > 0$ and that, relative to ∂, the function f is subdifferentially continuous and s-order σ-prox-regular at $\bar{x}$ for $\bar{x}^*$ with $\sigma \in [0, \gamma^{-1/\theta}[$. Then for any $\kappa \in {]}0, \gamma^{-1/\theta}[$ there are reals $\delta_0 > 0$ and $\eta_0 > 0$ along with a mapping $\zeta : B_{X^*}(\bar{x}^*, \delta_0) \to B_X(\bar{x}, \eta_0)$ with $\zeta(\bar{x}^*) = \bar{x}$ such that for all $u^* \in B_{X^*}(\bar{x}^*, \delta_0)$ and $x \in B_X(\bar{x}, \eta_0)$ one has

$$f(x) \geq f(\zeta(u^*)) + \langle u^*, x - \zeta(u^*) \rangle + \frac{\kappa}{s} \|x - \zeta(u^*)\|^s,$$

that is, $\bar{x}$ is a stable s-order local minimizer of $f - \langle \bar{x}^*, \cdot \rangle$ with constant κ.

PROOF. By subdifferential continuity and prox-regularity there is a real $\delta' > 0$ such that, for every $x' \in B(\bar{x}, \delta')$ and every $(x, x^*) \in \mathrm{gph}\, \partial f$ with $x \in B(\bar{x}, \delta')$ and $x^* \in B(\bar{x}^*, \delta')$ one has

$$\langle x^*, x' - x \rangle \leq f(x') - f(x) + \frac{\sigma}{s} \|x' - x\|^s.$$

By θ-order strong metric regularity and by Proposition 7.126(c) there are a positive real $\delta'' < \delta'$ and a mapping $\xi : B(\bar{x}^*, \delta'') \to X$ continuous at $\bar{x}^*$ with $\xi(\bar{x}^*) = \bar{x}$ and such that $\xi(u^*) \in (\partial f)^{-1}(u^*)$ for all $u^* \in B(\bar{x}^*, \delta'')$ and

$$\|x - \xi(u^*)\| \leq \gamma d\big(u^*, \partial f(x)\big)^\theta \quad \text{for all } x \in B(\bar{x}, \delta''), u^* \in B(\bar{x}^*, \delta'').$$

Choose $0 < \delta < \min\{\delta', \delta''\}$ such that $\xi(B(\bar{x}^*, \delta)) \subset B(\bar{x}, \delta')$ and denote by ζ the restriction of ξ to $B(\bar{x}^*, \delta)$. Define $\mathcal{Q} : X \rightrightarrows X$ by $\mathcal{Q}(x^*) := \{\zeta(x^*)\}$ if $x^* \in B(\bar{x}^*, \delta)$ and $\mathcal{Q}(x^*) = \emptyset$ otherwise. It ensues that

$$d(x, \mathcal{Q}(x^*)) \leq \gamma d\big(x^*, \partial f(x)\big)^\theta \quad \text{for all } x \in B(\bar{x}, \delta), x^* \in B(\bar{x}^*, \delta),$$

and that for every $x' \in B(\bar{x}, \delta)$ and every $(x^*, x) \in \mathrm{gph}\, \mathcal{Q}$ with $x \in B(\bar{x}, \delta)$ and $x^* \in B(\bar{x}^*, \delta)$

$$\langle x^*, x' - x \rangle \leq f(x') - f(x) + \frac{\sigma}{s} d\big(x', \mathcal{Q}(x^*)\big)^\theta.$$

Lemma 7.153 then yields reals $\delta_0 > 0$ and $r_0 > 0$ such that for all $u^* \in B(\bar{x}^*, \delta_0)$ and $x \in B(\bar{x}, r_0)$ one has

$$f(x) \geq f(\zeta(u^*)) + \langle u^*, x - \zeta(u^*) \rangle + \frac{\kappa}{s} \|x - \zeta(u^*)\|^s,$$

which finishes the proof. □

A similar result also holds when the point $\bar{x}$ at which ∂f is strongly metrically regular is a local minimizer.

PROPOSITION 7.156. *Let $f : X \to \mathbb{R} \cup \{+\infty\}$ be a proper lower semicontinuous function on a Banach space X and let $\theta \in]0, +\infty[$ and $s := 1 + 1/\theta$. Let $\bar{x} \in \mathrm{dom}\, f$ be a local minimizer of f. If the multimapping ∂f is θ-order strongly metrically regular at $\bar{x}$ for 0 with constant $\gamma > 0$, then for any $\kappa \in]0, \gamma^{-1/\theta}[$ the point $\bar{x}$ is a stable s-order local minimizer of f with constant κ.*

PROOF. Let $r > 0$ be such that $f(\bar{x}) = \inf_{B[\bar{x},r]} f$. By the θ-order strong metric regularity of ∂f and by Proposition 7.126(c) there are reals $\gamma > 0$, $\delta > 0$ and a mapping $\zeta : B_{X^*}(0, \delta) \to X$ continuous at 0 with $\zeta(0) = \bar{x}$ such that $\zeta(u^*) \in (\partial f)^{-1}(u^*)$ for all $u^* \in B_{X^*}(0, \delta)$ and

(7.51) $\qquad \|u - \zeta(u^*)\| \leq \gamma\, d(u^*, \partial f(u))^\theta \quad \text{for all } u \in B(\bar{x}, \delta),\ u^* \in B_{X^*}(0, \delta).$

This gives in particular (since $0 \in \partial f(\bar{x})$)

(7.52) $\qquad \|\bar{x} - \zeta(u^*)\| \leq \gamma\, d(u^*, \partial f(\bar{x}))^\theta \leq \gamma \|u^*\|^\theta.$

Taking $u^* = 0$ in (7.51) also gives

$$\|u - \bar{x}\| \leq \gamma\, d(0, \partial f(u))^\theta \quad \text{for all } u \in B(\bar{x}, \delta).$$

With $\sigma := 0$ and $\alpha := 1/2$ choose an integer $N \in \mathbb{N}$ such that, for κ_n defined in Lemma 7.132, one has $\kappa_N > \kappa$ (keep in mind that $\kappa_n \to \gamma^{-1/\theta}$ as $n \to \infty$). Put

$$\kappa' := \kappa_N \quad \text{and} \quad \eta := \frac{1}{(2-\alpha)^{p+N-1}} \min\{r, \delta\}.$$

Lemma 7.132 applied with $Q := \{\bar{x}\}$ gives

(7.53) $\qquad \dfrac{\kappa'}{s} \|x - \bar{x}\|^{1+1/\theta} \leq f(x) - f(\bar{x}) \quad \text{for all } x \in B(\bar{x}, \eta).$

Let $\delta' := \min\left\{\frac{\eta}{2}, \left(\frac{\eta}{2\gamma}\right)^{1/\theta}, \frac{\kappa'}{s}\left(\frac{\eta}{2}\right)^{1/\theta}\right\}$. Fix any $u^* \in B_{X^*}(0, \delta')$ and let us show the equality

$$(f - \langle u^*, \cdot \rangle)(\zeta(u^*)) = \inf_{B[\zeta(u^*), \eta/2]} (f - \langle u^*, \cdot \rangle).$$

By the inequality in (7.52) and by the definition of δ' we have

$$\|\bar{x} - \zeta(u^*)\| \leq \gamma \|u^*\|^\theta < \gamma(\delta')^\theta \leq \eta/2,$$

hence there exists some $\eta' \in]\eta/2, \eta[$ such that $B[\zeta(u^*), \eta/2] \subset B[\bar{x}, \eta']$. Let $\varepsilon_n \to 0$ with $\varepsilon_n > 0$. Fix any $n \in \mathbb{N}$ and choose $x_n \in B[\bar{x}, \eta']$ such that

(7.54) $\qquad (f - \langle u^*, \cdot \rangle)(x_n) < \inf_{B[\bar{x}, \eta']} (f - \langle u^*, \cdot \rangle) + \varepsilon_n^2.$

By the Ekeland variational principle there is $u_n \in B[\bar{x}, \eta']$ satisfying $\|u_n - x_n\| \leq \varepsilon_n$ and

(7.55) $\qquad (f - \langle u^*, \cdot \rangle)(u_n) \leq (f - \langle u^*, \cdot \rangle)(x) + \varepsilon_n \|x - u_n\|,\ \forall x \in B[\bar{x}, \eta'],$

hence in particular
$$f(u_n) - f(\bar{x}) \leq \langle u^*, u_n - \bar{x}\rangle + \varepsilon_n \|\bar{x} - u_n\|$$
$$\leq (\delta' + \varepsilon_n)\|u_n - \bar{x}\| \leq \left(\frac{\kappa'}{s}\left(\frac{\eta}{2}\right)^{1/\theta} + \varepsilon_n\right)\|u_n - \bar{x}\|.$$

This and (7.53) ensure that
$$\frac{\kappa'}{s}\|u_n - \bar{x}\|^{1/\theta} \leq \frac{\kappa'}{s}\left(\frac{\eta}{2}\right)^{1/\theta} + \varepsilon_n,$$

so for n large enough, say $n \geq n_0$, we have $s^{-1}\kappa'\|u_n - \bar{x}\|^{1/\theta} < s^{-1}\kappa'(\eta')^{1/\theta}$, that is, $\|u_n - \bar{x}\| < \eta'$. From this and the inequality in (7.55) it results, for $n \geq n_0$, that $u^* \in \partial f(u_n) + \varepsilon_n \mathbb{B}_{X^*}$, hence by (7.51) we obtain $\|u_n - \zeta(u^*)\| \leq \gamma \varepsilon_n^\theta$. This combined with the inequality $\|u_n - x_n\| \leq \varepsilon_n$ tells us that $x_n \to \zeta(u^*)$ as $n \to \infty$, which entails, by (7.54) and the lower semicontinuity of f along with the inclusion $\zeta(u^*) \in B[\bar{x}, \eta']$, that
$$(f - \langle u^*, \cdot\rangle)(\zeta(u^*)) = \inf_{B[\bar{x},\eta']}(f - \langle u^*, \cdot\rangle).$$

This and the inclusion $B[\zeta(u^*), \eta/2] \subset B[\bar{x}, \eta']$ yield the announced equality
$$(f - \langle u^*, \cdot\rangle)(\zeta(u^*)) = \inf_{B[\zeta(u^*),r'']}(f - \langle u^*, \cdot\rangle) \quad \text{with } r'' := \eta/2,$$

and for $\delta'' := \eta/2$ we also have
$$\|x - \zeta(u^*)\| \leq \gamma\, d(u^*, \partial f(x))^\theta \quad \text{for all } x \in B(\zeta(u^*), \delta'').$$

Since $u^* \in \partial f(\zeta(u^*))$, putting $\eta'' := (2-\alpha)^{-p-N+1}(\eta/2)$ (depending only on $\gamma, \theta, \kappa, r, \delta$) and applying Lemma 7.132 with $Q = \{\zeta(u^*)\}$ and $\sigma = 0$ we obtain for all $x \in B(\zeta(u^*), \eta'')$
$$\frac{\kappa'}{s}\|x - \zeta(u^*)\|^s \leq f(x) - \langle u^*, x\rangle - f(\zeta(u^*)) + \langle u^*, \zeta(u^*)\rangle.$$

Set $\eta_0 := \eta''/2$ and choose a real $\delta_0 \in\,]0, \delta'[$ such that for all $u^* \in B_{X^*}(0, \delta_0)$ one has $\|\zeta(u^*) - \bar{x}\| < \eta_0$, which entails $B_X(\bar{x}, \eta_0) \subset B_X(\zeta(u^*), \eta'')$. Consequently, $\zeta(B_{X^*}(0, \delta_0)) \subset B_X(\bar{x}, \eta_0)$ and for all $u^* \in B_{X^*}(0, \delta_0)$ and $x \in B_X(\bar{x}, \eta_0)$ one has
$$\frac{\kappa'}{s}\|x - \zeta(u^*)\|^s \leq f(x) - \langle u^*, x\rangle - f(\zeta(u^*)) + \langle u^*, \zeta(u^*)\rangle,$$

which translates (since $\kappa' > \kappa$) that $\bar{x}$ is a stable s-order local minimizer of f with constant κ. The proof is then finished. $\square$

EXERCISE 7.157. To prove the converse of (a) in the above proposition under the convexity of f, assume that $\bar{x}$ is a stable $(1+1/\theta)$-order local minimizer of f with constant $\kappa > 0$. By definition 7.140 let $\delta > 0$, $\eta > 0$ and $\zeta : B_{X^*}(0, \delta) \to B(\bar{x}, \eta)$ with $\zeta(0) = \bar{x}$ such that

(7.56)
$$\kappa\|x - \zeta(u^*)\|^{1+1/\theta} \leq f(x) - f(\zeta(u^*)) - \langle u^*, x - \zeta(u^*)\rangle, \ \forall u^* \in B_{X^*}(0, \delta), \forall x \in B(\bar{x}, \eta),$$

which entails in particular
$$\kappa\|x - \bar{x}\|^{1+1/\theta} \leq f(x) - f(\bar{x}) \quad \forall x \in B(\bar{x}; \eta)$$

and
$$\kappa\|\bar{x} - \zeta(u^*)\|^{1+1/\theta} \leq f(\bar{x}) - f(\zeta(u^*)) - \langle u^*, \bar{x} - \zeta(u^*)\rangle, \ \forall u^* \in B_{X^*}(0, \delta).$$

Since $\bar{x}$ is a global minimizer of f (by the convexity of f), we also have for any $u^* \in B_{X^*}(0, \delta)$
$$f(\bar{x}) - f(\zeta(u^*)) - \langle u^*, \bar{x} - \zeta(u^*)\rangle \leq -\langle u^*, \bar{x} - \zeta(u^*)\rangle \leq \eta \|u^*\|,$$
hence $\kappa \|\bar{x} - \zeta(u^*)\|^{1+1/\theta} \leq \eta \|u^*\|$, so $\zeta(u^*) \to \bar{x}$ as $u^* \to 0$.

Let $\delta' := \min\{\delta, \kappa\eta^{1/\theta} 2^{-1-1/\theta}\}$ and fix any $u^* \in B_{X^*}(0, \delta')$.

(a) Argue that $\|\bar{x} - \zeta(u^*)\| < \eta/2$ and $B(\zeta(u^*), \eta/2) \subset B(\bar{x}, \eta)$, and that with $\eta' := \eta/2$
$$\kappa \|x - \zeta(u^*)\|^{1+1/\theta} \leq f(x) - f(\zeta(u^*)) - \langle u^*, x - \zeta(u^*)\rangle, \ \forall x \in B(\zeta(u^*), \eta').$$

(b) For $\gamma := \kappa^{-\theta}$ and $r := \min\left\{\frac{\eta'}{2}, \left(\frac{\eta'}{2\kappa}\right)^{\theta}\right\}$, derive from Proposition 7.135 applied with $Q := \{\zeta(u^*)\}$ that
$$\|x - \zeta(u^*)\| \leq \gamma d(u^*, \partial f(x))^{\theta} \ \forall x \in B(\zeta(u^*), r).$$

(c) Keeping in mind that $\zeta(u^*) \to \bar{x}$ as $u^* \to 0$, choose $\delta_0 \in]0, \delta'[$ such that
$$\zeta(B_{X^*}(0, \delta_0)) \subset B(\bar{x}, \eta_0), \text{ where } \eta_0 := r/2.$$
Argue that $B(\bar{x}, \eta_0) \subset B(\zeta(u^*), r)$ for all $u^* \in B_{X^*}(0, \delta_0)$, and derive that ζ maps $B_{X^*}(0, \delta_0)$ into $B(\bar{x}, \eta_0)$ with $\zeta(0) = \bar{x}$ and
$$\|x - \zeta(u^*)\| \leq \gamma d(u^*, \partial f(x))^{\theta}, \ \forall u^* \in B_{X^*}(0, \delta_0), \forall x \in B(\bar{x}, \eta_0).$$
Conclude by applying Proposition 7.126(b). □

Proposition 7.156 furnishes every constant $\kappa \in]0, \gamma^{-1/\theta}[$ for the stable minimum property as Proposition 7.155 does, but the former uses the assumption that $\bar{x}$ is a local minimizer. Nevertheless, it allows us to obtain such constants with a significant relaxation of the prox-regularity assumption in Proposition 7.155.

PROPOSITION 7.158. Let $f : X \to \mathbb{R} \cup \{+\infty\}$ be a proper lower semicontinuous function on a Banach space X and let $\theta \in]0, +\infty[$ and $s := 1 + 1/\theta$. Let $(\bar{x}, \bar{x}^*) \in \text{gph} \, \partial f$ be such that f is subdifferentially continuous at $\bar{x}$ for $\bar{x}^*$. Assume that ∂f is θ-order strongly metrically regular at $\bar{x}$ for $\bar{x}^*$ with constant $\gamma > 0$ and that there are reals $\sigma \in [0, \gamma^{-1/\theta}[$ and $\delta > 0$ such that
$$f(x) \geq f(\bar{x}) + \langle \bar{x}^*, x - \bar{x}\rangle - \frac{\sigma}{s} \|x - \bar{x}\|^s, \ \forall x \in B_X(\bar{x}, \delta).$$
Then for any $\kappa \in]0, \gamma^{-1/\theta}[$ there are reals $\delta_0 > 0$ and $\eta_0 > 0$ along with a mapping $\zeta : B_{X^*}(\bar{x}^*, \delta_0) \to B_X(\bar{x}, \eta_0)$ with $\zeta(\bar{x}^*) = \bar{x}$ such that for all $u^* \in B_{X^*}(\bar{x}^*, \delta_0)$ and $x \in B_X(\bar{x}, \eta_0)$ one has
$$f(x) \geq f(\zeta(u^*)) + \langle u^*, x - \zeta(u^*)\rangle + \frac{\kappa}{s} \|x - \zeta(u^*)\|^s,$$
that is, $\bar{x}$ is a stable s-order local minimizer of $f - \langle \bar{x}^*, \cdot\rangle$ with constant κ.

PROOF. By s-order strong metric regularity of f with constant γ and by Proposition 7.126(d) there are reals $\delta_1, \eta_1 > 0$ and a mapping $\zeta_1 : B_{X^*}(\bar{x}^*, \delta_1) \to B_X(\bar{x}, \eta_1)$ such that for all $x \in B_X(\bar{x}, \eta_1)$ and $u^* \in B_{X^*}(\bar{x}^*, \delta_1)$
$$(7.57) \quad (\partial f)^{-1}(u^*) \cap B_X(\bar{x}, \eta_1) = \{\zeta_1(u^*)\} \text{ and } \|x - \zeta_1(u^*)\| \leq \gamma d(u^*, \partial f(x))^{\theta}.$$
By taking $u^* = \bar{x}^*$ in the above equality, it ensues that $\zeta_1(\bar{x}^*) = \bar{x}$.

On the other hand, the s-order strong metric regularity of ∂f at $\bar{x}$ for $\bar{x}^*$ is easily seen to be equivalent with that of $\partial(f - \langle \bar{x}^*, \cdot \rangle)$ at $\bar{x}$ for 0. The latter property and the inequality in the assumptions allow us to apply Corollary 7.146(a), which gives that the point $\bar{x}$ is a local minimizer of $f - \langle \bar{x}^*, \cdot \rangle$. Then, Proposition 7.156 furnishes some reals $\kappa_2, \delta_2, \eta_2 > 0$ and a mapping $\zeta_2 : B_{X^*}(0, \delta_2) \to B_X(\bar{x}, \eta_2)$ with $\zeta_2(0) = \bar{x}$ such that for all $v^* \in B_{X^*}(0, \delta_2)$ and $x \in B_X(\bar{x}, \eta_2)$

(7.58) $(f - \langle \bar{x}^*, \cdot \rangle)(x) \geq (f - \langle \bar{x}^*, \cdot \rangle)(\zeta_2(v^*)) + \langle v^*, x - \zeta_2(v^*) \rangle + \dfrac{\kappa_2}{s} \|x - \zeta_2(v^*)\|^s.$

This ensures that $\zeta_2(0) = \bar{x}$ is a minimizer of $f - \langle \bar{x}^*, \cdot \rangle$ on $B_X(\bar{x}, \eta_2)$ and also that, for all $v^* \in B_{X^*}(0, \delta_2)$

$$s^{-1} \kappa_2 \|\bar{x} - \zeta_2(v^*)\|^s \leq \langle v^*, \zeta_2(v^*) - \bar{x} \rangle \leq \eta_2 \|v^*\|,$$

so $\zeta_2(v^*) \to \bar{x}$ as $v^* \to 0$. Putting $\eta := \min\{\eta_1, \eta_2\}$ we can choose a positive real $\delta < \min\{\delta_1, \delta_2\}$ such that $\zeta_2(B_{X^*}(0, \delta)) \subset B_X(\bar{x}, \eta)$. Then, for any $v^* \in B_{X^*}(0, \delta)$ we have $\zeta_2(v^*) \in (\partial f - \bar{x}^*)^{-1}(v^*) \cap B_X(\bar{x}, \eta)$ by (7.58), or equivalently $\zeta_2(u^* - \bar{x}^*) \in (\partial f)^{-1}(u^*) \cap B_X(\bar{x}, \eta)$ for all $u^* \in B_{X^*}(\bar{x}^*, \delta)$. This inclusion according to the equality in (7.57) entails

$$\{\zeta_2(u^* - \bar{x}^*)\} = \{\zeta_1(u^*)\} = (\partial f)^{-1}(u^*) \cap B_X(\bar{x}, \eta).$$

Defining $\zeta_0 : B_{X^*}(\bar{x}^*, \delta) \to B_X(\bar{x}, \eta)$ by $\zeta_0(u^*) := \zeta_2(u^* - \bar{x}^*)$ for all $u^* \in B_{X^*}(\bar{x}^*, \delta)$, we see that $\{\zeta_0(u^*)\} = (\partial f)^{-1}(u^*) \cap B_X(\bar{x}, \eta)$ for all $u^* \in B_{X^*}(\bar{x}^*, \delta)$. This and the inequality in (7.58) tell us that f is s-order prox-regular at $\bar{x}$ for $\bar{x}^*$ with constant $\sigma' := 0$. Consequently, taking the s-order strong metric regularity assumption into account, Proposition 7.155 guarantees that, for every $\kappa \in]0, \gamma^{-1/\theta}[$ the point $\bar{x}$ is a stable s-order local minimizer of $f - \langle \bar{x}^*, \cdot \rangle$ with constant κ. □

A partial converse of Proposition 7.158 with a constant γ relied to both κ and σ holds true.

PROPOSITION 7.159. *Let $f : X \to \mathbb{R} \cup \{+\infty\}$ be a proper lower semicontinuous function on a Banach space X and let $\theta \in]0, +\infty[$ and $s := 1 + 1/\theta$. Let $\bar{x} \in (\partial f)^{-1}(0)$. Assume that $\bar{x}$ is a stable s-order local minimizer of f with constant $\kappa > 0$. Assume also that, relative to ∂, the function f is subdifferentially continuous and s-order σ-prox-regular at $\bar{x}$ for $\bar{x}^*$ with $\sigma \in [0, \kappa[$. Then, for $\gamma := s^\theta(\kappa - \sigma)^{-\theta}$ the multimapping ∂f is strongly metrically regular at $\bar{x}$ for $\bar{x}^* = 0$.*

EXERCISE 7.160 (Proof of Proposition 7.159). Keep notation in the proposition. (I) Argue (through either Remark 7.141 or Proposition 7.144) that there are reals $\delta > 0, \eta > 0$ and a mapping $\zeta : B_{X^*}(0, \delta) \to B_X(\bar{x}, \eta)$ continuous at 0 with $\zeta(0) = \bar{x}$ and such that

$$f(x) \geq f(\zeta(u^*)) + \langle u^*, x - \zeta(u^*) \rangle + \dfrac{\kappa}{s} \|x - \zeta(u^*)\|^s, \ \forall u^* \in B_{X^*}(0, \delta), x \in B_X(\bar{x}, \eta),$$

and

$$\langle x^*, x' - x \rangle \leq f(x') - f(x) + \dfrac{\sigma}{s} \|x' - x\|^s$$

for all $x' \in B_X(\bar{x}, \eta)$ and $(x, x^*) \in \partial f$ with $x \in B(\bar{x}, \eta)$ and $x^* \in B_{X^*}(0, \eta)$. Note that the first inequality entails $\zeta(u^*) \in (\partial f)^{-1}(u^*)$ for all $u^* \in B_{X^*}(0, \delta)$.

(II) Let $r := \min\left\{\dfrac{\eta}{4}, \left(\dfrac{\eta s}{2(\kappa - \sigma)}\right)^\theta\right\}$ and let $0 < \delta_0 < \delta$ be such that $\zeta(B_{X^*}(0, \delta_0)) \subset$

$B_X(\bar{x}, r/2)$. Let $u^* \in B_{X^*}(0, \delta_0)$ and $x \in B_X(\bar{x}, r/2)$.
(a) When $d(u^*, \partial f(x)) \geq \eta/2$, argue that $\|x - \zeta(u^*)\| < r$, and hence
$$d(u^*, \partial f(x)) \geq \frac{\eta}{2} \geq \frac{\kappa - \sigma}{s} r^{1/\theta} \geq \frac{\kappa - \sigma}{s} \|x - \zeta(u^*)\|^{1/\theta}.$$
(b) Suppose that $d(u^*, \partial f(x)) < \eta/2$. For any $x^* \in \partial f(x) \cap B_{X^*}(0, \eta/2)$, argue that
$$\|x^* - u^*\| \|x - \zeta(u^*)\| \geq \langle x^*, x - \zeta(u^*) \rangle - \langle u^*, x - \zeta(u^*) \rangle \geq \frac{\kappa - \sigma}{s} \|x - \zeta(u^*)\|^s,$$
and then that
$$\|x^* - u^*\| \geq \frac{\kappa - \sigma}{s} \|x - \zeta(u^*)\|^{1/\theta};$$
derive that
$$d(u^*, \partial f(x)) \geq \frac{\kappa - \sigma}{s} \|x - \zeta(u^*)\|^{1/\theta}.$$
(c) Deduce that for any $u^* \in B_{X^*}(0, \delta_0)$ and any $x \in B_X(\bar{x}, r/2)$
$$\|x - \zeta(u^*)\| \leq \gamma \, d(u^*, \partial f(x)),$$
and conclude by Proposition 7.126(b) that the multimapping ∂f is s-order strongly metrically regular at $\bar{x}$ for 0 with constant γ. $\square$

The next corollary from Proposition 7.159 gives an equivalent modification of definition of stable s-order local minimizers.

COROLLARY 7.161. *Let $f : X \to \mathbb{R} \cup \{+\infty\}$ be a proper lower semicontinuous function on a Banach space X and let $\theta \in]0, +\infty[$ and $s := 1 + 1/\theta$. Let $\bar{x} \in (\partial f)^{-1}(0)$ be a point at which f is ∂-subdifferentially continuous. The following assertions are equivalent:*
(a) *the point $\bar{x}$ is a stable s-order local minimizer of f with constant $\kappa > 0$ and, relative to ∂, the function f is s-order σ-prox-regular at $\bar{x}$ for 0 with a certain real $\sigma \in [0, \kappa[$;*
(b) *there exist reals $\delta, \eta > 0$ and a mapping $\zeta : B_{X^*}(0, \delta) \to B_X(\bar{x}, \eta)$ such that, for all $x \in B_X(\bar{x}, \eta)$ and $u^* \in B_{X^*}(0, \delta)$ one has $(\partial f)^{-1}(u^*) \cap B_X(\bar{x}, \eta) = \{\zeta(u^*)\}$ and*
$$f(x) \geq f(\zeta(u^*)) + \langle u^*, x - \zeta(u^*) \rangle + \frac{\kappa}{s} \|x - \zeta(u^*)\|^s.$$

PROOF. The equality of sets in (b) tells us in particular, with $u^* = 0$, that $\zeta(0) = \bar{x}$. Then, the assertion (b) obviously entails that $\bar{x}$ is a stable s-order local minimizer of f with constant κ (according to Definition 7.140(b)) as well as it implies that f is s-order prox-regular at $\bar{x}$ for 0 with constant $\sigma = 0$. The implication (b)$\Rightarrow$(a) then holds true.

Now suppose that the properties in (a) are satisfied, that is, f is σ-prox-regular and there are $\delta, \eta > 0$ and a mapping $\zeta : B_{X^*}(0, \delta) \to B_X(\bar{x}, \eta)$ with $\zeta(0) = \bar{x}$ such that, for all $x \in B_X(\bar{x}, \eta)$ and $u^* \in B_{X^*}(0, \delta)$
$$f(x) \geq f(\zeta(u^*)) + \langle u^*, x - \zeta(u^*) \rangle + \frac{\kappa}{s} \|x - \zeta(u^*)\|^s,$$
which gives in particular $u^* \in \partial f(\zeta(u^*))$. By the latter inequality and by Proposition 7.159 the multimapping ∂f is strongly metrically regular at $\bar{x}$ for 0, hence in particular there are reals $\delta', \eta' > 0$ such that $(\partial f)^{-1}(u^*) \cap B_X(\bar{x}, \eta')$ is a singleton for all $u^* \in B_{X^*}(0, \delta')$. From the equality $\zeta(0) = \bar{x}$ and the continuity of ζ at 0 (see Remark 7.141) we may choose some positive real $\delta_0 < \min\{\delta, \delta'\}$

such that $\zeta(B_{X^*}(0, \delta_0)) \subset B_X(\bar{x}, \eta_0)$, where $\eta_0 := \min\{\eta, \eta'\}$. Considering $\zeta_0 : B_{X^*}(0, \delta_0) \to B_X(\bar{x}, \eta_0)$ with $\zeta_0(x^*) = \zeta(x^*)$ for all $x^* \in B_{X^*}(0, \delta_0)$, we see that for all $u^* \in B_{X^*}(0, \delta_0)$ one has $\partial f^{-1}(u^*) \cap B_X(\bar{x}, \eta_0) = \{\zeta_0(u^*)\}$ along with

$$f(x) \geq f(\zeta_0(u^*)) + \langle u^*, x - \zeta_0(u^*)\rangle + \frac{\kappa}{s}\|x - \zeta_0(u^*)\|^s, \quad \forall x \in B_X(\bar{x}, \eta_0).$$

This justifies the implication (a)⇒(b). □

Our next step is to investigate the links with tilt-stable local minimizers. This type of minimizers is related to the property in Proposition 7.144 enjoyed by stable s-order local minimizers.

DEFINITION 7.162. Let $f : X \to \mathbb{R} \cup \{+\infty\}$ be a proper extended real-valued function on a normed space X and let $s \in\,]1, +\infty[$. An element $\bar{x} \in \operatorname{dom} f$ is called a *tilt-stable s-order local minimizer of f with constant* $\gamma \in [0, +\infty[$ if there are $\delta, \eta > 0$ and a mapping $\zeta : B_{X^*}(0, \delta) \to B_X[\bar{x}, \eta]$ with $\zeta(0) = \bar{x}$ such that

$$\zeta(u^*) \in \underset{B_X[\bar{x},\eta]}{\operatorname{Argmin}}(f - u^*) \quad \text{for all } u^* \in B_{X^*}(0, \delta),$$

and

$$\|\zeta(u^*) - \zeta(v^*)\| \leq \gamma \|u^* - v^*\|^{1/(s-1)} \quad \text{for all } u^*, v^* \in B_{X^*}(0, \delta).$$

When $s = 2$, so ζ is Lipschitz on $B_{X^*}(0, \delta)$, it is usual to say that $\bar{x}$ is a *tilt-stable local minimizer* of the function f.

First, we will study relationships between the tilt-stability of f at $\bar{x}$ and the Legendre-Fenchel conjugate of $f + \Psi_{B[\bar{x},\eta]}$, where η is as in the above definition. Given an extended real-valued function $\varphi : X \to \mathbb{R} \cup \{-\infty, +\infty\}$ on a normed space X, as already recalled in (4.36), its Legendre-Fenchel conjugate $\varphi^* : X^* \to \mathbb{R} \cup \{-\infty, +\infty\}$ is defined by

$$\varphi^*(x^*) := \sup_{x \in X}\left(\langle x^*, x\rangle - \varphi(x)\right), \quad \text{for all } x^* \in X^*,$$

and it is clearly a convex w^*-lower semicontinuous function. For any $x \in X$ and $x^* \in X^*$ we recall that the evident so-called Fenchel inequality

$$\langle x^*, x\rangle - \varphi^*(x^*) \leq \varphi(x)$$

holds, hence for the biconjugate $\varphi^{**} := (\varphi^*)^*$ we have

(7.59) $$\varphi^{**}(x) \leq \varphi(x) \quad \text{for all } x \in X.$$

Let $u \in X$ and $u^* \in X^*$ be such that $\langle u^*, u\rangle - \varphi(u) = \varphi^*(u^*)$. Then

$$\varphi^{**}(u) \geq \langle u^*, u\rangle - \varphi^*(u^*) = \varphi(u),$$

so by (7.59) we obtain in this case

$$\varphi^{**}(u) = \langle u^*, u\rangle - \varphi^*(u^*) = \varphi(u).$$

From the left equality and the definition of $(\varphi^*)^*(x)$ we derive

$$\varphi^{**}(u) - \langle u^*, u\rangle = -\varphi^*(u^*) \leq \varphi^{**}(x) - \langle u^*, x\rangle, \quad \forall x \in X.$$

If in addition $\varphi(u)$ is finite, then $u \in \partial \varphi^*(u^*)$ since for all $x^* \in X^*$

$$\langle u^*, u\rangle - \varphi^*(u^*) = \varphi(u) \geq \langle x^*, u\rangle - \varphi^*(x^*),$$

by definition of φ^*. Those properties justify in particular the following lemma.

LEMMA 7.163. Let $\varphi : X \to \mathbb{R}\cup\{+\infty\}$ be a proper function on a normed space X. Then, for any $u^* \in X^*$ one has
$$\operatorname*{Argmin}_X(\varphi - u^*) \subset \operatorname*{Argmin}_X(\varphi^{**} - \langle \cdot, u^* \rangle),$$
and for any $u \in \operatorname*{Argmin}_X(\varphi - u^*)$
$$\varphi(u) = \varphi^{**}(u) = \langle u^*, u \rangle - \varphi^*(u^*) \quad \text{and} \quad u \in \partial \varphi^*(u^*).$$

When φ is of class $\mathcal{C}^{1,\theta}$ on an open set, the Fenchel inequality can be improved as follows.

LEMMA 7.164. Let $\varphi : X \to \mathbb{R}\cup\{+\infty\}$ be a proper function on a normed space X. Assume that φ is differentiable on an open set $U \subset X$ and that there exist reals $\theta > 0$ and $\gamma \geq 0$ such that
$$\|D\varphi(u) - D\varphi(v)\| \leq \gamma \|u - v\|^\theta \quad \text{for all } u, v \in U.$$
Let $\bar{u} \in U$ and $\delta > 0$ be such that
$$B(\bar{u}, (1 + 2^{1/\theta})\delta) \subset U.$$
Then, for all $u \in B_X(\bar{u}, \delta)$ and $u^* \in B_{X^*}(D\varphi(\bar{u}), \delta^\theta \gamma)$ one has
$$\varphi^*(u^*) \geq \langle u^*, u \rangle - \varphi(u) + \frac{\theta}{(1+\theta)\gamma^{1/\theta}} \|u^* - D\varphi(u)\|^{1+1/\theta}.$$

PROOF. Put $\eta := (1 + 2^{1/\theta})\delta$ and fix any $u \in B(\bar{u}, \delta)$ and $u^* \in B_{X^*}(D\varphi(\bar{u}), \delta^\theta \gamma)$. Observing that, for any $v \in B(\bar{u}, \eta)$
$$\varphi(v) - \varphi(u) - \langle D\varphi(u), v - u \rangle = \int_0^1 \langle D\varphi(u + t(v - u)) - D\varphi(u), v - u \rangle\, dt$$
$$\leq \gamma \|v - u\|^{1+\theta} \int_0^1 t^\theta\, dt = \frac{\gamma}{1 + \theta} \|v - u\|^{1+\theta},$$
we derive that
$$\varphi^*(u^*) \geq \sup_{v \in B(\bar{u}, \eta)} \left(\langle u^*, v \rangle - \varphi(v) \right)$$
$$\geq \sup_{v \in B(\bar{u}, \eta)} \left(\langle u^*, v \rangle - \varphi(u) - \langle D\varphi(u), v - u \rangle - \frac{\gamma}{1 + \theta} \|v - u\|^{1+\theta} \right)$$
(7.60) $\quad = \langle u^*, u \rangle - \varphi(u) + \sup_{v \in B(\bar{u}, \eta)} \left(\langle u^* - D\varphi(u), v - u \rangle - \frac{\gamma}{1 + \theta} \|v - u\|^{1+\theta} \right).$

Consider now any $b \in \mathbb{B}_X$ and set $v_b := u + \gamma^{-1/\theta} \|u^* - D\varphi(u)\|^{1/\theta} b$. Note that
$$\|v_b - \bar{u}\| \leq \|u - \bar{u}\| + \gamma^{-1/\theta} \|u^* - D\varphi(u)\|^{1/\theta}$$
$$< \delta + \gamma^{-1/\theta} \left(\|u^* - D\varphi(\bar{u})\| + \|D\varphi(\bar{u}) - D\varphi(u)\| \right)^{1/\theta}$$
$$\leq \delta + \gamma^{-1/\theta} \left(\gamma \delta^\theta + \gamma \|\bar{u} - u\|^\theta \right)^{1/\theta},$$
so $\|v_b - \bar{u}\| < \delta(1 + 2^{1/\theta}) = \eta$. Setting
$$A := \sup_{v \in B(\bar{u}, \eta)} \left(\langle u^* - D\varphi(u), v - u \rangle - \frac{\gamma}{1 + \theta} \|v - u\|^{1+\theta} \right),$$

we obtain

$$A \geq \langle u^* - D\varphi(u), v_b - u \rangle - \frac{\gamma}{1+\theta}\|v_b - u\|^{1+\theta}$$

$$= \gamma^{-1/\theta}\|u^* - D\varphi(u)\|^{1/\theta}\langle u^* - D\varphi(u), b \rangle - \frac{\gamma^{-1/\theta}}{1+\theta}\|u^* - D\varphi(u)\|^{1+1/\theta}\|b\|^{1+\theta}$$

$$\geq \gamma^{-1/\theta}\|u^* - D\varphi(u)\|^{1/\theta}\langle u^* - D\varphi(u), b \rangle - \frac{\gamma^{-1/\theta}}{1+\theta}\|u^* - D\varphi(u)\|^{1+1/\theta},$$

which yields by taking the supremum over $b \in \mathbb{B}_X$

$$A \geq \gamma^{-1/\theta}\|u^* - D\varphi(u)\|^{1+1/\theta} - \frac{\gamma^{-1/\theta}}{1+\theta}\|u^* - D\varphi(u)\|^{1+1/\theta}$$

$$= \frac{\theta\gamma^{-1/\theta}}{1+\theta}\|u^* - D\varphi(u)\|^{1+1/\theta}.$$

This combined with (7.60) justifies the desired inequality of the lemma. □

LEMMA 7.165. *Let $f : X \to \mathbb{R} \cup \{+\infty\}$ be a proper function on a normed space X and let $\bar{x} \in \mathrm{dom}\, f$ and $\theta \in {]}0, +\infty[$. Assume that there exist $\delta, \eta > 0$ and a continuous mapping $\zeta : B_{X^*}(0, \delta) \to B_X[\bar{x}, \eta]$ with $\zeta(0) = \bar{x}$ such that*

$$\zeta(u^*) \in \underset{B_X[\bar{x},\eta]}{\mathrm{Argmin}}(f - u^*), \quad \text{for all } u^* \in B_{X^*}(0, \delta).$$

Then $(f + \Psi_{B_X[\bar{x},\eta]})^*$ is of class $\mathcal{C}^1$ on $B_{X^*}(0, \delta)$ and for all $u^* \in B_{X^*}(0, \delta)$

$$\underset{B_X[\bar{x},\eta]}{\mathrm{Argmin}}(f - u^*) = \{\zeta(u^*)\} \quad \text{and} \quad D\big((f + \Psi_{B_X[\bar{x},\eta]})^*\big)(u^*) = \zeta(u^*).$$

PROOF. Fix any $u^* \in B_{X^*}(0, \delta)$ and set $f_\eta := f + \Psi_{B_X[\bar{x},\eta]}$. For any $u \in \mathrm{Argmin}_{B[\bar{x},\eta]}(f - u^*) = \mathrm{Argmin}_X(f_\eta - u^*)$ we have $u \in \partial f_\eta^*(u^*)$ by Lemma 7.163. In particular $\zeta(u^*) \in \partial f_\eta^*(u^*)$, so ζ is a continuous selection of ∂f_η^* on $B_{X^*}(0, \delta)$. The convex function f_η^* is then of class $\mathcal{C}^1$ on $B_{X^*}(0, \delta)$ according to Corollary 3.110. All together ensure that, for all $u^* \in B_{X^*}(0, \delta)$

$$\underset{B_X[\bar{x},\eta]}{\mathrm{Argmin}}(f - u^*) = \{\zeta(u^*)\} \quad \text{and} \quad D(f_\eta^*)(u^*) = \zeta(u^*),$$

which are the desired equalities. □

LEMMA 7.166. *Let $f : X \to \mathbb{R} \cup \{+\infty\}$ be a proper function on a normed space X and let $\bar{x} \in \mathrm{dom}\, f$. Let also $\delta, \eta > 0$, a real $\gamma \geq 0$ and a continuous mapping $\zeta : B_{X^*}(0, \delta) \to B_X[\bar{x}, \eta]$ with $\zeta(0) = \bar{x}$. The following assertions are equivalent:*
(a) *for all $u^* \in B_{X^*}(0, \delta)$*

$$\underset{B_X[\bar{x},\eta]}{\mathrm{Argmin}}(f - u^*) = \{\zeta(u^*)\};$$

(b) *for all $u^* \in B_{X^*}(0, \delta)$*

$$\underset{X}{\mathrm{Argmin}}\big((f + \Psi_{B_X[\bar{x},\eta]})^{**} - \langle \cdot, u^* \rangle\big) = \{\zeta(u^*)\}$$

and

$$(f + \Psi_{B_X[\bar{x},\eta]})^{**}(x) = f(x) \quad \text{for all } x \in X \cap \partial(f + \Psi_{B_X[\bar{x},\eta]})^*\big(B_{X^*}(0, \delta)\big).$$

PROOF. Put $f_\eta := f + \Psi_{B_X[\bar{x},\eta]}$.
(a)$\Rightarrow$(b). Suppose that (a) holds and fix any $u^* \in B_{X^*}(0,\delta)$. By Lemma 7.163 we have $f(\zeta(u^*)) = f_\eta^{**}(\zeta(u^*))$. Further, Lemma 7.165 tells us that $\partial f_\eta^*(u^*) = \{\zeta(u^*)\}$, that is, $x \in \partial f_\eta^*(u^*)$ means $x = \zeta(u^*)$. This combined with what precedes yields $f_\eta^{**}(x) = f(x)$ for all $x \in X \cap \partial f_\eta^*(B_{X^*}(0,\delta))$, which is the second equality in (b). Concerning the first equality in (b), since

$$\operatorname*{Argmin}_X(f_\eta^{**} - \langle \cdot, u^* \rangle) = X \cap \partial f_\eta^*(u^*)$$

(as easily checked), we see that $\operatorname*{Argmin}_X(f_\eta^{**} - \langle \cdot, u^* \rangle) = \{\zeta(u^*)\}$ according to the above equality $\partial f_\eta^*(u^*) = \{\zeta(u^*)\}$.
(b)$\Rightarrow$(a). Suppose (b) and fix any $u^* \in B_{X^*}(0,\delta)$. By the first equality in (b) and by Lemma 7.163 we have

$$\operatorname*{Argmin}_{B[\bar{x},\eta]}(f - \langle \cdot, u^* \rangle) = \operatorname*{Argmin}_X(f_\eta - u^*) \subset \operatorname*{Argmin}_X(f_\eta^{**} - u^*) = \{\zeta(u^*)\}.$$

On the other hand, the first equality in (b) means $u^* \in \partial f_\eta^{**}(\zeta(u^*))$, which by the second equality in (b) entails $f(\zeta(u^*)) = f_\eta^{**}(\zeta(u^*))$. We deduce that, for every $x \in B_X[\bar{x},\eta]$ we have according to (7.59) and to the first equality in (b)

$$f(x) - \langle u^*, x \rangle = f_\eta(x) - \langle u^*, x \rangle \geq f_\eta^{**}(x) - \langle u^*, x \rangle$$
$$\geq f_\eta^{**}(\zeta(u^*)) - \langle u^*, \zeta(u^*) \rangle = f(\zeta(u^*)) - \langle u^*, \zeta(u^*) \rangle,$$

hence $\zeta(u^*) \in \operatorname*{Argmin}_{B_X[\bar{x},\eta]}(f - u^*)$. The implication (b)$\Rightarrow$(a) is then justified. $\square$

We can now prove that, in some sense, the concepts of stable s-order local minimizers and tilt-stable s-order local minimizers are equivalent.

PROPOSITION 7.167. *Let $f : X \to \mathbb{R} \cup \{+\infty\}$ be a proper function on a normed space X and let $\bar{x} \in \operatorname{dom} f$ and $\theta \in]0, +\infty[$. Let also $s := 1 + 1/\theta$. The following assertions hold.*
(a) *If $\bar{x}$ is a stable s-order local minimizer of f with constant $\kappa > 0$, then $\bar{x}$ is a tilt-stable s-order local minimizer of f with constant $(2\kappa)^{-\theta}$.*
(b) *If $\bar{x}$ is a tilt-stable s-order local minimizer of f with constant $\gamma > 0$, then $\bar{x}$ is a stable s-order local minimizer of f with constant $\gamma^{-1/\theta}$.*

PROOF. (a) Suppose that $\bar{x}$ is a stable s-order local minimizer of f with constant $\kappa > 0$. By Definition 7.140(b), by Remark 7.141(b) and by Proposition 7.144 there exist reals $\delta > 0$, $\eta > 0$ and a mapping $\zeta : B_{X^*}(0,\delta) \to B_X(\bar{x},\eta)$ with $\zeta(0) = \bar{x}$ such that for all $u^*, v^* \in B_{X^*}(0,\delta)$

$$\operatorname*{Argmin}_{B_X(\bar{x},\eta)}(f - u^*) = \{\zeta(u^*)\} \text{ and } \|\zeta(v^*) - \zeta(u^*)\| \leq (2\kappa)^{-\theta} \|v^* - u^*\|^{1/(s-1)}.$$

Fix $0 < \eta_0 < \eta$ and choose $0 < \delta_0 < \delta$ such that $\zeta(B_{X^*}(0,\delta_0)) \subset B_X[\bar{x},\eta_0]$ according to the continuity of ζ. Define $\zeta_0 : B_{X^*}(0,\delta_0) \to B_X[\bar{x},\eta_0]$ by $\zeta_0(x^*) = \zeta(x^*)$ for all $x^* \in B_{X^*}(0,\delta_0)$. Then for any $u^*, v^* \in B_{X^*}(0,\delta_0)$ we see that

$$\operatorname*{Argmin}_{B_X[\bar{x},\eta_0]}(f - u^*) = \{\zeta_0(u^*)\} \text{ and } \|\zeta_0(v^*) - \zeta_0(u^*)\| \leq (2\kappa)^{-\theta} \|v^* - u^*\|^{1/(s-1)},$$

where the equality is due to the equality $\operatorname*{Argmin}_{B_X(\bar{x},\eta)}(f - u^*) = \{\zeta(u^*)\}$ and to the fact

$$\zeta(u^*) = \zeta_0(u^*) \in B_X[\bar{x},\eta_0] \subset B_X(\bar{x},\eta).$$

The point $\bar{x}$ is then a tilt-stable s-order local minimizer of f with constant $(2\kappa)^{-\theta}$, that is, the statement (a) is justified.

(b) Concerning (b), let $\delta, \eta > 0$ and $\zeta : B_{X^*}(0, \delta) \to B_X[\bar{x}, \eta]$ be given by Definition 7.162. Put $f_\eta := f + \Psi_{B_X[\bar{x},\eta]}$. By Lemma 7.165 the function f_η^* is of class $\mathcal{C}^{1,\theta}$ on $B_{X^*}(0, \delta)$ with

$$\|Df_\eta^*(u^*) - Df_\eta^*(v^*)\| \leq \gamma \|u^* - v^*\|^\theta, \quad \forall u^*, v^* \in B_{X^*}(0, \delta).$$

Applying Lemma 7.164 with the function f_η^* we obtain $\delta_0, \eta_0 > 0$ such that for any fixed $u^* \in B(0, \delta_0)$ and $x \in B(\bar{x}, \eta_0)$

$$f_\eta^{**}(x) \geq \langle u^*, x \rangle - f_\eta^*(u^*) + \frac{\gamma^{-1/\theta}\theta}{1+\theta}\|x - \zeta(u^*)\|^{1+1/\theta}.$$

By (7.59) and by the equality $s = 1 + 1/\theta$ we see that

$$f(x) - \langle u^*, x \rangle \geq -f_\eta^*(u^*) + \frac{\gamma^{-1/\theta}}{s}\|x - \zeta(u^*)\|^s.$$

Since $f_\eta^*(u^*) = \langle u^*, \zeta(u^*) \rangle - f(\zeta(u^*))$ by the second equality of (b) in Lemma 7.166, it ensues that

$$f(x) - \langle u^*, x \rangle \geq f(\zeta(u^*)) - \langle u^*, \zeta(u^*) \rangle + \frac{\gamma^{-1/\theta}}{s}\|x - \zeta(u^*)\|^s,$$

which translates that $\bar{x}$ is a stable s-order local minimizer of f with constant $\gamma^{-1/\theta}$. $\square$

COROLLARY 7.168. *Let $f : X \to \mathbb{R} \cup \{+\infty\}$ be a proper function on a normed space X and let $\bar{x} \in \text{dom } f$ and $\theta \in \,]0, 1]$. Let also $s := 1 + 1/\theta$. The following assertions are equivalent:*
(a) *the point $\bar{x}$ is a stable s-order local minimizer of f;*
(b) *the point $\bar{x}$ is a tilt-stable s-order local minimizer of f;*
(c) *there are reals $\delta, \eta > 0$ such that, with $f_\eta := f + \Psi_{B_X[\bar{x},\eta]}$ the point $\bar{x}$ is a tilt-stable local minimizer of the convex function $f_\eta^{**}|_X$ (the restriction of f_η^{**} to X) and*

$$f_\eta^{**}(x) = f(x) \quad \text{for all } x \in X \cap \partial f_\eta^*(B_{X^*}(0, \delta)).$$

PROOF. The equivalence between (a) and (b) is obtained from Proposition 7.167 and the equivalence between (b) and (c) follows from Lemma 7.166. $\square$

THEOREM 7.169. *Let $f : X \to \mathbb{R} \cup \{+\infty\}$ be a proper lower semicontinuous function on a Banach space X which is ∂-subdifferentially continuous at $\bar{x} \in (\partial f)^{-1}(0)$. Let $\theta \in \,]0, +\infty[$ and $s := 1 + 1/\theta$. The following assertions are equivalent:*
(a) *the point $\bar{x}$ is a stable s-order local minimizer of f with some constant $\gamma > 0$ and f is s-order prox-regular at $\bar{x}$ for 0 with some constant $\sigma \in [0, \gamma^{-1/\theta}[$;*
(b) *the point $\bar{x}$ is a tilt-stable s-order local minimizer of f with some constant $\kappa > 0$ and f is s-order prox-regular at $\bar{x}$ with some constant σ;*
(c) *there are reals $\delta, \eta > 0$ such that, with $f_\eta := f + \Psi_{B_X[\bar{x},\eta]}$ the point $\bar{x}$ is a tilt-stable local minimizer of the convex function $f_\eta^{**}|_X$ (the restriction of f_η^{**} to X) and*

$$f_\eta^{**}(x) = f(x) \quad \text{for all } x \in X \cap \partial f_\eta^*(B_{X^*}(0, \delta)).$$

7.10. Comments

As said in the detailed paper [**521**] of A. D. Ioffe, the importance of metric inequalities in the line of (7.4) already appeared in the 1934 paper [**682**] by L. A. Lyusternik where he proved the famous theorem saying that the tangent manifold at a point $\bar{x}$ of the level set $f^{-1}(\bar{y})$ of a continuously differentiable mapping $f : X \to Y$ between two Banach spaces X and Y coincides with $\operatorname{Ker} f'(\bar{x})$ whenever the derivative $f'(\bar{x})$ is surjective. In the proof of that result Lyusternik established, for some real constant $\gamma \geq 0$, the inequality $d(x, f^{-1}(\bar{y})) \leq \gamma d(\bar{y}, f(x))$ for all $x \in \bar{x} + \operatorname{Ker} f'(\bar{x})$ close enough to $\bar{x}$. Through that result, Lyusternik proved in the same paper a Lagrange multiplier rule for the minimization of a continuously differentiable function $\varphi : X \to \mathbb{R}$ subject to the equality constraint $f(x) = 0$.

Almost fifteen years later the above paper of Lyusternik, two other fundamental results appeared in the celebrated 1950 paper [**448**] of L. M. Graves. The first result there can be stated, for a $\mathcal{C}^1$-mapping f between two Banach spaces X and Y whose derivative $f'(\bar{x})$ is surjective, as saying that there are real constants $c, r > 0$ such that $B[f(\bar{x}), crt] \subset f(B[\bar{x}, rt])$ for all $t \in [0, 1]$. Compared with Definition 7.13, the latter is a partial openness property with linear rate. Nevertheless, the second result fills a part of the gap, since it can be formulated as follows: If $f : X \to Y$ is a continuous mapping for which there exists a continuous linear mapping $A : X \to Y$ and two real constants $0 < \eta < \beta$ such that $\beta^{-1} \mathbb{B}_Y \subset A(\mathbb{B}_X)$ and

$$\|f(x) - f(u) - A(x - u)\| \leq \eta \|x - u\|,$$

for all x, u near a point $\bar{x}$, then the equation $f(x) = y$ (with unknown x) has a solution satisfying $\|x - \bar{x}\| \leq t$ whenever $\|y - f(\bar{x})\| < t(\beta - \eta)$ for all $t > 0$ sufficiently small. This entails that $d(\bar{x}, f^{-1}(y)) \leq (\beta - \eta) d(y, f(\bar{x}))$ for all y close enough to $f(\bar{x})$. The latter inequality, which has been first observed by A. D. Ioffe and V. M. Tikhomirov in their 1974 book [**529**], is different from the Lyusternik inequality above in the sense that $\bar{x}$ is fixed and y may vary near $f(\bar{x})$, while in the former $\bar{y} = f(\bar{x})$ is fixed and x may vary in $\bar{x} + \operatorname{Ker} f'(\bar{x})$ near $\bar{x}$. So, L. A. Lyusternik and L. M. Graves are the pioneers of *metric regularity*. Since the metric regularity is equivalent to the openness with linear rate, the names of S. Banach and J. Schauder have to be associated as well, because of their respective open mapping theorem for linear mappings in [**70**] and [**882**], generally called the *Banach-Schauder open mapping theorem*. In addition, both Lyusternik and Graves aforementioned papers used in their arguments the Banach-Schauder theorem as well as iterative constructions similar to those of Banach [**70**]. Two other names have to be added among the first people who recognized the interest of estimates in the line of metric regularity: A. J. Hoffman and S. M. Robinson because of their celebrated respective papers on the estimates for the distance to sets of solutions to generalized equations (that is, not reduced to the equality-level set of a mapping). Such a study probably began in 1952 with A. J. Hoffman [**499**] who established, in finite dimensions, estimates for the distance to the solution set of finitely many equality-inequality constraints. Several years later, the 1975 paper [**943**] of C. Ursescu and the 1976 paper [**839**] of S. M. Robinson generalized the Banach-Schauder theorem as well as the Hoffmann inequality to multimappings with closed convex graphs between Banach spaces.

Another crucial step was the 1976 paper [**840**] of S. M. Robinson which is probably the first paper in the nonlinear setting where the inequality (7.4) appears in its full statement. The seminal 1980 paper [**340**] by A. V. Dmitruk, A. A. Milyutin and N. P. Osmolovskii offered another study where the inequality (7.4) is examined in the nonconvex context. Note also that certain forms of estimates are involved in Pták papers [**823, 824**]. H. Federer noticed in his 1959 paper [**402**] the interest of metric regularity/subregularity in the calculus of tangent cones in showing in Theorem 4.10(3) there that

$$T^B(S_1 \cap S_2; \bar{x}) = T^B(S_1; \bar{x}) \cap T^B(S_2; \bar{x})$$

whenever S_1, S_2 are subsets in $\mathbb{R}^N$ for which there is a real constant $\kappa > 0$ such that

(7.61) $$d(x, S_1 \cap S_2) \leq \kappa\big(d(x, S_1) + d(x, S_2)\big)$$

for all x near $\bar{x} \in S_1 \cap S_2$; in [**402**] the Bouligand-Peano tangent cone $T^B(\cdot;\cdot)$ is denoted by $\mathrm{Tan}(\cdot,\cdot)$. In addition to this, by means of Peano's existence theorem for differential equations, Federer [**402**] proved in Theorem 4.10(2) that (7.61) holds true under the condition that the sets S_1, S_2 in $\mathbb{R}^N$ are prox-regular near $\bar{x}$ in the sense of Chapter 15 (positively reached in Federer's terminology) and

$$N^F(S_1; x) \cap -N^F(S_2; x) = \{0\} \quad \text{for all } x \in S_1 \cap S_2 \text{ near } \bar{x}.$$

The inequality of subregular type

(7.62) $$d(x, C \cap F^{-1}(D)) \leq \kappa\big(d(x, C) + d(F(x), D)\big) \quad \text{for } x \text{ near } \bar{x} \in C \cap F^{-1}(D)$$

was mainly used for establishing optimality conditions with the constraint system $C \cap F^{-1}(D)$ by A. D. Ioffe [**517, 521**], A. Jourani and L. Thibault [**587**], B. S. Mordukhovich [**723, 725**], R. T. Rockafellar [**860**], R. T. Rockafellar and R. J-B. Wets [**865**] etc. Under additive perturbation $f+F$, the preservation of either (7.62), or its variants, or its metric regularity stronger form (or the openness form with a linear rate) is largely studied by A. L. Dontchev, A. S. Lewis and R. T. Rockafellar [**351**], by A. L. Dontchev and R. T. Rockafellar [**353, 354**], by A. L. Dontchev and Hager. The Aubin-Lipschitz property in Definition 7.1 for multimappings has been introduced by J.-P. Aubin in his 1984 paper [**41**]. Many applications illustrating the usefulness of this Lipschitz type property were provided by Aubin in the same paper [**41**].

In addition to the aforementioned articles, diverse other contributions to metric regularity and Lipschitz-like properties of mappings and multimappings were offered in [**346, 347, 348, 349, 350, 352, 793, 861**], etc. In the literature there are also many works on conditions for metric regularity of mappings and multimappings by means of the concept of slope; see, for example, the articles [**56**] of D. Azé, [**57**] and the book [**526**] of A. D. Ioffe and the reference therein. B. Kummer in his papers [**638, 639**], and D. Klatte and B. Kummer in their book [**607**] and their paper [**608**] provided also useful conditions for Aubin-Lipschitz conditions for mappings and multimappings in terms of certain concepts of multiderivates.

Links between metric regularity/subregularity and the concepts *Conic Hull Intersection Property* and *Linear Regularity* can be found in A. Bakan, F. Deutsch and W. Li [**62**], H. H. Baushke and J. M. Borwein [**75**], H. H. Bauschke, J. M. Borwein and W. Li [**76**], H. H. Bauschke, J. M. Borwein and P. Tseng [**77**], K. F. Ng and W. H. Yang [**766**], X. Y. Zheng and K. F. Ng [**1004**], etc. The use of metric regularity in semi-infinite linear optimization problems can be found for example in the papers [**203**] by M. J. Cánovas, D. Klatte, M. A. Lopez and J. Parra and

[**204**] by M. J. Cánovas, M. A. Lopez, J. Parra and F. J. Toledo, and references in these papers.

The 1979 paper [**508**] of A. D. Ioffe was probably the first paper with the use of subdifferentials in the study of metric estimates.

Theorem 7.17 with $Q = Y$ was observed by J. M. Borwein and D. M. Zhuang [**153**], and by J.-P. Penot [**788**]. Ideas in that line previously appeared in the paper [**340**] of A. V. Dmitruk, A. A. Milyutin, and N. P. Osmolovskii. The proof in the text follows that given in R. T. Rockafellar and R. J-B. Wets [**865**] for $Q = Y$, and Proposition 7.5 with $P = X$ is also due to R. T. Rockafellar and R. J-B. Wets [**865**]. The concept of graphical metric regularity has been introduced by A. Jourani and L. Thibault [**581**] and its equivalence in Proposition 7.20 with the metric regularity has been observed and proved by L. Thibault [**919, 920**]. Proposition 7.21 was probably first noticed by R. Henrion in his Habilitation thesis [**476**]. Proposition 7.22 is essentially due to A. D. Ioffe [**521**].

Theorem 7.31 was proved by H. Frankowska in her 1990 paper [**418**]; the proof of the theorem and of its preparatory Lemma 7.30 follows the approach in [**418, 44**]. Another proof of Theorem 7.31 can be found in the 2020 paper [**549**] by M. Ivanov and N. Zlateva. Extended results in the line of Theorem 7.31 for Hölder metric regularity were provided by H. Frankowska and M. Quincampoix in their 2012 paper [**419**]. Theorem 7.32 is due to J.-P. Aubin; it was essentially contained in [**40**] (see also [**43**]). The statemennt and proof in the manuscript of Theorem 7.32 are taken from Aubin and Frankowska's book [**44**].

Theorem 7.33 is due to B. S. Mordukhovich and Y. Shao [**729**], and its proof is obtained in the text as a direct consequence of Theorem 7.29. Our statement and proof of Theorem 7.29 follow Theorem 6 in A. Y. Kruger [**633**]. Theorem 7.35 is due to B. S. Mordukhovich [**722, 723**] when X and Y are finite-dimensional. As stated with the coderivative compactness of M, Theorem 7.38 has a long story. The partially compactly epi-Lipschitz property of the multimapping M (as partially related to the compact epi-Lipschitzness of sets previously introduced and studied by J. M. Borwein and H. Strojwas [**142**]) has been introduced by A. Jourani and L. Thibault [**586, 588**] in order to obtain point-based sufficient conditions for the metric regularity of M; those papers [**586, 588**] were the natural continuation of [**581**]. The coderivative compactness property as stated in Definition 7.36 appeared in several papers as [**732, 588, 793**]; the cousin "*sequential normal compactness*" of that concept has been deeply studied and largely used in a series of papers by B. S. Mordukhovich and Y. Shao [**730, 731, 732**], B. S. Mordukhovich and N. M. Nam [**727**], and a comprehensive treatment and many comments are provided in Mordukhovich's book [**725**]. Connexions with the local compactness properties, as deeply initiated by P. D Loewen [**670**], can be found in details in [**588**]. Previously to all those papers, A. D. Ioffe probably was the first to provide point-based sufficient conditions under (what he called) the "*finite codimension property*" in a series of papers [**513, 516**], and other conditions of the same type were given by B. Ginsburg and A. D. Ioffe [**440**]. To complete the story, we shall say that it has been proved by A. Jourani and L. Thibault [**588**] that the partial compactness epi-Lipschitz property entails the full coderivative compactness (with nets in Definition 7.36), and a few years later A. D. Ioffe has shown the converse in the paper [**520**]; in the same paper A. D. Ioffe also proved that the full coderivative compactness is even equivalent to some form of the finite codimension property. As stated Lemma appeared in A. Jourani [**569**] and its statement and proof was inspired by similar similar statements and arguments from A. D Ioffe [**508**] and J. M. Borwein [**122**].

Lemma 7.47 and Theorem 7.48 as well as their proofs are due to A. Y. Kruger [**629**] where the expression of "*weak stationarity*" at $\bar{x}$ is used in place of "*metrically regular position*" at $\bar{x}$. The concept of "*weak stationarity*" is connected with the concept of "*extremality for sets*" introduced earlier by A. Y. Kruger and B. S. Mordukhovich [**634**]; various other notions of stationarity and extremality for systems of sets have been developed by A. Y. Kruger in a series of papers [**621, 623, 624, 625, 626, 627, 628, 630, 631, 632, 633**]; see also [**635**]; earlier concepts of extremality for systems of sets were introduced and studied by A. J. Dubovicki and A. A. Miljutin [**360**].

Other sufficient subdifferential conditions for metric subregularity and calmness of mappings and multimappings can be found in the papers [**314**] by A. Daniilidis and L. Thibault, [**393**] by M. Fabian, R. Henrion, A. Y. Kruger and J. V. Outrata, [**477**] by R. Henrion and A. Jourani, [**478**] by R. Henrion, A. Jourani and J. V. Outrata [**478**], [**527**] by A. D. Ioffe and J. V. Outrata, [**1005, 1006**] by X. Y. Zheng and K. F. Ng, etc.

The metric inequality in Corollary 7.66, known in the literature as Hoffman inequality, was first obtained by A. J. Hoffman in his 1952 paper [**499**]. Its extension in Theorem 7.65 to Banach spaces (that we called "Hoffman-Ioffe theorem") is due to A. D. Ioffe [**508**, Theorem 3]. Theorem 7.67 appeared in an equivalent form in the paper [**340**] by A. V. Dmitruk, A. A. Milyutin, and N. P. Osmolovskii. The arguments in the manuscript for Theorem 7.67 mainly follow those of Theorem 6 in the long 2000 survey [**521**] of A. D. Ioffe. Theorem 7.70 was established by S. M. Robinson [**840**].

The concept of fan in (7.34) has been introduced and developed by A. D. Ioffe in his 1981 paper [**510**]. The name "*adjacent cone*" for the cone defined in (7.35) appeared in particular in the book [**44**] of J.-P. Aubin and H. Frankowska.

Proposition 7.114 is taken from A. L. Dontchev and R. T. Rockafellar's book [**354**].

Concerning Subsection 7.9.2 the paper [**1002**] is one of the first related pioneering contributions. R. Zhang and J. Treiman showed in Theorem 4.2 of this 1995 paper [**1002**] that, if an extended real-valued proper lower semicontinuous function f on $\mathbb{R}^n$ is such that the multimapping $(\partial_L f)^{-1}$ is *upper Lipschitz semicontinuous* at 0 in the sense that for some real $c \geq 0$

$$(\partial_L f)^{-1}(v) \subset (\partial_L f)^{-1}(0) + c\|v\|\mathbb{B} \quad \text{for } v \text{ near } 0,$$

then for any set $C \subset (\partial f)^{-1}(0)$ there exist reals $\delta, \kappa > 0$ for which the (lower) quadratic-type growth condition

$$f(x) \geq \inf f + \kappa\, d\big(x, (C + 2\delta\,\mathbb{B}) \cap (\partial f)^{-1}(0)\big)^2, \quad \forall x \in C + \delta\,\mathbb{B}$$

is satisfied. In fact, for any lower semicontinuous function $f : X \to \mathbb{R} \cup \{+\infty\}$ on a Banach space X such that $(\partial f)^{-1}$ is upper Lipschitz semicontinuous at 0 with a subdifferential ∂ with a full sum rule, the proof in [**1002**] is still valid (with a very slight modification of (ii) therein) to produce reals $\delta, \kappa > 0$ such that

$$f(x) \geq \inf f + \kappa\, d\big(x, (C+2\delta\,\mathbb{B}) \cap (\partial f)^{-1}(0)\big)^2, \quad \forall x \in U_\delta(C) := \{x \in X : d_C(x) < \delta\}.$$

The first part (i) (which is the main part) in the proof of [**1002**, Theorem 4.2] establishes the (lower) quadratic-like growth behavior:

(7.63) $f(x) \geq \inf f + \kappa\, d\big(x, (\partial f)^{-1}(0)\big)^2$, $\forall x \in U_\delta(C) := \{x \in X : d_C(x) < \delta\}$,

which is similar (without precision of constant κ) with the statement of Corollary 7.138(a) with $s = 2$. A second fundamental step was the introduction of the concept of *tilt-stability* for a local minimizer by R. A. Poliquin and R. T. Rockafellar in their 1998 paper [**811**] where they provided characterizations of this concept in finite dimensions in terms of second-order coderivatives for prox-regular functions. In the 2000 book by J. F. Bonnans and A. S. Shapiro one can also find diverse other results for the uniform quadratic growth property. Coming back to the above result (7.63) by Zhang and Treiman, it is clear that the upper Lipschitz assumption of $(\partial f)^{-1}$ at 0 entails that $(\partial f)^{-1}$ is calm at $\bar{x}$ for 0, for any $\bar{x} \in (\partial f)^{-1}(0)$, or equivalently, ∂f is metrically subregular at $\bar{x}$ for 0. Motivated by this, F. J. Aragón Artacho and M. H. Geoffroy showed in [**20**, Theorem 3.5], for an extended real-valued lower semicontinuous convex function f on a Hilbert space, the equivalence in Corollary 7.139 between the metric subregularity of ∂f at $\bar{x}$ for $\bar{x}^* \in \partial f(\bar{x})$ and a growth condition like (7.46) with $s = 2$; they also showed in [**20**, Theorem 3.6 and Corollary 3.9] the similar equivalence in Corollary 7.150 with $\theta = 1$ between the strong metric regularity of the subdifferential of the convex function f and the (lower) uniform quadratic growth condition around $\bar{x}$ or stable second-order local minimizer property of $\bar{x}$. Aragón Artacho and Geoffroy extended those results in [**21**] to the setting of Banach spaces. With $s = 2$ (or equivalently $\theta = 1$) Propositions 7.158 and 7.159 related to the second-order uniform growth around $\bar{x}$ have been proved by B. S. Mordukhovich and T. T. Nghia in the equivalence in [**728**, Theorem 3.2]. Under the assumption that $\bar{x}$ is a local minimizer that equivalence (without exact constants) have been previously shown in the finite-dimensional framework by D. Drusvyatskiy and A. S. Lewis in [**358**, Theorem 3.3]. The study of tilt-stability of minimizers and its equivalence with uniform quadratic growth has been continued in finite dimensions by A. S. Lewis and and S. Zhang [**661**] for partially smooth functions and by D. Drusvyatskiy and A. S. Lewis [**358**] for prox-regular functions in Theorem 3.3 therein. The equivalence between uniform quadratic growth and tilt-stable minimizers of prox-regular functions has been extended to Asplund spaces by B. S. Mordukhovich and T. T. Nghia [**728**, Theorems 3.2 and 4.2]. With $\theta = 1$ or equivalently with $s = 2$, the exact constant κ of the growth-like condition in Corollary 7.137(a) and the exact constant γ in Corollary 7.137 for the metric subregularity of the subdifferential have been obtained by D. Drusvyatskiy, B. S. Mordukhovich and T. T. Nghia [**359**] in Asplund spaces, but their arguments (as those in [**728**]) still work with other subdifferentials with full sum rule in general Banach spaces. The study in this subsection of the link between θ-order metric subregularity (resp. regularity) of subdifferential and s-order local minimizer (resp. stable/tilt-stable local minimizer) has been initiated by X. Y. Zheng and K. F. Ng in their two 2015 papers [**1007, 1008**]. The study has even been carried out for local minimizer (resp. stable or tilt-stable local minimizer) with respect to a general admissible convex function $\varphi : [0, +\infty[\to [0, +\infty[$ in place of $[0, +\infty[\ni t \mapsto t^s$ by X. Y Zheng and J. Zhu [**1009**] and J. Yao, X. Y. Zheng and J. Zhu [**976**]. Various statements and proofs in this subsection are adaptations of corresponding results by Yao, Zheng and Zhu [**976**]. The basic result in Lemma 7.130 is a reproduction with $\varphi(t) = t^s$ of [**976**, Lemma 4.1] by Yao, Zheng and Zhu while Lemma 7.132 and Proposition 7.134 reproduce the statement and proof of [**976**, Theorem 4.1]. The proof of Proposition 7.156 follows Zhang and Treiman [**1002**, Theorem 4.2] and X. Y. Zheng and K. F. Ng [**1007**, Theorem 4.3]. The main ideas of Proposition 7.158 are from Yao, Zheng and Zhu [**976**, Theorem 5.2]. The proofs of Lemma 7.153 and Proposition 7.154 follow the arguments in the proof of [**976**, Theorem 5.1]. The statements and proofs of Lemma 7.164 and Proposition 7.167(b) are taken from [**1007**, Proposition 3.1 and Theorem 5.3]. The example in Remark 7.141(c) is taken from [**358**] and the example in Remark 7.152 is taken from [**359**]. Among other results it is worth mentioning that the equivalence between the strong metric subregularity of the subdifferential and second-order local minimizers has been also established for semialgebraic functions in finite dimensions by D. Drusvyatskiy and A. D. Ioffe [**357**]. For various results of other types concerning s-order metric regularity/subregularity we refer to [**419, 434, 662, 1008**] and the references in these papers.

APPENDIX A

Topology

Recall that a preorder $\preceq_J$ on a nonempty set J is a relation on J which is reflexive and transitive, and for such a relation the pair $(J, \preceq_J)$ is called a preordered set. The reflexivity amounts to saying that

$$j \preceq_J j \quad \text{for all } j \in J,$$

while the transivity means that given $j_1, j_2, j_3 \in J$

$$(j_1 \preceq_J j_2 \text{ and } j_2 \preceq_J j_3) \implies j_1 \preceq_J j_3.$$

It is often convenient to write $j_2 \succeq_J j_1$ instead of $j_1 \preceq_J j_2$. When the preorder $\preceq_J$ is such that for any $j_1, j_2 \in J$ there exists $j_3 \in J$ satisfying $j_1 \preceq_J j_3$ and $j_2 \preceq_J j_3$, one generally says that $(J, \preceq_J)$ is a *directed set*.

Given a nonempty set T, any family $(t_j)_{j \in J}$ indexed by a directed set $(J, \preceq_J)$ is called a *net*; one also writes $(t_j)_j$ when there is no ambiguity for the set J. Any sequence in T is clearly a particular net of T. Several other nets (which are not sequences) are already given in Example 1.15. Subsequences are known to be crucial in metric spaces to characterize compactness of sets. The same type of characterization also holds true in topological spaces through the concept of subnet. A *subnet* of the net $(t_j)_{j \in J}$ is a net of the form $(t_{s(i)})_{i \in I}$ where $(I, \preceq_I)$ is a directed set and $s : I \to J$ is a *directed mapping* in the sense that, for each $j_0 \in J$ there is $i_0 \in I$ such that $j_0 \preceq_J s(i)$ for all $i \in I$ satisfying $i_0 \preceq_I i$. Any subsequence of a sequence is obviously a subnet of this sequence, but there are obvious subnets of a sequence which are not subsequences. Particular subnets are achieved via the concept of cofinal set. A nonempty set $J_0 \subset J$ is called *cofinal* in the set $(J, \preceq_J)$ whenever for each $j \in J$ there is some $j' \in J_0$ such $j \preceq_J j'$. Endowing J_0 with the induced preorder we see that $(t_j)_{j \in J_0}$ is a subnet of the net $(t_j)_{j \in J}$.

If T is endowed with a topology τ, one says that the net $(t_j)_{j \in J}$ converges to some t in (T, τ_T) provided that for each neighborhood V of t there is some $j_V \in J$ such that $t_j \in V$ for all $j \in J$ with $j \succeq_J j_V$; in such a case one writes $t_j \xrightarrow[j \in J]{} t$ or $t_j \to t$. When the topology τ_T is Hausdorff, such an element (if any) is unique; in such a case it is called the limit of $(t_j)_{j \in J}$ and one writes $t = \lim_{j \in J} t_j$.

The following proposition summarizes some fundamental properties of nets and subnets. For its statement as well as for the foregoing concepts we refer, for example, to J. L. Kelley's book [**604**, Chapter 2, p. 65-72].

PROPOSITION A.1. *Let (T, τ_T) and (X, τ_X) be topological spaces, S be a subset of T and $f : T \to X$ be a mapping. The following holds.*
(a) *One has $t \in \operatorname{cl} S$ if and only if there is a net $(t_j)_j$ in S converging to t in (T, τ_T). The set S is thus τ_T closed if and only if S contains any element to which τ_T-converges a net in S.*
(b) *The mapping f is continuous at a point $\bar{t} \in T$ if and only if $f(t_j) \to f(\bar{t})$ in*

(X, τ_X) for any net $(t_j)_{j \in J}$ in T with $t_j \to \bar{t}$.

(c) A net $(t_j)_{j \in J}$ converges to t in (T, τ_T) if and only if everyone of its subnets converges to t.

(d) A net $(t_j)_{j \in J}$ in (T, τ_T) converges to some $t \in T$ if and only if every subnet admits itself a subnet converging to t.

(e) Given another topological space $(T', \tau_{T'})$ and endowing $T \times T'$ with the product topology, a net $(t_j, t'_j)_{j \in J}$ converges to (t, t') in $T \times T'$ if and only if $t_j \to t$ and $t'_j \to t'$.

(f) A subset $S \subset T$ is τ_T-compact if and only if every net in S admits a subnet τ_T-converging to an element in S.

Another remarkable feature of nets is the following property of iterated limits (see, for example, [**604**, Chapter 2, Theorem 4]):

PROPOSITION A.2. *Let $(\Lambda, \preceq_\Lambda)$ be a directed set and for each $\lambda \in \Lambda$ let $(I_\lambda, \preceq_\lambda)$ be also a directed set. Let $J := \Pi_{\lambda \in \Lambda} I_\lambda$ and let $\Lambda \times J = \Lambda \times \Pi_{\lambda \in \Lambda} I_\lambda$ be endowed with the product preorder. Assume that for each $\lambda \in \Lambda$ a net $(y_{\lambda,i})_{i \in I_\lambda}$ of a topological space Y converges to some z_λ in Y and that the net $(z_\lambda)_{\lambda \in \Lambda}$ converges to some $z \in Y$. Then putting $\xi_{\lambda,j} := y_{\lambda,j_\lambda}$ for every (λ, j) in $\Lambda \times J$, the net $(\xi_{\lambda,j})_{(\lambda,j) \in \Lambda \times J}$ converges to z.*

Let us turn now to paracompact spaces. Let (T, τ) be a Hausdorff topological space. A family $(U_j)_{j \in J}$ is called an *open cover* (or *open covering*) of T if $\bigcup_{j \in J} U_j = T$ and each U_j is an open set in T. The open cover $(U_j)_{j \in J}$ is *locally finite* when each $t \in T$ has a neighborhood intersecting only finitely many U_j. An open cover $(O_i)_{i \in I}$ of T is said to be an *open refinement* of an open cover $(U_j)_{j \in J}$ provided that for each $i \in I$ there is some $j \in J$ such that $O_i \subset U_j$. One says that T is *paracompact* if any open cover admits a locally finite open refinement.

Metric spaces and compact Hausdorff topological spaces are remarkable examples of paracompact spaces.

Recall that a family of real-valued functions $(\varphi_i)_{i \in I}$ on T is called a locally finite continuous partition of unity on T if

(i) each point $t \in T$ has a neighborhood on which all but a finite number of functions φ_i are null;

(ii) all the functions φ_i are non-negative and

$$\sum_{i \in I} \varphi_i(t) = 1 \quad \text{for every } t \in T.$$

If T is a Banach space X and each function φ_i is in addition of class $\mathcal{C}^p$ with $p \in \mathbb{N} \cup \{\infty\}$, one says that $(\varphi_i)_{i \in I}$ is a *locally finite $\mathcal{C}^p$-partition of unity* on X.

A continuous (resp. $\mathcal{C}^p$) partition $(\varphi_i)_{i \in I}$ of unity on T (resp. on X) is said to be *subordinated* to an open cover $(U_j)_{j \in J}$ of T (resp. of X) if each function φ_i is null outside some U_j.

Paracompact spaces enjoy the following fundamental property (see, for example, [**604**, p. 171]).

THEOREM A.3 ($\mathcal{C}^0$-*partition of unity under paracompactness*). *For any open cover $(U_j)_{j \in J}$ of a paracompact Hausdorff topological space T there is a locally finite continuous partition of unity on T which is subordinated to $(U_j)_{j \in J}$.*

Theorem 3 in the paper [**932**] of H. Toruńczyk that we recall below established $\mathcal{C}^\infty$-partition of unity in Hilbert spaces.

THEOREM A.4 (H. Toruńczyk). *For any open cover $(U_j)_{j\in J}$ of a Hilbert space H there is a locally finite $\mathcal{C}^\infty$-partition of unity on H which is subordinated to $(U_j)_{j\in J}$.*

APPENDIX B

Topological properties of convex sets

Let us recall first the classical analytic Hahn-Banach theorem. We continue with the setting that *all vector spaces will be considered over the field* $\mathbb{R}$.

THEOREM B.1 (Hahn-Banach theorem: analytic form). *Let X be a vector space and $s : X \to \mathbb{R}$ be a positive homogeneous convex function (also called sublinear function). Let $\ell_0 : X_0 \to \mathbb{R}$ be a linear functional on a vector subspace X_0 of X such that $\ell_0(u) \leq s(u)$ for all $u \in X_0$. Then ℓ_0 can be extended on a linear functional $\ell : X \to \mathbb{R}$ to the whole space X so that $\ell(x) \leq s(x)$ for all $x \in X$.*

From this theorem it is classical to derive the geometric Hahn-Banach theorems. The main lines of the proofs will be sketched.

THEOREM B.2 (Hahn-Banach theorem: geometric form I). *Let U be a nonempty open convex set in a topological vector space X and let a point $b \in X \setminus U$. Then there exists a continuous linear functional ℓ on X such that $\ell(x) < \ell(b)$ for all $x \in U$.*

PROOF. Fix $u \in U$ and observe that $V := -u + U$ is an open convex set containing zero and that $a := b - u \notin V$. Consider the linear functional $\zeta : \mathbb{R}a \to \mathbb{R}$ defined by $\zeta(ta) = tj_V(a)$ for all $t \in \mathbb{R}$, where j_V denotes the Minkowski gauge function of V. It is easy to verify that $\zeta(ta) \leq j_V(ta)$ for all $t \in \mathbb{R}$, hence the analytic Hahn-Banach theorem above gives that ζ can be extended to a linear functional $\ell : X \to \mathbb{R}$ such that $\ell(x) \leq j_V(x)$ for all $x \in X$. Since j_V is continuous and $V = \{x \in X : j_V(x) < 1\}$, the linear functional ℓ is continuous and $j_V(a) \geq 1$. It results that
$$\ell(v) \leq j_V(v) < 1 \leq j_V(a) = \zeta(a) = \ell(a) \quad \text{for all } v \in V,$$
which easily yields that $\ell(x) < \ell(b)$ for all $x \in U$. $\square$

THEOREM B.3 (Hahn-Banach theorem: geometric form II). *Let X be a topological vector space, C be a nonempty open convex set in X and D be a nonempty convex set in X such that $C \cap D = \emptyset$. Then there exist a nonzero continuous linear functional $\ell : X \to \mathbb{R}$ and a real α such that*
$$C \subset \{\ell(\cdot) < \alpha\} \quad \text{and} \quad D \subset \{\ell(\cdot) \geq \alpha\}.$$

PROOF. Since the open convex set $U := C - D$ does not contain zero, the previous theorem furnishes a continuous linear functional $\ell : X \to \mathbb{R}$ such that $\ell(u) < \ell(0) = 0$ for all $u \in U$. It ensues that $\ell(x) < \ell(y)$ for all $x \in C$ and $y \in D$, hence $\sup_{x \in C} \ell(x) \leq \inf_{y \in C} \ell(y)$ and both the supremum and infimum are finite. Choosing α between the two latter reals, we obtain $C \subset \{\ell(\cdot) \leq \alpha\}$ and $D \subset \{\ell(\cdot) \geq \alpha\}$. Since ℓ is nonzero, it is not difficult to deduce from the latter inclusion that $C \subset \{\ell(\cdot) < \alpha\}$ as desired. $\square$

THEOREM B.4 (Hahn-Banach theorem: geometric form III). *Let X be a locally convex vector space, C, D be nonempty disjoint convex subsets such that $C - D$ is closed, which holds in particular if one convex set is compact and the other is closed. Then there exist a nonzero continuous linear functional $\ell : X \to \mathbb{R}$ and reals α, ε with $\varepsilon > 0$ such that*

$$C \subset \{\ell(\cdot) \leq \alpha - \varepsilon\} \quad \text{and} \quad D \subset \{\ell(\cdot) \geq \alpha + \varepsilon\}.$$

PROOF. Since the convex set $C-D$ is closed and $0 \notin C-D$, there is a symmetric open convex neighborhood V of zero in X such that $(V + V) \cap (C - D) = \emptyset$, or equivalently $(C + V) \cap (D + V) = \emptyset$. Noting that $D + V$ is convex and $C + V$ is convex and open, there exists by Theorem B.3 a continuous linear functional $\ell : X \to \mathbb{R}$ and a real $\alpha > 0$ such that $\ell(x + u) \leq \alpha \leq \ell(y + v)$ for all $x \in C$, $y \in D$ and $u, v \in V$. From this the conclusion of the theorem easily follows. □

Convex sets admit well-known remarkable closure and interior properties.

PROPOSITION B.5. *Let X be a topological vector space and C be a convex set in X. The following hold.*
(a) *The topological closure* $\operatorname{cl} C$ *and interior* $\operatorname{int} C$ *of C are convex.*
(b) *If* $\operatorname{int} C \neq \emptyset$, *then*

$$[x, y[\subset \operatorname{int} C \quad \text{for all } x \in \operatorname{int} C \text{ and } y \in \operatorname{cl} C,$$

along with

$$\operatorname{cl}(\operatorname{int} C) = \operatorname{cl} C \quad \text{and} \quad \operatorname{int}(\operatorname{cl} C) = \operatorname{int} C.$$

PROOF. (b) Assume that $\operatorname{int} C \neq \emptyset$. Fix any $x \in \operatorname{int} C$. Take any $y \in \operatorname{cl} C$ and any $z \in]x, y[$ (if any). There are $s, t \in]0, 1[$ with $s + t = 1$ such that $z = sx + ty$. Choose a neighborhood U of zero such that $x+U+U \subset C$. Note that $y \in C+t^{-1}sU$ (since $y \in \operatorname{cl} C$), so there is $c \in C$ such that $y \in c + t^{-1}sU$, hence

$$z + sU = sx + ty + sU \subset s(x + U + U) + tc \subset sC + tC = C.$$

Since sU is a neighborhood of 0, it ensues that $z \in \operatorname{int} C$. It then results that $[x, y[\subset \operatorname{int} C$. The latter inclusion gives $sx + (1 - s)y \in \operatorname{int} C$ for all $s \in]0, 1[$, hence $y \in \operatorname{cl}(\operatorname{int} C)$. This justifies the equality $\operatorname{cl}(\operatorname{int} C) = \operatorname{cl} C$. To prove the other equality in (b), take any $u \in \operatorname{int}(\operatorname{cl} C)$. There is some $\alpha > 1$ such that $v := \alpha u + (1 - \alpha)x \in \operatorname{cl} C$. By what precedes we have $[x, v[\subset \operatorname{int} C$, hence $u \in \operatorname{int} C$. This justifies the equality $\operatorname{int}(\operatorname{cl} C) = \operatorname{int} C$.
(a) The convexity of $\operatorname{cl} C$ is obvious. Concerning $\operatorname{int} C$, its convexity is trivial if it is empty, and otherwise its convexity follows from the first property in (b) established above. □

It is known that the closure of intersection of convex sets does not coincide with the intersection of closures even for intervals in $\mathbb{R}$ as confirmed by $C_1 := [0, 1[$ and $C_2 := [1, 2]$. Nevertheless, we have the following proposition (see S. Dolecki [341]).

PROPOSITION B.6. *Let X be a topological vector space and C, D be two convex subsets in X such that $C \cap \operatorname{int} D \neq \emptyset$. Then one has*

$$\operatorname{cl}(C \cap \operatorname{int} D) = \operatorname{cl}(C \cap D) = (\operatorname{cl} C) \cap (\operatorname{cl} D).$$

PROOF. Choose $\bar{x} \in C \cap \operatorname{int} D$ and an open neighborhood U_0 of zero in X such that $\bar{x}+U_0 \subset D$. To show that the third member is included in the first one fix any $x \in (\operatorname{cl} C) \cap (\operatorname{cl} D)$. Consider any neighborhood U of zero and choose a balanced

neighborhood V of zero in X satisfying $V + V \subset U \cap U_0$. Choose a real $r \geq 1$ such that $\bar{x} - x \in rV$ and then choose a real $s > 0$ such that $s(1 + 2r)/(1 + 2s) < 1$. There exists $y \in (x + sV) \cap D$. Put $t := 2s/(1 + 2s)$ (so, $0 < t < 1$) and note by Proposition B.5 that
$$(1 - t)y + t(\bar{x} + U_0) \subset \operatorname{int} D.$$
Since
$$(1 - t)(x + sV) + t\bar{x} \subset (1 - t)(y + sV + sV) + t\bar{x}$$
$$= (1 - t)y + \frac{t}{2}V + \frac{t}{2}V + t\bar{x} \subset (1 - t)y + t(\bar{x} + V + V),$$
it ensues that $(1 - t)(x + sV) + t\bar{x} \subset \operatorname{int} D$. Noticing that $(x + sV) \cap C \neq \emptyset$ and using the inclusion $\bar{x} \in C$, we deduce that
(B.1) $$\left((1 - t)(x + sV) + t\bar{x}\right) \cap (C \cap \operatorname{int} D) \neq \emptyset.$$
Since $2rs/(1 + 2s) < 1$ and V is balanced, we observe that
$$t(\bar{x} - x) + s(1 - t)V \subset rtV + s(1 - t)V \subset \frac{2rs}{1 + 2s}V + \frac{s}{1 + 2s}V \subset V + V \subset U,$$
which combined with (B.1) yields $(x + U) \cap (C \cap \operatorname{int} D) \neq \emptyset$. It results that $x \in \operatorname{cl}(C \cap \operatorname{int} D)$, hence the third member of the proposition is included in the first. From this the equalities of the proposition are clear. □

Now let us recall that a subset F of a vector space X is *affine* if by definition it contains any *affine combination* of its elements, that is, for any $x_1, \cdots, x_p$ in F and any reals $t_1, \cdots, t_p$ with $t_1 + \cdots + t_p = 1$ one has
$$t_1 x_1 + \cdots + t_p x_p \in F.$$
When F is nonempty, taking any $u \in F$ we see that $F - u$ is a vector subspace of X. Consequently, any nonempty affine subset F of X is of the form $u + E$, where $u \in X$ and E is a vector subspace of X; such a vector space E is unique and its dimension is called the dimension of F, which is generally stated as $\dim F := \dim E$. Given any subset S of X, the intersection of all affine subsets of X containing S is an affine set, and it is the smallest affine subset of X containing S. It is called the *affine hull* of S and denoted by $\operatorname{aff} S$. When S is nonempty, it is easily seen that $\operatorname{aff} S$ coincides with the set of all affine combinations of elements in S, that is, $x \in \operatorname{aff} S$ if and only if for some integer p there are $x_1, \cdots, x_p \in S$ and $t_1, \cdots, t_p \in \mathbb{R}$ with $t_1 + \cdots + t_p = 1$ such that $x = t_1 x_1 + \cdots + t_p x_p$.

Assume now that X is a finite-dimensional normed space and let C be a convex set in X. We recall that the *relative interior* $\operatorname{rint} C$ of C is the interior of C relative to the affine hull $\operatorname{aff} C$ endowed with the induced topology, which can be written in the form
$$\operatorname{rint} C = \operatorname{int}_{\operatorname{aff} C}(C).$$
An affine subset F of X being closed, we see that
$$\operatorname{rint} F = F = \operatorname{cl} F.$$
At the opposite of the classical topological interior, it is worth noticing that, for two convex sets C_1, C_2 in X
$$C_1 \subset C_2 \not\Rightarrow \operatorname{rint} C_1 \subset \operatorname{rint} C_2;$$
indeed, for the subsets $C_1 := \{0\}$ and $C_2 := [0, 1[$ in $\mathbb{R}$, we have $C_1 \subset C_2$ but $\operatorname{rint} C_1 \not\subset \operatorname{rint} C_2$ since $\operatorname{rint} C_1 = \{0\}$ whereas $\operatorname{rint} C_2 =]0, 1[$.

Suppose for a moment that the convex set C in X is nonempty. We note that for any $u \in C$ the vector space $E := \text{Vect}\,(C - u)$ spanned by $C - u$ coincides with $\mathbb{R}(C - u)$, does not depend on $u \in C$ and satisfies the equality

(B.2) $$-u + \text{rint}\,C = \text{int}_E(C - u);$$

one defines the dimension of C as $\dim C := \dim E$. For $p := \dim E$, it is clear that there are $v_1, \cdots, v_p \in C - u$ linearly independent such that

$$Q := \{\lambda_1 v_1 + \cdots + \lambda_p v_p : \lambda_i \in\,]0,1[, \lambda_1 + \cdots + \lambda_p < 1\} \subset C - u \subset E.$$

Since the nonempty subset Q of E is open relative to E, it ensues that $\text{int}_E(C-u) \neq \emptyset$. This and (B.2) tell us that $\text{rint}\,C \neq \emptyset$. Further, taking $u \in \text{rint}\,C$ the same equality B.2 ensures that the vector space spanned by $-u + \text{rint}\,C$ coincides with E, hence $\text{int}_E(-u + \text{rint}\,C) = \text{int}_E(-u + C)$, which means by (B.2) again that $\text{rint}\,(\text{rint}\,C) = \text{rint}\,C$, which also holds true if $C = \emptyset$.

We summarize all those features as follows.

PROPOSITION B.7. *Let C, C_1, C_2 be convex sets in a finite-dimensional normed space X. One has $\text{rint}\,(\text{rint}\,C) = \text{rint}\,C$ along with*

$$C_1 \subset C_2 \not\Rightarrow \text{rint}\,C_1 \subset \text{rint}\,C_2$$

and

$$C \neq \emptyset \Rightarrow \text{rint}\,C \neq \emptyset.$$

From the second property of the proposition we can derive the following corollary.

COROLLARY B.8. *If C is a nonempty convex set of a finite-dimensional normed space X with $\text{int}_X C = \emptyset$, then for any $u \in C$, one has $\text{Vect}(C - u) \neq X$.*

PROOF. Fix any $u \in C$ and notice that $\text{int}_X(C - u) = \emptyset$ since $\text{int}_X C = \emptyset$. Denoting $E := \text{Vect}(C - u)$, we have $-u + \text{rint}\,C = \text{int}_E(C - u)$ by (B.2) above. Since $\text{rint}\,C \neq \emptyset$ by Proposition B.7, it ensues that $\text{int}_E(C - u) \neq \emptyset$. This and the above equality $\text{int}_X(C - u) = \emptyset$ entail that $E \neq X$. □

By Proposition B.5(a) we know that, for any convex set Q of a locally convex space X, any $x \in \text{int}\,Q$ and any $y \in \text{cl}\,Q$, one has $[x, y[\subset \text{int}\,Q$. This and (B.2) yield:

PROPOSITION B.9. *Let C be a nonempty convex set in a finite-dimensional normed space X. The following hold.*
(a) *For any $x \in \text{rint}\,C$ and any $y \in \text{cl}\,C$, one has $[x, y[\subset \text{rint}\,C$.*
(b) *One also has the equalities*

$$\text{cl}(\text{rint}\,C) = \text{cl}\,C \quad \text{and} \quad \text{rint}\,(\text{cl}\,C) = \text{rint}\,C.$$

As a corollary we have the additional equalities in finite dimensions for convex sets.

COROLLARY B.10. *If C is a convex set in a finite-dimensional normed space X, then*

$$\text{int}_X(\text{cl}\,C) = \text{int}_X(C) \quad \text{and} \quad \text{bdry}_X(\text{cl}\,C) = \text{bdry}_X C.$$

PROOF. We may suppose that $C \neq \emptyset$.

Case 1: $\operatorname{int}_X C \neq \emptyset$. In this case, the equality $\operatorname{int}_X(\operatorname{cl} C) = \operatorname{int}_X(C)$ follows from Proposition B.5(b).

Case 2: $\operatorname{int}_X C = \emptyset$. Suppose $\operatorname{int}_X(\operatorname{cl} C) \neq \emptyset$. Choose some $c \in \operatorname{int}_X(\operatorname{cl} C)$ and some real $r > 0$ with $B_X(c, r) \subset \operatorname{int}(\operatorname{cl} C)$. The fact that $\operatorname{int}_X(\operatorname{cl} C) \neq \emptyset$, entails that
$$\operatorname{int}_X(\operatorname{cl} C) = \operatorname{rint}(\operatorname{cl} C) = \operatorname{rint} C,$$
where the right equality is due to Proposition B.9(b). It ensues that
$$B_X(c, r) \subset \operatorname{rint} C \subset C,$$
hence $c \in \operatorname{int}_X C$, and this contradicts that $\operatorname{int}_X C = \emptyset$. Consequently, $\operatorname{int}_X(\operatorname{cl} C) = \emptyset$, thus $\operatorname{int}_X(\operatorname{cl} C) = \operatorname{int}_X C$.

In any case we obtain $\operatorname{int}_X(\operatorname{cl} C) = \operatorname{int}_X C$.

Finally, the second equality $\operatorname{bdry}_X(\operatorname{cl} C) = \operatorname{bdry}_X C$ is a consequence of the latter. □

In addition to Proposition B.6, we have the following situations where the closure of the intersection of finitely many convex sets coincides with the intersection of closures.

PROPOSITION B.11. Let X be a topological vector space and $(C_i)_{i \in I}$ be a family of convex sets in X. Assume that either $\bigcap_{i \in I} \operatorname{int} C_i \neq \emptyset$ or X is a finite-dimensional normed space and $\bigcap_{i \in I} \operatorname{rint} C_i \neq \emptyset$. Then one has $\operatorname{cl}\left(\bigcap_{i \in I} C_i\right) = \bigcap_{i \in I} \operatorname{cl} C_i$.

PROOF. The left side is obviously included in the right one. Conversely, fix z in $\bigcap_{i \in I} \operatorname{int} C_i$ (resp. in $\bigcap_{i \in I} \operatorname{rint} C_i$) and take any x in $\bigcap_{i \in I} \operatorname{cl} C_i$. For each $t \in {]0, 1[}$ putting $x_t := tz + (1-t)x$, by Proposition B.5(a) (resp. Proposition B.9(a)) we have $x_t \in C_i$ for every $i \in I$, and hence $x_t \in \bigcap_{i \in I} C_i$. Since $x_t \to x$ as $t \downarrow 0$, it follows that $x \in \operatorname{cl}\left(\bigcap_{i \in I} C_i\right)$, which finishes the proof. □

Via the above assertion (a) in Proposition B.9 the next proposition establishes a useful characterization of the relative interior of a convex set. The arguments that we follow here as well as those for Propositions B.14, B.15 and B.18 are taken from R. T. Rockafellar [852].

PROPOSITION B.12. Let C be a nonempty convex set in a finite-dimensional normed space X. An element $x \in C$ belongs to $\operatorname{rint} C$ if and only if for each $y \in C$ there exists some real $t > 1$ such that $tx + (1-t)y \in C$.

PROOF. By (B.2) the implication $\Rightarrow$ is obvious. Conversely, suppose that $x \in C$ fulfills the property in the statement. Since $\operatorname{rint} C \neq \emptyset$ according to Proposition B.7, we can choose $y \in \operatorname{rint} C$. There exist some real $t > 1$ such that $z := tx + (1-t)y \in C$. With $\theta := 1/t \in {]0, 1[}$ we see that $x = (1-\theta)y + \theta z$, hence by Proposition B.9(a) we conclude that $x \in \operatorname{rint} C$. □

The following corollary is easily derived from the latter proposition. (Notice that it could be also proved via the definition.)

COROLLARY B.13. If $C_1, \cdots, C_p$ are convex sets of finite-dimensional normed spaces $X_1, \cdots, X_p$ respectively, then one has
$$\mathrm{rint}(C_1 \times \cdots \times C_p) = (\mathrm{rint}\, C_1) \times \cdots \times (\mathrm{rint}\, C_p).$$

At the opposite of the usual topological interior, the finite intersection operation is not preserved by the relative interior even for convex sets. This is illustrated with the same above intervals $C_1 := [0,1]$ and $C_2 := [1,2]$ in $\mathbb{R}$. However, Proposition B.12 allows us to prove that this preservation holds true under the same condition in Proposition B.11.

PROPOSITION B.14. Let $C_1, \cdots, C_p$ be a finite family of convex sets in a finite dimensional normed space X such that $\bigcap_{i=1}^{p} \mathrm{rint}\, C_i \neq \emptyset$. Then one has
$$\mathrm{rint}\left(\bigcap_{i=1}^{p} C_i\right) = \bigcap_{i=1}^{p} \mathrm{rint}\, C_i.$$

PROOF. Denote by K_ℓ and K_r the left and right sides of the equality to prove. We note first by Proposition B.11 and Proposition B.9(b) that both convex sets $\bigcap_{i=1}^{p} C_i$ and $\bigcap_{i=1}^{p} \mathrm{rint}\, C_i$ have the same closure $\left(\text{equal to } \bigcap_{i=1}^{p} \mathrm{cl}\, C_i\right)$, hence the same relative interior by Proposition B.9(b). This gives
$$\mathrm{rint}\left(\bigcap_{i=1}^{p} C_i\right) = \mathrm{rint}\left(\bigcap_{i=1}^{p} \mathrm{rint}\, C_i\right) \subset \bigcap_{i=1}^{p} \mathrm{rint}\, C_i,$$
which justifies the inclusion $K_\ell \subset K_r$.

To prove the converse inclusion, fix any $x \in K_r$. Take any $y \in \bigcap_{i=1}^{p} C_i$. For each $i = 1, \cdots, p$ by Proposition B.12 there exits a real $t_i > 1$ such that $(1-t_i)x + t_i y \in C_i$. Putting $t := \min\{t_1, \cdots, t_p\}$, we see that $t > 1$ and $(1-t)x + ty \in C_i$ for all $i = 1, \cdots, p$ by convexity of C_i. Then $(1-t)x + ty \in \bigcap_{i=1}^{p} C_i$, so Proposition B.12 again tells us that $x \in \mathrm{rint}\left(\bigcap_{i=1}^{p} C_i\right)$. This translates the inclusion $K_r \subset K_\ell$ and finishes the proof. $\square$

The sequence of intervals $(J_n)_{n \in \mathbb{N}}$ with $J_n := [0, 1 + (1/n)]$ confirms that in general $\mathrm{rint}\left(\bigcap_{i \in I} C_i\right) \neq \bigcap_{i \in I} \mathrm{rint}\, C_i$ for infinite families of convex sets even when the condition $\bigcap_{i \in I} \mathrm{rint}\, C_i \neq \emptyset$ is fulfilled.

At the opposite of the topological interior, the relative interior is well-behaved with respect to images of convex sets under linear mappings.

PROPOSITION B.15. Let $A : X \to Y$ be an affine mapping between two finite-dimensional normed spaces. For any convex set C in X one has
$$\mathrm{rint}\,(A(C)) = A(\mathrm{rint}\, C).$$

PROOF. We may suppose that $C \neq \emptyset$. We note first by Proposition B.9(b) that
$$A(\mathrm{rint}\, C) \subset A(C) \subset A(\mathrm{cl}\, C) = A(\mathrm{cl}\,(\mathrm{rint}\, C)) \subset \mathrm{cl}(A(\mathrm{rint}\, C)),$$

where the latter inclusion is due to the continuity of A. This entails that the convex sets $A(C)$ and $A(\text{rint } C)$ have the same closure, hence the same relative interior by Proposition B.9(b). This ensures that

$$\text{rint}\,(A(C)) = \text{rint}\,(A(\text{rint } C)) \subset A(\text{rint } C).$$

To show the converse inclusion, fix any $v \in A(\text{rint } C)$, so $v = A(u)$ for some $u \in \text{rint } C$. For any $y \in A(C)$, choosing $x \in C$ with $y = A(x)$, Proposition B.12 gives some real $t > 1$ such that $tu + (1-t)x \in C$, hence $tv + (1-t)y \in A(C)$. Proposition B.12 again entails that $v \in \text{rint}\,(A(C))$, and the proof is finished. $\square$

Given convex sets $C_1, \cdots, C_p$ in finite-dimensional normed spaces $X_1, \cdots, X_p$ respectively, we saw in Corollary B.13 that

$$\text{rint}\,(C_1 \times \cdots \times C_p) = (\text{rint } C_1) \times \cdots \times (\text{rint } C_p).$$

Assuming $X_i = X$, for $i = 1, \cdots, p$, we have $C_1 + \cdots + C_p = A(C_1 \times \cdots \times C_p)$, where $A : X^p \to X$ is the linear mapping defined by $A(x_1, \cdots, x_p) := x_1 + \cdots + x_p$. The following corollary of the above proposition then directly follows.

COROLLARY B.16. *Let X be a finite-dimensional normed space.*
(a) *For any convex set C in X and any real $\alpha \in \mathbb{R}$ one has*

$$\text{rint}(\alpha C) = \alpha \,\text{rint}\, C.$$

(b) *For any convex sets $C_1, \cdots, C_p$ in X one has*

$$\text{rint}(C_1 + \cdots + C_p) = \text{rint}(C_1) + \cdots + \text{rint}(C_p).$$

Although the sets $C_1 := [-1, 1]$ and $C_2 := \{0\}$ in $\mathbb{R}$ tell us that the inclusion $0 \in \text{int}\,(C_1 - C_2)$ does not imply that $(\text{int } C_1) \cap (\text{int } C_2) \neq \emptyset$, the following equivalence for relative interiors is a direct corollary of Corollary B.16.

COROLLARY B.17. *Let C_1, C_2 be convex sets in a finite-dimensional normed space X. Then one has*

$$0 \in \text{rint}(C_1 - C_2) \iff (\text{rint } C_1) \cap (\text{rint } C_2) \neq \emptyset.$$

Given a linear mapping $A : X \to Y$ between finite-dimensional spaces, we saw in Proposition B.15 that $\text{rint}\,(A(C)) = A(\text{rint } C)$ for any convex set $C \subset X$. Via a certain condition, the relative interior is also well-behaved under inverse image of convex sets by linear mappings.

PROPOSITION B.18. *Let $A : X \to Y$ be an affine mapping between finite-dimensional normed spaces. For any convex set C in Y with $A^{-1}(\text{rint } C) \neq \emptyset$ one has*

$$\text{rint}\,(A^{-1}(C)) = A^{-1}(\text{rint } C).$$

PROOF. Considering the projector $\pi_X : X \times Y \to X$ (with $\pi_X(x, y) := x$) and the affine set $\text{gph } A$, we see that $A^{-1}(C) = \pi_X Q$, where $Q := (\text{gph } A) \cap (X \times C)$. Choosing by assumption $u \in A^{-1}(\text{rint } C)$, that is, $A(u) \in \text{rint } C$, we obtain $(u, A(u))$ in both $\text{gph } A = \text{rint}\,(\text{gph } A)$ and $\text{rint}\,(X \times C)$. We deduce by Proposition B.14 that $\text{rint } Q = (\text{gph } A) \cap (X \times \text{rint } C)$, which yields by Proposition B.15 that

$$\text{rint}\,(A^{-1}(C)) = \pi_X(\text{rint } Q) = A^{-1}(\text{rint } C),$$

which completes the proof. $\square$

In addition to the above features in finite dimensions, we present now the class of polyhedral convex sets which in particular enjoy the intrinsic closedness property. We maintain X as a finite-dimensional normed space. A set C in X is called *polyhedral convex* if it is the intersection of a finite collection $(\mathcal{H}_i)_{i \in I}$ of half-spaces in X, that is, there are a finite set I, elements $a_i^* \in X^*$ and $\beta_i \in \mathbb{R}$ such that
$$C = \{x \in X : \langle a_i^*, x \rangle \leq \beta_i,\ \forall i \in I\}.$$
The sets X (with $I = \emptyset$) is the greatest convex polyhedral set in X, and the empty set $\emptyset$ (as the intersection of two disjoint half-spaces) is the smallest. A basic result is the following *vertex/ray representation* for which a complete proof can be found, for example, in R. T. Rockafellar and R. J-B. Wets [**865**, Theorem 3.52 and Corollary 3.53].

THEOREM B.19 (Minkowski-Weyl theorem for convex polyhedra). A nonempty set C in the finite dimensional vector space X is polyhedral convex if and only if there are points $a_1, \cdots, a_p$ in X and directions associated with $a_{p+1}, \cdots, a_q$ in X such that $x \in C$ provided there are $t_1, \cdots, t_p \geq 0$ with $t_1 + \cdots + t_p = 1$ and $t_{p+1}, \cdots, t_q \geq 0$ such that

(B.3) $$x = t_1 a_1 + \cdots + t_p a_p + t_{p+1} a_{p+1} + \cdots + t_q a_q.$$

APPENDIX C

Functional analysis

Let us now recall certain fundamental basic facts concerning continuous linear operators/mappings. We start with a well-known invertibility result.

PROPOSITION C.1. Let $A : X \to X$ be a continuous linear operator from a Banach space X into itself. If $\|\text{id}_X - A\| < 1$, then A is invertible and $A^{-1} = \text{id}_X + \sum_{n=1}^{\infty} A^n$, where $A^n := A \circ \cdots \circ A$.

In the setting of Hilbert space it is also worth recalling the following property of positivity of a linear operator.

PROPOSITION C.2. Let $A : H \to H$ be a continuous linear operator from a Hilbert space H into itself. If $\|\text{id}_H - A\| \leq 1$, then the linear operator A is positive, that is $\langle Ax, x \rangle \geq 0$ for all $x \in H$.

PROOF. It suffices to notice that
$$\langle Ax, x \rangle = \langle x, x \rangle + \langle (A - \text{id}_H)x, x \rangle \geq \|x\|^2 - \|\text{id}_H - A\| \, \|x\|^2,$$
and hence $\langle Ax, x \rangle \geq 0$. $\square$

We continue now with the Banach-Schauder open mapping theorem (see, for example, Theorem 2.6 in H. Brezis' book [**178**]).

THEOREM C.3 (Banach-Schauder open mapping theorem). Let $A : X \to Y$ be a surjective continuous linear mapping between Banach spaces X, Y. Then there exits a real constant $c > 0$ such that $c\mathbb{B}_Y \subset A(\mathbb{B}_X)$.

The Krein-Šmulian theorem (also called Banach-Dieudonné theorem in the literature) furnishes a practical way to verify the weak* closedness of convex sets. A proof can be found, for example, in the book of N. Dunford and J. T. Schwartz [**366**, Theorem V.5.7].

THEOREM C.4 (Krein-Šmulian theorem). Let X be a Banach space and X^* its topological dual. A convex set in X^* is $w(X^*, X)$ closed if and only if its intersection with every closed ball in X^* centered at the origin is $w(X^*, X)$ closed.

Given a continuous linear operator/mapping $A : X \to Y$ between two Hausdorff locally convex spaces X and Y, the adjoint $A^* : Y^* \to X^*$ of A is defined by
$$A^*(y^*) := y^* \circ A \quad \text{for all } y^* \in X^*.$$
Consider the important situation when X and Y are two normed spaces. It is easily verified that A^* is norm-to-norm continuous and $\|A^*\| = \|A\|$; if A is an isomorphism (resp. isometric isomorphism), then so is A^* and

(C.1) $$(A^{-1})^* = (A^*)^{-1}.$$

Further, it is not difficult to see that
$$\text{Ker}(A^*) = (A(X))^\perp \quad \text{and} \quad \text{Ker}(A) = (A^*(Y^*))^{\perp_X},$$
where $(A^*(Y^*))^{\perp_X}$ denotes the orthogonal of $(A^*(Y^*))$ in X.

One also has the following theorem (see, for example, Theorem 2.19 in Brezis' book [178]) concerning the ranges of A and A^*.

THEOREM C.5 (closedness of the range of a continuous operator). Let $A : X \to Y$ be a continuous linear mapping between two Banach spaces. The following assertions are equivalent:
(a) $A(X)$ is closed in Y;
(b) $A^*(Y^*)$ is norm-closed in X^*;
(c) $A(X)$ coincides with the orthogonal in Y of the kernel of A^*;
(d) $A^*(Y^*)$ coincides with the orthogonal in X^* of the kernel of A.

The surjectivity of a continuous linear mapping between Banach spaces is known (see, for example, Theorem 2.20 in [178]) to be characterized with the help of its adjoint as follows.

THEOREM C.6 (surjectivity of continuous operator and its adjoint). Let $A : X \to Y$ be a continuous linear mapping between two Banach spaces X, Y. Then the following are equivalent:
(a) A is surjective;
(b) There exists a real constant $c \geq 0$ such that
$$\|y^*\| \leq c \|A^* y^*\| \quad \text{for all } y^* \in Y^*;$$
(c) $\text{Ker } A^* = \{0\}$ and $A^*(Y^*)$ is norm-closed in X^*.

Let us recall now the Goldstine theorem.

THEOREM C.7 (Goldstine theorem). For any normed space $(X, \|\cdot\|)$ the unit ball $\mathbb{B}_X$ of X is $w(X^{**}, X^*)$-dense in the unit ball $\mathbb{B}_{X^{**}}$ of the topological bidual space X^{**}.

The famous James theorem on weak compactness [552] is generally stated for subsets of Banach spaces in the following form (see, for example, J. Diestel's book [334, Chapter I, Theorem 3, (i) and (iv)] and see also the proof there).

THEOREM C.8 (James characterization theorem of weak compactness in Banach space). A nonempty weakly closed set C in a Banach space X is weakly compact if and only if every continuous linear functional on X attains it supremum on C.

In fact, as stated in Theorem C.10 below, a more general version holds true in certain locally convex spaces. Given a Hausdorff locally convex space (X, τ_X) and its topological dual X^*, we recall that the strong topology $\beta(X^*, X)$ on X^* is the topology generated by the uniform convergence over bounded sets of X. The bidual of X, denoted by X^{**}, is the topological dual of $(X^*, \beta(X^*, X))$. The locally convex space X is called *semi-reflexive* if the canonical embedding (or evaluation mapping) $X \ni x \mapsto \langle \cdot, x \rangle$ from X into X^{**} is surjective. In contrast, the Hausdorff locally convex space (X, τ_X) is called *reflexive* if the canonical embedding is a homeomorphism from (X, τ_X) onto $(X^{**}, \beta(X^{**}, X^*))$, where $\beta(X^{**}, X^*)$ is the topology on X^{**} of uniform convergence over bounded sets in $(X^*, \beta(X^*, X))$ (see

[**881**] for more details). It is worth mentioning that every semi-reflexive normed space is a reflexive Banach space (see [**881**, Corollary 2, IV p. 145]).

THEOREM C.9 (semi-reflexivity). *A Hausdorff locally convex space X is semi-reflexive if and only if every bounded subset of X is relatively weakly compact.*

We recall next the following James' theorem for which we refer to R. C. James' paper [**551**]; a proof can also be found in J. Diestel's book [**334**, Chapter I, Theorem 5, (i) and (iv)].

THEOREM C.10 (James characterization theorem of weak compactness in locally convex space). *Let X be a complete Hausdorff locally convex space. A nonempty weakly closed subset C in X is weakly compact if and only if every continuous linear functional on X attains it supremum on C.*

We recall now some renorming features for Banach spaces. Remind that a norm $\|\cdot\|$ on a Banach space X has the (sequential) *Kadec-Klee property* when for any sequence $(x_n)_n$ of X converging weakly to x with $\|x_n\| \to \|x\|$ one has $\|x_n - x\| \to 0$. The following renorming theorem for reflexive Banach spaces can be found for example in R. Deville, G. Godefroy and V. Zizler's book [**333**, Proposition VII-2-1].

THEOREM C.11 (F-differentiable LUC renorm under reflexivity). *Any reflexive Banach space can be renormed with an equivalent norm which is both Fréchet differentiable (off zero) and locally uniformly convex, so in particular the renorm is both Kadec-Klee and strictly convex.*

Recall that a Banach space $(X, \|\cdot\|)$ is *Asplund* provided that any continuous convex function $f: U \to \mathbb{R}$ on an open convex set U of X is Fréchet differentiable on a dense G_δ set in U. It is known (see, for example, the book of R. Deville, G. Godefroy and V. Zizler [**333**, Theorem I.5.7] and the book of R. R. Phelps [**801**, Theorem 2.34]) that a Banach space is Asplund if and only if the topological dual of any separable subspace is separable. It is also known (see, for example, [**333**, Corollary II.3.3]) that a separable Banach space admits an equivalent Fréchet differentiable (off zero) norm if and only if its topological dual is separable. Consequently, we can reformulate both above results in the following unified way.

THEOREM C.12 (Asplund property via separable subspaces). *For a Banach space X, the following are equivalent:*
(a) *The space X is Asplund;*
(b) *the topological dual of any separable subspace of X is separable;*
(c) *any separable subspace of X admits an equivalent norm which is Fréchet differentiable off zero.*

By the assertion (b) we have the following corollary with the example of reflexive Banach spaces as Asplund spaces.

COROLLARY C.13. *Any reflexive Banach space is an Asplund space.*

Note that the corollary can also be justified by Theorem C.11 and the Ekeland-Lebourg Theorem 4.49.

By reflexivity the Lebesgue and Sobolev spaces $L^p(\Omega, \mathbb{R})$ and $W^{m,p}(\Omega, \mathbb{R})$ with $p \in]1, \infty[$ (and Ω open in $\mathbb{R}^n$) are examples of Asplund spaces. However, these spaces with either $p = 1$ or $p = \infty$ as well as the Banach space of continuous functions $\mathcal{C}([0,1], \mathbb{R})$ are not Asplund spaces.

Bounded subsets of topological duals of Asplund spaces enjoy a remarkable weak* sequential compactness property. Given a Banach space X, recall that a subset Q of the topological dual X^* is said to be *weak* sequentially compact* whenever every sequence in Q admits a subsequence weak* converging to some element in Q. A key result related to such a sequential property is the famous Hagler-Johnson theorem (see [**460**]) whose proof can also be found, for example, in J. Diestel's book [**335**, Chapter XIII, Theorem 6].

THEOREM C.14 (J. Hagler and W.B. Johnson). *If the topological dual X^* of a Banach space X contains a bounded sequence without a weak* convergent subsequence, then X contains a separable subspace with nonseparable dual.*

As a direct consequence of the latter theorem and the characterization (b) in Theorem C.12, one has the following theorem for the situation of Asplund spaces.

THEOREM C.15 (sequential w*-compactness of balls in Asplund spaces). *If X is an Asplund space, then any closed ball of the topological dual X^* is weak* sequentially compact, or equivalently, any bounded sequence in X^* contains a weak* convergent subsequence.*

It is also worth stating a particular case of a result of J. Hagler and F. Sullivan in [**461**, Theorem 1] related to the weak* compactness of dual balls.

THEOREM C.16 (J. Hagler and F. Sullivan). *If a Banach space X admits an equivalent Gâteaux differentiable (off zero) norm, then any bounded sequence in X^* has a weak* convergent subsequence.*

Concerning uniformly convex (resp. smooth) norms defined in Definition 18.12 (resp. Definition 18.24) and presented in Subsections 18.1.2 and 18.1.3, some additional details are needed for certain renorming features utilized in the theory developed in Sections 18.5 and 18.6 for prox-regular sets in uniformly convex Banach spaces. The story of renorming uniformly convex Banach spaces with norms possessing improved properties can be briefly presented as follows.

Given a property $(\mathcal{P})$ for Banach spaces, one says that a Banach space $(X, \|\cdot\|_X)$ has super-$(\mathcal{P})$ whenever every Banach space $(Y, \|\cdot\|_Y)$ which is finitely representable in $(X, \|\cdot\|_X)$ possesses the property $(\mathcal{P})$. A Banach space $(Y, \|\cdot\|_Y)$ is *finitely representable* in $(X, \|\cdot\|_X)$ if for any finite dimensional vector subspace Y_0 of Y and any real $\lambda > 1$ there exists an isomorphism T from Y_0 onto a vector subspace of X such that
$$\frac{1}{\lambda}\|y\|_Y \leq \|T(y)\|_X \leq \lambda\|y\|_Y \quad \text{for every } y \in Y_0;$$
see James' paper [**554**, Definition 1]. Clearly, $(X, \|\cdot\|_X)$ possesses the property $(\mathcal{P})$ whenever it has super-$(\mathcal{P})$. The concept of Banach spaces with super-$(\mathcal{P})$ is mostly known when the property "$\mathcal{P}$" is the "reflexivity property". The Banach space $(X, \|\cdot\|_X)$ is called *super-reflexive* by R. C. James [**554**, Definition 3] provided that any Banach space which is finitely representable in $(X, \|\cdot\|_X)$ is reflexive. Many results of interest were established for super-reflexive spaces by James. In particular, he proved in [**554**, Theorem 2] the following result:

THEOREM C.17 (R.C. James). *A Banach space is super-reflexive if and only if its topological dual is.*

Evidently, by what precedes Theorem C.17 every super-reflexive Banach space is reflexive, but the converse fails as confirms Enflo's Theorem C.18 below. Via uniformly non-square Banach spaces, James also proved (see [**553**, Lemma C]) that any Banach space admitting an equivalent uniformly convex (resp. uniformly smooth) norm is super-reflexive. The converse implication of the latter result was achieved with the fundamental Enflo theorem [**385**].

THEOREM C.18 (P. Enflo). *A Banach space $(X, \|\cdot\|)$ is super-reflexive if and only if it admits an equivalent norm which is uniformly convex.*

For the development of that result of Enflo we refer also to the book of R. Deville, G. Godefroy and V. Zizler [**333**, p. 139-152]; as stated above the result corresponds to Corollary IV.4.6 in that book [**333**].

Let $(X, \|\cdot\|)$ be a Banach space and let $\|\cdot\|_i$, $i = 1, 2$, be two norms on X equivalent to $\|\cdot\|$. Suppose that the norm $\|\cdot\|_1$ possesses a certain degree of convexity $(\mathcal{C}_1)$ (of type "strict convexity", "uniform convexity", "uniform convexity with power type"q, "local uniform convexity"). Suppose also that the dual norm of $\|\cdot\|_2$ on X^* has a certain degree of convexity $(\mathcal{C}_2)$ (of one of the above types). Recall that E. Asplund averaging procedure (see, for example, J. Diestel's book [**334**, p. 113]) provides a norm $\|\cdot\|$ equivalent to $\|\cdot\|$ and which possesses the degree of convexity $(\mathcal{C}_1)$ and whose dual norm $\|\cdot\|_*$ on X^* possesses the degree of convexity $(\mathcal{C}_2)$. The Asplund's avering procedure is largely and well developed in Disetel's book [**334**, p. 106-113].

According to this Asplund's averaging procedure, to Enflo Theorem C.18 and to James Theorem C.17 recalled above, one can state:

THEOREM C.19 (P. Enflo). *Any uniformly convex Banach space admits an equivalent norm which is both uniformly convex and uniformly smooth.*

G. Pisier [**803**] extended the above Enflo Theorem C.18 in showing that the equivalent renorm can be required to be such that its modulus of convexity be of power type. Pisier's result can also be found, for example, in B. Beauzamy's book [**79**, p. 273] and a proof is developed in pages 273-290 of that book [**79**]. By the Asplund averaging procedure again and by Proposition 18.38 we can state:

THEOREM C.20 (G. Pisier). *Any uniformly convex Banach space admits an equivalent norm which is uniformly convex with modulus of convexity of power type $q \in [2, +\infty[$ and uniformly smooth with modulus of smoothness of power type $s \in]1, 2]$.*

APPENDIX D

Measure theory

Let λ denote the *outer Lebesgue measure* on the Euclidean space $\mathbb{R}^N$ given, for every subset $S \subset \mathbb{R}^N$, by $\lambda(S)$ equal to the infimum of $\sum_{n\in\mathbb{N}} \mathrm{meas}_N(P_n)$ over the sequences $(P_n)_n$ of N-rectangles in $\mathbb{R}^N$ with $S \subset \bigcup_{n\in\mathbb{N}} P_n$. Given a set S in $\mathbb{R}^N$, a point $\bar{x} \in S$ is called a λ-*density point* or an N-*dimensional outer Lebesgue density point* of S whenever
$$\lim_{r\downarrow 0} \frac{\lambda(S \cap B(\bar{x},r))}{\lambda(B(\bar{x},r))} = 1.$$
The classical theorem on density points (see, e.g., [389, Corollary 3, p. 45], [695, Corollary 2.14(1), p. 38]) is generally stated as: λ-almost every point of a Lebesgue measurable set A in $\mathbb{R}^N$ is a Lebesgue density point of A. Consider any nonempty bounded set S in $\mathbb{R}^N$ and take a Lebesgue measurable set $A \supset S$ with $\lambda(S) = \lambda(A)$. For any Lebesgue measurable set B, by outer measure property of λ we have
$$\lambda(A \cap B) + \lambda(A \cap B^c) = \lambda(A) = \lambda(S) \le \lambda(S \cap B) + \lambda(S \cap B^c),$$
where $B^c := \mathbb{R}^N \setminus B$. This and the inequality $\lambda(A \cap E) \ge \lambda(S \cap E)$ gives $\lambda(S \cap B) = \lambda(A \cap B)$ for any Lebesgue measurable set B in $\mathbb{R}^N$. Using this and the definition of λ-density point, as noticed in [695, Remark 2.15(2), p 39], it is easily seen that the theorem on Lebesgue density points can be reformulated as follows:

THEOREM D.1. *Lebesgue almost every point of a set S of the Euclidean space $\mathbb{R}^N$ is a Lebesgue density point of S.*

Let (X,d) be a metric space, $s \in [0, +\infty[$ and $0 < \alpha(s) < +\infty$ with $\alpha(0) = 1$. For any set $C \subset X$ and any real $\delta > 0$ let
$$\mathcal{H}_{s,\delta}(C) = \inf\left\{\sum_{j\in\mathbb{N}} \frac{\alpha(s)}{2^s} (\mathrm{diam}(Q_j))^s : C \subset \bigcup_{j\in\mathbb{N}} Q_j, \mathrm{diam}(Q_j) \le \delta\right\},$$
where we use the convention $0^0 = 0$ and $\mathrm{diam}(\emptyset) = 0$. One defines $\mathcal{H}_s(C)$ by
$$\mathcal{H}_s(C) = \lim_{\delta\downarrow 0} \mathcal{H}_{s,\delta}(C) = \sup_{\delta>0} \mathcal{H}_{s,\delta}(C).$$
It is known that for any subsets C and C_j, with $j \in \mathbb{N}$, of X satisfying $C \subset \bigcup_{j\in\mathbb{N}} C_j$ one has (see, for example, [389, Chapter 2] and [695, Chapter 4])
$$\mathcal{H}_s(C) \le \sum_{j\in\mathbb{N}} \mathcal{H}_s(C_j),$$
so $\mathcal{H}_s$ is an outer measure on the set of all subsets of X. It is called the s-*Hausdorff measure on X relative to $\alpha(\cdot)$*. For $s = 0$ the Hausorff measure $\mathcal{H}_0$ is the *counting measure*, that is,
$$\mathcal{H}_0(C) = \mathrm{card}\, C,$$

and for $0 \leq s < t < +\infty$ (see, for example, [**695**, Chapter 4])
$$\bigl(\mathcal{H}_s(C) < +\infty \Rightarrow \mathcal{H}_t(C) = 0\bigr) \quad \text{and} \quad \bigl(\mathcal{H}_t(C) > 0 \Rightarrow \mathcal{H}_s(C) = +\infty\bigr).$$
Further, if (Y, d_Y) is another metric space and $f : X \to Y$ is a Lipschitz continuous mapping with Lipschitz constant ℓ, one has

(D.1) $$\mathcal{H}_s(f(C)) \leq \ell^s \mathcal{H}_s(C).$$

When (X, d) is the Euclidean space $X = \mathbb{R}^N$ with its canonical Euclidean distance, one has (see, for example, [**389**, Chapter 2])

  (i) $\mathcal{H}_s \equiv 0$ for any real $s > N$;
  (ii) $\mathcal{H}_s(a + A(C)) = \mathcal{H}_s(C)$ for any $a \in \mathbb{R}^N$ and any isometric linear mapping $A : \mathbb{R}^N \to \mathbb{R}^N$;
  (iii) $\mathcal{H}_s(\rho C) = \rho^s \bigl(\mathcal{H}_s(C)\bigr)$ for any real $\rho > 0$.

Furthermore, in the same setting of $\mathbb{R}^N$ and with

(D.2) $$\alpha(s) = \frac{\pi^{s/2}}{\Gamma(\frac{s}{2} + 1)},$$

where $\Gamma(\cdot)$ is the usual Γ-function, it is known (see, for example, [**389**, Chapter 2, Theorem 2]) that the Hausdorff measure $\mathcal{H}_N$ relative to $\alpha(N)$ with $\alpha(\cdot)$ as in (D.2) coincides with the N-dimensional Lebesgue measure on $\mathbb{R}^N$. Accordingly, the Hausdorff measure $\mathcal{H}_s$ relative to $\alpha(s)$ in (D.2) is called the *normalized s-Hausdorff measure*. When $s = k$ is an integer $0 \leq k \leq N$ and $\alpha(k)$ is taken as in (D.2), one says that $\mathcal{H}_k$ is the *normalized k-dimensional Hausdorff measure* on $\mathbb{R}^N$.

APPENDIX E

Differential calculus and differentiable manifolds

We pass now to three pillars of differential calculus in Banach spaces: inverse mapping theorem, implicit function theorem and local submersion theorem. Let us begin first with the inverse mapping theorem. This requires to recall the concept of diffeomorphism.

Let $f : U \to V$ be a mapping from an open set U of a Banach space X into an open set V of a Banach space Y. One says that f is a $\mathcal{C}^p$-*diffeomorphism* for an integer $p \geq 1$ if f is a bijection from U onto V such that f and its inverse f^{-1} are of class $\mathcal{C}^p$ on U and V respectively. With this we can state the classical inverse mapping theorem which can be found, for example, in Theorem 2.5.2 of the book [1] of R. Abraham, J. E. Marsden and T. Ratiu.

THEOREM E.1 (inverse mapping theorem). *Let $f : O \to Y$ be a mapping from an open set O of a Banach space X into a Banach space Y and let $\bar{x} \in O$ be a point around which f is of class $\mathcal{C}^p$ (resp. let $\bar{x} \in O$ be a point at which f is strictly Fréchet differentiable) and such that $Df(\bar{x})$ is a bijection from X onto Y. Then there exist an open neighborhood $U \subset O$ of $\bar{x}$ such that $V := f(U)$ is an open neighborhood of $f(\bar{x})$ and the restriction $f_U : U \to V$ of f to U is a $\mathcal{C}^p$-diffeomorphism (resp. and the inverse f_U^{-1} is strictly Fréchet differentiable at $f(\bar{x})$). Further, one has*

$$D(f_U^{-1})(f(x)) = \bigl(Df(x)\bigr)^{-1} \quad \text{for all } x \in U,$$

(resp. $D(f_U^{-1})(f(\bar{x})) = \bigl(Df(\bar{x})\bigr)^{-1}$).

We state now the implicit function theorem (see, for example, [1, Theorem 2.5.5]).

THEOREM E.2 (implicit function theorem). *Let X, Y be Banach spaces, $O \subset X \times Y$ be an open set in $X \times Y$, and $f : O \to Z$ be a mapping from O into a Banach space Z. Assume that f is of class $\mathcal{C}^p$ around a point $(\bar{x}, \bar{y}) \in O$ (with an integer $p \geq 1$) and that the derivative with respect to the second variable $D_2 f(\bar{x}, \bar{y})$ is a bijection from Y onto Z. Then there exist open neighborhoods U of $\bar{x}$, V of $\bar{y}$ with $U \times V \subset O$, an open neighborhood W of $f(\bar{x}, \bar{y})$, and a unique $\mathcal{C}^p$ mapping $g : U \times W \to V$ such that*

$$f(x, g(x, z)) = z \quad \text{for all } (x, z) \in U \times W.$$

Given a Banach space X, a closed vector subspace X_1 admitting a complement closed vector space X_2, we recall (with $i = 1, 2$) the notation of the mapping $\pi_{X_i} : X \to X_i$ defined for every $x \in X = X_1 \oplus X_2$ by $x = \pi_{X_1}(x) \oplus \pi_{X_2}(x)$ with $\pi_{X_i}(x) \in X_i$. With this we can recall the local submersion theorem for which the reader can consult, for example, Theorem 2.5.13 in [1].

THEOREM E.3 (local submersion theorem). Let X, Y be Banach spaces, O be an open set in X and $f : O \to Y$ be a mapping of class $\mathcal{C}^p$ on O with an integer $p \geq 1$. Let $\overline{x} \in O$ be such that $Df(\overline{x})$ is surjective and $X_2 := \operatorname{Ker} Df(\overline{x})$ admits a topological complement vector space X_1 in X with $X = X_1 \oplus X_2$. Then there exist an open neighborhood $U \subset O$ of $\overline{x}$ in X, an open neighborhood V of $(f(\overline{x}), \pi_{X_2}(\overline{x}))$ in $Y \times X_2$, and a $\mathcal{C}^p$-diffeomorphism $\psi : V \to U$ from V onto U such that

$$f \circ \psi(v_1, v_2) = v_1 \quad \text{for all } (v_1, v_2) \in V$$

along with $\psi(v_1, v_2) = \psi_1(v_1, v_2) \oplus v_2$ with a $\mathcal{C}^p$-mapping $\psi_1 : V \to X_1$. Further, for each $(v_1, v_2) \in V$ one has that $D\psi(v_1, v_2)_{|Y \times \{0_{X_2}\}} : Y \times \{0_{X_2}\} \to X_1$ is an isomorphism.

We now recall some basic concepts concerning submanifolds in Banach spaces. To do so, we need first some features on locally uniformly continuous mappings. Given two normed spaces X, Y and a nonempty open set U of X, a mapping $f : U \to Y$ is said to be locally uniformly continuous when for each $\overline{x} \in U$ there exists some neighborhood $U_0 \subset U$ of $\overline{x}$ such that f is uniformly continuous on U_0; otherwise stated, for every real $\varepsilon > 0$ there is $\delta > 0$ such that

$$\|f(x) - f(x')\| \leq \varepsilon \quad \text{for all } x, x' \in U_0 \text{ with } \|x - x'\| \leq \delta.$$

While the class of continuous mappings from U into Y is usually denoted by $\mathcal{C}(U, Y)$ or $\mathcal{C}^0(U, Y)$, the class of locally uniformly continuous mappings from U into Y will be denoted by $\mathcal{C}^{0,0}(U, Y)$, and similarly $\mathcal{C}^{0,1}(U, Y)$ will stand for the class of locally Lipschitz continuous mappings from U into Y. Analogously, recalling (for $p \in \mathbb{N}$) that the class of p-continuously differentiable mappings from U into Y is denoted by $\mathcal{C}^p(U, Y)$, the class of mappings $f \in \mathcal{C}^p(U, Y)$ such that the p-th derivative $D^p f$ is locally uniformly continuous (resp. locally Lipschitz continuous) will be denoted by $\mathcal{C}^{p,0}(U, Y)$ (resp. $\mathcal{C}^{p,1}(U, Y)$).

When the normed space X is finite-dimensional, it is clear according to the compactness of closed balls in X that the classes $\mathcal{C}^{p,0}(U, Y)$ and $\mathcal{C}^p(U, Y)$ coincide for each $p \in \{0\} \cup \mathbb{N}$. Such a coincidence fails in infinite dimensions as shown in the following Izzo example [550].

EXAMPLE E.4 (A.J. Izzo's example). The function $f : \ell_2(\mathbb{N}) \to \mathbb{R}$, defined by $f(x) = \sum_{n=1}^{\infty} x_n^2 \cos(nx_n)$ for all $x := (x_n)_{n \in \mathbb{N}} \in \ell_2(\mathbb{N})$, is continuous but fails to be locally uniformly continuous. □

We present, in Lemma E.5 and Proposition E.6, two properties of locally uniformly continuous mappings.

LEMMA E.5. Let X_1, X_2 and X_3 be three normed spaces and let U and V be two nonempty open sets of X_1 and X_2, respectively. Let $f : U \to V$ and $g : V \to X_3$ be such that $f \in \mathcal{C}^{p,0}(U, X_2)$ and $g \in \mathcal{C}^{p,0}(V, X_3)$ (resp. $f \in \mathcal{C}^{p,1}(U, X_2)$ and $g \in \mathcal{C}^{p,1}(V, X_3)$), where p is an integer $p \geq 0$. Then, one has

$$g \circ f \in \mathcal{C}^{p,0}(U, X_3) \quad (\text{resp. } g \circ f \in \mathcal{C}^{p,1}(U, X_3)).$$

PROOF. We consider only the $\mathcal{C}^{p,0}$-case since the $\mathcal{C}^{p,1}$-case is similar. We proceed by induction. The result is clear for $p = 0$. Let us show it for $p = 1$. Suppose that f and g are of class $\mathcal{C}^{1,0}$. By chain rule, we know that for every $u \in U$,

$$D(g \circ f)(u) = Dg(f(u)) \circ Df(u) = \beta(Dg(f(u)), Df(u)),$$

where $\beta : \mathcal{L}(X_2, X_3) \times \mathcal{L}(X_1, X_2) \to \mathcal{L}(X_1, X_3)$ is the continuous bilinear operator given by $\beta(\Lambda_2, \Lambda_1) = \Lambda_2 \circ \Lambda_1$. Define
$$F : U \to X := \mathcal{L}(X_2, X_3) \times \mathcal{L}(X_1, X_2) \quad \text{by} \quad F(u) = (h(u), Df(u))$$
with $h := (Dg) \circ f$, and notice that F is locally uniformly continuous since both mappings h and Df are locally uniformly continuous. The mapping β being locally uniformly continuous (in fact, it is of class $\mathcal{C}^\infty$), the equality $D(g \circ f)(\cdot) = \beta \circ F(\cdot)$ ensures the local uniform continuity of $D(g \circ f)$, which justifies the result for $p = 1$.

Now assume that the result holds for $p \geq 1$ and let us show that it holds for $p + 1$. Suppose that $f \in \mathcal{C}^{p+1,0}(U, X_2)$ and $g \in \mathcal{C}^{p+1,0}(V, X_3)$. Then $Dg \in \mathcal{C}^{p,0}(V, \mathcal{L}(X_2, X_3))$ and by the induction assumption we also has $h := (Dg) \circ f \in \mathcal{C}^{p,0}(U, \mathcal{L}(X_2, X_3))$. Thus, the mapping $D^p F(\cdot) = (D^p h(\cdot), D^p f(\cdot))$ is locally uniformly continuous, which combined with the fact that β is of class $\mathcal{C}^{p,0}$ entails by the induction assumption again that
$$D(g \circ f)(\cdot) = \beta \circ F(\cdot) \in \mathcal{C}^{p,0}(U, \mathcal{L}(X_1, X_3)).$$
This means that $g \circ f \in \mathcal{C}^{p+1,0}(U, X_3)$, and the proof is finished. □

Consider now two nonempty open sets U and V of Banach spaces X and Y respectively, and a $\mathcal{C}^p$-diffeomorphism $F : U \to V$ with $p \in \mathbb{N}$, so by definition $F^{-1} : V \to U$ is also a $\mathcal{C}^p$-diffeomorphism. We have for every $v \in V$

(E.1) $$DF^{-1}(v) = \left(DF(F^{-1}(v))\right)^{-1} = J \circ DF \circ F^{-1}(v),$$

where $J : \mathrm{Iso}(X, Y) \to \mathrm{Iso}(Y, X)$ is the homeomorphism defined by $J(\Lambda) := \Lambda^{-1}$ and $\mathrm{Iso}(X, Y)$ denotes the subset in $\mathcal{L}(X, Y)$ of isomorphisms from X onto Y. It is known (see, for example, [409, Theorem 3.1.5]) that J is a $\mathcal{C}^\infty$-mapping, and hence it belongs to $\mathcal{C}^{p,0}(\mathrm{Iso}(X, Y), \mathcal{L}(Y, X))$. Note that F^{-1} is of class $\mathcal{C}^{p-1,0}$ since it is of class $\mathcal{C}^p$. If in fact $F \in \mathcal{C}^{p,0}(U, Y)$, then E.1 and Lemma E.5 give that $DF^{-1} \in \mathcal{C}^{p-1,0}(V, \mathcal{L}(Y, X))$, so $F^{-1} \in \mathcal{C}^{p,0}(V, X)$. Similar arguments are valid when F is a $\mathcal{C}^{p,1}$-diffeomorphism. Then, by changing the roles of F and F^{-1} we obtain:

PROPOSITION E.6. *Let U and V be two nonempty open sets of Banach spaces X and Y respectively. For any $\mathcal{C}^p$-diffeomorphism $F : U \to V$ (with $p \geq 1$), one has*
$$F \in \mathcal{C}^{p,0}(U, Y) \Leftrightarrow F^{-1} \in \mathcal{C}^{p,0}(V, X) \quad (\text{resp. } F \in \mathcal{C}^{p,1}(U, Y) \Leftrightarrow F^{-1} \in \mathcal{C}^{p,1}(V, X))$$

REMARK E.7. Through Proposition E.6 we can see, by the proof of Theorem 2.5.13 in [1], that the mapping $\psi : V \to U$ in Theorem E.3 above is a $\mathcal{C}^{p,0}$-diffeomorphism (resp. $\mathcal{C}^{p,1}$-diffeomorphism) whenever the mapping $f : O \to Y$ therein is of class $\mathcal{C}^{p,0}$ (resp. of class $\mathcal{C}^{p,1}$). □

Let us now recall the definition of the three concepts of $\mathcal{C}^p$-submanifolds, $\mathcal{C}^{p,0}$-submanifolds and $\mathcal{C}^{p,1}$-submanifolds.

DEFINITION E.8. *Let M be a subset of a Banach space X and let $m_0 \in M$. One says that M is a $\mathcal{C}^p$-submanifold with $p \in \mathbb{N}$ at/near m_0 if there exist a closed vector subspace E of X, an open neighborhood U of m_0 in X, an open neighborhood V of zero in X, and a mapping $\varphi : U \to V$ such that*

(a) *φ is a $\mathcal{C}^p$-diffeomorphism, that is, $\varphi : U \to V$ is bijective and both φ and its inverse are of class $\mathcal{C}^p$;*

(b) $\varphi(m_0) = 0$ and $\varphi(M \cap U) = E \cap \varphi(U)$.

In such a case, E is a *model space* and (U, φ) is a *local chart*.

If in addition the condition

(c) $\varphi \in \mathcal{C}^{p,0}(U, X)$ (resp. $\varphi \in \mathcal{C}^{p,1}(U, X)$)

is satisfied, one says that M is a $\mathcal{C}^{p,0}$-submanifold (resp. $\mathcal{C}^{p,1}$-submanifold) at/near $m_0 \in M$.

When M is a $\mathcal{C}^p$-submanifold (resp. $\mathcal{C}^{p,0}$-submanifold, or $\mathcal{C}^{p,1}$-submanifold) at/near each point in M with the same model vector space E, M is called a $\mathcal{C}^p$-submanifold (resp. $\mathcal{C}^{p,0}$-submanifold, or $\mathcal{C}^{p,1}$-submanifold).

If $M \subset X$ is a $\mathcal{C}^1$-submanifold at $m_0 \in M$, its *tangent space* at m_0 is

$$\text{(E.2)} \qquad T_{m_0} M := \left\{ h \in X : \begin{array}{c} \exists \gamma :]-1, 1[\to M \text{ a } \mathcal{C}^1\text{-curve with} \\ \gamma(0) = m_0 \text{ and } \gamma'(0) = h \end{array} \right\}.$$

For any local chart (U, φ) and any model (closed vector) space E representing M as a $\mathcal{C}^1$-submanifold at m_0, one has (see (17.14))

$$\text{(E.3)} \qquad T_{m_0} M = D\varphi(0)^{-1} E = T^C(M; m_0) = T^B(M; m_0),$$

where we recall that $T^C(\cdot; \cdot)$ (resp. $T^B(\cdot; \cdot)$) denotes the Clarke (resp. the Bouligand-Peano) tangent cone.

It is also worth noticing by Proposition E.6 that condition (c) in Definition E.8 guarantees that the inverse mapping φ^{-1} also belongs to $\mathcal{C}^{p,0}(\varphi(U), X)$ (resp. belongs to $\mathcal{C}^{p,1}(\varphi(U), X)$).

Assume that the model vector space E admits a topological vector complement E_c, so $X = E \oplus E_c$. With the local chart (U, φ) associated with E in Definition E.8, define $g : U \to E_c$ by $g(x) = \pi_{E_c} \circ \varphi(x)$. It is clear that g is of class $\mathcal{C}^p$ and that $U \cap M = \{x \in U : g(x) = 0\}$. A certain converse holds as shown in the following proposition.

PROPOSITION E.9. *Let U_0 be an open set of a Banach space X with $\bar{x} \in U_0$, let $p \in \mathbb{N}$ and let g be a mapping of class $\mathcal{C}^p$ (resp. $\mathcal{C}^{p,0}$, or $\mathcal{C}^{p,1}$) from U_0 into a Banach space Y such that $Dg(\bar{x})$ is surjective. Assume that $\text{Ker } Dg(\bar{x})$ admits a topological vector complement in X (which is the case, in particular, if X is a Hilbert space). Then, with $\bar{y} := g(\bar{x})$, the level set $M := \{x \in U_0 : g(x) = \bar{y}\}$ is a $\mathcal{C}^p$-submanifold (resp. $\mathcal{C}^{p,0}$-submanifold, or $\mathcal{C}^{p,1}$-submanifold) of X at $\bar{x}$.*

PROOF. Suppose (without loss of generality) that $\bar{x} = 0$ and $\bar{y} = 0$. We know by the local submersion theorem (see foregoing Theorem E.3 and Remark E.7) that, denoting by X_1 a topological vector complement in X of $X_2 := \text{Ker } A$ with $A := Dg(\bar{x})$, there exist an open neighborhood $U \subset U_0$ of $\bar{x} = 0$ in X, an open neighborhood V of $(0, 0)$ in $Y \times X_2$, and a $\mathcal{C}^p$-diffeomorphism (resp. $\mathcal{C}^{p,0}$-diffeomorphism, or $\mathcal{C}^{p,1}$-diffeomorphism) $\psi : V \to U$ from V onto U such that for all $(v_1, v_2) \in V$ one has $g \circ \psi(v_1, v_2) = v_1$ along with $\psi(v_1, v_2) = \psi_1(v_1, v_2) \oplus v_2$ with $\psi_1(v_1, v_2) \in X_1$. The continuous linear mapping $A_0 : X_1 \to Y$ with $A_0(x_1) := A(x_1)$ for all $x_1 \in X_1$ is bijective, hence an isomorphism from X_1 onto Y by the closed graph theorem, so the mapping $j : X_1 \oplus X_2 \to Y \times X_2$ defined by $j(x_1 \oplus x_2) = (A_0(x_1), x_2)$ is also an isomorphism. Let $j_V : j^{-1}(V) \to V$ the bijective restriction from $j^{-1}(V)$ onto V and consider the $\mathcal{C}^p$-diffeomorphism (resp. $\mathcal{C}^{p,0}$-diffeomorphism, or $\mathcal{C}^{p,1}$-diffeomorphism) $\varphi := j_V^{-1} \circ \psi^{-1}$ from U onto $j_V^{-1}(V)$.

Then, with $\pi_i := \pi_{X_i}$ we have
$$x \in \varphi(U \cap M) \Leftrightarrow \psi \circ j_V(x) \in U \text{ and } g \circ \psi(j_V(x)) = 0$$
$$\Leftrightarrow \psi \circ j_V(x) \in U \text{ and } g \circ \psi(A_0 \circ \pi_1(x), \pi_2(x)) = 0$$
$$\Leftrightarrow \psi \circ j_V(x) \in U \text{ and } A_0 \circ \pi_1(x) = 0.$$

From this we see that
$$x \in \varphi(U \cap M) \Leftrightarrow (\psi \circ j_V(x) \in U \text{ and } \pi_1 x = 0) \Leftrightarrow x \in \varphi(U) \cap X_2,$$
which means that $\varphi(U \cap M) = \varphi(U) \cap X_2$. Choose $(\bar{v}_1, \bar{v}_2) \in V$ such that $\psi(\bar{v}_1, \bar{v}_2) = 0$. We have $\bar{v}_1 = g \circ \psi(\bar{v}_1, \bar{v}_2) = 0$. Further, the equalities $0 = \psi(\bar{v}_1, \bar{v}_2) = \psi_1(\bar{v}_1, \bar{v}_2) \oplus \bar{v}_2$ entails that $\bar{v}_2 = 0$. Consequently, $\psi^{-1}(0) = (0, 0)$, so $\varphi(0) = j_V^{-1}(\psi^{-1}(0)) = 0$. Altogether, it results that M is a $\mathcal{C}^p$-submanifold (resp. $\mathcal{C}^{p,0}$-submanifold, or $\mathcal{C}^{p,1}$-submanifold) at $\bar{x} = 0$. $\square$

The next theorem shows that a submanifold as defined in Definition E.8 can be seen as the graph of a suitable mapping.

THEOREM E.10. *Let M be a subset of a Hilbert space H and $m_0 \in M$, and let Z be a closed vector subspace in H. Given an integer $p \in \mathbb{N}$, the set M is a $\mathcal{C}^p$-submanifold (resp. $\mathcal{C}^{p,0}$-submanifold, or $\mathcal{C}^{p,1}$-submanifold) at m_0 with Z as model vector space if and only if there exist an open neighborhood U of m_0 in H, an open neighborhood $V_Z \subset Z$ of zero in Z, and a mapping $\theta : V_Z \to Z^\perp$ such that*

(a) $\theta \in \mathcal{C}^p(V_Z, Z^\perp)$ (resp. $\theta \in \mathcal{C}^{p,0}(V_Z, Z^\perp)$, or $\theta \in \mathcal{C}^{p,1}(V_Z, Z^\perp)$);
(b) $\theta(0) = 0$ and $D\theta(0) = 0$;
(c) $M \cap U = (m_0 + L^{-1}(\text{gph}\,\theta)) \cap U$,

where $L : X \to Z \times Z^\perp$ is the canonic isomorphism given by the equality $L(x) = (\pi_Z(x), \pi_{Z^\perp}(x))$.

Further, under (a), (b) (c) one has $T_{m_0} M = Z$.

PROOF. We will consider only the situation when M is a $\mathcal{C}^{p,0}$-submanifold at m_0. The $\mathcal{C}^p$-submanifold is easier with similar arguments and the locally Lipschitz case is analogous.

To show the implication $\Leftarrow$, assume that (a), (b), (c) hold. Let us first prove that gph θ is a $\mathcal{C}^{p,0}$-submanifold at $(0,0)$. Indeed, consider the open set $U_1 := V_Z \times Z^\perp$ in $Z \times Z^\perp$ and the mapping $\varphi : U_1 \to Z \times Z^\perp$ defined by
$$\varphi(v, z_2) = (v, z_2 - \theta(v)) \quad \text{for all } (v, z_2) \in V_Z \times Z^\perp.$$
Obviously, we have $\varphi \in \mathcal{C}^{p,0}(U_1, Z \times Z^\perp)$ and $D\varphi(0,0) = \text{id}_{Z \times Z^\perp}$. By the local inverse mapping theorem (see Theorem E.1 above) there exists an open neighborhood $U_2 \subset U_1$ of $(0,0)$ in $Z \times Z^\perp$ such that $\phi := \varphi|_{U_2} : U_2 \to \varphi(U_2)$ is a $\mathcal{C}^p$-diffeomorphism, and clearly $\phi \in \mathcal{C}^{p,0}(U_2, Z \times Z^\perp)$. Further, since ϕ is the restriction of φ to U_2, we also have the equivalences
$$(v, z_2) \in U_2 \cap \text{gph}\,(\theta) \iff (v, z_2) \in U_2 \text{ and } z_2 = \theta(v)$$
$$\iff \phi(v, z_2) \in \phi(U_2) \text{ and } \pi_{Z^\perp}(\phi(v, z_2)) = 0$$
$$\iff \phi(v, z_2) \in \phi(U_2) \cap (Z \times \{0\}).$$

Thus, by Definition E.8 we see that gph θ is a $\mathcal{C}^{p,0}$-submanifold at $(0,0)$.

Now consider the affine mapping $\tilde{L} : X \to Z \times Z^\perp$ defined by $\tilde{L}(x) = L(x - m_0)$. Without loss of generality, we may suppose that $U_3 := \tilde{L}^{-1}(U_2) = L^{-1}(U_2) + m_0 \subset$

U. Consider the mapping $\widetilde{\phi} : U_3 \to L^{-1}(\phi(U_2))$ defined by $\widetilde{\phi} := L^{-1} \circ \phi \circ \widetilde{L}$. Since the bijective mappings L and $\widetilde{L}$ are of class $\mathcal{C}^\infty$ as well as their inverses, the mapping $\widetilde{\phi} : U_3 \to L^{-1}(\phi(U_2))$, defined by $\widetilde{\phi} := L^{-1} \circ \phi \circ \widetilde{L}$, satisfies the following properties:

(a) $\widetilde{\phi}$ is a $\mathcal{C}^p$-diffeomorphism;

(b) $\widetilde{\phi}(M \cap U_3) = \widetilde{\phi}(\widetilde{L}^{-1}(\operatorname{gph}\theta) \cap \widetilde{L}^{-1}(U_2))) = \widetilde{\phi}(\widetilde{L}^{-1}(\operatorname{gph}\theta \cap U_2)))$
$= L^{-1}(\phi(U_2) \cap Z \times \{0\}) = \widetilde{\phi}(U_3) \cap Z$;

(c) $\widetilde{\phi} \in \mathcal{C}^{p,0}(U_3, X)$ (by Lemma E.5).

This says that M is a $\mathcal{C}^{p,0}$-submanifold at m_0 with model space Z, so the desired implication $\Leftarrow$ is proved. Further, keeping in mind that $D\phi(0,0) = \operatorname{id}_{Z \times Z^\perp}$ the equality (E.3) gives that

$$T_0 M = D\widetilde{\phi}(0)^{-1} Z = (L^{-1} \circ D\phi(0,0))^{-1} \circ L)Z = Z,$$

which justifies the additional tangential equality in the statement of the theorem.

To prove the converse, assume that M is a $\mathcal{C}^{p,0}$-submanifold at m_0 with model space Z. Choose an open neighborhood W of m_0 in H and a $\mathcal{C}^p$-diffeomorphism $\varphi : W \to \varphi(W) \subset H$ such that $\varphi \in \mathcal{C}^{p,0}(W, H)$, $\varphi(m_0) = 0$ and that $\varphi(W \cap M) = \varphi(W) \cap Z$. Replacing φ by $D\varphi^{-1}(0) \circ \varphi$ if necessary and using the equality (E.3), we may and do suppose that $T_{m_0} M = Z$.

Consider the mapping $\phi : \varphi(W) \cap Z \to Z$ defined by $\phi(z) = \pi_Z(\varphi^{-1}(z) - m_0)$. Since $D\phi(0) = D\varphi^{-1}(0)|_Z$ is an isomorphism from Z to Z, the local inverse mapping theorem furnishes an open neighborhood O of zero in Z such that $\phi : O \to \phi(O)$ is a $\mathcal{C}^p$-diffeomorphism. Moreover, Lemma E.5 and Proposition E.6 give that $\phi \in \mathcal{C}^{p,0}(O, Z)$.

Choose $\delta > 0$ small enough such that

$$U := \varphi^{-1}(B_X(0,\delta)) \subset W \quad \text{and} \quad Z \cap B_X(0,\delta) \subset O.$$

Put $V := \phi(Z \cap B_X(0,\delta))$ and observe that

(E.4) $\qquad \varphi(M \cap U) = Z \cap \varphi(U) = Z \cap B_X(0,\delta) = \phi^{-1}(V).$

Now define $\theta : V \to Z^\perp$ by $\theta := \pi_{Z^\perp} \circ (\varphi^{-1} \circ \phi^{-1}(\cdot) - m_0)|_V$, and note that it belongs to $\mathcal{C}^{p,0}(V, Z^\perp)$ according to Lemma E.5. Further, we have $\theta(0) = 0$ and

$$D\theta(0) = \pi_{Z^\perp} \circ D\varphi^{-1}(0) \circ D\phi^{-1}(0) = 0,$$

since $D\varphi^{-1}(0) \circ D\phi^{-1}(0)Z = Z$. It remains to show that $U \cap M = (L^{-1}(\operatorname{gph}\theta) + m_0) \cap U$. For any $v = \phi(\varphi(m))$ with $m \in U \cap M$ noticing that

$$v_m = \phi(\varphi(m)) = \pi_Z(\varphi^{-1}(\varphi(m)) - m_0) = \pi_Z(m - m_0),$$

the definition of θ yields

(E.5) $\qquad L(m - m_0) = (\pi_Z(m - m_0), \pi_{Z^\perp}(\varphi^{-1} \circ \phi^{-1}(v) - m_0)) = (v, \theta(v)).$

This ensures that for any $m \in U \cap M$, we have $L(m - m_0) = (v_m, \theta(v_m))$ with $v_m := \phi(\varphi(m))$, while (E.4) gives $v_m \in V$. We deduce that $U \cap M \subset (L^{-1}(\operatorname{gph}\theta) + m_0) \cap U$. For the converse inclusion, take $v \in V$ such that $L^{-1}(v, \theta(v)) + m_0 \in U$. By (E.4) again there exists $m \in M \cap U$ such that $v = \phi \circ \varphi(m)$, so (E.5) entails the equality $L(m - m_0) = (v, \theta(v))$, hence $L^{-1}(v, \theta(v)) + m_0 \in M \cap U$. This justifies the desired converse inclusion and finishes the proof. $\square$

Bibliography

[1] R. Abraham, J. E. Marsden and T. Ratiu, *Manifolds, Tensor Analysis, and Applications*, Third Edition, Springer-Verlag, New York (2002).

[2] R. A. Adams, *Sobolev Spaces*, Academic Press, New York (1975).

[3] R. A. Adams and J. J. F. Fournier, *Sobolev Spaces*, Volume 140, Academic Press, Cambridge (2003).

[4] S. Adly, *A Variational Approach to Nonsmooth Dynamics, Applications in Unilateral Mechanics and Electronics*, Springer Briefs in Mathematics, Springer, New-York (2017).

[5] S. Adly, F. Nacry and L. Thibault, *Preservation of prox-regularity of sets with applications to constrained optimization*, SIAM J. Optim. 26 (2016), 448-473.

[6] S. Adly, F. Nacry and L. Thibault, *Discontinuous sweeping process with prox-regular sets*, ESAIM: COCV 23 (2017), 1293-1329

[7] S. Adly, F. Nacry and L. Thibault, *Prox-regularity approach to generalized equations and image projection*, ESAIM: COCV 24 (2018), 677-708.

[8] S. Adly, F. Nacry and L. Thibault, *New metric properties for prox-regular sets*, Math. Prog. 189 (2021), 7-36.

[9] N. I. Akhiezer and I. M. Glazman, *Theory of Linear Operators in Hilbert space*, Moscow, Gostekhizdat, (1950) (in Russian); English version Pitman Press (1980).

[10] P. Albano, *Some properties of semiconcave functions with general modulus*, J. Math. Anal. Appl. 271 (2002), 217-231.

[11] P. Albano and P. Cannarsa, *Singularities of semiconcave functions in Banach spaces*, in "Stochastic Analysis, Control, Optimization and Applications", Birkhäuser, Boston (1999), 171-190.

[12] P. Albano and P. Cannarsa, *Structural properties of singularities of semiconcave functions*, Ann. Scuola Norm. Sup. Pisa 28 (1999), 719–740.

[13] G. Alberti, *On the structure of singular points of convex functions*, Calc. Var. Partial Differential Equations, 2 (1994), 17-27.

[14] G. Alberti, L. Ambrosio and P. Cannarsa, *On the singularities of convex functions*, Manuscr. Math. 76 (1992), 421-435.

[15] O. Alvarez, P. Cardaliaguet and R. Monneau, *Existence and uniqueness for dislocation dynamics with positive velocity*, Interfaces Free Bound. 7 (2005), 415-434.

[16] C. Amara and M. Ciligot-Travain, *Lower CS-closed sets and functions*, J. Math. Anal. Appl. 239 (1999), 371-389.

[17] L. Ambrosio, P. Cannarsa and H. M. Soner, *On the propagation of singularities of semiconvex functions*, Ann. Scuola Norm. Sup. Pisa 20 (1993), 597–616.

[18] L. Ambrosio, N. Fusco, D. Pallara, *Functions of Bounded Variation and Free Discontinuity Problems*, Oxford Science Publications, Clarendon, Oxford (2000).

[19] R. D. Anderson and V. L. Klee, *Convex functions and upper semicontinuous collections*, Duke Math. J. 19 (1952), 349-357.

[20] F. J. Aragón Artacho and M. H. Geoffroy, *Characterizations of metric regularity of subdifferentials*, J. Convex Anal. 15 (2008), 365-380.

[21] F. J. Aragón Artacho and M. H. Geoffroy, *Metric regularity of the convex subdifferential in Banach spaces*, J. Nonlinear Convex Anal. 15 (2015), 35-47.

[22] G. Aronsson, *Extension of functions satisfying Lipschitz conditions*, Arkiv Math. 6 (1967), 551-561.

[23] N. Aronszajn, *Differentiability of Lipschitzian mappings between Banach spaces*, Studia Math. 57 (1976), 147-190.

[24] N. Aronszajn and K. T. Smith, *Functional spaces and functional completion*, Ann. Inst. Fourier, 6 (1956), 125-185.
[25] E. Asplund, *Farthest points in reflexive locally uniformly rotund Banach spaces*, Israel J. Math. 4 (1966), 213-216.
[26] E. Asplund, *Sets with unique farthest points*, Israel J. Math. 5 (1967), 201-209.
[27] E. Asplund, *Averaged norms*, Israel J. Math. 5 (1967), 227-233.
[28] E. Asplund, *Fréchet differentiability of convex functions*, Acta. Math. 121 (1968), 31-47.
[29] E. Asplund, *Čebyšev sets in Hilbert spaces*, Trans. Amer. Math. Soc. 144 (1969), 235-240.
[30] E. Asplund and R. T. Rockafellar, *Gradients of convex functions*, Trans. Amer. Math. Soc. 139 (1969), 443-467.
[31] H. Attouch, *Convergence de fonctions, des sous-différentiels et semi-groupes associés*, C. R. Acad. Sci. Paris 284 (1977), 539-542.
[32] H. Attouch, *Variational Convergence for Functions and Operators*, Pitman, Boston (1984).
[33] H. Attouch, J.-B. Baillon, M. Théra, *Variational sum of monotone operators*, J. Convex Anal. 1 (1994), 1-29.
[34] H. Attouch and G. Beer, *On the convergence of subdifferentials of convex functions*, Arch. Mat. 60 (1993), 389-400.
[35] H. Attouch, G. Buttazzo and G. Michaille, *Variational Analysis in Sobolev and BV Spaces: Applications to PDEs and Optimization*, (Second Edition) MPS-SIAM Book Series on Optimization 17, SIAM, Philadelphia (2014).
[36] H. Attouch and R. J-B. Wets, *Isometries for the Legendre-Fenchel transform*, Trans. Amer. Math. Soc. 296 (1986), 33-60.
[37] H. Attouch and R. J-B. Wets, *Quantitative stability of variational systems: I. The epigraphical distance*, Trans. Amer. Math. Soc. 328 (1991), 695-730.
[38] H. Attouch and R. J-B. Wets, *Quantitative stability of variational systems: II. A framework for nonlinear conditioning*, SIAM J. Optim. 3 (1993), 359-381.
[39] H. Attouch and R. J-B. Wets, *Quantitative stability of variational systems: III. ε-approximate solutions*, Math. Prog. 61 (1993), 197-214.
[40] J.-P. Aubin, *Contingent derivatives of set-valued maps and existence of solutions to nonlinear inclusions and differential inclusions*, in Mathematical Analysis and Applications, edited by L. Nachbin, pp. 159-229, Academics Press, New York (1981).
[41] J.-P. Aubin, *Lipschitz behavior of solutions to convex minimization problems*, Math. Oper. Res. 9 (1984), 87-111.
[42] J.-P. Aubin, A.Cellina, *Differential Inclusions. Set-Valued Maps and Viability Theory*, Springer, Berlin (1984).
[43] J.-P. Aubin and I. Ekeland, *Applied Nonlinear Analysis*, Wiley, New-York (1984).
[44] J.-P. Aubin and H. Frankowska, *Set-Valued Analysis*, Birkhäuser, Boston (1990).
[45] A. Auslender and M. Teboulle, *Asymptotic Cones and Functions in Optimization and Variational Inequalities*, Springer Monographs in Mathematics, Springer, New York (2003).
[46] D. Aussel, J.-N. Corvellec and M. Lassonde, *Subdifferential characterization of quasiconvexity and convexity*, J. Convex Anal. 1 (1994), 1-7.
[47] D. Aussel, J.-N. Corvellec and M. Lassonde, *Mean value property and subdifferential criteria for lower semicontinuous functions*, Trans. Amer. Math. Soc. 347 (1995), 4147-4161.
[48] D. Aussel, J.-N. Corvellec and M. Lassonde, *Nonsmooth constrained optimization and multidirectional mean value inequalities*, SIAM J. Optim. 9 (1999), 690-706.
[49] D. Aussel, A. Daniilidis and L. Thibault, *Subsmooth sets: functional characterizations and related concepts*, Trans. Amer. Math. Soc. 357 (2005), 1275-1301.
[50] D. Azagra, *Smooth negligibility and subdifferential calculus in Banach spaces, with applications*, Ph.D. dissertation thesis, Department of Mathematics, Universidad Complutense, Madrid (1997).
[51] D. Azagra and R. Deville, *Subdifferential Rolle's and mean value inequality theorems*, Bull. Austral. Math. Soc. 56 (1997), 317-329.
[52] D. Azagra, J. Ferrera and F. López-Mesas, *Approximate Rolle's theorem for the proximal subgradient and the generalized gradient*, J. Math. Anal. Appl. 283 (2003), 180-191.
[53] D. Azagra and M. Jiménez-Sevilla, *The failure of Rolle's theorem in infinite dimensional Banach spaces*, J. Funct. Anal. 182 (2001), 207-226.
[54] D. Azé, *Duality for the sum of convex functions in normed spaces*, Arch. Math. 62 (1994), 554-561.

[55] D. Azé, *Éléments d'Analyse Convexe et Variationnelle*, Ellipses, Paris (1997).
[56] D. Azé, *A unified theory for metric regularity of multifunctions*, J. Convex Anal. 13 (2006), 225-252.
[57] D. Azé and J.-N. Corvellec, *Characterizations of error bounds for lower semicontinuous functions on metric spaces*, ESAIM Control Optim. Calc. Var. 10 (2004), 409-425.
[58] D. Azé and J.-B. Hiriart-Urruty, *Sur un air de Rolle and Rolle*, Revue des Math. de l'Ens. Supérieur 3-4 (2000), 455-460.
[59] D. Azzam-Laouir, C. Castaing and M. D. P. Monteiro Marques, *Perturbed evolution problems with continuous bounded variation in time and applications*, Set-Valued Var. Anal. 26 (2018), 693-728.
[60] M. Bacák, J. M. Borwein, A. Eberhard and B. S. Mordukhovich *Infimal convolution and Lipschitzian properties of subdifferentials for prox-regular functions in Hilbert spaces*, J. Convex Anal. 17 (2010), 737-763.
[61] R. Baire, *Sur les Fonctions de Variables Réelles*, Thesis (Thèse de Doctorat ès Sciences Mathématiques), Université de Paris (1899).
[62] A. Bakan, F. Deutsch and W. Li, *Strong CHIP, normality, and linear regularity of convex sets*, Trans. Amer. Math. Soc. 357 (2005), 3831-3863.
[63] V. S. Balaganski and L. P. Vlasov, *The problem of convexity of Chebyshev sets*, Russ. Math. Surv. 51:6 (1996), 1127-1190.
[64] M. V. Balashov, *Weak convexity of the distance function*, J. Convex Anal. 20 (2013), 93-106.
[65] M. V. Balashov, *Proximal smoothness of a set with the Lipschitz metric projection*, J. Math. Anal. Appl. 406 (2013), 360-363.
[66] M. V. Balashov and M. O. Golubev, *About the Lipschitz property of the metric projection in the Hilbert space*, J. Math. Anal. Appl. 394 (2012), 545-551.
[67] M. V. Balashov and G. E. Ivanov, *Weakly convex and proximally smooth sets in Banach spaces*, Izvestiya RAN: Ser. Mat. 73:3 (2009), 23-66 (in Russian); English translation in Izvestiya: Mathematics 73:3 (2009), 455-499.
[68] M. V. Balashov and D. Repovš, *Weakly convex sets and modulus of nonconvexity*, J. Math. Anal. Appl. 371 (2010), 113-127.
[69] L. Ban and W. Song, *Duality gap of the conic convex constrained optimization problems in normed spaces*, Math. Program. 119 (2009), 195-214.
[70] S. Banach, *Théorie des Opérations Linéaires*, Monografje Matematyczne, Warsaw (1932).
[71] R. Foygel Barber and W. Ha, *Gradient descent with nonconvex constraints: local concavity determines convergence*, Information and Inference: A Journal of the IMA 4 (2018), 755-806.
[72] V. Barbu and T. Precupanu, *Convexity and Optimization in Banach Spaces*, 4th Edition, Springer Monographs in Mathematics (2012).
[73] M. S. Bazaraa, J. J. Goode and M. Z. Nashed, *On the cone of tangents with applications to mathematical programming*, J. Optim. Theory Appl. 13 (1974), 389-426.
[74] M. Bardi, I. Capuzzo-Dolcetta, *Optimal Control and Viscosity Solutions of Hamilton–Jacobi–Bellman Equations*, Birkhäuser, Boston (1997).
[75] H. H. Bauschke and J. M. Borwein, *On the convergence of von Neumann's alternating projection algorithm for two sets*, Set-Valued Anal. 1 (1993), 185-212.
[76] H. H. Bauschke, J. M. Borwein and W. Li, *Strong conical hull intersection property, bounded linear regularity, Jameson's property (G), and error bounds in convex optimization*, Math. Program., Ser. A, 86 (1999), 135-160.
[77] H. H. Bauschke, J. M. Borwein and P. Tseng, *Bounded linear regularity, strong CHIP, and CHIP are distinct properties*, J. Convex Anal., 7 (2000), 395-412.
[78] H. H. Bauschke and P. L. Combettes, *Convex Analysis and Monotone Operator Theory in Hilbert Spaces*, CMS Books in Mathematics, Springer (second edition), New York (2017).
[79] B. Beauzamy, *Introduction to Banach Spaces and their Geometry*, 2nd Edition, North-Holland, Amsterdam (1985).
[80] E. F. Beckenbach, *Convex functions*, Bull. Amer. Math. Soc. 54 (1948), 439-460.
[81] G. Beer, *Conjugate convex functions and the epi-distance topology*, Proc. Amer. Math. Soc. 108 (1990), 117-126.
[82] G. Beer, *Topologies on Closed and Closed Convex Sets*, Kluwer, Dordrecht (1993).
[83] R. Bellman and W. Karush, *On a new functional transform in analysis: the maximum transform*, J. Soc. Indust. Appl. Math. 67 (1961), 501-503.

[84] R. Bellman and W. Karush, *Mathematical programming and the maximum transform*, Bull. Amer. Math. Soc. 10 (1962), 550-567.

[85] R. Bellman and W. Karush, *On the maximum transform*, J. Math. Anal. Appl. 6 (1963), 67-74.

[86] H. Benabdellah, *Existence of solutions to nonconvex sweeping process*, J. Differential Equations 164 (2000), 286-295.

[87] H. Benabdellah, C. Castaing, A. Salvadori and A. Syam, *Nonconvex sweeping processes*, J. Appl. Anal. 2 (1996), 217-240.

[88] J. Benoist and J.-B. Hiriart-Urruty, *General squeeze theorems in nonsmooth analysis*, Canadian Mathematical Society Conference Proceedings, 24 (2000), 7-17.

[89] Y. Benyamini and J. Lindenstrauss, *Geometric Nonlinear Functional Analysis, Vol. 1*, American Matheamtical Society Colloquium Publications 48, American Mathematical Society, Providence, Rhode Island (2000).

[90] C. Berg, J. P. R. Christensen and P. Ressen, *Harmonic Analysis on Semigroups*, Springer-Verlag, New-York (1984).

[91] C. Berge, *Espaces Topologiques et Fonctions Multivoques*, Dunod, Paris (1959).

[92] E. R. Berkson, *Some metrics on the subspaces of a Banach space*, Pacific J. Math. 13 (1963), 7-22.

[93] F. Bernard and L. Thibault, *Prox-regularity of functions and sets in Banach spaces*, Set-valued Anal. 12 (2004), 25-47.

[94] F. Bernard and L. Thibault, *Prox-regular functions in Hilbert spaces*, J. Math. Anal. Appl. 303 (2005), 1-14.

[95] F. Bernard and L. Thibault, *Uniform prox-regularity of functions in Hilbert spaces*, Nonlinear Analysis 60 (2005), 187-207.

[96] F. Bernard, L. Thibault and D. Zagrodny, *Integration of primal lower nice functions in Hilbert spaces*, J. Optim. Theory Appl. 124 (2005), 561-579.

[97] F. Bernard, L. Thibault and N. Zlateva, *Characterization of prox-regular sets in uniformly convex Banach spaces*, J. Convex Anal. 13 (2006), 525-559.

[98] F. Bernard, L. Thibault and N. Zlateva, *Prox-regular sets in uniformly convex Banach space: Tangential and various other properties*, Trans. Amer. Math. Soc. 363 (2011), 2211-2247 (submitted in 2008).

[99] A. S. Besicovitch, *On tangents to general sets of points*, Fundam. Math. 22 (1934), 49-53.

[100] A. S. Besicovitch, *On singular points of convex surfaces*, Proceedings of Symposia in Pure Mathematics, vol VII, Providence, Rhode Island (1963).

[101] D. N. Bessis and F. H. Clarke, *Partial subdifferentials, derivates and Rademacher's theorem*, Trans. Amer. Math. Soc. 351 (1999), 2899-2926.

[102] Z. Birnbaum and W. Orlicz, *Über die Verallgemeinerung des begriffes der zueinander ko,jugierten Potenzen*, Studia Math. 3 (1931), 1-67.

[103] E. Bishop and R. R. Phelps, *A proof that every Banach space is subreflexive*, Bull. Amer. Math. Soc. 67 (1961), 97-98.

[104] E. Bishop and R. R. Phelps, *The support functional of convex sets*, in "Convexity", edited by V. Klee, Symposia Pure Maths VII, Amer. Math. Soc. Providence (1963), 27-35.

[105] J. Blatter, *Weiteste Punkte und nächste Punkte*, Rev. Roumaine Math. Pures Appl. 14 (1969), 615-621.

[106] V. I. Bogachev, *Measure Theory. Volume I*, Mathematics-Analysis, Springer, (2007).

[107] V. I. Bogachev, *Measure Theory. Volume II*, Mathematics-Analysis, Springer, (2007).

[108] J. Bolte and E. Pauwels, *Conservative set valued fields, automatic differentiation, stochastic gradient method and deep learning*, Math. Program.188 (2021), 19-51.

[109] D. Bongiorno, *Stepanoff's theorem in separable Banach spaces*, Comment. Math. Univ. Carolin. 39 (1998), 323-335.

[110] D. Bongiorno, *Radon-Nikodým property of the range of Lipschitz extensions*, Atti. Sem. Mat. Fis. Univ. Modena 48 (2000), 517-525.

[111] J. F. Bonnans and A. S. Shapiro, *Perturbation Analysis of Optimization Problems*, Springer-Verlag, New-York (2000).

[112] J.-M. Bonnisseau and B. Cornet, *Fixed point theorems and Morse's lemma for Lipschitzian functions*, J. Math. Anal. Appl. 146 (1990), 318-332.

[113] J.-M. Bonnisseau and B. Cornet, *Existence of marginal cost pricing equilibria: The nonsmooth case*, International Economic Review 31 (1990), 685-708.

[114] J.-M. Bonnisseau and B. Cornet, *Existence of equilibria with a tight marginal pricing rule*, J. Mathematical Economics 44 (2008), 613-624.
[115] J.-M. Bonnisseau, B. Cornet and M.-O. Czarnecki, *The marginal pricing rule revisited*, Economic Theory 33 (2007), 579-589.
[116] S. Boralugada and R. A. Poliquin, *Local integration of prox-regular functions in Hilbert spaces*, J. Convex Anal. 13 (2006), 27-36.
[117] Yu. G. Borisovich, B. D. Gelman, A. D. Myshkis and V. V. Obukhovskii, *Multi-valued mappings*, J. Soviet Math. 18 (1982), 719-791.
[118] J. M. Borwein, *A Lagrange multiplier theorem and a sandwich theorem for convex relations*, Math. Scand. 48 (1081), 189-204.
[119] J. M. Borwein, *Convex relations in analysis and optimization in generalized concavity in optimization and economics*, (Academic Press, London 1981), 335-371.
[120] J. M. Borwein, *A note on ε-subgradients and maximal monotonicity*, Pacific. J. Math. 103 (1982), 305-314.
[121] J. M. Borwein, *Continuity and differentiability properties of convex operators*, Proc. London Math. Soc. 44 (1982), 420-444.
[122] J. M. Borwein, *Stability and regular points of inequality systems*, J. Optim. Theory Appl. 48 (1986), 9-52.
[123] J. M. Borwein, *Epi-Lipschitz-like sets in Banach space: theorems and examples*, Nonlinear Anal. 11 (1987), 1207-1217.
[124] J. M. Borwein, *Minimal CUSCOS and subgradients of Lipschitz functions*, Fixed PointTheory and its Applications, Pitman Research Notes 252 (1991), 57-81.
[125] J. M. Borwein, *Proximality and Chebyshev sets*, Optimization letters, (2007), 21-32.
[126] J. M. Borwein, J. V. Burke and A. S. Lewis, *Differentiability of cone-monotone functions on separable Banach space*, Proc. Amer. Math. Soc. 132 (2004), 1067-1076.
[127] J. M. Borwein and M. Fabian, *A note on regularity of sets and of distance functions in Banach space*, J. Math. Anal. Appl. 182 (1994), 566-570.
[128] J. M. Borwein and S. Fitzpatrick, *Existence of nearest points in Banach spaces*, Canad. J. Math. 61 (1989), 702-720.
[129] J. M. Borwein and S. Fitzpatrick, *Weak* sequential compactness and bornological limit derivatives*, J. Convex Anal. 2 (1995), 59-67.
[130] J. M. Borwein and S. Fitzpatrick, *Duality inequalities and sandwiched functions* (Preprint, Simon Fraser University at Burnaby 1998), Nonlinear Anal. Th. Meth. Appl. 46 (2001), 365-380.
[131] J. M. Borwein, S. Fitzpatrick and J. R. Giles, *The differentiability of real functions on normed space using generalized gradients*, J. Math. Anal. Appl. 128 (1987), 512-534.
[132] J. M. Borwein, S. Fitzpatrick and R. Girgensohn, *Subdifferentials whose graphs are not norm$\times$weak* closed*, Canad. Math. Bull. 48 (2003), 538-545.
[133] J. M. Borwein and J. R. Giles, *The proximal normal formula in Banach spaces*, Trans. Amer. Math. Soc. 302 (1987), 371-381.
[134] J. M. Borwein and A. D. Lewis, *Convex Analysis and Nonlinear Optimization: Theory and Examples*, CMS Books in Mathematics, Springer (2005).
[135] J. M. Borwein and W. B. Moors, *Essentially smooth Lipschitz functions*, J. Func. Anal. 149 (1997), 305-351.
[136] J. M. Borwein and W. B. Moors, *A chain rule for essentially smooth Lipschitz functions*, SIAM J. Optim. 8 (1998), 300-308.
[137] J. M. Borwein and W. B. Moors, *Null sets and essentially smooth Lipschitz functions*, SIAM J. Optim. 8 (1998), 309-323.
[138] J. M. Borwein, W. B. Moors and X. Wang, *Lipschitz functions with prescribed derivatives and subderivatives*, Nonlinear Anal. 29 (1997), 53-64.
[139] J. M. Borwein, W. B. Moors and X. Wang, *Generalized subdifferentials: A Baire categorical approach*, Trans. Amer. Math. Soc. 353 (2001), 3875-3893.
[140] J. M. Borwein and D. Preiss, *A smooth variational principle with applications to subdifferentiability and to differentiability of convex functions*, Tran. Amer. Math. Soc. 303 (1987), 517-527.
[141] J. M. Borwein and H. Strójwas, *Directionally Lipschitzian mappings on Baire spaces*, Canad. J. Math. 36 (1984), 95-130.

[142] J. M. Borwein and H. Strójwas, *Tangential approximations*, Nonlinear Anal. 9 (1985), 1347-1366.

[143] J. M. Borwein and H. Strójwas, *Proximal analysis and boundaries of closed sets in Banach space. I: Theory*, Canad. J. Math. 38 (1986), 431-452 (submitted: November 1984).

[144] J. M. Borwein and H. Strójwas, *Proximal analysis and boundaries of closed sets in Banach space. II: Applications*, Canad. J. Math. 39 (1987), 428-472.

[145] J. M. Borwein and H. Strójwas, *The hypertangent cone*, Nonlinear Anal. 13 (1989), 125-144.

[146] J. M. Borwein and J. D. Vanderwerff, *Convex Functions: Constructions, Characterizations and Counterexamples*, Cambridge University Press (2010).

[147] J. M. Borwein and X. Wang, *Distinct differentiable functions may share the same Clarke subdifferential at all points*, Proc. Amer. Math. Soc. 125 (1997), 807–813.

[148] J. M. Borwein and X. Wang, *Lipschitz functions with maximal Clarke subdifferentials are generic*, Proc. Amer. Math. Soc. 128 (2000), 3221-3229.

[149] J. M. Borwein and X. Wang, *Cone-monotone functions, differentiability and continuity*, Canad. J. Math. 57 (2005), 961-982.

[150] J. M. Borwein and Q. J. Zhu, *Viscosity solutions and viscosity subderivatives in smooth Banach spaces with applications to metric regularity*, SIAM J. Control Optim. 34 (1996), 1568–1591.

[151] J. M. Borwein and Q. J. Zhu, *A survey of subdifferential calculus with applications*, Nonlinear Anal. 38 (1999), 687–773.

[152] J. M. Borwein and Q. J. Zhu, *Techniques of Variational Analysis*, CMS Books in Mathematics, Springer-Verlag, New-York (2005).

[153] J. M. Borwein and D. M. Zhuang, *Verifiable necessary and sufficient conditions for openness and regularity for set-valued and single-valued maps*, J. Math. Anal. Appl. 134 (1988), 441-459.

[154] U. Boscain, B. Piccoli, *Optimal Syntheses for Control Systems on 2-D Manifolds*, Springer, Berlin (2004).

[155] R. I. Bot, *Conjugate Duality in Convex Optimization*, Lecture Notes in Economics and Mathematical Systems, Vol. 637, Springer-Verlag, Berlin (2010).

[156] M. Bougeard, *Contributions à la théorie de Morse en dimension finie*, Thesis, Université Paris IX Dauphine, (1978).

[157] G. Bouligand, *Sur quelques points de topologie restreinte du premier ordre*, Bull. Soc. Math. France 56 (1928), 26-35.

[158] G. Bouligand, *Sur l'existence des demi-tangentes à une courbe de Jordan*, Fundamenta Mathematicae 15 (1930), 215-218.

[159] G. Bouligand, *Problèmes connexes de la notion d'enveloppe de M. Geeorges Durand*, C.R. Acad. Sci. Paris 189 (1929), p. 146.

[160] G. Bouligand, *Expression générale de la solidarité entre le problème du minimum d'une intégrale et l'équation correspondante d'Hamilton-Jacobi*, Rendiconti dei Lincei, (1930).

[161] G. Bouligand, *Sur quelques applications de la théorie des ensembles à la géométrie infinitésimale*, Bull. Intern. Acad. Polonaise Sc. L (1930), 407-420.

[162] G. Bouligand *Sur quelques points de méthodologie géométrique*, Revue Générale des Sciences Pures et Appliquées, 41 (1930), 39-43.

[163] G. Bouligand, *Sur une application du contingent à la théorie de la mesure*, Acta Math. 56 (1931), p. 371.

[164] G. Bouligand, *Introduction à la Géométrie Infinitésimale Directe*, Paris (1932).

[165] M. Bounkhel, *Régularité Tangentielle en Analyse Non Lisse*, Ph.D. dissertation thesis, Université de Montpellier (1999).

[166] M. Bounkhel, *On the distance function associated with a set-valued mapping*, J. Nonlinear Convex Anal. 2 (2001), 265-278.

[167] M. Bounkhel, *On arc-wise essentially smooth mappings between Banach spaces*, Optimization 51 (2002), 11-29.

[168] M. Bounkhel, *Scalarization of normal Fréchet regularity for set-valued mappings*, New Zealand J. Math. 33 (2004), 129-146.

[169] M. Bounkhel, *Regularity Concepts in Nonsmooth Analysi. Theory and Applications*, Springer Optimization and Its Applications, Vol. 59, Springer, New York (2011).

[170] M. Bounkhel and D.-L. Azzam, *Existence results on the second-order nonconvex sweeping processes with perturbations*, Set-Valued Anal. 12 (2004), 291-318.

[171] M. Boukhel and L. Thibault, *Scalarization of tangential regularity of set-valued mappings*, Set-Valued Anal. 7 (1999), 33-53.
[172] M. Bounkhel and L. Thibault, *On various notions of regularity of sets in nonsmooth analysis*, Nonlinear Anal. 48 (2002), 223-246.
[173] M. Bounkhel and L. Thibault, *Nonconvex sweeping process and prox-regularity in Hilbert space*, J. Nonlinear Convex Anal. 6 (2005), 359-374.
[174] N. Bourbaki, *Espaces Vectoriels Topologiques*, Fascicule XV, XVII, Hermann, Paris (1966).
[175] D. G. Bourgin, *Approximate isometries*, Bull. Amer. Math. Soc. 52 (1946), 704-714.
[176] H. Brezis, *Propriétés régularisantes de certains semi-groupes non linéaires*, Israel J. Math. 9 (1971), 513-534.
[177] H. Brezis, *Opérateurs Maximaux Monotones et Semi-groupes de Contractions dans les Espaces de Hilbert*, Math. Stud. 5, North-Holland, Amsterdam, 1973.
[178] H. Brezis, *Functional Analysis, Sobolev Spaces and Partial Differential Equations*, Springer, New-York (2011).
[179] A. Brønsted, *Conjugate convex functions in topological vector spaces*, Math. Fys. Medd. Dansk. Vid. Selsk. 34 (1964), 1-27.
[180] A. Brønsted, *On the subdifferential of the supremum of two convex functions*, Math. Scand. 31 (1972), 225-230.
[181] A. Brønsted and R. T. Rockafellar, *On the subdifferentiability of convex functions*, Proc. Amer. Math. Soc. 16 (1965), 605-611.
[182] J. V. Burke, M. C. Ferris and M. Qian, *On the Clarke subdifferential of the distance function of a closed set*, J. Math. Anal. Appl. 166 (1992), 199-213.
[183] J. V. Burke, A. S. Lewis, M. L. Overton, *Approximating subdifferentials by random sampling of gradients*, Math. Oper. Res. 27 (2002), 567-584.
[184] J. V. Burke and R. A. Poliquin, *Optimality conditions for non-finite convex composite functions*, Math. Program. 57 (1992), 103-120.
[185] L. N. H. Bunt, *Bijdrage tot de theorie de convexe puntverzamelingen*, Thesis, Univ. of Groningen, Amsterdam (1934).
[186] F. Cabello Sánchez, *Nearly convex functions, perturbations of norms and K-spaces*, Proc. Amer. Math. Soc. 129 (2001), 753-758.
[187] F. Cabello Sánchez, J.M.F. Castillo and P.L. Papini, *Seven views on approximate convexity and the geometry of K-spaces*, J. Lond. Math. Soc. 72 (2005), 457-477.
[188] A. Cabot, H. Engler and S. Gadat, *On the long time behavior of second order differential equations with asymptotically small dissipation*, Tran. Amer. Math. Soc. 361 (2009), 5983-6017.
[189] A. Cabot and L. Thibault, *Inclusion of subdifferentials, linear well-conditioning and steepest descent equation*, SIAM J. Optim. 23 (2013), 552-575.
[190] A. Cabot and L. Thibault, *Sequential formulae for the normal cone to sublevel sets*, Trans. Amer. Math. Soc. 366 (2014), 6591-6628.
[191] L. Caffarelli, *The regularity of free boundaries in higher dimensions*, Acta Math. 139 (1977), 155-184.
[192] A. Canino, *On p-convex sets and geodesics*, J. Differential Equations 75 (1988), 118-157; submitted 18 February 1987.
[193] A. Canino, *Existence of a closed geodesic on p-convex sets*, Ann. Inst. Henri Poincaré 5 (1988), 501-518.
[194] A. Canino, *Local properties of geodesics on p-convex sets*, Ann. Mat. Pura Appl. 159 (1991), 17-44.
[195] A. Canino, *Periodic solutions of quadratic Lagrangian systems on p-convex sets*, Ann. Fac. Sci. Toulouse Math. 12 (1991), 37-60.
[196] P. Cannarsa, *Regularitiy properties of solutions to Hamilton Jacobi equations in infinite dimensions and nonlinear optimal control*, Diff. Integral Equations 2 (1989), 479-493.
[197] P. Cannarsa and P. Cardialaguet, *Representation of equilibrium solutions to the table problem for growing sandpiles*, J. Eur. Math. Soc. 6 (2004), 435-464.
[198] P. Cannarsa and P. Cardaliaguet, *Perimeter estimates for reachable sets of control systems*, J. Convex Anal. 13 (2006), 253-267.
[199] P. Cannarsa and H. Frankowska, *Interior sphere property of attainable sets and time optimal control problems*, ESAIM: Control Optim. Calc. Var. 12 (2006), 350-370.
[200] P. Cannarsa and C. Sinestrari, *Convexity properties of the minimum time function*, Calc. Var. 3 (1995), 273-298.

[201] P. Cannarsa and C. Sinestrari, *Semiconcave Functions, Hamilton–Jacobi Equations, and Optimal Control*, Birkhäuser, Boston (2004).

[202] P. Cannarsa and H. M. Soner, *On the singularities of the viscosity solutions to Hamilton–Jacobi–Bellman equations*, Indiana Univ. Math. J. 36 (1987), 501–524.

[203] M. J. Cánovas, D. Klatte, M. A. Lopez, and J. Parra, *Metric regularity of convex semi-infinite programming problems under convex perturbations*, SIAM J. Optim. 16 (2007), 717-732.

[204] M. J. Cánovas, M. A. Lopez, J. Parra and F. J. Toledo, *Lipschitz continuity of the optimal value via bunds on the optimal set in linear semi-infinite optimization*, Math. Oper. Res; 31 (2006), 478-489.

[205] I. Capuzzo Dolcetta and I. Ishii, *Approximate solutions of the Bellman equation of deterministic control theory*, Appl. Math. Optim. 11 (1984), 161-181.

[206] A. Carioli and L. Vesely, *Normal cones and continuity of vector-valued convex functions*, J. Convex Anal. 25 (2013), 495-500.

[207] B. Cascales, J. Orihuela and A. Pérez, A., *One-sided James compactness theorem*, J. Math. Anal. Appl. 445 (2017), 1267-1283.

[208] E. Casini and P. L. Papini, *A counterexample to the infinity version of the Hyers-Ulam stability theorem*, Proc. Amer. Math. Soc., 118 (1993), 885-890.

[209] C. Castaing, *Sur les Multi-Applications Mesurables*, Thesis (Thése de Doctorat d'État ès Sciences Mathématiques), Université de Caen (1967).

[210] C. Castaing, *Sur les multi-applications mesurables*, Revue Francaise d'Informatique et de Recherche Opérationnelle, 1 (1967), 91-126.

[211] C. Castaing, A. G. Ibrahim and M. Yarou, *Existence problems in second order evolution inclusions: discretization and variational approach*, Taiwanese J. Math. 12 (2008), 1433-1475.

[212] C. Castaing, A. G. Ibrahim and M. Yarou, *Some contributions to nonconvex sweeping process*, J. Nonlinear Convex Anal. 10 (2009), 1-20.

[213] C. Castaing, A. Salvadori and L. Thibault, *Functional evolution equations governed by nonconvex sweeping process*, J. Nonlinear Convex Anal. 2 (2001), 217-241.

[214] C. Castaing and M. Valadier, *Convex Analysis and Measurable Multifunctions*, Lectures Notes in Mathematics, Vol. 580 springer-Verlag, Berlin (1977).

[215] A. Cellina, A. Marino and C. Olech, Eds, *Methods of Nonconvex Analysis*, Lecture Notes in Mathematics, Springer (Berlin) 1990.

[216] Y. Chabrillac and J.-P. Crouzeix, *Continuity and differentiability of monotone real functions of several variables*, in: Nonlinear Analysis and Optimization (Louvain-la-Neuve, 1983), Math. Prog. Study 30 (1987), 1-16.

[217] L. Chamard, *Sur les Propriétés de la Distance à un Ensemble Ponctuel*, Thesis (Thèse de Doctorat ès Sciences Mathématiques), Université de Poitiers (1933).

[218] T. Champion, *Duality gap in convex programming*, Math. Program. 99 (2004), 487-498.

[219] G. Chavent, *Quasiconvex sets and size × curvature condition, application to non linear inversion*, Appl. Math. Optim. 24 (1991), 129-169.

[220] G. Chavent, *New size × curvature conditions for strict quasiconvexity of sets*, SIAM J. Control Optim. 29 (1991), 1348-1372.

[221] G. Chavent, *On p-convex, proximally smooth, quasi-convex, strictly quasi-convex and approximately convex sets*, J. Convex Anal. 22 (2015), 427-446.

[222] N. Chemetov, M. D. P. Monteiro Marques, *Non-convex quasi-variational differential inclusions*, Set-Valued Anal. 15 (2007), 209-221.

[223] P. W. Cholewa, *Remarks on the stability of functional equations*, Aequationes Math. 27 (1984), 76-86.

[224] G. Choquet, *Sur les notions de filtre et de grille*, C. R. Acad. Sciences, 224 (1947), 171-173.

[225] G. Choquet, *Convergences*, Annales de l'université de Grenoble, 23 (1947-1948), 57-112.

[226] G. Choquet, *Ensembles et cônes convexes faiblement complets*, C. R. Acad. Sei. Paris 254 (1962), 1908-1910.

[227] J. P. R. Christensen, *On sets of Haar-measure zero in Abelian groups*, Israel J. Math. 13 (1972), 255-260.

[228] J. P. R. Christensen, *Topology and Borel Structure*, Math. Studies 10, Notas Mathematica, Amsterdam (1974).

[229] R. Cibulka and M. Fabian, *Attainment and (sub)differentiability of the infimal convolution of a function and the square of the norm*, J. Math. Anal. Appl. 368 (2010), 538-550.

[230] M. Ciligot-Travain, *On Lagrange-Kuhn-Tucker multipliers for Pareto optimization problems*, Numer. Funct. Anal. Optim. 15 (1994), 689-693.

[231] M. Ciligot-Travain, *An intersection formula for the normal cone associated with the hypertangent cone*, J. Appl. Anal. 5 (1999), 239-247.

[232] F. H. Clarke, *Necessary Conditions for Nonsmooth Problems in Optimal Control and the Calculus of Variations*, Ph.D. dissertation thesis, University of Washington, Seattle (1973).

[233] F. H. Clarke, *Generalized gradients and applications*, Trans. Amer. Math. Soc. 205 (1975), 247-262.

[234] F. H. Clarke, *The Euler-Lagrange differential inclusion*, J. Differential Equations 19 (1975), 80-90.

[235] F. H. Clarke, *Admissible relaxation in variational and control problems*, J. Math. Anal. Appl. 51 (1975), 557-576.

[236] F. H. Clarke, *The generalized problem of Bolza*, SIAM J. Control Optim. 14 (1976), 682-699.

[237] F. H. Clarke, *The maximum principle under minimal hypotheses*, SIAM J. Control Optim. 14 (1976), 1078-1091.

[238] F. H. Clarke, *Optimal solutions to differential inclusions*, J. Optim. Theory Appl. 19 (1976), 469-478.

[239] F. H. Clarke, *A new approach to Lagrange multipliers*, Math. Oper. Res. 1 (1976), 165-174.

[240] F. H. Clarke, *On the inverse function theorem*, Pac. J. Math. 64 (1976), 97-102.

[241] F. H. Clarke, *Extremal arcs and extended Hamiltonian systems*, Trans. Amer. Math. Soc. 231 (1977), 349-367.

[242] F. H. Clarke, *Inequality constraints in the calculus of variations*, Canad. J. Math. 3 (1977), 528-540.

[243] F. H. Clarke, *Optimization and Nonsmooth Analysis*, Wiley Intersciences, New York (1983). Second Edition: Classics in Applied Mathematics, 5, Society for Industrial and Applied Mathematics, Philadelphia (1990).

[244] F. H. Clarke, *Functional Analysis, Calculus of Variations and Optimal Control*, Graduate Texts in Mathematics, Springer, New York (2013).

[245] F. H. Clarke and Yu. S. Ledyaev *Mean value inequalities*, Proc. Amer. Math. Soc. 122 (1994), 1075-1083.

[246] F. H. Clarke and Yu. S. Ledyaev *Mean value inequalities in Hilbert space*, Trans. Amer. Math. Soc. 344 (1994), 307-324.

[247] F. H. Clarke, Yu. S. Ledyaev, R. J. Stern and P. R. Wolenski, *Nonsmooth Analysis and Control Theory*, Springer, Berlin (1998).

[248] F. H. Clarke and R. M. Redheffer, *The proximal subgradient and constancy*, Canad. Math. Bull. 36 (1993), 30-32.

[249] F. H. Clarke, R. J. Stern and P. R. Wolenski, *Subgradient criteria for monotonicity, the Lipschitz condition, and convexity*, Canad. J. Math. 45 (1993), 1167-1183; submitted 24 August 1992.

[250] F. H. Clarke, R. J. Stern and P. R. Wolenski, *Proximal smoothness and the lower C^2 property*, J. Convex Analysis 2 (1995), 117-144.

[251] J. A. Clarkson, *Uniformly convex spaces*, Trans. Amer. Math. Soc. 40 (1936), 396-414.

[252] S. Cobzas, *Antiproximinal sets in Banach spaces*, Math. Balkanica 4 (1974), 79-82.

[253] S. Cobzas, *Geometric properties of Banach spaces and the existence of nearest and farthest points*, Abst. Appl. Anal. 3 (2005), 259-285.

[254] G. Colombo and V. V. Goncharov, *The sweeping processes without convexity*, Set-Valued Anal. 7 (1999), 357–374.

[255] G. Colombo and V. V. Goncharov, *Variational inequalities and regularity properties of closed sets in Hilbert spaces*, J. Convex Anal. 8 (2001), 197–221.

[256] G. Colombo and V. V. Goncharov, *Continuous selections via geodesics*, Topological methods in Nonlin. Anal. 18 (2001), 171–182.

[257] G. Colombo, V. V. Goncharov and B. S. Mordukhovich, *Well-posedness of minimal time problem with constant dynamics in Banach space*, Set-Valued Var. Anal. 18 (2010), 349-372.

[258] G. Colombo and A. Marigonda, *Differentiability properties for a class of non-convex functions*, Calc. Var. 25 (2006), 1–31.

[259] G. Colombo and A. Marigonda, *Singularities for a class of non-convex sets and functions, and viscosity solutions of some Hamilton–Jacobi equations*, J. Convex Anal. 15 (2008), 105–129.

[260] G. Colombo, A. Marigonda and P. R. Wolenski, *Some new regularity properties for the minimal time function*, SIAM J. Control 44 (2006), 2285–2299.

[261] G. Colombo, A. Marigonda and P. R. Wolenski, *The Clarke generalized gradient for functions whose epigraph has positive reach*, Math. Oper. Res. 38 (2013), 451-468.

[262] G. Colombo and M. D. P. Monteiro Marques, *Sweeping by a continuous φ-convex set*, J. Differential Equations 187 (2003), 46–72.

[263] G. Colombo and T. K. Nguyen, *Quantitative isoperimetric inequalities for a class of nonconvex sets*, Calc. Var. 37 (2010), 141-166.

[264] G. Colombo and T. K. Nguyen, *On the structure of the minimum time function*, SIAM J. Control 48 (2010), 4776-4814.

[265] G. Colombo and L. Thibault, *Prox-regular sets and applications*, Handbook of Nonconvex Analysis and Applications, p. 99-182, D.Y. Gao, D. Motreanu Eds., International Press, Somerville (2010).

[266] G. Colombo and P. R. Wolenski, *The subgradient formula for the minimal time function in the case of constant dynamics in Hilbert space*, J. Global Optim. 28 (2004), 269-282.

[267] G. Colombo and P. R. Wolenski, *Variational analysis for a class of minimal time functions in Hilbert spaces*, J. Convex Anal. 11 (2004), 335-361.

[268] C. Combari, A. Elhilali Alauoi, A. Levy, R. Poliquin and L. Thibault, *Convex composite functions in Banach spaces and the primal lower-nice property*, Proc. Amer. Math. Soc. 126 (1998), 3701-3708.

[269] C. Combari, M. Laghdir and L. Thibault, *Sous-différentiels de fonctions composées*, Ann. Sci. Math. Québec 62 (1994), 119-148.

[270] C. Combari, M. Laghdir and L. Thibault, *A note on subdifferentials of convex composite functionals*, Arch. Math. 67 (1996), 239-252.

[271] C. Combari, M. Laghdir and L. Thibault, *On subdifferential calculus for convex functions defined on locally convex spaces*, Ann. Sci. Math. Québec 23 (1999), 23-36.

[272] C. Combari, S. Marcellin and L. Thibault, *On the graph convergence of ε-subdifferentials of convex functions*, J. Nonlin. Convex Anal. 4 (2003), 309-324.

[273] C. Combari, R. A. Poliquin and L. Thibault, *Convergence of subdifferentials of convexly composite functions*, Canad. J. Math. 51 (1999), 250-265.

[274] C. Combari and L. Thibault, *Epi-convergence of convexly composite functions in Banach spaces*, SIAM J. Optim. 13 (2003), 986–1003.

[275] R. Cominetti, *On Pseudo-differentiability*, Trans. Amer. Math. Soc. 324 (1991), 843-865.

[276] H. O. Cordes and J. P. Labrousse, *The invariance of the index in the metric space of closed operators*, J. Math. Mech. 12 (1963), 693-720.

[277] B. Cornet, *A remark on tangent cones*, CEREMADE Publication (1979), Univ. Paris-Dauphine.

[278] B. Cornet, *Regular properties of tangent and normal cones*, CEREMADE Publication 8130 (1981), Univ. Paris-Dauphine.

[279] B. Cornet, *Existence of equilibria in economies with increasing returns*, in Contributions to Operations Research and Economics: The twentieth anniversary of C.O.R.E., MIT Press, Cambridge, (1982).

[280] B. Cornet, *Existence of slow solutions for a class of differential inclusions*, J. Math. Anal. Appl. 96 (1983), 179-186.

[281] B. Cornet and M.-O. Czarnecki, *Smooth representations of epi-Lipschitzian subsets*, Nonlinear Anal. Th. Meth. Appl. 37 (1999), 139-160.

[282] B. Cornet and M.-O. Czarnecki, *Existence of generalized equilibria*, Nonlinear Anal. Th. Meth. Appl. 44 (2001), 555-574.

[283] R. Correa, P. Gajardo and L. Thibault, *Subdifferential representation formula and subdifferential criteria for the behavior of nonsmooth functions*, Nonlinear Anal. 65 (2006), 864-891.

[284] R. Correa, P. Gajardo and L. Thibault, *Links between directional derivatives through multidirectional mean value inequalities*, Math. Program. 116 Ser. B (2009), 57-77.

[285] R. Correa, P. Gajardo and L. Thibault, *Various Lipschitz-like properties for functions and sets. I. Directional derivative and tangential characterizations*, SIAM J. Optim. 20 (2010), 1766-1785.

[286] R. Correa, P. Gajardo and L. Thibault, *Various Lipschitz-like properties for functions and sets. II. Subdifferential and normal characterizations*, J. Convex Anal. (2020).

[287] R. Correa, A. Hantoute and A. Jourani, *Characterizations of convex approximate subdifferential calculus in Banach spaces*, Trans. Amer. Math. Soc. 368 (2016), 4831-4854.

[288] R. Correa, A. Hantoute, and M. A. López *Valadier-like formulas for the supremum function I* J. Convex Anal., 25 (2018), 1253-1278.

[289] R. Correa, A. Hantoute, M. A. López, *Valadier-like formulas for the supremum function II: the compactly indexed case*, J. Convex Anal. 26 (2019), 299-324.

[290] R. Correa and A. Jofre, *Tangentially continuous directional derivatives in nonsmooth analysis*, J. Optim. Theor. Appl. 61 (1989), 1-21.

[291] R. Correa, A. Jofre and L. Thibault, *Characterization of lower semicontinuous convex functions*, Proc. Amer. Math. Soc. 116 (1992), 67-72.

[292] R. Correa, A. Jofre and L. Thibault, *Subdifferential monotonicity as characterization of convex functions*, Numer. Funct. Anal. Optimiz. 15 (1994), 531-536.

[293] R. Correa, A. Jofre and L. Thibault, *Subdifferential characterization of convexity*, in Recent Advances in Nonsmooth Optimization, edited by D. Du, L. Qi and R. Womersley, World Scientific Publishing, Singapore, 1-23 (1995).

[294] R. Correa, D. Salas and L. Thibault, *Smoothness of the metric projection onto nonconvex bodies in Hilbert spaces*, J. Math. Anal. Appl., 457 (2018), 1307-1322.

[295] R. Correa and L. Thibault, *Subdifferential analysis of bivariate separately regular functions*, J. Math. Anal. Appl., 148 (1990), 157-174.

[296] R. Courant ans D. Hilbert, *Methods of Mathematical Physics*, Volume 2, 1961; German original edition 1938.

[297] M. G. Crandall, H. Ishii and P.-L. Lions, *User's guide to viscosity solutions of second-order partial differential equations*, Bull. Amer. Math. Soc. 27 (1992), 1-67.

[298] M. G. Crandall and P.-L. Lions, *Viscosity solutions of Hamilton-Jacobi equations*, Trans. Amer. Math. Soc. 277 (1983), 1-42; submitted: December 1, 1981.

[299] J.-P. Crouzeix, *Continuity and differentiability of quasi-convex functions*, Handbook of generalized convexity and generalized monotonicity, 121-149, Nonconvex Optim. Appl. 76 Springer, New-York (2005).

[300] M. Csörnyei, *Aronszajn null and Gaussian null sets coincide*, Israel J. Math. 111 (1999), 191-202.

[301] A. Cwiszewski and W. Kryszewski, *Equilibria of set-valued maps: a variational approach*, Nonlinear Anal. Th. Meth. Appl. 48 (2002), 707-746.

[302] A. Cwiszewski and W. Kryszewski, *The constrained degree and fixed point index theory for set-valued maps*, Nonlinear Anal. Th. Meth. Appl. 64 (2006), 2643-2664.

[303] M.-O. Czarnecki, Ph.D thesis, Université Paris 1 Panthéon Sorbonne, (1996).

[304] M.-O. Czarnecki, A. N. Gudovich, *Representation of epi-Lipschitzian sets*, Nonlinear Anal. Th. Meth. Appl. 73 (2010), 2361-2367.

[305] M.-O. Czarnecki and L. Thibault, *Sublevel representations of epi-Lipschitz sets and other properties*, Math. Program. 18 (2018), 555-569.

[306] J. Daneš, *A geometric theorem useful in nonlinear functional analysis*, Boll. Un. Mat. Ital. 6 (1972), 369-375.

[307] J. Daneš, *Equivalence of some geometric and related results of nonlinear functional analysis*, Commentationes Mathematicae Universitatis Carolinae 26 (1985), 443-454.

[308] J. W. Daniel, *The continuity of metric projections as functions of the data*, J. Approximation Theory 12 (1974), 234-239.

[309] A. Daniilidis and P. Georgiev, *Approximate convexity and submonotonicity*, J. Math. Anal. Appl. 291 (2004), 292-301.

[310] A. Daniilidis, W. L. Hare and J. Malick, *Geometrical interpretation of the predictor-corrector type algorithms in structured optimization problems*, Optimization 55 (2006), 481-503.

[311] A. Daniilidis, F. Jules and M. Lassonde, *Subdifferential characterization of approximate convexity: the lower semicontinuous case*, Math. Program. Ser. B 116 (2009), 115-127; submitted: September 18, 2005.

[312] A. Daniilidis, A. S. Lewis, J. Malick and H. Sendov, *Prox-regularity of spectral functions and spectral sets*, J. Convex Anal. 15 (2008), 547-560.

[313] A. Daniilidis and J. Malick, *Filling the gap between lower-C^1 and lower-C^2 functions*, J. Convex Anal. 12 (2005), 315-329.

[314] A. Daniilidis and L. Thibault, *Subsmooth sets and metrically subsmooth sets and functions in Banach space*, Unpublished paper, Univ. Montpellier II, 2008.

[315] J. M. Danskin, *The theory of max-min with applications*, SIAM J. Appl. Math. 14 (1966), 641-644.
[316] G. Dantzig, *Linear programming*, Operations Res. 50 (2002), 42-47.
[317] M. M. Day, *Reflexive Banach spaces not isomorphic to uniformly convex spaces*, Bull. Amer. Math. Soc. 47 (1941), 313-317.
[318] M. M. Day, *Uniform convexity in factor and conjugate spaces*, Ann. Math. 45 (1944), 375-385.
[319] J.-P. Dedieu, *Cône asymptote d'un ensemble non convexe. Application à l'optimisation*, C. R. Acad. Sci., Paris, Sér. A 285 (1977), 501-503.
[320] E. De Giorgi, M. Degiovanni, A. Marino and M. Tosques, *Evolution equations for a class of nonlinear operators*, Atti Accad. Naz. Lincei Rend. Cl. Sci. Fis. Mat. Natur. (8) 15 (1983), 1-8.
[321] M. Degiovanni, A. Marino and M. Tosques, *Evolution equations with lack of convexity*, Nonlinear Anal. 9 (1985), 1401-1443.
[322] E. De Giorgi, *Sulla proprietà isoperimetrica dell'ipersfera, nella classe degli insiemi aventi frontiera orientata di misura finita*, (Italian) Atti Accad. Naz. Lincei. Mem. Cl. Sci. Fis. Mat. Nat., Sez. I 8 (1958), 33-44; also (in English) in De Giorgi, *Selected papers*, L. Ambrosio, G. Dal Maso, M. Forti, M. Miranda, S. Spagnolo (Eds.), Springer, Berlin (2006).
[323] F. Delbaen, *Coherent Risk Measure*, Lectures at the Scuola Normale di Pisa, March 2000.
[324] M. C. Delfour and J.-P. Zolésio, *Shape analysis via oriented distance functions* J. Funct. Anal. 123 (1994), 129-201.
[325] M. C. Delfour and J.-P. Zolésio, *Shapes and Geometries: Metrics, Analysis, Differential Calculus, and Optimization*, SIAM series on Advances in Design and Control, Society for Industrial and Applied Mathematics, Philadelphia, 2nd ed. (2011).
[326] A. Denjoy, *Les quatre cas fondamentaux des nombres dérivés* C.R. Acad. Sci. Paris, 161 (1915), 124-127.
[327] A. Denjoy, *Mémoire sur les nombres dérivés des fonctions continues*, J. Math. Pures Appl. 1 (1915), 105-240.
[328] A. Denjoy, *Sur les fonctions dérivés sommables*, Bull. Soc. Math. France 43 (1915), 161-248.
[329] A. Denjoy, *Mémoire sur la totalisation des nombres dérivés non sommables*, Ann. École Norm. Sup. 33 (1916), 127-222.
[330] A. Denjoy, *Mémoire sur la totalisation des nombres dérivés non sommables*, Ann. École Norm. Sup. 34 (1916), 181-238.
[331] A. Denjoy, *Sur l'intégration des coefficients différentiels d'ordre supérieur*, Fundam. Math. 25 (1935), 273-326.
[332] R. Deville, *A mean value theorem for the non differentiable mappings*, Serdica Math. J. 21 (1995), 59-66.
[333] R. Deville, G. Godefroy and V. Zizler, *Smoothness and Renormings in Banach Spaces*, Longman and Wiley, New York (1993).
[334] J. Diestel, *Geometry of Banach Spaces-Selected Topics*, Springer-Verlag, Berlin (1975).
[335] J. Diestel, *Sequences and Series in Banach Spaces*, Springer-Verlag, Berlin (1984).
[336] S. J. Dilworth, R. Howard and J. W. Roberts, *Extremal approximately convex functions and estimating the size of convex hulls*, Advances in Math. 148 (1999), 1-43.
[337] S. J. Dilworth, R. Howard and J. W. Roberts, *On the size of approximately convex sets in normed spaces*, Studia Math. 140 (2000), 213-241.
[338] S. J. Dilworth, R. Howard and J. W. Roberts, *Extremal approximately convex functions and the best constants in a theorem of Hyers and Ulam*, Adv. Math. 172 (2002), 1-14.
[339] U. Dini, *Fondamenti per la Teorica delle Funzioni di Variabili Reali*, Pisa (1878).
[340] A. V. Dmitruk, A. A. Milyutin, and N. P. Osmolovskii, *Lyusternik's theorem and the theory of extrema*, Uspekhi Mat. Nauk 35:6 (1980), 11-46; English transl., Russian Math. Surveys 35:6 (1980), 11-51.
[341] S. Dolecki, *Tangency and differentiation: Some applications of convergence theory*, Ann. Math. Pura Appl. 130 (1982), 281-301.
[342] S. Dolecki and G. H. Greco, *Towards historical roots of necessary conditions of optimality: Regula of Peano*, Control and Cybernetics 36 (2007), 491-518.
[343] S. Dolecki and G. H. Greco, *Tangency vis-à-vis differentiability by Peano, Severi and Guareschi*, J. Convex Anal. 18 (2011), 301-339.

[344] S. Dolecki and S. Kurcyusz, *On Φ-convexity in extremal problems*, SIAM J. Cont. Optim. 16 (1978), 277-300.

[345] E. P. Dolzhenko, *Boundary properties of arbitrary functions* (in Russian), Izv. Akad. Nauk. SSSR ser. Mat. 31 (1967), 3-14.

[346] A. L. Dontchev, *The Graves theorem revisited*, J. Convex Anal. 3 (1996), 45-54.

[347] A. L. Dontchev and H. Frankowska, *Lyusternik-Graves theorem and fixed-points*, Proc. Amer. Math. Soc. 139 (2011), 521-534.

[348] A. L. Dontchev and H. Frankowska, *Lyusternik-Graves theorem and fixed-points 2*, J. Convex Anal. 19 (2012), 955-974.

[349] A. L. Dontchev and W. W. Hager, *An inverse mapping theorem for set-valued maps*, Proc. Amer. Math. Soc. 121 (1994), 481-489.

[350] A. L. Dontchev and W. W. Hager, *Lipschitz, functions, Lipschitz maps and stability in optimization*, Math. Oper. Res. 3 (1994), 753-768.

[351] A. L. Dontchev, A. S. Lewis and R. T. Rockafellar, *The radius of metric regularity*, Trans. Amer. Math. Soc. 355 (2003), 493-517.

[352] A. L. Dontchev, M. Quincampoix and N. Zlateva, *Aubin criterion for metric regularity*, J. Convex Anal. 13 (2006), 281-297.

[353] A. L. Dontchev and R. T. Rockafellar, *Regularity and conditioning of solution mappings in variational analysis*, Set-Valued Anal. 12 (2004), 79-109.

[354] A. L. Dontchev and R. T. Rockafellar, *Implicit Functions and Solution Mappings: A view From Variational Analysis*, Springer Monographs in Mathematics, 2nd edition, New York (2014).

[355] R. Douady, *Petites perturbations d'une suite exacte et d'une suite quasi-exacte*, Seminaire d'Analyse, Nice (1965-1966), 21-34.

[356] A. Douglis, *The continuous dependence of generalized solutions of non-linear partial differential equations upon initial data*, Comm. Pure Appl. Math. 14 (1961), 267-284.

[357] D. Drusvyatskiy and A. D. Ioffe, *Quadratic growth and critical point stability of semi-algebraic functions*, Math. Program. 153 (2015), 635-653.

[358] D. Drusvyatskiy and A. S. Lewis, *Tilt stability, uniform quadratic growth, and strong metric regularity of the subdifferential*, SIAM J. Optim. 23 (2013), 256-267.

[359] D. Drusvyatskiy, B. S. Mordukhovich and T. T. Nghia, *Second-order growth, tilt stability, and metric regularity of the subdifferential*, J. Convex Anal. 21 (2014), 1165-1192.

[360] A. J. Dubovicki and A. A. Milyutin, *Extremal problems with constraints*, U.S.S.R. Comp. Maths. Math. Phys. 5 (1965), 1-80.

[361] J. Duda, *On Gâteaux differentiability of pointwise Lipschitz mappings*, Canad. Math. Bull. 51 (2008), 205-216.

[362] J. Duda, *Cone monotone mapping: Continuity and differentiability*, Nonlinear Anal. 68 (2008), 1963-1972.

[363] J. Duda and L. Zajíček, *Semiconvex functions: representations as suprema of smooth functions and extensions*, J. Convex Anal. 16 (2009), 239-260.

[364] J. Duda and L. Zajíček, *Smallness of singular sets of semiconvex functions in separable Banach spaces*, J. Convex Anal. 20 (2013), 573-598.

[365] R. M. Dudley, *Real Analysis and Probability*, Cambridge University Press, Cambridge (2002).

[366] N. Dunford and J. T. Schwartz, *Linear Operators, part I: General Theory*, Interscience, New York (1958).

[367] G. Durand, *Sur un critère de dénombrabilité*, Acta. Math. 56, 1931, p. 363.

[368] G. Durand, *Sur une généralisation des surfaces convexes*, J. Math. Pures Appl. 10 (1931), 335-414.

[369] S. Dutta, *Generalized subdifferential of the distance function*, Proc. Amer. Math. Soc. 133 (2005), 2949-2955.

[370] M. Edelstein, *Farthest points of sets in uniformly convex Banach spaces*, Israel J. Math. 4 (1966), 171-176.

[371] M. Edelstein, *On nearest points of sets in uniformly convex Banach spaces*, J. Lond. Math. Soc. 43, (1968), 375–377.

[372] M. Edelstein, *A note on nearest points*, Quarterly J. Math. 21 (1970), 403-405.

[373] M. Edelstein, *Weakly proximinal sets*, J. Approx. Theory 18 (1976), 1-8.

[374] J. F. Edmond, *Delay perturbed sweeping processes*, Set-Valued AnaL. 14 (2006), 295-317.

[375] J. F. Edmond and L. Thibault, *Inclusions and integration of subdifferentials*, J. Nonlin. Convex Anal. 3 (2002), 411–434.
[376] J. F. Edmond and L. Thibault, *Relaxation of an optimal control problem involving a perturbed sweeping process*, Math. Program. ser B 104 (2005), 347-373.
[377] J. F. Edmond and L. Thibault, *BV solution of nonconvex sweeping process with perturbation*, J. Differential Equations 226 (2006), 135-179.
[378] A. V. Efimov, *Linear methods of approximating continuous periodic functions*, Amer. Math. Soc. Transl. 28 (1963), 221-268; translation of Russian original version in Mat.sb. 54 (1961), 51-90.
[379] N. V. Efimov and S. B. Stechkin, *Some supporting properties of sets in Banach spaces and Chebyshev sets*, Dokl. Akad. Nauk SSSR 127:2 (1959), 254–257. (in Russian)
[380] I. Ekeland, *On the variational principle*, J. Math. Anal. Appl. 47 (1974), 324-353.
[381] I. Ekeland, *Nonconvex minimization problems*, Bull. Amer. Math. Soc. 1 (1979), 443-474.
[382] I. Ekeland and J.-M. Lasry, *On the number of periodic trajectories for a Hamiltonian flow*, Ann. Math. 112 (1980), 293-319.
[383] I. Ekeland and G. Lebourg, *Generic Fréchet differentiability and perturbed optimization problems in Banach spaces*, Trans. Amer. Math. Soc. 224 (1976), 193-216.
[384] I. Ekeland and R. Temam, *Analyse Convexe et Problèmes Variationnels*, Dunod, Paris (1974).
[385] P. Enflo, *Banach spaces which can be given an equivalent uniformly convex norm*, Israel J. Math. 13 (1972), 281-288.
[386] P. Erdös, *On the Hausdorff dimension of some sets in Euclidean space*, Bull. Amer. Math. Soc. 52 (1946), 107–109.
[387] E. Ernst and M. Volle, *Generalized Courant-Beltrami penalty functions and zero duality gap for conic convex programs*, Positivity 17 (2013), 945-964.
[388] E. Ernst and M. Volle, *Zero duality gap and attainment with possibly non-convex data*, J. Convex Anal. 23 (2016), 615-629.
[389] L. C. Evans and R. F. Gariepy, *Measure Theory and Fine Properties of Functions*, Studies in Adv. Math., CRC Press, Boca Raton (1992).
[390] M. Fabian, *On classes of subdifferentiabiliy spaces of Ioffe*, Nonlinear Anal. Th. Meth. Appl. 12 (1988), 63-74.
[391] M. Fabian, *Subdifferentiability and trustworthiness in the light of a new variational principle of Borwein and Preiss*, Acta Univ. Carolinae 30 (1989), 51-56.
[392] M. Fabian, *Infimal convolution in Efimov-Stečkin Banach spaces*, J. Math. Anal. Appl. 339 (2008), 735-739.
[393] M. Fabian, R. Henrion, A. Y. Kruger and J. V. Outrata, *Error bounds: necessary and sufficient conditions*, Set-Valued Var. Anal. 18 (2010), 121-149.
[394] M. Fabian and A. D. Ioffe, *Separable reduction in the theory of Fréchet subdifferentiability*, Set-Valued Var. Anal. 21 (2013), 661-671.
[395] M. Fabian and A. D. Ioffe, *Separable reduction and rich family in the theory of Fréchet subdifferentials*, J. Convex Anal. 3 (2016), 631-648.
[396] M. Fabian and B. S. Mordukhovich, *Separable reduction and supporting properties of Fréchet-like normals in Banach spaces*, Canad. J. Math. 51 (1999), 51-56.
[397] M. Fabian and B. S. Mordukhovich, *Separable reduction and extremal principles in variational analysis*, Nonlinear Anal. Th. Meth. Appl. 49 (2002), 265-292.
[398] M. Fabian and D. Preiss, *On intermediate differentiability of Lipschitz functions on certain Banach spaces*, Proc. Amer. Math. Soc. 113 (1991), 733-740.
[399] M. Fabian and J. Revalski, *A variational principle in reflexive spaces with Kadec–Klee norm*, J. Convex Anal. 16 (2009) 211-226.
[400] M. Fabian and N. V. Zhivkov, *A characterization of Asplund spaces with the help of local ε-supports of Ekeland and Lebourg*, C. R. Acad. Bulg. Sci. 38 (1985), 671-674.
[401] E. Fabry, *Sur les rayons de convergence d'une série double*, C. R. Acad. Sci. Paris 134 (1902), 1190-1192.
[402] H. Federer, *Curvature measures*, Trans. Amer. Math. Soc. 93 (1959), 418-491.
[403] H. Federer, *Geometric Measure Theory*, Vol Grundlehren der mathematischen Wissenschaften, Springer, Berlin (1969).
[404] W. Fenchel, *On conjugate convex functions*, Canad. J. Math. 1 (1949), 73-77.

[405] W. Fenchel, *Convex Cones, Sets and Functions*, Lecture Notes, Princeton University, Princeton, New Jersey (1951).
[406] J. Ferrer, *Rolle's theorem fails in ℓ_2*, Amer. Math. Monthly (1996), 161-165.
[407] J. Ferrer, *On Rolle's theorem in spaces of infinite dimension*, Indian J. Math., 42 (2000), 21-36.
[408] J. Ferrer, *An approximate Rolle's theorem for polynomials of degree four in a Hilbert space*, Publ. RIMS, Kyoto Univ. 41 (2005), 375-384.
[409] M. Field, *Differential Calculus and Its Applications*, Dover Publications, Mineola (2012).
[410] S. Fitzpatrick, *Metric projections and the differentiability of distance functions*, Bull. Austral. Math. Soc. 22 (1980), 291-312.
[411] S. Fitzpatrick, *Differentiation of real-valued functions and continuity of metric projections*, Proc. Amer. Math. Soc. 91 (1984), 544-548.
[412] S. Fitzpatrick and R. R. Phelps, *Differentiability of the metric projection in Hilbert space*, Tran. Amer. Math. Soc. 170 (1982), 483-501.
[413] S. Fitzpatrick and S. Simons, *The conjugates, compositions and marginals of convex functions*, J. Convex Anal. 8 (2001), 423-446.
[414] S. Flåm, *Upward slopes and inf-convolution*, Math. Oper. Res. 31 (2006), 188-198.
[415] S. Flåm, J.-B. Hiriart-Urruty and A. Jourani, *Feasibility in finite time*, J. Dynamical Control Syst. 15 (2009), 537-555.
[416] A. Fougéres, *Coercivité des intégrandes convexes normales. Application à la minimisation des fonctionnelles intégrales et du clacul des variations*, Sémin. Anal. Convexe Montpellier, exposé 19 (1976).
[417] A. Fougères and A. Truffert, *Régularisation s.c.i. et épiconvergence: Approximations inf-convolutives associées à un référentiel*, Ann. Math. Pura Appl. 152 (1988), 21-51.
[418] H. Frankowska, *Some inverse mapping theorems*, Ann. Inst. H. Poincaré 7 (1990), 183-234.
[419] H. Frankowska and M. Quincampoix, *Hölder metric regularity of set-valued maps*, Math. Program. A 132 (2012), 333-354.
[420] M. Fréchet, *Sur la notion de différentielle*, C. R. Acad. Sci. Paris (1911), 845-847.
[421] M. Fréchet, *Sur la notion de différentielle*, C. R. Acad. Sci. Paris (1911), 1050-1051.
[422] M. Fréchet, Comptes Rendus du Congrès des Sociétés Savantes, Paris, (1912), p. 44.
[423] M. Fréchet, *Sur la notion de différentielle totale*, Nouvelles Annales de Mathématiques (4e série) 12 (1912), 385-433.
[424] M. Fréchet, *Sur la notion de différentielle totale*, Nouvelles Annales de Mathématiques (4e série) 12 (1912), 433-449.
[425] M. Fréchet, *Sur la notion de différentielle d'une fonction de ligne*, Tran. Amer. Math. Soc. 10 (1914), 135-161.
[426] M. Fréchet, *La notion de différentielle dans l'analyse générale*, Ann. Éc. Norm. Sup. 42 (1925), 293-323.
[427] J. H. G. Fu, *Tubular neighborhoods in Euclidean spaces*, Duke Math. J. 52 (1985), 1025-1046.
[428] B. Fuglede, *Stability in the isoperimetric problem for convex or nearly spherical domains in $\mathbb{R}^n$*, Trans. Am. Math. Soc. 314 (1989), 619-638.
[429] L. Gajek and D. Zagrodny, *Geometric mean value theorems for the Dini derivative*, J. Math. Anal. Appl. 191 (1995), 56-76.
[430] D. Gale, H. W. Kuhn and A. W. Tucker, *Linear programming and the theory of games*, in Activity Analysis of Production and Allocation, edited by T. C.Koopmans, Cowles Commission Monographs, Vol. 13 (1951), 317-329. Wiley, New York.
[431] R. Gâteaux, *Sur les fonctionnelles continues et les fonctionnelles analytiques*, C. R. Acad. Sci. Paris 157 (1913), 325-327.
[432] R. Gâteaux, *Fonctions d'une infinité de variables indépendantes*, Bull. Soc. Math. France 47 (1919), 70-96.
[433] R. Gâteaux, *Sur diverses questions de calcul fonctionnel*, Bull. Soc. Math. France 50 (1922), 1-37.
[434] M. Gaydu, M. H. Geoffroy and C. Jean-Alexis, *Metric subregularity of order q and the solving of inclusions*, Cent. Eur. J. Math., 9 (2011), 327-344.
[435] M. H. Geoffroy and M. Lassonde, *On a convergence of lower semicontinuous functions linked with the graph convergence of their subdifferentials*, Canadian Mathematical Society Conference Proceedings, 27 (2000), 93-109.

[436] E. Giner, *On the Clarke subdifferential of an integral functional on L_p, $1 \leq p < \infty$*, Canad. Math. Bull. 41 (1998), 41-48.

[437] E. Giner, *Subdifferential regularity and characterizations of Clarke subgradients of integral functionals*, J. Nonlin. Convex Anal. 9 (2008), 25-36.

[438] E. Giner, *Calmness properties and contingent subgradients of integral functionals on Lebesgue spaces L_p, $1 \leq p < \infty$*, Set-Valued Anal. 17 ((2009), 321-357.

[439] E. Giner, *Lagrange multipliers and lower bounds for integral functionals*, J. Convex Anal. 17 (2010), 301-308.

[440] B. Ginsburg and A. D. Ioffe, *The maximum principle in optimal control of systems governed by semilinear equations*, IMA Vol. Math. Appl. 78 (1996), 81-110.

[441] I. C. Gohberg and M. G. Krein, *The basic propositions on defect numbers, root numbers, and indices of linear operators*, Uspehi Mat. Nauk 12, 2 (74) (1957), 43-118 (in Russian).

[442] I. C. Gohberg and A. S. Markus, *Two theorems on the opening of subspaces of Banach space*, Uspehi Mat. Nauk 14, 5 (89) (1959), 135-140 (in Russian).

[443] G. Godefroy, *Some remarks on subdifferential calculus*, Revista matematica Complutense 11 (1998), 269-279.

[444] V. V. Goncharov and G. E. Ivanov, *Strong and weak convexity of closed sets in a Hilbert space*, in N. Daras and T. Rassias (eds) Operations Research, Engineering, and Cyber Security, Springer Optimization and its Applications, vol 113 (2017), 259-297,Springer.

[445] V. V. Goncharov, F. Pereira, *Neighbourhood retractions of nonconvex sets in a Hilbert space via sublinear functionals*, J. Convex Anal. 18 (2011), 1-36.

[446] L. Górniewicz, *Topological Fixed Point Theory for Multivalued Mappings*, Kluwer Academic Publishers, Dordrecht (1999).

[447] A. S. Granero, M. Jiménez-Sevilla, J. P. Moreno, *Intersections of closed balls and geometry of Banach spaces*, Extracta Mathematicae 19 (2004), 55-92.

[448] L. M. Graves, *Some mapping theorems*, Duke Math. J. 17 (1950), 111-114.

[449] J. W. Green, *Approximately convex functions*, Duke Math. J. 19 (1952), 499-504.

[450] S. Guillaume, *Evolution equations governed by the subdifferential of a convex composite function in finite dimensional equations*, Discrete Contin. Dynam. Systems 2 (1996), 23-52.

[451] V. I.Gurariŭ, *Openings and inclinations of subspaces of a Banach space*, Teor. Funktsii, Funktsional. Anal. i Prilozhen. 1 (1965), 194-204 (in Russian).

[452] J. Hadamard, *Étude sur les propriétés des fonctions entières et en particulier d'une fonction considérée par Riemann*, J. Math. Pures Appl. 58 (1893), 171-215.

[453] J. Hadamard, *Sur les fonctions entières*, Bull. Soc. Math. Fr. 24 (1896), 186-187.

[454] J. Hadamard, *Sur certaines propriétés des trajectoires en dynamique*, J. Math. Pures Appl. 62 (1897), 331-388.

[455] J. Hadamard, *La notion de différentielle dans l'enseignement*, Scripta Universitatis atque Bibliothecae Hierosolymitanarum (1923); Reprinted in the Mathematical Gazette 19, no. 236 (1935), 341–342.

[456] T. Haddad, A. Jourani and L. Thibault, *Reduction of sweeping process to unconstrained differential inclusion*, Pac. J. Optim. 4 (2008), 493-512.

[457] T. Haddad, J. Noel and L. Thibault, *Perturbed sweeping processes with a subsmooth set depending on the state*, Linear Nonlin. Anal. 2 (2016), 155-174.

[458] T. Haddad and L. Thibault, *Mixed upper semicontinuous perturbation of nonconvex sweeping process*, Math. Program. 123 (2010), 225–240.

[459] N. Hadjisavvas and D. T. Luc, *Second-order asymptotic directions of unbounded sets with application to optimization*, J. Convex Anal. 18 (2011), 181-202.

[460] J. Hagler and W. B. Johnson, *On Banach spaces whose dual balls are not weak* sequentially compact*, Israel J. Math. 28 (1977), 325–330.

[461] J. Hagler and F. Sullivan, *Smoothness and weak* sequential compactness*, Proc. Amer. Math. Soc. 78 (1980), 497-503.

[462] M. Hamamdjiev and M. Ivanov, *New multidirectional mean value inequality*, J. Convex Anal. 25 (2018), 1319-1334.

[463] A. Hantoute and M. A. López, *A new tour on the subdifferential of suprema. Highlighting the relationship between suprema and finite sums*, to appear.

[464] A. Hantoute, M. A. López, and C. Zalinescu, *Subdifferential calculus rules in convex analysis: A unifying approach via pointwise supremum functions*, SIAM J. Optim. 19 (2008), 863-882.

[465] A. Hantoute and A. Svensson, *A general representation of normal sets to sublevels of convex functions*, Set-Valued Var. Anal. 25 (2017), 651-678.

[466] A. Haraux, *How to differentiate the projection on a convex set in Hilbert space. Some applications to variational inequalities*, J. Math. Soc. Japan 29 (1977), 615-631 (submitted: September 1975).

[467] W. L. Hare and A. S. Lewis, *Identifying active constraints via partial smoothness and prox-regularity*, J. Convex Anal. 11 (2004), 251-266.

[468] W. L. Hare and R. A. Poliquin, *Prox-regularity and stability of the proximal mapping*, J. Convex Anal. 14 (2007), 589-606.

[469] W. L. Hare and C. Sagastizábal, *Computing proximal points of nonconvex functions*, Math. Program. 116 (2009), Ser. B, 221-258.

[470] U. S. Haslam-Jones, *Derivate planes and tangent plane of a measurable function*, Quart. J. Math. Oxford Ser. 3 (1932), 120-132.

[471] U. S. Haslam-Jones, *Tangential properties of a plane set of points*, Quart. J. Math. Oxford Ser. 7 (1936), 116-123.

[472] U. S. Haslam-Jones, *The discontinuities of an arbitrary function of two variables*, Quart. J. Math. Oxford Ser. 7 (1936), 184-190.

[473] F. Hausdorff, *Mengenlehre*, Walter de Gruyter, Berlin (1927).

[474] F. Hausdorff, *Set Theory*, Chelsea, English translation of the third (1937) German edition of *Mengenlehre*.

[475] J. Heinonen, *Lectures on Lipschitz Analysis*, Lectures at the 14th Jyväskylä Summer School in August 2004.

[476] R. Henrion, *The approximation subdifferential and parametric optimization*, Habilitation thesis, Humbold University, Berlin (1997).

[477] R. Henrion and A. Jourani, *Subdifferential conditions for calmness of convex constraints*, SIAM J. Optim. 13 (2002), 520-534.

[478] R. Henrion, A. Jourani and J. V. Outrata, *On the calmness of a class of multifunctions*, SIAM J. Optim. 13 (2002), 603-618.

[479] C. Henry, An existence theorem for a class of differential equations with multivalued right-hand side, J. Math. Anal. Appl. 41 (1973), 179-186.

[480] Ch. Hermite, *Sur deux limites d'une intégrale définie*, Mathesis 3 (1983), p. 82.

[481] J.-B. Hiriart-Urruty, *Contributions à la Programmation Mathématique: Cas Déterministe et Stochastique*, Thesis (Thèse de Doctorat d'État ès Sciences Mathhématiques), Université de Clermont-Ferrand (1977).

[482] J.-B. Hiriart-Urruty, *On optimality conditions in non-differentiable programming*, Math. Program. 14 (1978), 73-86.

[483] J.-B. Hiriart-Urruty, *Gradients généralisés de fonctions marginales*, SIAM J. Control Optim. 16 (1978), 301-316.

[484] J.-B. Hiriart-Urruty, *Tangent cones, generalized gradients and mathematical programming in Banach spaces*, Math. Oper. Res. 4 (1979), 79-97 (submitted: May, 1977).

[485] J.-B. Hiriart-Urruty, *New concepts in non differentiable programming*, Bull. Soc. Math. France Mémoire 60 (1979), 57-85.

[486] J.-B. Hiriart-Urruty, *Refinements of necessary optimality conditions in nondifferentiable programming. I*, Appl. Math. Optim. 5 (1979), 63-82.

[487] J.-B. Hiriart-Urruty, *A note on the mean value theorem for convex functions*, Boll. Un. Mat. Ital. B 17 (1980), 765-775.

[488] J.-B. Hiriart-Urruty, *Mean value theorems in nonsmooth analysis*, Numer. Funct. Anal. Optim. 2 (1980), 1-30.

[489] J.-B. Hiriart-Urruty, *Extension of Lipschitz functions*, J. Math. Anal. Appl. 77 (1980), 539-554.

[490] J.-B. Hiriart-Urruty, *Refinements of necessary optimality conditions in nondifferentiable programming. II*, Math. Program. Study 19 (1982), 120-139.

[491] J.-B. Hiriart-Urruty, *A short proof of the variational principle for approximate solutions of a minimization problem*, Amer. Math. Monthly 90 (1983), 206-207.

[492] J.-B. Hiriart-Urruty, *Images of connected sets by semicontinuous multifunctions*, J. Math. Anal. Appl. 111 (1985), 407-422.

[493] J.-B. Hiriart-Urruty, *Ensembles de Tchebychev vs. ensembles convexes: l'état de la situation vu par l'analyse non lisse*, Ann. Sci. Math. Que. 22 (1998), 47-62.

[494] J.-B. Hiriart-Urruty and C. Lemaréchal, *Convex Analysis and Minimization Algorithms. Part 1: Fundamentals*, Grundlehren der Mathematischen Wissenschaften, Springer-Verlag 305, Berlin (1993).

[495] J.-B. Hiriart-Urruty, M. Moussaoui, A. Seeger and M. Volle, *Subdifferential calculus without qualification conditions, using approximate subdifferentials: a survey*, Nonlinear Anal. th. Meth. Appl. 24 (1995), 1727-1754.

[496] J.-B. Hiriart-Urruty and R. R. Phelps, *Subdifferential calculus using ε-subdifferentials*, J. Funct. Analysis 118 (1993), 154-166.

[497] J.-B. Hiriart-Urruty and Ph. Plazanet, *Moreau's theorem revisited*, Ann. Inst. H. Poincaré 6 (1993), 544-555.

[498] J.-B. Hiriart-Urruty and L. Thibault, *Existence et caractérisation de différentielles généralisées d'applications localement lipschitziennes d'un espace de Banach séparable dans un espace de Banach réflexif séparable*, C. R. Acad. Sci. Paris Sér. A-B 290 (1980), 1091-1094.

[499] A. J. Hoffman, *On approximate solutions of systems of linear inequalities*, J. Res. Nat. Bur. Standards B 49 (1952), 263-265.

[500] M. O. Hölder, *Über einen Mittelwertsatz*, Nachr. Ges. Wiss. Göttingen, 1889, 38-47.

[501] R. B. Holmes, *Smoothness of certain metric projections on Hilbert space*, Trans. Amer. Math. Soc. 184 (1973), 87-100.

[502] L. Hörmander, *Sur la fonction d'appui des ensembles convexes dans un espace localement convexe*, Arkiv Math. 3 (1955), 181-186.

[503] M. M. Hrustalev, *Necessary and sufficient optimality conditions in the form of Bellman's equation*, Soviet Math. Dokl. 19 (1978), 1262-1266.

[504] D. H. Hyers, *On the stability of the linear functional equation*, Proc. Nat. Acad. Sei. U.S.A. vol. 27 (1941), 222-224.

[505] D. H. Hyers and S. M. Ulam, *On approximate isometries*, Bull. Amer. Math. Soc. 51 (1945), 288-292.

[506] D. H. Hyers and S. M. Ulam, *Approximate isometries of the space of continuous functions*, Ann. Math. 48 (1947), 285-289.

[507] D. H. Hyers and S. M. Ulam, *Approximately convex functions*, Proc. Amer. Math. Soc. 3 (1952), 821-828.

[508] A. D. Ioffe, *Regular points of Lipschitz functions*, Trans. Amer. Math. Soc. 251 (1979), 61-69.

[509] A. D. Ioffe, *Approximate subdifferentials of nonconvex functions*, CEREMADE Publication 8120 Université Paris IX Dauphine (1981).

[510] A. D. Ioffe, *Nonsmooth analysis: differential calculus and non-differentiable mappings*, Trans. Amer. Math. Soc. 266 (1981), 1-56.

[511] A. D. Ioffe, *On subdifferentiability spaces*, Ann. N. Y. Acad. Sci. 410 (1983), 107-121.

[512] A. D. Ioffe, *Calculus of Dini subdifferentials of functions and contingent derivatives of set-valued maps*, Nonlin. Anal. Th. Meth. Appl. 8 (1984), 517-539.

[513] A. D. Ioffe, *Necessary conditions in nonsmooth optimization*, Math. Oper. Res. 9 (1984), 159-189.

[514] A. D. Ioffe, *Approximate subdifferentials and applications, I: the finite dimensional theory*, Trans. Amer. Math. Soc. 281 (1984), 389-416; submitted: November 20, 1981.

[515] A. D. Ioffe, *Approximate subdifferentials and applications, II: functions on locally convex spaces*, Mathematika 33 (1986), 111-128.

[516] A. D. Ioffe, *On the local surjection property*, Nonlin. Anal. Th. Meth. Appl. 11 (1986), 565-592.

[517] A. D. Ioffe, *Approximate subdifferential and applications, III: the metric theory*, Mathematika 71 (1989), 1-38.

[518] A. D. Ioffe, *Proximal analysis and approximate subdifferentials*, J. London Math. Soc. 41 (1990), 175-192.

[519] A. D. Ioffe, *Fuzzy principles and characterization of trustworthiness*, Set-Valued Anal. 6 (1998), 265-276.

[520] A. D. Ioffe, *Codirectional compactness, metric regularity and subdifferential calculus*, Amer. Math. Soc., Providence, RI 2000.

[521] A. D. Ioffe, *Metric regularity and subdifferential calculus*, Russian Math. Surveys 55 (2000), 501-558.

[522] A. D. Ioffe, *Three theorems on subdifferentiation of convex integral functionals*, J. Convex Anal., 13 (2006), 759-772.

[523] A. D. Ioffe, *Typical convexity (concavity) of Dini-Hadamard upper (lower) directional derivatives of functions on separable Banach spaces*, J. Convex Anal. 17 (2010), 1019-1032.

[524] A. D. Ioffe, *Separable reduction revisited*, Optimization 60 (2011), 211-221.

[525] A. D. Ioffe, *On the general theory of subdifferentials*, Adv. Nonlinear Anal. 1 (2012), 47-120.

[526] A. D. Ioffe, *Variational Analysis of Regular Mappings: Theory and Applications*, Springer (2017).

[527] A. D. Ioffe, and J. V. Outrata, *On metric and calmness qualification conditions in subdifferential calculus*, Set-Valued Anal. 16 (2008), 199-227.

[528] A. D. Ioffe and Y. Sekiguchi, *Regularity estimates for convex multifunctions*, Math. Program. 117 (2009), 255–270 (submitted: November 2005).

[529] A. D. Ioffe and V. M. Tikhomirov, *Theory of Extremal Problems*, Nauka, Moscow (1974); English transl., Studies in Mathematics and its Applications, no. 6, North-Holland, Amsterdam, New York (1979).

[530] A. Iusem and A. Seeger, *Distances between closed convex cones: old and new results*, 17 (2010), 1033-1055.

[531] G. E. Ivanov, *Weak convexity in the sense of Vial and Efimov-Stechkin*, Izv. Ross. Akad. Nauk Ser. Mat. 69 (2005), 35-60 (in Russian); English tanslation in Izv. Math. (2005), 1113-1135.

[532] G. E. Ivanov, *Weakly convex sets and their properties*, Mat. Zametki 79 (2006), 60–86.

[533] G. E. Ivanov, *Weakly Convex Sets and Functions: Theory and Applications*, Fizmatlit, Moscow (2006) (in Russian).

[534] G. E. Ivanov, *On well posed best approximation problems for a nonsymmetric seminorm*, J. Convex Anal. 20 (2013), 501-529.

[535] G. E. Ivanov, *Weak convexity of sets and functions in Banach spaces*, J. Convex Anal. 22 (2015), 365-398.

[536] G. E. Ivanov, *Continuity and selections of the intersection operator applied to nonconvex sets*, J. Convex Anal. 22 (2015), 939-962.

[537] G. E. Ivanov, *Sharp estimates for the moduli of continuity of metric projections onto weakly convex sets*, Izvestiya Math. 79 (2015), 668-697.

[538] G. E. Ivanov, *Nonlinear images of sets. I: strong and weak convexity*, J. Convex Anal. 27 (2020), 361-380.

[539] G. E. Ivanov and M. V. Balashov, *Lipschitz parametrizations of set-valued maps with weakly convex images*, Izv. Ross. Akad. Nauk Ser. Mat. 71 (2007), 47-68 (in Russian); English translation in Izv. Math. 71 (2007), 1123-1143.

[540] G. E. Ivanov and L. Thibault, *Infimal convolution and optimal time control problem I: Fréchet and proximal subdifferentials*, Set-Valued Var. Anal. 26 (2018), 581-606.

[541] G. E. Ivanov and L. Thibault, *Infimal convolution and optimal time control problem II: Limiting subdifferential*, Set-Valued Var. Anal. 25 (2017), 517-542.

[542] G. E. Ivanov and L. Thibault, *Infimal convolution and optimal time control problem III: Minimal time projection set*, SIAM J. Optim. 28 (2018), 30-44.

[543] G. E. Ivanov and L. Thibault, *Well-posedness and subdifferentials of optimal value and infimal convolution*, Ser-Valued Var. Anal.

[544] M. Ivanov and N. Zlateva, *A Clarke-Ledyaev type inequality for certain nonconvex sets*, Serdica Math. J; 26 (2000), 277-286.

[545] M. Ivanov and N. Zlateva, *On primal lower nice property*, C.R. Acad. Bulgare Sci. 54 (2001), 5-10.

[546] M. Ivanov and N. Zlateva, *On nonconvex version of the inequality of Clarke and Ledyaev*, Nonlinear Anal. 49 (2002), 1023-1036.

[547] M. Ivanov and N. Zlateva, *Subdifferential characterization of primal lower nice functions on smooth spaces*, C.R. Acad. Bulgare Sci. 57 (2004), 13-18.

[548] M. Ivanov and N. Zlateva, *A new proof of the integrability of the subdifferential of a convex function on a Banach space*, Proc. Amer. Math. Soc. 136 (2008), 1787-1793.

[549] M. Ivanov and N. Zlateva, *On characterizations of metric regularity of multi-valued maps*, J. Convex Anal. 27 (2020), 381-388.

[550] A. J. Izzo, *Locally uniformly continuous functions*, Proc. Amer. Math. Soc. 122 (1994), 1095-1100.

[551] R. C. James, *Characterizations of reflexivity*, Studia Math. 23 (1963/1964), 205-216.
[552] R. C. James, *Weakly compact sets*, Trans. Amer. Math. Soc. 113 (1964), 129-140.
[553] R. C. James, *Some self-dual properties of normed linear spaces*, Symposium on Infinite Dimensional Topology, Annals of Mathematics Studies 69 (1972), 159-175.
[554] R. C. James, *Super-reflexive Banach spaces*, Can. J. Math. 24 (1972), 896-904.
[555] R. Janin, *Sur une classe de fonctions sous-linéarisables*, C. R. Acad. Sci. Paris 277 (1973), 265-267.
[556] R. Janin, *Sur la Dualité et la Sensibilité dans les Problèmes de Programmes Mathématiques*, Thesis (Thèse de Doctorat d'État ès Sciences Mathématiques), Université Paris VI, (1974).
[557] R. Janin, *Sur des multi-applications qui sont des gradients généralisés*, C. R. Acad. Sci. Paris 294 (1982), 115-117.
[558] S. Janiszewski, *Sur les continus irreductibles entre deux points*, Journal de l'Ecole Polytechnique 2e série - 16e Cahier (1912), pp. 79-170; (and Thèse de Doctorat ès Scienses Mathématiques, Universitè de Paris (1911)).
[559] J. L. W. V. Jensen *Om konvexe Funktioner og Uligheder mellem Middelvaerdier*, Nyt. Tidsskrift for Mathematik 16 B (1905), 49-69.
[560] J. L. W. V. Jensen *Sur les fonctions convexes et les inégalités entre les valeurs moyennes*, Acta Math. 30 (1906), 175-193.
[561] W. L. Jones, *On conjugate functionals*, Ph.D. dissertation thesis, Columbia University (1960).
[562] M. Jouak and L. Thibault, *Equicontinuity of families of convex and concave-convex operators*, Canad. J. Math. 36 (1984), 883-898.
[563] M. Jouak and L. Thibault, *Directional derivatives and almost everywhere differantiability of biconvex and concave-conve operators*, Math. Scand. 57 (1985), 215-224.
[564] A. Jofre and L. Thibault, *D-representation of subdifferentials of directionally Lipschitz functions*, Proc. Amer. Math. Soc. 110 (1990), 117-123.
[565] E. Jouini, *A remark on Clarke's normal cone and the marginal cost pricing rule*, J. Math. Econ. 18 (1989), 95-101.
[566] E. Jouini, *Functions with constant generalized gradients*, J. Math. Anal. Appl. 148 (1990), 121-130.
[567] E. Jouini, W. Schachermayer, and N. Touzi, *Law invariant risk measures have the Fatou property*, Adv. Math. Econ. 9 (2006), 49-71.
[568] A. Jourani, *On metric regularity of multifunctions*, Bull. Austral. Math. Soc. 44 (1991), 1-9.
[569] A. Jourani, *Regularity and strong sufficient optimality conditions in differentiable optimization problems*, Numer. Funct. Anal. optim. 14 (1993), 69-87.
[570] A. Jourani, *Compactly epi-Lipschitzian sets and A-subdifferentials in WT-spaces*, Optimization, 34 (1995), 1-17.
[571] A. Jourani, *Qualification conditions for multivalued functions in Banach spaces with applications to nonsmooth vector optimization problems*, Math. Program. 66 (1994), 1-23.
[572] A. Jourani, *Open mapping theorem and inversion theorem for γ-para-convex multivalued mappings and applications*, Studia Math. 117 (1996), 123-136.
[573] A. Jourani, *Subdifferentiability and subdifferential monotonicity of γ-para-convex functions*, Control Cybernetics, 25 (1996), 721-737.
[574] A. Jourani, *The role of locally compact cones in nonsmooth analysis*, Communications Appl. Nonlinear Anal. 5 (1998), 1-35.
[575] A. Jourani, *Variational sum of subdifferentials of convex functions*, Proceedings of the IV Catalan Days of Applied Mathematics, Tarragona, (1998), 71-79.
[576] A. Jourani, *Limits superior of subdifferentials of uniformly convergent functions in Banach spaces*, Positivity 3 (1999), 33-47.
[577] A. Jourani, *Weak regularity of functions and sets in Asplund spaces*, Nonlinear Analysis 45 (2006), 660-676.
[578] A. Jourani, *Radiality and semismoothness*, Control Cybernetics, 36 (2007), 669-680.
[579] A. Jourani and M. Sene, *Characterization of the Clarke regularity of subanalytic sets*, Proc. Amer. Math. Soc. 146 (2018), 1639-1649.
[580] A. Jourani and M. Sene, *Geometric characterizations of the strict Hadamard differentiability of sets*, Pure Appl. Funct. Anal., to appear.
[581] A. Jourani and L. Thibault, *Approximate subdifferential and metric regularity: the finite-dimensional case*, Math. Program. 47 (1990), 203-218.

[582] A. Jourani and L. Thibault, *Approximate subdifferentials of composite functions*, Bull. Aust. Math. Soc. 47 (1993), 443-455.

[583] A. Jourani and L. Thibault, *A note on approximate subdifferentials of composite functions*, Bull. Aust. Math. Soc. 49 (1994), 111-116.

[584] A. Jourani and L. Thibault, *Metric regularity and subdifferential calculus in Banach spaces*, Set-Valued Anal. 3 (1995), 87-100.

[585] A. Jourani and L. Thibault, *Metric regularity for strongly compactly Lipschitzian mappings*, Nonlinear Analysis 24 (1995), 229-240.

[586] A. Jourani and L. Thibault, *Verifiable conditions for openness and regularity of multivalued mappings*, Trans. Amer. Math. Soc. 347 (1995), 1255-1268.

[587] A. Jourani and L. Thibault, *Extensions of subdifferentials calculus rules in Banach spaces and applications*, Canadian J. Math. 48 (1996), 834-848.

[588] A. Jourani and L. Thibault, *Coderivatives of multivalued mappings, locally compact cones and metric regularity*, Nonlinear Anal. 35 (1999), 925-945.

[589] A. Jourani, L. Thibault and D. Zagrodny, $C^{1,\omega(\cdot)}$-*regularity and Lipschitz-like properties of subdifferential*, Proc. London Math. Soc. 105 (2012), 189-223.

[590] A. Jourani, L. Thibault and D. Zagrodny, *Differential properties of the Moreau envelope*, J. Funct. Anal. 266 (2014), 1185-1237.

[591] A. Jourani and E. Vilches, *Positively α-far sets and existence results for generalized perturbed sweeping processes*, J. Convex Anal. 23 (2016), 775-821.

[592] A. Jourani and T. Zakaryan, *The validity of the liminf formula and a characterization of Asplund spaces*, J. Math. Anal. Appl. 416 (2014), 824-838.

[593] F. Jules, *Sur la somme de sous-différentiels de fonctions semi-continues inférieurement*, Dissertationes Mathematicae 423 (2003), 62 pp.

[594] F. Jules and M. Lassonde, *Formulas for subdifferentials of sums of convex functions*, J. Convex Anal. 9 (2002), 519-533.

[595] F. Jules and M. Lassonde, *Subdifferential test for optimality*, J. Global Optim. 59 (2014), 101–106.

[596] W. Karush, *A queuing model for an inventory problem*, Operations Res. 5 (1957), 693-703.

[597] W. Karush, *A general algorithm f or the optimal distribution of effort*, Management Science 9 (1962), 50-72.

[598] T. Kato, *Perturbation theory for nullity, deficiency and other quantities of linear operators*, J. Analyse Math. 6 (1958), 261-322.

[599] T. Kato, *Perturbation Theory for Linear Operators*, Vol. 132 Grundlehren der mathematischen Wissenschaften, Springer-Verlag, Berlin, (2006).

[600] G. Katriel, *Are the approximate and the Ckarke subgradients generically equal?*, J. Math. Anal. Appl. 193 (1995), 588-593.

[601] I. Kecis and L. Thibault, *Subdifferential characterization of s-lower regular functions*, Appl. Anal. 94 (2015), 85-98.

[602] I. Kecis and L. Thibault, *Moreau envelopes of s-lower regular functions*, Nonlinear Anal. 127 (2015), 157-181.

[603] I. Kecis and L. Thibault, *Evolution differential inclusions associated to primal lower regular functions*, J. Nonlin. Convex Anal. 20 (2019), 1949-1979.

[604] J.L. Kelley, *General Topology*, D. Van Nostran Company, Princeton, New Jersey (1961).

[605] M.D. Kirszbraun, *Über die zusammenziehenden und Lipschitzchen tranformationen*, Fund. Math. 22 (1934), 77-108.

[606] T.H. Kjeldsen, *A Contextualized historical analysis of the Kuhn-Tucker theorem in nonlinear programming: the Impact of world war II*, Historia Mathematica 27 (2000), 331-361.

[607] D. Klatte and B. Kummer, *Nonsmooth Equations in Optimization: Regularity, Calculus, Methods, and Applications*, Kluwer, Boston, Massachusetts (2002)

[608] D. Klatte and B. Kummer, *Stability of inclusions: characterizations via suitable Lipschitz functions and algorithms*, Optimization 55(5-6) (2006), 627-660.

[609] V. Klee, *Convexity of Chebyshev sets*, Math. Annalen 142 (1961), 292-304.

[610] V. Klee, *On a question of Bishop and Phelps*, Amer. J. Math. 85 (1963), 95-98.

[611] A. Kolmogoroff and J. Verčenko, *Ueber Unstetigkeitspunkte von Funktionen zweier Veränderlichen*, C. R. Acad. Sci. URSS 1 (1934), 105-107.

[612] A. Kolmogoroff and J. Verčenko, *Weitere Untersuchungen über Unstetigkeitspunkte von Funktionen zweier Veränderlichen*, C. R. Acad. Sci. URSS 4 (1934), 361-364.

[613] S. V. Konjagin, *On approximation properties of closed sets in Banach spaces and the characterization of strongly convex spaces*, Soviet Math. Dokl. 21 (1980), 418-422.
[614] M. Konstantinov and N. Zlateva, *Epsilon subdifferential method and integrability*, J. Convex Anal. (2022), to appear.
[615] G. Köthe, *Topological Vector Spaces I*, Grundlehren der mathematischen Wissenschaften, Springer-Verlag, Berlin (1969) (English translation).
[616] S. G. Krantz and H. R. Parks, *Distance to C^k hypersurfaces*, J. Differential Equations, 40 (1981), 116-120.
[617] M. G. Krein and M. A. Krasnoselskiĭ, *Fundamental theorems concerning the extension of Hermitian operators and some of their applications to the theory of orthogonal polynomials and the moment problem*, Uspekhi Mat. Nauk. 2, 3 (1947), 60-106 (in Russian).
[618] M. G. Krein, M. A. Krasnoselskiĭ, and D. P. Milman, *Concerning the deficiency numbers of linear operators in Banach space and some geometric questions*, Sbornik Trudov Inst. A. N. Ukr. S. S. R., 11 (1948) (in Russian).
[619] A. Y. Kruger, *Epsilon-semidifferentials and epsilon-normal elements*, Deposited in VINITI no 1331-81, Minsk, 1981 (in Russian).
[620] A. Y. Kruger, *Generalized differentials of nonsmooth functions*, Deposited in VINITI no. 1332-81, Minsk, 1981 (in Russian).
[621] A. Y. Kruger, *Generalized differentials of nonsmooth functions and necessary conditions for an extremum*, Siberian Math. J. 26 (1985), 370-379.
[622] A. Y. Kruger, *Properties of generalized differentials*, Siberian Math. J. 26 (1985), 822-832.
[623] A. Y. Kruger, *A covering theorem for set-valued mappings*, Optimization 19 (1988), 763-780.
[624] A. Y. Kruger, *Strict ε-semidifferentials and extremality conditions*, Dokl. Nat. Akad. Nauk Belarus 41 (1997), 21-26 (in Russian).
[625] A. Y. Kruger, *On extremality of sets systems*, Dokl. Nat. Akad. Nauk Belarus 42 (1998), 24-28 (in Russian).
[626] A. Y. Kruger, *Strict (ε, δ)-semidifferentials and extremality of sets and functions*, Dokl. Nat. Akad. Nauk Belarus 44 (2000), 21-24 (in Russian).
[627] A. Y. Kruger, *Strict (ε, δ)-semidifferentials and extremality conditions*, Optimization 51 (2002), 539-554.
[628] A. Y. Kruger, *On Fréchet subdifferentials*, J. Math. Sci. 116 (2003), 3325-3358.
[629] A. Y. Kruger, *Weak stationarity: eliminating the gap between necessary and sufficient conditions*, Optimization 53(2) (2004), 147-164.
[630] A. Y. Kruger, *Stationarity and regularity of set systems*, Pac. J. Optim. 1(1) (2005), 101-126.
[631] A. Y. Kruger, *About regularity of collections of sets*, Set-Valued Anal. 14(2) (2006), 187-206.
[632] A. Y. Kruger, *Stationarity and regularity of real-valued functions*, Appl. Comput. Math. 5(1) (2006), 79-93.
[633] A. Y. Kruger, *About stationarity and regularity in variational analysis*, Taiwanese J. Math. 13 (2009), 1737-1785.
[634] A. Y. Kruger and B. S. Mordukhovich, *Extremal points and the Euler equation in nonsmooth problems of optimization*, Dokl. Akad. Nauk BSSR 24 (1980), 684-687 (in Russian).
[635] A. Y. Kruger and B. S. Mordukhovich, *Generalized normals and derivatives and necessary conditions for an extremum in problems of nondifferentiable programming*, Parts I and II, Deposited in VINITI, I - no. 408-80, II - no. 494-80, Minsk, 1980 (in Russian).
[636] S. N. Kruzhkov, *The Cauchy problem in the large for certain nonlinear first order differential equations*, Soviet. Math. Dokl. 1 (1960), 474-477.
[637] W. Kryszewski and S. Plaskacz, *Periodic solutions to impulsive differential inclusions with constraints*, Nonlin. Anal. Th. Meth. Apl. 65 (2006), 1974-1804.
[638] B. Kummer, *Metric regularity: Characterizations, nonsmooth variations and successive approximation*, Optimization 46 (1999), 247-281.
[639] B. Kummer, *Inverse functions of pseudo regular mappings and regularity conditions*, Math. Program., Ser. B 88, 2 (2000), 313-339.
[640] C. Kuratowski, *Les fonctions semi-continues dans l'espace des ensembles fermés*, Fund. Math. 18 (1932), 148-159; submitted in 1931.
[641] C. Kuratowski, *Topologie I*, Monografje Matematyczne (1933).
[642] C. Kuratowski, *Topologie II*, Monografje Matematyczne, (1933).
[643] A. G. Kusraev and S. S. Kutateladze, *Subdifferentials: Theory and Applications*, Mathematics and its Applications Vol. 323, Kluwer Academic Publishers, Dordrecht (1995).

[644] M. Laczkovich, *The local stability of convexity, affinity and of the Jensen equation*, Aequationes Math. 58/1-2 (1999), 135-142.

[645] A. Largillier, *A note on the gap convergence*, Appl. Math. Lett. 7 (1994), 67–71.

[646] M. Lassonde, *First-order rules for nonsmooth constrained optimization*, Nonlinear Anal. 44 (2001), 1031-1056.

[647] K.-S. Lau, *Farthest points in weakly compact sets*, Israel J. Math. 22 (1975), 168-174.

[648] K.-S. Lau, *Almost Chebyshev subsets in reflexive Banach spaces*, Indiana Univ. Math. J. 27 (1978), 791-795.

[649] P.-J. Laurent, *Approximation et Optimisation*, Hermann (1972).

[650] G. Lebourg, *Valeur moyenne pour gradients généralisés*, C. R. Acad. Sci. Paris Sér. A 281 (1975), 795-798.

[651] G. Lebourg, *Solutions en densité de problèmes d'optimisation paramétrés*, C. R. Acad. Sci. PAris Sér. A 289 (1979), 79-82.

[652] G. Lebourg, *Generic differentiability of Lipschitzian functions*, Trans. Amer. Math. Soc. 256 (1979), 123-144.

[653] A.-M. Legendre, *Mémoire sur l'intégration de quelques équations aux différences partielles*, Histoire de l'Académie royale des sciences (1787), 309-351.

[654] E. S. Levitin and B. T. Polyak, *Constrained minimization methods*, U.S.S.R. Computational Math. and Math. Phys. 6 (1966), 787-823.

[655] A. B. Levy, R. A. Poliquin and R. T. Rockafellar, *Stability of locally optimal solutions*, SIAM J. Optim. 10 (2000), 580-604.

[656] A. B. Levy, R. A. Poliquin and L. Thibault, *A partial extension of Attouch's theorem and its applications to second-order differentiation*, Trans. Amer. Math. Soc. 347 (1995), 1269-1294.

[657] A. B. Levy, R. A. Poliquin, R. T. Rockafellar, *Stability of locally optimal solutions*, SIAM J. Optim. 10 (2000), 580-604.

[658] A. S. Lewis and R. Lucchetti, *Nonsmooth duality, sandwich and squeeze theorems*, (Preprint, Università degli studi Milano 1998), SIAM J. Control Optim. 38 (2000), 613-626.

[659] A. S. Lewis, D. R. Luke and J. Malick, *Local linear convergence for alternating and averaged nonconvex projections*, Found. Comput. Math. 9 (2009), 485-513.

[660] A. S. Lewis and A. D. Ralph, *A nonlinear duality result equivalent to the Clarke-Ledyaev mean value inequality*, Nonlinear Anal. Th. Meth. Appl. 26 (1996), 343-350.

[661] A. S. Lewis and S. Zhang, *Partial smoothness, tilt stability, and generalized Hessians*, SIAM J. Optim. 23 (2013), 74-94.

[662] G. Li and B. S. Mordukhovich, *Hölder metric subregularity with applications to proximal point method*, SIAM J. Optim. 22 (2012), 1655-1684.

[663] J. Lindenstrauss, *On the modulus of smoothness and divergent series in Banach spaces*, Michigan Math. J. 10 (1963), 241-252.

[664] J. Lindenstrauss, *On operators which attain their norm*, Israel J. Math. 1 (1963), 139-148.

[665] J. Lindenstaruss, L. Tzafiri, *Classical Banach Spaces I. Sequence Spaces*, Springer, Berlin (1977).

[666] J. Lindenstaruss, L. Tzafiri, *Classical Banach Spaces II. Function Spaces*, Springer, Berlin (1979).

[667] V. I. Liokkoumovich, *The existence of B-spaces with non-convex modulus of convexity*, Izv. Vysš. Učebn. Zadev. Matematika 12 (1973), 13-50 (in Russian).

[668] P. D. Loewen, *The proximal normal formula in Hilbert space*, Nonlin. Anal. Th. Meth. AppL. 11 (1987), 979-995 (submitted: January 1986).

[669] P. D. Loewen, *The proximal subgradient formula in Banach spaces*, Canad. Math. Bull. 31 (1988), 353-361.

[670] P. D. Loewen, *Limit of Fréchet normals in nonsmooth analysis*, Optimization and Nonlinear Analysis (A.D. Ioffe, L. Marcus and S. Reich editors), Pitman Research Notes Math. Ser; 244 (1992), 178-188.

[671] P. D. Loewen, *Optimal Control via Nonsmooth Analysis*, Amer. Math. Soc., Providence, Rhode Island (1993).

[672] P. D. Loewen, *A mean value theorem for Fréchet subgradients*, Nonlin. Anal. Th. Meth. Appl. 23 (1994), 1365-1381.

[673] P. D. Loewen and R. T. Rockafellar, *Optimal control of unbounded differential inclusions*, SIAM J. Control Optim. 32 (1994), 442-470.

[674] P. D. Loewen and X. Wang, *A generalized variational principle*, Canad. J. Math. 53 (2001), 1174-1193.

[675] O. Lopez and L. Thibault, *Sequential formula for subdifferential of integral sum of convex functions*, J. Nonlinear Convex Anal., 9 (2008), 295-308.

[676] R. A. Lovaglia, *Locally uniformly convex Banach spaces*, Trans. Amer. Math. Soc. 78 (1955), 225-238.

[677] R. Lozano, B. Brogliato, O. Egeland and B. M. Maschke, *Dissipative Systems Analysis and Control*, Springer London, CCE Series (2000).

[678] Dinh The Luc, *Theory of Vector Optimization*, Lect. Notes Econ. Math. Sci. 319, Springer, Berlin (1989).

[679] Dinh The Luc, *A strong mean value theorem and applications*, Nonlinear Anal. 26 (1996), 915-923.

[680] Dinh The Luc and S. Swaminathan, *A characterization of convex functions*, Nonlinear Anal. Th. Meth. Appl. 20 (1993), 697-701.

[681] K.R. Lucas, *Submanifolds of dimension $n-1$ in $\mathcal{E}^n$ with normals satisfying a Lipschitz condition*, Studies In Eigenvalue Problems, Technical Report 18, Kansas University, Dept. of Mathematics, 1957. http://hdl.handle.net/2027/mdp.39015017419931

[682] L. A. Lyusternik, *On conditional extrema of functionals*, Mat. Sb. 41 (1934), 390-401.

[683] J. Makó and Z. Páles, *Strengthening of strong and approximate convexity*, Acta Math. Hung. 132 (2011), 78-91.

[684] J. Malý, *A simple proof of the Stepanov theorem on differentiabilty almost everywhere*, Exposition. Math. 17 (1999), 59-61.

[685] J. Malý and L. Zajíček, *On Stepanov type of differentiability theorems*, Acta Mathemaica Hungarica 145 (2015), 174–190.

[686] S. Mandelbrojt, *Sur les fonctions convexes*, C. R. Acad. Sci. Paris, 209 (1939), 977-978.

[687] P. Mankiewicz, *On the differentiabiliy of Lipschitz mappings in Fréchet spaces*, Studia Math. 45 (1973), 15-29.

[688] S. Marcellin and L. Thibault, *Integration of ε-subdifferential and maximal cyclic monotonicity*, J. Global Optim. 32 (2005), 83-91.

[689] S. Marcellin and L. Thibault, *Evolution problems associated with primal lower nice functions*, J. Convex Anal. 13 (2006), 385-421.

[690] J. Marcinkiewicz, *Sur les nombres dérivés*, Fundam. Math. 24 (1935), 305-308.

[691] J. Marcinkiewicz, *Sur les séries de Fourier*, Fundam. Math. 27 (1936), 38-69.

[692] J. Marcinkiewicz and A. Zygmund, *On the differentiability of functions and summability of trigonometrical series*, Fundam. Math. 26 (1936), 1-43.

[693] A. Marigonda, *Differentiability Properties for a Class of Non-Lipschitz Functions and Applications*, Ph.D. dissertation thesis, University of Padova, 2006.

[694] M. Marques-Alves and B. F. Svaiter, *A new proof for maximal monotonicity of subdifferential operators*, J. Convex Anal. 15 (2008), 345-348.

[695] P. Mattila, *Geometry of Sets and Measures in Euclidean Spaces*, Cambridge studies in advanced mathematics, Cambridge University Press (1995).

[696] B. Maury and J. Venel, *A mathematical framework for a crowd motion model*, C. R. Math. Acad. Sci. Paris 346 (2008), no. 23-24, 1245-1250.

[697] M. Mazade and L. Thibault, *Differential variational inequalities with locally prox regular sets*, J. Convex Anal. 19 (2012), 1109–1139.

[698] M. Mazade and L. Thibault, *Regularization of differential variational inequalities with locally prox-regular sets*, Math. Program. Ser. B 139 (2013), 243-269.

[699] M. Mazade and L. Thibault, *Primal lower nice functions and their Moreau envelopes*, In: Computational and Analytical Mathematics, New York (NY): Springer Proceedings in Mathematics Statistics (2013), pp. 521-553.

[700] S. Mazur, *Über konvexe Mengen in linearen normierten Räumen*, Studia Math. 4 (1933), 70-84.

[701] S. Mazur, *Über schwache Konvergentz in den Räumen L^p*, Studia Math. 4 (1933), 128-133.

[702] S. Mazurkiewicz, *Sur les nombres dérivés*, Fund. Math. 23 (1934), 9-10.

[703] E. J. Mc Shane, *Extension of range of functions*, Bull. Amer. Math. Soc. 40 (1934), 837-842 (submitted, June 1934).

[704] R. E. Megginson, *An Introduction to Banach Space Theory*, Graduate Texts in Mathematics, Springer, New York (1998).

[705] E. Michael, *An existence theorem for continuous functions*, Bull. Amer. Math. Soc. 59 (1953), p. 180.
[706] E. Michael, *Continuous selections. I*, Ann. Math. 63 (1956), 361-382 (submitted: December 1954).
[707] E. Michael, *Selected selection theorems*, Amer. Math. Monthly 63 (1956), 233-238.
[708] P. Michel and J.-P. Penot, *A generalized derivative for calm and stable functions*, Diff. Int. Equations 5 (1992), 433-454.
[709] R. Mifflin, *Semismooth and semiconvex functions in constrained optimization*, SIAM J. Control Optim. 15
[710] F. Mignot, *Contrôle dans les inèquations variationelles elliptiques*, J. Funct. Anal. 22 (1976), 130-185.
[711] D. Milman, *On some criteria for the regularity of spaces of the type (B)*, Comptes Rendus (Doklady) de l'Académie des Sciences de l'URSS, new series, vol. 20 (1938), 243-246.
[712] G. J. Minty, *On the monotonicity of the gradient of a convex function*, Pacific J. Math. 14 (1964), 243-247.
[713] G. J. Minty, *On the extension of Lipschitz, Lipschitz Hölder continuous, and monotone functions*, Bull. Amer. Math. Soc. 76 (1970), 334-339.
[714] J. Mirguet, *Nouvelles Recherches sur les Notions Infinitésimales Directes du Premier Ordre*, Thesis (Thèse de Doctorat ès Sciences Mathématiques) Université de Paris (1934).
[715] D. S. Mitrinović and I. B. Lacković, *Hermite and convexity*, Aequationes Mathematicae, 28 (1985), 229-232.
[716] M. D. P. Monteiro Marques, *Differential Inclusions in Nonsmooth Mechanical Problems– Shocks and Dry Friction*, Birkhäuser, Boston (1993).
[717] V. Montesinos, *Drop property equals reflexivity*, Studia Math. 87 (1987), 93-100.
[718] W. B. Moors, *Weak compactness of sublevel sets*, Proc. Amer. Math. Soc. 145 (2017), 3377-3379.
[719] W. B. Moors, *On a one-sided James' theorem*, J. Math. Anal. Appl. 449 (2017), 528-530.
[720] B. S. Mordukhovich, *Maximum principle in the problem of time optimal response with nonsmooth constraints*, J. Appl. Math. Mech. 40 (1976), 960-969.
[721] B. S. Mordukhovich, *Metric approximation and necessary optimality conditions for general classes of nonsmooth extremal problems*, Soviet Math. Dokl. 22 (1980), 526-530.
[722] B. S. Mordukhovich, *Nonsmooth analysis and nonconvex generalized differentials and adjoint mappings*, Dokl. Akad. Nauk BSSR 28 (1984), 976-979 (in Russian).
[723] B. S. Mordukhovich, *Approximation methods in problems of optimization and control*, Nauka, Moscow (1988) (in Russian).
[724] B. S. Mordukhovich, *Complete characterization of openness, metric regularity, and Lipschitzian properties of multifunctions*, Trans. Amer. Math. Soc. 340 (1993), 1-36.
[725] B. S. Mordukhovich, *Variational Analysis and Generalized Differentiation, I: Basic Theory*, Vol. 330 Grundlehren der mathematischen Wissenschaften, Springer-Verlag, Berlin, (2006).
[726] B. S. Mordukhovich, *Variational Analysis and Generalized Differentiation, II: Applications*, Vol. 331 Grundlehren der mathematischen Wissenschaften, Springer-Verlag, Berlin, (2006).
[727] B. S. Mordukhovich and N. M. Nam, *Subgradient of distance functions and applications to Lipschitzian stability*, Math. Program. Ser. B 104 (2005), 635-668.
[728] B. S. Mordukhovich and T. T. Nghia, *Second-order variational analysis and characterizations of tilt-stable optimal solutions in finite and infinite dimensions*, Nonlinear Anal. 86 (2013), 159-180.
[729] B. S. Mordukhovich and Y. Shao, *Differential characterizations of covering, metric regularity, and Lipschitzian properties of multifunctions between Banach spaces*, Nonlinear Anal. 25 (1995), 1401-1424.
[730] B. S. Mordukhovich and Y. Shao, *Nonsmooth sequential analysis in Asplund spaces*, Trans. Amer. Math. soc. 124 (1996), 1235-1280.
[731] B. S. Mordukhovich and Y. Shao, *Nonconvex differential calculus for infinite dimensional multifunctions*, Set-Valued Anal. 4 (1996), 205-256.
[732] B. S. Mordukhovich and Y. Shao, *Stability of set-valued mappings in infinite dimensions: point criteria and apllications*, SIAM J. Control Optim. 35 (1997), 285-314.
[733] J. J. Moreau, *Fonctions convexes en dualité*, (multigraphié), Séminaires de Mathématiques, Faculté des Sciences de Monpellier (1962), 18 pages.

[734] J. J. Moreau, *Fonctions convexes duales et points proximaux dans un espace hilbertien*, C.R. Acad. Sci. Paris 255 (1962), 2897-2899.

[735] J. J. Moreau, *Inf-concolution*, (multigraphié), Séminaires de Mathématiques, Faculté des Sciences de Montpellier (1963), 48 pages.

[736] J. J. Moreau, *Propriétés des applications 'prox'*, C.R. Acad. Sci. Paris, 256 (1963), 1069-1071.

[737] J. J. Moreau, *Inf-convolution des fonctions numériques sur un espace vectoriel*, C.R. Acad. Sci. Paris, 256 (1963), 5047-5049.

[738] J. J. Moreau, *Etude locale d'une fonctionnelle convexe*, (multigrapié) Séminaires de Mathématique, Faculté des Sciences de Montpellier (1963), 25 pages.

[739] J. J. Moreau, *Fonctionnelles sous-différentiables*, C. R. Acad. Sci. Paris 257 (1963), 4117-4119.

[740] J. J. Moreau, *Sur la fonction polaire d'une fonction semicontinue supérieurement*, C. R. Acad. Sci. Paris 258 (1964), 1128-1131.

[741] J. J. Moreau, *Proximité et dualité dans un espace hilbertien*, Bull. Soc. Math. France 93 (1965), 273-299.

[742] J. J. Moreau, *Semi-continuité du sous-gradient d'une fonctionnelle*, C. R. Acad. Sci. Paris 260 (1965), 1067-1070.

[743] J. J. Moreau, *Quadratic programming in mechanics: dynamics of one-sided constraints*, SIAM J. Control 4 (1966), 153-158.

[744] J. J. Moreau, *Fonctionnelles Convexes*, Lecture Notes, Collège de France (1966-1967); Second edition by Facoltà di ingegneria, Università di Roma "Tor Vergata" (2003).

[745] J. J. Moreau, *Inf-convolution, sous-additivité, convexité des fonctions numériques*, J. Math. Pures Appl. 49 (1970), 109-154.

[746] J. J. Moreau, *Rafle par un convexe variable I.* Sém. Anal. Convexe Montpellier, Exposé 15, (1971).

[747] J. J. Moreau, *Rafle par un convexe variable II.* Sém. Anal. Convexe Montpellier, Exposé 3, (1972).

[748] J. J. Moreau, *On unilateral constraints, friction and plasticity*, Corso tenuto a Bressanone dal 17 al 26 giugno 1973, Centro Internazionale Matematico Estivo (1973) p. 172-322.

[749] J. J. Moreau, *Multi-applications à rétraction finie*, Ann. Scuola Norm. Sup. Pisa 1 (1974), 169-203.

[750] J. J. Moreau, *On unilateral constraints, friction and plasticity*, in New Variational Techniques in Mathematical Physics, Proceedings from CIME (Capriz and Stampacchia eds.), Cremonese, Rome (1974), 173-322.

[751] J. J. Moreau, *Evolution problem associated with a moving convex set in a Hilbert space*, J. Differential Equations 26 (1977), 347-374.

[752] U. Mosco, *Convergence of convex sets and of solutions of variational inequalities*, Advances Math. 3 (1969), 510-585.

[753] T. S. Motzkin, *Sur quelques propriétés caractéristiques des ensembles convexes*, Att. R. Acad. Lincei Rend. 21 (1935), 562-567.

[754] F. Nacry, *Truncated nonconvex state-dependent sweeping process: implicit and semi-implicit adapted Moreau's catching-up algorithms*, J. Fixed Point Theory Appl. 20 (2018), Paper No 121.

[755] F. Nacry and L. Thibault, *Regularization of sweeping process: old and new*, Pure Appl. Funct. Anal. 4 (2019), 59-117.

[756] F. Nacry and L. Thibault, *BV prox-regular sweeping process with bounded truncated variation*, Optimization 69 (2020), 1391-1437.

[757] F. Nacry and L. Thibault, *Distance function associated to a prox-regular set*, Set-Valued Var. Anal., to appear.

[758] F. Nacry and L. Thibault, *Farthest points and strong convexity*, in progress.

[759] N. M. Nam, *Subdifferential formulas for a class of nonconvex infimal convolutions*, Optimization 64 (2015), 2213-2222.

[760] N. M. Nam and D. V. Cuong, *Generalized differentiation and characterizations for differentiability of infimal convolutions*, Set-Valued Var. Anal. 23 (2015), 333-353.

[761] J. Nash, *Real algebraic manifolds*, Ann. Math. 56 (1952), 405-421.

[762] J. Nečas, *Les Méthodes Directes en Théorie des Équations Elliptiques*, Masson, Paris (1967).

[763] A. Nekvinda and L. Zajíček, *A simple proof of the Rademacher theorem*, Časopis Pěst. Mat., 113 (1988), 337-341.

[764] A. Nekvinda and L. Zajíček, *Gâteaux differentuability of Lipschitz functions via directional derivatives*, Real Analysis Exchange 28 (2002/2003), 287-320.

[765] J. D. Newburg, *A topology for closed operators*, Ann. Math. 53 (1951), 250-255.

[766] K. F. Ng and W. H. Yang, *Regularities and their relations to error bounds*, Math. Program., Ser. A, 99 (2004), 521-538.

[767] H. V. Ngai, D. T. Luc and M. Théra, *Approximate convex functions*, J. Nonlinear Convex Anal. 1 (2000), 155-176.

[768] H. V. Ngai and J.-P. Penot, *Approximately convex sets*, J. Nonlinear Convex Anal. 8 (2007), 337-355.

[769] H. V. Ngai and J.-P. Penot, *Approximately convex functions and approximately monotone operators*, Nonlinear Anal. 66 (2007), 547-564; submitted 6 January 2005.

[770] H. V. Ngai and J.-P. Penot, *Subdifferentiation of regularized functions*, Set-Valued Var. Anal., 24 (2016), 167-189.

[771] H. V. Ngai and M. Théra, *φ-regular functions in Asplund spaces*, Control Cyber. 36 (2007), 755-774.

[772] T. K. Nguyen, *Hypographs satisfying an external sphere condition and the regularity of the minimum time function*, J. Math. Anal. Appl. 372 (2010), 611-628.

[773] G. Norlander, *The modulus of convexity in normed linear spaces*, Ark. Mat. 4 (1960), 15-17.

[774] C. Nour, R. J. Stern and J. Takche, *The union of uniform closed balls conjecture*, Control and Cybernetics 38 (2009), 1525-1534.

[775] C. Nour, R. J. Stern and J. Takche, *Proximal smoothness and the exterior sphere condition*, J. Convex Anal. 16 (2009), 501-514.

[776] C. Nour, R. J. Stern and J. Takche, *Validity of the union of uniform closed balls conjecture*, J. Convex Anal. 18 (2011), 589-600.

[777] J. Orihuela, *Conic James' compactness theorem*, J. Convex Anal. 25 (2018), 1335-1344.

[778] M. I. Ostrovskiĭ, *Topologies on the set of all subspaces of a Banach space and related questions of Banach space geometry*, Quaest. Math. 17 (1994), 259-319.

[779] P. Painlevé, *Sur les Lignes Singulières des Fonctions Analytiques*, Thesis (Thèse de Doctorat ès Sciences Mathématiques) Université de Paris (1887).

[780] P. Painlevé, *Sur les lignes singulières des fonctions analytiques*, Annales de la Faculté des Sciences de Toulouse 1re série, tome 2 (1888), p. B1-B130

[781] Z. Páles, *On approximately convex functions*, Proc. Amer. Math. Soc. 131 (2002), 243-252.

[782] G. Peano *Applicazioni Geometriche del Calcolo Infinitesimale*, Fratelli Bocca, Torino (1887).

[783] G. Peano *Lezioni di Analisi Infinitesimale*, Candeletti, Torino (1893).

[784] G. Peano *Formulario Mathematico*, Fratelli Bocca, Torino (1908).

[785] J.-P. Penot, *Calcul sous-différentiel et optimisation*, J. Funct. Anal. 27 (1978), 248-276.

[786] J.-P. Penot, *A characterization of tangential regularity*, Nonlinear Anal. Th. Meth. Appl. 5 (1981), 625-643.

[787] J.-P. Penot, *The drop theorem, the petal theorem and Ekeland's variational principle*, Nonlinear Anal. Th. Meth. Appl. 10 (1986), 459-468.

[788] J.-P. Penot, *Metric regularity, openness and Lipschitzian behavior of multifunctions*, Nonlin. Anal. Th. Meth. Appl. 13 (1989), 629-643.

[789] J.-P. Penot, *The cosmic Hausdorff topology, the bounded Hausdorff topology, and continuity of polarity*, Proc. Amer. Math. Soc. 113 (1991), 275-286.

[790] J.-P. Penot, *On the interchange of subdifferentiation and epi-convergence*, J. Math. Anal. Appl. 196 (1995), 676-698.

[791] J.-P. Penot, *Favorable classes of mappings and multimappings in nonlinear analysis and optimization*, J. Convex Anal. 3 (1996), 97-116.

[792] J.-P. Penot, *Subdifferential calculus without qualification assumptions*, J. Convex Anal. 3 (1996), 207-219.

[793] J.-P. Penot, *Compactness properties, openness criteria and coderivatives*, Set-Valued Anal. 6 (1998), 363-380.

[794] J.-P. Penot, *A short proof of the separable reduction theorem*, Demonstratio Math. 4 (2010), 653-663.

[795] J.-P. Penot, *Calculus Without Derivatives*, Graduate Texts in Mathematics, Springer, Berlin (2013).

[796] J.-P. Penot and M. Bougeard, *Approximation and decomposition properties of some classes of locally d.c. functions*, Math. Program. 41 (1988), 195-227.
[797] P. Pérez-Aros and L. Thibault, *Weak compactness of sublevel sets in complete locally convex spaces*, J. Convex Anal. 26 (2019), 739-751.
[798] P. Pérez-Aros and E. Vilches, *Moreau envelope of supremum functions with applications to infinite and stochastic programming*, to appear.
[799] B. J. Pettis, *A proof that every uniformly convex space is reflexive*, Duke Math. J. 5 (1939), 249-253.
[800] R. R. Phelps, *Gaussian null sets and differentiability of Lipschitz maps on Banach spaces*, Pacific J. Math. 18 (1978), 523-531.
[801] R. R. Phelps, *Convex Functions, Monotone Operators and Differentiability*, 2nd ed. Lectrure Notes in Mathematics, vol. 1364, Springer-Verlag, New York (1993).
[802] J. Pierpont, *Theory of Functions of Real Variables*, t. I, Boston, (1905).
[803] G. Pisier, *Martingales with values in uniformly convex spaces*, Israel J. Math. 20 (1975), 326-350.
[804] R. A. Poliquin, *Subgradient monotonicity and convex functions*, Nonlinear Anal. 14 (1990), 305-317.
[805] R. A. Poliquin, *Integration of subdifferentials of nonconvex functions*, Nonlinear Anal. 17 (1991), 385-398.
[806] R. A. Poliquin, *An extension of Attouch's theorem and its application to second-order epi-differentiation of convexly composite functions*, Trans. Amer. Math. Soc. 332 (1992), 861-874.
[807] R. A. Poliquin and R. T. Rockafellar, *Amenable functions in optimization*, in Nonsmooth Optimization Methods and Applications, edited by F. Gianessi, (1992) pp. 338-353, Gordon and Breach, Philadelphia.
[808] R. A. Poliquin and R. T. Rockafellar, *A calculus of epi-derivatives applicable to optimization*, Canad. J. Math. 45 (1993), 879-896.
[809] R. A. Poliquin and R. T. Rockafellar, *Prox-regular functions in variational analysis*, Trans. Amer. math. Soc. 348 (1996), 1805-1838.
[810] R. A. Poliquin and R. T. Rockafellar, *Generalized Hessian properties of regularized nonsmooth functions*, SIAM J. Optimization 6 (1996), 1121-1137.
[811] R. A. Poliquin, R. T. Rockafellar, *Tilt stability of a local minimum*, SIAM J. Optim. 8 (1998), 287-299.
[812] R. A. Poliquin and R. T. Rockafellar, *A calculus of prox-regularity*, J. Convex Anal. 17 (2010), 203-210.
[813] R. A. Poliquin, R. T. Rockafellar and L. Thibault, *Local differentiability of distance functions*, Trans. Amer. Math. Soc. 352 (2000), 5231-5249; submitted 17 June 1997.
[814] R. A. Poliquin, J. Vanderwerff and V. Zizler, *Renormings and convex composite representation of functions*, Bull. Polish Acad. Sci. Math. 42 (1994), 9-19.
[815] E. S. Polovinkin and M. V. Balashov, *Elements of Convex and Strongly Convex Analysis*, Fizmatlit, Moscow (2004). (in Russian)
[816] J.-B. Poly and G. Raby, *Fonction distance et sigularités*, Bull. Sc. Math. (2me Série) 108(2) (1984), 187-195.
[817] B. T. Polyak, *Existence theorems and convergence of minimizing sequences in extremum problems with restrictions*, Dokl. Akad. Nauk SSSR, 166 (1966), 287 - 290 (in Russian); English translation in Soviet Math. Dokl. 7 (1966), 72-75.
[818] D. Pompeiu, *Sur la continuité des fonctions complexes*, Annales de la Faculté des Sciences de Toulouse, tome 7, no 3 (1905), 265-315.
[819] D. Preiss, *Differentiability of Lipschitz functions on Banach spaces*, J. Func. Anal. 91 (1990), 312-345.
[820] D. Preiss and J. Tišer, *Two unexpected examples concerning differentiability of Lipschitz functions on Banach spaces*, in Collection: Geometric Aspects of Functional Analysis, Operator Theory: Advances and Applictions 77, Birkhäuser, Boston (1995), 219-238.
[821] D. Preiss and L. Zajíček, *Sigma-porous sets in products of metric spaces and sigma-directionally porous sets in Banach spaces*, Real Anal. Exchange 24 (1998/9), 295-314.
[822] D. Preiss and L. Zajíček, *Directional derivatives of Lipschitz functions*, Israel J. Math; 125 (2001), 1-27.
[823] V. Pták, *A quantitative refinement of the closed graph theorem*, Czechoslovak Math. J. 24 (1974), 503-506.

[824] V. Pták, *A nonlinear subtraction theorem*, Proc. Roy. Irish Acad. Sect. A 82 (1982), 47-53.
[825] M. Quincampoix, *Differential inclusions and target problems*, SIAM J. Control Optim. 30 (1992), 324-335.
[826] G. Rabaté, *Sur les Notions Originelles de la Géométrie Infinitésimale Directe*, Thesis (Thèse de Doctorat ès Sciences Mathématiques) Université de Toulouse (1931)
[827] P. J. Rabier, *Points of continuity of quasiconvex functions on topological vector spaces*, J. Convex Anal. 20 (2013), 701-721.
[828] P. J. Rabier, *Differentiability of quasiconvex functions on separable Banach spaces*, Israel J. Math. 207 (2015), 11-51.
[829] H. Rademacher, *Über partielle und totale Dijfferenzierbarkeit*, Math. Ann. 79 (1918), 340-359.
[830] H. Rådström, *An embedding theorem for spaces of convex sets*, Proc. Amer. Math. Soc. 3 (1952), 165-169.
[831] J. Rataj and L. Zajíček, *On the structure of sets with positive reach*, Math. Nachr. 290 (2017), 1806-1829.
[832] R. Redefer and W. Walter, *The subgradient in $\mathbb{R}^n$*, Nonlinear Anal. Th. Meth. Appl. 20 (1993), 1345-1348; submitted 1 January 1992.
[833] S. Reich, *On the asymptotic behavior of nonlinear semigroups and the range of accretive operators*, J. Math. Anal. Appl. 79 (1981), 113-126.
[834] P. Redont, Personal communication.
[835] Yu. G. Reshetnyak, *On a generalization of convex surfaces*, Mat. Sb. N.S. 40 (1956), 381-398 (in Russian).
[836] F. Riesz, *Sur l'existence de la dérivée des fonctions d'une variable réelle et des fonctions d'intervalle*, Verhandl. Internat. Math. Kongress Zurich, I (1932), 258-259.
[837] S. M. Robinson, *Normed convex processes*, Trans. Amer. Math. Soc. 174 (1972), 127-140.
[838] S. M. Robinson, *Stability theory for systems of inequalities in nonlinear programming I: linear systems*, SIAM J. Numer. Anal. 12 (1975), 754-769.
[839] S. M. Robinson, *Regularity and stability for convex multivalued functions*, Math. Oper. Res. 1 (1976), 130-143.
[840] S. M. Robinson, *Stability theory for systems of inequalities in nonlinear programming II: Differentiable nonlinear systems*, SIAM J. Numer. Anal. 13 (1976), 497-513.
[841] S. M. Robinson, *Strongly regular generalized equations*, Math. Oper. Res. 5 (1980), 43-62.
[842] S. M. Robinson, *Localized normal maps and the stability of variational conditions*, Set-Valued Anal. 12 (2004), 259-274.
[843] S. M. Robinson, *Aspects of the projector on prox-regular sets*, Variational analysis and applications, 963-973, Nonconvex Optim. Appl., 79, Springer, New York, (2005).
[844] R. T. Rockafellar, *Convex Functions and Dual Extremum Problems*, Ph.D. dissertation, Harvard University (1963).
[845] R. T. Rockafellar, *Duality theorems for convex functions*, Bull. Amer. Math. Soc. 70 (1964), 189-192.
[846] R. T. Rockafellar, *Characterization of the subdifferentials of convex functions*, Pac. J. Math. 17 (1966), 497-510.
[847] R. T. Rockafellar, *Level sets and continuity of conjugate convex functions*, Trans. Amer. math. Soc. 123 (1966), 46-63.
[848] R. T. Rockafellar, *Extension of Fenchel's duality theorem for convex functions*, Duke Math. J., 33 (1966), 81-90.
[849] R. T. Rockafellar, *Conjugates and Legendre transforms of convex functions*, Canadian J. Math., 19 (1967), 200-205.
[850] R. T. Rockafellar, *Duality and stability in extremum problems involving convex functions*, Pacific J. Math. 21 (1967), 167-187.
[851] R. T. Rockafellar, *On the maximal monotonicity of subdifferential mapping*, Pac. J. Math. 33 (1970), 209-216.
[852] R. T. Rockafellar, *Convex Analysis*, Princeton University Press, Princeton, New Jersey (1970).
[853] R. T. Rockafellar, *Conjugate Duality and Optimization*, Conference Board of Mathematical Sciences Series, 16, SIAM Publications, Philadelphia (1972).
[854] R. T. Rockafellar, *Clarke's tangent cone and the boundaries of closed sets in $\mathbb{R}^n$*, Nonlin. Anal. Th. Meth. Appl. 3 (1979), 145-154.

[855] R. T. Rockafellar, *Directional Lipschitzian functions and subdifferential calculus*, Proc. London Math. Soc. 39 (1980), 331-355.
[856] R. T. Rockafellar, *Generalized directional derivatives and subgradients of nonconvex functions*, Canad. J. Math. 32 (1980), 157-180 (submitted: March, 1978).
[857] R. T. Rockafellar, *The Theory of Subgradients and its Applications to Problems of Optimization: Convex and Nonconvex Functions*, Heldermann Verlag, Berlin (1981).
[858] R. T. Rockafellar, *Proximal subgradients, marginal values and augmented Lagrangians in nonconvex optimization*, Math. Oper. Res. 6 (1981), 424-436 (sumitted: july 1980).
[859] R. T. Rockafellar, *Favorable classes of Lipschitz continuous functions in subgradient optimization*, in Progress in Nondifferentiable Optimization, edited by E.A. Nurminskii, pp. 125-143, IIASA, Laxenburg, Austria (1982).
[860] R. T. Rockafellar, *Extensions of subgradient calculus with applications to optimization*, Nonlin. Anal. Th. Meth. Appl. 9 (1985), 665-698.
[861] R. T. Rockafellar, *Lipschitzian properties of multifunctions*, Nonlin. Anal. Th. Meth. Appl. 9 (1985), 867-895.
[862] R. T. Rockafellar, *First and second order epi-differentiability in nonlinear programming*, Trans. Amer. Math. Soc. 307 (1988), 75-108.
[863] R. T. Rockafellar, *Proto-differentiability of set-valued mappings and its applications in optimization*, Ann. Inst. H. Poincaré (1989), 449-482.
[864] R. T. Rockafellar, *Generalized second derivatives of convex functions and saddle functions*, Trans; Amer. Math. Soc. 322 (1990), 51-77.
[865] R. T. Rockafellar and R. J-B. Wets, *Variational Analysis*, Vol 317 Grundlehren der mathematischen Wissenschaften, Springer, Berlin (1998).
[866] B. Rodriguez and S. Simons, *Conjugate functions and subdifferentials in non-normed situations for operators with complete graphs*, Nonlinear Anal. 12 (1988), 1062-1078.
[867] F. Roger, *Les propriétés tangentielles des ensembles euclidiens de points*, 69 (1937), 99-133.
[868] F. Roger, *Sur l'extension à l'ordre n des théorèmes de M. Denjoy sur le nombres dérivés du premier ordre*, C. R. Acad. Sci. Paris 209 (1939), 11-14.
[869] S. Rolewicz, *On paraconvex multifunctions*, Oper. Res. Verfahren 31 (1979), 540-546.
[870] S. Rolewicz, *On γ-paraconvex multifunctions*, Math. Japon. 24 (1979), 293-300.
[871] S. Rolewicz, *On drop property*, Studia Math. 85 (1987), 27-35.
[872] S. Rolewicz, *On $\alpha(\cdot)$-paraconvex and strongly $\alpha(\cdot)$-paraconvex functions*, Control and Cybernetics 29 (2000), 367-377.
[873] S. Rolewicz, *Paraconvex Analysis*, Control and Cybernetics 34 (2005), 951-965.
[874] J. Saint Raymond, *Weak compactness and variational characterization of the convexity*, Mediterr. J. Math. 10 (2013), 927-940.
[875] S. Saks, *Sur quelques propriétés métriques d'ensembles*, Fundam. Math. 26 (1936), 234-240.
[876] S. Saks, *Theory of the integral*, Monographie Matematyczne, Tom VIII, Second revised edition, Hafner Publ. Comp., New-York (1937).
[877] S. Saks and A. Zygmund, *Sur les faisceaux de tangentes à une courbe*, Fundam. Math. 6 (1924), 117-121.
[878] D. Salas and L. Thibault, *On characterizations of submanifolds via smoothness of the distance function in Hilbert spaces*, J. Optim. Theory Appl. 182 (2019), 189-210.
[879] D. Salas and L. Thibault, *Quantified characterizations of bodies with smooth boundaries*, unpublished paper.
[880] D. Salas and L. Thibault, *Quantitative characterizations of nonconvex bodies with smooth boundaries via the metric projection in Hilbert spaces*, J. Math. Anal. Appl. 494 (2021), 1-21.
[881] H. H. Schaefer and M. P. Wolff, *Topological Vector Spaces*, Second edition. Graduate Texts in Mathematics, 3. Springer-Verlag, New York, (1999).
[882] J. Schauder, *Uber die Umkehrung linearer, stetiger Funktionaloperationen*, Studia Math. 2 (1930), 1-6.
[883] M. Sebbah and L. Thibault, *Metric projection and compatibly parameterized families of prox-regular sets in Hilbert space*, Nonlinear Anal. 75 (2012), 1547-1562.
[884] O. Serea, *On reflecting boundary problem for optimal control*, SIAM J. Control Optim. 42 (2003), 559-575.
[885] O. Serea and L. Thibault, *Primal lower nice property of value functions in optimization and control problems*, Set-Valued Var. Anal. 18 (2010), 569-600.

[886] F. Severi, *Su alcune questioni di topologia infinitesimale*, Ann. Polon. Math. 9 (1930), 97-108.
[887] A. Shapiro, *Existence and differentiability of metric projections in Hilbert spaces*, SIAM J. Optimization 4 (1994), 130-141.
[888] S. H. Shkarin, *On Rolle's theorem in infinite dimensional Banach spaces*, Math. Zam. 51 (1992), 128-136.
[889] V. Sierpinski and A. N. Singh, *On derivatives of discontinuous functions*, Fund. Math. 36 (1949), 283-287.
[890] S. Simons, *Minimax and Monotonicity*, Lecture Notes in Mathematics 1693, Springer, Berlin (1998).
[891] S. Simons, *A new proof of the maximal monotonicity of subdifferentials*, J. Convex Anal. 16 (2009), 165-168.
[892] A. N. Singh, *On infinite derivates*, Fund. Math. 33 (1945), 106-107.
[893] V. L. Šmulian, Rec. Math., 6(48):1, (1939).
[894] V. L. Šmulian, *On some geometrical properties of a sphere in the space of the type (B)*, C. R. Acad. Sci. U.R.S.S. XXIV (1939), 648-652
[895] V. L. Šmulian, *On some geometrical properties of the unit sphere in the space of the type (B)*, Mat. Sb. 6 (1939), 77-94 (in Russian).
[896] V. L. Šmulian, *Sur la dérivabilité de la norme dans l'espace de Banach*, C.R. Acad. Sci. U.R.S.S., XXVII, (1940).
[897] V. L. Šmulian, *Sur la structure de la sphère unitaire dans l'espace de Banach*, Rec. Math. [Mat. Sbornik] N.S. Vol. 9(51), No 3 (1941), 545-561.
[898] J. E. Spingarn, *Submonotone subdifferentials of Lipschitz functions*, Trans. Amer. Math. Soc. 264 (1981), 77-89 (submitted, November 1979).
[899] S. B. Stechkin, *Approximative properties of sets in normed linear spaces*, Revue Math. Pures Appl. 8 (1963), 5-18.
[900] C. Stegall, *Optimization of functions on certain subsets of Banach spaces*, Math. Ann. 236 (1978), 171–176.
[901] J. Steiner, *Über parallele Flächen*, Monatsber. Preuss. Akad. Wiss. (1840), 114–118, [Ges. Werke, Vol II (Reimer, Berlin, 1882), 245–308].
[902] E. Steinitz, *Bedingt konvergente Reihen und konvexe Systeme*, J. Reine Angew. Math. 143 (1913), 128-175.
[903] O. Stolz, *Grundzüge der Differential und Integral-Rechnung*, t. I, Leipzig, (1893).
[904] J. J. Stoker, *Unbounded convex point sets*, Amer. J. Math. 62 (1940), 165-179.
[905] Ch. Thäle, *50 Years Sets with Positive Reach - A Survey*, Surveys in Mathematics and its Applications, Vol. 3 (2008), pp. 123-165.
[906] L. Thibault, *Problème de Bolza dans un espace de Baanach séparable*, C. R. Acad. Sci. Paris Sér. I Math. 282 (1976), 1303-1306.
[907] L. Thibault, *Subdifferentials of compactly Lipschitz functions*, Sémin. Anal. Convexe Montpellier, Exposé 5 (1978).
[908] L. Thibault, *Mathematical programming and optimal control problems defined by compactly Lipschitzian mappings*, Sémin. Anal. Convexe Montpellier, Exposé 10 (1978).
[909] L. Thibault, *Subdifferentials of compactly Lipschitzian vector valued functions*, Ann. Mat. Pura Appl. 125 (1980), 157-192.
[910] L. Thibault, *On generalized differentials and subdifferentials of Lipschitz vector-valued functions*, Nonlinear Anal. 6 (1982), 1037-1053.
[911] L. Thibault, *Tangent cones and quasi-interiorly tangent cones to multifunctions*, Trans. Amer. Math. Soc. 277 (1983), 601-621.
[912] L. Thibault, *V-subdifferentials of convex operators*, J. Math. Anal. Appl. 115 (1986), 442-460.
[913] L. Thibault, *On subdifferentials of optimal value functions*, SIAM J. Control Optim. 29 (1991), 1019-1036.
[914] L. Thibault, *A generalized sequential formula for subdifferentials of sum of convex functions defined on Banach spaces*, in Recent Developments in Optimization, 7th French-German Conference on Optimization, Dijon, France, june 27-july 2, 1994, R. Durier, C. Michelot eds.; Lecture Notes Econ. Math. Syst. 429 (1995), 1434-1444, Springer-Verlag, Berlin.
[915] L. Thibault, *A note on the Zagrodny mean value theorem*, Optimization 35 (1995), 127-130.

[916] L. Thibault, *Sequential convex subdifferential calculus and sequential Lagrange multipliers*, (preprint 1994) SIAM J. Control Optim. 39 (1997), 331-355.

[917] L. Thibault, *A direct proof of a sequential formula for the subdifferential of the sum of two convex functions*, Unpublished paper, Univ. de Montpellier, Montpellier (1994).

[918] L. Thibault, *On compactly Lipschitzian mappings*, in Recent Advances in Optimization, Lecture Notes Econ. Math. Syst. 456 (1997) Springer-Verlag, Berlin.

[919] L. Thibault, *Equivalence between metric regularity and graphical metric regularity for set-valued mappings*, Unpublished note, Univ. de Montpellier, Montpellier (1999).

[920] L. Thibault, *Various forms of metric regularity*, Unpublished note, Univ. de Montpellier, Montpellier (1999).

[921] L. Thibault, *Limiting convex subdifferential calculus with applications to integration and maximal monotonicity of subdifferential*, Canadian Mathematical Society Conference Proceedings, 27 (2000), 279-289.

[922] L. Thibault, *Sweeping process with regular and nonregular sets*, J. Differential Equations 193 (2003), 1-26.

[923] L. Thibault, *Regularization of nonconvex sweeping process in Hilbert space*, Set-Valued Anal. 16 (2008), 319-333.

[924] L. Thibault, *Jean Jacques Moreau: a selected review of his mathematical works*, J. Convex AnaL. 23 (2016), 5-22.

[925] L. Thibault, *Subsmooth functions and sets*, Linear Nonlinear Anal. 4 (2018), 157-269.

[926] L. Thibault and D. Zagrodny, *Integration of subdifferentials of lower semicontinuous functions on Banach spaces*, J. Math. Anal. Appl. 189 (1995), 33-58; submitted 19 February 1993.

[927] L. Thibault and D. Zagrodny, *Enlarged inclusion of subdifferentials*, Canad. Math. Bull. 48 (2005), 283–301.

[928] L. Thibault and D. Zagrodny, *Subdifferential determination of essentially directionally smooth functions in Banach space*, SIAM J. Optim. 20 (2010), 2300-2326.

[929] L. Thibault and D. Zagrodny, *Determining functions by slopes*, to appear.

[930] L. Thibault and N. Zlateva, *Integrability of subdifferentials of directionally Lipschitz functions*, Proc. Amer. Math. Soc. 133 (2005), 2939–2948.

[931] A. A. Tolstonogov, *Differential Inclusions in a Banach Space*, Kluwer, Dordrecht (2000).

[932] H. Toruńczyk, *Smooth partitions of unity on some non-separable Banach spaces*, Studia Math. 46 (1973), 43-51.

[933] J. S. Treiman, Ph.D. dissertation, University of Washington, Seattle (1983).

[934] J. S. Treiman, *Characterization of Clarke's tangent and normal cones in finite and infinite dimensions*, Nonlinear Anal. Th. Meth. Appl. 7 (1983), 771-783.

[935] J. S. Treiman, *Generalized gradients, Lipschitz behavior and directional derivatives*, Canad. J. Math. 37 (1985), 1074-1084.

[936] J. S. Treiman, *Generalized gradients and paths of descent*, Optimization 17 (1986), 181-186.

[937] J. S. Treiman, *Clarke's gradients and epsilon-subgradients in Banach spaces*, Trans. Amer. Math. Soc. 294 (1986), 65-78.

[938] J. S. Treiman, *Shrinking generalized gradients*, Nonlinear Anal. Th. Meth. Appl. 12 (1988), 1429-1450.

[939] J. S. Treiman, *Finite dimensional optimality conditions: B-gradients*, J. Optim. Theory Appl. 62 (1989), 771-783.

[940] J. S. Treiman, *The linear nonconvex generalized gradients and Lagrange multipliers*, SIAM J. Optim. 5 (1995), 670-680.

[941] J. S. Treiman, *Lagrange multipliers for nonconvex generalized gradients with equality, inequality, and set constraints*, SIAM J. Control Optim. 37 (1999), 1313-1329.

[942] J. S. Treiman, *The linear generalized gradient in infinite dimensions*, Nonlinear Anal. Th. Meth. Appl. 48 (2002), 427-443.

[943] C. Ursescu, *Multifunctions with closed convex graphs*, Czech. Math. J. 25 (1975), 438-441.

[944] M. Valadier, *Sous-différentiel d'une borne supérieure et d'une somme continue de fonctions convexes*, C. R. Acad. Sci. Paris Sér. A-B, 268 (1969), A39-A42.

[945] M. Valadier, *Contribution à l'Analyse Convexe*, Thesis (Thèse de Doctorat d'État ès Sciences Mathématiques), Université de Paris (1970).

[946] M. Valadier, *Sous-différentiabilité de fonctions convexes dans un espace vectoriel ordonné*, Math. Scand. 30 (1972), 65-74.

[947] M. Valadier, *Quelques résultats de base concernant le processus de rafle*, Sémin. Anal. Convexe Montpellier (1988), exp. 3 (30 pages).
[948] M. Valadier, *Quelques problèmes d'entrainement unilatéral en dimension finie*, Sémin. Anal. Convexe Montpellier (1988), exp. 8 (21 pages).
[949] M. Valadier, *Lignes de descente de fonctions lipschitziennes non pathologiques*, Sémin. Anal. Convexe Montpellier (1988), exp. 9 (10 pages).
[950] M. Valadier, *Entrainement unilateral, lignes de descente, fonctions lipschitziennes non pathologiques*, C. R. Acad. Sc. Paris 308 (1989), 241-244.
[951] F. A. Valentine, *On the extension of a vector function so as to preserve a Lipschitz condition*, Bull. Amer. Math. Soc. 49 (1943), 100-108.
[952] F. A. Valentine, *A Lipschitz condition preserving extension for a vector function*, Amer. J. Math. 67 (1945), 83-93.
[953] F. Vasilesco, *Essai sur les Fonctions Multiformes de Variables Réelles*, Thesis (Thèse de Doctorat ès Sciences Mathématiques), Université de Paris (1925).
[954] L. Veselý, *On the multiplicity points of monotone operators ob separable Banach spaces*, Comment. Math. Univ. Carolinae 27 (1986), 551-570.
[955] L. Veselý, *On the multiplicity points of monotone operators ob separable Banach spaces II*, Comment. Math. Univ. Carolinae 28 (1987), 295-299.
[956] L. Veselý and L. Zajíček, *Delta-convex mappings between Banach spaces and applications*, Dissertationes Mathematicae 289, Warszawa (1989), 48 pp.
[957] J.-P. Vial, *Strong convexity of sets and functions* J. Math. Econom. 9 (1982), 187-205; submitted in 1978.
[958] J.-P. Vial, *Strong and weak convexity of sets and functions*, Math. Oper. Res. 8 (1983), 231-259; submitted in 1981.
[959] R. Vinter, *Optimal Control*, Birkhäuser, Boston, Massachusetts (2000).
[960] L.P. Vlasov, *On Chebyshev sets*, Soviet Math. Dokl. 8 (1967), 401-404.
[961] M. Volle, *On the subdifferential of an upper envelope of convex functions*, Acta Math. Vietnam. 19 (1994), 137-148.
[962] M. Volle, J.-B. Hiriart-Urruty and C. Zalinescu, *When some variational properties force convexity*, ESAIM: COCV 19 (2013), 701-709.
[963] J. von Neumann, *Zur Theorie der Gesellschaftsspiele*, Math. Ann. 100 (1928), 295-320.
[964] J. von Neumann, *Discussion of a maximum problem*, in John von Neumann Collected Works, edited by A. H. Taub, Vol. 6, pp. 89-95, Pergamon, Oxford.
[965] G. Wachsmuth, *A Guided Tour of Polyhedric Sets: Basic Properties, New Results on Intersections and Applications*, J. Convex Anal. 26 (2020), 153-188.
[966] G. Wachsmuth, *No-gap second-order conditions under n-polyhedric constraints and finitely many nonlinear constraints*, J. Convex Anal. 27 (2020), 733-751.
[967] A. J. Ward, *On the differential structure of real functions*, Proc. Lond. Math. Soc. 39 (1935), 339-362.
[968] D. W. Walkup and R. J-B. Wets, *Continuity of some convex-cone valued mappings*, Proc. Amer. Math. Soc. 18 (1967), 229-235.
[969] X. Wang, *On Chebyshev functions and Klee functions*, J. Math. Anal. Appl. 368 (2010), 293-310.
[970] V. Weckesser, *The subdifferential in Banach spaces*, Nonlin. Anal. Th. Meth. Appl. 20 (1993), 1349-1354; submitted 1 January 1992.
[971] H. Weyl, *On the volume of tubes*, Amer. J. Math. 61 (1939), 461-472.
[972] H. Whitney, *Analytic extension of differentiable functions*, Trans. Amer. Math. Soc. 36 (1934), 63-89 (submitted, December 1932).
[973] M. D. Wills, *Hausdorff distance and convex sets*, J. Convex Anal. 14 (2007), 109-117.
[974] P. R. Wolenski and Y. Zhuang, *Proximal analysis and the minimal time function*, SIAM J. Control Optim. 36 (1998), 1048-1072.
[975] Zong-Ben Xu and G. F. Roach, *Characteristic inequalities of uniformly convex and uniformly smooth Banach spaces*, J. Math. Anal. Appl. 157 (1991), 189-210.
[976] J. C. Yao, X. Y. Zheng and J. Zhu, *Stable minimizers of φ-regular functions*, SIAM J. Optim. 27 (2017), 1150-1170.
[977] G. C. Young, *On the derivates of a function*, Proc. London Math. Soc. 15 (1916), 360-384.
[978] W. H. Young, *The Fundamental Theorems of Differential Calculus*, Cambridge (1910).

[979] W. H. Young, *On classes of summable functions and their Fourier series*, Proc. Royal Society, A (1912), 225-229.
[980] W. H. Young, *La symétrie de structure des fonctions de variables réelles*, Bull. Sci. Math., 52 (1928), 265-280.
[981] D. Zagrodny, *Approximate mean value theorem for upper subderivatives*, Nonlinear Anal. 12 (1988), 1413-1428.
[982] D. Zagrodny, *A note on the equivalence between the mean value theorem for the Dini derivative and the Clarke-Rockafellar derivative*, Optimization 21 (1990), 179-183.
[983] D. Zagrodny, *Some recent mean value theorems in nonsmooth analysis*, Nonsmooth Optimization: Methods and Applications (Erice, 1991), Gordon and Breach, Montreux (1992), 421-428.
[984] L. Zajíček, *On the points of multiplicity of monotone operators*, Comment. Math. Univ. Carolinae 19 (1978), 179-189.
[985] L. Zajíček, *On the points of multivaluedness of metric projections in separable Banach spaces*, Comment. Math. Univ. Carolinae 19 (1978), 513-523.
[986] L. Zajíček, *On the differentiation of convex functions in finite and infinite dimensional spaces*, Czechoslovak Math. J. 29 (1979), 340-348.
[987] L. Zajíček, *Differentiability of the distance function and points of multivaluedness of the metric projection in Banach space*, Czechoslovak Math. J. 33 (1983), 292-308.
[988] L. Zajíček, *Strict differentiability via differentiability*, Acta Univ. Carolinae 28 (1987), 157-159.
[989] L. Zajíček, *Porosity and σ-porosity*, Real Analysis Exchange 13 (1987/88), 314-350.
[990] L. Zajíček, *Fréchet, strict differentiability and subdifferentiability*, Czechoslovak Math. J. 41 (1991), 471-489.
[991] L. Zajíček, *On differentiability properties of Lipschitz functions on a Banach space with a Lipschitz uniformly Gâteaux differentiable bump function*, Comment. Math. Univ. Carolinae 38 (1997), 329-336.
[992] L. Zajíček, *A note on intermediate differentiability of Lipschitz functions*, Comment. Math. Univ. Carolinae 40 (1999), 1501–1505.
[993] L. Zajíček, *On σ-porous sets in abstract spaces*, Abstr. Appl. Anal. (2005), 509-534.
[994] L. Zajíček, *On Lipschitz and D.C. surfaces of finite codimension in a Banach space*, Czechoslovak Math. J. 58 (2008), 849-864.
[995] L. Zajíček, *Differentiability of approximately convex, semiconcave and strongly paraconvex functions*, J. Convex Anal. 15 (2008), 1-15.
[996] L. Zajíček, *Singular points of order k of Clarke regular and arbitrary functions*, Comment. Math. Univ. Carolinae 53 (2012), 51-63.
[997] L. Zajíček, *Remarks on Fréchet differentiability of pointwise Lipschitz, cone-monotone and quasi-convex functions*, Comment. Math. Univ. Carolinae 55 (2014), 203-213.
[998] L. Zajíček, *Hadamard differentiability via Gâteaux differentiabilty*, Proc. Amer. Math. Soc. 143(2015), 279-288.
[999] L. Zajíček, *Properties of Hadamard directional derivatives: Denjoy-Young-Saks theorem for functions on Banach spaces*, J. Convex Anal. 22 (2015), 161-176.
[1000] C. Zalinescu, *Convex Analysis in General Vector Spaces*, World Scientific (2002).
[1001] E. H. Zarantonello, *Projections on convex sets in Hilbert space and spectral theory I and II*, in E. H. Zarantonello, ed., Contributions to Nonlinear Functional Analysis, Academic Press, New York (1971), pp. 237-424.
[1002] R. Zhang and J. S. Treiman, *Upper-Lipschitz multifunctions and inverse subdifferentials*, Nonlinear Anal. Theory Meth. Appl. 24 (1995), 273-286.
[1003] X. Y. Zheng and Q. H. He, *Characterization for metric regularity for σ-submooth multifunctions* Nonlinear Anal. 100 (2014), 111-121.
[1004] X. Y. Zheng and K. F. Ng, *Linear regularity for a collection of subsmooth sets in Banach spaces*, SIAM J. Optim., 19 (2008), 62-76.
[1005] X. Y. Zheng and K. F. Ng, *Calmness for L-subsmooth multifunctions in Banach spaces*, SIAM J. Optim. 19 (2009), 1648-1673.
[1006] X. Y. Zheng and K. F. Ng, *Metric subregularity for proximal generalized equations in Hilbert spaces*, Nonlinear Anal. 75 (2012) 1686-1699.
[1007] X. Y. Zheng and K. F. Ng, *Hölder stable minimizers, tilt stability and Hölder metric regularity of subdifferential*, SIAM J. Optim., 25 (2015), 416-438.

[1008] X. Y. Zheng and K. F. Ng, *Hölder weak sharp minimizers and Hölder tilt-stability*, Nonlinear Anal., 120 (2015), 186-201.

[1009] X. Y. Zheng and J. Zhu, *Stable well-posedness and tilt stability with respect to admissible functions*, ESAIM: COCV 23 (2017), 1397-1418.

[1010] N. V. Zhivkov, *Metric projections and antiprojections in strictly convex normed spaces*, C.R. Acad. Bulgare Sci. 31 (1978), 369-372.

[1011] N. V. Zhivkov, *Generic Gâteaux differentiability of locally Lipschitzian functions*, Collection: Constructive function theory 81 (Varna, 1981), 590-594.

[1012] N. V. Zhivkov, *Generic Gâteaux differentiability of directionally differentiable mappings*, Rev. Roumaine Math. Pures Appl., 32 (1987), 179-188.

[1013] Q. J. Zhu, *Clarke-Ledyaev mean value inequalities in smooth Banach spaces*, Nonlinear Anal., 32 (1998), 315-324.

[1014] Q. J. Zhu, *The equivalence of several basic theorems for subdifferentials*, Set-Valued Anal. 6 (1998), 171-185.

[1015] N. Zlateva, *Integrability through infimal regularization*, Compt. Rend. Acad. Bulg. Sci. 68 (2015), 551-560.

[1016] L. Zoretti, *Sur les fonctions analytiques uniformes qui possèdent un ensemble parfait discontinu de points singuliers*, J. Math. Pure. Appl., 6e série - tome I - Fasc. I (1905), p. 1-51.

Index

C^1-property under strict Fréchet differentiability on open set, 91
Y_+^*-scalarly lower semicontinuous, 441
ε-subdifferential chain rule for convex functions under qualification conditions, 440
ε-subdifferential chain rule: convex functions without qualification condition, 442
ε-subdifferential of $f + g \circ G$: convex functions without qualification condition, 444
ε-subdifferential of sum of convex functions under qualification conditions, 440
$\mathcal{B}$-differentiability, 304
$\mathcal{B}$-differentiability of continuous convex functions: intrinsic characterization, 307
(Minkowski) sum of sets, 4

ε-subdifferential of $f + g \circ G$: convex functions under qualification conditions, 438

A-norma cone: nonpositivity of second component of A-normal to epigraph, 624
A-normal cone: invariance with respect to equivalent renorm, 621
A-subdifferential of Lipschitz functions: link with L-subdifferential in Asplund spaces, 601
A-subdifferential of Lipschitz functions: sequential description in separable Banach spaces, 600
A-subdifferential: coincidence with L-subdifferential for Lipschitz functions on separable Asplund spaces, 601
A-subdifferential: coincidence with L-subdifferential for lsc functions in reflexive spaces, 626
A-subdifferential: coincidence with strict H-derivative, 608

A-subdifferential: description for Lipschitz functions with admissible classes of vector subspaces, 613
A-subdifferential: description for Lipschitz functions with Hadamard ε-subdifferential, 613
A-subdifferential: outer/upper semicontinuity for Lipschitz functions, 608
Argmin of a function, 2
Asplund space, 514
Asplund space: F-subdifferential characterization, 516
Asplund-Rockafellar theorem, 312
asymptotic cone, 319
asymptotic cone and support function, 321
asymptotic cone of the ε-subdifferential of a convex function, 323
asymptotic cone of the polar of a set, 321
asymptotic cone of the subdifferential of a convex function, 322
asymptotic function, 318
asymptotic function of a convex function f and support function of dom f^*, 325
asymptotic function: calculus, 324
Attouch-Brezis condition, 779
Attouch-Wets convergence, 40
Aubin theorem on openness and metric regularity of multimappings, 733
Aubin-Hölder property of θ-order for multimappings, 782
Aubin-Lipschitz property, 715

B-tangent and C-tangent cones to sublevels, 182
Banach-Steinhaus type theorem for families of convex mappings, 168
Berge theorem on semicontinuity of marginal functions I, 25
Berge theorem on semicontinuity of marginal functions II, 26
biconjugate, 295
biconjugate relative to X, 295
biconjugate relative to X^{**}, 295

biconvex function, 651
Bishop-Phelps variational principle for support functionals, 201, 494
Bishop-Phelps variational principle for support points, 201
Borwein's version of Brønsted-Rockafellar theorem, 327, 384
Borwein-Preiss variational principle, 505, 604, 702
Bouligand directional derivative, 132
Bouligand-Peano tangent cone, 128
Bouligand-Peano tangent cone to boundary set, 131
boundary point: L-normal characterization in finite dimensions, 540
boundary points of domains of convex functions in finite dimensions: subdifferential characterization, 363
boundedly weakly inf-compact, 371
boundedness of domain of conjugate, 403
boundedness of sublevels in normed space, 406
Brønsted-Rockafellar theorem, 327
Brønsted: subdifferential maximum rule, 448

C-directional derivative: expression via Dini directional derivative, 650
C-normal cone to inverse image, 359
C-normal cone via A-normals, 631
C-normal cone: via L-normals, 558
C-normal via distance function, 121
C-normal: to intersection, 73
C-normal: calculus, 778
C-normal: definition, 65
C-normal: norm estimate under interior tangent condition, 82
C-normals to sublevel sets, 184
C-optimality Fermat-type condition for minimization without constraint, 93
C-subdifferential chain rule for a non-Lipschitz function and a strictly differentiable mapping, 359
C-subdifferential of convex functions, 99
C-subdifferential of directionally Lipschitz function: closedness of the graph, 120
C-subdifferential of distance function by means of ε-nearest points, 192
C-subdifferential of distance function by means of nearest points, 193
C-subdifferential of Lipschitz function as w^*-closed convex hull of A-subdifferential, 614
C-subdifferential of Lipschitz functions: outer semicontinuity, 107
C-subdifferential of Lipschitz functions: uniqueness of C-subgradient as characterization of strict Hadamard differentiability, 109
C-subdifferential of Lipschitz functions: upper semicontinuity, 107
C-subdifferential via A-subgradients and horizon A-subgradients, 631
C-subdifferential: definition, 83
C-subdifferential: via L-subgradients and horizon L-subgradients, 558
C-subdifferential: weak* compactness under Lipschitz continuity, 103
C-subgradient, 83
C-subregular functions, 137
C-subregularity: Dini directional derivative characterization, 650
C-tangent cone to inverse image, 146
C-tangent cone versus limit inferior of B-tangent cone, 208
C-tangent cone: linearity at boundary points under interior tangent condition, 82
C-tangent via distance function, 121
C-tangent/normal cone: fitted sets, 69
C-tangent/normal cones: linearity property for graphs of Lipschitz mappings, 117
C-tangent: definition, 65
C-tangent: sequential characterization, 66
C-tangent: sequential characterization via boundary points, 70
C-tangent: to boundary set via C-tangent cones of the set and its complement, 71
C-tangent: to intersection, 73
calculus for normal/tangent cones to convex sets: locally convex spaces, 360
calmness of θ-order for multimappings, 783
Cartesian product: A-normal cone, 621
Cartesian product: L-normal, 527
chain rule for C-subdifferential: composition under strict Hadamard differentiability of inner mappings, 143
chain rule for C-subdifferential: composition with compactly Lipschitzian mapping, 177
chain rule for F-subdifferential, 482
chain rule: A-subdifferential, 628
chain rule: A/L-subdifferential, 769
chain rule: F-subdifferential, 554
chain rule: L-subdifferential, 531
characterization of maximal cyclically monotone family of multimappings, 450
characterization of pairs of solutions of primal and dual problems, 338
Clarke directional derivative, 104
Clarke normal cone, 65
Clarke subdifferential of a function, 83
Clarke tangent cone, 65

Clarke theorem of gradient representation of C-subdifferential, 114
closed (open) ball: definition, 2
closed convex hull of a function, 289
closed convex hull of a function: Legendre-Fenchel conjugate, 292
closed convex hull: definition, 2
closed convex hull: epigraph, 291
closed convex hull: subdifferential, 292
closedness of graph of C-normal cone of sets with the interior tangent property, 120
closedness of multimapping at a point, 20
closure of a function, 289
coderivative, 729
coderivative compactness of multimappings, 734
coderivatives of multimappings: calculus, 764
coercivity of convex functions: dual criteria, 301
coercivity of convex functions: primal criteria, 264
coincidence $N^F(S;x) = \bigl(T^B(S;x)\bigr)^\circ$ in finite dimensions, 478
compact tangent property/condition, 233
compactly epi-Lipschitz sets, 233
compactly epi-Lipschitz sets: property of A-normals, 637
compactly Lipschitzian mapping, 172
compactness of images of compact sets by upper semicontinuous multimappings, 22
compactness of the metric space of p-dimensional vector subspaces of a finite-dimensional space, 219
complement vector subspaces, 212
conjugate and subdifferential of $\mathbf{f} + \mathbf{g} \circ \mathbf{G}$: convex functions with g partially non-decreasing, 364
conjugate and subdifferential chain rule for convex functions in locally convex space, 357
conjugate and subdifferential of $f + g \circ G$: convex functions with g fully non-decreasing, 353
conjugate and subdifferential of sum of convex functions in locally convex spaces, 380
conjugate and subdifferential of sum of convex functions in locally convex space, 357
conjugate and subdifferential of sum of convex functions in locally convex spaces, 396, 432
conjugate function: global Lipschitz property on locally convex spaces, 302
conjugate of $f + g \circ G$, 778

conjugate of a sum by means of closure of sum of conjugates, 299
conjugate of infimal convolution, 287
conjugate of sum, 299
conjugate of supremum of family of functions, 300
conjugate via support function of epigraph, 283
conjugate: a list, 282
conjugate: definition, 281
connectedness of image of connected set under semicontinuous multimapping, 50
continuity of one-variable lsc convex functions at endpoints, 259
continuity of lsc convex functions relative to convex polyhedral subsets of the domain, 259
continuity of lsc convex functions relative to line segment of the domain, 259
continuity of metric projection in convex set-variable, 48
continuous selection: extension, 36
continuous selection: Michael theorem, 35
conjugate and subdifferential of sum of convex functions in locally convex spaces, 372
convergence for sets: Attouch-Wets convergence, 40
convergence for sets: Hausdorff-Pompeiu convergence, 37
convex functions f with gph ∂f not norm$\times w^*$ closed, 266
convex functions: continuity, 256
convex functions: equi-Lipschitz property, 157, 159
convex functions: equicontinuity on topological vector spaces, 257
convex functions: Lipschitz property, 157, 159, 161
convex functions: Lipschitz property on Fréchet spaces, 259
convex functions: Lipschitz property on locally convex spaces, 258
convex functions: Lipschitz property under growth conditions, 261
convex hull of a function, 289
convex hull of a function: analytic description, 290
convex hull of a function: Legendre-Fenchel conjugate, 292
convex hull: definition, 2
convex multimapping, 726
convex set: definition, 2
convex-concave functions: continuity property, 160
convexity of distance function, 96

convexity of functions: characterization via Dini directional derivative, 649
convexity of functions: characterization via lower Hadamard directional derivative, 691
convexity of functions: subdifferential characterization, 691
convexity of infimum function, 96
core of a set, 97
cyclically monotone family of multimappings, 449

Daneš drop theorem, 205
Danskin type formula for supremum of $\mathcal{C}^1$ functions, 187
Denjoy-Young-Saks theorem, 674
density of P-subdifferentiability points in Hilbert space, 577
density of F-subdifferentiability points, 516
density of ranges of subdifferentials, 680
density of set of F-subdifferentiability in Asplund space, 516
density of subdifferentiability points, 680
density property of H-subdifferentiability points, 598
derivates at end-points of one-variable lsc convex functions, 275
derivative of conjugate function, 520, 522
derivative-free characterization of differentiability of continuous convex functions, 512
differentiability along a subspace, 617
differentiability at subdifferentiability points: strict convexity of conjugate, 436
differentiability of convex functions in finite dimensions: subdifferential characterization, 363
differentiability of convex functions: continuous selection subgradient characterization of Gâteaux differentiability, 309
differentiability of convex functions: continuous selection subgradient characterization of $\mathcal{B}$-differentiability, 308
differentiability of convex functions: continuous selection subgradient characterization of Fréchet differentiability, 309
differentiability: characterization of F-differentiability via bounded sets, 85
differentiability: characterization of H-differentiability via compact sets, 85
differentiability: characterization of H-differentiability via composition with one real variable derivable mapping, 86

differentiability: Fréchet, Hadamard, Gâteaux, 84
Dini derivates/semiderivates, 647
Dini directional derivative, 99
Dini directional derivative: lower, 133
Dini directional derivatives, 647
Dini infinite semiderivate points, 673
Dini semiderivates: coincidence of suprema of four Dini semiderivates on a same open interval, 666
Dini subdifferential, 674
direct images of sets: F-subdifferential, 555
direct images of sets: L-subdifferential, 556
directed mapping, 12
directed set, 12
directional derivative, 99
directional derivative of convex functions, 99
directional Lipschitz property: characterization with lower Hadamard directional derivative, 686
directional Lipschitz property: subdifferential characterization, 686
directionally Lipschitz functions, 119
distance between vector subspaces, 215
distance function to images, 580
distance function to images: F-subdifferential, 580
distance function to images: L-subdifferential, 581
distance function: A-subdifferential in Banach space as outer limit of Hadamard ε-subdifferential, 620
distance function: A-subdifferential in separable Banach space as outer limit of H-subdifferential with inside points, 605
distance function: A-subdifferential via metric projection in Hilbert space, 627
distance function: F-subdifferential at outside point, 491
distance function: F-subdifferential and metric projection, 548
distance function: F-subdifferential at inside point, 490
distance function: F-subdifferential for ball-compact/w-closed sets via nearest points in Hilbert space, 548
distance function: F-subgradient and minimizing sequence, 492
distance function: L-subdifferential via metric projection in Hilbert space, 627
distance function: approximation of H-subgradients by Hadamard ε-subgradients at inside points, 618
distance function: approximation of its F-subgradients with F-subgradients at inside points, 537

distance function: approximation of its
   H-subgradients via H-subgradients at
   inside points, 603
distance function: C-subdifferential via
   metric projection, 560
distance function: definition, 2
distance function: differentiability of its
   square for convex set in reflexive space,
   481
distance function: L-subdifferential via
   F-subgradients, 535
distance function: L-subdifferential via
   Fréchet ε-subgradients, 544
distance function: L-subdifferential via
   P-subgradients, 578
distance function: lower Dini directional
   derivative and metric projection, 548
distance function: subdifferential at outside
   point under convexity, 418
distance function: variational proximal
   subdifferential at inside point, 574
distance function: variational proximal
   subdifferential at outside point, 575
distance to subdifferential of a function, 266
distances for sets: ρ-pseudo
   excess/distance, 41
distances for sets: Hausdorff-Pompeiu
   excess/distance, 37
distances for sets: truncated
   ρ-excess/distance, 39
dual problem, 334
duality gap, 335, 336
duality result I, 337
duality result II, 343

effective domain of a function, 1
Ekeland variational principle, 198, 200, 205,
   328, 372, 505, 517, 536, 561, 619, 640
Ekeland-Lebourg theorem on
   F-differentiability of continuous
   convex function, 514
enlargement of set, 417
epi-convergence, 18
epi-Lipschitz set: Cartesian product, 79
epi-Lipschitz set: non-nullity of C-normal
   cone at boundary points, 79
epi-Lipschitz property of sublevel set, 110
epi-Lipschitz property for convex sets with
   nonempty interior, 82
epi-Lipschitz property: characterization via
   signed distance, 154
epi-Lipschitz set: C-tangent to
   complement, 81
epi-Lipschitz set: closure of its interior, 79
epi-Lipschitz set: definition, 76
epi-Lipschitz set: epigraphical
   characterization, 77

epi-Lipschitz set: estimate of norm of
   C-normal, 82
epi-Lipschitz set: in Hilbert space, 78
epi-Lipschitz set: interior of its closure, 81
epi-Lipschitz set:
   interior/closure/boundary, 81
epi-Lipschitz set: linearity of the C-tangent
   cone at boundary point, 82
epi-Lipschitz sets: B-tangential
   characterization, 687
epi-Lipschitz sets: C-normals via
   C-subgradients of signed distance, 185
epi-Lipschitz sets: C-tangential
   characterization, 687
epi-Lipschitz sets: characterization via
   normal vectors, 687
epi-Lipschitz sets: sublevel representation,
   156
epigraph of a function, 1
equi-Lipschitz property for families of
   convex vector mappings, 165
excess and Hausdorff-Pompeiu distance
   between convex sets by means of
   support functions, 112
extension of Lipschitz mappings between
   subsets of Hilbert spaces, 181
extension of real-valued Lipschitz function,
   111

F-differentiability of conjugate function f^*
   at u^* and strong maximizer for
   $\langle u^*, \cdot \rangle - f$, 520
F-differentiability: Fréchet ε-subdifferential
   characterization, 543
F-normal, 473
F-normal and L-normal of sums of sets, 556
F-normal cone and F-subdifferential:
   closedness property, 478
F-subdifferential, 473
F-subdifferential of concave function, 481
F-subdifferential of the difference of two
   functions, 483
F-subdifferential: infimum/supremum of
   functions, 484
F-subgradient, 473
F-subgradient as F-gradient of lower
   function, 487
F-subgradient as gradient of minorizing
   C^1-function, 487
F-subgradient as gradient of minorizing
   function, 475
F-subgradient: approximation by
   P-subgradients, 577
F-subgradient: Singh's example of one
   variable functions with no
   subdifferentiability point, 517
Fabian and Ioffe separable reduction of
   F-subdifferentiability, 497

Fabian and Ioffe separable reduction of a nul F-subgradient for a sum, 500
failure of closedness of graph of C-normal cone, 108
failure of tangential regularity of distance function of tangentially regular set, 149
Fenchel duality, 346
Fenchel-Moreau theorem for biconjugate, 296
filter, filter basis, 8
Fréchet ε-normal, 541
Fréchet ε-subdifferential, 540
Fréchet differentiability through subdifferentiability, 481
Fréchet normal cone as polar of B-tangent cone in finite dimensions, 478
Frankowska constant, 731
Frankowska theorem on openness and metric regularity of multimappings, 733
fuzzy chain rule of F-subdifferential in Asplund space, 518
fuzzy chain rule: F-subdifferential in Asplund space, 518
fuzzy chain rule: H-subdifferential, 599
fuzzy chain rule: proximal subdifferential, 579
fuzzy optimality condition for sum: H-subdifferential, 598
fuzzy sum rule for F-subdifferential in Asplund space, 515
fuzzy sum rule for F-subdifferential under differentiable renorm, 511
fuzzy sum rule: F-subdifferential in F-smooth space, 511
fuzzy sum rule: F-subdifferential in Asplund space, 515
fuzzy sum rule: H-subdifferential, 598
fuzzy sum rule: proximal subdifferential, 576

G-differentiability: Dini-subdifferential characterization, 675
gauge function, 256
gradient $\mathcal{D}$-representation: C-subdifferential of directionally Lipschitz functions, 661
gradient $\mathcal{D}$-representation: C-subdifferential of locally Lipschitz subregular functions, 661
gradient $\mathcal{D}$-representation: subdifferential of continuous convex functions, 664
gradient $\mathcal{D}$-representation: subdifferential of convex functions, 663
gradient representation: C-subdifferential of Lipschitz functions in finite dimensions, 114
graphical metric regularity, 723

graphical metric subregularity, 723
Graves open mapping theorem, 144
grill of a filter basis, 8
grill of the collection of neighborhoods of a point, 7
growth condition under θ-order metric subregularity, 793
growth condition under θ-order metric subregularity of subdifferential, 791

Hadamard ε-subdifferential, 592
Hadamard differentiability under H-subdifferentiability, 596
Hadamard directional derivative: lower, 133
Hadamard subdifferential, 592
Hadamard subgradient, 592
Hausdorff continuity of subdifferential of a convex function and its Fréchet differentiability, 314
Hausdorff distance: Hölder continuity of $C \mapsto \operatorname{proj}_C(u)$ for closed convex sets C of a Hilbert space, 48
Hausdorff-Pompeiu convergence/continuity, 37
Hausdorff-Pompeiu distance, 37, 215
Hausdorff-Pompeiu distance Vs Hausdorff truncated distance for cones, 55
Hausdorff-Pompeiu distance: Theorem of coincidence $\operatorname{haus}(S, S') = \operatorname{haus}(\operatorname{bdry}(S), \operatorname{bdry}(S'))$ for closed convex sets, 51
Hausdorff-Pompeiu excess, 37, 215
Hausdorff-Pompeiu truncated ρ-excess/distance, 39
Hiriart-Urruty and Phelps: subdifferential chain rule, 446
Hiriart-Urruty and Phelps: subdifferential sum rule, 447
Hoffman inequality, 756
Hoffman-Ioffe metric inequality, 753
horizon A-subdifferential, 631
horizon A-subgradient, 631
horizon C-subdifferential, 116
horizon C-subgradient, 116
horizon function, 318
horizon L-subdifferential, 557
horizon L-subgradient, 557
horizon L-subgradient: analytic description, 563
horizon/singular subgradients of convex functions, 383

inf-convolution, 286, 488
infimal convolution, 286, 488
infimal convolution: ε-subdifferential, 288
infimal convolution: F-subdifferential, 488
infimal convolution: F-subdifferential of its opposite, 489

infimal convolution: Fréchet
  differentiability through continuous
  selection, 489
infimal convolution: lower Hadamard
  derivative, 489
infimal convolution: Moreau-Rockafellar
  subgradient under attainement, 288
infimal convolution: variational proximal
  subdifferential, 573
inner limit of multimappings, 5
inner limit: by means of grill, 7
inner/outer limit: sequential
  characterizations for sequences of sets
  in metric space, 10
inner/outer limit: along filter basis, 8
inner/outer limit: of multimappings via
  distance function, 10
inner/outer limit: of pointwise closure, 9
inner/outer limit: of sequences of sets, 5
inner/outer limit: product, 6
inner/outer limit: sequential
  characterizations for multimappings
  between metric spaces, 11
integration of subdifferentials: convex
  functions, 379
interior tangent cone, 72
interior tangent cone property for convex
  sets with nonempty interior, 82
interior tangent cone property: C-tangent
  to complement, 80
interior tangent property: characterization
  via normal vectors, 687
interior tangent property: characterization
  via signed distance, 154
interior tangent property: characterization
  via tangent cones, 687
interior tangent property: non-vacuity of
  interior of C-tangent cone as
  characterization for compactly
  epi-Lipschitz sets, 235
interior tangent property: non-vacuity of
  interior of C-tangent cone as
  characterization in finite dimensions,
  235
inverse image of sets: F-normals, 553, 554
inverse image of sets: L-normals, 555
inverse images of sets: variational proximal
  normals, 565
Ioffe and Tikhomirov theorem:
  subdifferential of suprema of convex
  functions, 272
Ioffe approximate normal cone: general
  Banach space case, 616
Ioffe approximate normal cone: separable
  Banach space case, 603
Ioffe approximate subdifferential: general
  Banach space case, 623

Ioffe approximate subdifferential: Lipschitz
  functions on general Banach spaces,
  607
Ioffe approximate subdifferential: Lipschitz
  functions on separable Banach spaces,
  599
isolated point, 210
isolated point: tangential characterization,
  210

Kruger theorem for metric regular
  transversality, 744

L-normal cone as outer limit of P-normal
  cone in Hilbert space, 578
L-normal cone: description with
  L-subdifferential of distance function,
  535
L-subdifferential as outer limit of
  P-subdifferential in Hilbert space, 577
L-subdifferential of continuous concave
  function, 525
L-subdifferential of distance function in
  Asplund space at inside point, 535
L-subdifferential of Lipschitz functions:
  non vacuity in Asplund space, 530
L-subdifferential regularity, L-normal
  regularity, 530
L-subdifferential: extension to normed
  spaces, 544
L-subdifferential: Nonvacuity for Lipschitz
  functions, 530
Lagrange multipliers: with
  A-subdifferential/L-subdifferential, 762
Lagrange multipliers: with
  C-subdifferential, 202
Lagrangian function, 349
Lebesgue density point, 220
Legendre-Fenchel conjugate, 102
Legendre-Fenchel transform: definition, 281
limit inferior of B-tangent cone and
  C-tangent cone for sets with the
  compact tangent property, 233
limit inferior of multimappings, 5
limit inferior/superior: of extended
  real-valued functions, 10
limit inferior/superior: of sequences of sets,
  5
limit superior of multimappings, 5
limit superior of multimappings:
  topological versus sequential, 14
limiting chain rule for a convex function
  and a linear mapping: locally convex
  space case, 376, 452
limiting chain rule for a convex function
  and vector mapping, 377
limiting descriptions of horizon
  subgradients of convex functions via
  true subgradients, 384

limiting descriptions of normals to convex sublevel sets: reflexive space, 385
limiting descriptions of normals to convex sublevels: general Banach space, 389
limiting descriptions of normals to intersections of convex sublevels, 391
limiting descriptions of normals to vector convex sublevels: general Banach space, 400
limiting descriptions of normals to vector convex sublevels: reflexive Banach space, 399
limiting normal, 524
limiting normal cone, 524
limiting subdifferential, 523
limiting subgradient, 523
limiting sum rule for convex functions, 376
limiting sum/chain subdifferential rule for convex functions, 373
linear minimization problem, 350
linear optimization: duality result, 351
linear optimization: primal and dual problems, 350
link between F-subgradient and F-normal to epigraph, 475
Lipschitz continuity of ε-subdifferential: topological vector space, 316
Lipschitz continuity of ε-subdifferential: normed space, 317
Lipschitz continuity: characterization via Dini directional derivative, 649
Lipschitz continuity: characterization with lower Hadamard directional derivative, 684, 688
Lipschitz continuity: horizon L-subdifferential characterization, 689
Lipschitz continuity: subdifferential characterization, 684, 688, 689
Lipschitz global property of conjugates: on normed spaces, 303
Lipschitz global property of convex functions, 302
Lipschitz property for vector convex mapping, 166
Lipschitz property under strict Hadamard differentiability, 88
Lipschitz surface, 213
Lipschitz surface: along a vector subspace, 213
Lipschitz surface: characterizations, 213
Lipschitz surface: finite codimension, 213
local boundedness from below of convex functions, 263
local boundedness of multimappings, 50
local minimizer of s-order, 793
lower Γ-limit, 18
lower epi-limit, 18
lower one-sided Lipschitz function, 591
lower semicontinuity of ε-directional derivative of $f \in \Gamma_0(X)$, 326
lower semicontinuity of the ε-directional derivative of a lsc convex function, 326
lower/upper limit: of extended real-valued functions, 10
Lyusternik tangential equality, 146
Lyusternik-Graves theorem, 761

Mackey topology, 342
Mangasarian-Fromovitz constraint qualification condition, 760, 763, 773, 775
marginal function, 333
maximal cyclically monotone family of multimappings, 449
maximal cyclically monotone multimapping, 449
maximal monotone multimapping, 275
maximal monotonicity of subdifferential of a convex function, 380
maximum of finitely many functions: A-subdifferential, 630
maximum of finitely many Lipschitz functions: L-subdifferential, 532
mean value inequality for images of sets: Dini semiderivate, 671
mean value inequality for Lipschitz functions: proximal subdifferential, 580
mean value inequality for Lipschitz functions: A-subdifferential, 630
mean value inequality for Lipschitz functions: F-subdifferential, 535
mean value inequality for Lipschitz functions: L-subdifferential, 535
mean value inequality with Dini semiderivate of images of sets, 671
mean value inequality: Dini subgradient, 677
mean value inequality: version with Dini directional derivative, 648
mean value theorem with C-subdifferential for Lipschitz functions, 170
measures of aperture between cones, 55
metric regular transversality, 738, 741
metric regularity, 719
metric regularity for constraint systems under strict F-differentiability, 759
metric regularity for convex constraints: infinitely many inequalities and equalities, 752
metric regularity for convex vector valued mappings, 750, 751
metric regularity for nonsmooth constraint systems, 766, 767
metric regularity of θ-order of multimappings, 782

metric regularity of a multimapping: equivalence with Aubin-Lipschitz property of its inverse, 722
metric regularity of a multimapping: equivalence with openness with a linear rate, 722
metric regularity of multimappings: A-coderivative condition, 734
metric regularity of multimappings: F-coderivative criteria, 733
metric regularity of multimappings: Mordukhovich criterion, 734
metric regularity: equivalence with graphical metric regularity, 724
metric regularity: inverse images, 761
metric regularity: persistence under Lipschitz perturbation, 756
metric subregular transversality, 738
metric subregularity, 720
metric subregularity of multimappings whose graph is finite union of closed polyhedral convex sets, 781
metric subregularity of θ-order of multimappings, 782
metric subregularity of subdifferential under θ-order growth condition, 791
metric subregularity under θ-order growth condition, 793
Michael continuous selection theorem, 35
Michel-Penot tangent cone, 229
minimizer: countability of set of strict local minimizers, 95
minimum of finitely many Lipschitz functions: L-subdifferential, 532
Minkowski-Weyl theorem for convex polyhedra, 828
Minkowski-Weyl theorem for convex polyhedra, 278
modulus of calmness, 716
modulus of metric regular transversality, 738
modulus of metric regularity, 719
modulus of metric subregular transversality, 738
modulus of uniform continuity, 508
monotonicity property: subdifferential characterization, 690
Mordukhovich limiting normal cone, L-normal cone, 524
Mordukhovich limiting subdifferential, L-subdifferential, 523
Moreau envelope, 419
Moreau envelope: characterizations, 433
Moreau envelope: convergence, 421
Moreau envelope: differentiability under convexity, 424
Moreau envelope: Lipschitz property, 420

Moreau Rockafellar theorem for sum of convex functions on normed spaces, 178
Moreau-Rockafellar (global) subdifferential, 250
Moreau-Rockafellar continuity condition in locally convex spaces, 357
Moreau-Rockafellar theorem for sum of convex functions, 372
Moreau-Rockafellar theorem for sum of convex functions on normed spaces, 125, 328, 494, 597, 605, 660
Moreau-Rockafellar theorem on subdifferential determination of convex functions, 379
Moreau: continuity of $f \in \Gamma_0(X)$ via f^*, 404
Moreau: weak compactness of sublevels, 405
multidirectional mean value inequality: case of bounded convex sets, 710
multidirectional mean value inequality: case of compact convex sets, 710
multidirectional mean value inequality: case of unbounded convex sets, 707
multimapping: composition, 4
multimapping: definition, 2
multimapping: effective domain, 3
multimapping: graph, 3
multimapping: inverse, 3
multimapping: range, 3

necessary optimality conditions: Lagrange multipliers, 767
necessary optimality conditions: Lagrange multipliers, 202, 204, 762, 768
net of points, 12
non Y_+-decreasing, 339
non closedness of L-normal cone: Fabian example, 529
non-boundedness of subdifferential of a convex function at subdifferentiability points of the boundary of int (dom f), 268
non-nullity of intersection of C-normal cones at an intersection point, 229
nondecreasing property of one real variable function: lower Dini semiderivate condition, 664
nonincreasing property: characterization via Dini directional derivative, 650
nonincreasing property: characterization via Hadamard directional derivative, 690
normal cones: calculus for A/L-normals, 769, 770
normals to a convex sublevel at a non continuity point, 382

normals to a convex sublevel at a non continuity point: via ε-subgradients, 445
normals to intersection of convex sublevels at a continuity point, 382
normals to intersection of convex sublevels at a non continuity point, 381
normals to intersection: A/L-normals, 770, 771

one-sided Aubin-Lipschitz property, 716
openness with a linear rate, 720
orthogonal projection, 217
outer limit of multimappings, 5
outer limit: iteration, 9

penalization: Clarke lemma, 194
performance/value function, 335
perpendicular vector to a set, 563
polar cones: duality equality $\mathrm{exc}(P \cap \mathbb{B}_X, Q) = \mathrm{exc}(Q^\circ \cap \mathbb{B}_{X^*}, P^\circ)$, 217
polar of a set, 29
polar of image of a set, 29
polar of sum of cones, 29
polar of sum of sets, 29
polyhedral convex functions: definition, 279
polyhedral convex functions: max-representation, 280
polyhedral convex sets, 278
polyhedral convex sets: images under linear mappings, 278
polyhedral convex sets: inverse images under linear mappings, 278
polyhedral convexity: infimum function, 279
polyhedral convexity: preservation by infimal convolution, 287
polyhedral convexity: preservation by Legendre-Fenchel transform, 283
polyhedral convexity: preservation by polarity, 280
polyhedral convexity: subdifferentiability of polyhedral convex functions, 280
polyhedral convexity: support functions of polyhedral convex sets, 280
positive dual cone, 341
primal problem, 334
principle: Borwein-Preiss principle, 505, 604
projector mapping parallel to a vector subspace, 212
properties of F-normal to epigraph, 485
prox-regularity of functions: s-order σ-prox-regularity, 796
proximal fuzzy chain rule in Hilbert space, 579
proximal fuzzy sum rule in Hilbert space, 576
proximal mapping, 419
proximal mapping: characterizations, 433
proximal mapping: contraction, 426
proximal mapping: pair of conjugates, 428
proximal normal in Hilbert space, 563
proximal normal: analytic characterizations in Hilbert space, 564
proximal subdifferential in Hilbert space, 564
proximal subgradient: analytic global space description, 573

quasi interiorly tangent cone to graph, 230
quasi-convexity, 441

Rademacher theorem, 113
rate of calmness, 716
rate of metric regular transversality, 738
rate of metric regularity, 719
rate of metric subregular transversality, 738
recession cone, 319
recession function, 318
relative weak compactness of sublevels, 407, 413, 414
Robinson qualification condition, 772
Robinson theorem for metric regularity of constraint systems, 759
Robinson-Ursescu theorem, 727
robust lower/upper semicontinuity, 642
Rockafellar directional derivative, 106
Rockafellar dual problem, 334
Rockafellar duality, 346
Rockafellar relative interior condition, 357
Rockafellar theorem for sum rule for C-subdifferential, 123
Rockafellar: boundedness of sublevels in locally convex space, 405
Rockafellar: maximal cyclic monotonicity of subdifferential, 451
Roger theorem, 227
Roger theorem for the set $\{x \in S : \exists F \in \mathrm{Svect}_p(X), F \cap T^B(S;x) = \{0\}\}$, 224
Rolle-type theorem: subdifferential approximate form in infinite dimensions, 696
Rolle-type theorem: subdifferential form in finite dimensions, 694
Rolle-type theorem: subdifferential sequential form, 698
rule for C-subdifferential of infinitely many Lipschitz functions: Clarke theorem, 185
rule for C-subdifferential of supremum of finitely many Lipschitz functions, 179
rule for subdifferential of supremum of infinitely many convex functions in locally convex space: case of compactness of index set, 272

rule for subdifferential of supremum of infinitely many convex functions in locally convex space: case of convergence property of index set, 271
rule for subdifferential of supremum of infinitely many convex functions in locally convex space: Valadier theorem II, 269
rule for subdifferential of supremum of infinitely many convex functions in locally convex space: Volle's description, 270
rule for subdifferential of supremum of infinitely many convex functions: Valadier theorem, 190
Rådström cancellation laws, 28

saddle function, 651
saddle function: C-subdifferential, 655
selection of a multimapping, 34, 308
semicontinuity for multimappings: lower semicontinuity, 21
semicontinuity for multimappings: Attouch-Wets semicontinuity, 45
semicontinuity for multimappings: coincidence between inner and lower semicontinuity, 21
semicontinuity for multimappings: Hausdorff semicontinuity, 45
semicontinuity for multimappings: inner(lower) semicontinuity of M at $\bar{t}$ for $\bar{x} \in M(\bar{t})$., 21
semicontinuity for multimappings: inner/outer semicontinuity, 20
semicontinuity for multimappings: lower/upper chain rule, 22
semicontinuity for multimappings: scalarization of upper semicontinuity via support function, 29
semicontinuity for multimappings: sequential characterization, 24
semicontinuity for multimappings: sum and convexification, 33
semicontinuity for multimappings: theorem of scalarization of lower semicontinuity via support function, 31
semicontinuity for multimappings: upper semicontinuity, 21
semicontinuity for multimappings: upper versus outer semicontinuity under normality, 23
semicontinuity for multimappings: upper Vs outer semicontinuity under compactness condition for images, 23
semilimit superior: a case of equality between topological and sequential, 14
semilimits of multimappings: characterizations via nets, 13

separability of the metric space of p-dimensional vector subspaces of a separable space, 219
separable reduction principle of F-subdifferentiability, 497
separable reduction principle of a nul F-subgradient for a sum, 501
separately convex functions: continuity property, 160
separately convex/concave functions: Lipschitz property under growth conditions, 262
sequential characterization of compactly Lipschitzian mapping, 174
sequential closedness of multimapping at a point, 20
sequential limit inferior/superior of multimappings, 13
sharp local minimizer, 793
sharp local minimizer: characterization via growth condition under convexity, 794
sharp local minimizer: subdifferential characterization under convexity, 794
signed distance function, 151
signed distance function and C-tangent cone, 154
signed distance function: convexity under the convexity of the set, 160
Singh's example of functions with no F-subdifferentiability point, 517
slope distance of functions: lower semicontinuity, 266
slope inequality for convex functions, 97
stable s-order local minimizer, 793
Stegall variational principle in Asplund space, 522
Stegall variational principle in weak* Asplund space, 522
strict $\mathcal{B}$-differentiability, 304
strict $\mathcal{B}$-differentiability of $\mathcal{B}$-differentiable continuous convex functions, 306
strict F-differentiability: C-subdifferential characterization in finite dimensions, 690
strict F-differentiability: L-subdifferential characterization in finite dimensions, 690
strict differentiability: characterization of strict F-differentiability via bounded sets, 86
strict differentiability: characterization of strict H-differentiability via compact sets, 86
strict differentiability: equivalence in finite dimensions, 88
strict differentiability: Fréchet, Hadamard, Gâteaux, 85

strict differentiability: generic strict
F-differentiability, 92
strict Fréchet differentiability:
C-subdifferential characterization, 171
strict Hadamard differentiability:
C-subdifferential characterization, 109
strong convexity of functions, 430
strong convexity of functions:
characterization via Hadamard
directional derivative, 694
strong convexity of functions:
characterizations via conjugate, 431
strong convexity of functions:
subdifferential characterization, 694
strong metric regularity, 719
strong metric regularity of θ-order for a
multimapping: various
characterization properties via a
selection of the inverse, 785
strong metric regularity of θ-order of
multimappings, 782
strong metric regularity of θ-order of
subdifferential, 799
strong metric subregularity, 720
strong metric subregularity of θ-order of
multimappings, 782
strong metric subregularity of θ-order of
subdifferential, 796
strong metric subregularity of θ-order of
subdifferential: convex functions, 797
strong metric subregularity of θ-order of
subdifferential: prox-regular functions,
797
strong minimizer/maximizer, 502
sub-sup regular function: C-directional
derivative, 652
sub-sup regular function: C-subdifferential,
655
sub-sup regularity of saddle functions, 652
sub-sup regularity: definition, 651
subdifferentiability of convex functions at
continuity points, 264
subdifferentiability of convex functions at
points in relative interior, 277
subdifferential and ε-subdifferential: norms,
256
subdifferential determination:
ε-subdifferential determination of
convex functions, 448
subdifferential determination: convex
functions, 379
subdifferential of $f + g \circ G$: convex
functions without qualification
condition, 444
subdifferential of $f + g \circ G$: convex
functions on Banach spaces, 778

subdifferential of $g \circ G$: convex mapping G
and convex function g of one real
variable, 366
subdifferential of a Gâteaux differentiable
convex function, 304
subdifferential of convex functions:
norm × weak* closedness, 265
subdifferential of norms, 100
subdifferential of positively convex
functions: case of normed cases, 100
subdifferential of sum of convex functions:
Attouch-Brezis, 779
Subdifferential of sum of convex functions:
Robinson-Rockafellar, 780
subdifferential regularity:
A-subdifferentially regular function,
615
subdifferentially continuous functions, 796
subdifferentially pathological function, 638
subdifferentials of one-variable lsc convex
functions, 273
subgradient $\mathcal{D}$-representation:
C-subdifferential of directionally
Lipschitz functions, 657
subgradient $\mathcal{D}$-representation:
C-subdifferential of locally Lipschitz
subregular functions, 661
subgradient $\mathcal{D}$-representation:
subdifferential of continuous convex
functions, 664
subgradient $\mathcal{D}$-representation:
subdifferential of convex functions, 662
subnet of a net, 12
sum of separate variable functions:
F-subdifferential, 484
sum of separate variable functions:
L-subdifferential, 527
sum rule for C-subdifferential, 123
sum rule: A-subdifferential, 632
sum rule: A-subdifferential of sums of
Lipschitz functions, 615
sum rule: L-subdifferential, 530
sums of sets: L-normals, 556
support function of ε-subdifferential of a
convex function and its directional
ε-derivative, 326
support function of C-subdifferential of
extended real-valued functions, 294
support function of a set, 27
support function of subdifferential of a
convex function and its Hadamard
directional derivative, 294
support function: characterizations, 295
support functional, 201
support of subdifferential of a lsc convex
function and its ε-directional
derivative, 326

tangent cone: $\mathcal{B}$-tangent cone at Lebesgue density point, 220
tangent cone: calculus, 774
tangent cone: countability of the set of points of nullity of symmetrized $\mathcal{B}$-tangent cone, 212
tangent cone: countable cover with F-Lipschitz surfaces of $\{x \in S : F \cap T^B(S;x) = \{0\}\}$, 214
tangent cone: countable cover with p-dimensional Lipschitz surfaces of $\{x \in S : \exists F \in \mathrm{Svect}_p(X), F \cap T^B(S;x) = \{0\}\}$, 219
tangent cone: Roger theorem, 224
tangent cone: Roger theorem for the set of $x \in S \subset \mathbb{R}^N$ with $T^B(S;x) \neq \mathbb{R}^N$, 227
tangent cones to direct images, 146
tangent cones to inverse images, 140
tangent/normal cone of convex sets: normed spaces, 67
tangential regularity of Lipschitz function, 136
tangential regularity of sets and functions, 134
tangential regularity of sets versus subregularity of distance functions, 149
tangential regularity under F-normal regularity, 478
test for differentiability of convex functions: Fréchet case, 513
test for differentiability of convex functions: Gâteaux case, 512
tilt-stable s-order local minimizer, 805
topological complement vector subspaces, 212
truncated Hausdorff pseudo-distance of vector subspaces via orthogonal projection, 217

uniform $\mathcal{B}$-differentiability, 304
uniform $\mathcal{B}$-differentiability of continuous convex functions: intrinsic characterization, 307
uniform F-differentiability, 512
uniform s-order growth condition, 793
upper Γ-limit, 18
upper epi-limit, 18
upper Lipschitz semicontinuity: multimappings with closed polyhedral convex graph, 781
upper semicontinuity for vector mappings, 164
upper-Lipschitz semicontinuity, 780

Valadier theorem: subdifferential of suprema of convex functions, 190, 269
value of optimization problem, 332
variational principle under differentiable norm, 508
variational principle: a general metric version, 502
variational proximal normal cone: nonclosedness, 567
variational proximal normal functional in normed space, 565
variational proximal subdifferential in normed space, 565
variational proximal subdifferential: analytic description, 566
variational proximal subgradient: analytic global space description, 573
Volle's description of subdifferential of supremum of convex functions, 270

weak topology, 26
weak* Asplund space, 522
weak* topology, 26
weakly-star compactness of sublevels in normed space, 406

Yao-Zheng-Zhu lemma, 790

Zagrodny mean value inequality theorem, 681